McGraw-Hill
Dictionary of
Science and
Engineering

McGraw-Hill
Dictionary of
Science and
Engineering

Sybil P. Parker Editor in Chief

McGraw-Hill Book Company

New York   St. Louis   San Francisco

Auckland   Bogotá   Guatemala   Hamburg
Lisbon   London   Madrid   Mexico
Montreal   New Delhi   Panama   Paris   San Juan
São Paulo   Singapore   Sydney   Tokyo   Toronto

5 6 7 8 9 0   DODO   9 1

ISBN 0-07-045483-3

Library of Congress Cataloging In Publication Data

McGraw-Hill dictionary of science and engineering.

1. Science—Dictionaries.   2. Technology—Dictionaries.
I. Parker, Sybil P.   II. McGraw-Hill Book Company.
III. Title: Science and engineering.
Q123.M338   1984   503'.21   83-26768
ISBN 0-07-045483-3

Preface

With the continuing proliferation of science and technology, the rapidly expanding scientific literature, and the growing emphasis on science education, there is an urgent need to understand the vocabulary of science, technology, and engineering. This need has reached beyond the professional community of scientists and engineers, and this Dictionary is based on an awareness of that fact.

The *McGraw-Hill Dictionary of Science and Engineering* is conceived as a reference tool for serious investigators of science, including students, teachers, librarians, writers, and general readers. It is derived from the comprehensive *McGraw-Hill Dictionary of Scientific and Technical Terms* (3d ed., 1984), which has served as a standard reference for professionals for 10 years, but which, in the opinion of the editors, is more than many nonprofessionals require.

The more than 35,000 terms in the *Dictionary of Science and Engineering* represent 102 fields of science and engineering, and constitute the basic vocabulary of these fields. The information was extracted from the parent work by combining the input of editorial skills with the output of magnetic tape. Terms were selected by editorial consensus from the master file of the larger dictionary, and the new product was generated electronically and set by computer composition.

This Dictionary retains the clarity, simplicity, and authoritativeness that have been the hallmark of the parent work. Each definition is preceded by an abbreviation which identifies the field of primary use; many terms are defined in multiple fields. When a definition applies equally to more than one field, a more general field is given for identification. For example, a definition that is applied in both electrical and mechanical engineering is assigned to engineering. Synonyms, acronyms, and abbreviations are given along with the defining entry, and are also listed in cross-reference entries in the alphabetical sequence.

An explanation of the alphabetization, cross-referencing, format, field abbreviations, and other information on how to use the Dictionary begins on page ix.

The *Dictionary of Science and Engineering* is intended to provide information that is not available in general dictionaries for anyone reading or writing about science or engineering at a subprofessional level. In serving this function, it will be an indispensable reference.

Sybil P. Parker
EDITOR IN CHIEF

Editorial Staff

Consulting and Contributing Editors

for the McGraw-Hill Dictionary of Scientific and Technical Terms grams and Budget, and Dean of the Graduate School, Rensselaer Polytechnic Institute. METALLURGICAL ENGINEERING.

Alvin W. Knoerr—Mining Engineer, U.S. Bureau of Mines, Washington, D.C. MINING ENGINEERING.

John Markus—Author and Consultant. ELECTRONICS.

Dr. Nathaniel Martin—Department of Mathematics, University of Virginia. MATHEMATICS.

Dr. Edward C. Monahan—Statutory Lecturer in Physical Oceanography, University College, Galway, Ireland. OCEANOGRAPHY.

Dr. N. Karle Mottet—Professor of Pathology, University of Washington School of Medicine. BIOCHEMISTRY; MEDICINE; PSYCHOLOGY.

Dr. Charles Oviatt—State Department of Education of Missouri. CHEMISTRY.

Dr. Guido Pontecorvo—Imperial Cancer Research Fund Laboratories, London. GENETICS AND EVOLUTION.

Dr. John Quick—Arthur D. Little, Inc., Cambridge, Massachusetts. ARMAMENTS; GRAPHIC ARTS.

Prof. Alan Saleski—Department of Mathematics, Loyola University of Chicago. MATHEMATICS.

Brig. Gen. Peter C. Sandretto—Formerly, Director, Engineering Management, International Telephone and Telegraph Corporation. NAVIGATION.

Prof. Frederic Schwab—Department of Geology, Washington and Lee University. GEOLOGY; PHYSICAL GEOGRAPHY.

Dr. W. R. Sistrom—Department of Biology, University of Oregon. MICROBIOLOGY.

Dr. Leonard Spero—Walter Reed Hospital Unit, Fort Dietrick, Maryland. CHEMISTRY.

Dr. Aaron Strauss—Department of Mathematics, University of Maryland. MATHEMATICS.

Dr. C. N. Touart—Senior Scientist, Air Force Geophysics Laboratory. GEOCHEMISTRY; GEOPHYSICS; METEOROLOGY.

Dr. Joachim Weindling—Professor of System Engineering and Operations Research, Polytechnic Institute of Brooklyn. INDUSTRIAL AND PRODUCTION ENGINEERING.

How to Use the Dictionary

I. ALPHABETIZATION

The terms in the *McGraw-Hill Dictionary of Science and Engineering* are alphabetized on a letter-by-letter basis; word spacing, hyphen, comma, solidus, and apostrophe in a term are ignored in the sequencing. For example, an ordering of terms would be:

air-earth current
air ejector
airflow
air gap
ALGOL

Also ignored in the sequencing of terms (usually, chemistry terms) are italic elements, numbers, small capitals, and Greek letters. For example, the following terms appear within alphabet letter "A":

γ-aminobutyric acid
2-aminoisovaleric acid
3′,5′-AMP

II. FORMAT

The basic format for a defining entry provides the term in boldface, the field in small capitals, and the single definition in lightface:

term [FIELD] Definition.

A field may be followed by multiple definitions, each introduced by a boldface number:

term [FIELD] **1.** Definition. **2.** Definition. **3.** Definition.

A term may have definitions in two or more fields:

term [BOT] Definition. [GEOL] Definition.

A simple cross-reference entry appears as:

term *See* another term.

A cross-reference may also appear in combination with definitions:

term [BOT] Definition. [GEOL] *See* another term.

III. CROSS-REFERENCING

A cross-reference entry directs the user to the defining entry. For example, the user looking up "asar" finds:

asar *See* esker.

The user then turns to the "E" terms for the definition.

Cross-references are also made from variant spellings, acronyms, abbreviations, and symbols.

aestival *See* estival.
ASCII *See* American Standard Code for Information Interchange.
at. wt *See* atomic weight.
Au *See* gold.

IV. ALSO KNOWN AS . . . , etc.

A definition may conclude with a mention of a synonym of the term, a variant spelling, an abbreviation for the term, or other such information, introduced by "Also known as . . . ," "Also spelled . . . ," "Abbreviated . . . ," "Symbolized . . . ," "Derived from . . . ." When a term has more than one definition, the positioning of any of these phrases conveys the extent of applicability. For example:

> **term** [BOT] **1.** Definition. Also known as synonym. **2.** Definition. Symbolized T.

In the above arrangement, "Also known as . . ." applies only to the first definition: "Symbolized . . ." applies only to the second definition.

> **term** [BOT] **1.** Definition. **2.** Definition. [GEOL] Definition. Also known as synonym.

In the above arrangement "Also known as . . ." applies only to the second field.

> **term** [BOT] Also known as synonym. **1.** Definition. **2.** Definition. [GEOL] Definition.

In the above arrangement, "Also known as . . ." applies to both definitions in the first field.

> **term** Also known as synonym. [BOT] **1.** Definition. **2.** Definition. [GEOL] Definition.

In the above arrangement, "Also known as . . ." applies to all definitions in both fields.

V. CHEMICAL FORMULAS

Chemistry definitions may include either an empirical formula (say, for abscisic acid, $C_{16}H_{20}O_4$) or a line formula (for acrylonitrile, CH_2CHCN), whichever is appropriate.

Field Abbreviations

ACOUS	acoustics		GEOGR	geography
AERO ENG	aerospace engineering		GEOL	geology
AGR	agriculture		GEOPHYS	geophysics
ANALY CHEM	analytical chemistry		GRAPHICS	graphic arts
ANAT	anatomy		HISTOL	histology
ANTHRO	anthropology		HOROL	horology
ARCH	architecture		HYD	hydrology
ARCHEO	archeology		IMMUNOL	immunology
ASTRON	astronomy		IND ENG	industrial engineering
ASTROPHYS	astrophysics		INORG CHEM	inorganic chemistry
ATOM PHYS	atomic physics		INV ZOO	invertebrate zoology
BIOCHEM	biochemistry		LAP	lapidary
BIOL	biology		MAP	mapping
BIOPHYS	biophysics		MATER	materials
BOT	botany		MATH	mathematics
BUILD	building construction		MECH	mechanics
CHEM	chemistry		MECH ENG	mechanical engineering
CHEM ENG	chemical engineering		MED	medicine
CIV ENG	civil engineering		MET	metallurgy
CLIMATOL	climatology		METEOROL	meteorology
COMMUN	communications		MICROBIO	microbiology
COMPUT SCI	computer science		MIN ENG	mining engineering
CONT SYS	control systems		MINERAL	mineralogy
CRYO	cryogenics		MOL BIO	molecular biology
CRYSTAL	crystallography		MYCOL	mycology
CYTOL	cytology		NAV	navigation
DES ENG	design engineering		NAV ARCH	naval architecture
ECOL	ecology		NUCLEO	nucleonics
ELEC	electricity		NUC PHYS	nuclear physics
ELECTR	electronics		OCEANOGR	oceanography
ELECTROMAG	electromagnetism		OPTICS	optics
EMBRYO	embryology		ORD	ordnance
ENG	engineering		ORG CHEM	organic chemistry
ENG ACOUS	engineering acoustics		PALEOBOT	paleobotany
EVOL	evolution		PALEON	paleontology
FL MECH	fluid mechanics		PARTIC PHYS	particle physics
FOOD ENG	food engineering		PATH	pathology
FOR	forestry		PETR	petrology
GEN	genetics		PETRO ENG	petroleum engineering
GEOCHEM	geochemistry		PHARM	pharmacology
GEOD	geodesy		PHYS	physics

PHYS CHEM	physical chemistry	STAT	statistics
PHYSIO	physiology	STAT MECH	statistical mechanics
PL PATH	plant pathology	SYS ENG	systems engineering
PL PHYS	plasma physics	SYST	systematics
PSYCH	psychology	TEXT	textiles
QUANT MECH	quantum mechanics	THERMO	thermodynamics
RELAT	relativity	VERT ZOO	vertebrate zoology
SCI TECH	science and technology	VET MED	veterinary medicine
SOLID STATE	solid-state physics	VIROL	virology
SPECT	spectroscopy	ZOO	zoology

Scope of Fields

acoustics—The science of the production, transmission, and effects of sound.

aerospace engineering—Engineering pertaining to the design and construction of aircraft and space vehicles and of power units and dealing with the special problems of flight in both the earth's atmosphere and space, such as in the flight of air vehicles and the launching, guidance, and control of missiles, earth satellites, and space vehicles and probes.

agriculture—The production of plants and animals useful to humans, involving soil cultivation and the breeding and management of crops and livestock.

analytical chemistry—Science and art of determining composition of materials in terms of elements and compounds which they contain.

anatomy—The branch of morphology concerned with the gross and microscopic structure of animals, especially humans.

anthropology—The study of the interrelations of biological, cultural, geographical, and historical aspects of the human race.

archeology—The scientific study of the material remains of the cultures of historical and prehistorical peoples.

architecture—The art or practice of designing structures, especially habitable structures in accordance with principles determined by esthetic and practical or material considerations.

astronomy—The science concerned with celestial bodies and with the observations and interpretation of radiation received in the vicinity of earth from the component parts of the universe.

astrophysics—A branch of astronomy that treats the physical properties of celestial bodies, such as luminosity, size, mass, density, temperature, and chemical composition, and their origin and evolution.

atomic physics—A branch of physics concerned with the structures of the atom, the characteristics of the electrons and other elementary particles of which the atom is composed, the arrangement of the atom's energy states, and the processes involved in the radiation of light and x-rays.

biochemistry—The study of the chemical substances that occur in living organisms, the processes by which these substances enter into or are formed in the organisms and react with each other and the environment, and the methods by which the substances and processes are identified, characterized, and measured.

biology—The science of living organisms, concerned with the study of embryology, anatomy, physiology, cytology, morphology, taxonomy, genetics, evolution, and ecology.

biophysics—The hybrid science involving the methods and ideas of physics and chemistry to study and explain the structures of living organisms and the mechanics of life processes.

botany—That branch of biological science which embraces the study of plants and plant life, including algae; deals with taxonomy, morphology, physiology, and other aspects.

building construction—The art of business of assembling materials into a structure, especially one designated for occupancy.

chemical engineering—A branch of engineering that deals with the development and application of manufacturing processes, such as refinery processes, which chemically convert raw materials into a variety of products, and that deals with the design and operation of plants and equipment to perform such work.

chemistry—The scientific study of the properties, composition, and structure of matter, the changes in structure and composition of matter, and accompanying energy changes.

civil engineering—The planning, design, construction, and maintenance of fixed structures and ground facilities for industry, for transportation, for use and control of water, for occupancy, and for harbor facilities.

climatology—That branch of meteorology concerned with the mean physical state of the atmosphere together with its statistical variations in both space and time as reflected in the weather behavior over a period of many years.

communications—The science and technology by which information is collected from an originating source, transformed into electric currents or fields, transmitted over electrical networks or space to another point, and reconverted into a form suitable for interpretation by a receiver.

computer science—The branch of knowledge concerned with information processes, the structures and procedures that represent these processes, and their implementation in information-processing systems.

control systems—The study of those systems in which one or more outputs are forced to change in a desired manner as time progresses.

cryogenics—The science of producing and maintaining very low temperatures, of phenomena at those temperatures, and of technical operations performed at very low temperatures.

crystallography—The branch of science that deals with the geometric description of crystals, their internal arrangement, and their properties.

cytology—The branch of biological science which deals with the structure, behavior, growth, and reproduction of cells and the function and chemistry of cells and cell components.

design engineering—A branch of engineering concerned with the design of a product or facility according to generally accepted uniform standards and procedures, such as the specification of a linear dimension, or a manufacturing practice, such as the consistent use of a particular size of screw to fasten covers.

ecology—The study of the interrelationships between organisms and their environment.

electricity—The science of physical phenomena involving electric charges and their effects when at rest and when in motion.

electromagnetism—The branch of physics dealing with the observations and laws relating electricity to magnetism, and with magnetism produced by an electric current.

electronics—The branch of science and technology relating to the conduction of electricity through gases or vacuum or through semiconducting materials; concerned with the design, manufacture, and application of electron tubes.

embryology—The study of the development of the organism from the zygote, or fertilized egg.

engineering—The science by which properties of matter and sources of power in nature are made useful to humans in structures, machines, and products.

engineering acoustics—A field of acoustics that deals with the production, detection, and control of sound by electrical devices, including the study, design, and construction of such things as microphones, loudspeakers, sound recorders and reproducers, and public address systems.

evolution—The processes of biological and organic change in organisms by which descendants come to differ from their ancestors, and a history of the sequence of such change.

fluid mechanics—The science concerned with fluids, either at rest or in motion, and dealing with pressures, velocities, and accelerations in the fluid, including fluid deformation and compression or expansion.

food engineering—Technical discipline involved in food manufacturing and processing.

forestry—The science of developing, cultivating, and managing forest lands for wood, forage, water, wildlife, and recreation; the management of growing timber.

genetics—The science concerned with biological inheritance, that is, with the causes of the resemblances and differences among related individuals.

geochemistry—The study of the chemical composition of the various phases of the earth and the physical and chemical processes which have produced the observed distribution of the elements and nuclides in these phases.

geodesy—A subdivision of geophysics which includes determinations of the size and shape of the earth, the earth's gravitational field, and the location of points fixed to the earth's crust in an earth-referred coordinate system.

geography—The science that deals with the description of land, sea, and air and the distribution of plant and animal life, including humans.

geology—The study or science of earth, its history, and its life as recorded in the rocks; includes the study of the geologic features of an area, such as the geometry of rock formations, weathering and erosion, and sedimentation.

geophysics—A branch of geology in which the principles and practices of physics are used to study the earth and its environment, that is, earth, air, and (by extension) space.

graphic arts—The fine and applied arts of representation, decoration, and writing or printing on flat surfaces together with the techniques and crafts associated with each; includes painting, drawing, engraving, etching, lithography, photography, and printing arts.

histology—The study of the structure and chemical composition of animal tissues as related to their function.

horology—Science of time measurement and the principles and technology of constructing time-measuring instruments.

hydrology—The science that treats of the surface and ground waters of the earth; their occurrence, circulation, and distribution; their chemical and physical properties; and their reaction with their environment.

immunology—The division of biological science concerned with the native or acquired resistance of higher animal forms and humans to infection with microorganisms.

industrial engineering—The application of engineering principles and training and the techniques of scientific management to the maintenance of a high level of productivity at optimum cost in industrial enterprises, as by analytical study, improvement, and installation of methods and systems, operating procedures, quantity and quality measurements and controls, safety measures, and personnel administration.

inorganic chemistry—A branch of chemistry that deals with reactions and properties of all chemical elements and their compounds, excluding hydrocarbons but usually including carbides and other simple carbon compounds (such as CO_2, CO, and HCN).

invertebrate zoology—A branch of zoology concerned with the taxonomy, behavior, and morphology of invertebrate animals.

lapidary—The study relating to precious stones or the art of cutting them.

mapping—The art and practice of making a drawing or other representation, usually on a flat surface, of the whole or part of an area (as the surface of the earth or some other planet), indicating relative position and size according to a specified scale or projection of selected features, as countries, cities, rock formations, or bodies of water.

materials—The study of admixtures of matter or the basic matter from which products are made; includes adhesives, building materials, fuels, paints, leathers, and so on.

mathematics—The deductive study of shape, quantity, and dependence; the two main areas are applied mathematics and pure mathematics, the former arising from the study of physical phenomena, the latter involving the intrinsic study of mathematical structures.

mechanical engineering—The branch of engineering concerned with the generation, transmission, and utilization of heat and mechanical power, and with the production and operation of tools, machinery, and their products.

mechanics—The branch of physics which seeks to formulate general rules for predicting the behavior of a physical system under the influence of any type of interaction with its environment.

medicine—The study of cause and treatment of human disease, including the healing arts dealing with diseases which are treated by a physician or a surgeon.

metallurgy—The branch of engineering concerned with the production of metals and alloys, their adaptation to use, and their performance in service; and the study of chemical reactions involved in the processes by which metals are produced, and of the laws governing the physical, chemical, and mechanical behavior of metallic materials.

meteorology—The science concerned primarily with the observation of the atmosphere and its phenomena, including temperature, density, winds, clouds, and precipitation.

microbiology—The science and study of microorganisms, especially bacteria and rickettsiae, and of antibiotic substances.

mineralogy—The science concerning the study of natural inorganic substances called minerals, including origin, description, and classification.

mining engineering—A branch of engineering concerned with the location and evaluation of coal and mineral deposits, the survey of mining areas, the layout and equipment of mines, the supervision of mining operations, and the cleaning, sizing, and dressing of the product.

molecular biology—That branch of biology which attempts to interpret biological events in terms of the molecules in the cell.

mycology—A branch of biological science concerned with the study of fungi.

naval architecture—The study of the physical characteristics and the design and construction of buoyant structures which operate in water, and of the construction and operation of the power plant and other mechanical equipment of these structures.

navigation—The science or art of conducting ships or aircraft from one place to another, especially the method of determining position, course, and distance traveled over the surface of the earth by the principles of geometry and astronomy and by reference to devices (radar, beacons, and instruments) designed as aids.

nuclear physics—The study of the characteristics, behavior, and internal structure of the atomic nucleus.

nucleonics—The technology based on phenomena of the atomic nucleus such as radioactivity, fission, and fusion; includes nuclear reactors, various applications of radioisotopes and radiation, particle accelerators, and radiation detection devices.

oceanography—The scientific study and exploration of the oceans and seas in all their aspects.

optics—The study of phenomena associated with the generation, transmission, and detection of electromagnetic radiation in the spectral range extending from the long-wave edge of the x-ray region to the short-wave edge of the radio region; and the science of light.

ordnance—That military area concerned with supplies, including weapons, ammunition, combat vehicles, and the necessary repair equipment; and with heavy firearms discharged from mounts, including cannons and artillery.

organic chemistry—The study of the composition, reactions, and properties of carbon compounds except CO_2, CO, and certain ionic compounds such as Na_2CO_3 and $NaCN$.

paleobotany—The study of fossil plants and vegetation of the geologic past.

paleontology—The study of life in the geologic past as recorded by fossil remains.

pathology—The branch of biological science which deals with the nature of disease, through study of its causes, its processes, and its effects, together with the associated alterations of structure and function; and the laboratory findings of disease, as distinguished from clinical signs and symptoms.

particle physics—The branch of physics concerned with understanding the properties, behavior, and structure of elementary particles, especially through study of collisions or decays involving energies of hundreds of MeV or more.

petroleum engineering—A branch of engineering concerned with the search for and extraction of oil, gas, and liquefiable hydrocarbons.

petrology—The branch of geology dealing with the origin, occurrence, structure, and history of rocks, especially igneous and metamorphic rocks.

pharmacology—The science of detection and measurement of the effects of drugs or other chemicals on biological systems; includes all chemicals used as drugs.

physical chemistry—The description and prediction of chemical behavior by means of physical theory, with extensive use of graphs and mathematical formulas; main subject areas are structure, thermodynamics, and kinetics.

physics—The science concerned with those aspects of nature which can be understood in terms of elementary principles and laws.

physiology—The branch of biological science concerned with the basic activities that occur in cells and tissues of living organisms and involving physical and chemical studies of these organisms.

plant pathology—The branch of botany concerned with diseases of plants.

plasma physics—The study of highly ionized gases.

psychology—The science of the function of the mind and the behavior of an organism, both animal and human, in relation to its environment.

quantum mechanics—The modern theory of matter, of electromagnetic radiation, and of the interaction between matter and radiation; it differs from classical physics, which it generalizes and supersedes, mainly in the realm of atomic and subatomic phenomena.

relativity—The study of physics theory which recognizes the universal character of the propagation speed of light and the consequent dependence of space, time, and other mechanical measurements on the motion of the observer performing the measurements; the two main divisions are special theory and general theory.

science and technology—The study of the natural sciences and the application of this knowledge for practical purposes.

solid-state physics—The branch of physics centering on the physical properties of solid materials; it is usually concerned with the properties of crystalline materials only, but it is sometimes extended to include the properties of glasses or polymers.

spectroscopy—The branch of physics concerned with the production, measurement, and interpretation of electromagnetic spectra arising from either emission or absorption of radiant energy by various substances.

statistical mechanics—That branch of physics which endeavors to explain and predict the macroscopic properties and behavior of a system on the basis of the known characteristics and interactions of the microscopic constituents of the system, usually when the number of such constituents is very large.

statistics—The science dealing with the collection, analysis, interpretation, and presentation of masses of numerical data.

systematics—The science of animal and plant classification.

systems engineering—The branch of engineering dealing with the design of a complex interconnection of many elements (a system) to maximize an agreed-upon measure of system performance.

textiles—Area of industry involving thhe production of fibers, filaments, or yarn, and the cloth made from these materials.

thermodynamics—The branch of physics which seeks to derive, from a few basic postulates, relations between properties of substances, especially those which are affected by changes in temperature, and a description of the conversion of energy from one form to another.

vertebrate zoology—A branch of zoology concerned with the taxonomy, behavior, and morphology of vertebrate animals.

veterinary medicine—That branch of medical practice which treats of the diseases and injuries of animals.

virology—The science that deals with the study of viruses.

zoology—The science that deals with the taxonomy, behavior, and morphology of animal life.

A

a *See* ampere; atto-.
aΩ *See* abohm.
A *See* ampere; angstrom.
Å *See* angstrom.
A+ *See* A positive.
aA *See* abampere.
AA *See* antiaircraft.
aa lava [GEOL] Type of lava with a rough, fragmental surface; consists of clinkers and scoria.
AAM *See* air-to-air missile.
A AND NOT B gate *See* AND NOT gate.
aapamoor [ECOL] A moor with elevated areas or mounds supporting dwarf shrubs and sphagnum, interspersed with low areas containing sedges and sphagnum, thus forming a mosaic.
aardvark [VERT ZOO] A nocturnal, burrowing, insectivorous mammal of the genus *Orycteropus* in the order Tubulidentata. Also known as earth pig.
aardwolf [VERT ZOO] *Proteles cristatus.* A hyenalike African mammal of the family Hyaenidae.
a axis [CRYSTAL] One of the crystallographic axes used as reference in crystal description, usually oriented horizontally, front to back. [GEOL] The direction of movement or transport in a tectonite.
ab- [ELECTROMAG] A prefix used to identify centimeter-gram-second electromagnetic units, as in abampere, abcoulomb, abfarad, abhenry, abmho, abohm, and abvolt.
abac *See* nomograph.
abaca [BOT] *Musa textilis.* A plant of the banana family native to Borneo and the Philippines, valuable for its hard fiber. Also known as Manila hemp.
abactinal [INV ZOO] In radially symmetrical animals, pertaining to the surface opposite the side where the mouth is located.
abacus [ARCH] A slab forming the topmost division of the capital of a column. [MATH] An instrument for performing arithmetical calculations manually by sliding markers on rods or in grooves.
abalone [INV ZOO] A gastropod mollusk composing the single genus *Haliotis* of the family Haliotidae. Also known as ear shell; ormer; paua.
abampere [ELEC] The unit of electric current in the electromagnetic centimeter-gram-second system; 1 abampere equals 10 amperes in the absolute meter-kilogram-second-ampere system. Abbreviated aA. Also known as Bi; biot.
abamurus [ARCH] A masonry block, in the form of a buttress, used to support a structure.
A band [HISTOL] The region between two adjacent I bands in a sarcomere; characterized by partial overlapping of actin and myosin filaments.
abandon [ENG] To stop drilling and remove the drill rig from the site of a borehole before the intended depth or target is reached.

abandoned channel *See* oxbow.
abapertural [INV ZOO] Away from the shell aperture, referring to mollusks.
abapical [BIOL] On the opposite side to, or directed away from, the apex.
abatement [ENG] **1.** The waste produced in cutting a timber, stone, or metal piece to a desired size and shape. **2.** A decrease in the amount of a substance or other quantity, such as atmospheric pollution.
abat-jour [BUILD] A device that is used to deflect daylight downward as it streams through a window.
abat-vent [BUILD] A series of sloping boards or metal strips, or some similar contrivance, to break the force of wind without being an obstruction to the passage of air or sound, as in a louver or chimney cowl.
abaxial [BIOL] On the opposite side to, or facing away from, the axis of an organ or organism.
Abbe condenser [OPTICS] A variable large-aperture lens system arranged substage to image a light source into the focal plane of a microscope objective.
Abbe number [OPTICS] A number which expresses the deviating effect of an optical glass on light of different wavelengths.
Abbe prism [OPTICS] A system used for image erection which is composed of two double right-angle prisms and involves four reflections.
Abbe refractometer [OPTICS] An optical instrument for the measurement of the refractive index of liquids.
Abbe's theory [OPTICS] The theory that for a lens to produce a true image, it must be large enough to transmit the entire diffraction pattern of the object.
ABC *See* automatic brightness control.
abcoulomb [ELEC] The unit of electric charge in the electromagnetic centimeter-gram-second system, equal to 10 coulombs. Abbreviated aC.
abdomen [ANAT] **1.** The portion of the vertebrate body between the thorax and the pelvis. **2.** The cavity of this part of the body. [INV ZOO] The elongate region posterior to the thorax in arthropods.
abdominal regions [ANAT] Nine theoretical areas delineated on the abdomen by two horizontal and two parasagittal lines: above, the right hypochondriac, epigastric, and left hypochondriac; in the middle, the right lateral, umbilical, and left lateral; and below, the right inguinal, hypogastric, and left inguinal.
abducens [ANAT] The sixth cranial nerve in vertebrates; a paired, somatic motor nerve arising from the floor of the fourth ventricle of the brain and supplying the lateral rectus eye muscles.
abduction [PHYSIO] Movement of an extremity or other body part away from the axis of the body.
abductor [PHYSIO] Any muscle that draws a part of the body or an extremity away from the body axis.

Abegg's rule [CHEM] An empirical rule, holding for a large number of elements, that the sum of the maximum positive and negative valencies of an element equals eight.

Abelian domain *See* Abelian field.

Abelian field [MATH] A set of elements a, b, c,... forming Abelian groups with addition and multiplication as group operations where $a(b + c) = ab + ac$. Also known as Abelian domain; domain.

Abelian group [MATH] A group whose binary operation is commutative; that is, $ab = ba$ for each a and b in the group.

abelite [MATER] A substance made of ammonium nitrate and a nitrated aromatic hydrocarbon and used as an explosive.

Abel's inequality [MATH] An inequality which states that the absolute value of the sum of n terms, each in the form ab, where the bs are positive numbers, is not greater than the product of the largest b with the largest absolute value of a partial sum of the as.

Abel theorem [MATH] **1.** A theorem stating that if a power series in z converges for $z = a$, it converges absolutely for $|z| < |a|$. **2.** A theorem stating that if a power series in z converges to $f(z)$ for $|z| < 1$ and to a for $z = 1$, then the limit of $f(z)$ as z approaches $1 = a$. **3.** A theorem stating that if the three series with nth term a_n, b_n, and $c_n = a_0 b_n + a_1 b_{n-1} + \cdots + a_n b_0$, respectively, converge, then the third series equals the product of the first two series.

abend [COMPUT SCI] An unplanned program termination that occurs when a computer is directed to execute an instruction or to process information that it cannot recognize. Also known as bomb; crash.

abenteric [MED] Involving abdominal organs and structures outside the intestine.

aberrant [BIOL] An atypical group, individual, or structure, especially one with an aberrant chromosome number.

aberration [ASTRON] The apparent angular displacement of the position of a celestial body in the direction of motion of the observer, caused by the combination of the velocity of the observer and the velocity of light. [OPTICS] *See* optical aberration.

abfarad [ELEC] A unit of capacitance in the electromagnetic centimeter-gram-second system equal to 10^9 far-ads. Abbreviated aF.

abhenry [ELEC] A unit of inductance in the electromagnetic centimeter-gram-second system of units which is equal to 10^{-9} henry. Abbreviated aH. Also known as centimeter.

abietic acid [ORG CHEM] $C_{20}H_{30}O_2$ A tricyclic, crystalline acid obtained from rosin; used in making esters for plasticizers.

abiotic [BIOL] Referring to the absence of living organisms.

ablastin [IMMUNOL] An antibodylike substance elicited by *Trypanosoma lewisi* in the blood serum of infected rats that inhibits reproduction of the parasite.

ablation [AERO ENG] The carrying away of heat, generated by aerodynamic heating, from a vital part by arranging for its absorption in a nonvital part, which may melt or vaporize and then pass away, taking the heat with it. Also known as ablative cooling. [GEOL] The wearing away of rocks, as by erosion or weathering. [HYD] The reduction in volume of a glacier due to melting and evaporation. [MED] The removal of tissue or a part of the body by surgery, such as by excision or amputation.

ablative agent [MATER] A material from which the surface layer is to be removed, often for the purpose of dissipating extreme heat energy, as in space vehicles reentering the earth's atmosphere. Also known as ablative material.

ablative cooling *See* ablation.

ablative shielding [AERO ENG] A covering of material designed to reduce heat transfer to the internal structure through sublimation and loss of mass.

ABM *See* antiballistic missile.

abmho [ELEC] A unit of conductance in the electromagnetic centimeter-gram-second system of units equal to 10^9 mhos. Abbreviated $(a\Omega)^{-1}$. Also known as absiemens (aS).

Abney effect [OPTICS] A shift in the apparent hue of a light which occurs as colored light is desaturated by the addition of white light.

Abney level *See* clinometer..

abnormal fold [GEOL] An anticlinorium in which there is an upward convergence of the axial surfaces of the subsidiary folds.

abnormal psychology [PSYCH] A branch of psychology that deals with behavior disorders and internal psychic conflict in addition to certain normal phenomena such as dreams, motivations, and anxiety.

abnormal statement [COMPUT SCI] An element of a FORTRAN V (UNIVAC) program which specifies that certain function subroutines must be called every time they are referred to.

ABO blood group [IMMUNOL] An immunologically distinct, genetically determined group of human erythrocyte antigens represented by two blood factors (A and B) and four blood types (A, B, AB, and O).

abohm [ELEC] The unit of electrical resistance in the centimeter-gram-second system; 1 abohm equals 10^{-9} ohm in the meter-kilogram-second system. Abbreviated $a\Omega$.

abomasum [VERT ZOO] The final chamber of the complex stomach of ruminants; has a glandular wall and corresponds to a true stomach.

A bomb *See* atomic bomb.

aboral [INV ZOO] Opposite to the mouth.

abort [AERO ENG] **1.** To cut short or break off an action, operation, or procedure with an aircraft, space vehicle, or the like, especially because of equipment failure. **2.** An aircraft, space vehicle, or the like which aborts. **3.** An act or instance of aborting.

abortion [MED] The spontaneous or induced expulsion of the fetus prior to the time of viability, most often during the first 20 weeks of the human gestation period.

abrasion [ENG] **1.** The removal of surface material from any solid through the frictional action of another solid, a liquid, or a gas or combination thereof. **2.** A surface discontinuity brought about by roughening or scratching. [GEOL] Wearing away of sedimentary rock chiefly by currents of water laden with sand and other rock debris and by glaciers. [MED] A spot denuded of skin, mucous membrane, or superficial epithelium by rubbing or scraping.

abrasive [GEOL] A small, hard, sharp-cornered rock fragment, used by natural agents in abrading rock material or land surfaces. Also known as abrasive ground. [MATER] **1.** A material used, usually as a grit sieved by a specified mesh but also as a solid shape or as a paste or slurry or air suspension, for grinding, honing, lapping, superfinishing, polishing, pressure blasting, or barrel tumbling. **2.** A material sintered or formed into a solid mass such as a hone or a wheel disk, cone, or burr for grinding or polishing other materials. **3.** Having qualities conducive to or derived from abrasion.

abrasive ground   See abrasive.

abrasiveness   [MATER] 1. The property of a material causing wear of a surface by friction. 2. The quality or characteristic of being able to scratch, abrade, or wear away another material.

abreast milling   [MECH ENG] A milling method in which parts are placed in a row parallel to the axis of the cutting tool and are milled simultaneously.

abreuvoir   [CIV ENG] A space between stones in masonry to be filled with mortar.

abrupt junction   [ELECTR] A pn junction in which the concentration of impurities changes suddenly from acceptors to donors.

abscess   [MED] A localized collection of pus surrounded by inflamed tissue.

abscisic acid   [BIOCHEM] $C_{16}H_{20}O_4$ A plant hormone produced by fruits and leaves that promotes abscission and dormancy and retards vegetative growth. Formerly known as abscisin.

abscisin   See abscisic acid.

abscissa   [MATH] The horizontal coordinate denoted on y, the vertical axis.

abscission   [BOT] A physiological process promoted by abscisic acid whereby plants shed a part, such as a leaf, flower, seed, or fruit.

absiemens   See abmho.

absolute address   [COMPUT SCI] The numerical identification of each storage location which is wired permanently into a computer by the manufacturer.

absolute age   [GEOL] The geologic age of a fossil, or a geologic event or structure expressed in units of time, usually years. Also known as actual age.

absolute alcohol   [ORG CHEM] Ethyl alcohol that contains no more than 1% water. Also known as anhydrous alcohol.

absolute altimeter   [ENG] An instrument which employs radio, sonic, or capacitive technology to produce on its indicator the measurement of distance from the aircraft to the terrain below. Also known as terrain clearance indicator.

absolute altitude   [ENG] Altitude above the actual surface, either land or water, of a planet or natural satellite.

absolute boiling point   [CHEM] The boiling point of a substance expressed in the unit of an absolute temperature scale.

absolute ceiling   [AERO ENG] The greatest altitude at which an aircraft can maintain level flight in a standard atmosphere and under specified conditions.

absolute code   [COMPUT SCI] A code used when the addresses in a program are to be written in machine language exactly as they will appear when the instructions are executed by the control circuits.

absolute convergence   [MATH] That property of an infinite series (or infinite product) of real or complex numbers if the series (product) of absolute values converges; absolute convergence implies convergence.

absolute coordinate system   [NAV] The inertial coordinate system which has its origin on the axis of the earth and is fixed with respect to the stars. Also known as absolute reference frame.

absolute density   See absolute gravity.

absolute deviation   [ORD] The shortest distance between the center of the target and the point where a projectile hits or bursts. [STAT] The difference, without regard to sign, between a variate value and a given value.

absolute electrometer   [ELEC] A very precise type of attracted disk electrometer in which the attraction

between two disks is balanced against the force of gravity.

absolute error   [MATH] In an approximate number, the numerical difference between the number and a number considered exact. [ORD] 1. Shortest distance between the center of impact or the center of burst of a group of shots and the point of impact or burst of a single shot within the group. 2. Error of a sight consisting of its error in relation to a master service sight with which it is tested and of the known error of the master service sight.

absolute expansion   [THERMO] The true expansion of a liquid with temperature, as calculated when the expansion of the container in which the volume of the liquid is measured is taken into account; in contrast with apparent expansion.

absolute gravity   [CHEM] Density or specific gravity of a fluid reduced to standard conditions; for example, with gases, to 760 mmHg pressure and 0°C temperature. Also known as absolute density.

absolute humidity   [PHYS] The ratio of the mass of water vapor in a sample of air to the volume of the sample.

absolute inequality   See unconditional inequality.

absolute magnitude   [ASTROPHYS] 1. A measure of the brightness of a star equal to the magnitude the star would have at a distance of 10 parsecs from the observer. 2. The stellar magnitude any meteor would have if placed in the observer's zenith at a height of 100 kilometers. [MATH] The absolute value of a number or quantity.

absolute manometer   [ENG] 1. A gas manometer whose calibration, which is the same for all ideal gases, can be calculated from the measurable physical constants of the instrument. 2. A manometer that measures absolute pressure.

absolute motion   [NAV] Motion relative to a point fixed on the earth,s surface or to an apparently fixed celestial point. [PHYS] Motion of an object described by its measurement in a frame of reference that is preferred over all other frames.

absolute parallax   See absolute stereoscopic parallax.

absolute permeability   [ELECTROMAG] The ratio of the magnetic flux density to the intensity of the magnetic field in a medium; measurement is in webers per square meter in the meter-kilogram-second system. Also known as induced capacity.

absolute pitch   [ACOUS] The pitch of a musical tone expressed as the frequency of the sound wave of that tone.

absolute pressure   [PHYS] The pressure above the absolute zero value of pressure that theoretically obtains in empty space or at the absolute zero of temperature, as distinguished from gage pressure.

absolute pressure gage   [ENG] A device that measures the pressure exerted by a fluid relative to a perfect vacuum; used to measure pressures very close to a perfect vacuum.

absolute programming   [COMPUT SCI] Programming with the use of absolute code.

absolute reference frame   See absolute coordinate system.

absolute scale   See absolute temperature scale.

absolute space-time   [PHYS] A concept underlying Newtonian mechanics which postulates the existence of a preferred reference system of time and spatial coordinates; replaced in relativistic mechanics by Einstein's equivalency principle. Also known as absolute time.

absolute specific gravity   [MECH] The ratio of the weight of a given volume of a substance in a vacuum at a given temperature to the weight of an equal volume of water in a vacuum at a given temperature.

absolute stereoscopic parallax [GRAPHICS] Considering a pair of aerial photographs of equal principal distance, the absolute stereoscopic parallax of a point is the algebraic difference of the distances of the two images from their respective photograph nadirs, measured in a horizontal plane and parallel to the air base. Also known as absolute parallax; horizontal parallax; linear parallax; parallax; stereoscopic parallax; x-parallax.

absolute system of units [PHYS] A set of units for measuring physical quantities, defined by interrelated equations in terms of arbitrary fundamental quantities of length, mass, time, and charge or current.

absolute temperature [THERMO] 1. The temperature measurable in theory on the thermodynamic temperature scale. 2. The temperature in Celsius degrees relative to the absolute zero at $-273.16°C$ (the Kelvin scale) or in Fahrenheit degrees relative to the absolute zero at $-459.69°F$ (the Rankine scale).

absolute temperature scale [THERMO] A scale with which temperatures are measured relative to absolute zero. Also known as absolute scale.

absolute threshold [PHYSIO] The minimum stimulus energy that an organism can detect.

absolute time [GEOL] Geologic time measured in years, as determined by radioactive decay of elements. [PHYS] *See* absolute space-time.

absolute unit [PHYS] A unit defined in terms of units of fundamental quantities such as length, time, mass, and charge or current.

absolute-value computer [COMPUT SCI] A computer that processes the values of the variables rather than their increments.

absolute value of a complex number [MATH] The modulus of a complex number; the square root of the sum of the squares of its real and imaginary part. Also known as magnitude of a complex number; modulus of a complex number.

absolute value of a real number [MATH] The number if it is nonnegative, and the negative of the number if it is negative. Also known as magnitude of a real number; numerical value of a real number.

absolute value of a vector [MATH] The length of a vector, disregarding its direction; the square root of the sum of the squares of its orthogonal components. Also known as magnitude of a vector.

absolute velocity [PHYS] The vector sum of the velocity of a fluid parcel relative to the earth and the velocity of the parcel due to the earth's rotation; the east-west component is the only one affected.

absolute viscosity [FL MECH] The tangential force per unit area of two parallel planes at unit distance apart when the space between them is filled with a fluid and one plane moves with unit velocity in its own plane relative to the other.

absolute vorticity [FL MECH] The vorticity of a fluid relative to an absolute coordinate system; especially, the vorticity of the atmosphere relative to axes not rotating with the earth.

absolute zero [THERMO] The temperature of $-273.16°C$, or $-459.69°F$, or 0 K, thought to be the temperature at which molecular motion vanishes and a body would have no heat energy.

absorbance [PHYS CHEM] The common logarithm of the reciprocal of the transmittance of a pure solvent. Also known as absorbancy; extinction.

absorbancy *See* absorbance.

absorbed dose [MED] The part of an administered medication which is not excreted by the recipient's

body. [NUCLEO] The amount of energy imparted by ionizing particles to a unit mass of irradiated material at a place of interest. Also known as dosage; dose.

absorbed dose rate [NUCLEO] The absorbed dose of ionizing radiation imparted at a given location per unit of time (second, minute, hour, or day).

absorbency index *See* absorptivity.

absorber [CHEM ENG] Equipment in which a gas is absorbed by contact with a liquid. [ELECTR] A material or device that takes up and dissipates radiated energy; may be used to shield an object from the energy, prevent reflection of the energy, determine the nature of the radiation, or selectively transmit one or more components of the radiation. [MECH ENG] 1. A device which holds liquid for the absorption of refrigerant vapor or other vapors. 2. That part of the low-pressure side of an absorption system used for absorbing refrigerant vapor. [NUCLEO] A material that absorbs neutrons or other ionizing radiation.

absorber control *See* absorption control.

absorbing rod *See* control rod.

absorptance [PHYS] The ratio of the total unabsorbed radiation to the total incident radiation; equal to one (unity) minus the transmittance.

absorptiometer [ANALY CHEM] 1. An instrument equipped with a filter system or other simple dispersing system to measure the absorption of nearly monochromatic radiation in the visible range by a gas or a liquid, and so determine the concentration of the absorbing constituents in the gas or liquid. 2. A device for regulating the thickness of a liquid in spectrophotometry.

absorption [CHEM] The taking up of matter in bulk by other matter, as in dissolving of a gas by a liquid. [ELEC] The property of a dielectric in a capacitor which causes a small charging current to flow after the plates have been brought up to the final potential, and a small discharging current to flow after the plates have been short-circuited, allowed to stand for a few minutes, and short-circuited again. Also known as dielectric soak. [ELECTROMAG] The taking up of energy from radiation by the medium through which the radiation is passing. [HYD] The entrance of surface water into the lithosphere. [NUCLEO] The process by which the quantity of particles entering a body is reduced by their interaction with the matter. [PHYSIO] Passage of a chemical substance, a pathogen, or radiant energy through a body membrane.

absorption band [PHYS] A range of wavelengths or frequencies in the electromagnetic spectrum within which radiant energy is absorbed by a substance.

absorption coefficient Also known as absorption factor; absorption ratio; coefficient of absorption. [ACOUS] The ratio of the sound energy absorbed by a surface of a medium or material to the sound energy incident on the surface. [PHYS] If a flux through a material decreases with distance x in proportion to e^{-ax}, then a is called the absorption coefficient.

absorption column *See* absorption tower.

absorption constant *See* absorptivity.

absorption control [ELECTR] *See* absorption modulation. [NUCLEO] Control of a nuclear reactor by a material that absorbs neutrons, such as cadmium or boron steel. Also known as absorber control.

absorption cooling [ENG] Refrigeration in which cooling is effected by the expansion of liquid ammonia into gas and the absorption of the gas by water, the ammonia is reused after the water evaporates.

absorption cycle [MECH ENG] In refrigeration, the process whereby a circulating refrigerant, for example, ammonia, is evaporated by heat from an aqueous solution at elevated pressure and subsequently reabsorbed at low pressure, displacing the need for a compressor.

absorption edge [SPECT] The wavelength corresponding to a discontinuity in the variation of the absorption coefficient of a substance with the wavelength of the radiation. Also known as absorption limit.

absorption factor See absorption coefficient.

absorption field [CIV ENG] Trenches containing coarse aggregate about distribution pipes permitting septic-tank effluent to seep into surrounding soil. Also known as disposal field.

absorption index [OPTICS] The complex index of refraction may be written as $n(1+ik)$; the coefficient k is the absorption index. Also known as index of absorption.

absorption lens [OPTICS] Glass which prevents selected wavelengths from passing through it; used in eyeglasses.

absorption limit See absorption edge.

absorption line [SPECT] A minute range of wavelength or frequency in the electromagnetic spectrum within which radiant energy is absorbed by the medium through which it is passing.

absorption loss [CIV ENG] The quantity of water that is lost during the initial filling of a reservoir because of absorption by soil and rocks. [COMMUN] That part of the transmission loss due to the dissipation or conversion of either sound energy or electromagnetic energy into other forms of energy, either within the medium or attendant upon a reflection.

absorption meter [ENG] An instrument designed to measure the amount of light transmitted through a transparent substance, using a photocell or other light detector.

absorption modulation [ELECTR] A system of amplitude modulation in which a variable-impedance device is inserted in or coupled to the output circuit of the transmitter. Also known as absorption control; loss modulation.

absorption nebula See dark nebula.

absorption peak [SPECT] A wavelength of maximum electromagnetic absorption by a chemical sample; used to identify specific elements, radicals, or compounds.

absorption process [CHEM ENG] A method in which light oil is introduced into an absorption tower so that it absorbs the gasoline in the rising wet gas; the light oil is then distilled to separate the gasoline.

absorption ratio See absorption coefficient.

absorption refrigeration [MECH ENG] Refrigeration in which cooling is effected by the expansion of liquid ammonia into gas and absorption of the gas by water; the ammonia is reused after the water evaporates.

absorption spectroscopy [SPECT] The study of spectra obtained by the passage of radiant energy from a continuous source through a cooler, selectively absorbing medium.

absorption spectrum [SPECT] The array of absorption lines and absorption bands which results from the passage of radiant energy from a continuous source through a cooler, selectively absorbing medium.

absorption system [MECH ENG] A refrigeration system in which the refrigerant gas in the evaporator is taken up by an absorber and is then, with the application of heat, released in a generator.

absorption tower [ENG] A vertical tube in which a rising gas is partially absorbed by a liquid in the form of falling droplets. Also known as absorption column.

absorption tube [CHEM] A tube filled with a solid absorbent and used to absorb gases and vapors.

absorption unit See sabin.

absorptivity [ANALY CHEM] The constant a in the Beer's law relation $A = abc$, where A is the absorbance, b the path length, and c the concentration of solution. Also known as absorptive power. Formerly known as absorbency index; absorption constant; extinction coefficient.

ABS resin See acrylonitrile butadiene styrene resin.

abstract algebra [MATH] The study of mathematical systems consisting of a set of elements, one or more binary operations by which two elements may be combined to yield a third, and several rules (axioms) for the interaction of the elements and the operations; includes group theory, ring theory, and number theory.

abstriction [MYCOL] In fungi, the cutting off of spores in hyphae by formation of septa followed by abscission of the spores, especially by constriction.

abT See gauss.

abterminal [BIOL] Referring to movement from the end toward the middle; specifically, describing the mode of electric current flow in a muscle.

abtesla See gauss.

abundance See abundance ratio.

abundance ratio [NUCLEO] The ratio of the number of atoms of one isotope to the total number of atoms in a mixture of isotopes. Also known as abundance.

abutment [CIV ENG] A surface or mass provided to withstand thrust; for example, end supports of an arch or a bridge.

abvolt [ELEC] The unit of electromotive force in the electromagnetic centimeter-gram-second system; 1 abvolt equals 10^{-8} volt in the absolute meter-kilogram-second system. Abbreviated aV.

abwatt [ELEC] The unit of electrical power in the centimeter-gram-second system; 1 abwatt equals 1 watt in the absolute meter-kilogram-second system.

abWb See maxwell.

abweber See maxwell.

abyssal [GEOL] See plutonic. [OCEANOGR] Pertaining to the abyssal zone.

abyssal cave See submarine fan.

abyssal fan See submarine fan.

abyssal floor [GEOL] The ocean floor, or bottom of the abyssal zone.

abyssal plain [GEOL] A flat, almost level area occupying the deepest parts of many of the ocean basins.

abyssal zone [OCEANOGR] The biogeographic realm of the great depths of the ocean beyond the limits of the continental shelf, generally below 1000 meters.

ac See alternating current.

aC See abcoulomb.

Ac See actinium; altocumulus cloud.

Acadian orogeny [GEOL] The period of formation accompanied by igneous intrusion that took place during the Middle and Late Devonian in the Appalachian Mountains.

Acalyptratae [INV ZOO] A large group of small, two-winged flies in the suborder Cyclorrhapha characterized by small or rudimentary calypters.

acantha [BIOL] A sharp spine; a spiny process, as on vertebrae.

Acanthaceae [BOT] A family of dicotyledonous plants in the order Scrophulariales distinguished by their usually herbaceous habit, irregular flowers, axile placentation, and dry, dehiscent fruits.

acanthaceous [BOT] Having sharp points or prickles; prickly.

Acantharia [INV ZOO] A subclass of essentially pelagic protozoans in the class Actinopodea characterized by skeletal rods constructed of strontium sulfate (celestite).

acanthite [MINERAL] Ag_2S A blackish to lead-gray silver sulfide mineral, crystallizing in the orthorhombic system.

Acanthocephala [INV ZOO] The spiny-headed worms, a phylum of helminths; adults are parasitic in the alimentary canal of vertebrates.

acanthocladous [BOT] Having spiny branches.

Acanthodidae [PALEON] A family of extinct acanthodian fishes in the order Acanthodiformes.

Acanthodiformes [PALEON] An order of extinct fishes in the class Acanthodii having scales of acellular bone and dentine, one dorsal fin, and no teeth.

Acanthodii [PALEON] A class of extinct fusiform fishes, the first jaw-bearing vertebrates in the fossil record.

Acanthometrida [INV ZOO] An order of marine protozoans in the subclass Acantharia with 20 or less skeletal rods.

Acanthophractida [INV ZOO] An order of marine protozoans in the subclass Acantharia; skeleton includes a latticework shell and skeletal rods.

Acanthopteri [VERT ZOO] An equivalent name for the Perciformes.

Acanthopterygii [VERT ZOO] An equivalent name for the Perciformes.

Acanthosomatidae [INV ZOO] A small family of insects in the order Hemiptera.

acanthozooid [INV ZOO] A specialized individual in a bryozoan colony that secretes tubules which project as spines above the colony's outer surface.

Acanthuridae [VERT ZOO] The surgeonfishes, a family of perciform fishes in the suborder Acanthuroidei.

Acanthuroidei [VERT ZOO] A suborder of chiefly herbivorous fishes in the order Perciformes.

acapnia [MED] Absence of carbon dioxide in the blood and tissues.

Acari [INV ZOO] The equivalent name for Acarina.

acariasis [MED] Any skin disease resulting from infestation with acarids or mites.

acaricide [MATER] A pesticide used to destroy mites on domestic animals, crops, and man. Also known as miticide.

Acaridiae [INV ZOO] A group of pale, weakly sclerotized mites in the suborder Sarcoptiformes, including serious pests of stored food products and skin parasites of warm-blooded vertebrates.

Acarina [INV ZOO] The ticks and mites, a large order of the class Arachnida, characterized by lack of body demarcation into cephalothorax and abdomen.

acarology [INV ZOO] A branch of zoology dealing with the mites and ticks.

acarpellous [BOT] Lacking carpels.

acarpous [BOT] Not producing fruit.

ACAS See airborne collision avoidance system.

acaulous [BOT] 1. Lacking a stem. 2. Being apparently stemless but having a short underground stem.

accelerated life test [ENG] Operation of a device, circuit, or system above maximum ratings to produce premature failure; used to estimate normal operating life.

accelerated test [ELEC] A test of the serviceability of an electric cable in use for some time by applying twice the voltage normally carried.

accelerating agent [MATER] 1. A substance which increases the speed of a chemical reaction. 2. A compound which hastens and improves the curing of rubber.

accelerating electrode [ELECTR] An electrode used in cathode-ray tubes and other electron tubes to increase the velocity of the electrons that contribute the space current or form a beam.

acceleration [MECH] The rate of change of velocity with respect to time.

acceleration of free fall See acceleration of gravity.

acceleration of gravity [MECH] The acceleration imparted to bodies by the attractive force of the earth; has an international standard value of 980.665 cm/sec^2 but varies with latitude and elevation. Also known as acceleration of free fall; apparent gravity.

acceleration potential [FL MECH] The sum of the potential of the force field acting on a fluid and the ratio of the pressure to the fluid density; the negative of its gradient gives the acceleration of a point in the fluid.

acceleration time [COMPUT SCI] The time required for a magnetic tape transport or any other mechanical device to attain its operating speed.

accelerator [CHEM ENG] In the manufacture of rubber, a substance that acts as a catalyst of vulcanization but undergoes a chemical change in the process. [MATER] 1. Any substance added to stucco, plaster, mortar, concrete, cement, and so on to hasten the set. 2. In the vulcanization process, a substance, added with a curing agent, to speed processing and enhance physical characteristics of a vulcanized material. [MECH ENG] A device for varying the speed of an automotive vehicle by varying the supply of fuel. [PHYS] See particle accelerator.

accelerator jet [MECH ENG] The jet through which the fuel is injected into the incoming air in the carburetor of an automotive vehicle with rapid demand for increased power output.

accelerometer [ENG] An instrument which measures acceleration or gravitational force capable of imparting acceleration.

accentuation [ELECTR] The enhancement of signal amplitudes in selected frequency bands with respect to other signals.

accentuator [ELECTR] A circuit that provides for the first part of a process for increasing the strength of certain audio frequencies with respect to others, to help these frequencies override noise or to reduce distortion. Also known as accentuator circuit. [MATER] A material that acts to increase the selectivity or intensity of a stain.

accentuator circuit See accentuator.

accept [COMPUT SCI] A data transmission statement which is used in FORTRAN when the computer is in conversational mode, and which enables the programmer to input, through the teletypewriter, data the programmer wishes stored in memory.

acceptable quality level [IND ENG] The maximum percentage of defects that has been determined tolerable as a process average for a sampling plan during inspection or test of a product with respect to economic and functional requirements of the item. Abbreviated AQL.

acceptable reliability level [IND ENG] The required level of reliability for a part, system, device, and so forth; may be expressed in a variety of terms, for example, number of failures allowable in 1000 hours of operating life. Abbreviated ARL.

acceptor [CHEM] 1. A chemical whose reaction rate with another chemical increases because the other substance undergoes another reaction. **2.** A species that accepts electrons, protons, electron pairs, or molecules such as dyes. [SOLID STATE] An impurity element that increases the number of holes in a semiconductor crystal such as germanium or silicon; aluminum, gallium, and indium are examples. Also known as acceptor impurity; acceptor material.

acceptor atom [SOLID STATE] An atom of a substance added to a semiconductor crystal to increase the number of holes in the conduction band.

acceptor circuit [ELECTR] A series-resonant circuit that has a low impedance at the frequency to which it is tuned and a higher impedance at all other frequencies.

acceptor impurity *See* acceptor.

acceptor material *See* acceptor.

access [CIV ENG] Freedom, ability, or the legal right to pass without obstruction from a given point on earth to some other objective, such as the sea or a public highway. [COMPUT SCI] The reading of data from storage or the writing of data into storage.

access arm [COMPUT SCI] The mechanical device which positions the read/write head on a magnetic storage unit.

access code [COMMUN] **1.** Numeric identification for internetwork or facility switching. **2.** The preliminary digits that a user must dial to be connected through an automatic PBX to the serving switching center, as in AUTOVON.

access-control register [COMPUT SCI] A storage device which controls the word-by-word transmission over a given channel.

access-control words [COMPUT SCI] Permanently wired instructions channeling transmitted words into reserved locations.

access gap *See* memory gap.

access mode [COMPUT SCI] A programming clause in COBOL which is required when using a random-access device so that a specific record may be read out of or written into a mass storage bin.

accessory bud [BOT] An embryonic shoot occurring above or to the side of an axillary bud. Also known as supernumerary bud.

accessory cell [BOT] A morphologically distinct epidermal cell adjacent to, and apparently functionally associated with, guard cells on the leaves of many plants.

accessory chromosome *See* supernumerary chromosome.

accessory element *See* trace element.

accessory gland [ANAT] A mass of glandular tissue separate from the main body of a gland. [INV ZOO] A gland associated with the male reproductive organs in insects.

accessory mineral [MINERAL] A minor mineral in an igneous rock that does not affect its general character.

accessory nerve [ANAT] The eleventh cranial nerve in tetrapods, a paired visceral motor nerve; the bulbar part innervates the larynx and pharynx, and the spinal part innervates the trapezius and sternocleidomastoid muscles.

access time [COMPUT SCI] The time period required for reading out of or writing into the computer memory.

accidental inclusion *See* xenolith.

Accipitridae [VERT ZOO] The diurnal birds of prey, the largest and most diverse family of the order Falconiformes, including hawks, eagles, and kites.

acclimation *See* acclimatization.

acclimatization [EVOL] Adaptation of a species or population to a changed environment over several generations. Also known as acclimation.

acclivity [GEOL] A slope that is ascending from a reference point.

accolade [ARCH] Decorative molding in which two ogee curves meet centrally over the top of a window or door.

accommodation [ECOL] A population's location within a habitat. [MAP] The limits or range within which a stereo-plotting instrument is capable of operating. [PHYSIO] A process in most vertebrates whereby the focal length of the eye is changed by automatic adjustment of the lens to bring images of objects from various distances into focus on the retina.

accordant [GEOL] Pertaining to topographic features that have nearly the same elevation.

accordant fold [GEOL] One of several folds that are similarly oriented.

accretion [ASTRON] A process in which a star gathers molecules of interstellar gas to itself by gravitational attraction. [CIV ENG] Artificial buildup of land due to the construction of a groin, breakwater, dam, or beach fill. [GEOL] **1.** Gradual buildup of land on a shore due to wave action, tides, currents, airborne material, or alluvial deposits. **2.** The process whereby stones or other inorganic masses add to their bulk by adding particles to their surfaces. Also known as aggradation. [METEOROL] The growth of a precipitation particle by the collision of a frozen particle (ice crystal or snowflake) with a supercooled liquid droplet which freezes upon contact.

accumbent [BOT] Describing an organ that leans against another; specifically referring to cotyledons having their edges folded against the hypocotyl.

accumulation point *See* cluster point.

accumulative error *See* cumulative error.

accumulator [AERO ENG] A device sometimes incorporated in the fuel system of a gas-turbine engine to store fuel and release it under pressure as an aid in starting. [CHEM ENG] An auxiliary ram extruder on blow-molding equipment used to store melted material between deliveries. [COMPUT SCI] A specific register, in the arithmetic unit of a computer, in which the result of an arithmetic or logical operation is formed; here numbers are added or subtracted, and certain operations such as sensing, shifting, and complementing are performed. Also known as accumulator register; counter. [ELEC] *See* storage battery. [MECH ENG] **1.** A device, such as a bag containing pressurized gas, which acts upon hydraulic fluid in a vessel, discharging it rapidly to give high hydraulic power, after which the fluid is returned to the vessel with the use of low hydraulic power. **2.** A device connected to a steam boiler to enable a uniform boiler output to meet an irregular steam demand. **3.** A chamber for storing low-side liquid refrigerant in a refrigeration system. Also known as surge drum; surge header.

accumulator battery *See* storage battery.

accumulator jump instruction [COMPUT SCI] An instruction which programs a computer to ignore the previously established program sequence depending on the status of the accumulator. Also known as accumulator transfer instruction.

accumulator register *See* accumulator.

accumulator shift instruction [COMPUT SCI] A computer instruction which causes the word in a register to be displaced a specified number of bit positions to the left or right.

accumulator transfer instruction   *See* accumulator jump instruction.

ac/dc motor   *See* universal motor.

acellular   [BIOL] Not composed of cells.

acellular slime mold   [MYCOL] The common name for members of the Myxomycetes.

acentric   [BIOL] Not oriented around a middle point. [GEN] A chromosome or chromosome fragment lacking a centromere.

acentrous   [VERT ZOO] Lacking vertebral centra and having the notochord persistent throughout life, as in certain primitive fishes.

Acephalina   [INV ZOO] A suborder of invertebrate parasites in the protozoan order Eugregarinida characterized by nonseptate trophozoites.

acephalous   [BOT] Having the style originate at the base instead of at the apex of the ovary.   [ZOO] Lacking a head.

acerate   [BOT] Needle-shaped, specifically referring to leaves.

Acerentomidae   [INV ZOO] A family of wingless insects belonging to the order Protura; the body lacks tracheae and spiracles.

acervate   [BIOL] Growing in heaps or dense clusters.

acervulus   [MYCOL] A cushion- or disk-shaped mass of hyphae, peculiar to the Melanconiales, on which there are dense aggregates of conidiophores.

acetabulum   [ANAT] A cup-shaped socket on the hipbone that receives the head of the femur.   [INV ZOO] **1.** A cavity on an insect body into which a leg inserts for articulation. **2.** The sucker of certain invertebrates such as trematodes and tapeworms.

acetal resins   [ORG CHEM] Linear, synthetic resins produced by the polymerization of formaldehyde (acetal homopolymers) or of formaldehyde with trioxane (acetal copolymers); hard, tough plastics used as substitutes for metals. Also known as polyacetals.

acetate   [ORG CHEM] One of two species derived from acetic acid, CH_3COOH; one type is the acetate ion, CH_3COO^-; the second type is a compound whose structure contains the acetate ion, such as ethyl acetate.   [TEXT] The official name for the textile fiber produced from partially hydrolyzed cellulose acetate. Formerly known as acetate rayon.

acetate film   [MATER] A cellulose acetate resin sheet that is transparent, airproof, hygienic, and resistant to grease, oil, and dust; used for photographic film, magnetic tapes, and packaging.

acetate process   [CHEM ENG] Acetylation of cellulose (wood pulp or cotton linters) with acetic acid or acetic anhydride and sulfuric acid catalyst to make cellulose acetate resin or fiber.

acetolysis   [ORG CHEM] Decomposition of an organic molecule through the action of acetic acid or acetic anhydride.

acetone body   *See* ketone body.

acetonemia   [MED] A condition characterized by large amounts of acetone bodies in the blood. Also known as ketonemia.

acetone number   [CHEM] A ratio used to estimate the degree of polymerization of materials such as drying oils; it is the weight in grams of acetone added to 100 grams of a drying oil to cause an insoluble phase to form.

acetyl   [ORG CHEM] CH_3CO A two-carbon organic radical containing a methyl group and a carbonyl group.

acetylase   [BIOCHEM] Any enzyme that catalyzes the formation of acetyl esters.

acetylation   [ORG CHEM] The process of bonding an acetyl group onto an organic molecule.

acetylcholine   [BIOCHEM] $C_7H_{17}O_3N$ A compound released from certain autonomic nerve endings which acts in the transmission of nerve impulses to excitable membranes.

acetylcholinesterase   [BIOCHEM] An enzyme found in excitable membranes that inactivates acetylcholine.

acetyl coenzyme A   [BIOCHEM] $C_{23}H_{39}O_{17}N_7P_3S$ A coenzyme, derived principally from the metabolism of glucose and fatty acids, that takes part in many biological acetylation reactions; oxidized in the Krebs cycle.

acetylene series   [ORG CHEM] A series of unsaturated aliphatic hydrocarbons, each containing at least one triple bond and having the general formula C_nH_{2n-2}.

acetylene torch   *See* oxyacetylene torch.

acetylide   [ORG CHEM] A compound formed from acetylene with the H atoms replaced by metals, as in cuprous acetylide (Cu_2C_2).

acetyl number   [ANALY CHEM] A measure of free hydroxyl groups in fats or oils determined by the amount of potassium hydroxide required to neutralize the acetic acid formed by saponification of acetylated fat or oil.

acetyl phosphate   [BIOCHEM] $C_2H_5O_5P$ The anhydride of acetic and phosphoric acids occurring in the metabolism of pyruvic acid by some bacteria; phosphate is used by some microorganisms, in place of ATP, for the phosphorylation of hexose sugars.

Achaenodontidae   [PALEON] A family of Eocene dichobunoids, piglike mammals belonging to the suborder Palaeodonta.

achalasia   [MED] Inability of a hollow muscular organ or ring of muscle (sphincter) to relax.

achene   [BOT] A small, dry, indehiscent fruit formed from a simple ovary bearing a single seed.

Achilles   [ASTRON] An asteroid; member of the group known as the Trojan planets.

Achilles jerk   [PHYSIO] A reflex action seen as plantar flection in response to a blow to the Achilles tendon. Also known as Achilles tendon reflex.

Achilles tendon   [ANAT] The tendon formed by union of the tendons of the calf muscles, the soleus and gastrocnemius, and inserted into the heel bone.

Acholeplasmataceae   [MICROBIO] A family of the order Mycoplasmatales; characters same as for the order and class (Mollicutes); do not require sterol for growth.

achondrite   [GEOL] A stony meteorite that contains no chondrules.

achondroplasia   [MED] A hereditary deforming disease of the skeletal system, inherited in humans as an autosomal dominant trait and characterized by insufficient growth of the long bones, resulting in reduced length. Also known as chondrodystrophy fetalis.

achordate   [VERT ZOO] Lacking a notochord.

achromat   *See* achromatic lens.

Achromatiaceae   [MICROBIO] A family of gliding bacteria of uncertain affiliation; cells are spherical to ovoid or cylindrical, movements are slow and jerky, and microcysts are not known.

achromatic   [OPTICS] Capable of transmitting light without decomposing it into its constituent colors.

achromatic color   [OPTICS] A color that has no hue or saturation but only brightness, such as white, black, and various shades of gray.

achromatic condenser   [OPTICS] A condenser designed to eliminate chromatic and spherical aberrations, usually through the use of four elements, two of which are achromatic lenses; used in microscopes having high magnification.

achromatic lens [OPTICS] A combination of two or more lenses having a focal length that is the same for two quite different wavelengths, thereby removing a major portion of chromatic aberration. Also known as achromat.

achromatic prism [OPTICS] A prism consisting of two or more prisms with different refractive indices combined so that light passing through the device is deviated but not dispersed.

achromatin [CYTOL] The portion of the cell nucleus which does not stain easily with basic dyes.

achromic [BIOL] Colorless; lacking normal pigmentation.

acicular [SCI TECH] Needlelike; slender and pointed.

acid [CHEM] **1.** Any of a class of chemical compounds whose aqueous solutions turn blue litmus paper red, react with and dissolve certain metals to form salts, and react with bases to form salts. **2.** A compound capable of transferring a hydrogen ion in solution. **3.** A substance that ionizes in solution to yield the positive ion of the solvent. **4.** A molecule or ion that combines with another molecule or ion by forming a covalent bond with two electrons from the other species.

acid alcohol [ORG CHEM] A compound containing both a carboxyl group (—COOH) and an alcohol group (—CH$_2$OH, =CHOH, or ≡COH).

acid anhydride [CHEM] An acid with one or more molecules of water removed; for example, SO$_3$ is the acid anhydride of H$_2$SO$_4$, sulfuric acid.

acid-base indicator [ANALY CHEM] A substance that reveals, through characteristic color changes, the degree of acidity or basicity of solutions.

acid-base pair [CHEM] A concept in the Brönsted theory of acids and bases; the pair consists of the source of the proton (acid) and the base generated by the transfer of the proton.

acid-base titration [ANALY CHEM] A titration in which an acid of known concentration is added to a solution of base of unknown concentration, or the converse.

acid blowcase See blowcase.

acid brittleness [MET] Low ductility of a metal due to its absorption of hydrogen gas, which may occur during an electrolytic process or during cleaning. Also known as hydrogen embrittlement.

acid bronze [MET] A copper-tin alloy containing lead and nickel; used in pumping equipment.

acid dye [ORG CHEM] Any of a group of sodium salts of sulfonic and carboxylic acids used to dye natural and synthetic fibers, leather, and paper.

acid egg See blowcase.

acid-fast bacteria [MICROBIO] Bacteria, especially mycobacteria, that stain with basic dyes and fluorochromes and resist decoloration by acid solutions.

acidic [CHEM] **1.** Pertaining to an acid or to its properties. **2.** Forming an acid during a chemical process.

acidification [CHEM] Addition of an acid to a solution until the pH falls below 7.

acidimeter [ANALY CHEM] An apparatus or a standard solution used to determine the amount of acid in a sample.

acidizing [PETRO ENG] Well-stimulation method to increase oil production by injecting hydrochloric acid into the oil-bearing formation; the acid dissolves rock to enlarge the porous passages through which the oil must flow.

acid lead [METAL] A 99.9% pure commercial lead made by adding copper to fully refined lead.

acid number [CHEM] See acid value. [ENG] A number derived from a standard test indicating the acid or base composition of lubricating oils; it in no way indicates the corrosive attack of the used oil in service. Also known as corrosion number.

acidolysis [ORG CHEM] A chemical reaction involving the decomposition of a molecule, with the addition of the elements of an acid to the molecule; the reaction is comparable to hydrolysis or alcoholysis, in which water or alcohol, respectively, is used in place of the acid. Also known as acyl exchange.

acidophil [BIOL] **1.** Any substance, tissue, or organism having an affinity for acid stains. **2.** An organism having a preference for an acid environment. [HISTOL] **1.** An alpha cell of the adenohypophysis. **2.** See eosinophil.

acidosis [MED] A condition of decreased alkali reserve of the blood and other body fluids.

acid phosphatase [BIOCHEM] An enzyme in blood which catalyzes the release of phosphate from phosphate esters; optimum activity at pH 5.

acid precipitation [METEOROL] Rain or snow with a pH of less than 5.6.

acid rain [METEOROL] Low-ph rainfall resulting from atmospheric reactions of aerosols containing chlorides and sulfates (or other negative ions).

acid slag [MET] Furnace slag in which there is more silica and silicates than lime and magnesia.

acid sludge [CHEM ENG] The residue left after treating petroleum oil with sulfuric acid for the removal of impurities.

acid soil [GEOL] A soil with pH less than 7; results from presence of exchangeable hydrogen and aluminum ions.

acid treating [CHEM ENG] A refining process in which unfinished petroleum products, such as gasoline, kerosine, and diesel oil, are contacted with sulfuric acid to improve their color, odor, and other properties.

acidulant [FOOD ENG] One of a class of chemicals added to food to increase either tartness or acidity, such as malic or citric acids for tartness and phosphoric acid for acidity.

acid value [CHEM] The acidity of a solution expressed in terms of normality. Also known as acid number.

aciniform [ZOO] Shaped like a berry or a bunch of grapes.

acinous gland [ANAT] A multicellular gland with sac-shaped secreting units. Also known as alveolar gland.

acinus [ANAT] The small terminal sac of an acinous gland, lined with secreting cells. [BOT] An individual drupelet of a multiple fruit.

Acipenseridae [VERT ZOO] The sturgeons, a family of actinopterygian fishes in the order Acipenseriformes.

Acipenseriformes [VERT ZOO] An order of the subclass Actinopterygii represented by the sturgeons and paddlefishes.

Acmaeidae [INV ZOO] A family of gastropod mollusks in the order Archaeogastropoda; includes many limpets.

acme screw thread [DES ENG] A standard thread having a profile angle of 29° and a flat crest; used on power screws in such devices as automobile jacks, presses, and lead screws on lathes. Also known as acme thread.

acme thread See acme screw thread.

acmite [MINERAL] NaFeSi$_2$O$_6$ A brown or green silicate mineral of the pyroxene group, often in long, pointed prismatic crystals; hardness is 6–6.5 on Mohs scale, and specific gravity is 3.50–3.55; found in igneous and metamorphic rocks.

acne [MED] A pleomorphic, inflammatory skin disease involving sebaceous follicles of the face, back, and chest and characterized by blackheads, whiteheads, papules, pustules, and nodules.

acne rosacea [MED] A form of acne occurring in older persons and seen as reddened inflamed areas on the forehead, nose, and cheeks.

Acnidosporidia [INV ZOO] An equivalent name for the Haplosporea.

Acoela [INV ZOO] An order of marine flatworms in the class Turbellaria characterized by the lack of a digestive tract and coelomic cavity.

Acoelea [INV ZOO] An order of gastropod mollusks in the subclass Opistobranchia; includes many sea slugs.

Acoelomata [INV ZOO] A subdivision of the animal kingdom; individuals are characterized by lack of a true body cavity.

acoelous [ZOO] **1.** Lacking a true body cavity or coelom. **2.** Lacking a true stomach or digestive tract.

Aconchulinida [INV ZOO] An order of protozoans in the subclass Filosia comprising a small group of naked amebas having filopodia.

aconitase [BIOCHEM] An enzyme involved in the Krebs citric acid cycle that catalyzes the breakdown of citric acid to *cis*-aconitic and isocitric acids.

aconitine [PHARM] $C_{34}H_{47}O_{11}N$ A poisonous, white, crystalline alkaloid compound obtained from aconites such as monkshood (*Aconitum napellus*).

acorn [BOT] The nut of the oak tree, usually surrounded at the base by a woody involucre.

acorn barnacle [INV ZOO] Any of the sessile barnacles that are enclosed in conical, flat-bottomed shells and attach to ships and near-shore rocks and piles.

acorn worm [INV ZOO] Any member of the class Enteropneusta, free-living animals that usually burrow in sand or mud. Also known as tongue worm.

acouchi [VERT ZOO] A hystricomorph rodent represented by two species in the family Dasyproctidae; believed to be a dwarf variety of the agouti.

acoustic [ACOUS] Relating to, containing, producing, arising from, actuated by, or carrying sound.

acoustic absorption coefficient *See* sound absorption coefficient.

acoustic absorptivity *See* sound absorption coefficient.

acoustical Doppler effect [ACOUS] The change in pitch of a sound observed when there is relative motion between source and observer.

acoustical holography [PHYS] A technique for using sound to form visible images, in which acoustic beams form an interference pattern of an object and a beam of light interacts with this pattern and is focused to form an optical image.

acoustical plaster [MATER] A low-density sound-absorbing plaster applied as a finish coat to provide a uniform finished surface.

acoustical stiffness [ACOUS] The acoustic reactance associated with the potential energy of a medium multiplied by 2π times the sound frequency.

acoustical tile [MATER] A sound-absorptive material, usually having unit dimensions of 24×24 inches (approximately 61×61 centimeters) or less, used to cover an acoustical ceiling.

acoustic array [ENG ACOUS] A sound-transmitting or -receiving system whose elements are arranged to give desired directional characteristics.

acoustic coupler [ENG ACOUS] A device used between the modem of a computer terminal and a standard telephone line to permit transmission of digital data in either direction without making direct connections.

acoustic delay [ENG ACOUS] A delay which is deliberately introduced in sound reproduction by having the sound travel a certain distance along a pipe before conversion into electric signals.

acoustic emission [ACOUS] The phenomenon of transient elastic-wave generation due to a rapid release of strain energy caused by a structural alteration in a solid material. Also known as stress-wave emission.

acoustic energy *See* sound energy.

acoustic feedback [ENG ACOUS] The reverberation of 2ound waves from a loudspeaker to a preceding part of an audio system, such as to the microphone, in such a manner as to reinforce, and distort, the original input. Also known as acoustic regeneration.

acoustic filter *See* filter.

acoustic hologram [ENG] The phase interference pattern, formed by acoustic beams, that is used in acoustical holography; when light is made to interact with this pattern, it forms an image of an object placed in one of the beams.

acoustic horn *See* horn.

acoustic image [ACOUS] The geometric space figure that is made up of the acoustic foci of an acoustic lens, mirror, or other acoustic optical system and is the acoustic counterpart of an extended source of sound. Also known as image.

acoustic imaging [ACOUS] The production of real-time images of the internal structure of a metallic or biological object that is opaque to light. Also known as ultrasonic imaging.

acoustic impedance [ACOUS] The complex ratio of the sound pressure on a given surface to the sound flux through that surface, expressed in acoustic ohms.

acoustic lens [MATER] Selected materials shaped to refract sound waves in accordance with the principles of geometrical optics, as is done for light. Also known as lens.

acoustic levitation [ACOUS] The use of a very intense sound wave to keep a body suspended above the device producing the sound wave.

acoustic microscope [OPTICS] An instrument which employs acoustic radiation at microwave frequencies to allow visualization of the microscopic detail exhibited in elastic properties of an object.

acoustic mode [SOLID STATE] The type of crystal lattice vibrations which for long wavelengths act like an acoustic wave in a continuous medium, but which for shorter wavelengths approach the Debye frequency, showing a dispersive decrease in phase velocity.

acoustic nerve *See* auditory nerve.

acoustic radiometer [ENG] An instrument for measuring sound intensity by determining the unidirectional steady-state pressure caused by the reflection or absorption of a sound wave at a boundary.

acoustic ratio [ENG ACOUS] The ratio of the intensity of sound radiated directly from a source to the intensity of sound reverberating from the walls of an enclosure, at a given point in the enclosure.

acoustic receiver [ELECTR] The complete equipment required for receiving modulated radio waves and converting them into sound.

acoustic reflection coefficient *See* acoustic reflectivity.

acoustic reflectivity [ACOUS] Ratio of the rate of flow of sound energy reflected from a surface, on the side of incidence, to the incident rate of flow. Also known as acoustic reflection coefficient; sound reflection coefficient.

acoustic reflex [PHYSIO] Brief, involuntary closure of the eyes due to stimulation of the acoustic nerve by a sudden sound.

acoustic regeneration *See* acoustic feedback.

acoustic resistance [ACOUS] The real component of the acoustic impedance.

acoustics [PHYS] **1.** The science of the production, transmission, and effects of sound. **2.** The characteristics of a room that determine the qualities of sound in it relevant to hearing.

acoustic stiffness [ACOUS] The product of the angular frequency and the acoustic stiffness reactance.

acoustic strain gage [ENG] An instrument used for measuring structural strains; consists of a length of fine wire mounted so its tension varies with strain; the wire is plucked with an electromagnetic device, and the resulting frequency of vibration is measured to determine the amount of strain.

acoustic transducer [ENG ACOUS] A device that converts acoustic energy to electrical or mechanical energy, such as a microphone or phonograph pickup.

acoustic transformer [ENG ACOUS] A device, such as a horn or megaphone, for increasing the efficiency of sound radiation.

acoustic transmission coefficient *See* sound transmission coefficient.

acoustic transmissivity *See* sound transmission coefficient.

acoustic transponder [NAV] A device used in underwater navigation which, on being interrogated by coded acoustic signals, emits acoustic reply.

acoustic wave [ACOUS] **1.** An elastic nonelectromagnetic wave that has a frequency which may extend into the gigahertz range; one type is a surface acoustic wave, and the other type is a bulk or volume acoustic wave. Also known as elastic wave. **2.** *See* sound.

acoustic well logging [ENG] A ground exploration method that uses a high-energy sound source and a receiver, both underground.

acoustoelectronics [ENG ACOUS] The branch of electronics that involves use of acoustic waves at microwave frequencies (above 500 megahertz), traveling on or in piezoelectric or other solid substrates. Also known as pretersonics.

acoustooptics [OPTICS] The science that deals with interactions between acoustic waves and light.

acquire [ELECTR] **1.** Of acquisition radars, the process of detecting the presence and location of a target in sufficient detail to permit identification. **2.** Of tracking radars, the process of positioning a radar beam so that a target is in that beam to permit the effective employment of weapons. Also known as target acquisition.

acquired immunity [IMMUNOL] Resistance to a microbial or other antigenic substance taken on by a naturally susceptible individual; may be either active or passive.

acrania [MED] Partial or complete absence of the cranium at birth.

Acrania [ZOO] A group of lower chordates with no cranium, jaws, vertebrae, or paired appendages; includes the Tunicata and Cephalochordata.

Acrasiales [BIOL] A group of microorganisms that have plant and animal characteristics; included in the phylum Myxomycophyta by botanists and Mycetozoia by zoologists.

Acrasida [MYCOL] An order of Mycetozoia containing cellular slime molds.

Acrasieae [BIOL] An equivalent name for the Acrasiales.

acrasin [BIOCHEM] The chemotactic substance thought to be secreted by, and to effect aggregation of, myxamebas during their fruiting phase.

acre [MECH] A unit of area, equal to 43,560 square feet, or to 4046.8564224 square meters.

acridine [ORG CHEM] $(C_6H_4)_2NCH$ A typical member of a group of organic heterocyclic compounds containing benzene rings fused to the 2,3 and 5,6 positions of pyridine; derivatives include dyes and medicines.

acridine orange [ORG CHEM] A dye with an affinity for nucleic acids; the complexes of nucleic acid and dye fluoresce orange with RNA and green with DNA when observed in the fluorescence microscope.

acriflavine [ORG CHEM] $C_{14}H_{14}N_3Cl$ A yellow acridine dye obtained from proflavine by methylation in the form of red crystals; used as an antiseptic in solution.

acroblast [CYTOL] A vesicular structure in the spermatid formed from Golgi material.

acrodont [ANAT] Having teeth fused to the edge of the supporting bone.

acrodynia [MED] A childhood syndrome associated with mercury ingestion and characterized by periods of irritability alternating with apathy, anorexia, pink itching hands and feet, photophobia, sweating, tachycardia, hypertension, and hypotonia.

acrolein test [ANALY CHEM] A test for the presence of glycerin or fats; a sample is heated with potassium bisulfate, and acrolein is released if the test is positive.

acromegaly [MED] A chronic condition in adults caused by hypersecretion of the growth hormone and marked by enlarged jaws, extremities, and viscera, accompanied by certain physiological changes.

acromere [HISTOL] The distal portion of a rod or cone in the retina.

acromion [ANAT] The flat process on the outer end of the scapular spine that articulates with the clavicle and forms the outer angle of the shoulder.

acrophobia [PSYCH] Abnormal fear of great heights.

Acrosaleniidae [PALEON] A family of Jurassic and Cretaceous echinoderms in the order Salenoida.

acrosin [BIOCHEM] A proteolytic enzyme located in the acrosome of a spermatozoon; thought to be involved in penetration of the egg.

acrosome [CYTOL] The anterior, crescent-shaped body of spermatozoon, formed from Golgi material of the spermatid. Also known as perforatorium.

acroterion [ARCH] **1.** A pedestal on a pediment to support an ornamental, such as a statue. **2.** An ornamental placed on such a pedestal.

Acrothoracica [INV ZOO] A small order of burrowing barnacles in the subclass Cirripedia that inhabit corals and the shells of mollusks and barnacles.

Acrotretacea [PALEON] A family of Cambrian and Ordovician inarticulate brachiopods of the suborder Acrotretidina.

Acrotretida [INV ZOO] An order of brachiopods in the class Inarticulata; representatives are known from Lower Cambrian to the present.

Acrotretidina [INV ZOO] A suborder of inarticulate brachiopods of the order Acrotretida; includes only species with shells composed of calcium phosphate.

acrylate [ORG CHEM] **1.** A salt or ester of acrylic acid. **2.** *See* acrylate resin.

acrylate resin [ORG CHEM] Acrylic acid or ester polymer with a $—CH_2—CH(COOR)—$ structure; used in paints, sizings and finishes for paper and textiles, adhesives, and plastics. Also known as acrylate.

acrylic fiber [TEXT] Any of numerous synthetic textile fibers made by polymerization of acrylonitrile.

acrylic resin [ORG CHEM] A thermoplastic synthetic organic polymer made by the polymerization of acrylic derivatives such as acrylic acid, methacrylic acid, ethyl acrylate, and methyl acrylate; used for adhesives, protective coatings, and finishes.

acrylic rubber [ORG CHEM] Synthetic rubber containing acrylonitrile; for example, nitrile rubber.

acrylonitrile [ORG CHEM] CH_2CHCN A colorless liquid compound used in the manufacture of acrylic rubber and fibers. Also known as vinylcyanide.

acrylonitrile butadiene styrene resin [ORG CHEM] A polymer made by blending acrylonitrile-styrene copolymer with a butadiene-acrylonitrile rubber or by interpolymerizing polybutadiene with styrene and acrylonitrile; combines the advantages of hardness and strength of the vinyl resin component with the toughness and impact resistance of the rubbery component. Abbreviated ABS resin.

Actaeonidae [INV ZOO] A family of gastropod mollusks in the order Tectibranchia.

Actaletidae [INV ZOO] A family of insects belonging to the order Collembola characterized by simple tracheal systems.

ACTH *See* adrenocorticotropic hormone.

actin [BIOCHEM] A muscle protein that is the chief constituent of the Z-band myofilaments of each sarcomere.

Actiniaria [INV ZOO] The sea anemones, an order of coelenterates in the subclass Zoantharia.

actinic [PHYS] Pertaining to electromagnetic radiation capable of initiating photochemical reactions, as in photography or the fading of pigments.

actinide series [CHEM] The group of elements of atomic number 89 through 103. Also known as actinoid elements.

actinium [CHEM] A radioactive element, symbol Ac, atomic number 89; its longest-lived isotope is Ac^{227} with a half-life of 21.7 years; the element is trivalent; chief use is, in equilibrium with its decay products, as a source of alpha rays.

actinocarpous [BOT] Having flowers and fruit radiating from one point.

Actinochitinosi [INV ZOO] A group name for two closely related suborders of mites, the Trombidiformes and the Sarcoptiformes.

actinolite [MINERAL] $Ca_2(Mg,Fe)_5Si_8O_{22}(OH)_2$ A green, monoclinic rock-forming amphibole; a variety of asbestos occurring in needlelike crystals and in fibrous or columnar forms; specific gravity 3–3.2.

actinometry [ASTROPHYS] The science of measurement of radiant energy, particularly that of the sun, in its thermal, chemical, and luminous aspects.

actinomorphic [BIOL] Descriptive of an organism, organ, or part that is radially symmetrical.

Actinomycetaceae [MICROBIO] A family of bacteria in the order Actinomycetales; gram-positive, diphtheroid cells which form filaments but not mycelia; chemoorganotrophs that ferment carbohydrates.

Actinomycetales [MICROBIO] An order of bacteria; cells form branching filaments which develop into mycelia in some families.

actinomycosis [MED] An infectious bacterial disease caused by *Actinomyces bovis* in cattle, hogs, and occasionally in man. Also known as lumpy jaw.

Actinomyxida [INV ZOO] An order of protozoan invertebrate parasites of the class Myxosporidea characterized by trivalved spores with three polar capsules.

actinophage [MICROBIO] A bacteriophage that infects and lyses members of the order Actinomycetales.

Actinophryida [INV ZOO] An order of protozoans in the subclass Heliozoia; individuals lack an organized test, a centroplast, and a capsule.

Actinoplanaceae [MICROBIO] A family of bacteria in the order Actinomycetales with well-developed mycelia and spores formed on sporangia.

Actinopodea [INV ZOO] A class of protozoans belonging to the superclass Sarcodina; most are free-floating, with highly specialized pseudopodia.

Actinopteri [VERT ZOO] An equivalent name for the Actinopterygii.

Actinopterygii [VERT ZOO] The ray-fin fishes, a subclass of the Osteichthyes distinguished by the structure of the paired fins, which are supported by dermal rays.

actinostele [BOT] A protostele characterized by xylem that is either star-shaped in cross section or has ribs radiating from the center.

actinotherapy *See* radiation therapy.

actinouranium [NUC PHYS] A naturally occurring radioactive isotope of the actinium series, emitting only alpha decay; symbol AcU; atomic number 92; mass number 235; half-life 7.1×10^8 years; isotopic symbol U^{235}.

actinula [INV ZOO] A larval stage of some hydrozoans that has tentacles and a mouth; attaches and develops into a hydroid in some species, or metamorphoses into a medusa.

action [MECH] An integral associated with the trajectory of a system in configuration space, equal to the sum of the integrals of the generalized momenta of the system over their canonically conjugate coordinates. Also known as phase integral. [ORD] The mechanism of a gun, usually breechloading, by which it is loaded, fired, and unloaded.

action current [PHYSIO] The electric current accompanying membrane depolarization and repolarization in an excitable cell.

action integral *See* action variable.

action potential [PHYSIO] A transient change in electric potential at the surface of a nerve or muscle cell occurring at the moment of stimulation.

action spectrum [PHYSIO] Graphic representation of the comparative effects of different wavelengths of light on living systems or their components.

action variable [PHYS] The integral $\int p\,dq$ over a cycle of a dynamical system; q is some coordinate, and p the conjugate momentum. Also known as action integral.

activate [ELEC] To make a cell or battery operative by addition of a liquid. [ELECTR] To treat the filament, cathode, or target of a vacuum tube to increase electron emission. [ENG] To set up conditions so that the object will function as designed or required. [NUCLEO] To induce radioactivity through bombardment by neutrons or by other types of radiation. [ORD] **1.** To bring into existence by official order a unit, post, camp, station, base, or shore activity which has previously been constituted and designated by name or number, or both, so that it can be organized to function in its assigned capacity. **2.** To prepare for active service a naval ship or craft which has been in an inactive or reserve status. [PHYS] To start activity or motion in a device or material.

activated alumina [MATER] Highly porous, granular aluminum oxide that preferentially absorbs liquids from gases and vapors, and moisture from some liquids; also used as a catalyst or catalyst carrier, as an absorbent to remove fluorides from drinking water, and in chromatography.

activated carbon [MATER] A powdered, granular, or pelleted form of amorphous carbon characterized by very large surface area per unit volume because of an enormous number of fine pores. Also known as activated charcoal.

activated cathode [ELECTR] A thermionic cathode consisting of a tungsten filament to which thorium has been added, and then brought to the surface, by a process such as heating in the absence of an electric field in order to increase thermionic emission.

activated charcoal See activated carbon.

activated sludge [CIV ENG] A semiliquid mass removed from the liquid flow of sewage and subjected to aeration and aerobic microbial action; the end product is dark to golden brown, partially decomposed, granular, and flocculent, and has an earthy odor when fresh.

activation [CHEM] Treatment of a substance by heat, radiation, or activating reagent to produce a more complete or rapid chemical or physical change. [ELEC] The process of adding liquid to a manufactured cell or battery to make it operative. [ELECTR] The process of treating the cathode or target of an electron tube to increase its emission. Also known as sensitization. [MET] 1. A process of facilitating the separation and collection of ore powders by the use of substances which change the response of the particle surfaces to a flotation fluid. 2. A process that increases the rate of pressing and heating a metal powder into cohesion. [MOL BIO] A change that is induced in an amino acid before it is utilized for protein synthesis. [NUCLEO] The process of inducing radioactivity by bombardment with neutrons or with other types of radiation. [PHYSIO] The designation for all changes in the ovum during fertilization, from sperm contact to the dissolution of nuclear membranes.

activation analysis [NUCLEO] A method of chemical analysis based on the detection of characteristic radionuclides following a nuclear bombardment. Also known as radioactivity analysis.

activation energy [PHYS CHEM] The energy, in excess over the ground state, which must be added to an atomic or molecular system to allow a particular process to take place.

activation record [COMPUT SCI] A variable part of a program module, such as data and control information, that may vary with different instances of execution.

activator [CHEM] 1. A substance that increases the effectiveness of a rubber vulcanization accelerator; for example, zinc oxide or litharge. 2. A trace quantity of a substance that imparts luminescence to crystals; for example, silver or copper in zinc sulfide or cadmium sulfide pigments.

active center [ASTRON] A localized, transient region of the solar atmosphere in which sunspots, faculae, plages, prominences, solar flares, and so forth are observed.

active component [ELEC] In the phasor representation of quantities in an alternating-current circuit, the component of current, voltage, or apparent power which contributes power, namely, the active current, active voltage, or active power. Also known as power component. [ELECTR] See active element.

active computer [COMPUT SCI] When two or more computers are installed, the one that is on-line and processing data.

active current [ELEC] The component of an electric current in a branch of an alternating-current circuit that is in phase with the voltage. Also known as watt current.

active element [ELECTR] Any generator of voltage or current in an impedance network. Also known as active component. [NUC PHYS] A chemical element which has one or more radioactive isotopes.

active filter [ELECTR] A filter that uses an amplifier with conventional passive filter elements to provide a desired fixed or tunable pass or rejection characteristic.

active front [METEOROL] A front, or portion thereof, which produces appreciable cloudiness and, usually, precipitation.

active galaxy [ASTRON] A galaxy whose central region exhibits strong emission activity, from radio to x-ray frequencies, probably as a result of gravitational collapse; this category includes M82 galaxies, Seyfert galaxies, N galaxies, and possibly quasars.

active immunity [IMMUNOL] Disease resistance in an individual due to antibody production after exposure to a microbial antigen following disease, inapparent infection, or inoculation.

active jamming See jamming.

active layer [GEOL] That part of the soil which is within the suprapermafrost layer and which usually freezes in winter and thaws in summer. Also known as frost zone.

active leg [ELECTR] An electrical element within a transducer which changes its electrical characteristics as a function of the application of a stimulus.

active location system [NAV] A navigation system in which the navigation satellite interrogates the craft, and the craft responds; useful for surveillance by a ground station, or for automated navigation if the satellite subsequently transmits data.

active logic [ELECTR] Logic that incorporates active components which provide such functions as level restoration, pulse shaping, pulse inversion, and power gain.

active material [ELEC] 1. A fluorescent material used in screens for cathode-ray tubes. 2. An energy-storing material, such as lead oxide, used in the plates of a storage battery. 3. A material, such as the iron of a core or the copper of a winding, that is involved in energy conversion in a circuit. [ELECTR] The material of the cathode of an electron tube that emits electrons when heated. [NUCLEO] A material capable of releasing substantial quantities of nuclear energy during fission.

active region [ASTRON] A localized, transient, non-uniform region on the sun's surface, penetrating well down into the lower chromosphere. [ELECTR] The region in which amplifying, rectifying, light emitting, or other dynamic action occurs in a semiconductor device.

active site [MOL BIO] The region of an enzyme molecule at which binding with the substrate occurs. Also known as binding site; catalytic site.

active sludge [CIV ENG] A sludge rich in destructive bacteria used to break down raw sewage.

active sonar [ENG] A system consisting of one or more transducers to send and receive sound, equipment for the generation and detection of the electrical impulses to and from the transducer, and a display or recorder system for the observation of the received signals.

active substrate [SOLID STATE] A semiconductor or ferrite material in which active elements are formed; also a mechanical support for the other elements of a semiconductor device or integrated circuit.

active system [ENG] In radio and radar, a system that requires transmitting equipment, such as a beacon or transponder.

active transducer [ELECTR] A transducer whose output is dependent upon sources of power, apart from that supplied by any of the actuating signals, which power is controlled by one or more of these signals.

active transport [PHYSIO] The pumping of ions or other substances across a cell membrane against an osmotic gradient, that is, from a lower to a higher concentration.

activity [COMPUT SCI] The use or modification of information contained in a file. [NUC PHYS] The intensity of a radioactive source. Also known as radioactivity. [PHYS CHEM] A thermodynamic function that correlates changes in the chemical potential with changes in experimentally measurable quantities, such as concentrations or partial pressures, through relations formally equivalent to those for ideal systems. [SYS ENG] The representation in a PERT or critical-path-method network of a task that takes up both time and resources and whose performance is necessary for the system to move from one event to the next.

activity coefficient [PHYS CHEM] A characteristic of a quantity expressing the deviation of a solution from ideal thermodynamic behavior; often used in connection with electrolytes.

activity level [COMPUT SCI] 1. The value assumed by a structural variable during the solution of a programming problem. 2. A measure of the number of times that use or modification is made of the information contained in a file.

activity ratio [COMPUT SCI] The ratio between used or modified records and the total number of records in a file. [GEOL] The ratio of plasticity index to percentage of clay-sized minerals in sediment.

actomyosin [BIOCHEM] A protein complex consisting of myosin and actin; the major constituent of a contracting muscle fibril.

actual age See absolute age.

actual instruction See effective instruction.

actual key [COMPUT SCI] A data item in COBOL computer language which can be used as an address.

actual power [MECH ENG] The power delivered at the output shaft of a source of power. Also known as actual horsepower.

actual pressure [METEOROL] The atmospheric pressure at the level of the barometer (elevation of ivory point), as obtained from the observed reading after applying the necessary corrections for temperature, gravity, and instrumental errors.

actual relative movement See slip.

actuator [CONT SYS] A mechanism to activate process control equipment by use of pneumatic, hydraulic, or electronic signals; for example, a valve actuator for opening or closing a valve to control the rate of fluid flow. [ENG ACOUS] An auxiliary external electrode used to apply a known electrostatic force to the diaphragm of a microphone for calibration purposes. Also known as electrostatic actuator. [MECH ENG] A device that produces mechanical force by means of pressurized fluid. [ORD] Part of the receiver mechanism in certain types of automatic weapons.

acuate [BIOL] 1. Having a sharp point. 2. Needle-shaped.

Aculeata [INV ZOO] A group of seven superfamilies that constitute the stinging forms of hymenopteran insects in the suborder Apocrita.

Aculognathidae [INV ZOO] The ant-sucking beetles, a family of coleopteran insects in the superfamily Cucujoidea.

acuminate [BOT] Tapered to a slender point, especially referring to leaves.

acupuncture [MED] The ancient Chinese art of puncturing the body with long, fine gold or silver needles to relieve pain and cure disease.

acute [BIOL] Ending in a sharp point. [MED] Referring to a disease or disorder of rapid onset, short duration, and pronounced symptoms.

acute angle [MATH] An angle of less than 90°.

acute benign lymphoblastosis See infectious mononucleosis.

acute triangle [MATH] A triangle each of whose angles is less than 90°.

acutifoliate [BOT] Having sharply pointed leaves.

acyclic [BOT] Having flowers arranged in a spiral instead of a whorl. [MATH] 1. A transformation on a set to itself for which no nonzero power leaves an element fixed. 2. A chain complex all of whose homology groups are trivial. [PHYS] Continually varying without a regularly repeated pattern.

acyclic compound [ORG CHEM] A chemical compound with an open-chain molecular structure rather than a ring-shaped structure; for example, the alkane series.

acyclic feeding [COMPUT SCI] A method employed by alphanumeric readers in which the trailing edge or some other document characteristic is used to activate the feeding of the succeeding document.

acyclic machine See homopolar generator.

acyclic motion See irrotational flow.

acyl [ORG CHEM] A radical formed from an organic acid by removal of a hydroxyl group; the general formula is RCO, where R may be aliphatic, alicyclic, or aromatic.

acylation [ORG CHEM] Any process whereby the acyl group is incorporated into a molecule by substitution.

AD See average deviation.

adamellite See quartz monzonite.

adamite [MINERAL] $Zn_2(AsO_4)(OH)$ A colorless, white, or yellow mineral consisting of basic zinc arsenate, crystallizing in the orthorhombic system; hardness is 3.5 on Mohs scale, and specific gravity is 4.34–4.35.

adapical [BOT] Near or toward the apex or tip.

adaptation [GEN] The occurrence of genetic changes in a population or species as the result of natural selection so that it adjusts to new or altered environmental conditions. [PHYSIO] The occurrence of physiological changes in an individual exposed to changed conditions; for example, tanning of the skin in sunshine, or increased red blood cell counts at high altitudes.

adaptation brightness See adaptation luminance.

adaptation illuminance See adaptation luminance.

adaptation level See adaptation luminance.

adaptation luminance [OPTICS] The average luminance, or brightness, of objects and surfaces in the immediate vicinity of an observer estimating the visual range. Also known as adaptation brightness; adaptation illuminance; adaptation level; brightness level; field brightness; field luminance.

adaptation syndrome [MED] Endocrine-mediated stress reaction of the body in response to systemic injury; involves an initial stage of shock, followed by resistance or adaptation and then healing or exhaustion.

adapter [COMPUT SCI] A device which converts bits of information received serially into parallel bit form for use in the inquiry buffer unit. [ENG] A device used to make electrical or mechanical connections between items not originally intended for use together. [MET] A connecting piece, usually made of fireclay, between a horizontal zinc retort and the condenser in which the molten zinc collects. [OPTICS] An attachment to a camera that permits its use in a manner for which it was not designed.

adaptive colitis *See* irritable colon.

adaptive control [CONT SYS] A control method in which one or more parameters are sensed and used to vary the feedback control signals in order to satisfy the performance criteria.

adaptive control function [CONT SYS] That level in the functional decomposition of a large-scale control system which updates parameters of the optimizing control function to achieve a best fit to current plant behavior, and updates parameters of the direct control function to achieve good dynamic response of the closed-loop system.

adaptive divergence [EVOL] Divergence of new forms from a common ancestral form due to adaptation to different environmental conditions.

adaptive enzyme [MICROBIO] Any bacterial enzyme formed in response to the presence of a substrate specific for that enzyme.

adaptive optics [OPTICS] The theory and design of optical systems that measure and correct wavefront aberrations in real time, that is, simultaneously with the operation of the system.

adaptive radiation [EVOL] Diversification of a dominant evolutionary group into a large number of subsidiary types adapted to more restrictive modes of life (different adaptive zones) within the range of the larger group.

adaxial [BIOL] On the same side as or facing toward the axis of an organ or organism.

Adcock antenna [ELECTROMAG] A pair of vertical antennas separated by a distance of one-half wavelength or less and connected in phase opposition to produce a radiation pattern having the shape of a figure eight.

ADCON *See* address constant.

addendum [DES ENG] The radial distance between two concentric circles on a gear, one being that whose radius extends to the top of a gear tooth (addendum circle) and the other being that which will roll without slipping on a circle on a mating gear (pitch line).

adder [COMPUT SCI] A computer device that can form the sum of two or more numbers or quantities. [ELECTR] A circuit in which two or more signals are combined to give an output-signal amplitude that is proportional to the sum of the input-signal amplitudes. Also known as adder circuit. [VERT ZOO] Any of the venomous viperine snakes included in the family Viperidae.

adder circuit *See* adder.

addiction [MED] Habituation to a specific practice, such as drinking alcoholic beverages or using drugs.

adding circuit [ELECTR] A circuit that performs the mathematical operation of addition.

adding tape [ENG] A surveyor's tape that is calibrated from 0 to 100 by full feet (or meters) in one direction, and has 1 additional foot (or meter) beyond the zero end which is subdivided in tenths or hundredths.

Addison's disease [MED] A primary failure or insufficiency of the adrenal cortex to secrete hormones.

addition [MATH] An operation by which two elements of a set are combined to yield a third; denoted $+$; usually reserved for the operation for an Abelian group or the group operation in a ring or vector space.

addition reaction [ORG CHEM] A type of reaction of unsaturated hydrocarbons with hydrogen, halogens, halogen acids, and other reagents, so that no change in valency is observed and the organic compound forms a more complex one.

additive [MATER] **1.** A substance added to another to improve, strengthen, or otherwise alter it; for example, tetraethyllead added to gasoline to prevent engine knock. **2.** *See* admixture. [MATH] Pertaining to addition. [STAT] That property of a process in which increments of the dependent variable are independent for nonoverlapping intervals of the independent variable.

additive function [MATH] Any function which preserves addition, such as $f(x+y) = f(x) + f(y)$.

address [COMPUT SCI] The number or name that uniquely identifies a register, memory location, or storage device in a computer.

address computation [COMPUT SCI] The modification by a computer of an address within an instruction, or of an instruction based on results obtained so far. Also known as address modification.

address constant [COMPUT SCI] A value, or its expression, used in the calculation of storage addresses from relative addresses for computers. Abbreviated ADCON. Also known as base address; presumptive address; reference address.

address conversion [COMPUT SCI] The use of an assembly program to translate symbolic or relative computer addresses.

address counter [COMPUT SCI] A counter which increments an initial memory address as a block of data is being transferred into the memory locations indicated by the counter.

address field [COMPUT SCI] The portion of a computer program instruction which specifies where a particular piece of information is located in the computer memory.

address format [COMPUT SCI] A description of the number of addresses included in a computer instruction.

address generation [COMPUT SCI] An addressing technique which facilitates addressing large storages and implementing dynamic program relocation; the effective main storage address is obtained by adding together the contents of the base register of the index register and of the displacement field.

addressing mode [COMPUT SCI] The specific technique by means of which a memory reference instruction will be spelled out if the computer word is too small to contain the memory address.

addressless instruction format *See* zero address instruction format.

address modification *See* address computation.

address register [COMPUT SCI] A register wherein the address part of an instruction is stored by a computer.

address sort routine [COMPUT SCI] A debugging routine which scans all instructions of the program being checked for a given address.

address track [COMPUT SCI] A path on a magnetic tape, drum, or disk on which are recorded addresses used in the retrieval of data stored on other tracks.

address translation [COMPUT SCI] The assignment of actual locations in a computer memory to virtual addresses in a computer program.

adduct [CHEM] **1.** A chemical compound that forms from chemical addition of two species; for example, reaction of butadiene with styrene forms an adduct,

4-phenyl-1-cyclohexene. **2.** The complex compound formed by association of an inclusion complex.

adduction [PHYSIO] Movement of one part of the body toward another or toward the median axis of the body.

adductor [ANAT] Any muscle that draws a part of the body toward the median axis.

Adeleina [INV ZOO] A suborder of protozoan invertebrate parasites in the order Eucoccida in which the sexual and asexual stages are in different hosts.

adelite [MINERAL] $CaMg(AsO_4)(OH,F)$ A colorless to gray, bluish-gray, yellowish-gray, yellow, or light green orthorhombic mineral consisting of a basic arsenate of calcium and magnesium; usually occurs in massive form.

adenine [BIOCHEM] $C_5H_5N_5$ A purine base, 6-aminopurine, occurring in ribonucleic acid and deoxyribonucleic acid and as a component of adenosinetriphosphate.

adenitis [MED] Inflammation of a gland or lymph node.

adenocarcinoma [MED] A malignant tumor originating in glandular or ductal epithelium and tending to produce acinic structures.

adenohypophysis [ANAT] The glandular part of the pituitary gland, composing the anterior and intermediate lobes.

adenoid [ANAT] **1.** A mass of lymphoid tissue. **2.** Lymphoid tissue of the nasopharynx. Also known as pharyngeal tonsil.

adenoma [MED] A benign tumor of glandular origin and structure.

adenomatosis [MED] A condition characterized by multiple adenomas within an organ or in several related organs.

adenomyoma [MED] A benign tumor of glandular and muscular elements occurring principally in the uterus and rectum.

adenopathy [MED] Any glandular disease; common usage limits the term to any abnormal swelling or enlargement of lymph nodes.

Adenophorea [INV ZOO] A class of unsegmented worms in the phylum Nematoda.

adenosine [BIOCHEM] $C_{10}H_{13}N_5O_4$ A nucleoside composed of adenine and D-ribose.

adenosine 3′,5′-cyclic monophosphate *See* cyclic adenylic acid.

adenosine 3′,5′-cyclic phosphate *See* cyclic adenylic acid.

adenosinediphosphate [BIOCHEM] $C_{10}H_{15}N_5O_{10}P_2$ A coenzyme composed of adenosine and two molecules of phosphoric acid that is important in intermediate cellular metabolism. Abbreviated ADP.

adenosinemonophosphate *See* adenylic acid.

adenosine 3′,5′-monophosphate *See* cyclic adenylic acid.

adenosinetriphosphate [BIOCHEM] $C_{10}H_{16}N_5O_{12}P_3$ A coenzyme composed of adenosinediphosphate with an additional phosphate group; an important energy compound in metabolism. Abbreviated ATP.

adenosis [MED] Any nonneoplastic glandular disease, especially one involving the lymph nodes.

adenovirus [VIROL] A group of animal viruses which cause febrile catarrhs and other respiratory diseases.

adenylic acid [BIOCHEM] **1.** A generic term for a group of isomeric nucleotides. **2.** The phosphoric acid ester of adenosine. Also known as adenosinemonophosphate (AMP).

Adephaga [INV ZOO] A suborder of insects in the order Coleoptera characterized by fused hind coxae that are immovable.

ader wax *See* ozocerite.

ADF *See* automatic direction finder.

ADH *See* vasopressin.

adhesion [BOT] Growing together of members of different and distinct whorls. [ELECTROMAG] Any mutually attractive force holding together two magnetic bodies, or two oppositely charged nonconducting bodies. [ENG] Intimate sticking together of metal surfaces under compressive stresses by formation of metallic bonds. [MECH] The force of static friction between two bodies, or the effects of this force. [MED] The abnormal union of an organ or part with some other part by formation of fibrous tissue. [PHYS] The tendency, due to intermolecular forces, for matter to cling to other matter.

adhesive bond [MECH] The forces such as dipole bonds which attract adhesives and base materials to each other.

adiabatic [THERMO] Referring to any change in which there is no gain or loss of heat.

adiabatic chart *See* Stuve chart.

adiabatic condensation pressure *See* condensation pressure.

adiabatic condensation temperature *See* condensation temperature.

adiabatic demagnetization [CRYO] A method of cooling paramagnetic salts to temperatures of 10^{-3} K; the sample is cooled to boiling point of helium in a strong magnetic field, thermally isolated, and then removed from the field to demagnetize it. Also known as Giaque-Debye method; magnetic cooling; paramagnetic cooling.

adiabatic engine [MECH ENG] A heat engine or thermodynamic system in which there is no gain or loss of heat.

adiabatic equilibrium [METEOROL] A vertical distribution of temperature and pressure in an atmosphere in hydrostatic equilibrium such that an air parcel displaced adiabatically will continue to possess the same temperature and pressure as its surroundings, so that no restoring force acts on a parcel displaced vertically. Also known as convective equilibrium.

adiabatic equivalent temperature *See* equivalent temperature.

adiabatic law [PHYS] The relationship which states that, for adiabatic expansion of gases, $P\rho^{-\gamma} = $ constant, where P = pressure, ρ = density, and γ ratio of specific heats C_P/C_V.

adiabatic process [THERMO] Any thermodynamic procedure which takes place in a system without the exchange of heat with the surroundings.

adiabatic saturation pressure *See* condensation pressure.

adiabatic saturation temperature *See* condensation temperature.

adiabatic system [SCI TECH] A body or system whose condition is altered without gaining heat from or losing heat to the surroundings.

Adimeridae [INV ZOO] An equivalent name for the Colydiidae.

adipocerite *See* hatchettite.

adipose [BIOL] Fatty; of or relating to fat.

A display [ELECTR] A radar oscilloscope display in cartesian coordinates; the targets appear as vertical deflection lines; their Y coordinates are proportional to signal intensity; their X coordinates are proportional to distance to targets.

adit [CIV ENG] An access tunnel used for excavation of the main tunnel. [MIN ENG] A nearly horizontal tunnel for access, drainage, or ventilation of a mine. Also known as side drift.

adjacency [COMPUT SCI] A condition in character recognition in which two consecutive graphic characters are separated by less than a specified distance.

adjoint of a matrix See adjugate of a matrix; Hermitian conjugate of a matrix.

adjoint operator [MATH] An operator B such that the inner products (Ax,y) and (x,By) are equal for a given operator A and for all elements x and y of a Hilbert space. Also known as associate operator; Hermitian conjugate operator.

adjoint vector space [MATH] The complete normed vector space constituted by a class of bounded, linear, homogeneous scalar functions defined on a normed vector space.

adjoint wave functions [QUANT MECH] Functions in the Dirac electron theory which are formed by applying the Dirac matrix B to the Hermitian conjugates of the original wave functions.

adjugate of a matrix [MATH] The matrix obtained by replacing each element with the cofactor of the transposed element. Also known as adjoint of a matrix.

adjustable transformer See variable transformer.

adjusted decibel [ELECTR] A unit used to show the relationship between the interfering effect of a noise frequency, or band of noise frequencies, and a reference noise power level of -85 dBm. Abbreviated dBa. Also known as decibel adjusted.

adjusted value [SCI TECH] A value of a quantity derived from observed data by some orderly process which eliminates discrepancies arising from errors in those data.

adlittoral [OCEANOGR] Of, pertaining to, or occurring in shallow waters adjacent to a shore.

Administrative Terminal System [COMPUT SCI] A system developed by the International Business Machine Corporation to enable the handling by computer of texts that would otherwise require copying by a typist. Abbreviated ATS.

admiralty brass [MET] An alloy containing 71% copper, 28% zinc, and 0.75–1.0% tin for additional corrosion resistance.

admittance [ELEC] A measure of how readily alternating current will flow in a circuit; the reciprocal of impedance, it is expressed in mhos.

admixture [GEOL] One of the lesser or subordinate grades of sediment. [MATER] A material (other than aggregate, cement, or water) added in small quantities to concrete to produce some desired change in properties. Also known as additive.

adnate [BIOL] United through growth; used especially for unlike parts. [BOT] Pertaining to growth with one side adherent to a stem.

adolescence [GEOL] Stage in the cycle of erosion following youth and preceding maturity. [PSYCH] The period of life from puberty to maturity.

adoral [ZOO] Near the mouth.

ADP See adenosinediphosphate; automatic data processing.

adrenal gland [ANAT] An endocrine organ located close to the kidneys of vertebrates and consisting of two morphologically distinct components, the cortex and medulla. Also known as suprarenal gland.

adrenaline See epinephrine.

adrenergic [PHYSIO] Describing the chemical activity of epinephrine or epinephrine-like substances.

adrenocorticotropic hormone [BIOCHEM] The chemical secretion of the adenohypophysis that stimulates the adrenal cortex. Abbreviated ACTH. Also known as adrenotropic hormone.

adrenotropic hormone See adrenocorticotropic hormone.

adsorbate [CHEM] A solid, liquid, or gas which is adsorbed as molecules, atoms, or ions by such substances as charcoal, silica, metals, water, and mercury.

adsorbent [CHEM] A solid or liquid that adsorbs other substances; for example, charcoal, silica, metals, water, and mercury.

adsorption [CHEM] The surface retention of solid, liquid, or gas molecules, atoms, or ions by a solid or liquid, as opposed to absorbtion, the penetration of substances into the bulk of the solid or liquid.

adsorption chromatography [ANALY CHEM] Separation of a chemical mixture (gas or liquid) by passing it over an adsorbent bed which adsorbs different compounds at different rates.

adularia [MINERAL] A weakly triclinic form of the mineral orthoclase occurring in transparent, colorless to milky-white pseudo-orthorhombic crystals.

advance [GEOL] 1. A continuing movement of a shoreline toward the sea. 2. A net movement over a specified period of time of a shoreline toward the sea. [HYD] The forward movement of a glacier. [MECH ENG] To effect the earlier occurrence of an event, for example, spark advance or injection advance. [NAV] 1. In making a turn, the distance a vessel moves in its initial direction from the point where the rudder is started over until the heading has changed 90°. 2. The distance a vessel moves in the initial direction for heading changes of less than 90°.

advance feed tape [COMPUT SCI] Computer tape punched so that the leading edges of its feed holes will line up with the leading edges of the data holes in the tape usage device.

advection [METEOROL] The process of transport of an atmospheric property solely by the mass motion of the atmosphere. [OCEANOGR] The process of transport of water, or of an aqueous property, solely by the mass motion of the oceans, most typically via horizontal currents.

advection fog [METEOROL] A type of fog caused by the horizontal movement of moist air over a cold surface and the consequent cooling of that air to below its dew point.

adventitia [ANAT] The external, connective-tissue covering of an organ or blood vessel. Also known as tunica adventitia.

adventitious [BIOL] Also known as adventive. 1. Acquired spontaneously or accidentally, not by heredity. 2. Arising, as a tissue or organ, in an unusual or abnormal place.

adventive [BIOL] 1. An organism that is introduced accidentally and is imperfectly naturalized; not native. 2. See adventitious.

adventive cone [GEOL] A volcanic cone that is on the flank of and subsidiary to a larger volcano. Also known as lateral cone; parasitic cone.

adz [DES ENG] A cutting tool with a thin arched blade, sharpened on the concave side, at right angles on the handle; used for rough dressing of timber.

Aechminidae [PALEON] A family of extinct ostracods in the order Paleocopa in which the hollow central spine is larger than the valve.

Aeduellidae [PALEON] A family of Lower Permian palaeoniscoid fishes in the order Palaeonisciformes.

Aegeriidae [INV ZOO] The clearwing moths, a family of lepidopteran insects in the suborder Heteroneura characterized by the lack of wing scales.

Aegialitidae [INV ZOO] An equivalent name for the Salpingidae.

Aegidae [INV ZOO] A family of isopod crustaceans in the suborder Flabellifera whose members are economically important as fish parasites.

aegirine [MINERAL] NaFe(SiO$_3$)$_2$ A brown or green clinopyroxene occurring in alkali-rich igneous rocks. Also known as aegirite.

aegirite *See* aegirine.

Aegothelidae [VERT ZOO] A family of small Australo-Papuan owlet-nightjars in the avian order Caprimulgiformes.

Aegypiinae [VERT ZOO] The Old World vultures, a subfamily of diurnal carrion feeders of the family Accipitridae.

Aelosomatidae [INV ZOO] A family of microscopic fresh-water annelid worms in the class Oligochaeta characterized by a ventrally ciliated prostomium.

aenigmatite *See* enigmatite.

aeolian *See* eolian.

aeolotropy *See* anisotropy.

Aepophilidae [INV ZOO] A family of bugs in the hemipteran superfamily Saldoidea.

Aepyornithidae [PALEON] The single family of the extinct avian order Aepyornithiformes.

Aepyornithiformes [PALEON] The elephant birds, an extinct order of ratite birds in the superorder Neognathae.

aeration [ENG] 1. Exposing to the action of air. 2. Causing air to bubble through. 3. Introducing air into a solution by spraying, stirring, or similar method. 4. Supplying or infusing with air, as in sand or soil. [FOOD ENG] Charging a liquid with some gas, such as water with carbon dioxide (soda water). [MIN ENG] The introduction of air into the pulp in a flotation cell to form air bubbles.

aerator [DES ENG] A tool having a roller equipped with hollow fins; used to remove cores of soil from turf. [ENG] 1. One who aerates. 2. Equipment used for aeration. 3. Any device for supplying air or gas under pressure, as for fumigating, welding, or ventilating. [MECH ENG] Equipment used to inject compressed air into sewage in the treatment process. [MET] A device which decreases the density of sand by mixing it with air, thus facilitating the movement of sand particles in packing.

aerial [BIOL] Of, in, or belonging to the air or atmosphere. [ELECTR] *See* antenna. [ORD] 1. Of or pertaining to operations in the air or to aircraft. 2. Of weapons or missiles used in aircraft or launched, dropped, or shot from aircraft.

aeroallergen [MED] Any airborne particulate matter that can induce allergic responses in sensitive persons.

aerobe [BIOL] An organism that requires air or free oxygen to maintain its life processes.

aerobiology [BIOL] The study of the atmospheric dispersal of airborne fungus spores, pollen grains, and microorganisms; and, more broadly, of airborne propagules of algae and protozoans, minute insects such as aphids, and pollution gases and particles which exert specific biologic effects.

aerodrome *See* airport.

aeroduct [AERO ENG] A ramjet type of engine designed to scoop up ions and electrons freely available in the outer reaches of the atmosphere or in the atmospheres of other spatial bodies and, by a metachemical process within the engine duct, to expel particles derived from the ions and electrons as a propulsive jetstream.

aerodynamic [FL MECH] Pertaining to forces acting upon any solid or liquid body moving relative to a gas (especially air).

aerodynamic center [AERO ENG] A point on a cross section of a wing or rotor blade through which the forces of drag and lift are acting and about which the pitching moment coefficient is practically constant.

aerodynamic chord [AERO ENG] A straight line intersecting or touching an airfoil profile at two points; specifically, that part of such a line between two points of intersection.

aerodynamic drag [FL MECH] A retarding force that acts upon a body moving through a gaseous fluid and that is parallel to the direction of motion of the body; it is a component of the total fluid forces acting on the body. Also known as aerodynamic resistance.

aerodynamic force [FL MECH] The force between a body and a gaseous fluid caused by their relative motion. Also known as aerodynamic load.

aerodynamic heating [FL MECH] The heating of a body produced by passage of air or other gases over its surface; caused by friction and by compression processes and significant chiefly at high speeds.

aerodynamic instability [AERO ENG] An unstable state caused by oscillations of a structure that are generated by spontaneous and more or less periodic fluctuations in the flow, particularly in the wake of the structure.

aerodynamic lift [FL MECH] That component of the total aerodynamic force acting on a body perpendicular to the undisturbed airflow relative to the body. Also known as lift.

aerodynamic load *See* aerodynamic force.

aerodynamic moment [AERO ENG] The torque about the center of gravity of a missile or projectile moving through the atmosphere, produced by any aerodynamic force which does not act through the center of gravity.

aerodynamic resistance *See* aerodynamic drag.

aerodynamics [FL MECH] The science that deals with the motion of air and other gaseous fluids and with the forces acting on bodies when they move through such fluids or when such fluids move against or around the bodies.

aerodynamic stability [AERO ENG] The property of a body in the air, such as an aircraft or rocket, to maintain its attitude, or to resist displacement, and if displaced, to develop aerodynamic forces and moments tending to restore the original condition.

aeroelasticity [MECH] The deformation of structurally elastic bodies in response to aerodynamic loads.

aeroembolism [MED] A condition marked by the presence of nitrogen bubbles in the blood and other body tissues resulting from a sudden fall in atmospheric pressure. Also known as air embolism.

aerofilter [CIV ENG] A filter bed for sewage treatment consisting of coarse material and operated at high speed, often with recirculation.

aerofoil *See* airfoil.

aerolite *See* stony meteorite.

aeromechanics [FL MECH] The science of air and other gases in motion or equilibrium; has two branches, aerostatics and aerodynamics.

aeromedicine *See* aerospace medicine.

aeronautical engineering [AERO ENG] The branch of engineering concerned primarily with the design and construction of aircraft structures and power units, and with the special problems of flight in the atmosphere.

aeronautical flutter [FL MECH] An aeroelastic, self-excited vibration in which the external source of energy is the airstream and which depends on the elastic, inertial, and dissipative forces of the system in addition to the aerodynamic forces. Also known as flutter.

aeronautical mile *See* air mile.

aeronautics [FL MECH] The science that deals with flight through the air.

aeronomy [GEOPHYS] The study of the atmosphere of the earth or other bodies, particularly in relation to composition, properties, relative motion, and radiation from outer space or other bodies.

aerootitis *See* barotitis.

aerophyte *See* epiphyte.

aeropulse engine *See* pulsejet engine.

aerosol [CHEM] A gaseous suspension of ultramicroscopic particles of a liquid or a solid.

aerospace *See* airspace.

aerospace engineering [ENG] Engineering pertaining to the design and construction of aircraft and space vehicles and of power units, and to the special problems of flight in both the earth's atmosphere and space, as in the flight of air vehicles and in the launching, guidance, and control of missiles, earth satellites, and space vehicles and probes.

aerospace medicine [MED] The branch of medicine dealing with the effects of flight in the atmosphere or space upon the human body and with the prevention or cure of physiological or psychological malfunctions arising from these effects. Also known as aeromedicine; aviation medicine.

aerospace vehicle [AERO ENG] A vehicle capable of flight both within and outside the sensible atmosphere.

aerostatics [FL MECH] The science of the equilibrium of gases and of solid bodies immersed in them when under the influence only of natural gravitational forces.

aerothermodynamics [FL MECH] The study of aerodynamic phenomena at sufficiently high gas velocities that thermodynamic properties of the gas are important.

aerothermoelasticity [FL MECH] The study of the response of elastic structures to the combined effects of aerodynamic heating and loading.

AES *See* Auger electron spectroscopy.

aesthesia *See* esthesia.

aestival *See* estival.

aestivation [BOT] The arrangement of floral parts in a bud. [PHYSIO] The condition of dormancy or torpidity.

aF *See* abfarad.

AFC *See* automatic frequency control.

afferent [PHYSIO] Conducting or conveying inward or toward the center, specifically in reference to nerves and blood vessels.

affine geometry [MATH] The study of geometry using the methods of linear algebra.

affine transformation [MATH] A function on a linear space to itself, which is the sum of a linear transformation and a fixed vector.

affinity chromatography [ANALY CHEM] A chromatographic technique that utilizes the ability of biological molecules to bend to certain ligands specifically and reversibly; used in protein biochemistry.

afforestation [FOR] Establishment of a new forest by seeding or planting on nonforested land.

aflatoxin [BIOCHEM] The toxin produced by some strains of the fungus *Aspergillus flavus*, the most potent carcinogen yet discovered.

afocal system [OPTICS] An optical system of zero convergent power, for example, a telescope.

African sleeping sickness [MED] A disease of man confined to tropical Africa, caused by the protozoans *Trypanosoma gambiense* or *T. rhodesiense;* symptoms include local reaction at the site of the bite, fever, enlargement of adjacent lymph nodes, skin rash, edema, and during the late phase, somnolence and emaciation. Also known as African trypanosomiasis; maladie du sommeil; sleeping sickness.

African trypanosomiasis *See* African sleeping sickness.

African violet [BOT] *Saintpaulia ionantha.* A flowering plant typical of the family Gesneriaceae.

afterbirth [EMBRYO] The placenta and fetal membranes expelled from the uterus following birth of offspring in viviparous mammals.

afterburner [AERO ENG] A device for augmenting the thrust of a jet engine by burning additional fuel in the uncombined oxygen in the gases from the turbine.

aftercondenser [MECH ENG] A condenser in the second stage of a two-stage ejector; used in steam power plants, refrigeration systems, and air conditioning systems.

aftercooling [MECH ENG] The cooling of a gas after its compression. [NUCLEO] The cooling of a reactor after it has been shut down.

afterdamp [MIN ENG] The mixture of gases which remains in a mine after a mine fire or an explosion of firedamp.

afterglow [ATOM PHYS] *See* phosphorescence. [METEOROL] A broad, high arch of radiance or glow seen occasionally in the western sky above the highest clouds in deepening twilight, caused by the scattering effect of very fine particles of dust suspended in the upper atmosphere. [PL PHYS] The transient decay of a plasma after the power has been turned off.

afterimage [PHYSIO] A visual sensation occurring after the stimulus to which it is a response has been removed.

aftershock [GEOPHYS] A small earthquake following a larger earthquake and originating at or near the larger earthquake's epicenter.

Ag *See* silver.

agamete [BIOL] An asexual reproductive cell that develops into an adult individual.

Agamidae [VERT ZOO] A family of Old World lizards in the suborder Sauria that have acrodont dentition.

agammaglobulinemia [MED] The condition characterized by lack of or extremely low levels of gamma globulin in the blood, together with defective antibody production and frequent infections; primary agammaglobulinemia occurs in three clinical forms: congenital, acquired, and transient.

Agaontidae [INV ZOO] A family of small hymenopteran insects in the superfamily Chalcidoidea; commonly called fig insects for their role in cross-pollination of figs.

agar [MATER] A gelatinous product extracted from certain red algae and used chiefly as a gelling agent in culture media.

Agaricales [MYCOL] An order of fungi in the class Basidiomycetes containing all forms of fleshy, gilled mushrooms.

agate [GRAPHICS] A type size in printing of about 5½ points, where 72 points equals 1 inch. [MINERAL] SiO_2 A fine-grained, fibrous variety of chalcedony with color banding or irregular clouding.

Agavaceae [BOT] A family of flowering plants in the order Liliales characterized by parallel, narrow-veined leaves, a more or less corolloid perianth, and an agavaceous habit.

AGC *See* automatic gain control.

age [BIOL] Period of time from origin or birth to a later time designated or understood; length of existence. [GEOL] **1.** Any one of the named epochs in the history of the earth marked by specific phases of physical conditions or organic evolution, such as the Age of Mammals. **2.** One of the smaller subdivisions of the epoch as geologic time, corresponding to the stage or the formation, such as the Lockport Age in the Niagara Epoch.

aged [GEOL] Of a ground configuration, having been reduced to base level.

age hardening [MET] Increasing the hardness of an alloy by a relatively low-temperature heat treatment that causes precipitation of components or phases of the alloy from the supersaturated solid solution.

Agena *See* β Centauri.

agenda [COMPUT SCI] **1.** The sequence of control statements required to carry out the solution of a computer problem. **2.** A collection of programs used for manipulating a matrix in the solution of a problem in linear programming.

agglomeration [FOOD ENG] A technique that combines powdered material to form larger, more soluble particles by intermingling in a humid atmosphere. [MET] Conversion of small pieces of low-grade iron ore into larger lumps by application of heat. [METEOROL] The process in which particles grow by collision with and assimilation of cloud particles or other precipitation particles. Also known as coagulation. [SCI TECH] An indiscriminately formed cluster of particles.

agglutinate cone *See* spatter cone.

agglutination reaction [IMMUNOL] Clumping of a particulate suspension of antigen by a reagent, usually an antibody.

agglutinin [IMMUNOL] An antibody from normal or immune serum that causes clumping of its complementary particulate antigen, such as bacteria or erythrocytes.

agglutinogen [IMMUNOL] An antigen that stimulates production of a specific antibody (agglutinin) when introduced into an animal body.

aggradation [GEOL] *See* accretion. [HYD] A process of shifting equilibrium of stream deposition, with upbuilding approximately at grade.

aggregate [GEOL] A collection of soil grains or particles gathered into a mass. [MATER] The natural sands, gravels, and crushed stone used for mixing with cementing material in making mortars and concretes.

aggression [PSYCH] Feelings and behavior of anger and hostility usually manifested by punitive or destructive actions; often associated with frustration.

aggressive device [COMPUT SCI] A unit of a computer that can initiate a request for communication with another device.

aging [BIOL] Growing older. [CHEM] All irreversible structural changes that occur in a precipitate after it has formed. [ELEC] Allowing a permanent magnet, capacitor, meter, or other device to remain in storage for a period of time, sometimes with a voltage applied, until the characteristics of the device become essentially constant. [ELECTROMAG] Change in the magnetic properties of iron with passage of time, for example, increase in the hysteresis. [ENG] **1.** The changing of the characteristics of a device due to its use. **2.** Operation of a product before shipment to stabilize characteristics or detect early failures. [MATER] **1.** Change in the properties of any substance with time. **2.** Change occurring in powders or slips with the passage of time. **3.** Curing of ceramic materials, such as clays and glazes, by a definite period of time under controlled storage conditions. [MET] **1.** Change in properties of an alloy or metal which

generally proceeds slowly at room temperatures and faster at elevated temperatures. **2.** Strain relief, occurring through long storage outdoors under varying temperatures, of iron castings intended for use as toolroom plates or lathe-bed supports. **3.** A second heat treatment of an alloy at a lower temperature, causing precipitation of the unstable phase and increasing hardness, strength, and electrical conductivity. [NUCLEO] The slowing down of neutrons.

agitator [MECH ENG] A device for keeping liquids and solids in liquids in motion by mixing, stirring, or shaking.

Aglaspida [PALEON] An order of Cambrian and Ordovician merostome arthropods in the subclass Xiphosurida characterized by a phosphatic exoskeleton and vaguely trilobed body form.

Aglossa [VERT ZOO] A suborder of anuran amphibians represented by the single family Pipidea and characterized by the absence of a tongue.

Agnatha [VERT ZOO] The most primitive class of vertebrates, characterized by the lack of true jaws.

agnathia [MED] Lack of the jaws.

agonic line [GEOPHYS] The imaginary line through all points on the earth's surface at which the magnetic declination is zero; that is, the locus of all points at which magnetic north and true north coincide.

Agonidae [VERT ZOO] The poachers, a small family of marine perciform fishes in the suborder Cottoidei.

agouti [VERT ZOO] A hystricomorph rodent, *Dasyprocta*, in the family Dasyproctidae, represented by 13 species.

agranular leukocyte [HISTOL] A type of white blood cell, including lymphocytes and monocytes, characterized by the absence of cytoplasmic granules and by a relatively large spherical or indented nucleus.

agranulocytosis [MED] An acute febrile illness, usually resulting from drug hypersensitivity, manifested as severe leukopenia, often with complete disappearance of granulocytes.

agricolite *See* eulytite.

agricultural chemistry [AGR] The science of chemical compositions and changes involved in the production, protection, and use of crops and livestock; includes all the life processes through which food and fiber are obtained for man and his animals, and control of these processes to increase yields, improve quality, and reduce costs.

agricultural engineering [AGR] A discipline concerned with developing and improving the means for providing food and fiber for mankind's needs.

agricultural geology [GEOL] A branch of geology that deals with the nature and distribution of soils, the occurrence of mineral fertilizers, and the behavior of underground water.

agricultural meteorology [AGR] The study and application of relationships between meteorology and agriculture, involving problems such as timing the planting of crops. Also known as agrometeorology.

agricultural science [AGR] A discipline dealing with the selection, breeding, and management of crops and domestic animals for more economical production.

agriculture [BIOL] The production of plants and animals useful to man, involving soil cultivation and the breeding and management of crops and livestock.

Agriochoeridae [PALEON] A family of extinct tylopod ruminants in the superfamily Merycoidodontoidea.

Agrionidae [INV ZOO] A family of odonatan insects in the suborder Zygoptera characterized by black or red markings on the wings.

agometeorology *See* agricultural meteorology.

Agromyzidae [INV ZOO] A family of myodarian cyclorrhaphous dipteran insects of the subsection Acalyptratae; commonly called leaf-miner flies because the larvae cut channels in leaves.

agronomy [AGR] The principles and procedures of soil management and of field crop and special-purpose plant improvement, management, and production.

agrotechnology [AGR] An innovative technology designed to render agricultural production more efficient and profitable.

aH *See* abhenry.

Ah *See* ampere-hour.

ahlfeldite [MINERAL] (Ni,Co)SeO$_3$·2H$_2$O A triclinic mineral identified as green to yellow crystals with a reddish-brown coating, consisting of a hydrous selenite of nickel.

aided tracking [ENG] A system of radar-tracking a target signal in bearing, elevation, or range, or any combination of these variables, in which the rate of motion of the tracking equipment is machine-controlled in collaboration with an operator so as to minimize tracking error.

aikinite [MINERAL] PbCuBiS$_3$ A mineral crystallizing in the orthorhombic system and occurring in massive and in gray needle-shaped crystals; hardness is 2 on Mohs scale, and specific gravity is 7.07. Also known as needle ore.

aileron [AERO ENG] The hinged rear portion of an aircraft wing moved differentially on each side of the aircraft to obtain lateral or roll control moments. [ARCH] A half gable, such as that which closes the end of a penthouse roof or of a church aisle.

aiming circle [ENG] An instrument for measuring angles in azimuth and elevation in connection with artillery firing and general topographic work; equipped with fine and coarse azimuth micrometers and a magnetic needle.

A indicator *See* A scope.

aiophyllous *See* evergreen.

air [CHEM] A predominantly mechanical mixture of a variety of individual gases forming the earth's enveloping atmosphere.

Air Almanac [NAV] **1.** A periodical publication of astronomical data useful to and designed primarily for air navigation. **2.** A joint publication of the U.S. Naval Observatory and the Royal Greenwich Observatory, listing the Greenwich hour angle and declination of various celestial bodies at 10-minute intervals, the time of sunrise, sunset, moonrise, and moonset, and other astronomical information arranged in a form convenient for navigators; each publication covers 4 months.

air-arc furnace [ENG] An arc furnace designed to power wind tunnels, the air being superheated to 20,000 K and expanded to emerge at supersonic speeds.

air armament [ORD] All equipment through which a combat aircraft can release destructive power on a target.

airblast circuit breaker [ELEC] An electric switch which, on opening, utilizes a high-pressure gas blast (air or sulfur hexafluoride) to break the arc.

airblasting [ENG] A blasting technique in which air at very high pressure is piped to a steel shell in a shot hole and discharged. Also known as air breaking.

air bleeder [MECH ENG] A device, such as a needle valve, for removing air from a hydraulic system.

airborne [AERO ENG] Of equipment and material, carried or transported by aircraft. [ORD] **1.** Of a force or organization, transported, or designed to be transported, by aircraft. **2.** Of an action or operation, carried out with transport aircraft.

airborne collision avoidance system [NAV] A navigation system for preventing collisions between aircraft that relies primarily on equipment carried on the aircraft itself, but which may make use of equipment already employed in the ground-based air-traffic control system. Abbreviated ACAS.

airborne profile [GEOD] Continuous terrain-profile data produced by an absolute altimeter in an aircraft which is making an altimeter-controlled flight along a prescribed course.

air brake [MECH ENG] An energy-conversion mechanism activated by air pressure and used to retard, stop, or hold a vehicle or, generally, any moving element.

air breaking *See* airblasting.

air-break switch *See* air switch.

air-breathing [MECH ENG] Of an engine or aerodynamic vehicle, required to take in air for the purpose of combustion.

airbrush [GRAPHICS] A pencil-shaped air gun that fine-sprays paint; used in retouching photographs and in shading drawings.

air capacitor [ELEC] A capacitor having only air as the dielectric material between its plates. Also known as air condenser.

air cell [ELECTR] A cell in which depolarization at the positive electrode is accomplished chemically by reduction of the oxygen in the air. [ZOO] A cavity or receptacle for air such as an alveolus, an air sac in birds, or a dilation of the trachea in insects.

air chamber [MECH ENG] A pressure vessel, partially filled with air, for converting pulsating flow to steady flow of water in a pipeline, as with a reciprocating pump.

air classifier [MECH ENG] A device to separate particles by size through the action of a stream of air.

air compressor [MECH ENG] A machine that increases the pressure of air by increasing its density and delivering the fluid against the connected system resistance on the discharge side.

air condenser [ELEC] *See* air capacitor. [MECH ENG] **1.** A steam condenser in which the heat exchange occurs through metal walls separating the steam from cooling air. Also known as air-cooled condenser. **2.** A device that removes vapors, such as of oil or water, from the airstream in a compressed-air line.

air conditioner [MECH ENG] A mechanism primarily for comfort cooling that lowers the temperature and reduces the humidity of air in buildings.

air conditioning [MECH ENG] The maintenance of certain aspects of the environment within a defined space to facilitate the function of that space; aspects controlled include air temperature and motion, radiant heat level, moisture, and concentration of pollutants such as dust, microorganisms, and gases. Also known as climate control. [TEXT] A chemical process by which small fibers are sealed into yarn.

air-cooled condenser *See* air condenser.

air-cooled engine [MECH ENG] An engine cooled directly by a stream of air without the interposition of a liquid medium.

air-cooled heat exchanger [MECH ENG] A finned-tube (extended-surface) heat exchanger with hot fluids inside the tubes, and cooling air that is fan-blown (forced draft) or fan-pulled (induced draft) across the tube bank.

air course *See* airway.

aircraft [AERO ENG] Any structure, machine, or contrivance, especially a vehicle, designed to be supported by the air, either by the dynamic action of the air upon the surfaces of the structure or object or by its own buoyancy. Also known as air vehicle.

aircraft carrier [NAV ARCH] A ship that carries aircraft, has a takeoff and landing deck, and is otherwise designed and equipped to serve as a base of operations for the aircraft. Also known as carrier.

aircraft ceiling [METEOROL] After United States weather observing practice, the ceiling classification applied when the reported ceiling value has been determined by a pilot while in flight within 1.5 nautical miles (2.8 kilometers) of any runway of the airport.

aircraft low-approach system [NAV] Any of the various means for furnishing guidance in the vertical and horizontal planes to aircraft during descent from an initial approach altitude to a point near the ground.

air current [FL MECH] Very generally, any moving stream of air. [GEOPHYS] *See* air-earth conduction current. [MIN ENG] The flow of air ventilating the workings of a mine. Also known as airflow.

air-cushion landing system [AERO ENG] A landing system based on the ground-effect principle whereby a stratum of air is utilized as the aircraft ground-contacting medium (in place of landing gear).

air-cushion vehicle [MECH ENG] A transportation device supported by low-pressure, low-velocity air capable of traveling equally well over water, ice, marsh, or relatively level land. Also known as ground-effect machine (GEM); hovercraft.

air cycle [MECH ENG] A refrigeration cycle characterized by the working fluid, air, remaining as a gas throughout the cycle rather than being condensed to a liquid; used primarily in airplane air conditioning.

air cylinder [MECH ENG] A cylinder in which air is compressed by a piston, compressed air is stored, or air drives a piston.

air discharge [GEOPHYS] **1.** A form of lightning discharge, intermediate in character between a cloud discharge and a cloud-to-ground discharge, in which the multibranching lightning channel descending from a cloud base does not reach the ground, but succeeds only in neutralizing the space charge distributed in the subcloud layer. **2.** A type of diffuse electrical discharge occasionally reported as occurring in the region above an active thunderstorm.

air-earth conduction current [GEOPHYS] That part of the air-earth current contributed by the electrical conduction of the atmosphere itself; represented as a downward movement of positive space charge in storm-free regions all over the world. Also known as air current.

air-earth current [GEOPHYS] The transfer of electric charge from the positively charged atmosphere to the negatively charged earth; made up of the air-earth conduction current, a precipitation current, a convection current, and miscellaneous smaller contributions.

air ejector [MECH ENG] A device that uses a fluid jet to remove air or other gases, as from a steam condenser.

air embolism *See* aeroembolism.

air entrainment [ENG] The inclusion of minute bubbles of air in cement or concrete through the addition of some material during grinding or mixing to reduce the surface tension of the water, giving improved properties for the end product.

air equivalent [NUCLEO] A measure of the effectiveness of an absorber of nuclear radiation, equal to the thickness of a layer of air at standard pressure and temperature that absorbs the same fraction of radiation or results in the same energy loss as does the absorber.

air filter [ENG] A device that reduces the concentration of solid particles in an airstream to a level that can be tolerated in a process or space occupancy; a component of most systems in which air is used for industrial processes, ventilation, or comfort air conditioning.

airflow [FL MECH] **1.** A flow or stream of air which may take place in a wind tunnel or, as a relative airflow, past the wing or other parts of a moving craft. Also known as airstream. **2.** A rate of flow, measured by mass or volume per unit of time. [MIN ENG] *See* air current.

airfoil [AERO ENG] A body of such shape that the force exerted on it by its motion through a fluid has a larger component normal to the direction of motion than along the direction of motion; examples are the wing of an airplane and the blade of a propeller. Also known as aerofoil.

airfoil profile [AERO ENG] The closed curve defining the external shape of the cross section of an airfoil. Also known as airfoil section; airfoil shape; wing section.

airfoil section *See* airfoil profile.

airfoil shape *See* airfoil profile.

airframe [AERO ENG] The basic assembled structure of any aircraft or rocket vehicle, except lighter-than-air craft, necessary to support aerodynamic forces and inertia loads imposed by the weight of the vehicle and contents.

air-fuel ratio [CHEM] The ratio of air to fuel by weight or volume which is significant for proper oxidative combustion of the fuel.

air furnace [MET] **1.** A furnace using a natural air draft. **2.** A furnace in which the metal is melted by a flame originating from fuel burned at one end, passing over the hearth in the middle, and exiting at the other end.

air gage [ENG] **1.** A device that measures air pressure. **2.** A device that compares the shape of a machined surface to that of a reference surface by measuring the rate of passage of air between the surfaces.

air gap [ELECTR] **1.** A gap or an equivalent filler of nonmagnetic material across the core of a choke, transformer, or other magnetic device. **2.** A spark gap consisting of two electrodes separated by air. [ENG] **1.** The distance between two components or parts. **2.** In plastic extrusion coating, the distance from the opening of the extrusion die to the nip formed by the pressure and chill rolls. **3.** The unobstructed vertical distance between the lowest opening of a faucet (or the like) which supplies a plumbing fixture (such as a tank or washbowl) and the level at which the fixture will overflow. [GEOL] *See* wind gap.

airglow [GEOPHYS] The quasi-steady radiant emission from the upper atmosphere over middle and low latitudes, as distinguished from the sporadic emission of auroras which occur over high latitudes. Also known as light-of-the-night-sky; night-sky light; night-sky luminescence; permanent aurora.

air gun *See* air rifle.

air-hardening steel [MET] A steel whose content of carbon and other alloying elements is sufficient for the steel to harden fully by cooling in air or any other atmosphere from a temperature above its transformation range. Also known as self-hardening steel.

air heater *See* air preheater.

air-heating system *See* air preheater.

air-injection system [MECH ENG] A device that uses compressed air to inject the fuel into the cylinder of an internal combustion engine.

air intake [AERO ENG] An open end of an air duct or similar projecting structure so that the motion of the aircraft is utilized in capturing air to be conducted to an engine or ventilator. [MIN ENG] A device for supplying a compressor with clean air at the lowest possible temperature.

air knife [ENG] A device that uses a thin, flat jet of air to remove the excess coating from freshly coated paper.

air layering [BOT] A method of vegetative propagation, usually of a wounded part, in which the branch or shoot is enclosed in a moist medium until roots develop, and then it is severed and cultivated as an independent plant.

air lift [MECH ENG] **1.** Equipment for lifting slurry or dry powder through pipes by means of compressed air. **2.** *See* air-lift pump.

air-lift hammer [MECH ENG] A gravity drop hammer used in closed die forging in which the ram is raised to its starting point by means of an air cylinder.

air-lift pump [MECH ENG] A device composed of two pipes, one inside the other, used to extract water from a well; the lower end of the pipes is submerged, and air is delivered through the inner pipe to form a mixture of air and water which rises in the outer pipe above the water in the well; also used to move corrosive liquids, mill tailings, and sand. Also known as air lift.

air lock [ENG] **1.** A chamber capable of being hermetically sealed that provides for passage between two places of different pressure, such as between an altitude chamber and the outside atmosphere, or between the outside atmosphere and the work area in a tunnel or shaft being excavated through soil subjected to water pressure higher than atmospheric. **2.** An air bubble in a pipeline which impedes liquid flow. [MIN ENG] A casing atop an upcast mine shaft to minimize surface air leakage into the fan.

air mass [METEOROL] An extensive body of the atmosphere which approximates horizontal homogeneity in its weather characteristics, particularly with reference to temperature and moisture distribution.

air meter [ENG] A device that measures the flow of air, or gas, expressed in volumetric or weight units per unit time. Also known as airometer.

air mile [NAV] A unit of length used in air navigation and equal, since 1954, to 1 international nautical mile (1852 meters). Also known as aeronautical mile.

air mileage [NAV] The number of miles flown relative to the air; true air speed multiplied by time.

air navigation [NAV] The process of directing and monitoring the progress of an aircraft between selected geographic points or with respect to some predetermined plan. Also known as avigation.

airometer [ENG] **1.** An apparatus for both holding air and measuring the quantity of air admitted into it. **2.** *See* air meter.

airplane [AERO ENG] A heavier-than-air vehicle designed to use the pressures created by its motion through the air to lift and transport useful loads.

air pocket [ENG] An air-filled space that is normally occupied by a liquid. Also known as air trap. [METEOROL] An expression used in the early days of aviation for a downdraft; such downdrafts were thought to be pockets in which there was insufficient air to support the plane.

air pollution [ECOL] The presence in the outdoor atmosphere of one or more contaminants such as dust, fumes, gas, mist, odor, smoke, or vapor in quantities and of characteristics and duration such as to be injurious to human, plant, or animal life or to property, or to interfere unreasonably with the comfortable enjoyment of life and property.

airport [CIV ENG] A terminal facility used for aircraft takeoff and landing and including facilities for handling passengers and cargo and for servicing aircraft. Also known as aerodrome.

airport engineering [CIV ENG] The planning, design, construction, and operation and maintenance of facilities providing for the landing and takeoff, loading and unloading, servicing, maintenance, and storage of aircraft.

airport surface detection equipment [NAV] A short-range radar using millimeter waves and giving a panoramic presentation of all aircraft and vehicles, moving or stationary, on the surface of an aerodrome; used by air traffic controllers for expeditious movement of surface aircraft on the ramp, taxiway, and runway.

airport traffic control tower [NAV] The terminal control point at an airport for all takeoff and landing operations, departure and approach operations, and ground movements of aircraft and airport vehicles. Abbreviated ATCT.

air-position indicator [NAV] An airborne computing system which presents a continuous indication of the aircraft position on the basis of aircraft heading, airspeed, and elapsed time. Abbreviated API.

air preheater [MECH ENG] A device used in steam boilers to transfer heat from the flue gases to the combustion air before the latter enters the furnace. Also known as air heater; air-heating system.

air pressure [PHYS] The force per unit area that the air exerts on any surface in contact with it, arising from the collisions of the air molecules with the surface.

airproof *See* airtight.

air propeller [AERO ENG] A hub-and-multiblade device for changing rotational power of an aircraft engine into thrust power for the purpose of propelling an aircraft through the air. [MECH ENG] A rotating fan for moving air.

air pump [MECH ENG] A device for removing air from an enclosed space or for adding air to an enclosed space.

Air Pump *See* Antlia.

air register [ENG] A device attached to an air-distributing duct for the purpose of controlling the discharge of air into the space to be heated, cooled, or ventilated.

air rifle [ORD] A low-powered rifle that shoots small metal pellets (BBs) by the action of compressed air. Also known as air gun; BB gun.

air route [NAV] The navigable airspace between two points, identified to the extent necessary for the application of flight rules.

air sac [GEOL] *See* vesicle. [INV ZOO] One of large, thin-walled structures associated with the tracheal system of some insects. [VERT ZOO] In birds, any of the small vesicles that are connected with the respiratory system and located in bones and muscles to increase buoyancy.

air separator [MECH ENG] A device that uses an air current to separate a material from another of greater density or particles from others of greater size.

air shaft [BUILD] An open space surrounded by the walls of a building or buildings to provide ventilation for windows. Also known as air well. [MIN ENG] A usually vertical earth bore or shaft to supply surface air to an underground facility such as a mine.

airship [AERO ENG] A propelled and steered aerial vehicle, dependent on gases for flotation.

air shower See cosmic-ray shower.

air-slaked [CHEM] Having the property of a substance, such as lime, that has been at least partially converted to a carbonate by exposure to air.

airspace [AERO ENG] 1. The space occupied by an aircraft formation or used in a maneuver. 2. The area around an airplane in flight that is considered an integral part of the plane in order to prevent collision with another plane; the space depends on the speed of the plane. [METEOROL] 1. Of or pertaining to both the earth's atmosphere and space. Also known as aerospace. 2. The portion of the atmosphere above a particular land area, especially a nation or other political subdivision.

airspeed [AERO ENG] The speed of an airborne object relative to the atmosphere; in a calm atmosphere, airspeed equals ground speed.

air spring [MECH ENG] A spring in which the energy storage element is air confined in a container that includes an elastomeric bellows or diaphragm.

air stack [AERO ENG] A group of planes flying at prescribed heights while waiting to land at an airport. [MIN ENG] A chimney for ventilating a coal mine.

air-standard cycle [THERMO] A thermodynamic cycle in which the working fluid is considered to be a perfect gas with such properties of air as a volume of 12.4 ft³/lb at 14.7 psi (approximately 0.7756 m³/kg at 101.36 kPa) and 492°R and a ratio of specific heats of 1:4.

airstream See airflow.

air-supply mask See air-tube breathing apparatus.

air-suspension system [MECH ENG] Parts of an automotive vehicle that are intermediate between the wheels and the frame, and support the car body and frame by means of a cushion of air to absorb road shock caused by passage of the wheels over irregularities.

air sweetening [CHEM ENG] A process in which air or oxygen is used to oxidize lead mercaptides to disulfides instead of using elemental sulfur.

air switch [ELEC] A switch in which the breaking of the electric circuit takes place in air. Also known as air-break switch.

air system [MECH ENG] A mechanical refrigeration system in which air serves as the refrigerant in a cycle comprising compressor, heat exchanger, expander, and refrigerating core.

air temperature [METEOROL] 1. The temperature of the atmosphere which represents the average kinetic energy of the molecular motion in a small region and is defined in terms of a standard or calibrated thermometer in thermal equilibrium with the air. 2. The temperature that the air outside of the aircraft is assumed to have as indicated on a cockpit instrument.

air terminal [CIV ENG] A facility providing a place of assembly and amenities for airline passengers and space for administrative functions. [ELEC] A structure, such as a tower, that serves as a lightning arrester.

air thermometer [ENG] A device that measures the temperature of an enclosed space by means of variations in the pressure or volume of air contained in a bulb placed in the space.

airtight [ENG] Not permitting the passage of air. Also known as airproof.

air-to-air missile [ORD] A missile launched from an aircraft at an airborne target. Abbreviated AAM.

air-to-ground missile [ORD] A missile launched from an aircraft at a ground target. Abbreviated AGM.

air-to-space missile [ORD] A missile launched from an aircraft at a space target, such as an earth satellite.

air-to-surface missile [ORD] A missile launched from aircraft at a ground or sea target. Abbreviated ASM.

air-to-underwater missile [ORD] A missile launched from an aircraft at an underwater target.

air-traffic control [NAV] 1. A service which promotes the safe and fast movement of aircraft operating in the air or on an airport surface by providing rules, procedures, and information and advisory services for pilots. Abbreviated ATC. 2. A system comprising enabling legislation, operating procedures, and navigation and communication equipment which is intended to make for the safe and expeditious movement of aircraft from the time that they leave the departure gates to arrival at the terminal gates.

air-traffic control center [NAV] A place for receiving information regarding aircraft movement, interpreting this information, and issuing instructions to aircraft to promote safe, orderly, expeditious flow of traffic; it directs traffic along controlled airspace.

air trap [CIV ENG] A U-shaped pipe filled with water that prevents the escape of foul air or gas from such systems as drains and sewers. [ENG] See air pocket.

air-tube breathing apparatus [ENG] A device consisting of a smoke helmet, mask, or mouthpiece supplied with fresh air by means of a flexible tube. Also known as air-supply mask.

air valve [MECH ENG] A valve that automatically lets air out of or into a liquid-carrying pipe when the internal pressure drops below atmospheric.

air vehicle See aircraft.

air volcano [GEOL] An eruptive opening in the earth from which large volumes of gas emanate, in addition to mud and stones; a variety of mud volcano.

air wall [NUCLEO] A wall of an ionization chamber designed so that its effect on ionizing radiation approximates that of air.

air washer [MECH ENG] 1. A device for cooling and cleaning air in which the entering warm, moist air is cooled below its dew point by refrigerated water so that although the air leaves close to saturation with water, it has less moisture per unit volume than when it entered. 2. Apparatus to wash particulates and soluble impurities from air by passing the airstream through a liquid bath or spray.

airwave [ELECTR] A radio wave used in radio and television broadcasting. [METEOROL] A wavelike oscillation in the pattern of wind flow aloft, usually with reference to the stronger portion of the westerly current.

airway [BUILD] A passage for ventilation between thermal insulation and roof boards. [MIN ENG] A passage for air in a mine. Also known as air course. [NAV] A designated route of passage for aircraft.

airways forecast See aviation weather forecast.

air well See air shaft.

Airy differential equation [MATH] The differential equation $(d^2f/dz^2) - zf = 0$, where z is the independent variable and f is the value of the function; used in studying the diffraction of light near caustic surface.

Airy disk [OPTICS] The bright, diffuse central spot of light formed by an optical system imaging a point source of light.

Airy function [MATH] Either of the solutions of the Airy differential equation.

aisle [ARCH] **1.** A passageway between or alongside blocks of seats, as in an auditorium. **2.** One of the parts of a basilica which are located at the sides of the nave, with each aisle separated from it by a row of columns.

Aistopoda [PALEON] An order of Upper Carboniferous amphibians in the subclass Lepospondyli characterized by reduced or absent limbs and an elongate, snakelike body.

Aitken nuclei [METEOROL] The microscopic particles in the atmosphere which serve as condensation nuclei for droplet growth during the rapid adiabatic expansion produced by an Aitken dust counter.

Aitken's formula [ASTRON] The expression used to determine the separation limit for true binary stars: $\log p'' = 2.5-0.2m$, where p'' = limit, m = magnitude.

Aizoaceae [BOT] A family of flowering plants in the order Caryophyllales; members are unarmed leaf-succulents, chiefly of Africa.

Ajax powder [MATER] A high-strength, high-density, gelatinous permitted explosive having good water resistance; contains nitroglycerine, potassium perchlorate, ammonium oxalate, and wood flour.

akerite [PETR] A rock composed of quartz syenite containing soda microcline, oligoclase, and augite.

akermanite [MINERAL] $Ca_2MgSi_2O_7$ Anhydrous calcium-magnesium silicate found in igneous rocks; a melilite.

akinesia [MED] **1.** Loss of or impaired motor function. **2.** Immobility from any cause.

akrochordite [MINERAL] $Mn_4Mg(AsO_4)_2(OH)_4 \cdot 4H_2O$ Mineral consisting of a hydrous basic manganese magnesium arsenate and occurring in reddish-brown rounded aggregates; hardness is 3 on Mohs scale, and specific gravity is 3.2.

alabandite [MINERAL] MnS A complex sulfide mineral that is a component of meteorites and usually occurs in iron-black massive or granular form. Also known as manganblende.

alabaster [MINERAL] **1.** $CaSO_4 \cdot 2H_2O$ A fine-grained, colorless gypsum. **2.** See onyx marble.

alabaster glass [MATER] A glass that contains small inclusions of different diffractive indexes and shows no color reaction to light.

alamosite [MINERAL] $PbSiO_3$ A white or colorless monoclinic mineral consisting of lead silicate and occurring in radiating fibers; hardness is 4.5 on Mohs scale, and specific gravity is 6.5.

alang-alang See cogon.

alanine [BIOCHEM] $C_3H_7NO_2$ A white, crystalline, nonessential amino acid of the pyruvic acid family.

alarm system [ENG] A system which operates a warning device after the occurrence of a dangerous or undesirable condition.

Alaska cedar [MATER] Wood from *Chamaecyparis nootkaensis* or *Cupressus sitkaensis*; has a fine, uniform, straight grain and is light and moderately hard; used for furniture and boat building. Also known as Sitka cypress; yellow cedar; yellow cypress.

Alaska Current [OCEANOGR] A current that flows northwestward and westward along the coasts of Canada and Alaska to the Aleutian Islands.

Alaudidae [VERT ZOO] The larks, a family of Oscine birds in the order Passeriformes.

albatross [VERT ZOO] Any of the large, long-winged oceanic birds composing the family Diomedeidae of the order Procellariiformes.

albedo [NUCLEO] The reflection factor a surface, such as paraffin, has for neutrons. [OPTICS] That fraction of the total light incident on a reflecting surface, especially a celestial body, which is reflected back in all directions.

albertite [MINERAL] Jet-black, brittle natural hydrocarbon with conchoidal fracture, hardness of 1–2, and specific gravity of approximately 1.1. Also known as asphaltite coal.

albinism [BIOL] The state of having colorless chromatophores, which results in the absence of pigmentation in animals that are normally pigmented. [MED] A hereditary, metabolic disorder transmitted as an autosomal recessive and characterized by the inability to form melanin in the skin, hair, and eyes due to tyrosinase deficiency.

albite [MINERAL] $NaAlSi_3O_8$ A colorless or milky-white variety of plagioclase of the feldspar group found in granite and various igneous and metamorphic rocks. Also known as sodaclase; sodium feldspar; white feldspar; white schorl.

albitite [PETR] A porphyritic dike rock that is coarse-grained and composed almost wholly of albite; common accessory minerals are muscovite, garnet, apatite, quartz, and opaque oxides.

Alboll [GEOL] A suborder of the soil order Mollisol with distinct horizons, wet for some part of the year; occurs mostly on upland flats and in shallow depressions.

albumen [CYTOL] The white of an egg, composed principally of albumin.

albumin [BIOCHEM] Any of a group of plant and animal proteins which are soluble in water, dilute salt solutions, and 50% saturated ammonium sulfate.

albuminoid [BIOCHEM] See scleroprotein. [BIOL] Having the characteristics of albumin.

albuminuria [MED] The presence of albumin in the urine; usually indicating renal disease.

alburnum See sapwood.

alcaptonuria See alkaptonuria.

Alcedinidae [VERT ZOO] The kingfishers, a worldwide family of colorful birds in the order Coraciiformes; characterized by large heads, short necks, and heavy, pointed bills.

alchemy [CHEM] A speculative chemical system having as its central aims the transmutation of base metals to gold and the discovery of the philosopher's stone.

Alcidae [VERT ZOO] A family of shorebirds, predominantly of northern coasts, in the order Charadriiformes, including auks, puffins, murres, and guillemots.

Alciopidae [INV ZOO] A pelagic family of errantian annelid worms in the class Polychaeta.

alcohol [ORG CHEM] **1.** C_2H_5OH A colorless, volatile liquid; boiling point of pure liquid is 78.3°C; it is soluble in water, chloroform, and methyl alcohol; used as solvent and in manufacture of many chemicals and medicines. Also known as ethanol; ethyl alcohol. **2.** Any of a class of organic compounds containing the hydroxyl group, OH.

alcoholic [MED] An individual who consumes excess amounts of alcoholic beverages to the extent of being addicted, habituated, or dependent.

alcoholic fermentation [MICROBIO] The process by which certain yeasts decompose sugars in the absence of oxygen to form alcohol and carbon dioxide; method for production of ethanol, wine, and beer.

alcoholimeter See alcoholometer.

alcoholmeter See alcoholometer.

alcoholometer [ENG] A device, such as a form of hydrometer, that measures the quantity of an alcohol contained in a liquid. Also known as alcoholimeter; alcoholmeter.

alcoholysis [ORG CHEM] The breaking of a carbon-to-carbon bond by addition of an alcohol.

alcove [ARCH] **1.** A recessed part of a room. **2.** A small room that opens into a larger one. **3.** An arched opening in a wall. [GEOL] A large niche formed by a stream in a face of horizontal strata.

Alcyonacea [INV ZOO] The soft corals, an order of littoral anthozoans of the subclass Alcyonaria.

Alcyonaria [INV ZOO] A subclass of the Anthozoa; members are colonial coelenterates, most of which are sedentary and littoral.

aldehyde [ORG CHEM] One of a class of organic compounds containing the CHO radical.

alder [BOT] The common name for several trees of the genus *Alnus*.

aldol condensation [ORG CHEM] Formation of a β-hydroxycarbonyl compound by the condensation of an aldehyde or a ketone in the presence of an acid or base catalyst. Also known as aldol reaction.

aldose [ORG CHEM] A class of monosaccharide sugars; the molecule contains an aldehyde group.

aldosterone [BIOCHEM] $C_{21}H_{28}O_5$ A steroid hormone extracted from the adrenal cortex that functions chiefly in regulating sodium and potassium metabolism.

ale [FOOD ENG] A fermented malt beverage, differing from beer in containing up to 8% alcohol by volume and being hopped more heavily.

alecithal [CYTOL] Referring to an egg without yolk, such as the eggs of placental mammals.

aleph null [MATH] The cardinal number of any set which can be put in one-to-one correspondence with the set of positive integers.

Alepocephaloidei [VERT ZOO] The slickheads, a suborder of deap-sea teleostean fishes of the order Salmoniformes.

Aleutian Current [OCEANOGR] A current setting southwestward along the southern coasts of the Aleutian Islands.

alewife [VERT ZOO] *Pomolobus pseudoharengus*. A food fish of the herring family that is very abundant on the Atlantic coast.

Alexandrian [GEOL] Lower Silurian geologic time.

alexandrite [MINERAL] A gem variety of chrysoberyl; emerald green in natural light but red in transmitted or artificial light.

Aleyrodidae [INV ZOO] The whiteflies, a family of homopteran insects included in the series Sternorrhyncha; economically important as plant pests.

alfalfa [BOT] *Medicago sativa*. A herbaceous perennial legume in the order Rosales, characterized by a deep taproot. Also known as lucerne.

alfenol [METAL] A permeability alloy that has 16% aluminum and 84% iron; it is brittle and at 572°F (300°C) can be rolled into thin sheets; used for transformer cores and tape recorder heads.

Alfisol [GEOL] An order of soils with gray to brown surface horizons, a medium-to-high base supply, and horizons of clay accumulation.

Alford loop [ELECTROMAG] An antenna utilizing multielements which usually are contained in the same horizontal plane and adjusted so that the antenna has approximately equal and in-phase currents uniformly distributed along each of its peripheral elements and produces a substantially circular radiation pattern in the plane of polarization; it is known for its purity of polarization.

Alfvén number [PHYS] The ratio of the speed of the Alfvén wave to the speed of the fluid at a point in the fluid.

Alfvén speed [PHYS] The speed of motion of the Alfvén wave, which is $v_a = B_0/\sqrt{\rho\mu}$, where B_0 is the magnetic field strength, ρ the fluid density, and μ the magnetic permeability (in meter-kilogram-second units).

Alfvén wave [PHYS] A hydromagnetic shear wave which moves along magnetic field lines; a major accelerative mechanism of charged particles in plasma physics and astrophysics.

algae [BOT] General name for the chlorophyll-bearing organisms in the plant subkingdom Thallobionta.

algae bloom [ECOL] A heavy growth of algae in and on a body of water as a result of high phosphate concentration from farm fertilizers and detergents.

algal limestone [PETR] A type of limestone either formed from the remains of calcium-secreting algae or formed when algae bind together the fragments of other lime-secreting organisms.

algebra [MATH] **1.** A method of solving practical problems by using symbols, usually letters, for unknown quantities. **2.** The study of the formal manipulations of equations involving symbols and numbers. **3.** An abstract mathematical system consisting of a vector space together with a multiplication by which two vectors may be combined to yield a third, and some axioms relating this multiplication to vector addition and scalar multiplication. Also known as hypercomplex system.

algebraic addition [MATH] The addition of algebraic quantities in the sense that adding a negative quantity is the same as subtracting a positive one.

algebraic curve [MATH] **1.** The set of points in the plane satisfying a polynomial equation in two variables. **2.** More generally, the set of points in n-space satisfying a polynomial equation in n variables.

algebraic equation [MATH] An equation in which zero is set equal to an algebraic expression.

algebraic expression [MATH] An expression which is obtained by performing a finite number of the following operations on symbols representing numbers: addition, subtraction, multiplication, division, raising to a power.

algebraic function [MATH] A function whose value is obtained by performing only the following operations to its argument: addition, subtraction, multiplication, division, raising to a rational power.

algebraic number [MATH] Any root of a polynomial with rational coefficients.

algebraic number theory [MATH] The study of properties of real numbers, especially integers, using the methods of abstract algebra.

algebraic topology [MATH] The study of topological properties of figures using the methods of abstract algebra; includes homotopy theory, homology theory, and cohomology theory.

Algerian onyx *See* onyx marble.

algin *See* sodium alginate.

alginic acid [ORG CHEM] $(C_6H_8O_6)_n$ An insoluble colloidal acid obtained from brown marine algae; it is hard when dry and absorbent when moist. Also known as algin.

algodonite [MINERAL] Cu_6As A steel gray to silver white mineral consisting of copper arsenide and occuring as minute hexagonal crystals or in massive and granular form.

ALGOL [COMPUT SCI] An algorithmic and procedure-oriented computer language used principally in the programming of scientific problems.

algorithm [MATH] A set of well-defined rules for the solution of a problem in a finite number of steps.

algorithmic language [COMPUT SCI] A language in which a procedure or scheme of calculations can be expressed accurately.

algorithm translation [COMPUT SCI] A step-by-step computerized method of translating one programming language into another programming language.

alias [COMPUT SCI] An alternative entry point in a computer subroutine at which its execution may begin, if so instructed by another routine.

alicyclic [ORG CHEM] 1. Having the properties of both aliphatic and cyclic substances. 2. Referring to a class of organic compounds containing only carbon and hydrogen atoms joined to form one or more rings. 3. Any one of the compounds of the alicyclic class. Also known as cyclane.

alidade [ENG] 1. An instrument for topographic surveying and mapping by the plane-table method. 2. Any sighting device employed for angular measurement.

alignment [ELECTR] The process of adjusting components of a system for proper interrelationship, including the adjustment of tuned circuits for proper frequency response and the time synchronization of the components of a system. [ENG] Placing of surveying points along a straight line. [MAP] Representing of the correct direction, character, and relationships of a line or feature on a map. [MIN ENG] The act of laying out a tunnel or regulating by line; adjusting to a line. [NUC PHYS] A population $p(m)$ of the $2I + 1$ orientational substates of a nucleus; $m = -I$ to $+I$, such that $p(m) = p(-m)$.

alignment chart See nomograph.

alignment pin [DES ENG] Pin in the center of the base of an octal, loctal, or other tube having a single vertical projecting rib that aids in correctly inserting the tube in its socket.

alignment wire See ground wire.

alimentary canal [ANAT] The tube through which food passes; in man, includes the mouth, pharynx, esophagus, stomach, and intestine.

alimentation [BIOL] Providing nourishment by feeding. [HYD] See accumulation.

aliphatic [ORG CHEM] Of or pertaining to any organic compound of hydrogen and carbon characterized by a straight chain of the carbon atoms; three subgroups of such compounds are alkanes, alkenes, and alkynes.

aliphatic series [ORG CHEM] A series of open-chained carbon-hydrogen compounds; the two major classes are the series with saturated bonds and with the unsaturated.

aliquot [MED] A representative sample of a larger quantity.

Alismataceae [BOT] A family of flowering plants belonging to the order Alismatales characterized by schizogenous secretory cells, a horseshoe-shaped embryo, and one or two ovules.

Alismatales [BOT] A small order of flowering plants in the subclass Alismatidae, including aquatic and semiaquatic herbs.

Alismatidae [BOT] A relatively primitive subclass of aquatic or semiaquatic herbaceous flowering plants in the class Liliopsida, generally having apocarpous flowers, and trinucleate pollen and lacking endosperm.

alive [ELEC] See energized. [MIN ENG] That portion of a lode that is productive.

alkali [CHEM] Any compound having highly basic qualities. [PETR] See alkalic.

alkalic Also known as alkali. [PETR] 1. Of igneous rock, containing more than average alkali (K_2O and Na_2O)

for that clan in which they are found. 2. Of igneous rock, having feldspathoids or other minerals, such as acmite, so that the molecular ratio of alkali to silica is greater than 1:6. 3. Of igneous rock, having a low alkali-lime index (51 or less).

alkali emission [GEOPHYS] Light emission from free lithium, potassium, and especially sodium in the upper atmosphere.

alkali feldspar [MINERAL] A feldspar composed of potassium feldspar and sodium feldspar, such as orthoclase, microcline, albite, and anorthoclase; all are considered alkali-rich.

alkali lake [HYD] A lake with large quantities of dissolved sodium and potassium carbonates as well as sodium chloride.

alkaline [CHEM] 1. Having properties of an alkali. 2. Having a pH greater than 7.

alkaline cell [ELEC] A primary cell that uses an alkaline electrolyte, usually potassium hydroxide, and delivers about 1.5 volts at much higher current rates than the common carbon-zinc cell. Also known as alkaline-manganese cell.

alkaline-earth metals [CHEM] The heaviest members of group IIa in the periodic table; usually calcium, strontium, magnesium, and barium.

alkaline-manganese cell See alkaline cell.

alkaline storage battery [ELEC] A storage battery in which the electrolyte consists of an alkaline solution, usually potassium hydroxide.

alkali soil [GEOL] A soil, with salts injurious to plant life, having a pH value of 8.5 or higher.

alkali vapor lamp [ELECTR] A vapor lamp in which light is produced by an electric discharge between electrodes in an alkali vapor at low or high pressures.

alkaloid [ORG CHEM] One of a group of nitrogenous bases of plant origin, such as nicotine, cocaine, and morphine.

alkane [ORG CHEM] A member of a series of saturated aliphatic hydrocarbons having the empirical formula C_nH_{2n+2}.

alkaptonuria [MED] A hereditary metabolic disorder transmitted as an autosomal recessive in man in which large amounts of homogentisic acid (alkapton) are excreted in the urine due to a deficiency of homogentisic acid oxidase. Also spelled alcaptonuria.

alkene [ORG CHEM] One of a class of unsaturated aliphatic hydrocarbons containing one or more carbon-to-carbon double bonds.

alkyl [ORG CHEM] A monovalent radical, C_nH_{2n+1}, which may be considered to be formed by loss of a hydrogen atom from an alkane; usually designated by R.

alkylation [CHEM ENG] A refinery process for chemically combining isoparaffin with olefin hydrocarbons. [ORG CHEM] A chemical process in which an alkyl radical is introduced into an organic compound by substitution or addition.

alkyne [ORG CHEM] One of a group of organic compounds containing a carbon-to-carbon triple bond.

allanite [MINERAL] $(Ca,Ce,La,Y)_2(Al,Fe)_3Si_3O_{12}(OH)$ Monoclinic mineral distinguished from all other members of the epidote group of silicates by a relatively high content of rare earths. Also known as bucklandite; cerine; orthite; treanorite.

allantoic acid [BIOCHEM] $C_4H_8N_4O_4$ A crystalline acid obtained by hydrolysis of allantoin; intermediate product in nucleic acid metabolism.

allantoin [BIOCHEM] $C_4H_6N_4O_3$ A crystallizable oxidation product of uric acid found in allantoic and amniotic fluids and in fetal urine.

allantois [EMBRYO] A fluid-filled, saclike, extraembryonic membrane lying between the chorion and amnion of reptilian, bird, and mammalian embryos.

all-channel tuning [COMMUN] The ability of a television set to receive ultra-high-frequency as well as very-high-frequency channels.

Alleculidae [INV ZOO] The comb claw beetles, a family of mostly tropical coleopteran insects in the superfamily Tenebrionoidea.

allele [GEN] One of a pair of genes, or of multiple forms of a gene, located at the same locus of homologous chromosomes. Also known as allelomorph.

allelomorph *See* allele.

allelopathy [PL PHYS] The harmful effect of one plant or microorganism on another owing to the release of secondary metabolic products into the environment.

allemontite [MINERAL] AsSb Rhombohedric, gray or reddish, native antimony arsenide occurring in reniform masses. Also known as arsenical antimony.

Allen screw [DES ENG] A screw or bolt which has an axial hexagonal socket in its head.

Allen wrench [DES ENG] A wrench made from a straight or bent hexagonal rod, used to turn an Allen screw.

allergen [IMMUNOL] Any antigen, such as pollen, a drug, or food, that induces an allergic state in humans or animals.

allergic reaction *See* allergy.

allergic rhinitis *See* hay fever.

allergy [MED] A type of antigen-antibody reaction marked by an exaggerated physiologic response to a substance that causes no symptoms in nonsensitive individuals. Also known as allergic reaction.

alligator [VERT ZOO] Either of two species of archosaurian reptiles in the genus *Alligator* of the family Alligatoridae.

alligator clip [ELEC] A long, narrow spring clip with meshing jaws; used with test leads to make temporary connections quickly. Also known as crocodile clip.

Alligatorinae [VERT ZOO] A subgroup of the crocodilian family Crocodylidae that includes alligators, caimans, *Melanosuchus*, and *Paleosuchus*.

alligatoring [MATER] Cracking of a film of paint or varnish, with broad, deep cracks through one or more coats. Also known as crocodiling. [MET] **1.** A splitting of an end of a rolled steel slab in which the plane of the split is parallel to the rolled surface. Also known as fishmouthing. **2.** The roughening of a sheet-metal surface during forming due to the coarse grain of the metal used.

alligator wrench [DES ENG] A wrench having fixed jaws forming a V, with teeth on one or both jaws.

allobar [NUC PHYS] A form of an element differing in its atomic weight from the naturally occurring form and hence being of different isotopic composition. [PHYS] A barometric pressure change.

allocate [COMPUT SCI] To place a portion of a computer memory or a peripheral unit under control of a computer program, through the action of an operator, program instruction, or executive program. [IND ENG] To assign a portion of a resource to an activity.

allochromy [PHYS] Emission of electromagnetic radiation that results from incident radiation at a different wavelength, as occurs in fluorescence or the Raman effect.

allochthon [GEOL] A rock that was transported a great distance from its original deposition by some tectonic process, generally related to overthrusting, recumbent folding, or gravity sliding.

allochthonous [PETR] Of rocks whose primary constituents have not been formed in situ.

Alloeocoela [INV ZOO] An order of platyhelminthic worms of the class Turbellaria distinguished by a simple pharynx and a diverticulated intestine.

allogeneic [IMMUNOL] Referring to a transplant made to a different genotype within the same species.

allograft [BIOL] Graft from a donor transplanted to a genetically dissimilar recipient of the same species.

Allogromiidae [INV ZOO] A little-known family of protozoans in the order Foraminiferida; adults are characterized by a chitinous test.

Allogromiina [INV ZOO] A suborder of marine and fresh-water protozoans in the order Foraminiferida characterized by an organic test of protein and acid mucopolysaccharide.

allometry [BIOL] **1.** The quantitative relation between a part and the whole or another part as the organism increases in size. Also known as heterauxesis; heterogony. **2.** The quantitative relation between the size of a part and the whole or another part, in a series of related organisms that differ in size.

allopelagic [ECOL] Relating to organisms living at various depths in the sea in response to influences other than temperature.

allophane [GEOL] $Al_2O_3 \cdot SiO_2 \cdot nH_2O$ A clay mineral composed of hydrated aluminosilicate gel of variable composition; P_2O_5 may be present in appreciable quantity.

allopolyploid [GEN] An organism or strain arising from a combination of genetically distinct chromosome sets from two diploid organisms.

allosteric enzyme [BIOCHEM] Any of the regulatory bacterial enzymes, such as those involved in end-product inhibition.

Allotheria [PALEON] A subclass of Mammalia that appeared in the Upper Jurassic and became extinct in the Cenozoic.

allotrioblast *See* xenoblast.

Allotriognathi [VERT ZOO] An equivalent name for the Lampridiformes.

allotriomorphism *See* allotropy.

allotropism *See* allotropy.

allotropy [CHEM] The assumption by an element or other substance of two or more different forms or structures which are most frequently stable in different temperature ranges, such as different crystalline forms of carbon as charcoal, graphite, or diamond. Also known as allotriomorphism; allotropism.

allotter [COMMUN] A telephone term referring to a distributor, associated with the finder control group relay assembly, which allots an idle line-finder in preparation for an additional call.

allotter relay [COMMUN] A telephone term referring to a relay of the line-finder circuit whose function is to preallot an idle line-finder to the next incoming call from the line, and to guard relays.

allotype [SYST] A paratype of the opposite sex to the holotype.

allowable load [MECH] The maximum force that may be safely applied to a solid, or is permitted by applicable regulators.

allowable stress [MECH] The maximum force per unit area that may be safely applied to a solid.

allowance [DES ENG] An intentional difference in sizes of two mating parts, allowing clearance usually for a film of oil, for running or sliding fits.

allowed transition [QUANT MECH] A transition between two states which is permitted by the selection rules and which consequently has a relatively high priority.

alloxan [BIOCHEM] $C_4H_2N_2O_4$ Crystalline oxidation product of uric acid; induces diabetes experimentally by selective destruction of pancreatic beta cells. Also known as mesoxalylurea.

alloy [MET] Any of a large number of substances having metallic properties and consisting of two or more elements; with few exceptions, the components are usually metallic elements.

alloy junction diode [ELECTR] A junction diode made by placing a pill of doped alloying material on a semiconductor material and heating until the molten alloy melts a portion of the semiconductor, resulting in a *pn* junction when the dissolved semiconductor recrystallizes. Also known as fused-junction diode.

alloy-junction transistor [ELECTR] A junction transistor made by placing pellets of a *p*-type impurity such as indium above and below an *n*-type wafer of germanium, then heating until the impurity alloys with the germanium to give a *pnp* transistor. Also known as fused-junction transistor.

alloy steel [MET] A steel whose distinctive properties are due to the presence of one or more elements other than carbon.

all-pass network [ELECTR] A network designed to introduce a phase shift in a signal without introducing an appreciable reduction in energy of the signal at any frequency.

allspice [BOT] The dried, unripe berries of a small, tropical evergreen tree, *Pimenta officinalis*, of the myrtle family; yields a pungent, aromatic spice.

alluvial [GEOL] 1. Of a placer, or its associated valuable mineral, formed by the action of running water. 2. Pertaining to or consisting of alluvium, or deposited by running water.

alluvial cone [GEOL] An alluvial fan with steep slopes formed of loose material washed down the slopes of mountains by ephemeral streams and deposited as a conical mass of low slope at the mouth of a gorge. Also known as cone delta; cone of dejection; cone of detritus; debris cone; dry delta; hemicone; wash.

alluvial deposit *See* alluvium.

alluvial fan [GEOL] A fan-shaped deposit formed by a stream either where it issues from a narrow mountain valley onto a plain or broad valley, or where a tributary stream joins a main stream.

alluvial mining [MIN ENG] The exploitation of alluvial deposits by dredging, hydraulicking, or drift mining.

alluvial plain [GEOL] A plain formed from the deposition of alluvium usually adjacent to a river that periodically overflows. Also known as aggraded valley plain; river plain; wash plain; waste plain.

alluvial terrace [GEOL] A terraced embankment of loose material adjacent to the sides of a river valley. Also known as built terrace; drift terrace; fill terrace; stream-built terrace; wave-built platform; wave-built terrace.

alluvion *See* alluvium.

alluvium [GEOL] The detrital materials eroded, transported, and deposited by streams; an important constituent of shelf deposits. Also known as alluvial deposit; alluvion.

all-wave receiver [ELECTR] A radio receiver capable of being tuned from about 535 kilohertz to at least 20 megahertz; some go above 100 megahertz and thus cover the FM band also.

allyl- [ORG CHEM] A prefix used in names of compounds whose structure contains an allyl cation.

allylic rearrangement [ORG CHEM] In a three-carbon molecule, the shifting of a double bond from the 1,2 carbon position to the 2,3 position, with the accompanying migration of an entering substituent or substituent group from the third carbon to the first.

allyl plastic *See* allyl resin.

allyl resin [ORG CHEM] Any of a class of thermosetting synthetic resins derived from esters of allyl alcohol or allyl chloride; used in making cast and laminated products. Also known as allyl plastic.

almandine [MINERAL] $Fe_3Al_2(SiO_4)_3$ A variety of garnet, deep red to brownish red, found in igneous and metamorphic rocks in many parts of world; used as a gemstone and an abrasive. Also known as almandite.

almandite *See* almandine.

almond [BOT] *Prunus amygdalus*. A small deciduous tree of the order Rosales; it produces a drupaceous edible fruit with an ellipsoidal, slightly compressed nutlike seed.

almucantar *See* parallel of altitude.

alnico [MET] One of a series of ferrous alloys containing aluminum, nickel, and cobalt, valued because of their highly retentive magnetic properties; usually designated with a roman-numeral number, such as alnico VII. Also known as aluminum-nickel-cobalt alloy.

aloe [PHARM] The dried resinous juice extracted from the leaves of the genus *Aloe*, especially *Aloe vulgaris* of the West Indies and *A. perryi* of Africa; used in purgative mixtures.

alongshore current *See* littoral current.

Alopiidae [VERT ZOO] A family of pelagic isurid elasmobranchs commonly known as thresher sharks because of their long, whiplike tail.

alpaca [VERT ZOO] *Lama pacos*. An artiodactyl of the camel family (Camelidae); economically important for its long, fine wool.

alphabet [SCI TECH] Any ordered set of unique graphics called characters, such as the 26 letters of the Roman alphabet.

alphabetic string *See* character string.

alpha brass [MET] An alloy of copper and zinc containing up to 36% zinc dissolved, rather than chemically combined, with the copper; ductile, easily cold-worked, and corrosion resistant; used for hot-water pipes.

alpha cell [HISTOL] Any of the acidophilic chromophiles in the anterior lobe of the adenohypophysis.

alpha decay [NUC PHYS] A radioactive transformation in which an alpha particle is emitted by a nuclide.

alpha globulin [BIOCHEM] A heterogeneous fraction of serum globulins containing the proteins of greatest electrophoretic mobility.

alpha helix [MOL BIO] A spatial configuration of the polypeptide chains of proteins in which the chain assumes a helical form, 0.54 nanometer in pitch, 3.6 amino acids per turn, presenting the appearance of a hollow cylinder with radiating side groups.

alphameric characters *See* alphanumeric characters.

alphanumeric characters [COMPUT SCI] All characters used by a computer, including letters, numerals, punctuation marks, and such signs as $, @, and #. Also known as alphameric characters.

alphanumeric grid *See* atlas grid.

alphanumeric instruction [COMPUT SCI] The name given to instructions which can be read equally well with alphabetic or numeric kinds of fields of data.

alpha particle [ATOM PHYS] A positively charged particle consisting of two protons and two neutrons, identical with the nucleus of the helium atom; emitted by several radioactive substances.

alpha position [ORG CHEM] In chemical nomenclature, the position of a substituting group of atoms in the main group of a molecule; for example, in a straight-chain compound such as α-hydroxypropionic acid ($CH_3CHOHCOOH$), the hydroxyl radical is in the alpha position.

alpha ray [NUCLEO] A stream of alpha particles.

alpha rhythm [PHYSIO] An electric current from the occipital region of the brain cortex having a pulse frequency of 8 to 13 per second; associated with a relaxed state in normal human adults.

alphascope [COMPUT SCI] An interactive alphanumerical input/output device that consists of a cathode-ray tube, keyboard, method of generating characters, method of refreshing the display, and communications equipment, and that forms part of a computer-based system requiring a short response time for retrieving answers to queries from a computer random-access memory.

Alpheidae [INV ZOO] The snapping shrimp, a family of decapod crustaceans included in the section Caridea.

alpine glacier [HYD] A glacier lying on or occupying a depression in mountainous terrain. Also known as mountain glacier.

alpine tundra [ECOL] Large, flat or gently sloping, treeless tracts of land above the timberline.

alstonite See bromlite.

alt See altitude.

altazimuth [ENG] An instrument equipped with both horizontal and vertical graduated circles, for the simultaneous observation of horizontal and vertical directions or angles. Also known as astronomical theodolite; universal instrument.

alteration switch [COMPUT SCI] A hand-operated switch mounted on the console of a computer, used to feed a single bit of information into a program. Also known as sense switch.

alternate [BOT] 1. Of the arrangement of leaves on opposite sides of the stem at different levels. 2. Of the arrangement of the parts of one whorl between members of another whorl.

alternate-channel interference [COMMUN] Interference that is caused in one communications channel by a transmitter operating in the next channel beyond an adjacent channel. Also known as second-channel interference.

alternating current [ELEC] Electric current that reverses direction periodically, usually many times per second. Abbreviated ac.

alternating-current generator [ELEC] A machine, usually rotary, which converts mechanical power into alternating-current electric power.

alternating-current motor [ELEC] A machine that converts alternating-current electrical energy into mechanical energy by utilizing forces exerted by magnetic fields produced by the current flow through conductors.

alternating-current resistance See high-frequency resistance.

alternating-current transmission [ELECTR] In television, that form of transmission in which a fixed setting of the controls makes any instantaneous value of signal correspond to the same value of brightness for only a short time.

alternating function [MATH] A function in which the interchange of two independent variables causes the dependent variable to change sign.

alternating gradient [ELECTROMAG] A magnetic field in which successive magnets have gradients of opposite sign, so that the field increases with radius in one magnet and decreases with radius in the next; used in synchrotrons and cyclotrons.

alternating series [MATH] Any series of real numbers in which consecutive terms have opposite signs.

alternation of generations See metagenesis.

alternator [ELEC] A mechanical, electrical, or electromechanical device which supplies alternating current.

altimeter [ENG] An instrument which determines the altitude of an object with respect to a fixed level, such as sea level; there are two common types: the aneroid altimeter and the radio altimeter.

altitude Abbreviated alt. [ENG] 1. Height, measured as distance along the extended earth's radius above a given datum, such as average sea level. 2. Angular displacement above the horizon measured by an altitude curve. [MATH] The perpendicular distance from the base to the top (a vertex or parallel line) of a geometric figure such as a triangle or parallelogram.

altitude circle [ASTRON] See parallel of altitude. [ELECTROMAG] A bright circle which surrounds the central dark portion of a plan position indicator display or photograph, and which results from ground clutter.

altitude curve [ENG] The arc of a vertical circle between the horizon and a point on the celestial sphere, measured upward from the horizon. [NAV] A graphical representation of the altitude of a celestial body as it would appear from a single assumed position or a series of assumed positions over a period of time; such curves are precomputed.

altitude delay [ELECTR] Synchronization delay introduced between the time of transmission of the radar pulse and the start of the trace on the indicator to eliminate the altitude/height hole on the plan position indicator–type display.

altitude difference [ENG] The difference between computed and observed altitudes, or between precomputed and sextant altitudes. Also known as altitude intercept; intercept.

altitude hole [ELECTR] The blank area in the center of a plan position indicator–type radarscope display caused by the time interval between transmission of a pulse and the receipt of the first ground return.

altitude intercept See altitude difference.

altocumulus cloud [METEOROL] A principal cloud type, white or gray or both white and gray in color; occurs as a layer or patch with a waved aspect, the elements of which appear as laminae, rounded masses, or rolls; frequently appears at different levels in a given sky. Abbreviated Ac.

altostratus cloud [METEOROL] A principal cloud type in the form of a gray or bluish (never white) sheet or layer of striated, fibrous, or uniform appearance; very often totally covers the sky and may cover an area of several thousand square miles; vertical extent may be from several hundred to thousands of meters. Abbreviated As.

altricial [VERT ZOO] Pertaining to young that are born or hatched immature and helpless, thus requiring extended development and parental care.

ALU See arithmetical unit.

alula [ZOO] 1. Digit of a bird wing homologous to the thumb. 2. See calypter.

alum [INORG CHEM] 1. Any of a group of double sulfates of trivalent metals such as aluminum, chromium, or iron and a univalent metal such as potassium or sodium. 2. See aluminum sulfate; ammonium aluminum sulfate; potassium aluminum sulfate. [MINERAL] $KAl(SO_4)_2 \cdot 12H_2O$ A colorless, white, astringent-tasting evaporite mineral.

alumina [INORG CHEM] Al_2O_3 The native form of aluminum oxide occurring as corundum or in hydrated forms, as a powder or crystalline substance.

alumina cement [MATER] A cement made with bauxite and containing a high percentage of aluminate, having the property of setting to high strength in 24 hours.

aluminite [MINERAL] $Al_2(SO_4)(OH)_4 \cdot 7H_2O$ Native monoclinic hydrous aluminum sulfate; used in tanning, papermaking, and water purification. Also known as websterite.

aluminium See aluminum.

aluminize [ENG] To apply a film of aluminum to a material, such as glass. [MET] To form a protective surface alloy on a metal by treatment at elevated temperature with aluminum or an aluminum compound.

aluminum [CHEM] A chemical element, symbol Al, atomic number 13, and atomic weight 26.9815. Also spelled aluminium.

aluminum brass [MET] 1. A casting brass to which aluminum has been added as a flux to improve the casting qualities and, with the addition of lead, the machining qualities. 2. A wrought brass to which aluminum has been added to improve the extruding and forging qualities and the oxidation resistance.

aluminum bronze [MET] A copper-aluminum alloy which may also contain iron, manganese, nickel, or zinc.

aluminum foil [MET] Aluminum in the form of a sheet of thickness not exceeding 0.005 inch (0.127 mm).

aluminum-nickel-cobalt alloy See alnico.

aluminum potassium sulfate See potassium aluminum sulfate.

aluminum soap [ORG CHEM] Any of various salts of higher carboxylic acids and aluminum that are insoluble in water and soluble in oils; used in lubricating greases, paints, varnishes, and waterproofing substances.

aluminum sulfate [INORG CHEM] $Al_2(SO_4)_3 \cdot 18H_2O$ A colorless salt in the form of monoclinic crystals that decompose in heat and are soluble in water; used in papermaking, water purification, and tanning, and as a mordant in dyeing. Also known as alum.

alumite See alunite.

alum rock See alunite.

alumstone See alunite.

alunite [MINERAL] $KAl_3(SO_4)_2(OH)_6$ A mineral composed of a basic potassium aluminum sulfate; it occurs as a hydrothermal-alteration product in feldspathic igneous rocks and is used in the manufacture of alum. Also known as alumite; alum rock; alumstone.

alveolar gland See acinous gland.

alveolus [ANAT] 1. A tiny air sac of the lung. 2. A tooth socket. 3. A sac of a compound gland.

Alydidae [INV ZOO] A family of hemipteran insects in the superfamily Coreoidea.

Am See americium; ammonium.

AM See amplitude modulation.

Amagat law See Amagat-Leduc rule.

Amagat-Leduc rule [PHYS] The rule which states that the volume taken up by a gas mixture equals the sum of the volumes each gas would occupy at the temperature and pressure of the mixture. Also known as Amagat law; Leduc law.

Amagat system [PHYS] A system of units in which the unit of pressure is the atmosphere and the unit of volume is the gram-molecular volume (22.4 liters at standard conditions).

amalgam [MET] An alloy of mercury. [MINERAL] A silver mercury alloy occurring in nature.

amalgamate [MET] 1. To unite a metal in an alloy with mercury. 2. To unite two dissimilar metals. 3. To cover the zinc elements of a galvanic battery with mercury.

amalgamation [MET] Also known as amalgam treatment. 1. The process of separating metal from ore by alloying the metal with mercury; formerly used for gold and silver recovery, where it has been superseded by the cyanide process. 2. The formation of an alloy of a metal with mercury.

amalgam treatment See amalgamation.

Amalthea [ASTRON] The innermost known satellite of Jupiter, orbiting at a mean distance of 1.13×10^5 miles (1.82×10^5 kilometers); it has a diameter of about 150 miles (240 kilometers). Also known as Jupiter V.

Amaranthaceae [BOT] The characteristic family of flowering plants in the order Caryophyllales; they have a syncarpous gynoecium, a monochlamydeous perianth that is more or less scarious, and mostly perfect flowers.

Amaryllidaceae [BOT] The former designation for a family of plants now included in the Liliaceae.

amateur radio [ELECTR] A radio used for two-way radio communications by private individuals as leisure-time activity. Also known as ham radio.

amatol [MATER] An explosive mixture composed of ammonium nitrate and trinitrotoluene; mixtures with 50% and 80% ammonium nitrate are used for small and large shells, respectively.

amazonite [MINERAL] An apple-green, bright-green, or blue-green variety of microcline found in the United States and Soviet Union; sometimes used as a gemstone. Also known as amazon stone.

amazon stone See amazonite.

amber [MINERAL] A transparent yellow, orange, or reddish-brown fossil resin derived from a coniferous tree; used for ornamental purposes; it is amorphous, has a specific gravity of 1.05–1.10, and a hardness of 2–2.5 on Mohs scale.

ambergris [PHYSIO] A fatty substance formed in the intestinal tract of the sperm whale; used in the manufacture of perfume.

ambient [ENG] Surrounding; especially, of or pertaining to the environment about a flying aircraft or other body but undisturbed or unaffected by it, as in ambient air or ambient temperature.

ambipolar [SCI TECH] Simultaneously operating in two opposite directions; for example, an electric current arising from the movement of positive and negative ions.

amblygonite [MINERAL] $(Li,Na)AlPO_4(F,OH)$ A mineral occurring in white or greenish cleavable masses and found in the United States and Europe; important ore of lithium.

amblyopia [MED] Dimness of vision, especially that not due to refractive errors or organic disease of the eye; may be congenital or acquired.

Amblyopsidae [VERT ZOO] The cave fishes, a family of actinopterygian fishes in the order Percopsiformes.

Amblyopsiformes [VERT ZOO] An equivalent name for the Percopsiformes.

Amblypygi [INV ZOO] An order of chelicerate arthropods in the class Arachnida, commonly known as the tailless whip scorpions.

amboceptor [IMMUNOL] According to P. Ehrlich, an antibody present in the blood of immunized animals which contains two specialized elements: a cytophil group that unites with a cellular antigen, and a complementophil group that joins with the complement.

ambulacrum [INV ZOO] In echinoderms, any of the radial series of plates along which the tube feet are arranged.

Ambystomatidae [VERT ZOO] A family of urodele amphibians in the suborder Salamandroidea; neoteny occurs frequently in this group.

Ambystomoidea [VERT ZOO] A suborder to which the family Ambystomatidae is sometimes elevated.

Amebelodontinae [PALEON] A subfamily of extinct elephantoid proboscideans in the family Gomphotheriidae.

amebiasis [MED] A parasitic disease of man caused by the ameba *Entamoeba histolytica*, characterized by clinical-pathological intestinal manifestations, including an acute dysentery phase. Also known as amebic dysentery.

amebic dysentery See amebiasis.

amebocyte [INV ZOO] One of the wandering ameboid cells in the tissues and fluids of many invertebrates that function in assimilation and excretion.

Ameiuridae [VERT ZOO] A family of North American catfishes belonging to the suborder Siluroidei.

ameloblast [EMBRYO] One of the columnar cells of the enamel organ that form dental enamel in developing teeth.

amendment record See change record.

amenorrhea [MED] Absence of menstruation due to either normal or abnormal conditions.

amentia [MED] Congenital subnormal intellectual development.

American bond [CIV ENG] A bond in which every fifth, sixth, or seventh course of a wall consists of headers and the other courses consist of stretchers. Also known as common bond; Scotch bond.

American caisson See box caisson.

American filter See disk filter.

American jade See californite.

American lion See puma.

American spotted fever See Rocky Mountain spotted fever.

American Standard Code for Information Interchange [COMMUN] Coded character set to be used for the general interchange of information among information-processing systems, communications systems, and associated equipment. Abbreviated ASCII.

American standard pipe thread [DES ENG] Taper, straight, or dryseal pipe thread whose dimensions conform to those of a particular series of specified sizes established as a standard in the United States. Also known as Briggs pipe thread.

American standard screw thread [DES ENG] Screw thread whose dimensions conform to those of a particular series of specified sizes established as a standard in the United States; used for bolts, nuts, and machine screws.

American system drill See churn drill.

American wire gage [MET] A particular series of specified diameters and thicknesses established as a standard in the United States and used for nonferrous sheets, rods, and wires. Abbreviated AWG. Also known as Brown and Sharp gage (B and S gage).

americium [CHEM] A chemical element, symbol Am, atomic number 95; the mass number of the isotope with the longest half-life is 243.

Amerosporae [MYCOL] A spore group of the Fungi Imperfecti characterized by one-celled or threadlike spores.

ametabolous metamorphosis [INV ZOO] A growth stage of certain insects characterized by an increase in size without distinct external changes.

A-metal [METAL] A type of permeability alloy containing 44% nickel and a small amount of copper; used to give nondistortion characteristics upon magnetization in transformers and loudspeakers.

amethyst [MINERAL] The transparent purple to violet variety of the mineral quartz; used as a jeweler's stone.

ametropia [MED] Any deficiency in the refractive ability of the eye that causes an unfocused image to fall on the retina.

Amici prism [OPTICS] A compound prism, used in direct-vision spectroscopes, that disperses a beam of light into a spectrum without causing the beam as a whole to undergo any net deviation; it is made up of alternate crown and flint glass components, refracting in opposite directions. Also known as direct-vision prism.

amictic lake [HYD] A lake that is perennially frozen.

amidation [ORG CHEM] The process of forming an amide; for example, in the laboratory benzyl reacts with methyl amine to form *N*-methylbenzamide.

amide [ORG CHEM] One of a class of organic compounds containing the $CONH_2$ radical.

Amiidae [VERT ZOO] A family of actinopterygian fishes in the order Amiiformes represented by a single living species, the bowfin (*Amia calva*).

Amiiformes [VERT ZOO] An order of actinopterygian fishes characterized by an abbreviate heterocercal tail, fusiform body, and median fin rays.

A min See ampere-minute.

amin- See amino-.

amine [ORG CHEM] One of a class of organic compounds which can be considered to be derived from ammonia by replacement of one or more hydrogens by organic radicals.

amino- [CHEM] Having the property of a compound in which the group NH_2 is attached to a radical other than an acid radical. Also spelled amin-.

amino acid [BIOCHEM] Any of the organic compounds that contain one or more basic amino groups and one or more acidic carboxyl groups and that are polymerized to form peptides and proteins; only 20 of the more than 80 amino acids found in nature serve as building blocks for proteins; examples are tyrosine and lysine.

aminoaciduria [MED] A group of disorders in which excess amounts of amino acids are excreted in the urine; caused by abnormal protein metabolism.

***para*-aminobenzoic acid** [BIOCHEM] $C_7H_7O_2N$ A yellow-red, crystalline compound that is part of the folic acid molecule; essential in metabolism of certain bacteria. Abbreviated PABA.

γ-aminobutyric acid [ORG CHEM] $H_2NCH_2CH_2$-CH_2COOH Crystals which are either leaflets or needles, with a melting point of 202°C; thought to be a central nervous system postsynaptic inhibitory transmitter. Abbreviated GABA. Also known as γ-amino-*n*-butyric acid; piperidic acid.

γ-amino-*n*-butyric acid See γ-aminobutyric acid.

α-aminohydrocinnamic acid See phenylalanine.

2-aminoisovaleric acid See valine.

α-aminoisovaleric acid See valine.

2-amino-3-methylbutyric acid See valine.

aminopeptidase [BIOCHEM] An enzyme which catalyzes the liberation of an amino acid from the end of a peptide having a free amino group.

α-amino-β-phenylpropionic acid See phenylalanine.

para-**aminosalicylic acid**   [PHARM] $C_7H_7NO_3$
White, crystalline drug used with other drugs in the
treatment of tuberculosis. Abbreviated PAS.

amino sugar   [BIOCHEM] A monosaccharide in which
a nonglycosidic hydroxyl group is replaced by an
amino or substituted amino group; an example is D-
glucosamine.

aminotransferase   *See* transaminase.

AML   *See* automatic modulation limiting.

ammeter   [ENG] An instrument for measuring the
magnitude of electric current flow.

Ammodiscacea   [INV ZOO] A superfamily of forami-
niferal protozoans in the suborder Textulariina,
characterized by a simple to labyrinthic test wall.

Ammodytoidei   [VERT ZOO] The sand lances, a suborder of marine actinopterygian fishes in the order
Perciformes, characterized by slender, eel-shaped
bodies.

ammonation   [INORG CHEM] A reaction in which ammonia is added to other molecules or ions by covalent bond formation utilizing the unshared pair of
electrons on the nitrogen atom, or through ion-di-
pole electrostatic interactions.

ammonia   [INORG CHEM] NH_3 A colorless gaseous
alkaline compound that is very soluble in water, has
a characteristic pungent odor, is lighter than air, and
is formed as a result of the decomposition of most
nitrogenous organic material; used as a fertilizer and
as a chemical intermediate.

ammoniac   [INORG CHEM] *See* ammoniacal.   [MATER]
A gum resin obtained from the stems of the ammoniac plant; used in medicine, perfume, plaster,
concrete, and adhesive. Also known as ammoniac
gum; ammoniacum; gum ammoniac; Persian ammoniac.

ammoniacal   [INORG CHEM] Pertaining to ammonia
or its properties. Also known as ammoniac.

ammoniac gum   *See* ammoniac.

ammonia clock   [HOROL] A time-measuring device
dependent on the pyramidal ammonia molecule's
property of turning inside out readily and oscillating
between the two extreme positions at the precise
frequency of 2.387013×10^{10} hertz.

ammoniacum   *See* ammoniac.

ammonia maser clock   [HOROL] A gas maser that
utilizes the transition of high-energy ammonia molecules to generate a stable microwave output signal
for use as a time standard.

ammonification   [CHEM] Addition of ammonia or
ammonia compounds, especially to the soil.

ammonite   [MATER] An explosive containing 70–95%
ammonium nitrate.   [PALEON] A fossil shell of the
cephalopod order Ammonoidea.

ammonium   [CHEM] The radical NH_4^+. Abbreviated Am.

ammonium aluminum sulfate   [INORG CHEM] NH_4Al-
$(SO_4)_2 \cdot 12H_2O$ Colorless, odorless crystals that are
soluble in water; used in manufacturing medicines
and baking powder, dyeing, papermaking, and tanning. Also known as alum; aluminum ammonium
sulfate; ammonia alum; ammonium alum.

ammonoid   [PALEON] A cephalopod of the order Ammonoidea.

Ammonoidea   [PALEON] An order of extinct cephalopod mollusks in the subclass Tetrabranchia; important as index fossils.

ammonolysis   [CHEM] 1. A dissociation reaction of
the ammonia molecule producing H^+ and NH_2^-
species. 2. Breaking of a bond by addition of ammonia.

Ammotheidae   [INV ZOO] A family of marine arthropods in the subphylum Pycnogonida.

ammunition   [ORD] 1. All kinds of missiles to be
thrown against an enemy. 2. Missiles not for direct
use against an enemy, with such purposes as illumination, signaling, and decelerating. 3. A complete round and all its components, that is, the material required for firing a weapon such as a pistol.

amnesia   [MED] The pathological loss or impairment
of memory brought about by psychogenic or physiological disturbances.

amniocentesis   [MED] A procedure during pregnancy by which the abdominal wall and fetal membranes are punctured with a cannula to withdraw
amniotic fluid.

amnion   [EMBRYO] A thin extraembryonic membrane
forming a closed sac around the embryo in birds,
reptiles, and mammals.

Amniota   [VERT ZOO] A collective term for the Reptilia, Aves, and Mammalia, all of which have an amnion during development.

amniotic fluid   [PHYSIO] A substance that fills the
amnion to protect the embryo from desiccation and
shock.

Amoebida   [INV ZOO] An order of rhizopod protozoans in the subclass Lobosia characterized by the
absence of a protective covering (test).

amorphous   [PHYS] Pertaining to a solid which is
noncrystalline, having neither definite form nor
structure.

amorphous semiconductor   [SOLID STATE] A semiconductor material which is not entirely crystalline,
having only short-range order in its structure.

amortisseur winding   *See* damper winding.

amortize   [IND ENG] To reduce gradually an obliga-
tion, such as a mortgage, by periodically paying a
part of the principal as well as the interest.

amp   *See* amperage; ampere.

AMP   *See* adenylic acid.

3′,5′-AMP   *See* cyclic adenylic acid.

ampangabeite   *See* samarskite.

Ampeliscidae   [INV ZOO] A family of tube-dwelling
amphipod crustaceans in the suborder Gammaridea.

amperage   [ELEC] The amount of electric current in
amperes. Abbreviated amp.

ampere   [ELEC] The unit of electric current in the
rationalized meter-kilogram-second system of units;
defined in terms of the force of attraction between
two parallel current-carrying conductors. Abbreviated a; A; amp.

ampere-hour   [ELEC] A unit for the quantity of electricity, obtained by integrating current flow in amperes over the time in hours for its flow; used as a
measure of battery capacity. Abbreviated Ah; amp-
hr.

Ampère law   [ELECTROMAG] 1. A law giving the
magnetic induction at a point due to given currents
in terms of the current elements and their positions
relative to the point. Also known as Laplace law. 2.
A law giving the line integral over a closed path of
the magnetic induction due to given currents in terms
of the total current linking the path.

ampere-minute   [ELEC] A unit of electrical charge,
equal to the charge transported in 1 minute by a
current of 1 ampere, or to 60 coulombs. Abbreviated A min.

Ampère rule   [ELECTROMAG] The rule which states
that the direction of the magnetic field surrounding
a conductor will be clockwise when viewed from the
conductor if the direction of current flow is away
from the observer.

Ampère theorem [ELECTROMAG] The theorem which states that an electric current flowing in a circuit produces a magnetic field at external points equivalent to that due to a magnetic shell whose bounding edge is the conductor and whose strength is equal to the strength of the current.

ampere-turn [ELECTROMAG] A unit of magnetomotive force in the meter-kilogram-second system defined as the force of a closed loop of one turn when there is a current of 1 ampere flowing in the loop. Abbreviated amp-turn.

amperometric titration [PHYS CHEM] A titration that involves measuring an electric current or changes in current during the course of the titration.

amperometry [PHYS CHEM] Chemical analysis by techniques which involve measuring electric currents.

Ampharetidae [INV ZOO] A large, deep-water family of polychaete annelids belonging to the Sedentaria.

Ampharetinae [INV ZOO] A subfamily of annelids belonging to the family Ampharetidae.

amphetamine [PHARM] $C_6H_5CH_2CHNHCH_3$ A volatile, colorless liquid used as a central nervous system stimulant. Also known as racemic 1-phenyl-2-aminopropane and by the trade name Benzedrine.

amphiarthrosis [ANAT] An articulation of limited movement in which bones are connected by fibrocartilage, such as that between vertebrae or that at the tibiofibular junction.

Amphibia [VERT ZOO] A class of vertebrate animals in the superclass Tetrapoda characterized by a moist, glandular skin, gills at some stage of development, and no amnion during the embryonic stage.

Amphibicorisae [INV ZOO] A subdivision of the insect order Hemiptera containing surface water bugs with exposed antennae.

Amphibioidei [INV ZOO] A family of tapeworms in the order Cyclophyllidea.

amphibiotic [ZOO] Being aquatic during the larval stage and terrestrial in the adult stage.

amphibious [BIOL] Capable of living both on dry or moist land and in water. [MECH ENG] Said of vehicles or equipment designed to be operated or used on either land or water. [ORD] A military operation conducted by coordinated action of land, sea, and air forces.

amphibole [MINERAL] Any of a group of rock-forming, ferromagnesian silicate minerals commonly found in igneous and metamorphic rocks; includes hornblende, anthophyllite, tremolite, and actinolite (asbestos minerals).

Amphibolidae [INV ZOO] A family of gastropod mollusks in the order Basommatophora.

amphibolite [PETR] A crystalloblastic metamorphic rock composed mainly of amphibole and plagioclase; quartz may be present in small quantities.

Amphichelydia [PALEON] A suborder of Triassic to Eocene anapsid reptiles in the order Chelonia; these turtles did not have a retractable neck.

Amphicoela [VERT ZOO] A small suborder of amphibians in the order Anura characterized by amphicoelous vertebrae.

amphicoelous [VERT ZOO] Describing vertebrae that have biconcave centra.

Amphicyonidae [PALEON] A family of extinct giant predatory carnivores placed in the infraorder Miacoidea by some authorities.

Amphidiscophora [INV ZOO] A subclass of sponges in the class Hexactinellida characterized by an anchoring tuft of spicules and no hexasters.

Amphidiscosa [INV ZOO] An order of hexactinellid sponges in the subclass Amphidiscophora characterized by amphidisc spicules, that is, spicules having a stellate disk at each end.

amphidromic [OCEANOGR] Of or pertaining to progression of a tide wave or bulge around a point or center of little or no tide.

amphigene See leucite.

Amphilestidae [PALEON] A family of Jurassic triconodont mammals whose subclass is uncertain.

Amphilinidea [INV ZOO] An order of tapeworms in the subclass Cestodaria characterized by a protrusible proboscis, anterior frontal glands, and no holdfast organ; they inhabit the coelom of sturgeon and other fishes.

Amphimerycidae [PALEON] A family of late Eocene to early Oligocene tylopod ruminants in the superfamily Amphimerycoidea.

Amphimerycoidea [PALEON] A superfamily of extinct ruminant artiodactyls in the infraorder Tylopoda.

amphimixis [PHYSIO] The union of egg and sperm in sexual reproduction.

Amphimonadidae [INV ZOO] A family of zoomastigophorean protozoans in the order Kinetoplastida.

amphimorphic [GEOL] A rock or mineral formed by two geologic processes.

Amphineura [INV ZOO] A class of the phylum Mollusca; members are bilaterally symmetrical, elongate marine animals, such as the chitons.

Amphinomidae [INV ZOO] The stinging or fire worms, a family of amphinomorphan polychaetes belonging to the Errantia.

Amphinomorpha [INV ZOO] Group name for three families of errantian polychaetes: Amphenomidae, Euphrosinidae, and Spintheridae.

amphiphatic molecule [ORG CHEM] A molecule having both hydrophilic and hydrophobic groups; examples are wetting agents and membrane lipids such as phosphoglycerides.

amphipneustic [VERT ZOO] Having both gills and lungs through all life stages, as in some amphibians.

Amphipoda [INV ZOO] An order of crustaceans in the subclass Malacostraca; individuals lack a carapace, bear unstalked eyes, and respire through thoracic branchiae or gills.

amphiprotic See amphoteric.

amphisarca [BOT] An indehiscent fruit characterized by many cells and seeds, pulpy flesh, and a hard rind; melon is an example.

Amphisbaenidae [VERT ZOO] A family of tropical snakelike lizards in the suborder Sauria.

Amphisopidae [INV ZOO] A family of isopod crustaceans in the suborder Phreatoicoidea.

Amphissitidae [PALEON] A family of extinct ostracods in the suborder Beyrichicopina.

Amphistaenidae [VERT ZOO] The worm lizards, a family of reptiles in the suborder Sauria; structural features are greatly reduced, particularly the limbs.

amphistome [INV ZOO] An adult type of digenetic trematode having a well-developed ventral sucker (acetabulum) on the posterior end.

amphistylar [ARCH] Having free columns in porticoes at both ends or at both sides and across the full ends of sides.

amphitene See zygotene.

amphitheater [ARCH] A structure or large room containing oval, circular, or semicircular tiers of seats facing an open space. [GEOGR] A valley or gulch having an oval or circular floor and formed by glacial action.

Amphitheriidae [PALEON] A family of Jurassic therian mammals in the infraclass Pantotheria.

amphitrichous [BIOL] Having flagella at both ends, as in certain bacteria.

Amphitritinae [INV ZOO] A subfamily of sedentary polychaete worms in the family Terebellidae.

amphitropous [BOT] Having a half-inverted ovule with the funiculus attached near the middle.

Amphiumidae [VERT ZOO] A small family of urodele amphibians in the suborder Salamandroidea composed of three species of large, eellike salamanders with tiny limbs.

Amphizoidae [INV ZOO] The trout stream beetles, a small family of coleopteran insects in the suborder Adephaga.

amphoteric [CHEM] Having both acidic and basic characteristics. Also known as amphiprotic.

amp-hr See ampere-hour.

amplexus [BOT] Having the edges of a leaf overlap the edges of a leaf above it in vernation. [VERT ZOO] The copulatory embrace of frogs and toads.

amplidyne [ELEC] A rotating magnetic amplifier having special windings and brush connections so that small changes in power input to the field coils produce large changes in power output.

amplification [GEN] 1. Treatment with an antibiotic or other agent to increase the relative proportion of plasmid to bacterial deoxyribonucleic acid. 2. Bulk replication of a gene library. [SCI TECH] The production of an output of greater magnitude than the input.

amplifier [ENG] A device capable of increasing the magnitude or power level of a physical quantity, such as an electric current or a hydraulic mechanical force, that is varying with time, without distorting the wave shape of the quantity.

amplitude [MATH] The angle between a vector representing a specified complex number on an Argand diagram and the positive real axis. Also known as argument. [NAV] Angular distance north or south of the prime vertical; the arc of the horizon, or the angle at the zenith between the prime vertical and a vertical circle, measured north or south from the prime vertical to the vertical circle. [PHYS] The maximum absolute value attained by the disturbance of a wave or by any quantity that varies periodically.

amplitude distortion See frequency distortion.

amplitude fading [COMMUN] Fading in which the amplitudes of all frequency components of a modulated carrier wave are uniformly attenuated.

amplitude-frequency distortion See frequency distortion.

amplitude-frequency response See frequency response.

amplitude gate [ELECTR] A circuit which transmits only those portions of an input signal which lie between two amplitude boundary level values. Also known as slicer; slicer amplifier.

amplitude limiter See limiter.

amplitude-limiting circuit See limiter.

amplitude-modulated indicator [ENG] A general class of radar indicators, in which the sweep of the electron beam is deflected vertically or horizontally from a base line to indicate the existence of an echo from a target. Also known as deflection-modulated indicator; intensity-modulated indicator.

amplitude modulation [ELECTR] Abbreviated AM. 1. Modulation in which the aplitude of a wave is the characteristic varied in accordance with the intelligence to be transmitted. 2. In telemetry, those systems of modulation in which each component frequency f of the transmitted intelligence produces a pair of sideband frequencies at carrier frequency plus f and carrier minus f.

amplitude-modulation radio [COMMUN] Also known as AM radio. 1. The system of radio communication employing amplitude modulation of a radio-frequency carrier to convey the intelligence. 2. A receiver used in such a system.

amplitude modulator [PHYS] Any device which imposes amplitude modulation upon a carrier wave in accordance with a desired program.

amplitude resonance [PHYS] The frequency at which a given sinusoidal excitation produces the maximum amplitude of oscillation in a resonant system.

amplitude response [ELECTR] The maximum output amplitude obtainable at various points over the frequency range of an instrument operating under rated conditions.

amplitude separator [ELECTR] A circuit used to isolate the portion of a waveform with amplitudes above or below a given value or between two given values.

amplitude shift keying [COMMUN] A method of transmitting binary coded messages in which a sinusoidal carrier is pulsed so that one of the binary states is represented by the presence of the carrier while the other is represented by its absence. Abbreviated ASK.

amplitude suppression ratio [ELECTR] Ratio, in frequency modulation, of the undesired output to the desired output of a frequency-modulated receiver when the applied signal has simultaneous amplitude and frequency modulation.

amp-turn See ampere-turn.

Ampulicidae [INV ZOO] A small family of hymenopteran insects in the superfamily Sphecoidea.

ampulla [ANAT] A dilated segment of a gland or tubule. [BOT] A small air bladder in some aquatic plants. [INV ZOO] The sac at the base of a tube foot in certain echinoderms.

AM radio See amplitude-modulation radio.

amu See atomic mass unit.

amygdaloid [GEOL] Lava rock containing amygdules. Also known as amygdaloidal lava.

amygdaloidal lava See amygdaloid.

amygdule [GEOL] 1. A mineral filling formed in vesicles (cavities) of lava flows; it may be chalcedony, opal, calcite, chlorite, or prehnite. 2. An agate pebble.

amyl [ORG CHEM] Any of the eight isomeric arrangements of the radical C_5H_{11} or a mixture of them. Also known as pentyl.

amylase [BIOCHEM] An enzyme that hydrolyzes reserve carbohydrates, starch in plants and glycogen in animals.

amyloid [PATH] An abnormal protein deposited in tissues, formed from the infiltration of an unknown substance, probably a carbohydrate.

amylolytic enzyme [BIOCHEM] A type of enzyme capable of denaturing starch molecules; used in textile manufacture to remove starch added to slash sizing agents.

amylopectin [BIOCHEM] A highly branched, high-molecular-weight carbohydrate polymer composed of about 80% corn starch.

amyloplast [BOT] A colorless cell plastid packed with starch grains and occurring in cells of plant storage tissue.

Amynodontidae [PALEON] A family of extinct hippopotamuslike perissodactyl mammals in the superfamily Rhinoceratoidea.

amyotrophic lateral sclerosis [MED] A degenerative disease of the pyramidal tracts and lower motor neurons characterized by motor weakness and a spastic condition of the limbs associated with muscular atrophy, fibrillary twitching, and final involvement of nuclei in the medulla. Also known as lateral sclerosis.

Anabantidae [VERT ZOO] A fresh-water family of actinopterygian fishes in the order Perciformes, including climbing perches and gourami.

Anabantoidei [VERT ZOO] A suborder of fresh-water labyrinth fishes in the order Perciformes.

anabatic wind [METEOROL] An upslope wind; usually applied only when the wind is blowing up a hill or mountain as the result of a local surface heating, and apart from the effects of the larger-scale circulation.

anabiosis [BIOL] State of suspended animation induced by desiccation and reversed by addition of moisture; can be achieved in rotifers.

anabolism [BIOCHEM] A part of metabolism involving the union of smaller molecules into larger molecules; the method of synthesis of tissue structure.

Anacanthini [VERT ZOO] An equivalent name for the Gadiformes.

Anacardiaceae [BOT] A family of flowering plants, the sumacs, in the order Sapindales; many species are allergenic to man.

anaclinal [GEOL] Having a downward inclination opposite to that of a stratum.

anaconda [VERT ZOO] *Eunectes murinus*. The largest living snake, an arboreal-aquatic member of the boa family (Boidae).

Anactinochitinosi [INV ZOO] A group name for three closely related suborders of mites and ticks: Onychopalpida, Mesostigmata, and Ixodides.

anadromous [VERT ZOO] Said of a fish, such as the salmon and shad, that ascends fresh-water streams from the sea to spawn.

Anadyomenaceae [BOT] A family of green marine algae in the order Siphonocladales characterized by the expanded blades of the thallus.

anaerobe [BIOL] An organism that does not require air or free oxygen to maintain its life processes.

anaerobic glycolysis [BIOCHEM] A metabolic pathway in plants by which, in the absence of oxygen, hexose is broken down to lactic acid and ethanol with some adenosinetriphosphate synthesis.

analbuminemia [MED] A disorder transmitted as an autosomal recessive, characterized by drastic reduction or absence of serum albumin.

analcime [MINERAL] $NaAlSi_2O_6 \cdot H_2O$ A white or slightly colored isometric zeolite found in diabase and in alkali-rich basalts. Also known as analcite.

analcite *See* analcime.

analeptic [PHARM] Any drug used to restore respiration and a wakeful state.

anal fin [VERT ZOO] An unpaired fin located medially on the posterior ventral part of the fish body.

analgesia [PHYSIO] Insensibility to pain with no loss of consciousness.

analgesic [PHARM] Any drug, such as salicylates, morphine, or opiates, used primarily for the relief of pain.

anal gland [INV ZOO] A gland in certain mollusks that secretes a purple substance. [VERT ZOO] A gland located near the anus or opening into the rectum in many vertebrates.

analog [ELECTR] A physical variable which remains similar to another variable insofar as the proportional relationships are the same over some specified range; for example, a temperature may be represented by a voltage which is its analog. [METEOROL] A past large-scale synoptic weather pattern which resembles a given (usually current) situation in its essential characteristics.

analog comparator [ELECTR] 1. A comparator that checks digital values to determine whether they are within predetermined upper and lower limits. 2. A comparator that produces high and low digital output signals when the sum of two analog voltages is positive and negative, respectively.

analog computer [COMPUT SCI] A computer is which quantities are represented by physical variables; problem parameters are translated into equivalent mechanical or electrical circuits as an analog for the physical phenomenon being investigated.

analog data [COMPUT SCI] Data represented in a continuous form, as contrasted with digital data having discrete values.

analog-digital computer *See* hybrid computer.

analogous [BIOL] Referring to structures that are similar in function and general appearance but not in origin, such as the wing of an insect and the wing of a bird.

analog recording [ELECTR] Any method of recording in which some characteristic of the recording signal, such as amplitude or frequency, is continuously varied in a manner analogous to the time variations of the original signal.

analog signal [ELECTR] A nominally continuous electrical signal that varies in amplitude or frequency in response to changes in sound, light, heat, position, or pressure.

analog simulation [COMPUT SCI] The representation of physical systems and phenomena by variables such as translation, rotation, resistance, and voltage.

analog switch [ELECTR] 1. A device that either transmits an analog signal without distortion or completely blocks it. 2. Any solid-state device, with or without a driver, capable of bilaterally switching voltages or current.

analog-to-digital converter [ELECTR] A device which translates continuous analog signals into proportional discrete digital signals.

analog-to-frequency converter [ELECTR] A converter in which an analog input in some form other than frequency is converted to a proportional change in frequency.

analysis [ANALY CHEM] The determination of the composition of a substance. [MATH] The branch of mathematics most explicitly concerned with the limit process or the concept of convergence; includes the theories of differentiation, integration and measure, infinite series, and analytic functions. Also known as mathematical analysis. [METEOROL] A detailed study in synoptic meteorology of the state of the atmosphere based on actual observations, usually including a separation of the entity into its component patterns and involving the drawing of families of isopleths for various elements.

analytical balance [ENG] A balance with a sensitivity of 0.1–0.01 milligram.

analytical chemistry [CHEM] The branch of chemistry dealing with techniques which yield any type of information about chemical systems.

analytical function generator [ELECTR] An analog computer device in which the dependence of an output variable on one or more input variables is given by a function that also appears in a physical law. Also known as natural function generator; natural law function generator.

analytic curve [MATH] A curve whose parametric equations are real analytic functions of the same real variable.

analytic function [MATH] A function which can be represented by a convergent Taylor series. Also known as holomorphic function.

analytic geometry [MATH] The study of geometric figures and curves using a coordinate system and the methods of algebra. Also known as cartesian geometry.

analytic psychology [PSYCH] The school of psychology that regards the libido not as an expression of the sex instinct, but of the will to live; the unconscious mind is thought to express certain archaic memories of race. Also known as Jungian psychology.

analyzer [COMPUT SCI] 1. A routine for the checking of a program. 2. One of several types of computers used to solve differential equations. [ENG] A multifunction test meter, measuring volts, ohms, and amperes. Also known as set analyzer. [MECH ENG] The component of an absorption refrigeration system where the mixture of water vapor and ammonia vapor leaving the generator meets the relatively cool solution of ammonia in water entering the generator and loses some of its vapor content. [OPTICS] A device, such as a Nicol prism, which passes only plane polarized light; used in the eyepiece of instruments such as the polariscope.

anamnestic response [IMMUNOL] A rapidly increased antibody level following renewed contact with a specific antigen, even after several years. Also known as booster response.

Anamnia [VERT ZOO] Vertebrate animals which lack an amnion in development, including Agnatha, Chondrichthyes, Osteichthyes, and Amphibia.

Anamniota [VERT ZOO] The equivalent name for Anamnia.

anamorphic lens [OPTICS] A lens that produces different magnifications along lines in different directions in the image plane.

anamorphic system [OPTICS] An optical system incorporating a cylindrical surface in which the image is distorted so that the angle of coverage in a direction perpendicular to the cylinder is different for the image than for the object.

anamorphic zone [GEOL] The zone of rock flow, as indicated by reactions that may involve decarbonation, dehydration, and deoxidation; silicates are built up, and the formation of denser minerals and of compact crystalline structure takes place.

anamorphism [EVOL] *See* anamorphosis. [GEOL] A kind of metamorphism at considerable depth in the earth's crust and under great pressure, resulting in the formation of complex minerals from simple ones.

anamorphosis [EVOL] Gradual increase in complexity of form and function during evolution of a group of animals or plants. Also known as anamorphism. [GRAPHICS] A drawing which appears to be distorted unless viewed from a particular angle or with a special device. [OPTICS] The production of a distorted image by an optical system.

Anancinae [PALEON] A subfamily of extinct proboscidean placental mammals in the family Gomphotheriidae.

Ananke [ASTRON] A small satellite of Jupiter with a diameter of about 14 miles (23 kilometers), orbiting with retrograde motion at a mean distance of 1.3×10^7 miles (2.1×10^7 kilometers). Also known as Jupiter XII.

anapaite [MINERAL] $Ca_2Fe(PO_4)_2 \cdot 4H_2O$ A pale-green or greenish-white triclinic mineral consisting of a ferrous iron hydrous phosphate and occurring in crystals and massive forms; hardness is 3–4 on Mohs scale, and specific gravity is 3.81.

anaphase [CYTOL] 1. The stage in mitosis and in the second meiotic division when the centromere splits and the chromatids separate and move to opposite poles. 2. The stage of the first meiotic division when the two halves of a bivalent chromosome separate and move to opposite poles.

anaphoresis [MED] Deficient functioning of sweat glands. [PHYS CHEM] Upon application of an electric field, the movement of positively charged colloidal particles or macromolecules suspended in a liquid toward the anode. [PHYSIO] Movement of positively charged ions into tissues under the influence of an electric current.

anaphylaxis [MED] Hypersensitivity following parenteral injection of an antigen; local or systemic allergenic reaction occurs when the antigen is reintroduced after a time lapse.

anaplasia [MED] Reversion of cells to an embryonic, immature, or undifferentiated state; degree usually corresponds to malignancy of a tumor.

Anaplasmataceae [MICROBIO] A family of the order Rickettsiales; obligate parasites, either in or on red blood cells or in the plasma of various vertebrates.

Anaplotheriidae [PALEON] A family of extinct tylopod ruminants in the superfamily Anaplotherioidea.

Anaplotherioidea [PALEON] A superfamily of extinct ruminant artiodactyls in the infraorder Tylopoda.

Anapsida [VERT ZOO] A subclass of reptiles characterized by a roofed temporal region in which there are no temporal openings.

Anasca [PALEON] A suborder of extinct bryozoans in the order Cheilostomata.

Anaspida [PALEON] An order of extinct fresh- or brackish-water vertebrates in the class Agnatha.

Anaspidacea [INV ZOO] An order of the crustacean superorder Syncarida.

Anaspididae [INV ZOO] A family of crustaceans in the order Anaspidacea.

anastatic process [GRAPHICS] Reproduction of a printed page, either type or pictures, by moistening it with dilute acid and pressing it against a zinc plate; the acid etches the zinc wherever in contact with unprinted portions; the plate can then be inked and printed.

anastatic water [HYD] That part of the subterranean water in the capillary fringe between the zone of aeration and the zone of saturation in the soil.

anastigmat *See* anastigmatic lens.

anastigmatic lens [OPTICS] A compound lens corrected for astigmatism and curvature of field. Also known as anastigmat.

anastomosis [SCI TECH] The union or intercommunication of branched systems in either two or three dimensions. Also known as inosculation.

anatase [MINERAL] The brown, dark-blue, or black tetragonal crystalline form of titanium dioxide, TiO_2; used to make a white pigment. Also known as octahedrite.

anatexis [GEOL] A high-temperature process of metamorphosis by which plutonic rock in the lowest levels of the crust is melted and regenerated as a magma.

Anatidae [VERT ZOO] A family of waterfowl, including ducks, geese, mergansers, pochards, and swans, in the order Anseriformes.

anatomical dead space *See* dead space.

anatomy [BIOL] A branch of morphology dealing with the structure of animals and plants.

anauxite [MINERAL] $Al_2(SiO_7)(OH)_4$ A clay mineral that is a mixture of kaolinite and quartz. Also known as ionite.

anchor [CIV ENG] A device connecting a structure to a heavy masonry or concrete object to a metal plate or to the ground to hold the structure in place. [ENG] A device, such as a metal rod, wire, or strap, for fixing one object to another, such as specially formed metal connectors used to fasten together timbers, masonry, or trusses. [INV ZOO] **1.** An anchor-shaped spicule in the integument of sea cucumbers. **2.** An anchor-shaped ossicle in echinoderms. [MECH ENG] A vehicle used in steam plowing and located on the side of the field opposite that of the engine while maintaining the tension on the endless wire by means of a pulley. [MET] A device that prevents the movement of sand cores in molds. [NAV ARCH] A device attached by cable to a ship and dropped overboard so that its hooks or flukes engage the bottom and hold the ship at that location.

anchorage [CIV ENG] **1.** An area where a vessel anchors or may anchor because of either suitability or designation. Also known as anchor station. **2.** A device which anchors tendons to the posttensioned concrete member. **3.** In pretensioning, a device used to anchor tendons temporarily during the hardening of the concrete. **4.** *See* deadman.

anchor and collar [DES ENG] A door or gate hinge whose socket is attached to an anchor embedded in the masonry.

anchor ball [NAV ARCH] **1.** A projectile with grappling hooks which is fired into the rigging of a wrecked vessel for lifesaving purposes. **2.** A black, circular shape hoisted between the bow and foremast of a vessel to indicate that it is anchored in or near a channel.

anchor block [BUILD] A block of wood, replacing a brick in a wall to provide a nailing or fastening surface. [CIV ENG] *See* deadman.

anchor bolt [CIV ENG] A bolt used with its head embedded in masonry or concrete and its threaded part protruding to hold a structure or machinery in place. Also known as anchor rod.

anchor escapement [HOROL] A clock escapement in which pallets of an anchor-shaped component cause the escape wheel to recoil slightly as the wheel is arrested. Also known as recoil escapement.

anchor ice [HYD] Ice formed beneath the surface of water, as in a lake or stream, and attached to the bottom or to submerged objects. Also known as bottom ice; ground ice.

anchor log [CIV ENG] A log, beam, or concrete block buried in the earth and used to hold a guy rope firmly. Also known as deadman.

anchor nut [DES ENG] A nut in the form of a tapped insert forced under steady pressure into a hole in sheet metal.

anchor pile [CIV ENG] A pile that is located on the land side of a bulkhead or pier and anchors it through such devices as rods, cables, and chains.

anchor plate [CIV ENG] A metal or wooden plate fastened to or embedded in a support, such as a floor, and used to hold a supporting cable firmly.

anchor rod *See* anchor bolt.

anchor station [CIV ENG] *See* anchorage. [OCEANOGR] An anchoring site by a research vessel for the purpose of making a set of scientific observations.

anchor tower [CIV ENG] **1.** A tower which is a part of a crane staging or stiffleg derrick and serves as an

anchor. **2.** A tower that supports and anchors an overhead transmission line.

anchor wall *See* deadman.

anchovy [VERT ZOO] Any member of the Engraulidae, a family of herringlike fishes harvested commercially for human consumption.

anchylosis *See* ankylosis.

ancon [ARCH] A bracket, elbow, or console at the top of a wall or window jamb to support a cornice.

Ancylostomidae [INV ZOO] A family of nematodes belonging to the group Strongyloidea.

And *See* Andromeda.

andalusite [MINERAL] Al_2SiO_5 A brown, yellow, green, red, or gray neosilicate mineral crystallizing in the orthorhombic system, usually found in metamorphic rocks.

AND circuit *See* AND gate.

Andept [GEOL] A suborder of the soil order Inceptisol, formed chiefly in volcanic ash or in regoliths with high components of ash.

Anderson bridge [ELECTR] A six-branch modification of the Maxwell-Wien bridge, used to measure self-inductance in terms of capacitance and resistance; bridge balance is independent of frequency.

andersonite [MINERAL] $Na_2Ca(UO_2)(CO_3)_3 \cdot 6H_2O$ Bright yellow-green secondary mineral consisting of a hydrous sodium calcium uranium carbonate.

andesine [MINERAL] A plagioclase feldspar with a composition ranging from $Ab_{70}An_{30}$ to $Ab_{50}An_{50}$, where $Ab = NaAlSi_3O_8$ and $An = CaAl_2Si_2O_8$; it is a primary constituent of intermediate igneous rocks, such as andesites.

andesite [PETR] Very finely crystalline extrusive rock of volcanic origin composed largely of plagioclase feldspar (oligoclase or andesine) with smaller amounts of dark-colored mineral (hornblende, biotite, or pyroxene); the extrusive equivalent of diorite.

AND function [MATH] An operation in logical algebra on statements P, Q, R, such that the operation is true if all the statements P, Q, R, ... are true, and the operation is false if at least one statement is false.

AND gate [ELECTR] A circuit which has two or more input signal ports and which delivers an output only if and when every input signal port is simultaneously energized. Also known as AND circuit; passive AND gate.

AND/NOR gate [ELECTR] A single logic element whose operation is equivalent to that of two AND gates with outputs feeding into a NOR gate.

AND NOT gate [ELECTR] A coincidence circuit that performs the logic operation AND NOT, under which a result is true only if statement A is true and statement B is not. Also known as A AND NOT B gate.

AND-OR circuit [ELECTR] Gating circuit that produces a prescribed output condition when several possible combined input signals are applied; exhibits the characteristics of the AND gate and the OR gate.

AND-OR-INVERT gate [ELECTR] A logic circuit with four inputs, a_1, a_2, b_1, and b_2, whose output is 0 only if either a_1 and a_2 or b_1 and b_2 are 1. Abbreviated A-O-I gate.

Andr *See* Andromeda.

Andreaeales [BOT] The single order of mosses of the subclass Andreaeobrya.

Andreaeaceae [BOT] The single family of the Andreaeales, an order of mosses.

Andreaeobrya [BOT] The granite mosses, a subclass of the class Bryopsida.

Andrenidae [INV ZOO] The mining or burrower bees, a family of hymenopteran insects in the superfamily Apoidea.

andrite [GEOL] A meteorite composed principally of augite with some olivine and troilite.

androecium [BOT] The aggregate of stamens in a flower.

androgen [BIOCHEM] A class of steroid hormones produced in the testis and adrenal cortex which act to regulate masculine secondary sexual characteristics.

androgenesis [EMBRYO] Development of an embryo from a fertilized irradiated egg, involving only the male nucleus.

androgyny [MED] A form of pseudohermaphroditism in humans in which the individual has female external sexual characteristics, but has undescended testes. Also known as male pseudohermaphroditism.

Andromeda [ASTRON] A constellation with a right ascension of 1 hour and a declination of 40°N. Abbreviated And; Andr.

Andromeda Galaxy [ASTRON] The spiral galaxy of type Sb nearest to the Milky Way. Also known as Andromeda Nebula.

Andromeda Nebula *See* Andromeda Galaxy.

andromerogony [EMBRYO] Development of an egg fragment following cutting, shaking, or centrifugation of a fertilized or unfertilized egg.

androstane [BIOCHEM] $C_{19}H_{32}$ The parent steroid hydrocarbon for all androgen hormones. Also known as etioallocholane.

androstenedione [BIOCHEM] $C_{19}H_{26}O_2$ Any one of three isomeric androgens produced by the adrenal cortex.

androsterone [BIOCHEM] $C_{19}H_{30}O_2$ An androgenic hormone occurring as a hydroxy ketone in the urine of men and women.

anechoic chamber [ENG] 1. A test room in which all surfaces are lined with a sound-absorbing material to reduce reflections of sound to a minimum. Also known as dead room; free-field room. 2. A room completely lined with a material that absorbs radio waves at a particular frequency or over a range of frequencies; used principally at microwave frequencies, such as for measuring radar beam cross sections.

anelasticity [MECH] Deviation from a proportional relationship between stress and strain.

Anelytropsidae [VERT ZOO] A family of lizards represented by a single Mexican species.

anemia [MED] A condition marked by significant decreases in hemoglobin concentration and in the number of circulating red blood cells. Also known as oligochromemia.

anemobiagraph [ENG] A recording pressure-tube anemometer in which the wind scale of the float manometer is linear through the use of springs; an example is the Dines anemometer.

anemochory [ECOL] Wind dispersal of plant and animal disseminules.

anemoclastic [GEOL] Referring to rock that was broken by wind erosion and rounded by wind action.

anemometer [ENG] A device which measures air speed.

anemometry [METEOROL] The study of measuring and recording the direction and speed (or force) of the wind, including its vertical component.

anemophilous [BOT] Pollinated by wind-carried pollen.

anencephalia [MED] A congenital malformation in which all or most of the brain and flat skull bones are absent.

Anepitheliocystidia [INV ZOO] A superorder of digenetic trematodes proposed by G. LaRue.

aneroid [ENG] 1. Containing no liquid or using no liquid. 2. *See* aneroid barometer.

aneroid barometer [ENG] A barometer which utilizes an aneroid capsule. Also known as anerroid.

aneroid capsule [ENG] A thin, disk-shaped box or capsule, usually metallic, partially evacuated and sealed, held extended by a spring, which expands and contracts with changes in atmospheric or gas pressure. Also known as bellows.

aneroid flowmeter [ENG] A mechanism to measure fluid flow rate by pressure of the fluid against a bellows counterbalanced by a calibrated spring.

anesthesia [PHYSIO] 1. Insensibility, general or local, induced by anesthetic agents. 2. Loss of sensation, of neurogenic or psychogenic origin.

anesthetic [PHARM] A drug, such as ether, that produces loss of sensibility.

anestrus [VERT ZOO] A prolonged period of inactivity between two periods of heat in cyclically breeding female mammals.

aneuploidy [GEN] Deviation from a normal haploid, diploid, or polyploid chromosome complement by the presence in excess of, or in defect of, one or more individual chromosomes.

aneurine *See* thiamine.

aneurysm [MED] Localized abnormal dilation of an artery due to weakening of the vessel wall.

angel echo [ENG] A radar echo from a region where there are no visible targets; may be caused by insects, birds, or refractive index variations in the atmosphere.

angina [MED] 1. A sore throat. 2. Any tense, constricting pain.

angina pectoris [MED] Constricting chest pain which may be accompanied by pain radiating down the arms, up into the jaw, or to other sites.

angiocardiography [MED] Roentgenographic visualization of the heart chambers and thoracic vessels following injection of a radiopaque material.

angiogram [MED] An x-ray photograph of blood vessels following injection of a radiopaque material.

angiography [MED] Roentgenographic visualization of blood vessels following injection of a radiopaque material.

angioma [MED] A tumor composed of lymphatic vessels or blood.

angiosperm [BOT] The common name for members of the plant division Magnoliophyta.

Angiospermae [BOT] An equivalent name for the .agnoliophyta.

angiotensin [BIOCHEM] A decapeptide hormone that influences blood vessel constriction and aldosterone secretion by the adrenal cortex. Also known as hypertensin.

angle [MATH] The geometric figure, arithmetic quantity, or algebraic signed quantity determined by two rays emanating from a common point or by two planes emanating from a common line.

angle bracket [ARCH] A bracket used in an angle or corner of a molded cornice. [GRAPHICS] Either of a pair of marks ⟨⟩ enclosing a mutilated passage or the explanation of an abbreviation in a text, or to enclose quotations or illustrations in a reference work. Also known as broken bracket; pointed bracket.

angle equation [ENG] A condition equation which expresses the relationship between the sum of the measured angles of a closed figure and the theoretical value of that sum, the unknowns being the corrections to the observed directions or angles, depending on which are used in the adjustment. Also known as triangle equation.

angle gear   *See* angular gear.

angle modulation   [ELECTR] The variation in the angle of a sine-wave carrier; particular forms are phase modulation and frequency modulation. Also known as sinusoidal angular modulation.

angle of approach   [CIV ENG] The maximum angle of an incline onto which a vehicle can move from a horizontal plane without interference. [MECH ENG] The angle that is turned through by either of paired wheels in gear from the first contact between a pair of teeth until the pitch points of these teeth fall together. [ORD] Angle between the line along which a moving target is traveling and the line along which the gun is pointed.

angle of arrival   [ELECTROMAG] A measure of the direction of propagation of electromagnetic radiation upon arrival at a receiver (the term is most commonly used in radio); it is the angle between the plane of the phase front and some plane of reference, usually the horizontal, at the receiving antenna.

angle of bite   *See* angle of nip.

angle of climb   [AERO ENG] The angle between the flight path of a climbing vehicle and the local horizontal.

angle of commutation   [ASTRON] The difference between the celestial longitudes of the sun and a planet, as observed from the earth.

angle of departure   [AERO ENG] The vertical angle, at the origin, between the line of site and the line of departure. [CIV ENG] The maximum angle of an incline from which a vehicle can move onto a horizontal plane without interference, such as from rear bumpers.

angle of depression   [ENG] The angle in a vertical plane between the horizontal and a descending line. Also known as depression angle; descending vertical angle; minus angle.

angle of descent   [AERO ENG] The angle between the flight path of a descending vehicle and the local horizontal.

angle of deviation   *See* deviation.

angle of dip   *See* dip.

angle of elevation   [ENG] The angle in a vertical plane between the local horizontal and an ascending line, as from an observer to an object; used in astronomy, surveying, and so on. Also known as ascending vertical angle; elevation angle. [ORD] **1.** The vertical angle above the line of sight through which the axis of the gun bore must be raised so that the bullet or projectile will carry to the target. **2.** In aerial gunnery, an acute angle between the bore axis of a gun and the horizontal; called quadrant elevation in ground gunnery.

angle of friction   *See* angle of repose.

angle of lag   *See* lag angle.

angle of lead   *See* lead angle.

angle of nip   [MECH ENG] The largest angle that will just grip a lump between the jaws, rolls, or mantle and ring of a crusher. Also known as angle of bite; nip. [MIN ENG] In a rock-crushing machine, the maximum angle subtended by its approaching jaws or roll surfaces at which a specified piece of ore can be gripped.

angle of reflection   [PHYS] The angle between the direction of propagation of a wave reflected by a surface and the line perpendicular to the surface at the point of reflection. Also known as reflection angle.

angle of refraction   [PHYS] The angle between the direction of propagation of a wave that is refracted by a surface and the line that is perpendicular to the surface at the point of refraction.

angle of repose   [ENG] *See* angle of rest. [MECH] The angle between the horizontal and the plane of contact between two bodies when the upper body is just about to slide over the lower. Also known as angle of friction.

angle of rest   [ENG] The maximum slope at which a heap of any loose or fragmented solid material will stand without sliding, or will come to rest when poured or dumped in a pile or on a slope. Also known as angle of repose.

angle of roll   [AERO ENG] The angle that the lateral body axis of an aircraft or similar body makes with a chosen reference plane in rolling; usually, the angle between the lateral axis and a horizontal plane.

angle of shear   [GEOL] The angle between the planes of maximum shear which is bisected by the axis of greatest compression.

angle of thread   [DES ENG] The angle occurring between the sides of a screw thread, measured in an axial plane.

angle of torsion   [MECH] The angle through which a part of an object such as a shaft or wire is rotated from its normal position when a torque is applied. Also known as angle of twist.

angle of twist   *See* angle of torsion.

angle of vertical   [ASTRON] The angle on the celestial sphere between a given vertical circle and the prime vertical circle.

angle of view   [OPTICS] The angle subtended by an image at the second nodal point of a lens.

angle plate   [DES ENG] An L-shaped plate or a plate having an angular section.

angle post   [BUILD] A railing support used at a landing or other break in the stairs.

anglerfish   [VERT ZOO] Any of several species of the order Lophiiformes characterized by remnants of a dorsal fin seen as a few rays on top of the head that are modified to bear a terminal bulb.

anglesite   [MINERAL] $PbSO_4$ A mineral occurring in white or gray, tabular or prismatic orthorhombic crystals or compact masses. Also known as lead spar; lead vitriol.

angle valve   [DES ENG] A manually operated valve with its outlet opening oriented at right angles to its inlet opening; used for regulating the flow of a fluid in a pipe.

Angoumian   [GEOL] Upper middle Upper Cretaceous (Upper Turonian) geologic time.

angstrom   [MECH] A unit of length, 10^{-10} meter, used primarily to express wavelengths of optical spectra. Abbreviated A; Å. Also known as tenthmeter.

Anguidae   [VERT ZOO] A family of limbless, snakelike lizards in the suborder Sauria, commonly known as slowworms or glass snakes.

Anguilliformes   [VERT ZOO] A large order of actinopterygian fishes containing the true eels.

Anguilloidei   [VERT ZOO] The typical eels, a suborder of actinopterygian fishes in the order Anguilliformes.

angular acceleration   [MECH] The time rate of change of angular velocity.

angular displacement   [PHYS] A vector measure of the rotation of an object about an axis; the vector points along the axis according to the right-hand rule; the length of the vector is the rotation angle, in degrees or radians.

angular frequency   [PHYS] For any oscillation, the number of vibrations per unit time, multiplied by 2π. Also known as angular velocity; radian frequency.

angular gear [MECH ENG] A gear that transmits motion between two rotating shafts that are not parallel. Also known as angle gear.

angular momentum [MECH] **1.** The cross product of a vector from a specified reference point to a particle, with the particle's linear momentum. Also known as moment of momentum. **2.** For a system of particles, the vector sum of the angular momenta (first definition) of the particles.

angular perspective [GRAPHICS] A form of plane linear perspective in which some of the principal lines of the picture are either parallel or perpendicular to the picture plane and some are oblique.

angular rate *See* angular speed.

angular resolution [ELECTROMAG] A measure of the ability of a radar to distinguish between two targets solely by the measurement of angles.

angular speed [MECH] Change of direction per unit time, as of a target on a radar screen, without regard to the direction of the rotation axis; in other words, the magnitude of the angular velocity vector. Also known as angular rate.

angular unconformity [GEOL] An unconformity in which the older strata dip at a different angle (usually steeper) than the younger strata.

angular velocity [MECH] The time rate of change of angular displacement. [PHYS] *See* angular frequency.

anharmonicity [PHYS] **1.** Mechanical vibration where the restoring force acting on a system does not vary linearly with displacement from equilibrium position. **2.** Variation from a linear relationship of dipole moment with internuclear distance in the infrared portion of the electromagnetic spectrum.

anharmonic oscillator [PHYS] An oscillating system in which the restoring force opposing a displacement from the position of equilibrium is a nonlinear function of the displacement.

Anhimidae [VERT ZOO] The screamers, a family of birds in the order Anseriformes characterized by stout bills, webbed feet, and spurred wings.

Anhingidae [VERT ZOO] The anhingas or snakebirds, a family of swimming birds in the order Pelecaniformes.

anhydride [CHEM] A compound formed from an acid by removal of water.

anhydrite [MINERAL] $CaSO_4$ A mineral that represents gypsum without its water of crystallization, occurring commonly in white and grayish granular to compact masses; the hardness is 3–3.5 on Mohs scale, and specific gravity is 2.90–2.99. Also known as cube spar.

anhydrous [CHEM] Being without water, especially water of crystallization.

anhydrous alcohol *See* absolute alcohol.

Aniliidae [VERT ZOO] A small family of nonvenomous, burrowing snakes in the order Squamata.

aniline ink [MATER] A fast-drying printing ink that is a solution of a coal-tar dye in an organic solvent or a solution of a pigment in an organic solvent or water.

aniline printing *See* flexography.

aniline process *See* flexography.

animal black [CHEM] Finely divided carbon made by calcination of animal bones or ivory; used for pigments, decolorizers, and purifying agents; varieties include bone black and ivory black.

animal pole [CYTOL] The region of an ovum which contains the least yolk and where the nucleus gives off polar bodies during meiosis.

animal virus [VIROL] A small infectious agent able to propagate only within living animal cells.

anion exchange [CHEM] A type of ion exchange in which the immobilized functional groups on the solid resin are positive.

anionic detergent [MATER] A class of detergents having a negatively charged surface-active ion, such as sodium alkylbenzene sulfonate.

Anisakidae [INV ZOO] A family of parasitic roundworms in the superfamily Ascaridoidea.

anise [BOT] The small fruit of the annual herb *Pimpinella anisum* in the family Umbelliferae; fruit is used for food flavoring, and oil is used in medicines, soaps, and cosmetics.

anisic acid [ORG CHEM] $CH_3OC_6H_4COOH$ White crystals or powder with a melting point of 184°C; soluble in alcohol and ether; used in medicine and as an insect repellent and ovicide. Also known as *para*-methoxybenzoic acid.

anisocarpous [BOT] Referring to a flower whose number of carpels is different from the number of stamens, petals, and sepals.

anisogamete *See* heterogamete.

anisogamy *See* heterogamy.

anisomerous [BOT] Referring to flowers that do not have the same number of parts in each whorl.

Anisomyaria [INV ZOO] An order of mollusks in the class Bivalvia containing the oysters, scallops, and mussels.

anisophyllous [BOT] Having leaves of two or more shapes and sizes.

Anisoptera [INV ZOO] The true dragonflies, a suborder of insects in the order Odonata.

anisostemonous [BOT] Referring to a flower whose number of stamens is different from the number of carpels, petals, and sepals.

Anisotomidae [INV ZOO] An equivalent name for Leiodidae.

anisotropy [ASTRON] The departure of the cosmic microwave radiation from equal intensity in all directions. [BOT] The property of a plant that assumes a certain position in response to an external stimulus. [PHYS] The characteristic of a substance for which a physical property, such as index of refraction, varies in value with the direction in or along which the measurement is made. Also known as aeolotropy; eolotropy. [ZOO] The property of an egg that has a definite axis or axes.

anisotropy constant [ELECTROMAG] In a ferromagnetic material, temperature-dependent parameters relating the magnetization in various directions to the anisotropy energy.

anisotropy energy [ELECTROMAG] Energy stored in a ferromagnetic crystal by virtue of the work done in rotating the magnetization of a domain away from the direction of easy magnetization.

anisotropy factor *See* dissymmetry factor.

ankaratrite *See* olivine nephelinite.

ankerite [MINERAL] $Ca(Fe,Mg,Mn)(CO_3)_2$ A white, red, or gray iron-rich carbonate mineral associated with iron ores and found in thin veins in coal seams; specific gravity is 2.95–3.1. Also known as cleat spar.

ankle [ANAT] The joint formed by the articulation of the leg bones with the talus, one of the tarsal bones.

Ankylosauria [PALEON] A suborder of Cretaceous dinosaurs in the reptilian order Ornithischia characterized by short legs and flattened, heavily armored bodies.

ankylosis Also spelled anchylosis. [MED] Stiffness or immobilization of a joint due to a surgical or pathologic process. [PHYS] The loss by a system of one or more degrees of freedom through development of one or more frictional constraints.

anlage [EMBRYO] Any group of embryonic cells when first identifiable as a future organ or body part. Also known as blastema; primordium.

annabergite [MINERAL] $(Ni,Co)_3(AsO_4)_2 \cdot 8H_2O$ A monoclinic mineral usually found as apple-green incrustations as an alteration product of nickel arsenides; it is isomorphous with erythrite. Also known as nickel bloom; nickel ocher.

annatto [BOT] *Bixa orellana*. A tree found in tropical America, characterized by cordate leaves and spinose, seed-filled capsules; a yellowish-red dye obtained from the pulp around the seeds is used as a food coloring.

anneal [ENG] To treat a metal, alloy, or glass with heat and then cool to remove internal stresses and to make the material less brittle. [GEN] To recombine strands of denatured bacterial deoxyribonucleic acid that were separated.

annealing point [THERMO] The temperature at which the viscosity of a glass is $10^{13.0}$ poises. Also known as annealing temperature; 13.0 temperature.

annealing temperature *See* annealing point.

Annedidae [VERT ZOO] A small family of limbless, snakelike, burrowing lizards of the suborder Sauria.

Annelida [INV ZOO] A diverse phylum comprising the multisegmented wormlike animals.

Anniellidae [VERT ZOO] A family of limbless, snakelike lizards in the order Squamata.

annihilation [PARTIC PHYS] A process in which an antiparticle and a particle combine and release their rest energies in other particles.

annihilation operator [QUANT MECH] An operator which reduces the occupation number of a single state by unity; for example, an annihilation operator applied to a state of one particle yields the vacuum.

annihilation radiation [PARTIC PHYS] Electromagnetic radiation arising from the collision, and resulting annihilation, of an electron and a positron, or of any particle and its antiparticle.

Annonaceae [BOT] A large family of woody flowering plants in the order Magnoliales, characterized by hypogynous flowers, exstipulate leaves, a trimerous perianth, and distinct stamens with a short, thick filament.

annual layer [GEOL] 1. A sedimentary layer deposited, or presumed to have been deposited, during the course of a year; for example, a glacial varve. 2. A dark layer in a stratified salt deposit containing disseminated anhydrite.

annual magnetic variation *See* magnetic annual variation.

annual parallax [ASTRON] The apparent displacement of a celestial body viewed from two separated observation points whose base line is the radius of the earth's orbit.

annual plant [BOT] A plant that completes its growth in one growing season and therefore must be planted annually.

annual ring [BOT] A line appearing on tree cross sections marking the end of a growing season and showing the volume of wood added during the year.

annual storage [HYD] The capacity of a reservoir that can handle a watershed's annual runoff but cannot carry over any portion of the water for longer than the year.

annular eclipse [ASTRON] An eclipse in which a thin ring of the source of light appears around the obscuring body.

annular effect [FL MECH] A phenomenon observed in the flow of fluid in a tube when its motion is alternating rapidly, as in the propagation of sound waves, in which the mean velocity rises progressing from the center of the tube toward the walls and then falls within a thin laminar boundary layer to zero at the wall itself.

annulus [ANAT] Any ringlike anatomical part. [BOT] 1. An elastic ring of cells between the operculum and the mouth of the capsule in mosses. 2. A line of cells, partly or entirely surrounding the sporangium in ferns, which constricts, thus causing rupture of the sporangium to release spores. 3. A whorl resembling a calyx at the base of the strobilus in certain horsetails. [MATH] The ringlike figure that lies between two concentric circles. [MYCOL] A ring of tissue representing the remnant of the veil around the stipe of some agarics.

annunciator [ENG] A signaling apparatus which operates electromagnetically and serves to indicate visually, or visually and audibly, whether a current is flowing, has flowed, or has changed direction of flow in one or more circuits.

Anobiidae [INV ZOO] The deathwatch beetles, a family of coleopteran insects of the superfamily Bostrichoidea.

anode [ELEC] The negative terminal of a primary cell or of a storage battery. [ELECTR] 1. The collector of electrons in an electron tube. Also known as plate; positive electrode. 2. In a semiconductor diode, the terminal toward which forward current flows from the external circuit. [PHYS CHEM] The positive terminal of an electrolytic cell.

anode circuit [ELECTR] Complete external electrical circuit connected between the anode and the cathode of an electron tube. Also known as plate circuit.

anode current [ELECTR] The electron current flowing through an electron tube from the cathode to the anode. Also known as plate current.

anode effect [PHYS CHEM] A condition produced by polarization of the anode in the electrolysis of fused salts and characterized by a sudden increase in voltage and a corresponding decrease in amperage.

anode fall [ELECTR] A very thin space-charge region in front of an anode surface, characterized by a steep potential gradient through the region.

anode resistance [ELECTR] The resistance value obtained when a small change in the anode voltage of an electron tube is divided by the resulting small change in anode current. Also known as plate resistance.

anode saturation [ELECTR] The condition in which the anode current of an electron tube cannot be further increased by increasing the anode voltage; the electrons are then being drawn to the anode at the same rate as they are emitted from the cathode. Also known as current saturation; plate saturation; saturation; voltage saturation.

anodic cleaning [MET] The removal of a foreign substance from a metallic surface by electrolysis with the metal as the anode. Also known as anodic pickling; reverse-current cleaning.

anodic pickling *See* anodic cleaning.

anodic protection [MET] Reduction of the corrosion rate in an anode by polarizing it into a potential region where the dissolution rates low.

anodic reaction [MET] The reaction in the mechanism of electrochemical corrosion in which the metal forming the anode dissolves in the electrolyte in the form of positively charged ions.

anodize [MET] The formation of a decorative or protective passive film on a metal part by making it the anode of a cell and applying electric current.

anole [VERT ZOO] Any arboreal lizard of the genus *Anolis*, characterized by flattened adhesive digits and a prehensile outer toe.

Anomalinacea [INV ZOO] A superfamily of marine and benthic sarcodinian protozoans in the order Foraminiferida.

anomalistic month [ASTRON] The average period of revolution of the moon from perigee to perigee, a period of 27 days 13 hours 18 minutes 33.2 seconds.

anomalistic period [ASTRON] The interval between two successive perigee passages of a satellite in orbit about a primary. Also known as perigee-to-perigee period.

anomalistic year [ASTRON] The period of one revolution of the earth about the sun from perihelion to perhihelion; 365 days 6 hours 13 minutes 53.0 seconds in 1900 and increasing at the rate of 0.26 second per century.

anomalon [NUC PHYS] A nuclear fragment, produced in the collision of a projectile nucleus at relativistic energy with a target nucleus at rest, that has an anomalously short mean free path, comparable to that of a uranium nucleus.

anomaloscope [OPTICS] An optical instrument for testing color vision, in which a yellow light whose intensity may be varied is matched against red and green lights whose intensity is fixed.

anomalous [SCI TECH] Deviating from the normal; irregular.

anomalous dispersion [OPTICS] Extraordinary behavior in the curve of refractive index versus wavelength which occurs in the vicinity of absorption lines or bands in the absorption spectrum of a medium.

anomalous viscosity *See* non-Newtonian viscosity.

anomalous Zeeman effect [SPECT] A type of splitting of spectral lines of a light source in a magnetic field which occurs for any line arising from a combination of terms of multiplicity greater than one; due to a nonclassical magnetic behavior of the electron spin.

Anomaluridae [VERT ZOO] The African flying squirrels, a small family in the order Rodentia characterized by the climbing organ, a series of scales at the root of the tail.

anomaly [ASTRON] In celestial mechanics, the angle between the radius vector to an orbiting body from its primary (the focus of the orbital ellipse) and the line of apsides of the orbit, measured in the direction of travel, from the point of closest approach to the primary (perifocus). Also known as true anomaly. [BIOL] An abnormal deviation from the characteristic form of a group. [GEOL] A local deviation from the general geological properties of a region. [MED] Any part of the body that is abnormal in its position, form, or structure. [METEOROL] The deviation of the value of an element (especially temperature) from its mean value over some specified interval. [OCEANOGR] The difference between conditions actually observed at a serial station and those that would have existed had the water all been of a given arbitrary temperature and salinity. [SCI TECH] A deviation beyond normal variations.

Anomocoela [VERT ZOO] A suborder of toadlike amphibians in the order Anura characterized by a lack of free ribs.

Anomphalacea [PALEON] A superfamily of extinct gastropod mollusks in the order Aspidobranchia.

Anomura [VERT ZOO] A section of the crustacean order Decapoda that includes lobsterlike and crablike forms.

Anopla [INV ZOO] A class or subclass of the phylum Rhynchocoela characterized by a simple tubular proboscis and by having the mouth opening posterior to the brain.

Anoplocephalidae [INV ZOO] A family of tapeworms in the order Cyclophyllidea.

Anoplura [INV ZOO] The sucking lice, a small group of mammalian ectoparasites usually considered to constitute an order in the class Insecta.

anorexia [MED] Loss of appetite.

anorthic crystal *See* triclinic crystal.

anorthite [MINERAL] The white, grayish, or reddish calcium-rich end member of the plagioclase feldspar series; composition ranges from $Ab_{10}An_{90}$ to Ab_0An_{100}, where $Ab = NaAlSi_3O_8$ and $An = CaAl_2Si_2O_8$. Also known as calciclase; calcium feldspar.

anorthoclase [MINERAL] A triclinic alkali feldspar having a chemical composition ranging from $Or_{40}Ab_{60}$ to $Or_{10}Ab_{90}$ to about 20 mole % An, where $Or = KAlSi_3O_8$, $Ab = NaAlSi_3O_8$, and $An = CaAl_2Si_2O_8$. Also known as anorthose; soda microcline.

anorthosite [PETR] A visibly crystalline plutonic rock composed almost entirely of plagioclase feldspar (andesine to anorthite) with minor amounts of pyroxene and olivine.

Anostraca [INV ZOO] An order of shrimplike crustaceans generally referred to the subclass Branchiopoda.

anoxia [MED] The failure of oxygen to gain access to, or to be utilized by, the body tissues.

Anseranatini [VERT ZOO] A subfamily of aquatic birds in the family Anatidae represented by a single species, the magpie goose.

Anseriformes [VERT ZOO] An order of birds, including ducks, geese, swans, and screamers, characterized by a broad, flat bill and webbed feet.

ant [INV ZOO] The common name for insects in the hymenopteran family Formicidae; all are social, and colonies exhibit a highly complex organization.

Ant *See* Antlia.

antacid [CHEM] Any substance that counteracts or neutralizes acidity.

antagonism [BIOL] 1. Mutual opposition as seen between organisms, muscles, physiologic actions, and drugs. 2. Opposing action between drugs and disease or drugs and functions.

Antarctic Circle [GEOD] The parallel of latitude approximately 66°32′ south of the Equator.

Antarctic Circumpolar Current [OCEANOGR] The ocean current flowing from west to east through all the oceans around the Antarctic Continent. Also known as West Wind Drift.

Antarctic Convergence [OCEANOGR] The oceanic polar front indicating the boundary between the subantarctic and subtropical waters. Also known as Southern Polar Front.

antarctic front [METEOROL] The semipermanent, semicontinuous front between the antarctic air of the Antarctic continent and the polar air of the southern oceans; generally comparable to the arctic front of the Northern Hemisphere.

Antarctic Intermediate Water [OCEANOGR] A water mass in the Southern Hemisphere, formed at the surface near the Antarctic Convergence between 45° and 55°S; it can be traced in the North Atlantic to about 25°N.

Antarctic Zone [GEOGR] The region between the Antarctic Circle (66°32′S) and the South Pole.

antebrachium *See* forearm.

antecedent stream [HYD] A stream that has retained its early course in spite of geologic changes since its course was assumed.

antecedent valley [GEOL] A stream valley that existed before uplift, faulting, or folding occurred and which has maintained itself during and after these events.

anteconsequent stream [HYD] A stream consequent to the form assumed by the earth's surface as the result of early movement of the earth but antecedent to later movement.

antelope [VERT ZOO] Any of the hollow-horned, hoofed ruminants assigned to the artiodactyl subfamily Antilopinae; confined to Africa and Asia.

ante meridian [ASTRON] 1. A section of the celestial meridian; it lies below the horizon, and the nadir is included. 2. Before noon, or the period of time between midnight (0000) and noon (1200).

antenna [ELECTROMAG] A device used for radiating or receiving radio waves. Also known as aerial; radio antenna. [INV ZOO] Any one of the paired, segmented, and movable sensory appendages on the heads of many arthropods.

antenna counterpoise *See* counterpoise.

antenna coupler [ELECTROMAG] A radio-frequency transformer, tuned line, or other device used to transfer energy efficiently from a transmitter to a transmission line or from a transmission line to a receiver.

antenna crosstalk [ELECTROMAG] The ratio or the logarithm of the ratio of the undesired power received by one antenna from another to the power transmitted by the other.

antenna effect [ELECTROMAG] A distortion of the directional properties of a loop antenna caused by an input to the direction-finding receiver which is generated between the loop and ground, in contrast to that which is generated between the two terminals of the loop. Also known as electrostatic error; vertical component effect.

antenna gain [ELECTROMAG] A measure of the effectiveness of a directional antenna as compared to a standard nondirectional antenna. Also known as gain.

antennal gland [INV ZOO] An excretory organ in the cephalon of adult crustaceans and best developed in the Malacostraca. Also known as green gland.

antenna pair [ELECTROMAG] Two antennas located on a base line of accurately surveyed length, sometimes arranged so that the array may be rotated around an axis at the center of the base line; used to produce directional patterns and in direction finding. [NAV] Two antennas located on a common base line to produce a directional pattern, often arranged so that the array may be rotated around an axis which is at their center.

antenna pattern *See* radiation pattern.

antenna resistance [ELECTROMAG] The power supplied to an entire antenna divided by the square of the effective antenna current measured at the point where power is supplied to the antenna.

Antennata [INV ZOO] An equivalent name for the Mandibulata.

antennaverter [COMMUN] Receiving antenna combined with a converter as a single unit, feeding directly into the intermediate-frequency amplifier of the receiver.

antepartum [MED] Pertaining to the period before delivery or birth.

anterior [ZOO] Situated near or toward the front or head of an animal body.

anthelic arc [ASTRON] A rare type of halo phenomenon appearing in an area 180° from the sun's azimuth and at the sun's elevation.

anthelion [ASTRON] A luminous white spot which occasionally appears on the parhelic circle 180° in azimuth away from the sun. Also known as counter sun.

antheridium [BOT] 1. The sex organ that produces male gametes in cryptogams. 2. A minute structure within the pollen grain of seed plants.

Anthicidae [INV ZOO] The antlike flower beetles, a family of coleopteran insects in the superfamily Tenebrionoidea.

Anthocerotae [BOT] A small class of the plant division Bryophyta, commonly known as hornworts or horned liverworts.

Anthocoridae [INV ZOO] The flower bugs, a family of hemipteran insects in the superfamily Cimicimorpha.

anthocyanin [BIOCHEM] Any of the intensely colored, sap-soluble glycoside plant pigments responsible for most scarlet, purple, mauve, and blue coloring in higher plants.

Anthocyathea [PALEON] A class of extinct marine organisms in the phylum Archaeocyatha characterized by skeletal tissue in the central cavity.

Anthomedusae [INV ZOO] A suborder of hydrozoan coelenterates in the order Hydroida characterized by athecate polyps.

Anthomyzidae [INV ZOO] A family of cyclorrhaphous myodarian dipteran insects belonging to the subsection Acalyptratae.

anthophyllite [MINERAL] A clove-brown orthorhombic mineral of the amphibole group. A variety of asbestos occurring as lamellae, radiations, fibers, or massive in metamorphic rocks. Also known as bidalotite.

Anthosomidae [INV ZOO] A family of fish ectoparasites in the crustacean suborder Caligoida.

Anthozoa [INV ZOO] A class of marine organisms in the phylum Coelenterata including the soft, horny, stony, and black corals, the sea pens, and the sea anemones.

anthracene [ORG CHEM] $C_{14}H_{10}$ A crystalline tricyclic aromatic hydrocarbon, colorless when pure, melting at 218°C and boiling at 342°C; obtained in the distillation of coal tar; used as an important source of dyestuffs, and in coating applications.

anthracite [MINERAL] A high-grade metamorphic coal having a semimetallic luster, high content of fixed carbon, and high density, and burning with a short blue flame and little smoke or odor. Also known as hard coal; Kilkenny coal; stone coal.

anthracnose [PL PATH] A fungus disease of plants caused by members of the Melanconiales and characterized by dark or black limited stem lesions.

Anthracosauria [PALEON] An order of Carboniferous and Permian labyrinthodont amphibians that includes the ancestors of living reptiles.

anthracosilicosis [MED] Chronic lung inflammation caused by inhalation of carbon and silicon particles.

anthracosis [MED] The accumulation of inhaled black coal dust particles in the lung accompanied by chronic inflammation. Also known as blacklung.

Anthracotheriidae [PALEON] A family of middle Eocene and early Pleistocene artiodactyl mammals in the superfamily Anthracotherioidea.

Anthracotherioidea [PALEON] A superfamily of extinct artiodactyl mammals in the suborder Paleodonta.

anthrax [VET MED] An acute, infectious bacterial disease of sheep and cattle caused by *Bacillus anthracis;* transmissible to man. Also known as splenic fever; wool-sorter's disease.

Anthribidae [INV ZOO] The fungus weevils, a family of coleopteran insects in the superfamily Curculionoidea.

anthropochory [ECOL] Dispersal of plant and animal disseminules by humans.

anthropodesoxycholic acid See chenodeoxycholic acid.

anthropogenic [ECOL] Referring to environmental alterations resulting from the presence or activities of humans.

anthropography [ANTHRO] A branch of anthropology that deals with the geographic distribution of divisions of humans based on physical character, language, customs, and institutions.

Anthropoidea [VERT ZOO] A suborder of mammals in the order Primates including New and Old World monkeys.

anthropology [BIOL] The study of the interrelations of biological, cultural, geographical, and historical aspects of humankind.

anthropometry [ANTHRO] Description of the physical variation in humankind by measurement; a basic technique of physical anthropology.

anthroposcopy [ANTHRO] The description of physical variation in humankind by visual inspection; a basic technique of physical anthropology.

Anthuridea [INV ZOO] A suborder of crustaceans in the order Isopoda characterized by slender, elongate, subcylindrical bodies, and by the fact that the outer branch of the paired tail appendage (uropod) arches over the base of the terminal abdominal segment, the telson.

antiaircraft [ORD] Used, or designed to be used, against airborne aircraft or missiles. Abbreviated AA.

Antiarchi [PALEON] A division of highly specialized placoderms restricted to fresh-water sediments of the Middle and Upper Devonian.

antiatom [ATOM PHYS] An atom made up of antiprotons, antineutrons, and positrons in the same way that an ordinary atom is made up of protons, neutrons, and electrons.

antiballistic missile [ORD] Any object thrown, dropped, fired, or otherwise projected with the purpose of intercepting a ballistic missile. Abbreviated ABM.

antibaryon [ATOM PHYS] One of a class of antiparticles, including the antinucleons and the antihyperons, with strong interactions, baryon number -1, and hypercharge and charge opposite to those for the particles.

antibiosis [ECOL] Antagonistic association between two organisms in which one is adversely affected.

antibiotic [MICROBIO] A chemical substance, produced by microorganisms and synthetically, that has the capacity in dilute solutions to inhibit the growth of, and even to destroy, bacteria and other microorganisms.

antibody [IMMUNOL] A protein, found principally in blood serum, originating either normally or in response to an antigen and characterized by a specific reactivity with its complementary antigen. Also known as immune body.

antibonding orbital [PHYS] An atomic or molecular orbital whose energy increases as atoms are brought closer together, indicating a net repulsion rather than a net attraction and chemical bonding.

anticathode [ELECTR] The anode or target of an x-ray tube, on which the stream of electrons from the cathode is focused and from which x-rays are emitted.

anticenter [GEOL] The point on the surface of the earth that is diametrically opposite the epicenter of an earthquake. Also known as antiepicenter.

anticholinesterase [BIOCHEM] Any agent, such as a nerve gas, that inhibits the action of cholinesterase and thereby destroys or interferes with nerve conduction.

anticipatory staging [COMPUT SCI] Moving blocks of data from one storage device to another prior to the actual request for them by the program.

anticlastic [MATH] Having the property of a surface or portion of a surface whose two principal curvatures at each point have opposite signs, so that one normal section is concave and the other convex.

anticlinal [BOT] Perpendicular to the surface or periphery of an organ. [GEOL] Folded as in an anticline.

anticline [GEOL] A fold in which layered strata are inclined down and away from the axes.

anticlinorium [GEOL] A series of anticlines and synclines that form a general arch or anticline.

anticlutter gain control [ELECTR] Device which automatically and smoothly increases the gain of a radar receiver from a low level to the maximum, within a specified period after each transmitter pulse, so that short-range echoes producing clutter are amplified less than long-range echoes.

anticoagulant [PHARM] An agent, such as sodium citrate, that prevents coagulation of a colloid, especially blood.

anticodon [GEN] A three-nucleotide sequence of transfer RNA that complements the codon in messenger RNA.

anticoincidence circuit [ELECTR] Circuit that produces a specified output pulse when one (frequently predesignated) of two inputs receives a pulse and the other receives no pulse within an assigned time interval.

anticollision radar [ENG] A radar set designed to give warning of possible collisions during movements of ships or aircraft.

anticommutator [MATH] The anticommutator of two operators, A and B, is the operator $AB + BA$.

anticorona [OPTICS] A diffraction phenomenon appearing at a point before an observer with the sun or moon directly behind him; consists of rings of colored lights complementary to the coronal rings. Also known as Brocken bow.

anticyclolysis [METEOROL] Any weakening of anticyclonic circulation in the atmosphere.

anticyclone [METEOROL] High-pressure atmospheric closed circulation whose relative direction of rotation is clockwise in the Northern Hemisphere, counterclockwise in the Southern Hemisphere, and undefined at the Equator. Also known as high-pressure area.

antidepressant [PHARM] A drug, such as imipramine and tranylcypromine, that relieves depression by increasing central sympathetic activity.

antiderivative See indefinite integral.

antidetonant See antiknock.

antidiuretic hormone See vasopressin.

antidote [PHARM] An agent that relieves or counteracts the action of a poison.

antiepicenter See anticenter.

antierythrite See erythritol.

antifading antenna [ELECTR] An antenna designed to confine radiation mainly to small angles of elevation to minimize the fading of radiation directed at larger angles of elevation.

antiferroelectric crystal [SOLID STATE] A crystalline substance characterized by a state of lower symmetry consisting of two interpenetrating sublattices with equal but opposite electric polarization, and a state of higher symmetry in which the sublattices are unpolarized and indistinguishable.

antiferromagnetic domain [SOLID STATE] A region in a solid within which equal groups of elementary atomic or molecular magnetic moments are aligned antiparallel.

antiferromagnetic resonance [ELECTROMAG] Magnetic resonance in antiferromagnetic materials which may be observed by rotating magnetic fields in either of two opposite directions.

antiferromagnetic susceptibility [ELECTROMAG] The magnetic response to an applied magnetic field of a substance whose atomic magnetic moments are aligned in antiparallel fashion.

antiferromagnetism [SOLID STATE] A property possessed by some metals, alloys, and salts of transition elements by which the atomic magnetic moments form an ordered array which alternates or spirals so as to give no net total moment in zero applied magnetic field.

antifertilizin [BIOCHEM] An immunologically specific substance produced by animal sperm to implement attraction by the egg before fertilization.

antifix [ARCH] In classical architecture, either an ornament at the eaves to hide the ends of the joint tiles of the roof or an ornament of the cymatium of a classic cornice, sometimes pierced so that water can flow through.

antifoaming agent [ORG CHEM] A substance, such as silicones, organic phosphates, and alcohols, that inhibits the formation of bubbles in a liquid during its agitation by reducing its surface tension.

antifogging compound [MATER] A compound of one or more basic chemicals with filler or extenders for preventing condensation of moisture on glass and other transparent material, such as lenses or windshields.

antifreeze [CHEM] A substance added to a liquid to lower its freezing point; the principal automotive antifreeze component is ethylene glycol.

antifriction [MECH] Making friction smaller in magnitude. [MECH ENG] Employing a rolling contact instead of a sliding contact.

antifriction bearing [MECH ENG] Any bearing having the capability of reducing friction effectively.

antifungal agent [MATER] A chemical compound that either destroys or inhibits the growth of fungi.

antigen [IMMUNOL] A substance which reacts with the products of specific humoral or cellular immunity, even those induced by related heterologous immunogens.

antigorite [MINERAL] $Mg_3Si_2O_5(OH)_4$ Brownish-green variety of the mineral serpentine. Also known as baltimorite; picrolite.

antigravity [PHYS] The repulsion of one body by another by means of a gravitational type of force; this has never been observed.

antihemorrhagic vitamin See vitamin K.

antihistamine [PHARM] A drug that prevents or diminishes the effect of histamine; used in treating allergic reactions and common-cold symptoms.

antihyperon [PARTIC PHYS] An antiparticle to a hyperon, having the same mass, lifetime, and spin as the hyperon, but with charge and magnetic moment reversed in sign.

anti-infective vitamin See vitamin A.

antiknock [MATER] 1. Resisting detonation or pinging in spark-ignited engines. 2. A substance, such as tetraethyllead, added to motor and aviation gasolines to increase the resistance of the fuel to knock

in spark-ignited engines. Also known as antidetonant.

Antilles Current [OCEANOGR] A current formed by part of the North Equatorial Current that flows along the northern side of the Greater Antilles.

Antilocapridae [VERT ZOO] A family of artiodactyl mammals in the superfamily Bovoidea; the pronghorn is the single living species.

antilog See antilogarithm of a number.

antilogarithm of a number [MATH] A second number whose logarithm is the first number. Abbreviated antilog. Also known as inverse logarithm of a number.

antilymphocyte serum [IMMUNOL] An immunosuppressive agent effective in prolonging the lives of homografts in experimental animals by reducing the circulating lymphocytes.

antimatter [PHYS] Material consisting of atoms which are composed of positrons, antiprotons, and antineutrons.

antimere [INV ZOO] Any one of the equivalent parts into which a radially symmetrical animal may be divided.

antimetabolite [PHARM] A substance, such as sulfanilamide or amethopterin, that inhibits utilization of an essential metabolite because it is an analog of the metabolite.

antimicrobial agent [MICROBIO] A chemical compound that either destroys or inhibits the growth of microscopic and submicroscopic organisms.

antimissile missile [ORD] An explosive missile designed to intercept and destroy another missile in flight. Also known as auntie.

antimisting fuel [MATER] A kerosine fuel which has an additive to reduce misting.

antimitotic drug [PHARM] A substance, such as colchicine, vincristine, or vinblastine, that interferes with mitotic cellular division; used in the chemotherapy of leukemia.

antimolecule [ATOM PHYS] A molecule made up of antiprotons, antineutrons, and positrons in the same way that an ordinary molecule is made up of protons, neutrons, and electrons.

antimonite [MINERAL] Sb_2S_3 A lead-gray antimony sulfide mineral, the primary source of antimony; sometimes contains gold or silver; has a brilliant metallic luster, and occurs as prismatic orthorhombic crystals in massive forms. Also known as antimony glance; gray antimony; stibium; stibnite.

antimony [CHEM] A chemical element, symbol Sb, atomic number 51, atomic weight 121.75. [MINERAL] A very brittle, tin-white, hexagonal mineral, the native form of the element.

antimony blende See kermesite.

antineoplastic drug [PHARM] An agent, such as mercaptopurine compounds, that is antagonistic to the growth of a neoplasm.

antineutrino [PARTIC PHYS] The antiparticle to the neutrino; it has zero mass, spin ½, and positive helicity; there are two antineutrinos, one associated with electrons and one with muons.

antineutron [PARTIC PHYS] The antiparticle to the neutron; a strongly interacting baryon which has no charge, mass of 939.6 MeV, spin ½, and mean life of about 10^3 seconds.

antinode [ASTRON] Either of the two points on an orbit where a line in the orbit plane, perpendicular to the line of nodes and passing through the focus, intersects the orbit. [PHYS] A point, line, or surface in a standing-wave system at which some characteristic of the wave has maximum amplitude. Also known as loop.

antinucleon [PARTIC PHYS] An antineutron or antiproton, that is, particles having the same mass as their nucleon counterparts but opposite charge or opposite magnetic moment.

antinucleus [NUC PHYS] A nucleus made up of antineutrons and antiprotons in the same way that an ordinary nucleus is made up of neutrons and protons.

antioxidant [CHEM] An inhibitor, such as ascorbic acid, effective in preventing oxidation by molecular oxygen.

antiparallel [GEN] Pertaining to parallel molecules that point in opposite directions, as the strands of deoxyribonucleic acid. [PHYS] Property of two displacements or other vectors which lie along parallel lines but point in opposite directions.

antiparticle [PARTIC PHYS] A counterpart to a particle having mass lifetime and spin identical to the particle but with charge and magnetic moment reversed in sign.

Antipatharia [INV ZOO] The black or horny corals, an order of tropical and subtropical coelenterates in the subclass Zoantharia.

antipersonnel weapon [ORD] All instruments of combat, either offensive or defensive, used to destroy, wound, obstruct, or threaten enemy personnel.

antiperthite [GEOL] Natural intergrowth of feldspars formed by separation of sodium feldspar (albite) and potassium feldspar (orthoclase) during slow cooling of molten mixtures; the potassium-rich phase is exolved in a plagioclase host, exactly the inverse of perthite.

antipetalous [BOT] Having stamens positioned opposite to, rather than alternating with, the petals.

antiproton [PARTIC PHYS] The antiparticle to the proton; a strongly interacting baryon which is stable, carries unit negative charge, has the same mass as the proton (983.3 MeV), and has spin ½.

antipyretic [PHARM] Any agent, such as aspirin, that reduces or prevents fever.

antiquark [PARTIC PHYS] The hypothetical antiparticle of a quark, having electric charge, baryon number, and strangeness opposite in sign to that of the corresponding quark.

antirachitic vitamin See vitamin D.

antiresonance [ELEC] See parallel resonance. [ENG] The condition for which the impedance of a given electric, acoustic, or dynamic system is very high, approaching infinity.

antiroll fin [NAV ARCH] A control surface protruding from either of the lower sides of a ship to generate lift opposed to the rolling of the ship.

antirolling gyroscope [NAV ARCH] A very large gyroscope having a vertical axis and mounted in a ship so that the axis can be tipped fore and aft by an electric motor, resulting in the gyroscope's exerting a large torque on the ship about the fore-and-aft axis.

antiroll tank [NAV ARCH] One of paired tanks at opposite sides within a ship which are partially filled with liquid whose flow from one tank to the other helps to offset ship rolls.

antiseptic [PHARM] A substance used to destroy or prevent the growth of infectious microorganisms on or in the body.

antiserum [IMMUNOL] Any immune serum that contains antibodies active chiefly in destroying a specific infecting virus or bacterium.

antispasmodic [PHARM] An agent, such as benzyl benzoate, that relieves convulsions and the pain of muscular spasms.

antisubmarine [ORD] Descriptive of equipment, mines, or missiles designed to attack or destroy submarines.

antitoxin [IMMUNOL] An antibody elaborated by the body in response to a bacterial toxin that will combine with and generally neutralize the toxin.

antitrades [METEOROL] A deep layer of westerly winds in the troposphere above the surface trade winds of the tropics.

anti-transmit-receive tube [ELECTR] A switching tube that prevents the received echo signal from being dissipated in the transmitter.

antivenin [IMMUNOL] An immune serum that neutralizes the venoms of certain poisonous snakes and black widow spiders.

antixerophthalmic vitamin See vitamin A.

antler [VERT ZOO] One of a pair of solid bony, usually branched outgrowths on the head of members of the deer family (Cervidae); shed annually.

antlerite [MINERAL] $Cu_3SO_4(OH)_4$ Emerald- to blackish-green mineral occurring in aggregates of needlelike crystals; an ore of copper. Also known as vernadskite.

Antlia [ASTRON] A constellation with a right ascension of 10 hours and declination of 35°S. Abbreviated Ant. Also known as Air Pump.

ant lion [INV ZOO] The common name for insects of the family Myrmeleontidae in the order Neuroptera; larvae are commonly called doodlebugs.

anvil [ENG] **1.** The part of a machine that absorbs the energy delivered by a sharp force or blow. **2.** The stationary end of a micrometer caliper. [MET] **1.** A heavy wrought-iron, cast-iron, or steel block upon which metal is hammered in smith forging. **2.** The base of the hammer, holding the die bed and lower die part in drop forging. [METEOROL] See incus.

anvil cloud [METEOROL] The popular name given to a cumulonimbus capillatus cloud, a thunderhead whose upper portion spreads in the form of an anvil with a fibrous or smooth aspect; it also refers to such an upper portion alone when it persists beyond the parent cloud.

anxiety neurosis [PSYCH] A psychoneurotic disorder characterized by diffuse anxious expectation not restricted to definite situations, persons, or objects, and by emotional instability, irritability, apprehensiveness, and a sense of fatigue; caused by incomplete resolution of repressed emotional problems, and frequently associated with somatic symptoms.

A-O-I gate See AND-OR-INVERT gate.

AOQL See average outgoing quality limit.

aorta [ANAT] The main vessel of systemic arterial circulation arising from the heart in vertebrates. [INV ZOO] The large dorsal or anterior vessel in many invertebrates.

aortic arch [ANAT] The portion of the aorta extending from the heart to the third thoracic vertebra; single in warm-blooded vertebrates and paired in fishes, amphibians, and reptiles.

aortic body See aortic paraganglion.

aortic paraganglion [ANAT] A structure in vertebrates belonging to the chromaffin system and found on the front of the abdominal aorta near the mesenteric arteries. Also known as aortic body; organs of Zuckerkandl.

Apatemyidae [PALEON] A family of extinct rodentlike insectivorous mammals belonging to the Proteutheria.

apatetic [ECOL] Pertaining to the imitative protective coloration of an animal subject to being preyed upon.

Apathornithidae [PALEON] A family of Cretaceous birds, with two species, belonging to the order Ichthyornithiformes.

apatite [MINERAL] A group of phosphate minerals that includes 10 mineral species and has the general formula $X_5(YO_4)_3Z$, where X is usually Ca^{2+} or Pb^{3+}, Y is P^{5+} or As^{5+}, and Z is F^-, Cl^-, or OH^-.

APC See automatic phase control.

ape [VERT ZOO] Any of the tailless primates of the families Hylobatidae and Pongidae in the same superfamily as humans.

aperiodic [PHYS] Of irregular occurrence; not periodic; not displaying resonant response.

aperiodic antenna [ELECTROMAG] Antenna designed to have constant impedance over a wide range of frequencies because of the suppression of reflections within the antenna system; includes terminated wave and rhombic antennas.

aperiodic damping [PHYS] Condition of a system in which the amount of damping is so large that, when the system is subjected to a single disturbance, either constant or instantaneous, the system comes to a position of rest without passing through that position; while an aperiodically damped system is not strictly an oscillating system, it has such properties that it should become an oscillating system if the damping were sufficiently reduced.

aperture [ELECTR] An opening through which electrons, light, radio waves, or other radiation can pass. [OPTICS] The diameter of the objective of a telescope or other optical instrument, usually expressed in inches, but sometimes as the angle between lines from the principal focus to opposite ends of a diameter of the objective.

aperture antenna [ELECTROMAG] Antenna in which the beam width is determined by the dimensions of a horn, lens, or reflector.

aperture mask See shadow mask.

aperture stop [OPTICS] That opening in an optical system that determines the size of the bundle of rays which traverse the system from a given point of the object to the corresponding point of the image.

apetalous [BOT] Lacking petals.

apex [ANAT] 1. The upper portion of a lung extending into the root. 2. The pointed end of the heart. 3. The tip of the root of a tooth. [BOT] The pointed tip of a leaf. [ENG] In architecture or construction, the highest point, peak, or tip of any structure. [GEOL] The part of a mineral vein nearest the surface of the earth. [MATH] 1. The vertex of a triangle opposite the side which is regarded as the base. 2. The vertex of a cone or pyramid.

aphagia [MED] Inability to swallow; may be organic or psychic in origin.

aphanite [PETR] 1. A general term applied to dense, homogeneous rocks whose constituents are too small to be distinguished by the unaided eye. 2. A rock having aphanitic texture.

aphanitic [PETR] Referring to the texture of an igneous rock in which the crystalline components are not distinguishable by the unaided eye.

aphasia [MED] Impairment in the use or comprehension of language caused by lesions of the cerebral cortex.

Aphasmidea [INV ZOO] An equivalent name for the Adenophorea.

Aphelenchoidea [INV ZOO] A superfamily of plant and insect-associated nematodes in the order Tylenchida.

Aphelocheiridae [INV ZOO] A family of hemipteran insects belonging to the superfamily Naucoroidea.

aphid [INV ZOO] The common name applied to the soft-bodied insects of the family Aphididae; they are phytophagous plant pests and vectors for plant viruses and fungal parasites.

Aphididae [INV ZOO] The true aphids, a family of homopteran insects in the superfamily Aphidoidea.

Aphidoidea [INV ZOO] A superfamily of sternorrhynchan insects in the order Homoptera.

aphotic zone [OCEANOGR] The deeper part of the ocean where sunlight is absent.

Aphredoderidae [VERT ZOO] A family of actinopterygian fishes in the order Percopsiformes containing one species, the pirate perch.

aphrodisiac [PHYSIO] Any chemical agent or odor that stimulates sexual desires.

Aphroditidae [INV ZOO] A family of scale-bearing polychaete worms belonging to the Errantia.

Aphrosalpingoidea [PALEON] A group of middle Paleozoic invertebrates classified with the calcareous sponges.

aphrosiderite See ripidolite.

Aphylidae [INV ZOO] An Australian family of hemipteran insects composed of two species; not placed in any higher toxonomic group.

API See air-position indicator; armor-piercing incendiary.

apiary [AGR] A place where bees are kept, especially for breeding and honey making.

apical bud See terminal bud.

apical dominance [BOT] Inhibition of lateral bud growth by the apical bud of a shoot, believed to be a response to auxins produced by the apical bud.

apical meristem [BOT] A region of embryonic tissue occurring at the tips of roots and stems.

Apidae [INV ZOO] A family of hymenopteran insects in the superfamily Apoidea including the honeybees, bumblebees, and carpenter bees.

Apioceridae [INV ZOO] A family of orthorrhaphous dipteran insects in the series Brachycera.

Apistobranchidae [INV ZOO] A family of spioniform annelid worms belonging to the Sedentaria.

Aplacophora [INV ZOO] A subclass of vermiform mollusks in the class Amphineura characterized by no shell and calcareous integumentary spicules.

aplanatic lens [OPTICS] A lens corrected for spherical abberation.

aplanatic points [OPTICS] Two points on the axis of an optical system which are located so that all the rays emanating from one converge to, or appear to diverge from, the other.

aplastic anemia [MED] A blood disorder in which lymphocytes predominate while there is a deficiency of erythrocytes, hemoglobin, and granulocytes.

aplite [PETR] Fine-grained granitic dike rock made up of light-colored mineral constituents, mostly quartz and feldspar; used to manufacture glass and enamel.

Apneumonomorphae [INV ZOO] A suborder of arachnid arthropods in the order Araneida characterized by the lack of book lungs.

apocenter See apofocus.

apochromatic lens [OPTICS] A lens with corrections for chromatic and spherical aberration.

apochromatic system [OPTICS] An optical system which is free from both spherical and chromatic aberration for two or more colors.

apocrine gland [PHYSIO] A multicellular gland, such as a mammary gland or an axillary sweat gland, that extrudes part of the cytoplasm with the secretory product.

Apocynaceae [BOT] A family of tropical and subtropical flowering trees, shrubs, and vines in the order Gentianales, characterized by a well-developed latex system, granular pollen, a poorly developed corona, and the carpels often united by the style and stigma; well-known members are oleander and periwinkle.

Apoda [VERT ZOO] The caecilians, a small order of wormlike, legless animals in the class Amphibia.

Apodacea [INV ZOO] A subclass of echinoderms in the class Holothuroidea characterized by simple or pinnate tentacles and reduced or absent tube feet.

Apodes [VERT ZOO] An equivalent name for the Anguilliformes.

Apodi [VERT ZOO] The swifts, a suborder of birds in the order Apodiformes.

Apodida [INV ZOO] An order of worm-shaped holothurian echinoderms in the subclass Apodacea.

Apodidae [VERT ZOO] The true swifts, a family of apodiform birds belonging to the suborder Apodi.

Apodiformes [VERT ZOO] An order of birds containing the hummingbirds and swifts.

apodization [ELECTR] A technique for modifying the response of a surface acoustic wave filter by varying the overlap between adjacent electrodes of the interdigital transducer. [OPTICS] The modification of the amplitude transmittance of the aperture of an optical system so as to reduce or suppress the energy of the diffraction rings relative to that of the central Airy disk.

apoenzyme [BIOCHEM] The protein moiety of an enzyme; determines the specificity of the enzyme reaction.

apofocus [ASTRON] The point on an elliptic orbit at the greatest distance from the principal focus. Also known as apocenter.

apogee [ASTRON] That point in an orbit at which the moon or an artificial satellite is most distant from the earth; the term is sometimes loosely applied to positions of satellites of other planets.

Apogonidae [VERT ZOO] The cardinal fishes, a family of tropical marine fishes in the order Perciformes; males incubate eggs in the mouth.

Apoidea [INV ZOO] The bees, a superfamily of hymenopteran insects in the suborder Apocrita.

Apollo [ASTRON] 1. To the Greeks, the planet Mercury when it was a morning star. 2. An asteroid with a very eccentric orbit and perihelion inside the orbit of Venus that passed about 3×10^6 kilometers from earth in 1932.

apomixis [EMBRYO] Parthenogenetic development of sex cells without fertilization.

aponeurosis [ANAT] A broad sheet of regularly arranged connective tissue that covers a muscle or serves to connect a flat muscle to a bone.

apophyllite [MINERAL] A hydrous calcium potassium silicate containing fluorine and occurring as a secondary mineral with zeolites with geodes and other igneous rocks; the composition is variable but approximates $KFCa_4(Si_2O_5)_4 \cdot 8H_2$. Also known as fisheye stone.

apophyllous [BOT] Having the parts of the perianth distinct.

apophysis [ANAT] An outgrowth or process on an organ or bone. [MYCOL] A swollen filament in fungi.

apoplexy [MED] 1. A symptom complex caused by an acute vascular lesion of the brain and characterized by unconsciousness with various degrees of paralysis and sensory impairment. 2. Sudden, severe hemorrhage into any organ.

Aporidea [INV ZOO] An order of tapeworms of uncertain composition and affinities; parasites of anseriform birds.

A positive [ELEC] Also known as A+. 1. Positive terminal of an A battery or positive polarity of other sources of filament voltage. 2. Denoting the terminal to which the positive side of the filament voltage source should be connected.

apostilb [OPTICS] A luminance unit equal to one ten-thousandth of a lambert. Also known as blandel.

Apostomatida [INV ZOO] An order of ciliated protozoans in the subclass Holotrichia; majority are commensals on marine crustaceans.

apothecaries' dram *See* dram.

apothecaries' measure [PHARM] A system of units of volume, usually of liquid drugs, in which 16 fluid ounces equals 1 pint.

apothecaries' ounce *See* ounce.

apothecaries' pound *See* pound.

apothecaries' weight [PHARM] A system of units of mass, usually of drugs, in which 1 pound equals 5760 grains or 1 troy pound.

apothem [MATH] The perpendicular distance from the center of a regular polygon to one of its sides.

A power supply *See* A supply.

apparatus [SCI TECH] A compound instrument designed to carry out a specific function.

apparent [ASTRON] A term used to designate certain measured or measurable astronomic quantities to refer them to real or visible objects, such as the sun or a star.

apparent expansion [THERMO] The expansion of a liquid with temperature, as measured in a graduated container without taking into account the container's expansion.

apparent force [MECH] A force introduced in a relative coordinate system in order that Newton's laws be satisfied in the system; examples are the Coriolis force and the centrifugal force incorporated in gravity.

apparent gravity *See* acceleration of gravity.

apparent horizon *See* horizon.

apparent libration in longitude *See* lunar libration.

apparent luminance [OPTICS] Luminance, created by air light, of that portion of the visual field subtended by a dark, distant object; that is, the light scattered into the eye by particles, including air molecules, lying along the optic path from eye to object.

apparent magnitude [ASTRON] An index of a star's brightness relative to that of the other stars; it does not take into account the difference in distance between the stars and is not an indication of the star's true luminosity.

apparent motion *See* relative motion.

apparent noon [ASTRON] Twelve o'clock apparent time, or the instant the apparent sun is over the upper branch of the meridian.

apparent place *See* apparent position.

apparent position [ASTRON] The position on the celestial sphere at which a heavenly body (or a space vehicle) would be seen from the center of the earth at a particular time. Also known as apparent place.

apparent power [ELEC] The product of the root-mean-square voltage and the root-mean-square current delivered in an alternating-current circuit, no account being taken of the phase difference between voltage and current.

apparent precession *See* apparent wander.

apparent solar day [ASTRON] The duration of one rotation of the earth on its axis with respect to the apparent sun. Also known as true solar day.

apparent solar time [ASTRON] Time measured by the apparent diurnal motion of the sun. Also known as apparent time; true solar time.

apparent vertical [GEOPHYS] The direction of the resultant of gravitational and all other accelerations. Also known as dynamic vertical.

apparent wander [GEOPHYS] Apparent change in the direction of the axis of rotation of a spinning body, such as a gyroscope, due to rotation of the earth. Also known as apparent precession; wander.

apparent weight [MECH] For a body immersed in a fluid (such as air), the resultant of the gravitational force and the buoyant force of the fluid acting on the body; equal in magnitude to the true weight minus the weight of the displaced fluid.

apparition [ASTRON] A period during which a planet, asteroid, or comet is observable, generally between two successive conjunctions of the body with the sun.

appendage [BIOL] Any subordinate or nonessential structure associated with a major body part. [NAV ARCH] Any fitting installed outside the underwater hull proper, such as a rudder, shaft, shaft strut, and bilge keel. [ZOO] Any jointed, peripheral extension, especially limbs, of arthropod and vertebrate bodies.

appendectomy [MED] Surgical removal of the vermiform appendix.

appendicitis [MED] Inflammation of the vermiform appendix.

appendicular skeleton [ANAT] The bones of the pectoral and pelvic girdles and the paired appendages in vertebrates.

apple [BOT] *Malus domestica*. A deciduous tree in the order Rosales which produces an edible, simple, fleshy, pome-type fruit.

apple of Peru *See* jimsonweed.

Appleton layer *See* F_2 layer.

application development language [COMPUT SCI] A very-high-level programming language that generates coding in a conventional programming language or provides the user of a data-base management system with a programming language that is easier to implement than conventional programming languages.

application package [COMPUT SCI] A combination of required hardware, including remote inputs and outputs, plus programming of the computer memory to produce the specified results.

applications program [COMPUT SCI] A program written to solve a specific problem, produce a specific report, or update a specific file.

applications technology satellite [AERO ENG] Any artificial satellite in the National Aeronautics and Space Administration program for the evaluation of advanced techniques and equipment for communications, meteorological, and navigation satellites. Abbreviated ATS.

application study [COMPUT SCI] The detailed process of determining a system or set of procedures for using a computer for definite functions of operations, and establishing specifications to be used as a base for the selection of equipment suitable to the specific needs.

apposition beach [GEOL] One of a series of parallel beaches formed on the seaward side of an older beach.

apposition eye [INV ZOO] A compound eye found in diurnal insects and crustaceans in which each ommatidium focuses on a small part of the whole field of light, producing a mosaic image.

apposition fabric [PETR] A primary orientation of the elements of a sedimentary rock that is developed or formed at time of deposition of the material; fabrics of most sedimentary rocks belong to this type. Also known as primary fabric.

approach [MECH ENG] The difference between the temperature of the water leaving a cooling tower and the wet-bulb temperature of the surrounding air. [NAV] In air operations, a maneuver executed by an aircraft in making its transit from high-altitude enroute flight to the point where it begins the landing approach; includes maneuvers (such as flying race-track pattern) required for traffic control.

approach path [NAV] That portion of the flight path which extends from the point where a descent is started to the point where the aircraft touches down on the runway.

approximate [MATH] 1. To obtain a result that is not exact but is near enough to the correct result for some specified purpose. 2. To obtain a series of results approaching the correct result. [SCI TECH] 1. Close to the correct value. Abbreviated approx. 2. To be close to.

approximate absolute temperature [PHYS] A temperature scale with the ice point at 273° and boiling point of water at 373°; it is intended to approximate the Kelvin temperature scale with sufficient accuracy for many sciences, notably meteorology, and is widely used in the meteorological literature. Also known as tercentesimal thermometric scale.

appulse [ASTRON] 1. The near approach of one celestial body to another on the celestial sphere, as in occultation or conjunction. 2. A penumbral eclipse of the moon.

apricot [BOT] *Prunus armeniaca*. A deciduous tree in the order Rosales which produces a simple fleshy stone fruit.

a priori [MATH] Pertaining to deductive reasoning from assumed axioms or supposedly self-evident principles, supposedly without reference to experience.

a priori probability *See* mathematical probability.

apron [AERO ENG] A protective device specially designed to cover an area surrounding the fuel inlet on a rocket or spacecraft. [BUILD] 1. A board on an interior wall beneath a windowsill. 2. The vertical rear panel of a sink attached to a wall. [CIV ENG] 1. A covering of a material such as concrete or timber over soil to prevent erosion by flowing water, as at the bottom of a dam. 2. A concrete or wooden shield that is situated along the bank of a river, along a sea wall, or below a dam. [GEOL] *See* outwash plain. [HYD] *See* ram. [MECH ENG] A plate serving to protect or cover a machine. [MIN ENG] A canvas-covered frame set at such an angle in the miner's rocker that the gravel and water passing over it are carried to the head of the machine. [ORD] 1. That portion of the superior slope of a parapet or the interior slope of a pit designed to protect the slopes against blast. 2. The hinged portion of a shield. 3. A removal screen of camouflage material placed over or in front of artillery guns. 4. A hard-surfaced area, usually paved, adjacent to a ship or the like, used to park, load, unload, or service vehicles.

apron conveyor [MECH ENG] A conveyor used for carrying granular or lumpy material and consisting of two strands of roller chain separated by overlapping plates, forming the carrying surface, with sides 2–6 inches (5–15 centimeters) high.

apse [ARCH] A semicircular (or nearly semicircular) or semipolygonal space, usually in a church, terminating an axis and intended to house an altar. [ASTRON] *See* apsis.

Apsidospondyli [VERT ZOO] A term used to include, as a subclass, amphibians in which the vertebral centra are formed from cartilaginous arches.

apsis [ASTRON] In celestial mechanics, either of the two orbital points nearest or farthest from the center of attraction. Also known as apse.

apterous [BIOL] Lacking wings, as in certain insects, or winglike expansions, as in certain seeds.

Apterygidae [VERT ZOO] The kiwis, a family of nocturnal ratite birds in the order Apterygiformes.

Apterygiformes [VERT ZOO] An order of ratite birds containing three living species, the kiwis, characterized by small eyes, limited eyesight, and nostrils at the tip of the bill.

Apterygota [INV ZOO] A subclass of the Insecta characterized by being primitively wingless.

Aptian [GEOL] Lower Cretaceous geologic time, between Barremian and Albian. Also known as Vectian.

Aqil *See* Aquila.

Aql *See* Aquila.

AQL *See* acceptable quality level.

aquaculture *See* aquiculture.

aquagene tuff *See* hyaloclastite.

aqualf [GEOL] A suborder of the soil order Alfisol, seasonally wet and marked by gray or mottled colors; occurs in depressions or on wide flats in local landscapes.

aqualung [ENG] A self-contained underwater breathing apparatus (scuba) of the demand or open-circuit type developed by J.Y. Cousteau.

aquamarine [MINERAL] A pale-blue or greenish-blue transparent gem variety of the mineral beryl.

aqua regia [INORG CHEM] A fuming, highly corrosive, volatile liquid with a suffocating odor made by mixing 1 part concentrated nitric acid and 3 parts concentrated hydrochloric acid; reacts with all metals, including silver and gold. Also known as chloroazotic acid; chloronitrous acid; nitrohydrochloric acid; nitromuriatic acid.

Aquarius [ASTRON] A constellation with a right ascension of 23 hours and declination of 15°S. Abbreviated Aqr. Also known as Water-Bearer.

aquatic [BIOL] Living or growing in, on, or near water; having a water habitat.

aquatint [GRAPHICS] An etching process that produces several tones by varying the etching time of different areas of a copper plate; the resulting print resembles an ink or wash drawing.

aquatone [GRAPHICS] An offset printing process utilizing a zinc plate that is gelatin-coated and hardened and sensitized to print type, line drawings, and fine-screen halftones.

aqueduct [CIV ENG] An artificial tube or channel for conveying water.

Aquent [GEOL] A suborder of the soil order Entisol, bluish gray or greenish gray in color; under water until very recent times; located at the margins of oceans, lakes, or seas.

aqueous humor [PHYSIO] The transparent fluid filling the anterior chamber of the eye.

Aquept [GEOL] A suborder of the soil order Inceptisol, wet or drained, which lacks silicate clay accumulation in the soil profiles; surface horizon varies in thickness.

aquiclude [GEOL] A porous formation that absorbs water slowly but will not transmit it fast enough to furnish an appreciable supply for a well or spring.

aquiculture [BIOL] Cultivation of natural faunal resources of water. Also spelled aquaculture.

aquifer [GEOL] A permeable body of rock capable of yielding quantities of groundwater to wells and springs. [HYD] A subsurface zone that yields economically important amounts of water to wells.

Aquifoliaceae [BOT] A family of woody flowering plants in the order Celastrales characterized by pendulous ovules, alternate leaves, imbricate petals, and drupaceous fruit; common members include various species of holly (*Ilex*).

aquifuge [GEOL] An impermeable body of rock which contains no interconnected openings or interstices and therefore neither absorbs nor transmits water.

Aquila [ASTRON] A constellation with a right ascension of 20 hours and declination of 5°N. Abbreviated Aqil; Aql.

Aquitanian [GEOL] Lower lower Miocene or uppermost Oligocene geologic time.

Aquitard [GEOL] A bed of low permeability adjacent to an aquifer; may serve as a storage unit for groundwater, although it does not yield water readily.

Aquod [GEOL] A suborder of the soil order Spodosol, with a black or dark brown horizon just below the surface horizon; seasonally wet, it occupies depressions or wide flats from which water cannot escape easily.

Aquoll [GEOL] A suborder of the soil order Mollisol, with thick surface horizons; formed under wet conditions, it may be under water at times, but is seasonally rather than continually wet.

Aquox [GEOL] A suborder of the soil order Oxisol, seasonally wet, found chiefly in shallow depressions; deeper soil profiles are predominantly gray, sometimes mottled, and contain nodules or sheets of iron and aluminum oxides.

Aquult [GEOL] A suborder of the soil order Ultisol; seasonally wet, it is saturated with water a significant part of the year unless drained; surface horizon of the soil profile is dark and varies in thickness, grading to gray in the deeper portions; it occurs in depressions or on wide upland flats from which water drains very slowly.

Ar *See* argon.

Arabellidae [INV ZOO] A family of polychaete worms belonging to the Errantia.

arabinose [BIOCHEM] $C_5H_{10}O_5$ A pentose sugar obtained in crystalline form from plant polysaccharides such as gums, hemicelluloses, and some glycosides.

Araceae [BOT] A family of herbaceous flowering plants in the order Arales; plants have stems, roots, and leaves, the inflorescence is a spadix, and the growth habit is terrestrial or sometimes more or less aquatic; well-known members include dumb cane (*Dieffenbachia*), jack-in-the-pulpit (*Arisaema*), and *Philodendron*.

Arachnida [INV ZOO] A class of arthropods in the subphylum Chelicerata characterized by four pairs of thoracic appendages.

arachnoid [ANAT] A membrane that covers the brain and spinal cord and lies between the pia mater and dura mater. [BOT] Of cobweblike appearance, caused by fine white hairs. Also known as araneose. [INV ZOO] Any invertebrate related to or resembling the Arachnida.

Arachnoidea [INV ZOO] The name used in some classification schemes to describe a class of primitive arthropods.

Aradidae [INV ZOO] The flat bugs, a family of hemipteran insects in the superfamily Aradoidea.

Aradoidea [INV ZOO] A small superfamily of hemipteran insects belonging to the subdivision Geocorisae.

Araeoscelidia [PALEON] A provisional order of extinct reptiles in the subclass Euryapsida.

aragonite [MINERAL] $CaCO_3$ A white, yellowish, or gray orthorhombic mineral species of calcium carbonate but with a crystal structure different from those of vaterite and calcite, the other two polymorphs of the same composition. Also known as Aragon spar.

Aragon spar See aragonite.

Arales [BOT] An order of monocotyledonous plants in the subclass Arecidae.

Araliaceae [BOT] A family of dicotyledonous trees and shrubs in the order Umbellales; there are typically five carpels and the fruit, usually a berry, is fleshy or dry; well-known members are ginseng (*Panax*) and English ivy (*Hedera helix*).

aralkyl [ORG CHEM] A radical in which an aryl group is substituted for an alkyl H atom. Derived from arylated alkyl.

Aramidae [VERT ZOO] The limpkins, a family of birds in the order Gruiformes.

Araneae [INV ZOO] An equivalent name for Araneida.

Araneida [INV ZOO] The spiders, an order of arthropods in the class Arachnida.

araneology [INV ZOO] The study of spiders.

Arbacioida [INV ZOO] An order of echinoderms in the superorder Echinacea.

arbiter [COMPUT SCI] A computer unit that determines the priority sequence in which two or more processor inputs are connected to a single functional unit such as a multiplier or memory.

arbitrary course computer See course-line computer.

arbitration [IND ENG] A semijudicial means of settling labor-management disputes in which both sides agree to be bound by the decision of one or more neutral persons selected by some method mutually agreed upon.

arbor [HOROL] The axle of a wheel in a watch or clock. [MECH ENG] 1. A cylindrical device positioned between the spindle and outer bearing of a milling machine and designed to hold a milling cutter. 2. A shaft or spindle used to hold a revolving cutting tool or the work to be cut. [MET] A device which supports sand cores in molds.

arboreal Also known as arboreous. [BOT] Relating to or resembling a tree. [ZOO] Living in trees.

arboreous [BOT] 1. Wooded. 2. See arboreal.

arboretum [BOT] An area where trees and shrubs are cultivated for educational and scientific purposes.

arbovirus [VIROL] Small, arthropod-borne animal viruses that are unstable at room temperature and inactivated by sodium deoxycholate; cause several types of encephalitis. Also known as arthropod-borne virus.

arc [ELEC] See electric arc. [ENG] The graduated scale of an instrument for measuring angles, as a marine sextant; readings obtained on that part of the arc beginning at zero and extending in the direction usually considered positive are popularly said to be on the arc, and those beginning at zero and extending in the opposite direction are said to be off the arc. [GEOL] A geologic or topographic feature that is repeated along a curved line on the surface of the earth. [MATH] 1. A continuous piece of the circumference of a circle. 2. See edge.

arcade [ARCH] 1. A series of arches supported on columns. 2. An arched passageway. [INV ZOO] A

type of cell associated with the pharyngeal region of nematodes and united with like cells by an arch.

arcback [ELECTR] The flow of a principal electron stream in the reverse direction in a mercury-vapor rectifier tube because of formation of a cathode spot on an anode; this results in failure of the rectifying action. Also known as backfire.

arc chute [ELEC] A collection of insulating barriers in a circuit breaker for confining the arc and preventing it from causing damage.

arc converter [ELECTR] A form of oscillator using an electric arc as the generator of alternating or pulsating current.

arc cutting [MET] A type of thermal cutting of metal using the temperature generated by an electric arc.

arc discharge [ELEC] A direct-current electrical current between electrodes in a gas or vapor, having high current density and relatively low voltage drop.

Arcellinida [INV ZOO] An order of rhizopodous protozoans in the subclass Lobosia characterized by lobopodia and a well-defined aperture in the test.

arc furnace [MET] A furnace used to heat materials by the energy from an electric arc. Also known as electric-arc furnace.

arch [CIV ENG] A structure curved and so designed that when it is subjected to vertical loads, its two end supports exert reaction forces with inwardly directed horizontal components; common uses for the arch are as a bridge, support for a roadway or railroad track, or part of a building.

Archaeoceti [PALEON] The zeuglodonts, a suborder of aquatic Eocene mammals in the order Cetacea; the oldest known cetaceans.

Archaeocidaridae [PALEON] A family of Carboniferous echinoderms in the order Cidaroida characterized by a flexible test and more than two columns of interambulacral plates.

Archaeocopida [PALEON] An order of Cambrian crustaceans in the subclass Ostracoda characterized by only slight calcification of the carapace.

Archaeogastropoda [INV ZOO] An order of gastropod mollusks that includes the most primitive snails.

Archaeopteridales [PALEOBOT] An order of Upper Devonian sporebearing plants in the class Polypodiopsida characterized by woody trunks and simple leaves.

Archaeopterygiformes [PALEON] The single order of the extinct avian subclass Archaeornithes.

Archaeornithes [PALEON] A subclass of Upper Jurassic birds comprising the oldest fossil birds.

Archangiaceae [MICROBIO] A family of bacteria in the order Myxobacterales; microcysts are rod-shaped, ovoid, or spherical and are not enclosed in sporangia, and fruiting bodies are irregular masses.

arch beam [CIV ENG] A curved beam, used in construction, with a longitudinal section bounded by two arcs having different radii and centers of curvature so that the beam cross section is larger at either end than at the center.

arc heating [MET] The heating of a material by the heat energy from an electric arc, which has a very high temperature and very high concentration of heat energy. Also known as electric-arc heating.

archegonium [BOT] The multicellular female sex organ in all plants of the Embryobionta except the Pinophyta and Magnoliophyta.

archencephalon [EMBRYO] The primitive embryonic forebrain from which the forebrain and midbrain develop.

archenteron [EMBRYO] The cavity of the gastrula formed by ingrowth of cells in vertebrate embryos. Also known as gastrocoele; primordial gut.

archeology [SCI TECH] The scientific study of the material remains of the cultures of historical or prehistorical peoples.

Archeozoic [GEOL] **1.** The era during which, or during the latter part of which, the oldest system of rocks was made. **2.** The last of three subdivisions of Archean time, when the lowest forms of life probably existed; as more physical measurements of geologic time are made, the usage is changing; it is now considered part of the Early Precambrian.

Archer See Sagittarius.

archerfish [VERT ZOO] The common name for any member of the fresh-water family Toxotidae in the order Perciformes; individuals eject a stream of water from the mouth to capture insects.

archetype [EVOL] A hypothetical ancestral type conceptualized by eliminating all specialized character traits.

Archiacanthocephala [INV ZOO] An order of worms in the phylum Acanthocephala; adults are endoparasites of terrestrial vertebrates.

Archiannelida [INV ZOO] A group name applied to three families of unrelated annelid worms: Nerillidae, Protodrilidae, and Dinophilidae.

archibenthic zone [OCEANOGR] The biogeographic realm of the ocean extending from a depth of about 200 meters to 800–1100 meters (665 feet to 2625–3610 feet).

Archigregarinida [INV ZOO] An order of telosporean protozoans in the subclass Gregarinia; endoparasites of invertebrates and lower chordates.

Archimedean principle [PHYS] The principle that a body immersed in a fluid undergoes an apparent loss in weight equal to the weight of the fluid it displaces.

Archimedes' screw [MECH ENG] A device for raising water by means of a rotating broad-threaded screw or spirally bent tube within an inclined hollow cylinder.

archipelago [GEOGR] **1.** A large group of islands. **2.** A sea that has a large group of islands within it.

architect [ARCH] A person who is skilled and knowledgeable in the design of buildings and whose qualifications are recognized by a college degree and licensing by an appropriate professional organization.

architect's scale [GRAPHICS] A rule with a scale on it so chosen that by placing the rule's edge on a reduced-scale drawing the scale of the drawing (say, in inches) may be converted directly into the dimensions of the object (say, in feet).

architectural engineering [CIV ENG] The branch of engineering dealing primarily with building materials and components and with the design of structural systems for buildings, in contrast to heavy construction such as bridges.

architecture [ENG] **1.** The art and science of designing buildings. **2.** The product of this art and science.

architrave [ARCH] **1.** The lowest division of an entablature that rests on the column capital. **2.** The molded band, group of moldings, or other architectural member around an opening, such as a door, especially if rectangular.

archiving [COMPUT SCI] The storage of files, in the form of punch cards, microfilm, or magnetic tape, for very long periods of time, in case it is necessary to regenerate the file due to subsequent introduction of errors.

Archosauria [VERT ZOO] A subclass of reptiles composed of five orders: Thecodontia, Saurischia, Ornithschia, Pterosauria, and Crocodilia.

Archostemata [INV ZOO] A suborder of insects in the order Coleoptera.

arch rib [CIV ENG] One of a set of projecting molded members subdividing the undersurface of an arch.

arcing time [ELEC] **1.** Interval between the parting, in a switch or circuit breaker, of the arcing contacts and the extension of the arc. **2.** Time elapsing, in a fuse, from the severance of the fuse link to the final interruption of the circuit under a specified condition.

arc jet engine [AERO ENG] An electromagnetic propulsion engine used to supply motive power for flight; hydrogen and ammonia are used as the propellant, and some plasma is formed as the result of electric-arc heating.

arc lamp [ELEC] An electric lamp in which the light is produced by an arc made when current flows through ionized gas between two electrodes. Also known as electric-arc lamp.

arcmin See minute.

arc navigation [NAV] A navigation system in which the position of an airplane or ship is maintained along an arc of a circle which has a radius measured from a control station by means of electronic distance-measuring equipment, such as shoran or oboe.

arc-over [ELEC] An unwanted arc resulting from the opening of a switch or the breakdown of insulation.

arc spectrum [SPECT] The spectrum of a neutral atom, as opposed to that of a molecule or an ion; it is usually produced by vaporizing the substance in an electric arc; designated by the roman numeral I following the symbol for the element, for example, HeI.

arctic air [METEOROL] An air mass whose characteristics are developed mostly in winter over arctic surfaces of ice and snow.

arctic anticyclone See arctic high.

Arctic Circle [GEOD] The parallel of latitude $66°32'N$ (often taken as $66\frac{1}{2}°N$).

arctic front [METEOROL] The semipermanent, semicontinuous front between the deep, cold arctic air and the shallower, basically less cold polar air of northern latitudes.

arctic high [METEOROL] A weak high that appears on mean charts of sea-level pressure over the Arctic Basin during late spring, summer, and early autumn. Also known as arctic anticyclone; polar anticyclone; polar high.

Arctiidae [INV ZOO] The tiger moths, a family of lepidopteran insects in the suborder Heteroneura.

Arctocyonidae [PALEON] A family of extinct carnivorelike mammals in the order Condylarthra.

Arctolepiformes [PALEON] A group of the extinct joint-necked fishes belonging to the Arthrodira.

arc triangulation [ENG] A system of triangulation in which an arc of a great circle on the surface of the earth is followed in order to tie in two distant points.

Arcturidae [INV ZOO] A family of isopod crustaceans in the suborder Valvifera characterized by an almost cylindrical body and extremely long antennae.

arcuate [ANAT] Arched or curved; bow-shaped.

arcus [METEOROL] A dense and horizontal roll-shaped accessory cloud, with more or less tattered edges, situated on the lower front part of the main cloud.

arc welding See electric-arc welding.

Arcyzonidae [PALEON] A family of Devonian paleocopan ostracods in the superfamily Kirkbyacea characterized by valves with a large central pit.

Ardeidae [VERT ZOO] The herons, a family of wading birds in the order Ciconiiformes.

ardennite [MINERAL] $Mn_5Al_5(VO_4)(SiO_4)_5(OH)_2 \cdot 2H_2O$ A yellow to yellowish-brown mineral consisting of a hydrous silicate vanadate and arsenate of manganese and aluminum.

arduinite   *See* mordenite.

area   [COMPUT SCI] A section of a computer memory assigned by a computer program or by the hardware to hold data of a particular type.   [MATH] A measure of the size of a two-dimensional surface, or of a region on such a surface.

areal pattern   [PETRO ENG] Distribution pattern of oil-production wells and water- or gas-injection wells over a given oil reservoir.

area meter   [ENG] A mechanism to measure fluid flow rate through a fixed-area conduit by the movement of a weighted piston or float supported by the flowing fluid; includes rotameters and piston-type meters.

area search   [COMPUT SCI] A computer search that examines only those records which satisfy some broad criteria.

areaway   [CIV ENG] An open space at subsurface level adjacent to a building, providing access to and utilities for a basement.

Arecaceae   [BOT] The palms, the single family of the order Arecales.

Arecales   [BOT] An order of flowering plants in the subclass Arecidae composed of the palms.

Arecidae   [BOT] A subclass of flowering plants in the class Liliopsida characterized by numerous, small flowers subtended by a prominent spathe and often aggregated into a spadix, and broad, petiolate leaves without typical parallel venation.

A register   *See* arithmetic register.

arenaceous   [GEOL] Of sediment or sedimentary rocks that have been derived from sand or that contain sand. Also known as arenarious; psammitic; sabulous.

arene   *See* aromatic hydrocarbon.

Arenicolidae   [INV ZOO] The lugworms, a family of mud-swallowing worms belonging to the Sedentaria.

arenicolite   [GEOL] A hole, groove, or other mark in a sedimentary rock, generally sandstone, interpreted as a burrow made by an arenicolous marine worm or a trail of a mollusk or crustacean.

Arenigian   [GEOL] A European stage including Lower Ordovician geologic time (above Tremadocian, below Llanvirnian). Also known as Skiddavian.

arenite   [PETR] Consolidated sand-texture sedimentary rock of any composition. Also known as arenyte; psammite.

Arent   [GEOL] A suborder of the soil order Entisol, consisting of soils formerly of other classifications that have been severely disturbed, completely disrupting the sequence of horizons.

arenyte   *See* arenite.

areola   [ANAT] **1.** The portion of the iris bordering the pupil of the eye. **2.** A pigmented ring surrounding a nipple, vesicle, or pustule. **3.** A small space, interval, or pore in a tissue.

areolar tissue   [HISTOL] A loose network of fibrous tissue and elastic fiber that connects the skin to the underlying structures.

Ares   [ASTRON] The planet Mars.

arête   [GEOL] Narrow, jagged ridge produced by the merging of glacial cirques. Also known as arris; crib; serrate ridge.

Arg   *See* Argo.

Argand diagram   [MATH] A two-dimensional cartesian coordinate system for representing the complex numbers, the number $x + iy$ being represented by the point whose coordinates are x and y.

Argasidae   [INV ZOO] The soft ticks, a family of arachnids in the suborder Ixodides; several species are important as ectoparasites and disease vectors for man and domestic animals.

Argentinoidei   [VERT ZOO] A family of marine deepwater teleostean fishes, including deep-sea smelts, in the order Salmoniformes.

argentite   [MINERAL] Ag_2S A lustrous, lead-gray ore of silver; it is a monoclinic mineral and is dimorphous with acanthite. Also known as argyrite; silver glance; vitreous silver.

Argid   [GEOL] A suborder of the soil order Aridisol, well drained, having a characteristically brown or red color and a silicate accumulation below the surface horizon; occupies older land surfaces in deserts.

Argidae   [INV ZOO] A small family of hymenopteran insects in the superfamily Tenthredinoidea.

argillaceous   [GEOL] Of rocks or sediments made of or largely composed of clay-size particles or clay minerals.

argillite   [PETR] A compact rock formed from siltstone, shale, or claystone but intermediate in degree of induration and structure between them and slate; argillite is more indurated than mudstone but lacks the fissility of shale.

arginine   [BIOCHEM] $C_6H_{14}N_4O_2$ A colorless, crystalline, water-soluble, essential amino acid of the α-ketoglutaric acid family.

Argo   [ASTRON] The large Ptolemy constellation; a southern constellation, now divided into four groups (Carina, Pupis, Vela, and Pyxis Nautica). Abbreviated Arg. Also known as Ship.

argon   [CHEM] A chemical element, symbol Ar, atomic number 18, atomic weight 39.998.

argon laser   [OPTICS] A gas laser using ionized argon; emits a 4880-angstrom line as well as infrared radiation.

Argovian   [GEOL] Upper Jurassic (lower Lusitanian), a substage of geologic time in Great Britain.

Arguloida   [INV ZOO] A group of crustaceans known as the fish lice; taxonomic status is uncertain.

argument   [ASTRON] An angle or arc, as in argument of perigee.   [COMPUT SCI] A value applied to a procedure, subroutine, or macroinstruction which is required in order to evaluate any of these.   [MATH] *See* amplitude; independent variable.

argyrodite   [MINERAL] Ag_8GeS_6 A steel-gray mineral, one of two germanium minerals and a source for germanium; crystallizes in the isometric system and is isomorphous with canfieldite.

Arhynchobdellae   [INV ZOO] An order of annelids in the class Hirudinea characterized by the lack of an eversible proboscis; includes most of the important leech parasites of man and warm-blooded animals.

Arhynchodina   [INV ZOO] A suborder of ciliophoran protozoans in the order Thigmotrichida.

arhythmicity   [BIOL] A condition characterized by the absence of an expected behavioral or physiologic rhythm.

Ari   *See* Aries.

Aridisol   [GEOL] A soil order characterized by pedogenic horizons; low in organic matter and nitrogen and high in calcium, magnesium, and more soluble elements; usually dry.

aridity   [CLIMATOL] The degree to which a climate lacks effective, life-promoting moisture.

Ariel   [AERO ENG] A series of artificial satellites launched for Britain by the United States.   [ASTRON] **1.** A satellite of the planet Uranus orbiting at a mean distance of 192,000 kilometers (119,000 miles). **2.** A constellation in the northern celestial hemisphere, seen in proximity to Taurus and Pisces.

Aries   [ASTRON] A constellation with a right ascension of 3 hours and declination of 20°N. Abbreviated Ari. Also known as Ram.

Ariidae [VERT ZOO] A family of tropical salt-water catfishes in the order Siluriformes.

Arikareean [GEOL] Lower Miocene geologic time.

aril [BOT] An outgrowth of the funiculus in certain seeds that either remains as an appendage or envelops the seed.

Arionidae [INV ZOO] A family of mollusks in the order Stylommatophora, including some of the pulmonate slugs.

Aristolochiaceae [BOT] The single family of the plant order Aristolochiales.

Aristolochiales [BOT] An order of dicotyledonous plants in the subclass Magnoliidae; species are herbaceous to woody, often climbing, with perigynous to epigynous, apetalous flowers, uniaperturate or nonaperturate pollen, and seeds with a small embryo and copious endosperm.

arithmetic [MATH] Addition, subtraction, multiplication, and division, usually of integers, rational numbers, real numbers, or complex numbers.

arithmetic address [COMPUT SCI] An address in a computer program that results from performing an arithmetic operation on another address.

arithmetical element See arithmetical unit.

arithmetical instruction [COMPUT SCI] An instruction in a computer program that directs the computer to perform an arithmetic operation (addition, subtraction, multiplication, or division) upon specified items of data.

arithmetical operation [COMPUT SCI] A digital computer operation in which numerical quantities are added, subtracted, multiplied, divided, or compared.

arithmetical unit [COMPUT SCI] The section of the computer which carries out all arithmetic and logic operations. Also known as arithmetical element; arithmetic-logic unit (ALU); arithmetic section; logic-arithmetic unit; logic section.

arithmetic check [COMPUT SCI] The verification of an arithmetical operation or series of operations by another such process; for example, the multiplication of 73 by 21 to check the result of multiplying 21 by 73.

arithmetic-logic unit See arithmetical unit.

arithmetic mean [MATH] The average of a collection of numbers obtained by dividing the sum of the numbers by the quantity of numbers. Also known as average (av).

arithmetic progression [MATH] A sequence of numbers for which there is a constant d such that the difference between any two successive terms is equal to d.

arithmetic register [COMPUT SCI] A specific memory location reserved for intermediate results of arithmetic operations. Also known as A register.

arithmetic section See arithmetical unit.

arithmetic shift [COMPUT SCI] A shift of the digits of a number, expressed in a positional notation system, in the register without changing the sign of the number.

arithmetic sum [MATH] 1. The result of the addition of two or more positive quantities. 2. The result of the addition of the absolute values of two or more quantities.

arizonite [MINERAL] $Fe_2Ti_3O_9$ A steel-gray mineral containing iron and titanium and found in irregular masses in pegmatite. [PETR] A dike rock composed of mostly quartz, some orthoclase, and accessory mica and apatite.

arkite [PETR] A feldspathoid-rich rock consisting largely of pseudoleucite and nepheline, subordinate melanite and pyroxene, and accessory orthoclase, apatite, and sphene.

arkose [PETR] A sedimentary rock composed of sand-size fragments that contain a high proportion of feldspar in addition to quartz and other detrital minerals.

arkosic [PETR] Having wholly or partly the character of arkose.

ARL See acceptable reliability level.

arm [ANAT] The upper or superior limb in humans which comprises the upper arm with one bone and the forearm with two bones. [ELEC] See branch. [GEOL] A ridge or a spur that extends from a mountain. [MATH] A side of an angle. [NAV ARCH] The part of an anchor extending from the crown to one of the flukes. [OCEANOGR] A long, narrow inlet of water extending from another body of water. [ORD] 1. A combat branch of a military force; specifically, a branch of the U.S. Army, such as the Infantry Armored Cavalry, the primary function of which is combat. 2. Often plural. Weapons for use in war. 3. To supply with arms. 4. To ready ammunition for detonation, as by removal of safety devices or alignment of the explosive elements in the explosive train of the fuse. [PHYS] The perpendicular distance from the line along which a force is applied to a reference point.

armadillo [VERT ZOO] Any of 21 species of edentate mammals in the family Dasypodidae.

armament [ORD] 1. The weapons of an airplane, tank, ship, or the like or of a unit or organized force. 2. Often plural. War equipment, weapons, and supplies.

armature [ELECTROMAG] 1. That part of an electric rotating machine that includes the main current-carrying winding in which the electromotive force produced by magnetic flux rotation is induced; it may be rotating or stationary. 2. The movable part of an electromagnetic device, such as the movable iron part of a relay, or the spring-mounted iron part of a vibrator or buzzer.

Armilliferidae [INV ZOO] A family of pentastomid arthropods belonging to the suborder Porocephaloidea.

arming [ORD] The changing of a fuse from a safe condition to a state of readiness for functioning.

armor [ELEC] Metal sheath enclosing a cable, primarily for mechanical protection. [ORD] 1. Any physical protective covering, such as metal, used on vehicles or persons against projectiles or fragments. 2. Armored units or forces. 3. The component of a weapon system that gives protection to the vehicle or weapon on its way to the target.

armored cable [ELEC] An electrical cable provided with a sheath of metal primarily for mechanical protection.

armor-piercing [ORD] Of ammunition, bombs, bullets, and projectiles, designed to penetrate armor and other resistant targets.

armor-piercing incendiary [ORD] Armor-piercing projectile designed to set fires after piercing armor. Abbreviated API.

armor plate [BUILD] A metal plate which protects the lower part of a door from kicks and scratches, covering the door to a height usually 39 inches (1 meter) or more. [MET] Heavy, flat steel, either surface-hardened or hardened throughout, used as a sheathing for warships, tanks, and so forth to resist penetration and deformation from heavy gunfire.

armyworm [INV ZOO] Any of the larvae of certain species of noctuid moths composing the family Phalaenidae; economically important pests of corn and other grasses.

Arodoidea [INV ZOO] A superfamily of hemipteran insects belonging to the subdivision Geocorisae.

aromatic [ORG CHEM] 1. Pertaining to or characterized by the presence of at least one benzene ring. 2. Describing those compounds having physical and chemical properties resembling those of benzene.

aromatic hydrocarbon [ORG CHEM] A member of the class of hydrocarbons, of which benzene is the first member, consisting of assemblages of cyclic conjugated carbon atoms and characterized by large resonance energies. Also known as arene.

aromatization [CHEM ENG] Conversion of any non-aromatic hydrocarbon structure to aromatic hydrocarbon, particularly petroleum.

aroyl [ORG CHEM] The radical RCO, where R is an aromatic (benzoyl, napthoyl) group.

aroylation [ORG CHEM] A reaction in which the aroyl group is incorporated into a molecule by substitution.

array [COMPUT SCI] A collection of data items with each identified by a subscript or key and arranged in such a way that a computer can examine the collection and retrieve data from these items associated with a particular subscript or key. [ELECTR] A group of components such as antennas, reflectors, or directors arranged to provide a desired variation of radiation transmission or reception with direction. [STAT] The arrangement of a sequence of items in statistics according to their values, such as from largest to smallest.

array processor [COMPUT SCI] A multiprocessor composed of a set of identical central processing units acting synchronously under the control of a common unit.

arrester [ELEC] See lightning arrester. [ENG] A wire screen at the top of an incinerator or chimney which prevents sparks or burning material from leaving the stack.

Arrhenius equation [PHYS CHEM] The relationship that the specific reaction rate constant k equals the frequency factor constant s times $\exp(-\delta H_{act}/RT)$, where δH_{act} is the heat of activation, R the gas constant, and T the absolute temperature.

arrhythmia [MED] Absence of rhythm, especially of heart beat or respiration.

Arridae [VERT ZOO] A family of catfishes in the suborder Siluroidei found from Cape Cod to Panama.

arris [ARCH] A short edge or angle at the junction of two surfaces, especially moldings and raised edges. [GEOL] See arête.

Arrow See Sagitta.

arrowroot [BOT] Any of the tropical American plants belonging to the genus *Maranta* in the family Marantaceae.

arrowroot starch [FOOD ENG] A nutritive carbohydrate obtained from the underground stems of arrowroot plants and used as a food for infants and invalids.

arrowworm [INV ZOO] Any member of the phylum Chaetognatha; useful indicator organism for identifying displaced masses of water.

arroyo [GEOL] Small, deep gully produced by flash flooding in arid and semiarid regions of the southwestern United State.

arsenal [ORD] 1. An installation whose primary mission is research, development, and manufacture pertaining to assigned items or components. 2. An installation having coequal missions of maintenance and supply for assigned items or components.

arsenate [INORG CHEM] 1. AsO_4^{-3} A negative ion derived from orthoarsenic acid, $H_3AsO_4 \cdot \frac{1}{2}H_2O$. 2. A salt or ester of arsenic acid.

arsenic [CHEM] A chemical element, symbol As, atomic number 33, atomic weight 74.9216. [MINERAL] A brittle, steel-gray hexagonal mineral, the native form of the element.

arsenical nickel See niccolite.

arsenic bloom See arsenolite.

arsenide [CHEM] A binary compound of negative, trivalent arsenic; for example, H_3As or GaAs.

arsenite [INORG CHEM] 1. AsO_3^{-3} A negative ion derived from aqueous solutions of As_4O_6. 2. A salt or ester of arsenious acid.

arsenolite [MINERAL] As_2O_3 A mineral crystallizing in the isometric system and usually occurring as a white bloom or crust. Also known as arsenic bloom.

arsenopyrite [MINERAL] FeAsS A white to steel-gray mineral crystallizing in the monoclinic system with pseudo-orthorhombic symmetry because of twinning; occurs in crystalline rock and is the principal ore of arsenic. Also known as mispickel.

arsonium [INORG CHEM] AsH_4 A radical; it may be considered analogous to the ammonium radical in that a compound such as AsH_4OH may form.

Artacaminae [INV ZOO] A subfamily of polychaete annelids in the family Terebellidae of the Sedentaria.

arteriogram [MED] A roentgenogram of an artery after injection with radiopaque material.

arteriography [MED] 1. Graphic presentation of the pulse. 2. Roentgenography of the arteries after the intravascular injection of a radiopaque substance.

arteriole [ANAT] An artery of small diameter that terminates in capillaries.

arteriolosclerosis [MED] Thickening of the lining of arterioles, usually due to hyalinization or fibromuscular hyperplasia.

arteriosclerosis [MED] A degenerative arterial disease marked by hardening and thickening of the vessel walls.

arterite [PETR] 1. A migmatite produced as a result of regional contact metamorphism during which residual magmas were injected into the host rock. 2. Gneisses characterized by veins formed from the solution given off by deep-seated intrusions of molten granite. 3. A veined gneiss in which the vein material was injected from a magma.

artery [ANAT] A vascular tube that carries blood away from the heart.

artesian aquifer [HYD] An aquifer that is bounded above and below by impermeable beds and that contains artesian water. Also known as confined aquifer.

artesian basin [HYD] A geologic structural feature or combination of such features in which water is confined under artesian pressure.

artesian spring [HYD] A spring whose water issues under artesian pressure, generally through some fissure or other opening in the confining bed that overlies the aquifer.

artesian water [HYD] Groundwater that is under sufficient pressure to rise above the level at which it encounters a well, but which does not necessarily rise to or above the surface of the ground.

artesian well [HYD] A well in which the water rises above the top of the water-bearing bed.

arthritis [MED] Any inflammatory process affecting joints or their component tissues.

arthrodesis [MED] Fusion of a joint by removing the articular surfaces and securing bony union. Also known as operative ankylosis.

arthrodia [ANAT] A diarthrosis permitting only restricted motion between a concave and a convex surface, as in some wrist and ankle articulations. Also known as gliding joint.

Arthrodira [PALEON] The joint-necked fishes, an Upper Silurian and Devonian order of the Placodermi.

Arthrodonteae [BOT] A family of mosses in the subclass Eubrya characterized by their thin, membranous peristome teeth composed of cell walls.

arthrogram [MED] A roentgenogram of a joint space after injection of radiopaque material.

arthrography [MED] Roentgenography of a joint space after the injection of radiopaque material.

Arthromitaceae [MICROBIO] Formerly a family of nonmotile bacteria in the order Caryophanales found in the intestine of millipedes, cockroaches, and toads.

arthropathy [MED] 1. Any joint disease. 2. A neurotrophic disorder of a joint, usually due to lack of pain sensation, found in association with tabes dorsalis, leprosy, syringomyelia, diabetic polyneuropathy, and occasionally multiple sclerosis and myelodysplasias.

arthroplasty [MED] 1. The making of an artificial joint. 2. Reconstruction of a new and functioning joint from an ankylosed one; a plastic operation upon a joint.

Arthropoda [INV ZOO] The largest phylum in the animal kingdom; adults typically have a segmented body, a sclerotized integument, and many-jointed segmental limbs.

arthropod-borne virus See arbovirus.

arthrosis [ANAT] An articulation or suture uniting two bones. [MED] Any degenerative joint disease.

Arthrotardigrada [INV ZOO] A suborder of microscopic invertebrates in the order Heterotardigrada characterized by toelike terminations on the legs.

artichoke [BOT] *Cynara scolymus*. A herbaceous perennial plant belonging to the order Asterales; the flower head is edible.

Articulata [INV ZOO] 1. A class of the Brachiopoda having hinged valves that usually bear teeth. 2. The only surviving subclass of the echinoderm class Crinoidea.

articulation [ANAT] See joint. [BOT] A joint between two parts of a plant that can separate spontaneously. [COMMUN] The pecentage of speech units understood correctly by a listener in a communications system; it generally applies to unrelated words, as in code messages, in distinction to intelligibility. [INV ZOO] A joint between rigid parts of an animal body, as the segments of an appendage in insects. [PHYSIO] The act of enunciating speech.

articulite See itacolumite.

artifact [ARCHEO] Any crafted object of common use that reflects the skills of humans in past cultures. [COMMUN] Any component of a signal that is extraneous to the variable represented by the signal. [HISTOL] A structure in a fixed cell or tissue formed by manipulation or by the reagent.

artificial delay line See delay line.

artificial horizon [NAV] 1. A gyro-operated flight instrument that shows the pitching and banking attitudes of an aircraft or spacecraft with respect to a reference line horizon, within limited degrees of movement, by means of the relative position of lines or marks on the face of the instrument representing the aircraft and the horizon. Also known as auto-

matic horizon. 2. A device, such as a spirit level or pendulum, that establishes a horizontal reference in a navigation instrument.

artificial insemination [MED] A process by which spermatozoa are collected from males and deposited in female genitalia by instruments rather than by natural service.

artificial intelligence [COMPUT SCI] The property of a machine capable of reason by which it can learn functions normally associated with human intelligence.

artificial language [COMPUT SCI] A computer language that is specifically designed to facilitate communication in a particular field, but is not yet natural to that field; opposite of a natural language, which evolves through long usage.

artificial respiration [MED] The maintenance of breathing by artificial ventilation, in the absence of normal spontaneous respiration; effective methods include mouth-to-mouth breathing and the use of a respirator.

artificial satellite [AERO ENG] Any man-made object placed in a near-periodic orbit in which it moves mainly under the gravitational influence of one celestial body, such as the earth, sun, another planet, or a planet's moon.

artillery [ORD] A gun or a rocket launcher with mounting too large or too heavy to be classed as a small arm. Abbreviated arty.

Artinskian [GEOL] A European stage of geologic time including Lower Permian (above Sakmarian, below Kungurian).

Artiodactyla [VERT ZOO] An order of terrestrial, herbivorous mammals characterized by having an even number of toes and by having the main limb axes pass between the third and fourth toes.

arty See artillery.

aryl [ORG CHEM] An organic radical derived from an aromatic hydrocarbon by removal of one hydrogen.

arylamine [ORG CHEM] An organic compound formed from an aromatic hydrocarbon that has at least one amine group joined to it, such as aniline.

aryl compound [ORG CHEM] Molecules with the six-carbon aromatic ring structure characteristic of benzene or compounds derived from aromatics.

arylide [ORG CHEM] A compound formed from a metal and an aryl radical, for example, PbR_4, where R is the aryl radical.

arytenoid [ANAT] Relating to either of the paired, pyramid-shaped, pivoting cartilages on the dorsal aspect of the larynx, in man and most other mammals, to which the vocal cords and arytenoid muscles are attached.

aS See abmho.

As See altostratus cloud; arsenic.

asar See esker.

asbestos [MINERAL] A general name for the useful, fibrous varieties of a number of rock-forming silicate minerals that are heat-resistant and chemically inert; two varieties exist: amphibole asbestos, the best grade of which approaches the composition Ca_2Mg_5-$(OH)_2Si_8O_{22}$ (tremolite), and serpentine asbestos, usually chrysotile, $Mg_3Si_2(OH)_4O_5$.

asbestos cement [MATER] A building material composed of a mixture of asbestos fiber, portland cement, and water made into plain sheets, corrugated sheets, tiles, and piping.

asbestosis [MED] A chronic lung inflammation caused by inhalation of asbestos dust.

asbolane See asbolite.

asbolite [MINERAL] A black, earthy mineral aggregate containing hydrated oxides of manganese and cobalt. Also known as asbolane; black cobalt; earthy cobalt.

A scan *See* A scope.

Ascaphidae [VERT ZOO] A family of amphicoelous frogs in the order Anura, represented by four living species.

ascariasis [MED] Any parasitic infection of humans or domestic mammals caused by species of *Ascaris*.

Ascaridata [INV ZOO] An equivalent name for the Ascaridina.

Ascaridida [INV ZOO] An order of parasitic nematodes in the subclass Phasmidia.

Ascarididae [INV ZOO] A family of parasitic nematodes in the superfamily Ascaridoidea.

Ascaridina [INV ZOO] A suborder of parasitic nematodes in the order Ascaridida.

Ascaridoidea [INV ZOO] A large superfamily of parasitic nematodes of the suborder Ascaridina.

Ascaroidea [INV ZOO] An equivalent name for Ascaridoidea.

ascending aorta [ANAT] The first part of the aorta, extending from its origin in the heart to the aortic arch.

ascending branch [MECH] The portion of the trajectory between the origin and the summit on which a projectile climbs and its altitude constantly increases.

ascending chromatography [ANALY CHEM] A technique for the analysis of mixtures of two or more compounds in which the mobile phase (sample and carrier) rises through the fixed phase.

ascending node [ASTRON] Also known as northbound node. **1.** The point at which a planet, planetoid, or comet crosses to the north side of the ecliptic. **2.** The point at which a satellite crosses to the north side of the equatorial plane of its primary.

ascending series [MATH] **1.** A series each of whose terms is greater than the preceding term. **2.** *See* power series.

ascending vertical angle *See* angle of elevation.

Ascheim-Zondek test [PATH] A human pregnancy test that uses the reaction of ovaries in immature white mice to an injection of urine from a woman.

Aschelminthes [INV ZOO] A heterogeneous phylum of small to microscopic wormlike animals; individuals are pseudocoelomate and mostly unsegmented and are covered with a cuticle.

Ascidiacea [INV ZOO] A large class of the phylum Tunicata; adults are sessile and may be solitary or colonial.

ASCII *See* American Standard Code for Information Interchange.

ascites [MED] An abnormal accumulation of serous fluid in the abdominal cavity.

Asclepiadaceae [BOT] A family of tropical and subtropical flowering plants in the order Gentianales characterized by a well-developed latex system; milkweed (*Asclepias*) is a well-known member.

Ascolichenes [BOT] A class of the lichens characterized by the production of asci similar to those produced by Ascomycetes.

Ascomycetes [MYCOL] A class of fungi in the subdivision Eumycetes, distinguished by the ascus.

A scope [ELECTR] A radarscope on which the trace appears as a horizontal or vertical range scale and the signals appear as vertical or horizontal deflections. Also known as A indicator; A scan.

ascorbic acid [BIOCHEM] $C_6H_8O_6$ A white, crystalline, water-soluble vitamin found in many plant materials, especially citrus fruit. Also known as vitamin C.

Ascothoracica [INV ZOO] An order of marine crustaceans in the subclass Cirripedia occurring as endo- and ectoparasites of echinoderms and coelenterates.

Aselloidea [INV ZOO] A group of free-living, freshwater isopod crustaceans in the suborder Asellota.

Asellota [INV ZOO] A suborder of morphologically and ecologically diverse aquatic crustaceans in the order Isopoda.

asepsis [MED] The state of being free from pathogenic microorganisms.

asexual [BIOL] **1.** Not involving sex. **2.** Exhibiting absence of sex or of functional sex organs.

ash [BOT] **1.** A tree of the genus *Fraxinus*, deciduous trees of the olive family (Oleaceae) characterized by opposite, pinnate leaflets. **2.** Any of various Australian trees having wood of great toughness and strength; used for tool handles and in work requiring flexibility. [CHEM] The incombustible matter remaining after a substance has been incinerated. [ENG] An undesirable constituent of diesel fuel whose quantitative measurement indicates degree of fuel cleanliness and freedom from abrasive material. [GEOL] Volcanic dust and particles less than 4 millimeters in diameter.

ash collector *See* dust chamber.

ash cone [GEOL] A volcanic cone built primarily of unconsolidated ash and generally shaped somewhat like a saucer, with a rim in the form of a wide circle and a broad central depression often nearly at the same elevation as the surrounding country.

ash content [MATER] The mass of incombustible material remaining after burning a given coal sample as a percentage of the orginal mass of the coal.

ash flow [GEOL] **1.** An avalanche of volcanic ash, generally a highly heated mixture of volcanic gases and ash, traveling down the flanks of a volcano or along the surface of the ground. Also known as glowing avalanche; incandescent tuff flow; **2.** A deposit of volcanic ash and other debris resulting from such a flow and lying on the surface of the ground.

Ashgillian [GEOL] A European stage of geologic time in the Upper Orodovician (above Upper Caradocian, below Llandoverian of Silurian).

ashtonite *See* mordenite.

asiderite *See* stony meteorite.

Asilidae [INV ZOO] The robber flies, a family of predatory, orthorrhaphous, dipteran insects in the series Brachycera.

ASK *See* amplitude shift keying.

ASM *See* air-to-surface missile.

Asopinae [INV ZOO] A family of hemipteran insects in the superfamily Pentatomoidea including some predators of caterpillars.

asparagine [BIOCHEM] $C_4H_8N_2O_3$ A white, crystalline amino acid found in many plant seeds.

asparagus [BOT] *Asparagus officinalis*. A dioecious, perennial monocot belonging to the order Liliales; the shoot of the plant is edible.

aspartic acid [BIOCHEM] $C_4H_7NO_4$ A nonessential, crystalline dicarboxylic amino acid found in plants and animals, especially in molasses from young sugarcane and sugarbeet.

aspartoyl [BIOCHEM] —COCH$_2$CH(NH$_2$)CO— A bivalent radical derived from aspartic acid.

aspect [ARCH] The direction which a building faces with respect to the points of a compass. [ASTRON] The apparent position of a celestial body relative to another; particularly, the apparent position of the moon or a planet relative to the sun. [CIV ENG] Of railway signals, what the engineer sees when viewing the blades or lights in their relative positions or colors. [ECOL] Seasonal appearance.

aspect card [COMPUT SCI] A card on which is entered the accession numbers of documents in an information retrieval system; the documents are judged to be related importantly to the concept for which the card is established.

aspect ratio [AERO ENG] The ratio of the square of the span of an airfoil to the total airfoil area, or the ratio of its span to its mean chord. [DES ENG] 1. The ratio of frame width to frame height in television; it is 4:3 in the United States and Britain. 2. In any rectangular configuration (such as the cross section of a rectangular duct), the ratio of the longer dimension to the shorter.

aspen [BOT] Any of several species of poplars (*Populus*) characterized by weak, flattened leaf stalks which cause the leaves to flutter in the slightest breeze.

aspergillosis [MED] A rare fungus infection of humans and animals caused by several species of *Aspergillus*.

asphalt [MATER] A brown to black, hard, brittle, or plastic bituminous material composed principally of hydrocarbons; formed in oil-bearing rocks near the Dead Sea, and in Trinidad; prepared by pyrolysis from coal tar, certain petroleums, and lignite tar; melts on heating, insoluble in water but soluble in gasoline; used for paving and roofing and in paints and varnishes.

asphaltene [MATER] Any of the dark, solid constituents of crude oils and other bitumens which are soluble in carbon disulfide but insoluble in paraffin naphthas; they hold most of the organic constituents of bitumens.

asphaltite [GEOL] Any of the dark-colored, solid, naturally occurring bitumens that are insoluble in water, but more or less completely soluble in carbon disulfide, benzol, and so on, with melting points between 250 and 600°F (121–316°C); examples are gilsonite and grahamite.

asphaltite coal *See* albertite.

asphalt roofing [MATER] A roofing material made by impregnating a dry roofing felt with a hot asphalt saturant, applying asphalt coatings to the weather and reverse sides, and embedding a mineral surfacing in the coating on the weather side.

aspheric surface [OPTICS] A lens or mirror surface which is altered slightly from a spherical surface in order to reduce aberrations.

asphyxia [MED] Suffocation due to oxygen deprivation, resulting in anoxia and carbon dioxide accumulation in the body.

Aspidiphoridae [INV ZOO] An equivalent name for the Sphindidae.

Aspidobothria [INV ZOO] An equivalent name for the Aspidogastrea.

Aspidobranchia [INV ZOO] An equivalent name for the Archaeogastropoda.

Aspidochirotacea [INV ZOO] A subclass of bilaterally symmetrical echinoderms in the class Holothuroidea characterized by tube feet and 10–30 shield-shaped tentacles.

Aspidochirotida [INV ZOO] An order of holothurioid echinoderms in the subclass Aspidochirotacea characterized by respiratory trees and dorsal tube feet converted into tactile warts.

Aspidocotylea [INV ZOO] An equivalent name for the Aspidogastrea.

Aspidodiadematidae [INV ZOO] A small family of deep-sea echinoderms in the order Diadematoida.

Aspidogastrea [INV ZOO] An order of endoparasitic worms in the class Trematoda having strongly developed ventral holdfasts.

Aspidogastridae [INV ZOO] A family of trematode worms in the order Aspidogastrea occurring as endoparasites of mollusks.

Aspidorhynchidae [PALEON] The single family of the Aspidorhynchiformes, an extinct order of holostean fishes.

Aspidorhynchiformes [PALEON] A small, extinct order of specialized holostean fishes.

Aspinothoracida [PALEON] The equivalent name for Brachythoraci.

aspirating burner [ENG] A burner in which combustion air at high velocity is drawn over an orifice, creating a negative static pressure and thereby sucking fuel into the stream of air; the mixture of air and fuel is conducted into a combustion chamber, where the fuel is burned in suspension.

aspiration [MED] The removal of fluids from a cavity by suction. [MICROBIOL] The use of suction to draw up a sample in a pipette. [SCI TECH] Act or the result of removing, carrying along, or drawing by suction.

aspirator [ENG] Any instrument or apparatus that utilizes a vacuum to draw up gases or granular materials. [MIN ENG] A device made of wire gauze, of cloth, or of a fibrous mass held between pieces of meshed material and used to cover the mouth and nose to keep dusts from entering the lungs.

Aspredinidae [VERT ZOO] A family of salt-water catfishes in the order Siluriformes found off the coast of South America.

assay [ANALY CHEM] Qualitative or quantitative determination of the components of a material, as an ore or a drug.

assemblage [ARCHEO] All related cultural traits and artifacts associated with one archeological manifestation. [ECOL] A group of organisms sharing a common habitant by chance. [ORD] A collection of items designed to accomplish one general function and identified and issued as a single item. [PALEON] A group of fossils occurring together at one stratigraphic level.

assemblage zone [PALEON] A biostratigraphic unit defined and identified by a group of associated fossils rather than by a single index fossil.

assembler [COMPUT SCI] A program designed to convert symbolic instruction into a form suitable for execution on a computer. Also known as assembly program; assembly routine.

assembly [COMPUT SCI] The automatic translation into machine language of a computer program written in symbolic language. [MECH ENG] A unit containing the component parts of a mechanism, machine, or similar device.

assembly drawing [GRAPHICS] A working-type engineering drawing depicting a complete unit, usually included with detail drawings of all parts in a set of working drawings.

assembly language [COMPUT SCI] A low-level computer language one step above the binary machine language.

assembly line [IND ENG] A mass-production arrangement whereby the work in process is progressively transferred from one operation to the next until the product is assembled.

assembly list [COMPUT SCI] A printed list which is the by-product of an assembly procedure; it lists in logical instruction sequence all details of a routine, showing the coded and symbolic notation next to the actual notations established by the assembly procedure; this listing is highly useful in the debugging of a routine.

assembly system [COMPUT SCI] An automatic programming software system with a programming language and machine-language programs that aid the programmer by performing different functions such as checkout and updating.

assembly unit [COMPUT SCI] **1.** A device which performs the function of associating and joining several parts or piecing together a program. **2.** A portion of a program which is capable of being assembled into a larger program.

assets [IND ENG] All the resources, rights, and property owned by a person or a company; the book value of these items as shown on the balance sheet.

assign [COMPUT SCI] A control statement in FORTRAN which assigns a computed value i to a variable k, the latter representing the number of the statement to which control is then transferred.

assignment problem [COMPUT SCI] A special case of the transportation problem in a linear program, in which the number of sources (assignees) equals the number of designations (assignments) and each supply and each demand equals 1.

assimilation [GEOL] Incorporation of solid or fluid material that was originally in the rock wall into a magma. [PHYSIO] Conversion of nutritive materials into protoplasm.

assisted takeoff [AERO ENG] A takeoff of an aircraft or a missile by using a supplementary source of power, usually rockets.

associate curve See Bertrand curve.

associated gas [PETRO ENG] Gaseous hydrocarbons occurring as a free-gas phase under original oil-reservoir conditions of temperature and pressure.

associate matrix See Hermitian conjugate of a matrix.

associate operator See adjoint operator.

association [CHEM] Combination or correlation of substances or functions. [ECOL] Major segment of a biome formed by a climax community, such as an oak-hickory forest of the deciduous forest biome. [PSYCH] A connection formed through learning.

association neuron [ANAT] A neuron, usually within the central nervous system, between sensory and motor neurons.

association trail [COMPUT SCI] A linkage between two or more documents or items of information, discerned during the process of their examination and recorded with the aid of an information retrieval system.

associative law [MATH] For a binary operation designated $\circ$, the relationship expressed by $a \circ (b \circ c) = (a \circ b) \circ c$.

associative learning [PSYCH] The principle that items experienced together are mentally linked so that they tend to reinforce one another.

associative memory [COMPUT SCI] A data-storage device in which a location is identified by its informational content rather than by names, addresses, or relative positions, and from which the data may be retrieved. Also known as associative storage. [PSYCH] Recalling a previously experienced item by thinking of something that is linked with it, thus invoking the association.

associative processor [COMPUT SCI] A digital computer that consists of a content-addressable memory and means for searching rapidly changing random digital data stored within, at speeds up to 1000 times faster than conventional digital computers.

associative storage See associative memory.

associative thinking [PSYCH] **1.** The mental process of making associations between a given subject and all pertinent present factors without drawing on past experience. **2.** Free association.

associator [COMPUT SCI] A device for bringing like entities into conjunction or juxtaposition.

assortative mating [GEN] Nonrandom mating with respect to phenotypes.

assumed decimal point [COMPUT SCI] For a decimal number stored in a computer or appearing on a printout, a position in the number at which place values change from positive to negative powers of 10, but to which no location is assigned or at which no printed character appears, as opposed to an actual decimal point. Also known as virtual decimal point.

assured mineral See developed reserves.

astable circuit [ELECTR] A circuit that alternates automatically and continuously between two unstable states at a frequency dependent on circuit constants; for example, a blocking oscillator.

astable multivibrator [ELECTR] A multivibrator in which each active device alternately conducts and is cut off for intervals of time determined by circuit constants, without use of external triggers. Also known as free-running multivibrator.

Astacidae [INV ZOO] A family of fresh-water crayfishes belonging to the section Macrura in the order Decapoda, occurring in the temperate regions of the Northern Hemisphere.

Astartian See Sequanian.

astatic [PHYS] Without orientation or directional characteristics; having no tendency to change position.

astatic governor See isochronous governor.

astatic pair [ELECTROMAG] A pair of parallel magnets, equal in strength and having polarities in opposite directions, and perpendicular to an axis which bisects both of them; there is no net force or torque on the pair in a uniform field.

astatine [CHEM] A radioactive chemical element, symbol At, atomic number 85, the heaviest of the halogen elements.

Asteidae [INV ZOO] A small, obscure family of cyclorrhaphous myodarian dipteran insects in the subsection Acalyptratae.

Asterales [BOT] An order of dicotyledonous plants in the subclass Asteridae, including aster, sunflower, zinnia, lettuce, artichoke, and dandelion; ovary is inferior, flowers are borne in involucrate, centripetally flowering heads, and the calyx, when present, is modified into a set of scale-, hair-, or bristlelike structures called the pappus.

astereognosis [MED] Loss of recognition of objects by touch, although recognition occurs through another sense, usually vision. Also known as tactile agnosia.

Asteridae [BOT] A large subclass of dicotyledonous plants in the class Magnoliopsida; plants are sympetalous, with unitegmic, tenuinucellate ovules and with the stamens usually as many as, or fewer than, the corolla lobes and alternate with them.

Asteriidae [INV ZOO] A large family of echinoderms in the order Forcipulatida, including many predatory sea stars.

Asterinidae [INV ZOO] The starlets, a family of echinoderms in the order Spinulosida.

asterism [ASTRON] A constellation or small group of stars. [OPTICS] A starlike optical phenomenon seen in gemstones called star stones; due to reflection of light by lustrous inclusions reduced to sharp lines of light by a domed cabochon style of cutting. [SPECT] A star-shaped pattern sometimes seen in x-ray spectrophotographs.

astern [ENG] To the rear of an aircraft, vehicle, or vessel; behind; from the back.

asteroid [ASTRON] One of the many small celestial bodies revolving around the sun, most of the orbits being between those of Mars and Jupiter. Also known as minor planet; planetoid.

asteroid belt [ASTRON] The region between 2.1 and 3.5 astronomical units from the sun where most of the asteroids are found.

Asteroidea [INV ZOO] The starfishes, a subclass of echinoderms in the subphylum Asterozoa characterized by five radial arms.

Asteroschematidae [INV ZOO] A family of ophiuroid echinoderms in the order Phrynophiurida with individuals having a small disk and stout arms.

Asterozoa [INV ZOO] A subphylum of echinoderms characterized by a star-shaped body and radially divergent axes of symmetry.

asthenosphere [GEOL] That portion of the upper mantle beneath the rigid lithosphere which is plastic enough for rock flowage to occur; extends from a depth of 50–100 kilometers to about 400 kilometers and is seismically equivalent to the low velocity zone.

asthma [MED] A pulmonary disease marked by labored breathing, wheezing, and coughing; cause may be emotional stress, chemical irritation, or exposure to an allergen.

Astian [GEOL] A European stage of geologic time: upper Pliocene, above Plaisancian, below the Pleistocene stage known as Villafranchian, Calabrian, or Günz.

astigmat See astigmatic lens.

astigmatic lens [OPTICS] A planocylindrical, spherocylindrical, or spherotoric lens used in eyeglasses to correct astigmatism. Also known as astigmat.

astigmatism [ELECTR] In an electron-beam tube, a focus defect in which electrons in different axial planes come to focus at different points. [MED] A defect of vision due to irregular curvatures of the refractive surfaces of the eye so that focal points of light are distorted. [OPTICS] The failure of an optical system, such as a lens or a mirror, to image a point as a single point; the system images the point on two line segments separated by an interval.

Astomatida [INV ZOO] An order of mouthless protozoans in the subclass Holotrichia; all species are invertebrate parasites, typically in oligochaete annelids.

Aston dark space [ELECTR] A dark region in a glow-discharge tube which extends for a few millimeters from the cathode up to the cathode glow.

Aston whole-number rule [PHYS] The rule which states that when expressed in atomic weight units, the atomic weights of isotopes are very nearly whole numbers, and the deviations found in samples of elements are due to the presence of several isotopes with different weights.

astragalus [ANAT] The bone of the ankle which articulates with the bones of the leg. Also known as talus.

astrakanite See bloedite.

Astrapotheria [PALEON] A relatively small order of large, extinct South American mammals in the infraclass Eutheria.

Astrapotheroidea [PALEON] A suborder of extinct mammals in the order Astrapotheria, ranging from early Eocene to late Miocene.

astroballistics [MECH] The study of phenomena arising out of the motion of a solid through a gas at speeds high enough to cause ablation; for example, the interaction of a meteoroid with the atmosphere.

astrochanite See bloedite.

astrocompass [NAV] A direction-determining instrument into which can be set the coordinates of any celestial body and the latitude of the observer and which will then give an indication of azimuth, true north, and heading.

astrodynamics [AERO ENG] The practical application of celestial mechanics, astroballistics, propulsion theory, and allied fields to the problem of planning and directing the trajectories of space vehicles. [ASTROPHYS] The dynamics of celestial objects.

astrogation See astronavigation.

astrogeodetic datum orientation [GEOD] Adjustment of the ellipsoid of reference for a particular datum so that the sum of the squares of deflections of the vertical at selected points throughout the geodetic network is made as small as possible.

astrogeodetic deflection [GEOD] The angle at a point between the normal to the geoid and the normal to the ellipsoid of an astrogeodetically oriented datum. Also known as relative deflection.

astrogeodetic leveling [GEOD] A concept whereby the astrogeodetic deflections of the vertical are used to determine the separation of the ellipsoid and the geoid in studying the figure of the earth. Also known as astronomical leveling.

astrogeology [ASTRON] The science that applies the principles of geology, geochemistry, and geophysics to the moon and planets other than the earth.

astrograph [ASTRON] A telescope designed to be used exclusively for astronomical photography. [MAP] A device for projecting a set of precomputed altitude curves onto a chart or plotting sheet, the curves moving with time such that if they are properly adjusted, they will remain in the correct position on the chart or plotting sheet; used in mapping the heavens.

astrographic position See astrometric position.

astrogravimetric leveling [GEOD] A concept whereby a gravimetric map is used for the interpolation of the astrogeodetic deflections of the vertical to determine the separation of the ellipsoid and the geoid in studying the figure of the earth.

astrogravimetric points [GEOD] Astronomical positions corrected for the deflection of the vertical by gravimetric methods.

astrometric position [ASTRON] The position of a heavenly body or space vehicle on the celestial sphere corrected for aberration but not for planetary aberration. Also known as astrographic position.

astrometry [ASTRON] The branch of astronomy dealing with the geometrical relations of the celestial bodies and their real and apparent motions.

astronaut [AERO ENG] In United States terminology, a person who rides in a space vehicle.

astronautical engineering [AERO ENG] The engineering aspects of flight in space.

astronautics [AERO ENG] 1. The art, skill, or activity of operating spacecraft. 2. The science of space flight.

astronavigation [NAV] The plotting and directing of the movement of a vehicle from within by means of observations on celestial bodies. Also known as astrogation; celestial navigation.

astronomic See astronomical.

astronomical [ASTRON] Of or pertaining to astronomy or to observations of the celestial bodies. Also known as astronomic.

astronomical almanac [ASTRON] A publication giving the tables of coordinates of a number of celestial bodies at a number of specific times during a given period.

astronomical azimuth [GEOD] The angle between the astronomical meridian plane of the observer and the plane containing the observed point and the true normal (vertical) of the observer, measured in the plane of the horizon, preferably clockwise from north.

astronomical constants [ASTROPHYS] The elements of the orbits of the bodies of the solar system, their masses relative to the sun, their size, shape, orientation, rotation, and inner constitution, and the velocity of light.

astronomical coordinate system [ASTRON] Any system of spherical coordinates serving to locate astronomical objects on the celestial sphere.

astronomical day [ASTRON] A mean solar day beginning at mean noon, 12 hours later than the beginning of the civil day of the same date; astronomers now generally use the civil day.

astronomical distance [ASTRON] The distance of a celestial body expressed in units such as the light-year, astronomical unit, and parsec.

astronomical eclipse *See* eclipse.

astronomical ephemeris *See* ephemeris.

astronomical equator [GEOD] An imaginary line on the surface of the earth connecting points having 0° astronomical latitude. Also known as terrestrial equator.

astronomical latitude [GEOD] Angular distance between the direction of gravity (plumb line) and the plane of the celestial equator; applies only to positions on the earth and is reckoned from the astronomical equator.

astronomical longitude [GEOD] The angle between the plane of the reference meridian and the plane of the local celestial meridian.

astronomical meridian [GEOD] A line on the surface of the earth connecting points having the same astronomical longitude. Also known as terrestrial meridian.

astronomical observatory [ASTRON] A building designed and equipped for making observations of astronomical phenomena.

astronomical position [GEOD] **1.** A point on the earth whose coordinates have been determined as a result of observation of celestial bodies. Also known as astronomical station. **2.** A point on the earth defined in terms of astronomical latitude and longitude.

astronomical spectroscopy [SPECT] The use of spectrographs in conjunction with telescopes to obtain observational data on the velocities and physical conditions of astronomical objects.

astronomical station *See* astronomical position.

astronomical surveying [GEOD] The celestial determination of latitude and longitude; separations are calculated by computing distances corresponding to measured angular displacements along the reference spheroid.

astronomical theodolite *See* altazimuth.

astronomical traverse [ENG] A survey traverse in which the geographic positions of the stations are obtained from astronomical observations, and lengths and azimuths of lines are obtained by computation.

astronomical triangle [ASTRON] A spherical triangle on the celestial sphere.

astronomical twilight [ASTRON] The period of incomplete darkness when the center of the sun is more than 6° but not more than 18° below the celestial horizon.

astronomical unit [ASTRON] Abbreviated AU. **1.** A measure for distance within the solar system equal to the mean distance between earth and sun, that is, 149,599,000 kilometers. **2.** The semimajor axis of the elliptical orbit of earth.

astronomy [SCI TECH] The science concerned with celestial bodies and the observation and interpretation of the radiation received in the vicinity of the earth from the component parts of the universe.

Astropectinidae [INV ZOO] A family of echinoderms in the suborder Paxillosina occurring in all seas from tidal level downward.

astrophysics [ASTRON] A branch of astronomy that treats of the physical properties of celestial bodies, such as luminosity, size, mass, density, temperature, and chemical composition, and with their origin and evaluation.

astrotracker [NAV] An automatic sextant which has the ability to sight on and continuously to track selected stars throughout the day and night, providing continuous heading and position data with no intervention on the part of the airman. Also known as star tracker.

A supply [ELECTR] Battery, transformer filament winding, or other voltage source that supplies power for heating filaments of vacuum tubes. Also known as A power supply.

asymmetrical-side-band transmission *See* vestigial-side-band transmission.

asymmetric synthesis [ORG CHEM] Chemical synthesis of a pure enantiomer, or of an enantiomorphic mixture in which one enantiomer predominates, without the use of resolution.

asymptote [MATH] **1.** A line approached by a curve in the limit as the curve approaches infinity. **2.** The limit of the tangents to a curve as the point of contact approaches infinity.

asymptotic expansion [MATH] A series of the form $a_0 + (a_1/x) + (a_2/x^2) + \cdots + (a_n/x_n) + \cdots$ is an asymptotic expansion of the function $f(x)$ if there exists a number N such that for all $n > N$ the quantity $x_n[f(x) - S_n(x)]$ approaches zero as x approaches infinity, where $S_n(x)$ is the sum of the first n terms in the series. Also known as asymptotic series.

asymptotic formula [MATH] A statement of equality between two functions which is not a true equality but which means the ratio of the two functions approaches 1 as the variable approaches some value, usually infinity.

asymptotic series *See* asymptotic expansion.

asynchronous [COMPUT SCI] Operating at a speed determined by the circuit functions rather than by timing signals. [PHYS] Not synchronous.

asynchronous computer [COMPUT SCI] A computer in which the performance of any operation starts as a result of a signal that the previous operation has been completed, rather than on a signal from a master clock.

asynchronous control [CONT SYS] A method of control in which the time allotted for performing an operation depends on the time actually required for the operation, rather than on a predetermined fraction of a fixed machine cycle.

asynchronous device [CONT SYS] A device in which the speed of operation is not related to any frequency in the system to which it is connected.

asynchronous transmission [COMMUN] Data transmission in which each character contains its own start and stop pulses and there is no control over the time between characters.

asynchronous working [COMPUT SCI] The mode of operation of a computer in which an operation is performed only at the end of the preceding operation.

At *See* astatine.

atactic [ORG CHEM] Of the configuration for a polymer, having the opposite steric configurations for the carbon atoms of the polymer chain occur in equal frequency and more or less at random.

atavism [EVOL] Appearance of a distant ancestral form of an organism or one of its parts due to reactivation of ancestral genes.

ataxite [GEOL] An iron meteorite that lacks the structure of either hexahedrite or octahedrite and contains more than 10% nickel. [PETR] A taxitic rock whose components are arranged in a breccialike manner, that is, there is no specific arrangement.

ATC *See* air-traffic control.

atelectasis [MED] 1. Total or partial collapsed state of the lung. 2. Failure of the lung to expand at birth.

Ateleopoidei [VERT ZOO] A family of oceanic fishes in the order Cetomimiformes characterized by an elongate body, lack of a dorsal fin, and an anal fin continuous with the caudal fin.

Atelopodidae [VERT ZOO] A family of small, brilliantly colored South and Central American frogs in the suborder Procoela.

Atelostomata [INV ZOO] A superorder of echinoderms in the subclass Euechinoidea characterized by a rigid, exocylic test and lacking a lantern, or jaw, apparatus.

Athalamida [INV ZOO] An order of naked amebas of the subclass Granuloreticulosia in which pseudopodia are branched and threadlike (reticulopodia).

Athecanephria [INV ZOO] An order of tube-dwelling, tentaculate animals in the class Pogonophora characterized by a saclike anterior coelom.

Atherinidae [VERT ZOO] The silversides, a family of actinopterygian fishes of the order Atheriniformes.

Atheriniformes [VERT ZOO] An order of actinopterygian fishes in the infraclass Teleostei, including flyingfishes, needlefishes, killifishes, silversides, and allied species.

atheroma [MED] 1. Fatty degeneration of the inner arterial walls. 2. A fatty cyst.

atherosclerosis [MED] Deposition of lipid with proliferation of fibrous connective tissue cells in the inner walls of the arteries.

Athiorhodaceae [MICROBIO] Formerly the nonsulfur photosynthetic bacteria, a family of small, gram-negative, nonsporeforming, motile bacteria in the suborder Rhodobacteriineae.

athwartship [NAV ARCH] Perpendicular to the fore and aft centerline of a ship.

Athyrididina [PALEON] A suborder of fossil articulate brachiopods in the order Spiriferida characterized by laterally or, more rarely, ventrally directed spires.

Atlantacea [INV ZOO] A superfamily of mollusks in the subclass Prosobranchia.

Atlantic time [ASTRON] A time zone; the fourth zone west of Greenwich. Also known as Atlantic standard time.

atlas [ANAT] The first cervical vertebra. [MAP] A collection of charts or maps kept loose or bound in a volume.

atlas grid [MAP] A reference system that permits the designation of the location of a point or an area on a map, photo, or other graphic in terms of numbers and letters. Also known as alphanumeric grid.

atmidometer *See* atmometer.

atmometer [ENG] The general name for an instrument which measures the evaporation rate of water into the atmosphere. Also known as atmidometer; evaporation gage; evaporimeter.

atmosphere [MECH] A unit of pressure equal to 1.013250×10^6 dynes/cm^2, which is the air pressure measured at mean sea level. Abbreviated atm. Also known as standard atmosphere. [METEOROL]

The gaseous envelope surrounding a planet or celestial body.

atmospheric boundary layer *See* surface boundary layer.

atmospheric chemistry [METEOROL] The study of the production, transport, modification, and removal of atmospheric constituents in the troposphere and stratosphere.

atmospheric distillation [CHEM ENG] Distillation operation conducted at atmospheric pressure, in contrast to vacuum distillation or pressure distillation.

atmospheric electric field [GEOPHYS] The atmosphere's electric field strength in volts per meter at any specified point in time and space; near the earth's surface, in fair-weather areas, a typical datum is about 100 and the field is directed vertically in such a way as to drive positive charges downward.

atmospheric electricity [GEOPHYS] The electrical processes occurring in the lower atmosphere, including both the intense local electrification accompanying storms and the much weaker fair-weather electrical activity over the entire globe produced by the electrified storms continuously in progress.

atmospheric entry [AERO ENG] The penetration of any planetary atmosphere by any object from outer space; specifically, the penetration of the earth's atmosphere by a crewed or uncrewed capsule or spacecraft.

atmospheric interference [GEOPHYS] Electromagnetic radiation, caused by natural electrical disturbances in the atmosphere, which interferes with radio systems. Also known as atmospherics; sferics; strays.

atmospheric ionization [GEOPHYS] The process by which neutral atmospheric molecules or atoms are rendered electrically charged chiefly by collisions with high-energy particles.

atmospheric layer *See* atmospheric shell.

atmospheric optics *See* meteorological optics.

atmospheric pressure [PHYS] The pressure at any point in an atmosphere due solely to the weight of the atmospheric gases above the point concerned. Also known as barometric pressure.

atmospheric refraction [GEOPHYS] 1. The angular difference between the apparent zenith distance of a celestial body and its true zenith distance, produced by refraction effects as the light from the body penetrates the atmosphere. 2. Any refraction caused by the atmosphere's normal decrease in density with height.

atmospheric region *See* atmospheric shell.

atmospherics *See* atmospheric interference.

atmospheric scattering [GEOPHYS] A change in the direction of propagation, frequency, or polarization of electromagnetic radiation caused by interaction with the atoms of the atmosphere.

atmospheric shell [METEOROL] Any one of a number of strata or layers of the earth's atmosphere; temperature distribution is the most common criterion used for denoting the various shells. Also known as atmospheric layer; atmospheric region.

atmospheric sounding [METEOROL] A measurement of atmospheric conditions aloft, above the effective range of surface weather observations.

atmospheric tide [GEOPHYS] Periodic global motions of the earth's atmosphere, produced by gravitational action of the sun and moon; amplitudes are minute except in the upper atmosphere.

atmospheric turbulence [METEOROL] Apparently random fluctuations of the atmosphere that often constitute major deformations of its state of fluid flow.

atoll [GEOGR] A ring-shaped coral reef that surrounds a lagoon without projecting land area and that is surrounded by open sea.

atom [COMPUT SCI] A primitive data element in a data structure. [CHEM] The individual structure which constitutes the basic unit of any chemical element. [MATH] An element, A, of a measure algebra, other than the zero element, which has the property that any element which is equal to or less than A is either equal to A or equal to the zero element.

atomic battery *See* nuclear battery.

atomic beam resonance [PHYS] Phenomenon in which an oscillating magnetic field, superimposed on a uniform magnetic field at right angles to it, causes transitions between states with different magnetic quantum numbers of the nuclei of atoms in a beam passing through the field; the transitions occur only when the frequency of the oscillating field assumes certain characteristic values.

atomic bomb [ORD] Also known as A bomb. **1.** A device for suddenly producing an explosively rapid neutron chain reaction in a fissile material such as uranium-235 or plutonium-239. Also known as fission bomb. **2.** Any explosive device which derives its energy from nuclear reactions, including a fusion bomb. Also known as nuclear bomb.

atomic charge [ATOM PHYS] The electric charge of an ion, equal to the number of electrons the atom has gained or lost in its ionization multiplied by the charge on one electron.

atomic clock [HOROL] An electronic clock whose frequency is supplied or governed by the natural resonance frequencies of atoms or molecules of suitable substances.

atomic cloud [NUCLEO] The cloud of hot gases, smoke, dust, and other matter that is carried aloft after the explosion of a nuclear weapon in the air or near the surface; frequently has a mushroom shape.

atomic constants [PHYS] The physical constants which play a fundamental role in atomic physics, including the electronic charge, electronic mass, speed of light, Avogadro number, and Planck's constant.

atomic device *See* atomic weapon.

atomic diamagnetism [ATOM PHYS] Diamagnetic ionic susceptibility, important in providing correction factors for measured magnetic susceptibilities; calculated theoretically by considering electron density distributions summed for each electron shell.

atomic energy *See* nuclear energy.

atomic fallout *See* fallout.

atomic fission *See* fission.

atomic form factor *See* atomic scattering factor.

atomic fusion *See* fusion.

atomic ground state [ATOM PHYS] The state of lowest energy in which an atom can exist. Also known as atomic unexcited state.

atomic hydrogen maser [PHYS] A maser in which dissociated hydrogen atoms from an electric discharge source are formed into a beam that undergoes selective magnetic processing; can be used as an atomic clock.

atomic mass [PHYS] The mass of a neutral atom usually expressed in atomic mass units.

atomic mass unit [PHYS] An arbitrarily defined unit in terms of which the masses of individual atoms are expressed; the standard is the unit of mass equal to $\frac{1}{12}$ the mass of the carbon atom, having as nucleus the isotope with mass number 12. Abbreviated amu. Also known as dalton.

atomic nucleus *See* nucleus.

atomic number [NUC PHYS] The number of protons in an atomic nucleus. Also known as proton number.

atomic orbital [ATOM PHYS] The space-dependent part of a wave function describing an electron in an atom.

atomic particle [ATOM PHYS] One of the particles of which an atom is constituted, as an electron, neutron, or proton.

atomic photoelectric effect *See* photoionization.

atomic physics [PHYS] The science concerned with the structure of the atom, the characteristics of the elementary particles of which the atom is composed, and the processes involved in the interactions of radiant energy with matter.

atomic pile *See* nuclear reactor.

atomic radius [PHYS CHEM] **1.** Half the distance of closest approach of two like atoms not united by a bond. **2.** The experimentally determined radius of an atom in a covalently bonded compound.

atomic reactor *See* nuclear reactor.

atomic scattering factor [PHYS] A quantity which expresses the efficiency with which x-rays of a stated wavelength are scattered into a given direction by a particular atom, measured in terms of the corresponding scattering by a point electron. Also known as atomic form factor.

atomic spectroscopy [SPECT] The branch of physics concerned with the production, measurement, and interpretation of spectra arising from either emission or absorption of electromagnetic radiation by atoms.

atomic spectrum [SPECT] The spectrum of radiations due to transitions between energy levels in an atom, either absorption or emission.

atomic stopping power [NUCLEO] For an ionizing particle passing through an element, the particle's energy loss per atom within a unit area normal to the particle's path; equal to the linear energy transfer (energy loss per unit path length) divided by the number of atoms per unit volume.

atomic structure [ATOM PHYS] The arrangement of the parts of an atom, which consists of a massive, positively charged nucleus surrounded by a cloud of electrons arranged in orbits describable in terms of quantum mechanics.

atomic theory [CHEM] The assumption that matter is composed of particles called atoms and that these are the limit to which matter can be subdivided.

atomic time [HOROL] Any time system standardized with reference to an atomic resonance, such as the international standard cesium-133 transition.

atomic unexcited state *See* atomic ground state.

atomic units *See* Hartree units.

atomic vibration [ATOM PHYS] Periodic, nearly harmonic changes in position of the atoms in a molecule giving rise to many properties of matter, including molecular spectra, heat capacity, and heat conduction.

atomic volume [PHYS CHEM] The volume occupied by 1 gram-atom of an element in the solid state.

atomic weapon [ORD] Any bomb, warhead, or projectile using active nuclear material to cause a chain reaction upon detonation. Also known as atomic device; nuclear weapon.

atomic weight [CHEM] The relative mass of an atom based on a scale in which a specific carbon atom (carbon-12) is assigned a mass value of 12. Abbreviated at. wt.

atomization [MECH ENG] The mechanical subdivision of a bulk liquid or meltable solid, such as certain metals, to produce drops, which vary in diameter depending on the process from under 10 to over 1000 micrometers.

atomizer [MECH ENG] A device that produces a mechanical subdivision of a bulk liquid, as by spraying, sprinkling, misting, or nebulizing.

atom smasher *See* particle accelerator.

atopic allergy [MED] A type of immediate hypersensitivity in humans resulting from spontaneous sensitization, usually by inhaled or ingested antigens; for example, asthma, hay fever, or hives.

ATP *See* adenosinetriphosphate.

ATR *See* attenuated total reflectance.

Atractidae [INV ZOO] A family of parasitic nematodes in the superfamily Oxyuroidea.

atran [NAV] An acronym for automatic terrain recognition and navigation, a system which depends upon the correlation of terrain images appearing on a radar cathode-ray tube with previously prepared maps or simulated radar images of the terrain.

atrial flutter [MED] A cardiac arrhythmia characterized by rapid, irregular atrial impulses and ineffective atrial contractions; the heartbeat varies from 60 to 180 per minute and is grossly irregular in intensity and rhythm. Also known as auricular flutter.

atrioventricular valve [ANAT] A structure located at the orifice between the atrium and ventricle which maintains a unidirectional blood flow through the heart.

atrium [ANAT] 1. The heart chamber that receives blood from the veins. 2. The main part of the tympanic cavity, below the malleus. 3. The external chamber to receive water from the gills in lancelets and tunicates. [ARCH] An open court located within a building.

atrophic arthritis *See* rheumatoid arthritis.

atropine [PHARM] $C_{17}H_{23}O_3N$ An alkaloid extracted from *Atropa belladonna* and related plants of the family Solanaceae; used to relieve muscle spasms and pain, and to dilate the pupil of the eye.

ATS *See* Administrative Terminal System; applications technology satellite.

attachment plug [ELEC] A device having an attached flexible cord containing conductors, and capable of being inserted in a receptacle so as to form an electrical connection between the conductors in the cord and conductors permanently connected to the receptacle.

attack director [COMPUT SCI] An electromechanical analog computer which is designed for surface antisubmarine use and which computes continuous solution of several lines of submarine attack; it is part of several antisubmarine fire control systems.

attapulgite [MINERAL] $(Mg,Al)_2Si_4O_{10}(OH)\cdot 4H_2O$ A clay mineral with a needlelike shape from Georgia and Florida; active ingredient in most fuller's earth, and used as a suspending agent, as an oil well drilling fluid, as a thickener in latex paint.

attended time [COMPUT SCI] The time in which a computer is either switched on and capable of normal operation (including time during which it is temporarily idle but still watched over by computer personnel) or out of service for maintenance work.

attention span [PSYCH] The period of time a person is able to concentrate his attentions on a given item, usually with respect to learning.

attenuate [ENG ACOUS] To weaken a signal by reducing its level.

attenuated total reflectance [SPECT] A method of spectrophotometric analysis based on the reflection of energy at the interface of two media which have different refractive indices and are in optical contact with each other. Abbreviated ATR. Also known as frustrated internal reflectance; internal reflectance spectroscopy.

attenuated vaccine [IMMUNOL] A suspension of weakened bacteria, viruses, or fractions thereof used to produce active immunity.

attenuation [BOT] Tapering, sometimes to a long point. [MICROBIO] Weakening or reduction of the virulence of a microorganism. [PHYS] The reduction in level of a quantity, such as the intensity of a wave, over an interval of a variable, such as the distance from a source.

attenuation constant [PHYS] A rating for a line or medium through which a plane wave is being transmitted, equal to the relative rate of decrease of an amplitude of a field component, voltage, or current in the direction of propagation, in nepers per unit length.

attenuation distortion [COMMUN] 1. In a circuit or system, departure from uniform amplification or attenuation over the frequency range required for transmission. 2. The effect of such departure on a transmitted signal.

attenuation equalizer [ELECTR] Corrective network which is designed to make the absolute value of the transfer impedance, with respect to two chosen pairs of terminals, substantially constant for all frequencies within a desired range.

attenuation factor *See* attenuation constant.

attenuation network [ELECTR] Arrangement of circuit elements, usually impedance elements, inserted in circuitry to introduce a known loss or to reduce the impedance level without reflections.

attenuator [ELECTR] An adjustable or fixed transducer for reducing the amplitude of a wave without introducing appreciable distortion.

attitude [AERO ENG] The position or orientation of an aircraft, spacecraft, and so on, either in motion or at rest, as determined by the relationship between its axes and some reference line or plane or some fixed system of reference axes. [GRAPHICS] The orientation of a camera or a photograph with respect to a given external reference system.

atto- [PHYS] A prefix representing 10^{-18}, which is 0.000 000 000 000 000 001, or one-millionth of a millionth of a millionth. Abbreviated a.

attribute [COMPUT SCI] A data item containing information about a variable.

attrition [GEOL] The act of wearing and smoothing of rock surfaces by the flow of water charged with sand and gravel, by the passage of sand drifts, or by the movement of glaciers. [MATER] Wear caused by rubbing or friction; for metal surfaces, also known as scoring; scouring.

attritus [GEOL] 1. Visible-to-ultramicroscopic particles of vegetable matter produced by microscopic and other organisms in vegetable deposits, particularly in swamps and bogs. 2. The dull gray to nearly black, frequently striped portion of material that makes up the bulk of some coals and alternate bands of bright anthraxylon in well-banded coals.

Atwood machine [MECH ENG] A device comprising a pulley over which is passed a stretch-free cord with a weight hanging on each end.

at. wt *See* atomic weight.

Atyidae [INV ZOO] A family of decapod crustaceans belonging to the section Caridea.

AU *See* astronomical unit.

Au *See* gold.

aubrite [GEOL] An enstatite achondrite (meteorite) consisting almost wholly of crystalline-granular enstatite (and clinoenstatite) poor in lime and practically free from ferrous oxide, with accessory oligoclase. Also known as bustite.

Auchenorrhyncha [INV ZOO] A group of homopteran families and one superfamily, in which the beak arises at the anteroventral extremity of the face and is not sheathed by the propleura.

audibility [ACOUS] **1.** The state or quality of being heard. **2.** The intensity of a received audio signal, usually expressed in decibels above or below 1 milliwatt using a stated single-frequency sine wave.

audibility curve [ACOUS] **1.** The limits of hearing represented graphically as an area by plotting the minimum audible intensity of a sine wave sound versus frequency. **2.** See equal loudness contour.

audibility threshold [ACOUS] The sound intensity at a given frequency which is the minimum perceptible by a normal human ear under specified standard conditions.

audible frequency See audible tone.

audible tone [ACOUS] Sound of a frequency which the average human can hear, ranging from 30 to 16,000 hertz. Also known as audible frequency.

audio [ACOUS] **1.** Of or pertaining to sound in the range of frequencies considered audible at reasonable listening intensities to the average young adult listener, approximately 15 to 20,000 hertz. **2.** Pertaining to equipment for the recording, transmission, reproduction, or amplification of such sound.

audio amplifier See audio-frequency amplifier.

audio-frequency amplifier [ELECTR] An electronic circuit for amplification of signals within, and in some cases above, the audible range of frequencies in equipment used to record and reproduce sound. Also known as audio amplifier.

audio-frequency range [ACOUS] The range of frequencies to which the human ear is sensitive, approximately 15 to 20,000 hertz. Also known as audio range.

audio-frequency shift modulation [COMMUN] System of facsimile transmission over radio, in which the frequency shift required is applied through an 800-hertz shift of an audio signal, rather than shifting the radio transmitter frequency; the radio signal is modulated by the shifting audio signal, usually at 1500 to 2300 hertz.

audiogram [ACOUS] A graph showing hearing loss, percent hearing loss, or percent hearing as a function of frequency.

audio masking See masking.

audiometer [ENG] An instrument composed of an oscillator, amplifier, and attenuator and used to measure hearing acuity for pure tones, speech, and bone conduction.

audiometry [ACOUS] The study of hearing ability by means of audiometers.

audio range See audio-frequency range.

audio response [COMMUN] A form of computer output in which prerecorded spoken syllables, words, or messages are selected and put together by a computer as the appropriate verbal response to a keyboarded inquiry on a time-shared on-line information system.

audio signal [ACOUS] An electric signal having the frequency of a mechanical wave that can be detected as a sound by the human ear.

audio system See sound-reproducing system.

audiphone [ENG ACOUS] A device that enables persons with certain types of deafness to hear, consisting of a plate or diaphragm that is placed against the teeth and transmits sound vibrations to the inner ear.

audit [COMPUT SCI] The operations developed to corroborate the evidence as regards authenticity and validity of the data that are introduced into the data-processing problem or system.

audition [PHYSIO] Ability to hear.

auditory [PHYSIO] Pertaining to the act or the organs of hearing.

auditory nerve [ANAT] The eighth cranial nerve in vertebrates; either of a pair of sensory nerves composed of two sets of nerve fibers, the cochlear nerve and the vestibular nerve. Also known as acoustic nerve; vestibulocochlear nerve.

audit trail [COMPUT SCI] A system that provides a means for tracing items of data from processing step to step, particularly from a machine-produced report or other machine output back to the original source data.

aufeis [HYD] Icing of ground or river water in arctic areas with continuous permafrost on which the water has continued to flow.

augen [PETR] Large, lenticular eye-shaped mineral grain or mineral aggregate visible in some metamorphic rocks.

auger [DES ENG] **1.** A wood-boring tool that consists of a shank with spiral channels ending in two spurs, a central tapered feed screw, and a pair of cutting lips. **2.** A large augerlike tool for boring into soil.

auger bit [DES ENG] A bit shaped like an auger but without a handle; used for wood boring and for earth drilling. [MIN ENG] Hard steel or tungsten carbide–tipped cutting teeth used in an auger running on a torque bar or in an auger-drill head running on a continuous-flight auger.

auger conveyor See screw conveyor.

Auger effect [ATOM PHYS] The radiationless transition of an electron in an atom from a discrete electronic level to an ionized continuous level of the same energy. Also known as autoionization.

Auger electron spectroscopy [SPECT] The energy analysis of Auger electrons produced when an excited atom relaxes by a radiationless process after ionization by a high-energy electron, ion, or x-ray beam. Abbreviated AES.

augite [MINERAL] $(Ca,Mg,Fe)(Mg,Fe,Al)(Al,Si)_2O_6$ A general name for the monoclinic pyroxenes; occurs as dark green to black, short, stubby, prismatic crystals, often of octagonal outline.

augitite [PETR] A volcanic rock consisting of abundant phenocrysts of augite in a glassy groundmass containing microlites of nepheline and plagioclase, with accessory biotite, apatite, and opaque oxides.

augmentation [ASTRON] The apparent increase in the semidiameter of a celestial body, as observed from the earth, as the body's altitude (angular distance above the horizon) increases, due to the reduced distance from the observer. The term is used principally in reference to the moon.

augmentation distance [NUC PHYS] The extrapolation distance, which is the distance between the time boundary of a nuclear reactor and its boundary calculated by extrapolation.

augmented matrix [MATH] The matrix of the coefficients, together with the constant terms, in a system of linear equations.

augmented operation code [COMPUT SCI] An operation code which is further defined by information from another portion of an instruction.

auk [VERT ZOO] Any of several large, short-necked diving birds (*Alca*) of the family Alcidae found along North Atlantic coasts.

Aulolepidae [PALEON] A family of marine fossil teleostean fishes in the order Ctenothrissiformes.

Auloporidae [PALEON] A family of Paleozoic corals in the order Tabulata.

A* unit [PHYS] An atomic standard unit of length, based on the tungsten $K\alpha_1$ line, approximately 10^{-11} centimeter; used for measurements of x-ray wavelengths and of crystal dimensions.

auntie *See* antimissile missile.

aural [BIOL] Pertaining to the ear or the sense of hearing.

aural masking *See* masking.

aureole [GEOL] A ring-shaped contact zone surrounding an igneous intrusion. Also known as contact aureole; contact zone; exomorphic zone; metamorphic aureole; metamorphic zone; thermal aureole. [METEOROL] A poorly developed corona in the atmosphere characterized by a bluish-white disk immediately around the luminous celestial body, as around the sun or moon in the fog.

aurichalcite [MINERAL] $(Zn,Cu)_5(CO_3)_2(OH)_6$ Pale-green or pale-blue mineral consisting of a basic copper zinc carbonate and occurring in crystalline incrustations. Also known as brass ore.

auricle [ANAT] 1. An ear-shaped appendage to an atrium of the heart. 2. Any ear-shaped structure. 3. *See* pinna.

auricular fibrillation [MED] Arrhythmic contractions of the auricles.

auricular flutter *See* atrial flutter.

aurora [ELEC] *See* corona discharge. [GEOPHYS] The most intense of the several lights emitted by the earth's upper atmosphere, seen most often along the outer realms of the Arctic and Antarctic, where it is called the aurora borealis and aurora australis, respectively; excited by charged particles from space.

aurora australis [GEOPHYS] The aurora of southern latitudes. Also known as southern lights.

aurora borealis [GEOPHYS] The aurora of northern latitudes. Also known as northern lights.

auroral line [SPECT] A prominent green line in the spectrum of the aurora at a wavelength of 5577 angstroms, resulting from a certain forbidden transition of oxygen.

auroral poles [GEOPHYS] The points on the earth's surface on which the auroral isochasms are centered; coincide approximately with the magnetic-axis poles of the geomagnetic field.

auroral region [GEOPHYS] The region within 30° geomagnetic latitude of each auroral pole.

auroral storm [GEOPHYS] A rapid succession of auroral substorms, occurring in a short period, of the order of a day, during a geomagnetic storm.

auroral substorm [GEOPHYS] A characteristic sequence of auroral intensifications and movements occurring around midnight, in which a rapid poleward movement of auroral arcs produces a bulge in the auroral oval.

auroral zone [GEOPHYS] A roughly circular band around either geomagnetic pole within which there is a maximum of auroral activity; lies about 10–15° geomagnetic latitude from the geomagnetic poles.

aurosmiridium [MINERAL] A brittle, silver-white, isometric mineral consisting of a solid solution of gold and osmium in iridium.

ausforging [MET] The forming of austenitic steel into required shapes by hammering or pressing after cooling.

ausrolling [MET] The working of austenitic steel by passing it, after cooling, between oppositely revolving rollers.

austausch coefficient *See* exchange coefficient.

austempering [MET] A process for the heat treatment of austenitic steel.

austenite [MET] Gamma iron with carbon in solution.

austenitic [MET] Composed mainly of austenite.

Australopithecinae [PALEON] The near-men, a subfamily of the family Hominidae composed of the single genus *Australopithecus*.

Austroastacidae [INV ZOO] A family of crayfish in the order Decapoda found in temperate regions of the Southern Hemisphere.

Austrodecidae [INV ZOO] A monogeneric family of marine arthropods in the subphylum Pycnogonida.

authalic latitude [MAP] A latitude based on a sphere having the same area as the spheroid, and such that areas between successive parallels of latitude are exactly equal to the corresponding areas on the spheroid. Also known as equal-area latitude.

authalic map projection *See* equal-area map projection.

authentication [COMMUN] Security measure designed to protect a communications system against fraudulent transmissions and establish the authenticity of a message by an authenticator within the transmission derived from certain predetermined elements of the message itself.

authigene [MINERAL] A mineral which has not been transported but has been formed in place. Also known as authigenic mineral.

authigenic [GEOL] Of constituents that came into existence with or after the formation of the rock of which they constitute a part; for example, the primary and secondary minerals of igneous rocks.

authigenic mineral *See* authigene.

authoring language [COMPUT SCI] A programming language designed to be convenient for authors of computer-based learning materials.

autism [PSYCH] A schizophrenic symptom characterized by absorption in fantasy to the exclusion of perceptual reality.

auto-abstract [COMPUT SCI] 1. To select key words from a document, commonly by an automatic or machine method, for the purpose of forming an abstract of the document. 2. The material abstracted from a document by machine methods.

autoagglutination [IMMUNOL] Agglutination of an individual's erythrocytes by his own serum. Also known as autohemagglutination.

autoantibody [IMMUNOL] An antibody formed by an individual against his own tissues; common in hemolytic anemias.

autoantigen [IMMUNOL] A tissue within the body which acquires the ability to incite the formation of complementary antibodies.

autochthon [GEOL] A succession of rock beds that have been moved comparatively little from their original site of formation, although they may be folded and faulted extensively. [PALEON] A fossil occurring where the organism once lived.

autochthonous [GEOL] Having been formed or occurring in the place where found.

autochthonous coal [GEOL] Coal believed to have originated from accumulations of plant debris at the place where the plants grew. Also known as indigenous coal.

autoclastic [GEOL] Of rock, fragmented in place by folding due to orogenic forces when the rock is not so heavily loaded as to render it plastic.

autoclave [ENG] An airtight vessel for heating and sometimes agitating its contents under high steam pressure; used for industrial processing, sterilizing, and cooking with moist or dry heat at high temperatures.

autocode [COMPUT SCI] The process of using a computer to convert automatically a symbolic code into a machine code. Also known as automatic code.

autocollimation [OPTICS] A procedure for collimating a telescope or other optical instrument with objective and crosshairs, in which the instrument is directed toward a plane mirror and the crosshairs and lens are adjusted so that the crosshairs coincide with their reflected image.

autocollimator [OPTICS] 1. A device by which a single lens collimates diverging light from a slit, and then focuses the light on an exit slit after it has passed through a prism to a mirror and been reflected back through the prism. 2. A telescope which has a graduated reticle, enabling an observer to read off the angles subtended by distant objects. 3. A convex mirror at the focus of the principal mirror of a reflecting telescope, which causes light to leave the telescope in a parallel beam. 4. A telescope equipped with an eyepiece designed for autocollimation.

autoconsequent stream [HYD] A stream in the process of building a fan or an alluvial plain, the course of which is guided by the slopes of the alluvium the stream itself has deposited.

autoconvection [METEOROL] The phenomenon of the spontaneous initiation of convection in an atmospheric layer in which the lapse rate is equal to or greater than the autoconvective lapse rate.

autocorrelator [ELECTR] A correlator in which the input signal is delayed and multiplied by the undelayed signal, the product of which is then smoothed in a low-pass filter to give an approximate computation of the autocorrelation function; used to detect a nonperiodic signal or a weak periodic signal hidden in noise.

autodecrement addressing [COMPUT SCI] An addressing mode of minicomputers in which the register is first decremented and then used as a pointer.

autodyne circuit [ELECTR] A circuit in which the same tube elements serve as oscillator and detector simultaneously.

autoecious See autoicous.

autogamy [BIOL] A process of self-fertilization that results in homozygosis; occurs in some flowering plants, fungi, and protozoans.

autogenous [SCI TECH] Self-generated; produced without external influence.

autogenous electrification [PHYS] The process by which net charge is built up on an object, such as an airplane, moving relative to air containing dust or ice crystals; produced by frictional effects (triboelectrification) accompanying contact between the object and the particulate matter.

autogenous ignition temperature See ignition temperature.

autogenous vaccine [IMMUNOL] A vaccine prepared from a culture of microorganisms taken directly from the infected person.

autogenous welding [MET] A fusion welding process using heat without the addition of filler metal to join two pieces of the same metal.

autogeosyncline [GEOL] A parageosyncline that subsides as an elliptical basin or trough nearly without associated highlands. Also known as intracratonic basin.

autogiro [AERO ENG] A type of aircraft which utilizes a rotating wing (rotor) to provide lift and a conventional engine-propeller combination to pull the vehicle through the air.

autograft [BIOL] A tissue transplanted from one part to another part of an individual's body.

autohemagglutination See autoagglutination.

autoicous [BOT] Having male and female organs on the same plant but on different branches. Also spelled autoecious.

autoignition [CHEM] See spontaneous ignition. [MECH ENG] Spontaneous ignition of some or all of the fuel-air mixture in the combustion chamber of an internal combustion engine.

autoimmune disease [IMMUNOL] An illness involving the formation of autoantibodies which appear to cause pathological damage to the host.

autoincrement addressing [COMPUT SCI] An addressing mode of minicomputers in which the operand address is gotten from the specified register which is then incremented.

autoindexing See automatic indexing.

autoionization See Auger effect.

autolith [PETR] 1. A fragment of igneous rock enclosed in another igneous rock of later consolidation, each being regarded as a derivative from a common parent magma. 2. A round, oval, or elongated accumulation of iron-magnesium minerals of uncertain origin in granitoid rock.

autologous [BIOL] Derived from or produced by the individual in question, such as an autologous protein or an autologous graft.

autoluminescence [ATOM PHYS] Luminescence of a material (such as a radioactive substance) resulting from energy originating within the material itself.

autolysis [GEOCHEM] Return of a substance to solution, as of phosphate removed from seawater by plankton and returned when these organisms die and decay. [PATH] Self-digestion by body cells following somatic or organ death or ischemic injury.

automata theory [ENG] The theory concerning the operation principles, application, and behavior characteristics of automatic devices.

automated tape library [COMPUT SCI] A computer storage system consisting of several thousand magnetic tapes and equipment under computer control which automatically brings the tapes from storage, mounts them on tape drives, dismounts the tapes when the job is completed, and returns them to storage.

automatic [ENG] Having a self-acting mechanism that performs a required act at a predetermined time or in response to certain conditions. [ORD] See automatic weapon.

automatic abstracting [COMPUT SCI] Techniques whereby, on the basis of statistical properties, a subset of the sentences in a document is selected as representative of the general content of that document.

automatic back bias [ELECTR] Radar technique which consists of one or more automatic gain control loops to prevent overloading of a receiver by large signals, whether jamming or actual radar echoes.

automatic background control See automatic brightness control.

automatic bass compensation [ELECTR] A circuit related to the volume control in some radio receivers and audio amplifiers to make bass notes sound properly balanced, in the audio spectrum, at low volume-control settings.

automatic brightness control [ELECTR] A circuit used in a television receiver to keep the average brightness of the reproduced image essentially constant. Abbreviated ABC. Also known as automatic background control.

automatic carriage [COMPUT SCI] Any mechanism designed to feed continuous paper or plastic forms through a printing or writing device, often using sprockets to engage holes in the paper.

automatic C bias   *See* self-bias.

automatic celestial navigation   *See* celestial-inertial guidance.

automatic character recognition   [COMPUT SCI] The technology of using special machine systems to identify human-readable symbols, most often alphanumeric, and then to utilize this data.

automatic check   [COMPUT SCI] An error-detecting procedure performed by a computer as an integral part of the normal operation of a device, with no human attention required unless an error is actually detected.

automatic check-out system   [CONT SYS] A system utilizing test equipment capable of automatically and simultaneously providing actions and information which will ultimately result in the efficient operation of tested equipment while keeping time to a minimum.

automatic chroma control   *See* automatic color control.

automatic chrominance control   *See* automatic color control.

automatic code   *See* autocode.

automatic coding   [COMPUT SCI] Any technique in which a computer is used to help bridge the gap between some intellectual and manual form of describing the steps to be followed in solving a given problem, and some final coding of the same problem for a given computer.

automatic color control   [ELECTR] A circuit used in a color television receiver to keep color intensity levels essentially constant despite variations in the strength of the received color signal; control is usually achieved by varying the gain of the chrominance band-pass amplifier. Also known as automatic chroma control; automatic chrominance control.

automatic contrast control   [ELECTR] A circuit that varies the gain of the radio-frequency and video intermediate-frequency amplifiers in such a way that the contrast of the television picture is maintained at a constant average level.

automatic-control block diagram   [CONT SYS] A diagrammatic representation of the mathematical relationships defining the flow of information and energy through the automatic control system, in which the components of the control system are represented as functional blocks in series and parallel arrangements according to their position in the actual control system.

automatic controller   [CONT SYS] An instrument that continuously measures the value of a variable quantity or condition and then automatically acts on the controlled equipment to correct any deviation from a desired preset value. Also known as automatic regulator; controller.

automatic control system   [CONT SYS] A control system having one or more automatic controllers connected in closed loops with one or more processes. Also known as regulating system.

automatic cutout   [ELEC] A device, usually operated by centrifugal force or by an electromagnet, that automatically shorts part of a circuit at a particular time.

automatic data processing   [ENG] The machine performance, with little or no human assistance, of any of a variety of tasks involving informational data; examples include automatic and responsive reading, computation, writing, speaking, directing artillery, and the running of an entire factory. Abbreviated ADP.

automatic data-processing system   [COMPUT SCI] The equipment, personnel program, and application operations involved in the utilization of electronic data-processing equipment, along with associated electric accounting machines, to solve business and logistics data-processing problems, with a minimum of human intervention.

automatic dialer   [ELECTR] A device in which a telephone number up to a maximum of 14 digits can be stored in a memory and then activated, directly into the line, by the caller's pressing a button. Also known as mechanical dialer.

automatic dictionary   [COMPUT SCI] Any table within a computer memory which establishes a one-to-one correspondence between two sets of characters.

automatic direction finder   [ELECTR] A direction finder that without manual manipulation indicates the direction of arrival of a radio signal. Abbreviated ADF. Also known as radio compass.

automatic drill   [DES ENG] A straight brace for bits whose shank comprises a coarse-pitch screw sliding in a threaded tube with a handle at the end; the device is operated by pushing the handle.

automatic error correction   [COMMUN] A technique, usually requiring the use of special codes or automatic retransmission, which detects and corrects errors occurring in transmission; the degree of correction depends upon coding and equipment configuration.

automatic exchange   [ELECTR] A telephone, teletypewriter, or data-transmission exchange in which communication between subscribers is effected, without the intervention of an operator, by devices set in operation by the originating subscriber's instrument. Also known as automatic switching system; machine switching system.

automatic-feed punch   [COMPUT SCI] A card punch having a hopper, a card track, and a stacker; movement of cards through the punch is automatic.

automatic fine-tuning control   [ELECTR] A circuit used in a color television receiver to maintain the correct oscillator frequency in the tuner for best color picture by compensating for drift and incorrect tuning.

automatic focus   [OPTICS] A device in a camera or enlarger which automatically keeps the objective lens in focus through a range of magnification.

automatic frequency control   [ELECTR] Abbreviated AFC. **1.** A circuit used to maintain the frequency of an oscillator within specified limits, as in a transmitter. **2.** A circuit used to keep a superheterodyne receiver tuned accurately to a given frequency by controlling its local oscillator, as in an FM receiver. **3.** A circuit used in radar superheterodyne receivers to vary the local oscillator frequency so as to compensate for changes in the frequency of the received echo signal. **4.** A circuit used in television receivers to make the frequency of a sweep oscillator correspond to the frequency of the synchronizing pulses in the received signal.

automatic gain control   [ELECTR] A control circuit that automatically changes the gain (amplification) of a receiver or other piece of equipment so that the desired output signal remains essentially constant despite variations in input signal strength. Abbreviated AGC.

automatic grid bias   *See* self-bias.

automatic gun   *See* automatic weapon.

automatic indexing   [COMPUT SCI] Selection of key words from a document by computer for use as index entries. Also known as autoindexing.

automatic interrupt   [COMPUT SCI] Interruption of a computer program brought about by a hardware device or executive program acting as a result of some event which has occurred independently of the interrupted program.

automatic landing system [NAV] The means for automatically guiding and controlling aircraft from an initial approach altitude to a point where safe contact is made with the landing surface.

automatic mathematical translator [COMPUT SCI] An automatic-programming computer capable of receiving a mathematical equation from a remote input and returning an immediate solution.

automatic message accounting See automatic toll ticketing.

automatic modulation limiting [COMMUN] A circuit that prevents overmodulation in some citizen-band radio transmitters by reducing the gain of one or more audio amplifier stages when the voice signal becomes stronger. Abbreviated AML.

automatic peak limiter See limiter.

automatic phase control [ELECTR] Abbreviated APC. 1. A circuit used in color television receivers to reinsert a 3.58-megahertz carrier signal with exactly the correct phase and frequency by synchronizing it with the transmitted color-burst signal. 2. An automatic frequency-control circuit in which the difference between two frequency sources is fed to a phase detector that produces the required control signal.

automatic picture control [ELECTR] A multiple-contact switch used in some color television receivers to disconnect one or more of the regular controls and make connections to corresponding preset controls.

automatic pilot [NAV] Also known as autopilot. 1. Equipment which automatically stabilizes the attitude of an aircraft about its pitch, roll, and yaw axes and keeps it on a predetermined heading. 2. Equipment for automatically steering ships.

automatic programming [COMPUT SCI] The preparation of machine-language instructions by use of a computer.

automatic ranging See autoranging.

automatic regulator See automatic controller.

automatic routine [COMPUT SCI] A routine that is executed independently of manual operations, but only if certain conditions occur within a program or record, or during some other process.

automatic sampler [MECH ENG] A mechanical device to sample process streams (gas, liquid, or solid) either continuously or at preset time intervals.

automatic sequences [COMPUT SCI] The characteristic of a computer that can perform successive operations without human intervention.

automatic stop [COMPUT SCI] An automatic halting of a computer processing operation as the result of an error detected by built-in checking devices.

automatic switching system See automatic exchange.

automatic tape punch [COMPUT SCI] A device that punches holes in a paper tape upon reception of electronic signals from a central processing unit.

automatic tint control [ELECTR] A circuit used in color television receivers to maintain correct flesh tones when a station changes cameras or switches to commercials, by correcting phase errors before the chroma signal is demodulated.

automatic titrator [ANALY CHEM] 1. Titration with quantitative reaction and measured flow of reactant. 2. Electrically generated reactant with potentiometric, amperometric, or colorimetric end-point or null-point determination.

automatic toll ticketing [COMMUN] System whereby toll calls are automatically recorded and timed, and toll tickets printed, under control of the calling telephone's dial pulses and without the intervention of an operator. Also known as automatic message accounting.

automatic tracking [NAV] 1. Tracking in which a servomechanism automatically follows some characteristic of the signal; specifically, a process by which tracking or data-acquisition systems are enabled to keep their antennas continuously directed at a moving target without manual operation. 2. An instrument which displays the actual course made good through the use of navigation derived from several sources.

automatic tuning system [CONT SYS] An electrical, mechanical, or electromechanical system that tunes a radio receiver or transmitter automatically to a predetermined frequency when a button or lever is pressed, a knob turned, or a telephone-type dial operated.

automatic typewriter [ENG] An electric typewriter that produces a punched paper tape or magnetic recording simultaneously with the conventional typed copy, for subsequent automatic retyping at high speed.

automatic video noise leveling [ELECTR] Constant false-alarm rate scheme in which the video noise level at the output of the receiver is sampled at the end of each range sweep and the receiver gain is readjusted accordingly to maintain a constant video noise level at the output.

automatic voltage regulator See voltage regulator.

automatic volume compressor See volume compressor.

automatic volume control [ELECTR] An automatic gain control that keeps the output volume of a radio receiver essentially constant despite variations in input-signal strength during fading or when tuning from station to station. Abbreviated AVC.

automatic volume expander See volume expander.

automatic weapon [ORD] A gun that fires, extracts, ejects, and reloads without application of power from an outside source, repeating the cycle as long as the firing mechanism is held in the proper position. Also known as automatic; automatic gun.

automation [ENG] 1. The replacement of human or animal labor by machines. 2. The operation of a machine or device automatically or by remote control.

automation source data [COMPUT SCI] The many methods of recording information in coded forms on paper tapes, punched cards, or tags that can be used over and over again to produce many other records without rewriting. Also known as source data automation (SDA).

automaton [COMPUT SCI] A robot which functions without step-by-step guidance by a human operator.

autometamorphism [PETR] Metamorphism of an igneous rock by the action of its own volatile fluids. Also known as autometasomatism.

autometasomatism See autometamorphism.

automobile [MECH ENG] A four-wheeled, trackless, self-propelled vehicle for land transportation of as many as eight people. Also known as car.

automorphic [PETR] Of minerals in igneous rock bounded by their own crystal faces. Also known as euhedral; idiomorphic.

automorphism [MATH] An isomorphism of an algebraic structure with itself.

autonavigator [NAV] A navigation system that automatically measures absolute vehicle motions and computes distance and direction from the departure point.

autonomic nervous system [ANAT] The visceral or involuntary division of the nervous system in vertebrates, which enervates glands, viscera, and smooth, cardiac, and some striated muscles.

autooxidation [CHEM] **1.** Oxidation caused by the atmosphere. **2.** An oxidation reaction that is self-catalyzed and spontaneous. **3.** An oxidation reaction begun only by an inductor.

autophagic vacuole [CYTOL] A membrane-bound cellular organelle that engulfs pieces of the substance of the cell itself. Also known as autolysosome.

autophagy [CYTOL] The cellular process of self-digestion.

autopilot See automatic pilot.

autoplotter [COMPUT SCI] A machine which automatically draws a graph from input data.

autopolarity [ELECTR] Automatic interchanging of connections to a digital meter when polarity is wrong; a minus sign appears ahead of the value on the digital display if the reading is negative.

autopolyploid [GEN] A cell or organism having three or more sets of chromosomes all derived from the same species.

autopositive [GRAPHICS] A film or paper which produces a positive image when exposed to a positive (or a negative image from a negative) and which is processed in a single development stage.

autoprotolysis [CHEM] Transfer of a proton from one molecule to another of the same substance.

autopsy [PATH] A postmortem examination of the body to determine cause of death.

autoradiography [ENG] A technique for detecting radioactivity in a specimen by producing an image on a photographic film or plate. Also known as radioautography.

autoranging [ENG] Automatic switching of a multirange meter from its lowest to the next higher range, with the switching process repeated until a range is reached for which the full-scale value is not exceeded. Also known as automatic ranging.

autorotation [MECH] **1.** Rotation about any axis of a body that is symmetrical and exposed to a uniform airstream and maintained only by aerodynamic moments. **2.** Rotation of a stalled symmetrical airfoil parallel to the direction of the wind.

autosexing [BIOL] Displaying differential sex characters at birth, noted particularly in fowl bred for sex-specific colors and patterns.

autosome [GEN] Any chromosome other than a sex chromosome.

autostability [CONT SYS] The ability of a device (such as a servomechanism) to hold a steady position, either by virtue of its shape and proportions, or by control by a servomechanism.

autostarter [ELEC] **1.** An automatic starting and switchover generating system consisting of a standby generator coupled to the station load through an automatic power transfer control unit. **2.** See autotransformer starter.

autosyndesis [CYTOL] The act of pairing of homologous chromosomes from the same parent during meiosis in polyploids.

autotomy [MED] Surgical removal of a part of one's own body. [ZOO] The process of self-amputation of appendages in crabs and other crustaceans and tails in some salamanders and lizards under stress.

autotransformer [ELEC] A power transformer having one continuous winding that is tapped; part of the winding serves as the primary and all of it serves as the secondary, or vice versa; small autotransformers are used to start motors.

autotransformer starter [ELEC] Motor starter having an autotransformer to furnish a reduced voltage for starting; includes the necessary switching mechanism. Also known as autostarter.

autotroph [BIOL] An organism capable of synthesizing organic nutrients directly from simple inorganic substances, such as carbon dioxide and inorganic nitrogen.

autumn [ASTRON] The season of the year which is the transition period from summer to winter, occurring as the sun approaches the winter solstice; beginning is marked by the autumnal equinox. Also known as fall.

autumnal equinox [ASTRON] The point on the celestial sphere at which the sun's rays at noon are 90° above the horizon at the Equator, or at an angle of 90° with the earth's axis, and neither North nor South Pole is inclined to the sun; occurs in the Northern Hemisphere on approximately September 23 and marks the beginning of autumn. Also known as first point of Libra.

autunite [MINERAL] $Ca(UO_2)_2(PO_4)_2 \cdot 10H_2O$ A common fluorescent mineral that occurs as yellow tetragonal plates in uranium deposits; minor ore of uranium.

Auversian See Ledian.

auxesis [PHYSIO] Growth resulting from increase in cell size.

auxiliary fault [GEOL] A branch fault; a minor fault ending against a major one.

auxiliary instruction buffer [COMPUT SCI] A section of storage in the instruction unit, 16 bytes in length, used to hold prefetched instructions.

auxiliary landing gear [AERO ENG] The part or parts of a landing gear, such as an outboard wheel, which is intended to stabilize the craft on the surface but which bears no significant part of the weight.

auxiliary mineral [MINERAL] A light-colored, relatively rare or unimportant mineral in an igneous rock; examples are apatite, muscovite, corundrum, fluorite, and topaz.

auxiliary plane [GEOL] A plane at right angles to the net slip on a fault plane as determined from analysis of seismic data for an earthquake.

auxiliary power plant [MECH ENG] Ancillary equipment, such as pumps, fans, and soot blowers, used with the main boiler, turbine, engine, waterwheel, or generator of a power-generating station.

auxiliary processor [COMPUT SCI] Any equipment which performs an auxiliary operation in a computer.

auxiliary routine [COMPUT SCI] A routine designed to assist in the operation of the computer and in debugging other routines.

auxiliary storage [COMPUT SCI] Storage device in addition to the main storage of a computer; for example, magnetic tape, disk, or magnetic drum.

auxiliary switch [ELEC] A switch actuated by the main device (such as a circuit breaker) for signaling, interlocking, or other purposes.

auxin [BIOCHEM] Any organic compound which promotes plant growth along the longitudinal axis when applied to shoots free from indigenous growth-promoting substances.

auxochrome [CHEM] Any substituent group such as —NH₂ and —OH which, by affecting the spectral regions of strong absorption in chromophores, enhance the ability of the chromogen to act as a dye.

av See arithmetic mean.

aV See abvolt.

availability [COMPUT SCI] Of data, data channels, and input-output devices in computers, the condition of being ready for use and not immediately committed to other tasks. [PHYS] The difference between the enthalpy per unit mass of substance and the product of entropy per unit mass multiplied by the lowest temperature available to the substance for heat discard; used in determining the ratio of actual work performed during a process by a working substance to that which theoretically should have been performed.

available draft [MECH ENG] The usable differential pressure in the combustion air in a furnace, used to sustain combustion of fuel or to transport products of combustion.

available energy [MECH ENG] Energy which can in principle be converted to mechanical work.

available heat [MECH ENG] The heat per unit mass of a working substance that could be transformed into work in an engine under ideal conditions for a given amount of heat per unit mass furnished to the working substance.

available line [ELECTR] Portion of the length of the scanning line which can be used specifically for picture signals in a facsimile system.

available machine time [COMPUT SCI] Time during which a computer has the power turned on, is not under maintenance, and is known or believed to be operating correctly.

available power gain [ELECTR] Ratio, in an electronic transducer, of the available power from the output terminals of the transducer, under specified input termination conditions, to the available power from the driving generator.

avalanche [ELECTR] 1. The cumulative process in which an electron or other charged particle accelerated by a strong electric field collides with and ionizes gas molecules, thereby releasing new electrons which in turn have more collisions, so that the discharge is thus self-maintained. Also known as avalanche effect; cascade; cumulative ionization; Townsend avalanche; Townsend ionization. 2. Cumulative multiplication of carriers in a semiconductor as a result of avalanche breakdown. Also known as avalanche effect. [HYD] A mass of snow or ice moving rapidly down a mountain slope or cliff.

avalanche breakdown [ELECTR] Nondestructive breakdown in a semiconductor diode when the electric field across the barrier region is strong enough so that current carriers collide with valence electrons to produce ionization and cumulative multiplication of carriers.

avalanche conduction [PHYSIO] Conduction of a nerve impulse through several neurons which converge, increasing the discharge intensity by summation.

avalanche diode [ELECTR] A semiconductor breakdown diode, usually made of silicon, in which avalanche breakdown occurs across the entire *pn* junction and voltage drop is then essentially constant and independent of current; the two most important types are IMPATT and TRAPATT diodes.

avalanche effect *See* avalanche.

avalanche oscillator [ELECTR] An oscillator that uses an avalanche diode as a negative resistance to achieve one-step conversion from direct-current to microwave outputs in the gigahertz range.

avalanche photodiode [ELECTR] A photodiode operated in the avalanche breakdown region to achieve internal photocurrent multiplication, thereby providing rapid light-controlled switching operation.

avalanche transistor [ELECTR] A transistor that utilizes avalanche breakdown to produce chain generation of charge-carrying hole-electron pairs.

AVC *See* automatic volume control.

aven [GEOL] *See* pothole. [MIN ENG] A vertical shaft leading upward from a cave passage, sometimes connecting with passages above.

aventurine [MINERAL] 1. A glass or mineral containing sparkling gold-colored particles, usually copper or chromic oxide. 2. A shiny red or green translucent quartz having small, but microscopically visible, exsolved hematite or included mica particles.

average *See* arithmetic mean.

average assay value [MIN ENG] The weighted result of assays obtained from a number of samples by multiplying the assay value of each sample by the width or thickness of the ore face over which it is taken and dividing the sum of these products by the total width of cross section sampled.

average-calculating operation [COMPUT SCI] A common or typical calculating operation longer than an addition and shorter than a multiplication; often taken as the mean of nine additions and one multiplication.

average deviation [MATH] In statistics, the average or arithmetic mean of the deviation, taken without regard to sign, from some fixed value, usually the arithmetic mean of the data. Abbreviated AD. Also known as mean deviation.

average-edge line [COMPUT SCI] The imaginary line which traces or smooths the shape of any written or printed character to be recognized by a computer through optical, magnetic, or other means.

average effectiveness level *See* effectiveness level.

average heading [NAV] The average heading flown for a given period; it should be the same value as desired heading if the drift was predicted accurately.

average molecular weight [ORG CHEM] The calculated number to average the molecular weights of the varying-length polymer chains present in a polymer mixture.

average outgoing quality limit [IND ENG] The average quality of all lots that pass quality inspection, expressed in terms of percent defective. Abbreviated AOQL.

average power output [ELECTR] Radio-frequency power, in an audio-modulation transmitter, delivered to the transmitter output terminals, averaged over a modulation cycle.

average wind [NAV] In air navigation, the resultant wind which would produce, or has produced, the same wind effect during a given period as the summation of the actual winds which will affect, or have affected, the flight of an aircraft.

Aves [VERT ZOO] A class of animals composed of the birds, which are warm-blooded, egg-laying vertebrates primarily adapted for flying.

avian leukosis [VET MED] A disease complex in fowl probably caused by viruses and characterized by autonomous proliferation of blood-forming cells.

aviation medicine *See* aerospace medicine.

aviation weather forecast [METEOROL] A forecast of weather elements of particular interest to aviation, such as the ceiling, visibility, upper winds, icing, turbulence, and types of precipitation or storms. Also known as airways forecast.

avigation *See* air navigation.

avionics [ENG] The design and production of airborne electrical and electronic devices; term is derived from aviation electronics.

avocado [BOT] *Persea americana*. A subtropical evergreen tree of the order Magnoliales that bears a pulpy pear-shaped edible fruit.

Avogadro's hypothesis *See* Avogadro's law.

Avogadro's law [PHYS] The law which states that under the same conditions of pressure and temperature, equal volumes of all gases contain equal numbers of molecules; for example, 359 cubic feet at

32°F and 1 atmosphere for a perfect gas. Also known as Avogadro's hypothesis.

Avogadro's number [PHYS] The number (6.02×10^{23}) of molecules in a gram-molecular weight of a substance.

avoidable delay [IND ENG] An interruption under the control of the operator during the normal operating time.

avoirdupois pound *See* pound.

avoirdupois weight [MECH] The system of units which has been commonly used in English-speaking countries for measurement of the mass of any substance except precious stones, precious metals, and drugs; it is based on the pound (approximately 453.6 grams) and includes the short ton (2000 pounds), long ton (2240 pounds), ounce (one-sixteenth pound), and dram (one-sixteenth ounce).

Avonian *See* Dinantian.

avulsion [HYD] A sudden change in the course of a stream by which a portion of land is cut off, as where a stream cuts across and forms an oxbow. [MED] Tearing one part away from the other, either by trauma or surgery.

AWG *See* American wire gage.

awl [DES ENG] A point tool with a short wooden handle used to mark surfaces and to make small holes, as in leather or wood.

awn [BOT] Any of the bristles at the ends of glumes or bracts on the spikelets of oats, barley, and some wheat and grasses. Also known as beard.

awning window [BUILD] A window consisting of a series of vertically arranged, top-hinged rectangular sections; designed to admit air while excluding rain.

axenic culture [BIOL] The growth of organisms of a single species in the absence of cells or living organisms of any other species.

axes of inertia [PHYS] The three principal axes of inertia, namely, one about which the moment of inertia is a maximum, one about which the moment of inertia is a minimum, and one perpendicular to both.

axial [SCI TECH] Of, pertaining to, or along an axis.

axial angle [CRYSTAL] 1. The acute angle between the two optic axes of a biaxial crystal. Also known as optic angle; optic-axial angle. 2. In air, the larger angle between the optic axes after refraction on leaving the crystal.

axial element [CRYSTAL] The lengths, length ratios, and angles which define a crystal's unit cell.

axial fan [MECH ENG] A fan whose housing confines the gas flow to the direction along the rotating shaft at both the inlet and outlet.

axial filament [CYTOL] The central microtubule elements of a cilium or flagellum. [INV ZOO] An organic fiber which serves as the core for deposition of mineral substance to form a ray of a sponge spicule.

axial flow [FL MECH] Flow of fluid through an axially symmetric device such that the direction of the flow is along the axis of symmetry. Also known as axisymmetric flow.

axial gradient [EMBRYO] In some invertebrates, a graded difference in metobolic activity along the anterior-posterior, dorsal-ventral, and medial-lateral embryonic axes.

axial load [MECH] A force with its resultant passing through the centroid of a particular section and being perpendicular to the plane of the section.

axial musculature [ANAT] The muscles that lie along the longitudinal axis of the vertebrate body.

axial plane [CRYSTAL] 1. A plane that includes two of the crystallographic axes. 2. The plane of the op-

tic axis of an optically biaxial crystal. [GEOL] A plane that intersects the crest or trough in such a manner that the limbs or sides of the fold are more or less symmetrically arranged with reference to it. Also known as axial surface.

axial ratio [CRYSTAL] The ratio obtained by comparing the length of a crystallographic axis with one of the lateral axes taken as unity. [ELECTR] The ratio of the major axis to the minor axis of the polarization ellipse of a waveguide. Also known as ellipticity.

axial skeleton [ANAT] The bones composing the skull, vertebral column, and associated structures of the vertebrate body.

axial stream [HYD] 1. The chief stream of an intermontane valley, the course of which is along the deepest part of the valley and is parallel to its longer dimension. 2. A stream whose course is along the axis of an anticlinal or a synclinal fold.

axial symmetry [MATH] Property of a geometric configuration which is unchanged when rotated about a given line.

axial trace [GEOL] The intersection of the axial plane of a fold with the surface of the earth or any other specified surface; sometimes such a line is loosely and incorrectly called the axis.

axial vector *See* pseudovector.

axial winding [MATER] A winding used in filament-wound fiberglass-reinforced plastic construction in which the filaments run along the axis at a zero helix angle.

Axiidae [INV ZOO] A family of decapod crustaceans, including the hermit crabs, in the suborder Reptantia.

axil [BIOL] The angle between a structure and the axis from which it arises, especially for branches and leaves.

axilla [ANAT] The depression between the arm and the thoracic wall; the armpit. [BOT] An axil.

axillary [ANAT] Of, pertaining to, or near the axilla or armpit. [BOT] Placed or growing in the axis of a branch or leaf.

Axinellina [INV ZOO] A suborder of sponges in the order Clavaxinellida.

axinite [MINERAL] $H_2(Ca,Fe,Mn)_4(BO)Al_2(SiO_4)_5$ Brown, blue, green, gray, or purplish gem mineral that commonly forms glassy triclinic crystals. Also known as glass schorl.

axiom [MATH] Any of the assumptions upon which a mathematical theory (such as geometry, ring theory, and the real numbers) is based. Also known as postulate.

axis [ANAT] 1. The second cervical vertebra in higher vertebrates; the first vertebra of amphibians. 2. The center line of an organism, organ, or other body part. [GEOL] 1. A line where a folded bed has maximum curvature. 2. The central portion of a mountain chain. [GRAPHICS] The locus of intersection of two pencils of lines in perspective position. [MATH] 1. In a coordinate system, the line determining one of the coordinates, obtained by setting all other coordinates to zero. 2. A line of symmetry for a geometric figure. [MECH] A line about which a body rotates.

axis cylinder [CYTOL] 1. The central mass of a nerve fiber. 2. The core of protoplasm in a medullated nerve fiber.

axis of abscissas [MATH] The horizontal or x axis of a two-dimensional cartesian coordinate system, parallel to which abscissas are measured.

axis of freedom [DES ENG] An axis in a gyro about which a gimbal provides a degree of freedom.

axis of ordinates [MATH] The vertical or y axis of a two-dimensional cartesian coordinate system, parallel to which ordinates are measured.

axis of rotation [MECH] A straight line passing through the points of a rotating rigid body that remain stationary, while the other points of the body move in circles about the axis.

axis of symmetry [MECH] An imaginary line about which a geometrical figure is symmetric. Also known as symmetry axis.

axis of thrust See thrust axis.

axisymmetric flow See axial flow.

axle [MECH ENG] A supporting member that carries a wheel and either rotates with the wheel to transmit mechanical power to or from it, or allows the wheel to rotate freely on it.

axometer [ENG] An instrument that locates the optical axis of a lens, particularly a lens used in eyeglasses.

axon [ANAT] The process or nerve fiber of a neuron that carries the unidirectional nerve impulse away from the cell body. Also known as neuraxon; neurite.

axoneme [CYTOL] A bundle of fibrils enclosed by a membrane that is continuous with the plasma membrane.

axonometric projection [GRAPHICS] A drawing that shows an object's inclined position with respect to the planes of projection. Also known as isometric projection.

axopodium [INV ZOO] A semipermanent pseudopodium composed of axial filaments surrounded by a cytoplasmic envelope.

aye-aye [VERT ZOO] Daubentonia madagascariensis. A rare prosimian primate indigenous to eastern Madagascar; the single species of the family Daubentoniidae.

Ayrton-Jones balance [ELEC] A type of balance with which force between current-carrying conductors is measured; uses single-layer solenoids as the fixed and movable coils.

Ayrton-Perry winding [ELEC] Winding of two wires in parallel but opposite directions to give better cancellation of magnetic fields than is obtained with a single winding.

Ayrton shunt [ELEC] A shunt used to increase the range of a galvanometer without changing the damping. Also known as universal shunt.

azeotrope See azeotropic mixture.

azeotropic mixture [CHEM] A solution of two or more liquids, the composition of which does not change upon distillation. Also known as azeotrope.

azide [ORG CHEM] One of several types of compounds containing the N_3 group and derived from hydrazoic acid, HN_3.

azimuth [ASTRON] Horizontal direction of a celestial point from a terrestrial point, expressed as the angular distance from a reference direction, usually measured from 0° at the reference direction clockwise through 360°. [GEOD] Horizontal direction on the earth's surface.

azimuthal map projection [MAP] The transformation of a spherical representation of the earth's surface to a tangent or intersecting plane established perpendicular to a right line passing through the center of the spherical representation.

azimuthal orthomorphic projection See stereographic projection.

azimuth bar See azimuth instrument.

azimuth circle [DES ENG] A ring calibrated from 0 to 360 degrees over a compass, compass repeater, radar plan position indicator, direction finder, and so on, which provides means for observing compass bearings and azimuths.

azimuth compass [NAV] A compass with vertical sights so that the magnetic azimuths of celestial bodies may be read.

azimuth error [ASTRON] The angle by which the east-west axis of a transit telescope deviates from being perpendicular to the plane of the meridian. [ENG] An error in the indicated azimuth of a target detected by radar.

azimuth instrument [ENG] An instrument for measuring azimuths, particularly a device which fits over a central pivot in the glass cover of a magnetic compass. Also known as azimuth bar; bearing bar.

azimuth line [ENG] A radial line from the principal point, isocenter, or nadir point of a photograph, representing the direction to a similar point of an adjacent photograph in the same flight line; used extensively in radial triangulation.

azimuth marker [ENG] 1. A scale encircling the plan position indicator scope of a radar on which the azimuth of a target from the radar may be measured. 2. Any of the reference limits inserted electronically at 10 or 15° intervals which extend radially from the relative position of the radar on an off-center plan position indicator scope. [NAV] See electronic azimuth marker.

azimuth resolution [ELECTROMAG] Angle or distance by which two targets must be separated in azimuth to be distinguished by a radar set, when the targets are at the same range.

azimuth-stabilized plan position indicator [ENG] A north-upward plan position indicator (PPI), a radarscope, which is stabilized by a gyrocompass so that either true or magnetic north is always at the top of the scope regardless of vehicle orientation.

azimuth traverse [ENG] A survey traverse in which the direction of the measured course is determined by azimuth and verified by back azimuth.

azoic dye [ORG CHEM] A water-insoluble azo dye that is formed by coupling of the components on a fiber. Also known as ice color; ingrain color.

azoic printing [GRAPHICS] The method whereby azoic compositions, that is, mixtures of napthols and diazotized products, temporarily inhibited from color development are printed on cloth; when the printed material is passed through steam containing formic acid vapor, the coupling reaction occurs and development takes place.

azole [ORG CHEM] One of a class of organic compounds with a five-membered N-heterocycle containing two double bonds; an example is 1,2,4-triazole.

azonal soil [GEOL] Any group of soils without well-developed profile characteristics, owing to their youth, conditions of parent material, or relief that prevents development of normal soil-profile characteristics.

Azotobacteraceae [MICROBIO] A family of large, bluntly rod-shaped, gram-negative, aerobic bacteria capable of fixing molecular nitrogen.

azurite [MINERAL] $Cu_3(CO_3)_2(OH)_2$ A blue monoclinic mineral consisting of a basic carbonate of copper; an ore of copper. Also known as blue copper ore; blue malachite; chessylite.

azygos vein [ANAT] A branch of the right precava which drains the intercostal muscles and empties into the superior vena cava.

azygote [BIOL] An individual produced by haploid parthenogenesis.

B

B *See* bel; boron; brewster.

Ba *See* barium.

babbitt metal [MET] Any of the white alloys composed primarily of tin or lead and of lesser amounts of antimony, copper, and perhaps other metals, and used for bearings.

babble [COMMUN] 1. Aggregate crosstalk from a large number of channels. 2. Unwanted disturbing sounds in a carrier or other multiple-channel system which result from the aggregate crosstalk or mutual interference from other channels.

Babesiidae [INV ZOO] A family of protozoans in the suborder Haemosporina containing parasites of vertebrate red blood cells.

Babinet compensator [OPTICS] A device for working with polarized light, made of two quartz prisms, assembled in a rhomb, to enable the optical retardation to be adjusted to positive or negative values.

Babinet's principle [OPTICS] The principle that the diffraction patterns produced by complementary screens are identical; two screens are said to be complementary when the opaque parts of one correspond to the transparent parts of the other.

Babinski reflex [MED] An abnormal reflex after infancy associated with a disturbance of the pyramidal tract, characterized by extension of the great toe with fanning of the other toes on sharply stroking the lateral aspect of the sole.

baboon [VERT ZOO] Any of five species of large African and Asian terrestrial primates of the genus *Papio*, distinguished by a doglike muzzle, a short tail, and naked callosities on the buttocks.

babs *See* blind approach beacon system.

bacciferous [BOT] Bearing berries.

Bacillaceae [MICROBIO] The single family of endospore-forming rods and cocci.

Bacillariophyceae [BOT] The diatoms, a class of algae in the division Chrysophyta.

Bacillariophyta [BOT] An equivalent name for the Bacillariophyceae.

bacillary [MICROBIO] 1. Rod-shaped. 2. Produced by, pertaining to, or resembling bacilli.

Bacillus Calmette-Guerin vaccine [IMMUNOL] A vaccine prepared from attenuated human tubercle bacilli and used to immunize humans against tuberculosis. Abbreviated BCG vaccine.

back [ANAT] The part of the human body extending from the neck to the base of the spine. [GRAPHICS] The part of a book where the binding and pages are stitched together. [MIN ENG] 1. The upper part of any mining cavity. 2. A joint, usually a strike joint, perpendicular to the direction of working.

backacter *See* backhoe.

back beach *See* backshore.

back bearing [NAV] A bearing along the reverse direction of a line. Also known as reciprocal bearing.

back bias [ELECTR] 1. Degenerative or regenerative voltage which is fed back to circuits before its originating point; usually applied to a control anode of a tube or other device. 2. Voltage applied to a grid of a tube (or tubes) or electrode of another device to reduce a condition which has been upset by some external cause.

back bond [SOLID STATE] A chemical bond between an atom in the surface layer of a solid and an atom in the second layer.

backbone [ANAT] *See* spine. [GEOL] 1. A ridge forming the principal axis of a mountain. 2. The principal mountain ridge, range, or system of a region.

backcast stripping [MIN ENG] A stripping method using two draglines; one strips and casts the overburden, and the other recasts a portion of the overburden.

backcross [GEN] A cross between an F_1 heterozygote and an individual of P_1 genotype.

backdigger *See* backhoe.

back draft [MET] A reversed taper given to a casting model or pattern to prevent its withdrawal from the mold.

back echo [ELECTROMAG] An echo signal produced on a radar screen by one of the minor back lobes of a search radar beam.

back electromotive force *See* counterelectromotive force.

back-emission electron radiography [ELECTR] A technique used in microradiography to visualize, among other things, the presence of material of different atomic numbers in the surface of the specimen being observed; the polished side of the specimen is facing and in close contact with the emulsion side of a fine-grain photographic plate; a light-tight cover holds the specimen and plate in place to be subjected to hardened x-rays.

backfill [CIV ENG] Earth refilling a trench or an excavation around a building, bridge abutment, and the like. [MIN ENG] Waste sand or rock used to support the mine roof after removal of ore.

backfire [ELECTR] *See* arcback. [ENG] Momentary backward burning of flame into the tip of a torch. Also known as flashback. [ORD] Rearward escapement of gases or cartridge fragments upon firing a gun.

backfire antenna [ELECTROMAG] An antenna which exhibits significant gain in a direction 180° from its principal lobe.

backflash [CHEM] Rapid combustion of a material occurring in an area that the reaction was not intended for.

backfolding [GEOL] Process in mountain forming in which the folds are overturned toward the interior of an orogenic belt. Also known as backward folding.

back furrow See esker.

back gearing [MECH ENG] The technique of using gears on machine tools to obtain an increase in the number of speed changes that can be gotten with cone belt drives.

background [COMMUN] 1. Picture white of the facsimile copy being scanned when the picture is black and white only. 2. Undesired printing in the recorded facsimile copy of the picture being transmitted, resulting in shading of the background area. 3. Noise heard during radio reception caused by atmospheric interference or the operation of the receiver at such high gain that inherent tube and circuit noises become noticeable.

background count [PHYS] Responses of the radiation counting system to radiation coming from sources other than the source to be measured.

background noise [ACOUS] The unwanted residual sound that is present whether or not the sound source being studied is in operation. [ENG] The undesired signals that are always present in an electronic or other system, independent of whether or not the desired signal is present.

background radiation [NUCLEO] The radiation in humans' natural environment, including cosmic rays and radiation from the naturally radioactive elements. Also known as natural radiation. [PHYS] Radiation which is due to sources other than the source of interest in a measurement of radiation and which is detected by the measuring apparatus.

background reflectance [COMPUT SCI] The reflectance, relative to a standard, of the surface on which a printed or handwritten character has been inscribed in optical character recognition.

background returns [ENG] 1. Signals on a radar screen from objects which are of no interest. 2. See clutter.

backhand welding [MET] A welding technique in which the flame is directed back against the completed weld. Also known as backward welding.

backhoe [MECH ENG] An excavator fitted with a hinged arm to which is rigidly attached a bucket that is drawn toward the machine in operation. Also known as backacter; backdigger; dragshovel; pullshovel.

backing [ELECTR] Flexible material, usually cellulose acetate or polyester, used on magnetic tape as the carrier for the oxide coating. [MET] See backing strip. [METEOROL] 1. Internationally, a change in wind direction in a counterclockwise sense (for example, south to east) in either hemisphere of the earth. 2. In United States usage, a change in wind direction in a counterclockwise sense in the Northern Hemisphere, clockwise in the Southern Hemisphere. [MIN ENG] 1. Timbers across the top of a level, supported in notches cut in the rock. 2. Rough masonry of a wall faced with finer work. 3. Earth placed behind a retaining wall.

backing storage [COMPUT SCI] A computer storage device whose capacity is larger, but whose access time is slower, than that of the computer's main storage or immediate access storage; usually slower than main storage. Also known as bulk storage.

backing strip [MET] A piece of metal, asbestos, or other nonflammable material placed behind a joint to facilitate welding. Also known as backing.

backlands [GEOL] A section of a river floodplain lying behind a natural levee.

backlash [DES ENG] The amount by which the tooth space of a gear exceeds the tooth thickness of the mating gear along the pitch circles. [ELECTR] A small reverse current in a rectifier tube caused by the motion of positive ions produced in the gas by the impact of thermoelectrons. [ENG] 1. Relative motion of mechanical parts cused by looseness. 2. The difference between the actual values of a quantity when a dial controlling this quantity is brought to a given position by a clockwise rotation and when it is brought to the same position by a counterclockwise rotation.

backlight [GRAPHICS] A spotlight that illuminates from behind so that the subject is separated from the background; used in photography.

back lobe [ELECTROMAG] The three-dimensional portion of the radiation pattern of a directional antenna that is directed away from the intended direction.

backlog [IND ENG] 1. An accumulation of orders promising future work and profit. 2. An accumulation of unprocessed materials or unperformed tasks.

back nut [DES ENG] 1. A threaded nut, one side of which is dished to retain a grommet; used in forming a watertight pipe joint. 2. A locking nut on the shank of a pipe fitting, tap, or valve.

back off [ENG] 1. To unscrew or disconnect. 2. To withdraw the drill bit from a borehole. 3. To withdraw a cutting tool or grinding wheel from contact with the workpiece.

back order [IND ENG] 1. An order held for future completion. 2. A new order placed for previously unavailable materials of an old order.

backplane [COMPUT SCI] A wiring board, usually constructed as a printed circuit, used in microcomputers and minicomputers to provide the required connections between logic, memory, and input/output modules.

backplate [BUILD] A plate, usually metal or wood, which serves as a backing for a structural member.

back porch [ELECTR] The period of time in a television circuit immediately following a synchronizing pulse during which the signal is held at the instantaneous amplitude corresponding to a black area in the received picture.

back pressure [MECH] Pressure due to a force that is operating in a direction opposite to that being considered, such as that of a fluid flow. [MECH ENG] Resistance transferred from rock into the drill stem when the bit is being fed at a faster rate than the bit can cut.

back radiation See backscattering; counterradiation.

back rush [OCEANOGR] Return of water seaward after the uprush of the waves.

backscattering Also known as back radiation; backward scattering. [COMMUN] Propagation of extraneous signals by F- or E-region reflection in addition to the desired ionospheric scatter mode; the undesired signal enters the antenna through the back lobes. [ELECTROMAG] 1. Radar echoes from a target. 2. Undesired radiation of energy to the rear by a directional antenna. [PHYS] The deflection of radiation or nuclear particles by scattering processes through angles greater than 90° with respect to the original direction of travel.

back-set bed [GEOL] Cross bedding that dips in a direction against the flow of a depositing current.

backshore [GEOL] The upper shore zone that is beyond the advance of the usual waves and tides. Also known as back beach; backshore beach.

backshore beach See backshore.

backshore terrace See berm.

backsight [ENG] **1.** A sight on a previously established survey point or line. **2.** Reading a leveling rod in its unchanged position after moving the leveling instrument to a different location. [NAV] A marine sextant observation of a celestial body made by facing 180° from the azimuth of the body and using the visible horizon in the direction in which the observer is facing.

back slope *See* dip slope.

backspace [COMPUT SCI] To move a recording medium one unit in the reverse or background direction. [MECH ENG] To move a typewriter carriage back one space by depressing a back space key.

backstay [ENG] **1.** A supporting cable that prevents a more or less vertical object from falling forward. **2.** A spring used to keep together the cutting edges of purchase shears. **3.** A rod that runs from either end of a carriage's rear axle to the reach. **4.** A leather strip that covers and strengthens a shoe's back seam. [GRAPHICS] A rope or strap that keeps the carriage of a hand printing press from moving too far forward. [NAV ARCH] A rope, wire, or cable that runs from the top of a mast to the side of a ship and slants a little aft. [TEXT] A bar having a glass rod on its top, that runs across a loom beneath the lowest motion of the warp yarns.

back stoping *See* shrinkage stoping.

backtalk [COMPUT SCI] Passage of information from a standby computer to the active computer.

backup [BUILD] That part of a masonry wall behind the exterior facing. [CIV ENG] Overflow in a drain or piping system, due to stoppage. [ENG] **1.** An item under development intended to perform the same general functions that another item also under development performs. **2.** A compressible material used behind a sealant to reduce its depth and to support the sealant against sag or indentation. [GRAPHICS] **1.** An image printed on the reverse side of a printed sheet. **2.** The printing of such an image. [MET] A support used to balance the upsetting force in the workpieces during flash welding. [PETRO ENG] During drilling, the holding of one section of pipe while another is screwed out of it or into it.

backup arrangement *See* cascade.

backward diode [ELECTR] A semiconductor diode similar to a tunnel diode except that it has no forward tunnel current; used as a low-voltage rectifier.

backward error analysis [COMPUT SCI] A form of error analysis which seeks to replace all errors made in the course of solving a problem by an equivalent perturbation of the original problem.

backward folding *See* backfolding.

backward scattering *See* backscattering.

backward wave [ELECTROMAG] An electromagnetic wave traveling opposite to the direction of motion of some other physical quantity in an electronic device such as a traveling-wave tube or mismatched transmission line.

backward-wave oscillator [ELECTR] An electronic device which amplifies microwave signals simultaneously over a wide band of frequencies and in which the traveling wave produced is reflected backward so as to sustain the wave oscillations. Abbreviated BWO. Also known as carcinotron.

backward-wave tube [ELECTR] A type of microwave traveling-wave electron tube in which electromagnetic energy on a slow-wave circuit flows opposite in direction to the travel of electrons in a beam.

backward welding *See* backhand welding.

backwash [CHEM ENG] **1.** In an ion-exchange resin system, an upward flow of water through a resin bed that cleans and reclassifies the resin particles after exhaustion. **2.** *See* blowback. [OCEANOGR] **1.** Water or waves thrown back by an obstruction such as a ship or breakwater. **2.** The seaward return of water after a rush of waves onto the beach foreshore.

backwater [HYD] **1.** A series of connected lagoons, or a creek parallel to a coast, narrowly separated from the sea and connected to it by barred outlets. **2.** Accumulation of water resulting from and held back by an obstruction. **3.** Water reversed in its course by an obstruction.

back wave [COMMUN] Signal emitted from a radio telegraph transmitter during spacing portions of the code characters. Also known as spacing wave.

back work [MIN ENG] Any kind of operation in a mine not immediately concerned with production or transport; literally, work behind the face, such as repairs to roads.

bacteria [MICROBIO] Extremely small, relatively simple prokaryotic microorganisms traditionally classified with the fungi as Schizomycetes.

Bacteriaceae [MICROBIO] A former designation for Brevibacteriaceae.

bacterial methanogenesis *See* methanogenesis.

bacteriochlorophyll [BIOCHEM] $C_{52}H_{70}O_6N_4Mg$ A tetrahydroporphyrin chlorophyll compound occurring in the forms *a* and *b* in photosynthetic bacteria; there is no evidence that *b* has the empirical formula given.

bacteriology [MICROBIO] The science and study of bacteria; a specialized branch of microbiology.

bacteriophage [VIROL] Any of the viruses that infect bacterial cells; each has a narrow host range. Also known as phage.

bacteriotoxin [MICROBIO] **1.** Any toxin that destroys or inhibits growth of bacteria. **2.** A toxin produced by bacteria.

bacterioviridin *See* chlorobium chlorophyll.

Bacteroidaceae [MICROBIO] The single family of gram-negative anaerobic bacteria; cells are nonsporeforming rods; some species are pathogenic.

Bactrian camel [VERT ZOO] *Camelus bactrianus.* The two-humped camel.

baculum [VERT ZOO] The penis bone, or os priapi, in lower mammals.

baddeleyite [MINERAL] ZrO_2 A colorless, yellow, brown, or black monoclinic zirconium oxide mineral found in Brazil and Ceylon; used as heat- and corrosion-resistant linings for furnaces and muffles.

badge meter *See* film badge.

badger [DES ENG] *See* badger plane. [ENG] A tool used inside a pipe or culvert to remove any excess mortar or deposits. [VERT ZOO] Any of eight species of carnivorous mammals in six genera comprising the subfamily Melinae of the weasel family (Mustelidae).

badge reader [COMPUT SCI] A device that can read data appearing in the form of holes in plastic badges or prepunched cards.

badlands [GEOGR] An erosive physiographic feature in semiarid regions characterized by sharp-edged, sinuous ridges separated by steep-sided, narrow, winding gullies.

baffle [ELEC] Device for deflecting oil or gas in a circuit breaker. [ELECTR] An auxiliary member in a gas tube used, for example, to control the flow of mercury particles or deionize the mercury following conduction. [ENG] A plate that regulates the flow of a fluid, as in a steam-boiler flue or a gasoline muffler. [ENG ACOUS] A cabinet or partition used with a loudspeaker to reduce interaction between sound waves produced simultaneously by the two surfaces of the diaphragm.

baffle plate [ELECTROMAG] Metal plate inserted in a waveguide to reduce the cross-sectional area for wave conversion purposes.

bagasse [FOOD ENG] Remains of sugarcane after the juice has been extracted by pressure between the rolls of a mill; used to supply the fuel requirements of raw-sugar mills.

bagasse disease *See* bagassosis.

bagassosis [MED] A pneumoconiosis caused by the inhalation of bagasse dust, a dry sugarcane residue. Also known as bagasse disease.

bag filter [ENG] Filtering apparatus with porous cloth or felt bags through which dust-laden gases are sent, leaving the dust on the inner surfaces of the bags.

bag molding [ENG] A method of molding plastic or plywood-plastic combinations into curved shapes, in which fluid pressure acting through a flexible cover, or bag, presses the material to be molded against a rigid die.

Bagridae [VERT ZOO] A family of semitropical catfishes in the suborder Siluroidei.

bahada *See* bajada.

bail [ENG] A loop of heavy wire snap-fitted around two or more parts of a connector or other device to hold the parts together.

bailer [ENG] A long, cylindrical vessel fitted with a bail at the upper end and a flap or tongue valve at the lower extremity; used to remove water, sand, and mud- or cuttings-laden fluids from a borehole. Also known as bailing bucket.

Bailey meter [ENG] A flowmeter consisting of a helical quarter-turn vane which operates a counter to record the total weight of granular material flowing through vertical or near-vertical ducts, spouts, or pipes.

bailing [ENG] Removal of the cuttings from a well during cable-tool drilling, or of the liquid from a well, by means of a bailer.

bailing bucket *See* bailer.

Bairdiacea [INV ZOO] A superfamily of ostracod crustaceans in the suborder Podocopa.

bajada [GEOL] An alluvial plain formed as a result of lateral growth of adjacent alluvial fans until they finally coalesce to form a continuous inclined deposit along a mountain front. Also spelled bahada.

bakerite [MINERAL] $8CaO \cdot 5B_2O_3 \cdot 6SiO_2 \cdot 6H_2O$ White mineral, occurring in fine-grained, nodular masses, resembling marble and unglazed porcelain, and consisting of hydrous calcium borosilicate.

Baker-Nunn camera [OPTICS] A large camera with a Schmidt-type lens system used to track earth satellites.

Baker-Schmidt telescope [OPTICS] A type of Schmidt telescope in which the light reflected from the near-spheroidal primary mirror is again reflected from a smaller, near-spheroidal secondary mirror, producing an image that is free of astigmatism and distortion.

bakers' yeast [FOOD ENG] An industrial yeast used for baking purposes because of maximum growth and low alcohol production; composed of dry cells of one or more strains of *Saccharomyces cerevisiae*.

baking powder [FOOD ENG] A yeast substitute of sodium bicarbonate plus potassium tartrate, tartaric acid, anhydrous sodium aluminum sulfate, monocalcium phosphate, or any combination of these acids so formulated as to release carbon dioxide from the sodium bicarbonate (baking soda) when moistened.

Balaenicipitidae [VERT ZOO] A family of wading birds composed of a single species, the shoebill stork (*Balaeniceps rex*), in the order Ciconiiformes.

Balaenidae [VERT ZOO] The right whales, a family of cetacean mammals composed of five species in the suborder Mysticeti.

balance [ACOUS] The condition in a stereo system wherein both speakers produce the same average sound levels. [AERO ENG] **1.** The equilibrium attained by an aircraft, rocket, or the like when forces and moments are acting upon it so as to produce steady flight, especially without rotation about its axes. **2.** The equilibrium about any specified axis that counterbalances something, especially on an aircraft control surface, such as a weight installed forward of the hinge axis to counterbalance the surface aft of the hinge axis. [CHEM] To bring a chemical equation into balance so that reaction substances and reaction products obey the laws of conservation of mass and charge. [ELEC] The state of an electrical network when it is adjusted so that voltage in one branch induces or causes no current in another branch. [ENG] An instrument for measuring mass or weight. [MIN ENG] The counterpoise or weight attached by cable to the drum of a winding engine to balance the weight of the cage and hoisting cable and thus assist the engine in lifting the load out of the shaft.

Balance *See* Libra.

balance bar *See* balance beam

balance beam [CIV ENG] A long beam, attached to a gate (or drawbridge, and such) so as to counterbalance the weight of the gate during opening or closing. Also known as balance bar.

balance coil [ELEC] An iron-core solenoid with adjustable taps near the center; used to convert a two-wire circuit to a three-wire circuit, the taps furnishing a neutral terminal for the latter.

balance control [ELECTR] A control used in a stereo sound system to vary the volume of one loudspeaker system relative to the other while maintaining their combined volume essentially constant.

balanced amplifier [ELECTR] An electronic amplifier in which there are two identical signal branches connected so as to operate with the inputs in phase opposition and with the output connections in phase, each balanced to ground.

balanced bridge [ELEC] Wheatstone bridge circuit which, when in a quiescent state, has an output voltage of zero.

balanced circuit [ELEC] A circuit whose two sides are electrically alike and symmetrical with respect to a common reference point, usually ground.

balanced converter *See* balun.

balanced currents [ELEC] Currents flowing in the two conductors of a balanced line which, at every point along the line, are equal in magnitude and opposite in direction. Also known as push-pull currents.

balanced detector [ELECTR] A detector used in frequency-modulation receivers; in one form the audio output is the rectified difference between voltages produced across two resonant circuits, one being tuned slightly above the carrier frequency and one slightly below.

balanced fertilizer [MATER] A material of varying composition added to soil so as to provide essential mineral elements at required levels, improve soil structure, or enhance microbial activity.

balanced gasoline [MATER] An automotive gasoline blended from petroleum hydrocarbons of varying volatilities to provide desired performance for engine starting, warm-up, acceleration, and mileage.

balanced line [ELEC] A transmission line consisting of two conductors capable of being operated so that the voltages of the two conductors at any transverse plane are equal in magnitude and opposite in polarity with respect to ground. [IND ENG] A production line for which the time cycles of the operators are made approximately equal so that the work flows at a desired steady rate from one operator to the next.

balanced oscillator [ELECTR] Any oscillator in which, at the oscillator frequency, the impedance centers of the tank circuits are at ground potential, and the voltages between either end and their centers are equal in magnitude and opposite in phase.

balanced polymorphism [GEN] Maintenance in a population of two or more alleles in equilibrium at frequencies too high to be explained, particularly for the rarer of them, by mutation balanced by selection; for example, the selective advantage of heterozygotes over both homozygotes.

balanced range of error [STAT] A range of error in which the maximum and minimum possible errors are opposite in sign and equal in magnitude.

balanced translocation [GEN] Positional change of one or more chromosome segments in cells or gametes without alteration of the normal diploid or haploid complement of genetic material.

balanced valve [ENG] A valve having equal fluid pressure in both the opening and closing directions.

balanced voltages [ELEC] Voltages that are equal in magnitude and opposite in polarity with respect to ground. Also known as push-pull voltages.

balance equation [MATH] An equation expressing a balance of quantities in the sense that the local or individual rates of change are zero. [METEOROL] A diagnostic equation expressing a balance between the pressure field and the horizontal field of motion of the atmosphere.

balance error [COMPUT SCI] An error voltage that arises at the output of analog adders in an analog computer and is directly proportional to the drift error.

balancer [ELEC] A mechanism for equalizing the loads on the outer lines of a three-wire system for electric power distribution, consisting of two similar shunt or compound machines which are coupled together with the armatures connected in series across the outer lines. [INV ZOO] See haltere. [VERT ZOO] Either of a pair of rodlike lateral appendages on the heads of some larval salamanders.

balance spring [HOROL] An oscillating spring of spiral or (in a chronometer) cylindrical shape which governs the movement of a balance wheel in a timepiece. Also known as hairspring.

balance wheel [MECH ENG] 1. A wheel which governs or stabilizes the movement of a mechanism. 2. See flywheel.

balancing unit [ELEC] 1. Antenna-matching device used to permit efficient coupling of a transmitter or receiver having an unbalanced output circuit to an antenna having a balanced transmission line. 2. Device for converting balanced to unbalanced transmission lines, and vice versa, by placing suitable discontinuities at the junction between the lines instead of using lumped components.

Balanidae [INV ZOO] A family of littoral, sessile barnacles in the suborder Balanomorpha.

Balanomorpha [INV ZOO] The symmetrical barnacles, a suborder of sessile crustaceans in the order Thoracica.

Balanopaceae [BOT] A small family of dioecious dicotyledonous plants in the order Fagales characterized by exstipulate leaves, seeds with endosperm, and the pistillate flower solitary in a multibracteate involucre.

Balanophoraceae [BOT] A family of dicotyledonous terrestrial plants in the order Santalales characterized by dry nutlike fruit, one to five ovules, unisexual flowers, attachment to the stem of the host, and the lack of chlorophyll.

Balanopsidales [BOT] An order in some systems of classification which includes only the Balanopaceae of the Fagales.

balantidiasis [MED] An intestinal infection of humans caused by the protozoan *Balantidium coli.*

baleen [VERT ZOO] A horny substance, growing in fringed filter plates suspended from the upper jaws of whalebone whales. Also known as whalebone.

baler [MECH ENG] A machine which takes large quantities of raw or finished materials and binds them with rope or metal straps or wires into a large package.

ball [GEOL] 1. A low sand ridge, underwater by high tide, which extends generally parallel with the shoreline; usually separated by an intervening trough from the beach. 2. A spheroidal mass of sedimentary material. 3. Common name for a nodule, especially of ironstone. [MECH ENG] In fine grinding, one of the crushing bodies used in a ball mill. [ORD] 1. A bullet for general use, as distinguished from bullets for special uses such as armor-piercing, incendiary, or high explosive. 2. A small-arms solid propellant which is oblate spheroidal in shape, generally a double-base propellant.

ball-and-race-type pulverizer [MECH ENG] A grinding machine in which balls rotate under an applied force between two races to crush materials, such as coal, to fine consistency. Also known as ball-bearing pulverizer.

ball-and-socket joint [ANAT] *See* enarthrosis. [GEOL] *See* cup-and-ball joint. [MECH ENG] A joint in which a member ending in a ball is joined to a member ending in a socket so that relative movement is permitted within a certain angle in all planes passing through a line. Also known as ball joint.

ballast [AERO ENG] A relatively dense substance that is placed in the car of a vehicle and can be thrown out to reduce the load or can be shifted to change the center of gravity. [CIV ENG] Crushed stone used in a railroad bed to support the ties, hold the track in line, and help drainage. [ELEC] A circuit element that serves to limit an electric current or to provide a starting voltage, as in certain types of lamps, such as in fluorescent ceiling fixtures. [MATER] Coarse gravel used as an ingredient in concrete. [NAV ARCH] 1. A relatively heavy material such as lead, iron, or water placed in a ship to ensure stability or to maintain the proper draft or trim. 2. To pump seawater into empty fuel tanks of a ship to ensure its stability or suitable draft and trim for seaworthiness.

ballast lamp [ELEC] A light-producing electrical resistance device which maintains nearly constant current by increasing in resistance as the current increases.

ballast line [NAV ARCH] A ship's water line when the ship is ballasted.

ballast resistor [ELEC] A resistor that increases in resistance as current through it increases, and decreases in resistance as current decreases. Also known as barretter (British usage).

ballast tank [NAV ARCH] 1. One of several tanks in the hold of a ship which may be pumped full of water as ballast. 2. One of the tanks in a submarine that are filled with water or air to submerge or surface.

ballast tube [ELEC] A ballast resistor mounted in an evacuated glass or metal envelope, like that of a vacuum tube, to reduce radiation of heat from the resistance element and thereby improve the voltage-regulating action.

ball bearing [MECH ENG] An antifriction bearing permitting free motion between moving and fixed parts by means of balls confined between outer and inner rings.

ball-bearing pulverizer *See* ball-and-race-type pulverizer.

ball breaker [ENG] 1. A steel or iron ball that is hoisted by a derrick and allowed to fall on blocks of waste stone to break them or to swing against old buildings to demolish them. Also known as skull cracker; wrecking ball. 2. A coring and sampling device consisting of a hollow glass ball, 3 to 5 inches (7.5 to 12.5 centimeters) in diameter, held in a frame attached to the trigger line above the triggering weight of the corer; used to indicate contact between corer and bottom.

ball bushing [MECH ENG] A type of ball bearing that allows motion of the shaft in its axial direction.

ball check valve [ENG] A valve having a ball held by a spring against a seat; used to permit flow in one direction only.

ball-float liquid-level meter [ENG] A float which rises and falls with liquid level, actuating a pointer adjacent to a calibrated scale in order to measure the level of a liquid in a tank or other container.

ball grinder *See* ball mill.

ballhead [MECH ENG] That part of the governor which contains flyweights whose force is balanced, at least in part, by the force of compression of a speeder spring.

ballistic body [ENG] A body free to move, behave, and be modified in appearance, contour, or texture by ambient conditions, substances, or forces, such as by the pressure of gases in a gun, by rifling in a barrel, by gravity, by temperature, or by air particles.

ballistic coefficient [MECH] The numerical measure of the ability of a missile to overcome air resistance; dependent upon the mass, diameter, and form factor.

ballistic curve [MECH] The curve described by the path of a bullet, a bomb, or other projectile as determined by the ballistic conditions, by the propulsive force, and by gravity.

ballistic galvanometer [ELEC] A galvanometer having a long period of swing so that the deflection may measure the electric charge in a current pulse or the time integral of a voltage pulse.

ballistic missile [ORD] A missile capable of guiding and propelling itself in a direction and to a velocity such that it will follow a ballistic trajectory to a desired point.

ballistic pendulum [ENG] A device which uses the deflection of a suspended weight to determine the momentum of a projectile.

ballistics [MECH] Branch of applied mechanics which deals with the motion and behavior characteristics of missiles, that is, projectiles, bombs, rockets, guided missiles, and so forth, and of accompanying phenomena.

ballistic trajectory [MECH] The trajectory followed by a body being acted upon only by gravitational forces and resistance of the medium through which it passes.

ballistite [MATER] A smokeless propellant containing nitrocellulose and nitroglycerin; used in some rocket, mortar, and small-arms ammunition.

ballistocardiograph [MED] A device to measure the volume of blood passing through the heart in a given period of time.

ball joint *See* ball-and-socket joint.

ball lightning [GEOPHYS] A relatively rare form of lightning, consisting of a reddish, luminous ball, of the order of 1 foot (30 centimeters) in diameter, which may move rapidly along solid objects or remain floating in midair. Also known as globe lightning.

ball mill [MECH ENG] A pulverizer that consists of a horizontal rotating cylinder, up to three diameters in length, containing a charge of tumbling or cascading steel balls, pebbles, or rods. Also known as ball grinder.

balloon [AERO ENG] A nonporous, flexible spherical bag, inflated with a gas such as helium that is lighter than air, so that it will rise and float in the atmosphere; a large-capacity balloon can be used to lift a payload suspended from it.

balloon astronomy [ASTRON] The observation of celestial objects from instruments mounted on balloons and carried to altitudes up to 30 kilometers, to detect electromagnetic radiation at wavelengths which do not penetrate to the earth's surface.

ball-peen hammer [ENG] A hammer with a ball at one end of the head; used in riveting and forming metal.

ball valve [MECH ENG] A valve in which the fluid flow is regulated by a ball moving relative to a spherical socket as a result of fluid pressure and the weight of the ball.

Balmer lines [SPECT] Lines in the hydrogen spectrum, produced by transitions between $n = 2$ and $n > 2$ levels either in emission or absorption; here n is the principal quantum number.

balsa [BOT] *Ochroma lagopus*. A tropical American tree in the order Malvales; its wood is strong and lighter than cork.

balsam [MATER] An exudate of the balsam tree; a mixture of resins, essential oils, cinnamic acid, and benzoic acid.

Balsaminaceae [BOT] A family of flowering plants in the order Geraniales, including touch-me-not (*Impatiens*); flowers are irregular with five stamens and five carpels, leaves are simple with pinnate venation, and the fruit is an elastically dehiscent capsule.

balun [ELEC] A device used for matching an unbalanced coaxial transmission line or system to a balanced two-wire line or system. Also known as balanced converter; bazooka; line-balance converter.

baluster [BUILD] A post which supports a handrail and encloses the open sections of a stairway.

balustrade [BUILD] The railing assembly of a stairway consisting of the handrail, balusters, and usually a bottom rail.

bamboo [BOT] The common name of various tropical and subtropical, perennial, ornamental grasses in five genera of the family Gramineae characterized by hollow woody stems up to 6 inches (15 centimeters) in diameter.

Bambusoideae [BOT] A subfamily of grasses, composed of bamboo species, in the family Gramineae.

Banach algebra [MATH] An algebra which is a Banach space satisfying the property that for every pair of vectors, the norm of the product of those vectors does not exceed the product of their norms.

Banach space [MATH] A real or complex vector space in which each vector has a non-negative length, or norm, and in which every Cauchy sequence converges to a point of the space. Also known as complete normed linear space.

banana [BOT] Any of the treelike, perennial plants of the genus *Musa* in the family Musaceae; fruit is a berry characterized by soft, pulpy flesh and a thin rind.

banana plug [ELEC] A plug having a spring-metal tip shaped like a banana and used on test leads or as terminals for plug-in components.

band [ANALY CHEM] The position and spread of a solute within a series of tubes in a liquid-liquid extraction procedure. Also known as zone. [BUILD] Any horizontal flat member or molding or group of moldings projecting slightly from a wall plane and usually marking a division in the wall. Also known as band course; band molding. [COMMUN] A range of electromagnetic-wave frequencies between definite limits, such as that assigned to a particular type of radio service. [COMPUT SCI] A set of circular or cyclic recording tracks on a storage device such as a magnetic drum, disk, or tape loop. [DES ENG] A strip or cord crossing the back of a book to which the sections are sewn. [GEOD] Any latitudinal strip, designated by accepted units of linear or angular measurement, which circumscribes the earth. [GEOL] A thin layer or stratum of rock that is noticeable because its color is different from the colors of adjacent layers. [ORD] A metal sleeve joining together the barrel and stock of a gun. [SOLID STATE] A restricted range in which the energies of electrons in solids lie, or from which they are excluded, as understood in quantum-mechanical terms. [SPECT] *See* band spectrum.

bandage [BUILD] A strap, band, ring, or chain placed around a structure to secure and hold its parts together, as around the springing of a dome. [ELEC] Rubber ribbon about 4 inches (10 centimeters) wide for temporarily protecting a telephone or coaxial splice from moisture. [MED] A strip of gauze, muslin, flannel, or other material, usually in the form of a roll, but sometimes triangular or tailed, used to hold dressing in place, to apply pressure, to immobilize a part, to support a dependent or injured part, to obliterate tissue cavities, or to check hemorrhage.

band brake [MECH ENG] A brake in which the frictional force is applied by increasing the tension in a flexible band to tighten it around the drum.

band chain [ENG] A steel or Invar tape, graduated in feet and at least 100 feet (30.5 meters) long, used for accurate surveying.

band clutch [MECH ENG] A friction clutch in which a steel band, lined with fabric, contracts onto the clutch rim.

banded [PETR] Pertaining to the appearance of rocks that have thin and nearly parallel bands of different textures, colors, and minerals.

banded coal [GEOL] A variety of bituminous and subbituminous coal made up of a sequence of thin lenses of highly lustrous coalified wood or bark interspersed with layers of more or less striated bright or dull coal.

banded structure [MET] The appearance of a metal showing light and dark parallel bands in the direction of rolling or working. [METEOROL] The appearance of precipitation echoes in the form of long bands as presented on radar plan position indicator (PPI) scopes. [PETR] An outcrop feature in igneous and metamorphic rocks due to alternation of layers, stripes, flat lenses, or streaks that obviously differ in mineral composition or texture.

band-elimination filter *See* band-stop filter.

band gap [SOLID STATE] An energy difference between two allowed bands of electron energy in a metal.

band lightning *See* ribbon lightning.

band-pass [ELECTR] A range, in hertz or kilohertz, expressing the difference between the limiting frequencies at which a desired fraction (usually half power) of the maximum output is obtained.

band-pass filter [ELECTR] An electric filter which transmits more or less uniformly in a certain band, outside of which the frequency components are attenuated. [OPTICS] *See* Christiansen filter.

band printer [COMPUT SCI] A line printer that uses a band of type characters as its printing mechanism.

band-rejection filter *See* band-stop filter.

band saw [MECH ENG] A power-operated woodworking saw consisting basically of a flexible band of steel having teeth on one edge, running over two vertical pulleys, and operated under tension.

band scheme [SOLID STATE] The identification of energy bands of a solid with the levels of independent atoms from which they arise as the atoms are brought together to form the solid, together with the width and spacing of the bands.

band selector [ELECTR] A switch that selects any of the bands in which a receiver, signal generator, or transmitter is designed to operate and usually has two or more sections to make the required changes in all tuning circuits simultaneously. Also known as band switch.

band spectrum [SPECT] A spectrum consisting of groups or bands of closely spaced lines in emission or absorption, characteristic of molecular gases and chemical compounds. Also known as band.

band spreading [COMMUN] Method of double-sideband transmission in which the frequency band of the modulating wave is shifted upward in frequency so that the sidebands produced by modulation are separated in frequency from the carrier by an amount at least equal to the bandwidth of the original modulating wave, and second order distortion products may be filtered from the demodulator output.

band-stop filter [ELECTR] An electric filter which transmits more or less uniformly at all frequencies of interest except for a band within which frequency components are largely attenuated. Also known as band-elimination filter; band-rejection filter.

band switch *See* band selector.

band theory of solids [SOLID STATE] A quantum-mechanical theory of the motion of electrons in solids that predicts certain restricted ranges or bands for the energies of these electrons.

bandwidth [COMMUN] The difference between the frequency limits of a band containing the useful frequency components of a signal. Abbreviated BW.

bandylite [MINERAL] $CuB_2O_4 \cdot CuCl_2 \cdot 4H_2O$ A tetragonal mineral that is deep blue with greenish lights and consists of a hydrated copper borate-chloride.

bang-bang circuit [ELECTR] An operational amplifier with double feedback limiters that drive a high-speed relay (1–2 milliseconds) in an analog computer; involved in signal-controlled programming.

bang-bang control [COMPUT SCI] Control of programming in an analog computer through a bang-bang circuit. [CONT SYS] A type of automatic control system in which the applied control signals assume either their maximum or minimum values.

Bangiophyceae [BOT] A class of red algae in the plant division Rhodophyta.

Bang's disease *See* contagious abortion.

bank [AERO ENG] The lateral inward inclination of an airplane when it rounds a curve. [CIV ENG] *See* embankment. [ELEC] **1.** A number of similar electrical devices, such as resistors, connected together for use as a single device. **2.** An assemblage of fixed contacts over which one or more wipers or brushes move in order to establish electrical connections in automatic switching. [GEOL] **1.** The edge of a waterway. **2.** The rising ground bordering a body of water. **3.** A steep slope or face, generally consisting of unconsolidated material. [IND ENG] The amount of material allowed to accumulate at a point on a production line where it is not employed or worked upon, to permit reasonable fluctuations in line speed before and after the point. Also known as float. [MIN ENG] **1.** The top of the shaft. **2.** The surface around the mouth of a shaft. **3.** The whole, or sometimes only one side or one end, of a working place underground. **4.** To manipulate materials such as coal, gravel, or sand on a bank. **5.** A terrace-like bench in open-pit mining. [OCEANOGR] A relatively flat-topped raised portion of the sea floor occurring at shallow depth and characteristically on the continental shelf or near an island.

bank and turn indicator [AERO ENG] A device used to advise the pilot that the aircraft is turning at a certain rate, and that the wings are properly banked to preclude slipping or sliding of the aircraft as it continues in flight. Also known as bank indicator.

bank-full stage [HYD] The flow stage of a river in which the stream completely fills its channel and the elevation of the water surface coincides with the bank margins.

bank indicator *See* bank and turn indicator.

banking pin [HOROL] One of the erect pins in the bottom plate of a watch that restrict the movement of the lever.

banner cloud [METEOROL] A cloud plume often observed to extend downwind from isolated mountain peaks, even on otherwise cloud-free days. Also known as cloud banner.

bar [GEOL] Any of the various submerged or partially submerged ridges, banks, or mounds of sand, gravel, or other unconsolidated sediment built up by waves or currents within stream channels, at estuary mouths, and along coasts. **2.** Any band of hard rock, for example, a vein or dike, that extends across a lode. [MECH] A unit of pressure equal to 10^5 pascals, or 10^5 newtons per square meter, or 10^6 dynes per square centimeter. [MET] An elongated piece of metal of simple uniform cross-section dimensions, usually rectangular, circular, or hexagonal, produced by forging or hot rolling. Also known as barstock. [MIN ENG] *See* bar drill.

bararite [MINERAL] $(NH_4)_2SiF_6$ A white, hexagonal mineral consisting of ammonium silicon fluoride; occurs in tabular, arborescent, and mammillary forms.

barb [METEOROL] A means of representing wind speed in the plotting of a synoptic chart, being a short, straight line drawn obliquely toward lower pressure from the end of a wind-direction shaft. Also known as feather. [VERT ZOO] A side branch on the shaft of a bird's feather.

bar beach [GEOL] A straight beach of offshore bars that are separated by shallow bodies of water from the mainland.

barbel [VERT ZOO] **1.** A slender, tactile process near the mouth in certain fishes, such as catfishes. **2.** Any European fresh-water fish in the genus *Barbus*.

barbertonite [MINERAL] $Mg_6Cr_2(OH)_{16}CO_3 \cdot 4H_2O$ A lilac to rose pink, hexagonal mineral consisting of a hydrated carbonate-hydroxide of magnesium and chromium; occurs in massive form or in masses of fibers or plates.

barbiturate [PHARM] Any of a group of ureides, such as phenobarbital, Amytal, or Seconal, that act as central nervous system depressants.

barbiturism [MED] Intoxication following an overdose of barbiturates; characterized by delirium, coma, and sometimes death.

bar buoy [NAV] A buoy marking the location of a bar at the mouth of a river or approach to a harbor.

barchan [GEOL] A crescent-shaped dune or drift of windblown sand or snow, the arms of which point downwind; formed by winds of almost constant direction and of moderate speeds. Also known as barchane; barkhan; crescentic dune.

barchane *See* barchan.

bar chart *See* bar graph.

bar code [COMPUT SCI] The representation of alphanumeric characters by series of adjacent stripes of various widths, for example, the universal product code.

bar-code reader *See* bar-code scanner.

bar-code scanner [COMPUT SCI] An optical scanning device that reads texts which have been converted into a special bar code. Also known as bar-code reader.

Bardeen-Cooper-Schrieffer theory [SOLID STATE] A theory of superconductivity that describes quantum-mechanically those states of the system in which conduction electrons cooperate in their motion so as to reduce the total energy appreciably below that of other states by exploiting their effective mutual attraction; these states predominate in a superconducting material. Abbreviated BCS theory.

bar drill [MIN ENG] A small diamond type or other type of rock drill mounted on a bar and used in an underground workplace. Also known as bar.

bare charm [PARTIC PHYS] Charm that is carried by a quark and is not canceled by the charm of the corresponding antiquark, so that the hadron of which the quark is a constituent has net charm different from zero.

bare electrode [MET] An uncoated electrode used in submerged arc automatic welding with a gas-shielded arc or a granular flux deposited in an elongated mound over a joint.

barefaced tenon [ENG] A tenon having a shoulder cut on one side only.

barge [NAV ARCH] A large cargo-carrying craft which is towed or pushed by a tug on both seagoing and inland waters.

bar generator [ELECTR] Generator of pulses or repeating waves which are equally separated in time; these pulses are synchronized by the synchronizing pulses of a television system, so that they can produce a stationary bar pattern on a television screen.

bar graph [STAT] A diagram of frequency-table data in which a rectangle with height proportional to the frequency is located at each value of a variate that takes only certain discrete values. Also known as bar chart.

baring *See* overburden.

barite [MINERAL] $BaSO_4$ A white, yellow, or colorless orthorhombic mineral occurring in tabular crystals, granules, or compact masses; specific gravity is 4.5; used in paints and drilling muds and as a source of barium chemicals; the principal ore of barium. Also known as baryte; barytine; cawk; heavy spar.

BARITT diode   *See* barrier injection transit-time diode.

barium   [CHEM] A chemical element, symbol Ba, with atomic number 56 and atomic weight of 137.34.

barium enema   [MED] A suspension of barium sulfate administered as an enema into the lower bowel to render it radiopaque.

barium glass   [MATER] Glass which differs from ordinary lime-soda glass in that barium oxide replaces part of the calcium oxide.

bar joist   [BUILD] A small steel truss with wire or rod web lacing used for roof and floor supports.

bark   [BOT] The tissues external to the cambium in a stem or root.   [MET] The decarburized layer formed beneath the scale on the surface of steel heated in air.   [NAV ARCH] A three-masted sailing ship whose foremast and mainmast are square-rigged and whose mizzenmast is fore-and-aft-rigged.

bar keel   [NAV ARCH] A solid keel with a rectangular cross section in an iron or steel ship.

barkhan   *See* barchan.

Barkhausen effect   [ELECTROMAG] The succession of abrupt changes in magnetization occurring when the magnetizing force acting on a piece of iron or other magnetic material is varied.

Barkhausen-Kurz oscillator   [ELECTR] An oscillator of the retarding-field type in which the frequency of oscillation depends solely on the transit time of electrons oscillating about a highly positive grid before reaching the less positive anode. Also known as Barkhausen oscillator; positive-grid oscillator.

Barkhausen oscillation   [ELECTR] Undesired oscillation in the horizontal output tube of a television receiver, causing one or more ragged dark vertical lines on the left side of the picture.

Barkhausen oscillator   *See* Barkhausen-Kurz oscillator.

barley   [BOT] A plant of the genus *Hordeum* in the order Cyperales that is cultivated as a grain crop; the seed is used to manufacture malt beverages and as a cereal.

bar linkage   [MECH ENG] A set of bars joined together at pivots by means of pins or equivalent devices; used to transmit power and information.

Barlow lens   [OPTICS] A lens with one plane surface and one concave surface that is placed between the objective and eyepiece of a telescope to decrease the convergence of the beam from the objective and thereby increase the effective focal length.

bar magnet   [ELECTROMAG] A bar of hard steel that has been strongly magnetized and holds its magnetism, thereby serving as a permanent magnet.

barney   [MIN ENG] A small car or truck, attached to a rope or cable, used to push cars up a slope or an inclined plane. Also known as bullfrog; donkey; groundhog; larry; mule; ram; truck.

baroclinic disturbance   [METEOROL] Any migratory cyclone associated with strong baroclinity of the atmosphere, evidenced on synoptic charts by temperature gradients in the constant-pressure surfaces, vertical wind shear, tilt of pressure troughs with height, and concentration of solenoids in the frontal surface near the ground. Also known as baroclinic wave.

baroclinic field   [METEOROL] A distribution of atmospheric pressure and mass such that the specific volume, or density, of air is a function not solely of pressure.

baroclinic instability   [METEOROL] A hydrodynamic instability arising from the existence of a meridional

temperature gradient (and hence of a thermal wind) in an atmosphere in quasi-geostrophic equilibrium and possessing static stability.

baroclinicity   *See* baroclinity.

baroclinic wave   *See* baroclinic disturbance.

baroclinity   [PHYS] The state of stratification in a fluid in which surfaces of constant pressure (isobaric surfaces) intersect surfaces of constant density (isosteric surfaces). Also known as baroclinicity; barocliny.

barocliny   *See* baroclinity.

barometer   [ENG] An absolute pressure gage specifically designed to measure atmospheric pressure.

barometric   [ENG] Pertaining to a barometer or to the results obtained by using a barometer.   [PHYS] Loosely, pertaining to atmospheric pressure; for example, barometric gradient (meaning pressure gradient).

barometric altimeter   *See* pressure altimeter.

barometric condenser   [MECH ENG] A contact condenser that uses a long, vertical pipe into which the condensate and cooling liquid flow to accomplish their removal by the pressure created at the lower end of the pipe.

barometric corrections   [PHYS] The corrections which must be applied to the reading of a mercury barometer in order that the observed value may be rendered accurate. Also known as barometric errors.

barometric errors   *See* barometric corrections.

barometric gradient   *See* pressure gradient.

barometric pressure   *See* atmospheric pressure.

barometric switch   *See* baroswitch.

barometric tide   [GEOPHYS] A daily variation in atmospheric pressure due to the gravitational attraction of the sun and moon.

barometric wave   [METEOROL] Any wave in the atmospheric pressure field; the term is usually reserved for short-period variations not associated with cyclonic-scale motions or with atmospheric tides.

barometry   [ENG] The study of the measurement of atmospheric pressure, with particular reference to ascertaining and correcting the errors of the different types of barometer.

baroscope   [ENG] An apparatus which demonstrates the equality of the weight of air displaced by an object and its loss of weight in air.

barostat   [ENG] A mechanism which maintains constant pressure inside a chamber.

baroswitch   [ENG] **1.** A pressure-operated switching device used in a radiosonde which determines whether temperature, humidity, or reference signals will be transmitted. **2.** Any switch operated by a change in barometric pressure. Also known as barometric switch.

barothermograph   [ENG] An instrument which automatically records pressure and temperature.

barothermohygrograph   [ENG] An instrument that produces graphs of atmospheric pressure, temperature, and humidity on a single sheet of paper.

barotitis   [MED] Inflammation of the ear, or a part of it, caused by changes in atmospheric pressure. Also known as aerootitis.

barotropic   [PHYS] Of, pertaining to, or characterized by a condition of barotropy.

barotropic disturbance   [METEOROL] Also known as barotropic wave. **1.** A wave disturbance in a two-dimensional nondivergent flow; the driving mechanism lies in the variation of either vorticity of the basic current or the variation of the vorticity of the earth about the local vertical. **2.** An atmospheric wave of cyclonic scale in which troughs and ridges are approximately vertical.

barotropic field [METEOROL] A distribution of atmospheric pressures and mass such that the specific volume, or density, of air is a function solely of pressure.

barotropic wave *See* barotropic disturbance.

barotropy [PHYS] The state of a fluid in which surfaces of constant density (or temperature) are coincident with surfaces of constant pressure; it is the state of zero baroclinity.

bar pattern [ELECTR] Pattern of repeating lines or bars on a television screen.

barracuda [VERT ZOO] The common name for about 20 species of fishes belonging to the genus *Sphyraena* in the order Perciformes.

barrage jamming [COMMUN] The simultaneous jamming of a number of radio or radar bands of frequencies.

barrage reception [COMMUN] A type of radio reception in which interference from certain directions is reduced by selecting from a system of directional antennas those which yield the greatest signal-to-interference ratio.

Barr body [CYTOL] A condensed, inactivated X chromosome inside the nuclear membrane in interphase somatic cells of women and most female mammals.

barre [TEXT] 1. A pattern of bars or stripes parallel to the weft of a fabric. 2. A defect in the production of a fabric that results in a streak parallel to the weft.

barred basin *See* restricted basin.

barred spiral galaxy [ASTRON] A spiral galaxy whose spiral arms originate at the ends of a bar-shaped structure centered at the nucleus of the galaxy.

barrel [DES ENG] 1. A container having a circular lateral cross section that is largest in the middle, and ends that are flat; often made of staves held together by hoops. 2. A piece of small pipe inserted in the end of a cartridge to carry the squib to the powder. 3. That portion of a pipe having a constant bore and wall thickness. [MECH] Abbreviated bbl. 1. The unit of liquid volume equal to 31.5 gallons (approximately 119 liters). 2. The unit of liquid volume for petroleum equal to 42 gallons (approximately 158 liters). 3. The unit of dry volume equal to 105 quarts (approximately 116 liters). 4. A unit of weight that varies in size according to the commodity being weighed. [NAV ARCH] The central part of a windlass or capstan about which the cable is wound. [OPTICS] A tapering cylindrical housing which contains the lenses of a camera and the iris diaphragm. [ORD] The cylindrical metallic part of a gun which controls the initial direction of a projectile.

barrel arch [ARCH] A plain arch with a barrellike cross section in which the length is greater than the span (diameter).

barrel distortion [OPTICS] A defect in an optical system whereby lateral magnification decreases with object size; the image of a square then appears barrel-shaped.

barrel printer [COMPUT SCI] A computer printer in which the entire set of characters is placed around a rapidly rotating cylinder at each print position; computer-controlled print hammers opposite each print position strike the paper and press it against an inked ribbon between the paper and the cylinder when the appropriate character reaches a position opposite the print hammer.

barrels per month [CHEM ENG] A unit measuring the rate at which petroleum is produced at the refinery. Abbreviated BM; bpm.

Barremian [GEOL] Lower Cretaceous geologic age, between Hauterivian and Aptian.

barren liquor [CHEM ENG] Liquid (liquor) from filter-cake washing in which there is little or no recovery value; for example, barren cyanide liquor from washing of gold cake slimes.

barrens [GEOGR] An area that because of adverse environmental conditions is relatively devoid of vegetation compared with adjacent areas.

barretter [ELEC] 1. Bolometer that consists of a fine wire or metal film having a positive temperature coefficient of resistivity, so that resistance increases with temperature; used for making power measurements in microwave devices. 2. *See* ballast resistor.

barrier [ECOL] Any physical or biological factor that restricts the migration or free movement of individuals or populations. [NAV] Anything which obstructs or prevents passage of a craft. [PHYS] *See* potential barrier.

barrier bar [GEOL] Ridges whose crests are parallel to the shore and which are usually made up of water-worn gravel put down by currents in shallow water at some distance from the shore.

barrier beach [GEOL] A single, long, narrow ridge of sand which rises slightly above the level of high tide and lies parallel to the shore, from which it is separated by a lagoon. Also known as offshore beach.

barrier flat [GEOL] An area which is relatively flat and frequently occupied by pools of water that separate the seaward edge of the barrier from a lagoon on the landward side.

barrier-grid storage tube *See* radechon.

barrier ice *See* shelf ice.

barrier injection transit-time diode [ELECTR] A microwave diode in which the carriers that traverse the drift region are generated by minority carrier injection from a forward-biased junction instead of being extracted from the plasma of an avalanche region. Abbreviated BARITT diode.

barrier island [GEOL] 1. An island similar to an offshore bar but differing from it in having multiple ridges, areas of vegetation, and swampy terraces extending toward the lagoon. 2. A detached portion of offshore bar between two inlets.

barrier layer *See* depletion layer.

barrier-layer cell *See* photovoltaic cell.

barrier-layer photocell *See* photovoltaic cell.

barrier reef [GEOL] A coral reef that runs parallel to the coast of an island or continent, from which it is separated by a lagoon.

barrier separation [CHEM ENG] The separation of a two-component gaseous mixture by selective diffusion of one component through a separative barrier (microporous metal or nonporous polymeric).

barrier shield [ENG] A wall or enclosure made of a material designed to absorb ionizing radiation, shielding the operator from an area where radioactive material is being used or processed by remote-control equipment.

Bartonellaceae [MICROBIO] A family of the order Rickettsiales; rod-shaped, coccoid, ring- or disk-shaped cells; parasites of human and other vertebrate red blood cells.

Bartonian [GEOL] A European stage: Eocene geologic time above Auversian, below Ludian. Also known as Marinesian.

Bart reaction [ORG CHEM] Formation of an aryl arsonic acid by treating the aryl diazo compound with trivalent arsenic compounds, such as sodium arsenite.

bar winding [ELEC] An armature winding made up of a series of metallic bars connected at their ends.

barycenter [ASTRON] The center of gravity of the earth-moon system. [MATH] The center of mass of a system of finitely many equal point masses distributed in euclidean space in such a way that their position vectors are linearly independent.

barycentric coordinates   [MATH] The coefficients in the representation of a point in a simplex as a linear combination of the vertices of the simplex.

baryon   [PARTIC PHYS] Any elementary particle which can be transformed into a nucleon and some number of mesons and lighter particles. Also known as heavy particle.

baryon number   [PARTIC PHYS] A conserved quantum number, equal to the number of baryons minus the number of antibaryons in a system; neutrons and protons have baryon number one; mesons and leptons have baryon number zero.

barysphere   *See* centrosphere.

baryta feldspar   *See* hyalophane.

baryte   *See* barite.

Barytheriidae   [PALEON] A family of extinct proboscidean mammals in the suborder Barytherioidea.

Barytherioidea   [PALEON] A suborder of extinct mammals of the order Proboscidea, in some systems of classification.

barytine   *See* barite.

barytocalcite   [MINERAL] $CaBa(CO_3)_2$ A colorless to white, grayish, greenish, or yellowish monoclinic mineral consisting of calcium and barium carbonate.

basal   [BIOL] Of, pertaining to, or located at the base. [PHYSIO] Being the minimal level for, or essential for maintenance of, vital activities of an organism, such as basal metabolism.

basal body   [CYTOL] A cellular organelle that induces the formation of cilia and flagella and is similar to and sometimes derived from a centriole. Also known as kinetosome.

basal-cell carcinoma   [MED] A locally invasive, rarely metastatic nevoid tumor of the epidermis. Also known as basal-cell epithelioma.

basal-cell epithelioma   *See* basal-cell carcinoma.

basal cleavage   [CRYSTAL] Cleavage parallel to the base of the crystal structure or to the lattice plane which is normal to one of the lattice axes.

basal conglomerate   [GEOL] A coarse gravelly sandstone or conglomerate forming the lowest member of a series of related strata which lie unconformably on older rocks; records the encroachment of the seabeach on dry land.

basal coplane   [GRAPHICS] The condition of exposure of a pair of photographs in which the two photographs lie in a common plane parallel to the air base.

basal ganglia   [ANAT] The corpus striatum, or the corpus striatum and the thalamus considered together as the important subcortical centers.

basal membrane   [ANAT] The tissue beneath the pigment layer of the retina that forms the outer layer of the choroid.

basal metabolic rate   [PHYSIO] The amount of energy utilized per unit time under conditions of basal metabolism; expressed as calories per square meter of body surface or per kilogram of body weight per hour. Abbreviated BMR.

basal metabolism   [PHYSIO] The sum total of anabolic and catabolic activities of an organism in the resting state providing just enough energy to maintain vital functions.

basal orientation   [CRYSTAL] A crystal orientation in which the surface is parallel to the base of the lattice or to the lattice plane which is normal to one of the lattice axes.

basal plane   [CRYSTAL] The plane perpendicular to the long, or c, axis in all crystals except those of the isometric system.

basalt   [PETR] An aphanitic crystalline rock of volcanic origin, composed largely of plagioclase feldspar (labradorite or bytownite) and dark minerals such as pyroxene and olivine; the extrusive equivalent of gabbro.

basalt glass   *See* tachylite.

basaltic dome   *See* shield volcano.

basalt obsidian   *See* tachylite.

basculating fault   *See* wrench fault.

bascule   [ENG] A structure that rotates about an axis, as a seesaw, with a counterbalance (for the weight of the structure) at one end.

bascule bridge   [CIV ENG] A movable bridge consisting primarily of a cantilever span extending across a channel; it rotates about a horizontal axis parallel with the waterway.

base   [CHEM] Any chemical species, ionic or molecular, capable of accepting or receiving a proton (hydrogen ion) from another substance; the other substance acts as an acid in giving of the proton; the hydroxyl ion is a base.   [CHEM ENG] The primary substance in solution in crude oil, and remaining after distillation.   [COMPUT SCI] *See* root.   [ELECTR] **1.** The region that lies between an emitter and a collector of a transistor and into which minority carriers are injected. **2.** The part of an electron tube that has the pins, leads, or other terminals to which external connections are made either directly or through a socket. **3.** The plastic, ceramic, or other insulating board that supports a printed wiring pattern. **4.** A plastic film that supports the magnetic powder of magnetic tape or the emulsion of photographic film.   [ENG] Foundation or part upon which an object or instrument rests.   [LAP] *See* pavilion.   [ORD] Station or installation from which military forces operate and from which supplies are obtained.

base address   *See* address constant.

base apparatus   [ENG] Any apparatus designed for use in measuring with accuracy and precision the length of a base line in triangulation, or the length of a line in first- or second-order traverse.

baseband   [COMMUN] The band of frequencies occupied by all transmitted signals used to modulate the radio wave that is produced by the transmitter in the absence of the signals.

base bias   [ELECTR] The direct voltage that is applied to the majority-carrier contact (base) of a transistor.

baseboard   [BUILD] A finish board covering the interior wall at the junction of the wall and the floor.

base-centered lattice   [CRYSTAL] A space lattice in which each unit cell has lattice points at the centers of each of two opposite faces as well as at the vertices; in a monoclinic crystal, they are the faces normal to one of the lattice axes.

base course   [BUILD] The lowest course or first course of a wall.   [CIV ENG] The first layer of material laid down in construction of a pavement.

base language   [COMPUT SCI] The component of an extensible language which provides a complete but minimal set of primitive facilities, such as elementary data types, and simple operations and control constructs.

base line   Abbreviated BL.   [ELECTR] The line traced on amplitude-modulated indicators which corresponds to the power level of the weakest echo detected by the radar; it is retraced with every pulse transmitted by the radar but appears as a nearly continuous display on the scope. [ENG] **1.** A surveyed line, established with more than usual care, to which surveys are referred for coordination and correlation. **2.** A cardinal line extending east and west along the astronomic parallel passing through the initial point, along which standard township, section, and quarter-section corners are established. [NAV] The geodesic line joining the two stations between which electrical phase or time is compared in determining navigational coordinates. [SCI TECH] A line drawn in the graphical representation of a varying physical quantity, such as a voltage or current, to indicate a reference value, such as the voltage value of a bias.

base-line extension [NAV] In the loran navigation system, the extension in both directions of a line passing through a slave station and its master; on this line the maximum time interval between the reception of signals from the master and slave stations may be received.

base-line technique [ANALY CHEM] A method for measurement of absorption peaks for quantitative analysis of chemical compounds in which a base line is drawn tangent to the spectrum background; the distance from the base line to the absorption peak is the absorbence due to the sample under study.

baseload [ELEC] Minimum load of a power generator over a given period of time.

base magnification [OPTICS] The ratio of the distance between the centers of the objectives of a pair of binoculars to the distance between the centers of the eyepieces.

base map [MAP] A map having essential outlines and onto which additional geographical or topographical data may be placed for comparison or correlation. Also known as mother map.

basement [BUILD] A building story which is wholly or less than half below ground; it is generally used for living space. [GEOL] **1.** A complex, usually of igneous and metamorphic rocks, that is overlain unconformably by sedimentary strata. **2.** A crustal layer beneath a sedimentary one and above the Mohorovičić discontinuity.

basement membrane [HISTOL] A delicate connective-tissue layer underlying the epithelium of many organs.

base metal [CHEM] Any of the metals on the lower end of the electrochemical series. [MET] **1.** The metal that is to be worked. **2.** The principal metal of an alloy. **3.** Any metal that will oxidize when heated in air. **4.** The metal of parts to be welded. **5.** Metal to which cladding or plating is applied. Also known as basis metal.

base modulation [ELECTR] Amplitude modulation produced by applying the modulating voltage to the base of a transistor amplifier.

base net [ENG] A system, in surveying, of quadrilaterals and triangles that include and are quite close to a base line in a triangulation system.

base notation See radix notation.

base of a logarithm [MATH] The number of which the logarithm is the exponent.

base of a number system [MATH] The number whose powers determine place value.

base of a topological space [MATH] A collection of sets, unions of which form all open sets.

base pairing [MOL BIO] The hydrogen bonding of complementary purine and pyrimidine bases—adenine with thymine, guanine with cytosine—in double-stranded deoxyribonucleic acids (DNA) or ribo-nucleic acids (RNA) or in DNA/RNA hybrid molecules.

base pin See pin.

base plate [DES ENG] The part of a theodolite which carries the lower ends of the three foot screws and attaches the theodolite to the tripod for surveying. [ENG] A metal plate that provides support or a foundation.

base pressure [FL MECH] The pressure exerted on the base or extreme aft end of a body, as of a cylindrical or boat-tailed body or of a blunt-trailing-edge wing in fluid flow. [MECH] A pressure used as a reference base, for example, atmospheric pressure.

base quantity [PHYS] One of a small number of physical quantities in a system of measurement that are defined, independent of other physical quantities, by means of a physical standard and by procedures for comparing the quantity to be measured with the standard. Also known as fundamental quantity.

base rate area [COMMUN] Area within the telephone exchange in which all types of service are given without mileage charges; the area does not necessarily follow the same lines or the same area as city limits; an extension of city limits, for example, would have no bearing on the base rate area boundary; customers located outside the area pay for the extra cost of providing service.

base register See index register.

base station [COMMUN] **1.** A land station, in the land mobile service, carrying on a service with land mobile stations, (a base station may secondarily communicate with other base stations incident to communications with land mobile stations). **2.** A station in a land mobile system which remains in a fixed location and communicates with the mobile stations. [ENG] The point from which a survey begins. [GEOD] A geographic position whose absolute gravity value is known.

base unit [PHYS] One of a small number of units in a system of measurement that are defined, independent of other units, by means of a physical standard; equivalently, a unit of a base quantity. Also known as fundamental unit.

basic [CHEM] Of a chemical species that has the properties of a base. [PETR] Of igneous rocks, having low silica content (generally less than 54%) and usually rich in iron, magnesium, or calcium.

BASIC [COMPUT SCI] A procedure-level computer language well suited for conversational mode on a terminal usually connected with a remotely operated computer. Derived from Beginners All-purpose Symbolic Instruction Code.

basic dye [MATER] Any of the dyes which are salts of the colored organic bases containing amino and imino groups, combined with a colorless acid, such as hydrochloric or sulfuric.

basic group [CHEM] A chemical group (for example, OH^-) which, when freed by ionization in solution, produces a pH greater than 7.

basic instruction [COMPUT SCI] An instruction in a computer program which is systematically changed by the program to obtain the instructions which are actually carried out. Also known as presumptive instruction; unmodified instruction.

basic linkage [COMPUT SCI] Computer coding that provides a standard means of connecting a given routine or program with other routines and that can be used repeatedly according to the same rules.

basic open-hearth process [MET] An open-hearth process for steelmaking under basic slag; used for pig iron and scrap with a phosphorus content too low for the Bessemer process and too high for the acid open-hearth process.

basic slag [MET] A slag resulting from the steelmaking process; rich in phosphorus, it is ground and used as a nutrient in grasslands.

basic software [COMPUT SCI] Software requirements that are taken into account in the design of the data-processing hardware and usually provided by the original equipment manufacturer.

basic steel [MET] Steel made by the basic process, in a furnace with a basic lining.

basic variables [COMPUT SCI] The m variables in a basic feasible solution for a linear programming model.

Basidiolichenes [BOT] A class of the Lichenes characterized by the production of basidia.

Basidiomycetes [MYCOL] A class of fungi in the subdivision Eumycetes; important as food and as causal agents of plant diseases.

basilar [BIOL] Of, pertaining to, or situated at the base.

basilar membrane [ANAT] A membrane of the mammalian inner ear supporting the organ of Corti and separating two cochlear channels, the scala media and scala tympani.

basin [CIV ENG] 1. A dock employing floodgates to keep water level constant during tidal variations. 2. A harbor for small craft. [DES ENG] An open-top vessel with relatively low sloping sides for holding liquids. [GEOL] 1. A low-lying area, wholly or largely surrounded by higher land, that varies from a small, nearly enclosed valley to an extensive, mountain-rimmed depression. 2. An entire area drained by a given stream and its tributaries. 3. An area in which the rock strata are inclined downward from all sides toward the center. 4. An area in which sediments accumulate. [MET] The mouth of a sprue in a gating system of castings into which the molten metal is first poured. [OCEANOGR] Deep portion of sea surrounded by shallower regions.

basis [MATH] A set of linearly independent vectors in a vector space such that each vector in the space is a linear combination of vectors from the set.

basis weight [GRAPHICS] The weight in pounds of 500 sheets of standard-size paper; certain sizes for a given class of paper are accepted as standard.

basket-handle arch [ARCH] An arch whose width is much greater than its height and which resembles an ellipse and is drawn from three or more centers. Also known as multicenter arch.

basket-weave [BUILD] A checkerboard pattern of bricks, flat or on edge. [TEXT] Pertaining to fabric in which two or more yarns are worked in warp and weft.

Basommatophora [INV ZOO] An order of mollusks in the subclass Pulmonata containing many aquatic snails.

basonym [SYST] The original, validly published name of a taxon.

basophil [HISTOL] A white blood cell with granules that stain with basic dyes and are water-soluble.

basophilia [BIOL] An affinity for basic dyes. [MED] An increase in the number of basophils in the circulating blood. [PATH] Stippling of the red cells with basic staining granules, representing a degenerative condition as seen in severe anemia, leukemia, malaria, lead poisoning, and other toxic states.

bass [ACOUS] Sounds having frequencies at the lower end of the audio range, below about 250 hertz. [VERT ZOO] The common name for a number of fishes assigned to two families, Centrarchidae and Serranidae, in the order Perciformes.

bass compensation [ELECTR] A circuit that emphasizes the low-frequency response of an audio amplifier at low volume levels to offset the lower sensitivity of the human ear to weak low frequencies.

bass control [ELECTR] A manual tone control that attenuates higher audio frequencies in an audio amplifier and thereby emphasizes bass frequencies.

bass reflex baffle [ENG ACOUS] A loudspeaker baffle having an opening of such size that bass frequencies from the rear of the loudspeaker emerge to reinforce those radiated directly forward.

bass response [ELECTR] A measure of the output of an electronic device or system as a function of an input of low audio frequencies.

bast See phloem.

bast fiber [BOT] Any fiber stripped from the inner bark of plants, such as flax, hemp, jute, and ramie; used in textile and paper manufacturing.

bat [VERT ZOO] The common name for all members of the mammalian order Chiroptera.

Batales [BOT] A small order of dicotyledonous plants in the subclass Caryophillidae of the class Magnoliopsida containing a single family with only one genus, *Batis*.

batch [COMPUT SCI] A set of items, records, or documents to be processed as a single unit. [ENG] 1. The quantity of material required for or produced by one operation. 2. An amount of material subjected to some unit chemical process or physical mixing process to make the final product substantially uniform.

batch distillation [CHEM ENG] Distillation where the entire batch of liquid feed is placed into the still at the beginning of the operation, in contrast to continuous distillation, where liquid is fed continuously into the still.

batcher [MECH ENG] A machine in which the ingredients of concrete are measured and combined into batches before being discharged to the concrete mixer.

batching [COMPUT SCI] Grouping records for the purpose of processing them in a computer. [ENG] Weighing or measuring the volume of the ingredients of a batch of concrete or mortar, and then introducing these ingredients into a mixer.

batch process [ENG] A process that is not in continuous or mass production; operations are carried out with discrete quantities of material or a limited number of items.

batch processing [COMPUT SCI] A technique that uses a single program loading to process many individual jobs, tasks, or requests for service.

batch terminal [COMPUT SCI] A computer terminal that provides access similar to that provided by the computer's input/output devices, but at a location convenient to the user, so that an entire task or job can be submitted at one time.

Batesian mimicry [ECOL] Resemblance of an innocuous species to one that is distasteful to predators.

batholith [GEOL] A body of igneous rock, 40 square miles (100 square kilometers) or more in area, emplaced at great or intermediate depth in the earth's crust.

bathometer [ENG] A mechanism which measures depths in water.

Bathonian [GEOL] A European stage of geologic time: Middle Jurassic, below Callovian, above Bajocian. Also known as Bathian.

Bathornithidae [PALEON] A family of Oligocene birds in the order Gruiformes.

bathyal zone [OCEANOGR] The biogeographic realm of the ocean depths between 100 and 1000 fathoms (180 and 1800 meters).

Bathyctenidae [INV ZOO] A family of bathypelagic coelenterates in the phylum Ctenophora.

Bathyergidae [VERT ZOO] A family of mammals, including the South African mole rats, in the order Rodentia.

Bathylaconoidei [VERT ZOO] A suborder of deep-sea fishes in the order Salmoniformes.

bathymetric chart [MAP] A topographic map of the floor of the ocean.

bathymetry [ENG] The science of measuring ocean depths in order to determine the sea floor topography.

Bathynellacea [INV ZOO] An order of crustaceans in the superorder Syncarida found in subterranean waters in England and central Europe.

Bathynellidae [INV ZOO] The single family of the crustacean order Bathynellacea.

bathypelagic zone [OCEANOGR] The biogeographic realm of the ocean lying between depths of 900 and 3700 meters.

Bathypteroidae [VERT ZOO] A family of benthic, deep-sea fishes in the order Salmoniformes.

bathyscaph [NAV ARCH] A free, crewed vehicle having a spherical cabin on the underside for exploring the deep ocean.

bathysphere [NAV ARCH] A spherical chamber in which persons are lowered for observation and study of ocean depths.

Bathysquillidae [INV ZOO] A family of mantis shrimps, with one genus (*Bathysquilla*) and two species, in the order Stomatopoda.

bathythermograph [ENG] A device for obtaining a record of temperature against depth (actually, pressure) in the ocean from a ship underway. Abbreviated BT. Also known as bathythermosphere.

bathythermosphere See bathythermograph.

batik [TEXT] **1.** A method of dyeing fabric in which parts of the cloth not intended to be dyed are covered with removable wax. **2.** The print so produced. **3.** The dyed cloth.

Batoidea [VERT ZOO] The skates and rays, an order of the subclass Elasmobranchii.

Batrachoididae [VERT ZOO] The single family of the order Batrachoidiformes.

Batrachoidiformes [VERT ZOO] The toadfishes, an order of teleostean fishes in the subclass Actinopterygii.

batt [MATER] A blanket of insulating material usually 16 inches (41 centimeters) wide and 3 to 6 inches (8 to 15 centimeters) thick, used to insulate building walls and roofs. [MIN ENG] A thin layer of coal occurring in the lower part of shale strata that lie above and close to a coal bed.

batten [AERO ENG] Metal, wood, or plastic panels laced to the envelope of a blimp in the nose cone to add rigidity to the nose and provide a good point of attachment for mooring. [BUILD] **1.** A sawed timber strip of specific dimension—usually 7 inches (18 centimeters) broad, less than 4 inches (10 centimeters) thick, and more than 6 feet (1.8 meters) long—used for outside walls of houses, flooring, and such. **2.** A strip of wood nailed across a door or other structure made of parallel boards to strengthen it and prevent warping.

batter [CIV ENG] A uniformly steep slope in a retaining wall or pier; inclination is expressed as 1 foot horizontally per vertical unit (in feet).

batter board [CIV ENG] Horizontal boards nailed to corner posts located just outside the corners of a proposed building to assist in the accurate layout of foundation and excavation lines.

battery [CHEM ENG] A series of distillation columns or other processing equipment operated as a single unit. [ELEC] A direct-current voltage source made up of one or more units that convert chemical, thermal, nuclear, or solar energy into electrical energy. [ORD] A group of guns or other weapons, such as mortars, machine guns, artillery pieces, or of searchlights, set up under one tactical commander in a certain area.

battery clip [ELEC] A terminal of a connecting wire having spring jaws that can be quickly snapped on a terminal of a device, such as a battery, to which a temporary wire connection is desired.

battery depolarizer See depolarizer.

battery eliminator [ELECTR] A device which supplies electron tubes with voltage from electric power supply mains.

battery separator [ELEC] An insulating plate inserted between the positive and negative plates of a battery to prevent them from touching.

baud [COMMUN] A unit of telegraph signaling speed equal to the number of code elements (pulses and spaces) per second or twice the number of pulses per second. Also known as unit pulse.

Baudot code [COMMUN] A teleprinter code that uses a combination of five or six marking and spacing intervals of equal duration for each character.

Baumé hydrometer scale [PHYS CHEM] A calibration scale for liquids that is reducible to specific gravity by the following formulas: for liquids heavier than water, specific gravity = $145 \div (145 - n)$(at 60°F); for liquids lighter than water, specific gravity = $140 \div (130 + n)$(at 60°F); n is the reading on the Baumé scale, in degrees Baumé. Baumé is abbreviated Bé.

bauxite [PETR] A whitish, grayish, brown, yellow, or reddish-brown rock composed of hydrous aluminum oxides and aluminum hydroxides and containing impurities such as free silica, silt, iron hydroxides, and clay minerals; the principal commercial source of aluminum.

bavenite See duplexite.

b axis [CRYSTAL] A crystallographic axis that is oriented horizontally, right to left. [PETR] A direction in the plane of movement that is at a right angle to the tectonic transport direction.

bay [AERO ENG] A space formed by structural partitions on an aircraft. [ARCH] Division of a building between adjacent beams or columns. [BOT] *Laurus nobilis.* An evergreen tree of the laurel family. [ELECTROMAG] One segment of an antenna array. [ENG] A housing used for equipment. [GEOGR] **1.** A body of water, smaller than a gulf and larger than a cove in a recess in the shoreline. **2.** A narrow neck of water leading from the sea between two headlands. [GEOPHYS] A simple transient magnetic disturbance, usually an hour in duration, whose appearance on a magnetic record has the shape of a V or a bay of the sea.

bay bar See baymouth bar.

bayberry [BOT] **1.** *Pimenta acris.* A West Indian tree related to the allspice; a source of bay oil. Also known as bay-rum tree; Jamaica bayberry; wild cinnamon. **2.** Any tree of the genus *Myrica.*

bay delta [GEOL] A usually triangular alluvial deposit formed at the point where the mouth of a stream enters the head of a drowned valley.

Bayer name [ASTRON] The Greek (or Roman) letter and the possessive form of the Latin name of a constellation, used as a star name; examples are α Cygni

(Deneb), β Orionis (Rigel), and η Ursae Majoris (Alkaid).

Bayer process   [MET] A method of producing alumina from bauxite by heating it in a sodium hydroxide solution.

Bayes decision rule   [STAT] A decision rule under which the strategy chosen from among several available ones is the one for which the expected value of payoff is the greatest.

bayldonite   [MINERAL] $Cu_3(AsO_4)_2(OH)_2$ An apple green to yellowish-green monoclinic mineral consisting of a basic arsenate of copper and lead; occurs in minute mammillary concretions, in massive form, and as crusts.

bayleyite   [MINERAL] $Mg_2(UO_2)(CO_3)_3 \cdot 18H_2O$ A sulfur yellow monoclinic mineral consisting of a hydrated carbonate of magnesium and uranium; occurs as minute, short-prismatic crystals.

baymouth bar   [GEOL] A bar extending entirely or partially across the mouth of a bay. Also known as bay bar.

bayonet base   [ELEC] A tube base or lamp base having two projecting pins on opposite sides of a smooth cylindrical surface to engage in corresponding slots in a bayonet socket and hold the base firmly in the socket.

bayonet socket   [DES ENG] A socket, having J-shaped slots on opposite sides, into which a bayonet base or coupling is inserted against a spring and rotated until its pins are seated firmly in the slots.

bayonet-tube exchanger   [MECH ENG] A dual-tube apparatus with heating (or cooling) fluid flowing into the inner tube and out of the annular space between the inner and outer tubes; can be inserted into tanks or other process vessels to heat or cool the liquid contents.

bayou   [HYD] A small, sluggish secondary stream or lake that exists often in an abandoned channel or a river delta.

bayside beach   [GEOL] A beach formed at the side of a bay by materials eroded from nearby headlands and deposited by longshore currents.

bay window   [ARCH] A window that projects outward from the wall of a building and forms a small indoor alcove.

bazooka   [ELECTR] See balun.   [ORD] Popular name applied to the 2.36-inch (60-millimeter) rocket launcher.

B battery   [ELECTR] The battery that furnishes required direct-current voltages to the plate and screen-grid electrodes of the electron tubes in a battery-operated circuit.

BBD   See bucket brigade device.

BB gun   See air rifle.

bbl   See barrel.

B box   See index register.

BCD system   See binary coded decimal system.

B cell   [IMMUNOL] One of a heterogeneous population of bone-marrow-derived lymphocytes which participates in the immune responses.

BCG vaccine   See Bacillus Calmette-Guerin vaccine.

BCS theory   See Bardeen-Cooper-Schrieffer theory.

Bdelloidea   [INV ZOO] An order of the class Rotifera comprising animals which resemble leeches in body shape and manner of locomotion.

Bdellomorpha   [INV ZOO] An order of ribbonlike worms in the class Enopla containing the single genus *Malacobdella*.

Bdellonemertini   [INV ZOO] An equivalent name for the Bdellomorpha.

bdft   See board-foot.

B display   [ELECTR] The presentation of radar output data in rectangular coordinates in which range and azimuth are plotted on the coordinate axes. Also known as range-bearing display.

Be   See beryllium.

Bé   See Baumé hydrometer scale.

BE   See binding energy.

beach   [GEOL] The zone of unconsolidated material that extends landward from the low-water line to where there is marked change in material or physiographic form or to the line of permanent vegetation.   [NAV] To intentionally run a craft ashore, as a landing ship.

beach cusp   See cusp.

beach cycle   [GEOL] Periodic retreat and outbuilding of beaches resulting from waves and tides.

beach gravel   [GEOL] Gravels in which most of the particles cluster about one size.

beach plain   [GEOL] Embankments of wave-deposited material added to a prograding shoreline.

beach profile   [GEOL] Intersection of a beach's ground surface with a vertical plane perpendicular to the shoreline.

beacon   [NAV] 1. A light, group of lights, electronic apparatus, or other device which emits identifying signals related to their positions so that the information so produced can be used by the navigator or pilots of aircraft and ships for guidance orientation or warning. 2. A structure where such a device is mounted or located.

beacon delay   [ELECTR] The amount of transponding delay within a beacon, that is, the time between the arrival of a signal and the response of the beacon.

beacon presentation   [ELECTR] The radarscope presentation resulting from radio-frequency waves sent out by a radar beacon.

bead   [DES ENG] A projecting rim or band.   [ELECTROMAG] A glass, ceramic, or plastic insulator through which passes the inner conductor of a coaxial transmission line and by means of which the inner conductor is supported in a position coaxial with the outer conductor.   [MET] 1. The drop of precious metal obtained by cupellation in fire assaying. 2. See weld bead.

beaded lake   See paternoster lake.

bead thermistor   [ELEC] A thermistor made by applying the semiconducting material to two wire leads as a viscous droplet, which cements the leads upon firing.

beak   [BOT] Any pointed projection, as on some fruits, that resembles a bird bill.   [INV ZOO] The tip of the umbo in bivalves.   [VERT ZOO] 1. The bill of a bird or some other animal, such as the turtle. 2. A projecting jawbone element of certain fishes, such as the sawfish and pike.

beaker   [NAV ARCH] A shipboard vessel, usually used for storing water.   [SCI TECH] A deep, open-mouthed, cylindrical vessel with thin walls, which usually has a projecting lip for pouring.

beam   [CIV ENG] A body, with one dimension large compared with the other dimensions, whose function is to carry lateral loads (perpendicular to the large dimension) and bending movements.   [NAV ARCH] The width of a ship at its widest point.   [PHYS] A concentrated, nearly unidirectional flow of particles, or a like propagation of electromagnetic or acoustic waves.   [TEXT] Spool-shaped holder, 8 to 12 feet (2.4 to 3.7 meters) in length, on which is wrapped yarn that is to be transferred to the warp holder of a loom.

beam angle   See beam width.

beam antenna [ELECTROMAG] An antenna that concentrates its radiation into a narrow beam in a definite direction.

beam approach beacon system *See* blind approach beacon system.

beam attenuator [SPECT]s15 An attachment to the spectrophotometer that reduces reference to beam energy to accommodate undersized chemical samples.

beam-balanced pump [PETRO ENG] An oil well pumping unit having a center-pivoted beam with the sucker rod plunger (pump) at the front end and a counterweight on the rearward extension.

beam bridge [CIV ENG] A fixed structure consisting of a series of steel or concrete beams placed parallel to traffic and supporting the roadway directly on their top flanges.

beam column [CIV ENG] A structural member subjected simultaneously to axial load and bending moments produced by lateral forces or eccentricity of the longitudinal load.

beam current [ELECTR] The electric current determined by the number and velocity of electrons in an electron beam.

beam-deflection amplifier [MECH ENG] A jet-interaction fluidic device in which the direction of a supply jet is varied by flow from one or more control jets which are oriented at approximately 90° to the supply jet.

beam-foil spectroscopy [ATOM PHYS] A method of studying the structure of atoms and ions in which a beam of ions energized in a particle accelerator passes through a thin carbon foil from which the ions emerge with various numbers of electrons removed and in various excited energy levels; the light or Auger electrons emitted in the deexcitation of these levels are then observed by various spectroscopic techniques. Abbreviated BFS.

beam hole [NUCLEO] A hole through the shield, and usually the reflector, of a nuclear reactor which allows a beam of radiation, especially fast neutrons, to escape for experimental purposes. Also known as glory hole.

beam lead [ELECTR] A flat thick-film lead, sometimes of gold, deposited on a semiconductor chip chemically or by evaporation, as a connecting lead for a semiconductor device or integrated circuit.

beam lobe switching [ELECTR] Method of determining the direction of a remote object by comparison of the signals corresponding to two or more successive beam angles, differing slightly from the direction of the object.

beam magnet *See* convergence magnet.

beam neutralization [PL PHYS] Neutralization of charged particles in a beam that takes place by means of charge exchange with a neutral gas.

beam pattern *See* directivity pattern.

beam power tube [ELECTR] An electron-beam tube which uses directed electron beams to provide most of its power-handling capability and in which the control grid and screen grid are essentially aligned. Also known as beam tetrode.

beam rider [AERO ENG] A missile for which the guidance system consists of standard reference signals transmitted in a radar beam which enable the missile to sense its location relative to the beam, correct its course, and thereby stay on the beam.

beam splitter [OPTICS] A mirror that reflects part of a beam of light falling on it and transmits part.

beam splitting [ELECTR] Process for increasing accuracy in locating targets by radar; by noting the azimuths at which one radar scan first discloses a target and at which radar data from it ceases, beam splitting calculates the mean azimuth for the target.

[OPTICS] The division of a beam of light into two beams by placing a special type of mirror in the path of the beam that reflects part of the light falling on it and transmits part.

beam spread [ENG] The angle of divergence from the central axis of an electromagnetic or acoustic beam as it travels through a material.

beam steering [ELECTR] Changing the direction of the major lobe of a radiation pattern, usually by switching antenna elements.

beam storage [COMPUT SCI] A magnetic storage device that employs electron beams to enter information into, or retrieve information from, storage cells; for example, a cathode-ray-tube storage.

beam switching [ELECTR] Method of obtaining more accurately the bearing or elevation of an object by comparing the signals received when the beam is in directions differing slightly in bearing or elevation; when these signals are equal, the object lies midway between the beam axes. Also known as lobe switching.

beam-switching tube [ELECTR] An electron tube which has a series of electrodes arranged around a central cathode and in which an electron beam is switched from one electrode to another. Also known as cyclophon.

beam tetrode *See* beam-power tube.

beam width [ELECTROMAG] The angle, measured in a horizontal plane, between the directions at which the intensity of an electromagnetic beam, such as a radar or radio beam, is one-half its maximum value. Also known as beam angle.

bean [BOT] The common name for various leguminous plants used as food for man and livestock; important commercial beans are true beans (*Phaseolus*) and California blackeye (*Vigna sinensis*).

bear [MIN ENG] To underhole or undermine; to drive in at the top or side of a working. [VERT ZOO] The common name for a few species of mammals in the family Ursidae.

Bear Driver *See* Boötes.

bearer [CIV ENG] Any horizontal beam, joist, or member which supports a load. [GRAPHICS] 1. In photoengraving, the metal left on a plate to protect the printing surface while molding. 2. In composition, one of the type-high slugs locked inside a chase to protect the printing surface. 3. In printing presses, one of the surface-to-surface ends of cylinders that come in contact.

bearing [CIV ENG] That portion of a beam, truss, or other structural member which rests on the supports. [MECH ENG] A machine part that supports another part which rotates, slides, or oscillates in or on it. [MIN ENG] The direction of a mine drivage, usually given in terms of the horizontal angle turned off a datum direction, such as the true north and south line. [NAV] The horizontal direction from one terrestrial point to another; basically synonymous with azimuth.

bearing angle [NAV] Horizontal direction measured from 0° at the reference direction clockwise or counterclockwise through 90° or 180°.

bearing bar [BUILD] A wrought-iron bar placed on masonry to provide a level support for floor joists. [CIV ENG] A load-carrying bar which supports a grating and which extends in the direction of the grating span. [ENG] *See* azimuth instrument.

bearing capacity [MECH] Load per unit area which can be safely supported by the ground.

bearing circle [ENG] A ring designed to fit snugly over a compass or compass repeater, and provided with vanes for observing compass bearings.

bearing cursor [ENG] Of a radar set, the radial line inscribed on a transparent disk which can be rotated manually about an axis coincident with the center of the plan position indicator; used for bearing determination. Also known as mechanical bearing cursor.

bearing partition [BUILD] A partition which supports a vertical load.

bearing pile [ENG] A vertical post or pile which carries the weight of a foundation, transmitting the load of a structure to the bedrock or subsoil without detrimental settlement.

bearing pressure [MECH] Load on a bearing surface divided by its area. Also known as bearing stress.

bearing strain [MECH] The deformation of bearing parts subjected to a load.

bearing strength [MECH] The maximum load that a column, wall, footing, or joint will sustain at failure, divided by the effective bearing area.

bearing stress *See* bearing pressure.

bearing wall [CIV ENG] A wall capable of supporting an imposed load. Also known as structural wall.

beat [PHYS] The periodic variation in amplitude of a wave that is the superposition of two simple harmonic waves of different frequencies.

beater [AGR] 1. A device for chopping or pulverizing unwanted parts of crops such as cornstalks or potato vines. 2. The part of a thresher that strikes the grains. [ENG] 1. A tool for packing in material to fill a blasthole containing a charge of powder. 2. A laborer who shovels or dumps asbestos fibers and sprays them with water in order to prepare them for the beating. [MECH ENG] A machine that cuts or beats paper stock. [TEXT] The section of a loom that drives the weft from the shed into the cloth.

beater mill *See* hammer mill.

beat frequency [ELECTR] The frequency of a signal equal to the difference in frequencies of two signals which produce the signal when they are combined in a nonlinear circuit.

beat-frequency oscillator [ELECTR] An oscillator in which a desired signal frequency, such as an audio frequency, is obtained as the beat frequency produced by combining two different signal frequencies, such as two different radio frequencies. Abbreviated BFO. Also known as heterodyne oscillator.

beating-in [ELECTR] Interconnecting two transmitter oscillators and adjusting one until no beat frequency is heard in a connected receiver; the oscillators are then at the same frequency.

beat reception *See* heterodyne reception.

Beattie and Bridgman equation [THERMO] An equation that relates the pressure, volume, and temperature of a real gas to the gas constant.

Beaufort wind scale [METEOROL] A system of code numbers from 0 to 12 classifying wind speeds into groups from 0–1 mile per hour or 0–1.6 kilometers per hour (Beaufort 0) to those over 75 miles per hour or 121 kilometers per hour (Beaufort 12).

beaver [VERT ZOO] The common name for two different and unrelated species of rodents, the mountain beaver (*Aplodontia rufa*) and the true or common beaver (*Castor canadensis*).

becquerel [NUCLEO] The International System unit of activity of a radionuclide, equal to the activity of a quantity of a radionuclide having one spontaneous nuclear transition per second. Symbolized Bq.

bed [CHEM] The ion-exchange resin contained in the column in an ion-exchange system. [CIV ENG] 1. In masonry and bricklaying, the side of a masonry unit on which the unit lies in the course of the wall; the underside when the unit is placed horizontally. 2. The layer of mortar on which a masonry unit is set. [GEOL] 1. The smallest division of a stratified rock series, marked by a well-defined divisional plane from its neighbors above and below. 2. An ore deposit, parallel to the stratification, constituting a regular member of the series of formations; not an intrusion. [GRAPHICS] The surface of a flatbed printing press on which the chase of composed type is secured for printing. [HYD] The bottom of a channel for the passage of water. [MECH ENG] The part of a machine having precisely machined ways or bearing surfaces which support or align other machine parts.

bedbug [INV ZOO] The common name for a number of species of household pests in the insect family Cimicidae that infest bedding, and by biting humans obtain blood for nutrition.

bedded [GEOL] Pertaining to rocks exhibiting depositional layering or bedding formed from consolidated sediments.

bedded vein [GEOL] A lode occupying the position of a bed that is parallel with the enclosing rock stratification.

bedding [CIV ENG] 1. Mortar, putty, or other substance used to secure a firm and even bearing, such as putty laid in the rabbet of a window frame, or mortar used to lay bricks. 2. A base which is prepared in soil or concrete for laying masonry or concrete. [GEOL] Condition where planes divide sedimentary rocks of the same or different lithology.

bedding fault [GEOL] A fault whose fault surface is parallel to the bedding plane of the constituent rocks. Also known as bedding-plane fault.

bedding plane [GEOL] Any of the division planes which separate the individual strata or beds in sedimentary or stratified rock.

bedding-plane fault *See* bedding fault.

bedding-plane slip *See* flexural slip.

Bedford limestone *See* spergenite.

bed load [GEOL] Particles of sand, gravel, or soil carried by the natural flow of a stream on or immediately above its bed.

Bedoulian [GEOL] Lower Cretaceous (lower Aptian) geologic time in Switzerland.

bedrock [GEOL] General term applied to the solid rock underlying soil or any other unconsolidated surficial cover.

bedsore *See* decubitus ulcer.

bee [INV ZOO] Any of the membranous-winged insects which compose the superfamily Apoidea in the order Hymenoptera characterized by a hairy body and by sucking and chewing mouthparts.

beech [BOT] Any of various deciduous trees of the genus *Fagus* in the beech family (Fagaceae) characterized by smooth gray bark, triangular nuts enclosed in burs, and hard wood with a fine grain.

Beehive *See* Praesepe.

beehive oven [ENG] An arched oven that carbonizes coal into coke by using the heat of combustion of gases that are formed, and of a small part of the coke that is formed, with no recovery of by-products.

Beer-Lambert-Bouguer law *See* Bouguer-Lambert-Beer law.

Beer's law [PHYS CHEM] The law which states that the absorption of light by a solution changes exponentially with the concentration, all else remaining the same.

beet [BOT] *Beta vulgaris*. The red or garden beet, a cool-season biennial of the order Caryophyllales grown for its edible, enlarged fleshy root.

beetle [INV ZOO] The common name given to members of the insect order Coleoptera. [MIN ENG] A powerful, cable-hauled propulsion unit, operated under remote control, for moving a train of wagons at the mine surface.

before the wind [NAV] In a direction approximating that toward which the wind is blowing; applies particularly to the situation where the wind is aft, and the craft is being aided by it.

Beggiatoaceae [MICROBIO] A family of bacteria in the order Cytophagales; cells are in chains in colorless, flexible, motile filaments; microcysts are not known.

Beggiatoales [MICROBIO] Formerly an order of motile, filamentous bacteria in the class Schizomycetes.

BEGIN [COMPUT SCI] An enclosing statement of ALGOL used to indicate the beginning of a block; any variable in a block enclosed by BEGIN and END is normally local to this block.

beginning-of-information marker [COMPUT SCI] A section of magnetic tape covered with reflective material that indicates the beginning of the area on which information is to be recorded.

Begoniaceae [BOT] A family of dicotyledonous plants in the order Violales characterized by an inferior ovary, unisexual flowers, stipulate leaves, and two to five carpels.

behaviorism [PSYCH] A school of psychology concerned with observable, tangible, and measurable data regarding behavior and human activities, but excluding ideas and emotions as purely subjective phenomena.

bel [PHYS] A dimensionless unit expressing the ratio of two powers or intensities, or the ratio of a power to a reference power, such that the number of bels is the common logarithm of this ratio. Symbolized b; B.

belaying pin [NAV ARCH] A pin-shaped metal rod about which ropes are secured on shipboard.

Belemnoidea [PALEON] An order of extinct dibranchiate mollusks in the class Cephalopoda.

Belgian block [MATER] 1. A stone block used for paving, having the shape of a truncated pyramid, with a depth of 7–8 inches (18–20 centimeters), a base of 5–6 inches (13–15 centimeters) square, and a face opposite the base that is 1 inch (2.5 centimeters) or less smaller than the base. 2. Any stone block used for paving.

B eliminator [ELECTR] Power pack that changes the alternating-current powerline voltage to the direct-current source required by plant circuits of vacuum tubes or semiconductor devices.

Belinuracea [PALEON] An extinct group of horseshoe crabs; arthropods belonging to the Limulida.

belite See larnite.

bell [ENG] 1. A hollow metallic cylinder closed at one end and flared at the other; it is used as a fixed-pitch musical instrument or signaling device and is set vibrating by a clapper or tongue which strikes the lip. 2. See bell tap. [MET] A conical device that seals the top of a blast furnace.

belladonna [BOT] Atropa belladonna. A perennial poisonous herb that belongs to the family Solanaceae; atropine is produced from the roots and leaves; used as an antispasmodic, as a cardiac and respiratory stimulant, and to check secretions. Also known as deadly nightshade.

bell buoy [NAV] A channel buoy with a bell which rings as the buoy swings in the waves, usually marking rocks or shoals.

belled caisson [CIV ENG] A type of drilled caisson with a flared bottom.

Bellerophontacea [PALEON] A superfamily of extinct gastropod mollusks in the order Aspidobranchia.

bell glass See bell jar.

bell jar [ENG] A bell-shaped vessel, usually made of glass, which is used for enclosing a vacuum, holding gases, or covering objects. Also known as bell glass.

bell-metal ore See stannite.

bellows [ENG] 1. A mechanism that expands and contracts, or has a rising and falling top, to suck in air through a valve and blow it out through a tube. 2. Any of several types of enclosures which have accordionlike walls, allowing one to vary the volume. 3. See aneroid capsule. [OPTICS] An accordionlike component of a camera which forms a passage between the lens and the film and allows one to vary the distance between them.

bell screw See bell tap.

bell socket See bell tap.

bell tap [ENG] A cylindrical fishing tool having an upward-tapered inside surface provided with hardened threads; when slipped over the upper end of lost, cylindrical, downhole drilling equipment and turned, the threaded inside surface of the bell tap cuts into and grips the outside surface of the lost equipment. Also known as bell; bell screw; bell socket; box bill; die; die collar; die nipple.

bell-type manometer [ENG] A differential pressure gage in which one pressure input is fed into an inverted cuplike container floating in liquid, and the other pressure input presses down upon the top of the container so that its level in the liquid is the measure of differential pressure.

Beloniformes [VERT ZOO] The former ordinal name for a group of fishes now included in the order Atheriniformes.

Belostomatidae [INV ZOO] The giant water bugs, a family of hemipteran insects in the subdivision Hydrocorisae.

belt [CIV ENG] In brickwork, a projecting row (or rows) of bricks, or an inserted row made of a different kind of brick. [ECOL] 1. Any altitudinal vegetation zone or band from the base to the summit of a mountain. 2. Any benthic vegetation zone or band from sea level to the ocean depths. 3. Any of the concentric vegetation zones around bodies of fresh water. [HYD] A long area or strip of pack ice, with a width of 1 kilometer to more than 100 kilometers. [MECH ENG] A flexible band used to connect pulleys or to convey materials by transmitting motion and power.

belt conveyor [MECH ENG] A heavy-duty conveyor consisting essentially of a head or drive pulley, a take-up pulley, a level or inclined endless belt made of canvas, rubber, or metal, and carrying and return idlers.

belt drive [MECH ENG] The transmission of power between shafts by means of a belt connecting pulleys on the shafts.

belting [MATER] 1. A sturdy fabric, usually of cotton, used in belts. 2. A heavy leather, made from hides of cattle, used in power transmission belts. Also known as belting leather.

belting leather See belting.

belt of cementation See zone of cementation.

bench [GEOL] A terrace of level earth or rock that is raised and narrow and that breaks the continuity of a declivity. [MIN ENG] 1. One of two or more divisions of a coal seam, separated by slate and so forth or simply separated by the process of cutting the coal, one bench or layer being cut before the adjacent one. 2. A long horizontal ledge of ore in an underground working place. **3.** A ledge in an open-pit mine from which excavation takes place at a constant level.

bench blasting [MIN ENG] A mining system used either underground or in surface pits whereby a thick ore or waste zone is removed by blasting a series of successive horizontal layers called benches.

bench gravel [GEOL] Gravel beds found on the sides of valleys above the present stream bottoms, representing parts of the bed of the stream when it was at a higher level.

bench mark [ENG] A relatively permanent natural or artificial object bearing a marked point whose elevation above or below an adopted datum—for example, sea level—is known. Abbreviated BM.

bend [DES ENG] **1.** The characteristic of an object, such as a machine part, that is curved. **2.** A section of pipe that is curved. **3.** A knot formed by a rope fastened to an object or another rope. [ELECTROMAG] A smooth change in the direction of the longitudinal axis of a waveguide. [GEOL] **1.** A curve or turn occurring in a stream course, bed, or channel which has not yet become a meander. **2.** The land area partly encircled by a bend or meander. [NAV] The departure of a defined navigation course or bearing from a straight line.

Ben Day process See Benday process.

Benday process [GRAPHICS] A process for printing shadings consisting of patterns of lines, dots, stipples, and so on, which involves inking a Benday screen (a rectangle of hardened gelatin with the pattern in relief), printing it on portions of the metal plate on which an outline drawing has been photoprinted, and then etching the metal as a line plate. Also spelled Ben Day process.

bending [ENG] **1.** The forming of a metal part, by pressure, into a curved or angular shape, or the stretching or flanging of it along a curved path. **2.** The forming of a wooden member to a desired shape by pressure after it has been softened or plasticized by heat and moisture. [HYD] Movement of sea ice up or down resulting from lateral pressure exerted by wind or tide. [OCEANOGR] The first stage in the formation of pressure ice caused by the action of current, wind, tide, or air temperature changes.

bending moment [MECH] Algebraic sum of all moments located between a cross section and one end of a structural member; a bending moment that bends the beam convex downward is positive, and one that bends it convex upward is negative.

Bendix-Weiss universal joint [MECH ENG] A universal joint that provides for constant angular velocity of the driven shaft by transmitting the torque through a set of four balls lying in the plane that contains the bisector of, and is perpendicular to, the plane of the angle between the shafts.

bends See caisson disease.

Benedict's solution [ANALY CHEM] A solution of potassium and sodium tartrates, copper sulfate, and sodium carbonate; used to detect reducing sugars.

beneficiation [MET] Improving the chemical or physical properties of an ore so that metal can be recovered at a profit. Also known as mineral dressing.

Benguela Current [OCEANOGR] A strong current flowing northward along the southwestern coast of Africa.

benign myalgic encephalomyelitis   See neuromyasthenia.

Benioff extensometer   [ENG] A linear strainmeter for measuring the change in distance between two

reference points separated by 20–30 meters or more; used to observe earth tides.

Benioff zone   [GEOPHYS] A zone of earthquake hypocenters distributed on well-defined planes that dips from a shallow depth into the earth's mantle to depths as great as 700 kilometers.

Bennettitales   [PALEOBOT] An equivalent name for the Cycadeoidales.

Bennettitatae   [PALEOBOT] A class of fossil gymnosperms in the order Cycadeoidales.

benthic   [OCEANOGR] Of, pertaining to, or living on the bottom or at the greatest depths of a large body of water. Also known as benthonic.

benthonic   See benthic.

benthos   [ECOL] Bottom-dwelling forms of marine life. Also known as bottom fauna. [OCEANOGR] The floor or deepest part of a sea or ocean.

bentonite   [GEOL] A clay formed from volcanic ash decomposition and largely composed of montmorillonite and beidellite. Also known as taylorite.

bent-tube boiler   [MECH ENG] A water-tube steam boiler in which the tubes terminate in upper and lower steam-and-water drums. Also known as drum-type boiler.

benzaldehyde cyanohydrin   See mandelonitrile.

benzene   [ORG CHEM] C_6H_6 A colorless, liquid, flammable, aromatic hydrocarbon that boils at 80.1°C and freezes at 5.4–5.5°C; used to manufacture styrene and phenol. Also known as benzol.

benzene ring   [ORG CHEM] The six-carbon ring structure found in benzene, C_6H_6, and in organic compounds formed from benzene by replacement of one or more hydrogen atoms by other chemical atoms or radicals.

benzene series   [ORG CHEM] A series of carbon-hydrogen compounds based on the benzene ring, with the general formula C_nH_{2n-6}, where n is 6 or more; examples are benzene, C_6H_6; toluene, C_7H_8; and xylene, C_8H_{10}.

benzol   See benzene.

benzyl   [ORG CHEM] The radical $C_6H_5CH_2^-$ found, for example, in benzyl alcohol, $C_6H_5CH_2OH$.

Berberidaceae   [BOT] A family of dicotyledonous herbs and shrubs in the order Ranunculales characterized by alternate leaves, perfect, well-developed flowers, and a seemingly solitary carpel.

Berenice's Hair   See Coma Berenices.

berg crystal   See rock crystal.

Bergius process   [CHEM ENG] Treatment of carbonaceous matter, such as coal or cellulosic materials, with hydrogen at elevated pressures and temperatures in the presence of a catalyst, to form an oil similar to crude petroleum. Also known as coal hydrogenation.

berg till   See floe till.

beriberi   [MED] A disorder resulting from the deficiency of vitamin B_1 and characterized by neurologic symptoms, cardiovascular abnormalities, edema, and cerebral manifestations.

Berkefeld filter   [MICROBIO] A diatomaceous-earth filter used for sterilization of heat-labile liquids, such as blood serum, enzyme solutions, and antibiotics.

berkelium   [CHEM] A radioactive element, symbol Bk, atomic number 97, the eighth member of the actinide series; properties resemble those of the rare-earth cerium.

berkeyite   See lazulite.

berm   [CIV ENG] A horizontal ledge cut between the foot and top of an embankment to stabilize the slope by intercepting sliding earth. [GEOL] **1.** A narrow terrace which originates from the interruption of an erosion cycle with rejuvenation of a stream in the mature stage of its development and renewed dissection. **2.** A horizontal portion of a beach or backshore formed by deposit of material as a result of wave action. Also known as backshore terrace; coastal berm.

Bermuda high [METEOROL] The semipermanent subtropical high of the North Atlantic Ocean, especially when it is located in the western part of that ocean area.

Bernoulli differential equation *See* Bernoulli equation.

Bernoulli distribution *See* binomial distribution.

Bernoulli equation [FL MECH] *See* Bernoulli theorem. [MATH] A nonlinear first-order differential equation of the form $(dy/dx) + yf(x) = y^n g(x)$, where n is a number different from unity and f and g are given functions. Also known as Bernoulli differential equation.

Bernoulli law *See* Bernoulli theorem.

Bernoulli number [MATH] Numerical value of the coefficient of $x^{2n}/(2n)!$ is the expansion of $xe^x/(e^x - 1)$.

Bernoulli polynomial [MATH] The nth one is

$$\sum_{k=0}^{n} \binom{n}{k} B_k Z^{n-k}$$

where $\binom{n}{k}$ is a binomial coefficient, and B_k is a Bernoulli number.

Bernoulli's lemniscate [MATH] A curve shaped like a figure eight whose equation in rectangular coordinates is expressed as $(x^2 + y^2)^2 = a^2(x^2 - y^2)$.

Bernoulli theorem [FL MECH] An expression of the conservation of energy in the steady flow of an incompressible, inviscid fluid; it states that the quantity $(p/\rho) + gz + (v^2/2)$ is constant along any streamline, where p is the fluid pressure, v is the fluid velocity, ρ is the mass density of the fluid, g is the acceleration due to gravity, and z is the vertical height. Also known as Bernoulli equation; Bernoulli law. [STAT] *See* law of large numbers.

Beroida [INV ZOO] The single order of the class Nuda in the phylum Ctenophora.

Berriasian [GEOL] Part of or the underlying stage of the Valanginian at the base of the Cretaceous.

berry [BOT] A usually small, simple, fleshy or pulpy fruit, such as a strawberry, grape, tomato, or banana.

Berthelot equation [PHYS CHEM] A form of the equation of state which relates the temperature, pressure, and volume of a gas with the gas constant.

berthonite *See* bournonite.

Bertrand curve [MATH] One of a pair of curves having the same principal normals. Also known as associate curve; conjugate curve.

bertrandite [MINERAL] $Be_4Si_2O_7(OH)_2$ A colorless or pale-yellow mineral consisting of a beryllium silicate occurring in prismatic crystals; hardness is 6–7 on Mohs scale, and specific gravity is 2.59–2.60.

Bertrand lens [OPTICS] An auxiliary lens that can be inserted in the tube of a polarizing microscope to obtain interference figures.

Beryciformes [VERT ZOO] An order of actinopterygian fishes in the infraclass Teleostei.

Berycomorphi [VERT ZOO] An equivalent name for the Beryciformes.

beryllium [CHEM] A chemical element, symbol Be, atomic number 4, atomic weight 9.0122. [MET] A rare metal, occurring naturally in combinations, with density about one-third of aluminum; used most commonly in the manufacture of beryllium-copper alloys which find numerous industrial and scientific applications.

beryllonite [MINERAL] $NaBe(PO_4)$ A colorless or yellow mineral occurring in short, prismatic or tabular, monoclinic crystals with two good pinacoidal cleavages at right angles; hardness is 5.5–6 on Mohs scale, and specific gravity is 2.85.

Berytidae [INV ZOO] The stilt bugs, a small family of hemipteran insects in the superfamily Pentatomorpha.

Bessel equation [MATH] The differential equation $z^2 f''(z) + z f'(z) + (z^2 - n^2) f(z) = 0$.

Bessel inequality [MATH] The statement that the sum of the squares of the inner product of a vector with the members of an orthonormal set is no larger than the square of the norm of the vector.

Bessel transform *See* Hankel transform.

Bessemer converter [MET] A pear-shaped, basic-lined, cylindrical vessel for producing steel by the Bessemer process.

Bessemer iron [MET] Pig iron with about 1% silicon and a sulfur and a phosphorus content below 0.10%; used to make steel by the Bessemer or the acid open-hearth process. Also known as Bessemer pig iron.

Bessemer pig iron *See* Bessemer iron.

Bessemer process [MET] A steelmaking process in which carbon, silicon, phosphorus, and manganese contained in molten pig iron are oxidized by a strong blast of air.

beta-absorption gage *See* beta gage.

beta blocker [PHYSIO] An adrenergic blocking agent capable of blocking nerve impulses to special sites (beta receptors) in the cerebellum; reduces the rate of heartbeats and the force of heart contractions.

beta brass [MET] A type of brass containing nearly equal proportions of copper and zinc.

beta cell [HISTOL] **1.** Any of the basophilic chromophiles in the anterior lobe of the adenohypophysis. **2.** One of the cells of the islets of Langerhans which produce insulin.

betacyanin [BIOCHEM] A group of purple plant pigments found in leaves, flowers, and roots of members of the order Caryophyllales.

beta decay [NUC PHYS] Radioactive transformation of a nuclide in which the atomic number increases or decreases by unity with no change in mass number; the nucleus emits or absorbs a beta particle (electron or positron). Also known as beta disintegration.

beta disintegration *See* beta decay.

beta distribution [STAT] The probability distribution of a random variable with density function $f(x) = [x^{\alpha-1}(1-x)^{\beta-1}]/B(\alpha,\beta)$, where B represents the beta function, α and β are positive real numbers, and $0<x<1$. Also known as Pearson Type I distribution.

betafite *See* ellsworthite.

beta function [MATH] A function of two positive variables, defined by

$$B(m,n) = \int_0^1 x^{m-1}(1-x)^{n-1} dx$$

beta gage [NUCLEO] A penetration-type thickness gage that measures the absorption of beta rays in the sample. Also known as beta-absorption gage.

beta globulin [BIOCHEM] A heterogeneous fraction of serum globulins containing transferrin and various complement components.

beta interaction *See* weak interaction.

betalain [BIOCHEM] The name for a group of 35 red or yellow compounds found only in plants of the family Caryophyllales, including red beets, red chard, and cactus fruits.

beta particle [NUC PHYS] An electron or positron emitted from a nucleus during beta decay.

beta ray [NUC PHYS] A stream of beta particles.

beta rhythm [PHYSIO] An electric current of low voltage from the brain, with a pulse frequency of 13–30/sec, encountered in a person who is aroused and anxious.

beta taxonomy [SYST] The level of taxonomic study dealing with the arrangement of species into lower and higher taxa.

betatron [NUCLEO] A device for accelerating electrons in an evacuated ring by means of a time-varying magnetic flux encircled by the ring. Also known as induction accelerator; rheotron.

betaxanthin [BIOCHEM] The name given to any of the yellow pigments found only in plants of the family Caryophyllales; they always occur with betacyanins.

Bethylidae [INV ZOO] A small family of hymenopteran insects in the superfamily Bethyloidea.

Bethyloidea [INV ZOO] A superfamily of hymenopteran insects in the suborder Apocrita.

Betti group See homology group.

Betti number See connectivity number.

Betulaceae [BOT] A small family of dicotyledonous plants in the order Fagales characterized by stipulate leaves, seeds without endosperm, and by being monoecious with female flowers mostly in catkins.

Betz cell [HISTOL] Any of the large conical cells composing the major histological feature of the precentral motor cortex in humans.

bevel [DES ENG] 1. The angle between one line or surface and another line or surface, or the horizontal, when this angle is not a right angle. 2. A sloping surface or line. [GRAPHICS] An instrument composed of two rules joined together, opening to any angle, which is used to draw angles and measure and lay off bevels.

beveling [GEOL] Planing by erosion of the outcropping edges of strata. [MECH ENG] See chamfering.

beyerite [MINERAL] $(Ca,Pb)Bi_2(CO_3)_2O_2$ A bright yellow to lemon yellow, tetragonal mineral consisting of bismuth and calcium carbonate; occurs as thin plates and compact earthy masses.

Beyrichacea [PALEON] A superfamily of extinct ostracods in the suborder Beyrichicopina.

Beyrichicopina [PALEON] A suborder of extinct ostracods in the order Paleocopa.

Beyrichiidae [PALEON] A family of extinct ostracods in the superfamily Beyrichacea.

bezel [DES ENG] 1. A grooved rim used to hold a transparent glass or plastic window or lens for a meter, tuning dial, or some other indicating device. 2. A sloping face on a cutting tool. [LAP] The oblique face of a cut gem, between the table and the girdle.

BFO See beat-frequency oscillator.

BFS See beam-foil spectroscopy.

B-H curve [ELECTROMAG] A graphical curve showing the relation between magnetic induction B and magnetizing force H for a magnetic material. Also known as magnetization curve.

Bi See abampere; bismuth.

bias [ELEC] 1. A direct-current voltage used on signaling or telegraph relays or electromagnets to secure desired time spacing of transitions from marking to spacing. 2. The restraint of a relay armature by spring tension to secure a desired time spacing of transitions from marking to spacing. 3. The effect on teleprinter signals produced by the electrical characteristics of the line and equipment. 4. The force applied to a relay to hold it in a given position.

[ELECTR] 1. A direct-current voltage applied to a transistor control electrode to establish the desired operating point. 2. See grid bias. [STAT] In estimating the value of a parameter of a probability distribution, the difference between the expected value of the estimator and the true value of the parameter.

bias cell [ELECTR] A small dry cell used singly or in series to provide the required negative bias for the grid circuit of an electron tube. Also known as grid-bias cell.

bias current [ELECTR] 1. An alternating electric current above about 40,000 hertz added to the audio current being recorded on magnetic tape to reduce distortion. 2. An electric current flowing through the base-emitter junction of a transistor and adjusted to set the operating point of the transistor.

bias distortion [COMMUN] See bias telegraph distortion. [ELECTR] Distortion resulting from the operation on a nonlinear portion of the characteristic curve of a vacuum tube or other device, due to improper biasing.

biased automatic gain control See delayed automatic gain control.

biased statistic [STAT] A statistic whose expected value, as obtained from a random sampling, does not equal the parameter or quantity being estimated.

bias error [STAT] A measurement error that remains constant in magnitude for all observations; a kind of systematic error.

bias meter [COMMUN] A meter used in teletypewriter work for measuring signal bias directly in percent; a positive reading indicates a marking signal bias; a negative reading, a spacing signal bias.

bias oscillator [ELECTR] An oscillator used in a magnetic recorder to generate the alternating-current signal that is added to the audio current being recorded on magnetic tape to reduce distortion.

bias resistor [ELECTR] A resistor used in the cathode or grid circuit of an electron tube to provide a voltage drop that serves as the bias.

bias telegraph distortion [COMMUN] A distortion that causes telegraph mark-and-space pulses to be lengthened or shortened; often caused by changes in the amplitude of incoming pulses. Also known as bias distortion; spacing bias.

bias voltage [ELECTR] A voltage applied or developed between two electrodes as a bias.

bias winding [ELEC] A control winding that carries a steady direct current which serves to establish desired operating conditions in a magnetic amplifier or other magnetic device.

biaxial crystal [CRYSTAL] A crystal of low symmetry in which the index ellipsoid has three unequal axes.

biaxial stress [MECH] The condition in which there are three mutually perpendicular principal stresses; two act in the same plane and one is zero.

bibb cock See bibcock.

bibcock [DES ENG] A faucet or stopcock whose nozzle is bent downward. Also spelled bibb cock.

bicameral [BIOL] Having two chambers, as the heart of a fish.

bicapsular [BIOL] 1. Having two capsules. 2. Having a capsule with two locules.

bicarbonate [INORG CHEM] A salt obtained by the neutralization of one hydrogen in carbonic acid.

biceps [ANAT] 1. A bicipital muscle. 2. The large muscle of the front of the upper arm that flexes the forearm; biceps brachii. 3. The thigh muscle that flexes the knee joint and extends the hip joint; biceps femoris.

bicipital [ANAT] 1. Pertaining to muscles having two origins. 2. Pertaining to ribs having double articulation with the vertebrae. [BOT] Having two heads or two supports.

bicompact set See compact set.

biconditional gate See equivalence gate.

bicontinuous function See homeomorphism.

bicornuate uterus [ANAT] A uterus with two horn-shaped processes on the superior aspect.

Bicosoecida [INV ZOO] An order of colorless, free-living protozoans, each member having two flagella, in the class Zoomastigophorea.

bicuspid [ANAT] Any of the four double-pointed premolar teeth in humans. [BIOL] Having two points or prominences.

bidentate liquid [INORG CHEM] A chelating agent having two groups capable of attachment to a metal ion.

bidirectional antenna [ELECTROMAG] An antenna that radiates or receives most of its energy in only two directions.

bidirectional clamping circuit [ELECTR] A clamping circuit that functions at the prescribed time irrespective of the polarity of the signal source at the time the pulses used to actuate the clamping action are applied.

bidirectional clipping circuit [ELECTR] An electronic circuit that prevents transmission of the portion of an electrical signal that exceeds a prescribed maximum or minimum voltage value.

bidirectional counter See forward-backward counter.

bidirectional microphone [ENG ACOUS] A microphone that responds equally well to sounds reaching it from the front and rear, corresponding to sound incidences of 0 and 180°.

bidirectional transducer [ELECTR] A transducer capable of measuring in both positive and negative directions from a reference position. Also known as bilateral transducer.

bidirectional transistor [ELECTR] A transistor that provides switching action in either direction of signal flow through a circuit; widely used in telephone switching circuits.

bieberite [MINERAL] $CoSO_4 \cdot 7H_2O$ A rose red or flesh red, monoclinic mineral consisting of cobalt sulfate heptahydrate; occurs as crusts and stalactites.

biennial plant [BOT] A plant that requires two growing seasons to complete its life cycle.

bifacial [BOT] Of a leaf, having dissimilar tissues on the upper and lower surfaces. [DES ENG] Of a tool, having both sides alike.

bifilar electrometer [ENG] An electrostatic voltmeter in which two conducting quartz fibers, stretched by a small weight or spring, are separated by their attraction in opposite directions toward two plate electrodes carrying the voltage to be measured.

bifilar micrometer See filar micrometer.

bifilar winding [ELEC] A winding consisting of two insulated wires, side by side, with currents traveling through them in opposite directions.

bifocal lens [OPTICS] 1. A lens with two parts having different focal lengths. 2. In particular, an eyeglass lens having one part that corrects for distant vision and one part for near vision.

bifurcation [SCI TECH] 1. Division into two branches, parts, or aspects. 2. Point at which division occurs.

big bang theory [ASTRON] A theory of the origin and evolution of the universe which holds that approximately 2×10^{10} years ago all the matter in the universe was packed into a small agglomeration of extremely high density and temperature which

exploded, sending matter in all directions and giving rise to the expanding universe. Also known as superdense theory.

Big Dipper [ASTRON] A group of stars that is part of the constellation Ursa Major. Also known as Charles' wain.

bight [GEOL] 1. A long, gradual bend or recess in the coastline which forms a large, open receding bay. 2. A bend in a river or mountain range. [OCEANOGR] An indentation in shelf ice, fast ice, or a floe.

Bignoniaceae [BOT] A family of dicotyledonous trees or shrubs in the order Scrophulariales characterized by a corolla with mostly five lobes, mature seeds with little or no endosperm and with wings, and opposite or whorled leaves.

biharmonic function [MATH] A solution to the partial differential equation $\Delta^2 u(x,y,z) = 0$, where Δ is the Laplacian operator; occurs frequently in problems in electrostatics.

bilateral [BIOL] Of or relating to both right and left sides of an area, organ, or organism. [ELECTR] Having a voltage current characteristic curve that is symmetrical with respect to the origin.

bilateral antenna [ELECTROMAG] An antenna having maximum response in exactly opposite directions, 180° apart, such as a loop.

bilateral cleavage [EMBRYO] The division pattern of a zygote that results in a bilaterally symmetrical embryo.

bilateral symmetry [BIOL] Symmetry such that the body can be divided by one median, or sagittal, dorsoventral plane into equivalent right and left halves, each a mirror image of the other.

bilateral transducer See bidirectional transducer.

bilayer [CHEM] A layer two molecules thick, such as that formed on the surface of the aqueous phase by phospholipids in aqueous solution.

bile [PHYSIO] An alkaline fluid secreted by the liver and delivered to the duodenum to aid in the emulsification, digestion, and absorption of fats. Also known as gall.

bile acid [BIOCHEM] Any of the liver-produced steroid acids, such as taurocholic acid and glycocholic acid, that appear in the bile as sodium salts.

bile pigment [BIOCHEM] Either of two colored organic compounds found in bile: bilirubin and biliverdin.

bilge [NAV ARCH] 1. Part of the underwater body of a ship between the flat of the bottom and the straight vertical sides. 2. Internally, the lowest part of the hull, next to the keelson.

biliary colic [MED] Severe abdominal pain caused by passage of a gallstone through the bile ducts into the duodenum.

biliary system [ANAT] The complex of canaliculi, or microscopic bile ducts, that empty into the larger intrahepatic bile ducts.

bilirubin [BIOCHEM] $C_{33}H_{36}N_4O_6$ An orange, crystalline pigment occurring in bile; the major metabolic breakdown product of heme.

biliverdin [BIOCHEM] $C_{33}H_{34}N_4O_6$ A green, crystalline pigment occurring in the bile of amphibians, birds, and humans; oxidation product of bilirubin in humans.

bill [DES ENG] One blade of a pair of scissors. [INV ZOO] A flattened portion of the shell margin of the broad end of an oyster. [NAV ARCH] The point at the end of an anchor fluke. [VERT ZOO] The jaws, together with the horny covering, of a bird. [ZOO] Any jawlike mouthpart.

billet [ENG] In a hydraulic extrusion press, a large cylindrical cake of plastic material placed within the pressing chamber. [MET] A semifinished, short, thick bar of iron or steel in the form of a cylinder or rectangular prism produced from an ingot; limited to 1.5 inches (3.8 centimeters) in width and thickness with a cross-sectional area up to 36 square inches (232 square centimeters).

Billingsellacea [PALEON] A group of extinct articulate brachiopods in the order Orthida.

billow cloud [METEOROL] Broad, nearly parallel lines of cloud oriented normal to the wind direction, with cloud bases near an inversion surface. Also known as undulatus.

bimetal [MATER] A laminate of two dissimilar metals, with different coefficients of thermal expansion, bonded together.

bimetallic thermometer [ENG] A temperature-measuring instrument in which the differential thermal expansion of thin, dissimilar metals, bonded together into a narrow strip and coiled into the shape of a helix or spiral, is used to actuate a pointer. Also known as differential thermometer.

bin [COMPUT SCI] A magnetic-tape memory in which a number of tapes are stored in a single housing. [ENG] An enclosed space, box, or frame for the storage of bulk substance.

binary [COMPUT SCI] Possessing a property for which there exists two choices or conditions, one choice excluding the other. [SCI TECH] Composed of or characterized by two parts or elements.

binary arithmetic operation [COMPUT SCI] An arithmetical operation in which the operands are in the form of binary numbers. Also known as binary operation.

binary cell [COMPUT SCI] An elementary unit of computer storage that can have one or the other of two stable states and can thus store one bit of information.

binary chain [COMPUT SCI] A series of binary circuit elements so arranged that each can change the state of the one following it.

binary code [COMPUT SCI] A code in which each allowable position has one of two possible states, commonly 0 and 1; the binary number system is one of many binary codes.

binary coded character [COMPUT SCI] One element of a notation system representing alphanumeric characters such as decimal digits, alphabetic letters, and punctuation marks by a predetermined configuration of consecutive binary digits.

binary coded decimal system [COMPUT SCI] A system of number representation in which each digit of a decimal number is represented by a binary number. Abbreviated BCD system.

binary coded decimal-to-decimal converter [COMPUT SCI] A computer circuit which selects one of ten outputs corresponding to a four-bit binary coded decimal input, placing it in the 0 state and the other nine outputs in the 1 state.

binary coded octal system [COMPUT SCI] Octal numbering system in which each octal digit is represented by a three-place binary number.

binary compound [CHEM] A compound that has two elements; it may contain two or more atoms; examples are KCl and $AlCl_3$.

binary dump [COMPUT SCI] The operation of copying the contents of a computer memory in binary form onto an external storage device.

binary fission [BIOL] A method of asexual reproduction accomplished by the splitting of a parent cell into two equal, or nearly equal, parts, each of which grows to parental size and form.

binary image [COMPUT SCI] A representation in a computer storage device of each of the holes in a punch card or paper tape (for example, by indicating the places where there are holes with a 1 and the places where there are no holes with a 0), to be differentiated from the characters represented by the combinations of holes.

binary incremental representation [COMPUT SCI] A type of incremental representation in which the value of change in a variable is represented by one binary digit which is set equal to 1 if there is an increase in the variable and to 0 if there is a decrease.

binary loader [COMPUT SCI] A computer program which transfers to main memory an exact image of the binary pattern of a program held in a storage or input device.

binary notation *See* binary number system.

binary number [MATH] A number expressed in the binary number system of positional notation.

binary number system [MATH] A representation for numbers using only the digits 0 and 1 in which successive digits are interpreted as coefficients of successive powers of the base 2. Also known as binary notation; binary system.

binary operation [COMPUT SCI] *See* binary arithmetic operation. [MATH] A rule for combining two elements of a set to obtain a third element of that set, for example, addition and multiplication.

binary point [COMPUT SCI] The character, or the location of an implied symbol, that separates the integral part of a numerical expression from its fractional part in binary notation.

binary search [COMPUT SCI] A dichotomizing search in which the set of items to be searched is divided at each step into two equal, or nearly equal, parts.

binary star [ASTRON] A pair of stars located sufficiently near each other in space to be connected by the bond of mutual gravitational attraction, compelling them to describe an orbit around their common center of gravity. Also known as binary system.

binary system [ASTRON] *See* binary star. [ENG] Any system containing two principal components. [MATH] *See* binary number system.

binary word [COMPUT SCI] A group of bits which occupies one storage address and is treated by the computer as a unit.

binaural [ACOUS] Pertaining to sound that reaches the listener over two paths, to give the effect of auditory perspective.

binaural hearing [PHYSIO] The perception of sound by stimulation of two ears.

binaural sound [ACOUS] The sound resulting from a reproduction system which has two channels, each fed into a different earphone or loudspeaker, so that a listener hears sounds coming from their original directions (with reference to the separated microphones used in recording the original sounds).

binder [MATER] A resin or other cementlike material used to hold particles together and provide mechanical strength or to ensure uniform consistency, solidification, or adhesion to a surface coating; typical binders are resin, glue, gum, and casein.

binder course [CIV ENG] Coarse aggregate with a bituminous binder between the foundation course and the wearing course of a pavement.

B indicator *See* B scope.

binding energy [PHYS] Abbreviated BE. Also known as total binding energy (TBE). **1.** The net energy required to remove a particle from a system. **2.** The net energy required to decompose a system into its constituent particles.

binding site *See* active site.

bind-seize *See* freeze.

binistor [ELECTR] A silicon *npn* tetrode that serves as a bistable negative-resistance device.

binnacle [NAV ARCH] A stand, case, or box for a ship's magnetic compass, which may contain nonmagnetic gear such as a lamp.

binocular [BIOL] **1.** Of, pertaining to, or used by both eyes. **2.** Of a type of visual perception which provides depth-of-field focus due to angular difference between the two retinal images. [OPTICS] Any optical instrument designed for use with both eyes to give enhanced views of distant objects, whose distinguishing performance feature is the depth perception obtainable.

binocular accommodation [PHYSIO] Automatic lens adjustment by both eyes simultaneously for focusing on distant objects.

binode [ELECTR] An electron tube with two anodes and one cathode used as a full-wave rectifier. Also known as double diode.

binomen [SYST] A binomial name assigned to species, as *Canis familiaris* for the dog.

binomial [MATH] A polynomial with only two terms.

binomial coefficient [MATH] A coefficient in the expansion of $(x + y)_n$, where n is a positive integer; the $(k + 1)$st coefficient is equal to the number of ways of choosing k objects out of n without regard for order. Symbolized $\binom{n}{k}$; $_nC_k$; $C(n,k)$; C_k^n.

binomial distribution [STAT] The distribution of a binomial random variable; the distribution (n,p) is given by $P(B = r) = \binom{n}{k} p^r q^{n-r}$, $p + q = 1$. Also known as Bernoulli distribution.

binomial equation [MATH] An equation of the form $x^n - a = 0$.

binomial expansion *See* binomial series.

binomial nomenclature [SYST] The Linnean system of classification requiring the designation of a binomen, the genus and species name, for every species of plant and animal.

binomial series [MATH] The expansion of $(x + y)^n$ when n is neither a positive integer nor zero. Also known as binomial expansion.

binomial trials [STAT] A sequence of trials, on each trial of which a certain result may or may not happen.

bioacoustics [BIOL] The study of the relation between living organisms and sound.

bioassay [ANALY CHEM] A method for quantitatively determining the concentration of a substance by its effect on the growth of a suitable animal, plant, or microorganism under controlled conditions.

biocenose *See* biotic community.

biochemical oxygen demand [MICROBIO] The amount of dissolved oxygen required to meet the metabolic needs of anaerobic microorganisms in water rich in organic matter, such as sewage. Abbreviated BOD. Also known as biological oxygen demand.

biochemistry [CHEM] The study of chemical substances occurring in living organisms and the reactions and methods for identifying these substances.

biochore [ECOL] A group of similar biotopes.

biochrome [BIOCHEM] Any naturally occurring plant or animal pigment.

biochronology [GEOL] The relative age dating of rock units based on their fossil content.

biociation [ECOL] A subdivision of a biome distinguished by the predominant animal species.

bioclastic rock [PETR] Rock formed from material broken or arranged by animals, humans, or sometimes plants; a rock composed of broken calcareous remains of organisms.

bioclimatology [ECOL] The study of the effects of the natural environment on living organisms.

bioconversion [CHEM ENG] The transformation of algae or other biomass materials in successive stages to aliphatic organic acids, to aliphatic hydrocarbons, or to diesel or other liquid fuels.

biocycle [ECOL] A group of similar biotopes composing a major division of the biosphere; there are three biocycles: terrestrial, marine, and fresh-water.

biodegradability [MATER] The characteristic of a substance that can be broken down by microorganisms.

biodynamics [BIOPHYS] The study of the effects of dynamic processes (motion, acceleration, weightlessness, and so on) on living organisms.

bioelectric current [PHYSIO] A self-propagating electric current generated on the surface of nerve and muscle cells by potential differences across excitable cell membranes.

bioelectrochemistry [PHYSIO] The study of the control of biological growth and repair processes by electrical stimulation.

bioengineering [ENG] The application of engineering knowledge to the fields of medicine and biology.

biofacies [GEOL] **1.** A rock unit differing in biologic aspect from laterally equivalent biotic groups. **2.** Lateral variation in the biologic aspect of a stratigraphic unit.

bioflavonoid [BIOCHEM] A group of compounds obtained from the rinds of citrus fruits and involved with the homeostasis of the walls of small blood vessels; in guinea pigs a marked reduction of bioflavonoids results in increased fragility and permeability of the capillaries; used to decrease permeability and fragility in capillaries in certain conditions. Also known as citrus flavonoid compound; vitamin P complex.

biogenic [BIOL] **1.** Essential to the maintenance of life. **2.** Produced by actions of living organisms.

biogeochemical cycle [GEOCHEM] The chemical interactions that exist between the atmosphere, hydrosphere, lithosphere, and biosphere.

biogeochemistry [GEOCHEM] A branch of geochemistry that is concerned with biologic materials and their relation to earth chemicals in an area.

biogeography [ECOL] The science concerned with the geographical distribution of animal and plant life.

bioherm [GEOL] A circumscribed mass of rock exclusively or mainly constructed by marine sedimentary organisms such as corals, algae, and stromatoporoids. Also known as organic mound.

biolith [PETR] A rock formed from or by organic material. Also known as biolite.

biological [BIOL] Of or pertaining to life or living organisms. [IMMUNOL] A biological product used to induce immunity to various infectious diseases or noxious substances of biological origin.

biological clock [PHYSIO] Any physiologic factor that functions in regulating body rhythms.

biological half-life [PHYSIO] The time required by the body to eliminate half of the amount of an administered substance through normal channels of elimination.

biological oxidation [BIOCHEM] Energy-producing reactions in living cells involving the transfer of hydrogen atoms or electrons from one molecule to another.

biological oxygen demand *See* biochemical oxygen demand.

biological productivity [ECOL] The quantity of organic matter or its equivalent in dry matter, carbon, or energy content which is accumulated during a given period of time.

biological specificity [BIOL] The principle that defines the orderly patterns of metabolic and developmental reactions giving rise to the unique characteristics of the individual and of its species.

biological warfare [ORD] Abbreviated BW. 1. Employment of living microorganisms, toxic biological products, and plant growth regulators to produce death or injury in man, animals, or plants. 2. Defense against such action.

biology [SCI TECH] A division of the natural sciences concerned with the study of life and living organisms.

bioluminescence [BIOL] The emission of visible light by living organisms.

biomagnetism [BIO PHYS] 1. Phenomenon involving magnetic fields surrounding parts or the whole of a living biological system. 2. The effects of magnetism on parts or the whole of a biological entity.

biomass [ECOL] The dry weight of living matter, including stored food, present in a species population and expressed in terms of a given area or volume of the habitat.

biome [ECOL] A complex biotic community covering a large geographic area and characterized by the distinctive life-forms of important climax species.

biomechanics [BIOPHYS] The study of the mechanics of living organisms.

biomedical engineering [ENG] The application of engineering technology to the solution of medical problems; examples are the development of prostheses such as artificial valves for the heart, various types of sensors for the blind, and automated artificial limbs.

biomere [ECOL] A biostratigraphic unit bounded by abrupt nonevolutionary changes in the dominant elements of a single phylum.

biometrics [STAT] The use of statistics to analyze observations of biological phenomena.

bionavigation [VERT ZOO] The ability of animals such as birds to find their way back to their roost, even if the landmarks on the outward-bound trip were effectively concealed from them.

bionics [ENG] The study of systems, particularly electronic systems, which function after the manner of living systems.

biopelite See black shale.

biophage See macroconsumer.

biophysics [SCI TECH] The hybrid science involving the application of physical principles and methods to study and explain the structures of living organisms and the mechanics of life processes.

biopotential [PHYSIO] Voltage difference measured between points in living cells, tissues, and organisms.

biopsy [PATH] The removal and examination of tissues, cells, or fluids from the living body for the purposes of diagnosis.

biorheology [BIOPHYS] The study of the deformation and flow of biological fluids, such as blood, mucus, and synovial fluid.

biosatellite [AERO ENG] An artificial satellite designed to contain and support humans, animals, or other living material in a reasonably normal manner for a period of time and to return safely to earth.

biosphere [ECOL] The life zone of the earth, including the lower part of the atmosphere, the hydrosphere, soil, and the lithosphere to a depth of about 2 kilometers.

biostratigraphic unit [GEOL] A stratum or body of strata that is defined and identified by one or more distinctive fossil species or genera without regard to lithologic or other physical features or relations.

biostratigraphy [PALEON] A part of paleontology concerned with the study of the conditions and deposition order of sedimentary rocks.

biostrome [GEOL] A bedded structure or layer (bioclastic stratum) composed of calcite and dolomitized calcarenitic fossil fragments distributed over the sea bottom as fine lentils, independent of or in association with bioherms or other areas of organic growth.

biosynthesis [BIOCHEM] Production, by synthesis or degradation, of a chemical compound by a living organism.

biot [ELEC] See abampere. [OPTICS] A unit of rotational strength in substances exhibiting circular dichroism, equal to 10^{-40} times the corresponding centimeter-gram-second unit.

biota [BIOL] 1. Animal and plant life characterizing a given region. 2. Flora and fauna, collectively.

biotar lens [OPTICS] A modern camera lens which is a modified Gauss objective with a large aperture and a field of about 24°.

biotechnology [ENG] The application of engineering and technological principles to the life sciences.

biotelemetry [ENG] The use of telemetry techniques, especially radio waves, to study behavior and physiology of living things.

biotic [BIOL] 1. Of or pertaining to life and living organisms. 2. Induced by the actions of living organisms.

biotic community [ECOL] An aggregation of organisms characterized by a distinctive combination of both animal and plant species in a particular habitat. Also known as biocenose.

biotic isolation [ECOL] The occurrence of organisms in isolation from others of their species.

biotin [BIOCHEM] $C_{10}H_{16}N_2O_3S$ A colorless, crystalline vitamin of the vitamin B complex occurring widely in nature, mainly in bound form.

biotite [MINERAL] A black, brown, or dark green, abundant and widely distributed species of rock-forming mineral of the mica group; its chemical composition is variable: $K_2[Fe(II),Mg]_{6-4}[Fe(III),Al,Ti]_{0-2}(Si_{6-5},Al_{2-3})O_{20-22}(OH,F)_{4-2}$. Also known as black mica; iron mica; magnesia mica; magnesium-iron mica.

biotope [ECOL] An area of uniform environmental conditions and biota.

Biot-Savart law [ELECTROMAG] A law that gives the intensity of the magnetic field due to a wire carrying a constant electric current.

Biot's law [OPTICS] The law that an optically active substance rotates plane-polarized light through an angle inversely proportional in its wavelength.

bioturbation [GEOL] The disruption of marine sedimentary structures by the activities of benthic organisms.

biotype [GEN] A group of organisms having the same genotype.

biparous [BOT] Bearing branches on dichotomous axes. [VERT ZOO] Bringing forth two young at a birth.

bipartite uterus [ANAT] A uterus divided into two parts almost to the base.

biped [VERT ZOO] 1. A two-footed animal. 2. Any two legs of a quadruped.

Biphyllidae [INV ZOO] The false skin beetles, a family of coleopteran insects in the superfamily Cucujoidea.

bipinnate [BOT] Pertaining to a leaf that is pinnate for both its primary and secondary divisions.

bipolar   [SCI TECH] 1. Having two poles. 2. Capable of assuming positive or negative values, such as an electric charge, or pertaining to a quantity with this property, such as a bipolar transistor.

bipolar circuit   [ELECTR] A logic circuit in which zeros and ones are treated in a symmetric or bipolar manner, rather than by the presence or absence of a signal; for example, a balanced arrangement in a square-loop-ferrite magnetic circuit.

bipolar memory   [COMPUT SCI] A computer memory employing integrated-circuit bipolar junction transistors as bistable memory cells.

bipolar signal   [COMMUN] A signal in which different logical states are represented by electrical voltages of opposite polarity.

bipotentiality   [BIOL] 1. Capacity to function either as male or female. 2. Hermaphroditism.

biprism   [OPTICS] A prism with apex angle only a little less than 180°, which produces a double image of a point source, giving rise to interference fringes on a nearby screen.

bipropellant   [MATER] A rocket propellant consisting of two unmixed or uncombined chemicals (fuel and oxidizer) fed to the combustion chamber separately.

biquadratic   [MATH] Any fourth-degree algebraic expression. Also known as quartic.

biquadratic equation   *See* quartic equation.

biquartic filter   [ELECTR] An active filter that uses operational amplifiers in combination with resistors and capacitors to provide infinite values of Q and simple adjustments for band-pass and center frequency.

biquinary notation   [MATH] A mixed-base notation system in which the first of each pair of digits counts 0 or 1 unit of five, and the second counts 0, 1, 2, 3, or 4 units. Also known as biquinary number system.

biquinary number system   *See* biquinary notation.

biradial symmetry   [BIOL] Symmetry both radial and bilateral. Also known as disymmetry.

biradical   [CHEM] A chemical species having two independent odd-electron sites.

biramous   [BIOL] Having two branches, such as an arthropod appendage.

birch   [BOT] The common name for all deciduous trees of the genus *Betula* that compose the family Betulaceae in the order Fagales.

bird   [VERT ZOO] Any of the warm-blooded vertebrates which make up the class Aves.

bird-foot delta   [GEOL] A delta formed by the outgrowth of fingers or pairs of natural levees at the mouth of river distributaries; an example is the Mississippi delta.

bird-hipped dinosaur   [PALEON] Any member of the order Ornithischia, distinguished by the birdlike arrangement of their hipbones.

birefringence   [OPTICS] Splitting of a light beam into two components, which travel at different velocities, by a material. Also known as double refraction.

birefringent filter   [OPTICS] A filter consisting of alternate layers of polarizing films and plates cut from a birefringent crystal; transmits light in a series of sharp, widely spaced wavelength bands. Also known as Lyot filter; monochromatic filter.

birth canal   [ANAT] The channel in mammals through which the fetus is expelled during parturition; consists of the cervix, vagina, and vulva.

birth rate   [BIOL] The ratio between the number of live births and a specified number of people in a population over a given period of time.

biscuit   [ENG ACOUS] *See* preform. [MATER] 1. A clay object that has been fired once prior to glaz-

ing. 2. Pottery that is unglazed in its final form. [MET] An upset blank for drop forging.

biscuit cutter   [MIN ENG] A short (6–8 inches or 15–20 centimeters) core barrel that is sharpened at the bottom and forced into the rocks by the jars.

bisector   [MATH] The ray dividing an angle into two equal angles.

bisectrix   [CRYSTAL] A line that is the bisector of the angle between the optic axes of a biaxial crystal.

biserial   [BIOL] Arranged in two rows or series.

bisexual   [BIOL] Of or relating to two sexes. [PSYCH] 1. Possessing mental and behavioral characteristics of both sexes. 2. Having sexual desires for members of both sexes.

bismuth   [CHEM] A metallic element, symbol Bi, of atomic number 83 and atomic weight 208.980. [MINERAL] The brittle, rhombohedral mineral form of the native element bismuth.

bismuth blende   *See* eulytite.

bismuth glance   *See* bismuthinite.

bismuthinite   [MINERAL] Bi_2S_3 A mineral consisting of bismuth trisulfide, which has an orthorhombic structure and is usually found in fibrous or leafy masses that are lead gray with a yellowish tarnish and a metallic luster. Also known as bismuth glance.

bismuth spar   *See* bismutite.

bismutite   [MINERAL] $(BiO)_2CO_3$ A dull-white, yellowish, or gray, earthy, amorphous mineral consisting of basic bismuth carbonate. Also known as bismuth spar.

bison   [VERT ZOO] The common name for two species of the family Bovidae in the order Artiodactyla; the wisent or European bison (*Bison bonasus*), and the American species (*Bison bison*).

bisphenoid   [CRYSTAL] A form apparently consisting of two sphenoids placed together symmetrically.

bistable circuit   [ELECTR] A circuit with two stable states such that the transition between the states cannot be accomplished by self-triggering.

bistable multivibrator   [ELECTR] A multivibrator in which either of the two active devices may remain conducting, with the other nonconducting, until the application of an external pulse. Also known as Eccles-Jordan circuit; Eccles-Jordan multivibrator; flip-flop circuit; trigger circuit.

bistable system   [CHEM] A chemical system with two relatively stable states which permits an oscillation between domination by one of these states to domination by the other.

bistatic radar   [ENG] Radar system in which the receiver is some distance from the transmitter, with separate antennas for each.

bistatic reflectivity   [OPTICS] The characteristic of a reflector which reflects energy along a line, or lines, different from or in addition to that of the incident ray.

bit   [COMPUT SCI] 1. A unit of information content equal to one binary decision or the designation of one of two possible and equally likely values or states of anything used to store or convey information. 2. A dimensionless unit of storage capacity specifying that the capacity of a storage device is expressed by the logarithm to the base 2 of the number of possible states of the device. [DES ENG] 1. A machine part for drilling or boring. 2. The cutting plate of a plane. 3. The blade of a cutting tool such as an ax. 4. A removable tooth of a saw. 5. Any cutting device which is attached to or part of a drill rod or drill string to bore or penetrate rocks. [MET] In soldering, the portion of the iron that transfers either heat or solder to the joint involved.

bit blank [DES ENG] A steel bit in which diamonds or other cutting media may be inset by hand peening or attached by a mechanical process such as casting, sintering, or brazing. Also known as bit shank; blank; blank bit; shank.

bit count appendage [COMPUT SCI] One of the two-byte elements replacing the parity bit stripped off each byte transferred from main storage to disk volume (the other element is the cyclic check); these two elements are appended to the block during the write operation; on a subsequent read operation these elements are calculated and compared to the appended elements for accuracy.

bit density [COMPUT SCI] Number of bits which can be placed, per unit length, area, or volume, on a storage medium; for example, bits per inch of magnetic tape. Also known as record density.

bitegmic [BOT] Having two integuments, especially in reference to ovules.

bit location [COMPUT SCI] Storage position on a record capable of storing one bit.

bit pattern [COMPUT SCI] A combination of binary digits arranged in a sequence.

bit position [COMPUT SCI] The position of a binary digit in a word, generally numbered from the least significant bit.

bit rate [COMMUN] Quantity, per unit time, of binary digits (or pulses representing them) which will pass a given point on a communications line or channel in a continuous stream.

bit shank See bit blank.

bit-sliced microprocessor [COMPUT SCI] A microprocessor in which the major logic of the central processor is partitioned into a set of large-scale-integration circuits, as opposed to being placed on a single chip.

bit stream [COMPUT SCI] **1.** A consecutive line of bits transmitted over a circuit in a transmission method in which character separation is accomplished by the terminal equipment. **2.** A binary signal without regard to grouping by character.

bit string [COMPUT SCI] A set of consecutive binary digits representing data in coded form, in which the significance of each bit is determined by its position in the sequence and its relation to the other bits.

bitumen [MATER] Naturally occurring or pyrolytically obtained substances of dark to black color consisting almost entirely of carbon and hydrogen, with very little oxygen, nitrogen, or sulfur.

bitumenite See torbanite.

bituminization See coalification.

bituminous [MATER] **1.** Containing much organic, or at least carbonaceous, matter, mostly in the form of tarry hydrocarbons which are usually described as bitumen. **2.** Similar to bitumen. **3.** Giving off volatile bituminous substances on heating, as in bituminous coal. [MINERAL] Of a mineral, having the odor of bitumen.

bituminous coal [GEOL] A dark brown to black coal that is high in carbonaceous matter and has 15–50% volatile matter. Also known as soft coal.

bituminous sand [GEOL] Sand containing bituminous-like material, such as the tar sands at Athabasca, Canada, from which oil is extracted commercially.

bit zone [COMPUT SCI] **1.** One of the two left-most bits in a commonly used system in which six bits are used for each character; related to overpunch. **2.** Any bit in a group of bit positions that are used to indicate a specific class of items; for example, numbers, letters, special signs, and commands.

bivalent chromosome [CYTOL] The structure formed following synapsis of a pair of homologous chromosomes from the zygotene stage of meiosis up to the beginning of anaphase.

bivalve [INV ZOO] The common name for a number of diverse, bilaterally symmetrical animals, including mollusks, ostracod crustaceans, and brachiopods, having a soft body enclosed in a calcareous two-part shell.

Bivalvia [INV ZOO] A large class of the phylum Mollusca containing the clams, oysters, and other bivalves.

Bk See berkelium.

BL See base line.

black [CHEM] Fine particles of impure carbon that are made by the incomplete burning of carbon compounds, such as natural gas, naphthas, acetylene, bones, ivory, and vegetables. [COMMUN] See black signal. [OPTICS] Quality of an object which uniformly absorbs large percentages of light of all visible wavelengths.

black amber See jet coal.

blackberry [BOT] Any of the upright or trailing shrubs of the genus *Rubus* in the order Rosales; an edible berry is produced by the plant.

blackbody [THERMO] An ideal body which would absorb all incident radiation and reflect none. Also known as hohlraum; ideal radiator.

blackbody radiation [THERMO] The emission of radiant energy which would take place from a blackbody at a fixed temperature; it takes place at a rate expressed by the Stefan-Boltzmann law, with a spectral energy distribution described by Planck's equation.

blackbody temperature [THERMO] The temperature of a blackbody that emits the same amount of heat radiation per unit area as a given object; measured by a total radiation pyrometer. Also known as brightness temperature.

black box [ENG] Any component, usually electronic and having known input and output, that can be readily inserted into or removed from a specific place in a larger system without knowledge of the component's detailed internal structure.

black cobalt See asbolite.

blackdamp [MIN ENG] A nonexplosive mixture of carbon dioxide with other gases, especially with 85–90% nitrogen, which is heavier than air and cannot support flame or life. Also known as chokedamp.

black dwarf [ASTRON] A star that cannot generate thermonuclear energy.

blackeye bean See cowpea.

black frost [HYD] A dry freeze with respect to its effects upon vegetation, that is, the internal freezing of vegetation unaccompanied by the protective formation of hoarfrost. Also known as hard frost.

black granite See diorite.

blackhead See comedo.

black hole [RELAT] A region of space-time from which nothing can escape, according to classical physics; quantum corrections indicate a black hole radiates particles with a temperature inversely proportional to the mass and directly proportional to Planck's constant.

black ice [HYD] A type of ice forming on lake or salt water; compact, and dark in appearance because of its transparency.

black lead See graphite.

blackleg [VET MED] An acute, usually fatal bacterial disease of cattle, and occasionally of sheep, goats, and swine, caused by *Clostridium chauvoei*.

black level [ELECTR] The level of the television picture signal corresponding to the maximum limit of black peaks.

black light [OPTICS] Invisible light, such as ultraviolet rays which fall on fluorescent materials and cause them to emit visible light.

blacklung See anthracosis.

black mica See biotite.

blackout [COMMUN] See radio blackout. [ELEC] The shutting off of power in an electrical power transmission system, either deliberately or through failure of the system.

black peak [COMMUN] A peak excursion of the television picture signal in the black direction.

black pepper [FOOD ENG] A spice; the dried unripe berries of *Piper nigrum*, a vine of the pepper family (Piperaceae) in the order Piperales.

black print [GRAPHICS] The film having the black component in a four-color separation process, or the comparable plate in the four-color printing process.

black sand [GEOL] Heavy, dark, sandlike minerals found on beaches and in stream beds; usually magnetite and ilmenite and sometimes gold, platinum, and monazite are present.

black scope [ELECTR] Cathode-ray tube operating at the threshold of luminescence when no video signals are being applied.

black shale [PETR] Very thinly bedded shale rich in sulfides such as pyrite and organic material deposited under barred basin conditions so that there was an anaerobic accumulation. Also known as biopelite.

black signal [COMMUN] Signal at any point in a facsimile system produced by the scanning of a maximum density area of the subject copy. Also known as black; picture black.

black silver See stephanite.

black tellurium See nagyagite.

blacktop [MATER] A black bituminous material that is used to pave roadways; it is spread over a layer of crushed rocks and packed down into a level surface; it may be spread over small areas of roadways in need of repair.

bladder [ANAT] Any saclike structure in humans and animals, such as a swimbladder or urinary bladder, that contains a gas or functions as a receptacle for fluid. [GEOL] See vesicle.

bladder press [MECH ENG] A machine which simultaneously molds and cures (vulcanizes) a pneumatic tire.

blade [BOT] The broad, flat portion of a leaf. [ELEC] A flat moving conductor in a switch. [ENG] 1. A broad, flat arm of a fan, turbine, or propeller. 2. The broad, flat surface of a bulldozer or snowplow by which the material is moved. 3. The part of a cutting tool, such as a saw, that cuts. [VERT ZOO] A single plate of baleen.

blade slap noise [AERO ENG] Impulsive noise (short high-pressure sound waves) of rotating blades, primarily helicopter blades.

blakeite [MINERAL] A deep reddish-brown to deep brown mineral consisting of anhydrous ferric tellurite; occurs in massive form, as microcrystalline crusts.

Blake jaw crusher [MECH ENG] A crusher with one fixed jaw plate and one pivoted at the top so as to give the greatest movement on the smallest lump.

Blancan [GEOL] Upper Pliocene or lowermost Pleistocene geologic time.

blanc fixe [INORG CHEM] $BaSO_4$ A commercial name for barium sulfate, with some use in pure form in the paint, paper, and pigment industries as a pigment extender.

blanching [FOOD ENG] A hot-water or steam direct-scalding treatment of raw foodstuffs of particulate type to inactivate enzymes which otherwise might cause quality deterioration, particularly of flavor, during processing or storage.

blandel See apostilb.

blank [DES ENG] See bit blank. [ELECTR] To cut off the electron beam of a television picture tube, camera tube, or cathode-ray oscilloscope tube during the process of retrace by applying a rectangular pulse voltage to the grid or cathode during each retrace interval. Also known as beam blank. [ENG] 1. The result of the final cutting operation on a natural crystal. 2. See blind. [MET] 1. A semifinished piece of metal to be stamped or forged into a tool or implement. 2. A semifinished, pressed, compacted mass of powdered metal. 3. Metal sheet prepared for a forming operation. [ORD] Ammunition which contains no projectile but which does contain a charge of low explosive, such as black powder, to produce a noise.

blank bit See bit blank.

blank carburizing [MET] A simulated carburizing procedure carried out without a carburizing medium. Also known as pseudocarburizing.

blank character [COMPUT SCI] A character, either printed or appearing as a blank, used to denote a blank space among printed characters. Also known as space character.

blanket [COMMUN] To blank out or obscure weak radio signals by a stronger signal. [GRAPHICS] In offset lithography, a rubber sheet covering the cylinder of an offset press that transfers the image from the plate to the paper. [MIN ENG] A textile material used in ore treatment plants for catching coarse free gold and sometimes associated minerals, for example, pyrite. [NUCLEO] A layer of fertile uranium-238 or thorium-232 material placed around or within the core of a nuclear reactor to breed new fuel.

blanketing [COMMUN] Interference due to a nearby transmitter whose signals are so strong that they override other signals over a wide band of frequencies. [MIN ENG] The material caught on the blankets that are used in concentrating gold-bearing sands or slimes.

blank form See blank medium.

blanking [ENG] 1. The closing off of flow through a liquid-containing process pipe by the insertion of solid disks at joints or unions; used during maintenance and repair work as a safety precaution. Also known as blinding. 2. Cutting of plastic or metal sheets into shapes by striking with a punch. Also known as die cutting.

blanking circuit [ELECTR] A circuit preventing the transmission of brightness variations during the horizontal and vertical retrace intervals in television scanning.

blanking level [ELECTR] The level that separates picture information from synchronizing information in a composite television picture signal; coincides with the level of the base of the synchronizing pulses. Also known as pedestal; pedestal level.

blanking time [ELECTR] The length of time that the electron beam of a cathode-ray tube is shut off.

blank medium [COMPUT SCI] An empty position on the medium concerned, such as a column without holes on a punch tape, used to indicate a blank character. Also known as blank form.

blast [COMPUT SCI] To release internal or external memory areas from the control of a computer program in the course of dynamic storage allocation, making these areas available for reallocation to other programs. [ENG] The setting off of a heavy explosive charge. [PHYS] **1.** The brief and rapid movement of air or other fluid away from a center of outward pressure, as in an explosion. **2.** The characteristic instantaneous rise in pressure, followed by a sudden decrease, that results from this movement, differentiated from less rapid pressure changes.

blast burner [ENG] A burner in which a controlled burst of air or oxygen under pressure is supplied to the illuminating gas used. Also known as blast lamp.

blast cell [HISTOL] An undifferentiated precursor of a human blood cell in the reticuloendothelial tissue.

blast furnace [MET] A tall, cylindrical smelting furnace for reducing iron ore to pig iron; the blast of air blown through solid fuel increases the combustion rate.

blast lamp *See* blast burner; blowtorch.

Blastobasidae [INV ZOO] A family of lepidopteran insects in the superfamily Tineoidea.

Blastocladiales [MYCOL] An order of aquatic fungi in the class Phycomycetes.

blastocoele [EMBRYO] The cavity of a blastula. Also known as segmentation cavity.

blastodisk [EMBRYO] The embryo-forming, protoplasmic disk on the surface of a yolk-filled egg, such as in reptiles, birds, and some fish.

Blastoidea [PALEON] A class of extinct pelmatozoan echinoderms in the subphylum Crinozoa.

blastoma [MED] **1.** A tumor whose parenchymal cells have certain embryonal characteristics. **2.** A true tumor.

blast roasting [MIN ENG] The roasting of finely divided ores by means of a blast to maintain internal combustion within the charge. Also known as roast sintering.

blastula [EMBRYO] A hollow sphere of cells characteristic of the early metazoan embryo.

blastulation [EMBRYO] Formation of a blastula from a solid ball of cleaving cells.

Blattidae [INV ZOO] The cockroaches, a family of insects in the order Orthoptera.

BLC *See* boundary-layer control.

bleaching [GRAPHICS] An afterprocess in the production of direct positive photographs, in which an oxidizing solution dissolves the negative silver. [TEXT] A process in which natural coloring matter is removed from a fiber, yarn, or fabric to make it white.

bleb [MED] A localized collection of fluid, as serum or blood, in the epidermis.

bleed [CHEM] Diffusion of coloring matter from a substance. [COMPUT SCI] In optical character recognition, the flow of ink in printed characters beyond the limits specified for their recognition by a character reader. [ENG] To let a fluid, such as air or liquid oxygen, escape under controlled conditions from a pipe, tank, or the like through a valve or outlet. [GRAPHICS] The extension of a photograph or other artwork to the very edge of the printed page. [MED] To exude blood from a wound.

bleeder [ENG] A connection located at a low place in an air line or a gasoline container so that, by means of a small valve, the condensed water or other liquid can be drained or bled off from the line or container without discharging the air or gas. [ELECTR] A high resistance connected across the dc output of a high-voltage power supply which serves to discharge the filter capacitors after the power supply has been turned off, and to provide a stabilizing load. [MED]

1. A person subject to frequent hemorrhages, as a hemophiliac. **2.** A blood vessel from which there is persistent uncontrolled bleeding. **3.** A blood vessel which has escaped closure by cautery or ligature during a surgical procedure.

bleeder current [ELEC] Current drawn continuously from a voltage source to lessen the effect of load changes or to provide a voltage drop across a resistor.

bleeder resistor [ELEC] A resistor connected across a power pack or other voltage source to improve voltage regulation by drawing a fixed current value continuously; also used to dissipate the charge remaining in filter capacitors when equipment is turned off.

bleeder turbine [MECH ENG] A multistage turbine where steam is extracted (bled) at pressures intermediate between throttle and exhaust, for process or feedwater heating purposes.

bleeding [ENG] Natural separation of a liquid from a liquid-solid or semisolid mixture; for example, separation of oil from a stored lubricating grease, or water from freshly poured concrete. [TEXT] Referring to a fabric in which the dye is not fast and therefore comes out when the fabric is wet.

bleeding cycle [MECH ENG] A steam cycle in which steam is drawn from the turbine at one or more stages and used to heat the feedwater. Also known as regenerative cycle.

bleeding time [PHYSIO] The time required for bleeding to stop after a small puncture wound.

bleed line [GRAPHICS] A line-width change usually due to overexposure or overdeveloping in photography.

bleed-through [GRAPHICS] Of records printed on both sides, the obtrusive show-through of printed matter from one side to the other.

bleed valve [ENG] A small-flow valve connected to a fluid process vessel or line for the purpose of bleeding off small quantities of contained fluid.

blende *See* sphalerite.

blended fuel oil [MATER] A mixture of petroleum residual and distillate fuel oils.

blended unconformity [GEOL] An unconformity that is not sharp because the original erosion surface was covered by a thick residual soil that graded downward into the underlying rock.

blended whiskey [FOOD ENG] Whiskey containing at least 20% by volume of 100 proof straight whiskey and mixed with other whiskey or neutral spirits, the mixture being not less than 80 proof at the time of bottling.

blending [TEXT] Producing uniform yarn and fabric by bringing together two or more distinctive types of fibers of different lengths.

blending inheritance [GEN] Inheritance in which the character of the offspring is a blend of those in the parents; a common feature for quantitative characters, such as stature, determined by large numbers of genes and affected by environmental variation.

Blenniidae [VERT ZOO] The blennies, a family of carnivorous marine fishes in the suborder Blennioidei.

Blennioidei [VERT ZOO] A large suborder of small marine fishes in the order Perciformes that live principally in coral and rock reefs.

Blephariceridae [INV ZOO] A family of dipteran insects in the suborder Orthorrhapha.

blight [PL PATH] Any plant disease or injury that results in general withering and death of the plant without rotting.

blimp [AERO ENG] A name originally applied to nonrigid, pressure-type airships, usually of small size; now applied to airships with volumes of approximately 1,500,000 cubic feet (42,000 cubic meters).

blind [ENG] A solid disk inserted at a pipe joint or union to prevent the flow of fluids through the pipe; used during maintenance and repair work as a safety precaution. Also known as blank. [GEOL] Referring to a mineral deposit with no surface outcrop.

blind approach beacon system [NAV] A pulse-type, ground-based navigation beacon used for runway approach at airports, which sends out signals that produce range and runway position information on the L-scan cathode-ray indicator of an aircraft making an instrument approach. Also known as beam approach beacon system (British usage). Abbreviated babs.

blind controller system [CONT SYS] A process control arrangement that separates the in-plant measuring points (for example, pressure, temperature, and flow rate) and control points (for example, a valve actuator) from the recorder or indicator at the central control panel.

blind drainage See closed drainage.

blind drift [MIN ENG] In a mine, a horizontal passage not yet connected with the other workings.

blind hole [DES ENG] A hole which does not pass completely through a workpiece. [ENG] A type of borehole that does not have the drilling mud or other circulating medium carry the cuttings to the surface.

blinding [ENG] See blanking. [MIN ENG] Interference with the functioning of a screen mesh by a matting of fine materials during screening. Also known as blocking; plugging.

blind landing [AERO ENG] Landing an aircraft solely by the use of instruments because of poor visibility.

blindness [MED] 1. Loss or absence of the ability to perceive visual images. 2. The condition of a person having less than 1/10 (20/200 on the Snellen test) normal vision.

blind rollers [OCEANOGR] Long, high swells which have increased in height, almost to the breaking point, as they pass over shoals or run in shoaling water. Also known as blind seas.

blind seas See blind rollers.

blind spot [ENG] An area on a filter screen where no filtering occurs. Also known as dead area. [PHYSIO] A place on the retina of the eye that is insensitive to light, where the optic nerve passes through the eyeball's inner surface.

blind trial [STAT] A trial in which experimenters and subjects are kept uninformed as to whether or not treatment has been given.

blind valley [GEOL] A valley that has been made by a spring from an underground channel which emerged to form a surface stream, and that is enclosed at the head of the stream by steep walls.

blind zone [COMMUN] Area from which echoes cannot be received; generally, an area shielded from the transmitter by some natural obstruction and therefore from which there can be no return.

B line See index register.

blinking [COMMUN] Method of providing information in pulse systems by modifying the signal at its source so that signal presentation on the display scope alternately appears and disappears; in loran, this indicates that a station is malfunctioning. [ELECTR] Electronic-countermeasures technique employed by two aircraft separated by a short distance and within the same azimuth resolution so as to appear as one target to a tracking radar; the two aircraft alternately spot-jam, causing the radar system to oscillate from one place to another, making an accurate solution of a fire control problem impossible. [NAV] Regular shifting right and left or alternate appearance and disappearance of a loran signal to indicate that the signals of a pair of stations are out of synchronization.

blip [ELECTR] 1. The display of a received pulse on the screen of a cathode-ray tube. Also known as pip. 2. An ideal infrared radiation detector that detects with unit quantum efficiency all of the radiation in the signal for which the detector was designed, and responds only to the background radiation noise that comes from the field of view of the detector.

blister [ENG] A raised area on the surface of a metallic or plastic object caused by the pressure of gases developed while the surface was in a partly molten state, or by diffusion of high-pressure gases from an inner surface. [GEOL] A domelike protuberance caused by the buckling of the cooling crust of a molten lava before the flowing mass has stopped. [MED] A local swelling of the skin resulting from the accumulation of serous fluid between the epidermis and true skin. [MIN ENG] A protrusion, more or less circular in plan, extending downward into a coal seam. [NUCLEO] A protuberance that sometimes develops on the surface of a nuclear-reactor fuel element during use, generally because of entrapped gases.

blister copper [MET] Copper having 96–99% purity and a blistered appearance and formed by forcing air through molten copper matte.

blizzard [METEOROL] A severe weather condition characterized by low temperatures and by strong winds bearing a great amount of snow (mostly fine, dry snow picked up from the ground).

bloat [VET MED] Distension of the rumen in cattle and other ruminants due to excessive gas formation following heavy fermentation of legumes eaten wet.

blob [METEOROL] In radar, oscilloscope evidence of a fairly small-scale temperature and moisture inhomogeneity produced by turbulence within the atmosphere.

Bloch equations [SOLID STATE] Approximate equations for the rate of change of magnetization of a solid in a magnetic field due to spin relaxation and gyroscopic precession.

Bloch function [SOLID STATE] A wave function for an electron in a periodic lattice, of the form $u(r)\cdot\exp[i\mathbf{k}\cdot\mathbf{r}]$ where $u(r)$ has the periodicity of the lattice.

Bloch theorem [QUANT MECH] The theorem that the lowest state of a quantum-mechanical system without a magnetic field can carry no current. [SOLID STATE] The theorem that, in a periodic structure, every electronic wave function can be represented by a Bloch function.

block [COMPUT SCI] 1. A group of information units (such as records, words, characters, or digits) that are transported or considered as a single unit by virtue of their being stored in successive storage locations; for example, a group of logical records constituting a physical record. 2. The section of a computer memory or storage device that stores such a group of information units. Also known as storage block. 3. To combine two or more information units into a single unit. [DES ENG] 1. A metal or wood case enclosing one or more pulleys; has a hook with which it can be attached to an object. 2. See cylinder block. [MIN ENG] A division of a mine, usually bounded by workings but sometimes by survey lines or other arbitrary limits. [PETRO ENG]

The subdivision of a sea area for the licensing of oil and gas exploration and production rights.

block brake [MECH ENG] A brake which consists of a block or shoe of wood bearing upon an iron or steel wheel.

block caving [MIN ENG] A method of caving where a block, 150–250 feet (46–77 meters) on a side and several hundred feet high, is induced to cave in after it is undercut; the broken ore is drawn off at a bell-shaped draw point.

block chaining *See* chained block encryption.

block cipher [COMMUN] A cipher that transforms a string of input bits of fixed length into a string of output bits of fixed length.

block coal [MIN ENG] **1.** A bituminous coal that breaks into large lumps or cubical blocks. **2.** Coal that passes over 5-, 6-, and 8-inch (127, 152, and 203 millimeter) block screens; used in smelting iron.

block data [COMPUT SCI] A statement in FORTRAN which declares that the program following is a data specification subprogram.

block diagram [ENG] A diagram in which the essential units of any system are drawn in the form of rectangles or blocks and their relation to each other is indicated by appropriate connecting lines.

blocked impedance [ELEC] The impedance at the input of a transducer when the impedance of the output system is made infinite, as by blocking or clamping the mechanical system.

blocked-out ore *See* developed reserves.

blocked resistance [ENG ACOUS] Resistance of an audio-frequency transducer when its moving elements are blocked so they cannot move; represents the resistance due only to electrical losses.

blockette [COMPUT SCI] A subdivision of a group of consecutive machine words transferred as a unit, particularly with reference to input and output.

block faulting [GEOL] A type of faulting in which fault blocks are displaced at different orientations and elevations.

block-grid keying [COMMUN] Method of keying a continuous-wave transmitter by operating the amplifier stage as an electronic switch; during the spacing interval when the key is open, the bias on the control grid becomes highly negative and prevents the flow of plate current so that the tube has no output; during the marking interval when the key is closed, this bias is removed and full plate current flows.

block head [COMPUT SCI] A list of declarations at the beginning of a computer program with block structure.

block identifier [COMPUT SCI] A means of identifying an area of storage in FORTRAN so that this area may be shared by a program and its subprograms.

block ignore character [COMPUT SCI] A character associated with a block which indicates the presence of errors in the block.

blocking [CHEM] Undesired adhesion of granular particles; often occurs with damp powders or plastic pellets in storage bins or during movement through conduits. [COMPUT SCI] Combining two or more computer records into one block. [ELECTR] **1.** Applying a high negative bias to the grid of an electron tube to reduce its anode current to zero. **2.** Overloading a receiver by an unwanted signal so that the automatic gain control reduces the response to a desired signal. **3.** Distortion occurring in a resistance-capacitance-coupled electron tube amplifier stage when grid current flows in the following tube. [ENG] Undesired adhesion between layers of plastic materials in contact during

storage or use. [HISTOL] **1.** The process of embedding tissue in a solid medium, such as paraffin. **2.** A histochemical process in which a portion of a molecule is treated to prevent it from reacting with some other agent. [MET] **1.** A preliminary hot-forging operation which imparts an approximate shape to the rough stock. **2.** Reducing the oxygen content of the bath in an open-hearth furnace. [MIN ENG] *See* blinding. [SOLID STATE] The hindering of motion of dislocations in a solid substance by small particles of a second substance included in the solid; this results in hardening of the substance. [STAT] The grouping of sample data into subgroups with similar characteristics.

blocking anticyclone *See* blocking high.

blocking capacitor *See* coupling capacitor.

blocking factor [COMPUT SCI] The largest possible number of records of a given size that can be contained within a single block.

blocking group [ORG CHEM] In peptide synthesis, a group that is reacted with a free amino or carboxyl group on an amino acid to prevent its taking part in subsequent formation of peptide bonds.

blocking high [METEOROL] Any high (or anticyclone) that remains nearly stationary or moves slowly compared to the west-to-east motion upstream from its location, so that it effectively blocks the movement of migratory cyclones across its latitudes. Also known as blocking anticyclone.

blocking layer *See* depletion layer.

blocking oscillator [ELECTR] A relaxation oscillator that generates a short-time-duration pulse by using a single transistor or electron tube and associated circuitry. Also known as squegger; squegging oscillator.

block input [COMPUT SCI] **1.** A block of computer words considered as a unit and intended or destined to be transferred from an internal storage medium to an external destination. **2.** *See* output area.

block lava [GEOL] Lava flows which occur in a tumultuous assemblage of angular blocks.

block length [COMPUT SCI] The total number of records, words, or characters contained in one block.

block loading [COMPUT SCI] A program loading technique in which the control sections of a program or program segment are loaded into contiguous positions in main memory.

block mark [COMPUT SCI] A special character that indicates the end of a block.

block mountain [GEOL] A mountain formed by the combined processes of uplifting, faulting, and tilting. Also known as fault-block mountain.

block plane [DES ENG] A small type of hand plane, designed for cutting across the grain of the wood and for planing end grains.

block polymer [ORG CHEM] A copolymer whose chain is composed of alternating sequences of identical monomer units.

block printing [GRAPHICS] The earliest form of printing, involving the cutting of crude pictures and lettering on blocks of wood.

block sort [COMPUT SCI] A method of sorting a file, usually with punched card sorters, in which the file is first sorted according to the value of the digit in the highest digit position of the key; the resulting collections of records can next be sorted independently in smaller operations, and the separate sections then joined.

block structure [COMPUT SCI] In computer programming, a conceptual tool used to group sequences of statements into single compound statements and to allow the programmer explicit control over the scope of the program variables.

block transfer [COMPUT SCI] The movement of data in blocks instead of by individual records.

blödite *See* bloedite.

bloedite [MINERAL] $MgSO_4 \cdot Na_2SO_4 \cdot 4H_2O$ A white or colorless monoclinic mineral consisting of magnessium sodium sulfate. Also spelled blödite. Also known as astrakanite; astrochanite.

blomstrandine *See* priorite.

blood [HISTOL] A fluid connective tissue consisting of the plasma and cells that circulate in the blood vessels.

blood agar [MICROBIO] A nutrient microbiologic culture medium enriched with whole blood and used to detect hemolytic strains of bacteria.

blood count [PATH] Determination of the number of white and red blood cells in a definite volume of blood.

blood dyscrasia [MED] Any abnormal condition of the formed elements of blood or of the constituents required for clotting.

blood group [IMMUNOL] An immunologically distinct, genetically determined class of human erythrocyte antigens, identified as A, B, AB, and O.

blood-plate hemolysis [MICROBIO] Destruction of red blood cells in a blood agar medium by a bacterial toxin.

blood platelet *See* thrombocyte.

blood poisoning *See* septicemia.

blood pressure [PHYSIO] Pressure exerted by blood on the walls of the blood vessels.

bloodstone [MINERAL] 1. A form of deep green chalcedony flecked with red jasper. Also known as heliotrope; oriental jasper. 2. *See* hematite.

blood sugar [BIOCHEM] The carbohydrate, principally glucose, of the blood.

blood test [PATH] 1. A serologic test for syphilis. 2. A blood count. 3. A test for detection of blood, usually one based on the peroxidase activity of blood, such as the benzidine test or guaiac test.

blood typing [IMMUNOL] Determination of an individual's blood group.

blood vessel [ANAT] A tubular channel for blood transport.

bloom [BOT] 1. An individual flower. Also known as blossom. 2. To yield blossoms. 3. The waxy coating that appears as a powder on certain fruits, such as plums, and leaves, such as cabbage. [ECOL] A colored area on the surface of bodies of water caused by heavy planktonic growth. [ENG] 1. Fluorescence in lubricating oils or a cloudy surface on varnished or enameled surfaces. 2. To apply an antireflection coating to glass. [GEOL] *See* blossom. [GRAPHICS] A milky or foggy defect that may appear on the surface of a varnished painting; caused by moisture. [MINERAL] *See* efflorescence. [MET] 1. A semifinished bar of metal formed from an ingot and having a rectangular cross section exceeding 36 square inches (232 square centimeters). 2. To hammer or roll metal in order to make its surface bright. [OPTICS] Color of oil in reflected light, differing from its color in transmitted light. Also known as fluorescence.

blooming mill [MET] A rolling mill for making blooms from ingots. Also known as cogging mill.

blossom [BOT] *See* bloom. [GEOL] The oxidized or decomposed outcrop of a vein or coal bed. Also known as bloom.

blow [ELEC] Opening of a circuit because of excess current, particularly when the current is heavy and a melting or breakdown point is reached.

blowback [CHEM ENG] 1. A continuous stream of liquid or gas bled through air lines from instruments and to the process line being monitored; prevents process fluid from backing up and contacting the instrument. 2. Reverse flow of fluid through a filter medium to remove caked solids. Also known as backwash. [GRAPHICS] To enlarge, or make an enlargement of, an image.

blowby [MECH ENG] Leaking of fluid between a cylinder and its piston during operation.

blowcase [CHEM ENG] A cylindrical or spherical corrosion- and pressure-resistant container from which acid is forced by compressed air to the agitator; used in manufacture of acids but largely superseded by centrifugal pumps. Also known as acid blowcase; acid egg.

blowdown [CHEM ENG] Removal of liquids or solids from a process vessel or storage vessel or a line by the use of pressure. [MECH ENG] The difference between the pressure at which the safety valve opens and the closing pressure. [METEOROL] A wind storm that causes trees or structures to be blown down.

blower [MECH ENG] A fan which operates where the resistance to gas flow is predominantly downstream of the fan.

blowhole [GEOL] A longitudinal tunnel opening in a sea cliff, on the upland side away from shore; columns of sea spray are thrown up through the opening, usually during storms. [MET] A pocket of air or gas formed in a metal during solidification. [VERT ZOO] The nostril on top of the head of cetacean mammals.

blowing [CHEM ENG] The introduction of compressed air near the bottom of a tank or other container in order to agitate the liquid therein. [ENG] *See* blow molding.

blow molding [ENG] A method of fabricating hollow plastic objects, such as bottles, by forcing a parison into a mold cavity and shaping by internal air pressure. Also known as blowing.

blown glass [ENG] Glassware formed by blowing air into a ball of liquefied glass until it reaches the desired shape.

blowoff valves [MECH ENG] Valves in boiler piping which facilitate removal of solid matter present in the boiler water.

blowout [ELEC] The melting of an electric fuse because of excessive current. [ELECTROMAG] The extinguishing of an electric arc by deflection in a magnetic field. Also known as magnetic blowout. [ENG] 1. The bursting of a container (such as a tube pipe, pneumatic tire, or dam) by the pressure of the contained fluid. 2. The rupture left by such bursting. 3. The abrupt escape of air from the working chamber of a pneumatic caisson. [GEOL] Any of the various trough-, saucer-, or cuplike hollows formed by wind erosion on a dune or other sand deposit. [HYD] A bubbling spring which bursts from the ground behind a river levee when water at flood stage is forced under the levee through pervious layers of sand or silt. Also known as sand boil. [PETRO ENG] A sudden, unplanned escape of oil or gas from a well during drilling.

blowout coil [ELECTROMAG] A coil that produces a magnetic field in an electrical switching device for the purpose of lengthening and extinguishing an electric arc formed as the contacts of the switching device part to interrupt the current.

blowpipe [BIOL] A small tube, tapering to a straight or slightly curved tip, used in anatomy and zoology to reveal or clean a cavity. [ENG] 1. A long, straight tube, used in glass blowing, on which molten glass is gathered and worked. **2.** A small, tapered, and frequently curved tube that leads a jet, usually of air, into a flame to concentrate and direct it; used in flame tests in analytical chemistry and in brazing and soldering of fine work. **3.** *See* blowtorch.

blowtorch [ENG] A small, portable blast burner which operates either by having air or oxygen and gaseous fuel delivered through tubes or by having a fuel tank which is pressured by a hand pump. Also known as blast lamp; blowpipe.

blue [OPTICS] The hue evoked in an average observer by monochromatic radiation having a wavelength in the approximate range from 455 to 492 nanometers; however, the same sensation can be produced in a variety of other ways.

blue asbestos *See* crocidolite.

blue baby [MED] An infant with congenital cyanosis due to cardiac or pulmonary defect, causing shunting of unoxygenated blood into the systemic circulation.

blueberry [BOT] Any of several species of plants in the genus *Vaccinium* of the order Ericales; the fruit, a berry, occurs in clusters on the plant.

bluefish [VERT ZOO] *Pomatomus saltatrix.* A predatory fish in the order Perciformes. Also known as skipjack.

blue gas [MATER] A gas consisting chiefly of carbon monoxide and hydrogen, formed by the action of steam upon hot coke; used mainly as a source of hydrogen and in synthesis of other chemical compounds. Also known as blue water gas.

blue glow [ELECTR] A glow normally seen in electron tubes containing mercury vapor, due to ionization of the mercury molecules. [MET] Luminescence emitted by certain metallic oxides when heated.

bluegrass [BOT] The common name for several species of perennial pasture and lawn grasses in the genus *Poa* of the order Cyperales.

blue-green algae [BOT] The common name for members of the Cyanophyceae.

blue ice [HYD] Pure ice in the form of large, single crystals that is blue owing to the scattering of light by the ice molecules; the purer the ice, the deeper the blue.

blueing *See* bluing.

blue iron earth *See* vivianite.

blue lead *See* galena.

blueline [GRAPHICS] A blue image or outline printed on paper or plastic sheeting and used as a guide for drafting, stripping, or layout; it does not reproduce when the finished work is photographed.

blue magnetism [GEOPHYS] The magnetism displayed by the south-seeking end of a freely suspended magnet; this is the magnetism of the earth's north magnetic pole.

blue malachite *See* azurite.

blueprint [GRAPHICS] **1.** A contact print, with white lines on a blue background, of a drawing; made on linen or on ferroprussiate paper and developed in water or a special solution. **2.** A photoprint used in offset lithography or photoengraving for use in checking positions of image elements.

blue rot [PL PATH] A fungus disease of conifers caused by members of the genus *Ceratostomella*, characterized by blue discoloration of the wood. Also known as bluing.

blueschist facies [PETROL] High-pressure, low-temperature metamorphism associated with subduction zones which produces a broad mineral associa-

tion including glaucophane, actinolite, jadeite, aegirine, lawsonite, and pumpellyite.

blue spar *See* lazulite.

bluestem grass [BOT] The common name for several species of tall, perennial grasses in the genus *Andropogon* of the order Cyperales.

bluestone [MINERAL] *See* chalcanthite. [PETR] **1.** A sandstone that is highly argillaceous and of even texture and bedding. **2.** The commercial name for a feldspathic sandstone that is dark bluish gray; it is easily split into thin slabs and used as flagstone.

blue vitriol [MATER] A hydrous solution of copper sulfate that is applied to the surface of a metal for layout purposes. [MINERAL] *See* chalcanthite.

blue water gas *See* blue gas.

bluing [MET] Also spelled blueing. **1.** Formation of a bluish oxide film on polished steel; improves appearance and provides some corrosion resistance. **2.** Heating of formed springs to reduce internal stress. **3.** A blue oxide film formed on the polished surface of a metal due to extremely high temperatures. [PL PATH] *See* blue rot.

BM *See* barrels per month; bench mark.

B meson [PARTIC PHYS] An elementary particle with strong nuclear interactions, baryon number $B = 0$, and mass 1237 MeV.

BMR *See* basal metabolic rate.

boa [VERT ZOO] Any large, nonvenomous snake of the family Boidae in the order Squamata.

board drop hammer [MECH ENG] A type of drop hammer in which the ram is attached to wooden boards which slide between two rollers; after the ram falls freely on the forging, it is raised by friction between the rotating rollers. Also known as board hammer.

board-foot [ENG] Unit of volume in measuring lumber; equals 144 cubic inches (2360 cubic centimeters), or the volume of a board 1 foot square and 1 inch thick. Abbreviated bdft.

board hammer *See* board drop hammer.

bobbin [ELECTROMAG] An insulated spool serving as a support for a coil. [TEXT] A cylinder with projecting edges at one or both ends and a hole along the axis, used for winding twisted strands of textile fiber, thread, or yarn.

bobbing [ELECTR] Fluctuation of the strength of a radar echo, or its indication on a radarscope, due to alternate interference and reinforcement of returning reflected waves.

bobierrite [MINERAL] $Mg_3(PO_4)_2 \cdot 8H_2O$ A transparent, colorless or white, monoclinic mineral consisting of octahydrated magnesium phosphate.

Bobillier's law [MECH] The law that, in general plane rigid motion, when *a* and *b* are the respective centers of curvature of points A and B, the angle between Aa and the tangent to the centrode of rotation (pole tangent) and the angle between Bb and a line from the centrode to the intersection of AB and ab (collineation axis) are equal and opposite.

BOD *See* biochemical oxygen demand.

Bode's law [ASTRON] An empirical law giving mean distances of planets to the sun by the formula $a = 0.4 + 0.3 \times 2^n$, where a is in astronomical units and n equals $-\infty$ for Mercury, 0 for Venus, 1 for Earth, and so on; the asteroids are included as planets. Also known as Titius-Bode law.

bodily tide *See* earth tide.

Bodonidae [INV ZOO] A family of protozoans in the order Kinetoplastida characterized by two unequally long flagella, one of them trailing.

body [AERO ENG] **1.** The main part or main central portion of an airplane, airship, rocket, or the like; a fuselage or hull. **2.** Any fabrication, structure, or other material form, especially one aerodynamically or ballistically designed; for example, an airfoil is a body designed to produce an aerodynamic reaction. [GEOGR] A separate entity or mass of water, such as an ocean or a lake. [GEOL] An ore body, or pocket of mineral deposit. [MATER] The consistency or viscosity of fluid materials, such as lubricating oils, paints, and cosmetics. [MECH ENG] The part of a drill which runs from the outer corners of the cutting lips to the shank or neck.

body-centered lattice [CRYSTAL] A space lattice in which the point at the intersection of the body diagonals is identical to the points at the corners of the unit cell.

body of revolution [MATH] A symmetrical body having the form described by rotating a plane curve about an axis in its plane.

body rhythm [PHYSIO] Any bodily process having some degree of regular periodicity.

body wave [GEOPHYS] A seismic wave that travels within the earth, as distinguished from one that travels along the surface.

boehmite [MINERAL] AlO(OH) Gray, brown, or red orthorhombic mineral that is a major constituent of some bauxites.

bog [ECOL] A plant community that develops and grows in areas with permanently waterlogged peat substrates. Also known as moor; quagmire.

bogey See bogie.

boghead coal [GEOL] Bituminous or subbituminous coal containing a large proportion of algal remains and volatile matter; similar to cannel coal in appearance and combustion.

bogie Also spelled bogey; bogy. [AERO ENG] A type of landing-gear unit consisting of two sets of wheels in tandem with a central strut. [ENG] **1.** A supporting and aligning wheel or roller on the inside of an endless track. **2.** A low truck or cart of solid build. **3.** A truck or axle to which wheels are fixed, which supports a railroad car, the leading end of a locomotive, or the end of a vehicle (such as a gun carriage) and which is allowed to swivel under it. **4.** A railroad car or locomotive supported by a bogie. [MECH ENG] The drive-wheel assembly and supporting frame comprising the four rear wheels of a six-wheel truck, mounted so that they can self-adjust to sharp curves and irregularities in the road. [MIN ENG] A small truck or trolley upon which a bucket is carried from the shaft to the spoil bank.

bog iron ore [MINERAL] A soft, spongy, porous deposit of impure hydrous iron oxides formed in bogs, marshes, swamps, peat mosses, and shallow lakes by precipitation from iron-bearing waters and by the oxidation action of algae, iron bacteria, or the atmosphere. Also known as lake ore; limnite; marsh ore; meadow ore; morass ore; swamp ore.

bog-mine ore See bog ore.

bog moss [ECOL] Moss of the genus *Sphagnum* occurring as the characteristic vegetation of bogs.

bog ore [MINERAL] A poorly stratified accumulation of earthy metallic mineral substances, consisting mainly of oxides, that are formed in bogs, marshes, swamps, and other low-lying moist places. Also known as bog-mine ore.

bogy See bogie.

Bohemian ruby See rose quartz.

Bohemian topaz See citrine.

Bohr atom [ATOM PHYS] An atomic model having the structure postulated in the Bohr theory.

Bohr's correspondence principle See correspondence principle.

Bohr-Sommerfeld theory [ATOM PHYS] A modification of the Bohr theory in which elliptical as well as circular orbits are allowed.

Bohr theory [ATOM PHYS] A theory of atomic structure postulating an electron moving in one of certain discrete circular orbits about a nucleus with emission or absorption of electromagnetic radiation necessarily accompanied by transitions of the electron between the allowed orbits.

Boidae [VERT ZOO] The boas, a family of nonvenomous reptiles of the order Squamata, having teeth on both jaws and hindlimb rudiments.

boil See furuncle.

boiler [MECH ENG] A water heater for generating steam.

boiler compound [CHEM] Any chemical used to treat boiler water to prevent corrosion, the fouling of heat-absorbing surfaces, foaming, and the contamination of steam.

boiler efficiency [MECH ENG] The ratio of heat absorbed in steam to the heat supplied in fuel, usually measured in percent.

boiler plate [GEOL] A fairly smooth surface on a cliff, consisting of flush or overlapping slabs of rock, having little or no foothold. [HYD] A crusty, frozen surface of snow. [MET] Flat-rolled steel, usually ¼ to ½ inch (6 to 13 millimeters) thick; used mainly for covering ships and making boilers and tanks. Also known as boiler steel.

boiler steel See boiler plate.

boiler superheater [MECH ENG] A boiler component, consisting of tubular elements, in which heat is added to high-pressure steam to increase its temperature and enthalpy.

boiler tube [MECH ENG] One of the tubes in a boiler that carry water (water-tube boiler) to be heated by the high-temperature gaseous products of combustion or that carry combustion products (fire-tube boiler) to heat the boiler water that surrounds them.

boiling [ASTRON] The telescopic appearance of the limbs of the sun and planets when the earth's atmosphere is turbulent, characterized by a constant rippling motion and lack of a clearly defined edge. [PHYS CHEM] The transition of a substance from the liquid to the gaseous phase, taking place at a single temperature in pure substances and over a range of temperatures in mixtures.

boiling point [PHYS CHEM] Abbreviated bp. **1.** The temperature at which the transition from the liquid to the gaseous phase occurs in a pure substance at fixed pressure. **2.** See bubble point.

boiling range [CHEM] The temperature range of a laboratory distillation of an oil from start until evaporation is complete.

boiling spring [HYD] **1.** A spring which emits water at a high temperature or at boiling point. **2.** A spring located at the head of an interior valley and rising from the bottom of a residual clay basin. **3.** A rapidly flowing spring that develops strong vertical eddies.

boiling-water reactor [NUCLEO] A nuclear reactor in which the coolant is water, maintained at such a pressure as to allow it to boil and form steam. Abbreviated BWR.

boil-off [TEXT] Removal of impurities from fabric by boiling the material in a solution. [THERMO] The vaporization of a liquid, such as liquid oxygen or liquid hydrogen, as its temperature reaches its boiling point under conditions of exposure, as in the tank of a rocket being readied for launch.

boldface [GRAPHICS] A type that has thick heavy lines so that it has a conspicuous black appearance.

bole [GEOL] Any of various red, yellow, or brown earthy clays consisting chiefly of hydrous aluminum silicates. Also known as bolus; terra miraculosa. [GRAPHICS] The foundation laid for gold leaf.

bolometer [ENG] An instrument that measures the energy of electromagnetic radiation in certain wavelength regions by utilizing the change in resistance of a thin conductor caused by the heating effect of the radiation. Also known as thermal detector.

bolster [MET] A block of steel to which drop-forging dies are attached.

bolster plate [MECH ENG] A plate fixed on the bed of a power press to locate and support the die assembly.

bolt [DES ENG] A rod, usually of metal, with a square, round, or hexagonal head at one end and a screw thread on the other, used to fasten objects together. [MIN ENG] See bolthole. [ORD] The sliding part in a breechloading weapon that pushes a cartridge into position and holds it there as the gun is fired. [TEXT] The entire length of cloth from a loom.

bolthole [MIN ENG] A short, narrow opening made to connect the main working with the airhead or ventilating drift of a coal mine. Also known as bolt.

bolting [ENG] A fastening system using screw-threaded devices such as nuts, bolts, or studs. [FOOD ENG] The process of refining or purifying, especially of sifting flour or meal through a sieve. [MIN ENG] The use of vibrating sieves to separate particles of different sizes.

Boltzmann constant [STAT MECH] The ratio of the universal gas constant to the Avogadro number.

Boltzmann distribution [STAT MECH] A function giving the probability that a molecule of a gas in thermal equilibrium will have generalized position and momentum coordinates within given infinitesimal ranges of values, assuming that the molecules obey classical mechanics.

Boltzmann statistics See Maxwell-Boltzmann statistics.

Boltzmann transport equation [STAT MECH] An equation used to study the nonequilibrium behavior of a collection of particles; it states that the rate of change of a function which specifies the probability of finding a particle in a unit volume of phase space is equal to the sum of terms arising from external forces, diffusion of particles, and collisions of the particles. Also known as Maxwell-Boltzmann equation.

bolus [GEOL] See bole. [PHARM] A pill of large size. [PHYSIO] The mass of food prepared by the mouth for swallowing.

bolus alba See kaolin.

bomb See abend.

Bombacaceae [BOT] A family of dicotyledonous tropical trees in the order Malvales with dry or fleshy fruit usually having woolly seeds.

bombard [NUCLEO] To direct a stream of particles or photons against a target. [ORD] To carry out a sustained attack upon a city, fort, or the like with bombs, projectiles, rockets, or other explosive missiles.

bombardment [ELECTR] The use of induction heating to heat electrodes of electron tubes to drive out gases during evacuation.

bombardment aircraft See bomber.

bomb calorimeter [ENG] A calorimeter designed with a strong-walled container constructed of a corrosion-resistant alloy, called the bomb, immersed in about 2.5 liters of water in a metal container; the sample, usually an organic compound, is ignited by electricity, and the heat generated is measured.

bomber [AERO ENG] An airplane specifically designed to carry and drop bombs. Also known as bombardment aircraft.

bombiccite See hartite.

Bombidae [INV ZOO] A family of relatively large, hairy, black and yellow bumblebees in the hymenopteran superfamily Apoidea.

Bombycidae [INV ZOO] A family of lepidopteran insects of the superorder Heteroneura that includes only the silkworms.

Bombyliidae [INV ZOO] The bee flies, a family of dipteran insects in the suborder Orthorrhapha.

bond [CHEM] The strong attractive force that holds together atoms in molecules and crystalline salts. Also known as chemical bond. [ELEC] The connection made by bonding electrically. [ENG] 1. A wire rope that fixes loads to a crane hook. 2. Adhesion between cement or concrete and masonry or reinforcement. [MET] 1. Material added to molding sand to impart bond strength. 2. Junction of the base metal and filler metal, or the base metal beads, in a welded joint.

bond angle [PHYS CHEM] The angle between bonds sharing a common atom. Also known as valence angle.

bond distance [PHYS CHEM] The distance separating the two nuclei of two atoms bonded to each other in a molecule. Also known as bond length.

bonded-phase chromatography [ANALY CHEM] A type of high-pressure liquid chromatography which employs a stable, chemically bonded stationary phase.

bonded strain gage [ENG] A strain gage in which the resistance element is a fine wire, usually in zig-zag form, embedded in an insulating backing material, such as impregnated paper or plastic, which is cemented to the pressure-sensing element.

bond energy [PHYS CHEM] The heat of formation of a molecule from its constituent atoms.

bonding [CHEM] The joining together of atoms to form molecules or crystalline salts. [ELEC] The use of low-resistance material to connect electrically a chassis, metal shield cans, cable shielding braid, and other supposedly equipotential points to eliminate undesirable electrical interaction resulting from high-impedance paths between them. [ENG] 1. The fastening together of two components of a device by means of adhesives, as in anchoring the copper foil of printed wiring to an insulating baseboard. 2. See cladding. [TEXT] The joining of two fabrics, usually a face fabric and a lining fabric.

bonding electron [PHYS CHEM] An electron whose orbit spans the entire molecule and so assists in holding it together.

bonding orbital [PHYS CHEM] A molecular orbital formed by a bonding electron whose energy decreases as the nuclei are brought closer together, resulting in a net attraction and chemical bonding.

bonding pad [ELECTR] A metallized area on the surface of a semiconductor device, to which connections can be made.

bonding strength [MECH] Structural effectiveness of adhesives, welds, solders, glues, or of the chemical bond formed between the metallic and ceramic components of a cermet, when subjected to stress loading, for example, shear, tension, or compression.

bond length See bond distance.

bond paper [MATER] A paper used for writing paper, business forms, and typewriter paper; the less expensive bond papers are made from wood sulfite pulps; rag-content bonds contain 25, 50, 75, or 100% of pulp made from rags, and offer greater permanence and strength.

bone [ANAT] One of the parts constituting a vertebrate skeleton. [HISTOL] A hard connective tissue that forms the major portion of the vertebrate skeleton.

bone black [MATER] A black substance made by carbonizing crushed, defatted bones in closed vessels; used as a paint and varnish pigment, as a decolorizing absorbent in clarifying shellac, in cementation, and in gas masks. Also known as animal black; bone char.

bone char See bone black.

bone conduction [BIOPHYS] Transmission of sound vibrations to the internal ear via the bones of the skull.

Bonellidae [INV ZOO] A family of wormlike animals belonging to the order Echiuroinea.

bone marrow [HISTOL] A vascular modified connective tissue occurring in the long bones and certain flat bones of vertebrates.

bone meal [MATER] A substance made by grinding animal bones; steamed meal, made from pressure-steamed bones, is used as a fertilizer; raw meal is used in animal feed.

Bononian [GEOL] Upper Jurassic (lower Portlandian) geologic time.

bony fish [VERT ZOO] The name applied to all members of the class Osteichthyes.

bony labyrinth [ANAT] The system of canals within the otic bones of vertebrates that houses the membranous labyrinth of the inner ear.

Boo See Boötes.

Boodleaceae [BOT] A family of green marine algae in the order Siphonocladales.

book capacitor [ELEC] A trimmer capacitor consisting of two plates which are hinged at one end; capacitance is varied by changing the angle between them.

book gill [INV ZOO] A type of gill in king crabs consisting of folds of membranous tissue arranged like the leaves of a book.

bookkeeping operation [COMPUT SCI] A computer operation which does not directly contribute to the result, that is, arithmetical, logical, and transfer operations used in modifying the address section of other instructions in counting cycles and in rearranging data. Also known as red-tape operation.

book louse [INV ZOO] A common name for a number of insects belonging to the order Psocoptera; important pests in herbaria, museums, and libraries.

book lung [INV ZOO] A saccular respiratory organ in many arachnids consisting of numerous membranous folds arranged like the pages of a book.

Boolean algebra [MATH] An algebraic system with two binary operations and one unary operation important in representing a two-valued logic.

Boolean search [COMPUT SCI] A search for selected information, that is, information satisfying conditions that can be expressed by AND, OR, and NOT functions.

boom [COMMUN] A movable mechanical support, usually in a television or motion picture studio, to suspend a microphone within range of the performers but above the field of view of the camera. [NAV ARCH] 1. A spar attached to a mast or kingpost of a ship carrying cargo-hoisting gear. 2. A spar upon which the lower side of a sail is bent.

Boopidae [INV ZOO] A family of lice in the order Mallophaga, parasitic on Australian marsupials.

boost [AERO ENG] 1. An auxiliary means of propulsion such as by a booster. 2. To supercharge. 3. To launch or push along during a portion of a flight. 4. See boost pressure. [ELECTR] To augment in relative intensity, as to boost the bass response in an audio system. [ENG] To bring about a more potent explosion of the main charge of an explosive by using an additional charge to set it off.

booster [AERO ENG] See booster engine; booster rocket; launch vehicle. [ELEC] A small generator inserted in series or parallel with a larger generator to maintain normal voltage output under heavy loads. [ELECTR] 1. A separate radio-frequency amplifier connected between an antenna and a television receiver to amplify weak signals. 2. A radio-frequency amplifier that amplifies and rebroadcasts a received television or communication radio carrier frequency for reception by the general public. [IMMUNOL] The dose of an immunizing agent given to stimulate the effects of a previous dose of the same agent. [ORD] An assembly of metal parts and explosive charge provided to augment the explosive component of a fuse, to cause detonation of the main explosive charge of the munition.

booster engine [AERO ENG] An engine, especially a booster rocket, that adds its thrust to the thrust of the sustainer engine. Also known as booster.

booster response See anamnestic response.

booster rocket [AERO ENG] Also known as booster. 1. A rocket motor, either solid- or liquid-fueled, that assists the normal propulsive system or sustainer engine of a rocket or aeronautical vehicle in some phase of its flight. 2. A rocket used to set a vehicle in motion before another engine takes over.

booster voltage [ELECTR] The additional voltage supplied by the damper tube to the horizontal output, horizontal oscillator, and vertical output tubes of a television receiver to give greater sawtooth sweep output.

boost pressure [AERO ENG] Manifold pressure greater than the ambient at atmospheric pressure, obtained by supercharging. Also known as boost.

boot [ELEC] A protective covering over any portion of a cable, wire, or connector. [MIN ENG] 1. A projecting portion of a reinforced concrete beam acting as a corbel to support the facing material, such as brick or stone. 2. The lower end of a bucket elevator. [PETRO ENG] See surge column.

Boot See Boötes.

boot button See bootstrap button.

Boötes [ASTRON] A constellation which lies south and east of Ursa Major. The star Arcturus is a member of the group. Abbreviated Boo; Boot. Also known as Bear Driver.

boothite [MINERAL] $CuSO_4 \cdot 7H_2O$ A blue, monoclinic mineral consisting of copper sulfate heptahydrate; usually occurs in massive or fibrous form.

bootstrap [ENG] A technique or device designed to bring itself into a desired state by means of its own action.

bootstrap button [COMPUT SCI] The first button pressed when a computer is turned on, causing the operating system to be loaded into memory. Also known as boot button; initial program load button; IPL button.

bootstrap circuit [ELECTR] A single-stage amplifier in which the output load is connected between the negative end of the anode supply and the cathode, while signal voltage is applied between grid and cathode; a change in grid voltage changes the input signal voltage with respect to ground by an amount equal to the output signal voltage.

bootstrap driver [ELECTR] Electronic circuit used to produce a square pulse to drive the modulator tube; the duration of the square pulse is determined by a pulse-forming line.

bootstrap instructor technique [COMPUT SCI] A technique permitting a system to bring itself into an operational state by means of its own action. Also known as bootstrap technique.

bootstrap integrator [ELECTR] A bootstrap sawtooth generator in which an integrating amplifier is used in the circuit. Also known as Miller generator.

bootstrap loader [COMPUT SCI] A very short program loading routine, used for loading other loaders in a computer; often implemented in a read-only memory.

bootstrap memory [COMPUT SCI] A device that provides for the automatic input of new programs without erasing the basic instructions in the computer.

bootstrapping [ELECTR] A technique for lifting a generator circuit above ground by a voltage value derived from its own output signal.

bootstrap program *See* loading program.

bootstrap sawtooth generator [ELECTR] A circuit capable of generating a highly linear positive sawtooth waveform through the use of bootstrapping.

bootstrap technique *See* bootstrap instructor technique.

Bopyridae [INV ZOO] A family of epicaridean isopods in the tribe Bopyrina known to parasitize decapod crustaceans.

Bopyrina [INV ZOO] A tribe of dioecious isopods in the suborder Epicaridea.

Bopyroidea [INV ZOO] An equivalent name for Epicaridea.

bora [METEOROL] A fall wind whose source is so cold that when the air reaches the lowlands or coast the dynamic warming is insufficient to raise the air temperature to the normal level for the region; hence it appears as a cold wind.

boracic acid *See* boric acid.

boracite [MINERAL] $Mg_3B_7O_{13}Cl$ A white, yellow, green, or blue orthorhombic borate mineral occurring in crystals which appear isometric in external form; it is strongly pyroelectric, has a hardness of 7 on Mohs scale, and a specific gravity of 2.9.

Boraginaceae [BOT] A family of flowering plants in the order Lamiales comprising mainly herbs and some tropical trees.

Boralf [GEOL] A suborder of the soil order Alfisol, dull brown or yellowish brown in color; occurs in cool or cold regions, chiefly at high latitudes or high altitudes.

borane [INORG CHEM] 1. A class of binary compounds of boron and hydrogen; boranes are used as fuels. Also known as boron hydride. 2. A substance which may be considered a derivative of a boron-hydrogen compound, such as BCl_3 and $B_{10}H_{12}I_2$.

borate mineral [MINERAL] Any of the large and complex group of naturally occurring crystalline solids in which boron occurs in chemical combination with oxygen.

borax [MINERAL] $Na_2B_4O_7 \cdot 10H_2O$ A white, yellow, blue, green, or gray borate mineral that is an ore of boron and occurs as an efflorescence or in monoclinic crystals; when pure it is used as a cleaning agent, antiseptic, and flux. Also known as diborate; pyroborate; sodium (1:2) borate; sodium tetraborate; tincal.

Borda mouthpiece [FL MECH] A reentrant tube in a hydraulic reservoir, whose contraction coefficient

(the ratio of the cross section of the issuing jet of liquid to that of the opening) can be calculated more simply than for other discharge openings.

Bordeaux mixture [MATER] A fungicide made from a mixture of lime, copper sulfate, and water.

borderline psychosis [PSYCH] The psychiatric diagnosis of an individual whose symptoms are severe but are not clearly psychotic or neurotic.

borderline syndrome [PSYCH] An impairment of ego function noted in individuals who have been angry most of their lives and who cannot relate meaningfully to or love other people, and who in consequence develop nonpsychotic but deviant mechanisms, such as withdrawal while expressing angry concern, or showing passive compliance at the price of real involvement, to maintain a precarious mental equilibrium.

bore [DES ENG] Inside diameter of a pipe or tube. [MECH ENG] 1. The diameter of a piston-cylinder mechanism as found in reciprocating engines, pumps, and compressors. 2. To penetrate or pierce with a rotary tool. 3. To machine a workpiece to increase the size of an existing hole in it. [MIN ENG] 1. A tunnel under construction. 2. To cut or drill a hole for blasting, water infusion, exploration, or water or firedamp drainage. [OCEANOGR] 1. A high, breaking wave of water, advancing rapidly up an estuary. Also known as eager; mascaret; tidal bore. 2. A submarine sand ridge, in very shallow water, whose crest may rise to intertidal level. [ORD] The interior of a gun barrel or tube.

borehole [ENG] A hole made by boring into the ground to study stratification, to obtain natural resources, or to release underground pressures.

borehole survey [ENG] Also known as drillhole survey. 1. Determining the course of and the target point reached by a borehole, using an azimuth-and-dip recording apparatus small enough to be lowered into a borehole. 2. The record of the information thereby obtained.

borer [INV ZOO] Any insect or other invertebrate that burrows into wood, rock, or other substances. [MECH ENG] An apparatus used to bore openings into the earth up to about 8 feet (2.4 meters) in diameter.

Borhyaenidae [VERT ZOO] A family of carnivorous mammals in the superfamily Borhyaenoidea.

Borhyaenoidea [VERT ZOO] A superfamily of carnivorous mammals in the order Marsupialia.

boric acid [INORG CHEM] H_3BO_3 An acid derived from boric oxide in the form of white, triclinic crystals, melting at 185°C, soluble in water. Also known as boracic acid; orthoboric acid.

borickite [MINERAL] $CaFe_5(PO_4)_2(OH)_{11} \cdot 3H_2O$ A reddish-brown, isotropic mineral consisting of a hydrated basic phosphate of calcium and iron; occurs in compact reniform masses.

boride [INORG CHEM] A binary compound of boron and a metal formed by heating a mixture of the two elements.

boring log *See* drill log.

boring machine [MECH ENG] A machine tool designed to machine internal work such as cylinders, holes in castings, and dies; types are horizontal, vertical, jig, and single.

boring sponge [INV ZOO] Marine sponge of the family Clionidae represented by species which excavate galleries in mollusks, shells, corals, limestone, and other calcareous matter.

Born approximation [QUANT MECH] A method used for the computation of cross sections in scattering problems; the interactions are treated as perturbations of free-particle systems.

Born equation [PHYS CHEM] An equation for determining the free energy of solvation of an ion in terms of the Avogadro number, the ionic valency, the ion's electronic charge, the dielectric constant of the electrolytic, and the ionic radius.

Born-Haber cycle [SOLID STATE] A sequence of chemical and physical processes by means of which the cohesive energy of an ionic crystal can be deduced from experimental quantities; it leads from an initial state in which a crystal is at zero pressure and 0 K to a final state which is an infinitely dilute gas of its constituent ions, also at zero pressure and 0 K.

bornite [MINERAL] Cu_5FeS_4 A primary mineral in many copper ore deposits; specific gravity 5.07; the metallic and brassy color of a fresh surface rapidly tarnishes upon exposure to air to an iridescent purple.

Born-Mayer equation [SOLID STATE] An equation for the cohesive energy of an ionic crystal which is deduced by assuming that this energy is the sum of terms arising from the Coulomb interaction and a repulsive interaction between nearest neighbors.

Born-Oppenheimer method [PHYS CHEM] A method for calculating the force constants between atoms by assuming that the electron motion is so fast compared with the nuclear motions that the electrons follow the motions of the nuclei adiabatically.

Born–von Kármán theory [SOLID STATE] A theory of specific heat which considers an acoustical spectrum for the vibrations of a system of point particles distributed like the atoms in a crystal lattice.

borolanite [PETR] A hypabyssal rock that is essentially orthoclase and melanite with subordinate nepheline, biotite, and pyroxene.

Boroll [GEOL] A suborder of the soil order Mollisol, characterized by a mean annual soil temperature of less than 8°C and by never being dry for 60 consecutive days during the 90-day period following the summer solstice.

boron [CHEM] A chemical element, symbol B, atomic number 5, atomic weight 10.811; it has three valence electrons and is nonmetallic.

boronatrocalcite *See* ulexite.

boron carbide [ORG CHEM] Any compound of boron and carbon, especially B_4C (used as an abrasive, alloying agent, and neutron absorber).

boron steel [MET] Alloy steel with a small amount (as little as 0.0005) of boron added to increase hardenability; can be used to replace other alloys in short supply.

borrow [CIV ENG] Earth material such as sand and gravel that is taken from one location to be used as fill at another. [MATH] An arithmetically negative carry; it occurs in direct subtraction by raising the low-order digit of the minuend by one unit of the next-higher-order digit; for example, when subtracting 67 from 92, a tens digit is borrowed from the 9, to raise the 2 to a factor of 12; the 7 of 67 is then subtracted from the 12 to yield 5 as the units digit of the difference; the 6 is then subtracted from 8, or 9 − 1, yielding 2 as the tens digit of the difference.

bort bit *See* diamond bit.

Bose distribution *See* Bose-Einstein distribution.

Bose-Einstein condensation [CRYO] A phenomenon that occurs in the study of systems of bosons; there is a critical temperature below which the ground state is highly populated. Also known as condensation; Einstein condensation.

Bose-Einstein distribution [STAT MECH] For an assembly of independent bosons, such as photons or helium atoms of mass number 4, a function that specifies the number of particles in each of the allowed energy states. Also known as Bose distribution.

Bose-Einstein statistics [STAT MECH] The statistical mechanics of a system of indistinguishable particles for which there is no restriction on the number of particles that may exist in the same state simultaneously. Also known as Einstein-Bose statistics.

bosh [MET] **1.** Tapering lower portion of a blast furnace, from the blast holes of the hearth up to the maximum internal diameter at the bottom of the stack. **2.** Quartz deposited on the furnace lining during the smelting of copper ore.

boson [STAT MECH] A particle that obeys Bose-Einstein statistics; includes photons, pi mesons, and all nuclei having an even number of particles and all particles with integer spin.

boss [DES ENG] Protuberance on a cast metal or plastic part to add strength, facilitate assembly, provide for fastenings, or so forth. [GEOL] A large, irregular mass of crystalline igneous rock that formed some distance below the surface but is now exposed by denudation. [NAV ARCH] *See* bossing.

bossing [NAV ARCH] **1.** A faired structural extension of the hull covering the propeller shaft where it emerges from the main hull and housing the main shaft bearing. **2.** The hub of a screw propeller. Also known as boss.

Bostrichidae [INV ZOO] The powder-post beetles, a family of coleopteran insects in the superfamily Bostrichoidea.

Bostrichoidea [INV ZOO] A superfamily of beetles in the coleopteran suborder Polyphaga.

botany [BIOL] A branch of the biological sciences which embraces the study of plants and plant life. [TEXT] *See* merino.

Bothriocephaloidea [INV ZOO] The equivalent name for the Pseudophyllidea.

Bothriocidaroida [PALEON] An order of extinct echinoderms in the subclass Perischoechinoidea in which the ambulacra consist of two columns of plates, the interambulacra of one column, and the madreporite is placed radially.

botryogen [MINERAL] $MgFe(SO_4)_2(OH)\cdot7H_2O$ An orange-red, monoclinic mineral consisting of a hydrated basic sulfate of magnesium and trivalent iron.

botryoid [GEOL] **1.** A mineral formation shaped like a bunch of grapes. **2.** Specifically, such a formation of calcium carbonate occurring in a cave. Also known as clusterite.

botryoidal [SCI TECH] Of formations and structures, shaped like a bunch of grapes.

bottled gas [MATER] Butane, propane, or butane-propane mixtures liquefied and bottled under pressure for use as a domestic cooking or heating fuel. Also known as bugas.

bottle thermometer [ENG] A thermoelectric thermometer used for measuring air temperature; the name is derived from the fact that the reference thermocouple is placed in an insulated bottle.

bottom [GEOL] **1.** The bed of a body of running or still water. **2.** *See* root.

bottom blow [ENG] A type of plastics blow molding machine in which air is injected into the parison from the bottom of the mold.

bottomed hole [ENG] A completed borehole, or a borehole in which drilling operations have been discontinued.

bottom fermentation [FOOD ENG] A slow alcoholic fermentation during which yeast cells accumulate at the bottom of the fermenting liquid; occurs during fermentation of lager beer and wines of low alcoholic content.

bottom flow [ENG] A molding apparatus that forms hollow plastic articles by injecting the blowing air at the bottom of the mold. [HYD] A density current that is denser than any section of the surrounding water and that flows along the bottom of the body of water. Also known as underflow.

bottom ice *See* anchor ice.

bottoming drill [DES ENG] A flat-ended twist drill designed to convert a cone at the bottom of a drilled hole into a cylinder.

bottomland [GEOL] A lowland formed by alluvial deposit about a lake basin or a stream.

bottom moraine *See* ground moraine.

bottom rot [PL PATH] 1. A fungus disease of lettuce, caused by *Pellicularia filamentosa*, that spreads from the base upward. 2. A fungus disease of tree trunks caused by pore fungi.

bottom sampler [ENG] Any instrument used to obtain a sample from the bottom of a body of water.

bottom sediment [PETRO ENG] A mixture of liquids and solids which form in the bottom of oil storage tanks.

bottomset beds [GEOL] Horizontal or gently inclined layers of finer material carried out and deposited on the bottom of a lake or sea in front of a delta.

bottom terrace [GEOL] A landform deposited by streams with moderate or small bottom loads of coarse sand and gravel, and characterized by a broad, sloping surface in the direction of flow and a steep escarpment facing downstream.

bottom water [HYD] Water lying beneath oil or gas in productive formations. [OCEANOGR] The water mass at the deepest part of a water column in the ocean.

botulin [MICROBIO] The neurogenic toxin which is produced by *Clostridium botulinum* and *C. parabotulinum* and causes botulism. Also known as botulinus toxin.

botulism [MED] Food poisoning due to intoxication by the exotoxin of *Clostridium botulinum* and *C. parabotulinum*.

Bouguer gravity anomaly [GEOPHYS] A value that corrects the observed gravity for latitude and elevation variations, as in the free-air gravity anomaly, plus the mass of material above some datum (usually sea level) within the earth and topography.

Bouguer-Lambert-Beer law [ANALY CHEM] The intensity of a beam of monochromatic radiation in an absorbing medium decreases exponentially with penetration distance. Also known as Beer-Lambert-Bouguer law; Lambert-Beer law.

Bouguer-Lambert law [ANALY CHEM] The law that the change in intensity of light transmitted through an absorbing substance is related exponentially to the thickness of the absorbing medium and a constant which depends on the sample and the wavelength of the light. Also known as Lambert's law.

boulder [GEOL] A worn rock with a diameter exceeding 256 millimeters. Also spelled bowlder.

boulder buster [ENG] A heavy, pyramidical- or conical-point steel tool which may be attached to the bottom end of a string of drill rods and used to break, by impact, a boulder encountered in a borehole. Also known as boulder cracker.

boulder clay *See* till.

boulder cracker *See* boulder buster.

boulder pavement [GEOL] A surface of till with boulders; the till has been abraded to flatness by glacier movement.

boulder train [GEOL] Glacial boulders derived from one locality and arranged in a right-angled line or lines leading off in the direction in which the drift agency operated.

boule [CRYSTAL] A pure crystal, such as silicon, having the atomic structure of a single crystal, formed synthetically by rotating a small seed crystal while pulling it slowly out of molten material in a special furnace.

boundary [ELECTR] An interface between *p*- and *n*-type semiconductor materials, at which donor and acceptor concentrations are equal. [GEOL] A line between areas occupied by rocks or formations of different type and age. [SCI TECH] A line or area which determines inclusion in a system.

boundary condition [MATH] A requirement to be met by a solution to a set of differential equations on a specified set of values of the independent variables.

boundary-layer control [FL MECH] Control over the development of a boundary layer by reduction of surface roughness and choice of surface contours. Abbreviated BLC.

boundary-layer flow [FL MECH] The flow of that portion of a viscous fluid which is in the neighborhood of a body in contact with the fluid and in motion relative to the fluid.

boundary-layer photocell *See* photovoltaic cell.

boundary-layer plasma [FL MECH] A plasma resulting from the frictional heat of hypersonic spacecraft entering the earth's atmosphere.

boundary-layer separation [FL MECH] That point where the boundary layer no longer continues to follow the contour of the boundary because the residual momentum of the fluid (left after overcoming viscous forces) may be insufficient to allow the flow to proceed into regions of increasing pressure. Also known as flow separation.

boundary-layer theory *See* film theory.

boundary point [MATH] In a topological space, a point of a set with the property that every neighborhood of the point contains points of both the set and its complement.

boundary survey [ENG] A survey made to establish or to reestablish a boundary line on the ground or to obtain data for constructing a map or plat showing a boundary line.

boundary wave [GEOPHYS] A seismic wave that propagates along a free surface or an interface between defined layers.

bounded function [MATH] 1. A function whose image is a bounded set. 2. A function of a metric space to itself which moves each point no more than some constant distance.

bounded set [MATH] 1. A collection of numbers whose absolute values are all smaller than some constant. 2. A set of points, the distance between any two of which is smaller than some constant.

bounds register [COMPUT SCI] A device which stores the upper and lower bounds on addresses in the memory of a given computer program in a time-sharing system.

bound variable [MATH] In logic, a variable that occurs within the scope of a quantifier, and cannot be replaced by a constant.

bound vector [MECH] A vector whose line of application and point of application are both prescribed, in addition to its direction.

bound water [CHEM] Water that is a portion of a system such as tissues or soil and does not form ice crystals until the material's temperature is lowered to about $-20°C$.

Bourdon pressure gage [ENG] A mechanical pressure-measuring instrument employing as its sensing element a curved or twisted metal tube, flattened in cross section and closed. Also known as Bourdon tube.

Bourdon tube *See* Bourdon pressure gage.

bournonite [MINERAL] $PbCuSbS_3$ Steel-gray to black orthorhombic crystals; mined as an ore of copper, lead, and antimony. Also known as berthonite; cogwheel ore.

boussingaultite [MINERAL] $(NH_4)_2Mg(SO_4)_2 \cdot 6H_2O$ A colorless to yellowish-pink, monoclinic mineral consisting of a hydrated sulfate of ammonium and magnesium; usually occurs in massive form, as crusts or stalactites.

bouton [ANAT] A club-shaped enlargement at the end of a nerve fiber. Also known as end bulb.

Bovidae [VERT ZOO] A family of pecoran ruminants in the superfamily Bovoidea containing the true antelopes, sheep, and goats.

bovine [VERT ZOO] **1.** Any member of the genus *Bos*. **2.** Resembling or pertaining to a cow or ox.

Bovoidea [VERT ZOO] A superfamily of pecoran ruminants in the order Artiodactyla comprising the pronghorns and bovids.

bow [AERO ENG] The forward section of an aircraft. [ARCH] A part of a building shaped as an arc or a polygon and projecting from a straight wall. [NAV ARCH] The forward part of a ship.

Bowen reaction series [MINERAL] A series of minerals wherein any early-formed phase will react with the melt later in the differentiation to yield a new mineral further in the series.

bowfin [VERT ZOO] *Amia calva*. A fish recognized as the only living species of the family Amiidae. Also known as dogfish; grindle; mudfish.

bowl classifier [CHEM ENG] A shallow bowl with a concave bottom so that a liquid-solid suspension can be fed to the center; coarse particles fall to the bottom, where they are raked to a central discharge point, and liquid and fine particles overflow the edges and are collected.

bowline [NAV ARCH] **1.** A rope attached to the vertical edge of a square sail near its midpoint, and used to keep the sail's weather edge taut forward when the vessel is close-hauled. **2.** A knot forming a loop that does not slip under tension, used particularly for mooring and hauling.

bowlingite *See* saponite.

bowl mill *See* bowl-mill pulverizer.

bowl-mill pulverizer [MECH ENG] A type of pulverizer which directly feeds a coal-fired furnace, in which springs press pivoted stationary rolls against a rotating bowl grinding ring, crushing the coal between them. Also known as a bowl mill.

bowl scraper [MECH ENG] A towed steel bowl hung within a fabricated steel frame, running on four or two wheels; transports soil, in addition to spreading and leveling it.

Bowman's capsule [ANAT] A two-layered membranous sac surrounding the glomerulus and constituting the closed end of a nephron in the kidneys of all higher vertebrates.

bowshock [ASTROPHYS] The shock wave set up by the interaction of the supersonic solar wind with a planet's magnetic field.

bowsprit [NAV ARCH] A large spar that projects forward from the forward end of a sailing ship, used to carry sails and support the masts by stays.

bowstring beam [CIV ENG] A steel, concrete, or timber beam or girder shaped in the form of a bow and string; the string resists the horizontal forces caused by loads on the arch.

bow wave [FL MECH] A shock wave occurring in front of a body, such as an airfoil, or apparently attached to the forward tip of the body.

box annealing [MET] Slow heating of metal sheets in a closed metal box to prevent oxidation, followed by cooling; usually limited to iron-base alloys.

box beam *See* box girder.

box bill *See* bell tap.

box caisson [CIV ENG] A floating steel or concrete box with an open top which will be filled and sunk at a foundation site in a river or seaway. Also known as American caisson; stranded caisson.

box camera [OPTICS] A camera that consists of a box, an arrangement for loading and winding film, a simple lens of fixed focus, and a simple shutter with a speed of about 1/30 second.

boxcar [COMMUN] One of a series of long signal-wave pulses which are separated by very short intervals of time. [ENG] A railroad car with a flat roof and vertical sides, usually with sliding doors, which carries freight that needs to be protected from weather and theft.

boxcar circuit [ELECTR] A circuit used in radar for sampling voltage waveforms and storing the latest value sampled; the term is derived from the flat, steplike segments of the output voltage waveform.

box girder [CIV ENG] A hollow girder or beam with a square or rectangular cross section. Also known as box beam.

box piles [CIV ENG] Pile foundations made by welding together two sections of steel sheet piling or combinations of beams, channels, and plates.

box wrench [ENG] A closed-end wrench designed to fit a variety of sizes and shapes of bolt heads and nuts.

Boyle's law [PHYS] The law that the product of the volume of a gas times its pressure is a constant at fixed temperature. Also known as Mariotte's law.

Boyle's temperature [THERMO] For a given gas, the temperature at which the virial coefficient B in the equation of state $Pv = RT\ [1 + (B/v) + (C/v^2) + \cdots]$ vanishes.

bp *See* boiling point.

bpm *See* barrels per month.

Bq *See* becquerel.

brace [DES ENG] A cranklike device used for turning a bit. [ENG] A diagonally placed structural member that withstands tension and compression, and often stiffens a structure against wind.

braced framing [CIV ENG] Framing a building with post and braces for stiffness.

braced-rib arch [CIV ENG] A type of steel arch, usually used in bridge construction, which has a system of diagonal bracing.

brace head [ENG] A cross handle attached at the top of a column of drill rods by means of which the rods and attached bit are turned after each drop in chop-and-wash operations while sinking a borehole through overburden. Also known as brace key.

brace key *See* brace head.

brachial artery [ANAT] An artery which originates at the axillary artery and branches into the radial and ulnar arteries; it distributes blood to the various muscles of the arm, the shaft of the humerus, the elbow joint, the forearm, and the hand.

brachial plexus [ANAT] A plexus of nerves located in the neck and axilla and composed of the anterior rami of the lower four cervical and first thoracic nerves.

Brachiata [INV ZOO] A phylum of deuterostomous, sedentary bottom-dwelling marine animals that live encased in tubes.

brachiate [BOT] Possessing widely divergent branches. [ZOO] Having arms.

Brachiopoda [INV ZOO] A phylum of solitary, marine, bivalved coelomate animals.

brachioradialis [ANAT] The muscle of the arm that flexes the elbow joint; origin is the lateral supracondylar ridge of the humerus, and insertion is the lower end of the radius.

brachium [ANAT] The upper arm or forelimb, from the shoulder to the elbow. [INV ZOO] 1. A ray of a crinoid. 2. A tentacle of a cephalopod. 3. Either of the paired appendages constituting the lophophore of a brachiopod.

brachyaxis [CRYSTAL] The shorter lateral axis, usually the *a* axis, of an orthorhombic or triclinic crystal. Also known as brachydiagonal.

brachydiagonal *See* brachyaxis.

brachydont [ANAT] Of teeth, having short crowns, well-developed roots, and narrow root canals; characteristic of man.

Brachygnatha [INV ZOO] A subsection of brachyuran crustaceans to which most of the crabs are assigned.

Brachypsectridae [INV ZOO] A family of coleopteran insects in the superfamily Cantharoidea represented by a single species.

Brachypteraciidae [VERT ZOO] The ground rollers, a family of colorful Madagascan birds in the order Coraciiformes.

brachypterous [INV ZOO] Having rudimentary or abnormally small wings, referring to certain insects.

Brachythoraci [PALEON] An order of the joint-neck fishes, now extinct.

Brachyura [INV ZOO] The section of the crustacean order Decapoda containing the true crabs.

bracket [BUILD] A vertical board to support the tread of a stair. [CIV ENG] A projecting support. [ORD] 1. The distance between two strikes or series of strikes, one of which is over the target and the other short of it, or one of which is to the right and the other to the left of the target. 2. A group of shots (or bombs) which fall both over and short of the target.

bracket fungus [MYCOL] A basidiomycete characterized by shelflike sporophores, sometimes seen on tree trunks.

brackish [HYD] 1. Of water, having salinity values ranging from approximately 0.50 to 17.00 parts per thousand. 2. Of water, having less salt than sea water, but undrinkable.

Braconidae [INV ZOO] The braconid wasps, a family of hymenopteran insects in the superfamily Ichneumonoidea.

bract [BOT] A modified leaf associated with plant reproductive structures.

bracteole [BOT] A small bract, especially if on the floral axis. Also known as bractlet.

bractlet *See* bracteole.

bradleyite [MINERAL] $Na_3Mg(PO_4)(CO_3)$ A light gray mineral consisting of a phosphate-carbonate of sodium and magnesium; occurs as fine-grained masses.

Bradyodonti [PALEON] An order of Paleozoic cartilaginous fishes (Chondrichthyes), presumably derived from primitive sharks.

Bradypodidae [VERT ZOO] A family of mammals in the order Edentata comprising the true sloths.

bradytely [EVOL] Evolutionary change that is either arrested or occurring at a very slow rate over long geologic periods.

Bragg angle [SOLID STATE] One of the characteristic angles at which x-rays reflect specularly from planes of atoms in a crystal.

Bragg curve [ATOM PHYS] 1. A curve showing the average number of ions per unit distance along a beam of initially monoenergetic ionizing particles, usually alpha particles, passing through a gas. Also known as Bragg ionization curve. 2. A curve showing the average specific ionization of an ionizing particle of a particular kind as a function of its kinetic energy, velocity, or residual range.

Bragg diffraction *See* Bragg scattering.

Bragg ionization curve *See* Bragg curve.

Bragg-Kleeman rule *See* Bragg rule.

Bragg reflection *See* Bragg scattering.

Bragg rule [ATOM PHYS] An empirical rule according to which the mass stopping power of an element for alpha particles is inversely proportional to the square root of the atomic weight. Also known as Bragg-Kleeman rule.

Bragg scattering [SOLID STATE] Scattering of x-rays or neutrons by the regularly spaced atoms in a crystal, for which constructive interference occurs only at definite angles called Bragg angles. Also known as Bragg diffraction; Bragg reflection.

Bragg's equation *See* Bragg's law.

Bragg's law [SOLID STATE] A statement of the conditions under which a crystal will reflect a beam of x-rays with maximum intensity. Also known as Bragg's equation; Bravais' law.

Bragg spectrometer [ENG] An instrument for x-ray analysis of crystal structure and measuring wavelengths of x-rays and gamma rays, in which a homogeneous beam of x-rays is directed on the known face of a crystal and the reflected beam is detected in a suitably placed ionization chamber. Also known as crystal diffraction spectrometer; crystal spectrometer; ionization spectrometer.

braided stream [HYD] A stream flowing in several channels that divide and reunite.

braided wire [ELEC] A tube of fine wires woven around a conductor or cable for shielding purposes or used alone in flattened form as a grounding strap.

Braille [COMMUN] A system of written communication for the blind in which letters are represented by raised dots over which the trained blind person moves the fingertips.

brain [ANAT] The portion of the vertebrate central nervous system enclosed in the skull. [ZOO] The enlarged anterior portion of the central nervous system in most bilaterally symmetrical animals.

braincase *See* cranium.

brain coral [INV ZOO] A reef-building coral resembling the human cerebrum in appearance.

brainstem [ANAT] The portion of the brain remaining after the cerebral hemispheres and cerebellum have been removed.

brainstorming [IND ENG] A procedure used to find a solution for a problem by collecting all the ideas, without regard for feasibility, which occur from a group of people meeting together.

brain wave [PHYSIO] A rhythmic fluctuation of voltage between parts of the brain, ranging from about 1 to 60 hertz and 10 to 100 microvolts.

brake [MECH ENG] A machine element for applying friction to a moving surface to slow it (and often, the containing vehicle or device) down or bring it to rest.

brake drum [MECH ENG] A rotating cylinder attached to a rotating part of machinery, which the brake band or brake shoe presses against.

brake fluid [MATER] The liquid in the cylinder of an automotive brake which, under the action of a piston, is forced through tubing into cylinders at each car wheel, moving a pair of pistons outward so that the brake shoes are thrust against the revolving brake drums.

brake lining [MECH ENG] A covering, riveted or molded to the brake shoe or brake band, which presses against the rotating brake drum; made of either fabric or molded asbestos material.

brake shoe [MECH ENG] The renewable friction element of a shoe brake. Also known as shoe.

braking rocket *See* retrorocket.

branch [BOT] A shoot or secondary stem on the trunk or a limb of a tree. [COMPUT SCI] **1.** Any one of a number of instruction sequences in a program to which computer control is passed, depending upon the status of one or more variables. **2.** *See* jump. [ELEC] A portion of a network consisting of one or more two-terminal elements in series. Also known as arm. [HYD] A small stream that merges into another, generally bigger, stream. [MATH] A complex function which is analytic in some domain and which takes on one of the values of a multiple-valued function in that domain. [NUC PHYS] A product resulting from one mode of decay of a radioactive nuclide that has two or more modes of decay. [ORG CHEM] A carbon side chain attached to a molecule's main carbon chain. [SCI TECH] An area of study representing an independent offshoot of a related basic discipline.

branch circuit [ELEC] A portion of a wiring system in the interior of a structure that extends from a final overload protective device to a plug receptacle or a load such as a lighting fixture, motor, or heater.

branched acinous gland [ANAT] A multicellular structure with saclike glandular portions connected to the surface of the containing organ or structure by a common duct.

branched tubular gland [ANAT] A multicellular structure with tube-shaped glandular portions connected to the surface of the containing organ or structure by a common secreting duct.

branchia *See* gill.

branchial [ZOO] Of or pertaining to gills.

branchial arch [VERT ZOO] One of the series of paired arches on the sides of the pharynx which support the gills in fishes and amphibians.

branchial cleft [EMBRYO] A rudimentary groove in the neck region of air-breathing vertebrate embryos. [VERT ZOO] One of the openings between the branchial arches in fishes and amphibians.

branchial pouch [ZOO] In cyclostomes and some sharks, one of the respiratory cavities occurring in the branchial clefts.

branchial sac [INV ZOO] In tunicates, the dilated pharyngeal portion of the alimentary canal that has vascular walls pierced with clefts and serves as a gill.

branchial segment [EMBRYO] Any of the paired pharyngeal segments indicating the visceral arches and clefts posterior to and including the third pair in air-breathing vertebrate embryos.

branching [COMPUT SCI] The selection, under control of a computer program, of one of two or more branches. [NUC PHYS] The occurrence of two or more modes by which a radionuclide can undergo radioactive decay. Also known as multiple decay; multiple disintegration.

branching adaptation *See* divergent adaptation.

branching bay *See* estuary.

branching process [STAT] A stochastic process in which the members of a population may have offspring and the lines of descent branch out as the new members are born.

branching ratio [NUC PHYS] The ratio of the number of parent atoms or particles decaying by one mode to the number decaying by another mode; the ratio of two specified branching fractions.

branch instruction [COMPUT SCI] An instruction that makes the computer choose between alternative subprograms, depending on the conditions determined by the computer during the execution of the program.

branchiocranium [VERT ZOO] The division of the fish skull constituting the mandibular and hyal regions and the branchial arches.

branchiomere [EMBRYO] An embryonic metamere that will differentiate into a visceral arch and cleft; a branchial segment.

Branchiopoda [INV ZOO] A subclass of crustaceans containing small or moderate-sized animals commonly called fairy shrimps, clam shrimps, and water fleas.

Branchiostoma [ZOO] A genus of lancelets formerly designated as amphioxus.

Branchiotremata [INV ZOO] The hemichordates, a branch of the subphylum Oligomera.

branchite *See* hartite.

Branchiura [INV ZOO] The fish lice, a subclass of fish ectoparasites in the class Crustacea.

branch point [COMPUT SCI] A point in a computer program at which there is a branch instruction. [ELEC] A terminal in an electrical network that is common to more than two elements or parts of elements of the network. Also known as junction point; node. [MATH] A point at which two or more sheets of a Riemann surface join together.

branch sewer [CIV ENG] A part of a sewer system that is larger in diameter than the lateral sewer system; receives sewage from both house connections and lateral sewers.

brass [GEOL] A British term for sulfides of iron (pyrites) in coal. Also known as brasses. [MET] A copper-zinc alloy of varying proportions but typically containing 67% copper and 33% zinc.

brasses *See* brass.

brassin [BIOCHEM] Any of a class of plant hormones characterized as long-chain fatty-acid esters; brassins act to induce both cell elongation and cell division in leaves and stems.

brass ore *See* aurichalcite.

Brathinidae [INV ZOO] The grass root beetles, a small family of coleopteran insects in the superfamily Staphylinoidea.

Braulidae [INV ZOO] The bee lice, a family of cyclorrhaphous dipteran insects in the section Pupipara.

braunite [MINERAL] $3Mn_2O_3 \cdot MnSiO_3$ Brittle mineral that forms tetragonal crystals; commonly found as steel-gray or brown-black masses in the United States, Europe, and South America; it is an ore of manganese.

Braun tube *See* cathode-ray tube.

Bravais indices [CRYSTAL] A modification of the Miller indices; frequently used for hexagonal and trigonal crystalline systems; they refer to four axes: the *c*-axis and three others at 120° angles in the basal plane.

Bravais lattice [CRYSTAL] One of the 14 possible arrangements of lattice points in space such that the arrangement of points about any chosen point is identical with that about any other point.

Bravais' law See Bragg's law.

Brayton cycle [THERMO] A thermodynamic cycle consisting of two constant-pressure processes interspersed with two constant-entropy processes. Also known as complete-expansion diesel cycle; Joule cycle.

braze [MET] To solder metals by melting a nonferrous filler metal, such as brass or brazing alloy (hard solder), with a melting point lower than that of the base metals, at the point of contact. Also known as hard-solder.

braze welding [MET] A method of welding in which coalescence is produced by heating above 800°F (427°C) and by using a nonferrous filler metal having a melting point below that of the base metals; in distinction to brazing, capillary attraction does not distribute the filler metal in the joint.

Brazil Current [OCEANOGR] The warm ocean current that flows southward along the Brazilian coast below Natal; the western boundary current in the South Atlantic Ocean.

Brazil nut [BOT] *Bertholettia excelsa*. A large broadleafed evergreen tree of the order Lecythidales; an edible seed is produced by the tree fruit.

breadboarding [ELECTR] Assembling an electronic circuit in the most convenient manner, without regard for final locations of components, to prove the feasibility of the circuit and to facilitate changes when necessary.

breadboard model [ENG] Uncased assembly of an instrument or other piece of equipment, such as a radio set, having its parts laid out on a flat surface and connected together to permit a check or demonstration of its operation.

breadfruit [BOT] *Artocarpus altilis*. An Indo-Malaysian tree, a species of the mulberry family (Moraceae). The tree produces a multiple fruit which is edible.

break [COMMUN] An interruption of a radio, telegraph, or telephone communication, as for sending in the opposite direction. [ELEC] **1.** A fault in a circuit. **2.** The minimum distance in a circuit-opening device between the stationary and movable contacts when these contacts are in the open position. [ELECTR] A reflected radar pulse which appears on a radarscope as a line perpendicular to the base line. [GEOGR] A significant variation of topography, such as a deep valley. [GEOL] *See* knickpoint. [METEOROL] **1.** A sudden change in the weather; usually applied to the end of an extended period of unusually hot, cold, wet, or dry weather. **2.** A hole or gap in a layer of clouds. [MIN ENG] **1.** A plane of discontinuity in the coal seam such as a slip, fracture, or cleat; the surfaces are in contact or slightly separated. **2.** A fracture or crack in the roof beds as a result of mining operations.

break-bulk cargo [IND ENG] Miscellaneous goods packed in boxes, bales, crates, cases, bags, cartons, barrels, or drums; may also include lumber, motor vehicles, pipe, steel, and machinery.

breakdown [ELEC] A large, usually abrupt rise in electric current in the presence of a small increase in voltage; can occur in a confined gas between two electrodes, a gas tube, the atmosphere (as lightning), an electrical insulator, and a reverse-biased semiconductor diode. [MET] The initial process of rolling and drawing, or a series of such processes, which reduce a casting or extruded shape before its final reduction to desired size.

breaker cam [MECH ENG] A rotating, engine-driven device in the ignition system of an internal combustion engine which causes the breaker points to open, leading to a rapid fall in the primary current.

breaker points [ELEC] Low-voltage contacts used to interrupt the current in the primary circuit of a gasoline engine's ignition system.

break-even point [IND ENG] The point at which a company neither makes a profit nor suffers a loss from the operations of the business, and at which total costs are equal to total sales volume.

break frequency [CONT SYS] The frequency at which a graph of the logarithm of the amplitude of the frequency response versus the logarithm of the frequency has an abrupt change in slope. Also known as corner frequency; knee frequency.

breaking load [MECH] The stress which, when steadily applied to a structural member, is just sufficient to break or rupture it.

breakoutput [COMPUT SCI] An ALGOL procedure which causes all bytes in a device buffer to be sent to the device rather than wait until the buffer is full.

breakpoint [CHEM ENG] *See* breakthrough. [COMPUT SCI] A point in a program where an instruction, instruction digit, or other condition enables a programmer to interrupt the run by external intervention or by a monitor routine.

breakpoint switch [COMPUT SCI] A manually operated switch which controls conditional operation at breakpoints, used primarily in debugging.

breakpoint symbol [COMPUT SCI] A symbol which may be optionally included in an instruction, as an indication, tag, or flag, to designate it as a breakpoint.

breakthrough [CHEM ENG] **1.** A localized break in a filter cake or precoat that permits fluid to pass through without being filtered. Also known as breakpoint. **2.** In an ion-exchange system, the first appearance of unadsorbed ions of the type which deplete the activity of the resin bed; this indicates that the bed must be regenerated. [COMPUT SCI] An interruption in the intended character stroke in optical character recognition. [MIN ENG] A passage cut through the pillar to allow the ventilating current to pass from one room to another; larger than a doghole. Also known as room crosscut.

breakup [HYD] The spring melting of snow, ice, and frozen ground; specifically, the destruction of the ice cover on rivers during the spring thaw.

breakwater [CIV ENG] A wall built into the sea to protect a shore area, harbor, anchorage, or basin from the action of waves.

breast [ANAT] The human mammary gland. [MIN ENG] **1.** In coal mines, a chamber driven in the seam from the gangway, for the extraction of coal. **2.** *See* face.

breast drill [DES ENG] A small, portable hand drill customarily used by handsetters to drill the holes in bit blanks in which diamonds are to be set; it includes a plate that is pressed against the worker's breast.

breather pipe [MECH ENG] A pipe that opens into a container for ventilation, as in a crankcase or oil tank.

breath-hold diving [ENG] A form of diving without the use of any artificial breathing mixtures.

breathing [ENG] **1.** Opening and closing of a plastics mold in orderto let gases escape during molding. Also known as degassing. **2.** Movement of gas, vapors, or air in and out of a storage-tank vent line as a result of liquid expansions and contractions induced by temperature changes. [MATER] Permeability of plastic sheeting to air, bubbles, voids, or trapped gas globules. [PHYSIO] Inhaling and exhaling.

breathing apparatus [ENG] An appliance that enables a person to function in irrespirable or poisonous gases or fluids; contains a supply of oxygen and a regenerator which removes the carbon dioxide exhaled.

breathing line [CIV ENG] A level of 5 feet (1.5 meters) above the floor; suggested temperatures for various occupancies of rooms and other chambers are usually given at this level.

breccia [PETR] A rock made up of very angular coarse fragments; may be sedimentary or may be formed by grinding or crushing along faults.

breccia pipe See pipe.

breech [ORD] The rear part of the bore of a gun, especially the opening that permits the projectile to be inserted at the rear of the bore.

breech bolt [ORD] A mechanism which opens and closes the breech in a carbine, machine gun, rifle, and the like; designed to push a cartridge into the chamber by sliding action.

breeches buoy [NAV ARCH] A device for carrying people from a stranded ship to shore or between ships.

breeder reactor [NUCLEO] A nuclear reactor that produces more fissionable material than it consumes.

breeding [AGR] The application of genetic principles to the improvement of farm animals and of cultivated plants. [GEN] Controlled mating and selection, or hybridization of plants and animals in order to improve the species. [NUCLEO] The production of nuclear fuel, by absorption of neutrons in a nuclear reactor, at a rate exceeding that at which fuel is being consumed.

breeding factor See breeding ratio.

breeding ratio [NUCLEO] The ratio of the number of fissionable atoms produced in a breeder reactor to the number of fissionable atoms consumed in the reactor; breeding gain is the breeding ratio minus 1. Also known as breeding factor.

breeze [METEROL] 1. A light, gentle, moderate, fresh wind. 2. In the Beaufort scale, a wind speed ranging from 4 to 31 miles (6.4 to 49.6 kilometers) per hour.

B register See index register.

Breit-Wigner formula [NUC PHYS] A formula which relates the cross section of a particular nuclear reaction with the energy of the incident particle, when the energy is near that required to form a discrete resonance level of the component nucleus.

bremsstrahlung [ELECTROMAG] Radiation that is emitted by an electron accelerated in its collision with the nucleus of an atom.

Brentidae [INV ZOO] The straight-snouted weevils, a family of coleopteran insects in the superfamily Curculionoidea.

Brevibacteriaceae [MICROBIO] Formerly a family of gram-positive, rod-shaped, schizomycetous bacteria in the order Eubacteriales.

brewers' yeast [FOOD ENG] Dried yeast cells recovered as a by-product of the brewing of beer and used as a natural source of vitamin B and protein.

brewing [FOOD ENG] The process of making beer, ale, and other malt beverages by boiling mashed malt to produce a wort, flavoring the wort with hops, fermenting this mixture with yeast, and drawing off the fermented wort for bottling.

brewster [OPTICS] A unit of stress optical coefficient of a material; it is equal to the stress optical coefficient of a material in which a stress of 1 bar produces a relative retardation between the components of a linearly polarized light beam of 1 angstrom when the light passes through a thickness of 1 millimeter in a direction perpendicular to the stress. Abbreviated B.

brewsterite [MINERAL] $Sr(Al_2Si_6O_{18})\cdot 5H_2O$ A member of the zeolite family of minerals; crystallizes in the monoclinic system and usually contains some calcium.

brick [MATER] A building material usually made from clay, molded as a rectangular block, and baked or burned in a kiln.

bridge [CIV ENG] A structure erected to span natural or artificial obstacles, such as rivers, highways, or railroads, and supporting a footpath or roadway for pedestrian, highway, or railroad traffic. [ELEC] 1. An electrical instrument having four or more branches, by means of which one or more of the electrical constants of an unknown component may be measured. 2. An electrical shunt path. [MATH] A line whose removal disconnects a component of a graph. Also known as isthmus. [MIN ENG] A piece of timber held above the cap of a set by blocks and used to facilitate the driving of spiling in soft or running ground. [NAV ARCH] An elevated structure extending across or over the weather deck of a vessel, containing stations for control and visual communications.

bridge cable [CIV ENG] Cable from which a roadway or truss is suspended in a suspension bridge; may be of pencil-thick wires laid parallel or strands of wire wound spirally.

bridge circuit [ELEC] An electrical network consisting basically of four impedances connected in series to form a rectangle, with one pair of diagonally opposite corners connected to an input device and the other pair to an output device.

bridge crane [MECH ENG] A hoisting machine in which the hoisting apparatus is carried by a bridgelike structure spanning the area over which the crane operates.

bridge deck [NAV ARCH] A partial deck above the main deck on merchant ships, usually near the middle of the ship.

bridged-T network [ELEC] A T network with a fourth branch connected between an input and an output terminal and across two branches of the network.

bridge graft [BOT] A plant graft in which each of several scions is grafted in two positions on the stock, one above and the other below an injury.

bridge house [NAV ARCH] A structure above the main deck near the middle of a ship whose top forms the bridge deck.

bridge hybrid See hybrid junction.

bridge pier [CIV ENG] The main support for a bridge, upon which the bridge superstructure rests; constructed of masonry, steel, timber, or concrete founded on firm ground below river mud.

bridge rectifier [ELECTR] A full-wave rectifier with four elements connected as a bridge circuit with direct voltage obtained from one pair of opposite junctions when alternating voltage is applied to the other pair.

bridging [ELEC] 1. Connecting one electric circuit in parallel with another. 2. The action of a selector switch whose movable contact is wide enough to touch two adjacent contacts so that the circuit is not broken during contact transfer. [MATH] The operation of carrying in addition or multiplication. [MET] 1. Formation of arched cavities in a powder compact. 2. Jamming of the charge in a blast or a cupola furnace due to adherence of fine ore particles to the inner walls. **3.** Formation of solidified metal over the top of the charge in a mold or crucible. [MIN ENG] The obstruction of the receiving opening in a material-crushing device by two or more pieces wedged together, each of which could easily pass through.

bridging amplifier [ELECTR] Amplifier with an input impedance sufficiently high so that its input may be bridged across a circuit without substantially affecting the signal level of the circuit across which it is bridged.

bridging connection [ELECTR] Parallel connection by means of which some of the signal energy in a circuit may be withdrawn frequently, with imperceptible effect on the normal operation of the circuit.

bridging ligand [ORG CHEM] A ligand in which an atom or molecular species which is able to exist independently is simultaneously bonded to two or more metal atoms.

bridging loss [ELECTR] Loss resulting from bridging an impedance across a transmission system; quantitatively, the ratio of the signal power delivered to that part of the system following the bridging point, and measured before the bridging, to the signal power delivered to the same part after the bridging.

Bridgman anvil [PHYS] A device for producing high static pressures using two large massive opposed pistons bearing on a small thin sample confined by a gasket material.

Bridgman effect [SOLID STATE] The phenomenon that when an electric current passes through an anisotropic crystal, there is an absorption or liberation of heat due to the nonuniformity in current distribution.

Bridgman relation [SOLID STATE] $P = QT\sigma$ in a metal or semiconductor, where P is the Ettingshausen coefficient, Q the Nernst-Ettingshausen coefficient, T the temperature, and σ the thermal conductivity in a transverse magnetic field.

bridled-cup anemometer [ENG] A combination cup anemometer and pressure-plate anemometer, consisting of an array of cups about a vertical axis of rotation, the free rotation of which is restricted by a spring arrangement; by adjustment of the force constant of the spring, an angular displacement can be obtained which is proportional to wind velocity.

bridled pressure plate [METEOROL] An instrument for measuring air velocity in which the pressure on a plate exposed to the wind is balanced by the force of a spring, and the deflection of the plate is measured by an inductance-type transducer.

brig [PHYS] A unit to express the ratio of two quantities, as a logarithm to the base 10; that is, a ratio of 10^x is equal to x brig; it is analogous to the bel, but the latter is restricted to power ratios. Also known as dex.

Briggs' logarithm *See* common logarithm.

Briggs pipe thread *See* American standard pipe thread.

bright [MATER] Referring to lubricating oils that are clear, or free from moisture. [OPTICS] Attribute of an area that appears to emit a large amount of light.

bright annealing [MET] Heating and cooling a metal in an inert atmosphere to inhibit oxidation; surface remains relatively bright.

bright band [METEOROL] The enhanced echo of snow as it melts to rain, as displayed on a range-height indicator scope.

bright-banded coal *See* bright coal.

bright coal [GEOL] A jet-black, pitchlike type of banded coal that is more compact than dull coal and

breaks with a shell-shaped fracture; microscopic examination shows a consistency of more than 5% anthraxyllon and less than 20% opaque matter. Also known as bright-banded coal; brights.

brightness [OPTICS] **1.** The characteristic of light that gives a visual sensation of more or less light. **2.** *See* luminance.

brightness control [ELECTR] A control that varies the luminance of the fluorescent screen of a cathode-ray tube, for a given input signal, by changing the grid bias of the tube and hence the beam current. Also known as brilliance control; intensity control.

brightness level *See* adaptation luminance.

brightness temperature *See* blackbody temperature.

bright points [ASTRON] Relatively small regions on the sun, distributed uniformly over the solar disk, from which there is increased x-ray and ultraviolet emission, having lifetimes on the order of 8 hours.

brights *See* bright coal.

bril [OPTICS] A unit of subjective luminance; 100 brils is the luminance level that corresponds to a luminance of 1 millilambert, and a doubling of luminance level corresponds to an increase of 1 bril.

brilliance [ELECTR] **1.** The degree of brightness and clarity of the display of a cathode-ray tube. **2.** The degree to which the higher audio frequencies of an input sound are reproduced by a radio receiver, by a public address amplifier, or by a sound-recording playback system.

brilliance control *See* brightness control.

brilliancy [LAP] A characteristic of gems that depends on the refractive index, transparency, polish, and proportions of the cut stone.

Brillouin function [SOLID STATE] A function of x with index (or parameter) n that appears in the quantum-mechanical theories of paramagnetism and ferromagnetism and is expressed as $[(2n+1)/2n] \cdot \coth[(2n + 1)x/2n] - (1/2n) \coth(x/2n)$.

Brillouin zone [SOLID STATE] A fundamental region of wave vectors in the theory of the propagation of waves through a crystal lattice; any wave vector outside this region is equivalent to some vector inside it.

brine [MATER] A liquid used in a refrigeration system, usually an aqueous solution of calcium chloride or sodium chloride, which is cooled by contact with the evaporator surface and then goes to the space to be refrigerated. [OCEANOGR] Sea water containing a higher concentration of dissolved salt than that of the ordinary ocean.

Brinell number [ENG] A hardness rating obtained from the Brinell test; expressed in kilograms per square millimeter.

Brinell test [ENG] A test to determine the hardness of a material, in which a steel ball 1 centimeter in diameter is pressed into the material with a standard force (usually 3000 kilograms); the spherical surface area of indentation is measured and divided into the load; the results are expressed as Brinell number.

briquet [MATER] A block of some compressed substance, such as coal dust, metal powder, or sawdust, used as a fuel. Also spelled briquette.

briquette *See* briquet.

briquetting [ENG] **1.** The process of binding together pulverized minerals, such as coal dust, into briquets under pressure, often with the aid of a binder, such as asphalt. **2.** A process or method of mounting mineral ore, rock, or metal fragments in an embedding or casting material, such as natural or artificial resins, waxes, metals, or alloys, to facilitate handling during grinding, polishing, and microscopic examination.

Brisingidae [INV ZOO] A family of deep-water echinoderms with as many as 44 arms, belonging to the order Forcipulatida.

bristle [BIOL] A short stiff hair or hairlike structure on an animal or plant.

Bristol board [MATER] Cardboard with a surface smooth enough for painting or writing, usually at least 0.006 inch (0.15 millimeter) thick.

britannia metal [MET] A silver-white tin alloy, similar to pewter, containing about 7% antimony, 2% copper, and often some zinc and bismuth; used in domestic utensils.

British absolute system of units [PHYS] A measurement system based on the foot, the second, and the pound mass; force unit is the poundal. Also known as foot-pound-second system of units (fps system of units).

British engineering system of units See British gravitational system of units.

British gravitational system of units [PHYS] A measurement system based on the foot, the second, and the slug mass; 1 slug weighs 32.174 pounds at sea level and 45° latitude, and equals 14.594 kilograms. Also known as British engineering system of units; engineer's system of units.

British imperial pound [MECH] The British standard of mass, of which a standard is preserved by the government.

British thermal unit [THERMO] Abbreviated Btu. **1.** A unit of heat energy equal to the heat needed to raise the temperature of 1 pound of air-free water from 60° to 61°F at a constant pressure of 1 standard atmosphere; it is found experimentally to be equal to 1054.5 joules. Also known as sixty degrees Fahrenheit British thermal unit ($Btu_{60/61}$). **2.** A unit of heat energy that is equal to 1/180 of the heat needed to raise 1 pound of air-free water from 32°F (0°C) to 212°F (100°C) at a constant pressure of 1 standard atmosphere; it is found experimentally to be equal to 1055.79 joules. Also known as mean British thermal unit (Btu_{mean}). **3.** A unit of heat energy whose magnitude is such that 1 British thermal unit per pound equals 2326 joules per kilogram; it is equal to exactly 1055.05585262 joules. Also known as international table British thermal unit (Btu_{it}).

brittle mica [MINERAL] Hydrous sodium, calcium, magnesium, and aluminum silicates; a group of more or less related minerals that resemble true micas but cleave to brittle flakes and contain calcium as the essential constituent.

brittleness [MECH] That property of a material manifested by fracture without appreciable prior plastic deformation.

brittle silver ore See stephanite.

brittle star [INV ZOO] The common name for all members of the echinoderm class Ophiuroidea.

brittle temperature [THERMO] The temperature point below which a material, especially metal, is brittle; that is, the critical normal stress for fracture is reached before the critical shear stress for plastic deformation.

Brix scale [CHEM ENG] A hydrometer scale for sugar solutions indicating the percentage by weight of sugar in the solution at a specified temperature.

broaching [ENG] **1.** The restoration of the diameter of a borehole by reaming. **2.** The breaking down of the walls between two contiguous drill holes. [MECH ENG] The machine-shaping of metal or plastic by pushing or pulling a broach across a surface or through an existing hole in a workpiece.

broad band [COMMUN] A band with a wide range of frequencies.

broad-band channel [COMMUN] A data transmission channel that can handle frequencies higher than the normal voice-grade line limit of 3 to 4 kilohertz; can carry many voice or data channels simultaneously or can be used for high-speed single-channel data transmission.

broadcast [COMMUN] A television or radio transmission intended for public reception.

broadcast band [COMMUN] The band of frequencies extending from 535 to 1605 kilohertz, corresponding to assigned carrier frequencies that increase in multiples of 10 kHz between 540 and 1600 kHz for the United States. Also known as standard broadcast band.

broadcaster [AGR] A machine that utilizes a rotating fanlike distributor for sowing grain, grass, and clover seed or for spreading fertilizer.

broadcast station [COMMUN] A television or radio station used for transmitting programs to the general public. Also known as station.

broadcloth [TEXT] A closely woven ribbed fabric having the rib running in the direction of the weft.

broadleaf tree [BOT] Any deciduous or evergreen tree having broad, flat leaves.

broadside array [ELECTROMAG] An antenna array whose direction of maximum radiation is perpendicular to the line or plane of the array.

broad tuning [ELECTR] Poor selectivity in a radio receiver, causing reception of two or more stations at a single setting of the tuning dial.

brocade [TEXT] Fabric made in a jacquard weave, usually with raised designs, and having a luxurious appearance; made of silk, polyester, or blends.

broccoli [BOT] *Brassica oleracea* var. *italica*. A biennial crucifer of the order Capparales which is grown for its edible stalks and buds.

brochanite See brochantite.

brochanthite See brochantite.

brochantite [MINERAL] $Cu_4(SO_4)(OH)_6$ A monoclinic copper mineral, emerald to dark green, commonly found with copper sulfide deposits; a minor copper ore. Also known as brochanite; brochanthite; warringtonite.

Brocken bow See anticorona.

Brocken specter [METEOROL] The illusory appearance of a gigantic figure (actually, the observer's shadow projected on cloud surfaces), observed on the Brocken peak in the Hartz Mountains of Saxony, but visible from other mountaintops under suitable conditions.

broken [METEOROL] Descriptive of a sky cover of from 0.6 to 0.9 (expressed to the nearest tenth).

broken-back transit [OPTICS] A type of transit telescope in which the light path is broken by insertion of a right-angled prism at the intersection of the optical and rotational axes. Also known as prism transit.

broken bracket See angle bracket.

broken line [MATH] A line which is composed of a series of line segments lying end to end, and which does not form a continuous line.

broken stone See crushed stone.

bromate [CHEM] **1.** BrO_3^- A negative ion derived from bromic acid, $HBrO_3$ **2.** A salt of bromic acid. **3.** $C_9H_9ClO_3$ A light brown solid with a melting point of 118–119°C; used as an herbicide to control weeds in crops such as flax, cereals, and legumes. Also known as 2-methyl-4-chlorophenoxyacetic acid.

bromegrass [BOT] The common name for a number of forage grasses of the genus *Bromus* in the order Cyperales.

Bromeliaceae [BOT] The single family of the flowering plant order Bromeliales.

Bromeliales [BOT] An order of monocotyledonous plants in the subclass Commelinidae, including terrestrial xerophytes and some epiphytes.

bromellite [MINERAL] BeO A white hexagonal mineral consisting of beryllium oxide; it is harder than zincite.

bromeosin *See* eosin.

bromide [CHEM] A compound derived from hydrobromic acid, HBr, with the bromine atom in the −1 oxidation state.

bromination [CHEM] The process of introducing bromine into a molecule.

bromine [CHEM] A chemical element, symbol Br, atomic number 35, atomic weight 79.904; used to make dibromide ethylene and in organic synthesis and plastics.

bromine number [ANALY CHEM] The amount of bromine absorbed by a fatty oil; indicates the purity of the oil and degree of unsaturation.

bromlite [MINERAL] $BaCa(CO_3)_2$ An orthorhombic mineral composed of a carbonate of barium and calcium. Also known as alstonite.

bromo- [CHEM] A prefix that indicates the presence of bromine in a molecule.

bromo acid *See* eosin.

bromouracil [BIOCHEM] $C_4H_3N_2O_2Br$ 5-Bromouracil, an analog of thymine that can react with deoxyribonucleic acid to produce a polymer with increased susceptibility to mutation.

bromyrite [MINERAL] AgBr A secondary ore of silver that occurs in the oxidized zone of silver deposits; exists in crusts and coatings resembling a wax.

bronchial asthma [MED] Asthma usually due to hypersensitivity to an inhaled or ingested allergen.

bronchiectasis [MED] Dilation of the bronchi and bronchioles following a chronic inflammatory process or an infection attended by pus formation.

bronchiole [ANAT] A small, thin-walled branch of a bronchus, usually terminating in alveoli.

bronchitis [MED] An inflammation of the bronchial tubes.

bronchodilator [MED] An instrument used to increase the caliber of the pulmonary air passages. [PHARM] Any drug which has the property of increasing the caliber of the pulmonary air passages.

bronchopneumonia [MED] Inflammation of the lungs which has spread from infected bronchi. Also known as lobular pneumonia.

bronchus [ANAT] Either of the two primary branches of the trachea or any of the bronchi's pulmonary branches having cartilage in their walls.

Brönsted acid [CHEM] A chemical species which can act as a source of protons. Also known as proton acid; protonic acid.

Brönsted-Lowry theory [CHEM] A theory that all acid-base reactions consist simply of the transfer of a proton from one base to another. Also known as Brönsted theory.

Brönsted theory *See* Brönsted-Lowry theory.

Brontotheriidae [PALEON] The single family of the extinct mammalian superfamily Brontotherioidea.

Brontotherioidea [PALEON] The titanotheres, a superfamily of large, extinct perissodactyl mammals in the suborder Hippomorpha.

bronze [MET] An alloy of copper and tin in varying proportions; other elements such as zinc, nickel, and lead may be added.

bronze mica *See* phlogopite.

bronzite [MINERAL] $(Mg,Fe)(SiO_3)$ An orthopyroxene mineral that forms metallic green orthorhombic crystals; a form of the enstatite-hypersthene series.

brood [BOT] Heavily infested by insects. [ZOO] **1.** The young of animals. **2.** To incubate eggs or cover the young for warmth. **3.** An animal kept for breeding.

brood pouch [VERT ZOO] A pouch of an animal body where eggs or embryos undergo certain stages of development.

brookite [MINERAL] TiO_2 A brown, reddish, or black orthorhombic mineral; it is trimorphous with rutile and anatase, has hardness of 5.5–6 on Mohs scale, and a specific gravity of 3.87–4.08. Also known as pyromelane.

Brooks variable inductometer [ELEC] An inductometer providing a nearly linear scale and consisting of two movable coils, side by side in a plane, sandwiched between two pairs of fixed coils.

brooming [CIV ENG] A method of finishing uniform concrete surfaces, such as the tops of pavement slabs or floor slabs, by dragging a broom over the surface to produce a grooved texture.

Brotulidae [VERT ZOO] A family of benthic teleosts in the order Perciformes.

brown algae [BOT] The common name for members of the Phaeophyta.

Brown and Sharpe gage *See* American wire gage.

brown blight [PL PATH] A virus disease of lettuce characterized by spots and streaks on the leaves, reduction in leaf size, and gradual browning of the foliage, beginning at the base.

brown blotch [PL PATH] **1.** A bacterial disease of mushrooms caused by *Pseudomonas tolaasi* and characterized by brown blotchy discolorations. **2.** A fungus disease of the pear characterized by brown blotches on the fruit.

brown canker [PL PATH] A fungus disease of roses caused by *Cryptosporella embrina* and characterized by lesions that are initially purple and gradually become buff.

brown clay *See* red clay.

brown fat cell [HISTOL] A moderately large, generally spherical cell in adipose tissue that has small fat droplets scattered in the cytoplasm.

brown hematite *See* limonite.

Brownian movement [STAT MECH] Random movements of small particles suspended in a fluid, caused by the statistical pressure fluctuations over the particle.

brown induration [MED] A pathologic condition marked by acute pulmonary congestion and edema with leakage of blood into the alveoli.

brown iron ore *See* limonite.

brown lignite [GEOL] A type of lignite with a fixed carbon content ranging from 30 to 55% and total carbon from 65 to 73.6%; contains 6300 Btu per pound (14.65 megajoules per kilogram). Also known as lignite B.

brownline [GRAPHICS] A print with a white background and brown lines made by contact-printing a negative on sensitized paper.

brown mica *See* phlogopite.

brownout [ELEC] **1.** A restriction of electrical power usage during a power shortage, especially for advertising and display purposes. **2.** An extinguishing of some of the lights in a city as a defensive measure against enemy bombardment.

brown patch [PL PATH] A fungus disease of grasses in golf greens and lawns caused by various soil-inhabiting species, typically producing brown circular areas surrounded by a band of grayish-black mycelia.

brownprint [GRAPHICS] A print with a brown background and white lines made by contact-printing a negative on a sensitized paper.

brown-ring test [ANALY CHEM] A common qualitative test for the nitrate ion; a brown ring forms at the juncture of a dilute ferrous sulfate solution layered on top of concentrated sulfuric acid if the upper layer contains nitrate ion.

brown root [PL PATH] A fungus disease of numerous tropical plants, such as coconut and rubber, caused by *Hymenochaete noxia* and characterized by defoliation and by incrustation of the roots with earth and stones held together by brown mycelia.

brown root rot [PL PATH] 1. A fungus disease of plants of the pea, cucumber, and potato families caused by *Thielavia basicola* and characterized by blackish discoloration and decay of the roots and stem base. 2. A disease of tobacco and other plants comparable to the fungus disease but believed to be caused by nematodes.

brown rot [PL PATH] Any fungus or bacterial plant disease characterized by browning and tissue decay.

brown soil [GEOL] Any of a zonal group of soils, with a brown surface horizon which grades into a lighter-colored soil and then into a layer of carbonate accumulation.

brown spot [PL PATH] Any fungus disease of plants, especially Indian corn, characterized by brown leaf spots.

brown stem rot [PL PATH] A fungus disease of soybeans caused by *Cephalosporium gregatum* in which there is brownish internal stem rot followed by discoloration and withering of leaves.

brownstone [PETR] Ferruginous sandstone with its grains coated with iron oxide.

browse [BIOL] 1. Twigs, shoots, and leaves eaten by livestock and other grazing animals. 2. To feed on this vegetation.

Brucellaceae [MICROBIO] Formerly a family of small, coccoid to rod-shaped, gram-negative bacteria in the order Eubacteriales.

brucellosis [MED] An infectious bacterial disease of man caused by *Brucella* species acquired by contact with diseased animals. Also known as Malta fever; Mediterranean fever; undulant fever. [VET MED] *See* contagious abortion.

Bruchidae [INV ZOO] The pea and bean weevils, a family of coleopteran insects in the superfamily Chrysomeloidea.

brucite [MINERAL] $Mg(OH)_2$ A hexagonal mineral; native magnesium hydroxide that appears gray and occurs in serpentines and impure limestones; hardness is 2.5 on Mohs scale, and specific gravity is 2.38–2.40.

brugnatellite [MINERAL] $Mg_6Fe(OH)_{13}CO_3 \cdot 4H_2O$ A flesh pink to yellowish- or brownish-white, hexagonal mineral consisting of a hydrated carbonate-hydroxide of magnesium and ferric iron; occurs in massive form.

Brunner's glands [ANAT] Simple, branched, tubular mucus-secreting glands in the submucosa of the duodenum in mammals. Also known as duodenal glands; glands of Brunner.

brush [ECOL] *See* tropical scrub. [ELEC] A conductive metal or carbon block used to make sliding electrical contact with a moving part.

brush border [CYTOL] A superficial protoplasic modification in the form of filiform processes or microvilli; present on certain absorptive cells in the intestinal epithelium and the proximal convolutions of nephrons.

brush discharge [ELEC] A luminous electric discharge that starts from a conductor when its potential exceeds a certain value but remains too low for the formation of an actual spark.

brushing [TEXT] A process for finishing fabrics by rubbing them against a rough surface, such as a wire-covered cylinder, usually to produce a thick nap. Also known as napping.

brushite [MINERAL] $CaHPO_4 \cdot 2H_2O$ A nearly colorless mineral that is a constituent of rock phosphates that crystallizes in slender or massive crystals.

brush station [COMPUT SCI] A location in a device where the holes in a punched card are sensed by brushes sweeping electrical contacts.

brussels sprouts [BOT] *Brassica oleracea* var. *gemmifera* A biennial crucifer of the order Capparales cultivated for its small, edible, headlike buds.

Bruxellian [GEOL] Lower middle Eocene geologic time.

Bryophyta [BOT] A small phylum of the plant kingdom, including mosses, liverworts, and hornworts, characterized by the lack of true roots, stems, and leaves.

Bryopsida [BOT] The mosses, a class of small green plants in the phylum Bryophyta. Also known as Musci.

Bryopsidaceae [BOT] A family of green algae in the order Siphonales.

Bryozoa [INV ZOO] The moss animals, a major phylum of sessile aquatic invertebrates occurring in colonies with hardened exoskeleton.

B scan *See* B scope.

B scope [ELECTR] A cathode-ray scope on which signals appear as spots, with bearing angle as the horizontal coordinate and range as the vertical coordinate. Also known as B indicator; B scan.

BSR *See* bulk shielding reactor.

B store *See* index register.

BT *See* bathythermograph.

Btu *See* British thermal unit.

bu *See* bushel.

bubble [METEOROL] *See* bubble high. [PHYS] 1. A small, approximately spherical body of fluid within another fluid or solid. 2. A thin, approximately spherical film of liquid inflated with air or other gas. [SOLID STATE] *See* magnetic bubble.

bubble cap [CHEM ENG] A metal cap covering a hole in the plate within a distillation tower; designed to permit vapors to rise from below the plate, pass through the cap, and make contact with liquid on the plate.

bubble cavitation [FL MECH] 1. Formation of vapor- or gas-filled cavities in liquids by mechanical forces. 2. The formation of vapor-filled cavities in the interior of liquids in motion when the pressure is reduced without change in ambient temperature.

bubble chamber [NUCLEO] A chamber in which the movements and interactions of charged particles can be observed as visible tracks in a superheated liquid, the tracks being gas bubbles that form along the paths of the moving particles.

bubble high [METEOROL] A small high, complete with anticyclonic circulation, of the order of 50 to 300 miles (80 to 480 kilometers) across, often induced by precipitation and vertical currents associated with thunderstorms. Also known as bubble.

bubble memory [COMPUT SCI] A computer memory in which the presence or absence of a magnetic bubble in a localized region of a thin magnetic film designates a 1 or 0; storage capacity can be well over 1 megabit per cubic inch. Also known as magnetic bubble memory.

bubble point [PHYS CHEM] In a solution of two or more components, the temperature at which the first bubbles of gas appear. Also known as boiling point.

bubble tray [CHEM ENG] A perforated, circular plate placed within a distillation tower at specific places to collect the fractions of petroleum produced in fractional distillation.

bubo [MED] An inflammatory enlargement of lymph nodes, usually of the groin or axilla; commonly associated with chancroid, lymphogranuloma venereum, and plague.

buccal cavity [ANAT] The space anterior to the teeth and gums in the mouths of all vertebrates having lips and cheeks. Also known as vestibule.

buccal gland [ANAT] Any of the mucous glands in the membrane lining the cheeks of mammals, except aquatic forms.

Buccinacea [INV ZOO] A superfamily of gastropod mollusks in the order Prosobranchia.

Buccinidae [INV ZOO] A family of marine gastropod mollusks in the order Neogastropoda containing the whelks in the genus *Buccinum*.

Bucconidae [VERT ZOO] The puffbirds, a family of neotropical birds in the order Piciformes.

Bucerotidae [VERT ZOO] The hornbills, a family of Old World tropical birds in the order Coraciiformes.

bucket [BOT] *See* calyx. [COMPUT SCI] A name usually reserved for a storage cell in which data may be accumulated. [ENG] 1. A cup on the rim of a Pelton wheel against which water impinges. 2. A reversed curve at the toe of a spillway to deflect the water horizontally and reduce erosiveness. 3. A container on a lift pump or chain pump. 4. A container on some bulk-handling equipment, such as a bucket elevator, bucket dredge, or bucket conveyor. 5. A water outlet in a turbine. 6. *See* calyx.

bucket brigade device [ELECTR] A semiconductor device in which majority carriers store charges that represent information, and minority carriers transfer charges from point to point in sequence. Abbreviated BBD.

bucket thermometer [ENG] A thermometer mounted in a bucket and used to measure the temperature of water drawn into the bucket from the surface of the ocean.

buckeye [BOT] The common name for deciduous trees composing the genus *Aesculus* in the order Sapindales; leaves are opposite and palmately compound, and the seed is large with a firm outer coat.

buckling [MECH] Bending of a sheet, plate, or column supporting a compressive load. [NUCLEO] The size-shape factor that appears in the general nuclear reactor equation and is a measure of the curvature of the neutron density distribution in the reactor.

buckwheat [BOT] The common name for several species of annual herbs in the genus *Fagopyrum* of the order Polygonales; used for the starchy seed.

buckwheat coal [GEOL] An anthracite coal that passes through 9/16-inch (14 millimeter) holes and over 5/16-inch (8 millimeter) holes in a screen.

Bucky diaphragm *See* Potter-Bucky grid.

bud [BOT] An embryonic shoot containing the growing stem tip surrounded by young leaves or flowers or both and frequently enclosed by bud scales.

budding [BIOL] A form of asexual reproduction in which a new individual arises as an outgrowth of an older individual. Also known as gemmation. [BOT]

A method of vegetative propagation in which a single bud is grafted laterally onto a stock.

budding bacteria [MICROBIO] Bacteria that reproduce by budding.

bud rot [PL PATH] Any plant disease or symptom involving bud decay.

bud scale [BOT] One of the modified leaves enclosing and protecting buds in perennial plants.

buetschliite [MINERAL] $K_6Ca_2(CO_3)_5 \cdot 6H_2O$ A mineral that is probably hexagonal and consists of a hydrated carbonate of potassium and calcium.

buffalo [VERT ZOO] The common name for several species of artiodactyl mammals in the family Bovidae, including the water buffalo and bison.

buffer [CHEM] A solution selected or prepared to minimize changes in hydrogen ion concentration which would otherwise occur as a result of a chemical reaction. Also known as buffer solution. [COMPUT SCI] *See* buffer storage. [ECOL] An animal that is introduced to serve as food for other animals to reduce the losses of more desirable animals. [ELEC] An electric circuit or component that prevents undesirable electrical interaction between two circuits or components. [ELECTR] 1. An isolating circuit in an electronic computer used to prevent the action of a driven circuit from affecting the corresponding driving circuit. 2. *See* buffer amplifier.

buffer amplifier [ELECTR] An amplifier used after an oscillator or other critical stage to isolate it from the effects of load impedance variations in subsequent stages. Also known as buffer; buffer stage.

buffer capacitor [ELECTR] A capacitor connected across the secondary of a vibrator transformer or between the anode and cathode of a cold-cathode rectifier tube to suppress voltage surges that might otherwise damage other parts in the circuit.

buffered computer [COMPUT SCI] A computer having a temporary storage device to compensate for differences in transmission speeds.

buffered I/O channel [COMPUT SCI] A storage device located between input/output (I/O) channels and main storage control to free the channels for use by other operations.

buffered terminal [COMPUT SCI] A computer terminal which contains storage equipment so that the rate at which it sends or receives data over its line does not need to agree exactly with the rate at which the data are entered or printed.

buffer solution *See* buffer.

buffer stage *See* buffer amplifier.

buffer storage [COMPUT SCI] A synchronizing element used between two different forms of storage in a computer; computation continues while transfers take place between buffer storage and the secondary or internal storage. Also known as buffer.

buffeting [AERO ENG] 1. The beating of an aerodynamic structure or surfaces by unsteady flow, gusts, and so forth. 2. The irregular shaking or oscillation of a vehicle component owing to turbulent air or separated flow.

buffing [ENG] The smoothing and brightening of a surface by an abrasive compound pressed against it by a soft wheel or belt.

Bufonidae [VERT ZOO] A family of toothless frogs in the suborder Procoela including the true toads (*Bufo*).

bug [COMPUT SCI] A defect in a program code or in designing a routine or a computer. [ELECTR] 1. A semiautomatic code-sending telegraph key in which movement of a lever to one side produces a series of correctly spaced dots and movement to the other side produces a single dash. 2. An electronic listening device, generally concealed, used for commercial or military espionage. [ENG] **1.** A defect or imperfection present in a piece of equipment. **2.** *See* bullet. [INV ZOO] Any insect in the order Hemiptera.

bugas *See* bottled gas.

buhrstone [PETR] A silicified fossiliferous limestone with abundant cavities previously occupied by fossil shells. Also known as millstone.

buhrstone mill [MECH ENG] A mill for grinding or pulverizing grain in which a flat siliceous rock (buhrstone), generally of cellular quartz, rotates against a stationary stone of the same material.

building code [CIV ENG] Local building laws to promote safe practices in the design and construction of a building.

building line [CIV ENG] A designated line beyond which a building cannot extend.

buildup [MET] Deposition of excess metal, either by electroplating or spraying, on worn or undersized machine components to restore required dimensions.

built-in check [COMPUT SCI] A hardware device which controls the accuracy of data either moved or stored within the computer system.

built-in function [COMPUT SCI] A function that is available through a simple reference and specification of arguments in a given higher-level programming language. Also known as built-in procedure; intrinsic procedure.

built-in procedure *See* built-in function.

built terrace *See* alluvial terrace.

built-up beam [ENG] A structural steel member that is fabricated by welding or riveting rather than being rolled.

built-up fraction [GRAPHICS] In typesetting, a fraction set with at least two blocks of type, with each successive block to the right of the preceding one; each block contains one numeral, and the slash is between two blocks.

built-up roof [BUILD] A roof constructed of several layers of felt and asphalt.

bulb [BOT] A short, subterranean stem with many overlapping fleshy leaf bases or scales, such as in the onion and tulip. [ELEC] *See* envelope.

bulbourethral gland [ANAT] Either of a pair of compound tubular glands, anterior to the prostate gland in males, which discharge into the urethra. Also known as Cowper's gland.

bulimia [MED] Excessive, insatiable appetite, seen in psychotic states; a symptom of diabetes mellitus and of certain cerebral lesions. Also known as hyperphagia.

Buliminacea [INV ZOO] A superfamily of benthic, marine foraminiferans in the suborder Rotaliina.

bulk acoustic wave [ACOUS] An acoustic wave that travels through a piezoelectric material, as in a quartz delay line. Also known as volume acoustic wave.

bulk cargo [IND ENG] Cargo which is loaded into a ship's hold without being boxed, bagged, or hand stowed, or is transported in large tank spaces.

bulk carrier [NAV ARCH] A vessel designed to transport dry or liquid bulk cargo.

bulk density [ENG] The mass of powdered or granulated solid material per unit of volume.

bulk diode [ELECTR] A semiconductor microwave diode that uses the bulk effect, such as Gunn diodes and diodes operating in limited space-charge-accumulation modes.

bulk effect [ELECTR] An effect that occurs within the entire bulk of a semiconductor material rather than in a localized region or junction.

bulk flow *See* convection.

bulk-handling machine [MECH ENG] Any of a diversified group of materials-handling machines designed for handling unpackaged, divided materials.

bulkhead [AERO ENG] A wall, partition, or such in a rocket, spacecraft, airplane fuselage, or similar structure, at right angles to the longitudinal axis of the structure and serving to strengthen, divide, or help give shape to the structure. [MIN ENG] A tightseal partition of wood, rock, and mud or concrete in mines that serves to protect against gas, fire, and water. [NAV ARCH] An upright partition separating compartments in a ship.

bulk insulation [ENG] A type of insulation that retards the flow of heat by the interposition of many air spaces and, in most cases, by opacity to radiant heat.

bulk memory [COMPUT SCI] A high-capacity memory used in connection with a computer for bulk storage of large quantities of data.

bulk modulus *See* bulk modulus of elasticity.

bulk modulus of elasticity [MECH] The ratio of the compressive or tensile force applied to a substance per unit surface area to the change in volume of the substance per unit volume. Also known as bulk modulus; compression modulus; hydrostatic modulus; modulus of compression; modulus of volume elasticity.

bulk shielding reactor [NUCLEO] The prototype swimming-pool reactor located at Oak Ridge, Tennessee; it uses heterogeneous enriched fuel and provides a combination of high thermal-neutron flux, ready accessibility, and versatility. Abbreviated BSR.

bulk storage *See* backing storage.

Bull *See* Taurus.

bulldozer [MECH ENG] A wheeled or crawler tractor equipped with a reinforced, curved steel plate mounted in front, perpendicular to the ground, for pushing excavated materials. [MET] A machine for bending, forging, and punching narrow plates and bars, in which a ram is pushed along a horizontal path by a pair of cranks that are linked to two bullwheels with eccentric pins.

bullet [ENG] **1.** A conical-nosed cylindrical weight, attached to a wire rope or line, either notched or seated to engage and attach itself to the upper end of a wire line core barrel or other retrievable or retractable device that has been placed in a borehole. Also known as bug; go-devil; overshot. **2.** A scraper with self-adjusting spring blades, inserted in a pipeline and carried forward by the fluid pressure, clearing away accumulations or debris from the walls of a pipe. Also known as go-devil. **3.** A bullet-shaped weight or small explosive charge dropped to explode a charge of nitroglycerin placed in a borehole. Also known as go-devil. **4.** An electric lamp covered by a conical metal case, usually at the end of a flexible metal shaft. **5.** *See* torpedo. [GRAPHICS] A hollow hemispherical shell, made of iron and filled with pitch, which holds small objects during the execution of artistic designs in metal. [MATER] A small, lustrous, nearly spherical industrial diamond. [ORD] The projectile fired, or intended to be fired, from a small arm.

bullfrog *See* barney.

bullhorn [COMMUN] A portable loudspeaker, generally with built-in amplifier and microphone, used for voice messages to crowds or from one ship to another at sea.

bullion [MET] Gold or silver in bulk in the shape of bars or ingots. [MIN ENG] A concretion found in some types of coal, composed of carbonate or silica stained by brown humic derivatives; often well-preserved plant structures form the nuclei.

bull nose [BUILD] A rounded external angle, as one used at window returns and doorframes.

bull wheel [MECH ENG] 1. The main wheel or gear of a machine, which is usually the largest and strongest. 2. A cylinder which has a rope wound about it for lifting or hauling. 3. A wheel attached to the base of a derrick boom which swings the derrick in a vertical plane.

bulwark [NAV ARCH] The part of a ship's side that extends above the main deck to protect it against heavy weather.

bumblebee [INV ZOO] The common name for several large, hairy social bees of the genus *Bombus* in the family Apidae.

bumping [AERO ENG] *See* chugging. [CHEM] Uneven boiling of a liquid caused by irregular rapid escape of large bubbles of highly volatile components as the liquid mixture is heated. [MECH ENG] *See* chugging. [MET] Forming a dish in metal by many repeated blows.

bunching [ELECTR] The flow of electrons from cathode to anode of a velocity-modulated tube as a succession of electron groups rather than as a continuous stream.

bunching voltage [ELECTR] Radio-frequency voltage between the grids of the buncher resonator in a velocity-modulated tube such as a klystron; generally, the term implies the peak value of this oscillating voltage.

bunchy top [PL PATH] Any viral disease of plants in which there is a shortening of the internodes with crowding of leaves and shoots at the stem apex.

bundle branch [ANAT] Either of the components of the atrioventricular bundle passing to the right and left ventricles of the heart.

bundled program [COMPUT SCI] A computer program written, maintained, and updated by the computer manufacturer, and included in the price of the hardware.

bundle scar [BOT] A mark within a leaf scar that shows the point of an abscised vascular bundle.

bundling machine [MECH ENG] A device that automatically accumulates cans, cartons, or glass containers for semiautomatic or automatic loading or for shipping cartons by assembling the packages into units of predetermined count and pattern which are then machine-wrapped in paper, film paperboard, or corrugated board.

Bundsandstein *See* Bunter.

Buniakowski's inequality *See* Cauchy-Schwarz inequality.

bunion [MED] A swelling of a bursa of the foot, especially of the metatarsophalangeal joint of the great toe; associated with thickening of the adjacent skin and a forcing of the great toe into adduction.

bunker fuel oil [MATER] A heavy residual petroleum oil used as fuel by ships, industry, and large-scale heating and power-production installations.

bunodont [ANAT] Having tubercles or rounded cusps on the molar teeth, as in humans.

Bunsen burner [ENG] A type of gas burner with an adjustable air supply.

Bunter [GEOL] Lower Triassic geologic time. Also known as Buntsandstein.

buoy [ENG] An anchored or moored floating object, other than a lightship, intended as an aid to navigation, to attach or suspend measuring instruments, or to mark the position of something beneath the water.

buoyage [NAV] A system of buoys; for example, one in which the buoys are assigned shape, color, and number distinction in accordance with location relative to the nearest obstruction is called a cardinal buoyage system.

buoyancy [FL MECH] The resultant vertical force exerted on a body by a static fluid in which it is submerged or floating.

buoyant density [PHYS] A technique that uses the sedimentation equilibrium in a density gradient to characterize a solute.

buoyant force [FL MECH] The force exerted vertically upward by a fluid on a body wholly or partly immersed in it; its magnitude is equal to the weight of the fluid displaced by the body.

Buprestidae [INV ZOO] The metallic wood-boring beetles, the large, single family of the coleopteran superfamily Buprestoidea.

Buprestoidea [INV ZOO] A superfamily of coleopteran insects in the suborder Polyphaga including many serious pests of fruit trees.

burden [ELEC] The amount of power drawn from the circuit connecting the secondary terminals of an instrument transformer, usually expressed in volt-amperes. [GEOL] All types of rock or earthy materials overlying bedrock. [MET] 1. The material which is melted in a direct arc furnace. 2. In an iron blast furnace, the ratio of iron and flux to coke and other fuels in the charge.

buret [CHEM] A graduated glass tube used to deliver variable volumes of liquid; usually equipped with a stopcock to control the liquid flow.

Burgers vector [CRYSTAL] A translation vector of a crystal lattice representing the displacement of the material to create a dislocation.

burglar alarm [ENG] An alarm in which interruption of electric current to a relay, caused, for example, by the breaking of a metallic tape placed at an entrance to a building, deenergizes the relay and causes the relay contacts to operate the alarm indicator. Also known as intrusion alarm.

Burhinidae [VERT ZOO] The thick-knees or stone curlews, a family of the avian order Charadriiformes.

buried soil *See* paleosol.

Burkett's lymphoma [MED] A malignant lymphoma of children, typically involving the retroperitoneal area and the mandible, but sparing the peripheral lymph nodes, bone marrow, and spleen.

burlap [TEXT] A coarse cloth woven from jute fiber. Also known as hessian.

burling [TEXT] Removing loops, knots, and vegetable matter in the finishing section of a woolen or worsted mill.

Burmanniaceae [BOT] A family of monocotyledonous plants in the order Orchidales characterized by regular flowers, three or six stamens opposite the petals, and ovarian nectaries.

burn [ENG] To consume fuel. [MED] An injury to tissues caused by heat, chemicals, electricity, or irradiation effects.

burner [CHEM ENG] A furnace where sulfur or sulfide ore are burned to produce sulfur dioxide and other gases. [ENG] 1. The part of a fluid-burning device at which the flame is produced. 2. Any burning device used to soften old paint to aid in its removal. 3. A worker who operates a kiln which burns brick or tile. 4. A worker who alters the properties of a mineral substance by burning. 5. A worker who uses a flame-cutting torch to cut metals. [MECH ENG] A unit of a steam boiler which mixes and directs the flow of fuel and air so as to ensure rapid ignition and complete combustion.

burner fuel oil [MATER] Any of the petroleum distillate and residual oils used to heat homes and buildings.

burn-in [ELECTR] Operation of electronic components before they are applied in order to stabilize their characteristics and reveal defects. [ENG] *See* freeze.

burning [ENG] The firing of clay products placed in a kiln. [MET] 1. Permanent damage to a metal caused by heating beyond temperature limits of the treatment. 2. Deep pitting of a metal caused by excessive pickling.

burning point [ENG] The lowest temperature at which a volatile oil in an open vessel will continue to burn when ignited by a flame held close to its surface; used to test safety of kerosine and other illuminating oils.

burning velocity [CHEM] The normal velocity of the region of combustion reaction (reaction zone) relative to nonturbulent unburned gas, in the combustion of a flammable mixture.

burnish [ENG] To polish or make shiny. [MET] To develop a smooth, lustrous surface finish by tumbling with steel balls or rubbing with a hard metal pad.

burnout [AERO ENG] 1. An act or instance of fuel or oxidant depletion or of depletion of both at once. 2. The time at which this depletion occurs. [ELEC] Failure of a device due to excessive heat produced by excessive current. [ENG] An instance of a device or a part overheating so as to result in destruction or damage. [NUCLEO] 1. To receive the greatest amount of radiation permissible during a given time. 2. The point at which the heat flux across a surface causes film-blanketing of the surface, resulting in a drop in the film heat-transfer coefficient, overheating, and possible surface failure.

burr [BOT] 1. A rough or prickly envelope on a fruit. 2. A fruit so characterized. [MET] A thin, ragged fin left on the edge of a piece of metal by a cutting or punching tool.

burr mill [FOOD ENG] A mill for grinding crops, in which two ribbed plates or disks rub or crush the material between them.

bursa [ANAT] A simple sac or cavity with smooth walls containing a clear, slightly sticky fluid and interposed between two moving surfaces of the body to reduce friction.

Burseraceae [BOT] A family of dicotyledonous plants in the order Sapindales characterized by an ovary of two to five cells, prominent resin ducts in the bark and wood, and an intrastaminal disk.

bursitis [MED] Inflammation of a bursa.

burst [COMMUN] 1. A sudden increase in the strength of a signal being received from beyond line-of-sight range. 2. *See* color burst. [COMPUT SCI] 1. To separate a continuous roll of paper into stacks of individual sheets by means of a burster. 2. The transfer of a collection of records in a storage device, leaving an interval in which data for other requirements can be obtained from or entered into the device. 3. A sequence of signals regarded as a unit in data transmission. [ELECTR] 1. An exceptionally large electric pulse in the circuit of an ionization chamber due to the simultaneous arrival of several ionizing particles. 2. A radar term for a single pulse of radio energy. [ORD] 1. Continuous fire from an automatic weapon, as from an aircraft machine gun, sometimes described as a

long or short burst. 2. The explosion of a projectile, bomb, or similar munition.

burst amplifier [ELECTR] An amplifier stage in a color television receiver that is keyed into conduction and amplification by a horizontal pulse at the instant of each arrival of the color burst. Also known as chroma band-pass amplifier.

burster [BOT] An abnormally double flower having the calyx split or fragmented. [COMPUT SCI] An off-line device in a computer system used to separate the continuous roll of paper produced as output from a printer into individual sheets, generally along perforations in the roll. [ORD] An explosive element used in chemical ammunition such as bombs, mines, and shells to open the container and disperse the contents.

burst mode [COMPUT SCI] A method of transferring data between a peripheral unit and a control processing unit in a computer system in which the peripheral unit sends the central processor a signal to receive data until the peripheral unit signals that the transfer is completed.

burst transmission [COMMUN] A radio transmission in which messages stored for a given time are sent at from 10 to more than 100 times the normal rate, recorded when received, and then slowed down to the normal rate for the user.

bus [ELEC] 1. A set of two or more electric conductors that serve as common connections between load circuits and each of the polarities (in direct-current systems) or phases (in alternating-current systems) of the source of electric power. 2. *See* busbar. [ELECTR] One or more conductors in a computer along which information is transmitted from any of several sources to any of several destinations. [ENG] A motor vehicle for carrying a large number of passengers.

busbar [ELEC] A heavy, rigid metallic conductor, usually uninsulated, used to carry a large current or to make a common connection between several circuits. Also known as bus.

bushbaby [VERT ZOO] Any of six species of African arboreal primates in two genera (*Galago* and *Euoticus*) of the family Lorisidae. Also known as galago; night ape.

bushel [MECH] Abbreviated bu. 1. A unit of volume (dry measure) used in the United States, equal to 2150.42 cubic inches or approximately 35.239 liters. 2. A unit of volume (liquid and dry measure) used in Britain, equal to 2219.36 cubic inches or 8 imperial gallons (approximately 36.369 liters).

bushing [MECH ENG] A removable piece of soft metal or graphite-filled sintered metal, usually in the form of a bearing, that lines a support for a shaft.

busway [ELEC] A prefabricated assembly of standard lengths of busbars rigidly supported by solid insulation and enclosed in a sheet-metal housing.

busy test [COMMUN] A test, in telephony, made to find out whether certain facilities which may be desired, such as a subscriber line or trunk, are available for use.

butane [ORG CHEM] C_4H_{10} An alkane of which there are two isomers, *n* and isobutane; occurs in natural gas and is produced by cracking petroleum.

butanol [ORG CHEM] Any one of four isomeric alcohols having the formula C_4H_9OH; colorless, toxic liquids soluble in most organic liquids. Also known as butyl alcohol.

Butomaceae [BOT] A family of monocotyledonous plants in the order Alismatales characterized by secretory canals, linear leaves, and a straight embryo.

butt [BUILD] The bottom or cover edge of a shingle. [DES ENG] The enlarged and squared-off end of a connecting rod or similar link in a machine. [MIN ENG] Coal exposed at right angles to the face and, in contrast to the face, generally having a rough surface. [ORD] Rear end of the stock of a rifle or other small arm.

butte [GEOGR] A detached hill or ridge which rises abruptly.

butterfat [FOOD ENG] A mixture of glycerides derived from fatty acids; the natural fat of milk and butter.

butterfly [INV ZOO] Any insect belonging to the lepidopteran suborder Rhopalocera, characterized by a slender body, broad colorful wings, and club-shaped antennae.

butterfly capacitor [ELEC] A variable capacitor having stator and rotor plates shaped like butterfly wings, with the stator plates having an outer ring to provide an inductance so that both capacitance and inductance may be varied, thereby giving a wide tuning range.

butterfly damper   *See* butterfly valve.

butterfly nut   *See* wing nut.

butterfly valve [ENG] A valve that utilizes a turnable disk element to regulate flow in a pipe or duct system, such as a hydraulic turbine or a ventilating system. Also known as butterfly damper.

buttermilk [FOOD ENG] A fermentation food product prepared by inoculating sweet milk with cultures of *Streptococcus lactis* and *Leuconostoc citrovorum*.

butter rock   *See* halotrichite.

Butterworth filter [ELECTR] An electric filter whose pass band (graph of transmission versus frequency) has a maximally flat shape.

butt fusion [ENG] The joining of two pieces of plastic or metal pipes or sheets by heating the ends until they are molten and then pressing them together to form a homogeneous bond.

butt joint [ELEC] A connection formed by placing the ends of two conductors together and joining them by welding, brazing, or soldering. [ELECTROMAG] A connection giving physical contact between the ends of two waveguides to maintain electrical continuity. [ENG] A joint in which the parts to be joined are fastened end to end or edge to edge with one or more cover plates (or other strengthening) generally used to accomplish the joining.

buttock lines [ENG] The lines of intersection of the surface of an aircraft or its float, or of the hull of a ship, with its longitudinal vertical planes. Also known as buttocks.

buttocks [ANAT] The two fleshy parts of the body posterior to the hip joints. [ENG] *See* buttock lines. [NAV ARCH] The convex part of the stern end of a ship above the water line. Also known as counter.

button [ELECTR] 1. A small, round piece of metal alloyed to the base wafer of an alloy-junction transistor. Also known as dot. 2. The container that holds the carbon granules of a carbon microphone. Also known as carbon button. [MET] 1. Mass of metal remaining in a crucible after fusion has been completed. 2. That part of a weld which tears out in the destructive testing of spot-, seam-, or projection-welded specimens.

buttress [ARCH] An upright projection that supports or resists lateral forces in a building. [CIV ENG] A pier constructed at right angles to a restraining wall on the side opposite to the restrained material; increases the strength and thrust resistance of the wall. [PALEON] A ridge on the inner surface of a pelecypod valve which acts as a support for part of the hinge.

butyl [ORG CHEM] Any of the four variations of the hydrocarbon radical C_4H_9: $CH_3CH_2CH_2CH_2$—, $(CH_3)_2CHCH_2$—, $CH_3CH_2CHCH_3$—, and $(CH_3)_3C$—.

butyl alcohol   *See* butanol.

butyl rubber [MATER] A synthetic rubber made by the polymerization of isoprene and isobutylene.

Buxbaumiales [BOT] An order of very small, atypical mosses (Bryopsida) composed of three genera and found on soil, rock, and rotten wood.

buzz [AERO ENG] Sustained oscillation of an aerodynamic control surface caused by intermittent flow separation on the surface, or by a motion of shock waves across the surface, or by a combination of flow separation and shock-wave motion on the surface. [CONT SYS] *See* dither. [ELECTR] The condition of a combinatorial circuit with feedback that has undergone a transition, caused by the inputs, from an unstable state to a new state that is also unstable. [FL MECH] In supersonic diffuser aerodynamics, a nonsteady shock motion and airflow associated with the shock system ahead of the inlet.

BW   *See* bandwidth; biological warfare.

BWO   *See* backward-wave oscillator.

BWR   *See* boiling-water reactor.

bypass [CIV ENG] A road which carries traffic around a congested district or temporary obstruction. [ELEC] A shunt path around some element or elements of a circuit. [ENG] An alternating, usually smaller, diversionary flow path in a fluid dynamic system to avoid some device, fixture, or obstruction.

bypass channel [CIV ENG] 1. A channel built to carry excess water from a stream. Also known as flood relief channel; floodway. 2. A channel constructed to divert water from a main channel.

bypass filter [ELECTR] Filter which provides a low-attenuation path around some other equipment, such as a carrier frequency filter used to bypass a physical telephone repeater station.

bypass ratio [AERO ENG] Ratio of the secondary to the primary inlet airflows for a turbofan engine.

bypass valve [ENG] A valve that opens to direct fluid elsewhere when a pressure limit is exceeded.

by-product [ENG] A product from a manufacturing process that is not considered the principal material.

Byrrhidae [INV ZOO] The pill beetles, the single family of the coleopteran insect superfamily Byrrhoidea.

Byrrhoidea [INV ZOO] A superfamily of coleopteran insects in the suborder Polyphaga.

byte [COMPUT SCI] A sequence of adjacent binary digits operated upon as a unit in a computer and usually shorter than a word.

byte mode [COMPUT SCI] A method of transferring data between a peripheral unit and a central processor in which one byte is transferred at a time.

bytownite [MINERAL] A plagioclase feldspar with a composition ranging from $Ab_{30}An_{70}$ to $Ab_{10}An_{90}$, where $Ab = NaAlSi_3O_8$ and $An = CaAl_2Si_2O_8$; occurs in basic and ultrabasic igneous rock.

Byturidae [INV ZOO] The raspberry fruitworms, a small family of coleopteran insects in the superfamily Cucujoidea.

C

c *See* calorie; centi-; charmed quark; curie.

C *See* capacitance; capacitor; carbon; coulomb.

Ca *See* calcium.

Ca^{45} *See* calcium-45.

caballing [OCEANOGR] The mixing of two water masses of identical in situ densities but different in situ temperatures and salinities, such that the resulting mixture is denser than its components and therefore sinks.

cabane [AERO ENG] The arrangement of struts used on early types of airplanes to brace the wings.

cabbage [BOT] *Brassica oleracea* var. *capitata*. A biennial crucifer of the order Capparales grown for its head of edible leaves.

cabbage yellows [PL PATH] A fungus disease of cabbage caused by *Fusarium conglutinans* and characterized by yellowing and dwarfing.

cabezon [VERT ZOO] *Scorpaenichthys marmoratus*. A fish that is the largest of the sculpin species, weighing as much as 25 pounds (11.3 kilograms) and reaching a length of 30 inches (76 centimeters).

Cabibbo theory [PARTIC PHYS] A theory describing baryon beta-decay processes, according to which the amplitude for such processes is given by

$$G\{\cos \Theta \, [V \, (\Delta s = 0) + A \, (\Delta s = 0)] \\ + \sin \Theta \, [V \, (\Delta s = +1) + A \, (\Delta s = +1)]\}$$

where Θ is the Cabibbo angle, Δs is the change in strangeness for the baryon, G is a universal beta-decay amplitude, and V and A are vector and axial vector amplitudes, respectively; it is experimentally determined that $\sin \Theta \approx 0.25$, so that $\cos \Theta = 0.97$.

cabinet file [DES ENG] A coarse-toothed file with flat and convex faces used for woodworking.

cabinet saw [DES ENG] A short saw, one edge used for ripping, the other for crosscutting.

cable [ARCH] A convex molding within one of the vertical grooves of a column or pilaster. [DES ENG] A stranded, ropelike assembly of wire or fiber. [ELEC] Strands of insulated electrical conductors laid together, usually around a central core, and surrounded by a heavy insulation. [OCEANOGR] *See* cable length.

cable buoy [ENG] A buoy used to mark one end of a submarine underwater cable during time of installation or repair.

cable conveyor [MECH ENG] A powered conveyor in which a trolley runs on a flexible, torque-transmitting cable that has helical threads.

cable delay [COMPUT SCI] The time required for one bit of data to go through a cable, about 1.5 nanoseconds per foot of cable.

cable-laid [DES ENG] Consisting of three ropes with a left-hand twist, each rope having three twisted strands.

cable length [OCEANOGR] A unit of distance, originally equal to the length of a ship's anchor cable, now variously considered to be 600 feet (183 meters), 608 feet (185.3 meters; one-tenth of a British nautical mile), or 720 feet or 120 fathoms (219.5 meters). Also known as cable.

cable messenger [ELEC] Stranded group of wires supported above the ground at intervals by poles or other structures and employed to furnish, within these intervals, frequent points of support for conductors or cables.

cable railway [MECH ENG] An inclined track on which rail cars travel, with the cars fixed to an endless steel-wire rope at equal spaces; the rope is driven by a stationary engine.

cable-stayed bridge [CIV ENG] A modification of the cantilever bridge consisting of girders or trusses cantilevered both ways from a central tower and supported by inclined cables attached to the tower at top or sometimes at several levels.

cable-system drill *See* churn drill.

cable television [COMMUN] A television program distribution system in which signals from all local stations and usually a number of distant stations are picked up by one or more high-gain antennas at elevated locations, amplified on individual channels, then fed directly to individual receivers of subscribers by overhead or underground coaxial cable. Also known as community antenna television (CATV).

cable tools [MIN ENG] The bits and other bottom-hole tools and equipment used to drill boreholes by percussive action, using a cable, instead of rods, to connect the drilling bit with the machine on the surface.

cable vault [CIV ENG] A manhole containing electrical cables. [ELEC] Vault in which the outside plant cables are spliced to the tipping cables.

cableway [MECH ENG] A transporting system consisting of a cable extended between two or more points on which cars are propelled to transport bulk materials for construction operations. [MIN ENG] A cable system of material handling in which carriers are supported by a cable and not detached from the operating span.

cabochon [LAP] One of two basic types of cutting styles used to fashion gemstones; the gem is cut in convex form, highly polished, but not faceted.

cacao [BOT] *Theobroma cacao*. A small tropical tree of the order Theales that bears capsular fruits which are a source of cocoa powder and chocolate. Also known as chocolate tree.

cache [COMPUT SCI] A small, fast storage buffer integrated in the central processing unit of some large computers.

cachexia [MED] Weight loss, weakness, and wasting of the body encountered in certain diseases or in terminal illnesses.

cacodyl [ORG CHEM] $(CH_3)_2As^-$ A radical found in, for example, cacodylic acid, $(CH_3)_2AsOOH$.

cacomistle [VERT ZOO] *Bassariscus astutus*. A raccoonlike mammal that inhabits the southern and southwestern United States; distinguished by a bushy black-and-white ringed tail. Also known as civet cat; ringtail.

caconym [SYST] A taxonomic name that is linguistically unacceptable.

cacoxenite [MINERAL] $Fe_4(PO_4)_3(OH)_3 \cdot 12H_2O$ Yellow or brownish mineral consisting of a hydrous basic iron phosphate occurring in radiated tufts.

Cactaceae [BOT] The cactus family of the order Caryophyllales; represented by the American stem succulents, which are mostly spiny with reduced leaves.

cactus [BOT] The common name for any member of the family Cactaceae, a group characterized by a fleshy habit, spines and bristles, and large, brightly colored, solitary flowers.

cadang-cadang [PL PATH] An infectious virus disease of the coconut palm characterized by yellow-bronzing of the leaves.

cadastral survey [CIV ENG] A survey made to establish property lines.

caddis fly [INV ZOO] The common name for all members of the insect order Trichoptera; adults are mothlike and the immature stages are aquatic.

cadmium [CHEM] A chemical element, symbol Cd, atomic number 48, atomic weight 112.40. [MET] A tin-white, malleable, ductile metal capable of high polish; principal use is in the plating of iron and steel, and to a much less extent of copper, brass, and other alloys, to protect them from corrosion and improve solderability and surface conductivity, and as a control absorber and shield in nuclear reactors.

cadmium blende *See* greenockite.

cadmium cell [ELEC] A standard cell used as a voltage reference; at 20°C its voltage is 1.0186 volts.

cadmium lamp [ELEC] A lamp containing cadmium vapor; wavelength (6438.4696 international angstroms, or 643.84696 nanometers) of light emitted is a standard of length.

cadmium-nickel storage cell *See* nickel-cadmium battery.

cadmium ocher *See* greenockite.

cadmium ratio [NUCLEO] The ratio of the response of an uncovered neutron detector to that of the same detector under identical conditions when it is covered with cadmium of a specified thickness.

cadmium red [MATER] A pigment composed of a mixture of cadmium sulfide, cadmium selenide, and barite; used as a red pigment.

cadmium silver oxide cell [ELEC] An alkaline-electrolyte cell that may be used without recharging in primary batteries or that may be recharged for secondary-battery use.

cadmium sulfide cell [ELECTR] A photoconductive cell in which a small wafer of cadmium sulfide provides an extremely high dark-light resistance ratio.

cadmium telluride detector [ELECTR] A photoconductive cell capable of operating continuously at ambient temperatures up to 750°F (400°C); used in solar cells and infrared, nuclear-radiation, and gamma-ray detectors.

Cae *See* Caelum.

caecum *See* cecum.

Caelum [ASTRON] A southern constellation, right ascension 5 hours, declination 40°S. Abbreviated Cae. Also known as Chisel.

Caenolestidae [PALEON] A family of extinct insectivorous mammals in the order Marsupialia.

Caenolestoidea [VERT ZOO] A superfamily of marsupial mammals represented by the single living family Caenolestidae.

caenostylic [VERT ZOO] Having the first two visceral arches attached to the cranium and functioning in food intake; a condition found in sharks, amphibians, and chimaeras.

Caesalpinoidea [BOT] A subfamily of dicotyledonous plants in the legume family, Leguminosae.

caffeine [ORG CHEM] $C_8H_{10}O_2N_4 \cdot H_2O$ An alkaloid found in a large number of plants, such as tea, coffee, cola, and mate.

cage [CRYSTAL] A void occurring in a crystal structure capable of trapping one or more foreign atoms. [MECH ENG] A frame for maintaining uniform separation between the balls or rollers in a bearing. Also known as separator. [MIN ENG] The car which carries men and materials in a mine hoist.

cage antenna [ELECTROMAG] Broad-band dipole antenna in which each pole consists of a cage of wires whose overall shape resembles that of a cylinder or a cone.

cage mill [MECH ENG] Pulverizer used to disintegrate clay, press cake, asbestos, packing-house by-products, and various tough, gummy, high-moisture-content or low-melting-point materials.

caging [NAV] The process of orienting and mechanically locking the spin axis of a gyroscope to an internal reference position.

cahnite [MINERAL] $Ca_2B(OH)_4(AsO_4)$ A tetragonal borate mineral occurring in white, sphenoidal crystals.

CAI *See* computer-assisted instruction.

Cailletet and Mathias law [PHYS CHEM] The law that describes the relationship between the mean density of a liquid and its saturated vapor at that temperature as being a linear function of the temperature.

caiman [VERT ZOO] Any of five species of reptiles of the genus *Caiman* in the family Alligatoridae, differing from alligators principally in having ventral armor and a sharper snout.

Cainotheriidae [PALEON] The single family of the extinct artiodactyl superfamily Cainotherioidea.

Cainotherioidea [PALEON] A superfamily of extinct, rabbit-sized tylopod ruminants in the mammalian order Artiodactyla.

cairn [ENG] An artificial mound of rocks, stones, or masonry, usually conical or pyramidal, whose purpose is to designate or to aid in identifying a point of surveying or of cadastral importance.

caisson [CIV ENG] 1. A watertight, cylindrical or rectangular chamber used in underwater construction to protect workers from water pressure and soil collapse. 2. A float used to raise a sunken vessel. 3. *See* dry-dock caisson. [ORD] A two-wheeled, horse-drawn vehicle used for transporting ammunition and other military equipment.

caisson disease [MED] A condition resulting from a rapid change in atmospheric pressure from high to normal, causing nitrogen bubbles to form in the blood and body tissues. Also known as bends; compressed-air illness.

caisson foundation [CIV ENG] A shaft of concrete placed under a building column or wall and extending down to hardpan or rock. Also known as pier foundation.

caking [ENG] Changing of a powder into a solid mass by heat, pressure, or water.

caking coal [GEOL] A type of coal which agglomerates and softens upon heating; after volatile material has been expelled at high temperature, a hard, gray cellular mass of coke remains. Also known as binding coal.

cal See calorie.

CAL [COMPUT SCI] A higher-level language, developed especially for time-sharing purposes, in which a user at a remote console typewriter is directly connected to the computer and can work out problems on-line with considerable help from the computer. Derived from Conversational Algebraic Language.

Calabrian [GEOL] Lower Pleistocene geologic time.

calaite See turquoise.

calamine [MET] An alloy composed of zinc, lead, and tin. [MINERAL] See hemimorphite; smithsonite. [PHARM] A powder mixture of zinc oxide and ferric oxide, used in skin lotions and ointments.

Calamitales [PALEOBOT] An extinct group of reedlike plants of the subphylum Sphenopsida characterized by horizontal rhizomes and tall, upright, grooved, articulated stems.

calandria [CHEM ENG] One of the tubes through which the heating fluid circulates in an evaporator.

calandria evaporator See short-tube vertical evaporator.

Calanoida [INV ZOO] A suborder of the crustacean order Copepoda, including the larger and more abundant of the pelagic species.

Calappidae [INV ZOO] The box crabs, a family of reptantian decapods in the subsection Oxystomata of the section Brachyura.

calaverite [MINERAL] $AuTe_2$ A yellowish or tin-white, monoclinic mineral commonly containing gold telluride and minor amounts of silver.

calcaneum [ANAT] 1. A bony projection of the metatarsus in birds. 2. See calcaneus.

calcaneus [ANAT] A bone of the tarsus, forming the heel bone in humans. Also known as calcaneum.

calcar [ZOO] A spur or spurlike process, especially on an appendage or digit.

Calcarea [INV ZOO] A class of the phylum Porifera, including sponges with a skeleton composed of calcium carbonate spicules.

calcarenite [PETR] A type of limestone or dolomite composed of coral or shell sand or of sand formed by erosion of older limestones, with particle size ranging from $\frac{1}{16}$ to 2 millimeters.

calcareous [SCI TECH] Resembling, containing, or composed of calcium carbonate.

calcareous algae [BOT] Algae that grow on limestone or in soil impregnated with lime.

calcareous crust See caliche.

calcareous duricrust See caliche.

calcareous ooze [GEOL] A fine-grained pelagic sediment containing undissolved sand- or silt-sized calcareous skeletal remains of small marine organisms mixed with amorphous clay-sized material.

calcareous schist [PETR] A coarse-grained metamorphic rock derived from impure calcareous sediment.

calcareous tufa See tufa.

calcarine sulcus [ANAT] A sulcus on the medial aspect of the occipital lobe of the cerebrum, between the lingual gyrus and the cuneus. Also known as calcarine fissure.

Calcaronea [INV ZOO] A subclass of sponges in the class Calcarea in which the larva is amphiblastulae.

calcemia See hypercalcemia.

calcic [SCI TECH] Derived from or containing calcium.

calciferous [BIOL] Containing or producing calcium or calcium carbonate.

calcification [GEOCHEM] Any process of soil formation in which the soil colloids are saturated to a high degree with exchangeable calcium, thus rendering them relatively immobile and nearly neutral in reaction. [PHYSIO] The deposit of calcareous matter within the tissues of the body.

calcilutite [PETR] 1. A dolomite or limestone formed of calcareous rock flour that is typically nonsiliceous. 2. A rock of calcium carbonate formed of grains or crystals with average diameter less than $\frac{1}{16}$ millimeter.

calcine [ENG] 1. To heat to a high temperature without fusing, as to heat unformed ceramic materials in a kiln, or to heat ores, precipitates, concentrates, or residues so that hydrates, carbonates, or other compounds are decomposed and the volatile material is expelled. 2. To heat under oxidizing conditions. [MATER] Product of calcining or roasting.

Calcinea [INV ZOO] A subclass of sponges in the class Calcarea in which the larvae are parenchymulae.

calcined gypsum See plaster of Paris.

calcining furnace [ENG] A heating device, such as a vertical-shaft kiln, that raises the temperature (but not to the melting point) of a substance such as limestone to make lime. Also known as calciner.

calcinosis [MED] Deposition of calcium salts in the skin, subcutaneous tissue, or other part of the body in certain pathologic conditions.

calciocarnotite See tyuyamunite.

calcioferrite [MINERAL] $Ca_2Fe_2(PO_4)OH \cdot 7H_2O$ A yellow or green mineral consisting of a hydrous basic calcium iron phosphate and occurring in nodular masses.

calciovolborthite [MINERAL] $CaCu(VO_4)(OH)$ Green, yellow, or gray mineral consisting of a basic vanadate of calcium and copper. Also known as tangeite.

calcirudite [PETR] Dolomite or limestone formed of worn or broken pieces of coral or shells or of limestone fragments coarser than sand; the interstices are filled with sand, calcite, or mud, the whole bound together with a calcareous cement.

calcite [MINERAL] $CaCO_3$ One of the commonest minerals, the principal constituent of limestone; hexagonal-rhombohedral crystal structure, dimorphous with aragonite. Also known as calcspar.

calcite compensation depth [GEOL] The depth in the ocean (about 5000 meters) below which solution of calcium carbonate occurs at a faster rate than its deposition. Abbreviated CCD.

calcitonin [BIOCHEM] A polypeptide, calcium-regulating hormone produced by the ultimobranchial bodies in vertebrates. Also known as thyrocalcitonin.

calcium [CHEM] A chemical element, symbol Ca, atomic number 20, atomic weight 40.08; used in metallurgy as an alloying agent for aluminum-bearing metal, as an aid in removing bismuth from lead, and as a deoxidizer in steel manufacture, and also used as a cathode coating in some types of photo tubes.

calcium-45 [NUC PHYS] A radioisotope of calcium having a mass number of 45, often used as a radioactive tracer in studying calcium metabolism in humans and other organisms; half-life is 165 days. Designated Ca^{45}.

calcium acetate [ORG CHEM] $Ca(C_2H_3O_2)_2$ A compound that crystallizes as colorless needles that are soluble in water; formerly used as an important source of acetone and acetic acid; now used as a mordant and as a stabilizer of plastics.

calcium hardness [CHEM] Presence of calcium ions in water, from dissolved carbonates and bicarbonates; treated in boiler water by introducing sodium phosphate.

calcium metabolism [BIOCHEM] Biochemical and physiological processes involved in maintaining the concentration of calcium in plasma at a constant level and providing a sufficient supply of calcium for bone mineralization.

calcium star [ASTRON] A term sometimes used to denote a star of spectral class F, which has prominent absorption bands of calcium.

calclacite [MINERAL] $CaCl_2Ca(C_2H_3O_2)\cdot10H_2O$ A white mineral consisting of a hydrated chloride-acetate of calcium; occurs as hairlike efflorescences.

Calclamnidae [PALEON] A family of Paleozoic echinoderms of the order Dendrochirotida.

calcrete [GEOL] A conglomerate of surficial gravel and sand cemented by calcium carbonate.

calcspar *See* calcite.

calculated address *See* generated address.

calculating machine *See* calculator.

calculating punch [COMPUT SCI] A calculator having a card reader and a card punch.

calculator [COMPUT SCI] A device that performs logic and arithmetic digital operations based on numerical data which are entered by pressing numerical and control keys. Also known as calculating machine.

calculus [ANAT] A small and cuplike structure. [MATH] The branch of mathematics dealing with differentiation and integration and related topics. [PATH] An abnormal, solid concretion of minerals and salts formed around organic materials and found chiefly in ducts, hollow organs, and cysts.

calculus of tensors [MATH] The branch of mathematics dealing with the differentiation of tensors.

calculus of variations [MATH] The study of problems concerning maximizing or minimizing a given definite integral relative to the dependent variables of the integrand function.

calculus of vectors [MATH] That branch of calculus concerned with differentiation and integration of vector-valued functions.

caldera [GEOL] A more or less circular volcanic depression whose diameter is many times greater than that of the volcanic vent.

Caledonian orogeny [GEOL] Deformation of the crust of the earth by a series of diastrophic movements beginning perhaps in Early Ordovician and continuing through Silurian, extending from Great Britain through Scandinavia.

caledonite [MINERAL] $Cu_2Pb_5(SO_4)_3CO_3(OH)_6$ A mineral occurring as green, orthorhombic crystals composed of basic copper lead sulfate; found in copper-lead deposits.

calendar [ASTRON] A system for everyday use in which time is divided into days and longer periods, such as weeks, months, and years, and a definite order for these periods and a correspondence between them are established.

calendar day [ASTRON] The period from midnight to midnight; it is 24 hours of mean solar time in length and coincides with the civil day.

calendar month [ASTRON] The month of the calendar, varying from 28 to 31 days in length.

calendar year [ASTRON] The year in the Gregorian calendar, common years having 365 days and leap years 366. Also known as civil year.

calender [ENG] **1.** To pass a material between rollers or plates to thin it into sheets or to make it smooth and glossy. **2.** The machine which performs this operation.

calendering [TEXT] Mechanical finishing process to produce a hard, shiny fabric.

calf [OCEANOGR] *See* calved ice. [VERT ZOO] The young of the domestic cow, elephant, rhinoceros, hippopotamus, moose, whale, and others.

caliber [ORD] The diameter of a projectile or the diameter of the bore of a gun or launching tube; for example, a caliber .22 cartridge has a diameter of approximately 0.22 inch (5.6 millimeters).

calibrate [SCI TECH] To determine, by measurement or comparison with a standard, the correct value of each scale reading on a meter or other device, or the correct value for each setting of a control knob.

calibrating tank [ENG] A tank having known capacity used to check the volumetric accuracy of liquid delivery by positive-displacement meters. Also known as meter-proving tank.

calibration curve [ENG] A plot of calibration data, giving the correct value for each indicated reading of a meter or control dial.

calibration reference [ANALY CHEM] Any of the standards of various types that indicate whether an analytical instrument or procedure is working within prescribed limits; examples are test solutions used with pH meters, and solutions with known concentrations (standard solutions) used with spectrophotometers.

caliche [GEOL] **1.** Conglomerate of gravel, rock, soil, or alluvium cemented with sodium salts in Chilean and Peruvian nitrate deposits; contains sodium nitrate, potassium nitrate, sodium iodate, sodium chloride, sodium sulfate, and sodium borate. **2.** A thin layer of clayey soil capping auriferous veins (Peruvian usage). **3.** Whitish clay in the selvage of veins (Chilean usage). **4.** A recently discovered mineral vein. **5.** A secondary accumulation of opaque, reddish brown to buff or white calcareous material occurring in layers on or near the surface of stony soils in arid and semiarid regions of the southwestern United States; called hardpan, calcareous duricrust, and kanker in different geographic regions. Also known as calcareous crust; croute calcaire; nari; sabach; tepetate.

Caliciaceae [BOT] A family of lichens in the order Caliciales in which the disk of the apothecium is borne on a short stalk.

Caliciales [BOT] An order of lichens in the class Ascolichenes characterized by an unusual apothecium.

California Current [OCEANOGR] The ocean current flowing southward along the western coast of the United States to northern Baja California.

californite [MINERAL] $Ca_{10}Al_4(Mg,Fe)_2Si_9O_{34}(OH,F)_4$ A variety of vesuvianite that resembles jade; it is dark-, yellowish-, olive-, or grass-green and occurs in translucent to opaque compact or massive form. Also known as American jade.

californium [CHEM] A chemical element, symbol Cf, atomic number 98; all isotopes are radioactive.

Caligidae [INV ZOO] A family of fish ectoparasites belonging to the crustacean suborder Caligoida.

Caligoida [INV ZOO] A suborder of the crustacean order Copepoda, including only fish ectoparasites and characterized by a sucking mouth with styliform mandibles.

caliper [DES ENG] An instrument with two legs or jaws that can be adjusted for measuring linear dimensions, thickness, or diameter.

caliper gage [DES ENG] An instrument, such as a micrometer, of fixed size for calipering.

caliper log [MIN ENG] A graphic record showing the diameter of a drilled hole at each depth; measurements are obtained by drawing a caliper upward through the hole and recording the diameter on quadrile paper.

calite [MET] A heat-resistant alloy of iron, nickel, and aluminum which resists oxidation up to 2200°F (1204°C) and is practically noncorrodible under ordinary conditions of exposure.

calk See caulk.

call [COMPUT SCI] To transfer control to a specified closed subroutine.

call announcer [ELECTR] Device for receiving pulses from an automatic telephone office and audibly reproducing the corresponding number in words, so that it may be heard by a manual operator.

call circuit [ELEC] Communications circuit between switching points used by traffic forces for transmitting switching instructions.

Callendar's equation [THERMO] 1. An equation of state for steam whose temperature is well above the boiling point at the existing pressure, but less than the critical temperature: $(V - b) = (RT/p) - (a/T^n)$, where V is the volume, R is the gas constant, T is the temperature, p is the pressure, n equals 10/3, and a and b are constants. 2. A very accurate equation relating temperature and resistance of platinum, according to which the temperature is the sum of a linear function of the resistance of platinum and a small correction term, which is a quadratic function of temperature.

Callendar's thermometer See platinum resistance thermometer.

Callichthyidae [VERT ZOO] A family of tropical catfishes in the suborder Siluroidei.

calligraphy [GRAPHICS] Elegant writing or penmanship as an art and as a profession.

call in [COMPUT SCI] To transfer control of a digital computer, temporarily, from a main routine to a subroutine that is inserted in the sequence of calculating operations, to fulfill an ancillary purpose.

call indicator [ELECTR] Device for receiving pulses from an automatic switching system and displaying the corresponding called number before an operator at a manual switchboard.

calling sequence [COMPUT SCI] A specific set of instructions to set up and call a given subroutine, make available the data required by it, and tell the computer where to return after the subroutine is executed.

Callionymoidei [VERT ZOO] A suborder of fishes in the order Perciformes, including two families of colorful marine bottom fishes known as dragonets.

Callipallenidae [INV ZOO] A family of marine arthropods in the subphylum Pycnogonida lacking palpi and having chelifores and 10-jointed ovigers.

Calliphoridae [INV ZOO] The blow flies, a family of myodarian cyclorrhaphous dipteran insects in the subsection Calypteratae.

Callisto [ASTRON] A satellite of Jupiter orbiting at a mean distance of 1,884,000 kilometers. Also known as Jupiter IV.

Callithricidae [VERT ZOO] The marmosets, a family of South American mammals in the order Primates.

call letters [COMMUN] Identifying letters, sometimes including numerals, assigned to radio and television stations by the Federal Communications Commission and other regulatory authorities throughout the world. Also known as call sign.

call number [COMPUT SCI] In computer operations, a set of characters identifying a subroutine, and containing information concerning parameters to be inserted in the subroutine, or information to be used in generating the subroutine, or information related to the operands.

Callorhinchidae [VERT ZOO] A family of ratfishes of the chondrichthyan order Chimaeriformes.

callose [BIOCHEM] A carbohydrate component of plant cell walls; associated with sieve plates where calluses are formed. [BIOL] Having hardened protuberances, as on the skin or on leaves and stems.

Callovian [GEOL] A stage in uppermost Middle or lowermost Upper Jurassic which marks a return to clayey sedimentation.

callus [BOT] 1. A thickened callose deposit on sieve plates. 2. A hard tissue that forms over a damaged plant surface. [MED] Hard, thick area on the surface of the skin.

calm [METEOROL] The absence of apparent motion of the air; in the Beaufort wind scale, smoke is observed to rise vertically, or the surface of the sea is smooth and mirrorlike; in U.S. weather observing practice, the wind has a speed under 1 mile per hour or 1 knot (1.6 kilometers per hour).

calomel [MINERAL] Hg_2Cl_2 A colorless, white, grayish, yellowish, or brown secondary, sectile, tetragonal mineral; used as a cathartic, insecticide, and fungicide. Also known as calomelene; calomelite; horn quicksilver; mercurial horn ore.

calomel electrode [PHYS CHEM] A reference electrode of known potential consisting of mercury, mercury chloride (calomel), and potassium chloride solution; used to measure pH and electromotive force. Also known as calomel half-cell; calomel reference electrode.

calomelene See calomel.

calomel half-cell See calomel electrode.

calomelite See calomel.

calomel reference electrode See calomel electrode.

calorescence [PHYS] The production of visible light by infrared radiation; the transformation is indirect, the light being produced by heat and not by any direct change of wavelength.

calorie [THERMO] Abbreviated cal; often designated c. 1. A unit of heat energy, equal to 4.1868 joules. Also known as International Table calorie (IT calorie). 2. A unit of energy, equal to the heat required to raise the temperature of 1 gram of water from 14.5° to 15.5°C at a constant pressure of 1 standard atmosphere; equal to 4.1855 ± 0.0005 joules. Also known as fifteen-degrees calorie; gram-calorie (g-cal); small calorie. 3. A unit of heat energy equal to 4.184 joules; used in thermochemistry. Also known as thermochemical calorie.

calorific value [ENG] Quantity of heat liberated on the complete combustion of a unit weight or unit volume of fuel.

calorifier [ENG] A device that heats fluids by circulating them over heating coils.

calorimeter [ENG] An apparatus for measuring heat quantities generated in or emitted by materials in processes such as chemical reactions, changes of state, or formation of solutions.

calorimetric titration See thermometric titration.

calorimetry [ENG] The measurement of the quantity of heat involved in various processes, such as chemical reactions, changes of state, and formations of solutions, or in the determination of the heat capacities of substances; fundamental unit of measurement is the joule or the calorie (4.184 joules).

calotype [GRAPHICS] An obsolete method of photography in which paper is treated with silver iodide, silver nitrate, and acetic and gallic acids; after exposure the paper is developed in a solution of silver nitrate and gallic acid.

calved ice [OCEANOGR] A piece of ice floating in a body of water after breaking off from a mass of land ice or an iceberg. Also known as calf.

calving [GEOL] The breaking off of a mass of ice from its parent glacier, iceberg, or ice shelf. Also known as ice calving. [VERT ZOO] Giving birth to a calf.

calyculate [BOT] Having bracts that imitate a second, external calyx.

Calymnidae [INV ZOO] A family of echinoderms in the order Holasteroida characterized by an ovoid test with a marginal fasciole.

calypter [INV ZOO] A scalelike or lobelike structure above the haltere of certain two-winged flies. Also known as alula; squama.

Calyptoblastea [INV ZOO] A suborder of coelenterates in the suborder Hydroida, including the hydroids with protective cups around the hydranths and gonozooids.

calyptra [BOT] **1.** A membranous cap or hoodlike covering, especially the remains of the archegonium over the capsule of a moss. **2.** Tissue surrounding the archegonium of a liverwort. **3.** Root cap.

Calyptratae [INV ZOO] A subsection of dipteran insects in the suborder Cyclorrhapha characterized by calypters associated with wings.

Calyssozoa [INV ZOO] The single class of the bryozoan subphylum Entoprocta.

calyx [BOT] The outermost whorl of a flower; composed of sepals. [ENG] A steel tube that is a guide rod and is also used to catch cuttings from a drill rod. Also known as bucket; sludge barrel; sludge bucket. [INV ZOO] A cup-shaped structure to which the arms are attached in crinoids. [MED] **1.** A cuplike structure. **2.** In the kidney, a collecting structure extending from the renal pelvis.

cam [MECH ENG] A plate or cylinder which communicates motion to a follower by means of its edge or a groove cut in its surface.

Cam *See* Camelopardalis.

CAM *See* computer-aided manufacturing.

Camallanida [INV ZOO] An order of phasmid nematodes in the subclass Spiruria, including parasites of domestic animals.

Camarodonta [INV ZOO] An order of Euechinoidea proposed by R. Jackson and abandoned in 1957.

Cambaridae [INV ZOO] A family of crayfishes belonging to the section Macrura in the crustacean order Decapoda.

Cambarinae [INV ZOO] A subfamily of crayfishes in the family Astacidae, including all North American species east of the Rocky Mountains.

camber [AERO ENG] The rise of the curve of an airfoil section, usually expressed as the ratio of the departure of the curve from a straight line joining the extremities of the curve to the length of this straight line. [DES ENG] Deviation from a straight line; the term is applied to a convex, edgewise sweep or curve, or to the increase in diameter at the center of rolled materials. [GEOL] **1.** A terminal, convex shoulder of the continental shelf. **2.** A structural feature that is caused by plastic clay beneath a bed flowing toward a valley so that the bed sags downward and seems to be draped over the sides of the valley. [NAV ARCH] *See* round of beam.

camber angle [MECH ENG] The inclination from the vertical of the steerable wheels of an automobile.

camber arch [ARCH] An arch with a slightly curved interior and horizontal exterior.

cambium [BOT] A layer of cells between the phloem and xylem of most vascular plants that is responsible for secondary growth and for generating new cells.

Cambrian [GEOL] The lowest geologic system that contains abundant fossils of animals, and the first (earliest) geologic period of the Paleozoic era from 570 to 500 million years ago.

camel [VERT ZOO] The common name for two species of artiodactyl mammals, the bactrian camel (*Camelus bactrianus*) and the dromedary camel (*C. dromedarius*), in the family Camelidae.

Camelidae [VERT ZOO] A family of tylopod ruminants in the superfamily Cameloidea of the order Artiodactyla, including four species of camels and llamas.

Cameloidea [VERT ZOO] A superfamily of tylopod ruminants in the order Artiodiodactyla.

Camelopardalis [ASTRON] Latin name for the Giraffe constellation of the northern hemisphere. Abbreviated Cam; Caml. Also known as Camelopardus; Giraffe.

Camelopardus *See* Camelopardalis.

cam engine [MECH ENG] A piston engine in which a cam-and-roller mechanism seems to convert reciprocating motion into rotary motion.

cameo [LAP] A type of carved gemstone in which the background is cut away to leave the subject in relief.

camera [ELECTR] *See* television camera. [OPTICS] A light-tight enclosure containing an aperture (usually provided with an optical lens or system of lenses) through which the light from an object passes and forms an image, often on a light-sensitive material, inside.

camera lucida [OPTICS] An instrument having a peculiarly shaped prism or a system of mirrors, and often a microscope, which causes a virtual image of an object to be produced on a plane surface, enabling the image's outline to be traced.

camera obscura [OPTICS] A primitive camera in which the real image of an object can be observed or traced on the wall of the enclosure opposite the aperture, rather than being recorded photographically.

camera-ready [GRAPHICS] A layout prepared for the offset printer; contains the actual material to be reproduced.

Camerata [PALEON] A subclass of extinct stalked echinoderms of the class Crinoidea.

camera tube [ELECTR] An electron-beam tube used in a television camera to convert an optical image into a corresponding charge-density electric image and to scan the resulting electric image in a predetermined sequence to provide an equivalent electric signal. Also known as pickup tube; television camera tube.

cam follower [MECH ENG] The output link of a cam mechanism.

Caml *See* Camelopardalis.

cam mechanism [MECH ENG] A mechanical linkage whose purpose is to produce, by means of a contoured cam surface, a prescribed motion of the output link.

cAMP *See* cyclic adenylic acid.

Campanulaceae [BOT] A family of dicotyledonous plants in the order Campanulales characterized by a style without an indusium but with well-developed collecting hairs below the stigmas, and by a well-developed latex system.

Campanulales [BOT] An order of dicotyledonous plants in the subclass Asteridae distinguished by a chiefly herbaceous habit, alternate leaves, and inferior ovary.

cam pawl [MECH ENG] A pawl which prevents a wheel from turning in one direction by a wedging action, while permitting it to rotate in the other direction.

Campbell bridge [ELEC] 1. A bridge designed for comparison of mutual inductances. 2. A circuit for measuring frequencies by adjusting a mutual inductance, until the current across a detector is zero.

Campbell-Stokes recorder [ENG] A sunshine recorder in which the time scale is supplied by the motion of the sun and which has a spherical lens that burns an image of the sun upon a specially prepared card.

camp ceiling [BUILD] A ceiling that is flat in the center portion and sloping at the sides.

Camp-Meidell condition [STAT] For determining the distribution of a set of numbers, the guideline stating that if the distribution has only one mode, if the mode is the same as the arithmetic mean, and if the frequencies decline continuously on both sides of the mode, then more than $1 - (1/2.25t^2)$ of any distribution will fall within the closed range $\bar{X} \pm t\sigma$, where t = number of items in a set, $\bar{X}$ = average, and σ = standard deviation.

Campodeidae [INV ZOO] A family of primarily wingless insects in the order Diplura which are most numerous in the Temperate Zone of the Northern Hemisphere.

camp-on system [COMMUN] A circuit control feature whereby a user attempting to establish a telephone call and encountering a busy station will hold the connection for a preset time, to the exclusion of other callers, in case the original conversation should terminate.

camptonite [PETR] A lamprophyre containing pyroxene, sodic hornblende, and olivine as dark constituents and labradorite as the light constituent; sodic orthoclase may be present.

campylotropous [BOT] Having the ovule symmetrical but half inverted, with the micropyle and funiculus at right angles to each other.

camshaft [MECH ENG] A rotating shaft to which a cam is attached.

can [DES ENG] A cylindrical metal vessel or container, usually with an open top or a removable cover. [NUCLEO] See jacket.

Canaceidae [INV ZOO] The seashore flies, a family of myodarian cyclorrhaphous dipteran insects in the subsection Acalypteratae.

Canadian Shield See Laurentian Shield.

canal [BIOL] A tubular duct or passage in bone or soft tissues. [CIV ENG] An artificial open waterway used for transportation, waterpower, or irrigation. [DES ENG] A groove on the underside of a corona. [GEOGR] A long, narrow arm of the sea extending far inland, between islands, or between islands and the mainland. [NUCLEO] A water-filled trench or conduit associated with a nuclear reactor, used for removing and sometimes storing radioactive objects taken from the reactor; the water acts as a shield against radiation.

canaliculus [HISTOL] 1. One of the minute channels in bone radiating from a Haversian canal and connecting lacunae with each other and with the canal. 2. A passage between the cells of the cell cords in the liver.

canalization [ENG] Any system of distribution canals or conduits for water, gas, electricity, or steam. [MED] Surgical method of wound drainage without tubes by forming channels. [PHYSIO] The formation of new channels in tissues, such as the formation of new blood vessels through a thrombus.

canal of Schlemm [ANAT] An irregular channel at the junction of the sclera and cornea in the eye that drains aqueous humor from the anterior chamber.

canal valve [ANAT] The semilunar valve in the right atrium of the heart between the orifice of the inferior vena cava and the right atrioventricular orifice. Also known as eustachian valve.

Canary Current [OCEANOGR] The prevailing southward flow of water along the northwestern coast of Africa.

Canastotan [GEOL] Lower Upper Silurian geologic time.

can buoy [NAV] Floating cylindrical unlighted buoy with a flat top, constructed of metal; depending on position, it may be painted black with or without an odd number on it, or painted with horizontal stripes.

Canc See Cancer.

cancellate [BIOL] Lattice-shaped. Also known as clathrate.

cancellation circuit [ELECTR] A circuit used in providing moving-target indication on a plan position indicator scope; cancels constant-amplitude fixed-target pulses by subtraction of successive pulse trains.

cancellation law [MATH] A rule which allows formal division by common factors in equal products, even in systems which have no division, as integral domains; $ab = ac$ implies that $b = c$.

cancellous [BIOL] Having a reticular or spongy structure.

cancellous bone [HISTOL] A form of bone near the ends of long bones having a cancellous matrix composed of rods, plates, or tubes; spaces are filled with marrow.

cancer [MED] Any malignant neoplasm, including carcinoma and sarcoma.

Cancer [ASTRON] A constellation with right ascension 9 hours, declination 20°N. Abbreviated Canc. Also known as Crab.

cancrinite [MINERAL] $Na_3CaAl_3Si_3O_{12}CO_3(OH)_2$ A feldspathoid tectosilicate occurring in hexagonal crystals in nepheline syenites, usually in compact or disseminated masses.

candela [OPTICS] A unit of luminous intensity, defined as $1/60$ of the luminous intensity per square centimeter of a blackbody radiator operating at the temperature of freezing platinum. Formerly known as candle. Also known as new candle.

candidiasis [MED] A fungus infection of the skin, lungs, mucous membranes, and viscera of humans caused by a species of *Candida*, usually *C. albicans*. Also known as moniliasis.

candite See ceylonite.

candle See candela.

candlepower [OPTICS] Luminous intensity expressed in candelas. Abbreviated cp.

cane [BOT] 1. A hollow, usually slender, jointed stem, such as in sugarcane or the bamboo grasses. 2. A stem growing directly from the base of the plant, as in most Rosaceae, such as blackberry and roses.

cane blight [PL PATH] A fungus disease affecting the canes of several bush fruits, such as currants and raspberries; caused by several species of fungi.

Canes Venatici [ASTRON] A northern constellation with right ascension 13 hours, declination 40°N, between Ursa Major and Boötes. Abbreviated CVn. Also known as Hunting Dogs.

canfieldite [MINERAL] Ag_8SnS_6 A black mineral of the argyrodite series consisting of silver thiostannate, with a specific gravity of 6.28; found in Germany and Bolivia.

Canidae [VERT ZOO] A family of carnivorous mammals in the superfamily Canoidea, including dogs and their allies.

canine [ANAT] A conical tooth, such as one located between the lateral incisor and first premolar in humans and many other mammals. Also known as cuspid. [VERT ZOO] Pertaining or related to dogs or to the family Canidae.

Canis Major [ASTRON] A constellation with right ascension 7 hours, declination 20°S. Abbreviated CMa. Also known as Greater Dog.

Canis Minor [ASTRON] A constellation with right ascension 8 hours, declination 5°N. Abbreviated CMi. Also known as Lesser Dog.

α Canis Minoris *See* Procyon.

canker [PL PATH] An area of necrosis on a woody stem resulting in shrinkage and cracking followed by the formation of callus tissue, ultimately killing the stem. [VET MED] A localized chronic inflammation of the ear in cats, dogs, foxes, ferrets, and others caused by the mite *Otodectes cynotis*.

canker sore [MED] Small ulceration of the mucous membrane of the mouth, sometimes caused by a food allergy.

canker stain [PL PATH] A fungus disease of plane trees caused by *Endoconidiophora fimbriata platani* and characterized by bluish-black or reddish-brown discolorations beneath blackened cankers on the trunk and sometimes the branches.

cankerworm [INV ZOO] Any of several lepidopteran insect larvae in the family Geometridae which cause severe plant damage by feeding on buds and foliage.

Cannabaceae [BOT] A family of dicotyledonous herbs in the order Urticales, including Indian hemp (*Cannabis sativa*) and characterized by erect anthers, two styles or style branches, and the lack of milky juice.

Cannaceae [BOT] A family of monocotyledonous plants in the order Zingiberales characterized by one functional stamen, a single functional pollen sac in the stamen, mucilage canals in the stem, and numerous ovules in each of the one to three locules.

canned motor [MECH ENG] A motor enclosed within a casing along with the liquid stream (that is, a pump) so that the motor bearings are lubricated by the same liquid that is being pumped.

canned program [COMPUT SCI] A program which has been written to solve a particular problem, is available to users of a computer system, and is usually fixed in form and capable of little or no modification.

canned pump [MECH ENG] A watertight pump that can operate under water.

cannel coal [GEOL] A fine-textured, highly volatile bituminous coal distinguished by a greasy luster and blocky, conchoidal fracture; burns with a steady luminous flame. Also known as cannelite.

cannelite *See* cannel coal.

canneloid [GEOL] **1.** Coal that resembles cannel coal. **2.** Coal intermediate between bituminous and cannel. **3.** Durain laminae in banded coal. **4.** Cannel coal of anthracite or semianthracite rank.

cannel shale [GEOL] A black shale formed by the accumulation of an aquatic ooze rich in bituminous organic matter in association with inorganic materials such as silt and clay.

Cannizzaro reaction [ORG CHEM] The reaction in which aldehydes that do not have a hydrogen attached to the carbon adjacent to the carbonyl group,

upon encountering strong alkali, readily form an alcohol and an acid salt.

Canoidea [VERT ZOO] A superfamily belonging to the mammalian order Carnivora, including all dogs and doglike species such as seals, bears, and weasels.

canonical [SCI TECH] Relating to the simplest or most significant form of a general function, equation, statement, rule, or expression.

canonical ensemble [STAT MECH] A hypothetical collection of systems of particles used to describe an actual individual system which is in thermal contact with a heat reservoir but is not allowed to exchange particles with its environment.

canonical equations of motion *See* Hamilton's equations of motion.

canonical form [CONT SYS] A specific type of dynamical system representation in which the associated matrices possess specific row-column structures. [ORG CHEM] A resonance structure for a cyclic compound in which the bonds do not intersect. [SCI TECH] A particularly clean and simple representation which usually follows or satisfies a general rule.

canonical momentum *See* conjugate momentum.

canonical time unit [ASTRON] For geocentric orbits, the time required by a hypothetical satellite to move one radian in a circular orbit of the earth's equatorial radius, that is, 13.447052 minutes.

canonical transformation [MATH] Any function which has a standard form, depending on the context. [MECH] A transformation which occurs among the coordinates and momenta describing the state of a classical dynamical system and which leaves the form of Hamilton's equations of motion unchanged. Also known as contact transformation.

cantaloupe [BOT] The fruit (pepo) of *Cucumis malo*, a small, distinctly netted, round to oval muskmelon of the family Cucurbitaceae in the order Violales.

cant file [DES ENG] A fine-tapered file with a triangular cross section, used for sharpening saw teeth.

Cantharidae [INV ZOO] The soldier beetles, a family of coleopteran insects in the superfamily Cantharoidea.

Cantharoidea [INV ZOO] A superfamily of coleopteran insects in the suborder Polyphaga.

canthus [ANAT] Either of the two angles formed by the junction of the eyelids, designated outer or lateral, and inner or medial.

cantilever [ENG] A beam or member securely fixed at one end and hanging free at the other end.

cantilever arch [ARCH] An arch supported by flat projections on opposing walls.

cantilever bridge [CIV ENG] A fixed bridge consisting of two spans projecting toward each other and joined at their ends by a suspended simple span.

cantilever spring [MECH ENG] A flat spring supported at one end and holding a load at or near the other end.

canting [MECH] Displacing the free end of a beam which is fixed at one end by subjecting it to a sideways force which is just short of that required to cause fracture.

Cantor function [MATH] A real-valued nondecreasing continuous function defined on the closed interval $[0,1]$ which maps the Cantor ternary set onto the interval $[0,1]$.

Cantor theorem [MATH] A theorem that there is no one-to-one correspondence between a set and the collection of its subsets.

cant strip [BUILD] **1.** A strip placed along the angle between a wall and a roof so that the roofing will not bend sharply. **2.** A strip placed under the edge of the lowest row of tiles on a roof to give them the same slope as the other tiles.

canyon [GEOGR] A chasm, gorge, or ravine cut in the surface of the earth by running water; the sides are steep and form cliffs.

Cap *See* Capricornus.

capacitance [ELEC] The ratio of the charge on one of the conductors of a capacitor (there being an equal and opposite charge on the other conductor) to the potential difference between the conductors. Symbolized C. Formerly known as capacity.

capacitance box [ELEC] An assembly of capacitors and switches which permits adjustment of the capacitance existing at the terminals in nominally uniform steps, from a minimum value near zero to the maximum which exists when all the capacitors are connected in parallel.

capacitance bridge [ELEC] A bridge for comparing two capacitances, such as a Schering bridge.

capacitance relay [ELECTR] An electronic relay that responds to a small change in capacitance, such as that created by bringing a hand near a pickup wire or plate.

capacitance standard *See* standard capacitor.

capacitive coupling [ELEC] Use of a capacitor to transfer energy from one circuit to another.

capacitive diaphragm [ELECTROMAG] A resonant window used in a waveguide to provide the equivalent of capacitive reactance at the frequency being transmitted.

capacitive divider [ELEC] Two or more capacitors placed in series across a source, making available a portion of the source voltage across each capacitor; the voltage across each capacitor will be inversely proportional to its capacitance.

capacitive feedback [ELECTR] Process of returning part of the energy in the plate (or output) circuit of a vacuum tube (or other device) to the grid (or input) circuit by means of a capacitance common to both circuits.

capacitive load [ELECTROMAG] A load in which the capacitive reactance exceeds the inductive reactance; the load draws a leading current.

capacitive reactance [ELECTROMAG] Reactance due to the capacitance of a capacitor or circuit, equal to the inverse of the product of the capacitance and the angular frequency.

capacitor [ELEC] A device which consists essentially of two conductors (such as parallel metal plates) insulated from each other by a dielectric and which introduces capacitance into a circuit, stores electrical energy, blocks the flow of direct current, and permits the flow of alternating current to a degree dependent on the capacitor's capacitance and the current frequency. Symbolized C. Also known as condenser; electrical condenser.

capacitor antenna [ELECTROMAG] Antenna consisting of two conductors or systems of conductors, the essential characteristic of which is its capacitance. Also known as condenser antenna.

capacitor box [ELECTR] A box-shaped structure in which a capacitor is submerged in a heat-absorbing medium, usually water. Also known as condenser box.

capacitor loudspeaker *See* electrostatic loudspeaker.

capacitor microphone [ENG ACOUS] A microphone consisting essentially of a flexible metal diaphragm and a rigid metal plate that together form a two-plate air capacitor; sound waves set the diaphragm in vibration, producing capacitance variations that are converted into audio-frequency signals by a suitable amplifier circuit. Also known as condenser microphone; electrostatic microphone.

capacitor motor [ELEC] **1.** A single-phase induction motor having a main winding connected directly to a source of alternating-current power and an auxiliary winding connected in series with a capacitor to the source of ac power. **2.** *See* capacitor-start motor.

capacitor pickup [ENG ACOUS] A phonograph pickup in which movements of the stylus in a record groove cause variations in the capacitance of the pickup.

capacitor-start motor [ELEC] A capacitor motor in which the capacitor is in the circuit only during the starting period; the capacitor and its auxiliary winding are disconnected automatically by a centrifugal switch or other device when the motor reaches a predetermined speed. Also known as capacitor motor.

capacity [ANALY CHEM] In chromatography, a measurement used in ion-exchange systems to express the adsorption ability of the ion-exchange materials. [COMPUT SCI] *See* storage capacity. [ELEC] *See* capacitance. [SCI TECH] Volume, especially in reference to merchandise or containers thereof.

cap cloud [METEOROL] An approximately stationary cloud, or standing cloud, on or hovering above an isolated mountain peak; formed by the cooling and condensation of humid air forced up over the peak. Also known as cloud cap.

cape [GEOGR] A prominent point of land jutting into a body of water. Also known as head; headland; mull; naze; ness; point; promontory.

Cape Horn Current [OCEANOGR] That part of the west wind drift flowing eastward in the immediate vicinity of Cape Horn, and then curving northeastward to continue as the Falkland Current.

caper [FOOD ENG] The buds and berries of the caper plant *(Capparis spinosa)* pickled and used as a condiment.

capillarity [FL MECH] The action by which the surface of a liquid where it contacts a solid is elevated or depressed, because of the relative attraction of the molecules of the liquid for each other and for those of the solid. Also known as capillary action.

capillary [ANAT] The smallest vessel of both the circulatory and lymphatic systems; the walls are composed of a single cell layer.

capillary angioma *See* hemangioma.

capillary attraction [FL MECH] The force of adhesion existing between a solid and a liquid in capillarity.

capillary depression [FL MECH] The depression of the meniscus of a liquid contained in a tube where the liquid does not wet the walls of the container, as in a mercury barometer; the meniscus has a convex shape, resulting in a depression.

capillary ejecta *See* Pele's hair.

capillary electrometer [ENG] An electrometer designed to measure a small potential difference between mercury and an electrolytic solution in a capillary tube by measuring the effect of this potential difference on the surface tension between the liquids. Also known as Lippmann electrometer.

capillary pyrites *See* millerite.

capillary ripple *See* capillary wave.

capillary rise [FL MECH] The rise of a liquid in a capillary tube times the radius of the tube.

capillary tube [ENG] A tube sufficiently fine so that capillary attraction of a liquid into the tube is significant.

capillary viscometer [ENG] A long, narrow tube used to measure the laminar flow of fluids.

capillary water [HYD] Soil water held by capillarity as a continuous film around soil particles and in interstices between particles above the phreactic line.

capillary wave [FL MECH] **1.** A wave occurring at the interface between two fluids, such as the interface between air and water on oceans and lakes, in which the principal restoring force is controlled by surface tension. **2.** A water wave of less than 1.7 centimeters. Also known as capillary ripple; ripple.

capital amount factor [IND ENG] Any of 20 common compound interest formulas used to calculate the equivalent uniform annual cost of all cash flows.

capital expenditure [IND ENG] Money spent for long-term additions or improvements and charged to a capital assets account.

capitate [BIOL] Enlarged and swollen at the tip. [BOT] Forming a head, as certain flowers of the Compositae.

Capitellidae [INV ZOO] A family of mud-swallowing annelid worms, sometimes called bloodworms, belonging to the Sedentaria.

capitellum [ANAT] A small head or rounded process of a bone.

Capitonidae [VERT ZOO] The barbets, a family of pantropical birds in the order Piciformes.

Caponidae [INV ZOO] A family of arachnid arthropods in the order Araneida characterized by having tracheae instead of book lungs.

Capparaceae [BOT] A family of dicotyledonous herbs, shrubs, and trees in the order Capparales characterized by parietal placentation; hypogynous, mostly regular flowers; four to many stamens; and simple to trifoliate or palmately compound leaves.

Capparales [BOT] An order of dicotyledonous plants in the subclass Dilleniidae.

capped fuse [ENG] A length of safety fuse with the cap or detonator crimped on before it is taken to the place of use.

capped steel [MET] Partially deoxidized steel cast in an open-top mold, which is capped to solidify the top metal and enforce internal pressure, resulting in a surface condition similar to that of rimming steel.

cappelenite [MINERAL] $(Ba,Ca,Na)(Y,La)_6B_6Si_{13}(O,OH)_{27}$ A greenish-brown hexagonal mineral consisting of a rare yttrium-barium borosilicate occurring in crystals.

capping [ENG] Preparation of a capped fuse. [GEOL] **1.** Consolidated barren rock overlying a mineral or ore deposit. **2.** *See* gossan. [MIN ENG] The attachment at the end of a winding rope. [PETRO ENG] **1.** The process of sealing or covering a borehole such as an oil or gas well. **2.** The material or device used to seal or cover a borehole.

Caprellidae [INV ZOO] The skeleton shrimps, a family of slender, cylindrical amphipod crustaceans in the suborder Caprellidea.

Caprellidea [INV ZOO] A suborder of marine and brackish-water animals of the crustacean order Amphipoda.

Capricornus [ASTRON] A constellation with right ascension 21 hours, declination 20°S. Abbreviated Cap. Also known as Sea Horse.

Caprifoliaceae [BOT] A family of dicotyledonous, mostly woody plants in the order Dipsacales, including elderberry and honeysuckle; characterized by distinct filaments and anthers, typically five stamens and five corolla lobes, more than one ovule per locule, and well-developed endosperm.

Caprimulgidae [VERT ZOO] A family of birds in the order Caprimulgiformes, including the nightjars, or goatsuckers.

Caprimulgiformes [VERT ZOO] An order of nocturnal and crepuscular birds, including nightjars, potoos, and frog-mouths.

cap rock [GEOL] **1.** An overlying, generally impervious layer or stratum of rock that overlies an oil- or gas-bearing rock. **2.** Barren vein matter, or a pinch in a vein, supposed to overlie ore. **3.** A hard layer of rock, usually sandstone, a short distance above a coal seam. **4.** An impervious body of anhydrite and gypsum in a salt dome.

Capsaloidea [INV ZOO] A superfamily of ectoparasitic trematodes in the subclass Monogenea characterized by a sucker-shaped holdfast with anchors and hooks.

cap screw [DES ENG] A screw which passes through a clear hole in the part to be joined, screws into a threaded hole in the other part, and has a head which holds the parts together.

capsicum [BOT] The fruit of a plant of the genus *Capsicum*, especially *C. frutescens*, cultivated in southern India and the tropics; a strong irritant to mucous membranes and eyes.

capstan [ENG] A shaft which pulls magnetic tape through a machine at constant speed. [NAV ARCH] A rotating vertical spindle-mounted drum on which cable is wound for raising an anchor or other heavy weight.

capstan nut [DES ENG] A nut whose edge has several holes, in one of which a bar can be inserted for turning it.

capstan screw [DES ENG] A screw whose head has several radial holes, in one of which a bar can be inserted for turning it.

capsular ligament [ANAT] A saclike ligament surrounding the articular cavity of a freely movable joint and attached to the bones.

capsule [ANAT] A membranous structure enclosing a body part or organ. [BOT] A closed structure bearing seeds or spores; it is dehiscent at maturity. [MICROBIO] A thick, mucous envelope, composed of polypeptide or carbohydrate, surrounding certain microorganisms. [PHARM] A soluble shell in which drugs are enclosed for oral administration.

capture [AERO ENG] The process in which a missile is taken under control by the guidance system. [ASTROPHYS] Of a central force field, as of a planet, to overcome by gravitational force the velocity of a passing body and bring the body under the control of the central force field, in some cases absorbing its mass. [GEOCHEM] In a crystal structure, the substitution of a trace element for a lower-valence common element. [HYD] The natural diversion of the headwaters of one stream into the channel of another stream having greater erosional activity and flowing at a lower level. Also known as piracy; river capture; river piracy; robbery; stream capture; stream piracy; stream robbery. [PHYS] A process in which an atomic or nuclear system acquires an additional particle; for example, the capture of electrons by positive ions, or capture of neutrons by nuclei.

capture area [ENG ACOUS] The effective area of the receiving surface of a hydrophone, or the available power of the acoustic energy divided by its equivalent plane-wave intensity.

capture ratio [COMMUN] A measure of the ability of a frequency-modulation tuner to reject the weaker of two stations that are on the same frequency; the lower the ratio in decibels of desired and undesired signals, the better the performance of the tuner.

capybara [VERT ZOO] *Hydrochoerus capybara.* An aquatic rodent (largest rodent in existence) found in

South America and characterized by partly webbed feet, no tail, and coarse hair.

car See automobile.

Car See Carina.

Carabidae [INV ZOO] The ground beetles, a family of predatory coleopteran insects in the suborder Adephaga.

Caracarinae [VERT ZOO] The caracaras, a subfamily of carrion-feeding birds in the order Falconiformes.

Caradocian [GEOL] Lower Upper Ordovician geologic time.

Carangidae [VERT ZOO] A family of perciform fishes in the suborder Percoidei, including jacks, scads, and pompanos.

carapace [GEOL] The upper normal limb of a fold having an almost horizontal axial plane. [INV ZOO] A dorsolateral, chitonous case covering the cephalothorax of many arthropods. [VERT ZOO] The bony, dorsal part of a turtle shell.

Carapidae [VERT ZOO] The pearlfishes, a family of sinuous, marine shore fishes in the order Gadiformes that live as commensals in the body cavity of holothurians.

carat [LAP] A unit of weight of gemstones, equal to 200 milligrams. Also known as metric carat.

Carathéodory's principle [THERMO] An expression of the second law of thermodynamics which says that in the neighborhood of any equilibrium state of a system, there are states which are not accessible by a reversible or irreversible adiabatic process. Also known as principle of inaccessibility.

caraway [BOT] *Carum carvi*. A white-flowered perennial herb of the family Umbelliferae; the fruit is used as a spice and flavoring agent.

carbanion [CHEM] One of the charged fragments which arise on heterolytic cleavage of a covalent bond involving carbon; the fragment carries an unshared pair of electrons and bears a negative charge.

carbene [ORG CHEM] A compound of carbon which exhibits two valences to a carbon atom; the two valence electrons are distributed in the same valence; an example is CH_2.

carbide [INORG CHEM] A binary compound of carbon with an element more electropositive than carbon; carbon-hydrogen compounds are excluded. [MATER] A cemented or compacted mixture of powdered carbides of heavy metals forming a hard material used in metal-cutting tools. Also known as cemented carbide.

carbocation [ORG CHEM] A positively charged ion whose charge resides, at least in part, on a carbon atom or group of carbon atoms.

carbocyclic compound [ORG CHEM] A compound with a homocyclic ring in which all the ring atoms are carbon, for example, benzene.

carbohumin See ulmin.

carbohydrase [BIOCHEM] Any enzyme that catalyzes the hydrolysis of disaccharides and more complex carbohydrates.

carbohydrate [BIOCHEM] Any of the group of organic compounds composed of carbon, hydrogen, and oxygen, including sugars, starches, and celluloses.

carbohydrate metabolism [BIOCHEM] The sum of the biochemical and physiological processes involved in the breakdown and synthesis of simple sugars, oligosaccharides, and polysaccharides and in the transport of sugar across cell membranes.

carbon [CHEM] A nonmetallic chemical element, symbol C, atomic number 6, atomic weight 12.01115; occurs freely as diamond, graphite, and coal.

carbon-12 [NUC PHYS] A stable isotope of carbon with mass number of 12, forming about 98.9% of natural carbon; used as the basis of the newer scale of atomic masses, having an atomic mass of exactly 12u (relative nuclidic mass unit) by definition.

carbon-13 [NUC PHYS] A heavy isotope of carbon having a mass number of 13.

carbon-14 [NUC PHYS] A naturally occurring radioisotope of carbon having a mass number of 14 and half-life of 5780 years; used in radiocarbon dating and in the elucidation of the metabolic path of carbon in photosynthesis. Also known as radiocarbon.

carbonaceous [SCI TECH] Relating to or composed of carbon.

carbon arc [ELEC] An electric arc between two electrodes, at least one of which is made of carbon; used in welding and high-intensity lamps, such as in searchlights and photography lamps.

carbon arc lamp [ELEC] An arc lamp in which an electric current flows between two electrodes of pure carbon, with incandescence at one or both electrodes and some light from the luminescence of the arc.

carbon arc welding [MET] Welding by maintaining an electric arc between a nonconsumable carbon electrode and the work.

carbonate [CHEM] 1. An ester or salt of carbonic acid. 2. A compound containing the carbonate (CO_3^{--}) radical. 3. Containing carbonates.

carbonate cycle [GEOCHEM] The biogeochemical carbonate pathways, involving the conversion of carbonate to CO_2 and HCO_3, the solution and deposition of carbonate, and the metabolism and regeneration of it in biological systems.

carbonation [CHEM] Conversion to a carbonate. [CHEM ENG] The process by which a fluid, especially a beverage, is impregnated with carbon dioxide. [GEOCHEM] A process of chemical weathering whereby minerals that contain soda, lime, potash, or basic oxides are changed to carbonates by the carbonic acid in air or water.

carbonatite [PETR] 1. Intrusive carbonate rock associated with alkaline igneous intrusive activity. 2. A sedimentary rock composed of at least 80% calcium or magnesium.

carbon black [CHEM] 1. An amorphous form of carbon produced commercially by thermal or oxidative decomposition of hydrocarbons and used principally in rubber goods, pigments, and printer's ink. 2. See gas black.

carbon brush [ELEC] A rod made of carbon that bears against a commutator, collector ring, or slip ring to provide passage for the electric current from a dynamo through an outside circuit or for an external current through a motor.

carbon button See button.

carbon cycle [GEOCHEM] The cycle of carbon in the biosphere, in which plants convert carbon dioxide to organic compounds that are consumed by plants and animals, and the carbon is returned to the biosphere in the form of inorganic compounds by processes of respiration and decay. [NUC PHYS] See carbon-nitrogen cycle.

carbon-14 dating [NUCLEO] Determining the approximate age of organic material associated with archeological or fossil artifacts by measuring the rate of radiation of the carbon-14 isotope. Also known as radioactive carbon dating; radiocarbon dating.

carbon electrode [MET] A nonfiller-metal electrode consisting of carbon or a graphite rod; sometimes contains copper powder for increased electrical conductivity; used in carbon arc welding.

carbon fiber [MATER] Commercial material made by pyrolyzing any spun, felted, or woven raw material to a char at temperatures from 700 to 1800°C.

Carboniferous [GEOL] A division of late Paleozoic rocks and geologic time including the Mississippian and Pennsylvanian periods.

carbonification *See* coalification.

carbon isotope ratio [GEOL] Ratio of carbon-12 to either of the less common isotopes, carbon-13 or carbon-14, or the reciprocal of one of these ratios; if not specified, the ratio refers to C12/C13. Also known as carbon ratio.

carbonite *See* natural coke.

carbonitriding [MET] Surface-hardening of low-carbon steel or other solid ferrous alloy by introducing carbon and nitrogen in a gaseous atmosphere containing carbon monoxide or hydrocarbons and ammonia at 800–875°C. Also known as gas cyaniding; nicarbing.

carbonium ion [ORG CHEM] A carbocation which has a positively charged carbon with a coordination number greater than 3.

carbonization [CHEM] The conversion of a carbon-containing substance to carbon or a carbon residue as the destructive distillation of coal by heat in the absence of air, yielding a solid residue with a higher percentage of carbon than the original coal; carried on for the production of coke and of fuel gas. [GEOCHEM] 1. In the coalification process, the accumulation of residual carbon by changes in organic material and their decomposition products. 2. Deposition of a thin film of carbon by slow decay of organic matter underwater. 3. A process of converting a carbonaceous material to carbon by removal of other components.

carbon knock [MECH ENG] Premature ignition resulting in knocking or pinging in an internal combustion engine caused when the accumulation of carbon produces overheating in the cylinder.

carbon lamp [ELEC] An arc lamp with carbon electrodes.

carbon microphone [ENG ACOUS] A microphone in which a flexible diaphragm moves in response to sound waves and applies a varying pressure to a container filled with carbon granules, causing the resistance of the microphone to vary correspondingly.

carbonmonoxyhemoglobin [BIOCHEM] A stable combination of carbon monoxide and hemoglobin formed in the blood when carbon monoxide is inhaled. Also known as carbonylhemoglobin and carboxyhemoglobin.

carbon-nitrogen cycle [NUC PHYS] A series of thermonuclear reactions, with release of energy, which presumably occurs in the sun and other stars; the net accomplishment is the synthesis of four hydrogen atoms into a helium atom, the emission of two positrons and much energy, and restoration of a carbon-12 atom with which the cycle began. Also known as carbon cycle; nitrogen cycle.

carbon paper [MATER] A paper, coated with dark waxy pigment, used to make duplicate copies while typewriting or handwriting; a sheet of carbon paper is sandwiched between two paper sheets, so that the impression made on the top sheet causes the carbon paper to transfer a pigmented impression onto the bottom sheet.

carbon pile [ELEC] A variable resistor consisting of a stack of carbon disks mounted between a fixed metal plate and a movable one that serve as the terminals of the resistor; the resistance value is reduced by applying pressure to the movable plate.

carbon ratio [GEOL] 1. The ratio of fixed carbon to fixed carbon plus volatile hydrocarbons in a coal. 2. *See* carbon isotope ratio.

carbon star [ASTRON] Any of a class of stars with an apparently high abundance ratio of carbon to hydrogen; a majority of these are low-temperature red giants of the C class.

carbon steel [MET] Steel containing carbon, to about 2%, as the principal alloying element.

carbon transducer [ENG] A transducer consisting of carbon granules in contact with a fixed electrode and a movable electrode, so that motion of the movable electrode varies the resistance of the granules.

carbonyl [ORG CHEM] A radical (CO) that is made up of one atom of carbon and one atom of oxygen connected by a double bond; found, for example, in aldehydes and ketones. Also known as carbonyl group.

carbonyl group *See* carbonyl.

carbonylhemoglobin *See* carbonmonoxyhemoglobin.

carbonyl process [MET] 1. A process in powder metallurgy for the production of iron, nickel, and iron-nickel alloy powders for magnetic applications. 2. A process used in putting a metallic coating on molybdenum tungsten and other metals.

carborundum [MATER] A manufactured crystalline material (silicon carbide), prepared by fusing coke and sand in an electric furnace; used as an abrasive in the grinding of low-tensile-strength materials, and as a semiconductor with a maximum operating temperature of 1300°C, to rectify and detect radio waves.

carboxy *See* carboxyl

carboxyhemoglobin *See* carbonmonoxyhemoglobin.

carboxyl [ORG CHEM] COOH The radical which determines the basicity of an organic acid. Also know as carboxy; oxatyl.

carboxylase [BIOCHEM] Any enzyme that catalyzes a carboxylation or decarboxylation reaction.

carboxylation [ORG CHEM] Addition of a carboxyl group into a molecule.

carbuncle [MED] A bacterial infection of subcutaneous tissue caused by *Staphylococcus aureus;* multiple sinuses are created by tissue destruction.

carburetion [CHEM ENG] The process of enriching a gas by adding volatile carbon compounds, such as hydrocarbons, to it, as in the manufacture of carbureted water gas. [MECH ENG] The process of mixing fuel with air in a carburetor.

carburetor [CHEM ENG] An apparatus for vaporizing, cracking, and enriching oils in the manufacture of carbureted water gas. [MECH ENG] A device that makes and controls the proportions and quantity of fuel-air mixture fed to a spark-ignition internal combustion engine.

carburize [MET] To surface-harden steel by converting the outer layer of low-carbon steel to high-carbon steel by heating the steel above the transformation range in contact with a carbonaceous material.

Carcharhinidae [VERT ZOO] A large family of sharks belonging to the charcharinid group of galeoids, including the tiger sharks and blue sharks.

Carchariidae [VERT ZOO] A family of shallow-water predatory sharks belonging to the isurid group of galeoids.

carcinogen [MED] Any agent that incites development of a carcinoma or any other sort of malignancy.

carcinotron *See* backward-wave oscillator.

card [COMPUT SCI] An information-carrying medium, common to practically all computers, for the introduction of data and instructions into the computers either directly or indirectly.

cardamom See cardamon.

cardamon [BOT] *Elettaria cardamomum.* A perennial herbaceous plant in the family Zingiberaceae; the seed of the plant is used as a spice. Also spelled cardamom.

Cardan joint See Hooke's joint.

Cardan motion [MECH ENG] The straight-line path followed by a moving centrode in a four-bar centrode linkage.

Cardan's suspension [DES ENG] An arrangement of rings in which a heavy body is mounted so that the body is fixed at one point; generally used in a gyroscope.

card bed [COMPUT SCI] The metal plate along which the card travels to the punching and reading stations.

card checking [COMPUT SCI] The verification carried out by the computer to ensure that all the data keypunched on a card has been correctly read into the memory.

card code [COMPUT SCI] The representation of characters on a punched card by means of punching one or more holes per column.

card face [COMPUT SCI] The printed side of a punched card or, if printing is on both sides, the side of chief importance.

card feed [COMPUT SCI] A device that inserts cards into a machine one at a time.

card field [COMPUT SCI] A specified group of card columns used for a particular category of data.

card fluff [COMPUT SCI] Small bits of paper that may be left attached to the edge of a hole punched in a punch card, which may cause misfeeding or misreading of the card.

card hopper [COMPUT SCI] A device that holds cards and makes them available to a card-feed mechanism. Also known as hopper.

cardia [ANAT] 1. The orifice where the esophagus enters the stomach. 2. The large, blind diverticulum of the stomach adjoining the orifice. [INV ZOO] Anterior enlargement of the ventriculus in some insects.

cardiac [ANAT] 1. Of, pertaining to, or situated near the heart. 2. Of or pertaining to the cardia of the stomach.

cardiac cycle [PHYSIO] The sequence of events in the heart between the start of one contraction and the start of the next.

cardiac electrophysiology [PHYSIO] The science that is concerned with the mechanism, spread, and interpretation of the electric currents which arise within heart muscle tissue and initiate each heart muscle contraction.

cardiac failure [MED] A complex of symptoms resulting from failure of the heart to pump sufficient quantities of blood. Also known as heart failure.

cardiac gland [ANAT] Any of the mucus-secreting, compound tubular structures near the esophagus or in the cardia of the stomach of vertebrates; capable of secreting digestive enzymes.

cardiac muscle [HISTOL] The principal tissue of the vertebrate heart; composed of a syncytium of striated muscle fibers.

cardiac pacemaker See pacemaker.

card image [COMPUT SCI] A one-to-one representation of the contents of a punched card, such as a matrix in which a 1 represents a hole and a 0 represents the absence of a hole.

cardinal number [MATH] The number of members of a set; usually taken as a particular well-ordered

set representative of the class of all sets which are in one-to-one correspondence with one another.

cardinal point [GEOD] Any of the four principal directions: north, east, south, or west of a compass. [OPTICS] Any one of six points in an optical system, namely, the two principal points, two nodal points, and two focal points. Also known as Gauss point.

cardinal system [NAV] In marine operations, a buoyage system in which the buoys are assigned distinctive shape, color, and number in accordance with location relative to the nearest obstruction and pertinent to the cardinal points.

carding [TEXT] Straightening or smoothing of raw fibers in a parallel fashion, with a carding machine.

cardiography [MED] Analysis of heart movements in the cardiac cycle by means of electronic instruments, especially by tracings.

cardioid [MATH] A heart-shaped curve generated by a point of a circle that rolls without slipping on a fixed circle of the same diameter.

cardioid condenser [OPTICS] A substage condenser that cuts off the direct light and allows only the light diffracted or dispersed from the object to enter the microscope; used in dark-field microscopes.

cardioid microphone [ENG ACOUS] A microphone having a heart-shaped, or cardioid, response pattern, so it has nearly uniform response for a range of about 180° in one direction and minimum response in the opposite direction.

cardioid pattern [ENG] Heart-shaped pattern obtained as the response or radiation characteristic of certain directional antennas, or as the response characteristic of certain types of microphones.

cardioscope [MED] 1. An instrument for the examination or visualization of the interior of the cardiac chambers. 2. An instrument which, by means of a cathode-ray oscillograph, projects an electrocardiographic record on a luminous screen.

cardiovascular system [ANAT] Those structures, including the heart and blood vessels, which provide channels for the flow of blood.

card jam [COMPUT SCI] A condition in which one or more punch cards become jammed along the card bed of a machine that is processing them. Also known as card wreck.

card loader [COMPUT SCI] A programming routine which permits a deck of cards to be read into a memory.

card machine [COMPUT SCI] 1. Any of several small computers which perform particular operations called for by instruction cards, concurrently with the reading of data cards. 2. Loosely, any type of peripheral equipment which reads or punches cards.

card punch [COMPUT SCI] A computer output device that punches holes in a punch card upon reception of signals from the central processing unit.

card reader [COMPUT SCI] A mechanism that senses information punched on cards, using wire brushes, metal feelers, or a photoelectric system. Also known as punched-card reader.

card row [COMPUT SCI] A row of punching positions parallel to the long edge of a punched card.

card sorter [COMPUT SCI] A machine used to arrange punched cards into an appropriate sequence for further processing. Also known as punched-card sorter.

card system [COMPUT SCI] A computer system whose only input unit is a card reader and whose only output units are a card punch and a printer.

card-to-card transceiving [COMPUT SCI] A system that makes possible instantaneous and accurate duplication of punched cards over telephone and telegraph networks between locations separated by either just a few miles or thousands of miles.

card-to-disk conversion [COMPUT SCI] A straightforward operation which consists in loading the data in a deck of cards onto a disk by means of a utility program.

card-to-print program [COMPUT SCI] A small computer program that uses punch cards in concert with a printer listing the information on the cards and that does not use either tape or disk storage equipment.

card-to-tape conversion [COMPUT SCI] A straightforward operation which consists in loading the data in a deck of cards onto a magnetic tape by means of a utility program.

card verifier [COMPUT SCI] An electromechanical device which allows the operator to check that a card has been properly key-punched.

card wreck See card jam.

Carettochelyidae [VERT ZOO] A family of reptiles in the order Chelonia containing only one species, the New Guinea plateless turtle (*Carettochelys insculpta*).

cargo liner See cargo ship.

cargo ship [NAV ARCH] A power-driven ship employed exclusively in commercial transportation of commodities on the ocean and large inland bodies of water. Also known as cargo liner.

Cariamidae [VERT ZOO] The long-legged cariamas, a family of birds in the order Gruiformes.

Caribbean Current [OCEANOGR] A water current flowing westward through the Caribbean Sea.

Caribosireninae [VERT ZOO] A subfamily of trichechiform sirenean mammals in the family Dugongidae.

Caridea [INV ZOO] A large section of decapod crustaceans in the suborder Natantia including many diverse forms of shrimps and prawns.

caries [MED] 1. Bone decay. 2. Tooth decay. Also known as dental caries.

carina [BIOL] A ridge or a keel-shaped anatomical structure. [VERT ZOO] See keel.

Carina [ASTRON] A constellation, right ascension 9 hours, declination 60°S. Abbreviated Car. Also known as Keel.

Carinomidae [INV ZOO] A monogeneric family of littoral ribbonlike worms in the order Palaeonemertini.

Carlsbad law [CRYSTAL] A feldspar twin law in which the twinning axis is the *c* axis, the operation is rotation of 180°, and the contact surface is parallel to the side pinacoid.

Carme [ASTRON] A small satellite of Jupiter with a diameter of about 19 miles (31 kilometers), orbiting with retrograde motion at a mean distance of 1.4×10^7 miles (2.3×10^7 kilometers) Also known as Jupiter XI.

carminite [MINERAL] $PbFe_2(AsO_4)_2(OH)_2$ A carmine to tile-red mineral consisting of a basic arsenate of lead and iron.

carnallite [MINERAL] $KMgCl_3 \cdot 6H_2O$ A milky-white or reddish mineral that crystallizes in the orthorhombic system and occurs in deliquescent masses; it is valuable as an ore of potassium.

carnassial [ANAT] Of or pertaining to molar or premolar teeth specialized for cutting and shearing.

Carnian [GEOL] Lower Upper Triassic geologic time. Also spelled Karnian.

Carnivora [VERT ZOO] A large order of placental mammals, including dogs, bears, and cats, that is primarily adapted for predation as evidenced by dentition and jaw articulation.

carnivorous [BIOL] Eating flesh or, as in plants, subsisting on nutrients obtained from animal protoplasm.

Carnosauria [PALEON] A group of large, predacious saurischian dinosaurs in the suborder Theropoda having short necks and large heads.

Carnot cycle [THERMO] A hypothetical cycle consisting of four reversible processes in succession: an isothermal expansion and heat addition, an isentropic expansion, an isothermal compression and heat rejection process, and an isentropic compression.

Carnot engine [MECH ENG] An ideal, frictionless engine which operates in a Carnot cycle.

carnotite [MINERAL] $K(UO_2)_2(VO_4)_2 \cdot nH_2O$ A canary-yellow, fine-grained hydrous vanadate of potassium and uranium having monoclinic microcrystals; an ore of radium and uranium.

Carnot's theorem [THERMO] 1. The theorem that all Carnot engines operating between two given temperatures have the same efficiency, and no cyclic heat engine operating between two given temperatures is more efficient than a Carnot engine. 2. The theorem that any system has two properties, the thermodynamic temperature T and the entropy S, such that the amount of heat exchanged in an infinitesimal reversible process is given by $dQ = TdS$; the thermodynamic temperature is a strictly increasing function of the empirical temperature measured on an arbitrary scale.

carotene [BIOCHEM] $C_{40}H_{56}$ Any of several red, crystalline, carotenoid hydrocarbon pigments occurring widely in nature, convertible in the animal body to vitamin A, and characterized by preferential solubility in petroleum ether. Also known as carotin.

carotenoid [BIOCHEM] A class of labile, easily oxidizable, yellow, orange, red, or purple pigments that are widely distributed in plants and animals and are preferentially soluble in fats and fat solvents.

carotid artery [ANAT] Either of the two principal arteries on both sides of the neck which supply blood to the head and neck. Also known as common carotid artery.

carotid body [ANAT] Either of two chemoreceptors sensitive to changes in blood chemistry which lie near the bifurcations of the carotid arteries. Also known as glomus caroticum.

carotid sinus [ANAT] An enlargement at the bifurcation of each carotid artery that is supplied with sensory nerve endings and plays a role in reflex control of blood pressure.

carotin See carotene.

carp [VERT ZOO] The common name for a number of fresh-water, cypriniform fishes in the family Cyprinidae, characterized by soft fins, pharyngeal teeth, and a suckerlike mouth.

carpel [BOT] The basic specialized leaf of the female reproductive structure in angiosperms; a megasporophyll.

carpenter's level [DES ENG] A bar, usually of aluminum or wood, containing a spirit level.

carpholite [MINERAL] $MnAl_2Si_2O_6(OH)_4$ A straw-yellow fibrous mineral consisting of a hydrous aluminum manganese silicate occurring in tufts; specific gravity is 2.93.

carphosiderite [MINERAL] A yellow mineral consisting of a basic hydrous iron sulfate occurring in masses and crusts.

Carpoidea [PALEON] Former designation for a class of extinct homalozoan echinoderms.

carpus [ANAT] 1. The wrist in humans or the corresponding part in other vertebrates. 2. The eight bones of the human wrist. [INV ZOO] The fifth segment from the base of a generalized crustacean appendage.

carrageen [BOT] *Chondrus crispus.* A cartilaginous red algae harvested in the Northern Atlantic as a source of carrageenin.

carriage [ENG] **1.** A device that moves in a predetermined path in a machine and carries some other part, such as a recorder head. **2.** A mechanism designed to hold a paper in the active portion of a printing or typing device, for example, a typewriter carriage. [ORD] Mobile or fixed support for a cannon; sometimes includes the elevating and traversing mechanisms.

carriage bolt [DES ENG] A round-head type of bolt with a square neck, used with a nut as a through bolt.

carriage tape *See* control tape.

carrier [COMMUN] **1.** The radio wave produced by a transmitter when there is no modulating signal, or any other wave, recurring series of pulses, or direct current capable of being modulated. Also known as carrier wave; signal carrier. **2.** A wave generated locally at a receiver that, when combined with the sidebands of a suppressed-carrier transmission in a suitable detector, produces the modulating wave. **3.** *See* carrier system. [GEN] An individual who is heterozygous for a recessive gene. [MECH ENG] Any machine for transporting materials or people. [MED] A person who harbors and eliminates an infectious agent and so transmits it to others, but who may not show signs of the disease. [NAV ARCH] *See* aircraft carrier. [NUCLEO] A substance that, when associated with a radioactive trace of another substance, will carry the trace with it through a chemical or physical process; an isotope is often used for this purpose. Also known as isotopic carrier. [SOLID STATE] *See* charge carrier.

carrier amplifier [ELECTR] A direct-current amplifier in which the dc input signal is filtered by a low-pass filter, then used to modulate a carrier so it can be amplified conventionally as an alternating-current signal; the amplified dc output is obtained by rectifying and filtering the rectified carrier signal.

carrier beat [COMMUN] An undesirable heterodyne of facsimile signals, each synchronous with a different stable reference oscillator, causing a pattern in received copy.

carrier channel [COMMUN] The equipment and lines that make up a complete carrier-current circuit between two or more points.

carrier chrominance signal *See* chrominance signal.

carrier current [COMMUN] A higher-frequency alternating current superimposed on ordinary telephone, telegraph, and power-line frequencies for communication and control purposes.

carrier density [SOLID STATE] The density of electrons and holes in a semiconductor.

carrier frequency [COMMUN] The frequency generated by an unmodulated radio, radar, carrier communication, or other transmitter, or the average frequency of the emitted wave when modulated by a symmetrical signal. Also known as center frequency; resting frequency.

carrier level [COMMUN] The strength or level of an unmodulated carrier signal at a particular point in a radio system, expressed in decibels in relation to some reference level.

carrier loading [ELECTROMAG] The addition of lumped inductances to the cable section of a transmission line specifically designed for carrier transmission; it serves to minimize impedance mismatch

between cable and open wire and to reduce the cable attenuation.

carrier mobility [SOLID STATE] The average drift velocity of carriers per unit electric field in a homogeneous semiconductor; the mobility of electrons is usually different from that of holes.

carrier noise [COMMUN] Noise produced by undesired variation of a radio-frequency signal in the absence of any intended modulation. Also known as residual modulation.

carrier repeater [ELECTR] Equipment designed to raise carrier signal levels to such a value that they may traverse a succeeding line section at such amplitude as to preserve an adequate signal-to-noise ratio; while the heart of a repeater is the amplifier, necessary adjuncts are filters, equalizers, level controls, and so on, depending upon the operating methods.

carrier shift [COMMUN] **1.** Transmission of radio-teletypewriter messages by shifting the carrier frequency in one direction for a marking signal and in the opposite direction for a spacing signal. **2.** Condition resulting from imperfect modulation whereby the positive and negative excursions of the envelope pattern are unequal, thus effecting a change in the power associated with the carrier.

carrier signaling [COMMUN] Method by which busy signals, ringing, or dial signaling relays are operated by the transmission of a carrier-frequency tone.

carrier suppression [COMMUN] **1.** Suppression of the carrier frequency after conventional modulation at the transmitter, with reinsertion of the carrier at the receiving end before demodulation. **2.** Suppression of the carrier when there is no modulation signal to be transmitted; used on ships to reduce interference between transmitters.

carrier system [COMMUN] A system permitting a number of simultaneous, independent communications over the same circuit. Also known as carrier.

carrier transmission [COMMUN] Transmission in which the transmitted electric wave is a wave resulting from the modulation of a single-frequency wave by a modulating wave.

carrier transport [SOLID STATE] The motion of conduction electrons or holes in semiconductors.

carrier wave *See* carrier.

carrot [BOT] *Daucus carota.* A biennial umbellifer of the order Umbellales with a yellow or orange-red edible root.

carry [MATH] An arithmetic operation that occurs in the course of addition when the sum of the digits in a given position equals or exceeds the base of the number system; a multiple m of the base is subtracted from this sum so that the remainder is less than the base, and the number m is then added to the next-higher-order digit.

carry-complete signal [COMPUT SCI] A signal generated by a digital parallel adder, indicating that all carries from an adding operation have been generated and propagated, and that the addition operation is completed.

carry flag [COMPUT SCI] A flip-flop circuit which indicates overflow in arithmetic operations.

carry lookahead [COMPUT SCI] A circuit which allows low-order carries to ripple through all the way to the highest-order bit to output a completed sum.

carry signal [COMPUT SCI] A signal produced in a computer when the sum of two digits in the same column equals or exceeds the base of the number system in use or when the difference between two digits is less than zero.

Carterinacea [INV ZOO] A monogeneric superfamily of marine, benthic foraminiferans in the suborder Rotaliina characterized by a test with monocrystal calcite spicules in a granular groundmass.

cartesian axis [MATH] One of a set of mutually perpendicular lines which all pass through a single point, used to define a cartesian coordinate system; the value of one of the coordinates on the axis is equal to the directed distance from the intersection of axes, while the values of the other coordinates vanish.

cartesian coordinates [MATH] The set of numbers which locate a point in space with respect to a collection of mutually perpendicular axes.

cartesian coordinate system [MATH] A coordinate system in n dimensions where n is any integer made by using n number axes which intersect each other at right angles at an origin, enabling any point within that rectangular space to be identified by the distances from the n lines. Also known as rectangular cartesian coordinate system.

cartesian geometry See analytic geometry.

cartesian surface [MATH] A surface obtained by rotating the curve

$$n_0(x^2 + y^2)^{1/2} \pm n_1[(x - a)^2 + y^2]^{1/2} = c$$

about the x axis.

cartesian tensor [MATH] The aggregate of the functions of position in a tensor field in an n-dimensional cartesian coordinate system.

cartilage [HISTOL] A specialized connective tissue which is bluish, translucent, and hard but yielding.

cartilage bone [HISTOL] Bone formed by ossification of cartilage.

cartilaginous fish [VERT ZOO] The common name for all members of the class Chondrichthyes.

cartogram [MAP] A type of single-factor or topical map that is often diagrammatic to show traffic flow, movement of people or goods, or value by area, where areas of the political subdivisions are distorted so that their size is proportional to their monetary value.

cartographer [GRAPHICS] An individual who makes charts or maps.

cartography [GRAPHICS] The making of maps and charts for the purpose of visualizing spatial distributions over various areas of the earth.

cartridge [ENG] A cylindrical, waterproof, paper shell filled with high explosive and closed at both ends; used in blasting. [ENG ACOUS] See phonograph pickup; tape cartridge. [NUCLEO] See jacket. [ORD] **1.** An assemblage of the components required to function a weapon once. **2.** Ammunition for a gun which contains in a unit assembly all components required to function the gun once, and which is loaded into the gun in one operation.

cartridge brass [MET] An alloy containing 70% copper and 30% zinc; uses include cartridge cases, automotive radiator cores and tanks, lighting fixtures, eyelets, rivets, springs, screws, and plumbing products.

cartridge filter [ENG] A filter for the clarification of process liquids containing small amounts of solids; turgid liquid flows between thin metal disks, assembled in a vertical stack, to openings in a central shaft supporting the disks, and solids are trapped between the disks.

cartridge fuse [ELEC] A type of electric fuse in which the fusible element is connected between metal ferrules at either end of an insulating tube.

cartridge lamp [ELEC] A pilot or dial lamp that has a tubular glass envelope with metal-ferrule terminals at each end.

cartridge tape drive [COMPUT SCI] A tape drive which will automatically thread the tape on the take-up reels without human assistance. Formerly known as hypertape drive.

car tunnel kiln [ENG] A long kiln with the fire located near the midpoint; ceramic ware is fired by loading it onto cars which are pushed through the kiln.

caruncle [ANAT] Any normal or abnormal fleshy outgrowth, such as the comb and wattles of fowl or the mass in the inner canthus of the eye. [BOT] A fleshy outgrowth developed from the seed coat near the hilum on some seeds, such as the castor bean.

Caryophanaceae [MICROBIO] Formerly a family of large, gram-negative bacteria belonging to the order Caryophanales and having disklike cells arranged in chains.

Caryophanales [MICROBIO] Formerly an order of bacteria in the class Schizomycetes occurring as trichomes which produce short structures that function as reproductive units.

Caryophyllaceae [BOT] A family of dicotyledonous plants in the order Caryophyllales differing from the other families in lacking betalains.

Caryophyllales [BOT] An order of dicotyledonous plants in the subclass Caryophyllidae characterized by free-central or basal placentation.

Caryophyllidae [BOT] A relatively small subclass of plants in the class Magnoliopsida characterized by trinucleate pollen, ovules with two integuments, and a multilayered nucellus.

caryopsis [BOT] A small, dry, indehiscent fruit having a single seed with such a thin, closely adherent pericarp that a single body, a grain, is formed.

Cas See Cassiopeia.

casaba melon [BOT] *Cucumis melo.* A winter muskmelon with a yellow rind and sweet flesh belonging to the family Cucurbitaceae of the order Violales.

cascade [ELEC] An electric-power circuit arrangement in which circuit breakers of reduced interrupting ratings are used in the branches, the circuit breakers being assisted in their protection function by other circuit breakers which operate almost instantaneously. Also known as backup arrangement. [ELECTR] See avalanche. [ENG] An arrangement of separation devices, such as isotope separators, connected in series so that they multiply the effect of each individual device. [GEOL] A landform structure formed by gravity collapse, consisting of a bed that buckles into a series of folds as it slides down the flanks of an anticline. [HYD] A small waterfall or series of falls descending over rocks.

cascade amplifier [ELECTR] A vacuum-tube amplifier containing two or more stages arranged in the conventional series manner. Also known as multistage amplifier.

cascade compensation [CONT SYS] Compensation in which the compensator is placed in series with the forward transfer function. Also known as series compensation; tandem compensation.

cascade connection [ELECTR] A series connection of amplifier stages, networks, or tuning circuits in which the output of one feeds the input of the next. Also known as tandem connection.

cascade control [CONT SYS] An automatic control system in which various control units are linked in sequence, each control unit regulating the operation of the next control unit in line.

cascade cooler [CHEM ENG] Fluid-cooling device through which the fluid flows in a series of horizontal tubes, one above the other; cooling water from a trough drips over each tube, then to a drain. Also known as serpentine cooler; trickle cooler.

cascaded carry [COMPUT SCI] A carry process in which the addition of two numerals results in a sum numeral and a carry numeral that are in turn added together, this process being repeated until no new carries are generated.

cascade hyperon See xi hyperon.

cascade image tube [ELECTR] An image tube having a number of sections stacked together, the output image of one section serving as the input for the next section; used for light detection at very low levels.

cascade junction [ELECTR] Two *pn* semiconductor junctions in tandem such that the condition of the first governs that of the second.

cascade limiter [ELECTR] A limiter circuit that uses two vacuum tubes in series to give improved limiter operation for both weak and strong signals in a frequency-modulation receiver. Also known as double limiter.

cascade liquefaction [CRYO] A method of liquefying gases in which a gas with a high critical temperature is liquefied by increasing its pressure; evaporation of this liquid cools a second liquid so that it can also be liquefied by compression, and so on.

cascade networks [ELEC] Two networks in tandem such that the output of the first feeds the input of the second.

cascade noise [ELECTR] The noise in a communications receiver after an input signal has been subjected to two tandem stages of amplification.

cascade particle See xi hyperon; xi-minus particle.

cascade shower [PARTIC PHYS] A cosmic-ray shower of electrons, positrons, and gamma rays which grows by pair production and bremsstralung events.

cascade transformer [ELEC] A source of high voltage that is made up of a collection of step-up transformers; secondary windings are in series, and primary windings, except the first, are supplied from a pair of taps on the secondary winding of the preceding transformer.

Cascadian orogeny [GEOL] Post-Tertiary deformation of the crust of the earth in western North America.

cascading [ELEC] An effect in which a failure of an electrical power system causes this system to draw excessive amounts of power from power systems which are interconnected with it, causing them to fail, and these systems cause adjacent systems to fail in a similar manner, and so forth. [MECH ENG] An effect in ball-mill rotating devices when the upper level of crushing bodies breaks clear and falls to the top of the crop load.

cascode amplifier [ELECTR] An amplifier consisting of a grounded-cathode input stage that drives a grounded-grid output stage; advantages include high gain and low noise; widely used in television tuners.

case [COMPUT SCI] In computers, a set of data to be used by a particular program. [ENG] An item designed to hold a specific item in a fixed position by virtue of conforming dimensions or attachments; the item which it contains is complete in itself for removal and use outside the container. [GRAPHICS] The cover of a hardbound book. [MET] Outer layer of a ferrous alloy which has been made harder than the core by case hardening. [MIN ENG] A small fissure admitting water

into the mine workings. [PETRO ENG] To line a borehole with steel tubing, such as casing or pipe.

case hardening [GEOL] Formation of a mineral coating on the surface of porous rock by evaporation of a mineral-bearing solution. [MET] Process of carburizing low-carbon steel or other ferrous alloy for making the outer layer (case) harder than the core.

casing [BUILD] A finishing member around the opening of a door or window. [MECH ENG] A fire-resistant covering used to protect part or all of a steam generating unit. [PETRO ENG] A special steel tubing welded or screwed together and lowered into a borehole to prevent entry of loose rock, gas, or liquid into the borehole or to prevent loss of circulation liquid into porous, cavernous, or crevassed ground.

casing hanger See hanger.

casinghead [PETRO ENG] A fitting at the head of an oil or gas well that allows the pumping operation to take place, as well as the separation of oil and gas.

casing joint [PETRO ENG] Joint or union that connects two lengths of pipe used to form an oil-well casing.

casing log [PETRO ENG] Recorded data of a downhole inspection of an oil or gas well made to determine some characteristic of the formations penetrated by the drill hole; types of logs include resistivity, induction, radioactivity, geologic, temperature, and acoustic.

Casparian strip [BOT] A thin band of suberin- or lignin-like deposition in the radial and transverse walls of certain plant cells during the primary development phase of the endodermis.

Cassadagan [GEOL] Middle Upper Devonian geologic time, above Chemungian.

cassava [BOT] *Manihot esculenta*. A shrubby perennial plant grown for its starchy, edible tuberous roots. Also known as manihot; manioc.

Cassegrain antenna [ELECTROMAG] A microwave antenna in which the feed radiator is mounted at or near the surface of the main reflector and aimed at a mirror at the focus; energy from the feed first illuminates the mirror, then spreads outward to illuminate the main reflector.

Cassegrain-Newtonian telescope See Newtonian-Cassegrain telescope.

Cassegrain telescope [OPTICS] A reflecting telescope in which a small hyperboloidal mirror reflects the convergent beam from the paraboloidal primary mirror through a hole in the primary mirror to an eyepiece in back of the primary mirror.

Casselian See Chattian.

cassette [ENG] A light-tight container designed to hold photographic film or plates. [ENG ACOUS] A small, compact container that holds a magnetic tape and can be readily inserted into a matching tape recorder for recording or playback; the tape passes from one hub within the container to the other hub.

cassette-cartridge system [COMPUT SCI] An input system often used in minicomputers; its low cost and ease in mounting often offset its slow access time.

cassette memory [COMPUT SCI] A removable magnetic tape cassette that stores computer programs and data.

Cassidulinacea [INV ZOO] A superfamily of marine, benthic foraminiferans in the suborder Rotaliina, characterized by a test of granular calcite with monolamellar septa.

Cassiduloida [INV ZOO] An order of exocyclic Euechinoidea possessing five similar ambulacra which form petal-shaped areas (phyllodes) around the mouth.

Cassiopeia [ASTRON] A constellation with right ascension 1 hour, declination 60°N. Abbreviated Cas.

cassiterite [MINERAL] SnO_2 A yellow, black, or brown mineral that crystallizes in the tetragonal system in prisms terminated by dipyramids; the most important ore of tin. Also known as tin stone.

cassowary [VERT ZOO] Any of three species of large, heavy, flightless birds composing the family Casuariidae in the order Casuariiformes.

cast [ENG] 1. To form a liquid or plastic substance into a fixed shape by letting it cool in the mold. 2. Any object which is formed by placing a castable substance in a mold or form and allowing it to solidify. Also known as casting. [MED] 1. A rigid dressing used to immobilize a part of the body. 2. See strabismus. [NAV] 1. To turn a ship in its own water. 2. To turn a ship to a desired direction without gaining either headway or sternway. 3. To take a sounding with the lead. [OPTICS] A change in a color because of the adding of a different hue. [PALEON] A fossil reproduction of a natural object formed by infiltration of a mold of the object by waterborne minerals. [PHYSIO] A mass of fibrous material or exudate having the form of the body cavity in which it has been molded; classified from its source, such as bronchial, renal, or tracheal.

caste [INV ZOO] One of the levels of mature social insects in a colony that carry out a specific function; examples are workers and soldiers.

castellanus [METEOROL] A cloud species with at least a fraction of its upper part presenting some vertically developed cumuliform protuberances (some of which are more tall than wide) which give the cloud a crenellated or turreted appearance. Previously known as castellatus.

castellated nut [DES ENG] A type of hexagonal nut with a cylindrical portion above through which slots are cut so that a cotter pin or safety wire can hold it in place.

castellatus See castellanus.

caster [ENG] 1. The inclination of the kingpin or its equivalent in automotive steering, which is positive if the kingpin inclines forward, negative if it inclines backward, and zero if it is vertical as viewed along the axis of the front wheels. 2. A wheel which is free to swivel about an axis at right angles to the axis of the wheel, used to support trucks, machinery, or furniture.

cast-film extrusion See chill-roll extrusion.

Castigliano's principle See Castigliano's theorem.

Castigliano's theorem [MECH] The theorem that the component in a given direction of the deflection of the point of application of an external force on an elastic body is equal to the partial derivative of the work of deformation with respect to the component of the force in that direction. Also known as Castigliano's principle.

casting See cast.

casting-out nines [MATH] A method of checking the correctness of elementary arithmetical operations, based on the fact that an integer yields the same remainder as the sum of its decimal digits, when divided by 9.

cast iron [MET] Any carbon-iron alloy cast to shape and containing 1.8–4.5% carbon, that is, in excess of the solubility in austenite at the eutectic temperature. Abbreviated C.I.

Castle's intrinsic factor See intrinsic factor.

Castner process [CHEM ENG] A process used industrially to make high-test sodium cyanide by reacting sodium, glowed charcoal, and dry ammonia gas to form sodamide, which is converted to cyanamide immediately; the cyanamide is converted to cyanide with charcoal.

Castniidae [INV ZOO] The castniids; large diurnal, butterflylike moths composing the single, small family of the lepidopteran superfamily Castnioidea.

Castnioidea [INV ZOO] A superfamily of neotropical and Indo-Australian lepidopteran insects in the suborder Heteroneura.

Casuariidae [VERT ZOO] The cassowaries, a family of flightless birds in the order Casuariiformes lacking head and neck feathers and having bony casques on the head.

Casuariiformes [VERT ZOO] An order of large, flightless, ostrichlike birds of Australia and New Guinea.

Casuarinaceae [BOT] The single, monogeneric family of the plant order Casuarinales characterized by reduced flowers and green twigs bearing whorls of scalelike, reduced leaves.

Casuarinales [BOT] A monofamilial order of dicotyledonous plants in the subclass Hamamelidae.

cat [NAV ARCH] 1. A sturdy tackle used for bringing an anchor up to a ship's cathead. 2. To hoist an anchor up to the cathead of a ship. [VERT ZOO] The common name for all members of the carnivoran mammalian family Felidae, especially breeds of the domestic species, Felis domestica.

catabolism [BIOCHEM] That part of metabolism concerned with the breakdown of large protoplasmic molecules and tissues, often with the liberation of energy.

cataclasis [GEOL] Deformation of rock by fracture and rotation of aggregates or mineral grains.

cataclasite See cataclastic rock.

cataclastic metamorphism [PETR] Local metamorphism restricted to a region of faults and overthrusts involving purely mechanical forces resulting in cataclasis.

cataclastic rock [PETR] Rock containing angular fragments formed by cataclasis. Also known as cataclasite.

cataclastic structure See mortar structure.

cataclysmic variable [ASTRON] A star showing a sudden increase in the magnitude of light, followed by a slow fading of light; examples are novae and supernovae. Also known as explosive variable.

catalase [BIOCHEM] An enzyme that catalyzes the decomposition of hydrogen peroxide into molecular oxygen and water.

catalepsy [PSYCH] Suspended animation with loss of voluntary motion associated with hysteria and the schizophrenic reactions in humans, and with organic nervous system disease in animals.

catalog [COMPUT SCI] 1. All the indexes to data sets or files in a system. 2. The index to all other indexes; the master index. 3. To add an entry to an index or to build an entire new index.

catalysis [CHEM] A phenomenon in which a relatively small amount of substance augments the rate of a chemical reaction without itself being consumed.

catalyst [CHEM] Substance that alters the velocity of a chemical reaction and may be recovered essentially unaltered in form and amount at the end of the reaction.

catalytic cracking [CHEM ENG] Conversion of high-boiling hydrocarbons into lower-boiling types by a catalyst.

catalytic reforming [CHEM ENG] Rearranging of hydrocarbon molecules in a gasoline boiling-range feedstock to form hydrocarbons having a higher antiknock quality. Abbreviated CR.

catalytic site See active site.

catamaran [NAV ARCH] 1. A sailing or powered boat having two rather slender hulls joined by a deck or other structure. 2. A rectangular raft resting on two parallel cylindrical floats.

catamorphism See katamorophism

catamount See puma.

cat-and-mouse engine [MECH ENG] A type of rotary engine, typified by the Tschudi engine, which is an analog of the reciprocating piston engine, except that the pistons travel in a circular motion. Also known as scissor engine.

cataphoresis See electrophoresis.

catapleiite [MINERAL] $(Na_2,Ca)ZrSi_3O_9 \cdot 2H_2O$ A yellow or yellowish-brown mineral crystallizing in the hexagonal system, consisting of a hydrous silicate of sodium, calcium, and zirconium, and occurring in thin tabular crystals; hardness is 6 on Mohs scale, and specific gravity is 2.8.

Catapochrotidae [INV ZOO] A monospecific family of coleopteran insects in the superfamily Cucujoidea.

catapult [AERO ENG] 1. A power-actuated machine or device for hurling an object at high speed, for example, a device which launches aircraft from a ship deck. 2. A device, usually explosive, for ejecting a person from an aircraft. [ORD] A mechanical device for hurling grenades or bombs.

cataract [HYD] A waterfall of considerable volume with the vertical fall concentrated in one sheer drop. [MED] An opacity in the crystalline lens or the lens capsule of the eye.

catarrhal conjunctivitis [MED] A usually acute inflammation of the conjunctiva with smarting of the eyes, heaviness of the lids, photophobia, and excessive mucous or mucopurulent secretion, caused by a variety of contagious organisms, but sometimes becoming chronic as a sequela of the acute form or because of irritation from polluted atmosphere or allergic factors. Also known as pinkeye.

catarrhal jaundice See infectious hepatitis.

catastrophism [GEOL] The theory that most features in the earth were produced by the occurrence of sudden, short-lived, worldwide events. [PALEON] The theory that the differences between fossils in successive stratigraphic horizons resulted from a general catastrophe followed by creation of the different organisms found in the next-younger beds.

catatonia [PSYCH] A type of schizophrenic reaction in which the individual remains speechless and motionless, assumes fixed postures, and lacks the will and resists attempts to activate speech and movement.

catch basin [CIV ENG] 1. A basin at the point where a street gutter empties into a sewer, built to catch matter that would not easily pass through the sewer. 2. A well or reservoir into which surface water may drain off.

catchment area See drainage basin.

catecholamine [BIOCHEM] Any one of a group of sympathomimetic amines containing a catechol moiety, including especially epinephrine, norepinephrine (levarterenol), and dopamine.

catena [COMPUT SCI] A series of data items that appears in a chained list. [GEOL] A group of soils derived from uniform or similar parent material which nonetheless show variations in type because of differences in topography or drainage.

catenary [MATER] In fiberglass-reinforced plastics, a measure of the sag in an assemblage of a number of strands, which have a minimal amount of twist, at the midpoint of a specified length. [MATH] The

curve obtained by suspending a uniform chain by its two ends; the graph of the hyperbolic cosine function.

catenate [COMPUT SCI] To arrange a collection of items in a chained list or catena.

catenation [CHEM] The property of an element to link to itself to form molecules as with carbon.

Catenulida [INV ZOO] An order of threadlike, colorless fresh-water rhabdocoeles with a simple pharynx and a single, median protonephridium.

caterpillar [INV ZOO] 1. The wormlike larval stage of a butterfly or moth. 2. The larva of certain insects, such as scorpion flies and sawflies. [MECH ENG] A vehicle, such as a tractor or army tank, which runs on two endless belts, one on each side, consisting of flat treads and kept in motion by toothed driving wheels.

catfish [VERT ZOO] The common name for a number of fishes which constitute the suborder Siluroidei in the order Cypriniformes, all of which have barbels around the mouth.

Cathartidae [VERT ZOO] The New World vultures, a family of large, diurnal predatory birds in the order Falconiformes that lack a voice and have slightly webbed feet.

catheter [MED] A hollow, tubular device for insertion into a cavity, duct, or vessel to permit injection or withdrawal of fluids or to establish patency of the passageway.

cathode [ELEC] The positively charged pole of a primary cell or a storage battery. [ELECTR] 1. The primary source of electrons in an electron tube; in directly heated tubes the filament is the cathode, and in indirectly heated tubes a coated metal cathode surrounds a heater. Designated K. Also known as negative electrode. 2. The terminal of a semiconductor diode that is negative with respect to the other terminal when the diode is biased in the forward direction. [PHYS CHEM] The electrode at which reduction takes place in an electrochemical cell, that is, a cell through which electrons are being forced.

cathode bias [ELECTR] Bias obtained by placing a resistor in the common cathode return circuit, between cathode and ground; flow of electrode currents through this resistor produces a voltage drop that serves to make the control grid negative with respect to the cathode.

cathode-coupled amplifier [ELECTR] A cascade amplifier in which the coupling between two stages is provided by a common cathode resistor.

cathode dark space [ELECTR] The relatively nonluminous region between the cathode glow and the negative flow in a glow-discharge cold-cathode tube. Also known as Crookes dark space; Hittorf dark space.

cathode drop [ELECTR] The voltage between the arc stream and the cathode of a glow-discharge tube. Also known as cathode fall.

cathode fall See cathode drop.

cathode follower [ELECTR] A vacuum-tube circuit in which the input signal is applied between the control grid and ground, and the load is connected between the cathode and ground. Also known as grounded-anode amplifier; grounded-plate amplifier.

cathode ray [ELECTR] A stream of electrons, such as that emitted by a heated filament in a tube, or that emitted by the cathode of a gas-discharge tube when the cathode is bombarded by positive ions.

cathode-ray oscilloscope [ELECTR] A test instrument that uses a cathode-ray tube to make visible on a fluorescent screen the instantaneous values and waveforms of electrical quantities that are rapidly varying as a function of time or another quantity. Abbreviated CRO. Also known as oscilloscope; scope.

cathode-ray tube [ELECTR] An electron tube in which a beam of electrons can be focused to a small area and varied in position and intensity on a surface. Abbreviated CRT. Originally known as Braun tube; also known as electron-ray tube.

cathode-ray tuning indicator [ELECTR] A small cathode-ray tube having a fluorescent pattern whose size varies with the voltage applied to the grid; used in radio receivers to indicate accuracy of tuning and as a modulation indicator in some tape recorders. Also known as electric eye; electron-ray indicator; magic eye; tuning eye.

cathode resistor [ELECTR] A resistor used in the cathode circuit of a vacuum tube, having a resistance value such that the voltage drop across it due to tube current provides the correct negative grid bias for the tube.

cathode sputtering *See* sputtering.

cathodic protection [MET] Protecting a metal from electrochemical corrosion by using it as the cathode of a cell with a sacrificial anode. Also known as electrolytic protection.

cathodoluminescence [ELECTR] Luminescence produced when high-velocity electrons bombard a metal in vacuum, thus vaporizing small amounts of the metal in an excited state, which amounts emit radiation characteristic of the metal. Also known as electronoluminescence.

cation [CHEM] A positively charged atom or group of atoms, or a radical which moves to the negative pole (cathode) during electrolysis.

cation exchange [CHEM] A chemical reaction in which hydrated cations of a solid are exchanged, equivalent for equivalent, for cations of like charge in solution.

cation exchange resin [ORG CHEM] A highly polymerized synthetic organic compound consisting of a large, nondiffusible anion and a simple, diffusible cation, which later can be exchanged for a cation in the medium in which the resin is placed.

cationic detergent [CHEM] A member of a group of detergents that have molecules containing a quaternary ammonium salt cation with a group of 12 to 24 carbon atoms attached to the nitrogen atom in the cation; an example is alkyltrimethyl ammonium bromide.

catkin [BOT] An indeterminate type of inflorescence that resembles a scaly spike and sometimes is pendant.

catoptric light [OPTICS] Light reflected from a mirror, for example, light from a filament, concentrated into a parallel beam by means of a reflector.

Catostomidae [VERT ZOO] The suckers, a family of cypriniform fishes in the suborder Cyprinoidei.

cat's eye [LAP] Any of several gems cut in convex form, principally from crysoberyl, exhibiting opalescent reflections. Also known as cymophane.

CATT *See* controlled avalanche transit-time triode.

CATV *See* cable television.

catwalk [ENG] A narrow, raised platform or pathway used for passage to otherwise inaccessible areas, such as a raised walkway on a ship permitting fore and aft passage when the main deck is awash, a walkway on the roof of a freight car, or a walkway along a vehicular bridge.

Cauchy boundary conditions [MATH] The conditions imposed on a surface in euclidean space which are to be satisfied by a solution to a partial differential equation.

Cauchy dispersion formula [OPTICS] A semiempirical formula for the index of refraction n of a medium as a function of wavelength λ, according to which $n = A + (B/\lambda^2)$, where A and B are constants.

Cauchy distribution [STAT] A distribution function having the form $M/[\pi M^2 + (x - a)^2]$, where x is the variable and M and a are constants. Also known as Cauchy frequency distribution.

Cauchy formula [MATH] An expression for the value of an analytic function f at a point z in terms of a line integral

$$f(z) = \frac{1}{2\pi i} \int_C \frac{f(\zeta)}{\zeta - z} d\zeta$$

where C is a simple closed curve containing z. Also known as Cauchy integral formula.

Cauchy frequency distribution *See* Cauchy distribution.

Cauchy inequality [MATH] The square of the sum of the products of two variables for a range of values is less than or equal to the product of the sums of the squares of these two variables for the same range of values.

Cauchy integral formula *See* Cauchy formula.

Cauchy integral test *See* Cauchy's test for convergence.

Cauchy mean value theorem [MATH] If f and g are functions satisfying certain conditions on an interval $[a,b]$, then there is a point x in the interval at which the ratio of derivatives $f'(x)/g'(x)$ equals the ratio of the net change in f, $f(b) - f(a)$, to that of g.

Cauchy problem [MATH] The problem of determining the solution of a system of partial differential equation of order m from the prescribed values of the solution and of its derivatives of order less than m on a given surface.

Cauchy relations [SOLID STATE] A set of six relations between the compliance constants of a solid which should be satisfied provided the forces between atoms in the solid depend only on the distances between them and act along the lines joining them, and provided that each atom is a center of symmetry in the lattice.

Cauchy-Riemann equations [MATH] A pair of partial differential equations that is satisfied by the real and imaginary parts of a complex function $f(z)$ if and only if the function is analytic: $\partial u/\partial x = \partial v/\partial y$ and $\partial u/\partial y = - \partial v/\partial x$, where $f(z) = u + iv$ and $z = x + iy$.

Cauchy-Schwarz inequality [MATH] The square of the inner-product of two vectors does not exceed the product of the squares of their norms. Also known as Buniakowski's inequality; Schwarz' inequality.

Cauchy's test for convergence [MATH] **1.** A series is absolutely convergent if the limit as n approaches infinity of its nth term raised to the $1/n$ power is less than unity. **2.** A series a_n is convergent if there exists a monotonically decreasing function f such that $f(n) = a_n$ for n greater than some fixed number N, and if the integral of $f(x)dx$ from N to ∞ converges. Also known as Cauchy integral test; Maclaurin-Cauchy test.

Caudata [VERT ZOO] An equivalent name for Urodela.

caudate [ZOO] **1.** Having a tail or taillike appendage. **2.** Any member of the Caudata.

Caulerpaceae [BOT] A family of green algae in the order Siphonales.

caulescent [BOT] Having an aboveground stem.

cauliflower [BOT] *Brassica oleracea* var. *botrytis*. A biennial crucifer of the order Capparales grown for its edible white head or curd, which is a tight mass of flower stalks.

cauliflower disease [PL PATH] **1.** A disease of the strawberry plant caused by the eelworm and manifested as clustered, puckered, and malformed leaves. **2.** A bacterial disease of the strawberry and some other plants caused by *Corynebacterium fascians*.

caulk Also spelled calk. [ENG] To make a seam or point airtight, watertight, or steamtight by driving in caulking compound, dry pack, lead wool, or other material. [MATER] Material used to caulk seams.

caulking compound [MATER] A heavy paste, such as a synthetic, containing a polysulfide rubber and lead peroxide curing agent, or a natural product such as oakum, used for caulking.

Caulobacteraceae [MICROBIO] Formerly a family of aquatic, stalked, gram-negative bacteria in the order Pseudomonadales.

causality [MECH] In classical mechanics, the principle that the specification of the dynamical variables of a system at a given time, and of the external forces acting on the system, completely determines the values of dynamical variables at later times. Also known as determinism. [PHYS] **1.** The principle that an event cannot precede its cause; in a relativistic theory, an event cannot have an effect outside its future light cone. **2.** In relativistic quantum field theory, the principle that the field operators at different space-time points commute (for boson fields; anticommute in the case of fermion fields) if the separation of the points is spacelike. [QUANT MECH] The principle that the specification of the dynamical state of a system at a given time, and of the interaction of the system with its environment, determines the dynamical state of the system at later times, from which a probability distribution for the observation of any dynamical variable may be determined. Also known as determinism. [SCI TECH] The existence of regularities which control natural phenomena.

caustic [CHEM] **1.** Burning or corrosive. **2.** A hydroxide of a light metal. [OPTICS] A curve or surface which is tangent to the rays after reflection or refraction in an optical system. [PHYS] A curve or surface which is tangent to adjacent orthogonals to waves that have been reflected or refracted from a curved surface.

cave [ENG] A pit or tunnel under a glass furnace for collecting ashes or raking the fire. [GEOL] A natural, hollow chamber or series of chambers and galleries beneath the earth's surface, or in the side of a mountain or hill, with an opening to the surface. [MIN ENG] **1.** Fragmented rock materials, derived from the sidewalls of a borehole, that obstruct the hole or hinder drilling progress. Also known as cavings. **2.** The partial or complete failure of borehole sidewalls or mine workings. Also known as cave-in. [NUCLEO] A heavily shielded compartment in which highly radioactive material can be handled, generally by remote control. Also known as hot cell.

cave formation See speleothem.

cave-in See cave.

Cavellinidae [PALEON] A family of Paleozoic ostracods in the suborder Platycopa.

Cavendish balance [ENG] An instrument for determining the constant of gravitation, in which one measures the displacement of two small spheres of mass m, which are connected by a light rod suspended in the middle by a thin wire, caused by bringing two large spheres of mass M near them.

Caviidae [VERT ZOO] A family of large, hystricomorph rodents distinguished by a reduced number of toes and a rudimentary tail.

caving [MIN ENG] A mining procedure, used when the surface is expendable, in which the ore body is undercut and allowed to fall, breaking into small pieces that are recovered by passages (drifts) driven for that purpose; sublevel caving, block caving, and top slicing are examples.

cavings See cave.

cavitation [CHEM] Emulsification produced by disruption of a liquid into a liquid-gas two-phase system, when the hydrodynamic pressure of the liquid is reduced to the vapor pressure. [ENG] Pitting of a solid surface such as metal or concrete. [FL MECH] Formation of gas- or vapor-filled cavities within liquids by mechanical forces; broadly includes bubble formation when water is brought to a boil and effervescence of carbonated drinks; specifically, the formation of vapor-filled cavities in the interior or on the solid boundaries of vaporized liquids in motion where the pressure is reduced to a critical value without a change in ambient temperature. [PATH] The formation of one or more cavities in an organ or tissue, especially as the result of disease.

cavitation noise [ACOUS] Noise resulting from the formation of vapor- or gas-filled cavities in liquids by mechanical forces, as occurs near a propeller.

cavitation number [FL MECH] The excess of the local static pressure head over the vapor pressure head divided by the velocity head.

cavity [BIOL] A hole or hollow space in an organ, tissue, or other body part. [ELECTROMAG] See cavity resonator.

cavity charge See shaped charge.

cavity filter [ELECTROMAG] A microwave filter that uses quarter-wavelength-coupled cavities inserted in waveguides or coaxial lines to provide band-pass or other response characteristics at frequencies in the gigahertz range.

cavity radiator [THERMO] A heated enclosure with a small opening which allows some radiation to escape or enter; the escaping radiation approximates that of a blackbody.

cavity resonator [ELECTROMAG] A space totally enclosed by a metallic conductor and excited in such a way that it becomes a source of electromagnetic oscillations. Also known as cavity; microwave cavity; microwave resonance cavity; resonant cavity; resonant chamber; resonant element; rhumbatron; tuned cavity; waveguide resonator.

cavity tuning [ELECTROMAG] Use of an adjustable cavity resonator as a tuned circuit in an oscillator or amplifier, with tuning usually achieved by moving a metal plunger in or out of the cavity to change the volume, and hence the resonant frequency of the cavity.

cavity wall [BUILD] A wall constructed in two separate thicknesses with an air space between; provides thermal insulation. Also known as hollow wall.

cavy [VERT ZOO] Any of the rodents composing the family Caviidae, which includes the guinea pig, rock cavies, mountain cavies, capybara, salt desert cavy, and mara.

CAW See channel address word.

c axis [CRYSTAL] A vertically oriented crystal axis, usually the principal axis; the unique symmetry axis in tetragonal and hexagonal crystals. [GEOL] The reference axis perpendicular to the plane of movement of rock or mineral strata.

cay [GEOL] 1. A flat coral island. 2. A flat mound of sand built up on a reef slightly above high tide. 3. A small, low coastal islet or emergent reef composed largely of sand or coral.

Cayley-Klein parameters [MATH] A set of four complex numbers used to describe the orientation of a rigid body in space, or equivalently, the rotation which produces that orientation, starting from some reference orientation.

C battery [ELEC] The battery that supplies the steady bias voltage required by the control-grid electrodes of electron tubes in battery-operated equipment. Also known as grid battery.

C bias *See* grid bias.

Cc *See* cirrocumulus cloud.

CCD *See* calcite compensation depth.

CCR process *See* cyclic catalytic reforming process.

CCTV *See* closed-circuit television.

Cd *See* cadmium.

CD *See* circular dichroism.

C display [ELECTR] In radar, a rectangular display in which targets appear as blips with bearing indicated by the horizontal coordinate, and angles of elevation by the vertical coordinate.

CDM *See* code-division multiplex.

CDMA *See* code-division multiple access.

CD-4 sound *See* compatible discrete four-channel sound.

Ce *See* cerium.

Cebidae [VERT ZOO] The New World monkeys, a family of primates in the suborder Anthropoidea including the capuchins and howler monkeys.

Cebochoeridae [PALEON] A family of extinct palaeodont artiodactyls in the superfamily Entelodontoidae.

cebollite [MINERAL] $H_2Ca_4Al_2Si_3O_{16}$ A greenish to white mineral consisting of hydrous calcium aluminum silicate occurring in fibrous aggregates; hardness is 5 on Mohs scale, and specific gravity is 3.

Cebrionidae [INV ZOO] The robust click beetles, a family of cosmopolitan coleopteran insects in the superfamily Elateroidea.

Cecropidae [INV ZOO] A family of crustaceans in the suborder Caligoida which are external parasites on fish.

cecum [ANAT] The blind end of a cavity, duct, or tube, especially the sac at the beginning of the large intestine. Also spelled caecum.

cedar [BOT] The common name for a large number of evergreen trees in the order Pinales having fragrant, durable wood.

ceiling [BUILD] The covering made of plaster, boards, or other material that constitutes the overhead surface in a room. [METEOROL] In the United States, the height ascribed to the lowest layer of clouds or of obscuring phenomena when it is reported as broken, overcast, or obscuration and not classified as thin or partial.

Celastraceae [BOT] A family of dicotyledonous plants in the order Celastrales characterized by erect and basal ovules, a flower disk that surrounds the ovary at the base, and opposite or sometimes alternate leaves.

Celastrales [BOT] An order of dicotyledonous plants in the subclass Rosidae marked by simple leaves and regular flowers.

celery [BOT] *Apium graveolens* var. *dulce.* A biennial umbellifer of the order Umbellales with edible petioles or leaf stalks.

celestial body [ASTRON] Any aggregation of matter in space constituting a unit for astronomical study, as the sun, moon, a planet, comet, star, or nebula. Also known as heavenly body.

celestial coordinates [ASTRON] Any set of coordinates, such as zenithal distance, altitude, celestial latitude, celestial longitude, local hour angle, azimuth and declination, used to define a point on the celestial sphere.

celestial equator [ASTRON] The primary great circle of the celestial sphere in the equatorial system, everywhere 90° from the celestial poles; the intersection of the extended plane of the equator and the celestial sphere. Also known as equinoctial.

celestial equator system of coordinates *See* equatorial system.

celestial geodesy [GEOD] The branch of geodesy which utilizes observations of near celestial bodies and earth satellites to determine the size and shape of the earth.

celestial horizon [ASTRON] That great circle of the celestial sphere which is formed by the intersection of the celestial sphere and a plane through the center of the earth and is perpendicular to the zenith-nadir line. Also known as rational horizon.

celestial-inertial guidance [NAV] The process of directing the movements of an aircraft or spacecraft by an inertial guidance system which receives correction inputs from observations of celestial bodies. Also known as automatic celestial navigation.

celestial latitude [ASTRON] Angular distance north or south of the ecliptic; the arc of a circle of latitude between the ecliptic and a point on the celestial sphere, measured northward or southward from the ecliptic through 90°, and labeled N or S to indicate the direction of measurement. Also known as ecliptic latitude.

celestial longitude [ASTRON] Angular distance east of the vernal equinox, along the ecliptic; the arc of the ecliptic or the angle at the ecliptic pole between the circle of latitude of the vernal equinox and the circle of latitude of a point on the celestial sphere, measured eastward from the circle of latitude of the vernal equinox, through 360°. Also known as ecliptic longitude.

celestial mechanics [ASTROPHYS] The calculation of motions of celestial bodies under the action of their mutual gravitational attractions. Also known as gravitational astronomy.

celestial meridian [ASTRON] A great circle on the celestial sphere, passing through the two celestial poles and the observer's zenith.

celestial navigation *See* astronavigation.

celestial pole [ASTRON] Either of the two points of intersection of the celestial sphere and the extended axis of the earth, labeled N or S to indicate the north celestial pole or the south celestial pole.

celestial sphere [ASTRON] An imaginary sphere of indefinitely large radius, which is described about an assumed center, and upon which positions of celestial bodies are projected along radii passing through the bodies.

celestine *See* celestite.

celestite [MINERAL] $SrSO_4$ A colorless or sky-blue mineral occurring in orthorhombic, tabular crystals and in compact forms; fracture is uneven and luster is vitreous; principal ore of strontium. Also known as celestine.

cell [BIOL] The microscopic functional and structural unit of all living organisms, consisting of a nucleus, cytoplasm, and a limiting membrane. [MATH] The homeomorphic image of the unit ball. [MIN ENG] A compartment in a flotation machine. [NUCLEO] One of a set of elementary regions in a heterogeneous reactor, all of which have the same geometrical form and the same neutron characteristics. [PHYS CHEM] A cup, jar, or vessel containing electrolyte solutions and metal electrodes to produce an electric current (conductiometric or potentiometric) or for electrolysis (electrolytic).

cellar *See* push-down storage.

cell constancy [BIOL] The condition in which the entire body, or a part thereof, consists of a fixed number of cells that is the same for all adults of the species.

cell division [CYTOL] The process by which living cells multiply; may be mitotic or a amitotic.

cell lineage [EMBRYO] The developmental history of individual blastomeres from their first cleavage division to their ultimate differentiation into cells of tissues and organs.

cell membrane [CYTOL] A thin layer of protoplasm, consisting mainly of lipids and proteins, which is present on the surface of all cells. Also known as plasmalemma; plasma membrane.

cell plate [CYTOL] A membrane-bound disk formed during cytokinesis in plant cells which eventually becomes the middle lamella of the wall formed between daughter cells.

cell theory [BIOL] **1.** A principle that describes the cell as the fundamental unit of all living organisms. **2.** A principle that describes the properties of an organism as the sum of the properties of its component cells.

cellular [BIOL] Characterized by, consisting of, or pertaining to cells. [PETR] Pertaining to igneous rock having a porous texture, usually with the cavities larger than pore size and smaller than caverns.

cellular cofferdam [CIV ENG] A cofferdam consisting of interlocking steel-sheet piling driven as a series of interconnecting cells; cells may be of circular type or of straight-wall diaphragm type; space between lines of pilings is filled with sand.

cellular convection [METEOROL] An organized, convective, fluid motion characterized by the presence of distinct convection cells or convective units, usually with upward motion (away from the heat source) in the central portions of the cell, and sinking or downward flow in the cell's outer regions.

cellular glass [MATER] Sheets or blocks of thermal insulating material for walls and roofs made from pulverized glass that is heated with a gas-forming chemical at the flow temperature of glass. Also known as cellulated glass; foamed glass.

cellular horn *See* multicellular horn.

cellular splitting [COMPUT SCI] A method of adding records to a file in which the records are grouped into cells and each cell is divided into two when it becomes full.

cellular striation [ENG] Stratum of cells inside a cellular-plastic object that differs noticeably from the cell structure of the remainder of the material.

cellulase [BIOCHEM] Any of a group of extracellular enzymes, produced by various fungi, bacteria, insects, and other lower animals, that hydrolyze cellulose.

cellulated glass *See* cellular glass.

cellulitis [MED] Diffuse inflammation of a connective tissue.

cellulose [BIOCHEM] $(C_6H_{10}O_5)_n$ The main polysaccharide in living plants, forming the skeletal structure of the plant cell wall; a polymer of β-D-glucose units linked together, with the elimination of water, to form chains comprising 2000–4000 units.

cell wall [CYTOL] A semirigid, permeable structure that is composed of cellulose, lignin, or other substances and that envelops most plant cells.

celo [MECH] A unit of acceleration equal to the acceleration of a body whose velocity changes uniformly by 1 foot per second (0.3048 meter per second) in 1 second.

Celor lens system [OPTICS] An anastigmatic lens system consisting of two air-spaced achromatic doublet lenses, one on each side of the stop. Also known as Gauss lens system; Gauss objective lens.

Celsius degree [THERMO] Unit of temperature interval or difference equal to the kelvin.

Celsius temperature scale [THERMO] Temperature scale in which the temperature Θ_c in degrees Celsius (°C) is related to the temperature T_k in kelvins by the formula $\Theta_c = T_k - 273.15$; the freezing point of water at standard atmospheric pressure is very nearly 0°C and the corresponding boiling point is very nearly 100°C. Formerly known as centigrade temperature scale.

Celyphidae [INV ZOO] A family of myodarian cyclorrhaphous dipteran insects in the subsection Acalypteratae.

cement [GEOL] Any chemically precipitated material, such as carbonates, gypsum, and barite, occurring in the interstices of clastic rocks. [HISTOL] Calcified tissue which fastens the roots of teeth to the alveolus. Also known as cementum. [INV ZOO] Any of the various adhesive secretions, produced by certain invertebrates, that harden on exposure to air and are used to bind objects. [MATER] **1.** A dry powder made from silica, alumina, lime, iron oxide, and magnesia which hardens when mixed with water; used as an ingredient in concrete. **2.** An adhesive for the assembling of surfaces which are not in close contact.

cementation [CHEM] The setting of a plastic material. [ENG] **1.** Plugging a cavity or drill hole with cement. Also known as dental work. **2.** Consolidation of loose sediments or sand by injection of a chemical agent or binder. [GEOL] The precipitation of a binding material around minerals or grains in rocks. [MET] **1.** High-temperature impregnation of a metal surface with another material. **2.** Conversion of wrought iron into steel by packing layers of bars in charcoal sealed with clay and heating to 1000°C for 7–10 days.

cemented carbide *See* carbide.

cemented lens *See* compound lens.

cementite [MET] Fe_3C A hard, brittle, crystalline compound occurring as lamellae or plates in steel. Also known as iron carbide.

Cen *See* Centaurus.

Cenomanian [GEOL] Lower Upper Cretaceous geologic time.

cenote *See* pothole.

Cenozoic [GEOL] The youngest of the eras, or major subdivisions of geologic time, extending from the end of the Mesozoic Era to the present, or Recent.

cental *See* hundredweight.

α Centauri [ASTRON] A double star, the brightest in the constellation Centaurus; apart from the sun, it is the nearest bright star to earth, about 4.3 light-years away; spectral classification G2. Also known as Rigil Kent.

β Centauri [ASTRON] A first-magnitude navigational star in the constellation Centaurus; 200 light-years from the sun; spectral classification B0. Also known as Agena; Hadar.

Centaurus [ASTRON] A constellation with right ascension 13 hours, declination 50°S. Abbreviated Cen.

center [MATH] 1. The point which is equidistant from all the points on a circle or sphere. 2. For an ellipsoid or hyperboloid, the point about which the surface is symmetrical.

centerboard [NAV ARCH] A metal or wooden slab in a casing along the centerline of a sailboat which may be lowered to increase the boat's resistance to lateral motion, and raised when the boat is in shallow water or is beached.

center-feed tape [COMPUT SCI] Punched tape in which the centers of the sprocket holes are in line with the centers of the holes carrying the data or message.

center frequency *See* carrier frequency.

center of a distribution [STAT] The expected value of any random variable which has the given distribution.

center of area [MATH] For a plane figure, the center of mass of a thin uniform plate having the same boundaries as the plane figure. Also known as center of figure; centroid.

center of attraction [MECH] A point toward which a force on a body or particle (such as gravitational or electrostatic force) is always directed; the magnitude of the force depends only on the distance of the body or particle from this point.

center of buoyancy [MECH] The point through which acts the resultant force exerted on a body by a static fluid in which it is submerged or floating; located at the centroid of displaced volume.

center of figure *See* center of area; center of volume.

center of force [MECH] The point toward or from which a central force acts.

center of gravity [MECH] A fixed point in a material body through which the resultant force of gravitational attraction acts.

center of inertia *See* center of mass.

center of mass [MECH] That point of a material body or system of bodies which moves as though the system's total mass existed at the point and all external forces were applied at the point. Also known as center of inertia; centroid.

center of pressure [AERO ENG] The point in the chord of an airfoil section which is at the intersection of the chord (prolonged if necessary) and the line of action of the combined air forces (resultant air force).

center of symmetry [SCI TECH] A point in an object through which any straight line encounters exactly similar points on opposite sides. Also known as symmetry center.

center of thrust *See* thrust axis.

center of twist [MECH] A point on a line parallel to the axis of a beam through which any transverse force must be applied to avoid twisting of the section. Also known as shear center.

center of volume [MATH] For a three-dimensional figure, the center of mass of a homogeneous solid having the same boundaries as the figures. Also known as center of figure; centroid.

center punch [DES ENG] A tool similar to a prick punch but having the point ground to an angle of about 90°; used to enlarge prick-punch marks or holes.

centesis [MED] Surgical puncture, or perforation as of a tumor or membrane.

centi- [SCI TECH] A prefix representing 10^{-2}, which is 0.01 or one-hundredth. Abbreviated c.

centigrade heat unit [THERMO] A unit of heat energy, equal to 0.01 of the quantity of heat needed to raise 1 pound of air-free water from 0 to 100°C at a constant pressure of 1 standard atmosphere; equal to 1900.44 joules. Symbolized CHU; (more correctly) CHU_{mean}.

centigrade temperature scale *See* Celsius temperature scale.

centihg *See* centimeter of mercury.

centimeter [ELEC] *See* abhenry; statfarad. [MECH] A unit of length equal to 0.01 meter. Abbreviated cm.

centimeter-candle *See* phot.

centimeter-gram-second system [PHYS] An absolute system of metric units in which the centimeter, gram mass, and the second are the basic units. Abbreviated cgs system.

centimeter of mercury [MECH] A unit of pressure equal to the pressure that would support a column of mercury 1 centimeter high, having a density of 13.5951 grams per cubic centimeter, when the acceleration of gravity is equal to its standard value (980.665 centimeters per second per second); it is equal to 1333. 22387415 pascals; it differs from the decatorr by less than 1 part in 7,000,000. Abbreviated cmHg. Also known as centihg.

centipede [INV ZOO] The common name for an arthropod of the class Chilopoda.

centipoise [FL MECH] A unit of viscosity which is equal to 0.01 poise. Abbreviated cp.

centistoke [FL MECH] A cgs unit of kinematic viscosity in customary use, equal to the kinematic viscosity of a fluid having a dynamic viscosity of 1 centipoise and a density of 1 gram per cubic centimeter. Abbreviated cs.

centner *See* hundredweight.

central canal [ANAT] The small canal running through the center of the spinal cord from the conus medullaris to the lower part of the fourth ventricle; represents the embryonic neural tube.

Centrales [BOT] An order of diatoms (Bacillariophyceae) in which the form is often circular and the markings on the valves are radial.

central heating [CIV ENG] The use of a single steam or hot-water heating plant to serve a group of buildings, facilities, or even a complete community through a system of distribution pipes.

centralized data processing [COMPUT SCI] The processing of all the data concerned with a given activity at one place, usually with fixed equipment within one building.

central nervous system [ANAT] The division of the vertebrate nervous system comprising the brain and spinal cord.

central office [COMMUN] A switching unit, installed in a telephone system serving the general public, having the necessary equipment and operating arrangements for terminating and interconnecting lines and trunks. Also known as telephone central office.

central office line *See* subscriber line.

central placentation [BOT] Having the ovules located in the center of the ovary.

central processing unit [COMPUT SCI] The part of a computer containing the circuits required to interpret and execute the instructions. Abbreviated CPU. Also known as frame; main frame.

central quadric [MATH] A quadric surface that has a point about which the surface is symmetrical; namely, a sphere, ellipsoid, or hyperboloid.

central sulcus [ANAT] A groove situated about the middle of the lateral surface of the cerebral hemisphere, separating the frontal from the parietal lobe.

central terminal [COMPUT SCI] A communication device which queues tellers' requests for processing and which channels answers to the consoles originating the transactions.

central valley *See* rift valley.

Centrarchidae [VERT ZOO] A family of fishes in the order Perciformes, including the fresh-water or black basses and several sunfishes.

centrifugal [MECH] Acting or moving in a direction away from the axis of rotation or the center of a circle along which a body is moving.

centrifugal casting [ENG] A method for casting metals or forming thermoplastic resins in which the molten material solidifies in and conforms to the shape of the inner surface of a heated, rapidly rotating container.

centrifugal classifier [MECH ENG] A machine that separates particles into size groups by centrifugal force.

centrifugal clutch [MECH ENG] A clutch operated by centrifugal force from the speed of rotation of a shaft, as when heavy expanding friction shoes act on the internal surface of a rim clutch, or a flyball-type mechanism is used to activate clutching surfaces on cones and disks.

centrifugal compressor [MECH ENG] A machine in which a gas or vapor is compressed by radial acceleration in an impeller with a surrounding casing, and can be arranged multistage for high ratios of compression.

centrifugal filtration [MECH ENG] The removal of a liquid from a slurry by introducing the slurry into a rapidly rotating basket, where the solids are retained on a porous screen and the liquid is forced out of the cake by the centrifugal action.

centrifugal force [MECH] 1. An outward pseudoforce, in a reference frame that is rotating with respect to an inertial reference frame, which is equal and opposite to the centripetal force that must act on a particle stationary in the rotating frame. 2. The reaction force to a centripetal force.

centrifugal molecular still [CHEM ENG] A device used for molecular distillation; material is fed to the center of a hot, rapidly rotating cone housed in a chamber at a high vacuum; centrifugal force spreads the material rapidly over the hot surface, where the evaporable material goes off as a vapor to the condenser.

centrifugal pump [MECH ENG] A machine for moving a liquid, such as water, by accelerating it radially outward in an impeller to a surrounding volute casing.

centrifugal separation [MECH ENG] The separation of two immiscible liquids in a centrifuge within a much shorter period of time than could be accomplished solely by gravity. [NUCLEO] Separation of isotopes by spinning a mixture in gas or vapor form at high speed.

centrifugal settler [CHEM ENG] Spinning container that separates solid particles from liquids; centrifugal force causes suspended solids to move toward or away from the center of rotation, thus concentrating them in one area for removal.

centrifugal switch [MECH ENG] A switch opened or closed by centrifugal force; used on some induction motors to open the starting winding when the motor has almost reached synchronous speed.

centrifugal tachometer [MECH ENG] An instrument which measures the instantaneous angular speed of a shaft by measuring the centrifugal force on a mass rotating with it.

centrifuge [MECH ENG] 1. A rotating device for separating liquids of different specific gravities or for separating suspended colloidal particles, such as clay particles in an aqueous suspension, according to particle-size fractions by centrifugal force. 2. A large

motor-driven apparatus with a long arm, at the end of which human and animal subjects or equipment can be revolved and rotated at various speeds to simulate the prolonged accelerations encountered in rockets and spacecraft.

centrifuge microscope [OPTICS] An instrument which permits magnification and observation of living cells being centrifuged; image of the material magnified by the objective which rotates near the periphery of the centrifuge head is brought to the axis of rotation where it is observed in a stationary ocular.

centriole [CYTOL] A complex cellular organelle forming the center of the centrosome in most cells; usually found near the nucleus in interphase cells and at the spindle poles during mitosis.

centripetal [MECH] Acting or moving in a direction toward the axis of rotation or the center of a circle along which a body is moving.

centripetal acceleration [MECH] The radial component of the acceleration of a particle or object moving around a circle, which can be shown to be directed toward the center of the circle.

centripetal force [MECH] The radial force required to keep a particle or object moving in a circular path, which can be shown to be directed toward the center of the circle.

centrobaric [MECH] 1. Pertaining to the center of gravity, or to some method of locating it. 2. Possessing a center of gravity.

centrode [MECH] The path traced by the instantaneous center of a plane figure when it undergoes plane motion.

Centrohelida [INV ZOO] An order of protozoans in the subclass Heliozoia lacking a central capsule and having axopodia or filopodia, and siliceous scales and spines.

centroid *See* center of area; center of mass; center of volume.

centrolecithal ovum [CYTOL] An egg cell having the yolk centrally located; occurs in arthropods.

Centrolenidae [VERT ZOO] A family of arboreal frogs in the suborder Procoela characterized by green bones.

centromere [CYTOL] A specialized chromomere to which the spindle fibers are attached during mitosis. Also known as kinomere; primary constriction.

Centronellidina [PALEON] A suborder of extinct articulate brachiopods in the order Terebratulida.

centrosome [CYTOL] A spherical hyaline region of the cytoplasm surrounding the centriole in many cells; plays a dynamic part in mitosis as the focus of the spindle pole.

Centrospermae [BOT] An equivalent name for the Caryophyllales.

Centrospermales [BOT] An equivalent name for the Caryophyllales.

centrosphere [CYTOL] The differentiated layer of cytoplasm immediately surrounding the centriole. [GEOL] The central core of the earth. Also known as the barysphere.

centrum [ANAT] The main body of a vertebra. [BOT] The central space in hollow-stemmed plants.

Cep *See* Cepheus.

Cephalaspida [PALEON] An equivalent name for the Osteostraci.

Cephalaspidomorphi [VERT ZOO] An equivalent name for Monorhina.

cephalic [ZOO] Of or pertaining to the head or anterior end.

cephalic vein [ANAT] A superficial vein located on the lateral side of the arm which drains blood from the radial side of the hand and forearm into the axillary vein.

cephalin [BIOCHEM] Any of several acidic phosphatides whose composition is similar to that of lecithin but having ethanolamine, serine, and inositol instead of choline; found in many living tissues, especially nervous tissue of the brain.

Cephalina [INV ZOO] A suborder of protozoans in the order Eugregarinida that are parasites of certain invertebrates.

Cephalobaenida [INV ZOO] An order of the arthropod class Pentastomida composed of primitive forms with six-legged larvae.

Cephalocarida [INV ZOO] A subclass of Crustacea erected to include the primitive crustacean *Hutchinsoniella macracantha*.

Cephalochordata [VERT ZOO] A subphylum of the Chordata comprising the lancelets, including *Branchiostoma*.

Cephaloidae [INV ZOO] The false longhorn beetles, a small family of coleopteran insects in the superfamily Tenebrionoidea.

Cephalopoda [INV ZOO] Exclusively marine animals constituting the most advanced class of the Mollusca, including squids, octopuses, and *Nautilus*.

cephalothorax [INV ZOO] The body division comprising the united head and thorax of arachnids and higher crustaceans.

Cephalothrididae [INV ZOO] A family of ribbonlike worms in the order Palaeonemertini.

cepheid [ASTRON] One of a subgroup of periodic variable stars whose brightness does not remain constant with time and whose period of variation is a function of intrinsic mean brightness.

Cepheus [ASTRON] A constellation with right ascension 22 hours, declination 70°N. Abbreviated Cep.

Cephidae [INV ZOO] The stem sawflies, composing the single family of the hymenopteran superfamily Cephoidea.

Cephoidea [INV ZOO] A superfamily of hymenopteran insects in the suborder Symphyta.

Ceractinomorpha [INV ZOO] A subclass of sponges belonging to the class Demospongiae.

ceramal *See* cermet.

Cerambycidae [INV ZOO] The longhorn beetles, a family of coleopteran insects in the superfamily Chrysomeloidea.

ceramet *See* cermet.

ceramic [MATER] **1.** A product made by the baking or firing of a nonmetallic mineral, such as tile, cement, plaster refractories, and brick. **2.** Consisting of such a product.

ceramic amplifier [ELECTR] An amplifier that utilizes the piezoelectric properties of semiconductors such as silicon.

ceramic capacitor [ELEC] A capacitor whose dielectric is a ceramic material such as steatite or barium titanate, the composition of which can be varied to give a wide range of temperature coefficients.

ceramic earphones *See* crystal headphones.

ceramic reactor [NUCLEO] A nuclear reactor in which the fuel and moderator assemblies are made from high-temperature-resistant ceramic materials such as metal oxides, carbides, or nitrides.

ceramic transducer *See* electrostriction transducer.

Ceramoporidae [PALEON] A family of extinct, marine bryozoans in the order Cystoporata.

Cerapachyinae [INV ZOO] A subfamily of predacious ants in the family Formicidae, including the army ant.

Ceraphronidae [INV ZOO] A superfamily of hymenopteran insects in the superfamily Proctotrupoidea.

cerargyrite [MINERAL] AgCl A colorless to pearl-gray mineral; crystallizes in the isometric system, but crystals, usually cubic, are rare; a secondary mineral that is an ore of silver. Also known as chlorargyrite; horn silver.

Ceratiomyxaceae [MYCOL] The single family of the fungal order Ceratiomyxales.

Ceratiomyxales [MYCOL] An order of myxomycetous fungi in the subclass Ceratiomyxomycetidae.

Ceratiomyxomycetidae [MYCOL] A subclass of fungi belonging to the Myxomycetes.

Ceratodontidae [PALEON] A family of Mesozoic lungfishes in the order Dipteriformes.

Ceratomorpha [VERT ZOO] A suborder of the mammalian order Perissodactyla including the tapiroids and the rhinoceratoids.

Ceratophyllaceae [BOT] A family of rootless, free-floating dicotyledons in the order Nymphaeales characterized by unisexual flowers and whorled, cleft leaves with slender segments.

Ceratopsia [PALEON] The horned dinosaurs, a suborder of Upper Cretaceous reptiles in the order Ornithischia.

cercaria [INV ZOO] The larval generation which terminates development of a digenetic trematode in the intermediate host.

Cercopidae [INV ZOO] A family of homopteran insects belonging to the series Auchenorrhyncha.

Cercopithecidae [VERT ZOO] The Old World monkeys, a family of primates in the suborder Anthropoidea.

cere [VERT ZOO] A soft, swollen mass of tissue at the base of the upper mandible through which the nostrils open in certain birds, such as parrots and birds of prey.

cereal [BOT] Any member of the grass family (Graminae) which produces edible, starchy grains usable as food by man and his livestock. Also known as grain.

cerebellum [ANAT] The part of the vertebrate brain lying below the cerebrum and above the pons, consisting of three lobes and concerned with muscular coordination and the maintenance of equilibrium.

cerebral cortex [ANAT] The superficial layer of the cerebral hemispheres, composed of gray matter and concerned with coordination of higher nervous activity.

cerebral hemisphere [ANAT] Either of the two lateral halves of the cerebrum.

cerebral palsy [MED] Any nonprogressive motor disorder in man caused by brain damage incurred during fetal development.

cerebral peduncle [ANAT] One of two large bands of white matter (containing descending axons of upper motor neurons) which emerge from the underside of the cerebral hemispheres and approach each other as they enter the rostral border of the pons.

cerebrose *See* galactose.

cerebroside [BIOCHEM] Any of a complex group of glycosides found in nerve tissue, consisting of a hexose, a nitrogenous base, and a fatty acid. Also known as galactolipid.

cerebrospinal fluid [PHYSIO] A clear liquid that fills the ventricles of the brain and the spaces between the arachnoid mater and pia mater.

cerebrum [ANAT] The enlarged anterior or upper part of the vertebrate brain consisting of two lateral hemispheres.

cerelose *See* glucose.

Cerenkov counter [NUCLEO] An apparatus for detecting high-energy charged particles by observation of the Cerenkov radiation produced.

Cerenkov radiation [ELECTROMAG] Light emitted by a high-speed charged particle when the particle passes through a transparent, nonconducting material at a speed greater than the speed of light in the material.

Ceres [ASTRON] The largest asteroid, with a diameter of about 960 kilometers, mean distance from the sun of 2.766 astronomical units, and C-type surface composition.

ceresin [MATER] 1. A hydrocarbon wax refined from veins of wax shale known as ozocerite; used in manufacture of candles, shoe polish, electrical insulation, and floor waxes because of its great compatibility with other substances. Also known as ceresine; ozocerite. 2. A mixture of paraffin wax and beeswax, or a mixture of ozocerite and paraffin.

ceresine *See* ceresin.

Ceriantharia [INV ZOO] An order of the Zoantharia distinguished by the elongate form of the anemone-like body.

Ceriantipatharia [INV ZOO] A subclass proposed by some taxonomists to include the anthozoan orders Antipatharia and Ceriantharia.

cerite [MINERAL] $(Ca,Fe)Ce_3Si_3O_{12}\cdot H_2O$ A brown rare-earth hydrous silicate of cerium and other metals found in gneiss; hardness is 5.5 on Mohs scale, and specific gravity is 4.86.

Cerithiacea [INV ZOO] A superfamily of gastropod mollusks in the order Prosobranchia.

cerium [CHEM] A chemical element, symbol Ce, atomic number 58, atomic weight 140.12; a rare-earth metal, used as a getter in the metal industry, as an opacifier and polisher in the glass industry, in Welsbach gas mantles, in cored carbon arcs, and as a liquid-liquid extraction agent to remove fission products from spent uranium fuel.

cerium-140 [NUC PHYS] An isotope of cerium with atomic mass number of 140, 88.48% of the known amount of the naturally occurring element.

cerium-142 [NUC PHYS] A radioactive isotope of cerium with atomic mass number of 142; emits α-particles and has a half-life of 5×10^{15} years.

cerium-144 [NUC PHYS] A radioactive isotope of the element cerium with atomic mass number of 144; a beta emitter with a half-life of 285 days.

cermet [MATER] Any of a group of composite materials made by mixing, pressing, and sintering metal with ceramic; examples are silicon-silicon carbide and chromium-alumina carbide. Also known as ceramal; ceramet; metal ceramic.

Cerophytidae [INV ZOO] A small family of coleopteran insects in the superfamily Elateroidea.

certified color *See* food color.

ceruloplasmin [BIOCHEM] The copper-binding serum protein in human blood.

cerumen [PHYSIO] The waxy secretion of the ceruminous glands of the external ear. Also known as earwax.

cerussite [MINERAL] $PbCO_3$ A yellow or white member of the aragonite group occurring in orthorhombic crystals; produced by the action of carbon dioxide on lead ore.

cervantite [MINERAL] Sb_2O_4 A white or yellow secondary mineral crystallizing in the orthorhombic system and formed by oxidation of antimony sulfide.

cervical [ANAT] Of or relating to the neck, a necklike part, or the cervix of an organ.

cervical vertebra [ANAT] Any of the bones in the neck region of the vertebral column; the transverse process has a characteristic perforation by a transverse foramen.

cervicitis [MED] Inflammation of the cervix uteri.

Cervidae [VERT ZOO] A family of pecoran ruminants in the superfamily Cervoidea, characterized by solid, deciduous antlers; includes deer and elk.

cervix uteri [ANAT] The narrow outer end of the uterus.

Cervoidea [VERT ZOO] A superfamily of tylopod ruminants in infraorder Pecora, including deer, giraffes, and related species.

cesarolite [MINERAL] $H_2PbMn_3O_8$ A steel-gray mineral consisting of a hydrous lead manganate occurring in spongy masses.

cesium [CHEM] A chemical element, symbol Cs, atomic number 55, atomic weight 132.905.

cesium-134 [NUC PHYS] An isotope of cesium, atomic mass number of 134; emits negative beta particles and has a half-life of 2.19 years; used in photoelectric cells and in ion propulsion systems under development.

cesium-137 [NUC PHYS] An isotope of cesium with atomic mass number of 137; emits negative beta particles and has a half-life of 30 years; offers promise as an encapsulated radiation source for therapeutic and other purposes. Also known as radiocesium.

cesium-beam atomic clock [NUCLEO] An instrument, used as the primary standard of frequency and time, in which a microwave oscillator, which generates radiation in a microwave cavity, is maintained at a frequency such that a hyperfine transition is induced in cesium atoms in a beam passing through the cavity. Also known as cesium-beam atomic oscillator.

cesium-beam atomic oscillator *See* cesium beam atomic clock.

cespitose [BOT] 1. Tufted; growing in tufts, as grass. 2. Having short stems forming a dense turf.

Cestida [INV ZOO] An order of ribbon-shaped ctenophores having a very short tentacular axis and an elongated pharyngeal axis.

Cestoda [INV ZOO] A subclass of tapeworms including most members of the class Cestoidea; all are endoparasites of vertebrates.

Cestodaria [INV ZOO] A small subclass of worms belonging to the class Cestoidea; all are endoparasites of primitive fishes.

Cestoidea [INV ZOO] The tapeworms, endoparasites composing a class of the phylum Platyhelminthes.

Cet *See* Cetus.

Cetacea [VERT ZOO] An order of aquatic mammals, including the whales, dolphins, and porpoises.

cetane number [CHEM ENG] The percentage by volume of cetane (cetane number 100) in a blend with α-methylnaphthalene (cetane number 0); indicates the ability of a fuel to ignite quickly after being injected into the cylinder of an engine.

Cetomimiformes [VERT ZOO] An order of rare oceanic, deepwater, soft-rayed fishes that are structurally diverse.

Cetomimoidei [VERT ZOO] The whalefishes, a suborder of the Cetomimiformes, including bioluminescent, deep-sea species.

Cetorhinidae [VERT ZOO] The basking sharks, a family of large, galeoid elasmobranchs of the isurid line.

Cetus [ASTRON] A constellation with right ascension 2 hours, declination 10°S. Abbreviated Cet. Also known as Whale.

ceylonite [MINERAL] A dark-green, brown, or black iron-bearing variety of spinel. Also known as candite; pleonaste; zeylanite.

Cf *See* californium.

cgs system *See* centimeter-gram-second system.

Cha *See* Chamaeleon.

chabazite [MINERAL] $CaAl_2Si_4O_{12} \cdot 6H_2O$ A white to yellow or red member of the zeolite group occurring in glassy rhombohedral crystals; hardness is 4–5 on Mohs scale, and specific gravity is 2.08–2.16.

chad [COMPUT SCI] The piece of material removed when forming a hole or notch in a punched tape or punched card. Also known as chip. [NUCLEO] **1.** A unit of neutron flux equal to 1 neutron per square centimeter per second. **2.** A unit of neutron flux equal to 10^{12} neutrons per square centimeter per second.

chadacryst See xenocryst.

chadless tape [COMPUT SCI] Paper tape in which the perforations for code characters are made by an incomplete circular cut, with the resulting flap of material folded aside.

chad tape [COMPUT SCI] Paper tape in which perforations for code characters are completely punched out.

chaeta See seta.

Chaetetidae [PALEON] A family of Paleozoic corals of the order Tabulata.

Chaetodontidae [VERT ZOO] The butterflyfishes, a family of perciform fishes in the suborder Percoidei.

Chaetognatha [INV ZOO] A phylum of abundant planktonic arrowworms.

Chaetonotoidea [INV ZOO] An order of the class Gastrotricha characterized by two adhesive tubes connected with the distinctive paired, posterior tail forks.

Chaetophoraceae [BOT] A family of algae in the order Ulotrichales characterized as branched filaments which taper toward the apices, sometimes bearing terminal setae.

Chaetopteridea [INV ZOO] A family of spioniform polychaete annelids belonging to the Sedentaria.

chafing corrosion See fretting corrosion.

Chagas' disease [MED] An acute and chronic protozoan disease of humans caused by the hemoflagellate *Trypanosoma (Schizotrypanum) cruzi*. Also known as South American trypanosomiasis.

chain [CHEM] A structure in which similar atoms are linked by bonds. [CIV ENG] See engineer's chain; Gunter's chain. [COMMUN] A network of radio, television, radar, navigation, or other similar stations connected by special telephone lines, coaxial cables, or radio relay links so all can operate as a group for broadcast purposes, communication purposes, or determination of position. [COMPUT SCI] **1.** A series of data or other items linked together in some way. **2.** A sequence of binary digits used to construct a code. [DES ENG] **1.** A flexible series of metal links or rings fitted into one another; used for supporting, restraining, dragging, or lifting objects or transmitting power. **2.** A mesh of rods or plates connected together, used to convey objects or transmit power. [GEOL] A series of interconnected or related natural features, such as lakes, islands, or seamounts, arranged in a longitudinal sequence. [MATH] See linearly ordered set.

chain code [COMPUT SCI] A binary code consisting of a cyclic sequence of some or all of the possible binary words at a given length such that each word is derived from the previous one by moving the binary digits one position to the left, dropping the leading bit, and inserting a new bit at the end, in such a way that no word recurs before the cycle is complete.

chain command [COMPUT SCI] Any input/output command in a sequence of input/output commands such as WRITE, READ, SENSE.

chain conveyor [MECH ENG] A machine for moving materials that carries the product on one or two endless linked chains with crossbars; allows smaller parts to be added as the work passes.

chain drive [MECH ENG] A flexible device for power transmission, hoisting, or conveying, consisting of an endless chain whose links mesh with toothed wheels fastened to the driving and driven shafts.

chained block encryption [COMMUN] The use of a block cipher in which the bits of a given output block depend not only on the bits in the corresponding input block and in the key, but also on any or all prior data bits, either inputted to or produced during the enciphering or deciphering process. Also known as block chaining.

chained list [COMPUT SCI] A collection of data items arranged in a sequence so that each item contains an address giving the location of the next item in a computer storage device.

chained records [COMPUT SCI] A file of records arranged according to the chaining method.

chain grate stoker [MECH ENG] A wide, endless chain used to feed, carry, and burn a noncoking coal in a furnace, control the air for combustion, and discharge the ash.

chain homomorphism [MATH] A sequence of homomorphisms $f_n: C_n \to D_n$ between the groups of two chain complexes such that $f_{n-1} d_n = \bar{d}_n f_n$ where d_n and $\bar{d}_n$ are the boundary homomorphisms of $\{C_n\}$ and $\{D_n\}$ respectively.

chaining [COMPUT SCI] A method of storing records which are not necessarily contiguous, in which the records are arranged in a sequence and each record contains means to identify its successor.

chaining search [COMPUT SCI] A method of searching for a data item in a chained list in which an initial key is used to obtain the location of either the item sought or another item in the list, and the search then progresses through the chain until the required item is obtained or the chain is completed.

chain isomerism [ORG CHEM] A type of molecular isomerism seen in carbon compounds; as the number of carbon atoms in the molecule increases, the linkage between the atoms may be a straight chain or branched chains producing isomers that differ from each other by possessing different carbon skeletons.

chain printer [COMPUT SCI] A high-speed printer in which the type slugs are carried by the links of a revolving chain.

chain radar system [ENG] A number of radar stations located at various sites on a missile range to enable complete radar coverage during a missile flight; the stations are linked by data and communication lines for target acquisition, target positioning, or data-recording purposes.

chain reaction [CHEM] A chemical reaction in which many molecules undergo chemical reaction after one molecule becomes activated. [NUCLEO] See nuclear chain reaction.

chain rule [MATH] A rule for differentiating a composition of functions: $(d/dx) f(g(x)) = f'(g(x)) \cdot g'(x)$.

chain saw [MECH ENG] A gasoline-powered saw for felling and bucking timber, operated by one person; has cutting teeth inserted in a sprocket chain that moves rapidly around the edge of an oval-shaped blade.

chair form [PHYS CHEM] A particular nonplanar conformation of a cyclic molecule with more than five atoms in the ring; for example, in the chair form of cyclohexane, the hydrogens are staggered and directed perpendicularly to the mean plane of the carbons (axial conformation, *a*) or equatorially to the center of the mean plane (equatorial conformation, *e*).

chalaza [BOT] The region at the base of the nucellus of an ovule; gives rise to the integuments. [CYTOL] One of the paired, spiral, albuminous bands in a bird's egg that attach the yolk to the shell lining membrane at the ends of the egg.

chalazion [MED] A small tumor of the eyelid formed by retention of tarsal gland secretions. Also known as a Meibomian cyst.

chalcanthite [MINERAL] $CuSO_4 \cdot 5H_2O$ A blue to bluish-green mineral which occurs in triclinic crystals or in massive fibrous veins or stalactites. Also known as bluestone; blue vitriol.

chalcedony [MINERAL] A cryptocrystalline variety of quartz; occurs as crusts with a rounded, mammillary, or botryoidal surface and as a major constituent of nodular and bedded cherts; varieties include carnelian and bloodstone.

Chalcididae [INV ZOO] The chalcids, a family of hymenopteran insects in the superfamily Chalcidoidea.

Chalcidoidea [INV ZOO] A superfamily of hymenopteran insects in the suborder Apocrita, including primarily insect parasites.

chalcoalumite [MINERAL] $CuAl_4(SO_4)(OH)_{12} \cdot 3H_2O$ A turquoise-green to pale-blue mineral consisting of a hydrous basic sulfate of copper and aluminum.

chalcocite [MINERAL] Cu_2S A fine-grained, massive mineral with a metallic luster which tarnishes to dull black on exposure; crystallizes in the orthorhombic system, the crystals being rare and small usually with hexagonal outline as a result of twinning; hardness is 2.5–3 on Mohs scale, and specific gravity is 5.5–5.8. Also known as beta chalcocite; chalcosine; copper glance; redruthite; vitreous copper.

chalcocyanite [MINERAL] $CuSO_4$ A white mineral consisting of copper sulfate. Also known as hydrocyanite.

chalcolite *See* torbernite.

chalcomenite [MINERAL] $CuSeO_3 \cdot 2H_2O$ A blue mineral consisting of copper selenite occurring in crystals.

chalcophanite [MINERAL] $(Zn,Mn,Fe)Mn_2O_5 \cdot nH_2O$ Black mineral with metallic luster consisting of hydrous manganese and zinc oxide.

chalcophile [GEOL] Having an affinity for sulfur and therefore massing in greatest concentration in the sulfide phase of a molten mass.

chalcophyllite [MINERAL] $Cu_{18}Al_2(AsO_4)_3(OH)_{27} \cdot 33H_2O$ A green mineral consisting of basic arsenate and sulfate of copper and aluminum occurring in tabular crystals or foliated masses. Also known as copper mica.

chalcopyrite [MINERAL] $CuFeS_2$ A major ore mineral of copper; crystallizes in the tetragonal crystal system, but crystals are generally small with diphenoidal faces resembling the tetrahedron; usually massive with a metallic luster and brass-yellow color; hardness is 3.5–4 on Mohs scale, and specific gravity is 4.1–4.3. Also known as copper pyrite; yellow pyrite.

chalcosiderite [MINERAL] $Cu(Fe,Al)_6(PO_4)_4(OH)_8 \cdot 4H_2O$ A green mineral, isomorphous with turquoise, consisting of a hydrous basic phosphate of copper, iron, and aluminum.

chalice cell *See* goblet cell.

Chalicotheriidae [PALEON] A family of extinct perissodactyl mammals in the superfamily Chalicotherioidea.

Chalicotherioidea [PALEON] A superfamily of extinct, specialized perissodactyls having claws rather than hooves.

chalk [MATER] Artificially prepared pure calcium carbonate; used as the basis for pastels. Also known as whiting. [PETR] A variety of limestone formed from pelagic organisms; it is very fine-grained, porous, and friable; white or very light-colored, it consists almost entirely of calcite.

challenger *See* interrogator.

challenging signal *See* interrogation.

chalybite *See* siderite.

Chamaeleon [ASTRON] A constellation, right ascension 11 hours, declination 80°S. Abbreviated Cha. Also spelled Chameleon.

Chamaeleontidae [VERT ZOO] The chameleons, a family of reptiles in the suborder Sauria.

Chamaemyiidae [INV ZOO] The aphid flies, a family of myodarian cyclorrhaphous dipteran insects of the subsection Acalypteratae.

Chamaesiphonales [BOT] An order of blue-green algae of the class Cyanophyceae; reproduce by cell division, colony fragmentation, and endospores.

chamber [CIV ENG] The space in a canal lock between the upper and lower gates. [GRAPHICS] A sleeve or channel of a transparent film jacket. [MIN ENG] 1. The working place of a miner. 2. A body of ore with definite boundaries apparently filling a preexisting cavern. [ORD] The part of the gun in which the charge is placed; in a revolver, the hole in the cylinder; in a cannon, the space between the obturator or breechblock and the forcing cone.

chameleon [VERT ZOO] The common name for about 80 species of small-to-medium-sized lizards composing the family Chamaeleontidae.

Chameleon *See* Chamaeleon.

chamfer [ENG] To bevel a sharp edge on a machined part.

chamfering [MECH ENG] Machining operations to produce a beveled edge. Also known as beveling.

chamois [VERT ZOO] *Rupicapra rupicapra.* A goatlike mammal included in the tribe Rupicaprini of the family Bovidae.

chamosite [MINERAL] A greenish-gray or black mineral consisting of silicate belonging to the chlorite group and having monoclinic crystals; found in many oolitic iron ores.

Champlainian [GEOL] Middle Ordovician geologic time.

chance-constrained programming [COMPUT SCI] Type of nonlinear programming wherein the deterministic constraints are replaced by their probabilistic counterparts.

chancre [MED] 1. A lesion or ulcer at the site of primary inoculation by an infecting organism. 2. The initial lesion of syphilis.

chancroid [MED] A lesion of the genitalia, usually of venereal origin, caused by *Hemophilus ducreyi.* Also known as soft chancre.

Chandler motion *See* polar wandering.

Chandler wobble [GEOPHYS] A movement in the earth's axis of rotation, the period of motion being about 14 months. Also known as Eulerian nutation.

change dump [COMPUT SCI] A type of dump in which only those locations in a computer memory whose contents have changed since some previous event are copied.

changed memory routine [COMPUT SCI] A selective memory dump routine in which only those words that have been changed in the course of running a program are printed.

change gear [MECH ENG] A gear used to change the speed of a driven shaft while the speed of the driving remains constant.

change of control   [COMPUT SCI]   **1.** A break in a series of records at which processing of the records may be interrupted and some predetermined action taken. **2.** *See* jump.

change record   [COMPUT SCI]   A record that is used to alter information in a corresponding master record. Also known as amendment record; transaction record.

change tape   [COMPUT SCI]   A paper tape or magnetic tape carrying information that is to be used to update filed information; the latter is often on a master tape. Also known as transaction tape.

Chanidae   [VERT ZOO]   A monospecific family of teleost fishes in the order Gonorynchiformes which contain the milkfish (*Chanos chanos*), distinguished by the lack of teeth.

channel   [CHEM ENG]   In percolation filtration, a portion of the clay bed where there is a preponderance of flow.   [CIV ENG]   A natural or artificial waterway connecting two bodies of water or containing moving water.   [COMMUN]   **1.** A band of radio frequencies allocated for a particular purpose; a standard broadcasting channel is 10 kilohertz wide, a television channel 6 megahertz wide.   **2.** A path for carrier-current signals.   [COMPUT SCI]   **1.** A path along which digital or other information may flow in a computer.   **2.** The section of a storage medium that is accessible to a given reading station in a computer, such as a path parallel to the edge of a magnetic tape or drum or a path in a delay-line memory.   **3.** One of the longitudinal rows of intelligence holes punched along the length of paper tape. Also known as level.   [ELECTR]   **1.** A path for a signal, as an audio amplifier may have several input channels.   **2.** The main current path between the source and drain electrodes in a field-effect transistor or other semiconductor device.   [ENG]   The forming of cavities in a gear lubricant at low temperatures because of congealing.   [HYD]   The deeper portion of a waterway carrying the main current.   [NAV]   Navigable portion of a body of water.   [NUCLEO]   A passage for fuel slugs or heat-transfer fluid in a reactor.   [PETRO ENG]   In a drilling operation, a cavity appearing behind the casing because of a defect in the cement.

channel address word   [COMPUT SCI]   A four-byte code containing the protection key and the main storage address of the first channel command word at the start of an input/output operation. Abbreviated CAW.

channel bank   [ELECTR]   Part of a carrier-multiplex terminal that performs the first step of modulation of the transmitting voice frequencies into a higher-frequency band, and the final step in the demodulation of the received higher-frequency band into the received voice frequencies.

channel black   *See* gas black.

channel command   [COMPUT SCI]   The step, equivalent to a program instruction, required to tell an input/output channel what operation is to be performed, and where the data are or should be located.

channel control command   [COMPUT SCI]   An order to a control unit to perform a nondata input/output operation.

channel design   [COMPUT SCI]   The type of channel, characterized by the tasks it can perform, available to a computer.

channel effect   [ELECTR]   A leakage current flowing over a surface path between the collector and emitter in some types of transistors.

channel-end condition   [COMPUT SCI]   A signal indicating that the use of an input/output channel is no longer required.

channeling   [ANALY CHEM]   In chromatography, furrows or breaks in an ion-exchange bed which permit a solution to run through without having contact with active groups elsewhere in the bed.   [COMMUN]   A type of multiplex transmission in which the separation between communication channels is accomplished through the use of carriers or subcarriers.   [NUCLEO]   The transmission of extra particles through a medium in a nuclear reactor due to the presence of voids in the medium.   [PHYS]   The steering of energetic charged particles by the atomic rows or atomic planes of a crystalline solid.

channelizing   [COMMUN]   The process of subdividing a wide-band transmission facility so as to handle a number of different circuits requiring comparatively narrow bandwidths.

channel mask   [COMPUT SCI]   A portion of a program status word indicating which channels may interrupt the task by their completion signals.

channel noise   [COMMUN]   Bursts of interruptive pulses caused mainly by contact closures in electromagnetic equipment or by transient voltages in electric cables during transmission of signals or data.

channel program   [COMPUT SCI]   The set of steps, called channel commands, by means of which an input/output channel is controlled.

channel pulse   [COMMUN]   Telemetering pulse representing intelligence on a channel by virtue of its time or modulation characteristics.

channel sense command   [COMPUT SCI]   A command commonly used to denote an unusual condition existing in an input/output device and requesting more information.

channel shifter   [ELECTR]   Radiotelephone carrier circuit that shifts one or two voice-frequency channels from normal channels to higher voice-frequency channels to reduce cross talk between channels; the channels are shifted back by a similar circuit at the receiving end.

channel spacing   [COMMUN]   The difference in frequency between successive radio or television channels.

channel spin   [NUC PHYS]   The vector sum of the spins of the particles involved in a nuclear reaction, either before or after the reaction takes place.

channel status table   [COMPUT SCI]   A table that is set up by an executive program to show the status of the various channels that connect the central processing unit with peripheral units, enabling the program to control input/output operations.

channel status word   [COMPUT SCI]   A storage register containing the status information of the input/output operation which caused an interrupt. Abbreviated CSW.

channel synchronizer   [ELECTR]   An electronic device providing the proper interface between the central processing unit and the peripheral devices.

Channidae   [VERT ZOO]   The snakeheads, a family of fresh-water perciform fishes in the suborder Anabantoidei.

Chaoboridae   [INV ZOO]   The phantom midges, a family of dipteran insects in the suborder Orthorrhapha.

chaparral   [ECOL]   A vegetation formation characterized by woody plants of low stature, impenetrable because of tough, rigid, interlacing branches, which have simple, waxy, evergreen, thick leaves.

Chapman-Enskog theory   [STAT MECH]   A method of solving the Boltzmann transport equation by successive approximations, essentially in powers of the mean free path. Also known as Enskog theory.

Chapman equation   [GEOPHYS] A theoretical relation describing the distribution of electron density with height in the upper atmosphere.   [STAT MECH] The relationship that the viscosity of a gas equals $(0.499)mv/[\sqrt{2}\pi\sigma^2(1 + C/T)]$, where m is the mass of a molecule, v its average speed, σ its collision diameter, C the Sutherland constant, and T the absolute temperature (Kelvin scale).

Chapman region   [GEOPHYS] A hypothetical region in the upper atmosphere in which the distribution of electron density with height can be described by Chapman's theoretical equation.

Characeae   [BOT] The single family of the order Charales.

Characidae   [VERT ZOO] The characins, the single family of the suborder Characoidei.

Characoidei   [VERT ZOO] A suborder of the order Cypriniformes including fresh-water fishes with toothed jaws and an adipose fin.

character   [COMPUT SCI] 1. An elementary mark used to represent data, usually in the form of a graphic spatial arrangement of connected or adjacent strokes, such as a letter or a digit. 2. A small collection of adjacent bits used to represent a piece of data, addressed and handled as a unit, often corresponding to a digit or letter.   [GEOPHYS] A distinctive aspect of a seismic event, for example, the waveform.

character adjustment   [COMPUT SCI] An address modification affecting a specific number of characters of the address part of the instruction.

character boundary   [COMPUT SCI] In character recognition, a real or imaginary rectangle which serves as the delimiter between consecutive characters or successive lines on a source document.

character code   [COMMUN] A bit pattern assigned to a particular character in a coded character set.

character density   [COMPUT SCI] The number of characters recorded per unit of length or area. Also known as record density.

character emitter   [COMPUT SCI] In character recognition, an electromechanical device which conveys a specimen character in the form of a time pulse or group of pulses.

character fill   [COMPUT SCI] To fill one or more locations in a computer storage device by repeated insertion of some particular character, usually blanks or zeros.

character group   [MATH] The set of all continuous homomorphisms of a topological group onto the group of all complex numbers with unit norm.

characteristic   [MATH] That part of the logarithm of a number which is the integral (the whole number) to the left of the decimal point in the logarithm.

characteristic curve   [GRAPHICS] In photography, a graph that shows how increases in exposure increase the density of the film.   [MATH] 1. One of a pair of conjugate curves in a surface with the property that the directions of the tangents through any point of the curve are the characteristic directions of the surface. 2. A curve plotted on graph paper to show the relation between two changing values.

characteristic distortion   [COMMUN] 1. Displacement of signal transitions resulting from the persistence of transients caused by preceding transitions. 2. In teletypewriter transmission-systems, repetitive displacement or disruption peculiar to specific portions of a teletypewriter signal; the two types of characteristic distortions are line and equipment.

characteristic equation   [MATH] 1. Any equation which has a solution, subject to specified boundary conditions, only when a parameter occurring in it has certain values. 2. Specifically, the equation $A\mathbf{u} = \lambda\,\mathbf{u}$, which can have a solution only when the parameter λ has certain values, where A can be a square matrix which multiplies the vector $\mathbf{u}$, or a linear differential or integral operator which operates on the function $\mathbf{u}$, or in general, any linear operator operating on the vector $\mathbf{u}$ in a finite or infinite dimensional vector space. Also known as eigenvalue equation. 3. An equation which sets the characteristic polynomial of a given linear transformation on a finite dimensional vector space, or of its matrix representation, equal to zero. 4. The number preceding the decimal of a common logarithm.   [PHYS] An equation relating a set of variables, such as pressure, volume, and temperature, whose values determine a substance's physical condition.   [PL PHYS] An equation whose solutions give the frequencies and modes of those perturbations of a hydromagnetic system which decay or grow exponentially in time, and indicate regions of stability of such a system.

characteristic function   [MATH] 1. The function χ_A defined for any subset A of a set by setting $\chi_A(x) = 1$ if x is in A and $\chi_A(x) = 0$ if x is not in A. Also known as indicator function. 2. See eigenfunction.   [PHYS] A function, such as the point characteristic function or the principal function, which is the integral of some property of an optical or mechanical system over time or over the path followed by the system, and whose value for a path actually followed by a system is a maximum or a minimum with respect to nearby paths with the same end points.   [STAT] A function that uniquely defines a probability distribution; it is equal to $\sqrt{2\pi}$ times the Fourier transform of the frequency function of the distribution.

characteristic loss spectroscopy   [SPECT] A branch of electron spectroscopy in which a solid surface is bombarded with monochromatic electrons, and backscattered particles which have lost an amount of energy equal to the core-level binding energy are detected. Abbreviated CLS.

characteristic number   See eigenvalue.

characteristic overflow   [COMPUT SCI] An error condition encountered when the characteristic of a floating point number exceeds the limit imposed by the hardware manufacturer.

characteristic polynomial   [MATH] The polynomial whose roots are the eigenvalues of a given linear transformation on a finite dimensional vector space.

characteristic radiation   [ATOM PHYS] Radiation originating in an atom following removal of an electron, whose wavelength depends only on the element concerned and the energy levels involved.

characteristic ray   [MATH] For a differential equation, an integral curve which generates all the others.

characteristic root   See eigenvalue.

characteristic temperature   See Debye temperature.

characteristic underflow   [COMPUT SCI] An error condition encountered when the characteristic of a floating point number is smaller than the smallest limit imposed by the hardware manufacturer.

characteristic value   See eigenvalue.

characteristic vector   See eigenvector.

character-oriented computer   [COMPUT SCI] A computer in which the locations of individual characters, rather than words, can be addressed.

character reader   [COMPUT SCI] In character recognition, any device capable of locating, identifying, and translating into machine code the handwritten or printed data appearing on a source document.

character recognition [COMPUT SCI] The technology of using a machine to sense and encode into a machine language the characters which are originally written or printed by human beings.

character set [COMMUN] A set of unique representations called characters, for example, the 26 letters of the English alphabet, the Boolean 0 and 1, the set of signals in Morse code, and the 128 characters of the USASCII.

character skew [COMPUT SCI] In character recognition, an improper appearance of a character to be recognized, in which it appears in a tilted condition with respect to a real or imaginary horizontal base line.

character string [COMPUT SCI] A sequence of characters in a computer memory or other storage device. Also known as alphabetic string.

character string constant [COMPUT SCI] An arbitrary combination of letters, digits, and other symbols which, in the processing of nonnumeric data involving character strings, performs a function analogous to that of a numeric constant in the processing of numeric data.

character stroke See stroke.

Charadrii [VERT ZOO] The shore birds, a suborder of the order Charadriiformes.

Charadriidae [VERT ZOO] The plovers, a family of birds in the superfamily Charadrioidea.

Charadriiformes [VERT ZOO] An order of cosmopolitan birds, most of which live near water.

Charadrioidea [VERT ZOO] A superfamily of the suborder Charadrii, including plovers, sandpipers, and phalaropes.

Charales [BOT] Green algae composing the single order of the class Charophyceae.

charcoal rot [PL PATH] A fungus disease of potato, corn, and other plants caused by *Macrophomina phaseoli;* tissues of the root and lower stem are destroyed and blackened.

Chareae [BOT] A tribe of green algae belonging to the family Characeae.

charge [ELEC] 1. A basic property of elementary particles of matter; the charge of an object may be a positive or negative number or zero; only integral multiples of the proton charge occur, and the charge of a body is the algebraic sum of the charges of its constituents; the value of the charge may be inferred from the Coulomb force between charged objects. Also known as electric charge. 2. To convert electrical energy to chemical energy in a secondary battery. 3. To feed electrical energy to a capacitor or other device that can store it. [ENG] 1. A unit of an explosive, either by itself or contained in a bomb, projectile, mine, or the like, or used as the propellant for a bullet or projectile. 2. To load a borehole with an explosive. 3. The material or part to be heated by induction or dielectric heating. 4. The measurement or weight of material, either liquid, preformed, or powder, used to load a mold at one time during one cycle in the manufacture of plastics or metal. [MECH ENG] 1. In refrigeration, the quantity of refrigerant contained in a system. 2. To introduce the refrigerant into a refrigeration system. [MET] Material introduced into a furnace for melting. [NUCLEO] The fissionable material or fuel placed in a reactor to produce a chain reaction.

charge carrier [SOLID STATE] A mobile conduction electron or mobile hole in a semiconductor. Also known as carrier.

charge conjugation conservation [PARTIC PHYS] The principle that the laws of motion are left unchanged by the charge conjugation operation; it is violated by the weak interactions, but no other violations have as yet been established.

charge conjugation parity See charge parity.

charge conservation See conservation of charge.

charge-coupled devices [ELECTR] Semiconductor devices arrayed so that the electric charge at the output of one provides the input stimulus to the next.

charge-coupled memory [COMPUT SCI] A computer memory that uses a large number of charge-coupled devices for data storage and retrieval.

charge coupling [COMPUT SCI] Transfer of all electric charges within a semiconductor storage element to a similar, nearby element by means of voltage manipulations.

charge density [ELEC] The charge per unit area on a surface or per unit volume in space.

charge-density wave [SOLID STATE] The ground state of a metal in which the conduction–electron charge density is sinusoidally modulated in space.

charged particle [PARTIC PHYS] A particle whose charge is not zero; the charge of a particle is added to its designation as a superscript, with particles of charge $+1$ and -1 (in terms of the charge of the proton) denoted by $+$ and $-$ respectively; for example, π^+, Σ^-.

charge efficiency [ELEC] The efficiency of electric cell recharging.

charge exchange [PHYS] The transfer of electric charge from one particle to another during a collision between the two particles.

charge-injection device [ELECTR] A charge-transfer device used as an image sensor in which the image points are accessed by reference to their horizontal and vertical coordinates. Abbreviated CID.

charge multiplet See isospin multiplet.

charge neutrality [PL PHYS] The near equality in the density of positive and negative charges throughout a volume, which is characteristic of a plasma. [SOLID STATE] The condition in which electrons and holes are present in equal numbers in a semiconductor.

charge parity [PARTIC PHYS] The eigenvalue of the charge conjugation operation; it exists only for a system which goes into itself under this operation. Also known as charge conjugation parity.

charge quantization [ELEC] The principle that the electric charge of an object must equal an integral multiple of a universal basic charge.

charge-transfer device [ELECTR] A semiconductor device that depends upon movements of stored charges between predetermined locations, as in charge-coupled and charge-injection devices.

Charles' law [PHYS] The law that at constant pressure the volume of a fixed mass or quantity of gas varies directly with the absolute temperature; a close approximation. Also known as Gay-Lussac law.

Charles' Wain See Big Dipper.

charm [PARTIC PHYS] A quantum number which has been proposed to account for an apparent lack of symmetry in the behavior of hadrons relative to that of leptons, to explain why certain reactions of elementary particles do not occur, and to account for the longevity of the J-1 and J-2 particles.

charmed particle [PARTIC PHYS] A particle whose total charm is not equal to zero.

charmed quark [PARTIC PHYS] A quark with an electric charge of $+\frac{2}{3}$, baryon number of $\frac{1}{3}$, zero strangeness, and charm of $+1$. Symbolized c.

Charmouthian [GEOL] Middle Lower Jurassic geologic time.

charnockite [PETR] Any of various faintly foliated, nearly massive varieties of quartzofeldspathic rocks containing hypersthene.

Charon [ASTRON] The only known satellite of Pluto, with an orbital period of 6.387 days, distance from Pluto of approximately 18,000 kilometers, and diameter of approximately 1500 kilometers.

Charophyceae [BOT] A class of green algae, division Chlorophyta.

Charophyta [BOT] A group of aquatic plants, ranging in size from a few inches to several feet in height, that live entirely submerged in water.

Charpy test [MET] An impact test to determine the ductility of a metal; a freely swinging pendulum is allowed to strike and break a notched specimen laid loosely on a support; the work done by the pendulum is obtained by comparing the position of the pendulum before release with the position to which it swings after breaking the specimen.

chart datum *See* datum plane.

chart house [NAV ARCH] A room, usually adjacent to or on the bridge, where charts and other navigational equipment are stored, and where navigational computations, plots, and so on may be made. Also known as chart room.

chart recorder [ENG] A recorder in which a dependent variable is plotted against an independent variable by an ink-filled pen moving on plain paper, a heated stylus on heat-sensitive paper, a light beam or electron beam on photosensitive paper, or an electrode on electrosensitive paper. The plot may be linear or curvilinear on a strip chart recorder, or polar on a circular chart recorder.

chart room *See* chart house.

chase [BUILD] A vertical passage for ducts, pipes, or wires in a building. [DES ENG] A series of cuts, each having a path that follows the path of the cut before it; an example is a screw thread. [ENG] **1.** The main body of the mold which contains the molding cavity or cavities. **2.** The enclosure used to shrink-fit parts of a mold cavity in place to prevent spreading or distortion, or to enclose an assembly of two or more parts of a split-cavity block. **3.** To straighten and clean threads on screws or pipes. [GRAPHICS] A rectangular metal frame in which type and plates are locked for letterpress printing. [ORD] The exposed part of a gun (artillery) in front on the trunnion band or cradle.

chassis [ENG] **1.** A frame on which the body of an automobile or airplane is mounted. **2.** A frame for mounting the working parts of a radio or other electronic device.

chatoyant [MINERAL] Of a mineral or gemstone, having a changeable luster or color marked by a band of light, resembling the eye of a cat in this respect.

chatter [ELEC] Prolonged undesirable opening and closing of electric contacts. Also known as contact chatter. [ENG] An irregular alternating motion of the parts of a relief valve due to the application of pressure where contact is made between the valve disk and the seat. [ENG ACOUS] Vibration of a disk-recorder cutting stylus in a direction other than that in which it is driven.

Chattian [GEOL] Upper Oligocene geologic time. Also known as Casselian.

Chautauquan [GEOL] Upper Devonian geologic time, below Bradfordian.

chE *See* cholinesterase.

Chebyshev approximation *See* min-max technique.

Chebyshev filter [ELECTR] A filter in which the transmission frequency curve has an equal-ripple shape, with very small peaks and valleys.

Chebyshev's inequality [STAT] Given a nonnegative random variable $f(x)$, and $k > 0$, the probability that $f(x) \geqslant k$ is less than or equal to the expected value of f divided by k.

check [COMPUT SCI] A test which is necessary to detect a mistake in computer programming or a computer malfunction. [MATER] A lengthwise crack in a board.

check indicator [COMPUT SCI] A console device, usually a light, informing the operator that an error has occurred.

checking program [COMPUT SCI] A computer program which detects and determines the nature of errors in other programs, particularly those that involve incorrect coding or punching of wrong characters. Also known as checking routine.

checking routine *See* checking program.

checkpoint [COMPUT SCI] That place in a routine at which the entire state of the computer (memory, registers, and so on) is written out on auxiliary storage (tape, disk, cards) from which it may be read back into the computer if the program is to be restarted later. [NAV] Geographical location on land or water above which the position of an aircraft in flight may be determined by observation or by electronic means.

check problem *See* check routine.

check rail [BUILD] A rail, thicker than the window, that spans the opening between the top and bottom sash; usually beveled and rabbeted. [CIV ENG] *See* guard rail.

check register [COMPUT SCI] A register in which transferred data are temporarily stored so that they may be compared with a second transfer of the same data, to verify the accuracy of the transfer.

check routine [COMPUT SCI] A routine or problem designed primarily to indicate whether a fault exists in a computer, without giving detailed information on the location of the fault. Also known as check problem; test program; test routine.

check symbol [COMPUT SCI] One or more digits generated by performing an arithmetic check or summation check on a data item which are then attached to the item and copied along with it through various stages of processing, allowing the check to be repeated to verify the accuracy of the copying processes.

check valve [ENG] A device for automatically limiting flow in a piping system to a single direction.

cheetah [VERT ZOO] *Acinonyx jubatus.* A doglike carnivoran mammal belonging to the cat family, having nonretractile claws and long legs.

cheilosis [MED] Cracking at the corners of the mouth and scaling of the lips, usually associated with riboflavin deficiency.

Cheilostomata [INV ZOO] An order of ectoproct bryozoans in the class Gymnolaemata possessing delicate erect or encrusting colonies composed of loosely grouped zooecia.

chela [INV ZOO] **1.** A claw or pincer on the limbs of certain crustaceans and arachnids. **2.** A sponge spicule with talonlike terminal processes.

chelate [ORG CHEM] A molecular structure in which a heterocyclic ring can be formed by the unshared electrons of neighboring atoms.

chelating agent [ORG CHEM] An organic compound in which atoms form more than one coordinate bonds with metals in solution.

chelation [ORG CHEM] A chemical process involving formation of a heterocyclic ring compound which contains at least one metal cation or hydrogen ion in the ring.

chelicera [INV ZOO] Either appendage of the first pair in arachnids, usually modified for seizing, crushing, or piercing.

Chelicerata [INV ZOO] A subphylum of the phylum Arthropoda; chelicerae are characteristically modified as pincers.

Chelidae [VERT ZOO] The side-necked turtles, a family of reptiles in the suborder Pleurodira.

cheliform [INV ZOO] Having a forcepslike organ formed by a movable joint closing against an adjacent segment; referring especially to a crab's claw.

cheliped [INV ZOO] Either of the paired appendages bearing chelae in decapod crustaceans.

Chelonariidae [INV ZOO] A family of coleopteran insects in the superfamily Dryopoidea.

Chelonethida [INV ZOO] An equivalent name for the Pseudoscorpionida.

Chelonia [VERT ZOO] An order of the Reptilia, subclass Anapsida, including the turtles, terrapins, and tortoises.

Cheloniidae [VERT ZOO] A family of reptiles in the order Chelonia including the hawksbill, loggerhead, and green sea turtles.

Cheluridae [INV ZOO] A family of amphipod crustaceans in the suborder Gammaridea.

Chelydridae [VERT ZOO] The snapping turtles, a small family of reptiles in the order Chelonia.

chemical bond *See* bond.

chemical compound *See* compound.

chemical dating [ANALY CHEM] The determination of the relative or absolute age of minerals and of ancient objects and materials by measurement of their chemical compositions.

chemical element *See* element.

chemical energy [PHYS CHEM] Energy of a chemical compound which, by the law of conservation of energy, must undergo a change equal and opposite to the change of heat energy in a reaction; the rearrangement of the atoms in reacting compounds to produce new compounds causes a change in chemical energy.

chemical engineering [ENG] That branch of engineering serving those industries that chemically convert basic raw materials into a variety of products, and dealing with the design and operation of plants and equipment to perform such work; all products are formed in chemical processes involving chemical reactions carried out under a wide range of conditions and frequently accompanied by changes in physical state or form.

chemical equilibrium [CHEM] A condition in which a chemical reaction is occurring at equal rates in its forward and reverse directions, so that the concentrations of the reacting substances do not change with time.

chemical etching [MET] Formation of characteristic surface features when a polished metal surface is etched by suitable reagents.

chemical exchange process [CHEM] A method of separating isotopes of the lighter elements by the repetition of a process of chemical change which involves exchange of the isotopes.

chemical formula [CHEM] A notation utilizing chemical symbols and numbers to indicate the chemical composition of a pure substance; examples are CH_4 for methane and HCl for hydrogen chloride.

chemical indicator [ANALY CHEM] A substance whose physical appearance is altered at or near the end point of a chemical titration.

chemical-ion pump [CHEM ENG] A vacuum pump whose pumping action is based on evaporation of a metal whose vapor then reacts with the chemically active molecules in the gas to be evacuated.

chemical kinetics [PHYS CHEM] That branch of physical chemistry concerned with the mechanisms and rates of chemical reactions. Also known as reaction kinetics.

chemical laser *See* chemically pumped laser.

chemically pumped laser [OPTICS] A laser in which pumping is achieved by using a chemical action rather than electrical energy to produce the required pulses of light. Also known as chemical laser.

chemically pure [CHEM] Without impurities detectable by analysis. Abbreviated cp.

chemical metallurgy [MET] The science and technology of extracting metals from ores and refining them. Also known as process metallurgy.

chemical microscopy [ANALY CHEM] Application of the microscope to the solution of chemical problems.

chemical pathology [PATH] The study of disease by using chemical methods.

chemical polarity [PHYS CHEM] Tendency of a molecule, or compound, to be attracted or repelled by electrical charges because of an asymmetrical arrangement of atoms around the nucleus.

chemical potential [PHYS CHEM] In a thermodynamic system of several constituents, the rate of change of the Gibbs function of the system with respect to the change in the number of moles of a particular constituent.

chemical precipitates [GEOL] A sediment formed from precipitated materials as distinguished from detrital particles that have been transported and deposited.

chemical pulping [CHEM ENG] Separation of wood fiber for paper pulp by chemical treatment of wood chips to dissolve the lignin that cements the fibers together.

chemical reaction [CHEM] A change in which a substance (or substances) is changed into one or more new substances; there is only a minute change, Δm, in the mass of the system, given by $\Delta E = \Delta mc^2$, where ΔE is the energy emitted or absorbed and c is the speed of light.

chemical release module [AERO ENG] A shuttle-launched, free-flying spacecraft containing canisters for injecting chemicals into the upper atmosphere and for the measurement of the reactions.

chemical remanent magnetization [GEOPHYS] Permanent magnetization of rocks acquired when a magnetic material, such as hematite, is grown at low temperature throught the oxidation of some other iron mineral, such as magnetite or goethite; the growing mineral becomes magnetized in the direction of any field which is present. Abbreviated CRM.

chemical sense [PHYSIO] A process of the nervous system for reception of and response to chemical stimulation by excitation of specialized receptors.

chemical shift [PHYS CHEM] Shift in a nuclear magnetic-resonance spectrum resulting from diamagnetic shielding of the nuclei by the surrounding electrons.

chemical shim [NUCLEO] A chemical, usually boric acid, that is placed in the coolant system of a nuclear reactor to serve as a neutron absorber and that compensates for fuel burnup during normal operation.

chemical shutdown [NUCLEO] Addition of a dissolved poison to the coolant of a nuclear reactor to achieve shutdown.

chemical symbol [CHEM] A notation for one of the chemical elements, consisting of letters; for example Ne, O, C, and Na represent neon, oxygen, carbon, and sodium.

chemical thermodynamics [PHYS CHEM] The application of thermodynamic principles to problems of chemical interest.

chemical weathering [GEOCHEM] A weathering process whereby rocks and minerals are transformed into new, fairly stable chemical combinations by such chemical reactions as hydrolysis, oxidation, ion exchange, and solution. Also known as decay; decomposition.

chemiluminescence [PHYS CHEM] Emission of light as a result of a chemical reaction without an apparent change in temperature.

chemistry [SCI TECH] The scientific study of the properties, composition, and structure of matter, the changes in structure and composition of matter, and accompanying energy changes.

chemoautotroph [MICROBIO] Any of a number of autotrophic bacteria and protozoans which do not carry out photosynthesis.

chemocline [HYD] The transition in a meromictic lake between the mixolimnion layer (at the top) and the monimolimnion layer (at the bottom).

chemodifferentiation [EMBRYO] The process of cellular differentiation at the molecular level by which embryonic cells become specialized as tissues and organs.

chemoheterotroph [BIOL] An organism that derives energy and carbon from the oxidation of preformed organic compounds.

chemoreception [PHYSIO] Reception of a chemical stimulus by an organism.

chemosphere [METEOROL] The vaguely defined region of the upper atmosphere in which photochemical reactions take place; generally considered to include the stratosphere (or the top thereof) and the mesosphere, and sometimes the lower part of the thermosphere.

chemostat [MICROBIO] An apparatus, and a principle, for the continuous culture of bacterial populations in a steady state.

chemosynthesis [BIOCHEM] The synthesis of organic compounds from carbon dioxide by microorganisms using energy derived from chemical reactions.

chemotaxis [BIOL] The orientation movement of a motile organism with reference to a chemical agent.

chemotherapy [MED] Administering chemical substances for treatment of disease, especially cancer and diseases caused by parasites.

chemotropism [BIOL] Orientation response of a sessile organism with reference to chemical stimuli.

Chemungian [GEOL] Middle Upper Devonian geologic time, below Cassodagan.

chemurgy [CHEM ENG] A branch of chemistry concerned with the profitable utilization of organic raw materials, especially agricultural products, for non-food purposes such as for paints and varnishes.

chenevixite [MINERAL] $Cu_2Fe_2(AsO_4)_2(OH)_4 \cdot H_2O$ A dark-green to greenish-yellow mineral consisting of a hydrous copper iron arsenate occurring in masses.

chenic acid *See* chenodeoxycholic acid.

chenodeoxycholic acid [BIOCHEM] $C_{24}H_{40}O_4$ A constituent of bile; needlelike crystals with a melting point of 119°C; soluble in alcohol, methanol, and acetic acid; used on an experimental basis to prevent and dissolve gallstones. Also known as anthropodesoxycholic acid; chenic acid; gallodesoxycholic acid.

Chenopodiaceae [BOT] A family of dicotyledonous plants in the order Caryophyllales having reduced, mostly greenish flowers.

Chermidae [INV ZOO] A small family of minute homopteran insects in the superfamily Aphidoidea.

Cherminae [INV ZOO] A subfamily of homopteran insects in the family Chermidae; all forms have a beak and an open digestive tract.

Chernozem [GEOL] One of the major groups of zonal soils, developed typically in temperate to cool, subhumid climate; the Chernozem soils in modern classification include Borolls, Ustolls, Udolls, and Xerolls. Also spelled Tchernozem.

cherry [BOT] **1.** Any trees or shrub of the genus *Prunus* in the order Rosales. **2.** The simple, fleshy, edible drupe or stone fruit of the plant.

cherry leaf spot [PL PATH] A fungus disease of the cherry caused by *Coccomyces hiemalis;* spotting and chlorosis of the leaves occurs, with consequent retardation of tree and fruit development.

cherry picker [AERO ENG] A crane used to remove the aerospace capsule containing astronauts from the top of the rocket in the event of a malfunction. [MECH ENG] Any of several small traveling cranes, especially one used to hoist a passenger on the end of a boom. [MIN ENG] A small hoist used to facilitate car changing near the loader in a mine tunnel.

chert [PETR] A hard, dense, micro- or cryptocrystalline rock composed of chalcedony and microcrystalline quartz. Also known as hornstone; phthanite.

chertification [GEOL] A process of replacement by silica in limestone in the form of fine-grained quartz or chalcedony.

chestnut [BOT] The common name for several species of large, deciduous trees of the genus *Castanea* in the order Fagales, which bear sweet, edible nuts.

chestnut blight [PL PATH] A fungus disease of the chestnut caused by *Endothia parasitica*, which attacks the bark and cambium, causing cankers that girdle the stem and kill the plant. Also known as chestnut canker.

chestnut canker *See* chestnut blight.

Chestnut soil [GEOL] One of the major groups of zonal soils, developed typically in temperate to cool, subhumid to semiarid climate; the Chestnut soils in modern classification include Ustolls, Borolls, and Xerolls.

chevrotain [VERT ZOO] The common name for four species of mammals constituting the family Tragulidae in the order Artiodactyla. Also known as mouse deer.

Cheyne-Stokes respiration [MED] Breathing characterized by periods of hyperpnea alternating with periods of apnea; rhythmic waxing and waning of respiration; occurs most commonly in older patients with heart failure and cerebrovascular disease.

Chézy formula [FL MECH] For the velocity V of open-channel flow which is steady and uniform, $V = \sqrt{8g/f} \cdot \sqrt{mS}$, where f is the Darcy-Weisbach friction coefficient, m the hydraulic radius, S the energy dissipation per unit length, and g the acceleration of gravity.

chiasma [ANAT] A cross-shaped point of intersection of two parts, especially of the optic nerves. [CYTOL] The point of junction and fusion between paired chromatids or chromosomes, first seen during diplotene of meiosis.

Chiasmodontidae [VERT ZOO] A family of deep-sea fishes in the order Perciformes.

chiastolite [MINERAL] A variety of andalusite whose crystals have a cross-shaped appearance in cross section due to the arrangement of carbonaceous impurities. Also known as macle.

Chicago caisson [CIV ENG] A cofferdam about 4 feet (1.2 meters) in diameter lined with planks and sunk in medium-stiff clays to hard ground for pier foundations. Also known as open-well caisson.

chicken [VERT ZOO] *Galus galus*. The common domestic fowl belonging to the order Galliformes.

chickenpox [MED] A mild, highly infectious viral disease of humans caused by a herpesvirus and characterized by vesicular rash. Also known as varicella.

chicle [MATER] A gummy exudate obtained from the bark of *Achras zapota*, an evergreen tree belonging to the sapodilla family (Sapotaceae); used as the principal ingredient of chewing gum.

chicory [BOT] *Cichorium intybus*. A perennial herb of the order Campanulales grown for its edible green leaves.

chief cell [HISTOL] 1. A parenchymal, secretory cell of the parathyroid gland. 2. A cell in the lumen of the gastric fundic glands.

chigger [INV ZOO] The common name for blood-sucking larval mites of the Trombiculidae which parasitize vertebrates.

Child-Langmuir equation *See* Child's law.

Child-Langmuir-Schottky equation *See* Child's law.

Child's law [ELECTR] A law stating that the current in a thermionic diode varies directly with the three-halves power of anode voltage and inversely with the square of the distance between the electrodes, provided the operating conditions are such that the current is limited only by the space charge. Also known as Child-Langmuir equation; Child-Langmuir-Schottky equation; Langmuir-Child equation.

Chile niter *See* Chile saltpeter.

Chile saltpeter [MINERAL] Also known as Chile niter. 1. Soda niter found in large quantities in caliche in arid regions of northern Chile. 2. Deposits of sodium nitrate.

chill [MET] 1. A metal plate inserted in the surface of a sand mold or placed in the mold cavity to rapidly cool and solidify the casting, producing a hard surface. 2. White or mottled iron occurring on the surface of a rapidly cooled gray iron casting.

chill-roll extrusion [ENG] Method of extruding plastic film in which the film is cooled while being drawn around two or more highly polished chill rolls, inside of which there is cooling water. Also known as cast-film extrusion.

Chilobolbinidae [PALEON] A family of extinct ostracods in the superfamily Hollinacea showing dimorphism of the velar structure.

Chilopoda [INV ZOO] The centipedes, a class of the Myriapoda that is exclusively carnivorous and predatory.

Chimaeridae [VERT ZOO] A family of the order Chimaeriformes.

Chimaeriformes [VERT ZOO] The single order of the chondrichthyan subclass Holocephali comprising the ratfishes, marine bottom-feeders of the Atlantic and Pacific oceans.

chimera [BIOL] An organism or a part made up of tissues or cells exhibiting chimerism.

chi meson [PARTIC PHYS] A meson resonance of mass 958 MeV/c^2, designated χ_0, which has 0 isospin and charge, negative parity, positive G parity, and spin probably equal to 0. Also known as eta-prime meson (η'). Also denoted η'_A (958).

chimney [BUILD] A vertical, hollow structure of masonry, steel, or concrete, built to convey gaseous products of combustion from a building. [ELECTR] A pipelike enclosure that is placed over a heat sink to improve natural upward convection of heat and thereby increase the dissipating ability of the sink. [GEOL] *See* pipe; spouting horn.

chimney rock [GEOL] 1. A chimney-shaped remnant of a rock cliff whose sides have been cut into and carried away by waves and the gravel beach. 2. A rock column rising above its surroundings. [MATER] A porous phosphate rock used principally in chimney construction.

chimpanzee [VERT ZOO] Either of two species of Primates of the genus *Pan* indigenous to central-west Africa.

China grass *See* ramie.

chinchilla [VERT ZOO] The common name for two species of rodents in the genus *Chinchilla* belonging to the family Chinchillidae.

Chinchillidae [VERT ZOO] A family of rodents comprising the chinchillas and viscachas.

Chinese binary *See* column binary.

chinook [METEOROL] The foehn on the eastern side of the Rocky Mountains.

Chionididae [VERT ZOO] The white sheathbills, a family of birds in the order Charadriiformes.

chip [COMPUT SCI] *See* chad. [ELECTR] 1. The shaped and processed semiconductor die that is mounted on a substrate to form a transistor, diode, or other semiconductor device. 2. An integrated microcircuit performing a significant number of functions and constituting a subsystem.

chipboard [MATER] A low-density paper board made from mixed waste paper and used where strength and quality are needed.

chip circuit *See* large-scale integrated circuit.

chipmunk [VERT ZOO] The common name for 18 species of rodents belonging to the tribe Marmotini in the family Sciuridae.

chirality [CHEM] The handedness of an asymmetric molecule. [PARTIC PHYS] The characteristic of particles of spin ½ $\hbar$ that are allowed to have only one spin state with respect to an axis of quantization parallel to the particle's momentum; if the particle's spin is always parallel to its momentum, it has positive chirality; antiparallel, negative chirality. [PHYS] The characteristic of an object that cannot be superimposed upon its mirror image.

chiral twinning *See* optical twinning.

Chiridotidae [INV ZOO] A family of holothurians in the order Apodida having six-spoked, wheel-shaped spicules.

Chirodidae [PALEON] A family of extinct chondrostean fishes in the suborder Platysomoidei.

Chirognathidae [PALEON] A family of conodonts in the suborder Neurodontiformes.

chiropodist [MED] One who treats minor ailments of the feet. Also known as podiatrist.

chiropractic [MED] A system of therapeutics based upon the theory that disease is caused by abnormal function of the nervous system; attempts to restore normal function are made through manipulation and treatment of the structures of the body, especially those of the spinal column.

chiropractor [MED] One who practices the chiropractic arts.

Chiroptera [VERT ZOO] The bats, an order of mammals having the front limbs modified as wings.

Chirotheuthidae [INV ZOO] A family of mollusks comprising several deep-sea species of squids.

chirp radar [ENG] Radar in which a swept-frequency signal is transmitted, received from a target, then compressed in time to give a narrow pulse called the chirp signal.

chisel [AGR] A strong, heavy tool with curved points used for tilling; drawn by a tractor, it stirs the soil at an appreciable depth without turning it. [DES ENG] A tool for working the surface of various materials, consisting of a metal bar with a sharp edge at one end and often driven by a mallet.

Chisel *See* Caelum.

chi-square distribution [STAT] The distribution of the sample variances indicated by

$$S^2 = \sum_{i=1}^{n} (x_i - \bar{x})^2 1/(n - 1)$$

where x_1, x_2, ... x_n are observations of a random sample n drawn from a normal population.

chi-square test [STAT] A generalization, and an extension, of a test for significant differences between a binomial population and a multinomial population, wherein each observation may fall into one of several classes and furnishes a comparison among several samples instead of just two.

chitin [BIOCHEM] A white or colorless amorphous polysaccharide that forms a base for the hard outer integuments of crustaceans, insects, and other invertebrates.

chitinase [BIOCHEM] An externally secreted digestive enzyme produced by certain microorganisms and invertebrates that hydrolyzes chitin.

Chitinozoa [PALEON] An extinct group of unicellular microfossils of the kingdom Protista.

chiton [INV ZOO] The common name for over 600 extant species of mollusks which are members of the class Polyplacophora.

chlamydeous [BOT] **1.** Pertaining to the floral envelope. **2.** Having a perianth.

Chlamydiaceae [MICROBIO] The single family of the order Chlamydiales; characterized by a developmental cycle from a small elementary body to a larger initial body which divides, with daughter cells becoming elementary bodies.

Chlamydiales [MICROBIO] An order of coccoid rickettsias; gram-negative, obligate, intracellular parasites of vertebrates.

Chlamydobacteriaceae [MICROBIO] Formerly a family of gram-negative bacteria in the order Chlamydobacteriales possessing trichomes in which false branching may occur.

Chlamydobacteriales [MICROBIO] Formerly an order comprising colorless, gram-negative, algae-like bacteria of the class Schizomycetes.

Chlamydomonadidae [INV ZOO] A family of colorless, flagellated protozoans in the order Volvocida considered to be close relatives of protozoans that possess chloroplasts.

Chlamydoselachidae [VERT ZOO] The frilled sharks, a family of rare deep-water forms having a combination of hybodont and modern shark characteristics.

chlamydospore [MYCOL] A thick-walled, unicellar resting spore developed from vegetative hyphae in almost all parasitic fungi.

Chlamydozoaceae [MICROBIO] A family of small, gram-negative, coccoid bacteria in the order Rickettsiales; members are obligate intracytoplasmic parasites, or saprophytes.

Chloracea [MICROBIO] The green sulfur bacteria, a family of photosynthetic bacteria in the suborder Rhodobacteriineae.

Chlorangiaceae [BOT] A primitive family of colonial green algae belonging to the Tetrasporales in which the cells are directly attached to each other.

chlorapatite [MINERAL] $Ca_5(PO_4)_3Cl$ An apatite mineral containing chlorine.

chlordan *See* chlordane.

chlordane [ORG CHEM] $C_{10}H_6Cl_8$ A volatile liquid insecticide; a chlorinated hexahydromethanoindene. Also spelled chlordan.

chlorenchyma [BOT] Chlorophyll-containing tissue in parts of higher plants, as in leaves.

chlorhydrin *See* chlorohydrin.

chloride [CHEM] **1.** A compound which is derived from hydrochloric acid and contains the chlorine atom in the -1 oxidation state. **2.** In general, any binary compound containing chlorine.

chloride shift [PHYSIO] The reversible exchange of chloride and bicarbonate ions between erythrocytes and plasma to effect transport of carbon dioxide and maintain ionic equilibrium during respiration.

chloridization [CHEM] *See* chlorination. [MET] Treatment of mineral ores with hydrochloric acid or chlorine to form the chloride of the main metal present.

chlorination [CHEM] **1.** Introduction of chlorine into a compound. Also known as chloridization. **2.** Water sterilization by chlorine gas. [TEXT] A process in which wool is treated with a solution of hypochlorite and an acid or similar mixture to reduce the tendency of the fiber to shrink by matting.

chlorine [CHEM] A chemical element, symbol Cl, atomic number 17, atomic weight 35.453; used in manufacture of solvents, insecticides, and many non-chlorine-containing compounds, and to bleach paper and pulp.

chlorine-36 [NUCLEO] A radioactive isotope of chlorine with atomic mass number of 36; a beta emitter with a half-life of 3×10^5 years.

chlorite [INORG CHEM] A salt of chlorous acid. [MINERAL] Any of a group of greenish, platyhydrous monoclinic silicates of aluminum, ferrous iron, and magnesium which are closely associated with and resemble the micas.

chlorite schist [PETR] A metamorphic rock whose composition is dominated by members of the chlorite group.

chloritoid [MINERAL] $FeAl_4Si_2O_{10}(OH)_4$ A micaceous mineral related to the brittle mica group; has both monoclinic and triclinic modifications, a gray to green color, and weakly pleochroic crystals.

chloro- [ORG CHEM] A prefix describing an organic compound which contains chlorine atoms substituted for hydrogen.

chloroazotic acid *See* aqua regia.

Chlorobiaceae [MICROBIO] A family of bacteria in the suborder Chlorobiineae; cells are nonmotile and contain bacteriochlorophylls *c*, *d*, or *e* in chlorobium vesicles attached to the cytoplasmic membrane.

Chlorobiineae [MICROBIO] The green sulfur bacteria, a suborder of the order Rhodospirillales; contains the families Chlorobiaceae and Chloroflexaceae.

chlorobium chlorophyll [BIOCHEM] $C_{51}H_{67}O_4N_4Mg$ Either of two spectral forms of chlorophyll occurring as esters of farnesol in certain (*Chlorobium*) photosynthetic bacteria. Also known as bacterioviridin.

chlorobromide paper [GRAPHICS] A paper with an emulsion composed of silver bromide and silver chloride; used in photography for fast-speed contact paper, and medium-speed enlarging paper.

chlorocarbon [ORG CHEM] A compound of chlorine and carbon only, such as carbon tetrachloride, CCl_4.

Chlorococcales [BOT] A large, highly diverse order of unicellular or colonial, mostly fresh-water green algae in the class Chlorophyceae.

Chlorodendrineae [BOT] A suborder of colonial green algae in the order Volvocales comprising some genera with individuals capable of detachment and motility.

Chloroflexaceae [MICROBIO] A family of phototrophic bacteria in the suborder Chlorobiineae; cells possess chlorobium vesicles and bacteriochlorophyll *a* and *c*, are filamentous, and show gliding motility; capable of anaerobic, phototrophic growth or aerobic chemotrophic growth.

chloroform [ORG CHEM] CHCl₃ A colorless, sweet-smelling, nonflammable liquid; used at one time as an anesthetic. Also known as trichloromethane.

chlorohydrin [ORG CHEM] Any of the compounds derived from a group of glycols or polyhydroxy alcohols by chlorine substitution for part of the hydroxyl groups. Also spelled chlorhydrin.

chlorohydrocarbon [ORG CHEM] A carbon- and hydrogen-containing compound with chlorine substituted for some hydrogen in the molecule.

Chloromonadida [INV ZOO] An order of flattened, grass-green or colorless, flagellated protozoans of the class Phytamastigophorea.

Chloromonadina [INV ZOO] The equivalent name for Chloromonadida.

Chloromonadophyceae [BOT] A group of algae considered by some to be a class of the division Chrysophyta.

Chloromonadophyta [BOT] A division of algae in the plant kingdom considered by some to be a class, Chloromonadophyceae.

chloronitrous acid *See* aqua regia.

chloropal *See* nontronite.

Chlorophyceae [BOT] A class of microscopic or macroscopic green algae, division Chlorophyta, composed of fresh- or salt-water, unicellular or multicellular, colonial, filamentous or sheetlike forms.

chlorophyll [BIOCHEM] The generic name for any of several oil-soluble green tetrapyrrole plant pigments which function as photoreceptors of light energy for photosynthesis.

chlorophyll a [BIOCHEM] C₅₅H₇₂O₅N₄Mg A magnesium chelate of dihydroporphyrin that is esterified with phytol and has a cyclopentanone ring; occurs in all higher plants and algae.

chlorophyll b [BIOCHEM] C₅₅H₇₀O₆N₄Mg An ester similar to cholorphyll *a* but with a —CHO substituted for a —CH₃; occurs in small amounts in all green plants and algae.

Chlorophyta [BOT] The green algae, a highly diversified plant division characterized by chloroplasts, having chlorophyll *a* and *b* as the predominating pigments.

Chloropidae [INV ZOO] The chloropid flies, a family of myodarian cyclorrhaphous dipteran insects in the subsection Acalypteratae.

chloroplast [BOT] A type of cell plastid occurring in the green parts of plants, containing chlorophyll pigments, and functioning in photosynthesis and protein synthesis.

chlorotrifluoroethylene polymer [ORG CHEM] A colorless, noninflammable, heat-resistant resin, soluble in most organic solvents, and with a high impact strength can be made into transparent filling and thin sheets; used for chemical piping, fitting, and insulation for wire and cables, and in electronic components. Also known as fluorothene; polytrifluorochloroethylene resin.

choana [ANAT] A funnel-shaped opening, especially the posterior nares. [INV ZOO] A protoplasmic collar surrounding the basal ends of the flagella in certain flagellates and in the choanocytes of sponges.

Choanichthyes [VERT ZOO] An equivalent name for the Sarcopterygii.

choanocyte [INV ZOO] Any of the choanate, flagellate cells lining the cavities of a sponge. Also known as collar cell.

Choanoflagellida [INV ZOO] An order of single-celled or colonial, colorless flagellates in the class Zoomastigophorea; distinguished by a thin protoplasmic collar at the anterior end.

chock [MIN ENG] A square pillar for supporting the roof in a mine, constructed of prop timber laid up in alternate cross layers, in log-cabin style, the center being filled with waste. [NAV ARCH] 1. An open or closed metal fitting through which ropes, wires, or cables are passed. 2. A block or wedge for supporting a boat that is being repaired.

chocolate spot [PL PATH] A fungus disease of legumes caused by species of *Botrytis* and characterized by brown spots on leaves and stems, with withering of shoots.

chocolate tree *See* cacao.

Choeropotamidae [PALEON] A family of extinct palaeodont artiodactyls in the superfamily Entelodontoidae.

choke [ELEC] An inductance used in a circuit to present a high impedance to frequencies above a specified frequency range without appreciably limiting the flow of direct current. Also known as choke coil. [ELECTROMAG] A groove or other discontinuity in a waveguide surface so shaped and dimensioned as to impede the passage of guided waves within a limited frequency range. [MECH ENG] 1. To increase the fuel feed to an internal combustion engine through the action of a choke valve. 2. *See* choke valve. [ORD] A narrowing toward the muzzle in the bore of gun, hence the choked bore; often applied to shotguns. [PETRO ENG] A removable nipple inserted in a flow line to control oil or gas flow.

choke coil *See* choke.

chokedamp *See* blackdamp

choked flow [FL MECH] Flow in a duct or passage such that the flow upstream of a certain critical section cannot be increased by a reduction of downstream pressure.

choke piston [ELECTROMAG] A piston in which there is no metallic contact with the walls of the waveguide at the edges of the reflecting surface; the short circuit for high-frequency currents is achieved by a choke system. Also known as noncontacting piston; noncontacting plunger.

choke valve [MECH ENG] A valve which supplies the higher suction necessary to give the excess fuel feed required for starting a cold internal combustion engine. Also known as choke.

choking [FL MECH] The condition prevailing in compressible fluid flow when the upper limit of mass flow is reached, or when the speed of sound is reached in a duct.

cholecystectomy [MED] Surgical removal of the gallbladder and cystic duct.

cholecystography [MED] Radiography of the gallbladder following injection or ingestion of a radiopaque substance excreted in bile. Also known as Graham-Cole test.

cholecystokinin [BIOCHEM] A hormone produced by the mucosa of the upper intestine which stimulates contraction of the gallbladder.

choleglobin [BIOCHEM] Combined native protein (globin) and open-ring iron-porphyrin, which is bile pigment hemoglobin; a precursor of biliverdin.

cholelithiasis [MED] The production of or the condition associated with gallstones in the gallbladder or bile ducts.

cholera [MED] 1. An acute, infectious bacterial disease of humans caused by *Vibrio comma;* characterized by diarrhea, delirium, stupor, and coma. 2.

Any condition characterized by profuse vomiting and diarrhea.

cholesterol [BIOCHEM] $C_{27}H_{46}O$ A sterol produced by all vertebrate cells, particularly in the liver, skin, and intestine, and found most abundantly in nerve tissue.

choline [BIOCHEM] $C_5H_{15}O_2N$ A basic hygroscopic substance constituting a vitamin of the B complex; used by most animals as a precursor of acetylcholine and a source of methyl groups.

cholinergic [PHYSIO] Liberating, activated by, or resembling the physiologic action of acetylcholine.

cholinesterase [BIOCHEM] An enzyme found in blood and in various other tissues that catalyzes hydrolysis of choline esters, including acetylcholine. Abbreviated chE.

Chondrichthyes [VERT ZOO] A class of vertebrates comprising the cartilaginous, jawed fishes characterized by the absence of true bone.

chondrification [PHYSIO] Formation of or conversion into cartilage.

chondroblast [HISTOL] A cell that produces cartilage.

Chondrobrachii [VERT ZOO] The equivalent name for Ateleopoidei.

chondroclast [HISTOL] A cell that absorbs cartilage.

chondrocranium [ANAT] The part of the adult cranium derived from the cartilaginous cranium. [EMBRYO] The cartilaginous, embryonic cranium of vertebrates.

chondrocyte [HISTOL] A cartilage cell.

chondrodite [MINERAL] $Mg_5(SiO_4)_2(F_7OH)_2$ A monoclinic mineral of the humite group; has a resinous luster, is yellow-red in color, and occurs in contact-metamorphosed dolomites.

chondrodystrophy fetalis See achondroplasia.

chondroitin [BIOCHEM] A nitrogenous polysaccharide occurring in cartilage in the form of condroitin-sulfuric acid.

chondroma [MED] A benign tumor of bone, cartilage, or other tissue which simulates the structure of cartilage in its growth.

Chondrophora [INV ZOO] A suborder of polymorphic, colonial, free-floating coelenterates of the class Hydrozoa.

chondroskeleton [ANAT] 1. The parts of the bony skeleton which are formed from cartilage. 2. Cartilaginous parts of a skeleton. [VERT ZOO] A cartilaginous skeleton, as in Chondrostei.

Chondrostei [PALEON] The most archaic infraclass of the subclass Actinopterygii, or rayfin fishes.

Chondrosteidae [PALEON] A family of extinct actinopterygian fishes in the order Acipenseriformes.

chondrule [GEOL] A spherically shaped body consisting chiefly of pyroxene or olivine minerals embedded in the matrix of certain stony meteorites.

Chonetidina [PALEON] A suborder of extinct articulate brachiopods in the order Strophomenida.

Chonotrichida [INV ZOO] A small order of vase-shaped ciliates in the subclass Holotrichia; commonly found as ectocommensals on marine crustaceans.

chopper [ENG] Any knife, axe, or mechanical device for chopping or cutting an object into segments. [PHYS] A device for interrupting an electric current, beam of light, or beam of infrared radiation at regular intervals, to permit amplification of the associated electrical quantity or signal by an alternating-current amplifier; also used to interrupt a continuous stream of neutrons to measure velocity.

chopper amplifier [ELECTR] A carrier amplifier in which the direct-current input is filtered by a low-pass filter, then converted into a square-wave alternating-current signal by either one or two choppers.

chopping [ELECTR] The removal, by electronic means, of one or both extremities of a signal at a predetermined level. [PHYS] The act of interrupting an electric current, beam of light, beam of infrared radiation, or stream of neutrons at regular intervals.

chord [ACOUS] A combination of two or more tones. [AERO ENG] 1. A straight line intersecting or touching an airfoil profile at two points. 2. Specifically, that part of such a line between two points of intersection. [CIV ENG] The top or bottom, generally horizontal member of a truss. [MATH] A line segment which intersects a curve or surface only at the endpoints of the segment.

chordamesoderm [EMBRYO] The portion of the mesoderm in the chordate embryo from which the notochord and related structures arise, and which induces formation of ectodermal neural structures.

Chordata [ZOO] The highest phylum in the animal kingdom, characterized by a notochord, nerve cord, and gill slits; includes the urochordates, lancelets, and vertebrates.

Chordodidae [INV ZOO] A family of worms in the order Gordioidea distinguished by a rough cuticle containing thickenings called areoles.

chordotomy k1 [MED] Surgical division of a spinal nerve tract to relieve severe intractable pain.

chorea [MED] A nervous disorder seen as part of a syndrome following an organic dysfunction or an infection and characterized by irregular, involuntary movements of the body, especially of the face and extremities.

chorioallantois [EMBRYO] A vascular fetal membrane that is formed by the close association or fusion of the chorion and allantois.

chorion [EMBRYO] The outermost of the extraembryonic membranes of amniotes, enclosing the embryo and all of its other membranes.

chorionitis See scleroderma.

choripetalous See polypetalous.

chorisepalous See polysepalous.

Choristida [INV ZOO] An order of sponges in the class Demospongiae in which at least some of the megascleres are tetraxons.

Choristodera [PALEON] A suborder of extinct reptiles of the order Eosuchia composed of a single genus, *Champsosaurus.*

C horizon [GEOL] The portion of the parent material in soils which has been penetrated with roots.

choroid [ANAT] The highly vascular layer of the vertebrate eye, lying between the sclera and the retina.

choroid plexus [ANAT] Any of the highly vascular, folded processes that project into the third, fourth, and lateral ventricles of the brain.

Christiansen filter [OPTICS] A type of color filter, a solid-in-liquid suspension, which scatters all incident energy except that of a narrow frequency range out of the direct beam. Also known as band-pass filter.

Christmas factor [BIOCHEM] A soluble protein blood factor involved in blood coagulation. Also known as factor IX; plasma thromboplastin component (PTC).

Christmas tree [PETRO ENG] An assembly of valves, tees, crosses, and other fittings at the wellhead, used to control oil or gas production and to give access to the well tubing.

Christoffel symbols [MATH] Symbols which represent particular functions of the coefficients and their first-order derivatives of a quadratic form.

chroma [OPTICS] **1.** The dimension of the Munsell system of color that corresponds most closely to saturation, which is the degree of vividness of a hue. Also known as Munsell chroma. **2.** *See* color saturation.

chroma control [ELECTR] The control that adjusts the amplitude of the carrier chrominance signal fed to the chrominance demodulators in a color television receiver, so as to change the saturation or vividness of the hues in the color picture. Also known as color control; color-saturation control.

chromadizing [MET] Treating the surface of aluminum or aluminum alloys with chromic acid to improve paint adhesion.

Chromadoria [INV ZOO] A subclass of nematode worms in the class Adenophorea.

Chromadorida [INV ZOO] An order of principally aquatic nematode worms in the subclass Chromadoria.

Chromadoridae [INV ZOO] A family of soil and freshwater, free-living nematodes in the superfamily Chromadoroidea; generally associated with algal substances.

Chromadoroidea [INV ZOO] A superfamily of small to moderate-sized, free-living nematodes with spiral, transversely ellipsoidal amphids and a striated cuticle.

chromaffin body *See* paraganglion.

chromaffin cell [HISTOL] Any cell of the suprarenal organs in lower vertebrates, of the adrenal medulla in mammals, of the paraganglia, or of the carotid bodies that stains with chromium salts.

chromaffin system [PHYSIO] The endocrine organs and tissues of the body that secrete epinephrine; characterized by an affinity for chromium salts.

chroma oscillator [ELECTR] A crystal oscillator used in color television receivers to generate a 3.579545-megahertz signal for comparison with the incoming 3.579545-megahertz chrominance subcarrier signal being transmitted. Also known as chrominance-subcarrier oscillator; color oscillator; color-subcarrier oscillator.

chromate [INORG CHEM] CrO_4^{--} **1.** An ion derived from the unstable acid H_2CrO_4. **2.** A salt or ester of chromic acid. [MINERAL] A mineral characterized by the cation CrO_4^{--}.

Chromatiaceae [MICROBIO] A family of bacteria in the suborder Rhodospirillineae; motile cells have polar flagella, photosynthetic membranes are continuous with the cytoplasmic membrane, all except one species are anaerobic, and bacteriochlorophyll *a* or *b* is present.

chromatic aberration [ELECTR] An electron-gun defect causing enlargement and blurring of the spot on the screen of a cathode-ray tube, because electrons leave the cathode with different initial velocities and are deflected differently by the electron lenses and deflection coils. [OPTICS] An optical lens defect causing color fringes, because the lens material brings different colors of light to focus at different points. Also known as color aberration.

chromatic diagram *See* chromaticity diagram.

chromaticity [OPTICS] The color quality of light that can be defined by its chromaticity coordinates; depends only on hue and saturation of a color, and not on its luminance (brightness).

chromaticity diagram [OPTICS] A triangular graph for specifying colors, whose ordinate is the y chromaticity coordinate and whose abscissa is the x chromaticity coordinate; the apexes of the triangle represent primary colors. Also known as chromatic diagram.

chromatic parallax [OPTICS] A type of optical parallax that arises from the dependence of the position of the focal plane on the wavelength of light.

chromatic resolving power [OPTICS] The difference between two equally strong spectral lines that can barely be separated by a spectroscopic instrument, divided into the average wavelength of these two lines; for prisms and gratings Rayleigh's criteria are used, and the term is defined as the width of the emergent beam times the angular dispersion.

chromatic vision [PHYSIO] Vision pertaining to the color sense, that is, the perception and evaluation of the colors of the spectrum.

chromatid [CYTOL] **1.** One of the pair of strands formed by longitudinal splitting of a chromosome which are joined by a single centromere in somatic cells during mitosis. **2.** One of a tetrad of strands formed by longitudinal splitting of paired chromosomes during diplotene of meiosis.

chromatin [BIOCHEM] The deoxyribonucleoprotein complex forming the major portion of the nuclear material and of the chromosomes.

chromatogram [ANALY CHEM] The pattern formed by zones of separated pigments and of colorless substance in chromatographic procedures.

chromatography [ANALY CHEM] A method of separating and analyzing mixtures of chemical substances by chromatographic adsorption.

chromatophore [HISTOL] A type of pigment cell found in the integument and certain deeper tissues of lower animals that contains color granules capable of being dispersed and concentrated.

chromatosis [MED] A pathologic process or pigmentary disease in which there is a deposit of coloring matter in a normally unpigmented site, or an excessive deposit in a normally pigmented area.

chromatron [ELECTR] A single-gun color picture tube having color phosphors deposited on the screen in strips instead of dots. Also known as Lawrence tube.

chrome spinel *See* picotite.

chrome steel *See* chromium steel.

chrome-vanadium steel *See* chromium-vanadium steel.

chrominance [OPTICS] The difference between any color and a specified reference color of equal brightness; in color television, this reference color is white having coordinates $x = 0.310$ and $y = 0.316$ on the chromaticity diagram.

chrominance carrier *See* chrominance subcarrier.

chrominance demodulator [ELECTR] A demodulator used in a color television receiver for deriving the I and Q components of the chrominance signal from the chrominance signal and the chrominance-subcarrier frequency. Also known as chrominance-subcarrier demodulator.

chrominance modulator [ELECTR] A modulator used in a color television transmitter to generate the chrominance signal from the video-frequency chrominance components and the chrominance subcarrier. Also known as chrominance-subcarrier modulator.

chrominance signal [COMMUN] One of the two components, called the I signal and Q signal, that add together to produce the total chrominance signal in a color television system. Also known as carrier chrominance signal.

chrominance subcarrier [COMMUN] The 3.579545-megahertz carrier whose modulation sidebands are added to the monochrome signal to convey color information in a color television receiver. Also known as chrominance carrier; color carrier; color subcarrier; subcarrier.

chrominance-subcarrier demodulator *See* chrominance demodulator.

chrominance-subcarrier modulator *See* chrominance modulator.

chrominance-subcarrier oscillator *See* chroma oscillator.

chromite [MINERAL] $FeCr_2O_4$ A mineral of the spinel group; crystals and pure form are rare, and it usually is massive; the only important ore mineral of chromium. Also known as chrome iron ore.

chromium [CHEM] A metallic chemical element, symbol Cr, atomic number 24, atomic weight 51.996. [MET] A blue-white, hard, brittle metal used in chrome plating, in chromizing, and in many alloys.

chromium-51 [NUC PHYS] A radioactive isotope with atomic mass 51 made by neutron bombardment of chromium; radiates gamma rays.

chromium dioxide tape [ELECTR] A magnetic recording tape developed primarily to improve quality and brilliance of reproduction when used in cassettes operated at 1⅞ inches per second (4.76 centimeters per second); requires special recorders that provide high bias.

chromium molybdenum steel [MET] Cast steel containing up to 1% carbon, 0.7–1.1% chromium, and 0.2–0.4% molybdenum; characterized by high strength and ductility.

chromium steel [MET] Hard, wear-resistant steel containing chromium as the predominating alloying element. Also known as chrome steel.

chromium-vanadium steel [MET] Any of several strong, hard alloy steels containing 0.15–0.25% vanadium, 0.50–1% chromium, and 0.45–0.55% carbon. Also known as chrome-vanadium steel.

chromizing [MET] Surface-alloying of metals in which an alloy is formed by diffusion of chromium into the base metal.

chromodynamics [PARTIC PHYS] A theory of the interaction between quarks carrying color in which the quarks exchange gluons in a manner analogous to the exchange of photons between charged particles in electrodynamics.

chromolipid *See* lipochrome.

chromolithography [GRAPHICS] Lithographic printing with several colors, requiring a stone for each color.

chromomere [CYTOL] Any of the linearly arranged chromatin granules in leptotene and pachytene chromosomes and in polytene nuclei.

chromometer *See* colorimeter.

chromonema [CYTOL] The coiled core of a chromatid; it is thought to contain the genes.

chromoplast [CYTOL] Any colored cell plastid, excluding chloroplasts.

chromosome [CYTOL] Any of the complex, thread-like structures seen in animal and plant nuclei during karyokinesis which carry the linearly arranged genetic units.

chromosome aberration [GEN] Modification of the normal chromosome complement due to deletion, duplication, or rearrangement of genetic material.

chromosome complement [GEN] The species-specific, normal diploid set of chromosomes in somatic cells.

chromosome map *See* genetic map.

chromosome puff [CYTOL] Chromatic material accumulating at a restricted site on a chromosome; thought to reflect functional activity of the gene at that site during differentiation.

chromosphere [ASTRON] A transparent, tenuous layer of gas that rests on the photosphere in the atmosphere of the sun.

chron [GEOL] The time unit equivalent to the stratigraphic unit, subseries, and geologic name of a division of geologic time.

chronic [MED] Long-continued; of long duration.

chronic infectious arthritis *See* rheumatoid arthritis.

chronoamperometry [ANALY CHEM] Electroanalysis by measuring at a working electrode the rate of change of current versus time during a titration; the potential is controlled.

chronocline [PALEON] A cline shown by successive morphological changes in the members of a related group, such as a species, in successive fossiliferous strata.

chronograph [ENG] An instrument used to register the time of an event or graphically record time intervals such as the duration of an event.

chronolith *See* time-stratigraphic unit.

chronolithologic unit *See* time-stratigraphic unit.

chronology [SCI TECH] The arrangement of data in order of time of appearance.

chronometer [HOROL] 1. Any extremely accurate watch. 2. A large, strongly built timepiece that beats half seconds and is especially designed for precise timekeeping on ships at sea.

chronopotentiometry [ANALY CHEM] Electroanalysis based on the measurement at a working electrode of the rate of change in potential versus time; the current is controlled.

chronostratic unit *See* time-stratigraphic unit.

chronostratigraphic unit *See* time-stratigraphic unit.

Chroococcales [BOT] An order of blue-green algae (Cyanophyceae) that reproduce by cell division and colony fragmentation only.

Chryomyidae [INV ZOO] A family of myodarian cyclorrhaphous dipteran insects in the subsection Acalypteratae.

Chrysididae [INV ZOO] The cuckoo wasps, a family of hymenopteran insects in the superfamily Bethyloidea having brilliant metallic blue and green bodies.

chrysoberyl [MINERAL] $BeAl_2O_4$ A pale green, yellow, or brown mineral that crystallizes in the orthorhombic system and is found most commonly in pegmatite dikes; used as a gem. Also known as chrysopal; gold beryl.

Chrysochloridae [PALEON] The golden moles, a family of extinct lipotyphlan mammals in the order Insectivora.

chrysocolla [MINERAL] $CuSiO_3 \cdot 2H_2O$ A silicate mineral ordinarily occurring in impure cryptocrystalline crusts and masses with conchoidal fracture; a minor ore of copper; luster is vitreous, and color is normally emerald green to greenish-blue.

chrysolite [MINERAL] 1. A gem characterized by light-yellowish-green hues, especially the gem varieties of olivine, but also including beryl, topaz, and spinel. 2. A variety of olivine having a magnesium-iron ratio of 0.90–0.70.

Chrysomelidae [INV ZOO] The leaf beetles, a family of coleopteran insects in the superfamily Chrysomeloidea.

Chrysomeloidea [INV ZOO] A superfamily of coleopteran insects in the suborder Polyphaga.

Chrysomonadida [INV ZOO] An order of yellow to brown, flagellated colonial protozoans of the class Phytamastigophorea.

Chrysomonadina [INV ZOO] The equivalent name for the Chrysomonadida.

Chrysopetalidae [INV ZOO] A small family of scale-bearing polychaete worms belonging to the Errantia.

Chrysophyceae [BOT] Golden-brown algae making up a class of fresh- and salt-water unicellular forms in the division Chrysophyta.

Chrysophyta [BOT] The golden-brown algae, a division of plants with a predominance of carotene and xanthophyll pigments in addition to chlorophyll.

chrysoprase [MINERAL] An apple-green variety of chalcedony that contains nickel; used as a gem. Also known as green chalcedony.

chrysotile [MINERAL] $Mg_3Si_2O_5(OH)_4$ A fibrous form of serpentine that constitutes one type of asbestos.

Chthamalidae [INV ZOO] A small family of barnacles in the suborder Thoracica.

CHU See centigrade heat unit.

CHU_mean See centigrade heat unit.

chuck [DES ENG] A device for holding a component of an instrument rigid, usually by means of adjustable jaws or set screws, such as the workpiece in a metalworking or woodworking machine, or the stylus or needle of a phonograph pickup. [MET] A small bar between flask bars to secure the sand in the upper box (cope) of a flask.

chuffing See chugging.

chugging [AERO ENG] Also known as bumping; chuffing. **1.** A form of combustion instability in a rocket engine, characterized by a pulsing operation at a fairly low frequency, sometimes defined as occurring between particular frequency limits. **2.** The noise that is made in this kind of combustion. [NUCLEO] An instability in a water-moderated reactor in which the formation of steam bubbles in the core and their subsequent collapse cause oscillations in the reactivity.

churchite See weinschenkite.

churn drill [MECH ENG] Portable drilling equipment, with drilling performed by a heavy string of tools tipped with a blunt-edge chisel bit suspended from a flexible cable, to which a reciprocating motion is imparted by its suspension from an oscillating beam or sheave, causing the bit to be raised and dropped. Also known as American system drill; cable-system drill.

churn hole See pothole.

churning [FOOD ENG] A mechanical mixing process used to separate the fat phase from a fat-water system; universally used in the manufacture of butter.

chute [ENG] A conduit for conveying free-flowing materials at high velocity to lower levels. [HYD] A short channel across a narrow land area which bypasses a bend in a river; formed by the river's breaking through the land.

chyle [PHYSIO] Lymph containing emulsified fat, present in the lacteals of the intestine during digestion of ingested fats.

chyme [PHYSIO] The semifluid, partially digested food mass that is expelled into the duodenum by the stomach.

chymosin See rennin.

chymotrypsin [BIOCHEM] A proteinase in the pancreatic juice that clots milk and hydrolyzes casein and gelatin.

chymotrypsinogen [BIOCHEM] An inactive proteolytic enzyme of pancreatic juice; converted to the active form, chymotrypsin, by trypsin.

Chytridiales [MYCOL] An order of mainly aquatic fungi of the class Phycomycetes having a saclike to rhizoidal thallus and zoospores with a single posterior flagellum.

Chytridiomycetes [MYCOL] A class of true fungi.

Ci See cirrus cloud; curie.

CI See color index; cropping index; temperature humidity index.

C.I. See cast iron.

Cicadellidae [INV ZOO] Large family of homopteran insects belonging to the series Auchenorrhyncha; includes leaf hoppers.

Cicadidae [INV ZOO] A family of large homopteran insects belonging to the series Auchenorrhyncha; includes the cicadas.

cicatrix [BIOL] A scarlike mark, usually caused by previous attachment of a part or organ. [MED] The connective-tissue scar formed at the site of a healing wound.

Cichlidae [VERT ZOO] The cichlids, a family of perciform fishes in the suborder Percoidei.

Cicindelidae [INV ZOO] The tiger beetles, a family of coleopteran insects in the suborder Adephaga.

Ciconiidae [VERT ZOO] The tree storks, a family of wading birds in the order Ciconiiformes.

Ciconiiformes [VERT ZOO] An order of predominantly long-legged, long-necked birds, including herons, storks, ibises, spoonbills, and their relatives.

CID See charge-injection device.

Cidaroida [INV ZOO] An order of echinoderms in the subclass Perischoechinoidea in which the ambulacra comprise two columns of simple plates.

Cifax [COMMUN] Enciphered facsimile communication in which the output of a keyed pulse generator is mixed with the output of the facsimile converter.

Ciidae [INV ZOO] The minute, tree-fungus beetles, a family of coleopteran insects in the superfamily Cucujoidea.

cilia [ANAT] Eyelashes. [CYTOL] Relatively short, centriole-based, hairlike processes on certain anatomical cells and motile organisms.

ciliary body [ANAT] A ring of tissue lying just anterior to the retinal margin of the eye.

Ciliata [INV ZOO] The single class of the protozoan subphylum Ciliophora.

Ciliophora [INV ZOO] The ciliated protozoans, a homogeneous subphylum of the Protozoa distinguished principally by a mouth, ciliation, and infraciliature.

Cimbicidae [INV ZOO] The cimbicid sawflies, a family of hymenopteran insects in the superfamily Tenthredinoidea.

Cimicidae [INV ZOO] The bat, bed, and bird bugs, a family of flattened, wingless, parasitic hemipteran insects in the superfamily Cimicimorpha.

Cimicimorpha [INV ZOO] A superfamily, or group according to some authorities, of hemipteran insects in the subdivision Geocorisae.

Cimicoidea [INV ZOO] A superfamily of the Cimicimorpha in some newer systems of classification.

cimolite [MINERAL] $2Al_2O_3 \cdot 9SiO_3 \cdot 6H_2O$ A white, grayish, or reddish mineral consisting of hydrous aluminum silicate occurring in soft, claylike masses.

cinching [COMPUT SCI] Creases produced in magnetic tape when the supply reel is wound at low tension and suddenly stopped during playback. [GRAPHICS] The tightening of successive loops of film on a roll.

Cincinnatian [GEOL] Upper Ordovician geologic time.

Cinclidae [VERT ZOO] The dippers, a family of insect-eating songbirds in the order Passeriformes.

cinder [GEOL] Fine-grained pyroclastic material ranging in diameter from 4 to 32 millimeters. [MATER] Slag from a metal furnace. [MET] Scale cast off in forging metal.

cinder block [MATER] A hollow block made of cinder concrete. [MET] A block which closes the front of a blast furnace, containing the cinder notch.

cinder concrete [MATER] A concrete containing cinders as the aggregate.

cinder cone [GEOL] A conical elevation formed by the accumulation of volcanic debris around a vent.

C indicator See C scope.

cinematography [GRAPHICS] Motion picture photography.

cinemicrography [GRAPHICS] The photography of objects formed by a microscope, using a motion picture camera.

cinetheodolite [ENG] A surveying theodolite in which 35-millimeter motion picture cameras with lenses of 60- to 240-inch (1.5 to 6.1 meter) focal length are substituted for the surveyor's eye and telescope; used for precise time-correlated observation of distant airplanes, missiles, and artificial satellites.

Cingulata [VERT ZOO] A group of xenarthran mammals in the order Edentata, including the armadillos.

cingulate [BIOL] Having a girdle of bands or markings.

cingulum [ANAT] 1. The ridge around the base of the crown of a tooth. 2. The tract of association nerve fibers in the brain, connecting the callosal and the hippocampal convolutions. [BOT] The part of a plant between stem and root. [INV ZOO] 1. Any girdlelike structure. 2. A band of color or a raised line on certain bivalve shells. 3. The outer zone of cilia on discs of certain rotifers. 4. The clitellum in annelids.

cinnabar [MINERAL] HgS A vermilion-red mineral that crystallizes in the hexagonal system, although crystals are rare, and commonly occurs in fine, granular, massive form; the only important ore of mercury. Also known as cinnabarite; vermilion.

cinnamon [BOT] *Cinnamomum zeylanicum.* An evergreen shrub of the laurel family (Lauraceae) in the order Magnoliales; a spice is made from the bark.

cipher [COMMUN] A transposition or substitution code for transmitting secret messages.

ciphony [COMMUN] A technique by which security is accomplished by converting speech into a series of on-off pulses and mixing these with the pulses supplied by a key generator; to recover the original speech, the identical key must be subtracted and the resultant on-off pulses reconverted into the original speech pattern; unauthorized listeners are unable to reconstruct the plain text unless they have an identical key generator and the daily key setting.

Cir See Circinus.

circadian rhythm [PHYSIO] A rhythmic process within an organism occurring independently of external synchronizing signals.

circinate [BIOL] Having the form of a flat coil with the apex at the center.

Circinus [ASTRON] A constellation, right ascension 15 hours, declination 60°S. Abbreviated Cir. Also known as Compasses.

circle [MATH] 1. The set of all points in the plane at a given distance from a fixed point. 2. A unit of angular measure, equal to one complete revolution, that is, to 2π radians or 360°. Also known as turn.

circle diagram [ELEC] A diagram which gives a graphical solution of equations for a transmission line, giving the input impedance of the line as a function of load impedance and electrical length of the line.

circle of declination See hour circle.

circle of equal altitude [GEOD] A circle on the surface of the earth, on every point of which the altitude of a given celestial body is the same at a given instant; the pole of this circle is the geographical position of the body, and the great-circle distance from this pole to the circle is the zenith distance of the body.

circle of latitude [ASTRON] A great circle of the celestial sphere passing through the ecliptic poles, and hence perpendicular to the plane of the ecliptic. Also known as parallel of latitude. [GEOD] A meridian of the terrestrial sphere along which latitude is measured.

circle of longitude [ASTRON] A circle of the celestial sphere, parallel to the ecliptic. [GEOD] See parallel.

circle of right ascension See hour circle.

circle of Willis [ANAT] A ring of arteries at the base of the cerebrum.

circuit [ELEC] See electric circuit. [ELECTROMAG] A complete wire, radio, or carrier communications channel. [MATH] See cycle.

circuit breaker [ELEC] An electromagnetic device that opens a circuit automatically when the current exceeds a predetermined value.

circuit diagram [ELEC] A drawing, using standardized symbols, of the arrangement and interconnections of the conductors and components of an electrical or electronic device or installation. Also known as schematic circuit diagram; wiring diagram.

circuit element See component.

circuit interrupter [ELEC] A device in a circuit breaker to remove energy from an arc in order to extinguish it.

circuit loading [ELEC] Power drawn from a circuit by an electric measuring instrument, which may alter appreciably the quantity being measured.

circuit noise [COMMUN] In telephone practice, the noise which is brought to the receiver electrically from a telephone system, excluding noise picked up acoustically by telephone transmitters.

circuit noise level [COMMUN] Ratio of the circuit noise at that point to some arbitrary amount of circuit noise chosen as a reference; usually expressed in decibels above reference noise, signifying the reading of a circuit noise meter, or in adjusted decibels, signifying circuit noise meter reading adjusted to represent interfering effect under specified conditions.

circuitry [ELEC] The complete combination of circuits used in an electrical or electronic system or piece of equipment.

circuit shift See cyclic shift.

circuit switching [COMMUN] The method of providing communication service through a switching facility, either from local users or from other switching facilities.

circuit theory [ELEC] The mathematical analysis of conditions and relationships in an electric circuit. Also known as electric circuit theory.

circuit determinant [MATH] A determinant in which the elements of each row are the same as those of the previous row moved one place to the right, with the last element put first.

circulant matrix [MATH] A matrix in which the elements of each row are those of the previous row moved one place to the right.

circular accelerator See circular particle accelerator.

circular antenna [ELECTROMAG] A folded dipole that is bent into a circle, so the transmission line and the abutting folded ends are at opposite ends of a diameter.

circular birefringence [OPTICS] The phenomenon in which an optically active substance transmits right circularly polarized light with a different velocity from left circularly polarized light.

circular buffering [COMPUT SCI] A technique for receiving data in an input-output control system which uses a single buffer that appears to be organized in a circle, with data wrapping around it.

circular chromatography See radial chromatography.

circular cylinder [MATH] A solid bounded by two parallel planes and a cylindrical surface whose intersections with planes perpendicular to the straight lines forming the surface are circles.

circular deoxyribonucleic acid [BIOCHEM] A single- or double-stranded ring of deoxyribonucleic acid found in certain bacteriophages and in human wart virus. Also known as ring deoxyribonucleic acid.

circular dichroism [OPTICS] A change from planar to elliptic polarization when an initially plane-polarized light wave traverses an optically active medium. Abbreviated CD.

circular flow method [FL MECH] A method to determine viscosities of Newtonian fluids by measuring the torque from viscous drag of sample material between a closely spaced rotating plate–stationary cone assembly.

circular functions See trigonometric functions.

circular motion [MECH] **1.** Motion of a particle in a circular path. **2.** Motion of a rigid body in which all its particles move in circles about a common axis, fixed with respect to the body, with a common angular velocity.

circular nomograph [MATH] A chart with concentric circular scales for three variables, laid out so that any straight line passes through values of the variables satisfying a given equation.

circular paper chromatography [ANALY CHEM] A paper chromatographic technique in which migration from a spot in the sheet takes place in 360° so that zones separate as a series of concentric rings.

circular particle accelerator [NUCLEO] A particle accelerator which utilizes a magnetic field to bend charged-particle orbits and confine the extent of particle motion. Also known as circular accelerator.

circular polarization [PHYS] Attribute of a transverse wave (either of electromagnetic radiation, or in an elastic medium) whose electric or displacement vector is of constant amplitude and, at a fixed point in space, rotates in a plane perpendicular to the propagation direction with constant angular velocity.

circular saw [MECH ENG] Any of several power tools for cutting wood or metal, having a thin steel disk with a toothed edge that rotates on a spindle.

circular shift See cyclic shift.

circular sweep generation [ELECTR] The use of electronic circuits to provide voltage or current which causes an electron beam in a device such as a cathode-ray tube to move in a circular deflection path at constant speed.

circular velocity [MECH] At any specific distance from the primary, the orbital velocity required to maintain a constant-radius orbit.

circulating memory [ELECTR] A digital computer device that uses a delay line to store information in the form of a pattern of pulses in a train; the output pulses are detected electrically, amplified, reshaped, and reinserted in the delay line at the beginning. Also known as circulating storage; delay-line memory; delay-line storage.

circulating reactor [NUCLEO] A nuclear reactor in which the fissionable material circulates through the core in fluid form or as small particles suspended in a fluid.

circulating register [COMPUT SCI] A shift register in which data move out of one end and reenter the other end, as in a closed loop.

circulating storage See circulating memory.

circulating system [CHEM ENG] Fluid system in which the process fluid is taken from and pumped back into the system, as in the circulation of distillation column bottoms through an external heater.

circulation [FL MECH] The flow or motion of fluid in or through a given area or volume. [MATH] For the circulation of a vector field around a closed path, the line integral of the field vector around the path. [METEOROL] For an air mass, in the line integral of the tangential component of the velocity field about a closed curve. [OCEANOGR] A water current flow occurring within a large area, usually in a closed circular pattern. [PHYSIO] The movement of blood through defined channels and tissue spaces; movement is through a closed circuit in vertebrates and certain invertebrates.

circulation area [BUILD] The area required for human traffic in a building, including permanent corridors, stairways, elevators, escalators, and lobbies.

circulation control rotor [AERO ENG] A configuration to provide STOL capability on high-performance aircraft by means of tangential blowing over a rounded trailing edge and mass flows characteristic of turbine engine bleed.

circulation index [METEOROL] A measure of the magnitude of one of several aspects of large-scale atmospheric circulation patterns; indices most frequently measured represent the strength of the zonal (east-west) or meridional (north-south) components of the wind, at the surface or at upper levels, usually averaged spatially and often averaged in time.

circulation pattern [METEOROL] The general geometric configuration of atmospheric circulation usually applied, in synoptic meteorology, to the large-scale features of synoptic charts and mean charts.

circulator [ELECTROMAG] A waveguide component having a number of terminals so arranged that energy entering one terminal is transmitted to the next adjacent terminal in a particular direction. Also known as microwave circulator.

circulatory system [ANAT] The vessels and organs composing the lymphatic and cardiovascular systems.

circumduction [ANAT] Movement of the distal end of a body part in the form of an arc; performed at ball-and-socket and saddle joints.

circumference [MATH] **1.** The length of a circle. **2.** For a sphere, the length of any great circle on the sphere.

circummeridian altitude See exmeridian altitude.

circumpolar [ASTRON] Revolving about the elevated pole without setting. [GEOGR] Located around one of the polar regions of earth.

circumpolar westerlies See westerlies.

circumscribed [MATH] **1.** A closed curve (or surface) is circumscribed about a polygon (or polyhedron) if every vertex of the polygon (or polyhedron) is incident upon the curve (or surface) and the polygon (or polyhedron) is contained in the curve (or surface). **2.** A polygon (or polyhedron) is circumscribed about a closed curve (or surface) if every side of the polygon (or face of the polyhedron) is tangent to the curve (or surface) and the curve (or surface) is contained within the polygon (or polyhedron).

Cirolanidae [INV ZOO] A family of isopod crustaceans in the suborder Flabellifera composed of actively swimming predators and scavengers with biting mouthparts.

cirque [GEOL] A steep elliptic to elongated enclave high on mountains in calcareous districts, usually forming the blunt end of a valley.

Cirratulidae [INV ZOO] A family of fringe worms belonging to the Sedentaria which are important detritus feeders in coastal waters.

cirrhosis [MED] A progressive, inflammatory disease of the liver characterized by a real or apparent increase in the proportion of hepatic connective tissue.

cirriform [METEOROL] Descriptive of clouds composed of small particles, mostly ice crystals, which are fairly widely dispersed, usually resulting in relative transparency and whiteness and often producing halo phenomena not observed with other cloud forms. [ZOO] Having the form of a cirrus; generally applied to a prolonged, slender process.

Cirripedia [INV ZOO] A subclass of the Crustacea, including the barnacles and goose barnacles; individuals are free-swimming in the larval stages but permanently fixed in the adult stage.

cirrocumulus cloud [METEOROL] A principal cloud type, appearing as a thin, white path of cloud without shadows, composed of very small elements in the form of grains, ripples, and so on. Abbreviated Cc.

Cirromorpha [INV ZOO] A suborder of cephalopod mollusks in the order Octopoda.

cirrostratus cloud [METEOROL] A principal cloud type, appearing as a whitish veil, usually fibrous but sometimes smooth, which may totally cover the sky and often produces halo phenomena, either partial or complete. Abbreviated Cs.

cirrus [INV ZOO] 1. The conical locomotor structure composed of fused cilia in hypotrich protozoans. 2. Any of the jointed thoracic appendages of barnacles. 3. Any hairlike tuft on insect appendages. 4. The male copulatory organ in some mollusks and trematodes. [VERT ZOO] Any of the tactile barbels of certain fishes. [ZOO] A tendrillike animal appendage.

cirrus cloud [METEOROL] A principal cloud type composed of detached cirriform elements in the form of white, delicate filaments, of white (or mostly white) patches, or narrow bands. Abbreviated Ci.

cis [ORG CHEM] A descriptive term indicating a form of isomerism in which atoms are located on the same side of an asymmetric molecule.

cislunar [ASTRON] Of or pertaining to phenomena, projects, or activity in the space between the earth and moon, or between the earth and the moon's orbit.

cissoid [MATH] A plane curve consisting of all points which lie on a variable line passing through a fixed point on a circle, and whose distance from the fixed point is equal to the distance from the line's intersection with the circle to its intersection with the tangent to the circle at the point diametrically opposite the fixed point; in cartesian coordinates the equation is $y^2(2a - x) = x^3$.

cistern [ANAT] A closed, fluid-filled sac or vesicle, such as the subarachnoid spaces or the vesicles comprising the dictyosomes of a Golgi apparatus. [CIV ENG] A tank for storing water or other liquid. [GEOL] A hollow that holds water.

cistern barometer [ENG] A pressure-measuring device in which pressure is read by the liquid rise in a vertical, closed-top tube as a result of system pressure on a liquid reservoir (cistern) into which the bottom, open end of the tube is immersed.

cis-trans isomerism [ORG CHEM] A type of geometrical isomerism found in alkenic systems in which it is possible for each of the doubly bonded carbons to carry two different atoms or groups; two similar atoms or groups may be on the same side (cis) or on opposite sides (trans) of a plane bisecting the alkenic carbons and perpendicular to the plane of the alkenic system.

cistron [MOL BIO] The genetic unit (deoxyribonucleic acid fragment) that codes for a particular polypeptide; mutants do not complement each other within a cistron.

Citheroniinae [INV ZOO] Subfamily of lepidopteran insects in the family Saturniidae, including the regal moth and the imperial moth.

citizens' band [COMMUN] A frequency band allocated for citizens' radio service (460–470 or 26.965–27.405 megahertz). Also known as citizens' waveband.

citizens' radio service [COMMUN] A radio communication service intended for private or personal radio communication, including radio signaling and control of objects by radio.

citizens' waveband *See* citizens' band.

citrate test [MICROBIO] A differential cultural test to identify genera within the bacterial family Enterobacteriaceae that are able to utilize sodium citrate as a sole source of carbon.

citric acid [BIOCHEM] $C_6H_8O_7 \cdot H_2O$ A colorless crystalline or white powdery organic, tricarboxylic acid occurring in plants, especially citrus fruits, and used as a flavoring agent, as an antioxidant in foods, and as a sequestering agent; the commercially produced form melts at 153°C.

citric acid cycle *See* Krebs cycle.

citrine [MINERAL] An important variety of crystalline quartz, yellow to brown in color and transparent. Also known as Bohemian topaz; false topaz; quartz topaz; topaz quartz; yellow quartz.

citron [BOT] *Citrus medica*. A shrubby, evergreen citrus tree in the order Sapindales cultivated for its edible, large, lemonlike fruit.

citronella [BOT] *Cymbopogon nardus*. A tropical grass; the source of citronella oil.

citrus anthracnose [PL PATH] A fungus disease of citrus plants caused by *Colletotrichum gloeosporioides* and characterized by tip blight, stains on the leaves, and spots, stains, or rot on the fruit.

citrus blast [PL PATH] A bacterial disease of citrus trees caused by *Pseudomonas syringae* and marked by drying and browning of foliage and twigs and black pitting of the fruit.

citrus canker [PL PATH] A bacterial disease of citrus plants caused by *Xanthomonas citri* and producing lesions on twigs, foliage, and fruit.

citrus flavanoid compound *See* bioflavanoid.

citrus fruit [BOT] Any of the edible fruits having a pulpy endocarp and a firm exocarp that are produced by plants of the genus *Citrus* and related genera.

citrus scab [PL PATH] A fungus disease of citrus plants caused by *Sphaceloma rosarum*, producing scablike lesions on all plant parts.

city plan [MAP] A large-scale, comprehensive map of a city delineating streets, important buildings, and other urban elements; relief is shown as important. Also known as town plan.

civet [PHYSIO] A fatty substance secreted by the civet gland; used as a fixative in perfumes. [VERT ZOO] Any of 18 species of catlike, nocturnal carnivores assigned to the family Viverridae, having a long head, pointed muzzle, and short limbs with nonretractile claws.

civet cat *See* cacomistle.

civil airway [NAV] An airway designated for air commerce.

civil day [ASTRON] A mean solar day beginning at midnight instead of at noon; may be based on either apparent solar time or mean solar time.

civil engineering [ENG] The planning, design, construction, and maintenance of fixed structures and ground facilities for industry, transportation, use and control of water, or occupancy.

civil time [ASTRON] Solar time in a day (civil day) that begins at midnight; may be either apparent solar time or mean solar time.

civil twilight [ASTRON] The interval of incomplete darkness between sunrise (or sunset) and the time when the center of the sun's disk is 6° below the horizon.

civil year *See* calendar year.

Cl *See* chlorine.

cladding [ENG] Process of covering one material with another and bonding them together under high pressure and temperature. Also known as bonding. [NUCLEO] An outer jacket, usually metallic, for a nuclear fuel element; prevents corrosion of fuel and release of fission products into the coolant.

Cladistia [VERT ZOO] The equivalent name for Polypteriformes.

Cladocera [INV ZOO] An order of small, fresh-water branchiopod crustaceans, commonly known as water fleas, characterized by a transparent bivalve shell.

Cladocopa [INV ZOO] A suborder of the order Myodocopida including marine animals having a carapace that lacks a permanent aperture when the two valves are closed.

Cladocopina [INV ZOO] The equivalent name for Cladocopa.

cladode *See* cladophyll.

cladogenesis [EVOL] Evolution associated with altered habit and habitat, usually in isolated species populations.

cladogenic adaptation *See* divergent adaptation.

Cladoniaceae [BOT] A family of lichens in the order Lecanorales, including the reindeer mosses and cup lichens, in which the main thallus is hollow.

Cladophorales [BOT] An order of coarse, wiry, filamentous, branched and unbranched algae in the class Chlorophyceae.

cladophyll [BOT] A branch arising from the axil of a true leaf and resembling a foliage leaf. Also known as cladode.

Cladoselachii [PALEON] An order of extinct elasmobranch fishes including the oldest and most primitive of sharks.

clairite *See* enargite.

Claisen condensation [ORG CHEM] **1.** Condensation, in the presence of sodium ethoxide, of esters or of esters and ketones to form β-dicarbonyl compounds. **2.** Condensation of arylaldehydes and acylphenones with esters or ketones in the presence of sodium ethoxide to yield unsaturated esters. Also known as Claisen reaction.

Claisen reaction *See* Claisen condensation.

clam [INV ZOO] The common name for a number of species of bivalve mollusks, many of which are important as food.

clamp [DES ENG] A tool for binding or pressing two or more parts together, by holding them firmly in their relative positions. [ELECTR] *See* clamping circuit.

clamper *See* direct-current restorer.

clamping [ELECTR] The introduction of a reference level that has some desired relation to a pulsed waveform, as at the negative or positive peaks. Also known as direct-current reinsertion; direct-current restoration.

clamping circuit [ELECTR] A circuit that reestablishes the direct-current level of a waveform; used in the dc restorer stage of a television receiver to restore the dc component to the video signal after its loss in capacitance-coupled alternating-current amplifiers, to reestablish the average light value of the reproduced image. Also known as clamp.

clamping diode [ELECTR] A diode used to clamp a voltage at some point in a circuit.

clamshell bucket [MECH ENG] A two-sided bucket used in a type of excavator to dig in a vertical direction; the bucket is dropped while its leaves are open and digs as they close. Also known as clamshell grab.

clamshell grab *See* clamshell bucket.

clan [ECOL] A very small community, perhaps a few square yards in area, in climax formation, and dominated by one species. [PETR] A category of igneous rocks defined in terms of similarities in mineralogical or chemical composition.

clapboard [MATER] A board, thicker at one edge than the other, used to cover exterior walls.

Clapeyron-Clausius equation *See* Clausius-Clapeyron equation.

Clapeyron equation *See* Clausius-Clapeyron equation.

Clapeyron's theorem [MECH] The theorem that the strain energy of a deformed body is equal to one-half the sum over three perpendicular directions of the displacement component times the corresponding force component, including deforming loads and body forces, but not the six constraining forces required to hold the body in equilibrium.

clarain [GEOL] A coal lithotype appearing as stratifications parallel to the bedding plane and usually having a silky luster and scattered or diffuse reflection. Also known as clarite.

Clarendonian [GEOL] Lower Pliocene or upper Miocene geologic time.

clarifier [COMMUN] A fine-tuning control on some single-band citizen band transceivers; adjusted for maximum naturalness of the received voice signals. [ENG] A device for filtering a liquid.

clarifying agents [FOOD ENG] Tannin, gelatin, albumin, methylcellulose, pectinases, and proteinases used to remove turbidity from fruit juices, vinegar, wine, beer, and soft drinks.

Clariidae [VERT ZOO] A family of Asian and African catfishes in the suborder Siluroidei.

clarite *See* clarain.

Clarkecarididae [PALEON] A family of extinct crustaceans in the order Anaspidacea.

clarkeite [MINERAL] $(Na,Ca,Pb)_2U_2(O,OH)_7$ A dark reddish-brown or dark brown mineral consisting of a hydrous or hydrated uranium oxide.

clarodurain [GEOL] A transitional lithotype of coal composed of vitrinite and other macerals, principally micrinite and exinite.

clarofusain [GEOL] A transitional lithotype of coal composed of fusinite and vitrinite and other macerals.

clarovitrain [GEOL] A transitional lithotype of coal rock composed primarily of the maceral vitrinite, with lesser amounts of other macerals.

clasper [VERT ZOO] A modified pelvic fin of male elasmobranchs and holocephalians used for the transmission of sperm.

clasp nut [DES ENG] A split nut that clasps a screw when closed around it.

class A amplifier [ELECTR] 1. An amplifier in which the grid bias and alternating grid voltages are such that anode current in a specific tube flows at all times. 2. A transistor amplifier in which each transistor is in its active region for the entire signal cycle.

class AB amplifier [ELECTR] 1. An amplifier in which the grid bias and alternating grid voltages are such that anode current in a specific tube flows for appreciably more than half but less than the entire electric cycle. 2. A transistor amplifier whose operation is class A for small signals and class B for large signals.

class B amplifier [ELECTR] 1. An amplifier in which the grid bias is approximately equal to the cutoff value, so that anode current is approximately zero when no exciting grid voltage is applied, and flows for approximately half of each cycle when an alternating grid voltage is applied. 2. A transistor amplifier in which each transistor is in its active region for approximately half the signal cycle.

class C amplifier [ELECTR] 1. An amplifier in which the bias on the control element is appreciably greater than the cutoff valve, so that the output current in each device is zero when no alternating control signal is applied, and flows for appreciably less than half of each cycle when an alternating control signal is applied. 2. A transistor amplifier in which each transistor is in its active region for significantly less than half the signal cycle.

classical field theory [PHYS] The study of distributions of energy, matter, and other physical quantities under circumstances where their discrete nature is unimportant, and they may be regarded as (in general, complex) continuous functions of position. Also known as c-number theory; continuum mechanics; continuum physics.

classical wave equation *See* wave equation.

classification [ENG] 1. Sorting out or categorizing of particles or objects by established criteria, such as size, function, or color. 2. Stratification of a mixture of various-sized particles (that is, sand and gravel), with the larger particles migrating to the bottom. [IND ENG] *See* grading. [ORD] Placing of military documents in special groups for safeguarding defense information. [SYST] A systematic arrangement of plants and animals into categories based on a definite plan, considering evolutionary, physiologic, cytogenetic, and other relationships.

classifier [MECH ENG] Any apparatus for separating mixtures of materials into their constituents according to size and density.

classify [SCI TECH] To sort into groups that have common properties.

class interval [STAT] One of several convenient intervals into which the values of the variate of a frequency distribution may be grouped.

class mark [STAT] The mid-value of a class interval, or the integral value nearest the midpoint of the interval.

classons [PARTIC PHYS] Massless bosons which are quanta of the two classical fields, gravitational and electromagnetic.

clast [GEOL] An individual grain, fragment, or constituent of detrital sediment or sedimentary rock produced by physical breakdown of a larger mass.

clastation *See* weathering.

clastic [GEOL] Rock or sediment composed of clasts which have been transported from their place of origin, as sandstone and shale.

clathrate [BIOL] *See* cancellate. [CHEM] A well-defined addition compound formed by inclusion of molecules in cavities formed by crystal lattices or present in large molecules; examples include hydroxyquinone, urea, and cyclodextrin. [PETR] Pertaining to a condition, chiefly in leucite rock, in which clear leucite crystals are surrounded by tangential leucite crystals to give the rock an appearance of a net or a section of sponge.

Clathrinida [INV ZOO] A monofamilial order of sponges in the subclass Calcinea with an asconoid structure and lacking a true dermal membrane or cortex.

Clathrinidae [INV ZOO] The single family of the order Clathrinida.

Claude process [CHEM ENG] A process of ammonia synthesis which uses high operating pressures and a train of converters. [CRYO] A method of liquefying air or other gases in stages, in which the gas is cooled by doing work in an expansion engine and then undergoing the Joule-Thomson effect as it passes through an expansion valve.

claudetite [MINERAL] As_2O_3 A mineral containing arsenic that is dimorphous with arsenolite; crystallizes in the monoclinic system.

clause [COMPUT SCI] A part of a statement in the COBOL language which may describe the structure of an elementary item, give initial values to items in independent and group work areas, or redefine data previously defined by another clause.

clausius [THERMO] A unit of entropy equal to the increase in entropy associated with the absorption of 1000 international table calories at a temperature of 1 K, or to 4186.8 joules per kelvin.

Clausius-Clapeyron equation [THERMO] An equation governing phase transitions of a substance, $dp/dT = \Delta H/(T\Delta V)$, in which p is the pressure, T is the temperature at which the phase transition occurs, ΔH is the change in heat content (enthalpy), and ΔV is the change in volume during the transition. Also known as Clapeyron-Clausius equation; Clapeyron equation.

Clausius equation [THERMO] An equation of state in reference to gases which applies a correction to the van der Waals equation: $\{P + (n^2a/[T(V + c)^2])\}(V - nb) = nRT$, where P is the pressure, T the temperature, V the volume of the gas, n the number of moles in the gas, R the gas constant, a depends only on temperature, b is a constant, and c is a function of a and b.

Clausius inequality [THERMO] The principle that for any system executing a cyclical process, the integral over the cycle of the infinitesimal amount of heat transferred to the system divided by its temperature is equal to or less than zero. Also known as Clausius theorem; inequality of Clausius.

Clausius law [THERMO] The law that an ideal gas's specific heat at constant volume does not depend on the temperature.

Clausius number [THERMO] A dimensionless number used in the study of heat conduction in forced fluid flow, equal to $V^3L\rho/k\Delta T$, where V is the fluid velocity, ρ is its density, L is a characteristic dimension, k is the thermal conductivity, and ΔT is the temperature difference.

Clausius range [STAT MECH] The condition in which the mean free path of molecules in a gas is much smaller than the dimensions of the container.

Clausius' statement [THERMO] A formulation of the second law of thermodynamics, stating it is not possible that, at the end of a cycle of changes, heat has been transferred from a colder to a hotter body without producing some other effect.

Clausius theorem *See* Clausius inequality.

clava [BIOL] A club-shaped structure, as the tip on the antennae of certain insects or the fruiting body of certain fungi.

Clavatoraceae [PALEOBOT] A group of middle Mesozoic algae belonging to the Charophyta.

Clavaxinellida [INV ZOO] An order of sponges in the class Demospongiae; members have monaxonid megascleres arranged in radial or plumose tracts.

clavicle [ANAT] A bone in the pectoral girdle of vertebrates with articulation occurring at the sternum and scapula.

clavus [INV ZOO] Any of several rounded or finger-like processes, such as the club of an insect antenna or the pointed anal portion of the hemelytron in hemipteran insects.

claw [ANAT] A sharp, slender, curved nail on the toe of an animal, such as a bird. [DES ENG] A fork for removing nails or spikes. [INV ZOO] A sharp-curved process on the tip of the limb of an insect.

clay [GEOL] 1. A natural, earthy, fine-grained material which develops plasticity when mixed with a limited amount of water; composed primarily of silica, alumina, and water, often with iron, alkalies, and alkaline earths. 2. The fraction of an earthy material containing the smallest particles, that is, finer than 3 micrometers. [MATER] A special grade of absorbent clay used as a filtering medium in refineries for removing solids or colorizing matter from lubricating oils.

clay mineral [MINERAL] One of a group of finely crystalline, hydrous silicates with a two- or three-layer crystal structure; the major components of clay materials; the most common minerals belong to the kaolinite, montmorillonite, attapulgite, and illite groups.

claypan [GEOL] A stratum of compact, stiff, relatively impervious noncemented clay; can be worked into a soft, plastic mass if immersed in water.

clay soil [GEOL] A fine-grained inorganic soil which forms hard lumps when dry and becomes sticky when wet.

clean bomb [ORD] A nuclear bomb that produces relatively little radioactive fallout.

clean room [ENG] A room in which elaborate precautions are employed to reduce dust particles and other contaminants in the air, as required for assembly of delicate equipment.

clear [COMPUT SCI] 1. To restore a storage device, memory device, or binary stage to a prescribed state, usually that denoting zero. Also known as reset. 2. A function key on calculators, to delete an entire problem or just the last keyboard entry. [METEOROL] 1. After United States weather observing practice, the state of the sky when it is cloudless or when the sky cover is less than 0.1 (to the nearest tenth). 2. To change from a stormy or cloudy weather condition to one of no precipitation and decreased cloudiness. [NAV] In marine navigation, to leave or pass safely, as to clear port or clear a shoal. [ORD] 1. To give a person a security clearance. 2. To operate a gun, so as to unload it or make certain no ammunition remains; to free a gun of stoppages.

clearance [ENG] Unobstructed space required for occasional removal of parts of equipment. [MECH ENG] 1. In a piston-and-cylinder mechanism, the space at the end of the cylinder when the piston is at dead-center position toward the end of the cylinder. 2. The ratio of the volume of this space to the piston displacement during a stroke. [MIN ENG] The space between the top or side of a car and the roof or wall. [NAV] 1. The clear space

between a vessel and an object such as a navigation light, hazard to navigation, or another vessel. 2. A specific message from air-traffic control to a pilot of an aircraft allowing him to proceed in accordance with the flight plan which the pilot had filed, or with some modification of the original plan. 3. In the instrument landing system, the difference in the depth of modulation which is required to produce a full-scale deflection of the course deviation indicator needle in any flight sector outside the on-course sectors. [ORD] Elevation of a gun at such an angle that a projectile will not strike an obstacle between the muzzle and the target. [PETRO ENG] The annular space between down-hole drill-string equipment, such as bits, core barrels, and casing, and the walls of the borehole with the down-hole equipment centered in the hole.

clear area [COMPUT SCI] In optical character recognition, any area designated to be kept free of printing or any other extraneous markings.

clear band [COMPUT SCI] In character recognition, a continuous horizontal strip of blank paper which must be obtained between consecutive code lines on a source document.

clear channel [COMMUN] A standard broadcast channel in which the dominant station or stations render service over wide areas; stations are cleared of objectionable interference within their primary service areas and over all or a substantial portion of their secondary.

clear ice [HYD] Generally, a layer or mass of ice which is relatively transparent because of its homogeneous structure and small number and size of air pockets.

clear-voice override [COMMUN] The ability of a speech scrambler to receive a clear message even when the scrambler is set for scrambler operation.

cleat [CIV ENG] A strip of wood, metal, or other material fastened across something to serve as a batten or to provide strength or support. [DES ENG] A fitting having two horizontally projecting horns around which a rope may be made fast. [GEOL] Vertical breakage planes found in coal. Also spelled cleet.

cleavage [CRYSTAL] Splitting, or the tendency to split, along planes determined by crystal structure and always parallel to a possible face. [EMBRYO] The subdivision of activated eggs into blastomeres. [GEOL] Splitting, or the tendency to split, along parallel, closely positioned planes in rock.

cleavage plane [CRYSTAL] Plane along which a crystalline substance may be split.

cleavelandite [MINERAL] A white, lamellar variety of albite that is almost pure $NaAlSi_3O_8$ and has a tabular habit, with individuals often showing mosaic developments and tending to occur in fan-shaped aggregates.

Clebsch-Gordan coefficient *See* vector coupling coefficient.

cleet *See* cleat.

cleft lip *See* harelip.

cleft palate [MED] A birth defect resulting from incomplete closure of the palate during embryogenesis.

cleistocarpous [BOT] Of mosses, having the capsule opening irregularly without an operculum. [MYCOL] Forming or having cleistothecia.

cleistogamy [BOT] Production of small closed flowers that are self-pollinating and contain numerous seeds. [PL ANAT] Production of a hermaphroditic floral type that remains closed and self-pollinates in the bud.

Cleridae [INV ZOO] The checkered beetles, a family of coleopteran insects in the superfamily Cleroidea.

Cleroidea [INV ZOO] A superfamily of coleopteran insects in the suborder Polyphaga.

clevis [DES ENG] A U-shaped metal fitting with holes in the open ends to receive a bolt or pin; used for attaching or suspending parts. [MIN ENG] A spring hook or snap hook which, in coal mining, is used to attach the bucket to the hoisting rope. Also known as clivvy.

cliff of displacement See fault scarp.

climacteric See menopause.

climate [CLIMATOL] The long-term manifestations of weather.

climate control [CLIMATOL] Schemes for artificially altering or controlling the climate of a region. [ENG] See air conditioning.

Climatiidae [PALEON] A family of archaic tooth-bearing fishes in the suborder Climatioidei.

Climatiiformes [PALEON] An order of extinct fishes in the class Acanthodii having two dorsal fins and large plates on the head and ventral shoulder.

Climatioidei [PALEON] A suborder of extinct fishes in the order Climatiiformes.

climatography [CLIMATOL] A quantitative description of climate, particularly with reference to the tables and charts which show the characteristic values of climatic elements at a station or over an area.

climatological forecast [METEOROL] A weather forecast based upon the climate of a region instead of upon the dynamic implications of current weather, with consideration given to such synoptic weather features as cyclones and anticyclones, fronts, and the jet stream.

climatology [METEOROL] That branch of meteorology concerned with the mean physical state of the atmosphere together with its statistical variations in both space and time as reflected in the weather behavior over a period of many years.

climax [ECOL] A mature, relatively stable community in an area, which community will undergo no further change under the prevailing climate; represents the culmination of ecological succession.

climax plant formation [ECOL] A mature, stable plant population in a climax community.

climb cutting [MET] A milling technique in which the teeth of a cutting tool advance into the work in the same direction as the feed. Also known as climb milling; down cutting; down milling.

climbing crane [MECH ENG] A crane used on top of a high-rise construction that ascends with the building as work progresses.

climbing irons [DES ENG] Spikes attached to a steel framework worn on shoes to climb wooden utility poles and trees.

climbing stem [BOT] A long, slender stem that climbs up a support or along the tops of other plants by using spines, adventitious roots, or tendrils for attachment.

climb milling See climb cutting.

cline [BIOL] A graded series of morphological or physiological characters exhibited by a natural group (as a species) of related organisms, generally along a line of environmental or geographic transition.

clinical pathology [PATH] A medical specialty encompassing the diagnostic study of disease by means of laboratory tests of material from the living patient.

clinical thermometer [ENG] A thermometer used to accurately determine the temperature of the human body; the most common type is a mercury-in-glass thermometer, in which the mercury expands from a bulb into a capillary tube past a constriction that prevents the mercury from receding back into the bulb, so that the thermometer registers the maximum temperature attained.

clinker [GEOL] Burnt or vitrified stony material, as ejected by a volcano or formed in a furnace. [MATER] An overburned brick.

clinoaxis [CRYSTAL] The inclined lateral axis that makes an oblique angle with the vertical axis in the monoclinic system. Also known as clinodiagonal.

clinochlore [MINERAL] $(Mg,Fe,Al)_3(Si,Al)_2O_5(OH)_4$ Green mineral of the chlorite group, occurring in monoclinic crystals, in folia or scales, or massive.

clinoclase [MINERAL] $Cu_3(AsO_4)(OH)_3$ A dark-green mineral consisting of basic copper arsenate occurring in translucent prismatic crystals or massive. Also known as clinoclasite.

clinoclasite See clinoclase.

clinodiagonal See clinoaxis.

clinoenstatite [MINERAL] $Mg_2(Si_2O_6)$ A monoclinic pyroxene consisting principally of magnesium silicate; occurs frequently in stony meteorites, but is rare in terrestrial environments.

clinograph [ENG] A type of directional surveying instrument that records photographically the direction and magnitude of deviations from the vertical of a borehole, well, or shaft; the information is obtained by the instrument in one trip into and out of the well.

clinographic projection [GRAPHICS] A method of representing objects, especially crystals, in which each point P of the object to be represented is projected onto the foot of a perpendicular from P to a plane which is located so that no place surface of the object is represented by a line.

clinohedral class [CRYSTAL] A rare class of crystals in the monoclinic system having a plane of symmetry but no axis of symmetry. Also known as domatic class.

clinohedrite [MINERAL] $CaZnSiO_3(OH)_2$ A colorless, white, or purplish monoclinic mineral consisting of a calcium zinc silicate occurring in crystals; hardness is 5.5 on Mohs scale, and specific gravity is 3.33.

clinometer [ENG] 1. A hand-held surveying device for measuring vertical angles; consists of a sighting tube surmounted by a graduated vertical arc with an attached level bubble; used in meteorology to measure cloud height at night, in conjunction with a ceiling light, and in ordnance for boresighting. 2. A device for measuring the amount of roll aboard ship.

clinopyroxene [MINERAL] The general term for any of those pyroxenes that crystallize in the monoclinic system; on occasion, these pyroxenes have large amounts of calcium with or without aluminum and the alkalies. Also known as monopyroxene clinoaugite.

clinozoisite [MINERAL] $Ca_2Al_3(SiO_4)_3(OH)$ A grayish-white, pink, or green monoclinic mineral of the epidote group.

Clintonian [GEOL] Lower Middle Silurian geologic time.

clintonite [MINERAL] $Ca(Mg,Al)_3(Al,Si)O_{10}(OH)_2$ A reddish-brown, copper-red, or yellowish monoclinic mineral of the brittle mica group occurring in crystals or foliated masses. Also known as seybertite; xanthophyllite.

Clionidae [INV ZOO] The boring sponges, a family of marine sponges in the class Demospongiae.

clipper See limiter.

clipper diode [ELECTR] A bidirectional breakdown diode that clips signal voltage peaks of either polarity when they exceed a predetermined amplitude.

clipper-limiter [ELECTR] A device whose output is a function of the instantaneous input amplitude for a range of values lying between two predetermined limits but is approximately constant, at another level, for input values above the range.

clipping [COMMUN] The perceptible mutilation of signals or speech syllables during transmission. [ELECTR] *See* limiting.

clipping circuit *See* limiter.

clitellum [INV ZOO] The thickened, glandular, saddlelike portion of the body wall of some annelid worms.

clitoris [ANAT] The homolog of the penis in females, located in the anterior portion of the vulva.

clivvy *See* clevis.

cloaca [INV ZOO] The chamber which functions as a respiratory, excretory, and reproductive duct in certain invertebrates. [VERT ZOO] The chamber which receives the discharges of the intestine, urinary tract, and reproductive canals in monotremes, amphibians, birds, reptiles, and many fish.

clock [ELECTR] A source of accurately timed pulses, used for synchronization in a digital computer or as a time base in a transmission system. [HOROL] A device for indicating the passage of time, usually containing a means for producing a regularly recurring action.

clock control system [CONT SYS] A system in which a timing device is used to generate the control function. Also known as time-controlled system.

clock frequency [ELECTR] The master frequency of the periodic pulses that schedule the operation of a digital computer.

clock motor *See* timing motor.

clock paradox [RELAT] The apparent contradiction between the principle of relativity, which asserts the equivalence of different observers, and the prediction, also part of the theory of relativity, that the clock of an observer who passes back and forth will be slower than the clock of an observer at rest. Also known as twin paradox.

clock pulses [COMPUT SCI] Electronic pulses which are emitted periodically, usually by a crystal device, to synchronize the operation of circuits in a computer. Also known as clock signals.

clock rate [ELECTR] The rate at which bits or words are transferred from one internal element of a computer to another. [HOROL] The amount of time which a clock gains or loses during a fixed period of time, usually a day.

clock signals *See* clock pulses.

clock track [COMPUT SCI] A track on a magnetic recording medium that generates clock pulses for the synchronization of read and write operations.

cloister vault [ARCH] A vault resembling a pyramid with outward-curving sides.

clone [BIOL] All individuals, considered collectively, produced asexually or by parthenogenesis from a single individual.

clonus [PHYSIO] Irregular, alternating muscular contractions and relaxations.

close coupling [ELEC] 1. The coupling obtained when the primary and secondary windings of a radiofrequency or intermediate-frequency transformer are close together. 2. A degree of coupling that is greater than critical coupling. Also known as tight coupling.

closed circuit [COMMUN] Program source that is not broadcast for general consumption but is fed to remote monitoring units by wire. [ELEC] A complete path for current.

closed-circuit communications system [COMMUN] A communications systems which is entirely self-contained, and does not exchange intelligence with other facilities and systems.

closed-circuit signaling [COMMUN] Signaling in which current flows in the idle condition, and a signal is initiated by increasing or decreasing the current.

closed-circuit television [COMMUN] Any application of television that does not involve broadcasting for public viewing; the programs can be seen only on specified receivers connected to the television camera by circuits, which include microwave relays and coaxial cables. Abbreviated CCTV.

closed cycle [THERMO] A thermodynamic cycle in which the thermodynamic fluid does not enter or leave the system, but is used over and over again.

closed drainage [HYD] Drainage in which the surface flow of water collects in sinks or lakes having no surface outlet. Also known as blind drainage.

closed file [COMPUT SCI] A file that cannot be accessed for reading or writing.

closed fold [GEOL] A fold whose limbs have been compressed until they are parallel, and whose structure contour lines form a closed loop. Also known as tight fold.

closed intervals [MATH] A closed interval of real numbers, denoted by $[a,b]$, consists of all numbers equal to or greater than a and equal to or less than b.

closed loop [COMPUT SCI] A loop whose execution continues indefinitely in the absence of any external intervention. [CONT SYS] A family of automatic control units linked together with a process to form an endless chain; the effects of control action are constantly measured so that if the controlled quantity departs from the norm, the control units act to bring it back.

closed-loop control system *See* feedback control system.

closed operator [MATH] A linear transformation f whose domain A is contained in a normed vector space X satisfying the condition that if $\lim x_n = x$ for a sequence x_n in A, and $\lim f(x_n) = y$, then x is in A and $f(x) = y$.

closed orthonormal set *See* complete orthonormal set.

closed pair [MECH] A pair of bodies that are subject to constraints which prevent any relative motion between them.

closed set [MATH] A set of points which contains all its cluster points. Also known as topologically closed set.

closed shell [PHYS] An atomic or nuclear shell containing the maximum number of electrons or nucleons allowed by the Pauli exclusion principle.

closed subroutine [COMPUT SCI] A subroutine that can be stored outside the main routine and can be connected to it by linkages at one or more locations.

closed system [THERMO] A system which is isolated so that it cannot exchange matter or energy with its surroundings and can therefore attain a state of thermodynamic equilibrium.

close-packed crystal [CRYSTAL] A crystal structure in which the lattice points are centers of spheres of equal radius arranged so that the volume of the interstices between the spheres is minimal.

close routine [COMPUT SCI] A computer program that changes the state of a file from open to closed.

closest approach [ASTRON] 1. The event that occurs when two planets or other celestial bodies are nearest to each other as they orbit about the sun or other primary. 2. The place or time of such an event.

close-talking microphone [ENG ACOUS] A microphone designed for use close to the mouth, so noise from more distant points is suppressed. Also known as noise-canceling microphone.

closure [GEOL] The vertical distance between the highest and lowest point on an anticline which is enclosed by contour lines. [MATH] The union of a set and its cluster points; the smallest closed set containing the set.

clothoid See Cornu's spiral.

clotting time [PHYSIO] The length of time required for shed blood to coagulate under standard conditions. Also known as coagulation time.

cloud [METEOROL] Suspensions of minute water droplets or of ice crystals produced by the condensation of water vapor. [NUC PHYS] The nucleons that are in the nucleus of an atom but not in closed shells. [SCI TECH] Any suspension of particulate matter, such as dust or smoke, dense enough to be seen.

cloudage See cloud cover.

cloud bank [METEOROL] A fairly well-defined mass of cloud observed at a distance; covers an appreciable portion of the horizon sky, but does not extend overhead.

cloud bar [METEOROL] 1. A heavy bank of clouds that appears on the horizon with the approach of an intense tropical cyclone (hurricane or typhoon); it is the outer edge of the central cloud mass of the storm. 2. Any long, narrow, unbroken line of cloud, such as a crest cloud or an element of billow cloud.

cloud base [METEOROL] For a given cloud or cloud layer, that lowest level in the atmosphere at which the air contains a perceptible quantity of cloud particles.

cloudburst [METEOROL] In popular terminology, any sudden and heavy fall of rain, usually of the shower type, and with a fall rate equal to or greater than 100 millimeters (3.94 inches) per hour. Also known as rain gush; rain gust.

cloud chamber [NUCLEO] A particle detector in which the path of a charged particle is made visible by the formation of liquid droplets along the trail of ions left by the particle as it passes through the gas of the chamber. Also known as expansion chamber; fog chamber.

cloud cover [METEOROL] That portion of the sky cover which is attributed to clouds, usually measured in tenths of sky covered. Also known as cloudage; cloudiness.

cloud height [METEOROL] The absolute altitude of the base of a cloud.

cloudiness See cloud cover.

cloud layer [METEOROL] An array of clouds, not necessarily all of the same type, whose bases are at approximately the same level; may be either continuous or composed of detached elements.

cloud point [CHEM ENG] The temperature at which paraffin wax or other solid substance begins to separate from a solution of petroleum oil; a cloudy appearance is seen in the oil at this point.

cloud seeding [METEOROL] Any technique carried out with the intent of adding to a cloud certain particles that will alter its natural development.

cloud system [METEOROL] An array of clouds and precipitation associated with a cyclonic-scale feature of atmospheric circulation, and displaying typical patterns and continuity. Also known as nephsystem.

cloudy [METEOROL] The character of a day's weather when the average cloudiness, as determined from frequent observations, is more than 0.7 for the 24-hour period.

clove [BOT] 1. The unopened flower bud of a small, conical, symmetrical evergreen tree, Eugenia caryophyllata, of the myrtle family (Myrtaceae); the dried buds are used as a pungent, strongly aromatic spice. 2. A small bulb developed within a larger bulb, as in garlic.

clover [BOT] 1. A common name designating the true clovers, sweet clovers, and other members of the Leguminosa. 2. A herb of the genus Trifolium.

cloverleaf [CIV ENG] A highway intersection resembling a clover leaf and designed to allow movement and interchange of traffic without direct crossings and left turns.

CLS See characteristic loss spectroscopy.

clubfoot [MED] Congenital malpositioning of a foot such that the forefoot is inverted and rotated with a shortened Achilles tendon.

club fungi [MYCOL] The common name for members of the class Basidiomycetes.

club moss [BOT] The common name for members of the class Lycopodiatae.

clubroot [PL PATH] A disease principally of crucifers, such as cabbage, caused by the slime mold Plasmodiophora brassicae in which roots become enlarged and deformed, leading to plant death.

Clupeidae [VERT ZOO] The herrings, a family of fishes in the suborder Clupoidea composing the most primitive group of higher bony fishes.

Clupeiformes [VERT ZOO] An order of teleost fishes in the subclass Actinopterygii, generally having a silvery, compressed body.

Clupoidea [VERT ZOO] A suborder of fishes in the order Clupeiformes comprising the herrings and anchovies.

cluster [COMPUT SCI] In a clustered file, one of the classes into which records with similar sets of content identifiers are grouped.

cluster analysis [STAT] A general approach to multivariate problems whose aim is to determine whether the individuals fall into groups or clusters.

cluster bean See guar.

cluster cepheids See RR Lyrae stars.

clustered file [COMPUT SCI] A collection of records organized so that items which exhibit similar sets of content identifiers are automatically grouped into common classes.

clustering algorithm [COMPUT SCI] A computer program that attempts to detect and locate the presence of groups of vectors, in a high-dimensional multivariate space, that share some property of similarity.

clusterite See botryoid.

cluster point [MATH] A cluster point of a set in a topological space is a point p whose neighborhoods all contain at least one point of the set other than p. Also known as accumulation point; limit point.

cluster variables See RR Lyrae stars.

clutch [MECH ENG] A machine element for the connection and disconnection of shafts in equipment drives, especially while running. [VERT ZOO] A nest of eggs or a brood of chicks.

clutter [ELECTROMAG] Unwanted echoes on a radar screen, such as those caused by the ground, sea, rain, stationary objects, chaff, enemy jamming transmissions, and grass. Also known as background return; radar clutter.

Clypeasteroida [INV ZOO] An order of exocyclic Euechinoidea having a monobasal apical system in which all the genital plates fuse together.

clypeus [INV ZOO] An anterior medial plate on the head of an insect, commonly bearing the labrum on its anterior margin. [MYCOL] A disk of black tissue about the mouth of the perithecia in certain ascomycetes.

Clythiidae [INV ZOO] The flat-footed flies, a family of cyclorrhaphous dipteran insects in the series Aschiza characterized by a flattened distal end on the hind tarsus.

cm *See* centimeter.

Cm *See* curium.

CMa *See* Canis Major.

cmHg *See* centimeter of mercury.

CMi *See* Canis Minor.

CMI *See* computer-managed instruction.

CMOS device [ELECTR] A device formed by the combination of a PMOS (p-type-channel metal oxide semiconductor device) with an NMOS (n-type-channel metal oxide semiconductor device). Derived from complementary metal oxide semiconductor device.

C network [ELECTR] Network composed of three impedance branches in series, the free ends being connected to one pair of terminals, and the junction points being connected to another pair of terminals.

Cnidaria [INV ZOO] The equivalent name for Coelenterata.

cnidoblast [INV ZOO] A cell that produces nematocysts. Also known as nettle cell; stinging cell.

Cnidospora [INV ZOO] A subphylum of spore-producing protozoans that are parasites in cells and tissues of invertebrates, fishes, a few amphibians, and turtles.

cnoidal wave [FL MECH] A finite-amplitude progressive wave in shallow water having its wave profile represented by the Jacobian elliptic function "cn."

c-number theory *See* classical field theory.

Co *See* cobalt.

^{60}Co *See* cobalt-60.

CoA *See* coenzyme A.

coacervation [CHEM] The separation, by addition of a third component, of an aqueous solution of a macromolecule colloid (polymer) into two liquid phases, one of which is colloid-rich (the coacervate) and the other an aqueous solution of the coacervating agent (the equilibrium liquid).

coagulation [CHEM] A separation or precipitation from a dispersed state of suspensoid particles resulting from their growth; may result from prolonged heating, addition of an electrolyte, or from a condensation reaction between solute and solvent; an example is the setting of a gel. [METEOROL] *See* agglomeration.

coagulation time *See* clotting time.

coal [GEOL] The natural, rocklike, brown to black derivative of forest-type plant material, usually accumulated in peat beds and progressively compressed and indurated until it is finally altered into graphite or graphite-like material.

coal ball [GEOL] A subspherical mass containing mineral matter embedded with plant material, found in coal seams and overlying beds of the late Paleozoic.

coal bed [GEOL] A seam or stratum of coal parallel to the rock stratification. Also known as coal rake; coal seam.

coalescence [MET] The bonding of welded materials into one body. [METEOROL] In cloud physics, merging of two or more water drops into a single

larger drop. [PHYS] The uniting by growth in one body, as particles, gas, or a liquid.

coalescent [CHEM] Chemical additive used in immiscible liquid-liquid mixtures to cause small droplets of the suspended liquid to unite, preparatory to removal from the carrier liquid.

coalescent pack [CHEM ENG] High-surface-area packing to consolidate liquid droplets for gravity separation from a second phase (for example, gas or immiscible liquid); packing must be wettable by the droplet phase; Berl saddles, Raschig rings, knitted wire mesh, excelsior, and similar materials are used.

coal gas [MATER] 1. Flammable gas derived from coal either naturally in place, or by induced methods of industrial plants and underground gasification. 2. Specifically, fuel gas obtained from carbonization of coal.

coal gasification [CHEM ENG] The conversion of coal, char, or coke to a gaseous product by reaction with air, oxygen, steam, carbon dioxide, or mixtures of these.

coal hydrogenation *See* Bergius process.

coalification [GEOL] Formation of coal from plant material by the processes of diagenesis and metamorphism. Also known as bituminization; carbonification; incarbonization; incoalation.

coal liquefaction [CHEM ENG] The process of preparing a liquid mixture of hydrocarbons by destructive distillation of coal.

coal paleobotany [PALEOBOT] A branch of the paleobotanical sciences concerned with the origin, composition, mode of occurrence, and significance of fossil plant materials that occur in or are associated with coal seams.

coal rake *See* coal bed.

Coalsack [ASTRON] An area in one of the brighter regions of the Southern Milky Way which to the naked eye appears entirely devoid of stars and hence dark with respect to the surrounding Milky Way region.

coal seam *See* coal bed.

coal tar [MATER] A tar obtained from carbonization of coal, usually in coke ovens or retorts, containing several hundred organic chemicals.

Coanda effect [FL MECH] The tendency of a gas or liquid coming out of a jet to travel close to the wall contour even if the wall's direction of curvature is away from the jet's axis; a factor in the operation of a fluidic element.

coarse-grained [MATER] Having a coarse texture. [PETR] *See* phaneritic.

coast [ENG] A memory feature on a radar which, when activated, causes the range and angle systems to continue to move in the same direction and at the same speed as that required to track an original target. [GEOGR] The general region of indefinite width that extends from the sea inland to the first major change in terrain features.

coastal berm *See* berm.

coastal engineering [CIV ENG] A branch of civil engineering pertaining to the study of the action of the seas on shorelines and to the design of structures to protect against this action.

coastal ice *See* fast ice.

coast ice *See* fast ice.

coastline [GEOGR] 1. The line that forms the boundary between the shore and the coast. 2. The line that forms the boundary between the water and the land.

coated filament [ELECTR] A vacuum-tube filament coated with metal oxides to provide increased electron emission.

coated lens [OPTICS] A lens whose surfaces have been coated with a thin, transparent film having an index of refraction that minimizes light loss by reflection.

coated paper [MATER] Paper with a surface coating of clay and other materials to produce a smooth, shiny surface; especially useful for fine, detailed, blur-free reproductions in color or black and white. Also known as enamel paper.

coati [VERT ZOO] The common name for three species of carnivorous mammals assigned to the raccoon family (Procyonidae) characterized by their elongated snout, body, and tail.

coax See coaxial cable.

coaxial [MECH] Sharing the same axes. [MECH ENG] Mounted on independent concentric shafts.

coaxial antenna [ELECTROMAG] An antenna consisting of a quarter-wave extension of the inner conductor of a coaxial line and a radiating sleeve that is in effect formed by folding back the outer conductor of the coaxial line for a length of approximately a quarter wavelength.

coaxial cable [ELECTROMAG] A transmission line in which one conductor is centered inside and insulated from an outer metal tube that serves as the second conductor. Also known as coax; coaxial line; coaxial transmission line; concentric cable; concentric line; concentric transmission line.

coaxial capacitor See cylindrical capacitor.

coaxial diode [ELECTR] A diode having the same outer diameter and terminations as a coaxial cable, or otherwise designed to be inserted in a coaxial cable.

coaxial filter [ELECTROMAG] A section of coaxial line having reentrant elements that provide the inductance and capacitance of a filter section.

coaxial isolator [ELECTROMAG] An isolator used in a coaxial cable to provide a higher loss for energy flow in one direction than in the opposite direction; all types use a permanent magnetic field in combination with ferrite and dielectric materials.

coaxial line See coaxial cable.

coaxial relay [ELECTROMAG] A relay designed for opening or closing a coaxial cable circuit without introducing a mismatch that would cause wave reflections.

coaxial speaker [ENG ACOUS] A loudspeaker system comprising two, or less commonly three, speaker units mounted on substantially the same axis in an integrated mechanical assembly, with an acoustic-radiation-controlling structure.

coaxial switch [ELEC] A switch that changes connections between coaxial cables going to antennas, transmitters, receivers, or other high-frequency devices without introducing impedance mismatch.

coaxial transistor [ELECTR] A point-contact transistor in which the emitter and collector are point electrodes making pressure contact at the centers of opposite sides of a thin disk of semiconductor material serving as base.

coaxial transmission line See coaxial cable.

cobalamin See vitamin B_{12}.

cobalt [CHEM] A metallic element, symbol Co, atomic number 27, atomic weight 58.93; used chiefly in alloys.

cobalt-60 [NUC PHYS] A radioisotope of cobalt, symbol ^{60}Co, having a mass number of 60; emits gamma rays and has many medical and industrial uses; the most commonly used isotope for encapsulated radiation sources.

cobalt-beam therapy [MED] Therapy involving the use of gamma radiation from a cobalt-60 source mounted in a cobalt bomb. Also known as cobalt therapy.

cobalt bloom See erythrite.

cobaltite [MINERAL] CoAsS A silver-white mineral with a metallic luster that crystallizes in the isometric system, resembling crystals of pyrite; one of the chief ores of cobalt. Also known as cobalt glance; gray cobalt; white cobalt.

cobalt ocher See asbolite; erythrite.

cobalt therapy See cobalt-beam therapy.

cobber [MIN ENG] **1.** A device used to reject waste materials from ore concentrates. **2.** A person who breaks fibers from asbestos rocks or chips low-grade material from ore.

cobblestone [MATER] A rounded stone 64–256 millimeters in diameter used in construction, especially for paving.

Cobb's disease [PL PATH] A bacterial disease of sugarcane caused by Xanthomonas vascularum and characterized by a slime in the vascular bundles, dwarfing, streaking of leaves, and decay. Also known as sugarcane gummosis.

cobinotron [ELECTROMAG] The combination of a corbino disk and a coil arranged to produce a magnetic field perpendicular to the disk.

Cobitidae [VERT ZOO] The loaches, a family of small fishes, many eel-shaped, in the suborder Cyprinoidei, characterized by barbels around the mouth.

Coblentzian [GEOL] Upper Lower Devonian geologic time.

COBOL [COMPUT SCI] A business data-processing language that can be given to a computer as a series of English statements describing a complete business operation. Derived from common business-oriented language.

cobra [VERT ZOO] Any of several species of venomous snakes in the reptilian family Elaphidae characterized by a hoodlike expansion of skin on the anterior neck that is supported by a series of ribs.

coca [BOT] Erythroxylon coca. A shrub in the family Erythroxylaceae; its leaves are the source of cocaine.

cocaine [PHARM] $C_{17}H_{21}O_4N$ An alkaloid obtained from coca leaves that is used for local anesthesia and as a tonic in digestive and nervous disorders. Also known as erythroxylon; methylbenzoylecgonine.

Coccidia [INV ZOO] A subclass of protozoans in the class Telosporea; typically intracellular parasites of epithelial tissue in vertebrates and invertebrates.

coccidioidomycosis [MED] An infectious fungus disease of humans and animals of either a pulmonary or a cutaneous nature; caused by Coccidioides immitis. Also known as San Joaquin Valley fever.

Coccinellidae [INV ZOO] The ladybird beetles, a family of coleopteran insects in the superfamily Cucujoidea.

coccobacillus [MICROBIO] A short, thick, oval bacillus, midway between the coccus and the bacillus in appearance.

Coccoidea [INV ZOO] A superfamily of homopteran insects belonging to the Sternorrhyncha; includes scale insects and mealy bugs.

coccolith [BOT] One of the small, interlocking calcite plates covering members of the Coccolithophorida.

Coccolithophora [INV ZOO] An order of phytoflagellates in the protozoan class Phytamastigophorea.

Coccolithophorida [BOT] A group of unicellular, biflagellate, golden-brown algae characterized by a covering of coccoliths.

Coccomyxaceae [BOT] A family of algae belonging to the Tetrasporales composed of elongate cells which reproduce only by vegetative means.

coccosphere [PALEOBOT] The fossilized remains of a member of Coccolithophorida.

Coccosteomorphi [PALEON] An aberrant lineage of the joint-necked fishes.

coccus [MICROBIO] A form of eubacteria which are more or less spherical in shape.

coccyx [ANAT] The fused vestige of caudal vertebrae forming the last bone of the vertebral column in man and certain other primates.

cochannel interference [COMMUN] Interference caused on one communication channel by a transmitter operating in the same channel.

cochlea [ANAT] The snail-shaped canal of the mammalian inner ear; it is divided into three channels and contains the essential organs of hearing.

cochlear nerve [ANAT] A sensory branch of the auditory nerve which receives impulses from the organ of Corti.

Cochliodontidae [PALEON] A family of extinct chondrichthian fishes in the order Bradyodonti.

cock [ENG] Any mechanism which starts, stops, or regulates the flow of liquid, such as a valve, faucet, or tap. [VERT ZOO] The adult male of the domestic fowl and of gallinaceous birds.

Cockcroft-Walton accelerator [NUCLEO] An electrostatic particle accelerator utilizing as a source of high voltage a transformer and an array of rectifiers and condensers, giving voltage multiplication.

cockle [INV ZOO] The common name for a number of species of marine mollusks in the class Bivalvia characterized by a shell having convex radial ribs.

cockroach See roach.

coconut [BOT] *Cocos nucifera*. A large palm in the order Arecales grown for its fiber and fruit, a large, ovoid, edible drupe with a fibrous exocarp and a hard, bony endocarp containing fleshy meat (endosperm).

cocoon [INV ZOO] 1. A protective case formed by the larvae of many insects, in which they pass the pupa stage. 2. Any of the various protective egg cases formed by invertebrates.

cocurrent line [OCEANOGR] A line through places having the same tidal current hour.

cod [VERT ZOO] The common name for fishes of the subfamily Gadidae, especially the Atlantic cod (*Gadus morrhua*).

Coddington lens [OPTICS] A magnifier consisting of a glass sphere with a deep groove cut around a great circle to serve as a stop.

Coddington shape factor See shape factor.

code [COMMUN] A system of symbols and rules for expressing information, such as the Morse code, EIA color code, and the binary and other machine languages used in digital computers.

codecarboxylase [BIOCHEM] The prosthetic component of the enzyme carboxylase which catalyzes decarboxylation of D-amino acids. Also known as pyridoxal phosphate.

codeclination [NAV] In celestial navigation, 90° minus the declination; when the declination and latitude are of the same name, codeclination is the same as polar distance measured from the elevated pole.

code converter [COMPUT SCI] A converter that changes coded information to a different code system.

coded character set [COMPUT SCI] A set of characters together with the code assigned to each character for computer use.

coded decimal See decimal-coded digit.

code-division multiple access [COMMUN] A multiple-access technique used in satellite communication systems, in which neither the satellite frequency nor the time domain is divided among the earth stations; each earth station has common usage of the full satellite bandwidth and time by transmitting with its own unique pseudo-noise-coded waveform; correlation-detection techniques are used at each receiver to recover the original information. Abbreviated CDMA.

code-division multiplex [COMMUN] Multiplex in which two or more communication links occupy the entire transmission channel simultaneously, with code signal structures designed so a given receiver responds only to its own signals and treats the other signals as noise. Abbreviated CDM.

code holes [COMPUT SCI] The informational holes in perforated tape, as opposed to the feed holes or other holes.

codehydrogenase I See diphosphopyridine nucleotide.

codehydrogenase II See triphosphopyridine nucleotide.

codeine [PHARM] $C_{18}H_{21}NO_3$ An alkaloid prepared from morphine; used as mild analgesic and cough suppressant.

code line [COMPUT SCI] In character recognition, the area reserved for the inscription of the printed or handwritten characters to be recognized.

code position [COMPUT SCI] A location in a data-recording medium at which data may be entered, such as the intersection of a column and a row on a punch card, at which a hole may be punched.

coder [COMMUN] A device that generates a code by generating pulses having varying lengths or spacings, as required for radio beacons and interrogators. Also known as moder; pulse coder; pulse-duration coder. [COMPUT SCI] A person who translates a sequence of computer instructions into codes acceptable to the machine.

code translation [COMMUN] Conversion of a directory code or number into a predetermined code for controlling the selection of an outgoing trunk or line.

Codiaceae [BOT] A family of green algae in the order Siphonales having macroscopic thalli composed of aggregates of tubes.

codimer [ORG CHEM] 1. A copolymer formed from the polymerization of two dissimilar olefin molecules. 2. The product of polymerization of isobutylene with one of the two normal butylenes.

coding [COMPUT SCI] 1. The process of converting a program design into an accurate, detailed representation of that program in some suitable language. 2. A list, in computer code, of the successive operations required to carry out a given routine or solve a given problem.

codistor [ELECTR] A multijunction semiconductor device which provides noise rejection and voltage regulation functions.

codon [COMMUN] A basic unit in a coded message, corresponding to a single character. [GEN] The basic unit of the genetic code, comprising three-nucleotide sequences of messenger ribonucleic acid, each of which is translated into one amino acid in protein synthesis.

coefficient of absorption See absorption coefficient.

coefficient of compressibility [MECH] The decrease in volume per unit volume of a substance resulting from a unit increase in pressure; it is the reciprocal of the bulk modulus.

coefficient of contraction [FL MECH] The ratio of the minimum cross-sectional area of a jet of liquid discharging from an orifice to the area of the orifice. Also known as contraction coefficient.

coefficient of coupling   *See* coupling constant.
coefficient of discharge   *See* discharge coefficient.
coefficient of eddy diffusion   *See* eddy diffusivity.
coefficient of elasticity   *See* modulus of elasticity.
coefficient of friction   [MECH] The ratio of the frictional force between two bodies in contact, parallel to the surface of contact, to the force, normal to the surface of contact, with which the bodies press against each other. Also known as friction coefficient.
coefficient of kinetic friction   [MECH] The ratio of the frictional force, parallel to the surface of contact, that opposes the motion of a body which is sliding or rolling over another, to the force, normal to the surface of contact, with which the bodies press against each other.
coefficient of linear expansion   [THERMO] The increment of length of a solid in a unit of length for a rise in temperature of 1° at constant pressure.
coefficient of performance   [THERMO] In a refrigeration cycle, the ratio of the heat energy extracted by the heat engine at the low temperature to the work supplied to operate the cycle; when used as a heating device, it is the ratio of the heat delivered in the high-temperature coils to the work supplied.
coefficient of reflection   *See* reflection coefficient.
coefficient of resistance   [FL MECH] The ratio of the loss of head of fluid, issuing from an orifice or passing over a weir, to the remaining head.
coefficient of rigidity   *See* modulus of elasticity in shear.
coefficient of strain   [MATH] Multiplier used in transformations to elongate or compress configurations in a direction parallel to an axis.   [MECH] For a substance undergoing a one-dimensional strain, the ratio of the distance along the strain axis between two points in the body, to the distance between the same points when the body is undeformed.
coefficient of variation   [STAT] The ratio of the standard deviation of a distribution to its arithmetic mean.
coefficient of velocity   *See* velocity coefficient.
coelacanth   [VERT ZOO] Any member of the Coelacanthiformes, an order of lobefin fishes represented by a single living genus, *Latimeria*.
Coelacanthidae   [PALEON] A family of extinct lobefin fishes in the order Coelacanthiformes.
Coelacanthiformes   [VERT ZOO] An order of lobefin fishes in the subclass Crossopterygii which were common fresh-water animals of the Carboniferous and Permian; one genus, *Latimeria*, exists today.
Coelacanthini   [VERT ZOO] The equivalent name for Coelacanthiformes.
Coelenterata   [INV ZOO] A phylum of the Radiata whose members typically bear tentacles and possess intrinsic nematocysts.
coelenteron   [INV ZOO] The internal cavity of coelenterates.
Coelolepida   [PALEON] An order of extinct jawless vertebrates (Agnatha) distinguished by skin set with minute, close-fitting scales of dentine, similar to placoid scales of sharks.
coelom   [ZOO] The mesodermally lined body cavity of most animals higher on the evolutionary scale than flatworms and nonsegmented roundworms.
Coelomomycetaceae   [MYCOL] A family of entomophilic fungi in the order Blastocladiales which parasitize primarily mosquito larvae.
Coelomycetes   [MYCOL] A group set up by some authorities to include the Sphaerioidaceae and the Melanconiales.

Coelopidae   [INV ZOO] The seaweed flies, a family of myodarian cyclorrhaphous dipteran insects in the subsection Acalypteratae whose larvae breed on decomposing seaweed.
coelostat   [ENG] A device consisting of a clockwork-driven mirror that enables a fixed telescope to continuously keep the same region of the sky in its field of view.
Coelurosauria   [PALEON] A group of small, lightly built saurischian dinosaurs in the suborder Theropoda having long necks and narrow, pointed skulls.
Coenagrionidae   [INV ZOO] A family of zygopteran insects in the order Odonata.
Coenobitidae   [INV ZOO] A family of terrestrial decapod crustaceans belonging to the Anomura.
coenobium   [INV ZOO] A colony of protozoans having a constant size, shape, and cell number, but with undifferentiated cells.
coenocyte   [BIOL] A multinucleate mass of protoplasm formed by repeated nucleus divisions without cell fission.
Coenomyidae   [INV ZOO] A family of orthorrhaphous dipteran insects in the series Brachycera.
Coenopteridales   [PALEOBOT] A heterogeneous group of fernlike fossil plants belonging to the Polypodiophyta.
Coenothecalia   [INV ZOO] An order of the class Alcyonaria that forms colonies; lacks spicules but has a skeleton composed of fibrocrystalline argonite.
coenotype   [BIOL] An organism having the characteristic structure of the group to which it belongs.
coenzyme   [BIOCHEM] The nonprotein portion of an enzyme; a prosthetic group which functions as an acceptor of electrons or functional groups.
coenzyme A   [BIOCHEM] $C_{21}H_{36}O_{16}N_7P_3S$ A coenzyme in all living cells; required by certain condensing enzymes to act in acetyl or other acyl-group transfer and in fatty-acid metabolism. Abbreviated CoA.
coenzyme I   *See* diphosphopyridine nucleotide.
coenzyme II   *See* triphosphopyridine nucleotide.
coercive force   [ELECTROMAG] The magnetic field *H* which must be applied to a magnetic material in a symmetrical, cyclically magnetized fashion, to make the magnetic induction *B* vanish. Also known as magnetic coercive force.
coercivity   [ELECTROMAG] The coercive force of a magnetic material in a hysteresis loop whose maximum induction approximates the saturation induction.
coesite   [MINERAL] A high-pressure polymorph of SiO_2 formed in nature only under unique physical conditions, requiring pressures of more than 20 kilobars (2×10^9 newtons per square meter); usually found in meteor impact craters.
coextrusion   [ENG] Extrusion-forming of plastic or metal products in which two or more compatible feed materials are used in physical admixture through the same extrusion die.
coffee   [BOT] Any of various shrubs or small trees of the genus *Coffea* (family Rubiaceae) cultivated for the seeds (coffee beans) of its fruit; most coffee beans are obtained from the Arabian species, *C. arabica*.
cofferdam   [CIV ENG] A temporary damlike structure constructed around an excavation to exclude water.
coffered ceiling   [BUILD] An ornamental ceiling constructed of panels that are sunken or recessed.
coffinite   [MINERAL] $USiO_4$ A black silicate important as a uranium ore; found in sandstone deposits and hydrothermal veins in New Mexico, Utah, and Wyoming.

cofinal [MATH] A subset C of a directed set D is cofinal if for each element of D there is a larger element in C.

cog [DES ENG] A tooth on the edge of a wheel.

cogging [ELECTROMAG] Variations in torque and speed of an electric motor due to variations in magnetic flux as rotor poles move past stator poles.

cognate [GEOL] Pertaining to contemporaneous fractures in a system with regard to time of origin and deformational type.

cognition [PSYCH] The conscious faculty or process of knowing, of becoming or being aware of thoughts or perceptions, including understanding and reasoning.

COGO [COMPUT SCI] A higher-level computer language oriented toward civil engineering, enabling one to write a program in a technical vocabulary familiar to engineers and feed it to the computer; several versions have been implemented. Derived from coordinated geometry.

cogon [BOT] *Imperate cylindrica*. A grass found in rainforests. Also known as alang-alang.

cog railway [CIV ENG] A steep railway that employs a cograil that meshes with a cogwheel on the locomotive to ensure traction.

cogwheel [DES ENG] A wheel with teeth around its edge.

cogwheel ore *See* bournonite.

coherence [PHYS] **1.** The existence of a correlation between the phases of two or more waves, so that interference effects may be produced between them. **2.** Property of moving in unison, such as is characteristic of the particles in a synchrotron.

coherence distance *See* coherence length.

coherence length [PHYS] For a beam of particles, the typical length of a wave packet along the beam; the more monochromatic the beam, the greater its coherence length. [SOLID STATE] A measure of the distance through which the effect of any local disturbance is spread out in a superconducting material. Also known as coherence distance.

coherent detector [ELECTR] A detector used in moving-target indicator radar to give an output-signal amplitude that depends on the phase of the echo signal instead of on its strength, as required for a display that shows only moving targets.

coherent echo [ELECTR] A radar echo whose phase and amplitude at a given range remain relatively constant.

coherent light [OPTICS] Radiant electromagnetic energy of the same, or almost the same, wavelength, and with definite phase relationships between different points in the field.

coherent moving-target indicator [ENG] A radar system in which the Doppler frequency of the target echo is compared to a local reference frequency generated by a coherent oscillator.

coherent oscillator [ELECTR] An oscillator used in moving-target indicator radar to serve as a reference by which changes in the radio-frequency phase of successively received pulses may be recognized. Abbreviated coho.

coherent-pulse radar [ELECTR] A radar in which the radio-frequency oscillations of recurrent pulses bear a constant phase relation to those of a continuous oscillation.

coherent radiation [PHYS] Radiation in which there are definite phase relationships between different points in a cross section of the beam.

coherent signal [ELECTR] In a pulsed radar system, a signal having a constant phase; it is mixed with the echo signal, whose phase depends upon the range

of the target, in order to detect the phase shift and measure the target's range.

coherent source [PHYS] A source in which there is a constant phase difference between waves emitted from different parts of the source.

coherent system [NAV] A navigation system in which the signal output is obtained by demodulating the received signal after mixing with a local signal having a fixed phase relation to that of the transmitted signal, to permit use of the information carrier by the phase of the received signal.

coherent units [PHYS] A system of units, such as the International System, in which the units of derived quantities are formed as products or quotients of units of the base quantities according to the algebraic relations linking these quantities.

cohesion [BOT] The union of similar plant parts or organs, as of the petals to form a corolla. [PHYS] The tendency of parts of a body of like composition to hold together, as a result of intermolecular attractive forces. [SCI TECH] The state or process of sticking together.

cohesive energy [SOLID STATE] The difference between the energy per atom of a system of free atoms at rest far apart from each other, and the energy of the solid.

cohesiveness [GEOL] Property of unconsolidated fine-grained sediments by which the particles stick together by surface forces.

cohesive strength [MECH] **1.** Strength corresponding to cohesive forces between atoms. **2.** Hypothetically, the stress causing tensile fracture without plastic deformation.

coho *See* coherent oscillator.

cohomology theory [MATH] A theory which uses algebraic groups to study the geometric properties of topological spaces; closely related to homology theory.

coil [ELECTROMAG] A number of turns of wire used to introduce inductance into an electric circuit, to produce magnetic flux, or to react mechanically to a changing magnetic flux; in high-frequency circuits a coil may be only a fraction of a turn. Also known as electric coil; inductance; inductance coil; inductor. [SCI TECH] An arrangement of flexible material into a spiral or helix.

coil spring [DES ENG] A helical or spiral spring, such as one of the helical springs used over the front wheels in an automotive suspension.

coincidence [GEN] A numerical value equal to the number of double crossovers observed, divided by the number expected.

coincidence amplifier [ELECTR] An electronic circuit that amplifies only that portion of a signal present when an enabling or controlling signal is simultaneously applied.

coincidence circuit [ELECTR] A circuit that produces a specified output pulse only when a specified number or combination of two or more input terminals receives pulses within an assigned time interval. Also known as coincidence counter; coincidence gate.

coincidence counter *See* coincidence circuit.

coincidence counting [NUCLEO] A method of distinguishing particular types of events from background events and of measuring the velocities or directions of particles, by registering the occurrence of counts in two or more particle detectors within a given time interval by means of coincidence circuits.

coincidence gate *See* coincidence circuit.

coincidence rangefinder [OPTICS] An optical rangefinder in which one-eyed viewing through a single eyepiece provides the basis for manipulation of the rangefinder adjustment to cause two images of the target or parts of each, viewed over different paths, to match or coincide.

coitus [ZOO] The act of copulation. Also known as intercourse.

coke [MATER] A coherent, cellular, solid residue remaining from the dry (destructive) distillation of a coking coal or of pitch, petroleum, petroleum residues, or other carbonaceous materials; contains carbon as its principal constituent, together with mineral matter and volatile matter.

coking [CHEM ENG] 1. Destructive distillation of coal to make coke 2. A process for thermally converting the heavy residual bottoms of crude oil entirely to lower-boiling petroleum products and by-product petroleum coke.

col [GEOL] A high, sharp-edged pass occurring in a mountain ridge, usually produced by the headward erosion of opposing cirques. [METEOROL] The point of intersection of a trough and a ridge in the pressure pattern of a weather map; it is the point of relatively lowest pressure between two highs and the point of relatively highest pressure between two lows. Also known as neutral point; saddle point.

Col See Columba.

cola [BOT] *Cola acuminata*. A tree of the sterculia family (Sterculiaceae) cultivated for cola nuts, the seeds of the fruit; extract of cola nuts is used in the manufacture of soft drinks.

colchicine [ORG CHEM] $C_{22}H_{25}O_6N$ An alkaloid extracted from the stem of the autumn crocus; used experimentally to inhibit spindle formation and delay centromere division, and medicinally in the treatment of gout.

cold [ELEC] Pertaining to electrical circuits that are disconnected from voltage supplies and at ground potential; opposed to hot, pertaining to carrying an electrical charge.

cold-air machine [MECH ENG] A refrigeration system in which air serves as the refrigerant in a cycle of adiabatic compression, cooling to ambient temperature, and adiabatic expansion to refrigeration temperature; the air is customarily reused in a closed superatmospheric pressure system. Also known as dense-air system.

cold anticyclone See cold high.

cold-blooded [PHYSIO] Having body temperature approximating that of the environment and not internally regulated.

cold cathode [ELECTR] A cathode whose operation does not depend on its temperature being above the ambient temperature.

cold-cathode discharge See glow discharge.

cold-cathode ionization gage See Philips ionization gage.

cold-cathode rectifier [ELECTR] A cold-cathode gas tube in which the electrodes differ greatly in size so electron flow is much greater in one direction than in the other. Also known as gas-filled rectifier.

cold-cathode tube [ELECTR] An electron tube containing a cold cathode, such as a cold-cathode rectifier, mercury-pool rectifier, neon tube, phototube, or voltage regulator.

cold color [GRAPHICS] In printing, a color that has bluish tones.

cold-core cyclone See cold low.

cold-core high See cold high.

cold-core low See cold low.

cold drawing [MET] Drawing a tube or wire through a series of successively smaller dies, without the ap-

plication of heat, to reduce its diameter. [TEXT] Drawing a textile, as nylon, when cold.

cold emission See field emission.

cold extrusion [MET] Shaping cold metal by striking a slug in a closed cavity with a punch so that the metal is forced up around the punch. Also known as cold forging; cold pressing; extrusion pressing; impact extrusion.

cold forging See cold extrusion.

cold front [METEOROL] Any nonoccluded front, or portion thereof, that moves so that the colder air replaces the warmer air; the leading edge of a relatively cold air mass.

cold high [METEOROL] At a given level in the atmosphere, any high that is generally characterized by colder air near its center than around its periphery. Also known as cold anticyclone; cold-core high.

cold light [PHYS] 1. Light emitted in luminescence. 2. Visible light which is accompanied by little or no infrared radiation, and therefore has little heating effect.

cold lime-soda process [CHEM ENG] A water-softening process in which water is treated with hydrated lime (sometimes in combination with soda ash), which reacts with dissolved calcium and magnesium compounds to form precipitates that can be removed as sludge.

cold low [METEOROL] At a given level in the atmosphere, any low that is generally characterized by colder air near its center than around its periphery. Also known as cold-core cyclone; cold-core low.

cold neutron [SOLID STATE] A very-low-energy neutron in a reactor, used for research into solid-state physics because it has a wavelength of the order of crystal lattice spacings and can therefore be diffracted by crystals.

cold pressing See cold extrusion.

cold rolling [MET] Rolling metal at room temperature to reduce thickness or harden the surface; results in a smooth finish and improved resistance to fatigue.

cold shot [MET] Intensely hard, globular portions of the surface of an ingot or casting formed by premature solidification upon first contact with the cold sand during pouring.

cold slug [ENG] The first material to enter an injection mold in plastics manufacturing.

cold-spot hygrometer See dew-point hygrometer.

cold storage [ENG] The storage of perishables at low temperatures produced by refrigeration, usually above freezing, to increase storage life.

cold test [CHEM ENG] A test to determine the temperature at which clouding or coagulation is first visible in a sample of oil, as the temperature of the sample is reduced.

cold-type composition [GRAPHICS] Any typesetting method which produces copy suitable for offset lithography; copy may be obtained from a typewriter or photocomposition equipment.

cold wave [METEOROL] A rapid fall in temperature within 24 hours to a level requiring substantially increased protection to agriculture, industry, commerce, and social activities.

cold welding [MET] Welding in which a molecular bond is obtained by a cold flow of metal under extremely high pressures, without heat; widely used for sealing transistors and quartz crystal holders.

Coleochaetaceae [BOT] A family of green algae in the suborder Ulotrichineae; all occur as attached, disklike, or parenchymatous thalli.

Coleodontidae [PALEON] A family of conodonts in the suborder Neurodontiformes.

Coleoidea [INV ZOO] A subclass of cephalopod mollusks including all cephalopods except *Nautilus*, according to certain systems of classification.

Coleophoridae [INV ZOO] The case bearers, moths with narrow wings composing a family of lepidopteran insects in the suborder Heteroneura; named for the silk-and-leaf shell carried by larvae.

Coleoptera [INV ZOO] The beetles, holometabolous insects making up the largest order of the animal kingdom; general features of the Insecta are found in this group.

Coleorrhyncha [INV ZOO] A monofamilial group of homopteran insects in which the beak is formed at the anteroventral extremity of the face and the propleura form a shield for the base of the beak.

Coleosporaceae [MYCOL] A family of parasitic fungi in the order Uredinales.

colic [MED] 1. Acute paroxysmal abdominal pain usually caused by smooth muscle spasm, obstruction, or twisting. 2. In early infancy, paroxysms of pain, crying, and irritability caused by swallowing air, overfeeding, intestinal allergy, and emotional factors.

colidar See ladar.

Coliidae [VERT ZOO] The colies or mousebirds, composing the single family of the avian order Coliiformes.

Coliiformes [VERT ZOO] A monofamilial order of birds distinguished by long tails, short legs, and long toes, all four of which are directed forward.

coliphage [VIROL] Any bacteriophage able to infect *Escherichia coli*.

colk See pothole.

collagen [BIOCHEM] A fibrous protein found in all multicellular animals, especially in connective tissue.

collapse structure [GEOL] A structure resulting from rock slides under the influence of gravity. Also known as gravity-collapse structure.

collar [DES ENG] A ring placed around an object to restrict its motion, hold it in place, or cover an opening. [MIN ENG] The mouth of a mine shaft. [NAV ARCH] 1. An opening in the end or bight of a rope or cable supporting a mast that goes over the masthead. 2. A ring or loop of metal, rope, or other material, used to secure a heart or deadeye. 3. A fitting over a structural part passing through a bulkhead or deck.

collar cell See choanocyte.

collard [BOT] *Brassica oleracea* var. *acephala*. A biennial crucifer of the order Capparales grown for its rosette of edible leaves.

collared hole [ENG] A started hole drilled sufficiently deep to confine the drill bit and prevent slippage of the bit from normal position.

collate [COMPUT SCI] To combine two or more similarly ordered sets of values into one set that may or may not have the same order as the original sets. [GRAPHICS] To assemble in proper sequence all the sheets, signatures, or insertions for a printed piece.

collateral series [NUC PHYS] A radioactive decay series, initiated by transmutation, that eventually joins into one of the four radioactive decay series encountered in natural radioactivity.

collating unit [COMPUT SCI] An electromechanical device capable of performing singly or simultaneously the merging, sequence-checking, selection, and matching of punched cards. Also known as collator.

collator See collating unit.

collecting tubule [ANAT] One of the ducts conveying urine from the renal tubules (nephrons) to the minor calyces of the renal pelvis.

collective [METEOROL] In aviation weather observations, a group of observations transmitted in prescribed order by stations on the same long-line teletypewriter circuit. Also known as sequence.

collector [ELECTR] 1. A semiconductive region through which a primary flow of charge carriers leaves the base of a transistor; the electrode or terminal connected to this region is also called the collector. 2. An electrode that collects electrons or ions which have completed their functions within an electron tube; a collector receives electrons after they have done useful work, whereas an anode receives electrons whose useful work is to be done outside the tube. Also known as electron collector. [ENG] A class of instruments employed to determine the electric potential at a point in the atmosphere, and ultimately the atmospheric electric field; all collectors consist of some device for rapidly bringing a conductor to the same potential as the air immediately surrounding it, plus some form of electrometer for measuring the difference in potential between the equilibrated collector and the earth itself; collectors differ widely in their speed of response to atmospheric potential changes.

collector junction [ELECTR] A semiconductor junction located between the base and collector electrodes of a transistor.

collector plate [ELEC] One of several metal inserts that are sometimes embedded in the lining of an electrolyte cell to make the resistance between the cell lining and the current leads as small as possible.

Collembola [INV ZOO] The springtails, an order of primitive insects in the subclass Apterygota having six abdominal segments.

collenchyma [BOT] A primary, or early-differentiated, subepidermal supporting tissue in leaf petioles and vein ribs formed before vascular differentiation.

collenchyme [INV ZOO] A loose mesenchyme that fills the space between ectoderm and endoderm in the body wall of many lower invertebrates, such as sponges.

collet [DES ENG] A split, coned sleeve to hold small, circular tools or work in the nose of a lathe or other type of machine. [ENG] 1. The glass neck remaining on a bottle after it is taken off the glass-blowing iron. 2. Pieces of glass, ordinarily discarded, that are added to a batch of glass. Also spelled cullet. [HOROL] A small, friction-tight collar on a balance staff which holds the inner end of a balance spring. [LAP] The small, horizontal face at the bottom of a brilliant-cut gemstone.

Colletidae [INV ZOO] The colletid bees, a family of hymenopteran insects in the superfamily Apoidea.

colliery [MIN ENG] A whole coal mining plant; generally the term is used in connection with anthracite mining but sometimes to designate the mine, shops, and preparation plant of a bituminous operation.

colligative properties [PHYS CHEM] Properties dependent on the number of molecules but not their nature.

collimate [PHYS] To render parallel to a certain line or direction; paths of electrons in a flooding beam, or paths of various rays of a scanning beam are collimated to cause them to become more nearly parallel as they approach the storage assembly of a storage tube.

collimating lens [OPTICS] A lens on a collimator used to focus light from a source near one of its focal points into a parallel beam.

collimation error [ASTRON] The amount by which the angle between the optical axis of a transit telescope and its east-west mechanical axis deviates from 90°. [ENG] 1. Angular error in magnitude and direction between two nominally parallel lines of sight. 2. Specifically, the angle by which the line of sight of a radar differs from what it should be.

collimator [OPTICS] An instrument which produces parallel rays of light. [PHYS] A device for confining the elements of a beam within an assigned solid angle.

collinear array See linear array.

collinear vectors [MATH] Two vectors, one of which is a nonzero scaler multiple of the other.

collineation [MATH] A mapping which transforms points into points, lines into lines, and planes into planes. Also known as collineatory transformation.

collineatory transformation See collineation.

collinsite [MINERAL] $Ca_2(Mg,Fe)(PO_4)_2$ A phosphate mineral occurring in concentric layers in phosphoric nodules; found in meteorites.

collision [PHYS] An interaction resulting from the close approach of two or more bodies, particles, or systems of particles, and confined to a relatively short time interval during which the motion of at least one of the particles or systems changes abruptly.

collision-avoidance system [ENG] Electronic devices and equipment used by a pilot to perform the functions of conflict detection and avoidance.

collision broadening See collision line-broadening.

collision bulkhead [NAV ARCH] A watertight partition in a ship, perpendicular to the fore and aft centerline of the ship, usually near the bow, for keeping out water in the event of a collision.

collision cross section See cross section.

collision excitation [ATOM PHYS] The excitation of a gas by collisions of moving charged particles.

collision ionization [ATOM PHYS] The ionization of atoms or molecules of a gas or vapor by collision with other particles.

collision line-broadening [SPECT] Spreading of a spectral line due to interruption of the radiation process when the radiator collides with another particle. Also known as collision broadening.

collision matrix See scattering matrix.

collision of the first kind [PHYS] An inelastic collision in which some of the kinetic energy of translational motion is converted to internal energy of the colliding systems. Also known as endoergic collision.

collision of the second kind [PHYS] An inelastic collision in which some of the internal energy of the colliding systems is converted to kinetic energy of translation. Also known as exoergic collision.

collision theory [PHYS CHEM] Theory of chemical reaction proposing that the rate of product formation is equal to the number of reactant-molecule collisions multiplied by a factor that corrects for low-energy-level collisions. [QUANT MECH] Theory to describe collisions of simple or complex particles, the derivation of collision cross sections from postulated interactions and the study of properties of collision amplitudes which follow from invariance principles such as conservation of probability and time-reversal invariance.

colloblast [INV ZOO] An adhesive cell on the tentacles of ctenophores.

colloform [GEOL] Pertaining to the rounded, globular texture of mineral formed by colloidal precipitation.

colloid [CHEM] The phase of a colloidal system made up of particles having dimensions of 10–10,000 angstroms (1–1000 nanometers) and which is dispersed in a different phase.

colloidal dispersion See colloidal system.

colloidal suspension See colloidal system.

colloidal system [CHEM] An intimate mixture of two substances, one of which, called the dispersed phase (or colloid), is uniformly distributed in a finely divided state through the second substance, called the dispersion medium (or dispersing medium); the dispersion medium or dispersed phase may be a gas, liquid, or solid. Also known as colloidal dispersion; colloidal suspension.

colloid chemistry [PHYS CHEM] The scientific study of matter whose size is approximately 10 to 10,000 angstroms (1 to 1000 nanometers), and which exists as a suspension in a continuous medium, especially a liquid, solid, or gaseous substance.

collophane [MINERAL] A massive, cryptocrystalline, carbonate-containing variety of apatite and a principal source of phosphates for fertilizers. Also known as collophanite.

collophanite See collophane.

Collothecacea [INV ZOO] A monofamilial suborder of mostly sessile rotifers in the order Monogonata; many species are encased in gelatinous tubes.

Collothecidae [INV ZOO] The single family of the Collothecacea.

colluvium [GEOL] Loose, incoherent deposits at the foot of a slope or cliff, brought there principally by gravity.

Collyritidae [PALEON] A family of extinct, small, ovoid, exocyclic Euechinoidea with fascioles or a plastron.

colog See cologarithm.

cologarithm [MATH] The cologarithm of a number is the logarithm of the reciprocal of that number. Abbreviated colog.

colon [ANAT] The portion of the human intestine extending from the cecum to the rectum; it is divided into four sections: ascending, transverse, descending, and sigmoid. Also known as large intestine.

colonnade [ARCH] A series of columns placed at regular intervals.

colony [BIOL] A localized population of individuals of the same species which are living either attached or separately. [MICROBIO] A cluster of microorganisms growing on the surface of or within a solid medium; usually cultured from a single cell.

color [OPTICS] A general term that refers to the wavelength composition of light, with particular reference to its visual appearance. [PARTIC PHYS] A hypothetical quantum number carried by quarks, so that each type of quark comes in three varieties which are identical in all measurable qualities but which differ in this additional property; this quantity determines the coupling of quarks to the gluon field.

color aberration See chromatic aberration.

color balance [ELECTR] Adjustment of the circuits feeding the three electron guns of a television color picture tube to compensate for differences in light-emitting efficiencies of the three color phosphors on the screen of the tube.

color-bar code [IND ENG] A code that uses one or more different colors of bars in combination with black bars and white spaces, to increase the density of binary coding of data printed on merchandise tags or directly on products for inventory control and other purposes.

color-blind [GRAPHICS] Of a photographic emulsion, sensitive only to blue, violet, and ultraviolet light.

color blindness [MED] Inability to perceive one or more colors.

color burst [ELECTR] The portion of the composite color television signal consisting of a few cycles of a sine wave of chrominance subcarrier frequency. Also known as burst; reference burst.

color carrier See chrominance subcarrier.

color center [SOLID STATE] A point lattice defect which produces optical absorption bands in an otherwise transparent crystal.

color circle [OPTICS] An arrangement of hues about the circumference of a circle in the order in which they appear in the electromagnetic spectrum, with pairs of complementary colors at opposite ends of diameters.

color code [ELEC] A system of colors used to indicate the electrical value of a component or to identify terminals and leads. [ENG] 1. Any system of colors used for purposes of identification, such as to identify dangerous areas of a factory. 2. A system of colors used to identify the type of material carried by a pipe; for example, dangerous materials, protective materials, extra valuable materials.

color coder See matrix.

color comparator [ANALY CHEM] A photoelectric instrument that compares an unknown color with that of a standard color sample for matching purposes. Also known as photoelectric color comparator.

color control See chroma control.

color correction [GRAPHICS] Any method used to improve color rendition; for example, masking, dot etching, reetching, and scanning. [OPTICS] The construction of an optical system so that the image positions of an object are the same for two or more wavelengths, and chromatic aberration is thus minimized.

color decoder See matrix.

color encoder See matrix.

colorfast [TEXT] Referring to a fabric that does not fade during normal wear.

color filter [OPTICS] An optical element that partially absorbs incident light, consisting of a pane of glass or other partially transparent material, or of films separated by narrow layers; the absorption may be either selective or nonselective with respect to wavelength. Also known as light filter.

colorimeter [ANALY CHEM] A device for measuring concentration of a known constituent in solution by comparison with colors of a few solutions of known concentration of that constituent. Also known as chromometer. [OPTICS] An instrument that measures color by determining the intensities of the three primary colors that will give that color.

colorimetry [OPTICS] Any technique by which an unknown color is evaluated in terms of standard colors; the technique may be visual, photoelectric, or indirect by means of spectrophotometry; used in chemistry and physics.

color index Abbreviated CI. [ASTRON] Of a star, the numerical difference between the apparent photographic magnitude and the apparent photovisual magnitude. [PATH] The amount of hemoglobin per erythrocyte relative to normal, equal to the percent normal hemoglobin concentration divided by percent normal erythrocyte count.

coloring agent [FOOD ENG] Any substance of natural origin, such as tumeric, annatto, caramel, carmine, and carotine, or a synthetic certified food color added to food to compensate for color changes during processing or to give an appetizing color.

color kinescope See color picture tube.

color lake See lake.

color oscillator See chroma oscillator.

color phase [COMMUN] The difference in phase between components (I or Q) of a chrominance signal and the chrominance-carrier reference in a color television receiver.

color picture tube [ELECTR] A cathode-ray tube having three different colors of phosphors, so that when these are appropriately scanned and excited in a color television receiver, a color picture is obtained. Also known as color kinescope; color television picture tube; tricolor picture tube.

color printing [GRAPHICS] The art and craft of embellishing designs, pictures, and typographic pages with color for a more pleasing effect than obtained in black and white; in addition, pictures more closely represent the original object or painting.

color purity [ELECTR] Absence of undesired colors in the spot produced on the screen by each beam of a television color picture tube.

color saturation [OPTICS] The degree to which a color is mixed with white; high saturation means little white, low saturation means much white. Also known as chroma; saturation.

color-saturation control See chroma control.

color separation [GRAPHICS] The process of preparing a separate drawing, engraving, or negative for each color required in the reproduction of a colored picture.

color signal [COMMUN] Any signal that controls the chromaticity values of a color television picture, such as the color picture signal and the chrominance signal.

color solid [OPTICS] A three-dimensional diagram which represents the relationship of three attributes of surface color: hue, saturation, and brightness.

color subcarrier See chrominance subcarrier.

color-subcarrier oscillator See chroma oscillator.

color television picture tube See color picture tube.

color temperature [STAT MECH] Of a solid surface, that temperature of a blackbody from which the radiant energy has essentially the same spectral distribution as that from the surface.

color test [ANALY CHEM] The quantitative analysis of a substance by comparing the intensity of the color produced in a sample by a reagent with a standard color produced similarly in a solution of known strength.

color transmission [COMMUN] In television, the transmission of a signal wave which represents both the brightness values and the chromaticity values in the picture.

color vision [PHYSIO] The ability to discriminate light on the basis of wavelength composition.

Colossendeidae [INV ZOO] A family of deep-water marine arthropods in the subphylum Pycnogonida, having long palpi and lacking chelifores, except in polymerous forms.

colostomy [MED] Surgical formation of an artificial anus by joining the colon to an opening in the anterior abdominal wall.

colostrum [PHYSIO] The first milk secreted by the mammary gland during the first days following parturition.

Colpitts oscillator [ELECTR] An oscillator in which a parallel-tuned tank circuit has two voltage-dividing capacitors in series, with their common connection going to the cathode in the electron-tube version and the emitter circuit in the transistor version.

Colubridae [VERT ZOO] A family of cosmopolitan snakes in the order Squamata.

Columba [ASTRON] A constellation, right ascension 6 hours, declination 35°S. Abbreviated Col. Also known as Dove.

Columbidae [VERT ZOO] A family of birds in the order Columbiformes composed of the pigeons and doves.

Columbiformes [VERT ZOO] An order of birds distinguished by a short, pointed bill, imperforate nostrils, and short legs.

columbite [MINERAL] $(Fe,Mn)(Cb,Ta)_2O_6$ An iron-black mineral with a submetallic luster that crystallizes in the orthorhombic system; the chief ore mineral of niobium (columbium); hardness is 6 on Mohs scale, and specific gravity is 5.4–6.5. Also known as dianite; greenlandite; niobite.

columbium See niobium.

columella [ANAT] See stapes. [BIOL] Any part shaped like a column. [BOT] A sterile axial body within the capsules of certain mosses, liverworts, and many fungi.

column [CHEM ENG] See tower. [COMPUT SCI] A vertical arrangement of characters or other expressions, usually referring to a specific print position on a printer or a vertical area on a card. [ENG] A vertical shaft designed to bear axial loads in compression. [MATH] See place. [NUCLEO] A hollow cylinder of water and spray thrown up from an underwater burst of an atomic weapon, through which hot, high-pressure gases are vented to the atmosphere; a somewhat similar column of dirt is formed in an underground explosion. Also known as plume.

columnar epithelium [HISTOL] Epithelium distinguished by elongated, columnar, or prismatic cells.

columnar jointing [GEOL] Parallel, prismatic columns that are formed as a result of contraction during cooling in basaltic flow and other extrusive and intrusive rocks. Also known as columnar structure; prismatic jointing; prismatic structure.

columnar stem [BOT] An unbranched, cylindrical stem bearing a set of large leaves at its summit, as in palms, or no leaves, as in cacti.

columnar structure [GEOL] See columnar jointing. [MINERAL] Mineral structure consisting of parallel columns of slender prismatic crystals. [PETR] A primary sedimentary structure consisting of columns arranged perpendicular to the bedding.

column binary [COMPUT SCI] The binary representation of data on punched cards in which adjacent positions in a column correspond to adjacent bits of data. Also known as Chinese binary.

column chromatography [ANALY CHEM] Chromatographic technique of two general types: packed columns usually contain either a granular adsorbent or a granular support material coated with a thin layer of high-boiling solvent (partitioning liquid); open-tubular columns contain a thin film of partitioning liquid on the column walls and have an opening so that gas can pass through the center of the column.

column operations [MATH] A set of rules for manipulating the columns of a matrix so that the image of the corresponding linear transformation remains unchanged.

column printer [COMPUT SCI] A small line printer used with some calculators to provide hard-copy printout of input and output data; typically consists

of 20 columns of numerals and a limited number of alphabetic or other identifying characters.

colure [ASTRON] A great circle of the celestial sphere through the celestial poles and either the equinoxes or solstices, called respectively the equinoctial colure or the solstitial colure.

colusite [MINERAL] $Cu_3(As,Sn,V,Fe,Te)S_4$ A bronze-colored mineral consisting of a sulfide of copper and arsenic with vanadium, iron, and telluride substituting for arsenic; usually occurs in massive form.

Com See Coma Berenices.

coma [ASTRON] The gaseous envelope that surrounds the nucleus of a comet. Also known as head. [ELECTR] A cathode-ray tube image defect that makes the spot on the screen appear comet-shaped when away from the center of the screen. [MED] Unconsciousness from which the patient cannot be aroused. [OPTICS] A manifestation of errors in an optical system, so that a point has an asymmetrical image (that is, appears as a pear-shaped spot).

Coma Berenices [ASTRON] A constellation, right ascension 13 hours, declination 20°N. Abbreviated Com. Also known as Berenice's Hair.

comagmatic province See petrographic province.

Comanchian [GEOL] Lower Cretaceous geologic time.

Comasteridae [INV ZOO] A family of radially symmetrical Crinozoa in the order Comatulida.

comatose [MED] In a condition of coma; resembling coma.

Comatulida [INV ZOO] The feather stars, an order of free-living echinoderms in the subclass Articulata.

comb [INV ZOO] **1.** A system of hexagonal cells constructed of beeswax by a colony of bees. **2.** A comb-like swimming plate in ctenophores. [VERT ZOO] A crest of naked tissue on the head of many male fowl.

comb antenna [ELECTROMAG] A broad-band antenna for vertically polarized signals, in which half of a fishbone antenna is erected vertically and fed against ground by a coaxial line.

combed cotton [TEXT] Cotton yarn that has been cleaned with wire brushes and roller cards after carding to remove short fibers and other impurities.

comb filter [ELECTR] A wave filter whose frequency spectrum consists of a number of equispaced elements resembling the teeth of a comb.

combination [MATH] A selection of one or more of the elements of a given set without regard to order.

combination coefficient [GEOPHYS] A measure of the specific rate of disappearance of small ions in the atmosphere due to either union with neutral Aitken nuclei to form new large ions, or union with large ions of opposite sign to form neutral Aitken nuclei.

combination mill [MET] A rolling mill arranged with continuous rolls for roughing and a guide or looping mill for shaping.

combination principle See Ritz's combination principle.

combination square [DES ENG] A square head and steel rule that when used together have both a 45 and 90° face to allow the testing of the accuracy of two surfaces intended to have these angles.

combination trap [GEOL] Underground reservoir structure closure, deformation, or fault where reservoir rock covers only part of the structure.

combination unit [CHEM ENG] A processing unit that combines more than one process, such as straight-run distillation together with selective cracking.

combination vibration [SPECT] A vibration of a polyatomic molecule involving the simultaneous excitation of two or more normal vibrations.

combinatorial analysis [MATH] 1. The determination of the number of possible outcomes in ideal games of chance by using formulas for computing numbers of combinations and permutations. 2. The study of large finite problems.

combinatorial theory [MATH] The branch of mathematics which studies the arrangements of elements into sets.

combinatorial topology [MATH] The study of polyhedrons, simplicial complexes, and generalizations of these. Also known as piecewise-linear topology.

combinatorics [MATH] Combinatorial topology which studies geometric forms by breaking them into simple geometric figures.

combined footing [CIV ENG] A footing, either rectangular or trapezoidal, that supports two columns.

combined head See read/write head.

combiner circuit [ELECTR] The circuit that combines the luminance and chrominance signals with the synchronizing signals in a color television camera chain.

combining-volumes principle [CHEM] The principle that when gases take part in chemical reactions the volumes of the reacting gases and those of the products (if gaseous) are in the ratio of small whole numbers, provided that all measurements are made at the same temperature and pressure. Also known as Gay-Lussac law.

combining weight [CHEM] The weight of an element that chemically combines with 8 grams of oxygen or its equivalent.

combustion [CHEM] The burning of gas, liquid, or solid, in which the fuel is oxidized, evolving heat and often light.

combustion chamber [AERO ENG] That part of the rocket engine in which the combustion of propellants takes place at high pressure. Also known as firing chamber. [ENG] Any chamber in which a fuel such as oil, coal, or kerosine is burned to provide heat. [MECH ENG] The space at the head end of an internal combustion engine cylinder where most of the combustion takes place.

combustion efficiency [CHEM] The ratio of heat actually developed in a combustion process to the heat that would be released if the combustion were perfect.

combustion engine [MECH ENG] An engine that operates by the energy of combustion of a fuel.

combustion furnace [ANALY CHEM] A heating device used in the analysis of organic compounds for elements. [ENG] A furnace whose source of heat is the energy released in the oxidation of fossil fuel.

combustion knock See engine knock.

combustion turbine See gas turbine.

combustion wave [CHEM] 1. A zone of burning propagated through a combustible medium. 2. The zoned, reacting, gaseous material formed when an explosive mixture is ignited.

combustor [MECH ENG] The combustion chamber together with burners, igniters, and injection devices in a gas turbine or jet engine.

comedo [MED] A collection of sebaceous material and keratin retained in the hair follicle and excretory duct of the sebaceous gland, whose surface is covered with a black dot caused by oxidation of sebum at the follicular orifice. Also known as blackhead.

comes [ASTRON] The smaller star in a binary system. Also known as companion.

Comesomatidae [INV ZOO] A family of free-living nematodes in the superfamily Chromadoroidea found as deposit feeders on soft bottom sediments.

comet [ASTRON] A nebulous celestial body having a fuzzy head surrounding a bright nucleus, one of two major types of bodies moving in closed orbits about the sun; in comparison with the planets, the comets are characterized by their more eccentric orbits and greater range of inclination to the ecliptic.

comfort control [ENG] Control of temperature, humidity, flow, and composition of air by using heating and air-conditioning systems, ventilators, or other systems to increase the comfort of people in an enclosure.

comfort curve [ENG] A line drawn on a graph of air temperature versus some function of humidity (usually wet-bulb temperature or relative humidity) to show the varying conditions under which the average sedentary person feels the same degree of comfort; a curve of constant comfort.

comfort index See temperature humidity index.

comfort standard See comfort zone.

comfort zone [ENG] The ranges of indoor temperature, humidity, and air movement, under which most persons enjoy mental and physical well-being. Also known as comfort standard.

COMIT [COMPUT SCI] A user-oriented, general-purpose, symbol-manipulation programming language for computers.

Comleyan [GEOL] Lower Cambrian geologic time.

comma [ACOUS] The difference between the larger and smaller whole tones in the just scale, corresponding to a frequency ratio of 81/80.

command [COMPUT SCI] A signal that initiates a predetermined type of computer operation that is defined by an instruction. [CONT SYS] An independent signal in a feedback control system, from which the dependent signals are controlled in a predetermined manner.

command code See operation code.

command control See command guidance.

command control program [COMPUT SCI] The interface between a time-sharing computer and its users by means of which they can create, edit, save, delete, and execute their programs.

command guidance [ENG] A type of electronic guidance of guided missiles or other guided aircraft wherein signals or pulses sent out by an operator cause the guided object to fly a directed path. Also known as command control.

command interpreter [COMPUT SCI] A program that processes commands and other input and output from an active terminal in a time-sharing system.

command language [COMPUT SCI] The language of an operating system, through which the users of a data-processing system describe the requirements of their tasks to that system. Also known as job control language.

command mode [COMPUT SCI] The status of a terminal in a time-sharing environment enabling the programmer to use the command control program.

command module [AERO ENG] The spacecraft module that carries the crew, the main communication and telemetry equipment, and the reentry capsule during cruising flight.

command pulses [ELECTR] The electrical representations of bit values of 1 or 0 which control input/output devices.

Commelinaceae [BOT] A family of monocotyledonous plants in the order Commelinales characterized by differentiation of the leaves into a closed sheath and a well-defined, commonly somewhat succulent blade.

Commelinales [BOT] An order of monocotyledonous plants in the subclass Commelinidae marked by having differentiated sepals and petals but lacking nectaries and nectar.

Commelinidae [BOT] A subclass of flowering plants in the class Liliopsida.

commensalism [ECOL] An interspecific, symbiotic relationship in which two different species are associated, wherein one is benefited and the other neither benefited nor harmed.

comment [COMPUT SCI] An expression identifying or explaining one or more steps in a routine, which has no effect on execution of the routine.

commercial lecithin *See* lecithin.

comminution [MECH ENG] Breaking up or grinding into small fragments. Also known as pulverization.

commissure [BIOL] A joint, seam, or closure line where two structures unite.

common area [COMPUT SCI] An area of storage which two or more routines share.

common-base connection *See* grounded-base connection.

common-base feedback oscillator [ELECTR] A bipolar transistor amplifier with a common-base connection and a positive feedback network between the collector (output) and the emitter (input).

common bile duct [ANAT] The duct formed by the union of the hepatic and cystic ducts.

common bond *See* American bond.

common branch [ELEC] A branch of an electrical network which is common to two or more meshes. Also known as mutual branch.

common business-oriented language *See* COBOL.

common carotid artery *See* carotid artery.

common carrier [IND ENG] A company recognized by an appropriate regulatory agency as having a vested interest in furnishing communications services or in transporting commodities or people.

common-collector connection *See* grounded-collector connection.

common control unit [COMPUT SCI] Control unit that is shared by more than one machine.

common declaration statement [COMPUT SCI] A nonexecutable statement in FORTRAN which allows specified arrays or variables to be stored in an area available to other programs.

common denominator [MATH] Any common multiple of the denominators of a collection of fractions.

common-drain amplifier [ELECTR] An amplifier using a field-effect transistor so that the input signal is injected between gate and drain, while the output is taken between the source and drain. Also known as source-follower amplifier.

common-emitter connection *See* grounded-emitter connection.

common feldspar *See* orthoclase.

common fraction [MATH] A fraction whose numerator and denominator are both integers.

common-gate amplifier [ELECTR] An amplifier using a field-effect transistor in which the gate is common to both the input circuit and the output circuit.

common hepatic duct *See* hepatic duct.

common iliac artery *See* iliac artery.

common impedance coupling [ELECTROMAG] The interaction of two circuits by means of an inductance or capacitance in a branch which is common to both circuits.

common-ion effect [CHEM] The lowering of the degree of ionization of a compound when another ionizable compound is added to a solution; the compound added has a common ion with the other compound.

common language [COMPUT SCI] A machine-readable language that is common to a group of computers and associated equipment.

common logarithm [MATH] The exponent in the representation of a number as a power of 10. Also known as Briggs' logarithm.

common mica *See* muscovite.

common mode [ELECTR] Having signals that are identical in amplitude and phase at both inputs, as in a differential operational amplifier.

common multiple [MATH] A quantity (polynomial number) divisible by all quantities in a given set.

common pyrite *See* pyrite.

common salt *See* halite; sodium chloride.

common-source amplifier [ELECTR] An amplifier stage using a field-effect transistor in which the input signal is applied between gate and source and the output signal is taken between drain and source.

common storage [COMPUT SCI] A section of memory in certain computers reserved for temporary storage of program outputs to be used as input for other programs.

common-user channel [COMMUN] Any of the communications channels which are available to all authorized agencies for transmission of command, administrative, and logistic traffic.

common user circuit [ELEC] A circuit designated to furnish a communications service to a number of users.

communication [COMMUN] The transmission of intelligence between two or more points over wires or by radio; the terms telecommunication and communication are often used interchangeably, but telecommunication is usually the preferred term when long distances are involved.

communication band [COMMUN] The band of frequencies effectively occupied by a radio transmitter for the type of transmission and the speed of signaling used.

communication cable [COMMUN] A uniform conductive circuit used in the telephone industry to connect customers to their local switching centers and to interconnect local and long-distance switching centers.

communication channel [COMMUN] The wire or radio channel that serves to convey intelligence between two or more terminals.

communication engineering [COMMUN] The design, construction, and operation of all types of equipment used for radio, wire, or other types of communication.

communication link *See* data link.

communications [ENG] The science and technology by which information is collected from an originating source, transformed into electric currents or fields, transmitted over electrical networks or space to another point, and reconverted into a form suitable for interpretation by a receiver.

communications network [COMMUN] Organization of stations capable of intercommunications but not necessarily on the same channel.

communications satellite [AERO ENG] An orbiting, artificial earth satellite that relays radio, television, and other signals between ground terminal stations thousands of miles apart. Also known as radio relay satellite; relay satellite.

communication system [COMMUN] A telephone, telegraph, teletypewriter, picture, data transmission, or other system in which electrical impulses originated at one place are faithfully reproduced at a distant point.

community [ECOL] Aggregation of organisms characterized by a distinctive combination of two or more ecologically related species; an example is a deciduous forest. Also known as ecological community.

community antenna television *See* cable television.

commutating capacitor [ELECTR] A capacitor used in gas-tube rectifier circuits to prevent the anode from going highly negative immediately after extinction.

commutating pole [ELECTROMAG] One of several small poles between the main poles of a direct-current generator or motor, which serves to neutralize the flux distortion in the neutral plane caused by armature reaction. Also known as compole; interpole.

commutating reactance [ELECTR] An inductive reactance placed in the cathode lead of a three-phase mercury-arc rectifier to ensure that tube current holds over during transfer of conduction from one anode to the next.

commutating zone [ELECTROMAG] The part of the armature of an electric machine that contains the windings which are short-circuited by the brush on the commutator at a particular instant.

commutation [COMMUN] The sampling of various quantities in a repetitive manner for transmission over a single channel in telemetering. [ELECTR] The transfer of current from one channel to another in a gas tube. [ELECTROMAG] The process of current reversal in the armature windings of a direct-current rotating machine to provide direct current at the brushes.

commutation rules [QUANT MECH] The specification of the commutators of operators corresponding to the dynamical variables of a system, which are equal to $i\hbar$ times the Poisson brackets of the classical variables to which the operators correspond.

commutative law [MATH] A rule which requires that the result of a binary operation be independent of order; that is, $ab = ba$.

commutator [ELECTROMAG] That part of a direct-current motor or generator which serves the dual function, in combination with brushes, of providing an electrical connection between the rotating armature winding and the stationary terminals, and of permitting reversal of the current in the armature windings. [MATH] The commutator of a and b is the element c of a group such that $bac = ab$. [QUANT MECH] The commutator of a and b is $[a,b] = ab - ba$.

commutator pulse [COMPUT SCI] One of a series of pulses indicating the beginning or end of a signal representing a single binary digit in a computer word. Also known as position pulse (P pulse).

commutator switch [ELEC] A switch, usually rotary and mechanically driven, that performs a set of switching operations in repeated sequential order, such as is required for telemetering many quantities. Also known as sampling switch; scanning switch.

compaction [COMPUT SCI] A technique for reducing the space required for data storage without losing any information content. [ENG] Increasing the dry density of a granular material, particularly soil, by means such as impact or by rolling the surface layers. [GEOL] Process by which soil and sediment mass loses pore space in response to the increasing weight of overlying material.

compact-open topology [MATH] A topology on the space of all continuous functions from one topological space into another; a subbase for this topology is given by the sets $W(K,U) = \{f:f(K)\subset U\}$, where K is compact and U is open.

compact operator [MATH] A linear transformation from one normed vector space to another, with the property that the image of every bounded set has a compact closure.

compact set [MATH] A set in a topological space with the property that every open cover has a finite subset which is also a cover. Also known as bicompact set.

compact space [MATH] A topological space which is a compact set.

companding [ELECTR] A process in which compression is followed by expansion; often used for noise reduction in equipment, in which case compression is applied before noise exposure and expansion after exposure.

compandor [ELECTR] A system for improving the signal-to-noise ratio by compressing the volume range of the signal at a transmitter or recorder by means of a compressor and restoring the normal range at the receiving or reproducing apparatus with an expander.

companion *See* comes.

companion cell [BOT] A specialized parenchyma cell occurring in close developmental and physiologic association with a sieve-tube member.

comparator [COMPUT SCI] A device that compares two transcriptions of the same information to verify the accuracy of transcription, storage, arithmetical operation, or some other process in a computer, and delivers an output signal of some form to indicate whether or not the two sources are equal or in agreement. [CONT SYS] A device which detects the value of the quantity to be controlled by a feedback control system and compares it continuously with the desired value of that quantity. [ENG] A device used to inspect a gaged part for deviation from a specified dimension, by mechanical, electrical, pneumatic, or optical means.

comparator circuit [ELECTR] An electronic circuit that produces an output voltage or current whenever two input levels simultaneously satisfy predetermined amplitude requirements; may be linear (continuous) or digital (discrete).

comparator probe [COMPUT SCI] A component of a hardware monitor that is used to sense the number of bits that appear in parallel, as in an address register.

comparing brushes [COMPUT SCI] Sets of metallic brushes which verify that all the cards in a gang-punching operation have been properly punched.

comparing control change *See* control change.

comparing unit [ELECTR] An electromechanical device which compares two groups of timed pulses and signals to establish either identity or nonidentity.

comparison [COMPUT SCI] A computer operation in which two numbers are compared as to identity, relative magnitude, or sign.

comparison bridge [ELECTR] A bridge circuit in which any change in the output voltage with respect to a reference voltage creates a corresponding error signal, which, by means of negative feedback, is used to correct the output voltage and thereby restore bridge balance.

comparison indicators [COMPUT SCI] Registers, one of which is activated during the comparison of two quantities to indicate whether the first quantity is lower than, equal to, or greater than the second quantity.

comparison spectrum [SPECT] A line spectrum whose wavelengths are accurately known, and which is matched with another spectrum to determine the wavelengths of the latter.

compass [ENG] An instrument for indicating a horizontal reference direction relative to the earth. [GRAPHICS] An instrument used for describing arcs or circles with pencil or pen; has two legs hinged together at the top.

compass bearing [NAV] Direction relative to north as indicated by a compass.

compass card [DES ENG] The part of a compass on which the direction graduations are placed, it is usually in the form of a thin disk or annulus graduated in degrees, clockwise from 0° at the reference direction to 360°, and sometimes also in compass points.

compass declinometer [ENG] An instrument used for magnetic distribution surveys; employs a thin compass needle 6 inches (15 centimeters) long, supported on a sapphire bearing and steel pivot of high quality; peep sights serve for aligning the compass box on an azimuth mark.

compass deviation [NAV] The difference between the readings of a compass which is without mechanical defects and is held motionless in space, and the same instrument when it is installed in the same geographic position but is mounted on a ship or aircraft; deviation is a systematic error which is compensated by placing iron bars in places about the compass; residual deviation errors are calibrated and noted on a card so it can be used by the pilot or navigator.

compass error [NAV] The angle by which a compass direction differs from the true geographical direction; assuming that there is no lubber line error, it is the algebraic sum of the declination, variation, and motion errors.

Compasses *See* Circinus.

compass north [NAV] The direction of north as indicated by a magnetic compass; the reference direction for measurement of compass directions.

compass points [GEOD] The 32 divisions of a compass at intervals of 11¼°, with each division further divided into quarter points. Also known as points of the compass.

compass rose [NAV] A graduated circle, usually marked in degrees, indicating directions north, south, east, and west, and inscribed on a chart; used for the calibration of compasses on crafts.

compatibility [COMPUT SCI] The ability of one device to accept data handled by another device without conversion of the data or modification of the code. [IMMUNOL] Ability of two bloods or other tissues to unite and function together. [ORD] In ammunition, the ability of a given material to exist unchanged under certain conditions of temperature and moisture when in the presence of some other specificmaterial. [PHARM] The capacity of two or more ingredients in a medicine to mix without chemical change or loss of therapeutic effectiveness. [SYS ENG] The ability of a new system to serve users of an old system.

compatible discrete four-channel sound [ENG ACOUS] A sound system in which a separate channel is maintained from each of the four sets of microphones at the recording studio or other input location to the four sets of loudspeakers that serve as the output of the system. Abbreviated CD-4 sound.

compatible monolithic integrated circuit [ELECTR] Device in which passive components are deposited by thin-film techniques on top of a basic silicon-substrate circuit containing the active components and some passive parts.

compensated amplifier [ELECTR] A broad-band amplifier in which the frequency range is extended by choice of circuit constants.

compensated pendulum [DES ENG] A pendulum made of two materials with different coefficients of expansion so that the distance between the point of suspension and center of oscillation remains nearly constant when the temperature changes.

compensated semiconductor [ELECTR] Semiconductor in which one type of impurity or imperfection (for example, donor) partially cancels the electrical effects on the other type of impurity or imperfection (for example, acceptor).

compensating eyepiece [OPTICS] A type of Huygens eyepiece in which the eye lens is achromatized to compensate for the color errors of the objective.

compensating network [CONT SYS] A network used in a low-energy-level method for suppression of excessive oscillations in a control system.

compensation [CONT SYS] Introduction of additional equipment into a control system in order to reshape its root locus so as to improve system performance. Also known as stabilization. [ELECTR] The modification of the amplitude-frequency response of an amplifier to broaden the bandwidth or to make the response more nearly uniform over the existing bandwidth. Also known as frequency compensation. [PSYCH] Counterbalancing a weakness or failure in one area by stressing or substituting a strength or success in another area.

compensator [CONT SYS] A device introduced into a feedback control system to improve performance and achieve stability. Also known as filter. [ELECTR] A component that offsets an error or other undesired effect. [OPTICS] A device, usually consisting of two quartz wedges, for determining the phase difference between the two components of elliptically polarized light.

competence [EMBRYO] The ability of a reacting system to respond to the inductive stimulus during early developmental stages. [GEOL] The ability of the wind to transport solid particles either by rolling, suspension, or saltation (intermittent rolling and suspension); usually expressed in terms of the weight of a single particle. [HYD] The ability of a stream, flowing at a given velocity, to move the largest particles.

competent beds [GEOL] Beds or strata capable of withstanding the pressures of folding without flowing or changing in original thickness.

competitive inhibition [BIOCHEM] Enzyme inhibition in which the inhibitor competes with the natural substrate for the active site of the enzyme; may be overcome by increasing substrate concentration.

compile [COMPUT SCI] To prepare a machine-language program automatically from a program written in a higher programming language, usually generating more than one machine instruction for each symbolic statement.

compiler [COMPUT SCI] A program to translate a higher programming language into machine language. Also known as compiling routine.

compiler system [COMPUT SCI] The set consisting of a higher-level language, such as FORTRAN, and its compiler which translates the program written in that language into machine-readable instructions.

compiling routine *See* compiler.

complement [IMMUNOL] A heat-sensitive, complex system in fresh human and other sera which, in combination with antibodies, is important in the host defense mechanism against invading microorganisms. [MATH] 1. The complement of a number A

is another number B such that the sum $A + B$ will produce a specified result. **2.** *See* radix complement. **3.** For a subset of a set, the collection of all members of the set which are not in the given subset.

complementarity [QUANT MECH] The principle that nature has complementary aspects, particle and wave; the two aspects are related by $p = h/\lambda$ and $E = h\nu$, where p and E are the momentum and energy of the particle, λ and ν are the wavelength and frequency of the wave, and h is Planck's constant.

complementary [ELECTR] Having pnp and npn or p- and n-channel semiconductor elements on or within the same integrated-circuit substrate or working together in the same functional amplifier state.

complementary angle [MATH] One of a pair of angles whose sum is $90°$.

complementary colors [OPTICS] Two colors which lie on opposite sides of the white point in the chromaticity diagram so that an additive mixture of the two, in appropriate proportions, can be made to yield an achromatic mixture.

complementary function [MATH] Any solution of the equation obtained from a given linear differential equation by replacing the inhomogeneous term with zero.

complementary genes [GEN] Nonallelic genes that complement one another's expression in a trait.

complementary logic switch [ELECTR] A complementary transistor pair which has a common input and interconnections such that one transistor is on when the other is off, and vice versa.

complementary metal oxide semiconductor device *See* CMOS device.

complementary symmetry [ELECTR] A circuit using both pnp and npn transistors in a symmetrical arrangement that permits push-pull operation without an input transformer or other form of phase inverter.

complementation [GEN] The action of complementary genes. [MATH] The act of replacing a set by its complement.

complementation law [STAT] The law that the probability of an event E is 1 minus the probability of the event not E.

complemented lattice [MATH] A lattice with distinguished elements a and b, and with the property that corresponding to each point x of the lattice, there is a y such that the greatest lower bound of x and y is a, and the least upper bound of x and y is b.

complement number system [COMPUT SCI] System of number handling in which the complement of the actual number is operated upon; used in some computers to facilitate arithmetic operations.

complete blood count [PATH] Differential and absolute determinations of the numbers of each type of blood cell in a sample and, by extrapolation, in the general circulation.

complete carry [COMPUT SCI] In parallel addition, an arrangement in which the carries that result from the addition of carry digits are allowed to propagate from place to place.

complete-expansion diesel cycle *See* Brayton cycle.

complete flower [BOT] A flower having all four floral parts, that is, having sepals, petals, stamens, and carpels.

complete induction *See* mathematical induction.

complete integral [MATH] **1.** A solution of an nth order ordinary differential equation which depends on n arbitrary constants as well as the independent

variable. Also known as complete primitive. **2.** A solution of a first-order partial differential equation with n independent variables which depends upon n arbitrary parameters as well as the independent variables.

complete leaf [BOT] A dicotyledon leaf consisting of three parts: blade, petiole, and a pair of stipules.

completely inelastic collision *See* perfectly inelastic collision.

completely normal space [MATH] A topological space with the property that any pair of sets with disjoint closures can be separated by open sets.

completely regular space [MATH] A topological space X where for every point x and neighborhood U of x there is a continuous function from X to $[0,1]$ with $f(x) = 1$ and $f(y) = 0$, if y is not in U. Also known as Tychonoff space.

complete normed linear space *See* Banach space.

complete operation [COMPUT SCI] An operation which includes obtaining all operands from storage, performing the operation, returning resulting operands to storage, and obtaining the next instruction.

complete orthonormal set [MATH] A set of mutually orthogonal unit vectors in a (possibly infinite dimensional) vector space which is contained in no larger such set, that is, no nonzero vector is perpendicular to all the vectors in the set. Also known as closed orthonormal set.

complete primitive *See* complete integral.

complete routine [COMPUT SCI] A routine, generally supplied by a computer manufacturer, which does not have to be modified by the user before being applied.

complex [GEOL] An assemblage of rocks that has been folded together, intricately mixed, involved, or otherwise complicated. [MATH] A space which is represented as a union of simplices which intersect only on their faces. [MED] *See* syndrome. [MINERAL] Composed of many ingredients. [PSYCH] A group of associated ideas with strong emotional tones, which have been transferred from the conscious mind into the unconscious and which influence the personality.

complexation *See* complexing.

complex compound [CHEM] Any of a group of chemical compounds in which a part of the molecular bonding is of the coordinate type.

complex conjugate [MATH] One of a pair of complex numbers with identical real parts and with imaginary parts differing only in sign. Also known as conjugate.

complex data type [COMPUT SCI] A scalar data type which contains two real fields representing the real and imaginary components of a complex number.

complex dune [GEOL] A dune of varying forms, often very large, and produced by variable, shifting winds and the merging of various dune types.

complex frequency [ENG] A complex number used to characterize exponential and damped sinusoidal motion in the same way that an ordinary frequency characterizes simple harmonic motion; designated by the constant s corresponding to a motion whose amplitude is given by Ae^{st}, where A is a constant and t is time.

complex impedance *See* electrical impedance; impedance.

complexing [CHEM] Formation of a complex compound. Also known as complexation.

complex notation [PHYS] The representation of a physical quantity by a complex number whose real component equals the instantaneous value of the physical quantity, a sinusoidally varying quantity thus being represented by a point rotating in a circle centered at the origin of the complex plane with uniform speed.

complex number [MATH] Any number of the form $a + bi$, where a and b are real numbers, and $i^2 = -1$.

complexometric titration [ANALY CHEM] A technique of volumetric analysis in which the formation of a colored complex is used to indicate the end point of a titration. Also known as chelatometry.

complex salt [INORG CHEM] A class of salts in which there are no detectable quantities of each of the metal ions existing in solution; an example is $K_3Fe(CN)_6$, which in solution has K^+ but no Fe^{3+} because Fe is strongly bound in the complex ion, $Fe(CN)_6^{3-}$.

complex wave [PHYS] A waveform which varies from instant to instant, but can be resolved into a number of sine-wave components, each of a different frequency and probably of a different amplitude.

compliance constant [MECH] Any one of the coefficients of the relations in the generalized Hooke's law used to express strain components as linear functions of the stress components. Also known as elastic constant.

compole *See* commutating pole.

component [CHEM] **1.** A part of a mixture. **2.** The smallest number of chemical substances able to form all the constituents of a system in whatever proportion they may be present. [ELEC] Any electric device, such as a coil, resistor, capacitor, generator, line, or electron tube, having distinct electrical characteristics and having terminals at which it may be connected to other components to form a circuit. Also known as circuit element; element. [SCI TECH] A constituent part of a system; examples are a vector term which when added to others gives a vector sum, an ingredient of a chemical system, or the mineral portion of a rock.

Compositae [BOT] The single family of the order Asterales; perhaps the largest family of flowering plants, it contains about 19,000 species.

composite [ENG ACOUS] A re-recording consisting of at least two elements. [MATER] A structural material composed of combinations of metal alloys or plastics, usually with the addition of strengthening agents.

composite beam [CIV ENG] Beam action of two materials joined to act as a unit, especially that developed by a concrete slab resting on a steel beam and joined by shear connectors.

composite cable [ELEC] Cable in which conductors of different gages or types are combined under one sheath.

composite circuit [ELECTR] A circuit used simultaneously for voice communication and telegraphy, with frequency-discriminating networks serving to separate the two types of signals.

composite color signal [COMMUN] The color television picture signal plus all blanking and synchronizing signals. Also known as composite picture signal.

composite column [CIV ENG] A concrete column having a structural-steel or cast-iron core with a maximum core area of 20%.

composite flash [GEOPHYS] A lightning discharge which is made up of a series of distinct lightning strokes with all strokes following the same or nearly the same channel, and with successive strokes occurring at intervals of about 0.05 second. Also known as multiple discharge.

composite fuel [MATER] A broad class of solid chemical fuels composed of a fuel and oxidizer and used as propellants in rockets; an example of a fuel is phenol formaldehyde, and an oxidizer is ammonium perchlorate. Also known as composite propellant.

composite joint [MET] A joint connected by welding in conjunction with one or more mechanical means.

composite nerve [PHYSIO] A nerve containing both sensory and motor fibers.

composite number [MATH] Any positive integer which is not prime. Also known as composite quantity.

composite picture signal *See* composite color signal.

composite propellant *See* composite fuel.

composite pulse [ELECTR] A pulse composed of a series of overlapping pulses received from the same source over several paths in a pulse navigation system.

composite quantity *See* composite number.

composite vein [GEOL] A large fracture zone composed of parallel ore-filled fissures and converging diagonals, whose walls and intervening country rock have been replaced to a certain degree.

composite wave filter [ELECTR] A combination of two or more low-pass, high-pass, band-pass, or band-elimination filters.

composition [CHEM] The elements or compounds making up a material or produced from it by analysis. [GRAPHICS] The act of composing or combining type for printing, either by hand or by machine. [SCI TECH] The elements or compounds making up a material or produced from it by analysis.

composition diagram [CHEM ENG] Graphical plots to show the solvent-solute concentration relationships during various stages of extraction operations (leaching, or solid-liquid extraction; and liquid-liquid extraction).

composition face *See* composition surface.

composition of forces [MECH] The determination of a force whose effect is the same as that of two or more given forces acting simultaneously; all forces are considered acting at the same point.

composition of velocities law [MECH] A law relating the velocities of an object in two references frames which are moving relative to each other with a specified velocity.

composition series [MATH] A normal series G_1, G_2, ..., of a group, where each G_i is a proper normal subgroup of G_{i-1} and no further normal subgroups both contain G_i and are contained in G_{i-1}.

composition surface [CRYSTAL] The surface uniting individuals of a crystal twin; may or may not be planar. Also known as composition face.

compound [CHEM] A substance whose molecules consist of unlike atoms and whose constituents cannot be separated by physical means. Also known as chemical compound.

compound curve [MATH] A curve made up of two arcs of differing radii whose centers are on the same side, connected by a common tangent; used to lay out railroad curves because curvature goes from nothing to a maximum gradually, and vice versa.

compound engine [MECH ENG] A multicylinder-type displacement engine, using steam, air, or hot gas, where expansion proceeds successively (sequentially).

compound eye [INV ZOO] An eye typical of crustaceans, insects, centipedes, and horseshoe crabs, constructed of many functionally independent photoreceptor units (ommatidia) separated by pigment cells.

compound generator [ELEC] A direct-current generator which has both a series field winding and a shunt field winding, both on the main poles with the shunt field winding on the outside.

compound gland [ANAT] A secretory structure with many ducts.

compound leaf [BOT] A type of leaf with the blade divided into two or more separate parts called leaflets.

compound lens [OPTICS] 1. A combination of two or more lenses in which the second surface of one lens has the same radius as the first surface of the following lens, and the two lenses are cemented together. Also known as cemented lens. 2. Any optical system consisting of more than one element, even when they are not in contact.

compound microscope [OPTICS] A microscope which utilizes two lenses or lens systems; one lens forms an enlarged image of the object, and the second magnifies the image formed by the first.

compound modulation See multiple modulation.

compound motor [ELEC] A direct-current motor with two separate field windings, one connected in parallel with the armature circuit, the other connected in series with the armature circuit.

compound nucleus [NUC PHYS] An intermediate state in a nuclear reaction in which the incident particle combines with the target nucleus and its energy is shared among all the nucleons of the system.

compound number [MATH] A quantity which is expressed as the sum of two or more quantities in terms of different units, for example, 3 feet 10 inches, or 2 pounds 5 ounces.

compound screw [DES ENG] A screw having different or opposite pitches on opposite ends of the shank.

compound sugar See oligosaccharide.

compound twins [CRYSTAL] Individuals of one mineral group united in accordance with two or more different twin laws.

compound volcano [GEOL] 1. A volcano consisting of a complex of two or more cones. 2. A volcano with an associated volcanic dome.

compound wave [FL MECH] A plane wave of finite amplitude in which neither the sum of the velocity potential and the component of velocity in the direction of wave motion, nor the difference of these two quantities, is constant.

compound winding [ELEC] A winding that is a combination of series and shunt winding.

compressed air [MECH] Air whose density is increased by subjecting it to a pressure greater than atmospheric pressure.

compressed-air illness See caisson disease.

compressibility [MECH] The property of a substance capable of being reduced in volume by application of pressure; quantitively, the reciprocal of the bulk modulus.

compressibility error [FL MECH] The error in the readings of a differential-pressure-type airspeed indicator due to compression of the air on the forward part of the pitot tube component moving at high speeds.

compressibility factor [THERMO] The product of the pressure and the volume of a gas, divided by the product of the temperature of the gas and the gas constant; this factor may be inserted in the ideal gas law to take into account the departure of true gases from ideal gas behavior. Also known as deviation

factor; gas-deviation factor; supercompressibility factor.

compressible flow [FL MECH] Flow in which the fluid density varies.

compressible-flow principle [FL MECH] The principle that when flow velocity is large, it is necessary to consider that the fluid is compressible rather than to assume that it has a constant density.

compression [COMPUT SCI] See data compression. [ELECTR] 1. Reduction of the effective gain of a device at one level of signal with respect to the gain at a lower level of signal, so that weak signal components will not be lost in background and strong signals will not overload the system. 2. See compression ratio. [GEOD] See flattening. [GEOL] A system of forces which tend to decrease the volume or shorten rocks. [MECH] Reduction in the volume of a substance due to pressure; for example in building, the type of stress which causes shortening of the fibers of a wooden member. [MECH ENG] See compression ratio.

compressional wave [PHYS] A disturbance traveling in an elastic medium; characterized by changes in volume and by particle motion parallel with the direction of wave movement. Also known as dilatational wave; irrotational wave; pressure wave; P wave.

compression coupling [MECH ENG] 1. A means of connecting two perfectly aligned shafts in which a slotted tapered sleeve is placed over the junction and two flanges are drawn over the sleeve so that they automatically center the shafts and provide sufficient contact pressure to transmit medium loads. 2. A type of tubing fitting.

compression ignition [MECH ENG] Ignition produced by compression of the air in a cylinder of an internal combustion engine before fuel is admitted.

compression-ignition engine See diesel engine.

compression machine See compressor.

compression modulus See bulk modulus of elasticity.

compression ratio [ELECTR] The ratio of the gain of a device at a low power level to the gain at some higher level, usually expressed in decibels. Also known as compression. [MECH ENG] The ratio in internal combustion engines between the volume displaced by the piston plus the clearance space, to the volume of the clearance space. Also known as compression. [MET] Ratio of the volume of loose metal powder to the volume of the compact made from it.

compression refrigeration [MECH ENG] The cooling of a gaseous refrigerant by first compressing it to liquid form (with resultant heat buildup), cooling the liquid by heat exchange, then releasing pressure to allow the liquid to vaporize (with resultant absorption of latent heat of vaporization and a refrigerative effect).

compression strength [MECH] Property of a material to resist rupture under compression.

compression stroke [MECH ENG] The phase of a positive displacement engine or compressor in which the motion of the piston compresses the fluid trapped in the cylinder.

compressive strength [MECH] The maximum compressive stress a material can withstand without failure.

compressive stress [MECH] A stress which causes an elastic body to shorten in the direction of the applied force.

compressor [ELECTR] The part of a compandor that is used to compress the intensity range of signals at the transmitting or recording end of a circuit. [MECH

ENG] A machine used for increasing the pressure of a gas or vapor. Also known as compression machine.

compromise network   [ELEC] **1.** Network employed in conjunction with a hybrid coil to balance a subscriber's loop; adjusted for an average loop length or an average subscriber's set, or both, to secure compromise (not precision) isolation between the two directional paths of the hybrid. **2.** Hybrid balancing network which is designed to balance the average of the impedances that may be connected to the switchboard side of a hybrid arrangement of a repeater.

Compton-Debye effect   *See* Compton effect.

Compton effect   [QUANT MECH] The increase in wavelength of electromagnetic radiation in the x-ray and gamma-ray region on being scattered by material objects; the scattering is due to the interaction of the photons with electrons that are effectively free. Also known as Compton-Debye effect.

Compton incoherent scattering   [NUC PHYS] Scattering of gamma rays by individual nucleons in a nucleus or electrons in an atom when the energy of the gamma rays is large enough so that binding effects may be neglected.

Compton meter   [NUCLEO] An ionization chamber having a balance chamber with a uranium source that is adjusted until it balances out normal cosmic radiation; variations in cosmic radiation are then shown on an electrometer.

Compton process   *See* Compton scattering.

Compton rule   [PHYS CHEM] An empirical law stating that the heat of fusion of an element times its atomic weight divided by its melting point in degrees Kelvin equals approximately 2.

Compton scattering   [QUANT MECH] The elastic scattering of photons by electrons. Also known as Compton process; gamma-ray scattering.

Compton wavelength   [QUANT MECH] A convenient unit of length that is characteristic of an elementary particle, equal to Planck's constant divided by the product of the particle's mass and the speed of light.

computational chemistry   [CHEM] The use of calculations to predict molecular structure, properties, and reactions.

computed go to   [COMPUT SCI] A control procedure in FORTRAN which allows the transfer of control to the ith label of a set of n labels used as statement numbers in the program.

compute mode   [COMPUT SCI] The operation of an analog computer in which input signals are used by the computing units to calculate a solution, in contrast to hold mode and reset mode.

computer   [COMPUT SCI] A device that receives, processes, and presents data; the two types are analog and digital. Also known as computing machine.

computer-aided design   [COMPUT SCI] The generation of computer automated designs for display on cathode-ray tubes. Abbreviated CAD.

computer-aided management of instruction   *See* computer-managed instruction.

computer-aided manufacturing   [COMPUT SCI] The use of computers to communicate work instructions to automatic machinery for the handling and processing needed to produce a workpiece. Abbreviated CAM.

computer analyst   [COMPUT SCI] A person who defines a problem, determines exactly what is required in the solution, and defines the outlines of the machine solution; generally, an expert in automatic data processing applications.

computer-assisted instruction   [COMPUT SCI] The use of computers to present drills, practice exercises, and tutorial sequences to the student, and sometimes to engage the student in a dialog about the substance of the instruction. Abbreviated CAI.

computer code   [COMPUT SCI] The code representing the operations built into the hardware of a particular computer.

computer control   [COMPUT SCI] *See* control.   [CONT SYS] Process control in which the process variables are fed into a computer and the output of the computer is used to control the process.

computer control register   *See* program register.

computer efficiency   [COMPUT SCI] **1.** The ratio of actual operating time to scheduled operating time of a computer. **2.** In time-sharing, the ratio of user time to the sum of user time plus system time.

computer entry punch   [COMPUT SCI] Combination card-reader and key punch that enters data directly onto a computer's memory drum.

computer graphics   [COMPUT SCI] The process of pictorial communication between human and computers, in which the computer input and output have the form of charts, drawings, or appropriate pictorial representation; such devices as cathode-ray tubes, mechanical plotting boards, curve tracers, coordinate digitizers, and light pens are employed.

computerized composition   [GRAPHICS] Type composition in which line-end hyphenation and other typographic work has been done by a computer working from unjustified tape.

computerized tomography   [MED] The process of producing a picture showing human body organs in cross section by first electronically detecting the variation in x-ray transmission through the body section at different angles, and then using this information in a digital computer to reconstruct the x-ray absorption of the tissues at an array of points representing the cross section.

computer-limited   [COMPUT SCI] Pertaining to a situation in which the time required for computation exceeds the time required to read inputs and write outputs.

computer-managed instruction   [COMPUT SCI] The use of computer assistance in testing, diagnosing, prescribing, grading, and record keeping. Abbreviated CMI. Also known as computer-aided management of instruction.

computer memory   *See* memory.

computer-output typesetting   [GRAPHICS] Production of graphic arts quality printout of computer information on photographic paper or film.

computer performance evaluation   [COMPUT SCI] The measurement and evaluation of the performance of a computer system, aimed at ensuring that a minimum amount of effort, expense, and waste is incurred in the production of data-processing services, and encompassing such tools as canned programs, source program optimizers, software monitors, hardware monitors, simulation, and benchmark problems. Abbreviated CPE.

computer programming   *See* programming.

computer storage device   *See* storage device.

computer system   [COMPUT SCI] **1.** A set of related but unconnected components (hardware) of a computer or data-processing system. **2.** A set of hardware parts that are related and connected, and thus form a computer.

computer theory   [COMPUT SCI] A discipline covering the study of circuitry, logic, microprogramming, compilers, programming languages, file structures, and system architectures.

computer utility [COMPUT SCI] A computer that provides service on a time-sharing basis, generally over telephone lines, to subscribers who have appropriate terminals.

computer vision [COMPUT SCI] Capability of computers to analyze and act on visual input.

computer word *See* word.

computing machine *See* computer.

conarium *See* pineal body.

concatenate [COMPUT SCI] To unite in a sequence, link together, or link to a chain.

concatenated codes [COMPUT SCI] Two or more codes which are encoded and decoded in series.

concatenation [COMPUT SCI] **1.** An operation in which a number of conceptually related components are linked together to form a larger, organizationally similar entity. **2.** In string processing, the synthesis of longer character strings from shorter ones. [ELEC] A method of speed control of induction motors in which the rotors of two wound-rotor motors are mechanically coupled together and the stator of the second motor is supplied with power from the rotor slip rings of the first motor.

concave [SCI TECH] Having a curved form which bulges inward resembling the interior of a sphere or cylinder or a section of these bodies.

concave function [MATH] A function $f(x)$ is said to be concave over the interval a,b if for any three points x_1, x_2, x_3 such that $a < x_1 < x_2 < x_3 < b$, $f(x_2) \geq L(x_2)$, where $L(x)$ is the equation of the straight line passing through the points $[x_1, f(x_1)]$ and $[x_3, f(x_3)]$.

concave grating [SPECT] A reflection grating which both collimates and focuses the light falling upon it, made by spacing straight grooves equally along the chord of a concave spherical or paraboloid mirror surface. Also known as Rowland grating.

concentrate [CHEM] To increase the amount of a dissolved substance by evaporation. [MIN ENG] **1.** To separate ore or metal from its containing rock or earth. **2.** The clean product recovered in froth flotation or other methods of mineral separation.

concentration [CHEM] In solutions, the mass, volume, or number of moles of solute present in proportion to the amount of solvent or total solution. [HYD] The ratio of the area of the sea covered by ice to the total area of sea surface. [MIN ENG] Separation and accumulation of economic minerals from gangue.

concentration cell [PHYS CHEM] **1.** Electrochemical cell for potentiometric measurement of ionic concentrations where the electrode potential electromotive force produced is determined as the difference in emf between a known cell (concentration) and the unknown cell. **2.** An electrolytic cell in which the electromotive force is due to a difference in electrolyte concentrations at the anode and the cathode.

concentration gradient [CHEM] The graded difference in the concentration of a solute throughout the solvent phase.

concentration polarization [PHYS CHEM] That part of the polarization of an electrolytic cell resulting from changes in the electrolyte concentration due to the passage of current through the solution.

concentration potential [CHEM] Tendency for a univalent electrolyte to concentrate in a specific region of a solution.

concentration scale [CHEM] Any of several numerical systems defining the quantitative relation of the components of a mixture; for solutions, concentration is expressed as the mass, volume, or number of moles of solute present in proportion to the amount of solvent or total solution.

concentrator [ELECTR] Buffer switch (analog or digital) which to reduces the number of trunks required. [ENG] **1.** An apparatus used to concentrate materials. **2.** A plant where materials are concentrated.

concentric [SCI TECH] Pertaining to the relationship between two different-sized circular, cylindrical, or spherical shapes when the smaller one is exactly centered within the larger one.

concentric cable *See* coaxial cable.

concentric fold [GEOL] A fold in which the original thickness of the strata is unchanged during deformation. Also known as parallel fold.

concentric line *See* coaxial cable.

concentric transmission line *See* coaxial cable.

concentric tube column [CHEM ENG] A carefully insulated distillation apparatus which is capable of very high separating power, and in which the outer vapor-rising annulus of the column is concentric around an inner, bottom-discharging reflux return.

concentric windings [ELEC] Transformer windings in which the low-voltage winding is in the form of a cylinder next to the core, and the high-voltage winding, also cylindrical, surrounds the low-voltage winding.

conceptacle [BOT] A cavity which is shaped like a flask with a pore opening to the outside, contains reproductive structures, and is bound in a thallus such as in the brown algae.

concept coordination [COMPUT SCI] The basic principles of various punched-card, aspect, and mechanized information retrieval systems in which independently assigned concepts are used to characterize the subject content of documents, and the latter are identified during searching by means of either such assigned concepts or their combination.

conceptual modeling [COMPUT SCI] Writing a program by means of which a given result will be obtained, although the result is incapable of proof. Also known as heuristic programming.

conch [INV ZOO] The common name for several species of large, colorful gastropod mollusks of the family Strombidae; the shell is used to make cameos and porcelain.

conchoid [MATH] A plane curve consisting of the locus of both ends of a line segment of constant length on a line which rotates about a fixed point, while the midpoint of the segment remains on a fixed line which does not contain the fixed point.

conchoidal [GEOL] Having a smoothly curved surface; used especially to describe the fracture surface of a mineral or rock.

Conchorhagae [INV ZOO] A suborder of benthonic wormlike animals in the class Kinorhyncha.

Conchostraca [INV ZOO] An order of mussellike crustaceans of moderate size belonging to the subclass Branchiopoda.

concordance [GEN] Similarity in appearance of members of a twin pair with respect to one or more specific traits.

concordant body [GEOL] An intrusive igneous body whose contacts are parallel to the bedding of the country rock. Also known as concordant injection; concordant pluton.

concordant coastline [GEOL] A coastline parallel to the land structures which form the margin of an ocean basin.

concordant injection *See* concordant body.

concordant pluton *See* concordant body.

concrete [MATER] A mixture of aggregate, water, and a binder, usually portland cement; hardens to stonelike condition when dry.

concrete retarder [MATER] A material added to concrete that decreases the hydration rate of cement, thereby increasing the setting time and decreasing the strengthening rate during the early age.

concretion [GEOL] A hard, compact mass of mineral matter in the pores of sedimentary or fragmental volcanic rock; represents a concentration of a minor constituent of the enclosing rock or of cementing material.

concurrency [COMPUT SCI] Referring to two or more tasks of a computer system which are in progress simultaneously.

concurrent conversion [COMPUT SCI] The transfer of data from one medium to another, such as card to tape, under computer control while programs are being run on the same computer.

concurrent input/output [COMPUT SCI] The simultaneous reading from and writing on different media by a computer.

concurrent operations control [COMPUT SCI] The supervisory capability required by a computer to handle more than one program at a time.

concurrent processing [COMPUT SCI] The capability of a computer to process more than one program at the time.

concurrent real-time processing [COMPUT SCI] The capability of a computer to process simultaneously several programs, each of which requires responses within a time span related to its particular time frame.

condensate [MATER] **1.** The liquid product from a condenser. Also known as condensate liquid. **2.** A light hydrocarbon mixture formed as a liquid product in a gas-recycling plant through expansion and cooling of the gas.

condensate field [GEOL] A petroleum field developed in predominantly gas-bearing reservoir rocks, but within which condensation of gas to oil commonly occurs with decreases in field pressure.

condensate flash [CHEM ENG] Partial evaporation (flash) of hot condensed liquid by a stepwise reduction in system pressure, the hot vapor supplying heat to a cooler evaporator step (stage).

condensate liquid See condensate.

condensate well [MECH ENG] A chamber into which condensed vapor falls for convenient accumulation prior to removal. [PETRO ENG] Well that produces a natural gas highly saturated with condensable hydrocarbons heavier than methane and ethane.

condensation [ACOUS] A measure of the increase in the instantaneous density at a given point owing to a sound wave, namely $(\rho - \rho_0)/\rho_0$, where ρ is the density and ρ_0 is the constant mean density at the point. [CHEM] Transformation from a gas to a liquid. [CRYO] See Bose-Einstein condensation. [ELEC] An increase of electric charge on a capacitor conductor. [MECH] An increase in density. [METEOROL] The process by which water vapor becomes a liquid such as dew, fog, or cloud or a solid like snow; condensation in the atmosphere is brought about by either of two processes: cooling of air to its dew point, or addition of enough water vapor to bring the mixture to the point of saturation (that is, the relative humidity is raised to 100%). [OPTICS] Focusing or collimation of light.

condensation cloud [METEOROL] A mist or fog of minute water droplets that temporarily surrounds the fireball following an atomic detonation in a comparatively humid atmosphere.

condensation nucleus [METEOROL] A particle, either liquid or solid, upon which condensation of water vapor begins in the atmosphere.

condensation polymer [ORG CHEM] A high-molecular-weight compound formed by condensation polymerization.

condensation polymerization [ORG CHEM] The formation of high-molecular-weight polymers from monomers by chemical reactions of the condensation type.

condensation pressure [METEOROL] The pressure at which moist air expanded dry adiabatically reaches saturation. Also called adiabatic condensation pressure; adiabatic saturation pressure.

condensation reaction [CHEM] One of a class of chemical reactions involving a combination between molecules or between parts of the same molecule.

condensation temperature [ANALY CHEM] In boiling-point determination, the temperature established on the bulb of a thermometer on which a thin moving film of liquid coexists with vapor from which the liquid has condensed, the vapor phase being replenished at the moment of measurement from a boiling-liquid phase. [METEOROL] The temperature at which a parcel of moist air expanded dry adiabatically reaches saturation. Also known as adiabatic condensation temperature; adiabatic saturation temperature.

condenser [ELEC] See capacitor. [MECH ENG] A heat-transfer device that reduces a thermodynamic fluid from its vapor phase to its liquid phase, such as in a vapor-compression refrigeration plant or in a condensing steam power plant. [OPTICS] A system of lenses or mirrors in an optical projection system, which gathers as much of the light from the source as possible and directs it through the projection lens.

condenser antenna See capacitor antenna.

condenser box See capacitor box.

condenser microphone See capacitor microphone.

condenser transducer See electrostatic transducer.

condenser tubes [MECH ENG] Metal tubes used in a heat-transfer device, with condenser vapor as the heat source and flowing liquid such as water as the receiver.

condensing flow [FL MECH] The flow and simultaneous condensation (partial or complete) of vapor through a cooled pipe or other closed conduit or container.

condensing routine [COMPUT SCI] A routine that converts a program format having one instruction per card to a program format having several instructions per card.

conditional [COMPUT SCI] Subject to the result of a comparison made during computation in a computer, or subject to human intervention.

conditional assembly [COMPUT SCI] A feature of some assemblers which suppresses certain sections of code if stated program conditions are not met at assembly time.

conditional branch See conditional jump.

conditional breakpoint [COMPUT SCI] A conditional jump that, if a specified switch is set, will cause a computer to stop; the routine may then be continued as coded or a jump may be forced.

conditional distribution [STAT] If W and Z are random variables with discrete values $w_1, w_2, \ldots$, and $z_1, z_2, \ldots$, the conditional distribution of W given $Z = z$ is the distribution which assigns to w_i, $i = 1, 2, \ldots$, the conditional probability of $W = w_i$ given $Z = z$.

conditional expression [COMPUT SCI] A COBOL language expression which is either true or false, depending upon the status of the variables within the expression.

conditional frequency [STAT] If r and s are possible outcomes of an experiment which is performed n times, the conditional frequency of s given that r has occurred is the ratio of the number of times both r and s have occurred to the number of times r has occurred.

conditional jump [COMPUT SCI] A computer instruction that will cause the proper one of two or more addresses to be used in obtaining the next instruction, depending on some property of a numerical expression that may be the result of some previous instruction. Also known as conditional branch; conditional transfer; decision instruction; discrimination.

conditional lethal mutant [GEN] A mutant gene that has lethal expression only under particular growth conditions, such as a specific temperature range.

conditionally compact set [MATH] A set whose closure is compact. Also known as relatively compact set.

conditional transfer *See* conditional jump.

condition code [COMPUT SCI] Portion of a program status word indicating the outcome of the most recently executed arithmetic or boolean operation.

conditioned reflex [PSYCH] Response of an organism to a stimulus which was inadequate to elicit the response until paired for one or more times with an adequate stimulus.

conditioned stop instruction [COMPUT SCI] A computer instruction which causes the execution of a program to stop if some given condition exists, such as the specific setting of a switch on a computer console.

conditioning [ELECTR] Equipment modifications or adjustments necessary to match transmission levels and impedances or to provide equalization between facilities. [SCI TECH] Substimulus so that it will respond in a uniform and desired manner to subsequent testing or processing.

condor [NAV] A continuous-wave navigation system, similar to benito, that automatically measures bearing and distance from a single ground station; the distance is determined by phase comparison and the bearing by automatic direction finding. [VERT ZOO] *Vultur gryphus.* A large American vulture having a bare head and neck, dull black plumage, and a white neck ruff.

conductance [ELEC] The real part of the admittance of a circuit; when the impedance contains no reactance, as in a direct-current circuit, it is the reciprocal of resistance, and is thus a measure of the ability of the circuit to conduct electricity. Also known as electrical conductance. Designated G. [FL MECH] For a component of a vacuum system, the amount of a gas that flows through divided by the pressure difference across the component. [THERMO] *See* thermal conductance.

conductimetry [CHEM] The scientific study of conductance measurements of solutions; to avoid electrolytic complications, conductance measurements are usually taken with alternating current.

conducting polymer [MATER] A plastic having high conductivity, approaching that of metals.

conduction [ELEC] The passage of electric charge, which can occur by a variety of processes, such as the passage of electrons or ionized atoms. Also known as electrical conduction. [PHYS] Transmission of energy by a medium which does not involve movement of the medium itself.

conduction band [SOLID STATE] An energy band in which electrons can move freely in a solid, producing net transport of charge.

conduction deafness [MED] Deafness involving an impairment of the mechanism that conducts sound to the sense organ.

conduction electron [SOLID STATE] An electron in the conduction band of a solid, where it is free to move under the influence of an electric field. Also known as outer-shell electron; valence electron.

conduction field [ELECTROMAG] Energy surrounding a conductor when an electric current is passed through the conductor, which, because of the difference in phase between the electrical field and magnetic field set up in the conductor, cannot be detached from the conductor.

conductive coupling [ELEC] Electric connection of two electric circuits by their sharing the same resistor.

conductive gasket [ELEC] A flexible metallic gasket used to reduce radio-frequency leakage at joints in shielding.

conductivity [ELEC] The ratio of the electric current density to the electric field in a material. Also known as electrical conductivity; specific conductance.

conductivity cell [ELEC] A glass vessel with two electrodes at a definite distance apart and filled with a solution whose conductivity is to be measured.

conductivity modulation [ELECTR] Of a semiconductor, the variation of the conductivity of a semiconductor through variation of the charge carrier density.

conductivity theory [STAT MECH] Theory which treats the system of electrons in a metal as a gas and uses the Boltzmann transport equation to calculate conductivity.

conductometer [ENG] An instrument designed to measure thermal conductivity; in particular, one that compares the rates at which different rods transmit heat.

conductometric titration [ANALY CHEM] A titration in which electrical conductance of a solution is measured during the course of the titration.

conductor [ELEC] A wire, cable, or other body or medium that is suitable for carrying electric current.

conductor skin effect *See* skin effect.

conduit [ELEC] Solid or flexible metal or other tubing through which insulated electric wires are run. [ENG] Any channel or pipe for conducting the flow of water or other fluid. [GEOL] A water-filled underground passage that is always under hydrostatic pressure.

Condylarthra [PALEON] A mammalian order of extinct, primitive, hoofed herbivores with five-toed plantigrade to semidigitigrade feet.

condyle [ANAT] A rounded bone prominence that functions in articulation. [BOT] The antheridium of certain stoneworts. [INV ZOO] A rounded, articular process on arthropod appendages.

condyloid articulation [ANAT] A joint, such as the wrist, formed by an ovoid surface that fits into an elliptical cavity, permitting all movement except rotation.

cone [BOT] The ovulate or staminate strobilus of a gymnosperm. [ENG ACOUS] The cone-shaped paper or fiber diaphragm of a loudspeaker. [GEOL] A mountain, hill, or other landform having relatively steep slopes and a pointed top. [HISTOL] A photoreceptor of the vertebrate retina that responds differentially to light across the visible spectrum, providing both color vision and visual acuity in bright light. [MATH] A solid bounded by a region enclosed in a closed curve on a plane and a surface formed by the segments joining each point of the closed curve to a point which is not in the plane. [MET] The part of an oxygen gas flame adjacent to the orifice of the tip. [TEXT] A bobbin on which yarn is wound for weaving.

cone antenna *See* conical antenna.

cone classifier [MECH ENG] Inverted-cone device for the separation of heavy particulates (such as sand, ore, or other mineral matter) from a liquid stream; feed enters the top of the cone, heavy particles settle to the bottom where they can be withdrawn, and liquid overflows the top edge, carrying the smaller particles or those of lower gravity over the rim; used in the mining and chemical industries.

cone clutch [MECH ENG] A clutch which uses the wedging action of mating conical surfaces to transmit friction torque.

cone key [DES ENG] A taper saddle key placed on a shaft to adapt it to a pulley with a too-large hole.

cone loudspeaker [ENG ACOUS] A loudspeaker employing a magnetic driving unit that is mechanically coupled to a paper or fiber cone. Also known as cone speaker.

Conemaughian [GEOL] Upper Middle Pennsylvanian geologic time.

cone of depression [HYD] The depression in the water table around a well defining the area of influence of the well. Also known as cone of influence.

cone of influence *See* cone of depression.

cone of revolution [MATH] The surface obtained by rotating a line around another line which it intersects, using the intersection point as a pivot.

cone speaker *See* cone loudspeaker.

cone valve [CIV ENG] A divergent valve whose cone-shaped head in a fixed cylinder spreads water around the wide, downstream end of the cone in spillways of dams or hydroelectric facilities. Also known as Howell-Bunger valve.

Conewangoan [GEOL] Upper Upper Devonian geologic time.

confidence [STAT] The degree of assurance that a specified failure rate is not exceeded.

confidence coefficient [STAT] The probability associated with a confidence interval; that is, the probability that the interval contains a given parameter or characteristic. Also known as confidence level.

confidence interval [STAT] An interval which has a specified probability of containing a given parameter or characteristic.

confidence level [IND ENG] The probability in acceptance sampling that the quality of accepted lots manufactured will be better than the rejectable quality level (RQL); 90% level indicates that accepted lots will be better than the RQL 90 times in 100. [STAT] *See* confidence coefficient.

confidence limit [STAT] One of the end points of a confidence interval.

configuration [AERO ENG] A particular type of specific aircraft, rocket, or such, which differs from others of the same model by the arrangement of its components or by the addition or omission of auxiliary equipment; for example, long-range configuration or cargo configuration. [CHEM] The three-dimensional spatial arrangement of atoms in a stable or isolable molecule. [ELEC] A group of components interconnected to perform a desired circuit function. [MATH] An arrangement of geometric objects. [MECH] The positions of all the particles in a system. [SYS ENG] A group of machines interconnected and programmed to operate as a system.

confinement [ENG] Physical restriction, or degree of such restriction, to passage of detonation wave or

reaction zone, for example, that of a resistant container which holds an explosive charge.

confluence [HYD] **1.** A stream formed from the flowing together of two or more streams. **2.** The place where such streams join.

confocal coordinates [MATH] Coordinates of a point in the plane with norm greater than 1 in terms of the system of ellipses and hyperbolas whose foci are at $(1,0)$ and $(-1,0)$.

confocal resonator [ELECTROMAG] A wavemeter for millimeter wavelengths, consisting of two spherical mirrors facing each other; changing the spacing between the mirrors affects propagation of electromagnetic energy between them, permitting direct measurement of free-space wavelength.

conformable [GEOL] **1.** Pertaining to the contact of an intrusive body when it is aligned with the internal structures of the intrusion. **2.** Referring to strata in which layers are formed above one another in an unbroken, parallel order.

conformal array [ELECTR] A circular, cylindrical, hemispherical, or other shaped array of electronically switched antennas; provides the special radiation patterns required for Tacan, IFF, and other air navigation, radar, and missile control applications.

conformality [MAP] The retention of angular relationships at each point on a map projection.

conformal map projection [MAP] A map projection on which the shape of any small area of the surface mapped is preserved unchanged. Also known as orthomorphic map projection.

conformation [ORG CHEM] In a molecule, a specific orientation of the atoms that varies from other possible orientations by rotation or rotations about single bonds; generally in mobile equilibrium with other conformations of the same structure. Also known as conformational isomer; conformer.

conformational analysis [PHYS CHEM] The determination of the arrangement in space of the constituent atoms of a molecule that may rotate about a single bond.

confusion jamming [ELECTR] An electronic countermeasure technique in which the signal from an enemy tracking radar is amplified and retransmitted with distortion to create a false echo that affects accuracy of target range, azimuth, and velocity data.

congelifraction [GEOL] The splitting or disintegration of rocks as the result of the freezing of the water contained. Also known as frost bursting; frost riving; frost shattering; frost splitting; frost weathering; frost wedging; gelifraction; gelivation.

congeliturbation [GEOL] The churning and stirring of soil as a result of repeated cycles of freezing and thawing; includes frost heaving and surface subsidence during thaws. Also known as cryoturbation; frost churning; frost stirring; geliturbation.

congener [CHEM] A chemical substance that is related to another substance, such as a derivative of a compound or an element belonging to the same family as another element in the periodic table.

congenital [MED] Dating from or existing before birth.

congestion [MED] An abnormal accumulation of fluid, usually blood, but occasionally bile or mucus, within the vessels of an organ or part.

congestive heart failure [MED] A state in which circulatory congestion exists as a result of heart failure.

conglomerate [GEOL] Cemented, rounded fragments of water-worn rock or pebbles, bound by a siliceous or argillaceous substance.

conglutination [IMMUNOL] The completion of an agglutinating system, or the enhancement of an incomplete one, by the addition of certain substances. [MED] Abnormal union of two contiguous surfaces or bodies.

conglutinin [IMMUNOL] A heat-stable substance in bovine and other serums that aids or causes agglomeration or lysis of certain sensitized cells or particles.

congruence [MATH] 1. The property of geometric figures that can be made to coincide by a rigid transformation. 2. The property of two integers having the same remainder on division by another integer.

Coniacian [GEOL] Lower Senonian geologic time.

conical antenna [ELECTROMAG] A wide-band antenna in which the driven element is conical in shape. Also known as cone antenna.

conical helimagnet [SOLID STATE] A helimagnet in which the directions of atomic magnetic moments all make the same angle (greater than 0° and less than 90°) with a specified axis of the crystal, moments of atoms in successive basal planes are separated by equal azimuthal angles, and all moments have the same magnitude.

conical refraction [OPTICS] Phenomenon in which a ray incident on the surface of a biaxial crystal at a certain direction splits into a family of rays which lie along a cone.

conical scanning [ELECTR] Scanning in radar in which the direction of maximum radiation generates a cone, the vertex angle of which is of the order of the beam width; may be either rotating or nutating, according to whether the direction of polarization rotates or remains unchanged.

conical surface [MATH] A surface formed by the lines which pass through each of the points of a closed plane curve and a fixed point which is not in the plane of the curve.

conichalcite [MINERAL] $CaCu(AsO_4)(OH)$ A grass green to yellowish-green or emerald green, orthorhombic mineral consisting of a basic arsenate of calcium and copper.

Coniconchia [PALEON] A class name proposed for certain extinct organisms thought to have been mollusks; distinguished by a calcareous univalve shell that is open at one end and by lack of a siphon.

conic projection [MAP] A map deformation pattern resulting from the transfer of the map to a tangent or intersecting cone.

Conidae [INV ZOO] A family of marine gastropod mollusks in the order Neogastropoda containing the poisonous cone shells.

conidiophore [MYCOL] A specialized aerial hypha that produces conidia in certain ascomycetes and imperfect fungi.

conidiospore *See* conidium.

conidium [MYCOL] Unicellular, asexual reproductive spore produced externally upon a conidiophore. Also known as conidiospore.

conifer [BOT] The common name for plants of the order Pinales.

Coniferales [BOT] The equivalent name for Pinales.

Coniferophyta [BOT] The equivalent name for Pinicae.

Conjugales [BOT] An order of fresh-water green algae in the class Chlorophyceae distinguished by the lack of flagellated cells, and conjugation being the method of sexual reproduction.

conjugase [BIOCHEM] Any of a group of enzymes which catalyze the breakdown of pteroylglutamic acid.

conjugate [GEOL] 1. Pertaining to fractures in which both sets of veins or joints show the same strike but opposite dip. 2. Pertaining to any two sets of veins or joints lying perpendicular. [MATH] *See* complex conjugate.

conjugate acid-base pair [CHEM] An acid and a base related by the ability of the acid to generate the base by loss of a proton.

conjugate branches [ELEC] Any two branches of an electrical network such that a change in the electromotive force in either branch does not result in a change in current in the other. Also known as conjugate conductors.

conjugate bridge [ELECTR] A bridge in which the detector circuit and the supply circuits are interchanged, as compared with a normal bridge of the given type.

conjugate conductors *See* conjugate branches.

conjugate convex functions [MATH] Two functions $f(x)$ and $g(y)$ are conjugate convex functions if the derivative of $f(x)$ is 0 for $x = 0$ and constantly increasing for $x > 0$, and the derivative of $g(y)$ is the inverse of the derivative of $f(x)$.

conjugate curve *See* Bertrand curve.

conjugated circuits [ELEC] Branches of an electrical network such that a change in the electromotive force in either branch does not result in a current change in the other.

conjugate diameters [MATH] 1. For a conic section, any pair of straight lines either of which bisects all the chords that are parallel to the other. 2. For an ellipsoid or hyperboloid, any three lines passing through the point of symmetry of the surface such that the plane containing the conjugate diameters (first definition) of one of the lines also contains the other two lines.

conjugate foci *See* conjugate points.

conjugate impedances [ELEC] Impedances having resistance components that are equal, and reactance components that are equal in magnitude but opposite in sign.

conjugate lines [MATH] 1. For a conic section, two lines each of which passes through the intersection of the tangents to the conic at its points of intersection with the other line. 2. For a quadric surface, two lines each of which intersects the polar line of the other.

conjugate momentum [MECH] If q_j $(j = 1,2, ...)$ are generalized coordinates of a classical dynamical system, and L is its Lagrangian, the momentum conjugate to q_j is $p_j = \partial L/\partial \dot{q}_j$. Also known as canonical momentum; generalized momentum.

conjugate points [MATH] For a conic section, two points either of which lies on the line that passes through the points of contact of the two tangents drawn to the conic from the other. [OPTICS] Any pair of points such that all rays from one are imaged on the other within the limits of validity of Gaussian optics. Also known as conjugate foci.

conjugation [BOT] Sexual reproduction by fusion of two protoplasts in certain thallophytes to form a zygote. [INV ZOO] Sexual reproduction by temporary union of cells with exchange of nuclear material between two individuals, principally ciliate protozoans. [MICROBIO] A process involving contact between two bacterial cells during which genetic material is passed from one cell to the other.

conjunction [ASTRON] 1. The situation in which two celestial bodies have either the same celestial longitude or the same sidereal hour angle. 2. The time at which this conjunction takes place. [MATH] The connection of two statements by the word "and."

conjunctiva [ANAT] The mucous membrane covering the eyeball and lining the eyelids.

conjunctivitis [MED] Inflammation of the conjunctiva.

connate [GEOL] Referring to materials involved in sedimentary processes that are contemporaneous with surrounding materials. [SCI TECH] Born, originated, or produced in a united or fused condition.

connate leaf [BOT] A leaf shaped as though the bases of two opposite leaves had fused around the stem.

connate water [HYD] Water entrapped in the interstices of igneous rocks when the rocks were formed; usually highly mineralized.

connected load [ELEC] The sum of the continuous power ratings of all load-consuming apparatus connected to an electric power distribution system or any part thereof.

connected set [MATH] A set in a topological space which is not the union of two nonempty sets A and B for which both the intersection of the closure of A with B and the intersection of the closure of B with A are empty; intuitively, a set with only one piece.

connect function [COMPUT SCI] A signal sent over a data line to a selected peripheral device to connect it with the central processing unit.

connecting bar See tombolo.

connecting circuit [ELECTR] A functional switching circuit which directly couples other functional circuit units to each other to exchange information as dictated by the momentary needs of the switching system.

connecting rod [MECH ENG] Any straight link that transmits motion or power from one linkage to another within a mechanism, especially linear to rotary motion, as in a reciprocating engine or compressor.

connection box [COMPUT SCI] A mechanical device for altering electrical connections between various terminals, used to control the operations of a punched-card machine; its function is similar to that of a plug board.

connective tissue [HISTOL] A primary tissue, distinguished by an abundance of fibrillar and nonfibrillar extracellular components.

connectivity number [MATH] 1. The number of points plus 1 which can be removed from a curve without separating the curve into more than one piece. 2. The number of closed cuts or cuts joining points of previous cuts (or joining points on the boundary) plus 1 which can be made on a surface without separating the surface. Also known as Betti number.

connector [ELECTR] A switch, or relay group system, which finds the telephone line being called as a result of digits being dialed; it also causes interrupted ringing voltage to be applied to the called line or of returning a busy tone to the calling party if the line is busy. [ENG] 1. A detachable device for connecting electrical conductors. 2. A metal part for joining timbers.

connect time [COMPUT SCI] The time that a user at a terminal is signed on to a computer.

connellite [MINERAL] $Cu_{19}(SO_4)Cl_4(OH)_{32}\cdot 3H_2O$ A deep-blue striated copper mineral; crystals are in the hexagonal system. Also known as footeite.

conning tower [NAV ARCH] 1. The raised observation post of a submarine, which is in addition usually used as an entrance or exit. 2. The armored pilothouse of a warship.

Conoclypidae [PALEON] A family of Cretaceous and Eocene exocyclic Euechinoidea in the order Holectypoida having developed aboral petals, internal partitions, and a high test.

Conocyeminae [PALEON] A subfamily of Mesozoan parasites in the family Dicyemidae.

conode See tie line.

conodont [PALEON] A minute, toothlike microfossil, composed of translucent amber-brown, fibrous or lamellar calcium phosphate; taxonomic identity is controversial.

Conodontiformes [PALEON] A suborder of conodonts from the Ordovician to the Triassic having a lamellar internal structure.

Conodontophoridia [PALEON] The ordinal name for the conodonts.

conoid [SCI TECH] Shaped somewhat like a cone, but not quite conical.

Conopidae [INV ZOO] The wasp flies, a family of dipteran insects in the suborder Cyclorrhapha.

conoplain See pediment.

conoscope [OPTICS] An instrument, essentially a wide-angle microscope, used for study and observation of interference figures and related phenomena of specially cut crystal plates, especially for measuring the axial angle. Also known as hodoscope.

consanguinity [GEN] Blood relationship arising from common parentage. [PETR] The genetic relationship between igneous rocks in a single petrographic province which are presumably derived from a common parent magma.

consciousness [PSYCH] State of being aware of one's own existence, of one's mental states, and of the impressions made upon one's senses.

consequent [GEOL] Of, pertaining to, or characterizing movements of the earth resulting from the external transfer of material in the process of gradation.

consequent poles [ELECTROMAG] Pairs of magnetic poles in a magnetized body that are in excess of the usual single pair.

consequent stream [GEOL] A stream whose course is determined by the slope of the land. Also known as superposed stream.

consequent valley [GEOL] 1. A valley whose direction depends on corrugation. 2. A valley formed by the widening of a trench cut by a consequent stream.

conservation law [PHYS] A law which states that some physical quantity associated with an isolated system is constant.

conservation of charge [ELEC] A law which states that the total charge of an isolated system is constant; no violation of this law has been discovered. Also known as charge conservation.

conservation of energy [PHYS] The principle that energy cannot be created or destroyed, although it can be changed from one form to another; no violation of this principle has been found. Also known as energy conservation.

conservation of mass [PHYS] The notion that mass can neither be created nor destroyed; it is violated by many microscopic phenomena.

conservation of matter [PHYS] The notion that matter can be neither created nor destroyed; it is violated by microscopic phenomena.

conservation of momentum [MECH] The principle that, when a system of masses is subject only to internal forces that masses of the system exert on one another, the total vector momentum of the system is constant; no violation of this principle has been found. Also known as momentum conservation.

conservation of parity [QUANT MECH] The law that, if the wave function describing the initial state of a system has even (odd) parity, the wave function describing the final state has even (odd) parity; it is violated by the weak interactions. Also known as parity conservation.

consistency routine [COMPUT SCI] A debugging routine which is used to determine whether the program being checked gives consistent results at specified check points; for example, consistent between runs or with values calculated by other means.

console [COMPUT SCI] The section of a computer that is used to control the machine manually, correct errors, manually revise the contents of storage, and provide communication in other ways between the operator or service engineer and the central processing unit. [ENG] 1. A main control desk for electronic equipment, as at a radar station, radio or television station, or airport control tower. Also known as control desk. 2. A large cabinet for a radio or television receiver, standing on the floor rather than on a table. 3. A grouping of controls, indicators, and similar items contained in a specially designed model cabinet for floor mounting; constitutes an operator's permanent working position.

console switch [COMPUT SCI] A switch on a computer console whose setting can be sensed by a computer, so that an instruction in the program can direct the computer to use this setting to determine which of various alternative courses of action should be followed.

consolidated ice [OCEANOGR] Ice which has been compacted into a solid mass by wind and ocean currents and covers an area of the ocean.

consolidation [GEOL] 1. Processes by which loose, soft, or liquid earth become coherent and firm. 2. Adjustment of a saturated soil in response to increased load; involves squeezing of water from the pores and a decrease in void ratio.

consolute [CHEM] Of or pertaining to liquids that are perfectly miscible in all proportions under certain conditions.

consonance [ACOUS] The interval between two tones whose frequencies are in a ratio approximately equal to the quotient of two whole numbers, each equal to or less than 6, or to such a quotient multiplied or divided by some power of 2.

consortism See symbiosis.

constancy See persistence.

constant [SCI TECH] A value that does not change during a particular process.

constant-bandwidth analyzer [ACOUS] A tunable sound analyzer which has a fixed pass band that is swept through the frequency range of interest. Also known as constant-bandwidth filter.

constant-bandwidth filter See constant-bandwidth analyzer.

constant-current characteristic [ELECTR] The relation between the voltages of two electrodes in an electron tube when the current to one of them is maintained constant and all other electrode voltages are constant.

constant-current electrolysis [CHEM] Electrolysis in which a constant current flows through the cell; used in electrodeposition analysis.

constant-current generator [ELECTR] A vacuum-tube circuit, generally containing a pentode, in which the alternating-current anode resistance is so high that anode current remains essentially constant despite variations in load resistance.

constant-current modulation [COMMUN] System of amplitude modulation in which output circuits of the signal amplifier and the carrier-wave generator or amplifier are connected via a common coil to a constant-current source. Also known as Heising modulation.

constant-current source [ELECTR] A circuit which produces a specified current, independent of the load resistance or applied voltage.

constant-current titration See potentiometric titration.

constant-current transformer [ELEC] A transformer that automatically maintains a constant current in its secondary circuit under varying loads, when supplied from a constant-voltage source.

constant field See stationary field.

constant-head meter [ENG] A flow meter which maintains a constant pressure differential but varies the orifice area with flow, such as a rotameter or piston meter.

constant-height chart [METEOROL] A synoptic chart for any surface of constant geometric altitude above mean sea level (a constant-height surface), usually containing plotted data and analyses of the distribution of such variables as pressure, wind, temperature, and humidity at that altitude. Also known as constant-level chart; fixed-level chart; isohypsic chart.

constant-k lens [ELECTROMAG] A microwave lens that is constructed as a solid dielectric sphere; a plane electromagnetic wave brought to a focus at one point on the sphere emerges from the opposite side of the sphere as a parallel beam.

constant-k network [ELECTR] A ladder network in which the product of the series and shunt impedances is independent of frequency within the operating frequency range.

constant-level chart See constant-height chart.

constant of gravitation See gravitational constant.

constant-pressure chart [METEOROL] The synoptic chart for any constant-pressure surface, usually containing plotted data and analyses of the distribution of height of the surface, wind temperature, humidity, and so on. Also known as isobaric chart; isobaric contour chart.

constant series See displacement series.

constant-speed drive [MECH ENG] A mechanism transmitting motion from one shaft to another that does not allow the velocity ratio of the shafts to be varied, or allows it to be varied only in steps.

constant-velocity recording [ENG ACOUS] A sound-recording method in which, for input signals of a given amplitude, the resulting recorded amplitude is inversely proportional to the frequency; the velocity of the cutting stylus is then constant for all input frequencies having that given amplitude.

constant-volume gas thermometer See gas thermometer.

Constellariidae [PALEON] A family of extinct, marine bryozoans in the order Cystoporata.

constellation [ASTRON] 1. Any one of the star groups interpreted as forming configurations in the sky; examples are Orion and Leo. 2. Any one of the definite areas of the sky.

constituent day [ASTRON] The duration of one rotation of the earth on its axis with respect to an astre fictif, that is, a fictitious star representing one of the periodic elements in the tidal forces; approximates the length of a lunar or solar day.

constituent number [OCEANOGR] One of the harmonic elements in a mathematical expression for the tide-producing force, and in corresponding formulas for the tide or tidal current.

constitution diagram [MET] Graphical representation of phase-stability relationships in an alloy system as a function of temperature. Also known as phase diagram.

constitutive enzyme   [BIOCHEM] An enzyme whose concentration in a cell is constant and is not influenced by substrate concentration.

constrained mechanism   [MECH ENG] A mechanism in which all members move only in prescribed paths.

constraint   [ENG] Anything that restricts the transverse contraction which normally occurs in a solid under longitudinal tension.   [MECH] A restriction on the natural degrees of freedom of a system; the number of constraints is the difference between the number of natural degrees of freedom and the number of actual degrees of freedom.   [SCI TECH] A condition imposed on a system which limits the freedom of the system; may be physical or mathematical, necessary or incidental.

constraint function   [MATH] A function defining one of the prescribed conditions in a nonlinear programming problem.

constraint matrix   [COMPUT SCI] The set of equations and inequalities defining the set of admissible solutions in linear programming.

constriction   [MED] *See* stricture.   [SCI TECH] Narrowing of a channel or cylindrical member.

construction engineering   [CIV ENG] A specialized branch of civil engineering concerned with the planning, execution, and control of construction operations for projects such as highways, dams, utility lines, and buildings.

constructive interference   [PHYS] Phenomenon in which the phases of waves arriving at a specified point over two or more paths of different lengths are such that the square of the resultant amplitude is greater than the sum of the squares of the component amplitudes.

consumable electrode   [MET] A metal electrode that supplies the filler for welding.

consumer   [ECOL] A nutritional grouping in the food chain of an ecosystem, composed of heterotrophic organisms, chiefly animals, which ingest other organisms or particulate organic matter.

consumption   *See* tuberculosis.

contact   [ELEC] *See* electric contact.   [ENG] Initial detection of an aircraft, ship, submarine, or other object on a radarscope or other detecting equipment.   [FL MECH] The surface between two immiscible fluids contained in a reservoir.   [GEOL] The surface between two different kinds of rocks.

contact adsorption   [CHEM ENG] Process for removal of minor constituents from fluids by stirring in direct contact with powdered or granulated adsorbents, or by passing the fluid through fixed-position adsorbent beds (activated carbon or ion-exchange resin); used to decolorize petroleum lubricating oils and to remove solvent vapors from air.

contact aerator   [CIV ENG] A tank in which sewage that is settled on a bed of stone, cement-asbestos, or other surfaces is treated by aeration with compressed air.

contact anemometer   [ENG] An anemometer which actuates an electrical contact at a rate dependent upon the wind speed. Also known as contact-cup anemometer.

contact aureole   *See* aureole.

contact breccia   [PETR] Angular rock fragments resulting from shattering of wall rocks around laccolithic and other igneous masses.

contact catalysis   [CHEM ENG] Process of change in the structure of gas molecules adsorbed onto solid surfaces; the basis of many industrial processes.

contact chatter   *See* chatter.

contact condenser   [MECH ENG] A device in which a vapor, such as steam, is brought into direct contact with a cooling liquid, such as water, and is condensed by giving up its latent heat to the liquid. Also known as direct-contact condenser.

contact-cup anemometer   *See* contact anemometer.

contact dermatitis   [MED] An acute or chronic inflammation of the skin resulting from irritation by or sensitization to some substance coming in contact with the skin.

contact electromotive force   *See* contact potential difference.

contact filtration   [CHEM ENG] A process in which finely divided adsorbent clay is mixed with oil to remove color bodies and to improve the oil's stability.

contact-mask read-only memory   *See* last-mask read-only memory.

contact metamorphism   [PETR] Metamorphism that is genetically related to the intrusion or extrusion of magmas and takes place in rocks at or near their contact.

contact metasomatism   [GEOL] One of the main local processes of thermal metamorphism that is related to intrusion of magmas; takes place in rocks or near their contact with a body of igneous rock.

contact microphone   [ENG ACOUS] A microphone designed to pick up mechanical vibrations directly and convert them into corresponding electric currents or voltages.

contact potential   *See* contact potential difference.

contact potential difference   [ELEC] The potential difference that exists across the space between two electrically connected materials. Also known as contact electromotive force; contact potential; Volta effect.

contact print   [GRAPHICS] A photographic image produced by the exposure of a sensitized emulsion in direct contact with a negative or positive transparency.

contact printer   [GRAPHICS] **1.** A device which provides a light source and a means for holding the negative and the sensitive material in contact during exposure. **2.** A specialized device for exposing diapositive plates at the same scale as that of the negative.

contact protection   [ELEC] Any method for suppressing the surge which results when an inductive circuit is suddenly interrupted; the break would otherwise produce arcing at the contacts, leading to their deterioration.

contact rectifier   *See* metallic rectifier.

contact resistance   [ELEC] The resistance in ohms between the contacts of a relay, switch, or other device when the contacts are touching each other.

contact thermography   [ENG] A method of measuring surface temperature in which a thin layer of luminescent material is spread on the surface of an object and is excited by ultraviolet radiation in a darkened room; the brightness of the coating indicates the surface temperature.

contact transformation   *See* canonical transformation.

contact twin   [CRYSTAL] Twinned crystals whose members are symmetrically arranged about a twin plane.

contact vein   [GEOL] **1.** A variety of fissure vein formed by deposition of minerals in a fault fissure at a rock contact. **2.** A replacement vein formed by mineralized solutions percolating along the more permeable surface areas of the contact.

contact zone   *See* aureole.

contagion [MED] 1. The process whereby disease spreads from one person to another, by direct or indirect contact. 2. The bacterium or virus which transmits disease.

contagious abortion [VET MED] Brucellosis in cattle caused by *Brucella abortus* and inducing abortion. Also known as Bang's disease; infectious abortion.

contagious distribution [STAT] A probability distribution which is dependent on a parameter that itself has a probability distribution.

containerization [IND ENG] The practice of placing cargo in large containers such as truck trailers to facilitate loading on and off ships and railroad flat cars.

container ship [NAV ARCH] A cargo ship which carries its cargo in weatherproof boxes (usually metal) of standard size, called containers, which need not be opened and are rapidly loaded or unloaded from the ship.

containment [NUCLEO] 1. Provision of a gastight enclosure around the highly radioactive components of a nuclear power plant, to contain the radioactivity released by a possible major accident. 2. The use of remote-control devices (slave apparatus) to remove spent cores from nuclear power plants or, in shielded laboratory hoods, to perform chemical studies of dangerous radioactive materials.

containment vessel [NUCLEO] A gas-tight shell or other enclosure around a reactor.

contaminate [SCI TECH] To render unfit or to soil by the introduction of foreign or unwanted material.

contamination [GEOL] A process in which the chemical composition of a magma changes due to the assimilation of country rocks. [HYD] The addition to water of any substance or property that prevents its use without further treatment. [MICROBIO] The process or act of soiling with bacteria. [NUCLEO] The deposit of radioactive materials, such as fission fragments or radiological warfare agents, on any objective or surface or in the atmosphere. [PSYCH] The fusion of words, resulting in a new word. [SCI TECH] Something that contaminates.

contemporaneous [GEOL] 1. Formed, existing, or originating at the same time. 2. Of a rock, developing during formation of the enclosing rock.

content indicator [COMPUT SCI] Display unit that indicates the content in a computer, and the program or mode being used.

contention [COMMUN] A method of operating a multiterminal communication channel in which any station may transmit if the channel is free; if the channel is in use, the queue of contention requests may be maintained in predetermined sequence. [COMPUT SCI] The condition arising when two or more units attempt to transmit over a time-division-multiplex channel at the same time.

continental air [METEOROL] A type of air whose characteristics are developed over a large land area and which therefore has relatively low moisture content.

continental climate [CLIMATOL] Climate characteristic of the interior of a landmass of continental size, marked by large annual, daily, and day-to-day temperature ranges, low relative humidity, and a moderate or small irregular rainfall; annual extremes of temperature occur soon after the solstices.

continental code [COMMUN] The code commonly used for manual telegraph communication, consisting of short (dot) and long (dash) symbols, but not the various-length spaces used in the original Morse code. Also known as international Morse code.

continental crust [GEOL] The basement complex of rock, that is, metamorphosed sedimentary and volcanic rock with associated igneous rocks mainly granitic, that underlies the continents and the continental shelves.

continental drift [GEOL] The concept of continent formation by the fragmentation and movement of land masses on the surface of the earth.

continentality [CLIMATOL] The degree to which a point on the earth's surface is in all respects subject to the influence of a land mass.

continental margin [GEOL] Those provinces between the shoreline and the deep-sea bottom; generally consists of the continental borderland, shelf, slope, and rise.

continental plateau *See* tableland.

continental platform *See* continental shelf.

continental rise [GEOL] A transitional part of the continental margin; a gentle slope with a generally smooth surface, built up by the shedding of sediments from the continental block, and located between the continental slope and the abyssal plain.

continental shelf [GEOL] The zone around a continent, that part of the continental margin extending from the shoreline and the continental slope; composes with the continental slope the continental terrace. Also known as continental platform; shelf.

continental shield [GEOL] Large areas of Precambrian rocks exposed within the cratons of continents.

continental slope [GEOL] The part of the continental margin consisting of the declivity from the edge of the continental shelf extending down to the continental rise.

continental terrace [GEOL] The continental shelf and slope together.

contingency interrupt [COMPUT SCI] A processing interruption due to an operator's action or due to an abnormal result from the system or from a program.

contingency table [STAT] A table for classifying elements of a population according to two variables, the rows corresponding to one variable and the columns to the other.

continue statement [COMPUT SCI] A nonexecutable statement in FORTRAN used principally as a target for transfers, particularly as the last statement in the range of a do statement.

continuity [CIV ENG] Joining of structural members to each other, such as floors to beams, and beams to beams and to columns, so they bend together and strengthen each other when loaded. Also known as fixity. [ELEC] Continuous effective contact of all components of an electric circuit to give it high conductance by providing low resistance.

continuity equation [PHYS] An equation obeyed by any conserved, indestructible quantity such as mass, electric charge, thermal energy, electrical energy, or quantum-mechanical probability, which is essentially a statement that the rate of increase of the quantity in any region equals the total current flowing into the region. Also known as equation of continuity.

continuous bridge [CIV ENG] A fixed bridge supported at three or more points and capable of resisting bending and shearing forces at all sections throughout its length.

continuous casting [MET] A technique in which an ingot, billet, tube, or other shape is continuously solidified and withdrawn while it is being poured, so that its length is not determined by mold dimensions.

continuous comparator   *See* linear comparator.

continuous contact coking   [CHEM ENG] A thermal conversion process using the mass-flow lift principle to give continuous coke circulation; oil-wetted particles of coke move downward into the reactor in which cracking, coking, and drying take place; pelleted coke, gas, gasoline, and gas oil are products of the process.

continuous control   [CONT SYS] Automatic control in which the controlled quantity is measured continuously and corrections are a continuous function of the deviation.

continuous countercurrent leaching   [CHEM ENG] Process of leaching by the use of continuous equipment in which the solid and liquid are both moved mechanically, and by the use of a series of leach tanks and the countercurrent flow of solvent through the tanks in reverse order to the flow of solid.

continuous distillation   [CHEM ENG] Separation by boiling of a liquid mixture with different component boiling points; feed is introduced continuously, with continuous removal of overhead vapors and high-boiling bottoms liquids.

continuous distribution   [STAT] Distribution of a continuous population, which is a class of pairs such that the second member of each pair is a value, and the first member of the pair is a proportion density for that value.

continuous equilibrium vaporization   *See* equilibrium flash vaporization.

continuous footing   [CIV ENG] A footing that supports a wall.

continuous forms   [COMPUT SCI] In character recognition, any batch of source information that exists in reel form, such as tally rolls or cash-register receipts.

continuous function   [MATH] A function which is continuous at each point of its domain. Also known as continuous transformation.

continuous leader   *See* dart leader.

continuously adjustable transformer   *See* variable transformer.

continuous miner   [MIN ENG] Machine designed to remove coal or other soft minerals from the face and to load it into cars or conveyors continuously, without the use of cutting machines, drills, or explosives.

continuous mining   [MIN ENG] A type of mining in which the continuous miner cuts or rips coal or other soft minerals from the face and loads it in a continuous operation.

continuous operation   [ENG] A process that operates on a continuous flow (materials or time) basis, in contrast to batch, intermittent, or sequenced operations.

continuous operator   [MATH] A linear transformation of Banach spaces which is continuous with respect to their topologies.

continuous precipitation   [MET] Precipitation that is characteristic of certain alloys, from a supersaturated solid solution, involving a gradual change of the lattice parameter of the matrix with aging time.

continuous profiling   [GEOL] A method of shooting in seismic exploration in which uniformly placed seismometer stations along a line are shot from holes spaced along the same line so that each hole records seismic ray paths geometrically identical with those from adjacent holes.

continuous reaction series   [MINERAL] A branch of Bowen's reaction series comprising the plagioclase mineral group in which reaction of early-formed crystals with water takes place continuously, without abrupt changes in crystal structure.

continuous stationery   [COMPUT SCI] A continuous ribbon of paper consisting of several hundred or more sheets separated by perforations and folded to form a pack, used to feed a computer printer and generally having sprocket holes along the margin for this purpose.

continuous tone   [GRAPHICS] An image which has not been screened and contains unbroken gradient tones from black to white, and may be either in negative or positive form.

continuous transformation   *See* continuous function.

continuous wave   [ELECTROMAG] A radio or radar wave whose successive sinusoidal oscillations are identical under steady-state conditions. Abbreviated CW. Also known as type A wave.

continuous-wave Doppler radar   *See* continuous-wave radar.

continuous-wave laser   [OPTICS] A laser in which the beam of coherent light is generated continuously, as required for communication and certain other applications. Abbreviated CW laser.

continuous-wave radar   [ENG] A radar system in which a transmitter sends out a continuous flow of radio energy; the target reradiates a small fraction of this energy to a separate receiving antenna. Also known as continuous-wave Doppler radar.

continuum mechanics   *See* classical field theory.

continuum physics   *See* classical field theory.

contour   [MAP] *See* contour line. [PHYS] A curve drawnup on a two-dimensional diagram through points which satisfy $f(x,y) = c$, where c is a constant and f is some function, such as the field strength for a transmitter. [SCI TECH] The periphery of a figure or body.

contour feather   [VERT ZOO] Any of the large flight feathers or long tail feathers of a bird. Also known as penna; vane feather.

contouring control   [COMPUT SCI] The guidance by a computer of a machine tool along a programmed path by interpolating many intermediate points between selected points.

contour interval   [MAP] The difference in elevation between adjacent contours.

contour line   [MAP] A map line representing a contour, that is, connecting points of equal elevation above or below a datum plane, usually mean sea level. Also known as contour; isoheight; isohypse. [METEOROL] A line on a weather map connecting points of equal atmospheric pressure, temperature, or such.

contour map   [MAP] A map displaying topographic or structural contour lines.

contourograph   [ELECTR] Device using a cathode-ray oscilloscope to produce imagery that has a three-dimensional appearance.

contour plowing   [AGR] Cultivation of land along lines connecting points of equal elevation, to prevent water erosion. Also known as terracing.

contour turning   [MECH ENG] Making a three-dimensional reproduction of the shape of a template by controlling the cutting tool with a follower that moves over the surface of a template.

contour value   [MAP] A numerical value placed upon a contour line to denote its elevation relative to a given datum, usually mean sea level.

contraceptive   [MED] Any mechanical device or chemical agent used to prevent conception.

contracted curvature tensor   [MATH] A symmetric tensor of second order, obtained by summation on two indices of the Riemann curvature tensor which are not antisymmetric. Also known as contracted Riemann-Christoffel tensor; Ricci tensor.

contracted Riemann-Christoffel tensor *See* contracted curvature tensor.

contractile vacuole [CYTOL] A tiny, intracellular, membranous bladder that functions in maintaining intra- and extracellular osmotic pressures in equilibrium, as well as excretion of water, such as occurs in protozoans.

contraction [MATH] A continuous function of a metric space to itself which moves each pair of points closer together. [MECH] The action or process of becoming smaller or pressed together, as a gas on cooling. [PHYSIO] Shortening of the fibers of muscle tissue.

contraction coefficient *See* coefficient of contraction.

contraction crack [ENG] A crack resulting from restriction of metal in a mold while contracting.

contraction joint [CIV ENG] A break designed in a structure to allow for drying and temperature shrinkage of concrete, brickwork, or masonry, thereby preventing the formation of cracks.

contraction loss [FL MECH] In fluid flow, the loss in mechanical energy in a stream flowing through a closed duct or pipe when there is a sudden contraction of the cross-sectional area of the passage.

contracture [ARCH] The narrowing of a section of a column. [MED] 1. Shortening, as of muscle or scar tissue, producing distortion or deformity or abnormal limitation of movement of a joint. 2. Retarded relaxation of muscle, as when it is injected with veratrine.

contracurrent system *See* katoptric system.

contraflexure point [CIV ENG] The point in a structure where bending occurs in opposite directions.

contraindication [MED] A symptom, indication, or condition in which a remedy or a method of treatment is inadvisable or improper.

CONTRAN [COMPUT SCI] Computer programming language in which instructions are written at the compiler level, thereby eliminating the need for translation by a compiling routine.

contrarotation [ENG] Rotation in the direction opposite to another rotation.

contrast [COMMUN] The degree of difference in tone between the lightest and darkest areas in a television or facsimile picture. [COMPUT SCI] In optical character recognition, the difference in color, reflectance, or shading between two areas of a surface, for example, a character and its background.

contrast control [ELECTR] A manual control that adjusts the range of brightness between highlights and shadows on the reproduced image in a television receiver.

contrast ratio [ELECTR] The ratio of the maximum to the minimum luminance values in a television picture.

contrast sensitivity *See* threshold contrast.

contrast threshold *See* threshold contrast.

contravariant index [MATH] A tensor index such that, under a transformation of coordinates, the procedure for obtaining a component of the transformed tensor for which this index has the value p involves taking a sum over q of the product of a component of the original tensor for which the index has the value q times the partial derivative of the pth transformed coordinate with respect to the qth original coordinate; it is written as a superscript.

contributory *See* tributary.

control [COMPUT SCI] 1. The section of a digital computer that carries out instructions in proper se-

quence, interprets each coded instruction, and applies the proper signals to the arithmetic unit and other parts in accordance with this interpretation. 2. A mathematical check used with some computer operations. [CONT SYS] A means or device to direct and regulate a process or sequence of events. [ELECTR] An input element of a cryotron. [STAT] 1. A test made to determine the extent of error in experimental observations or measurements. 2. A procedure carried out to give a standard of comparison in an experiment. 3. Observations made on subjects which have not undergone treatment, to use in comparison with observations made on subjects which have undergone treatment.

control bit [COMPUT SCI] A bit which marks either the beginning or the end of a character transmitted in asynchronous communication.

control block [COMPUT SCI] A storage area containing (in condensed, formalized form) the information required for the control of a task, function, operation, or quantity of information.

control board [ELEC] A panel at which one can make circuit changes, as in lighting a theater. [ENG] A panel in which meters and other indicating instruments display the condition of a system, and dials, switches, and other devices are used to modify circuits to control the system. Also known as control panel; panel board.

control break [COMPUT SCI] A key change which takes place in a control data field, especially in the execution of a report program.

control card [COMPUT SCI] A punched card containing input data or parameters which are necessary to begin or modify a program, or containing instructions needed for the specific application of a general routine.

control change [COMPUT SCI] A change of function that occurs when successive records (such as those entered on punched cards) differ in the data entered in the control field; for example, a punched-card tabulator may change from adding to printing at the end of a series of items. Also known as comparing control change.

control character [COMPUT SCI] A character whose occurrence in a particular context initiates, modifies, or stops a control operation in a computer or associated equipment.

control chart [IND ENG] A chart in which quantities of data concerning some property of a product or process are plotted and used specifically to determine the variation in the process.

control circuit [COMPUT SCI] One of the circuits that responds to the instructions in the program for a digital computer. [ELEC] A circuit that controls some function of a machine, device, or piece of equipment. [ELECTR] The circuit that feeds the control winding of a magnetic amplifier.

control computer [COMPUT SCI] A computer which uses inputs from sensor devices and outputs connected to control mechanisms to control physical processes.

control counter [COMPUT SCI] A counter providing data used to control the execution of a computer program.

control data [COMPUT SCI] Data used for identifying, selecting, executing, or modifying another set of data, a routine, a record, or the like.

control desk *See* console.

control diagram *See* flow chart.

control electrode [ELECTR] An electrode used to initiate or vary the current between two or more electrodes in an electron tube.

control element [CONT SYS] The portion of a feedback control system that acts on the process or machine being controlled.

control field [COMPUT SCI] A constant location where information for control purposes is placed, such as specified columns on punched cards.

control grid [ELECTR] A grid, ordinarily placed between the cathode and an anode, that serves to control the anode current of an electron tube.

control head gap [COMPUT SCI] The distance maintained between the read/write head of a disk drive and the disk surface.

control hole See designation punch.

control inductor See control winding.

control instructions [COMPUT SCI] Those instructions in a computer program which ensure proper sequencing of instructions so that a programmed task can be performed correctly.

controllability [AERO ENG] The quality of an aircraft or guided weapon which determines the ease of producing changes in flight direction or in altitude by operation of its controls. [CONT SYS] Property of a system for which, given any initial state and any desired state, there exists a time interval and an input signal which brings the system from the initial state to the desired state during the time interval.

controllable-pitch propeller [MECH ENG] An aircraft or ship propeller in which the pitch of the blades can be changed while the propeller is in motion; five types used for aircraft are two-position, variable-pitch, constant-speed, feathering, and reversible-pitch. Abbreviated CP propeller.

controlled airspace [NAV] An airspace of defined dimensions within which air-traffic control is provided.

controlled atmosphere [SCI TECH] A specified gas or mixture of gases at a predetermined temperature, and sometimes humidity, in which selected processes take place.

controlled avalanche device [ELECTR] A semiconductor device that has rigidly specified maximum and minimum avalanche voltage characteristics and is able to operate and absorb momentary power surges in this avalanche region indefinitely without damage.

controlled avalanche rectifier [ELECTR] A silicon rectifier in which carefully controlled, nondestructive internal avalanche breakdown across the entire junction area protects the junction surface, thereby eliminating local heating that would impair or destroy the reverse blocking ability of the rectifier.

controlled avalanche transit-time triode [ELECTR] A solid-state microwave device that uses a combination of IMPATT diode and *npn* bipolar transistor technologies; avalanche and drift zones are located between the base and collector regions. Abbreviated CATT.

controlled carrier modulation [COMMUN] System of modulation wherein the carrier is amplitude-modulated by the signal frequencies and, in addition, the carrier is amplitude-modulated according to the envelope of the signal so that the modulation factor remains constant regardless of the amplitude of the signal. Also known as floating carrier modulation; variable carrier modulation.

controlled fusion [NUCLEO] The use of thermonuclear fusion reactions in a controlled manner to generate power.

controlled rectifier [ELECTR] A rectifier that has provisions for regulating output current, such as with thyratrons, ignitrons, or silicon controlled rectifiers.

controlled thermonuclear reaction [NUCLEO] A fusion reaction generated in a controlled manner for research purposes or for production of useful power.

controlled thermonuclear reactor [NUCLEO] The heart of a fusion spacecraft propulsion system, based on the thermonuclear reaction of deuterium with a helium-3 isotope to produce helium-4 and protons. Abbreviated CTR.

controlled variable [CONT SYS] In process automatic-control work, that quantity or condition of a controlled system that is directly measured or controlled. [SCI TECH] The quantity or condition that is measured and controlled.

controller See automatic controller.

control limits [ELECTR] In radar evaluation, upper and lower control limits are established at those performance figures within which it is expected that 95% of quality-control samples will fall when the radar is performing normally. [IND ENG] In statistical quality control, the limits of of acceptability placed on control charts; parts outside the limits are defective.

control mark See tape mark.

control module [COMPUT SCI] The set of registers and circuitry required to carry out a specific function.

control panel [COMPUT SCI] **1.** An array of jacks or sockets in which wires (or other elements) may be plugged to control the action of an electromechanical device in a data-processing system such as a printer. Also known as plugboard; wiring board. **2.** See panel. [ENG] See control board.

control point [ADP] The numerical value of the controlled variable (speed, temperature, and so on) which, under any fixed set of operating conditions, an automatic controller operates to maintain. [MAP] Any station in a horizontal and vertical grid that is identified on a photograph and used for correlating the data shown on the photograph. [NAV] A position marked by a buoy, boat, aircraft, electronic device, conspicuous terrain feature, or other identifiable object which is given a name or number and used as an aid for navigation or control of ships, boats, or aircraft.

control program [COMPUT SCI] A program which carries on input/output operations, loading of programs, detection of errors, communication with the operator, and so forth.

control punch See designation punch.

control register [COMPUT SCI] Any one of the registers in a computer used to control the execution of a computer program.

control rod [NUCLEO] Any rod used to control the reactivity of a nuclear reactor; may be a fuel rod or part of the moderator; in a thermal reactor, commonly a neutron absorber. Also known as absorbing rod.

control room [COMMUN] A room from which engineers and production people control and direct a television or radio program or a sound-recording session; the room is adjacent to the main studios and separated from them by large soundproof glass windows. [ENG] A room from which space flights are directed.

control section [COMPUT SCI] The smallest integral subsection of a program, that is, the smallest unit of code that can be separately relocated during loading.

control sequence [COMPUT SCI] The order in which a set of executions are carried to perform a specific function.

control signal [COMPUT SCI] A set of pulses used to identify the channels to be followed by transferred data. [CONT SYS] The signal applied to the device that makes corrective changes in a controlled process or machine.

control state [COMPUT SCI] The operating mode of a system which permits it to override its normal sequence of operations.

control statement *See* job control statement.

control surface [AERO ENG] **1.** Any movable airfoil used to guide or control an aircraft, guided missile, or the like in the air, including the rudder, elevators, ailerons, spoiler flaps, and trim tabs. **2.** In restricted usage, one of the main control surfaces, such as the rudder, an elevator, or an aileron.

control switching point [COMMUN] A telephone office which is an important switching center in the routing of long-distance calls in the direct distance dialing system. Abbreviated CSP.

control symbol [COMPUT SCI] A symbol which, coded into the machine memory, controls certain steps in the mechanical translation process; since control symbols are not contextual symbols, they appear neither in the input nor in the output.

control synchro *See* control transformer.

control system feedback [CONT SYS] A signal obtained by comparing the output of a control system with the input, which is used to diminish the difference between them.

control tape [COMPUT SCI] Loop of paper tape to control the carriage operation of character printers. Also known as carriage tape.

control total [COMPUT SCI] The sum of the numbers in a specified record field of a batch of records, determined repetitiously during computer processing so that any discrepancy from the control indicates an error.

control track [ENG ACOUS] A supplementary sound track, usually containing tone signals that control the reproduction of the sound track, such as by changing feed levels to loudspeakers in a theater to achieve stereophonic effects.

control transformer [ELEC] A synchro in which the electrical output of the rotor is dependent on both the shaft position and the electric input to the stator. Also known as control synchro.

control unit [COMPUT SCI] An electronic device containing data buffers and logical circuitry, situated between the computer channel and the input/output device, and controlling data transfers and such operations as tape rewind.

control valve [ENG] A valve which controls pressure, volume, or flow direction in a fluid transmission system.

control vane [AERO ENG] A movable vane used for control, especially a movable air vane or jet vane on a rocket used to control flight altitude.

control variable [CONT SYS] One of the input variables of a control system, such as motor torque or the opening of a valve, which can be varied directly by the operator to maximize some measure of performance of the system.

control winding [ELECTR] A winding used on a magnetic amplifier or saturable reactor to apply control magnetomotive forces to the core. Also known as control inductor.

Conularida [PALEON] A small group of extinct invertebrates showing a narrow, four-sided, pyramidal-shaped test.

Conulata [INV ZOO] A subclass of free-living coelenterates in the class Scyphozoa; individuals are described as tetraramous cones to elongate pyramids having tentacles on the oral margin.

Conulidae [PALEON] A family of Cretaceous exocyclic Euechinoidea characterized by a flattened oral surface.

conus arteriosus [EMBRYO] The cone-shaped projection from which the pulmonary artery arises on the right ventricle of the heart in man and mammals.

convection [FL MECH] Diffusion in which the fluid as a whole is moving in the direction of diffusion. Also known as bulk flow. [METEOROL] Atmospheric motions that are predominantly vertical, resulting in vertical transport and mixing of atmospheric properties. [OCEANOGR] Movement and mixing of ocean water masses. [PHYS] Transmission of energy or mass by a medium involving movement of the medium itself.

convectional stability *See* static stability.

convection cell [GEOPHYS] A concept in plate tectonics that accounts for the lateral and the upward and downward movement of subcrustal mantle material as due to heat variation in the earth. [METEOROL] An atmospheric unit in which organized convective fluid motion occurs.

convection coefficient *See* film coefficient.

convection cooling [ENG] Heat transfer by natural, upward flow of hot air from the device being cooled.

convection current [ELECTR] The time rate at which the electric charges of an electron stream are transported through a given surface. [GEOPHYS] Mass movement of subcrustal or mantle material as the result of temperature variations. [METEOROL] Any current of air involved in convection; usually, the upward-moving portion of a convection circulation, such as a thermal or the updraft in cumulus clouds. Also known as convective current.

convection modulus [FL MECH] An intrinsic property of a fluid which is important in determining the Nusselt number, equal to the acceleration of gravity times the volume coefficient of thermal expansion divided by the product of the kinematic viscosity and the thermal diffusivity.

convective cloud [METEOROL] A cloud which owes its vertical development, and possibly its origin, to convection.

convective current *See* convection current.

convective discharge [ELECTR] The movement of a visible or invisible stream of charged particles away from a body that has been charged to a sufficiently high voltage. Also known as electric wind; static breeze.

convective equilibrium *See* adiabatic equilibrium.

convective instability [METEOROL] The state of an unsaturated layer or column of air in the atmosphere whose wet-bulb potential temperature (or equivalent potential temperature) decreases with elevation. Also known as potential instability.

convenience receptacle *See* outlet.

conventional grouping [COMPUT SCI] When a unit record containing a coding field is used for a single item code and the set of codes of the terms which describe the item, the grouping is conventional; for example, a personnel file in which each individual is represented by a card on which are punched codes for his or her age, sex, education, and salary.

conventional mining [MIN ENG] The cycle which includes cutting the coal, drilling the shot holes, charging and shooting the holes, loading the broken coal, and installing roof support. Also known as cyclic mining.

conventional programming [COMPUT SCI] The use of standard programming languages, as opposed to application development languages, financial planning languages, query languages, and report programs.

convergence [ANAT] The coming together of a group of afferent nerves upon a motoneuron of the ventral horn of the spinal cord. [ANTHRO] Independent development of similarities between unrelated cultures. [EVOL] Development of similarities between animals or plants of different groups resulting from adaptation to similar habitats. [ELECTR] A condition in which the electron beams of a multibeam cathode-ray tube intersect at a specified point, such as at an opening in the shadow mask of a three-gun color television picture tube; both static convergence and dynamic convergence are required. [GEOL] Diminution of the interval between geologic horizons. [HYD] The line of demarcation between turbid river water and clear lake water. [MATH] The property of having a limit for infinite series, sequences, products, and so on. [METEOROL] The increase in wind setup observed beyond that which would take place in an equivalent rectangular basin of uniform depth, caused by changes in platform or depth.

convergence coil [ELECTR] One of the coils used to obtain convergence of electron beams in a three-gun color television picture tube.

convergence control [ELECTR] A control used in a color television receiver to adjust the potential on the convergence electrode of the three-gun color picture tube to achieve convergence.

convergence magnet [ELECTR] A magnet assembly whose magnetic field converges two or more electron beams; used in three-gun color picture tubes. Also known as beam magnet.

convergence pressure [PHYS CHEM] The pressure at which the different constant-temperature K (liquid-vapor equilibrium) factors for each member of a two-component system converge to unity.

convergence zone [ACOUS] A sound transmission channel produced in sea water by a combination of pressure and temperature changes in the depth range between 2500 and 15,000 feet (750 and 4500 meters); utilized by sonar systems.

convergent integral [MATH] An improper integral which has a finite value.

converging lens [OPTICS] A lens that has a positive focal length, and therefore causes rays of light parallel to its axis to converge.

Conversational Algebraic Language *See* CAL.

conversational compiler [COMPUT SCI] A compiler which immediately checks the validity of each source language statement entered to the computer and informs the user if the next statement can be entered or if a mistake must be corrected. Also known as interpreter.

conversational mode [COMMUN] A computer operating mode that permits queries and responses between the computer and human operators at keyboard terminals.

conversational processing [COMPUT SCI] The operating mode of a computer system which enables a user to have each statement he keys into the system processed immediately.

converse [MATH] The converse of the statement "if *p*, then *q*" is the statement "if *q*, then *p*."

conversion [COMPUT SCI] *See* data conversion. [CHEM] Change of a compound from one isomeric form to another. [CHEM ENG] The chemical change from reactants to products in an industrial chemical process. [NAV] Determination of the rhumb-line direction of one point from another when the initial great-circle direction is known, or

vice versa, the difference between the two directions being the conversion angle; used in connection with radio bearings, Consol, Consolan, and in great-circle sailing. [NUC PHYS] Nuclear transformation of a fertile substance into a fissile substance. [PHYS] Change in a quantity's numerical value as a result of using a different unit of measurement. [PSYCH] A defense mechanism whereby unconscious emotional conflict is transformed into physical disability, the affected part always having symbolic meaning pertinent to the nature of the conflict.

conversion coefficient Also known as conversion fraction; internal conversion coefficient. [NUC PHYS] 1. The ratio of the number of conversion electrons emitted per unit time to the number of photons emitted per unit time in the de-excitation of a nucleus between two given states. 2. In older literature, the ratio of the number of conversion electrons emitted per unit time to the number of conversion electrons plus the number of photons emitted per unit time in the de-excitation of a nucleus between two given states.

conversion equipment [COMPUT SCI] Equipment used for conversion of data from one recording medium to another, as from card to tape.

conversion factor [MATH] The numerical factor by which one must multiply (or divide) a quantity that is expressed in terms of a certain unit to express the quantity in terms of another unit. [NUCLEO] *See* conversion ratio.

conversion fraction *See* conversion coefficient.

conversion gain [ELECTR] 1. Ratio of the intermediate-frequency output voltage to the input signal voltage of the first detector of a superheterodyne receiver. 2. Ratio of the available intermediate-frequency power output of a converter or mixer to the available radio-frequency power input. [NUCLEO] The conversion ratio minus one in a nuclear reactor.

conversion ratio [MATH] *See* conversion factor. [NUCLEO] The number of fissionable atoms produced per fissionable atom fissioned in a converter type of nuclear reactor. Also known as conversion factor.

conversion reaction [PSYCH] The form of hysterical neurosis in which the impulse causing anxiety is converted into functional symptoms of the special senses or voluntary nervous system.

conversion routine [COMPUT SCI] A flexible, self-contained, and generalized program used for data conversion, which only requires specifications about very few facts in order to be used by a programmer.

conversion table [SCI TECH] A list of equivalent values for converting from one set of units to another.

converter [COMPUT SCI] A computer unit that changes numerical information from one form to another, as from decimal to binary or vice versa, from fixed-point to floating-point representation, or from punched cards to magnetic tape. Also known as data converter. [ELEC] 1. Any device for changing alternating current to direct current, or direct current to alternating current. 2. *See* synchronous converter. [ELECTR] 1. The section of a superheterodyne radio receiver that converts the desired incoming radio-frequency signal to the intermediate-frequency value; the converter section includes the oscillator and the mixer-first detector. Also known as heterodyne conversion transducer; oscillator-mixer-first detector. 2. An auxiliary unit used with a television or radio receiver to permit reception of channels or frequencies for which the receiver was not originally designed. **3.** In facsimile, a device that changes the type of modulation delivered by the scanner. **4.** Unit of a radar system in which the mixer of a superheterodyne receiver and usually two stages of intermediate-frequency amplification are located; performs a preamplifying operation. **5.** *See* remodulator. [MET] A type of furnace in which impurities are oxidized out by blowing air through or across a path of molten metal or matte. [NUCLEO] Also known as nuclear converter. **1.** A nuclear reactor that converts fertile atoms into fuel by neutron capture, using one kind of fuel and producing another. **2.** A nuclear reactor that produces some fissionable fuel, but less than it consumes; the fuel produced may be the same as that consumed or different.

converter tube [ELECTR] An electron tube that combines the mixer and local-oscillator functions of a heterodyne conversion transducer.

convertiplane [AERO ENG] A hybrid form of heavier-than-air craft capable, because of one or more horizontal rotors or units acting as rotors, of taking off, hovering, and landing in a fashion similar to a helicopter; and once aloft and moving forward, capable, by means of a mechanical conversion, of flying purely as a fixed-wing aircraft, especially in higher speed ranges.

convex [SCI TECH] Having a curved form which bulges outward, resembling the exterior of a sphere or cylinder or a section of these bodies.

conveyor [MECH ENG] Any materials-handling machine designed to move individual articles such as solids or free-flowing bulk materials over a horizontal, inclined, declined, or vertical path of travel with continuous motion.

convolute [BIOL] Twisted or rolled together, specifically referring to leaves, mollusk shells, and renal tubules.

convolute bedding [GEOL] The extremely contorted laminae usually confined to a single layer of sediment, resulting from subaqueous slumping.

convolution [ANAT] A fold, twist, or coil of any organ, especially any one of the prominent convex parts of the brain, separated from each other by depressions or sulci. [GEOL] **1.** The process of developing convolute bedding. **2.** A structure resulting from a convolution process, such as a small-scale but intricate fold.

Convolvulaceae [BOT] A large family of dicotyledonous plants in the order Polemoniales characterized by internal phloem, the presence of chlorophyll, two ovules per carpel, and plicate cotyledons.

Cooke objective [OPTICS] A three-lens objective consisting of one biconcave lens, the dispersive component, between two biconvex lenses, the collective components; used in astronomical cameras.

coolant [MATER] **1.** A cutting fluid for machine operations, which keeps the tool cool to prevent reduction in hardness and resistance to abrasion, and prevents distortion of the work. **2.** A substance, ordinarily fluid, used for cooling any part of a reactor in which heat is generated. **3.** In general, any cooling agent, usually a fluid.

cool flame [CHEM] A faint, luminous phenomenon observed when, for example, a mixture of ether vapor and oxygen is slowly heated; it proceeds by diffusion of reactive molecules which initiate chemical processes as they go.

cooling curve [THERMO] A curve obtained by plotting time against temperature for a solid-liquid mixture cooling under constant conditions.

cooling degree day [MECH ENG] A unit for estimating the energy needed for cooling a building; one unit is given for each degree Fahrenheit that the daily mean temperature exceeds 75°F (24°C).

cooling load [MECH ENG] The total amount of heat energy that must be removed from a system by a cooling mechanism in a unit time, equal to the rate at which heat is generated by people, machinery, and processes, plus the net flow of heat into the system not associated with the cooling machinery.

cooling tower [ENG] A towerlike device in which atmospheric air circulates and cools warm water, generally by direct contact (evaporation).

cool star [ASTROPHYS] A low-temperature star, generally visible in the infrared range of the electromagnetic spectrum.

Coombs serum [IMMUNOL] An immune serum containing antiglobulin that is used in testing for Rh and other sensitizations.

cooperite [MINERAL] (Pt,Pd)S A steel-gray tetragonal mineral of metallic luster consisting of a sulfide of platinum, occurring in irregular grains in igneous rock.

Cooper pairs [SOLID STATE] Pairs of bound electrons which occur in a superconducting medium according to the Bardeen-Cooper-Schrieffer theory.

coordinate axes [MATH] One of a set of lines or curves used to define a coordinate system; the value of one of the coordinates uniquely determines the location of a point on the axis, while the values of the other coordinates vanish on the axis.

coordinate data receiver [ELECTR] A receiver specifically designed to accept the signal of a coordinate data transmitter and reconvert this signal into a form suitable for input to associated equipment such as a plotting board, computer, or radar set.

coordinate data transmitter [ELECTR] A transmitter that accepts two or more coordinates, such as those representing a target position, and converts them into a form suitable for transmission.

coordinated complex *See* coordination compound.

coordinated geometry *See* COGO.

coordinate indexing [COMPUT SCI] An indexing scheme in which equal-rank descriptors are used to describe a document, for information retrieval by a computer or other means.

coordinate plotter [GRAPHICS] An automated drafting device in which a transverse beam and drafting head are driven over a drawing surface by a computer and tape readers to produce highly precise drawings at high speed. Also known as mechanical plotting board; XY coordinate plotter; XY plotter.

coordinates [MAP] **1.** Linear or angular quantities which designate the position that a point occupies in a given reference frame or system. **2.** A general term to designate the particular kind of reference frame or system, such as plane rectangular coordinates or spherical coordinates. [MATH] A set of numbers which locate a point in space.

coordinate storage *See* matrix storage.

coordinate systems [MATH] A rule for designating each point in space by a set of numbers.

coordinate transformation [MATH] A mathematical or graphic process of obtaining a modified set of coordinates by performing some nonsingular operation on the coordinate axes, such as rotating or translating them.

coordinate valence [CHEM] A chemical bond between two atoms in which a shared pair of electrons forms the bond and the pair has been supplied by one of the two atoms. Also known as coordinate bond.

coordination chemistry [CHEM] The chemistry of metal ions in their interactions with other molecules or ions.

coordination compound [CHEM] A compound with a central atom or ion and a group of ions or molecules surrounding it. Also known as coordinated complex; Werner complex.

coordination number [PHYS] The number of nearest neighbors of a point in a space lattice, of an atom or an ion in a solid, or of an anion or cation in a solution.

coordination polymer [ORG CHEM] Organic addition polymer that is neither free-radical nor simply ionic; prepared by catalysts that combine an organometallic (for example, triethyl aluminum) and a transition metal compound (for example, $TiCl_4$).

Copeognatha [INV ZOO] An equivalent name for Psocoptera.

Copepoda [INV ZOO] An order of Crustacea commonly included in the Entomostraca; contains free-living, parasitic, and symbiotic forms.

copiapite [MINERAL] 1. $Fe_5(SO_4)_6(OH)_2 \cdot 20H_2O$ A yellow mineral occurring in granular or scalar aggregates. Also known as ihleite; knoxvillite; yellow copperas. 2. A group of minerals containing hydrous iron sulfates.

coping [BUILD] A covering course on a wall. [MECH ENG] Shaping stone or other nonmetallic substance with a grinding wheel. [MIN ENG] 1. Process of cutting and trimming the edges of stone slabs. 2. Process of cutting a stone slab into two pieces.

coplanar forces [MECH] Forces that act in a single plane; thus the forces are parallel to the plane and their points of application are in the plane.

Copodontidae [PALEON] An obscure family of Paleozoic fishes in the order Bradyodonti.

copolymer [ORG CHEM] A mixed polymer, the product of polymerization of two or more substances at the same time.

copper [CHEM] A chemical element, symbol Cu, atomic number 29, atomic weight 63.546. [MET] One of the most important nonferrous metals; a ductile and malleable metal found in various ores and used in industry, engineering, and the arts in both pure and alloyed form.

copper blight [PL PATH] A leaf spot disease of tea caused by the fungus *Guignardia camelliae*.

copperhead [VERT ZOO] *Agkistrodon contortrix*. A pit viper of the eastern United States; grows to about 3 feet (90 centimeters) in length and is distinguished by its coppery-brown skin with dark transverse blotches.

copper mica See chalcophyllite.

copper nickel See niccolite.

copperplate engraving [GRAPHICS] A thin, rigid plate of copper, with the lines of a picture cut, or engraved, into it, used for printing purposes; it is inked over, the ink is removed so that it is retained only in the engraved lines, and the plate is placed in a handpress where ink is transferred from engraved lines to overlying paper.

copper spot [PL PATH] A fungus disease of lawn grasses caused by *Gloeocercospora sorghi* and marked by coppery-red areas.

copper uranite See torbernite.

coppice [ECOL] A growth of small trees that are repeatedly cut down at short intervals; the new shoots are produced by the old stumps.

coprolite [GEOL] Petrified excrement.

copulation [ZOO] The sexual union of two individuals, resulting in insemination or deposition of the male gametes in close proximity to the female gametes.

copy [COMMUN] 1. To transcribe Morse code signals into written form. 2. To reproduce graphical material usually by an electrostatic device. 3. To reproduce information in a new location and possibly in a different form, leaving the source of the information unchanged. [COMPUT SCI] A string procedure in ALGOL by means of which a new byte string can be generated from an existing byte string. [GRAPHICS] See subject copy.

coquimbite [MINERAL] $Fe_2(SO_4)_3 \cdot 9H_2O$ A white mineral that crystallizes in the hexagonal system; it is dimorphous with paracoquimbite.

coquina [INV ZOO] A small marine clam of the genus *Donax*. [PETR] A coarse-grained, porous, easily crumbled variety of limestone composed principally of mollusk shell and coral fragments cemented together as rock.

Coraciidae [VERT ZOO] The rollers, a family of Old World birds in the order Coraciiformes.

Coraciiformes [VERT ZOO] An order of predominantly tropical and frequently brightly colored birds.

coracite See uraninite.

coracoid [ANAT] One of the paired bones on the posterior-ventral aspect of the pectoral girdle in vertebrates.

coral [INV ZOO] The skeleton of certain solitary and colonial anthozoan coelenterates; composed chiefly of calcium carbonate.

Corallanidae [INV ZOO] A family of sometimes parasitic, but often free-living, isopod crustaceans in the suborder Flabellifera.

Corallidae [INV ZOO] A family of dimorphic coelenterates in the order Gorgonacea.

Corallimorpharia [INV ZOO] An order of solitary sea anemones in the subclass Zoantharia resembling coral in many aspects.

Corallinaceae [BOT] A family of red algae, division Rhodophyta, having compact tissue with lime deposits within and between the cell walls.

coral reef [GEOL] A ridge or mass of limestone built up of detrital material deposited around a framework of skeletal remains of mollusks, colonial coral, and massive calcareous algae.

coral rock See reef limestone.

corbel [ARCH] An architectural bracket projecting from within a wall, formed by extensions of the masonry and wood beyond the wall surface, and supporting a burdensome weight above.

corbel arch [ARCH] A structure resembling an arch, in which successive courses project farther into a gap until they close.

Corbiculidae [INV ZOO] A family of fresh-water bivalve mollusks in the subclass Eulamellibranchia; an important food in the Orient.

corbino disk [ELECTROMAG] A variable-resistance device utilizing the effect of a magnetic field on the flow of carriers from the center to the circumference of a disk made of semiconducting or conducting material.

corbinotron [ENG] The combination of a corbino disk, made of high-mobility semiconductor material, and a coil arranged to produce a magnetic field perpendicular to the disk.

cord [ELEC] A small, very flexible insulated cable. [MATER] 1. A unit of measure for wood stacked for fuel or pulp; equals $4 \times 4 \times 8$, or 128 cubic feet (approximately 3.6246 cubic meters). 2. A long, flexible, cylindrical construction of natural or synthetic fibers twisted or woven together.

Cordaitaceae [PALEOBOT] A family of fossil plants belonging to the Cordaitales.

Cordaitales [PALEOBOT] An extensive natural grouping of forest trees of the late Paleozoic.

cordierite [MINERAL] $Mg_2(Al_4Si_5O_{18})$ A blue, orthorhombic magnesium aluminosilicate mineral frequently occurring associated with thermally metamorphosed rocks derived from argillaceous sediments.

cordillera [GEOGR] A mountain range or group of ranges, including valleys, plains, rivers, lakes, and so on, forming the main mountain axis of a continent.

Cordulegasteridae [INV ZOO] A family of anisopteran insects in the order Odonata.

cordylite [MINERAL] $(Ce,La)_2Ba(CO_3)_3F_2$ A colorless to wax-yellow mineral consisting of a carbonate and fluoride of cerium, lanthanum, and barium.

core [ELECTR] See magnetic core. [ELECTROMAG] See magnetic core. [ENG] The inner material of a wall, column, veneered door, or similar structure. [GEOL] 1. Center of the earth, beginning at a depth of 2900 kilometers. Also known as earth core. 2. A vertical, cylindrical boring of the earth from which composition and stratification may be determined; in oil or gas well exploration the presence of hydrocarbons or water are items of interest. [MET] A specially formed part of a mold used to form internal holes in a casting. [NUCLEO] The active portion of a nuclear reactor, containing the fissionable material. [OCEANOGR] That area within a layer of ocean water where parameters such as temperature, salinity, or velocity reach extreme values. [SCI TECH] The central part of a body or structure.

core drill [MECH ENG] A mechanism designed to rotate and to cause an annular-shaped rock-cutting bit to penetrate rock formations, produce cylindrical cores of the formations penetrated, and lift such cores to the surface, where they may be collected and examined.

Coreidae [INV ZOO] The squash bugs and leaf-footed bugs, a family of hemipteran insects belonging to the superfamily Coreoidea.

core-image library [COMPUT SCI] A collection of computer programs residing on mass-storage device in ready-to-run form.

core logging [GEOL] The analysis of the strata through which a borehole passes by the taking of core samples at predetermined depth intervals as the well is drilled.

core loss [ELECTROMAG] The rate of energy conversion into heat in a magnetic material due to the presence of an alternating or pulsating magnetic field. Also known as excitation loss; iron loss.

core memory See magnetic core storage.

core memory resident [COMPUT SCI] A control program which is in the main memory of a computer at all times to supervise the processing of the computer.

core sample [GEOL] A sample of rock, soil, snow, or ice obtained by driving a hollow tube into the undisturbed medium and withdrawing it with its contained sample or core.

core storage [COMPUT SCI] 1. The main memory of a computer. 2. See magnetic core storage.

core wall See cutoff wall.

coriander [BOT] *Coriandrum sativum.* A strong-scented perennial herb in the order Umbellales; the dried fruit is used as a flavoring.

coring [MET] A variable composition of individual crystals across a casting, due to nonequilibrium growth over a range of temperature; the purest material is near the center. [PETRO ENG] The use of a core barrel (hollow length of tubing) to take samples from the underground formation during the drilling operation; used for core analysis.

Coriolis acceleration [MECH] 1. An acceleration which, when added to the acceleration of an object relative to a rotating coordinate system and to its centripetal acceleration, gives the acceleration of the object relative to a fixed coordinate system. 2. A vector which is equal in magnitude and opposite in direction to that of the first definition.

Coriolis deflection See Coriolis effect.

Coriolis effect [MECH] Also known as Coriolis deflection. 1. The deflection relative to the earth's surface of any object moving above the earth, caused by the Coriolis force; an object moving horizontally is deflected to the right in the Northern Hemisphere, to the left in the Southern. 2. The effect of the Coriolis force in any rotating system. [PHYSIO] The physiological effects (nausea, vertigo, dizziness, and so on) felt by a person moving radially in a rotating system, as a rotating space station.

Coriolis force [MECH] A velocity-dependent pseudoforce in a reference frame which is rotating with respect to an inertial reference frame; it is equal and opposite to the product of the mass of the particle on which the force acts and its Coriolis acceleration.

Coriolis operator [SPECT] An operator which gives a large contribution to the energy of an axially symmetric molecule arising from the interaction between vibration and rotation when two vibrations have equal or nearly equal frequencies.

corium [ANAT] See dermis. [INV ZOO] Middle portion of the forewing of hemipteran insects.

Corixidae [INV ZOO] The water boatmen, the single family of the hemipteran superfamily Corixoidea.

Corixoidea [INV ZOO] A superfamily of hemipteran insects belonging to the subdivision Hydrocorisae that lack ocelli.

cork [BOT] A protective layer of cells that replaces the epidermis in older plant stems.

Corliss valve [MECH ENG] An oscillating type of valve gear with a trip mechanism for the admission and exhaust of steam to and from an engine cylinder.

corm [BOT] A short, erect, fleshy underground stem, usually broader than high and covered with membranous scales.

corn [BOT] *Zea mays.* A grain crop of the grass order Cyperales grown for its edible seeds (technically fruits).

Cornaceae [BOT] A family of dicotyledonous plants in the order Cornales characterized by perfect or unisexual flowers, a single ovule in each locule, as many stamens as petals, and opposite leaves.

Cornales [BOT] An order of dicotyledonous plants in the subclass Rosidae marked by a woody habit, simple leaves, well-developed endosperm, and fleshy fruits.

cornea [ANAT] The transparent anterior portion of the outer coat of the vertebrate eye covering the iris and the pupil. [INV ZOO] The outer transparent portion of each ommatidium of a compound eye.

corner frequency See break frequency.

corner reflector [ELECTROMAG] An antenna consisting of two conducting surfaces intersecting at an angle that is usually 90°, with a dipole or other antenna located on the bisector of the angle. [OPTICS] A reflector which returns a laser beam in the direction of its source, consisting of perpendicular reflecting surfaces; used to make precise determinations of distances in surveying.

cornetite [MINERAL] $Cu_3(PO_4)(OH)_3$ A peacock-blue mineral consisting of basic copper phosphate.

cornice [ARCH] The crowning, overhanging part of an architectural structure.

corniculate [BIOL] Possessing small horns or horn-like processes.

corn smut [PL PATH] A fungus disease of corn caused by *Ustilago maydis*.

corn sugar *See* dextrose.

Cornu quartz prism [OPTICS] A prism constructed of two 30° quartz prisms, left- and right-handed, used in conjunction with left- and right-handed lenses, so that the rotation of polarization occurring in one half of the optical path is exactly compensated by the reverse rotation in the other; used in a quartz spectrograph.

Cornu's spiral [MATH] The graph of the function

$$f(x,y) = \int_{-\infty}^{+\infty} \exp[-y^2z^2 - z - xe^{-z}]dz.$$

Also known as clothoid.

cornwallite [MINERAL] $Cu_5(AsO_4)_2(OH)_4 \cdot H_2O$ A verdigris green to blackish-green mineral consisting of a hydrated basic arsenate of copper; occurs as small botryoidal crusts.

corolla [BOT] A collective term for the petals of a flower.

corona [ARCH] The overhanging vertical member of a cornice. [ASTRON] *See* solar corona. [BOT] **1.** An appendage or series of fused appendages between the corolla and stamens of some flowers. **2.** The region where stem and root of a seed plant merge. Also known as crown. [ELEC] *See* corona discharge. [GEOL] A mineral zone that is usually radial about another mineral or at the area between two minerals. Also known as kelyphite. [INV ZOO] **1.** The anterior ring of cilia in rotifers. **2.** A sea urchin test. **3.** The calyx and arms of a crinoid. [MET] An area sometimes surrounding the nugget at the faying surfaces of a spot weld which provides a degree of bond strength. [METEOROL] A set of one or more prismatically colored rings of small radii, concentrically surrounding the disk of the sun, moon, or other luminary when veiled by a thin cloud; due to diffraction by numerous waterdrops. [MINERAL] An annular zone of minerals disposed either around another mineral or at the contact between two minerals.

Corona Australis [ASTRON] A constellation, right ascension 19 hours, declination 40°S. Abbreviated CrA. Also known as Southern Crown.

Corona Borealis [ASTRON] A constellation, right ascension 16 hours, declination 30°N. Abbreviated CrB. Also known as Northern Crown.

corona current [ELEC] The current of electricity equivalent to the rate of charge transferred to the air from an object experiencing corona discharge.

corona discharge [ELEC] A discharge of electricity appearing as a bluish-purple glow on the surface of and adjacent to a conductor when the voltage gradient exceeds a certain critical value; due to ionization of the surrounding air by the high voltage. Also known as aurora; corona; electric corona.

coronadite [MINERAL] $Pb(Mn^{2+},Mn^{4+})_8O_{16}$ A black mineral consisting of a lead and manganese oxide, occurring in massive form with fibrous structure; an important constituent of manganese ore.

coronagraph [ASTRON] An instrument for photographing the corona and prominences of the sun at times other than at solar eclipse.

coronal loop [ASTRON] A looplike structure revealed in soft x-ray images of the solar limb and believed to evolve from the introduction of energy and density perturbations at the top of an arched, cylindrical magnetic flux tube initially in equilibrium in the coronal plasma.

coronal suture [ANAT] The union of the frontal with the parietal bones transversely across the vertex of the skull.

corona radiata [HISTOL] The layer of cells immediately surrounding a mammalian ovum.

coronary artery [ANAT] Either of two arteries arising in the aortic sinuses that supply the heart tissue with blood.

coronary failure [MED] Prolonged precordial pain or discomfort without conventional evidence of myocardial infarction; subendocardial ischemia caused by a disparity between coronary blood flow and myocardial needs; this condition is more commonly referred to as coronary artery insufficiency.

coronary valve [ANAT] A semicircular fold of the endocardium of the right atrium at the orifice of the coronary sinus.

coronary vein [ANAT] **1.** Any of the blood vessels that bring blood from the heart and empty into the coronary sinus. **2.** A vein along the lesser curvature of the stomach.

Coronatae [INV ZOO] An order of the class Scyphozoa which includes mainly abyssal species having the exumbrella divided into two parts by a coronal furrow.

coronavirus [VIROL] A major group of animal viruses including avian infectious bronchitis virus and mouse hepatitis virus.

coronoid process [ANAT] **1.** A thin, flattened process projecting from the anterior portion of the upper border of the ramus of the mandible, and serving for the insertion of the temporal muscle. **2.** A triangular projection from the upper end of the ulna, forming the lower part of the radial notch.

Corophiidae [INV ZOO] A family of amphipod crustaceans in the suborder Gammaridea.

coroutine [COMPUT SCI] A program module for which the lifetime of a particular activation record is independent of the time when control enters or leaves the module, and in which the activation record maintains a local instruction counter so that, whenever control enters the module, execution begins at the point where it stopped when control last left that particular instance of execution.

corpora quadrigemina [ANAT] The inferior and superior colliculi collectively. Also known as quadrigeminal body.

cor pulmonale [MED] Hypertrophy and dilation of the right ventricle secondary to obstruction to the pulmonary blood flow and consequent pulmonary hypertension.

corpus allatum [INV ZOO] An endocrine structure near the brain of immature arthropods that secretes a juvenile hormone, neotenin.

corpus callosum [ANAT] A band of nerve tissue connecting the cerebral hemispheres in humans and higher mammals.

corpus cavernosum [ANAT] The cylinder of erectile tissue forming the clitoris in the female and the penis in the male.

corpuscle [ANAT] **1.** A small, rounded body. **2.** An encapsulated sensory-nerve end organ. [OPTICS] A particle of light in the corpuscular theory, corresponding to the photon in the quantum theory.

corpuscular radiation [PHYS] Radiation consisting of subatomic particles, such as electrons, protons, deuterons, and neutrons, as distinguished from electromagnetic radiation.

corpuscular theory of light [OPTICS] Theory that light consists of a stream of particles; now considered a limiting case of the quantum theory. Also known as Newton's theory of light.

corpus luteum [HISTOL] The yellow endocrine body formed in the ovary at the site of a ruptured Graafian follicle.

corrasion [GEOL] Mechanical wearing away of rock and soil by the action of solid materials moved along by wind, waves, running water, glaciers, or gravity. Also known as mechanical erosion.

correcting [NAV] The process of applying corrections, particularly that of converting compass to magnetic direction, or compass, magnetic, or gyro to true direction.

correcting plate *See* corrector plate.

corrective network [ELEC] An electric network inserted in a circuit to improve its transmission properties, impedance properties, or both. Also known as shaping circuit; shaping network.

corrector plate [OPTICS] A thin lens or system of lenses used to correct the spherical aberration of a spherical lens or the coma of a parabolic lens; used particularly in telescopes such as the Schmidt telescope. Also known as correcting plate.

correlation [GEOL] **1.** The determination of the equivalence or contemporaneity of geologic events in separated areas. **2.** As a step in seismic study, the selecting of corresponding phases, taken from two or more separated seismometer spreads, of seismic events seemingly developing at the same geologic formation boundary.

correlation coefficient [STAT] A measurement, which is unchanged by both addition and multiplication of the random variable by positive constants, of the tendency of two random variables X and Y to vary together; it is given by the ratio of the covariance of X and Y to the square root of the product of the variance of X and the variance of Y.

correlation curve *See* correlogram.

correlation distance [COMMUN] In tropospheric scatter propagation, the minimum spatial separation between antennas which will give rise to independent fading of the received signals.

correlogram [MATH] A curve showing the assumed correlation between two mathematical variables. Also known as correlation curve.

correspondence principle [QUANT MECH] The principle that quantum mechanics has a classical limit in which it is equivalent to classical mechanics. Also known as Bohr's correspondence principle.

corresponding states [PHYS CHEM] The condition when two or more substances are at the same reduced pressures, the same reduced temperatures, and the same reduced volumes.

corrosion [GEOCHEM] Chemical erosion by motionless or moving agents. [MET] Gradual destruction of a metal or alloy due to chemical processes such as oxidation or the action of a chemical agent.

corrosion border *See* corrosion rim.

corrosion fatigue [MET] Damage to or failure of a metal due to corrosion combined with fluctuating fatigue stresses.

corrosion number *See* acid number.

corrosion rim [MINERAL] A modification of the outlines of a porphyritic crystal due to the corrosive action of a magma on previously stable minerals. Also known as corrosion border.

corrosive [MATER] A substance that causes corrosion.

corrosive flux [MET] A soldering flux, usually composed of inorganic salts and acids, which provides oxide removal of the base metal upon application of solder; flux remaining on the base metal is corrosive and should be removed.

corrugated fastener [DES ENG] A thin corrugated strip of steel that can be hammered into a wood joint to fasten it.

corrugated lens [OPTICS] A lens having circular sections cut out from the surface to reduce its weight without lowering its focal power.

cortex [ANAT] The outer portion of an organ or structure, such as of the brain and adrenal glands. [BOT] A primary tissue in roots and stems of vascular plants that extends inward from the epidermis to the phloem. [INV ZOO] The peripheral layer of certain protozoans.

corticosteroid [BIOCHEM] **1.** Any steroid hormone secreted by the adrenal cortex of vertebrates. **2.** Any steroid with properties of an adrenal cortex steroid.

corticosterone [BIOCHEM] $C_{21}H_{30}O_4$ A steroid hormone produced by the adrenal cortex of vertebrates that stimulates carbohydrate synthesis and protein breakdown and is antagonistic to the action of insulin.

corticotrophic [PHYSIO] Having an effect on the adrenal cortex.

corticotropin [BIOCHEM] A hormonal preparation having adrenocorticotropic activity, derived from the adenohypophysis of certain domesticated animals.

cortisol *See* hydrocortisone.

cortisone [BIOCHEM] $C_{21}H_{28}O_5I$ A steroid hormone produced by the adrenal cortex of vertbrates that acts principally in carbohydrate metabolism.

cortlandite [PETR] A peridotite consisting of large crystals of hornblende with poikilitically included crystals of olivine. Also known as hudsonite.

corundum [MINERAL] Al_2O_3 A hard mineral occurring in various colors and crystallizing in the hexagonal system; crystals are usually prismatic or in rounded barrel shapes; gem varieties are ruby and sapphire.

Corvidae [VERT ZOO] A family of large birds in the order Passeriformes having stout, long beaks; includes the crows, jays, and magpies.

Corvus [ASTRON] A constellation, right ascension 12 hours, declination 20°S. Abbreviated Crv. Also known as Crow.

corvusite [MINERAL] $V_2V_{12}O_{34} \cdot nH_2O$ A blue-black to brown mineral consisting of a hydrous oxide of vanadium; occurs in massive form.

Corylophidae [INV ZOO] The equivalent name for Orthoperidae.

corymb [BOT] An inflorescence in which the flower stalks arise at different levels but reach the same height, resulting in a flat-topped cluster.

Corynebacteriaceae [MICROBIO] Formerly a family of nonsporeforming, usually nonmotile rod-shaped bacteria in the order Eubacteriales including animal and plant parasites and pathogens.

Coryphaenidae [VERT ZOO] A family of pelagic fishes in the order Perciformes characterized by a blunt nose and deeply forked tail.

Coryphodontidae [PALEON] The single family of the Coryphodontoidea, an extinct superfamily of mammals.

Coryphodontoidea [PALEON] A superfamily of extinct mammals in the order Pantodonta.

coryza [MED] Inflammation of the mucous membranes of the nose, usually marked by sneezing and discharge of watery mucus.

cos *See* cosine function.

cosecant [MATH] The reciprocal of the sine. Denoted csc.

cosine emission law [OPTICS] The law that the energy emitted by a radiating surface in any direction is proportional to the cosine of the angle which that direction makes with the normal.

cosine function [MATH] In a right triangle with an angle θ, the cosine function gives the ratio of adjacent side to hypotenuse; more generally, it is the function which assigns to any real number θ the abscissa of the point on the unit circle obtained by moving from $(1,0)$ counterclockwise θ units along the circle, or clockwise $|θ|$ units if θ is less than 0. Denoted cos.

cosmic [ASTRON] Pertaining to the cosmos, the vast extraterrestrial regions of the universe.

cosmic dust [ASTRON] Fine particles of solid matter forming clouds in interstellar space.

cosmic electrodynamics [ASTROPHYS] The science concerned with electromagnetic phenomena in ionized media encountered in interstellar space, in stars, and above the atmosphere.

cosmic microwave background See cosmic microwave radiation.

cosmic microwave radiation [ASTRON] A nearly uniform flux of microwave radiation that is believed to permeate all of space and to have originated in the big bang. Also known as cosmic microwave background; microwave background.

cosmic noise [COMMUN] Radio static caused by a phenomenon outside the earth's atmosphere, such as sunspots.

cosmic radiation See cosmic rays.

cosmic radio waves [ASTRON] Radio waves reaching the earth from interstellar or intergalactic sources.

cosmic rays [NUC PHYS] Electrons and the nuclei of atoms, largely hydrogen, that impinge upon the earth from all directions of space with nearly the speed of light. Also known as cosmic radiation; primary cosmic rays.

cosmic-ray shower [NUC PHYS] The simultaneous appearance of a number of downward-directed ionizing particles, with or without accompanying photons, caused by a single cosmic ray. Also known as air shower; shower.

cosmic year [ASTRON] The period of rotation of the Milky Way Galaxy, about 220 million years.

Cosmocercidae [INV ZOO] A group of nematodes assigned to the suborder Oxyurina by some authorities and to the suborder Ascaridina by others.

cosmochemistry [ASTROPHYS] The branch of science which treats of the chemical composition of the universe and its origin.

cosmogony [ASTROPHYS] Study of the origin and evolution of specific astronomical systems and of the universe as a whole.

cosmoid scale [VERT ZOO] A structure in the skin of primitive rhipidistians and dipnoans that is composed of enamel, a dentine layer (cosmine), and laminated bone.

cosmology [ASTRON] The study of the overall structure of the physical universe.

cosmopolitan [ECOL] Having a worldwide distribution wherever the habitat is suitable, with reference to the geographical distribution of a taxon.

cospectrum [PHYS] **1.** The spectral decomposition of the in-phase components of the covariance of two functions of time. **2.** The real part of the cross spectrum of two functions.

Cossidae [INV ZOO] The goat or carpenter moths, a family of heavy-bodied lepidopteran insects in the

superfamily Cossoidea having the abdomen extending well beyond the hindwings.

Cossoidea [INV ZOO] A monofamilial superfamily of lepidopteran insects belonging to suborder Heteroneura.

Cossuridae [INV ZOO] A family of fringe worms belonging to the Sedentaria.

Cossyphodidae [INV ZOO] The lively ant guest beetles, a small family of coleopteran insects in the superfamily Tenebrionoidea.

costa [BIOL] A rib or riblike structure. [BOT] The midrib of a leaf. [INV ZOO] The anterior vein of an insect's wing.

cost accounting [IND ENG] The branch of accounting in which one records, analyzes, and summarizes costs of material, labor, and burden, and compares these actual costs with predetermined budgets and standards.

Costaceae [BOT] A family of monocotyledonous plants in the order Zingiberales distinguished by having one functional stamen with two pollen sacs and spirally arranged leaves and bracts.

costal cartilage [ANAT] The cartilage occupying the interval between the ribs and the sternum or adjacent cartilages.

costal process [ANAT] An anterior or ventral projection on the lateral part of a cervical vertebra. [EMBRYO] An embryonic rib primordium, the ventrolateral outgrowth of the caudal, denser half of a sclerotome.

cost analysis [IND ENG] Analysis of the factors contributing to the costs of operating a business and of the costs which will result from alternative procedures, and of their effects on profits.

costate [BIOL] Having ribs or ridges.

cotangent [MATH] The reciprocal of the tangent. Denoted cot; ctn.

cotar [ENG] A passive system used for tracking a vehicle in space by determining the line of direction between a remote ground-based receiving antenna and a telemetering transmitter in the missile, using phase-comparison techniques. Derived from correlated orientation tracking and range.

cotat [ENG] A trajectory-measuring system using several antenna base lines, each separated by large distances, to measure direction cosines to an object; then the object's space position is computed by triangulation. Derived from correlation tracking and triangulation.

cotidal chart [MAP] A chart of cotidal lines that show approximate locations of high water at hourly intervals as measured from a reference meridian, usually Greenwich.

cotidal hour [ASTRON] The average interval expressed in solar or lunar hours between the moon's passage over the meridian of Greenwich and the following high water at a specified place.

cotidal line [MAP] A line on a chart passing through all points where high water occurs at the same time.

Cotingidae [VERT ZOO] The cotingas, a family of neotropical suboscine birds in the order Passeriformes.

cotter pin [DES ENG] A split pin, inserted into a hole, used to hold a nut or cotter securely to a bolt or shaft, or to hold a pair of hinge plates together.

Cottidae [VERT ZOO] The sculpins, a family of perciform fishes in the suborder Cottoidei.

Cottiformes [VERT ZOO] An order set up in some classification schemes to include the Cottoidei.

Cottoidei [VERT ZOO] The mail-cheeked fishes, a suborder of the order Perciformes characterized by the expanded third infraorbital bone.

cotton [BOT] Any plant of the genus *Gossypium* in the order Malvales; cultivated for the fibers obtained from its encapsulated fruits or bolls. [TEXT] The most economical natural fiber, obtained from plants of the genus *Gossypium*, used in making fabrics, cordage, and padding and for producing artificial fibers and cellulose.

cotton anthracnose [PL PATH] A fungus disease of cotton caused by *Glomerella gossypii* and characterized by reddish-brown to light-colored or necrotic spots.

cotton ball *See* ulexite.

Cotton effect [ANALY CHEM] The characteristic wavelength dependence of the optical rotatory dispersion curve or the circular dichroism curve or both in the vicinity of an absorption band.

cottonmouth *See* water moccasin.

Cotton-Mouton birefringence *See* Cotton-Mouton effect.

Cotton-Mouton constant [OPTICS] A constant giving the strength of the Cotton-Mouton effect in a liquid; when multiplied by the path length and the square of the magnetic field, it gives the phase difference between the components of light parallel and perpendicular to the field.

Cotton-Mouton effect [OPTICS] The double refraction (birefringence) of light in a liquid in a magnetic field at right angles to the direction of light propagation. Also known as Cotton-Mouton birefringence.

cotton root rot [PL PATH] A fungus disease of cotton caused by *Phymatotrichum omnivorum* and marked by bronzing of the foliage followed by sudden wilting and death of the plant.

cotton rust [PL PATH] A fungus disease of cotton caused by *Puccinia stakmanii* producing low, greenish-yellow or orange elevations on the undersurface of leaves.

cotton wilt [PL PATH] 1. A fungus disease of cotton caused by *Fusarium vasinfectum* growing in the water-conducting vessels and characterized by wilt, yellowing, blighting, and death. 2. A fungus blight of cotton caused by *Verticillium albo-atrum* and characterized by yellow mottling of the foliage.

cottony rot [PL PATH] A fungus disease of many plants, especially citrus trees, marked by fluffy white growth caused by *Sclerotinia sclerotiorum*, in which there is stem wilt and rot.

Cottrell hardening [SOLID STATE] Hardening of a material caused by locking of its dislocations when impurity atoms whose size differs from that of the solvent cluster around them.

Cottrell precipitator [ENG] A machine for removing dusts and mists from gases, in which the gas passes through a grounded pipe with a fine axial wire at a high negative voltage, and particles are ionized by the corona discharge of the wire and migrate to the pipe.

cotunnite [MINERAL] PbCl$_2$ An alteration product of galena; a soft, white to yellowish mineral that crystallizes in the orthorhombic crystal system.

cotyledon [BOT] The first leaf of the embryo of seed plants.

Cotylosauria [PALEON] An order of primitive reptiles in the subclass Anapsida, including the stem reptiles, ancestors of all of the more advanced Reptilia.

cotype *See* syntype.

coudé telescope [OPTICS] An instrument in which light is reflected along the polar axis to come to focus at a fixed place where it is viewed through a fixed eyepiece or where a spectrograph can be mounted.

Couette flow [FL MECH] Low-speed, steady motion of a viscous fluid between two infinite plates moving parallel to each other.

Couette viscometer [ENG] A viscometer in which the liquid whose viscosity is to be measured fills the space between two vertical coaxial cylinders, the inner one suspended by a torsion wire; the outer cylinder is rotated at a constant rate, and the resulting torque on the inner cylinder is measured by the twist of the wire. Also known as rotational viscometer.

cougar *See* puma.

Couinae [VERT ZOO] The couas, a subfamily of Madagascan birds in the family Cuculidae.

coulomb [ELEC] A unit of electric charge, defined as the amount of electric charge that crosses a surface in 1 second when a steady current of 1 absolute ampere is flowing across the surface; this is the absolute coulomb and has been the legal standard of quantity of electricity since 1950; the previous standard was the international coulomb, equal to 0.999835 absolute coulomb. Abbreviated coul. Symbolized C.

Coulomb attraction [ELEC] The electrostatic force of attraction exerted by one charged particle on another charged particle of opposite sign. Also known as electrostatic attraction.

Coulomb barrier [NUC PHYS] 1. The Coulomb repulsion which tends to keep positively charged bombarding particles out of the nucleus. 2. Specifically, the Coulomb potential associated with this force.

Coulomb excitation [NUC PHYS] Inelastic scattering of a positively charged particle by a nucleus and excitation of the nucleus, caused by the interaction of the nucleus with the rapidly changing electric field of the bombarding particle.

coulombmeter [ENG] A measuring instrument that measures quantity of electricity in coulombs by integrating a stored charge in a circuit which has very high input impedance.

Coulomb potential [ELEC] A scalar point function equal to the work per unit charge done against the Coulomb force in transferring a particle bearing an infinitesimal positive charge from infinity to a point in the field of a specific charge distribution.

Coulomb repulsion [ELEC] The electrostatic force of repulsion exerted by one charged particle on another charged particle of the same sign. Also known as electrostatic repulsion.

Coulomb's law [ELEC] The law that the attraction or repulsion between two electric charges acts along the line between them, is proportional to the product of their magnitudes, and is inversely proportional to the square of the distance between them. Also known as law of electrostatic attraction.

coulometer [PHYS CHEM] An electrolytic cell for the precise measurement of electrical quantities or current intensity by quantitative determination of chemical substances produced or consumed.

coulometric analysis [ANALY CHEM] A technique in which the amount of a substance is determined quantitatively by measuring the total amount of electricity required to deplete a solution of the substance.

coulometric titration [ANALY CHEM] The slow electrolytic generation of a soluble species which is capable of reacting quantitatively with the substance sought; some independent property must be observed to establish the equivalence point in the reaction.

coulometry [ANALY CHEM] A determination of the amount of an electrolyte released during electrolysis by measuring the number of coulombs used.

coulostatic analysis [PHYS CHEM] An electrochemical technique involving the application of a very short, large pulse of current to the electrode; the pulse charges the capacitive electrode-solution interface to a new potential, then the circuit is opened, and the return of the working electrode potential to its initial value is monitored; the current necessary to discharge the electrode interface comes from the electrolysis of electroactive species in solution; the change in electrode potential versus time results in a plot, the shape of which is proportional to concentration.

count [AERO ENG] 1. To proceed from one point to another in a countdown or plus count, normally by calling a number to signify the point reached. 2. To proceed in a countdown, for example, T minus 90 and counting. [CHEM] An ionizing event. [NUCLEO] 1. A single response of the counting system in a radiation counter. 2. The total number of events indicated by a counter. [TEXT] The number of warp and filling threads per square inch of fabric.

countable [MATH] Either finite or denumerable. Also known as enumerable.

countably infinite set *See* denumerable set.

counter [COMPUT SCI] 1. A register or storage location used to represent the number of occurrences of an event. 2. *See* accumulator. [ELECTR] *See* scaler. [ENG] A complete instrument for detecting, totalizing, and indicating a sequence of events. [NAV ARCH] *See* buttocks. [NUCLEO] *See* radiation counter.

counterbalance *See* counterweight.

counterbore [DES ENG] A flat-bottom enlargement of the mouth of a cylindrical bore to enlarge a borehole and give it a flat bottom. [ENG] To enlarge a borehole by means of a counterbore.

counter circuit *See* counting circuit.

counter coupling [COMPUT SCI] The technique of combining two or more counters into one counter of larger capacity in electromechanical devices by means of control panel wiring.

countercurrent extraction [CHEM ENG] A liquid-liquid extraction process in which the solvent and the process stream in contact with each other flow in opposite directions. Also known as countercurrent separation.

countercurrent flow [MECH ENG] A sensible heat-transfer system in which the two fluids flow in opposite directions.

countercurrent separation *See* countercurrent extraction.

counterelectromotive force [ELECTROMAG] The voltage developed in an inductive circuit by a changing current; the polarity of the induced voltage is at each instant opposite that of the applied voltage. Also known as back electromotive force.

counterfort wall [CIV ENG] A type of retaining wall that resembles a cantilever wall but has braces at the back; the toe slab is a cantilever and the main steel is placed horizontally.

counterglow *See* gegenschein.

counterimmunoelectrophoresis [IMMUNOL] Immunoelectrophoresis which uses two wells of application, one above the other, along the electrical axis—the anodal well filled with antibody and the cathodal with a negatively charged antigen; electrophoresis results in the antigen and antibody migrating cath-

odally and anodally, respectively, and a line of precipitation appears where the two meet.

counterlath [BUILD] 1. A strip placed between two rafters to support crosswise laths. 2. A lath placed between a timber and a sheet lath. 3. A lath nailed at a more or less random spacing between two precisely spaced laths. 4. A lath put on one side of a partition after the other side has been finished.

counterpoise [ELEC] A system of wires or other conductors that is elevated above and insulated from the ground to form a lower system of conductors for an antenna. Also known as antenna counterpoise. [MECH ENG] *See* counterweight.

counterradiation [GEOPHYS] The downward flux of atmospheric radiation passing through a given level surface, usually taken as the earth's surface. Also known as back radiation.

countersinking [MECH ENG] Drilling operation to form a flaring depression around the rim of a hole.

counterstain [BIOL] A second stain applied to a biological specimen to color elements other than those demonstrated by the principal stain.

counter sun *See* anthelion.

counter tube [ELECTR] An electron tube having one signal-input electrode and 10 or more output electrodes, with each input pulse serving to transfer conduction sequentially to the next output electrode; beam-switching tubes and cold-cathode counter tubes are examples. [NUCLEO] An electron tube that converts an incident particle or burst of incident radiation into a discrete electric pulse, generally by utilizing the current flow through a gas that is ionized by the radiation; used in radiation counters. Also known as radiation counter tube.

counterweight [MECH ENG] 1. A device which counterbalances the original load in elevators and skip and mine hoists, going up when the load goes down, so that the engine must only drive against the unbalanced load and overcome friction. 2. Any weight placed on a mechanism which is out of balance so as to maintain a static equilibrium. Also known as counterbalance; counterpoise.

counting chamber [MICROBIO] An accurately dimensioned chamber in a microslide which can hold a specific volume of fluid and which is usually ruled into units to facilitate the counting under the microscope of cells, bacteria, or other structures in the fluid.

counting circuit [ELECTR] A circuit that counts pulses by frequency-dividing techniques, by charging a capacitor in such a way as to produce a voltage proportional to the pulse count, or by other means. Also known as counter circuit.

counting-down circuit *See* frequency divider.

country rock [GEOL] 1. Rock that surrounds and is penetrated by mineral veins. 2. Rock that surrounds and is invaded by an igneous intrusion.

couple [CHEM] Joining of two molecules. [ELEC] To connect two circuits so signals are transferred from one to the other. [ELECTR] Two metals placed in contact, as in a thermocouple. [ENG] To connect with a coupling, such as of two belts or two pipes. [MECH] A system of two parallel forces of equal magnitude and opposite sense.

coupled circuits [ELEC] Two or more electric circuits so arranged that energy can transfer electrically or magnetically from one to another.

coupled oscillators [ELECTROMAG] A set of alternating-current circuits which interact with each other, for example, through mutual inductances or capacitances. [MECH] A set of particles subject to elastic restoring forces and also to elastic interactions with each other.

coupled reaction [CHEM] A reaction which involves two oxidants with a single reductant, where one reaction taken alone would be thermodynamically unfavorable.

coupled systems [PHYS] Mechanical, electrical, or other systems which are connected in such a way that they interact and exchange energy with each other.

coupled transistors [ELECTR] Transistors connected in series by transformers or resistance-capacitance networks, in much the same manner as electron tubes.

coupler [ELEC] A component used to transfer energy from one circuit to another. [ELECTROMAG] 1. A passage which joins two cavities or waveguides, allowing them to exchange energy. 2. A passage which joins the ends of two waveguides, whose cross section changes continuously from that of one to that of the other. [ENG] A device that connects two railroad cars. [NAV] The portion of a navigation system that receives signals of one type from a sensor and transmits signals of a different type to an actuator.

coupling [ELEC] 1. A mutual relation between two circuits that permits energy transfer from one to another, through a wire, resistor, transformer, capacitor, or other device. 2. A hardware device used to make a temporary connection between two wires. [ENG] Any device that serves to connect the ends of adjacent parts, as railroad cars. [MECH ENG] The mechanical fastening that connects shafts together for power transmission. Also known as shaft coupling.

coupling capacitor [ELECTR] A capacitor used to block the flow of direct current while allowing alternating or signal current to pass; widely used for joining two circuits or stages. Also known as blocking capacitor; stopping capacitor.

coupling coefficient [ELECTR] The ratio of the maximum change in energy of an electron traversing an interaction space to the product of the peak alternating gap voltage and the electronic charge. [PHYS] *See* coupling constant.

coupling constant [PARTIC PHYS] A measure of the strength of a type of interaction between particles, such as the strong interaction between mesons and nucleons, and the weak interaction between four fermions; analogous to the electric charge, which is the coupling constant between charged particles and electromagnetic radiation. [PHYS] 1. A measure of the strength of the coupling between two systems, especially electric circuits; maximum coupling is 1 and no coupling is 0. Also known as coefficient of coupling; coupling coefficient. 2. A measure of the dependence of one physical quantity on another.

coupling loop [ELECTROMAG] A conducting loop projecting into a waveguide or cavity resonator, designed to transfer energy to or from an external circuit.

course [CIV ENG] A row of stone, block, or brick of uniform height. [NAV] The intended direction of travel expressed as an angle in the horizontal plane between a reference line (true magnetic north) and the course line (the line connecting the point of origin and the point of destination), usually measured clockwise from the reference line. Also known as desired track. [TEXT] A row of stitches across a knitted fabric; corresponds to the filling in woven fabric.

course line [NAV] 1. A line of position plotted on a chart, parallel or substantially parallel to the intended course of a craft, showing whether the craft is to the right or the left of its course. 2. Any line representing a course.

course-line computer [NAV] An airborne computer that accepts bearing and distance information derived from ground facilities and uses this data to compute a course from the aircraft's present position to any other point which the pilot selects, providing only that it be within the coverage of the ground facilities; the steering information is displayed on a right-left indicator, while additional displays such as distance to go are also provided. Also known as arbitrary course computer; bearing distance computer; off-line computer; parallel course computer; rho-theta computer.

course made good [NAV] The resultant direction of actual travel of a vehicle, equivalent to its bearing from the point of departure.

Couvinian [GEOL] Lower Middle Devonian geologic time.

covalence [CHEM] The number of covalent bonds which an atom can form.

covalent bond [CHEM] A bond in which each atom of a bound pair contributes one electron to form a pair of electrons. Also known as electron pair bond.

covariance [STAT] A measurement of the tendency of two random variables, X and Y, to vary together, given by the expected value of the variable $(X - \bar{X})(Y - \bar{Y})$, where $\bar{X}$ and $\bar{Y}$ are the expected values of the variables X and Y, respectively.

covariant [RELAT] A scalar, vector, or higher-order tensor.

covariant components [MATH] Vector or tensor components which, in a transformation from one set of basis vectors to another, transform in the same manner as the basis vectors.

covariant index [MATH] A tensor index such that, under a transformation of coordinates, the procedure for obtaining a component of the transformed tensor for which this index has value p involves taking a sum over q of the product of a component of the original tensor for which the index has the value q times the partial derivative of the qth original coordinate with respect to the pth transformed coordinate; it is written as a subscript.

cove [GEOGR] 1. A small, narrow, sheltered bay, inlet, or creek on a coast. 2. A deep recess or hollow occurring in a cliff or steep mountainside.

covellite [MINERAL] CuS An indigo-blue mineral of metallic luster that crystallizes in the hexagonal system; it is usually massive or occurs in disseminations through other copper minerals and represents an ore of copper. Also known as indigo copper.

cover crops [AGR] Crops, especially grasses, grown for the express purpose of preventing and protecting a bare soil surface.

covered smut [PL PATH] A seed-borne smut of certain grain crops caused by *Ustilago hordei* in barley and *U. avenae* in oats.

cover plate [ENG] A pane of glass in a welding helmet or goggles which protects the colored lens excluding harmful light rays from damage by weld spatter.

covey [VERT ZOO] 1. A brood of birds. 2. A small flock of birds of one kind, used typically of partridge and quail.

covite [PETR] A rock of igneous origin composed of sodic orthoclase, hornblende, sodic pyroxene, nepheline, and accessory sphene, apatite, and opaque oxides.

cowling [AERO ENG] The streamlined metal cover of an aircraft engine. [ENG] A metal cover that houses an engine.

cowpea [BOT] *Vigna sinensis*. An annual legume in the order Rosales cultivated for its edible seeds. Also known as blackeye bean.

Cowper's gland *See* bulbourethral gland.

coxa [INV ZOO] The proximal or basal segment of the leg of insects and certain other arthropods which articulates with the body.

coxsackievirus [VIROL] A large subgroup of the enteroviruses in the picornavirus group including various human pathogens.

coyote [VERT ZOO] *Canis latrans*. A small wolf native to western North America but found as far eastward as New York State. Also known as prairie wolf.

cp *See* candlepower; centipoise; chemically pure.

CPE *See* computer performance evaluation.

CP invariance [PARTIC PHYS] The principle that the laws of physics are left unchanged by a combination of the operations of charge conjugation C and space inversion P; a small violation of this principle has been observed in the decay of neutral K-mesons.

cpm *See* cycle per minute.

CPM *See* critical path method.

CP propeller *See* controllable-pitch propeller.

cps *See* hertz.

CPT theorem [PARTIC PHYS] A theorem which states that a Lorentz invariant field theory is invariant to the product of charge conjugation C, space inversion P, and time reversal T.

CPU *See* central processing unit.

Cr *See* chromium.

CR *See* catalytic reforming.

CrA *See* Corona Australis.

crab [INV ZOO] **1.** The common name for a number of crustaceans in the order Decapoda having five pairs of walking legs, with the first pair modified as chelipeds. **2.** The common name for members of the Merostoma. [NAV] To drift sideways or to leeward, as a ship.

Crab Nebula [ASTRON] A gaseous nebula in the constellation Taurus; an amorphous mass which radiates a continuous spectrum involved in a mesh of filaments that radiate a bright-line spectrum.

Cracidae [VERT ZOO] A family of New World tropical upland game birds in the order Galliformes; includes the chachalacas, guans, and curassows.

crack closure [MATER] A phenomenon which occurs when the cyclic plasticity of a material gives rise to the development of residual plastic deformations in the vicinity of a crack tip, causing the fatigue crack to close at positive load.

cracked gasoline [MATER] Gasoline manufactured by heating crude petroleum distillation fractions or residues under pressure, or by heating with or without pressure in the presence of a catalyst, so that heavier hydrocarbons are broken into others, some of which distill in the gasoline range.

cracking [CHEM ENG] A process that is used to reduce the molecular weight of hydrocarbons by breaking the molecular bonds by various thermal, catalytic, or hydrocracking methods. [ENG] Presence of relatively large cracks extending into the interior of a structure, usually produced by overstressing the structural material.

cradle [ENG] A framework or other resting place for supporting or restraining objects. [ORD] The nonrecoiling structure of a weapon that houses the recoiling parts and rotates to elevate the gun. [TEXT] A device that catches the cards as they fall from a jacquard head.

crag [GEOL] A steep, rugged point or eminence of rock, as one projecting from the side of a mountain.

Crambiinae [INV ZOO] The snout moths, a subfamily of lepidopteran insects in the family Pyralididae containing small marshland and grassland forms.

cramp [DES ENG] A metal plate with bent ends used to hold blocks together. [MED] **1.** Painful, involuntary contraction of a muscle, such as a leg or foot cramp that may occur in normal individuals at night or in swimming. **2.** Any cramplike pain, as of the intestine, or that accompanying dysmenorrhea. **3.** Spasm of certain muscles, which may be intermittent or constant, from excessive use.

crampon [DES ENG] A device for holding heavy objects such as rock or lumber to be lifted by a crane or hoist; shaped like scissors, with points bent inward for grasping the load. Also spelled crampoon.

crampoon *See* crampon.

cranberry [BOT] Any of several plants of the genus *Vaccinium*, especially *V. macrocarpon*, in the order Ericales, cultivated for its small, edible berries.

Cranchiidae [INV ZOO] A family of cephalopod mollusks in the subclass Dibranchia.

crandallite [MINERAL] $CaAl_3(PO_4)_2(OH)_5 \cdot H_2O$ A white to light-grayish mineral consisting of a hydrous phosphate of calcium and aluminum occurring in fine, fibrous masses.

crane [MECH ENG] A hoisting machine with a power-operated inclined or horizontal boom and lifting tackle for moving loads vertically and horizontally. [VERT ZOO] The common name for the long-legged wading birds composing the family Gruidae of the order Gruiformes.

Crane *See* Grus.

Craniacea [INV ZOO] A family of inarticulate branchiopods in the suborder Craniidina.

cranial fossa [ANAT] Any of the three depressions in the floor of the interior of the skull.

cranial nerve [ANAT] Any of the paired nerves which arise in the brainstem of vertebrates and pass to peripheral structures through openings in the skull.

Craniata [VERT ZOO] A major subdivision of the phylum Chordata comprising the vertebrates, from cyclostomes to mammals, distinguished by a cranium.

craniectomy [MED] Surgical removal of strips or pieces of the cranial bones.

Craniidina [INV ZOO] A subdivision of inarticulate branchiopods in the order Acrotretida known to possess a pedicle; all forms are attached by cementation.

cranioplasty [MED] Surgical correction of defects in the cranial bones, usually by implants of metal, plastic material, or bone.

cranium [ANAT] That portion of the skull enclosing the brain. Also known as braincase.

crank [MECH ENG] A link in a mechanical linkage or mechanism that can turn about a center of rotation.

crankcase [MECH ENG] The housing for the crankshaft of an engine, where, in the case of an automobile, oil from hot engine parts is collected and cooled before returning to the engine by a pump.

crankshaft [MECH ENG] The shaft about which a crank rotates.

crash *See* abend.

Crassulaceae [BOT] A family of dicotyledonous plants in the order Rosales notable for their succulent leaves and resistance to desiccation.

crater [GEOL] **1.** A large, bowl-shaped topographic depression with steep sides. **2.** A rimmed structure at the summit of a volcanic cone; the floor is equal to the vent diameter. [MECH ENG] A depression in the face of a cutting tool worn down by chip contact. [MET] A depression at the end of the weld head or under the electrode during welding.

Crater [ASTRON] A constellation, right ascension 11 hours, declination 15°S. Abbreviated Crt. Also known as Cup.

crater cone [GEOL] A cone built around a volcanic vent by lava extruded from the vent.

crater lake [HYD] A fresh-water lake formed by the accumulation of rain and groundwater in a caldera or crater.

crater lamp [ELECTR] A glow-discharge tube used as a point source of light whose brightness is proportional to the signal current sent through the tube; used for photographic recording of facsimile signals.

craton [GEOL] The large, relatively immobile portion of continents consisting of both shield and platform areas.

crawler [MECH ENG] 1. One of a pair of an endless chain of plates driven by sprockets and used instead of wheels by certain power shovels, tractors, bulldozers, drilling machines, and such, as a means of propulsion. 2. Any machine mounted on such tracks.

crawl space [BUILD] 1. A shallow space in a building which workers can enter to gain access to pipes, wires, and equipment. 2. A shallow space located below the ground floor of a house and surrounded by the foundation wall.

crayfish [INV ZOO] The common name for a number of lobsterlike fresh-water decapod crustaceans in the section Astacura.

crazing [ENG] A network of fine cracks on or under the surface of a material such as enamel, glaze, metal, or plastic. [MET] Development of a network of cracks on a metal surface.

CrB See Corona Borealis.

cream ice See sludge.

crednerite [MINERAL] $CuMn_2O_4$ A steel-gray to iron-black foliated mineral consisting of copper, manganese, and oxygen.

creedite [MINERAL] $Ca_3Al_2(SO_4)(F,OH)_{10}\cdot 2H_2O$ A white or colorless monoclinic mineral consisting of hydrous calcium aluminum fluoride with calcium sulfate, occurring in grains and radiating crystalline masses; hardness is 2 on Mohs scale, and specific gravity is 2.7.

creep [ELECTR] A slow change in a characteristic with time or usage. [ENG] The tendency of wood to move while it is being cut, particularly when being mitered. [GEOL] A slow, imperceptible downward movement of slope-forming rock or soil under sheer stress. [GRAPHICS] A forward movement of the blanket during offset printing. [MECH] A time-dependent strain of solids caused by stress. [MIN ENG] See squeeze.

creeping flow [FL MECH] Fluid flow in which the velocity of flow is very small.

creep limit [MECH] The maximum stress a given material can withstand in a given time without exceeding a specified quantity of creep.

creep strength [MECH] The stress which, at a given temperature, will result in a creep rate of 1 percent deformation in 100,000 hours.

crenate [BIOL] Having a scalloped margin; used specifically for foliar structures, shrunken erythrocytes, and shells of certain mollusks.

crenation [PHYSIO] A notched appearance of shrunken erythrocytes; seen when they are exposed to the air or to strong saline solutions.

Crenotrichaceae [MICROBIO] Formerly a family of bacteria in the order Chlamydobacteriales having trichomes that are differentiated at the base and tip and attached to a firm substrate.

crenulation cleavage See slip cleavage.

Creodonta [PALEON] A group formerly recognized as a suborder of the order Carnivora.

crepitation [MED] A noise produced by the rubbing of fractured ends of bones, by cracking joints, and by pressure upon tissues containing abnormal amounts of air, as in cellular emphysema.

crepuscular [ZOO] Active during the hours of twilight or preceding dawn.

crepuscular rays [ASTRON] Streaks of light radiating from the sun shortly before and after sunset which shine through breaks in the clouds or through irregular spaces along the horizon.

crescent beach [GEOL] A crescent-shaped beach at the head of a bay or the mouth of a stream entering the bay, with the concave side facing the sea.

crescentic dune See barchan.

crescentic lake See oxbow lake.

crescent phase [ASTRON] A phase of the moon or an inferior planet in which less than half of the visible hemisphere is illuminated.

cress [BOT] Any of several prostrate crucifers belonging to the order Capparales and grown for their flavorful leaves; includes watercress (*Nasturtium officinale*), garden cress (*Lepidium sativum*), and upland or spring cress (*Barbarea verna*).

crest [DES ENG] The top of a screw thread. [SCI TECH] The highest point of a structure or natural formation, such as the top edge of a dam, the ridge of a roof, the highest point of a gravity wave, or the highest natural projection of a hill or mountain.

crestal injection See external gas injection.

crest gate [CIV ENG] A gate in the spillway of a dam which functions to maintain or change the water level.

crest stage [HYD] The highest stage reached at a point along a stream culminating a rise by waters of that stream.

Cretaceous [GEOL] The latest system of rocks or period of the Mesozoic Era, between 136 and 65 million years ago.

cretinism [MED] A type of dwarfism caused by hypothyroidism and associated with generalized body changes, including mental deficiency.

crevasse [GEOL] An open, nearly vertical fissure in a glacier or other mass of land ice or the earth, especially after earthquakes.

crevasse hoar [HYD] Ice crystals which form and grow in glacial crevasses and in other cavities where a large cooled space is formed and in which water vapor can accumulate under calm, still conditions.

crib [CIV ENG] The space between two successive ties along a railway track. [ENG] 1. Any structure composed of a layer of timber or steel joists laid on the ground, or two layers across each other, to spread a load. 2. Any structure composed of frames of timber placed horizontally on top of each other to form a wall. [GEOL] See arête.

crib death [MED] Sudden death of a sleeping infant without apparent cause. Also known as sudden death syndrome.

Cribrariaceae [BOT] A family of true slime molds in the order Liceales.

cribriform [BIOL] Perforated, like a sieve.

Cricetidae [VERT ZOO] A family of the order Rodentia including hamsters, voles, and some mice.

Cricetinae [VERT ZOO] A subfamily of mice in the family Cricetidae.

cricket [BUILD] A device that is used to divert water at the intersections of roofs or at the intersection of a roof and chimney. [INV ZOO] 1. The common name for members of the insect family Gryllidae.

2. The common name for any of several related species of orthopteran insects in the families Tettigoniidae, Gryllotalpidae, and Tridactylidae.

cricoid [ANAT] The signet-ring-shaped cartilage forming the base of the larynx in man and most other mammals.

crimp [ENG] **1.** To cause something to become wavy, crinkled, or warped, such as lumber. **2.** To pinch or press together, especially a tubular or cylindrical shape, in order to seal or unite. [TEXT] To give curl to synthetic fibers.

crimp contact [ELEC] A contact whose back portion is a hollow cylinder that will accept a wire; after a bared wire is inserted, a swaging tool is applied to crimp the contact metal firmly against the wire. Also known as solderless contact.

Crinoidea [INV ZOO] A class of radially symmetrical crinozoans in which the adult body is flower-shaped and is either carried on an anchored stem or is free-living.

Crinozoa [INV ZOO] A subphylum of the Echinodermata comprising radially symmetrical forms that show a partly meridional pattern of growth.

cripple [BUILD] A structural member, such as a stud above a window, that is cut less than full length.

cristobalite [MINERAL] SiO_2 A silicate mineral that is a high-temperature form of quartz; stable above 1470°C; crystallizes in the tetragonal system at low temperatures and the isometric system at high temperatures.

crit [NUCLEO] The mass of fissionable material that is critical under a given set of conditions; sometimes applied to the mass of an untamped critical sphere of fissionable material.

critical absorption wavelength [SPECT] The wavelength, characteristic of a given electron energy level in an atom of a specified element, at which an absorption discontinuity occurs.

critical altitude [AERO ENG] The maximum altitude at which a supercharger can maintain a pressure in the intake manifold of an engine equal to that existing during normal operation at rated power and speed at sea level without the supercharger. [ORD] The maximum altitude at which the propulsion system of a missile performs satisfactorily.

critical angle [PHYS] An angle associated with total reflection of electromagnetic or acoustic radiation back into a medium from the boundary with another medium in which the radiation has a higher phase velocity; it is the smallest angle with the normal to the boundary at which total reflection occurs.

critical area See picture element.

critical assembly [NUCLEO] An assembly of sufficient fissionable and moderator material to sustain a fission chain reaction at a low power level.

critical constant [PHYS CHEM] A characteristic temperature, pressure, and specific volume of a gas above which it cannot be liquefied.

critical coupling [ELEC] The degree of coupling that provides maximum transfer of signal energy from one radio-frequency resonant circuit to another when both are tuned to the same frequency. Also known as optimum coupling.

critical current [SOLID STATE] The current in a superconductive material above which the material is normal and below which the material is superconducting, at a specified temperature and in the absence of external magnetic fields.

critical damping [PHYS] Damping in a linear system on the threshold between oscillatory and exponential behavior.

critical field [ELECTR] The smallest theoretical value of steady magnetic flux density that would prevent an electron emitted from the cathode of a magnetron at zero velocity from reaching the anode. Also known as cutoff field.

criticality [NUCLEO] The condition in which a nuclear reactor is just self-sustaining.

critical mass [NUCLEO] The mass of fissionable material of a particular shape that is just sufficient to sustain a nuclear chain reaction.

critical moisture content [CHEM ENG] The average moisture throughout a solid material being dried, its value being related to drying rate, thickness of material, and the factors that influence the movement of moisture within the solid.

critical path method [SYS ENG] A systematic procedure for detailed project planning and control. Abbreviated CPM.

critical phenomena [PHYS CHEM] Physical properties of liquids and gases at the critical point (conditions at which two phases are just about to become one); for example, critical pressure is that needed to condense a gas at the critical temperature, and above the critical temperature the gas cannot be liquefied at any pressure.

critical point [MATH] A point at which the first derivative of a function is either 0 or does not exist. [PHYS CHEM] **1.** The temperature and pressure at which two phases of a substance in equilibrium with each other become identical, forming one phase. **2.** The temperature and pressure at which two ordinarily partially miscible liquids are consolute.

critical potential [ATOM PHYS] The energy needed to raise an electron to a higher energy level in an atom (resonance potential) or to remove it from the atom (ionization potential). [ELEC] A potential which results in sudden change in magnitude of the current.

critical pressure [FL MECH] For a nozzle whose cross section at each point is such that a fluid in isentropic flow just fills it, the pressure at the section of minimum area of the nozzle; if the nozzle is cut off at this point with no diverging section, decrease in the discharge pressure below the critical pressure (at constant admission pressure) does not result in increased flow. [THERMO] The pressure of the liquid-vapor critical point.

critical properties [PHYS CHEM] Physical and thermodynamic properties of materials at conditions of critical temperature, pressure, and volume, that is, at the critical point.

critical range [MET] The temperature range for the reversible change of austenite to ferrite, pearlite, and cementite.

critical ratio [STAT] The ratio of a particular deviation from the mean value to the standard deviation.

critical reactor [NUCLEO] A nuclear reactor in which the ratio of moderator to fuel is either subcritical or just critical; used to study the properties of the system and determine critical size.

critical speed [CRYO] See critical velocity. [FL MECH] See critical velocity. [MECH ENG] The angular speed at which a rotating shaft becomes dynamically unstable with large lateral amplitudes, due to resonance with the natural frequencies of lateral vibration of the shaft.

critical state [PHYS CHEM] Unique condition of pressure, temperature, and composition wherein all properties of coexisting vapor and liquid become identical.

critical temperature [PHYS CHEM] The temperature of the liquid-vapor critical point, that is, the temperature above which the substance has no liquid-vapor transition.

critical value [MATH] The value of the dependent variable at a critical point of a function. [STAT] A number which causes rejection of the null hypothesis if a given test statistic is this number or more, and acceptance of the null hypothesis if the test statistic is smaller than this number.

critical velocity [AERO ENG] In rocketry, the speed of sound at the conditions prevailing at the nozzle throat. Also known as throat velocity. [CRYO] The velocity of a superfluid in very narrow channels (on the order of 10^{-5} centimeter), which is nearly constant. Also known as critical speed. [MECH] 1. The speed of flow equal to the local speed of sound. Also known as critical speed. 2. The speed of fluid flow through a given conduit above which it becomes turbulent.

critical voltage [ELECTR] The highest theoretical value of steady anode voltage, at a given steady magnetic flux density, at which electrons emitted from the cathode of a magnetron at zero velocity would fail to reach the anode. Also known as cutoff voltage.

critical volume [PHYS] The volume occupied by one mole of a substance at the liquid-vapor critical point, that is, at the critical temperature and pressure.

CRM See chemical remanent magnetization.

CRO See cathode-ray oscilloscope.

Crocco's equation [FL MECH] A relationship, expressed as $v \times \omega = -T \text{ grad } S$, between vorticity and entropy gradient for the steady flow of an inviscid compressible fluid; v is the fluid velocity vector, ω (= curl v) is the vorticity vector, T is the fluid temperature, and S is the entropy per unit mass of the fluid.

crocidolite [MINERAL] A lavender-blue, indigo-blue, or leek-green asbestiform variety of riebeckite; occurs in fibrous, massive, and earthy forms. Also known as blue asbestos; krokidolite.

crocodile [ELEC] A unit of potential difference or electromotive force, equal to 10^6 volts; used informally at some nuclear physics laboratories. [VERT ZOO] The common name for about 12 species of aquatic reptiles included in the family Crocodylidae.

crocodile clip See alligator clip.

Crocodilia [VERT ZOO] An order of the class Reptilia which is composed of large, voracious, aquatic species, including the alligators, caimans, crocodiles, and gavials.

crocodiling See alligatoring.

Crocodylidae [VERT ZOO] A family of reptiles in the order Crocodilia including the true crocodiles, false gavial, alligators, and caimans.

Crocodylinae [VERT ZOO] A subfamily of reptiles in the family Crocodylidae containing the crocodiles, *Osteolaemus*, and the false gavial.

crocoite [MINERAL] $PbCrO_4$ A yellow to orange or hyacinth-red secondary mineral occurring as monoclinic, prismatic crystals; it is also massive granular. Also known as crocoisite; red lead ore.

crocus [BOT] A plant of the genus *Crocus*, comprising perennial herbs cultivated for their flowers. [MATER] Finely powdered oxide of iron, of dark red color, used for buffing and polishing.

Croixian [GEOL] Upper Cambrian geologic time.

Cro-Magnon man [PALEON] 1. A race of tall, erect Caucasoid men having large skulls; identified from skeletons found in southern France. 2. A general term to describe all fossils resembling this race that belong to the upper Paleolithic (35,000–8000 B.C.) in Europe.

Cromwell Current [OCEANOGR] An eastward-setting subsurface current that extends about 1½° north and south of the equator, and from about 150°E to 92°W.

cronstedtite [MINERAL] $Fe_4^{2+} Fe_2^{3+} (Fe_2^{3+} Si_2) O_{10}$ $(OH)_8$ A black to brownish-black mineral consisting of a hydrous iron silicate crystallizing in hexagonal prisms; specific gravity is 3.34–3.35.

Crookes dark space See cathode dark space.

crookesite [MINERAL] $(Cu,Tl,Ag)_2 Se$ An important selenium mineral occurring in lead-gray masses and having a metallic appearance.

crop [AGR] A plant or animal grown for its commercial value. [MET] Defective end portion of an ingot which is removed for scrap before rolling the ingot. [VERT ZOO] A distensible saccular diverticulum near the lower end of the esophagus of birds which serves to hold and soften food before passage into the stomach.

crop out See outcrop.

cropping index [AGR] The number of crops grown per year on a given land area times 100. Abbreviated CI.

crop rotation [AGR] A method of protecting the soil and replenishing its nutrition by planting a succession of different crops on the same land.

Cross See Crux.

crossband [COMMUN] Two-way communication in which one radio frequency is used in one direction and a frequency having different propagation characteristics is used in the opposite direction.

crossbar switch [ELEC] A switch having a three-dimensional arrangement of contacts and a magnet system that selects individual contacts according to their coordinates in the matrix.

crossbar system [COMMUN] Automatic telephone switching system which is generally characterized by the following features: selecting mechanisms are crossbar switches, common circuits select and test the switching paths and control the operation of the selecting mechanisms, and method of operations is one in which the switching information is received and stored by controlling mechanisms that determine the operations necessary in establishing a telephone connection.

cross-bedding [GEOL] The condition of having laminae lying transverse to the main stratification planes of the strata; occurs only in granular sediments. Also known as cross-lamination; cross-stratification.

crossbreed [BIOL] To propagate new individuals by breeding two distinctive varieties of a species.

cross-correlation [STAT] 1. Correlation between corresponding members of two or more series: if $q_1, \ldots, q_n$ and $r_1, \ldots, r_n$ are two series, correlation between q_i and r_i, or between q_i and r_{i+j} (for fixed j), is a cross correlation. 2. Correlation between or expectation of the inner product of two series of random variables, where the difference in indices between the corresponding values of the two series is fixed.

cross-correlator [ELECTR] A correlator in which a locally generated reference signal is multiplied by the incoming signal and the result is smoothed in a low-pass filter to give an approximate computation of the cross-correlation function. Also known as synchronous detector.

crosscurrent [FL MECH] A current that flows across or opposite to another current.

crosscut [ENG] A cut made through wood across the grain. [MIN ENG] 1. A small passageway driven at right angles to the main entry of a mine to connect it with a parallel entry of air course. **2.** A passageway in a mine that cuts across the geological structure.

crosscut file   [DES ENG] A file with a rounded edge on one side and a thin edge on the other; used to sharpen straight-sided saw teeth with round gullets.

crosscut saw   [DES ENG] A type of saw for cutting across the grain of the wood; designed with about eight teeth per inch.

crossed-field device   [ELECTR] Any instrument which uses the motion of electrons in perpendicular electric and magnetic fields to generate microwave radiation, either as an amplifier or oscillator.

crossed lens   [OPTICS] A lens designed with radii of curvature which give minimum spherical aberration for parallel incident rays.

crossed prisms   [OPTICS] A pair of Nicol prisms whose principal planes are perpendicular to each other, so that light passing through one is extinguished by the other.

cross-fade   [ENG ACOUS] In dubbing, the overlapping of two sound tracks, wherein the outgoing track fades out while the incoming track fades in.

cross fault   [GEOL] **1.** A fault whose strike is perpendicular to the general trend of the regional structure. **2.** A minor fault that intersects a major fault.

cross-fertilization   [BOT] Fertilization between two separate plants.   [ZOO] Fertilization between different kinds of individuals.

cross fire   [COMMUN] Interfering current in one telegraph or signaling channel resulting from telegraph or signaling currents in another channel.

cross fold   [GEOL] A secondary fold whose axis is perpendicular or oblique to the axis of another fold. Also known as subsequent fold; superimposed fold; transverse fold.

crossfoot   [COMPUT SCI] To add numbers in several different ways in a computer, for checking purposes.

cross hair   [ENG] An inscribed line or a strand of hair, wire, silk, or the like used in an optical sight, transit, or similar instrument for accurate sighting.

crosshatch generator   [ELECTR] A signal generator that generates a crosshatch pattern for adjusting color television receiver circuits.

crosshed   [MECH ENG] A block sliding between guides and containing a wrist pin for the conversion of reciprocating to rotary motion, as in an engine or compressor.   [MET] A device generally employed in wire coating which is attached to the discharge end of the extruder cylinder; designed to facilitate extruding material at an angle.   [MIN ENG] A runner or guide positioned just above a sinking bucket to restrict excessive swinging.

crossing over   [GEN] The exchange of genetic material between paired homologous chromosomes during meiosis. Also known as crossover.

cross joint   [GEOL] A fracture in igneous rock perpendicular to the lineation caused by flow magma. Also known as transverse joint.

cross-lamination   See cross-bedding.

cross-linking   [ORG CHEM] The setting up of chemical links between the molecular chains of polymers.

crossmarks   [GRAPHICS] Register marks used for the exact positioning of images in step-and-repeat, double, or multicolor printing; also used for superimposing overlays onto a base or onto each other.

cross matching   [IMMUNOL] Determination of blood compatibility for transfusion by mixing donor cells with recipient serum, and recipient cells with donor serum, and examining for an agglutination reaction.

cross modulation   [COMMUN] A type of interference in which the carrier of a desired signal becomes modulated by the program of an undesired signal on a different carrier frequency; the program of the undesired station is then heard in the background of the desired program.

cross multiplication   [MATH] Multiplication of the numerator of each of two fractions by the denominator of the other, as when eliminating fractions from an equation.

cross-neutralization   [ELECTR] Method of neutralization used in push-pull amplifiers, whereby a portion of the plate-cathode alternating-current voltage of each vacuum tube is applied to the grid-cathode circuit of the other vacuum tube through a neutralizing capacitor.

Crossopterygii   [PALEON] A subclass of the class Osteichthyes comprising the extinct lobefins or choanate fishes and represented by one extant species; distinguished by two separate dorsal fins.

Crossosomataceae   [BOT] A monogeneric family of xerophytic shrubs in the order Dilleniales characterized by perigynous flowers, seeds with thin endosperm, and small, entire leaves.

crossover   [CIV ENG] An S-shaped section of railroad track joining two parallel tracks.   [ELEC] A point at which two conductors cross, with appropriate insulation between them to prevent contact.   [GEN] See crossing over.

crossover frequency   [ENG ACOUS] **1.** The frequency at which a dividing network delivers equal power to the upper and lower frequency channels when both are terminated in specified loads. **2.** See transition frequency.

crossover network   [ENG ACOUS] A selective network used to divide the audio-frequency output of an amplifier into two or more bands of frequencies. Also known as dividing network; loudspeaker dividing network.

crossover voltage   [ELECTR] In a cathode-ray storage tube, the voltage of a secondary writing surface, with respect to cathode voltage, on which the secondary emission is unity.

cross-pollination   [BOT] Transfer of pollen from the anthers of one plant to the stigmata of another plant.

cross section   [GEOL] **1.** A diagram or drawing that shows the downward projection of surficial geology along a vertical plane, for example, a portion of a stream bed drawn at right angles to the mean direction of the flow of the stream. **2.** An actual exposure or cut which reveals geological features.   [GRAPHICS] A diagram or drawing representing a cut at right angles to an axis.   [MAP] A horizontal grid system that is laid out on the ground for determining contours, quantities of earthwork, and so on, by means of elevations of the grid points.   [MATH] **1.** The intersection of an n-dimensional geometric figure in some euclidean space with a lower dimensional hyperplane. **2.** A right inverse for the projection of a fiber bundle.   [PHYS] An area characteristic of a collision reaction between atomic or nuclear particles or systems, such that the number of reactions which occur equals the product of the number of target particles or systems and the number of incident particles or systems which would pass through this area if their velocities were perpendicular to it. Also known as collision cross section.

cross-stone   See harmotome; staurolite.

cross-stratification   See cross-bedding.

crosstalk   [COMMUN] **1.** The sound heard in a receiver along with a desired program because of cross modulation or other undesired coupling to another communication channel; it is also observed between adjacent pairs in a telephone cable. **2.** Interaction of audio and video signals in a television system, causing video modulation of the audio carrier or audio modulation of the video signal at some point. **3.** Interaction of the chrominance and luminance signals in a color television receiver. [ENG ACOUS] *See* magnetic printing.

crosstalk unit [COMMUN] A measure of the coupling between two circuits; the number of crosstalk units is 1 million times the ratio of the current or voltage at the observing point to the current or voltage at the origin of the disturbing signal, the impedances at these points being equal. Abbreviated cu.

cross valley *See* transverse valley.

crosswind [METEOROL] A wind which has a component directed perpendicularly to the course (or heading) of an exposed, moving object.

Crotalidae [VERT ZOO] A family of proglyphodont venomous snakes in the reptilian suborder Serpentes.

crotch [SCI TECH] The angular form made by the parting of two branches, legs, or members.

croup [MED] Any condition of upper-respiratory pathway obstruction in children, especially acute inflammation of the pharynx, larynx, and trachea, characterized by a hoarse, brassy, and stridulent cough and difficulties in breathing.

croute calcaire *See* caliche.

crow [VERT ZOO] The common name for a number of predominantly black birds in the genus *Corvus* comprising the most advanced members of the family Corvidae.

Crow *See* Corvus.

crowbar [DES ENG] An iron or steel bar that is usually bent and has a wedge-shaped working end; used as a lever and for prying. [ELEC] A device or action that in effect places a high overload on the actuating element of a circuit breaker or other protective device, thus triggering it.

crown [ANAT] **1.** The top of the skull. **2.** The portion of a tooth above the gum. [BOT] **1.** The topmost part of a plant or plant part. **2.** *See* corona. [CIV ENG] Center of a roadway elevated above the sides. [ENG] **1.** The part of a drill bit inset with diamonds. **2.** The vertex of an arch or arched surface. **3.** The top or dome of a furnace or kiln. [LAP] The portion of a faceted gem above the girdle. [MET] That part of the sheet or roll where the thickness or diameter increases from edge to center. [MIN ENG] A horizontal roof member of a timber up to 16 feet (4.9 meters) long and supported at each end by an upright.

crown ether [ORG CHEM] A macrocyclic polyether whose structure exhibits a conformation with a so-called hole capable of trapping cations by coordination with a lone pair of electrons on the oxygen atoms.

crown fire [FOR] A forest fire burning primarily in the tops of trees and shrubs.

crown gall [PL PATH] A bacterial disease of many plants induced by *Bacterium tumefaciens* and marked by abnormal enlargement of the stem near the root crown.

crown rot [PL PATH] Any plant disease or disorder marked by deterioration of the stem at or near ground level.

crown rust [PL PATH] A rust disease of oats and certain other grasses caused by varieties of *Puccinia coronata* and marked by light-orange masses of fungi on the leaves.

crown saw [DES ENG] A saw consisting of a hollow cylinder with teeth around its edge; used for cutting round holes. Also known as hole saw.

crown wheel [DES ENG] A gear that is light and crown-shaped. [HOROL] **1.** The horizontal escape wheel of a verge escapement timepiece. **2.** The wheel in the winding mechanism of a watch that drives the ratchet wheel and is itself driven by the winding-stem pinion.

Crt *See* Crater.

CRT *See* cathode-ray tube.

Cru *See* Crux.

cruciate [ANAT] Resembling a cross.

crucible [SCI TECH] A refractory vessel or pot, varying in size from a small laboratory utensil to large industrial equipment for melting or calcining.

Cruciferae [BOT] A large family of dicotyledonous herbs in the order Capparales characterized by parietal placentation; hypogynous, mostly regular flowers; and a two-celled ovary with the ovules attached to the margins of the partition.

cruciform [SCI TECH] Resembling or arranged like a cross.

crude lecithin *See* lecithin.

crude oil [GEOL] A comparatively volatile liquid bitumen composed principally of hydrocarbon, with traces of sulfur, nitrogen, or oxygen compounds; can be removed from the earth in a liquid state.

cruise missile [AERO ENG] A pilotless airplane that can be launched from a submarine, surface ship, ground vehicle, or another airplane; range can be up to 1500 miles (2400 kilometers), flying at a constant altitude that can be as low as 60 meters.

cruiser [NAV ARCH] A type of large warship, but smaller than a battleship, having a displacement of 6000 to 15,000 tons (5442–13,605 metric tons), moderately armed and armored, and capable of any naval duty except combat with battleships.

cruising radius [NAV] **1.** The maximum distance that an aircraft, starting with full fuel tanks, can cruise under given or specified conditions from its takeoff point before returning with a specified fuel reserve remaining. **2.** The distance a craft can travel at cruising speed without refueling. Also known as cruising range.

cruising range *See* cruising radius.

crunode [MATH] A point on a curve through which pass two branches of the curve with different tangents. Also known as node.

crus [ANAT] **1.** The shank of the hindleg, that portion between the femur and the ankle. **2.** Any of various parts of the body resembling a leg or root.

crush breccia [GEOL] A breccia formed in place by mechanical fragmentation of rock during movements of the earth's crust.

crush conglomerate [GEOL] Beds similar to a fault breccia, except that the fragments are rounded by attrition. Also known as tectonic conglomerate.

crushed stone [MIN ENG] Irregular fragments of rock crushed or ground to smaller sizes after quarrying. Also known as broken stone.

crush fold [GEOL] A fold of large dimensions that may involve considerable minor folding and faulting such as would produce a mountain chain or an oceanic deep.

crushing mill *See* stamping mill.

crushing strength [MECH] The compressive stress required to cause a solid to fail by fracture; in essence, it is the resistance of the solid to vertical pressure placed upon it.

crust [GEOL] The outermost solid layer of the earth, mostly consisting of crystalline rock and extending no more than a few miles from the surface to the Mohorovičić discontinuity. Also known as earth crust. [HYD] A hard layer of snow lying on top of a soft layer.

crustal plate *See* tectonic plate.

crustose [BOT] Of a lichen, forming a thin crustlike thallus which adheres closely to the substratum of rock, bark, or soil.

Crux [ASTRON] A constellation having four principal bright stars which form the figure of a cross; right ascension 12 hours, declination 60°S. Abbreviated Cru. Also known as Cross; Southern Cross.

Crv *See* Corvus.

cry-, cryo- [SCI TECH] Combining form meaning cold, freezing.

cryobiology [BIOL] The use of low-temperature environments in the study of living plants and animals.

cryochemistry [PHYS CHEM] The study of chemical phenomena in very-low-temperature environments.

cryoelectronics [ELECTR] A branch of electronics concerned with the study and application of superconductivity and other low-temperature phenomena to electronic devices and systems.

cryogen *See* cryogenic fluid.

cryogenic conductor *See* superconductor.

cryogenic device [CRYO] A device whose operation depends on superconductivity as produced by temperatures near absolute zero. Also known as superconducting device.

cryogenic engineering [ENG] A branch of engineering specializing in technical operations at very low temperatures (about 200 to 400°R, or -160 to $-50°C$).

cryogenic fluid [CRYO] A liquid which boils at temperatures of less than about 110 K at atmospheric pressure, such as hydrogen, helium, nitrogen, oxygen, air, or methane. Also known as cryogen.

cryogenic pump [CRYO] A high-speed vacuum pump that can produce an extremely low vacuum and has a low power consumption; to reduce the pressure, gases are condensed on surfaces within an enclosure at extremely low temperatures, usually attained by using liquid helium or liquid or gaseous hydrogen. Also known as cryopump.

cryogenics [PHYS] The production and maintenance of very low temperatures, and the study of phenomena at these temperatures.

cryogenic wind tunnel [ENG] A wind tunnel employing a cryogenic environment and utilizing independent control over Mach number, Reynolds number, aeroelastic effects, and model-tunnel interactions.

cryolaccolith *See* hydrolaccolith.

cryolite [MINERAL] Na_3AlF_6 A white or colorless mineral that crystallizes in the monoclinic system but has a pseudocubic aspect; found in masses of waxy luster; hardness is 2.5 on Mohs scale, and specific gravity is 2.95–3.0; used chiefly as a flux in producing aluminum from bauxite and for making salts of sodium and aluminum and porcelaneous glass. Also known as Greenland spar; ice stone.

cryology [HYD] The study of ice and snow. [MECH ENG] The study of low-temperature (about 200°R, or $-160°C$) refrigeration.

cryophilic *See* cryophilous.

cryophilous [ECOL] Having a preference for low temperatures. Also known as cryophilic.

cryophysics [CRYO] Physics as restricted to phenomena occurring at very low temperatures, approaching absolute zero.

cryophyte [ECOL] A plant that forms winter buds below the soil surface.

cryoplanation [GEOL] Land erosion at high latitudes or elevations due to processes of intensive frost action.

cryopreservation [ENG] Preservation of food, biologicals, and other materials at extremely low temperatures.

cryopump *See* cryogenic pump.

cryosar [ELECTR] A cryogenic, two-terminal, negative-resistance semiconductor device, consisting essentially of two contacts on a germanium wafer operating in liquid helium.

cryoscopy [ANALY CHEM] A phase-equilibrium technique to determine molecular weight and other properties of a solute by dissolving it in a liquid solvent and then ascertaining the solvent's freezing point.

cryosorption pump [MECH ENG] A high-vacuum pump that employs a sorbent such as activated charcoal or synthetic zeolite cooled by nitrogen or some other refrigerant; used to reduce pressure from atmospheric pressure to a few millitorr.

cryosphere [GEOL] That region of the earth in which the surface is perennially frozen.

cryostat [ENG] An apparatus used to provide low-temperature environments in which operations may be carried out under controlled conditions.

cryosurgery [MED] Selective destruction of tissue by freezing, as the use of a liquid nitrogen probe to the brain in parkinsonism.

cryotron [ELECTR] A switch that operates at very low temperatures at which its components are superconducting; when current is sent through a control element to produce a magnetic field, a gate element changes from a superconductive zero-resistance state to its normal resistive state.

cryotronics [ELECTR] The branch of electronics that deals with the design, construction, and use of cryogenic devices.

cryoturbation *See* congeliturbation.

Cryphaeaceae [BOT] A family of mosses in the order Isobryales distinguished by a rough calyptra.

crypt [ANAT] **1.** A follicle or pitlike depression. **2.** A simple glandular cavity.

cryptand [ORG CHEM] A macropolycyclic polyazopolyether, where the three-coordinate nitrogen atoms provide the vertices of a three-dimensional structure.

cryptic coloration [ZOO] A phenomenon of protective coloration by which an animal blends with the background through color matching or countershading.

Cryptobiidae [INV ZOO] A family of flagellate protozoans in the order Kinetoplastida including organisms with two flagella, one free and one with an undulating membrane.

Cryptobranchidae [VERT ZOO] The giant salamanders and hellbenders, a family of tailed amphibians in the suborder Cryptobranchoidea.

Cryptobranchoidea [VERT ZOO] A primitive suborder of amphibians in the order Urodela distinguished by external fertilization and aquatic larvae.

Cryptocerata [INV ZOO] A division of hemipteran insects in some systems of classification that includes the water bugs (Hydrocorisae).

Cryptochaetidae [INV ZOO] A family of myodarian cyclorrhaphous dipteran insects in the subsection Acalypteratae.

cryptoclastic [GEOL] Composed of extremely fine, almost submicroscopic, broken or fragmental particles.

cryptoclimate [ENG] The climate of a confined space, such as inside a house, barn, or greenhouse, or in an artificial or natural cave; a form of microclimate. Also spelled kryptoclimate.

Cryptococcaceae [MYCOL] A family of imperfect fungi in the order Moniliales in some systems of classification; equivalent to the Cryptococcales in other systems.

Cryptococcales [MYCOL] An order of imperfect fungi, in some systems of classification, set up to include the yeasts or yeastlike organisms whose perfect or sexual stage is not known.

cryptocrystalline [GEOL] Having a crystalline structure but of such a fine grain that individual components are not visible with a magnifying lens.

Cryptodira [VERT ZOO] A suborder of the reptilian order Chelonia including all turtles in which the cervical spines are uniformly reduced and the head folds directly back into the shell.

cryptography [COMMUN] The science of preparing messages in a form which cannot be read by those not privy to the secrets of the form.

cryptohalite [MINERAL] $(NH_4)_2SiF_6$ A colorless to white or gray, isometric mineral consisting of ammonium silicon fluoride; occurs in massive and arborescent forms.

cryptolite *See* monazite.

cryptology [COMMUN] The science of preparing messages in forms which are intended to be unintelligible to those not privy to the secrets of the form, and of deciphering such messages.

cryptomelane [MINERAL] $KMn_8O_{16}\cdot H_2O$ A usually massive mineral, common in manganese ores; contains an oxide of manganese and potassium and crystallizes in the monoclinic system.

Cryptomonadida [BIOL] An order of the class Phytamastigophorea considered to be protozoans by biologists and algae by botanists.

Cryptomonadina [BIOL] The equivalent name for Cryptomonadida.

cryptoperthite [MINERAL] A fine-grained, submicroscopic variety of perthite consisting of an intergrowth of potassic and sodic feldspar, detectable only by means of x-rays or with the aid of an electron microscope.

Cryptophagidae [INV ZOO] The silken fungus beetles, a family of coleopteran insects in the superfamily Cucujoidea.

Cryptophyceae [BOT] A class of algae of the Pyrrhophyta in some systems of classification; equivalent to the division Cryptophyta.

Cryptophyta [BOT] A division of the algae in some classification schemes; equivalent to the Cryptophyceae.

cryptophyte [BOT] A plant that produces buds either underwater or underground on corms, bulbs, or rhizomes.

Cryptopidae [INV ZOO] A family of epimorphic centipedes in the order Scolopendromorpha.

cryptorchidism *See* cryptorchism.

cryptorchism [MED] Failure of the testes to descend into the scrotum from the abdomen or inguinal canals. Also known as cryptorchidism.

Cryptostomata [PALEON] An order of extinct bryozoans in the class Gymnolaemata.

cryptotext [COMMUN] In cryptology, a text of visible writing which conveys no intelligible meaning in any language, or which apparently conveys an intelligible meaning that is not the real meaning.

crypts of Lieberkühn [ANAT] Simple, tubular glands which arise as evaginations into the mucosa of the small intestine.

crystal [CRYSTAL] A homogeneous solid made up of an element, chemical compound or isomorphous mixture throughout which the atoms or molecules are arranged in a regularly repeating pattern. [ELECTR] A natural or synthetic piezoelectric or semiconductor material whose atoms are arranged with some degree of geometric regularity. [MINERAL] *See* rock crystal.

crystal axis [CRYSTAL] A reference axis used for the vectoral properties of a crystal.

crystal cartridge [ENG ACOUS] A piezoelectric unit used with a stylus in a phonograph pickup to convert disk recordings into audio-frequency signals, or used with a diaphragm in a crystal microphone to convert sound waves into af signals.

crystal chemistry [CRYSTAL] The study of the crystalline structure and properties of a mineral or other solid.

crystal class [CRYSTAL] One of 32 categories of crystals according to the inversions, rotations about an axis, reflections, and combinations of these which leaves the crystal invariant. Also known as symmetry class.

crystal clock [HOROL] A clock which uses the mechanical resonance of a crystal plate coupled piezoelectrically into an electronic circuit.

crystal counter [NUCLEO] A particle detector in which the sensitive material is a dielectric (nonconducting) crystal mounted between two metallic electrodes.

crystal defect [CRYSTAL] Any departure from crystal symmetry caused by free surfaces, disorder, impurities, vacancies and interstitials, dislocations, lattice vibrations, and grain boundaries. Also known as lattice defect.

crystal detector [ELECTR] **1.** A crystal diode, or an equivalent earlier crystal-catwhisker combination, used to rectify a modulated radio-frequency signal to obtain the audio or video signal directly. **2.** A crystal diode used in a microwave receiver to combine an incoming radio-frequency signal with a local oscillator signal to produce an intermediate-frequency signal.

crystal diffraction [SOLID STATE] Diffraction by a crystal of beams of x-rays, neutrons, or electrons whose wavelengths (or de Broglie wavelengths) are comparable with the interatomic spacing of the crystal.

crystal diffraction spectrometer *See* Bragg spectrometer.

crystal diode *See* semiconductor diode.

crystal face [CRYSTAL] One of the outward planar surfaces which define a crystal and reflect its internal structure. Also known as face.

crystal field theory [PHYS CHEM] The theory which assumes that the ligands of a coordination compound are the sources of negative charge which perturb the energy levels of the central metal ion and thus subject the metal ion to an electric field analogous to that within an ionic crystalline lattice.

crystal filter [ELECTR] A highly selective tuned circuit employing one or more quartz crystals; sometimes used in intermediate-frequency amplifiers of communication receivers to improve the selectivity.

crystal growth [CRYSTAL] The growth of a crystal, which involves diffusion of the molecules of the crystallizing substance to the surface of the crystal, diffusion of these molecules over the crystal surface to special sites on the surface, incorporation of molecules into the surface at these sites, and diffusion of heat away from the surface.

crystal habit [CRYSTAL] The size and shape of the crystals in a crystalline solid. Also known as habit.

crystal headphones [ENG ACOUS] Headphones using Rochelle salt or other crystal elements to convert audio-frequency signals into sound waves. Also known as ceramic earphones.

crystal indices See Miller indices.

crystal laser [OPTICS] A laser that uses a pure crystal of ruby or other material for generating a coherent beam of output light.

crystal lattice [CRYSTAL] A lattice from which the structure of a crystal may be obtained by associating with every lattice point an assembly of atoms identical in composition, arrangement, and orientation.

crystalline [CRYSTAL] Of, pertaining to, resembling, or composed of crystals.

crystalline anisotropy [SOLID STATE] The tendency of crystals to have different properties in different directions; for example, a ferromagnet will spontaneously magnetize along certain crystallographic axes.

crystalline lens See lens.

crystalline rock [PETR] 1. Rock made up of minerals in a clearly crystalline state. 2. Igneous and metamorphic rock, as opposed to sedimentary rock.

crystallinity [CRYSTAL] The quality or state of being crystalline. [PETR] Degree of crystallization exhibited by igneous rock.

crystallite [GEOL] A small, rudimentary form of crystal which is of unknown mineralogic composition and which does not polarize light.

crystallization [CRYSTAL] The formation of crystalline substances from solutions or melts.

crystallization differentiation See fractional crystallization.

crystallizer [CHEM ENG] Process vessel within which dissolved solids in a supersaturated solution are forced out of solution by cooling or evaporation, and then recovered as solid crystals.

crystallographic axis [CRYSTAL] One of three lines (sometimes four, in the case of a hexagonal crystal), passing through a common point, that are chosen to have definite relation to the symmetry properties of a crystal, and are used as a reference in describing crystal symmetry and structure.

crystallography [PHYS] The branch of science that deals with the geometric description of crystals and their internal arrangement.

crystal loudspeaker [ENG ACOUS] A loudspeaker in which movements of the diaphragm are produced by a piezoelectric crystal unit that twists or bends under the influence of the applied audio-frequency signal voltage. Also known as piezoelectric loudspeaker.

crystal microphone [ENG ACOUS] A microphone in which deformation of a piezoelectric bar by the action of sound waves or mechanical vibrations generates the output voltage between the faces of the bar. Also known as piezoelectric microphone.

crystal mixer [ELECTR] A mixer that uses the nonlinear characteristic of a crystal diode to mix two frequencies; widely used in radar receivers to convert the received radar signal to a lower intermediate-frequency value by mixing it with a local oscillator signal.

crystal monochromator [SPECT] A spectrometer in which a collimated beam of slow neutrons from a reactor is incident on a single crystal of copper, lead, or other element mounted on a divided circle.

crystal optics [OPTICS] The study of the propagation of light, and associated phenomena, in crystalline solids.

crystal oscillator [ELECTR] An oscillator in which the frequency of the alternating-current output is determined by the mechanical properties of a piezoelectric crystal. Also known as piezoelectric oscillator.

crystal pickup [ENG ACOUS] A phonograph pickup in which movements of the needle in the record groove cause deformation of a piezoelectric crystal, thereby generating an audio-frequency output voltage between opposite faces of the crystal. Also known as piezoelectric pickup.

crystal rectifier See semiconductor diode.

crystal set [ELECTR] A radio receiver having a crystal detector stage for demodulation of the received signals, but no amplifier stages.

crystal spectrometer See Bragg spectrometer.

crystal structure [CRYSTAL] The arrangement of atoms or ions in a crystalline solid.

crystal symmetry [CRYSTAL] The existence of nontrivial operations, consisting of inversions, rotations around an axis, reflections, and combinations of these, which bring a crystal into a position indistinguishable from its original position.

crystal system [CRYSTAL] One of seven categories (cubic, hexagonal, tetragonal, trigonal, orthorhombic, monoclinic, and triclinic) into which a crystal may be classified according to the shape of the unit cell of its Bravais lattice, or according to the dominant symmetry elements of its crystal class.

crystal transducer [ELECTR] A transducer in which a piezoelectric crystal serves as the sensing element.

crystal whiskers [CRYSTAL] Single crystals that have grown in a filamentary form.

cs See centistoke.

Cs See cesium; cirrostratus cloud.

csc See cosecant.

C scan See C scope.

C scope [ELECTR] A cathode-ray scope on which signals appear as spots, with bearing angle as the horizontal coordinate and elevation angle as the vertical coordinate. Also known as C indicator; C scan.

CSP See control switching point.

CSW See channel status word.

ctenidium [INV ZOO] 1. The comb- or featherlike respiratory apparatus of certain mollusks. 2. A row of spines on the head or thorax of some fleas.

Ctenodrilidae [INV ZOO] A family of fringe worms belonging to the Sedentaria.

ctenoid scale [VERT ZOO] A thin, acellular structure composed of bonelike material and characterized by a serrated margin; found in the skin of advanced teleosts.

Ctenophora [INV ZOO] The comb jellies, a phylum of marine organisms having eight rows of comblike plates as the main locomotory structure.

Ctenostomata [INV ZOO] An order of bryozoans in the class Gymnolaemata recognized as inconspicuous, delicate colonies made up of relatively isolated, short, tubular zooecia with chitinous walls.

Ctenostomatida [INV ZOO] The equivalent name for Odontostomatida.

Ctenothrissidae [PALEON] A family of extinct teleostean fishes in the order Ctenothrissiformes.

Ctenothrissiformes [PALEON] A small order of extinct teleostean fishes; important as a group on the evolutionary line leading from the soft-rayed to the spiny-rayed fishes.

CTR See controlled thermonuclear reactor.

C-type asteroid [ASTRON] A type of asteroid whose surface is very dark and neutral-colored, and probably is of carbonaceous composition similar to primitive carbonaceous chondritic meteorites.

cu See crosstalk unit; cubic.

Cu See copper.

cubanite [MINERAL] $CuFe_2S_3$ Bronze-yellow mineral that crystallizes in the orthorhombic system. Also known as chalmersite.

cube [MATH] Regular polyhedron whose faces are all square.

cubeb [BOT] The dried, nearly ripe fruit (berries) of a climbing vine, *Piper cubeba*, of the pepper family (Piperaceae).

cube ore *See* pharmacosiderite.

cube root [MATH] Another number whose cube is the original number.

cubic [MECH] Denoting a unit of volume, so that if x is a unit of length, a cubic x is the volume of a cube whose sides have length $1x$; for example, a cubic meter, or a meter cubed, is the volume of a cube whose sides have a length of 1 meter. Abbreviated cu.

cubic crystal [CRYSTAL] A crystal whose lattice has a unit cell with perpendicular axes of equal length.

cubic determinant [MATH] A mathematical form analogous to an ordinary determinant, with the elements forming a cube instead of a square.

cubic equation [MATH] A polynomial equation with no exponent larger than 3.

cubic plane [CRYSTAL] A plane that is at right angles to any one of the three crystallographic axes of the cubic system.

cubic polynomial [MATH] A polynomial in which all exponents are no greater than 3.

cubic spline [MATH] One of a collection of cubic polynomials used in interpolating a function whose value is specified at each of a collection of distinct ordered values, X_i $(i = 1, ..., n)$, and whose slope is specified at X_1 and X_n; one cubic polynomial is found for each interval, such that the interpolating system has the prescribed values at each of the X_i, the prescribed slope at X_1 and X_n, and a continuous slope at each of the X_i.

cubic system *See* isometric system.

cuboid [ANAT] The outermost distal tarsal bone in vertebrates. [INV ZOO] Main vein of the wing in many insects, particularly the flies (Diptera). [SCI TECH] Nearly cubic in shape.

Cubomedusae [INV ZOO] An order of coelenterates in the class Scyphozoa distinguished by a cubic umbrella.

cuckoo [VERT ZOO] The common name for about 130 species of primarily arboreal birds in the family Culidae; some are social parasites.

Cucujidae [INV ZOO] The flat-back beetles, a family of predatory coleopteran insects in the superfamily Cucujoidea.

Cucujoidea [INV ZOO] A large superfamily of coleopteran insects in the suborder Polyphaga.

Cuculidae [VERT ZOO] A family of perching birds in the order Cuculiformes, including the cuckoos and the roadrunner, characterized by long tails, heavy beaks and conspicuous lashes.

Cuculiformes [VERT ZOO] An order of birds containing the cuckoos and allies, characterized by the zygodactyl arrangement of the toes.

Cucumariidae [INV ZOO] A family of dendrochirotacean holothurian echinoderms in the order Dendrochirotida.

cucumber [BOT] *Cucumis sativus*. An annual cucurbit, in the family Cucurbitaceae grown for its edible, immature fleshy fruit.

cucumber mildew [PL PATH] 1. A downy mildew of cucumbers and melons caused by *Peronoplasmopara cubensis*. 2. A powdery mildew of cucumbers and melons caused by *Erysiphe cichoracearum*.

cucumber mosaic [PL PATH] A virus disease of cucumbers and related fruits, producing mottling of terminal leaves and fruits and dwarfing of vines.

cucurbit wilt [PL PATH] A bacterial disease of cucumbers and related plants caused by *Erwinia tracheiphila*, characterized by sudden wilting of the plant.

cue circuit [ELECTR] A one-way communication circuit used to convey program control information.

cuesta [GEOGR] A gently sloping plain which terminates in a steep slope on one side.

cul-de-sac [ANAT] Blind pouch or diverticulum. [CIV ENG] A dead-end street with a circular area for turning around.

culdoscope [MED] An instrument used to visualize female pelvic organs, introduced through the vagina or a perforation into the retrouterine pouch.

Culicidae [INV ZOO] The mosquitoes, a family of slender, orthorrhaphous dipteran insects in the series Nematocera having long legs and piercing mouthparts.

Culicinae [INV ZOO] A subfamily of the dipteran family Culicidae.

cullet *See* collet.

culm [BOT] 1. A jointed and usually hollow grass stem. 2. The solid stem of certain monocotyledons, such as the sedges. [MIN ENG] Fine, refuse coal, screened and separated from larger pieces.

culmination [ASTRON] 1. The position of a heavenly body when at highest apparent altitude. 2. For a heavenly body which is continually above the horizon, the position of lowest apparent altitude. [GEOL] A high point on the axis of a fold.

cultigen [BIOL] A cultivated variety or species of organism for which there is no known wild ancestor. Also known as cultivar.

cultivar *See* cultigen.

cultural anthropology [ANTHRO] The division of anthropology dealing with the study of all aspects of culture.

culture [ANTHRO] The complex pattern of behavior that distinguishes a social, ethnic, or religious group. [BIOL] A growth of living cells or microorganisms in a controlled artificial environment.

culture medium [MICROBIO] The nutrients and other organic and inorganic materials used for the growth of microorganisms and plant and animal tissue in culture.

culvert [ENG] A covered channel or a large-diameter pipe that takes a watercourse below ground level.

Cumacea [INV ZOO] An order of the class Crustacea characterized by a well-developed carapace which is fused dorsally with at least the first three thoracic somites and overhangs the sides.

cumberlandite [PETR] A coarse-grained, ultramafic, ultrabasic rock composed principally of olivine crystals in a ground mass of magnetite and ilmenite with minor plagioclase.

cumengite [MINERAL] $Pb_4Cu_4Cl_8(OH)_8 \cdot H_2O$ A deep-blue or light-indigo-blue tetragonal mineral consisting of a basic lead-copper chloride occurring in crystals.

cumin [BOT] *Cuminum cyminum* An annual herb in the family Umbelliferae; the fruit is valuable for its edible, aromatic seeds.

cummingtonite [MINERAL] $(Fe,Mg)_7Si_8O_{22}(OH)_2$ A brownish mineral that crystallizes in the monoclinic system; usually occurs as lamellae or fibers in metamorphic rocks.

cumulants [STAT] A set of parameters k_h ($h = 1, ..., r$) of a one-dimensional probability distribution defined by

$$\ln \chi_x(q) = \sum_{h=1}^{r} k_h[(iq)^h/h!] + o(q^r)$$

where $\chi_x(q)$ is the characteristic function of the probability distribution of x. Also known as semi-invariants.

cumulative error [STAT] An error whose magnitude does not approach zero as the number of observations increases. Also known as accumulative error.

cumulative excitation [ATOM PHYS] Process by which the atom is raised from one excited state to a higher state by collision, for example, with an electron.

cumulative ionization [ATOM PHYS] Ionization of an excited atom in the metastable state by means of cumulative excitation. [ELECTR] See avalanche.

cumuliform cloud [METEOROL] A fundamental cloud type, showing vertical development in the form of rising mounds, domes, or towers.

cumulonimbus calvus cloud [METEOROL] A species of cumulonimbus cloud evolving from cumulus congestus: the protuberances of the upper portion have begun to lose the cumuliform outline; they loom and usually flatten, then transform into a whitish mass with a more or less diffuse outline and vertical striation; cirriform cloud is not present, but the transformation into ice crystals often proceeds with great rapidity.

cumulonimbus capillatus cloud [METEOROL] A species of cumulonimbus cloud characterized by the presence of distinct cirriform parts, frequently in the form of an anvil, a plume, or a vast and more or less disorderly mass of hair, and usually accompanied by a thunderstorm.

cumulonimbus cloud [METEOROL] A principal cloud type, exceptionally dense and vertically developed, occurring either as isolated clouds or as a line or wall of clouds with separated upper portions.

cumulus cloud [METEOROL] A principal type of cloud in the form of individual, detached elements which are generally dense and possess sharp nonfibrous outlines; these elements develop vertically, appearing as rising mounds, domes, or towers, the upper parts of which often resemble a cauliflower.

cumulus congestus cloud [METEOROL] A strongly sprouting cumulus species with generally sharp outline and sometimes a great vertical development, and with cauliflower or tower aspect.

cumulus humilis cloud [METEOROL] A species of cumulus cloud characterized by small vertical development and a generally flattened appearance, vertical growth is usually restricted by the existence of a temperature inversion in the atmosphere, which in turn explains the unusually uniform height of the cloud. Also known as fair-weather cumulus.

cumulus mediocris cloud [METEOROL] A cloud species unique to the species cumulus, of moderate vertical development, the upper protuberances or sproutings being not very marked; there may be a small cauliflower aspect; while this species does not give any precipitation, it frequently develops into cumulus congestus and cumulonimbus.

cumulus oophorus [HISTOL] The layer of gelatinous, follicle cells surrounding the ovum in a Graafian follicle.

cuneate [BIOL] Wedge-shaped with the acute angle near the base, as in certain insect wings and the leaves of various plants.

cuneiform [ANAT] 1. Any of three wedge-shaped tarsal bones. 2. Either of a pair of cartilages lying dorsal to the thyroid cartilage of the larynx. 3. Wedge-shaped, chiefly referring to skeletal elements.

Cup See Crater.

cup-and-ball joint [GEOL] A dish-shaped transverse fracture which divides a basalt column into segments. Also known as ball-and-socket joint.

cup anemometer [ENG] A rotation anemometer, usually consisting of three or four hemispherical or conical cups mounted with their diametral planes vertical and distributed symmetrically about the axis of rotation; the rate of rotation of the cups, which is a measure of the wind speed, is determined by a counter.

Cupedidae [INV ZOO] The reticulated beetles, the single family of the coleopteran suborder Archostemata.

cupellation [MET] 1. Method using a cupel for assaying precious metals. 2. Process for refining gold and silver by alloying them with lead and then oxidizing the molten lead to separate the base metal from the precious metal.

cupola [GEOL] An isolated, upward-projecting body of plutonic rock that lies near a larger body; both bodies are presumed to unite at depth. [MET] A vertical cylindrical furnace for melting gray iron for foundry use; the metal, coke, and flux are put into the top of the furnace onto a bed of coke through which air is blown. Also known as furnace cupola.

cupping [MET] 1. First operation of a deep-drawing process. 2. Fracture of a wire or rod in which one fracture surface is conical and the other concave.

cupric [CHEM] The divalent ion of copper.

cuprite [MINERAL] Cu_2O A red mineral that crystallizes in the isometric system and is found in crystals and fine-grained aggregates or is massive; a widespread supergene copper ore. Also known as octahedral copper ore; red copper ore; ruby copper ore.

cuprodescloizite See mottramite.

cupronickel [MET] A copper-base alloy with 10–30% nickel and small amounts of manganese and iron; used in industrial and marine installations as condenser and heat-exchanger tubing.

cuprouranite See torbernite.

cupule [BOT] 1. The cup-shaped involucre characteristic of oaks. 2. A cup-shaped corolla. 3. The gemmae cup of the Marchantiales. [INV ZOO] A small sucker on the feet of certain male flies.

Curculionidae [INV ZOO] The true weevils or snout beetles, a family of coleopteran insects in the superfamily Curculionoidea.

Curculionoidea [INV ZOO] A superfamily of coleopteran insects in the suborder Polyphaga.

Curcurbitaceae [BOT] A family of dicotyledonous herbs or herbaceous vines in the order Violales characterized by an inferior ovary, unisexual flowers, one to five stamens but typically three, and a sympetalous corolla.

Curcurbitales [BOT] The ordinal name assigned to the Cucurbitaceae in some systems of classification.

curd [BOT] The edible flower heads of members of the mustard family such as broccoli. [FOOD ENG] 1. The clotted portion of soured milk or milk treated with an acid or enzyme; used in making cheese. 2. Any food resembling milk curd.

cure [CHEM] To change the properties of a resin material by chemical polycondensation or addition reactions. [ENG] A process by which concrete is kept moist for its first week or month to provide enough water for the cement to harden. Also known as mature.

curettage [MED] Scraping of the inside of a body cavity or the hollow of an organ with a curet.

curie [NUCLEO] A unit of radioactivity, defined as that quantity of any radioactive nuclide which has 3.700×10^{10} disintegrations per second. Abbreviated c; Ci.

Curie constant [ELECTROMAG] The electric or magnetic susceptibility at some temperature times the difference of the temperature and the Curie temperature, which is a constant at temperatures above the Curie temperature according to the Curie-Weiss law.

Curie point See Curie temperature.

Curie principle [THERMO] The principle that a macroscopic cause never has more elements of symmetry than the effect it produces; for example, a scalar cause cannot produce a vectorial effect.

Curie's law [ELECTROMAG] The law that the magnetic susceptibilities of most paramagnetic substances are inversely proportional to their absolute temperatures.

Curie temperature [ELECTROMAG] The temperature marking the transition between ferromagnetism and paramagnetism, or between the ferroelectric phase and paraelectric phase. Also known as Curie point.

Curie-Weiss law [ELECTROMAG] A relation between magnetic or electric susceptibilities and the absolute temperatures which is followed by ferromagnets, antiferromagnets, nonpolar ferroelectrics, antiferroelectrics, and some paramagnets.

curing agent See hardener.

curing time [CHEM] The period of time in which a part is subjected to heat or pressure to cure the resin. [ENG] Time interval between the stopping of moving parts during thermoplastics molding and the release of mold pressure. Also known as molding time.

curite [MINERAL] $Pb_2U_5O_{17}\cdot4H_2O$ An orange-red radioactive mineral, occurring in acicular crystals, an alteration product of uraninite.

curium [CHEM] An element, symbol Cm, atomic number 96; the isotope of mass 244 is the principal source of this artificially produced element.

curium-242 [NUC PHYS] An isotope of curium, mass number 242; half-life is 165.5 days for α-particle emission; 7.2×10^6 years for spontaneous fission.

curium-244 [NUC PHYS] An isotope of curium, mass number 244; half-life is 16.6 years for α-particle emission; 1.4×10^7 years for spontaneous fission; potential use as compact thermoelectric power source.

curl [FOR] A block of timber cut from a crotch for cutting into veneers. [MATER] A defect of paper caused by unequal alteration in the dimensions of the top and underside of the sheet. [MATH] The curl of a vector function is a vector which is formally the cross product of the del operator and the vector. Also known as rotation (rot).

curly top [PL PATH] A virus disease of sugarbeets and certain other plants that is transmitted by a leafhopper; affected plants are dwarfed and have curled, upturned leaves.

currant [BOT] A shrubby, deciduous plant of the genus *Ribes* in the order Rosales; the edible fruit, a berry, is borne in clusters on the plant.

currant leaf spot [PL PATH] 1. An angular leaf spot of currants caused by the fungus *Cercospora angulata*. 2. An anthracnose of currants caused by *Pseudopeziza ribis* and characterized by brown or black spots.

current [ELEC] The net transfer of electric charge per unit time; a specialization of the physics definition. Also known as electric current. [PHYS] 1. The rate of flow of any conserved, indestructible quantity across a surface per unit time. 2. See current density.

current algebra [PARTIC PHYS] The application of algebraic relationships among currents derived from approximate symmetries, such as broken SU_3 symmetry, to the study of hadrons.

current amplification [ELECTR] The ratio of output-signal current to input-signal current for an electron tube, transistor, or magnetic amplifier, the multiplier section of a multiplier phototube, or any other amplifying device; often expressed in decibels by multiplying the common logarithm of the ratio by 20.

current amplifier [ELECTR] An amplifier capable of delivering considerably more signal current than is fed in.

current balance [ELEC] An apparatus with which force is measured between current-carrying conductors, with the purpose of assigning the value of the ampere.

current curve [OCEANOGR] In marine operations, a graphic representation of the flow of a current, consisting of a rectangular-coordinate graph on which speed is represented by the ordinates and time by the abscissas.

current density [ELEC] The current per unit cross-sectional area of a conductor; a specialization of the physics definition. Also known as electric current density. [PHYS] A vector quantity whose component perpendicular to any surface equals the rate of flow of some conserved, indestructible quantity across that surface per unit area per unit time. Also known as current.

current diagram [OCEANOGR] A graph showing the average speeds of flood and ebb currents throughout the current cycle for a considerable part of a tidal waterway.

current drain [ELEC] The current taken from a voltage source by a load. Also known as drain.

current efficiency [PHYS CHEM] The ratio of the amount of electricity, in coulombs, theoretically required to yield a given quantity of material in an electrochemical process, to the amount actually consumed.

current feedback circuit [ELECTR] A circuit used to eliminate effects of amplifier gain instability in an indirect-acting recording instrument, in which the voltage input (error signal) to an amplifier is the difference between the measured quantity and the voltage drop across a resistor.

current function See Lagrange stream function.

current gain [ELECTR] The fraction of the current flowing into the emitter of a transistor which flows through the base region and out the collector.

current generator [ELECTR] A two-terminal circuit element whose terminal current is independent of the voltage between its terminals.

current hour [OCEANOGR] The average time interval between the moon's transit over the meridian of Greenwich and the time of the following strength of flood current modified by the times of slack water and strength of ebb.

current-instruction register See instruction register.

current intensity [ELEC] The magnitude of an electric current. Also known as current strength.

current interrupter [ELEC] Mechanism connected into a current-carrying line to periodically interrupt current flow to allow no-current tests of system components.

current limiter [ELECTR] A device that restricts the flow of current to a certain amount, regardless of applied voltage. Also known as demand limiter.

current-limiting reactor *See* series reactor.

current-limiting resistor [ELEC] A resistor inserted in an electric circuit to limit the flow of current to some predetermined value; used chiefly to protect tubes and other components during warm-up.

current location reference [COMPUT SCI] A symbolic expression, such as a star, which indicates the current location reached by the program; a transfer to * + 2 would bring control to the second statement after the current statement.

current-mode logic [ELECTR] Integrated-circuit logic in which transistors are paralleled so as to eliminate current hogging.

current node [ELEC] A point at which current is zero along a transmission line, antenna, or other circuit element having standing waves.

current noise [ELECTR] Electrical noise of uncertain origin which is observed in certain resistances when a direct current is present, and which increases with the square of this current.

current pole [ENG] A pole used to determine the direction and speed of a current; the direction is determined by the direction of motion of the pole, and the speed by the amount of an attached current line paid out in a specified time.

current regulator [ELECTR] A device that maintains the output current of a voltage source at a predetermined, essentially constant value despite changes in load impedance.

current relay [ELEC] A relay that operates at a specified current value rather than at a specified voltage value.

current ripple [GEOL] A type of ripple mark having a long, gentle slope toward the direction from which the current flows, and a shorter, steeper slope on the lee side.

current rips [OCEANOGR] Small waves formed on the surface of water by the meeting of opposing ocean currents; vertical oscillation, rather than progressive waves, is characteristic of current rips.

current saturation *See* anode saturation.

current strength *See* current intensity.

current transformer [ELEC] An instrument transformer intended to have its primary winding connected in series with a circuit carrying the current to be measured or controlled; the current is measured across the secondary winding.

current-type flowmeter [ENG] A mechanical device to measure liquid velocity in open and closed channels; similar to the vane anemometer (where moving liquid turns a small windmill-type vane), but more rugged.

curry [FOOD ENG] A mixture of plant spices including turmeric, coriander, cinnamon, cumin, ginger, cardamon, cayenne pepper, cloves, and nutmeg.

cursor [COMPUT SCI] A movable spot of light that appears on the screen of a visual display terminal and can be positioned horizontally and vertically through keyboard controls to instruct the computer at what point a change is to be made. [DES ENG] A clear or amber-colored filter that can be placed over a radar screen and rotated until an etched diameter line on the filter passes through a target echo; the bearing from radar to target can then be read accurately on a stationary 360° scale surrounding the filter.

cursorial [VERT ZOO] Adapted for running.

curtain array [ELECTROMAG] An antenna array consisting of vertical wire elements stretched between two suspension cables.

curtain board [BUILD] A fire-retardant partition applied to a ceiling.

curtain wall [CIV ENG] An external wall that is not load-bearing.

curtate [COMPUT SCI] A group of adjacent rows on a punch card.

curvature [MATH] The reciprocal of the radius of the circle which most nearly approximates a curve at a given point; the rate of change of the unit tangent vector to a curve with respect to arc length of the curve.

curvature correction [ASTRON] A correction applied to the mean of a series of observations on a star or planet to take account of the divergence of the apparent path of the star or planet from a straight line. [GEOD] The correction applied in some geodetic work to take account of the divergence of the surface of the earth (spheroid) from a plane.

curvature of field [OPTICS] Error in the image of a plane object formed on a flat screen by an optical system when the best image lies on a curved surface.

curvature of space [RELAT] **1.** The deviation of a spacelike three-dimensional subspace of curved space-time from euclidean geometry. **2.** The Gaussian curvature of a spacelike three-dimensional subspace of curved space-time.

curvature tensor *See* Riemann-Christoffel tensor.

curved space-time [RELAT] A four-dimensional Riemannian space, in which there are no straight lines but only curves, which is a generalization of the Minkowski universe in the general theory of relativity.

curve fitting [STAT] The calculation of a curve of some particular character (as a logarithmic curve) that most closely approaches a number of points in a plane.

curve follower [COMPUT SCI] A device in which a photoelectric, capacitive or inductive pick-off guided by a servomechanism reads data in the form of a graph, such as a curve drawn on paper with suitable ink. Also known as graph follower.

curve tracing [MATH] The method of graphing a function by plotting points and analyzing symmetries, derivatives, and so on.

curvilinear [SCI TECH] Pertaining to curved lines, as in curvilinear coordinates or curvilinear motion.

curvilinear coordinates [MAP] Any linear coordinates which are not cartesian coordinates; frequently used curvilinear coordinates are polar coordinates and cylindrical coordinates.

curvilinear motion [MECH] Motion along a curved path.

curvilinear regression [STAT] Regression study of jointly distributed random variables where the function measuring their statistical dependence is analyzed in terms of curvilinear coordinates.

curvilinear transformation [MATH] A transformation from one coordinate system to another in which the coordinates in the new system are arbitrary twice-differentiable functions of the coordinates in the old system.

Cuscutaceae [BOT] A family of parasitic dicotyledonous plants in the order Polemoniales which lack internal phloem and chlorophyll, have capsular fruit, and are not rooted to the ground at maturity.

Cushing's syndrome [MED] A complex of symptoms including facial and truncal obesity, hypertension, edema, and osteoporosis, resulting from oversecretion of adrenocortical hormones.

cusp [ANAT] **1.** A pointed or rounded projection on the masticating surface of a tooth. **2.** One of the flaps of a heart valve. [ARCH] A pointed projection or peak created by the intersection of two arcs. [GEOL] One of a series of low, crescent-shaped mounds of beach material separated by smoothly curved, shallow troughs spaced at more or less regular intervals along and generally perpendicular to the beach face. Also known as beach cusp. [GEOPHYS] Any of the funnel-shaped regions in the magnetosphere extending from the front magnetopause to the polar ionosphere, and filled with solar wind plasma. [MATH] A singular point of a curve at which the limits of the tangents of the portions of the curve on either side of the point coincide. Also known as spinode.

cuspid *See* canine.

cuspidate [BIOL] Having a cusp; terminating in a point.

cut [CHEM ENG] A fraction obtained by a separation process. [CRYSTAL] A section of a crystal having two parallel major surfaces; cuts are specified by their orientation with respect to the axes of the natural crystal, such as X cut, Y cut, BT cut, and AT cut. [GRAPHICS] A photoengraving used in letterpress printing. [LAP] The style in which a gem is cut, such as brilliant cut, single cut, or rose cut. [MET] *See* fraction. [MIN ENG] **1.** To intersect a vein or a working. **2.** To excavate coal. **3.** To shear one side of an entry or a crosscut by digging out the coal from floor to roof with a pick. [NUCLEO] The fraction that is removed as product or advanced to the next separative element in an isotope separation process. [TEXT] The number of needles per inch in the cylinder or needle bed in a knitting frame.

cut and fill [CIV ENG] Construction of a road, a railway, or a canal which is partly embanked and partly below ground. [GEOL] **1.** Lateral corrosion of one side of a meander accompanied by deposition on the other. **2.** A sedimentary structure consisting of a small filled-in channel.

cutaneous anaphylaxis [IMMUNOL] Hypersensitivity that is marked by an intense skin reaction following parenteral contact with a sensitizing agent.

cutaneous sensation [PHYSIO] Any feeling originating in sensory nerve endings of the skin, including pressure, warmth, cold, and pain.

Cuterebridae [INV ZOO] The robust botflies, a family of myodarian cyclorrhaphous dipteran insects in the subsection Calypteratae.

cuticle [ANAT] The horny layer of the nail fold attached to the nail plate at its margin. [BIOL] A noncellular, hardened or membranous secretion of an epithelial sheet, such as the integument of nematodes and annelids, the exoskeleton of arthropods, and the continuous film of cutin on certain plant parts.

cutin [BIOCHEM] A mixture of fatty substances characteristically found in epidermal cell walls and in the cuticle of plant leaves and stems.

cutis *See* dermis.

cutoff [CIV ENG] **1.** A channel constructed to straighten a stream or to bypass large bends, thereby relieving an area normally subjected to flooding or channel erosion. **2.** An impermeable wall, collar, or other structure placed beneath the base or within the abutments of a dam to prevent or reduce losses by seepage along otherwise smooth surfaces or through porous strata. [ELECTR] **1.** The minimum value of negative grid bias that will prevent the flow of anode current in an electron tube. **2.** *See*

cutoff frequency. [ENG] **1.** A misfire in a round of shots because of severance of fuse owing to rock shear as adjacent charges explode. **2.** The line on a plastic object formed by the meeting of the two halves of a compression mold. Also known as flash groove; pinch-off. [GEOL] A new, relatively short channel formed when a stream cuts through the neck of an oxbow or horseshoe bend. [MECH ENG] **1.** The shutting off of the working fluid to an engine cylinder. **2.** The time required for this process. [MIN ENG] **1.** A quarryman's term for the direction along which the granite must be channeled, because it will not split. **2.** The number of feet a bit may be used in a particular type of rock (as specified by the drill foreman). **3.** Minimum percentage of mineral in an ore that can be mined profitably. [PHYS] Technique used when the contribution to the value of a physical quantity given by integration over a certain variable is absurd (in particular, when the contribution is infinite); involves cutting off the integral at some limit.

cutoff field *See* critical field.

cutoff frequency [ELECTR] A frequency at which the attenuation of a device begins to increase sharply, such as the limiting frequency below which a traveling wave in a given mode cannot be maintained in a waveguide, or the frequency above which an electron tube loses efficiency rapidly. Also known as critical frequency; cutoff.

cutoff lake *See* oxbow lake.

cutoff valve [MECH ENG] A valve used to stop the flow of steam to the cylinder of a steam engine.

cutoff voltage [ELECTR] **1.** The electrode voltage value that reduces the dependent variable of an electron-tube characteristic to a specified low value. **2.** *See* critical voltage.

cutoff wall [CIV ENG] A thin, watertight wall of clay or concrete built up from a cutoff trench to reduce seepage. Also known as core wall.

cutout [ELEC] **1.** Pairs brought out of a cable and terminated at some place other than at the end of the cable. **2.** An electrical device that is used to interrupt the flow of current through any particular apparatus or instrument, either automatically or manually. Also known as electric cutout. [GEOL] *See* horseback.

cutout box [ELEC] A fireproof cabinet or box with one or more hinged doors that contains fuses and switches for various leads in an electrical wiring system. Also known as fuse box.

cut platform *See* wave-cut platform.

cut-set [ELEC] A set of branches of a network such that the cutting of all the branches of the set increases the number of separate parts of the network, but the cutting of all the branches except one does not.

cutter [ENG ACOUS] An electromagnetic or piezoelectric device that converts an electric input to a mechanical output, used to drive the stylus that cuts a wavy groove in the highly polished wax surface of a recording disk. Also known as cutting head; head; phonograph cutter; recording head. [MECH ENG] *See* cutting tool. [MIN ENG] **1.** An operator of a coal-cutting or rock-cutting machine, or a worker engaged in underholing by pick or drill. **2.** A joint, usually a dip joint, running in the direction of working; usually in the plural.

cutting [BOT] A piece of plant stem with one or more nodes, which, when placed under suitable conditions, will produce roots and shoots resulting in a complete plant.

cutting angle [MECH ENG] The angle that the cutting face of a tool makes with the work surface back of the tool.

cutting edge [DES ENG] 1. The point or edge of a diamond or other material set in a drill bit. Also known as cutting point. 2. The edge of a lathe tool in contact with the work during a machining operation.

cutting fluid [MATER] A fluid flowed over the tool and work in metal cutting to reduce heat generated by friction, lubricate, prevent rust, and flush away chips.

cutting head *See* cutter.

cutting point *See* cutting edge.

cutting speed [MECH ENG] The speed of relative motion between the tool and workpiece in the main direction of cutting. Also known as feed rate; peripheral speed.

cutting tool [MECH ENG] The part of a machine tool which comes into contact with and removes material from the workpiece by the use of a cutting medium. Also known as cutter.

cutting torch [ENG] A torch that preheats metal while the surface is rapidly oxidized by a jet of oxygen issuing through the flame from an additional feed line.

cuttlefish [INV ZOO] An Old World decapod mollusk of the genus *Sepia;* shells are used to manufacture dentifrices and cosmetics.

cutwater [CIV ENG] A sharp-edged structure built around a bridge pier to protect it from the flow of water and material carried by the water.

CVn *See* Canes Venatici.

CW *See* continuous wave.

CW laser *See* continuous-wave laser.

cwt *See* hundredweight.

Cyamidae [INV ZOO] The whale lice, a family of amphipod crustaceans in the suborder Caprellidea that bear a resemblance to insect lice.

cyanalcohol *See* cyanohydrin.

cyanate [INORG CHEM] A salt or ester of cyanic acid containing the radical CNO.

cyanide [INORG CHEM] Any of a group of compounds containing the CN group and derived from hydrogen cyanide, HCN.

cyaniding [MET] Introduction of carbon and nitrogen simultaneously into a ferrous alloy by heating while in contact with molten cyanide; usually followed by quenching to produce a hardened case.

cyanite *See* kyanite.

cyanocarbon [ORG CHEM] A derivative of hydrocarbon in which all of the hydrogen atoms are replaced by the CN group.

cyanocobalamin *See* vitamin B_{12}.

cyanoethylation [ORG CHEM] A chemical reaction involving the addition of acrylonitrile to compounds with a reactive hydrogen.

cyanohydrin [ORG CHEM] A compound containing the radicals CN and OH. Also known as cyanalcohol.

cyanophilous [BIOL] Having an affinity for blue or green dyes.

Cyanophyceae [BOT] The blue-green algae, a class of plants in the division Schizophyta.

cyanophycin [BIOCHEM] A granular protein food reserve in the cells of blue-green algae, especially in the peripheral cytoplasm.

Cyanophyta [BOT] An equivalent name for the Cyanophyceae.

cyanosis [MED] A bluish coloration in the skin and mucous membranes due to deficient levels of oxygen in the blood.

cyanotrichite [MINERAL] $Cu_4Al_2(SO_4)(OH)_{12} \cdot 2H_2O$ A bright-blue or sky-blue mineral consisting of a hydrous basic copper aluminum sulfate.

Cyatheaceae [BOT] A family of tropical and pantropical tree ferns distinguished by the location of sori along the veins.

cyathium [BOT] An inflorescence in which the flowers arise from the base of a cuplike involucre.

Cyathoceridae [INV ZOO] The equivalent name for the Lepiceridae.

cybernetics [SCI TECH] 1. The science of control and communication in all of their manifestations within and between machines, animals, and organizations. 2. Specifically, the interaction between automatic control and living organisms, especially humans and animals.

Cycadales [BOT] An ancient order of plants in the class Cycadopsida characterized by tuberous or columnar stems that bear a crown of large, usually pinnate leaves.

Cycadeoidaceae [PALEOBOT] A family of extinct plants in the order Cycadeoidales characterized by sparsely branched trunks and a terminal crown of leaves.

Cycadeoidales [PALEOBOT] An order of extinct plants that were abundant during the Triassic, Jurassic, and Cretaceous periods.

Cycadicae [BOT] A subdivision of large-leaved gymnosperms with stout stems in the plant division Pinophyta; only a few species are extant.

Cycadofilicales [PALEOBOT] The equivalent name for the extinct Pteridospermae.

Cycadophyta [BOT] An equivalent name for Cycadecae elevated to the level of a division.

Cycadophytae [BOT] An equivalent name for Cycadicae.

Cycadopsida [BOT] A class of gymnosperms in the plant subdivision Cycadicae.

cyclamate [ORG CHEM] The calcium or sodium salt of cyclohexylsulfamate, an artificial sweetener.

Cyclanthaceae [BOT] The single family of the order Cyclanthales.

Cyclanthales [BOT] An order of monocotyledonous plants composed of herbs; or, seldom, composed of more or less woody plants with leaves that usually have a bifid, expanded blade.

cycle [ENG] To run a machine through an operating cycle. [FL MECH] A system of phases through which the working substance passes in an engine, compressor, pump, turbine, power plant, or refrigeration system. [MATH] 1. A member of the kernel of a boundary homomorphism. 2. In graph theory, a walk, $(v_0, v_1, ..., v_n)$, such that $v_0 = v_n$, and all other vertices are distinct. Also known as circuit. [SCI TECH] 1. One complete sequence of values of an alternating quantity, or of a sequence of process operations. 2. A set of operations that is repeated as a unit.

cycle count [COMPUT SCI] The operation of keeping track of the number of cycles a computer system goes through during processing time.

cycle criterion [COMPUT SCI] Total number of times a cycle in a computer program is to be repeated.

cycle delay selector [COMPUT SCI] An electromechanical device in a sorter which causes a cycle to be skipped so that the card out of sequence may be directed to a different pocket.

cycle index [COMPUT SCI] 1. The number of times a cycle has been carried out by a computer. 2. The difference, or its negative, between the number of executions of a cycle which are desired and the number which have actually been carried out.

cycle of erosion   *See* geomorphic cycle.

cycle of sedimentation   [GEOL] Also known as sedimentary cycle. **1.** A series of related processes and conditions appearing repeatedly in the same sequence in a sedimentary deposit. **2.** The sediments deposited from the beginning of one cycle to the beginning of a second cycle of the spread of the sea over a land area, consisting of the original land sediments, followed by those deposited by shallow water, then deep water, and then the reverse process of the receding water. **3.** *See* cyclothem.

cycle per minute   [PHYS] A unit of frequency of action, equal to 1/60 hertz. Abbreviated cpm.

cycle per second   *See* hertz.

cycle reset   [COMPUT SCI] The resetting of a cycle index to its initial or other specified value.

cycle stock   [CHEM ENG] The unfinished product taken from a stage of a refinery process and recharged to the process at an earlier stage in the operation.

cycle time   [COMPUT SCI] The shortest time elapsed between one store (or fetch) and the next store (or fetch) in the same memory unit. Also known as memory cycle. [PETRO ENG] In a drilling operation, the time needed for the pump to move the drilling fluid in a bore hole. [SCI TECH] The time required to carry out a cycle; used principally for time-and-motion studies.

cyclic   [SCI TECH] **1.** Pertaining to som cycle. **2.** Repeating itself in some manner in space or time.

cyclic adenylic acid   [BIOCHEM] $C_{10}H_{12}N_5O_6P$ An isomer of adenylic acid; crystal platelets with a melting point of 219–220°C; a key regulator which acts to control the rate of a number of cellular processes in bacteria, most animals, and some higher plants. Abbreviated cAMP. Also known as adenosine 3′,5′-cyclic monophosphate; adenosine 3′,5′-cyclic phosphate; adenosine 3′,5′-monophosphate; 3′,5′-AMP; cyclic AMP.

cyclic AMP   *See* cyclic adenylic acid.

cyclic anhydride   [ORG CHEM] A ring compound formed by the removal of water from a compound; an example is phthalic anhydride.

cyclic catalytic reforming process   [CHEM ENG] A method for the production of low-Btu reformed gas consisting of the conversion of carbureted water-gas sets by installing a bed of nickel catalyst in the superheater and using the carburetor as a combustion chamber and process steam superheater. Abbreviated CCR process.

cyclic check   [COMPUT SCI] One of the two-byte elements replacing the parity bit stripped off each byte transferred from main storage to disk volume (the other element is the bit count appendage); these two elements are appended to the block during the write operation; on a subsequent read operation these elements are calculated and compared to the appended elements for accuracy.

cyclic code   [COMPUT SCI] A code, such as a binary code, that changes only in one digit when going from one number to the number immediately following, and in that digit by only one unit.

cyclic compound   [ORG CHEM] A compound that contains a ring of atoms.

cyclic coordinate   [MECH] A generalized coordinate on which the Lagrangian of a system does not depend explicitly.

cyclic curve   [MATH] **1.** A curve (such as a cycloid, cardioid, or epicycloid) generated by a point of a circle that rolls (without slipping) on a given curve. **2.** The intersection of a quadric surface with a sphere. Also known as spherical cyclic curve. **3.** The stereo-graphic projection of a spherical cyclic curve. Also known as plane cyclic curve.

cyclic feeding   [COMPUT SCI] In character recognition, a system employed by character readers in which each input document is issued to the document transport in a predetermined and constant period of time.

cyclic mining   *See* conventional mining.

cyclic permutation   [MATH] A permutation of an ordered set of symbols which sends the first to the second, the second to the third, …, the last to the first.

cyclic shift   [COMPUT SCI] A computer shift in which the digits dropped off at one end of a word are returned at the other end of the word. Also known as circuit shift; circular shift; end-around shift; nonarithmetic shift; ring shift.

cyclic storage   [COMPUT SCI] A computer storage device, such as a magnetic drum, whose storage medium is arranged in such a way that information can be read into or extracted from individual locations at only certain fixed times in a basic cycle.

cyclic train   [MECH ENG] A set of gears, such as an epicyclic gear system, in which one or more of the gear axes rotates around a fixed axis.

cyclic transfer   [COMPUT SCI] The automatic transfer of data from some medium to memory or from memory to some medium until all the data are read.

cycling   [CHEM ENG] A series of operations in petroleum refining or natural-gas processing in which the steps are repeated periodically in the same sequence. [CONT SYST] A periodic change of the controlled variable from one value to another in an automatic control system. Also known as oscillation.

cycloalkene   [ORG CHEM] An unsaturated, monocyclic hydrocarbon having the formula C_nH_{2n-2}. Also known as cycloolefin.

cycloalkyne   [ORG CHEM] A cyclic compound containing one or more triple bonds between carbon atoms.

cycloconverter   [ELEC] A device that produces an alternating current of constant or precisely controllable frequency from a variable-frequency alternating-current input, with the output frequency usually one-third or less of the input frequency.

Cyclocystoidea   [PALEON] A class of small, disk-shaped, extinct echinozoans in which the lower surface of the body probably consisted of a suction cup.

cyclogenesis   [METEOROL] Any development or strengthening of cyclonic circulation in the atmosphere.

cyclograph   [ENG] An electronic instrument that produces on a cathode-ray screen a pattern which changes in shape according to core hardness, carbon content, case depth, and other metallurgical properties of a test sample of steel inserted in a sensing coil.

cycloid   [MATH] The curve traced by a point on the circumference of a circle as the circle rolls along a straight line.

cycloidal gear teeth   [DES ENG] Gear teeth whose profile is formed by the trace of a point on a circle rolling without slippage on the outside or inside of the pitch circle of a gear; now used only for clockwork and timer gears.

cycloidal pendulum   [MECH] A modification of a simple pendulum in which a weight is suspended from a cord which is slung between two pieces of metal shaped in the form of cycloids; as the bob swings, the cord wraps and unwraps on the cycloids;

the pendulum has a period that is independent of the amplitude of the swing.

cycloid scale [VERT ZOO] A thin, acellular structure which is composed of a bonelike substance and shows annual growth rings; found in the skin of soft-rayed fishes.

cyclolysis [METEOROL] The weakening or decay of cyclonic circulation in the atmosphere.

cyclone [CHEM ENG] A static reaction vessel in which fluids under pressure form a vortex. [MECH ENG] Any cone-shaped air-cleaning apparatus operated by centrifugal separation that is used in particle collecting and fine grinding operations. [METEOROL] A low-pressure region of the earth's atmosphere with roundish to elongated-oval ground plan, in-moving air currents, centrally upward air movement, and generally outward movement at various higher elevations in the troposphere.

cyclone classifier See cyclone separator.

cyclone furnace [ENG] A water-cooled, horizontal cylinder in which fuel is fired cyclonically and heat is released at extremely high rates.

cyclone separator [MECH ENG] A funnel-shaped device for removing particles from air or other fluids by centrifugal means; used to remove dust from air or other fluids, steam from water, and water from steam, and in certain applications to separate particles into two or more size classes. Also known as cyclone classifier.

cyclone wave [METEOROL] 1. A disturbance in the lower troposphere, of wavelength 1000–2500 kilometers; cyclone waves are recognized on synoptic charts as migratory high- and low-pressure systems. 2. A frontal wave at the crest of which there is a center of cyclonic circulation, that is, the frontal wave of a wave cyclone.

cyclonic [GEOPHYS] Having a sense of rotation about the local vertical that is the same as that of the earth's rotation: as viewed from above, counterclockwise in the Northern Hemisphere, clockwise in the Southern Hemisphere, undefined at the Equator.

cyclonic shear [METEOROL] Horizontal wind shear of such a nature that it contributes to the cyclonic vorticity of the flow; that is, it tends to produce cyclonic rotation of the individual air particles along the line of flow.

cycloolefin See cycloalkene.

cyclopean [MATER] Mass concrete with aggregate larger than 6 inches (15 centimeters); used for thick structures such as dams. [PETR] See mosaic.

cyclophon See beam-switching tube.

Cyclophoracea [INV ZOO] A superfamily of gastropod mollusks in the order Prosobranchia.

Cyclophoridae [INV ZOO] A family of land snails in the order Pectinibranchia.

Cyclophyllidea [INV ZOO] An order of platyhelminthic worms comprising most tapeworms of warm-blooded vertebrates.

cyclopia [MED] A congenital anomaly characterized by fusion of the eye sockets with various degrees of fusion of the eyes, to the occurrence of a single median eye.

Cyclopinidae [INV ZOO] A family of copepod crustaceans in the suborder Cyclopoida, section Gnathostoma.

Cyclopoida [INV ZOO] A suborder of small copepod crustaceans.

Cyclopteridae [VERT ZOO] The lumpfishes and snailfishes, a family of deep-sea forms in the suborder Cottoidei of the order Perciformes.

Cyclorhagae [INV ZOO] A suborder of benthonic, microscopic marine animals in the class Kinorhyncha of the phylum Aschelminthes.

Cyclorrhapha [INV ZOO] A suborder of true flies, order Diptera, in which developing adults are always formed in a puparium from which they emerge through a circular opening.

cyclosilicate [MINERAL] A silicate having the SiO_4 tetrahedra linked to form rings, with a silicon-oxygen ratio of 1:3, such as $Si_3O_9^{6-}$ or $Si_6O_{18}^{12-}$. Also known as ring silicate.

cyclosis [CYTOL] Massive rotational streaming of cytoplasm in certain vacuolated cells, such as the stonewort *Nitella* and *Paramecium*.

Cyclosporeae [BOT] A class of brown algae, division Phaeophyta, in which there is only a free-living diploid generation.

Cyclosteroidea [PALEON] A class of Middle Ordovician to Middle Devonian echinoderms in the subphylum Echinozoa.

Cyclostomata [INV ZOO] An order of bryozoans in the class Stenolaemata. [VERT ZOO] A subclass comprising the simplest and most primitive of living vertebrates characterized by the absence of jaws and the presence of a single median nostril and an uncalcified cartilaginous skeleton.

cyclostrophic flow [FL MECH] A form of gradient flow in which the centripetal acceleration exactly balances the horizontal pressure force.

cyclothem [GEOL] A rock stratigraphic unit associated with unstable shelf of interior basin conditions, in which the sea has repeatedly covered the land.

cyclotomy [MATH] Theory of dividing the circle into equal parts or constructing regular polygons or, analytically, of finding the nth roots of unity.

cyclotron [NUCLEO] An accelerator in which charged particles are successively accelerated by a constant-frequency alternating electric field that is synchronized with movement of the particles on spiral paths in a constant magnetic field normal to their path. Also known as phasotron.

cyclotron D See dee.

cyclotron emission See cyclotron radiation.

cyclotron frequency [ELECTROMAG] The frequency at which an electron traverses an orbit when moving subject to a uniform magnetic field, at right angles to the field. Also known as gyrofrequency.

cyclotron radiation [ELECTROMAG] The electromagnetic radiation emitted by charged particles as they orbit in a magnetic field, at a speed which is not close to the speed of light. Also known as cyclotron emission.

cyclotron resonance [PHYS] Resonance absorption of energy from an alternating-current electric field by electrons in a uniform magnetic field when the frequency of the electric field equals the cyclotron frequency, or the cyclotron frequency corresponding to the electron's effective mass if the electrons are in a solid. Also known as diamagnetic resonance.

Cydippidea [INV ZOO] An order of the Ctenophora having well-developed tentacles.

Cydnidae [INV ZOO] The ground or burrower bugs, a family of hemipteran insects in the superfamily Pentatomorpha.

Cyg See Cygnus.

Cygnus [ASTRON] A conspicuous northern summer constellation; the five major stars are arranged in the form of a cross, but the constellation is represented by a swan with spread wings flying southward; right ascension 21 hours, declination 40°N. Abbreviated Cyg. Also known as Northern Cross; Swan.

Cygnus A [ASTRON] A strong, discrete radio source in the constellation Cygnus, associated with two spiral galaxies in collision.

cylinder [CIV ENG] **1.** A steel tube 10–60 inches (25–152 centimeters) in diameter with a wall at least 1/8 inch (3 millimeters) thick that is driven into bedrock, excavated inside, filled with concrete, and used as a pile foundation. **2.** A domed, closed tank for storing hot water to be drawn off at taps. Also known as storage calorifier. [COMPUT SCI] **1.** The virtual cylinder represented by the tracks of equal radius of a set of disks on a disk drive. **2.** See seek area. [ENG] A container used to hold and transport compressed gas for various pressurized applications. [MATH] **1.** A solid bounded by a cylindrical surface and two parallel planes, or the surface of such a solid. **2.** See cylindrical surface. [MECH ENG] See engine cylinder.

cylinder block [DES ENG] The metal casting comprising the piston chambers of a multicylinder internal combustion engine. Also known as block; engine block.

cylinder bore [DES ENG] The internal diameter of the tube in which the piston of an engine or pump moves.

cylinder head [MECH ENG] The cap that serves to close the end of the piston chamber of a reciprocating engine, pump, or compressor.

cylinder press [GRAPHICS] A large printing press in which paper is rolled against a flat, reciprocating printing surface by a rotating cylinder.

cylindrical array [ELECTR] An electronic scanning antenna that may consist of several hundred columns of vertical dipoles mounted in cylindrical radomes arranged in a circle.

cylindrical cam [MECH ENG] A cam mechanism in which the cam follower undergoes translational motion parallel to the camshaft as a roller attached to it rolls in a groove in a circular cylinder concentric with the camshaft.

cylindrical capacitor [ELEC] A capacitor made of two concentric metal cylinders of the same length, with dielectric filling the space between the cylinders. Also known as coaxial capacitor.

cylindrical coordinates [MATH] A system of curvilinear coordinates in which the position of a point in space is determined by its perpendicular distance from a given line, its distance from a selected reference plane perpendicular to this line, and its angular distance from a selected reference line when projected onto this plane.

cylindrical map projection [MAP] A map projection produced by projecting the geographic meridians and parallels onto a cylinder which is tangent to (or intersects) the surface of a sphere, and then developing the cylinder into a plane.

cylindrical pinch See pinch effect.

cylindrical surface [MATH] A surface consisting of each of the straight lines which are parallel to a given straight line and pass through a given curve. Also known as cylinder.

cylindrite [MINERAL] $Pb_3Sn_4Sb_2S_{14}$ A blackish-gray mineral consisting of sulfur, lead, antimony, and tin, occurring in cylindrical forms that separate under pressure into distinct sheets or folia.

Cylindrocapsaceae [BOT] A family of green algae in the suborder Ulotrichineae comprising thick-walled, sheathed cells having massive chloroplasts.

cyme [BOT] An inflorescence in which each main axis terminates in a single flower; secondary and tertiary axes may also have flowers, but with shorter flower stalks.

cymophane See cat's eye.

Cymothoidae [INV ZOO] A family of isopod crustaceans in the suborder Flabellifera; members are fish parasites with reduced maxillipeds ending in hooks.

Cynipidae [INV ZOO] A family of hymenopteran insects in the superfamily Cynipoidea.

Cynipoidea [INV ZOO] A superfamily of hymenopteran insects in the suborder Apocrita.

Cynoglossidae [VERT ZOO] The tonguefishes, a family of Asiatic flatfishes in the order Pleuronectiformes.

Cyperaceae [BOT] The sedges, a family of monocotyledonous plants in the order Cyperales characterized by spirally arranged flowers on a spike or spikelet; a usually solid, often triangular stem; and three carpels.

Cyperales [BOT] An order of monocotyledonous plants in the subclass Commelinidae with reduced, mostly wind-pollinated or self-pollinated flowers that have a unilocular, two- or three-carpellate ovary bearing a single ovule.

Cypheliaceae [BOT] A family of typically crustose lichens with sessile apothecia in the order Caliciales.

Cyphophthalmi [INV ZOO] A family of small, mite-like arachnids in the order Phalangida.

Cypraecea [INV ZOO] A superfamily of gastropod mollusks in the order Prosobranchia.

Cypraeidae [INV ZOO] A family of colorful marine snails in the order Pectinibranchia.

cypress [BOT] The common name for members of the genus *Cupressus* and several related species in the order Pinales.

Cypridacea [INV ZOO] A superfamily of mostly freshwater ostracods in the suborder Podocopa.

Cypridinacea [INV ZOO] A superfamily of ostracods in the suborder Myodocopa characterized by a calcified carapace and having a round back with a downward-curving rostrum.

Cyprinidae [VERT ZOO] The largest family of fishes, including minnows and carps in the order Cypriniformes.

Cypriniformes [VERT ZOO] An order of actinopterygian fishes in the suborder Ostariophysi.

Cyprinodontidae [VERT ZOO] The killifishes, a family of actinopterygian fishes in the order Atheriniformes that inhabit ephemeral tropical ponds.

Cyprinoidei [VERT ZOO] A suborder of primarily fresh-water actinopterygian fishes in the order Cypriniformes having toothless jaws, no adipose fin, and faliciform lower pharyngeal bones.

cypris [INV ZOO] An ostracod-like, free-swimming larval stage in the development of Cirripedia.

Cyrtophorina [INV ZOO] The equivalent name for Gymnostomatida.

cyrtopia [INV ZOO] A type of crustacean larva (Ostracoda) characterized by an elongation of the first pair of antennae and loss of swimming action in the second pair.

cyrtosis [PL PATH] A virus disease of cotton characterized by stunting, distortion, and abnormal branching and coloration.

cyst [MED] A normal or pathologic sac with a distinct wall, containing fluid or other material.

cystectomy [MED] **1.** Excision of the gallbladder, or part of the urinary bladder. **2.** Removal of a cyst. **3.** Removal of a piece of the anterior capsule of the lens for the extraction of a cataract.

cysteine [BIOCHEM] $C_3H_7O_2NS$ A crystalline amino acid occurring as a constituent of glutathione and cystine.

cystic duct [ANAT] The duct of the gallbladder.

cysticercosis [MED] The infestation in humans by cysticerci of the genus *Taenia*.

cystic fibrosis [MED] A hereditary disease of the pancreas transmitted as an autosomal recessive; involves obstructive lesions, atrophy, and fibrosis of the pancreas and lungs, and the production of mucus of high viscosity. Also known as mucoviscidosis.

cystine [BIOCHEM] $C_6H_{12}N_2S_2$ A white, crystalline amino acid formed biosynthetically from cysteine.

cystinuria [MED] The presence in the urine of crystals of cystine together with some lysine, arginine, and ornithine.

cystitis [MED] Inflammation of a fluid-filled organ, especially the urinary bladder.

Cystobacteraceae [MICROBIO] A family of bacteria in the order Myxobacterales; vegetative cells are tapered, and microcysts are rod-shaped and enclosed in sporangia.

cystocele [MED] Herniation of the urinary bladder into the vagina.

Cystoidea [PALEON] A class of extinct crinozoans characterized by an ovoid body that was either sessile or attached by a short aboral stem.

Cystoporata [PALEON] An order of extinct, marine bryozoans characterized by cystopores and minutopores.

cystoscope [MED] An optical instrument for visual examination of the urinary bladder, ureters, and kidneys.

Cytheracea [INV ZOO] A superfamily of ostracods in the suborder Podocopa comprising principally crawling and digging marine forms.

Cytherellidae [INV ZOO] The family comprising all living members of the ostracod suborder Platycopa.

cytidine [BIOCHEM] $C_9H_{13}N_3O_5$ Cytosine riboside, a nucleoside composed of one molecule each of cytosine and D-ribose.

cytidylic acid [BIOCHEM] $C_9H_{14}O_8N_3P$ A nucleotide synthesized from the base cytosine and obtained by hydrolysis of nucleic acid. Also known as cytidine monophosphate.

cytochemistry [CYTOL] The science concerned with the chemistry of cells and cell components, primarily with the location of chemical constituents and enzymes.

cytochrome [BIOCHEM] Any of the complex protein respiratory pigments occurring within plant and animal cells, usually in mitochondia, that function as electron carriers in biological oxidation.

cytochrome a₃ *See* cytochrome oxidase.

cytochrome oxidase [BIOCHEM] Any of a family of respiratory pigments that react directly with oxygen in the reduced state. Also known as cytochrome a_3.

cytogenetics [CYTOL] The comparative study of the mechanisms and behavior of chromosomes in populations and taxa, and the effect of chromosomes on inheritance and evolution.

cytokinesis [CYTOL] Division of the cytoplasm following nuclear division.

cytokinin [BIOCHEM] Any of a group of plant hormones which elicit certain plant growth and development responses, especially by promoting cell division.

cytology [BIOL] A branch of the biological sciences which deals with the structure, behavior, growth, and reproduction of cells and the function and chemistry of cell components.

cytolysis [PATH] Disintegration or dissolution of cells, usually associated with a pathologic process.

cytomegalovirus [VIROL] An animal virus belonging to subgroup B of the herpesvirus group; causes cytomegalic inclusion disease and pneumonia.

Cytophagaceae [MICROBIO] A family of bacteria in the order Cytophagales; cells are rods or filaments, unsheathed cells are motile, filaments are not attached, and carotenoids are present.

Cytophagales [MICROBIO] An order of gliding bacteria; cells are rods or filaments and motile by gliding, and fruiting bodies are not produced.

cytoplasm [CYTOL] The protoplasm of an animal or plant cell external to the nucleus.

cytoplasmic inheritance [GEN] The control of genetic difference by hereditary units carried in cytoplasmic organelles. Also known as extrachromosomal inheritance.

cytoplasmic streaming [CYTOL] Intracellular movement involving irreversible deformation of the cytoplasm produced by endogenous forces.

cytosine [BIOCHEM] $C_4H_5ON_3$ A pyrimidine occurring as a fundamental unit or base of nucleic acids.

cytoskeleton [CYTOL] Protein fibers composing the structural framework of a cell.

cytosome [CYTOL] The cytoplasm of the cell, as distinct from the nucleus.

cytostome [INV ZOO] The mouth-like opening in many unicellular organisms, particularly Ciliophora.

cytotaxis [PHYSIO] Attraction of motile cells by specific diffusible stimuli emitted by other cells.

Czapek's agar [MICROBIO] A nutrient culture medium consisting of salt, sugar, water, and agar; used for certain mold cultures.

Czochralski process [CRYSTAL] A method of producing large single crystals by inserting a small seed crystal of germanium, silicon, or other semiconductor material into a crucible filled with similar molten material, then slowly pulling the seed up from the melt while rotating it.

D

D *See* dee; diopter.

dac *See* digital-to-analog converter.

dachiardite [MINERAL] $(Na_2Ca)_2(Al_4Si_{20}O_{48})\cdot 12H_2O$ A white to colorless mineral in the mordenite group of the zeolite family that crystallizes in the monoclinic system.

Dacian [GEOL] Lower upper Pliocene geologic time.

dacite [GEOL] Very fine crystalline or glassy rock of volcanic origin, composed chiefly of sodic plagioclase and free silica with subordinate dark-colored minerals.

Dacromycetales [MYCOL] An order of jelly fungi in the subclass Heterobasidiomycetidae having branched basidia with the appearance of a tuning fork.

Dactylochirotida [INV ZOO] An order of dendrochirotacean holothurians in which there are 8–30 digitate or digitiform tentacles, which sometimes bifurcate.

Dactylogyroidea [INV ZOO] A superfamily of trematodes in the subclass Monogenea; all are fish ectoparasites.

Dactylopteridae [VERT ZOO] The flying gurnards, the single family of the perciform suborder Dactylopteroidei.

Dactylopteroidei [VERT ZOO] A suborder of marine shore fishes in the order Perciformes, characterized by tremendously expansive pectoral fins.

Dactyloscopidae [VERT ZOO] The sand stargazers, a family of small tropical and subtropical perciform fishes in the suborder Blennioidei.

dado [ARCH] **1.** The lower portion of an interior wall set off by molding or other decoration. **2.** The portion of a pedestal between surbase and base. **3.** The portion of a wall basement between surbase and base course.

dagger board [NAV ARCH] A device used in a sailing vessel to increase the area of lateral resistance; it is a narrow metal flat or wooden board that slides up and down in the trunk, which extends from inside the hull to the bottom.

Dagor lens [OPTICS] An anastigmatic lens consisting of two lens systems that are nearly symmetrical with respect to the stop, each system containing three or more lenses.

daguerreotype [GRAPHICS] A photograph produced on a silver plate or a copper plate coated with silver sensitized by the action of iodine; after exposure of the plate in a camera, a latent image is developed by use of mercury vapor.

daily aberration *See* diurnal aberration.

dairy [AGR] A farm concerned with the production of milk. [FOOD ENG] An establishment where milk products are made.

daisy wheel printer [COMPUT SCI] A serial printer in which the printing element is a plastic hub that has a large number of flexible radial spokes, each spoke having one or more different raised printing characters; the wheel is rotated as it is moved horizontally step by step under computer control, and stops when a desired character is in a desired print position so a hammer can drive that character against an inked ribbon.

Dakotan [GEOL] Lower Upper Cretaceous geologic time.

Dalatiidae [VERT ZOO] The spineless dogfishes, a family of modern sharks belonging to the squaloid group.

d'Alembertian [MATH] A differential operator in four-dimensional space,

$$\frac{\partial^2}{\partial x^2} + \frac{\partial^2}{\partial y^2} + \frac{\partial^2}{\partial z^2} - \frac{1}{c^2}\frac{\partial^2}{\partial t^2}$$

which is used in the study of relativistic mechanics.

d'Alembert's paradox [FL MECH] The paradox that no forces act on a body moving at constant velocity in a straight line through a large mass of incompressible, inviscid fluid which was initially at rest, or in uniform motion.

d'Alembert's principle [MECH] The principle that the resultant of the external forces and the kinetic reaction acting on a body equals zero.

d'Alembert's wave equation *See* wave equation.

Dalitz pair [PARTIC PHYS] The electron and positron resulting from the decay of a neutral pion to these particles and a photon.

Dalitz plot [PARTIC PHYS] Pictorial representation for data on the distribution of certain three-particle configurations that result from elementary-particle decay processes or high-energy nuclear reactions.

Dallis grass [BOT] The common name for the tall perennial forage grasses composing the genus *Paspalum* in the order Cyperales.

dalton *See* atomic mass unit.

Dalton's atomic theory [CHEM] Theory forming the basis of accepted modern atomic theory, according to which matter is made of particles called atoms, reactions must take place between atoms or groups of atoms, and atoms of the same element are all alike but differ from atoms of another element.

Dalton's law [PHYS] The law that the pressure of a gas mixture is equal to the sum of the partial pressures of the gases composing it. Also known as law of partial pressures.

Dalton's temperature scale [THERMO] A scale for measuring temperature such that the absolute temperature T is given in terms of the temperature on the Dalton scale τ by $T = 273.15(373.15/273.15)^{\tau/100}$.

dam [CIV ENG] **1.** A barrier constructed to obstruct the flow of a watercourse. **2.** A pair of cast-steel plates with interlocking fingers built over an expansion joint in the road surface of a bridge.

damage assessment [ENG] Estimate of injury or loss to components, subsystems, or entire systems, as well as the cost of repairs or replacement to restore serviceability.

damp [ENG] To reduce the fire in a boiler or a furnace by putting a layer of damp coals or ashes on the fire bed. [MIN ENG] A poisonous gas in a coal mine. [PHYS] To gradually diminish the amplitude of a vibration or oscillation.

damp down [MET] To stop the blast in a blast furnace by closing the openings in the furnace.

damped harmonic motion [PHYS] Also known as damped oscillation; damped vibration. **1.** The linear motion of a particle subject both to an elastic restoring force proportional to its displacement and to a frictional force in the direction opposite to its motion and proportional to its speed. **2.** A similar variation in a quantity analogous to the displacement of a particle, such as the charge on a capacitor in a simple series circuit containing a resistance.

damped oscillation [PHYS] **1.** Any oscillation in which the amplitude of the oscillating quantity decreases with time. Also known as damped vibration. **2.** *See* damped harmonic motion.

damped vibration *See* damped harmonic motion; damped oscillation.

damped wave [PHYS] **1.** A wave whose amplitude drops exponentially with distance because of energy losses which are proportional to the square of the amplitude. **2.** A wave in which the amplitudes of successive cycles progressively diminish at the source.

damper [ELECTR] A diode used in the horizontal deflection circuit of a television receiver to make the sawtooth deflection current decrease smoothly to zero instead of oscillating at zero; the diode conducts each time the polarity is reversed by a current swing below zero. [MECH ENG] A valve or movable plate for regulating the flow of air or the draft in a stove, furnace, or fireplace.

damper winding [ELEC] A winding consisting of several conducting bars on the field poles of a synchronous machine, short-circuited by conducting rings or plates at their ends, and used to prevent pulsating variations of the position or magnitude of the magnetic field linking the poles. Also known as amortisseur winding.

damping [ENG] Reducing or eliminating reverberation in a room by placing sound-absorbing materials on the walls and ceiling. Also known as soundproofing. [PHYS] **1.** The dissipation of energy in motion of any type, especially oscillatory motion and the consequent reduction or decay of the motion. **2.** The extent of such dissipation and decay.

damping coefficient *See* damping factor; resistance.

damping constant *See* resistance.

damping factor [PHYS] **1.** The ratio of the logarithmic decrement of any underdamped harmonic motion to its period. Also known as damping coefficient. **2.** *See* decrement.

damping-off [PL PATH] A fungus disease of seedlings and cuttings in which the parasites invade the plant tissues near the ground level, causing wilting and rotting.

damping ratio [PHYS] The ratio of the actual resistance in damped harmonic motion to that necessary to produce critical damping. Also known as relative damping ratio.

Danaidae [INV ZOO] A family of large tropical butterflies, order Lepidoptera, having the first pair of legs degenerate.

danalite [MINERAL] $(Fe,Mn,Zn)_4Be_3(SiO_4)_3S$ A mineral consisting of a silicate and sulfide of iron and beryllium; it is isomorphous with helvite and genthelvite.

danburite [MINERAL] $CaB_2(SiO_4)_2$ An orange-yellow, yellowish-brown, grayish, or colorless transparent to translucent borosilicate mineral with a feldspar structure crystallizing in the orthorhombic system; it resembles topaz and is used as an ornamental stone.

Danian [GEOL] Lowermost Paleocene or uppermost Cretaceous geologic time.

Daniell cell [PHYS CHEM] A primary cell with a constant electromotive force of 1.1 volts, having a copper electrode in a copper sulfate solution and a zinc electrode in dilute sulfuric acid or zinc sulfate, the solutions separated by a porous partition or by gravity.

Daniell hygrometer [ENG] An instrument for measuring dew point; dew forms on the surface of a bulb containing ether which is cooled by evaporation into another bulb, the second bulb being cooled by the evaporation of ether on its outer surface.

dannemorite [MINERAL] $(Fe,Mn,Mg)_7Si_8O_{22}(OH)_2$ A yellowish-brown to greenish-gray monoclinic mineral consisting of a columnar or fibrous amphibole.

daphnite [MINERAL] $(MgFe)_3(Fe,Al)_3(Si,Al)_4O_{10}(OH)_8$ A mineral of the chlorite group consisting of a basic aluminosilicate of magnesium, iron, and aluminum.

Daphoenidae [PALEON] A family of extinct carnivoran mammals in the superfamily Miacoidea.

darapskite [MINERAL] $Na_3(NO_3)(SO_4) \cdot H_2O$ A naturally occurring hydrate mineral consisting of a hydrous nitrate and sulfate of sodium.

darcy [PHYS] A unit of permeability, equivalent to the passage of 1 cubic centimeter of fluid of 1 centipoise viscosity flowing in 1 second under a pressure of 1 atmosphere through a porous medium having a cross-sectional area of 1 square centimeter and a length of 1 centimeter.

Darcy's law [FL MECH] The law that the rate at which a fluid flows through a permeable substance per unit area is equal to the permeability, which is a property only of the substance through which the fluid is flowing, times the pressure drop per unit length of flow, divided by the viscosity of the fluid.

dark current *See* electrode dark current.

dark-eclipsing variables [ASTRON] A binary star system, comprising a bright star and an almost dark companion that revolve about each other.

dark-field illumination [OPTICS] A method of microscope illumination in which the illuminating beam is a hollow cone of light formed by an opaque stop at the center of the condenser large enough to prevent direct light from entering the objective; the specimen is placed at the concentration of the light cone, and is seen with light scattered or diffracted by it.

dark-line spectrum [SPECT] The absorption spectrum that results when white light passes through a substance, consisting of dark lines against a bright background.

dark nebula [ASTRON] A cloud of solid particles which absorbs or scatters away radiation directed toward an observer and becomes apparent when silhouetted against a bright nebula or rich star field. Also known as absorption nebula.

dark-red silver ore *See* pyrargyrite.

dark-ruby silver *See* pyrargyrite.

dark segment [METEOROL] A bluish-gray band appearing along the horizon opposite the rising or setting sun and lying just below the antitwilight arch. Also known as earth's shadow.

dark space [ELECTR] A region in a glow discharge that produces little or no light.

dark spot [ELECTR] A spot on a television receiver tube that results from a spurious signal generated in the television camera tube during rescan, generally from the redistribution of secondary electrons over the mosaic in the tube.

dark star [ASTRON] A star that is not visible but is a part of a binary star system; in particular, a star which causes, in an eclipsing variable, a primary eclipse.

d'Arsonval current [ELEC] A current consisting of isolated trains of heavily damped high-frequency oscillations of high voltage and relatively low current, used in diathermy.

d'Arsonval galvanometer [ENG] A galvanometer in which a light coil of wire, suspended from thin copper or gold ribbons, rotates in the field of a permanent magnet when current is carried to it through the ribbons; the position of the coil is indicated by a mirror carried on it, which reflects a light beam onto a fixed scale. Also known as light-beam galvanometer.

dart [INV ZOO] A small sclerotized structure ejected from the dart sac of certain snails into the body of another individual as a stimulant before copulation.

dart leader [GEOPHYS] The leader which, after the first stroke, initiates each succeeding stroke of a composite flash of lightning. Also known as continuous leader.

Darwinism [BIOL] The theory of the origin and perpetuation of new species based on natural selection of those offspring best adapted to their environment because of genetic variation and consequent vigor. Also known as Darwin's theory.

Darwin's finch [VERT ZOO] A bird of the subfamily Fringillidae; Darwin studied the variation of these birds and used his data as evidence for his theory of evolution by natural selection.

Darwin's theory See Darwinism.

Darwinulacea [INV ZOO] A small superfamily of nonmarine, parthenogenetic ostracods in the suborder Podocopa.

Darzen's procedure [ORG CHEM] Preparation of alkyl halides by refluxing a molecule of an alcohol with a molecule of thionyl chloride in the presence of a molecule of pyridine.

Darzen's reaction [ORG CHEM] Condensation of aldehydes and ketones with α-haloesters to produce glycidic esters.

Dasayatidae [VERT ZOO] The stingrays, a family of modern sharks in the batoid group having a narrow tail with a single poisonous spine.

Dascillidae [INV ZOO] The soft-bodied plant beetles, a family of coleopteran insects in the superfamily Dascilloidea.

Dascilloidea [INV ZOO] Superfamily of coleopteran insects in the suborder Polyphaga.

dasheen [BOT] *Colocasia esculenta*. A plant in the order Arales, grown for its edible corm.

dashkesanite [MINERAL] $(Na,K)Ca_2(Fe,Mg)_5(Si,Al)_8-O_{22}Cl_2$ A monoclinic mineral of the amphibole group consisting of a chloroaluminosilicate of sodium, potassium, iron, and magnesium.

dashpot [MECH ENG] A device used to dampen and control a motion, in which an attached piston is loosely fitted to move slowly in a cylinder containing oil.

Dasycladaceae [BOT] A family of green algae in the order Dasycladales comprising plants formed of a central stem from which whorls of branches develop.

Dasycladales [BOT] An order of lime-encrusted marine algae in the division Chlorophyta, characterized by a thallus composed of nonseptate, highly branched tubes.

Dasyonygidae [INV ZOO] A family of biting lice, order Mallophaga, that are confined to rodents of the family Procaviidae.

Dasypodidae [VERT ZOO] The armadillos, a family of edentate mammals in the infraorder Cingulata.

Dasytidae [INV ZOO] An equivalent name for Melyridae.

Dasyuridae [VERT ZOO] A family of mammals in the order Marsupialia characterized by five toes on each hindfoot.

Dasyuroidea [VERT ZOO] A superfamily of marsupial mammals.

data [COMPUT SCI] **1.** General term for numbers, letters, symbols, and analog quantities that serve as input for computer processing. **2.** Any representations of characters or analog quantities to which meaning, if not information, may be assigned. [SCI TECH] Numerical or qualitative values derived from scientific experiments.

data acquisition [COMMUN] The phase of data handling that begins with the sensing of variables and ends with a magnetic recording or other record of raw data; may include a complete radio telemetering link.

data automation [COMPUT SCI] The use of electronic, electromechanical, or mechanical equipment and associated techniques to automatically record, communicate, and process data and to present the resultant information.

data bank [COMPUT SCI] A complete collection of information such as contained in automated files, a library, or a set of computer disks. Also known as data base.

data base See data bank.

data-base machine [COMPUT SCI] A computer that handles the storage and retrieval of data into and out of a data base.

data-base management system [COMPUT SCI] A special data-processing system, or part of a data-processing system, which aids in the storage, manipulation, reporting, management, and control of data.

data carrier [COMPUT SCI] A medium on which data can be recorded, and which is usually easily transportable, such as cards, tape, paper, or disks.

data carrier storage [COMPUT SCI] Any type of storage in which the storage medium is outside the computer, such as tape, cards, or disks, in contrast to inherent storage.

data cell drive [COMPUT SCI] A large-capacity storage device consisting of strips of magnetic tape which can be individually transferred to the read-write head.

data center [COMPUT SCI] An organization established primarily to acquire, analyze, process, store, retrieve, and disseminate one or more types of data.

data chain [COMPUT SCI] Any combination of two or more data elements, data items, data codes, and data abbreviations in a prescribed sequence to yield meaningful information; for example, "date" consists of data elements year, month, and day.

data chaining [COMPUT SCI] A technique used in scatter reading or scatter writing in which new storage areas are defined for use as soon as the current data transfer is completed.

data channel [COMPUT SCI] A bidirectional data path between input/output devices and the main memory of a digital computer permitting one or more input/

output operations to proceed concurrently with computation.

data collection [COMPUT SCI] The process of sending data to a central point from one or more locations.

data communication network [COMPUT SCI] A set of nodes, consisting of computers, terminals, or some type of communication control units in various locations, connected by links consisting of communication channels providing a data path between the nodes.

data communications [COMMUN] The conveying from one location to another by electrical means of information that originates or is recorded in alphabetic, numeric, or pictorial form, or as a signal that represents a measurement; includes telemetering, telegraphy, and facsimile but not voice or television. Also known as data transmission.

data compression [COMPUT SCI] Any means of increasing the quantity of data that can be stored in a given space, or decreasing the space needed to store a given quantity of data. Also known as compression.

data conversion [COMPUT SCI] The changing of the representation of data from one form to another, as from binary to decimal, or from one physical recording medium to another, as from card to disk. Also known as conversion.

data converter *See* converter.

data descriptor [COMPUT SCI] A pointer indicating the memory location of a data item.

data dictionary [COMPUT SCI] A catalog containing the names and structures of all data types.

data display [COMPUT SCI] Visual presentation of processed data by specially designed electronic or electromechanical devices through interconnection (either on- or off-line) with digital computers or component equipments; although line printers and punch cards may display data, they are not usually categorized as displays but as output equipments.

data distribution [COMPUT SCI] Data transmission to one or more locations from a central point.

data division [COMPUT SCI] The section of a program (written in the COBOL language) which describes each data item used for input, output, and storage.

data-driven execution [COMPUT SCI] A mode of carrying out a program in a data flow system, in which an instruction is carried out whenever all its input values are present.

data element [COMPUT SCI] A set of data items pertaining to information of one kind, such as months of a year.

data encryption standard [COMMUN] A cryptographic algorithm of validated strength which is in the public domain and is accepted as a standard. Abbreviated DES.

data entry terminal [COMPUT SCI] A portable keyboard and small numeric display designed for interactive communication with a computer.

data field [COMPUT SCI] An area in the main memory of the computer in which a data record is contained.

data flow [COMPUT SCI] The transfer of data from an external storage device, through the processing unit and memory, and out to an external storage device.

data flow language [COMPUT SCI] A programming language used in a data flow system.

data formatting [COMPUT SCI] Structuring the presentation of data as numerical or alphabetic and specifying the size and type of each datum.

data generator [COMPUT SCI] A specialized word generator in which the programming is designed to test a particular class of device, the pulse parameters

and timing are adjustable, and selected words may be repeated, reinserted later in the sequence, omitted, and so forth.

data-handling system [COMPUT SCI] Automatically operated equipment used to interpret data gathered by instrument installations. Also known as data-reduction system.

data independence [COMPUT SCI] Separation of data from processing, either so that changes in the size or format of the data elements require no change in the computer programs processing them or so that these changes can be made automatically by the database management system.

data item [COMPUT SCI] A single member of a data element. Also known as datum.

data level [COMPUT SCI] The rank of a data element in a source language with respect to other elements in the same record.

data link [COMMUN] The physical equipment for automatic transmission and reception of information. Also known as communication link; information link; tie line; tie-link.

data logging [COMPUT SCI] Conversion of electrical impulses from process instruments into digital data to be recorded, stored, and periodically tabulated.

data manipulation [COMPUT SCI] The standard operations of sorting, merging, input/output, and report generation.

datamation [COMPUT SCI] A shortened term for automatic data processing; taken from data and automation.

data name [COMPUT SCI] A symbolic name used to represent an item of data in a source program, in place of the address of the data item.

data organization [COMPUT SCI] Any one of the data-management conventions for physical and spatial arrangement of the physical records of a data set. Also known as data-set organization.

data processing [COMPUT SCI] Any operation or combination of operations on data, including everything that happens to data from the time they are observed or collected to the time they are destroyed. Also known as information processing.

data-processing center [COMPUT SCI] A computer installation providing data-processing service for others, sometimes called customers, on a reimbursable or nonreimbursable basis.

data-processing inventory [COMPUT SCI] An identification of all major data-processing areas in an agency for the purpose of selecting and focusing upon those in which the use of automatic data-processing (ADP) techniques appears to be potentially advantageous, establishing relative priorities and schedules for embarking on ADP studies, and identifying significant relationships among areas to pinpoint possibilities for the integration of systems.

data processor [COMPUT SCI] **1.** Any device capable of performing operations on data, for instance, a desk calculator, an analog computer, or a digital computer. **2.** Person engaged in processing data.

data purification [COMPUT SCI] The process of removing as many inaccurate or incorrect items as possible from a mass of data before automatic data processing is begun.

data record [COMPUT SCI] A collection of data items related in some fashion and usually contiguous in location.

data reduction [COMPUT SCI] The transformation of raw data into a more useful form.

data-reduction system *See* data-handling system.

data retrieval [COMPUT SCI] The searching, selecting, and retrieving of actual data from a personnel file, data bank, or other file.

data set [COMPUT SCI] **1.** A named collection of similar and related data records recorded upon some computer-readable medium. **2.** A data file in IBM 360 terminology.

data-set organization *See* data organization.

data statement [COMPUT SCI] An instruction in a source program that identifies an item of data in the program and specifies its format.

data station [COMPUT SCI] A remote input/output device which handles a variety of transmissions to and from certain centralized computers.

data structure [COMPUT SCI] A collection of data components that are constructed in a regular and characteristic way.

data system [COMPUT SCI] The means, either manual or automatic, of converting data into action or decision information, including the forms, procedures, and processes which together provide an organized and interrelated means of recording, communicating, processing, and presenting information relative to a definable function or activity.

data tablet *See* electronic tablet.

data tracks [COMPUT SCI] Information storage positions on drum storage devices; information is stored on the drum surface in the form of magnetized or nonmagnetized areas.

data transfer [COMPUT SCI] The technique used by the hardware manufacturer to transmit data from computer to storage device or from storage device to computer; usually under specialized program control.

data transmission *See* data communications.

data transmission equipment [COMPUT SCI] The communications equipment used in direct support of data-processing equipment.

data transmission line [ELEC] A system of electrical conductors, such as a coaxial cable or pair of wires, used to send information from one place to another or one part of a system to another.

data type [COMPUT SCI] The manner in which a sequence of bits represents data in a computer program.

data under voice [COMMUN] A telephone digital data service that allows digital signals to travel on the lower portion of the frequency spectrum of existing microwave radio systems; digital channels initially available handled speeds of 2.4, 4.8, 9.6, and 56 kilobits per second. Abbreviated DUV.

data unit [COMPUT SCI] A set of digits or characters treated as a whole.

data validation [COMPUT SCI] The checking of data for correctness, or the determination of compliance with applicable standards, rules, and conventions.

data word [COMPUT SCI] A computer word that is part of the data which the computer is manipulating, in contrast with an instruction word. Also known as information word.

dating [SCI TECH] The use of methods and techniques to fix dates, assign periods of time, and determine age in archeology, biology, and geology.

datolite [MINERAL] $CaBSiO_4(OH)$ A mineral nesosilicate crystallizing in the monoclinic system; luster is vitreous, and crystals are colorless or white with a greenish tinge.

datum [COMPUT SCI] *See* data item. [ENG] **1.** A direction, level, or position from which angles, heights, speeds or distances are conveniently measured. **2.** Any numerical or geometric quantity or value that serves as a base reference for

other quantities or values (such as a point, line, or surface in relation to which others are determined). [GEOD] The latitude and longitude of an initial point; the azimuth of a line from this point. [GEOL] The top or bottom of a bed of rock on which structure contours are drawn.

datum plane [ENG] A permanently established horizontal plane, surface, or level to which soundings, ground elevations, water surface elevations, and tidal data are referred. Also known as chart datum; datum level; reference level; reference plane.

datum point [MAP] Any reference point of known or assumed coordinates from which calculation or measurements may be taken.

Daubentoniidae [VERT ZOO] A family of Madagascan prosimian primates containing a single species, the aye-aye.

daughter [NUC PHYS] The immediate product of radioactive decay of an element, such as uranium. Also known as decay product; radioactive decay product.

Dauphine law [CRYSTAL] A twin law in which the twinned parts are related by a rotation of 180° around the c axis.

Davian [GEOL] A subdivision of the Upper Cretaceous in Europe; a limestone formation with abundant hydrocorals, bryozoans, and mollusks in Denmark; marine limestone and nonmarine rocks in southeastern France; and continental formations in the Davian of Spain and Portugal.

davidite [MINERAL] A black primary pegmatite uranium mineral of the general formula $A_6B_{15}(O,OH)_{36}$, where $A = Fe^{2+}$, rare earths, uranium, calcium, zirconium, and thorium, and B = titanium, Fe^{3+}, vanadium, and chromium.

Davidson Current [OCEANOGR] A coastal countercurrent of the Pacific Ocean running north, inshore of the California Current, along the western coast of the United States (from northern California to Washington to at least latitude 48°N) during the winter months.

daviesite [MINERAL] An orthorhombic mineral consisting of a lead oxychloride, occurring in minute crystals.

davisonite [MINERAL] $Ca_3Al(PO_4)_2(OH)_3H_2O$ A white mineral consisting of a hydrous basic phosphate of calcium and aluminum.

davit [NAV ARCH] A fixed or movable shipboard crane projecting over the side or over a hatchway, which hoists and lowers boats, anchors, or cargo.

Davy lamp [MIN ENG] An early safety lamp with a mantle of wire gauze around the flame to dissipate the heat from the flame to below the ignition temperature of methane.

dawn [ASTRON] The first appearance of light in the eastern sky before sunrise, or the time of that appearance. Also known as daybreak.

Dawsoniales [BOT] An order of mosses comprising rigid plants with erect stems rising from a rhizomelike base.

dawsonite [MINERAL] $NaAl(OH)_2CO_3$ A white, bladed mineral found in certain oil shales that contains large quantities of alumina; specific gravity is 2.40.

day [ASTRON] One of various units of time equal to the period of rotation of the earth with respect to one or another direction in space; specific examples are the mean solar day and the sidereal day.

daybreak *See* dawn.

day clock [COMPUT SCI] An internal binary counter, with a resolution usually of a microsecond and a cycle measured in years, providing an accurate measure of elapsed time independent of system activity.

daylight saving time [ASTRON] A variation of zone time, usually 1 hour more advanced than standard time, frequently kept during the summer to make better use of daylight. Also known as summer time.

day neutral [BOT] Reaching maturity regardless of relative length of light and dark periods.

dB *See* decibel.

dBa *See* adjusted decibel.

dc *See* direct current.

DCTL *See* direct-coupled transistor logic.

dc-to-ac converter *See* inverter.

dc-to-ac inverter *See* inverter.

dc-to-dc converter [ELEC] An electronic circuit which converts one direct-current voltage into another, consisting of an inverter followed by a step-up or step-down transformer and rectifier.

DDA *See* digital differential analyzer.

DDBS *See* Digital Data Broadcast System.

D display [ENG] In radar, a C display in which the blips extend vertically to give a rough estimate of distance.

DDS *See* digital data service.

DDT [ORG CHEM] Common name for an insecticide; melting point 108.5°C, insoluble in water, very soluble in ethanol and acetone, colorless, and odorless; especially useful against agricultural pests, flies, lice, and mosquitoes. Also known as dichlorodiphenyltrichloroethane.

DDTA *See* derivative differential thermal analysis.

deactivation [CHEM] 1. Rendering inactive, as of a catalyst. 2. Loss of radioactivity. [MET] Chemical removal of the active constituents of a corrosive liquid. [MIN ENG] Treatment of one or more species of mineral particles to reduce floating during froth flotation.

dead [ELEC] Free from any electric connection to a source of potential difference from electric charge; not having a potential different from that of earth; the term is used only with reference to current-carrying parts which are sometimes alive or charged. [GEOL] In economic geology, designating a region with no economic value. [MIN ENG] An area of subsidence that has totally settled and is not likely to move.

dead area *See* blind spot.

dead arm [PL PATH] A fungus disease caused by *Cryptosporella viticola* in which the main lateral shoots are destroyed; common in grapes.

dead band [ELEC] The portion of a potentiometer element that is shortened by a tap; when the wiper traverses this area, there is no change in output. [ENG] The range of values of the measured variable to which an instrument will not effectively respond. Also known as dead zone; neutral zone.

deadbeat [MECH] Coming to rest without vibration or oscillation, as when the pointer of a meter moves to a new position without overshooting. Also known as deadbeat response.

deadbeat algorithm [CONT SYS] A digital control algorithm which attempts to follow set-point changes in minimum time, assuming that the controlled process can be modeled approximately as a first-order plus dead-time system.

deadbeat escapement [HOROL] A watch escapement without recoil, having arresting faces of the pallets described by a circular arc whose center is at the pivot point of the anchor; escape-wheel teeth are contoured to give impulses to these pallet faces, over which they slide without recoil.

deadbeat response *See* deadbeat.

dead bolt [DES ENG] A lock bolt that is moved directly by the turning of a knob or key, not by spring action.

dead end [ACOUS] The end of a sound studio that has the greater sound-absorbing characteristics. [ELEC] The portion of a tapped coil through which no current is flowing at a particular switch position. [SCI TECH] The end of a conduit, passage, power line, or similar system having no exit or continuation.

dead-end tower [CIV ENG] Antenna or transmission line tower designed to withstand unbalanced mechanical pull from all the conductors in one direction together with the wind strain and vertical loads.

deadeye [NAV ARCH] A flat, rounded wooden block usually pierced with three holes to receive the lanyard; used to fasten shrouds and stays.

dead halt *See* drop-dead halt.

dead load *See* static load.

deadly nightshade *See* belladonna.

deadman [CIV ENG] 1. A buried plate, wall, or block attached at some distance from and forming an anchorage for a retaining wall. Also known as anchorage; anchor block; anchor wall. 2. *See* anchor log.

deadman's brake [MECH ENG] An emergency device that automatically is activated to stop a vehicle when the driver removes his foot from the pedal.

deadman's handle [MECH ENG] A handle on a machine designed so that the operator must continuously press on it in order to keep the machine running.

dead reckoning [NAV] Determination of position of a craft by advancing a previous position to a new one on the basis of assumed distance and direction traveled.

dead room *See* anechoic chamber.

dead space [ANAT] The space in the trachea, bronchi, and other air passages which contains air that does not reach the alveoli during respiration, the amount of air being about 140 milliliters. Also known as anatomical dead space. [MED] A cavity left after closure of a wound. [PHYSIO] A calculated expression of the anatomical dead space plus whatever degree of overventilation or underperfusion is present; it is alleged to reflect the relationship of ventilation to pulmonary capillary perfusion. Also known as physiological dead space. [THERMO] A space filled with gas whose temperature differs from that of the main body of gas, such as the gas in the capillary tube of a constant-volume gas thermometer.

dead spot [COMMUN] A geographic location in which signals from a radio, television, or radar transmitter are received poorly or not at all. [ELECTR] A portion of the tuning range of a receiver in which stations are heard poorly or not at all, due to improper design of tuning circuits.

dead time [CONT SYS] The time interval between a change in the input signal to a process control system and the response to the signal. [ENG] The time interval, after a response to one signal or event, during which a system is unable to respond to another. Also known as insensitive time.

dead water [OCEANOGR] The mass of eddying water associated with formation of internal waves near the keel of a ship; forms under a ship of low propulsive power when it negotiates water which has a thin layer of fresher water over a deeper layer of more saline water.

deadweight capacity *See* deadweight tonnage.

deadweight tonnage [NAV ARCH] The total carrying capacity of a ship expressed in long tons (2240 pounds);

displacement of a fully loaded ship, less the weight of the ship itself. Abbreviated dwt. Also known as deadweight capacity.

dead zone *See* dead band.

dead zone unit [COMPUT SCI] An analog computer device that maintains an output signal at a constant value over a certain range of values of the input signal.

deaeration [ENG] Removal of gas or air from a substance, as from feedwater or food.

deafness [MED] Temporary or permanent impairment or loss of hearing.

deamidase [BIOCHEM] An enzyme that catalyzes the removal of an amido group from a compound.

deaminase [BIOCHEM] An enzyme that catalyzes the hydrolysis of amino compounds, removing the amino group.

death rate *See* mortality rate.

deblocking [COMPUT SCI] Breaking up a block of records into individual records.

deblooming [CHEM ENG] The process by which the fluorescence, or bloom, is removed from petroleum oils by exposing them in shallow tanks to the sun and atmospheric conditions or by using chemicals.

debris avalanche [GEOL] The sudden and rapid downward movement of incoherent mixtures of rock and soil on deep slopes.

debris cone [GEOL] 1. A mound of fine-grained debris piled atop certain boulders moved by a landslide. 2. A mound of ice or snow on a glacier covered with a thin layer of debris.

debris fall [GEOL] A relatively free downward or forward falling of unconsolidated or poorly consolidated earth or rocky debris from a cliff, cave, or arch.

debris flow [GEOL] A variety of rapid mass movement involving the downslope movement of high-density coarse clast-bearing mudflows, usually on alluvial fans.

debris glacier [HYD] A glacier formed from ice fragments that have fallen from a larger and taller glacier.

debris slide [GEOL] A type of landslide involving a rapid downward sliding and forward rolling of comparatively dry, unconsolidated earth and rocky debris.

de Broglie equation *See* de Broglie relation.

de Broglie relation [QUANT MECH] The relation in which the de Broglie wave associated with a free particle of matter, and the electromagnetic wave in a vacuum associated with a photon, has a wavelength equal to Planck's constant divided by the particle's momentum and a frequency equal to the particle's energy divided by Planck's constant. Also known as de Broglie equation.

de Broglie wave [QUANT MECH] The quantum-mechanical wave associated with a particle of matter. Also known as matter wave.

debug [COMPUT SCI] To test for, locate, and remove mistakes from a program or malfunctions from a computer. [ELECTR] To detect and remove secretly installed listening devices popularly known as bugs. [ENG] To eliminate from a newly designed system the components and circuits that cause early failures.

debugging routine [COMPUT SCI] A routine to aid programmers in the debugging of their routines; some typical routines are storage printout, tape printout, and drum printout routines.

debugging statement [COMPUT SCI] Temporary instructions inserted into a program being tested so as to pinpoint problem areas.

debug on-line [COMPUT SCI] 1. To detect and correct errors in a computer program by using only certain parts of the hardware of a computer, while other routines are being processed simultaneously. 2. To detect and correct errors in a program from a console distant from a computer in a multiaccess system.

debunching [ELECTR] A tendency for electrons in a beam to spread out both longitudinally and transversely due to mutual repulsion; the effect is a drawback in velocity modulation tubes.

Debye effect [ELECTROMAG] Selective absorption of electromagnetic waves by a dielectric, due to molecular dipoles.

Debye equation [SOLID STATE] The equation for the Debye specific heat, which satisfies the Dulong and Petit law at high temperatures and the Debye T^3 law at low temperatures.

Debye-Falkenhagen effect [PHYS CHEM] The increase in the conductance of an electrolytic solution when the applied voltage has a very high frequency.

Debye force *See* induction force.

Debye-Hückel theory [PHYS CHEM] A theory of the behavior of strong electrolytes, according to which each ion is surrounded by an ionic atmosphere of charges of the opposite sign whose behavior retards the movement of ions when a current is passed through the medium.

Debye potentials [ELECTROMAG] Two scalar potentials, designated Π_e and Π_m, in terms of which one can express the electric and magnetic fields resulting from radiation or scattering of electromagnetic waves by a distribution of localized sources in a homogeneous isotropic medium.

Debye relaxation time [PHYS CHEM] According to the Debye-Hückel theory, the time required for the ionic atmosphere of a charge to reach equilibrium in a current-carrying electrolyte, during which time the motion of the charge is retarded.

Debye specific heat [SOLID STATE] The specific heat of a solid under the assumption that the energy of the lattice arises entirely from acoustic lattice vibration modes which all have the same sound velocity, and that frequencies are cut off at a maximum such that the total number of modes equals the number of degrees of freedom of the solid.

Debye temperature [SOLID STATE] The temperature Θ arising in the computation of the Debye specific heat, defined by $k\Theta = h\nu$, where k is the Boltzmann constant, h is Planck's constant, and ν is the Debye frequency. Also known as characteristic temperature.

decade box [ELEC] An assembly of precision resistors, coils, or capacitors whose individual values vary in submultiples and multiples of 10; by appropriately setting a 10-position selector switch for each section, the decade box can be set to any desired value within its range.

decade bridge [ELECTR] Electronic apparatus for measurement of unknown values of resistances or capacitances by comparison with known values (bridge); one secondary section of the oscillator-driven transformer is tapped in decade steps, the other in 10 uniform steps.

decanning [NUCLEO] Removing the outer container of an enriched uranium fuel rod, in preparation for reprocessing of the fuel.

decantation [ENG] A method for mechanical dewatering of a wet solid by pouring off the liquid without disturbing underlying sediment or precipitate.

Decapoda [INV ZOO] 1. A diverse order of the class Crustacea including the shrimps, lobsters, hermit crabs, and true crabs; all members have a carapace, well-developed gills, and the first three pairs of thoracic appendages specialized as maxillipeds. **2.** An order of dibranchiate cephalopod mollusks containing the squids and cuttle fishes, characterized by eight arms and two long tentacles.

decarboxylase [BIOCHEM] An enzyme that hydrolyzes the carboxyl radical, COOH.

decarburize [MET] To remove carbon from the surface of a ferrous alloy, particularly steel, by heating in a medium that reacts with carbon.

decay [GEOCHEM] See chemical weathering. [MATER] To undergo decomposition. [NUC PHYS] See radioactive decay. [OCEANOGR] In ocean-wave studies, the loss of energy from wind-generated ocean waves after they have ceased to be acted on by the wind; this process is accompanied by an increase in length and a decrease in height of the wave. [PHYS] Gradual reduction in the magnitude of a quantity, as of current, magnetic flux, a stored charge, or phosphorescence.

decay area [OCEANOGR] The area into which ocean waves travel (as swell) after leaving the generating area.

decay chain See radioactive series.

decay coefficient See decay constant.

decay constant [PHYS] The constant c in the equation $I = I_0e^{-ct}$, for the time dependence of rate of decay of a radioactive species; here, I is the number of disintegrations per unit time. Also known as decay coefficient; disintegration constant; radioactive decay constant; transformation constant.

decay family See radioactive series.

decay product See daughter.

decay rate [NUC PHYS] The time rate of disintegration of radioactive material, generally accompanied by emission of particles or gamma radiation.

decay series See radioactive series.

decay time [PHYS] The time taken by a quantity to decay to a stated fraction of its initial value; the fraction is commonly $1/e$. Also known as storage time (deprecated).

Decca [NAV] A hyperbolic navigation system which establishes a line of position from measurement of the phase difference between two continuous-wave signals; the intersection of the two lines of position from two pairs of transmitting stations establishes a navigational fix, or location.

deceleration [MECH] The rate of decrease of speed of a motion.

deceleration parachute See drogue.

deceleration time [COMPUT SCI] For a storage medium, such as magnetic tape that must be physically moved in order for reading or writing to take place, the minimum time that must elapse between the completion of a reading or writing operation and the moment that motion ceases. Also known as stop time.

decentered lens [OPTICS] A lens whose optical center does not coincide with the geometrical center of the rim of the lens; has the effect of a lens combined with a weak prism.

decentralized data processing [COMPUT SCI] An arrangement comprising a data-processing center for each division or location of a single organization.

decibel [PHYS] A unit for describing the ratio of two powers or intensities, or the ratio of a power to a reference power; in the measurement of sound intensity, the pressure of the reference sound is usually taken as 2×10^{-4} dyne per square centimeter; equal to one-tenth bel; if P_1 and P_2 are two amounts of power, the first is said to be n decibels greater, where $n = 10 \log_{10}(P_1/P_2)$. Abbreviated dB.

decibel adjusted See adjusted decibel.

decidua [MED] The endometrium of pregnancy and associated fetal membranes which are cast off at parturition.

deciduous [BIOL] Falling off or being shed at the end of the growing period or season.

decigram [MECH] A unit of mass, equal to 0.1 gram.

decile [STAT] Any of the points which divide the total number of items in a frequency distribution into 10 equal parts.

deciliter [MECH] A unit of volume, equal to 0.1 liter, or 10^{-4} cubic meter.

decimal [MATH] A number expressed in the scale of tens.

decimal-binary switch [ELEC] A switch that connects a single input lead to appropriate combinations of four output leads (representing 1, 2, 4, and 8) for each of the decimal-numbered settings of its control knob; thus, for position 7, output leads 1, 2, and 4 would be connected to the input.

decimal code [COMPUT SCI] A code in which each allowable position has one of 10 possible states; the conventional decimal number system is a decimal code.

decimal-coded digit [COMPUT SCI] One of 10 arbitrarily selected patterns of 1 and 0 used to represent the decimal digits. Also known as coded decimal.

decimal fraction [MATH] Any number written in the form: an integer followed by a decimal point followed by a (possibly infinite) string of digits.

decimal number [MATH] A number signifying a decimal fraction by a decimal point to the left of the numerator with the number of figures to the right of the point equal to the power of 10 of the denominator.

decimal number system [MATH] A representational system for the real numbers in which place values are read in powers of 10.

decimal place [MATH] Reference to one of the digits following the decimal point in a decimal fraction; the kth decimal place registers units of 10^{-k}.

decimal point [MATH] A dot written either on or slightly above the line; used to mark the point at which place values change from positive to negative powers of 10 in the decimal number system.

decimal system [MATH] A number system based on the number 10; in theory, each unit is 10 times the next smaller one.

decimeter [MECH] A metric unit of length equal to one-tenth meter.

decision [COMPUT SCI] The computer operation of determining if a certain relationship exists between words in storage or registers, and taking alternative courses of action; this is effected by conditional jumps or equivalent techniques.

decision calculus [SYS ENG] A guide to the process of decision-making, often outlined in the following steps: analysis of the decision area to discover applicable elements; location or creation of criteria for evaluation; appraisal of the known information pertinent to the applicable elements and correction for bias; isolation of the unknown factors; weighting of the pertinent elements, known and unknown, as to relative importance; and projection of the relative impacts on the objective, and synthesis into a course of action.

decision element [ELECTR] A circuit that performs a logical operation such as "and," "or," "not," or "except" on one or more binary digits of input information representing "yes" or "no" and that expresses the result in its output. Also known as decision gate.

decision gate [ELECTR] *See* decision element. [NAV] In an instrument landing, that point along the path at which the pilot must decide to land or to execute a missed-approach procedure.

decision instruction *See* conditional jump.

decision mechanism [COMPUT SCI] In character recognition, that component part of a character reader which accepts the finalized version of the input character and makes an assessment as to its most probable identity.

decision rule [SYS ENG] In decision theory, the mathematical representation of a physical system which operates upon the observed data to produce a decision.

decision table [COMPUT SCI] 1. A table of contingencies to be considered in the definition of a problem, together with the actions to be taken; sometimes used in place of a flow chart for program documentation. 2. *See* DETAB.

decision theory [SYS ENG] A broad spectrum of concepts and techniques which have been developed to both describe and rationalize the process of decision making, that is, making a choice among several possible alternatives.

decision tree [IND ENG] Graphic display of the underlying decision process involved in the introduction of a new product by a manufacturer.

deck [COMPUT SCI] A set of punched cards. [CIV ENG] 1. A floor, usually of wood, without a roof. 2. The floor or roadway of a bridge. [ENG] A magnetic-tape transport mechanism. [NAV ARCH] Horizontal or cambered and sloping surfaces on a ship, corresponding to the floors of a building.

deckhouse [NAV ARCH] A low building or superstructure, such as a cabin, constructed on the top deck of a ship which may or may not extend to the edges of the deck.

deckle rod [ENG] A small rod inserted at each end of the extrusion coating die to adjust the die opening length.

declarative macroinstruction [COMPUT SCI] An instruction in an assembly language which directs the compiler to take some action or take note of some condition and which does not generate any instruction in the object program.

declarative statement [COMPUT SCI] Any program statement describing the data which will be used or identifying the memory locations which will be required.

declination [ASTRON] The angular distance of a celestial object north or south of the celestial equator. [GEOPHYS] The angle between the magnetic and geographical meridians, expressed in degrees and minutes east or west to indicate the direction of magnetic north from true north. Also known as magnetic declination; variation. [NAV] 1. In a system of polar or spherical coordinates, the angle at the origin between a line to a point and the equatorial plane, measured in a plane perpendicular to the equatorial plane. 2. The arc between the equator and the point measured on a great circle perpendicular to the equator.

declination circle [ENG] For a telescope with an equatorial mounting, a setting circle attached to the declination axis that shows the declination to which the telescope is pointing.

declination compass *See* declinometer.

declination variometer [ENG] An instrument that measures changes in the declination of the earth's magnetic field, consisting of a permanent bar magnet, usually about 1 centimeter long, suspended with a plane mirror from a fine quartz fiber 5–15 centimeters in length; a lens focuses to a point a beam of light reflected from the mirror to recording paper mounted on a rotating drum. Also known as D variometer.

declinometer [ENG] A magnetic instrument similar to a surveyor's compass, but arranged so that the line of sight can be rotated to conform with the needle or to any desired setting on the horizontal circle; used in determining magnetic declination. Also known as declination compass.

declivity [GEOL] 1. A slope descending downward from a point of reference. 2. A downward deviation from the horizontal.

decode [COMMUN] 1. To translate coded characters into a more understandable form. 2. *See* demodulate.

decoder [COMMUN] A device that decodes. [ELECTR] 1. A matrix of logic elements that selects one or more output channels, depending on the combination of input signals present. 2. *See* decoder circuit; matrix; tree.

decoder circuit [ELECTR] A circuit that responds to a particular coded signal while rejecting others. Also known as decoder.

decoding gate [COMPUT SCI] The use of combinatorial logic in circuitry to select a device identified by a binary address code. Also known as recognition gate.

decollator [COMPUT SCI] A device which separates the sheets of continuous stationery that form the output of a computer printer into separate stacks.

decolorize [CHEM ENG] To remove the color from, as from a liquid.

decommutator [ELECTR] The section of a telemetering system that extracts analog data from a time-serial train of samples representing a multiplicity of data sources transmitted over a single radio-frequency link.

decomposer [ECOL] A heterotrophic organism (including bacteria and fungi) which breaks down the complex compounds of dead protoplasm, absorbs some decomposition products, and releases substances usable by consumers. Also known as microcomposer; microconsumer; reducer.

decomposition [CHEM] The more or less permanent structural breakdown of a molecule into simpler molecules or atoms. [GEOCHEM] *See* chemical weathering.

decomposition potential [PHYS CHEM] The electrode potential at which the electrolysis current begins to increase appreciably. Also known as decomposition voltage.

decomposition voltage *See* decomposition potential.

decompression chamber [ENG] 1. A steel chamber fitted with auxiliary equipment to raise its air pressure to a value two to six times atmospheric pressure; used to relieve a diver who has decompressed too quickly in ascending. 2. Such a chamber in which conditions of high atmospheric pressure can be simulated for experimental purposes.

decoupling [ELEC] Preventing transfer or feedback of energy from one circuit to another.

decoupling filter [ELECTR] One of a number of low-pass filters placed between each of several amplifier stages and a common power supply.

decoupling network [ELEC] Any combination of resistors, coils, and capacitors placed in power supply leads or other leads that are common to two or more circuits, to prevent unwanted interstage coupling.

decrement [COMPUT SCI] A specific part of an instruction word in some binary computers, thus a set of digits. [HYD] See groundwater discharge. [MATH] The quantity by which a variable is decreased. [PHYS] The ratio of the amplitudes of an underdamped harmonic motion during two successive oscillations. Also known as damping factor; numerical decrement.

decrement field [COMPUT SCI] That part of an instruction word which is used to modify the contents of a storage location or register.

decrepitation [GEOPHYS] Breaking up of mineral substances when exposed to heat; usually accompanied by a crackling noise.

Dectra [NAV] A radio navigation aid that provides coverage over a specific section of a long ocean route, using equipment and techniques similar to Decca; a master station and a slave station are located at each end of the route; by comparing the phase difference of a master and its slave, guidance is provided to the right or left of the course; comparison of transmission with those of a high-precision quartz clock provides distance information along the course.

decubitus ulcer [MED] An ulcer of the skin and subcutaneous tissues following prolonged lying down, due to pressure on bony protuberances. Also known as bedsore; pressure ulcer.

decumbent [BOT] Lying down on the ground but with an ascending tip, specifically referring to a stem.

decussate [BOT] Of the arrangement of leaves, occurring in alternating pairs at right angles. [SCI TECH] To intersect in the form of an X.

decyl [ORG CHEM] An isomeric grouping of univalent radicals, all with formulas $C_{10}H_{21}$, and derived from the decanes by removing one hydrogen.

dedicated line [COMPUT SCI] A permanent communications link that is used solely to transmit information between a computer and a data-processing system.

dedicated terminal [COMPUT SCI] A computer terminal that is permanently connected to a data-processing system by a communications link that is used only to transmit information between the two.

deduction [MATH] The process of deriving a statement from certain assumed statements by applying the rules of logic.

dee [NUCLEO] A hollow accelerating-cyclotron electrode in the shape of the letter D. Also known as cyclotron D; D.

deep-draw [MET] To form shapes with large depth-diameter ratios in sheet or strip metal by considerable plastic distortion in dies.

deep easterlies See equatorial easterlies.

deep-etch [GRAPHICS] The etching of an offset printing plate so that the printing area becomes slightly recessed, thereby leading to sharper definition and longer life on press. [MET] Severe etching of a metal surface to reveal gross features, such as abnormal grain size, segregation, or cracks, at magnifications of 10 diameters or less. Also known as macroetching.

deep hibernation [PHYSIO] Profound decrease in metabolic rate and physiological function during winter, with a body temperature near 0°C, in certain warm-blooded vertebrates. Also known as hibernation.

deep inelastic transfer See quasi-fission.

deep scattering layer [OCEANOGR] The stratified populations of organisms which scatter sound in most oceanic waters.

deep-sea channel [GEOL] A trough-shaped valley of low relief beyond the continental rise on the deep-sea floor. Also known as mid-ocean canyon.

deep-seated See plutonic.

deep-sea trench [GEOL] A long, narrow depression of the deep-sea floor having steep sides and containing the greatest ocean depths; formed by depression, to several kilometers' depth, of the high-velocity crustal layer and the mantle.

deep sleep [PSYCH] The third and fourth stage of the sleep cycle, determined by electroencephalographic recording and characterized by slow brain waves. Also known as slow wave sleep (SWS).

deep space [ASTRON] Space beyond the gravitational influence of the earth.

Deep Space Network [AERO ENG] A spacecraft network operated by NASA which tracks, commands, and receives telemetry for all types of spacecraft sent to explore deep space, the moon, and solar system planets. Abbreviated DSN.

deep-submergence rescue vehicle [NAV ARCH] A small rescue and research submarine; intended primarily to rescue the crew of another submarine to depths of 5000 feet (1500 meters). Abbreviated DSRV.

deep trades See equatorial easterlies.

deep water [OCEANOGR] An ocean area where depth of the water layer is greater than one-half the wave length.

deep-well injection [ENG] Storage of liquid wastes, particularly chlorohydrocarbons, by injection into subsurface geologic strata for long-term isolation from the environment.

deer [VERT ZOO] The common name for 41 species of even-toed ungulates that compose the family Cervidae in the order Artiodactyla; males have antlers.

defect cluster [CRYSTAL] A macroscopic cluster of crystal defects which can arise from attraction among defects.

defect conduction [SOLID STATE] Electric conduction in a semiconductor by holes in the valence band.

defective number See deficient number.

defective track [COMPUT SCI] Any circular path on the surface of a magnetic disk which is detected by the system as unable to accept one or more bits of data.

defective virus [VIROL] A virus, such as adeno-associated satellite virus, that can grow and reproduce only in the presence of another virus.

defeminization [PHYSIO] Loss or reduction of feminine attributes, usually caused by ovarian dysfunction or removal. [PSYCH] Psychic process involving a deep and permanent change in the character of a woman, resulting in a giving up of feminine feelings, and the assumption of masculine qualities.

defense mechanism [PSYCH] Any psychic device, such as rationalization, denial, or repression, for concealing unacceptable feelings or for protecting oneself against unpleasant feelings, memories, or experiences.

deferred addressing [COMPUT SCI] A type of indirect addressing in which the address part of an instruction specifies a location containing an address, the latter in turn specifies another location containing an address, and so forth, the number of iterations being controlled by a preset counter.

deferred entry [COMPUT SCI] The passing of control of the central processing unit to a subroutine or to an entry point as the result of an asynchronous event.

deferred processing [COMPUT SCI] The making of computer runs which are postponed until nonpeak periods.

defervescence See lysis.

defibrillation [MED] Stopping a local quivering of muscle fibers, especially of the heart.

deficient number [MATH] A positive integer the sum of whose divisors, including 1 but excluding itself, is less than itself. Also known as defective number.

definite composition law [CHEM] The law that a given chemical compound always contains the same elements in the same fixed proportions by weight. Also known as definite proportions law.

definite proportions law *See* definite composition law.

definite Riemann integral [MATH] A number associated with a function defined on an interval $[a,b]$ which is

$$\lim_{N\to\infty} \sum_{k=0}^{N-1} f\left(a + \frac{k}{N}\right) \cdot \frac{b-a}{N},$$

if f is bounded and continuous; denoted by

$$\int_a^b f(x)dx;$$

if f is a positive function, the definite integral measures the area between the graph of f and the x axis.

definition [COMMUN] The fidelity with which a television or facsimile receiver forms an image. [ELECTR] The extent to which the fine-line details of a printed circuit correspond to the master drawing. [OPTICS] Lens image clarity or discernible detail.

definitive host [BIOL] The host in which a parasite reproduces sexually.

deflagration [CHEM] A chemical reaction accompanied by vigorous evolution of heat, flame, sparks, or spattering of burning particles.

deflation [GEOL] The sweeping erosive action of the wind over the ground.

deflection [ELECTR] The displacement of an electron beam from its straight-line path by an electrostatic or electromagnetic field. [ENG] **1.** Shape change or reduction in diameter of a conduit, produced without fracturing the material. **2.** Elastic movement or sinking of a loaded structural member, particularly of the mid-span of a beam. [ORD] **1.** Horizontal clockwise angle between the axis of the bore and the line of sighting. **2.** The setting on the scale to compensate for deflection.

deflection circuit [ELECTR] A circuit which controls the deflection of an electron beam in a cathode-ray tube.

deflection electrode [ELECTR] An electrode whose potential provides an electric field that deflects an electron beam. Also known as deflection plate.

deflection meter [ENG] A flowmeter that applies the differential pressure generated by a differential-producing primary device across a diaphragm or bellows in such a way as to create a deflection proportional to the differential pressure.

deflection-modulated indicator *See* amplitude-modulated indicator.

deflection plate *See* deflection electrode.

deflection polarity [ELECTR] Relationship between the direction of a displacement of the cathode beam and the polarity of the applied signal wave.

deflection voltage [ELECTR] The voltage applied between a pair of deflection electrodes to produce an electric field.

deflection yoke [ELECTR] An assembly of one or more electromagnets that is placed around the neck of an electron-beam tube to produce a magnetic field for deflection of one or more electron beams. Also known as scanning yoke; yoke.

deflector [ENG] A plate, baffle, or the like that diverts the flow of a forward-moving stream.

deflexed [BIOL] Turned sharply downward.

deflocculant [CHEM] An agent that causes deflocculation; examples are sodium carbonate and other basic materials used to deflocculate clay slips.

deflocculate [CHEM ENG] To break up and disperse agglomerates and form a stable colloid.

defocus [ENG] To make a beam of x-rays, electrons, light, or other radiation deviate from an accurate focus at the intended viewing or working surface.

defoliant [MATER] A chemical sprayed on plants that causes leaves to fall off prematurely.

deforestation [FOR] The act or process of removing trees from or clearing a forest.

deformation [MATH] A homotopy of the identity map to some other map. [MECH] Any alteration of shape or dimensions of a body caused by stresses, thermal expansion or contraction, chemical or metallurgical transformations, or shrinkage and expansions due to moisture change.

deformation bands *See* Lüder's lines.

deformation curve [MECH] A curve showing the relationship between the stress or load on a structure, structural member, or a specimen and the strain or deformation that results. Also known as stress-strain curve.

deformation ellipsoid *See* strain ellipsoid.

deformation energy [NUC PHYS] The energy which must be supplied to an initially spherical nucleus to give it a certain deformation in the Bohr-Wheeler theory.

deformation fabric [GEOL] The space orientation of rock elements produced by external stress on the rock.

deformation lamella [GEOL] A type of slipband in the crystalline grains of a material (particularly quartz) produced by intracrystalline slip during tectonic deformation.

deformation potential [SOLID STATE] The effective electric potential experienced by free electrons in a semiconductor or metal resulting from a local deformation in the crystal lattice.

degas [ELECTR] To drive out and exhaust the gases occluded in the internal parts of an electron tube or other gastight apparatus, generally by heating during evacuation. [ENG] To remove gas from a liquid or solid.

degasser *See* getter.

degassing *See* breathing.

degauss [ELECTR] To remove, erase, or clear information from a magnetic tape, disk, drum, or core. [ELECTROMAG] To neutralize (demagnetize) a magnetic field of, for example, a ship hull or television tube; a direct current of the correct value is sent through a cable around the ship hull; a current-carrying coil is brought up to and then removed from the television tube. Also known as deperm.

degeneracy [MATH] The condition in which two characteristic functions of an operator have the same characteristic value. [PHYS] The condition in which two or more modes of a vibrating system have the same frequency; a special case of the mathematics definition. [QUANT MECH] The condition in which two or more stationary states of the same system have the same energy even though their wave functions are not the same; a special case of the mathematics definition.

degenerate amplifier [ELECTR] Parametric amplifier with a pump frequency exactly twice the signal frequency, producing an idler frequency equal to that of the signal input; it is considered as a single-frequency device.

degenerate code [GEN] A genetic code in which more than one triplet sequence of nucleotides (codon) can specify the insertion of the same amino acid into a polypeptide chain.

degenerate conduction band [SOLID STATE] A band in which two or more orthogonal quantum states exist that have the same energy, the same spin, and zero mean velocity.

degenerate semiconductor [SOLID STATE] A semiconductor in which the number of electrons in the conduction band approaches that of a metal.

degeneration [ELECTR] The loss or gain in an amplifier through unintentional negative feedback. [MED] 1. Deterioration of cellular integrity with no sign of response to injury or disease. 2. General deterioration of a physical, mental, or moral state. [STAT MECH] A phenomenon which occurs in gases at very low temperatures when the molecular heat drops to less than ¾ the gas constant.

degenerative arthritis *See* degenerative joint disease.

degenerative joint disease [MED] A chronic joint disease characterized pathologically by degeneration of articular cartilage and hypertrophy of bone, clinically by pain on activity which subsides with rest. Also known as degenerative arthritis; hypertrophic arthritis; osteoarthritis; senescent arthritis.

Degeneriaceae [BOT] A family of dicotyledonous plants in the order Magnoliales characterized by laminar stamens; a solitary, pluriovulate, unsealed carpel; and ruminate endosperm.

deglitcher [ELECTR] A nonlinear filter or other special circuit used to limit the duration of switching transients in digital converters.

deglutition [PHYSIO] Act of swallowing.

degradation [COMPUT SCI] Condition under which a computer operates when some area of memory or some units of peripheral equipment are not available to the user. [GEOL] The wearing down of the land surface by processes of erosion and weathering. [HYD] 1. Lowering of a steam bed. 2. Shrinkage or disappearance of permafrost. [ORG CHEM] Conversion of an organic compound to one containing a smaller number of carbon atoms. [PHYS] Loss of energy of a particle, such as a neutron or photon, through a collision. [THERMO] The conversion of energy into forms that are increasingly difficult to convert into work, resulting from the general tendency of entropy to increase.

degree [CHEM] Any one of several units for measuring hardness of water, such as the English or Clark degree, the French degree, and the German degree. [FL MECH] One of the units in any of various scales of specific gravity, such as the Baumé scale. [MATH] 1. A unit for measurement of plane angles, equal to 1/360 of a complete revolution, or 1/90 of a right angle. Symbolized °. 2. For a vertex in a graph, the number of arcs which have that vertex as an end point. [THERMO] One of the units of temperature or temperature difference in any of various temperature scales, such as the Celsius, Fahrenheit, and Kelvin temperature scales (the Kelvin degree is now known as the kelvin).

degree-day [MECH ENG] A measure of the departure of the mean daily temperature from a given standard; one degree-day is recorded for each degree of departure above (or below) the standard during a single day; used to estimate energy requirements for building heating and, to a lesser extent, for cooling.

degree of curve [CIV ENG] A measure of the curvature of a railway or highway, equal to the angle subtended by a 100-foot (30-meter) chord (railway) or by a 100-foot arc (highway).

degree of degeneracy [MATH] The number of characteristic functions of an operator having the same characteristic value. Also known as order of degeneracy.

degree of freedom [MECH] Of a gyro, the number of orthogonal axes about which the spin axis is free to rotate, the spin axis freedom not being counted; this is not a universal convention; for example, the free gyro is frequently referred to as a three-degree-of-freedom gyro, the spin axis being counted. [MECH ENG] Any one of the number of ways in which the space configuration of a mechanical system may change. [PHYS CHEM] Any one of the variables, including pressure, temperature, composition, and specific volume, which must be specified to define the state of a system. [STAT] A number one less than the number of frequencies being tested with a chi-square test.

de Haas–Van Alphen effect [SOLID STATE] An effect occurring in many complex metals at low temperatures, consisting of a periodic variation in the diamagnetic susceptibility of conduction electrons with changes in the component of the applied magnetic field at right angles to the principal axis of the crystal.

dehiscence [BOT] Spontaneous bursting open of a mature plant structure, such as fruit, anther, or sporangium, to discharge its contents. [MED] A defect in the boundary of a bony canal or cavity.

dehrnite [MINERAL] $(Ca,Na,K)_5(PO_4)_3(OH)$ A colorless to pale green, greenish-white, or gray, hexagonal mineral consisting of a basic phosphate of calcium, sodium, and potassium; occurs as botryoidal crusts and minute hexagonal prisms.

dehumidifier [MECH ENG] Equipment designed to reduce the amount of water vapor in the ambient atmosphere.

dehydrase [BIOCHEM] An enzyme which catalyzes the removal of water from a substrate.

dehydrator [CHEM] A substance that removes water from a material; an example is sulfuric acid. [CHEM ENG] Vessel or process system for the removal of liquids from gases or solids by the use of heat, absorbents, or adsorbents.

dehydrogenase [BIOCHEM] An enzyme which removes hydrogen atoms from a substrate and transfers it to an acceptor other than oxygen.

deicing [ENG] The removal of ice deposited on any object, especially as applied to aircraft icing, by heating, chemical treatment, and mechanical rupture of the ice deposit.

Deimos [ASTRON] A satellite of Mars orbiting at a mean distance of 23,500 kilometers.

Deinotheriidae [PALEON] A family of extinct proboscidean mammals in the suborder Deinotherioidea; known only by the genus *Deinotherium*.

Deinotherioidea [PALEON] A monofamilial suborder of extinct mammals in the order Proboscidea.

deionization [CHEM] An ion-exchange process in which all charged species or ionizable organic and inorganic salts are removed from solution. [ELECTR] The return of an ionized gas to its neutral state after all sources of ionization have been removed, involving diffusion of ions to the container walls and volume recombination of negative and positive ions.

Delaborne prism [OPTICS] A special compound prism which, when rotated about an axis parallel to the reflecting face and lying in a plane perpendicular to the refracting faces, rotates the image through twice the angle. Also known as Dove prism.

delay [COMMUN] 1. Time required for a signal to pass through a device or a conducting medium. 2. Time which elapses between the instant at which any designated point of a transmitted wave passes any two designated points of a transmission circuit; such delay is primarily determined by the constants of the circuit. [IND ENG] Interruption of the normal tempo of an operation; may be avoidable or unavoidable.

delay circuit See time-delay circuit.

delay counter [COMPUT SCI] A counter which inserts a time delay in a sequence of events.

delay distortion [ELECTR] Phase distortion in which the rate of change of phase shift with frequency of a circuit or system is not constant over the frequency range required for transmission. Also called envelope delay distortion.

delayed automatic gain control [ELECTR] An automatic gain control system that does not operate until the signal exceeds a predetermined magnitude; weaker signals thus receive maximum amplification. Also known as biased automatic gain control; delayed automatic volume control; quiet automatic volume control.

delayed automatic volume control See delayed automatic gain control.

delayed coincidence [NUCLEO] Occurrence of a count in one detector at a short but measurable time later than a count in another detector, the two counts being due to successive events in the same nucleus.

delayed hypersensitivity [IMMUNOL] Abnormal reactivity in a sensitized individual beginning several hours after contact with the allergen.

delayed neutron [NUC PHYS] A neutron emitted spontaneously from a nucleus as a consequence of excitation left from a preceding radioactive decay event; in particular, a delayed fission neutron.

delayed proton [NUC PHYS] A proton emitted spontaneously from a nucleus as a consequence of excitation left from a previous radioactive decay event.

delayed speech [MED] A speech disorder characterized by a complete absence of vocalization or vocalization with no communicative value; speech is considered delayed when it fails to develop by the second year, caused by impaired hearing, severe childhood illness, or emotional disturbance.

delay equalizer [ELECTR] A corrective network used to make the phase delay or envelope delay of a circuit or system substantially constant over a desired frequency range.

delay line [ELECTR] A transmission line (as dissipationless as possible), or an electric network approximation of it, which, if terminated in its characteristic impedance, will reproduce at its output a waveform applied to its input terminals with little distortion, but at a time delayed by an amount dependent upon the electrical length of the line. Also known as artificial delay line.

delay-line memory See circulating memory.

delay-line storage See circulating memory.

delay multivibrator [ELECTR] A monostable multivibrator that generates an output pulse a predetermined time after it is triggered by an input pulse.

delay relay [ELEC] A relay having predetermined delay between energization and closing of contacts or between deenergization and dropout.

delay time [CONT SYS] The amount of time by which the arrival of a signal is retarded after transmission

through physical equipment or systems. [ELECTR] The time taken for collector current to start flowing in a transistor that is being turned on from the cutoff condition.

delay unit [COMPUT SCI] See transport delay unit. [ELECTR] Unit of a radar system in which pulses may be delayed a controllable amount.

deleted representation [COMPUT SCI] In paper tape codes, the superposition of a pattern of holes upon another pattern of holes representing a character, to effectively remove or obliterate the latter.

deletion record [COMPUT SCI] A record which removes and replaces an existing record when it is added to a file.

deliquescence [BOT] The condition of repeated divisions ending in fine divisions; seen especially in venation and stem branching. [PHYS CHEM] The absorption of atmospheric water vapor by a crystalline solid until the crystal eventually dissolves into a saturated solution.

delirium [MED] Severely disordered mental state associated with fever, intoxication, head trauma, and other encephalopathies.

delirium tremens [MED] Delirium associated with tremors, insomnia, and other physical and neurological symptoms frequently following chronic alcoholism.

dellenite See rhyodacite.

Delmontian [GEOL] Upper Miocene or lower Pliocene geologic time.

del operator [MATH] The rule which replaces the function f of three variables, x, y, z, by the vector valued function whose components in the x, y, z directions are the respective partial derivatives of f. Written ∇f. Also known as nabla.

Delphinidae [VERT ZOO] A family of aquatic mammals in the order Cetacea; includes the dolphins.

Delphinus [ASTRON] A northern constellation, right ascension, 21 hours, declination, 10° north. Also known as Dolphin. [VERT ZOO] A genus of cetacean mammals, including the dolphin.

delta [ELECTR] The difference between a partial-select output of a magnetic cell in a one state and a partial-select output of the same cell in a zero state. [GEOL] An alluvial deposit, usually triangular in shape, at the mouth of a river, stream, or tidal inlet.

delta baryon [PARTIC PHYS] 1. Any excited baryon state belonging to a multiplet having a total isospin of 3/2, a hypercharge of +1, a spin of 3/2, positive parity, and an approximate mass of 1236 MeV. Designated $\Delta(1236)$. 2. Any excited baryon state belonging to any multiplet having a total isospin of 3/2 and a hypercharge of +1.

Delta Cephei [ASTRON] A cepheid variable, from which the name of this type of star is derived; it has a period of 5.3 days.

delta connection [ELEC] A combination of three components connected in series to form a triangle like the Greek letter delta. Also known as mesh connection.

delta ferrite See delta iron.

delta function [MATH] A distribution δ such that

$$\int_{-\infty}^{\infty} f(t)\delta(x-t)dt \text{ is } f(x).$$

Also known as Dirac delta function.

delta geosyncline See exogeosyncline.

delta-gun tube [ELECTR] A color television picture tube in which three electron guns, arranged in a triangle, provide electron beams that fall on phosphor dots on the screen, causing them to emit light in three primary colors; a shadow mask located just behind the screen ensures that each beam excites only dots of one color.

delta iron [MET] The nonmagnetic polymorphic form of iron stable between about 1403°C and the melting point, about 1535°C. Also known as delta ferrite.

delta meson [PARTIC PHYS] Any scalar meson resonance, with positive charge conjugation parity, belonging to a multiplet with a total isospin of 1, a hypercharge of zero, a mass of 962 ± 5 MeV, and a width <5 MeV. Designated δ(962).

delta modulation [ELECTR] A pulse-modulation technique in which a continuous signal is converted into a binary pulse pattern, for transmission through low-quality channels.

delta network [ELEC] A set of three branches connected in series to form a mesh.

delta ray [ATOM PHYS] An electron or proton ejected by recoil when a rapidly moving alpha particle or other primary ionizing particle passes through matter.

delta rhythm [PHYSIO] An electric current generated in slow waves with frequencies of 0.5–3 per second from the forward portion of the brain of normal subjects when asleep.

Deltatheridia [PALEON] An order of mammals that includes the dominant carnivores of the early Cenozoic.

delta-Y transformation See Y-delta transformation.

deltoid [ANAT] The large triangular shoulder muscle; originates on the pectoral girdle and inserts on the humerus. [BIOL] Triangular in shape.

deltoid ligament [ANAT] The ligament on the medial side of the ankle joint; the fibers radiate from the medial malleolus to the talus, calcaneus, and navicular bones.

delusion [PSYCH] A conviction based on faulty perceptions, feelings, and thinking.

demagnetization [ELECTROMAG] **1.** The process of reducing or removing the magnetism of a ferromagnetic material. **2.** The reduction of magnetic induction by the internal field of a magnet. [MIN ENG] Deflocculation in dense-media process using ferrosilicon by passing the fluid through an alternating-current field.

demand See demand factor.

demand factor [ELEC] The ratio of the maximum demand of a building for electric power to the total connected load. Also known as demand.

demand limiter See current limiter.

demand meter [ENG] Any of several types of instruments used to determine a customer's maximum demand for electric power over an appreciable time interval; generally used for billing industrial users.

demand paging [COMPUT SCI] The characteristic of a virtual memory system which retrieves only that part of a user's program which is required during execution.

demand processing [COMPUT SCI] The processing of data by a computer system as soon as it is received, so that it is not necessary to store large amounts of raw data. Also known as immediate processing.

demand reading [COMPUT SCI] A method of carrying out input operations in which blocks of data are transmitted to the central processing unit as needed for processing.

demand staging [COMPUT SCI] Moving blocks of data from one storage device to another when programs request them.

demand writing [COMPUT SCI] A method of carrying out output operations in which blocks of data are transmitted from the central processing unit as they are needed by the user.

Dematiaceae [MYCOL] A family of fungi in the order Moniliales; sporophores are not grouped, hyphae are always dark, and the spores are hyaline or dark.

Dembowska [ASTRON] The only moderately large asteroid whose surface composition resembles that of ordinary chondritic meteorites; has a diameter of about 140 kilometers and a mean distance from the sun of 2.924 astronomical units.

deme [ECOL] A local population in which the individuals freely interbreed among themselves but not with those of other demes.

dementia [PSYCH] Deterioration of intellectual and other mental processes due to organic brain disease.

dementia praecox See schizophrenia.

demethylation [ORG CHEM] Removal of the methyl group from a compound.

demineralization [CHEM ENG] Removal of mineral constituents from water. [MED] Removal or loss of minerals and salts from the body, especially by disease.

Demodicidae [INV ZOO] The pore mites, a family of arachnids in the suborder Trombidiformes.

demodifier [COMPUT SCI] A data element used to restore part of an instruction which has been modified to its original value.

demodulate [COMMUN] To recover the modulating wave from a modulated carrier. Also known as decode; detect.

demodulator See detector.

demography [ECOL] The statistical study of populations with reference to natality, mortality, migratory movements, age, and sex, among other social, ethnic, and economic factors.

De Moivre's theorem [MATH] The nth power of the quantity $\cos \Theta + i \sin \Theta$ is $\cos n\Theta + i \sin n\Theta$ for any integer n.

demolition [CIV ENG] The act or process of tearing down a building or other structure. [ORD] Destroying a structure or an area by the use of explosives.

demon of Maxwell [THERMO] Hypothetical creature who controls a trapdoor over a microscopic hole in an adiabatic wall between two vessels filled with gas at the same temperature, so as to supposedly decrease the entropy of the gas as a whole and thus violate the second law of thermodynamics. Also known as Maxwell's demon.

De Morgan's rules [MATH] The complement of the union of two sets equals the intersection of their respective complements; the complement of the intersection of two sets equals the union of their complements.

De Morgan's test [MATH] A series with term u_n, for which $|u_{n+1}/u_n|$ converges to 1, will converge absolutely if there is $c > 0$ such that the limit superior of $n(|u_{n+1}/u_n| - 1)$ equals $-1 - c$.

demorphism See weathering.

Demospongiae [INV ZOO] A class of the phylum Porifera, including sponges with a skeleton of one- to four-rayed siliceous spicules, or of spongin fibers, or both.

demulsification [CHEM ENG] Prevention or breaking of liquid-liquid emulsions by chemical, mechanical or electrical demulsifiers.

demultiplexer [ELECTR] A device used to separate two or more signals that were previously combined by a compatible multiplexer and transmitted over a single channel.

demyelination [PATH] Destruction of myelin; loss of myelin from nerve sheaths or nerve tracts.

denaturant [CHEM] An inert, bad-tasting, or poisonous chemical substance added to a product such as ethyl alcohol to make it unfit for human consumption. [NUCLEO] A nonfissionable isotope that can be added to fissionable material to make it unsuitable for use in atomic weapons without extensive processing.

denature [CHEM] **1.** To change a protein by heating it or treating it with alkali or acid so that the original properties such as solubility are changed as a result of the protein's molecular structure being changed in some way. **2.** To add a denaturant, such as methyl alcohol, to grain alcohol to make the grain alcohol poisonous and unfit for human consumption.

denatured alcohol [CHEM] Ethyl alcohol containing a poisonous substance, such as methyl alcohol or benzene, which makes it unfit for human consumption.

dendrite [ANAT] The part of a neuron that carries the unidirectional nerve impulse toward the cell body. Also known as dendron. [CRYSTAL] A crystal having a treelike structure.

dendritic drainage [HYD] Irregular stream branching, with tributaries joining the main stream at all angles.

Dendrobatinae [VERT ZOO] A subfamily of anuran amphibians in the family Ranidae, including the colorful poisonous frogs of Central and South America.

dendrobranchiate gill [INV ZOO] A respiratory structure of certain decapod crustaceans, characterized by extensive branching of the two primary series.

Dendroceratida [INV ZOO] A small order of sponges of the class Demospongiae; members have a skeleton of spongin fibers or lack a skeleton.

Dendrochirotacea [INV ZOO] A subclass of echinoderms in the class Holothuroidea.

Dendrochirotida [INV ZOO] An order of dendrochirotacean holothurian echinoderms with 10–30 richly branched tentacles.

dendrochronology [GEOL] The science of measuring time intervals and dating events and environmental changes by reading and dating growth layers of trees as demarcated by the annual rings.

Dendrocolaptidae [VERT ZOO] The woodcreepers, a family of passeriform birds belonging to the suboscine group.

dendrogram [BIOL] A genealogical tree; the trunk represents the oldest common ancestor, and the branches indicate successively more recent divisions of a lineage for a group.

Dendroidea [PALEON] An order of extinct sessile, branched colonial animals in the class Graptolithina occurring among typical benthonic fauna.

dendrology [FOR] The division of forestry concerned with the classification, identification, and distribution of trees and other woody plants.

Dendromurinae [VERT ZOO] The African tree mice and related species, a subfamily of rodents in the family Muridae.

dendron See dendrite.

Deneb [ASTRON] A white star of spectral classification A2-Ia in the constellation Cygnus; the star α Cygni.

Denebola [ASTRON] A white star of stellar magnitude 2.2, spectral classification A2, in the constellation Leo; the star β Leonis.

denitration [CHEM] Removal of nitrates or nitrogen. Also known as denitrification.

denitrification [CHEM] See denitration. [MICROBIO] The reduction of nitrate or nitrite to gaseous products such as nitrogen, nitrous oxide, and nitric oxide; brought about by denitrifying bacteria.

denominator [MATH] In a fraction, the term that divides the other term (called the numerator), and is written below the line.

dense-air refrigeration cycle See reverse Brayton cycle.

dense-air system See cold-air machine.

dense binary code [COMPUT SCI] A code in which all possible states of the binary pattern are used.

dense connective tissue [HISTOL] A fibrous connective tissue with an abundance of enlarged collagenous fibers which tend to crowd out the cells and ground substance.

dense subset [MATH] A subset of a topological space whose closure is the entire space.

densimeter [ENG] An instrument which measures the density or specific gravity of a liquid, gas, or solid. Also known as densitometer; density gage; density indicator; gravitometer.

densitometer [ENG] **1.** An instrument which measures optical density by measuring the intensity of transmitted or reflected light; used to measure photographic density. **2.** See densimeter.

density [MATER] Closeness of texture or consistency. [MECH] The mass of a given substance per unit volume. [OPTICS] **1.** The degree of opacity of a translucent material. **2.** The common logarithm of opacity. [PHYS] The total amount of a quantity, such as energy, per unit of space.

density correction [AERO ENG] A correction made necessary because the airspeed indicator is calibrated only for standard air pressure; it is applied to equivalent airspeed to obtain true airspeed, or to calibrated airspeed to obtain density airspeed. [ENG] **1.** The part of the temperature correction of a mercury barometer which is necessitated by the variation of the density of mercury with temperature. **2.** The correction, applied to the indications of a pressure-tube anemometer or pressure-plate anemometer, which is necessitated by the variation of air density with temperature.

density current [METEOROL] Intrusion of a dense air mass beneath a lighter air mass; the usage applies to cold fronts. [OCEANOGR] See turbidity current.

density function [MATH] A density function for a measure m is a function which gives rise to m when it is integrated with respect to some other specified measure. [STAT] See probability density function.

density gage See densimeter.

density gradient centrifugation [ANALY CHEM] Separation of particles according to density by employing a gradient of varying densities; at equilibrium each particle settles in the gradient at a point equal to its density.

density indicator See densimeter.

density log [PETRO ENG] Radioactivity logging of reservoir structure densities down an oil-well bore by emission and detection of gamma rays.

density matrix [QUANT MECH] A matrix ρ_{mn} describing an ensemble of quantum-mechanical systems in a representation based on an orthonormal set of functions ϕ_n; for any operator G with representation G_{mn}, the ensemble average of the expectation value of G is the trace of ρG.

density modulation [ELECTR] Modulation of an electron beam by making the density of the electrons in the beam vary with time.

density of states [SOLID STATE] A function of energy E equal to the number of quantum states in the energy range between E and $E + dE$ divided by the product of dE and the volume of the substance.

density packing [COMPUT SCI] In computers, the number of binary digit magnetic pulses stored on tape or drum per linear inch on a single track by a single head.

density-wave theory [ASTROPHYS] A theory explaining the spiral structure of galaxies by a periodic variation in space in the density of matter which rotates with a fixed angular velocity while the angular velocity of the matter itself varies with distance from the galaxy's center.

dental [ANAT] Pertaining to the teeth.

dental calculi [MED] Calcareous deposits of organic and mineral matter on the teeth. Also known as tartar.

dental caries See caries.

dental coupling [MECH ENG] A type of flexible coupling used to join a steam turbine to a reduction-gear pinion shaft; consists of a short piece of shaft with gear teeth at each end, and mates with internal gears in a flange at the ends of the two shafts to be joined.

dental formula [VERT ZOO] An expression of the number and kind of teeth in each half jaw, both upper and lower, of mammals.

Dentaliidae [INV ZOO] A family of mollusks in the class Scaphopoda; members have pointed feet.

dental pad [VERT ZOO] A firm ridge that replaces incisors in the maxilla of cud-chewing herbivores.

dental papilla [EMBRYO] The mass of connective tissue located inside the enamel organ of a developing tooth, and forming the dentin and dental pulp of the tooth.

dental plate [INV ZOO] A flat plate that replaces teeth in certain invertebrates, such as some worms. [VERT ZOO] A flattened plate that represents fused teeth in parrot fishes and related forms.

dental pulp [HISTOL] The vascular connective tissue of the roots and pulp cavity of a tooth.

dental ridge [EMBRYO] An elevation of the embryonic jaw that forms a cusp or margin of a tooth.

dentate [BIOL] 1. Having teeth. 2. Having toothlike or conical marginal projections.

denticle [ZOO] A small tooth or toothlike projection, as the type of scale of certain elasmobranchs.

dentil [ARCH] One of a series of small rectangular blocks under a cornice.

dentil band [ARCH] A molding or band on a cornice resembling a row of dentils, but with the spaces between dentils filled.

dentin [HISTOL] A bonelike tissue composing the bulk of a vertebrate tooth; consists of 70% inorganic materials and 30% water and organic matter.

dentistry [MED] A branch of medical science concerned with the prevention, diagnosis, and treatment of diseases of the teeth and adjacent tissues and the restoration of missing dental structures.

dentition [VERT ZOO] The arrangement, type, and number of teeth which are variously located in the oral or in the pharyngeal cavities, or in both, in vertebrates.

denudation [GEOL] General wearing away of the land; laying bare of subjacent lands.

denumerable set [MATH] A set which may be put in one-to-one correspondence with the positive integers. Also known as countably infinite set.

deodorizing [CHEM ENG] A process for removing odor-creating substances from oil or fat, in which the oil or fat is held at high temperatures and low pressure while steam is blown through.

deoxidize [CHEM] 1. To remove oxygen by any of several processes. 2. To reduce from the state of an oxide. [MET] To remove an oxide film from a metal surface.

deoxygenation [CHEM] Removal of oxygen from a substance, such as blood or polluted water.

deoxyribonucleic acid [BIOCHEM] A linear polymer made up of deoxyribonucleotide repeating units (composed of the sugar 2-deoxyribose, phosphate, and a purine or pyrimidine base) linked by the phosphate group joining the 3′ position of one sugar to the 5′ position of the next; most molecules are double-stranded and antiparallel, resulting in a right-handed helix structure kept together by hydrogen bonds between a purine on one chain and a pyrimidine on another; carrier of genetic information, which is encoded in the sequence of bases; present in chromosomes and chromosomal material of cell organelles such as mitochondria and chloroplasts, and also present in some viruses. Abbreviated DNA.

deoxyribonucleoprotein [BIOCHEM] A protein containing molecules of deoxyribonucleic acid in close association with protein molecules.

deoxyribose [BIOCHEM] $C_5H_{10}O_4$ A pentose sugar in which the hydrogen replaces the hydroxyl groups of ribose; a major constituent of deoxyribonucleic acid.

deoxy sugar [BIOCHEM] A substance which has the characteristics of a sugar, but which shows a deviation from the required hydrogen-to-oxygen ratio.

dependence [STAT] The existence of a relationship between frequencies obtained from two parts of an experiment which does not arise from the direct influence of the result of the first part on the chances of the second part but indirectly from the fact that both parts are subject to influences from a common outside factor.

dependent variable [MATH] If y is a function of x, that is, if the function assigns a single value of y to each value of x, then y is the dependent variable.

deperm See degauss.

Depertellidae [PALEON] A family of extinct perissodactyl mammals in the superfamily Tapiroidea.

dephlegmation [CHEM ENG] In a distillation operation, the partial condensation of vapor to form a liquid richer in higher boiling constituents than the original vapor.

depleted material [NUCLEO] Material in which the amount of one or more isotopes of a constituent has been reduced by an isotope separation process or by a nuclear reaction.

depleted uranium [NUCLEO] Uranium having a smaller percentage of uranium-235 than the 0.7% found in natural uranium.

depletion [ECOL] Using a resource, such as water or timber, faster than it is replenished. [ELECTR] Reduction of the charge-carrier density in a semiconductor below the normal value for a given temperature and doping level. [NUCLEO] The percentage reduction in the quantity of fissionable atoms in the fuel assemblies or fuel mixture that occurs during operation of a nuclear reactor.

depletion layer [ELECTR] An electric double layer formed at the surface of contact between a metal and a semiconductor having different work functions, because the mobile carrier charge density is insufficient to neutralize the fixed charge density of donors and acceptors. Also known as barrier layer (deprecated); blocking layer (deprecated); space-charge layer.

depletion-mode field-effect transistor See junction-gate field-effect transistor.

depletion region [ELECTR] The portion of the channel in a metal oxide field-effect transistor in which there are no charge carriers.

depletion-type reservoir [PETRO ENG] Oil reservoir which is initially in (and during depletion remains in) a state of equilibrium between the gas and liquid phases; includes single-phase gas, two-phase bubble-point, and retrograde-gas-condensate (or dewpoint) reservoirs.

depolarization [ELEC] The removal or prevention of polarization in a substance (for example, through the use of a depolarizer in an electric cell) or of polarization arising from the field due to the charges induced on the surface of a dielectric when an external field is applied. [OPTICS] The resolution of polarized light in an optical depolarizer.

depolarizer [PHYS CHEM] A substance added to the electrolyte of a primary cell to prevent excessive buildup of hydrogen bubbles by combining chemically with the hydrogen gas as it forms. Also known as battery depolarizer.

depolymerization [ORG CHEM] Decomposition of macromolecular compounds into relatively simple compounds.

deposit [COMPUT SCI] To preserve the contents of a portion of a computer memory by copying it in a backing storage. [GEOL] Consolidated or unconsolidated material that has accumulated by a natural process or agent. [MATER] Any material applied to a base by means of vacuum, electrical, chemical, screening, or vapor methods. [SCI TECH] Any solid matter which is gradually laid down on a surface by a natural process.

deposit feeder [INV ZOO] Any animal that feeds on the detritus that collects on the substratum at the bottom of water. Also known as detritus feeder.

depositional dip *See* primary dip.

depositional fabric [PETR] Arrangement of detrital particles settled from suspension or of crystals from a differentiating magma determined by the plane of the surface on which they come to rest.

depositional remanent magnetization [GEOPHYS] Remanent magnetization occurring in sedimentary rock following the depositional alignment of previously magnetized grains. Abbreviated DRM.

depositional sequence [GEOL] A major but informal assemblage of formations or groups and supergroups, bounded by regionally extensive unconformities at both their base and top and extending over broad areas of continental cratons.

deposition potential [PHYS CHEM] The smallest potential which can produce electrolytic deposition when applied to an electrolytic cell.

deposition rate [MET] The amount of welding material deposited per unit of time, expressed in pounds per hour.

deprecated usage [SCI TECH] Word usage which is disapproved by experts in the pertinent field because the term in question has misleading connotations; for example, a capacitor has frequently been called a "condenser," but it does not condense anything.

depression [GEOL] 1. A hollow of any size on a plain surface having no natural outlet for surface drainage. 2. A structurally low area in the crust of the earth. [METEOROL] An area of low pressure; usually applied to a certain stage in the development of a tropical cyclone, to migratory lows and troughs, and to upper-level lows and troughs that are only weakly developed. Also known as low. [PSYCH] A mood provoked by conscious awareness of an idea or feeling that was previously pushed into the unconscious.

depression angle *See* angle of depression.

depression spring [HYD] A type of gravity spring that flows onto the land surface because the surface slopes down to the water table.

depressor [ANAT] A muscle that draws a part down. [CHEM ENG] An agent that prevents or retards a chemical reaction or process.

depth charge [ORD] A cylindrical or teardrop-shaped container holding a charge of TNT or other explosive, dropped from the deck of a ship, and detonated at a preset depth as an antisubmarine weapon.

depth contour *See* isobath.

depth curve *See* isobath.

depth gage [DES ENG] An instrument or tool for measuring the depth of depression to a thousandth inch.

depth hoar [HYD] A layer of ice crystals formed between the ground and snow cover by sublimation. Also known as sugar snow.

depth magnification [OPTICS] The ratio of the distance between two nearby points of the axis on the image side of an optical system to the distance between their conjugate points on the object side.

depth marker [ENG] A thin board or other lightweight substance used as a means of identifying the surface of snow or ice which has been covered by a more recent snowfall.

depth of field [OPTICS] The range of distances over which a camera gives satisfactory definition, its lens in the best focus for a certain specific distance.

depth perception [PHYSIO] Ability to judge spatial relationships.

depth sounder [ENG] An instrument for mechanically measuring the depth of the sea beneath a ship.

Derbyshire spar *See* fluorite.

derivation [MATH] 1. The process of deducing a formula. 2. A function D on an algebra which satisfies the equation $D(uv) = uD(v) + vD(u)$.

derivative [CHEM] A substance that is made from another substance. [MATH] The slope of a graph $y = f(x)$ at a given point c; more precisely, it is the limit as h approaches zero of $f(c+h) - f(c)$ divided by h. Also known as differential coefficient; rate of change.

derivative action [CONT SYS] Control action in which the speed at which a correction is made depends on how fast the system error is increasing. Also known as derivative compensation; rate action.

derivative compensation *See* derivative action.

derivative network [CONT SYS] A compensating network whose output is proportional to the sum of the input signal and its derivative. Also known as lead network.

derivative rock *See* sedimentary rock.

derivative thermometric titration [ANALY CHEM] The use of a special resistance-capacitance network to record first and second derivatives of a thermometric titration curve (temperature versus weight change upon heating) to produce a sharp end-point peak.

derived curve [MATH] A curve whose ordinate, for each value of the abscissa, is equal to the slope of some given curve. Also known as first derived curve.

derived quantity [PHYS] A physical quantity which, in a specified system of measurement, is defined by operations based on other physical quantities.

derived unit [PHYS] A unit that is formed, in a specified system of measurement, by combining base units and other derived units according to the algebraic relations linking the corresponding quantities.

dermal [ANAT] Pertaining to the dermis.

dermal bone [ANAT] A type of bone that ossifies directly from membrane without a cartilaginous predecessor; occurs only in the skull and shoulder region. Also known as investing bone; membrane bone.

dermal denticle [VERT ZOO] A toothlike scale composed mostly of dentine with a large central pulp cavity, found in the skin of sharks.

Dermaptera [INV ZOO] An order of small or medium-size, slender insects having incomplete metamorphosis, chewing mouthparts, short forewings, and cerci.

Dermatemydinae [VERT ZOO] A family of reptiles in the order Chelonia; includes the river turtles.

dermatitis [MED] Inflammation of the skin.

Dermatocarpaceae [BOT] A family of lichens in the order Pyrenulales having an umbilicate or squamulose growth form; most members grow on limestone or calcareous soils.

dermatoglyphics [ANAT] 1. The integumentary patterns on the surface of the fingertips, palms, and soles. 2. The study of these patterns.

dermatology [MED] The science of the structure, function, and diseases of the skin.

Dermatophilaceae [MICROBIO] A family of bacteria in the order Actinomycetales; cells produce mycelial filaments or muriform thalli; includes human and mammalian pathogens.

dermatophytosis [MED] A fungus infection, such as ringworm, of the skin of humans caused by the organism living in the keratinized tissues, characterized by vesicles, cracking, and scaling.

dermatosclerosis See scleroderma.

Dermestidae [INV ZOO] The skin beetles, a family of coleopteran insects in the superfamily Dermestoidea, including serious pests of stored agricultural grain products.

Dermestoidea [INV ZOO] A superfamily of coleopteran insects in the suborder Polyphaga.

dermis [ANAT] The deep layer of the skin, a dense connective tissue richly supplied with blood vessels, nerves, and sensory organs. Also known as corium; cutis.

Dermochelidae [VERT ZOO] A family of reptiles in the order Chelonia composed of a single species, the leatherback turtle.

dermoid cyst [MED] A benign cystic teratoma with skin, skin appendages, and their products as the most prominent components, usually involving the ovary or the skin.

Dermoptera [VERT ZOO] The flying lemurs, an ancient order of primatelike herbivorous and frugivorous gliding mammals confined to southeastern Asia and eastern India.

Derodontidae [INV ZOO] The tooth-necked fungus beetles, a small family of coleopteran insects in the superfamily Dermestoidea.

derrick [MECH ENG] A hoisting machine consisting usually of a vertical mast, a slanted boom, and associated tackle; may be operated mechanically or by hand.

derrick post See king post.

derris [BOT] Any of certain tropical shrubs in the genus *Derris* in the legume family (Leguminosae), having long climbing branches.

DES See data encryption standard.

desalination [CHEM ENG] Removal of salt, as from water or soil. Also known as desalting.

desalting [CHEM ENG] 1. The process of extracting inorganic salts from oil. 2. See desalination.

descaling [ENG] Removing scale, usually oxides, from the surface of a metal or the inner surface of a pipe, boiler, or other object.

Descartes laws of refraction See Snell laws of refraction.

Descartes ray [OPTICS] A ray of light incident on a sphere of transparent material, such as a water droplet, which after one internal reflection leaves the drop at the smallest possible angle of deviation from the direction of the incident ray; these rays make the primary rainbow.

Descartes' rule of signs [MATH] A polynomial with real coefficients has at most k real positive roots, where k is the number of sign changes in the polynomial.

Descemet's membrane [HISTOL] A layer of the cornea between the posterior surface of the stroma and the anterior surface of the endothelium which contains collagen arranged on a crystalline lattice.

descending [ANAT] Extending or directed downward or caudally, as the descending aorta. [PHYSIO] In the nervous system, efferent; conducting impulses or progressing down the spinal cord or from central to peripheral.

descending branch [MECH] That portion of a trajectory which is between the summit and the point where the trajectory terminates, either by impact or air burst, and along which the projectile falls, with altitude constantly decreasing. Also known as descent trajectory.

descending chromatography [ANALY CHEM] A type of paper chromatography in which the sample-carrying solvent mixture is fed to the top of the developing chamber, being separated as it works downward.

descending node [AERO ENG] That point at which an earth satellite crosses to the south side of the equatorial plane of its primary. Also known as southbound node. [ASTRON] The point at which a planet, planetoid, or comet crosses the ecliptic from north to south.

descending vertical angle See angle of depression.

descent [AERO ENG] Motion of a craft in which the path is inclined with respect to the horizontal.

descent trajectory See descending branch.

descriptor [COMPUT SCI] A word or phrase used to identify a document in a computer-based information storage and retrieval system.

desensitization [COMMUN] Reduction in receiver sensitivity due to the presence of a high-level off-channel signal overloading the radio-frequency amplifier or mixer stages, or causing automatic gain control action. [IMMUNOL] Loss or reduction of sensitivity to infection or an allergen accomplished by means of frequent, small doses of the antigen. Also known as hyposensitization. [PSYCH] Relief from or removal of a mental complex.

desert [GEOGR] 1. A wide, open, comparatively barren tract of land with few forms of life and little rainfall. 2. Any waste, uninhabited tract, such as the vast expanse of ice in Greenland.

desert crust See desert pavement.

desertification [ECOL] The creation of desiccated, barren, desertlike conditions due to natural changes in climate or possibly through mismanagement of the semiarid zone.

desert pavement [GEOL] A mosaic of pebbles and large stones which accumulate as the finer dust and sand particles are blown away by the wind. Also known as desert crust.

desert wind [METEOROL] A wind blowing off the desert, which is very dry and usually dusty, hot in summer but cold in winter, and with a large diurnal range of temperature.

desiccation [HYD] The permanent decrease or disappearance of water from a region, caused by a decrease of rainfall, a failure to maintain irrigation, or deforestation or overcropping. [SCI TECH] Thorough removal of water from a substance, often with the use of a desiccant.

desiccation breccia [GEOL] Fragments of a mud-cracked layer of sediment deposited with other sediments.

desiccation crack See mud crack.

desiccator [CHEM ENG] A closed vessel, usually made of glass and having an airtight lid, used for drying solid chemicals by means of a desiccant.

designation [COMPUT SCI] An item of data forming part of a computer record that indicates the type of record and thus determines how it is to be processed.

designation hole See designation punch.

designation punch [COMPUT SCI] A hole in a punched card indicating what is the nature of the data on the card or which functions are to be performed by the computer. Also known as control hole; control punch; designation hole; function hole.

design engineering [ENG] A branch of engineering concerned with the creation of systems, devices, and processes useful to and sought by society.

design load [DES ENG] The most stressful combination of weight or other forces a building, structure, or mechanical system or device is designed to sustain.

design pressure [CIV ENG] 1. The force exerted by a body of still water on a dam. 2. The pressure which the dam can withstand. [DES ENG] The pressure used in the calculation of minimum thickness or design characteristics of a boiler or pressure vessel in recognized code formulas; static head may be added where appropriate for specific parts of the structure.

design stress [DES ENG] A permissible maximum stress to which a machine part or structural member may be subjected, which is large enough to prevent failure in case the loads exceed expected values, or other uncertainties turn out unfavorably.

desilting basin [CIV ENG] A space or structure constructed just below a diversion structure of a canal to remove bed, sand, and silt loads. Also known as desilting works.

desilting works See desilting basin.

desired track See course.

de Sitter space [RELAT] A constant-curvature, vacuum solution to Einstein's equations of general relativity with cosmological term.

desize [TEXT] To remove size or sizing agents from warp yarns prior to weaving to protect them against the abrasive action of loom parts.

desk calculator [COMPUT SCI] A device that is used to perform arithmetic operations and is small enough to be conveniently placed on a desk.

desk check See dry run.

Desmidiaceae [BOT] A family of desmids, mostly unicellular algae in the order Conjugales.

desmine See stilbite.

Desmodonta [PALEON] An order of extinct bivalve, burrowing mollusks.

Desmodontidae [VERT ZOO] A small family of chiropteran mammals comprising the true vampire bats.

Desmodoroidea [INV ZOO] A superfamily of marine- and brackish-water-inhabiting nematodes with an annulated, usually smooth cuticle.

Des Moinesian [GEOL] Lower Middle Pennsylvanian geologic time.

Desmokontae [BOT] The equivalent name for Desmophyceae.

Desmophyceae [BOT] A class of rare, mostly marine algae in the division Pyrrhophyta.

Desmoscolecida [INV ZOO] An order of the class Nematoda.

Desmoscolecidae [INV ZOO] A family of nematodes in the superfamily Desmoscolecoidea; individuals resemble annelids in having coarse annulation.

Desmoscolecoidea [INV ZOO] A small superfamily of free-living nematodes characterized by a ringed body, an armored head set, and hemispherical amphids.

Desmostylia [PALEON] An extinct order of large hippopotamuslike, amphibious, gravigrade, shellfish-eating mammals.

Desmostylidae [PALEON] A family of extinct mammals in the order Desmostylia.

Desmothoracida [INV ZOO] An order of sessile and free-living protozoans in the subclass Heliozoia having a spherical body with a perforate, chitinous test.

desorption [PHYS CHEM] The process of removing a sorbed substance by the reverse of adsorption or absorption.

Desor's larva [INV ZOO] An oval, ciliated larva of certain nemertineans in which the gastrula remains inside the egg membrane.

desquamation [PHYSIO] Shedding; a peeling and casting off, as of the superficial epithelium, mucous membranes, renal tubules, and the skin.

destination address [COMPUT SCI] The location to which a jump instruction passes control in a program.

destination time [COMPUT SCI] The time involved in a memory access plus the time required for indirect addressing.

destraction [CHEM ENG] A high-pressure technique for separating high-boiling or nonvolatile material by dissolving it with application of supercritical gases.

destroyer [NAV ARCH] A small, fast, lightly armored warship capable of a variety of functions, usually armed with 5-inch (15 centimeter) guns, torpedoes, depth charges, and mines, and sometimes with guided missiles.

destructive distillation [ORG CHEM] Decomposition of organic compounds by heat without the presence of air.

destructive interference [OPTICS] The interaction of superimposed light from two different sources when the phase relationship is such as to reduce or cancel the resultant intensity to less than the sum of the individual lights.

destructive read [COMPUT SCI] Reading that partially or completely erases the stored information as it is being read.

desulfurization [CHEM ENG] The removal of sulfur, as from molten metals or petroleum oil.

desyl [ORG CHEM] The radical $C_6H_5COCH(C_6H_5)$—; may be formed from desoxybenzoin. Also known as α-phenyl phenacyl.

DETAB [COMPUT SCI] A programming language based on COBOL in which problems can be specified in the form of decision tables. Acronym for decision table.

detachable plugboard [COMPUT SCI] A control panel that can be removed from the computer or other system and exchanged for another without altering the positions of the plugs and cords. Also known as removable plugboard.

detached core [GEOL] The inner bed or beds of a fold that may become separated or pinched off from the main body of the strata due to extreme folding and compression.

detached-lever escapement [HOROL] A watch escapement whose regulating device is given an impulse during only a small part of its operating cycle.

detail chart [COMPUT SCI] A flow chart representing every single step of a program.

detail drawing [GRAPHICS] A large-scale drawing of a small part of a structure or machine.

detail file [COMPUT SCI] A file containing current or transient data used to update a master file or processed with the master file to obtain a specific result. Also known as transaction file.

detailing *See* screening.

detail printing [COMPUT SCI] The printing of information for each card as the card passes through the machine; the function is used to prepare reports that show complete detail about each card; during this listing operation, the machine adds, subtracts, cross-adds, or cross-subtracts, and prints many combinations of totals.

detect *See* demodulate.

detector [ELECTR] The stage in a receiver at which demodulation takes place; in a superheterodyne receiver this is called the second detector. Also known as demodulator; envelope detector. [SCI TECH] Apparatus or system used to detect the presence of an object, radiation, chemical compound, or such.

detent [MECH ENG] A catch or lever in a mechanism which initiates or locks movement of a part, especially in escapement mechanisms.

detention basin [CIV ENG] A reservoir without control gates for storing water over brief periods of time until the stream has the capacity for ordinary flow plus released water; used for flood regulation.

detergent [MATER] A synthetic cleansing agent resembling soap in the ability to emulsify oil and hold dirt, and containing surfactants which do not precipitate in hard water; may also contain protease enzymes and whitening agents.

determinant [CONT SYS] The product of the partial return differences associated with the nodes of a signal-flow graph. [MATH] A certain real-valued function of the column vectors of a square matrix which is zero if and only if the matrix is singular; used to solve systems of linear equations and to study linear transformations.

determinate cleavage [EMBRYO] A type of cleavage which separates portions of the zygote with specific and distinct potencies for development as specific parts of the body.

determinate growth [BOT] Growth in which the axis, or central stem, being limited by the development of the floral reproductive structure, does not grow or lengthen indefinitely.

determinate structure [MECH] A structure in which the equations of statics alone are sufficient to determine the stresses and reactions.

determinism *See* causality.

detonating fuse [ENG] A device consisting of a core of high explosive within a waterproof textile covering and set off by an electrical blasting cap fired from a distance by means of a fuse line; used in large, deep boreholes.

detonation [CHEM] An exothermic chemical reaction that propagates with such rapidity that the rate of advance of the reaction zone into the unreacted material exceeds the velocity of sound in the unreacted material; that is, the advancing reaction zone is preceded by a shock wave. [MECH ENG] Spontaneous combustion of the compressed charge after

passage of the spark in an internal combustion engine; it is accompanied by knock.

detonation wave [FL MECH] A shock wave that accompanies detonation and has a shock front followed by a region of decreasing pressure in which the reaction occurs.

detonator [ENG] A device, such as a blasting cap, employing a sensitive primary explosive to detonate a high-explosive charge.

detorsion [INV ZOO] Untwisting of the 180° visceral twist imposed by embryonic torsion on many gastropod mollusks. [MED] Untwisting of an abnormal torsion, as of a ureter or intestine.

detoxification [BIOCHEM] The act or process of removing a poison or the toxic properties of a substance in the body.

detrital minerals [MINERAL] Grains of heavy minerals found in sediment, resulting from mechanical disintegration of the parent rock.

detrital sediment [GEOL] Accumulations of the organic and inorganic fragmental products of the weathering and erosion of land transported to the place of deposition.

detritus [GEOL] Any loose material removed directly from rocks and minerals by mechanical means, such as disintegration or abrasion.

detritus feeder *See* deposit feeder.

deuteranopia [MED] Defective vision consisting of red-green color confusion, with no marked reduction in the brightness of any color.

deuteric [GEOL] Of or pertaining to alterations in igneous rock during the later stages and as a direct result of consolidation of magma or lava. Also known as epimagmatic; paulopost.

deuterium [CHEM] The isotope of the element hydrogen with one neutron and one proton in the nucleus; atomic weight 2.0144. Designated D, d, H^2, or ^2H.

deuterium cycle *See* proton-proton chain.

deuterium oxide *See* heavy water.

Deuteromycetes [MYCOL] The equivalent name for Fungi Imperfecti.

deuteron [NUC PHYS] The nucleus of a deuterium atom, consisting of a neutron and a proton. Designated d. Also known as deuton.

Deuterophlebiidae [INV ZOO] A family of dipteran insects in the suborder Cyclorrhapha.

Deuterostomia [ZOO] A division of the animal kingdom which includes the phyla Echinodermata, Chaetognatha, Hemichordata, and Chordata.

deuton *See* deuteron.

deutoplasm [EMBRYO] The nutritive yolk granules in egg cells.

developed dye [CHEM] A direct azo dye that can be further diazotized by a developer after application to the fiber; it couples with the fiber to form colorfast shades. Also known as diazo dye.

developed ore *See* developed reserves.

developed reserves [MIN ENG] Ore that is exposed on three sides and for which tonnage yield and quality estimates have been made. Also known as assured mineral; blocked-out ore; developed ore; measured ore; ore in sight.

developer [CHEM] An organic compound which interacts on a textile fiber to develop a dye. [GRAPHICS] A chemical solution used to develop exposed photographic materials by reducing silver salts to metallic silver.

development [GEOL] The progression of changes in fossil groups which have succeeded one another during deposition of the strata of the earth. [METEOROL] The process of intensification of an atmospheric disturbance, most commonly applied to cyclones and anticyclones. [MIN ENG] Opening of a coal seam or ore body by sinking shafts or driving levels, as well as installing equipment, for proving ore reserves and exploiting them. [SCI TECH] The work required to determine the best production techniques to bring a new process or piece of equipment to the production stage.

developmental psychology [PSYCH] The branch of psychology that deals with changes in behavior occurring with changes in age.

development index [METEOROL] An index used as an aid in forecasting cyclogenesis; the development index I is defined most frequently as the difference in divergence between two well-separated, tropospheric, constant-pressure surfaces. Also known as relative divergence.

deviation [ENG] The difference between the actual value of a controlled variable and the desired value corresponding to the set point. [EVOL] Evolutionary differentiation involving interpolation of new stages in the ancestral pattern of morphogenesis. [OPTICS] The angle between the incident ray on an object or optical system and the emergent ray, following reflection, refraction, or diffraction. Also known as angle of deviation. [STAT] The difference between any given number in a set and the mean average of those numbers.

deviation factor *See* compressibility factor.

device [COMPUT SCI] A general-purpose term used, often indiscriminately, to refer to a computer component or the computer itself. [ELECTR] An electronic element that cannot be divided without destroying its stated function; commonly applied to active elements such as transistors and transducers. [ENG] A mechanism, tool, or other piece of equipment designed for specific uses.

device address [COMPUT SCI] The binary code which corresponds to a unique device, referred to when selecting this specific device.

device assignment [COMPUT SCI] The use of a logical device number used in conjunction with an input/output instruction, and made to refer to a specific device.

device driver [COMPUT SCI] A subroutine which handles a complete input/output operation.

device-end condition [COMPUT SCI] The completion of an input/output operation, such as the transfer of a complete data block, recognized by the hardware in the absence of a byte count.

device flag [COMPUT SCI] A flip-flop output which indicates the ready status of an input/output device.

device number [COMPUT SCI] The physical or logical number which refers to a specific input/output device.

device selector [COMPUT SCI] A circuit which gates data-transfer or command pulses to a specific input/output device.

devillite [MINERAL] $Cu_4Ca(SO_4)_2(OH)_6 3H_2O$ A dark-green mineral consisting of a hydrous basic sulfate of copper and calcium, occurring in six-sided platy crystals.

devitrified glass [MATER] A glassy material which has been changed from a vitreous to a brittle crystalline state during manufacture.

Devonian [GEOL] The fourth period of the Paleozoic Era.

dew [HYD] Water condensed onto grass and other objects near the ground, the temperatures of which have fallen below the dew point of the surface air because of radiational cooling during the night but are still above freezing.

Dewar flask [PHYS] A vessel having double walls, the space between being evacuated to prevent the transfer of heat and the surfaces facing the vacuum being heat-reflective; used to hold liquid gases and to study low-temperature phenomena.

dewatering [ENG] **1.** Removal of water from solid material by wet classification, centrifugation, filtration, or similar solid-liquid separation techniques. **2.** Removing or draining water from an enclosure or a structure, such as a riverbed, caisson, or mine shaft, by pumping or evaporation.

dew cell [ENG] An instrument used to determine the dew point, consisting of a pair of spaced, bare electrical wires wound spirally around an insulator and covered with a wicking wetted with a water solution containing an excess of lithium chloride; an electrical potential applied to the wires causes a flow of current through the lithium chloride solution, which raises the temperature of the solution until its vapor pressure is in equilibrium with that of the ambient air.

dewclaw [VERT ZOO] **1.** A vestigial digit on the foot of a mammal which does not reach the ground. **2.** A claw or hoof terminating such a digit.

dewetting [MET] Flow of solder away from the soldered surface during reheating following initial soldering.

dewlap [ANAT] A fleshy or fatty fold of skin on the throat of some humans. [BOT] One of a pair of hinges at the joint of a sugarcane leaf blade. [VERT ZOO] A fold of skin hanging from the neck of some reptiles and bovines.

dew point [CHEM] The temperature at which water vapor begins to condense. [METEOROL] The temperature at which air becomes saturated when cooled without addition of moisture or change of pressure; any further cooling causes condensation. Also known as dew-point temperature.

dew-point composition [CHEM ENG] The water vapor–air composition at saturation, that is, at the temperature at which water exerts a vapor pressure equal to the partial pressure of water vapor in the air-water mixture.

dew-point curve [CHEM ENG] On a PVT phase diagram, the line that separates the two-phase (gas-liquid) region from the one-phase (gas) region, and indicates the point at a given gas temperature or pressure at which the first dew or liquid phase occurs.

dew-point depression [CHEM ENG] Reduction of the liquid-vapor dew point of a gas by removal of a portion of the liquid (such as water) from the gas (such as air). [METEOROL] The number of degrees the dew point is found to be lower than the temperature.

dew-point hygrometer [CHEM ENG] An instrument for determining the dew point by measuring the temperature at which vapor being cooled in a silver vessel begins to condense. Also known as cold-spot hygrometer.

dew-point temperature *See* dew point.

dex *See* brig.

Dexaminidae [INV ZOO] A family of amphipod crustaceans in the suborder Gammeridea.

dextran [BIOCHEM] Any of the several polysaccharides, $(C_5H_{10}O_5)_n$, that yield glucose units on hydrolysis.

dextrin [BIOCHEM] A polymer of D-glucose which is intermediate in complexity between starch and maltose.

dextro *See* dextrorotatory.

dextrorotatory [OPTICS] Rotating clockwise the plane of polarization of a wave traveling through a medium in a clockwise direction, as seen by an eye observing the light. Abbreviated dextro.

dextrorse [BOT] Twining toward the right.

dextrose [BIOCHEM] $C_6H_{12}O_6 \cdot H_2O$ A dextrorotatory monosaccharide obtained as a white, crystalline, odorless, sweet powder, which is soluble in about one part of water; an important intermediate in carbohydrate metabolism; used for nutritional purposes, for temporary increase of blood volume, and as a diuretic. Also known as corn sugar; grape sugar.

D horizon [GEOL] A soil horizon sometimes occurring below a B or C horizon, consisting of unweathered rock.

Di See didymium.

diabantite [MINERAL] $(Mg,Fe^{2+},Al)_6(Si,Al)_4O_{10}(OH)_8$ Mineral of the chlorite group consisting of a basic silicate of magnesium, iron, and aluminum, occurring in cavities in basic igneous rock.

diabase [PETR] An intrusive rock consisting principally of labradorite and pyroxene.

diabasic [PETR] Denoting igneous rock in which the inter-stices between the feldspar crystals are filled with discrete crystals or grains of pyroxene.

diabatic [THERMO] Involving a thermodynamic change of state of a system in which there is a transfer of heat across the boundaries of the system.

diabetes [MED] Any of various abnormal conditions characterized by excessive urinary output, thirst, and hunger; usually refers to diabetes mellitus.

diabetes insipidus [MED] A form of diabetes due to a disfunction of the hypothalamus.

diabetes mellitus [MED] A metabolic disorder arising from a defect in carbohydrate utilization by the body, related to inadequate or abnormal insulin production by the pancreas.

diachronous [GEOL] Of a rock unit, varying in age in different areas or cutting across time planes or biostratigraphic zones. Also known as time-transgressive.

diacid [CHEM] An acid that has two acidic hydrogen atoms; an example is oxalic acid.

Diacodectidae [PALEON] A family of extinct artiodactyl mammals in the suborder Palaeodonta.

diactor [ELEC] Direct-acting automatic regulator for control of shunt generator voltage output.

diadelphous stamen [BOT] A stamen that has its filaments united into two sets.

Diadematacea [INV ZOO] A superorder of Euechinoidea having a rigid or flexible test, perforate tubercles, and branchial slits.

Diadematidae [INV ZOO] A family of large euechinoid echinoderms in the order Diadematoida having crenulate tubercles and long spines.

Diadematoida [INV ZOO] An order of echinoderms in the superorder Diadematacea with hollow primary radioles and crenulate tubercles.

diadochy [CRYSTAL] Replacement or ability to be replaced of one atom or ion by another in a crystal lattice.

diadromous [BOT] Having venation in the form of fanlike radiations. [VERT ZOO] Of fish, migrating between salt and fresh waters.

Diadumenidae [INV ZOO] A family of anthozoans in the order Actiniaria.

diafocal point [OPTICS] For a ray of light refracted by a lens, a point on the ray which lies on a plane passing through the axis of the lens which is parallel to the ray on the opposite side of the lens.

diagenesis [GEOL] Chemical and physical changes occurring in sediments during and after their deposition but before consolidation.

diagnosis [COMPUT SCI] The process of locating and explaining detectable errors in a computer routine or hardware component. [MED] Identification of a disease from its signs and symptoms. [SYST] In taxonomic study, a statement of the characters that distinguish a taxon from coordinate taxa.

diagnostic check See diagnostic routine.

diagnostic routine [COMPUT SCI] A routine designed to locate a computer malfunction or a mistake in coding. Also known as diagnostic check; diagnostic subroutine; diagnostic test; error detection routine.

diagnostic subroutine See diagnostic routine.

diagnostic test See diagnostic routine.

diagonal [CIV ENG] A sloping structural member, under compression or tension or both, of a truss or bracing system. [MATH] **1.** The set of points all of whose coordinates are equal to one another in an *n*-dimensional coordinate system. **2.** A line joining opposite vertices of a polygon with an even number of sides. [OPTICS] A plane mirror or prism face mounted near the eyepiece of a telescope at an angle to the light path, to redirect the light for convenience of observation or to reduce the intensity of the image of the sun so that it can be observed directly. [TEXT] A heavy twilled fabric.

diagonal fault [GEOL] A fault whose strike is diagonal or oblique to the strike of the adjacent strata. Also known as oblique fault.

diagonal horn antenna [ELECTROMAG] Horn antenna in which all cross sections are square and the electric vector is parallel to one of the diagonals; the radiation pattern in the far field has almost perfect circular symmetry.

diagonal joint [GEOL] A joint having its strike oblique to the strike of the strata of the sedimentary rock, or to the cleavage plane of the metamorphic rock in which it occurs. Also known as oblique joint.

diagonal pitch [ENG] In rows of staggered rivets, the distance between the center of a rivet in one row to the center of the adjacent rivet in the next row.

diagram [COMPUT SCI] A schematic representation of a sequence of subroutines designed to solve a problem; it is a coarser and less symbolic representation than a flow chart, frequently including descriptions in English words. [GRAPHICS] **1.** A line drawing that represents an object or area according to a scale. **2.** A graph which shows the relation between two variables or which plots the occurrence of events or objects as a function of two variables.

diakinesis [CYTOL] The last stage of meiotic prophase, when the chromatids attain maximum contraction and the bivalents move apart and position themselves against the nuclear membrane.

dial [COMMUN] In automatic telephone switching, a type of calling device which, when wound up and released, generates pulses required for establishing connections. [DES ENG] A separate scale or other device for indicating the value to which a control is set.

dialdehyde [ORG CHEM] A molecule that has two aldehyde groups, such as dialdehyde starch.

dial exchange [COMMUN] A telephone exchange area in which all subscribers originate their calls by dialing.

dial indicator [DES ENG] Meter or gage with a calibrated circular face and a pivoted pointer to give readings.

dialing key [COMMUN] Method of dialing in which a set of numerical keys is used to originate dial pulses instead of a dial; generally used in connection with voice-frequency dialing.

dial jacks [ELEC] Strip of jacks associated with and bridged to a regular out-going trunk jack circuit to provide a connection between the dial cords and the outgoing trunks.

dial key [ELEC] Key unit of the subscriber's cord circuit used to connect the dial into the line.

dial telephone system [COMMUN] A telephone system in which telephone connections between customers are ordinarily established by electronic and mechanical apparatus, controlled by manipulations of dials operated by calling parties.

dial tone [COMMUN] A tone employed in a dial telephone system to indicate that the equipment is ready for dialing operation.

dial-up [COMMUN] **1.** The service whereby a dial telephone can be used to initiate and effect station-to-station telephone calls. **2.** In computer networks, pertaining to terminals which must dial up to receive service, as contrasted with those hand-wired or permanently connected into the network.

dialysis [PHYS CHEM] A process of selective diffusion through a membrane; usually used to separate low-molecular-weight solutes which diffuse through the membrane from the colloidal and high-molecular-weight solutes which do not.

dialyzate [CHEM] The material that does not diffuse through the membrane during dialysis; alternatively, it may be considered the material that has diffused.

dialyzer [CHEM ENG] **1.** The semipermeable membrane used for dialyzing liquid. **2.** The container used in dialysis; it is separated into compartments by membranes.

diamagnet [ELECTROMAG] A substance which is diamagnetic, such as the alkali and alkaline earth metals, the halogens, and the noble gases.

diamagnetic [ELECTROMAG] Having a magnetic permeability less than 1; materials with this property are repelled by a magnet and tend to position themselves at right angles to magnetic lines of force.

diamagnetic Faraday effect [OPTICS] Faraday effect at frequencies near an absorption line which is split due to the splitting of the upper level only.

diamagnetic resonance See cyclotron resonance.

diamagnetic susceptibility [ELECTROMAG] The susceptibility of a diamagnetic material, which is always negative and usually on the order of -10^{-5} cm^3/mole.

diamagnetism [ELECTROMAG] The property of a material which is repelled by magnets.

diameter [MATH] **1.** A line segment which passes through the center of a circle, and whose end points lie on the circle. **2.** The length of such a line.

diametral curve [MATH] A curve that passes through the midpoints of a family of parallel chords of a given curve.

diametral plane [MATH] **1.** A plane that passes through the center of a sphere. **2.** A plane that passes through the midpoints of a family of parallel chords of a quadric surface that are parallel to a given chord.

diametral surface [MATH] A surface that passes through the midpoints of a family of parallel chords of a given surface that are parallel to a given chord.

diamictite [PETR] A calcareous, terrigenous sedimentary rock that is not sorted or poorly sorted and

contains particles of many sizes. Also known as mixtite.

diamide [ORG CHEM] A molecule that has two amide ($-CONH_2$) groups.

diamine [ORG CHEM] Any compound containing two amino groups.

diamino [ORG CHEM] A term used in chemical nomenclature to indicate the presence in a molecule of two amino ($-NH_2$) groups.

diamond [MINERAL] A colorless mineral composed entirely of carbon crystallized in the isometric system as octahedrons, dodecahedrons, and cubes; the hardest substance known; used as a gem and in cutting tools.

diamond antenna See rhombic antenna.

diamond bit [DES ENG] A rotary drilling bit crowned with bort-type diamonds, used for rock boring. Also known as bort bit.

diamond canker [PL PATH] A virus disease that affects the bark of certain stone-fruit trees, resulting in weakening of the trunk and limbs.

diamond circuit [ELECTR] A gate circuit that provides isolation between input and output terminals in its off state, by operating transistors in their cutoff region; in the on state the output voltage follows the input voltage as required for gating both analog and digital signals, while the transistors provide current gain to supply output current on demand.

diamond drill [DES ENG] A drilling machine with a hollow, diamond-set bit for boring rock and yielding continuous and columnar rock samples.

diamond indenter [ENG] An instrument that measures hardness by indenting a material with a diamond point.

diamond-pyramid hardness number [MET] The quotient of the load applied in the diamond-pyramid hardness test divided by the pyramidal area of the impression.

diamond-pyramid hardness test [MET] An indentation hardness test in which a diamond-pyramid indenter, with a 136° angle between opposite faces, is forced under variable loads into the surface of a test specimen. Also known as Vickers hardness test.

diandrous [BOT] Having two stamens.

Dianemaceae [MICROBIO] A family of slime molds in the order Trichales.

Dianulitidae [PALEON] A family of extinct, marine bryozoans in the order Cystoporata.

diapause [PHYSIO] A period of spontaneously suspended growth or development in certain insects, mites, crustaceans, and snails.

diapedesis [MED] Hemorrhage of blood cells, especially erythrocytes, through an intact vessel wall into the tissues.

Diapensiaceae [BOT] The single family of the Diapensiales, an order of flowering plants.

Diapensiales [BOT] A monofamilial order of dicotyledonous plants in the subclass Dilleniidae comprising certain herbs and dwarf shrubs in temperate and arctic regions of the Northern Hemisphere.

Diaphanocephalidae [INV ZOO] A family of parasitic roundworms belonging to the Strongyloidea; snakes are the principal host.

diaphorite [MINERAL] Pb$_2$Ag$_3$Sb$_3$S$_8$ A gray-black orthorhombic mineral consisting of sulfide of lead, silver, and antimony, occurring in crystals. Also known as ultrabasite.

diaphragm [ANAT] The dome-shaped partition composed of muscle and connective tissue that separates the abdominal and thoracic cavities in mammals. [ELECTROMAG] See iris. [ENG] A thin sheet placed between parallel parts of a member of structural steel to increase its rigidity. [ENG ACOUS] A thin, flexible sheet that can be moved by sound waves, as in a microphone, or can produce sound waves when moved, as in a loudspeaker. [OPTICS] Any opening in an optical system which controls the cross section of a beam of light passing through it, to control light intensity, reduce aberration, or increase depth of focus. Also known as lens stop. [PHYS] **1.** A separating wall or membrane, especially one which transmits some substances and forces but not others. **2.** In general, any opening, sometimes adjustable in size, which is used to control the flow of a substance or radiation.

diaphragmatic hernia [MED] Protrusion of an abdominal organ through the diaphragm into the thoracic cavity.

diaphragm cell [CHEM ENG] An electrolytic cell used to produce sodium hydroxide and chlorine from sodium chloride brine; porous diaphragm separates the anode and cathode compartments.

diaphragm meter [ENG] A flow meter which uses the movement of a diaphragm in the measurement of a difference in pressure created by the flow, such as a force-balance-type or a deflection-type meter.

diaphragm pump [MECH ENG] A metering pump which uses a diaphragm to isolate the operating parts from pumped liquid in a mechanically actuated diaphragm pump, or from hydraulic fluid in a hydraulically actuated diaphragm pump.

diaphragm valve [ENG] A fluid valve in which the open-close element is a flexible diaphragm; used for fluids containing suspended solids, but limited to low-pressure systems.

diaphthoresis *See* retrograde metamorphism.

diaphthorite [PETR] Schistose rocks in which minerals have formed by retrograde metamorphism.

diaphysis [ANAT] The shaft of a long bone.

diapir [GEOL] A dome or anticlinal fold in which a mobile plastic core has ruptured the more brittle overlying rock. Also known as diapiric fold; piercement dome; piercing fold.

diapiric fold *See* diapir.

diapophysis [ANAT] The articular portion of a transverse process of a vertebra.

Diapriidae [INV ZOO] A family of hymenopteran insects in the superfamily Proctotrupoidea.

diarthrosis [ANAT] A freely moving articulation, characterized by a synovial cavity between the bones.

diaspore [MINERAL] AlO(OH) A mineral composed of some bauxites occurring in white, lamellar masses; crystallizes in the orthorhombic system.

diastase [BIOCHEM] An enzyme that catalyzes the hydrolysis of starch to maltose. Also known as vegetable diastase.

diastasis [MED] Any simple separation of parts normally joined together, as the separation of an epiphysis from the body of a bone without true fracture, or the dislocation of an amphiarthrosis. [PHYSIO] The final phase of diastole, the phase of slow ventricular filling.

diastem [GEOL] A temporal break between adjacent geologic strata that represents nondeposition or local erosion but not a change in the general regimen of deposition.

diastereoisomer [ORG CHEM] One of a pair of optical isomers which are not mirror images of each other. Also known as diastereomer.

diastereomer *See* diastereoisomer.

diastole [PHYSIO] The rhythmic relaxation and dilation of a heart chamber, especially a ventricle.

diastolic pressure [PHYSIO] The lowest arterial blood pressure during the cardiac cycle; reflects relaxation and dilation of a heart chamber.

diastrophism [GEOL] **1.** The general process or combination of processes by which the earth's crust is deformed. **2.** The results of this deforming action.

diathermy [MED] The therapeutic use of high-frequency electric currents to produce localized heat in body tissues.

diatom [INV ZOO] The common name for algae composing the class Bacillariophyceae; noted for the symmetry and sculpturing of the siliceous cell walls.

diatomaceous earth [GEOL] A yellow, white, or light-gray, siliceous, porous deposit made of the opaline shells of diatoms; used as a filter aid, paint filler, adsorbent, abrasive, and thermal insulator. Also known as kieselguhr; tripolite.

diatomaceous ooze [GEOL] A pelagic, siliceous sediment composed of more than 30% diatom tests, up to 40% calcium carbonate, and up to 25% mineral grains.

diatonic scale [ACOUS] A musical scale in which the octave is divided into intervals of two different sizes, five of one and two of the other, with adjustments in tuning systems other than equal temperament.

diatreme [GEOL] A circular volcanic vent produced by the explosive energy of gas-charged magmas.

Diatrymiformes [PALEON] An order of extinct large, flightless birds having massive legs, tiny wings, and large heads and beaks.

diazoate [ORG CHEM] A salt with molecular formula of the type $C_6H_5N=NOOM$, where M is a nonvalent metal.

diazo compound [ORG CHEM] An organic compound containing the radical $-N=N-$.

diazo dye *See* developed dye.

diazole [ORG CHEM] A cyclic hydrocarbon with five atoms in the ring, two of which are nitrogen atoms and three are carbon.

diazonium salts [ORG CHEM] Compounds of the type $R\cdot X\cdot N:N$, with X being an acid radical such as chlorine.

diazo oxide [ORG CHEM] An organic molecule or a grouping of organic molecules that have a diazo group and an oxygen atom joined to ortho positions of an aromatic nucleus. Also known as diazophenol.

diazophenol *See* diazo oxide.

diazo process *See* diazotization.

diazotization [ORG CHEM] Reaction between a primary aromatic amine and nitrous acid to give a diazo compound. Also known as diazo process.

Dibamidae [VERT ZOO] The flap-legged skinks, a small family of lizards in the suborder Sauria comprising three species confined to southeastern Asia.

dibasic [CHEM] **1.** Compounds containing two hydrogens that may be replaced by a monovalent metal or radical. **2.** An alcohol that has two hydroxyl groups, for example, ethylene glycol.

dibit [COMPUT SCI] A pair of binary digits, used to specify one of four values.

diborate *See* borax.

Dibranchia [INV ZOO] A subclass of the Cephalopoda containing all living cephalopods except *Nautilus;* members possess two gills and, when present, an internal shell.

dibromo- [CHEM] A prefix indicating two bromine atoms.

dibutyl [ORG CHEM] Indicating the presence of two butyl groupings bonded through a third atom or group in a molecule.

dicarboxylic acid [ORG CHEM] A compound with two carboxyl groups.

dicaryon See dikaryon.

Dichelesthiidae [INV ZOO] A family of parasitic copepods in the suborder Caligoida; individuals attach to the gills of various fishes.

dichloride [CHEM] Any inorganic salt or organic compound that has two chloride atoms in its molecule.

dichlorodiphenyltrichloroethane See DDT.

Dichobunidae [PALEON] A family of extinct artiodactyl mammals in the superfamily Dichobunoidea.

Dichobunoidea [PALEON] A superfamily of extinct artiodactyl mammals in the suborder Paleodonta composed of small- to medium-size forms with tri- to quadritubercular bunodont upper teeth.

dichotomizing search [COMPUT SCI] A procedure for searching an item in a set, in which, at each step, the set is divided into two parts, one part being then discarded if it can be logically shown that the item could not be in that part.

dichotomy [ASTRON] The phase of the moon or an inferior planet at which exactly half of its disk is illuminated and the terminator is a straight line. [BIOL] 1. Divided in two parts. 2. Repeated branching or forking. [COMPUT SCI] A division into two subordinate classes; for example, all white and all nonwhite, or all zero and all nonzero.

dichroic mirror [OPTICS] A glass surface coated with a special metal film that reflects certain colors of light while allowing others to pass through.

dichroism [OPTICS] In certain anisotropic materials, the property of having different absorption coefficients for light polarized in different directions.

dichromatic [BIOL] Having or exhibiting two color phases independently of age or sex.

dichromatic dye [CHEM] Dye or indicator in which different colors are seen, depending upon the thickness of the solution.

dichromatism [MED] Partial color blindness in which vision is apparently based on two primary colors rather than the normal three.

Dickinsoniidae [PALEON] A family that comprises extinct flat-bodied, mutisegmented coelomates; identified as ediacaran fauna.

dickinsonite [MINERAL] $H_2Na_6(Mn,Fe,Ca,Mg)_{14}$-$(PO_4)_{12} \cdot H_2O$ A green mineral consisting of foliated hydrous acid phosphate, chiefly of manganese, iron, and sodium, and is isostructural with arrojadite; specific gravity is 3.34.

dickite [MINERAL] $Al_2Si_2O_5(OH)_4$ A mineral of the kaolin group found crystallized in clay in hydrothermal veins; it is polymorphous with kaolinite and nacrite.

Dicksoniaceae [BOT] A family of tree ferns characterized by marginal sori which are terminal on the veins and protected by a bivalved indusium.

Dick test [IMMUNOL] A skin test to determine immunity to scarlet fever; *Streptococcus pyogenes* toxin is injected intracutaneously and produces a reaction if there is no circulating antitoxin.

dicotyledon [BOT] Any plant of the class Magnoliopsida, all having two cotyledons.

Dicotyledoneae [BOT] The equivalent name for Magnoliopsida.

Dicranales [BOT] An order of mosses having erect stems, dichotomous branching, and dense foliation.

dictionary [COMPUT SCI] A table establishing the correspondence between specific words and their code representations.

dictionary code [COMPUT SCI] An alphabetical arrangement of English words and terms, associated with their code representations.

Dictyoceratida [INV ZOO] An order of sponges of the class Demospongiae; includes the bath sponges of commerce.

Dictyonellidina [PALEON] A suborder of extinct articulate brachiopods.

dictyosome [CYTOL] A stack of two or more cisternae; a component of the Golgi apparatus.

Dictyospongiidae [PALEON] A family of extinct sponges in the subclass Amphidiscophora having spicules resembling a one-ended amphidisc (paraclavule).

Dictyosporae [MYCOL] A spore group of the imperfect fungi characterized by multicelled spores with cross and longitudinal septae.

dictyospore [MYCOL] A multicellular spore in certain fungi characterized by longitudinal walls and cross septa.

dictyostele [BOT] A modified siphonostele in which the vascular tissue is dissected into a network of distinct strands; found in certain fern stems.

Dictyosteliaceae [MICROBIO] A family of microorganisms belonging to the Acrasiales and characterized by strongly differentiated fructifications.

Dicyemida [INV ZOO] An order of mesozoans comprising minute, wormlike parasites of the renal organs of cephalopod mollusks.

didelphic [ANAT] Having a double uterus or genital tract.

Didelphidae [VERT ZOO] The opossums, a family of arboreal mammals in the order Marsupialia.

Didolodontidae [PALEON] A family consisting of extinct medium-sized herbivores in the order Condylarthra.

Didymiaceae [MICROBIO] A family of slime molds in the order Physarales.

didymium [CHEM] A mixture of the rare-earth elements praeseodymium and neodymium. Abbreviated Di.

didymolite [MINERAL] $Ca_2Al_6Si_9O_{29}$ A dark-gray monoclinic mineral consisting of a calcium aluminum silicate, occurring in twinned crystals.

Didymosporae [MYCOL] A spore group of the imperfect fungi characterized by two-celled spores.

didynamous [BOT] Having four stamens occurring in two pairs, one pair long and the other short.

die [DES ENG] A tool or mold used to impart shapes to, or to form impressions on, materials such as metals and ceramics. [ELECTR] The tiny, sawed or otherwise machined piece of semiconductor material used in the construction of a transistor, diode, or other semiconductor device; plural is dice. [MED] To pass from physical life. [MIN ENG] See bell tap.

dieback [ECOL] A large area of exposed, unprotected swamp or marsh deposits resulting from the salinity of a coastal lagoon. [PL PATH] Of a plant, to die from the top or peripheral parts.

die block [ENG] 1. A tool-steel block which is bolted to the bed of a punch press and into which the desired impressions are machined. 2. The part of an extrusion mold die holding the forming bushing and core.

die casting [ENG] A metal casting process in which molten metal is forced under pressure into a permanent mold; the two types are hot-chamber and cold-chamber.

Dieckman condensation [CHEM ENG] Any condensation of esters of dicarboxylic acids which produce cyclic β-ketoesters.

die clearance [ENG] The distance between die members that meet during an operation.

die collar See bell tap.

die cutting [ENG] *See* blanking. [GRAPHICS] Cutting special shapes, such as labels, from printed sheets by using sharp steel rules; often done at the same time as the printing.

die down [BOT] Normal seasonal death of aboveground parts of herbaceous perennials.

die drawing [MET] Reducing the diameter of wire or tubing by pulling it through a die.

die forming [MET] Shaping metal by means of a die under pressure.

Diego blood group [IMMUNOL] A genetically determined, immunologically distinct group of human erythrocyte antigens recognized by reaction with a specific antibody.

die holder [ENG] A plate or block on which the die block is mounted; it is fastened to the bolster or press bed.

dielectric [MATER] A material which is an electrical insulator or in which an electric field can be sustained with a minimum dissipation in power.

dielectric absorption [ELEC] The persistence of electric polarization in certain dielectrics after removal of the electric field. [ELECTROMAG] *See* dielectric loss.

dielectric amplifier [ELECTR] An amplifier using a ferroelectric capacitor whose capacitance varies with applied voltage so as to give signal amplification.

dielectric constant [ELEC] **1.** For an isotropic medium, the ratio of the capacitance of a capacitor filled with a given dielectric to that of the same capacitor having only a vacuum as dielectric. **2.** More generally, $1 + \gamma\chi$, where γ is 4π in Gaussian and cgs electrostatic units or 1 in rationalized mks units, and χ is the electric susceptibility tensor. Also known as relative dielectric constant; relative permittivity; specific inductive capacity (SIC).

dielectric crystal [ELEC] A crystal which is electrically nonconducting.

dielectric current [ELEC] The current flowing at any instant through a surface of a dielectric that is located in a changing electric field.

dielectric displacement *See* electric displacement.

dielectric flux density *See* electric displacement.

dielectric heating [ELEC] Heating of a nominally electrical insulating material due to its own electrical (dielectric) losses, when the material is placed in a varying electrostatic field.

dielectric hysteresis *See* ferroelectric hysteresis.

dielectric lens [ELECTROMAG] A lens made of dielectric material so that it refracts radio waves in the same manner that an optical lens refracts light waves; used with microwave antennas.

dielectric-lens antenna [ELECTROMAG] An aperture antenna in which the beam width is determined by the dimensions of a dielectric lens through which the beam passes.

dielectric loss [ELECTROMAG] The electric energy that is converted into heat in a dielectric subjected to a varying electric field. Also known as dielectric absorption.

dielectric polarization *See* polarization.

dielectric soak *See* absorption.

dielectric strength [ELEC] The maximum electrical potential gradient that a material can withstand without rupture; usually specified in volts per millimeter of thickness. Also known as electric strength.

dielectric susceptibility *See* electric susceptibility.

dielectric test [ELEC] A test involving application of a voltage higher than the rated value for a specified time, to determine the margin of safety against later failure of insulating materials.

dielectric waveguide [ELEC] A waveguide consisting of a dielectric cylinder surrounded by air.

dielectric wedge [ELECTROMAG] A wedge-shaped piece of dielectric used in a waveguide to match its impedance to that of another waveguide.

Diels-Alder reaction [ORG CHEM] The 1,4 addition of a conjugated diolefin to a compound, known as a dienophile, containing a double or triple bond; the dienophile may be activated by conjugation with a second double bond or with an electron acceptor.

diencephalon [EMBRYO] The posterior division of the embryonic forebrain in vertebrates.

diene [ORG CHEM] One of a class of organic compounds containing two ethylenic linkages (carbon-to-carbon double bonds) in the molecules. Also known as diolefin.

die nipple *See* bell tap.

die opening [MET] The distance between electrodes in flash or upset welding; it is measured with parts in contact but before the beginning or immediately after completion of the weld cycle.

diesel cycle [THERMO] An internal combustion engine cycle in which the heat of compression ignites the fuel.

diesel engine [MECH ENG] An internal combustion engine operating on a thermodynamic cycle in which the ratio of compression of the air charge is sufficiently high to ignite the fuel subsequently injected into the combustion chamber. Also known as compression-ignition engine.

diesel fuel [MATER] Fuel used for internal combustion in diesel engines; usually that fraction of crude oil that distills after kerosine.

diesel index [CHEM ENG] An empirical expression for the correlation between the aniline number of a diesel fuel and its ignitability. [MECH ENG] Diesel fuel rating based on ignition qualities; high-quality fuel has a high index number.

die set [ENG] A tool or tool holder consisting of a die base for the attachment of a die and a punch plate for the attachment of a punch.

die shoe [MECH ENG] A block placed beneath the lower part of a die upon which the die holder is mounted; spreads the impact over the die bed, thereby reducing wear.

diester [ORG CHEM] A compound containing two ester groupings.

diesterase [BIOCHEM] An enzyme such as a nuclease which splits the linkages binding individual nucleotides of a nucleic acid.

diestrus [PHYSIO] The long, quiescent period following ovulation in the estrous cycle in mammals; the stage in which the uterus prepares for the reception of a fertilized ovum.

diether [ORG CHEM] A molecule that has two oxygen atoms with ether bonds.

dietrichite [MINERAL] $(Zn,Fe,Mn)Al_2(SO_4)_4 \cdot 22H_2O$ Mineral consisting of a hydrous sulfate of aluminum and one or more of the metals zinc, iron, and manganese.

dietzeite [MINERAL] $Ca_2(IO_3)_2(CrO_4)$ A dark-golden-yellow iodate mineral commonly in fibrous or columnar form as a component of caliche.

die welding [MET] Forge welding in which the weld is completed under pressure between dies.

difference [MATH] The result of subtracting one number from another.

difference amplifier *See* differential amplifier.

difference channel [ENG ACOUS] An audio channel that handles the difference between the signals in the left and right channels of a stereophonic sound system.

difference detector [ELECTR] A detector circuit in which the output is a function of the difference between the amplitudes of the two input waveforms.

difference equation [MATH] An equation expressing a functional relationship of one or more independent variables, one or more functions dependent on these variables, and successive differences of these functions.

difference limen *See* just-noticeable difference.

difference threshold *See* just-noticeable difference.

differentiable function [MATH] A function which has a derivative at each point of its domain.

differentiable manifold [MATH] A topological space with a maximal differentiable atlas; roughly speaking, a smooth surface.

differential [CONT SYS] The difference between levels for turn-on and turn-off operation in a control system. [MATH] 1. The differential of a real-valued function $f(x)$, where x is a vector, evaluated at a given vector c, is the linear, real-valued function whose graph is the tangent hyperplane to the graph of $f(x)$ at $x=c$; if x is a real number, the usual notation is $df = f'(c)dx$. 2. *See* total differential. [MECH ENG] Any arrangement of gears forming an epicyclic train in which the angular speed of one shaft is proportional to the sum or difference of the angular speeds of two other gears which lie on the same axis; allows one shaft to revolve faster than the other, the speed of the main driving member being equal to the algebraic mean of the speeds of the two shafts. Also known as differential gear.

differential amplifier [ELECTR] An amplifier whose output is proportional to the difference between the voltages applied to its two inputs. Also called difference amplifier.

differential analyzer [COMPUT SCI] A mechanical or electromechanical device designed primarily to solve differential equations.

differential calculus [MATH] The study of the manner in which the value of a function changes as one changes the value of the independent variable; includes maximum-minimum problems and expansion of functions into Taylor series.

differential calorimetry [THERMO] Technique for measurement of and comparison (differential) of process heats (reaction, absorption, hydrolysis, and so on) for a specimen and a reference material.

differential capacitance [ELECTR] The derivative with respect to voltage of a charge characteristic, such as an alternating charge characteristic or a mean charge characteristic, at a given point on the characteristic.

differential capacitor [ELEC] A two-section variable capacitor having one rotor and two stators so arranged that as capacitance is reduced in one section it is increased in the other.

differential centrifugation [CYTOL] The separation of mixtures such as cellular particles in a medium at various centrifugal forces to separate particles of different density, size, and shape from each other.

differential coefficient *See* derivative.

differential comparator [ELECTR] A comparator having at least two high-gain differential-amplifier stages, followed by level-shifting and buffering stages, as required for converting a differential input to single-ended output for digital logic applications.

differential compound motor [ELEC] A direct-current motor whose speed may be made nearly constant or may be adjusted to increase with increasing load.

differential correction [ASTRON] A method for finding from the observed residuals minus the computed residuals $(O - C)$ small corrections which, when applied to the orbital elements or constants, will reduce the deviations from the observed motion to a minimum.

differential cross section [PHYS] The cross section for a collision process resulting in the emission of particles or photons at a specified angle relative to the direction of the incident particles, per unit angle or per unit solid angle.

differential delay [COMMUN] The difference between the maximum and minimum frequency delays occurring across a band.

differential discriminator [ELECTR] A discriminator that passes only pulses whose amplitudes are between two predetermined values, neither of which is zero.

differential equation [MATH] An equation expressing a relationship between functions and their derivatives.

differential fault *See* scissor fault.

differential frequency circuit [ELEC] A circuit that provides a continuous output frequency equal to the absolute difference between two continuous input frequencies.

differential frequency meter [ENG] A circuit that converts the absolute frequency difference between two input signals to a linearly proportional direct-current output voltage that can be used to drive a meter, recorder, oscilloscope, or other device.

differential gain control [ELECTR] Device for altering the gain of a radio receiver according to expected change of signal level, to reduce the amplitude differential between the signals at the output of the receiver. Also known as gain sensitivity control.

differential galvanometer [ELEC] A galvanometer having a magnetic needle which is free to rotate in the magnetic field produced by currents flowing in opposite directions through two separate identical coils, so that there is no deflection when the currents are equal.

differential gear *See* differential.

differential heat of solution [THERMO] The partial derivative of the total heat of solution with respect to the molal concentration of one component of the solution, when the concentration of the other component or components, the pressure, and the temperature are held constant.

differential input [ELECTR] Amplifier input circuit that rejects voltages that are the same at both input terminals and amplifies the voltage difference between the two input terminals.

differential instrument [ENG] Galvanometer or other measuring instrument having two circuits or coils, usually identical, through which currents flow in opposite directions; the difference or differential effect of these currents actuates the indicating pointer.

differential ionization chamber [NUCLEO] A two-section ionization chamber in which electrode potentials are such that output current is equal to the difference between the separate ionization currents of the two sections.

differential keying [ELECTR] Method for obtaining chirp-free break-in keying of continuous wave transmitters by using circuitry that arranges to have the oscillator turn on fast before the keyed amplifier stage can pass any signal, and turn off fast after the keyed amplifier stage has cut off.

differential leukocyte count [PATH] The percentage of each variety of leukocytes in the blood, usually based on counting 100 leukocytes. Also known as differential blood count.

differential leveling [ENG] A surveying process in which a horizontal line of sight of known elevation is intercepted by a graduated standard, or rod, held vertically on the point being checked.

differentially coherent phase-shift keying *See* differential phase-shift keying.

differential manometer [ENG] An instrument in which the difference in pressure between two sources is determined from the vertical distance between the surfaces of a liquid in two legs of an erect or inverted U-shaped tube when each of the legs is connected to one of the sources.

differential microphone *See* double-button microphone.

differential-mode signal [ELECTR] A signal that is applied between the two ungrounded terminals of a balanced three-terminal system.

differential modulation [COMMUN] Modulation in which the choice of the significant condition for any signal element is dependent on the choice for the previous signal element.

differential operator [MATH] An operator on a space of functions which maps a function f into a linear combination of higher-order derivatives of f.

differential phase [ELECTR] Difference in output phase of a small high-frequency sine-wave signal at two stated levels of a low-frequency signal on which it is superimposed in a video transmission system.

differential phase-shift keying [COMMUN] Form of phase-shift keying in which the reference phase for a given keying interval is the phase of the signal during the preceding keying interval. Also known as differentially coherent phase-shift keying.

differential polarography [ANALY CHEM] Technique of polarographic analysis which measures the difference in current flowing between two identical dropping-mercury electrodes at the same potential but in different solutions.

differential pressure [PHYS] The difference in pressure between two points of a system, such as between the well bottom and wellhead or between the two sides of an orifice.

differential pressure gage [ENG] Apparatus to measure pressure differences between two points in a system; it can be a pressured liquid column balanced by a pressured liquid reservoir, a formed metallic pressure element with opposing force, or an electrical-electronic gage (such as strain, thermal-conductivity, or ionization).

differential pulse-code modulation [COMMUN] A type of pulse-code modulation proposed for television transmission, in which only the differences between the continuous picture elements on the scanning lines are transmitted, enabling the bandwidth of the signal to be reduced. Abbreviated DPCM.

differential reaction rate [PHYS CHEM] The order of a chemical reaction expressed as a differential equation with respect to time; for example, $dx/dt = k(a - x)$ for first order, $dx/dt = k(a - x)(b - x)$ for second order, and so on, where k is the specific rate constant, a is the concentration of reactant A, b is the concentration of reactant B, and dx/dt is the rate of change in concentration for time t.

differential relay [ELEC] A two-winding relay that operates when the difference between the currents in the two windings reaches a predetermined value.

differential screw [MECH ENG] A type of compound screw which produces a motion equal to the difference in motion between the two component screws.

differential selection [STAT] A biased selection of a conditioned sample.

differential separation [CHEM ENG] Release of gas (vapor) from liquids by a reduction in pressure that allows the vapor to come out of the solution, so that the vapor can be removed from the system; differs from flash separation, in which the vapor and liquid are kept in contact following pressure reduction.

differential spectrophotometry [SPECT] Spectrophotometric analysis of a sample when a solution of the major component of the sample is placed in the reference cell; the recorded spectrum represents the difference between the sample and the reference cell.

differential synchro *See* synchro differential receiver; synchro differential transmitter.

differential thermal analysis [THERMO] A method of determining the temperature at which thermal reactions occur in a material undergoing continuous heating to elevated temperatures; also involves a determination of the nature and intensity of such reactions.

differential thermometer *See* bimetallic thermometer.

differential thermometric titration [ANALY CHEM] Thermometric titration in which titrant is added simultaneously to the reaction mixture and to a blank in identically equipped cells.

differential topology [MATH] The branch of mathematics dealing with differentiable manifolds.

differential transducer [ELEC] A transducer that simultaneously senses two separate sources and provides an output proportional to the difference between them.

differential transformer [ELEC] A transformer used to join two or more sources of signals to a common transmission line.

differential voltmeter [ELEC] A voltmeter that measures only the difference between a known voltage and an unknown voltage.

differential winding [ELEC] A winding whose magnetic field opposes that of a nearby winding.

differentiating circuit [ELEC] A circuit whose output voltage is proportional to the rate of change of the input voltage. Also known as differentiating network.

differentiating network *See* differentiating circuit.

differentiation [MATH] The act of taking a derivative.

differentiator [ELECTR] A device whose output function is proportional to the derivative, or rate of change, of the input function with respect to one or more variables.

diffluence [FL MECH] A region of fluid flow in which the fluid is diverging from the direction of flow.

diffracted wave [PHYS] A wave whose front has been changed in direction by an obstacle or other nonhomogeneity in a medium, other than by reflection or refraction.

diffraction [PHYS] Any redistribution in space of the intensity of waves that results from the presence of an object causing variations of either the amplitude or phase of the waves; found in all types of wave phenomena.

diffraction grating [SPECT] An optical device consisting of an assembly of narrow slits or grooves which produce a large number of beams that can interfere to produce spectra. Also known as grating.

diffraction instrument *See* diffractometer.

diffraction pattern [PHYS] Pattern produced on a screen or plate by waves which have undergone diffraction.

diffraction ring [OPTICS] Circular light pattern which appears to surround particles in a microscope field.

diffraction spectrum [SPECT] Parallel light and dark or colored bands of light produced by diffraction.

diffraction symmetry [CRYSTAL] Any symmetry in a crystal lattice which causes the systematic annihilation of certain beams in x-ray diffraction.

diffraction velocimeter [OPTICS] A velocity-measuring instrument that uses a continuous-wave laser to send a beam of coherent light at objects moving at right angles to the beam; the needlelike diffraction lobes reflected by the moving objects sweep past the optical grating in the receiver, thereby generating in a photomultiplier a series of impulses from which velocity can be determined and read out. Also known as optical diffraction velocimeter.

diffraction zone [ELECTROMAG] The portion of a radio propagation path which lies outside a line-of-sight path.

diffractometer [PHYS] An instrument used to study the structure of matter by means of the diffraction of x-rays, electrons, neutrons, or other waves. Also known as diffraction instrument.

diffractometry [CRYSTAL] The science of determining crystal structures by studying the diffraction of beams of x-rays or other waves.

diffuse-cutting filter [OPTICS] A color filter that gradually changes in absorption with wavelength.

diffused-alloy transistor [ELECTR] A transistor in which the semiconductor wafer is subjected to gaseous diffusion to produce a nonuniform base region, after which alloy junctions are formed in the same manner as for an alloy-junction transistor; it may also have an intrinsic region, to give a *pnip* unit. Also known as drift transistor.

diffused-base transistor [ELECTR] A transistor in which a nonuniform base region is produced by gaseous diffusion; the collector-base junction is also formed by gaseous diffusion, while the emitter-base junction is a conventional alloy junction.

diffused junction [ELECTR] A semiconductor junction that has been formed by the diffusion of an impurity within a semiconductor crystal.

diffused-junction rectifier [ELECTR] A semiconductor diode in which the *pn* junction is produced by diffusion.

diffused-junction transistor [ELECTR] A transistor in which the emitter and collector electrodes have been formed by diffusion by an impurity metal into the semiconductor wafer without heating.

diffused-mesa transistor [ELECTR] A diffused-junction transistor in which an *n*-type impurity is diffused into one side of a *p*-type wafer; a second *pn* junction, required for the emitter, is produced by alloying or diffusing a *p*-type impurity into the newly formed *n*-type surface; after contacts have been applied, undesired diffused areas are etched away to create a flat-topped peak called a mesa.

diffused resistor [ELECTR] An integrated-circuit resistor produced by a diffusion process in a semiconductor substrate.

diffuse front [METEOROL] A front across which the characteristics of wind shift and temperature change are weakly defined.

diffuse nebula [ASTRON] A type of nebula ranging from huge masses presenting relatively high surface brightness down to faint, milky structures that are detectable only with long exposures and special filters; may contain both dust and gas or may be purely gaseous.

diffuser [ENG] A duct, chamber, or section in which a high-velocity, low-pressure stream of fluid (usually air) is converted into a low-velocity, high-pressure flow.

diffuse radiation [PHYS] Radiant energy propagating in many different directions through a given small volume of space.

diffuse reflector [OPTICS] Any surface whose irregularities are so large compared to the wavelength of the incident radiation that the reflected rays are sent back in a multiplicity of directions.

diffuse series [SPECT] A series occurring in the spectra of many atoms having one, two, or three electrons in the outer shell, in which the total orbital angular momentum quantum number changes from 2 to 1.

diffuse skylight See diffuse sky radiation.

diffuse sky radiation [ASTROPHYS] Solar radiation reaching the earth's surface after having been scattered from the direct solar beam by molecules or suspensoids in the atmosphere. Also known as diffuse skylight; skylight; sky radiation.

diffuse spectrum [SPECT] Any spectrum having lines which are very broad even when there is no possibility of line broadening by collisions.

diffuse transmission [PHYS] Transmission of electromagnetic or acoustic radiation in all directions by a transmitting body.

diffusion [ELECTR] A method of producing a junction by diffusing an impurity metal into a semiconductor at a high temperature. [MECH ENG] The conversion of air velocity into static pressure in the diffuser casing of a centrifugal fan, resulting from increases in the radius of the air spin and in area. [METEOROL] The exchange of fluid parcels (and hence the transport of conservative properties) between regions in space, in the apparently random motions of the parcels on a scale too small to be treated by the equations of motion; the diffusion of momentum (viscosity), vorticity, water vapor, heat (conduction), and gaseous components of the atmospheric mixture have been studied extensively. [OPTICS] **1.** The distribution of incident light by reflection. **2.** Transmission of light through a translucent material. [PHYS] The spontaneous movement and scattering of particles (atoms and molecules), of liquids, gases, and solids. [SOLID STATE] **1.** The actual transport of mass, in the form of discrete atoms, through the lattice of a crystalline solid. **2.** The movement of carriers in a semiconductor.

diffusion barrier [CHEM ENG] Porous barrier through which gaseous mixtures are passed for enrichment of the lighter-molecular-weight constituent of the diffusate; used as a many-stage cascade system for the recovery of $U^{235}F_6$ isotopes from a $U^{238}F_6$ stream.

diffusion bonding [MET] A solid-state process for joining metals by using only heat and pressure to achieve atomic bonding.

diffusion brazing [MET] A process which produces bonding of the faying surfaces by heating them to suitable temperatures; the filler metal is diffused with the base metal and approaches the properties of the base metal.

diffusion capacitance [ELECTR] The rate of change of stored minority-carrier charge with the voltage across a semiconductor junction.

diffusion coating [MET] An alloy coating produced by allowing the coating material to diffuse into the base at high temperature.

diffusion coefficient [PHYS] The weight of a material, in grams, diffusing across an area of 1 square centimeter in 1 second in a unit concentration gradient. Also known as diffusivity.

diffusion constant [SOLID STATE] The diffusion current density in a homogeneous semiconductor divided by the charge carrier concentration gradient.

diffusion current [ANALY CHEM] In polarography with a dropping-mercury electrode, the flow that is controlled by the rate of diffusion of the active solution species across the concentration gradient produced by the removal of ions or molecules at the electrode surface.

diffusion equation [PHYS] 1. An equation for diffusion which states that the rate of change of the density of the diffusing substance, at a fixed point in space, equals the sum of the diffusion coefficient times the Laplacian of the density, the amount of the quantity generated per unit volume per unit time, and the negative of the quantity absorbed per unit volume per unit time. 2. More generally, any equation which states that the rate of change of some quantity, at a fixed point in space, equals a positive constant times the Laplacian of that quantity.

diffusion gradient [PHYS] The graphed distance of penetration (diffusion) versus concentration of the material (or effect) diffusing through a second material; applies to heat, liquids, solids, or gases.

diffusion hygrometer [ENG] A hygrometer based upon the diffusion of water vapor through a porous membrane; essentially, it consists of a closed chamber having porous walls and containing a hygroscopic compound, whose absorption of water vapor causes a pressure drop within the chamber that is measured by a manometer.

diffusion length [PHYS] The average distance traveled by a particle, such as a minority carrier in a semiconductor or a thermal neutron in a nuclear reactor, from the point at which it is formed to the point at which it is absorbed.

diffusion pump [ENG] A vacuum pump in which a stream of heavy molecules, such as mercury vapor, carries gas molecules out of the volume being evacuated; also used for separating isotopes according to weight, the lighter molecules being pumped preferentially by the vapor stream.

diffusion theory [ELEC] The theory that in semiconductors, where there is a variation of carrier concentration, a motion of the carriers is produced by diffusion in addition to the drift determined by the mobility and the electric field.

diffusion transistor [ELECTR] A transistor in which current flow is a result of diffusion of carriers, donors, or acceptors, as in a junction transistor.

diffusion velocity [FL MECH] 1. The relative mean molecular velocity of a selected gas undergoing diffusion in a gaseous atmosphere, commonly taken as a nitrogen (N_2) atmosphere; a molecular phenomenon that depends upon the gaseous concentration as well as upon the pressure and temperature gradients present. 2. The velocity or speed with which a turbulent diffusion process proceeds as evidenced by the motion of individual eddies.

diffusion welding [MET] A welding process which utilizes high temperatures and pressures to coalesce the faying surfaces by solid-state bonding; there is no physical movement, visible deformation of the parts involved, or melting.

diffusivity [PHYS] *See* diffusion coefficient. [THERMO] The quantity of heat passing normally through a unit area per unit time divided by the product of specific heat, density, and temperature gradient. Also known as thermometric conductivity.

diffusivity analysis [ANALY CHEM] Analysis of difficult-to-separate materials in solution by diffusion effects, using, for example, dialysis, electrodialysis, interferometry, amperometric titration, polarography, or voltammetry.

digastric [ANAT] Of a muscle, having a fleshy part at each end and a tendinous part in the middle.

Digenea [INV ZOO] A group of parasitic flatworms or flukes constituting a subclass or order of the class Trematoda and having two types of generations in the life cycle.

digester [CHEM ENG] A vessel used to produce cellulose pulp from wood chips by cooking under pressure. [CIV ENG] A sludge-digestion tank containing a system of hot water or steam pipes for heating the sludge.

digestion [CHEM ENG] 1. Preferential dissolving of mineral constituents in concentrations of ore. 2. Liquefaction of organic waste materials by action of microbes. 3. Separation of fabric from tires by the use of hot sodium hydroxide. 4. Removing lignin from wood in manufacture of chemical cellulose paper pulp. [CIV ENG] The process of sewage treatment by the anaerobic decomposition of organic matter. [PHYSIO] The process of converting food to an absorbable form by breaking it down to simpler chemical compounds.

digestive system [ANAT] A system of structures in which food substances are digested.

digestive tract [ANAT] The alimentary canal.

digicom [COMMUN] A wire communication system that transmits speech signals in the form of corresponding trains of pulses and transmits digital information directly from computers, radar, tape readers, teleprinters, and telemetering equipment.

digit [COMPUT SCI] In a decimal digital computer, the space reserved for storage of one digit of information. [MATH] A character used to represent one of the nonnegative integers smaller than the base of a system of positional notation. Also known as numeric character.

digital [COMPUT SCI] Pertaining to data in the form of digits.

digital circuit [ELECTR] A circuit designed to respond at input voltages at one of a finite number of levels and, similarly, to produce output voltages at one of a finite number of levels.

digital communications [COMMUN] System of telecommunications employing a nominally discontinuous signal that changes in frequency, amplitude, or polarity.

digital comparator [ELECTR] A comparator circuit operating on input signals at discrete levels. Also known as discrete comparator.

digital computer [COMPUT SCI] A computer operating on discrete data by performing arithmetic and logic processes on these data.

digital converter [ELECTR] A device that converts voltages to digital form; examples include analog-to-digital converters, pulse-code modulators, encoders, and quantizing encoders.

digital counter [ELECTR] A discrete-state device (one with only a finite number of output conditions) that responds by advancing to its next output condition.

Digital Data Broadcast System [NAV] A system that will provide information aiding air-traffic control; digital data to aircraft over vortac channels will carry information on the geographic location, elevation, magnetic variation, and related data of the vortac station being received. Abbreviated DDBS.

digital data recorder [COMPUT SCI] Electronic device that converts continuous electrical analog signals into number (digital) values and records these values onto a data log via a high-speed typewriter.

digital data service [COMMUN] A telephone communication system developed specifically for digital data, using existing local digital lines combined with data-under-voice microwave transmission facilities. Abbreviated DDS.

digital differential analyzer [COMPUT SCI] A differential analyzer which uses numbers to represent analog quantities. Abbreviated DDA.

digital display [COMPUT SCI] A display in which the result is indicated in directly readable numerals.

digital filter [ELECTR] An electrical filter that responds to an input which has been quantified, usually as pulses.

digital incremental plotter [COMPUT SCI] A device for converting digital signals in the output of a computer into graphical form, in which the digital signals control the motion of a plotting pen and of a drum that carries the paper on which the graph is drawn.

digital integrator [COMPUT SCI] A device for computing definite integrals in which increments in the input variables and output variable are represented by digital signals.

digitalis [PHARM] The dried leaf of the purple foxglove plant (*Digitalis purpurea*), containing digitoxin and gitoxin; constitutes a powerful cardiac stimulant and diuretic.

digital modulation [COMMUN] A method of placing digital traffic on a microwave system without use of modems, by transmitting the information in the form of discrete phase or frequency states determined by the digital signal.

digital multiplier [ELECTR] A multiplier that accepts two numbers in digital form and gives their product in the same digital form, usually by making repeated additions; the multiplying process is simpler if the numbers are in binary form wherein digits are represented by a 0 or 1.

digital phase shifter [ELECTR] Device which provides a signal phase shift by the application of a control pulse; a reversal or phase shift requires a control pulse of opposite polarity.

digital plotter [ELECTR] A recorder that produces permanent hard copy in the form of a graph from digital input data.

digital printer [COMPUT SCI] A printer that provides a permanent readable record of binary-coded decimal or other coded data in a digital form that may include some or all alphanumeric characters and special symbols along with numerals. Also known as digital recorder.

digital recorder See digital printer.

digital recording [ELECTR] Magnetic recording in which the information is first coded in a digital form, generally with a binary code that uses two discrete values of residual flux.

digital representation [COMPUT SCI] The use of discrete impulses or quantities in coded patterns to represent variables or other data in the form of numbers or characters.

digital simulation [COMPUT SCI] The representation of a system in a form acceptable to a digital computer as opposed to an analog computer.

digital system [COMPUT SCI] Any of the levels of operation for a digital computer, including the wires and mechanical parts, the logical elements, and the functional units for reading, writing, storing, and manipulating information.

digital television [COMMUN] Television in which picture redundancy is reduced or eliminated by transmitting only the data needed to define motion in the picture, as represented by changes in the areas of continuous white or continuous black.

digital-to-analog converter [ELECTR] A converter in which digital input signals are changed to essentially proportional analog signals. Abbreviated dac.

digital transducer [ELECTR] A transducer that measures physical quantities and transmits the information as coded digital signals rather than as continuously varying currents or voltages.

digit-coded voice [COMPUT SCI] A limited, spoken vocabulary, each word of which corresponds to a code and which, upon keyed inquiry, can be strung in meaningful sequence and can be outputted as audio response to the inquiry.

digit compression [COMPUT SCI] Any process which increases the number of digits stored at a given location.

digitigrade [VERT ZOO] Pertaining to animals, such as dogs and cats, which walk on the digits with the posterior part of the foot raised from the ground.

digitize [COMPUT SCI] To convert an analog measurement of a quantity into a numerical value.

digitoxin [ORG CHEM] $C_{41}H_{64}O_{13}$ A poisonous steroid glycoside found as the most active principle of digitalis, from the foxglove leaf.

digit plane [COMPUT SCI] In a computer memory consisting of magnetic cores arranged in a three-dimensional array, a plane containing elements for a particular digit position in various words.

digit pulse [ELECTR] An electrical pulse which induces a magnetizing force in a number of magnetic cores in a computer storage, all corresponding to a particular digit position in a number of different words.

digit selector [COMPUT SCI] A device which separates a card column into individual pulses corresponding to punched row positions.

digoxin [ORG CHEM] $C_{41}H_{64}O_{14}$ A crystalline steroid obtained from a foxglove leaf (*Digitalis lanata*); similar to digitalis in pharmacological effects.

dihalide [CHEM] A molecule containing two atoms of halogen combined with a radical or element.

dihedral [AERO ENG] The upward or downward inclination of an airplane's wing or other supporting surface in respect to the horizontal; in some contexts, the upward inclination only. [MATH] *See* dihedron.

dihedron [MATH] A geometric figure formed by two half planes that are bounded by the same straight line. Also known as dihedral.

dihexagonal [CRYSTAL] Of crystals, having a symmetrical form with 12 sides.

dihexahedron [CRYSTAL] A type of crystal that has 12 faces, such as a double six-sided pyramid.

dihydrate [CHEM] A compound with two molecules of water of hydration.

dihydroxy [CHEM] A molecule containing two hydroxyl groups.

dihydroxyphenylalanine [BIOCHEM] $C_9H_{11}NO_4$ An amino acid that can be formed by oxidation of tyrosine; it is converted by a series of biochemical transformations, utilizing the enzyme dopa oxidase, to melanins. Also known as dopa.

dikaryon [MYCOL] Also spelled dicaryon. 1. A pair of distinct, unfused nuclei in the same cell brought together by union of plus and minus hyphae in certain mycelia. 2. Mycelium containing dikaryotic cells.

dike [CIV ENG] An embankment constructed on dry ground along a riverbank to prevent overflow of lowlands and to retain floodwater. [GEOL] A tabular body of igneous rock that cuts across adjacent rocks or cuts massive rocks.

diketone [ORG CHEM] A molecule containing two ketone carbonyl groups.

diktoma *See* neuroepithelioma.

dilatancy [CHEM] The property of a viscous suspension which sets solid under the influence of pressure. [GEOL] Expansion of deformed masses of granular material, such as sand, due to rearrangement of the component grains.

dilatation [PHYS] The increase in volume per unit volume of any continuous substance, caused by deformation.

dilatational wave *See* compressional wave.

dilation [MATH] A transformation which changes the size, and only the size, of a geometric figure. [SCI TECH] The act or process of stretching or expanding.

dilatometry [PHYS] The measurement of changes in the volume of a liquid or dimensions of a solid which occur in phenomena such as allotropic transformations, thermal expansion, compression, creep, or magnetostriction.

dill [BOT] *Anethum graveolens*. A small annual or biennial herb in the family Umbelliferae; the aromatic leaves and seeds are used for food flavoring.

Dilleniaceae [BOT] A family of dicotyledonous trees, woody vines, and shrubs in the order Dilleniales having hypogynous flowers and mostly entire leaves.

Dilleniales [BOT] An order of dicotyledonous plants in the subclass Dilleniidae characterized by separate carpels and numerous stamens.

Dilleniidae [BOT] A subclass of plants in the class Magnoliopsida distinguished by being syncarpous, having centrifugal stamens, and usually having bitegmic ovules and binucleate pollen.

dilute [CHEM] To make less concentrated.

dilute phase [CHEM ENG] In liquid-liquid extraction, the liquid phase that is dilute with respect to the material being extracted.

dilution [CHEM] Increasing the proportion of solvent to solute in any solution and thereby decreasing the concentration of the solute per unit volume. [MET] The use of a welding filler metal deposit with a base metal or a previously deposited weld material having a lower alloy content. [OPTICS] Reducing the intensity of a color by adding white.

dilution gene [GEN] Any modifier gene that acts to reduce the effect of another gene.

dilution method [MICROBIO] A technique in which a series of cultures is tested with various concentrations of an antibiotic to determine the minimum inhibiting concentration of antibiotic.

dimension [GRAPHICS] In a mechanical drawing, a labeled measure in a straight line of the breadth, height, or thickness of a part, the angular position of a line, or the location of a detail such as a hole or boss.

dimensional analysis [PHYS] A technique that involves the study of dimensions of physical quantities, used primarily as a tool for obtaining information about physical systems too complicated for full mathematical solutions to be feasible.

dimensional constant [PHYS] A physical quantity whose numerical value depends on the units chosen for fundamental quantities but not on the system being considered.

dimension declaration statement [COMPUT SCI] A FORTRAN statement identifying arrays and specifying the number and bounds of the subscripts.

dimensioning [GRAPHICS] Assigning of dimensions to a mechanical drawing.

dimensionless group [PHYS] Any combination of dimensional or dimensionless quantities possessing zero overall dimensions; an example is the Reynolds number.

dimensionless number [MATH] A ratio of various physical properties (such as density or heat capacity) and conditions (such as flow rate or weight) of such nature that the resulting number has no defining units of weight, rate, and so on. Also known as nondimensional parameter.

dimension theory [MATH] The study of abstract notions of dimension, which are topological invariants of a space.

dimer [CHEM] A condensation product consisting of two molecules.

dimeric water [INORG CHEM] Water in which pairs of molecules are joined by hydrogen bonds.

dimethyl [ORG CHEM] A compound that has two methyl groups.

dimictic lake [HYD] A lake which circulates twice a year.

dimorphism [CHEM] Having crystallization in two forms with the same chemical composition. [SCI TECH] Existing in two distinct forms, with reference to two members expected to be identical.

dimuon event [PARTIC PHYS] An inelastic collision of a neutrino or antineutrino with a nucleus in which there are two muons among the products of the collision.

Dimylidae [PALEON] A family of extinct lipotyphlan mammals in the order Insectivora; a side branch in the ancestry of the hedgehogs.

Dinantian [GEOL] Lower Carboniferous geologic time. Also known as Avonian.

D indicator *See* D scope.

Dines anemometer [ENG] A pressure-tube anemometer in which the pressure head on a weather vane is kept facing into the wind, and the suction head, near the bearing which supports the vane, develops a suction independent of wind direction; the pressure difference between the heads is proportional to the square of the wind speed and is measured by a float manometer with a linear wind scale.

dineutron [NUC PHYS] **1.** A hypothetical bound state of two neutrons, which probably does not exist. **2.** A combination of two neutrons which has a transitory existence in certain nuclear reactions.

Dinidoridae [INV ZOO] A family of hemipteran insects in the superfamily Pentatomoidea.

dinitrate [CHEM] A molecule that contains two nitrate groups.

dinitrogen fixation *See* nitrogen fixation.

Dinocerata [PALEON] An extinct order of large, herbivorous mammals having semigraviportal limbs and hoofed, five-toed feet; often called uintatheres.

Dinoflagellata [INV ZOO] The equivalent name for Dinoflagellida.

Dinoflagellida [INV ZOO] An order of flagellate protozoans in the class Phytamastigophorea; most members have fixed shapes determined by thick covering plates.

Dinophilidae [INV ZOO] A family of annelid worms belonging to the Archiannelida.

Dinophyceae [BOT] The dinoflagellates, a class of thallophytes in the division Pyrrhophyta.

Dinornithiformes [PALEON] The moas, an order of extinct birds of New Zealand; all had strong legs with four-toed feet.

dinosaur [PALEON] The name, meaning terrible lizard, applied to the fossil bones of certain large, ancient bipedal and quadripedal reptiles placed in the orders Saurischia and Ornithischia.

DIN system [GRAPHICS] A system in photography used to find the speed of photographic emulsions; it is stated in terms of the logarithm of the reciprocal of the exposure needed to obtain a density of 0.1 above fog density.

dioctahedral [CRYSTAL] Pertaining to a crystal structure in which only two of the three available octahedrally coordinated positions are occupied. [MATH] Having 16 faces.

Dioctophymatida [INV ZOO] An order of parasitic nematode worms in the subclass Enoplia.

Dioctophymoidea [INV ZOO] An order or superfamily of parasitic nematodes characterized by the peculiar structure of the copulatory bursa of the male.

dioctyl [ORG CHEM] A compound that has two octyl groups.

diode [ELECTR] 1. A two-electrode electron tube containing an anode and a cathode. 2. *See* semiconductor diode.

diode amplifier [ELECTR] A microwave amplifier using an IMPATT, TRAPATT, or transferred-electron diode in a microwave circulator providing the input/output isolation required for amplification; center frequencies are in the gigahertz range, from about 1 to 100 gigahertz, and power outputs are up to 20 watts continuous-wave or more than 200 watts pulsed, depending on the diode used.

diode characteristic [ELECTR] The composite electrode characteristic of an electron tube when all electrodes except the cathode are connected together.

diode clamp *See* diode clamping circuit.

diode clamping circuit [ELECTR] A clamping circuit in which a diode provides a very low resistance whenever the potential at a certain point rises above a certain value in some circuits or falls below a certain value in others. Also known as diode clamp.

diode clipping circuit [ELECTR] A clipping circuit in which a diode is used as a switch to perform the clipping action.

diode demodulator [ELECTR] A demodulator using one or more crystal or electron tube diodes to provide a rectified output whose average value is proportional to the original modulation. Also known as diode detector.

diode detector *See* diode demodulator.

diode drop *See* diode forward voltage.

diode forward voltage [ELECTR] The voltage across a semiconductor diode that is carrying current in the forward direction; it is usually approximately constant over the range of currents commonly used. Also known as diode drop; diode voltage; forward voltage drop.

diode function generator [ELECTR] A function generator that uses the transfer characteristics of resistive networks containing biased diodes; the desired function is approximated by linear segments.

diode gate [ELECTR] An AND gate that uses diodes as switching elements.

diode laser *See* semiconductor laser.

diode limiter [ELECTR] A peak-limiting circuit employing a diode that becomes conductive when signal peaks exceed a predetermined value.

diode modulator [ELECTR] A modulator using one or more diodes to combine a modulating signal with a carrier signal; used chiefly for low-level signaling because of inherently poor efficiency.

diode rectifier-amplifier meter [ELECTR] The most widely used vacuum tube voltmeter for measurement of alternating-current voltage; has separate tubes for rectification and direct-current amplification, permitting an optimum design for each.

diode-switch [ELECTR] Diode which is made to act as a switch by the successive application of positive and negative biasing voltages to the anode (relative to the cathode), thereby allowing or preventing, respectively, the passage of other applied waveforms within certain limits of voltage.

diode theory [ELEC] The theory that in a semiconductor, when the barrier thickness is comparable to or smaller than the mean free path of the carriers, then the carriers cross the barrier without being scattered, much as in a vacuum tube diode.

diode transistor logic [ELECTR] A circuit that uses diodes, transistors, and resistors to provide logic functions. Abbreviated DTL.

diode voltage *See* diode forward voltage.

diodide [CHEM] A molecule that contains two iodine atoms bonded to an element or radical.

dioecious [BIOL] Having the male and female reproductive organs on different individuals. Also known as dioic.

diogenite [MINERAL] An achondritic stony meteorite composed essentially of iron-rich pyroxene minerals. Also known as rodite.

dioic *See* dioecious.

diolefin *See* diene.

Diomedeidae [VERT ZOO] The albatrosses, a family of birds in the order Procellariiformes.

Dione [ASTRON] A satellite of Saturn that orbits at a mean distance of 377×10^2 kilometers and has a diameter of about 800 kilometers.

diophantine equations [MATH] Equations with more than one independent variable and with integer coefficients for which integer solutions are desired.

Diopsidae [INV ZOO] The stalk-eyed flies, a family of myodarian cyclorrhaphous dipteran insects in the subsection Acalypteratae.

diopside [MINERAL] $CaMg(SiO_3)_2$ A white to green monoclinic pyroxene mineral which forms gray to white, short, stubby, prismatic, often equidimensional crystals. Also known as malacolite.

dioptase [MINERAL] $CuSiO_2(OH)_2$ A rare emerald-green mineral that forms hexagonal, hydrous crystals.

diopter [OPTICS] A measure of the power of a lens or a prism, equal to the reciprocal of its focal length in meters. Abbreviated D.

dioptrics [OPTICS] The branch of optics that treats of the refraction of light, especially by the transparent medium of the eye, and by lenses.

diorite [PETR] A phaneritic plutonic rock with granular texture composed largely of plagioclase feldspar with smaller amounts of dark-colored minerals; used occasionally as ornamental and building stone. Also known as black granite.

Dioscoreaceae [BOT] A family of monocotyledonous, leafy-stemmed, mostly twining plants in the order Liliales, having an inferior ovary and septal nectaries and lacking tendrils.

dioxygenase [BIOCHEM] Any of a group of enzymes which catalyze the insertion of both atoms of an oxygen molecule into an organic substrate according to the generalized formula $AH_2 + O_2 \rightarrow A(OH)_2$.

dip [ENG] The vertical angle between the sensible horizon and a line to the visible horizon at sea, due to the elevation of the observer and to the convexity of the earth's surface. Also known as dip of horizon. [GEOL] 1. The angle that a stratum or fault plane makes with the horizontal. Also known as angle of dip; true dip. 2. A pronounced depression in the land surface.

dipeptidase [BIOCHEM] An enzyme that hydrolyzes a dipeptide.

dip fault [GEOL] A type of fault that strikes parallel with the dip of the strata involved.

diphenol [ORG CHEM] A compound that has two phenol groups, for example, resorcinol.

diphosphopyridine nucleotide [BIOCHEM] $C_{21}H_{27}O_{14}N_7P_2$ An organic coenzyme that functions in enzymatic systems concerned with oxidation-reduction reactions. Abbreviated DPN. Also known as codehydrogenase 1; coenzyme 1; nicotinamide adenine dinucleotide (NAD).

diphtheria [MED] A communicable bacterial disease of man caused by the growth of *Corynebacterium diphtheriae* on any mucous membrane, especially of the throat.

diphycercal [VERT ZOO] Pertaining to a tail fin, having symmetrical upper and lower parts, and with the vertebral column extending to the tip without upturning.

diphyletic [EVOL] Originating from two lines of descent.

Diphyllidea [INV ZOO] A monogeneric order of tapeworms in the subclass Cestoda; all species live in the intestine of elasmobranch fishes.

diphyllous [BOT] Having two leaves.

diphyodont [ANAT] Having two successive sets of teeth, deciduous followed by permanent, as in humans.

dip inductor *See* earth inductor.

dip joint [GEOL] A joint that strikes approximately at right angles to the cleavage or bedding of the constituent rock.

Diplacanthidae [PALEON] A family of extinct acanthodian fishes in the suborder Diplacanthoidei.

Diplacanthoidei [PALEON] A suborder of extinct acanthodian fishes in the order Climatiiformes.

diplacusis [MED] A difference in the pitch perceptions of the two ears when stimulated by the same sound frequency.

Diplasiocoela [VERT ZOO] A suborder of amphibians in the order Anura typically having the eighth vertebra biconcave.

diplegia [MED] Paralysis of similar parts on the two sides of the body.

diplexer [ELECTR] A coupling system that allows two different transmitters to operate simultaneously or separately from the same antenna.

Diplobathrida [PALEON] An order of extinct, camerate crinoids having two circles of plates beneath the radials.

diploblastic [ZOO] Having two germ layers, referring to embryos and certain lower invertebrates.

diploglossate [VERT ZOO] Pertaining to certain lizards, having the ability to retract the end of the tongue into the basal portion.

diploid [CRYSTAL] A crystal form in the isometric system having 24 similar quadrilateral faces arranged in pairs.

diploid state [GEN] A condition in which a chromosome set is present in duplicate in a nucleus (2N).

Diplomonadida [INV ZOO] An order of small, colorless protozoans in the class Zoomastigophorea, having a bilaterally symmetrical body with four flagella on each side.

Diplomystidae [VERT ZOO] A family of catfishes in the suborder Siluroidei confined to the waters of Chile and Argentina.

diplopia [MED] A disorder characterized by double vision.

Diplopoda [INV ZOO] The millipeds, a class of terrestrial tracheate, oviparous arthropods; each body segment except the first few bears two pairs of walking legs.

Diploporita [PALEON] An extinct order of echinoderms in the class Cystoidea in which the thecal canals were associated in pairs.

Diplorhina [VERT ZOO] The subclass of the class Agnatha that includes the jawless vertebrates with paired nostrils.

diplotene [CYTOL] The stage of meiotic prophase during which pairs of nonsister chromatids of each bivalent repel each other and are kept from falling apart by the chiasmata.

Diplura [INV ZOO] An order of small, primarily wingless insects of worldwide distribution.

dipmeter [ENG] 1. An instrument used to measure the direction and angle of dip of geologic formations. 2. An absorption wavemeter in which bipolar or field-effect transistors replace the electron tubes used in older grid-dip meters.

Dipneumonomorphae [INV ZOO] A suborder of the order Araneida comprising the spiders common in the United States, including grass spiders, hunting spiders, and black widows.

Dipneusti [VERT ZOO] The equivalent name for Dipnoi.

Dipnoi [VERT ZOO] The lungfishes, a subclass of the Osteichthyes having lungs that arise from a ventral connection in the gut.

Dipodidae [VERT ZOO] The Old World jerboas, a family of mammals in the order Rodentia.

dip of horizon *See* dip.

dipolar ion [CHEM] An ion carrying both a positive and a negative charge. Also known as zwitterion.

dipole [ELECTROMAG] Any object or system that is oppositely charged at two points, or poles, such as a magnet or a polar molecule; more precisely, the limit as either charge goes to infinity, the separation distance to zero, while the product remains constant. Also known as doublet; electric doublet.

dipole antenna [ELECTROMAG] An antenna approximately one-half wavelength long, split at its electrical center for connection to a transmission line whose radiation pattern has a maximum at right angles to the antenna. Also known as doublet antenna; half-wave dipole.

dipole moment *See* electric dipole moment; magnetic dipole moment.

dipole relaxation [ELEC] The process, occupying a certain period of time after a change in the applied electric field, in which the orientation polarization of a substance reaches equilibrium.

dipole transition [ATOM PHYS] A transition of an atom or nucleus from one energy state to another in which dipole radiation is emitted or absorbed.

dip plating *See* immersion plating.

dip pole *See* magnetic pole.

dip reversal *See* reversal of dip.

Diprionidae [INV ZOO] The conifer sawflies, a family of hymenopteran insects in the superfamily Tenthredinoidea.

dipropyl [ORG CHEM] A compound containing two propyl groups.

Diprotodonta [VERT ZOO] A proposed order of marsupial mammals to include the phalangers, wombats, koalas, and kangaroos.

Diprotodontidae [PALEON] A family of extinct marsupial mammals.

Dipsacales [BOT] An order of dicotyledonous herbs and shrubs in the subclass Asteridae characterized by an inferior ovary and usually opposite leaves.

dip slip [GEOL] The component of a fault parallel to the dip of the fault. Also known as normal displacement.

dip slope [GEOL] A slope of the surface of the land determined by and conforming approximately to the dip of the underlying rocks. Also known as back slope; outface.

Dipsocoridae [INV ZOO] A family of hemipteran insects in the superfamily Dipsocoroidea; members are predators on small insects under bark or in rotten wood.

Dipsocoroidea [INV ZOO] A superfamily of minute, ground-inhabiting hemipteran insects belonging to the subdivision Geocorisae.

dip stream [HYD] A consequent stream that flows in the direction of the dip of the strata it traverses.

Diptera [INV ZOO] The true flies, an order of the class Insecta characterized by possessing only two wings and a pair of balancers.

Dipteriformes [VERT ZOO] The single order of the subclass Dipnoi, the lungfishes.

Dipterocarpaceae [BOT] A family of dicotyledonous plants in the order Theales having mostly stipulate, alternate leaves, a prominently exserted connective, and a calyx that is mostly winged in fruit.

dipterous [BIOL] **1.** Of, related to, or characteristic of Diptera. **2.** Having two wings or winglike structures.

dipulse [COMMUN] Transmission of a binary code in which the presence of one cycle of a sine-wave tone represents a binary "1" and the absence of one cycle represents a binary "0."

Dirac delta function See delta function.

Dirac electron theory See Dirac theory.

Dirac equation [QUANT MECH] A relativistic wave equation for an electron in an electromagnetic field, in which the wave function has four components corresponding to four internal states specified by a two-valued spin coordinate and an energy coordinate which can have a positive or negative value.

Dirac h See h-bar.

Dirac matrix [QUANT MECH] Any one of four matrices, designated γ_μ (μ = 1, 2, 3, 4), each having four rows and four columns and satisfying $\gamma_\mu \gamma_\nu + \gamma_\nu \gamma_\mu = \delta_{\mu\nu}$, where $\delta_{\mu\nu}$ is the Kronecker delta function, which matrices operate on the four-component wave function in the Dirac equation. Also known as gamma matrix.

Dirac moment [QUANT MECH] Magnetic moment of the electron according to the Dirac theory, equal to $e\hbar/2mc$, where e and m are the charge and mass of the positron respectively, $\hbar$ is Planck's constant divided by 2π, and c is the speed of light.

Dirac monopole [QUANT MECH] A magnetic monopole whose magnetic charge is an integral multiple of $\hbar c/2e$, where $\hbar$ is Planck's constant divided by 2π, c is the speed of light, and e is the charge of the electron.

Dirac particle [PARTIC PHYS] A particle behaving according to the Dirac theory, which describes the behavior of electrons and muons except for radiative corrections, and is envisaged as describing a central core of a hadron of spin $\frac{1}{2}\hbar$ which remains when the effects of nuclear forces are removed.

Dirac quantization [QUANT MECH] The condition, arising from conservation of angular momentum, that for any electric charge q and magnetic monopole with magnetic charge m, one has $2qm = n\hbar c$, where n is an integer, $\hbar$ is Planck's constant divided by 2π, and c is the speed of light (gaussian units).

Dirac spinor See spinor.

Dirac theory [QUANT MECH] Theory of the electron based on the Dirac equation, which accounts for its spin angular momentum and gives its magnetic moment and its behavior in an electromagnetic field (except for higher-order corrections). Also known as Dirac electron theory.

Dirac wave function [QUANT MECH] A function appropriate for describing a spin ½ particle and antiparticle; it is a column matrix with four entries, each of which is a function of the space and time coordinates; the four-components form two first-rank Lorentz spinors.

direct access See random access.

direct-access memory See random-access memory.

direct-access storage See random-access memory.

direct-acting pump [MECH ENG] A displacement reciprocating pump in which the steam or power piston is connected to the pump piston by means of a rod, without crank motion or flywheel.

direct address [COMPUT SCI] Any address specifying the location of an operand.

direct allocation [COMPUT SCI] A system in which the storage locations and peripheral units to be assigned to use by a computer program are specified when the program is written, in contrast to dynamic allocation.

direct-arc furnace [ENG] A furnace in which a material in a refractory-lined shell is rapidly heated to pour temperature by an electric arc which goes directly from electrodes to the material.

direct code [COMPUT SCI] A code in which instructions are written in the basic machine language.

direct-connected [MECH ENG] The connection between a driver and a driven part, as a turbine and an electric generator, without intervening speed-changing devices, such as gears.

direct-contact condenser See contact condenser.

direct control [COMPUT SCI] The control of one machine in a data-processing system by another, without human intervention.

direct cost [IND ENG] The cost in goods and labor to produce a product which would not be spent if the product were not made.

direct-coupled amplifier [ELECTR] A direct-current amplifier in which a resistor or a direct connection provides the coupling between stages, so small changes in direct currents can be amplified.

direct-coupled transistor logic [ELECTR] Integrated-circuit logic using only resistors and transistors, with direct conductive coupling between the transistors; speed can be up to 1 megahertz. Abbreviated DCTL.

direct coupling [ELEC] Coupling of two circuits by means of a non-frequency-sensitive device, such as a wire, resistor, or battery, so both direct and alternating current can flow through the coupling path. [MECH ENG] The direct connection of the shaft of a prime mover (such as a motor) to the shaft of a rotating mechanism (such as a pump or compressor).

direct current [ELEC] Electric current which flows in one direction only, as opposed to alternating current. Abbreviated dc.

direct-current circuit [ELEC] Any combination of dc voltage or current sources, such as generators and batteries, in conjunction with transmission lines, resistors, and power converters such as motors.

direct-current component [COMMUN] The average value of a signal; in television, it represents the average luminance of the picture being transmitted; in radar, the level from which the transmitted and received pulses rise.

direct-current dump [ELECTR] Removal of all direct-current power from a computer system or component intentionally, accidentally, or conditionally;

in some types of storage, this results in loss of stored information.

direct-current inserter [ELECTR] A television transmitter stage that adds to the video signal a dc component known as the pedestal level.

direct-current picture transmission [COMMUN] Television transmission in which the signal contains a dc component that represents the average illumination of the entire scene. Also known as direct-current transmission.

direct-current quadruplex system [COMMUN] Direct-current telegraph system which affords simultaneous transmission of two messages in each direction over the same line, achieved by superimposing neutral telegraph upon polar telegraph.

direct-current reinsertion *See* clamping.

direct-current restoration *See* clamping.

direct-current restorer [ELECTR] A clamp circuit used to establish a dc reference level in a signal without modifying to any important degree the waveform of the signal itself. Also known as clamper; reinserter.

direct-current transducer [ELECTR] A transducer that requires dc excitation and provides a dc output that varies with the parameter being sensed.

direct-current transmission *See* direct-current picture transmission.

direct-current vacuum-tube voltmeter [ELECTR] The amplifying and indicating portions of the diode rectifier-amplifier meter, which are usually designed so that the diode rectifier can be disconnected for dc measurements.

direct-current voltage *See* direct voltage.

direct-cycle reactor [NUCLEO] A nuclear power plant in which the heat-transfer fluid circulates through the reactor and then passes directly to the turbine in a continuous cycle.

direct digital control [CONT SYS] The use of a digital computer generally on a time-sharing or multiplexing basis, for process control in petroleum, chemical, and other industries.

direct distance dialing [COMMUN] A telephone exchange service that allows a telephone user to dial subscribers outside the local area using a standard routing pattern from the local or end office.

direct drive [MECH ENG] A drive in which the driving part is directly connected to the driven part.

direct dye [MATER] A group of coal tar dyes that act without mordants, for example, benzidine dyes. Also known as substantive dye.

directed number [MATH] A number together with a sign.

directed set [MATH] A partially ordered set with the property that for every pair of elements a,b in the set, there is a third element which is larger than both a and b.

direct-entry terminal [COMPUT SCI] A device from which data are received into a computer immediately, and which edits data at the time of receipt, allowing computer files to be accessed to validate the information entered, and allowing the terminal operator to be notified immediately of any errors.

direct-expansion coil [MECH ENG] A finned coil, used in air cooling, inside of which circulates a cold fluid or evaporating refrigerant. Abbreviated DX coil.

direct feedback system [CONT SYS] A system in which electrical feedback is used directly, as in a tachometer.

direct-fired evaporator [CHEM ENG] An evaporator in which the flame and combustion gases are separated from the boiling liquid by a metal wall, or other heating surface.

direct-gap semiconductor [SOLID STATE] A semiconductor in which the minimum of the conduction band occurs at the same wave vector as the maximum of the valence band, and recombination radiation consequently occurs with relatively large intensity.

direct grid bias *See* grid bias.

direct hierarchy control [COMPUT SCI] A method of manipulating data in a computer storage hierarchy in which data transfer is completely under the control of built-in algorithms and the user or programmer is not concerned with the various storage subsystems.

direct-insert subroutine [COMPUT SCI] A body of coding or a group of instructions inserted directly into the logic of a program, often in multiple copies, whenever required.

direct instruction [COMPUT SCI] An instruction containing the address of the operand on which the operation specified in the instruction is to be performed.

direct inward dialing [COMMUN] The capability for dialing individual telephone extensions in a large organization directly from outside, without going through a central switchboard.

direction [ENG] The position of one point in space relative to another without reference to the distance between them; may be either three-dimensional or two-dimensional, the horizontal being the usual plane of the latter; usually indicated in terms of its angular distance from a reference direction. [GEOL] *See* trend.

directional antenna [ELECTROMAG] An antenna that radiates or receives radio waves more effectively in some directions than others.

directional coupler [ELECTR] A device that couples a secondary system only to a wave traveling in a particular direction in a primary transmission system, while completely ignoring a wave traveling in the opposite direction. Also known as directive feed.

directional filter [ELECTR] A low-pass, band-pass, or high-pass filter that separates the bands of frequencies used for transmission in opposite directions in a carrier system. Also known as directional separation filter.

directional gain *See* directivity index.

directional gyro [AERO ENG] A flight instrument incorporating a gyro that holds its position in azimuth and thus can be used as a directional reference. Also known as direction indicator. [MECH] A two-degrees-of-freedom gyro with a provision for maintaining its spin axis approximately horizontal.

directional microphone [ENG ACOUS] A microphone whose response varies significantly with the direction of sound incidence.

directional pattern *See* radiation pattern.

directional response pattern *See* directivity pattern.

directional separation filter *See* directional filter.

directional structure [GEOL] Any sedimentary structure having directional significance; examples are cross-bedding and ripple marks. Also known as vectorial structure.

directional well [PETRO ENG] A well drilled at an angle up to 70° from the vertical to avoid obstacles over the reservoir, such as towns, beaches, or bodies of water.

direction finder *See* radio direction finder.

direction indicator *See* directional gyro.

direction of propagation [PHYS] **1.** The normal to a surface of constant phase, in a propagating wave. **2.** The direction of the group velocity. **3.** The direction of time-average energy flow. (In a homogeneous isotropic medium, these three directions coincide.)

direction rectifier [ELECTR] A rectifier that supplies a direct-current voltage whose magnitude and polarity vary with the magnitude and relative polarity of an alternating-current synchro error voltage.

directive [COMPUT SCI] An instruction in a source program that guides the compiler in making the translation to machine language, and is usually not translated into instructions in the object program. [INV ZOO] Any of the dorsal and ventral paired mesenteries of certain anthozoan coelenterates.

directivity [ELECTR] The ability of a logic circuit to ensure that the input signal is not affected by the output signal. [ELECTROMAG] **1.** The value of the directive gain of an antenna in the direction of its maximum value. **2.** The ratio of the power measured at the forward-wave sampling terminals of a directional coupler, with only a forward wave present in the transmission line, to the power measured at the same terminals when the direction of the forward wave in the line is reversed; the ratio is usually expressed in decibels.

directivity factor [ENG ACOUS] **1.** The ratio of radiated sound intensity at a remote point on the principal axis of a loudspeaker or other transducer, to the average intensity of the sound transmitted through a sphere passing through the remote point and concentric with the transducer; the frequency must be stated. **2.** The ratio of the square of the voltage produced by sound waves arriving parallel to the principal axis of a microphone or other receiving transducer, to the mean square of the voltage that would be produced if sound waves having the same frequency and mean-square pressure were arriving simultaneously from all directions with random phase; the frequency must be stated.

directivity index [ENG ACOUS] The directivity factor expressed in decibels; it is 10 times the logarithm to the base 10 of the directivity factor. Also known as directional gain.

directivity pattern [ENG ACOUS] A graphical or other description of the response of a transducer used for sound emission or reception as a function of the direction of the transmitted or incident sound waves in a specified plane and at a specified frequency. Also known as beam pattern; directional response pattern.

direct keying device [COMPUT SCI] A computer input device which enables direct entry of information by means of a keyboard.

direct labor standard *See* standard time.

directly heated cathode *See* filament.

direct memory access [COMPUT SCI] The use of special hardware for direct transfer of data to or from memory to minimize the interruptions caused by program-controlled data transfers. Abbreviated dma.

direct motion [ASTRON] Eastward, or counterclockwise, motion of a planet or other object as seen from the North Pole (motion in the direction of increasing right ascension).

direct nuclear reaction [NUC PHYS] A nuclear reaction which is completed in the time required for the incident particle to transverse the target nucleus, so that it does not combine with the nucleus as a whole but interacts only with the surface or with some individual constituent.

direct numerical control [COMPUT SCI] The use of a computer to program, service, and log a process such as a machine-tool cutting operation.

director [ELECTR] Telephone switch which translates the digits dialed into the directing digits actually used to switch the call. [ELECTROMAG] A parasitic element placed a fraction of a wavelength ahead of a dipole receiving antenna to increase the gain of the array in the direction of the major lobe. [ORD] Electromechanical equipment which is used to track a moving target in azimuth and angular height and which, with the addition of other necessary information from an outside source, such as a radar set or a range finder, continuously computes firing data and transmits them to the guns.

direct organization [COMPUT SCI] A type of processing in which records within data sets stored on direct-access devices may be fetched directly if their physical locations are known.

directory [COMPUT SCI] The listing and description of all the fields of the records making up a file.

direct outward dialing [COMMUN] A private automatic branch telephone exchange that permits all local stations to dial outside numbers. Abbreviated DOD.

direct positive [GRAPHICS] A photographic positive made by exposure to light and developed without a negative.

direct-power generator [ENG] Any device which converts thermal or chemical energy into electric power by methods more direct than the conventional thermal cycle.

direct recording [ENG ACOUS] Recording in which a record is produced immediately, without subsequent processing, in response to received signals.

direct-reduction process [MET] Any of several methods for extracting iron ore below the melting point of iron, to produce solid reduced iron that may be converted to steel with little further refining.

direct reflection *See* specular reflection.

direct resistance-coupled amplifier [ELECTR] Amplifier in which the plate of one stage is connected either directly or through a resistor to the control grid of the next stage, with the plate-load resistor being common to both stages; used to amplify small changes in direct current.

directrix [MATH] A fixed line used in one method of defining a conic; the distance from this line divided by the distance from a fixed point (called the focus) is the same for all points on the conic.

direct route [ELEC] In wire communications, the trunks that connect a pair of switching centers, regardless of the geographical direction the actual trunk facilities may follow. [NAV] The shortest navigational distance between two points on the earth's surface, for example, the great circle.

direct solar radiation [ASTROPHYS] That portion of the radiant energy received at the actinometer direct from the sun, as distinguished from diffuse sky radiation, effective terrestrial radiation, or radiation from any other source.

direct sum [MATH] If each of the sets in a finite direct product of sets has a group structure, this structure may be imposed on the direct product by defining the composition "componentwise"; the resulting group is called the direct sum.

direct tide [GEOPHYS] A gravitational solar or lunar tide in the ocean or atmosphere which is in phase with the apparent motions of the attracting body, and consequently has its local maxima directly under the tide-producing body, and on the opposite side of the earth.

direct-view storage tube [ELECTR] A cathode-ray tube in which secondary emission of electrons from a storage grid is used to provide an intensely bright display for long and controllable periods of time. Also known as display storage tube; viewing storage tube.

direct-vision prism   *See* Amici prism.

direct voltage   [ELEC] A voltage that forces electrons to move through a circuit in the same direction continuously, thereby producing a direct current. Also known as direct-current voltage.

Dirichlet conditions   [MATH] The requirement that a function be bounded, and have finitely many maxima, minima, and discontinuities on the closed interval $[-\pi, \pi]$.

Dirichlet problem   [MATH] To determine a solution to Laplace's equation which satisfies certain conditions in a region and on its boundary.

Dirichlet series   [MATH] A series whose nth term is a complex number divided by n to the zth power.

dirty bomb   [ORD] An explosive based on nuclear fission that emits many long-lived radioactive isotopes.

disability glare   *See* glare.

disaccharide   [BIOCHEM] Any of the class of compound sugars which yield two monosaccharide units upon hydrolysis.

Disasteridae   [PALEON] A family of extinct burrowing, exocyclic Euechinoidea in the order Holasteroida comprising mainly small, ovoid forms without fascioles or a plastron.

disc   *See* disk.

Discellaceae   [MYCOL] A family of fungi of the order Sphaeropsidales, including saprophytes and some plant pathogens.

discharge   [ELEC] To remove a charge from a battery, capacitor, or other electric-energy storage device.   [ELECTR] The passage of electricity through a gas, usually accompanied by a glow, arc, spark, or corona. Also known as electric discharge.   [FL MECH] The flow rate of a fluid at a given instant expressed as volume per unit of time.

discharge coefficient   [FL MECH] In a nozzle or other constriction, the ratio of the mass flow rate at the discharge end of the nozzle to that of an ideal nozzle which expands an identical working fluid from the same initial conditions to the same exit pressure. Also known as coefficient of discharge.

discharged solids   *See* residue.

discharge head   [MECH ENG] Vertical distance between the intake level of a water pump and the level at which it discharges water freely to the atmosphere.

discharge lamp   [ELECTR] A lamp in which light is produced by an electric discharge between electrodes in a gas (or vapor) at low or high pressure. Also known as electric-discharge lamp; gas-discharge lamp; vapor lamp.

discharge liquor   [CHEM ENG] Liquid that has passed through a processing operation. Also known as effluent; product.

discharge tube   [ELECTR] An evacuated enclosure containing a gas at low pressure, through which current can flow when sufficient voltage is applied between metal electrodes in the tube. Also known as electric-discharge tube.   [MECH ENG] A tube through which steam and water are released into a boiler drum.

discharging arch   [CIV ENG] A support built over, and not touching, a weak structural member, such as a wooden lintel, to carry the main load. Also known as relieving arch.

Discinacea   [INV ZOO] A family of inarticulate brachiopods in the suborder Acrotretidina.

discoctaster   [INV ZOO] A type of spicule with eight rays terminating in discs in hexactinellid sponges.

Discoglossidae   [VERT ZOO] A family of anuran amphibians in and typical of the suborder Opisthocoela.

discoid   [BIOL] **1.** Being flat and circular in form. **2.** Any structure shaped like a disc.

Discoidiidae   [PALEON] A family of extinct conical or globular, exocyclic Euechinoidea in the order Holectypoida distinguished by the rudiments of internal skeletal partitions.

Discolichenes   [BOT] The equivalent name for Lecanorales.

Discolomidae   [INV ZOO] The tropical log beetles, a family of coleopteran insects in the superfamily Cucujoidea.

discomfort glare   *See* glare.

discomfort index   *See* temperature-humidity index.

discomposition   [NUCLEO] The process in which an atom is knocked out of its position in a crystal lattice by direct nuclear impact, as by fast neutrons or by fast ions that have been previously knocked out of their lattice positions.

discomposition effect   [NUCLEO] Changes in physical or chemical properties of a substance caused by discomposition. Also known as Wigner effect.

Discomycetes   [MYCOL] A group of fungi in the class Ascomycetes in which the surface of the fruiting body is exposed during maturation of the spores.

discone antenna   [ELECTROMAG] A biconical antenna in which one of the cones is spread out to 180° to form a disk; the center conductor of the coaxial line terminates at the center of the disk, and the cable shield terminates at the vertex of the cone.

disconformity   [GEOL] Unconformity between parallel beds or strata.

disconnect   [ELEC] To open a circuit by removing wires or connections, as distinguished from opening a switch to stop current flow.   [ENG] To sever a connection.

discontinuity   [ELECTROMAG] An abrupt change in the shape of a waveguide.   [GEOL] **1.** An interruption in sedimentation. **2.** A surface that separates unrelated groups of rocks.   [GEOPHYS] A boundary at which the velocity of seismic waves changes abruptly.   [MATH] A point at which a function is not continuous.   [MET] The place where the structural nature of a weldment is interfered with because of the materials involved or where the mechanical, physical, or metallurgical aspects are not homogeneous.   [PHYS] A break in the continuity of a medium or material at which a reflection of wave energy can occur.

discontinuous amplifier   [ELECTR] Amplifier in which the input waveform is reproduced on some type of averaging basis.

discontinuous phase   *See* disperse phase.

discontinuous reaction series   [GEOL] The branch of Bowen's reaction series that include olivine, pyroxene, amphibole, and biotite; each change in the series represents an abrupt change in phase.

Discorbacea   [INV ZOO] A superfamily of foraminiferan protozoans in the suborder Rotaliina characterized by a radial, perforate, calcite test and a monolamellar septa.

discord   *See* dissonance.

discordance   [GEOL] An unconformity characterized by lack of parallelism between strata which touch without fusion.

discrete   [SCI TECH] **1.** Composed of separate and distinct parts. **2.** Having an individually distinct identity.

discrete comparator   *See* digital comparator.

discrete sound system [ENG ACOUS] A quadraphonic sound system in which the four input channels are preserved as four discrete channels during recording and playback processes; sometimes referred to as a 4-4-4 system.

discrete spectrum [SPECT] A spectrum in which the component wavelengths constitute a discrete sequence of values rather than a continuum of values.

discrete system [CONT SYS] A control system in which signals at one or more points may change only at discrete values of time. Also known as discrete-time system.

discrete-time system See discrete system.

discrete variable [MATH] A variable for which the possible values form a discrete set.

discrete-word intelligibility [COMMUN] The percent of intelligibility obtained when the speech units under consideration are words, usually presented so as to minimize the contextual relation between them.

discriminant [MATH] The quantity $b^2 - 4ac$ where a,b,c are coefficients of a given quadratic polynomial: $ax^2 + bx + c$.

discrimination [COMMUN] 1. In frequency-modulated systems, the detection or demodulation of the imposed variations in the frequency of the carriers. 2. In a tuned circuit, the degree of rejection of unwanted signals. 3. Of any system or transducer, the difference between the losses at specified frequencies with the system or transducer terminated in specified impedances. [COMPUT SCI] See conditional jump.

discriminator [ELECTR] A circuit in which magnitude and polarity of the output voltage depend on how an input signal differs from a standard or from another signal.

dish See parabolic reflector.

disilicate [CHEM] A silicate compound that has two silicon atoms in the molecule.

disilicide [CHEM] A compound that has two silicon atoms joined to a radical or another element.

disintegration [NUC PHYS] Any transformation of a nucleus, whether spontaneous or induced by irradiation, in which particles or photons are emitted.

disintegration chain See radioactive series.

disintegration constant See decay constant.

disintegration energy [NUC PHYS] The energy released, or the negative of the energy absorbed, during a nuclear or particle reaction. Designated Q. Also known as Q value; reaction energy.

disintegration family See radioactive series.

disintegration rate [NUCLEO] 1. The absolute rate of decay of a radioactive substance, usually expressed in terms of disintegrations per unit of time. 2. The absolute rate of transformation of a nuclide under bombardment.

disintegration series See radioactive series.

disintegration voltage [ELECTR] The lowest anode voltage at which destructive positive-ion bombardment of the cathode occurs in a hot-cathode gas tube.

disjunction [CYTOL] Separation of chromatids or homologous chromosomes during anaphase.

disk Also spelled disc. [BIOL] Any of various rounded and flattened animal and plant structures. [COMPUT SCI] A rotating circular plate having a magnetizable surface on which information may be stored as a pattern of polarized spots on concentric recording tracks. [ENG ACOUS] See phonograph record. [MATH] The region in the plane consisting of all points with norm less than 1 (sometimes less than or equal to 1).

disk armature [ELEC] The armature in a motor that has a disk winding or is made up of a metal disk.

disk brake [MECH ENG] A type of brake in which disks attached to a fixed frame are pressed against disks attached to a rotating axle or against the inner surfaces of a rotating housing.

disk cam [MECH ENG] A disk with a contoured edge which rotates about an axis perpendicular to the disk, communicating motion to the cam follower which remains in contact with the edge of the disk.

disk cartridge [COMPUT SCI] A removable module that contains a single magnetic disk platter which remains attached to the housing when placed into the disk drive.

disk centrifuge [MECH ENG] A centrifuge with a large bowl having a set of disks that separate the liquid into thin layers to create shallow settling chambers.

disk clutch [MECH ENG] A clutch in which torque is transmitted by friction between friction disks with specially prepared friction material riveted to both sides and contact plates keyed to the inner surface of an external hub.

disk drive [COMPUT SCI] The physical unit that holds, spins, reads, and writes the magnetic disks. Also known as disk unit.

disk engine [MECH ENG] A rotating engine in which the piston is a disk.

diskette See floppy disk.

disk file [COMPUT SCI] An organized collection of records held on a magnetic disk.

disk filter [ENG] A filter in which the substance to be filtered is drawn through membranes stretched on segments of revolving disks by a vacuum inside each disk; the solids left on the membrane are lifted from the tank and discharged. Also known as American filter.

disk galaxy [ASTRON] A galaxy consisting of a central bulge of a spheroidal aggregation of stars and a surrounding disk of stars fanning outward in a thin layer.

disk memory See disk storage.

disk meter [ENG] A positive displacement meter to measure flow rate of a fluid; consists of a disk that wobbles or nutates within a chamber so that each time the disk nutates a known volume of fluid passes through the meter.

disk pack [COMPUT SCI] A set of magnetic disks that can be removed from a disk drive as a unit.

disk recording [ENG ACOUS] 1. The process of inscribing suitably transformed acoustical or electrical signals on a phonograph record. 2. See phonograph record.

disk sander [MECH ENG] A machine that uses a circular disk coated with abrasive to smooth or shape surfaces.

disk spring [MECH ENG] A mechanical spring that consists of a disk or washer supported by one force (distributed by a suitable chuck or holder) at the periphery and by an opposing force on the center or hub of the disk.

disk storage [ELECTR] An external computer storage device consisting of one or more disks spaced on a common shaft, and magnetic heads mounted on arms that reach between the disks to read and record information on them. Also known as disk memory; magnetic disk storage.

disk unit See disk drive.

dislocation [CRYSTAL] A defect occurring along certain lines in the crystal structure and present as a closed ring or a line anchored at its ends to other dislocations, grain boundaries, the surface, or other structural feature. Also known as line defect. [GEOL] Relative movement of rock on opposite sides of a fault. Also known as displacement. [MED] Displacement of one or more bones of a joint.

dislocation breccia *See* fault breccia.

disodium ethylene-bis-dithiocarbamate *See* nabam.

Disomidae [INV ZOO] A family of spioniform annelid worms belonging to the Sedentaria.

dispatching [COMPUT SCI] The control of priorities in a queue of requests in a multiprogramming or multitasking environment.

dispermy [PHYSIO] Entrance of two spermatozoa into an ovum.

dispersal pattern [GEOCHEM] Distribution pattern of metals in soil, rock, water, or vegetation.

disperse [COMPUT SCI] A data-processing operation in which grouped input items are distributed among a larger number of groups in the output.

dispersed elements [GEOCHEM] Elements which form few or no independent minerals but are present as minor ingredients in minerals of abundant elements.

disperse dye [MATER] A very slightly water-soluble, colored material for use on cellulose acetate and other synthetic fibers; color is transferred to the fiber as extremely finely divided particles, resulting in a solution of the dye in the solid fiber.

disperse phase [CHEM] The phase of a disperse system consisting of particles or droplets of one substance distributed through another system. Also known as discontinuous phase; internal phase.

disperse system [CHEM] A two-phase system consisting of a dispersion medium and a disperse phase.

dispersing prism [OPTICS] An optical prism which deviates light of different wavelengths by different amounts and can therefore be used to separate white light into its monochromatic parts.

dispersion [AERO ENG] Deviation from a prescribed flight path; specifically, circular dispersion especially as applied to missiles. [CHEM] A distribution of finely divided particles in a medium. [COMMUN] The entropy of the output of a communications channel when the input is known. [ELECTROMAG] Scattering of microwave radiation by an obstruction. [MINERAL] In optical mineralogy, the constant optical values at different positions on the spectrum. [PHYS] 1. The separation of a complex of electromagnetic or sound waves into its various frequency components. 2. Quantitatively, the rate of change of refractive index with wavelength or frequency at a given wavelength or frequency. 3. The rate of change of deviation with wavelength or frequency. 4. In general, any process separating radiation into components having different frequencies, energies, velocities, or other characteristics, such as the sorting of electrons according to velocity in a magnetic field.

dispersion equation *See* dispersion formula.

dispersion formula [PHYS] Any formula which gives the refractive index as a function of wavelength of electromagnetic radiation. Also known as dispersion equation.

dispersion mill [MECH ENG] Size-reduction apparatus that disrupts clusters or agglomerates of solids, rather than breaking down individual particles; used for paint pigments, food products, and cosmetics.

dispersion of a random variable [STAT] The spread of a random variable's distribution about its mean.

dispersion relation [NUC PHYS] A relation between the cross section for a given effect and the de Broglie wavelength of the incident particle, which is similar to a classical dispersion formula. [PHYS] An integral formula relating the real and imaginary parts of some function of frequency or energy, such as a refractive index or scattering amplitude, based on the causality principle and the Cauchy integral

formula. [PL PHYS] A relation between the radian frequency and the wave vector of a wave motion or instability in a plasma.

disphenoid [CRYSTAL] 1. A crystal form with four similar triangular faces combined in a wedge shape; can be tetragonal or orthorhombic. 2. A crystal form with eight scalene triangles combined in pairs.

displacement [CHEM] A chemical reaction in which an atom, radical, or molecule displaces and sets free an element of a compound. [ELEC] *See* electric displacement. [FL MECH] 1. The weight of fluid which is displaced by a floating body, equal to the weight of the body and its contents; the displacement of a ship is generally measured in long tons (1 long ton = 2240 pounds). 2. The volume of fluid which is displaced by a floating body. [GEOL] *See* dislocation. [MECH] 1. The linear distance from the initial to the final position of an object moved from one place to another, regardless of the length of path followed. 2. The distance of an oscillating particle from its equilibrium position. [MECH ENG] The volume swept out in one stroke by a piston moving in a cylinder as for an engine, pump, or compressor.

displacement chromatography [ANALY CHEM] Variation of column-development or elution chromatography in which the solvent is sorbed more strongly than the sample components; the freed sample migrates down the column, pushed by the solvent.

displacement current [ELECTROMAG] The rate of change of the electric displacement vector, which must be added to the current density to extend Ampère's law to the case of time-varying fields (meter-kilogram-second units). Also known as Maxwell's displacement current.

displacement engine *See* piston engine.

displacement law *See* radioactive displacement law; Wien's displacement law.

displacement meter [ENG] A water meter that measures water flow quantitatively by recording the number of times a vessel of known capacity is filled and emptied.

displacement pump [MECH ENG] A pump that develops its action through the alternate filling and emptying of an enclosed volume as in a piston-cylinder construction.

displacement series [CHEM] The elements in decreasing order of their negative potentials. Also known as constant series; electromotive series; Volta series.

display [ELECTR] 1. A visible representation of information, in words, numbers, or drawings, as on the cathode-ray tube screen of a radar set, navigation system, or computer console. 2. The device on which the information is projected. Also known as display device. 3. The image of the information.

display console [COMPUT SCI] A cathode-ray tube or other display unit on which data being processed or stored in a computer can be presented in graphical or character form; sometimes equipped with a light pen with which the user can alter the information displayed.

display control [COMPUT SCI] A unit in a computer system consisting of channels and associated control circuitry that connect a number of visual display units with a central processor.

display device *See* display.

display packing [COMPUT SCI] An efficient means of transmitting the x and y coordinates of a point packed in a single word to halve the time required to freshen the spot on a cathode-ray tube display.

display primary [COMMUN] One of the primary colors produced in a television receiver that, when mixed in proper proportions, serve to produce the other desired colors. Also known as receiver primary.

display processor [COMPUT SCI] A section of a computer, or a minicomputer which handles the routines required to display an output on a cathode-ray tube.

display storage tube See direct-view storage tube.

display terminal [COMPUT SCI] A computer output device in which characters, and sometimes graphic information, appear on the screen of a cathode-ray tube. Also known as display unit; video display terminal (VDT).

display tube [ELECTR] A cathode-ray tube used to provide a visual display. Also known as visual display unit.

display unit See display terminal.

disposal field See absorption field.

disruptive discharge [ELEC] A sudden and large increase in current through an insulating medium due to complete failure of the medium under electrostatic stress.

dissecting microscope [OPTICS] Either of two types of optical microscope used to magnify materials undergoing dissection.

dissection [GEOL] Destruction of the continuity of the land surface by erosive cutting of valleys or ravines into a relatively even surface.

dissector tube [ELECTR] Camera tube having a continuous photo cathode on which is formed a photoelectric emission pattern which is scanned by moving its electron-optical image over an aperture.

disseminule [BIOL] An individual organism or part of an individual adapted for the dispersal of a population of the same organisms.

dissipation [PHYS] Any loss of energy, generally by conversion into heat; quantitatively, the rate at which this loss occurs. Also known as energy dissipation.

dissipation coefficient See scattering coefficient.

dissipative muffler [ENG] A device which absorbs sound energy as the gas passes through it; a duct lined with sound-absorbing material is the most common type.

dissipative tunneling [SOLID STATE] Quantum-mechanical tunneling of individual electrons, rather than pairs, across a thin insulating layer separating two superconducting metals when there is a voltage across this layer, resulting in partial disruption of cooperative motion.

dissipator See heat sink.

dissociation [MED] Independent, uncoordinated functioning of the atria and ventricles. [MICROBIO] The appearance of a novel colony type on solid media after one or more subcultures of the microorganism in liquid media. [PHYS CHEM] Separation of a molecule into two or more fragments (atoms, ions, radicals) by collision with a second body or by the absorption of electromagnetic radiation. [PSYCH] The segregation of ideas from their affects or feelings, resulting in independent functioning of these components of a person's mental processes.

dissociation constant [PHYS CHEM] A constant whose numerical value depends on the equilibrium between the undissociated and dissociated forms of a molecule; a higher value indicates greater dissociation.

dissolution [CHEM] Dissolving of a material.

dissolve [CHEM] 1. To cause to disperse. 2. To cause to pass into solution. [GRAPHICS] A superimposing of one television or motion picture shot upon another, the emergent shot gradually brightening and the overlapped shot gradually darkening, so that as one scene disappears, another gradually appears. Also known as lap dissolve.

dissolved air floatation [CHEM ENG] A liquid-solid separation process wherein the main mechanism of suspended-solids removal is the change of apparent specific gravity of those suspended solids in relation to that of the suspending liquid by the attachment of small gas bubbles formed by the release of dissolved gas to the solids.

dissolved gas See solution gas.

dissonance [ACOUS] An unpleasant combination of harmonics heard when certain musical tones are played simultaneously. Also known as discord.

dissymmetry factor [OPTICS] A quantity which expresses the strength of circular dichroism, equal to the difference in the absorption indices for left and right circularly polarized light divided by the absorption index for ordinary light of the same wavelength. Also known as anisotropy factor.

Distacodidae [PALEON] A family of conodonts in the suborder Conodontiformes characterized as simple curved cones with deeply excavated attachment scars.

distance-luminosity relation [ASTRON] The relation in which the light intensity from a star is inversely proportional to the square of its distance.

distance marker [ENG] One of a series of concentric circles, painted or otherwise fixed on the screen of a plan position indicator, from which the distance of a target from the radar antenna can be read directly; used for surveillance and navigation where the relative distances between a number of targets are required simultaneously. Also known as radar range marker; range marker.

distance-measuring equipment [NAV] A radio aid to navigation that provides distance information by measuring total round-trip time of transmission from an airborne interrogator to a ground-based transponder and return. Abbreviated DME.

distemper [VET MED] Any of several contagious virus diseases of mammals, especially the form occurring in dogs, marked by fever, respiratory inflammation, and destruction of myelinated nerve tissue.

disthene See kyanite.

distillate [CHEM] The products of distillation formed by condensing vapors.

distillation [CHEM] The process of producing a gas or vapor from a liquid by heating the liquid in a vessel and collecting and condensing the vapors into liquids.

distilled water [CHEM] Water that has been freed of dissolved or suspended solids and organisms by distillation.

distortion [ELECTR] Any undesired change in the waveform of an electric signal passing through a circuit or other transmission medium. [OPTICS] A defect of an optical system in which the magnification varies with angular distance from the axis, causing straight lines to appear curved.

distortional wave See S wave.

distortion meter [ENG] An instrument that provides a visual indication of the harmonic content of an audio-frequency wave.

distributary [HYD] An irregular branch flowing out from a main stream and not returning to it, as in a delta. Also known as distributary channel.

distributary channel See distributary.

distributed amplifier [ELECTR] A wide-band amplifier in which tubes are distributed along artificial delay lines made up of coils acting with the input and output capacitances of the tubes.

distributed capacitance [ELEC] Capacitance that exists between the turns in a coil or choke, or between adjacent conductors or circuits, as distinguished from the capacitance concentrated in a capacitor.

distributed communications [COMMUN] Information transfer beyond the local level that may involve the originating source to transmit information to all communications centers on any one network, and may also cause an interchange of communictions among several whole networks.

distributed control system [CONT SYS] A collection of modules, each with its own specific function, interconnected tightly to carry out an integrated data acquisition and control application.

distributed fault See fault zone.

distributed inductance [ELECTROMAG] The inductance that exists along the entire length of a conductor, as distinguished from inductance concentrated in a coil.

distributed intelligence [COMPUT SCI] The existence of processing capability in terminals and other peripheral devices of a computer system.

distributed network [COMMUN] A communications network in which there exist alternative routings between the various nodes.

distributed-parameter system See distributed system.

distributed processing system [COMPUT SCI] An information processing system consisting of two or more programmable devices, connected so that information can be exchanged.

distributed system [SYS ENG] A system whose behavior is governed by partial differential equations, and not merely ordinary differential equations. Also known as distributed-parameter system.

distribution [MATH] An abstract object which generalizes the idea of function; used in applied mathematics, quantum theory, and probability theory; the delta function is an example. Also known as generalized function.

distribution amplifier [ELECTR] A radio-frequency power amplifier used to feed television or radio signals to a number of receivers, as in an apartment house or a hotel. [ENG ACOUS] An audio-frequency power amplifier used to feed a speech or music distribution system and having sufficiently low output impedance so changes in load do not appreciably affect the output voltage.

distribution cable [ELEC] Cable extending from a feeder cable into a specific area for the purpose of providing service to that area.

distribution center [ELEC] In an alternating-current power system, the point at which control and routing equipment is installed.

distribution coefficient [OPTICS] One of the tristimulus values of monochromatic radiations having equal power, usually denoted by $\bar{x}, \bar{y}, \bar{z}$. [PHYS CHEM] The ratio of the amounts of solute dissolved in two immiscible liquids at equilibrium.

distribution curve [STAT] The graph of the distribution function of a random variable.

distribution deck [COMPUT SCI] A card file which duplicates all or part of a master card file and used for disseminating or decentralizing.

distribution function See distribution of a random variable.

distribution law [ANALY CHEM] The law stating that if a substance is dissolved in two immiscible liquids, the ratio of its concentration in each is constant. [STAT MECH] A law which gives a density function specifying the probability of finding a particle in a unit volume of phase space, or the number of particles in each of the states which a particle may occupy, or the number of particles per unit volume of phase space.

distribution of a random variable [STAT] For a discrete random variable, a function (or table) which assigns to each possible value of the random variable the probability that this value will occur; for a continuous random variable x, the monotone nondecreasing function which assigns to each real t the probability that x is less than or equal to t. Also known as distribution function; probability distribution; statistical distribution.

distribution reservoir [CIV ENG] A service reservoir connected with the conduits of a primary water supply; used to supply water to consumers according to fluctuations in demand over short time periods and serves for local storage in case of emergency.

distribution substation [ELEC] An electric power substation associated with the distribution system and the primary feeders for supply to residential, commercial, and industrial loads.

distribution system [ELEC] Circuitry involving high-voltage switchgear, step-down transformers, voltage dividers, and related equipment used to receive high-voltage electricity from a primary source and redistribute it at lower voltages. Also known as electric distribution system.

distributive fault See step fault.

distributive lattice [MATH] A lattice in which "greatest lower bound" obeys a distributive law with respect to "least upper bound," and vice versa.

distributive law [MATH] A rule which stipulates how two binary operations on a set shall behave with respect to one another; in particular, if $+$, $\circ$ are two such operations then $\circ$ distributes over $+$ means $a \circ (b + c) = (a \circ b) + (a \circ c)$ for all a,b,c in the set.

distributor [ELEC] 1. Any device which allocates a telegraph line to each of a number of channels, or to each row of holes on a punched tape, in succession. 2. A rotary switch that directs the high-voltage ignition current in the proper firing sequence to the various cylinders of an internal combustion engine. [ELECTR] The electronic circuitry which acts as an intermediate link between the accumulator and drum storage.

district heating [MECH ENG] The supply of heat, either in the form of steam or hot water, from a central source to a group of buildings.

disturbance [COMMUN] An undesired interference or noise signal affecting radio, television, or facsimile reception. [CONT SYS] An undesired command signal in a control system. [GEOL] Folding or faulting of rock or a stratum from its original position. [METEOROL] 1. Any low or cyclone, but usually one that is relatively small in size and effect. 2. An area where weather, wind, pressure, and so on show signs of the development of cyclonic circulation. 3. Any deviation in flow or pressure that is associated with a disturbed state of the weather, such as cloudiness and precipitation. 4. Any individual circulatory system within the primary circulation of the atmosphere.

disulfate [CHEM] A compound that has two sulfate radicals.

disulfide [CHEM] 1. A compound that has two sulfur atoms bonded to a radical or element. 2. One of a group of organosulfur compounds RSSR′ that may be symmetrical (R = R′) or unsymmetrical (R and R′ different).

disulfonate [CHEM] A molecule that has two sulfonate groups.

disulfonic acid [CHEM] A molecule that has two sulfonic acid groups.

ditch [CIV ENG] A small artificial channel cut through earth or rock to carry water for irrigation or drainage.

dither [COMMUN] A technique for representing the entire gray scale of a picture by picture elements with only one of two levels ("white" and "black"), in which a multilevel input image signal is compared with a position-dependent set of thresholds, and picture elements are set to "white" only where the image input signal exceeds the threshold. [CONT SYS] A force having a controlled amplitude and frequency, applied continuously to a device driven by a servomotor so that the device is constantly in small-amplitude motion and cannot stick at its null position. Also known as buzz.

dither matrix [COMMUN] A square matrix of threshold values that is repeated as a regular array to provide a threshold pattern for an entire image in the dither method of image representation.

diuretic [PHARM] Any agent that increases the volume and flow of urine.

diuretic hormone [BIOCHEM] A neurohormone that promotes water loss in insects by increasing the volume of fluid secreted into the Malpighian tubules.

diurnal [BIOL] Active during daylight hours. [METEOROL] Pertaining to meteorological actions which are completed within 24 hours and which recur every 24 hours.

diurnal aberration [ASTRON] Aberration caused by the rotation of the earth; its value varies with the latitude of the observer and ranges from zero at the poles to 0.31 second of arc. Also known as daily aberration.

diurnal arc [ASTRON] That part of a celestial body's diurnal circle which lies above the horizon of the observer.

diurnal circle [ASTRON] The apparent daily path of a celestial body, approximating a parallel of declination.

diurnal inequality [OCEANOGR] The difference between the heights of the two high waters or the two low waters of a lunar day.

diurnal motion [ASTRON] The apparent daily motion of a celestial body as observed from a rotating body.

diurnal parallax See geocentric parallax.

diurnal tide [OCEANOGR] A tide in which there is only one high water and one low water each lunar day.

diurnal variation [GEOPHYS] Daily variations of the earth's magnetic field at a given point on the surface, with both solar and lunar periods having their source in the horizontal movements of air in the ionosphere.

divaricate [BIOL] Broadly divergent and spread apart.

divaricator [ZOO] A muscle that causes separation of parts, as of brachiopod shells.

dive [AERO ENG] A rapid descent by an aircraft or missile, nose downward, with or without power or thrust. [ENG] To submerge into an underwater environment so that it may be studied or utilized; includes the use of specialized equipment such as scuba, diving helmets, diving suits, diving bells, and underwater research vessels. [NAV] To submerge a submarine under power.

divergence [ELECTR] The spreading of a cathode-ray stream due to repulsion of like charges (electrons). [FL MECH] The ratio of the area of any section of fluid emerging from a nozzle to the area of

the throat of the nozzle. [METEOROL] The two-dimensional horizontal divergence of the velocity field. [NUCLEO] In a nuclear reactor, the condition wherein the number of neutrons produced increases in each succeeding generation. [OCEANOGR] A horizontal flow of water, in different directions, from a common center or zone.

divergence theorem See Gauss' theorem.

divergent adaptation [EVOL] Adaptation to different kinds of environment that results in divergence from a common ancestral form. Also known as branching adaptation; cladogenic adaptation.

divergent die [ENG] A die with the internal channels that lead to the orifice diverging, such as the dies used for manufacture of hollow-body plastic items.

divergent integral [MATH] An improper integral which does not have a finite value.

divergent nozzle [DES ENG] A nozzle whose cross section becomes larger in the direction of flow.

divergent series [MATH] An infinite series whose sequence of partial sums does not converge.

diverging lens [OPTICS] A lens whose focal length is negative, so that light incident parallel to its axis diverges after passing through it. Also known as negative lens.

diverging meniscus lens See negative meniscus lens.

diversion dam [CIV ENG] A fixed dam for diverting stream water away from its course.

diversion gate [CIV ENG] A gate which may be closed to divert water from the main conduit or canal to a lateral or some other channel.

diversity [COMMUN] Method of signal extraction by which an optimum resultant signal is derived from a combination of, or selection from, a plurality of transmission paths, channels, techniques, or physical arrangements; the system may employ space diversity, polarization diversity, frequency diversity, or any other arrangement by which a choice can be made between signals.

diversity radar [ENG] A radar that uses two or more transmitters and receivers, each pair operating at a slightly different frequency but sharing a common antenna and video display, to obtain greater effective range and reduce susceptibility to jamming.

diverter [ELEC] A low resistance which is connected in parallel with the series or compole winding of a direct-current machine and diverts current from it, causing the magnetomotive force produced by the winding to vary.

diverticulosis [MED] Presence of many diverticula in the intestine.

diverticulum [MED] An abnormal outpocketing or sac on the wall of a hollow organ.

divide [GEOGR] A ridge or section of high ground between drainage systems. [MATH] One object (integer, polynomial) divides another if their quotient is an object of the same type. [SCI TECH] A point or line of division.

divide check [COMPUT SCI] An error signal indicating that an illegal division (such as dividing by zero) was attempted.

divided slit scan [COMPUT SCI] In optical character recognition, a device consisting of a narrow column of photoelectric cells which scans an input character at given intervals for the purpose of obtaining its horizontal and vertical components.

dividend [MATH] A quantity which is divided by another quantity in the operation of division.

division [MATH] The inverse operation of multiplication; the number a divided by the number b is the number c such that b multiplied by c is equal to a.

division circle [ASTRON] A large circular structure attached to the horizontal axis of a transit circle with accurately calibrated markings; used to determine the inclination of the instrument.

division plate [MECH ENG] A diaphragm which surrounds the piston rod of a crosshead-type engine and separates the crankcase from the lower portion of the cylinder.

division subroutine [COMPUT SCI] A built-in program which achieves division by methods such as repetitive subtraction.

divisor [MATH] The quantity by which another quantity is divided in the operation of division.

dixenite [MINERAL] $Mn_5(SiO_3)(AsO_3)(OH)_2$ A black hexagonal mineral consisting of a manganese arsenite and silicate, occurring in scales.

Dixidae [INV ZOO] A family of orthorrhaphous dipteran insects in the series Nematocera.

dizygotic twins [BIOL] Twins derived from two eggs. Also known as fraternal twins.

djalmaite *See* microlite.

Djulfian [GEOL] Upper upper Permian geologic time.

D layer [GEOL] The lower mantle of the earth, between a depth of 1000 and 2900 km. [GEOPHYS] The lowest layer of ionized air above the earth, occurring in the D region only in the daytime hemisphere; reflects frequencies below about 50 kilohertz and partially absorbs higher-frequency waves.

D line [SPECT] The yellow line that is the first line of the major series of the sodium spectrum; the doublet in the Fraunhofer lines whose almost equal components have wavelengths of 5895.93 and 5889.96 angstroms respectively.

dma *See* direct memory access.

DME *See* distance-measuring equipment.

D meson [PARTIC PHYS] 1. A neutral pseudovector meson resonance having a mass of 1285 ± 4 MeV, width of about 30 MeV, and positive charge conjugation parity; the only singlet state consistent with the $(1^+)^+$ nonet. 2. Collective name for three charmed mesons that form an isotopic spin triplet, have masses of approximately 1865 MeV, and are pseudoscalar particles.

DNA *See* deoxyribonucleic acid.

dock [CIV ENG] 1. The slip or waterway that is between two piers or cut into the land for the berthing of ships. 2. A basin or enclosure for reception of vessels, provided with means for controlling the water level.

docking [AERO ENG] The mechanical coupling of two or more orbiting spacecraft.

dock landing ship [NAV ARCH] An amphibious warfare ship which transports and supports landing craft and can quickly unload troops and equipment onto a beach. Designated LSD.

Docodonta [PALEON] A primitive order of Jurassic mammals of North America and England.

doctor blade [ENG] A device for regulating the amount of liquid material on the rollers of a spreader. Also known as doctor knife.

doctor knife *See* doctor blade.

document [COMPUT SCI] 1. Any record, printed or otherwise, that can be read by man or machine. 2. To prepare a written text and charts describing the purpose, nature, usage, and operation of a program or a system of programs.

document alignment [COMPUT SCI] The phase of the reading process in which a transverse force is

applied to a document to line up its reference edge with that of the reading station.

document flow [COMPUT SCI] The path taken by documents as they are processed through a record handling system.

document handling [COMPUT SCI] In character recognition, the process of loading, feeding, transporting, and unloading a cut-form document that has been submitted for character recognition.

document leading edge [COMPUT SCI] In character recognition, that edge which is the foremost one encountered during the reading process and whose relative position defines the document's direction of travel.

document misregistration [COMPUT SCI] In character recognition, the improper state of appearance of a document, on site in a character reader, with respect to real or imaginary horizontal baselines.

document number [COMPUT SCI] The number given to a document by its originators to be used as a means for retrieval; this number will follow any one of various systems, such as chronological, subject area, or accession.

document reader [COMPUT SCI] An optical character reader which reads a limited amount of information (one to five lines) and generally operates from a predetermined format.

document reference edge [COMPUT SCI] In character recognition, that edge of a source document which provides the basis of all subsequent reading processes, insofar as it indicates the relative position of registration marks, and the impending text.

docuterm [COMPUT SCI] A word or phrase descriptive of the subject matter or concept of an item of information and considered important for later retrieval of information.

DOD *See* direct outward dialing.

dodo [ENG] A rectangular groove cut across the grain of a board. [VERT ZOO] *Raphus calcullatus*. A large, flightless, extinct bird of the family Raphidae.

dogfish *See* bowfin.

Dolby system [ELECTR] A noise-reduction system to reduce hiss and other high-frequency noise originating in magnetic tape by providing a predetermined amount of extra amplification for low levels of the higher audio frequencies during recording, with corresponding attenuation during playback to restore the music to its correct level while reducing tape noise.

doldrums [METEOROL] A nautical term for the equatorial trough, with special reference to the light and variable nature of the winds. Also known as equatorial calms.

dolerophanite [MINERAL] $Cu_2(SO_4)O$ A brown, monoclinic mineral consisting of a basic copper sulfate, occurring in crystals.

Dolichopodidae [INV ZOO] The long-legged flies, a family of orthorrhaphous dipteran insects in the series Brachycera.

Dolichothoraci [PALEON] A group of joint-necked fishes assigned to the Arctolepiformes in which the pectoral appendages are represented solely by large fixed spines.

dolioform [BIOL] Barrel-shaped.

Doliolida [INV ZOO] An order of pelagic tunicates in the class Thaliacea; transparent forms, partly or wholly ringed by muscular bands.

dollar [NUCLEO] A unit of reactivity, equal to the difference between the reactivities for delayed critical and prompt critical conditions in a given nuclear reactor.

dollar spot [PL PATH] A fungus disease of lawn grasses caused by *Sclerotinia homeocarpa* and characterized by small, round, brownish areas which gradually coalesce.

dolly [ENG] Any of several types of industrial hand trucks consisting of a low platform or specially shaped carrier mounted on rollers or combinations of fixed and swivel casters; used to carry such things as furniture, milk cans, paper rolls, machinery weighing up to 80 tons, and television cameras short distances.

dolomite [MINERAL] $CaMg(CO_3)_2$ The carbonate mineral; white or colorless with hexagonal symmetry and a structure similar to that of calcite, but with alternate layers of calcium ions being completely replaced by magnesium.

dolomite rock See dolomitic limestone.

dolomitic limestone [PETR] A limestone whose carbonate fraction contains more than 50% dolomite. Also known as dolomite rock; dolostone.

dolomitic marble See magnesian marble.

dolostone See dolomitic limestone.

dolphin [CIV ENG] 1. A group of piles driven close and tied together to provide a fixed mooring in the open sea or a guide for ships coming into a narrow harbor entrance. 2. A mooring post on a wharf. [VERT ZOO] The common name for about 33 species of cetacean mammals included in the family Delphinidae and characterized by the pronounced beak-shaped mouth.

Dolphin See Delphinus.

domain [MATH] 1. A nonempty open connected set in euclidean space. Also known as region. 2. See Abelian field. [SOLID STATE] A region in a solid within which elementary atomic or molecular magnetic or electric moments are uniformly arrayed.

domain theory [SOLID STATE] A theory of the behavior of ferromagnetic and ferroelectric crystals according to which changes in the bulk magnetization and polarization arise from changes in size and orientation of domains that are each polarized to saturation but which point in different directions.

domain-tip memory [COMPUT SCI] A computer memory in which the presence or absence of a magnetic domain in a localized region of a thin magnetic film designates a 1 or 0. Abbreviated DOT memory. Also known as magnetic domain memory.

domatic class See clinohedral class.

dome [ARCH] A hemispherical roof. [ASTRON] A shallow raised structure on the moon's surface with a smooth convex cross section and a diameter anywhere from a few kilometers up to about 80 kilometers. [CRYSTAL] An open crystal form consisting of two faces astride a symmetry plane. [ENG ACOUS] An enclosure for a sonar transducer, projector, or hydrophone and associated equipment; designed to have minimum effect on sound waves traveling underwater. [GEOL] 1. A circular or elliptical, almost symmetrical upfold or anticlinal type of structural deformation. 2. A large igneous intrusion whose surface is convex upward. [ORD] The mound of water spray created in air when the shock wave from an underwater detonation of an atomic weapon reaches the surface.

Domerian [GEOL] Upper Charmouthian geologic time.

domestication [BIOL] The adaptation of an animal or plant through breeding in captivity to a life intimately associated with and advantageous to humans.

domestic coke [MATER] Coke for residential heating, which must have as low an ash content and as high a softening temperature (preferably above 2300°F, or 1260°C) for the ash as possible.

domestic satellite [AERO ENG] A satellite in stationary orbit 22,300 miles (35,680 kilometers) above the equator for handling up to 12 separate color television programs, up to 14,000 private-line telephone calls, or an equivalent number of channels for other communication services within the United States. Abbreviated DOMSAT.

dominance [ECOL] The influence that a controlling organism has on numerical composition or internal energy dynamics in a community. [GEN] The expression of a heritable trait in the heterozygote such as to make it phenotypically indistinguishable from the homozygote.

dominant allele [GEN] The member of a pair of alleles which is phenotypically indistinguishable in both the homozygous and heterozygous condition.

dominant hemisphere [PHYSIO] The cerebral hemisphere which controls certain motor activities; usually the left hemisphere in right-handed individuals.

dominating integral [MATH] An improper integral whose nonnegative, nonincreasing integrand function has the property that its value for all sufficiently large positive integers n is no smaller than the nth term of a given series of positive terms; used in the integral test for convergence.

DOMSAT See domestic satellite.

donkey [MIN ENG] See barney. [VERT ZOO] A domestic ass (*Equus asinus*); a perissodactyl mammal in the family Equidae.

donkey engine [MECH ENG] A small auxiliary engine which is usually portable or semiportable and powered by steam, compressed air, or other means, particularly one used to power a windlass to lift cargo on shipboard or to haul logs.

Donnan equilibrium [PHYS CHEM] The particular equilibrium set up when two coexisting phases are subject to the restriction that one or more of the ionic components cannot pass from one phase into the other; commonly, this restriction is caused by a membrane which is permeable to the solvent and small ions but impermeable to colloidal ions or charged particles of colloidal size. Also known as Gibbs-Donnan equilibrium.

donor [SOLID STATE] An impurity that is added to a pure semiconductor material to increase the number of free electrons. Also known as donor impurity; electron donor.

donor impurity See donor.

donut See doughnut.

doodlebug [GEOL] Also known as douser. 1. Any unscientific device or apparatus, such as a divining rod, used to locate subsurface water, minerals, gas, or oil. 2. A scientific instrument used for locating minerals. [INV ZOO] The larva of an ant lion. [MECH ENG] 1. A small tractor. 2. A motor-driven railcar used for maintenance and repair work. [MIN ENG] The treatment plant or washing unit of a dredge which is mounted on a pontoon and can be floated in an excavation dug by a dragline. [ORD] 1. A small military tank or utility truck. 2. An airborne, magnetic submarine-detecting device.

dopa See dihydroxyphenylalanine.

dopant See doping agent.

dope [ELECTR] See doping agent. [MATER] A cellulose ester lacquer used as an adhesive or a coating.

doped junction [ELECTR] A junction produced by adding an impurity to the melt during growing of a semiconductor crystal.

doping [ELECTR] The addition of impurities to a semiconductor to achieve a desired characteristic, as in producing an n-type or p-type material. Also known as semiconductor doping. [ENG] Coating the mold or mandrel with a substance which will prevent the molded plywood part from sticking to it and will facilitate removal.

doping agent [ELECTR] An impurity element added to semiconductor materials used in crystal diodes and transistors. Also known as dopant; dope.

Doppler broadening [SPECT] Frequency spreading that occurs in single-frequency radiation when the radiating atoms, molecules, or nuclei do not all have the same velocity and may each give rise to a different Doppler shift.

Doppler effect [PHYS] The change in the observed frequency of an acoustic or electromagnetic wave due to relative motion of source and observer.

Doppler frequency *See* Doppler shift.

dopplerite [GEOL] A naturally occurring gel of humic acids found in peat bags or where an aqueous extract from a low-rank coal can collect.

Doppler navigation [NAV] Dead reckoning performed automatically by a device which gives a continuous indication of position by integrating the speed and the crab angle of the aircraft as derived from measurement of the Doppler effect of echoes from directed beams of radiant energy transmitted from the craft.

Doppler radar [ENG] A radar that makes use of the Doppler shift of an echo due to relative motion of target and radar to differentiate between fixed and moving targets and measure target velocities.

Doppler shift [PHYS] The amount of the change in the observed frequency of a wave due to Doppler effect, usually expressed in hertz. Also known as Doppler frequency.

Doppler VOR [NAV] A ground-based navigational aid operating at very high frequency and using a wide-aperture radiation system to reduce azimuth errors caused by reflection from terrain and other obstacles; makes use of the Doppler principle to solve the problem of ambiguity that arises from the use of a radiation system with apertures that exceed one-half wavelength; the system is so designed that its signals may be received on the equipment used for the narrow-aperture VOR (very-high-frequency omnidirectional radio range).

Doradidae [VERT ZOO] A family of South American catfishes in the suborder Siluroidei.

Dorado [ASTRON] A constellation of the southern hemisphere, right ascension 5 hours, declination 65° south. Also known as Swordfish.

Dorilaidae [INV ZOO] The big-headed flies, a family of cyclorrhaphous dipteran insects in the series Aschiza.

Dorippidae [INV ZOO] The mask crabs, a family of brachyuran decapods in the subsection Oxystomata.

dormancy [BOT] A state of quiescence during the development of many plants characterized by their inability to grow, though continuing their morphological and physiological activities.

dormouse [VERT ZOO] The common name applied to members of the family Gliridae; they are Old World arboreal rodents intermediate between squirrels and rats.

dorsal [ANAT] Located near or on the back of an animal or one of its parts.

dorsal aorta [ANAT] The portion of the aorta extending from the left ventricle to the first branch. [INV ZOO] The large, dorsal blood vessel in many invertebrates.

dorsal column [ANAT] A column situated dorsally in each lateral half of the spinal cord which receives the terminals of some afferent fibers from the dorsal roots of the spinal nerves.

dorsal fin [VERT ZOO] A median longitudinal vertical fin on the dorsal aspect of a fish or other aquatic vertebrate.

dorsum [ANAT] 1. The entire dorsal surface of the animal body. 2. The upper part of the tongue, opposite the velum.

Dorvilleidae [INV ZOO] A family of minute errantian annelids in the superfamily Euniceae.

Dorylaimoidea [INV ZOO] An order or superfamily of nematodes inhabiting soil and fresh water.

Dorylinae [INV ZOO] A subfamily of predacious ants in the family Formicidae, including the army ant (*Eciton hamatum*).

Dorypteridae [PALEON] A family of Permian palaeonisciform fishes sometimes included in the suborder Platysomoidei.

dosage [GEN] The number of genes with a similar action that control a given character. [MED] The prescribed or correct amount of medicine or other therapeutic agent administered to treat a given illness. Also known as dose. [NUCLEO] *See* absorbed dose.

dose [MED] 1. The measure, expressed in number of roentgens, of a property of x-rays at a particular place; used in radiology. 2. *See* dosage. [NUCLEO] *See* absorbed dose.

dosemeter *See* dosimeter.

dosimeter [NUCLEO] An instrument that measures the total dose of nuclear radiation received in a given period. Also spelled dosemeter.

dot [ELECTR] *See* button. [GRAPHICS] A subdivision of the printing surface into minute units, formed by a halftone screen and separated by the etching process.

dot character printer [COMPUT SCI] A computer printer that uses the dot matrix technique to generate characters.

dot etching [GRAPHICS] A technique in correcting the color of a positive or halftone negative employing the chemical reduction of halftone dots.

dot generator [ELECTR] A signal generator that produces a dot pattern on the screen of a three-gun color television picture tube, for use in convergence adjustments.

double-acting [MECH ENG] Acting in two directions, as with a reciprocating piston in a cylinder with a working chamber at each end.

double-amplitude-modulation multiplier [ELECTR] A multiplier in which one variable is amplitude-modulated by a carrier, and the modulated signal is again amplitude-modulated by the other variable; the resulting double-modulated signal is applied to a balanced demodulator to obtain the product of the two variables.

double arcing [MET] An occurrence in plasma arc welding and cutting where a secondary electric arc displaces the main arc at the outlet of a welding nozzle.

double-base diode *See* unijunction transistor.

double-base junction diode *See* unijunction transistor.

double-base junction transistor [ELECTR] A tetrode transistor that is essentially a junction triode transistor having two base connections on opposite sides of the central region of the transistor. Also known as tetrode junction transistor.

double beta decay [NUC PHYS] A nuclear transformation in which the atomic number changes by 2 and the mass number does not change; either two electrons are emitted or two orbital electrons are captured.

double blossom [PL PATH] A fungus disease of dewberry and blackberry caused by *Fusarium rubi* and characterized by witches'-brooms and enlargement and malformation of the flowers.

double bond [PHYS CHEM] A type of linkage between atoms in which two pair of electrons are shared equally.

double bridge *See* Kelvin bridge.

double-buffered data transfer [COMPUT SCI] The transmission of data into the buffer register and from there into the device register proper.

double-button microphone [ENG ACOUS] A carbon microphone having two carbon-filled buttonlike containers, one on each side of the diaphragm, to give twice the resistance change obtainable with a single button. Also known as differential microphone.

double circulation [PHYSIO] A circulatory system in which blood flows through two separate circuits, as pulmonary and systemic.

double cluster [ASTRON] A pair of globular clusters that are physically close to each other, near the northern boundary of the constellation Perseus.

double-diffused transistor [ELECTR] A transistor in which two *pn* junctions are formed in the semiconductor wafer by gaseous diffusion of both *p*-type and *n*-type impurities; an intrinsic region can also be formed.

double diode *See* binode; duodiode.

double-diode limiter [ELECTR] Type of limiter which is used to remove all positive signals from a combination of positive and negative pulses, or to remove all the negative signals from such a combination of positive and negative pulses.

double-doped transistor [ELECTR] The original grown-junction transistor, formed by successively adding *p*-type and *n*-type impurities to the melt during growing of the crystal.

double-entry method [MIN ENG] A mining arrangement involving twin entries in flat or gently dipping coal, so that rooms can be extended from both entryways.

double exposure [GRAPHICS] The act of recording two images on top of each other completely or in part.

double fertilization [BOT] In most seed plants, fertilization involving fusion between the egg nucleus and one sperm nucleus, and fusion between the other sperm nucleus and the polar nuclei.

double frequency shift keying [COMMUN] Multiplex system in which two telegraph signals are combined and transmitted simultaneously by a method of frequency shifting between four radio frequencies.

double group [QUANT MECH] A type of group useful in studying systems of half-integral spin; it is formed by modifying a finite point group by introducing an element which is a rotation through an angle of 2π about an arbitrary axis and which is not the unit element but gives the unit element when applied twice.

double-hump fission barrier [NUC PHYS] Two separated maxima in a plot of potential energy against nuclear deformation of an actinide nucleus, which inhibit spontaneous fission of the nucleus and give rise to isomeric states in the valley between the two maxima.

double-hung [BUILD] Of a window, having top and bottom sashes which are counterweighted or equipped with a spring on each side for easier raising and lowering.

double image [ELECTR] A television picture consisting of two overlapping images due to reception of the signal over two paths of different length so that signals arrive at slightly different times.

double layer *See* electric double-layer.

double-length number [COMPUT SCI] A number having twice as many digits as are ordinarily used in a given computer. Also known as double-precision number.

double limiter *See* cascade limiter.

double-list sorting [COMPUT SCI] A method of internal sorting in which the entire unsorted list is first placed in one portion of main memory and sorting action then takes place, creating a sorted list, generally in another area of memory.

double modulation [COMMUN] A method of modulation in which a subcarrier is first modulated with the desired intelligence, and the modulated subcarrier is then used to modulate a second carrier having a higher frequency.

double point [MATH] A point on a curve at which a curve crosses or touches itself, or has a cusp; that is, a point at which the curve has two tangents (which may be coincident).

double-pole switch [ELEC] A switch that operates simultaneously in two separate electric circuits or in both lines of a single circuit.

double precision [COMPUT SCI] The use of two computer words to represent a double-length number.

double-precision hardware [COMPUT SCI] Special arithmetic units in a computer designed to handle double-length numbers, employed in operations in which greater accuracy than normal is desired.

double-precision number *See* double-length number.

double-pulse recording [COMPUT SCI] A technique for recording binary digits in magnetic cells in which each cell consists of two regions that can be magnetized in opposite directions and the value of each bit (0 or 1) is determined by the order in which the regions occur.

doubler *See* frequency doubler; voltage doubler.

double refraction *See* birefringence.

double root [MATH] A number a such that $(x - a)^2 p(x) = 0$ where $p(x)$ is a polynomial of which a is not a root.

double sampling [IND ENG] Inspecting one sample and then deciding whether to accept or reject the lot or to defer action until a second sample is inspected.

double-sideband modulation [COMMUN] Amplitude modulation in which the modulated wave is composed of a carrier, an upper sideband whose frequency is the sum of the carrier and modulation frequencies, and a lower sideband whose frequency is the difference between the carrier and modulation frequencies. Abbreviated DSB.

double-sideband transmission [COMMUN] The transmission of a modulated carrier wave accompanied by both of the sidebands resulting from modulation; the upper sideband corresponds to the sum of the carrier and modulation frequencies, whereas the lower sideband corresponds to the difference between the carrier and modulation frequencies.

double-slider coupling *See* slider coupling.

double star [ASTRON] A star which appears as a single point of light to the eye but which can be resolved into two points by a telescope.

doublet [ATOM PHYS] Two stationary states which have the same orbital and spin angular momentum but which have different total angular momenta, and therefore have slightly different energies due to spin-orbit coupling. [ELECTROMAG] *See* dipole. [FL MECH] A source and a sink separated by an infinitesimal distance, each having an infinitely large strength so that the product of this strength and the separation is finite. [OPTICS] A lens made up of two components, especially an achromat. [PARTIC PHYS] Two elementary particles which have slightly differing masses and the same baryon number, spin, parity, and charge conjugation parity (if self-conjugate), but have different charges. [PHYS CHEM] Two electrons which are shared between two atoms and give rise to a nonpolar valence bond. [SPECT] Two closely separated spectral lines arising from a transition between a single state and a pair of states forming a doublet as described in the atomic physics definition.

doublet antenna *See* dipole antenna.

doublet flow [FL MECH] The motion of a fluid in the vicinity of a doublet; can be superposed with uniform flow to yield flow around a cylinder or a sphere.

double-track tape recorder [ENG ACOUS] A tape recorder with a recording head that covers half the tape width, so two parallel tracks can be recorded on one tape. Also known as dual-track tape recorder; half-track tape recorder.

double-tuned amplifier [ELECTR] Amplifier of one or more stages in which each stage uses coupled circuits having two frequencies of resonance, to obtain wider bands than those obtainable with single tuning.

double-tuned circuit [ELECTR] A circuit that is resonant to two adjacent frequencies, so that there are two approximately equal values of peak response, with a dip between.

double-tuned detector [ELECTR] A type of frequency-modulation discriminator in which the limiter output transformer has two secondaries, one tuned above the resting frequency and the other tuned an equal amount below.

double word [COMPUT SCI] A unit containing twice as many bits as a word.

double-word addressing [COMPUT SCI] An addressing mode in computers with short words (less than 16 bits) in which the second of two consecutive instruction words contains the address of a location.

doughnut [NUCLEO] Also spelled donut. **1.** The toroidal vacuum chamber in which electrons are accelerated in a betatron or synchrotron. Also known as toroid. **2.** An assembly of enriched fissionable material, often doughnut-shaped, used in a thermal reactor to provide a local increase in fast neutron flux for experimental purposes. [PETRO ENG] A ring of wedges or a threaded, tapered ring that supports a pipe string.

Douglas fir [BOT] *Pseudotsuga menziesii*. A large coniferous tree in the order Pinales; cones are characterized by bracts extending beyond the scales. Also known as red fir.

douglasite [MINERAL] $K_2FeCl_4 \cdot 2H_2O$ Ore from Stassfurt, Germany; a member of the erythrosiderite group; orthorhombic; in the isomorphous series.

douser *See* doodlebug.

Dove *See* Columba.

Dove prism *See* Delaborne prism.

dovetail joint [DES ENG] A joint consisting of a flaring tenon in a fitting mortise.

dovetail molding [ARCH] A molding in a zig-zag pattern resembling a series of dovetails.

down cutting *See* climb cutting.

downdip [GEOL] Pertaining to a position parallel to or in the direction of the dip of a stratum or bed.

downdraft [PHYS] A current of air or other gas that travels downward, as during a thunderstorm or in a mine shaft.

downdraft carburetor [MECH ENG] A carburetor in which the fuel is fed into a downward current of air.

downhole drill [MIN ENG] A hammer or percussive drill in which a reciprocating pneumatic piston is located immediately behind the drill bit and can follow and enter the bit down the hole, for minimizing energy losses.

downhole equipment *See* drill fittings.

down-lead *See* lead-in.

downlink [COMMUN] The ratio or optical transmission path downward from a communications satellite to the earth or an aircraft, or from an aircraft to the earth.

down-load [COMPUT SCI] To transfer a program or data file from a central computer to a remote computer or to the memory of an intelligent terminal.

down milling *See* climb cutting.

Downs cell [CHEM ENG] A brick-lined steel vessel with four graphite anodes projecting upward from the bottom, with cathodes in the form of steel cylinders concentric with the anodes, containing an electrolyte which is 40% NaCl and 60% $CaCl_2$ at 590°C; used to make sodium.

Down's process [CHEM ENG] A method for producing sodium and chlorine from sodium chloride; potassium chloride and fluoride are added to the sodium chloride to reduce the melting point; the fused mixture is electrolyzed, with sodium forming at the cathode and chlorine at the anode.

Down's syndrome [MED] A syndrome of congenital defects, especially mental retardation, typical facies responsible for the term mongoloid idiocy, or mongolism, and cytogenetic abnormality consisting of trisomy 21 or its equivalent in the form of an unbalanced translocation. Also known as mongolism; trisomy 21 syndrome.

downstream [CHEM ENG] Portion of a product stream that has already passed through the system; that portion located after a specific process unit. [HYD] In the direction of flow, as a current or waterway.

downtime [IND ENG] The lost production time during which a piece of equipment is not operating correctly due to a breakdown, maintenance, necessities, or power failure.

downwind [NAV] In the direction toward which the wind is blowing; applies particularly to moving downwind, whether desired or not.

downy mildew [PL PATH] A fungus disease of higher plants caused by members of the family Peronosporaceae and characterized by a white, downy growth on the diseased plant parts.

Dow oscillator *See* electron-coupled oscillator.

Dowtonian [GEOL] Uppermost Silurian or lowermost Devonian geologic time.

DPCM *See* differential pulse-code modulation.

DPN *See* diphosphopyridine nucleotide.

dr *See* dram.

Dra *See* Draco.

Drac *See* Draco.

drachm *See* dram.

Draco [ASTRON] A long, serpentine constellation that surrounds half of the Little Dipper in the north. Abbreviated Dra; Drac. Also known as Dragon.

Draconids [ASTRON] Several meteor showers whose radiants lie in the constellation Draco.

Dracunculoidea [INV ZOO] An order or superfamily of parasitic nematodes characterized by their habitat in host tissues and by the way larvae leave the host through a skin lesion.

draft Also spelled draught. [CIV ENG] A line of a traverse survey. [ENG] 1. In molds, the degree of taper on a side wall or the angle of clearance present to facilitate removal of cured or hardened parts from a mold. 2. The area of a water discharge opening. [FL MECH] 1. An air current in a confined space, such as that in a cooling tower or chimney. 2. The difference between atmospheric pressure and some lower pressure in a confined space that causes air to flow, such as exists in the furnace or gas passages of a steam-generating unit or in a chimney. [MET] 1. The act or process of drawing, with dies. 2. The work or quantity of work drawn. [NAV ARCH] The vertical distance from the top of the keel plate or bar keel to the load waterline.

drafting [GRAPHICS] The making of drawings of objects, structures, or systems that have been visualized by engineers, scientists, and others. [TEXT] 1. The process of lengthening raw fibers, in the form of slubbing, sliver, or roving, to make the stock look more like yarn. 2. Plotting directions for weaving on cross-section paper, showing the movement of the threads.

drag [FL MECH] Resistance caused by friction in the direction opposite to that of the motion of the center of gravity of a moving body in a fluid. [MET] 1. The bottom part of a flask used in casting. 2. In thermal cutting, the distance deviating from the theoretical vertical line of cutting measured along the bottom surface of the material. [MIN ENG] Movement of the hanging wall with respect to the foot wall due to the weight of the arch block in an inclined slope.

drag-chain conveyor [MECH ENG] A conveyor in which the open links of a chain drag material along the bottom of a hard-faced concrete or cast iron trough. Also known as dragline conveyor.

drag coefficient [FL MECH] A characteristic of a body in a flowing inviscous fluid, equal to the ratio of twice the force on the body in the direction of flow to the product of the density of the fluid, the square of the flow velocity, and the effective cross-sectional area of the body.

drag conveyor See flight conveyor.

drag-cup generator [ENG] A type of tachometer which uses eddy currents and functions in control systems; it consists of two stationary windings, positioned so as to have zero coupling, and a nonmagnetic metal cup, which is revolved by the source whose speed is to be measured; one of the windings is used for excitation, inducing eddy currents in the rotating cup. Also known as drag-cup tachometer.

drag-cup tachometer See drag-cup generator.

drag fold [GEOL] A minor fold formed in an incompetent bed by movement of a competent bed so as to subject it to couple; the axis is at right angles to the direction in which the beds slip.

drag force [PL PHYS] A force on an electrically conducting fluid arising from inelastic collisions of electrons and ions and proportional to the fluid velocity.

dragline [MECH ENG] An excavator operated by pulling a bucket on ropes towards the jib from which it is suspended. Also known as dragline excavator.

dragline conveyor See drag-chain conveyor.

dragline excavator See dragline.

Dragon See Draco.

dragonfly [INV ZOO] Any of the insects composing six families of the suborder Anisoptera and having four large, membranous wings and compound eyes that provide keen vision.

drag-type tachometer See eddy-current tachometer.

drain [CIV ENG] 1. A channel which carries off surface water. 2. A pipe which carries off liquid sewage. [ELEC] See current drain. [ELECTR] One of the electrodes in a thin-film transistor.

drainage basin [HYD] An area in which surface run-off collects and from which it is carried by a drainage system, as a river and its tributaries. Also known as catchment area; drainage area; feeding ground; gathering ground; hydrographic basin.

drainage canal [CIV ENG] An artificial canal built to drain water from an area having no natural outlet for precipitation accumulation.

drainage pattern [HYD] The configuration of a natural or artificial drainage system; stream patterns reflect the topography and rock patterns of the area.

drain tile [BUILD] A cylindrical tile with holes in the walls used at the base of a building foundation to carry away groundwater.

dram [MECH] 1. A unit of mass, used in the apothecaries' system of mass units, equal to $\frac{1}{8}$ apothecaries' ounce or 60 grains or 3.8879346 grams. Also known as apothecaries' dram (dram ap); drachm (British). 2. A unit of mass, formerly used in the United Kingdom, equal to $\frac{1}{16}$ ounce (avoirdupois) or approximately 1.77185 grams. Abbreviated dr.

dram ap See dram.

Draper catalog [ASTRON] A nine-volume catalog of stars completed in 1924; it gives positions, magnitudes, and spectral classes of 225,300 stars.

draught See draft.

draught stop See fire stop.

draw [ENG] To haul a load. [MET] 1. A fissure or pocket in a casting formed when the supply of molten metal is inadequate during solidification. 2. To remove a pattern from a foundry flask. [MIN ENG] 1. To remove timber supports, allowing overhanging coal to fall down for collection. 2. To allow ore to run down chutes from stopes, chambers, or ore bins. 3. To collect broken coal in trucks. 4. To hoist coal, rock, ore, or other materials to the surface. 5. The horizontal distance to which creep extends on the surface beyond the stopes.

drawbridge [CIV ENG] Any bridge that can be raised, lowered, or drawn aside to provide clear passage for ships.

drawdown [GRAPHICS] In inkmaking, a procedure for obtaining a rough estimation of a color shade in which a small sample of ink is placed on a piece of paper and spread with a spatula to yield a thin film of ink. [HYD] The magnitude of the change in water surface level in a well, reservoir, or natural body of water resulting from the withdrawal of water. [PETRO ENG] The difference between the static and the flowing bottom-hole pressure.

drawing [CHEM ENG] Removing ceramic ware from a kiln after it has been fired. [GRAPHICS] A surface portrayal of a form or figure in line. [MET] 1. Pulling a wire or tube through a die to reduce the cross section. 2. Forcing plastic deformation of metal in a die to form recessed parts.

dredge [ENG] A cylindrical or rectangular device for collecting samples of bottom sediment and benthic fauna. [MECH ENG] A floating excavator used for widening or deepening channels, building canals, constructing levees, raising material from stream or harbor bottoms to be used elsewhere as fill, or mining.

D region [GEOPHYS] The region of ionosphere up to about 60 miles (97 kilometers) above the earth, below the E and F regions, in which the D layer forms.

Drepanellacea [PALEON] A monomorphic superfamily of extinct paleocopan ostracods in the suborder Beyrichicopina having a subquadrate carapace, many with a marginal rim.

Drepanellidae [PALEON] A monomorphic family of extinct ostracods in the superfamily Drepanellacea.

Drepanidae [INV ZOO] The hooktips, a small family of lepidopteran insects in the suborder Heteroneura.

Dresbachian [GEOL] Lower Croixan geologic time.

dress [ELECTR] The arrangement of connecting wires in a circuit to prevent undesirable coupling and feedback. [MECH ENG] 1. To shape a tool. 2. To restore a tool to its original shape and sharpness. [MIN ENG] To sort, grind, clean, and concentrate ore.

dressing [AGR] Manure or compost used as a fertilizer. [ENG] The sharpening, repairing, and replacing of parts, notably drilling bits and tool joints, to ready equipment for reuse. [MED] 1. Application of various materials for protecting a wound and encouraging healing. 2. Material so applied.

drier [ENG] A device to remove water. [MATER] 1. A substance that absorbs water. 2. A substance that is used to hasten solidification. 3. Material, such as salts of lead, manganese, and cobalt, which facilitates the oxidation of oils; used in paints and varnishes to speed drying.

drift [ENG] 1. A gradual deviation from a set adjustment, such as frequency or balance current, or from a direction. 2. The deviation, or the angle of deviation, of a borehole from the vertical or from its intended course. [GEOL] 1. Rock material picked up and transported by a glacier and deposited elsewhere. 2. Detrital material moved and deposited on a beach by waves and currents. [MECH ENG] The water lost in a cooling tower as mist or droplets entrained by the circulating air, not including the evaporative loss. [MIN ENG] A horizontal mine opening which follows a vein or lies within the trend of an ore body. [NAV] 1. The movement of a craft caused by the action of wind or current. 2. To move gradually from a set position without control. [OCEANOGR] See drift current. [SOLID STATE] The movement of current carriers in a semiconductor under the influence of an applied voltage.

drift angle [NAV] 1. The horizontal angle between the axis of a ship and the tangent to its path. Also known as drift correction angle. 2. The angle between the longitudinal axis of an aircraft and its path relative to the ground.

drift bottle [OCEANOGR] A bottle which is released into the sea for studying currents; contains a card, identifying the date and place of release, to be returned by the finder with date and place of recovery. Also known as floater.

drift correction angle See drift angle.

drift current [OCEANOGR] 1. A wide, slow-moving ocean current principally caused by winds. Also known as drift; wind drift; wind-driven current. 2. Current determined from the differences between dead reckoning and a navigational fix. [PL PHYS] A current of free charged particles in perpendicular electric and magnetic fields that results from an average motion of the particles in a direction perpendicular to both fields.

drifter [MECH ENG] A rock drill, similar to but usually larger than a jack hammer, mounted for drilling holes up to 4½ inches (11.4 centimeters) in diameter. [MIN ENG] 1. A person who excavates mine drifts. 2. An air-driven rock drill used for excavating mine drifts and crosscuts.

drift error [COMPUT SCI] An error arising in the use of an analog computer due to gradual changes in the output of circuits (such as amplifiers) in the computer.

drift ice [OCEANOGR] Sea ice that has drifted from its place of formation.

drifting snow [METEOROL] Wind-driven snow raised from the surface of the earth to a height of less than 6 feet (1.8 meters).

drift mobility [SOLID STATE] The average drift velocity of carriers per unit electric field in a homogeneous semiconductor. Also known as mobility.

driftpin [DES ENG] A round, tapered metal rod that is driven into matching rivet holes of two metal parts for stretching the parts and bringing them into alignment.

drift transistor [ELECTR] 1. A transistor having two plane parallel junctions, with a resistivity gradient in the base region between the junctions to improve the high-frequency response. 2. See diffused-alloy transistor.

drift tube [NUCLEO] A tubular electrode placed in the vacuum chamber of a circular accelerator, to which radio-frequency voltage is applied to accelerate the particles.

drift velocity [SOLID STATE] The average velocity of a carrier that is moving under the influence of an electric field in a semiconductor, conductor, or electron tube.

drift wave [PL PHYS] An oscillation in a magnetically confined plasma which arises in the presence of density gradients, for example, at the plasma's surface, and which resembles the waves that propagate at the interface of two fluids of different density in a gravity field.

Drilidae [INV ZOO] The false firefly beetles, a family of coleopteran insects in the superfamily Cantharoidea.

drill [ENG] A rotating-end cutting tool for creating or enlarging holes in a solid material. Also known as drill bit. [TEXT] Strong twilled carded cotton cloth.

drill bit See drill.

drill cable [ENG] A cable used to pull up drill rods, casing, and other drilling equipment used in making a borehole.

drill carriage [MECH ENG] A platform or frame on which several rock drills are mounted and which moves along a track, for heavy drilling in large tunnels. Also known as jumbo.

drill collar [DES ENG] A ring which holds a drill bit and gives it radial location with respect to a bearing.

drill column [MIN ENG] A steel pipe that can be wedged across an underground opening in a vertical or horizontal position to serve as a base on which to mount a diamond or rock drill.

drill cuttings [ENG] Cuttings of rock and other subterranean materials brought to the surface during the drilling of wellholes.

drilled caisson [CIV ENG] A drilled hole filled with concrete and lined with a cylindrical steel casing if needed.

driller [ENG] A person who operates a drilling machine. [MECH ENG] See drilling machine.

drill fittings [ENG] All equipment used in a borehole during drilling. Also known as downhole equipment.

drill gage [DES ENG] A thin, flat steel plate that has accurate holes for many sizes of drills; each hole, identified as to drill size, enables the diameter of a drill to be checked. [ENG] Diameter of a borehole.

drillhole survey *See* borehole survey.

drilling fluid *See* drilling mud.

drilling machine [MECH ENG] A device, usually motor-driven, fitted with an end cutting tool that is rotated with sufficient power either to create a hole or to enlarge an existing hole in a solid material. Also known as driller.

drilling mud [MATER] A suspension of finely divided heavy material, such as bentonite and barite, pumped through the drill pipe during rotary drilling to seal off porous zones and flush out chippings, and to lubricate and cool the bit. Also known as drilling fluid.

drilling platform [ENG] The structural base upon which the drill rig and associated equipment is mounted during the drilling operation.

drill log [ENG] 1. A record of the events and features of the formations penetrated during boring. Also known as boring log. 2. A record of all occurrences during drilling that might help in a complete logging of the hole or in determining the cost of the drilling.

drill pipe [MIN ENG] A pipe used for driving a revolving drill bit, used especially in drilling wells; consists of a casing within which tubing is run to conduct oil or gas to ground level; drilling mud flows in the annular space between casing and tubing during the drilling operation.

drill press [MECH ENG] A drilling machine in which a vertical drill moves into the work, which is stationary.

drill stem *See* drill string.

drill string [MECH ENG] The assemblage of drill rods, core barrel, and bit, or of drill rods, drill collars, and bit in a borehole, which is connected to and rotated by the drill collar of the borehole. Also known as drill stem.

drip edge [BUILD] A metal strip that extends beyond the other parts of the roof and is used to direct rainwater off.

dripstone cave [GEOL] Any cave of calcium carbonate or other mineral formed by the action of dripping water.

drive [ELECTR] *See* excitation. [MECH ENG] The means by which a machine is given motion or power (as in steam drive, diesel-electric drive), or by which power is transferred from one part of a machine to another (as in gear drive, belt drive). [MIN ENG] 1. To excavate in a horizontal or inclined plane. 2. A horizontal underground tunnel along or parallel to a lode, vein, or ore body. [PSYCH] A strong impetus to behavior or active striving.

drive control *See* horizontal drive control.

drive fit [DES ENG] A fit in which the larger (male) part is pressed into a smaller (female) part; the assembly must be effected through the application of an external force.

driven array [ELECTROMAG] An antenna array consisting of a number of driven elements, usually half-wave dipoles, fed in phase or out of phase from a common source.

driven blocking oscillator *See* monostable blocking oscillator.

driven caisson [CIV ENG] A caisson formed by driving a cylindrical steel shell into the ground with a pile-driving hammer and then placing concrete inside; the shell may be removed when concrete sets.

driven gear [MECH ENG] The member of a pair of gears to which motion and power are transmitted by the other.

driven snow [METEOROL] Snow which has been moved by wind and collected into snowdrifts.

drivepipe [ENG] A thick-walled casing pipe that is driven through overburden or into a deep drill hole to prevent caving.

drive pulley [MECH ENG] The pulley that drives a conveyor belt.

drive pulse [ELECTR] An electrical pulse which induces a magnetizing force in an element of a magnetic core storage, reversing the polarity of the core.

driver [ELECTR] The amplifier stage preceding the output stage in a receiver or transmitter. [ENG ACOUS] The portion of a horn loudspeaker that converts electrical energy into acoustical energy and feeds the acoustical energy to the small end of the horn.

drivescrew [DES ENG] A screw that is driven all the way in, or nearly all the way in, with a hammer.

drive shaft [MECH ENG] A shaft which transmits power from a motor or engine to the rest of a machine.

drive winding [ELECTR] A coil of wire that is inductively coupled to an element of a magnetic memory. Also known as drive wire.

drive wire *See* drive winding.

drizzle [METEOROL] Very small, numerous, and uniformly dispersed water drops that may appear to float while following air currents; unlike fog droplets, drizzle falls to the ground; it usually falls from low stratus clouds and is frequently accompanied by low visibility and fog.

DRM *See* detrital remanent magnetization.

drogue [AERO ENG] 1. A small parachute attached to a body for stabilization and deceleration. Also known as deceleration parachute. 2. A funnel-shaped device at the end of the hose of a tanker aircraft in flight, to receive the probe of another aircraft that will take on fuel. [ENG] 1. A device, such as a sea anchor, usually shaped like a funnel or cone and dragged or towed behind a boat or seaplane for deceleration, stabilization, or speed control. 2. A current-measuring assembly consisting of a weighted current cross, sail, or parachute and an attached surface buoy. Also known as drag anchor; sea anchor.

Dromadidae [VERT ZOO] A family of the avian order Charadriiformes containing a single species, the crab plover (*Dromas ardeola*).

dromedary [VERT ZOO] *Camelus dromedarius* The Arabian camel, distinguished by a single hump.

Dromiacea [INV ZOO] The dromiid crabs, a subsection of the Brachyura in the crustacean order Decapoda.

Dromiceidae [VERT ZOO] The emus, a monospecific family of flightless birds in the order Casuariiformes.

drone [AERO ENG] A pilotless aircraft usually subordinated to the controlling influences of a remotely located command station, but occasionally preprogrammed. [INV ZOO] A haploid male bee or ant; one of the three castes in a colony.

drooped airfoil [AERO ENG] A base-line airfoil with an abrupt change in cross section at about midspan from the fuselage; the outboard portion of the wing has a cross-section with a nearly flat bottom and a drooped (downward) leading edge in relation to the inboard base-line wing.

droop governor [MECH ENG] A governor whose equilibrium speed decreases as the load on the machinery controlled by the governor increases.

drop [FL MECH] The quantity of liquid that coalesces into a single globule; sizes vary according to physical conditions and the properties of the fluid itself. [HYD] The difference in water-surface elevations that is measured up-and downstream from a narrowing in the stream. [MET] A casting defect due to the falling of a portion of sand from an overhanging section of the mold. [MINERAL] A funnel-shaped downward intrusion of sedimentary rock into the roof of a coal seam. [PL PATH] A fungus disease of various vegetables caused by *Sclerotinia sclerotiorum* and characterized by wilt and stem rot.

drop-dead halt [COMPUT SCI] A machine halt from which there is no recovery; such a halt may occur through a logical error in programming; examples in which a drop-dead halt could occur are division by zero and transfer to a nonexistent instruction word. Also known as dead halt.

drop forging [MET] Plastic deformation of hot metal under a falling weight, such as a drop hammer.

drop hammer [MECH ENG] *See* pile hammer. [MET] A hammer used in forging that is raised and then dropped on the metal resting on an anvil or on a die.

drop-in [COMPUT SCI] The accidental appearance of an unwanted bit, digit, or character on a magnetic recording surface or during reading from or writing to a magnetic storage device.

drop model of nucleus *See* liquid-drop model of nucleus.

dropout [ELEC] Of a relay, the maximum current, voltage, power, or such, at which it will release from its energized position. [ELECTR] A reduction in output signal level during reproduction of recorded data, sufficient to cause a processing error. [GRAPHICS] A halftone negative, print, or plate on which some of the original image has been removed by masking or opaquing.

dropping fraction [COMPUT SCI] In punched cards, the chance that a given sorting operation will cause a card taken at random to be selected.

dropping-mercury electrode [PHYS CHEM] An electrode consisting of a fine-bore capillary tube above which a constant head of mercury is maintained; the mercury emerges from the tip of the capillary at the rate of a few milligrams per second and forms a spherical drop which falls into the solution at the rate of one every 2–10 seconds.

drop press *See* punch press.

dropsonde [ENG] A radiosonde dropped by parachute from a high-flying aircraft to measure weather conditions and report them back to the aircraft.

dropsy *See* edema.

Droseraceae [BOT] A family of dicotyledonous plants in the order Sarraceniales, distinguished by leaves that do not form pitchers, parietal placentation, and several styles.

Drosophilidae [INV ZOO] The vinegar flies, a family of myodarian cyclorrhaphous dipteran insects in the subsection Acalyptratae, including the fruit fly (*Drosophila melanogaster*).

dross [MET] An impurity, usually an oxide, formed on the surface of a molten metal.

drought [CLIMATOL] A period of abnormally dry weather sufficiently prolonged so that the lack of water causes a serious hydrologic imbalance (such as crop damage, water supply shortage, and so on) in the affected area; in general, the term should be reserved for relatively extensive time periods and areas.

drowned coast [GEOL] A shoreline transformed from a hilly land surface to an archipelago of small islands by inundation by the sea.

drowned river mouth *See* estuary.

drug resistance [MICROBIO] A decreased reactivity of living organisms to the injurious actions of certain drugs and chemicals.

drug tolerance [MED] Condition that may follow repeated ingestion of a drug in so that the effect produced by the original dose no longer occurs.

drum [CHEM ENG] Tower or vessel in a refinery into which heated products are conducted so that volatile portions can separate. [DES ENG] **1.** A hollow, cylindrical container. **2.** A metal cylindrical shipping container for liquids having a capacity of 12–110 gallons (45–416 liters). [ELECTR] A computer storage device consisting of a rapidly rotating cylinder with a magnetizable external surface on which data can be read or written by many read/write heads floating a few millionths of an inch off the surface. Also known as drum memory; drum storage; magnetic drum; magnetic drum storage. [MECH ENG] A horizontal cylinder about which rope or wire rope is wound in a hoisting mechanism. Also known as hoisting drum.

drum brake [MECH ENG] A brake in which two curved shoes fitted with heat- and wear-resistant linings are forced against the surface of a rotating drum.

drum dryer [MECH ENG] A machine for removing water from substances such as milk, in which a thin film of the product is moved over a turning steam-heated drum and a knife scrapes it from the drum after moisture has been removed.

drum filter [MECH ENG] A cylindrical drum that rotates through thickened ore pulp, extracts liquid by a vacuum, and leaves solids, in the form of a cake, on a permeable membrane on the drum end. Also known as rotary filter; rotary vacuum filter.

drum gate [CIV ENG] A movable crest gate in the form of an arc hinged at the apex and operated by reservoir pressure to open and close a spillway.

drumlin [GEOL] A hill of glacial drift or bedrock having a half-ellipsoidal streamline form like the inverted bowl of a spoon, with its long axis paralleling the direction of movement of the glacier that fashioned it.

drum memory *See* drum.

drum plotter [ENG] A graphics output device that draws lines with a continuously moving pen on a sheet of paper rolled around a rotating drum that moves the paper in a direction perpendicular to the motion of the pen.

drum printer [COMPUT SCI] An impact printer in which a complete set of characters for each print position on a line is on a continuously rotating drum behind an inked ribbon, with paper in front of the ribbon; identical characters are printed simultaneously at all required positions on a line, on the fly, by signal-controlled hammers.

drum recorder [ELECTR] A facsimile recorder in which the record sheet is mounted on a rotating drum or cylinder.

drum storage *See* drum.

drum-type boiler *See* bent-tube boiler.

drum winding [ELEC] A type of winding in electric machines in which coils are housed in long, narrow gaps either in the outer surface of a cylindrical core or in the inner surface of a core with a cylindrical bore.

drupaceous [BOT] Of, pertaining to, or characteristic of a drupe.

drupe [BOT] A fruit, such as a cherry, having a thin or leathery exocarp, a fleshy mesocarp, and a single seed with a stony endocarp. Also known as stone fruit.

drupelet [BOT] An individual drupe of an aggregate fruit. Also known as grain.

dry-back boiler *See* scotch boiler.

dry battery [ELEC] A battery made up of a series, parallel, or series-parallel arrangement of dry cells in a single housing to provide desired voltage and current values.

dry-bone ore *See* smithsonite.

dry-bulb thermometer [ENG] An ordinary thermometer, especially one with an unmoistened bulb; not dependent upon atmospheric humidity.

dry cell [ELEC] A voltage-generating cell having an immobilized electrolyte.

dry-chemical fire extinguisher [CHEM ENG] A dry powder, consisting principally of sodium bicarbonate, which is used for extinguishing small fires, especially electrical fires.

dry circuit [ELEC] A relay circuit in which open-circuit voltages are very low and closed-circuit currents extremely small, so there is no arcing to roughen the contacts.

dry cooling tower [MECH ENG] A structure in which water is cooled by circulation through finned tubes, transferring heat to air passing over the fins; there is no loss of water by evaporation because the air does not directly contact the water.

dry corrosion [MET] Destruction of a metal or alloy by chemical processes resulting from attack by gases in the atmosphere above the dew point.

dry course [BUILD] An initial roofing course of felt or paper not bedded in tar or asphalt.

dry delta *See* alluvial fan.

dry-disk rectifier *See* metallic rectifier.

dry dock [CIV ENG] A dock providing support for a vessel and a means for removing the water so that the bottom of the vessel can be exposed.

dry-dock caisson [CIV ENG] The floating gate to a dry dock. Also known as caisson.

dry gangrene [MED] Local death of a part caused by arterial obstruction without associated venous obstruction or infection.

dry gas [MATER] A gas that does not contain fractions which may easily condense under normal atmospheric conditions, for example, natural gas with methane and ethane.

dry hole [ENG] A hole driven without the use of water. [PETRO ENG] A well in which no oil or gas is found.

dry ice [INORG CHEM] Carbon dioxide in the solid form, usually made in blocks to be used as a coolant; changes directly to a gas at $-78.5°C$ as heat is absorbed.

drying [CHEM] 1. An operation in which a liquid, usually water, is removed from a wet solid in equipment termed a dryer. 2. A process of oxidation whereby a liquid such as linseed oil changes into a solid film.

drying oil [MATER] Relatively highly unsaturated oil, such as cottonseed, soybean, and linseed oil, that is easily oxidized and polymerized to form a hard, dry film on exposure to air; used in paints and varnish.

Dryinidae [INV ZOO] A family of hymenopteran insects in the superfamily Bethyloidea.

dry offset *See* letterset.

Dryomyzidae [INV ZOO] A family of myodarian cyclorrhaphous dipteran insects in the subsection Acalypteratae.

Dryopidae [INV ZOO] The long-toed water beetles, a family of coleopteran insects in the superfamily Dryopoidea.

Dryopoidea [INV ZOO] A superfamily of coleopteran insects in the suborder Polyphaga, including the nonpredatory aquatic beetles.

dry pint *See* pint.

dry plate [GRAPHICS] A photographic plate that has a sensitized coating of an emulsion of silver halide in gelatin which is dried before exposure to light in the photographic process.

dry-plate rectifier *See* metallic rectifier.

drypoint etching [GRAPHICS] Etching in which a sharp tool (an etching needle) scratches through only the etching ground that is placed on the surface of the copper plate; the plate is then placed in an acid bath, and the chemical action produces a line deep enough to hold ink.

dry pt *See* pint.

dry rot [MICROBIO] A rapid decay of seasoned timber caused by certain fungi which cause the wood to be reduced to a dry, friable texture. [PL PATH] Any of various rot diseases of plants characterized by drying of affected tissues.

dry run [COMPUT SCI] A check of the logic and coding of a computer program in which the program's operations are followed from a flow chart and written instructions, and the results of each step are written down, before the program is run on a computer. Also known as desk check. [ENG] Any practice test or session. [ORD] Any simulated firing practice, particularly a dive-bombing approach made without the release of a bomb.

dry season [CLIMATOL] In certain types of climate, an annually recurring period of one or more months during which precipitation is at a minimum for the region.

dry socket [MED] Inflammation of the dental alveolus, especially the inflamed condition following the removal of a tooth. Also known as alveolitis.

dry wall [BUILD] A wall covered with wallboard, in contrast to plaster. [ENG] A wall constructed of rock without cementing material.

dry well [CIV ENG] 1. A well that has been completely drained. 2. An excavated well filled with broken stone and used to receive drainage when the water percolates into the soil. 3. Compartment of a pumping station in which the pumps are housed. [NUCLEO] The first containment tank surrounding a water-cooled nuclear reactor that uses the pressure-suppressing containment system.

DSB *See* double-sideband modulation.

D scan *See* D scope.

D scope [ELECTR] A cathode-ray scope which combines the features of B and C scopes, the signal appearing as a spot with bearing angle as the horizontal coordinate and elevation angle as the vertical coordinate, but with each spot expanded slightly in a vertical direction to give a rough range indication. Also known as D indicator; D scan.

DSN *See* Deep Space Network.

DSRV *See* deep-submergence rescue vehicle.

Dst [GEOPHYS] The "storm-time" component of variation of the terrestrial magnetic field, that is, the component which correlates with the interval of time since the onset of a magnetic storm; used as an index of intensity of the ring current.

DTL *See* diode transistor logic.

dual-channel amplifier [ENG ACOUS] An audio-frequency amplifier having two separate amplifiers for the two channels of a stereophonic sound system, usually operating from a common power supply mounted on the same chassis.

dual completion well [PETRO ENG] Single well casing containing two production tubing strings, each in a different zone of the reservoir (one higher, one lower) and each separately controlled.

dual coordinates [MATH] Point coordinates and plane coordinates are dual in geometry since an equation about one determines an equation about the other.

dual-cycle boiling-water reactor [NUCLEO] A boiling-water reactor in which part of the steam used to run the steam turbine is generated in the reactor core and part is generated in an external heat exchanger. Also known as dual-cycle reactor system.

dual-cycle reactor system *See* dual-cycle boiling-water reactor.

dual graph [MATH] A planar graph corresponding to a planar map obtained by replacing each country with its capital and each common boundary by an arc joining the two countries.

dual group [MATH] The group of all homomorphisms of an Abelian group G into the cyclic group of order n, where n is the smallest integer such that g_n is the identity element of G.

duality principle Also known as principle of duality. [ELEC] The principle that for any theorem in electrical circuit analysis there is a dual theorem in which one replaces quantities with dual quantities; current and voltage, impedance and admittance, and meshes and nodes are examples of dual quantities. [ELECTR] The principle that analogies may be drawn between a transistor circuit and the corresponding vacuum tube circuit. [ELECTROMAG] The principle that one can obtain new solutions of Maxwell's equations from known solutions by replacing $\mathbf{E}$ with $\mathbf{H}$, $\mathbf{H}$ with $-\mathbf{E}$, ϵ with μ, and μ with ϵ. [MATH] A principle that if a theorem is true, it remains true if each object and operation is replaced by its dual; important in projective geometry and Boolean algebra. [QUANT MECH] *See* wave-particle duality.

dual linear programming [MATH] Linear programming in which the maximum and minimum number are the same number.

dual-mode control [CONT SYS] A type of control law which consists of two distinct types of operation; in linear systems, these modes usually consist of a linear feedback mode and a bang-bang-type mode.

dual modulation [COMMUN] The process of modulating a common carrier wave or subcarrier with two different types of modulation, each conveying separate information.

dual network [ELEC] A network which has the same number of terminal pairs as a given network, and whose open-circuit impedance network is the same as the short-circuit admittance matrix of the given network, and vice versa.

dual space [MATH] The vector space consisting of all linear transformations from a given vector space into its scalar field.

dual tensor [MATH] The product of a given tensor, covariant in all its indices, with the contravariant form of the determinant tensor, contracting over the indices of the given tensor.

dual-trace oscilloscope [ELECTR] An oscilloscope which can compare two waveforms on the face of a single cathode-ray tube, using any one of several methods.

dual-track tape recorder *See* double-track tape recorder.

Duane-Hunt law [QUANT MECH] The law that the frequency of x-rays resulting from electrons striking a target cannot exceed eV/h, where e is the charge of the electron, V is the exciting voltage, and h is Planck's constant.

dub [ENG ACOUS] 1. To transfer recorded material from one recording to another, with or without the addition of new sounds, background music, or sound effects. 2. To combine two or more sources of sound into one record. 3. To add a new sound track or new sounds to a motion picture film, or to a recorded radio or television production.

duck-billed dinosaur [PALEON] Any of several herbivorous, bipedal ornithopods having the front of the mouth widened to form a ducklike beak.

duckbill platypus *See* platypus.

ducod punched card [COMPUT SCI] A punched card that has 12 rows of punching positions in each column, designated 0 through 9 and X and Y, and in which numbers from 0 through 99 are punched as two holes in positions 0 through 9, while digit order or digit duplication is indicated by punching or not punching in positions X and Y.

Ducrey test [IMMUNOL] A skin test to determine past or present infection with *Hemophilus ducreyi*.

duct [ANAT] An enclosed tubular channel for conducting a glandular secretion or other body fluid. [COMMUN] An enclosed runway for cables. [GEOPHYS] The space between two air layers, or between an air layer and the earth's surface, in which microwave beams are trapped in ducting. Also known as radio duct; tropospheric duct. [MECH ENG] A fluid flow passage which may range from a few inches in diameter to many feet in rectangular cross section, usually constructed of galvanized steel, aluminum, or copper, through which air flows in a ventilation system or to a compressor, supercharger, or other equipment at speeds ranging to thousands of feet per minute.

ducted fan [MECH ENG] A propeller or multibladed fan inside a coaxial duct or cowling. Also known as ducted propeller; shrouded propeller.

ducted propeller *See* ducted fan.

ducted rocket *See* rocket ramjet.

ductile iron *See* nodular cast iron.

ductility [MATER] The ability of a material to be plastically deformed by elongation, without fracture.

ducting [GEOPHYS] An atmospheric condition in the troposphere in which temperature inversions cause microwave beams to refract up and down between two air layers, so that microwave signals travel 10 or more times farther than the normal line-of-sight limit. Also known as superrefraction; tropospheric ducting.

ductless gland *See* endocrine gland.

ductus arteriosus [EMBRYO] Blood shunt between the pulmonary artery and the aorta of the mammalian embryo.

ductus deferens *See* vas deferens.

ductus venosus [EMBRYO] Blood shunt between the left umbilical vein and the right sinus venosus of the heart in the mammalian embryo.

dufrenite [MINERAL] A blackish-green, fibrous ferric phosphate mineral; commonly massive or in nodules.

dufrenoysite [MINERAL] $Pb_2As_2S_5$ A lead gray to steel gray, monoclinic mineral consisting of lead arsenic sulfide.

duftite [MINERAL] $PbCu(AsO_4)(OH)$ Orthorhombic mineral that is composed of a basic arsenate of lead and copper.

Dugongidae [VERT ZOO] A family of aquatic mammals in the order Sirenia comprising two species, the dugong and the sea cow.

Dugonginae [VERT ZOO] The dugongs, a subfamily of sirenian mammals in the family Dugongidae characterized by enlarged, sharply deflected premaxillae and the absence of nasal bones.

dull coal [GEOL] A component of banded coal with a grayish color and dull appearance, consisting of small anthraxylon constituents in addition to cuticles and barklike constituents embedded in the attritus.

Dulong-Petit law [THERMO] The law that the product of the specific heat per gram and the atomic weight of many solid elements at room temperature has almost the same value, about 6.3 calories (264 joules) per degree Celsius.

dulse [BOT] Any of several species of red algae of the genus *Rhodymenia* found below the intertidal zone in northern latitudes; an important food plant.

dumb terminal [COMPUT SCI] A computer input/output device that lacks the capability to process or formate data, and is thus entirely dependent on the main computer for these activities.

dumdum [ORD] A bullet that flattens excessively on contact, or one especially designed to flatten excessively.

dummy [COMMUN] Telegraphy network simulating a customer's loop for adjusting a telegraph repeater; the dummy side of the repeater is that toward the customer. [COMPUT SCI] An artificial address, instruction, or other unit of information inserted in a digital computer solely to fulfill prescribed conditions (such as word length or block length) without affecting operations. [ENG] Simulating device with no operating features, as a dummy heat coil. [GRAPHICS] A preliminary layout which shows the placement of illustrations and text as they will appear in the final printing. [MET] A cathode that undergoes electroplating at low current densities. [ORD] **1.** A nonexplosive bomb, projectile, or the like, or an object made to appear as one of these. **2.** An object made to appear as an airplane, gun emplacement, or the like from the air.

dummy argument [COMPUT SCI] The variable appearing in the definition of a macro or function which will be replaced by an address at call time.

dummy deck [COMPUT SCI] Complete set of tabulating cards containing only punched, coded information (nonaperture); used as machine-handling set for sorting, reproducing, and interpreting in conjunction with an aperture card system.

dummy instruction [COMPUT SCI] An artificial instruction or address inserted in a list to serve a purpose other than the execution as an instruction.

dumontite [MINERAL] $Pb_2(UO_2)_3(PO_4)_2(OH)_4 \cdot 3H_2O$ Yellow orthorhombic mineral consisting of a hydrated phosphate of uranium and lead, occurring in crystals.

dumortierite [MINERAL] $Al_8BSi_3O_{19}(OH)$ A pink, green, blue, or violet mineral that crystallizes in the orthorhombic system but commonly occurs in parallel or radiating fibrous aggregates; mined for the manufacture of high-grade porcelain.

dump [COMPUT SCI] To copy the contents of all or part of a storage, usually from an internal storage device into an external storage device. [ELECTR] To withdraw all power from a system or component accidentally or intentionally. [ORD] A temporary storage area, usually in the open, for bombs, ammunition, equipment, or supplies.

dump bucket [MECH ENG] A large bucket with movable discharge gates at the bottom; used to move soil or other construction materials by a crane or cable.

dump car [MECH ENG] Any of several types of narrow-gage rail cars with bodies which can easily be tipped to dump material.

dump check [COMPUT SCI] A computer check that usually consists of adding all the digits during dumping, and verifying the sum when retransferring.

dump truck [ENG] A motor or hand-propelled truck for hauling and dumping loose materials, equipped with a body that discharges its contents by gravity.

dumpy level [ENG] A surveyor's level which has the telescope with its level tube rigidly attached to a vertical spindle and is capable only of horizontal rotary movement.

dundasite [MINERAL] $PbAl_2(CO_3)_2(OH)_4 \cdot 2H_2O$ A white mineral consisting of a basic lead aluminum carbonate, occurring in spherical aggregates.

dune [GEOL] A mound or ridge of unconsolidated granular material, usually of sand size and of durable composition (such as quartz), capable of movement by transfer of individual grains entrained by a moving fluid.

dunite [PETR] An ultrabasic rock consisting almost solely of a magnesium-rich olivine with some chromite and picotite; an important source of chromium.

duodecimal number system [MATH] A representation system for real numbers using 12 as the base.

duodenal glands *See* Brunner's glands.

duodenum [ANAT] The first section of the small intestine of mammals, extending from the pylorus to the jejunum.

duodiode [ELECTR] An electron tube having two diodes in the same envelope, with either a common cathode or separate cathodes. Also known as double diode.

duolateral coil *See* honeycomb coil.

duoplasmatron [ELECTR] An ion-beam source in which electrons from a hot filament are accelerated sufficiently to ionize a gas by impact; the resulting positive ions are drawn out by high-voltage electrons and focused into a beam by electrostatic lens action.

duoprimed word [COMPUT SCI] A computer word containing a representation of the sixth, seventh, eighth, and ninth rows of information from an 80-column card.

duotone [GRAPHICS] A process in which two halftone cuts, one with a screen angle 30° different from the other, are made from the same black-and-white photograph, and the picture is printed in two tones, usually black and a color such as blue or green.

duotype [GRAPHICS] A process in which two halftone plates, for letterpress, are produced from a black-and-white original, each plate being etched differently; one plate is etched for detail and printed in a dark color, and the other is etched for a flat effect and printed in a light color.

duplex [ENG] Consisting of two parts working together or in a similar fashion.

duplexer [ELECTR] A switching device used in radar to permit alternate use of the same antenna for both transmitting and receiving; other forms of duplexers serve for two-way radio communication using a single antenna at lower frequencies. Also known as duplexing assembly.

duplexing *See* duplex operation; duplex process.

duplexing assembly *See* duplexer.

duplexite [MINERAL] $Ca_4BeAl_2Si_9O_{24}(OH)_2$ A white fibrous mineral consisting of hydrous beryllium calcium aluminosilicate. Also known as bavenite.

duplex operation [COMMUN] The operation of associated transmitting and receiving apparatus concurrently, as in ordinary telephones, without manual switching between talking and listening periods. Also known as duplexing; duplex transmission. [ENG] In radar, a condition of operation when two identical and interchangeable equipments are provided, one in an active state and the other immediately available for operation.

duplex practice See duplex process.

duplex process [MET] A two-step procedure in which steel is refined by one process (usually the Bessemer process) and finished by another process (usually open-hearth or electric-furnace). Also known as duplexing; duplex practice.

duplex transmission See duplex operation.

duplex uterus [ANAT] A condition in certain primitive mammals, such as rodents and bats, that have two distinct uteri opening separately into the vagina.

duplicate key [COMPUT SCI] A key on the card punch which, when depressed, will copy a card in the reading station onto a card in the write station.

duplicate record [COMPUT SCI] An unwanted record that has the same key as another record in the same file.

duplication check [COMPUT SCI] A check based on the identity in results of two independent performances of the same task.

durain [GEOL] A hard, granular ingredient of banded coal which occurs in lenticels and shows a close, firm texture. Also known as durite.

dura mater [ANAT] The fibrous membrane forming the outermost covering of the brain and spinal cord. Also known as endocranium.

durangite [MINERAL] NaAlF(AsO₄) An orange-red, monoclinic mineral consisting of a fluoarsenate of sodium and aluminum; occurs in crystals.

Durargid [GEOL] A great soil group constituting a subdivision of the Argids, indicating those soils with a hardpan cemented by silica and called a duripan.

durinite [GEOL] The principal maceral of durain; a heterogeneous material, semiopaque in section (including all parts of plants); micrinite, exinite, cutinite, resinite, collinite, xylinite, suberinite, and fusinite may be present.

durite See durain.

dusk [ASTRON] That part of either morning or evening twilight between complete darkness and civil twilight.

dust [GEOL] Dry solid matter of silt and clay size (less than 1/16 millimeter). [PHYS] A loose term applied to solid particles predominantly larger than colloidal size and capable of temporary gas suspension.

dust chamber [ENG] A chamber through which gases pass to permit deposition of solid particles for collection. Also known as ash collector; dust collector.

dust collector See dust chamber.

dust counter [ENG] A photoelectric apparatus which measures the size and number of dust particles per unit volume of air. Also known as Kern counter.

dust devil [METEOROL] A small but vigorous whirlwind, usually of short duration, rendered visible by dust, sand, and debris picked up from the ground; diameters range from about 10–100 feet (3–30 meters), and average height is about 600 feet (180 meters).

Dutch elm disease [PL PATH] A lethal fungus disease of elm trees caused by *Graphium ulmi*, which releases a toxic substance that destroys vascular tissue; transmitted by a bark beetle.

Dutchman's log [ENG] A buoyant object thrown overboard to determine the speed of a vessel; the time required for a known length of the vessel to pass the object is measured, and the speed can then be computed.

duty cycle [COMMUN] The product of the pulse duration and pulse frequency of a pulse carrier, equal to the time per second that pulse power is applied. Also known as duty factor. [ELECTR] See duty ratio. [ENG] 1. The time intervals devoted to starting, running, stopping, and idling when a device is used for intermittent duty. 2. The ratio of working time to total time for an intermittently operating device, usually expressed as a percent. Also known as duty factor. [MET] The percentage of time that current flows in equipment over a specific period during electric resistance welding. [NUCLEO] The fraction of time during which a pulsed accelerator beam is on target, usually expressed as a percent. Also known as duty factor.

duty factor See duty cycle.

duty ratio [ELECTR] In a pulse radar or similar system, the ratio of average to peak pulse power. Also known as duty cycle.

DUV See data under voice.

D variometer See declination variometer.

dwarf [BIOL] Being an atypically small form or variety of something. [MED] An abnormally small individual; especially one whose bodily proportions are altered.

dwarf disease [PL PATH] A virus disease marked by the inhibition of fruit production; common in plum trees.

dwarf galaxy [ASTRON] A galaxy with low luminosity.

dwarf star [ASTRON] A star that typically has surface temperature of 5730 K, radius of 690,000 kilometers, mass of 2×10^{33}g, and luminosity of 4×10^{33}ergs/sec. Also known as main sequence star.

dwell [DES ENG] That part of a cam that allows the cam follower to remain at maximum lift for a period of time. [ENG] A pause in the application of pressure to a mold.

dwt See deadweight tonnage; pennyweight.

DX coil See direct-expansion coil.

Dy See dysprosium.

dyad [CYTOL] Either of the two pair of chromatids produced by separation of a tetrad during the first meiotic division. [MATH] An abstract object which is a pair of vectors **AB** in a given order on which certain operations are defined.

dye [CHEM] A colored substance which imparts more or less permanent color to other materials. Also known as dyestuff.

dyeing [CHEM ENG] The application of color-producing agents to material, usually fibrous or film, in order to impart a degree of color permanence demanded by the projected end use.

dye laser [OPTICS] A type of tunable laser in which the active material is a dye such as acridine red or esculin, with very large molecules, and laser action takes place between the first excited and ground electronic states, each of which comprises a broad vibrational-rotational continuum.

dynamic address translator [COMPUT SCI] A hardware device used in a virtual memory system to automatically identify a virtual address inquiry in terms of segment number, page number within the segment, and position of the record with reference to the beginning of the page.

dynamical friction [PHYS] 1. The drag force between electrons and ions drifitng with respect to each other. 2. Sliding friction, in contrast to static friction.

dynamic algorithm [COMPUT SCI] An algorithm whose operation is, to some extent, unpredictable in advance, generally because it contains logical decisions that are made on the basis of quantities computed during the course of the algorithm. Also known as heuristic algorithm.

dynamical parallax   [ASTRON] A parallax of binary stars that is computed from the sum of the masses of the binary system.

dynamical similarity   [MECH] Two flow fields are dynamically similar if one can be transformed into the other by a change of length and velocity scales. All dimensionless numbers of the flows must be the same.

dynamical variable   [MECH] One of the quantities used to describe a system in classical mechanics, such as the coordinates of a particle, the components of its velocity, the momentum, or functions of these quantities.

dynamic analogies   [PHYS] Analogies that make it possible to convert the differential equations for mechanical and acoustical systems to equivalent electrical equations that can be represented by electric networks and solved by circuit theory.

dynamic balance   [MECH] The condition which exists in a rotating body when the axis about which it is forced to rotate, or to which reference is made, is parallel with a principal axis of inertia; no products of inertia about the center of gravity of the body exist in relation to the selected rotational axis.

dynamic braking   [MECH] A technique of electric braking in which the retarding force is supplied by the same machine that originally was the driving motor.

dynamic breccia   See tectonic breccia.

dynamic capillary pressure   [PETRO ENG] Capillary-pressure saturation curves of a core sample determined by the simultaneous steady-state flow of two fluids through the sample; capillarity pressures are determined by the difference in the pressures of the two fluids.

dynamic characteristic   See load characteristic.

dynamic check   [ENG] Check used to ascertain the correct performance of some or all components of equipment or a system under dynamic or operating conditions.

dynamic circuit   [ELECTR] An MOS circuit designed to make use of its high input impedance to store charge temporarily at certain nodes of the circuit and thereby increase the speed of the circuit.

dynamic climatology   [CLIMATOL] The climatology of atmospheric dynamics and thermodynamics, that is, a climatological approach to the study and explanation of atmospheric circulation.

dynamic debugging routine   [COMPUT SCI] A debugging routine which operates in conjunction with the program being checked and interacts with it while the program is running.

dynamic equilibrium   Also known as kinetic equilibrium. [MECH] The condition of any mechanical system when the kinetic reaction is regarded as a force, so that the resultant force on the system is zero according to d'Alembert's principle. [PHYS] A condition in which several processes act simultaneously to maintain a system in an overall state that does not change with time.

dynamic focusing   [ELECTR] The process of varying the focusing electrode voltage for a color picture tube automatically so the electron-beam spots remain in focus as they sweep over the flat surface of the screen.

dynamic forecasting   See numerical forecasting.

dynamic impedance   [ELEC] The impedance of a circuit having an inductance and a capacitance in parallel at the frequency at which this impedance has a maximum value. Also known as rejector impedance.

dynamic instability   See inertial instability.

dynamicizer   [COMPUT SCI] A device that converts a collection of data represented by a spatial arrangement of bits in a computer storage device into a series of signals occurring in time.

dynamic load   [AERO ENG] With respect to aircraft, rockets, or spacecraft, a load due to an acceleration of craft, as imposed by gusts, by maneuvering, by landing, by firing rockets, and so on. [CIV ENG] A force exerted by a moving body on a resisting member, usually in a relatively short time interval. Also known as energy load.

dynamic loudspeaker   [ENG ACOUS] A loudspeaker in which the moving diaphragm is attached to a current-carrying voice coil that interacts with a constant magnetic field to give the in-and-out motion required for the production of sound waves. Also known as dynamic speaker; moving-coil loudspeaker.

dynamic memory   See dynamic storage.

dynamic memory allocation   See dynamic storage allocation.

dynamic metamorphism   [GEOL] Metamorphism resulting exclusively or largely from rock deformation, principally faulting and folding. Also known as dynamometamorphism.

dynamic meteorology   [METEOROL] The study of atmospheric motions as solutions of the fundamental equations of hydrodynamics or other systems of equations appropriate to special situations, as in the statistical theory of turbulence.

dynamic microphone   [ENG ACOUS] A moving-conductor microphone in which the flexible diaphragm is attached to a coil positioned in the fixed magnetic field of a permanent magnet. Also known as moving-coil microphone.

dynamic pickup   [ELECTR] A pickup in which the electric output is due to motion of a coil or conductor in a constant magnetic field. Also known as dynamic reproducer; moving-coil pickup.

dynamic pressure   [FL MECH] **1.** The pressure that a moving fluid would have if it were brought to rest by isentropic flow against a pressure gradient. Also known as impact pressure; stagnation pressure; total pressure. **2.** The difference between the quantity in the first definition and the static pressure.

dynamic printout   [COMPUT SCI] A printout of data which occurs during the machine run as one of the sequential operations.

dynamic problem check   [COMPUT SCI] Any dynamic check used to ascertain that the computer solution satisfies the given system of equations in an analog computer operation.

dynamic programming   [MATH] A mathematical technique, more sophisticated than linear programming, for solving a multidimensional optimization problem, which transforms the problem into a sequence of single-stage problems having only one variable each.

dynamic program relocation   [COMPUT SCI] The act of moving a partially executed program to another location in main memory, without hindering its ability to finish processing normally.

dynamic reproducer   See dynamic pickup.

dynamics   [MECH] That branch of mechanics which deals with the motion of a system of material particles under the influence of forces, especially those which originate outside the system under consideration.

dynamic sequential control   [COMPUT SCI] Method of operation of a digital computer through which it can alter instructions as the computation proceeds, or the sequence in which instructions are executed, or both.

dynamic similarity [MECH ENG] A relation between two mechanical systems (often referred to as model and prototype) such that by proportional alterations of the units of length, mass, and time, measured quantities in the one system go identically (or with a constant multiple for each) into those in the other; in particular, this implies constant ratios of forces in the two systems.

dynamic speaker *See* dynamic loudspeaker.

dynamic stability [MECH] The characteristic of a body, such as an aircraft, rocket, or ship, that causes it, when disturbed from an original state of steady motion in an upright position, to damp the oscillations set up by restoring moments and gradually return to its original state. Also known as stability.

dynamic stop [COMPUT SCI] A loop in a computer program which is created by a branch instruction in the presence of an error condition, and which signifies the existence of this condition.

dynamic storage [COMPUT SCI] 1. Computer storage in which information at a certain position is not always available instantly because it is moving, as in an acoustic delay line or magnetic drum. Also known as dynamic memory. 2. Computer storage consisting of capacitively charged circuit elements which must be continually refreshed or recharged at regular intervals.

dynamic storage allocation [COMPUT SCI] A computer system in which memory capacity is made available to a program on the basis of actual, momentary need during program execution, and areas of storage may be reassigned at any time. Also known as dynamic allocation; dynamic memory allocation.

dynamic subroutine [COMPUT SCI] Subroutine that involves parameters, such as decimal point position or item size, from which a relatively coded subroutine is derived by the computer itself.

dynamic symmetry [PHYS] A symmetry law related, not to the geometric structure of the constituents of matter, but to the laws which govern the dynamic behavior of these constituents.

dynamic test [ENG] A test conducted under active or simulated load.

dynamic vertical *See* apparent vertical.

dynamo *See* generator.

dynamoelectric [PHYS] Pertaining to the conversion of mechanical energy to electric energy, or vice versa.

dynamometamorphism *See* dynamic metamorphism.

dynamometer [ENG] 1. An instrument in which current, voltage, or power is measured by the force between a fixed coil and a moving coil. 2. A special type of electric rotating machine used to measure the output torque or driving torque of rotating machinery by the elastic deformation produced.

dynamotor [ELEC] A rotating electric machine having two or more windings on a single armature containing a commutator for direct-current operation and slip rings for alternating-current operation; when one type of power is fed in for motor operation, the other type is delivered by generator action. Also known as rotary converter; synchronous inverter.

dynatron [ELECTR] A screen-grid tube in which secondary emission of electrons from the anode causes the anode current to decrease as anode voltage increases, resulting in a negative resistance characteristic. Also known as negatron.

dyne [MECH] The unit of force in the centimeter-gram-second system of units, equal to the force which imparts an acceleration of 1 cm/sec^2 to a 1 gram mass.

dynode [ELECTR] An electrode whose primary function is secondary emission of electrons; used in multiplier phototubes and some types of television camera tubes. Also known as electron mirror.

dysarthria [MED] Impairment of articulation caused by any disorder or lesion affecting the tongue or speech muscles.

dysarthrosis [MED] 1. Deformity, dislocation, or disease of a joint. 2. A false joint.

dysentery [MED] Inflammation of the intestine characterized by pain, intense diarrhea, and the passage of mucus and blood.

dyskinesia [MED] 1. Disordered movements of voluntary or involuntary muscles, particularly those seen in disorders of the extrapyramidal system. 2. Impaired voluntary movements.

dysmenorrhea [MED] Difficult or painful menstruation.

Dysodonta [PALEON] An order of extinct bivalve mollusks with a nearly toothless hinge and a ligament in grooves or pits.

dyspepsia [MED] Disturbed digestion.

dysphagia [MED] Difficulty in swallowing, or inability to swallow, of organic or psychic causation.

dysplasia [PATH] Abnormal development or growth, especially of cells.

dyspnea [MED] Difficult or labored breathing.

dysprosium [CHEM] A metallic rare-earth element, symbol Dy, atomic number 66, atomic weight 162.50.

dystrophy [MED] 1. Defective nutrition. 2. Defective or abnormal development or degeneration.

dysuria [MED] Painful urination.

Dytiscidae [INV ZOO] The predacious diving beetles, a family of coleopteran insects in the suborder Adephaga.

E

e [MATH] The base of the natural logarithms; the number defined by the equation

$$\int_1^e \frac{1}{x}\, dx = 1.$$

E *See* electric field vector.

eager *See* bore.

eagle [VERT ZOO] Any of several large, strong diurnal birds of prey in the family Accipitridae.

EAM *See* electric accounting machine.

ear [ANAT] The receptor organ that sends both auditory information and space orientation information to the brain in vertebrates.

eardrum *See* tympanic membrane.

earlandite [MINERAL] $Ca_3(C_6H_5O_7)_2 \cdot 4H_2O$ A mineral consisting of a hydrous citrate of calcium; found in sediments in the Weddell Sea.

early-warning radar [ORD] A line of air defense radar units along the perimeter of a defended area to provide the earliest possible warning of approaching aircraft.

EAROM *See* electrically alterable read-only memory.

earphone [ENG ACOUS] **1.** An electroacoustical transducer, such as a telephone receiver or a headphone, actuated by an electrical system and supplying energy to an acoustical system of the ear, the waveform in the acoustical system being substantially the same as in the electrical system. **2.** A small, lightweight electroacoustic transducer that fits inside the ear, used chiefly with hearing aids.

ear protector [ENG] A device, such as a plug or ear muff, used to protect the human ear from loud noise that may be injurious to hearing, such as that of jet engines.

ear rot [PL PATH] Any of several fungus diseases of corn, occurring both in the field and in storage and marked by decay and molding of the ears.

ear shell *See* abalone.

earth [ASTRON] The third planet in the solar system, lying between Venus and Mars; sometimes capitalized. [ELEC] *See* ground. [GEOL] **1.** Solid component of the globe, distinct from air and water. **2.** Soil; loose material composed of disintegrated solid matter.

earth connection *See* ground.

earth core *See* core.

earth crust *See* crust.

earth current [ELEC] Return, fault, leakage, or stray current passing through the earth from electrical equipment. Also known as ground current. [GEOPHYS] A current flowing through the ground and due to natural causes, such as the earth's mag-

netic field or auroral activity. Also known as telluric current.

earthenware [ENG] Ceramic products of natural clay, fired at 1742–2129°F (950–1165°C), that is slightly porous, opaque, and usually covered with a nonporous glaze.

earthflow [GEOL] A variety of mass movement involving the downslope slippage of soil and weathered rock in a series of subparallel sheets.

earth hummock [GEOL] A small, dome-shaped uplift of soil caused by the pressure of groundwater. Also known as earth mound.

earth inductor [ENG] A type of inclinometer that has a coil which rotates in the earth's field and in which a voltage is induced when the rotation axis does not coincide with the field direction; used to measure the dip angle of the earth's magnetic field. Also known as dip inductor; earth inductor compass; induction inclinometer.

earth inductor compass *See* earth inductor.

earthlight [ASTRON] The illumination of the dark part of the moon's disk, produced by sunlight reflected onto the moon from the earth's surface and atmosphere. Also known as earthshine.

earth mound *See* earth hummock.

earth movements [GEOPHYS] Movements of the earth, comprising revolution about the sun, rotation on the axis, precession of equinoxes, and motion of the surface of the earth relative to the core and mantle.

earthmover [MECH ENG] A machine used to excavate, transport, or push earth.

earth oscillations [GEOPHYS] Any rhythmic deformations of the earth as an elastic body; for example, the gravitational attraction of the moon and sun excite the oscillations known as earth tides.

earth pig *See* aardvark.

earthquake [GEOPHYS] A series of suddenly generated elastic waves in the earth occurring in shallow depths to about 700 kilometers.

earthquake tremor *See* tremor.

earthquake zone [GEOL] An area of the earth's crust in which movements, sometimes with associated volcanism, occur. Also known as seismic area.

earth radiation *See* terrestrial radiation.

earth resources technology satellite [AERO ENG] One of a series of satellites designed primarily to measure the natural resources of the earth; functions include mapping, cataloging water resources, surveying crops and forests, tracing sources of water and air pollution, identifying soil and rock formations, and acquiring oceanographic data. Abbreviated ERTS.

earth science [SCI TECH] The science that deals with the earth or any part thereof; includes the disciplines of geology, geography, oceanography, and meteorology, among others.

earthshine *See* earthlight.

earth's shadow *See* dark segment.

earthstar [MYCOL] A fungus of the genus *Geastrum* that resembles a puffball with a double peridium, the outer layer of which splits into the shape of a star.

earth tide [GEOPHYS] The periodic movement of the earth's crust caused by forces of the moon and sun. Also known as bodily tide.

earth tremor *See* tremor.

earth wax *See* ozocerite.

earthwork [CIV ENG] 1. Any operation involving the excavation or construction of earth embankments. 2. Any construction made of earth. [ORD] A temporary or permanent fortification for attack or defense, made chiefly of earth.

earthworm [INV ZOO] The common name for certain terrestrial members of the class Oligochaeta, especially forms belonging to the family Lumbricidae.

earthy cobalt *See* asbolite.

earwig [INV ZOO] The common name for members of the insect order Dermaptera.

easement [CIV ENG] The right held by one person over another person's land for a specific use; rights of tenants are excluded.

east [GEOD] The direction 90° to the right of north.

East Africa Coast Current [OCEANOGR] A current that is influenced by the monsoon drifts of the Indian Ocean, flowing southwestward along the Somalia coast in the Northern Hemisphere winter and northeastward in the Northern Hemisphere summer. Also known as Somali Current.

East Australia Current [OCEANOGR] The current which is formed by part of the South Equatorial Current and flows southward along the eastern coast of Australia.

eastern equine encephalitis [MED] A mosquito-borne virus infection of horses and mules in the eastern and southern United States caused by a member of arbovirus group A.

Eastern Hemisphere [GEOGR] The half of the earth lying mostly to the east of the Atlantic Ocean, including Europe, Africa, and Asia.

East Greenland Current [OCEANOGR] A current setting south along the eastern coast of Greenland and carrying water of low salinity and low temperature.

eastonite [MINERAL] $K_2Mg_5AlSi_5Al_3O_{20}(OH_4)$ A mineral consisting of basic silicate of potassium, magnesium, and aluminum; it is an end member of the biotite system.

east point [GEOD] That intersection of the prime vertical with the horizon which lies to the right of the observer when facing north.

east-west effect [ASTRON] The phenomenon due to the fact that a greater number of cosmic-ray particles approach the earth from a westerly direction than from an easterly.

eave [BUILD] The border of a roof overhanging a wall.

EBAM *See* electron-beam memory.

ebb current [OCEANOGR] The tidal current associated with the decrease in the height of a tide.

ebb tide [OCEANOGR] The portion of the tide cycle between high water and the following low water. Also known as falling tide.

EBCDIC *See* extended binary-coded decimal interchange code.

Ebenaceae [BOT] A family of dicotyledonous plants in the order Ebenales, in which a latex system is absent and flowers are mostly unisexual with the styles separate, at least distally.

Ebenales [BOT] An order of woody, sympetalous dicotyledonous plants in the subclass Dilleniidae, having axile placentation and usually twice as many stamens as corolla lobes.

EBM *See* electron-beam machining.

ebonite *See* hard rubber.

ebony [BOT] Any of several African and Asian trees of the genus *Diospyros*, providing a hard, durable wood.

Ebriida [INV ZOO] An order of flagellate protozoans in the class Phytamastigophorea characterized by a solid siliceous skeleton.

ebulliometry [PHYS CHEM] The precise measurement of the absolute or differential boiling points of solutions.

ebullition [PHYS] The process or state of a liquid bubbling up or boiling.

eccentric [SCI TECH] Situated to one side with reference to a center.

eccentric angle [MATH] For an ellipse having semimajor and semiminor angles of lengths a and b respectively, lying along the x and y axes of a coordinate system respectively, and for a point (x,y) on the ellipse, the angle arc cos (x/a) = arc sin (y/b).

eccentricity [MATH] The ratio of the distance of a point on a conic from the focus to the distance from the directrix. [MECH] The distance of the geometric center of a revolving body from the axis of rotation.

eccentric load [ENG] A load imposed on a structural member at some point other than the centroid of the section.

eccentric rotor engine [MECH ENG] A rotary engine, such as the Wankel engine, wherein motion is imparted to a shaft by a rotor eccentric to the shaft.

eccentric valve [ENG] A rubber-lined slurry or fluid valve with an eccentric rotary cut-off body to reduce corrosion and wear on mechanical moving valve parts.

ecchymosis [MED] A subcutaneous hemorrhage marked by purple discoloration of the skin.

Eccles-Jordan circuit *See* bistable multivibrator.

Eccles-Jordan multivibrator *See* bistable multivibrator.

eccrine gland [PHYSIO] One of the small sweat glands distributed all over the human body surface; they are tubular coiled merocrine glands that secrete clear aqueous sweat.

ecdemite [MINERAL] $Pb_6As_2O_7Cl_4$ A greenish-yellow to yellow, tetragonal mineral consisting of an oxychloride of lead and arsenic; occurs as coatings of small tabular crystals and as coarsely foliated masses.

ecdysis [INV ZOO] Molting of the outer cuticular layer of the body, as in insects and crustaceans.

ecdysone [BIOCHEM] The molting hormone of insects.

ECG *See* electrocardiogram.

echelette grating [SPECT] A diffraction grating with coarse groove spacing, designed for the infrared region; has grooves with comparatively flat sides and concentrates most of the radiation by reflection into a small angular coverage.

echelle grating [SPECT] A diffraction grating designed for use in high orders and at angles of illumination greater than 45° to obtain high dispersion and resolving power by the use of high orders of interference.

echelon grating [SPECT] A diffraction grating which consists of about 20 plane-parallel plates about 1 centimeter thick, cut from one sheet, each plate extending beyond the next by about 1 millimeter, and which has a resolving power on the order of 10^6.

Echeneidae [VERT ZOO] The remoras, a family of perciform fishes in the suborder Percoidei.

echidna [VERT ZOO] A spiny anteater; any member of the family Tachyglossidae.

Echinacea [INV ZOO] A suborder of echinoderms in the order Euechinoidea; individuals have a rigid test, keeled teeth, and branchial slits.

echinate [ZOO] Having a dense covering of spines or bristles.

Echinidae [INV ZOO] A family of echinacean echinoderms in the order Echinoida possessing trigeminate or polyporous plates with the pores in a narrow vertical zone.

Echiniscoidea [INV ZOO] A suborder of tardigrades in the order Heterotardigrada characterized by terminal claws on the legs.

echinococcosis [MED] Infestation by the larva (hydatid) of *Echinococcus granulosis* in man, and in some canines and herbivores. Also known as hydatid disease; hydatidosis.

Echinocystitoida [PALEON] An order of extinct echinoderms in the subclass Perischoechinoidea.

Echinodera [INV ZOO] The equivalent name for Kinorhyncha.

Echinodermata [INV ZOO] A phylum of exclusively marine coelomate animals distinguished from all others by an internal skeleton composed of calcite plates, and a water-vascular system to serve the needs of locomotion, respiration, nutrition, or perception.

Echinoida [INV ZOO] An order of Echinacea with a camarodont lantern, smooth test, and imperforate noncrenulate tubercles.

Echinoidea [INV ZOO] The sea urchins, a class of Echinozoa having a compact body enclosed in a hard shell, or test, formed by regularly arranged plates which bear movable spines.

Echinometridae [INV ZOO] A family of echinoderms in the order Echinoida, including polyporous types with either an oblong or a spherical test.

Echinosteliaceae [MYCOL] A family of slime molds in the order Echinosteliales.

Echinosteliales [MYCOL] An order of slime molds in the subclass Myxogastromycetidae.

Echinothuriidae [INV ZOO] A family of deep-water echinoderms in the order Echinothurioida in which the large, flexible test collapses into a disk at atmospheric pressure.

Echinothurioida [INV ZOO] An order of echinoderms in the superorder Diadematacea with solid or hollow primary radioles, diademoid ambulacral plates, noncrenulate tubercles, and the anus within the apical system.

Echinozoa [INV ZOO] A subphylum of free-living echinoderms having the body essentially globoid with meridional symmetry and lacking appendages.

Echiurida [INV ZOO] A small group of wormlike organisms regarded as a separate phylum of the animal kingdom; members have a saclike or sausage-shaped body with an anterior, detachable prostomium.

Echiuridae [INV ZOO] A small family of the order Echiurinea characterized by a flaplike prostomium.

Echiuroidea [INV ZOO] A phylum of schizocoelous animals.

Echiurinea [INV ZOO] An order of the Echiurida.

echo [ELECTR] **1.** The signal reflected by a radar target, or the trace produced by this signal on the screen of the cathode-ray tube in a radar receiver. Also known as radar echo; return. **2.** *See* ghost signal. [PHYS] A wave packet that has been reflected or otherwise returned with sufficient delay and

magnitude to be perceived as a signal distinct from that directly transmitted.

echo attenuation [ELECTR] The power transmitted at an output terminal of a transmission line, divided by the power reflected back to the same output terminal.

echocardiography [MED] A diagnostic technique for the heart that uses a transducer held against the chest to send high-frequency sound waves which pass harmlessly into the heart; as they strike structures within the heart, they are reflected back to the transducer and recorded on an oscilloscope.

echo chamber [ACOUS] A reverberant room or enclosure used in a studio to add echo effects to sounds for radio or television programs.

echo check [COMPUT SCI] A method of ascertaining the accuracy of transmission of data in which the transmitted data are returned to the sending end for comparison with original data. Also known as readback check.

echoencephalograph [MED] An instrument that uses ultrasonic pulses and echo-ranging techniques to give a pictorial representation of intracranial structure. Also known as sonoencephalograph.

echogram [ENG] The graphic presentation of echo soundings recorded as a continuous profile of the sea bottom. [MED] The pictorial display of anatomical structures using pulse-echo techniques.

echograph [ENG] An instrument used to record an echogram.

echolalia [MED] The purposeless, often seemingly involuntary repetition of words spoken by another person; a disorder seen in certain psychotic states and in certain organic brain syndromes. Also known as echophrasia.

echo location *See* echo ranging.

echo matching [ENG] Rotating an antenna to a position in which the pulse indications of an echo-splitting radar are equal.

echophrasia *See* echolalia.

echo ranging [ENG] Active sonar, in which underwater sound equipment generates bursts of ultrasonic sound and picks up echoes reflected from submarines, fish, and other objects within range, to determine both direction and distance to each target. Also known as echo location. [VERT ZOO] An auditory feedback mechanism in bats, porpoises, seals, and certain other animals whereby reflected ultrasonic sounds are utilized in orientation.

Echo satellite [AERO ENG] An aluminized-surface, Mylar balloon about 100 feet (30 meters) in diameter, placed in orbit as a passive communications satellite for reflecting microwave signals from a transmitter to receivers beyond the horizon.

echo signal *See* target signal.

echo sounder *See* sonic depth finder.

echo sounding [ENG] Determination of the depth of water by measuring the time interval between emission of a sonic or ultrasonic signal and the return of its echo from the sea bottom.

echo-splitting radar [ENG] Radar in which the echo is split by special circuits associated with the antenna lobe–switching mechanism, to give two echo indications on the radarscope screen; when the two echo indications are equal in height, the target bearing is read from a calibrated scale.

echo suppressor [ELECTR] **1.** A circuit that desensitizes radar navigation equipment for a fixed period after the reception of one pulse, for the purpose of rejecting delayed pulses arriving from longer, indirect reflection paths. **2.** A relay or other device used on a transmission line to prevent a reflected wave from returning to the sending end of the line.

echo talker [COMPUT SCI] The interference created by the retransmission of a message back to its source while the source is still transmitting.

echovirus [VIROL] A division of enteroviruses in the picornavirus group; the name is derived from the group designation enteric cytopathogenic human orphan virus.

eckermannite [MINERAL] $Na_3(Mg,Li)_4(Al,Fe)Si_8O_{22}(OH,F)_2$ Mineral of the amphibole group containing magnesium, lithium, iron, and fluorine.

ECL See emitter-coupled logic.

eclampsia [MED] A disorder occurring during the latter half of pregnancy, characterized by elevated blood pressure, edema, proteinuria, and convulsions or coma.

eclipse [ASTRON] 1. The reduction in visibility or disappearance of a body by passing into the shadow cast by another body. 2. The apparent cutting off, wholly or partially, of the light from a luminous body by a dark body coming between it and the observer. Also known as astronomical eclipse.

eclipsed conformation [PHYS CHEM] A particular arrangement of constituent atoms that may rotate about a single bond in a molecule; for ethane it is such that when viewed along the axis of the carbon-carbon bond the hydrogen atoms of one methyl group are exactly in line with those of the other methyl group.

eclipse period [VIROL] A phase in the proliferation of viral particles during which the virus cannot be detected in the host cell.

eclipsing binary See eclipsing variable star.

eclipsing variable star [ASTRON] A binary star whose orbit is such that every time one star passes between the observer and its companion an eclipse results. Also known as eclipsing binary.

ecliptic [ASTRON] The apparent annual path of the sun among the stars; the intersection of the plane of the earth's orbit with the celestial sphere.

ecliptic coordinate system [ASTRON] A celestial coordinate system in which the ecliptic is taken as the primary and the great circles perpendicular to it are then taken as secondaries.

ecliptic latitude See celestial latitude.

ecliptic longitude See celestial longitude.

ecliptic pole [ASTRON] On the celestial sphere, either of two points 90° from the ecliptic.

eclogite [PETR] A class of metamorphic rocks distinguished by their composition, consisting essentially of omphacite and pyrope with small amounts of diopside, enstatite, olivine, kyanite, rutile, and rarely, diamond.

eclogite facies [PETR] A type of facies composed of eclogite and formed by regional metamorphism at extremely high temperature and pressure.

ECM See electrochemical machining.

eco See electron-coupled oscillator.

ecocline [ECOL] A genetic gradient of adaptability to an environmental gradient; formed by the merger of ecotypes.

ecological association [ECOL] A complex of communities, such as an elm-hackberry association, which develops in accord with variations in physiography, soil, and successional history within the major subdivision of a biotic realm.

ecological community See community.

ecological interaction [ECOL] The relation between species that live together in a community; specifically, the effect an individual of one species may exert on an individual of another species.

ecological pyramid [ECOL] A pyramid-shaped diagram representing quantitatively the numbers of organisms, energy relationships, and biomass of an ecosystem; numbers are high for the lowest trophic levels (plants) and low for the highest trophic level (carnivores).

ecological succession [ECOL] A gradual process incurred by the change in the number of individuals of each species of a community and by establishment of new species populations that may gradually replace the original inhabitants.

ecological system See ecosystem.

ecology [BIOL] A study of the interrelationships which exist between organisms and their environment. Also known as environmental biology.

econometrics [IND ENG] The application of mathematical and statistical techniques to the estimation of mathematical relationships for testing of economic theories and the solution of economic problems.

economic entomology [BIOL] A branch of entomology concerned with the study of economic losses of commercially important animals and plants due to insect predation.

economic geography [GEOGR] A branch of geography concerned with the relations of physical environment and economic conditions to the manufacture and distribution of commodities.

economic geology [GEOL] 1. Application of geologic knowledge to materials usage and principles of engineering. 2. The study of metallic ore deposits.

economic life [IND ENG] The number of years after which a capital good should be replaced in order to minimize the long-run annual cost of operation, repair, depreciation, and capital.

economic order quantity [IND ENG] The number of orders required to fulfill the economic lot size.

economic purchase quantity [IND ENG] The economic lot size for a purchased quantity.

economizer [ENG] A reservoir in a continuous-flow oxygen system in which oxygen exhaled by the user is collected for recirculation in the system. [MECH ENG] A forced-flow, once-through, convection-heat-transfer tube bank in which feedwater is raised in temperature on its way to the evaporating section of a steam boiler, thus lowering flue gas temperature, improving boiler efficiency, and saving fuel.

economy [COMPUT SCI] The ratio of the number of characters to be coded to the maximum number available with the code; for example, binary-coded decimal using 4 bits provides 16 possible characters but uses only 10 of them.

ecosystem [ECOL] A functional system which includes the organisms of a natural community together with their environment. Derived from ecological system.

ecotone [ECOL] A zone of intergradation between ecological communities.

ecotype [ECOL] A subunit, race, or variety of a plant ecospecies that is restricted to one habitat; equivalent to a taxonomic subspecies.

ECR See electronic cash register.

Ecterocoelia [INV ZOO] The equivalent name for Protostomia.

ectocommensal [ECOL] An organism living on the outer surface of the body of another organism, without affecting its host.

ectoderm [EMBRYO] The outer germ layer of an animal embryo. Also known as epiblast. [INV ZOO] The outer layer of a diploblastic animal.

ectogenesis [EMBRYO] Development of an embryo or of embryonic tissue outside the body in an artificial environment.

ectohumus [GEOL] An accumulation of organic matter on the soil surface with little or no mixing with mineral material. Also known as mor; raw humus.

ectoparasite [ECOL] A parasite that lives on the exterior of its host.

ectophagous [INV ZOO] The larval stage of a parasitic insect which is in the process of development externally on a host.

ectopia [MED] A congenital or acquired positional abnormality of an organ or other part of the body.

ectopic pairing [CYTOL] Pairing between nonhomologous segments of the salivary gland chromosomes in *Drosophila*, presumably involving mainly heterochromatic regions.

ectoplasm [CYTOL] The outer, gelled zone of the cytoplasmic ground substance in many cells. Also known as ectosarc.

Ectoprocta [INV ZOO] A subphylum of colonial bryozoans having eucoelomate visceral cavities and the anus opening outside the circlet of tentacles.

ectosarc *See* ectoplasm.

ectotherm [PHYSIO] An animal that obtains most of its heat from the environment and therefore has a body temperature very close to that of its environment.

ectotrophic [BIOL] Obtaining nourishment from outside; applied to certain parasitic fungi that live on and surround the roots of the host plant.

Ectrephidae [INV ZOO] An equivalent name for Ptinidae.

eczema [MED] Any skin disorder characterized by redness, thickening, oozing from blisters or papules, and occasional formation of fissures and crusts.

ED₅₀ *See* effective dose 50.

edaphic community [ECOL] A plant community that results from or is influenced by soil factors such as salinity and drainage.

Edaphosuria [PALEON] A suborder of extinct, lowland, terrestrial, herbivorous reptiles in the order Pelycosauria.

eddy [FL MECH] A vortexlike motion of a fluid running contrary to the main current.

eddy coefficient *See* exchange coefficient.

eddy current [ELECTROMAG] An electric current induced within the body of a conductor when that conductor either moves through a nonuniform magnetic field or is in a region where there is a change in magnetic flux. Also known as Foucault current.

eddy-current brake [MECH ENG] A control device or dynamometer for regulating rotational speed, as of flywheels, in which energy is converted by eddy currents into heat.

eddy-current clutch [MECH ENG] A type of electromagnetic clutch in which torque is transmitted by means of eddy currents induced by a magnetic field set up by a coil carrying direct current in one rotating member.

eddy-current heating *See* induction heating.

eddy-current loss [ELECTROMAG] Energy loss due to undesired eddy currents circulating in a magnetic core.

eddy-current tachometer [ENG] A type of tachometer in which a rotating permanent magnet induces currents in a spring-mounted metal cylinder; the resulting torque rotates the cylinder and moves its attached pointer in proportion to the speed of the rotating shaft. Also known as drag-type tachometer.

eddy-current test [ELECTROMAG] A nondestructive test in which the change of impedance of a test coil brought close to a conducting specimen indicates the eddy currents induced by the coil, and thereby indicates certain properties or defects of the specimen.

eddy diffusion [FL MECH] Diffusion which occurs in turbulent flow, by the rapid process of mixing of the swirling eddies of fluid. Also known as turbulent diffusion.

eddy diffusion coefficient *See* eddy diffusivity.

eddy diffusivity [FL MECH] The exchange coefficient for the diffusion of a conservative property by eddies in a turbulent flow. Also known as coefficient of eddy diffusion; eddy diffusion coefficient.

eddy flux [FL MECH] The rate of transport (or flux) of fluid properties such as momentum, mass heat, or suspended matter by means of eddies in a turbulent motion; the rate of turbulent exchange. Also known as moisture flux; turbulent flux.

eddy mill *See* pothole.

eddy stress *See* Reynolds stress.

eddy velocity [FL MECH] The difference between the mean velocity of fluid flow and the instantaneous velocity at a point. Also known as fluctuation velocity.

eddy viscosity [FL MECH] The turbulent transfer of momentum by eddies giving rise to an internal fluid friction, in a manner analogous to the action of molecular viscosity in laminar flow, but taking place on a much larger scale.

edema [MED] An excessive accumulation of fluid in the cells, tissue spaces, or body cavities due to a disturbance in the fluid exchange mechanism. Also known as dropsy.

Edenian [GEOL] Lower Cincinnatian geologic stage in North America, above the Mohawkian and below Maysvillian.

Edentata [VERT ZOO] An order of mammals characterized by the absence of teeth or the presence of simple prismatic, unspecialized teeth with no enamel.

edentate [VERT ZOO] 1. Lacking teeth. 2. Any member of the Edentata.

edge [MATH] 1. A line along which two plane faces of a solid intersect. 2. A line segment connecting nodes or vertices in a graph (a geometric representation of the relation among situations). Also known as arc.

edge dislocation [CRYSTAL] A dislocation which may be regarded as the result of inserting an extra plane of atoms, terminating along the line of the dislocation. Also known as Taylor-Orowan dislocation.

edge focusing [ELECTROMAG] Axial focusing of a stream of ions which occurs when it crosses a fringe magnetic field obliquely; used in mass spectrometers and cyclotrons.

edge-notched card [COMPUT SCI] A card with a series of holes along one or more edges, and notches which open one or more holes, so that long needles inserted in specific holes in a deck of such cards let fall only cards with a desired type of data.

edge-punched card [COMPUT SCI] A card with one or more rows of holes, representing binary data, punched along the edges.

Ediacaran fauna [PALEON] The oldest known assemblage of fossil remains of soft-bodied marine animals; first discovered in the Ediacara Hills, Australia.

edingtonite [MINERAL] $BaAl_2Si_3O_{10}\cdot4H_2O$ Gray zeolite mineral that forms rhombic crystals; sometimes contains large amounts of calcium.

Edison battery [ELEC] A storage battery composed of cells having nickel and iron in an alkaline solution. Also known as nickel-iron battery.

Edison effect *See* thermionic emission.

E display [ELECTR] A rectangular radar display in which targets appear as blips with distance indicated by the horizontal coordinate, and elevation by the vertical coordinate.

edit [COMPUT SCI] **1.** To modify the form or format of an output or input by inserting or deleting characters such as page numbers or decimal points. **2.** A computer instruction directing that this step be performed.

edit capability [COMPUT SCI] The degree of sophistication available to the programmer to modify his statements while in the time-sharing mode.

edit mask [COMPUT SCI] The receiving word through which a source word is filtered, allowing for the suppression of leading zeroes, the insertion of floating dollar signs and decimal points, and other such formatting.

editor program [COMPUT SCI] A special program by means of which a user can easily perform corrections, insertions, modifications, or deletions in an existing program or data file.

EDM *See* electron discharge machining.

EDP *See* electronic data processing.

Edrioasteroidea [PALEON] A class of extinct Echinozoa having ambulacral radial areas bordered by tube feet, and the mouth and anus located on the upper side of the theca.

eductor [ENG] **1.** An ejectorlike device for mixing two fluids. **2.** *See* ejector.

EEG *See* electroencephalogram.

eel [VERT ZOO] The common name for a number of unrelated fishes included in the orders Anguilliformes and Cypriniformes; all have an elongate, serpentine body.

eel grass *See* tape grass.

eff *See* efficiency.

effective address [COMPUT SCI] The address that is obtained by applying any specified indexing or indirect addressing rules to the specified address; the effective address is then used to identify the current operand.

effective air path [NAV] A straight line on a navigation chart connecting two air positions, commonly used between the air positions of two pressure soundings in order to determine effective true airspeed between the two soundings.

effective aperture [OPTICS] The diameter of the image of the aperture stop of an optical system, as viewed from the object.

effective area [CHEM ENG] Absolute or cross-sectional area of process media involved in the process, such as the actual area of filter media through which a fluid passes, or the available surface area of absorbent contacted by a gas or liquid. [ELECTROMAG] Of an antenna in any specified direction, the square of the wavelength multiplied by the power gain (or directive gain) in that direction and divided by 4π (12.57).

effective bandwidth [ELECTR] The bandwidth of an assumed rectangular band-pass having the same transfer ratio at a reference frequency as a given actual band-pass filter, and passing the same mean-square value of a hypothetical current having even distribution of energy throughout that bandwidth.

effective capacitance [ELEC] Total capacitance existing between any two given points of an electric circuit.

effective current [ELEC] The value of alternating current that will give the same heating effect as the corresponding value of direct current. Also known as root-mean-square current.

effective dose 50 [PHARM] The amount of a drug required to produce a response in 50% of the sub-

jects to whom the drug is given. Abbreviated ED_{50}. Also known as median effective dose.

effective field intensity [ELECTROMAG] Root-mean-square value of the inverse distance fields at a distance of 1 mile (1.6 kilometers) from the transmitting antenna in all directions in the horizontal plane.

effective force *See* inertial force.

effective half-life [NUCLEO] The half-life of a radioisotope in a biological organism, resulting from a combination of radioactive decay and biological elimination.

effective height [ELECTROMAG] The height of the center of radiation of a transmitting antenna above the effective ground level.

effective horizon [COMMUN] A horizon whose distance at a given height above sea level is the distance to the horizon of a fictitious earth, having a radius 4/3 times the earth's true radius; used to estimate ranges of antennas, taking atmospheric refraction into account.

effective horsepower [NAV ARCH] The power necessary to overcome the resistance of water to motion of a ship towed at a given speed, measured in horsepower. Abbreviated ehp. Also known as towrope horsepower.

effective instruction [COMPUT SCI] The computer instruction that results from changing a basic instruction during program modification. Also known as actual instruction.

effective mass [SOLID STATE] A parameter with the dimensions of mass that is assigned to electrons in a solid; in the presence of an external electromagnetic field the electrons behave in many respects as if they were free, but with a mass equal to this parameter rather than the true mass.

effectiveness level [COMPUT SCI] A measure of the effectiveness of data-processing equipment, equal to the ratio of the operational use time to the total performance period, expressed as a percentage. Also known as average effectiveness level.

effective porosity [GEOL] A property of earth containing interconnecting interstices, expressed as a percent of bulk volume occupied by the interstices.

effective precipitation [HYD] **1.** The part of precipitation that reaches stream channels as runoff. Also known as effective rainfall. **2.** In irrigation, the portion of the precipitation which remains in the soil and is available for consumptive use.

effective radiated power [ELECTROMAG] The product of antenna input power and antenna power gain, expressed in kilowatts. Abbreviated ERP.

effective rainfall *See* effective precipitation.

effective resistance *See* high-frequency resistance.

effective sound pressure [ACOUS] The root-mean-square value of the instantaneous sound pressure at a point during a complete cycle, expressed in dynes per square centimeter. Also known as root-mean-square sound pressure; sound pressure.

effective temperature [ASTROPHYS] A measure of the temperature of a star, deduced by means of the Stefan-Boltzmann law, from the total energy that is emitted per unit area. [METEOROL] The temperature at which motionless, saturated air would induce, in a sedentary worker wearing ordinary indoor clothing, the same sensation of comfort as that induced by the actual conditions of temperature, humidity, and air movement.

effective thermal resistance [ELECTR] Of a semiconductor device, the effective temperature rise per unit power dissipation of a designated junction above the temperature of a stated external reference point under conditions of thermal equilibrium. Also known as thermal resistance.

effective time [COMPUT SCI] The time during which computer equipment is in actual use and produces useful results.

effective value *See* root-mean-square value.

effector [BIOCHEM] An activator of an allosteric enzyme. [CONT SYS] A motor, solenoid, or hydraulic piston that turns commands to a teleoperator into specific manipulatory actions. [PHYSIO] A structure that is sensitive to a stimulus and causes an organism or part of an organism to react to the stimulus, either positively or negatively.

effector organ [PHYSIO] Any muscle or gland that mediates overt behavior, that is, movement or secretion.

efferent [PHYSIO] Carrying or conducting away, as the duct of an exocrine gland or a nerve.

effervescence [CHEM] The bubbling of a solution of an element or chemical compound as the result of the emission of gas without the application of heat; for example, the escape of carbon dioxide from carbonated water.

efficiency Abbreviated eff. [CHEM] In an ion-exchange system, a measurement of the effectiveness of a system expressed as the amount of regenerant required to remove a given unit of adsorbed material. [ENG] **1.** Measure of the degree of heat output per unit of fuel when all available oxidizable materials in the fuel have been burned. **2.** Ratio of useful energy provided by a dynamic system to the energy supplied to it during a specific period of operation. [NUCLEO] The probability that a count will be produced in a counter tube by a specified particle or quantum incident. [PHYS] The ratio, usually expressed as a percentage, of the useful power output to the power input of a device. [THERMO] The ratio of the work done by a heat engine to the heat energy absorbed by it. Also known as thermal efficiency.

efficiency expert [IND ENG] An individual who analyzes procedures, productivity, and jobs in order to recommend methods for achieving maximum utilization of resources and equipment.

efflorescence [BOT] The period or process of flowering. [CHEM] The property of hydrated crystals to lose water of hydration and crumble when exposed to air. [MINERAL] A whitish powder, consisting of one or several minerals produced as an encrustation on the surface of a rock in an arid region. Also known as bloom.

effluent [CHEM ENG] *See* discharge liquor. [CIV ENG] The liquid waste of sewage and industrial processing. [HYD] **1.** Flowing outward or away from. **2.** Liquid which flows away from a containing space or a main waterway.

effusion [MED] A pouring out of any fluid into a body cavity or tissue. [PHYS CHEM] The movement of a gas through an opening which is small as compared with the average distance which the gas molecules travel between collisions. [SCI TECH] **1.** The act or process of leaking or pouring out. **2.** Any material that is effused.

EFL *See* error frequency limit.

e format [COMPUT SCI] A decimal, normalized form of a floating point number in FORTRAN in which a number such as 18.756 appears as .18756E + 02, which stands for .18756 × 10^2.

egest [PHYSIO] **1.** To discharge indigestible matter from the digestive tract. **2.** To rid the body of waste.

egg [CYTOL] **1.** A large, female sex cell enclosed in a porous, calcareous or leathery shell, produced by birds and reptiles. **2.** *See* ovum.

egg apparatus [BOT] A group of three cells, consisting of the egg and two synergid cells, in the micropylar end of the embryo sac in seed plants.

egg capsule *See* egg case.

egg case [INV ZOO] **1.** A protective capsule containing the eggs of certain insects and mollusks. Also known as egg capsule. **2.** A silk pouch in which certain spiders carry their eggs. Also known as egg sac. [VERT ZOO] A soft, gelatinous (amphibians) or strong, horny (skates) envelope containing the egg of certain vertebrates.

eggplant [BOT] *Solanum melongena*. A plant of the order Polemoniales grown for its edible egg-shaped, fleshy fruit.

egg raft [ZOO] A floating mass of eggs; produced by a variety of aquatic organisms.

egg sac [INV ZOO] **1.** The structure containing the eggs of certain microcrustaceans. **2.** *See* egg case.

eggstone *See* oolite.

egg tooth [VERT ZOO] A toothlike prominence on the tip of the beak of a bird embryo and the tip of the nose of an oviparous reptile, which is used to break the eggshell.

eglestonite [MINERAL] Hg_4Cl_2O Rare mercuric oxide mineral; forms yellow-brown isometric crystals upon exposure to air.

ego [PSYCH] **1.** The self. **2.** The conscious part of the personality that is in contact with reality.

egress [ASTRON] The departure of the moon from the shadow of the earth in an eclipse, or of a planet from the disk of the sun, or of a satellite (or its shadow) from the disk of the parent planet.

Egyptian henna *See* henna.

EHF *See* extremely high frequency.

ehp *See* effective horsepower.

Ehrenfest's adiabatic law [QUANT MECH] The law that, if the Hamiltonian of a system undergoes an infinitely slow change, and if the system is initially in an eigenstate of the Hamiltonian, then at the end of the change it will be in the eigenstate of the new Hamiltonian that derives from the original state by continuity, provided certain conditions are met. Also known as Ehrenfest's theorem.

Ehrenfest's equations [THERMO] Equations which state that for the phase curve $P(T)$ of a second-order phase transition the derivative of pressure P with respect to temperature T is equal to $(C_p{}^f - C_p{}^i)/TV(\gamma^f - \gamma^i) = (\gamma^f - \gamma^i)/(K^f - K^i)$, where i and f refer to the two phases, γ is the coefficient of volume expansion, K is the compressibility, C_p is the specific heat at constant pressure, and V is the volume.

Ehrenfest's theorem [QUANT MECH] **1.** The theorem that a quantum-mechanical wave packet obeys the equations of motion of the corresponding classical particle when the position, momentum, and force acting on the particle are replaced by the expectation values of these quantities. **2.** *See* Ehrenfest's adiabatic law.

Ehrlichieae [MICROBIO] A tribe of the family Rickettsiaceae; spherical and occasionally pleomorphic cells; pathogenic for some mammals, not including humans.

ehv *See* extra-high voltage.

eigenfunction [MATH] Also known as characteristic function. **1.** An eigenvector for a linear operator on a vector space whose vectors are functions. Also known as proper function. **2.** A solution to the Sturm-Liouville partial differential equation.

eigenmatrix [MATH] Corresponding to a diagonalizable matrix or linear transformation, this is the matrix all of whose entries are 0 save those on the principal diagonal where appear the eigenvalues.

eigenstate [QUANT MECH] 1. A dynamical state whose state vector (or wave function) is an eigenvector (or eigenfunction) of an operator corresponding to a specified physical quantity. 2. *See* energy state.

eigenvalue [MATH] That one of the scalars λ such that $T(v) = \lambda v$, where T is a linear operator on a vector space, and v is an eigenvector. Also known as characteristic number; characteristic root; characteristic value; latent root; proper value.

eigenvalue equation *See* characteristic equation.

eigenvalue problem *See* Sturm-Liouville problem.

eigenvector [MATH] A nonzero vector v whose direction is not changed by a given linear transformation T; that is, $T(v) = \lambda v$ for some scalar λ. Also known as characteristic vector.

eight-level code [COMMUN] A teletypewriter code that uses eight impulses, in addition to the start and stop impulses, to define a character.

eighty-column card [COMPUT SCI] A card with 80 columns of punch positions, and 12 punch positions in each column.

Eimeriina [INV ZOO] A suborder of coccidian protozoans in the order Eucoccida in which there is no syzygy and the microgametocytes produce a large number of microgametes.

E indicator *See* E scope.

einstein [PHYS] A unit of light energy used in photochemistry, equal to Avogadro's number times the energy of one photon of light of the frequency in question.

Einstein-Bohr equation [QUANT MECH] In a system undergoing a transition between two states so that it emits or absorbs radiation, that equation indicating that the radiation frequency equals the difference in energy between the two states divided by Planck's constant.

Einstein-Bose statistics *See* Bose-Einstein statistics.

Einstein condensation *See* Bose-Einstein condensation.

Einstein–de Haas effect [ELECTROMAG] A freely suspended body consisting of a ferromagnetic material acquires a rotation when its magnetization changes.

Einstein–de Haas method [ELECTROMAG] Method of measuring the gyromagnetic ratio of a ferromagnetic substance; one measures the angular displacement induced in a ferromagnetic cylinder suspended from a torsion fiber when magnetization of the object is reversed, and the magnetization change is measured with a magnetometer.

Einstein–de Sitter model [RELAT] A model of the universe in which ordinary euclidean geometry holds good, the distribution of matter extends infinitely at all times, and the universe expands from an infinitely condensed state at such a rate that the density is inversely proportional to the square of the time elapsed since the beginning of the expansion.

Einstein displacement *See* Einstein shift.

Einstein elevator [RELAT] A windowless elevator freely falling in its shaft, inside of which conditions resemble interstellar space; used to elucidate the principle of equivalence.

Einstein equations [STAT MECH] Equations for the density and pressure of a Bose-Einstein gas in terms of power series in a parameter which appears in the Bose-Einstein distribution law.

Einstein frequency [SOLID STATE] Single frequency with which each atom vibrates independently of other atoms, in a model of lattice vibrations; equal to frequency observed in infrared absorption studies.

einsteinium [CHEM] Synthetic radioactive element, symbol Es, atomic number 99; discovered in debris of 1952 hydrogen bomb explosion; now made in cyclotrons.

Einstein mass-energy relation [RELAT] The relation in which the energy of a system is equivalent to its mass times the square of the speed of light.

Einstein photoelectric law [QUANT MECH] The law that the energy of an electron emitted from a system in the photoelectric effect is $h\nu - W$, where h is Planck's constant, ν is the frequency of the incident radiation, and W is the energy needed to remove the electron from the system; if $h\nu$ is less than W, no electrons are emitted.

Einstein-Planck law [QUANT MECH] The law that the energy of a photon is given by Planck's constant times the frequency. [RELAT] The equation of motion of a charged particle in an electromagnetic field, according to which its rate of change of momentum is equal to the Lorentz force, where the magnitude of the momentum is $mv/(1 - v^2/c^2)^{1/2}$, where m and v are the particle's mass and velocity, and c is the speed of light.

Einstein relation [PHYS] The relation in which the mobility of charges in an ionic solution or semiconductor is equal to the magnitude of the charge times the diffusion coefficient divided by the product of the Boltzmann constant and the absolute temperature.

Einstein's absorption coefficient [ATOM PHYS] The proportionality constant governing the absorption of electromagnetic radiation by atoms, equal to the number of quanta absorbed per second divided by the product of the energy of radiation per unit volume per unit wave number and the number of atoms in the ground state.

Einstein's equivalency principle *See* equivalence principle.

Einstein's field equations [RELAT] Those equations relevant to the relationship in which the Einstein tensor equals -8π times the energy momentum tensor times the gravitational constant divided by the square of the speed of light. Also known as Einstein's law of gravitation.

Einstein shift [RELAT] A shift toward longer wavelengths of spectral lines emitted by atoms in strong gravitational fields. Also known as Einstein displacement.

Einstein's law of gravitation *See* Einstein's field equations.

Einstein's principle of relativity [RELAT] The principle that all the laws of physics must assume the same mathematical form in any inertial frame of reference; thus, it is impossible to determine the absolute motion of a system by any means.

Einstein's summation convention [MATH] A notational convenience used in tensor analysis whereupon it is agreed that any term in which an index appears twice will stand for the sum of all such terms as the index assumes all of a preassigned range of values.

Einstein's unified field theories [RELAT] A series of theories attempting to express a general unifying principle underlying electromagnetism and gravity.

Einstein universe [RELAT] A model of the universe which is a four-dimensional cylindrical surface in a five-dimensional space.

Einthoven galvanometer *See* string galvanometer.

ejaculation [PHYSIO] The act or process of suddenly discharging a fluid from the body; specifically, the ejection of semen during orgasm.

ejaculatory duct [ANAT] The terminal part of the ductus deferens after junction with the duct of a seminal vesicle, embedded in the prostate gland and opening into the urethra.

ejecta [GEOL] Material which is discharged by a volcano. [PHYSIO] Excrement. [SCI TECH] Material which is cast out.

ejection [ENG] The process of removing a molding from a mold impression by mechanical means, by hand, or by compressed air. [ORD] The expelling, by the ejector, of the empty cartridge case from small arms and rapid-fire guns.

ejector [ENG] **1.** Any of various types of jet pumps used to withdraw fluid materials from a space. Also known as eductor. **2.** A device that ejects the finished casting from a mold. [ORD] A device in the breech mechanism of a gun, rifle, or other firearm which automatically throws out an empty cartridge case, or unfired cartridge, from the breech or receiver.

ejector pin [ENG] A pin driven into the rear of a mold cavity to force the finished piece out. Also known as knockout pin.

ejector plate [ENG] The plate backing up the ejector pins and holding the ejector assembly together.

ejector rod [ENG] A rod that activates the ejector assembly of a mold when it is opened.

EKG See electrocardiogram.

Ekman convergence [OCEANOGR] A zone of convergence of warm surface water caused by Ekman transport, creating a marked depression of the ocean's thermocline in the affected area.

Ekman current meter [ENG] A mechanical device for measuring ocean current velocity which incorporates a propeller and a magnetic compass and can be suspended from a moored ship.

Ekman layer [METEOROL] The layer of transition between the surface boundary layer of the atmosphere, where the shearing stress is constant, and the free atmosphere, which is treated as an ideal fluid in approximate geostrophic equilibrium. Also known as spiral layer.

Ekman spiral [METEOROL] A theoretical representation that a wind blowing steadily over an ocean of unlimited depth and extent and uniform viscosity would cause, in the Northern Hemisphere, the immediate surface water to drift at an angle of 45° to the right of the wind direction, and the water beneath to drift further to the right, and with slower and slower speeds, as one goes to greater depths.

Ekman transport [OCEANOGR] The movement of ocean water caused by wind blowing steadily over the surface; occurs at right angles to the wind direction.

Elaeagnaceae [BOT] A family of dicotyledonous plants in the order Proteales, noted for peltate leaf scales which often give the leaves a silvery-gray appearance.

Elaphomycetaceae [MYCOL] A family of underground, saprophytic or mycorrhiza-forming fungi in the order Eurotiales characterized by ascocarps with thick, usually woody walls.

Elapidae [VERT ZOO] A family of poisonous reptiles, including cobras, kraits, mambas, and coral snakes; all have a pteroglyph fang arrangement.

Elara [ASTRON] A small satellite of Jupiter with a diameter of about 20 miles (32 kilometers), orbiting at a mean distance of 7.29×10^6 miles (11.73×10^6 kilometers). Also known as Jupiter VII.

Elasipodida [INV ZOO] An order of deep-sea aspidochirotacean holothurians in which there are no respiratory trees and bilateral symmetry is often quite conspicuous.

Elasmidae [INV ZOO] A family of hymenopteran insects in the superfamily Chalcidoidea.

Elasmobranchii [VERT ZOO] The sharks and rays, a subclass of the class Chondrichthyes distinguished by separate gill openings, amphistylic or hyostylic jaw suspension, and ampullae of Lorenzini in the head region.

Elassomatidae [VERT ZOO] The pygmy sunfishes, a family of the order Perciformes.

elastic [MECH] Capable of sustaining deformation without permanent loss of size or shape.

elastic buckling [MECH] An abrupt increase in the lateral deflection of a column at a critical load while the stresses acting on the column are wholly elastic.

elastic cartilage [HISTOL] A type of cartilage containing elastic fibers in the matrix.

elastic center [MECH] That point of a beam in the plane of the section lying midway between the flexural center and the center of twist in that section.

elastic collision [MECH] A collision in which the sum of the kinetic energies of translation of the participating systems is the same after the collision as before.

elastic constant See compliance constant; stiffness constant.

elastic deformation [MECH] Reversible alteration of the form or dimensions of a solid body under stress or strain.

elastic fiber [HISTOL] A homogeneous, fibrillar connective tissue component that is highly refractile and appears yellowish when arranged in bundles.

elasticity [MECH] **1.** The property whereby a solid material changes its shape and size under action of opposing forces, but recovers its original configuration when the forces are removed. **2.** The existence of forces which tend to restore to its original position any part of a medium (solid or fluid) which has been displaced.

elasticity modulus See modulus of elasticity.

elastic limit [MECH] The maximum stress a solid can sustain without undergoing permanent deformation.

elastic modulus See modulus of elasticity.

elasticoviscosity [FL MECH] That property of a fluid whose rate of deformation under stress is the sum of a part corresponding to a viscous Newtonian fluid and a part obeying Hooke's law.

elastic potential energy [MECH] Capacity that a body has to do work by virtue of its deformation.

elastic ratio [MECH] The ratio of the elastic limit to the ultimate strength of a solid.

elastic rebound theory [GEOL] A theory which attributes faulting to stresses (in the form of potential energy) which are being built up in the earth and which, at discrete intervals, are suddenly released as elastic energy; at the time of rupture the rocks on either side of the fault spring back to a position of little or no strain.

elastic wave [ACOUS] See acoustic wave. [PHYS] A wave propagated by a medium having inertia and elasticity (the existence of forces which tend to restore any part of a medium to its original position), in which displaced particles transfer momentum to adjoining particles, and are themselves restored to their original position.

elastin [BIOCHEM] An elastic protein composing the principal component of elastic fibers.

elastomer [MATER] A material, such as a synthetic rubber or plastic, which at room temperature can be stretched under low stress to at least twice its original length and, upon immediate release of the stress, will return with force to its approximate original length.

elater [BOT] A spiral, filamentous structure that functions in the dispersion of spores in certain plants, such as liverworts and slime molds.

Elateridae [INV ZOO] The click beetles, a large family of coleopteran insects in the superfamily Elateroidea; many have light-producing organs.

elaterite [GEOL] A light-brown to black asphaltic pyrobitumen that is moderately soft and elastic. Also known as elastic bitumen; mineral caoutchouc.

Elateroidea [INV ZOO] A superfamily of coleopteran insects in the suborder Polyphaga.

E layer [GEOPHYS] A layer of ionized air occurring at altitudes between 100 and 120 kilometers in the E region of the ionosphere, capable of bending radio waves back to earth. Also known as Heaviside layer; Kennelly-Heaviside layer.

elbow [ANAT] The arm joint formed at the junction of the humerus, radius, and ulna. [DES ENG] **1.** A fitting that connects two pipes at an angle, often of 90°. **2.** A sharp corner in a pipe. [ELECTRO-MAG] In a waveguide, a bend of comparatively short radius, normally 90°, and sometimes for acute angles down to 15°. [GEOGR] A sharp change in direction of a coast line, channel, bank, or so on.

elbow meter [ENG] Pipe elbow used as a liquids flowmeter; flow rate is measured by determining the differential pressure developed between the inner and outer radii of the bend by means of two pressure taps located midway on the bend.

elective culture [MICROBIO] A type of microorganism grown selectively from a mixed culture by culturing in a medium and under conditions selective for only one type of organism.

electret [ELEC] A solid dielectric possessing persistent electric polarization, by virtue of a long time constant for decay of a charge instability.

electret headphone [ENG ACOUS] A headphone consisting of an electret transducer, usually in the form of a push-pull transducer.

electret microphone [ENG ACOUS] A microphone consisting of an electret transducer in which the foil electret diaphragm is placed next to a perforated, ridged, metal or metal-coated backplate, and output voltage, taken between diaphragm and backplate, is proportional to the displacement of the diaphragm.

electret transducer [ELECTR] An electroacoustic or electromechanical transducer in which a foil electret, stretched out to form a diaphragm, is placed next to a metal or metal-coated plate, and motion of the diaphragm is converted to voltage between diaphragm and plate, or vice versa.

electric [ELEC] Containing, producing, arising from, or actuated by electricity; often used interchangeably with electrical.

electric accounting machine [COMPUT SCI] Data-processing equipment that is predominantly electromechanical in nature, such as sorters, collectors, and tabulators. Abbreviated EAM.

electrical [ELEC] Related to or associated with electricity, but not containing it or having its properties or characteristics; often used interchangeably with electric.

electrical angle [ELEC] An angle that specifies a particular instant in an alternating-current cycle or expresses the phase difference between two alternating quantities; usually expressed in electrical degrees.

electrical axis [SOLID STATE] The x axis in a quartz crystal; there are three such axes in a crystal, each

parallel to one pair of opposite sides of the hexagon; all pass through and are perpendicular to the optical, or z, axis.

electrical code [ELEC] A systematic body of rules governing the practical application and installation of electrically operated equipment and devices and electric wiring systems.

electrical condenser See capacitor.

electrical conductance See conductance.

electrical conduction See conduction.

electrical conductivity See conductivity.

electrical degree [ELEC] A unit equal to $\frac{1}{360}$ cycle of an alternating quantity.

electrical discharge machining See electron discharge machining.

electrical engineering [ENG] Engineering that deals with practical applications of electricity; generally restricted to applications involving current flow through conductors, as in motors and generators.

electrical equivalent [ANALY CHEM] In conductometric analyses of electrolyte solutions, an outside, calibrated current source as compared to (equivalent to) the current passing through the sample under analysis; for example, a Wheatstone-bridge balanced reading.

electrical fault See fault.

electrical impedance Also known as impedance. [ELEC] **1.** The total opposition that a circuit presents to an alternating current, equal to the complex ratio of the voltage to the current in complex notation. Also known as complex impedance. **2.** The ratio of the maximum voltage in an alternating-current circuit to the maximum current; equal to the magnitude of the quantity in the first definition.

electrical impedance meter [ELEC] An instrument which measures the complex ratio of voltage to current in a given circuit at a given frequency. Also known as impedance meter.

electrical instability [ELEC] A persistent condition of unwanted self-oscillation in an amplifier or other electric circuit.

electrical insulation See insulation.

electrical insulator See insulator.

electrical interference See interference.

electrical loading See loading.

electrical log [ENG] Recorded measurement of the conductivities and resistivities down the length of an uncased borehole; gives a complete record of the formations penetrated.

electrical logging [ENG] The recording in uncased sections of a borehole of the conductivities and resistivities of the penetrated formations; used for geological correlation of the strata and evaluation of possibly productive horizons. Also known as electrical well logging.

electrically alterable read-only memory [COMPUT SCI] A read-only memory that can be reprogrammed electrically in the field a limited number of times, after the entire memory is erased by applying an appropriate electric field. Abbreviated EAROM.

electrical noise [ELEC] Noise generated by electrical devices, for example, motors, engine ignition, power lines, and so on, and propagated to the receiving antenna direct from the noise source.

electrical potential energy [ELEC] Energy possessed by electric charges by virtue of their position in an electrostatic field.

electrical pressure transducer See pressure transducer.

electrical resistance See resistance.

electrical resistance thermometer See resistance thermometer.

electrical resistivity [ELEC] The electrical resistance offered by a material to the flow of current, times the cross-sectional area of current flow and per unit length of current path; the reciprocal of the conductivity. Also known as resistivity; specific resistance.

electrical resistor *See* resistor.

electrical resonator *See* tank circuit.

electrical symbol [ELEC] A simple geometrical symbol used to represent a component of a circuit in a schematic circuit diagram.

electrical system [ELEC] System of wiring, switches, relays, and other equipment associated with receiving and distributing electricity.

electrical tape *See* insulating tape.

electrical transcription *See* transcription.

electrical unit [ELEC] A standard in terms of which some electrical quantity is evaluated.

electrical well logging *See* electrical logging.

electric arc [ELEC] A discharge of electricity through a gas, normally characterized by a voltage drop approximately equal to the ionization potential of the gas. Also known as arc.

electric-arc furnace *See* arc furnace.

electric-arc heating *See* arc heating.

electric-arc lamp *See* arc lamp.

electric-arc welding [MET] Welding in which the joint is heated to fusion by an electric arc or by a large electric current. Also known as arc welding.

electric brake [MECH ENG] An actuator in which the actuating force is supplied by current flowing through a solenoid, or through an electromagnet which is thereby attracted to disks on the rotating member, actuating the brake shoes; this force is counteracted by the force of a compression spring. Also known as electromagnetic brake.

electric calamine *See* hemimorphite.

electric cell [ELEC] 1. A single unit of a primary or secondary battery that converts chemical energy into electric energy. 2. A single unit of a device that converts radiant energy into electric energy, such as a nuclear, solar, or photovoltaic cell.

electric charge *See* charge.

electric chopper [ELECTROMAG] A chopper in which an electromagnet driven by a source of alternating current sets into vibration a reed carrying a moving contact that alternately touches two fixed contacts in a signal circuit, thus periodically interrupting the signal.

electric circuit [ELEC] Also known as circuit. 1. A path or group of interconnected paths capable of carrying electric currents. 2. An arrangement of one or more complete, closed paths for electron flow.

electric circuit theory *See* circuit theory.

electric coil *See* coil.

electric connector [ELEC] A device that joins electric conductors mechanically and electrically to other conductors and to the terminals of apparatus and equipment.

electric contact [ELEC] A physical contact that permits current flow between conducting parts. Also known as contact.

electric control [ELEC] The control of a machine or device by switches, relays, or rheostats, as contrasted with electronic control by electron tubes or by devices that do the work of electron tubes.

electric converter *See* synchronous converter.

electric corona *See* corona discharge.

electric current *See* current.

electric current density *See* current density.

electric cutout *See* cutout.

electric delay line [ELECTR] A delay line using properties of lumped or distributed capacitive and inductive elements; can be used for signal storage by recirculating information-carrying wave patterns.

electric dipole [ELEC] A localized distribution of positive and negative electricity, without net charge, whose mean positions of positive and negative charges do not coincide.

electric dipole moment [ELEC] A quantity characteristic of a charge distribution, equal to the vector sum over the electric charges of the product of the charge and the position vector of the charge.

electric discharge *See* discharge.

electric-discharge lamp *See* discharge lamp.

electric-discharge tube *See* discharge tube.

electric displacement [ELEC] The electric field intensity multiplied by the permittivity. Symbolized D. Also known as dielectric displacement; dielectric flux density; displacement; electric displacement density; electric flux density; electric induction.

electric displacement density *See* electric displacement.

electric distribution system *See* distribution system.

electric double layer [PHYS CHEM] A phenomenon found at a solid-liquid interface; it is made up of ions of one charge type which are fixed to the surface of the solid and an equal number of mobile ions of the opposite charge which are distributed through the neighboring region of the liquid; in such a system the movement of liquid causes a displacement of the mobile counterions with respect to the fixed charges on the solid surface. Also known as double layer.

electric doublet *See* dipole.

electric eel [VERT ZOO] *Electrophorus electricus*. An eellike cypriniform electric fish of the family Gymnotidae.

electric energy [ELECTROMAG] 1. Energy of electric charges by virtue of their position in an electric field. 2. Energy of electric currents by virtue of their position in a magnetic field.

electric engine [AERO ENG] A rocket engine in which the propellant is accelerated by some electric device. Also known as electric propulsion system; electric rocket.

electric eye *See* cathode-ray tuning indicator; photocell; phototube.

electric field [ELEC] 1. One of the fundamental fields in nature, causing a charged body to be attracted to or repelled by other charged bodies; associated with an electromagnetic wave or a changing magnetic field. 2. Specifically, the electric force per unit test charge.

electric field effect *See* Stark effect.

electric field intensity *See* electric field vector.

electric field strength *See* electric field vector.

electric field vector [ELEC] The force on a stationary positive charge per unit charge at a point in an electric field. Designated E. Also known as electric field intensity; electric field strength; electric vector.

electric filter [ELECTR] 1. A network that transmits alternating currents of desired frequencies while substantially attenuating all other frequencies. Also known as frequency-selective device. 2. *See* filter.

electric fish [VERT ZOO] Any of several fishes capable of producing electric discharges from an electric organ.

electric flowmeter [ELEC] Fluid-flow measurement device relying on an inductance or impedance bridge or on electrical-resistance rod elements to sense flow-rate variations.

electric flux [ELEC] 1. The integral over a surface of the component of the electric displacement perpendicular to the surface; equal to the number of electric lines of force crossing the surface. **2.** The electric lines of force in a region.

electric flux density *See* electric displacement.

electric flux line *See* electric line of force.

electric forming [ELECTR] The process of applying electric energy to a semiconductor or other device to modify permanently its electrical characteristics.

electric furnace [ENG] A furnace which uses electricity as a source of heat.

electric fuse *See* fuse.

electric generator *See* generator.

electric heating [ENG] Any method of converting electric energy to heat energy by resisting the free flow of electric current.

electric hygrometer [ENG] An instrument for indicating by electrical means the humidity of the ambient atmosphere; usually based on the relation between the electric conductance of a film of hygroscopic material and its moisture content.

electric hysteresis *See* ferroelectric hysteresis.

electric ignition [MECH ENG] Ignition of a charge of fuel vapor and air in an internal combustion engine by passing a high-voltage electric current between two electrodes in the combustion chamber.

electric image [ELEC] A fictitious charge used in finding the electric field set up by fixed electric charges in the neighborhood of a conductor; the conductor, with its distribution of induced surface charges, is replaced by one or more of these fictitious charges. Also known as image.

electric induction *See* electric displacement.

electricity [PHYS] Physical phenomenon involving electric charges and their effects when at rest and when in motion.

electric lamp [ELEC] A lamp in which light is produced by electricity, as the incandescent lamp, arc lamp, glow lamp, mercury-vapor lamp, and fluorescent lamp.

electric line of force [ELEC] An imaginary line drawn so that each segment of the line is parallel to the direction of the electric field or of the electric displacement at that point, and the density of the set of lines is proportional to the electric field or electrical displacement. Also known as electric flux line.

electric main *See* power transmission line.

electric meter [ENG] An electricity-measuring device that totalizes with time, such as a watthour meter or ampere-hour meter, in contrast to an electric instrument.

electric moment [ELEC] One of a series of quantities characterizing an electric charge distribution; an l-th moment is given by integrating the product of the charge density, the l-th power of the distance from the origin, and a spherical harmonic Y^*_{lm} over the charge distribution.

electric monopole [ELEC] A distribution of electric charge which is concentrated at a point or is spherically symmetric.

electric motor *See* motor.

electric multipole [ELECTROMAG] One of a series of types of static or oscillating charge distributions; the multipole of order 1 is a point charge or a spherically symmetric distribution, and the electric and magnetic fields produced by an electric multipole of order 2^n are equivalent to those of two electric multipoles of order 2^{n-1} of equal strengths, but opposite sign, separated from each other by a short distance.

electric network *See* network.

electric organ [VERT ZOO] An organ consisting of rows of electroplaques which produce an electric discharge.

electric outlet *See* outlet.

electric polarizability [ELEC] Induced dipole moment of an atom or molecule in a unit electric field.

electric polarization *See* polarization.

electric potential [ELEC] The work which must be done against electric forces to bring a unit charge from a reference point to the point in question; the reference point is located at an infinite distance, or, for practical purposes, at the surface of the earth or some other large conductor. Also known as electrostatic potential; potential.

electric power [ELEC] The rate at which electric energy is converted to other forms of energy, equal to the product of the current and the voltage drop.

electric power line *See* power line.

electric power meter [ENG] A device that measures electric power consumed, either at an instant, as in a wattmeter, or averaged over a time interval, as in a demand meter. Also known as power meter.

electric power plant [MECH ENG] A power plant that converts a form of raw energy into electricity, for example, a hydro, steam, diesel, or nuclear generating station for stationary or transportation service.

electric power station [ELEC] A generating station or an electric power substation.

electric power substation [ELEC] An assembly of equipment in an electric power system through which electric energy is passed for transmission, transformation, distribution, or switching. Also known as substation.

electric probe [PL PHYS] A device used to measure electron temperatures, electron and ion densities, space and wall potentials, and random electron currents in a plasma; consists substantially of one or two small collecting electrodes to which various potentials are applied, with the corresponding collection currents being measured. Also known as electrostatic probe.

electric propulsion system *See* electric engine.

electric protective device [ELEC] A particular type of equipment used in electric power systems to detect abnormal conditions and to initiate appropriate corrective action. Also known as protective device.

electric quadrupole [ELEC] A charge distribution that produces an electric field equivalent to that produced by two electric dipoles whose dipole moments have the same magnitude but point in opposite directions and which are separated from each other by a small distance.

electric quadrupole lens [ELECTR] A device for focusing beams of charged particles which has four electrodes with alternately positive and negative polarity; used in electron microscopes and particle accelerators.

electric quadrupole moment [ELEC] A quantity characterizing an electric charge distribution, obtained by integrating the product of the charge density, the second power of the distance from the origin, and a spherical harmonic Y^*_{2m} over the charge distribution.

electric raceway *See* raceway.

electric reactor *See* reactor.

electric relay *See* relay.

electric resistance *See* resistance.

electric resistance furnace *See* resistance furnace.

electric rocket *See* electric engine.

electric rotating machinery [ELEC] Any form of apparatus which has a rotating member and generates, converts, transforms, or modifies electric power, such as a motor, generator, or synchronous converter.

electric scanning [ELECTR] Scanning in which the required changes in radar beam direction are produced by variations in phase or amplitude of the currents fed to the various elements of the antenna array.

electric shielding [ELECTROMAG] Any means of avoiding pickup of undesired signals or noise, suppressing radiation of undesired signals, or confining wanted signals to desired paths or regions, such as electrostatic shielding or electromagnetic shielding. Also known as screening; shielding.

electric shock [PHYSIO] The sudden pain, convulsion, unconsciousness, or death produced by the passage of electric current through the body.

electric shunt *See* shunt.

electric solenoid *See* solenoid.

electric spark *See* spark.

electric spark machining *See* electron discharge machining.

electric steel [MET] Steel melted in an electric furnace which permits close control and the addition of alloying elements directly into the furnace.

electric strength *See* dielectric strength.

electric susceptibility [ELEC] A dimensionless parameter measuring the ease of polarization of a dielectric, equal (in meter-kilogram-second units) to the ratio of the polarization to the product of the electric field strength and the vacuum permittivity. Also known as dielectric susceptibility.

electric switch *See* switch.

electric switchboard *See* switchboard.

electric tank *See* electrolytic tank.

electric terminal *See* terminal.

electric transducer [ELECTR] A transducer in which all of the waves are electric.

electric twinning [SOLID STATE] A defect occurring in natural quartz crystals, in which adjacent regions of quartz have their electric axes oppositely poled.

electric vector *See* electric field vector.

electric wave [ELECTROMAG] An electromagnetic wave, especially one whose wavelength is at least a few centimeters. Also known as Hertzian wave.

electric-wave filter *See* filter.

electric wind *See* convective discharge.

electric wire *See* wire.

electric wiring *See* wiring.

electrification [ELEC] 1. The process of establishing a charge in an object. 2. The generation, distribution, and utilization of electricity.

electroacoustics [ENG ACOUS] The conversion of acoustic energy and waves into electric energy and waves, or vice versa.

electroacoustic transducer [ENG ACOUS] A transducer that receives waves from an electric system and delivers waves to an acoustic system, or vice versa. Also known as sound transducer.

electrocardiogram [MED] A graphic recording of the electrical manifestations of the heart action as obtained from the body surfaces. Abbreviated ECG; EKG.

electrocardiography [MED] The medical specialty concerned with the production and interpretation of electrocardiograms.

electrocauterization [MED] The application of a direct galvanic current to tissues to cause destruction or coagulation.

electrochemical cell [PHYS CHEM] A combination of two electrodes arranged so that an overall oxidation-reduction reaction produces an electromotive force; includes dry cells, wet cells, standard cells, fuel cells, solid-electrolyte cells, and reserve cells.

electrochemical corrosion [MET] Corrosion of a metal associated with the flow of electric current in an electrolyte. Also known as electrolytic corrosion.

electrochemical effect [PHYS CHEM] Conversion of chemical to electric energy, as in electrochemical cells; or the reverse process, used to produce elemental aluminum, magnesium, and bromine from compounds of these elements.

electrochemical emf [PHYS CHEM] Electrical force generated by means of chemical action, in manufactured cells (such as dry batteries) or by natural means (galvanic reaction).

electrochemical equivalent [PHYS CHEM] The weight in grams of a substance produced or consumed by electrolysis with 100% current efficiency during the flow of a quantity of electricity equal to 1 faraday ($96,487.0 \pm 1.6$ coulombs).

electrochemical machining [MET] Removing excess metal by electrolytic dissolution, effected by the tool acting as the cathode against the workpiece acting as the anode. Abbreviated ECM. Also known as electrolytic machining.

electrochemical potential [PHYS CHEM] The difference in potential that exists when two dissimilar electrodes are connected through an external conducting circuit and the two electrodes are placed in a conducting solution so that electrochemical reactions occur.

electrochemical process [PHYS CHEM] 1. A chemical change accompanying the passage of an electric current, especially as used in the preparation of commercially important quantities of certain chemical substances. 2. The reverse change, in which a chemical reaction is used as the source of energy to produce an electric current, as in a battery.

electrochemical series [PHYS CHEM] A series in which the metals and other substances are listed in the order of their chemical reactivity or electrode potentials, the most reactive at the top and the less reactive at the bottom. Also known as electromotive series.

electrochemical transducer [ENG] A device which uses a chemical change to measure the input parameter; the output is a varying electrical signal proportional to the measurand.

electrochemistry [PHYS CHEM] A branch of chemistry dealing with chemical changes accompanying the passage of an electric current; or with the reverse process, in which a chemical reaction is used to produce an electric current.

electrochromatography [ANALY CHEM] Type of chromatography that utilizes application of an electric potential to produce an electric differential. Also known as electropherography.

electrochromic display [ELECTR] A solid-state passive display that uses organic or inorganic insulating solids which change color when injected with positive or negative charges.

electroconvulsive shock [MED] The technique of eliciting convulsions by applying an electric current through the brain for a brief period.

electrode [ELEC] 1. An electric conductor through which an electric current enters or leaves a medium, whether it be an electrolytic solution, solid, molten mass, gas, or vacuum. 2. One of the terminals used in dielectric heating or diathermy for applying the high-frequency electric field to the material being heated.

electrode admittance [ELECTR] Quotient of dividing the alternating component of the electrode current by the alternating component of the electrode voltage, all other electrode voltages being maintained constant.

electrodecantation [PHYS CHEM] A modification of electrodialysis in which a cell is divided into three sections by two membranes and electrodes are placed in the end sections; colloidal matter is concentrated at the sides and bottom of the middle section, and the liquid that floats to the top is drawn off.

electrode capacitance [ELECTR] Capacitance between one electrode and all the other electrodes connected together.

electrode characteristic [ELECTR] Relation between the electrode voltage and the current to an electrode, all other electrode voltages being maintained constant.

electrode conductance [ELECTR] Quotient of the inphase component of the electrode alternating current by the electrode alternating voltage, all other electrode voltage being maintained constant; this is a variational and not a total conductance. Also known as grid conductance.

electrode current [ELECTR] Current passing to or from an electrode, through the interelectrode space within a vacuum tube.

electrode dark current [ELECTR] The electrode current that flows when there is no radiant flux incident on the photocathode in a phototube or camera tube. Also known as dark current.

electrode drop [ELECTR] Voltage drop in the electrode due to its resistance.

electrode force [MET] The force that occurs between electrodes during seam, spot, and projection welding. Also known as welding force.

electrode impedance [ELECTR] Reciprocal of the electrode admittance.

electrodeposition [MET] Electrolytic process in which a metal is deposited at the cathode from a solution of its ions; includes electroplating and electroforming. Also known as electrolytic deposition.

electrodeposition analysis [ANALY CHEM] An electroanalytical technique in which an element is quantitatively deposited on an electrode.

electrode potential Also known as electrode voltage. [ELECTR] The instantaneous voltage of an electrode with respect to the cathode of an electron tube. [PHYS CHEM] The voltage existing between an electrode and the solution or electrolyte in which it is immersed; usually, electrode potentials are referred to a standard electrode, such as the hydrogen electrode.

electrode resistance [ELECTR] Reciprocal of the electrode conductance; this is the effective parallel resistance and is not the real component of the electrode impedance.

electrode voltage See electrode potential.

electrodialysis [PHYS CHEM] Dialysis that is conducted with the aid of an electromotive force applied to electrodes adjacent to both sides of the membrane.

electrodynamic ammeter [ENG] Instrument which measures the current passing through a fixed coil and a movable coil connected in series by balancing the torque on the movable coil (resulting from the magnetic field of the fixed coil) against that of a spiral spring.

electrodynamic drift [GEOPHYS] Motion of charged particles in the upper atmosphere due to the combined effect of electric and magnetic fields; in the ionospheric F region and above, the drift velocity is perpendicular to both the electric and magnetic fields.

electrodynamic instrument [ENG] An instrument that depends for its operation on the reaction between the current in one or more movable coils and the current in one or more fixed coils. Also known as electrodynamometer.

electrodynamic loudspeaker [ENG ACOUS] Dynamic loudspeaker in which the magnetic field is produced by an electromagnet, called the field coil, to which a direct current must be furnished.

electrodynamic machine [ELEC] An electric generator or motor in which the output load current is produced by magnetomotive currents generated in a rotating armature.

electrodynamics [ELECTROMAG] The study of the relations between electrical, magnetic, and mechanical phenomena.

electrodynamic shaker See shaker.

electrodynamic wattmeter [ENG] An electrodynamic instrument connected as a wattmeter, with the main current flowing through the fixed coil, and a small current proportional to the voltage flowing through the movable coil. Also known as moving-coil wattmeter.

electrodynamometer See electrodynamic instrument.

electroencephalogram [MED] A graphic recording of the electric discharges of the cerebral cortex as detected by electrodes on the surface of the scalp. Abbreviated EEG.

electroencephalography [MED] The medical specialty concerned with the production and interpretation of electroencephalograms.

electroendosmosis [PHYS] The production of an endosmosis effect by an electrical potential; that is, the use of electricity to cause diffusion of a liquid through an organic membrane.

electroepitaxy [CRYSTAL] A crystal growth process achieved by passing an electric current through the substrate solution.

electroerosive machining See electron discharge machining.

electrofocusing See isoelectric focusing.

electroforming [MET] Shaping components by electrodeposition of the metal on a pattern.

electrogalvanizing [MET] Coating of a metal, especially iron or steel, with zinc by electroplating.

electrogasdynamics [PHYS] Conversion of the kinetic energy of a moving gas to electricity, for such applications as high-voltage electric power generation, air-pollution control, and paint spraying.

electrogram [ELECTR] A record of an image of an object made by sparking, usually on paper. [METEOROL] A record, usually automatically produced, which shows the time variations of the atmospheric electric field at a given point. [PHYSIO] The graphic representation of electric events in living tissues; commonly, an electrocardiogram or electroencephalogram.

electrogravimetry [ANALY CHEM] Electrodeposition analysis in which the quantities of metals deposited may be determined by weighing a suitable electrode before and after deposition.

electrojet [GEOPHYS] A stream of intense electric current moving in the upper atmosphere around the equator and in polar regions.

electrokinetic phenomena [PHYS CHEM] The phenomena associated with movement of charged particles through a continuous medium or with the movement of a continuous medium over a charged surface.

electrokinetic potential See zeta potential.

electrokinetics [ELECTROMAG] The study of the motion of electric charges, especially of steady currents in electric circuits, and of the motion of electrified particles in electric or magnetic fields.

electrokinetic transducer [ELEC] An instrument which converts dynamic physical forces, such as vibration and sound, into corresponding electric signals by measuring the streaming potential generated by passage of a polar fluid through a permeable refractory-ceramic or fritted-glass member between two chambers.

electroless plating [MET] Deposition of a metal coating by immersion of a metal or nonmetal in a suitable bath containing a chemical reducing agent.

electroluminescence [ELECTR] The emission of light, not due to heating effects alone, resulting from application of an electric field to a material, usually solid.

electroluminescent cell *See* electroluminescent panel.

electroluminescent display [ELECTR] A display in which various combinations of electroluminescent segments may be activated by applying voltages to produce any desired numeral or other character.

electroluminescent lamp *See* electroluminescent panel.

electroluminescent panel [ELECTR] A surface-area light source employing the principle of electroluminescence; consists of a suitable phosphor placed between sheet-metal electrodes, one of which is essentially transparent, with an alternating current applied between the electrodes. Also known as electroluminescent cell; electroluminescent lamp; light panel; luminescent cell.

electroluminescent phosphor [MATER] Zinc sulfide powder, with small additions of copper or manganese, which emits light when suspended in an insulator in an intense alternating electric field. Also known as electroluminor.

electroluminor *See* electroluminescent phosphor.

electrolysis [PHYS CHEM] A method by which chemical reactions are carried out by passage of electric current through a solution of an electrolyte or through a molten salt.

electrolyte [PHYS CHEM] A chemical compound which when molten or dissolved in certain solvents, usually water, will conduct an electric current.

electrolyte-activated battery [ELEC] A reserve battery in which an aqueous electrolyte is stored in a separate chamber, and a mechanism, which may be operated from a remote location, drives the electrolyte out of the reservoir and into the cells of the battery for activation.

electrolytic brightening *See* electropolishing.

electrolytic capacitor [ELEC] A capacitor consisting of two electrodes separated by an electrolyte; a dielectric film, usually a thin layer of gas, is formed on the surface of one electrode. Also known as electrolytic condenser.

electrolytic cell [PHYS CHEM] A cell consisting of electrodes immersed in an electrolyte solution, for carrying out electrolysis.

electrolytic condenser *See* electrolytic capacitor.

electrolytic conductance [PHYS CHEM] The transport of electric charges, under electric potential differences, by charged particles (called ions) of atomic or larger size.

electrolytic conductivity [PHYS CHEM] The conductivity of a medium in which the transport of electric charges, under electric potential differences, is by particles of atomic or larger size.

electrolytic corrosion *See* electrochemical corrosion.

electrolytic deposition *See* electrodeposition.

electrolytic dissociation [CHEM] The ionization of a compound in a solution.

electrolytic grinding [MECH ENG] A combined grinding and machining operation in which the abrasive, cathodic grinding wheel is in contact with the anodic workpiece beneath the surface of an electrolyte.

electrolytic interrupter [ELEC] An interrupter that consists of two electrodes in an electrolytic solution; bubbles formed in the solution continually interrupt the passage of current between the electrodes.

electrolytic photocopying [GRAPHICS] Photocopying process in which an image is projected on a sheet consisting of a paper support, a thin aluminum laminate, and a coating of a white photoconductive substance in contact with an electrolyte, and electrolysis takes place in the exposed areas when a direct current is applied across the electrolyte and the aluminum underlayer.

electrolytic polishing *See* electropolishing.

electrolytic potential [PHYS CHEM] Difference in potential between an electrode and the immediately adjacent electrolyte, expressed in terms of some standard electrode difference.

electrolytic process [PHYS CHEM] An electrochemical process involving the principles of electrolysis, especially as relating to the separation and deposition of metals.

electrolytic protection *See* cathodic protection.

electrolytic rectifier [ELEC] A rectifier consisting of metal electrodes in an electrolyte, in which rectification of alternating current is accompanied by electrolytic action; polarizing film formed on one electrode permits current flow in one direction but not the other.

electrolytic refining *See* electrorefining.

electrolytic rheostat [ELEC] A rheostat that consists of a tank of conducting liquid in which electrodes are placed, and resistance is varied by changing the distance between the electrodes, the depth of immersion of the electrodes, or the resistivity of the solution. Also known as water rheostat.

electrolytic separation [PHYS CHEM] Separation of isotopes by electrolysis, based on differing rates of discharge at the electrode of ions of different isotopes.

electrolytic switch [ELEC] A switch having two electrodes projecting into a chamber partly filled with electrolyte, leaving an air bubble of predetermined width; the bubble shifts position and changes the amount of electrolyte in contact with the electrodes when the switch is tilted from true horizontal.

electrolytic tank [ENG] A tank in which voltages are applied to an enlarged scale model of an electron-tube system or a reduced scale model of an aerodynamic system immersed in a poorly conducting liquid, and equipotential lines between electrodes are traced; used as an aid to electron-tube design or in computing ideal fluid flow; the latter application is based on the fact that the velocity potential in ideal flow and the stream function in planar flow satisfy the same equation, Laplace's equation, as an electrostatic potential. Also known as electric tank; potential flow analyzer.

electromachining [MECH ENG] The application of electric or ultrasonic energy to a workpiece to effect removal of material.

electromagnet [ELECTROMAG] A magnet consisting of a coil wound around a soft iron or steel core; the core is strongly magnetized when current flows through the coil, and is almost completely demagnetized when the current is interrupted.

electromagnetic [PHYS] Pertaining to phenomena in which electricity and magnetism are related.

electromagnetic brake *See* electric brake.

electromagnetic cathode-ray tube [ELECTR] A cathode-ray tube in which electromagnetic deflection is used on the electron beam.

electromagnetic clutch [MECH ENG] A clutch based on magnetic coupling between conductors, such as a magnetic fluid and powder clutch, an eddy-current clutch, or a hysteresis clutch.

electromagnetic compatibility [ELECTR] The capability of electronic equipment or systems to be operated in the intended electromagnetic environment at design levels of efficiency.

electromagnetic constant See speed of light.

electromagnetic coupling [ELECTROMAG] Coupling that exists between circuits when they are mutually affected by the same electromagnetic field.

electromagnetic current [ELECTR] Motion of charged particles (for example, in the ionosphere) giving rise to electric and magnetic fields.

electromagnetic damping [ELEC] Retardation of motion that results from the reaction between eddy currents in a moving conductor and the magnetic field in which it is moving.

electromagnetic energy [ELECTROMAG] The energy associated with electric or magnetic fields.

electromagnetic field [ELECTROMAG] An electric or magnetic field, or a combination of the two, as in an electromagnetic wave.

electromagnetic field equations See Maxwell field equations.

electromagnetic flowmeter [ENG] A flowmeter that offers no obstruction to liquid flow; two coils produce an electromagnetic field in the conductive moving fluid; the current induced in the liquid, detected by two electrodes, is directly proportional to the rate of flow. Also known as electromagnetic meter.

electromagnetic focusing [ELECTR] Focusing the electron beam in a telelvision picture tube by means of a magnetic field parallel to the beam; the field is produced by sending an adjustable value of direct current through a focusing coil mounted on the neck of the tube.

electromagnetic horn See horn antenna.

electromagnetic induction [ELECTROMAG] The production of an electromotive force either by motion of a conductor through a magnetic field so as to cut across the magnetic flux or by a change in the magnetic flux that threads a conductor. Also known as induction.

electromagnetic interference [ELEC] Interference, generally at radio frequencies, that is generated inside systems, as contrasted to radio-frequency interference coming from sources outside a system. Abbreviated emi.

electromagnetic log [ENG] A log containing an electromagnetic sensing element extended below the hull of the vessel; this device produces a voltage directly proportional to speed through the water.

electromagnetic logging [ENG] A method of well logging in which a transmitting coil sets up an alternating electromagnetic field, and a receiver coil, placed in the drill hole above the transmitter coil, measures the secondary electromagnetic field induced by the resulting eddy currents within the formation. Also known as electromagnetic well logging.

electromagnetic meter See electromagnetic flowmeter.

electromagnetic potential [ELECTROMAG] Collective name for a scalar potential, which reduces to the electrostatic potential in a time-independent system, and the vector potential for the magnetic field; the electric and magnetic fields can be written in terms of these potentials.

electromagnetic properties [ELECTROMAG] The response of materials or equipment to electromagnetic fields, and their ability to produce such fields.

electromagnetic prospecting See electromagnetic surveying.

electromagnetic pump [ELEC] A pump in which a conductive liquid is made to move through a pipe by sending a large current transversely through the liquid; this current reacts with a magnetic field that is at right angles to the pipe and to current flow, to move the current-carrying liquid conductor.

electromagnetic radiation [ELECTROMAG] Electromagnetic waves and, especially, the associated electromagnetic energy.

electromagnetic relay [ELECTROMAG] A relay in which current flow through a coil produces a magnetic field that results in contact actuation.

electromagnetic scattering [PHYS] The process in which energy is removed from a beam of electromagnetic radiation and reemitted without appreciable changes in wavelength.

electromagnetic spectrum [ELECTROMAG] The total range of wavelengths or frequencies of electromagnetic radiation, extending from the longest radio waves to the shortest known cosmic rays.

electromagnetic surveying [ENG] Underground surveying carried out by generating electromagnetic waves at the surface of the earth; the waves penetrate the earth and induce currents in conducting ore bodies, thereby generating new waves that are detected by instruments at the surface or by a receiving coil lowered into a borehole. Also known as electromagnetic prospecting.

electromagnetic susceptibility [ELECTR] The tolerance of circuits and components to all sources of interfering electromagnetic energy.

electromagnetic system of units [ELECTROMAG] A centimeter-gram-second system of electric and magnetic units in which the unit of current is defined as the current which, if maintained in two straight parallel wires having infinite length and being 1 centimeter apart in vacuum, would produce between these conductors a force of 2 dynes per centimeter of length; other units are derived from this definition by assigning unit coefficients in equations relating electric and magnetic quantities. Also known as electromagnetic units (emu).

electromagnetic transducer See electromechanical transducer.

electromagnetic units See electromagnetic system of units.

electromagnetic wave [ELECTROMAG] A disturbance which propagates outward from any electric charge which oscillates or is accelerated; far from the charge it consists of vibrating electric and magnetic fields which move at the speed of light and are at right angles to each other and to the direction of motion.

electromagnetic well logging See electromagnetic logging.

electromagnetism [PHYS] 1. Branch of physics relating electricity to magnetism. 2. Magnetism produced by an electric current rather than by a permanent magnet.

electromechanical [MECH ENG] Pertaining to a mechanical device, system, or process which is electrostatically or electromagnetically actuated or controlled.

electromechanical circuit [ELEC] A circuit containing both electrical and mechanical parameters of consequence in its analysis.

electromechanical dialer [ELECTR] A telephone dialer which activates one of a set of desired numbers, precoded into it, when the user selects and presses a start button.

electromechanical relay [ELECTROMAG] A protective relay operating on the principle of electromagnetic attraction, as a plunger relay, or of electromagnetic induction.

electromechanical transducer [ELECTR] A transducer for receiving waves from an electric system and delivering waves to a mechanical system, or vice versa. Also known as electromagnetic transducer.

electromechanics [MECH ENG] The technology of mechanical devices, systems, or processes which are electrostatically or electromagnetically actuated or controlled.

electrometallurgy [MET] Industrial recovery and processing of metals by electrical and electrolytic procedures.

electrometer [ENG] An instrument for measuring voltage without drawing appreciable current.

electromodulation [SPECT] Modulation spectroscopy in which changes in transmission or reflection spectra induced by a perturbing electric field are measured.

electromotance *See* electromotive force.

electromotive force [PHYS CHEM] 1. The difference in electric potential that exists between two dissimilar electrodes immersed in the same electrolyte or otherwise connected by ionic conductors. 2. The resultant of the relative electrode potential of the two dissimilar electrodes at which electrochemical reactions occur. Abbreviated emf. Also known as electromotance.

electromotive series *See* electrochemical series.

electromyogram [MED] 1. A graphic recording of the electrical response of a muscle to electrical stimulation. 2. A graphic recording of eye movements during reading.

electromyography [MED] A medical specialty concerned with the production and study of electromyograms.

electron [PHYS] 1. A stable elementary particle which is the negatively charged constituent of ordinary matter, having a mass of about 9.11×10^{-28}g (equivalent to 0.511 MeV), a charge of about -1.602×10^{-19} coulomb, and a spin of ½. Also known as negative electron; negatron. 2. Collective name for the electron, as in the first definition, and the positron.

electron acceptor [PHYS CHEM] An atom or part of a molecule joined by a covalent bond to an electron donor. [SOLID STATE] *See* acceptor.

electron-acoustic microscopy [PHYS] A technique for producing images that show variations in an object's thermal and elastic properties; an electron beam generates ultrasonic waves in the specimen which are detected by a piezoelectric transducer whose output controls the brightness of a spot sweeping a cathode-ray tube in synchronism with the electron beam.

electron affinity [ATOM PHYS] The work needed in removing an electron from a negative ion, thus restoring the neutrality of an atom or molecule.

electron attachment *See* electron capture.

electron beam [ELECTR] A narrow stream of electrons moving in the same direction, all having about the same velocity.

electron-beam-accessed memory *See* electron-beam memory.

electron-beam laser [OPTICS] A semiconductor laser in which the electron beam that provides pumping action in a thin plate of cadmium sulfide or other material is swept electrically in two dimensions by a deflection yoke, much as in a cathode-ray tube.

electron-beam machining [MET] A machining process in which heat is produced by a focused electron beam at a sufficiently high temperature to volatilize and thereby remove metal in a desired manner; takes place in a vacuum. Abbreviated EBM.

electron-beam memory [COMPUT SCI] A memory that uses a high-resolution electron beam to store information on a target in a vacuum tube. Also known as electron-beam-accessed memory (EBAM).

electron-beam parametric amplifier [ELECTR] A parametric amplifier in which energy is pumped from an electrostatic field into a beam of electrons traveling down the length of the tube, and electron couplers impress the input signal at one end of the tube and translate spiraling electron motion into electric output at the other.

electron-beam pumping [ELECTR] The use of an electron beam to produce excitation for population inversion and lasing action in a semiconductor laser.

electron-beam tube [ELECTR] An electron tube whose performance depends on the formation and control of one or more electron beams.

electron-beam welding [MET] A technique for joining materials in which highly collimated electron beams are used at a pressure below 10^{-3} mmHg (0.1333 pascal) to produce a highly concentrated heat source; used in outer space.

electron capture [ATOM PHYS] The process in which an atom or ion passing through a material medium either loses or gains one or more orbital electrons. [NUC PHYS] A radioactive transformation of nuclide in which a bound electron merges with its nucleus. Also known as electron attachment.

electron charge [PHYS] The charge carried by an electron, equal to about -1.602×10^{-19} coulomb, or -4.803×10^{-10} statcoulomb.

electron cloud [ATOM PHYS] Picture of an electron state in which the charge is thought of as being smeared out, with the resulting charge density distribution corresponding to the probability distribution function associated with the Schrödinger wave function.

electron collector *See* collector.

electron compound [MET] Alloy of two metals in which a progressive change in composition is accompanied by a progression of phases, differing in crystal structure. Also known as Hume-Rothery compound; intermetallic compound.

electron conduction [ELEC] Conduction of electricity resulting from motion of electrons, rather than from ions in a gas or solution, or holes in a solid.

electron configuration [ATOM PHYS] The orbital and spin arrangement of an atom's electrons, specifying the quantum numbers of the atom's electrons in a given state.

electron-coupled oscillator [ELECTR] An oscillator employing a multigrid tube in which the cathode and two grids operate as an oscillator; the anode-circuit load is coupled to the oscillator through the electron stream. Abbreviated eco. Also known as Dow oscillator.

electron coupling [ELECTR] A method of coupling two circuits inside an electron tube, used principally with multigrid tubes; the electron stream passing between electrodes in one circuit transfers energy to electrodes in the other circuit. Also known as electronic coupling.

electron diffraction [PHYS] The phenomenon associated with the interference processes which occur when electrons are scattered by atoms in crystals to form diffraction patterns.

electron dipole moment *See* electron magnetic moment.

electron discharge machining [MET] A process by which materials that conduct electricity can be removed from a metal by an electric spark; used to form holes of varied shapes in materials of poor machinability. Abbreviated EDM. Also known as electrical discharge machining; electric spark machining; electroerosive machining; electrospark machining.

electron donor [PHYS CHEM] An atom or part of a molecule which supplies both electrons of a duplet forming a covalent bond. [SOLID STATE] *See* donor.

electron efficiency [ELECTR] The power which an electron stream delivers to the circuit of an oscillator or amplifier at a given frequency, divided by the direct power supplied to the stream. Also known as electronic efficiency.

electronegative [ELEC] 1. Carrying a negative electric charge. 2. Capable of acting as the negative electrode in an electric cell. [PHYS CHEM] Pertaining to an atom or group of atoms that has a relatively great tendency to attract electrons to itself.

electronegative potential [PHYS CHEM] Potential of an electrode expressed as negative with respect to the hydrogen electrode.

electron emission [ELECTR] The liberation of electrons from an electrode into the surrounding space, usually under the influence of heat, light, or a high electric field.

electron energy level [ATOM PHYS] A quantum-mechanical concept for energy levels of electrons about the nucleus; electron energies are functions of each particular atomic species.

electroneutrality principle [PHYS CHEM] The principle that in an electrolytic solution the concentrations of all the ionic species are such that the solution as a whole is neutral.

electron gun [ELECTR] An electrode structure that produces and may control, focus, deflect, and converge one or more electron beams in an electron tube.

electron hole *See* hole.

electron hole droplets [SOLID STATE] A form of electronic excitation observed in germanium and silicon at sufficiently low cryogenic temperatures; it is associated with a liquid-gas phase transition of the charge carriers, and consists of regions of conducting electron-hole Fermi liquid coexisting with regions of insulating exciton gas.

electronic [ELECTR] Pertaining to electron devices or to circuits or systems utilizing electron devices, including electron tubes, magnetic amplifiers, transistors, and other devices that do the work of electron tubes.

electronic altimeter *See* radio altimeter.

electronic angular momentum [ATOM PHYS] The total angular momentum associated with the orbital motion of the spins of all the electrons of an atom.

electronic azimuth marker [ELECTR] On an airborne radar plan position indicator (PPI) a bright rotatable radial line used for bearing determination. Also known as azimuth marker.

electronic band spectrum [SPECT] Bands of spectral lines associated with a change of electronic state of a molecule; each band corresponds to certain vibrational energies in the initial and final states and consists of numerous rotational lines.

electronic calculator [ELECTR] A calculator in which integrated circuits perform calculations and show results on a digital display; the displays usually use either seven-segment light-emitting diodes or liquid crystals.

electronic cash register [ENG] A system for automatically checking out goods from retail food stores, consisting of a device that scans packages and reads symbols imprinted on the label, and a computer that converts the symbol information to tell a cash register the price of the item; the computer can also keep records of sales and inventories. Abbreviated ECR.

electronic circuit [ELECTR] An electric circuit in which the equilibrium of electrons in some of the components (such as electron tubes, transistors, or magnetic amplifiers) is upset by means other than an applied voltage.

electronic commutator [ELECTR] An electron-tube or transistor circuit that switches one circuit connection rapidly and successively to many other circuits, without the wear and noise of mechanical switches.

electronic component [ELECTR] A component which is able to amplify or control voltages or currents without mechanical or other nonelectrical command, or to switch currents or voltages without mechanical switches; examples include electron tubes, transistors, and other solid-state devices.

electronic composition [GRAPHICS] Typesetting in which characters are generated by electron or laser beams at speeds above about 6000 words per minute.

electronic controller [ELECTR] Electronic device incorporating vacuum tubes or solid-state devices and used to control the action or position of equipment; for example, a valve operator.

electronic counter [ELECTR] A circuit using electron tubes or equivalent devices for counting electric pulses. Also known as electronic tachometer.

electronic countermeasure [ELECTR] An offensive or defensive tactic or device using electronic and reflecting apparatus to reduce the military effectiveness of enemy equipment involving electromagnetic radiation, such as radar, communication, guidance, or other radio-wave devices. Abbreviated ECM. Also known as electromagnetic countermeasure.

electronic coupling *See* electron coupling.

electronic data processing [COMPUT SCI] Processing data by using equipment that is predominantly electronic in nature, such as an electronic digital computer. Abbreviated EDP.

electronic distance-measuring equipment [NAV] A navigation system consisting of airborne devices that transmit microsecond pulses to special ground beacons, which retransmit the signals to the aircraft; the length of expired time between transmission and reception is measured, converted to kilometers or miles, and presented to the pilot.

electronic efficiency [ELECTR] Ratio of the power at the desired frequency, delivered by the electron stream to the circuit in an oscillator or amplifier, to the average power supplied to the stream.

electronic emission spectrum [SPECT] Spectrum resulting from emission of electromagnetic radiation by atoms, ions, and molecules following excitations of their electrons.

electronic energy curve [PHYS CHEM] A graph of the energy of a diatomic molecule in a given electronic state as a function of the distance between the nuclei of the atoms.

electronic engineering [ENG] Engineering that deals with practical applications of electronics.

electronic heating [ENG] Heating by means of radio-frequency current produced by an electron-tube oscillator or an equivalent radio-frequency power source. Also known as high-frequency heating; radio-frequency heating.

electronic intelligence [ORD] Electronic systems, apparatus, and operations for obtaining information concerning an enemy's capabilities, intentions, plans, and order of battle. Abbreviated elint.

electronic interference [ELECTR] Any electrical or electromagnetic disturbance that causes undesirable response in electronic equipment.

electronic jammer *See* jammer.

electronic jamming *See* jamming.

electronic listening device [ELECTR] A device used to capture the sound waves of conversation originating in an ostensibly private setting in a form, usually as a magnetic tape recording, which can be used against the target by adverse interests.

electronic magnetic moment [ATOM PHYS] The total magnetic dipole moment associated with the orbital motion of all the electrons of an atom and the electron spins; opposed to nuclear magnetic moment.

electronic navigation [NAV] Navigation by means of any electronic device or instrument.

electronic phase-angle meter [ELECTR] A phasemeter that makes use of electronic devices, such as amplifiers and limiters, that convert the alternating-current voltages being measured to square waves whose spacings are proportional to phase.

electronic polarization [ELEC] Polarization arising from the displacement of electrons with respect to the nuclei with which they are associated, upon application of an external electric field.

electronic position indicator [NAV] A radio navigation system used in hydrographic surveying which provides circular lines of position. Abbreviated EPI.

electronic power supply *See* power supply.

electronic pumping *See* pumping.

electronic-raster scanning *See* electronic scanning.

electronic recording [ELECTR] The process of making a graphical record of a varying quantity or signal (or the result of such a process) by electronic means, involving control of an electron beam by electric or magnetic fields, as in a cathode-ray oscillograph, in contrast to light-beam recording.

electronics [PHYS] Study, control, and application of the conduction of electricity through gases or vacuum or through semiconducting or conducting materials.

electronic scanning [ELECTR] Scanning in which an electron beam, controlled by electric or magnetic fields, is swept over the area under examination, in contrast to mechanical or electromechanical scanning. Also known as electronic-raster scanning.

electronic sculpturing [COMPUT SCI] Procedure for constructing a model of a system by using an analog computer, in which the model is devised at the console by interconnecting components on the basis of analogous configuration with real system elements; then, by adjusting circuit gains and reference voltages, dynamic behavior can be generated that corresponds to the desired response, or is recognizable in the real system.

electronic spectrum [SPECT] Spectrum resulting from emission or absorption of electromagnetic radiation during changes in the electron configuration of atoms, ions, or molecules, as opposed to vibrational, rotational, fine-structure, or hyperfine spectra.

electronic state [QUANT MECH] The physical state of electrons of a system, as specified, for example, by a Schrödinger-Pauli wave function of the positions and spin orientations of all the electrons.

electronic structure [PHYS] The arrangement of electrons in an atom, molecule, or solid, specified by their wave functions, energy levels, or quantum numbers.

electronic switch [ELECTR] **1.** Vacuum tube, crystal diodes, or transistors used as an on and off switching device. **2.** Test instrument used to present two wave shapes on a single gun cathode-ray tube.

electronic switching [COMMUN] Telephone switching using a computer with a storage containing program switching logic, whose output actuates reed switches that set up telephone connections automatically. [ELECTR] The use of electronic circuits to perform the functions of a high-speed switch.

electronic tablet [COMPUT SCI] A data-entry device consisting of stylus, writing surface, and circuitry that produces a pair of digital coordinate values corresponding continuously to the position of the stylus upon the surface. Also known as data tablet.

electronic tachometer *See* electronic counter.

electronic tuning [ELECTR] Tuning of a transmitter, receiver, or other tuned equipment by changing a control voltage rather than by adjusting or switching components by hand.

electronic video recording [ELECTR] The recording of black and white or color television visual signals on a reel of photographic film as coded black and white images. Abbreviated EVR.

electronic voltage regulator [ELECTR] A device which maintains the direct-current power supply voltage for electronic equipment nearly constant in spite of input alternating-current line voltage variations and output load variations.

electronic voltmeter [ENG] Voltmeter which uses the rectifying and amplifying properties of electron devices and their associated circuits to secure desired characteristics, such as high-input impedance, wide-frequency range, crest indications, and so on.

electronic work function [SOLID STATE] The energy required to raise an electron with the Fermi energy in a solid to the energy level of an electron at rest in vacuum outside the solid.

electronic writing [ELECTR] The use of electronic circuits and electron devices to reproduce symbols, such as an alphabet, in a prescribed order on an electronic display device for the purpose of transferring information from a source to a viewer of the display device.

electron image tube *See* image tube.

electron injection [ELECTR] **1.** The emission of electrons from one solid into another. **2.** The process of injecting a beam of electrons with an electron gun into the vacuum chamber of a mass spectrometer, betatron, or other large electron accelerator.

electron lens [ELECTR] An electric or magnetic field, or a combination thereof, which acts upon an electron beam in a manner analogous to that in which an optical lens acts upon a light beam. Also known as lens.

electron magnetic moment [ATOM PHYS] The magnetic dipole moment which an electron possesses by virtue of its spin. Also known as electron dipole moment.

electron mass [PHYS] The mass of an electron, equal to about 9.11×10^{-28}g, equivalent to 0.511 MeV. Also known as electron rest mass.

electron microprobe   [PHYS] An x-ray machine in which electrons emitted from a hot-filament source are accelerated electrostatically, then focused to an extremely small point on the surface of a specimen by an electromagnetic lens; nondestructive analysis of the specimen can then be made by measuring the backscattered electrons, the specimen current, the resulting x-radiation, or any other resulting phenomenon. Also known as electron probe.

electron microscope   [ELECTR] A device for forming greatly magnified images of objects by means of electrons, usually focused by electron lenses.

electron mirror   See dynode.

electron mobility   [SOLID STATE] The drift mobility of electrons in a semiconductor, being the electron velocity divided by the applied electric field.

electron multiplier   [ELECTR] An electron-tube structure which produces current amplification; an electron beam containing the desired signal is reflected in turn from the surfaces of each of a series of dynodes, and at each reflection an impinging electron releases two or more secondary electrons, so that the beam builds up in strength. Also known as multiplier.

electron-multiplier phototube   See multiplier phototube.

electron nuclear double resonance   [SPECT] Type of electron paramagnetic resonance (EPR) spectroscopy permitting greatly enhanced resolution, in which a material is simultaneously irradiated at one of its EPR frequencies and by a second oscillatory field whose frequency is swept over the range of nuclear frequencies. Abbreviated ENDOR.

electron number   [ATOM PHYS] The number of electrons in an ion or atom.

electron optics   [ELECTR] The study of the motion of free electrons under the influence of electric and magnetic fields.

electron orbit   [PHYS] The path described by an electron.

electron pair   [PHYS CHEM] A pair of valence electrons which form a nonpolar bond between two neighboring atoms.

electron pair bond   See covalent bond.

electron paramagnetic resonance   [PHYS] Magnetic resonance arising from the magnetic moment of unpaired electrons in a paramagnetic substance or in a paramagnetic center in a diamagnetic substance. Abbreviated EPR. Also known as electron spin resonance (ESR); paramagnetic resonance.

electron paramagnetism   [PHYS] Paramagnetism in a substance whose atoms or molecules possess a net electronic magnetic moment; arises because of the tendency of a magnetic field to orient the electronic magnetic moments parallel to itself.

electron-positron storage ring   [NUCLEO] An annular vacuum chamber, enclosed by bending and focusing magnets, in which counterrotating beams of electrons and positrons are stored for several hours and can be made to collide with each other.

electron probe   See electron microprobe.

electron probe microanalysis   [ANALY CHEM] A technique in analytical chemistry in which a finely focused beam of electrons is used to excite an x-ray spectrum characteristic of the elements in the sample; can be used with samples as small as 10^{-11} cubic centimeter.

electron radius   [PHYS] The classical value r of 2.81777×10^{-13} centimeter for the radius of an electron; obtained by equating mc^2 for the electron to e^2/r, where e and m are the charge and mass of the electron respectively; any classical model for an electron will have approximately this radius.

electron-ray indicator   See cathode-ray tuning indicator.

electron-ray tube   See cathode-ray tube.

electron rest mass   See electron mass.

electron runaway   [PL PHYS] High acceleration of electrons in a collisional plasma caused by a suddenly applied electric field (which greatly reduces the collision cross secion of the electrons).

electron shell   [ATOM PHYS] 1. The collection of all the electron states in an atom which have a given principal quantum number. 2. The collection of all the electron states in an atom which have a given principal quantum number and a given orbital angular momentum quantum number.

electron spectroscopy   [SPECT] The study of the energy spectra of photoelectrons or Auger electrons emitted from a substance upon bombardment by electromagnetic radiation, electrons, or ions; used to investigate atomic, molecular, or solid-state structure, and in chemical analysis.

electron spectroscopy for chemical analysis   See x-ray photoelectron spectroscopy.

electron spin   [QUANT MECH] That property of an electron which gives rise to its angular momentum about an axis within the electron.

electron spin density   [PHYS] The vector sum of the spin angular momenta of electrons at each point in a substance per unit volume.

electron spin resonance   See electron paramagnetic resonance.

electron synchrotron   [NUCLEO] A circular electron accelerator in which the frequency of the accelerating system is constant, the strength of the magnetic guide field increases, and the electrons move in orbits of nearly constant radius.

electron transfer   [PHYS] The passage of an electron from one constituent of a system to another.

electron transport system   [BIOCHEM] The components of the final sequence of reactions in biological oxidations; composed of a series of oxidizing agents arranged in order of increasing strength and terminating in oxygen.

electron trap   [SOLID STATE] A defect or chemical impurity in a semiconductor or insulator which captures mobile electrons in a special way.

electron tube   [ELECTR] An electron device in which conduction of electricity is provided by electrons moving through a vacuum or gaseous medium within a gastight envelope. Also known as radio tube; tube; valve (British usage).

electron-tube heater   See heater.

electron tunneling   [QUANT MECH] The passage of electrons through a potential barrier which they would not be able to cross according to classical mechanics, such as a thin insulating barrier between two superconductors.

electronvolt   [PHYS] A unit of energy which is equal to the energy acquired by an electron when it passes through a potential difference of 1 volt in a vacuum; it is equal to $(1.602192 \pm 0.000007) \times 10^{-19}$ volt. Abbreviated eV.

electron wave function   [QUANT MECH] Function of the spin orientation and position of one or more electrons, specifying the dynamical state of the electrons; the square of the function's modulus gives the probability per unit volume of finding electrons at a given position.

electron wavelength   [QUANT MECH] The de Broglie wavelength of an electron, given by Planck's constant divided by the momentum.

electrooptical birefringence   See electrooptical Kerr effect.

electrooptical character recognition   See optical character recognition.

electrooptical Kerr effect   [OPTICS] Birefringence induced by an electric field. Also known as electrooptical birefringence; Kerr effect.

electrooptical modulator   [COMMUN] An optical modulator in which a Kerr cell, an electrooptical crystal, or other signal-controlled electrooptical device is used to modulate the amplitude, phase, frequency, or direction of a light beam.

electrooptical material   [OPTICS] A material in which the indices of refraction are changed by an applied electric field.

electrooptics   [OPTICS] The study of the influence of an electric field on optical phenomena, as in the electrooptical Kerr effect and the Stark effect. Also known as optoelectronics.

electroosmosis   [PHYS CHEM] The movement in an electric field of liquid with respect to colloidal particles immobilized in a porous diaphragm or a single capillary tube.

electroosmotic driver   [ELECTR] A type of solion for converting voltage into fluid pressure, which uses depolarizing electrodes sealed in an electrolyte and operates through the streaming potential effect. Also known as micropump.

electropherography   See electrochromatography.

electrophilic   [PHYS CHEM] Any chemical process in which electrons are acquired from or shared with other molecules or ions.

electrophoresis   [PHYS CHEM] An electrochemical process in which colloidal particles or macromolecules with a net electric charge migrate in a solution under the influence of an electric current. Also known as cataphoresis.

electrophorus   [ELEC] A device used to produce electric charges; it consists of a hard-rubber disk, which is negatively charged by rubbing with fur, and a metal plate, held by an insulating handle, which is placed on the disk; the plate is then touched with a grounded conductor, so that negative charge is removed and the plate has net positive charge.

electrophotography   [GRAPHICS] An electrostatic image-forming process in which light, x-rays, or gamma rays form an electrostatic image on a photoconductive, insulating medium; the charged image areas attract and hold a fine powder called a toner, and the powder image is then transferred to paper or fused there by heat.

electrophysiology   [PHYSIO] The branch of physiology concerned with determining the basic mechanisms by which electric currents are generated within living organisms.

electroplating   [MET] Electrodeposition of a metal or alloy from a suitable electrolyte solution; the article to be plated is connected as the cathode in the electrolyte solution; direct current is introduced through the anode which consists of the metal to be deposited.

electroplax   [VERT ZOO] One of the structural units of an electric organ of some fishes, composed of thin, flattened plates of modified muscle that appear as two large, waferlike, roughly circular or rectangular surfaces.

electropolishing   [MET] Smoothing and enhancing the appearance of a metal surface by making it an anode in a suitable electrolyte. Also known as electrolytic brightening; electrolytic polishing.

electropositive   [ELEC] 1. Carrying a positive electric charge. 2. Capable of acting as the positive electrode in an electric cell.   [PHYS CHEM] Pertaining to elements, ions, or radicals that tend to give up or lose electrons.

electropulse engine   [AERO ENG] An engine, for propelling a flight vehicle, that is based on the use of spark discharges through which intense electric and magnetic fields are established for periods ranging from microseconds to a few milliseconds; a resulting electromagnetic force drives the plasma along the leads and away from the spark gap.

electrorefining   [CHEM ENG] Petroleum refinery process for light hydrocarbon streams in which an electrostatic field is used to assist in separation of chemical treating agents (acid, caustic, doctor) from the hydrocarbon phase.   [MET] Purifying metals by electrolysis using an impure metal as anode from which the pure metal is dissolved and subsequently deposited at the cathode. Also known as electrolytic refining.

electroreflectance   [SPECT] Electromodulation in which reflection spectra are studied. Abbreviated ER.

electroretinogram   [MED] A graphic recording of the electric discharges of the retina. Abbreviated ERG.

electroscope   [ENG] An instrument for detecting an electric charge by means of the mechanical forces exerted between electrically charged bodies.

electroshock therapy   [MED] Treatment of mental patients by passing an electric current of 85–110 volts through the brain.

electroslag welding   [MET] A welding process in which consumable electrodes are fed into a joint containing flux; the current melts the flux, and the flux in turn melts the faces of the joint and the electrodes, allowing the weld metal to form a continuously cast ingot between the joint faces.

electrospark machining   See electron discharge machining.

electrostatic   [ELEC] Pertaining to electricity at rest, such as an electric charge on an object.

electrostatic accelerator   [ELECTR] Any instrument which uses an electrostatic field to accelerate charged particles to high velocities in a vacuum.

electrostatic actuator   See actuator.

electrostatic analyzer   [ELECTR] A device which filters an electron beam, permitting only electrons within a very narrow velocity range to pass through.

electrostatic bond   [PHYS CHEM] A valence bond in which two atoms are kept together by electrostatic forces caused by transferring one or more electrons from one atom to the other.

electrostatic copying   See electrostatography.

electrostatic error   See antenna effect.

electrostatic field   [ELEC] A time-independent electric field, such as that produced by stationary charges.

electrostatic force   [ELEC] Force on a charged particle due to an electrostatic field, equal to the electric field vector times the charge of the particle.

electrostatic generator   [ELEC] Any machine which produces electric charges by friction or (more commonly) electrostatic induction.

electrostatic induction   [ELEC] The process of charging an object electrically by bringing it near another charged object, then touching it to ground. Also known as induction.

electrostatic lens   [ELECTR] An arrangement of electrostatic fields which acts upon beams of charged particles similar to the way a glass lens acts on light beams.

electrostatic loudspeaker   [ENG ACOUS] A loudspeaker in which the mechanical forces are produced by the action of electrostatic fields; in one type the fields are produced between a thin metal diaphragm and a rigid metal plate. Also known as capacitor loudspeaker.

electrostatic memory   *See* electrostatic storage.

electrostatic microphone   *See* capacitor microphone.

electrostatic octupole lens   [ELECTR] A device for controlling beams of electrons or other charged particles, consisting of eight electrodes arranged in a circular pattern with alternating polarities; commonly used to correct aberrations of quadrupole lens systems.

electrostatic potential   *See* electric potential.

electrostatic precipitator   [ENG] A device which removes dust or other finely divided particles from a gas by charging the particles inductively with an electric field, then attracting them to highly charged collector plates. Also known as precipitator.

electrostatic printer   [GRAPHICS] A line printer in which high-intensity lamps project images of characters onto a sensitized drum to form electrostatic patterns that attract ink powder; the images are then transferred to paper and fused.

electrostatic probe   *See* electric probe.

electrostatic quadrupole lens   [ELECTR] A device for focusing beams of electrons or other charged particles, consisting of four electrodes arranged in a circular pattern with alternating polarities.

electrostatics   [ELEC] The study of electric charges at rest, their electric fields, and potentials.

electrostatic separator   [ENG] A separator in which a finely pulverized mixture falls through a powerful electric field between two electrodes; materials having different specific inductive capacitances are deflected by varying amounts and fall into different sorting chutes.

electrostatic storage   [ELECTR] A storage in which information is retained as the presence or absence of electrostatic charges at specific spot locations, generally on the screen of a special type of cathode-ray tube known as a storage tube. Also known as electrostatic memory.

electrostatic storage tube   *See* storage tube.

electrostatic transducer   [ENG ACOUS] A transducer consisting of a fixed electrode and a movable electrode, charged electrostatically in opposite polarity; motion of the movable electrode changes the capacitance between the electrodes and thereby makes the applied voltage change in proportion to the amplitude of the electrode's motion. Also known as condenser transducer.

electrostatic tweeter   [ENG ACOUS] A tweeter loudspeaker in which a flat metal diaphragm is driven directly by a varying high voltage applied between the diaphragm and a fixed metal electrode.

electrostatic units   [ELEC] A centimeter-gram-second system of electric and magnetic units in which the unit of charge is that charge which exerts a force of 1 dyne on another unit charge when separated from it by a distance of 1 centimeter in vacuum; other units are derived from this definition by assigning unit coefficients in equations relating electric and magnetic quantities. Abbreviated esu.

electrostatography   [GRAPHICS] A generic term covering all processes involving the forming and use of electrostatic charged patterns for recording and reproducing images; the field is divided into electrophotography and electrography. Also known as electrostatic copying.

electrostriction   [MECH] A form of elastic deformation of a dielectric induced by an electric field, associated with those components of strain which are independent of reversal of field direction, in contrast to the piezoelectric effect. Also known as electrostrictive strain.

electrostriction transducer   [ENG ACOUS] A transducer which depends on the production of an elastic strain in certain symmetric crystals when an electric field is applied, or, conversely, which produces a voltage when the crystal is deformed. Also known as ceramic transducer.

electrostrictive strain   *See* electrostriction.

electrosurgery   [MED] The use of electricity to perform surgical procedures, as the use of electricity to simultaneously cut tissue and arrest bleeding.

electrotherapy   [MED] The therapeutic use of electricity.

electrothermal   [PHYS] 1. Pertaining to both heat and electricity. 2. In particular, pertaining to conversion of electrical energy into heat energy.

electrothermal ammeter   *See* thermoammeter.

electrotonus   [PHYSIO] The change of condition in a nerve or a muscle during the passage of a current of electricity.

electrotype   [GRAPHICS] A duplicate printing surface prepared by making a mold of the type page or half-tone plate, then suspending this mold in a bath of copper sulfate and sulfuric acid where, by electrolytic action, a thin shell of copper is deposited on it, and finally pouring molten type metal into this shell to strengthen it for use on the press.

electrovalence   [PHYS CHEM] The valence of an atom that has formed an ionic bond.

electrovalent bond   *See* ionic bond.

electrowinning   [MET] Extracting metal from solutions by electrochemical processes.

element   [CHEM] A substance made up of atoms with the same atomic number; common examples are hydrogen, gold, and iron. Also known as chemical element.   [COMPUT SCI] A circuit or device performing some specific elementary data-processing function.   [ELEC] 1. A part of an electron tube, semiconductor device, or antenna array that contributes directly to the electrical performance. 2. *See* component.   [ELECTROMAG] Radiator, active or parasitic, that is a part of an antenna.   [IND ENG] A brief, relatively homogeneous part of a work cycle that can be described and identified.   [MATH] 1. In an array such as a matrix or determinant, a quantity identified by the intersection of a given row or column. 2. In network topology, an edge.

element 104   [CHEM] The first element beyond the actinide series, and the twelfth transuranium element; the atoms of element 104, of mass number 260, were first produced by irradiating plutonium-242 with neon-22 ions in a heavy-ion cyclotron.

element 105   [CHEM] An artificial element whose isotope of mass number 260 was discovered by bombarding californium-249 with nitrogen-15 ions in a heavy-ion linear accelerator.

element 106   [CHEM] An artificial element whose isotope of mass number 263 was discovered by bombarding californium-249 with oxygen-18 ions in a heavy-ion linear accelerator, and whose isotope of mass number 259 was discovered by bombarding lead-207 and lead-208 with chromium-54 ions in a heavy-ion cyclotron.

element 107   [CHEM] An artificial element whose isotope of mass number 261 has been tentatively identified as a reaction product in the bombardment of bismuth-209 with chromium-54 ions and lead-208 with manganese-55 ions in a heavy-ion cyclotron.

elemental area   *See* picture element.

elementary charge [PHYS] An electric charge such that the electric charge of any body is an integral multiple of it, equal to the electron charge.

elementary excitation [QUANT MECH] The quantum of energy of some vibration or wave, such as a photon, phonon, plasmon, magnon, polaron, or exciton.

elementary item [COMPUT SCI] An item considered to have no subordinate item in the COBOL language.

elementary particle [PARTIC PHYS] A particle which, in the present state of knowledge, cannot be described as compound, and is thus one of the fundamental constituents of all matter. Also known as fundamental particle; particle.

elements of an orbit [ASTRON] A set of quantities specifying the orbit of a member of the solar system or of a binary star system, used to calculate the body's position at any time.

eleolite *See* nepheline.

elephant [METEOROL] *See* elephanta. [VERT ZOO] The common name for two living species of proboscidean mammals in the family Elephantidae; distinguished by the elongation of the nostrils and upper lip into a sensitive, prehensile proboscis.

elephanta [METEOROL] A strong southeasterly wind on the Malabar coast of southwest India in September and October, at the end of the southwest monsoon, bringing thundersqualls and heavy rain. Also known as elephant; elephanter.

elephanter *See* elephanta.

elephantiasis [MED] A parasitic disease of man caused by the filarial nematode *Wuchereria bancrofti;* characterized by cutaneous and subcutaneous tissue enlargement due to lymphatic obstruction.

Elephantidae [VERT ZOO] A family of mammals in the order Proboscidea containing the modern elephants and extinct mammoths.

elevating machine *See* elevator.

elevation [ENG] Vertical distance to a point or object from sea level or some other datum. [GRAPHICS] A graphic projection of a machine or structure on a vertical plane without perspective. [ORD] In antiaircraft artillery, a term sometimes applied to the angular height.

elevation angle [ELECTROMAG] The angle that a radio, radar, or other such beam makes with the horizontal. [ENG] *See* angle of elevation.

elevator [AERO ENG] The hinged rear portion of the longitudinal stabilizing surface or tail plane of an aircraft, used to obtain longitudinal or pitch-control moments. [MECH ENG] Also known as elevating machine. **1.** Vertical, continuous-belt, or chain device with closely spaced buckets, scoops, arms, or trays to lift or elevate powders, granules, or solid objects to a higher level. **2.** Pneumatic device in which air or gas is used to elevate finely powdered materials through a closed conduit. **3.** An enclosed platform or car that moves up and down in a shaft for transporting people or materials. Also known as lift. [PETRO ENG] A clamp gripping a stand or column of casing tubing, drill pipe, or sucker rods so that it can be moved up or down in a borehole being drilled.

eleven punch *See* X punch.

elevon [AERO ENG] The hinged rear portion of an aircraft wing, moved in the same direction on each side of the aircraft to obtain longitudinal control and differentially to obtain lateral control; elevon is a combination of the words elevator and aileron to denote that an elevon combines the functions of aircraft elevators and ailerons.

ELF *See* extremely low frequency.

elimination [MATH] A process of deriving from a system of equations a new system with fewer variables, but with precisely the same solutions.

elimination factor [COMPUT SCI] In information retrieval, the ratio obtained in dividing the number of documents that have not been retrieved by the total number of documents in the file.

elimination reaction [ORG CHEM] A chemical reaction involving elimination of some portion of a reactant compound, with the production of a second compound.

elint *See* electronic intelligence.

elixir [PHARM] A sweetened, aromatic solution, usually hydroalcoholic, sometimes containing soluble medicants; intended for use only as a flavor or vehicle.

elk [VERT ZOO] *Alces alces.* A mammal (family Cervidae) in Europe and Asia that resembles the North American moose but is smaller; it is the largest living deer.

ell [BUILD] A wing built perpendicular to the main section of a building.

ellestadite [MINERAL] A pale rose, hexagonal mineral consisting of an apatite-like calcium sulfate-silicate; occurs in granular massive form.

ellipse [MATH] The locus of all points in the plane at which the sum of the distances from a fixed pair of points, the foci, is a given constant.

ellipsoid [MATH] A surface whose intersection with every plane is an ellipse (or circle).

ellipsoidal coordinates [MATH] Coordinates in space determined by confocal quadrics.

ellipsoid of revolution [MATH] An ellipsoid generated by rotation of an ellipse about one of its axes. Also known as spheroid.

ellipsoid of wave normals *See* index ellipsoid.

ellipsometer [OPTICS] An instrument for determining the degree of ellipticity of polarized light; used to measure the thickness of very thin transparent films by observing light reflected from the film.

elliptical galaxy [ASTRON] A galaxy whose overall shape ranges from a spheroid to an ellipsoid, without any noticeable structural features. Also known as spheroidal galaxy.

elliptical orbit [MECH] The path of a body moving along an ellipse, such as that described by either of two bodies revolving under their mutual gravitational attraction but otherwise undisturbed.

elliptical polarization [ELECTROMAG] Polarization of an electromagnetic wave in which the electric field vector at any point in space describes an ellipse in a plane perpendicular to the propagation direction.

elliptical projection [MAP] A map of the surface of the earth formed on an ellipse's interior.

elliptical system [ENG] A tracking or navigation system where ellipsoids of position are determined from time or phase summation relative to two or more fixed stations which are the focuses for the ellipsoids.

elliptic coordinates [MATH] The coordinates of a point in the plane determined by confocal ellipses and hyperbolas.

elliptic function [MATH] An inverse function of an elliptic integral; alternatively, a doubly periodic, meromorphic function of a complex variable.

elliptic geometry [MATH] The geometry obtained from euclidean geometry by replacing the parallel line postulate with the postulate that infinitely many lines may be drawn through a given point, parallel to a given line. Also known as Riemannian geometry.

elliptic integral [MATH] An integral over x whose integrand is a rational function of x and the square root of $p(x)$, where $p(x)$ is a third- or fourth-degree polynomial without multiple roots.

ellipticity See axial ratio.

ellsworthite [MINERAL] $(Ca,Na,U)_2(Nb,Ta)_2O_6$-(O,OH) A yellow, brown, greenish or black mineral of the pyrochlore group occurring in isometric crystals and consisting of an oxide of niobium, titanium, and uranium. Also known as betafite; hatchettolite.

elm [BOT] The common name for hardwood trees composing the genus *Ulmus*, characterized by simple, serrate, deciduous leaves.

Elmidae [INV ZOO] The drive beetles, a small family of coleopteran insects in the superfamily Dryopoidea.

El Niño [OCEANOGR] A warm current setting south along the coast of Peru; generally develops during February and March concurrently with a southerly shift in the tropical rain belt.

elongation [ASTRON] The difference between the celestial longitude of the moon or a planet, as measured from the earth, and that of the sun. [COMMUN] The extension of the envelope of a signal due to delayed arrival of multipath components. [MECH] The fractional increase in a material's length due to stress in tension or to thermal expansion.

Elopidae [VERT ZOO] A family of fishes in the order Elopiformes, including the tarpon, ladyfish, and machete.

Elopiformes [VERT ZOO] A primitive order of actinopterygian fishes characterized by a single dorsal fin composed of soft rays only, cycloid scales, and toothed maxillae.

elpasolite [MINERAL] K_2NaAlF_6 Mineral composed of sodium potassium aluminum fluoride.

elpidite [MINERAL] $Na_2ZrSi_6O_{15} \cdot 3H_2O$ A white to brick-red mineral composed of hydrated sodium zirconium silicate.

eluant [CHEM] A liquid used to extract one material from another, as in chromatography.

eluate [CHEM] The solution that results from the elution process.

elution [CHEM] The removal of adsorbed species from a porous bed or chromatographic column by means of a stream of liquid or gas.

elutriate [ENG] To separate or purify; for example, to separate ore by washing, decanting, and settling.

elutriation [CHEM ENG] The process of removing substances from a mixture through washing and decanting. [ENG] In a mixture, the separation of finer lighter particles from coarser heavier particles through a slow stream of fluid moving upward so that the lighter particles are carried with it. [GEOL] The washing away of the lighter or finer particles in a soil, especially by the action of raindrops.

elutriator [ENG] An apparatus used to separate suspended solid particles according to size by the process of elutriation.

eluvial [GEOL] Of, composed of, or relating to eluvium.

eluviation [HYD] The process of transporting dissolved or suspended materials in the soil by lateral or downward water flow when rainfall exceeds evaporation.

eluvium [GEOL] Disintegrated rock material formed and accumulated in situ or moved by the wind alone.

elytron [INV ZOO] 1. One of the two sclerotized or leathery anterior wings of beetles which serve to cover and protect the membranous hindwings. 2. A dorsal scale of certain Polychaeta.

em [GRAPHICS] A unit of linear measurement used in printing which is equal to the point size of the type; for example, an em in 6-point type is 6 points wide.

Emballonuridae [VERT ZOO] The sheath-tailed bats, a family of mammals in the order Chiroptera.

embankment [CIV ENG] 1. A ridge constructed of earth, stone, or other material to carry a roadway or railroad at a level above that of the surrounding terrain. 2. A ridge of earth or stone to prevent water from passing beyond desirable limits. Also known as bank.

embayed [GEOGR] Formed into a bay. [NAV] Pertaining to a vessel in a bay unable to put to sea or to put to sea safely because of wind, current, or sea.

embayment [GEOGR] Indentation in a shoreline forming a bay. [GEOL] 1. Act or process of forming a bay. 2. A reentrant of sedimentary rock into a crystalline massif.

embed Also spelled imbed. [BIOL] To prepare a specimen for sectioning for microscopic examination by infiltrating with or enclosing in paraffin or other supporting material. [SCI TECH] 1. To enclose in a matrix. 2. To closely surround.

Embiidina [INV ZOO] An equivalent name for Embioptera.

Embioptera [INV ZOO] An order of silk-spinning, orthopteroid insects resembling the grasshoppers; commonly called the embiids or web spinners.

Embiotocidae [VERT ZOO] The surfperches, a family of perciform fishes in the suborder Percoidei.

embolism [MED] The blocking of a blood vessel by an embolus.

embolite [MINERAL] $Ag(Cl,Br)$ A yellow-green mineral resembling cerargyrite; composed of native silver chloride and silver bromide.

Embolomeri [PALEON] An extinct side branch of slender-bodied, fish-eating aquatic anthracosaurs in which intercentra as well as centra form complete rings.

embolus [MED] A clot or other mass of particulate matter foreign to the bloodstream which lodges in a blood vessel and causes obstruction.

embrasure [ARCH] 1. Opening in a wall or parapet, especially one through which a gun is fired. 2. An opening such as for a door or window with sloping or beveled sides.

Embrithopoda [PALEON] An order established for the unique Oligocene mammal *Arsinoitherium*, a herbivorous animal that resembled the modern rhinoceros.

embrittlement [MECH] Reduction or loss of ductility or toughness in a metal or plastic with little change in other mechanical properties.

embryo [BOT] The young sporophyte of a seed plant. [EMBRYO] 1. An early stage of development in multicellular organisms. 2. The product of conception up to the third month of human pregnancy.

Embryobionta [BOT] The land plants, a subkingdom of the Plantae characterized by having specialized conducting tissue in the sporophyte (except bryophytes), having multicellular sex organs, and producing an embryo.

embryogenesis [EMBRYO] The formation and development of an embryo from an egg.

embryology [BIOL] The study of the development of the organism from the zygote, or fertilized egg.

embryonal-cell lipoma See liposarcoma.

embryonated egg culture [VIROL] Embryonated hen's eggs inoculated with animal viruses for the purpose of identification, isolation, titration, or for quantity cultivation in the production of viral vaccines.

embryonic differentiation [EMBRYO] The process by which specialized and diversified structures arise during embryogenesis.

embryonic inducer [EMBRYO] The acting system in embryos, which contributes to the formation of specialized tissues by controlling the mode of development of the reacting system.

embryonic induction [EMBRYO] The influence of one cell group (inducer) over a neighboring cell group (induced) during embryogenesis. Also known as induction.

Embryophyta [BOT] The equivalent name for Embryobionta.

embryo sac [BOT] The female gametophyte of a seed plant, containing the egg, synergids, and polar and antipodal nuclei; fusion of the antipodals and a pollen generative nucleus forms the endosperm.

emerald [MINERAL] $Al_2(Be_3Si_6O_{18})$ A brilliant-green to grass-green gem variety of beryl that crystallizes in the hexagonal system; green color is caused by varying amounts of chromium. Also known as smaragd.

emergence [GEOL] **1.** Dry land which was part of the ocean floor. **2.** The act or process of becoming an emergent land mass. [HYD] See resurgence.

emergency broadcast system [COMMUN] A system of broadcast stations and interconnecting facilities authorized by the U.S. Federal Communications Commission to operate in a controlled manner during a war, threat of war, state of public peril or disaster, or other national emergency.

emergency power supply [ELEC] A source of power that becomes available, usually automatically, when normal power line service fails.

emergency recorder plot [NAV] An emergency substitute for a tactical range recorder, consisting of a plastic sheet representing space and time coordinates on which speed and distance data are marked with a grease pencil as they are called off verbally. Abbreviated ERP.

emersion [ASTRON] The reappearance of a celestial body after an eclipse or occultation.

emery [MATER] An abrasive which is composed of pulverized, impure corundum; used in polishing and grinding. [MINERAL] A fine, granular, gray-black, impure variety of corundum containing iron oxides, either hematite or magnetite; occurs as masses in limestone and as segregations in igneous rock.

emesis [MED] The act of vomiting.

emetic [PHARM] Any agent that induces emesis.

emf See electromotive force.

emi See electromagnetic interference.

emission [ELECTROMAG] Any radiation of energy by means of electromagnetic waves, as from a radio transmitter.

emission electron microscope [ELECTR] An electron microscope in which thermionic, photo, secondary, or field electrons emitted from a metal surface are projected on a fluorescent screen, with or without focusing.

emission flame photometry [ANALY CHEM] A form of flame photometry in which a sample solution to be analyzed is aspirated into a hydrogen-oxygen or acetylene-oxygen flame; the line emission spectrum is formed, and the line or band of interest is isolated with a monochromator and its intensity measured photoelectrically.

emission lines [SPECT] Spectral lines resulting from emission of electromagnetic radiation by atoms, ions, or molecules during changes from excited states to states of lower energy.

emission nebula [ASTRON] A type of bright diffuse nebula whose luminosity results from the excitation and ionization of its gas atoms by ultraviolet radiation from a nearby O- or B-type star.

emission spectrometer [SPECT] A spectrometer that measures percent concentrations of preselected elements in samples of metals and other materials; when the sample is vaporized by an electric spark or arc, the characteristic wavelengths of light emitted by each element are measured with a diffraction grating and an array of photodetectors.

emission spectrum [SPECT] Electromagnetic spectrum produced when radiations from any emitting source, excited by any of various forms of energy, are dispersed.

emissive power See emittance.

emissivity [THERMO] The ratio of the radiation emitted by a surface to the radiation emitted by a perfect blackbody radiator at the same temperature. Also known as thermal emissivity.

emittance [THERMO] The power radiated per unit area of a radiating surface. Also known as emissive power; radiating power.

emitter [COMPUT SCI] A time pulse generator found in some equipment, such as a card punch. [ELECTR] A transistor region from which charge carriers that are minority carriers in the base are injected into the base, thus controlling the current flowing through the collector; corresponds to the cathode of an electron tube. Symbolized E. Also known as emitter region.

emitter barrier [ELECTR] One of the regions in which rectification takes place in a transistor, lying between the emitter region and the base region.

emitter-coupled logic [ELECTR] A form of current-mode logic in which the emitters of two transistors are connected to a single current-carrying resistor in such a way that only one transistor conducts at a time. Abbreviated ECL.

emitter follower [ELECTR] A grounded-collector transistor amplifier which provides less than unity voltage gain but high input resistance and low output resistance, and which is similar to a cathode follower in its operations.

emitter junction [ELECTR] A transistor junction normally biased in the low-resistance direction to inject minority carriers into a base.

emitter pulse [COMPUT SCI] In a punch-card machine, one of a set of pulses associated with a particular row of punch positions on a punch card.

emitter region See emitter.

emmonsite [MINERAL] $FE_2Te_3O_9 \cdot 2H_2O$ Yellow-green mineral composed of a hydrous oxide of iron and tellurium.

E mode See transverse magnetic mode.

empennage [AERO ENG] The assembly at the rear of an aircraft; it comprises the horizontal and vertical stabilizers. Also known as tail assembly.

Empididae [INV ZOO] The dance flies, a family of orthorrhaphous dipteran insects in the series Nematocera.

empirical [SCI TECH] Based on actual measurement, observation, or experience, rather than on theory.

empirical curve [MATH] A smooth curve drawn through or close to points representing measured values of two variables on a graph.

empirical formula [CHEM] A chemical formula indicating the variety and relative proportions of the atoms in a molecule but not showing the manner in which they are linked together.

empirical probability [STAT] The ratio of the number of times an event has occurred to the total number of trials performed. Also known as a posteriori probability.

empirical rule [SCI TECH] A rule which is derived from measurements or observations, and is not based on any theory.

emplacement [GEOL] Intrusion of igneous rock or development of an ore body in older rocks. [ORD] **1.** Prepared poition for one or more weapons or pieces of equipment, protecting against hostile fire or bombardment but permitting execution of the mission of the weapons. **2.** Act of fixing a gun in a prepared position from which it may be fired.

empty medium [COMPUT SCI] A material which has been prepared to have data recorded on it by the entry of some preliminary data, such as feed holes punched on a paper tape or header labels written on a magnetic tape; in contrast to a virgin medium.

empyema [MED] The presence of pus in the body cavity, hollow organ, or tissue space; when the term is used without qualification, it generally refers to pus in the pleural space.

emu [ELECTROMAG] *See* electromagnetic system of units. [VERT ZOO] *Dromiceius novae-hollandiae*. An Australian ratite bird, the second largest living bird, characterized by rudimentary wings and a feathered head and neck without wattles.

emulation mode [COMPUT SCI] A method of operation in which a computer actually executes the instructions of a different (simpler) computer, in contrast to normal mode.

emulator [COMPUT SCI] The microprogram-assisted macroprogram which allows a computer to run programs written for another computer.

emulsification [CHEM] The process of dispersing one liquid in a second immiscible liquid; the largest group of emulsifying agents are soaps, detergents, and other compounds, whose basic structure is a paraffin chain terminating in a polar group.

emulsion [CHEM] A stable dispersion of one liquid in a second immiscible liquid, such as milk (oil dispersed in water). [GRAPHICS] In photography, the photosensitized material on film, plates, and various photographic papers.

emulsion speed [GRAPHICS] Sensitivity of a photographic emulsion to light, under standard conditions of exposure and development.

Emydidae [VERT ZOO] A family of aquatic and semiaquatic turtles in the suborder Cryptodira.

enable [COMPUT SCI] To authorize an activity which would otherwise be suppressed, such as to write on a tape.

enabled instruction [COMPUT SCI] An instruction in a program in data flow language, all of whose input values are present, so that the instruction may be carried out.

Enaliornithidae [PALEON] A family of extinct birds assigned to the order Hesperornithiformes, having well-developed teeth found in grooves in the dentary and maxillary bones of the jaws.

enamel [MATER] **1.** A finely ground, resin-containing oil paint that dries relatively harder, smoother, and glossier than ordinary paint. **2.** *See* glaze.

enamel organ [EMBRYO] The epithelial ingrowth from the dental lamina which covers the dental papilla, furnishes a mold for the shape of a developing tooth, and forms the dental enamel.

enamel paper *See* coated paper.

enantiomer *See* enantiomorph.

enantiomorph [CHEM] One of an isomeric pair of crystalline forms or compounds whose molecules are nonsuperimposable mirror images. Also known as enantiomer; optical antipode; optical isomer.

enantiomorphism [CHEM] A phenomenon of mirror-image relationship exhibited by right-handed and left-handed crystals or by the molecular structures of two stereoisomers.

enantiotopic ligand [ORG CHEM] A ligand whose replacement or addition gives rise to enantiomers.

enantiotropy [CHEM] The relation of crystal forms of the same substance in which one form is stable above the transition-point temperature and the other stable below it, so that the forms can change reversibly one into the other.

Enantiozoa [INV ZOO] The equivalent name for Parazoa.

enargite [MINERAL] A lustrous, grayish-black mineral which is found in orthorhombic crystals but is more commonly columnar, bladed, or massive; hardness is 3 on Mohs scale, specific gravity is 4.44; in some places enargite is a valuable copper ore. Also known as clairite; luzonite.

enarthrosis [ANAT] A freely movable joint that allows a wide range of motion on all planes. Also known as ball-and-socket joint.

encapsulate [SCI TECH] To surround, encase, or enclose as if in a capsule; for example, the formation of a protective coating around a bacterium, or the enclosure of an item such as an electronic component in plastic.

encephalitis [MED] Inflammation of the brain.

encephalitis lethargica [MED] Epidemic encephalitis, probably of viral etiology, characterized by lethargy, ophthalmoplegia, hyperkinesia, and at times residual neurologic disability, particularly parkinsonism with oculogyric crisis. Also known as epidemic encephalitis; sleeping sickness; von Economo's disease.

encephalogram [MED] A roentgenogram of the brain made in encephalography.

encephalography [MED] Roentgenography of the brain following removal of cerebrospinal fluid, by lumbar or cisternal puncture, and its replacement by air or other gas.

encephalomyelitis [MED] Inflammation of the brain and spinal cord.

enchondroma [MED] A benign tumor composed of dysplastic cartilage cells, occurring in the metaphysis of cylindric bones, especially of the hands and feet.

encipher [COMMUN] To convert a plain-text message into unintelligible language by means of a cryptosystem. Also known as encrypt.

Encke roots [MATH] For any two numbers a_1 and a_2, the numbers $-x_1$ and $-x_2$, where x_1 and x_2 are the roots of the equation $x^2 + a_1x + a_2 = 0$, with $|x_1| < |x_2|$.

Encke's Comet [ASTRON] A very faint comet with the shortest period of any known comet, 3.3 years.

encode [COMMUN] To express given information by means of a code. [COMPUT SCI] To prepare a routine in machine language for a specific computer.

encoder [COMPUT SCI] In character recognition, that class of printer which is usually designed for the specific purpose of printing a particular type font in predetermined positions on certain size forms. [ELECTR] **1.** In an electronic computer: a network or system in which only one input is excited at a time and each input produces a combination of outputs. **2.** *See* matrix.

encoding strip [COMPUT SCI] In character recognition, the area reserved for the inscription of magnetic-ink characters, as in bank checks.

encrypt *See* encipher.

encryption [COMPUT SCI] The coding of a clear text message by a transmitting unit so as to prevent unauthorized eavesdropping along the transmission line; the receiving unit uses the same algorithm as the transmitting unit to decode the incoming message.

Encyrtidae [INV ZOO] A family of hymenopteran insects in the superfamily Chalcidoidea.

encystment [BIOL] The process of forming or becoming enclosed in a cyst or capsule.

end See warp.

end-around carry [COMPUT SCI] A carry from the most significant digit place to the least significant digit place.

end-around shift See cyclic shift.

endarteritis [MED] Inflammation of the lining (tunica intima) of an artery.

end-bearing pile [CIV ENG] A bearing pile that is driven down to hard ground so that it carries the full load at its point. Also known as a point-bearing pile.

end bulb See bouton.

end bulb of Krause See Krause's corpuscle.

end cell [ELEC] One of a group of cells in series with a storage battery, which can be switched in to maintain the output voltage of the battery when it is not being charged.

Endeidae [INV ZOO] A family of marine arthropods in the subphylum Pycnogonida.

endellite [MINERAL] $Al_2Sl_2O_5(OH)_4 \cdot 4H_2O$ Term used in the United States for a clay mineral, the more hydrous form of halloysite. Also known as hydrated halloysite; hydrohalloysite; hydrokaolin.

endemic [MED] Peculiar to a certain region, specifically referring to a disease which occurs more or less constantly in any locality.

endergonic [BIOCHEM] Of or pertaining to a biochemical reaction in which the final products possess more free energy than the starting materials; usually associated with anabolism.

endite [INV ZOO] 1. One of the appendages on the inner aspect of an arthropod limb. 2. A ridgelike chewing surface on the inner part of the pedipalpus or maxilla of many arachnids.

endless loop [COMPUT SCI] A sequence of instructions in a computer program that is repeated over and over without end, due to a mistake in the programming.

end mark [COMPUT SCI] A mark which signals the end of a unit of information.

end member [MINERAL] One of the two or more pure chemical compounds that enters into solid solution with other pure chemical compounds to make up a series of minerals of similar crystal structure (that is, an isomorphous, solid-solution series).

endo- [ORG CHEM] Prefix that denotes inward-directed valence bonds of a six-membered ring in its boat form. [SCI TECH] Prefix denoting within or inside.

endocarditis [MED] Inflammation of the endocardium.

endocardium [ANAT] The membrane lining the heart.

endocarp [BOT] The inner layer of the wall of a fruit or pericarp.

endochondral ossification [PHYSIO] The conversion of cartilage into bone. Also known as intracartilaginous ossification.

endocommensal [ECOL] A commensal that lives within the body of its host.

endocranium [ANAT] 1. The inner surface of the cranium. 2. See dura mater. [INV ZOO] The processes on the inner surface of the head capsule of certain insects.

endocrine gland [PHYSIO] A ductless structure whose secretion (hormone) is passed into adjacent tissue and then to the bloodstream either directly or by way of the lymphatics. Also known as ductless gland.

endocrine system [PHYSIO] The chemical coordinating system in animals, that is, the endocrine glands that produce hormones.

endocrinology [PHYSIO] The study of the endocrine glands and the hormones that they synthesize and secrete.

endocytosis [CYTOL] 1. An active process in which extracellular materials are introduced into the cytoplasm of cells by either phagocytosis or pinocytosis. 2. The process by which animal cells internalize large molecules and large collections of fluid.

endoderm [EMBRYO] The inner, primary germ layer of an animal embryo; sometimes referred to as the hypoblast. Also known as entoderm; hypoblast.

endodermis [BOT] The innermost tissue of the cortex of most plant roots and certain stems consisting of a single layer of at least partly suberized or cutinized cells; functions to control the movement of water and other substances into and out of the stele.

endoenzyme [BIOCHEM] An intracellular enzyme, retained and utilized by the secreting cell.

endoergic collision See collision of the first kind.

end-of-data-mark [COMPUT SCI] A character or word signaling the end of all data held in a particular storage unit.

end-of-field-mark [COMPUT SCI] A data item signaling the end of a field of data, generally a variable-length field.

end of file [COMPUT SCI] 1. Termination or point of completion of a quantity of data; end of file marks are used to indicate this point. 2. Automatic procedures to handle tapes when the end of an input or output tape is reached; a reflective spot, called a record mark, is placed on the physical end of the tape to signal the end.

end-of-file gap [COMPUT SCI] A gap of precise dimension to indicate the end of a file on tape. Abbreviated EOF gap.

end-of-file indicator [COMPUT SCI] 1. A device that indicates the end of a file on tape. 2. See end-of-file mark.

end-of-file mark [COMPUT SCI] A control character which signifies that the last record of a file has been read. Also known as end-of-file indicator.

end-of-file routine [COMPUT SCI] A program which checks that the contents of a file read into the computer were correctly read; may also start the rewind procedure.

end-of-file spot [COMPUT SCI] A reflective piece of tape indicating the end of the tape.

end-of-job control card [COMPUT SCI] The last card in a deck, normally punched with a distinctive code indicating that no additional cards are required for the job.

end-of-record gap [COMPUT SCI] A gap of precise dimension (shorter than the end-of-file gap) which indicates the physical end of a record on a magnetic tape. Abbreviated EOR gap.

end-of-record word [COMPUT SCI] The last word in a record, usually written in a special format that enables identification of the end of the record.

end-of-run routine [COMPUT SCI] A routine that carries out various housekeeping operations such as rewinding tapes and printing control totals before a run is completed.

end-of-tape routine [COMPUT SCI] A program which is brought into play when the end of a tape is reached;

may involve a series of validity checks and initiate the tape rewind.

end-of-transmission card [COMMUN] Last card of each message; used to signal the end of a transmission and contains the same information as the header card, plus additional data for traffic analysis.

endogamy [BIOL] Sexual reproduction between organisms which are closely related. [BOT] Pollination of a flower by another flower of the same plant.

endogenetic See endogenic.

endogenic [GEOL] Of or pertaining to a geologic process, or its resulting feature such as a rock, that originated within the earth. Also known as endogenetic; endogenous.

endogenous [BIOCHEM] Relating to the metabolism of nitrogenous tissue elements. [GEOL] See endogenic. [MED] Pertaining to diseases resulting from internal causes. [PSYCH] Pertaining to mental disorders caused by hereditary or constitutional factors.

endogenous pyrogen [BIOCHEM] A fever-inducing substance (protein) produced by cells of the host body, such as leukocytes and macrophages.

endogenous variables [MATH] In a mathematical model, the dependent variables; their values are to be determined by the solution of the model equations.

endolecithal [INV ZOO] A type of egg found in turbellarians with yolk granules in the cytoplasm of the egg. Also spelled entolecithal.

endolymph [PHYSIO] The lymph fluid found in the membranous labyrinth of the ear.

endometrium [ANAT] The mucous membrane lining the uterus.

endomitosis [CYTOL] Division of the chromosomes without dissolution of the nuclear membrane; results in polyploidy or polyteny.

endomixis [INV ZOO] Periodic division and reorganization of the nucleus in certain ciliated protozoans.

endomorphism [MATH] A function from a set with some structure (such as a group, ring, vector space, or topological space) to itself which preserves this structure.

Endomycetales [MICROBIO] Former designation for Saccharomycetales.

Endomycetoideae [MICROBIO] A subfamily of ascosporogenous yeasts in the family Saccharomycetaceae.

Endomychidae [INV ZOO] The handsome fungus beetles, a family of coleopteran insects in the superfamily Cucujoidea.

endomysium [HISTOL] The connective tissue layer surrounding an individual skeletal muscle fiber.

endoneural fibroma See neurofibroma.

endonuclease [BIOCHEM] Any of a group of enzymes which degrade deoxyribonucleic acid or ribonucleic acid molecules by attaching nucleotide linkages within the polynucleotide chain.

endoparasite [ECOL] A parasite that lives inside its host.

endoplasmic reticulum [CYTOL] A vacuolar system of the cytoplasm in differentiated cells that functions in protein synthesis and sequestration. Abbreviated ER.

endopleurite [INV ZOO] 1. The portion of a crustacean apodeme which develops from the interepimeral membrane. 2. One of the laterally located parts on the thorax of an insect which fold inward, extending into the body cavity.

endopodite [INV ZOO] The inner branch of a biramous crustacean appendage.

Endoprocta [INV ZOO] The equivalent name for Entoprocta.

Endopterygota [INV ZOO] A division of the insects in the subclass Pterygota, including those orders which undergo a holometabolous metamorphosis.

ENDOR See electron nuclear double resonance.

end organ [ANAT] The expanded termination of a nerve fiber in muscle, skin, mucous membrane, or other structure.

β-endorphin [BIOCHEM] A 31–amino acid peptide fragment of pituitary β-lipotropic hormone having morphinelike activity.

endorser [COMPUT SCI] A special feature available on most magnetic-ink character-recognition readers that imprints a bank's endorsement on successful document reading.

endoscope [MED] An instrument used to visualize the interior of a body cavity or hollow organ.

endosmosis [PHYSIO] The passage of a liquid inward through a cell membrane.

endosperm [BOT] 1. The nutritive protein material within the embryo sac of seed plants. 2. Storage tissue in the seeds of gymnosperms.

endotheliochorial placenta [EMBRYO] A type of placenta in which the maternal blood is separated from the chorion by the maternal capillary endothelium; occurs in dogs.

endothelium [HISTOL] The epithelial layer of cells lining the heart and vessels of the circulatory system.

Endotheriidae [PALEON] A family of Cretaceous insectivores from China belonging to the Proteutheria.

endotherm [PHYS CHEM] In differential thermal analysis, a graph of the temperature difference between a sample compound and a thermally inert reference compound (commonly aluminum oxide) as the substances are simultaneously heated to elevated temperatures at a predetermined rate, and the sample compound undergoes endothermal or exothermal processes. [PHYSIO] An animal that produces enough heat from its own metabolism and employs devices to retard heat loss so that it is able to keep its body temperature higher than that of its environment.

Endothyracea [PALEON] A superfamily of extinct benthic marine foraminiferans in the suborder Fusulinina, having a granular or fibrous wall.

endotoxin [MICROBIO] A toxin that is produced within a microorganism and can be isolated only after the cell is disintegrated.

end point [ANALY CHEM] That stage in the titration at which an effect, such as a color change, occurs, indicating that a desired point in the titration has been reached. [CHEM ENG] In the distillation analysis of crude petroleum and its products, the highest reading of a thermometer when a specified proportion of the liquid has boiled off. Also known as final boiling point.

end product [PHYS] The final product of a chemical or nuclear reaction or process.

end-to-end data system [COMPUT SCI] A comprehensive data system which demonstrates the processing of sensor data to the user, thus reducing data fragmentation.

endurance limit See fatigue limit.

endurance ratio See fatigue ratio.

endurance strength See fatigue strength.

en echelon [GEOL] Referring to an overlapped or staggered arrangement of geologic features.

energized [ELEC] Electrically connected to a voltage source. Also known as alive; hot; live.

energy [PHYS] The capacity for doing work.

energy absorption [PHYS] Conversion of mechanical or radiant energy into the internal potential energy or heat energy of a system.

energy balance [PHYS] The arithmetic balancing of energy inputs versus outputs for an object, reactor, or other processing system; it is positive if energy is released, and negative if it is absorbed. [PHYSIO] The relation of the amount of utilizable energy taken into the body to that which is employed for internal work, external work, and the growth and repair of tissues.

energy conservation See conservation of energy.

energy dissipation See dissipation.

energy eigenstate See energy state.

energy flux [PHYS] A vector quantity whose component perpendicular to any surface equals the energy transported across that surface by some medium per unit area per unit time.

energy gap [SOLID STATE] A range of forbidden energies in the band theory of solids.

energy head [FL MECH] The elevation of the hydraulic grade line at any section of a waterway plus the velocity head of the mean velocity of the water in that section.

energy integral [MECH] A constant of integration resulting from integration of Newton's second law of motion in the case of a conservative force; equal to the sum of the kinetic energy of the particle and the potential energy of the force acting on it.

energy level [GEOL] The kinetic energy supplied by waves or current action in an aqueous sedimentary environment either at the interface of deposition or several meters above. [QUANT MECH] An allowed energy of a physical system; there may be several allowed states at one level.

energy load See dynamic load.

energy metabolism [BIOCHEM] The chemical reactions involved in energy transformations within cells.

energy operator [QUANT MECH] The operator corresponding to the energy or Hamiltonian of a classical system. Also known as Hamiltonian operator.

energy pyramid [ECOL] An ecological pyramid illustrating the energy flow within an ecosystem.

energy spectrum [PHYS] Any plot, display, or photographic record of the intensity of some type of radiation as a function of its energy.

energy state [QUANT MECH] An eigenstate of the energy (Hamiltonian) operator, so that the energy has a definite stationary value. Also known as eigenstate; energy eigenstate; quantum state; stationary state.

engine [MECH ENG] A machine in which power is applied to do work by the conversion of various forms of energy into mechanical force and motion.

engine block See cylinder block.

engine cycle [THERMO] Any series of thermodynamic phases constituting a cycle for the conversion of heat into work; examples are the Otto cycle, Stirling cycle, and Diesel cycle.

engine cylinder [MECH ENG] A cylindrical chamber in an engine in which the energy of the working fluid, in the form of pressure and heat, is converted to mechanical force by performing work on the piston. Also known as cylinder.

engine displacement [MECH ENG] Volume displaced by each piston moving from bottom dead center to top dead center multiplied by the number of cylinders.

engineer [ENG] An individual who specializes in one of the branches of engineering.

engineering [SCI TECH] The science by which the properties of matter and the sources of power in nature are made useful to humans in structures, machines, and products.

engineering geology [CIV ENG] The application of education and experience in geology and other geosciences to solve geological problems posed by civil engineering structures.

engineer's chain [CIV ENG] A surveyor's measuring instrument consisting of 1-foot (30.48-centimeter) steel links joined together by rings, 100 (30.5 meters) or 50 feet long. Also known as chain.

engineer's system of units See British gravitational system of units.

engine knock [MECH ENG] In spark ignition engines, the sound and other effects associated with ignition and rapid combustion of the last part of the charge to burn, before the flame front reaches it. Also known as combustion knock.

engine performance [MECH ENG] Relationship between power output, revolutions per minute, fuel or fluid consumption, and ambient conditions in which an engine operates.

engine starter [ELEC] The electric motor in the electric system of an automobile that cranks the engine for starting. Also known as starter; starting motor.

englishite [MINERAL] $K_2Ca_4Al_8(PO_4)_8(OH)_{10} \cdot 9H_2O$ A white mineral composed of hydrous basic phosphate of potassium, calcium, and aluminum.

engram [PHYSIO] A memory imprint; the alteration that has occurred in nervous tissue as a result of an excitation from a stimulus, which hypothetically accounts for retention of that experience. Also known as memory trace.

Engraulidae [VERT ZOO] The anchovies, a family of herringlike fishes in the suborder Clupoidea.

engraving [GRAPHICS] A photomechanical process in which lines are scribed on line negatives by means of a needle that produces an even transparent printing line by removal of a thin layer of photographic emulsion.

enhancement [ELECTR] An increase in the density of charged carriers in a particular region of a semiconductor.

enhancer gene [GEN] Any modifier gene that acts to enhance the action of a nonallelic gene.

Enicocephalidae [INV ZOO] The gnat bugs, a family of hemipteran insects in the superfamily Enicocephaloidea.

Enicocephaloidea [INV ZOO] A superfamily of the Hemiptera in the subdivision Geocorisae containing a single family.

enigmatite [MINERAL] $Na_2Fe_5TiSi_6O_{20}$ A black amphibole mineral occurring in triclinic crystals; specific gravity is 3.14–3.80. Also spelled aenigmatite.

enkephalin [BIOCHEM] A mixture of two polypeptides isolated from the brain; central mode of action is an inhibition of neurotransmitter release.

enlarger [OPTICS] An optical projector used to project an enlarged image of a photograph's negative onto photosensitized film or paper. Also known as photoenlarger.

enol [ORG CHEM] An organic compound with a hydroxide group adjacent to a double bond; varies with a ketone form in the effect known as enol-keto tautomerism; an example is the compound $CH_3COH=CHCO_2C_2H_5$.

enolase [BIOCHEM] An enzyme that catalyzes the reversible dehydration of phosphoglyceric acid to phosphopyruvic acid.

enol-keto tautomerism [ORG CHEM] The tautomeric migration of a hydrogen atom from an adjacent carbon atom to a carbonyl group of a keto compound to produce the enol form of the compound; the reverse process of hydrogen atom migration also occurs.

Enopla [INV ZOO] A class or subclass of ribbonlike worms of the phylum Rhynchocoela.

Enoplia [INV ZOO] A subclass of nematodes in the class Adenophorea.

Enoplida [INV ZOO] An order of nematodes in the subclass Enoplia.

Enoplidae [INV ZOO] A family of free-living marine nematodes in the superfamily Enoploidea, characterized by a complex arrangement of teeth and mandibles.

Enoploidea [INV ZOO] A superfamily of small to very large free-living marine nematodes having pocketlike amphids opening to the exterior via slitlike apertures.

Enoploteuthidae [INV ZOO] A molluscan family of deep-sea squids in the class Cephalopoda.

enriched material [NUC ENG] Material in which the amount of one or more isotopes has been increased above that occurring in nature, such as uranium in which the abundance of ^{235}U is increased.

enrichment [NUCLEO] A process that changes the isotopic ratio in a material; for uranium, for example, the ratio of ^{235}U to ^{238}U may be increased by gaseous diffusion of uranium hexafluoride.

enrichment culture [MICROBIO] A medium of known composition and specific conditions of incubation which favors the growth of a particular type or species of bacterium.

enrichment factor [NUCLEO] The ratio of the abundance of a particular isotope in an enriched material to its abundance in the original material.

Enskog theory See Chapman-Enskog theory.

enstatite [MINERAL] MgOSiO$_2$ A member of the pyroxene mineral group that crystallizes in the orthorhombic system; usually yellowish gray but becomes green when a little iron is present.

Enteletacea [PALEON] A group of extinct articulate brachiopods in the order Orthida.

Entelodontidae [PALEON] A family of extinct palaeodont artiodactyls in the superfamily Entelodontoidea.

Entelodontoidea [PALEON] A superfamily of extinct piglike mammals in the suborder Palaeodonta having huge skulls and enlarged incisors.

enteric bacilli [MICROBIO] Microorganisms, especially the gram-negative rods, found in the intestinal tract of man and animals.

Enterobacteriaceae [MICROBIO] A family of gram-negative, facultatively anaerobic rods; cells are nonsporeforming and may be nonmotile or motile with peritrichous flagella; includes important human and plant pathogens.

enterocolitis [MED] Inflammation of the small intestine and colon.

Enteropneusta [INV ZOO] The acorn worms or tongue worms, a class of the Hemichordata; free-living solitary animals with no exoskeleton and with numerous gill slits and a straight gut.

enterovirus [VIROL] One of the two subgroups of human picornaviruses; includes the polioviruses, the coxsackieviruses, and the echoviruses.

Enterozoa [ZOO] Animals with a digestive tract or cavity; includes all animals except Protozoa, Mesozoa, and Parazoa.

enthalpimetric analysis [ANALY CHEM] Generic designation for a group of modern thermochemical methodologies such as thermometric enthalpy titrations which rely on monitoring the temperature changes produced in adiabatic calorimeters by heats of reaction occurring in solution; in contradistinction, classical methods of thermoanalysis such as thermogravimetry focus primarily on changes occurring in solid samples in response to externally imposed programmed alterations in temperature.

enthalpy [THERMO] The sum of the internal energy of a system plus the product of the system's volume multiplied by the pressure exerted on the system by its surroundings. Also known as heat content; sensible heat; total heat.

enthalpy of vaporization See heat of vaporization.

enthalpy titration See thermometric titration.

entire function [MATH] A function of a complex variable which is analytic throughout the entire complex plane. Also known as integral function.

entire ring See integral domain.

entire series [MATH] A power series which converges for all values of its variable; a power series with an infinite radius of convergence.

Entisol [GEOL] An order of soil having few or faint horizons.

entoderm See endoderm.

Entodiniomorphida [INV ZOO] An order of highly evolved ciliated protozoans in the subclass Spirotrichia, characterized by a smooth, firm pellicle and the lack of external ciliature.

entolecithal See endolecithal.

Entomoconchacea [PALEON] A superfamily of extinct marine ostracods in the suborder Myodocopa that are without a rostrum above the permanent aperture.

entomology [INV ZOO] A branch of the biological sciences that deals with the study of insects.

entomophagous [ZOO] Feeding on insects.

Entomophthoraceae [MYCOL] The single family of the order Entomophthorales.

Entomophthorales [MYCOL] An order of mainly terrestrial fungi in the class Phycomycetes having a hyphal thallus and nonmotile sporangiospores, or conidia.

Entomostraca [INV ZOO] A group of Crustacea comprising the orders Cephalocarida, Branchiopoda, Ostracoda, Copepoda, Branchiura, and Cirripedia.

Entoniscidae [INV ZOO] A family of isopod crustaceans in the tribe Bopyrina that are parasitic in the visceral cavity of crabs and porcellanids.

Entoprocta [INV ZOO] A group of bryozoans, sometimes considered to be a subphylum, having a pseudocoelomate visceral cavity and the anus opening inside the circlet of tentacles.

entrainment [CHEM ENG] A process in which the liquid boils so violently that suspended droplets of liquid are carried in the escaping vapor. [HYD] The pickup and movement of sediment as bed load or in suspension by current flow. [METEOROL] The mixing of environmental air into a preexisting organized air current so that the environmental air becomes part of the current. [OCEANOGR] The transfer of fluid by friction from one water mass to another, usually occurring between currents moving in respect to each other.

entrance [CIV ENG] The seaward end of a channel, harbor, and so on. [COMPUT SCI] The location of a program or subroutine at which execution is to start. Also known as entry point. [ENG] A place of physical entering, such as a door or passage. [NAV ARCH] The part of a ship's underwater hull which is forward of the amidships.

entrapment [GEOL] The underground trapping of oil or gas reserves by folds, faults, domes, asphaltic seals, unconformities, and such.

entrenched stream [HYD] A stream that flows in a valley or narrow trench cut into a plain or relatively level upland. Also spelled intrenched stream.

entropy [COMMUN] A measure of the absence of information about a situation, or, equivalently, the uncertainty associated with the nature of a situation. [MATH] In a mathematical context, this concept is attached to dynamical systems, transformations between measure spaces, or systems of events with probabilities; it expresses the amount of disorder inherent or produced. [STAT MECH] Measure of the disorder of a system, equal to the Boltzmann constant times the natural logarithm of the number of microscopic states corresponding to the thermodynamic state of the system; this statistical-mechanical definition can be shown to be equivalent to the thermodynamic definition. [THERMO] Function of the state of a thermodynamic system whose change in any differential reversible process is equal to the heat absorbed by the system from its surroundings divided by the absolute temperature of the system. Also known as thermal charge.

entry [COMPUT SCI] Input data fed during the execution of a program by means of a terminal.

entry ballistics [MECH] That branch of ballistics which pertains to the entry of a missile, spacecraft, or other object from outer space into and through an atmosphere.

entry block [COMPUT SCI] The area of main memory reserved for the data which will be introduced at execution time.

entry condition [COMPUT SCI] A requirement that must be met before a program or routine can be entered by a computer program. Also known as initial condition.

entry instruction [COMPUT SCI] The first instruction to be executed in a subroutine.

entry point See entrance.

entry sorting [COMPUT SCI] A method of internal sorting in which records or blocks of records are placed, one at a time, in a buffer area and then integrated into the sorted list before the next record is placed in the buffer.

enuresis [MED] Urinary incontinence, especially in the absence of organic cause.

envelope [COMMUN] A curve drawn to pass through the peaks of a graph, such as that of a moduated radio-frequency carrier signal. [ENG] The glass or metal housing of an electron tube or the glass housing of an incandescent lamp. Also known as bulb.

envelope delay distortion See delay distortion.

envelope detector See detector.

environment [ECOL] The sum of all external conditions and influences affecting the development and life of organisms. [ENG] The aggregate of all natural, operational, or other conditions that affect the operation of equipment or components. [PHYS] The aggregate of all the conditions and the influences that determine the behavior of a physical system.

environmental biology See ecology.

environmental chemistry [CHEM] The complex chemical relationships involving the atmosphere, climatology, air and water pollution, fuels, pesticides, energy, biochemistry, geochemistry, and so on.

environmental engineering [ENG] The technology concerned with the reduction of pollution, contamination, and deterioration of the surroundings in which humans live.

environmental impact statement [ENG] A report of the potential effect of plans for land use in terms of the environmental, engineering, esthetic, and economic aspects of the proposed objective.

environmental protection [ENG] The protection of humans and equipment against stresses of climate and other elements of the environment.

environmental survey satellite [AERO ENG] One of a series of meteorological satellites which completely photographs the earth each day. Abbreviated ESSA.

environmental test [ENG] A laboratory test conducted to determine the functional performance of a component or system under conditions that simulate the real environment in which the component or system is expected to operate.

environment division [COMPUT SCI] The section of a program written in COBOL which defines the hardware and files to be used by the program.

enzootic [VET MED] 1. A disease affecting animals in a limited geographic region. 2. Pertaining to such a disease.

enzyme [BIOCHEM] Any of a group of catalytic proteins that are produced by living cells and that mediate and promote the chemical processes of life without themselves being altered or destroyed.

enzyme induction [MICROBIO] The process by which a microbial cell synthesizes an enzyme in response to the presence of a substrate or of a substance closely related to a substrate in the medium.

enzyme inhibition [BIOCHEM] Prevention of an enzymic process as a result of the interaction of some substance with the enzyme so as to decrease the rate of reaction.

enzyme repression [BIOCHEM] The process by which the rate of synthesis of an enzyme is reduced in the presence of a metabolite, often the end product of a chain of reactions in which the enzyme in question operates near the beginning.

Eocambrian [GEOL] Pertaining to the thick sequences of strata conformably underlying Lower Cambrian fossils. Also known as Infracambrian.

Eocanthocephala [INV ZOO] An order of the Acanthocephala characterized by the presence of a small number of giant subcuticular nuclei.

Eocene [GEOL] The next to the oldest of the five major epochs of the Tertiary period (in the Cenozoic era).

Eocrinoidea [PALEON] A class of extinct echinoderms in the subphylum Crinozoa that had biserial brachioles like those of cystoids combined with a theca like that of crinoids.

EOF gap See end-of-file gap.

Eogene See Paleogene.

eolian [METEOROL] Pertaining to the action or the effect of the wind, as in eolian sounds or eolian deposits (of dust). Also spelled aeolian.

eolian dune [GEOL] A dune resulting from entrainment of grains by the flow of moving air.

eolian soil [GEOL] A type of soil ranging from sand dunes to loess deposits whose particles are predominantly of silt size.

eolotropy See anisotropy.

Eomoropidae [PALEON] A family of extinct perissodactyl mammals in the superfamily Chalicotherioidea.

EOR gap See end-of-record gap.

Eosentomidae [INV ZOO] A family of primitive wingless insects in the order Protura that possess spiracles and tracheae.

eosin [ORG CHEM] $C_{20}H_8O_5Br_4$ 1. A red fluorescent dye in the form of triclinic crystals that are insoluble in water; used chiefly in cosmetics and as a toner.

Also known as bromeosin; bromo acid; eosine; tetra-bromofluorescein. **2.** The red to brown crystalline sodium or potassium salt of this dye; used in organic pigments, as a biological stain, and in pharmaceuticals. Also known as eosine; eosine G; eosine Y; eosine yellowish.

eosine See eosin.

eosine G See eosin.

eosine Y See eosin.

eosine yellowish See eosin.

eosinophil [HISTOL] A granular leukocyte having cytoplasmic granules that stain with acid dyes and a nucleus with two lobes connected by a thin thread of chromatin. Also known as acidophil.

Eosuchia [PALEON] The oldest, most primitive, and only extinct order of lepidosaurian reptiles.

eötvös [GEOPHYS] A unit of horizontal gradient of gravitational acceleration, equal to a change in gravitational acceleration of 10^{-9} galileo over a horizontal distance of 1 centimeter.

Eötvös constant [PHYS] A constant that appears in an expression for the behavior of the surface tension γ of a liquid as the temperature T drops to a critical temperature T_c at which the surface tension disappears, equal to $\gamma(M/\rho)^{2/3}/(T_c - T)$, where M is the molecular weight and ρ the density of the liquid.

Eötvös experiment [RELAT] An experiment which tests the equality of inertial mass and gravitational mass by balancing on a given body the earth's gravitational attraction against the kinetic reaction arising from the rotation of the earth.

Eötvös rule [THERMO] The rule that the rate of change of molar surface energy with temperature is a constant for all liquids; deviations are encountered in practice.

Eötvös torsion balance [ENG] An instrument which records the change in the acceleration of gravity over the horizontal distance between the ends of a beam; used to measure density variations of subsurface rocks.

Epacridaceae [BOT] A family of dicotyledonous plants in the order Ericales, distinguished by palmately veined leaves, and stamens equal in number with the corolla lobes.

epaxial [BIOL] Above or dorsal to an axis.

epaxial muscle [ANAT] Any of the dorsal trunk muscles of vertebrates.

epeiric sea See epicontinental sea.

ependyma [HISTOL] The layer of epithelial cells lining the cavities of the brain and spinal cord. Also known as ependymal layer.

ependymal layer See ependyma.

ephaptic transmission [PHYSIO] Electrical transfer of activity to a postephaptic unit by the action current of a preephaptic cell.

Ephedrales [BOT] A monogeneric order of gymnosperms in the subdivision Gneticae.

ephemeral plant [BOT] An annual plant that completes its life cycle in one short moist season; desert plants are examples.

ephemeral stream [HYD] A stream channel which carries water only during and immediately after periods of rainfall or snowmelt.

Ephemerida [INV ZOO] An equivalent name for the Ephemeroptera.

ephemeris [ASTRON] A periodical publication tabulating the predicted positions of celestial bodies at regular intervals, such as daily, and containing other data of interest to astronomers. Also known as astronomical ephemeris.

ephemeris day [ASTRON] A unit of time equal to 86,400 ephemeris seconds (International System of Units).

ephemeris second [ASTRON] The fundamental unit of time of the International System of Units of 1960, equal to 1/31556925.9747 of the tropical year defined by the mean motion of the sun in longitude at the epoch 1900 January 0 day 12 hours.

ephemeris time [ASTRON] The uniform measure of time defined by the laws of dynamics and determined in principle from the orbital motions of the planets, specifically the orbital motion of the earth as represented by Newcomb's Tables of the Sun. Abbreviated E.T.

Ephemeroptera [INV ZOO] The mayflies, an order of exopterygote insects in the subclass Pterygota.

Ephydridae [INV ZOO] The shore flies, a family of myodarian cyclorrhaphous dipteran insects in the subsection Acalypteratae.

epi- [ORG CHEM] A prefix used in naming compounds to indicate the presence of a bridge or intramolecular connection. [SCI TECH] Prefix denoting upon, beside, near to, over, outer, anterior, prior to, or after.

EPI See electronic position indicator.

epiblast See ectoderm.

epiboly [EMBRYO] The growing or extending of one part, such as the upper hemisphere of a blastula, over and around another part, such as the lower hemisphere, in embryogenesis.

epicardium [ANAT] The inner, serous portion of the pericardium that is in contact with the heart. [INV ZOO] A tubular prolongation of the branchial sac in certain ascidians which takes part in the process of budding.

Epicaridea [INV ZOO] A suborder of the Isopoda whose members are parasitic on various marine crustaceans.

epicenter [GEOL] A point on the surface of the earth which is directly above the seismic focus of an earthquake and where the earthquake vibrations reach first.

epicontinental sea [OCEANOGR] That portion of the sea lying upon the continental shelf, and the portions which extend into the interior of the continent with similar shallow depths. Also known as epeiric sea; inland sea.

epicotyl [BOT] The embryonic plant stem above the cotyledons.

epicranium [INV ZOO] The dorsal wall of an insect head. [VERT ZOO] The structures covering the cranium in vertebrates.

epicuticle [INV ZOO] The outer, waxy layer of an insect cuticle or exoskeleton.

epicycle [MATH] The circle which generates an epicycloid or hypocycloid.

epicyclic gear [MECH ENG] A system of gears in which one or more gears travel around the inside or the outside of another gear whose axis is fixed.

epicycloid [MATH] The curve traced by a point on a circle as it rolls along the outside of a fixed circle.

epidemic [MED] A sudden increase in the incidence rate of a disease to a value above normal, affecting large numbers of people and spread over a wide area.

epidemic encephalitis See encephalitis lethargica.

epidemic hepatitis See infectious hepatitis.

epidemic jaundice See infectious hepatitis.

epidemic roseola See rubella.

epidemiology [MED] The study of the mass aspects of disease.

epidermis [BOT] The outermost layer (sometimes several layers) of cells on the primary plant body. [HISTOL] The outer nonsensitive, nonvascular portion of the skin comprising two strata of cells, the stratum corneum and the stratum germinativum.

epidiascope [OPTICS] 1. An optical projection system for forming an enlarged real image of a flat opaque object, in which light is reflected from the object and then from a mirror before being focused by a projection lens. Also known as episcope. 2. An optical projection system which can easily be altered to project either transparent or opaque objects.

epididymis [ANAT] The convoluted efferent duct lying posterior to the testis and connected to it by the efferent ductules of the testis.

epidiorite [PETR] A dioritic rock formed by alteration of pyroxenic igneous rocks.

epidosite [PETR] A rare metamorphic rock composed of epidote and quartz.

epidote [MINERAL] A pistachio-green to blackish-green calcium aluminum sorosilicate mineral that crystallizes in the monoclinic system; the luster is vitreous, hardness is 6½ on Mohs scale, and specific gravity is 3.35–3.45.

epidote-amphibolite facies [PETR] Metamorphic rocks formed under pressures of 3000–7000 bars and temperatures of 250–450°C with conditions intermediate between those that formed greenschist and amphibolite, or with characteristics intermediate.

epidotization [GEOL] The introduction of epidote into, or the formation of epidote from, rocks.

epigastric region [ANAT] The upper and middle part of the abdominal surface between the two hypochondriac regions. Also known as epigastrium.

epigastrium [ANAT] *See* epigastric region. [INV ZOO] The ventral side of mesothorax and metathorax in insects.

epigenesis [EMBRYO] Development in gradual stages of differentiation. [GEOL] Alteration of the mineral content of rock due to outside influences.

epigenite [MINERAL] $(Cu,Fe)_5AsS_6$ A steel-gray, orthorhombic mineral consisting of copper and iron arsenic sulfide.

epiglottis [ANAT] A flap of elastic cartilage covered by mucous membrane that protects the glottis during swallowing.

epigynous [BOT] Having the perianth and stamens attached near the top of the ovary; that is, the ovary is inferior.

epilepsy [MED] A condition characterized by the paroxysmal recurrence of transient, uncontrollable episodes of abnormal neurological or mental function, or both.

epilimnion [HYD] A fresh-water zone of relatively warm water in which mixing occurs as a result of wind action and convection currents.

epimagmatic *See* deuteric.

epimer [ORG CHEM] A type of isomer in which the difference between the two compounds is the relative position of the H (hydrogen) group and OH (hydroxyl) group on the last asymmetric C (carbon) atom of the chain, as in the sugars D-glucose and D-mannose.

epimerase [BIOCHEM] A type of enzyme that catalyzes the rearrangement of hydroxyl groups on a substrate.

epimere [ANAT] The dorsal muscle plate of the lining of a coelomic cavity. [EMBRYO] The dorsal part of a mesodermal segment in the embryo of chordates.

epimerization [ORG CHEM] A process in which an optically active compound that contains two or more asymmetric centers, only one of these is altered by some reaction to form an epimer.

epimeron [INV ZOO] 1. The posterior plate of the pleuron in insects. 2. The portion of a somite between the tergum and the insertion of a limb in arthropods.

epimysium [ANAT] The connective-tissue sheath surrounding a skeletal muscle.

epinasty [BOT] Growth changes in which the upper surface of a leaf grows, thus bending the leaf downward.

epinephrine [BIOCHEM] $C_9H_{13}O_3N$ A hormone secreted by the adrenal medulla that acts to increase blood pressure due to stimulation of heart action and constriction of peripheral blood vessels. Also known as adrenaline.

epipelagic [OCEANOGR] Of or pertaining to the portion of oceanic zone into which enough light penetrates to allow photosynthesis.

epiphyseal arch [EMBRYO] The arched structure in the third ventricle of the embryonic brain, which marks the site of development of the pineal body.

epiphyseal plate [ANAT] 1. The broad, articular surface on each end of a vertebral centrum. 2. The thin layer of cartilage between the epiphysis and the shaft of a long bone. Also known as metaphysis.

epiphysis [ANAT] 1. The end portion of a long bone in vertebrates. 2. *See* pineal body.

epiphyte [ECOL] A plant which grows nonparasitically on another plant or on some nonliving structure, such as a building or telephone pole, deriving moisture and nutrients from the air. Also known as aerophyte.

epipodite [INV ZOO] A branch of the basal joint of the protopodite of thoracic limbs of many arthropods.

epipodium [BOT] The apical portion of an embryonic phyllopodium. [INV ZOO] 1. A ridge or fold on the lateral edges of each side of the foot of certain gastropod mollusks. 2. The elevated ring on an ambulacral plate in Echinoidea.

Epipolasina [INV ZOO] A suborder of sponges in the order Clavaxinellida having radially arranged monactinal or diactinal megascleres.

epiproct [INV ZOO] A plate above the anus forming the dorsal part of the tenth or eleventh somite of certain insects.

episiotomy [MED] Medial or lateral incision of the vulva during childbirth, to avoid undue laceration.

episome [GEN] A circular genetic element in bacteria, presumably a deoxyribonucleic acid fragment, which is not necessary for survival of the organism and which can be integrated in the bacterial chromosome or remain free.

epispadias [MED] A congenital defect of the anterior urethra in which the canal terminates on the dorsum of the penis and posterior to its normal opening.

epistasis [GEN] The suppression of the effect of one gene by another. [MED] A checking or stoppage of a hemorrhage or other discharge. [PATH] A scum or film of substance floating on the surface of urine.

epistilbite [MINERAL] $CaAl_2Si_6O_{16}\cdot5H_2O$ A mineral of the zeolite family that contains calcium and aluminosilicate and crystallizes in the monoclinic system; occurs in white prismatic crystals or granular forms.

epistome [INV ZOO] 1. The area between the mouth and the second antennae in crustaceans. 2. The plate covering this region. 3. The area between the labrum and the epicranium in many insects. 4. A flap covering the mouth of certain bryozoans. 5. The area just above the labrum in certain dipterans.

epitaxial diffused-junction transistor [ELECTR] A junction transistor produced by growing a thin, high-purity layer of semiconductor material on a heavily doped region of the same type.

epitaxial layer [SOLID STATE] A semiconductor layer having the same crystalline orientation as the substrate on which it is grown.

epitaxial transistor [ELECTR] Transistor with one or more epitaxial layers.

epitaxy [CRYSTAL] Growth of one crystal on the surface of another crystal in which the growth of the deposited crystal is oriented by the lattice structure of the substrate.

epitheliochorial placenta [EMBRYO] A type of placenta in which the maternal epithelium and fetal epithelium are in contact. Also known as villous placenta.

epithelioma [MED] A tumor derived from epithelium; usually a skin cancer, occasionally cancer of a mucous membrane.

epithelium [HISTOL] A primary animal tissue, distinguished by cells being close together with little intercellular substance; covers free surfaces and lines body cavities and ducts.

epithermal [GEOL] Pertaining to mineral veins and ore deposits formed from warm waters at shallow depth, at temperatures ranging from 50–200°C, and generally at some distance from the magnetic source.

epithermal reactor [NUCLEO] A nuclear reactor in which a substantial fraction of fissions is induced by neutrons having more than thermal energy.

epitrochoid [MATH] A curve traced by a point rigidly attached to a circle at a point other than the center when the circle rolls without slipping on the outside of a fixed circle.

epizone [GEOL] 1. The zone of metamorphism characterized by moderate temperature, low hydrostatic pressure, and powerful stress. 2. The outer depth zone of metamorphic rocks.

epizootic [VET MED] 1. Affecting many animals of one kind in one region simultaneously; widely diffuse and rapidly spreading. 2. An extensive outbreak of an epizootic disease.

epoch [ASTRON] A particular instant for which certain data are valid; for example, star positions in an astronomical catalog, epoch 1950.0. [GEOL] A major subdivision of a period of geologic time. [PHYS] *See* time.

epoxidation [ORG CHEM] Reaction yielding an epoxy compound, such as the conversion of ethylene to ethylene oxide.

epoxide [ORG CHEM] A reactive group in which an oxygen atom is joined to each of two carbon atoms which are already bonded.

epoxy- [ORG CHEM] A prefix indicating presence of an epoxide group in a molecule.

epoxy adhesive [MATER] An adhesive material made of epoxy resin.

epoxy matrix composite [MATER] Any of the high-strength compositions consisting of epoxy resin and a reinforcing matrix of filaments or fibers of glass, metal, or other materials.

epoxy resin [ORG CHEM] A polyether resin formed originally by the polymerization of bisphenol A and epichlorohydrin, having high strength, and low shrinkage during curing; used as a coating, adhesive, casting, or foam.

EPR *See* electron paramagnetic resonance.

EPROM *See* erasable programmable read-only memory.

epsilon meson [PARTIC PHYS] Neutral, scalar, meson resonance having positive charge conjugation parity and G-parity, a mass of about 730 MeV, and a width of about 600 MeV; decays to two pions.

epsilon structure [SOLID STATE] The hexagonal close-packed structure of the ε-phase of an electron compound.

epsomite [MINERAL] $MgSO_4 \cdot 7H_2O$ A mineral that occurs in clear, needlelike, orthorhombic crystals; commonly it is massive or fibrous; luster varies from vitreous to milky, hardness is 2–2.5 on Mohs scale, and specific gravity is 1.68; it has a salty bitter taste and is soluble in water. Also known as epsom salt.

epsom salt *See* epsomite.

Epstein-Barr virus [VIROL] Herpeslike virus particles first identified in cultures of cells from Burkett's malignant lymphoma.

equal [MATH] Being the same in some sense determined by context.

equal-area latitude *See* authalic latitude.

equal-area map projection [MAP] A map projection having a constant area scale; it is not conformal and is not used for navigation. Also known as authalic map projection; equivalent map projection.

equal-arm balance [MECH] A simple balance in which the distances from the point of support of the balance-arm beam to the two pans at the end of the beam are equal.

equality gate *See* equivalence gate.

equalization [ELECTR] The effect of all frequency-discriminating means employed in transmitting, recording, amplifying, or other signal-handling systems to obtain a desired overall frequency response. Also known as frequency-response equalization.

equalizer [ELECTR] A network designed to compensate for an undesired amplitude-frequency or phase-frequency response of a system or component; usually a combination of coils, capacitors, and resistors. Also known as equalizing circuit. [MECH ENG] 1. A bar to which one attaches a vehicle's whiffletrees to make the pull of draft animals equal. Also known as equalizing bar. 2. A bar which joins a pair of axle springs on a railway locomotive or car for equalization of weight. Also known as equalizing bar. 3. A device which distributes braking force among independent brakes of an automotive vehicle. Also known as equalizer brake. 4. A machine which saws wooden stock to equal lengths. [ORD] A device attached to the carriage of those artillery weapons that, when emplaced, rest on two wheels and two trail ends; it is a compensating mechanism to transmit equally the weapon weight and firing shock.

equalizer brake *See* equalizer.

equalizing bar *See* equalizer.

equalizing circuit *See* equalizer.

equalizing current [ELEC] Current that circulates between two parallel-connected compound generators to equalize their output.

equalizing line [CHEM ENG] A pipe or tubing interconnection between two closed vessels, containers, or process systems to allow pressure equalization.

equalizing pulses [ELECTR] In television, pulses at twice the line frequency, occurring just before and after the vertical synchronizing pulses, which minimize the effect of line frequency pulses on the interlace.

equal loudness contour [ACOUS] A curve on a graph of sound intensity in decibels versus frequency at each point along which sound appears to be equally loud to a listener. Also known as Fletcher-Munson contour.

equal tails test [STAT] A technique for choosing two critical values for use in a two-sided test; it consists of selecting critical values c and d so that the probability of acceptance of the null hypothesis if the test statistic does not exceed c is the same as the probability of acceptance of the null hypothesis if the test statistic is not smaller than d.

equate [MATH] To state algebraically that two expressions are equal to one another.

equation [CHEM] A symbolic expression that represents in an abbreviated form the laboratory observations of a chemical change; an equation (such as $2H_2 + O_2 \rightarrow 2H_2O$) indicates what reactants are consumed (H_2 and O_2) and what products are formed (H_2O), the correct formula of each reactant and product, and satisfies the law of conservation of atoms in that the symbols for the number of atoms reacting equals the number of atoms in the products. [MATH] A statement that each of two expressions is equal to the other.

equation of continuity *See* continuity equation.

equation of motion [FL MECH] One of a set of hydrodynamical equations representing the application of Newton's second law of motion to a fluid system; the total acceleration on an individual fluid particle is equated to the sum of the forces acting on the particle within the fluid. [MECH] **1.** Equation which specifies the coordinates of particles as functions of time. **2.** A differential equation, or one of several such equations, from which the coordinates of particles as functions of time can be obtained if the initial positions and velocities of the particles are known. [QUANT MECH] A differential equation which enables one to predict the statistical distribution of the results of any measurement upon a system at any time if the initial dynamical state of the system is known.

equation of state [PHYS CHEM] A mathematical expression which defines the physical state of a homogeneous substance (gas, liquid, or solid) by relating volume to pressure and absolute temperature for a given mass of the material.

equation of the center [ASTRON] The angle between the actual longitude of the moon and the longitude of an imaginary body that moves with constant angular velocity with the same period as the moon.

equation of time [ASTRON] The addition of a quantity to mean solar time to obtain apparent solar time; formerly, when apparent solar time was in common use, the opposite convention was used; apparent solar time has annual variation as a result of the sun's inclination in the ecliptic and the eccentricity of the earth's elliptical orbit.

equator [GEOD] The great circle around the earth, equally distant from the North and South poles, which divides the earth into the Northern and Southern hemispheres; the line from which latitudes are reckoned.

equatorial axis [GEOD] The diameter of the earth described between two points on the equator.

equatorial bulge [GEOD] The excess of the earth's equatorial diameter over the polar diameter.

equatorial calms *See* doldrums.

equatorial convergence zone *See* intertropical convergence zone.

Equatorial Countercurrent [OCEANOGR] An ocean current flowing eastward (counter to and between the westward-flowing North Equatorial Current and South Equatorial Current) through all the oceans.

Equatorial Current *See* North Equatorial Current; South Equatorial Current.

equatorial easterlies [METEOROL] The trade winds in the summer hemisphere when they are very deep, extending at least 8 to 10 kilometers in altitude, and generally not topped by upper westerlies; if upper

westerlies are present, they are too weak and shallow to influence the weather. Also known as deep easterlies; deep trades.

equatorial front *See* intertropical front.

equatorial horizontal parallax [ASTRON] The parallax of a member of the solar system measured from positional observations made at the same time at two stations on earth, whose distance apart is the earth's equatorial radius.

equatorial mounting [ENG] The mounting of an equatorial telescope; it has two perpendicular axes, the polar axis (parallel to the earth's axis) that turns on fixed bearings, and the declination axis, supported by the polar axis.

equatorial orbit [ASTRON] An orbit in the plane of the earth's equator.

equatorial plane [ASTRON] The plane passing through the equator of the earth, or of another celestial body, perpendicular to its axis of rotation and equidistant from its poles. [CYTOL] The plane in a cell undergoing mitosis that is midway between the centrosomes and perpendicular to the spindle fibers. [MECH] A plane perpendicular to the axis of rotation of a rotating body and equidistant from the intersections of this axis with the body's surface, provided that the body is symmetric about the axis of rotation and is symmetric under reflection through this plane.

equatorial projection [MAP] A map projection centered on the equator.

equatorial system [ASTRON] A set of celestial coordinates based on the celestial equator as the primary great circle; usually declination and hour angle or sidereal hour angle. Also known as celestial equator system of coordinates; equinoctial system of coordinates.

equatorial tide [OCEANOGR] **1.** A lunar fortnightly tide. **2.** A tidal component with a period of 328 hours.

Equatorial Undercurrent [OCEANOGR] **1.** A subsurface current flowing from west to east in the Indian Ocean near the 150-meter depth at the equator during the time of the Northeast Monsoon. **2.** A permanent subsurface current in the equatorial region of the Atlantic and Pacific oceans.

equatorial westerlies [METEOROL] The westerly winds occasionally found in the equatorial trough and separated from the mid-latitude westerlies by the broad belt of easterly trade winds.

Equidae [VERT ZOO] A family of perissodactyl mammals in the superfamily Equoidea, including the horses, zebras, and donkeys.

equidistant [MATH] Being the same distance from some given object.

equilateral arch [ARCH] An arch described by two circular curves intersecting at the peak of the arch, each curve having a chord equal to the span.

equilateral polygon [MATH] A polygon all of whose sides are the same length.

equilateral polyhedron [MATH] A polyhedron all of whose faces are identical.

equilibrium [MECH] Condition in which a particle, or all the constituent particles of a body, are at rest or in unaccelerated motion in an inertial reference frame. Also known as static equilibrium. [PHYS] Condition in which no change occurs in the state of a system as long as its surroundings are unaltered. [STAT MECH] Condition in which the distribution function of a system is time-independent.

equilibrium constant [CHEM] A constant at a given temperature such that when a reversible chemical reaction $cC + bB = gG + hH$ has reached equilibrium, the value of this constant K^0 is equal to

$$\frac{a_G^g a_H^h}{a_C^c a_B^b}$$

where a_G, a_H, a_C, and a_B represent chemical activities of the species G, H, C, and B at equilibrium.

equilibrium diagram [PHYS CHEM] A phase diagram of the equilibrium relationship between temperature, pressure, and composition in any system.

equilibrium flash vaporization [CHEM ENG] Process in which a continuous liquid-mixture feed stream is partly vaporized in a column or vessel, with continuous withdrawal of vapor and liquid portions, the vapor and liquid in equilibrium. Also known as continuous equilibrium vaporization; simple continuous distillation.

equilibrium potential [PHYS CHEM] A point in which forward and reverse reaction rates are equal in an electrolytic solution, thereby establishing the potential of an electrode.

equilibrium ratio [PHYS CHEM] 1. In any system, relation of the proportions of the various components (gas, liquid) at equilibrium conditions. 2. See equilibrium vaporization ratio.

equilibrium vaporization ratio [PHYS CHEM] In a liquid-vapor equilibrium mixture, the ratio of the mole fraction of a component in the vapor phase (y) to the mole fraction of the same component in the liquid phase (x), or $y/x = K$ (the K factor). Also known as equilibrium ratio.

equine [VERT ZOO] 1. Resembling a horse. 2. Of or related to the Equidae.

equine encephalitis [MED] A disease of equines and humans caused by one of three viral strains: eastern, western, and Venezuelan equine viruses. Also known as equine encephalomyelitis.

equine encephalomyelitis See equine encephalitis.

equinoctial See celestial equator.

equinoctial point See equinox.

equinoctial system of coordinates See equatorial system.

equinox [ASTRON] 1. Either of the two points of intersection of the ecliptic and the celestial equator, occupied by the sun when its declination is 0°. Also known as equinoctial point. 2. That instant when the sun occupies one of the equinoctial points.

equipartition [CHEM] 1. The condition in a gas where under equal pressure the molecules of the gas maintain the same average distance between each other. 2. The equal distribution of a compound between two solvents. 3. The distribution of the atoms in an orderly fashion, such as in a crystal.

equipartition law [STAT MECH] In a classical ideal gas, the average kinetic energy per molecule associated with any degree of freedom which occurs as a quadratic term in the expression for the mechanical energy, is equal to half of Boltzmann's constant times the absolute temperature.

equipment augmentation [COMPUT SCI] 1. Procuring additional automatic data-processing equipment capability to accommodate increased work load within an established data system. 2. Obtaining additional sites or locations.

equipment compatibility [COMPUT SCI] The ability of a device to handle data prepared or handled by other equipment, without alteration of the code or of the form of the data.

equipotential surface [ELEC] A surface on which the electric potential is the same at every point.

[GEOPHYS] A surface characterized by the potential being constant everywhere on it for the attractive forces concerned. [MECH] A surface which is always normal to the lines of force of a field and on which the potential is everywhere the same.

Equisetales [BOT] The horsetails, a monogeneric order of the class Equisetopsida; the only living genus is *Equisetum*.

Equisetatae See Equisetopsida.

Equisetineae [BOT] The equivalent name for the Equisetophyta.

Equisetophyta [BOT] A division of the subkingdom Embryobionta represented by a single living genus, *Equisetum*.

Equisetopsida [BOT] A class of the division Equisetophyta whose members made up a major part of the flora, especially in moist or swampy places, during the Carboniferous Period.

equivalence [MAP] In an equal-area map projection, the property of having the ratio between areas on the map the same as the ratio between corresponding areas on the earth's surface. [MATH] A logic operator having the property that if P, Q, R, etc., are statements, then the equivalence of P, Q, R, etc., is true if and only if all statements are true or all statements are false.

equivalence classes [MATH] The collection of pairwise disjoint subsets determined by an equivalence relation on a set; two elements are in the same equivalence class if and only if they are equivalent under the given relation.

equivalence element See equivalence gate.

equivalence gate [COMPUT SCI] A logic circuit that produces a binary output signal of 1 if its two binary input signals are the same, and an output signal of 0 if the input signals differ. Also known as biconditional gate; equality gate; equivalence element; exclusive-NOR gate; match gate.

equivalence point [CHEM] The point in a titration where the amounts of titrant and material being titrated are equivalent chemically.

equivalence principle [RELAT] In general relativity, the principle that the observable local effects of a gravitational field are indistinguishable from those arising from acceleration of the frame of reference. Also known as Einstein's equivalency principle; principle of equivalence.

equivalent airspeed [AERO ENG] The product of the true airspeed and the square root of the density ratio; used in structural design work to designate various design conditions.

equivalent bending moment [MECH] A bending moment which, acting alone, would produce in a circular shaft a normal stress of the same magnitude as the maximum normal stress produced by a given bending moment and a given twisting moment acting simultaneously.

equivalent binary digits [COMPUT SCI] The number of binary positions required to enumerate the elements of a given set.

equivalent circuit [ELEC] A circuit whose behavior is identical to that of a more complex circuit or device over a stated range of operating conditions.

equivalent conductance [PHYS CHEM] Property of an electrolyte, equal to the specific conductance divided by the number of gram equivalents of solute per cubic centimeter of solvent.

equivalent electrons [ATOM PHYS] Electrons in an atom which have the same principal and orbital quantum numbers, but not necessarily the same magnetic orbital and magnetic spin quantum numbers.

equivalent focal length [OPTICS] The focal length of a thin lens which forms images that most nearly duplicate those of a given compound lens, thick lens, or system of lenses.

equivalent height *See* virtual height.

equivalent map projection *See* equal-area map projection.

equivalent nuclei [PHYS CHEM] A set of nuclei in a molecule which are transformed into each other by rotations, reflections, or combinations of these operations, leaving the molecule invariant.

equivalent temperature [METEOROL] **1.** The temperature that an air parcel would have if all water vapor were condensed out at constant pressure, the latent heat released being used to heat the air. Also known as isobaric equivalent temperature. **2.** The temperature that an air parcel would have after undergoing the following theoretical process: dry-adiabatic expansion until saturated, pseudoadiabatic expansion until all moisture is precipitated out, and dry adiabatic compression to the initial pressure; this is the equivalent temperature as read from a thermodynamic chart and is always greater than the isobaric equivalent temperature. Also known as adiabatic equivalent temperature; pseudoequivalent temperature. [THERMO] A term used in British engineering for that temperature of a uniform enclosure in which, in still air, a sizable blackbody at 75°F (23.9°C) would lose heat at the same rate as in the environment.

equivalent twisting moment [MECH] A twisting moment which, if acting alone, would produce in a circular shaft a shear stress of the same magnitude as the shear stress produced by a given twisting moment and a given bending moment acting simultaneously.

equivalent weight [CHEM] The number of parts by weight of an element or compound which will combine with or replace, directly or indirectly, 1.008 parts by weight of hydrogen, 8.00 parts of oxygen, or the equivalent weight of any other element or compound.

equivoluminal wave *See* S wave.

Equoidea [VERT ZOO] A superfamily of perissodactyl mammals in the suborder Hippomorpha comprising the living and extinct horses and their relatives.

Equuleus [ASTRON] A northern constellation near Aquarius, right ascension 21 hours, declination 10° north. Abbreviated Equl. Also known as Little Horse.

Er *See* erbium.

ER *See* electroreflectance; endoplasmic reticulum.

era [GEOL] A division of geologic time of the highest order, comprising one or more periods.

eradiation *See* terrestrial radiation.

erasable programmable read-only memory [COMPUT SCI] A read-only memory in which stored data can be erased by ultraviolet light or other means and reprogrammed bit by bit with appropriate voltage pulses. Abbreviated EPROM.

erasable storage [COMPUT SCI] Any storage medium which permits new data to be written in place of the old, such as magnetic disk or tape, but not punched card or punched tape.

erase [COMPUT SCI] To change all the binary digits in a digital computer storage device to binary zeros. [ELECTR] **1.** To remove recorded material from magnetic tape by passing the tape through a strong, constant magnetic field (dc erase) or through a high-frequency alternating magnetic field (ac erase). **2.** To eliminate previously stored information in a

charge-storage tube by charging or discharging all storage elements.

erase character *See* ignore character.

erasing head [ELECTR] A magnetic head used to obliterate material previously recorded on magnetic tape.

erbium [CHEM] A trivalent metallic rare-earth element, symbol Er, of the yttrium subgroup, found in euxenite, gadolinite, fergusonite, and xenotine; atomic number 68, atomic weight 167.26, specific gravity 9.051; insoluble in water, soluble in acids; melts at 1400–1500°C.

erect image [OPTICS] An image in which directions are the same as those in the object, in contrast to an inverted image.

erection [CIV ENG] Positioning and fixing the frame of a structure. [PHYSIO] The enlarged state of erectile tissue when engorged with blood, as of the penis or clitoris.

erection tower [CIV ENG] A temporary framework built at a construction site for hoisting equipment.

Eremascoideae [BOT] A monogeneric subfamily of ascosporogenous yeasts characterized by mostly septate mycelia, and spherical asci with eight oval to round ascospores.

Erethizontidae [VERT ZOO] The New World porcupines, a family of rodents characterized by sharply pointed, erectile hairs and four functional digits.

ERG *See* electroretinogram.

Ergasilidae [INV ZOO] A family of copepod crustaceans in the suborder Cyclopoida in which the females are parasitic on aquatic animals, while the males are free-swimming.

ergodic [STAT] **1.** Property of a system or process in which averages computed from a data sample over time converge, in a probabilistic sense, to ensemble or special averages. **2.** Pertaining to such a system or process.

ergodic theory [MATH] The study of measure-preserving transformations. [STAT MECH] Mathematical theory which attempts to show that the various possible microscopic states of a system are equally probable, and that the system is therefore ergodic.

ergodic transformation [MATH] A measure-preserving transformation on X with the property that whenever X is written as a union of two disjoint invariant subsets, one of these must have measure zero.

ergosterin *See* ergosterol.

ergosterol [BIOCHEM] $C_{28}H_{44}O$ A crystalline, water-insoluble, unsaturated sterol found in ergot, yeast, and other fungi, and which may be converted to vitamin D_2 on irradiation with ultraviolet light or activation with electrons. Also known as ergosterin.

ergot [MYCOL] The dark purple or black sclerotium of the fungus *Claviceps purpurea*. [ORG CHEM] Any of the five optically isomeric pairs of alkaloids obtained from this fungus; only the levorotatory isomers are physiologically active.

ergotism [MED] Acute or chronic intoxication resulting from ingestion of grain infected with ergot fungus, or from chronic use of drugs containing ergot.

Eri *See* Eridanus.

Erian [GEOL] Middle Devonian geologic time; a North American provincial series.

Erian orogeny [GEOL] One of the orogenies during Phanerozoic geologic time, at the end of the Silurian; the last part of the Caledonian orogenic era. Also known as Hibernian orogeny.

Ericaceae [BOT] A large family of dicotyledonous plants in the order Ericales distinguished by having twice as many stamens as corolla lobes.

Ericales [BOT] An order of dicotyledonous plants in the subclass Dilleniidae; plants are generally sympetalous with unitegmic ovules and they have twice as many stamens as petals.

Ericsson cycle [THERMO] An ideal thermodynamic cycle consisting of two isobaric processes interspersed with processes which are, in effect, isothermal, but each of which consists of an infinite number of alternating isentropic and isobaric processes.

Erid See Eridanus.

Eridanus [ASTRON] A southern constellation made up of a long, crooked line of stars beginning near Rigel in the foot of Orion, and winding west and south to the first-magnitude star Achernar. Abbreviated Eri; Erid. Also known as River Po.

erikite [MINERAL] A brown mineral consisting of a silicate and phosphate of cerium metals; occurs in orthorhombic crystals.

Erinaceidae [VERT ZOO] The hedgehogs, a family of mammals in the order Insectivora characterized by dorsal and lateral body spines.

erineum [PL PATH] An abnormal growth of hairs induced on the epidermis of a leaf by certain mites.

erinite [MINERAL] $Cu_5(OH)_4(AsO_4)_2$ Emerald-green mineral composed of basic copper arsenate.

Erinnidae [INV ZOO] A family of orthorrhaphous dipteran insects in the series Brachycera.

Eriocaulaceae [BOT] The single family of the order Eriocaulales.

Eriocaulales [BOT] An order of monocotyledonous plants in the subclass Commelinidae, having a perianth reduced or lacking and having unisexual flowers aggregated on a long peduncle.

Eriococcinae [INV ZOO] A family of homopteran insects in the superfamily Coccoidea; adult females and late instar nymphs have an anal ring.

Eriocraniidae [INV ZOO] A small family of lepidopteran insects in the superfamily Eriocranioidea.

Eriocranioidea [INV ZOO] A superfamily of lepidopteran insects in the suborder Homoneura comprising tiny moths with reduced, untoothed mandibles.

erionite [MINERAL] A chabazite mineral of the zeolite family that contains calcium ions and crystallizes in the hexagonal system.

Eriophyidae [INV ZOO] The bud mites or gall mites, a family of economically important plant-feeding mites in the suborder Trombidiformes.

Erlenmeyer synthesis [ORG CHEM] Preparation of cyclic ethers by the condensation of an aldehyde with an α-acylamino acid in the presence of acetic anhydride and sodium acetate.

Eros [ASTRON] An asteroid that is about 32 kilometers in diameter; its closest approach to the earth is every 44 years.

erosion [GEOL] 1. The loosening and transportation of rock debris at the earth's surface. 2. The wearing away of the land, chiefly by rain and running water. [MED] 1. Surgical removal of tissues by scraping. 2. Excision of a joint.

erosion-corrosion [MET] Attack on a metal surface resulting from the combined effects of erosion and corrosion.

erosion pavement [GEOL] A layer of pebbles and small rocks that prevents the soil underneath from eroding.

erosion platform See wave-cut platform.

erosive burning [CHEM] Combustion of solid propellants accompanied by nonsteady, high-velocity flows of product gases across burning propellant surfaces.

Erotylidae [INV ZOO] The pleasing fungus beetles, a family of coleopteran insects in the superfamily Cucujoidea.

ERP See effective radiated power; emergency recorder plot.

Errantia [INV ZOO] A group of 34 families of polychaete annelids in which the anterior region is exposed and the linear body is often long and is dorsoventrally flattened.

erratic [GEOL] A rock fragment that has been transported a great distance, generally by glacier ice or floating ice, and differs from the bedrock on which it rests.

error [COMPUT SCI] An incorrect result arising from approximations used in numerical methods, rather than from a human mistake or computer malfunction. [SCI TECH] Any discrepancy between a computed, observed, or measured quantity and the true, specified, or theoretically correct value of that quantity.

error burst [COMPUT SCI] The condition when more than one bit is in error in a given number of bits.

error character [COMPUT SCI] A character that indicates the existence of an error in the data being processed or transmitted, and usually specifies that a certain amount of preceding or following data is to be ignored.

error checking and recovery [COMPUT SCI] An automatic procedure which checks for parity and will proceed with the execution after error correction.

error-checking code See self-checking code.

error code [COMPUT SCI] A specific character punched into a card or tape to indicate that a conscious error was made in the associated block of data; machines reading the error code may be programmed to throw out the entire block automatically.

error coefficient [CONT SYS] The steady-state value of the output of a control system, or of some derivative of the output, divided by the steady-state actuating signal. Also known as error constant.

error constant See error coefficient.

error-correcting code [COMPUT SCI] Data representation that allows for error detection and error correction if the error is of a specific kind.

error correction [COMMUN] Any system for reducing errors in an incoming message, such as sending redundant signals as a check. [COMPUT SCI] Computer device for automatically locating and correcting a machine error of dropping a bit or picking up an extraneous bit, without stopping the machine or having it go to a programmed recovery routine. [ELEC] Correction of time errors in interconnected alternating-current power systems resulting from deviations from normal frequency, in order to make all areas synchronous.

error-detecting code See self-checking code.

error-detecting system [COMPUT SCI] An automatic system which detects an error due to a lack of data, or erroneous data during transmission.

error detection and feedback system [COMPUT SCI] An automatic system which retransmits a piece of data detected by the computer as being in error.

error detection routine See diagnostic routine.

error diagnostic [COMPUT SCI] A computer printout of an instruction or data statement, pinpointing an error in the instruction or statement and spelling out the type of error involved.

error frequency limit [COMPUT SCI] The maximum number of single bit errors per unit of time that a computer will accept before a machine check interrupt is initiated. Abbreviated EFL.

error-indicating system [COMPUT SCI] Built-in circuits designed to indicate automatically that certain computational errors have occurred.

error interrupt [COMPUT SCI] The halt in execution of a program because of errors which the computer is not capable of correcting.

error list [COMPUT SCI] A list generated by a compiler showing invalid or erroneous instructions in a source program.

error message [COMPUT SCI] A message indicating detection of an error.

error range [COMPUT SCI] A range of values such that an error condition will result if a specified data item falls within it. [STAT] The difference between the highest and lowest error values; a measure of the uncertainty associated with a number.

error routine [COMPUT SCI] A routine which takes control of a program and initiates corrective actions when an error is detected.

error signal [CONT SYS] In an automatic control device, a signal whose magnitude and sign are used to correct the alignment between the controlling and the controlled elements. [ELEC] *See* error voltage. [ELECTR] A voltage that depends on the signal received from the target in a tracking system, having a polarity and magnitude dependent on the angle between the target and the center of the scanning beam.

error voltage [ELEC] A voltage, usually obtained from a selsyn, that is proportional to the difference between the angular positions of the input and output shafts of a servosystem; this voltage acts on the system to produce a motion that tends to reduce the error in position. Also known as error signal.

ERTS *See* earth resources technology satellite.

eruption [GEOL] The ejection of solid, liquid, or gaseous material from a volcano.

eruptive star [ASTRON] A star that has a rapid change in its intensity because of the physical change it undergoes; examples are flare stars, recurrent novae, novae, supernovae, and nebular variables.

Erwinieae [MICROBIO] Formerly a tribe of phytopathogenic bacteria in the family Enterobacteriaceae, including the single genus *Erwinia*.

Erysiphaceae [MYCOL] The powdery mildews, a family of ascomycetous fungi in the order Erysiphales with light-colored mycelia and conidia.

Erysiphales [MYCOL] An order of ascomycetous fungi which are obligate parasites of seed plants, causing powdery mildew and sooty mold.

erythema [MED] Localized redness of the skin in areas of variable size.

erythremia *See* erythrocytosis; polycythemia vera.

erythrine *See* erythrite.

erythrite [MINERAL] $Co_3(AsO_4)_2 \cdot 8H_2O$ A crimson, peach, or pink-red oxidized cobalt mineral that occurs in monoclinic crystals, in globular and reniform masses, or in earthy forms. Also known as cobalt bloom; cobalt ocher; erythrine; peachblossom ore; red cobalt. [ORG CHEM] *See* erythritol.

erythritol [ORG CHEM] $H(CHOH)_4H$ A tetrahydric alcohol; occurs as tetragonal prisms, melting at 121°C, soluble in water; used in medicine as a vasodilator. Also known as antierythrite; erythrite; erythrol; ethroglucin; phycite; tetrahydroxy butane.

erythroblast [HISTOL] A nucleated cell occurring in bone marrow as the earliest recognizable cell of the erythrocytic series.

erythroblastosis fetalis [MED] A form of hemolytic anemia affecting the fetus and newborn infant when a mother is Rh-negative and has developed antibodies against an Rh-positive fetus. Also known as hemolytic disease of newborn.

erythrocyte [HISTOL] A type of blood cell that contains a nucleus in all vertebrates but humans and that has hemoglobin in the cytoplasm. Also known as red blood cell.

erythrocytopoiesis *See* erythropoiesis.

erythrocytosis [MED] An increase in the number of circulating erythrocytes of more than two standard deviations above the mean normal, usually occurring secondary to hypoxia. Also known as erythremia; polycythemia.

erythrol *See* erythritol.

erythropoiesis [PHYSIO] The process by which erythrocytes are formed. Also known as erythrocytopoiesis.

Erythroxylaceae [BOT] A homogeneous family of dicotyledonous woody plants in the order Linales characterized by petals that are internally appendiculate, three carpels, and flowers without a disk.

erythroxylon *See* cocaine.

Es *See* einsteinium.

Esaki tunnel diode *See* tunnel diode.

ESCA *See* x-ray photoelectron spectroscopy.

E scan *See* E scope.

escape character [COMPUT SCI] A character used to indicate that the succeeding character or characters are expressed in a code different from the code currently in use.

escapement [HOROL] A device in a timepiece consisting of a toothed wheel which engages alternate pallets attached to an oscillating member. [MECH ENG] A ratchet device that permits motion in one direction slowly.

escape velocity [AERO ENG] In space flight, the speed at which an object is able to overcome the gravitational pull of the earth.

escar *See* esker.

escarpment [GEOL] A cliff or steep slope of some extent, generally separating two level or gently sloping areas, and produced by erosion or by faulting. Also known as scarp. [ORD] The ground surrounding a fortified place which has been cut away nearly vertically to prevent an enemy's approach.

Escherichieae [MICROBIO] Formerly a tribe of bacteria in the family Enterobacteriaceae defined by the ability to ferment lactose, with the rapid production of acid and visible gas.

eschynite [MINERAL] $(Ce,Ca,Fe,Th)(Ti,Cb)_2O_6$ A black mineral, occurring in prismatic crystals; a rare oxide of cesium, titanium, and other metals, which is isomorphous with priorite.

E scope [ELECTR] A cathode-ray scope on which signals appear as spots, with range as the horizontal coordinate and elevation angle or height as the vertical coordinate. Also known as E indicator; E scan.

escutcheon [DES ENG] An ornamental shield, flange, or border used around a dial, window, control knob, or other panel-mounted part. Also known as escutcheon plate.

escutcheon plate *See* escutcheon.

ESD *See* external symbol dictionary.

eskar *See* esker.

esker [GEOL] A sinuous ridge of constructional form, consisting of stratified accumulations, glacial sand, and gravel. Also known as asar; back furrow; escar; eschar; eskar; osar; serpent kame.

Esocidae [VERT ZOO] The pikes, a family of fishes in the order Clupeiformes characterized by an elongated beaklike snout and sharp teeth.

Esocoidei [VERT ZOO] A small suborder of fresh-water fishes in the order Salmoniformes; includes the pikes, mudminnows, and pickerels.

esophagus [ANAT] The tubular portion of the alimentary canal interposed between the pharynx and the stomach. Also known as gullet.

ESP See extrasensory perception.

ESR See electron paramagnetic resonance.

ESSA See environmental survey satellite.

essential amino acid [BIOCHEM] Any of eight of the 20 naturally occurring amino acids that are indispensable for optimum animal growth but cannot be formed in the body and must be supplied in the diet.

essential hypertension [MED] Elevation of the systemic blood pressure, of unknown origin. Also known as primary hypertension.

essential oil [MATER] Any of the odoriferous oily products of plant origin which are distillable; the principal constituents are terpenes, but benzenoid and aliphatic compounds may also be present. Also known as ethereal oil.

essexite [PETR] A rock of igneous origin composed principally of plagioclase hornblende, biotite, and titanaugite.

established flow [FL MECH] The flow when the boundary layer of a fluid flowing in a duct completely fills the duct; that is, when the effect of the wall shearing stress extends completely across the duct.

ester [ORG CHEM] The compound formed by the elimination of water and the bonding of an alcohol and an organic acid.

esterase [BIOCHEM] Any of a group of enzymes that catalyze the synthesis and hydrolysis of esters.

esterification [ORG CHEM] A chemical reaction whereby esters are formed.

esthesia [PHYSIO] The capacity for sensation, perception, or feeling. Also spelled aesthesia.

esthesioneuroblastoma See neuroepithelioma.

esthesioneuroepithelioma See neuroepithelioma.

estival [ASTRON] Of or pertaining to the summer. Also spelled aestival.

estivation [PHYSIO] 1. The adaptation of certain animals to the conditions of summer, or the taking on of certain modifications, which enables them to survive a hot, dry summer. 2. The dormant condition of an organism during the summer.

estradiol [BIOCHEM] $C_{18}H_{24}O_2$ An estrogenic hormone produced by follicle cells of the vertebrate ovary; provokes estrus and proliferation of the human endometrium, and stimulates ICSH (interstitial-cell-stimulating hormone) secretion.

estrogen [BIOCHEM] Any of various natural or synthetic substances possessing the biologic activity of estrus-producing hormones.

estrous cycle [PHYSIO] The physiological changes that take place between periods of estrus in the female mammal.

estrus [PHYSIO] The period in female mammals during which ovulation occurs and the animal is receptive to mating.

estuarine oceanography [OCEANOGR] The study of the chemical, physical, biological, and geological properties of estuaries.

estuary [GEOGR] A semienclosed coastal body of water which has a free connection with the open sea and within which sea water is measurably diluted with fresh water. Also known as branching bay; drowned river mouth; firth.

esu See electrostatic units.

E.T. See ephemeris time.

etalon [OPTICS] 1. Two adjustable parallel mirrors mounted so that either one may serve as one of the mirrors in a Michelson interferometer; used to measure distances in terms of wavelengths of spectral lines. 2. An instrument similar to the Fabry-Perot interferometer, except that the distance between the plates is fixed. Also known as Fabry-Perot etalon.

eta meson [PARTIC PHYS] Neutral pseudoscalar meson having zero isotopic spin and hypercharge, positive charge parity and G parity, and a mass of about 549 MeV; decays via electromagnetic interactions.

eta-prime meson See chi meson.

etch [GRAPHICS] To incise lines on a plate of metal, glass, or other material by covering it with an acid-resistant coating, scratching through the coating, and then permitting an acid bath to erode exposed parts of the plate. [MET] To corrode the surface of a metal in order to reveal its composition and structure.

etch cracks [MET] Shallow cracks in the surface of hardened steel that result from reaction with an acid, causing hydrogen cracking.

etched circuit [ENG] A printed circuit formed by chemical or electrolytic removal of unwanted portions of a layer of conductive material bonded to an insulating base.

ethanol [ORG CHEM] C_2H_5OH A colorless liquid, miscible with water, boiling point 78.32°C; used as a reagent and solvent. Also known as alcohol; ethyl alcohol; grain alcohol.

ether [ELECTROMAG] The medium postulated to carry electromagnetic waves, similar to the way a gas carries sound waves. [ORG CHEM] 1. One of a class of organic compounds characterized by the structural feature of an oxygen linking two hydrocarbon groups (such as R—O—R). 2. $(C_2H_5)_2O$ A colorless liquid, slightly soluble in water; used as a reagent, intermediate, anesthetic, and solvent. Also known as ethyl ether.

ether drag [ELECTROMAG] The hypothesis, advanced unsuccessfully to account for results of the Michelson-Morley experiment, that ether is dragged along with matter.

ethereal oil See essential oil.

etherification [ORG CHEM] The process of making an ether from an alcohol.

etherize [MED] To produce anesthesia by administration of ether.

ethmoid bone [ANAT] An irregularly shaped cartilage bone of the skull, forming the medial wall of each orbit and part of the roof and lateral walls of the nasal cavities.

ethology [VERT ZOO] The study of animal behavior in a natural context.

ethoxide [ORG CHEM] A compound formed from ethanol by replacing the hydrogen of the hydroxy group by a monovalent metal. Also known as ethylate.

ethoxylation [CHEM ENG] A catalytic process which involves the direct addition of ethylene oxide to an alkyl phenol or to an aliphatic alcohol.

ethroglucin See erythritol.

ethyl alcohol See ethanol.

ethylate See ethoxide.

ethylation [ORG CHEM] Formation of a new compound by introducing the ethyl radical (C_2H_{5-}).

ethylene resin [ORG CHEM] A thermoplastic material composed of polymers of ethylene; the resin is synthesized by polymerization of ethylene at elevated temperatures and pressures in the presence of catalysts. Also known as polyethylene; polyethylene resin.

ethyl ether *See* ether.

ethylic compound [ORG CHEM] Generic term for ethyl compounds.

ethynylation [ORG CHEM] Production of an acetylenic derivative by the condensation of acetylene with a compound such as an aldehyde; for example, production of butynediol from the union of formaldehyde with acetylene.

etioallocholane *See* androstane.

etiolation [BOT] The yellowing or whitening of green plant parts grown in darkness.

etiology [MED] Any factors which cause disease. [SCI TECH] A branch of science dealing with the causes of phenomena.

Ettingshausen effect [PHYS] The phenomenon that, when a metal strip is placed with its plane perpendicular to a magnetic field and an electric current is sent longitudinally through the strip, corresponding points on opposite edges of the strip have different temperatures.

Ettingshausen-Nernst effect [PHYS] The phenomenon that, when a conductor or semiconductor is subjected to a temperature gradient and to a magnetic field perpendicular to the temperature gradient, an electric field arises perpendicular to both the temperature gradient and the magnetic field. Also known as Nernst effect.

ettringite [MINERAL] $Ca_6Al_2(SO_4)_3(OH)_{12}\cdot26H_2O$ A mineral composed of hydrous basic calcium and aluminum sulfate.

Eu *See* europium.

EU *See* expected value.

Eubacteriales [MICROBIO] Formerly an order of the class Schizomycetes; considered the true bacteria and characterized by simple, undifferentiated, rigid cells of either spherical or straight, rod-shaped form.

Eubasidiomycetes [MYCOL] An equivalent name for Homobasidiomycetidae.

Eubrya [BOT] A subclass of the mosses (Bryopsida); the leafy gametophytes arise from buds on the protonemata, which are nearly always filamentous or branched green threads attached to the substratum by rhizoids.

Eubryales [BOT] An order of mosses (Bryatae); plants have the sporophyte at the end of a stem, vary in size from small to robust, and generally grow in tufts.

eucairite [MINERAL] CuAgSe A white, native selenide that crystallizes in the isometric crystal system.

Eucarida [INV ZOO] A large superorder of the decapod crustaceans, subclass Malacostraca, including shrimps, lobsters, hermit crabs, and crabs; characterized by having the shell and thoracic segments fused dorsally and the eyes on movable stalks.

Eucaryota [BIOL] Primitive, unicellular organisms having a well-defined nuclear membrane, chromosomes, and mitotic cell division.

eucaryote *See* eukaryote.

Eucestoda [INV ZOO] The true tapeworms, a subclass of the class Cestoda.

Eucharitidae [INV ZOO] A family of hymenopteran insects in the superfamily Chalcidoidea.

euchlorin [MINERAL] $(K,Na)_8Cu_9(SO_4)_{10}(OH)_6$ An emerald-green mineral consisting of a basic sulfate of potassium, sodium, and copper; found in lava at Vesuvius.

euchroite [MINERAL] $Cu_2(AsO_4)(OH)\cdot3H_2O$ An emerald green or leek green, orthorhombic mineral consisting of a hydrated basic copper arsenate.

euchromatin [CYTOL] The portion of the chromosomes that stains with low intensity, uncoils during interphase, and condenses during cell division.

Eucinetidae [INV ZOO] The plate thigh beetles, a family of coleopteran insects in the superfamily Dascilloidea.

euclase [MINERAL] $BeAlSiO_4(OH)$ A brittle, pale green, blue, yellow, or violet monoclinic mineral, occurring as prismatic crystals.

Euclasterida [INV ZOO] An order of asteroid echinoderms in which the arms are sharply distinguished from a small, central disk-shaped body.

Eucleidae [INV ZOO] The slug moths, a family of lepidopteran insects in the suborder Heteroneura.

euclidean geometry [MATH] The study of the properties preserved by isometries of two- and three-dimensional euclidean space.

Euclymeninae [INV ZOO] A subfamily of annelids in the family Maldonidae of the Sedentaria, having well-developed plaques and an anal pore within the plaque.

Eucnemidae [INV ZOO] The false click beetles, a family of coleopteran insects in the superfamily Elateroidea.

Eucoccida [INV ZOO] An order of parasitic protozoans in the subclass Coccidia characterized by alternating sexual and asexual phases; stages of the life cycle occur intracellularly in vertebrates and invertebrates.

Eucoelomata [ZOO] A large sector of the animal kingdom including all forms in which there is a true coelom or body cavity; includes all phyla above Aschelminthes.

Eucommiales [BOT] A monotypic order of dicotyledonous plants in the subclass Hamamelidae; plants have two, unitegmic ovules and lack stipules.

eucrite [MINERAL] An olivine-bearing gabbro containing unusually calcic plagiocase; a meteorite component.

eucryptite [MINERAL] $LiAlSiO_4$ A colorless or white lithium aluminum silicate mineral, crystallizing in the hexagonal system; specific gravity is 2.67.

Eudactylinidae [INV ZOO] A family of parasitic copepod crustaceans in the suborder Caligoida; found as ectoparasites on the gills of sharks.

eudialite [MINERAL] $(Na,Ca,Fe)_6ZrSi_6O_{18}(OH,Cl)$ Hexagonal-crystalline silicate chloride mineral; color is red to brown.

eudidymite [MINERAL] $NaBeSi_3O_7(OH)$ A glassy white mineral composed of sodium beryllium silicate.

Euechinoidea [INV ZOO] A subclass of echinoderms in the class Echinoidea; distinguished by the relative stability of ambulacra and interambulacra.

eugenics [GEN] The use of practices that influence the hereditary qualities of future generations, with the aim of improving the genetic future of humanity.

eugeosyncline [GEOL] The internal volcanic belt of an orthogeosyncline.

Euglenida [INV ZOO] An order of protozoans in the class Phytamastigophorea, including the largest green, noncolonial flagellates.

Euglenidae [INV ZOO] The antlike leaf beetles, a family of coleopteran insects in the superfamily Tenebrionoidea.

Euglenoidina [INV ZOO] The equivalent name for Euglenida.

Euglenophyceae [BOT] The single class of the plant division Euglenophyta.

Euglenophyta [BOT] A division of the plant kingdom including one-celled, chiefly aquatic flagellate organisms having a spindle-shaped or flattened body, naked or with a pellicle.

euglobulin [BIOCHEM] True globulin; a simple protein that is soluble in distilled water and dilute salt solutions.

Eugregarinida [INV ZOO] An order of protozoans in the subclass Gregarinia; parasites of certain invertebrates.

euhedral *See* automorphic.

eukaryote [BIOL] A cell with a definitive nucleus. Also spelled eucaryote.

Eulamellibranchia [INV ZOO] The largest subclass of the molluscan class Bivalvia, having a heterodont shell hinge, leaflike gills, and well-developed siphons.

Euler angles [MECH] Three angular parameters that specify the orientation of a body with respect to reference axes.

Euler diagram [MATH] A diagram consisting of closed curves, used to represent relations between logical propositions or sets; similar to a Venn diagram.

Euler equation [MECH] Expression for the energy removed from a gas stream by a rotating blade system (as a gas turbine), independent of the blade system (as a radial- or axial-flow system).

Euler equations of motion [MECH] A set of three differential equations expressing relations between the force moments, angular velocities, and angular accelerations of a rotating rigid body.

Eulerian coordinates [FL MECH] Any system of coordinates in which properties of a fluid are assigned to points in space at each given time, without attempt to identify individual fluid parcels from one time to the next; a sequence of synoptic charts is a Eulerian representation of the data.

Eulerian equation [FL MECH] A mathematical representation of the motions of a fluid in which the behavior and the properties of the fluid are described at fixed points in a coordinate system.

Eulerian nutation *See* Chandler wobble.

Euler-Lagrange equation [MATH] A partial differential equation arising in the calculus of variations, which provides a necessary condition that $y(x)$ minimize the integral over some finite interval of $f(x,y,y')dx$, where $y' = dy/dx$; the equation is $(\partial f(x,y,y')/\partial y) - (d/dx)(\partial f(x,y,y')/\partial y') = 0$. Also known as Euler's equation.

Euler's equation *See* Euler-Lagrange equation.

Euler's expansion [FL MECH] The transformation of a derivative (d/dt) describing the behavior of a moving particle with respect to time, into a local derivative $(\delta/\delta t)$ and three additional terms that describe the changing motion of a fluid as it passes through a fixed point.

Euler's formula [MATH] The formula $e^{ix} = \cos x + i \sin x$, where $i = \sqrt{-1}$.

Euler transformation [MATH] A method of obtaining from a given convergent series a new series which converges faster to the same limit, and for defining sums of certain divergent series; the transformation carries the series $a_0 - a_1 + a_2 - a_3 + \ldots$ into a series whose nth term is

$$\sum_{r=0}^{n-1} (-1)^r \binom{n-1}{r} a_r/2^n.$$

eulittoral [OCEANOGR] A subdivision of the benthic division of the littoral zone of the marine environment, extending from high-tide level to about 60 meters, the lower limit for abundant growth of attached plants.

Eulophidae [INV ZOO] A family of hymenopteran insects in the superfamily Chalcidoidea including species that are parasitic on the larvae of other insects.

eulytine *See* eulytite.

eulytite [MINERAL] $Bi_4Si_3O_{12}$ A bismuth silicate mineral usually found as minute dark-brown or gray tetrahedral crystals; specific gravity is 6.11. Also known as agricolite; bismuth blende; eulytine.

Eumalacostraca [INV ZOO] A series of the class Crustacea comprising shrimplike crustaceans having eight thoracic segments, six abdominal segments, and a telson.

Eumetazoa [ZOO] A section of the animal kingdom that includes the phyla above the Porifera; contains those animals which have tissues or show some tissue formation and organ systems.

Eumycetes [MYCOL] The true fungi, a large group of microorganisms characterized by cell walls, lack of chlorophyll, and mycelia in most species; includes the unicellular yeasts.

Eumycetozoida [INV ZOO] An order of protozoans in the subclass Mycetozoia; includes slime molds which form a plasmodium.

Eumycophyta [MYCOL] An equivalent name for the Eumycetes.

Eumycota [MYCOL] An equivalent name for Eumycetes.

Eunicea [INV ZOO] A superfamily of polychaete annelids belonging to the Errantia.

Eunicidae [INV ZOO] A family of polychaete annelids in the superfamily Eunicea having characteristic pharyngeal armature consisting of maxillae and mandibles.

Euomphalacea [PALEON] A superfamily of extinct gastropod mollusks in the order Aspidobranchia characterized by shells with low spires, some approaching bivalve symmetry.

eupelagic *See* pelagic.

Eupelmidae [INV ZOO] A family of hymenopteran insects in the superfamily Chalcidoidea.

Euphausiacea [INV ZOO] An order of planktonic malacostracans in the class Crustacea possessing photophores which emit a brilliant blue-green light.

euphenics [GEN] The production of a satisfactory phenotype by means other than eugenics.

Eupheterochlorina [INV ZOO] A suborder of flagellate protozoans in the order Heterochlorida.

Euphorbiaceae [BOT] A family of dicotyledonous plants in the order Euphorbiales characterized by dehiscent fruit having more than one seed and by epitropous ovules.

Euphorbiales [BOT] An order of dicotyledonous plants in the subclass Rosidae having simple leaves and unisexual flowers that are aggregated and reduced.

euphotic [OCEANOGR] Of or constituting the upper levels of the marine environment down to the limits of effective light penetration for photosynthesis.

Euphrosinidae [INV ZOO] A family of amphinomorphan polychaete annelids with short, dorsolaterally flattened bodies.

Euplexoptera [INV ZOO] The equivalent name for Dermaptera.

euploid [GEN] Having a chromosome complement that is an exact multiple of the haploid complement.

Eupodidae [INV ZOO] A family of mites in the suborder Trombidiformes.

Euproopacea [PALEON] A group of Paleozoic horseshoe crabs belonging to the Limulida.

Europa [ASTRON] **1.** A satellite of Jupiter with a mean distance from Jupiter of 4.17×10^5 miles (6.71×10^5 kilometers), orbital period of 3.6 days, and diameter of about 1950 miles (3100 kilometers). Also known as Jupiter II. **2.** An asteroid with a diameter of about 280 kilometers, mean distance from the sun of 3.096 astronomical units, and C-type surface composition.

European canker [PL PATH] **1.** A fungus disease of apple, pear, and other fruit and shade trees caused by *Nectria galligena* and characterized by cankers with concentric rings of callus on the trunk and branches. **2.** A fungus disease of poplars caused by *Dothichiza populea*.

europium [CHEM] A member of the rare-earth elements in the cerium subgroup, symbol Eu, atomic number 63, atomic weight 151.96, steel gray and malleable, melting at 1100–1200°C.

Eurotiaceae [MYCOL] A family of ascomycetous fungi of the order Eurotiales in which the asci are borne in cleistothecia or closed fruiting bodies.

Eurotiales [MYCOL] An order of fungi in the class Ascomycetes bearing ascospores in globose or broadly oval, delicate asci which lack a pore.

Euryalae [INV ZOO] The basket fishes, a family of echinoderms in the subclass Ophiuroidea.

Euryalina [INV ZOO] A suborder of ophiuroid echinoderms in the order Phrynophiurida characterized by a leathery integument.

Euryapsida [PALEON] A subclass of fossil reptiles distinguished by an upper temporal opening on each side of the skull.

eurybathic [ECOL] Living at the bottom of a body of water.

Eurychilinidae [PALEON] A family of extinct dimorphic ostracods in the superfamily Hollinacea.

Eurylaimi [VERT ZOO] A monofamilial suborder of suboscine birds in the order Passeriformes.

Eurylaimidae [VERT ZOO] The broadbills, the single family of the avian suborder Eurylaimi.

Eurymylidae [PALEON] A family of extinct mammals presumed to be the ancestral stock of the order Lagomorpha.

Euryphoridae [INV ZOO] A family of copepod crustaceans in the order Caligoida; members are fish ectoparasites.

Eurypterida [PALEON] A group of extinct aquatic arthropods in the subphylum Chelicerata having elongate-lanceolate bodies encased in a chitinous exoskeleton.

Eurypygidae [VERT ZOO] The sun bitterns, a family of tropical and subtropical New World birds belonging to the order Gruiformes.

Eurytomidae [INV ZOO] The seed and stem chalcids, a family of hymenopteran insects in the superfamily Chalcidoidea.

Eusiridae [INV ZOO] A family of pelagic amphipod crustaceans in the suborder Gammaridea.

eustachian tube [ANAT] A tube composed of bone and cartilage that connects the nasopharynx with the middle ear cavity.

eustachian valve *See* canal valve.

eustele [BOT] A modified siphonostele containing collateral or bicollateral vascular bundles; found in most gymnosperm and angiosperm stems.

Eusuchia [VERT ZOO] The modern crocodiles, a suborder of the order Crocodilia characterized by a fully developed secondary palate and procoelous vertebrae.

Eusyllinae [INV ZOO] A subfamily of polychaete annelids in the family Syllidae having a thick body and unsegmented cirri.

Eutardigrada [INV ZOO] An order of tardigrades which lack both a sensory cephalic appendage and a club-shaped appendage.

eutectic [MET] The microstructure that results when a metal of eutectic composition solidifies. [PHYS CHEM] An alloy or solution that has the lowest possible constant melting point.

eutectic crystallization [MET] Simultaneous crystallization of the constituents of a eutectic alloy during cooling of the melt.

eutectic point [PHYS CHEM] The point in the constitutional diagram indicating the composition and temperature of the lowest melting point of a eutectic.

eutectic system [PHYS CHEM] The particular composition and temperature of materials at the eutectic point.

eutectic temperature [PHYS CHEM] The temperature at the lowest melting point of a eutectic.

eutectoid [MET] A mixture of phases whose composition is determined by the eutectoid point in the solid region of an equilibrium diagram and whose constituents are formed by the eutectoid reaction. [PHYS CHEM] The point in an equilibrium diagram for a solid solution at which the solution on cooling is converted to a mixture of solids.

Euthacanthidae [PALEON] A family of extinct acanthodian fishes in the order Climatiiformes.

Eutheria [VERT ZOO] An infraclass of therian mammals including all living forms except the monotremes and marsupials.

Eutrichosomatidae [INV ZOO] Small family of hymenopteran insects in the superfamily Chalcidoidea.

eutrophic [HYD] Pertaining to a lake containing a high concentration of dissolved nutrients; often shallow, with periods of oxygen deficiency.

eutrophication [ECOL] Of bodies of water, the process of becoming better nourished either naturally by processes of maturation or artificially by fertilization.

euxenite [MINERAL] A brownish-black rare-earth mineral that crystallizes in the orthorhombic system, contains oxide of calcium, cerium, columbium, tantalum, titanium, and uranium, and has a metallic luster; hardness is 6.5 on Mohs scale, and specific gravity is 4.7–5.0.

euxinic [HYD] Of or pertaining to an environment of restricted circulation and stagnant or anaerobic conditions.

eV *See* electron volt.

EV *See* expected value.

Evaniidae [INV ZOO] The ensign flies, a family of hymenopteran insects in the superfamily Proctotrupoidea.

evansite [MINERAL] $Al_3(PO_4)(OH)_6 \cdot 6H_2O$ A colorless to milky white mineral consisting of a hydrated basic aluminum phosphate; occurs in massive form and as stalactites.

evaporation [PHYS] Conversion of a liquid to the vapor state by the addition of latent heat.

evaporation gage *See* atmometer.

evaporative condenser [MECH ENG] An apparatus in which vapor is condensed within tubes that are cooled by the evaporation of water flowing over the outside of the tubes.

evaporative cooling [ENG] **1.** Lowering the temperature of a large mass of liquid by utilizing the latent heat of vaporization of a portion of the liquid. **2.** Cooling air by evaporating water into it. **3.** *See* vaporization cooling.

evaporative cooling tower *See* wet cooling tower.

evaporative heat regulation [PHYSIO] The composite process by which an animal body is cooled by evaporation of sensible perspiration; this avenue of heat loss serves as a physical means of regulating the body temperature.

evaporator [CHEM ENG] A device used to vaporize part or all of the solvent from a solution; the valuable product is usually either a solid or concentrated solution of the solute. [MECH ENG] Any of many devices in which liquid is changed to the vapor state by the addition of heat, for example, distiller, still, dryer, water purifier, or refrigeration system element where evaporation proceeds at low pressure and consequent low temperature.

evaporimeter See atmometer.

evaporite [GEOL] Deposits of mineral salts from sea water or salt lakes due to evaporation of the water.

evapotranspiration [HYD] Discharge of water from the earth's surface to the atmosphere by evaporation from lakes, streams, and soil surfaces and by transpiration from plants. Also known as fly-off; total evaporation; water loss.

even-even nucleus [NUC PHYS] A nucleus which has an even number of neutrons and an even number of protons.

even function [MATH] A function with the property that $f(x) = f(-x)$ for each number x.

even number [MATH] An integer which is a multiple of 2.

even-odd nucleus [NUC PHYS] A nucleus which has an even number of protons and an odd number of neutrons.

even parity check [COMPUT SCI] A parity check in which the number of 0's or 1's in each word is expected to be even.

event [COMPUT SCI] The moment of time at which a specified change of state occurs; usually marks the completion of an asynchronous input/output operation. [GEOL] An incident of probable tectonic significance, but whose full implications are unknown. [PHYS] A point in space-time. [STAT] A mathematical model of the result of a conceptual experiment; this model is a measurable subset of a probability space.

Eventognathi [VERT ZOO] The equivalent name for Cypriniformes.

evergreen [BOT] Pertaining to a perennially green plant. Also known as aiophyllous.

Evjen method [SOLID STATE] Method of calculating lattice sums in which groups of charges whose total charge is zero are taken together, so that the contribution of each group is small and the series rapidly converges.

evjite [PETR] A gabbro of hornblende in which the only light-colored mineral is labradorite or bytownite; hornblende must be primary, not uralitic.

evoked potential [PHYSIO] Electrical response of any neuron to stimuli.

evolute [MATH] The locus of the centers of curvature of a curve.

evolution [BIOL] The processes of biological and organic change in organisms by which descendants come to differ from their ancestors.

evorsion hollow See pothole.

EVR See electronic video recording.

Ewald method [SOLID STATE] Method of calculating lattice sums in which certain mathematical techniques are employed to make series converge rapidly.

E wave See transverse magnetic wave.

EXAFS See extended x-ray absorption fine structure.

exalted-carrier receiver [ELECTR] Receiver that counteracts selective fading by maintaining the carrier at a high level at all times; this minimizes the second harmonic distortion that would otherwise occur when the carrier drops out while leaving most of the sidebands at their normal amplitudes.

exanthem subitum [MED] A mild, sometimes epidemic viral disease of young children, with abrupt onset, high fever, and rash. Also known as roseola infantum.

except [MATH] A logical operator which has the property that if P and Q are two statements, then the statement "P except Q" is true only when P alone is true; it is false for the other three combinations (P false Q false, P false Q true, and P true Q true).

except gate [ELECTR] A gate that produces an output pulse only for a pulse on one or more input lines and the absence of a pulse on one or more other lines.

exceptional space [QUANT MECH] A space used to describe a system with a finite number of degrees of freedom in a generalization of quantum mechanics; this generalization is achieved by reformulating quantum mechanics in terms of a Jordan algebra of observables and states, and then generalizing this to the exceptional Jordan algebra realized by the algebra of 3×3 Hermitian matrices over the Cayley numbers.

exception-item encoding [COMPUT SCI] A technique which allows the uninterrupted flow of a process by the automatic shunting of erroneous records to an error tape for later corrections.

exception-principle system [COMPUT SCI] A technique which assumes no printouts except when an error is encountered.

exception reporting [COMPUT SCI] A form of programming in which only values that are outside predetermined limits, representing significant changes, are selected for printout at the output of a computer.

Excepulaceae [MYCOL] The equivalent name for Discellaceae.

excess electron [SOLID STATE] Electron introduced into a semiconductor by a donor impurity and available for conduction.

excess-fifty code [COMPUT SCI] A number code in which the number n is represented by the binary equivalent of $n + 50$.

excess-three code [COMPUT SCI] A number code in which the decimal digit n is represented by the four-bit binary equivalent of $n + 3$.

exchange [COMMUN] 1. A unit established by a telephone company for the administration of telephone service in a specified area, usually a town, a city, or a village and its environs, and consisting of one or more central offices together with the associated plant used in furnishing telephone service in that area. Also known as local exchange. 2. Room or building equipped so telephone lines terminating there may be interconnected as required; equipment may include a switchboard or automatic switching apparatus. [COMPUT SCI] The interchange of contents between two locations. [QUANT MECH] 1. Operation of exchanging the space and spin coordinates in a Schrödinger-Pauli wave function representing two identical particles; this operation must leave the wave function unchanged, except possibly for sign. 2. Process of exchanging a real or virtual particle between two other particles.

exchangeable disk storage [COMPUT SCI] A type of disk storage, used as a backing storage, in which the disks come in capsules, each containing several disks; the capsules can be replaced during operation of the computer and can be stored until needed.

exchange broadening [SPECT] The broadening of a spectral line by some type of chemical or spin exchange process which limits the lifetime of the absorbing or emitting species and produces the broadening via the Heisenberg uncertainty principle.

exchange buffering [COMPUT SCI] An input/output buffering technique that avoids the internal moving of data.

exchange coefficient [FL MECH] A coefficient of eddy flux in turbulent flow, defined in analogy to those coefficients of the kinetic theory of gases. Also known as austausch coefficient; eddy coefficient; interchange coefficient.

exchange current [ELEC] The magnitude of the current which flows through a galvanic cell when it is operating in a reversible manner.

exchange message [COMPUT SCI] A device, placed between a communication line and a computer, in order to take care of certain communication functions and thereby free the computer for other work.

exchange narrowing [SPECT] The phenomenon in which, when a spectral line is split and thereby broadened by some variable perturbation, the broadening may be narrowed by a dynamic process that exchanges different values of the perturbation.

exchange operator [QUANT MECH] An operator which exchanges the spatial coordinates of the particles in a wave function, or their spins, or both positions and spins.

exchanger *See* heat exchanger.

exchange reaction [CHEM] Reaction in which two atoms or ions exchange places either in two different molecules or in the same molecule.

Excipulaceae [MYCOL] The equivalent name for Discellaceae.

excitable [BIOL] Describing a tissue or organism that exhibits irritability.

excitation [ATOM PHYS] A process in which an atom or molecule gains energy from electromagnetic radiation or by collision, raising it to an excited state. [CONT SYS] The application of energy to one portion of a system or apparatus in a manner that enables another portion to carry out a specialized function; a generalization of the electricity and electronics definitions. [ELEC] The application of voltage to field coils to produce a magnetic field, as required for the operation of an excited-field loudspeaker or a generator. [ELECTR] 1. The signal voltage that is applied to the control electrode of an electron tube. Also known as drive. 2. Application of signal power to a transmitting antenna. [QUANT MECH] The addition of energy to a particle or system of particles at ground state to produce an excited state.

excitation curve [NUC PHYS] A curve showing the relative yield of a specified nuclear reaction as a function of the energy of the incident particles or photons. Also known as excitation function.

excitation energy [QUANT MECH] The minimum energy required to change a system from its ground state to a particular excited state.

excitation function [ATOM PHYS] 1. The cross section for an incident electron to excite an atom to a particular excited state expressed as a function of the electron energy. 2. *See* excitation curve.

excitation index [SPECT] In emission spectroscopy, the ratio of intensities of a pair of extremely nonhomologous spectra lines; used to provide a sensitive indication of variation in excitation conditions.

excitation loss *See* core loss.

excitation potential [QUANT MECH] Electric potential which gives the excitation energy when multiplied by the magnitude of the electron charge.

excitation spectrum [SPECT] The graph of luminous efficiency per unit energy of the exciting light absorbed by a photoluminescent body versus the frequency of the exciting light.

excited state [QUANT MECH] A stationary state of higher energy than the lowest stationary state or ground state of a particle or system of particles.

exciter [ELEC] 1. A small auxiliary generator that provides field current for an alternating-current generator. 2. *See* exciter lamp. [ELECTR] A crystal oscillator or self-excited oscillator used to generate the carrier frequency of a transmitter. [ELECTROMAG] 1. The portion of a directional transmitting antenna system that is directly connected to the transmitter. 2. A loop or probe extending into a resonant cavity or waveguide.

exciter lamp [ELEC] A bright incandescent lamp having a concentrated filament, used to excite a phototube or photocell in sound movie and facsimile systems. Also known as exciter.

exciting line [SPECT] The frequency of electromagnetic radiation, that is, the spectral line from a non-continuous source, which is absorbed by a system in connection with some particular process.

exciton [SOLID STATE] An excited state of an insulator or semiconductor which allows energy to be transported without transport of electric charge; may be thought of as an electron and a hole in a bound state.

exclusion principle [QUANT MECH] The principle that no two fermions of the same kind may simultaneously occupy the same quantum state. Also known as Pauli exclusion principle.

exclusive-NOR gate *See* equivalence gate.

exclusive or [COMPUT SCI] An instruction which performs the "exclusive or" operation on a bit-by-bit basis for its two operand words, usually storing the result in one of the operand locations. [MATH] A logic operator which has the property that if P is a statement and Q is a statement, then P exclusive or Q is true if either but not both statements are true, false if both are true or both are false.

exclusive segments [ADP] Parts of an overlay program structure that cannot be resident in main memory simultaneously.

Excorallanidae [INV ZOO] A family of free-living and parasitic isopod crustaceans in the suborder Flabellifera which have mandibles and first maxillae modified as hooklike piercing organs.

excoriation [MED] Abrasion of a portion of the skin.

excrescence [BIOL] 1. Abnormal or excessive increase in growth. 2. An abnormal outgrowth.

excretion [PHYSIO] The removal of unusable or excess material from a cell or a living organism.

excretory system [ANAT] Those organs concerned with solid, fluid, or gaseous excretion.

excurrent [BIOL] Flowing out. [BOT] 1. Having an undivided main stem or trunk. 2. Having the midrib extending beyond the apex.

excursion [NUCLEO] A sudden, very rapid rise in the power level of a nuclear reactor caused by supercriticality.

execute [COMPUT SCI] Usually, to run a compiled or assembled program on the computer; by extension, to compile or assemble and to run a source program.

execution control program [COMPUT SCI] The program delivered by the manufacturer which permits the computer to handle the programs fed to it.

execution cycle [COMPUT SCI] The time during which an elementary operation takes place.

execution error detection [COMPUT SCI] The detection of errors which become apparent only during execution time.

execution time [COMPUT SCI] The time during which actual work, such as addition or multiplication, is carried out in the execution of a computer instruction.

executive communications [COMPUT SCI] The routine information transmitted to the operator on the status of programs being executed and of the requirements made by these programs of the various components of the system.

executive-control language [COMPUT SCI] The generic term for a finite set of instructions which enables the programmer to run a program more efficiently.

executive file-control system [COMPUT SCI] The assignment of intermediate storage devices performed by the computer, and over which the programmer has no control.

executive instruction [COMPUT SCI] Instruction to determine how a specially written computer program is to operate.

executive routine [COMPUT SCI] A digital computer routine designed to process and control other routines. Also known as master routine; monitor routine.

executive supervisor [COMPUT SCI] The component of the computer system which controls the sequencing, setup, and execution of the jobs presented to it.

executive system [COMPUT SCI] A set of programs and routines which guides a computer in the performance of its tasks, assists the programs (and programmers) with certain supporting functions, and increases the usefulness of the computer's hardware. Also known as monitor system; operating system.

exergonic [BIOCHEM] Of or pertaining to a biochemical reaction in which the end products possess less free energy than the starting materials; usually associated with catabolism.

exfoliation [GEOL] *See* sheeting. [MED] **1.** The separation of bone or other tissue in thin layers. **2.** A peeling and shedding of the horny layer of the skin. [MET] Peeling off or separation of metal at its surface in the form of thin, parallel scales or lamellae. [PETR] The breaking off of thin concentric shells, sheets, scales, plates, and so on, from a rock mass; measuring less than a centimeter to several meters in thickness, the loosened rock is spalled, peeled, or stripped. [SCI TECH] Flaking away or peeling off in scales.

exfoliation joint *See* sheeting structure.

exfoliative cytology [PATH] The study of cells shed spontaneously from the body surfaces; used principally in the diagnosis of cancer.

exhalation [GEOPHYS] The process by which radioactive gases escape from the surface layers of soil or loose rock, where they are formed by decay of radioactive salts. [PHYSIO] The giving off or sending forth in the form of vapor; expiration.

exhaust [MECH ENG] **1.** The working substance discharged from an engine cylinder or turbine after performing work on the moving parts of the machine. **2.** The phase of the engine cycle concerned with this discharge. **3.** A duct for the escape of gases, fumes, and odors from an enclosure, sometimes equipped with an arrangement of fans. [SCI TECH] *See* evacuate.

exhaustion region [ELECTR] A layer in a semiconductor, adjacent to its contact with a metal, in which there is almost complete ionization of atoms in the lattice and few charge carriers, resulting in a space-charge density.

exhaust valve [MECH ENG] The valve on a cylinder in an internal combustion engine which controls the discharge of spent gas.

exhaust velocity [FL MECH] The velocity of gaseous or other particles in the exhaust stream of the nozzle of a reaction engine, relative to the nozzle.

exine *See* exosporium.

exinite [GEOL] A hydrogen-rich maceral group consisting of spore exines, cuticular matter, resins, and waxes; includes sporinite, cutinite, alginite, and resinite. Also known as liptinite.

exit [COMPUT SCI] **1.** A way of terminating a repeated cycle of operations in a computer program. **2.** A place at which such a cycle can be stopped. [ENG] A door, passage, or place of egress.

exite [INV ZOO] A movable appendage or lobe located on the external side of the limb of a generalized arthropod.

exit pupil [OPTICS] The image of the aperture stop of an optical system formed in the image space by rays emanating from a point on the optical axis in the object space.

exmeridian altitude [ASTRON] An altitude of a celestial body near the celestial meridian of the observer to which a correction is to be applied to determine the meridian altitude. Also known as circummeridian altitude.

exmeridian observation [ASTRON] **1.** Measurement of the altitude of a celestial body near the celestial meridian of the observer, for conversion to a meridian altitude. **2.** The altitude so measured.

exo- [ORG CHEM] A conformation of carbon bonds in a six-membered ring such that the molecule is boat-shaped with one or more substituents directed outward from the ring. [SCI TECH] A prefix denoting outside or outer.

Exocoetidae [VERT ZOO] The halfbeaks, a family of actinopterygian fishes in the order Atheriniformes.

exocrine gland [PHYSIO] A structure whose secretion is passed directly or by ducts to its exterior surface, or to another surface which is continuous with the external surface of the gland.

exodermis *See* hypodermis.

exoelectrons [PHYS] Electrons emitted from the surfaces of metals and certain ceramics after these surfaces have been freshly formed by a process such as abrasion or fracture; electrons obtain energy required for emission from processes such as establishment of surface films and rearrangement of disturbed atoms.

exoenzyme [BIOCHEM] An enzyme that functions outside the cell in which it was synthesized.

exoergic *See* exothermic.

exoergic collision *See* collision of the second kind.

exogamy [GEN] Union of gametes from organisms that are not closely related. Also known as outbreeding.

exogenote [GEN] The genetic fragment transferred from the donor to the recipient cell during the process of recombination in bacteria.

exogenous [BIOL] **1.** Due to an external cause; not arising within the organism. **2.** Growing by addition to the outer surfaces. [PHYSIO] Pertaining to those factors in the metabolism of nitrogenous substances obtained from food.

exogenous inclusion *See* xenolith.

exogenous variables [MATH] In a mathematical model, the independent variables, which are predetermined and given outside the model.

exogeosyncline [GEOL] A parageosyncline that lies along the cratonal border and obtains its clastic sediments from erosion of the adjacent orthogeosynclinal belt outside the craton. Also known as deltageosyncline; foredeep; transverse basin.

Exogoninae [INV ZOO] A subfamily of polychaete annelids in the family Syllidae having a short, small body of few segments.

exomorphic zone *See* aureole.

exon [GEN] That portion of deoxyribonucleic acid which codes for the final messenger ribonucleic acid.

exonuclease [BIOCHEM] Any of a group of enzymes which catalyze hydrolysis of single nucleotide residues from the end of a deoxyribonucleic acid chain.

exopeptidase [BIOCHEM] An enzyme that acts on the terminal peptide bonds of a protein chain.

exophthalmic goiter *See* hyperthyroidism.

exophthalmos [MED] Abnormal protrusion of the eyeball from the orbit.

exopodite [INV ZOO] The outer branch of a biramous crustacean appendage.

Exopterygota [INV ZOO] A division of the insect subclass Pterygota including those insects which undergo a hemimetabolous metamorphosis.

exoskeleton [INV ZOO] The external supportive covering of certain invertebrates, such as arthropods. [VERT ZOO] Bony or horny epidermal derivatives, such as nails, hoofs, and scales.

exosphere [METEOROL] An outermost region of the atmosphere, estimated at 500–1000 kilometers, where the density is so low that the mean free path of particles depends upon their direction with respect to the local vertical, being greatest for upward-traveling particles. Also known as region of escape.

exosporium [BOT] The outer of two layers forming the wall of spores such as pollen and bacterial spores. Also known as exine.

exothermic [PHYS] Indicating liberation of heat. Also known as exoergic.

exotic stream [HYD] A stream that crosses a desert as it flows to the sea, or any stream which derives most of its water from the drainage system of another region.

exotoxin [MICROBIO] A toxin that is excreted by a microorganism.

expanded plastic [MATER] A light, spongy plastic made by introducing pockets of air or gas. Also known as foamed plastic; plastic foam.

expander [ELECTR] A transducer that, for a given input amplitude range, produces a larger output range.

expanding brake [MECH ENG] A brake that operates by moving outward against the inside rim of a drum or wheel.

expanding universe [ASTROPHYS] Explanation of the red shift observed in spectral lines from distant galaxies as due to a mutual recession of galaxies away from each other. [RELAT] A model of the universe describing the process defined in the astronomy definition, in which the universe is nonstatic, homogeneous, and isotropic; based on Einstein's field equations with a nonvanishing cosmical constant.

expandor [ELECTR] The part of a compandor that is used at the receiving end of a circuit to return the compressed signal to its original form; attenuates weak signals and amplifies strong signals.

expansion [ELECTR] A process in which the effective gain of an amplifier is varied as a function of signal magnitude, the effective gain being greater for large signals than for small signals; the result is greater volume range in an audio amplifier and greater contrast range in facsimile. [MECH ENG] Increase in volume of working material with accompanying drop in pressure of a gaseous or vapor fluid, as in an internal combustion engine or steam engine cylinder. [PHYS] Process in which the volume of a constant mass of a substance increases.

expansion bolt [DES ENG] A bolt having an end which, when embedded into masonry or concrete, expands under a pull on the bolt, thereby providing anchorage.

expansion chamber *See* cloud chamber.

expansion ellipsoid [SOLID STATE] An ellipsoid whose axes have lengths which are proportional to the coefficient of linear expansion in the corresponding direction in a crystal.

expansion fissures [GEOL] A system of fissures which radiate randomly and pass through feldspars and other minerals adjacent to olivine crystals that have been replaced by serpentine.

expansion fit [DES ENG] A condition of optimum clearance between certain mating parts in which the cold inner member is placed inside the warmer outer member and the temperature is allowed to equalize.

expansion joint [GEOL] *See* sheeting structure. [MECH ENG] 1. A joint between parts of a structure or machine to avoid distortion when subjected to temperature change. 2. A pipe coupling which, under temperature change, allows movement of a piping system without hazard to associated equipment.

expansion ratio [FL MECH] For the calculation of the mass flow of a gas out of a nozzle or other expanding duct, the ratio of the nozzle exit section area to the nozzle throat area, or the ratio of final to initial volume. [MECH ENG] In a reciprocating piston engine, the ratio of cylinder volume with piston at bottom dead center to cylinder volume with piston at top dead center.

expectation *See* expected value.

expected utility *See* expected value.

expected value [MATH] 1. For a random variable x with probability density function $f(x)$, this is the integral from $-\infty$ to ∞ of $xf(x)dx$. Also known as expectation. 2. For a random variable x on a probability space (Ω, P), the integral of x with respect to the probability measure P. [SYS ENG] In decision theory, a measure of the value or utility expected to result from a given strategy, equal to the sum over states of nature of the product of the probability of the state times the consequence or outcome of the strategy in terms of some value or utility parameter. Abbreviated EV. Also known as expected utility (EU).

experimental design [STAT] A pattern for setting up experiments and making observations about the relationship between several variables in which one attempts to obtain as much information as possible for a fixed expenditure level.

exploded view [GRAPHICS] A drawing or picture of any article or piece of equipment in which the component parts are separated but so arranged to show their relationship to the whole.

exploration [MIN ENG] The search for economic deposits of minerals, ore, gas, oil, or coal by geological surveys, geophysical prospecting, boreholes and trial pits, or surface or underground headings, drifts, or tunnels.

exploring coil [ELECTROMAG] A small coil used to measure a magnetic field or to detect changes produced in a magnetic field by a hidden object; the coil is connected to an indicating instrument either directly or through an amplifier. Also known as magnetic test coil; search coil.

explosion [CHEM] A chemical reaction or change of state which is effected in an exceedingly short space of time with the generation of a high temperature and generally a large quantity of gas.

explosion welding [MET] A solid-state process wherein bonding is produced by a controlled detonation, resulting in rapid movement together of the members to be joined.

explosive [MATER] A substance, such as trinitrotoluene, or a mixture, such as gunpowder, that is characterized by chemical stability but may be made to undergo rapid chemical change without an outside source of oxygen, whereupon it produces a large quantity of energy generally accompanied by the evolution of hot gases.

explosive bolt [AERO ENG] A bolt designed to contain a remote-initiated explosive charge which, upon detonation, will shear the bolt or cause it to fail otherwise; applicable to such uses as stage separation of rockets, jettison of expended fuel tanks, and ejection of parachutes.

explosive echo ranging [ENG] Sonar in which a charge is exploded underwater to produce a shock wave that serves the same purpose as an ultrasonic pulse; the elapsed time for return of the reflected wave gives target range.

explosive evolution [EVOL] Rapid diversification of a group of fossil organisms in a short geological time.

explosive forming [MET] Shaping metal parts in dies by using an explosive charge to generate forming pressure.

explosive limits [CHEM ENG] The upper and lower limits of percentage composition of a combustible gas mixed with other gases or air within which the mixture explodes when ignited.

explosive rivet [ENG] A rivet holding a charge of explosive material; when the charge is set off, the rivet expands to fit tightly in the hole.

explosive variable *See* cataclysmic variable.

exponent [MATH] A number or symbol placed to the right and above some given mathematical expression.

exponential curve [MATH] A graph of the function $y = a^x$, where a is a positive constant.

exponential density function [MATH] A probability density function obtained by integrating a function of the form $\exp(-|x - m|/\sigma)$, where m is the mean and σ the standard deviation.

exponential distribution [STAT] A continuous probability distribution whose density function is given by $f(x) = ae^{-ax}$, where $a > 0$ for $x > 0$, and $f(x) = 0$ for $x \leq 0$; the mean and standard deviation are both $1/a$.

exponential law [MATH] *See* law of exponents. [PHYS] The principle that growth or decay of some physical quantity is at a rate such that its value at a certain time or place is the initial value times e raised to a power equal to a constant times some convenient coordinate, such as the elapsed time or the distance traveled by a wave; there is growth if the constant is positive, decay if it is negative.

exponential pulse [PHYS] Variation of some quantity with time similar to the displacement of a critically damped harmonic oscillator which is initially given an impulse in its equilibrium position.

exponential smoothing [IND ENG] A mathematical-statistical method of forecasting used in industrial engineering which assumes that demand for the following period is some weighted average of the demands for the past periods.

exposure [BUILD] The distance from the butt of one shingle to the butt of the shingle above it, or the

amount of a shingle that is seen. [GRAPHICS] The act of permitting light to fall upon a photosensitive material. [MED] The state of being open to some action or influence that may affect detrimentally, as cold, disease, or wetness. [METEOROL] The general surroundings of a site, with special reference to its openness to winds and sunshine. [NUCLEO] 1. The total quantity of radiation at a given point, measured in air. 2. The cumulative amount of radiation exposure to which nuclear fuel has been subjected in a nuclear reactor; usually expressed in terms of the thermal energy produced by the reactor per ton of fuel initially present, as megawatt days per ton. [OPTICS] *See* light exposure; radiant exposure.

exposure factor [NUCLEO] A quantity f used to specify radiographic exposure equal to st/d^2, where s is the intensity of radioactive source, t is the time, and d is the source to film distance.

expression [CHEM ENG] Separation of liquid from a two-phase solid-liquid system by compression under conditions that permit liquid to escape while the solid is retained between the compressing surfaces. Also known as mechanical expression. [COMPUT SCI] A mathematical or logical statement written in a source language, consisting of a collection of operands connected by operations in a logical manner.

exsecant [MATH] The trigonometric function defined by subtracting unity from the secant, that is exsec θ = sec θ − 1.

exserted [BIOL] Protruding beyond the enclosing structure, such as stamens extending beyond the margin of the corolla.

exsolution [GEOL] A phenomenon during which molten rock solutions separate when cooled.

extended-area service [COMMUN] Telephone exchange service, without toll charges, that extends over an area where there is a community of interest, in return for a somewhat higher exchange service rate.

extended binary-coded decimal interchange code [COMPUT SCI] A computer code that uses eight binary positions to represent a single character, giving a possible maximum of 256 characters. Abbreviated EBCDIC.

extended dislocation [CRYSTAL] A dislocation in a close-packed structure consisting of a strip of stacking fault edged by two lines across which slip through a fraction of a lattice constant, into one of the alternative stacking positions, has occurred.

extended forecast [METEOROL] In general, a forecast of weather conditions for a period extending beyond 2 days from the day of issue. Also known as long-range forecast.

extended-precision word [COMPUT SCI] A piece of data of 16 bytes in floating-point arithmetic when additional precision is required.

extended-range forecast *See* medium-range forecast.

extended state [QUANT MECH] A state of motion in which an electron may be found anywhere within a region of a material of linear extent equal to that of the material itself.

extended x-ray absorption fine structure [PHYS] A variation in the x-ray absorption of a substance as a function of energy, at energies just above that required for photons to liberate core electrons into the continuum; it is due to interference between the outgoing photoelectron waves and electron waves backscattered from atoms adjacent to the absorbing atoms. Abbreviated EXAFS.

extender [CHEM] A material used to dilute or extend or change the properties of resins, ceramics, paints, rubber, and so on. [ELEC] A male or female receptacle connected by a short cable to make a test point more conveniently accessible to a test probe.

extend flip-flop [COMPUT SCI] A special flag set when there is a carry-out of the most significant bit in the register after an addition or a subtraction.

extensibility [MECH] The amount to which a material can be stretched or distorted without breaking.

extensible language [COMPUT SCI] A programming language which can be modified by adding new features or changing existing ones.

extensible system [COMPUT SCI] A computer system in which users may extend the basic system by implementing their own languages and subsystems and making them available for others to use.

extensional fault See tension fault.

extension bolt [DES ENG] A vertical bolt that can be slid into place by a long extension rod; used at the top of doors.

extension joints [GEOL] Fractures that form parallel to a compressive force.

extension of a field [MATH] Any field containing the original field.

extensometer [ENG] 1. A strainometer that measures the change in distance between two reference points separated 20–30 meters or more; used in studies of displacements due to seismic activities. 2. An instrument designed to measure minute deformations of small objects subjected to stress.

exterior angle [MATH] 1. An angle between one side of a polygon and the prolongation of an adjacent side. 2. An angle made by a line (the transversal) that intersects two other lines, and either of the latter on the outside.

exterior ballistics [MECH] The science concerned with behavior of a projectile after leaving the muzzle of the firing weapon.

exterior of a set [MATH] The largest open set contained in the complement of a given set.

extern [COMPUT SCI] A pseudoinstruction found in several assembly languages which explicitly tells an assembler that a symbol is external, that is, not defined in the program module.

external angle [MATH] The angle defined by an arc around the boundaries of an internal angle or included angle.

external auditory meatus [ANAT] The external passage of the ear, leading to the tympanic membrane in reptiles, birds, and mammals.

external beam [NUCLEO] A beam of particles which originate in a particle accelerator and are directed outside the accelerator so that they can be used for experiments with external apparatus.

external brake [MECH ENG] A brake that operates by contacting the outside of a brake drum.

external combustion engine [MECH ENG] An engine in which the generation of heat is effected in a furnace or reactor outside the engine cylinder.

external delay [COMPUT SCI] Time during which a computer cannot be operated due to circumstances beyond the reasonable control of the operators and maintenance engineers, such as a failure of the public power supply.

external ear [ANAT] The portion of the ear that receives sound waves, including the pinna and external auditory meatus.

external error [COMPUT SCI] An error sensed by the computer when this error occurs in a device such as a disk drive.

external galaxy [ASTRON] Any galaxy known to exist, besides the Milky Way.

external gas injection [PETRO ENG] Pressure-maintenance gas injection with wells located in the structurally higher positions of the reservoir, usually in the primary or secondary gas cap. Also known as crestal injection; gas-cap injection.

external interrupt [COMPUT SCI] Any interrupt caused by the operator or by some external device such as a tape drive.

external-interrupt status word [COMPUT SCI] The content of a special register which indicates, among other things, the source of the interrupt.

external memory [COMPUT SCI] Any storage device not an integral part of a computer system, such as a magnetic tape or a deck of cards.

external photoelectric effect See photoemission.

external Q [ELECTR] The inverse of the difference between the loaded and unloaded Q values of a microwave tube.

external respiration [PHYSIO] The processes by which oxygen is carried into living cells from the outside environment and by which carbon dioxide is carried in the reverse direction.

external signal [COMPUT SCI] Any message to an operator for which no printout is required but which is self-explanatory, such as a light condition indicating whether the equipment is on or off.

external sorting [COMPUT SCI] The sorting of a list of items by a computer in which the list is too large to be brought into the memory at one time, and instead is brought into the memory a piece at a time so as to produce a collection of ordered sublists which are subsequently reordered by the computer to produce a single list.

external storage [COMPUT SCI] Large-capacity, slow-access data storage attached to a digital computer and used to store information that exceeds the capacity of main storage.

external symbol dictionary [COMPUT SCI] A list of external symbols and their relocatable addresses which allows the linkage editor to resolve interprogram references. Abbreviated ESD.

external wave [FL MECH] 1. A wave in fluid motion having its maximum amplitude at an external boundary such as a free surface. 2. Any surface wave on the free surface of a homogeneous incompressible fluid is an external wave.

extinction [ASTRON] The reduction in the apparent brightness of a celestial object due to absorption and dispersion of its light by the atmosphere; it is greater at low altitudes. [EVOL] The worldwide death and disappearance of a specific organism or group of organisms. [HYD] The drying up of lake by either water loss or destruction of the lake basin. [OPTICS] Phenomenon in which plane polarized light is almost completely absorbed by a polarizer whose axis is perpendicular to the plane of polarization. [PHYS CHEM] See absorbance. [PSYCH] Decrease in frequency and elimination of a conditioned response if reinforcement of the response is withheld.

extinction coefficient See absorptivity.

extrachromosomal inheritance See cytoplasmic inheritance.

extract [CHEM] Material separated from liquid or solid mixture by a solvent. [COMPUT SCI] 1. To form a new computer word by extracting and putting together selected segments of given words. 2. To remove from a computer register or memory all items that meet a specified condition. [PHARM] 1. A pharmaceutical preparation obtained by dissolving the active constituents of a drug with a suitable menstruum, evaporating the solvent, and adjusting to prescribed standards. **2.** A preparation, usually in a concentrated form, obtained by treating plant or animal tissue with a solvent to remove desired odiferous, flavorful, or nutritive components of the tissue.

extract instruction [COMPUT SCI] An instruction that requests the formation of a new expression from selected parts of given expressions.

extraction [CHEM] A method of separation in which a solid or solution is contacted with a liquid solvent (the two being essential mutually insoluble) to transfer one or more components into the solvent. [MED] The act or process of pulling out a tooth.

extraction column [CHEM ENG] Vertical-process vessel in which a desired product is separated from a liquid by countercurrent contact with a solvent in which the desired product is preferentially soluble.

extractive distillation [CHEM ENG] A distillation process to separate components from eutectic mixtures; a solution of the mixture is cooled, causing one component to crystallize out and the other to remain in solution; used to separate p-xylene and m-xylene, using n-pentane as the solvent.

extractor [CHEM ENG] An apparatus for solvent-contact with liquids or solids for removal of specified components. [COMPUT SCI] *See* mask. [ENG] **1.** A machine for extracting a substance by a solvent or by centrifugal force, squeezing, or other action. **2.** An instrument for removing an object. [ORD] A device in the breech mechanism of a gun, rifle, or the like, for pulling an empty cartridge case or an unfired cartridge out of the chamber of a gun, rifle, or the like.

extraembryonic membrane *See* fetal membrane.

extragalactic [ASTRON] Beyond the Milky Way.

extra-high voltage [ELEC] A voltage above 345 kilovolts used for power transmission. Abbreviated ehv.

extraneous response [ELECTR] Any undesired response of a receiver, recorder, or other susceptible device, due to the desired signals, undesired signals, or any combination or interaction among them.

extraordinary component *See* extraordinary wave.

extraordinary ray [OPTICS] One of two rays into which a ray incident on an anisotropic uniaxial crystal is split; its deviation at the crystal's surface depends on the orientation of the crystal, and it is deviated even in the case of normal incidence.

extraordinary wave [GEOPHYS] Magnetoionic wave component which, when viewed below the ionosphere in the direction of propagation, has clockwise or counterclockwise elliptical polarization respectively, accordingly as the earth's magnetic field has a positive or negative component in the same direction. Also known as X wave. [OPTICS] Component of electromagnetic radiation propagating in an anisotropic uniaxial crystal whose electric displacement vector lies in the plane containing the optical axis and the direction normal to the wavefront; it gives rise to the extraordinary ray. Also known as extraordinary component.

extrapolation [MATH] Estimating a function at a point which is larger than (or smaller than) all the points at which the value of the function is known.

extrapyramidal system [ANAT] Descending tracts of nerve fibers arising in the cortex and subcortical motor areas of the brain.

extrasensory perception [PSYCH] The alleged phenomenon of perception or awareness of external events in the absence of any sensory stimulation arising from the events. Abbreviated ESP.

extravasation [GEOL] The eruption of lava from a vent in the earth. [MED] The pouring out or eruption of a body fluid from its proper channel or vessel into the surrounding tissue.

extreme [CLIMATOL] The highest, and in some cases the lowest, value of a climatic element observed during a given period or during a given month or season of that period; if this is the whole period for which observations are available, it is the absolute extreme. [MATH] *See* extremum.

extremely high frequency [COMMUN] The frequency band from 30,000 to 300,000 megahertz in the radio spectrum. Abbreviated EHF.

extremely low frequency [COMMUN] A frequency below 300 hertz in the radio spectrum. Abbreviated ELF.

extreme point [MATH] **1.** A maximum or minimum value of a function. **2.** A point in a convex subset K of a vector space is called extreme if it does not lie on the interior of any line segment contained in K.

extreme ultraviolet radiation *See* vacuum ultraviolet radiation.

extremum [MATH] A maximum or minimum value of a function. Also known as extreme.

extrinsic factor *See* vitamin B_{12}.

extrinsic semiconductor [ELECTR] A semiconductor whose electrical properties are dependent on impurities added to the semiconductor crystal, in contrast to an intrinsic semiconductor, whose properties are characteristic of an ideal pure crystal.

extrinsic variable star [ASTRON] A variable star, such as an eclipsing variable, whose variation in apparent brightness is due to some external cause, rather than to actual variation in the amount of radiation emitted.

extrorse [BIOL] Directed outward or away from the axis of growth.

extruder [ENG] A device that forces ductile or semisoft solids through the die openings of appropriate shape to produce a continuous film, strip, or tubing.

extrusion [ENG] A process in which a hot or cold semisoft solid material, such as metal or plastic, is forced through the orifice of a die to produce a continuously formed piece in the shape of the desired product. [GEOL] Emission of magma or magmatic materials at the surface of the earth. [TEXT] A process for making continuous-filament synthetic fibers by forcing a syruplike liquid through minute holes of a spinneret.

extrusion coating [ENG] A process of placing resin on a substrate by extruding a thin film of molten resin and pressing it onto or into the substrates, or both, without the use of adhesives.

extrusion pressing *See* cold extrusion.

extrusive rock *See* volcanic rock.

exudate [MED] **1.** A proteinaceous, cellular material that passes through blood vessel walls into the surrounding tissue in inflammation or a superficial lesion. **2.** Any substance that is exuded.

exumbrella [INV ZOO] The outer, convex surface of the umbrella of jellyfishes.

eye [FOOD ENG] A hole formed in certain cheeses during ripening, such as in swiss cheese. [ZOO] A photoreceptive sense organ that is capable of forming an image in vertebrates and in some invertebrates such as the squids and crayfishes.

eye of the storm [METEOROL] The center of a tropical cyclone, marked by relatively light winds, confused seas, rising temperature, lowered relative humidity, and often by clear skies.

eyepiece [OPTICS] A lens or optical system which offers to the eye the image originating from another system (the objective) at a suitable viewing distance. Also known as ocular.

eye socket   *See* orbit.

eyespot   [BOT] 1. A small photosensitive pigment body in certain unicellular algae. 2. A dark area around the hilum of certain seeds, as some beans.   [INV ZOO] A simple organ of vision in many invertebrates consisting of pigmented cells overlying a sensory termination.   [PL PATH] A fungus disease of sugarcane and certain other grasses which is caused by *Helminthosporium sacchari* and characterized by yellowish oval lesions on the stems and leaves.

eyestalk   [INV ZOO] A movable peduncle bearing a terminal eye in decapod crustaceans.

Eykman formula   [OPTICS] An empirical formula which relates the molal refraction of a liquid at a given optical frequency to its index of refraction, density, and molecular weight.

Eyring equation   [PHYS CHEM] An equation, based on statistical mechanics, which gives the specific reaction rate for a chemical reaction in terms of the heat of activation, entropy of activation, the temperature, and various constants.

E zone   [COMMUN] One of the three zones into which the earth is divided to show the variations of the F_2 layer in respect of longitude when making frequency predictions; it roughly covers Asia, Australia, the Philippines, and Japan.

F

F *See* farad; fluorine.

F₁ [GEN] Notation for the first filial generation resulting from a cross.

F₂ [GEN] Notation for the progeny produced by intercrossing members of the F_1. Also known as second generation.

fabric [ARCH] The framework of a building. [GEOL] The spatial orientation of the elements of a sedimentary rock. [PETR] The sum of all the structural and textural features of a rock. Also known as petrofabric; rock fabric; structural fabric. [SCI TECH] **1.** Arrangement or pattern of constituent parts. **2.** Materials used in fabrication. [TEXT] A thin, flexible material made of any combination of cloth, fiber, polymeric film, sheet, or foam.

fabric analysis *See* structural petrology.

fabric diagram [PETR] In structural petrology, a graphic representation of the data of fabric elements. Also known as petrofabric diagram.

fabric element [PETR] A surface or line of structural discontinuity in a rock fabric.

Fabriciinae [INV ZOO] A subfamily of small to minute, colonial, sedentary polychaete annelids in the family Sabellidae.

Fabry-Perot etalon *See* etalon.

Fabry-Perot interferometer [OPTICS] An interferometer having two parallel glass plates (whose separation of a few centimeters may be varied), silvered on their inner surfaces so that the incoming wave is multiply reflected between them and ultimately transmitted.

facade [ARCH] The front of a building or a face of a building, given special architectural treatment.

face [ANAT] The anterior portion of the head, including the forehead and jaws. [CRYSTAL] *See* crystal face. [ELECTR] *See* faceplate. [GEOL] **1.** The main surface of a landform. **2.** The original surface of a layer of rock. [GRAPHICS] **1.** A particular style or size of letter as distinguished from another style or size. Also known as typeface. **2.** The printing surface of a printing plate or the front surface of a piece of paper. [MIN ENG] A surface on which mining operations are being performed. Also known as breast. [TEXT] The side of a fabric which is more attractive than the other side because of features such as weave, luster, or finish.

face angle [MATH] An angle between two successive edges of a polyhedral angle.

face-bonding [ELECTR] Method of assembling hybrid microcircuits wherein semiconductor chips are provided with small mounting pads, turned facedown, and bonded directly to the ends of the thin-film conductors on the passive substrate.

face brick [MATER] A brick of some esthetic quality to be used on the exposed surface of a building wall or other structure.

face-centered cubic lattice [CRYSTAL] A lattice whose unit cells are cubes, with lattice points at the center of each face of the cube, as well as at the vertices. Abbreviated fcc lattice.

face-centered orthorhombic lattice [CRYSTAL] An orthorhombic lattice which has lattice points at the center of each face of a unit cell, as well as at the vertices.

facellite *See* kaliophillite.

faceplate [ELECTR] The transparent or semitransparent glass front of a cathode-ray tube, through which the image is viewed or projected; the inner surface of the face is coated with fluorescent chemicals that emit light when hit by an electron beam. Also known as face. [ENG] **1.** A disk fixed perpendicularly to the spindle of a lathe and used for attachment of the workpiece. **2.** A protective plate used to cover holes in machines or other devices. **3.** In scuba or skin diving, a glass or plastic window positioned over the face to provide an air space between the diver's eyes and the water.

facet [ANAT] A small plane surface, especially on a bone or a hard body; may be produced by wear, as a worn spot on the surface of a tooth. [GEOGR] Any part of an intersecting surface that constitutes a unit of geographic study, for example, a flat or a slope. [INV ZOO] The surface of a simple eye in the compound eye of arthropods and certain other invertebrates. [MATER] The plane surface of a crystal, a cut precious stone, or other fractured surface.

facework [CIV ENG] Ornamental or otherwise special material on the front side or outside of a wall.

facial bone [ANAT] The bone comprising the nose and jaws, formed by the maxilla, zygoma, nasal, lacrimal, palatine, inferior nasal concha, vomer, mandible, and parts of the ethmoid and sphenoid.

facial nerve [ANAT] The seventh cranial nerve in vertebrates; a paired composite nerve, with motor elements supplying muscles of facial expression and with sensory fibers from the taste buds of the anterior two-thirds of the tongue and from other sensory endings in the anterior part of the throat.

facies [ANAT] Characteristic appearance of the face in association with a disease or abnormality. [ECOL] The makeup or appearance of a community or species population. [GEOL] Any observable attribute or attributes of a rock or stratigraphic unit, such as overall appearance or composition, of one part of the rock or unit as contrasted with other parts of the same rock or unit.

facility assignment [COMPUT SCI] The allocation of core memory and external devices by the executive as required by the program being executed.

facing [CIV ENG] A covering or casting of some material applied to the outer face of embankments, buildings, and other structures. [MECH ENG] Machining the end of a flat rotating surface by applying a tool perpendicular to the axis of rotation in a spiral planar path. [MET] A fine molding sand applied to the face of a mold.

facsimile [COMMUN] **1.** A system of communication in which a transmitter scans a photograph, map, or other fixed graphic material and converts the information into signal waves for transmission by wire or radio to a facsimile receiver at a remote point. Also known as fax; phototelegraphy; radiophoto; telephoto; telephotography; wirephoto. **2.** A photograph transmitted by radio to a facsimile receiver. Also known as radiophoto. [GRAPHICS] An exact copy of a book, document, painting, or other material.

facsimile modulation [COMMUN] Process in which the amplitude, frequency, or phase of a transmitted wave is varied with time in accordance with a facsimile transmission signal.

facsimile posting [COMPUT SCI] The process of transferring by a duplicating process a printed line of information from a report, such as a listing of transactions prepared on an accounting machine, to a ledger or other recorded sheet.

factor I *See* fibrinogen.

factor II *See* prothrombin.

factor III *See* thromboplastin.

factor IX *See* Christmas factor.

factor analysis [MATH] Given sets of variables which are related linearly, factor analysis studies techniques of approximating each set relative to the others; usually the variables denote numbers.

factorial design [STAT] A design for an experiment that allows the experimenter to find out the effect levels of each factor on levels of all the other factors.

factoring [MATH] Finding the factors of an integer or polynomial.

factor of an integer [MATH] Any integer which when multiplied by another integer gives the original integer.

factor of a polynomial [MATH] Any polynomial which when multiplied by another polynomial gives the original polynomial.

facula [ASTRON] Any of the large patches of bright material forming a veined network in the vicinity of sunspots; faculae appear to be more permanent than sunspots and are probably due to elevated clouds of luminous gas.

facultative aerobe [MICROBIO] An anaerobic microorganism which can grow under aerobic conditions.

facultative anaerobe [MICROBIO] A microorganism that grows equally well under aerobic and anaerobic conditions.

fade-in [COMMUN] A gradual increase in signal strength, as at the start of a radio or television program or when changing to a new scene, to make sound volume and picture brightness increase gradually. [GRAPHICS] In motion pictures, the gradual emergence of a screen image from black.

fade-out [COMMUN] A gradual and temporary loss of a received radio or television signal due to magnetic storms, atmospheric disturbances, or other conditions along the transmission path.

fader [ELECTR] A multiple-unit level control used for gradual changeover from one microphone, audio channel, or television camera to another.

fading [COMMUN] Variations in the field strength of a radio signal, usually gradual, that are caused by changes in the transmission medium.

Fagaceae [BOT] A family of dicotyledonous plants in the order Fagales characterized by stipulate leaves, seeds without endosperm, female flowers generally not in catkins, and mostly three styles and locules.

Fagales [BOT] An order of dicotyledonous woody plants in the subclass Hamamelidae having simple leaves and much reduced, mostly unisexual flowers.

fahlore *See* tetrahedrite.

Fahrenheit scale [THERMO] A temperature scale; the temperature in degrees Fahrenheit (°F) is the sum of 32 plus 9/5 the temperature in degrees Celsius; water at 1 atmosphere (101,325 newtons per square meter) pressure freezes very near 32°F and boils very near 212°F.

fail-safe system [ENG] A system designed so that failure of power, control circuits, structural members, or other components will not endanger people operating the system or other people in the vicinity.

failsafe tape *See* incremental dump tape.

failure [MECH] Condition caused by collapse, break, or bending, so that a structure or structural element can no longer fulfill its purpose.

failure logging [COMPUT SCI] The automatic recording of the state of various components of a computer system following detection of a machine fault; used to initiate corrective procedures, such as repeating attempts to read or write a magnetic tape, and to aid customer engineers in diagnosing errors.

fair [METEOROL] Generally descriptive of pleasant weather conditions, with regard for location and time of year; it is subject to popular misinterpretation, for it is a purely subjective description; when this term is used in forecasts of the U.S. Weather Bureau, it is meant to imply no precipitation, less than 0.4 sky cover of low clouds, and no other extreme conditions of cloudiness or windiness.

faired cable [DES ENG] A trawling cable covered by streamlined surfaces to reduce hydrodynamic drag.

fairfieldite [MINERAL] $Ca_2Mn(PO_4)_2 \cdot 2H_2O$ A white or pale-yellow mineral composed of hydrous calcium manganese phosphate and occurring in foliated or fibrous form.

fairing [AERO ENG] A structure or surface on an aircraft or rocket that functions to reduce drag, such as the streamlined nose of a satellite-launching rocket.

fairlead [AERO ENG] A guide through which an airplane antenna or control cable passes. [MECH ENG] A group of pulleys or rollers used in conjunction with a winch or similar apparatus to permit the cable to be reeled from any direction. [NAV ARCH] A block, ring, or other fitting through which passes a line or the running rigging on a ship to prevent chafing.

fair-weather cumulus *See* cumulus humilis cloud.

fairy ring spot [PL PATH] A fungus disease of carnations caused by *Heterosporium echinulatum*, producing bleached spots with concentric dark zones on the leaves.

fairy stone *See* staurolite.

falcate [ASTRON] Crescent-shaped; applied usually to the appearance of the moon, Venus, and Mercury during their crescent phases. [BIOL] Shaped like a sickle.

falcon [VERT ZOO] Any of the highly specialized diurnal birds of prey composing the family Falconidae; these birds have been captured and trained for hunting.

Falconidae [VERT ZOO] The falcons, a family consisting of long-winged predacious birds in the order Falconiformes.

Falconiformes [VERT ZOO] An order of birds containing the diurnal birds of prey, including falcons, hawks, vultures, and eagles.

falculate [ZOO] Curved and with a sharp point.

Falkland Current [OCEANOGR] An ocean current flowing northward along the Argentine coast.

fall [ASTRON] 1. Of a spacecraft or spatial body, to drop toward a spatial body under the influence of its gravity. 2. *See* autumn. [MECH ENG] The rope or chain of a hoisting tackle. [MIN ENG] A mass of rock, coal, or ore which has fallen from the roof or side in any subterranean working or gallery.

fallback [COMPUT SCI] The system, electronic or manual, which is substituted for the computer system in case of breakdown. [GEOL] Fragmented ejecta from an impact or explosion crater during formation which partly refills the true crater almost immediately. [NUCLEO] That part of the material carried into the air by an atomic explosion which ultimately drops back to the earth or water at the site of the explosion.

fall block [MECH ENG] A pulley block that rises and falls with the load on a lifting tackle.

falling-ball viscometer *See* falling-sphere viscometer.

falling body [MECH] A body whose motion is accelerated toward the center of the earth by the force of gravity, other forces acting on it being negligible by comparison.

falling-drop method [PHYS] Technique for measurement of liquid densities in which the time of fall of a drop of the sample liquid through a reference liquid is measured.

falling film [FL MECH] A theoretical liquid film that moves downward in even flow on a vertical surface in laminar flow; the concept is used for heat- and mass-transfer calculations.

falling-sphere viscometer [ENG] A viscometer which measures the speed of a spherical body falling with constant velocity in the fluid whose viscosity is to be determined. Also known as falling-ball viscometer.

falling tide *See* ebb tide.

fall line [GEOL] 1. The zone or boundary between resistant rocks of older land and weaker strata of plains. 2. The line indicated by the edge over which a waterway suddenly descends, as in waterfalls.

falloff curve [PETRO ENG] Graphical representation of bottom-hole pressure falloff for a shut-in well as the reservoir drainage area expands.

Fallopian tube [ANAT] Either of the paired oviducts that extend from the ovary to the uterus for conduction of the ovum in mammals.

fallout [NUCLEO] The material that descends to the earth or water well beyond the site of a surface or subsurface nuclear explosion. Also known as atomic fallout; radioactive fallout.

fall velocity *See* settling velocity.

false acceptance [STAT] Accepting on the basis of a statistical test a hypothesis which is wrong.

false blossom [PL PATH] 1. A fungus disease of the cranberry caused by *Exobasidium oxycocci;* erect flower buds are formed which produce malformed flowers that set no fruit. Also known as rosebloom. 2. A similar virus disease of the cranberry transmitted by the leafhopper, *Scleroracus vaccinii.* Also known as Wisconsin false blossom.

false bottom [CIV ENG] A temporary bottom installed in a caisson to add to its buoyancy. [MET] An insert put in either member of a die set to increase the strength and improve the life of the

die. [MIN ENG] A flat, hexagonal or cylindrical iron die upon which ore is crushed in a stamp mill.

false drop *See* false retrieval.

false form *See* pseudomorph.

false galena *See* sphalerite.

false lapis *See* lazulite.

false retrieval [COMPUT SCI] An item retrieved in an automatic library search which is unrelated or vaguely related to the subject of the search. Also known as false drop.

false rib [ANAT] A rib that is not attached to the sternum directly; any of the five lower ribs on each side in humans.

false smut [PL PATH] 1. A fungus disease of palm caused by *Graphiola phoenicis* and characterized by small cylindrical protruding pustules, often surrounded by yellowish leaf tissue. 2. *See* green smut.

false sorts [COMPUT SCI] Entries irrelevant to the subject sought which are retrieved in a search.

false white rainbow *See* fogbow.

familial [BIOL] Of, pertaining to, or occurring among the members of a family.

family [CHEM] A group of elements whose chemical properties, such as valence, solubility of salts, and behavior toward reagents, are similar. [SYST] A taxonomic category based on the grouping of related genera.

fan [AGR] A mechanical device used for winnowing grain. [BIOL] Any structure, such as a leaf or the tail of a bird, resembling an open fan. [ELECTROMAG] Volume of space periodically energized by a radar beam (or beams) repeatedly traversing an established pattern. [GEOL] A gently sloping, fan-shaped feature usually found near the lower termination of a canyon. [MECH ENG] 1. A device, usually consisting of a rotating paddle wheel or an airscrew, with or without a casing, for producing currents in order to circulate, exhaust, or deliver large volumes of air or gas. 2. A vane to keep the sails of a windmill facing the direction of the wind.

fan antenna [ELECTROMAG] An array of folded dipoles of different length forming a wide-band ultra-high-frequency or very-high-frequency antenna.

fan beam [ELECTROMAG] 1. A radio beam having an elliptically shaped cross section in which the ratio of the major to the minor axis usually exceeds 3 to 1; the beam is broad in the vertical plane and narrow in the horizontal plane. 2. A radar beam having the shape of a fan.

Fanconi's anemia [MED] An infantile anemia that resembles pernicious anemia; related to excessive chromosomal breakage and associated with the risk of developing leukemia.

fan fold [GEOL] A fold of strata in which both limbs are overturned, forming a syncline or anticline.

fanjet [AERO ENG] A turbojet engine whose performance has been improved by the addition of a fan which operates in an annular duct surrounding the engine.

fanning mill [FOOD ENG] A device consisting of two vibrating screens and utilizing an air blast to clean and separate grain.

Fanning's equation [FL MECH] The equation expressing that frictional pressure drop of fluid flowing in a pipe is a function of the Reynolds number, rate of flow, acceleration due to gravity, and length and diameter of the pipe.

Fanno flow [FL MECH] An ideal flow used to study the flow of fluids in long pipes; the flow obeys the same simplifying assumptions as Rayleigh flow except that the assumption there is no friction is replaced by the requirement the flow be adiabatic.

fan-out [ELECTR] The number of parallel loads that can be driven from one output mode of a logic circuit.

fantail [NAV ARCH] The area of the upper deck of a ship which is nearest the stern.

farad [ELEC] The unit of capacitance in the meter-kilogram-second system, equal to the capacitance of a capacitor which has a potential difference of 1 volt between its plates when the charge on one of its plates is 1 coulomb, there being an equal and opposite charge on the other plate. Symbolized F.

faraday [PHYS] The electric charge required to liberate 1 gram-equivalent of a substance by electrolysis; experimentally equal to $96,487.0 \pm 1.6$ coulombs. Also known as Faraday constant.

Faraday birefringence [OPTICS] Difference in the indices of refraction of left and right circularly polarized light passing through matter parallel to an applied magnetic field; it is responsible for the Faraday effect.

Faraday constant *See* faraday.

Faraday dark space [ELECTR] The relatively nonluminous region that separates the negative glow from the positive column in a cold-cathode glow-discharge tube.

Faraday effect [OPTICS] Rotation of polarization of a beam of linearly polarized light when it passes through matter in the direction of an applied magnetic field; it is the result of Faraday birefringence. Also known as Faraday rotation; Kundt effect; magnetic rotation.

Faraday rotation *See* Faraday effect.

Faraday's law of electromagnetic induction [ELECTROMAG] The law that the electromotive force induced in a circuit by a changing magnetic field is equal to the negative of the rate of change of the magnetic flux linking the circuit. Also known as law of electromagnetic induction.

Faraday's laws of electrolysis [PHYS CHEM] **1.** The amount of any substance dissolved or deposited in electrolysis is proportional to the total electric charge passed. **2.** The amounts of different substances dissolved or deposited by the passage of the same electric charge are proportional to their equivalent weights.

far-end crosstalk [COMMUN] Crosstalk that travels along the disturbed circuit in the same direction as desired signals in that circuit.

farinaceous [BIOL] Having a mealy surface covering. [FOOD ENG] **1.** Containing starch or flour. **2.** Having the texture of meal. [GEOL] Of a rock or sediment, having a texture that is mealy, soft, and friable, for example, a limestone or a pelagic ooze.

Farinales [BOT] An order that includes several groups regarded as orders of the Commelinidae in other systems of classification.

far-infrared radiation [ELECTROMAG] Infrared radiation the wavelengths of which are the longest of those in the infrared region, about 50–1000 micrometers; requires diffraction gratings for spectroscopic analysis.

Farinosae [BOT] The equivalent name for Farinales.

farringtonite [MINERAL] $Mg_3(PO_4)_2$ Colorless, waxwhite, or yellow phosphate mineral known only in meteorites.

farsightedness *See* hypermetropia.

far-ultraviolet radiation [ELECTROMAG] Ultraviolet radiation in the wavelength range of 200–300 nanometers; germicidal effects are greatest in this range.

fascia [BUILD] A wide board fixed vertically on edge to the rafter ends or wall which carries the gutter around the eaves of a roof. [HISTOL] Layers of areolar connective tissue under the skin and between muscles, nerves, and blood vessels.

fasciculus [ANAT] A bundle or tract of nerve, muscle, or tendon fibers isolated by a sheath of connective tissues and having common origins, innervation, and functions.

fast-access storage [COMPUT SCI] The section of a computer storage from which data can be obtained most rapidly.

fast breeder reactor [NUCLEO] A type of fast reactor using highly enriched fuel in the core, fertile material in the blanket, and a liquid-metal coolant, such as sodium; high-speed neutrons fission the fuel in the compact core, and the excess neutrons convert fertile material to fissionable isotopes; the breeding ratio is 1.0 or larger. Abbreviated FBR.

fast chemical reaction [PHYS CHEM] A reaction with a half-life of milliseconds or less; such reactions occur so rapidly that special experimental techniques are required to observe their rate.

fast coupling [MECH ENG] A flexible geared coupling that uses two interior hubs on the shafts with circumferential gear teeth surrounded by a casing having internal gear teeth to mesh and connect the two hubs.

fastener [DES ENG] **1.** A device for joining two separate parts of an article or structure. **2.** A device for holding closed a door, gate, or similar structure.

fast ice [HYD] Any type of sea, river, or lake ice attached to the shore (ice foot, ice shelf), beached (shore ice), stranded in shallow water, or frozen to the bottom of shallow waters (anchor ice). Also known as landfast ice. [OCEANOGR] Sea ice generally remaining in the position where originally formed and sometimes attaining a considerable thickness; it is attached to the shore or over shoals where it may be held in position by islands, grounded icebergs, or polar ice. Also known as coastal ice; coast ice.

fast neutron [NUCLEO] A neutron having energy much greater than some arbitrary lower limit (that may be only a few thousand electron volts).

fast-neutron spectrometry [NUC PHYS] Neutron spectrometry in which nuclear reactions are produced by or yield fast neutrons; such reactions are more varied than in the slow-neutron case.

fast time constant [ELEC] An electric circuit which combines resistance and capacitance to give a short time constant for capacitor discharge through the resistor. [ELECTR] Circuit with short time constant used to emphasize signals of short duration to produce discrimination against low-frequency components of clutter in radar.

fast time scale [COMPUT SCI] In simulation by an analog computer, a scale in which the time duration of a simulated event is less than the actual time duration of the event in the physical system under study.

fat [ANAT] Pertaining to an obese person. [BIOCHEM] Any of the glyceryl esters of fatty acids which form a class of neutral organic compounds. [PHYSIO] The chief component of fat cells and other animal and plant tissues.

fat cell [HISTOL] The principal component of adipose connective tissue; two types are yellow fat cells and brown fat cells.

fate map [EMBRYO] A graphic scheme indicating the definite spatial arrangement of undifferentiated embryonic cells in accordance with their destination to become specific tissues.

fathom [OCEANOGR] The common unit of depth in the ocean, equal to 6 feet (1.8288 meters).

fathom curve *See* isobath.

fatigue [ELECTR] The decrease of efficiency of a luminescent or light-sensitive material as a result of excitation. [MECH] Failure of a material by cracking resulting from repeated or cyclic stress. [PHYSIO] Exhaustion of strength or reduced capacity to respond to stimulation following a period of activity.

fatigue limit [MECH] The maximum stress that a material can endure for an infinite number of stress cycles without breaking. Also known as endurance limit.

fatigue ratio [MECH] The ratio of the fatigue limit or fatigue strength to the static tensile strength. Also known as endurance ratio.

fatigue strength [MECH] The maximum stress a material can endure for a given number of stress cycles without breaking. Also known as endurance strength.

fatty acid [ORG CHEM] An organic monobasic acid of the general formula $C_nH_{2n+1}COOH$ derived from the saturated series of aliphatic hydrocarbons; examples are palmitic acid, stearic acid, and oleic acid; used as a lubricant in cosmetics and nutrition, and for soaps and detergents.

fauces [ANAT] The passage in the throat between the soft palate and the base of the tongue. [BOT] The throat of a calyx, corolla, or similar part.

faucial tonsil See palatine tonsil.

fault [ELEC] A defect, such as an open circuit, short circuit, or ground, in a circuit, component, or line. Also known as electrical fault; faulting. [ELECTR] Any physical condition that causes a component of a data-processing system to fail in performance. [GEOL] A fracture in rock along which the adjacent rock surfaces are differentially displaced.

fault basin [GEOL] A region depressed in relation to surrounding regions and separated from them by faults.

fault block [GEOL] A rock mass that is bounded by faults; the faults may be elevated or depressed and not necessarily the same on all sides.

fault-block mountain See block mountain.

fault breccia [GEOL] The assembly of angular fragments found frequently along faults. Also known as dislocation breccia.

fault cliff See fault scarp.

fault current See fault electrode current.

fault electrode current [ELEC] The current to an electrode under fault conditions, such as during arc-backs and load short circuits. Also known as fault current; surge electrode current.

fault escarpment See fault scarp.

faulting [ELEC] See fault. [GEOL] The fracturing and displacement processes which produce a fault.

fault ledge See fault scarp.

fault line [GEOL] Intersection of the fault surface with the surface of the earth or any other horizontal surface of reference. Also known as fault trace.

fault masking [COMPUT SCI] Any type of hardware redundancy in which faults are corrected immediately and the operations of fault detection, location, and correction are indistinguishable.

fault scarp [GEOL] A steep cliff formed by movement along one side of a fault. Also known as cliff of displacement; fault cliff; fault escarpment; fault ledge.

fault strike [GEOL] The angular direction, with respect to north, of the intersection of the fault surface with a horizontal plane.

fault system [GEOL] Two or more fault sets which interconnect.

fault tolerance [SYS ENG] The capability of a system to perform its functions in accord with design specifications even in the presence of component failures.

fault trace See fault line.

fault zone [GEOL] A fault expressed as an area of numerous small fractures. Also known as distributed fault.

fauna [ZOO] 1. Animals. 2. The animal life characteristic of a particular region or environment.

Favositidae [PALEON] A family of extinct Paleozoic corals in the order Tabulata.

fax See facsimile.

fayalite [MINERAL] Fe_2SiO_4 A brown to black mineral of the olivine group, consisting of iron silicate and found either massive or in crystals; specific gravity is 4.1.

FBR See fast breeder reactor.

fcc lattice See face-centered cubic lattice.

F center [SOLID STATE] A color center consisting of an electron trapped by a negative ion vacancy in an ionic crystal, such as an alkali halide or an alkaline-earth fluoride or oxide.

F corona [ASTRON] The outer layer of the sun's corona. Also known as Fraunhofer corona.

FD&C color See food color.

F display [ELECTR] A rectangular display in which a target appears as a centralized blip when the radar antenna is aimed at it; horizontal and vertical aiming errors are respectively indicated by the horizontal and vertical displacement of the blip.

Fe See iron.

feasibility study [SYS ENG] 1. A study of applicability or desirability of any management or procedural system from the standpoint of advantages versus disadvantages in any given case. 2. A study to determine the time at which it would be practicable or desirable to install such a system when determined to be advantageous. 3. A study to determine whether a plan is capable of being accomplished successfully.

feather [MECH ENG] To change the pitch on a propeller. [METEOROL] See barb. [VERT ZOO] An ectodermal derivative which is a specialized keratinous outgrowth of the epidermis of birds; functions in flight and in providing insulation and protection.

feather alum See alunogen; halotrichite.

feathering [FOOD ENG] Flocculation of the cream (fat) in homogenized milk when added to hot coffee or tea due to a defect in the chemistry of the cream. [MECH ENG] A pitch position in a controllable-pitch propeller; it is used in the event of engine failure to stop the windmilling action, and occurs when the blade angle is about 90° to the plane of rotation. Also known as full feathering. [VERT ZOO] Plumage.

feather joint [ENG] A joint made by cutting a mating groove in each of the pieces to be joined and inserting a feather in the opening formed when the pieces are butted together. Also known as ploughed-and-tongued joint. [GEOL] One of a series of joints in a fault zone formed by shear and tension. Also known as pinnate joint.

feather ore See jamesonite.

feather rot [PL PATH] A fungus rot of both dead and living tree trunks caused by *Poria subacida* and characterized by the white stringy or spongy nature of the rotted tissue.

Fechner law [PHYSIO] The intensity of a sensation produced by a stimulus varies directly as the logarithm of the numerical value of that stimulus.

fecundity [BIOL] The innate potential reproductive capacity of the individual organism, as denoted by its ability to form and separate from the body the mature germ cells.

feed [AGR] Any crops or other food substances for livestock. [COMPUT SCI] **1.** To supply the material to be operated upon to a machine. **2.** A device capable of so feeding. [ELECTR] To supply a signal to the input of a circuit, transmission line, or antenna. [ELECTROMAG] The part of a radar antenna that is connected to or mounted on the end of the transmission line and serves to radiate radiofrequency electromagnetic energy to the reflector or receive energy therefrom. [ENG] **1.** Process or act of supplying material to a processing unit for treatment. **2.** The material supplied to a processing unit for treatment. [FOOD ENG] The fermenting wort that is removed from the yeast troughs during brewing processes. [MECH ENG] Forward motion imparted to the cutters or drills of cutting or drilling machinery.

feedback [ELECTR] The return of a portion of the output of a circuit or device to its input. [SCI TECH] The control of input as a function of output by returning a portion of the output to the input.

feedback amplifier [ELECTR] An amplifier in which a passive network is used to return a portion of the output signal to its input so as to change the performance characteristics of the amplifier.

feedback circuit [ELECTR] A circuit that returns a portion of the output signal of an electronic circuit or control system to the input of the circuit or system.

feedback compensation [CONT SYS] Improvement of the response of a feedback control system by placing a compensator in the feedback path, in contrast to cascade compensation. Also known as parallel compensation.

feedback control loop *See* feedback loop.

feedback control system [CONT SYS] A system in which the value of some output quantity is controlled by feeding back the value of the controlled quantity and using it to manipulate an input quantity so as to bring the value of the controlled quantity closer to a desired value. Also known as closed-loop control system.

feedback inhibition [BIOCHEM] A cellular control mechanism by which the end product of a series of metabolic reactions inhibits the activity of the first enzyme in the sequence.

feedback loop [CONT SYS] A closed transmission path or loop that includes an active transducer and consists of a forward path, a feedback path, and one or more mixing points arranged to maintain a prescribed relationship between the loop input signal and the loop output signal. Also known as feedback control loop.

feedback oscillator [ELECTR] An oscillating circuit, including an amplifier, in which the output is fed back in phase with the input; oscillation is maintained at a frequency determined by the values of the components in the amplifier and the feedback circuits.

feedback regulator [CONT SYS] A feedback control system that tends to maintain a prescribed relationship between certain system signals and other predetermined quantities.

feeder [ELEC] **1.** A transmission line used between a transmitter and an antenna. **2.** A conductor, or several conductors, connecting generating stations, substations, or feeding points in an electric power distribution system. **3.** A group of conductors in an interior wiring system which link a main distribution center with secondary or branch-circuit distribution

centers. [GEOL] A small ore-bearing vein which merges with a larger one. [HYD] *See* tributary. [MECH ENG] **1.** A conveyor adapted to control the rate of delivery of bulk materials, packages, or objects, or a control device which separates or assembles objects. **2.** A device for delivering materials to a processing unit. [MET] A runner or riser so placed that it can feed molten metal to the contracting mass of the casting as it cools in its flask, therefore preventing formation of cavities or porous structure. [ORD] A device that supplies ammunition to a weapon, usually actuated by an automatic or semiautomatic mechanism.

feeder cable [COMMUN] In communications practice, a cable extending from the central office along a primary route (main feeder cable) or from a main feeder cable along a secondary route (branch feeder cable) and providing connections to one or more distribution cables.

feedforward control [CONT SYS] Process control in which changes are detected at the process input and an anticipating correction signal is applied before process output is affected.

feedhead [MET] A reservoir of molten metal that is left above a casting in order to supply additional metal as the casting solidifies and shrinks. Also known as riser; sinkhead.

feed ratio [MECH ENG] The number of revolutions a drill stem and bit must turn to advance the drill bit 1 inch when the stem is attached to and rotated by a screw- or gear-feed type of drill swivel head with a particular pair of the set of gears engaged. Also known as feed speed.

feed speed *See* feed ratio.

feedstock [ENG] The raw material furnished to a machine or process.

feedthrough [ELEC] A conductor that connects patterns on opposite sides of a printed circuit board. Also known as interface connection.

feedthrough insulator *See* feedthrough terminal.

feedthrough terminal [ELEC] An insulator designed for mounting in a hole in a panel, wall, or bulkhead, with a conductor in the center on the insulator to permit feeding electricity through the partition. Also known as feedthrough insulator.

feedwater [MECH ENG] The water supplied to a boiler or still.

Fehling's reagent [ANALY CHEM] A solution of cupric sulfate, sodium potassium tartrate, and sodium hydroxide, used to test for the presence of reducing compounds such as sugars.

feldspar [MINERAL] A group of silicate minerals that make up about 60% of the outer 15 kilometers of the earth's crust; they are silicates of aluminum with the metals potassium, sodium, and calcium, and rarely, barium.

feldspathic graywacke [PETR] Sandstone containing less than 75% quartz and chert and 15–75% detrital clay matrix, and having feldspar grains in greater abundance than rock fragments. Also known as arkosic wacke; high-rank graywacke.

feldspathic sandstone [PETR] Sandstone rich in feldspar; intermediate in composition between arkosic sandstone and quartz sandstone, made up of 10–25% feldspar and less than 20% matrix material.

feldspathization [GEOL] Formation of feldspar in a rock usually as a result of metamorphism leading toward granitization.

feldspathoid [GEOL] Aluminosilicates of sodium, potassium, or calcium that are similar in composition to feldspars but contain less silica than the corresponding feldspar.

Felidae [VERT ZOO] The cats and saber-toothed cats, a family of mammals in the superfamily Feloidea.

feline [VERT ZOO] 1. Of or relating to the genus *Felis*. 2. Catlike.

Feloidea [VERT ZOO] A superfamily of catlike mammals in the order Carnivora.

felsic [MINERAL] A light-colored mineral. [PETR] Of an igneous rock, having a mode containing light-colored minerals.

felsite [PETR] 1. A light-colored, fine-grained igneous rock composed chiefly of quartz or feldspar. 2. A rock characterized by felsitic texture.

female connector [ELEC] A connector having one or more contacts set into recessed openings; jacks, sockets, and wall outlets are examples.

female fitting [DES ENG] In a paired pipe or an electrical or mechanical connection, the portion (fitting) that receives, contrasted to the male portion (fitting) that inserts.

female pseudohermaphroditism *See* gynandry.

feminizing syndrome [MED] Any of a number of symptom complexes in which males tend to take on feminine characteristics due to alterations of adrenocortiocotropin output.

femoral artery [ANAT] The principal artery of the thigh; originates as a continuation of the external iliac artery.

femoral hernia [MED] A hernia that occurs at the passage of the arteries and veins from the abdomen into the legs below the inguinal ligament.

femoral nerve [ANAT] A mixed nerve of the leg; the motor portion innervates muscles of the thigh, and the sensory portion innervates portions of the skin of the thigh, leg, hip, and knee.

femto- [SCI TECH] A prefix representing 10^{-15}, which is 0.000 000 000 000 001, or one-thousandth of a millionth of a millionth.

femur [ANAT] 1. The proximal bone of the hind or lower limb in vertebrates. 2. The thigh bone in humans, articulating with the acetabulum and tibia.

fen [GEOGR] Peat land covered by water, especially in the upper regions of old estuaries and around lakes, that can be drained only artificially.

fence [AERO ENG] A stationary plate or vane projecting from the upper surface of an airfoil, substantially parallel to the airflow, used to prevent spanwise flow. [COMPUT SCI] *See* fence cell. [ENG] 1. A line of data-acquisition or tracking stations used to monitor orbiting satellites. 2. A line of radar or radio stations for detection of satellites or other objects in orbit. 3. A line or network of early-warning radar stations. 4. A concentric steel fence erected around a ground radar transmitting antenna to serve as an artificial horizon and suppress ground clutter that would otherwise drown out weak signals returning at a low angle from a target. 5. An adjustable guide on a tool.

fender [CIV ENG] A timber, cluster of piles, or bag of rope placed along dock or bridge pier to prevent damage by docking ships or floating objects. [ENG] A cover over the upper part of a wheel of an automobile or other vehicle. [MIN ENG] A thin pillar of coal adjacent to the gob, left for protection while driving a lift through the mine pillar. [NAV ARCH] A padded device acting as a buffer to prevent damage between two ships or between a ship and dock.

Fenestellidae [PALEON] A family of extinct fenestrated, cryptostomatous bryozoans which abounded during the Silurian.

fenestration [ARCH] The arrangement of openings, especially windows, in the wall of a building. [BIOL] 1. A transparent or windowlike break or opening in the surface. 2. The presence of windowlike openings.

fennel [BOT] *Foeniculum vulgare*. A tall perennial herb of the family Umbelliferae; a spice is derived from the fruit.

fenster *See* window.

ferberite [MINERAL] $FeNO_4$ A black mineral of the wolframite solid-solution series occurring as monoclinic, prismatic crystals and having a submetallic luster; hardness is 4.5 on Mohs scale, and specific gravity is 7.5.

fergusonite [MINERAL] $Y_2O_3 \cdot (Nb,Ta)_2O_5$ Brownishblack rare-earth mineral with a tetragonal crystal form; it is isomorphous with formanite.

Fermat's principle [OPTICS] The principle that an electromagnetic wave will take a path that involves the least travel time when propagating between two points. Also known as least-time principle; stationary time principle.

fermentation [MICROBIO] An enzymatic transformation of organic substrates, especially carbohydrates, generally accompanied by the evolution of gas; a physiological counterpart of oxidation, permitting certain organisms to live and grow in the absence of air; used in various industrial processes for the manufacture of products such as alcohols, acids, and cheese by the action of yeasts, molds, and bacteria; alcoholic fermentation is the best-known example. Also known as zymosis.

Fermi age [NUCLEO] The value calculated for the slowing-down area in the Fermi age model; it has the dimensions of area, not time. Also known as age; neutron age; symbolic age of neutrons.

Fermi age model [NUCLEO] A model used in studying the slowing down of neutrons by elastic collisions; it is assumed that the slowing down takes place by a very large number of very small energy changes.

Fermi constant [NUC PHYS] A universal constant, introduced in beta-disintegration theory, that expresses the strength of the interaction between the transforming nucleon and the electron-neutrino field.

Fermi-Dirac distribution function [STAT MECH] A function specifying the probability that a member of an assembly of independent fermions, such as electrons in a semiconductor or metal, will occupy a certain energy state when thermal equilibrium exists.

Fermi-Dirac statistics [STAT MECH] The statistics of an assembly of identical half-integer spin particles; such particles have wave functions antisymmetrical with respect to particle interchange and satisfy the Pauli exclusion principle.

Fermi distribution [SOLID STATE] Distribution of energies of electrons in a semiconductor or metal as given by the Fermi-Dirac distribution function; nearly all energy levels below the Fermi level are filled, and nearly all above this level are empty.

Fermi hole [SOLID STATE] A region surrounding an electron in a solid in which the energy band theory predicts that the probability of finding other electrons is less than the average over the volume of the solid.

fermion [QUANT MECH] A particle, such as the electron, proton, or neutron, which obeys the rule that the wave function of several identical particles changes sign when the coordinates of any pair are interchanged; it therefore obeys the Pauli exclusion principle.

Fermi plot *See* Kurie plot.

Fermi resonance [PHYS CHEM] In a polyatomic molecule, the relationship of two vibrational levels that have in zero approximation nearly the same energy; they repel each other, and the eigenfunctions of the two states mix.

Fermi selection rules [NUC PHYS] Selection rules for beta decay in a Fermi transition; that is, there is no change in total angular momentum or parity of the nucleus in an allowed transition.

Fermi surface [SOLID STATE] A constant-energy surface in the space containing the wave vectors of states of members of an assembly of independent fermions, such as electrons in a semiconductor or metal, whose energy is that of the Fermi level.

Fermi temperature [STAT MECH] The energy of the Fermi level of an assembly of fermions divided by Boltzmann's constant, which appears as a parameter in the Fermi-Dirac distribution function.

fermium [CHEM] A synthetic radioactive element, symbol Fm, with atomic number 100; discovered in debris of the 1952 hydrogen bomb explosion, and now made in nuclear reactors.

fermorite [MINERAL] $(Ca,Sr)_5[(As,P)O_4]_3$ A white mineral composed of arsenate, phosphate, and fluoride of calcium and strontium, occurring in crystalline masses.

fern [BOT] Any of a large number of vascular plants composing the division Polypodiophyta.

ferric [INORG CHEM] The term for a compound of trivalent iron, for example, ferric bromide, $FeBr_3$.

ferrichrome [MICROBIO] A cyclic hexapeptide that is a microbial hydroxamic acid and is involved in iron transport and metabolism in microorganisms.

ferric oxide [INORG CHEM] Fe_2O_3 Red, hexagonal crystals or powder, insoluble in water and soluble in acids, melting at $1565°C$; used as a catalyst and pigment for metal polishing, in metallurgy, and in medicine. Also known as ferric oxide red; jeweler's rouge; red ocher.

ferric oxide red See ferric oxide.

ferrihemoglobin [BIOCHEM] Hemoglobin in the oxidized state. Also known as methemoglobin.

ferrimagnet See ferrimagnetic material.

ferrimagnetic material [SOLID STATE] A material displaying ferrimagnetism; the ferrites are the principal example. Also known as ferrimagnet.

ferrimagnetism [SOLID STATE] A type of magnetism in which the magnetic moments of neighboring ions tend to align nonparallel, usually antiparallel, to each other, but the moments are of different magnitudes, so there is an appreciable resultant magnetization.

ferrimolybdite [MINERAL] $Fe_2(MoO_4)_3·8H_2O$ A colorless to canary yellow, probably orthorhombic mineral consisting of hydrated ferric molybdate; occurs in massive form, as crusts or aggregates.

ferrinatrite [MINERAL] $Na_3Fe(SO_4)_3·3H_2O$ A greenish or white mineral composed of sodium ferric iron double sulfate; usually occurs in spherical forms.

ferrisicklerite [MINERAL] $(Li,Fe,Mn)(PO_4)$ Mineral composed of phosphate of lithium, ferric iron, and manganese, more iron being present than manganese; it is isomorphous with sicklerite.

ferristor [ELECTR] A miniature, two-winding, saturable reactor that operates at a high carrier frequency and may be connected as a coincidence gate, current discriminator, free-running multivibrator, oscillator, or ring counter.

ferrite [INORG CHEM] An unstable compound of a strong base and ferric oxide which exists in alkaline solution, such as $NaFeO_2$. [MET] Iron that has not combined with carbon in pig iron or steel. [PETR] Grains or scales of unidentifiable, generally transparent amorphous iron oxide in the matrix of a porphyritic rock. [SOLID STATE] Any ferrimag-

netic material having high electrical resistivity which has a spinel crystal structure and the chemical formula XFe_2O_4, where X represents any divalent metal ion whose size is such that it will fit into the crystal structure.

ferrite attenuator See ferrite limiter.

ferrite-core memory [ELECTR] A magnetic memory consisting of a matrix of tiny toroidal cores molded from a square-loop ferrite, through which are threaded the pulse-carrying wires and the sense wire.

ferrite device [ELEC] An electrical device whose principle of operation is based upon the use of ferrites in powdered, compressed, sintered form, making use of their ferrimagnetism and their high electrical resistivity, which makes eddy-current losses extremely low at high frequencies.

ferrite limiter [ELECTROMAG] A passive, low-power microwave limiter having an insertion loss of less than 1 decibel when operating in its linear range, with minimum phase distortion; the input signal is coupled to a single-crystal sample of either yttrium iron garnet or lithium ferrite, which is biased to resonance by a magnetic field. Also known as ferrite attenuator.

ferrite switch [ELECTROMAG] A ferrite device that blocks the flow of energy through a waveguide by rotating the electric field vector 90°; the switch is energized by sending direct current through its magnetizing coil; the rotated electromagnetic wave is then reflected from a reactive mismatch or absorbed in a resistive card.

ferritic stainless steel [MET] Any magnetic iron alloy containing more than 12% chromium having a body-centered cubic structure. Also known as stainless iron.

ferritungstite [MINERAL] $Fe_2(WO_4)(OH)_4·4H_2O$ A yellow ocher mineral composed of hydrous ferric tungstate, occurring as a powder.

ferroalloy [MET] Any alloy containing iron, usually in major amount. Also known as ferrous alloy.

ferroan dolomite [MINERAL] A species of ankerite having less than 20% of the manganese positions occupied by iron.

Ferrod [GEOL] A suborder of the soil order Spodosol that is well drained and contains an iron accumulation with little organic matter.

ferrodolomite [MINERAL] $CaFe(CO_3)_2$ A mineral composed of calcium iron carbonate, isomorphous with dolomite, and occurring in ankerite.

ferroelectric [SOLID STATE] A crystalline substance displaying ferroelectricity, such as barium titanate, potassium dihydrogen phosphate, and Rochelle salt; used in ceramic capacitors, acoustic transducers, and dielectric amplifiers. Also known as seignette-electric.

ferroelectric domain [SOLID STATE] A region of a ferroelectric material within which the spontaneous polarization is constant.

ferroelectric hysteresis [ELEC] The dependence of the polarization of ferroelectric materials not only on the applied electric field but also on their previous history; analogous to magnetic hysteresis in ferromagnetic materials. Also known as dielectric hysteresis; electric hysteresis.

ferroelectricity [SOLID STATE] Spontaneous electric polarization in a crystal; analogous to ferromagnetism.

ferromagnetic amplifier [ELECTR] A parametric amplifier based on the nonlinear behavior of ferromagnetic resonance at high radio-frequency power levels; incorrectly known as garnet maser.

ferromagnetic domain [SOLID STATE] A region of a ferromagnetic material within which atomic or molecular magnetic moments are aligned parallel. Also known as magnetic domain.

ferromagnetic film *See* magnetic thin film.

ferromagnetic material [SOLID STATE] A material displaying ferromagnetism, such as the various forms of iron, steel, cobalt, nickel, and their alloys.

ferromagnetics [ELECTR] The science that deals with the storage of binary information and the logical control of pulse sequences through the utilization of the magnetic polarization properties of materials.

ferromagnetism [SOLID STATE] A property, exhibited by certain metals, alloys, and compounds of the transition (iron group) rare-earth and actinide elements, in which the internal magnetic moments spontaneously organize in a common direction; gives rise to a permeability considerably greater than that of vacuum, and to magnetic hysteresis.

ferrosilite [MINERAL] A mineral in the orthopyroxene group; the iron analog of enstatite; occurs in hypersthene, but is not found separately in nature.

ferrous [CHEM] The term or prefix used to denote compounds of iron in which iron is in the divalent $(2+)$ state.

ferrous alloy *See* ferroalloy.

ferruginous [SCI TECH] 1. Pertaining to or containing iron. 2. Having the appearance or color of iron rust (ferric oxide).

ferrule [DES ENG] 1. A metal ring or cap attached to the end of a tool handle, post, or other device to strengthen and protect it. 2. A bushing inserted in the end of a boiler flue to spread and tighten it. [ENG] *See* stabilizer.

fersmanite [MINERAL] $(Na,Ca)_2(TI,Cb)Si(O,F)_6$ A brown mineral composed of a silicate fluoride of sodium, calcium, titanium, and columbium.

fersmite [MINERAL] $(Ca,Ce)(Cb,Ti)_2(O,F)_6$ A black mineral composed of an oxide and fluoride of calcium and columbium with cerium and titanium.

fertility [BIOL] The state of or capacity for abundant productivity.

fertility factor [GEN] An episomal bacterial sex factor which determines the role of a bacterium as either a male donor or as a female recipient of genetic material. Also known as F factor; sex factor.

fertilization [PHYSIO] The physicochemical processes involved in the union of the male and female gametes to form the zygote.

fertilization membrane [CYTOL] A membrane that separates from the surface of and surrounds many eggs following activation by the sperm; prevents multiple fertilization.

fertilizer [MATER] Material that is added to the soil to supply chemical elements needed for plant nutrition.

fervanite [MINERAL] $Fe_4V_4O_{16} \cdot 5H_2O$ Golden-brown mineral composed of a hydrated iron vanadate; although itself not radioactive, it occurs with radioactive minerals.

fescue [BOT] A group of grasses of the genus *Festuca*, used for both hay and pasture.

FET *See* field-effect transistor.

fetal fat-cell lipoma *See* liposarcoma.

fetal hemoglobin [BIOCHEM] A normal embryonic hemoglobin having alpha chains identical to those of normal adult human hemoglobin, and gamma chains similar to adult beta chains.

fetal membrane [EMBRYO] Any one of the membranous structures which surround the embryo during its development period. Also known as extraembryonic membrane.

fetch [COMPUT SCI] To locate and load into main memory a requested load module, relocating it as necessary and leaving it in a ready-to-execute condition. [OCEANOGR] 1. The distance traversed by waves without obstruction. 2. An area of the sea surface over which seas are generated by a wind having a constant speed and direction. 3. The length of the fetch area, measured in the direction of the wind in which the seas are generated. Also known as generating area.

fetch ahead *See* instruction lookahead.

fetch bit [COMPUT SCI] The fifth bit in a storage key; the value of the fetch bit can protect a stored block from destruction or from being accessed by unauthorized programs.

fetus [EMBRYO] 1. The unborn offspring of viviparous mammals in the later stages of development. 2. In human beings, the developing body in utero from the beginning of the ninth week after fertilization through the fortieth week of intrauterine gestation, or until birth.

Feulgen reaction [ANALY CHEM] An aldehyde specific reaction based on the formation of a purple-colored compound when aldehydes react with fuchsin-sulfuric acid; deoxyribonucleic acid gives this reaction after removal of its purine bases by acid hydrolysis; used as a nuclear stain.

Feyliniidae [VERT ZOO] The limbless skinks, a family of reptiles in the suborder Sauria represented by four species in tropical Africa.

Feynman diagram [QUANT MECH] A diagram which gives an intuitive picture of a term in a perturbation expansion of a scattering matrix element or other physical quantity associated with interactions of particles; each line represents a particle, each vertex an interaction.

Feynman integral [QUANT MECH] A term in a perturbation expansion of a scattering matrix element; it is an integral over the Minkowski space of various particles (or over the corresponding momentum space) of the product of propagators of these particles and quantities representing interactions between the particles.

F factor *See* fertility factor.

F format [COMPUT SCI] 1. In data management, a fixed-length logical record format. 2. In FORTRAN, a real variable formatted as $Fw.d$, where w is the width of the field and d represents the number of digits to appear after the decimal point.

fiber [BOT] 1. An elongate, thick-walled, tapering plant cell that lacks protoplasm and has a small lumen. 2. A very slender root. [MET] 1. The characteristic of wrought metal that indicates directional properties as revealed by etching or by fracture appearance. 2. The pattern of preferred orientation of metal crystals after a deformation process, usually wiredrawing. [OPTICS] A transparent threadlike object made of glass or clear plastic, used to conduct light along selected paths. [TEXT] An extremely long, pliable, cohesive natural or manufactured threadlike object from which yarns are spun to be woven into textiles.

fiber bundle [OPTICS] A flexible bundle of glass or other transparent fibers, parallel to each other, used in fiber optics to transmit a complete image from one end of the bundle to the other.

fiber crops [AGR] Plants, such as flax, hemp, jute, and sisal, cultivated for their content or yield of fibrous material.

fiber flax [BOT] The flax plant grown in fertile, well-drained, well-prepared soil and cool, humid climate; planted in the early spring and harvested when half the seed pods turn yellow; used in the manufacture of linen.

fiber optics [OPTICS] The technique of transmitting light through long, thin, flexible fibers of glass, plastic, or other transparent materials; bundles of parallel fibers can be used to transmit complete images.

fiberscope [OPTICS] An arrangement of parallel glass fibers with an objective lens on one end and an eyepiece at the other; the assembly can be bent as required to view objects that are inaccessible for direct viewing.

fiber waveguide See optical waveguide.

fibril [BIOL] A small thread or fiber, as a root hair or one of the structural units of a striated muscle. [MATER] One of the minute threadlike elements of a natural or synthetic fiber.

fibrillation [PHYSIO] An independent, spontaneous, local twitching of muscle fibers. [TEXT] A process in which yarn is extruded as a ribbon or a tape rather than as a fine filament.

fibrin [BIOCHEM] The fibrous, insoluble protein that forms the structure of a blood clot; formed by the action of thrombin.

fibrinogen [BIOCHEM] A plasma protein synthesized by the parenchymal cells of the liver; the precursor of fibrin. Also known as factor I.

fibrinolysin See plasmin.

fibroadenoma [MED] A benign tumor containing both fibrous and glandular elements.

fibroblast [HISTOL] A stellate connective tissue cell found in fibrous tissue. Also known as a fibrocyte.

fibroblastic [PETR] Of a metamorphic rock, having a texture that is homeoblastic as a result of the development of minerals with a fibrous habit during recrystallization.

fibrocartilage [HISTOL] A form of cartilage rich in dense, closely opposed bundles of collagen fibers; occurs in intervertebral disks, in the symphysis pubis, and in certain tendons.

fibrocyte See fibroblast.

fibroferrite [MINERAL] $Fe(SO_4)(OH)\cdot5H_2O$ A yellowish mineral composed of a hydrous basic ferric sulfate, occurring in fibrous form.

fibroid tumor See fibroma.

fibrolite See sillimanite.

fibroma [MED] A benign tumor composed primarily of fibrous connective tissue. Also known as fibroid tumor.

fibromatosis [MED] 1. The occurrence of multiple fibromas. 2. Localized proliferation of fibroblasts without apparent cause.

fibroplasia [MED] The growth of fibrous tissue, as in the second phase of wound healing.

fibrosarcoma [MED] A sarcoma composed of spindle cells that produce collagenous fibrils.

fibrosis [MED] Growth of fibrous connective tissue in an organ or part in excess of that naturally present.

fibrous protein [BIOCHEM] Any of a class of highly insoluble proteins representing the principal structural elements of many animal tissues.

fibula [ANAT] The outer and usually slender bone of the hind or lower limb below the knee in vertebrates; it articulates with the tibia and astragalus in humans, and is ankylosed with the tibia in birds and some mammals.

Fick's law [PHYS] The law that the rate of diffusion of matter across a plane is proportional to the negative of the rate of change of the concentration of the diffusing substance in the direction perpendicular to the plane.

fidelity [COMMUN] The degree to which a system accurately reproduces at its output the essential characteristics of the signal impressed on its input.

fiducial point [OPTICS] A mark, or one of several marks, visible in the field of view of an optical instrument, used as a reference or for measurement. Also known as fiduciary point.

fiducial temperature [METEOROL] That temperature at which, in a specified latitude, the reading of a particular barometer does not require temperature or latitude correction. [THERMO] Any of the temperatures assigned to a number of reproducible equilibrium states on the International Practical Temperature Scale; standard instruments are calibrated at these temperatures.

fiduciary point See fiducial point.

fiedlerite [MINERAL] $Pb_3(OH)_2Cl_4$ A colorless mineral composed of a hydroxychloride of lead, occurring as monoclinic crystals.

field [COMPUT SCI] A specified area, such as a group of card columns or a set of bit locations in a computer word, used for a particular category of data. [ELEC] That part of an electric motor or generator which produces the magnetic flux which reacts with the armature, producing the desired machine action. [ELECTR] One of the equal parts into which a frame is divided in interlaced scanning for television; includes one complete scanning operation from top to bottom of the picture and back again. [GEOL] A region or area with a particular mineral resource, for example, a gold field. [GEOPHYS] That area or space in which a particular geophysical effect, such as gravity or magnetism, occurs and can be measured. [MATH] An algebraic system possessing two operations which have all the properties that addition and multiplication of real numbers have. [MED] The area in which surgery is taking place, bounded on all sides by sterilized tissue or drapes. Also known as sterile field. [OPTICS] See field of view. [PHYS] 1. An entity which acts as an intermediary in interactions between particles, which is distributed over part or all of space, and whose properties are functions of space coordinates and, except for static fields, of time; examples include gravitational field, sound field, and the strain tensor of an elastic medium. 2. The quantum-mechanical analog of this entity, in which the function of space and time is replaced by an operator at each point in space-time.

field artillery [ORD] Artillery mounted on carriages, and mobile enough to accompany infantry or armored units in the field.

field brightness See adaptation luminance.

field coil [ELECTROMAG] A coil used to produce a constant-strength magnetic field in an electric motor, generator, or excited-field loudspeaker; depending on the type of motor or generator, the field core may be on the stator or the rotor. Also known as field winding.

field delimiter [COMPUT SCI] Any symbol, such as a slash, colon, tab, or space, which enables an assembler to recognize the end of a field.

field-desorption [SOLID STATE] A technique which tears atoms from a surface by an electric field applied at a sharp dip to produce very well-ordered, clean, plane surfaces of many crystallographic orientations.

field-desorption mass spectroscopy [SPECT] A technique for analysis of nonvolatile molecules in which a sample is deposited on a thin tungsten wire containing sharp microneedles of carbon on the surface; a voltage is applied to the wire, thus producing high electric-field gradients at the points of the needles, and moderate heating then causes desorption from the surface or molecular ions, which are focused into a mass spectrometer.

field-effect device [ELECTR] A semiconductor device whose properties are determined largely by the effect of an electric field on a region within the semiconductor.

field-effect display [OPTICS] A type of numerical display device in which a liquid-crystal cell is sandwiched between polarizers; the cell is treated so that it normally rotates light 90°, but ceases to rotate light when an electric field is applied to it, altering the transmission of the device.

field-effect transistor [ELECTR] A transistor in which the resistance of the current path from source to drain is modulated by applying a transverse electric field between grid or gate electrodes; the electric field varies the thickness of the depletion layer between the gates, thereby reducing the conductance. Abbreviated FET.

field-effect-transistor resistor [ELECTR] A field-effect transistor in which the gate is generally tied to the drain; the resultant structure is used as a resistance load for another transistor.

field emission [ELECTR] The emission of electrons from the surface of a metallic conductor into a vacuum (or into an insulator) under influence of a strong electric field; electrons penetrate through the surface potential barrier by virtue of the quantum-mechanical tunnel effect. Also known as cold emission.

field-emission microscope [ELECTR] A device that uses field emission of electrons or of positive ions (field-ion microscope) to produce a magnified image of the emitter surface on a fluorescent screen.

field engineer [COMPUT SCI] A professional who installs computer hardware on customers' premises, performs routine preventive maintenance, and repairs equipment when it is out of order. Also known as field service representative. [ENG] **1.** An engineer who is in charge of directing civil, mechanical, and electrical engineering activities in the production and transmission of petroleum and natural gas. **2.** An engineer who operates at a construction site.

field excitation [MECH ENG] Control of the speed of a series motor in an electric or diesel-electric locomotive by changing the relation between the armature current and the field strength, either through a reduction in field current by shunting the field coils with resistance, or through the use of field taps.

field frequency [ELECTR] The number of fields transmitted per second in television; equal to the frame frequency multiplied by the number of fields that make up one frame. Also known as field repetition rate.

field geology [GEOL] The study of rocks and rock materials in their environment and in their natural relations to one another.

field gradient [PHYS] **1.** A vector obtained by applying the del operator to a scalar field. **2.** A tensor obtained by dyadic multiplication of the del operator with a vector field.

field intensity [COMMUN] In Federal Communications Commission regulations, the electric field intensity in the horizontal direction. [PHYS] *See* field strength.

field-ion microscope [ELECTR] A microscope in which atoms are ionized by an electric field near a sharp tip; the field then forces the ions to a fluorescent screen, which shows an enlarged image of the tip, and individual atoms are made visible; this is

the most powerful microscope yet produced. Also known as ion microscope.

field length [COMPUT SCI] The number of columns, characters, or bits in a specified field.

field luminance *See* adaptation luminance.

field of view [OPTICS] The area or solid angle which can be viewed through an optical instrument. Also known as field.

field operator [QUANT MECH] An operator function of space and time for the annihilation or creation of a particle.

field pattern *See* radiation pattern.

field-programmable logic array [ELECTR] A programmed logic array in which the internal connections of the logic gates can be programmed once in the field by passing high current through fusible links, by using avalanche-induced migration to short base-emitter junctions at desired interconnections, or by other means. Abbreviated FPLA. Also known as programmable logic array.

field quenching [MET] The quench cooling and tempering of a heated metal object at the site of construction or operation by using portable equipment rather than fixed manufacturing facilities. [SOLID STATE] Decrease in the emission of light of a phosphor excited by ultraviolet radiation, x-rays, alpha particles, or cathode rays when an electric field is simultaneously applied.

field repetition rate *See* field frequency.

field section [COMPUT SCI] A portion of a field, such as the section formed by the second and third character of a 10-character field.

field service representative *See* field engineer.

field strength [PHYS] A vector characterizing a field. Also known as field intensity.

field test [SCI TECH] A nonformal experiment, that is, one with fewer controls than a laboratory experiment, conducted under field conditions.

field theory [MATH] The study of fields and their extensions. [PHYS] A theory in which the basic quantities are fields; classically the equations governing the fields may be given; in quantum field theory the commutation rules satisfied by the field operators also are specified. [PSYCH] A psychological theory that emphasizes the importance of interactions between events in an individual's environment.

field winding *See* field coil.

FIFO *See* first-in, first-out.

fifteen-degrees calorie *See* calorie.

fifth sound [CRYO] A temperature oscillation which propagates in helium II contained in a porous material such as a tightly packed powder, where the normal component is immobilized by its viscosity.

fig [BOT] *Ficus carica.* A deciduous tree of the family Moraceae cultivated for its edible fruit, which is a syconium, consisting of a fleshy hollow receptacle lined with pistillate flowers.

fighter aircraft [AERO ENG] A military aircraft designed primarily to destroy other aircraft in the air; may also be used to bomb military targets; it is maneuverable and has a high rate of climb.

fighter bomber [AERO ENG] A fighter aircraft that is designed to have bombs, or rockets, added to it so that it may be used as a bomber.

Figitidae [INV ZOO] A family of hymenopteran insects in the superfamily Cynipoidea.

figure of merit [ELECTR] A performance rating that governs the choice of a device for a particular application; for example, the figure of merit of a magnetic amplifier is the ratio of usable power gain to the control time constant.

filament [BOT] 1. The stalk of a stamen which supports the anther. 2. A chain of cells joined end to end, as in certain algae. [ELEC] Metallic wire or ribbon which is heated in an incandescent lamp to produce light, by passing an electric current through the filament. [ELECTR] A cathode made of resistance wire or ribbon, through which an electric current is sent to produce the high temperature required for emission of electrons in a thermionic tube. Also known as directly heated cathode; filamentary cathode; filament-type cathode. [INV ZOO] A single silk fiber in the cocoon of a silkworm. [MET] A long, flexible metal wire drawn very fine. [SCI TECH] A long, flexible object with a small cross section. [TEXT] A single continuous manufactured fiber which is extruded from a spinneret and joined with others to make a thread.

filamentary cathode *See* filament.

filament lamp *See* incandescent lamp.

filamentous bacteria [MICROBIO] Bacteria, especially in the order Actinomycetales, whose cells resemble filaments and are often branched.

filament saturation *See* temperature saturation.

filament-type cathode *See* filament.

filament winding [ELECTR] The secondary winding of a power transformer that furnishes alternating-current heater or filament voltage for one or more electron tubes. [ENG] A process for fabricating a composite structure in which continuous fiber reinforcement (glass, boron, silicon carbide), either previously impregnated with a matrix material or impregnated during winding, are wound under tension over a rotating core.

Filarioidea [INV ZOO] An order of the class Nematoda comprising highly specialized parasites of humans and domestic animals.

filar micrometer [DES ENG] An instrument used to measure small distances in the field of an eyepiece by using two parallel wires, one of which is fixed while the other is moved at right angles to its length by means of an accurately cut screw. Also known as bifilar micrometer.

filbert [BOT] Either of two European plants belonging to the genus *Corylus* and producing a thick-shelled, edible nut. Also known as hazelnut.

file [COMPUT SCI] A collection of related records treated as a unit. [DES ENG] A steel bar or rod with cutting teeth on its surface; used as a smoothing or forming tool.

file gap [COMPUT SCI] An area in a data storage medium which is used mainly to indicate the end of a file and sometimes the beginning of another.

file header [COMPUT SCI] A set of words comprising the file name and various characteristics of the file, found at the beginning of a file stored on magnetic tape or disk.

file identification [COMPUT SCI] A device, such as a label or tag, used to identify, describe, or name a physical medium, such as a reel of digital magnetic tape or a box of punched cards, which contains data.

file maintenance [COMPUT SCI] Data-processing operation in which a master file is updated on the basis of one or more transaction files.

file name [COMPUT SCI] The name given by the programmer to a specific set of data.

file opening [COMPUT SCI] The process, carried out by computer software, of identifying a file and comparing the file header with specifications in the program being run to ensure that the file corresponds.

file organization [COMPUT SCI] The structure of a file meeting two requirements: to minimize the run-

ning time of the program, and to simplify the work involved in modifying the contents of the file.

file-oriented system [COMPUT SCI] A computer configuration which considers a heavy, or exclusive, usage of data files.

file search [COMPUT SCI] An operation involving looking through the file for information on all items falling in a specified category, extracting the information for any item where the information recorded meets certain criteria, and determining whether or not there exists a specified pattern of information anywhere in the file.

Filicales [BOT] The equivalent name for Polypodiales.

Filicineae [BOT] The equivalent name for Polypodiatae.

Filicornia [INV ZOO] A group of hyperiid amphipod crustaceans in the suborder Genuina having the first antennae inserted anteriorly.

filiform [BIOL] Threadlike or filamentous.

filiform lapilli *See* Pele's hair.

fill [CIV ENG] Earth used for embankments or as backfill. [MIN ENG] *See* pack.

fill characters [COMPUT SCI] Nondata characters or bits which are used to fill out a field on the left if data are right-justified or on the right if data are left-justified.

filled band [SOLID STATE] An energy band, each of whose energy levels is occupied by an electron.

filled stopes [MIN ENG] Stopes filled with barren stone, low-grade ore, sand, or tailings (mill waste) after the ore has been extracted.

filled-system thermometer [ENG] A thermometer which has a bourdon tube connected by a capillary tube to a hollow bulb; the deformation of the bourdon tube depends on the pressure of a gas (usually nitrogen or helium) or on the volume of a liquid filling the system. Also known as filled thermometer.

filled thermometer *See* filled-system thermometer.

filled thermoplastic [MATER] A thermoplastic resin material that has been extended (filled) with an inert filler powder or fibers before curing.

filler [MATER] 1. An inert material added to paper, resin, bituminous material, and other substances to modify their properties and improve quality. 2. A material used to fill holes in wood, plaster, or other surfaces before applying a coating such as paint or varnish. [MET] The rod used to deposit metal in a joint in brazing, soldering, or welding.

fillet [BUILD] A flat molding that separates rounded or angular moldings. [DES ENG] A concave transition surface between two otherwise intersecting surfaces. [ENG] 1. Any narrow, flat metal or wood member. 2. A corner piece at the juncture of perpendicular surfaces to lessen the danger of cracks, as in core boxes for castings. [FOOD ENG] A boneless slice of meat or fish.

fillet lightning *See* ribbon lightning.

fillet weld [MET] A weld joining two edges at right angles; cross-sectional configuration is approximately triangular.

film [BIOL] A thin, membranous skin, such as a pellicle. [ELEC] The layer adjacent to the valve metal in an electrochemical valve, in which is located the high voltage drop when current flows in the direction of high impedance. [GRAPHICS] Plastic material, such as cellulose acetate or cellulose nitrate, coated with a light-sensitive emulsion, used to make negatives or transparencies in radiography or photography. [MATER] A flat section of a thermoplastic resin, a regenerated cellulose derivative, or other material that is extremely thin in comparison to its length and breadth and has a nominal maximum thickness of 0.25 millimeter. [MED] A pathological opacity, as of the cornea. [MET] Oxide coating on a metal.

film badge [NUCLEO] A device worn for the purpose of indicating the absorbed dose of radiation received by the wearer; usually made of metal, plastic, or paper and loaded with one or more pieces of x-ray film. Also known as badge meter.

film coefficient [THERMO] For a fluid confined in a vessel, the rate of flow of heat out of the fluid, per unit area of vessel wall divided by the difference between the temperature in the interior of the fluid and the temperature at the surface of the wall. Also known as convection coefficient.

film pressure [PHYS] The difference between the surface tension of a pure liquid and the surface tension of the liquid with a unimolecular layer of a given substance adsorbed on it. Also known as surface pressure.

film reader [ELECTR] A device for converting a pattern of transparent or opaque spots on a photographic film into a series of electric pulses. [OPTICS] A device for projecting or displaying microfilm so that an operator can read the data on the film; usually provided with equipment for moving or holding the film.

film scanning [ELECTR] The process of converting motion picture film into corresponding electric signals that can be transmitted by a television system.

film strength [MATER] 1. The measurement of a lubricant's ability to keep an unbroken film over surfaces. 2. The resistance to disruption by films of all types, such as plastic films or surface-coating films.

filmstrip [GRAPHICS] A continuous length of 35-millimeter film containing a number of still photographs, drawings, or charts, which are projected on a screen one at a time.

film theory [PHYS] A theory of the transfer of material or heat across a phase boundary, where one or both of the phases are flowing fluids, the main controlling factor being resistance to heat conduction or mass diffusion through a relatively stagnant film of the fluid next to the surface. Also known as boundary-layer theory.

film transport [MECH ENG] 1. The mechanism for moving photographic film through the region where light strikes it in recording film tracks or sound tracks of motion pictures. 2. The mechanism which moves the film print past the area where light passes through it in reproduction of picture and sound.

filoplume [VERT ZOO] A specialized feather that may be decorative, sensory, or both; it is always associated with papillae of contour feathers.

filopodia [INV ZOO] Filamentous pseudopodia.

Filosia [INV ZOO] A subclass of the class Rhizopodea characterized by slender filopodia which rarely anastomose.

filosus See fibratus.

filter [COMPUT SCI] A device or program that separates data or signals in accordance with specified criteria. [CONT SYS] See compensator. [ELECTR] Any transmission network used in electrical systems for the selective enhancement of a given class of input signals. Also known as electric filter; electric-wave filter. [ENG] A porous article or material for separating suspended particulate matter from liquids by passing the liquid through the pores in the filter and sieving out the solids. [ENG ACOUS] A device employed to reject sound in a particular range of frequencies while passing sound in an-

other range of frequencies. Also known as acoustic filter. [MATH] A family of subsets of a set S: it does not include the empty set, the intersection of any two members of the family is also a member, and any subset of S containing a member is also a member. [OPTICS] An optical element that partially absorbs incident electromagnetic radiation in the visible, ultraviolet, or infrared spectra, consisting of a pane of glass or other partially transparent material, or of films separated by narrow layers; the absorption may be either selective or nonselective with respect to wavelength. Also known as optical filter. [SCI TECH] In general, a selective device that transmits a desired range of matter or energy while substantially attenuating all other ranges.

filterable virus [VIROL] Virus particles that remain in a fluid after passing through a diatomite or glazed porcelain filter with pores too minute to allow the passage of bacterial cells.

filter bed [CIV ENG] A fill of pervious soil that provides a site for a septic field. [ENG] A contact bed used for filtering purposes.

filter cake [MATER] A concentrated solid or semisolid material that is separated from a liquid and remains on the filter after pressure filtration.

filter cake washing [CHEM ENG] An operation performed at the end of a filtration, in which residual liquid impurities are washed out of the cake by the flow of another liquid through the cake.

filter capacitor [ELEC] A capacitor used in a power-supply filter system to provide a low-reactance path for alternating currents and thereby suppress ripple currents, without affecting direct currents.

filter factor [OPTICS] The number of times the exposure must be increased when a filter is used on a camera, because the filter absorbs some of the light.

filter paper [MATER] Porous cellulose paper used for filtering, especially for quantitative purposes.

filter pass band See filter transmission band.

filter photometry [ANALY CHEM] 1. Colorimetric analysis of solution colors with a filter applied to the eyepiece of a conventional colorimeter. 2. Inspection of a pair of Nessler tubes through a filter.

filter press [ENG] A metal frame on which iron plates are suspended and pressed together by a screw device; liquid to be filtered is pumped into canvas bags between the plates, and the screw is tightened so that pressure is furnished for filtration.

filter transmission band [ELECTR] Frequency band of free transmission; that is, frequency band in which, if dissipation is neglected, the attenuation constant is zero. Also known as filter pass band.

filtrate [SCI TECH] The discharge liquor in filtration. Also known as mother liquor; strong liquor.

filtration [SCI TECH] A process of separating particulate matter from a fluid, such as air or a liquid, by passing the fluid carrier through a medium that will not pass the particulates.

fimbria See pilus.

fin [AERO ENG] A fixed or adjustable vane or airfoil affixed longitudinally to an aerodynamically or ballistically designed body for stabilizing purposes. [DES ENG] A projecting flat plate or structure, such as a cooling fin. [ENG] Material which remains in the holes of a molded part and which must be removed. [VERT ZOO] A paddle-shaped appendage on fish and other aquatic animals that is used for propulsion, balance, and guidance.

final boiling point See end point.

finder [COMMUN] 1. An optical or electronic device that shows the field of action covered by a television camera. **2.** Switch or relay group in telephone switching systems that selects the path which the call is to take through the system; operates under the control of the calling station's dial. [OPTICS] A small telescope having a wide-angle lens and low power, which is attached to a larger telescope and points in the same direction; used to locate objects that are to be viewed in the larger telescope.

F indicator *See* F scope.

fine gold [MET] Almost pure gold; the value of bullion gold depends on its percentage of fineness. [MIN ENG] In placer mining, gold in exceedingly small particles.

fines [MATER] **1.** Particles smaller than average in a mixture of particles varying in size. **2.** Fine material which passes through a standard screen on which coarser fragments are retained. [MET] That portion of a metal powder consisting of particles smaller than a specified size.

fine-screen halftone [GRAPHICS] An illustration (photograph or artwork) reproduced for printing in a continuous tone by photographing the illustration with a 120-line to the inch screen between the illustration and the camera.

fine structure [ATOM PHYS] The splitting of spectral lines in atomic and molecular spectra caused by the spin angular momentum of the electrons and the coupling of the spin to the orbital angular momentum.

finger [ANAT] Any of the four digits on the hand other than the thumb. [GEOL] The tendency for gas which is displacing liquid hydrocarbons in a heterogeneous reservoir rock system to move forward irregularly (in fingers), rather than on a uniform front. [PETRO ENG] A pair or set of bracketlike projections placed at a strategic point in a drill tripod or derrick to keep a number of lengths of drill rods or casing in place when they are standing in the tripod or derrick.

finger lake [HYD] A long, comparatively narrow lake, generally glacial in origin; may occupy a rock basin in the floor of a glacial trough or be confined by a morainal dam across the lower end of the valley.

finial [ARCH] An ornamental terminating or capping feature on an upper extremity, as over a door or on a pinnacle or gable.

finish [MATER] **1.** A chemical or other material applied to surfaces to protect them, to alter their appearances, or to modify their physical properties; finishes can be physically, chemically, or electrolytically applied and have value for fabrics and fibers, metals, paper products, plastics, woods, and so on. **2.** The ultimate quality, condition, or appearance of the surface of a material.

finisher [CIV ENG] A construction machine used to smooth the freshly placed surface of a roadway, or to prepare the foundation for a pavement.

finishing hardware [BUILD] Items, such as hinges, door pulls, and strike plates, made in attractive shapes and finishes, and usually visible on the completed structure.

finish turning [MECH ENG] The operation of machining a surface to accurate size and producing a smooth finish.

finite difference [MATH] The difference between the values of a function at two discrete points, used to approximate the derivative of the function.

finite-element method [ENG] An approximation method for studying continuous physical systems, used in structural mechanics, electrical field theory, and fluid mechanics; the system is broken into discrete elements interconnected at discrete node points.

finite mathematics [MATH] **1.** Those parts of mathematics which deal with finite sets. **2.** Those fields of mathematics which make no use of the concept of limit.

finite set [MATH] A set whose elements can be indexed by integers $1,2,3, \ldots, n$ inclusive.

fin keel [NAV ARCH] A metal plate or thin fairing attached to the keel of sailing craft to give resistance to lateral motion, and frequently having a cigar-shaped lead bulb at the bottom to give transverse stability. Also known as ballast fin.

fin spine [VERT ZOO] A bony process that supports the fins of certain fishes.

fiord *See* fjord.

fiorite *See* siliceous sinter.

fir [BOT] The common name for any tree of the genus *Abies* in the pine family; needles are characteristically flat.

fire [CHEM] The manifestation of rapid combustion, or combination of materials with oxygen. [ENG] To blast with gunpowder or other explosives. [MIN ENG] A warning that a shot is being fired. [ORD] **1.** The discharge of a gun, launching of a missile, or the like. **2.** The projectiles or missiles fired. **3.** To discharge a weapon.

fire assay [MET] Analysis of a metal-bearing material, especially gold and silver, by assaying a sample with a suitable flux and measuring the content of resulting metal leads by weighing or atomic absorption techniques.

fireball [ASTRON] A bright meteor with luminosity equal to or exceeding that of the brightest planets. [NUCLEO] The luminous sphere of hot gases that forms a few millionths of a second after a nuclear explosion.

fire blight [PL PATH] A bacterial disease of apple, pear, and related pomaceous fruit trees caused by *Erwinia amylovora*; leaves are blackened, cankers form on the trunk, and flowers and fruits become discolored.

firebreak [FOR] A cleared area of land intended to check the spread of forest or prairie fire. [MIN ENG] A strip across an area in which either no combustible material is employed or in which, if timber supports are used, sand is filled and packed tightly around them.

firebrick [MATER] A refractory brick, often made of fireclay, that is able to withstand high temperature (up to 1500–1600°C) without fusion; used to line furnaces, fireplaces, and chimneys.

fireclay [GEOL] **1.** A clay that can resist high temperatures without becoming glassy. **2.** Soft, embedded, white or gray clay rich in hydrated aluminum silicates or silica and deficient in alkalies and iron.

firedamp [MIN ENG] **1.** A gas formed in mines by decomposition of coal or other carbonaceous matter; consists chiefly of methane and is combustible. **2.** An airtight stopping to isolate an underground fire and to prevent the inflow of fresh air and the outflow of foul air. Also known as fire wall.

fire door [ENG] **1.** The door or opening through which fuel is supplied to a furnace or stove. **2.** A door that can be closed to prevent the spreading of fire, as through a building or mine.

fired state [ELECTR] The "on" state of a silicon controlled rectifier or other semiconductor switching device, occurring when a suitable triggering pulse is applied to the gate.

firefly [INV ZOO] Any of various flying insects which produce light by bioluminescence.

fire load [CIV ENG] The load of combustible material per square foot of floor space.

fire opal [MINERAL] A translucent or transparent, orangy-yellow, brownish-orange, or red variety of opal that gives out fiery reflections in bright light and that may have a play of colors. Also known as pyrophane; sun opal.

fire point [CHEM] The lowest temperature at which a volatile combustible substance vaporizes rapidly enough to form above its surface an air-vapor mixture which burns continuously when ignited by a small flame.

fireproof [BUILD] Having noncombustible walls, stairways, and stress-bearing members, and having all steel and iron structural members which could be damaged by heat protected by refractory materials. [MATER] The property of being relatively resistant to combustion.

fireproofing compound See fire retardant.

fire-resistant [CIV ENG] Of a structural element, able to resist combustion for a specified time under conditions of standard heat intensity without burning or failing structurally.

fire retardant [MATER] A chemical used as a coating for or a component of a combustible material to reduce or eliminate a tendency to burn; used with textiles, plastics, rubbers, paints, and other materials. Also known as fireproofing compound.

firestone See flint.

fire stop [BUILD] An incombustible, horizontal or vertical barrier, as of brick across a hollow wall or across an open room, to stop the spread of fire. Also known as draught stop.

fire-tube boiler [MECH ENG] A steam boiler in which hot gaseous products of combustion pass through tubes surrounded by boiler water.

fire wall [CIV ENG] **1.** A fire-resisting wall separating two parts of a building from the lowest floor to several feet above the roof to prevent the spread of fire. **2.** A fire-resisting wall surrounding an oil storage tank to retain oil that may escape and to confine fire. [MIN ENG] See firedamp.

fire welding See forge welding.

firing [ELECTR] **1.** The gas ionization that initiates current flow in a gas-discharge tube. **2.** Excitation of a magnetron or transmit-receive tube by a pulse. **3.** The transition from the unsaturated to the saturated state of a saturable reactor. [ENG] **1.** The act or process of adding fuel and air to a furnace. **2.** Igniting an explosive mixture. **3.** Treating a ceramic product with heat.

firing chamber See combustion chamber.

firing circuit [ELECTR] **1.** Circuit used with an ignitron to deliver a pulse of current of 5–50 amperes in the forward direction, from the igniter to the mercury, to start a cathode spot and to control the time of firing. **2.** By analogy, a similar control circuit of silicon-controlled rectifiers and like devices.

firmware [COMPUT SCI] A computer program or instruction, such as a microprogram, used so often that it is stored in a read-only memory instead of being included in software; often used in computers that monitor production processes.

firn [HYD] Material transitional between snow and glacier ice; it is formed from snow after existing through one summer melt season and becomes glacier ice when its permeability to liquid water drops to zero. Also known as firn snow.

firn ice See iced firn.

firn snow See firn; old snow.

first derived curve See derived curve.

first detector See mixer.

first filial generation [GEN] The first generation resulting from a cross with all members being heterozygous for characters which differ from those of the parents. Symbolized F_1.

first-generation [COMPUT SCI] Denoting electronic hardware, logical organization, and software characteristic of a first-generation computer.

first-generation computer [COMPUT SCI] A computer from the earliest stage of computer development, ending in the early 1960s, characterized by the use of vacuum tubes, the performance of one operation at a time in strictly sequential fashion, and elementary software, usually including a program loader, simple utility routines, and an assembler to assist in program writing.

first harmonic See fundamental.

first-in, first-out [IND ENG] An inventory cost evaluation method which transfers costs of material to the product in chronological order. Abbreviated FIFO.

first-item list [COMPUT SCI] A series of records that is printed with descriptive information from only the first record of each group.

first law of motion See Newton's first law.

first law of thermodynamics [THERMO] The law that heat is a form of energy, and the total amount of energy of all kinds in an isolated system is constant; it is an application of the principle of conservation of energy.

first level address [COMPUT SCI] The location of a referenced operand.

first-order spectrum [SPECT] A spectrum, produced by a diffraction grating, in which the difference in path length of light from adjacent slits is one wavelength.

first-order subroutine [COMPUT SCI] A subroutine which is entered directly from a main routine or program and which leads back to that program. Also known as first-remove subroutine.

first-order theory [OPTICS] See Gaussian optics. [PHYS] A theory which takes into account only the most important terms, such as the term proportional to the independent variable in the series expansion of a function appearing in the theory.

first point of Aries See vernal equinox.

first point of Cancer See summer solstice.

first point of Capricorn See winter solstice.

first point of Libra See autumnal equinox.

first-remove subroutine See first-order subroutine.

first selector [ELECTR] Selector which immediately follows a line finder in a switch train and which responds to dial pulses of the first digit of the called telephone number.

firth See estuary.

Fischer's distribution [STAT] Given data from a normal population with S_1^2 and S_2^2 two independent estimates of variance, the distribution ½ log (S_1^2/S_2^2).

Fischer-Tropsch process [CHEM ENG] A catalytic process to synthesize hydrocarbons and their oxygen derivatives by the controlled reaction of hydrogen and carbon monoxide.

fish [VERT ZOO] The common name for the cold-blooded aquatic vertebrates belonging to the groups Cyclostomata, Chondrichthyes, and Osteichthyes.

fish-bone antenna [ELECTROMAG] **1.** Antenna consisting of a series of coplanar elements arranged in collinear pairs, loosely coupled to a balanced transmission line. **2.** Directional antenna in the form of a plane array of doublets arranged transversely along both sides of a transmission line.

fisher [VERT ZOO] Martes pennanti. An arboreal, carnivorous mammal of the family Mustelidae; a relatively large weasellike animal with dark fur, found in northern North America.

fishery [ECOL] A place for harvesting fish or other aquatic life, particularly in sea waters.

Fishes See Pisces.

fisheye [MATER] A small globular mass which has not blended completely into the surrounding material and is particularly evident in a transparent or translucent material, such as a plastic coating or surface coating. [MET] See flake.

fish gelatin See isinglass.

fishing [ENG] In drilling, the operation by which lost or damaged tools are secured and brought to the surface from the bottom of a well or drill hole.

fish lice [INV ZOO] The common name for all members of the crustacean group Arguloida.

fish plate [CIV ENG] One of a pair of steel plates bolted to the sides of a rail or beam joint, to secure the joint.

fishpole antenna See whip antenna.

Fissidentales [BOT] An order of the Bryopsida having erect to procumbent, simple or branching stems and two rows of leaves arranged in one plane.

fissile [GEOL] Capable of being split along the line of the grain or cleavage plane. [NUCLEO] See fissionable.

fission [BIOL] A method of asexual reproduction among bacteria, algae, and protozoans by which the organism splits into two or more parts, each part becoming a complete organism. [NUC PHYS] The division of an atomic nucleus into parts of comparable mass; usually restricted to heavier nuclei such as isotopes of uranium, plutonium, and thorium. Also known as atomic fission; nuclear fission.

fissionable [NUCLEO] 1. A property of material whose nuclei are capable of undergoing fission. Also known as fissile. 2. A material capable of fission.

fission barrier [NUC PHYS] One or more maxima in the plot of potential energy against nuclear deformation of a heavy nucleus, which inhibits spontaneous fission of the nucleus.

fission chamber [NUCLEO] An ionization chamber used to detect slow neutrons; the inside wall has a thin coating of uranium, in which a slow neutron produces a fission; the resulting highly ionizing fission fragments produce a count in the chamber. Also known as fission counter.

fission counter See fission chamber.

fission fuel See nuclear fuel.

fission fungi [MICROBIO] A misnomer once used to describe the Schizomycetes.

fission-fusion bomb [NUCLEO] An explosive device which derives its energy in comparable amounts from nuclear fission and nuclear fusion.

fission reactor See nuclear reactor.

fission threshold [NUC PHYS] The minimum kinetic energy of a bombarding neutron required to induce fission of a nucleus.

fission-track dating [GEOL] A method of dating geological specimens by counting the radiation-damage tracks produced by spontaneous fission of uranium impurities in minerals and glasses.

Fissipeda [VERT ZOO] Former designation for a suborder of the Carnivora.

fissure [GEOL] 1. A high, narrow cave passageway. 2. An extensive crack in a rock. [MET] A small cracklike discontinuity with a slight opening or displacement of the fracture surfaces.

Fissurellidae [INV ZOO] The keyhole limpets, a family of gastropod mollusks in the order Archeogastropoda.

fistula [MED] An abnormal congenital or acquired communication between two surfaces or between a viscus or other hollow structure and the exterior.

Fistuliporidae [PALEON] A diverse family of extinct marine bryozoans in the order Cystoporata.

Fittig's synthesis [ORG CHEM] The synthesis of aromatic hydrocarbons by the condensation of aryl halides with alkyl halides, using sodium as a catalyst.

fitting [ENG] A small auxiliary part of standard dimensions used in the assembly of an engine, piping system, machine, or other apparatus.

FitzGerald-Lorentz contraction [RELAT] The contraction of a moving body in the direction of its motion when its speed is comparable to the speed of light. Also known as Lorentz contraction; Lorentz-FitzGerald contraction.

five-dimensional space [MATH] A vector space whose basis has five vectors.

five-level code [COMPUT SCI] A code which uses five bits to specify each character.

fix [BIOL] To kill, harden, or preserve a tissue, organ, or organism by immersion in dilute acids, alcohol, or solutions of coagulants. [COMPUT SCI] A piece of coding that is inserted in a computer program to correct an error. [NAV] A position of a vessel or craft determined by its master, pilot, or navigator through the use of some or all of the equipments and techniques available.

fixative [MATER] 1. A chemical or a mixture of chemicals used to treat biological specimens before preservation so as to retain a reasonable facsimile of their appearance when alive. 2. A substance used to increase the durability of another substance; used to fix dye mordants, hold textile dyes and pigments, and slow the rate of perfume evaporation. Also known as fixing agent.

fixed arch [CIV ENG] A stiff arch having rotation prevented at its supports.

fixed area [COMPUT SCI] That portion of the main storage occupied by the resident portion of the control program.

fixed attenuator See pad.

fixed-bed operation [CHEM ENG] An operation in which the additive material (catalyst, absorbent, filter media, ion-exchange resin) remains stationary in the chemical reactor.

fixed bias [ELECTR] A constant value of bias voltage, independent of signal strength.

fixed-block [COMPUT SCI] Pertaining to an arrangement of data in which all the blocks of data have the same number of words or characters, as determined by either the hardware requirements of the computer or the programmer.

fixed bridge [CIV ENG] A bridge having permanent horizontal or vertical alignment.

fixed capacitor [ELEC] A capacitor having a definite capacitance value that cannot be adjusted.

fixed-cycle operation [COMPUT SCI] An operation completed in a specified number of regularly timed execution cycles.

fixed disk [COMPUT SCI] A disk drive that permanently holds the disk platters.

fixed end moment See fixing moment.

fixed field [COMPUT SCI] A field in computers, film selection devices, or punched cards, or a given number of holes along the edge of a marginal punched card, set aside, or "fixed," for the recording of a given type of characteristic.

fixed-field method [COMPUT SCI] A method of data storage in which the same type of data is always placed in the same relative position.

fixed-focus lens [OPTICS] A lens whose focus is invariable, as on inexpensive cameras with no mechanism for adjusting focus but so designed that all objects from a few feet away to infinity are tolerably in focus.

fixed form coding [COMPUT SCI] Any method of coding a source language in which each part of the instruction appears in a fixed field.

fixed-head disk [COMPUT SCI] A disk storage device in which the read-write heads are fixed in position, one to a track, and the arms to which they are attached are immovable.

fixed-length field [COMPUT SCI] A field that always has the same number of characters, regardless of its content.

fixed-length record [COMPUT SCI] One of a file of records, each of which must have the same specified number of data units, such as blocks, words, characters, or digits.

fixed-level chart See constant-height chart.

fixed logic [COMPUT SCI] Circuit logic of computers or peripheral devices that cannot be changed by external controls; connections must be physically broken to arrange the logic.

fixed memory [COMPUT SCI] Of a computer, a nondestructive readout memory that is only mechanically alterable.

fixed point [ENG] A reproducible value, as for temperature, used to standardize measurements; derived from intrinsic properties of pure substances. [MATH] For a function f mapping a set S to itself, any element of S which f sends to itself.

fixed-point arithmetic [COMPUT SCI] 1. A method of calculation in which the computer does not consider the location of the decimal or radix point because the point is given a fixed position. 2. A type of arithmetic in which the operands and results of all arithmetic operations must be properly scaled so as to have a magnitude between certain fixed values.

fixed-point computer [COMPUT SCI] A computer in which numbers in all registers and storage locations must have an arithmetic point which remains in the same fixed location.

fixed-point part See mantissa.

fixed-point representation [COMPUT SCI] Any method of representing a number in which a fixed-point convention is used.

fixed-point system [COMPUT SCI] A number system in which the location of the point is fixed with respect to one end of the numerals, according to some convention.

fixed-position addressing [COMPUT SCI] Direct access to an item in a data file on disk or drum, as opposed to a sequential search for this item starting with the first item in the file.

fixed-product area [COMPUT SCI] The area in core memory where multiplication takes place for certain types of computers.

fixed-program computer [COMPUT SCI] A special-purpose computer having a program permanently wired in.

fixed storage [COMPUT SCI] A storage for data not alterable by computer instructions, such as magnetic-core storage with a lockout feature.

fixed word length [COMPUT SCI] The length of a computer machine word that always contains the same number of characters or digits.

fixing agent See fixative.

fixing moment [MECH] The bending moment at the end support of a beam necessary to fix it and prevent rotation. Also known as fixed end moment.

fixity See continuity.

Fizeau fringes [OPTICS] 1. Interference fringes of monochromatic light from interference in a geometrical situation other than plane parallel plates. Also known as fringes of equal thickness. 2. Interference fringes in light from a Fizeau interferometer.

Fizeau interferometer [OPTICS] Interferometer in which light from a point source is collimated and multiply reflected between a plane mirror and the partially silvered inner surface of a parallel plane plate, and is viewed in reflection.

fjord [GEOGR] A narrow, deep inlet of the sea between high cliffs or steep slopes. Also spelled fiord.

Flabellifera [INV ZOO] The largest and morphologically most generalized suborder of isopod crustaceans; the biramous uropods are attached to the sides of the abdomen and may form, with the last abdominal fragment, a caudal fan.

Flabelligeridae [INV ZOO] The cage worms, a family of spioniform worms belonging to the Sedentaria; the anterior part of the body is often concealed by a cage of setae arising from the first few segments.

flaccid [BOT] Deficient in turgor. [PHYSIO] Soft, flabby, or relaxed.

Flacourtiaceae [BOT] A family of dicotyledonous plants in the order Violales having the characteristics of the more primitive members of the order.

flag [COMPUT SCI] Any of various types of indicators used for identification, such as a work mark, or a character that signals the occurrence of some condition, such as the end of a word. [ELECTR] A small metal tab that holds the getter during assembly of an electron tube. [ENG] 1. A piece of fabric used as a symbol or as a signaling or marking device. 2. A large sheet of metal or fabric used to shield television camera lenses from light when not in use.

flagella [BIOL] Relatively long, whiplike, centriole-based locomotor organelles on some motile cells.

Flagellata [INV ZOO] The equivalent name for Mastigophora.

flag operand [COMPUT SCI] A part of the instruction of some assembly languages denoting which elements of the object instruction will be flagged.

flag smut [PL PATH] A smut affecting the leaves and stems of cereals and other grasses, characterized by formation of sori within the tissues, which rupture releasing black spore masses and causing fraying of the infected area.

flagstone [GEOL] 1. A hard, thin-bedded sandstone, firm shale, or other rock that splits easily along bedding planes or joints into flat slabs. 2. A piece of flagstone used for making pavement or covering the side of a house.

flake [MATER] 1. Dry, unplasticized, cellulosic plastics base. 2. Plastic chip used as feed in molding operations. [MET] 1. Discontinuous, internal cracks formed in steel during cooling due usually to the release of hydrogen. Also known as fisheye; shattercrack; snowflake. 2. Fish-scale, flat particles in powder metallurgy. Also known as flake powder.

flake powder See flake.

flaking [CHEM ENG] Continuous process operation to remove heat from material in the liquid state to cause its solidification. [ENG] 1. Reducing or separating into flakes. 2. See frosting. [MIN ENG] Breaking small chips from the face of a refractory, particularly chrome ore containing refractories.

flame arc lamp [ELEC] An arc lamp in which carbon electrodes are impregnated with chemicals, such as calcium, barium, or titanium, which are more volatile than the carbon and radiate light when driven into the arc.

flame cell [INV ZOO] A hollow cell that contains the terminal branches of excretory vessels in certain flatworms and rotifers and some other invertebrates.

flame deflector [AERO ENG] 1. In a vertical launch, any of variously designed obstructions that intercept the hot gases of the rocket engine so as to deflect them away from the ground or from a structure. 2. In a captive test, an elbow in the exhaust conduit or flame bucket that deflects the flame into the open.

flame emission spectroscopy [SPECT] A flame photometry technique in which the solution containing the sample to be analyzed is optically excited in an oxyhydrogen or oxyacetylene flame.

flame excitation [SPECT] Use of a high-temperature flame (such as oxyacetylene) to excite spectra emission lines from alkali and alkaline-earth elements and metals.

flame hardening [MET] A method for local surface hardening of steel by passing an oxyacetylene or similar flame over the work at a predetermined rate.

flame laser [OPTICS] A molecular gas laser in which gases such as carbon disulfide and oxygen are mixed at low pressures and ignited; the flame is then self-sustaining and produces carbon monoxide laser emission.

flame photometer [SPECT] One of several types of instruments used in flame photometry, such as the emission flame photometer and the atomic absorption spectrophotometer, in each of which a solution of the chemical being analyzed is vaporized; the spectral lines resulting from the light source going through the vapors enters a monochromator that selects the band or bands of interest.

flame photometry [SPECT] A branch of spectrochemical analysis in which samples in solution are excited to produce line emission spectra by introduction into a flame.

flame plate [ENG] One of the plates on a boiler firebox which are subjected to the maximum furnace temperature.

flame spectrometry [SPECT] A procedure used to measure the spectra or to determine wavelengths emitted by flame-excited substances.

flame spectrophotometry [SPECT] A method used to determine the intensity of radiations of various wavelengths in a spectrum emitted by a chemical inserted into a flame.

flame spectrum [SPECT] An emission spectrum obtained by evaporating substances in a nonluminous flame.

flamingo [VERT ZOO] Any of various long-legged and long-necked aquatic birds of the family Phoenicopteridae characterized by a broad bill resembling that of a duck but abruptly bent downward and rosy-white plumage with scarlet coverts.

flammability [CHEM] A measure of the extent to which a material will support combustion. Also known as inflammability.

flange [SCI TECH] A projecting rim of an organism or mechanical part.

flank [CIV ENG] The outer edge of a carriageway. [DES ENG] 1. The end surface of a cutting tool, adjacent to the cutting edge. 2. The side of a screw thread. [GEOL] See limb. [VERT ZOO] The part of a quadruped mammal between the ribs and the pelvic girdle.

flare [AERO ENG] To descend in a smooth curve, making a transition from a relatively steep descent to a direction substantially parallel to the surface, when landing an aircraft. [ASTRON] A bright eruption from the sun's chromosphere; flares may appear within minutes and fade within an hour, cover a wide range of intensity and size, and tend to occur between sunspots or over their penumbrae. [CHEM ENG] A device for disposing of combustible gases from refining or chemical processes by burning in the open, in contrast to combustion in a furnace or closed vessel or chamber. [DES ENG] An expansion at the end of a cylindrical body, as at the base of a rocket. [ELECTR] A radar screen target indication having an enlarged and distorted shape due to excessive brightness. [ELECTROMAG] See horn antenna. [ENG] A pyrotechnic item designed to produce a single source of intense light for such purposes as target or airfield illumination. [NAV ARCH] A concave curve of a boat's or ship's sides away from the center line, above the waterline, normally at the bow.

flareback [ORD] A rearward escapement of flame or gas from a gun.

flare spot [OPTICS] A small, diffuse, brightly illuminated region produced by multiple reflections of light from the various surfaces of an optical system.

flare stars See UV Ceti stars.

flaser [GEOL] Streaky layer of parallel, scaly aggregates that surrounds the lenticular bodies of granular material in flaser structure; caused by pressure and shearing during metamorphism.

flaser structure [GEOL] 1. A metamorphic structure in which small lenses and layers of granular material are surrounded by a matrix of sheared, crushed material, resembling a crude flow structure. Also known as pachoidal structure. 2. A primary sedimentary structure consisting of fine-sand or silt lenticles that are aligned and cross-bedded.

flash [ENG] In plastics or rubber molding or in metal casting, that portion of the charge which overflows from the mold cavity at the joint line. [MET] A fin of excess metal along the mold joint line of a casting, occurring between mating die faces of a forging or expelled from a joint in resistance welding.

flashback See backfire.

flash barrier [ELEC] A fireproof structure between conductors of an electric machine, designed to minimize flashover or the damage caused by flashover.

flash boiler [MECH ENG] A boiler with hot tubes of small capacity; designed to immediately convert small amounts of water to superheated steam.

flash butt welding [MET] Resistance welding to produce a butt joint by passing an electric current through two pieces of metal in light contact to create an arc which causes flashing and consequent heating; the weld is completed by applying pressure at the joint.

flash chamber [CHEM ENG] A conventional oil-and-gas separator operated at low pressure, with the liquid from a higher-pressure vessel being flashed into it. Also known as flash trap; flash vessel.

flash factor [OPTICS] In photography using a photoflash lamp, a number dependent on the lamp and the film speed, equal to the product of the distance of the lamp from the subject and the correct f-number for that distance.

flash groove [ENG] 1. A groove in a casting die so that excess material can escape during casting. 2. See cutoff.

flashing [BUILD] A strip of sheet metal placed at the junction of exterior building surfaces to render the joint watertight. [CHEM ENG] Vaporization of volatile liquids by either heat or vacuum. [ENG] Burning brick in an intermittent air supply in order to impart irregular color to the bricks. [MET]

The violent expulsion of small metal particles due to arcing during flash butt welding.

flashing over [ELEC] Accidental formation of an arc over the surface of a rotating commutator from brush-to-brush; usually caused by faulty insulation between commutator segments.

flash lamp [ELECTR] A gaseous-discharge lamp used in a photoflash unit to produce flashes of light of short duration and high intensity for stroboscopic photography. Also known as stroboscopic lamp.

flash mold [ENG] A mold which permits excess material to escape during closing.

flashover [ELEC] An electric discharge around or over the surface of an insulator.

flashover voltage [ELECTR] The voltage at which an electric discharge occurs between two electrodes that are separated by an insulator; the value depends on whether the insulator surface is dry or wet. Also known as sparkover voltage.

flash pasteurization [MICROBIO] A pasteurization method in which a heat-labile liquid, such as milk, is briefly subjected to temperatures of 230°F (110°C).

flash photolysis [PHYS CHEM] A method of studying fast photochemical reactions in gas molecules; a powerful lamp is discharged in microsecond flashes near a reaction vessel holding the gas, and the products formed by the flash are observed spectroscopically.

flash point [CHEM] The lowest temperature at which vapors from a volatile liquid will ignite momentarily upon the application of a small flame under specified conditions; test conditions can be either open- or closed-cup.

flash separation [CHEM ENG] Process for separation of gas (vapor) from liquid components under reduced pressure; the liquid and gas remain in contact as the gas evolves from the liquid.

flash spectroscopy [SPECT] The study of the electronic states of molecules after they absorb energy from an intense, brief light flash.

flash trap *See* flash chamber.

flash vaporization [CHEM ENG] Rapid vaporization achieved by passing a volatile liquid through continuously heated coils. [ENG] A method used for withdrawing liquefied petroleum gas from storage in which liquid is first flashed into a vapor in an intermediate pressure system, and then a second stage regulator provides the low pressure required to use the gas in appliances.

flash vessel *See* flash chamber.

flash welding [MET] A form of resistance butt welding used to weld wide, thin members or members with irregular faces, and tubing to tubing.

flat [ACOUS] A musical note that is a half step lower than a specified note. [ENG] A nonglossy painted surface. [GEOGR] A level tract of land. [GEOL] *See* mud flat. [GRAPHICS] **1.** The sheet of glass on which negative films are placed close together for printing on sensitized metal in the photoengraving process. **2.** An assemblage of negative or positive films used in preparing a photo-offset plate. [MINERAL] An inferior grade of rough diamonds. [NAV] **1.** A place covered with water too shallow for ordinary navigation. **2.** The area between high- and low-water marks along the edge of an arm of the sea, a bay, or tidal river; the term is usually used in the plural. [NAV ARCH] A partial deck below the main deck, constructed without any camber. [SCI TECH] **1.** A smooth, even surface. **2.** An object with a broad, shallow or thin form.

flat arch [ARCH] **1.** A straight horizontal arch consisting of mutually supportive wedge-shaped blocks. **2.** Any arch with a small rise-to-span ratio.

flatbed press [GRAPHICS] A press whose printing plates for type and cuts are flat; it and the rotary press are the two major kinds of presses.

flat-bottom crown *See* flat-face bit.

flat cable [ELEC] A cable made of round or rectangular, parallel copper wires arranged in a plane and laminated or molded into a ribbon of flexible insulating plastic.

flat-face bit [DES ENG] A diamond core bit whose face in cross section is square. Also known as flat-bottom crown; flat-nose bit; square-nose bit.

flatfish [VERT ZOO] Any of a number of asymmetrical fishes which compose the order Pleuronectiformes; the body is laterally compressed, and both eyes are on the same side of the head.

flat-nose bit *See* flat-face bit.

flatpack [ELECTR] Semiconductor network encapsulated in a thin, rectangular package, with the necessary connecting leads projecting from the edges of the unit.

flats [GRAPHICS] Stage constructions, used in series, to produce a painted background, usually architectural details; they are canvas-covered frames and are used for all flat surfaces on the stage, such as room interiors or exterior walls of buildings.

flat space-time [RELAT] Space-time in which the Riemann-Christoffel tensor vanishes; geometry is then equivalent to that of the Minkowski universe used in special relativity.

flat spring *See* leaf spring.

flattening [GEOD] The ratio of the difference between the equatorial and polar radii of the earth; the flattening of the earth is the ellipticity of the spheroid; the magnitude of the flattening is sometimes expressed as the numerical value of the reciprocal of the flattening. Also known as compression. [MET] Straightening of metal sheet by passing it through special rollers which flatten it without changing its thickness. Also known as roll flattening.

flat-top antenna [ELECTROMAG] An antenna having two or more lengths of wire parallel to each other and in a plane parallel to the ground, each fed at or near its midpoint.

flat tuning [ELECTR] Tuning of a radio receiver in which a change in frequency of the received waves produces only a small change in the current in the tuning apparatus.

flatworm [INV ZOO] The common name for members of the phylum Platyhelminthes; individuals are dorsoventrally flattened.

flavescence [PL PATH] Yellowing or blanching of green plant parts due to diminution of chlorophyll accompanying certain virus disease.

flavin [BIOCHEM] **1.** A yellow dye obtained from the bark of quercitron trees. **2.** Any of several water-soluble yellow pigments occurring as coenzymes of flavoproteins.

flavoprotein [BIOCHEM] Any of a number of conjugated protein dehydrogenases containing flavin that play a role in biological oxidations in both plants and animals; a yellow enzyme.

flavor [PARTIC PHYS] A label used to distinguish different types of leptons (the electron, electron neutrino, muon, muon neutrino, and possibly others) and different color triplets of quarks (the up, down, strange, and charmed quarks, and possibly others).

flax [BOT] *Linum usitatissimum.* An erect annual plant with linear leaves and blue flowers; cultivated as a source of flaxseed and fiber.

flaxseed [BOT] The seed obtained from the seed flax plant; a source of linseed oil.

flax wilt [PL PATH] A fungus disease of flax caused by *Fusarium oxysporum lini;* diseased plants wilt, yellow, and die.

F layer [GEOPHYS] An ionized layer in the F region of the ionosphere which consists of the F_1 and F_2 layers in the day hemisphere, and the F_2 layer alone in the night hemisphere; it is capable of reflecting radio waves to earth at frequencies up to about 50 megahertz.

F_1 layer [GEOPHYS] The ionosphere layer beneath the F_2 layer during the day, at a virtual height of 200–300 kilometers, being closest to earth around noon; characterized by a distinct maximum of free-electron density, except at high latitudes during winter, when the layer is not detectable.

F_2 layer [GEOPHYS] The highest constantly observable ionosphere layer, characterized by a distinct maximum of free-electron density at a virtual height from about 225 kilometers in the polar winter to more than 400 kilometers in daytime near the magnetic equator. Also known as Appleton layer.

flea [INV ZOO] Any of the wingless insects composing the order Siphonaptera; most are ectoparasites of mammals and birds.

Fleming's rule *See* left-hand rule; right-hand rule.

Fleming's solution [MATER] A tissue fixative made up of a mixture of osmic, chromic and acetic acids.

Flemish bond [CIV ENG] A masonry bond consisting of alternating stretchers and headers in each course, laid with broken joints.

fleshy fruit [BOT] A fruit having a fleshy pericarp that is usually soft and juicy, but sometimes hard and tough.

Fletcher-Munson contour *See* equal loudness contour.

Flexibilia [PALEON] A subclass of extinct stalked or creeping Crinoidea; characteristics include a flexible tegmen with open umbulacral grooves, uniserial arms, a cylindrical stem, and five conspicuous basals and radials.

flexible circuit [ELECTR] A printed circuit made on a flexible plastic sheet that is usually die-cut to fit between large components.

flexible coupling [ELECTROMAG] A coupling designed to allow a limited angular movement between the axes of two waveguides. [MECH ENG] A coupling used to connect two shafts and to accommodate their misalignment.

flexible pavement [CIV ENG] A road or runway made of bituminous material which has little tensile strength and is therefore flexible.

flexible resistor [ELEC] A wire-wound resistor having the appearance of a flexible lead; made by winding the Nichrome resistance wire around a length of asbestos or other heat-resistant cord, then covering the winding with asbestos and braided insulating covering.

flexible shaft [MECH ENG] **1.** A shaft that transmits rotary motion at any angle up to about 90°. **2.** A shaft made of flexible material or of segments. **3.** A shaft whose bearings are designed to accommodate a small amount of misalignment.

flexible spacecraft [AERO ENG] A space vehicle (usually a space structure or rotating satellite) whose surfaces or appendages may be subject to elastic flexural deformations (vibrations).

flexion [BIOL] Act of bending, especially of a joint.

flexional symbols [COMPUT SCI] Symbols in which the meaning of each component digit is dependent on those which precede it.

flexion reflex [PHYSIO] An unconditioned, segmental reflex elicited by noxious stimulation and consisting of contraction of the flexor muscles of all joints on the same side. Also known as the nocioceptive reflex.

flexographic printing *See* flexography.

flexography [GRAPHICS] Relief printing with plates fastened to a cylinder and with a single inking roller supplied with aniline ink from two rollers in the ink fountain. Also known as aniline printing; aniline process; flexographic printing.

flexural modulus [MECH] A measure of the resistance of a beam of specified material and cross section to bending, equal to the product of Young's modulus for the material and the square of the radius of gyration of the beam about its neutral axis.

flexural slip [GEOL] The slipping of sedimentary strata along bedding planes during folding, producing disharmonic folding and, when extreme, décollement. Also known as bedding-plane slip.

flexural strength [MECH] Strength of a material in blending, that is, resistance to fracture.

flexure [EMBRYO] A sharp bend of the anterior part of the primary axis of the vertebrate embryo. [GEOL] **1.** A broad, domed structure. **2.** A fold. [MECH] **1.** The deformation of any beam subjected to a load. **2.** Any deformation of an elastic body in which the points originally lying on any straight line are displaced to form a plane curve. [VERT ZOO] The last joint of a bird's wing.

flicker [OPTICS] A visual sensation produced by periodic fluctuations in light at rates ranging from a few cycles per second to a few tens of cycles per second.

flicker photometer [OPTICS] A photometer in which a single field of view is alternately illuminated by the light sources to be compared, and the rate of alternation is such that color flicker is absent but brightness flicker is not; disappearance of flicker signifies equality of luminance.

flight conveyor [MECH ENG] A conveyor in which paddles, attached to single or double strands of chain, drag or push pulverized or granulated solid materials along a trough. Also known as drag conveyor.

flight deck [AERO ENG] In certain airplanes, an elevated compartment occupied by the crew for operating the airplane in flight. [NAV ARCH] The topmost complete deck of an aircraft carrier, used mainly for takeoff and landing of planes.

flight envelope [AERO ENG] The boundary depicting, for a specific aircraft, the limits of speed, altitude, and acceleration which that aircraft cannot safely exceed.

flight feather [VERT ZOO] Any of the long contour feathers on the wing of a bird. Also known as remex.

flight forecast [METEOROL] An aviation weather forecast for a specific flight.

flight instrument [AERO ENG] An aircraft instrument used in the control of the direction of flight, attitude, altitude, or speed of an aircraft, for example, the artificial horizon, airspeed indicator, altimeter, compass, rate-of-climb indicator, accelerometer, turn-and-bank indicator, and so on.

flight log [NAV] **1.** A complete written record of a flight, normally showing flight planning information together with actual data recorded during the flight. **2.** A device that automatically records on a screen or map the flight path flown by an aircraft.

flight path [AERO ENG] The path made or followed in the air or in space by an aircraft, rocket, or such.

flight-path computer [COMPUT SCI] A computer that includes all of the functions of a course-line computer and also provides means for controlling the altitude of an aircraft in accordance with a desired plan of flight.

flight plan [NAV] Information provided to air-traffic service units, giving in detail the proposed plan of flight, including times, altitudes, way points, and so on, and submitted for approval.

flight recorder [ENG] Any instrument or device that records information about the performance of an aircraft in flight or about conditions encountered in flight, for future study and evaluation.

flight rules [NAV] Rules established by competent authority to govern flights; the type of flight involved determines whether instrument flight rules or visual flight rules apply.

flight track *See* track.

flint [MINERAL] A black or gray, massive, hard, somewhat impure variety of chalcedony, breaking with a conchoidal fracture. Also known as firestone.

flint glass [MATER] **1.** Heavy, colorless, brilliant glass that contains lead oxide. **2.** Any high-quality glass.

flip chip [ELECTR] A tiny semiconductor die having terminations all on one side in the form of solder pads or bump contacts; after the surface of the chip has been passivated or otherwise treated, it is flipped over for attaching to a matching substrate.

flip coil [ELECTROMAG] A small coil used to measure the strength of a magnetic field; it is placed in the field, connected to a ballistic galvanometer or other instrument, and suddenly flipped over 180°; alternatively, the coil may be held stationary and the magnetic field reversed.

flip-flop circuit *See* bistable multivibrator.

flip-over process *See* Umklapp process.

flipper [VERT ZOO] A broad, flat appendage used for locomotion by aquatic mammals and sea turtles.

float [AGR] A device consisting of one or more blades used to level a seedbed. [BIOL] An air-filled sac in many pelagic flora and fauna that serves to buoy up the body of the organism. [DES ENG] A file which has a single set of parallel teeth. [ENG] **1.** A flat, rectangular piece of wood with a handle, used to apply and smooth coats of plaster. **2.** A mechanical device to finish the surface of freshly placed concrete paving. **3.** A marble-polishing block. **4.** Any structure that provides positive buoyancy such as a hollow, watertight unit that floats or rests on the surface of a fluid. **5.** *See* plummet. [GEOL] An isolated, displaced rock or ore fragment. [IND ENG] *See* bank. [TEXT] **1.** A thread used to create patterns in fabric by passing over other threads. **2.** A fabric defect caused by passing a thread over other threads where it should be interwoven.

float chamber [ENG] A vessel in which a float regulates the level of a liquid.

float gage [ENG] Any one of several types of instruments in which the level of a liquid is determined from the height of a body floating on its surface, by using pullies, levers, or other mechanical devices.

floating address [COMPUT SCI] The symbolic address used prior to its conversion to a machine address.

floating arithmetic *See* floating-point arithmetic.

floating axle [MECH ENG] A live axle used to turn the wheels of an automotive vehicle; the weight of the vehicle is borne by housings at the ends of a fixed axle.

floating carrier modulation *See* controlled carrier modulation.

floating-decimal arithmetic *See* floating-point arithmetic.

floating dollar sign [COMPUT SCI] A dollar sign used with an edit mask, allowing the sign to be inserted before the nonzero leading digit of a dollar amount.

floating floor [BUILD] A floor constructed so that the wearing surface is separated from the supporting structure by an insulating layer of mineral wool, resilient quilt, or other material to provide insulation against impact sound.

floating foundation [CIV ENG] **1.** A reinforced concrete slab that distributes the concentrated load from columns; used on soft soil. **2.** A foundation mat several meters below the ground surface when it is combined with external walls.

floating grid [ELECTR] Vacuum-tube grid that is not connected to any circuit; it assumes a negative potential with respect to the cathode. Also known as free grid.

floating-point arithmetic [MATH] A method of performing arithmetical operations, used especially by automatic computers, in which numbers are expressed as integers multiplied by the radix raised to an integral power, as 87×10^{-4} instead of 0.0087. Also known as floating arithmetic; floating-decimal arithmetic.

floating-point coefficient *See* mantissa.

floating-point package [COMPUT SCI] A program which enables a computer to perform arithmetic operations when such capabilities are not wired into the computer. Also known as floating-point routine.

floating-point routine *See* floating-point package.

floating-point system [COMPUT SCI] A number system in which the location of the point does not remain fixed with respect to one end of the numerals.

floating rib [ANAT] One of the last two ribs in humans which have the anterior end free.

floating roof [ENG] A type of tank roof (steel, plastic, sheet, or microballoons) which floats upon the surface of the stored liquid; used to decrease the vapor space and reduce the potential for evaporation.

float valve [ENG] A valve whose on-off action is controlled directly by the fall or rise of a float concurrent with the fall or rise of liquid level in a liquid-containing vessel.

floc [CHEM] Small masses formed in a fluid through coagulation, agglomeration, or biochemical reaction of fine suspended particles.

flocculate [BIOL] Having small tufts of hairs. [CHEM] To cause to aggregate or coalesce into a flocculent mass.

flocculus [ANAT] A prominent lobe of the cerebellum situated behind and below the middle cerebellar peduncle on each side of the median fissure. [ASTRON] A patch in the sun's surface seen in the light of calcium or hydrogen; the patch may be bright or dark and is usually in the vicinity of sunspots.

floccus [BOT] A tuft of woolly hairs. [METEOROL] A cloud species in which each element is a small tuft with a rounded top and a ragged bottom.

flock [TEXT] **1.** Pulverized wool, cotton, silk, or rayon fiber used to form velvety patterns on cloth. **2.** Woolen or cotton refuse reduced by machinery and used to stuff furniture.

floc test [ANALY CHEM] A quantitative test applied to kerosine and other illuminating oils to detect substances rendered insoluble by heat.

floe [OCEANOGR] A piece of floating sea ice other than fast ice or glacier ice; may consist of a single fragment or of many consolidated fragments, but is larger than an ice cake and smaller than an ice field. Also known as ice floe.

floe till [GEOL] 1. A glacial till resulting from the intact deposition of a grounded iceberg in a lake bordering an ice sheet. 2. A lacustrine clay with boulders, stones, and other glacial matter dropped into it by melting icebergs. Also known as berg till.

flokite *See* mordenite.

flood [ELECTR] To direct a large-area flow of electrons toward a storage assembly in a charge storage tube. [ENG] To cover or fill with fluid. [HYD] The condition that occurs when water overflows the natural or artificial confines of a stream or other body of water, or accumulates by drainage over low-lying areas. [MECH ENG] To supply an excess of fuel to a carburetor so that the level rises above the nozzle. [OCEANOGR] The highest point of a tide.

flood current [OCEANOGR] The tidal current associated with the increase in the height of a tide.

floodgate [CIV ENG] 1. A gate used to restrain a flow or, when opened, to allow a flood flow to pass. 2. The lower gate of a lock.

flood icing *See* icing.

flooding [AGR] Filling of ditches or covering of land with water during the raising of crops; rice, for example, must have occasional flooding to grow properly. [CHEM ENG] Condition in a liquid-vapor counterflow device (such as a distillation column) in which the rate of vapor rise is such as to prevent liquid downflow, causing a buildup of the liquid (flooding) within the device. [PETRO ENG] Technique of increasing recovery of oil (secondary recovery) from a reservoir by injection of water into the formation to drive the oil toward producing wellholes. Also known as waterflooding.

flooding ice *See* icing.

floodlight [ELEC] A light projector used for outdoor lighting of buildings, parking lots, sports fields, and the like, usually having a filament lamp or mercury-vapor lamp and a parabolic reflector.

floodplain [GEOL] The relatively smooth valley floors adjacent to and formed by alluviating rivers which are subject to overflow.

flood relief channel *See* bypass channel.

flood stage [HYD] The stage, on a fixed river gage, at which overflow of the natural banks of the stream begins to cause damage in any portion of the reach for which the gage is used as an index.

flood tide [OCEANOGR] 1. That period of tide between low water and the next high water. 2. A tide at its highest point.

flood tuff *See* ignimbrite.

floodway *See* bypass channel.

floor [ENG] The bottom, horizontal surface of an enclosed space. [GEOL] 1. The rock underlying a stratified or nearly horizontal deposit, corresponding to the footwall of more steeply dipping deposits. 2. A horizontal, flat ore body. [MIN ENG] Boards laid at the heading to receive blasted rocks and to facilitate ore loading. [NAV ARCH] One of a series of vertical plates extending across the bottom of a ship at right angles to the center line and forming part of the bottom framing of the hull.

flooring [MATER] Material suitable for use as a floor.

floor outlet [ELEC] An electrical outlet whose face is level with or recessed into a floor. Also known as floor plug.

floor plan [ARCH] A diagram of a floor showing partitions, doors, windows, and other features.

floor plug *See* floor outlet.

floor system [CIV ENG] The structural floor assembly between supporting beams or girders in buildings and bridges.

flopover [ELECTR] A defect in television reception in which a series of frames move vertically up or down the screen, caused by lack of synchronization between the vertical and horizontal sweep frequencies.

floppy disk [COMPUT SCI] A flexible plastic disk about 7½ inches (19 centimeters) in diameter, coated with magnetic oxide and used for data entry to a computer; a slot in its protective envelope or housing, which remains stationary while the disk rotates, exposes the track positions for the magnetic read/write head of the drive unit. Also known as diskette.

florentium *See* promethium-147.

floret [BOT] A small individual flower that is part of a compact group of flowers, such as the head of a composite plant or inflorescence.

floriculture [AGR] A segment of horticulture concerned with commercial production, marketing, and retail sale of cut flowers and potted plants, as well as home gardening and flower arrangement.

Florida Current [OCEANOGR] A fast current that sets through the Straits of Florida to a point north of Grand Bahama Island, where it joins the Antilles Current to form the Gulf Stream.

Florideophyceae [BOT] A class of red algae, division Rhodophyta, having prominent pit connections between cells.

Flosculariacea [INV ZOO] A suborder of rotifers in the order Monogononta having a malleoramate mastax.

Flosculariidae [INV ZOO] A family of sessile rotifers in the suborder Flosculariacea.

flospinning [MET] Power-spinning or flowing metal over a rotating bar for shaping into cylindrical, conical, and curvilinear parts.

flotation [ENG] A process used to separate particulate solids by causing one group of particles to float; utilizes differences in surface chemical properties of the particles, some of which are entirely wetted by water, others are not; the process is primarily applied to treatment of minerals but can be applied to chemical and biological materials; in mining engineering it is referred to as froth flotation.

flotation collar [ENG] A buoyant bag carried by a spacecraft and designed so that it inflates and surrounds part of the outer surface if the spacecraft lands in the sea.

flotsam [ENG] Floating articles, particularly those that are thrown overboard to lighten a vessel in distress.

flounder [VERT ZOO] Any of a number of flatfishes in the families Pleuronectidae and Bothidae of the order Pleuronectiformes.

flow [COMPUT SCI] The sequence in which events take place or operations are carried out. [ENG] A forward movement in a continuous stream or sequence of fluids or discrete objects or materials, as in a continuous chemical process or solids-conveying or production-line operations. [FL MECH] The forward continuous movement of a fluid, such as gases, vapors, or liquids, through closed or open channels or conduits. [GEOL] Any rock deformation that is not instantly recoverable without permanent loss of cohesion. Also known as flowage; rock flowage. [PHYS] The movement of electric charges, gases, liquids, or other materials or quantities.

flow banding [GEOL] An igneous rock structure resulting from flowing of magmas or lavas and characterized by alternation of mineralogically unlike layers.

flow breccia [GEOL] A breccia formed with the movement of lava flow while the flow is still in motion.

flow chart [ENG] A graphical representation of the progress of a system for the definition, analysis, or solution of a data-processing or manufacturing problem in which symbols are used to represent operations, data or material flow, and equipment, and lines and arrows represent interrelationships among the components. Also known as control diagram; flow diagram; flow sheet.

flow coefficient [FL MECH] An experimentally determined proportionality constant, relating the actual velocity of fluid flow in a pipe, duct, or open channel to the theoretical velocity expected under certain assumptions. [MECH ENG] A dimensionless number used in studying the power required by fans, equal to the volumetric flow rate through the fan divided by the product of the rate of rotation of the fan and the cube of the impeller diameter.

flow control valve [ENG] A valve whose flow opening is controlled by the rate of flow of the fluid through it; usually controlled by differential pressure across an orifice at the valve. Also known as rate-of-flow control valve.

flow counter *See* gas-flow counter tube.

flow diagram *See* flow chart.

flow equation [FL MECH] Equation for the calculation of fluid (gas, vapor, liquid) flow through conduits or channels; consists of an interrelation of fluid properties (such as density or viscosity), environmental conditions (such as temperature or pressure), and conduit or channel geometry and conditions (such as diameter, cross-sectional shape, or surface roughness).

flower [BOT] The characteristic reproductive structure of a seed plant, particularly if some or all of the parts are brightly colored.

flow fold [GEOL] Folding in beds, composed of relatively plastic rock, that assume any shape impressed upon them by the more rigid surrounding rocks or by the general stress pattern of the deformed zone; there are no apparent surfaces of slip.

flowing well [PETRO ENG] Oil reservoir in which gas-drive pressure is sufficient to force oil flow up through and out of a wellhole.

flow line [ENG] 1. The connecting line or arrow between symbols on a flow chart or block diagram. 2. Mark on a molded plastic or metal article made by the meeting of two input-flow fronts during molding. Also known as weld mark. [HYD] A contour of the water level around a body of water. [PETR] In an igneous rock, any internal structure produced by parallel orientation of crystals, mineral streaks, or inclusions. [PETRO ENG] A pipeline that takes oil from a single well or a series of wells to a gathering center.

flowmeter [ENG] An instrument used to measure pressure, flow rate, and discharge rate of a liquid, vapor, or gas flowing in a pipe. Also known as fluid meter.

flow nozzle [ENG] A flowmeter in a closed conduit, consisting of a short flared nozzle of reduced diameter inset into the inner diameter of a pipe; used to cause a temporary pressure drop in flowing fluid to determine flow rate via measurement of static pressures before and after the nozzle.

flow pattern [FL MECH] Pattern of two-phase flow in a conduit or channel pipe, taking into consideration the ratio of gas to liquid and conditions of flow resistance and liquid holdup.

flow rate [FL MECH] Also known as rate of flow. 1. Time required for a given quantity of flowable material to flow a measured distance. 2. Weight or volume of flowable material flowing per unit time.

flow reactor [CHEM ENG] A dynamic reactor system in which reactants flow continuously into the vessel and products are continuously removed, in contrast to a batch reactor.

flow resistance [FL MECH] 1. Any factor within a conduit or channel that impedes the flow of fluid, such as surface roughness or sudden bends, contractions, or expansions. 2. *See* viscosity.

flow separation *See* boundary-layer separation.

flow sheet *See* flow chart.

flow soldering [ENG] Soldering of printed circuit boards by moving them over a flowing wave of molten solder in a solder bath; the process permits precise control of the depth of immersion in the molten solder and minimizes heating of the board. Also known as wave soldering.

flowstone [GEOL] Deposits of calcium carbonate that accumulated against the walls of a cave where water flowed on the rock.

flow stress [MECH] The stress along one axis at a given value of strain that is required to produce plastic deformation.

flow valve [ENG] A valve that closes itself when the flow of a fluid exceeds a particular value.

flow welding [MET] A welding process in which coalescence is produced by heating with molten filler metal, which is poured over the joint until the welding temperature is attained and the required amount of filler metal is added.

fl oz *See* fluid ounce.

fluctuation [OCEANOGR] 1. Wavelike motion of water. 2. The variations of water-level height from mean sea level that are not due to tide-producing forces. [SCI TECH] 1. Variation, especially back and forth between successive values in a series of observations. 2. Variation of data points about a smooth curve passing among them.

fluctuation noise *See* random noise.

fluctuation velocity *See* eddy velocity.

flue [ENG] A channel or passage for conveying combustion products from a furnace, boiler, or fireplace to or through a chimney.

fluid [PHYS] An aggregate of matter in which the molecules are able to flow past each other without limit and without fracture planes forming.

fluid-bed process [CHEM ENG] A type of process based on the tendency of finely divided powders to behave in a fluidlike manner when supported and moved by a rising gas or vapor stream; used mainly for catalytic cracking of petroleum distillates.

fluid clutch *See* fluid drive.

fluid computer [COMPUT SCI] A digital computer constructed entirely from air-powered fluid logic elements; it contains no moving parts and no electronic circuits; all logic functions are carried out by interaction between jets of air.

fluid coupling [MECH ENG] A device for transmitting rotation between shafts by means of the acceleration and deceleration of a fluid such as oil. Also known as hydraulic coupling.

fluid drive [MECH ENG] A power coupling operated on a hydraulic turbine principle in which the engine flywheel has a set of turbine blades which are connected directly to it and which are driven in oil, thereby turning another set of blades attached to the transmission gears of the automobile. Also known as fluid clutch; hydraulic clutch.

fluid dynamics [FL MECH] The science of fluids in motion.

fluid-film bearing [MECH ENG] An antifriction bearing in which rubbing surfaces are kept apart by a film of lubricant such as oil.

fluid friction [FL MECH] Conversion of mechanical energy in fluid flow into heat energy.

fluidics [ENG] A control technology that employs fluid dynamic phenomena to perform sensing, control, information, processing, and actuation functions without the use of moving mechanical parts.

fluidity [FL MECH] The reciprocal of viscosity; expresses the ability of a substance to flow.

fluidization [CHEM ENG] A roasting process in which finely divided solids are suspended in a rising current of air (or other fluid), producing a fluidized bed; used in the calcination of various minerals, in Fischer-Tropsch synthesis, and in the coal industry.

fluidized bed [ENG] A cushion of air or hot gas blown through the porous bottom slab of a container which can be used to float a powdered material as a means of drying, heating, quenching, or calcining the immersed components.

fluidized-bed combustion [MECH ENG] A method of burning particulate fuel, such as coal, in which the amount of air required for combustion far exceeds that found in conventional burners; the fuel particles are continually fed into a bed of mineral ash in the proportions of 1 part fuel to 200 parts ash, while a flow of air passes up through the bed, causing it to act like a turbulent fluid.

fluidized-bed reactor See fluidized reactor.

fluidized reactor [NUCLEO] A nuclear reactor in which the fuel has been given the properties of a quasi-fluid, such as by suspension of fine fuel particles in a carrying gas or liquid. Also known as fluidized-bed reactor.

fluid mechanics [MECH] The science concerned with fluids, either at rest or in motion, and dealing with pressures, velocities, and accelerations in the fluid, including fluid deformation and compression or expansion.

fluid meter See flow meter.

fluid ounce [MECH] Abbreviated fl oz. **1.** A unit of volume that is used in the United States for measurement of liquid substances, equal to 1/16 liquid pint, or 231/128 (approximately 1.804) cubic inches, or $2.95735295625 \times 10^{-5}$ cubic meter. **2.** A unit of volume used in the United Kingdom for measurement of liquid substances, and occasionally of solid substances, equal to 1/20 pint or approximately 2.84130×10^{-5} cubic meter.

fluid resistance [FL MECH] The force exerted by a gas or liquid opposing the motion of a body through it. Also known as resistance.

fluid statics [FL MECH] The determination of pressure intensities and forces exerted by fluids at rest.

fluke [INV ZOO] The common name for more than 40,000 species of parasitic flatworms that form the class Trematoda. [NAV ARCH] The broad end of each arm of an anchor. [VERT ZOO] A flatfish, especially summer flounder.

flume [ENG] **1.** An open channel constructed of steel, reinforced concrete, or wood and used to convey water to be utilized for power, to transport logs, and so on. **2.** To divert by a flume, as the waters of a stream, in order to lay bare the auriferous sand and gravel forming the bed. [GEOL] A ravine with a stream flowing through it.

fluoborate [INORG CHEM] **1.** Any of a group of compounds related to the borates in which one or more oxygens have been replaced by fluorine atoms. **2.**

The BF_4^- ion, which is derived from fluoboric acid, HBF_4.

fluoborite [MINERAL] $Mg_3(BO_3)(F,OH)_3$ A colorless mineral composed of magnesium fluoborate; occurs in hexagonal prisms. Also known as nocerite.

fluolite See pitchstone.

fluophor See luminophor.

fluor See fluorite.

fluorapatite [MINERAL] **1.** $Ca_5(PO_4)_3F$ A mineral of the solid-solution series of the apatite group; common accessory mineral in igneous rocks. **2.** An apatite mineral in which the fluoride member dominates.

fluorescence [ATOM PHYS] **1.** Emission of electromagnetic radiation that is caused by the flow of some form of energy into the emitting body and which ceases abruptly when the excitation ceases. **2.** Emission of electromagnetic radiation that is caused by the flow of some form of energy into the emitting body and whose decay, when the excitation ceases, is temperature-independent. [NUC PHYS] Gamma radiation scattered by nuclei which are excited to and radiate from an excited state. [OPTICS] See bloom.

fluorescence analysis See fluorometric analysis.

fluorescence microscope [OPTICS] A variation of the compound laboratory light microscope which is arranged to admit ultraviolet, violet, and sometimes blue radiations to a specimen, which then fluoresces.

fluorescence spectra [SPECT] Emission spectra of fluorescence in which an atom or molecule is excited by absorbing light and then emits light of characteristic frequencies.

fluorescent antibody test [IMMUNOL] A clinical laboratory test based on the antigen used in the diagnosis of syphilis and lupus erythematosus and for identification of certain bacteria and fungi, including the tubercle bacillus.

fluorescent lamp [ELECTR] A tubular discharge lamp in which ionization of mercury vapor produces radiation that activates the fluorescent coating on the inner surface of the glass.

fluoridation [ENG] The addition of the fluorine ion (F^-) to municipal water supplies in a final concentration of 0.8–1.6 ppm (parts per million) to help prevent dental caries in children. [GEOCHEM] Formation in rocks of fluorine-containing minerals such as fluorite or topaz.

fluorine [CHEM] A gaseous or liquid chemical element, symbol F, atomic number 9, atomic weight 18.998; a member of the halide family, it is the most electronegative and the most chemically energetic of the nonmetallic elements; highly toxic, corrosive, and flammable; used in rocket fuels and as a chemical intermediate.

fluorite [MINERAL] CaF_2 A transparent to translucent, often blue or purple mineral, commonly found in crystalline cubes in veins and associated with lead, tin, and zinc ores; hardness is 4 on Mohs scale; the principal ore of fluorine. Also known as Derbyshire spar; fluor; fluorspar.

fluorocarbon [ORG CHEM] A hydrocarbon such as Freon in which part or all hydrogen atoms have been replaced by fluorine atoms; can be liquid or gas and is nonflammable and heat-stable; used as refrigerant, aerosol propellant, and solvent. Also known as fluorohydrocarbon.

fluorocarbon resin [ORG CHEM] Polymeric material made up of carbon and fluorine with or without other halogens (such as chlorine) or hydrogen; the resin is extremely inert and more dense than corresponding fluorocarbons such as Teflon.

fluorography [GRAPHICS] Photography of an image produced on a fluorescent screen. Also known as photofluorography.

fluorohydrocarbon See fluorocarbon.

fluorometric analysis [ANALY CHEM] A method of chemical analysis in which a sample, exposed to radiation of one wavelength, absorbs this radiation and reemits radiation of the same or longer wavelength in about 10^{-9} second; the intensity of reemited radiation is almost directly proportional to the concentration of the fluorescing material. Also known as fluorescence analysis; fluorometry.

fluorometry See fluorometric analysis.

fluoroscope [ENG] A fluorescent screen designed for use with an x-ray tube to permit direct visual observation of x-ray shadow images of objects interposed between the x-ray tube and the screen.

fluorothene See chlorotrifluoroethylene polymer.

fluorspar See fluorite.

flurry [METEOROL] A brief shower of snow accompanied by a gust of wind, or a sudden, brief wind squall.

flush [ECOL] An evergreen herbaceous or nonflowering vegetation growing in habitats where seepage water causes the surface to be constantly wet but rarely flooded. [ENG] Pertaining to separate surfaces that are on the same level. [GRAPHICS] A printing term that means no indention; headings are often run flush left, that is, they align at the left margin; flush-right lines align at the right.

flushing [CIV ENG] The removal or reduction to a permissible level of dissolved or suspended contaminants in an estuary or harbor. [ENG] Removing lodged deposits of rock fragments and other debris by water flow at high velocity; used to clean water conduits and drilled boreholes.

flute [DES ENG] A groove having a curved section, especially when parallel to the main axis, as on columns, drills, and other cylindrical or conical shaped pieces. [GEOL] 1. A natural groove running vertically down the face of a rock. 2. A groove in a sedimentary structure formed by the scouring action of a turbulent, sediment-laden water current, and having a steep upcurrent end.

fluted coupling See stabilizer.

flutter [ACOUS] Distortion that occurs in sound reproduction as a result of undesired speed variations during the recording, duplicating, or reproducing process. [ELECTROMAG] A fast-changing variation in received signal strength, such as may be caused by antenna movements in a high wind or interaction with a signal or another frequency. [ENG] The irregular alternating motion of the parts of a relief valve due to the application of pressure where no contact is made between the valve disk and the seat. [FL MECH] See aeronautical flutter. [MED] Rapid, regular contraction of the atrial muscle of the heart.

flutter echo [ACOUS] A multiple echo in which the reflections rapidly follow each other. [ELECTROMAG] A radar echo consisting of a rapid succession of reflected pulses resulting from a single transmitted pulse.

flutter valve [ENG] A valve that is operated by fluctuations in pressure of the material flowing over it; used in carburetors.

Fluvent [GEOL] A suborder of the soil order Entisol that is well-drained with visible marks of sedimentation and no identifiable horizons; occurs in recently deposited alluvium along streams or in fans.

fluvial [HYD] 1. Pertaining to or produced by the action of a stream or river. 2. Existing, growing, or living in or near a river or stream.

fluvial cycle of erosion See normal cycle.

flux [ELECTROMAG] The electric or magnetic lines of force in a region. [MATER] 1. In soldering, welding, and brazing, a material applied to the pieces to be united to reduce the melting point of solders and filler metals and to prevent the formation of oxides. 2. A substance used to promote the fusing of minerals or metals. 3. Additive for plastics composition to improve flow during physical processing. 4. In enamel work, a substance composed of silicates and other materials that forms a colorless, transparent glass when fired. Also know as fondant. [NUCLEO] The product of the number of particles per unit volume and their average velocity; a special case of the physics definition. Also known as flux density. [PHYS] 1. The integral over a given surface of the component of a vector field (for example, the magnetic flux density, electric displacement, or gravitational field) perpendicular to the surface; by definition, it is proportional to the number of lines of force crossing the surface. 2. The amount of some quantity flowing across a given area (often a unit area perpendicular to the flow) per unit time; the quantity may be, for example, mass or volume of fluid, electromagnetic energy, or number of particles.

flux density [NUCLEO] See flux. [PHYS] Any vector field whose flux is a significant physical quantity; examples are magnetic flux density, electric displacement, gravitational field, and the Poynting vector.

flux gate [ENG] A detector that gives an electric signal whose magnitude and phase are proportional to the magnitude and direction of the external magnetic field acting along its axis; used to indicate the direction of the terrestrial magnetic field.

fluxional compound [ORG CHEM] 1. Any of a group of molecules which undergo rapid intramolecular rearrangements in which the component atoms are interchanged among equivalent structures. 2. Molecules in which bonds are broken and reformed in the rearrangement process.

flux jumping See Meissner effect.

flux line See line of force.

flux linkage [ELECTROMAG] The product of the number of turns in a coil and the magnetic flux passing through the coil. Also known as linkage.

flux refraction [ELECTROMAG] The abrupt change in direction of magnetic flux lines at the boundary between two media having different permeabilities, or of the electric flux lines at the boundary between two media having different dielectric constants, when these lines are oblique to the boundary.

fly [INV ZOO] The common name for a number of species of the insect order Diptera characterized by a single pair of wings, antennae, compound eyes, and hindwings modified to form knoblike balancing organs, the halters. [MECH ENG] A fan with two or more blades used in timepieces or light machinery to govern speed by air resistance.

Fly See Musca.

flyback [ELECTR] The time interval in which the electron beam of a cathode-ray tube returns to its starting point after scanning one line or one field of a television picture or after completing one trace in an oscilloscope. Also known as retrace; return trace. [HOROL] The return to zero of the timing hand of a stopwatch or chronograph.

flyback transformer *See* horizontal output transformer.

fly-by-tube control [AERO ENG] Fluidic flight control for aircraft in which a hydraulic control signal link connects the pilot's controls to the control surface actuators.

fly-by-wire system [AERO ENG] A flight control system that uses electric wiring instead of mechanical or hydraulic linkages to control the actuators for the ailerons, flaps, and other control surfaces of an aircraft.

flying-aperture scanner [ELECTR] An optical scanner, used in character recognition, in which a document is flooded with light, and light is collected sequentially spot by spot from the illuminated image.

flying bridge [NAV ARCH] A narrow walkway or platform built at the level of the top of the pilothouse and containing a duplicate set of controls for steering gear and engine-room signals and for navigating instruments; the platform extends from one side of the ship to the other. Also known as navigating bridge.

flying buttress [ARCH] A buttress connected to the building it supports by an arch.

flying fish [VERT ZOO] Any of about 65 species of marine fishes which form the family Exocoetidae in the order Atheriniformes; characteristic enlarged pectoral fins are used for gliding.

Flying Fish *See* Volan.

flying head [ELECTR] A read/write head used on magnetic disks and drums, so designed that it flies a microscopic distance off the moving magnetic surface and is supported by a film of air.

flying spot [ELECTR] A small point of light, controlled mechanically or electrically, which moves rapidly in a rectangular scanning pattern in a flying-spot scanner.

flying-spot scanner [ELECTR] A scanner used for television film and slide transmission, electronic writing, and character recognition, in which a moving spot of light, controlled mechanically or electrically, scans the image field, and the light reflected from or transmitted by the image field is picked up by a phototube to generate electric signals. Also known as optical scanner.

fly-off *See* evapotranspiration.

flysch [GEOL] Deposits of dark, fine-grained, thinly bedded sandstone shales and of clay, thought to be deposited by turbidity currents and originally defined as rock formations on the northern and southern borders of the Alps.

flyway [VERT ZOO] A geographic migration route for birds, including the breeding and wintering areas that it connects.

flywheel [MECH ENG] A rotating element attached to the shaft of a machine for the maintenance of uniform angular velocity and revolutions per minute. Also known as balance wheel.

Fm *See* fermium.

FM *See* frequency modulation.

FM/AM multiplier [ELECTR] Multiplier in which the frequency deviation from the central frequency of a carrier is proportional to one variable, and its amplitude is proportional to the other variable; the frequency-amplitude-modulated carrier is then consecutively demodulated for frequency modulation (FM) and for amplitude modulation (AM); the final output is proportional to the product of the two variables.

fnp *See* fusion point.

f number [OPTICS] A lens rating obtained by dividing the lens's focal length by its effective maximum diameter; the larger the f number, the less exposure

is given. Also known as focal ratio. Also known as stop number.

foam [CHEM] An emulsionlike two-phase system where the dispersed phase is gas or air. [FL MECH] A froth of bubbles on the surface of a liquid, often stabilized by organic contaminants, as found at sea or along shore. [GEOL] *See* pumice.

foamed plastic *See* expanded plastic.

foam glass [MATER] A light, black, opaque, cellular glass made by adding powdered carbon to crushed glass and firing the mixture.

foam-in-place [ENG] The deposition of reactive foam ingredients onto the surface to be covered, allowing the foaming reaction to take place upon that surface, as with polyurethane foam; used in applying thermal insulation for homes and industrial equipment.

focal distance *See* focal length.

focal infection [MED] Infection in a limited area, such as the tonsils, teeth, sinuses, or prostate.

focal length [OPTICS] The distance from the focal point of a lens or curved mirror to the principal point; for a thin lens it is approximately the distance from the focal point to the lens. Also known as focal distance.

focal plane [OPTICS] A plane perpendicular to the axis of an optical system and passing through the focal point of the system.

focal-plane shutter [OPTICS] A camera shutter consisting of a blind containing a slot; the blind is pulled rapidly across the film, exposing it through the slot.

focal point [OPTICS] The point to which rays that are initially parallel to the axis of a lens, mirror, or other optical system are converged or from which they appear to diverge. Also known as principal focus.

focal ratio *See* f number.

focus [ELECTR] To control convergence or divergence of the electron paths within one or more beams, usually by adjusting a voltage or current in a circuit that controls the electric or magnetic fields through which the beams pass, in order to obtain a desired image or a desired current density within the beam. [GEOPHYS] The center of an earthquake and the origin of its elastic waves within the earth. [MATH] A point in the plane which together with a line (directrix) defines a conic section. [NUCLEO] To guide particles along a desired path in a particle accelerator by means of electric or magnetic fields. [OPTICS] **1.** The point or small region at which rays converge or from which they appear to diverge. **2.** To move an optical lens toward or away from a screen or film to obtain the sharpest possible image of a desired object.

focus control [ELECTR] A control that adjusts spot size at the screen of a cathode-ray tube to give the sharpest possible image; it may vary the current through a focusing coil or change the position of a permanent magnet. [OPTICS] A device to adjust a lens system to produce a sharp image.

focusing coil [ELECTR] A coil that produces a magnetic field parallel to an electron beam for the purpose of focusing the beam.

focusing electrode [ELECTR] An electrode to which a potential is applied to control the cross-sectional area of the electron beam in a cathode-ray tube.

focusing magnet [ELECTR] A permanent magnet used to produce a magnetic field for focusing an electron beam.

foehn [METEOROL] A warm, dry wind on the lee side of a mountain range, the warmth and dryness being due to adiabatic compression as the air descends the mountain slopes. Also spelled föhn.

fog [GRAPHICS] A dark, hazy deposit or veil of uniform density over all or parts of a piece of film or paper; can be caused by light other than that forming the image, lens flare, aged materials, or chemical impurities. [METEOROL] Water droplets or, rarely, ice crystals suspended in the air in sufficient concentration to reduce visibility appreciably.

fogbank [METEOROL] A fairly well-defined mass of fog observed in the distance, most commonly at sea.

fogbow [OPTICS] A faintly colored circular arc similar to a rainbow but formed on fog layers containing drops whose diameters are of the order of 100 micrometers or less. Also known as false white rainbow; mistbow; white rainbow.

fog chamber *See* cloud chamber.

föhn *See* foehn.

foil [MET] A thin sheet of metal, usually less than 0.006 inch (0.15 millimeter) thick.

foil dosimeter [NUCLEO] A device for measuring the amount of radiation exposure by means of the degree of activation created in a metal foil inserted in the radiation field.

foil electret [ELEC] A thin film of strongly insulating material capable of trapping charge carriers, such as polyfluoroethylenepropylene, that is electrically charged to produce an external electric field; in the conventional design, charge carriers of one sign are injected into one surface, and a compensation charge of opposite sign forms on the opposite surface or an adjacent electrode.

Fokker-Planck equation [STAT MECH] An equation for the distribution function of a gas, analogous to the Boltzmann equation but applying where the forces are long-range and the collisions are not binary.

fold [ANAT] A plication or doubling, as of various parts of the body such as membranes and other flat surfaces. [GEOL] A bend in rock strata or other planar structure, usually produced by deformation; folds are recognized where layered rocks have been distorted into wavelike form. [MET] *See* lap.

fold belt *See* orogenic belt.

folded dipole *See* folded-dipole antenna.

folded-dipole antenna [ELECTROMAG] A dipole antenna whose outer ends are folded back and joined together at the center; the impedance is about 300 ohms, as compared to 70 ohms for a single-wire dipole; widely used with television and frequency-modulation receivers. Also known as folded dipole.

foliaceous [BOT] Consisting of or having the form or texture of a foliage leaf. [GEOL] Having a leaflike or platelike structure composed of thin layers of minerals. [ZOO] Resembling a leaf in growth form or mode.

foliage [BOT] The leaves of a plant.

foliation [BOT] **1.** The process of developing into a leaf. **2.** The state of being in leaf. [GEOL] A laminated structure formed by segregation of different minerals into layers that are parallel to the schistosity. [MET] Beating metal into thin sheets.

folic acid [BIOCHEM] $C_{19}H_{19}N_7O_6$ A yellow, crystalline vitamin of the B complex; it is slightly soluble in water, usually occurs in conjugates containing glutamic acid residues, and is found especially in plant leaves and vertebrate livers. Also known as pteroylglutamic acid (PGA).

Folist [GEOL] A suborder of the soil order Histosol, consisting of wet forest litter resting on rock or rubble.

follicle [BIOL] A deep, narrow sheath or a small cavity. [BOT] A type of dehiscent fruit composed of one carpel opening along a single suture.

follicle-stimulating hormone [BIOCHEM] A protein hormone released by the anterior pituitary of vertebrates which stimulates growth and secretion of the Graafian follicle and also promotes spermatogenesis. Abbreviated FSH.

follow current [ELEC] The current at power frequency that passes through a surge diverter or other discharge path after a high-voltage surge has started the discharge.

follower [ENG] A drill used for making all but the first part of a hole, the first part being made with a drill of larger gage.

following limb [ASTRON] The half of the limb of a celestial body with an observable disk that appears to follow the body in its apparent motion across the field of view of a fixed telescope.

following wind [METEOROL] **1.** A wind blowing in the direction of ocean-wave advance. **2.** *See* tailwind. [NAV] Wind blowing in the general direction of a vessel's course.

fondant *See* flux.

font [GRAPHICS] A particular typeface and size, including all the uppercase and lowercase letters, punctuation marks, numerals, and so forth.

fontanelle [ANAT] A membrane-covered space between the bones of a fetal or young skull. [INV ZOO] A depression on the head of termites.

Fontéchevade man [PALEON] A fossil man representing the third interglacial *Homo sapiens* and having browridges and a cranial vault similar to those of modern *Homo sapiens*.

food chain [ECOL] The scheme of feeding relationships by trophic levels which unites the member species of a biological community.

food color [MATER] A colorant, either a dye (soluble) or a lake (insoluble), permitted by the Food and Drug Administration for use in foods, drugs, and cosmetics. Also known as certified color; FD&C color.

food engineering [ENG] The technical discipline involved in food manufacturing and processing.

food infection [MED] A type of bacterial food poisoning in which the host is infected by organisms carried by food.

food poisoning [MED] Poisoning due to intake of food contaminated with bacteria or poisonous substances produced by bacteria.

food science [FOOD ENG] The applied science which deals with the chemical, biochemical, physical, physiochemical, and biological properties of foods.

food technology [FOOD ENG] The application of science and engineering to the refining, manufacturing, and handling of foods; many food technologists are food scientists rather than engineers.

food vacuole [CYTOL] A membrane-bound organelle in which digestion occurs in cells capable of phagocytosis. Also known as heterophagic vacuole; phagocytic vacuole.

fool's gold *See* pyrite.

foot [ANAT] Terminal portion of a vertebrate leg. [INV ZOO] An organ for locomotion or attachment. [MECH] The unit of length in the British systems of units, equal to exactly 0.3048 meter. Abbreviated ft.

foot-and-mouth disease [VET MED] A highly contagious virus disease of cattle, pigs, sheep, and goats that is transmissible to man; characterized by fever, salivation, and formation of vesicles in the mouth and pharynx and on the feet. Also known as hoof-and-mouth disease.

footcandle [OPTICS] A unit of illumination, equal to the illumination of a surface, 1 square foot in area, on which there is a luminous flux of 1 lumen uniformly distributed, or equal to the illumination of a surface all points of which are at a distance of 1 foot from a uniform point source of 1 candela; equal to approximately 10.7639 lux. Abbreviated ftc.

foot-pound [MECH] **1.** Unit of energy or work in the English gravitational system, equal to the work done by 1 pound of force when the point at which the force is applied is displaced 1 foot in the direction of the force; equal to approximately 1.355818 joule. Abbreviated ft-lb; ft-lbf. **2.** Unit of torque in the English gravitational system, equal to the torque produced by 1 pound of force acting at a perpendicular distance of 1 foot from an axis of rotation. Also known as pound-foot. Abbreviated lbf-ft.

foot-poundal [MECH] **1.** A unit of energy or work in the English absolute system, equal to the work done by a force of magnitude 1 poundal when the point at which the force is applied is displaced 1 foot in the direction of the force; equal to approximately 0.04214011 joules. Abbreviated ft-pdl. **2.** A unit of torque in the English absolute system, equal to the torque produced by a force of magnitude 1 poundal acting at a perpendicular distance of 1 foot from the axis of rotation. Also known as poundal-foot. Abbreviated pdl-ft.

foot-pound-second system of units See British absolute system of units.

foot rot [PL PATH] Any disease that involves rotting of the stem or trunk of a plant. [VET MED] See foul foot.

footwall [GEOL] The mass of rock that lies beneath a fault, an ore body, or a mine working. Also known as heading side; heading wall; lower plate.

foramen [BIOL] A small opening, orifice, pore, or perforation.

foramen magnum [ANAT] A large oval opening in the occipital bone at the base of the cranium that allows passage of the spinal cord, accessory nerves, and vertebral arteries.

foramen ovale [ANAT] An opening in the sphenoid for the passage of nerves and blood vessels. [EMBRYO] An opening in the fetal heart partition between the two atria.

Foraminiferida [INV ZOO] An order of dominantly marine protozoans in the subclass Granuloreticulosia having a secreted or agglutinated shell enclosing the ameboid body.

forbesite [MINERAL] $H(Ni,Co)AsO_4\cdot3\frac{1}{2}H_2O$ A grayish-white mineral composed of hydrous nickel cobalt arsenate; occurs in fibrocrystalline form.

forbidden band [SOLID STATE] A range of unallowed energy levels for an electron in a solid.

forbidden-character code [COMPUT SCI] A bit code which exists only when an error occurs in the binary coding of characters.

forbidden-combination check [COMPUT SCI] A test for the occurrence of a nonpermissible code expression in a computer; used to detect computer errors.

forbidden line [ATOM PHYS] A spectral line associated with a transition forbidden by selection rules; optically this might be a magnetic dipole or electric quadrupole transition.

forbidden transition [QUANT MECH] A transition between two states of a quantum-mechanical system which is considerably less probable than a competing allowed transition.

force [COMPUT SCI] To intervene manually in a computer routine and cause the computer to execute a jump instruction. [MECH] That influence on a body which causes it to accelerate; quantitatively it is a vector, equal to the body's time rate of change of momentum.

force constant [MECH] The ratio of the force to the deformation of a system whose deformation is proportional to the applied force. [PHYS CHEM] An expression for the force acting to restrain the relative displacement of the nuclei in a molecule.

forced-air heating [MECH ENG] A warm-air heating system in which positive air circulation is provided by means of a fan or a blower.

forced convection [THERMO] Heat convection in which fluid motion is maintained by some external agency.

forced draft [MECH ENG] Air under positive pressure produced by fans at the point where air or gases enter a unit, such as a combustion furnace.

forced oscillation [MECH] An oscillation produced in a simple oscillator or equivalent mechanical system by an external periodic driving force. Also known as forced vibration.

forced programming See minimum-access programming.

forced ventilation [MECH ENG] A system of ventilation in which air is forced through ventilation ducts under pressure.

forced vibration See forced oscillation.

force feedback [CONT SYS] A method of error detection in which the force exerted on the effector is sensed and fed back to the control, usually by mechanical, hydraulic, or electric transducers.

force fit See press fit.

force gage [ENG] An instrument which measures the force exerted on an object.

force plug [ENG] A mold member that fits into the cavity block, exerting pressure on the molding compound. Also known as piston; plunger.

forceps [DES ENG] A pincerlike instrument for grasping objects. [INV ZOO] A pair of curved, hard, movable appendages at the end of the abdomen of certain insects, for example, the earwig. [MED] A device with two blades or limbs opposite each other which is operated by handles or by direct force on the blades; used in surgery to grasp, compress, and hold tissue, a body part, or surgical substances.

forcipate [BIOL] Shaped like forceps; deeply forked.

Forcipulatida [INV ZOO] An order of echinoderms in the subclass Asteroidea characterized by crossed pedicellariae.

fore [NAV ARCH] **1.** The front part of a ship. **2.** In the direction of or toward the bow.

forearm [ANAT] The part of the upper extremity between the wrist and the elbow. Also known as antebrachium.

forebay [CIV ENG] **1.** A small reservoir at the head of the pipeline that carries water to the consumer; it is the last free water surface of a distribution system. **2.** A reservoir feeding the penstocks of a hydro-power plant.

forebrain [EMBRYO] The most anterior expansion of the neural tube of a vertebrate embryo. [VERT ZOO] The part of the adult brain derived from the embryonic forebrain; includes the cerebrum, thalamus, and hypothalamus.

forecasting [COMMUN] The prediction of conditions of radio propagation for a period extending anywhere from a few hours to a few months. [METEOROL] Procedures for extrapolation of the future characteristics of weather on the basis of present and past conditions.

foredeep [GEOL] **1.** A long, narrow depression that borders an orogenic belt, such as an island arc, on the convex side. **2.** See exogeosyncline.

foreground [COMPUT SCI] A program or process of high priority that utilizes machine facilities as needed, with less critical, background work performed in otherwise unused time.

foreland [GEOGR] An extensive area of land jutting out into the sea. [GEOL] 1. A lowland area onto which piedmont glaciers have moved from adjacent mountains. 2. A stable part of a continent bordering an orogenic or mobile belt.

forelimb [ANAT] An appendage (as a wing, fin, or arm) of a vertebrate that is, or is homologous to, the foreleg of a quadruped.

forensic chemistry [CHEM] The application of chemistry to the study of materials or problems in cases where the findings may be presented as technical evidence in a court of law.

forensic medicine [MED] Application of medical evidence or medical opinion for purposes of civil or criminal law.

forensic science [SCI TECH] The application of science for discussion, debate, argumentative or legal purposes.

foreset bed [GEOL] One of a series of inclined symmetrically arranged layers of a cross-bedding unit formed by deposition of sediments that rolled down a steep frontal slope of a delta or dune.

foreshock [GEOPHYS] A tremor which precedes a larger earthquake or main shock.

forest [ECOL] An ecosystem consisting of plants and animals and their environment, with trees as the dominant form of vegetation.

forestry [ECOL] The management of forest lands for wood, forages, water, wildlife, and recreation.

forge [MET] 1. To form a metal, usually hot, into desirable shapes by employing compressive forces. 2. A machine or place in which metal is formed hot, or where iron is produced from its ore.

forge welding [MET] A group of welding processes in which the parts to be joined, usually iron, are heated to about 1000°C and then hammered or pressed together. Also known as fire welding.

forging brass [MET] Brass composed of 60% copper, 38% zinc, and 2% lead, used for hot forgings, hardware, and plumbing supplies; it is extremely plastic when hot, is corrosion-resistant, and has excellent mechanical properties.

forked lightning [GEOPHYS] A common form of lightning, in a cloud-to-ground discharge, which exhibits downward-directed branches from the main lightning channel.

forklift [MECH ENG] A machine, usually powered by hydraulic means, consisting of two or more prongs which can be raised and lowered and are inserted under heavy materials or objects for hoisting and moving them.

forklift truck See fork truck.

fork truck [MECH ENG] A vehicle equipped with a forklift. Also known as forklift truck.

formal language [COMPUT SCI] An abstract mathematical object used to model the syntax of a programming or natural language.

formanite [MINERAL] A mineral composed of an oxide of uranium, zirconium, thorium, calcium, tantalum, and niobium with some rare-earth metals.

format [COMPUT SCI] The specific arrangement of data on a printed page, punched card, or such to meet established presentation requirements.

formation [GEOL] Any assemblage of rocks which have some common character and are mappable as a unit.

formation water [HYD] Water present with petroleum or gas in reservoirs. Also known as oil-reservoir water.

formatted tape [COMPUT SCI] A magnetic tape which employs a prerecorded timing track by means of which blocks of data can be found after reference to a directory table.

form factor [ELEC] 1. The ratio of the effective value of a periodic function, such as an alternating current, to its average absolute value. 2. A factor that takes the shape of a coil into account when computing its inductance. Also known as shape factor. [MECH] The theoretical stress concentration factor for a given shape, for a perfectly elastic material. [PHYS] A function which describes the internal structure of a particle, allowing calculations to be made even though the structure is unknown. [QUANT MECH] An expression used in studying the scattering of electrons or radiation from atoms, nuclei, or elementary particles, which gives the deviation from point particle scattering due to the distribution of charge and current in the target.

form feed character [COMPUT SCI] A control character that determines when a printer or display device moves to the next page, form, or equivalent unit of data.

form feeding [COMPUT SCI] The positioning of documents in order to move them past printing or sensing devices, either singly or in continuous rolls.

Formicariidae [VERT ZOO] The antbirds, a family of suboscine birds in the order Passeriformes.

Formicidae [INV ZOO] The ants, social insects composing the single family of the hymenopteran superfamily Formicoidea.

Formicoidea [INV ZOO] A monofamilial superfamily of hymenopteran insects in the suborder Apocrita, containing the ants.

forming [ELEC] Application of voltage to an electrolytic capacitor, electrolytic rectifier, or semiconductor device to produce a desired permanent change in electrical characteristics as a part of the manufacturing process. [ENG] A process for shaping or molding sheets, rods, or other pieces of hot glass, ceramic ware, plastic, or metal by the application of pressure.

form line [MAP] An approximation of a contour line without a definite elevation value, as derived by visual observation, sometimes supplemented by measured elevations but not in sufficient quantity to produce accurate results; used principally to indicate the appearance of terrain which has not been accurately surveyed.

forms control buffer [COMPUT SCI] A reserved storage containing coordinates for a page position on the printer; earlier printers utilized a carriage control tape, allowing the page to be set at a specific position.

form stop [COMPUT SCI] A device which stops a machine when its supply of paper has run out.

formula [CHEM] 1. A combination of chemical symbols that expresses a molecule's composition. 2. A reaction formula showing the interrelationship between reactants and products. [MATH] An equation or rule relating mathematical objects or quantities.

formulation [CHEM] The particular mixture of base chemicals and additives required for a product.

formula weight [CHEM] 1. The gram-molecular weight of a substance. 2. In the case of a substance of uncertain molecular weight such as certain proteins, the molecular weight calculated from the composition, assuming that the element present in the smallest proportion is represented by only one atom.

formwork [CIV ENG] A temporary wooden casing used to contain concrete during its placing and hardening. Also known as shuttering.

for-next loop [COMPUT SCI] In computer programming, a high-level logic statement which defines a part of a computer program that will be repeated a certain number of times.

fors See G; gram-force.

forsterite [MINERAL] Mg_2SiO_4 A whitish or yellowish, magnesium-rich variety of olivine. Also known as white olivine.

fortnightly tide [OCEANOGR] A tide occurring at intervals of one-half the period of oscillation of the moon, approximately 2 weeks.

FORTRAN [COMPUT SCI] A family of procedure-oriented languages used mostly for scientific or algebraic applications; derived from formula translation.

forward-backward counter [COMPUT SCI] A counter that has both an add and a subtract input so as to count in either an increasing or a decreasing direction. Also known as bidirectional counter.

forward bias [ELECTR] A bias voltage that is applied to a *pn*-junction in the direction that causes a large current flow; used in some semiconductor diode circuits.

forward error analysis [COMPUT SCI] A method of error analysis based on the assumption that small changes in the input data lead to small changes in the results, so that bounds for the errors in the results caused by rounding or truncation errors in the input can be calculated.

forward extrusion [MET] A cold extrusion process in which a formed blank is placed in a die cavity and struck by a punch; the metal is extruded through an annular space between the die and the end of the punch, moving in the same direction as the punch.

forward scatter [COMMUN] 1. Propagation of electromagnetic waves at frequencies above the maximum usable high frequency through use of the scattering of a small portion of the transmitted energy when the signal passes from an unionized medium into a layer of the ionosphere. 2. Collectively, the very-high-frequency forward propagation by ionospheric scatter and ultra-high-frequency forward propagation by tropospheric scatter communications techniques. [GEOPHYS] The scattering of radiant energy into the hemisphere of space bounded by a plane normal to the direction of the incident radiation and lying on the side toward which the incident radiation was advancing.

forward voltage drop See diode forward voltage.

fossa [ANAT] A pit of depression. [VERT ZOO] *Cryptoprocta ferox*. A Madagascan carnivore related to the civets.

fossil [PALEON] The organic remains, traces, or imprint of an organism preserved in the earth's crust since some time in the geologic past.

fossil fuel [GEOL] Any hydrocarbon deposit that may be used for fuel; examples are petroleum, coal, and natural gas.

fossil man [PALEON] Ancient man identified from prehistoric skeletal remains which are archeologically earlier than the Neolithic.

fossil soil See paleosol.

fossil wax See ozocerite.

Foucault current See eddy current.

Foucault mirror [OPTICS] Experiment for measuring the speed of light in which light is reflected from a rapidly rotating mirror to a distant mirror and back, and the speed of light is deduced from the displacement of the beam after its second reflection from the rotating mirror, the angular speed of the rotating mirror, and the distance the light travels.

Foucault pendulum [MECH] A swinging weight supported by a long wire, so that the wire's upper support restrains the wire only in the vertical direction, and the weight is set swinging with no lateral or circular motion; the plane of the pendulum gradually changes, demonstrating the rotation of the earth on its axis.

foul bottom [CIV ENG] A hard, uneven, rocky or obstructed bottom having poor holding qualities for anchors, or one having rocks or wreckage that would endanger an anchored vessel. [NAV ARCH] Referring to the underwater portion of the hull when covered with foreign matter such as barnacles or grass.

foul foot [VET MED] A feedlot disease of cattle and sheep marked by inflammation and ulceration of the feet; common in wet feedlots. Also called foot rot.

Foulger's test [ANALY CHEM] A test for fructose in which urea, sulfuric acid, and stannous chloride are added to the solution to be tested, the solution is boiled, and in the presence of fructose a blue coloration forms.

foundation [CIV ENG] 1. The ground that supports a building or other structure. 2. The portion of a structure which transmits the building load to the ground.

foundation mat See raft foundation.

foundry [ENG] A building where metal or glass castings are produced.

fountain effect [FL MECH] The effect occurring when two containers of superfluid helium are connected by a capillary tube and one of them is heated, so that helium flows through the tube in the direction of higher temperature.

four-address [COMPUT SCI] Pertaining to an instruction address which contains four address parts.

four-ball tester [ENG] A machine designed to measure the efficiency of lubricants by driving one ball against three stationary balls clamped together in a cup filled with the lubricant; performance is evaluated by measuring wear-scar diameters on the stationary balls.

four-bar linkage [MECH ENG] A plane linkage consisting of four links pinned tail to head in a closed loop with lower, or closed, joints.

four-channel sound system See quadraphonic sound system.

four-color printing [GRAPHICS] A method of reproducing full-color originals, such as paintings and color photographs, by overprinting a series of four plates in yellow, magenta, cyan, and black ink.

four-color separation process [GRAPHICS] Conversion of a color illustration into four negative films from which the four printing plates [yellow, magenta, cyan (blue) and black] that will be used in the printing process are made; the negative film for yellow is made by photographing the illustration through a blue filter, the magenta through a green filter, the cyan through a red filter, and the black through a yellow filter.

Fourier analysis [MATH] The study of convergence of Fourier series and when and how a function is approximated by its Fourier series or transform.

Fourier-Bessel integrals [MATH] Given a function $F(v,\theta)$ independent of θ where v,θ are the polar coordinates in the plane, these integrals have the form

$$\int_0^\infty u\,du \int_0^\infty F(r)J_m(ur)r\,dr,$$

where J_m is a Bessel function order m.

Fourier-Bessel transform See Hankel transform.

Fourier expansion   *See* Fourier series.
Fourier heat equation   *See* Fourier law of heat conduction; heat equation.
Fourier integrals   [MATH] For a function $f(x)$ the Fourier integrals are

$$\frac{1}{\pi} \int_0^\infty du \int_{-\infty}^\infty f(t) \cos u(x - t)dt,$$

$$\frac{1}{\pi} \int_0^\infty du \int_{-\infty}^\infty f(t) \sin u(x - t)dt.$$

Fourier law of heat conduction   [THERMO] The law that the rate of heat flow through a substance is proportional to the area normal to the direction of flow and to the negative of the rate of change of temperature with distance along the direction of flow. Also known as Fourier heat equation.
Fourier-Legendre series   [MATH] Given a function $f(x)$, the series from $n = 0$ to infinity of $a_n P_n(x)$, where $P_n(x)$, $n = 0,1,2, \ldots$ are the Legendre polynomials, and a_n is the product of $(2n+1)/2$ and the integral over x from -1 to 1 of $f(x)P_n(x)$.
Fourier number   [FL MECH] A dimensionless number used in unsteady-state flow problems, equal to the product of the dynamic viscosity and a characteristic time divided by the product of the fluid density and the square of a characteristic length. Symbolized Fo_f. [PHYS] A dimensionless number used in the study of unsteady-state mass transfer, equal to the product of the diffusion coefficient and a characteristic time divided by the square of a characteristic length. Symbolized N_{Fo_m}. [THERMO] A dimensionless number used in the study of unsteady-state heat transfer, equal to the product of the thermal conductivity and a characteristic time, divided by the product of the density, the specific heat at constant pressure, and the distance from the midpoint of the body through which heat is passing to the surface. Symbolized N_{Fo_h}.
Fourier series   [MATH] The Fourier series of a function $f(x)$ is

$$\frac{1}{2} a_0 + \sum_{n=1}^\infty (a_n \cos nx + b_n \sin nx)$$

with

$$a_n = \frac{1}{\pi} \int_{-\pi}^\pi f(x) \cos nxdx.$$

$$b_n = \frac{1}{\pi} \int_{-\pi}^\pi f(x) \sin nxdx,$$

Also known as Fourier expansion.
Fourier-Stieltjes transform   [MATH] For a function $f(y)$ of bounded variation on the interval $(-\infty, \infty)$, the function $F(x)$ equal to $1/\sqrt{2\pi}$ times the integral from $y = -\infty$ to $y = \infty$ of exp $(-ixy)df(y)$.
Fourier transform   [MATH] For a function $f(t)$, the function $F(x)$ equal to $1/\sqrt{2\pi}$ times the integral over t from $-\infty$ to ∞ of $f(t)$ exp (itx).
four-layer diode   [ELECTR] A semiconductor diode having three junctions, terminal connections being made to the two outer layers that form the junctions; a Shockley diode is an example.
four-layer transistor   [ELECTR] A junction transistor having four conductivity regions but only three terminals; a thyristor is an example.
four-pole double-throw   [ELEC] A 12-terminal switch or relay contact arrangement that simultaneously connects two pairs of terminals to either of two other pairs of terminals. Abbreviated 4PDT.

four-quadrant multiplier   [COMPUT SCI] A multiplier in an analog computer in which both the reference signal and the number represented by the input may be bipolar, and the multiplication rules for algebraic sign are obeyed. Also known as quarter-square multiplier.
four-stroke cycle   [MECH ENG] An internal combustion engine cycle completed in four piston strokes; includes a suction stroke, compression stroke, expansion stroke, and exhaust stroke.
four-tape   [COMPUT SCI] To sort input data, supplied on two tapes, into incomplete sequences alternately on two output tapes; the output tapes are used for input on the succeeding pass, resulting in longer and longer sequences after each pass, until the data are all in one sequence on one output tape.
fourth dimension   [RELAT] Time in the theory of relativity, in which space and time are conceived as particular aspects of a four-dimensional world.
fourth-power law   *See* Stefan-Boltzmann law.
fourth sound   [CRYO] A pressure wave which propagates in helium II contained in a porous material such as a tightly packed powder, and which results entirely from motion of the superfluid component, the normal component being immobilized by its viscosity.
four-track tape   [ENG ACOUS] Magnetic tape on which two tracks are recorded for each direction of travel, to provide stereo sound reproduction or to double the amount of source material that can be recorded on a given length of 1/4-inch (0.635 centimeter) tape.
four-vector   [RELAT] A set of four quantities which transform under a Lorentz transformation in the same way as the three space coordinates and the time coordinate of an event. Also known as Lorentz four-vector.
four-way switch   [ELEC] An electric switch employed in house wiring, that makes it possible to turn a light on or off at three or more places.
four-way valve   [MECH ENG] A valve at the junction of four waterways which allows passage between any two adjacent waterways by means of a movable element operated by a quarter turn.
four-wheel drive   [MECH ENG] An arrangement in which the drive shaft acts on all four wheels of the automobile.
four-wire circuit   [COMMUN] A two-way circuit using two paths so arranged that communication currents are transmitted in one direction only on one path, and in the opposite direction on the other path; the transmission path may or may not employ four wires.
four-wire repeater   [ELECTR] Telephone repeater for use in a four-wire circuit and in which there are two amplifiers, one serving to amplify the telephone currents in one side of the four-wire circuit, and the other serving to amplify the telephone currents in the other side of the four-wire circuit.
fovea   [BIOL] A small depression or pit.
fovea centralis   [ANAT] A small, rodless depression of the retina in line with the visual axis, which affords acute vision.
foveal vision   [PHYSIO] Vision achieved by looking directly at objects in the daylight so that the image falls on or near the fovea centralis. Also known as photopic vision.
Fowler-DuBridge theory   [SOLID STATE] Theory of photoelectric emission from a metal based on the Sommerfeld model, which takes into account the thermal agitation of electrons in the metal and predicts the photoelectric yield and the energy spectrum of photoelectrons as functions of temperature and the frequency of incident radiation.
fp   *See* freezing point.

FPLA See field-programmable logic array.

fps system of units See British absolute system of units.

Fr See francium; statcoulomb.

fractal [MATH] A geometrical shape whose structure is such that magnification by a given factor reproduces the original object.

fraction [CHEM] One of the portions of a volatile liquid within certain boiling point ranges, such as petroleum naphtha fractions or gas-oil fractions. [MATH] An expression which is the product of a real number or complex number with the multiplicative inverse of a real or complex number. [MET] In powder metallurgy, that portion of sample that lies between two stated particle sizes. Also known as cut. [SCI TECH] A portion of a mixture which represents a discrete unit and can be isolated from the whole system.

fractional crystallization [PETR] Separation of a cooling magma into multiple minerals as the different minerals cool and congeal at progressively lower temperatures. Also known as crystallization differentiation; fractionation.

fractional distillation [CHEM] A method to separate a mixture of several volatile components of different boiling points; the mixture is distilled at the lowest boiling point, and the distillate is collected as one fraction until the temperature of the vapor rises, showing that the next higher boiling component of the mixture is beginning to distill; this component is then collected as a separate fraction.

fractional equation [MATH] 1. Any equation that contains fractions. 2. An equation in which the unknown variable appears in the denominator of one or more terms.

fractionating column [CHEM] An apparatus used widely for separation of fluid (gaseous or liquid) components by vapor-liquid fractionation or liquid-liquid extraction or liquid-solid adsorption.

fractionation [CHEM] Separation of a mixture in successive stages, each stage removing from the mixture some proportion of one of the substances, as by differential solubility in water-solvent mixtures. [NUCLEO] Alterations in the isotopic composition of substances found in nature or in radioactive weapon debris, which result from small differences in the physical and chemical properties of isotopes of an element. [PETR] See fractional crystallization.

fracture [GEOL] A crack, joint, or fault in a rock due to mechanical failure by stress. Also known as rupture. [MED] The breaking of bone, cartilage, or teeth. [MINERAL] A break in a mineral other than along a cleavage plane. [SCI TECH] 1. The act, process, or state of being broken. 2. The surface appearance of a freshly broken material. 3. The break produced by fracturing.

fracture cleavage [GEOL] Cleavage that occurs in deformed but only slightly metamorphosed rocks along closely spaced, parallel joints and fractures.

fracture dome [MIN ENG] The zone of loose or semiloose rock which exists in the immediate hanging or footwall of a stope.

fracture strength See fracture stress.

fracture stress [MECH] The minimum tensile stress that will cause fracture. Also known as fracture strength.

fracture test [ENG] 1. Macro- or microscopic examination of a fractured surface to determine characteristics such as grain pattern, composition, or the presence of defects. 2. A test designed to evaluate fracture stress.

fracture wear [MECH] The wear on individual abrasive grains on the surface of a grinding wheel caused by fracture.

fracture zone [GEOL] An elongate zone on the deep-sea floor that is of irregular topography and often separates regions of different depths; frequently crosses and displaces the midoceanic ridge by faulting.

fragility test [PATH] A measure of the resistance of red blood cells to osmotic hemolysis in hypotonic salt solutions of graded dilutions.

fragmentation [CYTOL] Amitotic division; a type of asexual reproduction. [MIN ENG] The blasting of coal, ore, or rock into pieces small enough to load, handle, and transport without the need for hand-breaking or secondary blasting. [PSYCH] Disordered behavior and mental processes.

fragmentation bomb [ORD] An item designed to be dropped from aircraft to produce many small, high-velocity fragments when detonated.

Frahm frequency meter See vibrating-reed frequency meter.

frame [BUILD] The skeleton structure of a building. Also known as framing. [COMMUN] One cycle of a regularly recurring series of pulses. [COMPUT SCI] 1. A row of recording or punch positions extending across a magnetic or paper tape in a direction at right angles to its motion. 2. See central processing unit. [ELECTR] 1. One complete coverage of a television picture. 2. A rectangular area representing the size of copy handled by a facsimile system. [GRAPHICS] A single complete picture on motion picture film.

frame set [MIN ENG] The arrangement of the legs and cap or crossbar so as to provide support for the roof of an underground passage. Also known as framing; set.

framework silicate See tectosilicate.

framing [BUILD] See frame. [ELECTR] 1. Adjusting a television picture to a desired position on the screen of the picture tube. 2. Adjusting a facsimile picture to a desired position in the direction of line progression. Also known as phasing. [MIN ENG] See frame set.

framing camera [OPTICS] A motion picture camera that automatically controls the position of successive still photographs on the film so that, when the film is subsequently projected, the image will appear steady on the screen.

francium [CHEM] A radioactive alkali-metal element, symbol Fr, atomic number 87, atomic weight distinguished by nuclear instability; exists in short-lived radioactive forms, the chief isotope being francium-223.

Franck-Condon principle [PHYS CHEM] The principle that in any molecular system the transition from one energy state to another is so rapid that the nuclei of the atoms involved can be considered to be stationary during the transition.

franckeite [MINERAL] A dark-gray or black massive mineral composed of lead antimony tin sulfide.

Frankiaceae [MICROBIO] A family of bacteria in the order Actinomycetales; filamentous cells form true mycelia; they are symbiotic and found in active, nitrogen-fixing root nodules.

franklin See statcoulomb.

franklinite [MINERAL] $ZnFe_2O_4$ Black, slightly magnetic mineral member of the spinel group; usually possesses extensive substitution of divalent manganese and iron for the divalent zinc, and limited trivalent manganese for the trivalent iron.

Frasch process [MIN ENG] A process to remove sulfur from sulfur beds; superheated water is forced under pressure into the sulfur bed, and the molten sulfur is thus forced to the surface.

Fraunhofer corona *See* F corona.

Fraunhofer diffraction [OPTICS] Diffraction of a beam of parallel light observed at an effectively infinite distance from the diffracting object, usually with the aid of lenses which collimate the light before diffraction and focus it at the point of observation.

Fraunhofer spectrum [SPECT] The absorption lines in sunlight, due to the cooler outer layers of the sun's atmosphere.

frazil [HYD] Ice crystals which form in supercooled water that is too turbulent to permit coagulation of the crystals into sheet ice.

freckle [MED] A pigmented macule resulting from focal increase in melanin, usually associated with exposure to sunlight, commonly on the face.

free-air anomaly [GEOPHYS] A gravity anomaly calculated as the difference between the measured gravity and the theoretical gravity at sea level and a free-air coefficient determined by the elevation of the measuring station.

free-air temperature [METEOROL] Temperature of the atmosphere, obtained by a thermometer located so as to avoid as completely as practicable the effects of extraneous heating.

free atom [ATOM PHYS] An atom, as in a gas, whose properties, such as spectrum and magnetic moment, are not significantly affected by other atoms, ions, or molecules nearby.

freeboard [ANALY CHEM] The space provided above the resin bed in an ion-exchange column to allow for expansion of the bed during backwashing. [CIV ENG] The height between normal water level and the crest of a dam or the top of a flume. [NAV ARCH] The vertical distance from the intersection of the top of the freeboard deck amidships with the outer surface of the side plating to the upper edge of the summer load line, or in general terms, the distance from the waterline to the deck.

free convection *See* natural convection.

free electron [PHYS] An electron that is not constrained to remain in a particular atom, and is therefore able to move in matter or in a vacuum when acted on by external electric or magnetic fields.

free-electron laser [OPTICS] A multifrequency laser utilizing optical radiation amplification by a beam of free electrons passing through a vacuum in a transverse periodic magnetic field.

free-electron theory of metals [SOLID STATE] A model of a metal in which the free electrons, that is, those giving rise to the conductivity, are regarded as moving in a potential (due to the metal ions in the lattice and to all the remaining free electrons) which is approximated as constant everywhere inside the metal. Also known as Sommerfeld model; Sommerfeld theory.

free energy [THERMO] 1. The internal energy of a system minus the product of its temperature and its entropy. Also known as Helmholtz free energy; Helmholtz function; Helmholtz potential; thermodynamic potential at constant volume; work function. 2. *See* Gibbs free energy.

free enthalpy *See* Gibbs free energy.

free fall [MECH] The ideal falling motion of a body acted upon only by the pull of the earth's gravitational field. [PETRO ENG] In deep drilling, an arrangement by which the bit is permitted to fall freely to the bottom at each drop or down stroke.

free field [ACOUS] An isotropic, homogeneous sound field that is free from all bounding surfaces. [COMPUT SCI] A property of information retrieval devices which permits recording of information in the search medium without regard to preassigned fixed fields. [PHYS] A field in empty space not interacting with other fields or sources.

free-field room *See* anechoic chamber.

free-field storage [COMPUT SCI] Data storage that allows recording of the data without regard for fixed or preassigned fields.

free fit [DES ENG] A fit between mating pieces where accuracy is not essential or where large variations in temperature may occur.

free grid *See* floating grid.

free gyroscope [ENG] A gyroscope that uses the property of gyroscopic rigidity to sense changes in altitude of a machine, such as an airplane; the spinning wheel or rotor is isolated from the airplane by gimbals; when the plane changes from level flight, the gyro remains vertical and gives the pilot an artificial horizon reference.

free ion [PHYS CHEM] An ion, such as found in an ionized gas, whose properties, such as spectrum and magnetic moment, are not significantly affected by other atoms, ions, or molecules nearby.

freemartin [VERT ZOO] An intersexual, usually sterile female calf twinborn with a male.

free molecule [PHYS CHEM] A molecule, as in a gas, whose properties, such as spectrum and magnetic moment, are not affected by other atoms, ions, and molecules nearby.

free-running multivibrator *See* astable multivibrator.

free space [COMMUN] A region high enough so that the radiation pattern of an antenna is not affected by surrounding objects such as buildings, trees, hills, and the earth.

freestone [BOT] A fruit stone to which the fruit does not cling, as in certain varieties of peach. [GEOL] Stone, particularly a thick-bedded, even-textured, fine-grained sandstone, that breaks freely and is able to be cut and dressed with equal facility in any direction without tending to split.

free surface [FL MECH] A boundary between two homogeneous fluids.

free symbol [COMPUT SCI] A contextual symbol preceded and followed by a space; it is always meaningful and always used to symbolize both grammatical and nongrammatical meaning; an example is the English "I."

free-water elevation *See* water table.

free-water surface *See* water table.

freeze [ENG] 1. To permit drilling tools, casing, drivepipe, or drill rods to become lodged in a borehole by reason of caving walls or impaction of sand, mud, or drill cuttings, to the extent that they cannot be pulled out. Also known as bind-seize. 2. To burn in a bit. Also known as burn-in. 3. The premature setting of cement, especially when cement slurry hardens before it can be ejected fully from pumps or drill rods during a borehole cementation operation. 4. The act or process of drilling a borehole by utilizing a drill fluid chilled to minus 30–40°F, (minus 34–40°C) as a means of consolidating, by freezing, the borehole wall materials or core as the drill penetrates a water-saturated formation, such as sand or gravel. [PHYS CHEM] To solidify a liquid by removal of heat.

freeze drying [ENG] A method of drying materials, such as certain foods, that would be destroyed by the loss of volatile ingredients or by drying temperatures above the freezing point; the material is frozen under high vacuum so that ice or other frozen solvent will quickly sublime and a porous solid remain.

freeze etching [CRYO] A method using cryogenics to prepare specimens for study with a microscope.

freezing drizzle [METEOROL] Drizzle that falls in liquid form but freezes upon impact with the ground to form a coating of glaze.

freezing mixture [PHYS CHEM] A mixture of substances whose freezing point is lower than that of its constituents.

freezing point [PHYS CHEM] The temperature at which a liquid and a solid may be in equilibrium. Abbreviated fp.

freezing precipitation [METEOROL] Any form of liquid precipitation that freezes upon impact with the ground or exposed objects; that is, freezing rain or freezing drizzle.

Fregatidae [VERT ZOO] Frigate birds or man-o'-war birds, a family of fish-eating birds in the order Pelecaniformes.

F region [GEOPHYS] The general region of the ionosphere in which the F_1 and F_2 layers tend to form.

freibergite [MINERAL] A steel-gray, silver-bearing variety of tetrahedrite.

freieslebenite [MINERAL] $Pb_3Ag_5Sb_5S_{12}$ A steel-gray to dark mineral composed of a sulfide of antimony, lead, and silver.

freighter [ENG] A ship or aircraft used mainly for carrying freight.

freight ton See ton.

freirinite [MINERAL] $Na_3Cu_3(AsO_4)_2(OH)_3 \cdot H_2O$ A lavender to turquoise-blue mineral composed of a basic hydrous arsenate of sodium and copper.

fremontite See natromontebrasite.

french curve [GRAPHICS] A guide, usually made of clear plastic, used for making regular, irregular, and reverse curves in mechanical drawings and illustrations.

French drain [CIV ENG] An underground passage for water, consisting of loose stones covered with earth.

French measles See rubella.

Frenkel defect [SOLID STATE] A crystal defect consisting of a vacancy and an interstitial which arise when an atom is plucked out of a normal lattice site and forced into an interstitial position. Also known as Frenkel pair.

Frenkel pair See Frenkel defect.

frenulum [ANAT] 1. A small fold of integument or mucous membrane. 2. A small ridge on the upper part of the anterior medullary velum. [INV ZOO] A spine on most moths that projects from the hindwings and is held to the forewings by a clasp, thus coupling the wings together.

frenum [ANAT] A fold of tissue that restricts the movements of an organ.

frequency [PHYS] The number of cycles completed by a periodic quantity in a unit time.

frequency analyzer [ELECTR] A device which measures the intensity of many different frequency components in some oscillation, as in a radio band; used to identify transmitting sources.

frequency band [PHYS] A continuous range of frequencies extending between two limiting frequencies.

frequency bridge [ELECTR] A bridge in which the balance varies with frequency in a known manner, such as the Wien bridge; used to measure frequency.

frequency changer See frequency converter.

frequency characteristic See frequency-response curve.

frequency compensation See compensation.

frequency converter [ELEC] A circuit, device, or machine that changes an alternating current from one frequency to another, with or without a change in voltage or number of phases. Also known as frequency changer; frequency translator.

frequency counter [ELECTR] An electronic counter used to measure frequency by counting the number of cycles in an electric signal during a preselected time interval.

frequency curve [STAT] A graphical representation of a continuous frequency distribution; the value of the variable is the abscissa and the frequency is the ordinate.

frequency deviation [COMMUN] The peak difference between the instantaneous frequency of a frequency-modulated wave and the carrier frequency.

frequency discriminator [ELECTR] A discriminator circuit that delivers an output voltage which is proportional to the deviations of a signal from a predetermined frequency value.

frequency distortion [ELECTR] Distortion in which the relative magnitudes of the different frequency components of a wave are changed during transmission or amplification. Also known as amplitude distortion; amplitude-frequency distortion; waveform-amplitude distortion.

frequency distribution [MATH] A function which measures the relative frequency or probability that a variable can take on a set of values.

frequency divider [ELECTR] A harmonic conversion transducer in which the frequency of the output signal is an integral submultiple of the input frequency. Also known as counting-down circuit.

frequency-division multiplexing [COMMUN] A multiplex system for transmitting two or more signals over a common path by using a different frequency band for each signal. Abbreviated fdm. Also known as frequency multiplexing.

frequency domain [COMMUN] A plane on which signal strength can be represented graphically as a function of frequency, instead of a function of time. [CONT SYS] Pertaining to a method of analysis, particularly useful for fixed linear systems in which one does not deal with functions of time explicitly, but with their Laplace or Fourier transforms, which are functions of frequency.

frequency doubler [ELECTR] An amplifier stage whose resonant anode circuit is tuned to the second harmonic of the input frequency; the output frequency is then twice the input frequency. Also known as doubler.

frequency function See probability density function.

frequency meter [ENG] 1. An instrument for measuring the frequency of an alternating current; the scale is usually graduated in hertz, kilohertz, and megahertz. 2. A device calibrated to indicate frequency of a radio wave.

frequency-modulated cyclotron See synchrocyclotron.

frequency modulation [COMMUN] Modulation in which the instantaneous frequency of the modulated wave differs from the carrier frequency by an amount proportional to the instantaneous value of the modulating wave. Abbreviated FM.

frequency-modulation broadcast band [COMMUN] The band of frequencies extending from 88 to 108 megahertz; used for frequency-modulation radio broadcasting in the United States.

frequency modulator [ELECTR] A circuit or device for producing frequency modulation.

frequency monitor [ELECTR] An instrument for indicating the amount of deviation of the carrier frequency of a transmitter from its assigned value.

frequency multiplexing See frequency-division multiplexing.

frequency multiplier [ELECTR] A harmonic conversion transducer in which the frequency of the output signal is an exact integral multiple of the input frequency. Also known as multiplier.

frequency optimum traffic See optimum working frequency.

frequency polygon [STAT] A graph obtained from a frequency distribution by joining with straight lines points whose abscissae are the midpoints of successive class intervals and whose ordinates are the corresponding class frequencies.

frequency response [ENG] A measure of the effectiveness with which a circuit, device, or system transmits the different frequencies applied to it; it is a phasor whose magnitude is the ratio of the magnitude of the output signal to that of a sine-wave input, and whose phase is that of the output with respect to the input. Also known as amplitude-frequency response; sine-wave response.

frequency-response curve [ENG] A graph showing the magnitude or the phase of the freqency response of a device or system as a function of frequency. Also known as frequency characteristic.

frequency-response equalization See equalization.

frequency-selective device See electric filter.

frequency separator [ELECTR] The circuit that separates the horizontal and vertical synchronizing pulses in a monochrome or color television receiver.

frequency shift [ELECTR] A change in the frequency of a radio transmitter or oscillator. Also known as radio-frequency shift.

frequency-shift keying [COMMUN] A form of frequency modulation used especially in telegraph and facsimile transmission, in which the modulating wave shifts the output frequency between predetermined values corresponding to the frequencies of correlated sources. Abbreviated FSK. Also known as frequency-shift modulation; frequency-shift transmission.

frequency-shift modulation See frequency-shift keying.

frequency-shift transmission See frequency-shift keying.

frequency spectrum [PHYS] A plot of the distribution of the intensity of some type of electromagnetic or acoustic radiation as a function of frequency. [SYS ENG] In the analysis of a random function of time, such as the amplitude of noise in a system, the limit as T approaches infinity of $1/2\pi T$ times the ensemble average of the squared magnitude of the amplitude of the Fourier transform of the function from $-T$ to T. Also known as power-density spectrum; power spectrum; spectral density.

frequency splitting [ELECTR] One condition of operation of a magnetron which causes rapid alternating from one mode of operation to another; this results in a similar rapid change in oscillatory frequency and consequent loss in power at the desired frequency.

frequency standard [ELECTR] A stable oscillator, usually controlled by a crystal or tuning fork, that is used primarily for frequency calibration.

frequency synthesizer [ELECTR] A device that provides a choice of a large number of different frequencies by combining frequencies selected from groups of independent crystals, frequency dividers, and frequency multipliers.

frequency translator See frequency converter.

fresco [GRAPHICS] Painting of two types, buon fresco and fresco secco; in buon fresco, dry pigments are ground with water and applied on a wet lime plaster wall, and as the plaster dries, the pigment is permanently bound to it; fresco secco utilizes pigments ground in glue, casein, or polymer emulsions and applied on a dry plaster wall that has been wetted with limewater.

freshet [HYD] **1.** The annual spring rise of streams in cold climates as a result of melting snow. **2.** A flood resulting from either rain or melting snow; usually applied only to small streams and to floods of minor severity. **3.** A small fresh-water stream.

fresh water [HYD] Water containing no significant amounts of salts, such as in rivers and lakes.

Fresnel biprism [OPTICS] A very flat triangular prism which has two very acute angles and one very obtuse angle; used to observe the interference of light from a slit passing through the two halves of the prism.

Fresnel diffraction [OPTICS] Diffraction in which the source of light or the observing screen are at a finite distance from the aperture or obstacle.

Fresnel lens [OPTICS] A thin lens constructed with stepped setbacks so as to have the optical properties of a much thicker lens.

Fresnel mirrors [OPTICS] Two plane mirrors which are inclined to each other on the order of a degree and used to observe the interference of light which originates from a slit and is reflected from both mirrors.

Fresnel rhomb [OPTICS] A glass rhomb which has an acute angle of about 52°; light which is incident normal to the end of the rhomb undergoes two internal reflections, and if it is initially linearly polarized at an angle of 45° to the plane of incidence, it emerges circularly polarized.

Fresnel zones [ELECTROMAG] Circular portions of a wavefront transverse to a line between an emitter and a point where the disturbance is being observed; the nth zone includes all paths whose lengths are between $n - 1$ and n half-wavelengths longer than the line-of-sight path. Also known as half-period zones.

fretting corrosion [MET] Surface damage usually in an air environment between two surfaces, one or both of which are metals, in close contact under pressure and subject to a slight relative motion. Also known as chafing corrosion.

Freudianism [PSYCH] The psychoanalytic school of psychiatry founded by Sigmund Freud.

friable [MATER] Referring to the property of a substance capable of being easily rubbed, crumbled, or pulverized into powder.

friction [MECH] A force which opposes the relative motion of two bodies whenever such motion exists or whenever there exist other forces which tend to produce such motion.

friction bearing [MECH ENG] A solid bearing that directly contacts and supports an axle end.

friction bonding [ENG] Soldering of a semiconductor chip to a substrate by vibrating the chip back and forth under pressure to create friction that breaks up oxide layers and helps alloy the mating terminals.

friction coefficient See coefficient of friction.

friction damping [MECH] The conversion of the mechanical vibrational energy of solids into heat energy by causing one dry member to slide on another.

friction factor [FL MECH] Any of several dimensionless numbers used in studying fluid friction in pipes, equal to the Fanning friction factor times some dimensionless constant.

friction flow [FL MECH] Fluid flow in which a significant amount of mechanical energy is dissipated into heat by action of viscosity.

friction gear [MECH ENG] Gearing in which motion is transmitted through friction between two surfaces in rolling contact.

friction head [FL MECH] The head lost by the flow in a stream or conduit due to frictional disturbances set up by the moving fluid and its containing conduit and by intermolecular friction.

friction layer See surface boundary layer.

frictionless flow See inviscid flow.

friction loss [MECH] Mechanical energy lost because of mechanical friction between moving parts of a machine.

friction tape [MATER] Cotton tape impregnated with a sticky moisture-repelling compound; used chiefly to hold rubbertape insulation in position over a joint or splice.

Friedel-Crafts reaction [ORG CHEM] A substitution reaction, catalyzed by aluminum chloride in which an alkyl (R—) or an acyl (RCO—) group replaces a hydrogen atom of an aromatic nucleus to produce hydrocarbon or a ketone.

friedelite [MINERAL] $Mn_8Si_6O_{18}(OH,Cl)_4 \cdot 3H_2O$ A rose-red mineral composed of manganese silicate with chlorine.

Friedlander synthesis [ORG CHEM] A synthesis of quinolines; the method is usually catalyzed by bases and consists of condensation of an aromatic o-aminocarbonyl derivative with a compound containing a methylene group in the alpha position to the carbonyl.

frieze [ARCH] A decorated band immediately below the cornice on an interior wall. [TEXT] Thick, heavyweight coating and upholstery fabric, with a rough, raised fibrous surface and a more or less hard feel.

frigidoreceptor [PHYSIO] A cutaneous sense organ which is sensitive to cold.

fringe [OPTICS] One of the light or dark bands produced by interference or diffraction of light.

fringe multiplication [MATH] The duplicating effect of a family of curves superimposed on another family of curves so that the curves intersect at angles less than 45°, a new family of curves appears which pass through intersection of the original curves.

fringes of equal thickness See Fizeau fringes.

Fringillidae [VERT ZOO] The finches, a family of oscine birds in the order Passeriformes.

fringing reef [GEOL] A coral reef attached directly to or bordering the shore of an island or continental landmass.

frit [MATER] Fusible ceramic mixture used to make glazes and enamels for dinnerware and metallic surfaces, as on stoves and metal-base basins and tubs.

fritillary [BOT] The common name for plants of the genus *Fritillaria*. [INV ZOO] The common name for butterflies in several genera of the subfamily Nymphalinae.

frog [DES ENG] A hollow on one or both of the larger faces of a brick or block; reduces weight of the brick or block; may be filled with mortar. [ENG] A device which permits the train or tram wheels on one rail of a track to cross the rail of an intersecting track. [VERT ZOO] The common name for a number of tailless amphibians in the order Anura; most have hindlegs adapted for jumping, scaleless skin, and large eyes.

frohbergite [MINERAL] $FeTe_2$ A mineral composed of iron telluride; it is isomorphous with marcasite.

frond [BOT] 1. The leaf of a palm or fern. 2. A foliaceous thallus or thalloid shoot.

frondelite [MINERAL] $MnFe_4(PO_4)_5(OH)_5$ A mineral composed of basic phosphate of manganese and iron; it is isomorphous with rockbridgeite.

front [METEOROL] A sloping surface of discontinuity in the troposphere, separating air masses of different density or temperature.

frontal apron See outwash plain.

frontal bone [ANAT] Either of a pair of flat membrane bones in vertebrates, and a single bone in humans, forming the upper frontal portion of the cranium; the forehead bone.

frontal cyclone [METEOROL] Any cyclone associated with a front; often used synonymously with wave cyclone or with extratropical cyclone (as opposed to tropical cyclones, which are nonfrontal).

frontal inversion [METEOROL] A temperature inversion in the atmosphere, encountered upon vertical ascent through a sloping front (or frontal zone).

frontal lobe [ANAT] The anterior portion of a cerebral hemisphere, bounded behind by the central sulcus and below by the lateral cerebral sulcus.

frontal occlusion See occluded front.

frontal plain See outwash plain.

frontal thunderstorm [METEOROL] A thunderstorm associated with a front; limited to thunderstorms resulting from the convection induced by frontal lifting.

frontal wave [METEOROL] A horizontal, wavelike deformation of a front in the lower levels, commonly associated with a maximum of cyclonic circulations in the adjacent flow; it may develop into a wave cyclone.

frontal zone [METEOROL] The three-dimensional zone or layer of large horizontal density gradient, bounded by frontal surfaces and surface front.

front-end [COMPUT SCI] Of a minicomputer, under programmed instructions, performing data transfers and control operations to relieve a larger computer of these routines.

front-end loader [MECH ENG] An excavator consisting of an articulated bucket mounted on a series of movable arms at the front of a crawler or rubber-tired tractor.

front-end processor [COMPUT SCI] A computer which connects to the main computer at one end and communications channels at the other, and which directs the transmitting and receiving of messages, detects and corrects transmission errors, assembles and disassembles messages, and performs other processing functions so that the main computer receives pure information.

frontogenesis [METEOROL] 1. The initial formation of a frontal zone or front. 2. The increase in the horizontal gradient of an air mass property, mainly density, and the formation of the accompanying features of the wind field that typify a front.

frost [HYD] A covering of ice in one of its several forms, produced by the sublimation of water vapor on objects colder than 32°F (0°C).

frost action [GEOL] 1. The weathering process caused by cycles of freezing and thawing of water in surface pores, cracks, and other openings. 2. Alternate or repeated cycles of freezing and thawing of water contained in materials; the term is especially applied to disruptive effects of this action.

frostbite [MED] Injury to skin and subcutaneous tissues, and in severe cases to deeper tissues also, from exposure to extreme cold.

frost bursting   See congelifraction.

frost churning   See congeliturbation.

frosted glass   [MATER] Glass that has been etched with sand, or appears to have been so treated.

frost flakes   See ice fog.

frost flowers   See ice flowers.

frost fog   See ice fog.

frost heaving   [GEOL] The lifting and distortion of a surface due to internal action of frost resulting from subsurface ice formation; affects soil, rock, pavement, and other structures.

frosting   [ENG] Decorating a scraped metal surface with a handscraper. Also known as flaking.

frost line   [GEOL] 1. The maximum depth of frozen ground during the winter. 2. The lower limit of the permafrost. [MATER] In polyethylene film extrusion, a ring-shaped area with a frosty appearance at the point where the film reaches its final diameter.

frost mound   [GEOL] A hill and knoll associated with frozen ground in a permafrost region, containing a core of ice. Also known as soffosian knob; soil blister.

frost point   [METEOROL] The temperature to which atmospheric moisture must be cooled to reach the point of saturation with respect to ice.

frost ring   [BOT] A false annual growth ring in the trunk of a tree due to out-of-season defoliation by frost and subsequent regrowth of foliage.

frost riving   See congelifraction.

frost smoke   [METEOROL] 1. A rare type of fog formed in the same manner as a steam fog, but at colder temperatures so that it is composed of ice particles instead of water droplets. 2. See steam fog.

frost splitting   See congelifraction.

frosty mildew   [PL PATH] A leaf spot caused by fungi of the genus Cercosporella and characterized by pale to white lesions.

frost zone   See seasonally frozen ground.

froth flotation   [ENG] A process for recovery of particles of ore or other material, in which the particles adhere to bubbles and can be removed as part of the froth.

frozen fog   See ice fog.

frozen ground   [GEOL] Soil having a temperature below freezing, generally containing water in the form of ice. Also known as gelisol; merzlota; taele; tjaele.

frozen section   [BIOL] A thin slice of material cut from a frozen sample of tissue or organ.

fructification   [BOT] 1. The process of producing fruit. 2. A fruit and its appendages. [MYCOL] A sporogenous structure.

fructivorous   See frugivorous.

D-fructopyranose   See fructose.

fructose   [BIOCHEM] $C_6H_{12}O_5$ The commonest of ketoses and the sweetest of sugars, found in the free state in fruit juices, honey, and nectar of plant glands. Also known as D-fructopyranose.

frugivorous   [ZOO] Fruit-eating. Also known as fructivorous.

fruit   [BOT] A fully matured plant ovary with or without other floral or shoot parts united with it at maturity. [NAV] Radar-beacon-system video display of a synchronous beacon return which results when several interrogator stations are located within the same general area; each interrogator receives its own interrogated reply as well as many synchronous replies resulting from interrogation of the airborne transponders by other ground stations.

fruit fly   [INV ZOO] 1. The common name for those acalypterate insects composing the family Tephritidae. 2. Any insect whose larvae feed on fruit or decaying vegetable matter.

frustrated internal reflectance   See attenuated total reflectance.

frutescent   [BIOL] See fruticose. [BOT] Shrublike in habit.

fruticose   [BIOL] Resembling a shrub; applied especially to lichens. Also known as frutescent.

F scan   See F scope.

F scope   [ELECTR] A cathode-ray scope on which a single signal appears as a spot with bearing error as the horizontal coordinate and elevation angle error as the vertical coordinate, with cross hairs on the scope face to assist in bringing the system to bear on the target. Also known as F indicator; F scan.

FSH   See follicle-stimulating hormone.

FSK   See frequency-shift keying.

F star   [ASTRON] A star whose spectral type is F; surface temperature is 7000K, and color is yellowish.

f stop   [OPTICS] An aperture setting for a camera lens; indicated by the f number.

ft   See foot.

ftc   See footcandle.

F test   See variance ratio test.

ft-lb   See foot-pound.

ft-lbf   See foot-pound.

ft-pdl   See foot-poundal.

Fucales   [BOT] An order of brown algae composing the class Cyclosporeae.

fuchsite   [MINERAL] A bright-green variety of muscovite rich in chromium.

Fucophyceae   [BOT] A class of brown algae.

fuel   [MATER] A material that is burnt to release heat energy, for example, coal, oil, or uranium.

fuel assembly   [NUCLEO] A combination of fuel and structural materials, used in some nuclear reactors to facilitate assembly of the core.

fuel cell   [ELEC] A cell that converts chemical energy directly into electric energy, with electric power being produced as a part of a chemical reaction between the electrolyte and a fuel such as kerosine or industrial fuel gas.

fuel cycle   See reactor fuel cycle.

fuel element   [NUCLEO] A rod, tube, plate, or other geometrical form into which nuclear fuel is fabricated for use in a reactor.

fuel gas   [MATER] A gaseous fuel used to provide heat energy when burned with oxygen.

fuel injection   [MECH ENG] The delivery of fuel to an internal combustion engine cylinder by pressure from a mechanical pump.

fuel oil   [MATER] A liquid product burned to generate heat, exclusive of oils with a flash point below 100°F (38°C); includes heating oils, stove oils, furnace oils, bunker fuel oils.

fuel pump   [MECH ENG] A pump for drawing fuel from a storage tank and delivering it to an engine or furnace.

fugacious   [BOT] Lasting a short time; used principally to describe plant parts that fall soon after being formed.

fugacity   [THERMO] A function used as an analog of the partial pressure in applying thermodynamics to real systems; at a constant temperature it is proportional to the exponential of the ratio of the chemical potential of a constituent of a system divided by the product of the gas constant and the temperature, and it approaches the partial pressure as the total pressure of the gas approaches zero.

fulcrum   [MECH] The rigid point of support about which a lever pivots.

Fulgoroidea [INV ZOO] The lantern flies, a superfamily of homopteran insects in the series Auchenorrhyncha distinguished by the anterior and middle coxae being of equal length and joined to the body at some distance from the median line.

fulgurite [GEOL] A glassy, rootlike tube formed when a lightning stroke terminates in dry sandy soil; the intense heating of the current passing down into the soil along an irregular path fuses the sand.

full adder [ELECTR] A logic element which operates on two binary digits and a carry digit from a preceding stage, producing as output a sum digit and a new carry digit.

full duplex [COMMUN] Of a telegraph or other data channel, able to operate in both directions simultaneously. [COMPUT SCI] The complete duplication of any data-processing facility.

fuller's earth [GEOL] A natural, fine-grained earthy material, such as a clay, with high adsorptive power; consists principally of hydrated aluminum silicates; used as an adsorbent in refining and decolorizing oils, as a catalyst, and as a bleaching agent.

full-face tunneling [CIV ENG] A system of tunneling in which the tunnel opening is enlarged to desired diameter before extension of the tunnel face.

full feathering *See* feathering.

full load [ELEC] The greatest load that a circuit or piece of equipment is designed to carry under specified conditions.

full moon [ASTRON] The moon at opposition, with a phase angle of 0°, when it appears as a round disk to an observer on the earth because the illuminated side is toward the observer.

full-wave bridge [ELECTR] A circuit having a bridge with four diodes, which provides full-wave rectification and gives twice as much direct-current output voltage for a given alternating-current input voltage as a conventional full-wave rectifier.

full-wave rectifier [ELECTR] A double-element rectifier that provides full-wave rectification; one element functions during positive half cycles and the other during negative half cycles.

full-wave vibrator [ELEC] A vibrator having an armature that moves back and forth between two fixed contacts so as to change the direction of direct-current flow through a transformer at regular intervals and thereby permit voltage stepup by the transformer; used in battery-operated power supplies for mobile and marine radio equipment.

full-word boundary [COMPUT SCI] In the IBM 360 system, any address which ends in 00, and is therefore a natural boundary for a four-byte machine word.

fulmar [VERT ZOO] Any of the oceanic birds composing the family Procellariidae; sometimes referred to as foul gulls because of the foul-smelling substance spat at intruders upon their nests.

fumarase [BIOCHEM] An enzyme that catalyzes the hydration of fumaric acid to malic acid, and the reverse dehydration.

Fumariaceae [BOT] A family of dicotyledonous plants in the order Papaverales having four or six stamens, irregular flowers, and no latex system.

fumarole [GEOL] A hole, usually found in volcanic areas, from which vapors or gases escape.

fumigant [CHEM] A chemical compound which acts in the gaseous state to destroy insects and their larvae and other pests; examples are dichlorethyl ether, *p*-dichlorobenzene, and ethylene oxide.

fuming nitric acid [INORG CHEM] Concentrated nitric acid containing dissolved nitrogen dioxide; may be prepared by adding formaldehyde to concentrated nitric acid.

Funariales [BOT] An order of mosses; plants are usually annual, are terrestrial, and have stems that are erect, short, simple, or sparingly branched.

function [COMPUT SCI] In FORTRAN, a subroutine of a particular kind which returns a computational value whenever it is called. [MATH] A mathematical rule between two sets which assigns to each member of the first, exactly one member of the second.

functional [COMPUT SCI] In a linear programming problem involving a set of variables x_j, $j = 1, 2, \ldots, n$, a function of the form $c_1x_1 + c_2x_2 + \ldots + c_nx_n$ (where the c_j are constants) which one wishes to optimize (maximize or minimize, depending on the problem) subject to a set of restrictions. [MATH] Any function from a vector space into its scalar field.

functional analysis [MATH] A branch of analysis which studies the properties of mappings of classes of functions from one topological vector space to another.

functional design [COMPUT SCI] A level of the design process in which subtasks are specified and the relationships among them defined, so that the total collection of subsystems performs the entire task of the system.

functional diagram [COMPUT SCI] A diagram that indicates the functions of the principal parts of a total system and also shows the important relationships and interactions among these parts.

functional generator *See* function generator.

functional group [ORG CHEM] An atom or group of atoms, acting as a unit, that has replaced a hydrogen atom in a hydrocarbon molecule and whose presence imparts characteristic properties to this molecule; frequently represented as R—.

functional interleaving [COMPUT SCI] Alternating the parts of a number of sequences in a cyclic fashion, such as a number of accesses to memory followed by an access to a data channel.

functional paralysis *See* hysterical paralysis.

functional requirement [COMPUT SCI] The documentation which accompanies a program and states in detail what is to be performed by the system.

functional switching circuit [ELECTR] One of a relatively small number of types of circuits which implements a Boolean function and constitutes a basic building block of a switching system; examples are the AND, OR, NOT, NAND, and NOR circuits.

functional unit [COMPUT SCI] The part of the computer required to perform an elementary process such as an addition or a pulse generation.

function code [COMPUT SCI] Special code which appears on a medium such as a paper tape and which controls machine functions such as a carriage return.

function generator [ELECTR] Also known as functional generator. **1.** An analog computer device that indicates the value of a given function as the independent variable is increased. **2.** A signal generator that delivers a choice of a number of different waveforms, with provisions for varying the frequency over a wide range.

function hole *See* designation punch.

function key [COMPUT SCI] A special key on a keyboard to control a mechanical function, initiate a specific computer operation, or transmit a signal that would otherwise require multiple key strokes.

function switch [ELECTR] A network having a number of inputs and outputs so connected that input signals expressed in a certain code will produce output signals that are a function of the input information but in a different code.

function table [COMPUT SCI] 1. Sets of computer information arranged so an entry in one set selects one or more entries in the other sets. 2. A computer device that converts multiple inputs into a single output or encodes a single input into multiple outputs. [MATH] A table that lists the values of a function for various values of the variable.

function unit [COMPUT SCI] In computer systems, a device which can store a functional relationship and release it continuously or in increments.

functor [COMPUT SCI] *See* logic element. [MATH] A function between categories which associates objects with objects and morphisms with morphisms.

fundamental [PHYS] The lowest frequency component of a complex wave. Also known as first harmonic; fundamental component.

fundamental circle *See* primary great circle.

fundamental component *See* fundamental.

fundamental frequency [PHYS] 1. The lowest frequency at which a system vibrates freely. 2. The lowest frequency in a complex wave.

fundamental group [COMMUN] In wire communications, a group of trunks that connect each local or trunk switching center to a trunk switching center of higher rank on which it homes; the term also applies to groups that interconnect zone centers.

fundamental interaction [PARTIC PHYS] One of the fundamental forces that act between the elementary particles of matter.

fundamental jelly *See* ulmin.

fundamental particle *See* elementary particle.

fundamental series [SPECT] A series occurring in the line spectra of many atoms and ions having one, two, or three electrons in the outer shell, in which the total orbital angular momentum quantum number changes from 3 to 2.

fundamental substance *See* ulmin.

fundamental tensor *See* metric tensor.

fundamental unit *See* base unit.

fundus [ANAT] The bottom of a hollow organ.

fungi [MYCOL] Nucleated, usually filamentous, sporebearing organisms devoid of chlorophyll.

fungicide [MATER] An agent that kills or destroys fungi.

Fungi Imperfecti [MYCOL] A class of the subdivision Eumycetes; the name is derived from the lack of a sexual stage.

fungistat [MATER] A compound that inhibits or prevents growth of fungi.

Fungivoridae [INV ZOO] The fungus gnats, a family of orthorrhaphous dipteran insects in the series Nematocera; the larvae feed on fungi.

fungivorous [ZOO] Feeding on or in fungi.

funicle *See* funiculus.

funiculus [ANAT] Also known as funicle. 1. Any structure in the form of a chord. 2. A column of white matter in the spinal cord. [BOT] The stalk of an ovule. [INV ZOO] A band of tissue extending from the adoral end of the coelom to the adoral body wall in bryozoans.

funnel cloud [METEOROL] The popular term for the tornado cloud, often shaped like a funnel with the small end nearest the ground.

furca [INV ZOO] A forked process as the last abdominal segment of certain crustaceans, and as part of the spring in collembolans.

Furipteridae [VERT ZOO] The smoky bats, a family of mammals in the order Chiroptera having a vestigial thumb and small ears.

furlong [MECH] A unit of length, equal to 1/8 mile, 660 feet, or 201.168 meters.

furnace [ENG] An apparatus in which heat is liberated and transferred directly or indirectly to a solid or fluid mass for the purpose of effecting a physical or chemical change.

furnace cupola *See* cupola.

Furnariidae [VERT ZOO] The oven birds, a family of perching birds in the superfamily Furnarioidea.

Furnarioidea [VERT ZOO] A superfamily of birds in the order Passeriformes characterized by a predominance of gray, brown, and black plumage.

furring [BUILD] Thin strips of wood or metal applied to the joists, studs, or wall of a building to level the surface, create an air space, or add thickness.

furuncle [MED] A small cutaneous abscess, usually resulting from infection of a hair follicle by *Staphylococcus aureus*. Also known as boil.

fuse [ELEC] An expandable device for opening an electric circuit when the current therein becomes excessive, containing a section of conductor which melts when the current through it exceeds a rated value for a definite period of time. Also known as electric fuse. [ENG] Also spelled fuze. 1. A device with explosive components designed to initiate a train of fire or detonation in an item of ammunition by an action such as hydrostatic pressure, electrical energy, chemical energy, impact, or a combination of these. 2. A nonexplosive device designed to initiate an explosion in an item of ammunition by an action such as continuous or pulsating electromagnetic waves or acceleration.

fuse box *See* cutout box.

fused-electrolyte battery *See* thermal battery.

fused-junction diode *See* alloy junction diode.

fused-junction transistor *See* alloy-junction transistor.

fused-salt electrolysis [PHYS CHEM] Electrolysis with use of purified fused salts as raw material and as an electrolyte.

fused-salt reactor *See* molten-salt reactor.

fused semiconductor [ELECTR] Junction formed by recrystallization on a base crystal from a liquid phase of one or more components and the semiconductor.

fuselage [AERO ENG] In an airplane, the central structure to which wings and tail are attached; it accommodates flight crew, passengers, and cargo.

fusibility [THERMO] The quality or degree of being capable of being liquefied by heat.

fusible alloy [MET] A low melting alloy, usually of bismuth, tin, cadmium, and lead, which melts at temperatures as low as 70°C (160°F).

fusible resistor [ELEC] A resistor designed to protect a circuit against overload; its resistance limits current flow and thereby protects against surges when power is first applied to a circuit; its fuse characteristic opens the circuit when current drain exceeds design limits.

fusiform [BIOL] Spindle-shaped; tapering toward the ends.

fusion [NUC PHYS] Combination of two light nuclei to form a heavier nucleus (and perhaps other reaction products) with release of some binding energy. Also known as atomic fusion; nuclear fusion. [PHYS CHEM] A change of the state of a substance from the solid phase to the liquid phase. Also known as melting.

fusion bomb [ORD] A bomb that depends upon nuclear fusion for release of energy. Also known as fusion weapon.

fusion piercing [ENG] A method of producing vertical blastholes by virtually burning holes in rock. Also known as piercing.

fusion point [NUCLEO] The temperature of a plasma above which the rate of energy generation by nuclear fusion reactions exceeds the rate of energy loss from the plasma, so that the fusion reaction can be self-sustaining. Abbreviated fnp.

fusion reactor [NUCLEO] A proposed device in which controlled, self-sustaining nuclear fusion reactions would be carried out in order to produce useful power.

fusion weapon *See* fusion bomb.

fusion welding [MET] Any welding operation involving melting of the base or parent metal.

Fusulinacea [PALEON] A superfamily of large, marine extinct protozoans in the order Foraminiferida characterized by a chambered calcareous shell.

Fusulinidae [PALEON] A family of extinct protozoans in the superfamily Fusulinacea.

Fusulinina [PALEON] A suborder of extinct rhizopod protozoans in the order Foraminiferida having a monolamellar, microgranular calcite wall.

future address patch [COMPUT SCI] A computer output containing the address of a symbol and the address of the last reference to that symbol.

future label [COMPUT SCI] An address referenced in the operand field of an instruction, but which has not been previously defined.

future of an event [RELAT] Those events which can be reached by a signal emitted at the event and which move at a speed less than or equal to the speed of light in a vacuum.

fuze *See* fuse.

G

g *See* gram.
G [ELEC] *See* conductance. [MECH] A unit of acceleration equal to the standard acceleration of gravity, 9.80665 meters per second per second, or approximately 32.1740 feet per second per second. Also known as fors; grav. [SCI TECH] *See* giga-.
Ga *See* gallium.
GABA *See* γ-aminobutyric acid.
gabbro [PETR] A group of dark-colored, intrusive igneous rocks with granular texture, composed largely of basic plagioclase and clinopyroxene.
gable [ARCH] The upper, triangular portion of the terminal wall of a building under the ridge of a sloped roof.
Gadidae [VERT ZOO] A family of fishes in the order Gadiformes, including cod, haddock, pollock, and hake.
Gadiformes [VERT ZOO] An order of actinopterygian fishes that lack fin spines and a swim bladder duct and have cycloid scales and many-rayed pelvic fins.
gadolinite [MINERAL] $Be_2FeY_2Si_2O_{10}$ A black, greenish-black, or brown rare-earth mineral; hardness is 6.5–7 on Mohs scale, and specific gravity is 4–4.5.
gadolinium [CHEM] A rare-earth element, symbol Gd, atomic number 64, atomic weight 157.25; highly magnetic, especially at low temperatures.
gage Also spelled gauge. [CIV ENG] The distance between the inner faces of the rails of railway track; standard gage in the United States is 4 feet 8 ½ inches (1.44 meters). [DES ENG] **1.** A device for determining the relative shape or size of an object. **2.** The thickness of a metal sheet, a rod, or a wire. [ENG] The minimum sieve size through which most (95% or more) of an aggregate will pass. [ORD] The interior diameter of the barrel of a shotgun expressed by the number of spherical lead bullets fitting it that are required to make a pound. [TEXT] A measure of the density of knit cloth, given in the number of stitches in 1.5 inches (3.7 centimeters).
gage block [DES ENG] A chrome steel block having two flat, parallel surfaces with the parallel distance between them being the size marked on the block to a guaranteed accuracy of a few millionths of an inch; used as the standard of precise lineal measurement for most manufacturing processes. Also known as precision block; size block.
gage invariance [ELECTROMAG] The invariance of electric and magnetic fields and electrodynamic interactions under gage transformations.
gageite [MINERAL] $(Mn,Mg,Zn)_8Si_3O_{14}\cdot2H_2O(or\ 3H_2O)$ A mineral composed of a hydrous silicate of manganese, magnesium, and zinc.
gage theory [PHYS] Any field theory in which, as the result of the conservation of some quantity, it is possible to perform a transformation in which the phase of the fields is altered by a function of space and time without altering any measurable physical quantity, so that the fields obtained by any such transformation give a valid description of a given physical situation.
gage transformation [ELECTROMAG] The addition of the gradient of some function of space and time to the magnetic vector potential, and the addition of the negative of the partial derivative of the same function with respect to time, divided by the speed of light, to the electric scaler potential; this procedure gives different potentials but leaves the electric and magnetic fields unchanged.
gaging [NUCLEO] The measurement of the thickness, density, or quantity of material by the amount of radiation it absorbs; this is the most common use of radioactive isotopes in industry. Also spelled gauging.
gahnite [MINERAL] $ZnAl_2O_4$ A usually dark-green, but sometimes yellow, gray, or black spinel mineral consisting of an oxide of zinc and aluminum. Also known as zinc spinel.
gain [ELECTR] **1.** The increase in signal power that is produced by an amplifier; usually given as the ratio of output to input voltage, current, or power, expressed in decibels. Also known as transmission gain. **2.** *See* antenna gain. [ENG] A cavity in a piece of wood prepared by notching or mortising so that a hinge or other hardware or another piece of wood can be placed on the cavity.
gain sensitivity control *See* differential gain control.
gal [MECH] **1.** The unit of acceleration in the centimeter-gram-second system, equal to 1 centimeter per second squared; commonly used in geodetic measurement. Formerly known as galileo. Symbolized Gal. **2.** *See* gallon.
Gal *See* gal.
galactic center [ASTRON] The gravitational center of the Milky Way Galaxy; the sun and other stars of the Galaxy revolve about this center.
galactic circle *See* galactic equator.
galactic cluster *See* open cluster.
galactic coordinates *See* galactic system.
galactic disk [ASTRON] The flat distribution of stars and interstellar matter in the spiral arms and plane of the Milky Way Galaxy.
galactic equator [ASTRON] A great circle of the celestial sphere, inclined 62° to the celestial equator, coinciding approximately with the center line of the Milky Way, and constituting the primary great circle for the galactic system of coordinates; it is everywhere 90° from the galactic poles. Also known as galactic circle.
galactic latitude [ASTRON] Angular distance north or south of the galactic equator; the arc of a great circle

through the galactic poles, between the galactic equator and a point on the celestial sphere, measured northward or southward from the galactic equator through 90° and labeled N or S to indicate the direction of measurement.

galactic longitude [ASTRON] Angular distance east of sidereal hour angle 94°.4 along the galactic equator; the arc of the galactic equator or the angle at the galactic pole between the great circle through the intersection of the galactic equator and the celestial equator in Sagittarius (SHA 94°.4) and a great circle through the galactic poles, measured eastward from the great circle through SHA 94°.4 through 360°.

galactic nebula [ASTRON] A nebula that is in or near the galactic system known as the Milky Way.

galactic nova [ASTRON] One of the novae that are concentrated largely in a band 10° on each side of the plane of the galaxy and are most frequent toward the center of the galaxy.

galactic nucleus [ASTRON] The center area in the galaxy about which there is a large spherical distribution of stars and from which the spiral arms emanate.

galactic plane [ASTRON] The plane that may be drawn through the galactic equator; the plane of the Milky Way Galaxy.

galactic pole [ASTRON] On the celestial sphere, either of the two points 90° from the galactic equator.

galactic system [ASTRON] An astronomical coordinate system using latitude measured north and south from the galactic equator, and longitude measured in the sense of increasing right ascension from 0 to 360°. Also known as galactic coordinates.

galactic windows [ASTROPHYS] The regions near the equator of the Milky Way where there is low absorption of light by interstellar clouds so that some distant external galaxies may be seen through them.

galactokinase [BIOCHEM] An enzyme which reacts D-galactose with adenosinetriphosphate to give D-galactose-1-phosphate and adenosinediphosphate.

galactolipid *See* cerebroside.

galactose [BIOCHEM] $C_6H_{12}O_6$ A monosaccharide occurring in both levo and dextro forms as a constituent of plant and animal oligosaccharides (lactose and raffinose) and polysaccharides (agar and pectin). Also known as cerebrose.

galactosemia [MED] A congenital metabolic disorder caused by an enzyme deficiency and marked by high blood levels of galactose.

Galatheidea [INV ZOO] A group of decapod crustaceans belonging to the Anomura and having a symmetrical abdomen bent upon itself and a well-developed tail fan.

Galaxioidei [VERT ZOO] A suborder of mostly small, fresh-water fishes in the order Salmoniformes.

galaxy [ASTRON] A large-scale aggregate of stars, gas, and dust; the aggregate is a separate system of stars covering a mass range from 10^7 to 10^{12} solar masses and ranging in diameter from 1500 to 300,000 light-years.

Galaxy *See* Milky Way Galaxy.

galaxy cluster [ASTRON] A collection of from two to several hundred galaxies which are much more densely distributed than the average density of galaxies in space.

Galbulidae [VERT ZOO] The jacamars, a family of highly iridescent birds of the order Piciformes that resemble giant hummingbirds.

gale [METEOROL] 1. An unusually strong wind. 2. In storm-warning terminology, a wind of 28–47 knots (52–87 kilometers per hour). 3. In the Beaufort wind

scale, a wind whose speed is 28–55 knots (52–102 kilometers per hour).

galea [ANAT] The epicranial aponeurosis linking the occipital and frontal muscles. [BIOL] A helmet-shaped structure. [BOT] A helmet-shaped petal near the axis. [INV ZOO] 1. The endopodite of the maxilla of certain insects. 2. A spinning organ on the movable digit of chelicerae of pseudoscorpions.

galena [MINERAL] PbS A bluish-gray to lead-gray mineral with brilliant metallic luster, specific gravity 7.5, and hardness 2.5 on Mohs scale; occurs in cubic or octahedral crystals, in masses, or in grains. Also known as blue lead; lead glance.

Galeritidae [PALEON] A family of extinct exocyclic Euechinoidea in the order Holectypoida, characterized by large ambulacral plates with small, widely separated pore pairs.

Galilean glass *See* Galilean telescope.

Galilean satellites [ASTRON] The four largest and brightest satellites of Jupiter (Io, Europa, Ganymede, and Callisto).

Galilean telescope [OPTICS] A refracting telescope whose objective is a converging (convex) lens and whose eyepiece is a diverging (concave) lens; it forms erect images. Also known as Galilean glass.

Galilean transformation [MECH] A mathematical transformation used to relate the space and time variables of two uniformly moving (inertial) reference systems in nonrelativistic kinematics.

galileo *See* gal.

Galileo's law of inertia *See* Newton's first law.

gall [MED] A sore on the skin that is caused by chafing. [PHYSIO] *See* bile. [PL PATH] A large swelling on plant tissues caused by the invasion of parasites, such as fungi or bacteria, following puncture by an insect; insect oviposit and larvae of insects are found in galls.

gallbladder [ANAT] A hollow, muscular organ in humans and most vertebrates which receives dilute bile from the liver, concentrates it, and discharges it into the duodenum.

galleria forest [ECOL] A modified tropical deciduous forest occurring along stream banks.

Galleriinae [INV ZOO] A monotypic subfamily of lepidopteran insects in the family Pyralididae; contains the bee moth or wax worm (*Galleria mellonella*), which lives in beehives and whose larvae feed on beeswax.

gallery [GEOL] 1. A horizontal, or nearly horizontal, underground passage, either natural or artificial. 2. A subsidiary passage in a cave at a higher level than the main passage. [MIN ENG] A level or drift.

galley [ENG] The kitchen of a ship, airplane, or trailer. [GRAPHICS] A flat, oblong, open-ended tray into which the letters assembled by hand in a composing stick are transferred after the composing stick is full.

galley proof [GRAPHICS] A reproduction taken from type while the type is still in the galley; the reproduction is reviewed for errors.

Galliformes [VERT ZOO] An order of birds that includes important domestic and game birds, such as turkeys, pheasants, and quails.

gallinaceous [VERT ZOO] Of, pertaining to, or resembling birds of the order Galliformes.

gallium [CHEM] A chemical element, symbol Ga, atomic number 31, atomic weight 69.72. [MET] A silvery-white metal, melting at 29.7°C, boiling at 1983°C.

gallodesoxycholic acid *See* chenodeoxycholic acid.

gallon [MECH] Abbreviated gal. 1. A unit of volume used in the United States for measurement of liquid substances, equal to 231 cubic inches, or to 3.785 411 784 × 10^{-3} cubic meter, or to 3.785 411 784 liters; equal to 128 fluid ounces. **2.** A unit of volume used in the United Kingdom for measurement of liquid and solid substances, usually the former; equal to the volume occupied by 10 pounds of weight of water of density 0.998859 gram per milliliter in air of density 0.001217 gram per milliliter against weights of density 8.136 grams per milliliter (the milliliter here has its old definition of 1.000028 × 10^{-6} cubic meter); approximately equal to 277.420 cubic inches, or to 4.54609 × 10^{-3} cubic meter, or to 4.54609 liters; equal to 160 fluid ounces.

gallstone [PATH] A nodule formed in the gallbladder or biliary tubes and composed of calcium, cholesterol, or bilirubin, or a combination of these.

galmei See hemimorphite.

Galumnidae [INV ZOO] A family of oribatid mites in the suborder Sarcoptiformes.

galvanic [ELEC] Pertaining to electricity flowing as a result of chemical action.

galvanic battery [ELEC] A galvanic cell, or two or more such cells electrically connected to produce energy.

galvanic cell [ELEC] An electrolytic cell that is capable of producing electric energy by electrochemical action.

galvanic corrosion [MET] Electrochemical corrosion associated with the current in a galvanic cell, caused by dissimilar metals in an electrolyte because of the difference in potential (emf) of the two metals.

galvanic series [CHEM] The relative hierarchy of metals arranged in order from magnesium (least noble) at the anodic, corroded end through platinum (most noble) at the cathodic, protected end.

galvanic skin response [PHYSIO] The electrical reactions of the skin to any stimulus as detected by a sensitive galvanometer; most often used experimentally to measure the resistance of the skin to the passage of a weak electric current.

galvanism [BIOL] The use of a galvanic current for medical or biological purposes.

galvanize [MET] To deposit zinc on the surface of metal by the processes of hot dipping, sherardizing, or sometimes electroplating.

galvanometer [ENG] An instrument for indicating or measuring a small electric current by means of a mechanical motion derived from electromagnetic or electrodynamic forces produced by the current.

gambrel roof [BUILD] A roof with two sloping sides stepped at different angles on each side of the center ridge; the lower slope is steeper than the upper slope.

game [MATH] A mathematical model expressing a contest between two players under specified rules.

gamete [BIOL] A mature germ cell.

game theory [MATH] The mathematical study of games or abstract models of conflict situations from the viewpoint of determining an optimal policy or strategy. Also known as theory of games.

gametogenesis [BIOL] The formation of gametes, or reproductive cells such as ova or sperm.

gametophyte [BOT] **1.** The haploid generation producing gametes in plants exhibiting metagenesis. **2.** An individual plant of this generation.

gamma-absorption gage See gamma gage.

gamma counter [ENG] A device for detecting gamma radiation, primarily through the detection of fast electrons produced by the gamma rays; it either yields information about integrated intensity within a time interval or detects each photon separately.

gamma decay See gamma emission.

gamma distribution [STAT] A normal distribution whose frequency function involves a gamma function.

gamma emission [NUC PHYS] A quantum transition between two energy levels of a nucleus in which a gamma ray is emitted. Also known as gamma decay.

gamma function [MATH] The complex function given by the integral with respect to t from 0 to ∞ of $e^{-t}t^{-l}$; this function helps determine the general solution of Gauss' hypergeometric equation.

gamma gage [NUCLEO] A penetration-type thickness gage that measures the thickness or density of a sample by measuring its absorption of gamma rays. Also known as gamma-absorption gage.

gamma matrix See Dirac matrix.

gamma ray [NUC PHYS] A high-energy photon, especially as emitted by a nucleus in a transition between two energy levels.

gamma-ray astronomy [ASTRON] The study of gamma rays from extraterrestrial sources, especially gamma-ray bursts.

gamma-ray bursts [ASTRON] Intense blasts of soft gamma rays of unknown origin, which range in duration from a tenth of a second to tens of seconds and occur several times a year from sources widely distributed over the sky.

gamma-ray detector [ENG] An instrument used on ships to identify and measure abnormal concentrations of gamma rays in the oceans.

gamma-ray laser [PHYS] A hypothetical device which would generate coherent radiation in the range 0.005–0.5 nanometer by inducing isomeric radiative transitions between isomeric nuclear states. Also known as graser.

gamma-ray scattering See Compton scattering.

gamma-ray spectrometry [NUCLEO] **1.** Determination of the energy distribution of gamma rays emitted by nuclei. Also known as gamma-ray spectroscopy. **2.** In particular, a variation of neutron activation analysis in which the induced radiation from the sample is gamma rays instead of neutrons.

gamma-ray spectroscopy See gamma-ray spectrometry.

gamma-ray telescope [ASTRON] Any device for detecting and determining the directions of extraterrestrial gamma rays, using coincidence or anticoincidence circuits with scintillation or semiconductor detectors to obtain directional discrimination.

Gammaridea [INV ZOO] The scuds or sand hoppers, a suborder of amphipod crustaceans; individuals are usually compressed laterally, are poor walkers, and lack a carapace.

gamma taxonomy [SYST] The level of taxonomic study concerned with biological aspects of taxa, including intraspecific populations, speciation, and evolutionary rates and trends.

gamma transition See glass transition.

gamogony [INV ZOO] Spore formation by multiple fission in sporozoans. [ZOO] Sexual reproduction.

gamopetalous [BOT] Having petals united at their edges. Also known as sympetalous.

gamophyllous [BOT] Having the leaves of the perianth united.

gamosepalous [BOT] Having sepals united at their edges. Also known as synsepalous.

Gamow-Condon-Gurney theory [NUC PHYS] An early quantum-mechanical theory of alpha-particle decay according to which the alpha particle penetrates a potential barrier near the surface of the nucleus by a tunneling process.

Gamow-Teller interaction [NUC PHYS] Interaction between a nucleon source current and a lepton field which has an axial vector or tensor form.

Gamow-Teller selection rules [NUC PHYS] Selection rules for beta decay caused by the Gamow-Teller interaction; that is, in an allowed transition there is no parity change of the nuclear state, and the spin of the nucleus can either remain unchanged or change by $\pm$ 1; transitions from spin 0 to spin 0 are excluded, however.

Gampsonychidae [PALEON] A family of extinct crustaceans in the order Palaeocaridacea.

ganged control [ELECTR] Controls of two or more circuits mounted on a common shaft to permit simultaneous control of the circuits by turning a single knob.

ganglioma [MED] A form of ganglioneuroma in which neuronal and glial elements appear in about equal proportions.

ganglion [ANAT] A group of nerve cell bodies, usually located outside the brain and spinal cord.

ganglioside [BIOCHEM] One of a group of glycosphingolipids found in neuronal surface membranes and spleen; they contain an N-acyl fatty acid derivative of sphingosine linked to a carbohydrate (glucose or galactose); they also contain N-acetylglucosamine or N-acetylgalactosamine, and N-acetylneuraminic acid.

gang-punch [COMPUT SCI] To punch identical or constant information into all of a group of punched cards.

gangrene [MED] A form of tissue death usually occurring in an extremity due to insufficient blood supply.

gangue [GEOL] The valueless rock or aggregates of minerals in an ore.

ganoid scale [VERT ZOO] A structure having several layers of enamellike material (ganoin) on the upper surface and laminated bone below.

ganophyllite [MINERAL] $(Na,K)(Mn,Fe,Al)_5(Si,Al)_6$-$O_{15}(OH)_5 \cdot 2H_2O$ A brown, prismatic crystalline or foliated mineral composed of a hydrous silicate of manganese and aluminum.

gantry [ENG] A frame erected on side supports so as to span an area and support and hoist machinery and heavy materials.

Gantt chart [IND ENG] In production planning and control, a type of bar chart depicting the work planned and done in relation to time; each division of space represents both a time interval and the amount of work to be done during that interval.

Ganymede [ASTRON] A satellite of Jupiter orbiting at a mean distance of 1,071,000 kilometers. Also known as Jupiter III.

gap [COMMUN] A region not adequately covered by the main lobes of a radar antenna. [COMPUT SCI] A uniformly magnetized area in a magnetic storage device (tape, disk), used to indicate the end of an area containing information. [ELEC] The spacing between two electric contacts. [ELECTROMAG] A break in a closed magnetic circuit, containing only air or filled with a nonmagnetic material. [GEN] A short region that is missing in one strand of a double-stranded deoxyribonucleic acid. [GEOGR] Any sharp, deep notch in a mountain ridge or between hills. [MET] An opening at the point of closest approach between faces of members in a weld joint.

gap digit [COMPUT SCI] A digit in a machine word that does not represent data or instructions, such as a parity bit or a digit included for engineering purposes.

gape [ANAT] The margin to margin distance between open jaws. [INV ZOO] The space between the margins of a closed mollusk valve.

gap filling [ELECTROMAG] Electrical or mechanical rearrangement of an antenna array, or the use of a supplementary array, to produce lobes where gaps previously occurred.

gapless tape [COMPUT SCI] A magnetic tape upon which raw data is recorded in a continuous manner; the data are streamed onto the tape without the word gaps; the data still may contain signs and end-of-record marks in the gapless form.

gapped tape [COMPUT SCI] A magnetic tape upon which blocked data has been recorded; it contains all of the flag bits and format to be read directly into a computer for immediate use.

gap scatter [COMPUT SCI] The deviation from the exact distance required between read/write heads and the magnetized surface.

gar [VERT ZOO] The common name for about seven species of bony fishes in the order Semionotiformes having a slim form, an elongate snout, and close-set ganoid scales.

garbage See hash.

garbage collection [COMPUT SCI] In a computer program with dynamic storage allocation, the automatic process of identifying those memory cells whose contents are no longer useful for the computation in progress and then making them available for some other use.

garbage in, garbage out [COMPUT SCI] A phrase often stressed during introductory courses in computer utilization as a reminder that, regardless of the correctness of the logic built into the program, no answer can be valid if the input is erroneous. Abbreviated GIGO.

garble [COMMUN] To alter a message intentionally or unintentionally so that it is difficult to understand.

garboard strake [NAV ARCH] The strake of shell plating adjacent to the keel; this row of plates acts in conjunction with the keel, and the plates are made heavier than the other bottom plates.

garlic [BOT] *Allium sativum*. A perennial plant of the order Liliales grown for its pungent, edible bulbs.

garnet [MINERAL] A generic name for a group of mineral silicates that are isometric in crystallization and have the general chemical formula $A_3B_2(SiO_4)_3$, where A is Fe^{2+}, Mn^{2+}, Mg, or Ca, and B is Al, Fe^{3+}, Cr^{3+}, or Ti^{3+}; used as a gemstone and as an abrasive.

garnierite [MINERAL] $(Ni,Mg)_3Si_2O_5(OH)_4$ An apple-green or pale-green, monoclinic serpentine; a gemstone and an ore of nickel. Also known as nepuite; noumeite.

garronite [MINERAL] $Na_2Ca_5Al_{12}Si_{20}O_{64} \cdot 27H_2O$ A zeolite mineral belonging to the phillipsite group; cyrstallizes in the tetragonal system.

gas [MATER] See gasoline. [ORD] To expose to a war gas. [PHYS] A phase of matter in which the substance expands readily to fill any containing vessel; characterized by relatively low density.

gas absorption operation [CHEM ENG] The recovery of solute gases present in gaseous mixtures of noncondensables; this recovery is generally achieved by contacting the gas stream with a liquid that offers specific or selective solubility for the solute gas to be recovered, or with an adsorbent (for example, synthetic or natural zeolite) that accepts only specific molecule sizes or shapes.

gas analysis [ANALY CHEM] Analysis of the constituents or properties of a gas (either pure or mixed); composition can be measured by chemical adsorption, combustion, electrochemical cells, indicator papers, chromatography, mass spectroscopy, and so on; properties analyzed for include heating value, molecular weight, density, and viscosity.

gas black [CHEM] Fine particles of carbon formed by partial combustion or thermal decomposition of natural gas; used to reinforce rubber products such as tires. Also known as carbon black; channel black.

gas cap [GEOPHYS] The gas immediately in front of a meteoroid as it travels through the atmosphere. [PETRO ENG] Gas occurring above liquid hydrocarbons in a reservoir under such trap conditions as the presence of water which prevents downward migration or the abutment of an impermeable formation against the reservoir.

gas capacitor [ELEC] A capacitor consisting of two or more electrodes separated by a gas, other than air, that serves as a dielectric.

gas-cap injection See external gas injection.

gas chromatograph [ANALY CHEM] The instrument used in gas chromatography to detect volatile compounds present; also used to determine certain physical properties such as distribution or partition coefficients and adsorption isotherms, and as a preparative technique for isolating pure components or certain fractions from complex mixtures.

gas chromatography [ANALY CHEM] A separation technique involving passage of a gaseous moving phase through a column containing a fixed adsorbent phase; it is used principally as a quantitative analytical technique for volatile compounds.

gas-compression cycle [MECH ENG] A refrigeration cycle in which hot, compressed gas is cooled in a heat exchanger, then passes into a gas expander which provides an exhaust stream of cold gas to another heat exchanger that handles the sensible-heat refrigeration effect and exhausts the gas to the compressor.

gas compressor [MECH ENG] A machine that increases the pressure of a gas or vapor by increasing the gas density and delivering the fluid against the connected system resistance.

gas-condensate reservoir [GEOL] Hydrocarbon reservoir in which conditions of temperature and pressure have resulted in the condensation of the heavier hydrocarbon constituents from the reservoir gas.

gas-condensate well [PETRO ENG] A well producing hydrocarbons from a gas-condensate reservoir.

gas constant [THERMO] The constant of proportionality appearing in the equation of state of an ideal gas, equal to the pressure of the gas times its molar volume divided by its temperature. Also known as gas-law constant.

gas-cooled reactor [NUCLEO] A nuclear reactor in which a gas, such as air, carbon dioxide, or helium, is used as a coolant.

gas current [ELECTR] A positive-ion current produced by collisions between electrons and residual gas molecules in an electron tube. Also known as ionization current.

gas cyaniding See carbonitriding.

gas cycle [THERMO] A sequence in which a gaseous fluid undergoes a series of thermodynamic phases, ultimately returning to its original state.

gas-deviation factor See compressibility factor.

gas discharge [ELECTR] Conduction of electricity in a gas, due to movements of ions produced by collisions between electrons and gas molecules.

gas-discharge lamp See discharge lamp.

gas-discharge laser [OPTICS] A gas laser in which optical pumping is caused by nonequilibrium processes in a gas discharge.

gas doping [ELECTR] The introduction of impurity atoms into a semiconductor material by epitaxial growth, by using streams of gas that are mixed before being fed into the reactor vessel.

gas dynamics [PHYS] The study of the motion of gases, and of its causes, which takes into account thermal effects generated by the motion.

gas embolus [MED] An embolus composed of a gas resulting from trauma or other causes.

gas engine [MECH ENG] An internal combustion engine that uses gaseous fuel.

gaseous diffusion [CHEM ENG] 1. Pressure-induced free-molecular transfer of gas through microporous barriers as in the process of making fissionable fuel. 2. Selective solubility diffusion of gas through nonporous polymers by absorption and solution of the gas in the polymer matrix.

gaseous nebulae [ASTRON] Clouds of gas, such as the Network Nebula in Cygnus, that are members of the Milky Way galactic system and are small compared with its overall dimensions.

gas field [PETRO ENG] An area underlain with little or no interruption by one or more reservoirs of commercially valuable gas.

gas-filled rectifier See cold-cathode rectifier.

gas-flow counter tube [NUCLEO] A radiation-counter tube in which an appropriate atmosphere is maintained by a flow of gas through the tube. Also known as flow counter; gas-flow radiation counter.

gas-flow radiation counter See gas-flow counter tube.

gas furnace [ENG] An enclosure in which a gaseous fuel is burned.

gas gangrene [MED] A localized, but rapidly spreading, necrotizing bacterial wound infection characterized by edema, gas production, and discoloration; caused by several species of *Clostridium*.

gasification [CHEM ENG] Any chemical or heat process used to convert a substance to a gas; coal is converted by the Hygas process to a gaseous fuel. [MIN ENG] A method for exploiting poor-quality coal and thin coal seams by burning the coal in place to produce combustible gas which can be burned to generate power or processed into chemicals and fuels. Also known as underground gasification.

gas injection [MECH ENG] Injection of gaseous fuel into the cylinder of an internal combustion engine at the appropriate part of the cycle. [PETRO ENG] The injection of gas into a reservoir to maintain formation pressure and to drive liquid hydrocarbons toward the wellbores.

gas-injection well [PETRO ENG] A well hole in a reservoir into which pressurized gas is injected to maintain formation pressure or to drive liquid hydrocarbons toward other well holes.

gas ionization [ELECTR] Removal of the planetary electrons from the atoms of gas filling an electron tube, so that the resulting ions participate in current flow through the tube.

gasket [ENG] A packing made of deformable material, usually in the form of a sheet or ring, used to make a pressure-tight joint between stationary parts. Also known as static seal.

gas kinematics [FL MECH] The motion of a gas considered by itself, without regard for the causes of motion.

gas laser [OPTICS] A laser in which the active medium is a discharge in a gas contained in a glass or quartz tube with a Brewster-angle window at each end; the gas can be excited by a high-frequency oscillator or direct-current flow between electrodes inside the tube; the function of the discharge is to pump the medium, to obtain population inversion.

gas law [THERMO] Any law relating the pressure, volume, and temperature of a gas.

gas-law constant *See* gas constant.

gas lift [CHEM ENG] Solids movement operation in which an upward-flowing gas stream in a closed conduit or vessel is used to lift and move powdered or granular solid material. [PETRO ENG] The injection of gas near the bottom of an oil well to aerate and lighten the column of oil to increase oil production from the well.

gas-liquid chromatography [ANALY CHEM] A form of gas chromatography in which the fixed phase (column packing) is a liquid solvent distributed on an inert solid support. Abbreviated GLC. Also known as gas-liquid partition chromatography.

gas-liquid partition chromatography *See* gas-liquid chromatography.

gas logging [PETRO ENG] Hot-wire-detector or gas-chromatographic analysis and record of gas contained in the mud stream and cuttings for a well being drilled; a common way of detecting subsurface oil and gas shows.

gas maser [PHYS] A maser in which the microwave electromagnetic radiation interacts with the molecules of a gas such as ammonia; used chiefly in highly stable oscillator applications, as in atomic clocks.

gas mask [ENG] A device to protect the eyes and respiratory tract from noxious gases, vapors, and aerosols, by removing contamination with a filter and a bed of adsorbent material.

gas meter [ENG] An instrument for measuring and recording the amount of gas flow through a pipe.

gasohol [MATER] Synthetic fuel consisting of a mixture of gasoline and grain alcohol (ethanol).

gas oil [MATER] A petroleum distillate boiling within the general range 450–800°F (232–426°C); usually includes kerosine, diesel fuel, heating oils, and light fuel oils.

gas-oil separator [PETRO ENG] An oil-field stock tank or series of tanks in which wellhead pressure is reduced so that the dissolved gas associated with reservoir oil is flashed off or separated as a separate phase. Also known as gas separator; oil-field separator; oil-gas separator; oil separator; separator.

gasoline [MATER] A fuel for internal combustion engines consisting essentially of volatile flammable liquid hydrocarbons; derived from crude petroleum by processes such as distillation reforming, polymerization, catalytic cracking, and alkylation; the common name is gas. Also known as petrol.

gasoline engine [MECH ENG] An internal combustion engine that uses a mixture of air and gasoline vapor as a fuel.

gaspeite [MINERAL] NaCO$_3$ An anhydrous normal carbonate mineral with calcite structure.

gas pocket [GEOL] A gas-filled cavity in rocks, especially above an oil pocket. [MET] A cavity in a metal which contains trapped gases.

gas reservoir [GEOL] An accumulation of natural gas found with or near accumulations of crude oil in the earth's crust.

gas seal [ENG] A seal which prevents gas from leaking to or from a machine along a shaft.

gas separator *See* gas-oil separator.

gas-shielded arc welding [MET] Use of a gas atmosphere to shield the molten metal from air in arc welding.

gassing [ELEC] The evolution of gas in the form of small bubbles in a storage battery when charging continues after the battery has been completely charged. [ENG] **1.** Absorption of gas by a material. **2.** Formation of gas pockets in a material. **3.** Evolution of gas from a material during a process or procedure. [TEXT] *See* singeing.

gas-solid chromatography [ANALY CHEM] A form of gas chromatography in which the moving phase is a gas and the stationary phase is a surface-active sorbent (charcoal, silica gel, or activated alumina). Abbreviated GSC.

Gasteromycetes [MYCOL] A group of basidiomycetous fungi in the subclass Homobasidiomycetidae with enclosed basidia and with basidiospores borne symmetrically on long sterigmata and not forcibly discharged.

Gasterophilidae [INV ZOO] The horse bots, a family of myodarian cyclorrhaphous dipteran insects in the subsection Calypteratae, including individuals that cause myiasis in horses.

Gasterosteidae [VERT ZOO] The sticklebacks, a family of actinopterygian fishes in the order Gasterosteiformes.

Gasterosteiformes [VERT ZOO] An order of actinopterygian fishes characterized by a ductless swim bladder, a pelvic fin that is abdominal to subthoracic in position, and an elongate snout.

Gasteruptiidae [INV ZOO] A family of hymenopteran insects in the superfamily Proctotrupoidea.

gas thermometer [ENG] A device to measure temperature by measuring the pressure exerted by a definite amount of gas enclosed in a constant volume; the gas (preferably hydrogen or helium) is enclosed in a glass or fused-quartz bulb connected to a mercury manometer. Also known as constant-volume gas thermometer.

gas thermometry [ENG] Measurement of temperatures with a gas thermometer; used with helium down to about 1 K.

gas trap [CIV ENG] A bend or chamber in a drain or sewer pipe that prevents sewer gas from escaping.

gastrectomy [MED] Surgical removal of all or part of the stomach.

gastric acid [BIOCHEM] Hydrochloric acid secreted by parietal cells in the fundus of the stomach.

gastric juice [PHYSIO] The digestive fluid secreted by gastric glands; contains gastric acid and enzymes.

gastric ulcer [MED] An ulcer of the mucous membrane of the stomach.

gastrin [BIOCHEM] A polypeptide hormone secreted by the pyloric mucosa which stimulates the pancreas to release pancreatic fluid and the stomach to release gastric acid.

gastrocnemius [ANAT] A large muscle of the posterior aspect of the leg, arising by two heads from the posterior surfaces of the lateral and medial condyles of the femur, and inserted with the soleus muscle into the calcaneal tendon, and through this into the back of the calcaneus.

gastroenterology [MED] The branch of medicine concerned with study of the stomach and intestine.

gastrointestinal system [ANAT] The portion of the digestive system including the stomach, intestine, and all accessory organs.

Gastromyzontidae [VERT ZOO] A small family of actinopterygian fishes of the suborder Cyprinoidei found in southeastern Asia.

Gastropoda [INV ZOO] A large, morphologically diverse class of the phylum Mollusca, containing the snails, slugs, limpets, and conchs.

gastroscope [MED] A hollow, tubular instrument used to examine the inside of the stomach by passage through the mouth and esophagus.

gastrostomy [MED] The establishment of a fistulous opening into the stomach, with an external opening in the skin; used for artificial feeding.

Gastrotricha [INV ZOO] A group of microscopic, pseudocoelomate animals considered either to be a class of the Aschelminthes or to constitute a separate phylum.

gastrula [EMBRYO] The stage of development in animals in which the endoderm is formed and invagination of the blastula has occurred.

gastrulation [EMBRYO] The process by which the endoderm is formed during development.

gas tube [ELECTR] An electron tube into which a small amount of gas or vapor is admitted after the tube has been evacuated; ionization of gas molecules during operation greatly increases current flow.

gas-tube boiler See waste-heat boiler.

gas turbine [MECH ENG] A heat engine that converts energy of fuel into work by using compressed, hot gas as the working medium and that usually delivers its mechanical output through a rotating shaft. Also known as combustion turbine.

gas vacuole [BIOL] A membrane-bound, gas-filled cavity in some algae and protozoans; thought to control buoyancy.

gas welding [MET] A welding process in which metals are joined by the heat of an oxyacetylene flame.

gas well [PETRO ENG] A well drilled for extraction of natural gas from a gas reservoir.

gas zone [GEOL] A rock formation containing gas under a pressure large enough to force the gas out if tapped from the surface.

GAT See Greenwich apparent time.

gate [CIV ENG] A movable barrier across an opening in a large barrier, a fence, or a wall. [ELECTR] **1.** A circuit having an output and a multiplicity of inputs and so designed that the output is energized only when a certain combination of pulses is present at the inputs. **2.** A circuit in which one signal, generally a square wave, serves to switch another signal on and off. **3.** One of the electrodes in a field-effect transistor. **4.** An output element of a cryotron. **5.** To control the passage of a pulse or signal. **6.** In radar, an electric waveform which is applied to the control point of a circuit to alter the mode of operation of the circuit at the time when the waveform is applied. Also known as gating waveform. [ENG] **1.** A device, such as a valve or door, for controlling the passage of materials through a pipe, channel, or other passageway. **2.** A metallization that determine the exact functionn of each cell and interconnect the cells to form a specific network when the customer orders the device.

gate-array device [ELECTR] An integrated logic circuit that is manufactured by first fabricating a two-dimensional array of logic cells, each of which is equivalent to one or a few logic gates, and then adding final layers of metallization that determine the exact function of each cell and interconnect the cells to form a specific network when the customer orders the device.

gate-controlled rectifier [ELECTR] A three-terminal semiconductor device, such as a silicon controlled rectifier, in which the unidirectional current flow between the rectifier terminals is controlled by a signal applied to a third terminal called the gate.

gate-controlled switch [ELECTR] A semiconductor device that can be switched from its nonconducting or "off" state to its conducting or "on" state by applying a negative pulse to its gate terminal and that can be turned off at any time by applying reverse drive to the gate. Abbreviated GCS.

gate equivalent circuit [ELECTR] A unit of measure for specifying relative complexity of digital circuits, equal to the number of individual logic gates that would have to be interconnected to perform the same function as the digital circuit under evaluation.

gate pulse [ELECTR] A pulse that triggers a gate circuit so it will pass a signal.

gate valve [DES ENG] A valve with a disk-shaped closing element that fits tightly over an opening through which water passes.

gateway [COMMUN] A point of entry and exit to another system, such as the connection point between a local-area network and an external-communications network.

gather [GRAPHICS] In bookbinding, to arrange in sequence the folded sheets or signatures of a book. [MIN ENG] **1.** To assemble loaded cars from several production points and deliver them to main haulage for transport to the surface or pit bottom. **2.** To drive a heading through disturbed or faulty ground so as to meet the seam of coal at a convenient level or point on the opposite side.

gather write [COMPUT SCI] An operation that creates a single output record from data items gathered from nonconsecutive locations in main memory.

gating [ELECTR] The process of selecting those portions of a wave that exist during one or more selected time intervals or that have magnitudes between selected limits. [ENG] A network of connecting channels, including sprues, runners, gates, and cavities, which conduct molten metal to the mold.

gating waveform See gate.

Gatterman reaction [ORG CHEM] **1.** Reaction of a phenol or phenol ester, and hydrogen chloride or hydrogen cyanide, in the presence of a metallic chloride such as aluminum chloride to form, after hydrolysis, an aldehyde. **2.** Reaction of an aqueous ethanolic solution of diazonium salts with precipitated copper powder or other reducing agent to form diaryl compounds.

gauge [ELECTROMAG] One of the family of possible choices for the electric scalar potential and magnetic vector potential, given the electric and magnetic fields.

gauging See gaging.

gauss [ELECTROMAG] Unit of magnetic induction in the electromagnetic and Gaussian systems of units, equal to 1 maxwell per square centimeter, or 10^{-4} weber per square meter. Also known as abtesla (abt).

Gauss' error curve See normal distribution.

Gauss eyepiece [OPTICS] A Ramsden eyepiece which has a thin glass plate between the two lenses, making an angle of 45° with the optical axis; used to set a telescope perpendicular to a plane reflecting surface.

Gauss formulas [MATH] Formulas dealing with the sine and cosine of angles in a spherical triangle.

Gaussian curvature [MATH] The invariant of a surface specified by Gauss' theorem. Also known as total curvature.

Gaussian curve [STAT] The bell-shaped curve corresponding to a population which has a normal distribution. Also known as normal curve.

Gaussian distribution See normal distribution.

Gaussian noise [COMMUN] Noise that has a frequency distribution which follows the Gaussian curve. [MATH] *See* Wiener process.

Gaussian optics [OPTICS] An approximation which describes rays which are very close to the axis of an optical system and are nearly parallel to this axis, so that only the linear terms of Taylor series for the distance of a point from the axis or the angle which a ray makes with the axis need be considered. Also known as first-order theory.

Gaussian reduction [MATH] A procedure of simplification of the rows of a matrix which is based upon the notion of solving a system of simultaneous equations. Also known as Gauss-Jordan elimination.

Gaussian system [ELECTROMAG] A combination of the electrostatic and electromagnetic systems of units (esu and emu), in which electrostatic quantities are expressed in esu and magnetic and electromagnetic quantities in emu, with appropriate use of the conversion constant c (the speed of light) between the two systems. Also known as Gaussian units.

Gaussian units *See* Gaussian system.

Gauss-Jordan elimination *See* Gaussian reduction.

Gauss' law of the arithmetic mean [MATH] The law that a harmonic function can attain its maximum value only on the boundary of its domain of definition, unless it is a constant.

Gauss-Legendre rule [MATH] An approximation technique of definite integrals by a finite series which uses the zeros and derivatives of the Legendre polynomials.

Gauss lens system *See* Celor lens system.

Gauss' mean value theorem [MATH] The value of a harmonic function at a point in a planar region is equal to its integral about a circle centered at the point.

Gauss objective lens *See* Celor lens system.

Gauss point *See* cardinal point.

Gauss' principle of least constraint [MECH] The principle that the motion of a system of interconnected material points subjected to any influence is such as to minimize the constraint on the system; here the constraint, during an infinitesimal period of time, is the sum over the points of the product of the mass of the point times the square of its deviation from the position it would have occupied at the end of the time period if it had not been connected to other points.

Gauss' theorem [MATH] 1. The assertion, under certain light restrictions, that the volume integral through a volume V of the divergence of a vector function is equal to the surface integral of the exterior normal component of the vector function over the boundary surface of V. Also known as divergence theorem. 2. At a point on a surface the product of the principal curvatures is an invariant of the surface, called the Gaussian curvature.

gavial [VERT ZOO] The name for two species of reptiles composing the family Gavialidae.

Gavialidae [VERT ZOO] The gavials, a family of reptiles in the order Crocodilia distinguished by an extremely long, slender snout with an enlarged tip.

Gaviidae [VERT ZOO] The single, monogeneric family of the order Gaviiformes.

Gaviiformes [VERT ZOO] The loons, a monofamilial order of diving birds characterized by webbed feet, compressed, bladelike tarsi, and a heavy, pointed bill.

Gay-Lussac's law *See* Charles' law; combining-volumes principle.

gaylussite [MINERAL] $Na_2Ca(CO_3)_2 \cdot 5H_2O$ A translucent, yellowish-white hydrous carbonate mineral,

with a vitreous luster, crystallizing in the monoclinic system; found in dry lakes.

g-cal *See* calorie.

GCA radar *See* ground-controlled approach radar.

GCI radar *See* ground-controlled intercept radar.

GCS *See* gate-controlled switch.

Gd *See* gadolinium.

G display [ELECTR] A rectangular radar display in which horizontal and vertical aiming errors are indicated by horizontal and vertical displacement, respectively, and range is indicated by the length of wings appearing on the blip, with length increasing as range decreases.

Ge *See* germanium.

gear [DES ENG] A toothed machine element used to transmit motion between rotating shafts when the center distance of the shafts is not too large. [MECH ENG] 1. A mechanism performing a specific function in a machine. 2. An adjustment device of the transmission in a motor vehicle which determines mechanical advantage, relative speed, and direction of travel.

gearbox *See* transmission.

gear down [MECH ENG] To arrange gears so that the driven part rotates at a slower speed than the driving part.

gear drive [MECH ENG] Transmission of motion or torque from one shaft to another by means of direct contact between toothed wheels.

gearing [MECH ENG] A set of gear wheels.

gearksutite [MINERAL] $CaAl(OH)F_4 \cdot H_2O$ A clayey mineral composed of hydrous calcium aluminum fluoride, occurring with cryolite.

gear level [MECH ENG] To arrange gears so that the driven part and driving part turn at the same speed.

gear ratio [MECH ENG] The ratio of the angular speed of the driving member of a gear train or similar mechanism to that of the driven member; specifically, the number of revolutions made by the engine per revolution of the rear wheels of an automobile.

gearshift [MECH ENG] A device for engaging and disengaging gears.

gear teeth [DES ENG] Projections on the circumference or face of a wheel which engage with complementary projections on another wheel to transmit force and motion.

gear train [MECH ENG] A combination of two or more gears used to transmit motion between two rotating shafts or between a shaft and a slide.

gear up [MECH ENG] To arrange gears so that the driven part rotates faster than the driving part.

gear wheel [MECH ENG] A wheel that meshes gear teeth with another part.

Gecarcinidae [INV ZOO] The true land crabs, a family of decapod crustaceans belonging to the Brachygnatha.

gecko [VERT ZOO] The common name for more than 300 species of arboreal and nocturnal reptiles composing the family Gekkonidae.

geepound *See* slug.

gegenschein [ASTRON] A round or elongated, faint, ill-defined spot of light in the sky at a point 180° from the sun. Also known as counterglow; zodiacal counterglow.

gehlenite [MINERAL] Ca_2Al_2SiO A mineral of the melilite group that crystallizes in the tetragonal crystal system and is isomorphous with akermanite; a green, resinous material found with spinel.

Geiger counter *See* Geiger-Müller counter.

Geiger counter tube *See* Geiger-Müller tube.

Geiger-Müller counter [NUCLEO] 1. A radiation counter that uses a Geiger-Müller tube in appropriate circuits to detect and count ionizing particles; each particle crossing the tube produces ionization of gas in the tube which is roughly independent of the particle's nature and energy, resulting in a uniform discharge across the tube. Abbreviated GM counter. Also known as Geiger counter. **2.** See Geiger-Müller tube.

Geiger-Müller counter tube See Geiger-Müller tube.

Geiger-Müller tube [NUCLEO] A radiation-counter tube operated in the Geiger region; it usually consists of a gas-filled cylindrical metal chamber containing a fine-wire anode at its axis. Also known as Geiger counter tube; Geiger-Müller counter; Geiger-Müller counter tube.

Geiger-Nutall rule [NUC PHYS] The rule that the logarithm of the decay constant of an alpha emitter is linearly related to the logarithm of the range of the alpha particles emitted by it.

geikielite [MINERAL] $MgTiO_3$ A bluish-black or brownish-black mineral that crystallizes in the rhombohedral system and occurs in the form of rolled pebbles; it is isomorphous with ilmenite.

Gekkonidae [VERT ZOO] The geckos, a family of small lizards in the order Squamata distinguished by a flattened body, a long sensitive tongue, and adhesive pads on the toes of many species.

gel [CHEM] A two-phase colloidal system consisting of a solid and a liquid in more solid form than a sol.

Gelastocoridae [INV ZOO] The toad bugs, a family of tropical and subtropical hemipteran insects in the subdivision Hydrocorisae.

gelatin [MATER] See gelatin dynamite. [ORG CHEM] A protein derived from the skin, white connective tissue, and bones of animals; used as a food and in photography, the plastics industry, metallurgy, and pharmaceuticals.

gelatin dynamite [MATER] A high explosive consisting mainly of a jellylike mass of nitroglycerin, with sodium nitrate, meal, collodion cotton, and sodium carbonate; commonly used by drillers to shatter boulders encountered in driving pipe through overburden, especially in water-filled or saturated ground. Also known as gelatin; gelignite; nitrogelatin.

Gelechiidae [INV ZOO] A large family of minute to small moths in the lepidopteran superfamily Tineoidea, generally having forewings and trapezoidal hindwings.

gel filtration [ANALY CHEM] A type of column chromatography which separates molecules on the basis of size; higher-molecular-weight substances pass through the column first. Also known as molecular exclusion chromatography; molecular sieve chromatography.

gelifluction [GEOL] The slow, continuous downslope movement of rock debris and water-saturated soil that occurs above frozen ground, as in most polar regions and in many high mountain ranges.

gelignite See gelatin dynamite.

gelisol See frozen ground.

Gelocidae [PALEON] A family of extinct pecoran ruminants in the superfamily Traguloidea.

gelose See ulmin.

gel permeation chromatography [ANALY CHEM] Analysis by chromatography in which the stationary phase consists of beads of porous polymeric material such as a cross-linked dextran carbohydrate derivative sold under the trade name Sephadex; the moving phase is a liquid.

gem [MINERAL] A natural or artificially produced mineral or other material that has sufficient beauty and durability for use as a personal adornment.

Gemini [ASTRON] A northern constellation; right ascension 7 hours, declination 20°N. Also known as Twins.

gemmule [ANAT] A minute dendritic process functioning as a synaptic contact point. [BIOL] Any bud formed by gemmation. [INV ZOO] A cystlike, asexual reproductive structure of many Porifera that germinates when proper environmental conditions exist; it is a protective, overwintering structure which germinates the following spring.

Gemolite [OPTICS] A binocular magnifier with dark-field illumination, used to distinguish natural from synthetic gem materials.

gemology [MINERAL] The science concerned with the identification, grading, evaluation, fashioning, and other aspects of gemstones.

Gempylidae [VERT ZOO] The snake mackerels, a family of the suborder Scombroidei comprising compressed, elongate, or eel-shaped spiny-rayed fishes with caniniform teeth.

Gemuendinoidei [PALEON] A suborder of extinct raylike placoderm fishes in the order Rhenanida.

gene [GEN] The basic unit of inheritance.

gene action [GEN] The functioning of a gene in determining the phenotype of an individual.

gene flow [GEN] The passage and establishment of genes characteristic of a breeding population into the gene complex of another population through hybridization and backcrossing.

gene frequency [GEN] The number of occurrences of a specific gene within a population.

gene penetrance See penetrance.

gene pool [GEN] The totality of the genes of a specific population at a given time.

general anesthesia [MED] Loss of sensation with loss of consciousness, produced by administration of anesthetic drugs.

generalized coordinates [MECH] A set of variables used to specify the position and orientation of a system, in principle defined in terms of cartesian coordinates of the system's particles and of the time in some convenient manner; the number of such coordinates equals the number of degrees of freedom of the system Also known as Lagrangian coordinates.

generalized function See distribution.

generalized momentum See conjugate momentum.

generalized routine [COMPUT SCI] A routine which can process a wide variety of jobs; for example, a generalized sort routine which will sort in ascending or descending order on any number of fields whether alphabetic or numeric, or both, and whether binary coded decimals or pure binaries.

general paresis [MED] An inflammatory and degenerative disease of the brain caused by infection with *Treponema pallidum*. Also known as syphilitic meningoencephalitis.

general program [COMPUT SCI] A computer program designed to solve a specific type of problem when values of appropriate parameters are supplied.

general register See local register.

general relativity [RELAT] The theory of Einstein which generalizes special relativity to noninertial frames of reference and incorporates gravitation, and in which events take place in a curved space.

general routine [COMPUT SCI] In computers, a routine, or program, applicable to a class of problems; it provides instructions for solving a specific problem when appropriate parameters are supplied.

generate [COMPUT SCI] To create a particular program by selecting parts of a general-program skeleton (or outline) and specializing these parts into a cohesive entity.

generated address [COMPUT SCI] An address calculated or determined by instructions contained in a computer program for subsequent use by that program. Also known as calculated address; synthetic address.

generating area *See* fetch.

generating flow [FL MECH] For a liquid allowed to flow smoothly into a duct, the flow while the boundary layer, which starts at the entrance and grows until it fills the duct, is growing.

generating routine *See* generator.

generation [BIOL] A group of organisms having a common parent or parents and comprising a single level in line of descent. [COMPUT SCI] **1.** Any one of three groups used to historically classify computers according to their electronic hardware components, logical organization and software, or programming techniques; computers are thus known as first-, second-, or third-generation; a particular computer may possess characteristics of all generations simultaneously. **2.** One of a family of data sets, related to one another in that each is a modification of the next most recent data set.

generation number [COMPUT SCI] A number contained in the file label of a reel of magnetic tape that indicates the generation of the data set of the tape.

generator [COMPUT SCI] A program that produces specific programs as directed by input parameters. Also known as generating routine. [ELEC] A machine that converts mechanical energy into electrical energy; in its commonest form, a large number of conductors are mounted on an armature that is rotated in a magnetic field produced by field coils. Also known as dynamo; electric generator. [ELECTR] **1.** A vacuum-tube oscillator or any other nonrotating device that generates an alternating voltage at a desired frequency when energized with direct-current power or low-frequency alternating-current power. **2.** A circuit that generates a desired repetitive or nonrepetitive waveform, such as a pulse generator. [MATH] **1.** One of the set of elements of an algebraic system such as a group, ring, or module which determine all other elements when all admissible operations are performed upon them. **2.** *See* generatrix.

generator set [ENG] The aggregate of one or more generators together with the equipment and plant for producing the energy that drives them.

generatrix [MATH] The straight line generating a ruled surface. Also known as generator.

genet [VERT ZOO] The common name for nine species of small, arboreal African carnivores in the family Viverridae.

genetic code [MOL BIO] The genetic information in the nucleotide sequences in deoxyribonucleic acid represented by a four-letter alphabet that makes up a vocabulary of 64 three-nucleotide sequences, or codons; a sequence of such codons (averaging about 100 codons) constructs a message for a polypeptide chain.

genetic drift [GEN] The random fluctuation of gene frequencies from generation to generation that occurs in small populations.

genetic engineering [GEN] The intentional production of new genes and alteration of genomes by the substitution or addition of new genetic material.

genetic fingerprinting [MOL BIO] Identification of chemical entities in animal tissues as indicative of the presence of specific genes.

genetic homeostasis [GEN] The tendency of Mendelian populations to maintain a constant genetic composition.

genetic map [GEN] A graphic presentation of the linear arrangement of genes on a chromosome; gene positions are determined by percentages of recombination in linkage experiments. Also known as chromosome map.

genetics [BIOL] The science that is concerned with the study of biological inheritance.

geniculate [SCI TECH] Bent abruptly at an angle, as a bent knee.

Geniohyidae [PALEON] A family of extinct ungulate mammals in the order Hyracoidea; all members were medium to large-sized animals with long snouts.

genitalia [ANAT] The organs of reproduction, especially those which are external.

genital ridge [EMBRYO] A medial ridge or fold on the ventromedial surface of the mesonephros in the embryo, produced by growth of the peritoneum; the primordium of the gonads and their ligaments.

genital tract [ANAT] The ducts of the reproductive system.

genitourinary system *See* urogenital system.

genome [GEN] **1.** The genetic endowment of a species. **2.** The haploid set of chromosomes.

genotype [GEN] The genetic constitution of an organism, usually in respect to one gene or a few genes relevant in a particular context. [SYST] The type species of a genus.

Gentianaceae [BOT] A family of dicotyledonous herbaceous plants in the order Gentianales distinguished by lacking stipules and having parietal placentation.

Gentianales [BOT] A family of dicotyledonous plants in the subclass Asteridae having well-developed internal phloem and opposite, simple, mostly entire leaves.

genu *See* knee.

genus [MATH] An integer associated to a surface which measures the number of holes in the surface. [SYST] A taxonomic category that includes groups of closely related species; the principal subdivision of a family.

geobotanical prospecting [GEOL] The use of the distribution, appearance, and growth anomalies of plants in locating ore deposits.

geocentric coordinates [ASTRON] Coordinates that define the position of a point with respect to the center of the earth; can be either cartesian (x, y, and z) or spherical (latitude, longitude, and radial distance). Also known as geocentric position; geocentric coordinate system.

geocentric coordinate system *See* geocentric coordinates.

geocentric latitude [ASTRON] The latitude of a celestial body from the center of the earth. [GEOD] Of a position on the earth's surface, the angle between a line to the center of the earth and the plane of the equator.

geocentric longitude [ASTRON] The celestial longitude of the position of a body projected on the celestial sphere when the body is viewed from the center of the earth. [GEOD] At a position on the earth's surface, the angle between the plane of the reference meridian and a plane through the polar axis and a line from the position in question to the center of mass of the earth.

geocentric parallax [ASTRON] The difference in the apparent direction or position of a celestial body, measured in seconds of arc, as determined from the center of the earth and from a point on its surface; this varies with the body's altitude and distance from the earth. Also known as diurnal parallax.

geocentric position *See* geocentric coordinates.

geocerite [MINERAL] A white, waxy mineral composed of carbon, oxygen, and hydrogen, occurring in brown coal.

geochemical cycle [GEOCHEM] During geologic changes, the sequence of stages in the migration of elements between the lithosphere, hydrosphere, and atmosphere.

geochemical prospecting [ENG] The use of geochemical and biogeochemical principles and data in the search for economic deposits of minerals, petroleum, and natural gases.

geochemistry [GEOL] The study of the chemical composition of the various phases of the earth and the physical and chemical processes which have produced the observed distribution of the elements and nuclides in these phases.

geochronology [GEOL] **1.** The dating of the events in the earth's history. **2.** A system of dating developed for the purposes of study of the earth's history.

geochronometry [GEOL] The study of the absolute age of the rocks of the earth based on the radioactive decay of isotopes, such as ^{238}U, ^{235}U, ^{232}Th, ^{87}Rb, ^{40}K, and ^{14}C, present in minerals and rocks.

Geocorisae [INV ZOO] A subdivision of hemipteran insects containing those land bugs with conspicuous antennae and an ejaculatory bulb in the male.

geode [GEOL] A roughly spheroidal, hollow body lined inside with inward-projecting, small crystals; found frequently in limestone beds but may occur in shale.

geodesic [MATH] A curve joining two points in a Riemannian manifold which has minimum length.

geodesic coordinates [RELAT] Coordinates in the neighborhood of a point P such that the gradient of the metric tensor is zero at P.

geodesic dome [ARCH] A dome constructed of many light, straight structural elements in tension, arranged in a framework of triangles to reduce stress and weight.

geodesy [GEOPHYS] A subdivision of geophysics which includes determination of the size and shape of the earth, the earth's gravitational field, and the location of points fixed to the earth's crust in an earth-referred coordinate system.

geodetic coordinates [GEOD] The quantities latitude, longitude, and elevation which define the position of a point on the surface of the earth with respect to the reference spheroid.

geodetic equator [GEOD] The great circle midway between the poles of revolution of the earth, connecting points of 0° geodetic latitude.

geodetic latitude [GEOD] Angular distance between the plane of the equator and a normal to the spheroid; a geodetic latitude differs from the corresponding astronomical latitude by the amount of the meridional component of station error. Also known as geographic latitude; topographical latitude.

geodetic longitude [GEOD] The angle between the plane of the reference meridian and the plane through the polar axis and the normal to the spheroid; a geodetic longitude differs from the corresponding astronomical longitude by the amount of the prime-vertical component of station error divided by the cosine of the latitude. Also known as geographic longitude.

geodetic meridian [GEOD] A line on a spheroid connecting points of equal geodetic longitude. Also known as geographic meridian.

geodetic parallel [GEOD] A line connecting points of equal geodetic latitude. Also known as geographic parallel.

geodetic satellite [AERO ENG] An artificial earth satellite used to obtain data for geodetic triangulation calculations.

geodetic survey [ENG] A survey in which the figure and size of the earth are considered; it is applicable for large areas and long lines and is used for the precise location of basic points suitable for controlling other surveys.

geoelectricity *See* terrestrial electricity.

geographical coordinates [GEOGR] Spherical coordinates, designating both astronomical and geodetic coordinates, defining a point on the surface of the earth, usually latitude and longitude. Also known as terrestrial coordinates.

geographical cycle *See* geomorphic cycle.

geographical position [ASTRON] That point on the earth at which a given celestial body is in the zenith at a specified time. [GEOGR] Any position on the earth defined by means of its geographical coordinates, either astronomical or geodetic.

geographic latitude *See* geodetic latitude.

geographic longitude *See* geodetic longitude.

geographic meridian *See* geodetic meridian.

geographic parallel *See* geodetic parallel.

geographic position [GEOGR] The position of a point on the surface of the earth expressed in terms of geographical coordinates either geodetic or astronomical.

geography [SCI TECH] The study of all aspects of the earth's surface, comprising its natural and political divisions, the differentiation of areas, and, sometimes people in relationship to the environment.

geoid [GEOD] The figure of the earth considered as a sea-level surface extended continuously over the entire earth's surface.

geoisotherm [GEOPHYS] The locus of points of equal temperature in the interior of the earth; a line in two dimensions or a surface in three dimensions. Also known as geotherm; isogeotherm.

geolith *See* rock-stratigraphic unit.

geologic age [GEOL] **1.** Any time period in the earth's history marked by special phases of physical conditions or organic development. **2.** A formal geologic unit of time that corresponds to a stage. **3.** An informal geologic time unit that corresponds to any stratigraphic unit.

geologic column [GEOL] **1.** The vertical sequence of strata of various ages found in an area or region. **2.** The geologic time scale as represented by rocks.

geologic thermometry *See* geothermometry.

geologic time [GEOL] The period of time covered by historical geology, from the end of the formation of the earth as a separate planet to the beginning of written history.

geologic time scale [GEOL] The relative age of various geologic periods and the absolute time intervals.

geology [SCI TECH] The study or science of the earth, its history, and its life as recorded in the rocks; includes the study of geologic features of an area, such as the geometry of rock formations, weathering and erosion, and sedimentation.

geomagnetic coordinates [GEOPHYS] A system of spherical coordinates based on the best fit of a centered dipole to the actual magnetic field of the earth.

geomagnetic equator [GEOPHYS] That terrestrial great circle which is 90° from the geomagnetic poles.

geomagnetic field reversal [GEOPHYS] Reversed magnetization in sedimentary and igneous rock, that is, polarized opposite to the mean geomagnetic field.

geomagnetic latitude [GEOPHYS] The magnetic latitude that a location would have if the field of the earth were to be replaced by a dipole field closely approximating it.

geomagnetic longitude [GEOPHYS] Longitude that is determined around the geomagnetic axis instead of around the rotation axis of the earth.

geomagnetic meridian [GEOPHYS] A semicircle connecting the geomagnetic poles.

geomagnetic pole [GEOPHYS] Either of two antipodal points marking the intersection of the earth's surface with the extended axis of a powerful bar magnet assumed to be located at the center of the earth and having a field approximating the actual magnetic field of the earth.

geomagnetic reversal [GEOPHYS] Reversed magnetization of the earth's magnetic dipole.

geomagnetic secular variation *See* secular variation.

geomagnetic storm [GEOPHYS] A large disturbance of the earth's magnetic field.

geomagnetic variation [GEOPHYS] Temporal changes in the geomagnetic field, both long-term (secular) and short-term (transient).

geomagnetism [GEOPHYS] 1. The magnetism of the earth. Also known as terrestrial magnetism. 2. The branch of science that deals with the earth's magnetism.

geometrical acoustics *See* ray acoustics.

geometrical optics [OPTICS] The geometry of paths of light rays and their imagery through optical systems.

geometric average *See* geometric mean.

geometric mean [MATH] The geometric mean of n given quantities is the nth root of their product. Also known as geometric average.

geometric programming [SYS ENG] A nonlinear programming technique in which the relative contribution of each of the component costs is first determined; only then are the variables in the component costs determined.

geometric progression [MATH] A sequence which has the form a, ar, ar^2, ar^3, ...

Geometridae [INV ZOO] A large family of lepidopteran insects in the superfamily Geometroidea that have slender bodies and relatively broad wings; includes measuring worms, loopers, and cankerworms.

Geometroidea [INV ZOO] A superfamily of lepidopteran insects in the suborder Heteroneura comprising small to large moths with reduced maxillary palpi and tympanal organs at the base of the abdomen.

geomorphic cycle [GEOL] The cycle of change in the surface configuration of the earth. Also known as cycle of erosion; geographical cycle.

geomorphology [GEOL] The study of the origin of secondary topographic features which are carved by erosion in the primary elements and built up of the erosional debris.

Geomyidae [VERT ZOO] The pocket gophers, a family of rodents characterized by fur-lined cheek pouches which open outward, a stout body with short legs, and a broad, blunt head.

Geophilomorpha [INV ZOO] An order of centipedes in the class Chilopoda including specialized forms that are blind, epimorphic, and dorsoventrally flattened.

geophone [ELECTR] A transducer, used in seismic work, that responds to motion of the ground at a location on or below the surface of the earth.

geophysical engineering [ENG] A branch of engineering that applies scientific methods for locating mineral deposits.

geophysics [GEOL] The physics of the earth and its environment, that is, earth, air, and (by extension) space.

geopotential [PHYS] The potential energy of a unit mass relative to sea level, numerically equal to the work that would be done in lifting the unit mass from sea level to the height at which the mass is located, against the force of gravity.

georgiadesite [MINERAL] $Pb_3(AsO_4)Cl_3$ A white or brownish-yellow mineral composed of lead chloroarsenate, occurring in orthorhombic crystals.

Georyssidae [INV ZOO] The minute mud-loving beetles, a family of coleopteran insects belonging to the Polyphaga.

Geosiridaceae [BOT] A monotypic family of monocotyledonous plants in the order Orchidales characterized by regular flowers with three stamens that are opposite the sepals.

geosphere [GEOL] 1. The solid mass of earth, as distinct from the atmosphere and hydrosphere. 2. The lithosphere, hydrosphere, and atmosphere combined.

Geospizinae [VERT ZOO] Darwin finches, a subfamily of perching birds in the family Fringillidae.

geostatic pressure *See* ground pressure.

geostationary satellite [AERO ENG] A satellite that orbits the earth from west to east at such a speed as to remain fixed over a given place on the earth's equator at approximately 35,900 kilometers altitude; makes one revolution in 24 hours, synchronous with the earth's rotation. Also known as synchronous satellite.

geostrophic [GEOPHYS] Pertaining to deflecting force resulting from the earth's rotation.

geostrophic approximation [GEOPHYS] The assumption that the geostrophic current can represent the actual horizontal current. Also known as geostrophic assumption.

geostrophic assumption *See* geostrophic approximation.

geostrophic current [GEOPHYS] A current defined by assuming the existence of an exact balance between the horizontal pressure gradient force and the Coriolis force.

geostrophic wind [METEOROL] That horizontal wind velocity for which the Coriolis acceleration exactly balances the horizontal pressure force.

geosynclinal couple *See* orthogeosyncline.

geosynclinal cycle *See* tectonic cycle.

geosyncline [GEOL] A part of the crust of the earth that sank deeply through time.

geotaxis [PHYSIO] Movement of a free-living organism in response to the stimulus of gravity.

geotechnics [CIV ENG] The application of scientific methods and engineering principles to civil engineering problems through acquiring, interpreting, and using knowledge of materials of the crust of the earth.

geotectonics *See* tectonics.

geotextile [MATER] A woven or nonwoven fabric manufactured from synthetic fibers or yarns that is designed to serve as a continuous membrane between soil and aggregate in a variety of earth structures.

geotherm *See* geoisotherm.

geothermal gradient [GEOPHYS] The change in temperature with depth of the earth.

geothermal well logging [ENG] Measurement of the change in temperature of the earth by means of well logging.

geothermometry [GEOL] Measurement of the temperatures at which geologic processes occur or occurred. Also known as geologic thermometry.

Gephyrea [INV ZOO] A class of burrowing worms in the phylum Annelida.

Geraniaceae [BOT] A family of dicotyledonous plants in the order Geraniales in which the fruit is beaked, styles are usually united, and the leaves have stipules.

Geraniales [BOT] An order of dicotyledonous plants in the subclass Rosidae comprising herbs or soft shrubs with a superior ovary and with compound or deeply cleft leaves.

Gerardiidae [INV ZOO] A family of anthozoans in the order Zoanthidea.

gerbil [VERT ZOO] The common name for about 100 species of African and Asian rodents composing the subfamily Gerbillinae.

Gerbillinae [VERT ZOO] The gerbils, a subfamily of rodents in the family Muridae characterized by hindlegs that are longer than the front ones, and a long, slightly haired, usually tufted tail.

geriatrics [MED] The study of the biological and physical changes and the diseases of old age.

germ [BIOL] A primary source, especially one from which growth and development are expected. [MICROBIO] General designation for a microorganism.

germanite [MINERAL] $Cu_3(Ge,Ga,Fe)(S,As)_4$ Reddish-gray mineral occurring in massive form; an important source of germanium.

germanium [CHEM] A brittle, water-insoluble, silvery-gray metallic element in the carbon family, symbol Ge, atomic number 32, atomic weight 72.59, melting at 959°C. [MET] A rare metal used in semiconductors, alloys, and glass.

germanium diode [ELECTR] A semiconductor diode that uses a germanium crystal pellet as the rectifying element. Also known as germanium rectifier.

germanium rectifier See germanium diode.

German measles See rubella.

germ cell [BIOL] An egg or sperm or one of their antecedent cells.

germination [BOT] The beginning or the process of development of a spore or seed. [PETR] See grain growth.

germ layer [EMBRYO] One of the primitive cell layers which appear in the early animal embryo and from which the embryo body is constructed.

germ theory [MED] The theory that contagious and infectious diseases are caused by microorganisms.

gerontology [PHYSIO] The scientific study of aging processes in biological systems, particularly in humans.

Gerrhosauridae [VERT ZOO] A small family of lizards in the suborder Sauria confined to Africa and Madagascar.

Gerridae [INV ZOO] The water striders, a family of hemipteran insects in the subdivision Amphibicorisae having long middle and hind legs and a median scent gland opening on the metasternum.

Gerroidea [INV ZOO] The single superfamily of the hemipteran subdivision Amphibicorisae; all members have conspicuous antennae and hydrofuge hairs covering the body.

gersdorffite [MINERAL] NiAsS A silver-white to steel-gray mineral, crystallizing in the isometric sys-

tem; resembles cobaltite and may contain some iron and cobalt. Also known as nickel glance.

Gesneriaceae [BOT] A family of dicotyledonous plants in the order Scrophulariales characterized by parietal placentation, mostly opposite or whorled leaves, and a well-developed embryo.

gestate [EMBRYO] To carry the young in the uterus from conception to delivery.

gestation period [EMBRYO] The period in mammals from fertilization to birth.

getter [PHYS CHEM] 1. A substance, such as thallium, that binds gases on its surface and is used to maintain a high vacuum in a vacuum tube. 2. A special metal alloy that is placed in a vacuum tube during manufacture and vaporized after the tube has been evacuated; when the vaporized metal condenses, it absorbs residual gases. Also known as degasser.

getter-ion pump [ENG] A high-vacuum pump that employs chemically active metal layers which are continuously or intermittently deposited on the wall of the pump, and which chemisorb active gases while inert gases are "cleaned up" by ionizing them in an electric discharge and drawing the positive ions to the wall, where the neutralized ions are buried by fresh deposits of metal. Also known as sputter-ion pump.

geyser [HYD] A natural spring or fountain which discharges a column of water or steam into the air at more or less regular intervals.

geyserite See siliceous sinter.

gf See gram-force.

g factor See Landé g factor.

g force [PHYS] A force such that a body subjected to it would have the acceleration of gravity at sea level; used as a unit of measurement for bodies undergoing the stress of acceleration.

GH See growth hormone.

GHA See Greenwich hour angle.

ghost crystal See phantom crystal.

ghost image [ELECTR] 1. An undesired duplicate image at the right of the desired image on a television receiver; due to multipath effect, wherein a reflected signal traveling over a longer path arrives slightly later than the desired signal. 2. See ghost pulse.

ghost pulse [ELECTR] An unwanted signal appearing on the screen of a radar indicator and caused by echoes which have a basic repetition frequency differing from that of the desired signals. Also known as ghost image; ghost signal.

ghost signal [ELECTR] 1. The reflection-path signal that produces a ghost image on a television receiver. Also known as echo. 2. See ghost pulse.

ghost spot [PL PATH] A disease of tomato characterized by small white rings on the fruit.

giant granite See pegmatite.

giant planets [ASTRON] The planets Jupiter, Saturn, Uranus, and Neptune.

giant pulse laser See Q-switched laser.

giant star [ASTRON] One of a class of stars that is 20 or 30 or more times larger than the sun and over 100 times more luminous.

Giaque-Debye method See adiabatic demagnetization.

gib [ENG] A removable plate designed to hold other parts in place or act as a bearing or wear surface. [MIN ENG] 1. A temporary support at the face to prevent coal from falling before the cut is complete, either by hand or by machine. 2. A prop put in the holing of a seam while being undercut. 3. A piece of metal often used in the same hole with a wedge-shaped key for holding pieces together.

gibberellin [BIOCHEM] Any member of a family of naturally derived compounds which have a gibbane skeleton and a broad spectrum of biological activity but are noted as plant growth regulators.

gibberish *See* hash.

gibbon [VERT ZOO] The common name for seven species of large, tailless primates belonging to the genus *Hylobates*; the face and ears are hairless, and the arms are longer than the legs.

gibbous moon [ASTRON] The shape of the moon's visible surface when the sun is illuminating more than half of the side facing the earth.

Gibbs-Donnan equilibrium *See* Donnan equilibrium.

Gibbs free energy [THERMO] The thermodynamic function $G = H - TS$, where H is enthalpy, T absolute temperature, and S entropy. Also known as free energy; free enthalpy; Gibbs function.

Gibbs function *See* Gibbs free energy.

Gibbs-Helmholtz equation [PHYS CHEM] An expression for the influence of temperature upon the equilibrium constant of a chemical reaction, $(d \ln K^0/dT)_P = \Delta H^0/RT^2$, where K^0 is the equilibrium constant, ΔH^0 the standard heat of the reaction at the absolute temperature T, and R the gas constant. [THERMO] **1.** Either of two thermodynamic relations that are useful in calculating the internal energy U or enthalpy H of a system; they may be written $U = F - T(\partial F/\partial T)_V$ and $H = G - T(\partial G/\partial T)_P$, where F is the free energy, G is the Gibbs free energy, T is the absolute temperature, V is the volume, and P is the pressure. **2.** Any of the similar equations for changes in thermodynamic potentials during an isothermal process.

gibbsite [MINERAL] $Al(OH)_3$ A white or tinted mineral, crystallizing in the monoclinic system; a principal constituent of bauxite. Also known as hydrargillite.

Gibbs phase rule [PHYS CHEM] A relation describing the nature of a heterogeneous chemical system at equilibrium, $F = C + 2 - P$, where F is the degrees of freedom, P the number of phases, and C the number of components. Also known as Gibbs rule.

Gibbs rule *See* Gibbs phase rule.

Gibraltar stone *See* onyx marble.

giga- [SCI TECH] A prefix representing 10^9, which is 1,000,000,000, or a billion. Abbreviated G. Also known as kilomega- (deprecated usage).

gigantism [MED] Abnormal largeness of the body due to hypersecretion of growth hormone.

Giganturoidei [VERT ZOO] A suborder of small, mesopelagic actinopterygian fishes in the order Cetomimiformes having large mouths and strong teeth.

GIGO *See* garbage in, garbage out.

Gila monster [VERT ZOO] The common name for two species of reptiles in the genus *Heloderma* (Helodermatidae) distinguished by a rounded body that is covered with multicolored beaded tubercles, and a bifid protrusible tongue.

gilbert [ELECTROMAG] The unit of magnetomotive force in the electromagnetic system, equal to the magnetomotive force of a closed loop of one turn in which there is a current of 1/4 abamp.

gilding [GRAPHICS] Overlaying material with a thin layer of gold.

gill [MECH] **1.** A unit of volume used in the United States for the measurement of liquid substances, equal to 1/4 U.S. liquid pint, or to $1.1829411825 \times 10^{-4}$ cubic meter. **2.** A unit of volume used in the United

Kingdom for the measurement of liquid substances, and occasionally of solid substances, equal to 1/4 U.K. pint, or to approximately 1.42065×10^{-4} cubic meter. [VERT ZOO] The respiratory organ of water-breathing animals. Also known as branchia.

gillespite [MINERAL] $BaFeSi_4O_{10}$ A micalike mineral composed of barium and iron silicate.

gilsonite [MINERAL] A variety of asphalt; it has black color, brilliant luster, brown streaks, and conchoidal fracture.

gimbal [ENG] **1.** A device with two mutually perpendicular and intersecting axes of rotation, thus giving free angular movement in two directions, on which an engine or other object may be mounted. **2.** In a gyro, a support which provides the spin axis with a degree of freedom. **3.** To move a reaction engine about on a gimbal so as to obtain pitching and yawing correction moments. **4.** To mount something on a gimbal.

gimlet [DES ENG] A small tool consisting of a threaded tip, grooved shank, and a cross handle; used for boring holes in wood.

gin [AGR] **1.** A machine used to separate cotton fiber from the seed and waste. **2.** To thus separate cotton fiber. [FOOD ENG] An alcoholic beverage made from distilled spirits flavored with an extract of the juniperberry or other flavoring botanicals. [MECH ENG] A hoisting machine in the form of a tripod with a windlass, pulleys, and ropes.

G indicator *See* G scope.

ginger [BOT] *Zingiber officinale.* An erect perennial herb of the family Zingiberaceae having thick, scaly branched rhizomes; a spice oleoresin is made by an organic solvent extraction of the ground dried rhizome.

gingiva [ANAT] The mucous membrane surrounding the teeth sockets.

gingivitis [MED] Inflammation of the gingiva.

Ginglymodi [VERT ZOO] An equivalent name for Semionotiformes.

ginglymus [ANAT] A type of diarthrosis permitting motion only in one plane. Also known as hinge joint.

Ginkgoales [BOT] An order of gymnosperms composing the class Ginkgoopsida with one living species, the dioecious maidenhair tree (*Ginkgo biloba*).

Ginkgoatae *See* Ginkgoopsida.

Ginkgoopsida [BOT] A class of the subdivision Pinicae containing the single, monotypic order Ginkgoales.

Ginkgophyta [BOT] The equivalent name for Ginkgoopsida.

ginorite [MINERAL] $Ca_2B_{14}O_{23} \cdot 8H_2O$ A white monoclinic mineral composed of hydrous borate of calcium.

ginseng [BOT] The common name for plants of the genus *Panax*, a group of perennial herbs in the family Araliaceae; the aromatic root of the plant has been used medicinally in China.

Ginzburg-Landau theory [CRYO] A phenomenological theory of superconductivity which accounts for the coherence length; the ordered state of a superconductor is described by a complex order parameter which is similar to a Schrödinger wave function, but describes all the condensed superelectrons, rather than a single charged particle. Also known as Landau-Ginzburg theory.

Ginzburg-London superconductivity theory [SOLID STATE] A modification of the London superconductivity theory to take into account the boundary energy.

giobertite *See* magnesite.

Giorgi system See meter-kilogram-second-ampere system.

giraffe [VERT ZOO] *Giraffa camelopardalis.* An artiodactyl mammal in the family Giraffidae characterized by extreme elongation of the neck vertebrae, and two prominent horns on the head.

Giraffe See Camelopardalis.

Giraffidae [VERT ZOO] A family of pecoran ruminants in the superfamily Bovoidea including giraffe, okapi, and relatives.

girder [CIV ENG] A large beam made of metal or concrete, and sometimes of wood.

girdle [ANAT] Either of the ringlike groups of bones supporting the forelimbs (arms) and hindlimbs (legs) in vertebrates. [INV ZOO] 1. Either of the hooplike bands constituting the sides of the two valves of a diatom. 2. The peripheral portion of the mantle in chitons. [LAP] The periphery of a cut gemstone that is usually grasped by the setting or mounting. [PETR] With reference to a fabric diagram or equal-area projection net, a belt showing concentration of points which is approximately coincident with a great circle of the net and which represents orientation of the fabric elements.

girt [CIV ENG] 1. A timber in the second-floor corner posts of a house to serve as a footing for roof rafters. 2. A horizontal member to stiffen the framework of a building frame or trestle. [ENG] A brace member running horizontally between the legs of a drill tripod or derrick. [MIN ENG] In square-set timbering, a horizontal brace running parallel to the drift.

gismondite [MINERAL] $CaAl_2Si_2O_8 \cdot 4H_2O$ A light-colored mineral composed of hydrous calcium aluminum silicate, occurring in pyramidal crystals.

gizzard [VERT ZOO] The muscular portion of the stomach of most birds where food is ground with the aid of ingested pebbles.

glabrous [BIOL] Having a smooth surface; specifically, having the epidermis devoid of hair or down.

glacial [GEOL] Pertaining to an interval of geologic time which was marked by an equatorward advance of ice during an ice age; the opposite of interglacial; these intervals are variously called glacial periods, glacial epochs, glacial stages, and so on. [HYD] Pertaining to ice, especially in great masses such as sheets of land ice or glaciers.

glacial epoch [GEOL] 1. Any of the geologic epochs characterized by an ice age; thus, the Pleistocene epoch may be termed a glacial epoch. 2. Generally, an interval of geologic time which was marked by a major equatorward advance of ice; the term has been applied to an entire ice age or (rarely) to the individual glacial stages which make up an ice age.

glacial erosion [GEOL] Movement of soil or rock from one point to another by the action of the moving ice of a glacier. Also known as ice erosion.

glacial geology [GEOL] The study of land features resulting from glaciation.

glacial lake [GEOL] A lake that exists because of the effects of the glacial period.

glacial mill See moulin.

glacial outwash See outwash.

glacial period [GEOL] 1. Any of the geologic periods which embraced an ice age; for example, the Quaternary period may be called a glacial period. 2. Generally, an interval of geologic time which was marked by a major equatorward advance of ice.

glacial scour [GEOL] Erosion resulting from glacial action, whereby the surface material is removed and the rock fragments carried by the glacier abrade, scratch, and polish the bedrock. Also known as scouring.

glacial till See till.

glacial varve See varve.

glaciated terrain [GEOL] A region that once bore great masses of glacial ice; a distinguishing feature is marks of glaciation.

glaciation [GEOL] Alteration of any part of the earth's surface by passage of a glacier, chiefly by glacial erosion or deposition. [METEOROL] The transformation of cloud particles from waterdrops to ice crystals, as in the upper portion of a cumulonimbus cloud.

glacier [HYD] A mass of land ice, formed by the further recrystallization of firn, flowing slowly (at present or in the past) from an accumulation area to an area of ablation.

glacier ice [HYD] Any ice that is or was once a part of a glacier, consolidated from firn by further melting and refreezing and by static pressure; for example, an iceberg.

glacier mill See moulin.

glacier pothole See moulin.

glacier well See moulin.

glaciology [GEOL] The study of existing or modern glaciers in their entirety.

gladite [MINERAL] $PbCuBi_5S_9$ A lead gray mineral consisting of lead and copper bismuth sulfide; occurs as prismatic crystals.

gland [ANAT] A structure which produces a substance essential and vital to the existence of the organism. [ENG] 1. A device for preventing leakage at a machine joint, as where a shaft emerges from a vessel containing a pressurized fluid. 2. A movable part used in a stuffing box to compress the packing.

glands of Brunner See Brunner's glands.

glandular fever See infectious mononucleosis.

glans [ANAT] The conical body forming the distal end of the clitoris or penis.

glare [OPTICS] 1. Discomfort produced in an observer by one or more visible sources of light. Also known as discomfort glare. 2. Visual disability caused by visible sources or areas of luminance which are in an observer's field of view but do not assist in viewing. Also known as disability glare. 3. Dazzling brightness of the atmosphere, caused by excessive reflection and scattering of light by particles in the line of sight.

Glareolidae [VERT ZOO] A family of birds in the order Charadriiformes including the ploverlike coursers and the swallowlike pratincoles.

glass [MATER] A hard, amorphous, inorganic, usually transparent, brittle substance made by fusing silicates, sometimes borates and phosphates, with certain basic oxides and then rapidly cooling to prevent crystallization. [METEOROL] In nautical terminology, a contraction for "weather glass" (a mercury barometer).

glass-plate capacitor [ELEC] High-voltage capacitor in which the metal plates are separated by sheets of glass serving as the dielectric, with the complete assembly generally immersed in oil.

glass porphyry See virtophyre.

glass sand [MATER] High-quartz sand used in glassmaking; contains small amounts of aluminum oxide, iron oxide, calcium oxide, and magnesium oxide.

glass sponge [INV ZOO] A siliceous sponge belonging to the class Hyalospongiae.

glass switch [ELECTR] An amorphous solid-state device used to control the flow of electric current. Also known as ovonic device.

glass transition [PHYS CHEM] The change in an amorphous region of a partially crystalline polymer from a viscous or rubbery condition to a hard and relatively brittle one; usually brought about by changing the temperature. Also known as gamma transition; glassy transition.

glass-tube manometer [ENG] A manometer for simple indication of difference of pressure, in contrast to the metallic-housed mercury manometer, used to record or control difference of pressure or fluid flow.

glass wool [MATER] A mass of glass fibers resembling wool and used as insulation, packing, and air filters.

glassy feldspar See sanidine.

glassy state See vitreous state.

glassy transition See glass transition.

glauberite [MINERAL] $Na_2Ca(SO_4)_2$ A brittle, gray-yellow monoclinic mineral having vitreous luster and saline taste.

glaucocerinite [MINERAL] A mineral composed of a hydrous basic sulfate of copper, zinc, and aluminum.

glaucochroite [MINERAL] $CaMnSiO_4$ A bluish-green mineral that is related to monticellite, is composed of calcium manganese silicate, and occurs in prismatic crystals.

glaucodot [MINERAL] $(Co,Fe)AsS$ A grayish-white, metallic-looking mineral composed of cobalt iron sulfarsenide, occurring in orthorhombic crystals.

glaucoma [MED] A disease of the eye characterized by increased fluid pressure within the eyeball.

glauconite [MINERAL] $K_{1.5}(Fe,Mg,Al)_{4-6}(Si,Al)_8O_{20}(OH)_4$ A type of clay mineral; it is dioctahedral and occurs in flakes and as pigmentary material.

glaucophane [MINERAL] $Na_2Mg_3Al_2Si_8$ A blue to black monoclinic sodium amphibole; blue to black coloration with marked pleochroism.

glaze [ENG] A glossy coating. Also known as enamel. [HYD] A coating of ice, generally clear and smooth but usually containing some air pockets, formed on exposed objects by the freezing of a film of supercooled water deposited by rain, drizzle, or fog, or possibly condensed from supercooled water vapor. Also known as glazed frost; glaze ice; verglas.

glazed frost See glaze.

glaze ice See glaze.

glazier's point [ENG] A small piece of sheet metal, usually shaped like a triangle, used to hold a pane of glass in place.

glazing [ENG] 1. Cutting and fitting panes of glass into frames. 2. Smoothing the lead of a wiped pipe joint by passing a hot iron over it.

glazing compound [MATER] A caulking compound used to seal and support a glass pane in place.

GLC See gas-liquid chromatography.

glide [AERO ENG] Descent of an aircraft at a normal angle of attack, with little or no thrust. [CRYSTAL] See slip.

glide angle See gliding angle.

glide bomb [ORD] A bomb fitted with airfoils to provide lift, released in the direction of a target by an airplane.

glide fold See shear fold.

glide path [AERO ENG] 1. The flight path of an aeronautical vehicle in a glide, seen from the side. Also known as glide trajectory. 2. The path used by an aircraft or spacecraft in a landing approach procedure.

glide path indicator [NAV] A system which provides signals for indicating vertical guidance of an aircraft along an inclined surface.

glide plane [CRYSTAL] A lattice plane in a crystal on which translation or twin gliding occurs. Also known as slip plane. [NAV] See glide slope.

glide slope [AERO ENG] See gliding angle. [NAV] An inclined electromagnetic surface which is generated by instrument-landing approach facilities and which includes a glide path supplying guidance in the vertical plane. Also known as glide plane.

glide trajectory See glide path.

gliding angle [AERO ENG] The angle between the horizontal and the glide path of an aircraft. Also known as glide angle; glide slope.

gliding bacteria [MICROBIO] The descriptive term for members of the orders Beggiatoales and Myxobacterales; they are motile by means of creeping movements.

gliding joint See arthrodia.

glime [HYD] An ice coating with a consistency intermediate between glaze and rime.

glint [ELECTR] 1. Pulse-to-pulse variation in amplitude of a reflected radar signal, owing to the reflection of the radar from a body that is rapidly changing its reflecting surface, for example, a spinning airplane propeller. 2. The use of this effect to degrade tracking or seeking functions of an enemy weapons system. [OPTICS] A small region designed to strongly reflect light from a target.

Gliridae [VERT ZOO] The dormice, a family of mammals in the order Rodentia.

glitch [ASTRON] A sudden change in the period of a pulsar, believed to result from a phenomenon analogous to an earthquake that changes the pulsar's moment of inertia. [ELECTR] 1. An undesired transient voltage spike occurring on a signal being processed. 2. A minor technical problem arising in electronic equipment.

glitter [MATER] A group of decorative materials consisting of flakes large enough so that each flake produces a plainly visible sparkle or reflection; incorporated into plastic during compounding. [OPTICS] The spots of light reflected from a point source by the surface of the sea or wave facets, that is, specular reflection.

Global Positioning System [NAV] A satellite navigation system which will use many (up to 24) satellites in three sets of orbits to supply the navigator with a precise time standard and three-dimensional information of position and velocity.

global search and replace [COMPUT SCI] A text-editing function of a word-processing system in which text is scanned for a given combination of characters, and each such combination is repalced by another set of characters.

global variable [COMPUT SCI] A variable which can be accessed (used or changed) throughout a computer program and is not confined to a single block.

globe [MAP] A sphere on the surface of which is a map of the world.

globe lightning See ball lightning.

globe valve [ENG] A device for regulating flow in a pipeline, consisting of a movable disk-type element and a stationary ring seat in a generally spherical body.

Globigerinacea [INV ZOO] A superfamily of foraminiferan protozoans in the suborder Rotaliina characterized by a radial calcite test with bilamellar septa and a large aperture.

globular protein [BIOCHEM] Any protein that is readily soluble in aqueous solvents.

globular star cluster [ASTRON] A group of many thousands of stars that are much closer to each other than the stars around the group and that are traveling through space together; a globular cluster has a slightly flattened spheroidal shape.

globule [ASTRON] A black volume of cosmic dust viewed against the brighter background of bright nebulae.

globulin [BIOCHEM] A heat-labile serum protein precipitated by 50% saturated ammonium sulfate and soluble in dilute salt solutions.

glockerite [MINERAL] A brown, ocher yellow, black, or dull green mineral consisting of a hydrated basic sulfate of ferric iron; occurs in stalactitic, encrusting, or earthy forms.

glomerulonephritis [MED] Inflammation of the kidney, primarily involving the glomeruli.

glomerulus [ANAT] A tuft of capillary loops projecting into the lumen of a renal corpuscle.

glomus [ANAT] **1.** A fold of the mesothelium arising near the base of the mesentery in the pronephros and containing a ball of blood vessels. **2.** A prominent portion of the choroid plexus of the lateral ventricle of the brain.

glory hole [CIV ENG] A funnel-shaped, fixed-crest spillway. [ENG] A furnace for resoftening or fire polishing glass during working, or an entrance in such a furnace. [MIN ENG] An opening formed by the removal of soft or broken ore through an underground passage. [NUCLEO] *See* beam hole.

gloss [OPTICS] The ratio of the light specularly reflected from a surface to the total light reflected.

glossimeter [ENG] An instrument, often photoelectric, for measuring the ratio of the light reflected from a surface in a definite direction to the total light reflected in all directions. Also known as glossmeter.

Glossinidae [INV ZOO] The tsetse flies, a family of cyclorrhaphous dipteran insects in the section Pupipara.

Glossiphoniidae [INV ZOO] A family of small leeches with flattened bodies in the order Rhynchobdellae.

glossitis [MED] Inflammation of the tongue.

glossmeter *See* glossimeter.

glossopharyngeal nerve [ANAT] The ninth cranial nerve in vertebrates; a paired mixed nerve that supplies autonomic innervation to the parotid gland and contains sensory fibers from the posterior one-third of the tongue and the anterior pharynx.

glossy [OPTICS] Property of a surface from which much more light is specularly reflected than is diffusely reflected.

glossy print [GRAPHICS] A photograph dried on a ferrotype plate or drum; the surface has a glazed appearance and is not easily scratched or soiled in handling.

glottis [ANAT] The opening between the margins of the vocal folds.

glow discharge [ELECTR] A discharge of electricity through gas at relatively low pressure in an electron tube, characterized by several regions of diffuse, luminous glow and a voltage drop in the vicinity of the cathode that is much higher than the ionization voltage of the gas. Also known as cold-cathode discharge.

glow-discharge cold-cathode tube *See* glow-discharge tube.

glow-discharge tube [ELECTR] A gas tube that depends for its operation on the properties of a glow discharge. Also known as glow-discharge cold-cathode tube; glow tube.

glow-discharge voltage regulator [ELECTR] Gas tube that varies in resistance, depending on the value of the applied voltage; used for voltage regulation.

glow lamp [ELECTR] A two-electrode electron tube containing a small quantity of an inert gas, in which light is produced by a negative glow close to the negative electrode when voltage is applied between the electrodes.

glow plug [MECH ENG] A small electric heater, located inside a cylinder of a diesel engine, that preheats the air and aids the engine in starting.

glow tube *See* glow-discharge tube.

glucagon [BIOCHEM] The protein hormone secreted by α-cells of the pancreas which plays a role in carbohydrate metabolism. Also known as hyperglycemic factor; hyperglycemic glycogenolytic factor.

glucocorticoid [BIOCHEM] A corticoid that affects glucose metabolism; secreted principally by the adrenal cortex.

glucogenesis [BIOCHEM] Formation of glucose within the animal body from products of glycolysis.

glucokinase [BIOCHEM] An enzyme that catalyzes the phosphorylation of D-glucose to glucose-6-phosphate.

gluconeogenesis [BIOCHEM] Formation of glucose within the animal body from substances other than carbohydrates, particularly proteins and fats.

D-glucopyranose *See* glucose.

glucose [BIOCHEM] $C_6H_{12}O_6$ A monosaccharide; occurs free or combined and is the most common sugar. Also known as cerelose; D-glucopyranose.

glucose tolerance test [PATH] A test to measure the ability of the liver to convert glucose to glycogen.

glucosidase [BIOCHEM] An enzyme that hydrolyzes glucosides.

glucosyltransferase [BIOCHEM] An enzyme that catalyzes the glucosylation of hydroxymethyl cytosine; a constituent of bacteriophage deoxyribonucleic acid.

glucuronidase [BIOCHEM] An enzyme that catalyzes hydrolysis of glucuronides. Also known as glycuronidase.

glue [MATER] A crude, impure, amber-colored form of commercial gelatin of unknown detailed composition produced by the hydrolysis of animal collagen; gelatinizes in aqueous solutions and dries to form a strong, adhesive layer.

Glumiflorae [BOT] An equivalent name for Cyperales.

gluon [PARTIC PHYS] One of eight hypothetical massless particles with spin quantum number and negative parity that mediate strong interactions between quarks.

glutelin [BIOCHEM] A class of simple, heat-labile proteins occurring in seeds of cereals; soluble in dilute acids and alkalies.

gluten [BIOCHEM] **1.** A mixture of proteins found in the seeds of cereals; gives dough elasticity and cohesiveness. **2.** An albuminous element of animal tissue.

Glyceridae [INV ZOO] A family of polychaete annelids belonging to the Errantia and characterized by an enormous eversible proboscis.

glycogen [BIOCHEM] A nonreducing, white, amorphous polysaccharide found as a reserve carbohydrate stored in muscle and liver cells of all higher animals, as well as in cells of lower animals.

glycogenolysis [BIOCHEM] The metabolic breakdown of glycogen.

glycolipid [BIOCHEM] Any of a class of complex lipids which contain carbohydrate residues.

glycolysis [BIOCHEM] The enzymatic breakdown of glucose or other carbohydrate, with the formation of lactic acid or pyruvic acid and the release of energy in the form of adenosinetriphosphate.

glycolytic pathway [BIOCHEM] The principal series of phosphorylative reactions involved in pyruvic acid production in phosphorylative fermentations. Also known as Embden-Meyerhof pathway; hexose diphosphate pathway.

glyconeogenesis [BIOCHEM] The metabolic process of glycogen formation from noncarbohydrate precursors.

glycopeptide *See* glycoprotein.

glycoprotein [BIOCHEM] Any of a class of conjugated proteins containing both carbohydrate and protein units. Also known as glycopeptide.

glycuronidase *See* glucuronidase.

Glyphocyphidae [PALEON] A family of extinct echinoderms in the order Temnopleuroida comprising small forms with a sculptured test, perforate crenulate tubercles, and diademoid ambulacral plates.

Glyptocrinina [PALEON] A suborder of extinct crinoids in the order Monobathrida.

gm *See* gram.

GM counter *See* Geiger-Müller counter.

gmelinite [MINERAL] $(Na_2Ca)Al_2Si_4O_{12} \cdot 6H_2O$ Zeolite mineral that is colorless or lightly colored and crystallizes in the hexagonal system.

GMT *See* Greenwich mean time.

gnat [INV ZOO] The common name for a large variety of biting insects in the order Diptera.

Gnathiidea [INV ZOO] A suborder of isopod crustaceans characterized by a much reduced second thoracomere, short antennules and antennae, and a straight pleon.

Gnathobdellae [INV ZOO] A suborder of leeches in the order Arhynchobdellae having jaws and a conspicuous posterior sucker; contains most of the important blood-sucking leeches of humans and other warm-blooded animals.

Gnathobelodontinae [PALEON] A subfamily of extinct elephantoid proboscideans containing the shovel-jawed forms of the family Gomphotheriidae.

Gnathodontidae [PALEON] A family of extinct conodonts having platforms with large, cup-shaped attachment scars.

Gnathostomata [INV ZOO] A suborder of echinoderms in the order Echinoidea characterized by a rigid, exocyclic test and a lantern or jaw apparatus. [VERT ZOO] A group of the subphylum Vertebrata which possess jaws and usually have paired appendages.

Gnathostomidae [INV ZOO] A family of parasitic nematodes in the superfamily Spiruroidea; sometimes placed in the superfamily Physalopteroidea.

Gnathostomulida [INV ZOO] Microscopic marine worms of uncertain systematic relationship, mainly characterized by cuticular structures in the pharynx and a monociliated skin epithelium.

gnd *See* ground.

gneiss [PETR] A variety of rocks with a banded or coarsely foliated structure formed by regional metamorphism.

Gnetales [BOT] A monogeneric order of the subdivision Gneticae; most species are lianas with opposite, oval, entire-margined leaves.

Gnetatae *See* Gnetopsida.

Gneticae [BOT] A subdivision of the division Pinophyta characterized by vessels in the secondary wood, ovules with two integuments, opposite leaves, and an embryo with two cotyledons.

Gnetophyta [BOT] The equivalent name for Gnetopsida.

Gnetopsida [BOT] A class of gymnosperms comprising the subdivision Gneticae.

gnomon [ENG] On a sundial, the inclined plate or pin that casts a shadow. Also known as style. [MATH] A geometric figure formed by removing from a parallelogram a similar parallelogram that contains one of its corners.

gnomonic chart [MAP] A chart on the gnomonic projection where great circles project as straight lines. Also known as great-circle chart.

gnomonic projection [CRYSTAL] A projection for displaying the poles of a crystal in which the poles are projected radially from the center of a reference sphere onto a plane tangent to the sphere. [MAP] A projection on a plane tangent to the surface of a sphere having the point of projection at the center of the sphere.

Gnostidae [INV ZOO] An equivalent name for Ptinidae.

gnotobiology [BIOL] That branch of biology dealing with known living forms; the study of higher organisms in the absence of all demonstrable, viable microorganisms except those known to be present.

gnu [VERT ZOO] Any of several large African antelopes of the genera *Connochaetes* and *Gorgon* having a large oxlike head with horns that characteristically curve downward and outward and then up, with the bases forming a frontal shield in older individuals.

goat [VERT ZOO] The common name for a number of artiodactyl mammals in the genus *Capra*; closely related to sheep but differing in having a lighter build and hollow, swept-back, sometimes spiral or twisted horns.

Gobiatheriinae [PALEON] A subfamily of extinct herbivorous mammals in the family Uintatheriidae known from one late Eocene genus; characterized by extreme reduction of anterior dentition and by lack of horns.

Gobiesocidae [VERT ZOO] The single family of the order Gobiesociformes.

Gobiesociformes [VERT ZOO] The clingfishes, a monofamilial order of scaleless bony fishes equipped with a thoracic sucking disk which serves for attachment.

Gobiidae [VERT ZOO] A family of perciform fishes in the suborder Gobioidei characterized by pelvic fins united to form a sucking disk on the breast.

Gobioidei [VERT ZOO] The gobies, a suborder of morphologically diverse actinopterygian fishes in the order Perciformes; all lack a lateral line.

goblet cell [HISTOL] A unicellular, mucus-secreting intraepithelial gland that is distended on the free surface. Also known as chalice cell. [INV ZOO] Any of the unicellular choanocytes of the genus *Monosiga*.

go-devil [ENG] **1.** A device inserted in a pipe or hole for purposes such as cleaning or for detonating an explosive. **2.** A sled for moving logs or cultivating. **3.** A large rake for gathering hay. **4.** A small railroad car used for transporting workers and materials.

goethite [MINERAL] FeO(OH) A yellow, red, or dark-brown mineral crystallizing in the orthorhombic system, although it is usually found in radiating fibrous aggregates; a common constituent of natural rust or limonite. Also known as xanthosiderite.

go gage [DES ENG] A test device that just fits a part if it has the proper dimensions (often used in pairs with a "no go" gage to establish maximum and minimum dimensions).

goiter [MED] An enlargement of all or part of the thyroid gland; may be accompanied by a hormonal dysfunction.

gold [CHEM] A chemical element, symbol Au, atomic number 79, atomic weight 196.967; soluble in aqua regia; melts at 1065°C. [MET] The native metallic element; a deep-yellow, very dense, soft, isometric metal, usually found alloyed with silver or copper; used in jewelry, dentistry, gilding, anodes, alloys, and solders.

gold-198 [NUC PHYS] The radioisotope of gold, atomic mass number 198 and half-life 2.7 days; used in medical treatment of tumors by injecting it in colloidal form directly into tumor tissue.

goldbeater's skin [MATER] The treated outside membrane of the large intestine of cattle; used between leaves of metal in goldbeating, and sometimes in hygrometers.

golden algae [BOT] The common name for members of the class Chrysophyceae.

golden-brown algae [BOT] The common name for members of the division Chrysophyta.

golden section search [COMPUT SCI] A dichotomizing search in which, in each step, the remaining items are divided as closely as possible according to the golden section.

goldfish [VERT ZOO] *Crassius auratus.* An orange cypriniform fish of the family Cyprinidae that can grow to over 18 inches (46 centimeters); closely related to the carps.

Goldhaber triangle [PARTIC PHYS] A plot describing a high-energy reaction leading to four or more particles; its coordinates are the invariant masses of two intermediate-state quasi-particle composites, and its kinematical limits form a right-angled isosceles triangle; resonances in the quasi-particle composites appear as horizontal and vertical bands.

gold leaf [MET] Gold beaten or rolled into extremely thin sheets or leaves (10^{-6} inch or 25 nanometers thick); leaves are stored in books (a book consists of 25 leaves), the paper of which is rubbed with chalk to keep the leaves from sticking.

gold-leaf electroscope [ELEC] An electroscope in which two narrow strips of gold foil or leaf suspended in a glass jar spread apart when charged; the angle between the strips is related to the charge.

gold number [ANALY CHEM] A measure of the amount of protective colloid which must be added to a standard red gold sol mixed with sodium chloride solution to prevent the solution from causing the sol to coagulate, as manifested by a change in color from red to blue.

gold plate [MET] Gold which has been electroplated on a material in a thin layer of controlled thickness; used on electric contacts for corrosion resistance and solderability and on jewelry and oraments.

gold point [THERMO] The temperature of the freezing point of gold at a pressure of 1 standard atmosphere (101,325 newtons per square meter); used to define the International Practical Temperature Scale of 1968, on which it is assigned a value of 1337.58 K or 1064.43°C.

goldschmidtine See stephanite.

goldschmidtite See sylvanite.

Golgi apparatus [CYTOL] A cellular organelle that is part of the cytoplasmic membrane system; it is composed of regions of stacked cisternae and it functions in secretory processes.

Golgi cell [ANAT] **1.** A nerve cell with long axons. **2.** A nerve cell with short axons that branch repeatedly and terminate near the cell body.

Golgi tendon organ [PHYSIO] Any of the kinesthetic receptors situated near the junction of muscle fibers and a tendon which act as muscle-tension recorders.

Gomphidae [INV ZOO] A family of dragonflies belonging to the Anisoptera.

gomphosis [ANAT] An immovable articulation, as that formed by the insertion of teeth into the bony sockets.

Gomphotheriidae [PALEON] A family of extinct proboscidean mammals in the suborder Elephantoidea consisting of species with shoveling or digging specializations of the lower tusks.

Gomphotheriinae [PALEON] A subfamily of extinct elephantoid proboscideans in the family Gomphotheriidae containing species with long jaws and bunomastodont teeth.

gon See grade.

gonad [ANAT] A primary sex gland; an ovary or a testis.

gonadotropic hormone [BIOCHEM] Either of two adenohypophyseal hormones, FSH (follicle-stimulating hormone) or ICSH (interstitial-cell-stimulating hormone), that act to stimulate the gonads.

gonadotropin [BIOCHEM] A substance that acts to stimulate the gonads.

Gondwanaland [GEOL] The ancient continent that is supposed to have fragmented and drifted apart to form eventually the present continents.

Goniadidae [INV ZOO] A family of marine polychaete annelids belonging to the Errantia.

goniometer [ELECTROMAG] An instrument for determining the direction of maximum response to a received radio signal, or selecting the direction of maximum radiation of a transmitted radio signal; consists of two fixed perpendicular coils, each attached to one of a pair of loop antennas which are also perpendicular, and a rotatable coil which bears the same space relationship to the coils as the direction of the signal to the antennas. [ENG] **1.** An instrument used to measure the angles between crystal faces. **2.** An instrument which uses x-ray diffraction to measure the angular positions of the axes of a crystal. **3.** Any instrument for measuring angles.

gonnardite [MINERAL] $Na_2CaAl_4Si_6O_{20}\cdot7H_2O$ Zeolite mineral occurring in fibrous, radiating spherules; specific gravity is 2.3.

Gonodactylidae [INV ZOO] A family of mantis shrimp in the order Stomatopoda.

Gonorhynchiformes [VERT ZOO] A small order of soft-rayed teleost fishes having fusiform or moderately compressed bodies, single short dorsal and anal fins, a forked caudal fin, and weak toothless jaws.

gonorrhea [MED] A bacterial infection of man caused by the gonococcus (*Neisseria gonorrhoeae*) which invades the mucous membrane of the urogenital tract.

goose [VERT ZOO] The common name for a number of waterfowl in the subfamily Anatinae; they are intermediate in size and features between ducks and swans.

gooseberry [BOT] The common name for about six species of thorny, spreading bushes of the genus *Ribes* in the order Rosales, producing small, acidic, edible fruit.

gooseberry stone See grossularite.

gooseneck barnacle [INV ZOO] Any stalked barnacle, especially of the genus *Lepas*.

gopher [VERT ZOO] The common name for North American rodents composing the family Geomyidae. Also known as pocket gopher.

gorceixite [MINERAL] $BaAl_3(PO_4)_2(OH)_5\cdot H_2O$ A brown mineral composed of a hydrous basic phosphate of barium and aluminum.

Gordiidae [INV ZOO] A monogeneric family of worms in the order Gordioidea distinguished by a smooth cuticle.

Gordioidea [INV ZOO] An order of worms belonging to the Nematomorpha in which there is one ventral epidermal cord, a body cavity filled with mesenchymal tissue, and paired gonads.

gordonite [MINERAL] $MgAl_2(PO_4)_2(OH)_2 \cdot 8H_2O$ A colorless mineral composed of a hydrous basic phosphate of magnesium and aluminum.

gorge [ARCH] The entrance to a bastion. [GEOGR] A narrow passage between mountains or the walls of a canyon, especially one with steep, rocky walls. [OCEANOGR] A collection of solid matter obstructing a channel or a river, as an ice gorge.

Gorgonacea [INV ZOO] The horny corals, an order of the coelenterate subclass Alcyonaria; colonies are fanlike or featherlike with branches spread radially or oppositely in one plane.

Gorgonocephalidae [INV ZOO] A family of ophiuroid echinoderms in the order Phrynophiurida in which the individuals often have branched arms.

gorilla [VERT ZOO] *Gorilla gorilla.* An anthropoid ape, the largest living primate; the two African subspecies are the lowland gorilla and the mountain gorilla.

goslarite [MINERAL] $ZnSO_4 \cdot 7H_2O$ A white mineral composed of hydrous zinc sulfate.

gossan [GEOL] A rusty, ferruginous deposit filling the upper regions of mineral veins and overlying a sulfide deposit; formed by oxidation of pyrites. Also known as capping; gozzan; iron hat.

Gotlandian [GEOL] A geologic time period recognized in Europe to include the Ordovician; it appears before the Devonian.

gouge [DES ENG] A curved chisel for wood, bone, stone, and so on. [GEOL] Soft, pulverized mixture of rock and mineral material found along shear (fault) zones and produced by the differential movement across the plane of slippage. [MIN ENG] A layer of soft material along the wall of a vein which favors miners by enabling them, after gouging it out with a pick, to attack the solid vein from the side.

gout [MED] A condition of purine metabolism resulting in increased blood levels of uric acid with ultimate deposition as urates in soft tissues around joints.

governor [MECH ENG] A device, especially one actuated by the centrifugal force of whirling weights opposed by gravity or by springs, used to provide automatic control of speed or power of a prime mover.

goyazite [MINERAL] $SrAl_3(PO_4)_2(OH)_5 \cdot H_2O$ A granular, yellowish-white mineral composed of a hydrous strontium aluminum phosphate.

gozzan *See* gossan.

gr *See* grain.

Graafian follicle [HISTOL] The mature mammalian ovum with its surrounding epithelial cells.

grab bucket [MECH ENG] A bucket with hinged jaws or teeth that is hung from cables on a crane or excavator and is used to dig and pick up materials.

graben [GEOL] A block of the earth's crust, generally with a length much greater than its width, that has dropped relative to the blocks on either side.

Gracilariidae [INV ZOO] A family of small moths in the superfamily Tineoidea; both pairs of wings are lanceolate and widely fringed.

gracilis [ANAT] A long slender muscle on the medial aspect of the thigh.

grade [CIV ENG] **1.** To prepare a roadway or other land surface of uniform slope. **2.** A surface prepared for the support of rails, a road, or a conduit. [COMMUN] One of two types of television service, designated grade A and grade B, each having a specified signal strength, that of grade A being several times as large as B. [ENG] The degree of strength of a high explosive. [GEOL] The slope of the bed of a stream, or of a surface over which water flows, upon which the current can just transport its load without either eroding or depositing. [MATH] A unit of plane angle, equal to 0.01 right angle, or $\pi/200$ radians, or $0.9°$. Also known as gon. [MIN ENG] **1.** A classification of ore according to recoverable amount of a valuable metal. **2.** To sort and classify diamonds.

grade crossing [CIV ENG] The intersection of roadways, railways, pedestrian walks, or combinations of these at grade.

graded bedding [GEOL] A stratification in which each stratum displays a gradation in the size of grains from coarse below to fine above.

graded-junction transistor *See* rate-grown transistor.

graded stream [HYD] A stream in which, over a period of years, slope is adjusted to yield the velocity required for transportation of the load supplied from the drainage basin.

grader [MECH ENG] A high-bodied, wheeled vehicle with a leveling blade mounted between the front and rear wheels; used for fine-grading relatively loose and level earth.

gradient current [OCEANOGR] A current defined by assuming that the horizontal pressure gradient in the sea is balanced by the sum of the Coriolis and bottom frictional forces; at some distance from the bottom the effect of friction becomes negligible, and above this the gradient and geostrophic currents are equivalent. Also known as slope current.

gradient index optics [OPTICS] Optical systems with components whose refractive indexes vary continuously within the material used for the optical elements.

gradient microphone [ENG ACOUS] A microphone whose electrical response corresponds to some function of the difference in pressure between two points in space.

gradient wind [METEOROL] A wind for which Coriolis acceleration and the centripetal acceleration exactly balance the horizontal pressure force.

grading [GEOL] The gradual reduction of the land to a level surface; for example, erosion of land to base level by streams. [IND ENG] Segregating a product into a number of adjoining categories which often form a spectrum of quality. Also known as classification.

graduate [CHEM] A cylindrical vessel that is calibrated in fluid ounces or milliliters or both; used to measure the volume of liquids.

graft [BIOL] **1.** To unite to form a graft. **2.** A piece of tissue transplanted from one individual to another or to a different place on the same individual. **3.** An individual resulting from the grafting of parts. [BOT] To unite a scion to an understock so that the two grow together and continue development as a single plant without change in scion or stock.

graftonite [MINERAL] $(Fe,Mn,Ca)_3(PO_4)_2$ A salmon-pink mineral, crystallizing in the monoclinic system, and found as laminated intergrowths of triphylite; hardness is 5 on Mohs scale, and specific gravity is 3.7.

Graham-Cole test *See* cholecystography.

grahamite [GEOL] *See* mesosiderite. [MINERAL] A solid, jet-black hydrocarbon that occurs in veinlike masses; soluble in carbon disulfide and chloroform.

grain [BOT] **1.** A rounded, granular prominence on the back of a sepal. **2.** *See* cereal. **3.** *See* drupelet. [GEOL] The particles or discrete crystals that make up a sediment or rock. [GRAPHICS] A small particle of metallic silver remaining in a photographic emulsion after developing and fixing; these grains together form the dark areas of a photographic image. [HYD] The particles which make up settled snow, firn, and glacier ice. [MATER] **1.** The appearance and texture of wood due to the arrangement of constituent fibers. **2.** The woodlike appearance or texture of a rock, metal, or other material. **3.** The direction in which most fibers lie in a sample of paper, which corresponds with the way the paper was made on the manufacturing machine. [MECH] A unit of mass in the United States and United Kingdom, common to the avoirdupois, apothecaries', and troy systems, equal to 1/7000 of a pound, or to 6.479891×10^{-5} kilogram. Abbreviated gr. [ORD] A single piece of solid propellant, regardless of size or shape, used in a gun or rocket; a rocket grain is often very large and shaped to fit its requirements. [TEXT] The direction in a piece of fabric which is parallel with the selvage.

grain alcohol *See* ethanol.

grain boundary [MET] Surface between individual grains in a metal.

grain direction [COMPUT SCI] In character recognition, the arrangement of paper fibers in relation to a document's travel through a character reader.

grain growth [MET] Enlargement of grains in a metal, usually through heat treatment. [PETR] Enlargement of some individual crystals in a monomineralic rock, producing a coarser texture. Also known as germination.

grain size [GEOL] Average size of mineral particles composing a rock or sediment. [MET] Average size of grains in a metal expressed as average diameter, or grains per unit area or volume.

grain sorghum [BOT] *Sorghum bicolor.* A grass plant cultivated for its grain and to a lesser extent for forage. Also known as nonsaccharine sorghum.

gram [MECH] The unit of mass in the centimeter-gram-second system of units, equal to 0.001 kilogram. Abbreviated g; gm.

gram-atomic weight [CHEM] The atomic weight of an element expressed in grams, that is, the atomic weight on a scale on which the atomic weight of carbon-12 isotope is taken as 12 exactly.

gram-calorie *See* calorie.

gramenite *See* nontronite.

gram-equivalent weight [CHEM] The equivalent weight of an element or compound expressed in grams on a scale in which carbon-12 has an equivalent weight of 3 grams in those compounds in which its formal valence is 4.

gram-force [MECH] A unit of force in the centimeter-gram-second gravitational system, equal to the gravitational force on a 1-gram mass at a specified location. Abbreviated gf. Also known as fors; gram-weight; pond.

Graminales [BOT] The equivalent name for Cyperales.

Gramineae [BOT] The grasses, a family of monocotyledonous plants in the order Cyperales characterized by distichously arranged flowers on the axis of the spikelet.

graminicolous [ECOL] Living upon grass.

graminivorous [ZOO] Feeding on grasses.

gram-molecular volume [CHEM] The volume occupied by a gram-molecular weight of a chemical in the gaseous state at 0°C and 760 millimeters of pressure (101,325 newtons per square meter).

gram-molecular weight [CHEM] The molecular weight of compound expressed in grams, that is, the molecular weight on a scale on which the atomic weight of carbon-12 isotope is taken as 12 exactly.

gram-negative [MICROBIO] Of bacteria, decolorizing and staining with the counterstain when treated with Gram's stain.

gram-positive [MICROBIO] Of bacteria holding the color of the primary stain when treated with Gram's stain.

Gram's stain [MICROBIO] A differential bacteriological stain; a fixed smear is stained with a slightly alkaline solution of basic dye, treated with a solution of iodine in potassium iodide, and then with a neutral decolorizing agent, and usually counterstained; bacteria stain either blue (gram-positive) or red (gram-negative).

gram-weight *See* gram-force.

grand canonical ensemble [STAT MECH] A collection of systems of particles used to describe an individual system which is allowed to exchange both energy and particles with its environment.

grandfather [COMPUT SCI] A data set that is two generations earlier than the data set under consideration.

grandfather cycle [COMPUT SCI] The period during which records are kept but not used except to reconstruct other records which are accidentally lost.

grand unified field theory [PARTIC PHYS] A theory in which the strong, electromagnetic, and weak interactions become aspects of one interaction.

granite [PETR] A visibly crystalline plutonic rock with granular texture; composed of quartz and alkali feldspar with subordinate plagioclase and biotite and hornblende.

granite pegmatite *See* pegmatite.

granite porphyry *See* quartz porphyry.

granitic layer *See* sial.

granoblastic fabric [PETR] The texture of metamorphic rocks composed of equidimensional elements formed during recrystallization.

granodiorite [PETR] A visibly crystalline plutonic rock composed chiefly of sodic plagioclase, alkali feldspar, quartz, and subordinate dark-colored minerals.

granogabbro [PETR] Plutonic rock composed of quartz, basic plagioclase, potash-feldspar, and at least one ferromagnesian mineral; intermediate between a granite and a gabbro, and in a strict sense, a granodiorite with more than 50% boric plagioclase.

Grantiidae [INV ZOO] A family of calcareous sponges in the order Sycettida.

granular [SCI TECH] Having a grainy texture.

granular ice [HYD] Ice composed of many tiny, opaque, white or milky pellets or grains frozen together and presenting a rough surface; this is the type of ice deposited as rime and compacted as névé.

granular leukocyte *See* granulocyte.

granulation [ASTRON] The small "rice grain" markings on the sun's photosphere. Also known as photospheric granulation. [MED] **1.** Tiny red granules made of capillary loops in the base of an ulcer. **2.** Process of granular tissue formation around a focus of inflammation. [PL PATH] Dry, tasteless condition of citrus fruit due to hardening of the juice sacs when fruit is left on trees too late in the season. [SCI TECH] The state or process of reducing a material to grains or small particles.

granulocyte [HISTOL] A leukocyte containing granules in the cytoplasm. Also known as granular leukocyte; polymorph; polymorphonuclear leukocyte.

grape [BOT] The common name for plants of the genus *Vitis* characterized by climbing stems with cylindrical-tapering tendrils and polygamodioecious flowers; grown for the edible, pulpy berries.

grapefruit [BOT] *Citrus paradisi.* An evergreen tree with a well-rounded top cultivated for its edible fruit, a large, globose citrus fruit characterized by a yellow rind and white, pink, or red pulp.

grape sugar *See* dextrose.

graph [MATH] **1.** The planar object, formed from points and line segments between them, used in the study of circuits and networks. **2.** The graph of a function *f* is the set of all ordered pairs $[x, f(x)]$, where *x* is in the domain of *f*. **3.** *See* graphical representation.

graph follower *See* curve follower.

graphical representation [MATH] The plot of the points in the plane which constitute the graph of a given real function or a pictorial diagram depicting interdependence of variables. Also known as graph.

graphic arts [GRAPHICS] Those methods of applied arts used to form a visual end product that conveys information or is a decoration; methods include drawing, painting, photography, all types of printing, and bookmaking.

graphic display [ELECTR] The display of data in graphical form on the screen of a cathode-ray tube.

graphic recording instrument [ENG] An instrument that makes a graphic record of one or more quantities as a function of another variable, usually time.

graphics [COMMUN] **1.** In communications systems, an information mode in which a graphic system is used to reproduce intelligence; a variation of facsimile. **2.** Nonvoice analog information devices and modes such as facsimile, photographics, and television. [SCI TECH] **1.** The graphic media. **2.** The art of drawing a three-dimensional object on a two-dimensional surface according to mathematical rules of projection.

graphic tellurium *See* sylvanite.

graphic terminal [ELECTR] A cathode-ray-tube or other type of computer terminal capable of producing some form of line drawing based on data being processed by or stored in a computer.

Graphidaceae [BOT] A family of mosses formerly grouped with lichenized Hysteriales but now included in the order Lecanorales; individuals have true paraphyses.

Graphiolaceae [MYCOL] A family of parasitic fungi in the order Ustilaginales in which teleutospores are produced in a cuplike fruiting body.

graphite [MINERAL] A mineral consisting of a low-pressure allotropic form of carbon; it is soft, black, and lustrous and has a greasy feeling; it occurs naturally in hexagonal crystals or massive or can be synthesized from petroleum coke; hardness is 1–2 on Mohs scale, and specific gravity is 2.09–2.23; used in pencils, crucibles, lubricants, paints, and polishes. Also known as black lead; plumbago.

graphite-epoxy composite [MATER] A structural material composed of epoxy resins reinforced with graphite.

graphite-polyimide composite [MATER] A composite material utilizing graphite reinforcing fibers in a resin matrix.

graphitization [ORG CHEM] The formation of graphitelike material from organic compounds.

graphitizing [MET] Annealing cast iron to convert all or some of the combined carbon to graphitic carbon.

graphoepitaxy [CRYSTAL] The use of artificial surface relief structures to induce crystallographic orientation in thin films.

graph theory [MATH] **1.** The mathematical study of the structure of graphs and networks. **2.** The body of techniques used in graphing functions in the plane.

grapnel [DES ENG] An implement with claws used to recover a lost core, drill fittings, and junk from a borehole or for other grappling operations. Also known as grapple. [NAV ARCH] An anchor with four or five hooks used for dragging the bottom or for anchorage.

grapple *See* grapnel.

Grapsidae [INV ZOO] The square-backed crabs, a family of decapod crustaceans in the Brachyura.

Graptolithina [PALEON] A class of extinct colonial animals believed to be related to the class Pterobranchia of the Hemichordata.

Graptoloidea [PALEON] An order of extinct animals in the class Graptolithina including branched, planktonic forms described from black shales.

Graptozoa [PALEON] The equivalent name for Graptolithina.

graser *See* gamma-ray laser.

grass [BOT] The common name for all members of the family Gramineae; monocotyledonous plants having leaves that consist of a sheath which fits around the stem like a split tube, and a long, narrow blade. [ELECTR] Clutter due to circuit noise in a radar receiver, seen on an A scope as a pattern resembling a cross section of turf. Also known as hash.

grasserie [INV ZOO] A polyhedrosis disease of silkworms characterized by spotty yellowing of the skin and internal liquefaction. Also known as jaundice.

grasshopper [INV ZOO] The common name for a number of plant-eating orthopteran insects composing the subfamily Saltatoria; individuals have hindlegs adapted for jumping, and mouthparts adapted for biting and chewing.

grasshopper linkage [MECH ENG] A straight-line mechanism used in some early steam engines.

grassland [ECOL] Any area of herbaceous terrestrial vegetation dominated by grasses and graminoid species.

grate [ENG] A support for burning solid fuels; usually made of closely spaced bars to hold the burning fuel, while allowing combustion air to rise up to the fuel from beneath, and ashes to fall away from the burning fuel.

graticule [MAP] A network of lines representing the earth's parallels of latitude and meridians of longitude on a map, chart, or plotting sheet. [OPTICS] A scale at the focal plane of an optical instrument to aid in the measurement of objects.

grating [ELECTROMAG] **1.** An arrangement of fine, parallel wires used in waveguides to pass only a certain type of wave. **2.** An arrangement of crossed metal ribs or wires that acts as a reflector for a microwave antenna and offers minimum wind resistance. [SPECT] *See* diffraction grating.

grating spectrograph [SPECT] A grating spectroscope provided with a photographic camera or other device for recording the spectrum.

grating spectroscope [SPECT] A spectroscope which employs a transmission or reflection grating to disperse light, and usually also has a slit, a mirror or lenses to collimate the light sent through the slit and to focus the light dispersed by the grating into spectrum lines, and an eyepiece for viewing the spectrum.

grav *See* G.

gravel [GEOL] A loose or unconsolidated deposit of rounded pebbles, cobbles, or boulders.

Grave's disease *See* hyperthyroidism.

gravid [ZOO] **1.** Of the uterus, containing a fetus. **2.** Pertaining to female animals when carrying young or eggs.

Gravigrada [VERT ZOO] The sloths, a group of herbivorous xenarthran mammals in the order Edentata; members are completely hairy and have five upper and four lower prismatic teeth without enamel.

gravimeter [ENG] A highly sensitive weighing device used for relative measurement of the force of gravity by detecting small weight differences of a constant mass at different points on the earth. Also known as gravity meter.

gravimetric analysis [ANALY CHEM] That branch of quantitative analytical chemistry in which a desired constituent is converted, usually by precipitation or combustion, to a pure compound or element, of definite known composition, and is weighed; in a few cases a compound or element is formed which does not contain the constituent but bears a definite mathematical relationship to it.

gravimetric geodesy [GEOD] The science that utilizes measurements and characteristics of the earth's gravity field, as well as theories regarding this field, to deduce the shape of the earth and, in combination with arc measurements, the earth's size.

gravitation [PHYS] The mutual attraction between all masses in the universe. Also known as gravitational attraction.

gravitational acceleration [PHYS] The acceleration imparted to a body by the attraction of the earth; approximately equal to 980.7 cm/sec^2, or 32.2 ft/sec^2.

gravitational astronomy *See* celestial mechanics.

gravitational attraction *See* gravitation.

gravitational collapse [ASTRON] The implosion of a star or other astronomical body from an initial size to a size hundreds or thousands of times smaller.

gravitational constant [MECH] The constant of proportionality in Newton's law of gravitation, equal to the gravitational force between any two particles times the square of the distance between them, divided by the product of their masses. Also known as constant of gravitation.

gravitational convection *See* thermal convection.

gravitational field [MECH] The field in a region in space in which a test particle would experience a gravitational force; quantitatively, the gravitational force per unit mass on the particle at a particular point.

gravitational-field theory [RELAT] A theory in which gravity is treated as a field, as opposed to a theory in which the force acts instantaneously at a distance.

gravitational force [MECH] The force on a particle due to its gravitational attraction to other particles.

gravitational lens [ASTRON] A massive galaxy whose gravitational field focuses light from a distant quasar near or along its line of sight, giving a double or multiple image of the quasar.

gravitational potential [MECH] The amount of work which must be done against gravitational forces to move a particle of unit mass to a specified position from a reference position, usually a point at infinity.

gravitational radiation *See* gravitational wave.

gravitational red shift [RELAT] A displacement of spectral lines toward the red when the gravitational potential at the observer of the light is greater than at its source.

gravitational systems of units [MECH] Systems in which length, force, and time are regarded as fundamental, and the unit of force is the gravitational force on a standard body at a specified location on the earth's surface.

gravitational tide [OCEANOGR] An atmospheric tide due to gravitational attraction of the sun and moon.

gravitational wave [RELAT] A propagating gravitational field predicted by general relativity, which is produced by some change in the distribution of matter; it travels at the speed of light, exerting forces on masses in its path. Also known as gravitational radiation.

gravitometer *See* densimeter.

graviton [PHYS] A theoretically deduced particle postulated as the quantum of the gravitational field, having a rest mass and charge of zero and a spin of 2.

gravity [MECH] The gravitational attraction at the surface of a planet or other celestial body.

gravity chute [ENG] A gravity conveyor in the form of an inclined plane, trough, or framework that depends on sliding friction to control the rate of descent.

gravity-collapse structure *See* collapse structure.

gravity conveyor [ENG] Any unpowered conveyor such as a gravity chute or a roller conveyor, which uses the force of gravity to move materials over a downward path.

gravity dam [CIV ENG] A dam which depends on its weight for stability.

gravity fault *See* normal fault.

gravity flow [HYD] A form of glacier movement in which the flow of the ice results from the downslope gravitational component in an ice mass resting on a sloping floor.

gravity meter [ENG] **1.** U-tube-manometer type of device for direct reading of solution specific gravities in semimicro quantities. **2.** An electrical device for measuring variations in gravitation through different geologic formations; used in mineral exploration. **3.** *See* gravimeter.

gravity pendulum *See* pendulum.

gravity railroad [ENG] A cable railroad in which cars descend a slope by gravity and are hauled back up the slope by a stationary engine, or there may be two tracks with cars so connected that cars going down may help to raise the cars going up and thus conserve energy.

gravity separation [ENG] Separation of immiscible phases (gas-solid, liquid-solid, liquid-liquid, solid-solid) by allowing the denser phase to settle out under the influence of gravity; used in ore dressing and various industrial chemical processes.

gravity settling chamber [ENG] Chamber or vessel in which the velocity of heavy particles (solids or liquids) in a fluid stream is reduced to allow them to settle downward by gravity, as in the case of a dust-laden gas stream.

gravity tide [GEOPHYS] Cyclic motion of the earth's surface caused by interaction of gravitational forces of the moon, sun, and earth.

gravity wave [FL MECH] **1.** A wave at a gas-liquid interface which depends primarily upon gravitational forces, surface tension and viscosity being of secondary importance. **2.** A wave in a fluid medium in which restoring forces are provided primarily by buoyancy (that is, gravity) rather than by compression.

gravity yard *See* hump yard.

gravure printing [GRAPHICS] A process that prints from sunken or depressed surfaces or cups on a plate, in contrast to the raised printing surfaces of letterpress; the depth (and the area) of the depressed areas varies, thus yielding more or less ink on the paper.

gray antimony *See* antimonite; jamesonite.

gray blight [PL PATH] A fungus disease of tea caused by *Pestalotia (Pestalozzia) theae*, which invades the tissues and causes the formation of black dots on the leaves.

graybody [THERMO] An energy radiator which has a blackbody energy distribution, reduced by a constant factor, throughout the radiation spectrum or within a certain wavelength interval. Also known as nonselective radiator.

gray copper ore *See* tetrahedrite.

gray hematite *See* specularite.

gray iron [MET] Pig or cast iron in which the carbon not contained in pearlite is present in the form of graphitic carbon.

gray leaf spot [PL PATH] A fungus disease of tomatoes caused by *Stemphylium solani* and characterized by water-soaked brown spots on the leaves that become gray with age.

gray manganese ore *See* manganite.

gray matter [HISTOL] The part of the central nervous system composed of nerve cell bodies, their dendrites, and the proximal and terminal unmyelinated portions of axons.

gray scab [PL PATH] A fungus disease of willow caused by *Sphaceloma murrayae* and characterized by irregular raised leaf spots having grayish-white centers and dark-brown margins.

gray scale [OPTICS] A series of achromatic tones having varying proportions of white and black, to give a full range of grays between white and black; a gray scale is usually divided into 10 steps.

gray speck [PL PATH] A manganese-deficiency disease of oats characterized by light-green to grayish leaf spots, and later by the buff or light-brown discoloration of the blades.

graywacke [PETR] An argillaceous sandstone characterized by an abundance of unstable mineral and rock fragments and a fine-grained clay matrix binding the larger, sand-size detrital fragments.

graywall [PL PATH] A disease of tomatoes thought to be caused by excess sunlight and characterized by translucent grayish-brown streaks or blotches on the fruit and browning of the vascular strands.

grease [MATER] **1.** Rendered, inedible animal fat that is soft at room temperature and is obtained from lard, tallow, bone, raw animal fat, and other waste products. **2.** A lubricant in the form of a solid to semisolid dispersion of a thickening agent in a fluid lubricant, such as petroleum oil thickened with metallic soap.

grease spot [PL PATH] A fungus disease of turf grasses caused by *Pythium aphanidermatum* and characterized by spots that have a greasy border of blackened leaves and intermingled cottony mycelia. Also known as spot blight.

grease trap [CIV ENG] A trap in a drain or waste pipe to stop grease from entering a sewer system.

greasewood [BOT] Any plant of the genus *Sarcobatus*, especially *S. vermiculatus*, which is a low shrub that grows in alkali soils of the western United States.

greasy quartz *See* milky quartz.

great circle [GEOD] A circle, or near circle, described on the earth's surface by a plane passing through the center of the earth. [MATH] The circle on the Riemann sphere produced by a plane passing through the center of the sphere.

great-circle chart *See* gnomonic chart.

Greater Dog *See* Canis Major

greater omentum [ANAT] A fold of peritoneum that is attached to the greater curvature of the stomach and hangs down over the intestine and fuses with the mesocolon.

greatest common divisor [MATH] The greatest common divisor of integers $n_1, n_2, \ldots, n_k$ is the largest of all integers that divide each n_i. Abbreviated gcd. Also known as highest common factor (hcf).

Great Nebula of Orion *See* Orion Nebula.

grebe [VERT ZOO] The common name for members of the family Podicipedidae; these birds have legs set far posteriorly, compressed bladelike tarsi, individually broadened and lobed toes, and a rudimentary tail.

Greco-Latin square [STAT] An arrangement of combinations of two sets of letters (one set Greek, the other Roman) in a square array, in such a way that no letter occurs more than once in the array. Also known as orthogonal Latin square.

Greeffiellidae [INV ZOO] A superfamily of free-living nematodes in the order Desmoscolecoidea.

green [MET] Pertaining to an unsintered powder. [OPTICS] The hue evoked in an average observer by monochromatic radiation having a wavelength in the approximate range from 492 to 577 nanometers; however, the same sensation can be produced in a variety of other ways.

green algae [BOT] The common name for members of the plant division Chlorophyta.

green gland *See* antennal gland.

green gold [MET] A greenish alloy of gold obtained by using silver, silver and cadmium, or silver and copper as the alloying metal.

greenhouse effect [METEOROL] The effect of the earth's atmosphere in trapping heat from the sun; the atmosphere acts like a greenhouse.

Greenland spar *See* cryolite.

green laser [OPTICS] A gas laser using mercury and argon to generate a green line at 5225 angstroms, corresponding to the wavelength that is most readily transmitted through seawater.

green lead ore *See* pyromorphite.

green lumber [MATER] Freshly sawed lumber, before drying.

green manure [AGR] Herbaceous plant material plowed into the soil while still green.

green mold [MYCOL] Any fungus, especially *Penicillium* and *Aspergillus* species, that is green or has green spores.

greenockite [MINERAL] CdS A green or orange mineral that crystallizes in the hexagonal system; occurs as an earthy encrustation and is dimorphous with hawleyite. Also known as cadmium blende; cadmium ocher; xanthochroite.

green rosette [PL PATH] A virus disease of the peanut characterized by bunching and yellowing of the leaves with severe stunting of the plant.

green rot [PL PATH] A decay of fallen deciduous trees in which the wood is colored a malachite green by the fungus *Peziza aeruginosa*.

greensand [GEOL] A greenish sand consisting principally of grains of glauconite and found between the low-water mark and the inner mud line. [PETR] Sandstone composed of greensand with little or no cement.

greenschist [PETR] A schistose metamorphic rock with abundant chlorite, epidote, or actinolite present, giving it a green color.

green smut [PL PATH] A fungus disease of rice characterized by enlarged grains covered with a green powder consisting of conidia, and caused by *Ustilaginoidea virens*. Also known as false smut.

greenstick fracture [MED] An incomplete fracture of a long bone, seen in children; the bone is bent but splintered only on the convex side.

greenstone [MINERAL] *See* nephrite. [PETR] Any altered basic igneous rock which is green due to the presence of chlorite, hornblende, or epidote.

Greenwich apparent time [ASTRON] Local apparent time at the Greenwich meridian. Abbreviated GAT.

Greenwich civil time *See* Greenwich mean time.

Greenwich hour angle [ASTRON] Angular distance west of the Greenwich celestial meridian; the arc of the celestial equator, or the angle at the celestial pole, between the upper branch of the Greenwich celestial meridian and the hour circle of a point on the celestial sphere, measured westward from the Greenwich celestial meridian through 360°. Abbreviated GHA.

Greenwich mean time [ASTRON] Mean solar time at the meridian of Greenwich. Abbreviated GMT. Also known as Greenwich civil time; universal time; Z time; zulu time.

Greenwich meridian [GEOD] The meridian passing through Greenwich, England, and serving as the reference for Greenwich time; it also serves as the origin of measurement of longitude.

Gregarinia [INV ZOO] A subclass of the protozoan class Telosporea occurring principally as extracellular parasites of invertebrates.

Gregorian calendar [ASTRON] The calendar used for civil purposes throughout the world, replacing the Julian calendar and closely adjusted to the tropical year.

Gregorian telescope [OPTICS] A reflecting telescope having a paraboloidal mirror with a hole in the center and a small secondary (concave ellipsoidal) mirror placed beyond the focus of the primary mirror; light is reflected to the secondary mirror and back to an eyepiece at the hole; the telescope produces an erect image but a small field of view.

greisen [PETR] A pneumatolytically altered granite consisting of mainly quartz and a light-green mica.

grenatite *See* leucite; staurolite.

grid [COMPUT SCI] In optical character recognition, a system of two groups of parallel lines, perpendicular to each other, used to measure or specify character images. [DES ENG] A network of equally spaced lines forming squares, used for determining permissible locations of holes on a printed circuit board or a chassis. [ELEC] **1.** A metal plate with holes or ridges, used in a storage cell or battery as a conductor and a support for the active material. **2.** Any systematic network, such as of telephone lines or power lines. [ELECTR] An electrode located between the cathode and anode of an electron tube, which has one or more openings through which electrons or ions can pass, and serves to control the flow of electrons from cathode to anode. [MAP] A system of uniformly spaced perpendicular lines and horizontal lines running north and south, and east and west on a map, chart, or aerial photograph; used in locating points. [MED] *See* Potter-Bucky grid. [MIN ENG] Imaginary line used to divide the surface of an area when following a checkerboard placement of boreholes.

grid-anode transconductance *See* transconductance.

grid azimuth [NAV] In grid navigation, the angle in the plane of projection measured clockwise between a straight line and the central meridian of a plane-rectangular coordinate system.

grid battery *See* C battery.

grid bearing [NAV] In grid navigation, the angle in the plane of the projection between a line and a north-south grid line.

grid bias [ELECTR] The direct-current voltage applied between the control grid and cathode of an electron tube to establish the desired operating point. Also known as bias; C bias; direct grid bias.

grid-bias cell *See* bias cell.

grid blocking [ELECTR] **1.** Method of keying a circuit by applying negative grid bias several times cutoff value to the grid of a tube during key-up conditions; when the key is down, the blocking bias is removed and normal current flows through the keyed circuit. **2.** Blocking of capacitance-coupled stages in an amplifier caused by the accumulation of charge on the coupling capacitors due to grid current passed during the reception of excessive signals.

grid blocking capacitor *See* grid capacitor.

grid capacitor [ELECTR] A small capacitor used in the grid circuit of an electron tube to pass signal current while blocking the direct-currentanode voltage of the preceding stage. Also known as grid blocking capacitor; grid condenser.

grid circuit [ELECTR] The circuit connected between the grid and cathode of an electron tube.

grid condenser *See* grid capacitor.

grid conductance *See* electrode conductance.

grid-controlled mercury-arc rectifier [ELECTR] A mercury-arc rectifier in which one or more electrodes are employed exclusively to control the starting of the discharge. Also known as grid-controlled rectifier.

grid-controlled rectifier *See* grid-controlled mercury-arc rectifier.

grid current [ELECTR] Electron flow to a positive grid in an electron tube.

grid element [ELEC] A sinuous resistor used to heat a furnace, made of heavy wire, strap, or casting and suspended from refractory or stainless supports built into the furnace walls, floor, and roof.

grid heading [NAV] In grid navigation, the heading relative to grid north.

gridistor [ELECTR] Field-effect transistor which uses the principle of centripetal striction and has a multichannel structure, combining advantages of both field effect transistors and minority carrier injection transistors.

grid leak [ELECTR] A resistor used in the grid circuit of an electron tube to provide a discharge path for the grid capacitor and for charges built up on the control grid.

grid limiter [ELECTR] Limiter circuit which operates by limiting positive grid voltages by means of a large ohmic value resistor; as the exciting signal moves in a positive direction with respect to the cathode, current through the resistor causes an *IR* drop which holds the grid voltage essentially at cathode potential; during negative excursions no current flows in the grid circuit, so no voltage drop occurs across the resistor.

grid locking [ELECTR] Defect of tube operation in which the grid potential becomes continuously positive due to excessive grid emission.

grid meridian [NAV] One of the grid lines extending in a grid north-south direction.

grid navigation [NAV] A navigation system related to a reference grid instead of the true north for the measurement of angles.

grid north [NAV] In grid navigation, the northerly or zero direction indicated by the grid datum of directional reference.

grid-plate transconductance *See* transconductance.

grid pulsing [ELECTR] Circuit arrangement of a radio-frequency oscillator in which the grid of the oscillator is biased so negatively that no oscillation takes place even when full plate voltage is applied; pulsing is accomplished by removing this negative bias through the application of a positive pulse on the grid.

grid spectrometer [SPECT] A grating spectrometer in which a large increase in light flux without loss of resolution is achieved by replacing entrance and exit slits with grids consisting of opaque and transparent areas, patterned to have large transmittance only when the entrance grid image coincides with that of the exit grid.

grid transformer [ELECTR] Transformer to supply an alternating voltage to a grid circuit or circuits.

grid voltage [ELECTR] The voltage between a grid and the cathode of an electron tube.

Grignard reaction [ORG CHEM] A reaction between an alkyl or aryl halide and magnesium metal in a suitable solvent, usually absolute ether, to form an organometallic halide.

Grignard reagent [ORG CHEM] RMgX The organometallic halide formed in the Grignard reaction; an example is C_2H_5MgCl; it is useful in organic synthesis.

grillage [CIV ENG] A footing that consists of two or more tiers of closely spaced structural steel beams resting on a concrete block, each tier being at right angles to the one below.

grille [ENG ACOUS] An arrangement of wood, metal, or plastic bars placed across the front of a loudspeaker in a cabinet for decorative and protective purposes.

Grimmiales [BOT] An order of mosses commonly growing upon rock in dense tufts or cushions and having hygroscopic, costate, usually lanceolate leaves arranged in many rows on the stem.

grinding [ELECTR] 1. A mechanical operation performed on silicon substrates of semiconductors to provide a smooth surface for epitaxial deposition or diffusion of impurities. 2. A mechanical operation performed on quartz crystals to alter their physical size and hence their resonant frequencies. [MECH ENG] 1. Reducing a material to relatively small particles. 2. Removing material from a workpiece with a grinding wheel. [MIN ENG] The act or process of continuing to drill after the bit or core barrel is blocked, thereby crushing and destroying any core that might have been produced.

grinding mill [LAP] A lathe designed for lapidary work. [MECH ENG] A machine consisting of a rotating cylindrical drum, that reduces the size of particles of ore or other materials fed into it; three main types are ball, rod, and tube mills.

grinding ratio [MECH ENG] Ratio of the volume of ground material removed from the workpiece to the volume removed from the grinding wheel.

grinding stress [MECH] Residual tensile or compressive stress, or a combination of both, on the surface of a material due to grinding.

grinding wheel [DES ENG] A wheel or disk having an abrasive material such as alumina or silicon carbide bonded to the surface.

grindle *See* bowfin.

grindstone [ENG] A stone disk on a revolving axle, used for grinding, smoothing, and shaping.

griphite [MINERAL] $(Na,Al,Ca,Fe)_6Mn_4(PO_4)_5(OH)_4$ Mineral composed of a basic phosphate of sodium, calcium, iron, aluminum, and manganese.

grit [GEOL] 1. A hard, sharp granule, as of sand. 2. A coarse sand. [MATER] An abrasive material composed of angular grains. [PETR] A sandstone composed of angular grains of different sizes.

grit blasting [ENG] Surface treatment in which steel grit, sand, or other abrasive material is blown against an object to produce a roughened surface or to remove dirt, rust, and scale. Also known as sandblasting.

grit chamber [CIV ENG] A chamber designed to remove sand, gravel, or other heavy solids that have subsiding velocities or specific gravities substantially greater than those of the organic solids in waste water.

grizzly [ENG] 1. A coarse screen used for rough sizing and separation of ore, gravel, or soil. 2. A grating to protect chutes, manways, and winzes, in mines, or to prevent debris from entering a water inlet.

grizzly bear [VERT ZOO] The common name for a number of species of large carnivorous mammals in the genus *Ursus*, family Ursidae.

groin [ANAT] Depression between the abdomen and the thigh. [ARCH] The projecting edge at the intersection of two vaults. [CIV ENG] A barrier built out from a seashore or riverbank to protect the land from erosion and sand movements, among other functions. Also known as groyne; jetty; spur dike; wing dam.

Gromiida [INV ZOO] An order of protozoans in the subclass Filosia; the test, which is chitinous in some species and thin and somewhat flexible in others, is reinforced with sand grains or siliceous particles.

grommet [ENG] 1. A metal washer or eyelet. 2. A piece of fiber soaked in a packing material and used under bolt and nut heads to preserve tightness. [ORD] Device made of rope, plastic, rubber, or metal to protect the rotating band of projectiles.

gross anatomy [ANAT] Anatomy that deals with the naked-eye appearance of tissues.

gross errors [STAT] Errors that occur when a measurement process is subject occasionally to being very far off.

gross index [COMPUT SCI] The first of two indexes consulted to gain access to a record.

gross information content [COMMUN] Measure of the total information, redundant or otherwise, contained in a message; expressed as the number of bits, nits, or Hartleys required to transmit the message with specified accuracy over a noiseless medium without coding.

gross ton *See* ton.

gross tonnage [NAV ARCH] The total volume of the interior of a ship, measured in tons (units of 100 cubic feet).

grossular *See* grossularite.

grossularite [MINERAL] $Ca_3Al_2(SiO_4)_3$ The colorless or green, yellow, brown, or red end member of the garnet group, often occurring in contact-metamorphosis impure limestones. Also known as gooseberry stone; grossular.

gross weight [IND ENG] The weight of a vehicle or container when it is loaded with goods. Abbreviated gr wt.

grothite *See* sphene.

ground [AERO ENG] To forbid (an aircraft or individual) to fly, usually for a short time. [ELEC] 1. A conducting path, intentional or accidental, between an electric circuit or equipment and the earth, or some conducting body serving in place of the earth. Abbreviated gnd. Also known as earth (British usage); earth connection. 2. To connect electrical equipment to the earth or to some conducting body which serves in place of the earth.

[GEOL] 1. Any rock or rock material. 2. A mineralized deposit. 3. Rock in which a mineral deposit occurs. [NAV] To touch bottom or run aground; in a serious grounding, the vessel is said to strand.

ground cable [ELEC] A heavy cable connected to earth for the purpose of grounding electric equipment.

ground circuit [ELEC] A telephone or telegraph circuit part of which passes through the ground.

ground clutter [ELECTROMAG] Clutter on a ground or airborne radar due to reflection of signals from the ground or objects on the ground. Also known as ground flutter; ground return; land return; terrain echoes.

ground control [CIV ENG] Supervision or direction of all airport surface traffic, except an aircraft landing or taking off. [ENG] The marking of survey, triangulation, or other key points or system of points on the earth's surface so that they may be recognized in aerial photographs.

ground-controlled approach [NAV] Technique or procedures by which a ground controller directs an aircraft to reach a point from which a landing may be made; the controller utilizes information from various sensors, including radar.

ground-controlled approach radar [ENG] Ground radar system providing information by which aircraft approaches may be directed by radio communications. Abbreviated GCA radar.

ground-controlled intercept radar [ENG] A radar system by means of which a controller may direct an aircraft to make an interception of another aircraft. Abbreviated GCI radar.

ground controller [ENG] Aircraft controller stationed on the ground; a generic term, applied to the controller in ground-controlled approach, ground-controlled interception, and so on.

ground data equipment [ENG] Any device located on the ground that aids in obtaining space-position or tracking data (including computation function); reads out data telemetry, video, and so on, from payload instrumentation, or is capable of transmitting command and control signals to a satellite or space vehicle.

grounded-anode amplifier *See* cathode follower.

grounded-base amplifier [ELECTR] An amplifier that uses a transistor in a grounded-base connection.

grounded-base connection [ELECTR] A transistor circuit in which the base electrode is common to both the input and output circuits; the base need not be directly connected to circuit ground. Also known as common-base connection.

grounded-cathode amplifier [ELECTR] Electron-tube amplifier with a cathode at ground potential at the operating frequency, with input applied between control grid and ground, and with the output load connected between plate and ground.

grounded-collector connection [ELECTR] A transistor circuit in which the collector electrode is common to both the input and output circuits; the collector need not be directly connected to circuit ground. Also known as common-collector connection.

grounded-emitter amplifier [ELECTR] An amplifier that uses a transistor in a grounded-emitter connection.

grounded-emitter connection [ELECTR] A transistor circuit in which the emitter electrode is common to both the input and output circuits; the emitter need not be directly connected to circuit ground. Also known as common-emitter connection.

grounded-gate amplifier [ELECTR] Amplifier that uses thin-film transistors in which the gate electrode is connected to ground; the input signal is fed to the source electrode and the output is obtained from the drain electrode.

grounded-grid amplifier [ELECTR] An electron-tube amplifier circuit in which the control grid is at ground potential at the operating frequency; the input signal is applied between cathode and ground, and the output load is connected between anode and ground.

grounded ice *See* stranded ice.

grounded-plate amplifier *See* cathode follower.

ground-effect machine *See* air-cushion vehicle.

ground electrode [ELEC] A conductor buried in the ground, used to maintain conductors connected to it at ground potential and dissipate current conducted to it into the earth, or to provide a return path for electric current in a direct-current power transmission system.

ground fault interrupter [ELEC] A fast-acting circuit breaker that also senses very small ground fault currents such as might flow through the body of a person standing on damp ground while touching a hot alternating-current line wire.

ground flutter *See* ground clutter.

ground fog [METEOROL] A fog that hides less than 0.6 of the sky and does not extend to the base of any clouds that may lie above it.

ground glass [OPTICS] A sheet of matte-surfaced glass on the back of a view camera or process camera so that the image of the subject can be focused on it; it is exactly in the film plane.

ground handling equipment *See* ground support equipment.

groundhog *See* barney.

ground ice [HYD] 1. A body of clear ice in frozen ground, most commonly found in more or less permanently frozen ground (permafrost), and may be of sufficient age to be termed fossil ice. Also known as stone ice; subsoil ice; subterranean ice; underground ice. 2. *See* anchor ice.

grounding [ELEC] Intentional electrical connection to a reference conducting plane, which may be earth, but which more generally consists of a specific array of interconnected electrical conductors referred to as the grounding conductor.

grounding receptacle [ELEC] A receptacle which has an extra contact that accepts the third round or U-shaped prong of a grounding attachment plug and is connected internally to a supporting strap, providing a ground both through the outlet box and the grounding conductor, armor, or raceway of the wiring system.

ground junction *See* grown junction.

ground layer *See* surface boundary layer.

ground loop [COMMUN] Return currents or magnetic fields from relatively high-powered circuits or components which generate unwanted noisy signals in the common return of relatively low-level signal circuits. [ELEC] Potentially detrimental loop formed when two or more points in an electric system that are nominally at ground potential are connected by a conducting path.

groundmass *See* matrix.

ground moraine [GEOL] Rock material carried and deposited in the base of a glacier. Also known as bottom moraine; subglacial moraine.

ground noise [ENG ACOUS] The residual system noise in the absence of the signal in recording and reproducing; usually caused by inhomogeneity in the recording and reproducing media, but may also include tube noise and noise generated in resistive elements in the amplifier system. [GEOPHYS] In seismic exploration, disturbance of the ground due to some cause other than the shot.

ground outlet [ELEC] Outlet equipped with a receptacle of the polarity type having, in addition to the current-carrying contacts, one grounded contact which can be used for the connection of an equipment-grounding conductor.

ground plane antenna [ELECTROMAG] Vertical antenna combined with a grounded horizontal disk, turnstile element, or similar ground plane simulation; such antennas may be mounted several wavelengths above the ground, and provide a low radiation angle.

ground-position indicator [NAV] An airborne computing mechanism that provides a continuous indication of position through data derived from instruments which give heading, airspeed, elapsed time, and drift.

ground pressure [GEOPHYS] The pressure to which a rock formation is subjected by the weight of the superimposed rock and rock material or by diastrophic forces created by movements in the rocks forming the earth's crust. Also known as geostatic pressure; lithostatic pressure; rock pressure.

ground protection [ELEC] Protection provided a circuit by a device which opens the circuit when a fault to ground occurs.

ground return [ELEC] Use of the earth as the return path for a transmission line. [ELECTROMAG] 1. An echo received from the ground by an airborne radar set. 2. *See* ground clutter.

ground rod [ELEC] A rod that is driven into the earth to provide a good ground connection.

groundscatter propagation [COMMUN] Multihop ionospheric radio propagation along other than the great-circle path between transmitting and receiving stations; radiation from the transmitter is first reflected back to earth from the ionosphere, then scattered in many directions from the earth's surface.

ground streamer [METEOROL] An upward advancing column of high-ion density which rises from a point on the surface of the earth toward which a stepped leader descends at the start of a lightning discharge.

ground substance *See* matrix.

ground support equipment [AERO ENG] That equipment on the ground, including all implements, tools, and devices (mobile or fixed), required to inspect, test, adjust, calibrate, appraise, gage, measure, repair, overhaul, assemble, disassemble, transport, safeguard, record, store, or otherwise function in support of a rocket, space vehicle, or the like, either in the research and development phase or in an operational phase, or in support of the guidance system used with the missile, vehicle, or the like. Abbreviated GSE. Also known as ground handling equipment.

ground surveillance radar [ENG] A surveillance radar operated at a fixed point on the earth's surface for observation and control of the position of aircraft or other vehicles in the vicinity.

ground-to-cloud discharge [GEOPHYS] A lightning discharge in which the original streamer processes start upward from an object located on the ground.

ground-up read-only memory [COMPUT SCI] A read-only memory which is designed from the bottom up, and for which all fabrication masks used in the multiple mask process are custom-generated.

ground visibility [METEOROL] In aviation terminology, the horizontal visibility observed at the ground, that is, surface visibility or control-tower visibility.

groundwater [HYD] All subsurface water, especially that part that is in the zone of saturation.

groundwater decrement *See* groundwater discharge.

groundwater discharge [HYD] 1. Water released from the zone of saturation. 2. Release of such water. Also known as decrement; groundwater decrement; phreatic-water discharge.

groundwater increment *See* recharge.

groundwater level [HYD] 1. The level below which the rocks and subsoil are full of water. 2. *See* water table.

groundwater recharge *See* recharge.

groundwater replenishment *See* recharge.

groundwater surface *See* water table.

groundwater table *See* water table.

ground wave [COMMUN] A radio wave that is propagated along the earth and is ordinarily affected by the presence of the ground and the troposphere; includes all components of a radio wave over the earth except ionospheric and tropospheric waves. Also known as surface wave. [ORD] One of the waves formed in the ground by an explosion.

ground wire [CIV ENG] A small-gage, high-strength steel wire used to establish line and grade for air-blown mortar or concrete. Also known as alignment wire; screed wire. [ELEC] A conductor used to connect electric equipment to a ground rod or other grounded object.

group [CHEM] 1. A family of elements with similar chemical properties. 2. A combination of bonded atoms that behave as a single unit under certain conditions. [COMMUN] A subdivision containing a number of voice channels, either within a supergroup or separately, normally comprised of up to 12 voice channels occupying the frequency band 60–108 kilohertz; each voice channel may be multiplexed for teletypewriter operation, if required; the number of voice channels which may be simultaneously multiplexed for teletypewriter operation will vary according to equipment design. [MATH] A set G with an associative binary operation where $g_1 \cdot g_2$ always exists and is an element of G; each g has an inverse element g^{-1}, and G contains an identity element.

group bus [ELEC] A scheme of electrical connections for a generating station in which more than two feeder lines are supplied by two bus-selector circuit breakers which lead to a main bus and an auxiliary bus.

group busy tone [COMMUN] High tone connected to the jack sleeves of an outgoing trunk group as an indication that all trunks in the group are busy.

group code *See* systematic error-checking code.

group dynamics [PSYCH] A branch of social psychology which studies problems involving the structure of a group.

grouped records [COMPUT SCI] Two or more records placed together and identified by a single key, to save storage space or reduce access time.

group-indicate [COMPUT SCI] To print indicative information from only the first record of a group.

grouping circuits [COMMUN] Circuits used to interconnect two or more switchboard positions together, so that one operator may handle the several switchboard positions from one operator's set.

grouping of records [COMPUT SCI] Placing records together in a group to either conserve storage space or reduce access time.

group mark [COMPUT SCI] A character signaling the beginning or end of a group of data.

group modulation [COMMUN] Process by which a number of channels, already separately modulated to a specific frequency range, are again modulated to shift the group to another range.

group therapy [PSYCH] A technique of psychotherapy, consisting of a group of persons who discuss personal problems under the guidance of a therapist.

grouse [VERT ZOO] Any of a number of game birds in the family Tetraonidae having a plump body and strong, feathered legs.

grouser [ENG] A temporary pile or a heavy, iron-shod pole driven into the bottom of a stream to hold a drilling or dredging boat or other floating object in position. Also known as spud.

grout [MATER] 1. A fluid mixture of cement and water, or a mixture of cement, sand, and water. 2. Waste material of all sizes obtained in quarrying stone.

grout curtain [ENG] A row of vertically drilled holes filled with grout under pressure to form the cutoff wall under a dam, or to form a barrier around an excavation through which water cannot seep or flow.

groutite [MINERAL] $HMnO_2$ A mineral of the diaspore group, composed of manganese, hydrogen, and oxygen; it is polymorphous with manganite.

grove cell [ELEC] Primary cell, having a platinum electrode in an electrolyte of nitric acid within a porous cup, outside of which is a zinc electrode in an electrolyte of sulfuric acid; it normally operates on a closed circuit.

growler [ELEC] An electromagnetic device consisting essentially of two field poles arranged as in a motor, used for locating short-circuited coils in the armature of a generator or motor and for magnetizing or demagnetizing objects; a growling noise indicates a short-circuited coil. [OCEANOGR] A small piece of floating sea ice, usually a fragment of an iceberg or floeberg; it floats low in the water, and its surface often is heavily pitted; it often appears greenish in color.

grown-diffused transistor [ELECTR] A junction transistor in which the final junctions are formed by diffusion of impurities near a grown junction.

grown junction [ELECTR] A junction produced by changing the types and amounts of donor and acceptor impurities that are added during the growth of a semiconductor crystal from a melt. Also known as ground junction.

growth curve [MICROBIO] A graphic representation of the growth of a bacterial population in which the log of the number of bacteria or the actual number of bacteria is plotted against time. [NUCLEO] A curve showing how some quantity associated with a radioactive transformation or induced nuclear reaction increases with time.

growth hormone [BIOCHEM] 1. A polypeptide hormone secreted by the anterior pituitary which promotes an increase in body size. Abbreviated GH. 2. Any hormone that regulates growth in plants and animals.

growth spiral [CRYSTAL] A structure on a crystal surface, observed after growth, consisting of a growth step winding downward and outward in an Archimedean spiral which may be distorted by the crystal structure.

growth step [CRYSTAL] A ledge on a crystal surface, one or more lattice spacings high, where crystal growth can take place.

groyne *See* groin.

grub screw [DES ENG] A headless screw with a slot at one end to receive a screwdriver.

Gruidae [VERT ZOO] The cranes, a family of large, tall, cosmopolitan wading birds in the order Gruiformes.

Gruiformes [VERT ZOO] A heterogeneous order of generally cosmopolitan birds including the rails, coots, limpkins, button quails, sun grebes, and cranes.

Grüneisen constant [SOLID STATE] Three times the bulk modulus of a solid times its linear expansion coefficient, divided by its specific heat per unit volume; it is reasonably constant for most cubic crystals. Also known as Grüneisen gamma.

Grüneisen gamma *See* Grüneisen constant.

Grüneisen relation [SOLID STATE] The relation stating that the electrical resistivity of a very pure metal is proportional to a mathematical function which depends on the ratio of the temperature to a characteristic temperature.

Grus [ASTRON] A constellation, right ascension 22 hours, declination 45°S. Also known as Crane.

gr wt *See* gross weight.

Gryllidae [INV ZOO] The true crickets, a family of orthopteran insects in which individuals are dark-colored and chunky with long antennae and long, cylindrical ovipositors.

Grylloblattidae [INV ZOO] A monogeneric family of crickets in the order Orthoptera; members are small, slender, wingless insects with hindlegs not adapted for jumping.

Gryllotalpidae [INV ZOO] A family of North American insects in the order Orthoptera which live in sand or mud; they eat the roots of seedlings growing in moist, light soils.

GSC *See* gas-solid chromatography.

G scan *See* G scope.

G scope [ELECTR] A cathode-ray scope on which a single signal appears as a spot on which wings grow as the distance to the target is decreased, with bearing error as the horizontal coordinate and elevation angle error as the vertical coordinate. Also known as G indicator; G scan.

GSE *See* ground support equipment.

G star [ASTRON] A star of spectral type G; many metallic lines are seen in the spectra, with hydrogen and potassium being strong; G stars are yellow stars, with surface temperatures of 5500–4200 K for giants, 6000–5000 K for dwarfs.

guanine [BIOCHEM] $C_5H_5ON_5$ A purine base; occurs naturally as a fundamental component of nucleic acids.

guanophore *See* iridocyte.

guanylic acid [BIOCHEM] A nucleotide composed of guanine, a pentose sugar, and phosphoric acid and formed during the hydrolysis of nucleic acid. Abbreviated GMP. Also known as guanosine monophosphate; guanosine phosphoric acid.

guar [AGR] *Cyanopsis tetragonaloba*. A leguminous crop adapted to semiarid regions of the southwestern United States and Mexico. Also known as cluster bean.

guard band [ELECTR] A narrow frequency band provided between adjacent channels in certain portions of the radio spectrum to prevent interference between stations.

guard cell [BOT] Either of two specialized cells surrounding each stoma in the epidermis of plants; functions in regulating stoma size.

guardrail [CIV ENG] 1. A handrail. 2. A rail made of posts and a metal strip used on a road as a divider between lines of traffic in opposite directions or used as a safety barrier on curves. 3. A rail fixed close to the outside of the inner rail on railway curves to hold the inner wheels of a railway car on the rail. Also known as check rail; safety rail; slide rail.

guard relay [ELEC] Used in the linefinder circuit to make sure that only one linefinder can be connected to any line circuit when two or more line relays are operated simultaneously.

guard ring [ELEC] A ring-shaped auxiliary electrode surrounding one of the plates of a parallel-plate capacitor to reduce edge effects. [ELECTR] A ring-shaped auxiliary electrode used in an electron tube or other device to modify the electric field or reduce insulator leakage; in a counter tube or ionization chamber a guard ring may also serve to define the sensitive volume. [THERMO] A device used in heat flow experiments to ensure an even distribution of heat, consisting of a ring that surrounds the specimen and is made of a similar material.

guard shield [ELECTR] Internal floating shield that surrounds the entire input section of an amplifier; effective shielding is achieved only when the absolute potential of the guard is stabilized with respect to the incoming signal.

guard signal [COMPUT SCI] A signal used in digital-to-analog converters, analog-to-digital converters, or other converters which permits values to be read or converted only when the values are not changing, usually to avoid ambiguity error.

guard wire [ELEC] A grounded conductor placed beneath an overhead transmission line in order to ground the line, in case it breaks, before reaching the ground.

guava [BOT] *Psidium guajava.* A shrub or low tree of tropical America belonging to the family Myrtaceae; produces an edible, aromatic, sweet, juicy berry.

guayule [BOT] *Parthenium argentatum.* A subshrub of the family Compositae that is native to Mexico and the southwestern United States; it has been cultivated as a source of rubber.

gubernaculum [ANAT] A guiding structure, as the fibrous cord extending from the fetal testes to the scrotal swellings. [INV ZOO] **1.** A posterior flagellum of certain protozoans. **2.** A sclerotized structure associated with the copulatory spicules of certain nematodes.

gudmundite [MINERAL] FeSbS A silver-white to steel-gray orthorhombic mineral composed of a sulfide and antimonide of iron.

guest element *See* trace element.

Guiana Current [OCEANOGR] A current flowing northwestward along the northeastern coast of South America.

guidance [NAV] The process of directing the movements of any vehicle, especially an aeronautical vehicle or space vehicle, with particular reference to the selection of a course or flight path.

guidance system [AERO ENG] The control devices used in guidance of an aircraft or spacecraft. [NAV] Apparatus for generating and detecting the path along which a vehicle or craft is guided, often remotely and automatically.

guidance tape [COMPUT SCI] A magnetic tape or punched paper tape that is placed in a missile or its computer to program desired events during flight.

guide edge [COMPUT SCI] The edge of a paper or magnetic tape, punch card, printed sheet, or other medium used to properly align its position.

guided missile [ORD] An unmanned self-propelled vehicle, with or without a warhead, which is designed to move in a trajectory or flight path all or partially above the earth's surface and whose trajectory or course, while in flight, is capable of being directed by remote control, by homing systems, or by inertial or programmed guidance from within; excludes drones, torpedoes, and rockets and other vehicles whose trajectory or course cannot be controlled while in flight.

guided propagation [COMMUN] Type of radio-wave propagation in which radiated rays are bent excessively by refraction in the lower layers of the atmosphere; this bending creates an effect much as if a duct or waveguide has been formed in the atmosphere to guide part of the radiated energy over distances far beyond the normal range. Also known as trapping.

guided wave [ELECTROMAG] A wave whose energy is concentrated near a boundary or between substantially parallel boundaries separating materials of different properties and whose direction of propagation is effectively parallel to these boundaries; waveguides transmit guided waves.

guide holes [COMPUT SCI] One or more rows of holes prepunched around the margins of hand-sorted cards.

guide margin [COMPUT SCI] The distance between the guide edge of a paper tape and the center line of the track of holes that is closest to this edge.

guide number [GRAPHICS] A number that relates the output of a light source (such as a flashbulb) to the sensitivity of a particular film; when the number is divided by the distance in feet to the subject, it gives the T stop at which the lens should be set.

guide rail [CIV ENG] A track or rail that serves to guide movement, as of a sliding door, window, or similar element.

guides [MECH ENG] **1.** Pulleys to lead a driving belt or rope in a new direction or to keep it from leaving its desired direction. **2.** Tracks that support and determine the path of a skip bucket and skip bucket bail. **3.** Tracks guiding the chain or buckets of a bucket elevator. **4.** The runway paralleling the path of the conveyor which limits the conveyor or parts of a conveyor to movement in a defined path. [MIN ENG] **1.** Steel, wood, or steel-wire rope conductors in a mine shaft to guide the movement of the cages. **2.** Timber, rope, or metal tracks in a hoisting shaft, which are engaged by shoes on the cage or skip so as to steady it in transit. **3.** The holes in a crossbeam through which the stems of the stamps in a stamp mill rise and fall.

guildite [MINERAL] $(Cu,Fe)_3(Fe,Al)_4(SO_4)_7(OH)_4 \cdot 15H_2O$ A dark-brown mineral composed of a basic hydrated sulfate of copper, iron, and aluminum.

Guillain-Barré syndrome *See* Landry-Guillain-Barré syndrome.

Guinea Current [OCEANOGR] A current flowing eastward along the southern coast of northwestern Africa into the Gulf of Guinea.

guinea fowl [VERT ZOO] The common name for plump African game birds composing the family Numididae; individuals have few feathers on the head and neck, but may have a crest of feathers and various fleshy appendages.

guinea pig [VERT ZOO] The common name for several species of wild and domestic hystricomorph rodents in the genus *Cavia,* family Caviidae; individuals are stocky, short-eared, short-legged, and nearly tailless.

guinea worm [INV ZOO] *Dracunculus medinensis.* A parasitic nematode that infects the subcutaneous tissues of humans and other mammals.

guitarfish [VERT ZOO] The common name for fishes composing the family Rhinobatidae.

guitermanite [MINERAL] $Pb_{10}Ar_6S_{19}$ A bluish-gray mineral composed of lead, arsenic, and sulfur, occurring in compact masses.

gular [ANAT] Of, pertaining to, or situated in the gula or upper throat. [VERT ZOO] A horny shield on the plastron of turtles.

gulch [GEOGR] A gulley, sometimes occupied by a torrential stream.

Gulf Stream [OCEANOGR] A relatively warm, well-defined, swift, relatively narrow, northward-flowing ocean current which originates north of Grand Bahama Island where the Florida Current and the Antilles Current meet, and which eventually becomes the eastward-flowing North Atlantic Current.

Gulf Stream Countercurrent [OCEANOGR] 1. A surface current opposite to the Gulf Stream, one current component on the Sargasso Sea side and the other component much weaker, on the inshore side. 2. A predicted, but as yet unobserved, large current deep under the Gulf Stream but opposite to it.

Gulf Stream meander [OCEANOGR] One of the changeable, winding bends in the Gulf Stream; such bends intensify as the Gulf Stream merges into North Atlantic Drift and break up into detached eddies at times, at about 40°N.

Gulf Stream system [OCEANOGR] The Florida Current, Gulf Stream, and North Atlantic Current, collectively.

gull [VERT ZOO] The common name for a number of long-winged swimming birds in the family Laridae having a stout build, a thick, somewhat hooked bill, a short tail, and webbed feet.

gullet [ANAT] See esophagus. [INV ZOO] A canal between the cytostome and reservoir that functions in food intake in ciliates.

gum ammoniac See ammoniac.

gumbo [BOT] See okra. [GEOL] A soil that forms a sticky mud when wet.

gummite [MINERAL] Any of various yellow, orange, red, or brown secondary minerals containing hydrous oxides of uranium, thorium, and lead. Also known as uranium ocher.

gummosis [PL PATH] Production of gummy exudates in diseased plants as a result of cell degeneration.

gum resin [MATER] A group of oleoresinous substances obtained from plants; mixtures of true gums and resins less soluble in alcohol than natural resins; examples are rubber, gutta-percha, gamboge, myrrh, and olibanum; used to make certain pharmaceuticals.

gunmetal [MET] 1. Bronze composed of copper and tin in proportions of 9:1, formerly used to make cannons. 2. Any metal or alloy from which guns are made. 3. Any metal or alloy treated to give the appearance of black, tarnished copper-alloy gunmetal.

Gunn amplifier [ELECTR] A microwave amplifier in which a Gunn oscillator functions as a negative-resistance amplifier when placed across the terminals of a microwave source.

Gunn diode See Gunn oscillator.

Gunn effect [ELECTR] Development of a rapidly fluctuating current in a small block of a semiconductor (perhaps n-type gallium arsenide) when a constant voltage above a critical value is applied to contacts on opposite faces.

gunnel See gunwale.

Gunneraceae [BOT] A family of dicotyledonous terrestrial herbs in the order Haloragales, distinguished by two to four styles, a unilocular bitegmic ovule, large inflorescences with no petals, and drupaceous fruit.

Gunn oscillator [ELECTR] A microwave oscillator utilizing the Gunn effect. Also known as Gunn diode.

Gunter's chain [ENG] A chain 66 feet (20.1168 meters) long, consisting of 100 steel links, each 7.92 inches (20.1168 centimeters) long, joined by rings, which is used as the unit of length for surveying public lands in the United States. Also known as chain.

gunwale [NAV ARCH] The upper edge of the side of a boat. Also spelled gunnel.

Günz [GEOL] A European stage of geologic time, in the Pleistocene (above Astian of Pliocene, below Mindel); it is the first stage of glaciation of the Pleistocene in the Alps.

Günz-Mindel [GEOL] The first interglacial stage of the Pleistocene in the Alps, between Günz and Mindel glacial stages.

Gurney-Mott theory [CHEM] A theory of the photographic process that proposes a two-stage mechanism; in the first stage, a light quantum is absorbed at a point within the silver halide gelatin, releasing a mobile electron and a positive hole; these mobile defects diffuse to trapping sites (sensitivity centers) within the volume or on the surface of the grain; in the second stage, trapped (negatively charged) electron is neutralized by an interstitial (positively charged) silver ion, which combines with the electron to form a silver atom; the silver atom is capable of trapping a second electron, after which the process repeats itself, causing the silver speck to grow.

gust [METEOROL] A sudden, brief increase in the speed of the wind; it is of a more transient character than a squall and is followed by a lull or slackening in the wind speed.

gustation [PHYSIO] The act or the sensation of tasting.

gust tunnel [AERO ENG] A type of wind tunnel that has an enclosed space and is used to test the effect of gusts on an airplane model in free flight to determine how atmospheric gusts affect the flight of an airplane.

gut [ANAT] The intestine. [EMBRYO] The embryonic digestive tube. [GEOL] 1. A narrow water passage such as a strait. 2. A channel deeper that the surrounding water; generally formed by water in motion.

gutter [BUILD] A trough along the edge of the eaves of a building to carry off rainwater. [CIV ENG] A shallow trench provided beside a canal, bordering a highway, or elsewhere, for surface drainage. [GRAPHICS] In the pages of a book, the unprinted space or inner margin between the printed area and the binding. [MET] A groove along the periphery of a die impression to allow for excess flash during forging. [MIN ENG] A drainage trench cut along the side of a mine shaft to conduct the water back into a lodge or sump.

Guttiferae [BOT] A family of dicotyledonous plants in the order Theales characterized by extipulate leaves and conspicuous secretory canals or cavities in all organs.

Guttulinaceae [MICROBIO] A family of microorganisms in the Acrasiales characterized by simple fruiting structures with only slightly differentiated component cells containing little or no cellulose.

guy [ENG] A rope or wire securing a pole, derrick, or similar temporary structure in a vertical position.

guyot [GEOL] A seamount, usually deeper than 100 fathoms (180 meters), having a smooth platform top. Also known as tablemount.

Gymnarchidae [VERT ZOO] A monotypic family of electrogenic fishes in the order Osteoglossiformes in which individuals lack pelvic, anal, and caudal fins.

Gymnarthridae [PALEON] A family of extinct lepospondylous amphibians that have a skull with only a single bone representing the tabular and temporal elements of the primitive skull roof.

Gymnoascaceae [MYCOL] A family of ascomycetous fungi in the order Eurotiales including dermatophytes and forms that grow on dung, soil, and feathers.

Gymnoblastea [INV ZOO] A suborder of coelenterates in the order Hydroida comprising hydroids without protective cups around the hydranths and gonozooids.

Gymnocerata [INV ZOO] An equivalent name for Hydrocorisae.

Gymnocodiaceae [PALEOBOT] A family of fossil red algae.

Gymnodinia [INV ZOO] A suborder of flagellate protozoans in the order Dinoflagellida that are naked or have thin pellicles.

Gymnolaemata [INV ZOO] A class of ectoproct bryozoans possessing lophophores which are circular in basal outline and zooecia which are short, wide, and vaselike or boxlike.

Gymnonoti [VERT ZOO] An equivalent name for Cypriniformes.

Gymnophiona [VERT ZOO] An equivalent name for Apoda.

Gymnopleura [INV ZOO] A subsection of brachyuran decapod crustaceans including the primitive burrowing crabs with trapezoidal or elongate carapaces, the first pereiopods subchelate, and some or all of the remaining pereiopods flattened and expanded.

gymnosperm [BOT] The common name for members of the division Pinophyta; seed plants having naked ovules at the time of pollination.

Gymnospermae [BOT] The equivalent name for Pinophyta.

Gymnostomatida [INV ZOO] An order of the protozoan subclass Holotrichia containing the most primitive ciliates, distinguished by the lack of ciliature in the oral area.

Gymnotidae [VERT ZOO] The single family of the suborder Gymnotoidei; eel-shaped fishes having numerous vertebrae, and anus located far forward, and lacking pelvic and developed dorsal fins.

Gymnotoidei [VERT ZOO] A monofamilial suborder of actinopterygian fishes in the order Cypriniformes.

gynander [BIOL] A mosaic individual composed of diploid female portions derived from both parents and haploid male portions derived from an extra egg or sperm nucleus.

gynandry [PHYSIO] A form of pseudohermaphroditism in which the external sexual characteristics are partly or wholly of the male aspect, but internal female genitalia are present. Also known as female pseudohermaphroditism; virilism.

gynecology [MED] The branch of the medical science dealing with diseases of women, particularly those affecting the sex organs.

gynecomastia [MED] Abnormal enlargement of the mammary glands in the male.

gynogenesis [EMBRYO] Development of a fertilized egg through the action of the egg nucleus, without participation of the sperm nucleus.

gypsite [GEOL] A variety of gypsum consisting of dirt and sand; found as an efflorescent deposit in arid regions, overlying gypsum. Also known as gypsum earth.

gypsum [MINERAL] $CaSO_4 \cdot 2H_2O$ A mineral, the commonest sulfate mineral; crystals are monoclinic, clear, white to gray, yellowish, or brownish in color, with well-developed cleavages; luster is subvitreous to pearly, hardness is 2 on Mohs scale, and specific gravity is 2.3; it is calcined at 190–200°C to produce plaster of paris.

gypsum earth *See* gypsite.

gypsy moth [INV ZOO] *Porthetria dispar.* A large lepidopteran insect of the family Lymantriidae that was accidentally imported into New England from Europe in the late 19th century; larvae are economically important as pests of deciduous trees.

Gyracanthidae [PALEON] A family of extinct acanthodian fishes in the suborder Diplacanthoidei.

gyration tensor [SOLID STATE] A tensor characteristic of an optically active crystal, whose product with a unit vector in the direction of propagation of a light ray gives the gyration vector.

gyration vector [OPTICS] For light propagating in an optically active medium, a vector whose cross product with the time derivative of the electric displacement vector gives a negative contribution to the electric field.

gyrator [ELECTROMAG] A waveguide component that uses a ferrite section to give zero phase shift for one direction of propagation and 180° phase shift for the other direction; in other words, it causes a reversal of signal polarity for one direction of propagation but not for the other direction. Also known as microwave gyrator.

gyrator filter [ELECTR] A highly selective active filter that uses a gyrator which is terminated in a capacitor so as to have an inductive input impedance.

gyratory breaker *See* gyratory crusher.

gyratory crusher [MECH ENG] A primary breaking machine in the form of two cones, an outer fixed cone and a solid inner erect cone mounted on an eccentric bearing. Also known as gyratory breaker.

gyre [OCEANOGR] A closed circulatory system, but larger than a whirlpool or eddy.

Gyrinidae [INV ZOO] The whirligig beetles, a family of large coleopteran insects in the suborder Adephaga.

Gyrinocheilidae [VERT ZOO] A monogeneric family of cypriniform fishes in the suborder Cyprinoidei.

gyro *See* gyroscope.

gyrocompass [NAV] A north-seeking form of gyroscope used as a vehicle's or craft's directional reference. Also known as gyroscopic compass.

Gyrocotylidae [INV ZOO] An order of tapeworms of the subclass Cestodaria; species are intestinal parasites of chimaeroid fishes and are characterized by an anterior eversible proboscis and a posterior ruffled adhesive organ.

Gyrocotyloidea [INV ZOO] A class of trematode worms according to some systems of classification.

Gyrodactyloidea [INV ZOO] A superfamily of ectoparasitic trematodes in the subclass Monogenea; the posterior holdfast is solid and is armed with central anchors and marginal hooks.

gyrodamper [AERO ENG] A single-gimbal control moment gyro actively controlled to extract the structural vibratory energy through the local rotational deformations of a structure; used for large space structures.

gyrofrequency *See* cyclotron frequency.

gyromagnetic compass [NAV] A magnetic compass in which gyroscopic stabilization is used to indicate direction.

gyromagnetic coupler [ELECTR] A coupler in which a single-crystal yig (yttrium iron garnet) resonator provides coupling at the required low signal levels between two crossed strip-line resonant circuits.

gyromagnetic effect [ELECTROMAG] The rotation induced in a body by a change in its magnetization, or the magnetization resulting from a rotation.

gyromagnetic ratio [PHYS] **1.** The ratio of the magnetic dipole moment to the angular momentum for a classical, atomic, or nuclear system. **2.** Occasionally, the reciprocal of the quantity in the first definition.

gyromagnetics [ELECTROMAG] The study of the relation between the angular momentum and the magnetization of a substance as exhibited in the gyromagnetic effect.

gyropendulum [MECH ENG] A gravity pendulum attached to a rapidly spinning gyro wheel.

Gyropidae [INV ZOO] A family of biting lice in the order Mallophaga; members are ectoparasites of South American rodents.

gyroplane [AERO ENG] A rotorcraft whose rotors are not power-driven.

gyroscope [ENG] An instrument that maintains an angular reference direction by virtue of a rapidly spinning, heavy mass; all applications of the gyroscope depend on a special form of Newton's second law, which states that a massive, rapidly spinning body rigidly resists being disturbed and tends to react to a disturbing torque by precessing (rotating slowly) in a direction at right angles to the direction of torque. Also known as gyro.

gyroscopic compass *See* gyrocompass.

gyroscopic couple [MECH ENG] The turning moment which opposes any change of the inclination of the axis of rotation of a gyroscope.

gyroscopic precession [MECH] The turning of the axis of spin of a gyroscope as a result of an external torque acting on the gyroscope; the axis always turns toward the direction of the torque.

gyrotron [ELECTR] **1.** A device that detects motion of a system by measuring the phase distortion that occurs when a vibrating tuning fork is moved. **2.** A type of microwave tube in which microwave amplification or generation results from cyclotron resonance coupling between microwave fields and an electron beam in vacuum. Also known as cyclotron resonance maser.

gyrus [ANAT] One of the convolutions (ridges) on the surface of the cerebrum.

gyttja [GEOL] A fresh-water anaerobic mud containing an abundance of organic matter; capable of supporting aerobic life.

H

H *See* henry; hydrogen.

ha *See* hectare.

habit [CRYSTAL] *See* crystal habit. [PSYCH] A repetitious behavior pattern.

habit plane [CRYSTAL] The crystallographic plane or system of planes along which certain phenomena such as twinning occur.

habituation [MED] **1.** A condition of tolerance to the effects of a drug or a poison, acquired by its continued use; marked by a psychic craving for it when the drug is withdrawn. **2.** Mild drug addiction in which withdrawal symptoms are not severe.

hachure [MAP] A short line used to denote slopes of the ground, as on maps.

hackberry [BOT] **1.** *Celtis occidentalis.* A tree of the eastern United States characterized by corky or warty bark, and by alternate, long-pointed serrate leaves unequal at the base; produces small, sweet, edible drupaceous fruit. **2.** Any of several other trees of the genus *Celtis.*

hackmanite [MINERAL] A mineral of the sodalite family containing a small amount of sulfur; fluoresces orange or red in ultraviolet light.

Hadamard's inequality [MATH] An inequality that gives an upper bound for the square of the absolute value of the determinant of a matrix in terms of the squares of the matrix entries; the upper bound is the product, over the rows of the matrix, of the sum of the squares of the absolute values of the entries in a row.

Hadar *See* β Centauri.

haddock [VERT ZOO] *Melanogrammus aeglefinus.* A fish of the family Gadidae characterized by a black lateral line and a dark spot behind the gills.

Hadley cell [METEOROL] A direct, thermally driven, and zonally symmetric circulation first proposed by George Hadley as an explanation for the trade winds; it consists of the equatorward movement of the trade winds between about latitude 30° and the equator in each hemisphere, with rising wind components near the equator, poleward flow aloft, and finally descending components at about latitude 30° again.

Hadromerina [INV ZOO] A suborder of sponges in the class Clavaxinellida having monactinal megascleres, usually with a terminal knob at one end.

hadron [PARTIC PHYS] An elementary particle which has strong interactions.

hadronic atom [ATOM PHYS] An atom consisting of a negatively charged, strongly interacting particle orbiting around an ordinary nucleus.

Haemosporina [INV ZOO] A suborder of sporozoan protozoans in the subclass Coccidia; all are parasites of vertebrates, and human malarial parasites are included.

hafnium [CHEM] A metallic element, symbol Hf, atomic number 72, atomic weight 178.49; melting point 2000°C, boiling point above 5400°C.

hagfish [VERT ZOO] The common name for the jawless fishes composing the order Myxinoidea.

Haidinger brushes [OPTICS] Faint yellow, brushlike patterns that are observed when a bright surface is viewed through a polarizer such as a rotating Nicol prism or sheet of Polaroid film; believed to be caused by birefringence of fibers at the fovea of the eye.

Haidinger fringes [OPTICS] Interference fringes produced by nearly normal incidence of light on thick, flat plates. Also known as constant-angle fringes; constant-deviation fringes.

haidingerite [MINERAL] $HCaAsO_4 \cdot H_2O$ A white mineral composed of hydrous calcium arsenate.

hail [METEOROL] Precipitation in the form of balls or irregular lumps of ice, always produced by convective clouds, nearly always cumulonimbus.

hailstone [METEOROL] A single unit of hail, ranging in size from that of a pea to that of a grapefruit, or from less than ¼ inch (6 millimeters) to more than 5 inches (13 centimeters) diameter; may be spheroidal, conical, or generally irregular in shape.

hair [ZOO] **1.** A threadlike outgrowth of the epidermis of animals, especially a keratinized structure in mammalian skin. **2.** The hairy coat of a mammal, or of a part of the animal.

hair cell [HISTOL] The basic sensory unit of the inner ear of vertebrates; a columnar, polarized structure with specialized cilia on the free surface.

hair cracks [MATER] Fine, random cracks in the surface of the top coat of paint or other coating material.

hair follicle [ANAT] An epithelial ingrowth of the dermis that surrounds a hair.

hair gland [ANAT] Sebaceous gland associated with hair follicles.

hair hygrometer [ENG] A hygrometer in which the sensing element is a bundle of human hair, which is held under slight tension by a spring and which expands and contracts with changes in the moisture of the surrounding air or gas.

hair pyrites *See* millerite.

hairspring *See* balance spring.

hairworm [INV ZOO] The common name for about 80 species of worms composing the class Nematomorpha.

Halacaridae [INV ZOO] A family of marine arachnids in the order Acarina.

halation [ELECTR] An area of glow surrounding a bright spot on a fluorescent screen, due to scattering by the phosphor or to multiple reflections at front and back surfaces of the glass faceplate. [OPTICS] A halo on a photographic image of a bright object caused by light reflected from the back of the film or plate.

Haldane's rule [GEN] The rule that if one sex in a first generation of hybrids is rare, absent, or sterile, then it is the heterogametic sex.

Halecomorphi [VERT ZOO] The equivalent name for Amiiformes.

Halecostomi [VERT ZOO] The equivalent name for Pholidophoriformes.

half-adder [ELECTR] A logic element which operates on two binary digits (but no carry digits) from a preceding stage, producing as output a sum digit and a carry digit.

half-adjust [COMPUT SCI] A rounding process in which the least significant digit is dropped and, if the least significant digit is one-half or more of the number base, one is added to the next more significant digit and all carries are propagated.

half-angle formulas [MATH] In trigonometry, formulas that express the trigonometric functions of half an angle in terms of trigonometric functions of the angle.

half block [COMPUT SCI] The unit of transfer between main storage and the buffer control unit; it consists of a column of 128 elements, each element 16 bytes long.

half-bridge [ELEC] A bridge having two power supplies, located in two of the bridge arms, to replace the single power supply of a conventional bridge.

half-cell [PHYS CHEM] A single electrode immersed in an electrolyte.

half-cell potential [PHYS CHEM] In electrochemical cells, the electrical potential developed by the overall cell reaction; can be considered, for calculation purposes, as the sum of the potential developed at the anode and the potential developed at the cathode, each being a half-cell.

half-life [CHEM] The time required for one-half of a given material to undergo chemical reactions. [NUCLEO] The average time interval required for one-half of any quantity of identical radioactive atoms to undergo radioactive decay. Also known as half-value period; radioactive half-life.

half line *See* ray.

half-moon [ASTRON] The moon as seen in the first quarter and the last, or third, quarter.

half nut [DES ENG] A nut split lengthwise so that it can be clamped around a screw.

half-period zones *See* Fresnel zones.

half-shift register [ELECTR] Logic circuit consisting of a gated input storage element, with or without an inverter.

half tide [OCEANOGR] The condition when the tide is at the level between any given high tide and the following or preceding low tide. Also known as mean tide.

half time [NUCLEO] The time during which half the radioactive material resulting from a nuclear explosion remains in the atmosphere.

halftone [GRAPHICS] An engraving used in printing to reproduce photographs and drawings that contain continuous tones, that is, grays (middle tones or half-tones) in addition to black and white; preparation involves photographing the artwork through a screen.

half-track tape recorder *See* double-track tape recorder.

half-value period *See* half-life.

half-wave [ELEC] Pertaining to half of one cycle of a wave. [ELECTROMAG] Having an electrical length of a half wavelength.

half-wave amplifier [ELECTR] A magnetic amplifier whose total induced voltage has a frequency equal to the power supply frequency.

half-wave antenna [ELECTROMAG] An antenna whose electrical length is half the wavelength being transmitted or received.

half-wave plate [OPTICS] A thin section of a doubly refracting crystal, of a thickness such that the ordinary and extraordinary components of normally incident light emerge from it with a phase difference corresponding to an odd number of half wavelengths.

half-word I/O buffer [COMPUT SCI] A buffer, the upper half being used to store the upper half of a word for both input and output characters, the lower half of the buffer being used for purposes such as the storage of constants.

halibut [VERT ZOO] Either of two large species of flatfishes in the genus *Hippoglossus;* commonly known as a right-eye flounder.

Halichondrida [INV ZOO] A small order of sponges in the class Demospongiae with a skeleton of diactinal or monactinal, siliceous megascleres (or both), a skinlike dermis, and small amounts of spongin.

Halictidae [INV ZOO] The halictid and sweat bees, a family of hymenopteran insects in the superfamily Apoidea.

halide [CHEM] A compound of the type MX, where X is fluorine, chlorine, iodine, bromine, or astatine, and M is another element or organic radical.

Haliplidae [INV ZOO] The crawling water beetles, a family of coleopteran insects in the suborder Adephaga.

halite [MINERAL] NaCl Native salt; an evaporite mineral occurring as isometric crystals or in massive, granular, or compact form. Also known as common salt; rock salt.

Halitheriinae [PALEON] A subfamily of extinct sirenian mammals in the family Dugongidae.

Hall coefficient [ELECTROMAG] A measure of the Hall effect, equal to the transverse electric field (Hall field) divided by the product of the current density and the magnetic induction. Also known as Hall constant.

Hall constant *See* Hall coefficient.

Hall effect [ELECTROMAG] The development of a transverse electric field in a current-carrying conductor placed in a magnetic field; ordinarily the conductor is positioned so that the magnetic field is perpendicular to the direction of current flow and the electric field is perpendicular to both.

Halley's Comet [ASTRON] A member of the solar system, with an orbit and a period of about 76 years; its nucleus has been estimated to be about 15 kilometers in radius; next due to appear in 1985–1986.

halliard *See* halyard.

Hall mobility [SOLID STATE] The product of conductivity and the Hall constant for a conductor or semiconductor; a measure of the mobility of the electrons or holes in a semiconductor.

halloysite [MINERAL] $Al_2Si_2O_5(OH)_4 \cdot 2H_2O$ Porcelainlike clay mineral whose composition is like that of kaolinite but contains more water and is structurally distinct; varieties are known as metahalloysites.

hallucination [PSYCH] A perception without an appropriate stimulus.

hallucinogen [PHARM] A substance, such as LSD, that induces hallucinations.

hallux [ANAT] The first digit of the hindlimb; the big toe of human.

halo [ASTRON] A type of ray system in which many short, filamentary streaks form a complex network of bright matter surrounding the lunar crater. Also known as nimbus. [ELECTR] An undesirable bright or dark ring surrounding an image on the fluorescent screen of a television cathode-ray tube; generally due to overloading or maladjustment of the camera tube. [GEOL] A ring or cresent surrounding an area of opposite sign; it is a diffusion of a high concentration of the sought mineral into surrounding ground or rock; it is encountered in mineral prospecting and in magnetic and geochemical surveys. [METEOROL] Any one of a large class of atmospheric optical phenomena which appear as colored or whitish rings and arcs about the sun or moon when seen through an ice crystal cloud or in a sky filled with falling ice crystals. [OPTICS] A ring around the photographic image of a bright source caused by light scattering in any one of a number of possible ways.

Halobacteriaceae [MICROBIO] A family of gram-negative, aerobic rods and cocci which require high salt (sodium chloride) concentrations for maintenance and growth.

halo blight [PL PATH] A bacterial blight of beans and sometimes other legumes caused by *Pseudomonas phaseolicola* and characterized by water-soaked lesions surrounded by a yellow ring on the leaves, stems, and pods.

halocarbon [ORG CHEM] A compound of carbon and a halogen, sometimes with hydrogen.

halocline [OCEANOGR] A well-defined vertical gradient of salinity in the oceans and seas.

Halocypridacea [INV ZOO] A superfamily of ostracods in the suborder Myodocopa; individuals are straight-backed with a very thin, usually calcified carapace.

haloform reaction [ORG CHEM] Halogenation of acetaldehyde or a methyl ketone in aqueous basic solution; the reaction is characteristic of compounds containing a CH_3CO group linked to a hydrogen or to another carbon.

halogen [CHEM] Any of the elements of the halogen family, consisting of fluorine, chlorine, bromine, iodine, and astatine.

halogenated hydrocarbon [ORG CHEM] One of a group of halogen derivatives of organic hydrogen- and carbon-containing compounds; the group includes monohalogen compounds (alkyl or aryl halides) and polyhalogen compounds that contain the same or different halogen atoms.

halogenation [ORG CHEM] A chemical process or reaction in which a halogen element is introduced into a substance, generally by the use of the element itself.

halogen mineral [MINERAL] Any of the naturally occurring compounds containing a halogen as the sole or principal anionic constituent.

halophile [BIOL] An organism that requires high salt concentrations for growth and maintenance.

halophyte [ECOL] A plant or microorganism that grows well in soils having a high salt content.

Haloragaceae [BOT] A family of dicotyledonous plants in the order Haloragales distinguished by an apical ovary of 2–4 loculi, small inflorescences, and small, alternate or opposite or whorled, exstipulate leaves.

Haloragales [BOT] An order of dicotyledonous plants in the subclass Rosidae containing herbs with perfect or often unisexual, more or less reduced flowers, and a minute or vestigial perianth.

Halosauridae [VERT ZOO] A family of mostly extinct deep-sea teleost fishes in the order Notacanthiformes.

halosere [ECOL] The series of communities succeeding one another, from the pioneer stage to the climax, and commencing in salt water or on saline soil.

halotrichite [MINERAL] 1. $FeAl_2(SO_4)_4 \cdot 22H_2O$ A mineral composed of hydrous sulfate of iron and aluminum. Also known as butter rock; feather alum; iron alum; mountain butter. 2. Any sulfate mineral resembling halotrichite in structure and habit.

halt [COMPUT SCI] The cessation of the execution of the sequence of operations in a computer program resulting from a halt instruction, hang-up, or interrupt.

haltere [INV ZOO] Either of a pair of capitate filaments representing rudimentary hindwings in Diptera. Also known as balancer.

halyard [NAV ARCH] A rope or tackle used to raise or lower a flag, sail, or spar. Also spelled halliard.

Halysitidae [PALEON] A family of extinct Paleozoic corals of the order Tabulata.

Hamamelidaceae [BOT] A family of dicotyledonous trees or shrubs in the order Hamamelidales characterized by united carpels, alternate leaves, perfect or unisexual flowers, and free filaments.

Hamamelidae [BOT] A small subclass of plants in the class Magnoliopsida having strongly reduced, often unisexual flowers with poorly developed or no perianth.

Hamamelidales [BOT] A small order of dicotyledonous plants in the subclass Hamamelidae characterized by vessels in the wood and a gynoecium consisting either of separate carpels or of united carpels that open at maturity.

hambergite [MINERAL] Be_2BO_3OH A grayish-white or colorless mineral composed of beryllium borate and occurring as prismatic crystals; hardness is 7.5 on Mohs scale, and specific gravity is 2.35.

Hamiltonian function [MECH] A function of the generalized coordinates and momenta of a system, equal in value to the sum over the coordinates of the product of the generalized momentum corresponding to the coordinate, and the coordinate's time derivative, minus the Lagrangian of the system; it is numerically equal to the total energy if the Lagrangian does not depend on time explicitly; the equations of motion of the system are determined by the functional dependence of the Hamiltonian on the generalized coordinates and momenta.

Hamiltonian operator *See* energy operator.

Hamilton-Jacobi equation [MATH] A particular partial differential equation useful in studying certain systems of ordinary equations arising in the calculus of variations, dynamics, and optics: $H(q_1, \ldots, q_n, \partial\phi/\partial q_1, \ldots, \partial\phi/\partial q_n, t) + \partial\phi/\partial t = 0$, where $q_1, \ldots, q_n$ are generalized coordinates, t is the time coordinate, H is the Hamiltonian function, and ϕ is a function that generates a transformation by means of which the generalized coordinates and momenta may be expressed in terms of new generalized coordinates and momenta which are constants of motion.

Hamilton-Jacobi theory [MATH] The study of the solutions of the Hamilton-Jacobi equation and the information they provide concerning solutions of the related systems of ordinary differential equations. [MECH] A theory that provides a means for discussing the motion of a dynamic system in terms of a single partial differential equation of the first order, the Hamilton-Jacobi equation.

Hamilton's equations of motion [MECH] A set of first-order, highly symmetrical equations describing the motion of a classical dynamical system, namely $\dot{q}_j = \partial H/\partial \dot{p}_j$, $p_j = -\partial H/\partial q_j$; here q_j ($j = 1, 2, \ldots$) are generalized coordinates of the system, p_j is the momentum conjugate to q_j, and H is the Hamiltonian. Also known as canonical equations of motion.

Hamilton's principle [MECH] A variational principle which states that the path of a conservative system in configuration space between two configurations is such that the integral of the Lagrangian function over time is a minimum or maximum relative to nearby paths between the same end points and taking the same time.

hammarite [MINERAL] $Pb_2Cu_2Bi_4S_9$ A monoclinic mineral whose color is a steel gray with red tone; consists of lead and copper sulfide.

hammer [DES ENG] 1. A hand tool used for pounding and consisting of a solid metal head set crosswise on the end of a handle. 2. An arm with a striking head for sounding a bell or gong. [MECH ENG] A power tool with a metal block or a drill for the head. [ORD] A metallic pivoted item in a firing mechanism designed to strike a firing pin or percussion cap and thus fire a gun.

hammer code See Hamming code.

hammer drill [MECH ENG] Any of three types of fast-cutting, compressed-air rock drills (drifter, sinker, and stoper) in which a hammer strikes rapid blows on a loosely held piston, and the bit remains against the rock in the bottom of the hole, rebounding slightly at each blow, but does not reciprocate.

hammer mill [MECH ENG] 1. A type of impact mill or crusher in which materials are reduced in size by hammers revolving rapidly in a vertical plane within a steel casing. Also known as beater mill. 2. A grinding machine which pulverizes feed and other products by several rows of thin hammers revolving at high speed.

hammertoe [MED] A condition of the toe, usually the second, in which the proximal phalanx is extremely extended while the two distal phalanges are flexed.

Hamming code [COMMUN] An error-correcting code used in data transmission. Also known as hammer code.

hamming distance See signal distance.

Hamoproteidae [INV ZOO] A family of parasitic protozoans in the suborder Haemosporina; only the gametocytes occur in blood cells.

ham radio See amateur radio.

hamster [VERT ZOO] The common name for any of 14 species of rodents in the family Cricetidae characterized by scent glands in the flanks, large cheek pouches, and a short tail.

hamstring muscles [ANAT] The biceps femoris, semitendinosus, and semimembranosus collectively.

hancockite [MINERAL] A complex silicate mineral containing lead, calcium, strontium, and other minerals; it is isomorphous with epidote.

hand [ANAT] The terminal part of the upper extremity modified for grasping. [TEXT] The quality or feel of a fabric.

handedness [PHYS] A division of objects, such as coordinate systems, screws, and circularly polarized light beams, into two classes (right and left), which distinguishes an object from a mirror image but not from a rotated object.

hand feed [ENG] A drill machine in which the rate at which the bit is made to penetrate the rock is controlled by a hand-operated ratchet and lever or a hand-turned wheel meshing with a screw mechanism.

hand-feed punch See hand punch.

hand forging [MET] Plastic deformation of a metal by manual force.

handhole [ENG] A shallow access hole large enough for a hand to be inserted for maintenance and repair of machinery or equipment.

hand lens See simple microscope.

hand level [ENG] A hand-held surveyor's level, basically a telescope with a bubble tube attached so that the position of the bubble can be seen when looking through the telescope.

hand punch [COMPUT SCI] A device which punches holes in punch cards and moves the cards as they are punched as a direct result of pressure on the keys of a keyboard, and which requires feeding and removal of cards by hand, one at a time. Also known as hand-feed punch. [DES ENG] A hand-held device for punching holes in paper or cards.

handrail [ENG] A narrow rail to be grasped by a person for support.

hand rule [ELECTROMAG] The rule that, when grasping the conductor in the right hand with the thumb pointing in the direction of the current, the fingers will then point in the direction of the lines of flux.

handset [DES ENG] A combination of a telephone-type receiver and transmitter, designed for holding in one hand.

hand time [IND ENG] The time necessary to complete a manual element. Also known as manual time.

hand truck [ENG] 1. A manually operated, two-wheeled truck consisting of a rectangular frame with handles at the top and a plate at the bottom to slide under the load. 2. Any of various small, manually operated, multiwheeled platform trucks for transporting materials.

hangar [CIV ENG] A building at an airport specially designed in height and width to enable aircraft to be stored or maintained in it.

hangar deck [NAV ARCH] A deck, below the flight deck of a carrier, where aircraft are parked and serviced.

hanger [CIV ENG] An iron strap which lends support to a joist beam or pipe. [GEOL] See hanging wall. [PETRO ENG] 1. A device to seat in the bowl of a lowermost casing head to suspend the next-smaller casing string and form a seal between the two. Also known as casing hanger. 2. A device to provide a seal between the tubing and the tubing head. Also known as tubing hanger.

hanging [GEOL] See hanging wall. [MET] Sticking or wedging of part of the charge in a blast furnace.

hanging-drop atomizer [MECH ENG] An atomizing device used in gravitational atomization; functions by quasi-static emission of a drop from a wetted surface. Also known as pendant atomizer.

hanging glacier [HYD] A glacier lying above a cliff or steep mountainside; as the glacier advances, calving can cause ice avalanches.

hanging load [MECH ENG] 1. The weight that can be suspended on a hoist line or hook device in a drill tripod or derrick without causing the members of the derrick or tripod to buckle. 2. The weight suspended or supported by a bearing.

hanging scaffold [CIV ENG] A movable platform suspended by ropes and pulleys; used by workers for above-ground building construction and maintenance.

hanging side See hanging wall.

hanging valley [GEOL] A valley whose floor is higher than the level of the shore or other valley to which it leads.

hanging wall [GEOL] The rock mass above a fault plane, vein, lode, ore body, or other structure. Also known as hanger; hanging; hanging side.

hangover [COMMUN] 1. In television, overlapping and blurring of successive frames opposite to direction of subject motion, due to improper adjustment of transient response. **2.** In facsimile, distortion produced when the signal changes from maximum to minimum conditions at a slower rate than required, resulting in tailing on the lines in the recorded copy. [MED] Aftereffect following excessive intake of alcohol or certain drugs, such as barbiturates.

hang-up [COMPUT SCI] A nonprogrammed stop in a computer routine caused by a human mistake or a computer malfunction. [ENG] A virtual leak resulting from the release of entrapped tracer gas from a leak detector vacuum system. [MIN ENG] Blockage of the movement of ore by rock in an underground chute.

Hankel functions [MATH] The Bessel functions of the third kind, occurring frequently in physical studies.

Hankel transform [MATH] The Hankel transform of order m of a real function $f(t)$ is the function $F(s)$ given by the integral from 0 to ∞ of $f(t)tJ_m(st)dt$, where J_m denotes the mth-order Bessel function. Also known as Bessel transform; Fourier-Bessel transform.

hanksite [MINERAL] $Na_{22}K(SO_4)_9(CO_3)_2Cl$ A white or yellow mineral crystallizing in the hexagonal system; found in California.

hannayite [MINERAL] $Mg_3(NH_4)_2H_2(PO_4)_4 \cdot 8H_2O$ Mineral composed of hydrous acid ammonium magnesium phosphate; occurs as yellow crystals in guano.

Hansen's disease [MED] An infectious disease of humans thought to be caused by *Mycobacterium leprae*; common manifestations are cutaneous and neural lesions. Also known as leprosy.

Haplodoci [VERT ZOO] The equivalent name for Batrachoidiformes.

haploid [GEN] Having half of the diploid or full complement of chromosomes, as in mature gametes.

Haplolepidae [PALEON] A family of Carboniferous chondrostean fishes in the suborder Palaeoniscoidei having a reduced number of fin rays and a vertical jaw suspension.

Haplomi [VERT ZOO] An equivalent name for Salmoniformes.

Haplosclerida [INV ZOO] An order of sponges in the class Demospongiae including species with a skeleton made up of siliceous megascleres embedded in spongin fibers or spongin cement.

Haplosporea [INV ZOO] A class of Protozoa in the subphylum Sporozoa distinguished by the production of spores lacking polar filaments.

Haplosporida [INV ZOO] An order of Protozoa in the class Haplosporea distinguished by the production of uninucleate spores that lack both polar capsules and filaments.

haplostele [BOT] A type of protostele with the core of xylem characterized by a smooth outline.

hapten [IMMUNOL] A simple substance that reacts like an antigen in vitro by combining with antibody; may function as an allergen when linked to proteinaceous substances of the tissue.

Haptophyceae [BOT] A class of the phylum Chrysophyta that contains the Coccolithophorida.

haptor [INV ZOO] The posterior organ of attachment in certain monogenetic trematodes characterized by multiple suckers and the presence of hooks.

harbor [GEOGR] Any body of water of sufficient depth for ships to enter and find shelter from storms or other natural phenomena. Also known as port.

hard coal See anthracite.

hard copy [COMPUT SCI] Human-readable typewritten or printed characters produced on paper at the same time that information is being keyboarded in a coded machine language, as when punching cards

or paper tape. [GRAPHICS] The typewritten or paper copy of material keyboarded into a computer, word processor, or typesetter.

hard data [SCI TECH] Data in the form of numbers or graphs, as opposed to qualitative information.

hard disk [COMPUT SCI] A magnetic disk made of rigid material.

hard-drawn wire [MET] Cold-drawn metal wire, usually of high tensile strength.

hardenability [MET] In a ferrous alloy, the property that determines the depth and distribution of hardness induced by quenching from elevated temperatures.

hardened steel [MET] Steel hardened by quenching from high temperatures.

hardener [MET] A master alloy added to a melt to control hardness. [ORG CHEM] Compound reacted with a resin polymer to harden it, such as the amines or anhydrides that react with epoxides to cure or harden them into plastic materials. Also known as curing agent.

hardening [MET] **1.** Imparting hardness to carbon steel by abrupt cooling (quenching) through a critical temperature range. **2.** Heat-treating an age-hardening or precipitation-hardening alloy at intermediate temperatures.

hard-face [MET] To apply a layer of hard, abrasion-resistant metal to a less resistant metal part by plating, welding, spraying, or other techniques. Also known as hard-surface.

hard freeze [HYD] A freeze in which seasonal vegetation is destroyed, the ground surface is frozen solid underfoot, and heavy ice is formed on small water surfaces such as puddles and water containers.

hard frost See black frost.

hardness [CHEM] The amount of calcium carbonate dissolved in water, usually expressed as parts of calcium carbonate per million parts of water. [ELECTROMAG] That quality which determines the penetrating ability of x-rays; the shorter the wavelength, the harder and more penetrating the rays. [ENG] Property of an installation, facility, transmission link, or equipment that will prevent an unacceptable level of damage. [MATER] Resistance of a metal or other material to indentation, scratching, abrasion, or cutting.

hardness number [ENG] A number representing the relative hardness of a mineral, metal, or other material as determined by any of more than 30 different hardness tests.

hardpan See caliche.

hard radiation [PHYS] Radiation whose particles or photons have a high energy and, as a result, readily penetrate all kinds of materials, including metals.

hard rime [HYD] Opaque, granular masses of rime deposited chiefly on vertical surfaces by a dense supercooled fog; it is more compact and amorphous than soft rime, and may build out into the wind as glazed cones or feathers.

hard rot [PL PATH] **1.** Any plant disease characterized by lesions with hard surfaces. **2.** A fungus disease of gladiolus caused by *Septoria gladioli* which produces hard-surfaced lesions on the leaves and corms.

hard rubber [MATER] Rubber that has been vulcanized at high temperatures and pressures to give hardness; used as an electrical insulating material and in tool handles. Also known as ebonite.

hard-solder See braze.

hard superconductor [CRYO] A superconductor that requires a strong magnetic field, over 1000 oersteds

(79,577 amperes per meter), to destroy superconductivity; niobium and vanadium are examples.

hard-surface [CIV ENG] To treat a ground surface in order to prevent muddiness. [MET] See hard-face.

hardware [COMPUT SCI] The physical, tangible, and permanent components of a computer or a data-processing system. [ENG] Items made of metal, such as tools, fittings, fasteners, and appliances. [ORD] Metal military items for use in combat.

hardware compatibility [COMPUT SCI] Property of two computers such that the object code from one machine can be loaded and executed on the other to produce exactly the same results.

hardware diagnostic [COMPUT SCI] A computer program designed to determine whether the components of a computer are operating properly.

hardware monitor [COMPUT SCI] A system used to evaluate the performance of computer hardware; it collects information such as central processing unit usage from voltage level sensors that are attached to the circuitry and measure the length of time or the number of times various signals occur, and displays this information or stores it on a medium that is then fed into a special data-reduction program.

hard water [CHEM] Water that contains certain salts, such as those of calcium or magnesium, which form insoluble deposits in boilers and form precipitates with soap.

hard-wire [ELEC] To connect electric components with solid, metallic wires as opposed to radio links, and the like.

hard-wired [COMPUT SCI] Having a fixed wired program or control system built in by the manufacturer and not subject to change by programming.

hardwood [MATER] Dense, close-grained wood of an angiospermous tree, such as oak, walnut, cherry, and maple.

hardystonite [MINERAL] $Ca_2ZnSi_2O_7$ A white mineral composed of zinc calcium silicate.

hare [VERT ZOO] The common name for a number of lagomorphs in the family Leporidae; they differ from rabbits in being larger with longer ears, legs, and tails.

Hare See Lepus.

harelip [MED] A congenital defect, sometimes hereditary, marked by an abnormal cleft between the upper lip and the base of the nose. Also known as cleft lip.

Harlechian [GEOL] A European stage of geologic time: Lower Cambrian.

harmonic [ACOUS] One of a series of sounds, each of which has a frequency which is an integral multiple of some fundamental frequency. [MATH] A solution of Laplace's equation which is separable in a specified coordinate system. [PHYS] A sinusoidal component of a periodic wave, having a frequency that is an integral multiple of the fundamental frequency. Also known as harmonic component.

harmonic analysis [MATH] A study of functions by attempting to represent them as infinite series or integrals which involve functions from some particular well-understood family; it subsumes studying a function via its Fourier series. [PHYS] Any method of identifying and evaluating the harmonics that make up a complex waveform of sound pressure, voltage, current, or some other varying quantity.

harmonic analyzer [ELECTR] An instrument that measures the strength of each harmonic in a complex wave. Also known as harmonic wave analyzer.

harmonic component See harmonic.

harmonic distortion [ELECTR] Nonlinear distortion in which undesired harmonics of a sinusoidal input signal are generated because of circuit nonlinearity.

harmonic folding [GEOL] Folding in the earth's surface, with no sharp changes with depth in the form of the folds.

harmonic frequency [PHYS] An integral multiple of the fundamental frequency of a periodic wave.

harmonic function [MATH] 1. A function of two real variables which is a solution of Laplace's equation in two variables. 2. A function of three real variables which is a solution of Laplace's equation in three variables.

harmonic generator [ELECTR] A generator operated under conditions such that it generates strong harmonics along with the fundamental frequency.

harmonic loss [ELECTROMAG] Energy loss in a generator due to space harmonics of the magnetomotive force produced by armature current, especially losses resulting from the fifth and seventh harmonics.

harmonic mean [MATH] For n positive numbers $x_1, x_2, \ldots, x_n$ their harmonic mean is the number $n/(1/x_1 + 1/x_2 + \ldots + 1/x_n)$.

harmonic motion [MECH] A periodic motion that is a sinusoidal function of time, that is, motion along a line given by the equation $x = a \cos(kt + \theta)$, where t is the time parameter, and a, k, and θ are constants. Also known as harmonic vibration; simple harmonic motion (SHM).

harmonic oscillator [ELECTR] See sinusoidal oscillator. [MECH] Any physical system that is bound to a position of stable equilibrium by a restoring force or torque proportional to the linear or angular displacement from this position. [PHYS] Anything which has equations of motion that are the same as the system in the mechanics definition. Also known as linear oscillator; simple oscillator.

harmonic progression [MATH] A sequence of numbers whose reciprocals form an arithmetic progression.

harmonic speed changer [MECH ENG] A mechanical-drive system used to transmit rotary, linear, or angular motion at high ratios and with positive motion.

harmonic vibration See harmonic motion.

harmonic wave [PHYS] A transverse waveform obtained by mapping onto a time base the periodic up and down excursions of simple harmonic motion.

harmonic wave analyzer See harmonic analyzer.

harmotome [MINERAL] $(K,Ba)(Al,Si)_2(Si_6O_{16}) \cdot 6H_2O$ A zeolite mineral with ion-exchange properties that forms cruciform twin crystals. Also known as cross-stone.

Harpacticoida [INV ZOO] An order of minute copepod crustaceans of variable form, but generally being linear and more or less cylindrical.

Harpidae [INV ZOO] A family of marine gastropod mollusks in the order Neogastropoda.

harrow [AGR] An implement that is pulled over plowed soil to break clods, level the surface, and destroy weeds.

harstigite [MINERAL] $Be_2Ca_3Si_3O_{11}$ A mineral composed of silicate of beryllium and calcium.

hartite [GEOL] A white, crystalline, fossil resin that is found in lignites. Also known as bombiccite; branchite; hofmannite; josen.

hartley [COMMUN] A unit of information content, equal to the designation of 1 of 10 possible and equally likely values or states of anything used to store or convey information.

Hartley oscillator [ELECTR] A vacuum-tube oscillator in which the parallel-tuned tank circuit is connected between grid and anode; the tank coil has an intermediate tap at cathode potential, so the grid-cathode portion of the coil provides the necessary feedback voltage.

Hartley principle [COMMUN] The principle that the total number of bits of information that can be transmitted over a channel in a given time is proportional to the product of channel bandwidth and transmission time.

Hartmann lines *See* Lüders lines.

Hartmann test [OPTICS] A test for telescope mirrors in which the mirror is covered with a screen with regularly spaced holes, and a photographic plate is placed near the focus; for a perfect mirror, this results in regularly spaced dots on the plate. [SPECT] A test for spectrometers in which light is passed through different parts of the entrance slit; any resulting changes of the spectrum indicate a fault in the instrument.

hartree [ATOM PHYS] A unit of energy used in studies of atomic spectra and structure, equal (in centimeter-gram-second units) to $4\pi^2 me^4/h^2$, where e and m are the charge and mass of the electron, and h is Planck's constant; equal to approximately 27.21 electron volts or 4.360×10^{-18} joule.

Hartree equation [ELECTR] An equation which gives the lowest anode voltage at which it is theoretically possible to maintain oscillation in the different modes of a magnetron.

Hartree-Fock approximation [QUANT MECH] A refinement of the Hartree method in which one uses determinants of single-particle wave functions rather than products, thereby introducing exchange terms into the Hamiltonian.

Hartree method [QUANT MECH] An iterative variational method of finding an approximate wave function for a system of many electrons, in which one attempts to find a product of single-particle wave functions, each one of which is a solution of the Schrödinger equation with the field deduced from the charge density distribution due to all the other electrons. Also known as self-consistent field method.

Hartree units [ATOM PHYS] A system of units in which the unit of angular momentum is Planck's constant divided by 2π, the unit of mass is the mass of the electron, and the unit of charge is the charge of the electron. Also known as atomic units.

harvest moon [ASTRON] A full moon that is seen nearest the autumnal equinox.

harzburgite [PETR] A peridotite consisting principally of olivine and orthopyroxene.

Hasche process [CHEM ENG] A thermal reforming process for hydrocarbon fuels; it is a noncatalytic regenerative method in which a mixture of hydrocarbon gas or vapor and air is passed through a regenerative mass that is progressively hotter in the direction of the gas flow; partial combustion occurs, liberating heat to crack the remaining hydrocarbons in a combustion zone.

hash [COMPUT SCI] Data which are obviously meaningless, caused by human mistakes or computer malfunction. Also known as garbage; gibberish. [ELEC] Electric noise produced by the contacts of a vibrator or by the brushes of a generator or motor. [ELECTR] *See* grass.

hashing [COMPUT SCI] 1. A direct addressing technique which derives the required address from a random number table. 2. Any computer operation which transforms one or more fields into a different arrangement which is usually more compact and easily manipulated. Also known as hash coding.

hash total [COMPUT SCI] A sum obtained by adding together numbers having different meanings; the sole purpose is to ensure that the correct number of data have been read by the computer.

hastate [BIOL] Shaped like an arrowhead with divergent barbs.

hastingsite [MINERAL] $NaCa_2(Fe,Mg)_5Al_2Si_6O_{22}(OH)_2$ A mineral of the amphibole group crystallizing in the monoclinic system and composed chiefly of sodium, calcium, and iron, but usually with some potassium and magnesium.

hatch [ENG] A door or opening, especially on an airplane, spacecraft, or ship.

hatchettine *See* hatchettite.

hatchettite [MINERAL] $C_{38}H_{78}$ A yellow-white mineral paraffin wax, melting at 55–65°C in the natural state and 79°C in the pure state; occurs in masses in ironstone nodules or in cavities in limestone. Also known as adipocerite; adipocire; hatchettine; mineral tallow; mountain tallow; naphthine.

hatchettolite *See* ellsworthite.

hatching [GRAPHICS] Parallel lines drawn on sections of plans for buildings or machines to distinguish between different materials.

hatchite [MINERAL] A lead-gray mineral composed of sulfide of lead and arsenic; occurs in triclinic crystals.

hauerite [MINERAL] MnS_2 A reddish-brown or brownish-black mineral composed of native manganese sulfide; occurs massive or in octahedral or pyritohedral crystals.

haughtonite [PETR] A black variety of biotite that is rich in iron.

haulage [MIN ENG] The movement, in cars or otherwise, of men, supplies, ore, and waste, underground and on the surface.

Hausdorff space [MATH] A topological space where each pair of distinct points can be enclosed in disjoint open neighborhoods.

hausmannite [MINERAL] Mn_3O_4 Brownish-black, opaque mineral composed of manganese tetroxide.

haustorium [BOT] 1. An outgrowth of certain parasitic plants which serves to absorb food from the host. 2. Food-absorbing cell of the embryo sac in nonparasitic plants.

Hauterivian [GEOL] A European stage of geologic time, in the Lower Cretaceous, above Valanginian and below Barremian.

haüyne [MINERAL] $(Na,Ca)_{4-8}(Al_6Si_6O_{24})(SO_4,S)_{1-2}$ An isometric silicate mineral of the sodalite group occurring as grains embedded in various igneous rocks; hardness is 5.5–6 on Mohs scale, and specific gravity is 2.4–2.5. Also known as haüynite.

haüynite *See* haüyne.

Hauzeur furnace [MET] A double furnace for the distillation of zinc wherein waste heat from one set of retorts is utilized for heating the second set.

Haversian canal [HISTOL] The central, longitudinal channel of an osteon containing blood vessels and connective tissue.

HA virus *See* hemadsorption virus.

hawk [ENG] A board with a handle underneath used by a workman to hold mortar. [VERT ZOO] Any of the various smaller diurnal birds of prey in the family Accipitridae; some species are used for hunting hare and partridge in India and other parts of Asia.

hawse [NAV ARCH] 1. The area in the bow of the ship where the hawsepipes are located. 2. The distance between the bow of a ship and the anchors.

hawsepipe [NAV ARCH] A pipe, made of heavy cast iron or steel, through which the anchor chain runs; placed in the ship's bow on each side of the stem, or in some cases also at the stern when a stern anchor is used.

hawser [NAV ARCH] A large rope or cable, usually over 5 inches (13 centimeters) in diameter, generally used to tow or moor a ship or secure it at a dock.

hay fever [MED] An allergic disorder of the nasal membranes and related structures due to sensitization by certain plant pollens. Also known as allergic rhinitis.

haze [METEOROL] Fine dust or salt particles dispersed through a portion of the atmosphere; the particles are so small that they cannot be felt, or individually seen with the naked eye, but they diminish horizontal visibility and give the atmosphere a characteristic opalescent appearance that subdues all colors. [OPTICS] The degree of cloudiness in a solution, cured plastic material, or coating material.

hazelnut See filbert.

h-bar [QUANT MECH] A fundamental constant equal to $h/2\pi$, where h is Planck's constant. Symbolized $\hbar$. Also known as Dirac h; h-line.

H beam [CIV ENG] A beam similar to the I beam but with longer flanges.

H bomb See hydrogen bomb.

hcf See greatest common divisor.

hcp structure See hexagonal close-packed structure.

He See helium.

head [ANAT] The region of the body consisting of the skull, its contents, and related structures. [ASTRON] See coma. [BUILD] The upper part of the frame on a door or window. [COMPUT SCI] A device that reads, records, or erases data on a storage medium such as a drum or tape; examples are a small electromagnet or a sensing or punching device. [ELECTR] The photoelectric unit that converts the sound track on motion picture film into corresponding audio signals in a motion picture projector. [ENG] The end section of a plastics blow-molding machine in which a hollow parison is formed from the melt. [ENG ACOUS] See cutter. [FL MECH] See pressure head. [GEOGR] See headland.

headboard [MIN ENG] 1. A wooden wedge placed against the hanging wall, and against which one end of the stull is jammed. 2. A board in the roof of a heading, contacting the earth above and supported by a headtree on each side.

header [BUILD] A framing beam positioned between trimmers and supported at each end by a tail beam. [CIV ENG] Brick or stone laid in a wall with its narrow end facing the wall. [COMMUN] The first section of a message, which contains information such as the addressee, routing, and origination time. [ELEC] A mounting plate through which the insulated terminals or leads are brought out from a hermetically sealed relay, transformer, transistor, tube, or other device. [ENG] A pipe, conduit, or chamber which distributes fluid from a series of smaller pipes or conduits; an example is a manifold. [MECH ENG] A machine used for gathering or upsetting materials; used for screw, rivet, and bolt heads. [MIN ENG] 1. An entry-boring machine that bores the entire section of the entry in one operation. 2. A rock that heads off or delays progress. 3. A blasthole at or above the head.

header card [COMPUT SCI] A card that contains supplemental information related to the data on the succeeding cards.

header record [COMPUT SCI] Computer input record containing common, constant, or identifying information for records that follow.

header-type boiler See straight-tube boiler.

headframe [MIN ENG] 1. The frame at the top of a shaft, on which is mounted the hoisting pulley. Also known as gallows; gallows frame; headstock; hoist frame. 2. The shaft frame, sheaves, hoisting arrangements, dumping gear, and connected works at the top of a shaft. Also known as headgear.

head gap [COMPUT SCI] The space between the read/write head and the recording medium, such as a disk in a computer.

head gate [CIV ENG] 1. A gate on the upstream side of a lock or conduit. 2. A gate at the starting point of an irrigation ditch.

headgear See headframe.

heading [NAV] 1. The horizontal direction in which a ship actually points or heads at any instant, expressed in angular units from a reference direction, usually from 0° at the reference direction clockwise through 360°. 2. In air navigation, the horizontal direction in which an aircraft points or heads, that is the direction of the longitudinal axis, measured as in the first definition.

heading side See footwall.

heading wall See footwall.

headland [GEOGR] 1. A high, steep-faced promontory extending into the sea. Also known as head; mull. 2. High ground surrounding a body of water.

head loss [FL MECH] The drop in the sum of pressure head, velocity head, and potential head between two points along the path of a flowing fluid, due to causes such as fluid friction.

head-per-track [COMPUT SCI] An arrangement having one read/write head for each magnetized track on a disk or drum to eliminate the need to move a single head from track to track.

headphone [ENG ACOUS] An electroacoustic transducer designed to be held against an ear by a clamp passing over the head, for private listening to the audio output of a communications, radio, or television receiver or other source of audio-frequency signals. Also known as phone.

heads [MIN ENG] 1. Material removed from the ore in the treatment plant and containing the valuable metallic constituents. 2. The feed to a concentrating system in ore dressing.

headset [ENG ACOUS] A single headphone or a pair of headphones, with a clamping strap or wires holding them in position.

head smut [PL PATH] A fungus disease of corn and sorghum caused by *Sphacelotheca reiliana* which destroys the head of the plant.

headstock [MECH ENG] 1. The device on a lathe for carrying the revolving spindle. 2. The movable head of certain measuring machines. 3. The device on a cylindrical grinding machine for rotating the work. [MIN ENG] See headframe.

headwall [CIV ENG] A retaining wall at the outlet of a drain or culvert. [GEOL] The steep cliff at the back of a cirque.

headwaters [HYD] The source and upstream waters of a stream.

hearing aid [ENG ACOUS] A miniature, portable sound amplifier for persons with impaired hearing, consisting of a microphone, audio amplifier, earphone, and battery.

heart [ANAT] The hollow muscular pumping organ of the cardiovascular system in vertebrates.

heartbeat [PHYSIO] Pulsation of the heart coincident with ventricular systole.

heart failure   *See* cardiac failure.
hearth   [BUILD] 1. The floor of a fireplace or brick oven. 2. The projection in front of a fireplace, made of brick, stone, or cement.   [MET] The floor of a reverberatory, open-hearth, cupola, or blast furnace; it is made of refractory material able to support the charge and to collect the molten products.
heart-lung machine   [MED] A machine through which blood is shunted to maintain circulation during heart surgery.
heart pacemaker   *See* pacemaker.
heartrot   [PL PATH] 1. A rot involving disintegration of the heartwood of a tree. 2. A fungus disease of beets and rutabagas caused by *Mycosphaerella tabifica* which results in decay of the central tissues of the plant. 3. A boron-deficiency disease of sugarbeets that causes rot. 4. A fatal disease of palms associated with a trypanosomatid flagellate *Phytomonas*. Also known as cedros wilt; marchitez sorpresiva.
heart valve   [ANAT] Flaps of tissue that prevent reflux of blood from the ventricles to the atria or from the pulmonary arteries or aorta to the atria.
heart worm   [INV ZOO] *Dirofilaria immitis*. A filarial nematode parasitic on dogs and other carnivores.
heat   [THERMO] Energy in transit due to a temperature difference between the source from which the energy is coming and a sink toward which the energy is going; other types of energy in transit are called work.
heat-activated battery   *See* thermal battery.
heat balance   [GEOPHYS] The equilibrium which exists on the average between the radiation received by the earth and atmosphere from the sun and that emitted by the earth and atmosphere.   [MET] The calculation used in fluidization roasting so that the addition or removal of heat can be controlled to maintain the optimum temperature in the reacting vessel.   [THERMO] The equilibrium which is known to exist when all sources of heat gain and loss for a given region or body are accounted for.
heat barrier   *See* thermal barrier.
heat budget   [GEOPHYS] Amount of heat needed to raise a lake's water from the winter temperature to the maximum summer temperature.   [THERMO] The statement of the total inflow and outflow of heat for a planet, spacecraft, biological organism, or other entity.
heat capacity   [THERMO] The quantity of heat required to raise a system one degree in temperature in a specified way, usually at constant pressure or constant volume. Also known as thermal capacity.
heat conduction   [THERMO] The flow of thermal energy through a substance from a higher- to a lower-temperature region.
heat conductivity   *See* thermal conductivity.
heat content   *See* enthalpy.
heat convection   [THERMO] The transfer of thermal energy by actual physical movement from one location to another of a substance in which thermal energy is stored. Also known as thermal convection.
heat cramps   [MED] Painful voluntary-muscle spasm and cramps following strenuous exercise, usually in persons in good physical condition, due to loss of sodium chloride and water from excessive sweating.
heat cycle   *See* thermodynamic cycle.
heat dump   *See* heat sink.
heat engine   [MECH ENG] A machine that converts heat into work (mechanical energy).   [THERMO] A thermodynamic system which undergoes a cyclic process during which a positive amount of work is done by the system; some heat flows into the system and a smaller amount flows out in each cycle.
heat equation   [THERMO] A parabolic second-order differential equation for the temperature of a substance in a region where no heat source exists: $\partial t/\partial \tau = (k/\rho c)(\partial^2 t/\partial x^2 + \partial^2 t/\partial y^2 + \partial t^2/\partial z^2)$, where x, y, and z are space coordinates, τ is the time, $t(x,y,z,\tau)$ is the temperature, k is the thermal conductivity of the body, ρ is its density, and c is its specific heat; this equation is fundamental to the study of heat flow in bodies. Also known as Fourier heat equation; heat flow equation.
heater   [ELECTR] An electric heating element for supplying heat to an indirectly heated cathode in an electron tube. Also known as electron-tube heater.   [ENG] A contrivance designed to give off heat.
heater oil   *See* heating oil.
heat exchange   [CHEM ENG] A unit operation based on heat transfer which functions in the heating and cooling of fluids with or without phase change.
heat exchanger   [ENG] Any device, such as an automobile radiator, that transfers heat from one fluid to another or to the environment. Also known as exchanger.
heat flow   [THERMO] Heat thought of as energy flowing from one substance to another; quantitatively, the amount of heat transferred in a unit time. Also known as heat transmission.
heat flow equation   *See* heat equation.
heating fuel   *See* heating oil.
heating load   [CIV ENG] The quantity of heat per unit time that must be provided to maintain the temperature in a building at a given level.
heating oil   [MATER] No. 2 fuel oil; used in domestic heating units. Also known as heater oil; heating fuel.
heating plant   [CIV ENG] The whole system for heating an enclosed space. Also known as heating system.
heating system   *See* heating plant.
heating value   *See* heat of combustion.
heat lamp   [ELEC] An infrared lamp used for brooders in farming, for drying paint or ink, for keeping food warm, and for therapeutic and other applications requiring heat with or without some visible light.
heat low   *See* thermal low.
heat of ablation   [THERMO] A measure of the effective heat capacity of an ablating material, numerically the heating rate input divided by the mass loss rate which results from ablation.
heat of activation   [PHYS CHEM] The increase in enthalpy when a substance is transformed from a less active to a more reactive form at constant pressure.
heat of adsorption   [THERMO] The increase in enthalpy when 1 mole of a substance is adsorbed upon another at constant pressure.
heat of association   [PHYS CHEM] Increase in enthalpy accompanying the formation of 1 mole of a coordination compound from its constituent molecules or other particles at constant pressure.
heat of combustion   [PHYS CHEM] The amount of heat released in the oxidation of 1 mole of a substance at constant pressure, or constant volume. Also known as heat value; heating value.
heat of compression   [THERMO] Heat generated when air is compressed.
heat of condensation   [THERMO] The increase in enthalpy accompanying the conversion of 1 mole of vapor into liquid at constant pressure and temperature.

heat of crystallization [THERMO] The increase in enthalpy when 1 mole of a substance is transformed into its crystalline state at constant pressure.

heat of dissociation [PHYS CHEM] The increase in enthalpy at constant pressure, when molecules break apart or valence linkages rupture.

heat of evaporation *See* heat of vaporization.

heat of formation [PHYS CHEM] The increase in enthalpy resulting from the formation of 1 mole of a substance from its elements at constant pressure.

heat of fusion [THERMO] The increase in enthalpy accompanying the conversion of 1 mole, or a unit mass, of a solid to a liquid at its melting point at constant pressure and temperature. Also known as latent heat of fusion.

heat of hydration [PHYS CHEM] The increase in enthalpy accompanying the formation of 1 mole of a hydrate from the anhydrous form of the compound and from water at constant pressure.

heat of ionization [PHYS CHEM] The increase in enthalpy when 1 mole of a substance is completely ionized at constant pressure.

heat of solidification [THERMO] The increase in enthalpy when 1 mole of a solid is formed from a liquid or, less commonly, a gas at constant pressure and temperature.

heat of solution [PHYS CHEM] The enthalpy of a solution minus the sum of the enthalpies of its components. Also known as integral heat of solution; total heat of solution.

heat of sublimation [THERMO] The increase in enthalpy accompanying the conversion of 1 mole, or unit mass, of a solid to a vapor at constant pressure and temperature. Also known as latent heat of sublimation.

heat of transformation [THERMO] The increase in enthalpy of a substance when it undergoes some phase change at constant pressure and temperature.

heat of vaporization [THERMO] The quantity of energy required to evaporate 1 mole, or a unit mass, of a liquid, at constant pressure and temperature. Also known as enthalpy of vaporization; heat of evaporation; latent heat of vaporization.

heat of wetting [THERMO] **1.** The heat of adsorption of water on a substance. **2.** The additional heat required, above the heat of vaporization of free water, to evaporate water from a substance in which it has been absorbed.

heat pipe [ENG] A heat-transfer device consisting of a sealed metal tube with an inner lining of wicklike capillary material and a small amount of fluid in a partial vacuum; heat is absorbed at one end by vaporization of the fluid and is released at the other end by condensation of the vapor.

heat pump [MECH ENG] A device which transfers heat from a cooler reservoir to a hotter one, expending mechanical energy in the process, especially when the main purpose is to heat the hot reservoir rather than refrigerate the cold one.

heat radiation [THERMO] The energy radiated by solids, liquids, and gases in the form of electromagnetic waves as a result of their temperature. Also known as thermal radiation.

heat resistance *See* thermal resistance.

heat seal [ENG] A union between two thermoplastic surfaces by application of heat and pressure to the joint.

heatseeker [ORD] A guided missile incorporating an infrared device for homing on heat-radiating machines or installations, such as an aircraft engine or a blast furnace.

heatsink [AERO ENG] **1.** A type of protective device capable of absorbing heat and used as a heat shield. **2.** In nuclear propulsion, any thermodynamic device, such as a radiator or condenser, that is designed to absorb the excess heat energy of the working fluid. Also known as heat dump. [ELEC] A mass of metal that is added to a device for the purpose of absorbing and dissipating heat; used with power transistors and many types of metallic rectifiers. Also known as dissipator. [THERMO] Any (gas, solid, or liquid) region where heat is absorbed.

heatstroke [MED] A heat-exposure syndrome characterized by hyperpyrexia and prostration due to diminution or cessation of sweating, occurring most commonly in persons with underlying disease.

heat transmission *See* heat flow.

heat treatment [MET] Heating and cooling a metal or alloy to obtain desired properties or conditions.

heat value *See* heat of combustion.

heave [GEOL] **1.** The horizontal component of the slip, measured at right angles to the strike of the fault. [MIN ENG] A rising of the floor of a mine caused by its being too soft to resist the weight on the pillars. **2.** A predominantly upward movement of the surface of the soil due to expansion or displacement. [OCEANOGR] The motion imparted to a floating body by wave action.

heavenly body *See* celestial body.

Heaviside calculus [MATH] A type of operational calculus that is used to completely analyze a linear dynamical system which represents some vibrating physical system.

Heaviside layer *See* E layer.

heavy artillery [ORD] Artillery other than antiaircraft artillery; consists of howitzers and longer-barreled cannon not classified as medium artillery.

heavy-duty oil [MATER] Lubricating oil with good oxidation stability and corrosion-preventive and detergent-dispersant characterisitics; used in high-speed diesel and gasoline engines under heavy-duty service conditions.

heavy ends [MATER] The highest boiling portion of a petroleum fraction.

heavy hydrogen [NUC PHYS] Hydrogen consisting of isotopes whose mass number is greater than one, namely deuterium or tritium.

heavy-ion linac *See* heavy-ion linear accelerator.

heavy-ion linear accelerator [NUCLEO] A linear accelerator which produces a beam of heavy particles of high intensity and sharp energy; used to produce transuranic elements and short-lived isotopes, and to study nuclear reactions, nuclear spectroscopy, and the absorption of heavy ions in matter. Also known as heavy-ion linac; hilac.

heavy oxygen *See* oxygen-18.

heavy particle *See* baryon.

heavy water [INORG CHEM] A compound of hydrogen and oxygen containing a higher proportion of the hydrogen isotope deuterium than does naturally occurring water. Also known as deuterium oxide.

heavy-water reactor [NUCLEO] A nuclear reactor in which heavy water serves as moderator and sometimes also as coolant.

Hebrovellidae [INV ZOO] A family of hemipteran insects in the subdivision Amphicorisae.

hecatolite *See* moonstone.

hectare [MECH] A unit of area in the metric system equal to 100 ares or 10,000 square meters. Abbreviated ha.

hecto- [SCI TECH] A prefix representing 10^2 or 100.

hectorite [MINERAL] $(Mg,Li)_3Si_4O_{10}(OH)_2$ A trioctohedral clay mineral of the montmorillonite group composed of a hydrous silicate of magnesium and lithium.

hedenbergite [MINERAL] $CaFeSi_2O_6$ A black mineral consisting of calcium-iron pyroxene and occurring at the contacts of limestone with granitic masses.

hedgehog [ORD] 1. A portable obstacle, made of crossed poles laced with barbed wire, in the general shape of an hourglass. 2. A beach obstacle, usually made of steel bars or channel iron, imbedded in concrete and used to interfere with beach landings. 3. A concentration of troops securely entrenched or fortified, with arms and defenses facing all directions. [VERT ZOO] The common name for members of the insectivorous family Erinaceidae characterized by spines on their back and sides.

Hedwigiaceae [BOT] A family of mosses in the order Isobryales.

hedyphane [MINERAL] $(Ca,Pb)_5Cl(AsO_4)_3$ Yellowish-white mineral composed of lead and calcium arsenate and chloride; occurs in monoclinic crystals.

HEED See high-energy electron diffraction.

heel post [CIV ENG] A post to which are secured the hinges of a gate or door.

Heidelberg man [PALEON] An early type of European fossil man known from an isolated lower jaw; considered a variant of *Homo erectus* or an early stock of Neanderthal man.

height [MATH] The perpendicular distance between horizontal lines or planes passing through the top and bottom of an object.

height-position indicator [ELECTR] Radar display which shows simultaneously angular elevation, slant range, and height of objects detected in the vertical sight plane.

height-range indicator [ELECTR] 1. Radar display which shows an echo as a bright spot on a rectangular field, slant range being indicated along the X axis, height above the horizontal plane being indicated (on a magnified scale) along the Y axis, and height above the earth being shown by a cursor. 2. Cathode-ray tube from which altitude and range measurements of flightborne objects may be viewed.

heiligenschein [OPTICS] A diffuse white ring surrounding the shadow cast by the observer's head upon a dew-covered lawn when the solar elevation is low and, therefore, the distance from observer to shadow is great.

Heine-Medin disease See poliomyelitis.

Heisenberg equation of motion [QUANT MECH] An equation which gives the rate of change of an operator corresponding to a physical quantity in the Heisenberg picture.

Heisenberg picture [QUANT MECH] A mode of description of a system in which dynamic states are represented by stationary vectors and physical quantities are represented by operators which evolve in the course of time. Also known as Heisenberg representation.

Heisenberg representation See Heisenberg picture.

Heisenberg uncertainty principle See uncertainty principle.

Heisenberg uncertainty relation See uncertainty relation.

Heising modulation See constant-current modulation.

Helaletidae [PALEON] A family of extinct perissodactyl mammals in the superfamily Tapiroidea.

Helcionellacea [PALEON] A superfamily of extinct gastropod mollusks in the order Aspidobranchia.

Helderbergian [GEOL] A North American stage of geologic time, in the lower Lower Devonian.

Heleidae [INV ZOO] The biting midges, a family of orthorrhaphous dipteran insects in the series Nematocera.

Heliasteridae [INV ZOO] A family of echinoderms in the subclass Asteroidea lacking pentameral symmetry but structurally resembling common asteroids.

helical [MATH] Pertaining to a cylindrical spiral, for example, a screw thread.

helical angle [MECH] In the study of torsion, the angular displacement of a longitudinal element, originally straight on the surface of an untwisted bar, which becomes helical after twisting.

helical gear [MECH ENG] Gear wheels running on parallel axes, with teeth twisted oblique to the gear axis.

helical scanning [COMMUN] A method of facsimile scanning in which a single-turn helix rotates against a stationary bar to give horizontal movement of an elemental area. [ENG] A method of radar scanning in which the antenna beam rotates continuously about the vertical axis while the elevation angle changes slowly from horizontal to vertical, so that a point on the radar beam describes a distorted helix.

helical traveling-wave tube See helix tube.

helicity [QUANT MECH] The component of the spin of a particle along its momentum.

helicoid [INV ZOO] Of a gastropod shell, shaped like a flat coil or flattened spiral. [MATH] A surface generated by a curve which is rotated about a straight line and also is translated in the direction of the line at a rate that is a constant multiple of its rate of rotation.

Heliconiaceae [BOT] A family of monocotyledonous plants in the order Zingiberales characterized by perfect flowers with a solitary ovule in each locule, schizocarpic fruit, and capitate stigma.

Helicoplacoidea [PALEON] A class of free-living, spindle- or pear-shaped, plated echinozoans known only from the Lower Cambrian of California.

helicopter [AERO ENG] An aircraft fitted to sustain itself by motor-driven horizontal rotating blades (rotors) that accelerate the air downward, providing a reactive lift force, or accelerate the air at an angle to the vertical, providing lift and thrust.

Helicosporae [MYCOL] A spore group of the Fungi Imperfecti characterized by spirally coiled, septate spores.

Helicosporida [INV ZOO] An order of protozoans in the class Myxosporidea characterized by production of spores with a relatively thick, single intrasporal filament and three uninucleate sporoplasms.

helicotrema [ANAT] The opening at the apex of the cochlea through which the scala tympani and the scala vestibuli communicate with each other.

Heligmosomidae [INV ZOO] A family of parasitic roundworms belonging to the Strongyloidea.

helimagnetism [SOLID STATE] A property possessed by some metals, alloys, and salts of transition elements or rare earths, in which the atomic magnetic moments, at sufficiently low temperatures, are arranged in ferromagnetic planes, the direction of the magnetism varying in a uniform way from plane to plane.

heliocentric [ASTRON] Relative to the sun as a center.

Heliodinidae [INV ZOO] A family of lepidopteran insects in the suborder Heteroneura.

heliogram [COMMUN] A message transmitted on a heliograph.

heliograph [COMMUN] An instrument for sending telegraphic messages by reflecting the sun's rays from a mirror. [ENG] An instrument that records the duration of sunshine and gives a qualitative measure of its amount by action of sun's rays on blueprint paper.

heliolite *See* sunstone.

Heliolitidae [PALEON] A family of extinct corals in the order Tabulata.

heliometer [OPTICS] A split-lens telescope used to measure the sun's diameter as well as small distances between stars or other celestial bodies.

heliophyllite [MINERAL] $Pb_6As_2O_7Cl_4$ A yellow to greenish-yellow, orthorhombic mineral consisting of an oxychloride of lead and arsenic; occurs in massive and tabular form and as crystals.

Heliornithidae [VERT ZOO] The lobed-toed sun grebes, a family of pantropical birds in the order Gruiformes.

helioscope [OPTICS] A telescope for observing the sun that protects the observer's eyes from the sun's glare.

heliostat [ENG] A clock-driven instrument mounting which automatically and continuously points in the direction of the sun; it is used with a pyrheliometer when continuous direct solar radiation measurements are required.

heliotaxis [BIOL] Orientation movement of an organism in response to the stimulus of sunlight.

heliotrope [BOT] A plant whose flower or stem turns toward the sun. [ENG] An instrument that reflects the sun's rays over long distances; used in geodetic surveys. [MINERAL] *See* bloodstone.

heliotropic wind [METEOROL] A subtle, diurnal component of the wind velocity leading to a diurnal shift of the wind or turning of the wind with the sun, produced by the east-to-west progression of daytime surface heating.

heliotropism [BIOL] Growth or orientation movement of a sessile organism or part, such as a plant, in response to the stimulus of sunlight.

Heliozoia [INV ZOO] A subclass of the protozoan class Actinopodea; individuals lack a central capsule and have either axopodia or filopodia.

helitron [ELECTR] An electrostatically focused, low-noise backward-wave oscillator; the microwave output signal frequency can be swept rapidly over a wide range by varying the voltage applied between the cathode and the associated radio-frequency circuit.

helium [CHEM] A gaseous chemical element, symbol He, atomic number 2, and atomic weight 4.0026; one of the noble gases in group 0 of the periodic table.

helium I [CRYO] The phase of liquid helium-4 which is stable at temperatures above the lambda point (about 2.2 K) and has the properties of a normal liquid, except low density.

helium II [CRYO] The phase of liquid helium-4 which is stable at temperatures between absolute zero and the lambda point (about 2.2 K), and has many remarkable properties such as vanishing viscosity, extremely high heat conductivity, and the fountain effect.

helium-3 [NUC PHYS] The isotope of helium with mass number 3, constituting approximately 1.3 parts per million of naturally occurring helium.

helium-4 [NUC PHYS] The isotope of helium with mass number 4, constituting nearly all naturally occurring helium.

helium-cadmium laser [OPTICS] A metal-vapor ion laser in which cadmium vapor, produced by heat or other means, migrates through a high-voltage glow discharge in helium, generating a continuous laser beam at wavelengths in the ultraviolet and blue parts of the spectrum (about 0.3 to 0.5 micrometers).

helix [ELEC] A spread-out, single-layer coil of wire, either wound around a supporting cylinder or made of stiff enough wire to be self-supporting. [MATH] A curve traced on a cylindrical or conical surface where all points of the surface are cut at the same angle; this screwlike winding curve is parametrically given by equations $x = a \cos \theta$, $y = a \sin \theta$, $z = b\theta$, where θ is the parameter and a and b are constants.

helix angle [DES ENG] That angle formed by the helix of the thread at the pitch-diameter line and a line at right angles to the axis. [MATH] The constant angle between the tangent to a helix and a generator of the cylinder upon which the helix lies.

helix tube [ELECTR] A traveling-wave tube in which the electromagnetic wave travels along a wire wound in a spiral about the path of the beam, so that the wave travels along the tube at a velocity approximately equal to the beam velocity. Also known as helical traveling-wave tube.

hellandite [MINERAL] Mineral composed of silicate of metals in the cerium group with aluminum, iron, manganese, and calcium.

hellbender [VERT ZOO] *Cryptobranchus alleganiensis*. A large amphibian of the order Urodela which is the most primitive of the living salamanders, retaining some larval characteristics.

helmholtz [ELEC] A unit of dipole moment per unit area, equal to 1 Debye unit per square angstrom, or approximately 3.335×10^{-10} coulomb per meter.

Helmholtz coils [ELECTROMAG] A pair of flat, circular coils having equal numbers of turns and equal diameters, arranged with a common axis, and connected in series; used to obtain a magnetic field more nearly uniform than that of a single coil.

Helmholtz double layer [PHYS] An electrical double layer of positive and negative charges one molecule thick which occurs at a surface where two bodies of different materials are in contact, or at the surface of a metal or other substance capable of existing in solution as ions and immersed in a dissociating solvent.

Helmholtz equation [MATH] A partial differential equation obtained by setting the Laplacian of a function equal to the function multiplied by a negative constant. [OPTICS] An equation which relates the linear and angular magnifications of a spherical refracting interface. Also known as Lagrange-Helmholtz equation. [PHYS CHEM] The relationship stating that the emf (electromotive force) of a reversible electrolytic cell equals the work equivalent of the chemical reaction when charge passes through the cell plus the product of the temperature and the derivative of the emf with respect to temperature.

Helmholtz free energy *See* free energy.

Helmholtz function *See* free energy.

Helmholtz instability [FL MECH] The hydrodynamic instability arising from a shear, or discontinuity, in current speed at the interface between two fluids in two-dimensional motion; the perturbation gains kinetic energy at the expense of that of the basic currents. Also known as shearing instability.

Helmholtz potential *See* free energy.

Helmholtz resonator [ENG ACOUS] An enclosure having a small opening consisting of a straight tube of such dimensions that the enclosure resonates at a single frequency determined by the geometry of the resonator.

Helmholtz's theorem [ELEC] *See* Thévenin's theorem. [FL MECH] The theorem that in the isentropic flow of a nonviscous fluid which is not subject to body forces, individual vortices always consist of the same fluid particles. [MATH] The theorem determining a general class of vector fields as being everywhere expressible as the sum of an irrotational vector with a divergence-free vector.

Helmholtz theory *See* Young-Helmholtz theory.

Helmholtz wave [FL MECH] An unstable wave in a system of two homogeneous fluids with a velocity discontinuity at the interface.

Helobiae [BOT] The equivalent name for Helobiales.

Helobiales [BOT] An order embracing most of the Alismatidae in certain systems of classification.

Helodermatidae [VERT ZOO] A family of lizards in the suborder Sauria.

Helodidae [INV ZOO] The marsh beetles, a family of coleopteran insects in the superfamily Dascilloidea.

Helodontidae [PALEON] A family of extinct ratfishes conditionally placed in the order Bradyodonti.

Helomyzidae [INV ZOO] The sun flies, a family of myodarian cyclorrhaphous dipteran insects in the subsection Acalypteratae.

Heloridae [INV ZOO] A family of hymenopteran insects in the superfamily Proctotrupoidea.

Helotiales [MYCOL] An order of fungi in the class Ascomycetes.

Helotidae [INV ZOO] The metallic sap beetles, a family of coleopteran insects in the superfamily Cucujoidea.

Helotrephidae [INV ZOO] A family of true aquatic, tropical hemipteran insects in the subdivision Hydrocorisae.

helper virus [VIROL] A virus that, by its infection of a cell, enables a defective virus to multiply by supplying one or more functions that the defective virus lacks.

help screen [COMPUT SCI] Instructions that explain how to use the software of a computer system and that can be presented on the screen of a video display terminal at any time.

helvine *See* helvite.

helvite [MINERAL] $(Mn,Fe,Zn)_4Be_3(SiO_4)_3S$ A silicate mineral isomorphous with danalite and genthelvite. Also known as helvine.

hemacytometer *See* hemocytometer.

hemadsorption virus [VIROL] A descriptive term for myxoviruses that agglutinate red blood cells and cause the cells to adsorb to each other. Abbreviated HA virus.

hemafibrite [MINERAL] $Mn_3(AsO_4)(OH)_3 \cdot H_2O$ A brownish to garnet-red mineral composed of basic manganese arsenate.

hemal arch [ANAT] 1. A ventral loop on the body of vertebrate caudal vertebrae surrounding the blood vessels. 2. In man, the ventral vertebral process formed by the centrum together with the ribs.

hemangioma [MED] A tumor composed of blood vessels. Also known as capillary angioma.

hematite [MINERAL] Fe_2O_3 An iron mineral crystallizing in the rhombohedral system; the most important ore of iron, it is dimorphous with maghemite, occurs in black metallic-looking crystals, in reniform masses or fibrous aggregates, or in reddish earthy forms. Also known as bloodstone; red hematite; red iron ore; red ocher; rhombohedral iron ore.

hematocrit [PATH] The volume, after centrifugation, occupied by the cellular elements of blood, in relation to the total volume.

hematolite [MINERAL] $(Mn,Mg)_4Al(AsO_4)(OH)_8$ A brownish-red mineral composed of aluminum manganese arsenate; occurs in rhombohedral crystals.

hematology [MED] The science of the blood, its nature, functions, and diseases.

hematoma [MED] A localized mass of blood in tissue; usually it clots and becomes encapsulated by connective tissue.

hematophanite [MINERAL] $Pb_5Fe_4O_{10}(Cl,OH)_2$ A mineral composed of oxychloride lead and iron.

hematopoiesis [PHYSIO] The process by which the cellular elements of the blood are formed. Also known as hemopoiesis.

hematopoietic system *See* reticuloendothelial system.

hematopoietin [BIOCHEM] A substance which is produced by the juxtaglomerular apparatus in the kidney and controls the rate of red cell production. Also known as hemopoietin.

hematuria [MED] A pathological condition in which the urine contains blood.

heme [BIOCHEM] $C_{34}H_{32}O_4N_4Fe$ An iron-protoporphyrin complex associated with each polypeptide unit of hemoglobin.

hemi- [BIOL] 1. Prefix for half. 2. Prefix denoting one side of the body.

Hemiascomycetes [MYCOL] The equivalent name for Hemiascomycetidae.

Hemiascomycetidae [MYCOL] A subclass of fungi in the class Ascomycetes.

hemiazygous vein [ANAT] A vein on the left side of the vertebral column which drains blood from the left ascending lumbar vein to the azygos vein.

Hemibasidiomycetes [MYCOL] The equivalent name for Heterobasidiomycetidae.

Hemichordata [SYST] A group of marine animals categorized as either a phylum of deuterostomes or a subphylum of chordates; includes the Enteropneusta, Pterobranchia, and Graptolithina.

Hemicidaridae [PALEON] A family of extinct Echinacea in the order Hemicidaroida distinguished by a stirodont lantern, and ambulacra abruptly widened at the ambitus.

Hemicidaroida [PALEON] An order of extinct echinoderms in the superorder Echinacea characterized by one very large tubercle on each interambulacral plate.

hemicrystalline *See* hypocrystalline.

Hemidiscosa [INV ZOO] An order of sponges in the subclass Amphidiscophora distinguished by birotulates that are hemidiscs with asymmetrical ends.

hemihedral symmetry [CRYSTAL] The possession by a crystal of only half of the elements of symmetry which are possible in the crystal system to which it belongs.

Hemileucinae [INV ZOO] A subfamily of lepidopteran insects in the family Saturnidae consisting of the buck moths and relatives.

Hemimetabola [INV ZOO] A division of the insect subclass Pterygota; members are characterized by hemimetabolous metamorphosis.

hemimetabolous metamorphosis [INV ZOO] An incomplete metamorphosis; gills are present in aquatic larvae, or naiads.

hemimorphic crystal [CRYSTAL] A crystal with no transverse plane of symmetry and no center of symmetry; composed of forms belonging to only one end of the axis of symmetry.

hemimorphite [MINERAL] $Zn_4Si_2O_7(OH)_2 \cdot H_2O$ A white, colorless, pale-green, blue, or yellow mineral having an orthorhombic crystal structure; an ore of zinc. Also known as calamine; electric calamine; galmei.

hemin [BIOCHEM] $C_{34}H_{32}O_4N_4FeCl$ The crystalline salt of ferriheme, containing iron in the ferric state.

hemipenis [VERT ZOO] Either of a pair of nonerectile, evertible sacs that lie on the floor of the cloaca in snakes and lizards; used as intromittent organs.

Hemipeplidae [INV ZOO] An equivalent name for Cucujidae.

Hemiprocnidae [VERT ZOO] The crested swifts, a family comprising three species of perching birds found only in southeastern Asia.

Hemiptera [INV ZOO] The true bugs, an order of the class Insecta characterized by forewings differentiated into a basal area and a membranous apical region.

Hemisphaeriales [MYCOL] A group of ascomycetous fungi characterized by the wall of the fruit body being a stroma.

hemisphere [GEOGR] A half of the earth divided into north and south sections by the equator, or into an east section containing Europe, Asia, and Africa, and a west section containing the Americas. [MATH] One of the two pieces of a sphere divided by a great circle.

hemispheroid [MATH] One of the halves into which a spheroid is divided by a plane of symmetry.

Hemist [GEOL] A suborder of the soil order Histosol, consisting of partially decayed plant residues and saturated with water most of the time.

hemitropic [CRYSTAL] Pertaining to a twinned structure in which, if one part were rotated 180°, the two parts would be parallel.

Hemizonida [PALEON] A Paleozoic order of echinoderms of the subclass Asteroidea having an ambulacral groove that is well defined by adambulacral ossicles, but with restricted or undeveloped marginal plates.

hemizygous [GEN] Pertaining to the condition or state of having a gene present in a single dose; for instance, in the X chromosome of male mammals.

hemlock [BOT] The common name for members of the genus *Tsuga* in the pine family characterized by two white lines beneath the flattened, needlelike leaves.

hemocoel [INV ZOO] An expanded portion of the blood system in arthropods that replaces a portion of the coelom.

hemocyanin [BIOCHEM] A blue respiratory pigment found only in mollusks and in arthropods other than insects.

hemocytometer [PATH] A specifically designed, ruled and calibrated glass slide used with a microscope to count red and white blood cells. Also spelled hemacytometer.

hemoglobin [BIOCHEM] The iron-containing, oxygen-carrying molecule of the red blood cells of vertebrates comprising four polypeptide subunits in a heme group.

hemoglobinuria [MED] A pathological condition in which the urine contains hemoglobin.

hemolymph [INV ZOO] The circulating fluid of the open circulatory systems of many invertebrates.

hemolysin [IMMUNOL] A substance that lyses erythrocytes.

hemolytic anemia [MED] A decrease in the blood concentration of hemoglobin and the number of erythrocytes, due to the inability of the mature erythrocytes to survive in the circulating blood.

hemolytic disease of newborn *See* erythroblastosis fetalis.

hemophilia [MED] A rare, hereditary blood disorder marked by a tendency toward bleeding and hemorrhages due to a deficiency of factor VIII.

hemophilic bacteria [MICROBIO] Bacteria of the genera *Hemophilus*, *Bordetella*, and *Moraxella*; all are small, gram-negative, nonmotile, parasitic rods, dependent upon blood factors for growth.

hemopoiesis *See* hematopoiesis.

hemopoietin *See* hematopoietin.

hemorrhoid [MED] A varicosity of the external hemorrhoidal veins, causing painful swelling in the anal region.

hemostat [MED] An instrument to compress a bleeding vessel.

hemp-core cable *See* standard wire rope.

henbane [BOT] *Hyoscyamus niger*. A poisonous herb containing the toxic alkaloids hyoscyamine and hyoscine; extracts have properties similar to belladonna.

henequen [MATER] A hard plant fiber, obtained from the leaves of the American agave (*Agave fourcroydes*) and other agave species; used to make rope, twine, and cord.

Henicocephalidae [INV ZOO] A family of hemopteran insects of uncertain affinities.

Henle's loop *See* loop of Henle.

henna [BOT] *Lawsonia inermis*. An Old World plant having small opposite leaves and axillary panicles of white flowers; a reddish-brown dye extracted from the leaves is used in hair dyes. Also known as Egyptian henna.

henry [ELECTROMAG] The mks unit of self and mutual inductance, equal to the self-inductance of a circuit or the mutual inductance between two circuits if there is an induced electromotive force of 1 volt when the current is changing at the rate of 1 ampere per second. Symbolized H.

heparin [BIOCHEM] An acid mucopolysaccharide acting as an antithrombin, antithromboplastin, and antiplatelet factor to prolong the clotting time of whole blood; occurs in a variety of tissues, most abundantly in liver.

Hepaticae [BOT] The equivalent name for Marchantiatae.

hepatic artery [ANAT] A branch of the celiac artery that carries blood to the stomach, pancreas, great omentum, liver, and gallbladder.

hepatic duct [ANAT] The common duct draining the liver. Also known as common hepatic duct.

hepatic portal system [ANAT] A system of veins in vertebrates which collect blood from the digestive tract and spleen and pass it through capillaries in the liver.

hepatic vein [ANAT] A blood vessel that drains blood from the liver into the inferior vena cava.

hepatitis [MED] Inflammation of the liver; commonly of viral origin but also occurring in association with syphilis, typhoid fever, malaria, toxemias, and parasitic infestations.

Hepialidae [INV ZOO] A family of lepidopteran insects in the superfamily Hepialoidea.

Hepialoidea [INV ZOO] A superfamily of lepidopteran insects in the suborder Homoneura including medium- to large-sized moths which possess rudimentary mouthparts.

Hepsogastridae [INV ZOO] A family of parasitic insects in the order Mallophaga.

heptode [ELECTR] A seven-electrode electron tube containing an anode, a cathode, a control electrode, and four additional electrodes that are ordinarily grids. Also known as pentagrid.

heptose [BIOCHEM] Any member of the group of monosaccharides containing seven carbon atoms.

herb [BOT] **1.** A seed plant that lacks a persistent, woody stem aboveground and dies at the end of the season. **2.** An aromatic plant or plant part used medicinally or for food flavoring.

herbaceous [BOT] **1.** Resembling or pertaining to a herb. **2.** Pertaining to a stem with little or no woody tissue.

herbarium [BOT] A collection of plant specimens, pressed and mounted on paper or placed in liquid preservatives, and systematically arranged.

Hercules [ASTRON] A constellation with no stars brighter than third magnitude; right ascension 17 hours, declination 30° north.

Hercules stone *See* lodestone.

Hercynian geosyncline [GEOL] A principal area of geosynclinal sediment accumulation in Devonian time; found in south-central and southern Europe and northern Africa.

hercynite [MINERAL] (Fe, Mg)Al$_2$O$_4$ A black mineral of the spinel group; crystallizes in the isometric system. Also known as ferrospinel; iron spinel.

herderite [MINERAL] CaBe(PO$_4$)(F,OH) A colorless to pale-yellow or greenish-white mineral consisting of phosphate and fluoride of calcium and beryllium; hardness is 7.5–8 on Mohs scale, and specific gravity is 3.92.

heredity [GEN] The sum of genetic endowment obtained from the parents.

Hering theory [PHYSIO] A theory of color vision which assumes that three qualitatively different processes are present in the visual system, and that each of the three is capable of responding in two opposite ways.

hermaphrodite [BIOL] An individual (animal, plant, or higher vertebrate) exhibiting hermaphroditism. [PHYSIO] An abnormal condition, especially in humans and other higher vertebrates, in which both male and female reproductive organs are present in the individual.

hermaphrodite caliper [DES ENG] A layout tool having one leg pointed and the other like that of an inside caliper; used to locate the center of irregularly shaped stock or to lay out a line parallel to an edge.

hermaphroditic *See* monoecious.

hermaphroditic connector [ELEC] A connector in which both mating parts are exactly alike at their mating surfaces. Also known as sexless connector.

hermetic seal [ENG] An airtight seal.

hermit crab [INV ZOO] The common name for a number of marine decapod crustaceans of the families Paguridae and Parapaguridae; all lack right-sided appendages and have a large, soft, coiled abdomen.

Hermitian conjugate of a matrix [MATH] The transpose of the complex conjugate of a matrix. Also known as adjoint of a matrix; associate matrix.

Hermitian conjugate operator *See* adjoint operator.

Hermitian matrix [MATH] A matrix which equals its conjugate transpose matrix, that is, is self-adjoint.

Hermitian operator [MATH] A linear operator A on vectors in a Hilbert space, such that if x and y are in the range of A then the inner products (Ax,y) and (x,Ay) are equal.

hernia [MED] Abnormal protrusion of an organ or other body part through its containing wall. Also called rupture.

herniated disk [MED] An intervertebral disk in which the pulpy center has pushed through the fibrocartilage. Also known as slipped disk.

heroin [PHARM] C$_{21}$H$_{23}$O$_5$N A white, crystalline powder made from morphine; the hydrochloride compound is used as a sedative and narcotic.

heron [VERT ZOO] The common name for wading birds composing the family Ardeidae characterized by long legs and neck, a long tapered bill, large wings, and soft plumage.

herpes [MED] An acute inflammation of the skin or mucous membranes, characterized by the development of groups of vesicles on an inflammatory base.

herpes simplex [MED] An acute vesicular eruption of the skin or mucous membranes caused by a virus, commonly seen as cold sores or fever blisters.

herpesvirus [VIROL] A major group of deoxyribonucleic acid–containing animal viruses, distinguished by a cubic capsid, enveloped virion, and affinity for the host nucleus as a site of maturation.

herpes zoster [MED] A systemic virus infection affecting spinal nerve roots, characterized by vesicular eruptions distributed along the course of a cutaneous nerve. Also known as shingles; zoster.

herring [VERT ZOO] The common name for fishes composing the family Clupeidae; fins are soft-rayed and have no supporting spines, there are usually four gill clefts, and scales are on the body but absent on the head.

Herschel-Cassegrain telescope [OPTICS] A modification of a Cassegrain telescope in which the primary paraboloidal mirror is slightly inclined to the optical axis, and both the secondary hyperboloidal mirror and the eyepiece are located off the axis, so that it is not necessary to pierce the primary.

hertz [PHYS] Unit of frequency; a periodic oscillation has a frequency of n hertz if in 1 second it goes through n cycles. Also known as cycle per second (cps). Symbolized Hz.

Hertzian oscillator [ELECTROMAG] **1.** A generator of electric dipole radiation; consists of two capacitors joined by a conducting rod having a small spark gap; an oscillatory discharge occurs when the two halves of the oscillator are raised to a sufficiently high potential difference. **2.** A dumbbell-shaped conductor in which electrons oscillate from one end to the other, producing electric dipole radiation.

Hertzian wave *See* electric wave.

Hertz's law [MECH] A law which gives the radius of contact between a sphere of elastic material and a surface in terms of the sphere's radius, the normal force exerted on the sphere, and Young's modulus for the material of the sphere.

Hertzsprung-Russell diagram [ASTRON] A plot showing the relation between the luminosity and surface temperature of stars; other related quantities frequently used in plotting this diagram are the absolute magnitude for luminosity, and spectral type or color index for the surface temperatures. Abbreviated H-R diagram.

Hertz vector *See* polarization potential.

hervidero *See* mud volcano.

Hesionidae [INV ZOO] A family of small polychaete worms belonging to the Errantia.

hesitation [COMPUT SCI] A brief automatic suspension of the operations of a main program in order to perform all or part of another operation, such as rapid transmission of data to or from a peripheral unit.

Hesperiidae [INV ZOO] The single family of the superfamily Hesperioidea comprising butterflies known as skippers because of their rapid, erratic flight.

Hesperioidea [INV ZOO] A monofamilial superfamily of lepidopteran insects in the suborder Heteroneura including heavy-bodied, mostly diurnal insects with clubbed antennae that are bent, curved, or reflexed at the tip.

Hesperornithidae [PALEON] A family of extinct North American birds in the order Hesperornithiformes.

Hesperornithiformes [PALEON] An order of ancient extinct birds; individuals were large, flightless, aquatic diving birds with the shoulder girdle and wings much reduced and the legs specialized for strong swimming.

hessian *See* burlap.

hessite [MINERAL] Ag_2Te A lead-gray sectile mineral crystallizing in the isometric system; usually massive and often auriferous.

Hess's law [PHYS CHEM] The law that the evolved or absorbed heat in a chemical reaction is the same whether the reaction takes one step or several steps. Also known as the law of constant heat summation.

Heteractinida [PALEON] A group of Paleozoic sponges with calcareous spicules; probably related to the Calcarea.

Heterakidae [INV ZOO] A group of nematodes assigned either to the suborder Oxyurina or the suborder Ascaridina.

heterauxesis *See* allometry.

hetero- [CHEM] Chemical prefix meaning different; for example, a heterocyclic compound is one in which the ring is made of more than one kind of atom.

Heterobasidiomycetidae [MYCOL] A class of fungi in which the basidium either is branched or is divided into cells by cross walls.

heteroblastic [EMBRYO] Arising from different tissues or germ layers, in referring to similar organs in different species. [PETR] Pertaining to rocks in which the essential constituents are of two distinct orders of magnitude of size.

Heterocapsina [BOT] An order of green algae in the class Xanthophyceae. [INV ZOO] A suborder of yellow-green to green flagellate protozoans in the order Heterochlorida.

heterocarpous [BOT] Producing two distinct types of fruit.

Heterocera [INV ZOO] A formerly recognized suborder of Lepidoptera including all forms without clubbed antennae.

heterocercal [VERT ZOO] Pertaining to the caudal fin of certain fishes and indicating that the upper lobe is larger, with the vertebral column terminating in this lobe.

Heteroceridae [INV ZOO] The variegated mud-loving beetles, a family of coleopteran insects in the superfamily Dryopoidea.

Heterocheilidae [INV ZOO] A family of parasitic roundworms in the superfamily Ascaridoidea.

Heterochlorida [INV ZOO] An order of yellow-green to green flagellate oraganisms of the class Phytamastigophorea.

heterochromatin [CYTOL] Specialized chromosome material which remains tightly coiled even in the nondividing nucleus and stains darkly in interphase.

heterocoelous [ANAT] Pertaining to vertebrae with centra having saddle-shaped articulations.

Heterocorallia [PALEON] An extinct small, monofamilial order of fossil corals with elongate skeletons; found in calcareous shales and in limestones.

Heterocotylea [INV ZOO] The equivalent name for Monogenea.

heterocyclic compound [ORG CHEM] Compound in which the ring structure is a combination of more than one kind of atom; for example, pyridine, C_5H_5N.

Heterodera [INV ZOO] The cyst nematodes, a genus of phytoparasitic worms that live in the internal root systems of many plants.

heterodont [ANAT] Having teeth that are variable in shape and differentiated into incisors, canines, and molars. [INV ZOO] In bivalves, having two types of teeth on one valve which fit into depressions on the other valve.

Heterodonta [INV ZOO] An order of bivalve mollusks in some systems of classification; hinge teeth are few in number and variable in form.

Heterodontoidea [VERT ZOO] A suborder of sharks in the order Selachii which is represented by the single living genus *Heterodontus*.

heterodyne [ELECTR] To mix two alternating-current signals of different frequencies in a nonlinear device for the purpose of producing two new frequencies, the sum of and difference between the two original frequencies.

heterodyne conversion transducer *See* converter.

heterodyne detector [ELECTR] A detector in which an unmodulated carrier frequency is combined with the signal of a local oscillator having a slightly different frequency, to provide an audio-frequency beat signal that can be heard with a loudspeaker or headphones; used chiefly for code reception.

heterodyne frequency [COMMUN] Either of the two new frequencies resulting from heterodyne action between the two input frequencies of a heterodyne detector.

heterodyne modulator *See* mixer.

heterodyne oscillator [ELECTR] 1. A separate variable-frequency oscillator used to produce the second frequency required in a heterodyne detector for code reception. 2. *See* beat-frequency oscillator.

heterodyne reception [ELECTR] Radio reception in which the incoming radio-frequency signal is combined with a locally generated rf signal of different frequency, followed by detection. Also known as beat reception.

heterodyne repeater [ELECTR] A radio repeater in which the received radio signals are converted to an intermediate frequency, amplified, and reconverted to a new frequency band for transmission over the next repeater section.

heteroecious [BIOL] Pertaining to forms that pass through different stages of a life cycle in different hosts.

heteroerotism [PSYCH] Sexual desire directed away from one's own self or sex.

heterogamete [BIOL] A gamete that differs in size, appearance, structure, or sex chromosome content from the gamete of the opposite sex. Also known as anisogamete.

heterogamy [BIOL] 1. Alternation of a true sexual generation with a parthenogenetic generation. 2. Sexual reproduction by fusion of unlike gametes. Also known as anisogamy. [BOT] Condition of producing two kinds of flowers.

heterogeneous [CHEM] Pertaining to a mixture of phases such as liquid-vapor, or liquid-vapor-solid. [MATH] Pertaining to quantities having different degrees or dimensions. [SCI TECH] Composed of dissimilar or nonuniform constituents.

heterogeneous catalysis [CHEM] Catalysis occurring at a phase boundary, usually a solid-fluid interface.

Heterogeneratae [BOT] A class of brown algae distinguished by a heteromorphic alteration of generations.

heterogenetic antigen *See* heterophile antigen.

heterogenite [MINERAL] $CoO(OH)$ A black cobalt mineral, sometimes with some copper and iron, found in mammillary masses. Also known as stainierite.

Heterognathi [VERT ZOO] An equivalent name for Cypriniformes.

heterogony [BIOL] 1. *See* allometry. 2. Alteration of generations in a complete life cycle, especially of a dioecious and hermaphroditic generation. [BOT] Having heteromorphic perfect flowers with respect to lengths of the stamens or styles.

heterograft [IMMUNOL] A tissue or organ obtained from an animal of one species and transplanted to the body of an animal of another species. Also known as heterologous graft.

heterojunction [ELECTR] The boundary between two different semiconductor materials, usually with a negligible discontinuity in the crystal structure.

heterolecithal [CYTOL] Of an egg, having the yolk distributed unevenly throughout the cytoplasm.

heterologous graft *See* heterograft.

Heteromera [INV ZOO] The equivalent name for Tenebrionoidea.

Heteromi [VERT ZOO] An equivalent name for Notacanthiformes.

heteromorphic [CYTOL] Having synoptic or sex chromosomes that differ in size or form. [MED] Differing from the normal in size or morphology. [ZOO] Having a different form at each stage of the life history.

heteromorphite [MINERAL] $Pb_7Sb_8S_{19}$ An iron black, monoclinic mineral consisting of lead antimony sulfide.

heteromorphosis [BIOL] Regeneration of an organ or part that differs from the original structure at the site. [EMBRYO] Formation of an organ at an abnormal site. Also known as homeosis.

Heteromyidae [VERT ZOO] A family of the mammalian order Rodentia containing the North American kangaroo mice and the pocket mice.

Heteromyinae [VERT ZOO] The spiny pocket mice, a subfamily of the rodent family Heteromyidae.

Heteromyota [INV ZOO] A monospecific order of wormlike animals in the phylum Echiurida.

Heterodontidae [VERT ZOO] The Port Jackson sharks, a family of aberrant modern elasmobranchs in the suborder Heterodontoidea.

Heteronemertini [INV ZOO] An order of the class Anopla; individuals have a middorsal blood vessel and a body wall composed of three muscular layers.

Heteroneura [INV ZOO] A suborder of Lepidoptera; individuals are characterized by fore- and hindwings that differ in shape and venation and by sucking mouthparts.

heterophagic vacuole *See* food vacuole.

heterophile antibody [IMMUNOL] Substance that will react with heterophile antigen; found in the serum of patients with infectious mononucleosis.

heterophile antigen [IMMUNOL] A substance that occurs in unrelated species of animals but has similar serologic properties among them. Also known as heterogenetic antigen.

Heterophyllidae [PALEON] The single family of the extinct coelenterate order Heterocorallia.

Heteropiidae [INV ZOO] A family of calcareous sponges in the suborder Sycettida.

heteroploidy [GEN] The condition of a chromosome complement in which one or more chromosomes, or parts of chromosomes, are present in number different from the numbers of the rest.

heteropolar generator [ELECTROMAG] A generator whose active conductors successively pass through magnetic fields of opposite direction.

Heteroporidae [INV ZOO] A family of trepostomatous-like bryozoans in the order Cyclostomata.

Heteroptera [INV ZOO] The equivalent name for Hemiptera.

heterosexuality [PSYCH] Having sexual feeling toward members of the opposite sex.

heterosis [GEN] The increase in size, yield, and performance found in some hybrids, especially of inbred parents. Also known as hybrid vigor.

heterosite [MINERAL] A mineral composed of phosphate of iron and manganese; it is isomorphous with purpurite.

Heterosomata [VERT ZOO] The equivalent name for Pleuronectiformes.

Heterosoricinae [PALEON] A subfamily of extinct insectivores in the family Soricidae distinguished by a short jaw and hedgehoglike teeth.

Heterospionidae [INV ZOO] A monogeneric family of spioniform worms found in shallow and abyssal depths of the Atlantic and Pacific oceans.

Heterostraci [PALEON] An extinct group of ostracoderms, or armored, jawless vertebrates; armor consisted of bone lacking cavities for bone cells.

Heterotardigrada [INV ZOO] An order of the tardigrades exhibiting wide morphologic variations.

Heterotrichida [INV ZOO] A large order of large ciliates in the protozoan subclass Spirotrichia; buccal ciliature is well developed and some species are pigmented.

Heterotrichina [INV ZOO] A suborder of the protozoan order Heterotrichida.

heterotroph [BIOL] An organism that obtains nourishment from the ingestion and breakdown of organic matter.

heterotropia *See* strabismus.

heterozygote [GEN] An individual that has different alleles at one or more loci and therefore produces gametes of two or more different kinds.

heulandite [MINERAL] $CaAl_2Si_6O_{16} \cdot 5H_2O$ A zeolite mineral that crystallizes in the monoclinic system; often occurs as foliated masses or in crystal form in cavities of decomposed basic igneous rocks.

heuristic algorithm *See* dynamic algorithm.

heuristic method [MATH] A method of solving a problem in which one tries each of several approaches or methods and evaluates progress toward a solution after each attempt.

heuristic program [COMPUT SCI] A program in which a computer tries each of several methods of solving a problem and judges whether the program is closer to solution after each attempt. Also known as heuristic routine.

heuristic routine *See* heuristic program.

hewettite [MINERAL] $CaV_6O_{16} \cdot 9H_2O$ A deep-red mineral composed of hydrated calcium vanadate; found in silky orthorhombic crystal aggregates in Colorado, Utah, and Peru.

Hexactinellida [INV ZOO] A class of the phylum Porifera which includes sponges with a skeleton made up basically of hexactinal siliceous spicules.

Hexactinosa [INV ZOO] An order of sponges in the subclass Hexasterophora; parenchymal megascleres form a rigid framework and consist of simple hexactins.

hexadecimal [MATH] Pertaining to a number system using the base 16. Also known as sexadecimal.

hexadecimal notation [COMPUT SCI] A notation in the scale of 16, using decimal digits 0 to 9 and six more digits that are sometimes represented by *A*, *B*, *C*, *D*, *E*, and *F*.

hexadecimal number system [MATH] A digital system based on powers of 16, as compared with the use of powers of 10 in the decimal number system. Also known as sexadecimal number system.

hexagonal close-packed structure [CRYSTAL] A close-packed crystal structure characterized by the regular alternation of two layers; the atoms in each layer lie at the vertices of a series of equilateral triangles, and the atoms in one layer lie directly above the centers of the triangles in neighboring layers. Abbreviated hcp structure.

hexagonal lattice [CRYSTAL] A Bravais lattice whose unit cells are right prisms with hexagonal bases and whose lattice points are located at the vertices of the unit cell and at the centers of the bases.

hexagonal nut [DES ENG] A plain nut in hexagon form.

hexagonal system [CRYSTAL] A crystal system that has three equal axes intersecting at 120° and lying in one plane; a fourth, unequal axis is perpendicular to the other three.

hexahedron [MATH] A polyhedron with six faces.

Hexanchidae [VERT ZOO] The six- and seven-gill sharks, a group of aberrant modern elasmobranchs in the suborder Notidanoidea.

Hexapoda [INV ZOO] An equivalent name for Insecta.

Hexasterophora [INV ZOO] A subclass of sponges of the class Hexactinellida in which parenchymal microscleres are typically hexasters.

hexenbesen See witches'-broom disease.

hex nut [DES ENG] A nut in the shape of a hexagon.

hexoctahedron [CRYSTAL] A cubic crystal form that has 48 equal triangular faces, each of which cuts the three crystallographic axes at different distances.

hexode [ELECTR] A six-electrode electron tube containing an anode, a cathode, a control electrode, and three additional electrodes that are ordinarily grids.

hexose diphosphate pathway See glycolytic pathway.

Hf See hafnium.

HF See high frequency.

hfs See hyperfine structure.

Hg See mercury.

hiatus hernia [MED] Hernia through the esophageal hiatus, usually of a portion of the stomach.

hibernaculum [BIOL] A winter shelter for plants or dormant animals. [BOT] A winter bud or other winter plant part. [INV ZOO] A winter resting bud produced by a few fresh-water bryozoans which grows into a new colony in the spring.

hibernation [PHYSIO] 1. Condition of dormancy and torpor found in cold-blooded vertebrates and invertebrates. 2. See deep hibernation.

Hibernian orogeny See Erian orogeny.

hickory [BOT] The common name for species of the genus *Carya* in the order Fagales; tall·deciduous tree with pinnately compound leaves, solid pith, and terminal, scaly winter buds.

hiddenite [MINERAL] A transparent green or yellowish-green spodumene mineral containing chromium and valued as a gem.

hidrosis [MED] Abnormally profuse sweat. [PHYSIO] The formation and excretion of sweat.

hierarchical distributed processing system [COMPUT SCI] A type of distributed processing system in which processing functions are distributed outward from a central computer to intelligent terminal controllers or satellite information processors. Also known as host-centered system; host/satellite system.

hieratite [MINERAL] K_2SiF_6 A grayish mineral composed of potassium fluosilicate; occurs as deposits in volcanic holes.

hi-fi See high fidelity.

Higgs bosons [PARTIC PHYS] Massive scalar mesons whose existence is predicted by certain unified gage

theories of the weak and electromagnetic interactions; they are not eliminated by the Higgs mechanism.

Higgs mechanism [PARTIC PHYS] The feature of the spontaneously broken gage symmetries that the Goldstone bosons do not appear as physical particles, but instead constitute the zero helicity states of gage vector bosons of nonzero mass (such as the intermediate vector boson).

high [METEOROL] An area of high pressure, referring to a maximum of atmospheric pressure in two dimensions (closed isobars) in the synoptic surface chart, or a maximum of height (closed contours) in the constant-pressure chart; since a high is, on the synoptic chart, always associated with anticyclonic circulation, the term is used interchangeably with anticyclone.

high-alloy steel [MET] Steel containing large percentages of elements other than carbon.

high-altitude radio altimeter See radar altimeter.

high-alumina brick [MATER] Refractory brick made from raw materials rich in alumina, such as diaspore and bauxite; when well fired, they contain a large amount of mullite; used for unusually severe temperature or load conditions.

high band [COMMUN] The television band extending from 174 to 216 megahertz, which includes channels 7 to 13.

high boost See high-frequency compensation.

high brass [MET] The most common commercial wrought brass containing about 35% zinc and 65% copper.

high-carbon steel [MET] A cast or forged steel containing more than 0.5% carbon.

high clouds [METEOROL] Types of clouds whose mean lower level is above 20,000 feet (6100 meters); principal clouds in this group are cirrus, cirrocumulus, and cirrostratus.

high contrast [GRAPHICS] That area where the degree of difference between black and white approaches the maximum.

high-current rectifier [ELECTR] A solid-state device, gas tube, or vacuum tube used to convert alternating to direct current for powering low-impedance loads.

high-current switch [ELEC] A switch used to redirect heavy current flow; usually has a make-before-break feature to prevent excessive arcing.

high-energy bond [PHYS CHEM] Any chemical bond yielding a decrease in free energy of at least 5 kilocalories per mole.

high-energy electron diffraction [PHYS] The diffraction of electrons with high energies, usually in the range of 30,000–70,000 electron volts, mainly to study the structure of atoms and molecules in gases and liquids. Abbreviated HEED.

high-energy particle [PARTIC PHYS] An elementary particle having an energy of hundreds of MeV (megaelectronvolts) or more.

high-energy physics See particle physics.

higher-level language See high-level language.

higher-order language See high-level language.

highest common factor See greatest common divisor.

high explosive [MATER] An explosive with a nitroglycerin base requiring a detonator; the explosion is violent and practically instantaneous.

high fidelity [ENG ACOUS] Audio reproduction that closely approximates the sound of the original performance. Also known as hi-fi.

high frequency [COMMUN] Federal Communications Commission designation for the band from 3 to 30 megahertz in the radio spectrum. Abbreviated HF.

high-frequency compensation [ELECTR] Increasing the amplification at high frequencies with respect to that at low and middle frequencies in a given band, such as in a video band or an audio band. Also known as high boost.

high-frequency heating *See* electronic heating.

high-frequency propagation [COMMUN] Propagation of radio waves in the high-frequency band, which depends entirely on reflection from the ionosphere.

high-frequency resistance [ELEC] The total resistance offered by a device in an alternating-current circuit, including the direct-current resistance and the resistance due to eddy current, hysteresis, dielectric, and corona losses. Also known as alternating-current resistance; effective resistance; radio-frequency resistance.

high-frequency transformer [ELECTR] A transformer which matches impedances and transmits a frequency band in the carrier (or higher) frequency ranges.

high-frequency welding [MET] Resistance welding in which the heat is produced by the current flow induced by a high-frequency electromagnetic field. Also known as radio-frequency welding.

high heat [THERMO] Heat absorbed by the cooling medium in a calorimeter when products of combustion are cooled to the initial atmospheric (ambient) temperature.

high-impedance voltmeter [ELEC] A voltage-measuring device with a high-impedance input to reduce load on the unit under test; a vacuum-tube voltmeter is one type.

high-level language [COMPUT SCI] A computer language whose instructions or statements each correspond to several machine language instructions, designed to make coding easier. Also known as higher-level language; higher-order language.

high-level modulation [COMMUN] Modulation produced at a point in a system where the power level approximates that at the output of the system.

highlights [ELECTR] Bright areas occurring in a television image. [GRAPHICS] The bright parts of a subject that show up as the densest part of the negative and the lightest or whitest part of a print.

high-octane [MATER] Having an octane number in the middle or high 90s and therefore having good antiknock properties.

high-order [COMPUT SCI] Pertaining to a digit location in a numeral, the leftmost digit being the highest-order digit.

high-pass filter [ELECTR] A filter that transmits all frequencies above a given cutoff frequency and substantially attenuates all others.

high polymer [ORG CHEM] A large molecule (of molecular weight greater than 10,000) usually composed of repeat units of low-molecular-weight species; for example, ethylene or propylene.

high-positive indicator [COMPUT SCI] A component in some computers whose status is "on" if the number tested is positive and nonzero.

high-pressure area *See* anticyclone.

high-pressure cloud chamber [NUCLEO] A cloud chamber in which the gas is maintained at high pressure to reduce the range of high-energy particles and thereby increase the probability of observing events.

high-pressure gas injection [PETRO ENG] Oil reservoir pressure maintenance by injection of gas at pressures higher than those used in conventional equilibrium gas drives.

high-pressure phenomena [PHYS] Natural conditions and processes occurring at high pressures, and their duplication in the laboratory.

high-pressure physics [PHYS] The study of the effects of high pressure on the properties of matter.

high-pressure process [CHEM ENG] A chemical process operating at elevated pressure; for example, phenol manufacture at 330 atmospheres (1 atmosphere $= 1.01325 \times 10^5$ newtons per square meter), ethylene polymerization at 2000 atm, ammonia synthesis at 100–1000 atm, and synthetic-diamond manufacture up to 100,000 atm.

high Q [ELECTR] A characteristic wherein a component has a high ratio of reactance to effective resistance, so that its Q factor is high.

high-rank coal [GEOL] Coal consisting of less than 4% moisture when air-dried, or more than 84% carbon.

high-rank graywacke *See* feldspathic graywacke.

high-speed carry [COMPUT SCI] A technique in parallel addition to speed up the propagation of carries.

high-speed photography [GRAPHICS] Photography to record movement or events that occur too quickly to be observed by usual visual or photographic means; motion pictures may be shot at high speeds (50–500 frames per second) and projected at normal rates, so that the action of the subject is slowed to a point where it can be observed; or a series of individual still photos may be produced.

high-speed printer [COMPUT SCI] A printer which can function at a high rate, relative to the state of the art; 600 lines per minute is considered high speed. Abbreviated HSP.

high-speed reader [COMPUT SCI] The fastest input device existing at a particular time in the state of the technology.

high-speed storage *See* rapid storage.

high-strength alloy *See* high-tensile alloy.

high-strength low-alloy steel [MET] Steel containing small amounts of niobium or vanadium and having higher strength, better low-temperature impact toughness, and, in some grades, better atmospheric corrosion resistance than carbon steel. Abbreviated HSLA steel.

high-strength steel *See* high-tensile steel.

high-temperature alloy [MET] An alloy suitable for use at temperatures of 500°C and above.

high-temperature chemistry [PHYS CHEM] The study of chemical phenomena occurring above about 500 K.

high-temperature gas-cooled reactor [NUCLEO] A prototype gas-cooled reactor in which the coolant is pressurized helium gas with an inlet temperature of about 325°C and an outlet temperature of about 750°C, and the fuel consists of fully enriched uranium and thorium. Abbreviated HTGR.

high-temperature phenomena [PHYS] Phenomena occurring at temperatures above about 500K.

high-temperature reactor [NUCLEO] A power reactor in which the temperature is high enough for efficient generation of mechanical power.

high-tensile alloy [MET] An alloy having a high tensile strength. Also known as high-strength alloy.

high-tensile steel [MET] Low-alloy steel having a yield strength range of 50,000–100,000 pounds per square inch (3.4×10^8 to 6.9×10^8 newtons per square meter). Also known as high-strength steel.

high tension *See* high voltage.

high tide [OCEANOGR] The maximum height reached by a rising tide. Also known as high water.

high vacuum [PHYS] A vacuum with a pressure between 1×10^{-3} and 1×10^{-6} mmHg (0.1333224 and 0.0001333 newton per square meter).

high voltage [ELEC] A voltage on the order of thousands of volts. Also known as high tension.

high-voltage direct current [ELEC] A long-distance direct-current power transmission system that uses direct-current voltages up to about 1 megavolt to keep transmission losses down. Abbreviated HVDC.

highwall [MIN ENG] The unexcavated face of exposed overburden and coal or ore in an opencast mine or the face or bank of the uphill side of a contour strip-mine excavation.

high water See high tide.

high-water line [OCEANOGR] The intersection of the plane of mean high water with the shore.

high-water lunitidal interval [GEOPHYS] The interval of time between the transit (upper or lower) of the moon and the next high water at a place.

highway engineering [CIV ENG] A branch of civil engineering dealing with highway planning, location, design, and maintenance.

hilac See heavy-ion linear accelerator.

Hilbert space [MATH] A Banach space which also is an inner-product space with the inner product of a vector with itself being the same as the square of the norm of the vector.

hilgardite [MINERAL] $Ca_8(B_6O_{11})_3Cl_4 \cdot H_2O$ Colorless mineral composed of hydrous borate and chloride of calcium; occurs as monoclinic domatic crystals.

hill [GEOGR] A land surface feature characterized by strong relief; it is a prominence smaller than a mountain.

hill-climbing [MATH] Any numerical procedure for finding the maximum or maxima of a function. [MECH ENG] Adjustment, either continuous or periodic, of a self-regulating system to achieve optimum performance.

hillebrandite [MINERAL] $Ca_2SiO_3(OH)_2$ A white mineral composed of hydrous calcium silicate; occurs in masses.

Hill reaction [BIOCHEM] The release of molecular oxygen by isolated chloroplasts in the presence of a suitable electron receptor, such as ferricyanide. [ORG CHEM] Production of substituted phenylacetic acids by the oxidation of the corresponding alkylbenzene by potassium permanganate in the presence of acetic acid.

hilum [ANAT] See hilus. [BOT] Scar on a seed marking the point of detachment from the funiculus.

hilus [ANAT] An opening or recess in an organ, usually for passage of a vessel or duct. Also known as hilum.

Himalia [ASTRON] A small satellite of Jupiter with a diameter of about 35 miles (56 kilometers), orbiting at a mean distance of 7.12×10^6 miles (11.46×10^6 kilometers). Also known as Jupiter VI.

Himantandraceae [BOT] A family of dicotyledonous plants in the order Magnoliales characterized by several, uniovulate carpels and laminar stamens.

Himantopterinae [INV ZOO] A subfamily of lepidopteran insects in the family Zygaenidae including small, brightly colored moths with narrow hindwings, ribbonlike tails, and long hairs covering the body and wings.

hindbrain See rhombencephalon.

hindgut [EMBRYO] The caudal portion of the embryonic digestive tube in vertebrates.

H indicator See H scope.

hinged arch [CIV ENG] A structure that can rotate at its supports or in the center or at both places.

hinge fault [GEOL] A fault whose movement is an angular or rotational one on a side of an axis that is normal to the fault plane.

hinge joint See ginglymus.

hinge line [GEOL] 1. The line separating the region in which a beach has been thrust upward from that in which it is horizontal. 2. A line in the plane of a hinge fault separating the part of a fault along which thrust or reverse movement occurred from that having normal movement.

hinge plate [INV ZOO] 1. In bivalve mollusks, the portion of a valve that supports the hinge teeth. 2. The socket-bearing part of the dorsal valve in brachiopods.

hinge tooth [INV ZOO] A projection on a valve of a bivalve mollusk near the hinge line.

hinsdalite [MINERAL] $(Pb,Sr)Al_3(PO_4)(SO_4)(OH)_6$ A dark-gray or greenish rhombohedral mineral composed of basic lead and strontium aluminum sulfate and phosphate; occurs in coarse crystals and masses.

hinterland [GEOL] 1. The region behind the coastal district. 2. The terrain on the back of a folded mountain chain. 3. The moving block which forces geosynclinal sediments toward the foreland.

Hiodontidae [VERT ZOO] A family of tropical, freshwater actinopterygian fishes in the order Osteoglossiformes containing the mooneyes of North America.

hip [ANAT] 1. The region of the junction of thigh and trunk. 2. The hip joint, formed by articulation of the femur and hipbone.

Hippidea [INV ZOO] A group of decapod crustaceans belonging to the Anomura and including cylindrical or squarish burrowing crustaceans in which the abdomen is symmetrical and bent under the thorax.

Hippoboscidae [INV ZOO] The louse flies, a family of cyclorrhaphous dipteran insects in the section Pupipara.

hippocampus [ANAT] A ridge that extends over the floor of the descending horn of each lateral ventricle of the brain.

Hippocrateaceae [BOT] A family of dicotyledonous plants in the order Celastrales distinguished by an extrastaminal disk, mostly opposite leaves, seeds without endosperm, and a well-developed latex system.

Hippoglossidae [VERT ZOO] A family of actinopterygian fishes in the order Pleuronectiformes composed of the flounders and plaice.

Hippomorpha [VERT ZOO] A suborder of the mammalian order Perissodactyla containing horses, zebras, and related forms.

Hippopotamidae [VERT ZOO] The hippopotamuses, a family of palaeodont mammals in the superfamily Anthracotherioidea.

hippopotamus [VERT ZOO] The common name for two species of artiodactyl ungulates composing the family Hippopotamidae.

Hipposideridae [VERT ZOO] The Old World leaf-nosed bats, a family of mammals in the order Chiroptera.

hip rafter [BUILD] A diagonal rafter extending from the plate to the ridge of a roof.

hip roof [ARCH] A roof having four slopes; the two shorter ends are triangular.

hirsutism [MED] An abnormal condition characterized by growth of hair in unusual places and in unusual amounts.

Hirudinea [INV ZOO] A class of parasitic or predatory annelid worms commonly known as leeches; all have 34 body segments and terminal suckers for attachment and locomotion.

Hirudinidae [VERT ZOO] The swallows, a family of passeriform birds in the suborder Oscines.

hisingerite [MINERAL] $Fe_2^{3+}Si_2O_5(OH)_4 \cdot 2H_2O$ A black, amorphous mineral composed of hydrous ferric silicate; an iron ore.

histidine [BIOCHEM] $C_6H_9O_2N_3$ A crystalline basic amino acid present in large amounts in hemoglobin and resulting from the hydrolysis of most proteins.

histiocyte See macrophage.

histochemistry [BIOCHEM] A science that deals with the distribution and activities of chemical components in tissues.

histocompatibility [IMMUNOL] The capacity to accept or reject a tissue graft.

histogram [STAT] A graphical representation of a distribution function by means of rectangles whose widths represent intervals into which the range of observed values is divided and whose heights represent the number of observations occurring in each interval.

histology [ANAT] The study of the structure and chemical composition of animal and plant tissues as related to their function.

Histosol [GEOL] An order of wet soils consisting mostly of organic matter, popularly called peats and mucks.

Histriobdellidae [INV ZOO] A small family of errantian polychaete worms that live as ectoparasites on crayfishes.

hit [COMPUT SCI] The obtaining of a correct answer in a mechanical information-retrieval system. [ELEC] A momentary electrical disturbance on a transmission line. [ORD] 1. A blow or impact on a target by a bullet, bomb, or other projectile. 2. An instance of striking something with a bomb or the like.

hitch [GEOL] 1. A fault of strata common in coal measures, accompanied by displacement. 2. A minor dislocation of a vein or stratum not exceeding in extent the thickness of the vein or stratum. [MIN ENG] 1. A step cut in the rock face to hold timber support in an underground working. 2. A hole cut in side rock solid enough to hold the cap of a set of timbers, permitting the leg to be dispensed with.

hit rate [COMPUT SCI] The ratio of the number of records accessed in a run to the number of records in the file, expressed as a percentage.

Hittorf dark space See cathode dark space.

Hittorf method [PHYS CHEM] A procedure for determining transference numbers in which one measures changes in the composition of the solution near the cathode and near the anode of an electrolytic cell, due to passage of a known amount of electricity.

Hittorf principle [ELECTR] The principle that a discharge between electrodes in a gas at a given pressure does not necessarily occur between the closest points of the electrodes if the distance between these points lies to the left of the minimum on a graph of spark potential versus distance. Also known as shortpath principle.

hives See urticaria.

h-line See h-bar.

H mode See transverse electric mode.

Ho See holmium.

hoarfrost [HYD] A deposit of interlocking ice crystals formed by direct sublimation on objects. Also known as white frost.

hobber See hobbing machine.

hobbing machine [MECH ENG] A machine for cutting gear teeth in gear blanks or for cutting worm, spur, or helical gears. Also known as hobber.

hobnail [DES ENG] A short, large-headed, sharp-pointed nail; used to attach soles to heavy shoes.

hod [CIV ENG] A tray fitted with a handle by which it can be carried on the shoulder for transporting bricks or mortar.

Hodgkin's disease [MED] A disease characterized by a neoplastic proliferation of atypical histiocytes in one or several lymph nodes. Also known as lymphogranulomatosis.

hodgkinsonite [MINERAL] $MnZnSiO_5 \cdot H_2O$ A pink to reddish-brown mineral composed of hydrous zinc manganese silicate; occurs as crystals.

hodograph [PHYS] 1. The curve traced out in the course of time by the tip of a vector representing some physical quantity. 2. In particular, the path traced out by the velocity vector of a given particle.

hodoscope [NUCLEO] An array of small Geiger counters, scintillation counters, or other radiation counters used in tracing paths of high-energy particles in experiments with particle accelerators or in cosmic rays. [OPTICS] See conoscope.

Hodotermitidae [INV ZOO] A family of lower (primitive) termites in the order Isoptera.

hoegbomite [MINERAL] $Mg(Al,Fe,Ti)_4O_7$ A black mineral composed of an oxide of magnesium, aluminum, iron, and titanium. Also spelled högbomite.

hoernesite [MINERAL] $Mg_3As_2O_8 \cdot H_2O$ A white, monoclinic mineral composed of hydrous magnesium arsenate; occurs as gypsumlike crystals.

hofmannite See hartite.

Hofmann reaction [ORG CHEM] A reaction in which amides are degraded by treatment with bromine and alkali (caustic soda) to amines containing one less carbon; used commercially in the production of nylon.

Hofmann rearrangement [ORG CHEM] A chemical rearrangement of the hydrohalides of N-alkylanilines upon heating to give aminoalkyl benzenes.

högbomite See hoegbomite.

hoghorn antenna See horn antenna.

hohlraum See blackbody.

hohmannite [MINERAL] $Fe_2(SO_4)_2(OH)_2 \cdot 7H_2O$ A chestnut brown to burnt orange and amaranth red, triclinic mineral consisting of a hydrated basic sulfate of iron.

hoist [MECH ENG] 1. To move or lift something by a rope-and-pulley device. 2. A power unit for a hoisting machine, designed to lift from a position directly above the load and therefore mounted to facilitate mobile service.

hoist frame See headframe.

hoisting drum See drum.

hoisting machine [MECH ENG] A mechanism for raising and lowering material with intermittent motion while holding the material freely suspended.

Holasteridae [INV ZOO] A family of exocyclic Euechinoidea in the order Holasteroida; individuals are oval or heart-shaped, with fully developed pore pairs.

Holasteroida [INV ZOO] An order of exocyclic Euechinoidea in which the apical system is elongated along the anteroposterior axis and teeth occur only in juvenile stages.

hold [COMPUT SCI] To retain information in a computer storage device for further use after it has been initially utilized. [ELECTR] To maintain storage elements at equilibrium voltages in a charge storage tube by electron bombardment. [ENG] The interior of a ship or plane, especially the cargo compartment. [IND ENG] A therblig, or basic operation, in time-and-motion study in which the hand or other body member maintains an object in a fixed position and location.

hold circuit [ELECTR] A circuit in a sampled-data control system that converts the series of impulses, generated by the sampler, into a rectangular function, in order to smooth the signal to the motor or plant.

hold control [ELECTR] A manual control that changes the frequency of the horizontal or vertical sweep oscillator in a television receiver, so that the frequency more nearly corresponds to that of the incoming synchronizing pulses.

holdenite [MINERAL] A red, orthorhombic mineral composed of basic manganese zinc arsenate with a small amount of calcium, magnesium, and iron.

holdfast [BOT] 1. A suckerlike base which attaches the thallus of certain algae to the support. 2. A disklike terminal structure on the tendrils of various plants used for attachment to a flat surface. [INV ZOO] An organ by which parasites such as tapeworms attach themselves to the host.

holding anode [ELECTR] A small auxiliary anode used in a mercury-pool rectifier to keep a cathode spot energized during the intervals when the main-anode current is zero.

holding beam [ELECTR] A diffused beam of electrons used to regenerate the charges stored on the dielectric surface of a cathode-ray storage tube.

holding coil [ELECTR] A separate relay coil that is energized by contacts which close when a relay pulls in, to hold the relay in its energized position after the orginal operating circuit is opened.

holding current [ELECTR] The minimum current required to maintain a switching device in a closed or conducting state after it is energized or triggered.

holding time [COMMUN] Period of time a trunk or circuit is in use on a call, including operator's time in connecting and subscriber's or user's conversation time.

hold lamp [ELEC] Indicating lamp which remains lighted while a telephone connection is being held.

hold mode [COMPUT SCI] The state of an analog computer in which its operation is interrupted without altering the values of the variables it is handling, so that computation can continue when the interruption is over. Also known as interrupt mode.

hold-over command [COMPUT SCI] A command punched at the end of each card to cause machines to treat the several cards as if they were one continuous record.

hole [SOLID STATE] A vacant electron energy state near the top of an energy band in a solid; behaves as though it were a positively charged particle. Also known as electron hole.

hole conduction [ELECTR] Conduction occurring in a semiconductor when electrons move into holes under the influence of an applied voltage and thereby create new holes.

Holectypidae [PALEON] A family of extinct exocyclic Euechinoidea in the order Holectypoida; individuals are hemispherical.

Holectypoida [INV ZOO] An order of exocyclic Euechinoidea with keeled, flanged teeth, with distinct genital plates, and with the ambulacra narrower than the interambulacra on the adoral side.

hole injection [ELECTR] The production of holes in an *n*-type semiconductor when voltage is applied to a sharp metal point in contact with the surface of the material.

hole mobility [ELECTR] A measure of the ability of a hole to travel readily through a semiconductor, equal to the average drift velocity of holes divided by the electric field.

hole saw *See* crown saw.

hole site [COMPUT SCI] The area on a punch card where a hole may be punched.

hole trap [ELECTR] A semiconductor impurity capable of releasing electrons to the conduction or valence bands, equivalent to trapping a hole.

holism [BIOL] The view that the whole of a complex system, such as a cell or organism, is functionally greater than the sum of its parts. Also known as organicism.

holistic masks [COMPUT SCI] In character recognition, that set of characters which resides within a character reader and theoretically represents the exact replicas of all possible input characters.

hollandite [MINERAL] $Ba(Mn^{2+},Mn^{4+})_8O_{16}$ A silvery-gray to black mineral composed of manganate of barium and manganese; occurs as crystals.

Hollerith code [COMPUT SCI] A code used to represent letters, numbers, or special symbols to be punched in standard 80-column punch cards.

Hollerith string [COMPUT SCI] A sequence of characters preceded by an H and a character count in FORTRAN, as 4HSTOP.

Hollinacea [PALEON] A dimorphic superfamily of extinct ostracods in the suborder Beyrichicopina including forms with sulci, lobation, and some form of velar structure.

Hollinidae [PALEON] An extinct family of ostracods in the superfamily Hollinacea distinguished by having a bulbous third lobe on the valve.

hollow cathode [ELECTR] A cathode which is hollow and closed at one end in a discharge tube filled with inert gas, designed so that radiation is emitted from the cathode glow inside the cathode.

hollow drill [DES ENG] A drill rod or stem having an axial hole for the passage of water or compressed air to remove cuttings from a drill hole. Also known as hollow rod; hollow stem.

hollow-pipe waveguide [ELECTROMAG] A waveguide consisting of a hollow metal pipe; electromagnetic waves are transmitted through the interior and electric currents flow on the inner surfaces.

hollow rod *See* hollow drill.

hollow stem *See* hollow drill.

hollow wall [BUILD] A masonry wall provided with an air space between the inner and outer wythes.

holly [BOT] The common name for the trees and shrubs composing the genus *Ilex*; distinguished by spiny leaves and small berries.

hollywood lignumvitae *See* lignumvitae.

holmium [CHEM] A rare-earth element belonging to the yttrium subgroup, symbol Ho, atomic number 67, atomic weight 164.93, melting point 1400–1525°C.

holmquisite [MINERAL] $(Na,K,Ca)Li(Mg,Fe)_3Al_2Si_8O_{22}(OH)_2$ A bluish-black, orthorhombic mineral composed of alkali and silicate of iron, magnesium, lithium, and aluminum.

Holobasidiomycetes [MYCOL] An equivalent name for Homobasidiomycetidae.

holoblastic [EMBRYO] Pertaining to eggs that undergo total cleavage due to the absence of a mass of yolk.

Holocentridae [VERT ZOO] A family of nocturnal beryciform fishes found in shallow tropical and subtropical reefs; contains the squirrelfishes and soldierfishes.

Holocephali [VERT ZOO] The chimaeras, a subclass of the Chondrichthyes, distinguished by four pairs of gills and gill arches, an erectile dorsal fin and spine, and naked skin.

holoclastic [PETR] Being or belonging to ordinary (sedimentary) clastic rock.

holocrine gland [PHYSIO] A structure whose cells undergo dissolution and are entirely extruded, together with the secretory product.

holoenzyme [BIOCHEM] A complex, fully active enzyme, containing an apoenzyme and a coenzyme.

hologamy [BIOL] Condition of having gametes similar in size and form to somatic cells. [BOT] Condition of having the whole thallus develop into a gametangium.

hologram [OPTICS] The special photographic plate used in holography; when this negative is developed and illuminated from behind by a coherent gas-laser beam, it produces a three-dimensional image in space. Also known as hologram interferometer.

hologram interferometer See hologram.

holographic interferometry [OPTICS] The study of the formation and interpretation of the fringe pattern which appears when a wave, generated at some earlier time and stored in a hologram, is later reconstructed and caused to interfere with a comparison wave.

holographic memory [COMPUT SCI] A memory in which information is stored in the form of holographic images on thermoplastic or other recording films.

holography [PHYS] A technique for recording, and later reconstructing, the amplitude and phase distributions of a wave disturbance; widely used as a method of three-dimensional optical image formation, and also with acoustical and radio waves; in optical image formation, the technique is accomplished by recording on a photographic plate the pattern of interference between coherent light reflected from the object of interest, and light that comes directly from the same source or is reflected from a mirror.

holohedral [CRYSTAL] Pertaining to a crystal structure having the highest symmetry in each crystal class. Also known as holosymmetric; holosystemic.

holohedron [CRYSTAL] A crystal form of the holohedral class, having all the faces needed for complete symmetry.

Holometabola [INV ZOO] A division of the insect subclass Pterygota whose members undergo holometabolous metamorphosis during development.

holometabolous metamorphosis [INV ZOO] Complete metamorphosis, during which there are four stages; the egg, larva, pupa, and imago or adult.

holomictic lake [HYD] A lake whose water circulates completely from top to bottom.

holomorphic function See analytic function.

holonephros [VERT ZOO] Type of kidney having one nephron beside each somite along the entire length of the coelom; seen in larvae of myxinoid cyclostomes.

holonomic constraints [MECH] An integrable set of differential equations which describe the restrictions on the motion of a system; a function relating several variables, in the form $f(x_1, ..., x_n) = 0$, in optimization or physical problems.

holonomic system [MECH] A system in which the constraints are such that the original coordinates can be expressed in terms of independent coordinates and possibly also the time.

Holoptychidae [PALEON] A family of extinct lobefin fishes in the order Osteolepiformes.

Holostei [VERT ZOO] An infraclass of fishes in the subclass Actinopterygii descended from the Chondrostei and ancestral to the Teleostei.

holostratotype [GEOL] The originally defined stratotype.

holosymmetric See holohedral.

holosystemic See holohedral.

Holothuriidae [VERT ZOO] A family of aspidochirotacean echinoderms in the order Aspidochirotida possessing tentacular ampullae and only the left gonad.

Holothuroidea [INV ZOO] The sea cucumbers, a class of the subphylum Echinozoa characterized by a cylindrical body and smooth, leathery skin.

Holothyridae [INV ZOO] The single family of the acarine suborder Holothyrina.

Holothyrina [INV ZOO] A suborder of mites in the order Acarina which are large and hemispherical with a deep-brown, smooth, heavily sclerotized cuticle.

Holotrichia [INV ZOO] A major subclass of the protozoan class Ciliatea; body ciliation is uniform with cilia arranged in longitudinal rows.

holotype [SYST] A nomenclatural type for the single specimen designated as "the type" by the orginal author at the time of publication of the original description.

Holuridae [PALEON] A group of extinct chondrostean fishes in the suborder Palaeoniscoidei distinguished in having lepidotrichia of all fins articulated but not bifurcated, fins without fulcra, and the tail not cleft.

Homacodontidae [PALEON] A family of extinct palaeodont mammals in the superfamily Dichobunoidea.

Homalopteridae [VERT ZOO] A small family of cypriniform fishes in the suborder Cyprinoidei.

Homalorhagae [INV ZOO] A suborder of the class Kinorhyncha having a single dorsal plate covering the neck and three ventral plates on the third zonite.

Homalozoa [INV ZOO] A subphylum of echinoderms characterized by the complete absence of radial symmetry.

Homaridae [INV ZOO] A family of marine decapod crustaceans containing the lobsters.

home address [COMPUT SCI] A technique used to identify each disk track uniquely by means of a 9-byte record immediately following the index marker; the record contains a flag (good or defective track), cylinder number, head number, cyclic check, and bit count appendage.

homeoblastic [PETR] Of a metamorphic crystalloblastic texture, having constituent minerals of approximately the same size.

homeomorphic spaces [MATH] Two topological spaces with a homeomorphism existing between them; intuitively one can be obtained from the other by stretching, twisting, or shrinking.

homeomorphism [MATH] A continuous map between topological spaces which is one-to-one, onto, and its inverse function is continuous. Also known as bicontinuous function; topological mapping.

homeosis See heteromorphosis.

homeostasis [BIOL] In higher animals, the maintenance of an internal constancy and an independence of the environment.

homeothermia [PHYSIO] The condition of being warm-blooded.

home record [COMPUT SCI] The first record in the chaining method of file organization.

hometaxial-base transistor [ELECTR] Transistor manufactured by a single-diffusion process to form both emitter and collector junctions in a uniformly doped silicon slice; the resulting homogeneously doped base region is free from accelerating fields in the axial (collector-to-emitter) direction, which could cause undesirable high current flow and destroy the transistor.

holosystemic See holohedral.

homilite [MINERAL] $(Ca_2(Fe,Mg)B_2Si_2O_{10}$ A black or blackish brown mineral composed of iron calcium borosilicate.

homing beacon [NAV] A radio beacon, either airborne or on the ground, toward which an aircraft can fly if equipped with a radio compass or homing adapter. Also known as radio homing beacon.

homing device [ELECTR] A control device that automatically starts in the correct direction of motion or rotation to achieve a desired change, as in a remote-control tuning motor for a television receiver. [ENG] A device incorporated in a guided missile or the like to home it on a target. [NAV] A transmitter, receiver, or adapter used for homing aircraft or used by aircraft for homing purposes.

homing guidance [ENG] A guidance system in which a missile directs itself to a target by means of a self-contained mechanism that reacts to a particular characteristic of the target.

homing relay [ELEC] A stepping relay that returns to a specified starting position before each operating cycle.

homing torpedo [ORD] A torpedo having homing guidance, designed for homing on a surface vessel or a submerged submarine.

homing transponder [NAV] A small acoustic transponder used in navigation of a submersible vehicle, which can be carried by the vehicle and quickly dropped when an area of interest is reached; a single transponder and a dead reckoning system are used.

Hominidae [VERT ZOO] A family of primates in the superfamily Hominoidea containing one living species, *Homo sapiens*.

Hominoidea [VERT ZOO] A superfamily of the order Primates comprising apes and humans.

homo- [ORG CHEM] **1.** Indicating the homolog of a compound differing in formula from the latter by an increase of one CH_2 group. **2.** Indicating a homopolymer made up of a single type of monomer, such as polyethylene from ethylene. **3.** Indicating that a skeletal atom has been added to a well-known structure. [SCI TECH] Prefix indicating the same or similar.

Homobasidiomycetidae [MYCOL] A subclass of basidiomycetous fungi in which the basidium is not divided by cross walls.

homocentric [OPTICS] Pertaining to rays which have the same focal point, or which are parallel. Also known as stigmatic.

homocercal [VERT ZOO] Pertaining to the caudal fin of certain fishes which has almost equal lobes, with the vertebral column terminating near the middle of the base.

homochlamydeous [BIOL] Having all members of the perianth similar or not differentiated into calyx or corolla.

homocysteine [BIOCHEM] $C_4H_9O_2NS$ An amino acid formed in animals by demethylation of methionine.

homodont [VERT ZOO] Having all teeth similar in form; characteristic of nonmammalian vertebrates.

homogametic sex [GEN] The sex of a species in which the paired sex chromosomes are of equal size and which therefore produces homogametes.

homogamety [GEN] The production of homogametes by one sex of a species.

homogamy [BIOL] Inbreeding due to isolation. [BOT] Condition of having all flowers alike.

homogeneous [CHEM] Pertaining to a substance having uniform composition or structure. [MATH] Pertaining to a group of mathematical symbols of uniform dimensions or degree. [SCI TECH] Uniform in structure or composition.

homogeneous catalysis [CHEM] Catalysis occurring within a single phase, usually a gas or liquid.

homogeneous coordinates [MATH] To a point in the plane with cartesian coordinates (x,y) there corresponds the homogeneous coordinates (x_1,x_2,x_3), where $x_1/x_3 = x$, $x_2/x_3 = y$; any polynomial equation in cartesian coordinates becomes homogeneous if a change into these coordinates is made.

homogeneous differential equation [MATH] A differential equation where every scalar multiple of a solution is also a solution.

homogeneous equation [MATH] An equation that can be rewritten into the form having zero on one side of the equal sign and a homogeneous function of all the variables on the other side.

homogeneous polynomial [MATH] A polynomial all of whose terms have the same total degree; equivalently it is a homogenous function of the variables involved.

homogeneous radiation [PHYS] Radiation having an extremely narrow band of frequencies, or a beam of monoenergetic particles of a single type, so that all components of the radiation are alike.

homogeneous reactor [NUCLEO] A nuclear reactor in which fissionable material and moderator (if used) are intimately mixed to form an effectively homogeneous medium for neutrons.

homogeneous space [MATH] A topological space having a group of transformations acting upon it, that is, a transformation group, where for any two points x and y some transformation from the group will send x to y.

homogeneous transformation See linear transformation.

homogenizer [MECH ENG] A machine that blends or emulsifies a substance by forcing it through fine openings against a hard surface.

homograft [BIOL] Graft from a donor transplanted to a genetically dissimilar recipient of the same species. Also known as allograft.

Homoistela [PALEON] A class of extinct echinoderms in the subphylum Homalozoa.

homolecithal [CYTOL] Referring to eggs having small amounts of evenly distributed yolk. Also known as isolecithal.

homologous [BIOL] Pertaining to a structural relation between parts of different organisms due to evolutionary development from the same or a corresponding part, such as the wing of a bird and the pectoral fin of a fish. [GEOL] **1.** Referring to strata, in separated areas, that are correlatable (contemporaneous) and are of the same general character or facies, or occupy analogous structural positions along the strike. **2.** Pertaining to faults, in separated areas, that have the same relative position or structure.

homology [CHEM] The relation among elements of the same group, or family, in the periodic table. [ORG CHEM] That state, in a series of organic compounds that differ from each other by a CH_2 such as the methane series C_nH_{2n+2}, in which there is a similarity between the compounds in the series and a graded change of their properties.

homology group [MATH] Associated to a topological space X, one of a sequence of Abelian groups $H_n(X)$ that reflect how n-dimensional simplicial complexes can be used to fill up X and also help determine the presence of n-dimensional holes appearing in X. Also known as Betti group.

homology theory [MATH] Theory attempting to compare topological spaces and investigate their structures by determining the algebraic nature and interrelationships appearing in the various homology groups.

homomorphism [BOT] Having perfect flowers consisting of only one type. [MATH] A function between two algebraic systems of the same type which preserves the algebraic operations.

Homoneura [INV ZOO] A suborder of the Lepidoptera with mandibulate mouthparts, and fore- and hindwings that are similar in shape and venation.

homopolar [ELEC] 1. Electrically symmetrical. 2. Having equal distribution of charge.

homopolar bond [PHYS CHEM] A covalent bond whose total dipole moment is zero.

homopolar generator [ELECTR] A direct-current generator in which the poles presented to the armature are all of the same polarity, so that the voltage generated in active conductors has the same polarity at all times; a pure direct current is thus produced, without commutation. Also known as acyclic machine; homopolar machine; unipolar machine.

homopolar machine See homopolar generator.

Homoptera [INV ZOO] An order of the class Insecta including a large number of sucking insects of diverse forms.

Homosclerophorida [INV ZOO] An order of primitive sponges of the class Demospongiae with a skeleton consisting of equirayed, tetraxonid, siliceous spicules.

homoserine [BIOCHEM] $C_4H_9O_3N$ An amino acid formed as an intermediate product in animals in the metabolic breakdown of cystathionine to cysteine.

homosexual [BIOL] Of, pertaining to, or being the same sex. [PSYCH] 1. Of, pertaining to, or exhibiting homosexuality. 2. One who practices homosexuality.

homosexuality [PSYCH] 1. State of being sexually attracted to members of the same sex. 2. A form of homoerotism involving sexual interest without genital expression.

homotopy [MATH] Between two mappings of the same topological spaces, a continuous function representing how, in a step-by-step fashion, the image of one mapping can be continuously deformed onto the image of the other.

homotopy theory [MATH] The study of the topological structure of a space by examining the algebraic properties of its various homotopy groups.

homotype [SYST] A taxonomic type for a specimen which has been compared with the holotype by another than the author of the species and determined by him to be conspecific with it.

homozygote [GEN] An individual that has identical alleles at one or more loci and therefore produces gametes which are all identical.

honeycomb coil [ELECTROMAG] A coil wound in a crisscross manner to reduce distributed capacitance. Also known as duolateral coil; lattice-wound coil.

honeycomb radiator [MECH ENG] A heat-exchange device utilizing many small cells, shaped like a bees' comb, for cooling circulating water in an automobile.

honeydew [INV ZOO] The viscous secretion deposited on leaves by many aphids and scale insects; an attractant for ants.

Honey Dew melon [BOT] A variety of muskmelon (Cucumis melo) belonging to the Violales; fruit is large, oval, smooth, and creamy yellow to ivory, without surface markings.

honing [MECH ENG] The process of removing a relatively small amount of material from a cylindrical surface by means of abrasive stones to obtain a desired finish or extremely close dimensional tolerance.

hood [DES ENG] An opaque shield placed above or around the screen of a cathode-ray tube to eliminate extraneous light. [ENG] 1. Close-fitting, rubber head covering that leaves the face exposed; used in scuba diving. 2. A protective covering, usually providing special ventilation to carry away objectionable fumes, dusts, and gases, in which dangerous chemical, biological, or radioactive materials can be safely handled.

hoof-and-mouth disease See foot-and-mouth disease.

hook [DES ENG] A piece of hard material, especially metal, formed into a curve for catching, holding, or pulling something. [ELECTR] A circuit phenomenon occurring in four-zone transistors, wherein hole or electron conduction can occur in opposite directions to produce voltage drops that encourage other types of conduction. [GEOGR] The end of a spit of land that is turned toward shore. Also known as recurved spit.

hook collector transistor [ELECTR] A transistor in which there are four layers of alternating n- and p-type semiconductor material and the two interior layers are thin compared to the diffusion length. Also known as hook transistor; pn hook transistor.

hooked spit See hook.

Hooker diaphragm cell [CHEM ENG] A device used in industry for the electrolysis of brine (sodium chloride) to make chlorine and caustic soda (sodium hydroxide) or caustic potash (potassium hydroxide); saturated purified brine fed around the anode passes through the diaphragm to the cathode; chlorine is formed at the anode and hydrogen released at the cathode, leaving sodium hydroxide and residual sodium chloride in the cell liquor; the diaphragm prevents the products from mixing.

Hookeriales [BOT] An order of the mosses with irregularly branched stems and leaves that appear to be in one plane.

Hooke's joint [MECH ENG] A simple universal joint; consists of two yokes attached to their respective shafts and connected by means of a spider. Also known as Cardan joint.

hook transistor See hook collector transistor.

hookup [ELEC] An arrangement of circuits and apparatus for a particular purpose.

hookworm [INV ZOO] The common name for parasitic roundworms composing the family Ancylostomidae.

hoot stop [COMPUT SCI] A closed loop that generates an audible signal; usually employed to signal an error or for operating convenience.

hop [BOT] Humulus lupulus. A dioecious liana of the order Urticales distinguished by herbaceous vines produced from a perennial crown; the inflorescence, a catkin, of the female plant is used commercially for beer production. [COMMUN] A single reflection of a radio wave from the ionosphere back to the earth in traveling from one point to another.

hopeite [MINERAL] $Zn_3(PO_4)_2 \cdot 4H_2O$ A gray, orthorhombic mineral composed of hydrous phosphate of zinc; specific gravity is 2.76–2.85; dimorphous with parahopeite.

hophornbeam [BOT] Any tree of the genus Ostrya in the birch family recognized by its very scaly bark and the fruit which closely resembles that of the hopvine.

Hoplocarida [INV ZOO] A superorder of the class Crustacea with the single order Stomatopoda.

Hoplonemertini [INV ZOO] An order of unsegmented, ribbonlike worms in the class Enopla; all species have an armed proboscis.

Hoplophoridae [INV ZOO] A family of prawns containing numerous bathypelagic representatives.

hopper [COMPUT SCI] *See* card hopper. [ENG] A funnel-shaped receptacle with an opening at the top for loading and a discharge opening at the bottom for bulk-delivering material such as grain or coal.

hopperburn [PL PATH] A disease of potato and peanut plants caused by a leafhopper which secretes a toxic substance on the leaves, causing browning and shriveling.

hordeolum [MED] A furuncular inflammation of the connective tissue of the eyelids near a hair follicle. Also known as sty.

horizon [ASTRON] The apparent boundary line between the sky and the earth or sea. Also known as apparent horizon. [GEOL] **1.** The surface separating two beds. **2.** One of the layers, each of which is a few inches to a foot thick, that make up a soil.

horizon system of coordinates [ASTRON] A set of celestial coordinates based on the celestial horizon as the primary great circle.

horizontal [SCI TECH] Being in a plane perpendicular to the gravitational field, that is, perpendicular to a plumb line, at a given point on the earth's surface.

horizontal blanking pulse [ELECTR] The rectangular pulse that forms the pedestal of the composite television signal between active horizontal lines and causes the beam current of the picture tube to be cut off during retrace. Also known as line-frequency blanking pulse.

horizontal centering control [ELECTR] The centering control provided in a television receiver or cathode-ray oscilloscope to shift the position of the entire image horizontally in either direction on the screen.

horizontal chromatography [ANALY CHEM] Paper chromatography in which the chromatogram is horizontal instead of vertical.

horizontal convergence control [ELECTR] The control that adjusts the amplitude of the horizontal dynamic convergence voltage in a color television receiver.

horizontal definition *See* horizontal resolution.

horizontal deflection electrode [ELECTR] One of a pair of electrodes that move the electron beam horizontally from side to side on the fluorescent screen of a cathode-ray tube employing electrostatic deflection.

horizontal deflection oscillator [ELECTR] The oscillator that produces, under control of the horizontal synchronizing signals, the sawtooth voltage waveform that is amplified to feed the horizontal deflection coils on the picture tube of a television receiver. Also known as horizontal oscillator.

horizontal distributed processing system [COMPUT SCI] A type of distributed system in which two or more computers which are logically equivalent are connected together, with no hierarchy or master/slave relationship.

horizontal drive control [ELECTR] The control in a television receiver, usually at the rear, that adjusts the output of the horizontal oscillator. Also known as drive control.

horizontal feed [COMPUT SCI] A card feed in which punch cards are placed in a hopper and enter and pass through a card track, all in a horizontal position.

horizontal flow chart [COMPUT SCI] A graphical representation of the movement of forms, punch cards, and other recording media through an organization, showing the movement of each medium from the time it is first used to the time it is destroyed.

horizontal frequency *See* line frequency.

horizontal hold control [ELECTR] The hold control that changes the free-running period of the horizontal deflection oscillator in a television receiver, so that the picture remains steady in the horizontal direction.

horizontal instruction [COMPUT SCI] An instruction in machine language to carry out independent operations on various operands in parallel or in a well-defined time sequence.

horizontal intensity variometer [ENG] Essentially a declination variometer with a larger, stiffer fiber than in the standard model; there is enough torsion in the fiber to cause the magnet to turn 90° out of the magnetic meridian; the magnet is aligned with the magnetic prime vertical to within 0.5° so it does not respond appreciably to changes in declination. Also known as H variometer.

horizontal linearity control [ELECTR] A linearity control that permits narrowing or expanding of the width of the left half of a television receiver image, to give linearity in the horizontal direction so that circular objects appear as true circles.

horizontal line frequency *See* line frequency.

horizontal magnetometer [ENG] A measuring instrument for ascertaining changes in the horizontal component of the magnetic field intensity.

horizontal oscillator *See* horizontal deflection oscillator.

horizontal output stage [ELECTR] The television receiver stage that feeds the horizontal deflection coils of the picture tube through the horizontal output transformer; may also include a part of the second-anode power supply for the picture tube.

horizontal output transformer [ELECTR] A transformer used in a television receiver to provide the horizontal deflection voltage, the high voltage for the second-anode power supply of the picture tube, and the filament voltage for the high-voltage rectifier tube. Also known as flyback transformer; horizontal sweep transformer.

horizontal parallax [ASTRON] The geocentric parallax of a celestial object when it is rising or setting. [GRAPHICS] *See* absolute stereoscopic parallax.

horizontal pendulum [MECH] A pendulum that moves in a horizontal plane, such as a compass needle turning on its pivot.

horizontal projection [GRAPHICS] A drawing of a structure as it would appear if projected on a horizontal plane.

horizontal resolution [ELECTR] The number of individual picture elements or dots that can be distinguished in a horizontal scanning line of a television or facsimile image. Also known as horizontal definition.

horizontal return tubular boiler [MECH ENG] A fire-tube boiler having tubes within a cylindrical shell that are attached to the end closures; products of combustion are transported under the lower half of the shell and back through the tubes.

horizontal scanning [ENG] In radar scanning, rotating the antenna in azimuth around the horizon or in a sector. Also known as searching lighting.

horizontal sweep [ELECTR] The sweep of the electron beam from left to right across the screen of a cathode-ray tube.

horizontal sweep transformer *See* horizontal output transformer.

horizontal synchronizing pulse [ELECTR] The rectangular pulse transmitted at the end of each line in a television system, to keep the receiver in line-by-line synchronism with the transmitter. Also known as line synchronizing pulse.

horizontal system [COMPUT SCI] A programming system in which instructions are written horizontally, that is, across the page.

hormone [BIOCHEM] **1.** A chemical messenger produced by endocrine glands and secreted directly into the bloodstream to exert a specific effect on a distant part of the body. **2.** An organic compound that is synthesized in minute quantities in one part of a plant and translocated to another part, where it influences a specific physiological process.

horn [ELECTROMAG] *See* horn antenna. [ENG ACOUS] A tube whose cross-sectional area increases from one end to the other, used to radiate or receive sound waves and to intensify and direct them. Also known as acoustic horn. [GEOL] A topographically high, sharp, pyramid-shaped mountain peak produced by the headward erosion of mountain glaciers; the Matterhorn is the classic example. [MET] Holding arm for the electrode of a resistance spot-welding machine.

horn angle [MATH] A geometric figure formed by two tangent plane curves that lie on the same side of their mutual tangent line in the neighborhood of the point of tangency.

horn antenna [ELECTROMAG] A microwave antenna produced by flaring out the end of a circular or rectangular waveguide into the shape of a horn, for radiating radio waves directly into space. Also known as electromagnetic horn; flare (British usage); hoghorn antenna (British usage); horn; horn radiator.

hornbeam [BOT] Any tree of the genus *Carpinus* in the birch family distinguished by doubly serrate leaves and by small, pointed, angular winter buds with scales in four rows.

hornblende [MINERAL] A general name given to the monoclinic calcium amphiboles that form an extensive solid-solution series between the various metals in the generalized formula $(Ca,Na)_2(Mg,Fe,Al)_5$-$(Al,Si)_8O_{22}(OH,F)_2$.

horned dinosaur [PALEON] Common name for extinct reptiles of the suborder Ceratopsia.

horned liverwort [BOT] The common name for bryophytes of the class Anthocerotae. Also known as hornwort.

horned toad [VERT ZOO] The common name for any of the lizards of the genus *Phrynosoma;* a reptile that resembles a toad but is less bulky.

hornet [INV ZOO] The common name for a number of large wasps in the hymenopteran family Vespidae.

hornfels [PETR] A common name for a class of metamorphic rocks produced by contact metamorphism and characterized by equidimensional grains without preferred orientation.

hornfels facies [PETR] Rock formed at depths in the earth's crust not exceeding 10 kilometers at temperatures of 250–800°C; includes albite-epidote hornfels facies, pyroxene-hornfels facies, and hornblende-hornfels facies.

horn quicksilver *See* calomel.

horn radiator *See* horn antenna.

horn silver *See* cerargyrite.

hornwort *See* horned liverwort.

horseback [GEOL] A low and sharp ridge of sand, gravel, or rock. [MIN ENG] **1.** Shale or sandstone occurring in a channel that was cut by flowing water in a coal seam. Also known as cutout; horse; roll; swell; symon fault; washout. **2.** To move or raise

a heavy piece of machinery or timber by using a pinch bar as a lever. Also known as pinch.

horse chestnut [BOT] *Aesculus hippocastanum.* An ornamental buckeye tree in the order Sapindales, usually with seven leaflets per leaf and resinous buds.

horsehair blight [PL PATH] A fungus disease of tea and certain other tropical plants caused by *Marasmius equicrinis* and characterized by black festoons of mycelia hanging from the branches.

horsehead [MIN ENG] Timbers or steel joists used to support planks in tunneling through loose ground.

Horsehead Nebula [ASTRON] A cloud of obscuring particles between the earth and a gaseous emission nebula in the constellation Orion.

horse latitudes [METEOROL] The belt of latitudes over the oceans at approximately 30–35°N and S where winds are predominantly calm or very light and weather is hot and dry.

horsepower [MECH] The unit of power in the British engineering system, equal to 550 foot-pounds per second, approximately 745.7 watts. Abbreviated hp.

horseradish [BOT] *Armoracia rusticana.* A perennial crucifer belonging to the order Capparales and grown for its pungent roots, used as a condiment.

horseshoe bend *See* oxbow.

horseshoe crab [INV ZOO] The common name for arthropods composing the subclass Xiphosurida, especially the subgroup Limulida.

horseshoe lake *See* oxbow lake.

horsetail [BOT] The common name for plants of the genus *Equisetum* composing the order Equisetales.

horst [GEOL] **1.** A block of the earth's crust uplifted along faults relative to the rocks on either side. **2.** A mass of the earth's crust limited by faults and standing in relief. **3.** One of the older mountain masses limiting the Alps to the west and north. **4.** A knobby ledge of limestone beneath a thin soil mantle.

hortonolite [MINERAL] $(Fe,Mg,Mn)_2SiO_4$ A dark mineral composed of silicate of iron, magnesium, and manganese; a member of the olivine series.

host [BIOL] **1.** An organism on or in which a parasite lives. **2.** The dominant partner of a symbiotic or commensal pair.

host-centered system *See* hierarchical distributed processing system.

host computer [COMPUT SCI] The computer upon which depends a specialized computer handling the input/output functions in a real-time system.

host processor [COMPUT SCI] The central computer in a hierarchical distributed processing system, which is typically located at some central site where it serves as a focal point for the collection of data, and often for the provision of services which cannot economically be distributed.

host/satellite system *See* hierarchical distributed processing system.

hot [ELEC] *See* energized. [NUCLEO] Being highly radioactive.

hot-air engine [MECH ENG] A heat engine in which air or other gases, such as hydrogen, helium, or nitrogen, are used as the working fluid, operating on cycles such as the Stirling or Ericsson.

hot-air furnace [MECH ENG] An encased heating unit providing warm air to ducts for circulation by gravity convection or by fans.

hot-air sterilization [ENG] A method of sterilization using dry heat for glassware and other heat-resistant materials which need to be dry after treatment; temperatures of 160–165°C are generated for at least 2 hours.

hot atom [NUCLEO] An atom that has high internal or kinetic energy as a result of a nuclear process such as beta decay or neutron capture.

hot-bulb [MECH ENG] Pertaining to an ignition method used in semidiesel engines in which the fuel mixture is ignited in a separate chamber kept above the ignition temperature by the heat of compression.

hot carrier [ELECTR] A carrier, which may be either an electron or a hole, that has relatively high energy with respect to the carriers normally found in majority-carrier devices such as thin-film transistors.

hot-carrier diode See Schottky barrier diode.

hot cathode [ELECTR] A cathode in which electron or ion emission is produced by heat. Also known as thermionic cathode.

hot-cathode gas-filled tube See thyratron.

hot-cathode tube See thermionic tube.

hot cell See cave.

hot chamber die casting [ENG] A die-casting process in which a piston is driven through a reservoir of molten metal and thereby delivers a quantity of molten metal to the die cavity.

hot dipping [MET] Coating metal components by immersion in a molten metal bath, such as tin or zinc.

hot electron [ELECTR] An electron that is in excess of the thermal equilibrium number and, for metals, has an energy greater than the Fermi level; for semiconductors, the energy must be a definite amount above that of the edge of the conduction band.

hot-filament ionization gage [ELECTR] An ionization gage in which electrons emitted by an incandescent filament, and attracted toward a positively charged grid electrode, collide with gas molecules to produce ions which are then attracted to a negatively charged electrode; the ion current is a measure of the number of gas molecules.

hot forming [MET] Shaping operations performed at temperatures above the recrystallization temperature of the metal.

hot-gas welding [ENG] Joining of thermoplastic materials by softening first with a jet of hot air, then joining at the softened points.

hot hole [ELECTR] A hole that can move at much greater velocity than normal holes in a semiconductor.

hot junction [ELECTR] The heated junction of a thermocouple.

hot pressing [ENG] Forming a metal-powder compact or a ceramic shape by applying pressure and heat simultaneously at temperatures high enough for sintering to occur.

hot-press printing See hot stamping.

hot-quenching [MET] Quenching from high temperatures into a medium of lower but still high temperature.

hot spot [CHEM ENG] An area or point within a reaction system at which the temperature is appreciably higher than in the bulk of the reactor; usually locates the reaction front. [FOR] A forest region where fires occur at frequent intervals. [GRAPHICS] A region of excessive illumination on a photo. [MOL BIO] A site in a gene at which there is an unusually high frequency of mutation. [NUCLEO] **1.** A surface area of higher than average radioactivity. **2.** A part of a reactor fuel surface element that has become overheated. [PHYS] A localized region with temperature higher than the surroundings.

hot spring [HYD] A thermal spring whose water temperature is above 98°F (37°C).

hot stamping [GRAPHICS] A method of printing in which heated type or stamping dies are pressed against a thin leaf of gold or other metal, or against pigment upon a surface such as paper, board, plastic, or leather; the thin leaf contains a sizing agent that is set by heat; this system is sometimes used for making movie titles and artwork for slides. Also known as hot-press printing.

hot strength See tensile strength.

hot-water heating [MECH ENG] A heating system for a building in which the heat-conveying medium is hot water and the heat-emitting means are radiators, convectors, or panel coils. Also known as hydronic heating.

hot-wire ammeter [ENG] An ammeter which measures alternating or direct current by sending it through a fine wire, causing the wire to heat and to expand or sag, deflecting a pointer. Also known as thermal ammeter.

hot-wire anemometer [ENG] An anemometer used in research on air turbulence and boundary layers; the resistance of an electrically heated fine wire placed in a gas stream is altered by cooling by an amount which depends on the fluid velocity.

hot-wire microphone [ENG ACOUS] A velocity microphone that depends for its operation on the change in resistance of a hot wire as the wire is cooled by varying particle velocities in a sound wave.

hour [MECH] A unit of time equal to 3600 seconds. Abbreviated hr.

hour angle [ASTRON] Angular distance west of a celestial meridian or hour circle; the arc of the celestial equator, or the angle at the celestial pole, between the upper branch of a celestial meridian or hour circle and the hour circle of a celestial body or the vernal equinox, measured westward through 360°.

hour circle [ASTRON] An imaginary great circle passing through the celestial poles on the celestial sphere above which declination is measured. Also known as circle of declination; circle of right ascension.

hourglass [HOROL] An instrument for measuring time, consisting of two somewhat funnel-shaped vessels joined at their narrow points by a thin, hollow neck; one hour is required for a given quantity of sand to fall from the upper to the lower vessel.

housefly [INV ZOO] *Musca domestica*. A dipteran insect with lapping mouthparts commonly found near human habitations; a vector in the transmission of many disease pathogens.

housekeeping [COMPUT SCI] Those operations or routines which do not contribute directly to the solution of a computer program, but rather to the organization of the program.

housekeeping run [COMPUT SCI] The performance of a program or routine to maintain the structure of files, such as sorting, merging, addition of new records, or deletion or modification of existing records.

hovercraft See air-cushion vehicle.

howardite [GEOL] An achondritic stony meteorite composed chiefly of calcic plagioclase and orthopyroxene.

Howell-Bunger valve See cone valve.

Howell-Jolly bodies [PATH] Small, round basophilic inclusions of nuclear material in erythrocytes of splenectomized persons.

Howe truss [CIV ENG] A truss for spans up to 80 feet (24 meters) having both vertical and diagonal members; made of steel or timber or both.

howl [ENG ACOUS] Undesirable prolonged sound produced by a radio receiver or audio-frequency amplifier system because of either electric or acoustic feedback.

howler [COMMUN] In telephone practice, an associated unit by which the test desk operator may connect a high tone of varying loudness to a subscriber's line to call the subscriber's attention to the fact that the phone receiver is off the hook. [ELECTR] An audio device used to warn a radar operator that signals are appearing on a radar screen.

howlite [MINERAL] $Ca_2Bi_5SiO_9(OH)_5$ A white mineral occurring in nodular or earthy form.

Hoyer method of prestressing *See* pretensioning.

hp *See* horsepower.

H plane [ELECTROMAG] The plane of an antenna in which lies the magnetic field vector of linearly polarized radiation.

hr *See* hour.

H-R diagram *See* Hertzsprung-Russell diagram.

H scan *See* H scope.

H scope [NAV] A modified form of B scope on which the signal appears as two dots, the slope of the line joining them indicating elevation angle; the horizontal coordinate indicates bearing angle and the vertical coordinate range, with respect to the left-hand dot. Also known as H indicator; H scan.

HSLA steel *See* high-strength low-alloy steel.

HSP *See* high-speed printer.

HTGR *See* high-temperature gas-cooled reactor.

hub [COMPUT SCI] An electric socket in a plugboard into which one may insert or connect leads or may plug wires. [DES ENG] 1. The cylindrical central part of a wheel, propeller, or fan. 2. A piece in a lock that is turned by the knob spindle, causing the bolt to move. 3. A short coupling that joins plumbing pipes. [MET] A steel punch used in making a working die for a coin or medal.

hubble [ASTRON] A unit of astronomical distance equal to 10^9 light-years or 9.4605×10^{24} meters.

Hubble constant [ASTROPHYS] The rate at which the velocity of recession of the galaxies increases with distance; the value is about 30 kilometers per second per million light-years (or 3.2×10^{-18} sec^{-1}) with an uncertainty of about ± 15 km/sec.

Hubble effect *See* red shift.

huckleberry [BOT] The common name for shrubs of the genus *Gaylussacia* in the family Ericaceae distinguished by an ovary with 10 locules and 10 ovules; the dark-blue berries are edible.

hue [OPTICS] The name of a color, such as red, yellow, green, blue, or purple, as perceived subjectively.

huebnerite [MINERAL] $MnWO_4$ A brownish-red to black manganese member of the wolframite series, occurring in short, monoclinic, prismatic crystals; isomorphous with ferberite.

hue control [ELECTR] A control that varies the phase of the chrominance signals with respect to that of the burst signal in a color television receiver, in order to change the hues in the image. Also known as phase control.

Hugoniot function [PHYS] A function specifying the locus of states which are possible immediately after the passage of a shock front; gives the state's pressure as a function of its specific volume.

hühnerkobelite [MINERAL] $(Na,Ca)(Fe,Mn)_2(PO_4)_2$ A mineral composed of phosphate of sodium, calcium, iron, and manganese; it is isomorphous with varulite.

hull [BOT] The outer, usually hard, covering of a fruit or seed. [FOOD ENG] 1. To remove husks from fruits and seeds, as from ears of corn, nuts, or peas. 2. To remove the shell of a crustacean or mollusk, as an oyster. [NAV ARCH] The body or shell of a ship. [ORD] 1. The outer casing of a rocket, guided missile, or the like. 2. Massive ar-

mored body of a tank, exclusive of tracks, motor, turret, and armament.

hulsite [MINERAL] $(Fe^{2+},Mg)_2(Fe^{3+},Sn)(BO_3)O_2$ A black mineral composed of iron calcium magnesium tin borate.

hum [ELEC] A sound produced by an iron core of a transformer due to loose laminations or magnetostrictive effects; the frequency of the sound is twice the power line frequency. [ELECTR] An electrical disturbance occurring at the power supply frequency or its harmonics, usually 60 or 120 hertz in the United States.

human ecology [ECOL] The branch of ecology that considers the relations of individual persons and of human communities with their particular environment.

human-factors engineering [ENG] The area of knowledge dealing with the capabilities and limitations of human performance in relation to design of machines, jobs, and other modifications of the human's physical environment.

Humboldt Current *See* Peru Current.

humboldtine [MINERAL] $FeC_2O_4 \cdot 2H_2O$ A mineral composed of hydrous ferrous oxalate. Also known as humboldtite; oxalite.

humboldtite *See* humboldtine.

hum-bucking coil [ENG ACOUS] A coil wound on the field coil of an excited-field loudspeaker and connected in series opposition with the voice coil, so that hum voltage induced in the voice coil is canceled by that induced in the hum-bucking coil.

humectant [CHEM] A substance which absorbs or retains moisture; examples are glycerol, propylene glycol, and sorbitol; used in preparing confectioneries and dried fruit.

humeral [ANAT] Of or pertaining to the humerus or the shoulder region.

Hume-Rothery compound *See* electron compound.

Hume-Rothery rule [SOLID STATE] The rule that the ratio of the number of valence electrons to the number of atoms in a given phase of an electron compound depends only on the phase, and not on the elements making up the compounds.

humerus [ANAT] The proximal bone of the forelimb in vertebrates; the bone of the upper arm in humans, articulating with the glenoid fossa and the radius and ulna.

humic coal [GEOL] A coal whose attritus is composed mainly of transparent humic degradation material.

humidification [ENG] The process of increasing the water vapor content of a gas.

humidifier [MECH ENG] An apparatus for supplying moisture to the air and for maintaining desired humidity conditions.

humidistat [ENG] An instrument that measures and controls relative humidity. Also known as hydrostat.

humidity [METEOROL] Atmospheric water vapor content, expressed in any of several measures, such as relative humidity.

humidity index [CLIMATOL] An index of the degree of water surplus over water need at any given station; it is calculated as humidity index = $100s/n$, where s (the water surplus) is the sum of the monthly differences between precipitation and potential evapotranspiration for those months when the normal precipitation exceeds the latter, and where n (the water need) is the sum of monthly potential evapotranspiration for those months of surplus.

humin [GEOL] *See* ulmin. [ORG CHEM] An insoluble pigment formed in the acid hydrolysis of a protein that contains tryptophan.

Humiriaceae [BOT] A family of dicotyledonous plants in the order Linales characterized by exappendiculate petals, usually five petals, flowers with an intrastaminal disk, and leaves lacking stipules.

humite [MINERAL] 1. A humic coal mineral. 2. A series of magnesium neosilicate minerals closely related in crystal structure and chemical composition.

humivore [ECOL] An organism that feeds on humus.

hummingbird [VERT ZOO] The common name for members of the family Trochilidae; fast-flying, short-legged, weak-footed insectivorous birds with a tubular, pointed bill and a fringed tongue.

hummocked ice [OCEANOGR] Pressure ice, characterized by haphazardly arranged mounds or hillocks; it has less definite form, and show the effects of greater pressure, than either rafted ice or tented ice, but in fact may develop from either of those.

hummocky [GEOL] Any topographic surface characterized by rounded or conical mounds.

Humod [GEOL] A suborder of the soil order Spodosol having an accumulation of humus, and of aluminum but not iron.

humogelite See ulmin.

humor [PHYSIO] A fluid or semifluid part of the body.

Humox [GEOL] A suborder of the soil order Oxisol that is high in organic matter, well drained but moist all or nearly all year, and restricted to relatively cool climates and high altitudes for Oxisols.

humpback See kyphosis.

hump yard [CIV ENG] A switch yard in a railway system that has a hump or steep incline down which freight cars can coast to prescheduled locations. Also known as gravity yard.

Humult [GEOL] A suborder of the soil order Ultisol, well drained with a moderately thick surface horizon; formed under conditions of high rainfall distributed evenly over the year; common in southeastern Brazil.

humus [GEOL] The amorphous, ordinarily dark-colored, colloidal matter in soil; a complex of the fractions of organic matter of plant, animal, and microbial origin that are most resistant to decomposition.

hundredweight [SCI TECH] Abbreviated cwt. 1. A unit of weight, in common use in the United States, equal to 100 pounds or the weight of 45.359237 kilograms. Also known as cental; centner; kintal; quintal; short hundredweight. 2. A unit of weight in common use in the United Kingdom equal to 112 pounds or to the weight of 50.80234544 kilograms. Also known as long hundredweight. 3. A unit of weight in troy measure equal to 100 troy pounds or the weight of 37.32417216 kilograms.

Hund rules [ATOM PHYS] Two rules giving the order in energy of atomic states formed by equivalent electrons: of the terms given by equivalent electrons, the ones with greatest multiplicity have the least energy, and of these the one with greatest orbital angular momentum is lowest; the state of a multiplet with lowest energy is that in which th total angular momentum is the least possible, if the shell is less than half-filled, and the greatest possible, if more than half filled.

hunting [CONT SYS] Undesirable oscillation of an automatic control system, wherein the controlled variable swings on both sides of the desired value. [ELECTR] Operation of a selector in moving from terminal to terminal until one is found which is idle. [MECH ENG] Irregular engine speed resulting from instability of the governing device.

Hunting Dogs See Canes Venatici.

Huntington's chorea [MED] A rare hereditary disease of the basal ganglia and cerebral cortex resulting in choreiform (dancelike) movements, intellectual deterioration, and psychosis.

hunting tooth [DES ENG] An extra tooth on the larger of two gear wheels so that the total number of teeth will not be an integral multiple of the number on the smaller wheel.

huntite [MINERAL] $CaMg_3(CO_3)_4$ A white mineral consisting of calcium magnesium carbonate.

hureaulite [MINERAL] $Mn_5H_2(PO_4)_4 \cdot 4H_2O$ A monoclinic mineral of varying colors consisting of a hydrated acid phosphate of manganese.

Huronian [GEOL] The lower system of the restricted Proterozoic.

hurricane [METEOROL] A tropical cyclone of great intensity; any wind reaching a speed of more than 73 miles per hour (117 kilometers per hour) is said to have hurricane force.

hurricane-force wind [METEOROL] In the Beaufort wind scale, a wind whose speed is 64 knots (117 kilometers per hour) or higher.

hurricane surge See hurricane wave.

hurricane tide See hurricane wave.

hurricane wave [OCEANOGR] As experienced on islands and along a shore, a sudden rise in the level of the sea associated with a hurricane. Also known as hurricane surge; hurricane tide.

hurricane wind [METEOROL] In general, the severe wind of an intense tropical cyclone (hurricane or typhoon); the term has no further technical connotation, but is easily confused with the strictly defined hurricane-force wind. Also known as typhoon wind.

husk [BOT] The outer coat of certain seeds, particularly if it is a dry, membranous structure.

hutchinsonite [MINERAL] $(Pb,Tl)_2(Cu,Ag)As_5S_{10}$ Red mineral composed of sulfide of lead, copper, and arsenic, with varying amounts of thallium and silver, occurring in small orthorhombic crystals.

Huttig equation [THERMO] An equation which states that the ratio of the volume of gas adsorbed on the surface of a nonporous solid at a given pressure and temperature to the volume of gas required to cover the surface completely with a unimolecular layer equals $(1 + r) c^r/(1 + c^r)$, where r is the ratio of the equilibrium gas pressure to the saturated vapor pressure of the adsorbate at the temperature of adsorption, and c is the product of a constant and the exponential of $(q - q_l)/RT$, where q is the heat of adsorption into a first layer molecule, q_l is the heat of liquefaction of the adsorbate, T is the temperature, and R is the gas constant.

huttonite [MINERAL] $ThSiO_4$ A colorless to pale-green monoclinic mineral composed of silicate of thorium; it is dimorphous with thorite.

Huygens' approximation [MATH] The length of a small circular arc is approximately $\frac{1}{3}(8c' - c)$, where c is the chord of the arc and c' is the chord of half the arc.

Huygens eyepiece [OPTICS] An eyepiece in which there are two plano-convex lenses, and the plane sides of both lenses face the eye.

Huygens' principle [OPTICS] The principle that each point on a light wavefront may be regarded as a source of secondary waves, the envelope of these secondary waves determining the position of the wavefront at a later time.

H variometer See horizontal intensity variometer.

HVDC See high-voltage direct current.

H vector [ELECTROMAG] A vector that is the magnetic field. For a plane wave in free space, it is perpendicular to the E vector and to the direction of propagation.

H wave See transverse electric wave.

hyacinth *See* zircon.

Hyades [ASTRON] A V-shaped open star cluster about 150 light-years from the sun, which appears in the constellation Taurus near the star Aldebaran.

Hyaenidae [VERT ZOO] A family of catlike carnivores in the superfamily Feloidea including the hyenas and aardwolf.

Hyaenodontidae [PALEON] A family of extinct carnivorous mammals in the order Deltatheridia.

Hyalellidae [INV ZOO] A family of amphipod crustaceans in the suborder Gammaridea.

hyaline [BIOCHEM] A clear, homogeneous, structureless material found in the matrix of cartilage, vitreous body, mucin, and glycogen.   [GEOL] Transparent and resembling glass.

hyaline membrane [HISTOL] 1. A basement membrane. 2. A membrane of a hair follicle between the inner fibrous layer and the outer root sheath.

hyaline test [INV ZOO] A translucent wall or shell of certain foraminiferans composed of layers of calcite interspersed with separating membranes.

hyalite [MINERAL] A colorless, clear or translucent variety of opal occurring as globular concretions or botryoidal crusts in cavities or cracks of rocks. Also known as Müller's glass; water opal.

hyalobasalt *See* tachylite.

hyaloclastite [GEOL] A tufflike deposit formed by the flowing of basalt under water and ice and its consequent fragmentation. Also known as aquagene tuff.

Hyalodictyae [MYCOL] A subdivision of the spore group Dictyosporae characterized by hyaline spores.

Hyalodidymae [MYCOL] A subdivision of the spore group Didymosporae characterized by hyaline spores.

Hyalohelicosporae [MYCOL] A subdivision of the spore group Helicosporae characterized by hyaline spores.

hyaloophitic [PETR] Of the texture of igneous rocks, being composed principally of a glassy ground mass with little interstitial texture.

hyalophane [MINERAL] $BaAl_2Si_2O_8$ A colorless feldspar mineral crystallizing in the monoclinic system; isomorphous with adularia. Also known as baryta feldspar.

Hyalophragmiae [MYCOL] A subdivision of the spore group Phragmosporae characterized by hyaline spores.

hyaloplasm [CYTOL] The optically clear, viscous to gelatinous ground substance of cytoplasm in which formed bodies are suspended.

hyalopsite *See* obsidian.

Hyaloscolecosporae [MYCOL] A subdivision of the spore group Scalecosporae characterized by hyaline spores.

Hyalospongia [PALEON] A class of extinct glass sponges, equivalent to the living Hexactinellida, having siliceous spicules made of opaline silica.

Hyalosporae [MYCOL] A subdivision of the spore group Amerosporae characterized by hyaline spores.

Hyalostaurosporae [MYCOL] A subdivision of the spore group Staurosporae characterized by hyaline spores.

hyalotekite [MINERAL] $(Pb,Ca,Ba)_4BSi_6O_{17}(OH,F)$ A white gray mineral composed of borosilicate and fluoride of lead, barium, and calcium, occurring in crystalline masses.

hyaluronate lyase *See* hyaluronidase.

hyaluronic acid [BIOCHEM] A polysaccharide found as an integral part of the gellike substance of animal connective tissue.

hyaluronidase [BIOCHEM] Any one of a family of enzymes which catalyze the breakdown of hyalu-

ronic acid. Also known as hyaluronate lyase; spreading factor.

Hybodontoidea [PALEON] An ancient suborder of extinct fossil sharks in the order Selachii.

hybrid [GEN] The offspring of genetically dissimilar parents.   [PETR] Pertaining to a rock formed by the assimilation of two magmas.   [SCI TECH] Having two or more different characteristics or types of structure.

hybrid circuit [ELEC] A circuit in which two or more basically different types of components, such as tubes and transistors, performing similar functions are used together.

hybrid coil *See* hybrid transformer.

hybrid computer [COMPUT SCI] A computer designed to handle both analog and digital data. Also known as analog-digital computer.

hybrid input/output [COMPUT SCI] The routines required to handle inputs to and outputs from a computer system comprising digital and analog computers.

hybrid integrated circuit [ELECTR] A circuit in which one or more discrete components are used in combination with integrated-circuit construction.

hybrid interface [COMPUT SCI] A device that joins a digital to an analog computer, converting digital signals transmitted serially by the digital computer into analog signals that are transmitted simultaneously to the various units of the analog computer, and vice versa.

hybridized orbital [PHYS CHEM] A molecular orbital which is a linear combination of two or more orbitals of comparable energy (such as $2s$ and $2p$ orbitals), is concentrated along a certain direction in space, and participates in formation of a directed valence bond.

hybrid junction [ELECTR] A transformer, resistor, or waveguide circuit or device that has four pairs of terminals so arranged that a signal entering at one terminal pair divides and emerges from the two adjacent terminal pairs, but is unable to reach the opposite terminal pair. Also known as bridge hybrid.

hybrid network [COMMUN] Nonhomogeneous communications network required to operate with signals of dissimilar characteristics (such as analog and digital modes).

hybridoma [IMMUNOL] A hybrid myeloma formed by fusing myeloma cells with lymphocytes that produce a specific antibody; the individual cells can be cloned, and each clone produces large amounts of identical (monoclonal) antibody.

hybrid relay [ELEC] A relay in which solid-state elements are combined with moving contacts.

hybrid repeater *See* hybrid transformer.

hybrid rocket [AERO ENG] A rocket with an engine utilizing a liquid propellant with a solid propellant in the same rocket engine.

hybrid set [ELEC] Two or more transformers interconnected to form a hybrid junction. Also known as transformer hybrid.

hybrid sterility [GEN] Inability to form functional gametes in a hybrid due to disturbances in sex-cell development or in meiosis, caused by incompatible genetic constitution.

hybrid thin-film circuit [ELECTR] Microcircuit formed by attaching discrete components and semiconductor devices to networks of passive components and conductors that have been vacuum-deposited on glazed ceramic, sapphire, or glass substrates.

hybrid transformer [ELEC] A single transformer that performs the essential functions of a hybrid set. Also known as bridge transformer; hybrid coil; hybrid repeater.

hybrid vigor   See heterosis.

hybrid wave function   [QUANT MECH] A linear combination of wave functions of one problem used as an approximation to the wave function in another problem; for example, a linear combination of atomic orbitals used to represent a molecular bond.

hydatid disease   See echinococcosis.

hydatidosis   See echinococcosis.

hydatiform mole   [MED] A benign placental tumor formed as a cystic growth of the chorionic villi. Also known as hydatiform tumor.

hydatiform tumor   See hydatiform mole.

hydra   [INV ZOO] Any species of *Hydra* or related genera, consisting of a simple, tubular body with a mouth at one end surrounded by tentacles, and a foot at the other end for attachment.

Hydra   [ASTRON] A large constellation of the Southern Hemisphere, right ascension 10 hours, declination 20° south. Also known as Water Monster. [INV ZOO] A common genus of coelenterates in the suborder Anthomedusae.

Hydrachnellae   [INV ZOO] A family of generally freshwater predacious mites in the suborder Trombidiformes, including some parasitic forms.

Hydraenidae   [INV ZOO] The equivalent name for Limnebiidae.

hydranencephaly   [MED] A congenital anomaly in which there are vestiges of a cerebellum, occipital lobes, and basal nuclei; frontal and parietal lobes are replaced by a cyst; and the neurocranium is undeveloped.

hydrargillite   See gibbsite.

hydrase   [BIOCHEM] An enzyme that catalyzes removal or addition of water to a substrate without hydrolyzing it.

hydrate   [CHEM] **1.** A form of a solid compound which has water in the form of H_2O molecules associated with it; for example, anhydrous copper sulfate is a white solid with the formula $CuSO_4$, but when crystallized from water a blue crystalline solid with formula $CuSO_4 \cdot 5H_2O$ results, and the water molecules are an integral part of the crystal. **2.** A crystalline compound resulting from the combination of water and a gas; frequently a constituent of natural gas that is under pressure.

hydrated cellulose   See hydrocellulose.

hydrated halloysite   See endellite.

hydration   [CHEM] The incorporation of molecular water into a complex molecule with the molecules or units of another species; the complex may be held together by relatively weak forces or may exist as a definite compound.

hydraulic   [ENG] Operated or effected by the action of water or other fluid of low viscosity.

hydraulic accumulator   [MECH ENG] A hydraulic flywheel that stores potential energy by accumulating a quantity of pressurized hydraulic fluid in a suitable enclosed vessel.

hydraulic actuator   [MECH ENG] A cylinder or fluid motor that converts hydraulic power into useful mechanical work; mechanical motion produced may be linear, rotary, or oscillatory.

hydraulic amplifier   [CONT SYS] A device which increases the power of a signal in a hydraulic servomechanism or other system through the use of fixed and variable orifices. Also known as hydraulic intensifier.

hydraulic analogy   [FL MECH] The analogy between the flow of a shallow liquid and the flow of a compressible gas; various phenomena such as shock waves

occur in both systems; the analogy requires neglect of vertical accelerations in the liquid, and restrictions on the ratio of specific heats for the gas.

hydraulic brake   [MECH ENG] A brake in which the retarding force is applied through the action of a hydraulic press.

hydraulic circuit   [MECH ENG] A circuit whose operation is analogous to that of an electric circuit except that electric currents are replaced by currents of water or other fluids, as in a hydraulic control.

hydraulic clutch   See fluid drive.

hydraulic coupling   See fluid coupling.

hydraulic engineering   [CIV ENG] A branch of civil engineering concerned with the design, erection, and construction of sewage disposal plants, waterworks, dams, water-operated power plants, and such.

hydraulic excavation   See hydraulicking.

hydraulic extraction   See hydraulicking.

hydraulic fluid   [MATER] A low-viscosity fluid used in operating a hydraulic mechanism.

hydraulic fracturing   [PETRO ENG] A method in which sand-water mixtures are forced into underground wells under pressure; the pressure splits the petroleum-bearing sandstone, thereby allowing the oil to move toward the wells more freely.

hydraulic friction   [FL MECH] Resistance to flow which is exerted on the surface of contact between a stream and its conduit and which induces a loss of energy.

hydraulic gradient   [FL MECH] With regard to an aquifer, the rate of change of pressure head per unit of distance of flow at a given point and in a given direction. [HYD] The slope of the hydraulic grade line of a stream.

hydraulic intensifier   See hydraulic amplifier.

hydraulic jump   [FL MECH] A steady-state, finite-amplitude disturbance in a channel, in which water passes turbulently from a region of (uniform) low depth and high velocity to a region of (uniform) high depth and low velocity; when applied to hydraulic jumps, the usual hydraulic formulas governing the relations of velocity and depth do not conserve energy.

hydraulicking   [MIN ENG] Excavating alluvial or other mineral deposits by means of high-pressure water jets. Also known as hydraulic excavation; hydraulic extraction; hydroextraction.

hydraulic limestone   [MATER] Limestone, containing silica and alumina, which produces lime that hardens in water.

hydraulic machine   [MECH ENG] A machine powered by a motor activated by the confined flow of a stream of liquid, such as oil or water under pressure.

hydraulic nozzle   [MECH ENG] An atomizing device in which fluid pressure is converted into fluid velocity.

hydraulic packing   [ENG] Packing material that resists the effects of water even under high pressure.

hydraulic power system   [MECH ENG] A power transmission system comprising machinery and auxiliary components which function to generate, transmit, control, and utilize hydraulic energy.

hydraulic radius   [FL MECH] The ratio of the cross-sectional area of a conduit in which a fluid is flowing to the inner perimeter of the conduit.

hydraulics   [FL MECH] The branch of science and technology concerned with the mechanics of fluids, especially liquids.

hydride   [INORG CHEM] A compound containing hydrogen and another element; examples are H_2S, which is a hydride although it may be properly called hydrogen sulfide, and lithium hydride, LiH.

Hydrobatidae [VERT ZOO] The storm petrels, a family of oceanic birds in the order Procellariiformes.

hydrobiotite [MINERAL] A light-green, trioctahedral clay mineral of mixed layers of biotite and vermiculite.

hydroboracite [MINERAL] $CaMgB_6O_{11} \cdot 6H_2O$ A white mineral composed of hydrous calcium magnesium borate, occurring in fibrous and foliated masses.

hydroboration [ORG CHEM] The process of producing organoboranes by the addition of a compound with a B-H bond to an unsaturated hydrocarbon; for example, the reaction of diborane ion with a carbonyl compound.

hydrocalumite [MINERAL] $Ca_2Al(OH)_7 \cdot 3H_2O$ A colorless to light-green mineral composed of a hydrous hydroxide of calcium and aluminum.

hydrocarbon [ORG CHEM] One of a very large group of chemical compounds composed only of carbon and hydrogen; the largest source of hydrocarbons is from petroleum crude oil.

hydrocarbon resins [ORG CHEM] Brittle or gummy materials prepared by the polymerization of several unsaturated constituents of coal tar, rosin, or petroleum; they are inexpensive and find uses in rubber and asphalt formulations and in coating and caulking compositions.

hydrocellulose [MATER] A gelatinous mass formed from the reaction of cellulose with water either by grinding cellulose and mixing with water, or by using strong salt solutions, acids, or alkalies; used in the manufacture of artifical fibers such as rayon, mercerized cotton, paper, and vulcanized fiber. Also known as hydrated cellulose.

hydrocephaly [MED] Increased volume of cerebrospinal fluid in the skull.

hydrocerussite [MINERAL] $Pb_3(OH)_2(CO_3)_2$ A colorless mineral composed of basic lead carbonate, occurring as crystals in thin hexagonal plates.

Hydrocharitaceae [BOT] The single family of the order Hydrocharitales, characterized by an inferior, compound ovary with laminar placentation.

Hydrocharitales [BOT] A monofamilial order of aquatic monocotyledonous plants in the subclass Alismatidae.

Hydrocorallina [INV ZOO] An order in some systems of classification set up to include the coelenterate groups Milleporina and Stylasterina.

Hydrocorisae [INV ZOO] A subdivision of the Hemiptera containing water bugs with concealed antennae and without a bulbus ejaculatorius in the male.

hydrocortisone [BIOCHEM] $C_{21}H_{30}O_5$ The generic name for 17-hydroxycorticosterone; an adrenocortical steroid occurring naturally and prepared synthetically; its effects are similar to cortisone, but it is more active. Also known as cortisol.

hydrocracking [CHEM ENG] A catalytic, high-pressure petroleum refinery process that is flexible enough to produce either high-octane gasoline or aviation jet fuel; the two main reactions are the adding of hydrogen to petroleum-derived molecules too massive and complex for gasoline and then the cracking of them to the required fuels; the catalyst is an acidic solid and a hydrogenating metal component.

hydrocyanite See chalcocyanite.

Hydrodamalinae [VERT ZOO] A monogeneric subfamily of sirenian mammals in the family Dugongidae.

hydrodesulfurization [CHEM ENG] A catalytic process in which the petroleum feedstock is reacted with hydrogen to reduce the sulfur content in the oil.

hydrodynamic equations [FL MECH] Three equations which express the net acceleration of a unit water particle as the sum of the partial accelerations due to pressure gradient force, frictional force, earth's deflecting force, gravitational force, and other factors.

hydrodynamic pressure [FL MECH] The difference between the pressure of a fluid and the hydrostatic pressure; this concept is useful chiefly in problems of the steady flow of an incompressible fluid in which the hydrostatic pressure is constant for a given elevation (as when the fluid is bounded above by a rigid plate), so that the external force field (gravity) may be eliminated from the problem.

hydrodynamics [FL MECH] The study of the motion of a fluid and of the interactions of the fluid with its boundaries, especially in the incompressible inviscid case.

hydroelectric generator [MECH ENG] An electric rotating machine that transforms mechanical power from a hydraulic turbine or water wheel into electric power.

hydroelectricity [ELEC] Electric power produced by hydroelectric generators.

hydroextraction See hydraulicking.

hydrofining [CHEM ENG] A fixed-bed catalytic process to desulfurize and hydrogenate a wide range of charge stocks, from gases through waxes; the catalyst comprises cobalt oxide and molybdenum oxide on an extruded alumina support and may be regenerated in place by air and steam or flue gas.

hydrofoil [NAV ARCH] **1.** A flat or airfoil-shaped plate attached to a ship to stabilize it in roll. **2.** Foils attached by struts to the bottom of a hydrofoil boat to lift the hull out of the water as its speed is increased.

hydroforming [CHEM ENG] A petroleum-refinery process in which naphthas are passed over a catalyst at elevated temperatures and moderate pressures in the presence of added hydrogen or hydrogen-containing gases, to form high-octane BTX aromatics for motor fuels or chemical manufacture.

hydroformylation [CHEM ENG] The reaction of adding hydrogen and the —CHO group to the carbon atoms across a double bond to yield oxygenated derivatives; an example is in the oxo process where the term hydroformylation applies to those reactions brought about by treating olefins with a mixture of hydrogen and carbon monoxide in the presence of a cobalt catalyst.

hydrogasification [CHEM ENG] A technique to manufacture synthetic pipeline gas from coal; pulverized coal is reacted with hot, raw, hydrogen-rich gas containing a substantial amount of steam at 1000 psig (6.9×10^6 newtons per square meter, gage) to form methane.

hydrogel [CHEM] The formation of a colloid in which the disperse phase (colloid) has combined with the continuous phase (water) to produce a viscous jellylike product; for example, coagulated silicic acid.

hydrogen [CHEM] The first chemical element, symbol H, in the periodic table, atomic number 1, atomic weight 1.00797; under ordinary conditions it is a colorless, odorless, tasteless gas composed of diatomic molecules, H_2; used in manufacture of ammonia and methanol, for hydrofining, for desulfurization of petroleum products, and to reduce metallic oxide ores.

hydrogenase [BIOCHEM] Enzyme that catalyzes the oxidation of hydrogen.

hydrogenation [CHEM ENG] Saturation of diolefin impurities in gasolines to form a stable product. [ORG CHEM] Catalytic reaction of hydrogen with other compounds, usually unsaturated; for example, unsaturated cottonseed oil is hydrogenated to form solid fats.

hydrogen bomb [ORD] A device in which heavy hydrogen nuclei, under intense heat and pressure, undergo an uncontrolled, self-sustaining fusion reaction to produce an explosion. Also known as H bomb.

hydrogen bond [PHYS CHEM] A type of bond formed when a hydrogen atom bonded to atom A in one molecule makes an additional bond to atom B either in the same or another molecule; the strongest hydrogen bonds are formed when A and B are highly electronegative atoms, such as fluorine, oxygen, or nitrogen.

hydrogen burning [ASTROPHYS] Thermonuclear reactions occurring in the cores of main-sequence stars, in which nuclei of hydrogen fuse to form helium nuclei.

hydrogen electrode [PHYS CHEM] A noble metal (such as platinum) of large surface area covered with hydrogen gas in a solution of hydrogen ion saturated with hydrogen gas; metal is used in a foil form and is welded to a wire sealed in the bottom of a hollow glass tube, which is partially filled with mercury; used as a standard electrode with a potential of zero to measure hydrogen ion activity.

hydrogen embrittlement See acid brittleness.

hydrogen ion See hydronium ion.

hydrogen laser [OPTICS] A molecular gas laser in which hydrogen is used to generate coherent wavelengths near 0.6 micrometer in the vacuum ultraviolet region.

hydrogen line [SPECT] A spectral line emitted by neutral hydrogen having a frequency of 1420 megahertz and a wavelength of 21 centimeters; radiation from this line is used in radio astronomy to study the amount and velocity of hydrogen in the Galaxy.

hydrogen maser [PHYS] A maser in which hydrogen gas is the basis for providing an output signal with a high degree of stability and spectral purity.

hydrogenolysis [CHEM] A reaction in which hydrogen gas causes a chemical change that is similar to the role of water in hydrolysis.

hydrogen peroxide [INORG CHEM] H_2O_2 Unstable, colorless, heavy liquid boiling at 158°C; soluble in water and alcohol; used as a bleach, chemical intermediate, rocket fuel, and antiseptic. Also known as peroxide.

hydrogen star [ASTROPHYS] A star of spectral class A, a white star with a surface temperature of 8000 to 11,000 K.

hydrogeology [HYD] The science dealing with the occurrence of surface and ground water, its utilization, and its functions in modifying the earth, primarily by erosion and deposition.

hydrograph [HYD] A graphical representation of stage, flow, velocity, or other characteristics of water at a given point as a function of time.

hydrographic survey [OCEANOGR] Survey of a water area with particular reference to tidal currents, submarine relief, and any adjacent land.

hydrography [GEOGR] Science which deals with the measurement and description of the physical features of the oceans, lakes, rivers, and their adjoining coastal areas, with particular reference to their control and utilization. [NAV] Measurement of the tides and currents as an aid to navigation.

hydrohalite [MINERAL] $Na_2Cl \cdot 2H_2O$ A mineral composed of hydrated sodium chloride, formed only from salty water cooled below 0°C.

hydrohalloysite See endellite.

hydrohetaerolite [MINERAL] $Zn_2Mn_4O_8 \cdot H_2O$ A dark brown to brownish-black mineral consisting of a hydrated oxide of zinc and manganese; occurs in massive form.

hydroid [INV ZOO] 1. The polyp form of a hydrozoan coelenterate. Also known as hydroid polyp; hydropolyp. 2. Any member of the Hydroida.

Hydroida [INV ZOO] An order of coelenterates in the class Hydrozoa including usually colonial forms with a well-developed polyp stage.

hydroid polyp See hydroid.

hydrokaolin See endellite.

hydrokinematics [FL MECH] The study of the motion of a liquid apart from the cause of motion.

hydrokinetics [FL MECH] The study of the forces produced by a liquid as a consequence of its motion.

hydrolaccolith [GEOL] A frost mound, 0.1–6 meters in height, having a core of ice and resembling a laccolith in section. Also known as cryolaccolith.

hydrolase [BIOCHEM] Any of a class of enzymes which catalyze the hydrolysis of proteins, nucleic acids, starch, fats, phosphate esters, and other macromolecular substances.

hydrolith [PETR] 1. A chemically precipitated aqueous rock, such as rock salt. 2. A rock that is free of organic material.

hydrologic cycle [HYD] The complete cycle through which water passes, from the oceans, through the atmosphere, to the land, and back to the ocean. Also known as water cycle.

hydrology [GEOPHYS] The science that treats the occurrence, circulation, distribution, and properties of the waters of the earth, and their reaction with the environment.

hydrolysis [CHEM] 1. Decomposition or alteration of a chemical substance by water. 2. In aqueous solutions of electrolytes, the reactions of cations with water to produce a weak base or of anions to produce a weak acid.

hydrolytic enzyme [BIOCHEM] A catalyst that acts like a hydrolase.

hydromagnesite [MINERAL] $Mg_4(OH)_2(CO_3)_3 \cdot 3H_2O$ White, earthy mineral crystallizing in the monoclinic system and found in small crystals, amorphous masses, or chalky crusts.

hydromagnetic instability See magnetohydrodynamic instability.

hydromagnetics See magnetohydrodynamics.

hydromagnetic stability See magnetohydrodynamic stability.

hydrometallurgy [MET] Treatment of metals and metal-containing materials by wet processes.

hydrometamorphism [GEOL] Alteration of rocks by material carried in solution by water without the influence of high temperature or pressure.

hydrometeorology [METEOROL] That part of meteorology of direct concern to hydrologic problems, particularly to flood control, hydroelectric power, irrigation, and similar fields of engineering and water resources.

hydrometer [ENG] A direct-reading instrument for indicating the density, specific gravity, or some similar characteristic of liquids.

Hydrometridae [INV ZOO] The marsh treaders, a family of hemipteran insects in the subdivision Amphibicorisae.

hydrometry [FL MECH] The science and technology of measuring specific gravities, particularly of liquids.

hydronephrosis [MED] Accumulation of urine in and distension of the renal pelvis and calyces due to obstructed outflow.

hydronic heating See hot-water heating.

hydronium ion [INORG CHEM] H_3O^+ A proton combined with a molecule of water; found in pure water and in all aqueous solutions. Also known as hydrogen ion; oxonium ion.

Hydrophiidae [VERT ZOO] A family of proglyphodont snakes in the suborder Serpentes found in Indian-Pacific oceans.

hydrophilic [CHEM] Having an affinity for, attracting, adsorbing, or absorbing water.

Hydrophilidae [INV ZOO] The water scavenger beetles, a large family of coleopteran insects in the superfamily Hydrophiloidea.

Hydrophiloidea [INV ZOO] A superfamily of coleopteran insects in the suborder Polyphaga.

hydrophobia [MED] See rabies. [PSYCH] An abnormal fear of water.

hydrophobic [CHEM] Lacking an affinity for, repelling, or failing to adsorb or absorb water. [MED] Of, pertaining to, or suffering from hydrophobia.

hydrophone [ENG ACOUS] A device which receives underwater sound waves and converts them to electric waves.

Hydrophyllaceae [BOT] A family of dicotyledonous plants in the order Polemoniales distinguished by two carpels, parietal placentation, and generally imbricate corolla lobes in the bud.

hydropneumatic [ENG] Operated by both water and air power.

hydropolyp See hydroid.

hydroponics [BOT] Growing of plants in a nutrient solution with the mechanical support of an inert medium such as sand.

Hydroscaphidae [INV ZOO] The skiff beetles, a small family of coleopteran insects in the suborder Myxophaga.

hydroskeleton [INV ZOO] Water contained within the coelenteron and serving a skeletal function in most coelenterate polyps.

hydrosol [CHEM] A colloidal system in which the dispersion medium is water, and the dispersed phase may be a solid, a gas, or another liquid.

hydrosphere [HYD] The water portion of the earth as distinguished from the solid part (lithosphere) and from the gaseous outer envelope (atmosphere).

hydrostat See humidistat.

hydrostatic modulus See bulk modulus of elasticity.

hydrostatics [FL MECH] The study of liquids at rest and the forces exerted on them or by them.

hydrostatic stability See static stability.

hydrotalcite [MINERAL] $Mg_6Al_2(OH)_{16}(CO_3)\cdot4H_2O$ Pearly-white mineral composed of hydrous aluminum and magnesium hydroxide and carbonate.

hydrothermal [GEOL] Of or pertaining to heated water, to its action, or to the products of such action.

hydrothermal deposit [GEOL] A mineral deposit precipitated from a hot, aqueous solution.

hydrothermal solution [GEOL] Hot, residual watery fluids derived from magmas during the later stages of their crystallization and commonly containing large amounts of dissolved metals which are deposited as ore veins in fissures along which the solutions often move.

hydrothorax [MED] Collection of serous fluid in the pleural spaces.

hydrotreating [CHEM ENG] Oil refinery catalytic process in which hydrogen is contacted with petroleum intermediate or product streams to remove impurities, such as oxygen, sulfur, nitrogen, or unsaturated hydrocarbons.

hydrotroilite [MINERAL] $FeS\cdot NH_2O$ A black, finely divided colloidal material reported in many muds and clays; thought to be formed by bacteria on bottoms of marine basins.

hydrotropism [BIOL] Orientation involving growth or movement of a sessile organism or part, especially plant roots, in response to the presence of water.

hydrotungstite [MINERAL] $H_2WO_4\cdot H_2O$ A mineral composed of hydrous tungstic acid.

hydrous [CHEM] Indicating the presence of an indefinite amount of water. [MINERAL] Indicating a definite proportion of combined water.

hydroxide [CHEM] Compound containing the OH^- group; the hydroxides of metals are usually bases and those of nonmetals are usually acids; a hydroxide can be organic or inorganic.

hydroximino See nitroso.

hydroxy- [ORG CHEM] Chemical prefix indicating the OH^- group in an organic compound, such as hydroxybenzene for phenol, C_6H_5OH; the use of just oxy- for the prefix is incorrect. Also spelled hydroxyl-.

hydroxyl- See hydroxy-.

hydroxylase [BIOCHEM] Any of several enzymes that catalyze certain hydroxylation reactions involving atomic oxygen.

hydroxylation reaction [ORG CHEM] One of several types of reactions used to introduce one or more hydroxyl groups into organic compounds; an oxidation reaction as opposed to hydrolysis.

5-hydroxytryptamine See serotonin.

hydrozincite [MINERAL] $Zn_5(OH)_5(CO_3)_2$ A white, grayish, or yellowish mineral composed of basic zinc carbonate, occurring as masses or crusts.

Hydrozoa [INV ZOO] A class of the phylum Coelenterata which includes the fresh-water hydras, the marine hydroids, many small jellyfish, a few corals, and the Portuguese man-of-war.

Hydrus [ASTRON] A southern constellation, right ascension 2 hours, declination 75°S. Also known as Water Snake.

hyena [VERT ZOO] An African carnivore represented by three species of the family Hyaenidae that resemble dogs but are more closely related to cats.

Hyeniales [PALEOBOT] An order of Devonian plants characterized by small, dichotomously forked leaves borne in whorls.

Hyeniopsida [PALEOBOT] An extinct class of the division Equisetophyta.

Hygrobiidae [INV ZOO] The squeaker beetles, a small family of coleopteran insects in the suborder Adephaga.

hygrology [METEOROL] The study which deals with the water vapor content (humidity) of the atmosphere.

hygrometer [ENG] An instrument for giving a direct indication of the amount of moisture in the air or other gas, the indication usually being in terms of relative humidity as a percentage which the moisture present bears to the maximum amount of moisture that could be present at the location temperature without condensation taking place.

hygroscopic [BOT] Being sensitive to moisture, such as certain tissues. [CHEM] 1. Possessing a marked ability to accelerate the condensation of water vapor; applied to condensation nuclei composed of salts which yield aqueous solutions of a very low equilibrium vapor pressure compared with that of pure water at the same temperature. 2. Pertaining to a substance whose physical characteristics are appreciably altered by effects of water vapor. 3. Pertaining to water absorbed by dry soil minerals from the atmosphere; the amounts depend on the physicochemical character of the surfaces, and increase with rising relative humidity.

hygroscopic water   [HYD] The component of soil water that is held adsorbed on the surface of soil particles and is not available to vegetation.

Hylidae   [VERT ZOO] The tree frogs, a large amphibian family in the suborder Procoela; many are adapted to arboreal life, having expanded digital disks.

Hylobatidae   [VERT ZOO] A family of anthropoid primates in the superfamily Hominoidea including the gibbon and the siamang of southeastern Asia.

hymen   [ANAT] A mucous membrane partly closing off the vaginal orifice. Also known as maidenhead.

Hymenomycetes   [MYCOL] A group of the Homobasidiomycetidae including forms such as mushrooms and pore fungi in which basidia are formed in an exposed layer (hymenium) and basidiospores are borne asymmetrically on slender stalks.

Hymenoptera   [INV ZOO] A large order of insects including ants, wasps, bees, sawflies, and related forms; head, thorax and abdomen are clearly differentiated; wings, when present, and legs are attached to the thorax.

Hymenostomatida   [INV ZOO] An order of ciliated protozoans in the subclass Holotrichia having fairly uniform ciliation and a definite buccal ciliature.

Hynobiidae   [VERT ZOO] A family of salamanders in the suborder Cryptobranchoidea.

Hyocephalidae   [INV ZOO] A monospecific family of hemipteran insects in the superfamily Pentatomorpha.

hyoglossus   [ANAT] An extrinsic muscle of the tongue arising from the hyoid bone.

hyoid   [ANAT] **1.** A bone or complex of bones at the base of the tongue supporting the tongue and its muscles. **2.** Of or pertaining to structures derived from the hyoid arch.

hyomandibular cleft   [EMBRYO] The space between the hyoid arch and the mandibular arch in the vertebrate embryo.

hyomandibular pouch   [EMBRYO] A portion of the endodermal lining of the pharyngeal cavity which separates the paired hyoid and mandibular arches in vertebrate embryos.

Hyopssodontidae   [PALEON] A family of extinct mammalian herbivores in the order Condylarthra.

hypabyssal rock   [PETR] Those igneous rocks that rose from great depths as magmas but solidified as minor intrusions before reaching the surface.

hypandrium   [INV ZOO] A plate covering the genitalia on the ninth abdominal segment of certain male insects.

hypaxial musculature   [ANAT] The ventral portion of the axial musculature of vertebrates including subvertebral flank and ventral abdominal muscle groups.

hypengyophobia   [PSYCH] An abnormal fear of responsibility.

hyperacid   [PHYSIO] Containing more than the normal concentration of acid in the gastric juice.

hyperbaric   [MED] Pertaining to an anesthetic solution with a specific gravity greater than that of the cerebrospinal fluid.

hyperbaric chamber   [ENG] A specially equipped pressure vessel used in medicine and physiological research to administer oxygen at elevated pressures.

hyperbola   [MATH] The plane curve obtained by intersecting a circular cone of two nappes with a plane parallel to the axis of the cone.

hyperbolic antenna   [ELECTROMAG] A radiator whose reflector in cross section describes a half hyperbola.

hyperbolic functions   [MATH] The real or complex functions sinh (x), cosh (x), tanh (x), coth (x), sech (x), csch (x); they are related to the hyperbola in somewhat the same fashion as the trigonometric functions are related to the circle, and have properties analogous to those of the trigonometric functions.

hyperbolic logarithm   *See* logarithm.

hyperbolic navigation   [NAV] Navigation by maintaining constant the indication of two parameters; the parameters can have any reasonable ratio to each other.

hyperbolic point   [FL MECH] A singular point in a streamline field which constitutes the intersection of a convergence line and a divergence line; it is analogous to a col in the field of a single-valued scalar quantity. Also known as neutral point. [MATH] A point on a surface where the Gaussian curvature is strictly negative.

hyperbolic space   [MATH] A space described by hyperbolic rather than cartesian coordinates.

hyperboloid of revolution   [MATH] A surface generated by rotating a hyperboloid about one of its axes.

hypercalcemia   [MED] Excessive amounts of calcium in the blood. Also known as calcemia.

hypercharge   [PARTIC PHYS] A quantum number conserved by strong interactions, equal to twice the average of the charges of the members of an isospin multiplet.

hypercholesteremia   [MED] Elevated cholesterol levels in the blood.

hypercomplex number   *See* quaternion.

hypercomplex system   *See* algebra.

hyperconjugation   [PHYS CHEM] An arrangement of bonds in a molecule that is similar to conjugation in its formulation and manifestations, but the effects are weaker; it occurs when a CH_2 or CH_3 group (or in general, an AR_2 or AR_3 group where A may be any polyvalent atom and R any atom or radical) is adjacent to a multiple bond or to a group containing an atom with a lone π-electron, π-electron pair or quartet, or π-electron vacancy; it can be sacrificial (relatively weak) or isovalent (stronger).

hyperdisk   [COMPUT SCI] A mass-storage technique which uses a large-capacity storage and a disk for overflow.

hyperemia   [MED] An excess of blood within an organ or tissue caused by blood vessel dilation or impaired drainage, especially of the skin.

hypereutectic alloy   [MET] Any binary alloy whose composition lies to the right of the eutectic on an equilibrium diagram and which contains some eutectic structure.

hyperfine structure   [SPECT] A splitting of spectral lines due to the spin of the atomic nucleus or to the occurrence of a mixture of isotopes in the element. Abbreviated hfs.

hyperfocal distance   [OPTICS] The distance from the camera lens to the nearest object in acceptable focus when the lens is focused on infinity.

hypergammaglobulinemia   [MED] Increased blood levels of gamma globulin, usually associated with hepatic disease.

hypergene   *See* supergene.

hypergeometric function   [MATH] A function which is a solution to the hypergeometric equation and obtained as an infinite series expansion.

hyperglycemia   [MED] Excessive amounts of sugar in the blood.

hyperglycemic factor   *See* glucagon.

hyperglycemic glycogenolytic factor   *See* glucagon.

Hyperiidea [INV ZOO] A suborder of amphipod crustaceans distinguished by large eyes which cover nearly the entire head.

Hyperion [ASTRON] A satellite of Saturn approximately 300 miles (480 kilometers) in diameter.

hyperkinesia [MED] Excessive and usually uncontrollable muscle movement.

hyperlipemia [MED] Excessive amounts of fat in the blood.

Hypermastigida [INV ZOO] An order of the multiflagellate protozoans in the class Zoomastigophorea; all inhabit the alimentary canal of termites, cockroaches, and woodroaches.

hypermetropia [MED] A defect of vision resulting from too short an eyeball so that unaccommodated rays focus behind the retina. Also known as farsightedness; hyperopia.

Hyperoartii [VERT ZOO] A superorder in the subclass Monorhina distinguished by the single median dorsal nasal opening leading into a blind hypophyseal sac.

hyperon [PARTIC PHYS] 1. An elementary particle which has baryon number $B = +1$, that is, which can be transformed into a nucleon and some number of mesons or lighter particles, and which has nonzero strangeness number. 2. A hyperon (as in the first definition) which is semistable (the lifetime is much longer than 10^{-22} second).

hyperopia See hypermetropia.

Hyperotreti [VERT ZOO] A suborder in the subclass Monorhina distinguished by the nasal opening which is located at the tip of the snout and communicates with the pharynx by a long duct.

hyperparasite [ECOL] An organism that is parasitic on other parasites.

hyperphagia See bulimia.

hyperpituitarism [MED] Any abnormal condition resulting from overactivity of the anterior pituitary.

hyperplasia [MED] Increase in cell number causing an increase in the size of a tissue or organ.

hyperploid [GEN] Having one or more chromosomes or parts of chromosomes in excess of the haploid number, or of a whole multiple of the haploid number.

hypersensitivity [IMMUNOL] The state of being abnormally sensitive, especially to allergens; responsible for allergic reactions.

hypersonic [ACOUS] Pertaining to frequencies above 500 megahertz. [FL MECH] Pertaining to hypersonic speeds, or air currents moving at hypersonic speeds.

hypersonic flight [AERO ENG] Flight at speeds well above the local velocity of sound; by convention, hypersonic regime starts at about five times the speed of sound and extends upward indefinitely.

hypersonics [ACOUS] Production and utilization of sound waves of frequencies above 500 megahertz.

hypersonic speed [FL MECH] A speed of an object greater than about five times the speed of sound in the fluid through which the object is moving.

hypersthene [MINERAL] $(Mg,Fe)SiO_3$ A grayish, greenish, black, or dark-brown rock-forming mineral of the orthopyroxene group, with bronzelike luster on the cleavage surface.

hypersthenfels See norite.

hypertape control unit See tape control unit.

hypertape drive See cartridge tape drive.

hypertely [EVOL] An extreme overdevelopment of an organ or body part during evolution that is disadvantageous to the organism. [ZOO] An extreme degree of imitative coloration, beyond the aspect of utility.

hypertensin See angiotensin.

hypertension [MED] Abnormal elevation of blood pressure, generally regarded to be levels of 165 systolic and 95 diastolic.

hyperthyroidism [MED] The constellation of signs and symptoms caused by excessive thyroid hormone in the blood, either from exaggerated functional activity of the thyroid gland or from excessive administration of thyroid hormone, and manifested by thyroid enlargement, emaciation, sweating, tachycardia, exophthalmos, and tremor. Also known as exophthalmic goiter; Grave's disease; thyrotoxicosis; toxic goiter.

hypertonic [PHYSIO] 1. Excessive or above normal in tone or tension, as a muscle. 2. Having an osmotic pressure greater than that of physiologic salt solution or of any other solution taken as a standard.

Hypertragulidae [PALEON] A family of extinct chevrotainlike pecoran ruminants in the superfamily Traguloidea.

hypertrophic arthritis See degenerative joint disease.

hypertrophy [PATH] Increase in cell size causing an increase in the size of an organ or tissue.

hypervalent atom [CHEM] A central atom in a single-bonded structure that imparts more than eight valence electrons in forming covalent bonds.

hypervelocity [MECH] 1. Muzzle velocity of an artillery projectile of 3500 feet per second (1067 meters per second) or more. 2. Muzzle velocity of a small-arms projectile of 5000 feet per second (1524 meters per second) or more. 3. Muzzle velocity of a tank-cannon projectile in excess of 3350 feet per second (1021 meters per second).

hyperventilation [MED] Increase in air intake or of the rate or depth of respiration.

hypha [MYCOL] One of the filaments composing the mycelium of a fungus.

Hyphochytriales [MYCOL] An order of aquatic fungi in the class Phycomycetes having a saclike to limited hyphal thallus and zoospores with two flagella.

Hyphochytridiomycetes [MYCOL] A class of the true fungi; usually grouped with other classes under the general term Phycomycetes.

Hyphomicrobiaceae [MICROBIO] Formerly a family of bacteria in the order Hyphomicrobiales; cells occurring in free-floating groups with individual cells attached to each other by a slender filament.

Hyphomicrobiales [MICROBIO] Formerly an order of bacteria in the class Schizomycetes containing forms that multiply by budding.

Hypnineae [BOT] A suborder of the Hypnobryales characterized by complanate, glossy plants with ecostate or costate leaves and paraphyllia rarely present.

Hypnobryales [BOT] An order of mosses composed of procumbent and pleurocumbent plants with usually symmetrical leaves arranged in more than two rows.

hypnosis [PSYCH] An altered state of consciousness in which the individual is more susceptible to suggestion and in which regressive behavior may spontaneously occur.

hypoallergenic [PHARM] Having a low tendency to induce allergic reactions; used particularly for formulated dermatologic preparations.

hypobaric [MED] Pertaining to an anesthetic solution of specific gravity lower than the cerebrospinal fluid. [PHYS] Having less weight or pressure.

hypoblast See endoderm.

hypobranchial musculature [ANAT] The ventral musculature in vertebrates extending from the pectoral girdle forward to the hyoid arch, chin, and tongue.

Hypochilidae [INV ZOO] A family of true spiders in the order Araneida.

Hypochilomorphae [INV ZOO] A monofamilial suborder of spiders in the order Araneida.

hypochondriasis [PSYCH] A chronic condition in which the patient is morbidly concerned with his own health and believes himself suffering from grave bodily diseases.

Hypocopridae [INV ZOO] A small family of coleopteran insects in the superfamily Cucujoidea.

hypocotyl [BOT] The portion of the embryonic plant axis below the cotyledon.

Hypocreales [MYCOL] An order of fungi belonging to the Ascomycetes and including several entomophilic fungi.

hypocrystalline [PETR] Pertaining to the texture of igneous rock characterized by crystalline components in an amorphous groundmass. Also known as hemicrystalline; hypohyaline; merocrystalline; miocrystalline; semicrystalline.

hypocycloid [MATH] The curve which is traced in the plane as a given point fixed on a circle moves while this circle rolls along the inside of another circle.

Hypodermatidae [INV ZOO] The warble flies, a family of myodarian cyclorrhaphous dipteran insects in the subsection Calypteratae.

hypodermic needle [MED] A hollow needle, with a slanted open point, used for subcutaneous and intramuscular injections of fluid.

hypodermis [BOT] The outermost cell layer of the cortex of plants. Also known as exodermis. [INV ZOO] The layer of cells that underlies and secretes the cuticle in arthropods and some other invertebrates.

hypoeutectic alloy [MET] Any binary alloy whose composition lies to the left of the eutectic.

hypogammaglobulinemia [MED] Reduced blood levels of gamma globulin.

hypogene [GEOL] 1. Of minerals or ores, formed by ascending waters. 2. Of geologic processes, originating within or below the crust of the earth.

hypoglossal nerve [ANAT] The twelfth cranial nerve; a paired motor nerve in tetrapod vertebrates innervating tongue muscles; corresponds to the hypobranchial nerve in fishes.

hypoglycemia [MED] Condition caused by low levels of sugar in the blood.

hypogynous [BOT] Having all flower parts attached to the receptacle below the pistil and free from it.

hypohyaline See hypocrystalline.

hypoid gear [MECH ENG] Gear wheels connecting nonparallel, nonintersecting shafts, usually at right angles.

hypolimnion [HYD] The lower level of water in a stratified lake, characterized by a uniform temperature that is generally cooler than that of other strata in the lake.

hypophysis [ANAT] A small rounded endocrine gland which lies in the sella turcica of the sphenoid bone and is attached to the floor of the third ventricle of the brain in all craniate vertebrates. Also known as pituitary gland.

hypopituitarism [MED] Condition caused by insufficient secretion of pituitary hormones, especially of the adenohypophysis. Also known as panhypopituitarism.

hypoplasia [MED] Failure of a tissue or organ to achieve complete development.

hypoplastic dwarf [MED] A normally proportioned individual of subnormal size.

hypoploid [GEN] Having one or more less chromosomes, or parts of chromosomes, than a whole multiple of the haploid number.

hyposensitization See desensitization.

hypospadias [MED] 1. Congenital anomaly in which the urethra opens on the ventral surface of the penis or in the perineum. 2. Congenital anomaly in which the urethra opens into the vagina.

hypotension [MED] Abnormally low blood pressure, commonly considered to be levels below 100 diastolic and 40 systolic.

hypothalamus [ANAT] The floor of the third brain ventricle; site of production of several substances that act on the adenohypophysis.

hypothermal [GEOL] Referring to the high-temperature (300–500°C) environment of hypothermal deposits.

hypothermia [PHYSIO] Condition of reduced body temperature in homeotherms.

hypothesis [SCI TECH] 1. A proposition which is assumed to be true in proving another proposition. 2. A proposition which is thought to be true because its consequences are found to be true. [STAT] A statement which specifies a population or distribution, and whose truth can be tested by sample evidence.

hypothyroidism [MED] Condition caused by deficient secretion of the thyroid hormone.

hypotonic [PHYSIO] 1. Pertaining to subnormal muscle strength or tension. 2. Referring to a solution with a lower osmotic pressure than physiological saline.

Hypotrichida [INV ZOO] An order of highly specialized protozoans in the subclass Spirotrichia characterized by cirri on the ventral surface and a lack of ciliature on the dorsal surface.

hypotrochoid [MATH] A curve traced by a point rigidly attached to a circle at a point other than the center when the circle rolls without slipping on the inside of a fixed circle.

hypotype [SYST] A specimen of a species, which, though not a member of the original type series, is known from a published description or listing.

hypovolemic shock [MED] Shock caused by reduced blood volume which may be due to loss of blood or plasma as in burns, the crush syndrome, perforating gastrointestinal wounds, or other trauma. Also known as wound shock.

hypoxemia See hypoxia.

hypoxia [MED] Oxygen deficiency; any state wherein a physioloically inadequate amount of oxygen is available to or is utilized by tissue, without respect to cause or degree. Also known as hypoxemia.

hypsography [GEOGR] The science of measuring or describing elevations of the earth's surface with reference to a given datum, usually sea level.

hypsometer [ENG] 1. An instrument for measuring atmospheric pressure to ascertain elevations by determining the boiling point of liquids. 2. Any of several instruments for determining tree heights by triangulation.

hypsometric map [MAP] In topographic surveying, a map giving elevations by contours, or sometimes by means of shading, tinting, or batching.

hypsometry [ENG] The measuring of elevation with reference to sea level.

Hyracodontidae [PALEON] The running rhinoceroses, an extinct family of perissodactyl mammals in the superfamily Rhinoceratoidea.

Hyracoidea [VERT ZOO] An order of ungulate mammals represented only by the conies of Africa, Arabia, and Syria.

hyster- [MED] A combining form that denotes either a relation to or a connection with the uterus, or hysteria.

hysterectomy [MED] Surgical removal of all or part of the uterus.

hysteresis [ELECTR] An oscillator effect wherein a given value of an operating parameter may result in multiple values of output power or frequency. [ELECTROMAG] *See* magnetic hysteresis. [NUCLEO] A temporary change in the counting-rate-voltage characteristic of a radiation counter tube, caused by its previous operation. [PHYS] The dependence of the state of a system on its previous history, generally in the form of a lagging of a physical effect behind its cause.

hysteresis clutch [MECH ENG] A clutch in which torque is produced by attraction between induced poles in a magnetized iron ring and the control field.

hysteresis loop [PHYS] The closed curve followed by a material displaying hysteresis (such as a ferromagnet or ferroelectric) on a graph of a driven variable (such as magnetic flux density or electric polarization) versus the driving variable (such as magnetic field or electric field).

hysteresis motor [ELEC] A synchronous motor without salient poles and without direct-current excitation which utilizes the hysteresis and eddy-current losses induced in its hardened-steel rotor to produce rotor torque.

hysteretic damping [MECH] Damping of a vibrating system in which the retarding force is proportional to the velocity and inversely proportional to the frequency of the vibration.

hysteria [PSYCH] A type of neurosis characterized by extreme emotionalism involving disorders of somatic and psychological functions; the conversion type is associated with neuromuscular and sensory symptoms such as paralysis, tremors, seizures, or blindness, whereas the dissociative displays disorders of consciousness such as amnesia, somnolence, and multiple personalities.

Hysteriales [BOT] An order of lichens in the class Ascolichenes including those species with an ascolocular structure.

hysterical paralysis [MED] Muscle weakness or paralysis without loss of reflex activity, in which no organic nerve lesion can be demonstrated, but which is due to psychogenic factors. Also known as functional paralysis.

Hystrichospherida [PALEON] A group of protistan microfossils.

Hystricidae [VERT ZOO] The Old World porcupines, a family of Rodentia ranging from southern Europe to Africa and eastern Asia and into the Philippines.

Hystricomorpha [VERT ZOO] A superorder of the class Rodentia.

Hz *See* hertz.

I

I *See* iodine.

IA *See* international angstrom.

ianthinite [MINERAL] $2UO_2 \cdot 7H_2O$ A violet mineral composed of hydrous uranium dioxide, occurring as orthorhombic crystals.

IAS *See* indicated airspeed.

IAT *See* international atomic time.

Iballidae [INV ZOO] A small family of hymenopteran insects in the superfamily Cynipoidea.

ibis [VERT ZOO] The common name for wading birds making up the family Threskiornithidae and distinguished by a long, slender, downward-curving bill.

IC *See* integrated circuit.

Icacinaceae [BOT] A family of dicotyledonous plants in the order Celastrales characterized by haplostemonous flowers, pendulous ovules, stipules wanting or vestigial, a polypetalous corolla, valvate petals, and usually one (sometimes three) locules.

ICBM *See* intercontinental ballistic missile.

ice [PHYS CHEM] 1. The dense substance formed by the freezing of water to the solid state; has a melting point of 32°F (0°C) and commonly occurs in the form of hexagonal crystals. 2. A layer or mass of frozen water.

ice age [GEOL] 1. A major interval of geologic time during which extensive ice sheets (continental glaciers) formed over many parts of the world. 2. *(Capitalized) See* Pleistocene.

ice apron [CIV ENG] A wedge-shaped structure which protects a bridge pier from floating ice. [HYD] 1. The snow and ice attached to the walls of a cirque. 2. The ice that is flowing from an ice sheet over the edge of a plateau. 3. A piedmont glacier's lobe. 4. Ice that adheres to a wall of a valley below a hanging glacier.

iceberg [OCEANOGR] A large mass of detached land ice floating in the sea or stranded in shallow water.

ice blink [METEOROL] A relatively bright, usually yellowish-white glare on the underside of a cloud layer, produced by light reflected from an ice-covered surface such as pack ice; used in polar regions with reference to the sky map; ice blink is not as bright as snow blink, but much brighter than water sky or land sky.

icebreaker [NAV ARCH] A vessel designed for operating in and breaking heavy ice, having a special-shaped, reinforced bow, protected propellers, and powerful engines.

ice calving *See* calving.

ice canopy *See* pack ice.

ice cap [HYD] 1. A perennial cover of ice and snow in the shape of a dome or plate on the summit area of a mountain through which the mountain peaks emerge. 2. A perennial cover of ice and snow on a flat land mass such as an Arctic island.

ice cascade *See* icefall.

ice cave [GEOL] A cave that is cool enough to hold ice through all or most of the warm season. [HYD] A cave in ice such as a glacier formed by a stream of melted water.

ice color *See* azoic dye.

ice concentration [OCEANOGR] In sea ice reporting, the ratio of the areal extent of ice present and the total areal extent of ice and water. Also known as ice cover.

ice cover *See* ice concentration.

ice crust [HYD] A type of snow crust; a layer of ice, thicker than a film crust, upon a snow surface, formed by the freezing of meltwater or rainwater which has flowed onto it.

ice-crystal fog *See* ice fog.

iced firn [HYD] A mixture of glacier ice and firn; firn permeated with meltwater and then refrozen. Also known as firn ice.

ice erosion [GEOL] 1. Erosion due to freezing of water in rock fractures. 2. *See* glacial erosion.

icefall [HYD] That portion of a glacier where a sudden steepening of descent causes a chaotic breaking up of the ice. Also known as ice cascade.

ice field [HYD] A mass of land ice resting on a mountain region and covering all but the highest peaks. [OCEANOGR] A flat sheet of sea ice that is more than 5 miles (8 kilometers) across.

ice floe *See* floe.

ice flowers [HYD] 1. Formations of ice crystals on the surface of a quiet, slowly freezing body of water. 2. Delicate tufts of hoarfrost that occasionally form in great abundance on an ice or snow surface. Also known as frost flowers. 3. Frost crystals resembling a flower, formed on salt nuclei on the surface of sea ice as a result of rapid freezing of sea water. Also known as salt flowers.

ice fog [METEOROL] A type of fog composed of suspended particles of ice, partly ice crystals 20–100 micrometers in diameter but chiefly, especially when dense, droxtals 12–20 micrometers in diameter; occurs at very low temperatures and usually in clear, calm weather in high latitudes. Also known as frost flakes; frost fog; frozen fog; ice-crystal fog; pogonip; rime fog.

ice island [OCEANOGR] A large tabular fragment of shelf ice found in the Arctic Ocean and having an irregular surface, thickness of 15–50 meters, and an area between a few thousand square meters and 500 square kilometers or more.

ice jam [HYD] 1. An accumulation of broken river ice caught in a narrow channel, frequently producing local floods during a spring breakup. 2. Fields of lake or sea ice thawed loose from the shores in early spring, and blown against the shore, sometimes exerting great pressure.

ice-laid drift *See* till.

Iceland agate   *See* obsidian.

Iceland crystal   *See* Iceland spar.

Iceland spar   [MINERAL] A pure, transparent form of calcite found particularly in Iceland; easily cleaved to form rhombohedral crystals that are doubly refracting. Also known as Iceland crystal.

ice mantle   *See* ice sheet.

ice needle   [PHYS CHEM] A long, thin ice crystal whose cross section perpendicular to its long dimension is typically hexagonal. Also called ice spicule.

ice nucleus   [METEOROL] Any particle which may act as a nucleus in formation of ice crystals in the atmosphere.

ice pack   *See* pack ice.

ice point   [PHYS CHEM] The true freezing point of water; the temperature at which a mixture of air-saturated pure water and pure ice may exist in equilibrium at a pressure of 1 standard atmosphere (101,325 newtons per square meter).

ice rind   [HYD] A thin but hard layer of sea ice, river ice, or lake ice, which is either a new encrustation upon old ice or a single layer of ice usually found in bays and fiords, where fresh water freezes on top of slightly colder sea water.

ice sheet   [HYD] A thick glacier, more than 50,000 square kilometers in area, forming a cover of ice and snow that is continuous over a land surface and moving outward in all directions. Also known as ice mantle.

ice spar   *See* sanidine.

ice spicule   *See* ice needle.

ice storm   [METEOROL] A storm characterized by a fall of freezing precipitation, forming a glaze on terrestrial objects that creates many hazards. Also known as silver storm.

ice tongue   [HYD] Any narrow extension of a glacier or ice shelf, such as a projection floating in the sea or an outlet glacier of an ice cap.

ichneumon   [INV ZOO] The common name for members of the family Ichneumonidae.

Ichneumonidae   [INV ZOO] The ichneumon flies, a large family of cosmopolitan, parasitic wasps in the superfamily Ichneumonoidea.

Ichneumonoidea   [INV ZOO] A superfamily of hymenopteran insects; members are parasites of other insects.

ichnofossil   *See* trace fossil.

ichthyism   [MED] Food poisoning caused by eating spoiled fish.

Ichthyobdellidae   [INV ZOO] A family of leeches in the order Rhynchobdellae distinguished by cylindrical bodies with conspicuous, powerful suckers.

ichthyocolla   *See* isinglass.

Ichthyodectidae   [PALEON] A family of Cretaceous marine osteoglossiform fishes.

ichthyology   [VERT ZOO] A branch of vertebrate zoology that deals with the study of fishes.

Ichthyopterygia   [PALEON] A subclass of extinct Mesozoic reptiles composed of predatory fish-finned and sea-swimming forms with short necks and a porpoiselike body.

Ichthyornithes   [PALEON] A superorder of fossil birds of the order Ichthyornithiformes according to some systems of classification.

Ichthyornithidae   [PALEON] A family of extinct birds in the order Ichthyornithiformes.

Ichthyornithiformes   [PALEON] An order of ancient fossil birds including strong flying species from the Upper Cretaceous that possessed all skeletal characteristics of modern birds.

Ichthyosauria   [PALEON] The only order of the reptilian subclass Ichthyopterygia, comprising the extinct predacious fish-lizards; all were adapted to a sea life in having tail flukes, paddles, and dorsal fins.

ichthyosis   [MED] A congenital skin disease characterized by dryness and scales, especially on the extensor surfaces of the extremities.

Ichthyostegalia   [PALEON] An extinct Devonian order of labyrinthodont amphibians, the oldest known representatives of the class.

Ichthyotomidae   [INV ZOO] A monotypic family of errantian annelids in the superfamily Eunicea.

icicle   [HYD] Ice shaped like a narrow cone, hanging point downward from a roof, fence, or other sheltered or heated source from which water flows and freezes in below-freezing air.

icing   [HYD] **1.** Any deposit or coating of ice on an object, caused by the impingement and freezing of liquid (usually supercooled) hydrometeors. **2.** A mass or sheet of ice formed on the ground surface during the winter by successive freezing of sheets of water that may seep from the ground, from a river, or from a spring. Also known as flood icing; flooding ice.

iconoscope   [ELECTR] A television camera tube in which a beam of high-velocity electrons scans a photoemissive mosaic that is capable of storing an electric charge pattern corresponding to an optical image focused on the mosaic. Also known as storage camera; storage-type camera tube.

icosahedron   [MATH] A 20-sided polyhedron.

icositetrahedron   *See* trapezohedron.

Icosteidae   [VERT ZOO] The ragfishes, a family of perciform fishes in the suborder Stromateoidei found in high seas.

icotype   [SYST] A typical, accurately identified specimen of a species, but not the basis for a published description.

ICSH   *See* luteinizing hormone.

ICS system   *See* intercarrier sound system.

ICT   *See* International Critical Tables.

Icteridae   [VERT ZOO] The troupials, a family of New World perching birds in the suborder Oscines.

icterus   *See* jaundice.

Ictidosauria   [PALEON] An extinct order of mammal-like reptiles in the subclass Synapsida including small carnivorous and herbivorous terrestrial forms.

ID   *See* inside diameter.

ID$_{50}$ *See* infective dose 50.

iddingsite   [MINERAL] A reddish-brown mixture of silicates, forming patches in basic igneous rocks.

ideal aerodynamics   [FL MECH] A branch of aerodynamics that deals with simplifying assumptions that help explain some airflow problems and provide approximate answers. Also known as ideal fluid dynamics.

ideal dielectric   [ELEC] Dielectric in which all the energy required to establish an electric field in the dielectric is returned to the source when the field is removed. Also known as perfect dielectric.

ideal flow   [FL MECH] **1.** Fluid flow which is incompressible, two-dimensional, irrotational, steady, and nonviscous. **2.** *See* inviscid flow.

ideal fluid   [FL MECH] **1.** A fluid which has ideal flow. **2.** *See* inviscid fluid.

ideal fluid dynamics   *See* ideal aerodynamics.

ideal gas   [THERMO] Also known as perfect gas. **1.** A gas whose molecules are infinitely small and exert no force on each other. **2.** A gas that obeys Boyle's law (the product of the pressure and volume is constant at constant temperature) and Joule's law (the internal energy is a function of the temperature alone).

ideal gas law   [THERMO] The equation of state of an ideal gas which is a good approximation to real gases at sufficiently high temperatures and low pressures;

that is, $PV = RT$, where P is the pressure, V is the volume per mole of gas, T is the temperature, and R is the gas constant.

ideal line [MATH] The collection of all ideal points, each corresponding to a given family of parallel lines. Also known as line at infinity.

ideal point [MATH] In projective geometry, all lines parallel to a given line are hypothesized to meet at a point at infinity, called an ideal point. Also known as point at infinity.

ideal radiator *See* blackbody.

ideal solution [CHEM] A solution that conforms to Raoult's law over all ranges of temperature and concentration and shows no internal energy change on mixing and no attractive force between components.

ideation [PSYCH] Conceptualization of an idea.

idempotent [MATH] **1.** An element x of an algebraic system satisfying the equation $x^2 = x$. **2.** An algebraic system in which every element x satisfies $x^2 = x$.

identical twins *See* monozygotic twins.

identifier [COMPUT SCI] A symbol whose purpose is to specify a body of data.

identifier word [COMPUT SCI] A full-length computer word associated with a search function.

identity [MATH] An equation satisfied for all possible choices of values for the variables involved.

identity crisis [PSYCH] The critical period in emotional maturation and personality development, occurring usually during adolescence, which involves the reworking and abandonment of childhood identifications and the integration of new personal and social identifications.

identity gate *See* identity unit.

identity unit [COMPUT SCI] A logic element with several binary input signals and a single binary output signal whose value is 1 if all the input signals have the same value and 0 if they do not. Also known as identity gate.

ideogenous *See* syngenetic.

ideokinetic apraxia *See* ideomotor apraxia.

ideomotor [PHYSIO] **1.** Pertaining to involuntary movement resulting from or accompanying some mental activity, as moving the lips while reading. **2.** Pertaining to both ideation and motor activity.

ideotype [SYST] A specimen identified as belonging to a specific taxon but collected from other than the type locality.

idiomorphic *See* automorphic.

idiopathy [MED] **1.** A primary disease; one not a result of any other disease, but of spontaneous origin. **2.** Disease for which no cause is known.

Idiostolidae [INV ZOO] A small family of hemipteran insects in the superfamily Pentatomorpha.

I display [ELECTR] A radarscope display in which a target appears as a complete circle when the radar antenna is correctly pointed at it, the radius of the circle being proportional to target distance; when the antenna is not aimed at the target, the circle reduces to a circle segment.

idle current *See* reactive current.

idler gear [MECH ENG] A gear situated between a driving gear and a driven gear to transfer motion, without any change of direction or of gear ratio.

idler pulley [MECH ENG] A pulley used to guide and tighten the belt or chain of a conveyor system.

idler wheel [MECH ENG] **1.** A wheel used to transmit motion or to guide and support something. **2.** A roller with a rubber surface used to transfer power by frictional means in a sound-recording or sound-reproducing system.

idle time [COMPUT SCI] The time during which a piece of hardware in good operating condition is unused. [IND ENG] A period of time during a regular work cycle when a worker is not active because of waiting for materials or instruction. Also known as waiting time.

idling system [MECH ENG] A system to obtain adequate metering forces at low airspeeds and small throttle openings in an automobile carburetor in the idling position.

idocrase *See* vesuvianite.

Idoteidae [INV ZOO] A family of isopod crustaceans in the suborder Valvifera having a flattened body and seven pairs of similar walking legs.

IDP *See* integrated data processing.

idrialite [MINERAL] A mineral composed of crystalline hydrocarbon, $C_{22}H_{14}$.

i-f *See* intermediate frequency.

i-f amplifier *See* intermediate-frequency amplifier.

IFR *See* instrument flight rules.

if then else [COMPUT SCI] A logic statement in a high-level programming language that defines the data to be compared and the actions to be taken as the result of a comparison.

i-f transformer *See* intermediate-frequency transformer.

igneous [PETR] Pertaining to rocks which have congealed from a molten mass.

igneous facies [PETR] A part of an igneous rock differing in structure, texture, or composition from the main mass.

igneous province *See* petrographic province.

ignimbrite [PETR] A silicic volcanic rock that forms thick, compact, lavalike sheets over a wide area of New Zealand. Also known as flood tuff.

igniter [ENG] **1.** A device for igniting a fuel mixture. **2.** A charge, as of black powder, to facilitate ignition of a propelling or bursting charge.

ignition [CHEM] The process of starting a fuel mixture burning, or the means for such a process.

ignition coil [ELECTROMAG] A coil in an ignition system which stores energy in a magnetic field relatively slowly and releases it suddenly to ignite a fuel mixture.

ignition delay *See* ignition lag.

ignition lag [MECH ENG] In the internal combustion engine, the time interval between the passage of the spark and the inflammation of the air-fuel mixture. Also known as ignition delay.

ignition point *See* ignition temperature.

ignition system [MECH ENG] The system in an internal combustion engine that initiates the chemical reaction between fuel and air in the cylinder charge by producing a spark.

ignition temperature [CHEM] The lowest temperature at which combustion begins and continues in a substance when it is heated in air. Also known as autogenous ignition temperature; ignition point. [PL PHYS] The lowest temperature at which the fusion energy generated in a plasma exceeds the energy lost through bremsstrahlung radiation.

ignitor [ELECTR] **1.** An electrode used to initiate and sustain the discharge in a switching tube. Also known as keep-alive electrode (deprecated). **2.** A pencil-shaped electrode, made of carborundum or some other conducting material that is not wetted by mercury, partly immersed in the mercury-pool cathode of an ignitron and used to initiate conduction at the desired point in each alternating-current cycle.

ignitron [ELECTR] A single-anode pool tube in which an ignitor electrode is employed to initiate the cathode spot on the surface of the mercury pool before each conducting period.

ignore character [COMPUT SCI] Also known as erase character. **1.** A character indicating that no action whatever is to be taken, that is, a character to be ignored; often used to obliterate an erroneous character. **2.** A character indicating that the preceding or following character is to be ignored, as specified. **3.** A character indicating that some specified action is not to be taken.

iguana [VERT ZOO] The common name for a number of species of herbivorous, arboreal reptiles in the family Iguanidae characterized by a dorsal crest of soft spines and a dewlap; there are only two species of true iguanas.

iguanid [VERT ZOO] The common name for members of the reptilian family Iguanidae.

Iguanidae [VERT ZOO] A family of reptiles in the order Squamata having teeth fixed to the inner edge of the jaws, a nonretractile tongue, a compressed body, five clawed toes, and a long but rarely prehensile tail.

ihleite *See* copiapite.

ihp *See* indicated horsepower.

I indicator *See* I scope.

I²L *See* integrated injection logic.

ileocecal valve [ANAT] A muscular structure at the junction of the ileum and cecum which prevents reflex of the cecal contents.

ileocolitis [MED] Inflammation of both the ileum and the colon.

ileostomy [MED] Surgical formation of an artificial anus through the abdominal wall into the ileum.

ilesite [MINERAL] $(Mn,Zn,Fe)SO_4 \cdot 4H_2O$ A green mineral composed of hydrous manganese zinc iron sulfate.

ileum [ANAT] The last portion of the small intestine, extending from the jejunum to the large intestine.

ileus [MED] Acute intestinal obstruction of neurogenic origin.

ILF *See* infralow frequency.

iliac artery [ANAT] Either of the two large arteries arising by bifurcation of the abdominal aorta and supplying blood to the lower trunk and legs (or hind limbs in quadrupeds). Also known as common iliac artery.

iliac region *See* inguinal region.

iliac vein [ANAT] Any of the three veins on each side of the body which correspond to and accompany the iliac artery.

ilium [ANAT] Either of a pair of bones forming the superior portion of the pelvis bone in vertebrates.

illegal character [COMPUT SCI] A character or combination of bits that is not accepted by a computer or by a specific routine; commonly detected and used as an indication of a machine malfunction.

illinium *See* promethium-147.

Illinoian [GEOL] The third glaciation of the Pleistocene in North America, between the Yarmouth and Sangamon interglacial stages.

illite [MINERAL] A group of gray, green, or yellowish-brown micalike clay minerals found in argillaceous sediments; intermediate in composition and structure between montmorillonite and muscovite.

illuminance [OPTICS] The density of the luminous flux on a surface. Also known as illumination; luminous flux density.

illumination [ELECTROMAG] **1.** The geometric distribution of power reaching various parts of a dish reflector in an antenna system. **2.** The power distribution to elements of an antenna array. [OPTICS]

1. The science of the application of lighting. **2.** *See* illuminance.

illuminometer [OPTICS] A portable photometer which is used in the field or outside the laboratory and yields results of lower accuracy than a laboratory photometer.

illusion [PSYCH] A false interpretation of a real sensation; a perception that misinterprets the object perceived.

illuvial [GEOL] Pertaining to a region or material characterized by the accumulation of soil by the illuviation of another zone or material.

illuvium [GEOL] Material leached by chemical or other processes from one soil horizon and deposited in another.

ilmenite [MINERAL] $FeTiO_3$ An iron-black, opaque, rhombohedral mineral that is the principal ore of titanium. Also known as mohsite; titanic iron ore.

ILS *See* instrument landing system.

ilsemannite [MINERAL] A black, blue-black, or blue mineral composed of hydrous molybdenum oxide or perhaps sulfate, occurring in earthy massive form.

image [ACOUS] *See* acoustic image. [COMMUN] **1.** One of two groups of side bands generated in the process of modulation; the unused group is referred to as the unwanted image. **2.** The scene reproduced by a television or facsimile receiver. [COMPUT SCI] A copy of the information contained in a punched card (or other data record) recorded on a different data medium. [ELEC] *See* electric image. [ELECTROMAG] The input reflection coefficient corresponding to the reflection coefficient of a specified load when the load is placed on one side of a waveguide junction and a slotted line is placed on the other. [OPTICS] An optical counterpart of a self-luminous or illuminated object formed by the light rays that traverse an optical system; each point of the object has a corresponding point in the image from which rays diverge or appear to diverge. [PHYS] Any reproduction of an object produced by means of focusing light, sound, electron radiation, or other emanations coming from the object or reflected by the object. [PSYCH] A representation of a sensory experience, occurring in the brain.

image card [COMPUT SCI] A representation in storage of the holes punched in a card, in such a manner that the holes are represented by one binary digit, and the unpunched spaces are represented by the other binary digit.

image converter [ELECTR] *See* image tube. [OPTICS] A converter that uses a fiber optic bundle to change the form of an image, for more convenient recording and display or for the coding of secret messages.

image dissector [COMPUT SCI] In optical character recognition, a device that optically examines an input character for the purpose of breaking it down into its prescribed elements.

image effect [ELECTROMAG] Effect produced on the field of an antenna due to the presence of the earth; electromagnetic waves are reflected from the earth's surface, and these reflections often are accounted for by an image antenna at an equal distance below the earth's surface.

image intensifier *See* light amplifier.

image isocon [ELECTR] A television camera tube which is similar to the image orthicon but whose return beam consists of scanning beam electrons that are scattered by positive stored charges on the target.

image orthicon [ELECTR] A television camera tube in which an electron image is produced by a photoemitting surface and focused on one side of a separate storage tube that is scanned on its opposite side by a beam of low-velocity electrons; electrons that are reflected from the storage tube, after positive stored charges are neutralized by the scanning beam, form a return beam which is amplified by an electron multiplier.

image plane [OPTICS] The plane in which an image produced by an optical system is formed; if the object plane is perpendicular to the optical axis, the image plane will ordinarily also be perpendicular to the axis.

image processing [COMPUT SCI] A technique in which the data from an image are digitized and various mathematical operations are applied to the data, generally with a digital computer, in order to create an enhanced image that is more useful or pleasing to a human observer, or to perform some of the interpretation and recognition tasks usually performed by humans. Also known as picture processing.

image ratio [ELECTR] In a heterodyne receiver, the ratio of the image frequency signal input at the antenna to the desired signal input for identical outputs.

image restoration [COMPUT SCI] Operation on a picture with a digital computer to make it more closely resemble the original object.

image-storage array [ELECTR] A solid-state panel or chip in which the image-sensing elements may be a metal oxide semiconductor or a charge-coupled or other light-sensitive device that can be manufactured in a high-density configuration.

image tube [ELECTR] An electron tube that reproduces on its fluorescent screen an image of the optical image or other irradiation pattern incident on its photosensitive surface. Also known as electron image tube; image converter.

imaginary number [MATH] A complex number of the form $a + bi$, with b not equal to zero, where a and b are real numbers, and $i = \sqrt{-1}$; some writers require also that $a = 0$. Also known as imaginary quantity.

imaginary quantity See imaginary number.

imaging radar [ENG] Radar carried on aircraft which forms images of the terrain.

imago [INV ZOO] The sexually mature, usually winged stage of insect development. [PSYCH] An unconscious mental picture, usually idealized, of a parent or loved person important in the early development of an individual and carried into adulthood.

imbecile [PSYCH] A person of middle-grade mental deficiency; the individual's mental age is between 3 and 7 years.

imbed See embed.

imbibition [PHYS CHEM] Absorption of liquid by a solid or a semisolid material.

imbricate structure [GEOL] 1. A sedimentary structure characterized by shingling of pebbles all inclined in the same direction with the upper edge of each leaning downstream or toward the sea. Also known as shingle structure. 2. Tabular masses that overlap one another and are inclined in the same direction. Also known as schuppen structure; shingle-block structure.

imerinite [MINERAL] $Na_2(Mg,Fe)_6Si_8O_{22}(O,OH)_2$ A colorless to blue mineral composed of a basic silicate of sodium, iron, and magnesium, occurring as acicular crystals.

Imhoff tank [CIV ENG] A sewage treatment tank in which digestion and settlement take place in separate compartments, one below the other.

imide [ORG CHEM] 1. A compound derived from acid anhydrides by replacing the oxygen (O) with the $=$NH group. 2. A compound that has either the $=$NH group or a secondary amine in which R is an acyl radical, as R_2NH.

imine [ORG CHEM] A class of compounds that are the product of condensation reactions of aldehydes or ketones with ammonia or amines; they have the NH radical attached to the carbon with the double bond, as R—HC$=$NH; an example is benzaldimine.

immediate-access [COMPUT SCI] 1. Pertaining to an access time which is relatively brief, or to a relatively fast transfer of information. 2. Pertaining to a device which is directly connected with another device.

immediate address [COMPUT SCI] The value of an operand contained in the address part of an instruction and used as data by this instruction.

immediate hypersensitivity [IMMUNOL] A type of hypersensitivity in which the response rapidly occurs following exposure of a sensitized individual to the antigen.

immediate processing See demand processing.

immersion [ASTRON] The disappearance of a celestial body either by passing behind another or passing into another's shadow. [MATH] A mapping f of a topological space X into a topological space Y such that for every $x \in X$ there exists a neighborhood N of x, such that f is a homeomorphism of N onto $f(N)$. [SCI TECH] Placement into or within a fluid, usually water.

immersion electron microscope [ELECTR] An emission electron microscope in which the specimen is a flat conducting surface which may be heated, illuminated, or bombarded by high-velocity electrons or ions so as to emit low-velocity thermionic, photo-, or secondary electrons; these are accelerated to a high velocity in an immersion objective or cathode lens and imaged as in a transmission electron microscope.

immersion lens See immersion objective.

immersion objective [OPTICS] A high-power microscope objective designed to work with the space between the objective and the cover glass over the object filled with an oil whose index of refraction is nearly the same as that of the objective and the cover glass, in order to reduce reflection losses and increase the index of refraction of the object space. Also known as immersion lens.

immersion plating [MET] Applying an adherent layer of more-noble metal to the surface of a metal object by dipping in a solution of more-noble metal ions; a replacement reaction. Also known as dip plating; metal replacement.

immiscible [CHEM] Pertaining to liquids that will not mix with each other.

immittance [ELEC] A term used to denote both impedance and admittance, as commonly applied to transmission lines, networks, and certain types of measuring instruments.

immittance bridge [ELECTROMAG] A modification of an admittance bridge which compares the output current of a four-terminal device with admittance standards in a T configuration in order to measure transfer admittance by a null method.

immune [IMMUNOL] 1. Safe from attack; protected against a disease by an innate or an acquired immunity. 2. Pertaining to or conferring immunity.

immune body See antibody.

immune serum [IMMUNOL] Blood serum obtained from an immunized individual and carrying antibodies.

immunity [IMMUNOL] The condition of a living organism whereby it resists and overcomes an infection or a disease. [MET] The ability of metal to resist corrosion as a result of thermodynamic stability.

immunization [IMMUNOL] Rendering an organism immune to a specific communicable disease.

immunochemistry [IMMUNOL] A branch of science dealing with the chemical changes associated with immunity factors.

immunodiffusion [IMMUNOL] A serological procedure in which antigen and antibody solutions are permitted to diffuse toward each other through a gel matrix; interaction is manifested by a precipitin line for each system.

immunoelectrophoresis [IMMUNOL] A serological procedure in which the components of an antigen are separated by electrophoretic migration and then made visible by immunodiffusion of specific antibodies.

immunofluorescence [IMMUNOL] Fluorescence as the result of, or identifying, an immune response; a specifically stained antigen fluoresces in ultraviolet light and can thus be easily identified with a homologous antigen.

immunogen [IMMUNOL] A substance which stimulates production of specific antibody or of cellular immunity, and which can react with these products.

immunogenetics [MED] A branch of immunology dealing with the relationships between immunity and genetic factors in disease.

immunologic suppression [IMMUNOL] The use of x-irradiation, chemicals, corticosteroid hormones, or antilymphocyte antisera to suppress antibody production, particularly in graft transplants.

immunologic tolerance [IMMUNOL] **1.** A condition in which an animal will accept a homograft without rejection. **2.** A state of specific unresponsiveness to an antigen or antigens in adult life as a consequence of exposure to the antigen in utero or in the neonatal period.

immunology [BIOL] A branch of biological science concerned with the native or acquired resistance of higher animal forms and humans to infection with microorganisms.

immunopathology [MED] The study of various human and animal diseases in which humoral and cellular immune factors seem important in causing pathological damage to cells, tissues, and the host.

immunosuppression [IMMUNOL] Suppression of an immune response by the use of drugs or radiation.

impact [MECH] A forceful collision between two bodies which is sufficient to cause an appreciable change in the momentum of the system on which it acts. Also known as impulsive force.

impact avalanche and transit time diode See IMPATT diode.

impact energy [MECH] The energy necessary to fracture a material. Also known as impact strength.

impact excitation [ELEC] Starting of damped oscillations in a radio circuit by a sudden surge, such as that produced by a spark discharge.

impact extrusion [MET] A cold extrusion process for producing tubular components by striking a slug of the metal, which has been placed in the cavity of the die, with a punch moving at high velocity.

impaction [MED] **1.** The state of being lodged and retained in a body part. **2.** Confinement of a tooth in the jaw so that its eruption is prevented. **3.** A condition in which one fragment of bone is firmly driven into another fragment so that neither can move against the other.

impact ionization [ELECTR] Ionization produced by the impact of a high-energy charge carrier on an atom of semiconductor material; the effect is an increase in the number of charge carriers.

impact load [ENG] A force delivered by a blow, as opposed to a force applied gradually and maintained over a long period.

impact loss [FL MECH] Loss of head in a flowing stream due to the impact of water particles upon themselves or some bounding surface.

impactometer See impactor.

impactor [ENG] A general term for instruments which sample atmospheric suspensoids by impaction; such instruments consist of a housing which constrains the air flow past a sensitized sampling plate. Also known as impactometer. [MECH ENG] A machine or part whose operating principle is striking blows. [MIN ENG] A rotary hammermill which crushes ore by impacting it against crushing plates or elements.

impact pressure See dynamic pressure.

impact printer [GRAPHICS] A line printer that has one or more character fonts, a ribbon or other inking device, a paper transport, and some means of impacting desired characters or character elements on the paper.

impact strength [MECH] **1.** Ability of a material to resist shock loading. **2.** See impact energy.

impact test [ENG] Determination of the degree of resistance of a material to breaking by impact, under bending, tension, and torsion loads; the energy absorbed is measured in breaking the material by a single blow.

impact tube See pitot tube.

IMPATT diode [ELECTR] A *pn* junction diode that has a depletion region adjacent to the junction, through which electrons and holes can drift, and is biased beyond the avalanche breakdown voltage. Derived from impact avalanche and transit time diode.

impedance [ELEC] See electrical impedance. [PHYS] **1.** The ratio of a sinusoidally varying quantity to a second quantity which measures the response of a physical system to the first, both being considered in complex notation; examples are electrical impedance, acoustic impedance, and mechanical impedance. Also known as complex impedance. **2.** The ratio of the greatest magnitude of a sinusoidally varying quantity to the greatest magnitude of a second quantity which measures the response of a physical system to the first; equal to the magnitude of the quantity in the first definition.

impedance bridge [ELEC] A device similar to a Wheatstone bridge, used to compare impedances which may contain inductance, capacitance, and resistance.

impedance compensator [ELEC] Electric network designed to be associated with another network or a line with the purpose of giving the impedance of the combination a desired characteristic with frequency over a desired frequency range.

impedance drop [ELEC] The total voltage drop across a component or conductor of an alternating-current circuit, equal to the phasor sum of the resistance drop and the reactance drop.

impedance match [ELEC] The condition in which the external impedance of a connected load is equal to the internal impedance of the source or to the surge impedance of a transmission line, thereby giving maximum transfer of energy from source to load, minimum reflection, and minimum distortion.

impedance-matching network [ELEC] A network of two or more resistors, coils, or capacitors used to couple two circuits in such a manner that the impedance of each circuit will be equal to the impedance into which it locks. Also known as line-building-out network.

impedance meter *See* electrical impedance meter.

impeller [MECH ENG] The rotating member of a turbine, blower, fan, axial or centrifugal pump, or mixing apparatus. Also known as rotor.

Impennes [VERT ZOO] A superorder of birds for the order Sphenisciformes in some systems of classification.

imperfect flower [BOT] A flower lacking either stamens or carpels.

imperfect gas *See* real gas.

imperforate [BIOL] Lacking a normal opening. [GRAPHICS] In printing, no perforations between repeated designs, for example, stamps or labels, which are to be used separately.

imperial gallon *See* gallon.

imperial pint *See* pint.

impetigo [MED] An acute, contagious, inflammatory skin disease caused by streptococcal or staphylococcal infections and characterized by vesicular or pustular lesions.

impingement [ENG] Removal of liquid droplets from a flowing gas or vapor stream by causing it to collide with a baffle plate at high velocity, so that the droplets fall away from the stream. Also known as liquid knockout.

impinger [ENG] A device used to sample dust in the air that draws in a measured volume of dusty air and directs it through a jet to impact on a wetted glass plate; the dust particles adhering to the plate are counted.

implant [MED] 1. A quantity of radioactive material in a suitable container, intended to be embedded in a tissue or tumor for therapeutic purposes. 2. A tissue graft placed in depth in the body.

implanted atom [ELECTR] An atom introduced into semiconductor material by ion implantation.

implanted device [ELECTR] A resistor or other device that is fabricated within a silicon or other semiconducting substrate by ion implantation. [MED] A heart pacemaker or other medical electronic device that is surgically placed in the body.

implosion [CHEM] The sudden reduction of pressure by chemical reaction or change of state which causes an inrushing of the surrounding medium. [PHYS] A bursting inward, as in the inward collapse of an evacuated container (such as the glass envelope of a cathode-ray tube) or the compression of fissionable material by ordinary explosives in a nuclear weapon.

imposition [GRAPHICS] The pattern of arranging pages for a signature of a book so that the pages will be in sequence when folding occurs.

impotence [MED] 1. Inability in the male to perform the sexual act. 2. Lack of sexual vigor.

impounding reservoir [CIV ENG] A reservoir with outlets controlled by gates that release stored surface water as needed in a dry season; may also store water for domestic or industrial use or for flood control. Also known as storage reservoir.

impregnate [ENG] To force a liquid substance into the spaces of a porous solid in order to change its properties, as the impregnation of turquoise gems with plastic to improve color and durability, the impregnation of porous tungsten with a molten barium

compound to manufacture a dispenser cathode, or the impregnation of wood with creosote to preserve its integrity against water damage. [MED] To fertilize or cause to become pregnant.

impression [GEOL] A form left on a soft soil surface by plant parts; the soil hardens and usually the imprint is a concave feature. [GRAPHICS] 1. A print made from an engraved plate. 2. A press run or printing of a book. [MET] A machined cavity in a forging die for production of a specific geometric shape in the workpiece.

impression cylinder [GRAPHICS] A cylinder onto which an inked image is pressed, so that this image can be transferred to paper in the offset duplicating process.

imprinter [GRAPHICS] Any device for entering markings onto a form, including, but not limited to, printing presses, typewriters, pressure imprinting devices such as those used with credit cards and address plates, pencils, pens, cash registers, adding machines, and bookkeeping machines.

imprinting [PSYCH] The very rapid development of a response or learning pattern to a stimulus at an early and usually critical period of development; particularly characteristic of some species of birds.

improper fraction [MATH] 1. In arithmetic, the quotient of two integers in which the numerator is greater than or equal to the denominator. 2. In algebra, the quotient of two polynomials in which the degree of the numerator is greater than or equal to that of the denominator.

impsonite [GEOL] A black, asphaltic pyrobitumen with a high fixed-carbon content derived from the metamorphosis of petroleum.

impulse [MECH] The integral of a force over an interval of time. [MET] A single pulse or several pulses in welding current used in resistance welding. [PHYS] A pulse which lasts for so short a time that its duration can be thought of as infinitesimal. [PSYCH] A sudden psychogenic urge to act.

impulse excitation *See* shock excitation.

impulse generator [ELEC] An apparatus which produces very short surges of high-voltage or high-current power by discharging capacitors in parallel or in series. Also known as pulse generator.

impulse modulation [CONT SYS] Modulation of a signal in which it is replaced by a series of impulses, equally spaced in time, whose strengths (integrals over time) are proportional to the amplitude of the signal at the time of the impulse.

impulse strength [ELEC] Voltage breakdown of insulation under voltage surges on the order of microseconds in duration.

impulse turbine [MECH ENG] A prime mover in which fluid under pressure enters a stationary nozzle where its pressure (potential) energy is converted to velocity (kinetic) energy and absorbed by the rotor.

impulse voltage [ELEC] A unidirectional voltage that rapidly rises to a peak value and then drops to zero more or less rapidly. Also known as pulse voltage.

impulsive force *See* impact.

impurity [SCI TECH] An undesirable foreign material in a pure substance. [SOLID STATE] A substance that, when diffused into semiconductor metal in small amounts, either provides free electrons to the metal or accepts electrons from it.

impurity semiconductor [SOLID STATE] A semiconductor whose properties are due to impurity levels produced by foreign atoms.

In *See* indium.

in. *See* inch.

Inadunata [PALEON] An extinct subclass of stalked Paleozoic Crinozoa characterized by branched or simple arms that were free and in no way incorporated into the calyx.

Inarticulata [INV ZOO] A class of the phylum Brachiopoda; valves are typically not articulated and are held together only by soft tissue of the living animal.

inborn [BIOL] Of or pertaining to a congenital or hereditary characteristic.

inbreeding [GEN] Breeding of closely related individuals; self-fertilization, as in some plants, is the most extreme form.

incandescent lamp [ELEC] An electric lamp that produces light when a metallic filament is heated white-hot in a vacuum by passing an electric current through the filament. Also known as filament lamp; light bulb.

incarbonization See coalification.

Inceptisol [GEOL] A soil order characterized by soils that are usually moist, with pedogenic horizons of alteration of parent materials but not of illuviation.

incertae sedis [SYST] Placed in an uncertain taxonomic position.

inch [MECH] A unit of length in common use in the United States and Great Britain, equal to $\frac{1}{12}$ foot or 2.54 centimeters. Abbreviated in.

inching See jogging.

inch of mercury [MECH] The pressure exerted by a 1-inch-high (2.54-centimeter-high) column of mercury that has a density of 13.5951 grams per cubic centimeter when the acceleration of gravity has the standard value of 9.80665 m/sec^2 or approximately 32.17398 ft/sec^2; equal to 3386.388640341 newtons per square meter; used as a unit in the measurement of atmospheric pressure.

incidence matrix [MATH] In a graph the p by q matrix (b_{ij}) for which $b_{ij} = 1$ if the ith vertex is an end point of the jth edge, and $b_{ij} = 0$ otherwise.

incident power [ELEC] Product of the outgoing current and voltage, from a transmitter, traveling down a transmission line to the antenna.

incident wave [ELECTR] A current or voltage wave that is traveling through a transmission line in the direction from source to load. [PHYS] A wave that impinges on a discontinuity, particle, or body, or on a medium having different propagation characteristics.

incineration [CHEM] The process of burning a material so that only ashes remain.

incinerator [ENG] A furnace or other container in which materials are burned.

Incirrata [INV ZOO] A suborder of cephalopod mollusks in the order Octopoda.

incised [BIOL] Having a deeply and irregularly notched margin. [MED] Made by cutting, as a wound.

incision [MED] A cut or wound of the body tissue, as an abdominal incision or a vertical or oblique incision.

incisor [ANAT] A tooth specialized for cutting, especially those in front of the canines on the upper jaw of mammals.

inclination [GEOL] The angle at which a geological body or surface deviates from the horizontal or vertical; often used synonymously with dip. [GEOPHYS] In magnetic inclination, the dip angle of the earth's magnetic field. Also known as magnetic dip. [MATH] **1.** The inclination of a line in a plane is the angle made with the positive x axis. **2.** The inclination of a line in space with respect to a plane is the smaller angle the line makes with its orthogonal projection in the plane. **3.** The inclina-

tion of a plane with respect to a given plane is the smaller of the dihedral angles which it makes with the given plane. [SCI TECH] **1.** Angular deviation of a direction or surface from the true vertical or horizontal. **2.** The angle which a direction or surface makes with the vertical or horizontal. **3.** A surface which deviates from the vertical or horizontal.

inclination of axis [ASTRON] The angle between a planet's axis of rotation and the perpendicular to the plane of its orbit.

inclined bedding [GEOL] A type of bedding in which the strata dip in the direction of current flow.

inclined extinction [OPTICS] Extinction in which the vibration directions are inclined to a crystal axis or direction of cleavage. Also known as oblique extinction.

inclined orbit [AERO ENG] A satellite orbit which is inclined with respect to the earth's equator.

inclined plane [MECH] A plane surface at an angle to some force or reference line.

inclined-tube manometer [ENG] A glass-tube manometer with the leg inclined from the vertical to extend the scale for more minute readings.

inclinometer [ENG] **1.** An instrument that measures the attitude of an aircraft with respect to the horizontal. **2.** An instrument for measuring the angle between the earth's magnetic field vector and the horizontal plane. **3.** An apparatus used to ascertain the direction of the magnetic field of the earth with reference to the plane of the horizon.

inclusion [CRYSTAL] **1.** A crystal or fragment of a crystal found in another crystal. **2.** A small cavity filled with gas or liquid in a crystal. [CYTOL] A visible product of cellular metabolism within the protoplasm. [MET] An impure particle, such as sand, trapped in molten metal during solidification. [PETR] A fragment of older rock enclosed in an igneous rock.

inclusion blennorrhea See inclusion conjunctivitis.

inclusion body [VIROL] Any of the abnormal structures appearing within the cell nucleus or the cytoplasm during the course of virus multiplication.

inclusion conjunctivitis [MED] An acute inflammation of the conjunctiva with pus formation caused by a virus and identified from epithelial-cell inclusion bodies in conjunctival scrapings. Also known as inclusion blennorrhea; paratrachoma; swimmer's conjunctivitis; swimming pool conjunctivitis.

inclusive or See or.

incoalation See coalification.

incoherence [MED] Lack of coherence, relevance, or continuity of ideas or language.

incoherent light [OPTICS] Electromagnetic radiant energy not all of the same phase, and possibly also consisting of various wavelengths.

incoherent scattering [PHYS] Scattering of particles or photons in which the scattering elements act independently of one another, so that there are no definite phase relationships among the different parts of the scattered beam.

incompetent bed [GEOL] A bed not combining sufficient firmness and flexibility to transmit a thrust and to lift a load by bending.

incompetent rock [ENG] Soft or fragmented rock in which an opening, such as a borehole or an underground working place, cannot be maintained unless artificially supported by casing, cementing, or timbering.

incomplete flower [BOT] A flower lacking one or more modified leaves, such as petals, sepals, pistils, or stamens.

incomplete fusion   *See* quasi-fission.
incompressibility   [MECH] Quality of a substance which maintains its original volume under increased pressure.
incompressible flow   [FL MECH] Fluid motion without any change in density.
incompressible fluid   [FL MECH] A fluid which is not reduced in volume by an increase in pressure.
incongruous   [GEOL] Of a drag fold, having an axis and axial surface not parallel to the axis and axial surface of the main fold to which it is related.
incontinence   [MED] Inability to control the natural evacuations, as the feces or the urine; specifically, involuntary evacuation from organic causes.
increment   [HYD] *See* recharge.   [MATH] A change in the argument or values of a function, usually restricted to being a small positive or negative quantity.   [SCI TECH] A small change in the value of a variable.
incremental compiler   [COMPUT SCI] A compiler that generates code for a statement, or group of statements, which is independent of the code generated for other statements.
incremental computer   [COMPUT SCI] A special-purpose computer designed to process changes in variables as well as absolute values; for instance, a digital differential analyzer.
incremental digital recorder   [COMPUT SCI] Magnetic tape recorder in which the tape advances across the recording head step by step, as in a punched-paper-tape recorder; used for recording an irregular flow of data economically and reliably.
incremental dump tape   [COMPUT SCI] A safety technique used in time-sharing which consists in copying all files (created or modified by a user during a day) on a magnetic tape; in case of system failure, the file storage can then be reconstructed. Also known as failsafe tape.
incremental mode   [COMPUT SCI] The plotting of a curve on a cathode-ray tube by illuminating a fixed number of points at a time.
incremental representation   [COMPUT SCI] A way of representing variables used in incremental computers, in which changes in the variables are represented instead of the values of the variables themselves.
incubation   [CHEM] Maintenance of chemical mixtures at specified temperatures for varying time periods to study chemical reactions, such as enzyme activity.   [MED] The phase of an infectious disease process between infection by the pathogen and appearance of symptoms.   [VERT ZOO] The act or process of brooding or incubating.
incubator   [AGR] A device for the artificial hatching of eggs.   [MED] A small chamber with controlled oxygen, temperature, and humidity for newborn infants requiring special care.   [MICROBIO] A laboratory cabinet with controlled temperature for the cultivation of bacteria, or for facilitating biologic tests.
incumbent   [BIOL] Lying on or down.   [GEOL] Lying above, said of a stratum that is superimposed or overlies another stratum.
incurrent siphon   *See* inhalant siphon.
Incurvariidae   [INV ZOO] A family of lepidopteran insects in the superfamily Incurvarioidea; includes yucca moths and relatives.
Incurvarioidea   [INV ZOO] A monofamilial superfamily of lepidopteran insects in the suborder Heteroneura having wings covered with microscopic spines, a single genital opening in the female, and venation that is almost complete.

incus   [ANAT] The middle one of three ossicles in the middle ear. Also known as anvil.   [METEOROL] A supplementary cloud feature peculiar to cumulonimbus capillatus; the spreading of the upper portion of cumulonimbus when this part takes the form of an anvil with a fibrous or smooth aspect. Also known as anvil; thunderhead.
indefinite integral   [MATH] An indefinite integral of a function $f(x)$ is a function $F(x)$ whose derivative equals $f(x)$. Also known as antiderivative; integral.
indehiscent   [BOT] Remaining closed at maturity, as certain fruits. **2.** Not splitting along regular lines.
indentation hardness   [MET] The resistance of a metal surface to indention when subjected to pressure by a hard pointed or rounded tool. Also known as penetration hardness.
independent assortment   [GEN] The random assortment of the alleles at two or more loci on different chromosome pairs or far apart on the same chromosome pair which occurs at meiosis; first discovered by G. Mendel.
independent events   [STAT] Two events in probability such that the occurrence of one of them does not affect the probability of the occurrence of the other.
independent footing   [CIV ENG] A footing that supports a concentrated load, such as a single column.
independent functions   [MATH] A set of functions such that knowledge of the values obtained by all but one of them at a point is insufficient to determine the value of the remaining function.
independent random variables   [STAT] the discrete random variables $X_1, X_2, \ldots , X_n$ are independent if for arbitrary values $x_1, x_2, \ldots , x_n$ of the variables the probability that $X_1 = x_1$ and $X_2 = x_2$, etc., is equal to the product of the probabilities that $X_1 = x_i$ for $i = 1, 2, \ldots , n$; random variables which are unrelated.
independent sector   [COMPUT SCI] A device on some punched-card tabulators that allows only the first of a series of similar data items to be printed and prevents printing of the rest.
independent-sideband modulation   [COMMUN] Modulation in which the radio-frequency carrier is reduced or eliminated and two channels of information are transmitted, one on an upper and one on a lower sideband. Abbreviated ISB modulation.
independent suspension   [MECH ENG] In automobiles, a system of springs and guide links by which wheels are mounted independently on the chassis.
independent variable   [MATH] In an equation $y = f(x)$, the input variable x. Also known as argument.
inderborite   [MINERAL] $CaMgB_6O_{11} \cdot 11H_2O$ A monoclinic mineral composed of hydrous calcium and magnesium borate.
inderite   [MINERAL] $Mg_2B_6O_{11} \cdot 15H_2O$ A hydrated borate mineral.
indeterminacy principle   *See* uncertainty principle.
indeterminate cleavage   [EMBRYO] Cleavage in which all the early cells have the same potencies with respect to development of the entire zygote.
indeterminate equations   [MATH] A set of equations possessing an infinite number of solutions.
indeterminate growth   [BOT] Growth of a plant in which the axis is not limited by development of a reproductive structure, and therefore growth continues indefinitely.
index   [COMPUT SCI] **1.** A list of record surrogates arranged in order of some attribute expressible in machine-orderable form. **2.** To produce a machine-

orderable set of record surrogates, as in indexing a book. **3.** To compute a machine location by indirection, as is done by index registers. **4.** The portion of a computer instruction which indicates what index register (if any) is to be used to modify the address of an instruction. [MATH] **1.** Unity of a logarithmic scale, as the C scale of a slide rule. **2.** A subscript or superscript used to indicate a specific element of a set or sequence. [PHYS] A numerical quantity, usually dimensionless, denoting the magnitude of some physical effect, such as the refractive index.

index array [COMPUT SCI] A group of registers corresponding one-for-one with the buffer registers, each register containing the main storage address of the data contained in the corresponding buffer register.

index bed *See* key bed.

index center [MECH ENG] One of two machine-tool centers used to hold work and to rotate it by a fixed amount.

index counter [ENG] A counter indicating revolutions of the tape supply reel, making it possible to index selections within a reel of tape.

indexed address [COMPUT SCI] An address which is modified, generally by means of index registers, before or during execution of a computer instruction.

index ellipsoid [OPTICS] An ellipsoid whose three perpendicular axes are proportional in length to the principal values of the index of refraction of light in an anisotropic medium and point in the direction of the corresponding electric vector. Also known as ellipsoid of wave normals; indicatrix; optical indicatrix; polarizability ellipsoid; reciprocal ellipsoid.

index fossil [PALEON] The ancient remains and traces of an organism that lived during a particular geologic time period and that geologically date the containing rocks.

index line *See* isopleth.

index marker [COMPUT SCI] The beginning (and end) of each track in a disk, which is recognized by a special sensing device within the disk mechanism.

index mineral [PETR] A mineral whose first appearance in passing from low to higher grades of metamorphism indicates the outer limit of a zone.

index number [STAT] A number indicating change in magnitude, as of cost or of volume of production, as compared with the magnitude at a specified time, usually taken as 100; for example, if production volume in 1970 was two times as much as the volume in 1950 (taken as 100), its index number is 200.

index of absorption *See* absorption index.

index of refraction [OPTICS] The ratio of the phase velocity of light in a vacuum to that in a specified medium. Also known as refractive index; refracture index.

index point [COMPUT SCI] A hardware reference mark on a disk or drum for use in timing.

index register [COMPUT SCI] A hardware element which holds a number that can be added to (or, in some cases, subtracted from) the address portion of a computer instruction to form an effective address. Also known as base register; B box; B line; B register; B store; modifier register.

index thermometer [ENG] A thermometer in which steel index particles are carried by mercury in the capillary and adhere to the capillary wall in the high and low positions, thus indicating minimum and maximum inertial scales.

index word *See* modifier.

indialite [MINERAL] $Mg_2Al_4Si_5O_{18}$ A hexagonal cordierite mineral; it is isotypic with beryl.

Indian *See* Indus.

indianaite [MINERAL] A white porcelainlike clay mineral; a variety of halloysite found in Indiana.

Indiana limestone *See* spergenite.

Indian summer [CLIMATOL] A period, in mid- or late autumn, of abnormally warm weather, generally clear skies, sunny but hazy days, and cool nights; in New England, at least one killing frost and preferably a substantial period of normally cool weather must precede this warm spell in order for it to be considered a true Indian summer; it does not occur every year, and in some years there may be two or three Indian summers; the term is most often heard in the northeastern United States, but its usage extends throughout English-speaking countries.

indicated airspeed [AERO ENG] The airspeed as shown by a differential-pressure airspeed indicator, uncorrected for instrument and installation errors; a simple computation for altitude and temperature converts indicated airspeed to true airspeed. Abbreviated IAS.

indicated horsepower [MECH ENG] The horsepower delivered by an engine as calculated from the average pressure of the working fluid in the cylinders and the displacement. Abbreviated ihp.

indicated ore [MIN ENG] A known mineral deposit for which quantitative estimates are made partly from inference and partly from specific sampling. Also known as probable ore.

indicator [COMPUT SCI] A device announcing an error or failure. [ELECTR] A cathode-ray tube or other device that presents information transmitted or relayed from some other source, as from a radar receiver. [ENG] An instrument for obtaining a diagram of the pressure-volume changes in a running positive-displacement engine, compressor, or pump cylinder during the working cycle.

indicator card [ENG] A chart on which an indicator diagram is produced by an instrument called an engine indicator which traces the real-performance cycle diagram as the machine is running.

indicator diagram [ENG] A pressure-volume diagram representing and measuring the work done by or on a fluid while performing the work cycle in a reciprocating engine, pump, or compressor cylinder.

indicator element [ELECTR] A component whose variability under conditions of manufacture or use is likely to cause the greatest variation in some measurable parameter.

indicator function *See* characteristic function.

indicator gate [ELECTR] Rectangular voltage waveform which is applied to the grid or cathode circuit of an indicator cathode-ray tube to sensitize or desensitize it during a desired portion of the operating cycle.

indicator lamp [ELEC] A neon lamp whose on-off condition is used to convey information.

indicator plant [BOT] A plant used in geobotanical prospecting as an indicator of a certain geological phenomenon.

indicator tube [ELECTR] An electron-beam tube in which useful information is conveyed by the variation in cross section of the beam at a luminescent target.

indicatrix *See* index ellipsoid.

indicolite [MINERAL] An indigo-blue variety of tourmaline that is used as a gemstone. Also known as indigolite.

indigenous [SCI TECH] Existing and having originated naturally in a particular region or environment.

indigenous coal *See* autochthonous coal.

indigo [ORG CHEM] 1. A blue dye extracted from species of the *Indigofera* bush. 2. *See* indigo blue.

indigolite *See* indicolite.

indirect address [COMPUT SCI] An address in a computer instruction that indicates a location where the address of the referenced operand is to be found. Also known as multilevel address.

indirect addressing [COMPUT SCI] A programming device whereby the address part of an instruction is not the address of the operand but rather the location in storage where the address of the operand may be found.

indirect control [COMPUT SCI] The control of one peripheral unit by another through some sequence of events that involves human intervention.

indirect cost [IND ENG] A cost that is not readily indentifiable with or chargeable to a specific product or service.

indirect cycle [NUCLEO] A nuclear reactor cycle in which a heat exchanger transfers heat from the reactor coolant to a second fluid, which then drives a prime mover.

indirect extrusion [MET] An extrusion process in which the billet remains stationary while a hollow die stand forces the die back into the cylinder.

indirect stratification *See* secondary stratification.

indirect vision *See* peripheral vision.

indium [CHEM] A metallic element, symbol In, atomic number 49, atomic weight 114.82; soluble in acids; melts at 156°C, boils at 1450°C. [MET] A ductile, silver-white, shiny metal that resists tarnishing and is used in precious-metal alloys for jewelry and dentistry, in glass-sealing alloys, lubricants, and bearing metals, and as an atomic-pile neutron indicator.

individual line [COMMUN] Subscriber line arranged to serve only one main station, although additional stations may be connected to the line as extensions; an individual line is not arranged for discriminatory ringing with respect to the stations on that line.

Indriidae [VERT ZOO] A family of Madagascan prosimians containing wholly arboreal vertical clingers and leapers.

induced capacity *See* absolute permeability.

induced current [ELECTROMAG] A current produced in a conductor by a time-varying magnetic field, as in induction heating.

induced draft [MECH ENG] A mechanical draft produced by suction stream jets or fans at the point where air or gases leave a unit.

induced drag [FL MECH] That part of the drag caused by the downflow or downwash of the airstream passing over the wing of an aircraft, equal to the lift times the tangent of the induced angle of attack.

induced electromotive force [ELECTROMAG] An electromotive force resulting from the motion of a conductor through a magnetic field, or from a change in the magnetic flux that threads a conductor.

induced fission [NUC PHYS] Fission which takes place only when a nucleus is bombarded with neutrons, gamma rays, or other carriers of energy.

induced magnetization [GEOPHYS] That component of a rock's magnetization which is proportional to, and has the same direction as, the ambient magnetic field.

induced moment [ELEC] The average electric dipole moment per molecule which is produced by the action of an electric field on a dielectric substance.

induced potential *See* induced voltage.

induced voltage [ELECTROMAG] A voltage produced by electromagnetic or electrostatic induction. Also known as induced potential.

inducer [EMBRYO] The cell group that functions as the acting system in embryonic induction by controlling the mode of development of the reacting system. Also known as inductor.

inducible enzyme [BIOCHEM] An enzyme which is present in trace quantities within a cell but whose concentration increases dramatically in the presence of substrate molecules.

inductance [ELECTROMAG] 1. That property of an electric circuit or of two neighboring circuits whereby an electromotive force is generated (by the process of electromagnetic induction) in one circuit by a change of current in itself or in the other. 2. Quantitatively, the ratio of the emf (electromotive force) to the rate of change of the current. 3. *See* coil.

inductance bridge [ELECTROMAG] 1. A device, similar to a Wheatstone bridge, for comparing inductances. 2. A four-coil alternating-current bridge circuit used for transmitting a mechanical movement to a remote location over a three-wire circuit; half of the bridge is at each location.

inductance coil *See* coil.

induction [ELEC] *See* electrostatic induction. [ELECTROMAG] *See* electromagnetic induction. [EMBRYO] *See* embryonic induction. [MED] The period from administration of the anesthetic to loss of consciousness by the patient. [SCI TECH] The act or process of causing.

induction accelerator *See* betatron.

induction brazing [MET] A brazing process in which coalescence is produced by heat generated within the work by an induced electric current.

induction coil [ELECTROMAG] A device for producing high-voltage alternating current or high-voltage pulses from low-voltage direct current, in which interruption of direct current in a primary coil, containing relatively few turns of wire, induces a high voltage in a secondary coil, containing many turns of wire wound over the primary.

induction force [PHYS CHEM] A type of van der Waals force resulting from the interaction of the dipole moment of a polar molecule and the induced dipole moment of a nonpolar molecule. Also known as Debye force.

induction furnace [ENG] An electric furnace in which heat is produced in a metal charge by electromagnetic induction.

induction generator [ELEC] A nonsynchronous alternating-current generator whose construction is identical to that of an ac motor, and which is driven above synchronous speed by external sources of mechanical power.

induction heating [ENG] Increasing the temperature in a material by induced electric current. Also known as eddy-current heating.

induction inclinometer *See* earth inductor.

induction log [ENG] An electric log of the conductivity of rock with depth obtained by lowering into an uncased borehole a generating coil that induces eddy currents on the rocks and these are detected by a receiver coil.

induction machine [ELEC] An asynchronous alternating-current machine, such as an induction motor or induction generator, in which the windings of two electric circuits rotate with respect to each other and power is transferred from one circuit to the other by electromagnetic induction.

induction motor [ELEC] An alternating-current motor in which a primary winding on one member (usually the stator) is connected to the power source, and a secondary winding on the other member (usually the rotor) carries only current induced by the magnetic field of the primary.

induction pump [MECH ENG] Any pump operated by electromagnetic induction.

induction valve See inlet valve.

induction voltage regulator [ELECTROMAG] A type of transformer having a primary winding connected in parallel with a circuit and a secondary winding in series with the circuit; the relative positions of the primary and secondary windings are changed to vary the voltage or phase relations in the circuit.

induction welding [MET] A process of welding by means of heat generated within the work by induced electric currents.

inductive circuit [ELEC] A circuit containing a higher value of inductive reactance than capacitive reactance.

inductive coupling [ELEC] Coupling of two circuits by means of the mutual inductance provided by a transformer. Also known as transformer coupling.

inductive filter [ELECTR] A low-pass filter used for smoothing the direct-current output voltage of a rectifier; consists of one or more sections in series, each section consisting of an inductor on one of the pair of conductors in series with a capacitor between the conductors. Also known as LC filter.

inductive load [ELEC] A load that is predominantly inductive, so that the alternating load current lags behind the alternating voltage of the load. Also known as lagging load.

inductive pressure transducer [ELECTROMAG] A type of pressure transducer in which changes in pressure cause a bourdon tube or other pressure-sensing element to move a magnetic core, and this results in a change in inductance of one or more windings of a coil surrounding the core.

inductometer [ELECTROMAG] A coil of wire of known inductance; the inductance may be fixed as in the case of primary standards, adjustable by means of switches, or continuously variable by means of a movable-coil construction.

inductor See coil; inducer.

inductor generator [ELEC] An alternating-current generator in which all the windings are fixed, and the flux linkages are varied by rotating an appropriately toothed ferromagnetic rotor; sometimes used for generating high power at frequencies up to several thousand hertz for induction heating.

inductor tachometer [ENG] A type of impulse tachometer in which the rotating member, consisting of a magnetic material, causes the magnetic flux threading a circuit containing a magnet and a pickup coil to rise and fall, producing pulses in the circuit which are rectified for a permanent-magnet, movable-coil instrument.

induplicate [BOT] Having the edges turned or rolled inward without twisting or overlapping; applied to the leaves of a bud.

induration [BIOL] The process of hardening, especially by increasing the fibrous elements. [GEOL] **1.** The hardening of a rock material by the application of heat or pressure or by the introduction of a cementing material. **2.** A hardened mass formed by such processes. **3.** The hardening of a soil horizon by chemical action to form a hardpan. [MED] Hardening of a tissue or organ due to an accumulation of blood, inflammation, or neoplastic growth.

Indus [ASTRON] A constellation, right ascension 21 hours, declination 55° south. Also known as Indian.

indusium [ANAT] A covering membrane such as the amnion. [BOT] An epidermal outgrowth covering the sori in many ferns. [MYCOL] The annulus of certain fungi.

industrial climatology [CLIMATOL] A type of applied climatology which studies the effect of climate and weather on industry's operations; the goal is to provide industry with a sound statistical basis for all administrative and operational decisions which involve a weather factor.

industrial engineering [ENG] A branch of engineering concerned with the design, improvement, and installation of integrated systems of people, materials, and equipment. Also known as management engineering.

industrial jewel [MINERAL] A hard stone, such as ruby or sapphire, used for bearings and impulse pins in instruments and for recording needles.

industrial microbiology [MICROBIO] The study, utilization, and manipulation of those microorganisms capable of economically producing desirable substances or changes in substances, and the control of undesirable microorganisms.

industrial psychology [PSYCH] Psychology applied to problems in industry, dealing chiefly with the selection, efficiency, and mental health of personnel.

ineffective time [COMPUT SCI] Time during which a computer can operate normally but which is not used effectively because of mistakes or inefficiency in operating the installation or for other reasons.

inelastic [MECH] Not capable of sustaining a deformation without permanent change in size or shape.

inelastic buckling [MECH] Sudden increase of deflection or twist in a column when compressive stress reaches the elastic limit but before elastic buckling develops.

inelastic collision [MECH] A collision in which the total kinetic energy of the colliding particles is not the same after the collision as before it.

inelastic stress [MECH] A force acting on a solid which produces a deformation such that the original shape and size of the solid are not restored after removal of the force.

inequality [MATH] A statement that one quantity is less than, less than or equal to, greater than, or greater than or equal to another quantity.

inequality of Clausius See Clausius inequality.

inert [SCI TECH] Lacking an activity, reactivity, or effect.

inert gas See noble gas.

inertia [MECH] That property of matter which manifests itself as a resistance to any change in the momentum of a body. [MED] Sluggishness, especially of muscular activity.

inertia ellipsoid [MECH] An ellipsoid used in describing the motion of a rigid body; it is fixed in the body, and the distance from its center to its surface in any direction is inversely proportional to the square root of the moment of inertia about the corresponding axis. Also known as momental ellipsoid; Poinsot ellipsoid.

inertial coordinate system See inertial reference frame.

inertial flow [FL MECH] Flow in which no external forces are exerted on a fluid. [GEOPHYS] Frictionless flow in a geopotential surface in which there is no pressure gradient; the centrifugal and Coriolis accelerations must therefore be equal and opposite, and the constant inertial wind speed V_i is given by $V_i = fR$, where f is the Coriolis parameter and R the radius of curvature of the path.

inertial force [MECH] The fictitious force acting on a body as a result of using a noninertial frame of reference; examples are the centrifugal and Coriolis forces that appear in rotating coordinate systems. Also known as effective force.

inertial instability [FL MECH] 1. Generally, instability in which the only form of energy transferred between the steady state and the disturbance in the fluid is kinetic energy. 2. The hydrodynamic instability arising in a rotating fluid mass when the velocity distribution is such that the kinetic energy of a disturbance grows at the expense of kinetic energy of the rotation. Also known as dynamic instability.

inertial mass [MECH] The mass of an object as determined by Newton's second law, in contrast to the mass as determined by the proportionality to the gravitational force.

inertial reference frame [MECH] A coordinate system in which a body moves with constant velocity as long as no force is acting on it. Also known as inertial coordinate system.

inertia of energy [RELAT] The principle that the inertial properties of matter both determine and are determined by its total energy content.

inertia welding [MET] A form of friction welding which utilizes kinetic energy stored in a flywheel system to supply the power required for all of the heating and much of the forging.

inertinite [GEOL] A carbon-rich maceral group, which includes micrinite, sclerotinite, fusinite, and semifusinite.

inesite [MINERAL] $Ca_2Mn_7Si_{10}O_{28}(OH)_2\cdot5H_2O$ A pale-red mineral composed of hydrous manganese calcium silicate, occurring in small prismatic crystals or massive.

infantile amaurotic familial idiocy See Tay-Sachs disease.

infantile eczema [MED] An allergic inflammation of the skin in young children, usually due to common antigens such as food or inhalants.

infantile genitalia [ANAT] The genital organs of an infant. [MED] Underdevelopment of the adult genitals.

infantile paralysis See poliomyelitis.

infantilism [MED] Persistence of physical, behavioral, or mental infantile characteristics into childhood, adolescence, or adult life.

infarct [MED] Localized death of tissue caused by obstructed inflow of arterial blood. Also known as infarction.

infarction [MED] 1. Condition or process leading to the development of an infarct. 2. See infarct.

infection [MED] 1. Invasion of the body by a pathogenic organism, with or without disease manifestation. 2. Pathologic condition resulting from invasion of a pathogen.

infectious abortion See contagious abortion.

infectious hepatitis [MED] Type A viral hepatitis, an acute infectious virus disease of the liver associated with hepatic inflammation and characterized by fever, liver enlargement, and jaundice. Also known as catarrhal jaundice; epidemic hepatitis; epidemic jaundice; virus hepatitis.

infectious mononucleosis [MED] A disorder of unknown etiology characterized by irregular fever, pathology of lymph nodes, lymphocytosis, and high serum levels of heterophil antibodies against sheep erythrocytes. Also known as acute benign lymphoblastosis; glandular fever; kissing disease; lymphocytic angina; monocytic angina; Pfeiffer's disease.

infective dose 50 [MICROBIO] The dose of microorganisms required to cause infection in 50% of the experimental animals; a special case of the median effective dose. Abbreviated ID_{50}. Also known as median infective dose.

inferior [BIOL] The lower of two structures.

inferior conjunction [ASTRON] A type of configuration in which two celestial bodies have their least apparent separation; the smaller body is nearer the observer than the larger body, about which it orbits; for example, Venus is closest to the earth at its inferior conjunction.

inferior planet [ASTRON] A planet that circles the sun in an orbit that is smaller than the earth's.

inferior vena cava [ANAT] A large vein which drains blood from the iliac veins, lower extremities, and abdomen into the right atrium.

infertility [MED] Involuntary reduction in reproductive ability.

infiltration [GEOL] Deposition of mineral matter among the pores or grains of a rock by permeation of water carrying the matter in solution. [HYD] Movement of water through the soil surface into the ground. [MET] 1. Filling the pores of a metal powder compact with metal having a lower melting point. 2. Movement of molten metal into the pores of a fiber or foam metal.

infiltration gallery [CIV ENG] A large, horizontal underground conduit of porous material or with openings on the sides for collecting percolating water by infiltration.

infinite [MATH] Larger than any fixed number.

infinite sequence See sequence.

infinite set [MATH] A set with more elements than any fixed integer; such a set can be put into a one to one correspondence with a proper subset of itself.

infinity [COMPUT SCI] Any number larger than the maximum number that a computer is able to store in any register. [MATH] The concept of a value larger than any finite value.

infinity method [OPTICS] Method of adjusting two lines of sight to make them parallel; lines are adjusted on an object at great distance, for example, a star.

infix operation [COMPUT SCI] An operation carried out within an operation, as the addition of a and b prior to the multiplication by c or division by d in the operation $(a + b)c/d$.

inflammability See flammability.

inflammation [MED] Local tissue response to injury characterized by redness, swelling, pain, and heat.

inflected [BOT] Curved or bent sharply inward, downward, or toward the axis. Also known as inflexed.

inflected arch See inverted arch.

inflection point See point of inflection.

inflexed See inflected.

influence line [MECH] A graph of the shear, stress, bending moment, or other effect of a movable load on a structural member versus the position of the load.

influent [ECOL] An organism that disturbs the ecological balance of a community. [SCI TECH] An input stream of a fluid, as water into a reservoir, or liquid into a process vessel.

influenza [MED] An acute virus disease of the respiratory system characterized by headache, muscle pain, fever, and prostration.

influenza virus [VIROL] Any of three immunological types, designated A, B, and C, belonging to the myxovirus group which cause influenza.

influx See mouth.

information [COMMUN] Data which has been recorded, classified, organized, related, or interpreted within a framework so that meaning emerges.

information bit [COMMUN] Bit that is generated by the data source but is not used by the data-transmission system.

information channel [COMMUN] A facility used to transmit information between data-processing terminals separated by large distances.

information flow [COMPUT SCI] The graphic representation of data collection, data processing, and report distribution throughout an organization.

information link See data link.

information processing [COMPUT SCI] 1. The manipulation of data so that new data (implicit in the original) appear in a useful form. 2. See data processing.

information retrieval [COMPUT SCI] The technique and process of searching, recovering, and interpreting information from large amounts of stored data.

information selection systems [COMPUT SCI] A class of information processing systems which carry out a sequence of operations necessary to locate in storage one or more items assumed to have certain specified characteristics and to retrieve such items directly or indirectly, in whole or in part.

information separator [COMPUT SCI] A character that separates items or fields of information in a record, especially a variable-length record.

information system [COMMUN] Any means for communicating knowledge from one person to another, such as by simple verbal communication, punched-card systems, optical coincidence systems based on coordinate indexing, and completely computerized methods of storing, searching, and retrieving of information.

information theory [COMMUN] A branch of communications theory which is devoted to problems in coding, and which provides criteria for comparing different communications systems on the basis of signaling rate, using a numerical measure of the amount of information gained when the content of a message is learned. [MATH] The branch of probability theory concerned with the likelihood of the transmission of messages, accurate to within specified limits, when the bits of information composing the message are subject to possible distortion.

information unit [COMMUN] A unit of information content, equal to a bit, nit, or hartley, according to whether logarithms are taken to base 2, e, or 10.

information word See data word.

infrabasal [BIOL] Inferior to a basal structure.

Infracambrian See Eocambrian.

infraciliature [INV ZOO] The neuromotor apparatus, silverline system, or neuroneme system of ciliates.

infraclass [SYST] A subdivision of a subclass; equivalent to a superorder.

infradyne receiver [ELECTR] A superheterodyne receiver in which the intermediate frequency is higher than the signal frequency, so as to obtain high selectivity.

infralow frequency [COMMUN] A designation for the band from 0.3 to 3 kilohertz in the radio spectrum. Abbreviated ILF.

infrared astronomy [ASTROPHYS] The study of electromagnetic radiation in the spectrum between 0.75 and 1000 micrometers emanating from astronomical sources.

infrared detector [ELECTR] A device responding to infrared radiation, used in detecting fires, or overheating in machinery, planes, vehicles, and people,

and in controlling temperature-sensitive industrial processes.

infrared-emitting diode [ELECTR] A light-emitting diode that has maximum emission in the near-infrared region, typically at 0.9 micrometer for pn gallium arsenide.

infrared filter [OPTICS] A substance or device which is highly transparent to infrared radiation at certain wavelengths while absorbing other types of electromagnetic radiation.

infrared galaxy [ASTRON] A galaxy or quasar whose nucleus emits enormous amounts of infrared radiation, in some cases more than 1000 times the output of the entire Milky Way Galaxy at all wavelengths.

infrared image converter [ELECTR] A device for converting an invisible infrared image into a visible image, consisting of an infrared-sensitive, semitransparent photocathode on one end of an evacuated envelope and a phosphor screen on the other, with an electrostatic lens system between the two. Also known as infrared image tube.

infrared image tube See infrared image converter.

infrared lamp [ELEC] An incandescent lamp which operates at reduced voltage with a filament temperature of 4000°F (2200°C) so that it radiates electromagnetic energy primarily in the infrared region.

infrared laser [PHYS] A laser which emits infrared radiation, especially in the near- and intermediate-infrared regions.

infrared mapping [MAP] Mapping in which a sensitive infrared detector is mounted on an infrared scanner, and the resulting thermal image is translated into varying shades of gray on photographic film, to give a line-pattern image much like that seen on a television screen.

infrared maser [PHYS] A laser which emits infrared radiation, especially in the far-infrared region, or which is pumped with radiation at infrared frequencies and emits radiation at millimeter wavelengths.

infrared microscope [OPTICS] A type of reflecting microscope which uses radiation of wavelengths greater than 700 nanometers and is used to reveal detail in materials that are opaque to light, such as molybdenum, wood, corals, and many red-dyed materials.

infrared photography [GRAPHICS] Photography in which an infrared optical system projects an image directly on infrared film, to provide a record of point-to-point variations in temperature of a scene.

infrared radiation [ELECTROMAG] Electromagnetic radiation whose wavelengths lie in the range from 0.75 to 1000 micrometer (the long-wavelength limit of visible red light) to 1000 micrometers (the shortest microwaves).

infrared spectrometer [SPECT] Device used to identify and measure the concentrations of heteroatomic compounds in gases, in many nonaqueous liquids, and in some solids by arc or spark excitation and subsequent measurement of the electromagnetic emissions in the wavelength range of 0.78 to 300 micrometers.

infrared spectroscopy [SPECT] The study of the properties of material systems by means of their interaction with infrared radiation; ordinarily the radiation is dispersed into a spectrum after passing through the material.

infrared spectrum [ELECTROMAG] 1. The range of wavelengths of infrared radiation. 2. A display or graph of the intensity of infrared radiation emitted or absorbed by a material as a function of wavelength or some related parameter.

infrared star [ASTROPHYS] A star that emits a large amount of radiant energy in the infrared portion of the electromagnetic spectrum.

infrared telescope [OPTICS] An instrument that converts an invisible infrared image into a visible image and enlarges this image, consisting of an infrared image converter tube, an objective lens for imaging the scene to be viewed onto the photocathode of the tube, and an ocular for viewing the phosphor screen of the tube.

infrared window [GEOPHYS] A frequency region in the infrared where there is good transmission of electromagnetic radiation through the atmosphere.

infrasonic [ACOUS] Pertaining to signals, equipment, or phenomena involving frequencies below the range of human hearing, hence below about 15 hertz. Also known as subsonic (deprecated usage).

infrasound [ACOUS] Vibrations of the air at frequencies too low to be perceived as sound by the human ear, below about 15 hertz.

infundibular process [ANAT] The distal portion of the neural lobe of the pituitary.

infundibulum [ANAT] 1. A funnel-shaped passage or part. 2. The stalk of the neurohypophysis.

infusion [CHEM] The aqueous solution of a soluble constituent of a substance as the result of the substance's steeping in the solvent for a period of time. [MED] The slow injection of a solution into a vein or into subcutaneous or other tissue of the body.

in-gate See gate.

ingestion [BIOL] The act or process of taking food and other substances into the animal body.

Ingolfiellidea [INV ZOO] A suborder of amphipod crustaceans in which both abdomen and maxilliped are well developed and the head often bears a separate ocular lobe lacking eyes.

ingot [MET] 1. A solid metal casting suitable for remelting or working. 2. A bar of gold or silver.

ingrain color See azoic dye.

ingress [ASTRON] The entrance of the moon into the shadow of the earth in an eclipse, of a planet into the disk of the sun, or of a satellite (or its shadow) onto the disk of the parent planet. [SCI TECH] The act of entering, as of air into the lungs or a liquid into an orifice.

inguinal canal [ANAT] A short, narrow passage between the abdominal ring and the inguinal ring in which lies the spermatic cord in males and the round ligament in females.

inguinal hernia [MED] Protrusion of the abdominal viscera through the inguinal canal.

inguinal region [ANAT] The abdominal region occurring on each side of the body as a depression between the abdomen and the thigh. Also known as iliac region.

inhalant canal [INV ZOO] The incurrent canal in sponges and mollusks.

inhalant siphon [INV ZOO] A channel for water intake in the mantle of bivalve mollusks. Also known as incurrent siphon.

inhalator [MED] A device for facilitating the inhalation of a gas or spray, as for providing oxygen or oxygen–carbon dioxide mixtures for respiration in resuscitation.

inhaler [MED] 1. A device containing a solid medication through which air is drawn into the air passages. 2. An atomizer containing a liquid medication. [MIN ENG] An apparatus, of different forms, for permitting the supply of fresh air to a miner.

inherent damping [MECH ENG] A method of vibration damping which makes use of the mechanical hysteresis of such materials as rubber, felt, and cork.

inherent storage [COMPUT SCI] Any type of storage in which the storage medium is part of the hardware of the computer medium.

inheritance [GEN] 1. The acquisition of characteristics by transmission of germ plasm from ancestor to descendant. 2. The sum total of characteristics dependent upon the constitution of the fertilized ovum.

inherited error [COMPUT SCI] The error existing in the data supplied at the beginning of a step in a step-by-step calculation as executed by a program.

inhibit-gate [ELECTR] Gate circuit whose output is energized only when certain signals are present and other signals are not present at the inputs.

inhibiting signal [ELECTR] A signal, which when entered into a specific circuit will prevent the circuit from exercising its normal function; for example, an inhibit signal fed into an AND gate will prevent the gate from yielding an output when all normal input signals are present.

inhibitor [CHEM] A substance which is capable of stopping or retarding a chemical reaction; to be technically useful, it must be effective in low concentration.

inhour [NUCLEO] A unit of reactivity of a reactor; 1 inhour is the reactivity that will give the reactor a period of 1 hour. Derived from inverse hour.

Iniomi [VERT ZOO] An equivalent name for Salmoniformes.

initial condition [COMPUT SCI] See entry condition. [METEOROL] A prescription of the state of a dynamical system at some specified time; for all subsequent times the equations of motion and boundary conditions determine the state of the system; the appropriate synoptic weather charts, for example, constitute a (discrete) set of initial conditions for a forecast; in many contexts, initial conditions are considered as boundary conditions in the dimension of time.

initial condition mode See reset mode.

initial detention See surface storage.

initial dip See primary dip.

initial instructions [COMPUT SCI] A routine stored in a computer to aid in placing a program in memory. Also known as initial orders.

initialize [COMPUT SCI] 1. To set counters, switches, and addresses to zero or other starting values at the beginning of, or at prescribed points in, a computer routine. 2. To begin an operation, and more specifically, to adjust the environment to the required starting configuration.

initial orders See initial instructions.

initial program load button See bootstrap button.

initiate See trigger.

injected [PETR] Pertaining to intrusive igneous rock or other mobile rock that has erupted through rock walls to neighboring older rocks.

injection [AERO ENG] The process of placing a spacecraft into a specific trajectory, such as an earth orbit or an encounter trajectory to Mars. [ELECTR] 1. The method of applying a signal to an electronic circuit or device. 2. The process of introducing electrons or holes into a semiconductor so that their total number exceeds the number present at thermal equilibrium. [GEOL] Also known as intrusion; sedimentary injection. 1. A process by which sedimentary material is forced under abnormal pressure into a preexisting rock or deposit. 2. A structure formed by an injection process. [MATH] A mapping f from a set A into a set B which has the property that for any element b of B there is at most one element a of A for which

$f(a) = b$. Also known as injective mapping; one-to-one mapping. [MECH ENG] The introduction of fuel, fuel and air, fuel and oxidizer, water, or other substance into an engine induction system or combustion chamber. [MED] **1.** Introduction of a fluid into the skin, vessels, muscle, subcutaneous tissue, or any cavity of the body. **2.** The substance injected. [MIN ENG] The introduction under pressure of a liquid or plastic material into cracks, cavities, or pores in a rock formation.

injection carburetor [MECH ENG] A carburetor in which fuel is delivered under pressure into a heated part of the engine intake system. Also known as pressure carburetor.

injection fluid [PETRO ENG] Gas or water, depending on the nature of the reservoid and its fluid content, for injection into the formation to increase hydrocarbon production.

injection gneiss [PETR] A composite rock with banding entirely or partly caused by layer-by-layer injection of granitic magma into rock layers.

injection grid [ELECTR] Grid introduced into a vacuum tube in such a way that it exercises control over the electron stream without causing interaction between the screen grid and control grid.

injection laser [OPTICS] A laser in which a forward-biased gallium arsenide diode converts direct-current input power directly into coherent light, without optical pumping.

injection locking [ELECTR] The capture or synchronization of a free-running oscillator by a weak injected signal at a frequency close to the natural oscillator frequency or to one of its subharmonics; used for frequency stabilization in IMPATT or magnetron microwave oscillators, gas-laser oscillators, and many other types of oscillators.

injection luminescent diode [ELECTR] Gallium arsenide diode, operating in either the laser or the noncoherent mode, that can be used as a visible or near-infrared light source for triggering such devices as light-activated switches.

injection molding [ENG] Molding metal, plastic, or nonplastic ceramic shapes by injecting a measured quantity of the molten material into dies.

injection pump [MECH ENG] A pump that forces a measured amount of fuel through a fuel line and atomizing nozzle in the combustion chamber of an internal combustion engine.

injection well [HYD] See recharge well. [PETRO ENG] In secondary recovery of petroleum, a well in which a fluid such as gas or water is injected to provide supplemental energy to drive the oil remaining in the reservoir to the vicinity of production wells.

injective mapping See injection.

injector [ELECTR] An electrode through which charge carriers (holes or electrons) are forced to enter the high-field region in a spacistor. [MECH ENG] **1.** An apparatus containing a nozzle in an actuating fluid which is accelerated and thus entrains a second fluid, so delivering the mixture against a pressure in excess of the actuating fluid. **2.** A plug with a valved nozzle through which fuel is metered to the combustion chambers in diesel- or full-injection engines. **3.** A jet through which feedwater is injected into a boiler, or fuel is injected into a combustion chamber.

ink bleed [COMPUT SCI] In character recognition, the capillary extension of ink beyond the original edges of a printed or handwritten character.

ink-jet printer [GRAPHICS] A nonimpact printer that uses electrostatic acceleration and deflection of ink particles emerging from nozzles to form characters on plain paper in a dot-matrix format.

ink knife [GRAPHICS] An instrument resembling a large spatula, with a handle and a blade with square or rounded end; used for handling paste inks. Also known as ink slice.

ink sac [INV ZOO] An organ attached to the rectum in many cephalopods which secretes and ejects an inky fluid.

ink slice See ink knife.

ink smudge [COMPUT SCI] In character recognition, the overflow of ink beyond the original edges of a printed or handwritten character.

inland sea See epicontinental sea.

inlay [GRAPHICS] A picture or ornament made by inserting a material such as metal into a space in metal, stone, or wood; the material (such as wire) may be burnished, heated, or fused.

inlet [ENG] An entrance or orifice for the admission of fluid. [GEOGR] **1.** A short, narrow waterway connecting a bay or lagoon with the sea. **2.** A recess or bay in the shore of a body of water. **3.** A waterway flowing into a larger body of water.

inlet valve [MECH ENG] The valve through which a fluid is drawn into the cylinder of a positive-displacement engine, pump, or compressor. Also known as induction valve.

in line [ENG] **1.** Over the center of a borehole and parallel with its long axis. **2.** Of a drill motor, mounted so that its drive shaft and the drive rod in the drill swivel head are parallel, or mounted so that the shaft driving the drill-swivel-head bevel gear and the drill-motor drive shaft are centered in a direct line and parallel with each other. **3.** Having similar units mounted together in a line.

in-line coding [COMPUT SCI] Any group of instructions within the main body of a program.

in-line linkage [MECH ENG] A power-steering linkage which has the control valve and actuator combined in a single assembly.

in-line procedure [COMPUT SCI] A short body of coding or instruction which accomplishes some purpose.

in-line processing [COMPUT SCI] The processing of data in random order, not subject to preliminary editing or sorting.

in-line subroutine [COMPUT SCI] A subroutine which is an integral part of a program.

innate [BIOL] Pertaining to a natural or inborn character dependent on genetic constitution. [BOT] Positioned at the apex of a supporting structure. [MYCOL] Embedded in, especially of an organ such as the fruiting body embedded in the thallus of some fungi.

inner ear [ANAT] The part of the vertebrate ear concerned with labyrinthine sense and sound reception; consists generally of a bony and a membranous labyrinth, made up of the vestibular apparatus, three semicircular canals, and the cochlea. Also known as internal ear.

inner mantle See lower mantle.

inner planet [ASTRON] Any of the four planets (Mercury, Venus, Earth, and Mars) in the solar system whose orbits are closest to the sun.

innervation [ANAT] The distribution of nerves to a part. [PHYSIO] The amount of nerve stimulation received by a part.

innominate artery [ANAT] The first artery branching from the aortic arch; distributes blood to the head, neck, shoulder, and arm on the right side of the body.

inoculation [BIOL] Introduction of a disease agent into an animal or plant to produce a mild form of disease and render the individual immune. [MET] Treating a molten material with another material before casting in order to nucleate crystals. [MICROBIO] Introduction of microorganisms onto or into a culture medium.

inoculum [MICROBIO] A small amount of substance containing bacteria from a pure culture which is used to start a new culture or to infect an experimental animal.

inorganic [INORG CHEM] Pertaining to or composed of chemical compounds that do not contain carbon as the principal element (excepting carbonates, cyanides, and cyanates), that is, matter other than plant or animal.

inorganic acid [INORG CHEM] A compound composed of hydrogen and a nonmetal element or radical; examples are hydrochloric acid, HCl, sulfuric acid, H_2SO_4, and carbonic acid, H_2CO_3.

inorganic chemistry [CHEM] The study of chemical reactions and properties of all the elements and their compounds, with the exception of hydrocarbons, and usually including carbides, oxides of carbon, metallic carbonates, carbon-sulfur compounds, and carbon-nitrogen compounds.

inosculation *See* anastomosis.

inositol [ORG CHEM] $C_6H_6(OH)_6·2H_2O$ A water-soluble alcohol often grouped with the vitamins; there are nine stereoisomers of hexahydroxycyclohexane, and the only one of biological importance is optically inactive *meso*-inositol, comprising white crystals, widely distributed in animals and plants; it serves as a growth factor for animals and microorganisms.

in phase [PHYS] Having waveforms that are of the same frequency and that pass through corresponding values at the same instant.

input [COMPUT SCI] The information that is delivered to a data-processing device from the external world, the process of delivering this data, or the equipment that performs this process. [ELECTR] 1. The power or signal fed into an electrical or electronic device. 2. The terminals to which the power or signal is applied. [SCI TECH] Those resources and other environmental factors converted by a system.

input area [COMPUT SCI] A section of internal storage reserved for storage of data or instructions received from an input unit such as cards or tape. Also known as input block; input storage.

input block [COMPUT SCI] 1. A block of data read or transferred into a computer. 2. *See* input area.

input equipment [COMPUT SCI] 1. The equipment used for transferring data and instructions into an automatic data-processing system. 2. The equipment by which an operator transcribes original data and instructions to a medium that may be used in an automatic data-processing system.

input-limited [COMPUT SCI] Pertaining to a system or operation whose speed or efficiency depends mainly on the speed of input into the machine rather than the speed of the machine itself.

input magazine [COMPUT SCI] A part of a card-handling device which supplies the cards to the processing portion of the machine. Also known as magazine.

input/output [COMPUT SCI] Pertaining to all equipment and activity that transfers information into or out of a computer. Abbreviated I/O.

input/output adapter [COMPUT SCI] A circuitry which allows input/output devices to be attached directly to the central processing unit.

input/output buffer [COMPUT SCI] An area of a computer memory used to temporarily store data and instructions transferred into and out of a computer, permitting several such transfers to take place simultaneously with processing of data.

input/output control unit [COMPUT SCI] The piece of hardware which controls the operation of one or more of a type of devices such as tape drives or disk drives; this unit is frequently an integral part of the input/output device itself.

input/output device [COMPUT SCI] A unit that accepts new data, sends it into the computer for processing, receives the results, and translates them into a usable medium.

input/output interrupt identification [COMPUT SCI] The ascertainment of the device and channel taking part in the transfer of information into or out of a computer that causes a particular input/output interrupt, and of the status of the device and channel.

input/output interrupt indicator [COMPUT SCI] A device which registers an input/output interrupt associated with a particular input/output channel; it can be used in input/output interrupt identification.

input/output referencing [COMPUT SCI] The use of symbolic names in a computer program to indicate data on input/output devices, the actual devices allocated to the program being determined when the program is executed.

input/output register [COMPUT SCI] Computer register that provides the transfer of information from inputs to the central computer, or from it to output equipment.

input/output switching [COMPUT SCI] A technique in which a number of channels can connect input and output devices to a central processing unit; each device may be assigned to any available channel, so that several different channels may service a particular device during the execution of a program.

input/output wedge [COMPUT SCI] The characteristic shape of a Kiviat graph of a system which is approaching complete input/output boundedness.

input record [COMPUT SCI] 1. A record that is read from an input device into a computer memory during the performance of a program or routine. 2. A record that has been stored in an input area and is ready to be processed.

input register [COMPUT SCI] A register that accepts input information from a computer at one speed and supplies the information to the central processing unit at another speed, usually much greater.

input routine [COMPUT SCI] A routine which controls the loading and reading of programs, data, and other routines into a computer for storage or immediate use. Also known as loading routine.

input station [COMPUT SCI] A terminal in an in-plant communications system at which data can be entered into the system directly as events take place, enabling files to be immediately updated.

input storage *See* input area.

inquiline [ZOO] An animal that inhabits the nest of another species.

inquiry [COMPUT SCI] A request for the retrieval of a particular item or set of items from storage.

inscribe [COMPUT SCI] To rewrite data on a document in a form which can be read by an optical or magnetic ink character recognition machine.

inscribed [MATH] A polygon is inscribed in a circle or some curve if every vertex of the polygon lies on the circle or curve.

insect [INV ZOO] 1. A member of the Insecta. 2. An invertebrate that resembles an insect, such as a spider, mite, or centipede.

Insecta [INV ZOO] A class of the Arthropoda typically having a segmented body with an external, chitinous covering, a pair of compound eyes, a pair of antennae, three pairs of mouthparts, and two pairs of wings.

insecticide [MATER] A chemical agent that destroys insects.

Insectivora [VERT ZOO] An order of mammals including hedgehogs, shrews, moles, and other forms, most of which have spines.

insectivorous [BIOL] Feeding on a diet of insects.

insemination [BIOL] The planting of seed. [PHYSIO] 1. The introduction of sperm into the vagina. 2. Impregnation.

insensitive time See dead time.

insequent stream [HYD] A stream that has developed on the present surface, but not consequent upon it, and seemingly not controlled or adjusted by the rock structure and surface features.

inserted [BIOL] United or attached to the supporting structure by natural growth.

insertion gain [ELECTR] The ratio of the power delivered to a part of the system following insertion of an amplifier, to the power delivered to that same part before insertion of the amplifier; usually expressed in decibels.

insertion loss [ELECTR] The loss in load power due to the insertion of a component or device at some point in a transmission system; generally expressed as the ratio in decibels of the power received at the load before insertion of the apparatus, to the power received at the load after insertion.

insertion switch [COMPUT SCI] Process by which information is inserted into the computer by an operator who manually operates switches.

insert pump See rod pump.

inshore current [OCEANOGR] The horizontal movement of water inside the surf zone, including longshore and rip currents.

inshore zone [GEOL] The zone of variable width extending from the shoreline at low tide through the breaker zone.

inside caliper [DES ENG] A caliper that has two legs with feet that turn outward; used to measure inside dimensions, as the diameter of a hole.

inside diameter [DES ENG] The length of a line which passes through the center of a hollow cylindrical or spherical object, and whose end points lie on the inner surface of the object. Abbreviated ID.

in situ [SCI TECH] In the original location.

insol See insoluble.

insolation [ASTRON] 1. Exposure of an object to the sun. 2. Solar energy received, often expressed as a rate of energy per unit horizontal surface.

insoluble [CHEM] Incapable of being dissolved in another material; usually refers to solid-liquid or liquid-liquid systems. Abbreviated insol.

inspissation [CHEM] The process of thickening a liquid by evaporation. [GEOCHEM] Thickening of an oil deposit by evaporation or oxidation, resulting, for example, after long exposure in pitch or gum formation.

instability [CONT SYS] A condition of a control system in which excessive positive feedback causes persistent, unwanted oscillations in the output of the system. [PHYS] A property of the steady state of a system such that certain disturbances or perturbations introduced into the steady state will increase in magnitude, the maximum perturbation amplitude always remaining larger than the initial amplitude.

installation tape number [COMPUT SCI] A number that is permanently assigned to a reel of magnetic tape to identify it.

installed capacity [ELEC] The maximum runoff of a hydroelectric facility that can be constantly maintained and utilized by equipment.

instantaneous automatic gain control [ELECTR] In a radar system, the portion that automatically adjusts the gain of an amplifier for each pulse to obtain a substantially constant output-pulse peak amplitude with different input-pulse peak amplitudes; the circuit is fast enough to act during the time a pulse is passing through the amplifier.

instantaneous axis [MECH] The axis about which a rigid body is carrying out a pure rotation at a given instant in time.

instantaneous center [MECH] A point about which a rigid body is rotating at a given instant in time. Also known as instant center.

instantaneous power [ELEC] The product of the instantaneous voltage and the instantaneous current for a circuit or component.

instant center See instantaneous center.

instant-on switch [ELECTR] A switch that applies a reduced filament voltage to all tubes in a television receiver continuously, so the picture appears almost instantaneously after the set is turned on.

instar [INV ZOO] A stage between molts in the life of arthropods, especially insects.

instinct [PSYCH] A primary tendency or inborn drive, as toward life, sexual reproduction, and death. [ZOO] A precise form of behavior in which there is an invariable association of a particular series of responses with specific stimuli; an unconditioned compound reflex.

instruction [COMPUT SCI] A pattern of digits which signifies to a computer that a particular operation is to be performed and which may also indicate the operands (or the locations of operands) to be operated on.

instruction address [COMPUT SCI] The address of the storage location in which a given instruction is stored.

instruction address register [COMPUT SCI] A special storage location, forming part of the program controller, in which addresses of instructions are stored in order to control their sequential retrieval from memory during the execution of a program.

instruction code [COMPUT SCI] That part of an instruction which distinguishes it from all other instructions and specifies the action to be performed.

instruction counter [COMPUT SCI] A counter that indicates the location of the next computer instruction to be interpreted. Also known as location counter; program counter; sequence counter.

instruction format [COMPUT SCI] Any rule which assigns various functions to the various digits of an instruction.

instruction length [COMPUT SCI] The number of bits or bytes (eight bits per byte) which defines an instruction.

instruction lookahead [COMPUT SCI] A technique for speeding up the process of fetching and decoding instructions in a computer program, and of computing addresses of required operands and fetching them, in which the control unit fetches any unexecuted instructions on hand, to the extent this is feasible. Also known as fetch ahead.

instruction modification [COMPUT SCI] A change, carried out by the program, in an instruction so that, upon being repeated, this instruction will perform a different operation.

instruction register [COMPUT SCI] A hardware element that receives and holds an instruction as it is extracted from memory; the register either contains or is connected to circuits that interpret the instruction (or discover its meaning). Also known as current-instruction register.

instruction repertory *See* instruction set.

instruction set [COMPUT SCI] Also known as instruction repertory. **1.** The set of instructions which a computing or data-processing system is capable of performing. **2.** The set of instructions which an automatic coding system assembles.

instruction transfer [COMPUT SCI] An instruction which transfers control to one or another subprogram, depending upon the value of some operation.

instrument approach system [NAV] An aircraft navigation system that furnishes guidance in the vertical and horizontal planes to aircraft during descent from an initial-approach altitude to a point near the landing area; completion of a landing requires guidance to touchdown by visual or other means.

instrumentation [ENG] Designing, manufacturing, and utilizing physical instruments or instrument systems for detection, observation, measurement, automatic control, automatic computation, communication, or data processing.

instrumentation amplifier [ELECTR] An amplifier that accepts a voltage signal as an input and produces a linearly scaled version of this signal at the output; it is a closed-loop fixed-gain amplifier, usually differential, and has high input impedance, low drift, and high common-mode rejection over a wide range of frequencies.

instrument flight [NAV] A flight in which the navigation of the aircraft is controlled solely by reference to instruments.

instrument flight rules [NAV] Regulations governing flying when weather conditions are below the minimum for visual flight rules. Abbreviated IFR.

instrument landing [NAV] Landing made through the use of a system of electronic beacons and radar.

instrument landing system [NAV] A system of radio navigation which provides lateral and vertical guidance, as well as other navigational parameters required by a pilot in a low approach or a landing. Abbreviated ILS.

instrument multiplier [ELEC] A highly accurate resistor used in series with a voltmeter to extend its voltage range. Also known as voltage multiplier; voltage-range multiplier.

instrument panel [ENG] A panel or board containing indicating meters.

instrument reading time [ENG] The time, after a change in a measured quantity, which it takes for the indication of an instrument to come and remain within a specified percentage of its final value.

insulated [ELEC] Separated from other conducting surfaces by a nonconducting material.

insulated-gate field-effect transistor *See* metal oxide semiconductor field-effect transistor.

insulating compound [MATER] A liquid, at low temperatures, which is poured into joint boxes and allowed to solidify; as a poor conductor of heat and electricity, it provides good insulation.

insulating strength [ELEC] Measure of the ability of an insulating material to withstand electric stress without breakdown; it is defined as the voltage per unit thickness necessary to initiate a disruptive discharge; usually measured in volts per centimeter.

insulating tape [MATER] Tape impregnated with insulating material, and usually adhesive; used to cover joints in insulated wires or cables. Also known as electrical tape.

insulation [BUILD] Material used in walls, ceilings, and floors to retard the passage of heat and sound. [ELEC] A material having high electrical resistivity and therefore suitable for separating adjacent conductors in an electric circuit or preventing possible future contact between conductors. Also known as electrical insulation.

insulator [ELEC] A device having high electrical resistance and used for supporting or separating conductors to prevent undesired flow of current from them to other objects. Also known as electrical insulator. [MATER] A material that is a poor conductor of heat, sound, or electricity. [SOLID STATE] A substance in which the normal energy band is full and is separated from the first excitation band by a forbidden band that can be penetrated only by an electron having an energy of several electron volts, sufficient to disrupt the substance.

insulin [BIOCHEM] A protein hormone produced by the beta cells of the islets of Langerhans which participates in carbohydrate and fat metabolism.

insulin shock [MED] Clinical manifestation of hypoglycemia due to excess amounts of insulin in the blood.

intaglio [LAP] A type of carved gemstone in which the figure is engraved on the surface of the stone rather than left in relief by cutting away the background, as in a cameo.

intaglio printing [GRAPHICS] A printing method in which the printing elements are all below the plate surface, having been cut, scratched, engraved, or etched into the metal to form ink-retaining grooves or cups; surplus ink on the surface must be wiped or scraped off after each inking and before each printing impression.

intake [ENG] **1.** An entrance for air, water, fuel, or other fluid, or the amount of such fluid taken in. **2.** A main passage for air in a mine. [HYD] *See* recharge.

intake manifold [MECH ENG] A system of pipes which feeds fuel to the various cylinders of a multicylinder internal combustion engine.

intake valve [MECH ENG] The valve which opens to allow air or an air-fuel mixture to enter an engine cylinder.

integer [MATH] Any positive or negative counting number or zero.

integer constant [COMPUT SCI] A constant that uses the values 0, 1, ..., 9 with no decimal point in FORTRAN.

integer programming [SYS ENG] A series of procedures used in operations research to find maxima or minima of a function subject to one or more constraints, including one which requires that the values of some or all of the variables be whole numbers.

integer spin [QUANT MECH] Property of a particle whose spin angular momentum is a whole number times Planck's constant divided by 2π; bosons have this property; in contrast, fermions have half-integer spin.

integral [MATH] **1.** A solution of a differential equation is sometimes called an integral of the equation. **2.** *See* definite Riemann integral; indefinite integral.

integral absorbed dose *See* integral dose.

integral calculus [MATH] The study of integration and its applications to finding areas, volumes, or solutions of differential equations.

integral control [CONT SYS] Use of a control system in which the control signal changes at a rate proportional to the error signal.

integral domain [MATH] A commutative ring with identity where the product of nonzero elements is never zero. Also known as entire ring.

integral dose [NUCLEO] The total energy imparted to an irradiated body by an ionizing radiation; usually expressed in gram-rads or gram-roentgens. Also known as integral absorbed dose; volume dose.

integral equation [MATH] An equation where the unknown function occurs under an integral sign.

integral function [MATH] 1. A function taking on integer values. 2. See entire function.

integral heat of solution See heat of solution.

integral network [CONT SYS] A compensating network which produces high gain at low input frequencies and low gain at high frequencies, and is therefore useful in achieving low steady-state errors. Also known as lagging network; lag network.

integral operator [MATH] A rule for transforming one function into another function by means of an integral; this often is in context a linear transformation on some vector space of functions.

integral transform See integral transformation.

integral transformation [MATH] A transform of a function $F(x)$ given by the function

$$f(y) = \int_a^b K(x,y)F(x) \, dx$$

Where $K(x,y)$ is some function. Also known as integral transform.

integrand [MATH] The function which is being integrated in a given integral.

integrated circuit [ELECTR] An interconnected array of active and passive elements integrated with a single semiconductor substrate or deposited on the substrate by a continuous series of compatible processes, and capable of performing at least one complete electronic circuit function. Abbreviated IC. Also known as integrated semiconductor.

integrated-circuit memory See semiconductor memory.

integrated communications system [COMMUN] Communications system on either a unilateral or joint basis in which a message can be filed at any communications center in that system and be delivered to the addressee by any other appropriate communications center in that system without reprocessing enroute.

integrated data processing [COMPUT SCI] Data processing that has been organized and carried out as a whole, so that intermediate outputs may serve as inputs for subsequent processing with no human copying required. Abbreviated IDP.

integrated data retrieval system [COMPUT SCI] A section of a data-processing system that provides facilities for simultaneous operation of several videodata interrogations in a single line and performs required communications with the rest of the system; it provides storage and retrieval of both data subsystems and files and standard formats for data representation.

integrated drainage [HYD] Drainage resulting after folding and faulting of a surface under arid conditions; the streams by working headward have joined basins across intervening mountains or ridges.

integrated electronics [ELECTR] A generic term for that portion of electronic art and technology in which the interdependence of material, device, circuit, and system-design consideration is especially significant; more specifically, that portion of the art dealing with integrated circuits.

integrated information processing [COMPUT SCI] System of computers and peripheral systems arranged and coordinated to work concurrently or independently on different problems at the same time.

integrated injection logic [ELECTR] Integrated-circuit logic that uses a simple and compact bipolar transistor gate structure which makes possible large-scale integration on silicon for logic arrays, memories, watch circuits, and various other analog and digital applications. Abbreviated I²L. Also known as merged-transistor logic.

integrated optics [OPTICS] A thin-film device containing tiny lenses, prisms, and switches to transmit very thin laser beams, and serving the same purposes as the manipulation of electrons in thin-film devices of integrated electronics.

integrated semiconductor See integrated circuit.

integrating amplifier [ELECTR] An operational amplifier with a shunt capacitor such that mathematically the waveform at the output is the integral (usually over time) of the input.

integrating detector [ELECTR] A frequency-modulation detector in which a frequency-modulated wave is converted to an intermediate-frequency pulse-rate modulated wave, from which the original modulating signal can be recovered by use of an integrator.

integrating filter [ELECTR] A filter in which successive pulses of applied voltage cause cumulative buildup of charge and voltage on an output capacitor.

integrating frequency meter [ENG] An instrument that measures the total number of cycles through which the alternating voltage of an electric power system has passed in a given period of time, enabling this total to be compared with the number of cycles that would have elapsed if the prescribed frequency had been maintained. Also known as master frequency meter.

integrating meter [ENG] An instrument that totalizes electric energy or some other quantity consumed over a period of time.

integrating network [ELECTR] A circuit or network whose output waveform is the time integral of its input waveform. Also known as integrator.

integrator [ELECTR] 1. A computer device that approximates the mathematical process of integration. 2. See integrating network.

integrity [COMPUT SCI] Property of data which can be recovered in the event of its destruction through failure of the recording medium, user carelessness, program malfunction, or other mishap.

integument [ANAT] An outer covering, especially the skin, together with its various derivatives.

intelligence [COMMUN] Data, information, or messages that are to be transmitted. [PSYCH] 1. The intellect or astuteness of the mind. 2. Ability to recognize and understand qualities and attributes of the physical world and of mankind. 3. Ability to solve problems and engage in abstract thought processes.

intelligence quotient [PSYCH] The numerical designation for intelligence expressed as a ratio of an individual's performance on a standardized test to the average performance according to age. Abbreviated IQ.

intelligent terminal [COMPUT SCI] A computer input/output device with its own memory and logic circuits which can perform certain operations normally carried out by the computer. Also known as smart terminal.

intelligibility [COMMUN] The percentage of speech units understood correctly by a listener in a communications system; customarily used for regular messages where the context aids the listener, in distinction to articulation. Also known as speech intelligibility.

Intelsat [COMMUN] A satellite network under international control, used for global communication by more than 80 countries; the system requires stationary satellites over the Atlantic, Pacific, and Indian oceans and highly directional antennas at earth stations. Derived from international telecommunications satellite.

intensifier [GRAPHICS] In photography, a means used to strengthen the image of either a negative or a positive; usually, metal is added to the silver image, and the increase in density is proportional to the existing density; thus when a negative is intensified, the highlights are always intensified more than the shadows. [PETRO ENG] Hydrofluoric acid added to hydrochloric acid for oil well acidizing; the fluoride destroys silica films that are insoluble by hydrochloric acid.

intensifier image orthicon [ELECTR] An image orthicon combined with an image intensifier that amplifies the electron stream originating at the photocathode before it strikes the target.

intensity [PHYS] **1.** The strength or amount of a quantity, as of electric field, current, magnetization, radiation, or radioactivity. **2.** The power transmitted by a light or sound wave across a unit area perpendicular to the wave.

intensity control See brightness control.

intensity level [PHYS] The logarithm of the ratio of two intensities, powers or energies, usually expressed in decibels.

intensity modulation [ELECTR] Modulation of electron beam intensity in a cathode-ray tube in accordance with the magnitude of the received signal.

intensity of magnetization See intrinsic induction.

interaction [FL MECH] With respect to wave components, the nonlinear action by which properties of fluid flow (such as momentum, energy, vorticity), are transferred from one portion of the wave spectrum to another, or viewed in another manner, between eddies of different size-scales. [PHYS] A process in which two or more bodies exert mutual forces on each other. [STAT] The phenomenon which causes the response to applying two treatments not to be the simple sum of the responses to each treatment.

interactive graphical input [COMPUT SCI] Information which is delivered to a computer by using hand-held devices, such as writing styli used with electronic tablets and light-pens used with cathode-ray tube displays, to sketch a problem description in an on-line interactive mode in which the computer acts as a drafting assistant with unusual powers, such as converting rough freehand motions of a pen or stylus to accurate picture elements.

interactive information system [COMPUT SCI] An information system in which the user communicates with the computing facility through a terminal and receives rapid responses which can be used to prepare the next input.

interactive processing [COMPUT SCI] Computer processing in which the user can modify the operation appropriately while observing results at critical steps.

interactive terminal [COMPUT SCI] A computer terminal designed for two-way communication between operator and computer.

interambulacrum [INV ZOO] In echinoderms, an area between two ambulacra.

interbedded [GEOL] Having beds lying between other beds with different characteristics.

interblock [COMPUT SCI] A device or system that prevents one part of a computing system from interfering with another.

interblock gap [COMPUT SCI] A space separating two blocks of data on a magnetic tape.

intercalary [BOT] Referring to growth occurring between the apex and leaf. [SCI TECH] Inserted between two original components.

intercardinal point [GEOD] Any of the four directions midway between the cardinal points, that is, northeast, southeast, southwest, and northwest. Also known as quadrantal point.

intercarrier channel [COMMUN] A carrier telegraph channel in the available frequency spectrum between carrier telephone channels.

intercarrier noise suppression [ELECTR] Means of suppressing the noise resulting from increased gain when a high-gain receiver with automatic volume control is tuned between stations; the suppression circuit automatically blocks the audio-frequency input of the receiver when no signal exists at the second detector. Also known as interstation noise suppression.

intercarrier sound system [ELECTR] A television receiver arrangement in which the television picture carrier and the associated sound carrier are amplified together by the video intermediate-frequency amplifier and passed through the second detector, to give the conventional video signal plus a frequency-modulated sound signal whose center frequency is the 4.5 megahertz difference between the two carrier frequencies. Abbreviated ICS system.

intercellular [HISTOL] Of or pertaining to the region between cells.

intercept [CRYSTAL] One of the distances cut off a crystal's reference axis by planes. [MAP] See altitude difference. [MATH] The point where a straight line crosses one of the axes of a cartesian coordinate system.

interception [COMMUN] Tapping or tuning in to a telephone or radio message not intended for the listener. [HYD] **1.** The process by which precipitation is caught and retained on vegetation or structures and subsequently evaporated without reaching the ground. **2.** That part of the precipitation intercepted by vegetation. [METEOROL] **1.** The loss of sunshine, a part of which may be intercepted by hills, trees, or tall buildings. **2.** The depletion of part of the solar spectrum by atmospheric gases and suspensoids; this commonly refers to the absorption of ultraviolet radiation by ozone and dust. [ORD] Meeting or interrupting the course of a moving vessel, aircraft, or missile.

intercept station [COMMUN] Provides service for subscribers whereby calls to disconnected stations or dead lines are either routed to an intercept operator for explanation, or the calling party receives a distinctive tone that informs him that he has made such a call.

interchange [CIV ENG] A junction of two or more highways at a number of separate levels so that traffic can pass from one highway to another without the crossing at grade of traffic streams. [ELEC] The current flowing into or out of a power system which is interconnected with one or more other power systems.

interchangeability [ENG] The ability to replace the components, parts, or equipment of one manufacturer with those of another, without losing function or suitability.

interchange coefficient *See* exchange coefficient.

intercolumnation [ARCH] Distance between columns, measured between the bottoms of shafts, just above the apophyge, and expressed in terms of the lower diameter of the column.

intercom *See* intercommunicating system.

intercommunicating system Also known as intercom. [COMMUN] **1.** A telephone system providing direct communication between telephones on the same premises. **2.** A two-way communication system having a microphone and loudspeaker at each station and providing communication within a limited area.

intercontinental ballistic missile [ORD] A missile flying a ballistic trajectory after guided powered flight, usually over ranges in excess of 4000 miles (6500 kilometers). Abbreviated ICBM.

intercontinental sea [GEOGR] A large body of salt water extending between two continents.

intercostal [ANAT] Situated or occurring between adjacent ribs. [NAV ARCH] Situated or fitted between adjacent members of a ship's frame.

intercourse *See* coitus.

intercropping [AGR] A form of multiple cropping in which two or more crops simultaneously occupy the same field.

intercycle [COMPUT SCI] A cycle of operation of a punched-card tabulator or other punched-card machine during which card feeding is stopped to permit calculation and printing of control totals or to effect a change in control.

interdendritic attack *See* interdendritic corrosion.

interdendritic corrosion [MET] Preferential corrosion of the metal immediately surrounding dendrites in unworked or slightly worked alloys caused by composition gradients. Also known as interdendritic attack.

interface [COMPUT SCI] **1.** Some form of electronic device that enables one piece of gear to communicate with or control another. **2.** A device linking two otherwise incompatible devices, such as an editing terminal of one manufacturer to typesetter of another. [GEOPHYS] *See* seismic discontinuity. [PHYS CHEM] The boundary between any two phases: among the three phases (gas, liquid, and solid), there are five types of interfaces: gas-liquid, gas-solid, liquid-liquid, liquid-solid, and solid-solid. [SCI TECH] A shared boundary; it may be a piece of hardware used between two pieces of equipment, a portion of computer storage accessed by two or more programs, or a surface that forms the boundary between two types of materials.

interface connection *See* feedthrough.

interfacial energy [PHYS] The free energy of the surfaces at an interface, resulting from differences in the tendencies of each phase to attract its own molecules; equal to the surface tension. Also known as surface energy.

interfacial force *See* interfacial tension.

interfacial polarization [ELEC] *See* space-charge polarization. [OPTICS] Polarization of light by reflection from the surface of a dielectric at Brewster's angle.

interfacial tension [PHYS] A kind of surface tension, occurring at the interface between two liquids. Also known as interfacial force.

interference [COMMUN] Any undesired energy that tends to interfere with the reception of desired signals. Also known as electrical interference; radio interference. [PHYS] The variation with distance or time of the amplitude of a wave which results from the superposition (algebraic or vector addition) of two or more waves having the same, or nearly the same, frequency. Also known as wave interference.

interference colors [OPTICS] Colors formed by interference of a beam of light passed through a thin section of a mineral placed in a polarizing microscope.

interference fading [COMMUN] Fading of the signal produced by different wave components traveling slightly different paths in arriving at the receiver.

interference figure [OPTICS] A pattern of light and dark areas observed with a conoscope when a birefringent crystal is placed in a convergent beam of linearly polarized light.

interference filter [ELECTR] **1.** A filter used to attenuate artificial interference signals entering a receiver through its power line. **2.** A filter used to attenuate unwanted carrier-frequency signals in the tuned circuits of a receiver. [OPTICS] An optical filter in which the wavelengths that are not transmitted are removed by interference phenomena rather then by absorbtion or scattering.

interference microscope [OPTICS] A microscope used for visualizing and measuring differences in phase or optical paths in transparent or reflecting specimens; it differs from the phase contrast microscope in that the incident and diffracted waves are not separated, but interference is produced between the transmitted wave and another wave which originates from the same source.

interference pattern [ELECTR] Pattern produced on a radarscope by interference signals. [PHYS] Resulting space distribution of pressure, particle density, particle velocity, energy density, or energy flux when progressive waves of the same frequency and kind are superimposed.

interference wave [COMMUN] A radio wave reflected by the lower atmosphere which produces an interference pattern when combined with the direct wave.

interferometry [OPTICS] The design and use of optical inferometers; uses include precise measurement of wavelength, measurement of very small distances and thicknesses, study of hyperfine structure of spectral lines, precise measurement of indices of refraction, and determination of separations of binary stars and diameters of very large stars.

interferon [BIOCHEM] A protein produced by intact animal cells when infected with viruses; acts to inhibit viral reproduction and to induce resistance in host cells.

interfix [COMPUT SCI] A technique for describing relationships of key words in an item or document in a way which prevents crosstalk from causing false retrievals when very specific entries are made.

intergenic suppression [GEN] The restoration of a suppressed function or character by a second mutation that is located in a different gene than the original or first mutation.

interglacial [GEOL] Pertaining to or formed during a period of geologic time between two successive glacial epochs or between two glacial stages.

intergranular [SCI TECH] Occurring between grains.

interhemispheric integration [PSYCH] The process by which information is exchanged between the cerebral hemispheres.

interior angle [MATH] 1. An angle between two adjacent sides of a polygon that lies within the polygon. 2. For a line (called the transversal) that intersects two other lines, an angle between the transversal and one of the two lines that lies within the space between the two lines.

interior ballistics [MECH] The science concerned with the combustion of powder, development of pressure, and movement of a projectile in the bore of a gun.

interior of a set [MATH] The set of all interior points of a set in a topological space.

interior point [MATH] A point p in a topological space is an interior point of a set S if there is some open neighborhood of p which is contained in S.

interkinesis *See* interphase.

interlace [COMPUT SCI] To assign successive memory location numbers to physically separated locations on a storage tape or magnetic drum of a computer, usually to reduce access time.

interlace operation [COMPUT SCI] System of computer operation where data can be read out or copied into memory without interfering with the other activities of the computer.

interlacing arches [ARCH] Arches with intersecting curved moldings that appear to be interlaced.

interleave [COMPUT SCI] 1. To alternate parts of one sequence with parts of one or more other sequences in a cyclic fashion such that each sequence retains its identity. 2. To arrange the members of a sequence of memory addresses in different memory modules of a computer system, in order to reduce the time taken to access the sequence.

interlobate moraine *See* intermediate moraine.

interlock [COMPUT SCI] 1. A mechanism, implemented in hardware or software, to coordinate the activity of two or more processes within a computing system, and to ensure that one process has reached a suitable state such that the other may proceed. 2. *See* deadlock. [ENG] A switch or other device that prevents activation of a piece of equipment when a protective door is open or some other hazard exists.

interlock relay [ELEC] A relay composed of two or more coils, each with its own armature and associated contacts, so arranged that movement of one armature or the energizing of its coil is dependent on the position of the other armature.

interlock switch [ELEC] A switch designed for mounting on a door, drawer, or cover so that it opens automatically when the door or other part is opened.

interlude [COMPUT SCI] A small routine or program which is designed to carry out minor preliminary calculations or housekeeping operations before the main routine begins to operate, and which can usually be overwritten after it has performed its function.

intermediary metabolism [BIOCHEM] Intermediate steps in the chemical synthesis and breakdown of foodstuffs within body cells.

intermediate [CHEM] A precursor to a desired product; ethylene is an intermediate for polyethylene, and ethane is an intermediate for ethylene. [GRAPHICS] That print which is used as a master for further reproduction.

intermediate control data [COMPUT SCI] Control data at a level which is neither the most nor the least significant, or which is used to sort records into groups that are neither the largest nor the smallest used; for example, if control data are used to specify state, town, and street, then the data specifying town would be intermediate control data.

intermediate distributing frame [ELEC] Frame in a local telephone central office, the primary purpose of which is to cross-connect the subscriber line multiple to the subscriber line circuit; in a private exchange, the intermediate distributing frame is for similar purposes.

intermediate frequency [ELECTR] The frequency produced by combining the received signal with that of the local oscillator in a superheterodyne receiver. Abbreviated i-f.

intermediate-frequency amplifier [ELECTR] The section of a superheterodyne receiver that amplifies signals after they have been converted to the fixed intermediate-frequency value by the frequency converter. Abbreviated i-f amplifier.

intermediate-frequency response ratio [ELECTR] In a superheterodyne receiver, the ratio of the intermediate-frequency signal input at the antenna to the desired signal input for identical outputs. Also known as intermediate-interference ratio.

intermediate-frequency transformer [ELECTR] The transformer used at the input and output of each intermediate-frequency amplifier stage in a superheterodyne receiver for coupling purposes and to provide selectivity. Abbreviated i-f transformer.

intermediate host [BIOL] The host in which a parasite multiplies asexually.

intermediate-interference ratio *See* intermediate-frequency response ratio.

intermediate layer *See* sima.

intermediate memory storage [COMPUT SCI] An electronic device for holding working figures temporarily until needed and for releasing final figures to the output.

intermediate moraine [GEOL] A type of lateral moraine formed at the junction of two adjacent glacial lobes. Also known as interlobate moraine.

intermediate-range ballistic missile [ORD] A missile flying a ballistic trajectory after guided powered flight and having a range of about 200 to 1500 miles (300 to 2500 kilometers). Abbreviated IRBM.

intermediate result [COMPUT SCI] A quantity or value derived from an operation performed in the course of a program or subroutine which is itself used as an operand in further operations.

intermediate storage [COMPUT SCI] The portion of the computer storage facilities that usually stores information in the processing stage.

intermediate total [COMPUT SCI] A sum that is produced when there is a change in the value of control data at a level that is neither the most nor the least significant.

intermediate value theorem [MATH] If $f(x)$ is a continuous real-valued function on the closed interval from a to b, then, for any y between the least upper bound and the greatest lower bound of the values of f, there is an x between a and b with $f(x) = y$.

intermediate vector boson [PARTIC PHYS] One of the hypothetical particles with spin quantum number 1 and negative parity, which would interact with weak currents and mediate the weak interactions in the same way that photons interact with electromagnetic currents and mediate the electromagnetic interactions.

intermedin [BIOCHEM] A hormonal substance produced by the intermediate portion of the hypophysis of certain animal species which influences pigmentation; similar to melanocyte-stimulating hormone in humans.

intermenstrual flow *See* metrorrhagia.

intermetallic compound See electron compound.

intermitotic [CYTOL] Of or pertaining to a stage of the cell cycle between two successive mitoses.

intermittent current [ELEC] A unidirectional current that flows and ceases to flow at irregular or regular intervals. [OCEANOGR] A unidirectional current interrupted at intervals.

intermittent stream [HYD] A stream which carries water a considerable portion of the time, but which ceases to flow occasionally or seasonally because bed seepage and evapotranspiration exceed the available water supply.

intermittent weld [MET] A weld in which the continuity is broken by recurring unwelded spaces.

intermodulation [ELECTR] Modulation of the components of a complex wave by each other, producing new waves whose frequencies are equal to the sums and differences of integral multiples of the component frequencies of the original complex wave.

intermolecular force [PHYS CHEM] The force between two molecules; it is that negative gradient of the potential energy between the interacting molecules, if energy is a function of the distance between the centers of the molecules.

intermontane [GEOL] Located between or surrounded by mountains.

internal acoustic meatus [ANAT] An opening in the hard portion of the temporal bone for passage of the facial and acoustic nerves and internal auditory vessels.

internal arithmetic [COMPUT SCI] Arithmetic operations carried out in a computer's arithmetic unit within the central processing unit.

internal brake [MECH ENG] A friction brake in which an internal shoe follows the inner surface of the rotating brake drum, wedging itself between the drum and the point at which it is anchored; used in motor vehicles.

internal carotid [ANAT] A main division of the common carotid artery, distributing blood through three sets of branches to the cerebrum, eye, forehead, nose, internal ear, trigeminal nerve, dura mater, and hypophysis.

internal combustion engine [MECH ENG] A prime mover in which the fuel is burned within the engine and the products of combustion serve as the thermodynamic fluid, as with gasoline and diesel engines.

internal conversion [NUC PHYS] A nuclear de-excitation process in which energy is transmitted directly from an excited nucleus to an orbital electron, causing ejection of that electron from the atom.

internal conversion coefficient See conversion coefficient.

internal ear See inner ear.

internal elastic membrane [HISTOL] A sheet of elastin found between the tunica intima and the tunica media in medium-and small-caliber arteries.

internal energy [THERMO] A characteristic property of the state of a thermodynamic system, introduced in the first law of thermodynamics; it includes intrinsic energies of individual molecules, kinetic energies of internal motions, and contributions from interactions between molecules, but excludes the potential or kinetic energy of the system as a whole; it is sometimes erroneously referred to as heat energy.

internal fertilization [PHYSIO] Fertilization of the egg within the body of the female.

internal friction [FL MECH See viscosity. [MECH] Conversion of mechanical strain energy to heat within a material subjected to fluctuating stress.

internal gear [DES ENG] An annular gear having teeth on the inner surface of its rim.

internal iliac artery [ANAT] The medial terminal division of the common iliac artery.

internalization [PSYCH] A mental mechanism operating outside of and beyond conscious awareness by which certain external attributes, attitudes, or standards are taken within oneself.

internal loss See loss.

internally stored program [COMPUT SCI] A sequence of instructions, stored inside the computer in the same storage facilities as the computer data, as opposed to external storage on tape, disk, drum, or cards.

internal memory See internal storage.

internal phase See disperse phase.

internal photoelectric effect [SOLID STATE] A process in which the absorption of a photon in a semiconductor results in the excitation of an electron from the valence band to the conduction band.

internal pressure See intrinsic pressure.

internal reflectance spectroscopy See attenuated total reflectance.

internal sorting [COMPUT SCI] The sorting of a list of items by a computer in which the entire list can be brought into the main computer memory and sorted in memory.

internal storage [COMPUT SCI] The total memory or storage that is accessible automatically to a computer without human intervention. Also known as internal memory.

internal stress [MECH] A stress system within a solid that is not dependent on external forces. Also known as residual stress.

internal thread [DES ENG] A screw thread cut on the inner surface of a hollow cylinder.

internal wave [FL MECH] A wave motion of a stably stratified fluid in which the maximum vertical motion takes place below the surface of the fluid.

international ampere [ELEC] The current that, when flowing through a solution of silver nitrate in water, deposits silver at a rate of 0.001118 gram per second; it has been superseded by the ampere as a unit of current, and is equal to approximately 0.999850 ampere.

international angstrom [PHYS] A unit of length, equal to 1/6438.4696 of the wavelength of the red cadmium line in dry air at standard atmospheric pressure, at a temperature of 15°C containing 0.03% by volume of carbon dioxide; equal to 1.0000002 angstroms. Abbreviated IA.

international atomic time [HOROL] Time based on atomic clocks operating in conformity with the definition of the second as the International System unit of time. Abbreviated IAT.

international broadcasting [COMMUN] International radio broadcasting for public entertainment, on frequency bands between 5950 and 21,750 kilohertz, that are assigned by international agreement.

international call sign [COMMUN] Call sign assigned according to the provisions of the International Telecommunication Union to identify a radio station; the nationality of the radio station is identified by the first or the first two characters.

international candle [OPTICS] A unit of luminous intensity, now replaced by the candela; as defined in the United States, it was a specified fraction of the average luminous intensity radiated in a horizontal direction by a group of 45 carbon-filament lamps preserved at the National Bureau of Standards when the lamps were operated at a specified voltage. Also known as standard candle.

international code signal [COMMUN] Code adopted by many nations for international communications; it uses combinations of letters in lieu of words, phrases, and sentences; the letters are transmitted by the hoisting of international alphabet flags or by transmitting their dot and dash equivalents in the international Morse code. Also known as international signal code.

International Critical Tables [PHYS] A seven-volume series of tables of numerical data in physics, chemistry, and technology, published in 1926–1930, prepared by experts who gave the "best" value which could be derived from all the data available at the time. Abbreviated ICT.

international date line [ASTRON] A jagged arbitrary line, roughly equal to the 180° meridian, where a date change occurs: if the line is crossed from east to west a day is skipped, if from west to east the same day is repeated.

international ellipsoid of reference [GEOD] The reference ellipsoid, based upon the Hayford spheroid, the semimajor axis of which is 6,378,388 meters; the flattening or ellipticity equals 1/297; by computation the semiminor axis is 6,356,911.946 meters. Also known as international spheroid.

international henry [ELECTROMAG] A unit of electrical inductance which has been superseded by the henry, and is equal to 1.00049 henry. Also known as quadrant; secohm.

international Morse code See continental code.

international nautical mile [NAV] A unit of length equal to 1852 meters.

international ohm [ELEC] A unit of resistance, equal to that of a column of mercury of uniform cross section that has a length of 160.3 centimeters and a mass of 14.4521 grams at the temperature of melting ice; it has been superseded by the ohm, and is equal to 1.00049 ohms.

international practical temperature scale [THERMO] Temperature scale based on six points; the water triple point, the boiling points of oxygen, water, sulfur, and the solidification points of silver and gold; designated as °C, degrees Celsius, or t_{int}.

international radio silence [COMMUN] Three-minute periods of radio silence, on the frequency of 500 kilohertz only, commencing 15 and 45 minutes after each hour, during which all marine radio stations must listen on that frequency for distress signals of ships and aircraft.

international signal code See international code signal.

international spheroid See international ellipsoid of reference.

international synoptic code [METEOROL] A synoptic code approved by the World Meteorological Organization in which the observable meteorological elements are encoded and transmitted in words of five numerical digits length.

international system of electrical units [ELEC] System of electrical units based on agreed fundamental units for the ohm, ampere, centimeter, and second, in use between 1893 and 1947, inclusive; in 1948, the Giorgi, or meter-kilogram-second-absolute system, was adopted for international use.

International System of Units [PHYS] A system of physical units in which the fundamental quantities are length, time, mass, electric current, temperature, luminous intensity, and amount of substance, and the corresponding units are the meter, second, kilogram, ampere, kelvin, candela, and mole; it has been given official status and recommended for universal use by the General Conference on Weights and Measures. Also known (in French) as Système International d'Unités. Abbreviated SI (in all languages).

international table British thermal unit See British thermal unit.

international table calorie See calorie.

international volt [ELEC] A unit of potential difference or electromotive force, equal to 1/1.01858 of the electromotive force of a Weston cell at 20°C; it has been superseded by the volt, and is equal to 1.00034 volts.

internode [BIOL] The interval between two nodes, as on a stem or along a nerve fiber.

interoceptor [PHYSIO] A sense receptor located in visceral organs and yielding diffuse sensations.

interoffice trunk [COMMUN] A direct trunk between local central offices in the same exchange.

interpenetration twin [CRYSTAL] Two or more individual crystals so twinned that they appear to have grown through one another. Also known as penetration twin.

interphase [CYTOL] Also known as interkinesis. **1.** The period between succeeding mitotic divisions. **2.** The period between the first and second meiotic divisions in those organisms where nuclei are reconstituted at the end of the first division.

interphase transformer [ELECTR] Autotransformer or a set of mutually coupled reactors used in conjunction with three-phase rectifier transformers to modify current relations in the rectifier system to increase the number of anodes of different phase relations which carry current at any instant.

interphone [COMMUN] An intercommunication system using headphones and microphones for communication between adjoining or nearby studios or offices, or between crew locations on an aircraft, vessel, or tank or other vehicle. Also known as talkback circuit.

interplanetary probe [AERO ENG] An instrumented spacecraft that flies through the region of space between the planets.

interplanetary spacecraft [AERO ENG] A spacecraft designed for interplanetary flight.

interpolation [MATH] A process used to estimate an intermediate value of one (dependent) variable which is a function of a second (independent) variable when values of the dependent variable corresponding to several discrete values of the independent variable are known.

interpole See commutating pole.

interposition trunk [COMMUN] Trunk which connects two positions of a large switchboard so that a line on one position can be connected to a line on another position.

interpret [COMPUT SCI] To print on a punched card the infomation punched in that card.

interpreter [COMPUT SCI] **1.** A program that translates and executes each source program statement before proceeding to the next one. Also known as interpretive routine. **2.** A machine that senses a punched card and prints the punched information on that card. Also known as punched-card interpreter. **3.** See conversational compiler.

interpretive code See interpretive language.

interpretive language [COMPUT SCI] A computer programming language in which each instruction is immediately translated and acted upon by the computer, as opposed to a compiler which decodes a whole program before a single instruction can be executed. Also known as interpretive code; pseudocode.

interpretive programming [COMPUT SCI] The writing of computer programs in an interpretive language, which generally uses mnemonic symbols to represent operations and operands and must be translated into machine language by the computer at the time the instructions are to be executed.

interpretive trace program [COMPUT SCI] An interpretive routine that provides a record of the machine code into which the source program is translated and of the result of each step, or of selected steps, of the program.

interquartile range [STAT] The distance between the top of the lower quartile and the bottom of the upper quartile of a distribution.

interrecord gap See record gap.

interrogating typewriter [COMPUT SCI] A typewriter designed to insert data into a computer program in main memory or receive output from the program.

interrogation [COMMUN] The transmission of a radio-frequency pulse, or combination of pulses, intended to trigger a transponder or group of transponders, a racon system, or an IFF system, in order to elicit an electromagnetic reply. Also known as challenging signal.

interrogator [ELECTR] 1. A radar transmitter which sends out a pulse that triggers a transponder; usually combined in a single unit with a responsor, which receives the reply from a transponder and produces an output suitable for actuating a display of some navigational parameter. Also known as challenger; interrogator-transmitter. 2. See interrogator-responsor.

interrogator-responsor [ELECTR] A transmitter and receiver combined, used for sending out pulses to interrogate a radar beacon and for receiving and displaying the resulting replies. Also known as interrogator.

interrogator-transmitter See interrogator.

interrupt [COMPUT SCI] 1. To stop a running program in such a way that it can be resumed at a later time, and in the meanwhile permit some other action to be performed. 2. The action of such a stoppage.

interrupted current [ELEC] A current produced by opening and closing at regular intervals a circuit that would otherwise carry a steady current or one that varied continuously with time.

interrupted dc tachometer [ENG] A type of impulse tachometer in which the frequency of pulses generated by the interrupted direct current of an ignition-circuit primary of an internal combustion engine is used to measure the speed of the engine.

interrupted screw [DES ENG] A screw with longitudinal grooves cut into the thread, and which locks quickly when inserted into a similar mating part.

interrupter [ELEC] An electric, electronic, or mechanical device that periodically interrupts the flow of a direct current so as to produce pulses. [ORD] A barrier in a fuse which prevents transmission of an explosive effect to some element beyond the interrupter.

interrupt mask [COMPUT SCI] A technique of suppressing certain interrupts and allowing the control program to handle these masked interrupts at a later time.

interrupt mode See hold mode.

interrupt system [COMPUT SCI] The means of interrupting a program and proceeding with it at a later time; this is accomplished by saving the contents of the program counter and other specific registers, storing them in reserved areas, starting the new in-

struction sequence, and upon completion, reloading the program counter and registers to return to the original program, and reenabling the interrupt.

intersect [ENG] To find a position by the triangulation method. [SCI TECH] To pass through or across.

intersection [CIV ENG] 1. A point of junction or crossing of two or more roadways. 2. A surveying method in which a plane table is used alternately at each end of a measured baseline. [MATH] The intersection of two sets is the set consisting of all elements common to both of the two sets.

intersection data [COMPUT SCI] Data which are meaningful only when associated with the concatenation of two segments.

intersegmental reflex [PHYSIO] An unconditioned reflex arc connecting input and output by means of afferent pathways in the dorsal spinal roots and efferent pathways in the ventral spinal roots.

intersex [PHYSIO] An individual who is intermediate in sexual constitution between male and female.

interspersion [ECOL] 1. An intermingling of different organisms within a community. 2. The level or degree of intermingling of one kind of organism with others in the community.

interstation noise suppression See intercarrier noise suppression.

interstellar [ASTRON] Between the stars.

interstice [SOLID STATE] A space or volume between atoms of a lattice, or between groups of atoms 'or grains of a solid structure.

interstitial [CRYSTAL] A crystal defect in which an atom occupies a position between the regular lattice positions of a crystal. [SCI TECH] Of, pertaining to, or situated in a space between two things.

interstitial cell [HISTOL] A cell that is not peculiar to or characteristic of a particular organ or tissue but which comprises fibrous tissue binding other cells and tissue elements; examples are neuroglial cells and Leydig cells.

interstitial cell–stimulating hormone See luteinizing hormone.

interstitial water [HYD] Subsurface water contained in pore spaces between the grains of rock and sediments.

intersystem communications [COMPUT SCI] The ability of two or more computer systems to share input, output, and storage devices, and to send messages to each other by means of shared input and output channels or by channels that directly connect central processors.

intertidal zone [OCEANOGR] The part of the littoral zone above low-tide mark.

intertropical convergence zone [METEOROL] The axis, or a portion thereof, of the broad trade-wind current of the tropics; this axis is the dividing line between the southeast trades and the northeast trades (of the Southern and Northern hemispheres, respectively). Also known as equatorial convergence zone; meteorological equator.

intertropical front [METEOROL] The interface or transition zone occurring within the equatorial trough between the Northern and Southern hemispheres. Also known as equatorial front; tropical front.

interval [ACOUS] The spacing in pitch or frequency between two sounds; the grequency interval is the ratio of the frequencies or the logarithm of this ratio. [MATH] A set of numbers which consists of those numbers that are greater than one fixed number and less than another, and that may also include one or both of the end numbers. [PHYS] The time separating two events, or the distance between two objects. [RELAT] 1. in special relativity, the Lorentz invariant quantity $c^2(\Delta t)^2 - (\Delta x)^2 - (\Delta y)^2 - (\Delta z)^2$, where c is the speed of light, Δt is the difference in the time coordinates of two specified events, and Δx, Δy, and Δz are the differences in their x, y, and z coordinates, respectively. **2.** In general relativity, a generalization of this concept, namely the sum over the indices μ and ν of $g_{\mu\nu}dx^\mu dx^\nu$, where dx^μ and dx^ν are the differences in the x^μ and x^ν coordinates of two specified neighboring events, and $g_{\mu\nu}$ is an element of the metric tensor.

interval scale [STAT] A rule or system for assigning numbers to objects in such a way that the difference between any two objects is reflected in the difference in the numbers assigned to them; used in interval measurement.

interval timer [ENG] A device which operates a set of contacts during a preset time interval and, at the end of the interval, returns the contacts to their normal positions. Also known as timer.

interventricular septum [ANAT] The muscular wall between the heart ventricles. Also known as ventricular septum.

intestinal juice [PHYSIO] An alkaline fluid composed of the combined secretions of all intestinal glands.

intestine [ANAT] The tubular portion of the vertebrate digestive tract, usually between the stomach and the cloaca or anus.

in-the-seam mining [MIN ENG] The usual method of mining characterized by the driving of development shafts into the coal seam.

in the wind [NAV] In the direction from which the wind is blowing; windward; used particularly in reference to a heading or to the position of an object.

intoxication [MED] **1.** Poisoning. **2.** The state produced by overindulgence in alcohol.

intracartilaginous ossification See endochondral ossification.

intracistron complementation [GEN] The process whereby two different mutant alleles, each of which determines in homozygotes an inactive enzyme, determine the formation of the active enzyme when present in the same nucleus.

Intracoastal Waterway [NAV] An inside protected route extending through New Jersey; from Norfolk, Virginia to Key West, Florida; across Florida from St. Lucie Inlet to Fort Myers, Charlotte Harbor, Tampa Bay, and Tarpon Springs; and from Carabelle, Florida, to Brownsville, Texas.

intraformational breccia [PETR] A rock resulting from cracking and desiccation-shrinking of a mud after withdrawal of water followed by almost contemporaneous sedimentation.

intraformational conglomerate [GEOL] **1.** A conglomerate in which clasts and the matrix are contemporaneous in origin. **2.** A conglomerate formed in the midst of a geologic formation.

intragenic [MOL BIO] Within a gene, in referring to certain events.

intragenic recombination [MOL BIO] Recombination occurring between the mutons of one cistron.

intragenic suppression [GEN] The restoration of a suppressed function or character as a consequence of a second mutation located within the same gene as the original or first mutation.

intramembranous ossification [HISTOL] Formation of bone tissue directly from connective tissue without a preliminary cartilage stage.

intramuscular [ANAT] Lying within or going into the substance of a muscle.

intraocular [ANAT] Occurring within the globe of the eye.

intraocular pressure [PHYSIO] The hydrostatic pressure within the eyeball.

intraspinal block [MED] Anesthesia of the spinal column by injection of an anesthetic into the spinal canal.

intravenous [ANAT] Located within, or going into, the veins.

intrinsic-barrier diode [ELECTR] A *pin* diode, in which a thin region of intrinsic material separates the *p*-type region and the *n*-type region.

intrinsic-barrier transistor [ELECTR] A *pnip* or *npin* transistor, in which a thin region of intrinsic material separates the base and collector.

intrinsic conductivity [SOLID STATE] The conductivity of a semiconductor or metal in which impurities and structural defects are absent or have a very low concentration.

intrinsic factor [BIOCHEM] A substance, produced by the stomach, which combines with the extrinsic factor (vitamin B_{12}) in food to yield an antianemic principle; lack of the intrinsic factor is believed to be a cause of pernicious anemia. Also known as Castle's intrinsic factor.

intrinsic flux density See intrinsic induction.

intrinsic induction [ELECTROMAG] The vector difference between the magnetic flux density at a given point and the magnetic flux density which would exist there, for the same magnetic field strength, if the point were in a vacuum. Symbol B_i. Also known as intensity of magnetization; intrinsic flux density; magnetic polarization.

intrinsic layer [ELECTR] A layer of semiconductor material whose properties are essentially those of the pure undoped material.

intrinsic parity [PARTIC PHYS] A quantum number, equal to $+1$ or -1, which is assigned to particles so that the product of the intrinsic parities of the particles composing a system times the parity of the system's wave function yields the total parity.

intrinsic photoemission [SOLID STATE] Photoemission which can occur in an ideally pure and perfect crystal, in contrast to other types of photoemission which are associated with crystal defects.

intrinsic pressure [PHYS] Pressure in a fluid resulting from inward forces on molecules near the fluid surface, caused by attraction between molecules. Also known as internal pressure.

intrinsic procedure See built-in function.

intrinsic semiconductor [SOLID STATE] A semiconductor in which the concentration of charge carriers is characteristic of the material itself rather than of the content of impurities and structural defects of the crystal. Also known as *i*-type semiconductor.

introitus [ANAT] An opening or entryway, especially the opening into the vagina.

intromission [ZOO] The act or process of inserting one body into another, specifically, of the penis into the vagina.

introvert [PSYCH] An individual whose interests are self-directed, and not directed toward the outside world. [ZOO] **1.** A structure capable of introversion. **2.** To turn inward.

intrusion [GEOL] **1.** The process of emplacement of magma in preexisting rock. Also known as injection; invasion; irruption. **2.** A large-scale sedimentary injection. Also known as sedimentary intrusion. **3.** Any rock mass formed by an intrusive process.

intubation [MED] The introduction of a tube into a hollow organ to keep it open, especially into the larynx to ensure the passage of air.

intumescence [MATER] The property of a material to swell when heated; intumescent materials in bulk and sheet form are used as fireproofing agents.

invagination [EMBRYO] The enfolding of a part of the wall of the blastula to form a gastrula. [PHYSIO] **1.** The act of ensheathing or becoming ensheathed. **2.** The process of burrowing or enfolding to form a hollow space within a previously solid structure, as the invagination of the nasal mucosa within a bone of the skull to form a paranasal sinus.

invariance [MATH] *See* invariant property. [PHYS] The property of a physical quantity or physical law of being unchanged by certain transformations or operations, such as reflection of spatial coordinates, time reversal, charge conjugation, rotations, or Lorentz transformations. Also known as symmetry.

invariance principle [PHYS] Any principle which states that a physical quantity or physical law possesses invariance under certain transformations. Also known as symmetry law; symmetry principle. [RELAT] In general relativity, the principle that the laws of motion are the same in all frames of reference, whether accelerated or not.

invariant property [MATH] A mathematical property of some space unchanged after the application of any member from some given family of transformations. Also known as invariance.

invasion [GEOL] **1.** The movement of one material into a porous reservoir area that has been occupied by another material. **2.** *See* intrusion; transgression. [MED] **1.** The phase of an infectious disease during which the pathogen multiplies and is distributed; precedes signs and symptoms. **2.** The process by which microorganisms enter the body.

inventory control [IND ENG] Systematic management of the balance on hand of inventory items, involving the supply, storage, distribution, and recording of items.

inverse beta decay [NUC PHYS] A reaction providing evidence for the existence of the neutrino, in which an antineutrino (or neutrino) collides with a proton (or neutron) to produce a neutron (or proton) and a positron (or electron).

inverse Compton effect [QUANT MECH] A process in which relativistic particles give up some of their energy to long-wavelength radiation, converting it to shorter-wavelength radiation.

inverse cylindrical orthomorphic projection *See* transverse Mercator projection.

inverse direction [ELECTR] The direction in which the electron flow encounters greater resistance in a rectifier, going from the positive to the negative electrode; the opposite of the conducting direction. Also known as reverse direction.

inverse feedback *See* negative feedback.

inverse function [MATH] An inverse function for a function f is a function g whose domain is the range of f and whose range is the domain of f with the property that both f composed with g and g composed with f give the identity function.

inverse hour *See* inhour.

inverse limiter [ELECTR] A transducer, the output of which is constant for input of instantaneous values within a specified range and a linear or other prescribed function of the input for inputs above and below that range.

inverse logarithm of a number *See* antilogarithm of a number.

inverse Mercator projection *See* transverse Mercator projection.

inverse network [ELEC] Two two-terminal networks are said to be inverse when the product of their impedances is independent of frequency within the range of interest.

inverse of a number [MATH] The additive inverse of a real or complex number a is the number which when added to a gives 0; the multiplicative inverse of a is the number which when multiplied with a gives 1.

inverse piezoelectric effect [SOLID STATE] The contraction or expansion of a piezoelectric crystal under the influence of an electric field, as in crystal headphones; also occurs at pn junctions in some semiconductor materials.

inverse points [MATH] A pair of points lying on a diameter of a circle or sphere such that the product of the distances of the points from the center equals the square of the radius.

inverse-square law [PHYS] Any law in which a physical quantity varies with distance from a source inversely as the square of that distance.

inverse Zeeman effect [SPECT] A splitting of the absorption lines of atoms or molecules in a static magnetic field; it is the Zeeman effect observed with absorption lines.

inversion [CHEM] Change of a compound into an isomeric form. [COMMUN] The process of scrambling speech for secrecy by beating the voice signal with a fixed, higher audio frequency and using only the difference frequencies. [CRYSTAL] A change from one crystal polymorph to another. Also known as transformation. [ELEC] The solution of certain problems in electrostatics through the use of the transformation in Kelvin's inversion theorem. [GEN] A type of chromosomal rearrangement in which two breaks take place in a chromosome and the fragment between breaks rotates 180° before rejoining. [GEOL] **1.** Development of inverted relief through which anticlines are transformed into valleys and synclines are changed into mountains. **2.** The occupancy by a lava flow of a ravine or valley that occurred in the side of a volcano. **3.** A diagenetic process in which unstable minerals are converted to a more stable form without a change in chemical composition. [MECH ENG] The conversion of basic four-bar linkages to special motion linkages, such as parallelogram linkage, slider-crank mechanism, and slow-motion mechanism by successively holding fast, as ground link, members of a specific linkage (as drag link). [MED] The act or process of turning inward or upside down. [METEOROL] A departure from the usual decrease or increase with altitude of the value of an atmospheric property, most commonly temperature. [OPTICS] The formation of an inverted image by an optical system. [PHYS] The simultaneous reflection of all three directions in space, so that each coordinate is replaced by the negative of itself. Also known as space inversion. [SOLID STATE] The production of a layer at the surface of a semiconductor which is of opposite type from that of the bulk of the semiconductor, usually as the result of an applied electric field. [THERMO] A reversal of the usual direction of a variation or process, such as the change in sign of the expansion coefficient of water at 4°C, or a change in sign in the Joule-Thomson coefficient at a certain temperature.

inversion axis *See* rotation-inversion axis.

inversion temperature [ENG] The temperature to which one junction of a thermocouple must be raised in order to make the thermoelectric electromotive force in the circuit equal to zero, when the other junction of the thermocouple is held at a constant low temperature. [THERMO] The temperature at hich the Joule-Thomson effect of a gas changes sign.

invertase See saccharase.

Invertebrata [INV ZOO] A division of the animal kingdom including all animals which lack a spinal column; has no taxonomic status.

invertebrate [INV ZOO] An animal lacking a backbone and internal skeleton.

invertebrate zoology [ZOO] A branch of biology that deals with the study of Invertebrata.

inverted amplifier [ELECTR] A two-tube amplifier in which the control grids are grounded and the input signal is applied between the cathodes; the grid then serves as a shield between the input and output circuits.

inverted arch [CIV ENG] An arch with the crown downward, below the line of the springings; commonly used in tunnels and foundations. Also known as inflected arch.

inverted file [COMPUT SCI] 1. A file, or method of file organization, in which labels indicating the locations of all documents of a given type are placed in a single record. 2. A file whose usual order has been inverted.

inverted image [OPTICS] An image in which up and down, as well as left and right, are interchanged; that is, an image that results from rotating the object 180° about a line from the object to the observer; such images are formed by most astronomical telescopes. Also known as reversed image.

inverted vee [ELECTROMAG] 1. A directional antenna consisting of a conductor which has the form of an inverted V, and which is fed at one end and connected to ground through an appropriate termination at the other. 2. A center-fed horizontal dipole antenna whose arms have ends bent downward 45°.

inverter [ELEC] A device for converting direct current into alternating current; it may be electromechanical, as in a vibrator or synchronous inverter, or electronic, as in a thyratron inverter circuit. Also known as dc-to-ac converter; dc-to-ac inverter. [ELECTR] See phase inverter.

inverter circuit See NOT circuit.

invertin See saccharase.

inverting amplifier [ELECTR] Amplifier whose output polarity is reversed as compared to its input; such an amplifier obtains its negative feedback by a connection from output to input, and with high gain is widely used as an operational amplifier.

inverting terminal [ELECT] The negative input terminal of an operational amplifier; a positive-going voltage at the inverting terminal gives a negative-going output voltage.

invert sugar [FOOD ENG] A mixture consisting of 50% glucose and 50% fructose, obtained by hydrolysis of sucrose, which absorbs water readily; used in the food industry.

investing bone See dermal bone.

investment casting [MET] A casting method designed to achieve high dimensional accuracy for small castings by making a mold of refractory slurry, which sets at room temperature, surrounding a wax pattern which is then melted out to leave a mold without joints.

inviscid flow [FL MECH] Flow of an inviscid fluid. Also known as frictionless flow; ideal flow; nonviscous flow.

inviscid fluid [FL MECH] A fluid which has no viscosity; it therefore can support no shearing stress,

and flows without energy dissipation. Also known as ideal fluid; nonviscous fluid; perfect fluid.

in vitro [BIOL] Pertaining to a biological reaction taking place in an artificial apparatus.

in vivo [BIOL] Pertaining to a biological reaction taking place in a living cell or organism.

involucre [BOT] Bracts forming one or more whorls at the base of an inflorescence or fruit in certain plants.

involuntary muscle [PHYSIO] Muscle not under the control of the will; usually consists of smooth muscle tissue and lies within a vescus, or is associated with skin.

involute [BIOL] Being coiled, curled, or rolled in at the edge. [MATH] A curve produced by any point of a perfectly flexible inextensible thread that is kept taut as it is wound upon or unwound from another curve.

involution [BIOL] A turning or rolling in. [EMBRYO] Gastrulation by ingrowth of blastomeres around the dorsal lip. [MED] 1. The retrogressive change to their normal condition that organs undergo after fulfilling their functional purposes, as the uterus after pregnancy. 2. The period of regression or the processes of decline or decay which occur in the human constitution after middle life.

inward-outward dialing system [COMMUN] Dialing system whereby calls within the local exchange area may be dialed directly to and from base private branch exchange telephone stations without the assistance of the base private branch exchange operator; CENTREX, a service offered by some telephone companies, is a form of inward-outward dialing.

inyoite [MINERAL] $Ca_2B_6O_{11} \cdot 13H_2O$ A colorless, monoclinic mineral consisting of a hydrous calcium borate; hardness is 2 on Mohs scale, and specific gravity is 2.

Io [ASTRON] A satellite of Jupiter; its diameter is 2300 miles (3700 kilometers). Also known as Jupiter I. [NUC PHYS] See ionium.

I/O See input/output.

iodargyrite [MINERAL] AgI A yellowish or greenish hexagonal mineral composed of native silver iodide, usually occurring in thin plates. Also known as iodyrite.

iodide [CHEM] 1. A compound which contains the iodine atom in the −1 oxidation state and which may be considered to be derived from hydriodic acid (HI); examples are KI and NaI. 2. A compound of iodine, such as CH_3CH_2I, in which the iodine has combined with a more electropositive group.

iodine [CHEM] A nonmetallic halogen element, symbol I, atomic number 53, atomic weight 126.9044; melts at 114°C, boils at 184°C; the poisonous, corrosive, dark plates or granules are readily sublimed; insoluble in water, soluble in common solvents; used as germicide and antiseptic, in dyes, tinctures, and pharmaceuticals, in engraving lithography, and as a catalyst and analytical reagent.

iodized salt [FOOD ENG] Common table salt which has been treated with iodide to provide iodine as nutritional supplement.

iodyrite See iodargyrite.

ion [CHEM] An isolated electron or positron or an atom or molecule which by loss or gain of one or more electrons has acquired a net electric charge.

ion accelerator [NUCLEO] A linear accelerator in which ions are accelerated by an electric field in a standing-wave pattern that is set up in a resonant cavity by external oscillators or amplifiers.

ion-acoustic wave [PL PHYS] A longitudinal compression wave in the ion density of a plasma which can occur at high electron temperatures and low frequencies, caused by a combination of ion inertia and electron pressure.

ion atmosphere *See* ion cloud.

ion burn *See* ion spot.

ion chamber *See* ionization chamber.

ion cloud [GEOPHYS] An inhomogeneity or patch of unusually great ion density in one of the regular regions of the ionosphere; such patches occur quite often in the E region. [PHYS CHEM] A slight preponderance of negative ions around a positive ion in an electrolyte, and vice versa, according to the Debye-Hückel theory. Also known as ion atmosphere.

ion column [GEOPHYS] The trail of ionized gases in the trajectory of a meteoroid entering the upper atmosphere; a part of the composite phenomenon known as a meteor. Also known as meteor trail.

ion exchange [PHYS CHEM] A chemical reaction in which mobile hydrated ions of a solid are exchanged, equivalent for equivalent, for ions of like charge in solution; the solid has an open, fishnetlike structure, and the mobile ions neutralize the charged, or potentially charged, groups attached to the solid matrix; the solid matrix is termed the ion exchanger.

ion-exchange chromatography [ANALY CHEM] A chromatographic procedure in which the stationary phase consists of ion-exchange resins which may be acidic or basic.

ion-exchange resin [MATER] A synthetic resin that can combine or exchange ions with a solution; such a resin produces the exchange of sodium for calcium ions in the softening of hard water.

ion exclusion [CHEM] Ion-exchange resin system in which the mobile ions in the resin-gel phase electrically neutralize the immobilized charged functional groups attached to the resin, thus preventing penetration of solvent electrolyte into the resin-gel phase; used in separations where electrolyte is to be excluded from the resin, but not nonpolar materials, as the separation of salt from nonpolar glycerin.

ion-exclusion chromatography [ANALY CHEM] Chromatography in which the adsorbent material is saturated with the same mobile ions (cationic or anionic) as are present in the sample-carrying eluent (solvent), thus repelling the similar sample ions.

ion gage *See* ionization gage.

ionic bond [PHYS CHEM] A type of chemical bonding in which one or more electrons are transferred completely from one atom to another, thus converting the neutral atoms into electrically charged ions; these ions are approximately spherical and attract one another because of their opposite charge. Also known as electrovalent bond.

ionic conductance [PHYS CHEM] The contribution of a given type of ion to the total equivalent conductance in the limit of infinite dilution.

ionic conduction [SOLID STATE] Electrical conduction of a solid due to the displacement of ions within the crystal lattice.

ionic crystal [CRYSTAL] A crystal in which the lattice-site occupants are charged ions held together primarily by their electrostatic interaction.

ionic equilibrium [PHYS CHEM] The condition in which the rate of dissociation of nonionized molecules is equal to the rate of combination of the ions.

ionic membrane [CHEM ENG] Semipermeable membrane that conducts electricity; the application of an electric field to the membrane achieves an electrophoretic movement of ions through the membrane; used in electrodialysis.

ionic semiconductor [SOLID STATE] A solid whose electrical conductivity is due primarily to the movement of ions rather than that of electrons and holes.

ion implantation [ENG] A process of introducing impurities into the near-surface regions of solids by directing a beam of ions at the solid.

ionium [NUC PHYS] A naturally occurring radioisotope, symbol Io, of thorium, atomic weight 230.

ionization [CHEM] A process by which a neutral atom or molecule loses or gains electrons, thereby acquiring a net charge and becoming an ion; occurs as the result of the dissociation of the atoms of a molecule in solution ($NaCl \rightarrow Na^+ + Cl^-$) or of a gas in an electric field ($H_2 \rightarrow 2H^+$).

ionization chamber [NUCLEO] A particle detector which measures the ionization produced in the gas filling the chamber by the fast-moving charged particles as they pass through. Also known as ion chamber.

ionization constant [PHYS CHEM] Analog of the dissociation constant, where $k = [H^+][A^-]/[HA]$; used for the application of the law of mass action to ionization; in the equation HA represents the acid, such as acetic acid.

ionization current *See* gas current.

ionization energy [ATOM PHYS] The amount of energy needed to remove an electron from a given kind of atom or molecule to an infinite distance; usually expressed in electron volts, and numerically equal to the ionization potential in volts.

ionization gage [ELECTR] An instrument for measuring low gas densities by ionizing the gas and measuring the ion current. Also known as ion gage; ionization vacuum gage.

ionization potential [ATOM PHYS] The energy per unit charge needed to remove an electron from a given kind of atom or molecule to an infinite distance; usually expressed in volts. Also known as ion potential.

ionization radiation *See* ionizing radiation.

ionization spectrometer *See* Bragg spectrometer.

ionization vacuum gage *See* ionization gage.

ionizing radiation [NUCLEO] 1. Particles or photons that have sufficient energy to produce ionization directly in their passage through a substance. Also known as ionization radiation. 2. Particles that are capable of nuclear interactions in which sufficient energy is released to produce ionization.

ion laser [OPTICS] A gas laser in which stimulated emission takes place between two energy levels of an ion; gases used include argon, krypton, neon, and xenon; examples include helium-cadmium lasers and metal vapor lasers.

ion microprobe *See* secondary ion mass spectrometer.

ion microprobe mass spectrometer [ENG] A type of secondary ion mass spectrometer in which primary ions are focused on a spot 1–2 micrometers in diameter, mass-charge separation of secondary ions is carried out by a double focusing mass spectrometer or spectrograph, and a magnified image of elemental or isotopic distributions on the sample surface is produced using synchronous scanning of the primary ion beam and an oscilloscope.

ion microscope *See* field-ion microscope.

ionosonde [ENG] A radar system for determining the vertical height at which the ionosphere reflects signals back to earth at various frequencies; a pulsed vertical beam is swept periodically through a frequency range from 0.5 to 20 megahertz, and the variation of echo return time with frequency is photographically recorded.

ionosphere [GEOPHYS] That part of the earth's upper atmosphere which is sufficiently ionized by solar ultraviolet radiation so that the concentration of free electrons affects the propagation of radio waves; its base is at about 70 or 80 kilometers and it extends to an indefinite height.

ionospheric propagation [COMMUN] Propagation of radio waves over long distances by reflection from the ionosphere, useful at frequencies up to about 25 megahertz.

ionospheric scatter [COMMUN] A form of scatter propagation in which radio waves are scattered by the lower E layer of the ionosphere to permit communication over distances from 600 to 1400 miles (1000 to 2250 kilometers) when using the frequency range of about 25 to 100 megahertz.

ionospheric storm [GEOPHYS] A turbulence in the F region of the ionosphere, usually due to a sudden burst of radiation from the sun; it is accompanied by a decrease in the density of ionization and an increase in the virtual height of the region.

ion pair [NUCLEO] A positive ion and an equal-charge negative ion, usually an electron, that are produced by the action of radiation on a neutral atom or molecule.

ion-permeable membrane [MATER] A film or sheet of a substance which is preferentially permeable to some species or types of ions.

ion potential *See* ionization potential.

ion probe *See* secondary ion mass spectrometer.

ion propulsion [AERO ENG] Vehicular motion caused by reaction from the high-speed discharge of a beam of electrically equally charged minute particles ejected behind the vehicle.

ion pump [ELECTR] A vacuum pump in which gas molecules are first ionized by electrons that have been generated by a high voltage and are spiraling in a high-intensity magnetic field, and the molecules are then attracted to a cathode, or propelled by electrodes into an auxiliary pump or an ion trap.

ion spot [ELECTR] Of a cathode-ray tube screen, an area of localized deterioration of luminescence caused by bombardment with negative ions. Also known as ion burn.

iontophoresis [MED] A medical treatment used to drive positive or negative ions into a tissue, in which two electrodes are placed in contact with tissue, one of the electrodes being a pad of absorbent material soaked with a solution of the material to be administered, and a voltage is applied between the electrodes.

ion trap [ELECTR] **1.** An arrangement whereby ions in the electron beam of a cathode-ray tube are prevented from bombarding the screen and producing an ion spot, usually employing a magnet to bend the electron beam so that it passes through the tiny aperture of the electron gun, while the heavier ions are less affected by the magnetic field and are trapped inside the gun. **2.** A metal electrode, usually of titanium, into which ions in an ion pump are absorbed.

Iospilidae [INV ZOO] A small family of pelagic polychaetes assigned to the Errantia.

Iowan glaciation [GEOL] The earliest substage of the Wisconsin glacial stage; occurred more than 30,000 years ago.

ipecac [BOT] Any of several low, perennial, tropical South American shrubs or half shrubs in the genus *Cephaelis* of the family Rubiaceae; the dried rhizome and root, containing emetine, cephaeline, and other alkaloids, is used as an emetic and expectorant.

IPL button *See* bootstrap button.

ipsonite [GEOL] The final stage of weathered asphalt; a black, infusible substance, only slightly soluble in carbon disulfide, containing 50–80% fixed carbon and very little oxygen.

IQ *See* intelligence quotient.

Ir *See* iridium.

IRBM *See* intermediate-range ballistic missile.

IR drop *See* resistance drop.

IRG *See* record gap.

Iridaceae [BOT] A family of monocotyledonous herbs in the order Liliales distinguished by three stamens and an inferior ovary.

iridescence [OPTICS] A rainbow color effect exhibited in various bodies as a result of interference in a thin film (as of soap bubbles or mother of pearl) or of diffraction of light reflected from a ribbed surface (as of the plumage of some birds).

iridium [CHEM] A metallic element, symbol Ir, atomic number 77, atomic weight 192.2, in the platinum group; insoluble in acids, melting at 2454°C. [MET] A silver-white, brittle, hard metal used in jewelry, electric contacts, electrodes, resistance wires, and pen tips.

iridocyte [HISTOL] A specialized cell in the integument of certain animal species which is filled with iridescent crystals of guanine and a variety of lipophores. Also known as guanophore; iridophore.

iridophore *See* iridocyte.

iris [ANAT] A pigmented diaphragm perforated centrally by an adjustable pupil which regulates the amount of light reaching the retina in vertebrate eyes. [BOT] Any plant of the genus *Iris*, the type genus of the family Iridaceae, characterized by linear or sword-shaped leaves, erect stalks, and brightcolored flowers with the three inner perianth segments erect and the outer perianth segments drooping. [ELECTROMAG] A conducting plate mounted across a waveguide to introduce impedance; when only a single mode can be supported, an iris acts substantially as a shunt admittance and may be used for matching the waveguide impedance to that of a load. Also known as diaphragm; waveguide window. [OPTICS] A circular mechanical device, whose diameter can be varied continuously, which controls the amount of light reaching the film of a camera. Also known as iris diaphragm.

iris diaphragm *See* iris.

Irish potato *See* potato.

Irminger Current [OCEANOGR] An ocean current that is one of the terminal branches of the Gulf Stream system, flowing west off the southern coast of Iceland.

iron [CHEM] A silvery-white metallic element, symbol Fe, atomic number 26, atomic weight 55.847, melting at 1530°C. [MET] A heavy, magnetic, malleable and ductile metal occurring in meteorites and combined in a wide range of ores and most igneous rocks; it is one of the most widely used metals, and plays a role in biological processes.

Iron Age [ARCHEO] Period characterized by production and widespread use of iron, starting approximately 1000 B.C.

iron alum *See* halotrichite.

iron carbide *See* cementite.

iron core [ELECTROMAG] A core made of solid or laminated iron, or some other magnetic material which may contain very little iron.

iron-deficiency anemia [MED] Hypochromic microcytic anemia due to excessive loss, deficient intake, or poor absorption of iron. Also known as nutritional hypochromic anemia.

iron glance *See* specularite.

iron hat *See* gossan.

iron loss *See* core loss.

iron mica *See* lepidomelane.

iron pyrite *See* pyrite.

iron spar *See* siderite.

ironwood [BOT] Any of a number of hardwood trees in the United States, including the American hornbeam, the buckwheat, and the eastern hophornbeam.

irradiance *See* radiant flux density.

irradiation [BIOPHYS] Subjection of a biological system to sound waves of sufficient intensity to modify their structure or function. [ENG] The exposure of a material, object, or patient to x-rays, gamma rays, ultraviolet rays, or other ionizing radiation. [OPTICS] An optical illusion which makes bright objects appear larger than they really are.

irreducible equation [MATH] An equation that is equivalent to one formed by setting an irreducible polynomial equal to zero.

irreducible function *See* irreducible polynomial.

irreducible polynomial [MATH] A polynomial is irreducible over a field K if it cannot be written as the product of two polynomials of lesser degree whose coefficients come from K. Also known as irreducible function.

irregular [BOT] Lacking symmetry, as of a flower having petals unlike in size or shape.

irregular cleavage [EMBRYO] Division of a zygote into random masses of cells, as in certain coelenterates.

irregular connective tissue [HISTOL] A loose or dense connective tissue with fibers irregularly woven and irregularly distributed; collagen is the dominant fiber type.

Irregularia [INV ZOO] An artificial assemblage of echinoderms in which the anus and periproct lie outside the apical system, the ambulacral plates remain simple, the primary radioles are hollow, and the rigid test shows some degree of bilateral symmetry.

irreversible process [THERMO] A process which cannot be reversed by an infinitesimal change in external conditions.

irreversible thermodynamics *See* nonequilibrium thermodynamics.

irrigation [CIV ENG] Artificial application of water to arable land for agricultural use. [MED] Therapeutic washing out by means of a continuous stream of water.

irritability [PHYSIO] 1. A condition or quality of being excitable; the power of responding to a stimulus. 2. A condition of abnormal excitability of an organism, organ, or part, when it reacts excessively to a slight stimulation.

irritable colon [MED] Any of several disturbed colonic functions associated with anxiety or emotional stress. Also known as adaptive colitis; mucous colitis; spastic colon; unstable colon.

irrotational flow [FL MECH] Fluid flow in which the curl of the velocity function is zero everywhere, so that the circulation of the velocity about any closed curve vanishes. Also known as acyclic motion; irrotational motion.

irrotational motion *See* irrotational flow.

irrotational strain [GEOL] Strain in which the orientation of the axes of strain does not change. Also known as nonrotational strain.

irrotational vector field [MATH] A vector field whose curl is identically zero; every such field is the gra-

dient of a scalar function. Also known as lamellar vector field.

irrotational wave *See* compressional wave.

irruption *See* intrusion.

Irvngtonian [GEOL] A stage of geologic time in southern California, in the lower Pleistocene, below the Rancholabrean.

isarithm *See* isopleth.

ISB modulation *See* independent-sideband modulation.

I scan *See* I scope.

ischemia [MED] Localized tissue anemia as a result of obstruction of the blood supply or to vasoconstriction.

ischiopodite [INV ZOO] The segment nearest the basipodite of walking legs in certain crustaceans. Also known as ischium.

ischium [ANAT] Either of a pair of bones forming the dorsoposterior portion of the vertebrate pelvis; the inferior part of the human pelvis upon which the body rests in sitting. [INV ZOO] *See* ischiopodite.

Ischnacanthidae [PALEON] The single family of the acanthodian order Ischnacanthiformes.

Ischnacanthiformes [PALEON] A monofamilial order of extinct fishes of the order Acanthodii; members were slender, lightly armored predators with sharp teeth, deeply inserted fin spines, and two dorsal fins.

I scope [ELECTR] A cathode-ray scope on which a single signal appears as a circular segment whose radius is proportional to the range and whose circular length is inversely proportional to the error of aiming the antenna, true aim resulting in a complete circle; the position of the arc, relative to the center, indicates the position of the target relative to the beam axis. Also known as broken circle indicator; I indicator; I scan.

Isectolophidae [PALEO] A family of extinct ceratomorph mammals in the superfamily Tapiroidea.

isenthalpic expansion [THERMO] Expansion which takes place without any change in enthalpy.

isentropic [THERMO] Having constant entropy; at constant entropy.

isentropic expansion [THERMO] Expansion which occurs without any change in entropy.

isentropic flow [THERMO] Fluid flow in which the entropy of any part of the fluid does not change as that part is carried along with the fluid.

isentropic mixing [METEOROL] Any atmospheric mixing process which occurs within an isentropic surface; the fact that many atmospheric motions are reversible adiabatic processes renders this type of mixing important, and exchange coefficients have been computed therefor.

Ising coupling [SOLID STATE] A model of coupling between two atoms in a lattice, used to study ferromagnetism, in which the spin component of each atom along some axis is taken to be $+1$ or -1, and the energy of interaction is proportional to the negative of the product of the spin components along this axis.

isinglass [MATER] A gelatin made from the dried swim bladders of sturgeon and other fishes; used in glues, cements, and printing inks. Also known as fish gelatin; ichthyocolla. [MINERAL] Sheet mica, usually in the form of single cleavage plates; used in furnace and stove doors.

Ising model [SOLID STATE] A crude model of a ferromagnetic material or an analogous system, used to study phase transitions, in which atoms in a one-, two-, or three-dimensional lattice interact via Ising coupling between nearest neighbors, and the spin components of the atoms are coupled to a uniform magnetic field.

island [GEOGR] A tract of land smaller than a continent and surrounded by water; normally in an ocean, sea, lake, or stream.

island arc [GEOGR] A group of islands usually with a curving archlike pattern, generally convex toward the open ocean, having a deep trench or trough on the convex side and usually enclosing a deep basin on the concave side.

island of Langerhans *See* islet of Langerhans.

islet of Langerhans [HISTOL] A mass of cell cords in the pancreas that is of an endocrine nature, secreting insulin and a minor hormone like lipocaic. Also known as island of Langerhans; islet of the pancreas.

islet of the pancreas *See* islet of Langerhans.

isoagglutinin [IMMUNOL] An agglutinin which acts upon the red blood cells of members of the same species. Also known as isohemagglutinin.

isoallele [GEN] An allele that carries mutational alterations at the same site.

isoantigen [IMMUNOL] An antigen in an individual capable of stimulating production of a specific antibody in another member of the same species. Also known as alloantigen.

isobar [METEOROL] A line drawn through all points of equal atmospheric pressure along a given reference surface, such as a constant-height surface (notably mean sea level on surface charts), an isentropic surface, or the vertical plan of a synoptic cross section. [NUC PHYS] One of two or more nuclides having the same number of nucleons in their nuclei but differing in their atomic numbers and chemical properties. [PHYS] **1.** A line connecting points of equal pressure along a given surface in a physical system. **2.** A line connecting points of equal pressure on a graph plotting thermodynamic variables.

isobaric [THERMO] Of equal or constant pressure, with respect to either space or time.

isobaric chart *See* constant-pressure chart.

isobaric contour chart *See* constant-pressure chart.

isobaric equivalent temperature *See* equivalent temperature.

isobaric spin *See* isotopic spin.

isobath [OCEANOG] A contour line connecting points of equal water depths on a chart. Also known as depth contour; depth curve; fathom curve.

isobathytherm [OCEANOGR] A line or surface showing the depth in oceans or lakes at which points have the same temperatures.

isobits [COMPUT SCI] Binary digits having the same value.

Isobryales [BOT] An order of mosses in which the plants are slender to robust and up to 90 centimeters in length.

isochor *See* isochore.

isochore [PHYS] A graph that shows the variation of one quantity with another; for example, the variation of pressure with temperature, when the volume of the substance is held constant. Also known as isochor; isometric.

isochromatic [OPTICS] **1.** Pertaining to a variation of certain quantities related to light (such as density of the medium through which the light is passing, index of refraction), in which the color or wavelength of the light is held constant. **2.** Pertaining to lines connecting points of the same color.

isochromosome [CYTOL] An abnormal chromosome with a medial centromere and identical arms formed as a result of transverse, rather than longitudinal, splitting of the centromere.

isochrone [PHYS] A line on a chart connecting all points having the same time of occurrence of particular phenomena or of a particular value of a quantity.

isochronous governor [MECH ENG] A governor that keeps the speed of a prime mover constant at all loads. Also known as astatic governor.

isoclasite [MINERAL] $Ca_2(PO_4)(OH) \cdot 2H_2O$ A white mineral composed of a basic hydrous calcium phosphate; occurring in small crystals or columnar forms.

isoclinal *See* isoclinic line.

isocline [GEOL] A fold of strata so tightly compressed that parts on each side dip in the same direction.

isoclinic line [GEOPHYS] A line connecting points on the earth's surface which have the same magnetic dip. Also known as isoclinal. [SOLID STATE] A line joining points in a plate at which the principal stresses have parallel directions.

Isocrinida [INV ZOO] An order of stalked articulate echinoderms with nodal rings of cirri.

isodynamic [MECH] Pertaining to equality of two or more forces or to constancy of a force.

isoelectric [ELEC] Pertaining to a constant electric potential.

isoelectric focusing [PHYS CHEM] Protein separation technique in which a mixture of protein molecules is resolved into its components by subjecting the mixture to an electric field in a supporting gel having a previously established pH gradient. Also know as electrofocusing.

isoelectric point [PHYS CHEM] The pH value of the dispersion medium of a colloidal suspension at which the colloidal particles do not move in an electric field.

isoelectric precipitation [CHEM] Precipitation of materials at the isoelectric point (the pH at which the net charge on a molecule in solution is zero); proteins coagulate best at this point.

isoelectronic [ATOM PHYS] Pertaining to atoms having the same number of electrons outside the nucleus of the atom.

isoenzyme [BIOCHEM] Any of the electrophoretically distinct forms of an enzyme, representing different polymeric states but having the same function. Also known as isozyme.

Isoetaceae [BOT] The single family assigned to the order Isoetales in some systems of classification.

Isoetales [BOT] A monotypic order of the class Isoetopsida containing the single genus *Isoetes*, characterized by long, narrow leaves with a spoonlike base, spirally arranged on an underground cormlike structure.

Isoetatae *See* Isoetopsida.

Isoetopsida [BOT] A class of the division Lycopodiophyta; members are heterosporous and have a distinctive appendage, the ligule, on the upper side of the leaf near the base.

isofacies map [GEOL] A stratigraphic map showing the distribution of one or more facies within a particular stratigraphic unit.

isogamete [BIOL] A reproductive cell that is morphologically similar in both male and female and cannot be distinguished on form alone.

Isogeneratae [BOT] A class of brown algae distinguished by having an isomorphic alternation of generations.

isogeotherm *See* geoisotherm.

isogonic line [GEOPHYS] **1.** Any of the lines on a chart or map showing the same direction of the wind vector. **2.** Any of the lines on a chart or map connecting points of equal magnetic variation.

isogram *See* isopleth.

isohaline [OCEANOGR] 1. Of equal or constant salinity. 2. A line on a chart connecting all points of equal salinity.

isoheight *See* contour line.

isohemagglutinin *See* isoagglutinin.

isohypse *See* contour line.

isohypsic chart *See* constant-height chart.

isokinetic *See* isotach.

isolate [CHEM ENG] To separate two portions of a process system by means of valving or line blanks; used as safety measure during maintenance or repair, or to redirect process flows. [ELEC] To disconnect a circuit or piece of equipment from an electric supply system.

isolated camera [ELECTR] 1. A television camera that views a particular portion of a scene of action and produces a tape which can then be used either immediately for instant replay or for video replay at a later time. 2. The technique of video replay involving such a camera.

isolated footing [CIV ENG] A concrete slab or block under an individual load or column.

isolated location [COMPUT SCI] A location in a computer memory which is protected by some hardware device so that it cannot be addressed by a computer program and its contents cannot be accidentally altered.

isolated point [MATH] A point p in a topological space is an isolated point of a set if p is in the set and there is a neighborhood of p which contains no other points of the set.

isolation [CHEM] Separation of a pure chemical substance from a compound or mixture; as in distillation, precipitation, or absorption. [COMPUT SCI] The ability of a logic circuit having more than one input to ensure that each input signal is not affected by any of the others. [MED] Separation of an individual with a communicable disease from other, healthy individuals. [MICROBIO] Separation of an individual or strain from a natural, mixed population. [PHYSIO] Separation of a tissue, organ, system, or other part of the body for purposes of study. [PSYCH] Dissociation of a memory or thought from the emotions or feelings associated with it.

isolation network [ELEC] A network inserted in a circuit or transmission line to prevent interaction between circuits on each side of the insertion point.

isolator [ELECTR] A passive attenuator in which the loss in one direction is much greater than that in the opposite direction; a ferrite isolator for waveguides is an example. [ENG] Any device that absorbs vibration or noise, or prevents its transmission.

isolecithal *See* homolecithal.

isoleucine [BIOCHEM] $C_6H_{13}O_2$ An essential monocarboxylic amino acid occurring in most dietary proteins.

isolith [ELECTR] Integrated circuit of components formed on a single silicon slice, but with the various components interconnected by beam leads and with circuit parts isolated by removal of the silicon between them. [GEOL] A line on a contour-type map that denotes the aggregate thickness of a single lithology in a stratigraphic succession composed of one or more lithologies.

isomer [CHEM] One of two or more chemical substances having the same elementary percentage composition and molecular weight but differing in structure, and therefore in properties; there are many ways in which such structural differences occur; one example is provided by the compounds *n*-butane, $CH_3(CH_2)_2CH_3$, and isobutane, $CH_3CH(CH_3)_2$.

[NUC PHYS] One of two or more nuclides having the same mass number and atomic number, but existing for measurable times in different quantum states with different energies and radioactive properties.

isomerase [BIOCHEM] An enzyme that catalyzes isomerization reactions.

isomerism [BIOL] The condition of having two or more comparable parts made up of identical numbers of similar segments. [CHEM] The phenomenon whereby certain chemical compounds have structures that are different although the compounds possess the same elemental composition. [NUC PHYS] The occurrence of nuclear isomers.

isomerization [CHEM] A process whereby a compound is changed into an isomer; for example, conversion of butane into isobutane.

isometric *See* isochore.

isometric projection *See* axonometric projection.

isometric system [CRYSTAL] The crystal system in which the forms are referred to three equal, mutually perpendicular axes. Also known as cubic system.

isometry [MATH] A mapping f from a metric space X to a metric space Y where the distance between any two points of X equals the distance between their images under f in Y.

isomorph *See* isomorphic mineral.

isomorphic mineral [MINERAL] Any two or more crystalline mineral compounds having different chemical composition but identical structure, such as the garnet series or the feldspar group Also known as isomorph.

isomorphism [MATH] A one to one function of an algebraic structure (for example, group, ring, module, vector space) onto another of the same type, preserving all algebraic relations; its inverse function behaves likewise. [PHYS CHEM] A condition present when an ion at high dilution is incorporated by mixed crystal formation into a precipitate, even though such formation would not be predicted on the basis of crystallographic and ionic radii; an example is coprecipitation of lead with potassium chloride. [SCI TECH] The quality or state of being identical or similar in form, shape, or structure, such as between organisms resulting from evolutionary convergence, or crystalline forms of similar composition.

isopleth [MATH] The straight line which cuts the three scales of a nomograph at values satisfying some equation. Also known as index line. [METEOROL] 1. A line of equal or constant value of a given quantity with respect to either space or time. Also known as isogram. 2. More specifically, a line drawn through points on a graph at which a given quantity has the same numerical value (or occurs with the same frequency) as a function of the two coordinate variables. Also known as isarithm.

Isopoda [INV ZOO] An order of malacostracan crustaceans characterized by a cephalon bearing one pair of maxillipeds in addition to the antennae, mandibles, and maxillae.

isopotential map [PETRO ENG] A contour-line map to show the initial or calculated daily rate of oil well production in a multiwell field.

Isoptera [INV ZOO] An order of Insecta containing morphologically primitive forms characterized by gradual metamorphosis, lack of true larval and pupal stages, biting and prognathous mouthparts, two pairs of subequal wings, and the abdomen joined broadly to the thorax.

isopulse system [COMMUN] In adaptive communications, a pulse coding system wherein the number of information pulses transmitted is indicated by special inserted pulses.

isopycnic [METEOROL] A line on a chart connecting all points of equal or constant density. [PHYS] Of equal or constant density, with respect to either space or time.

isosceles triangle [MATH] A triangle with two sides of equal length.

isospin See isotopic spin.

isospin multiplet [PARTIC PHYS] A collection of elementary particles which have approximately the same mass and the same quantum numbers except for charge, but have a sequence of charge values, $Y/2 - I$, $Y/2 - I + 1$, ..., $Y/2 + I$ times the proton charge, where Y is an integer known as the hypercharge, and I is an integer or half-integer known as the isospin; examples are the pions ($Y = 0$, $I = 1$) and the nucleons ($Y = 1$, $I = \frac{1}{2}$). Also known as charge multiplet; particle multiplet.

Isospondyli [VERT ZOO] A former equivalent name for Clupeiformes.

isostatics [MECH] In photoelasticity studies of stress analyses, those curves, the tangents to which represent the progressive change in principal-plane directions. Also known as stress trajectories. Also known as stress lines.

isostemonous [BOT] Having the number of stamens of a flower equal to the number of perianth divisions.

isotach [METEOROL] A line in a given surface connecting points with equal wind speed. Also known as isokinetic; isovel.

isotherm [GEOPHYS] A line on a chart connecting all points of equal or constant temperature. [THERMO] A curve or formula showing the relationship between two variables, such as pressure and volume, when the temperature is held constant. Also known as isothermal.

isothermal [THERMO] 1. Having constant temperature; at constant temperature. 2. See isotherm.

isothermal chart [GEOPHYS] A map showing the distribution of air temperature (or sometimes sea-surface or soil temperature) over a portion of the earth or at some level in the atmosphere; places of equal temperature are connected by lines called isotherms.

isotonic [PHYSIO] 1. Having uniform tension, as the fibers of a contracted muscle. 2. Of a solution, having the same osmotic pressure as the fluid phase of a cell or tissue.

isotope [NUC PHYS] One of two or more atoms having the same atomic number but different mass number.

isotope dilution [NUCLEO] The introduction of a radioisotope into stable isotopes of an element in order to make volume, mass, and age measurements of the element.

isotope effect [PHYS CHEM] The effect of difference of mass between isotopes of the same element on nonnuclear physical and chemical properties, such as the rate of reaction or position of equilibrium, of chemical reactions involving the isotopes. [SOLID STATE] Variation of the transition temperatures of the isotopes of a superconducting element in inverse proportion to the square root of the atomic mass.

isotope exchange [NUCLEO] 1. Exchange of places by two atoms, but different isotopes, of the same element in two different molecules, or in different locations of the same molecule. 2. The transfer of isotopically tagged atoms from one chemical form or valence state to another, without net chemical reaction.

isotope exchange reaction [CHEM] A chemical reaction in which interchange of the atoms of a given element between two or more chemical forms of the element occurs, the atoms in one form being isotopically labeled so as to distinguish them from atoms in the other form.

isotope shift [SPECT] A displacement in the spectral lines due to the different isotopes of an element.

isotopic age determination See radiometric dating.

isotopic carrier See carrier.

isotopic enrichment [NUCLEO] The process by which the relative abundances of the isotopes of a given element are altered in a batch, thus producing a form of the element enriched in a particular isotope.

isotopic indicator See isotopic tracer.

isotopic label See isotopic tracer.

isotopic spin [NUC PHYS] A quantum-mechanical variable, resembling the angular momentum vector in algebraic structure whose third component distinguished between members of groups of elementary particles, such as the nucleons, which apparently behave in the same way with respect to strong nuclear forces, but have different charges. Also known as isobaric spin; isospin; i-spin.

isotopic tracer [CHEM] An isotope of an element, either radioactive or stable, a small amount of which may be incorporated into a sample material (the carrier) in order to follow the course of that element through a chemical, biological, or physical process, and also follow the larger sample. Also known as isotopic indicator; isotopic label; label; tag.

isotropic [BIOL] Having a tendency for equal growth in all directions. [CYTOL] An ovum lacking any predetermined axis. [PHYS] Having identical properties in all directions.

isotropic antenna See unipole.

isotropic dielectric [ELEC] A dielectric whose polarization always has a direction that is parallel to the applied electric field, and a magnitude which does not depend on the direction of the electric field.

isotropic fluid [FL MECH] A fluid whose properties are not dependent on the direction along which they are measured.

isotropy [PHYS] The quality of a property which does not depend on the direction along which it is measured, or of a medium or entity whose properties do not depend on the direction along which they are measured.

isotypes [GEN] A series of antigens, for example, blood types, common to all members of a species but differentiating classes and subclasses within the species.

isotypic [CRYSTAL] Pertaining to a crystalline substance whose chemical formula is analogous to, and whose structure is like, that of another specified compound.

isovalent conjugation [PHYS CHEM] An arrangement of bonds in a conjugated molecule such that alternative structures with an equal number of bonds can be written; an example occurs in benzene.

isovalent hyperconjugation [PHYS CHEM] An arrangement of bonds in a hyperconjugated molecule such that the number of bonds is the same in the two resonance structures but the second structure is energetically less favorable than the first structure; examples are $H_3C\!\!\equiv\!\!C\text{—}C^+H_2$ and $H_3C\!\!\equiv\!\!C\text{—}CH_2$.

isovel See isotach.

isozyme See isoenzyme.

i-spin See isotopic spin.

isthmus [BIOL] A passage or constricted part connecting two parts of an organ. [GEOGR] A narrow strip of land having water on both sides and connecting two large land masses. [MATH] *See* bridge.

Istiophoridae [VERT ZOO] The billfishes, a family of oceanic perciform fishes in the suborder Scombroidei.

Isuridae [VERT ZOO] The mackerel sharks, a family of pelagic, predacious galeoids distinguished by a heavy body, nearly symmetrical tail, and sharp, awl-like teeth.

itacolumite [PETR] A fine-grained, thin-bedded sandstone or a schistose quartzite that contains mica, chlorite, and talc and that exhibits flexibility when split into slabs. Also known as articulite.

IT calorie *See* calorie.

itch [PHYSIO] An irritating cutaneous sensation allied to pain.

item [COMPUT SCI] A set of adjacent digits, bits, or characters which is treated as a unit and conveys a single unit of information.

iteration *See* iterative method.

iteration process [COMPUT SCI] The process of repeating a sequence of instructions with minor modifications between successive repetitions.

iterative array [COMPUT SCI] In a computer, an array of a large number of interconnected identical processing modules, used with appropriate driver and control circuits to permit a large number of simultaneous parallel operations.

iterative impedance [ELECTR] Impedance that, when connected to one pair of terminals of a four-terminal transducer, will cause the same impedance to appear between the other two terminals.

iterative method [MATH] Any process of successive approximation used in such problems as numerical solution of algebraic equations, differential equations, or the interpolation of the values of a function. Also known as iteration.

iterative process [MATH] A process for calculating a desired result by means of a repeated cycle of operations, which comes closer and closer to the desired result; for example, the arithmetical square root of N may be approximated by an iterative process using additions, subtractions, and divisions only.

iterative routine [COMPUT SCI] A computer program that obtains a result by carrying out a series of operations repetitiously until some specified condition is met.

Ithomiinae [INV ZOO] The glossy-wings, a subfamily of weak-flying lepidopteran insects having on the wings broad, transparent areas in which the scales are reduced to short hairs.

Itonididae [INV ZOO] The gall midges, a family of orthorrhaphous dipteran insects in the series Nematocera; most are plant pests.

i-type semiconductor *See* intrinsic semiconductor.

Ixodides [INV ZOO] The ticks, a suborder of the Acarina distinguished by spiracles behind the third or fourth pair of legs.

Izod test [MET] An impact test in which a falling pendulum strikes a fixed, usually notched specimen with 120 foot-pounds (163 joules) of energy at a velocity of 11.5 feet (3.5 meters) per second; the height of the pendulum swing after striking is a measure of the energy absorbed and thus indicates impact strength.

J

J *See* joule.

Jacanidae [VERT ZOO] The jacanas or lily-trotters, the single family of the avian superfamily Jacanoidea.

Jacanoidea [VERT ZOO] A monofamilial superfamily of colorful marshbirds distinguished by greatly elongated toes and claws, long legs, a short tail, and a straight bill.

jacinth *See* zircon.

jack [ELEC] A connecting device into which a plug can be inserted to make circuit connections; may also have contacts that open or close to perform switching functions when the plug is inserted or removed. [MECH ENG] A portable device for lifting heavy loads through a short distance, operated by a lever, a screw, or a hydraulic press. [MINERAL] *See* sphalerite. [TEXT] **1.** A frame in lace-manufacturing equipment that has horizontal bars to support fixed vertical wires, against which bobbins containing the yarn can freely revolve. **2.** An oscillating lever that raises the harness of a dobby loom.

jackal [VERT ZOO] **1.** *Canis aureus.* A wild dog found in southeastern Europe, southern Asia, and northern Africa. **2.** Any of various similar Old World wild dogs; they resemble wolves but are smaller and more yellowish.

jacket [MECH ENG] The space around an engine cylinder through which a cooling liquid circulates. [NUCLEO] A thin container for one or more fuel slugs, used to prevent the fuel from escaping into the coolant of a reactor. Also known as can; cartridge. [ORD] **1.** Cylinder of steel covering and strengthening the breech end of a gun or howitzer tube. **2.** The water jacket on some machine guns. [PETRO ENG] The support structure of a steel offshore production platform; it is fixed to the seabed by piling, and the superstructure is mounted on it.

jackhammer [MECH ENG] A hand-held rock drill operated by compressed air.

jack ladder [ENG] A V-shaped trough holding a toothed endless chain, and used to move logs from pond to sawmill. [NAV ARCH] *See* Jacob's ladder.

jackscrew [MECH ENG] **1.** A jack operated by a screw mechanism. Also known as screw jack. **2.** The screw of such a jack.

jackstay [NAV ARCH] A rope, rod, or pipe rove through the eyebolts fitted on a yard or mast for the purpose of attaching sails to the yard or mast.

Jacobian [MATH] The Jacobian of functions $f_i(x_1, x_2, \ldots, x_n)$, $i = 1, 2, \ldots, n$, of real variables x_i is the determinant of the matrix whose ith row lists all the first-order partial derivatives of the function $f_i(x_1, x_2, \ldots, x_n)$. Also known as Jacobian determinant.

Jacobian determinant *See* Jacobian.

Jacobian matrix [MATH] The matrix used to form the Jacobian.

jacobsite [MINERAL] $MnFe_2O_4$ A black magnetic mineral composed of an oxide of manganese and iron; a member of the magnetite series.

Jacob's ladder [NAV ARCH] A portable ladder, having rope, chain, or wire sides, and wooden or iron rungs, slung over the ship's side for temporary use. Also known as jack ladder.

Jacobsoniidae [INV ZOO] The false snout beetles, a small family of coleopteran insects in the superfamily Dermestoidea.

Jacobson radical *See* radical.

jacquard [TEXT] **1.** A loom or knitting machine for weaving figured fabrics and whose apparatus is controlled by punched cards. **2.** A fabric of jacquard weave.

jade [MINERAL] A hard, compact, dark-green or greenish-white gemstone composed of either jadeite or nephrite. Also known as jadestone.

jadeite [MINERAL] $NaAl(SiO_3)_2$ A clinopyroxene mineral occurring as green, fibrous monoclinic crystals; the most valuable variety of jade.

jadestone *See* jade.

jaguar [VERT ZOO] *Felis onca.* A large, wild cat indigenous to Central and South America; it is distinguished by a buff-colored coat with black spots, and has a relatively large head and short legs.

Jahn-Teller effect [PHYS CHEM] The effect whereby, except for linear molecules, degenerate orbital states in molecules are unstable.

jalousie [BUILD] A window that consists of a number of long, narrow panels, each hinged at the top.

jam [COMPUT SCI] In punched-card equipment, a feed malfunction causing blockage of passages with crumpled cards.

jamb [BUILD] The vertical member on the side of an opening, as a door or window.

jamesonite [MINERAL] $Pb_4FeSb_6S_{14}$ A lead-gray to gray-black mineral that crystallizes in the orthorhombic system, occurs in acicular crystals with fibrous or featherlike forms, and has a metallic luster. Also known as feather ore; gray antimony.

jammer [ELECTR] A transmitter used in jamming of radio or radar transmissions. Also known as electronic jammer.

jamming [ELECTR] Radiation or reradiation of electromagnetic waves so as to impair the usefulness of a specific segment of the radio spectrum that is being used by the enemy for communication or radar. Also known as active jamming; electronic jamming.

jam nut *See* locknut.

Janus [ASTRON] The innermost satellite of Saturn, which orbits at a mean distance of 159.5×10^3 kilometers, very near Saturn's rings, and is judged to have a diameter of approximately 220 kilometers.

japan [MATER] A glossy, black baking paint or varnish that consists primarily of a hard asphalt base.

Japan Current *See* Kuroshio Current.

japanning [MET] The finishing of metal objects with japan.

Japygidae [INV ZOO] A family of wingless insects in the order Diplura with forcepslike anal appendages; members attack and devour small soil arthropods.

jarlite [MINERAL] $NaSr_3Al_3F_{16}$ A colorless to brownish mineral composed of aluminofluoride of sodium and strontium.

jarosite [MINERAL] $KFe_3(SO_4)_2(OH)_6$ An ocher-yellow or brown alunite mineral having rhombohedral crystal structure. Also known as utahite.

jasper [PETR] A dense, opaque to slightly translucent cryptocrystalline quartz containing iron oxide impurities; characteristically red. Also known as jasperite; jasperoid; jaspis.

jasperite *See* jasper.

jasperoid *See* jasper.

jaspilite [PETR] A compact siliceous rock resembling jasper and containing iron oxides in bands.

jaspis *See* jasper.

jaspoid *See* tachylite.

jaundice [INV ZOO] *See* grasserie. [MED] Yellow coloration of the skin, mucous membranes, and secretions resulting from hyperbile-rubinemia. Also known as icterus.

Java black rot [PL PATH] A fungus disease of stored sweet potatoes caused by *Diplodia tubericola*; the inside of the root becomes black and brittle.

Java man [PALEON] An overspecialized, apelike form of *Homo sapiens* from the middle Pleistocene having a small brain capacity, low cranial vault, and massive browridges.

jaw [ANAT] Either of two bones forming the skeleton of the mouth of vertebrates: the upper jaw or maxilla, and the lower jaw or mandible. [ENG] A notched part that permits a railroad-car axle box to move vertically. [GEOL] The side of a narrow passage such as a gorge.

jawbreaker *See* jaw crusher.

jaw clutch [MECH ENG] A clutch that provides positive connection of one shaft with another by means of interlocking faces; may be square or spiral; the most common type of positive clutch.

jaw crusher [MECH ENG] A machine for breaking rock between two steel jaws, one fixed and the other swinging. Also known as jawbreaker.

jawless vertebrate [VERT ZOO] The common name for members of the Agnatha.

J box *See* junction box.

J-carrier system [COMMUN] Broad-band carrier system, providing 12 telephone channels, which uses frequencies up to about 140 kilohertz by means of effective four-wire transmission on a single open-wire pair.

J display [ELECTR] A modified radarscope A display in which the time base is a circle; the target signal appears as an outward radial deflection from the time base.

jeffersonite [MINERAL] $Ca(Mn,Zn,Fe)Si_2O_6$ A dark-green or greenish-black mineral composed of pyroxene.

jejunum [ANAT] The middle portion of the small intestine, extending between the duodenum and the ileum.

jelly *See* ulmin.

jellyfish [INV ZOO] Any of various free-swimming marine coelenterates belonging to the Hydrozoa or Scyphozoa and having a bell- or bowl-shaped body. Also known as medusa.

jelly fungus [MYCOL] The common name for many members of the Heterobasidiomycetidae, especially the orders Tremallales and Dacromycetales, distinguished by a jellylike appearance or consistency.

jenny [VERT ZOO] **1.** A female animal, as a jenny wren. **2.** A female donkey.

jerboa [VERT ZOO] The common name for 25 species of rodents composing the family Dipodidae; all are adapted for jumping, having extremely long hindlegs and feet.

jeremejevite [MINERAL] $AlBO_3$ A colorless or yellowish mineral composed of aluminum borate that occurs in hexagonal crystals.

jersey [TEXT] A knitted wool, cotton, polyester, rayon, or other fabric with a slight rib on one side.

jet [FL MECH] A strong, well-defined stream of compressible fluid, either gas or liquid, issuing from an orifice or nozzle or moving in a contracted duct.

jet coal [GEOL] A hard, lustrous, pure black variety of lignite, occurring in isolated masses in bituminous shale; thought to be derived from waterlogged driftwood. Also known as black amber.

jet drilling [MECH ENG] A drilling method that utilizes a chopping bit, with a water jet run on a string of hollow drill rods, to chop through soils and wash the cuttings to the surface. Also known as wash boring.

jet engine [AERO ENG] An aircraft engine that derives all or most of its thrust by reaction to its ejection of combustion products (or heated air) in a jet and that obtains oxygen from the atmosphere for the combustion of its fuel. [MECH ENG] Any engine that ejects a jet or stream of gas or fluid, obtaining all or most of its thrust by reaction to the ejection.

jet fuel [MATER] Special grade of kerosine with a flash point of 125°F (52°C), used for jet aircraft; may have methane or naphthene added to produce a 110°F (43°C) flash point, for military aircraft.

jet lag [PHYSIO] Desynchronization of biological rhythms because of transmeridian flight.

jet membrane process [CHEM ENG] A method for separating or enriching isotopes of the same element by using a condensable vapor as the carrier fluid, a process gas containing the isotopes enters a chamber into which a heavy condensable gas (the jet) flows, the lighter of the two isotopes is enriched relative to the heavier species and is collected by a probe downstream for further enrichment or analysis.

jet nozzle [DES ENG] A nozzle, usually specially shaped, for producing a jet, such as the exhaust nozzle on a jet or rocket engine.

jetsam [ENG] Articles that sink when thrown overboard, particularly those jettisoned for the purpose of lightening a vessel in distress.

jet stream [AERO ENG] The stream of gas or fluid expelled by any reaction device, in particular the stream of combustion products expelled from a jet engine, rocket engine, or rocket motor. [METEOROL] A relatively narrow, fast-moving wind current flanked by more slowly moving currents; observed principally in the zone of prevailing westerlies above the lower troposphere, and in most cases reaching maximum intensity with regard to speed and concentration near the troposphere.

jettison [ENG] The throwing overboard of objects, especially to lighten a craft in distress.

jetty *See* groin.

jewel [ENG] **1.** A bearing usually made of synthetic corundum and used in precision timekeeping devices, gyros, and other instruments. **2.** A bearing lining of soft metal, used in railroad cars, for example.

jeweler's rouge *See* ferric oxide.

jezekite *See* morinite.

JFET *See* junction field-effect transistor.

jib [NAV ARCH] A triangular sail bent to a foremast stay.

jig [ENG] A machine for dyeing piece goods by moving the cloth at full width (open width) through the dye liquor on rollers. [MECH ENG] A device used to position and hold parts for machining operations and to guide the cutting tool. [MIN ENG] A vibrating device in which coal is cleaned and ore is concentrated in water.

jigging [MIN ENG] A gravity method which separates mineral from gangue particles by utilizing an effective difference in settling rate through a periodically dilated bed.

jigsaw [MECH ENG] A tool with a narrow blade suitable for cutting intricate curves and lines.

jimsonweed [BOT] *Datura stramonium.* A tall, poisonous annual weed having large white or violet trumpet-shaped flowers and globose prickly fruits. Also known as apple of Peru.

J indicator *See* J scope.

jird [VERT ZOO] Any one of the diminutive rodents composing related species of the genus *Meriones* which are inhabitants of northern Africa and southwestern Asia; they serve as experimental hosts for studies of schistosomiasis.

jitter [COMMUN] In facsimile, distortion in the received copy caused by momentary errors in synchronism between the scanner and recorder mechanisms; does not include slow errors in synchronism due to instability of the frequency standards used in the facsimile transmitter and recorder. [ELECTR] Small, rapid variations in a waveform due to mechanical vibrations, fluctuations in supply voltages, control-system instability, and other causes.

JND *See* just-noticeable difference.

joaquinite [MINERAL] $NaBa_2Ce_2Fe(Ti,Nb)_2$-$Si_8O_{26}(OH,F)_2$ A honey-yellow mineral composed of sodium iron titanium silicate, occurring in orthorhombic crystals.

job [COMPUT SCI] A unit of work to be done by the computer; it is a single entity from the standpoint of computer installation management, but may consist of one or more job steps. [IND ENG] **1.** The combination of duties, skills, knowledge, and responsibilities assigned to an individual employee. **2.** A work order.

job control block [COMPUT SCI] A group of data containing the execution-control data and the job identification when the job is initiated as a unit of work to the operating system.

job control language *See* command language.

job control statement [COMPUT SCI] Any of the statements used to direct an operating system in its functioning, as contrasted to data, programs, or other information needed to process a job but not intended directly for the operating system itself. Also known as control statement.

job flow control [COMPUT SCI] Control over the order in which jobs are handled by a computer in order to use the central processing units and the units under the computer's control as efficiently as possible.

job library [COMPUT SCI] A partitioned data set, or a concatenation of partitioned data sets, used as the primary source of object programs (load modules) for a particular job, and more generally, as a source

of runnable programs from which all or most of the programs for a given job will be selected.

job-oriented terminal [COMPUT SCI] A terminal, such as a point-of-sale terminal, at which data taken directly from a source can enter a communication network directly.

job processing control [COMPUT SCI] The section of the control program responsible for initiating operations, assigning facilities, and proceeding from one job to the next.

job stacking [COMPUT SCI] The presentation of jobs to a computer system, each job followed by another.

jogging [ELEC] Quickly repeated opening and closing of a circuit to produce small movements of the driven machine. Also known as inching.

johannite [MINERAL] $Cu(UO_2)_2(SO_4)_2 \cdot 6H_2O$ An emerald green to apple green, triclinic mineral consisting of a hydrated basic copper and uranium sulfate.

johannsenite [MINERAL] $CaMnSi_2O_6$ A clove-brown, grayish, or greenish clinopyroxene mineral composed of a silicate of calcium and manganese; a member of the pyroxene group.

Johnson noise *See* thermal noise.

johnstrupite [MINERAL] A mineral composed of a complex silicate of cerium and other metals, approximately $(Ca,Na)_3(Ce,Ti,Zr)(SiO_4)_2F$; occurs in prismatic crystals.

joint [ANAT] A contact surface between two individual bones. Also known as articulation. [ELEC] A juncture of two wires or other conductive paths for current. [ENG] The surface at which two or more mechanical or structural components are united. [GEOL] A fracture that traverses a rock and does not show any discernible displacement of one side of the fracture relative to the other.

jointer [ENG] **1.** Any tool used to prepare, make, or simulate joints, such as a plane for smoothing wood surfaces prior to joining them, or a hand tool for inscribing grooves in fresh cement. **2.** A file for making sawteeth the same height. **3.** An attachment to a plow that covers discarded material. **4.** A worker who makes joints, particularly a construction worker who cuts stone to proper fit.

joist [CIV ENG] A steel or wood beam providing direct support for a floor.

jojoba [AGR] *Simmondsia californica.* A shrub adapted to the arid portions of the southwestern United States and Mexico; liquid extracted from the seed may be used as a substitute for sperm oil.

Joppeicidae [INV ZOO] A monospecific family of hemipteran insects included in the Pentatomorpha; found in the Mediterranean regions.

Jordan algebra [MATH] A nonassociative algebra over a field in which the products satisfy the Jordan identity $(xy)x^2 = x(yx^2)$.

Jordan curve [MATH] A simple closed curve in the plane, that is, a curve that is closed, connected, and does not cross itself.

jordanite [MINERAL] $(Pb,Tl)_{13}As_7S_{23}$ A lead-gray mineral composed of lead arsenic sulfide, occurring as monoclinic crystals.

joseite [MINERAL] $Bi_3Te(Si,S)$ A mineral composed of telluride of bismuth containing sulfur and selenium.

josen *See* hartite.

josephinite [MINERAL] A mineral consisting of an alloy of iron and nickel; occurs naturally in stream gravel.

Josephson effect [CRYO] The tunneling of electron pairs through a thin insulating barrier between two superconducting materials. Also known as Josephson tunneling.

Josephson junction [CRYO] A thin insulator separating two superconducting materials; it displays the Josephson effect.

Josephson tunneling *See* Josephson effect.

joule [MECH] The unit of energy or work in the meter-kilogram-second system of units, equal to the work done by a force of 1 newton magnitude when the point at which the force is applied is displaced 1 meter in the direction of the force. Symbolized J. Also known as newton-meter of energy.

Joule cycle *See* Brayton cycle.

Joule effect [PHYS] **1.** The heating effect produced by the flow of current through a resistance. **2.** A change in the length of a ferromagnetic substance which occurs parallel to an applied magnetic field. Also known as Joule magnetostriction; longitudinal magnetostriction.

Joule equivalent [THERMO] The numerical relation between quantities of mechanical energy and heat; the present accepted value is 1 fifteen-degrees calorie equals 4.1855 ± 0.0005 joules. Also known as mechanical equivalent of heat.

Joule heat [ELEC] The heat which is evolved when current flows through a medium having electrical resistance, as given by Joule's law.

Joule magnetostriction *See* Joule effect.

Joule's law [ELEC] The law that when electricity flows through a substance, the rate of evolution of heat in watts equals the resistance of the substance in ohms times the square of the current in amperes. [THERMO] The law that at constant temperature the internal energy of a gas tends to a finite limit, independent of volume, as the pressure tends to zero.

journal [MECH ENG] That part of a shaft or crank which is supported by and turns in a bearing.

journal bearing [MECH ENG] A cylindrical bearing which supports a rotating cylindrical shaft.

Jovian planet [ASTRON] Any of the four major planets (Jupiter, Saturn, Uranus, and Neptune) that are at a greater distance from the sun than the terrestrial planets (Mercury, Venus, Earth, and Mars).

joystick [AERO ENG] A lever used to control the motion of an aircraft; fore-and-aft motion operates the elevators while lateral motion operates the ailerons. [ENG] A two-axis displacement control operated by a lever or ball, for XY positioning of a device or an electron beam.

J particle [PARTIC PHYS] A neutral meson which has a mass of 3095 megaelectronvolts, spin quantum number 1, and negative parity and charge parity; it has an anomalously long lifetime of approximately 10^{-20} second (corresponding to a width of approximately 70 kiloelectronvolts). Also known as psi particle (symbolized ψ).

J scan *See* J scope.

J scope [ELECTR] A modification of an A scope in which the trace appears as a circular range scale near the circumference of the cathode-ray tube face, the signal appearing as a radial deflection of the range scale; no bearing indication is given. Also known as J indicator; J scan.

Jugatae [INV ZOO] The equivalent name for Homoneura.

Juglandaceae [BOT] A family of dicotyledonous plants in the order Juglandales having unisexual flowers, a solitary basal ovule in a unilocular inferior ovary, and pinnately compound, exstipulate leaves.

Juglandales [BOT] An order of dicotyledonous plants in the subclass Hamamelidae distinguished by compound leaves; includes hickory, walnut, and butternut.

jugular [ANAT] Pertaining to the region of the neck above the clavicle.

jugular vein [ANAT] The vein in the neck which drains the brain, face, and neck into the innominate.

Julian calendar [ASTRON] A calendar replaced by the Gregorian calendar; the Julian year was 365.25 days, the fraction allowed for the extra day every fourth year (leap year). There are 12 months, each 30 or 31 days except for February which has 28 days or in leap year 29.

julienite [MINERAL] $Na_2Co(SCN)_4 \cdot 8H_2O$ A blue, tetragonal mineral consisting of a hydrated sodium cobalt thiocyanate.

jumbo *See* drill carriage.

jump [COMPUT SCI] A transfer of control which terminates one sequence of instructions and begins another sequence at a different location. Also known as branch; transfer.

jumper [ELEC] A short length of conductor used to make a connection between two points or terminals in a circuit or to provide a path around a break in a circuit.

jump fire [FOR] A fire carried ahead of a forest fire by wind-borne burning material.

jump function [MATH] A function used to represent a sampled data sequence arising in the numerical study of linear difference equations.

jumping trace routine [COMPUT SCI] A trace routine which is primarily concerned with providing a record of jump instructions in order to show the sequence of program steps that the computer followed.

jump phenomenon [CONT SYS] A phenomenon occurring in a nonlinear system subjected to a sinusoidal input at constant frequency, in which the value of the amplitude of the forced oscillation can jump upward or downward as the input amplitude is varied through either of two fixed values, and the graph of the forced amplitude versus the input amplitude follows a hysteresis loop.

Juncaceae [BOT] A family of monocotyledonous plants in the order Juncales characterized by an inflorescence of diverse sorts, vascular bundles with abaxial phloem, and cells without silica bodies.

Juncales [BOT] An order of monocotyledonous plants in the subclass Commelinidae marked by reduced flowers and capsular fruits with one too many anatropous ovules per carpel.

junction [CIV ENG] A point of intersection of roads or highways, especially where one terminates. [ELEC] *See* major node. [ELECTR] A region of transition between two different semiconducting regions in a semiconductor device, such as a *pn* junction, or between a metal and a semiconductor. [ELECTROMAG] A fitting used to join a branch waveguide at an angle to a main waveguide, as in a tee junction. Also known as waveguide junction.

junction box [ENG] A protective enclosure into which wires or cables are led and connected to form joints. Also known as J box.

junction capacitor [ELECTR] An integrated-circuit capacitor that uses the capacitance of a reverse-biased *pn* junction.

junction detector [NUCLEO] A reverse-biased semiconductor junction functioning as a solid ionization chamber to produce an electric output pulse whose amplitude is linearly proportional to the energy deposited in the junction depletion layer by the incident ionizing radiation.

junction diode [ELECTR] A semiconductor diode in which the rectifying characteristics occur at an alloy, diffused, electrochemical, or grown junction between n-type and p-type semiconductor materials. Also known as junction rectifier.

junction field-effect transistor [ELECTR] A field-effect transistor in which there is normally a channel of relatively low-conductivity semiconductor joining the source and drain, and this channel is reduced and eventually cut off by junction depletion regions, reducing the conductivity, when a voltage is applied between the gate electrodes. Abbreviated JFET. Also known as depletion-mode field-effect transistor.

junction loss [COMMUN] In telephone circuits, that part of the repetition equivalent assignable to interaction effects arising at trunk terminals.

junction point *See* branch point.

junction rectifier *See* junction diode.

junction transistor [ELECTR] A transistor in which emitter and collector barriers are formed between semiconductor regions of opposite conductivity type.

June solstice [ASTRON] Summer solstice in the Northern Hemisphere.

Jungermanniales [BOT] The leafy liverworts, an order of bryophytes in the class Marchantiatae characterized by chlorophyll-containing, ribbonlike or leaflike bodies and an undifferentiated thallus.

Jungian psychology *See* analytic psychology.

jungle [ECOL] An impenetrable thicket of second-growth vegetation replacing tropical rain forest that has been disturbed; lower growth layers are dense.

Jupiter [ASTRON] The largest planet in the solar system, and the fifth in order of distance from the sun; semimajor axis = 485×10^6 miles (780×10^6 kilometers); sidereal revolution period = 11.86 years; mean orbital velocity = 8.2 miles per second (13.2 kilometers per second); inclination of orbital plane to ecliptic = 1.03; equatorial diameter = 88,700 miles (142,700 kilometers); polar diameter = 82,800 miles (133,300 kilometers); mass = about 318.4 (earth = 1).

Jupiter I *See* Io.

Jupiter II *See* Europa.

Jupiter III *See* Ganymede.

Jupiter IV *See* Callisto.

Jupiter V *See* Amalthea.

Jupiter VI *See* Himalia.

Jupiter VII *See* Elara.

Jupiter VIII *See* Pasiphae.

Jupiter IX *See* Sinope.

Jupiter X *See* Lysithea.

Jupiter XI *See* Carme.

Jupiter XII *See* Ananke.

Jupiter XIII *See* Leda.

Jura *See* Jurassic.

Jurassic [GEOL] Also known as Jura. **1.** The second period of the Mesozoic era of geologic time. **2.** The corresponding system of rocks.

jurupaite [MINERAL] $(Ca,Mg)_2(Si_2O_5)(OH)_2$ A mineral composed of hydrous calcium magnesium silicate.

jury rudder [NAV ARCH] A temporary device used to steer a boat when the rudder is out of commission.

justification [GRAPHICS] In type composition, the adjustment of spacing in each line of type so that all lines are filled out to the same desired length.

justify [COMPUT SCI] To shift data so that they assume a particular position relative to one or more reference points, lines, or marks in a storage medium.

just-noticeable difference [PSYCH] A subjective scale used in psychophysical tests, defined by C. Fechner as the subjective unit, and the absolute threshold as the zero point of the subjective scale; for example, the subjective intensity of a particular brightness of light would be specified when it was given as 100 just-noticeable differences above the absolute threshold. Abbreviated JND. Also known as difference limen; difference threshold.

just ton *See* ton.

just tuning [ACOUS] A tuning system generated by octave rearrangements of the notes of three consecutive triads, each having the frequency ratio 4:5:6, with the highest note of one triad serving as the lowest note of the next.

jute [BOT] Either of two Asiatic species of tall, slender, half-shrubby annual plants, *Corchorus capsularis* and *C. olitorius*, in the family Malvaceae, useful for their fiber.

juvenile hormone *See* neotenin.

juvenile-onset diabetes [MED] A form of diabetes mellitus which develops early in life and presents much more severe symptoms than the more common maturity-onset diabetes.

Jynginae [VERT ZOO] The wrynecks, a family of Old World birds in the order Piciformes; a subfamily of the Picidae in some systems of classification.

K

k *See* kilo-.

K *See* cathode; potassium.

K-A decay [NUC PHYS] Radioactive decay of potassium-40 (K^{40}) to argon-40 (A^{40}), as the nucleus of potassium captures an orbital electron and then decays to argon-40; the ratio of K^{40} to A^{40} is used to determine the age of rock (K-A age).

kainite [MINERAL] $MgSO_4 \cdot KCl \cdot 3H_2O$ A white, gray, pink, or black monoclinic mineral, occurring in irregular granular masses; used as a fertilizer and as a source of potassium and magnesium compounds.

kainosite [MINERAL] $Ca_2(Ce,Y)_2(SiO_4)_3CO_3 \cdot H_2O$ A yellowish-brown mineral composed of a hydrous silicate and carbonate of calcium, cerium, and yttrium.

kaki [BOT] *Diospyros kaki.* The Japanese persimmon; it provides a type of ebony wood that is black with gray, yellow, and brown streaks, has a close, even grain, and is very hard.

kala azar [MED] Visceral leishmaniasis due to the protozoan *Leishmania donovani*, transmitted by certain sandflies (*Phlebotomus*); characterized by chronic, irregular fever, enlargement of the spleen and liver, emaciation, anemia, and leukopenia.

kale [BOT] Either of two biennial crucifers, *Brassica oleracea* var. *acephala* and *B. fimbriata*, in the order Capparales, grown for the nutritious curled green leaves.

kaleidoscope [OPTICS] An optical toy consisting of a tube containing two plane mirrors placed at an angle of 60° and mounted so that a symmetrical pattern produced by multiple reflection is observed through a peephole at one end when objects (such as pieces of colored glass) at the other end are suitably illuminated.

kaliborite [MINERAL] $HKMg_2B_{12}O_{21} \cdot 9H_2O$ A colorless to white mineral composed of a hydrous borate of potassium and magnesium. Also known as paternoite.

kalicinite [MINERAL] $KHCO_3$ A colorless to white or yellowish, monoclinic mineral consisting of potassium bicarbonate; occurs in crystalline aggregates.

kalinite [MINERAL] $KAl(SO_4)_2 \cdot 11H_2O$ A birefringent mineral of the alum group composed of a hydrous sulfate of potassium and aluminum, occurring in fibrous form. Also known as potash alum.

kaliophilite [MINERAL] $KAlSiO_4$ A rare hexagonal tectosilicate mineral found in volcanic rocks; high in potassium and low in silica, it is dimorphous with kalsilite. Also known as facellite; phacellite.

kalium *See* potassium.

kalkowskite [MINERAL] $Fe_2Ti_3O_9$ A rare, brownish or black mineral composed of an oxide of iron and titanium, usually with small amounts of rare-earth elements, niobium, and tantalum.

Kalman filter [CONT SYS] A linear system in which the mean squared error between the desired output and the actual output is minimized when the input is a random signal generated by white noise.

Kalotermitidae [INV ZOO] A family of relatively primitive, lower termites in the order Isoptera.

kalsilite [MINERAL] $KAlSiO_4$ A rare mineral from volcanic rocks in southwestern Uganda; the crystal system is hexagonal; kalsilite is dimorphous with kaliophilite and sometimes contains sodium.

kamacite [MINERAL] A mineral composed of a nickel-iron alloy and comprising with taenite the bulk of most iron meteorites.

kame [GEOL] A low, long, steep-sided mound of glacial drift, commonly stratified sand and gravel, deposited as an alluvial fan or delta at the terminal margin of a melting glacier.

kame terrace [GEOL] A terracelike ridge deposited along the margins of glaciers by meltwater streams flowing adjacent to the valley walls.

Kamptozoa [INV ZOO] An equivalent name for Entoprocta.

kangaroo [VERT ZOO] Any of various Australian marsupials in the family Macropodidae generally characterized by a long, thick tail that is used as a balancing organ, short forelimbs, and enlarged hindlegs adapted for jumping.

kaolin [MINERAL] Any of a group of clay minerals, including kaolinite, nacrite, dickite, and anauxite, with a two-layer crystal in which silicon-oxygen and aluminum-hydroxyl sheets alternate; approximate composition is $Al_2O_3 \cdot 2SiO_2 \cdot 2H_2O$. [PETR] A soft, nonplastic white rock composed principally of kaolin minerals. Also known as bolus alba; white clay.

kaolinite [MINERAL] $Al_2Si_2O_5(OH)_4$ The principal mineral of the kaolin group of clay minerals; a white, gray, or yellowish high-alumina mineral consisting of sheets of tetrahedrally coordinated silicon linked by an oxygen shared with octahedrally coordinated aluminum.

kaon *See* K meson.

kapok tree [BOT] *Ceiba pentandra.* A tree of the family Bombacaceae which produces pods containing seeds covered with silk cotton. Also known as silk cotton tree.

kappa particle [CYTOL] A self-duplicating nucleoprotein particle found in various strains of *Paramecium* and thought to function as an infectious agent; classed as an intracellular symbiont, occupying a position between the viruses and the bacteria and organelles.

karat [MET] A unit for designating the fineness of gold in an alloy; represents a twenty-fourth part; thus, 18-karat gold is 18/24 or 75% pure.

Kármán vortex street [FL MECH] A double row of line vortices in a fluid which, under certain conditions, is shed in the wake of cylindrical bodies when the relative fluid velocity is perpendicular to the axis of the cylinder.

Karnian *See* Carnian.

karst [GEOL] A topography formed over limestone, dolomite, or gypsum and characterized by sinkholes, caves, and underground drainage.

Karumiidae [INV ZOO] The termitelike beetles, a small family of coleopteran insects in the superfamily Cantharoidea distinguished by having a tenth tergum.

karyotype [CYTOL] The normal diploid or haploid complement of chromosomes, with respect to size, form, and number, characteristic of an individual, species, genus, or other grouping.

kasolite [MINERAL] $Pb(UO_2)SiO_4 \cdot H_2O$ Yellow-ocher mineral composed of a hydrous lead uranium silicate, occurring in monoclinic crystals.

katamorphism [GEOL] A process of metamorphism occurring at or near the earth's surface and resulting in production of simpler minerals from complex minerals through oxidation, hydration, solution, and other processes. Also spelled catamorphism.

Kata thermometer [ENG] An alcohol thermometer used to measure low velocities in air circulation, by heating the large bulb of the thermometer above 100°F (38°C) and noting the time it takes to cool from 100 to 95°F (38 to 35°C) or some other interval above ambient temperature, the time interval being a measure of the air current at that location.

Kathlaniidae [INV ZOO] A family of nematodes assigned to the Ascaridina by some authorities and to the Oxyurina by others.

katoptric system [OPTICS] An optical system such that, when the object is displaced in a direction parallel to the axis, the image is displaced in the opposite direction (in contrast to a dioptric system). Also known as contracurrent system.

kay *See* key.

kayser [SPECT] A unit of reciprocal length, especially wave number, equal to the reciprocal of 1 centimeter. Also known as rydberg.

Kazanian [GEOL] A European stage of geologic time: Upper Permian (above Kungurian, below Tatarian).

K band [COMMUN] A band of radio frequencies extending from 10,900 to 36,000 megahertz, corresponding to wavelengths of 2.75 to 0.834 centimeters. [SOLID STATE] An optical absorption band which appears together with an F-band and has a lower intensity and shorter wavelength than the latter.

K capture [NUC PHYS] A type of beta interaction in which a nucleus captures an electron from the K shell of atomic electrons (the shell nearest the nucleus) and emits a neutrino.

K display [ELECTR] A modified radarscope A display in which a target appears as a pair of vertical deflections instead of as a single deflection; when the radar antenna is correctly pointed at the target in azimuth, the deflections are of equal height; when the antenna is not correctly pointed, the difference in pulse heights is an indication of direction and magnitude of azimuth pointing error.

keel [NAV ARCH] A steel beam or timber, or a series of steel beams and plates or timbers joined together, extending along the center of the bottom of a ship from stem to stern and often projecting below the bottom, to which the frames and hull plating are attached. [VERT ZOO] The median ridge on the breastbone in certain birds. Also known as carina.

Keel *See* Carina.

keel block [CIV ENG] docking block used to support a ship's keel. [MET] A simple shape from which a test casting is made in the form of a large head, which is removed and discarded, with a keel on the bottom.

keelson [NAV ARCH] A structure of timbers or steel beams which are bolted to the top of a keel to increase its strength. Also spelled kelson.

keep-alive electrode *See* ignitor.

keeper [ELECTROMAG] A bar of iron or steel placed across the poles of a permanent magnet to complete the magnetic circuit when the magnet is not in use, to avoid the self-demagnetizing effect of leakage lines. Also known as magnet keeper.

Keewatin [GEOL] A division of the Archeozoic rocks of the Canadian Shield.

kehoeite [MINERAL] An amorphous mineral composed of a basic hydrous calcium aluminum zinc phosphate, occurring in massive form.

K electron [ATOM PHYS] An electron in the K shell.

keloid [MED] A firm, elevated fibrous formation of tissue at the site of a scar.

kelson *See* keelson.

kelvin [ELEC] A name formerly given to the kilowatt-hour. Also known as thermal volt. [THERMO] A unit of absolute temperature equal to 1/273.16 of the absolute temperature of the triple point of water. Symbolized K. Formerly known as degree Kelvin.

Kelvin absolute temperature scale [THERMO] A temperature scale in which the ratio of the temperatures of two reservoirs is equal to the ratio of the amount of heat absorbed from one of them by a heat engine operating in a Carnot cycle to the amount of heat rejected by this engine to the other reservoir; the temperature of the triple point of water is defined as 273.16 K. Also known as Kelvin temperature scale.

Kelvin bridge [ELEC] A specialized version of the Wheatstone bridge network designed to eliminate, or greatly reduce, the effect of lead and contact resistance, and thus permit accurate measurement of low resistance. Also known as double bridge; Kelvin network; Thomson bridge.

Kelvin equation [THERMO] An equation giving the increase in vapor pressure of a substance which accompanies an increase in curvature of its surface; the equation describes the greater rate of evaporation of a small liquid droplet as compared to that of a larger one, and the greater solubility of small solid particles as compared to that of larger particles.

Kelvin-Helmholtz contraction [ASTROPHYS] A contraction of a star once it is formed and before it is hot enough to ignite its hydrogen; the contraction converts gravitational potential energy into heat, some of which is radiated, with the remainder used to raise the internal temperature of the star.

Kelvin network *See* Kelvin bridge.

Kelvin relations *See* Thomson relations.

Kelvin scale [THERMO] The basic scale used for temperature definition; the triple point of water (comprising ice, liquid, and vapor) is defined as 273.16 K; given two reservoirs, a reversible heat engine is built operating in a cycle between them, and the ratio of their temperatures is defined to be equal to the ratio of the heats transferred.

Kelvin's circulation theorem [FL MECH] The theorem that, if the external forces acting on an inviscid fluid are conservative and if the fluid density is a function of the pressure only, then the circulation along a closed curve which moves with the fluid does not change with time.

Kelvin skin effect   *See* skin effect.

Kelvin temperature scale   [THERMO]   **1.** An International Practical Temperature Scale which agrees with the Kelvin absolute temperature scale within the limits of experimental determination. **2.** *See* Kelvin absolute temperature scale.

kelyphite   *See* corona.

kempite   [MINERAL] $Mn_2(OH)_3Cl$ An emerald-green orthorhombic mineral composed of a basic manganese oxychloride, occurring in small crystals.

Kendall effect   [COMMUN]   A spurious pattern or other distortion in a facsimile record caused by unwanted modulation products arising from the transmission of a carrier signal; occurs principally when the width of one side band is greater than half the facsimile carrier frequency.

Kennelly-Heaviside layer   *See* E layer.

kentrolite   [MINERAL] $Pb_2Mn_2Si_2O_9$ A dark reddish-brown mineral composed of a lead manganese silicate.

Kentucky coffee tree   [BOT]   *Gymnocladus dioica.* An extremely tall, dioecious tree of the order Rosales readily recognized when in fruit by its leguminous pods containing heavy seeds, once used as a coffee substitute.

Kenyapithecus   [PALEON]   An early member of Hominidae from the Miocene.

Keplerian motion   [ASTRON]   Orbital movement of a body about another that is not disturbed by the presence of a third celestial body.

Keplerian telescope   [OPTICS]   A telescope that forms a real intermediate image in the focal plane and can be used for introducing a reticle or a scale into the focal plane.

Kepler's equations   [ASTRON]   The mathematical relationship between two different systems of angular measurements of the position of a body in an ellipse.

Kepler's laws   [ASTRON]   Three laws, determined by Johannes Kepler, that describe the motions of planets in their orbits: the orbits of the planets are ellipses with the sun at a common focus; the line joining a planet and the sun sweeps over equal areas during equal intervals of time; the squares of the periods of revolution of any two planets are proportional to the cubes of their mean distances from the sun.

kerabitumen   *See* kerogen.

keratin   [BIOCHEM]   Any of various albuminoids characteristic of epidermal derivatives, such as nails and feathers, which are insoluble in protein solvents, have a high sulfur content, and generally contain cystine and arginine as the predominating amino acids.

keratitis   [MED]   Inflammation of the cornea.

keratoconjunctivitis   [MED]   Concurrent inflammation of the cornea and the conjunctiva. Also known as shipyard eye.

keratosis   [MED]   Any disease of the skin characterized by an overgrowth of the cornified epithelium.

kermesite   [MINERAL] Sb_2S_2O A cherry-red mineral occurring as tufts of capillary crystals, and formed from an alteration of stibnite. Also known as antimony blende; purple blende; pyrostibite; red antimony.

Kern counter   *See* dust counter.

kernel   [ATOM PHYS]   An atom that has been stripped of its valence electrons, or a positively charged nucleus lacking the outermost orbital electrons. [BOT] **1.** The inner portion of a seed. **2.** A whole grain or seed of a cereal plant, such as corn or barley. [MATH] *See* null space.

kernel blight   [PL PATH]   Any of several fungus diseases of barley caused chiefly by *Gibberella zeae,* *Helminthosporium sativum,* and *Alternaria* species shriveling and discoloring the grain.

kernel of a homomorphism   [MATH]   For a homomorphism h from a group G to a group H, this consists of all elements of G which h sends to the identity element of H.

kernel of a linear transformation   [MATH]   All those vectors which a linear transformation maps to the zero vector.

kernel of an integral transform   [MATH]   The function $K(x, t)$ in the transformation which sends the function $f(x)$ to the function $\int K(x,t)f(t)dt = F(x)$.

kernel spot   [PL PATH]   A fungus disease of the pecan kernel caused by *Coniothyrium caryogenum* and characterized by dull-brown roundish spots.

Kernig's sign   [MED]   In meningeal irritation, with the patient lying face up and the thigh flexed at the hip, the pain and spasm of the hamstring muscles when an attempt is made to completely extend the leg at the knee.

kernite   [MINERAL] $Na_2B_4O_7 \cdot 4H_2O$ A colorless to white hydrous borate mineral crystallizing in the monoclinic system and having vitreous luster; an important source of boron. Also known as rasorite.

kerogen   [GEOL]   The complex, fossilized organic material present in sedimentary rocks, especially in shales; converted to petroleum products by distillation. Also known as kerabitumen; petrologen.

kerogen shale   *See* oil shale.

kerosine   [MATER]   A refined petroleum fraction used as a fuel for heating and cooking, jet engines, lamps, and weed burning and as a base for insecticides; specific gravity is about 0.8; components are mostly paraffinic and naphthenic hydrocarbons in the C_{10} to C_{14} range. Also known as lamp oil.

kerosine shale   *See* torbanite.

Kerr cell   [OPTICS]   A glass cell containing a dielectric liquid that exhibits the Kerr effect, such as nitrobenzene, in which is inserted the two plates of a capacitor, used to observe the Kerr effect on light passing through the cell.

Kerr effect   *See* electrooptical Kerr effect.

Kerr magnetooptical effect   *See* magnetooptic Kerr effect.

ketal   [ORG CHEM]   **1.** Former term for the $=CO$ group, as in dimethyl ketal (acetone). **2.** Any of the ketone acetates from condensation of alkyl orthoformates with ketones in the presence of alcohols.

keto-   [ORG CHEM]   Organic chemical prefix for the keto or carbonyl group, C:O, as in a ketone.

ketoacidosis   *See* ketosis.

ketone   [ORG CHEM]   One of a class of chemical compounds of the general formula RR'CO, where R and R' are alkyl, aryl, or heterocyclic radicals; the groups R and R' may be the same or different, or incorporated into a ring; the ketones, acetone, and methyl ethyl ketone are used as solvents, and ketones in general are important intermediates in the synthesis of organic compounds.

ketone body   [BIOCHEM]   Any of various ketones which increase in blood and urine in certain conditions, such as diabetic acidosis, starvation, and pregnancy. Also known as acetone body.

ketonemia   *See* acetonemia.

ketosis   [MED]   Excess amounts of ketones in the body, especially associated with diabetes mellitus. Also known as ketoacidosis.

kettle   [GEOL]   **1.** A bowl-shaped depression with steep sides in glacial drift deposits that is formed by the melting of glacier ice left behind by the retreating glacier and buried in the drift. Also known as kettle basin; kettle hole. **2.** *See* pothole.

kettle basin *See* kettle.
kettle hole *See* kettle.
Keuper [GEOL] A European stage of geologic time, especially in Germany; Upper Triassic.
Keweenawan [GEOL] The younger of two Precambrian time systems that constitute the Proterozoic period in Michigan and Wisconsin.
key [BUILD] 1. Plastering that is forced between laths to secure the rest of the plaster in place. 2. The roughening on a surface to be glued or plastered to increase adhesiveness. [CIV ENG] A projecting portion that serves to prevent movement of parts at a construction joint. [COMPUT SCI] A data item that serves to uniquely identify a data record. [DES ENG] 1. An instrument that is inserted into a lock to operate the bolt. 2. A device used to move in some manner in order to secure or tighten. 3. One of the levers of a keyboard. 4. *See* machine key. [ELEC] 1. A hand-operated switch used for transmitting code signals. Also known as signaling key. 2. A special lever-type switch used for opening or closing a circuit only as long as the handle is depressed. Also known as switching key. [ENG] The pieces of core causing a block in a core barrel, the removal of which allows the rest of the core in the barrel to slide out. [GEOL] A cay, especially one of the islets off the south of Florida. Also spelled kay. [PETRO ENG] A hooklike wrench fitted to the square of a sucker rod to pull and run each sucker rod of a pumping oil well. [SYST] An arrangement of the distinguishing features of a taxonomic group to serve as a guide for establishing relationships and names of unidentified members of the group.
key bed [GEOL] Also known as index bed; key horizon; marker bed. 1. A stratum or body of strata that has distinctive characteristics so that it can be easily identified. 2. A bed whose top or bottom is employed as a datum in the drawing of structure contour maps.
keyboard [ENG] A set of keys or control levers having a systematic arrangement and used to operate a machine or other piece of equipment such as a typewriter, typesetter, processing unit of a computer, or piano.
keyboard entry [COMPUT SCI] A piece of information fed manually into a computing system by means of a set of keys, such as a typewriter.
keyboard inquiry [COMPUT SCI] A question asked a computer concerning the status of a program being run, or concerning the value achieved by a specific variable, by means of a console typewriter.
keyboard lockout [COMPUT SCI] An arrangement for preventing transmission from a particular keyboard while other transmissions are taking place on the same circuit.
keyboard printer [COMPUT SCI] A computer input device that includes a keyboard and a printer that prints the keyed-in data and often also prints computer output information.
key change [COMPUT SCI] The occurrence, in a file of records which have been sorted according to their keys and are being read into a computer, of a record whose key differs from that of its immediate predecessor.
key click [COMMUN] Transient signal sometimes produced when a radiotelegraph sending key is opened or closed; it is heard in a loudspeaker or headphone as a click or chirp.
key-disk machine [COMPUT SCI] A keyboard machine used to record data directly on a magnetic disk.

key-driven calculator [COMPUT SCI] A mechanical desk calculator in which numeric selection keys, arranged in columns, are coupled directly to accumulator dials on which the results are visible.
keyed clamp [ELECTR] Clamping circuit in which the time of clamping is determined by a control signal.
keyer [ELECTR] Device which changes the output of a transmitter from one condition to another according to the intelligence to be transmitted.
keyhole [DES ENG] A hole or a slot for receiving a key. [MET] A welding method wherein the heat source, because of its concentration, causes a hole through the surface immediately ahead of the molten weld metal in the direction of welding; the hole is filled as the welding progresses, ensuring complete joint penetration. [ORD] Of a bullet, to strike a target after tumbling in flight so that the long axis of the bullet and the line of flight are not the same; usually caused by failure of the bullet to receive sufficient spin from the rifling in the barrel.
key horizon *See* key bed.
keying [CIV ENG] Establishing a mechanical bond in a construction joint. [ELEC] The forming of signals, such as for telegraph transmission, by modulating a direct-current or other carrier between discrete values of some characteristic.
keying frequency [COMMUN] In facsimile, the maximum number of times a second that a black-line signal occurs when scanning the subject copy.
keying interval [COMMUN] In a periodically keyed transmission system, one of the set of intervals starting from a change in state and equal in length to the shortest time between changes of state.
keylock switch [ELEC] A switch that can be operated only by inserting and turning a key such as that used in ordinary locks.
key punch [COMPUT SCI] A keyboard-actuated device that punches holes in a card; it may be a hand-feed punch or an automatic feed punch.
key seat *See* keyway.
keystone [ARCH] Wedge-shaped stone at the crown of an arch. [MATER] Small crushed stone used as filler for the large aggregate in bituminous bound roads.
keyswitch [COMPUT SCI] A switch that is operated by depressing a key on the keyboard of a data entry terminal.
key telephone system [COMMUN] A telephone system consisting of phones with several keys, connecting cables, and relay switching apparatus, which does not need a special operator to handle incoming or outgoing calls and which generally permits users to select one of several possible lines and to hold calls.
key-to-disk system [COMPUT SCI] A data-entry system in which information entered on several keyboards is collected on different sections of a magnetic disk, and the data are extracted from the disk when complete, and are copied onto a magnetic tape or another disk for further processing on the main computer.
key-to-tape system [COMPUT SCI] A data-entry system consisting of several keyboards connected to a central controlling unit, typically a minicomputer, which collects information from each keyboard and then directs it to a magnetic tape.
key verify [COMPUT SCI] The use of a key punch verifier to ascertain that the data punched on a card corresponds to the data of the original document.

keyway [DES ENG] 1. An opening in a lock for passage of a flat metal key. 2. The pocket in the driven element to provide a driving surface for the key. 3. A groove or channel for a key in any mechanical part. Also known as key seat. [ENG] An interlocking channel or groove in a cement or wood joint to provide reinforcement.

key-word-in-context index [COMPUT SCI] A computer-generated listing of titles of documents, produced on a line printer, with the key words lined up vertically in a fixed position within the title and arranged in alphabetical order. Abbreviated KWIC index.

K feldspar *See* potassium feldspar.

kgf-m *See* meter-kilogram.

khibinite *See* mosandrite.

kickdown [MECH ENG] 1. Shifting to lower gear in an automotive vehicle. 2. The device for shifting.

kick over [MECH ENG] To start firing; applied to internal combustion engines.

kickplate [BUILD] A plate used on the bottom of doors and cabinets or on the risers of steps to protect them from shoe marks.

Kick's law [ENG] The law that the energy needed to crush a solid material to a specified fraction of its original size is the same, regardless of the original size of the feed material.

kidney [ANAT] Either of a pair of organs involved with the elimination of water and waste products from the body of vertebrates; in humans they are bean-shaped, about 5 inches (12.7 centimeters) long, and are located in the posterior part of the abdomen behind the peritoneum.

kidney ore [MINERAL] A form of hematite found in compact masses, concretions, or nodules that are kidney-shaped.

kidney stone *See* nephrite.

kieserite [MINERAL] $MgSO_4 \cdot H_2O$ A white mineral that crystallizes in the monoclinic system, is composed of hydrous magnesium sulfate, and occurs in saline residues.

Kilkenny coal *See* anthracite.

killed steel [MET] Thoroughly deoxidized steel, for example, by additions of aluminum or silicon, in which the reaction between carbon and oxygen during solidification is suppressed.

killed vaccine [IMMUNOL] A suspension of killed microorganisms used as antigens to produce immunity.

killer circuit [ELECTR] Vacuum tube or tubes and associated circuits in which are generated the blanking pulses used to temporarily disable a radar set.

killer whale [VERT ZOO] *Orcinus orca.* A predatory, cosmopolitan cetacean mammal, about 30 feet (9 meters) long, found only in cold waters.

kiln [ENG] A heated enclosure used for drying, burning, or firing materials such as ore or ceramics.

kilo- [SCI TECH] A prefix representing 10^3 or 1000. Abbreviated k.

kilogram-meter *See* meter-kilogram.

kilomega- *See* giga-.

kimberlite [PETR] A form of mica periodite that is formed mainly of phenocrysts, olivine, phlogopite, and subordinate melilite with minor amounts of pyroxene, apatite, perovskite, and opaque oxides.

Kimmeridgian [GEOL] A European stage of geologic time; middle Upper Jurassic, above Oxfordian, below Portlandian.

kinase [BIOCHEM] Any enzyme that catalyzes phosphorylation reactions.

Kinderhookian [GEOL] Lower Mississippian geologic time, above the Chautauquan of Devonian, below Osagian.

K indicator *See* K scope.

kinematics [MECH] The study of the motion of a system of material particles without reference to the forces which act on the system.

kinescope *See* picture tube.

kinescope recording [COMMUN] A motion picture film made by photographing images on the face of the picture tube in a television monitor or receiver, to permit repeating the same television program later and at different stations. Also known as television recording.

kinesis [PHYSIO] The general term for physical movement, including that induced by stimulation, for example, light.

kinesthesis [PHYSIO] The system of sensitivity present in the muscles and their attachments.

kinetic energy [MECH] The energy which a body possesses because of its motion; in classical mechanics, equal to one-half of the body's mass times the square of its speed.

kinetic equilibrium *See* dynamic equilibrium.

kinetic potential *See* Lagrangian.

kinetic theory [STAT MECH] A theory which attempts to explain the behavior of physical systems on the assumption that they are composed of large numbers of atoms or molecules in vigorous motion; it is further assumed that energy and momentum are conserved in collisions of these particles, and that statistical methods can be applied to deduce the particles' average behavior. Also known as molecular theory.

kinetochore [CYTOL] Within the centromere, the granule upon which the spindle fibers attach.

kinetoplast [CYTOL] A genetically autonomous, membrane-bound organelle associated with the basal body at the base of flagella in certain flagellates, such as the trypanosomes. Also known as parabasal body.

Kinetoplastida [INV ZOO] An order of colorless protozoans in the class Zoomastigophorea having pliable bodies and possessing one or two flagella in some stage of their life.

kinetosome *See* basal body.

kingdom [SYST] One of the primary divisions that include all living organisms: most authorities recognize two, the animal kingdom and the plant kingdom, while others recognize three or more, such as Protista, Plantae, Animalia, and Mychota.

kingfisher [VERT ZOO] The common name for members of the avian family Alcedinidae; most are tropical Old World species characterized by short legs, long bills, bright plumage, and short wings.

king post [NAV ARCH] A short, strong post for supporting a cargo boom on a cargo ship. Also known as derrick post; samson post.

king post truss [BUILD] A wooden roof truss having two principal rafters held by a horizontal tie beam, a king post upright between tie beam and ridge, and usually two struts to the rafters from a thickening at the king post foot.

kinin [PHARM] Any of several pharmacologically active polypeptides that act as hypotensives, contracting isolated smooth muscles and increasing capillary permeability; an example is bradykinin.

kink instability [PL PHYS] A type of hydromagnetic instability in which the ionized gas and its magnetic confining field tend to form a loop or kink, which then grows steadily larger. Also known as sausage instability.

kinocilium [CYTOL] A type of cilium containing one central pair of microfibrils and nine peripheral pairs; they extend from the apex of hair cells in all vertebrate ears except mammals.

kinomere *See* centromere.

Kinorhyncha [INV ZOO] A class of the phylum Aschelminthes consisting of superficially segmented microscopic marine animals lacking external ciliation.

Kinosternidae [VERT ZOO] The mud and musk turtles, a family of chelonian reptiles in the suborder Cryptodira found in North, Central, and South America.

kintal *See* hundredweight.

kinzigite [PETR] A coarse-grained metamorphic rock that is formed principally of garnet and biotite, with K feldspar, quartz, mica, cordierite, and sillimanite.

Kirchhoff's current law [ELEC] The law that at any given instant the sum of the instantaneous values of all the currents flowing toward a point is equal to the sum of instantaneous values of all the currents flowing away from the point. Also known as Kirchhoff's first law.

Kirchhoff's equations [THERMO] Equations which state that the partial derivative of the change of enthalpy (or of internal energy) during a reaction, with respect to temperature, at constant pressure (or volume) equals the change in heat capacity at constant pressure (or volume).

Kirchhoff's first law *See* Kirchhoff's current law.

Kirchhoff's law [ELEC] Either of the two fundamental laws dealing with the relation of currents at a junction and voltages around closed loops in an electric network; they are known as Kirchhoff's current law and Kirchhoff's voltage law. [THERMO] The law that the ratio of the emissivity of a heat radiator to the absorptivity of the same radiator is the same for all bodies, depending on frequency and temperature alone, and is equal to the emissivity of a blackbody. Also known as Kirchhoff's principle.

Kirchhoff's principle *See* Kirchhoff's law.

Kirchhoff's second law *See* Kirchhoff's voltage law.

Kirchhoff's voltage law [ELEC] The law that at each instant of time the algebraic sum of the voltage rises around a closed loop in a network is equal to the algebraic sum of the voltage drops, both being taken in the same direction around the loop. Also known as Kirchhoff's second law.

Kirchhoff theory [OPTICS] A theory of diffraction of light which gives a mathematical formulation of Huygens' principle, based on the wave equation and Green's theorem, and enables quantitative determination of the amplitude and phase at any point to a very close approximation.

Kirkbyacea [PALEON] A monomorphic superfamily of extinct ostracods in the suborder Beyrichicopina, all of which are reticulate.

Kirkbyidae [PALEON] A family of extinct ostracods in the superfamily Kirkbyacea in which the pit is reduced and lies below the middle of the valve.

Kirkwood gaps [ASTRON] Regions in the main zone of asteroids where almost no asteroids are found.

kirovite [MINERAL] $(Fe,Mg)SO_4 \cdot 7H_2O$ A mineral composed of a hydrous sulfate of iron and magnesium; it is isomorphous with malanterite and pisanite.

kissing disease *See* infectious mononucleosis.

kiwi [VERT ZOO] The common name for three species of nocturnal ratites of New Zealand composing the family Apterygidae; all have small eyes, vestigial wings, a long slender bill, and short powerful legs.

Kjeldahl method [ANALY CHEM] Quantitative analysis of organic compounds to determine nitrogen content by interaction with concentrated sulfuric acid; ammonia is distilled from the NH_4SO_4 formed.

klebelsbergite [MINERAL] A mineral composed of basic antimony sulfate, occurring between crystals of stibnite.

Klein-Gordon equation [QUANT MECH] A wave equation describing a spinless particle which is consistent with the special theory of relativity. Also known as Schrödinger-Klein-Gordon equation.

kleinite [MINERAL] A yellow to orange mineral composed of a basic oxide, sulfate, and chloride of mercury and ammonium.

Klinefelter's syndrome [MED] A complex of symptoms associated with hypogonadism in males as an accompaniment of an anomaly of the sex chromosomes; somatic cells are found to have a Y chromosome and more than one X chromosome.

klinotaxis [BIOL] Positive orientation movement of a motile organism induced by a stimulus.

klint [GEOL] An exhumed coral reef or bioherm that is more resistant to the processes of erosion than the rocks that enclose it so that the core remains in relief as hills and ridges.

klintite [GEOL] The dense, hard dolomite composing a klint; gives to the core a strength and resistance to erosion.

Kloedenellacea [PALEON] A dimorphic superfamily of extinct ostracods in the suborder Kloedenellocopina having the posterior part of one dimorph longer and more inflated than the other dimorph.

Kloedenellocopina [PALEON] A suborder of extinct ostracods in the order Paleocopa characterized by a relatively straight dorsal border with a gently curved or nearly straight ventral border.

klystron [ELECTR] An evacuated electron-beam tube in which an initial velocity modulation imparted to electrons in the beam results subsequently in density modulation of the beam; used as an amplifier in the microwave region or as an oscillator.

klystron generator [ELECTR] Klystron tube used as a generator, with its second cavity or catcher directly feeding waves into a waveguide.

klystron oscillator *See* velocity-modulated oscillator.

klystron repeater [ELECTR] Klystron tube operated as an amplifier and inserted directly in a waveguide in such a way that incoming waves velocity-modulate the electron stream emitted from a heated cathode; a second cavity converts the energy of the electron clusters into waves of the original type but of greatly increased amplitude and feeds them into the outgoing guide.

K meson [PARTIC PHYS] **1.** Collective name for four pseudo-scalar mesons, having masses of about 495 MeV (megaelectronvolts) and decaying via weak interactions: K^+, K^-, K_S^0, and K_L^0; they consist of two isotopic spin doublets, the (K^+, K^0) doublet and its antiparticle doublet $(K^-, \bar{K}^0)$, having hypercharge or strangeness of $+1$ and -1 respectively, where K^0 and $\bar{K}^0$ are certain combinations of K_L^0 and K_S^0 states. Also known as kaon. **2.** Collective name for any meson resonance belonging to an isotopic doublet with hypercharge $+1$ or -1, denoted $K_{JP}(m)$ or $\bar{K}_{JP}(m)$ respectively, where m is the mass, and J and P are the spin and parity.

knee [ANAT] **1.** The articulation between the femur and the tibia in humans. Also known as genu. **2.** The corresponding articulation in the hindlimb of a quadrupedal vertebrate. [MECH ENG] In a knee-and-column type of milling machine, the part which supports the saddle and table and which can move vertically on the column. [MET] The lower supporting structure for an arm in a resistance welding machine.

kneecap *See* patella.

knee frequency *See* break frequency.

knee rafter [BUILD] A brace placed diagonally between a principal rafter and a tie beam.

knee wall [BUILD] A partition that forms a side wall or supports roof rafters under a pitched roof.

Kneriidae [VERT ZOO] A small family of tropical African fresh-water fishes in the order Gonorynchiformes.

knick *See* knickpoint.

knickpoint [GEOL] A point of sharp change of slope, especially in the longitudinal profile of a stream or of its valley. Also known as break; knick; nick; nickpoint; rejuvenation head; rock step.

knife-edge bearing [MECH ENG] A balance beam or lever arm fulcrum in the form of a hardened steel wedge; used to minimize friction.

knife-edge cam follower [DES ENG] A cam follower having a sharp narrow edge or point like that of a knife; useful in developing cam profile relationships.

knife switch [ELEC] An electric switch consisting of a metal blade hinged at one end to a stationary jaw, so that the blade can be pushed over to make contact between spring clips.

knock intensity [ENG] The intensity of knock (detonation) recorded when testing a motor gasoline for octane or knock rating.

knock-off [MECH ENG] 1. The automatic stopping of a machine when it is operating improperly. 2. The device that causes automatic stopping.

knockout [ENG] A partially cutout piece in metal or plastic that can be forced out when a hole is needed.

knock rating [ENG] Rating of gasolines according to knocking tendency.

Knoop hardness [MET] The relative microhardness of a material, such as metal, determined by the Knoop indentation test.

Knoop indentation test [MET] A diamond pyramid hardness test employing the Knoop indenter; hardness is determined by the depth to which the Knoop indenter penetrates.

Knorr synthesis [ORG CHEM] A condensation reaction carried out in either glacial acetic acid or an aqueous alkali in which an α-aminoketone combines with an α-carbonyl compound to form a pyrrole; possibly the most versatile pyrrole synthesis.

knot [COMPUT SCI] *See* deadlock. [MATER] A scar on lumber marking a place where a branch grew out of the tree truck. [MATH] In the general case, a knot consists of an embedding of an n-dimensional sphere in an $(n+2)$-dimensional sphere; classically, it is an interlaced closed curve, homeomorphic to a circle. [ORG CHEM] A chiral structure in which rings containing 50 or more members have a knotlike configuration. [PHYS] A speed unit of 1 nautical mile (1.852 kilometers) per hour, equal to approximately 0.51444 meters per second.

knoxvillite *See* copiapite.

knuckle joint [DES ENG] A hinge joint between two rods in which an eye on one piece fits between two flat projections with eyes on the other piece and is retained by a round pin.

Knudsen cell [PHYS CHEM] A vessel used to measure very low vapor pressures by measuring the mass of vapor which escapes when the vessel contains a liquid in equilibrium with its vapor.

Knudsen gage [ENG] An instrument for measuring very low pressures, which measures the force of a gas on a cold plate beside which there is an electrically heated plate.

knurl [ENG] To provide a surface, usually a metal, with small ridges or knobs to ensure a firm grip or as a decorative feature.

koala [VERT ZOO] *Phascolarctos cinereus*. An arboreal marsupial mammal of the family Phalangeridae having large hairy ears, gray fir, and two clawed toes opposing three others on each limb.

kobellite [MINERAL] $Pb_2(Bi,Sb)_2S_5$ A blackish-gray mineral composed of antimony bismuth lead sulfide.

Kochab [ASTRON] The brighter of the two stars called the Guardian of the Pole in the constellation Ursa Minor.

Koch's postulates [MICROBIO] A set of laws elucidated by Robert Koch: the microorganism identified as the etiologic agent must be present in every case of the disease; the etiologic agent must be isolated and cultivated in pure culture; the organism must produce the disease when inoculated in pure culture into susceptible animals; a microorganism must be observed in and recovered from the experimentally diseased animal. Also known as law of specificity of bacteria.

koenenite [MINERAL] $Mg_5Al_2(OH)_{12}Cl_4$ A very soft mineral composed of a basic magnesium aluminum chloride.

kohlrabi [BOT] A biennial crucifer, designated *Brassica caulorapa* and *B. oleracea* var. *caulo-rapa*, of the order Capparales grown for its edible turniplike, enlarged stem.

koktaite [MINERAL] $(NH_4)_2Ca(SO_4)_2 \cdot H_2O$ A mineral composed of a hydrous calcium ammonium sulfate.

kolbeckite [MINERAL] A blue to gray mineral composed of a hydrous beryllium aluminum calcium silicate and phosphate. Also known as sterrettite.

Kolbe hydrocarbon synthesis [ORG CHEM] The production of an alkane by the electrolysis of a water-soluble salt of a carboxylic acid.

Komodo dragon [VERT ZOO] *Varanus komodoensis*. A predatory reptile of the family Varanidae found only on the island of Komodo; it is the largest living lizard and may grow to 10 feet (3 meters).

kona [METEOROL] A stormy, rain-bringing wind from the southwest or south-southwest in Hawaii; it blows about five times a year on the southwest slopes, which are in the lee of the prevailing northeast trade winds.

Kondo effect [MET] The large anomalous increase in the resistance of certain dilute alloys of magnetic materials in nonmagnetic hosts as the temperature is lowered.

Kondo temperature [MET] The temperature below which the Kondo effect predominates for a specified magnetic impurity and host material.

kongsbergite [MINERAL] A silver-rich variety of a native amalgam composed of silver (95%) and mercury (5%).

Koonungidae [INV ZOO] A family of Australian crustaceans in the order Anaspidacea with sessile eyes and the first thoracic limb modified for digging.

Koplik's sign [MED] Small red spots surrounded by white areas seen in the mucous membrane of the mouth in the prodromal stage of measles. Also known as Koplik's spots.

Koplik's spots *See* Koplik's sign.

koppite [MINERAL] Mineral composed of a form of pyrochlore containing cerium, iron, and potassium.

kornelite [MINERAL] $Fe_2(SO_4)_3 \cdot 7H_2O$ A colorless to brown mineral composed of hydrous ferric sulfate.

Korner's method [ORG CHEM] A method for determining the absolute position of substituents for positional isomers in benzene by the experimental production of positional isomers from a given disubstituted benzene.

kornerupine [MINERAL] $(Mg,Fe,Al)_{20}(Si,B)_9O_{43}$ A colorless, yellow, brown, or sea-green mineral composed of magnesium iron borosilicate.

Korsakoff's neurosis *See* Korsakoff's syndrome.

Korsakoff's psychosis *See* Korsakoff's syndrome.

Korsakoff's syndrome [PSYCH] A form of amnesic-confabulatory syndrome characterized by confusion, loss of memory, retrograde amnesia with compensatory confabulation, and polyneuritis; seen in chronic alcoholism and other causes of vitamin B deficiency. Also known as Korsakoff's neurosis; Korsakoff's psychosis.

kotoite [MINERAL] $Mg_3(BO_3)_2$ An orthorhombic borate mineral; it is isostructural with jimboite.

Kr *See* krypton.

kraft paper [MATER] A strong paper or cardboard made from sulfate-process wood pulp; unbleached varieties are used for wrapping paper and shipping artons.

Kramers-Kronig relation [OPTICS] A relation between the real and imaginary parts of the index of refraction of a substance, based on the causality principle and Cauchy's theorem.

Krause's corpuscle [ANAT] One of the spheroid nerve-end organs resembling lamellar corpuscles, but having a more delicate capsule; found especially in the conjunctiva, the mucosa of the tongue, and the external genitalia; they are believed to be cold receptors. Also known as end bulb of Krause.

krausite [MINERAL] $KFe(SO_4)_2 \cdot H_2O$ Yellowish-green mineral composed of hydrous potassium iron sulfate.

Krebs cycle [BIOCHEM] A sequence of enzymatic reactions involving oxidation of a two-carbon acetyl unit to carbon dioxide and water to provide energy for storage in the form of high-energy phosphate bonds. Also known as citric acid cycle; tricarboxylic acid cycle.

kremersite [MINERAL] $[NH_4,K]_2FeCl_5 \cdot H_2O$ A red mineral composed of hydrous potassium ammonium iron chloride, occurring in octahedral crystals.

krennerite [MINERAL] $AuTe_2$ A silver-white to pale-yellow mineral composed of gold telluride and often containing silver. Also known as white tellurium.

kribergite [MINERAL] $Al_5(PO_4)_3(SO_4)(OH)_4 \cdot 2H_2O$ White, chalklike mineral composed of hydrous basic aluminum sulfate and phosphate.

krill [INV ZOO] A name applied to planktonic crustaceans that constitute the diet of many whales, particularly whalebone whales.

krohnkite [MINERAL] $Na_2Cu(SO_4)_2 \cdot 2H_2O$ An azure-blue monoclinic mineral composed of hydrous copper sodium sulfate, occurring in massive form.

krokidolite *See* crocidolite.

Kronig-Penney model [SOLID STATE] An idealized one-dimensional model of a crystal in which the potential energy of an electron is an infinite sequence of periodically spaced square wells.

kryolithionite [MINERAL] $Na_3Li_3(AlF_6)_2$ Variety of spodumene found in Greenland; has a crystal structure resembling that of garnet.

kryptoclimate *See* cryptoclimate.

krypton [CHEM] A colorless, inert gaseous element, symbol Kr, atomic number 36, atomic weight 83.80; it is odorless and tasteless; used to fill luminescent electric tubes.

K scan *See* K scope.

K scope [ELECTR] A modified form of A scope on which one signal appears as two pips, the relative amplitudes of which indicate the error of aiming the antenna. Also known as K indicator; K scan.

K shell [ATOM PHYS] The innermost shell of electrons surrounding the atomic nucleus, having electrons characterized by the principal quantum number 1.

K truss [BUILD] A building truss in the form of a K due to the orientation of the vertical member and two oblique members in each panel.

kudzu [BOT] Any of various perennial vine legumes of the genus *Pueraria* in the order Rosales cultivated principally as a forage crop.

Kuehneosauridae [PALEON] The gliding lizards, a family of Upper Triassic reptiles in the order Squamata including the earliest known aerial vertebrates.

kumquat [BOT] A citrus shrub or tree of the genus *Fortunella* in the order Sapindales grown for its small, flame- to orange-colored edible fruit having three to five locules filled with an acid pulp, and a sweet, pulpy rind.

Kundt effect *See* Faraday effect.

Kundt rule [SPECT] The rule that the optical absorption bands of a solution are displaced toward the red when its refractive index increases because of changes in composition or other causes.

Kungurian [GEOL] A European stage of geologic time; Middle Permian, above Artinskian, below Kazanian.

Kurie plot [NUC PHYS] Graph used in studying beta decay, in which the square root of the number of beta particles whose momenta (or energy) lie within a certain narrow range, divided by a function worked out by Fermi, is plotted against beta-particle energy; it is a straight line for allowed transitions and some forbidden transitions, in accord with the Fermi beta-decay theory. Also known as Fermi plot.

kurnakovite [MINERAL] $Mg_2B_6O_{11} \cdot 13H_2O$ A white mineral composed of hydrous magnesium borate.

Kuroshio Countercurrent [OCEANOGR] A component of the Kuroshio system flowing south and southwest between latitudes 155° and 160°E about 70 kilometers from the coast of Japan on the right-hand side of the Kuroshio Current.

Kuroshio Current [OCEANOGR] A fast ocean current (2–4 knots) flowing northeastward from Taiwan to the Ryukyu Islands and close to the coast of Japan to about 150°E. Also known as Japan Current.

Kuroshio system [OCEANOGR] A system of ocean currents which includes part of the North Equatorial Current, the Tsushima Current, the Kuroshio Current, and the Kuroshio extension.

Kurtoidei [VERT ZOO] A monogeneric suborder of perciform fishes having a unique ossification that encloses the upper part of the swim bladder, and an occipital hook in the male for holding eggs during brooding.

kurtosis [STAT] The extent to which a frequency distribution is concentrated about the mean or peaked; it is sometimes defined as the ratio of the fourth moment of the distribution to the square of the second moment.

kutnahorite [MINERAL] $Ca(Mn,Mg,Fe)(CO_3)_2$ A rare carbonate of calcium and manganese, found with some magnesium and iron substituting for manganese; forms rhombohedral crystals and is isomorphous with dolomite.

Kutorginida [PALEON] An order of extinct brachiopod mollusks that is unplaced taxonomically.

Kutta-Joukowski equation [FL MECH] An equation which states that the lift force exerted on a body by an ideal fluid, per unit length of body perpendicular to the flow, is equal to the product of the mass density of the fluid, the linear velocity of the fluid relative to the body, and the fluid circulation. Also known as Kutta-Joukowski theorem.

Kutta-Joukowski theorem *See* Kutta-Joukowski equation.

kwashiorkor [MED] A nutritional deficiency disease in infants and young children, mainly in the tropics, caused primarily by a diet low in proteins and rich in carbohydrates. Also known as nutritional dystrophy.

KWIC index *See* key-word-in-context index.

kyanite [MINERAL] Al_2SiO_5 A blue or light-green neosilicate mineral; crystallizes in the triclinic system, and luster is vitreous to pearly; occurs in long, thin bladed crystals and crystalline aggregates. Also known as cyanite; disthene; sappare.

kyphosis [MED] Angular curvature of the spine, usually in the thoracic region. Also known as humpback; hunchback.

L

l *See* liter.

L *See* lambert.

La *See* lanthanum.

label [COMPUT SCI] A data item that serves to identify a data record (much in the same way as a key is used), or a symbolic name used in a program to mark the location of a particular instruction or routine. [NUCLEO] *See* isotopic tracer.

label constant *See* location constant.

labellate [BIOL] Having a labellum.

labellum [BOT] The median membrane of the corolla of an orchid often differing in size and morphology from the other two petals. [INV ZOO] **1.** A prolongation of the labrum in certain beetles and true bugs. **2.** In Diptera, either of a pair of sensitive fleshy lobes consisting of the expanded end of the labium.

label record [COMPUT SCI] A tape record containing information concerning the file on that tape, such as format, record length, and block size.

labial gland [ANAT] Any of the small, tubular mucous and serous glands underneath the mucous membrane of mammalian lips. [INV ZOO] A salivary gland, or modification thereof, opening at the base of the labium in certain insects.

labial palp [INV ZOO] **1.** Either of a pair of fleshy appendages on either side of the mouth of certain bivalve mollusks. **2.** A jointed appendage attached to the labium of certain insects.

Labiatae [BOT] A large family of dicotyledonous plants in the order Lamiales; members are typically aromatic and usually herbaceous or merely shrubby.

labiate [ANAT] Having liplike margins that are thick and fleshy. [BIOL] Having lips. [BOT] Having the limb of a tubular calyx or corolla divided into two unequal overlapping parts.

labile [PSYCH] Unstable in mood. [SCI TECH] Also known as metastable. **1.** Readily changed, as by heat, oxidation, or other processes. **2.** Moving from place to place.

labium [BIOL] **1.** A liplike structure. **2.** The lower lip, as of a labiate corolla or of an insect.

labium majus [ANAT] Either of the two outer folds surrounding the vulva in the female.

labium minus [ANAT] Either of the two inner folds, at the inner surfaces of the labia majora, surrounding the vulva in the female.

Laboulbeniales [MYCOL] An order of ascomycetous fungi made up of species that live primarily on the external surfaces of insects.

Labrador Current [OCEANOGR] A current that flows southward from Baffin Bay, through the Davis Strait, and southwestward along the Labrador and Newfoundland coasts.

labradorite [MINERAL] A gray, blue, green, or brown plagioclase feldspar with composition ranging from $Ab_{50}An_{50}$ to $Ab_{30}An_{70}$, where $Ab = NaAlSi_3O_8$ and $An = CaAl_2Si_2O_8$; in the course of formation when the natural material cools, the feldspar sometimes exhibits a variously colored luster. Also known as Labrador spar.

Labrador spar *See* labradorite.

Labridae [VERT ZOO] The wrasses, a family of perciform fishes in the suborder Percoidei.

labrum [INV ZOO] **1.** The upper lip of certain arthropods, lying in front of or above the mandibles. **2.** The outer edge of a gastropod shell.

labyrinth [ANAT] **1.** Any body structure full of intricate cavities and canals. **2.** The inner ear. [ENG ACOUS] A loudspeaker enclosure having air chambers at the rear that absorb rearward-radiated acoustic energy, to prevent it from interfering with the desired forward-radiated energy.

labyrinthine syndrome *See* Ménière's syndrome.

labyrinthitis [MED] Inflammation of the labyrinth of the inner ear.

Labyrinthodontia [PALEON] A subclass of fossil amphibians descended from crossopterygian fishes, ancestral to reptiles, and antecedent to at least part of other amphibian types.

Labyrinthulia [INV ZOO] A subclass of the protozoan class Rhizopoda containing mostly marine, ovoid to spindle-shaped, uninucleate organisms that secrete a network of filaments (slime tubes) along which they glide.

Labyrinthulida [INV ZOO] The single order of the protozoan subclass Labyrinthulia.

Lacciferinae [INV ZOO] A subfamily of scale insects in the superfamily Coccoidea in which the male lacks compound eyes, the abdomen is without spiracles in all stages, and the apical abdominal segments of nymphs and females do not form a pygidium.

laccolith [GEOL] A body of igneous rock intruding into sedimentary rocks so that the overlying strata have been notably lifted by the force of intrusion.

lace [COMPUT SCI] To punch all the holes in some area of a punch card, such as a card row or card column. [TEXT] A patterned, openwork fabric made by hand with needles or hooks, or by machinery.

lacerate [MED] To inflict a wound by tearing.

lacerated [BIOL] Having a deeply and irregularly incised margin or apex.

Lacerta [ASTRON] A small northern constellation lying between Cygnus and Andromeda, and adjoining the northern boundary of Pegasus. Also known as Lizard.

Lacertidae [VERT ZOO] A family of reptiles in the suborder Sauria, including all typical lizards, characterized by movable eyelids, a fused lower jaw, homodont dentition, and epidermal scales.

lacing [CIV ENG] **1.** A lightweight metallic piece that is fixed diagonally to two channels or four angle sections, forming a composite strut. **2.** A course of brick, stone, or tiles in a wall of rubble to give strength. **3.** A course of upright bricks forming a bond between two or more arch rings. **4.** Distribution steel in a slab of reinforced concrete. **5.** A light timber fastened to pairs of struts or walings in the timbering of excavations (including mines). [COMPUT SCI] Extra multiple punching in a card column to signify the end of a specific card run; the term is derived from the lacework appearance of the card. [ELEC] Tying insulated wires together to support each other and form a single neat cable, with separately laced branches.

lacmus *See* litmus.

lacquer [MATER] A material which contains a substantial quantity of a cellulose derivative, most commonly nitrocellulose but sometimes a cellulose ester, such as cellulose acetate or cellulose butyrate, or a cellulose ether such as ethyl cellulose; used to give a glossy finish, especially on brass and other bright metals.

lacquer tree *See* varnish tree.

lacrimal [ANAT] Pertaining to tears, tear ducts, or tear-secreting organs.

lacrimal bone [ANAT] A small bone located in the anterior medial wall of the orbit, articulating with the frontal, ethmoid, maxilla, and inferior nasal concha.

lacrimal gland [ANAT] A compound tubuloalveolar gland that secretes tears. Also known as tear gland.

lacroixite [MINERAL] A pale yellowish-green mineral composed of basic phosphate of aluminum, calcium, manganese, and sodium (often with fluorine), occurring as crystals.

lactalbumin [BIOCHEM] A simple protein contained in milk which resembles serum albumin and is of high nutritional quality.

lactam [ORG CHEM] An internal (cyclic) amide formed by heating gamma (γ) and delta (δ) amino acids; thus γ-aminobutyric acid readily forms γ-butyrolactam lactam (pyrrolidone); many lactams have physiological activity.

lactase [BIOCHEM] An enzyme that catalyzes the hydrolysis of lactose to dextrose and galactose.

lactation [PHYSIO] Secretion of milk by the mammary glands.

lactim [ORG CHEM] A tautomeric enol form of a lactam with which it forms an equilibrium whenever the lactam nitrogen carries a free hydrogen.

lactin *See* lactose.

Lactobacillaceae [MICROBIO] The single family of gram-positive, asporogenous, rod-shaped bacteria; they are saccharoclastic, and produce lactate from carbohydrate metabolism.

lactoflavin *See* riboflavin.

lactogenic hormone *See* prolactin.

lactose [BIOCHEM] $C_{12}H_{22}O_{11}$ A disaccharide composed of D-glucose and D-galactose which occures in milk. Also known as lactin; milk sugar.

lacuna [BIOL] A small space or depression. [HISTOL] A cavity in the matrix of bone or cartilage which is occupied by the cell body.

lacustrine [GEOL] Belonging to or produced by lakes.

Lacydonidae [INV ZOO] A benthic family of pelagic errantian polychaetes.

ladar [OPTICS] A missile-tracking system that uses a visible light beam in place of a microwave radar beam to obtain measurements of speed, altitude, direction, and range of missiles. Derived from laser detecting and ranging. Also known as colidar; laser radar.

ladder attenuator [ELECTR] A type of ladder network designed to introduce a desired, adjustable loss when working between two resistive impedances, one of which has a shunt arm that may be connected to any of various switch points along the ladder.

ladder network [ELECTR] A network composed of a sequence of H, L, T, or pi networks connected in tandem; chiefly used as an electric filter. Also known as series-shunt network.

ladder road *See* ladderway.

ladderway [MIN ENG] Also known as ladder road; manway. **1.** Mine shaft between two main levels, equipped with ladders. **2.** The particular shaft, or compartment of a shaft, containing ladders.

Ladinian [GEOL] A European stage of geologic time: upper Middle Triassic (above Anisian, below Carnian).

Laemobothridae [INV ZOO] A family of lice in the order Mallophaga including parasites of aquatic birds, especially geese and coots.

lag [ELECTR] A persistence of the electric charge image in a camera tube for a small number of frames. [PHYS] **1.** The difference in time between two events or values considered together. **2.** *See* lag angle.

lag angle [PHYS] The negative of phase difference between a sinusoidally varying quantity and a reference quantity which varies sinusoidally at the same frequency, when this phase difference is negative. Also known as angle of lag; lag.

lag coefficient *See* time constant.

Lagenidiales [MYCOL] An order of aquatic fungi belonging to the class Phycomycetales characterized by a saclike to limited hyphal thallus and zoospores having two flagella.

lag fault [GEOL] A minor low-angle thrust fault occurring within an overthrust; it develops when one part of the mass is thrust farther than an adjacent higher or lower part.

lagging [CIV ENG] **1.** Horizontal wooden strips fastened across an arch under construction to transfer weight to the centering form. **2.** Wooden members positioned vertically to prevent cave-ins in earthworking. [MATER] Asbestos and magnesia plaster that is used as a thermal insulation on process equipment and piping.

lagging load *See* inductive load.

lagging network *See* integral network.

lag network *See* integral network.

Lagomorpha [VERT ZOO] The order of mammals including rabbits, hares, and pikas; differentiated from rodents by two pairs of upper incisors covered by enamel, vertical or lateral jaw motion, three upper and two lower premolars, fused tibia and fibula, and a spiral valve in the cecum.

lagoon [GEOGR] **1.** A shallow sound, pond, or lake generally near but separated from or communicating with the open sea. **2.** A shallow fresh-water pond or lake generally near or communicating with a larger body of fresh water.

Lagoon Nebula [ASTRON] A patchy, luminous gaseous nebula that appears to be surrounded by a much larger region of cold, neutral hydrogen.

lag phase [MICROBIO] The period of physiological activity and diminished cell division following the addition of inoculum of bacteria to a new culture medium.

Lagrange function *See* Lagrangian.

Lagrange-Hamilton theory [MECH] The formalized study of continuous systems in terms of field variables where a Lagrangian density function and Hamiltonian density function are introduced to produce equations of motion.

Lagrange-Helmholtz equation   *See* Helmholtz equation.

Lagrange's equations   [MECH] Equations of motion of a mechanical system for which a classical (nonquantum-mechanical) description is suitable, and which relate the kinetic energy of the system to the generalized coordinates, the generalized forces, and the time. Also known as Lagrangian equations of motion.

Lagrange's theorem   [MATH] In a group of finite order, the order of any subgroup must divide the order of the entire group.

Lagrange stream function   [FL MECH] A scalar function of position used to describe steady, incompressible two-dimensional flow; constant values of this function give the streamlines, and the rate of flow between a pair of streamlines is equal to the difference between the values of this function on the streamlines. Also known as current function; stream function.

Lagrangian   [MECH] 1. The difference between the kinetic energy and the potential energy of a system of particles, expressed as a function of generalized coordinates and velocities from which Lagrange's equations can be derived. Also known as kinetic potential; Lagrange function. 2. For a dynamical system of fields, a function which plays the same role as the Lagrangian of a system of particles; its integral over a time interval is a maximum or a minimum with respect to infinitesimal variations of the fields, provided the initial and final fields are held fixed.

Lagrangian coordinates   *See* generalized coordinates.

Lagrangian equations of motion   *See* Lagrange's equations.

Lagrangian function   [MECH] The function which measures the difference between the kinetic and potential energy of a dynamical system.

Lagriidae   [INV ZOO] The long-jointed bark beetles, a family of coleopteran insects in the superfamily Tenebrionoidea.

Laguerre polynomials   [MATH] A sequence of orthogonal polynomials which solve Laguerre's differential equation for positive integral values of the parameter.

Laguerre's differential equation   [MATH] The equation $xy'' + (1 - x)y' + \alpha y = 0$, where α is a constant.

Lagynacea   [INV ZOO] A superfamily of foraminiferan protozoans in the suborder Allogromiina having a free or attached test that has a membranous to tectinous wall and a single, ovoid, tubular, or irregular chamber.

lake   [HYD] An inland body of water, small to moderately large, with its surface water exposed to the atmosphere. [MATER] Any of a large group of dyes that have been combined with or adsorbed by salts of calcium, barium, chromium, aluminum, phosphotungstic acid, or phosphomolybdic acid; used for textile dyeing. Also known as color lake.

lake breeze   [METEOROL] A wind, similar in origin to the sea breeze but generally weaker, blowing from the surface of a large lake onto the shores during the afternoon; it is caused by the difference in surface temperature of land and water, as in the land and sea breeze system.

lake ore   *See* bog iron ore.

Lalande cell   [ELEC] A type of wet cell that uses a zinc anode and cupric oxide cathode cast as flat plates or hollow cylinders, and an electrolyte of sodium hydroxide in aqueous solution (caustic soda).

Lamarckism   [EVOL] The theory that organic evolution takes place through the inheritance of modifications caused by the environment, and by the effects of use and disuse of organs.

lambda hyperon   [PARTIC PHYS] 1. A quasi-stable baryon, forming an isotopic singlet, having zero charge and hypercharge, a spin of ½, positive parity and mass of 1115.5 MeV (megaelectronvolts). Designated Λ. Also known as lambda particle. 2. Any baryon resonance having zero hypercharge and total isotopic spin; designated $\Lambda_J P(m)$, where m is the mass of the baryon in MeV, and J and P are its spin and parity (if known).

lambda particle   *See* lambda hyperon.

lambda point   [CRYO] The temperature (2.1780 K), at atmospheric pressure, at which the transformation between the liquids helium I and helium II takes place; a special case of the thermodynamics definition. [THERMO] A temperature at which the specific heat of a substance has a sharply peaked maximum, observed in many second-order transitions.

lambert   [OPTICS] A unit of luminance (photometric brightness) that is equal to $1/\pi$ candela per square centimeter, or to the uniform luminance of a perfectly diffusing surface emitting or reflecting light at the rate of 1 lumen per square centimeter. Abbreviated L.

Lambert-Beer law   *See* Bouguer-Lambert-Beer law.

Lambert conformal projection   [MAP] A conformal conic projection with two standard parallels, or a conformal conic map projection in which the surface of a sphere or spheroid, such as the earth, is conceived as developed on a cone which intersects the sphere or spheroid at two standard parallels; the cone is then spread out to form a plane which is the map.

Lambert's law   [OPTICS] 1. The law that the illumination of a surface by a light ray varies as the cosine of the angle of incidence between the normal to the surface and the incident ray. 2. The law that the luminous intensity in a given direction radiated or reflected by a perfectly diffusing plane surface varies as the cosine of the angle between that direction and the normal to the surface. 3. *See* Bouguer-Lambert law.

Lamb shift   [ATOM PHYS] A small shift in the energy levels of a hydrogen atom, and of hydrogenlike ions, from those predicted by the Dirac electron theory, in accord with principles of quantum electrodynamics.

lamella   [ANAT] A thin scale or plate.

lamella arch   [CIV ENG] An arch consisting basically of a series of intersecting skewed arches made up of relatively short straight members; two members are bolted, riveted, or welded to a third piece at its center.

lamella roof   [BUILD] A large span vault built of members connected in a diamond pattern.

lamellar vector field   *See* irrotational vector field.

Lamellibranchiata   [INV ZOO] An equivalent name for Bivalvia.

Lamellisabellidae   [INV ZOO] A family of marine animals in the order Thecanephria.

Lamé's equations   [MATH] A general collection of second-order differential equations which have five regular singularities.

Lamé's relations   [MATH] Six independent relations which when satisfied by the covariant metric tensor of a three-dimensional space provide necessary and sufficient conditions for the space to be euclidean.

Lamiaceae   [BOT] An equivalent name for Labiatae.

Lamiales   [BOT] An order of dicotyledonous plants in the subclass Asteridae marked by its characteristic gynoecium, consisting of usually two biovulate carpels, with each carpel divided between the ovules by a false partition, or with the two halves of the carpel seemingly wholly separate.

lamina [ANAT] A thin sheet or layer of tissue; a scalelike structure. [GEOL] A thin, clearly differentiated layer of sedimentary rock or sediment, usually less than 1 centimeter thick. [MATER] A flat or curved arrangement of unidirectional or woven fibers in a matrix.

laminar [SCI TECH] 1. Arranged in thin layers. 2. Pertaining to viscous streamline flow without turbulence.

laminar flow [FL MECH] Streamline flow of an incompressible, viscous Newtonian fluid; all particles of the fluid move in distinct and separate lines.

Laminariophyceae [BOT] A class of algae belonging to the division Phaeophyta.

laminate [MATER] A sheet of material made of several different bonded layers.

lamination [GRAPHICS] A plastic protective film on a printed sheet that has been bonded by heat and pressure. [MATER] One of the thin punchings of iron or steel used in building up a laminated core for a magnetic circuit. [MED] An operation in embryotomy in which the skull is cut in slices. [SCI TECH] Arrangement in layers.

laminography See sectional radiography.

lampadite [MINERAL] A mineral composed chiefly of hydrous manganese oxide with as much as 18% copper oxide and often cobalt oxide.

lamp bank [ELEC] A number of incandescent lamps connected in parallel or series to serve as a resistance load for full-load tests of electric equipment.

lampblack [MATER] A grayish-black amorphous, practically pure form of carbon made by burning oil, coal tar, resin, or other carbonaceous substance in an insufficient supply of air; used in making paints, lead pencils, metal polishes, electric brush carbons, crayons, and carbon papers.

lampbrush chromosome [CYTOL] An exceptionally large chromosome characterized by fine lateral projections which are associated with active ribonucleic acid and protein synthesis.

lamphouse [ENG] 1. The light housing in a motion picture projector, located behind the projector head ordinarily consisting of a carbon arc lamp operating on direct current at about 60 volts, a concave reflector behind the arc which collects the light and concentrates it on the film, and cooling devices. 2. A box with a small hole containing an electric lamp and a concave mirror behind it, used as a concentrated source of light in a microscope, photographic enlarger, or other instrument.

lamp oil See kerosine.

lamprey [VERT ZOO] The common name for all members of the order Petromyzonida.

Lampridiformes [VERT ZOO] An order of teleost fishes characterized by a compressed, often ribbonlike body, fins composed of soft rays, a ductless swim bladder, and protractile maxillae among other distinguishing features.

lamprophyllite [MINERAL] $Na_2SrTiSi_2O_8$ A mineral composed of titanium strontium sodium silicate.

lamprophyre [PETR] Any of a group of igneous rocks characterized by a porphyritic texture in which abundant, large crystals of dark-colored minerals appear set in a not visibly crystalline matrix.

Lampyridae [INV ZOO] The firefly beetles, a large cosmopolitan family of coleopteran insects in the superfamily Cantharoidea.

lanarkite [MINERAL] Pb_2OSO_4 A white, greenish, or gray monoclinic mineral consisting of basic lead sulfate, with specific gravity of 6.92; formed by action of heat and air on galena.

lancelet [ZOO] The common name for members of the subphylum Cephalochordata.

Lanceolidae [INV ZOO] A family of bathypelagic amphipod crustaceans in the suborder Hyperiidea.

land [AERO ENG] Of an aircraft, to alight on land or a ship deck. [DES ENG] The top surface of the tooth of a cutting tool, behind the cutting edge. [ELECTR] See terminal area. [ENG] 1. In plastics molding equipment, the horizontal bearing surface of a semipositive or flash mold to allow excess material to escape; or the bearing surface along the top of the screw flight in a screw extruder; or the surface of an extrusion die that is parallel to the direction of melt flow. 2. The surface between successive grooves of a diffraction grating or phonograph record. [GEOGR] The portion of the earth's surface that stands above sea level. [ORD] One of the raised ridges in the bore of a rifled gun barrel.

Landau damping [PL PHYS] Damping of a plasma oscillation wave which occurs in situations where the particles of the plasma are able to increase their average energy at the expense of the wave, and thus to damp it out, even in cases where the dissipative effects of collisions are unimportant.

Landau-Ginzburg theory See Ginzburg-Landau theory.

land breeze [METEOROL] A coastal breeze blowing from land to sea, caused by the temperature difference when the sea surface is warmer than the adjacent land; therefore, the land breeze usually blows by night and alternates with a sea breeze which blows in the opposite direction by day.

land bridge [GEOGR] A strip of land linking two landmasses, often subject to temporary submergence, but permitting intermittent migration of organisms.

Landé g factor [ATOM PHYS] Also known as g factor. 1. The negative ratio of the magnetic moment of an electron or atom, in units of the Bohr magneton, to its angular momentum, in units of Planck's constant divided by 2π. 2. The ratio of the difference in energy between two energy levels which differ only in magnetic quantum number to the product of the Bohr magneton, the applied magnetic field, and the difference between the magnetic quantum numbers of free atoms; identical to the first definition for free atoms. Also known as Landé splitting factor; spectroscopic splitting factor. [NUC PHYS] The ratio of the magnetic moment of a nucleon, in units of the nuclear magneton, to its angular momentum in units of Planck's constant divided by 2π.

Landenian [GEOL] A European stage of geologic time: upper Paleocene (above Montian, below Ypresian of Eocene).

Landé splitting factor See Landé g factor.

landfast ice See fast ice.

landform [GEOGR] All the physical, recognizable, naturally formed features of land, having a characteristic shape; includes major forms such as a plain, mountain, or plateau, and minor forms such as a hill, valley, or alluvial fan.

land ice [HYD] Any part of the earth's seasonal or perennial ice cover which has formed over land as the result, principally, of the freezing of precipitation.

landing [CIV ENG] A place where boats receive or discharge passengers, freight, and so on. [MIN ENG]

1. Level stage in a shaft at which cages are loaded and discharged. **2.** The top or bottom of a slope, shaft, or inclined plane. [NAV] The termination of an aircraft's flight or of a ship's voyage.

landing chart [NAV] An aeronautical chart showing obstructions in the immediate vicinity of an aerodrome and the layout of the runways or landing area for use in landing and taxiing.

landing flap [AERO ENG] A movable airfoil-shaped structure located aft of the rear beam or spar of the wing; extends about two-thirds of the span of the wing and functions to substantially increase the lift, permitting lower takeoff and landing speeds.

landing gear [AERO ENG] Those components of an aircraft or spacecraft that support and provide mobility for the craft on land, water, or other surface.

land mile *See* mile.

land return *See* ground clutter.

Landry-Guillain-Barré syndrome [MED] A diffuse motor-neuron paresis, rapid in onset, and usually ascending and symmetrical in distribution, with proximal involvement greater than distal, and motor deficits greater than sensory. Also known as Guillain-Barré syndrome; Landry's paralysis.

Landry's paralysis *See* Landry-Guillain-Barré syndrome.

landscape architecture [CIV ENG] The art of arranging and fitting land for human use and enjoyment.

landscape engineer [CIV ENG] A person who applies engineering principles and methods to planning, design, and construction of natural scenery arrangements on a tract of land.

landslide [GEOL] The perceptible downward sliding or falling of a relatively dry mass of earth, rock, or combination of the two under the influence of gravity. Also known as landslip.

landslip *See* landslide.

land tie [CIV ENG] A rod or chain connecting an outside structure such as a retaining wall to a buried anchor plate.

langbanite [MINERAL] An ironblack hexagonal mineral composed of silicate and oxide of manganese, iron, and antimony, occurring in prismatic crystals.

langbeinite [MINERAL] $K_2Mg_2(SO_4)_3$ Colorless, yellowish, reddish, or greenish hexagonal mineral with vitreous luster, used in the salt industry in the fertilizer industry as a source of potassium sulfate.

Langevin-Debye formula [STAT MECH] A formula for the polarizability of a dielectric material or the paramagnetic susceptibility of a magnetic material, in which these quantities are the sum of a temperature-independent contribution and a contribution arising from the partial orientation of permanent electric or magnetic dipole moments which varies inversely with the temperature. Also known as Langevin-Debye law.

Langevin-Debye law *See* Langevin-Debye formula.

Langevin function [ELECTROMAG] A mathematical function, $L(x)$, which occurs in the expressions for the paramagnetic susceptibility of a classical (non-quantum-mechanical) collection of magnetic dipoles, and for the polarizability of molecules having a permanent electric dipole moment; given by $L(x) = \coth x - 1/x$.

langite [MINERAL] A blue to green mineral composed of basic hydrous copper sulfate.

langley [PHYS] A unit of energy per unit area commonly employed in radiation theory; equal to 1 gram-calorie per square centimeter.

Langmuir-Child equation *See* Child's law.

language [COMPUT SCI] The set of words and rules used to construct sentences with which to express and process information for handling by computers and associated equipment.

language converter [COMPUT SCI] A device which translates a form of data (such as that on microfilm) into another form of data (such as that on magnetic tape).

language translator [COMPUT SCI] **1.** Any assembler or compiler that accepts human-readable statements and produces equivalent outputs in a form closer to machine language. **2.** A program designed to convert one computer language to equivalent statements in another computer language, perhaps to be executed on a different computer. **3.** A routine that performs or assists in the performance of natural language translations, such as Russian to English, or Chinese to Russian.

Languriidae [INV ZOO] The lizard beetles, a cosmopolitan family of coleopteran insects in the superfamily Cucujoidea.

Laniatores [INV ZOO] A suborder of arachnids in the order Phalangida having flattened, often colorful bodies and found chiefly in tropical areas.

lansfordite [MINERAL] $MgCO_3 \cdot 5H_2O$ A mineral composed of hydrous basic carbonate of magnesium when extracted from the earth, changing to nesquehovite after exposure to the air.

lantern fish [VERT ZOO] The common name for the deep-sea teleost fishes composing the family Myctophidae and distinguished by luminous glands that are widely distributed upon the body surface.

lantern ring [DES ENG] A ring or sleeve around a rotating shaft; an opening in the ring provides for forced feeding of oil or grease to bearing surfaces; particularly effective for pumps handling liquids.

lanthanide contraction [ATOM PHYS] A phenomenon encountered in the rare-earth elements; the radii of the atoms of the members of the series decrease slightly as the atomic numbers increase; starting with element 58 in the periodic table, the balancing electron fills in an inner incomplete $4f$ shell as the charge on the nucleus increases.

lanthanide series [CHEM] Rare-earth elements of atomic numbers 57 through 71; their chemical properties are similar to those of lanthanum, atomic number 57.

lanthanite [MINERAL] $(La,Ce)_2(CO_3)_3 \cdot 8H_2O$ A colorless, white, pink, or yellow mineral composed of hydrous lanthanum carbonate, occurring in crystals or in earthy form.

lanthanum [CHEM] A chemical element, symbol La, atomic number 57, atomic weight 138.91; it is the second most abundant element in the rare-earth group. [MET] A white, soft, malleable metal; tarnishes in moist air; a major component of misch metal.

Lanthonotidae [VERT ZOO] A family of lizards (Sauria) belonging to the Anguimorpha line; restricted to North Borneo.

lanugo [ANAT] A downy covering of hair, especially that seen on the fetus or persisting on the adult body.

lap [CIV ENG] The length by which a reinforcing bar must overlap the bar it will replace. [MATER] An abrasive material used for lapping. [MET] A defect caused by folding and then rolling or forging a hot metal fin or corner onto a surface without welding. Also known as fold.

laparoscopy [MED] A method of visually examining the peritoneal cavity by means of a long slender endoscope equipped with sheath, obturator, biopsy forceps, a sphygmomanometer bulb and tubing, scissors, and a syringe; the endoscope is introduced into the peritoneal cavity through a small incision in the abdominal wall. Also known as peritoneoscopy.

laparotomy [MED] A surgical incision through the abdominal wall into the abdominal cavity.

lap dissolve [ELECTR] Changeover from one television scene to another so that the new picture appears gradually as the previous picture simultaneously disappears. [GRAPHICS] See dissolve.

lapis lazuli [PETR] An azure-blue, violet-blue, or greenish-blue, translucent to opaque crystalline rock used as a semiprecious stone; composed chiefly of lazurite and calcite with some haüyne, sodalite, and other minerals. Also known as lazuli.

lap joint [ENG] A simple joint between two members made by overlapping the ends and fastening them together with bolts, rivets, or welding.

Laplace irrotational motion [FL MECH] Irrotational flow of an inviscid, incompressible fluid.

Laplace operator [MATH] The linear operator defined on differentiable functions which gives for each function the sum of all its nonmixed second partial derivatives. Also known as Laplacian.

Laplace's equation [ACOUS] An equation for the speed c of sound in a gas; it may be written $c = \sqrt{\gamma p/\rho}$, where p is the pressure, ρ is the density, and γ is the ratio of specific heats. [MATH] The partial differential equation which states that the sum of all the nonmixed second partial derivatives equals 0; the potential functions of many physical systems satisfy this equation.

Laplace transform [MATH] For a function $f(x)$ its Laplace transform is the function $F(y)$ defined as the integral over x from 0 to ∞ of the function $e^{-yx}f(x)$.

Laplacian See Laplace operator.

lapping [ELECTR] Moving a quartz, semiconductor, or other crystal slab over a flat plate on which a liquid abrasive has been poured, to obtain a flat polished surface or to reduce the thickness a carefully controlled amount. [MET] Polishing with a material such as cloth, lead, plastic, wood, iron, or copper having fine abrasive particles incorporated or rubbed into the surface.

lapstrake [NAV ARCH] A method of hull construction where each continuous band of hull planking (strake) is lapped on the outside of the one beneath.

lap winding [ELEC] A two-layer winding in which each coil is connected in series to the adjacent coil.

Laramic orogeny See Laramidian orogeny.

Laramide orogeny See Laramidian orogeny.

Laramide revolution See Laramidian orogeny.

Laramidian orogeny [GEOL] An orogenic era typically developed in the eastern Rocky Mountains; phases extended from Late Cretaceous until the end of the Paleocene. Also known as Laramic orogeny; Laramide orogeny; Laramide revolution.

larch [BOT] The common name for members of the genus *Larix* of the pine family, having deciduous needles and short, spurlike branches which annually bear a crown of needles.

larderillite [MINERAL] $(NH_4)B_5O_8 \cdot 2H_2O$ A white mineral composed of hydrous ammonium borate, occurring as a crystalline powder.

large dyne See newton.

large intestine See colon.

Large Magellanic Cloud [ASTRON] An irregular cloud of stars in the constellation Doradus; it is 150,000 light-years away and nearly 30,000 light-years in diameter. Abbreviated LMC. Also known as Nubecula Major.

large nuclei [OCEANOGR] Particles of concentrated seawater or crystalline salt in the marine atmosphere having radii larger than 10^{-5} centimeter.

large scale [MAP] A scale of sufficient size to permit the plotting of much detail with exactness. [METEOROL] A scale such that the curvature of the earth may not be considered negligible; this scale is applicable to the high tropospheric long-wave patterns, with four or five waves around the hemisphere in the middle latitudes.

large-scale integrated circuit [ELECTR] A very complex integrated circuit, which contains well over 100 interconnected individual devices, such as basic logic gates and transistors, placed on a single semiconductor chip. Abbreviated LSI circuit. Also known as chip circuit; multiple-function circuit.

large-scale integrated memory See semiconductor memory.

Largidae [INV ZOO] A family of hemipteran insects in the superfamily Pentatomorpha.

Laridae [VERT ZOO] A family of birds in the order Charadriiformes composed of the gulls and terns.

Larinae [VERT ZOO] A subfamily of birds in the family Laridae containing the gulls and characterized by a thick, slightly hooked beak, a square tail, and a stout white body, with shades of gray on the back and the upper wing surface.

Larmor frequency [ELECTROMAG] The angular frequency of the Larmor precession, equal in esu (electrostatic units) to the negative of a particle's charge times the magnetic induction divided by the product of twice the particle's mass and the speed of light.

Larmor precession [ELECTROMAG] A common rotation superposed upon the motion of a system of charged particles, all having the same ration of charge to mass, by a magnetic field.

larnite [MINERAL] β-Ca_2SiO_4 A gray mineral that is a metastable monoclinic phase of calcium orthosilicate, stable from 520 to 670°C. Also known as belite.

larry [MIN ENG] 1. A car with a hopper bottom and adjustable chutes for feeding coke ovens. Also known as lorry. 2. See barney.

larsenite [MINERAL] $PbZnSiO_4$ A colorless or white mineral composed of lead zinc silicate, occurring in orthorhombic crystals.

Larvacea [INV ZOO] A class of the subphylum Tunicata consisting of minute planktonic animals in which the tail, with dorsal nerve cord and notochord, persists throughout life.

Larvaevoridae [INV ZOO] The tachina flies, a large family of dipteran insects in the suborder Cyclorrhapha distinguished by a thick covering of bristles on the body; most are parasites of arthropods.

laryngitis [MED] Inflammation of the larynx.

laryngophone [ENG ACOUS] A microphone designed to be placed against the throat of a speaker, to pick up voice vibrations directly without responding to background noise.

laryngoscope [MED] A tubular instrument, combining a light system and a telescopic system, used in the visualization of the interior larynx and adaptable for diagnostic, therapeutic, and surgical procedures.

larynx [ANAT] The complex of cartilages and related structures at the opening of the trachea into the pharynx in vertebrates; functions in protecting the entrance of the trachea, and in phonation in higher forms.

laser [OPTICS] An active electron device that converts input power into a very narrow, intense beam of coherent visible or infrared light; the input power excites the atoms of an optical resonator to a higher energy level, and the resonator forces the excited atoms to radiate in phase. Derived from light amplification by stimulated emission of radiation.

laser amplifier   [ELECTR] A laser which is used to increase the output of another laser. Also known as light amplifier.

laser annealing   [MET] The rapid heating of metals or alloys with the use of lasers.

laser-beam printer   [GRAPHICS] A nonimpact printer that operates at well over 10,000 lines per minute, using a low-power laser to produce image-forming charges a line at a time on the photoconductive surface of a drum; dry powder that adheres only to charged areas is applied to the drum, transferred to plain paper, and fused by heat.

laser camera   [OPTICS] An airborne camera system for night photography in which a laser beam is split into two beams; one beam, which is almost invisible, scans the ground, while the second beam is modulated by a detector of light reflected from the ground area being scanned, and is in turn swept back and forth over a moving film by the same scanner.

laser communication   [COMMUN] Optical communication in which the light source is a laser whose beam is modulated for voice, video, or data communication over information bandwidths up to 1 gigahertz.

laser diode   *See* semiconductor laser.

laser indused fusion   *See* laser fusion.

laser infrared radar   *See* lidar.

laser memory   [COMPUT SCI] A computer memory in which a controlled laser beam acts on individual and extremely small areas of a photosensitive or other type of surface, for storage and subsequent readout of digital data or other types of information.

laser propulsion   [AERO ENG] The use of high-power lasers for aircraft, rocket, or spacecraft propulsion by indirect conversion of laser-heated propellants or working fluids to produce thrust; for direct thrust generation with laser light pressure on the vehicle; or for direct conversion of laser energy into electricity for propulsion.

laser pumping   [OPTICS] The application of a laser beam of appropriate frequency to a laser medium so that absorption of the radiation increases the population of atoms or molecules in higher energy states.

laser radar   *See* lidar.

laserscope   [ENG] A pulsed high-power laser used with appropriate scanning and imaging devices to sense objects over the sea at night or in fog and provide three-dimensional images on a viewing screen.

laser spectroscopy   [SPECT] A branch of spectroscopy in which a laser is used as an intense, monochromatic light source; in particular, it includes saturation spectroscopy, as well as the application of laser sources to Raman spectroscopy and other techniques.

laser velocimeter   [OPTICS] Any velocity measuring instrument that makes use of a laser.

Lasiocampidae   [INV ZOO] The tent caterpillars and lappet moths, a family of cosmopolitan (except New Zealand) lepidopteran insects in the suborder Heteroneura.

last in, first out   [IND ENG] A method of determining the inventory costs by transferring the costs of material to the product in reverse chronological order. Abbreviated LIFO.

last-mask read-only memory   [COMPUT SCI] A read-only memory in which the final mask used in the fabrication process determines the connections to the internal transistors, and these connections in turn determine the data pattern that will be read out when the cell is accessed. Also known as contact-mask read-only memory.

last quarter   [ASTRON] The phase of the moon at western quadrature, half of the illuminated hemisphere being visible from the earth; has the characteristic half-moon shape.

latch-up   [ELECTR] A self-sustaining low-impedance state in a $p^r p^n$ device, which is a type of electronic malfunction.

latency   [COMPUT SCI] The waiting time between the order to read/write some information from/to a specified place and the beginning of the data–read/write operation.   [MED] The stage of an infectious disease, other than the incubation period, in which there are neither clinical signs nor symptoms.   [PHYSIO] The period between the introduction of and the response to a stimulus.   [PSYCH] The phase between the Oedipal period and adolescence, characterized by an apparent cessation of psychosexual development.

latent defect   [IND ENG] A flaw or other imperfection in any article which is discovered after delivery; usually, latent defects are inherent weaknesses which normally are not detected by examination or routine tests, but which are present at time of manufacture and are aggravated by use.

latent heat   [THERMO] The amount of heat absorbed or evolved by 1 mole, or a unit mass, of a substance during a change of state (such as fusion, sublimation or vaporization) at constant temperature and pressure.

latent heat of fusion   *See* heat of fusion.

latent heat of sublimation   *See* heat of sublimation.

latent heat of vaporization   *See* heat of vaporization.

latent image   [GRAPHICS] An invisible image produced by the physical or chemical effects of light on the individual crystals (usually silver halide) of photographic emulsions; the development process makes the image visible, in the negative.

latent period   [MED] Any stage of an infectious disease in which there are no clinical signs of symptoms of the infection.   [PHYSIO] The period between the introduction of a stimulus and the response to it.   [VIROL] The initial period of phage growth after infection during which time virus nucleic acid is manufactured by the host cell.

latent root   *See* eigenvalue.

lateral aberration   [OPTICS] 1. The distance from the axis of an optical system at which a ray intersects a plane perpendicular to the axis through the focus of paraxial rays. 2. The difference between the reciprocals of the image distances for paraxial and rim rays. 3. For chromatic aberration, the difference in sizes of the images of an object for two different colors.

lateral bud   [BOT] Any bud that develops on the side of a stem.

lateral cone   *See* adventive cone.

lateral line   [INV ZOO] A longitudinal lateral line along the sides of certain oligochaetes consisting of cell bodies of the layer of circular muscle.   [VERT ZOO] A line along the sides of the body of most fishes, often distinguished by differently colored scales, which marks the lateral line organ.

lateral-line organ   [VERT ZOO] A small, pear-shaped sense organ in the skin of many fishes and amphibians that is sensitive to pressure changes in the surrounding water.

lateral magnification   [OPTICS] The ratio of some linear dimension, perpendicular to the optical axis, of an image formed by an optical system, to the corresponding linear dimension of the object. Also known as magnification.

lateral meristem [BOT] Strips or cylinders of dividing cells located parallel to the long axis of the organ in which they occur; the lateral meristem functions to increase the diameter of the organ.

lateral moraine [GEOL] Drift material, usually thin, that was deposited by a glacier in a valley after the glacier melted.

lateral parity check [COMPUT SCI] The number of one bits counted across the width of the magnetic tape; this number plus a one or a zero must always be odd (or even), depending upon the manufacturer.

lateral sclerosis *See* amyotrophic lateral sclerosis; primary lateral sclerosis.

laterite [GEOL] Weathered material composed principally of the oxides of iron, aluminum, titanium, and manganese; laterite ranges from soft, earthy, porous soil to hard, dense rock.

latex [MATER] **1.** Milky colloid in which natural or synthetic rubber or plastic is suspended in water. **2.** An elastomer product made from latex.

latex paint [MATER] A paint consisting of a water suspension or emulsion of latex combined with pigments and additives such as binders and suspending agents. Also known as latex water paint.

latex water paint *See* latex paint.

lath [CIV ENG] **1.** A narrow strip of wood used in making a level base, as for plaster or tiles, or in constructing a light framework, as a trellis. **2.** A sheet of material used as a base for plaster.

lathe [MECH ENG] A machine for shaping a workpiece by gripping it in a holding device and rotating it under power against a suitable cutting tool for turning, boring, facing, or threading.

Lathridiidae [INV ZOO] The minute brown scavenger beetles, a large cosmopolitan family of coleopteran insects in the superfamily Cucujoidea.

Latimeridae [VERT ZOO] A family of deep-sea lobefin fishes (Coelacanthiformes) known from a single living species, *Latimeria chalumnae.*

Latin square [MATH] An $n \times n$ square array of n different symbols, each symbol appearing once in each row and once in each column; these symbols prove useful in ordering the observations of an experiment.

latite [PETR] A not visibly crystalline rock of volcanic origin composed chiefly of sodic plagioclase and alkali feldspar with subordinate quantities of dark-colored minerals in a finely crystalline to glassy groundmass.

latitude [GEOD] Angular distance from a primary great circle or plane, as on the celestial sphere or the earth.

lattice [CIV ENG] A network of crisscrossed strips of metal or wood. [CRYSTAL] A regular periodic arrangement of points in three-dimensional space; it consists of all those points P for which the vector from a given fixed point has the form $n_1 \mathbf{a} + n_2 \mathbf{b} + n_3 \mathbf{c}$, where n_1, n_2, and n_3 are integers, and $\mathbf{a}$, $\mathbf{b}$, and $\mathbf{c}$ are fixed, linearly independent vectors. Also known as periodic lattice; space lattice. [MATH] A partially ordered set in which each pair of elements has both a greatest lower bound and least upper bound. [NAV] A pattern formed by two or more families of intersecting lines of position, such as the hyperbolic lines of position from two or more loran stations. [NUCLEO] An orderly array or pattern of nuclear fuel elements and moderator in a reactor or critical assembly.

lattice constant [CRYSTAL] A parameter defining the unit cell of a crystal lattice, that is, the length of one of the edges of the cell or an angle between edges. Also known as lattice parameter.

lattice defect *See* crystal defect.

lattice girder [CIV ENG] An open girder, beam, or column built from members joined and braced by intersecting diagonal bars. Also known as open-web girder.

lattice network [ELEC] A network that is composed of four branches connected in series to form a mesh; two nonadjacent junction points serve as input terminals, and the remaining two junction points serve as output terminals.

lattice parameter *See* lattice constant.

lattice vibration [SOLID STATE] A periodic oscillation of the atoms in a crystal lattice about their equilibrium positions.

lattice-wound coil *See* honeycomb coil.

lattissimus dorsi [ANAT] The widest muscle of the back; a broad, flat muscle of the lower back that adducts and extends the humerus, is used to pull the body upward in climbing, and is an accessory muscle of respiration.

Lattorfian *See* Tongrian.

Laue condition [CRYSTAL] **1.** The condition for a vector to lie in a Laue plane: its scalar product with a specified vector in the reciprocal lattice must be one-half of the scalar product of the latter vector with itself. **2.** *See* Laue equations.

Laue equations [CRYSTAL] Three equations which must be satisfied for an x-ray beam of specified wavelength to be diffracted through a specified angle by a crystal; they state that the scaler products of each of the crystallographic axial vectors with the difference between unit vectors in the directions of the incident and scattered beams, are integral multiples of the wavelength. Also known as Laue condition.

Laue method [CRYSTAL] A method of studying crystalline structures by x-ray diffraction, in which a finely collimated beam of polychromatic x-rays falls on a single crystal whose orientation can be set as desired, and diffracted beams are recorded on a photographic film.

Laugiidae [PALEON] A family of Mesozoic fishes in the order Coelacanthiformes.

laumonite *See* laumontite.

laumontite [MINERAL] $CaAl_2Si_4O_{12}\cdot 4H_2O$ A white zeolite mineral crystallizing in the monoclinic system; loses water on exposure to air, eventually becoming opaque and crumbling. Also known as laumonite; lomonite; lomontite.

launch [AERO ENG] **1.** To send off a rocket vehicle under its own rocket power, as in the case of guided aircraft rockets, artillery rockets, and space vehicles. **2.** To send off a missile or aircraft by means of a catapult or by means of inertial force, as in the release of a bomb from a flying aircraft. **3.** To give a space probe an added boost for flight into space just before separation from its launch vehicle.

launch pad [AERO ENG] The load-bearing base or platform from which a rocket vehicle is launched. Also known as pad.

launch vehicle [AERO ENG] A rocket or other vehicle used to launch a probe, satellite, or the like. Also known as booster.

launch window [AERO ENG] The time period during which a spacecraft or missile must be launched in order to achieve a desired encounter, rendezvous, or impact.

lauoho o pele *See* Pele's hair.

Lauraceae [BOT] The laurel family of the order Magnoliales distinguished by definite stamens in series of three, a single pistil, and the lack of petals.

Laurasia [GEOL] A continent theorized to have existed in the Northern Hemisphere; supposedly it broke up to form the present northern continents about the end of the Pennsylvanian period.

Lauratonematidae [INV ZOO] A family of marine nematodes of the superfamily Enoploidea; many females possess a cloaca.

Laurentian Plateau *See* Laurentian Shield.

Laurentian Shield [GEOL] A Precambrian plateau extending over half of Canada from Labrador southwest along Hudson Bay and northwest to the Arctic Ocean. Also known as Canadian Shield; Laurentian Plateau.

laurionite [MINERAL] $Pb(OH)Cl$ A colorless mineral composed of basic lead chloride, occurring in prismatic crystals; it is dimorphous with paralaurionite.

laurite [MINERAL] RuS_2 A black mineral composed of ruthenium sulfide (often with osmium), occurring as small crystals or grains.

lausenite [MINERAL] $Fe_2(SO_4)_3 \cdot 6H_2O$ A white, monoclinic mineral consisting of hydrated ferric sulfate; occurs in lumpy aggregates of fibers.

lautarite [MINERAL] $Ca(IO_3)_2$ A monoclinic mineral composed of calcium iodate that occurs in prismatic crystals.

Lauxaniidae [INV ZOO] A family of myodarian cyclorrhaphous dipteran insects in the subsection Acalypteratae; larvae are leaf miners.

lava [GEOL] **1.** Molten extrusive material that reaches the earth's surface through volcanic vents and fissures. **2.** The rock mass formed by consolidation of molten rock issuing from volcanic vents and fissures, consisting chiefly of magnesium silicate; used for insulators.

lava cone [GEOL] A volcanic cone that was formed of lava flows.

lava dome *See* shield volcano.

lava field [GEOL] A wide area of lava flow; it is commonly several square kilometers in area and forms along the base of a large compound volcano or on the flanks of shield volcanoes.

lava tube [GEOL] A long, tubular opening under the crust of solidified lava.

lavenite [MINERAL] $(Na,Ca)_3Zr(Si_2O_7)(O,OH,F)_2$ A mineral composed of complex silicate, occurring in prismatic crystals.

law [SCI TECH] A regularity which applies to all members of a broad class of phenomena.

law of action and reaction *See* Newton's third law.

law of constant heat summation *See* Hess's law.

law of corresponding states [CHEM] The law that when, for two substances, any two ratios of pressure, temperature, or volume to their respective critical properties are equal, the third ratio must equal the other two.

law of definite composition *See* law of definite proportion.

law of definite proportion [CHEM] The law that a given chemical compound always contains the same elements in the same fixed proportion by weight. Also known as law of definite composition.

law of electromagnetic induction *See* Faraday's law of electromagnetic induction.

law of exponents [MATH] Any of the laws $a^m a^n = a^{m+n}$, $a^m/a^n = a^{m-n}$, $(a^m)^n = a^{mn}$, $(ab)^n = a^n b^n$, $(a/b)^n = a^n/b^n$; these laws are valid when m and n are any integers, or when a and b are positive and m and n are any real numbers. Also known as exponential law.

law of gravitation *See* Newton's law of gravitation.

law of large numbers [STAT] The law that if, in a collection of independent identical experiments, $N(B)$

represents the number of occurrences of an event B in n trials, and p is the probability that B occurs at any given trial, then for large enough n it is unlikely that $N(B)/n$ differs from p by very much. Also known as Bernoulli's theorem.

law of partial pressures *See* Dalton's law.

law of reflection *See* reflection law.

law of signs [MATH] The product or quotient of two numbers is positive if the numbers have the same sign, negative if they have opposite signs.

law of specificity of bacteria *See* Koch's postulate.

law of superposition [GEOL] The law that strata underlying other strata must be the older if there has been neither overthrust nor inversion.

Lawrence tube *See* chromatron.

lawrencite [MINERAL] $(Fe,Ni)Cl_2$ A brown or green mineral composed of ferrous chloride and found as an abundant accessory mineral in iron meteorites.

lawrencium [CHEM] A chemical element, symbol Lr, atomic number 103; two isotopes have been discovered, mass number 257 or 258 and mass number 256.

laws of refraction *See* Snell laws of refraction.

Lawson criterion [PL PHYS] The requirement for the energy produced by fusion in a plasma to exceed that required to produce the confined plasma; it states that for a mixture of deuterium and tritium in the temperature range from 1×10^8 to 5×10^8 degrees Celsius, the product of the ionic density and the confinement time must be about 10^{14} seconds per cubic centimeter.

lawsonite [MINERAL] $CaAl_2(Si_2O_7)(OH)_2 \cdot H_2O$ A colorless or grayish-blue mineral crystallizing in the orthorhombic system; found in gneisses and schists.

layer [GEOL] A tabular body of rock, ice, sediment, or soil lying parallel to the supporting surface and distinctly limited above and below. [GEOPHYS] One of several strata of ionized air, some of which exist only during the daytime, occurring at altitudes between 50 and 400 kilometers; the layers reflect radio waves at certain frequencies and partially absorb others. [MET] The stratum of weld metal consisting of one or more passes and lying parallel to the welding surface.

layering [BOT] A propagation method by which root formation is induced on a branch or a shoot attached to the parent stem by covering the part with soil. [ECOL] A stratum of plant forms in a community, such as mosses, shrubs, or trees in a bog area.

layer lattice *See* layer structure.

layer silicate *See* phyllosilicate.

layer structure [CRYSTAL] A crystalline structure found in substances such as graphites and clays, in which the atoms are largely concentrated in a set of parallel planes, with the regions between the planes comparatively vacant. Also known as layer lattice.

layout [GRAPHICS] A design drawing or graphical statement of the overall form of a component, system, or device, which is usually prepared during innovative stages of a design.

lazuli *See* lapis lazuli.

lazulite [MINERAL] $(Mg,Fe)Al_2(OH)_2(PO_4)_2$ A violet-blue or azure-blue mineral with vitreous luster; composed of basic aluminum phosphate and occurring in small masses or monoclinic crystals; hardness is 5-6 on Mohs scale, and specific gravity is 3.06–3.12. Also known as berkeyite; blue spar; false lapis.

lazurite [MINERAL] $(Na,Ca)_8(Al,Si)O_{24}(S,SO_4)$ A blue or violet-blue feldspathoid mineral crystallizing in the isometric system; the chief mineral constituent of lapis lazuli.

lb *See* pound.

L band [COMMUN] A band of radio frequencies extending from 390 to 1550 megahertz, corresponding to wavelengths of 76.9 to 19.37 centimeters.

lb ap *See* pound.

lb apoth *See* pound.

lbf *See* pound.

lbf-ft *See* foot-pound.

lb t *See* pound.

lb tr *See* pound.

L capture [NUC PHYS] A type of generalized beta interaction in which a nucleus captures an electron from the *L* shell of atomic electrons (the shell second closest to the nucleus).

LCD *See* liquid crystal display.

LC filter *See* inductive filter.

L display [ELECTR] A radarscope display in which the target appears as two horizontal pulses or blips, one extending to the right and one to the left from a central vertical time base; when the radar antenna is correctly aimed in azimuth at the target, both blips are of equal amplitude; when not correctly aimed, the relative blip amplitudes indicate the pointing error; the position of the signal along the baseline indicates target distance; the display may be rotated 90° when used for elevation aiming instead of azimuth aiming.

leaching [CHEM ENG] The dissolving, by a liquid solvent, of soluble material from its mixture with an insoluble solid; leaching is an industrial separation operation based on mass transfer; examples are the washing of a soluble salt from the surface of an insoluble precipitate, and the extraction of sugar from sugarbeets. [GEOCHEM] The separation or dissolving out of soluble constituents from a rock or ore body by percolation of water. [MIN ENG] Dissolving soluble minerals or metals out of the ore, as by the use of percolating solutions such as cyanide or chlorine solutions, acids, or water. Also known as lixiviation.

lead [CHEM] A chemical element, symbol Pb, atomic number 82, atomic weight 207.19. [DES ENG] The distance that a screw will advance or move into a nut in one complete turn. [ELEC] A wire used to connect two points in a circuit. [ENG] A mass of lead attached to a line, as used for sounding at sea. [GEOL] A small, narrow passage in a cave. [GRAPHICS] A thin strip of metal used during the composition process to space lines of type. [MET] A soft, heavy metal with a silvery-bluish color; when freshly cut it is malleable and ductile; occurs naturally, mostly in combination; used principally in alloys in pipes, cable sheaths, type metal, and shields against radioactivity. [ORD] **1.** The action of aiming ahead of a moving target with a gun, bomb, rocket, or torpedo so as to hit the target, including whatever action is necessary to correct for deflection. **2.** The distance between the moving target and the point at which the gun or missile is aimed. **3.** The number of diameters for one complete turn of the rifling. **4.** An explosive train component which consists of a column of high explosive, usually small in diameter, used to transmit detonation from one detonating component to a succeeding high explosive component; it is generally used to transmit the detonation from a detonator to a booster charge. [PHYS] *See* lead angle.

lead-acid battery [ELEC] A storage battery in which the electrodes are grids of lead containing lead oxides that change in composition during charging and discharging, and the electrolyte is dilute sulfuric acid.

lead angle [DES ENG] The angle that the tangent to a helix makes with the plane normal to the axis of the helix. [MET] The angle at the point of welding between an electrode and a line perpendicular to the weld axis. [ORD] **1.** The angle between the line of sight to a moving target and the line of sight to a point ahead of the target. **2.** A dropping angle. [PHYS] The phase difference between a sinusoidally varying quantity and a reference quantity which varies sinusoidally at the same frequency, when this phase difference is positive. Also known as angle of lead; lead; phase lead.

lead bronze [MET] An alloy of 60–70% copper, up to 2% nickel, and up to 15% tin with the balance lead; used as a bearing metal.

leaded gasoline [MATER] Motor gasoline into which a small amount of TEL (tetraethyllead) has been added to increase octane number or rating.

leader [COMPUT SCI] A record which precedes a group of detail records, giving information about the group not present in the detail records; for example, "beginning of batch 17." [ENG] The unrecorded length of magnetic tape that enables the operator to thread the tape through the drive and onto the take-up reel without losing data or recorded music, speech, or such. [GEOPHYS] The streamer which initiates the first phase of each stroke of a lightning discharge; it is a channel of very high ion density which propagates through the air by the continual establishment of an electron avalanche ahead of its tip. Also known as leader streamer. [GRAPHICS] **1.** A short piece of blank film at each end of a strip of photographic film or reel of motion picture film, used to thread the film through the mechanism of a camera or projector. **2.** Lines or rows of dashes or dots used to guide the reader's eye across a printed page.

leader label [COMPUT SCI] A record appearing at the beginning of a magnetic tape to uniquely identify the tape as one required by the system.

leader streamer *See* leader.

leader stroke [GEOPHYS] The entire set of events associated with the propagation of any leader between cloud and ground in a lightning discharge.

lead glance *See* galena.

lead glass [MATER] Glass into which lead oxide is incorporated to give high refractive index, optical dispersion, and surface brilliance; used in optical glass.

leadhillite [MINERAL] $Pb_4(SO_4)(CO_3)_2(OH)_2$ A yellowish or greenish- or grayish-white monoclinic mineral consisting of basic sulfate and carbonate of lead; dimorphous with susanite.

lead-in [ELEC] A single wire used to connect a single-terminal outdoor antenna to a receiver or transmitter. Also known as down-lead.

leading [GRAPHICS] Space inserted between lines of type to open them up vertically.

leading edge [COMPUT SCI] The edge of a punched card or document which enters a machine first. [AERO ENG] The front edge of an airfoil or wing. [DES ENG] The surfaces or inset cutting points on a bit that face in the same direction as the rotation of the bit. [PHYS] The major portion of the rise of a pulse.

leading end [COMPUT SCI] The end of a paper or magnetic tape that is read first, or has data entered on it first.

lead-in groove [DES ENG] A blank spiral groove at the outside edge of a disk recording, generally of a pitch much greater than that of the recorded grooves, provided to bring the pickup stylus quickly to the first recorded groove. Also known as lead-in spiral.

leading stone   *See* lodestone.
lead-in spiral   *See* lead-in groove.
lead marcasite   *See* sphalerite.
lead network   *See* derivative network.
leaf   [BOT] A modified aerial appendage which develops from a plant stem at a node, usually contains chlorophyll, and is the principal organ in which photosynthesis and transpiration occur.   [BUILD] **1.** A separately movable division of a folding or sliding door. **2.** One of a pair of doors or windows. **3.** One of the two halves of a cavity wall.   [COMPUT SCI] *See* terminal vertex.
leaf blight   [PL PATH] Any of various blight diseases which cause browning, death, and falling of the leaves.
leaf blotch   [PL PATH] A plant disease characterized by discolored areas in the leaves with indistinct or diffuse margins.
leaf curl   [PL PATH] A fungus or viral disease of plants marked by the curling of leaves.
leaf drop   [PL PATH] Premature falling of leaves, associated with disease.
leafhopper   [INV ZOO] The common name for members of the homopteran family Cicadellidae.
leaflet   [BOT] **1.** A division of a compound leaf. **2.** A small or young foliage leaf.
leaf miner   [INV ZOO] Any of the larvae of various insects which burrow into and eat the parenchyma of leaves.
leaf mottle   [PL PATH] A fungus disease characterized by chlorotic mottling of the leaves; for example, caused by *Verticillium dahliae* in sunflower.
leaf roll   [PL PATH] Any of several virus diseases characterized by upward or inward rolling of the leaf margins.
leaf rot   [PL PATH] Any plant disease characterized by break-down of leaf tissues; for example, caused by *Pellicularia koleroga* in coffee.
leaf rust   [PL PATH] Any rust disease that primarily affects leaves; common in coffee, alfalfa, and wheat, barley, and other cereals.
leaf scald   [PL PATH] A bacterial disease of sugarcane caused by *Bacterium albilineans* which invades the vascular tissues, causing creamy or grayish streaking and withering of the leaves.
leaf scar   [BOT] A mark on a stem, formed by secretion of suberin and a gumlike substance, showing where a leaf has abscised.
leaf scorch   [BOT] Any of several disorders and fungus diseases marked by a burned appearance of the leaves; for example, caused by the fungus *Diplocarpon earliana* in strawberry.
leaf spot   [PL PATH] Any of various diseases or disorders characterized by the appearance of well-defined discolored spots on the leaves.
leaf spring   [DES ENG] A beam of cantilever design, firmly anchored at one end and with a large deflection under a load. Also known as flat spring.
leaf stripe   [PL PATH] Any of various plant diseases characterized by striped discolorations on the foliage.
leak   [PL PATH] A watery rot of fruits and vegetables caused by various fungi, such as *Rhizopus nigricans* in strawberry.
leakage   [ENG] Undesired and gradual escape or entry of a quantity, such as loss of neutrons by diffusion from the core of a nuclear reactor, escape of electromagnetic radiation through joints in shielding, flow of electricity over or through an insulating material, and flow of magnetic lines of force beyond the working region.   [MIN ENG] An unintentional diversion of ventilation air from its designed path.   [PHYS CHEM] A phenomenon occurring in an ion-exchange process in which some influent ions are not adsorbed by the ion-exchange bed and appear in the effluent.
leakage current   [ELEC] **1.** Undesirable flow of current through or over the surface of an insulating material or insulator. **2.** The flow of direct current through a poor dielectric in a capacitor.   [ELECTR] The alternating current that passes through a rectifier without being rectified.
leakage halo   [GEOCHEM] The dispersion of elements along channels and paths followed by mineralizing solutions leading into and away from the central focus of mineralization.
lean fuel mixture   *See* lean mixture.
lean mixture   [MECH ENG] A fuel-air mixture containing a low percentage of fuel and a high percentage of air, as compared with a normal or rich mixture. Also known as lean fuel mixture.
leapfrog test   [COMPUT SCI] A computer test using a special program that performs a series of arithmetical or logical operations on one group of storage locations, transfers itself to another group, checks the correctness of the transfer, then begins the series of operations again; eventually, all storage positions will have been tested.
least-action principle   *See* principle of least action.
least common denominator   [MATH] The least common multiple of the denominators of a collection of fractions.
least common multiple   [MATH] The least common multiple of a set of quantities (for example, numbers or polynomials) is the smallest quantity divisible by each of them.
least-energy principle   [MECH] The principle that the potential energy of a system in stable equilibrium is a minimum relative to that of nearby configurations.
least significant bit   [COMPUT SCI] The bit that carries the lowest value or weight in binary notation for a numeral; for example, when 13 is represented by binary 1101, the 1 at the right is the least significant bit. Abbreviated LSB.
least significant character   [COMPUT SCI] The character in the rightmost position in a number or word.
least squares method   [STAT] A technique of fitting a curve close to some given points which minimizes the sum of the squares of the deviations of the given points from the curve.
least-time principle   *See* Fermat's principle.
Lebesgue integral   [MATH] The generalization of Riemann integration of real valued functions, which allows for integration over more complicated sets, existence of the integral even though the function has many points of discontinuity, and convergence properties which are not valid for Riemann integrals.
Lebesgue number   [MATH] The Lebesgue number of an open cover of a compact metric space X is a positive real number so that any subset of X whose diameter is less than this number must be completely contained in a member of the cover.
Lebesgue-Stieltjes integral   [MATH] A Lebesgue integral of the form

$$\int_b^a f(x)\,d\phi(x)$$

where ϕ is of bounded variation; if $\phi(x) = x$, it reduces to the Lebesgue integral of $f(x)$; if $\phi(x)$ is differentiable, it reduces to the Lebesgue integral of $f(x)\phi'(x)$.
Lecanicephaloidea   [INV ZOO] An order of tapeworms of the subclass Cestoda distinguished by hav-

ing the scolex divided into two portions; all species are intestinal parasites of elasmobranch fishes.

Lecanoraceae [BOT] A temperate and boreal family of lichens in the order Lecanorales characterized by a crustose thallus and a distinct thalloid rim on the apothecia.

Lecanorales [BOT] An order of the Ascolichenes having open, discoid apothecia with a typical hymenium and hypothecium.

lechatelierite [MINERAL] A natural silica glass, occurring in fulgurites and impact craters and formed by the melting of quartz sand at high temperatures generated by lightning or by the impact of a meteorite.

Le Chatelier's principle [PHYS] The principle that when an external force is applied to a system at equilibrium, the system adjusts so as to minimize the effect of the applied force.

Lecher line See Lecher wires.

Lecher wires [ELECTROMAG] Two parallel wires that are several wavelengths long and a small fraction of a wavelength apart, used to measure the wavelength of a microwave source that is connected to one end of the wires; a shorting bar which slides along the wires is used to determine the position of standing-wave nodes. Also known as Lecher line; Lecher wire wavemeter.

Lecher wire wavemeter See Lecher wires.

Lecideaceae [BOT] A temperate and boreal family of lichens in the order Lecanorales; members lack a thalloid rim around the apothecia.

lecithin [BIOCHEM] Any of a group of phospholipids having the general composition $CH_2OR_1 \cdot CHOR_2 \cdot CH_2OPO_2OHR_3$, in which R_1 and R_2 are fatty acids and R_3 is choline, and with emulsifying, wetting, and antioxidant properties. [MATER] 1. A mixture of phosphatides and oil obtained by drying the separate gums from the degumming of soybean oil; consists of the phosphatides (lecithin), cephalin, other fatlike phosphorus-containing compounds, and 30–35% entrained soybean oil; may be treated to produce more refined grades; used in foods, cosmetics, and paints. Also known as commercial lecithin; crude lecithin; soybean lecithin; soy lecithin. 2. A waxy mixture of phosphatides obtained by refining commercial lecithin to remove the soybean oil and other materials; used in pharmaceuticals. Also known as refined lecithin.

lecithinase [BIOCHEM] An enzyme that catalyzes the breakdown of a lecithin into its constituents.

Leclanché cell [ELEC] The common dry cell, which is a primary cell having a carbon positive electrode and a zinc negative electrode in an electrolyte of sal ammoniac and a depolarizer.

lecontite [MINERAL] $Na(NH_4,K)SO_4 \cdot 2H_2O$ A colorless mineral composed of a hydrous sodium potassium ammonium sulfate; found in bat guano.

le cri du chat syndrome [MED] A complex of congenital malformations resulting from a deletion in chromosome 4 or 5 and characterized by mental retardation and the production of a catlike cry.

Lecythidaceae [BOT] The single family of the order Lecythidales.

Lecythidales [BOT] A monofamilial order of dicotyledonous tropical woody plants in the subclass Dilleniidae; distinguished by entire leaves, valvate sepals, separate petals, numerous centrifugal stamens, and a syncarpous, inferior ovary with axile placentation.

LED See light-emitting diode.

Leda [ASTRON] A small satellite of Jupiter with a diameter probably less than 8 kilometers, orbiting at a mean distance of 1.11×10^7 kilometers. Also known as Jupiter XIII.

Ledian [GEOL] Lower upper Eocene geologic time. Also known as Auversian.

Leduc current [ELEC] An asymmetrical alternating current obtained from, or similar to that obtained from, the secondary winding of an induction coil; used in electrobiology.

Leduc law See Amagat-Leduc rule.

lee [SCI TECH] The side of an object, such as an island or a ship, away from the direction in which the wind is coming, and sheltered from wind or waves.

Leeaceae [BOT] A family of dicotyledonous plants in the order Rhamnales distinguished by solitary ovules in each locule, simple to compound leaves, a small embryo, and hypogynous flowers.

leech [INV ZOO] The common name for members of the annelid class Hirudinea.

LEED See low-energy electron diffraction.

leek [BOT] *Allium porrum.* A biennial herb known only by cultivation; grown for its mildly pungent succulent leaves and thick cylindrical stalk.

left-hand [DES ENG] Of drilling and cutting tools, screw threads, and other threaded devices, designed to rotate clockwise or cut to the left.

left-hand limit See limit on the left.

left-hand rule [ELECTROMAG] 1. For a current-carrying wire, the rule that if the fingers of the left hand are placed around the wire so that the thumb points in the direction of electron flow, the fingers will be pointing in the direction of the magnetic field produced by the wire. 2. For a current-carrying wire in a magnetic field, such as a wire on the armature of a motor, the rule that if the thumb, first, and second fingers of the left hand are extended at right angles to one another, with the first finger representing the direction of magnetic lines of force and the second finger representing the direction of current flow, the thumb will be pointing in the direction of force on the wire. Also known as Fleming's rule.

leg [ANAT] The lower extremity of a human limb, between the knee and the ankle. [COMPUT SCI] The sequence of instructions that is followed in a computer routine from one branch point to the next. [ENG] 1. Anything that functionally or structurally resembles an animal leg. 2. One of the branches of a forked or jointed object. 3. One of the main upright members of a drill derrick or tripod. [GEOPHYS] A single cycle of more or less periodic motion in a wave train on a seismogram. [MECH ENG] The case that encloses the vertical part of the belt carrying the buckets within a grain elevator. [MET] In a fillet weld, the distance between the root and the toe. [MIN ENG] 1. In mine timbering, a prop or upright member of a set or frame. 2. A stone that has to be wedged out from beneath a larger one. [NAV] 1. One part of a craft's track, consisting of a single course line. 2. A track identified by an aid to navigation. [ZOO] An appendage or limb used for support and locomotion.

Legendre contact transformation See Legendre transformation.

Legendre equation [MATH] The second-order linear homogeneous differential equation $(1 - x^2)y'' - 2xy' + \nu(\nu + 1)y = 0$, where ν is real and nonnegative.

Legendre polynomials [MATH] A collection of orthogonal polynomials which provide solutions to the

Legendre equation for nonnegative integral values of the parameter.

Legendre transformation [MATH] A mathematical procedure in which one replaces a function of several variables with a new function which depends on partial derivatives of the original function with respect to some of the original independent variables. Also known as Legendre contact transformation.

legrandite [MINERAL] $Zn_{14}(OH)(AsO_4)_9 \cdot 12H_2O$ A yellow to nearly colorless mineral composed of basic hydrous zinc arsenate.

legume [BOT] A dry, dehiscent fruit derived from a single simple pistil; common examples are alfalfa, beans, peanuts, and vetch.

Leguminosae [BOT] The legume family of the plant order Rosales characterized by stipulate, compound leaves, 10 or more stamens, and a single carpel; many members harbor symbiotic nitrogen-fixing bacteria in their roots.

leightonite [MINERAL] $K_2Ca_2Cu(SO_4)_4 \cdot 2H_2O$ A pale-blue mineral composed of hydrous sulfate of copper, calcium, and potassium.

Leiodidae [INV ZOO] The round carrion beetles, a cosmopolitan family of coleopteran insects in the superfamily Staphylinoidea; commonly found under decaying bark.

Leitneriales [BOT] A monofamilial order of flowering plants in the subclass Hamamelidae; members are simple-leaved, dioecious shrubs with flowers in catkins, and have a superior, pseudomonomerous ovary with a single ovule.

Lelapiidae [INV ZOO] A family of calcaronean sponges in the order Sycettida characterized by a rigid skeleton composed of tracts or bundles of modified triradiates.

LEM See lunar excursion module.

lemma [BOT] Either of the pair of bracts that are borne above the glumes and enclose the flower of a grass spikelet. [MATH] A mathematical fact germane to the proof of some theorem.

lemming [VERT ZOO] The common name for the small burrowing rodents composing the subfamily Microtinae.

Lemnaceae [BOT] The duckweeds, a family of monocotyledonous plants in the order Arales; members are small, free-floating, thalloid aquatics with much reduced flowers that form a miniature spadix.

lemniscate of Bernoulli [MATH] The locus of points (x,y) in the plane satisfying the equation $(x^2 + y^2)^2 = a^2(x^2 - y^2)$ or, in polar coordinates (r,θ), the equation $r^2 = a^2 \cos 2\theta$.

lemniscus [ANAT] A secondary sensory pathway of the central nervous system, usually terminating in the thalamus.

lemon [BOT] *Citrus limon*. A small evergreen tree belonging to the order Sapindales cultivated for its acid citrus fruit which is a modified berry called a hesperidium.

lemur [VERT ZOO] The common name for members of the primate family Lemuridae; characterized by long tails, foxlike faces, and scent glands on the shoulder region and wrists.

Lemuridae [VERT ZOO] A family of prosimian primates of Madagascar belonging to the Lemuroidea; all members are arboreal forest dwellers.

Lemuroidea [VERT ZOO] A suborder or superfamily of Primates including the lemurs, tarsiers, and lorises, or sometimes simply the lemurs.

length block [COMPUT SCI] The total number of records, words, or characters contained in one block.

lenitic See lentic.

lens [ANAT] A transparent, encapsulated, nearly spherical structure located behind the pupil of vertebrate eyes, and in the complex eyes of many invertebrates, that focuses light rays on the retina. Also known as crystalline lens. [COMMUN] A dielectric or metallic structure that is transparent to radio waves and can bend them to produce a desired radiation pattern; used with antennas for radar and microwave relay systems. [ELECTR] See electron lens. [ELECTROMAG] See magnetic lens. [GEOL] A geologic deposit that is thick in the middle and converges toward the edges, resembling a convex lens. [MATER] See acoustic lens. [OPTICS] A curved piece of ground and polished or molded material, usually glass, used for the refraction of light, its two surfaces having the same axis; or two or more such surfaces cemented together. Also known as optical lens.

lens antenna [ELECTROMAG] A microwave antenna in which a dielectric lens is placed in front of the dipole or horn radiator to concentrate the radiated energy into a narrow beam or to focus received energy on the receiving dipole or horn.

lens stop See diaphragm.

lentic [ECOL] Of or pertaining to still waters such as lakes, reservoirs, ponds, and bogs. Also spelled lenitic.

lenticel [BOT] A loose-structured opening in the periderm beneath the stomata in the stem of many woody plants that facilitates gas transport.

lenticular [OPTICS] Of or pertaining to a lens. [SCI TECH] Having the shape of a lentil or double convex lens.

lenticular cloud See lenticularis.

lenticularis [METEOR] A cloud species, the elements of which have the form of more or less isolated, generally smooth lenses; the outlines are sharp. Also known as lenticular cloud.

lentil [BOT] *Lens esculenta*. A seminiviny annual legume having pinnately compound, vetchlike leaves; cultivated for its thin, lens-shaped, edible seed. [GEOL] 1. A rock body that is lens-shaped and enclosed in a stratum of different material. 2. A rock stratigraphic unit that is a subdivision of a formation and has limited geographic extent; it thins out in all directions.

lentor See stoke.

Lenz's law [ELECTROMAG] The law that whenever there is an induced electromotive force (emf) in a conductor, it is always in such a direction that the current it would produce would oppose the change which causes the induced emf.

Leo [ASTRON] A northern constellation, right ascension 11 hours, declination 15° north. Also known as Lion.

Leo Minor [ASTRON] A northern constellation, right ascension 10 hours, declination 35° north. Also known as Lesser Lion.

Leonardian [GEOL] A North American provincial series: Lower Permian (above Wolfcampian, below Guadalupian).

Leonids [ASTRON] A meteor shower, the radiant of which lies in the constellation Leo; it is visible between November 10 and 15.

leonite [MINERAL] $K_2Mg(SO_4)_2 \cdot 4H_2O$ A colorless, white, or yellowish mineral composed of hydrous magnesium potassium sulfate, occurring in monoclinic crystals.

leopard [VERT ZOO] *Felis pardus*. A species of wildcat in the family Felidae found in Africa and Asia; the coat is characteristically buff-colored with black spots.

leopoldite *See* sylvite.

Leotichidae [INV ZOO] A small Oriental family of hemipteran insects in the superfamily Leptopodoidea.

Lepadomorpha [INV ZOO] A suborder of barnacles in the order Thoracica having a peduncle and a capitulum which is usually protected by calcareous plates.

Leperditicopida [PALEON] An order of extinct ostracods characterized by very thick, straight-backed valves which show unique muscle scars and other markings.

Leperditillacea [PALEON] A superfamily of extinct paleocopan ostracods in the suborder Kloedenellocopina including the unisulcate, nondimorphic forms.

Lepiceridae [INV ZOO] Horn's beetle, a family of Central American coleopteran insects composed of two species.

Lepidocentroida [INV ZOO] The name applied to a polyphyletic assemblage of echinoids now regarded as members of the Echinocystitoida and Echinothurioida.

lepidocrocite [MINERAL] $\alpha FeO(OH)$ A ruby- or blood-red mineral crystallizing in the orthorhombic system; it is associated with limonite in iron ores and is a component of meteorites.

Lepidodendrales [PALEOBOT] The giant club mosses, an order of extinct lycopods (Lycopodiopsida) consisting primarily of arborescent forms characterized by dichotomous branching, small amounts of secondary vascular tissue, and heterospory.

lepidolite [MINERAL] $K(Li,Al)_3(Si,Al)_4O_{10}(F,OH)_2$ A rose-colored mineral of the mica group crystallizing in the monoclinic system. Also known as lithionite; lithium mica.

lepidomelane [MINERAL] A black variety of biotite that is characterized by the presence of large amounts of ferric iron. Also known as iron mica.

lepidophyllous [BOT] Having scaly leaves.

Lepidoptera [INV ZOO] Large order of scaly-winged insects, including the butterflies, skippers, and moths; adults are characterized by two pairs of membranous wings and sucking mouthparts, featuring a prominent, coiled proboscis.

Lepidorsirenidae [VERT ZOO] A family of slender, obligate air-breathing, eellike fishes in the order Dipteriformes having small thin scales, slender ribbonlike paired fins, and paired ventral lungs.

Lepidosaphinae [INV ZOO] A family of homopteran insects in the superfamily Coccoidea having dark-colored, noncircular scales.

Lepidosauria [VERT ZOO] A subclass of reptiles in which the skull structure is characterized by two temporal openings on each side which have reduced bony arcades, and by the lack of an antorbital opening in front of the orbit.

Lepidotrichidae [INV ZOO] A family of wingless insects in the order Thysanura.

Lepismatidae [INV ZOO] A family of silverfish in the order Thysanura characterized by small or missing compound eyes.

Lepisostei [VERT ZOO] An equivalent name for Semionotiformes.

Lepisosteidae [VERT ZOO] A family of fishes in the order Semionotiformes.

Lepisosteiformes [VERT ZOO] An equivalent name for Semionotiformes.

Leporidae [VERT ZOO] A family of mammals in the order Lagomorpha including the rabbits and hares.

Lepospondyli [PALEON] A subclass of extinct amphibians including all forms in which the vertebral centra are formed by ossification directly around the notochord.

lepospondylous [VERT ZOO] Having the notochord enclosed by cylindrical vertebrae shaped like an hourglass in longitudinal section.

leprosy *See* Hansen's disease.

Leptaleinae [INV ZOO] A subfamily of the Formicidae including largely arboreal ant forms which inhabit plants in tropical and subtropical regions.

Leptictidae [PALEON] A family of extinct North American insectivoran mammals belonging to the Proteutheria which ranged from the Cretaceous to middle Oligocene.

Leptinidae [INV ZOO] The mammal nest beetles, a small European and North American family of coleopteran insects in the superfamily Staphylinoidea.

Leptocardii [ZOO] The equivalent name for Cephalochordata.

leptocercal [VERT ZOO] Of the tail of a fish, tapering to a long, slender point.

Leptochoeridae [PALEON] An extinct family of palaeodont artiodactyl mammals in the superfamily Dichobunoidea.

Leptodactylidae [VERT ZOO] A large family of frogs in the suborder Procoela found principally in the American tropics and Australia.

leptodactylous [VERT ZOO] Having slender toes, as certain birds.

Leptodiridae [INV ZOO] The small carrion beetles, a cosmopolitan family of coleopteran insects in the superfamily Staphylinoidea.

leptogeosyncline [GEOL] A deep oceanic trough that has not been filled with sedimentation and is associated with volcanism.

leptokurtic distribution [STAT] A distribution in which the ratio of the fourth moment to the square of the second moment is greater than 3, which is the value for a normal distribution; it appears to be more heavily concentrated about the mean, or more peaked, than a normal distribution.

Leptolepidae [PALEON] An extinct family of fishes in the order Leptolepiformes representing the first teleosts as defined on the basis of the advanced structure of the caudal skeleton.

Leptolepiformes [PALEON] An extinct order of small, ray-finned teleost fishes characterized by a relatively strong, ossified axial skeleton, thin cycloid scales, and a preopercle with an elongated dorsal portion.

Leptomedusae [INV ZOO] A suborder of hydrozoan coelenterates in the order Hydroida characterized by the presence of a hydrotheca.

Leptomitales [MYCOL] An order of aquatic Phycomycetes characterized by a hyphal thallus, or basal rhizoids and terminal hyphae, and zoospores with two flagella.

lepton [PARTIC PHYS] A fermion having a mass smaller than the proton mass; leptons interact with electromagnetic and gravitational fields, but beyond this they interact only through weak interactions.

Leptopodidae [INV ZOO] A tropical and subtropical family of hemipteran insects in the superfamily Leptopodoidea distinguished by the spiny body and appendages.

Leptopodoidea [INV ZOO] A superfamily of hemipteran insects in the subdivision Geocorisae.

Leptosomatidae [VERT ZOO] The cuckoo rollers, a family of Madagascan birds in the order Coraciiformes composed of a single species distinguished by the downy covering on the newly hatched young.

Leptostraca [INV ZOO] A primitive group of crustaceans considered as one of a series of Malacostraca distinguished by an additional abdominal somite that lacks appendages, and a telson bearing two movable articulated prongs.

Leptostromataceae [MYCOL] A family of fungi of the order Sphaeropsidales; pycnidia are black and shield-shaped, circular or oblong, and slightly asymmetrical; included are some fruit-tree pathogens.

leptotene [CYTOL] The first stage of meiotic prophase, when the chromosomes appear as thin threads having well-defined chromomeres.

Leptotyphlopidae [VERT ZOO] A family of small, harmless, burrowing circumtropical snakes (Serpentes) in the order Squamata; teeth are present only on the lower jaw and are few in number.

Lepus [ASTRON] A southern constellation, right ascension 5.5 hours, declination 20°S. Also known as Hare. [VERT ZOO] The type genus of the family Leporidae, comprising the typical hares.

Lernaeidae [INV ZOO] A family of copepod crustaceans in the suborder Caligoida; all are fixed ectoparasites, that is, they penetrate the skin of freshwater fish.

Lernaeopodidae [INV ZOO] A family of ectoparasitic crustaceans belonging to the Lernaeopodoida; individuals are attached to the walls of the fishes' gill chambers by modified second maxillae.

Lernaeopodoida [INV ZOO] The fish maggots, a group of ectoparasitic crustaceans characterized by a modified postembryonic development reduced to two or three stages, a free-swimming larva, and the lack of external signs of physical maturity in adults.

Leskeineae [BOT] A suborder of mosses in the order Hypnobryales; plants are not complanate, paraphyllia are frequently present, leaves are costate, and alar cells are not generally differentiated.

Lesser Dog *See* Canis Minor.

Lesser Lion *See* Leo Minor.

Lestidae [INV ZOO] A family of odonatan insects belonging to the Zygoptera; distinguished by the wings being held in a V position while at rest.

Lestoniidae [INV ZOO] A monospecific family of hemipteran insects in the superfamily Pentatomorpha found only in Australia.

lethal gene [GEN] A gene mutation that causes premature death in heterozygotes if dominant, and in homozygotes if recessive. Also known as lethal mutation.

lethal mutation *See* lethal gene.

letter code [COMPUT SCI] A Baudot code function which cancels errors by causing the receiving terminal to print nothing.

letterpress [GRAPHICS] A method of printing by impressing paper on a raised inked surface; the oldest printing process, it employs type or plates cast or engraved in relief on materials such as metal, wood, or rubber. Also known as relief press.

letterset [GRAPHICS] A method of printing that uses a blanket for transferring an image from plate to paper; unlike offset lithography, it uses a relief plate and requires no dampening system. Also known as dry offset.

letter spacing [GRAPHICS] Space inserted between letters of a word to open it up horizontally.

lettuce [BOT] *Lactuca sativa*. An annual plant of the order Asterales cultivated for its succulent leaves; common varieties are head lettuce, leaf or curled lettuce, romaine lettuce, and iceberg lettuce.

Leucaltidae [INV ZOO] A family of calcinean sponges in the order Leucettida having numerous small, interstitial, flagellated chambers.

Leucascidae [INV ZOO] A family of calcinean sponges in the order Leucettida having a radiate arrangement of flagellated chambers.

Leucettida [INV ZOO] An order of calcareous sponges in the subclass Calcinea having a leuconoid structure and a distinct dermal membrane or cortex.

leucine [BIOCHEM] $C_6H_{13}O_2N$ A monocarboxylic essential amino acid obtained by hydrolysis of protein-containing substances such as milk.

leucite [MINERAL] $KAlSi_2O_6$ A white or gray rock-forming mineral belonging to the feldspathoid group; at ordinary temperatures the mineral exists as aggregates of trapezohedral crystals with glassy fracture; hardness is 5.5–6.0 on Mohs scale, and specific gravity is 2.45–2.50. Also known as amphigene; grenatite; vesuvian; Vesuvian garnet; white garnet.

leucitohedron *See* trapezohedron.

leucochalcite *See* olivenite.

Leucodontineae [BOT] A family of mosses in the order Isobryales with foliated branches, often bearing catkins.

leucophanite [MINERAL] $(Na,Ca)_2BeSi_7(O,F,OH)_7$ Greenish mineral composed of beryllium sodium calcium silicate containing fluorine and occurring in glassy, tabular crystals.

leucopyrite *See* loellingite.

Leucosiidae [INV ZOO] The purse crabs, a family of true crabs belonging to the Oxystomata.

Leucosoleniida [INV ZOO] An order of calcareous sponges in the subclass Calcaronea characterized by an asconoid structure and the lack of a true dermal membrane or cortex.

leucosphenite [MINERAL] $Na_4BaTi_2Si_{10}O_{27}$ A white mineral composed of sodium barium silicotitanate and occurring as wedge-shaped crystals.

Leucospidae [INV ZOO] A small family of hymenopteran insects in the superfamily Chalcidoidea distinguished by a longitudinal fold in the forewings.

Leucothoidae [INV ZOO] A family of amphipod crustaceans in the suborder Gammaridea including semiparasitic and commensal species.

Leucotrichaceae [MICROBIO] A family of bacteria in the order Cytophagales; long, colorless, unbranched filaments having conspicuous cross-walls and containing cylindrical or ovoid cells; filaments attach to substrate.

leukemia [MED] Any of several diseases of the hemopoietic system characterized by uncontrolled leukocyte proliferation. Also known as leukocythemia.

leukocyte [HISTOL] A colorless, ameboid blood cell having a nucleus and granular or nongranular cytoplasm. Also known as white blood cell; white corpuscle.

leukocythemia *See* leukemia.

leukocytopenia *See* leukopenia.

leukocytosis [MED] Elevation of the leukocyte count to values above the normal limit.

leukoderma [MED] Defective skin pigmentation, especially the congenital absence of pigment in patches or bands.

leukopenia [MED] A reduction in the leukocyte count to values below the normal limit. Also known as leukocytopenia.

leukoplakia [MED] Formation of thickened white patches on mucous membranes, particularly of the mouth and vulva.

leukotomy *See* lobotomy.

leuneburgite [MINERAL] $Mg_3B_2(PO_4)_2(OH)_6 \cdot 5H_2O$ A colorless mineral consisting of a hydrous basic phosphate of magnesium and boron.

levator [MED] An instrument used for raising a depressed portion of the skull. [PHYSIO] Any muscle that raises or elevates a part.

levee [CIV ENG] 1. A dike for confining a stream. 2. A pier along a river. [GEOL] 1. An embankment bordering one or both sides of a sea channel or the low-gradient seaward part of a canyon or valley. **2.** A low ridge sometimes deposited by a stream on its sides.

level [CIV ENG] **1.** A surveying instrument with a telescope and bubble tube used to take level sights over various distances, commonly 100 feet (30 meters). **2.** To make the earth surface horizontal. [COMMUN] A specified position on an amplitude scale applied to a signal waveform, such as reference white level and reference black level in a standard television signal. [COMPUT SCI] **1.** The status of a data item in COBOL language indicating whether this item includes additional items. **2.** See channel. [DES ENG] A device consisting of a bubble tube that is used to find a horizontal line or plane. Also known as spirit level. [ELEC] A single bank of contacts, as on a stepping relay. [ELECTR] **1.** The difference between a quantity and an arbitrarily specified reference quantity, usually expressed as the logarithm of the ratio of the quantities. **2.** A charge value that can be stored in a given storage element of a charge storage tube and distinguished in the output from other charge values. [MIN ENG] **1.** Mine workings that are at the same elevation. **2.** A gutter for the water to run in.

level compensator [ELECTR] **1.** Automatic transmission-regulating feature or device used to minimize the effects of variations in amplitude of the received signal. **2.** Automatic gain control device used in the receiving equipment of a telegraph circuit.

level error [ASTRON] **1.** The difference between the apparent altitude of a celestial object above the apparent horizon and its true altitude above the celestial horizon. **2.** The angle between the east-west mechanical axis of a transit telescope and the horizontal plane.

level indicator [ENG] An instrument that indicates liquid level. [ENG ACOUS] An indicator that shows the audio voltage level at which a recording is being made; may be a volume-unit meter, neon lamp, or cathode-ray tuning indicator.

leveling [ENG] Adjusting any device, such as a launcher, gun mount, or sighting equipment, so that all horizontal or vertical angles will be measured in the true horizontal and vertical planes. [IND ENG] A method of performance rating which seeks to rate the principal factors that cause the speed of motions rather than speed itself; it considers that the level at which the operator works is influenced by effort and skill. [MET] Flattening rolled sheet by evening out irregularities, using a roller or tensile straining. [MIN ENG] Measurement of rises and falls, heights, and contour lines.

level of saturation See water table.

level of significance of a test [STAT] The probability of false rejection of the null hypothesis. Also known as significance level.

level shifting [ELECTR] Changing the logic level at the interface between two different semiconductor logic systems.

lever [ENG] A rigid bar, pivoted about a fixed point (fulcrum), used to multiply force or motion; used for raising, prying, or dislodging an object.

leverage [MECH] The multiplication of force or motion achieved by a lever.

lever escapement [HOROL] A clock movement in which the balance wheel is connected to the escapement by a lever attached to a roller; the wheel

swings through a much larger angle than does a pendulum.

levigate [BOT] See glabrous. [CHEM] To separate a finely divided powder from a coarser material by suspending in a liquid in which both substances are insoluble. Also known as elutriation. **2.** To grind a moist solid to a fine powder.

levo form [PHYS CHEM] An optical isomer which induces levorotation in a beam of plane polarized light.

levorotation [OPTICS] Rotation of the plane of polarization of plane polarized light in a counterclockwise direction, as seen by an observer facing in the direction of light propagation. Also known as levulorotation.

levulorotation See levorotation.

levulose [BIOCHEM] Levorotatory D-fructose.

levyine See levynite.

levyite See levynite.

levyne See levynite.

levynite [MINERAL] $NaCa_3Al_7Si_{11}O_{36}\cdot15H_2O$ A white or light-colored mineral of the zeolite group, composed of hydrous silicate of aluminum, sodium, and calcium, and occurring in rhombohedral crystals. Also known as levyine; levyite; levyne.

Lewis acid [CHEM] A substance that can accept an electron pair from a base; thus, $AlCl_3$, BF_3, and SO_3 are acids.

Lewis base [CHEM] A substance that can donate an electron pair; examples are the hydroxide ion, OH^-, and ammonia, NH_3.

Lewis blood group system [IMMUNOL] An antigen, designated by Le^a, first recognized in a Mrs. Lewis, occurring in about 22% of the population, detected by anti-Le^a antibodies; primarily composed of soluble antigens of serum and body fluids like saliva, with secondary absorption by erythrocytes.

lewistonite [MINERAL] $(Ca,K,Na)_5(PO_4)_3(OH)$ White mineral composed of basic calcium potassium sodium phosphate.

Leydig cell [HISTOL] One of the interstitial cells of the testes; thought to produce androgen.

LF See low-frequency.

LGV See lymphogranuloma venereum.

lherzolite [PETR] Peridotite composed principally of olivine with orthopyroxene and clinopyroxene.

l'Hospital's rule [MATH] A rule useful in evaluating indeterminate forms: if both the functions $f(x)$ and $g(x)$ and all their derivatives up to order $(n-1)$ vanish at $x = a$, but the nth derivatives both do not vanish or both become infinite at $x = a$, then

$$\lim_{x \to a} f(x)/g(x) = f^{(n)}(a)/g^{(n)}(a),$$

$f^{(n)}$ denoting the nth derivative.

l'Huilier's equation [MATH] An equation used in the solution of a spherical triangle, involving tangents of various functions of its angles and sides.

Li See lithium.

Liapunov function See Lyapunov function.

Lias See Liassic.

Liassic [GEOL] The Lower Jurassic period of geologic time. Also known as Lias.

Libellulidae [INV ZOO] A large family of odonatan insects belonging to the Anisoptera and distinguished by a notch on the posterior margin of the eyes and a foot-shaped anal loop in the hindwing.

libethenite [MINERAL] $Cu_2(PO_4)OH$ An olive-green mineral composed of basic copper sulfate, occurring as small prismatic crystals or in masses.

libido [PSYCH] 1. Sexual desire. 2. The sum total of all instinctual forces; psychic energy or drive usually associated with the sexual instinct.

Libra [ASTRON] A southern constellation, right ascension 15 hours, declination 15° south. Also known as Balance.

librarian [COMPUT SCI] The program which maintains and makes available all programs and routines composing the operating system.

library [COMPUT SCI] 1. A computerized facility containing a collection of organized information used for reference. 2. An organized collection of computer programs together with the associated program listings, documentation, users' directions, decks, and tapes.

libration [PHYS] Any oscillatory rotational motion, such as that of the moon, or of a molecule in a solid which does not have enough energy to make full rotations.

libration in latitude See lunar libration.

Libytheidae [INV ZOO] The snout butterflies, a family of cosmopolitan lepidopteran insects in the suborder Heteroneura distinguished by long labial palps; represented in North America by a single species.

Liceaceae [MYCOL] A family of plasmodial slime molds in the order Liceales.

Liceales [MYCOL] An order of plasmodial slime molds in the subclass Myxogastromycetidae.

lichen [BOT] The common name for members of the Lichenes.

Lichenberger figures See Lichtenberg figures.

lichen blue See litmus.

Lichenes [BOT] A group of organisms consisting of fungi and algae growing together symbiotically.

Lichenes Imperfecti [BOT] A class of the Lichenes containing species with no known method of sexual reproduction.

Lichnophorina [INV ZOO] A suborder of ciliophoran protozoans belonging to the Heterotrichida.

Lichtenberg figures [ELEC] Patterns produced on a photographic emulsion, or in fine powder spread over the surface of a solid dielectric, by an electric discharge produced by a high transient voltage. Also known as Lichenberger figures.

lidar [OPTICS] An instrument in which a ruby laser generates intense infrared pulses in beam widths as small as 30 seconds of arc; beam reflections and scattering effects of clouds, smog layers, and some atmospheric discontinuities are measured by radar techniques; it can also be used for tracking weather balloons, smoke puffs, and rocket trails. Derived from laser infrared radar.

Lie algebra [MATH] The algebra of vector fields on a manifold with additive operation given by pointwise sum and multiplication by the Lie bracket.

liebigite [MINERAL] $Ca_2U(CO_3)_4 \cdot 10H_2O$ An apple- or yellow-green mineral composed of hydrous uranium calcium carbonate; occurs as a coating or concretion in rock.

lie detector [ENG] An instrument that indicates or records one or more functional variables of a person's body while the person undergoes the emotional stress associated with a lie. Also known as polygraph; psychointegroammeter.

Lie group [MATH] A topological group which is also a differentiable manifold in such a way that the group operations are themselves analytic functions.

Liesegang banding [GEOL] Colored or compositional rings or bands in a fluid-saturated rock due to rhythmic precipitation. Also known as Liesegang rings.

Liesegang rings See Liesegang banding.

life form [ECOL] The form characteristically taken by a plant at maturity.

life zone [ECOL] A portion of the earth's land area having a generally uniform climate and soil, and a biota showing a high degree of uniformity in species composition and adaptation.

LIFO See last in, first out.

lift [FL MECH] See aerodynamic lift. [MECH ENG] See elevator. [MIN ENG] 1. The vertical height traveled by a cage in a shaft. 2. The distance between the first level and the surface or between any two levels. 3. Any of the various gangways from which coal is raised at a slope colliery.

lift bridge [CIV ENG] A drawbridge whose movable spans are raised vertically.

lift coefficient [AERO ENG] The quantity $C_L = 2L/\rho V^2 S$, where L is the lift of a whole airplane wing, ρ is the mass density of the air, V is the free-stream velocity, and S is the wing area; this is also applicable to other airfoils.

lift-drag ratio [AERO ENG] The lift of an aerodynamic form, such as an airplane wing, divided by the drag.

lifting reentry [AERO ENG] A reentry into the atmosphere by a space vehicle where aerodynamic lift is used, allowing a more gradual descent, greater accuracy in landing at a predetermined spot; it can accommodate greater errors in the guidance system and greater temperature control.

lift pump [MECH ENG] A pump for lifting fluid to the pump's own level.

lift valve [MECH ENG] A valve that moves perpendicularly to the plane of the valve seat.

ligament [ENG] The section of solid material in a tube sheet or shell between adjacent holes. [HISTOL] A flexible, dense white fibrous connective tissue joining, and sometimes encapsulating, the articular surfaces of bones.

ligamentum nuchae See nuchal ligament.

ligand [CHEM] The molecule, ion, or group bound to the central atom in a chelate or a coordination compound; an example is the ammonia molecules in $[Co(NH_3)_6]^{3+}$.

ligase [BIOCHEM] An enzyme that catalyzes the union of two molecules, involving the participation of a nucleoside triphosphate which is converted to a nucleoside diphosphate or monophosphate. Also known as synthetase.

light [OPTICS] 1. Electromagnetic radiation with wavelengths capable of causing the sensation of vision, ranging approximately from 400 (extreme violet) to 770 nanometers (extreme red). Also known as light radiation; visible radiation. 2. More generally, electromagnetic radiation of any wavelength; thus, the term is sometimes applied to infrared and ultraviolet radiation.

light adaptation [PHYSIO] The disappearance of dark adaptation; the chemical processes by which the eyes, after exposure to a dim environment, become accustomed to bright illumination, which initially is perceived as quite intense and uncomfortable.

light amplification by stimulated emission of radiation See laser.

light amplifier [ELECTR] 1. Any electronic device which, when actuated by a light image, reproduces a similar image of enhanced brightness, and which is capable of operating at very low light levels without introducing spurious brightness variations (noise) into the reproduced image. Also known as image intensifier. 2. See laser amplifier.

light-beam galvanometer See d'Arsonval galvanometer.

light-beam oscillograph [ELECTROMAG] An oscillograph in which a beam of light, focused to a point by a lens, is reflected from a tiny mirror attached to the moving coil of a galvanometer onto a photographic film moving at constant speed.

light bulb *See* incandescent lamp.

light chopper [ELECTR] A rotating fan or other mechanical device used to interrupt a light beam that is aimed at a phototube, to permit alternating-current amplification of the phototube output and to make its output independent of strong, steady ambient illumination.

light curve [ASTROPHYS] A graph showing the variations in brightness of a celestial object; the stellar magnitude is usually shown on the vertical axis, and time is the horizontal coordinate.

light-emitting diode [ELECTR] A semiconductor diode that converts electric energy efficiently into spontaneous and noncoherent electromagnetic radiation at visible and near-infrared wavelengths by electroluminescence at a forward-biased *pn* junction. Abbreviated LED. Also known as solid-state lamp.

light ends [MATER] The lower-boiling components of a mixture of hydrocarbons, such as those evaporated or distilled off easily in comparison to the bulk of the mixture; for hydrocarbon mixtures, usually considered to be butane and lighter.

lighter-than-air craft [AERO ENG] An aircraft, such as a dirigible, that weighs less than the air it displaces.

light exposure [OPTICS] A measure of the total amount of light falling on a surface; equal to the integral over time of the luminance of the surface. Also known as exposure.

light filter *See* color filter.

light frost [HYD] A thin and more or less patchy deposit of hoarfrost on surface objects and vegetation.

light guide *See* optical fiber.

light intensity *See* luminous intensity.

light meter [ENG] A small, portable device for measuring illumination; an exposure meter is a specific application, being calibrated to give photographic exposures.

light microscope *See* optical microscope.

light modulator [ELECTR] The combination of a source of light, an appropriate optical system, and a means for varying the resulting light beam to produce an optical sound track on motion picture film.

lightning [GEOPHYS] The large spark produced by an abrupt discontinuous discharge of electricity through the air, resulting most often from the creation and separation of electric charge in cumulonimbus clouds.

lightning arrester [ELEC] A protective device designed primarily for connection between a conductor of an electrical system and ground to limit the magnitude of transient overvoltages on equipment. Also known as arrester; surge arrester.

lightning rod [ELEC] A metallic rod set up on an exposed elevation of a structure and connected to a low-resistance ground to intercept lightning discharges and to provide a direct conducting path to ground for them.

light-of-the-night-sky *See* airglow.

light panel *See* electroluminescent panel.

light pen [ELECTR] A tiny photocell or photomultiplier, mounted with or without fiber or plastic light pipe in a pen-shaped housing; it is held against a cathode-ray screen to make measurements from the screen or to change the nature of the display.

light pillar *See* sun pillar.

light quantum *See* photon.

light radiation *See* light.

light-red silver ore *See* proustite.

light reflex [PHYSIO] The postural orientation response of certain aquatic forms stimulated by the source of light; receptors may be on the ventral or dorsal surface.

light relay *See* photoelectric relay.

light-ruby silver *See* proustite.

light-sensitive cell *See* photodetector.

light-sensitive detector *See* photodetector.

light-sensitive tube *See* phototube.

light sensor photodevice *See* photodetector.

lightship [NAV] A distinctively marked vessel anchored or moored at a charted point, to serve as an aid to navigation.

light transport aircraft [AERO ENG] A multiengine airplane having a maximum passenger capacity of 30 seats and a gross weight of about 35,000 pounds (16,000 kilograms).

light-year [ASTROPHYS] A unit of measurement of astronomical distance; it is the distance light travels in one sidereal year and is equivalent to 9.461×10^{12} kilometers or 5.879×10^{12} miles.

Ligiidae [INV ZOO] A family of primitive terrestrial isopods in the suborder Oniscoidea.

lignify [BOT] To convert cell wall constituents into wood or woody tissue by chemical and physical changes.

lignin [BIOCHEM] A substance that together with cellulose forms the woody cell walls of plants and cements them together; a colorless to brown substance removed from paper-pulp sulfite liquor.

lignite [GEOL] Coal of relatively recent origin, intermediate between peat and bituminous coal; often contains patterns from the wood from which it formed. Also known as brown coal.

lignumvitae [BOT] *Guaiacum sanctum.* A medium-sized evergreen tree in the order Sapindales that yields a resin or gum known as gum guaiac or resin of guaiac. Also known as hollywood lignumvitae.

ligule [BOT] **1.** A small outgrowth in the axis of the leaves in Selaginellales. **2.** A thin outgrowth of a foliage leaf or leaf sheath. [INV ZOO] A small lobe on the parapodium of certain polychaetes.

likelihood [MATH] The likelihood of a sample of independent values of $x_1, x_2, \ldots, x_n$, with $f(x)$ the probability function, is the product $f(x_1) \circ f(x_2) \circ \ldots \circ f(x_n)$.

likelihood ratio [STAT] The probability of a random drawing of a specified sample from a population, assuming a given hypothesis about the parameters of the population, divided by the probability of a random drawing of the same sample, assuming that the parameters of the population are such that this probability is maximized.

Liliaceae [BOT] A family of the order Liliales distinguished by six stamens, typically narrow, parallel-veined leaves, and a superior ovary.

Liliales [BOT] An order of monocotyledonous plants in the subclass Liliidae having the typical characteristics of the subclass.

Liliatae *See* Liliopsida.

Liliidae [BOT] A subclass of the Liliopsida; all plants are syncarpous and have a six-membered perianth, with all members petaloid.

Liliopsida [BOT] The monocotyledons, making up a class of the Magnoliophyta; characterized generally by a single cotyledon, parallel-veined leaves, and stems and roots lacking a well-defined pith and cortex.

lily [BOT] **1.** Any of the perennial bulbous herbs with showy unscented flowers constituting the genus *Lilium*. **2.** Any of various other plants having similar flowers.

lily-pad ice *See* pancake ice.

Limacidae [INV ZOO] A family of gastropod mollusks containing the slugs.

limb [ANAT] An extremity or appendage used for locomotion or prehension, such as an arm or a leg. [ASTRON] The circular outer edge of a celestial body; the half with the greater altitude is called the upper limb, and the half with the lesser altitude, the lower limb. [BOT] A large primary tree branch. [DES ENG] **1.** The graduated margin of an arc or circle in an instrument for measuring angles, as that part of a marine sextant carrying the altitude scale. **2.** The graduated staff of a leveling rod. [GEOL] One of the two sections of an anticline or syncline on either side of the axis. Also known as flank.

limbic system [ANAT] The inner edge of the cerebral cortex in the medial and temporal regions of the cerebral hemispheres.

limburgite [PETR] A dark, glass-rich igneous rock with abundant large crystals of olivine and pyroxene and with little or no feldspar.

limbus [BIOL] A border clearly defined by its color or structure, as the margin of a bivalve shell or of the cornea of the eye.

lime [BOT] *Citrus aurantifolia*. A tropical tree with elliptic oblong leaves cultivated for its acid citrus fruit which is a hesperidium.

lime mortar [MATER] A mixture of hydrated lime, sand, and water having a compressive strength up to 400 pounds per square inch (2.8×10^6 newtons per square meter); used for interior non-load-bearing walls in buildings.

limestone [PETR] **1.** A sedimentary rock composed dominantly (more than 95%) of calcium carbonate, principally in the form of calcite; examples include chalk and travertine. **2.** Any rock containing 80% or more of calcium carbonate or magnesium carbonate.

liminal contrast *See* threshold contrast.

limited-access data [COMPUT SCI] Data to which only authorized users have access.

limited-entry decision table [COMPUT SCI] A decision table in which the condition stub specifies exactly the condition or the value of the variable.

limited integrator [ELECTR] A device used in analog computers that has two input signals and one output signal whose value is proportional to the integral of one of the input signals with respect to the other as long as this output signal does not exceed specified limits.

limiter [ELECTR] An electronic circuit used to prevent the amplitude of an electronic waveform from exceeding a specified level while preserving the shape of the waveform at amplitudes less than the specified level. Also known as amplitude limiter; amplitude-limiting circuit; automatic peak limiter; clipper; clipping circuit; limiter circuit; peak limiter. [NUC PHYS] A material aperture in a fusion power reactor which collects particles from the outer surfaces of the plasmas to control their transport to regions of low density.

limiter circuit *See* limiter.

limit governor [MECH ENG] A mechanical governor that takes over control from the main governor to shut the machine down when speed reaches a predetermined excess above the allowable rated. Also known as topping governor.

limit inferior [MATH] **1.** The limit inferior of a sequence whose nth term is a_n is the limit as N approaches infinity of the greatest lower bound of the terms a_n for which n is greater than N; denoted by $\liminf_{n \to \infty} a_n$ or $\underline{\lim}_{n \to \infty} a_n$. **2.** The limit inferior of a function f at a point c is the limit as ϵ approaches zero of the greatest lower bound of $f(x)$ for $|x - c| < \epsilon$ and $x \neq c$; denoted by $\liminf_{x \to c} f(x)$ or $\underline{\lim}_{x \to c} f(x)$.

limiting [ELECTR] A desired or undesired amplitude-limiting action performed on a signal by a limiter. Also known as clipping.

limit-load design *See* ultimate-load design.

limit of a function [MATH] A function $f(x)$ has limit L as x tends to c if given any positive number ϵ (no matter how small) there is a positive number δ such that if x is in the domain of f, x is not c, and $|x - c| < \delta$, then $|f(x) - L| < \epsilon$; written $\lim_{x \to c} f(x) = L$.

limit on the left [MATH] The limit on the left of the function f at a point c is the limit of f at c which would be obtained if only values of x less than c were taken into account; more precisely, it is the number L which has the property that for any positive number ϵ, there is a positive number δ so that if x is in the domain of f and $0 < c - x < \delta$ then $|f(x) - L| < \epsilon$; denoted by $\lim_{x \to c^-} f(x) = L$, or $f(c^-) = L$. Also known as left-hand limit.

limit on the right [MATH] The limit on the right of the function $f(x)$ at a point c is the limit of f at c which would be obtained if only values of x greater than c were taken into account; more precisely, it is the number L which has the property that for any positive number ϵ there is a positive number δ so that if x is in the domain of f and $0 < x - c < \delta$, then $|f(x) - L| < \epsilon$; denoted by $\lim_{x \to c^+} f(x) = L$ or $f(c^+) = L$. Also known as right-hand limit.

limit point *See* cluster point.

limit superior [MATH] **1.** The limit superior of a sequence whose nth term is a_n is the limit as N approaches infinity of the least upper bound of the terms a_n for which n is greater than N; denoted by $\limsup_{x \to \infty} a_n$ or $\overline{\lim}_{x \to \infty} a_n$. **2.** The limit superior of a function f at a point c is the limit as ϵ approaches zero of the least upper bound of $f(x)$ for $|x - c| < \epsilon$ and $x \neq c$; denoted by $\limsup_{x \to c} f(x)$ or $\overline{\lim}_{x \to c} f(x)$.

limit switch [ELEC] A switch designed to cut off power automatically at or near the limit of travel of a moving object controlled by electrical means.

limivorous [ZOO] Feeding on mud, as certain annelids, for the organic matter it contains.

Limnebiidae [INV ZOO] The minute moss beetles, a family of coleopteran insects in the superfamily Hydrophiloidea.

limnetic [ECOL] Of, pertaining to, or inhabiting the pelagic region of a body of fresh water.

Limnichidae [INV ZOO] The minute false water beetles, a cosmopolitan family of coleopteran insects in the superfamily Dryopoidea.

limnite *See* bog iron ore.

Limnocharitaceae [BOT] A family of monocotyledonous plants in the order Alismatales characterized by schizogenous secretory canals, multiaperturate pollen, several or many ovules, and a horseshoe-shaped embryo.

limnology [ECOL] The science of the life and conditions for life in lakes, ponds, and streams.

Limnomedusae [INV ZOO] A suborder of hydrozoan coelenterates in the order Hydroida characterized by naked hydroids.

Limnoriidae [INV ZOO] The gribbles, a family of isopod crustaceans in the suborder Flabellifera that burrow into submerged marine timbers.

limonite [MINERAL] A group of brown or yellowish-brown, amorphous, naturally occurring ferric oxides of variable composition; commonly formed secondary material by oxidation of iron-bearing minerals; a minor ore of iron. Also known as brown hematite; brown iron ore.

limpet [INV ZOO] Any of several species of marine gastropod mollusks composing the families Patellidae and Acmaeidae which have a conical and tentlike shell with ridges extending from the apex to the border.

Limulacea [INV ZOO] A group of horseshoe crabs belonging to the Limulida.

Limulida [INV ZOO] A subgroup of Xiphosurida including all living members of the subclass.

Limulodidae [INV ZOO] The horseshoe crab beetles, a family of coleopteran insects in the superfamily Staphylinoidea.

linac See linear accelerator.

Linaceae [BOT] A family of herbaceous or shrubby dicotyledonous plants in the order Linales characterized by mostly capsular fruit, stipulate leaves, and exappendiculate petals.

Linales [BOT] An order of dicotyledonous plants in the subclass Orsidae containing simple-leaved herbs or woody plants with hypogynous, regular, syncarpous flowers having five to many stamens which are connate at the base.

lindackerite [MINERAL] $Cu_6Ni_3(AsO_4)_4(SO_4)(OH)_4$ A light-green or apple-green mineral composed of hydrous basic sulfate and arsenate of nickel and copper; occurs in tabular crystals or massive.

Linde's rule [SOLID STATE] The rule that the increase in electrical resistivity of a monovalent metal produced by a substitutional impurity per atomic percent impurity is equal to $a + b(v - 1)^2$, where a and b are constants for a given solvent metal and a given row of the periodic table for the impurity, and v is the valence of the impurity.

L indicator See L scope.

line [BOT] A unit of length, equal to $\frac{1}{12}$ inch, or approximately 2.117 millimeters; it is most frequently used by botanists in describing the size of plants. [ELECTR] 1. The path covered by the electron beam of a television picture tube in one sweep from left to right across the screen. 2. One horizontal scanning element in a facsimile system. 3. See trace. [MATH] The set of points $(x_1,...,x_n)$ in euclidean space, each of whose coordinates is a linear function of a single parameter t; $x_i = f_i(t)$.

linea alba [ANAT] A tendinous ridge extending in the median line of the abdomen from the pubis to the tiphoid process and formed by the blending of aponeuroses of the oblique and transverse muscles of the abdomen.

lineage [GEN] Descent from a common progenitor.

lineament [ASTRON] A prominent linear feature on the lunar surface. [GEOL] A straight or gently curved, lengthy topographic feature expressed as depressions or lines of depressions. Also known as linear. [GRAPHICS] A structurally controlled line on an aerial photograph; applied to lines representing beds, veins, faults, rock boundaries, and such.

linear [CONT SYS] Having an output that varies in direct proportion to the input. [GEOL] See linea-

ment. [SCI TECH] 1. Of or relating to a line. 2. Having a single dimension.

linear accelerator [NUCLEO] A particle accelerator which accelerates electrons, protons, or heavy ions in a straight line by the action of alternating voltages. Also known as linac.

linear actuator [MECH ENG] A device that converts some kind of power, such as hydraulic or electric power, into linear motion.

linear algebra [MATH] The study of vector spaces and linear transformations.

linear amplifier [ELECTR] An amplifier in which changes in output current are directly proportional to changes in applied input voltage.

linear array [ELECTROMAG] An antenna array in which the dipole or other half-wave elements are arranged end to end on the same straight line. Also known as collinear array.

linear circuit See linear network.

linear comparator [ELECTR] A comparator circuit which operates on continuous, or nondiscrete, waveforms. Also known as continuous comparator.

linear control system [CONT SYS] A linear system whose inputs are forced to change in a desired manner as time progresses.

linear differential equation [MATH] A differential equation in which all derivatives occur linearly, and all coefficients are functions of the independent variable.

linear distortion [ELECTR] Amplitude distortion in which the output signal envelope is not proportional to the input signal envelope and no alien frequencies are involved.

linear equation [MATH] A linear equation in the variables $x_1, \ldots, x_n$ and y is any equation of the form $a_1x_1 + a_2x_2 + \ldots + a_nx_n = y$.

linear flow structure See platy flow structure.

linear function See linear transformation.

linearity [MATH] The property whereby a mathematical system is well behaved (in the context of the given system) with regard to addition and scalar multiplication. [PHYS] The relationship that exists between two quantities when a change in one of them produces a directly proportional change in the other.

linearity control [ELECTR] A cathode-ray-tube control which varies the distribution of scanning speed throughout the trace interval. Also known as distribution control.

linearization [CONT SYS] 1. The modification of a system so that its outputs are approximately linear functions of its inputs, in order to facilitate analysis of the system. 2. The mathematical approximation of a nonlinear system, whose departures from linearity are small, by a linear system corresponding to small changes in the variables about their average values.

linearly ordered set [MATH] A set with an ordering $\leq$ such that for any two elements a and b either $a \leq b$ or $b \leq a$. Also known as chain; serially ordered set; simply ordered set.

linear manifold [MATH] A subset of a vector space which is itself a vector space with the induced operations of addition and scalar multiplication.

linear molecule [PHYS CHEM] A molecule whose atoms are arranged so that the bond angle between each is 180°; an example is carbon dioxide, CO_2.

linear momentum See momentum.

linear motion See rectilinear motion.

linear network [ELEC] A network in which the parameters of resistance, inductance, and capacitance are constant with respect to current or voltage, and in which the voltage or current of sources is independent of or directly proportional to other voltages and currents, or their derivatives, in the network. Also known as linear circuit.

linear operator   *See* linear transformation.

linear oscillator   *See* harmonic oscillator.

linear parallax   *See* absolute stereoscopic parallax.

linear polarization   [OPTICS] Polarization of an electromagnetic wave in which the electric vector at a fixed point in space remains pointing in a fixed direction, although varying in magnitude. Also known as plane polarization.

linear polymer   [ORG CHEM] A polymer whose molecule is arranged in a chainlike fashion with few branches or bridges between the chains.

linear programming   [MATH] The study of maximizing or minimizing a linear function $f(x_1, \ldots, x_n)$ subject to given constraints which are linear inequalities involving the variables x_i.

linear rectifier   [ELECTR] A rectifier, the output current of voltage of which contains a wave having a form identical with that of the envelope of an impressed signal wave.

linear space   *See* vector space.

linear-sweep delay circuit   [ELECTR] A widely used form of linear time-delay circuit in which the input signal initiates action by a linear sawtooth generator, such as the bootstrap or Miller integrator, whose output is then compared with a calibrated direct-current reference voltage level.

linear-sweep generator   [ELECTR] An electronic circuit that provides a voltage or current that is a linear function of time; the waveform is usually recurrent at uniform periods of time.

linear system   [CONT SYS] A system in which the outputs are components of a vector which is equal to the value of a linear operator applied to a vector whose components are the inputs.   [MATH] A system where all the interrelationships among the quantities involved are expressed by linear equations which may be algebraic, differential, or integral.

linear transformation   [MATH] A function T defined in a vector space E and having its values in another vector space over the same field, such that if f and g are vectors in E, and c is a scalar, then $T(f+g) = Tf + Tg$ and $T(cf) = c(Tf)$. Also known as homogeneous transformation; linear function; linear operator.

linear variable-differential transformer   [ELECTR] A transformer in which a diaphragm or other transducer sensing element moves an armature linearly inside the coils of a differential transformer, to change the output voltage by changing the inductances of the coils in equal but opposite amounts. Abbreviated LVDT.

linear velocity   *See* velocity.

line at infinity   *See* ideal line.

line balance   [ELEC] 1. Degree of electrical similarity of the two conductors of a transmission line. 2. Matching impedance, equaling the impedance of the line at all frequencies, that is used to terminate a two-wire line.

line-balance converter   *See* balun.

line block   *See* line cut.

line-building-out network   *See* impedance-matching network.

line circuit   [ELEC] 1. Relay equipment associated with each station connected to a dial or manual switchboard. 2. A circuit to interconnect an individual telephone and a channel terminal.

line code   [COMPUT SCI] The single instruction required to solve a specific type of problem on a special-purpose computer.

line-controlled blocking oscillator   [ELECTR] A circuit formed by combining a monostable blocking oscillator with an open-circuit transmission line in the regenerative circuit; it is capable of generating pulses with large amounts of power.

line cord   [ELEC] A two-wire cord terminating in a two-prong plug at one end and connected permanently to a radio receiver or other appliance at the other end; used to make connections to a source of power. Also known as power cord.

line cut   [GRAPHICS] A relief printing plate made by photographing a design and then transferring the negative onto a zinc or copper plate that is then developed, with the lines that will form the printing surface being protected and the rest of the plate etched down; used exclusively for the reproduction of materials executed in black (or a color) and white, with no intermediate shades of gray (or tones). Also known as line block; line engraving; line etching; line plate.

line defect   *See* dislocation.

line dot matrix   [COMPUT SCI] A line printer that uses the dot matrix printing technique. Also known as parallel dot character printer.

line drop   [ELEC] The voltage drop existing between two points on a power line or transmission line, due to the impedance of the line.

line editor   [COMPUT SCI] A text-editing system that stores a file of discrete lines of text to be printed out on the console (or displayed) and manipulated on a line-by-line basis, so that editing operations are limited and are specified for lines identified by a specific number.

line engraving   *See* line cut.

line etching   *See* line cut.

line fault   [ELEC] A defect, such as an open circuit, short circuit, or ground, in an electric line for transmission or distribution of power or of speech, music, or other content.

line feed   [COMPUT SCI] 1. Signal that causes a printer to feed the paper up a discrete number of lines. 2. Rate at which paper is fed through a printer.

line filter   [ELEC] 1. A filter inserted between a power line and a receiver, transmitter, or other unit of electric equipment to prevent passage of noise signals through the power line in either direction. Also known as power-line filter. 2. A filter inserted in a transmission line or high-voltage power line for carrier communication purposes.

line frequency   [ELECTR] The number of times per second that the scanning spot sweeps across the screen in a horizontal direction in a television system. Also known as horizontal frequency; horizontal line frequency.

line-frequency blanking pulse   *See* horizontal blanking pulse.

line graph   [MATH] A graph in which successive points representing the value of a variable at selected values of the independent variable are connected by straight lines.

line map   *See* planimetric map.

line microphone   [ENG ACOUS] A highly directional microphone consisting of a single straight-line element or an array of small parallel tubes of different lengths, with one end of each abutting a microphone element. Also known as machine-gun microphone.

line of apsides   [ASTRON] 1. The line connecting the two points of an orbit that are nearest and farthest from the center of attraction, as the perigee and apogee of the moon or the perihelion and aphelion of a planet. **2.** The length of this line.

line of collimation [OPTICS] In a surveying telescope, the imaginary line through the optical center of the object glass and the cross-hair intersection in the diaphragm.

line of flight [MECH] The line of movement, or the intended line of movement, of an aircraft, guided missile, or projectile in the air.

line of flux *See* line of force.

line of force [PHYS] An imaginary line in a field of force (such as an electric, magnetic, or gravitational field) whose tangent at any point gives the direction of the field at that point; the lines are spaced so that the number through a unit area perpendicular to the field represents the intensity of the field. Also known as flux line; line of flux.

line of magnetic induction *See* maxwell.

line of position [NAV] A line indicating a series of possible positions of a craft, determined by observation or measurement. Also known as position line.

line of sight [ELECTROMAG] The straight line for a transmitting radar antenna in the direction of the beam. [SCI TECH] A straight, unobstructed path or line between two points, as between an observer's eye and a target.

line-of-sight communication [COMMUN] Electromagnetic wave propagation, usually microwaves, in a straight line between the transmitter and receiver; the useful transmission distance is generally limited to the horizon as sighted from the elevation of the transmitter.

line-of-sight velocity *See* radial velocity.

line of strike *See* strike.

line pair [SPECT] In spectrographic analysis, a particular spectral line and the internal standard line with which it is compared to determine the concentration of a substance.

line plate *See* line cut.

line printer [COMPUT SCI] A device that prints an entire line in a single operation, without necessarily printing one character at a time.

liner [DES ENG] A replaceable tubular sleeve inside a hydraulic or pump-pressure cylinder in which the piston travels. [ENG] A string of casing in a borehole. [MET] **1.** The cylindrical chamber that holds the billet for extrusion. **2.** The slab of coating metal that is placed on the core alloy and is subsequently rolled down to form a clad composite. [MIN ENG] **1.** A foot piece for uprights in timber sets. **2.** Timber supports erected to reinforce existing sets which are beginning to collapse due to heavy strata pressure. **3.** A bar put up between two other bars to assist in carrying the roof. **4.** Replaceable facings inside a grinding mill. [NAV ARCH] A merchant vessel engaged in regular, usually high-speed service.

line skew [COMPUT SCI] In character recognition, a form of line misregistration, when the string of characters to be recognized appears in a uniformly slanted condition with respect to a real or imaginary baseline.

line spectrum [SPECT] **1.** A spectrum of radiation in which the quantity being studied, such as frequency or energy, takes on discrete values. **2.** Conventionally, the spectra of atoms, ions, and certain molecules in the gaseous phase at low pressures, distinguished from band spectra of molecules, which consist of a pattern of closely spaced spectral lines which could not be resolved by early spectroscopes.

line speed [COMMUN] Maximum rate at which signals may be transmitted over a given channel, usually in bauds or bits per second.

line switching [COMMUN] A telephone switching system in which a switch attached to a subscriber line connects an originating call to an idle part of the switching apparatus. [ELECTR] Connecting or disconnecting the line voltage from a piece of electronic equipment.

line synchronizing pulse *See* horizontal synchronizing pulse.

line transformer [ELEC] Transformer connecting a transmission line to terminal equipment; used for isolation, line balance, impedance matching, or additional circuit connections.

line voltage [ELEC] The voltage provided by a power line at the point of use.

Linguatulida [INV ZOO] The equivalent name for Pentastomida.

Linguatuloidea [INV ZOO] A suborder of pentastomid arthropods in the order Porocephalida; characterized by an elongate, ventrally flattened, annulate, posteriorly attenuated body, simple hooks on the adult, and binate hooks in the larvae.

linguistics [COMMUN] The study of human speech in its various aspects, especially units of language, phonetics, syntax, accent, semantics, and grammar.

Lingulacea [INV ZOO] A superfamily of inarticulate brachiopods in the order Lingulida characterized by an elongate, biconvex calcium phosphate shell, with the majority having a pedicle.

lingulate [BIOL] Tongue- or strap-shaped.

Lingulida [INV ZOO] An order of inarticulate brachiopods represented by two living genera, *Lingula* and *Glottidia*.

link [CIV ENG] A standardized part of a surveyor's chain, which is 7.92 inches (20.1168 centimeters) in the Gunter's chain and 1 foot (30.48 centimeters) in the engineer's chain. [COMMUN] General term used to indicate the existence of communications facilities between two points. [DES ENG] **1.** One of the rings of a chain. **2.** A connecting piece in the moving parts of a machine.

linkage [COMPUT SCI] In programming, coding that connects two separately coded routines. [ELECTROMAG] *See* flux linkage. [GEN] Failure of nonallelic genes to recombine at random as a result of their being located within the same chromosome or chromosome fragment. [MECH ENG] A mechanism that transfers motion in a desired manner by using some combination of bar links, slides, pivots, and rotating members.

Linofilm typesetter [GRAPHICS] A photographic typesetting machine consisting of a keyboard, photographic unit, corrector, and composer; the keyboard produces a perforated paper tape containing information for operating the photographic unit, which produces right-reading positive type on film or on photographic paper; the corrector affixes correct lines in position automatically; and the composer produces a made-up page.

lintel [BUILD] A horizontal member over an opening, such as a door or window, usually carrying the wall load.

linters [MATER] Short residual fibers that adhere to ginned cottonseed; used for making fabrics that do not require long fibers, as plastic fillers, and in the manufacture of cellulosic plastics.

LIOCS [COMPUT SCI] Set of routines handling buffering, blocking, label checking, and overlap of input/output with processing. Derived from logical input/output control system.

lion [VERT ZOO] *Felis leo.* A large carnivorous mammal of the family Felidae distinguished by a tawny coat and blackish tufted tail, with a heavy blackish or dark-brown mane in the male.

Lion *See* Leo.

Liopteridae [INV ZOO] A small family of hymenopteran insects in the superfamily Cynipoidea.

Liouville equation [STAT MECH] An equation which states that the density of points representing an ensemble of systems in phase space which are in the neighborhood of some given system does not change with time.

Liouville's theorem [MATH] Every function of a complex variable which is bounded and analytic in the entire complex plane must be constant.

lip [ANAT] A fleshy fold above and below the entrance to the mouth of mammals. [CIV ENG] A parapet placed on the downstream margin of a millrace or apron in order to minimize scouring of the river bottom. [DES ENG] Cutting edge of a fluted drill formed by the intersection of the flute and the lip clearance angle, and extending from the chisel edge at the web to the circumference. [MED] The margin of an open wound. [SCI TECH] The edge of a hollow cavity or container.

Lipalian [GEOL] A hypothetical geologic period that supposedly antedated the Cambrian.

Liparidae [INV ZOO] The equivalent name for Lymantriidae.

lipase [BIOCHEM] An enzyme that catalyzes the hydrolysis of fats or the breakdown of lipoproteins.

Liphistiidae [INV ZOO] A family of spiders in the suborder Liphistiomorphae in which the abdomen shows evidence of true segmentation by the presence of tergal and sternal plates.

Liphistiomorphae [INV ZOO] A suborder of arachnids in the order Araneida containing families with a primitively segmented abdomen.

lipid [BIOCHEM] One of a class of compounds which contain long-chain aliphatic hydrocarbons and their derivatives, such as fatty acids, alcohols, amines, amino alcohols, and aldehydes; includes waxes, fats, and derived compounds. Also known as lipin; lipoid.

lipid metabolism [BIOCHEM] The physiologic and metabolic processes involved in the assimilation of dietary lipids and the synthesis and degradation of lipids.

lipid storage disease [MED] Any of various rare diseases characterized by the accumulation of large histiocytes containing lipids throughout reticuloendothelial tissues; examples are Goucher's disease, Niemann-Pick disease, and amaurotic familial idiocy.

lipin [BIOCHEM] **1.** A compound lipid, such as a cerebroside. **2.** *See* lipid.

lipoblastoma *See* liposarcoma.

lipochrome [BIOCHEM] Any of various fat-soluble pigments, such as carotenoid, occurring in natural fats. Also known as chromolipid.

lipoid [BIOCHEM] **1.** A fatlike substance. **2.** *See* lipid.

lipoma [MED] A benign tumor composed of fat cells.

Lipomycetoideae [MICROBIO] A subfamily of oxidative yeasts in the family Saccharomycetaceae characterized by budding cells and a saclike appendage which develops into an ascus.

lipomyxoma *See* liposarcoma.

lipopolysaccharide [BIOCHEM] Any of a class of conjugated polysaccharides consisting of a polysaccharide combined with a lipid.

lipoprotein [BIOCHEM] Any of a class of conjugated proteins consisting of a protein combined with a lipid.

liposarcoma [MED] A sarcoma originating in adipose tissue. Also known as embryonal-cell lipoma; fetal fat-cell lipoma; infiltrating lipoma; lipoblastoma; lipomyxoma; myxolipoma; myxoma lipomatodes.

Lipostraca [PALEON] An order of the subclass Branchiopoda erected to include the single fossil species *Lepidocaris rhyniensis.*

lipotropic hormone [BIOCHEM] Any hormone having lipolytic activity on adipose tissue.

Lipotyphla [VERT ZOO] A group of insectivoran mammals composed of insectivores which lack an intestinal cecum and in which the stapedial artery is the major blood supply to the brain.

Lippmann electrometer *See* capillary electrometer.

liptinite *See* exinite.

liq pt *See* pint.

liquefaction [PHYS] A change in the phase of a substance to the liquid state; usually, a change from the gaseous to the liquid state, especially of a substance which is a gas at normal pressure and temperature.

liquefied gas [MATER] A gaseous compound or mixture converted to the liquid phase by cooling or compression; examples are liquefied petroleum gas (LPG), liquefied natural gas (LNG), liquid oxygen, and liquid ammonia.

liquefied natural gas [MATER] A product of natural gas which consists primarily of methanes; its critical temperature is about $-100°F$ ($-73°C$), and thus it must be liquefied by cooling to cryogenic temperatures and must be well insulated to be held in the liquid state; used as a domestic fuel. Abbreviated LNG.

liquefied petroleum gas [MATER] A product of petroleum gases; principally propane and butane, it must be stored under pressure to keep it in a liquid state; it is often stored in metal cylinders (bottled gas) and used as fuel for tractors, trucks, and buses, and as a domestic cooking or heating fuel in rural areas. Abbreviated LPG.

liqueur [FOOD ENG] An alcoholic beverage prepared by combining a spirit, usually brandy, with certain flavorings and sugar.

liquid [PHYS] A state of matter intermediate between that of crystalline substances and gases in which a substance has the capacity to flow under extremely small shear stresses and conforms to the shape of a confining vessel, but is relatively incompressible, lacks the capacity to expand without limit, and can possess a free surface.

liquid chromatography [ANALY CHEM] A form of chromatography employing a liquid as the moving phase and a solid or a liquid on a solid support as the stationary phase; techniques include column chromatography, gel permeation chromatography, and partition chromatography.

liquid-column gage *See* U-tube manometer.

liquid cooling [ENG] Use of circulating liquid to cool process equipment and hermetically sealed components such as transistors.

liquid crystal [PHYS CHEM] A liquid which is not isotropic; it is birefringent and exhibits interference patterns in polarized light; this behavior results from the orientation of molecules parallel to each other in large clusters.

liquid crystal display [ELECTR] A digital display that consists of two sheets of glass separated by a sealed-in, normally transparent, liquid crystal material; the outer surface of each glass sheet has a transparent conductive coating such as tin oxide or indium oxide, with the viewing-side coating etched into character-forming segments that have leads going to the edges of the display; a voltage applied between front and back electrode coatings disrupts the orderly arrangement of the molecules, darkening the liquid enough to form visible characters even though no light is generated. Abbreviated LCD.

liquid-drop model of nucleus   [NUC PHYS] A model of the nucleus in which it is compared to a drop of incompressible liquid, and the nucleons are analogous to molecules in the liquid; used to study binding energies, fission, collective motion, decay, and reactions. Also known as drop model of nucleus.

liquid extraction   See solvent extraction.

liquid fuel   [MATER] A rocket fuel which is liquid under the conditions in which it is utilized in the rocket. Also known as liquid propellant.

liquid gas   [PHYS] A gas in the liquid state.

liquid helium   [CRYO] The state of helium which exists at atmospheric pressure at temperatures below $-268.95°C$ (4.2 K), and for temperatures near absolute zero at pressures up to about 25 atmospheres (2.53 megapascals); has two phases, helium I and helium II.

liquid-in-glass thermometer   [ENG] A thermometer in which the thermally sensitive element is a liquid contained in a graduated glass envelope; the indication of such a thermometer depends upon the difference between the coefficients of thermal expansion of the liquid and the glass; mercury and alcohol are liquids commonly used in meteorological thermometers.

liquid knockout   See impingement.

liquid-phase epitaxy   [CRYSTAL] A liquid-phase transformation during crystal growth.

liquid pint   See pint.

liquid piston rotary compressor   [MECH ENG] A rotary compressor in which a multiblade rotor revolves in a casing partly filled with liquid, for example, water.

liquid propellant   See liquid fuel.

liquidus line   [THERMO] For a two-component system, a curve on a graph of temperature versus concentration which connects temperatures at which fusion is completed as the temperature is raised.

liquor   [CHEM ENG] 1. Supernatant liquid decanted from a liquid-solids mixture in which the solids have settled. 2. Liquid overflow from a liquid-liquid extraction unit. [FOOD ENG] 1. Sugarcane sap before it is crystallized into sugar. 2. A strong distilled alcoholic beverage. [PHARM] A solution of a medicinal substance in water.

Liriopeidae   [INV ZOO] The phantom craneflies, a family of dipteran insects in the suborder Orthorrhapha distinguished by black and white banded legs.

LISP   [COMPUT SCI] An interpretive language developed for the manipulation of symbolic strings of recursive data; can also be used to manipulate mathematical and arithmetic logic. Derived from list processing language.

Lissajous figure   [PHYS] The path of a particle moving in a plane when the components of its position along two perpendicular axes each undergo simple harmonic motions and the ratio of their frequencies is a rational number.

Lissamphibia   [VERT ZOO] A subclass of Amphibia including all living amphibians; distinguished by pedicellate teeth and an operculum-plectrum complex of the middle ear.

list   [COMPUT SCI] 1. A last-in, first-out storage organization, usually implemented by software, but sometimes implemented by hardware. 2. In FORTRAN, a set of data items to be read or written. [ENG] To lean to one side, or deviate from the vertical.

list processing   [COMPUT SCI] A programming technique in which list structures are used to organize memory.

list processing language   See LISP.

list structure   [COMPUT SCI] A set of data items, connected together because each element contains the address of a successor element (and sometimes of a predecessor element).

litchi   See lychee.

liter   [MECH] A unit of volume or capacity, equal to 1 decimeter cubed, or 0.001 cubic meter, or 1000 cubic centimeters. Abbreviated l.

lithifaction   See lithification.

lithification   [GEOL] 1. Conversion of a newly deposited sediment into an indurated rock. Also known as lithifaction. 2. Compositional change of coal to bituminous shale or other rock.

lithionite   See lepidolite.

lithiophilite   [MINERAL] $Li(Mn,Fe)PO_4$ A salmon-pink or clove-brown mineral crystallizing in the orthorhombic system; isomorphous with triphylite.

Lithistida   [PALEON] An order of fossil sponges in the class Demospongia having a reticulate skeleton composed of irregular and knobby siliceous spicules.

lithium   [CHEM] A chemical element, symbol Li, atomic number 3, atomic weight 6.939; an alkali metal.

lithium cell   [CHEM] An electrolytic cell for the production of metallic lithium.   [ELEC] A primary cell for producing electrical energy by using lithium metal for one electrode immersed in usually an organic electrolyte.

lithium mica   See lepidolite.

lithoclase   [GEOL] A naturally produced rock fracture.

Lithodidae   [INV ZOO] The king crabs, a family of anomuran decapods in the superfamily Paguridea distinguished by reduced last pereiopods and by the asymmetrical arrangement of the abdominal plates in the female.

lithofacies   [GEOL] A subdivision of a specified stratigraphic unit distinguished on the basis of lithologic features.

lithogenesis   [PATH] The process of formation of calculi or stones.   [PETR] The branch of science dealing with the formation of rocks, especially the formation of sedimentary rocks.

lithograph   [GRAPHICS] Originally, a reproduction of a writing sample or a drawing made from a litho stone onto which the writing or drawing had been drawn with a greasy ink or crayon; now, a reproduction from litho metal plates produced by photolithography and run on an offset press.

lithography   [GRAPHICS] A printing process in which a design is sketched with an oily ink or a litho crayon on a flat, smooth stone; in printing, the entire surface of the stone is wetted, and the design areas repel the water, but accept a greasy ink; a clean impression is then made by pressing a sheet of paper against the surface of the stone and running the whole through a press.

lithologic unit   See rock-stratigraphic unit.

lithology   [GEOL] The description of the physical character of a rock as determined by eye or with a low-power magnifier, and based on color, structures, mineralogic components, and grain size.

lithosphere   [GEOL] 1. The rigid outer crust of rock on the earth about 50 miles (80 kilometers) thick, above the asthenosphere. Also known as oxysphere. 2. Since the development of plate tectonics theory, a term referring to the rigid, upper 100 kilometers of the crust and upper mantle, above the asthenosphere.

lithostatic pressure   *See* ground pressure.

lithostratic unit   *See* rock-stratigraphic unit.

lithostratigraphic unit   *See* rock-stratigraphic unit.

litmus   [MATER] Blue, water-soluble powder from various lichens, especially *Variolaria lecanora* and *V. rocella*; turns red in solutions at pH 4.5, and blue at pH 8.3; used as an acid-base indicator. Also known as lacmus; lichen blue.

litmus paper   [MATER] White, unsized paper saturated by litmus in water; used as a pH indicator.

Litopterna   [PALEON] An order of hoofed, herbivorous mammals confined to the Cenozoic of South America; characterized by a skull without expansion of the temporal or squamosal sinuses, a postorbital bar, primitive dentition, and feet that were three-toed or reduced to a single digit.

Little Bear   *See* Ursa Minor.

little cherry disease   [PL PATH] A virus disease of sweet cherries characterized by small, angular pointed fruits which retain the bright red color of immaturity and never reach mature size.

Little Dipper   *See* Ursa Minor.

Little Ice Age   [GEOL] A period of expansion of mountain glaciers, marked by climatic deterioration, that began about 5500 years ago and extended to as late as A.D. 1550–1850 in some regions, as the Alps, Norway, Iceland, and Alaska.

little leaf   [PL PATH] Any of various plant diseases and disorders characterized by chlorotic, underdeveloped, and sometimes distorted leaves.

little peach disease   [PL PATH] A virus disease of the peach tree in which the fruit is dwarfed and delayed in ripening, the leaves yellow, and the tree dies.

littoral current   [OCEANOGR] A current, caused by wave action, that sets parallel to the shore; usually in the nearshore region within the breaker zone. Also known as alongshore current; longshore current.

littoral zone   [ECOL] Of or pertaining to the biogeographic zone between the high- and low-water marks.

Littorinacea   [PALEON] An extinct superfamily of gastropod mollusks in the order Prosobranchia.

Littorinidae   [INV ZOO] The periwinkles, a family of marine gastropod mollusks in the order Pectinibranchia distinguished by their spiral, globular shells.

Littrow grating spectrograph   [SPECT] A spectrograph having a plane grating at an angle to the axis of the instrument, and a lens in front of the grating which both collimates and focuses the light.

Littrow prism   [OPTICS] A prism having angles of 30, 60, and 90°, silvered on the side opposite the 60° angle; a lens used with it can serve both as a telescope and as a collimator.

Littrow quartz spectrograph   [SPECT] A spectrograph in which dispersion is accomplished by a Littrow quartz prism with a rear reflecting surface that reverses the light; a lens in front of the prism acts as both collimator and focusing lens.

Lituolacea   [INV ZOO] A superfamily of benthic marine foraminiferans in the suborder Textulariina having a multilocular, rectilinear, enrolled or uncoiled test with a simple to labyrinthic wall.

live   [COMMUN] Being broadcast directly at the time of production, instead of from recorded or filmed program material. [ELEC] *See* energized.

live axle   [MECH ENG] An axle to which wheels are rigidly fixed.

live load   [MECH] A moving load or a load of variable force acting upon a structure, in addition to its own weight.

liver   [ANAT] A large vascular gland in the body of vertebrates, consisting of a continuous parenchymal mass covered by a capsule; secretes bile, manufactures certain blood proteins and enzymes, and removes toxins from the systemic circulation. [MATER] Intermediate layer of dark-colored, oily material formed by hydrolyzation of acid sludge from sulfuric acid treatment of petroleum oil; insoluble in weak acid and oil.

liverwort   [BOT] The common name for members of the Marchantiatae.

live-virus vaccine   [IMMUNOL] A suspension of attenuated live viruses injected to produce immunity.

living fossil   [BIOL] A living species belonging to an ancient stock otherwise known only as fossils.

livingstonite   [MINERAL] $HgSb_4S_7$ A lead-gray mineral with red streak and metallic luster; a source of mercury.

livor mortis   [PATH] The reddish-blue discoloration of the cadaver that occurs in the dependent portions of the body due to gradual gravitational flow of unclotted blood.

lizard-hipped dinosaur   [PALEON] The name applied to members of the Saurichia because of the comparatively unspecialized three-pronged pelvis.

llama   [VERT ZOO] Any of three species of South American artiodactyl mammals of the genus *Lama* in the camel family; differs from the camel in being smaller and lacking a hump.

Llandellian   [GEOL] Upper Middle Ordovician geologic time.

Llandoverian   [GEOL] Lower Silurian geologic time.

LLL circuit   *See* low-level logic circuit.

lm   *See* lumen.

LM   *See* lunar excursion model.

LMC   *See* Large Magellanic Cloud.

lm-hr   *See* lumen-hour.

lm-sec   *See* lumen-second.

lm/w   *See* lumen per watt.

L network   [ELECTR] A network composed of two branches in series, with the free ends connected to one pair of terminals; the junction point and one free end are connected to another pair of terminals.

LNG   *See* liquefied natural gas.

load   [COMPUT SCI] **1.** To place data into an internal register under program control. **2.** To place a program from external storage into central memory under operator (or program) control, particularly when loading the first program into an otherwise empty computer. **3.** An instruction, or operator control button, which causes the computer to initiate the load action. [ELEC] **1.** A device that consumes electric power. **2.** The amount of electric power that is drawn from a power line, generator, or other power source. **3.** The material to be heated by an induction heater or dielectric heater. Also known as work. [ELECTR] The device that receives the useful signal output of an amplifier, oscillator, or other signal source. [ENG] **1.** To place ammunition in a gun, bombs on an airplane, explosives in a missile or borehole, fuel in a fuel tank, cargo or passengers into a vehicle, and the like. **2.** The quantity of gas delivered or required at any particular point on a gas supply system; develops primarily at gas-consuming equipment. [MECH] **1.** The weight that is supported by a structure. **2.** Mechanical force that is applied to a body. **3.** The burden placed on any machine, measured by units such as horsepower, kilowatts, or tons. [MIN ENG] Unit of weight of ore used in the South African diamond mines; equal to 1600 pounds (725 kilograms); the equivalent of about 16 cubic feet (0.453 cubic meter) of broken ore.

load-and-go [COMPUT SCI] An operating technique with no stops between the loading and execution phases of a program; may include assembling or compiling.

load-break switch [ELEC] An electric switch in a circuit with several hundred thousand volts, designed to carry a large amount of current without overheating the open position, having enough insulation to isolate the circuit in closed position, and equipped with arc interrupters to interrupt the load current.

load characteristic [ELECTR] Relation between the instantaneous values of a pair of variables such as an electrode voltage and an electrode current, when all direct electrode supply voltages are maintained constant. Also known as dynamic characteristic.

load compensation [CONT SYS] Compensation in which the compensator acts on the output signal after it has generated feedback signals. Also known as load stabilization.

loaded Q [ELEC] The Q factor of an impedance which is connected or coupled under working conditions. Also known as working Q. [ELECTROMAG] The Q factor of a specific mode of resonance of a microwave tube or resonant cavity when there is external coupling to that mode.

loader [COMPUT SCI] A computer program that takes some other program from an input or storage device and places it in memory at some predetermined address. [MECH ENG] A machine such as a mechanical shovel used for loading bulk materials. [ORD] Mechanical device which loads guns with cartridges.

loading [CHEM ENG] Condition of vapor overcapacity in a liquid-vapor-contact tower, in which rising vapor lifts or holds falling liquid. [ELEC] The addition of inductance to a transmission line to improve its transmission characteristics throughout a given frequency band. Also known as electrical loading. [ENG] 1. Buildup on a cutting tool of the material removed in cutting. 2. Filling the pores of a grinding wheel with material removed in the grinding process. [ENG ACOUS] Placing material at the front or rear of a loudspeaker to change its acoustic impedance and thereby alter its radiation. [MET] Filling of a die cavity with powdered metal. [NUCLEO] Placing fuel in a nuclear reactor.

loading program [COMPUT SCI] Program used to load other programs into computer memory. Also known as bootstrap program.

loading routine See input routine.

load line [ELECTR] A straight line drawn across a series of tube or transistor characteristic curves to show how output signal current will change with input signal voltage when a specified load resistance is used. [NAV ARCH] A line, painted or cut on the outside of a ship, which marks the maximum waterline when the ship is loaded with the greatest cargo which it can carry safely.

load loss [ELEC] The sum of the copper loss of a transformer, due to resistance in the windings, plus the eddy current loss in the winding, plus the stray loss.

load module [COMPUT SCI] A program in a form suitable for loading into memory and executing.

load stabilization See load compensation.

loadstone See lodestone.

loam [GEOL] Soil mixture of sand, silt, clay, and humus. [MET] Molding material consisting of sand, silt, and clay used over backup material for producing massive castings, usually of iron or steel.

lobate [BIOL] Having lobes. [VERT ZOO] Of a fish, having the skin of the fin extend onto the bases of the fin rays.

lobe [BIOL] A rounded projection on an organ or body part. [DES ENG] A projection on a cam wheel or a noncircular gear wheel. [ELECTROMAG] A part of the radiation pattern of a directional antenna representing an area of stronger radio-signal transmission. Also known as radiation lobe. [ENG ACOUS] A portion of the directivity pattern of a transducer representing an area of increased emission or response. [HYD] A curved projection on the margin of a continental ice sheet.

lobed impeller meter [ENG] A type of positive displacement meter in which a fluid stream is separated into discrete quantities by rotating, meshing impellers driven by interlocking gears.

lobefin fish [VERT ZOO] The common name for members composing the subclass Crossopterygii.

lobe switching See beam switching.

loblolly pine [BOT] Pinus taeda. A hard yellow pine of the central and southeastern United States having a reddish-brown fissured bark, needles in groups of three, and a full bushy top.

lobopodia [INV ZOO] Broad, thick pseudopodia.

Lobosia [INV ZOO] A subclass of the protozoan class Rhizopodea generally characterized by lobopodia.

lobotomy [MED] An operative section of the fibers between the frontal lobes of the brain. Also known as leukotomy; prefrontal lobotomy.

lobster [INV ZOO] The common name for several bottom-dwelling decapod crustaceans making up the family Homaridae which are commercially important as a food item.

lobular pneumonia See bronchopneumonia.

local anesthetic [PHARM] A drug which induces loss of sensation only in the region to which it is applied.

local apparent noon [ASTRON] Twelve o'clock local apparent time, or the instant the apparent sun is over the upper branch of the local meridian.

local apparent time [ASTRON] The arc of the celestial equator, or the angle at the celestial pole, between the lower branch of the local celestial meridian and the hour circle of the apparent or true sun, measured westward from the lower branch of the local celestial meridian through 24 hours.

local-area network [COMPUT SCI] A communications network connecting various hardware devices together within a building by means of a continuous cable or an in-house voice-data telephone system.

local exchange See exchange.

local group [ASTRON] A group of at least 20 known galaxies in the vicinity of the sun; the Andromeda Spiral is the largest of the group, and the Milky Way Galaxy is the second largest.

local hour angle [ASTRON] Angular distance west of the local celestial meridian.

local invariance [PHYS] The property of physical laws which remain unchanged under a specified set of symmetry transformations even when these transformations are chosen independently at every point of space and time.

localization [COMPUT SCI] Imposing some physical order upon a set of objects, so that a given object has a greater probability of being in some particular regions of space than in others.

local lunar time [ASTRON] The arc of the celestial equator, or the angle at the celestial pole, between the lower branch of the local celestial meridian and the hour circle of the moon, measured westward from the lower branch of the local celestial meridian through 24 hours; local hour angle of the moon, expressed in time units, plus 12 hours; local lunar time at the Greenwich meridian is called Greenwich lunar time.

local mean time [ASTRON] The arc of the celestial equator, or the angle at the celestial pole, between the lower branch of the local celestial meridian and the hour circle of the mean sun, measured westward from the lower branch of the local celestial meridian through 24 hours.

local meridian [ASTRON] The meridian through any particular position which serves as the reference for local time.

local oscillator [ELECTR] The oscillator in a superheterodyne receiver, whose output is mixed with the incoming modulated radio-frequency carrier signal in the mixer to give the frequency conversions needed to produce the intermediate-frequency signal.

local register [COMPUT SCI] One of a relatively small number (usually less than 32) of high-speed storage elements in a computer system which may be directly referred to by the instructions in the programs. Also known as general register.

local sidereal time [ASTRON] The arc of the celestial equator, or the angle at the celestial pole which is between the upper branch of the local celestial meridian and the hour circle of the vernal equinox.

local storm [METEOROL] A storm of mesometeorological scale; thus, thunderstorms, squalls, and tornadoes are often put in this category.

local time [ASTRON] 1. Time based upon the local meridian as reference, as contrasted with that based upon a zone meridian, or the meridian of Greenwich. 2. Any time kept locally.

local variable [COMPUT SCI] A variable which can be accessed (used or changed) only in one block of a computer program.

locate mode [COMPUT SCI] A method of communicating with an input/output control system (IOCS), in which the address of the data involved, but not the data themselves, is transferred between the IOCS routine and the program.

location [COMPUT SCI] Any place in which data may be stored; usually expressed as a number.

location counter *See* instruction counter.

location fit [DES ENG] The characteristic wherein mechanical sizes of mating parts are such that, when assembled, the parts are accurately positioned in relation to each other.

lock [CIV ENG] A chamber with gates on both ends connecting two sections of a canal or other waterway, to raise or lower the water level in each section. [DES ENG] A fastening device in which a releasable bolt is secured. [ELECTR] To fasten onto and automatically follow a target by means of a radar beam. [MET] A condition in forging in which the flash line is in more than one plane. [ORD] 1. Position of a safety mechanism which prevents a weapon from being fired. 2. Fastening device used to secure against accidental movement, as on a control surface. 3. To secure or make safe, as to set the safety on a weapon.

locked oscillator [ELECTR] A sine-wave oscillator whose frequency can be locked by an external signal to the control frequency divided by an integer.

lock-in [ELECTR] Shifting and automatic holding of one or both of the frequencies of two oscillating systems which are coupled together, so that the two frequencies have the ratio of two integral numbers.

lock-in amplifier [ELECTR] An amplifier that uses some form of automatic synchronization with an external reference signal to detect and measure very weak electromagnetic radiation at radio or optical wavelengths in the presence of very high noise levels.

lockjaw *See* tetanus.

locknut [DES ENG] 1. A nut screwed down firmly against another or against a washer to prevent loosening. Also known as jam nut. 2. A nut that is self-locking when tightened. 3. A nut fitted to the end of a pipe to secure it and prevent leakage.

lock-on [ELECTR] 1. The procedure wherein a target-seeking system (such as some types of radars) is continuously and automatically following a target in one or more coordinates (for example, range, bearing, elevation). 2. The instant at which radar begins to track a target automatically.

lockout [COMMUN] 1. In a telephone circuit controlled by two voice-operated devices, the inability of one or both subscribers to get through, because of either excessive local circuit noise or continuous speech from one or both subscribers. Also known as receiver lockout system. 2. In mobile communications, an arrangement of control circuits whereby only one receiver can feed the system at one time to avoid distortion. Also known as receiver lockout system. [COMPUT SCI] 1. In computer communications, the inability of a remote terminal to achieve entry to a computer system until project programmer number, processing authority code, and password have been validated against computer-stored lists. 2. The precautions taken to ensure that two or more programs executing simultaneously in a computer system do not access the same data at the same time, make unauthorized changes in shared data, or otherwise interfere with each other.

lock washer [DES ENG] A solid or split washer placed underneath a nut or screw to prevent loosening by exerting pressure.

locus [GEN] The fixed position of a gene in a chromosome. [MATH] A collection of points in a euclidean space whose coordinates satisfy one or more algebraic conditions.

locust [BOT] Either of two species of commercially important trees, black locust (*Robinia pseudoacacia*) and honey locust (*Gladitsia triacanthos*), in the family Leguminosae. [INV ZOO] The common name for various migratory grasshoppers of the family Locustidae.

Locustidae [INV ZOO] A family of insects in the order Orthoptera; antennae are usually less than half the body length, hindlegs are adapted for jumping, and the ovipositor is multipartite.

lode [GEOL] A fissure in consolidated rock filled with mineral; usually applied to metalliferous deposits.

lodestone [MINERAL] The naturally occurring magnetic iron oxide, or magnetite, possessing polarity, and attracting iron objects to itself. Also known as Hercules stone; leading stone; loadstone.

loellingite [MINERAL] FeAs$_2$ A silver-white to steel-gray mineral composed of iron arsenide with some cobalt, nickel, antimony, and sulfur; isomorphous with arsenopyrite; a source of arsenic. Also known as leucopyrite; löllingite.

loess [GEOL] An essentially unconsolidated, unstratified calcareous silt; commonly it is homogeneous, permeable, and buff to gray in color, and contains calcareous concretions and fossils.

log [COMMUN] A written record of radio and television station operating data, required by law. [COMPUT SCI] A record of computer operating runs, including tapes used, control settings, halts, and other pertinent data. [ENG] The record of, or the act or process of recording, events or the type and characteristics of the rock penetrated in drilling a borehole as evidenced by the cuttings, core recovered, or information obtained from electronic devices. [MATER] Unshaped timber either rough or squared. [NAV] **1.** An instrument for measuring the speed or distance or both traveled by a vessel. **2.** A written record of the movements of a craft, with regard to courses, speeds, positions, and other information of interest to navigators, and of important happenings aboard the craft. **3.** A written record of specific related information, such as that concerning performance of an instrument.

Loganiaceae [BOT] A family of mostly woody dicotyledonous plants in the order Gentianales; members lack a latex system and have fully united carpels and axile placentation.

logarithm [MATH] **1.** The real-valued function log u defined by log $u = v$ if $e^v = u$, e^v denoting the exponential function. Also known as hyperbolic logarithm; Naperian logarithm; natural logarithm. **2.** An analog in complex variables relative to the function e^z.

logarithmic coordinates [MATH] In the plane, logarithmic coordinates are defined by two coordinate axes, each marked with a scale where the distance between two points is the difference of the logarithms of the two numbers.

logarithmic curve [MATH] A curve whose equation in cartesian coordinates is $y = \log ax$, where a is greater than 1.

logarithmic multiplier [ELECTR] A multiplier in which each variable is applied to a logarithmic function generator, and the outputs are added together and applied to an exponential function generator, to obtain an output proportional to the product of two inputs.

logarithmic scale [MATH] A scale in which the distances that numbers are at from a reference point are proportional to their logarithms.

logarithmic transformation [STAT] The replacement of a variate y with a new variate $z = \log y$ or $z = \log (y + c)$, where c is a constant; this operation is often performed when the resulting distribution is normal, or if the resulting relationship with another variable is linear.

loggia [ARCH] A roofed open arcade on the side of a building.

logging [ENG] Continuous recording versus depth of some characteristic datum of the formations penetrated by a drill hole; for example, resistivity, spontaneous potential, conductivity, fluid content, radioactivity, or density. [FOR] The cutting and removal of the woody stem portions of forest trees.

logic [ELECTR] **1.** The basic principles and applications of truth tables, interconnections of on/off circuit elements, and other factors involved in mathematical computation in a computer. **2.** General term for the various types of gates, flip-flops, and other on/off circuits used to perform problem-solving functions in a digital computer.

logical comparison [COMPUT SCI] The operation of comparing two items in a computer and producing a one output if they are equal or alike, and a zero output if not alike.

logical connectives [MATH] Symbols which link mathematical statements; these symbols represent the terms "and," "or," "implication," and "negation."

logical construction [COMPUT SCI] A simple logical property that determines the type of characters which a particular code represents; for example, the first

two bits can tell whether a character is numeric or alphabetic.

logical data independence [COMPUT SCI] A data base structured so that changing the logical structure will not affect its accessibility by the program reading it.

logical decision [COMPUT SCI] The ability to select one of many paths, depending upon intermediate programming data.

logical expression [COMPUT SCI] Two arithmetic expressions connected by a relational operator indicating whether an expression is greater than, equal to, or less than the other, or connected by a logical variable, logical constant (true or false), or logical operator.

logical gate *See* switching gate.

logical instruction [COMPUT SCI] A digital computer instruction which forms a logical combination (on a bit-by-bit basis) of its operands and leaves the result in a known location.

logical shift [COMPUT SCI] A shift operation that treats the operand as a set of bits, not as a signed numeric value or character representation.

logical sum [COMPUT SCI] A computer addition in which the result is 1 when either one or both input variables is 1, and the result is 0 when the input variables are both 0.

logic-arithmetic unit *See* arithmetical unit.

logic circuit [COMPUT SCI] A computer circuit that provides the action of a logic function or logic operation.

logic design [COMPUT SCI] The design of a computer at the level which considers the operation of each functional block and the relationships between the functional blocks.

logic diagram [COMPUT SCI] A graphical representation of the logic design or a portion thereof; displays the existence of functional elements and the paths by which they interact with one another.

logic element [COMPUT SCI] A hardware circuit that performs a simple, predefined transformation on its input and presents the resulting signal as its output. Occasionally known as functor.

logic operation [COMPUT SCI] A nonarithmetical operation in a computer, such as comparing, selecting, making references, matching, sorting, and merging, where logical yes-or-no quantities are involved.

logic operator [COMPUT SCI] A rule which assigns, to every combination of the values "true" and "false" among one or more independent variables, the value "true" or "false" to a dependent variable.

logic section *See* arithmetical unit.

logic switch [ELECTR] A diode matrix or other switching arrangement that is capable of directing an input signal to one of several outputs.

logic unit [COMPUT SCI] A separate unit which exists in some computer systems to carry out logic (as opposed to arithmetic) operations.

logic word [COMPUT SCI] A machine word which represents an arbitrary set of digitally encoded symbols.

log-on [COMPUT SCI] The procedure for users to identify themselves to a computer system for authorized access to their programs and information.

log-periodic antenna [ELECTROMAG] A broad-band antenna which consists of a sheet of metal with two wedge-shaped cutouts, each with teeth cut into its radii along circular arcs; characteristics are repeated at a number of frequencies that are equally spaced on a logarithmic scale.

löllingite *See* loellingite.

lomonite   See laumontite.

lomontite   See laumontite.

Lonchaeidae   [INV ZOO] A family of minute myodarian cyclorrhaphous dipteran insects in the subsection Acalyptreatae.

London dispersion force   See van der Waals force.

London equations   [SOLID STATE] Equations for the time derivative and the curl of the current in a superconductor in terms of the electric and magnetic field vectors respectively, derived in the London superconductivity theory.

London superconductivity theory   [SOLID STATE] An extension of the two-fluid model of superconductivity, in which it is assumed that superfluid electrons behave as if the only force acting on them arises from applied electric fields, and that the curl of the superfluid current vanishes in the absence of a magnetic field.

London superfluidity theory   [CRYO] A theory, based on the fact that helium-4 obeys Bose-Einstein statistics, in which helium-4 is treated as an ideal Bose-Einstein gas, and its superfluid component is equated with the finite fraction of the atoms of such a gas which are in the ground state at very low temperatures.

long bone   [ANAT] A bone in which the length markedly exceeds the width, as the femur or the humerus.

long discharge   [ELEC] 1. A capacitor or other electrical charge accumulator which takes a long time to leak off. 2. A gaseous electrical discharge in which the length of the discharge channel is very long compared with its diameter; lightning discharges are natural examples of long discharges. Also known as long spark.

long hundredweight   See hundredweight.

Longidorinae   [INV ZOO] A subfamily of nematodes belonging to the Dorylaimoidea including economically important plant parasites.

Longipennes   [VERT ZOO] An equivalent name for Charadriiformes.

longitude   [GEOD] Angular distance, along the Equator, between the meridian passing through a position and, usually, the meridian of Greenwich.

longitudinal   [SCI TECH] Pertaining to the lengthwise dimension.

longitudinal magnetorestriction   See Joule effect.

long-period tide   [OCEANOGR] A tide or tidal current constituent with a period which is independent of the rotation of the earth but which depends upon the orbital movement of the moon or of the earth.

long-playing record   [ENG ACOUS] A 10- or 12-inch (25.4- or 30.48-centimeter) phonograph record that operates at a speed of 33⅓ rpm (revolutions per minute) and has closely spaced grooves, to give playing times up to about 30 minutes for one 12-inch side. Also known as LP record; microgroove record.

long-range forecast   [METEOROL] A weather forecast covering periods from 48 hours to a week in advance (medium-range forecast), and ranging to even longer forecasts over periods of a month, a season, and so on.

longshore bar   [GEOL] A ridge of sand, gravel, or mud built on the seashore by waves and currents, generally parallel to the shore and submerged by high tides. Also known as offshore bar.

longshore current   See littoral current.

long spark   See long discharge.

long string   [PETRO ENG] The last string of casing that is set in a well, through the producing zone. Also known as oil string.

long ton   See ton.

longwall system   [MIN ENG] A method of mining in which the faces are advanced from the shaft toward the boundary, and the roof is allowed to cave in behind the miners as work progresses.

long wave   [COMMUN] An electromagnetic wave having a wavelength longer than the longest broadcast-band wavelength of about 545 meters, corresponding to frequencies below about 550 kilohertz. [METEOROL] With regard to atmospheric circulation, a wave in the major belt of westerlies which is characterized by large length (thousands of kilometers) and significant amplitude; the wavelength is typically longer than that of the rapidly moving individual cyclonic and anticyclonic disturbances of the lower troposphere. Also known as major wave; planetary wave.

long-wave radio   [COMMUN] A radio which can receive frequencies below the lowest broadcast frequency of 550 kilohertz.

look-up   [COMPUT SCI] An operation or process in which a table of stored values is scanned (or searched) until a value equal to (or sometimes, greater than) a specified value is found.

loon   [VERT ZOO] The common name for birds composing the family Gaviidae, all of which are fish-eating diving birds.

loop   [AERO ENG] A flight maneuver in which an airplane flies a circular path in an approximately vertical plane, with the lateral axis of the airplane remaining horizontal, that is, an inside loop. [COMPUT SCI] A sequence of computer instructions which are executed repeatedly, but usually with address modifications changing the operands of each iteration, until a terminating condition is satisfied. [ELEC] 1. A closed path or circuit over which a signal can circulate, as in a feedback control system. 2. Commercially, the portion of a connection from central office to subscriber in a telephone system. 3. See mesh. [ELECTROMAG] See coupling loop; loop antenna. [ENG] 1. A reel of motion picture film or magnetic tape whose ends are spliced together, so that it may be played repeatedly without interruption. 2. A closed circuit of pipe in which materials and components may be placed to test them under different conditions of temperature, irradiation, and so forth. [MATH] A line which begins and ends at the same point of the graph. [PHYS] 1. A closed curve on a graph, such as a hysteresis loop. 2. That part of a standing wave where the vertical motion is greatest and the horizontal velocities are least. 3. See antinode.

loop antenna   [ELECTROMAG] A directional-type antenna consisting of one or more complete turns of a conductor, usually tuned to resonance by a variable capacitor connected to the terminals of the loop. Also known as loop.

loop lake   See oxbow lake.

loop network   See ring network.

loop of Henle   [ANAT] The U-shaped portion of a renal tubule formed by a descending and an ascending tubule. Also known as Henle's loop.

loop stop   [COMPUT SCI] A small closed loop that is entered to stop the progress of a computer program, usually when some condition occurs that requires intervention by the operator or that should be brought to the operator's attention. Also known as stop loop.

loose fit   [DES ENG] A fit with enough clearance to allow free play of the joined members.

loose kernel smut   [PL PATH] A type of kernel smut disease distinguished by the ruptured spore-containing gall.

Lopadorrhynchidae [INV ZOO] A small family of pelagic polychaete annelids belonging to the Errantia.

Lophialetidae [PALEON] A family of extinct perissodactyl mammals in the superfamily Tapiroidea.

Lophiiformes [VERT ZOO] A modified order of actinopterygian fishes distinguished by the reduction of the first dorsal fin to a few flexible rays, the first of which is on the head and bears a terminal bulb; includes anglerfish and allies.

Lophiodontidae [PALEON] An extinct family of perissodactyl mammals in the superfamily Tapiroidea.

lophocercous [VERT ZOO] Having a ridgelike caudal fin that lacks rays.

lophodont [VERT ZOO] Having molar teeth whose grinding surfaces have transverse ridges.

Lophogastrida [INV ZOO] A suborder of free-swimming marine crustaceans in the order Mysidacea characterized by imperfect fusion of the sixth and seventh abdominal somites, seven pairs of gills and brood lamellae, and natatory, biramous pleopods.

lophophore [INV ZOO] A food-gathering organ consisting of a fleshy basal ridge or lobe, from which numerous ciliated tentacles arise; found in Bryozoa, Phoronida, and Brachiopoda.

loran [NAV] The designation of a family of radio navigation systems by which hyperbolic lines of position are determined by measuring the difference in the times of reception of synchronized pulse signals from two or more fixed transmitters. Derived from long-range navigation.

Loranthaceae [BOT] A family of dicotyledonous plants in the order Santalales in which the ovules have no integument and are embedded in a large, central placenta.

Lorentz-Boltzmann equation [STAT MECH] An approximation to the Boltzmann transport equation for states that are near equilibrium, which shows that the Maxwell-Boltzmann distribution applies at equilibrium.

Lorentz contraction See FitzGerald-Lorentz contraction.

Lorentz equation [ELECTROMAG] The equation of motion for a charged particle, which sets the rate of change of its momentum equal to the Lorentz force.

Lorentz-FitzGerald contraction See FitzGerald-Lorentz contraction.

Lorentz four-vector See four-vector.

Lorentz frame [RELAT] Any of the family of inertial coordinate systems, with three space coordinates and one time coordinate, used in the special theory of relativity; each frame is in uniform motion with respect to all the other Lorentz frames, and the interval between any two events is the same in all frames.

Lorentz-Lorenz molar refraction See molar refraction.

Lorentz relation See Wiedemann-Franz law.

Lorentz transformation [MATH] Any linear transformation of euclidean four space which preserves the quadratic form $q(x,y,z,t) = t^2 - x^2 - y^2 - z^2$. [RELAT] Any of the family of mathematical transformations used in the special theory of relativity to relate the space and time variables of different Lorentz frames.

Lorentz curve [STAT] A graph for showing the concentration of ownership of economic quantities such as wealth and income; it is formed by plotting the cumulative distribution of the amount of the variable concerned against the cumulative frequency distribution of the individuals possessing the amount.

lorica [INV ZOO] A hard shell or case in certain invertebrates, as in many rotifers and protozoans; functions as an exoskeleton.

Loricaridae [VERT ZOO] A family of catfishes in the suborder Siluroidei found in the Andes.

Loricata [INV ZOO] An equivalent name for Polyplacophora.

loris [VERT ZOO] Either of two slow-moving, nocturnal, arboreal primates included in the family Lorisidae: the slender loris (*Loris tardigradus*) and slow loris (*Nycticebus coucang*).

Lorisidae [VERT ZOO] A family of prosimian primates comprising the lorises of Asia and the galagos of Africa.

lorry See larry.

Loschmidt number [PHYS] The number of molecules in 1 cubic centimeter of an ideal gas at 1 atmosphere pressure and 0°C, equal to approximately 2.687×10^{19}.

loss [COMMUN] See transmission loss. [ENG] Power that is dissipated in a device or system without doing useful work. Also known as internal loss.

loss current [ELEC] The current which passes through a capacitor as a result of the conductivity of the dielectric and results in power loss in the capacitor. [ELECTROMAG] The component of the current across an inductor which is in phase with the voltage (in phasor notation) and is associated with power losses in the inductor.

loss modulation See absorption modulation.

loss of head [FL MECH] Energy decrease between two points in a hydraulic system due to such causes as friction, bends, obstructions, or expansions.

loss of information See walk-down.

lotrite See pumpellyite.

loudness [ACOUS] The magnitude of the physiological sensation produced by a sound, which varies directly with the physical intensity of sound but also depends on frequency of sound and waveform.

loudness unit [ACOUS] A unit of loudness equal to the loudness of a sound having a loudness level of 0 phon; the loudness unit has been replaced by the sone.

loudspeaker [ENG ACOUS] A device that converts electrical signal energy into acoustical energy, which it radiates into a bounded space, such as a room, or into outdoor space. Also known as speaker.

loughlinite [MINERAL] $Na_2Mg_3Si_6O_{16} \cdot 8H_2O$ A pearly-white mineral that resembles asbestos, consisting of a hydrous silicate of sodium and magnesium.

louping ill [VET MED] A virus disease of sheep, similar to encephalomyelitis, transmitted by the tick *Ixodes racinus*. Also known as ovine encephalomyelitis; trembling ill.

louse [INV ZOO] The common name for the apterous ectoparasites composing the orders Anoplura and Mallophaga.

louver [BUILD] An opening in a wall or ceiling with slanted or sloping slats to allow sun and ventilation and exclude rain; may be fixed or adjustable, and may be at the opening of a ventilating duct. [ENG] Any arrangement of fixed or adjustable slatlike openings to provide ventilation. [ENG ACOUS] An arrangement of concentric or parallel slats or equivalent grille members used to conceal and protect a loudspeaker while allowing sound waves to pass.

lovchorrite See mosandrite.

low See depression.

low-angle thrust See overthrust.

low-approach system [NAV] A means for furnishing guidance in the vertical and horizontal planes to aircraft during descent from an initial approach altitude to a point near the ground.

low-carbon steel [MET] Steel containing 0.15% or less of carbon.

low clouds [METEOROL] Types of clouds, the mean level of which is between the surface and 6500 feet (1980 meters); the principal clouds in this group are stratocumulus, stratus, and nimbostratus.

low-energy electron diffraction [SOLID STATE] A technique for studying the atomic structure of single crystal surfaces, in which electrons of uniform energy in the approximate range 5–500 electron volts are scattered from a surface, and those scattered electrons that have lost no energy are selected and accelerated to a fluorescent screen where the diffraction pattern from the surface can be observed. Abbreviated LEED.

low-energy physics [PHYS] That part of physics which studies microscopic phenomena involving energies of several million electronvolts or less, such as the arrangement of electrons in an atom or a solid, and the arrangement of protons and neutrons within the atomic nucleus, and the nature of forces between these particles.

lower atmosphere [METEOROL] That part of the atmosphere in which most weather phenomena occur (that is, the troposphere and lower stratosphere); in other contexts, the term implies the lower troposphere.

Lower Cambrian [GEOL] The earliest epoch of the Cambrian period of geologic time, ending about 540,000,000 years ago.

Lower Cretaceous [GEOL] The earliest epoch of the Cretaceous period of geologic time, extending from about 140- to 120,000,000 years ago.

Lower Devonian [GEOL] The earliest epoch of the Devonian period of geologic time, extending from about 400- to 385,000,000 years ago.

Lower Jurassic [GEOL] The earliest epoch of the Jurassic period of geologic time, extending from about 185- to 170,000,000 years ago.

lower mantle [GEOL] The portion of the mantle below a depth of about 1000 kilometers. Also known as inner mantle; mesosphere; pallasite shell.

Lower Mississippian [GEOL] The earliest epoch of the Mississippian period of geologic time, beginning about 350,000,000 years ago.

Lower Ordovician [GEOL] The earliest epoch of the Ordovician period of geologic time, extending from about 490- to 460,000,000 years ago.

Lower Pennsylvanian [GEOL] The earliest epoch of the Pennsylvanian period of geologic time, beginning about 310,000,000 years ago.

Lower Permian [GEOL] The earliest epoch of the Permian period of geologic time, extending from about 275- to 260,000,000 years ago.

lower plate See footwall.

Lower Silurian [GEOL] The earliest epoch of the Silurian period of geologic time, beginning about 420,000,000 years ago.

Lower Triassic [GEOL] The earliest epoch of the Triassic period of geologic time, extending from about 230- to 215,000,000 years ago.

low-frequency [COMMUN] A Federal Communications Commission designation for the band from 30 to 300 kilohertz in the radio spectrum. Abbreviated LF.

low-level counting [NUCLEO] The measurement of very small amounts of radioactivity, such as that generated by long-lived natural radioactive isotopes, and isotopes produced by cosmic rays and nuclear explosions.

low-level language [COMPUT SCI] A computer language consisting of mnemonics that directly correspond to machine language instructions; for example, an assembler that converts the interpreted code of a higher-level language to machine language.

low-level logic circuit [ELECTR] A modification of a diode-transistor logic circuit in which a resistor and capacitor in parallel are replaced by a diode, with the result that a relatively small voltage swing is required at the base of the transistor to switch it on or off. Abbreviated LLL circuit.

low-pass filter [ELEC] A filter that transmits alternating currents below a given cutoff frequency and substantially attenuates all other currents.

low-temperature physics [CRYO] A study of the properties of gross matter at low temperature's especially at temperatures so low that the quantum character of the substance becomes observable in effects such as superconductivity, superfluid liquid helium, magnetic cooling, and nuclear orientation.

low-temperature thermometry [CRYO] The assignment of numbers on the Kelvin absolute temperature scale to achievable and reproducible low-temperature states, and the choice and calibration of suitable instruments for the practical measurement of low temperatures, such as thermocouples, and resistance, vapor-pressure, gas, and magnetic thermometers.

low tide See low water.

low-velocity layer [GEOPHYS] A layer in the solid earth in which seismic wave velocity is lower than the layers immediately below or above.

low voltage [ELEC] **1.** Voltage which is small enough to be regarded as safe for indoor use, usually 120 volts in the United States. **2.** Voltage which is less than that needed for normal operation; a result of low voltage may be burnout of electric motors due to loss of electromotive force.

low water [OCEANOGR] The lowest limit of the surface water level reached by the lowering tide. Also known as low tide.

loxodont [VERT ZOO] Having molar teeth with shallow hollows between the ridges.

loxodrome See rhumb line.

loxolophodont [VERT ZOO] Having crests on the molar teeth that connect three of the tubercles and with the fourth or posterior inner tubercle being rudimentary or absent.

Loxonematacea [PALEON] An extinct superfamily of gastropod mollusks in the order Prosobranchia.

LPG See liquefied petroleum gas.

LP record See long-playing record.

Lr See lawrencium.

LSB See least significant bit.

L scan See L scope.

L scope [ELECTR] A cathode-ray scope on which a trace appears as a vertical or horizontal range scale, the signals appearing as left and right horizontal (or up and down vertical) deflections as echoes are received by two antennas, the left and right (or up and down) deflections being proportional to the strength of the echoes received by the two antennas. Also known as L indicator; L scan.

LSD See dock landing ship; lysergic acid diethylamide.

LSD-25 See lysergic acid diethylamide.

L shell [ATOM PHYS] The second shell of electrons surrounding the nucleus of an atom, having electrons whose principal quantum number is 2.

LSI circuit See large-scale integrated circuit.

Lu See lutetium.

lubricant [MATER] A substance used to reduce friction between parts or objects in relative motion.

Lucanidae [INV ZOO] The stag beetles, a cosmopolitan family of coleopteran insects in the superfamily Scarabaeoidea.

lucerne *See* alfalfa.

Luciocephalidae [VERT ZOO] A family of fresh-water fishes in the suborder Anabantoidei.

Lüders' lines [MET] Surface markings on a metal caused by flow of the material strained beyond its elastic limit. Also known as deformation bands; Hartmann lines; Lüders' bands; Piobert lines; stretcher strains.

Ludlovian [GEOL] A European stage of geologic time; Upper Silurian, below Gedinnian of Devonian, above Wenlockian.

ludwigite [MINERAL] $(Mg,Fe)_2FeBO_5$ Blackish-green mineral that crystallizes in the monoclinic system and occurs in fibrous masses; isomorphous with ronsenite.

lueneburgite [MINERAL] $Mg_3B_2(OH)_6(PO_4)_2 \cdot 6H_2O$ A colorless mineral composed of hydrous basic phosphate of magnesium and boron.

lug [DES ENG] A projection or head on a metal part to serve as a cap, handle, support, or fitting connection.

Luidiidae [INV ZOO] A family of echinoderms in the suborder Paxillosina.

Luisian [GEOL] A North American stage of geologic time: Miocene (above Relizian, below Mohnian).

Lukasiewicz notation *See* Polish notation.

lumbar artery [ANAT] Any of the four or five pairs of branches of the abdominal aorta opposite the lumbar region of the spine; supplies blood to loin muscles, skin on the sides of the abdomen, and the spinal cord.

lumbar nerve [ANAT] Any of five pairs of nerves arising from lumbar segments of the spinal cord; characterized by motor, visceral sensory, somatic sensory, and sympathetic components; they innervate the skin and deep muscles of the lower back and the lumbar plexus.

lumbar vertebrae [ANAT] Those vertebrae located between the lowest ribs and the pelvic girdle in all vertebrates.

Lumbricidae [INV ZOO] A family of annelid worms in the order Oligochaeta; includes the earthworm.

Lumbriclymeninae [INV ZOO] A subfamily of mud-swallowing sedentary worms in the family Maldanidae.

Lumbriculidae [INV ZOO] A family of aquatic annelids in the order Oligochaeta.

Lumbrineridae [INV ZOO] A family of errantian polychaetes in the superfamily Eunicea.

lumen [OPTICS] The unit of luminous flux, equal to the luminous flux emitted within a unit solid angle (1 steradian) from a point source having a uniform intensity of 1 candela, or to the luminous flux received on a unit surface, all points of which are at a unit distance from such a source. Symbolized lm. [SCI TECH] The space within a tube.

lumen-hour [OPTICS] A unit of quantity of light (luminous energy), equal to the quantity of light radiated or received for a period of 1 hour by a flux of 1 lumen. Abbreviated lm-hr.

lumen per watt [OPTICS] The unit of luminosity factor and of luminous efficacy. Abbreviated lm/w.

lumen-second [OPTICS] A unit of quantity of light (luminous energy), equal to the quantity of light radiated or received for a period of 1 second by a flux of 1 lumen. Abbreviated lm-sec.

luminaire [ELEC] An electric lighting fixture, wall bracket, portable lamp, or other complete lighting unit designed to contain one or more electric lighting sources and associated reflectors, refractors, housing, and such support for those items as necessary.

luminance [OPTICS] The ratio of the luminous intensity in a given direction of an infinitesimal element of a surface containing the point under consideration, to the orthogonally projected area of the element on a plane perpendicular to the given direction. Formerly known as brightness.

luminescence [PHYS] Light emission that cannot be attributed merely to the temperature of the emitting body, but results from such causes as chemical reactions at ordinary temperatures, electron bombardment, electromagnetic radiation, and electric fields.

luminescent cell *See* electroluminescent panel.

luminescent center [SOLID STATE] A point-lattice defect in a transparent crystal that exhibits luminescence.

luminophor [PHYS] A luminescent material that converts part of the absorbed primary energy into emitted luminescent radiation. Also known as fluophor; fluor; phosphor.

luminosity *See* luminosity factor.

luminosity classes [ASTRON] A classification of stars in an orderly sequence according to their absolute brightness.

luminosity curve *See* luminosity function.

luminosity factor [OPTICS] The ratio of luminous flux in lumens emitted by a source at a particular wavelength to the corresponding radiant flux in watts at the same wavelength; thus this is a measure of the visual sensitivity of the eye. Also known as luminosity.

luminosity function [ASTRON] The functional relationship between stellar magnitude and the number and distribution of stars of each magnitude interval. Also known as relative luminosity factor. [OPTICS] A standard measure of the response of an eye to monochromatic light at various wavelengths; the function is normalized to unity at its maximum value. Also known as luminosity curve; spectral luminous efficiency; visibility function.

luminous cloud *See* sheet lightning.

luminous coefficient [OPTICS] A measure of the fraction of the radiant power of a light source which contributes to its luminous properties, equal to the average of the luminosity function at various wavelengths, weighted according to the spectral intensity of the source. Also known as luminous efficiency.

luminous efficacy [OPTICS] 1. The ratio of the total luminous flux in lumens emitted by a light source over all wavelengths to the total radiant flux in watts. Formerly known as luminous efficiency. 2. The ratio of the total luminous flux emitted by a light source to the power input of the source; expressed in lumens per watt.

luminous efficiency *See* luminous coefficient; luminous efficacy.

luminous flux density *See* illuminance.

luminous intensity [OPTICS] The luminous flux incident on a small surface which lies in a specified direction from a light source and is normal to this direction, divided by the solid angle (in steradians) which the surface subtends at the source of light. Also known as light intensity.

lunar appulse [ASTRON] An eclipse of the moon in which the penumbral shadow of the earth falls on the moon. Also known as penumbral eclipse.

lunar day [ASTRON] The time interval between two successive crossings of the meridian by the moon.

lunar eclipse [ASTRON] Obscuration of the full moon when it passes through the shadow of the earth.

lunar ephemeris [ASTRON] A computed list of positions the moon will occupy in the sky on certain dates.

lunar excursion module [AERO ENG] A manned spacecraft designed to be carried on top of the Apollo service module and having its own power plant for making a manned landing on the moon and a return from the moon to the orbiting Apollo spacecraft. Abbreviated LEM. Also known as lunar module (LM).

lunar geology *See* selenology.

lunar inequality [ASTRON] Variation in the moon's motion in its orbit, due to attraction by other bodies of the solar system. [GEOPHYS] A minute fluctuation of a magnetic needle from its mean position, caused by the moon.

lunar interval [ASTRON] The difference in time between the transit of the moon over the Greenwich meridian and a local meridian; the lunar interval equals the difference between the Greenwich and local intervals of a tide or current phase.

lunar libration [ASTRON] **1.** The effect wherein the face of the moon appears to swing east and west about 8° from its central position each month. Also known as apparent libration in longitude. **2.** The state wherein the inclination of the moon's polar axis allows an observer on earth to see about 59% of the moon's surface. Also known as libration in latitude. **3.** The small oscillation with which the moon rocks back and forth about its mean rotation rate. Also known as physical libration of the moon.

lunar module *See* lunar excursion module.

lunar month [ASTRON] The period of revolution of the moon about the earth, especially a synodical month.

lunar nutation [ASTRON] A nodding motion of the earth's axis caused by the inclination of the moon's orbit to the ecliptic; it can displace the celestial pole by 9 seconds of arc from its mean position and has a period of 18.6 years.

lunar tide [OCEANOGR] The portion of a tide produced by forces of the moon.

lunar time [ASTRON] **1.** Time based upon the rotation of the earth relative to the moon; it may be designated as local or Greenwich, as the local or Greenwich meridian is used as the reference. **2.** Time on the moon.

lunar year [ASTRON] A time interval comprising 12 lunar (synodic) months.

Luneberg lens [ELECTROMAG] A type of antenna consisting of a dielectric sphere whose index of refraction varies with distance from the center of the sphere so that a beam of parallel rays falling on the lens is focused at a point on the lens surface diametrically opposite from the direction of incidence, and, conversely, energy emanating from a point on the surface is focused into a plane wave. Also spelled Luneburg lens.

lung [ANAT] Either of the paired air-filled sacs, usually in the anterior or anteroventral part of the trunk of most tetrapods, which function as organs of respiration.

lungfish [VERT ZOO] The common name for members of the Dipnoi; all have lungs that arise from a ventral connection with the gut.

lunitidal interval [OCEANOGR] The period between the moon's upper or lower transit over a specified meridian and a specified phase of the tidal current following the transit.

lupine [BOT] A leguminous plant of the genus *Lupinus* with an upright stem, leaves divided into sev-

eral digitate leaflets, and terminal racemes of pea-shaped blossoms.

Lupus [ASTRON] A southern constellation lying between Centaurus and Scorpius. Also known as Wolf.

lupus erythematosus [MED] An acute or subacute febrile collagen disease characterized by a butterfly-shaped rash over the cheeks and perilingual erythema.

luster [OPTICS] The appearance of a surface dependent on reflected light; types include metallic, vitreous, resinous, adamantine, silky, pearly, greasy, dull, and earthy; applied to minerals, textiles, and many other materials.

luteinizing hormone [BIOCHEM] A glycoprotein hormone secreted by the adenohypophysis of vertebrates which stimulates hormone production by interstitial cells of gonads. Also known as interstitial-cell-stimulating hormone (ICSH).

luteotropic hormone *See* prolactin.

lutetium [CHEM] A chemical element, symbol Lu, atomic number 71, atomic weight 174.97; a very rare metal and the heaviest member of the rare-earth group.

Lutheran blood group [IMMUNOL] The erythrocyte antigens defined by reactions with an antibody designated anti-Lu^a, initially detected in the serum of a multiply transfused patient with lupus erythematosus, who developed antibodies against erythrocytes of a donor named Lutheran, and by anti-Lu^b.

Lutjanidae [VERT ZOO] The snappers, a family of perciform fishes in the suborder Percoidei.

lux [OPTICS] A unit of illumination, equal to the illumination on a surface 1 square meter in area on which there is a luminous flux of 1 lumen uniformly distributed, or the illumination on a surface all points of which are at a distance of 1 meter from a uniform point source of 1 candela. Symbolized lx. Also known as meter-candle.

luxon *See* troland.

luzonite *See* enargite.

LVDT *See* linear variable-differential transformer.

lx *See* lux.

Lyapunov function [MATH] A function of a vector and of time which is positive-definite and has a negative-definite derivative with respect to time for nonzero vectors, is identically zero for the zero vector, and approaches infinity as the norm of the vector approaches infinity; used in determining the stability of control systems. Also spelled Liapunov function.

lyase [BIOCHEM] An enzyme that catalyzes the non-hydrolytic cleavage of its substrate with the formation of a double bond; examples are decarboxylases.

Lycaenidae [INV ZOO] A family of heteroneuran lepidopteran insects in the superfamily Papilionoidea including blue, gossamer, hairstreak, copper, and metalmark butterflies.

Lycaeninae [INV ZOO] A subfamily of the Lycaenidae distinguished by functional prothoracic legs in the male.

lychee [BOT] A tree of the genus *Litchi* in the family Sapindaceae, especially *L. chinensis* which is cultivated for its edible fruit, a one-seeded berry distinguished by the thin, leathery, rough pericarp that is red in most varieties. Also spelled litchi.

Lychniscosa [INV ZOO] An order of sponges in the subclass Hexasterophora in which parenchymal megascleres form a rigid framework and are all or in part lychniscs.

Lycopodiales [BOT] The type order of Lycopodiopsida.

Lycopodiatae *See* Lycopodiopsida.

Lycopodineae [BOT] The equivalent name for Lycopodiopsida.

Lycopodiophyta [BOT] A division of the subkingdom Embryobionta characterized by a dominant independent sporophyte, dichotomously branching roots and stems, a single vascular bundle, and small, simple, spirally arranged leaves.

Lycopodiopsida [BOT] The lycopods, the type class of Lycophodiophyta.

Lycopsida [BOT] Former subphylum of the Embryophyta now designated as the division Lycopodiophyta.

Lycoriidae [INV ZOO] A family of small, dark-winged dipteran insects in the suborder Orthorrhapha.

Lycosidae [INV ZOO] A family of hunting spiders in the suborder Dipneumonomorphae that actively pursue their prey.

Lycoteuthidae [INV ZOO] A family of squids.

Lyctidae [INV ZOO] The large-winged beetles, a large cosmopolitan family of coleopteran insects in the superfamily Bostrichoidea.

lye [INORG CHEM] **1.** A solution of potassium hydroxide or sodium hydroxide used as a strong alkaline solution in industry. **2.** The alkaline solution that is obtained from the leaching of wood ashes.

Lygaeidae [INV ZOO] The lygaeid bugs, a large family of phytophagous hemipteran insects in the superfamily Lygaeoidea.

Lygaeoidea [INV ZOO] A superfamily of pentatomorphan insects having four-segmented antennae and ocelli.

Lyginopteridaceae [PALEOBOT] An extinct family of the Lyginopteridales including monostelic pteridosperms having one or two vascular traces entering the base of the petiole.

Lyginopteridales [PALEOBOT] An order of the Pteridospermae.

Lyginopteridatae [PALEOBOT] The equivalent name for Pteridospermae.

Lyman band [SPECT] A band in the ultraviolet spectrum of molecular hydrogen, extending from 125 to 161 nanometers.

Lyman series [SPECT] A group of lines in the ultraviolet spectrum of hydrogen covering the wavelengths of 121.5–91.2 nanometers.

Lymantriidae [INV ZOO] The tussock moths, a family of heteroneuran lepidopteran insects in the superfamily Noctuoidea; the antennae of males is broadly pectinate and there is a tuft of hairs on the end of the female abdomen.

Lymexylonidae [INV ZOO] The ship timber beetles composing the single family of the coleopteran superfamily Lymexylonoidea.

Lymexylonoidea [INV ZOO] A monofamilial superfamily of wood-boring coleopteran insects in the suborder Polyphaga characterized by a short neck and serrate antennae.

lymph [HISTOL] The colorless fluid which circulates through the vessels of the lymphatic system.

lymphadenoma [MED] A tumorlike enlargement of a lymph node.

lymphangioma [MED] An abnormal mass of lymphatic vessels.

lymphatic system [ANAT] A system of vessels and nodes conveying lymph in the vertebrate body, beginning with capillaries in tissue spaces and eventually forming the thoracic ducts which empty into the subclavian veins.

lymph gland See lymph node.

lymph node [ANAT] An aggregation of lymphoid tissue surrounded by a fibrous capsule; found along the course of lymphatic vessels. Also known as lymph gland.

lymphocyte [HISTOL] An agranular leukocyte formed primarily in lymphoid tissue; occurs as the principal cell type of lymph and composes 20–30% of the blood leukocytes.

lymphocytic angina See infectious mononucleosis.

lymphocytic leukemia [MED] A type of leukemia in which lymphocytic cells predominate.

lymphocytic sarcoma See lymphosarcoma.

lymphocytopenia [MED] Reduction of the absolute number of lymphocytes per unit volume of peripheral blood. Also known as lymphopenia.

lymphocytosis [MED] An abnormally high lymphocyte count in the blood.

lymphogranuloma inguinale See lymphogranuloma venereum.

lymphogranulomatosis See Hodgkin's disease.

lymphogranuloma venereum [MED] A systemic infectious venereal disease caused by a microorganism belonging to the PLT-Bedsonia group, characterized by enlargement of inguinal lymph nodes and genital ulceration. Abbreviated LGV. Also known as lymphogranuloma inguinale; lymphopathia venereum; venereal bubo.

lymphoma [MED] Any neoplasm, usually malignant, of the lymphoid tissues.

lymphopathia venereum See lymphogranuloma venereum.

lymphopenia See lymphocytopenia.

lymphosarcoma [MED] A malignant lymphoma composed of anaplastic lymphoid cells resembling lymphocytes or lymphoblasts, according to the degree of differentiation. Also known as lymphocytic sarcoma.

lynx [VERT ZOO] Any of several wildcats of the genus *Lynx* having long legs, short tails, and usually tufted ears; differs from other felids in having 28 instead of 30 teeth.

Lyomeri [VERT ZOO] The equivalent name for Saccopharyngiformes.

lyophilic [CHEM] Referring to a substance which will readily go into colloidal suspension in a liquid.

lyophilization [CHEM ENG] Rapid freezing of a material, especially biological specimens for preservation, at a very low temperature followed by rapid dehydration by sublimation in a high vacuum.

lyophobic [CHEM] Referring to a substance in a colloidal state that has a tendency to repel liquids.

Lyopomi [VERT ZOO] An equivalent name for Notacanthiformes.

Lyot filter See birefringent filter.

Lyra [ASTRON] A northern constellation; right ascension 19 hours, declination 40° north; its first-magnitude star, Vega, is a navigational star and the most brilliant star in this part of the sky.

Lysaretidae [INV ZOO] A family of errantian polychaete worms in the superfamily Eunicea.

lysergic acid diethylamide [ORG CHEM] $C_{15}H_{15}N_2\text{-}CON(C_2H_5)_2$ A psychotomimetic drug synthesized from compounds derived from ergot. Abbreviated LSD; LSD-25.

lysin [IMMUNOL] A substance, particularly antibodies, capable of lysing a cell.

lysis [CYTOL] Dissolution of a cell or tissue by the action of a lysin. [MED] **1.** Gradual decline in the manifestations of a disease, especially an infectious disease. Also known as defervescence. **2.** Gradual fall of fever.

Lysithea [ASTRON] A small satellite of Jupiter with a diameter of about 15 miles (24 kilometers), orbiting at a mean distance of about 7.30 × 10⁶ miles (11.75 × 10⁶ kilometers). Also known as Jupiter X.

lysosome [CYTOL] A specialized cell organelle surrounded by a single membrane and containing a mixture of hydrolytic (digestive) enzymes.

lysozyme [BIOCHEM] An enzyme present in certain body secretions, principally tears, which acts to hydrolyze certain bacterial cell walls.

Lyssacinosa [INV ZOO] An order of sponges in the subclass Hexasterophora in which parenchymal megascleres are typically free and unconnected but are sometimes secondarily united.

lytic phage [VIROL] Any phage that causes host cells to lyse.

lytic reaction [CYTOL] A reaction that leads to lysis of a cell.

m *See* meter; milli-.

M *See* mega-; molarity.

macadam [CIV ENG] Uniformly graded stones consolidated by rolling to form a road surface; may be bound with water or cement, or coated with tar or bitumen.

macaluba *See* mud volcano.

macaque [VERT ZOO] The common name for 12 species of Old World monkeys composing the genus *Macaca*, including the Barbary ape and the rhesus monkey.

macaw [VERT ZOO] The common name for large South and Central American parrots of the genus *Ara* and related genera; individuals are brilliantly colored with a long tail, a hooked bill, and a naked area around the eyes.

macedonite [MINERAL] $PbTiO_3$ A mineral composed of an oxide of lead and titanium. [PETR] A basaltic rock that contains orthoclase, sodic plagioclase, biotite, olivine, and rare pyriboles.

maceral [GEOL] The microscopic organic constituents found in coal.

Machaeridea [INV ZOO] A class of homolozoan echinoderms in older systems of classification.

Mach cone [FL MECH] 1. The cone-shaped shock wave theoretically emanating from an infinitesimally small particle moving at supersonic speed through a fluid medium; it is the locus of the Mach lines. 2. The cone-shaped shock wave generated by a sharp-pointed body, as at the nose of a high-speed aircraft.

Machilidae [INV ZOO] A family of insects belonging to the Thysanura having large compound eyes and ocelli and a monocondylous mandible of the scraping type.

machinability [MET] 1. The ability of a metal to be machined. 2. The difficulty or ease with which a metal can be machined.

machinable *See* machine-sensible.

Mach indicator *See* Machmeter.

machine [COMPUT SCI] 1. A mechanical, electric, or electronic device, such as a computer, tabulator, sorter, or collator. 2. A simplified, abstract model of an internally programmed computer, such as a Turing machine. [MECH ENG] A combination of rigid or resistant bodies having definite motions and capable of performing useful work.

machine address [COMPUT SCI] The actual and unique internal designation of the location at which an instruction or datum is to be stored or from which it is to be retrieved.

machine bolt [DES ENG] A heavy-weight bolt with a square, hexagonal, or flat head used in the automotive, aircraft, and machinery fields.

machine code [COMPUT SCI] 1. A computer representation of a character, digit, or action command in internal form. 2. A computer instruction in internal

format, or that part of the instruction which identifies the action to be performed. 3. The set of all instruction types that a particular computer can execute.

machine conditions [COMPUT SCI] A component of a task descriptor that specifies the contents of all programmmable registers in the processor, such as arithmetic and index registers.

machine controlled time [IND ENG] The time necessary for a machine to complete the automatic portion of a work cycle. Also known as independent machine time; machine element; machine time.

machine cycle [COMPUT SCI] 1. The shortest period of time at the end of which a series of events in the operation of a computer is repeated. 2. The series of events itself.

machine element [DES ENG] Any of the elementary mechanical parts, such as gears, bearings, fasteners, screws, pipes, springs, and bolts used as essentially standardized components for most devices, apparatus, and machinery. [IND ENG] *See* machine controlled time.

machine-gun microphone *See* line microphone.

machine-independent [COMPUT SCI] Referring to programs and procedures which function in essentially the same manner regardless of the machine on which they are carried out.

machine instruction [COMPUT SCI] A set of digits, binary bits, or characters that a computer can recognize and act upon, and that, when interpreted or decoded, indicates the action to be performed and which operand is to be involved in the action.

machine interruption [COMPUT SCI] A halt in computer operations followed by the beginning of a diagnosis procedure, as a result of an error detection.

machine key [DES ENG] A piece inserted between a shaft and a hub to prevent relative rotation. Also known as key.

machine language [COMPUT SCI] The set of instructions available to a particular digital computer, and by extension the format of a computer program in its final form, capable of being executed by a computer.

machine logic [COMPUT SCI] The structure of a computer, the operation it performs, and the type and form of data used internally.

machine operator [COMPUT SCI] The person who manipulates the computer controls, brings up and closes down the computer, and can override a number of computer decisions.

machine-oriented programming system [COMPUT SCI] A system written in assembly language (or macro code) directly oriented toward the computer's internal language.

machine-readable *See* machine-sensible.

machine-recognizable *See* machine-sensible.

machine run *See* run.

machine screw [DES ENG] A blunt-ended screw with a standardized thread and a head that may be flat, round, fillister, or oval, and may be slotted, or constructed for wrenching; used to fasten machine parts together.

machine-sensible [COMPUT SCI] Capable of being read or sensed by a device, usually by one designed and built specifically for this task. Also known as machinable; machine-readable; machine-recognizable; mechanized.

machine switching system *See* automatic exchange.

machine time *See* machine controlled time.

machine tool [MECH ENG] A stationary power-driven machine for the shaping, cutting, turning, boring, drilling, grinding, or polishing of solid parts, especially metals.

machine translation *See* mechanical translation.

machine word [COMPUT SCI] The fundamental unit of information in a word-organized digital computer, consisting of a fixed number of binary bits, decimal digits, characters, or bytes.

Mach line [FL MECH] **1.** A line representing a Mach wave. **2.** *See* Mach wave.

Machmeter [ENG] An instrument that measures and indicates speed relative to the speed of sound, that is, indicates the Mach number. Also known as Mach indicator.

Mach number [FL MECH] The ratio of the speed of a body or of a point on a body with respect to the surrounding air or other fluid, or the ratio of the speed of a fluid, to the speed of sound in the medium. Symbolized N_{Ma}. Also known as relative Mach number.

Mach wave [FL MECH] Also known as Mach line. **1.** A shock wave theoretically occurring along a common line of intersection of all the pressure disturbances emanating from an infinitesimally small particle moving at supersonic speed through a fluid medium, with such a wave considered to exert no changes in the condition of the fluid passing through it. **2.** A very weak shock wave appearing, for example, at the nose of a very sharp body, where the fluid undergoes no substantial change in direction.

mackerel [VERT ZOO] The common name for perciform fishes composing the subfamily Scombroidei of the family Scombridae, characterized by a long slender body, pointed head, and large mouth.

mackerel shark [VERT ZOO] The common name for isurid galeoid elasmobranchs making up the family Isuridae; heavy-bodied fish with sharp-edged, awllike teeth and a nearly symmetrical tail.

Maclaurin-Cauchy test *See* Cauchy's test for convergence.

macle [CRYSTAL] A twinned crystal. [LAP] A twin structure in a diamond. [MINERAL] **1.** A dark or discolored spot in a mineral specimen. **2.** *See* chiastolite.

Macleod equation [FL MECH] An equation which states that the fourth root of the surface tension of a liquid is proportional to the difference between the densities of the liquid and of its vapor.

Macraucheniidae [PALEON] A family of extinct herbivorous mammals in the order Litopterna; members were proportioned much as camels are, and eventually lost the vertebral arterial canal of the cervical vertebrae.

Macristiidae [VERT ZOO] A family of oceanic teleostean fishes assigned by some zoologists to the order Ctenothrissiformes.

macro- [SCI TECH] Prefix meaning large.

macroanalysis [ANALY CHEM] Qualitative or quantitative analysis of chemicals that are in quantities of the order of grams.

macroassembler [COMPUT SCI] A program made up of one or more sequences of assembly language statements, each sequence represented by a symbolic name.

macroassembly program [COMPUT SCI] A set of assembly languages for the IBM 7090 and 7040 series of computers, and the assemblers for these languages; the assemblers operate under the IBSYS systems and prepare relocatable or absolute binary output in computer language. Abbreviated MAP.

macrocode [COMPUT SCI] A coding and programming language that assembles groups of computer instructions into single instructions.

macroconsumer [ECOL] A large consumer which ingests other organisms or particulate organic matter. Also known as biophage.

macrocrystalline [PETR] **1.** Pertaining to the texture of holocrystalline rock in which the constituents are visible without magnification. **2.** Pertaining to the texture of a rock with grains or crystals greater than 0.75 millimeter in diameter in recrystallized sediment.

Macrocypracea [INV ZOO] A superfamily of marine ostracods in the suborder Podocopa having all thoracic legs different from each other, greatly reduced furcae, and long, thin Zenker's organs.

Macrodasyoidea [INV ZOO] An order of wormlike invertebrates of the class Gastrotricha characterized by distinctive, cylindrical adhesive tubes in the cuticle which are moved by delicate muscle strands.

macrodefinition [COMPUT SCI] A statement that defines a macroinstruction and the set of ordinary instructions which it replaces.

macroetching *See* deep-etching.

macroevolution [EVOL] The larger course of evolution by which the categories of animal and plant classification above the species level have been evolved from each other and have differentiated into the forms within each.

macrofacies [GEOL] A collection of sedimentary facies that are related genetically.

macrogamete [BIOL] The larger, usually female gamete produced by a heterogamous organism.

macrogenerator *See* macroprocessor.

macroinstruction [COMPUT SCI] An instruction in a higher-level language which is equivalent to a specific set of one or more ordinary instructions in the same language.

macrolanguage [COMPUT SCI] A computer language that manipulates stored strings in which particular sites of the string are marked so that other strings can be inserted in these sites when the stored string is brought forth.

Macrolepidoptera [INV ZOO] A former division of Lepidoptera that included the larger moths and butterflies.

macrolibrary [COMPUT SCI] A collection of prewritten specialized but unparticularized routines (or sets of statements) which reside in mass storage.

macromere [EMBRYO] Any of the large blastomeres composing the vegetative hemisphere of telolecithal morulas and blastulas.

macrometeorology [METEOROL] The study of the largest-scale aspects of the atmosphere, such as the general circulation, and weather types.

macromolecule [ORG CHEM] A large molecule in which there is a large number of one or several relatively simple structural units, each consisting of several atoms bonded together.

macronutrient [BIOCHEM] An element, such as potassium and nitrogen, essential in large quantities for plant growth.

macrophage [HISTOL] A large phagocyte of the reticuloendothelial system. Also known as a histiocyte.

Macropodidae [VERT ZOO] The kangaroos, a family of Australian herbivorous mammals in the order Marsupialia.

macroprocessor [COMPUT SCI] A piece of software which replaces each macroinstruction in a computer program by the set of ordinary instructions which it stands for. Also known as macrogenerator.

macroprogramming [COMPUT SCI] The process of writing machine procedure statements in terms of macroinstructions.

Macroscelidea [VERT ZOO] A monofamilial order of mammals containing the elephant shrews and their allies.

Macroscelididae [VERT ZOO] The single, African family of the mammalian order Macroscelidea.

macroscopic [SCI TECH] Large enough to be observed with the naked eye.

macroskeleton [COMPUT SCI] A definition of a macroinstruction in a precise but content-free way, which can be particularized by a processor as directed by macroinstruction parameters. Also known as model.

macrospore [BOT] The larger of two spore types produced by heterosporous plants; the female gamete. Also known as megaspore. [INV ZOO] The larger gamete produced by certain radiolarians; the female gamete.

Macrostomida [INV ZOO] An order of rhabdocoels having a simple pharynx, paired protonephridia, and a single pair of longitudinal nerves.

Macrostomidae [INV ZOO] A family of rhabdocoels in the order Macrostomida; members are broad and flattened in shape and have paired sex organs.

Macrotermitinae [INV ZOO] A subfamily of termites in the family Termitidae.

Macrouridae [VERT ZOO] The grenadiers, a family of actinopterygian fishes in the order Gadiformes in which the body tapers to a point, and the dorsal, caudal, and anal fins are continuous.

Macroveliidae [INV ZOO] A family of hemipteran insects in the subdivision Amphibicorisae.

Macrura [INV ZOO] A group of decapod crustaceans in the suborder Reptantia including eryonids, spiny lobsters, and true lobsters; the abdomen is extended and bears a well-developed tail fan.

macula [ANAT] Any anatomical structure having the form of a spot or stain.

macula lutea [ANAT] A yellow spot on the retina; the area of maximum visual acuity, being made up almost entirely of retinal cones.

MAD See magnetic anomaly detector.

Madreporaria [INV ZOO] The equivalent name for Scleractinia.

madreporite [INV ZOO] A delicately perforated sieve plate at the distal end of the stone canal in echinoderms.

Madsen impedance meter [ENG] An instrument for measuring the acoustic impedance of normal and deaf ears, based on the principle of the Wheatstone bridge.

maelstrom [OCEANOGR] A powerful and often destructive water current caused by the combined effects of high, wind-generated waves and a strong, opposing tidal current.

Maestrichtian [GEOL] A European stage of geologic time: Upper Cretaceous (above Menevian, below Fastiniogian).

mafic mineral [MINERAL] 1. A mineral that is composed predominantly of the ferromagnesian rock-forming silicates. 2. In general, any dark mineral.

MAG See maximum available gain.

magamp See magnetic amplifier.

magazine [COMPUT SCI] 1. A holder of microfilm or magnetic recording media strips. 2. See input magazine. [ENG] 1. A storage area for explosives. 2. A building, compartment, or structure constructed and located for the storage of explosives or ammunition. [ORD] That part of a gun or firearm that holds ammunition ready for chambering.

Magellanic Clouds [ASTRON] Two irregular clouds of stars that are the nearest galaxies to the galactic system; both the Large and Small Magellanic Clouds are identified as Irregular in the classification of E.P. Hubble. Also known as Nubeculae.

Magelonidae [INV ZOO] A monogeneric family of spioniform annelid worms belonging to Sedentaria.

maghemite [MINERAL] γ-Fe_2O_3 A mineral form of iron oxide that is strongly magnetic and a member of the magnetite series.

magic eye See cathode-ray tuning indicator.

magic numbers [NUC PHYS] The integers 8, 20, 28, 50, 82, 126; nuclei in which the number of protons, neutrons, or both is magic have a stability and binding energy which is greater than average, and have other special properties.

magic square [MATH] A square array of integers where the sum of the entries of each row, each column, and each diagonal is the same.

magma [GEOL] The molten rock material from which igneous rocks are formed.

magma province See petrographic province.

magmatism [PETR] The formation of igneous rock from magma.

magnesian calcite [MINERAL] $(Ca,Mg)CO_3$ A variety of calcite consisting of randomly substituted magnesium carbonate in a disordered calcite lattice. Also known as magnesium calcite.

magnesian limestone [PETR] Limestone with at least 90% calcite, a maximum of 10% dolomite, an approximate magnesium oxide equivalent of 1.1–2.1%, and an approximate magnesium carbonate equivalent of 2.3–4.4%.

magnesian marble [PETR] A type of magnesian limestone that has been metamorphosed; contains some dolomite. Also known as dolomitic marble.

magnesiochromite [MINERAL] $MgCr_2O_4$ A mineral of the spinel group composed of magnesium chromium oxide; it is isomorphous with chromite. Also known as magnochromite.

magnesiocopiapite [MINERAL] $MgFe_4(SO_4)_6(OH)_2 \cdot 20H_2O$ A mineral of the copiapite group composed of hydrous basic magnesium and iron sulfate; it is isomorphous with copiapite and cuprocopiapite.

magnesioferrite [MINERAL] $(Mg,Fe)Fe_2O_4$ A black, strongly magnetic mineral of the magnetite series in the spinel group. Also known as magnoferrite.

magnesite [MINERAL] $MgCO_3$ The mineral form of magnesium carbonate, usually massive and white, with hexagonal symmetry; specific gravity is 3, and hardness is 4 on Mohs scale. Also known as giobertite.

magnesium [CHEM] A metallic element, symbol Mg, atomic number 12, atomic weight 24.312. [MET] A silvery-white, lightweight, malleable, ductile metal, used in metallurgical and chemical processes, photography, pyrotechny, and light alloys.

magnesium calcite See magnesian calcite.

magnesium-iron mica See biotite.

magnet [ELECTROMAG] A piece of ferromagnetic or ferrimagnetic material whose domains are sufficiently aligned so that it produces a net magnetic field outside itself and can experience a net torque when placed in an external magnetic field.

magnetic amplifier [ELECTR] A device that employs saturable reactors to modulate the flow of alternating-current electric power to a load in response to a lower-energy-level direct-current input signal. Abbreviated magamp. Also known as transductor.

magnetic annual variation [GEOPHYS] The small, systematic temporal variation in the earth's magnetic field which occurs after the trend for secular change has been removed from the average monthly values. Also known as annual magnetic variation.

magnetic anomaly detector [ELECTROMAG] A sensitive magnetometer carried at the end of a boom projecting from the tail of a patrol plane which can detect very small changes in the earth's magnetic field caused by a ferrous object, such as a submerged submarine; used to pinpoint a submarine's location, for effective deployment of suitable weapons. Abbreviated MAD.

magnetic azimuth [NAV] Azimuth relative to magnetic north.

magnetic bearing [NAV] Bearing relative to magnetic north, with the compass bearing corrected for deviation.

magnetic bias [ELECTROMAG] A steady magnetic field applied to the magnetic circuit of a relay or other magnetic device.

magnetic blowout [ELECTROMAG] **1.** A permanent magnet or electromagnet used to produce a magnetic field that lengthens the arc between opening contacts of a switch or circuit breaker, thereby helping to extinguish the arc. **2.** See blowout.

magnetic brake [MECH ENG] A friction brake under the control of an electromagnet.

magnetic bubble [SOLID STATE] A cylindrical stable (nonvolatile) region of magnetization produced in a thin-film magnetic material by an external magnetic field; direction of magnetization is perpendicular to the plane of the material. Also known as bubble.

magnetic bubble memory See bubble memory.

magnetic card [COMPUT SCI] A card with a magnetic surface on which data can be stored by selective magnetization.

magnetic card file [COMPUT SCI] A direct-access storage device in which units of data are stored on magnetic cards contained in one or more magazines from which they are withdrawn, when addressed, to be carried at high speed past a read/write head.

magnetic character [COMPUT SCI] A character printed with magnetic ink, as on bank checks, for reading by machines as well as by humans.

magnetic character reader [COMPUT SCI] A character reader that reads special type fonts printed in magnetic ink, such as those used on bank checks, and feeds the character data directly to a computer for processing.

magnetic character sorter [COMPUT SCI] A device that reads documents printed with magnetic ink; all data read are stored, and records are sorted on any required field. Also known as magnetic document sorter-reader.

magnetic circuit [ELECTROMAG] A group of magnetic flux lines each forming a closed path, especially when this circuit is regarded as analogous to an electric circuit because of the similarity of its magnetic field equations to direct-current circuit equations.

magnetic coercive force See coercive force.

magnetic compass [NAV] A compass depending for its directive force upon the attraction of the horizontal component of the earth's magnetic field for a magnetized needle or sensing element free to turn with a minimum of friction in any horizontal direction.

magnetic compression [PL PHYS] The force exerted by a magnetic field on an electrically conducting fluid or on a plasma.

magnetic confinement [PL PHYS] The containment of a plasma within a region of space by the forces of magnetic fields on the charged particles in the gas.

magnetic constant [ELECTROMAG] The absolute permeability of empty space, equal to 1 electromagnetic unit in the centimeter-gram-second system, and to $4\pi \times 10^{-7}$ henry per meter or, numerically, to 1.25664×10^{-6} henry per meter in International System units. Symbolized μ_0.

magnetic cooling See adiabatic demagnetization.

magnetic core Also known as core. [ELECTR] A configuration of magnetic material, usually a mixture of iron oxide or ferrite particles mixed with a binding agent and formed into a tiny doughnutlike shape, that is placed in a spatial relationship to current-carrying conductors, and is used to maintain a magnetic polarization for the purpose of storing data, or for its nonlinear properties as a logic element. Also known as memory core. [ELECTROMAG] A quantity of ferrous material placed in a coil or transformer to provide a better path than air for magnetic flux, thereby increasing the inductance of the coil and increasing the coupling between the windings of a transformer.

magnetic core storage [COMPUT SCI] A computer storage system in which each of thousands of magnetic cores stores one bit of information; current pulses are sent through wires threading through the cores to record or read out data. Also known as core memory; core storage.

magnetic coupling [ELECTROMAG] For a pair of particles or systems, the effect of the magnetic field created by one system on the magnetic moment or angular momentum of the other.

magnetic Curie temperature [SOLID STATE] The temperature below which a magnetic material exhibits ferromagnetism, and above which ferromagnetism is destroyed and the material is paramagnetic.

magnetic daily variation See magnetic diurnal variation.

magnetic damping [ELECTROMAG] Damping of a mechanical motion by means of the reaction between a magnetic field and the current generated by the motion of a coil through the magnetic field.

magnetic declination See declination.

magnetic delay line [ELECTR] Delay line, used for the storage of data in a computer, consisting essentially of a metallic medium along which the velocity of the propagation of magnetic energy is small compared to the speed of light; storage is accomplished by the recirculation of wave patterns containing information, usually in binary form.

magnetic dip See inclination.

magnetic dipole [ELECTROMAG] An object, such as a permanent magnet, current loop, or particle with angular momentum, which experiences a torque in a magnetic field, and itself gives rise to a magnetic field, as if it consisted of two magnetic poles of opposite sign separated by a small distance.

magnetic dipole density See magnetization.

magnetic dipole moment [ELECTROMAG] A vector associated with a magnet, current loop, particle, or such, whose cross product with the magnetic induction (or alternatively, the magnetic field strength) of a magnetic field is equal to the torque exerted on the system by the field. Also known as dipole moment; magnetic moment.

magnetic disk storage *See* disk storage.

magnetic displacement *See* magnetic induction.

magnetic diurnal variation [GEOPHYS] Oscillations of the earth's magnetic field which have a periodicity of about a day and which depend to a close approximation only on local time and geographic latitude. Also known as magnetic daily variation.

magnetic document sorter-reader *See* magnetic character sorter.

magnetic domain *See* ferromagnetic domain.

magnetic-domain memory *See* domain-tip memory.

magnetic drum *See* drum.

magnetic drum storage *See* drum.

magnetic field [ELECTROMAG] **1.** One of the elementary fields in nature; it is found in the vicinity of a magnetic body or current-carrying medium and, along with electric field, in a light wave; charges moving through a magnetic field experience the Lorentz force. **2.** *See* magnetic field strength.

magnetic field intensity *See* magnetic field strength.

magnetic field strength [ELECTROMAG] An auxiliary vector field, used in describing magnetic phenomena, whose curl, in the case of static charges and currents, equals (in meter-kilogram-second units) the free current density vector, independent of the magnetic permeability of the material. Also known as magnetic field; magnetic field intensity; magnetic force; magnetic intensity; magnetizing force.

magnetic film *See* magnetic thin film.

magnetic flux [ELECTROMAG] **1.** The integral over a specified surface of the component of magnetic induction perpendicular to the surface. **2.** *See* magnetic lines of force.

magnetic flux density *See* magnetic induction.

magnetic focusing [ELECTROMAG] Focusing of a beam of electrons or other charged particles by using the action of a magnetic field.

magnetic force *See* magnetic field strength.

magnetic hysteresis [ELECTROMAG] Lagging of changes in the magnetization of a substance behind changes in the magnetic field as the magnetic field is varied. Also known as hysteresis.

magnetic induction [ELECTROMAG] A vector quantity that is used as a quantitative measure of magnetic field; the force on a charged particle moving in the field is equal to the particle's charge times the cross product of the particle's velocity with the magnetic induction (mks units). Also known as magnetic displacement; magnetic flux density; magnetic vector.

magnetic-ink character recognition [COMPUT SCI] That branch of character recognition which involves the sensing of magnetic-ink characters for the purpose of determining the character's most probable identity. Abbreviated MICR.

magnetic intensity *See* magnetic field strength.

magnetic iron ore *See* magnetite.

magnetic lens [ELECTROMAG] A magnetic field with axial symmetry, capable of converging beams of charged particles of uniform velocity and of forming images of objects placed in the path of such beams; the field may be produced by solenoids, electromagnets, or permanent magnets. Also known as lens.

magnetic lines of flux *See* magnetic lines of force.

magnetic line of force [ELECTROMAG] Lines used to represent the magnetic induction in a magnetic field, selected so that they are parallel to the magnetic induction at each point, and so that the number of lines per unit area of a surface perpendicular to the induction is equal to the induction. Also known as magnetic flux; magnetic lines of flux.

magnetic memory *See* magnetic storage.

magnetic meridian [GEOPHYS] A line which is at any point in the direction of horizontal magnetic force of the earth; a compass needle without deviation lies in the magnetic meridian.

magnetic microphone [ENG ACOUS] A microphone consisting of a diaphragm acted upon by sound waves and connected to an armature which varies the reluctance in a magnetic field surrounded by a coil. Also known as reluctance microphone; variable-reluctance microphone.

magnetic monopole [ELECTROMAG] A hypothetical particle carrying magnetic charge; it would be a source for magnetic field in the same way that a charged particle is a source for electric field. Also known as monopole.

magnetic multipole [ELECTROMAG] One of a series of types of static or oscillating distributions of magnetization, which is a magnetic multipole of order 2; the electric and magnetic fields produced by a magnetic multipole of order 2^n are equivalent to those of two magnetic multipoles of order 2^{n-1} of equal strength but opposite sign, separated from each other by a short distance.

magnetic north [GEOPHYS] At any point on the earth's surface, the horizontal direction of the earth's magnetic lines of force (direction of a magnetic meridian) toward the north magnetic pole; a particular direction indicated by the needle of a magnetic compass.

magnetic nuclear resonance *See* nuclear magnetic resonance.

magnetic pinch *See* pinch effect.

magnetic polarization *See* intrinsic induction.

magnetic pole [ELECTROMAG] **1.** One of two regions located at the ends of a magnet that generate and respond to magnetic fields in much the same way that electric charges generate and respond to electric fields. **2.** A particle which generates and responds to magnetic fields in exactly the same way that electric charges generate and respond to electric fields; the particle probably does not have physical reality, but it is often convenient to imagine that a magnetic dipole consists of two magnetic poles of opposite sign, separated by a small distance. [GEOPHYS] In geomagnetism, either of the two points on the earth's surface where the magnetic meridians converge, that is, where the magnetic field is vertical. Also known as dip pole.

magnetic pressure [PL PHYS] A function, proportional to the square of the magnetic induction, such that the force exerted by a magnetic field on an electrically conducting fluid (excluding the force associated with curvature of magnetic flux lines) is the same as the force that would be exerted by a hydrostatic pressure equal to this function.

magnetic pressure transducer [ENG] A type of pressure transducer in which a change of pressure is converted into a change of magnetic reluctance or inductance when one part of a magnetic circuit is moved by a pressure-sensitive element, such as a bourdon tube, bellows, or diaphragm.

magnetic prospecting [ENG] Carrying out airborne or ground surveys of variations in the earth's magnetic field, using a magnetometer or other equipment, to locate magnetic deposits of iron, nickel, or titanium, or nonmagnetic deposits which either contain magnetic gangue minerals or are associated with magnetic structures.

magnetic quadrupole lens [ELECTROMAG] A magnetic field generated by four magnetic poles of alternating sign arranged in a circle; used to focus beams of charged particles in devices such as electron microscopes and particle accelerators.

magnetic recorder [ELECTR] An instrument that records information, generally in the form of audio-frequency or digital signals, on magnetic tape or magnetic wire as magnetic variations in the medium.

magnetic reluctance *See* reluctance.

magnetic reluctivity *See* reluctivity.

magnetic reversal [GEOPHYS] A reversal of the polarity of the earth's magnetic field that has occurred at irregular intervals on the order of 1,000,000 years.

magnetic rotation *See* Faraday effect.

magnetics [ELECTROMAG] The study of magnetic phenomena, comprising magnetostatics and electromagnetism.

magnetic saturation [ELECTROMAG] The condition in which, after a magnetic field strength becomes sufficiently large, further increase in the magnetic field strength produces no additional magnetization in a magnetic material. Also known as saturation.

magnetic shift register [COMPUT SCI] A shift register in which the pattern of settings of a row of magnetic cores is shifted one step along the row by each new input pulse; diodes in the coupling loops between cores prevent backward flow of information.

magnetic spectrometer [NUCLEO] A device for measuring the momentum of charged particles, or their distribution of intensity versus momentum, by passing the particles through a magnetic field which bends their paths in proportion to their momentum.

magnetic stepping motor *See* stepper motor.

magnetic storage [COMPUT SCI] A device utilizing magnetic properties of materials to store data; may be roughly divided into two categories, moving (drum, disk, tape) and static (core, thin film). Also known as magnetic memory.

magnetic storm [GEOPHYS] A worldwide disturbance of the earth's magnetic field; frequently characterized by a sudden onset, in which the magnetic field undergoes marked changes in the course of an hour or less, followed by a very gradual return to normalcy, which may take several days.

magnetic survey [GEOPHYS] 1. Magnetometer map of variations in the earth's total magnetic field; used in petroleum exploration to determine basement-rock depths and geologic anomalies. 2. Measurement of a component of the geomagnetic field at different locations.

magnetic susceptibility [ELECTROMAG] The ratio of the magnetization of a material to the magnetic field strength; it is a tensor when these two quantities are not parallel; otherwise it is a simple number. Also known as susceptibility.

magnetic tape [ELECTR] A plastic, paper, or metal tape that is coated or impregnated with magnetizable iron oxide particles; used in magnetic recording.

magnetic test coil *See* exploring coil.

magnetic thin film [SOLID STATE] A sheet or cylinder of magnetic material less than 5 micrometers thick, usually possessing uniaxial magnetic anisotropy; used mainly in computer storage and logic elements. Also known as ferromagnetic film; magnetic film.

magnetic transducer [ELECTROMAG] A device for transforming mechanical into electrical energy, which consists of a magnetic field including a variable-reluctance path and a coil surrounding all or a part of

this path, so that variation in reluctance leads to a variation in the magnetic flux through the coil and a corresponding induced emf (electromotive force).

magnetic vector *See* magnetic induction.

magnetic vector potential *See* vector potential.

magnetic wave [SOLID STATE] The spread of magnetization from a small portion of a substance where an abrupt change in the magnetic field has taken place.

magnetism [PHYS] Phenomena involving magnetic fields and their effects upon materials.

magnetite [MINERAL] An opaque iron-black and streak-black isometric mineral and member of the spinel structure type, usually occurring in octahedrals or in granular to massive form; hardness is 6 on Mohs scale, and specific gravity is 5.20. Also known as magnetic iron ore; octahedral iron ore.

magnetization [ELECTROMAG] 1. The property and in particular, the extent of being magnetized; quantitatively, the magnetic moment per unit volume of a substance. Also known as magnetic dipole density; magnetization intensity. 2. The process of magnetizing a magnetic material.

magnetization curve *See* B-H curve; normal magnetization curve.

magnetization intensity *See* magnetization.

magnetizing force *See* magnetic field strength.

magnet keeper *See* keeper.

magneto [ELEC] An alternating-current generator that uses one or more permanent magnets to produce its magnetic field; frequently used as a source of ignition energy on tractor, marine, industrial, and aviation engines. Also known as magnetoelectric generator.

magnetoacoustics [PHYS] The study of the effects of magnetic fields on acoustical phenomena, such as various oscillations in the attenuation of ultrasonic sound waves by a crystal placed in a magnetic field at a very low temperature, as the magnetic field strength or sound frequency is varied.

magnetoaerodynamics [PL PHYS] Study of the properties and characteristics of, and the forces exerted by, highly ionized air and other gases; applied principally to study of reentering ballistic missiles and spacecraft.

magnetocaloric effect [THERMO] The reversible change of temperature accompanying the change of magnetization of a ferromagnetic material.

magnetoelectric generator *See* magneto.

magnetofluid [PHYS CHEM] A Newtonian or shear-thinning fluid whose flow properties become viscoplastic when it is modulated by a magnetic field.

magnetofluid dynamics [PHYS] 1. The study of the motion of an electrically conducting metal, such as mercury, in the presence of electric and magnetic fields. 2. *See* magnetohydrodynamics.

magnetogas dynamics [PL PHYS] The science of motion in a plasma under the influence of mechanical, electric, and magnetic forces.

magnetograph [ELECTROMAG] A set of three variometers attached to a suitable recording unit, which records the components of the magnetic field vector in each of three perpendicular directions.

magnetohydrodynamic generator [ELEC] A system for generating electric power in which the kinetic energy of a flowing conducting fluid is converted to electric energy by a magnetohydrodynamic interaction. Abbreviated MHD generator.

magnetohydrodynamic instability [PL PHYS] An instability of a plasma in which the plasma expands while moving into a region of weaker magnetic field, until it is expelled from the field. Also known as hydromagnetic instability.

magnetohydrodynamics [PHYS] The study of the dynamics or motion of an electrically conducting fluid, such as an ionized gas or liquid metal, interacting with a magnetic field. Abbreviated MHD. Also known as hydromagnetics; magnetofluid dynamics.

magnetohydrodynamic stability [PL PHYS] The condition of a plasma in which fluctuations in density, pressure, velocity, or the distribution of particles in phase space, die out rather than increase. Also known as hydromagnetic stability.

magneto ignition system [ELECTROMAG] An ignition system in which the voltage required to cause a flow of current in the primary winding of the ignition coil is generated by a set of permanent magnets, instead of being supplied by a battery.

magnetometer [ENG] An instrument for measuring the magnitude and sometimes also the direction of a magnetic field, such as the earth's magnetic field.

magnetomotive force [ELECTROMAG] The work that would be required to carry a magnetic pole of unit strength once around a magnetic circuit. Abbreviated mmf.

magneton [PHYS] A unit of magnetic moment used for atomic, molecular, or nuclear magnets, such as the Bohr magneton, Weiss magneton, or nuclear magneton.

magnetooptical shutter [OPTICS] A device in which light passes through crossed Nicol prisms and a glass cell containing a liquid displaying the Faraday effect between the prisms; light can pass through the system only when a magnetic field is applied to the cell at an angle of 45° to the polarization planes of both prisms.

magnetooptic Kerr effect [OPTICS] Changes produced in the optical properties of a reflecting surface of a ferromagnetic substance when the substance is magnetized; this applies especially to the elliptical polarization of reflected light, when the ordinary rules of metallic reflection would give only plane polarized light. Also known as Kerr magnetooptical effect.

magnetooptics [OPTICS] The study of the effect of a magnetic field on light passing through a substance in the field.

magnetopause [GEOPHYS] A boundary that marks the transition from the earth's magnetosphere to the interplanetary medium.

magnetoplumbite [MINERAL] $(PbMn)_2Fe_6O_{11}$ Black mineral consisting of a ferric oxide of plumbite and manganese, and occurring in acute metallic hexagonal crystals.

magnetoresistance [ELECTROMAG] The change in electrical resistance produced in a current-carrying conductor or semiconductor on application of a magnetic field.

magnetoresistivity [ELECTROMAG] The change in resistivity produced in a current-carrying conductor or semiconductor on application of a magnetic field.

magnetoresistor [ELECTR] Magnetic field–controlled variable resistor.

magnetosheath [GEOPHYS] The relatively thin region between the earth's magnetopause and the shock front in the solar wind.

magnetosphere [GEOPHYS] The region of the earth in which the geomagnetic field plays a dominant part in controlling the physical processes that take place; it is usually considered to begin at an altitude of about 100 kilometers and to extend outward to a distant boundary that marks the beginning of interplanetary space.

magnetostatics [ELECTROMAG] The study of magnetic fields that remain constant with time. Also known as static magnetism.

magnetostriction [ELECTROMAG] The dependence of the state of strain (dimensions) of a ferromagnetic sample on the direction and extent of its magnetization.

magnetotail [GEOPHYS] The portion of the magnetosphere extending from earth in the direction away from the sun for a variable distance of the order of 1000 earth radii.

magnetron [ELECTR] One of a family of crossed-field microwave tubes, wherein electrons, generated from a heated cathode, move under the combined force of a radial electric field and an axial magnetic field in such a way as to produce microwave radiation in the frequency range 1–40 gigahertz; a pulsed microwave radiation source for radar, and continuous source for microwave cooking.

magnetron sputtering [ELECTR] A method for depositing a thin layer of metal in which a microwave tube is utilized to confine a plasma magnetically to produce high deposition rates and a low working-gas partial pressure.

magnet wire [ELEC] The insulated copper or aluminum wire used in the coils of all types of electromagnetic machines and devices.

magnification [OPTICS] 1. A measure of the effectiveness of an optical system in enlarging or reducing an image; the magnification may be lateral, longitudinal, or angular. 2. See lateral magnification.

magnifier See simple microscope.

magnifying glass [OPTICS] 1. Any device that uses a simple lens which enlarges the object being viewed. 2. See simple microscope.

magnifying power [OPTICS] The ratio of the tangent of the angle subtended at the eye by an image formed by an optical system, to the tangent of the angle subtended at the eye by the corresponding object at a distance for convenient viewing.

magnitude [ASTRON] The relative luminance of a celestial body; the smaller (algebraically) the number indicating magnitude, the more luminous the body. Also known as stellar magnitude. [GEOPHYS] A measure of the amount of energy released by an earthquake.

magnitude of a complex number See absolute value of a complex number.

magnochromite See magnesiochromite.

magnoferrite See magnesioferrite.

Magnoliaceae [BOT] A family of dicotyledonous plants of the order Magnoliales characterized by hypogynous flowers with few to numerous stamens, stipulate leaves, and uniaperturate pollen.

Magnoliales [BOT] The type order of the subclass Magnoliidae; members are woody plants distinguished by the presence of spherical ethereal oil cells and by a well-developed perianth of separate tepals.

Magnoliatae See Magnoliopsida.

Magnoliidae [BOT] A primitive subclass of flowering plants in the class Magnoliopsida generally having a well-developed perianth, numerous centripetal stamens, and bitegmic, crassinucellate ovules.

Magnoliophyta [BOT] The angiosperms, a division of vascular seed plants having the ovules enclosed in an ovary and well-developed vessels in the xylem.

Magnoliopsida [BOT] The dicotyledons, a class of flowering plants in the division Magnoliophyta generally characterized by having two cotyledons and net-veined leaves, with vascular bundles borne in a ring enclosing a pith.

magnon [SOLID STATE] A quasi-particle which is introduced to describe small departures from complete ordering of electronic spins in ferro-, ferri-, antiferro-, and helimagnetic arrays. Also known as quantized spin wave.

Magnus effect [FL MECH] A force on a rotating cylinder in a fluid flowing perpendicular to the axis of the cylinder; the force is perpendicular to both flow direction and cylinder axis. Also known as Magnus force.

Magnus force *See* Magnus effect.

mag-slip *See* synchro.

mahogany [BOT] Any of several tropical trees in the family Meliaceae of the Geraniales. [MATER] The hard wood of these trees, especially the red or yellow-brown wood of the West Indies mahogany tree (*Swietenia mahagoni*).

maidenhead *See* hymen.

main [CIV ENG] A duct or pipe that supplies or drains ancillary branches. [ELEC] **1.** One of the conductors extending from the service switch, generator bus, or converter bus to the main distribution center in interior wiring. **2.** *See* power transmission line.

main clock *See* master clock.

main distributing frame [ELEC] Frame which terminates the permanennt outside line entering the central office building on one side and the subscriber-line multiple cabling, trunk multiple cabling, and so on, used for associating an outside line with any desired terminal on the other side; it usually carries the control-office protective devices, and functions as a test point between line and office. Also known as main frame.

Maindroniidae [INV ZOO] A family of wingless insects belonging to the Thysanura proper.

main frame *See* central processing unit; main distributing frame.

main lobe *See* major lobe.

main memory *See* main storage.

main path [COMPUT SCI] The principal branch of a routine followed by a computer in the process of carrying out the routine.

main program [COMPUT SCI] **1.** The central part of a computer program, from which control may be transferred to various subroutines and to which control is eventually returned. Also known as main routine. **2.** *See* executive routine.

main routine *See* executive routine; main program.

main sequence [ASTRON] The band in the spectrum luminosity diagram which has the great majority of stars; their energy derives from core burning of hydrogen into helium.

main sequence star [ASTRON] **1.** Any of those stars in the smooth curve termed the main sequence in a Hertzsprung-Russell diagram. **2.** *See* dwarf star.

main shaft [MECH ENG] The line of shafting receiving its power from the engine or motor and transmitting power to other parts.

main storage [COMPUT SCI] A digital computer's principal working storage, from which instructions can be executed or operands fetched for data manipulation. Also known as main memory.

main stroke *See* return streamer.

maintenance routine [COMPUT SCI] A computer program designed to detect conditions which may give rise to a computer malfunction in order to assist a service engineer in performing routine preventive maintenance.

Majidae [INV ZOO] The spider, or decorator, crabs, a family of decapod crustaceans included in the Brachyura; members are slow-moving animals that often conceal themselves by attaching seaweed and sessile animals to their carapace.

Majorana force [NUC PHYS] A force between two nucleons postulated to explain various phenomena, which can be derived from a potential containing an operator which exchanges the nucleons' positions but not their spins.

Majorana neutrino [PARTIC PHYS] A particle described by a wave function that satisfies the Dirac equation with mass equal to zero, and that is self–charge-conjugate.

major axis [MATH] The longer of the two axes with respect to which an ellipse is symmetric.

major cycle [COMPUT SCI] The time interval between successive appearances of a given storage position in a serial-access computer storage.

major diatonic scale [ACOUS] A diatonic scale in which the relative sizes of the sequence of intervals are approximately $2,2,1,2,2,2,1$.

majority [MATH] A logic operator having the property that if P, Q, R are statements, then the function (P, Q, R, ...) is true if more than half the statements are true, or false if half or less are true.

majority carrier [ELECTR] The type of carrier, that is, electron or hole, that constitutes more than half the carriers in a semiconductor.

majority element *See* majority gate.

majority emitter [ELECTR] Of a transistor, an electrode from which a flow of minority carriers enters the interelectrode region.

majority gate [COMPUT SCI] A logic circuit which has one output and several inputs, and whose output is energized only if a majority of its inputs are energized. Also known as majority element; majority logic.

majority logic *See* majority gate.

major lobe [ELECTROMAG] Antenna lobe indicating the direction of maximum radiation or reception. Also known as main lobe.

major node [ELEC] A point in an electrical network at which three or more elements are connected together. Also known as junction.

major planet [ASTRON] Any of the four planets that are larger than earth: Jupiter, Saturn, Neptune, and Uranus.

major wave *See* long wave.

makeready [GRAPHICS] The careful leveling of relief printing plates on the bed of the press so that they yield the best possible impression.

Malachiidae [INV ZOO] An equivalent name for Malyridae.

malachite [MINERAL] $Cu_2(OH)_2(CO_3)$ A bright-green monoclinic mineral consisting of a basic carbonate of copper and usually occurring in massive forms or in bundles of radiating fibers; specific gravity is 4.05, and hardness is 3.5–4 on Mohs scale.

Malacobothridia [INV ZOO] A subclass of worms in the class Trematoda; members typically have one or two soft, flexible suckers and are endoparasitic in vertebrates and invertebrates.

Malacocotylea [INV ZOO] The equivalent name for Digenea.

malacology [INV ZOO] The study of mollusks.

malacoplakia [MED] The accumulation of modified histiocytes (malacoplakia cells) to produce soft, pale, elevated plaques, usually in the urinary bladder of middle-aged women.

Malacopoda [INV ZOO] A subphylum of invertebrates in the phylum Oncopoda.

Malacopterygii [VERT ZOO] An equivalent name for Clupeiformes in older classifications.

Malacostraca [INV ZOO] A large, diversified subclass of Crustacea including shrimps, lobsters, crabs, sow bugs, and their allies; generally characterized by having a maximum of 19 pairs of appendages and trunk limbs which are sharply differentiated into thoracic and abdominal series.

maladie du sommeil *See* African sleeping sickness.

Malapteruridae [VERT ZOO] A family of African catfishes in the suborder Siluroidei.

malar bone *See* zygomatic bone.

malchite [PETR] A fine-grained lamprophyre with small, rare phenocrysts or hornblende, labradorite, and sometimes biotite embedded in a matrix of hornblende, andesine, and some quartz.

Malcidae [INV ZOO] A small family of Ethiopian and Oriental hemipternan insects in the superfamily Pentatomorpha.

Malcodermata [INV ZOO] The equivalent name for Cantharoidea.

Maldanidae [INV ZOO] The bamboo worms, a family of mud-swallowing annelids belonging to the Sedentaria.

Maldaninae [INV ZOO] A subfamily of the Maldanidae distinguished by cephalic and anal plaques with the anal aperture located dorsally.

male [BOT] A flower lacking pistils. [ZOO] 1. Of or pertaining to the sex that produces spermatozoa. 2. An individual of this sex.

male pseudohermaphroditism *See* androgyny.

malfunction routine [COMPUT SCI] A program used in troubleshooting.

malignant [MED] 1. Endangering the life or health of an individual. 2. Of or pertaining to the growth and proliferation of certain neoplasms which terminate in death if not checked by treatment.

malladrite [MINERAL] Na_2SiF_6 A hexagonal mineral composed of sodium fluosilicate, occurring as small crystals in volcanic holes in Vesuvius.

mallardite [MINERAL] $MnSO_4 \cdot 7H_2O$ A pale-rose, monoclinic mineral composed of hydrous manganese sulfate.

malleable [MET] Capable of undergoing plastic deformation without rupture; a property characteristic of metals.

mallee *See* tropical scrub.

malleolus [ANAT] A projection on the distal end of the tibia and fibula at the ankle.

malleus [ANAT] The outermost, hammer-shaped ossicle of the middle ear; attaches to the tympanic membrane and articulates with the incus.

Mallophaga [INV ZOO] Biting lice, a comparatively small order of wingless insects characterized by five-segmented antennae, distinctly developed mandibles, one or two terminal claws on each leg, and a prothorax developed as a distinct segment.

Malm [GEOL] The Upper Jurassic geologic series, above Dogger and below Cretaceous.

Malpighiaceae [BOT] A family of dicotyledonous plants in the order Polygalales distinguished by having three carpels, several fertile stamens, five petals that are commonly fringed or toothed, and indehiscent fruit.

Malpighian corpuscle [ANAT] 1. A lymph nodule of the spleen. 2. *See* renal corpuscle.

Malpighian tubule [INV ZOO] Any of the blind tubes that open into the posterior portion of the gut in most insects and certain other arthropods and excrete matter or secrete substances such as silk.

malt [FOOD ENG] A nutrient material made from grain, commonly barley, which has been soaked, allowed to germinate, and dried.

Malta fever *See* brucellosis.

maltase [BIOCHEM] An enzyme that catalyzes the conversion of maltose to dextrose.

maltobiose *See* maltose.

maltose [BIOCHEM] $C_{12}H_{22}O_{11}$ A crystalline disaccharide that is a product of the enzymatic hydrolysis of starch, dextrin, and glycogen; does not appear to exist free in nature. Also known as maltobiose; malt sugar.

malt sugar *See* maltose.

Malus *See* Pyxis.

Malvaceae [BOT] A family of herbaceous dicotyledons in the order Malvales characterized by imbricate or contorted petals, mostly unilocular anthers, and minutely spiny, multiporate pollen.

Malvales [BOT] An order of flowering plants in the subclass Dilleniidae having hypogynous flowers with valvate calyx, mostly separate petals, numerous centrifugal stamens, and a syncarpous pistil.

malysite [MINERAL] $FeCl_3$ A halogen mineral deposited by sublimation; found most commonly at Mount Vesuvius, Italy.

mamma [ANAT] A milk-secreting organ characterizing all mammals.

mammal [VERT ZOO] A member of Mammalia.

Mammalia [VERT ZOO] A large class of warm-blooded vertebrates containing animals characterized by mammary glands, a body covering of hair, three ossicles in the middle ear, a muscular diaphragm separating the thoracic and abdominal cavities, red blood cells without nuclei, and embryonic development in the allantois and amnion.

mammary gland [PHYSIO] A highly modified sebaceous gland that secretes milk; a unique anatomical feature of mammals.

mammary ridge [EMBRYO] An ectodermal thickening forming a longitudinal elevation on the chest between the limbs from which the mammary glands develop.

mammary-stimulating hormone [BIOCHEM] 1. Estrogen and progesterone considered together as the hormones which induce proliferation of the mammary ductile and acinous elements respectively. 2. *See* prolactin.

mammectomy *See* mastectomy.

mammillary [ANAT] 1. Of or pertaining to the nipple. 2. Breast- or nipple-shaped. [MINERAL] Of or pertaining to an aggregate of crystals in the form of a rounded mass.

mammillary structure *See* pillow structure.

mammogen *See* prolactin.

mammogenic hormone [BIOCHEM] 1. Any hormone that stimulates or induces development of the mammary gland. 2. *See* prolactin.

mammography [MED] Radiographic examination of the breast, performed with or without injection of the glandular ducts with a contrast medium.

mammoplasty [MED] Plastic surgery performed to alter the shape of the breast.

mammoth [PALEON] Any of various large Pleistocene elephants having long, upcurved tusks and a heavy coat of hair.

mammotropin *See* prolactin.

Mammutinae [PALEON] A subfamily of extinct proboscidean mammals in the family Mastodontidae.

management engineering *See* industrial engineering.

management information system [COMMUN] A communication system in which data are recorded and processed to form the basis for decisions by top management of an organization.

manandonite [MINERAL] $Li_4Al_{14}B_4Si_6O_{29}(OH)_{24}$ A white mineral composed of basic borosilicate of lithium and aluminum.

manasseite [MINERAL] $MgAl_2(OH)_{16}(CO_3)\cdot 4H_2O$ A hexagonal mineral composed of basic hydrous magnesium and aluminum carbonate; it is dimorphous with hydrotalcite.

mandarin [BOT] A large and variable group of citrus fruits in the species *Citrus reticulata* and some of its hybrids; many varieties of the trees are compact with willowy twigs and small, narrow, pointed leaves; includes tangerines, King oranges, Temple oranges, and tangelos.

mandelic acid nitrile *See* mandelonitrile.

mandelonitrile [ORG CHEM] $C_6H_5CH(OH)CN$ A liquid used to prepare bitter almond water. Also known as benzaldehyde cyanohydrin; mandelic acid nitrile.

Mandelstam representation [PARTIC PHYS] For a reaction in which there are two particles both before and after scattering: an expression, containing several integrals, for a function related to the scattering amplitude; the arguments of the function are the center-of-mass energy and scattering angle, extended to complex values; the function is conjectured to be analytic in these variables except for certain cuts and to have values along these cuts which give the scattering amplitude of the reaction, and of the two reactions derivable from it by the crossing principle.

mandible [ANAT] **1.** The bone of the lower jaw. **2.** The lower jaw. [INV ZOO] Any of various mouthparts in many invertebrates designed to hold or bite into food.

mandibular gland *See* submandibular gland.

Mandibulata [INV ZOO] A subphylum of Arthropoda; members possess a pair of mandibles which characterize the group.

mandrel [MECH ENG] A shaft inserted through a hole in a component to support the work during machining. [MET] A metal bar serving as a core around which other metals are cast, forged, or extruded, forming a true central hole.

mandrill [VERT ZOO] *Mandrillus sphinx*. An Old World cercopithecoid monkey found in West-central Africa and characterized by large red callosities near the ischium and by blue ridges on each side of the nose in males.

manganate [INORG CHEM] **1.** Salts that have manganese in the anion. **2.** In particular, a salt of manganic acid formed by fusion of manganese dioxide with an alkali.

manganese [CHEM] A metallic element, symbol Mn, atomic weight 54.938, atomic number 25; a transition element whose properties fall between those of chromium and iron. [MET] A hard, brittle, grayish-white metal used chiefly in making steel.

manganese epidote *See* piemontite.

manganese nodule [GEOL] Small, irregular black to brown concretions consisting chiefly of manganese salts and manganese oxide minerals; formed in oceans as a result of pelagic sedimentation or precipitation.

manganite [MINERAL] $MnO(OH)$ A brilliant steel-gray or black polymorphous mineral; crystallizes in the orthorhombic system. Also known as gray manganese ore.

manganosite [MINERAL] MnO An emerald-green isometric mineral occurring in small octahedrons that blacken on exposure; hardness is 5–6 on Mohs scale, and specific gravity is 5.18.

Manger *See* Praesepe.

mangle gearing [MECH ENG] Gearing for producing reciprocating motion; a pinion rotating in a single direction drives a rack with teeth at the ends and on both sides.

mango [BOT] *Mangifera indica*. A large evergreen tree of the sumac family (Anacardiaceae), native to southeastern Asia, but now cultivated in Africa, tropical America, Florida, and California for its edible fruit, a thick-skinned, yellowish-red, fleshy drupe.

mangrove [BOT] A tropical tree or shrub of the genus *Rhizophora* characterized by an extensive, impenetrable system of prop roots which contribute to land building. [MATER] Liquid derived from the mangrove tree *Rhizophora mucronata* and used in the leather tanning industry.

mania [PSYCH] Excessive enthusiasm or excitement; a violent desire or passion; manifestation of a psychotic disorder.

manic-depressive psychosis [PSYCH] A severe disturbance of affect characterized by extreme and pathological elation alternating with severe dejection, both of which may last for months or years.

Manidae [VERT ZOO] The pangolins, a family of mammals comprising the order Pholidota.

manifold [ENG] The branch pipe arrangement which connects the valve parts of a multicylinder engine to a single carburetor or to a muffler. [MATH] A topological space which is locally euclidean; there are four types: topological, piecewise linear, differentiable, and complex, depending on whether the local coordinate systems are obtained from continuous, piecewise linear, differentiable, or complex analytic functions of those in euclidean space; intuitively, a surface.

manihot *See* cassava.

Manila hemp *See* abaca.

manioc *See* cassava.

man-machine system [ENG] A system in which the functions of the worker and the machine are interrelated and necessary for the operation of the system.

mannan [BIOCHEM] Any of a group of polysaccharides composed chiefly or entirely of D-mannose units.

mannose [BIOCHEM] $C_6H_{12}O_6$ A fermentable monosaccharide obtained from manna.

manometer [ENG] A double-leg liquid-column gage used to measure the difference between two fluid pressures.

mansfieldite [MINERAL] $Al(AsO_4)\cdot 2H_2O$ A white to pale-gray orthorhombic mineral composed of hydrous aluminum arsenate; it is isomorphous with scorodite.

Mantidae [INV ZOO] A family of predacious orthopteran insects characterized by a long, slender prothorax bearing a pair of large, grasping legs, and a freely moving head with large eyes.

mantis [INV ZOO] The common name for insects composing the family Mantidae.

mantissa [COMPUT SCI] A fixed point number composed of the most significant part of a given floating point number. Also known as fixed-point part; floating-point coefficient. [MATH] The positive decimal part of a common logarithm.

mantle [ANAT] Collectively, the convolutions, corpus callosum, and fornix of the brain. [BIOL] An enveloping layer, as the external body wall lining the shell of many invertebrates, or the external meristematic layers in a stem apex. [ENG] A lace-like hood or envelope (sack) of refractory material which, when positioned over a flame and heated to incandescence, gives light. [GEOL] The intermediate shell zone of the earth below the crust and above the core (to a depth of 3480 kilometers). [MET] That part of the outer wall and casing of a blast furnace located above the hearth. [VERT ZOO]A bird's back and wing plumage if distinguished from the rest of the plumage by a uniform color.

mantle rock   *See* regolith.

Mantodea   [INV ZOO] An order equivalent to the family Mantidae in some systems of classification.

manual input   [COMPUT SCI] The entry of data by hand into a device at the time of processing.

manual number generator   *See* manual word generator.

manual system   [COMPUT SCI] A system involving data processing which does not make use of stored-program computing equipment; by this somewhat arbitrary definition, systems using other types of tabulating equipment, such as the card-programmed calculator, are considered to be manual.

manual telephone system   [COMMUN] A telephone system in which connections between customers are ordinarily established manually by telephone operators in accordance with orders given verbally by calling parties.

manual time   *See* hand time.

manual word generator   [COMPUT SCI] A device into which an operator can enter a computer word by hand, either for direct insertion into memory or to be held until it is read during the execution of a program. Also known as manual number generator.

manubrium   [ANAT] 1. The triangular cephalic portion of the sternum in humans and certain other mammals. 2. The median anterior portion of the sternum in birds. 3. The process of the malleus. [BOT] A cylindrical cell that projects inward from the middle of each shield composing the antheridium in stoneworts. [INV ZOO] The elevation bearing the mouth in hydrozoan polyps.

manus   [ANAT] The hand of a human or the forefoot of a quadruped. [INV ZOO] The proximal enlargement of the propodus of the chela of arthropods.

many-body problem   [MECH] The problem of predicting the motions of three or more objects obeying Newton's laws of motion and attracting each other according to Newton's law of gravitation. Also known as n-body problem.

many-body theory   [PHYS] A scheme for calculating physical quantities for systems with large numbers of particles, without finding details of each particle's motion, often at temperatures close to absolute zero.

manyplies   *See* omasum.

map   [COMPUT SCI] 1. An output produced by an assembler, compiler, linkage editor, or relocatable loader which indicates the (absolute or relocatable) locations of such elements as programs, subroutines, variables, or arrays. 2. By extension, an index of the storage allocation on a magnetic disk or drum. [GRAPHICS] A representation, usually on a plane surface, of all or part of the surface of the earth, celestial sphere, or other area; shows relative size and position, according to a given projection, of the physical features represented and such other information as may be applicable to the purpose intended. [MATH] *See* mapping.

MAP   *See* macroassembly program.

maple   [BOT] Any of various broad-leaved, deciduous trees of the genus *Acer* in the order Sapindales characterized by simple, opposite, usually palmately lobed leaves and a fruit consisting of two long-winged samaras. [MATER] The hard, light-colored, close-grained wood, especially from sugar maple (*A. saccharum*).

mapping   [GRAPHICS] Preparation of a map or engaging in a mapping operation. [MATH] 1. Any function or multiple-valued relation. Also known as map. 2. In topology, a continuous function.

map projection   *See* projection.

maquis   [ECOL] A type of vegetation composed of shrubs, or scrub, usually not exceeding 3 meters in height, the majority having small, hard, leathery, often spiny or needlelike drought-resistant leaves and occurring in areas with a Mediterranean climate.

Marantaceae   [BOT] A family of monocotyledonous plants in the order Zingiberales characterized by one functional stamen with a single functional pollen sac, solitary ovules in each locule, and mostly arillate seeds.

Marattiaceae   [BOT] A family of ferns coextensive with the order Marattiales.

Marattiales   [BOT] An ancient order of ferns having massive eusporangiate sporangia in sori on the lower side of the circinate leaves.

marble   [PETR] 1. Metamorphic rock composed of recrystallized calcite or dolomite. 2. Commercially, any limestone or dolomite taking polish.

marcasite   [MINERAL] FeS_2 A pale bronze-yellow to nearly white mineral, crystallizing in the orthorhombic system; hardness is 6–6.5 on Mohs scale, and specific gravity is 4.89.

marcescent   [BOT] Withering without falling off.

Marcgraviaceae   [BOT] A family of dicotyledonous shrubs or vines in the order Theales having exstipulate leaves with scanty or no endosperm, two integuments, and highly modified bracts.

Marchantiales   [BOT] The thallose liverworts, an order of the class Marchantiopsida having a flat body composed of several distinct tissue layers, smooth-walled and tuberculate-walled rhizoids, and male and female sex organs borne on stalks on separate plants.

Marchantiatae   *See* Marchantiopsida.

Marchantiopsida   [BOT] The liverworts, a class of lower green plants; the plant body is usually a thin, prostrate thallus with rhizoids on the lower surface.

March equinox   *See* vernal equinox.

mare   [ASTRON] 1. One of the large, dark, flat areas on the lunar surface. 2. One of the less well-defined areas on Mars. [VERT ZOO] A mature female horse or other equine.

margarine   [FOOD ENG] A plastic food fat product composed of processed vegetable oils or animal fats or both, cultured milk, salt, and emulsifiers.

margarite   [GEOL] A string of beadlike globulites; commonly found in glassy igneous rocks. [MINERAL] $CaAl_2(Al_2Si_2)O_{10}(OH)_2$ A pink, reddish, or yellow, brittle mineral.

Margaritiferidae   [INV ZOO] A family of gastropod mollusks with nacreous shells that provide an important source of commercial pearls.

Margarodinae   [INV ZOO] A subfamily of homopteran insects in the superfamily Coccoidea in which abdominal spiracles are present in all stages of development.

margarosanite   [MINERAL] $PbCa_2(SiO_3)_3$ A colorless or snow-white triclinic mineral composed of lead calcium silicate, occurring in lamellar masses.

margin   [GEOGR] The boundary around a body of water. [GRAPHICS] The blank area at the vertical and horizontal edges of a printed page. [SCI TECH] An outside limit.

marginal blight   [PL PATH] A bacterial disease of lettuce caused by *Pseudomonas marginalis*, characterized by brownish marginal discoloration of the foliage.

marginal chlorosis   [PL PATH] A virus disease characterized by yellowing or blanching of leaf margins; common disease of peanut plants.

marginal moraine   *See* terminal moraine.

marginal plain   *See* outwash plain.

marginal probability [STAT] Probability expressed by the two conditional probability distributions which arise from the joint distribution of two random variables.

marginal sea [GEOGR] A semiclosed sea adjacent to a continent and connected with the ocean at the water surface.

margin of safety [DES ENG] A design criterion, usually the ratio between the load that would cause failure of a member or structure and the load that is imposed upon it in service.

mariculture [AGR] The cultivation of marine organisms, plant and animal, for purposes of human consumption.

marihuana See marijuana.

marijuana [BOT] The Spanish name for the dried leaves and flowering tops of the hemp plant (*Cannabis sativa*), which have narcotic ingredients and are smoked in cigarettes. Also spelled marihuana.

marine climate [CLIMATOL] A regional climate which is under the predominant influence of the sea, that is, a climate characterized by oceanity; the antithesis of a continental climate. Also known as maritime climate; oceanic climate.

marine engineering [ENG] The design, construction, installation, operation, and maintenance of main power plants, as well as the associated auxiliary machinery and equipment, for the propulsion of ships.

marine forecast [METEOROL] A forecast, for a specified oceanic or coastal area, of weather elements of particular interest to maritime transportation, including wind, visibility, the general state of the weather, and storm warnings.

marine stack See stack.

marine terrace [GEOL] A seacoast terrace formed by the merging of a wave-built terrace and a wave-cut platform. Also known as sea terrace; shore terrace.

marine transgression See transgression.

Mariotte's law See Boyle's law.

maritime air [METEOROL] A type of air whose characteristics are developed over an extensive water surface and which, therefore, has the basic maritime quality of high moisture content in at least its lower levels.

maritime climate See marine climate.

maritime frequency bands [COMMUN] In the United States, a collection of radio frequencies allocated for communication between coast stations and ships or between ships.

maritime mobile service [COMMUN] A mobile service between coast stations and ship stations, or between ship stations, in which survival craft stations may also participate.

marjoram [BOT] Any of several perennial plants of the genera *Origanum* and *Majorana* in the mint family, Labiatae; the leaves are used as a food seasoning.

mark [COMMUN] The closed-circuit condition in telegraphic communication, during which the signal actuates the printer; the opposite of space. [COMPUT SCI] A distinguishing feature used to signal some particular location or condition. [NAV] 1. A charted conspicuous object, structure, or light serving as an indicator for guidance or warning to craft; a beacon; it may be a day-beacon or sea-mark depending upon its location, or a day-mark or lighted beacon depending upon its period of usefulness. 2. Fathoms marked on a lead line. [ORD] A designation followed by a serial number, used to identify models of military equipment.

marker beacon [NAV] A low-power radio beacon transmitting a signal to designate a small area, as an aid to navigation.

marker bed [GEOL] 1. A stratified unit with distinctive characteristics making it an easily recognized geologic horizon. 2. A rock layer which accounts for a characteristic portion of a seismic refraction time-distance curve. 3. See key bed.

marker buoy [NAV] 1. A temporary buoy used in surveying to mark a location of particular interest, such as a shoal or reef. 2. See station buoy.

Markov chain [MATH] A Markov process whose state space is finite or countably infinite.

Markov process [MATH] A stochastic process which assumes that in a series of random events the probability of an occurrence of each event depends only on the immediately preceding outcome.

mark reading [COMPUT SCI] In character recognition, that form of mark detection which employs a photoelectric device to locate and convey intended information; the information appears as special marks on sites (windows) within the document coding area.

mark sensing [COMPUT SCI] In character recognition, that form of mark detection which depends on the conductivity of graphite pencil marks to locate and convey intended information; the information appears as special marks on sites (windows) within the document coding area.

marl [GEOL] A deposit of crumbling earthy material composed principally of clay with magnesium and calcium carbonate; used as a fertilizer for lime-deficient soils. [TEXT] Two yarns of different colors or kinds twisted around each other.

marlite See marlstone.

marlstone [PETR] 1. A consolidated rock that has about the same composition as marl; considered to be an earthy or impure argillaceous limestone. Also known as marlite. 2. A hard ferruginous rock of the Middle Lias in England.

marmoset [VERT ZOO] Any of 10 species of South American primates belonging to the family Callithricidae; individuals are primitive in that they have claws rather than nails and a nonprehensile tail.

marmot [VERT ZOO] Any of several species of stout-bodied, short-legged burrowing rodents of the genus *Marmota* in the squirrel family Sciuridae.

Mars [ASTRON] The planet fourth in distance from the sun; it is visible to the naked eye as a bright red star, except for short periods when it is near its conjunction with the sun; its diameter is about 6700 kilometers.

marsh [ECOL] A transitional land-water area, covered at least part of the time by estuarine or coastal waters, and characterized by aquatic and grasslike vegetation, especially without peatlike accumulation.

marsh gas [GEOCHEM] Combustible gas, consisting chiefly of methane, produced as a result of decay of vegetation in stagnant water.

marshite [MINERAL] CuI A reddish, oil-brown isometric mineral composed of cuprous iodide and occurring as crystals; hardness is 2.5 on Mohs scale, and specific gravity is 5.6.

marsh ore See bog iron ore.

Marsileales [BOT] A small monofamilial order of heterosporous, leptosporangiate ferns (Polypodiophyta); leaves arise on long stalks from the rhizome, and sporangia are enclosed in modified folded leaves or leaf segments called sporocarps.

Marsipobranchii [VERT ZOO] An equivalent name for Cyclostomata.

marsupial [VERT ZOO] 1. A member of the Marsupialia. 2. Having a marsupium. 3. Of, pertaining to, or constituting a marsupium.

Marsupialia [VERT ZOO] The single order of the mammalian infraclass Metatheria, characterized by the presence of a marsupium in the female.

Marsupicarnivora [VERT ZOO] An order proposed to include the polydactylous and polyprotodont carnivorous superfamilies of Marsupialia.

marsupium [VERT ZOO] A fold of skin that forms a pouch enclosing the mammary glands on the abdomen of most marsupials.

martempering [MET] Quenching austenitized steel to a temperature just above, or in the upper part of, the martensite range, holding it at this point until the temperature is equalized throughout, and then cooling in air to room temperature.

marten [VERT ZOO] Any of seven species of carnivores of the genus *Martes* in the family Mustelidae which resemble the weasel but are larger and of a semiarboreal habit.

martensite [MET] A metastable transitional structure formed by a shear process during a phase transformation, characterized by an acicular or needlelike pattern; in carbon steel it is a hard, supersaturated solid solution of carbon in a body-centered tetragonal lattice of iron.

martingale [STAT] A sequence of random variables $x_1, x_2, \ldots$, where the conditional expected value of x_{n+1} given $x_1, x_2, \ldots, x_n$, equals x_n.

mA s See milliampere-second.

mascagnite [MINERAL] $(NH_4)_2SO_4$ A yellowish-gray mineral found in guano, near burning coal beds, or as lava incrustation; specific gravity is 1.77; hardness is 2–2.5 on Mohs scale.

mascaret See bore.

maser [PHYS] A device for coherent amplification or generation of electromagnetic waves in which an ensemble of atoms or molecules, raised to an unstable energy state, is stimulated by an electromagnetic wave to radiate excess energy at the same frequency and phase as the stimulating wave. Derived from microwave amplification by stimulated emission of radiation. Also known as paramagnetic amplifier.

mask [COMPUT SCI] A pattern of characters used to control the retention or elimination of portions of another pattern of characters. Also known as extractor. [DES ENG] A frame used in front of a television picture tube to conceal the rounded edges of the screen. [ELECTR] A thin sheet of metal or other material containing an open pattern, used to shield selected portions of a semiconductor or other surface during a deposition process. [GRAPHICS] 1. In color separation photography, an intermediate negative or positive that is used to correct color. 2. In offset lithography, opaque material that protectively covers open or selected areas of a printing plate during the exposure process. [MET] A protective device in thermal spraying against blasting or coating effects which are reflected from the substrate surface.

masking [ACOUS] The amount by which the threshold of audibility of a sound is raised by the presence of another sound; the unit customarily used is the decibel. Also known as audio masking; aural masking. [COMPUT SCI] 1. Replacing specific characters in one register by corresponding characters in another register. 2. Extracting certain characters from a string of characters. [ELECTR] 1. Using a covering or coating on a semiconductor surface to provide a masked area for selective deposition or etching. 2. A programmed procedure for eliminating radar coverage in areas where such transmissions may be of use to the enemy for navigation purposes, by weakening the beam in appropriate directions or by use of additional transmitters on the same frequency at suitable sites to interfere with homing; also used to suppress the beam in areas where it would interfere with television reception. [ENG] Preventing entrance of a tracer gas into a vessel by covering the leaks.

masonry [CIV ENG] A construction of stone or similar materials such as concrete or brick.

mass [MECH] A quantitative measure of a body's resistance to being accelerated; equal to the inverse of the ratio of the body's acceleration to the acceleration of a standard mass under otherwise identical conditions.

mass absorption coefficient [PHYS] The linear absorption coefficient divided by the density of the medium.

mass-data multiprocessing [COMPUT SCI] The basic concept of time sharing, with many inquiry stations to a central location capable of on-line data retrieval.

mass defect [NUC PHYS] The difference between the mass of an atom and the sum of the masses of its individual components in the free (unbound) state.

mass-energy conservation [RELAT] The principle that energy cannot be created or destroyed; however, one form of energy is that which a particle has because of its rest mass, equal to this mass times the square of the speed of light.

mass-energy relation [RELAT] The relation whereby the total energy content of a body is equal to its inertial mass times the square of the speed of light.

masseter [ANAT] The masticatory muscle, arising from the zygomatic arch and inserted into the lower jaw.

mass flow [FL MECH] The mass of a fluid in motion which crosses a given area in a unit time.

massif [GEOL] A massive block of rock within an erogenic belt, generally more rigid than the surrounding rocks, and commonly composed of crystalline basement or younger plutons.

massive [GEOL] Of a mineral deposit, having a large concentration of ore in one place. [MINERAL] Of a mineral, lacking an internal structure. [PALEON] Of corallum, composed of closely packed corallites. [PETR] 1. Of a competent rock, being homogeneous, isotropic, and elastically perfect. 2. Of a metamorphic rock, having constituents which do not show parallel orientation and are not arranged in layers. 3. Of igneous rocks, being homogeneous over wide areas and lacking any layering, foliation, cleavage, or similar features.

mass-luminosity relation [ASTROPHYS] A relation between stellar magnitudes and mass of the stars; when the absolute magnitudes of stars are plotted versus the logarithms of their masses, the points fall closely along a smooth curve.

mass movement [GEOL] Movement of a portion of the land surface as a unit.

mass number [NUC PHYS] The sum of the numbers of protons and neutrons in the nucleus of an atom or nuclide. Also known as nuclear number; nucleon number.

mass spectrograph [ENG] A mass spectroscope in which the ions fall on a photographic plate which after development shows the distribution of particle masses.

mass spectrometer [ENG] A mass spectroscope in which a slit moves across the paths of particles with various masses, and an electrical detector behind it records the intensity distribution of masses.

mass spectroscope [ENG] An instrument used for determining the masses of atoms or molecules, in which a beam of ions is sent through a combination of electric and magnetic fields so arranged that the ions are deflected according to their masses.

mass spectrum [PARTIC PHYS] A plot of masses of elementary particles, including unstable states. Also known as particle spectrum. [PHYS] A display, record, or plot of the distribution in mass, or in mass-to-charge ratio, of ionized atoms, molecules, or molecular fragments.

mass storage [COMPUT SCI] A computer storage with large capacity, especially one whose contents are directly accessible to a computer's central processing unit.

mass-storage system [COMPUT SCI] A computer system containing a large number of storage devices, with one of these devices containing the master file of the operating system, routines, and library routines.

mass-transfer rate [PHYS] The measurement of the movement of matter as a function of time.

mass transport [FL MECH] 1. Carrying of loose materials in a moving medium such as water or air. 2. The movement of fluid, especially water, from one place to another.

mass wasting [GEOL] Dislodgement and downslope transport of loose rock and soil material under the direct influence of gravitational body stresses.

mast [ENG] 1. A vertical metal pole serving as an antenna or antenna support. 2. A slender vertical pole which must be held in position by guy lines. 3. A drill, derrick, or tripod mounted on a drill unit, which can be raised to operating position by mechanical means. 4. A single pole, used as a drill derrick, supported in its upright or operating position by guys. [NAV ARCH] A long wooden or metal pole or spar, usually vertical, on the deck or keel of a ship, to support other spars which in turn support or are attached to sails, as well as derricks.

Mastacembeloidei [VERT ZOO] The spiny eels, a suborder of perciform fishes that are eellike in shape and have the pectoral girdle suspended from the vertebral column.

mast cell [HISTOL] A connective-tissue cell with numerous large, basophilic, metachromatic granules in the cytoplasm.

mastectomy [MED] Surgical removal of the breast. Also known as mammectomy.

master [ENG] 1. A device which controls subsidiary devices. 2. A precise workpiece through which duplicates are made. [ENG ACOUS] See master phonograph record. [NAV] See master station.

master card [COMPUT SCI] A computer card that contains information about a group of computer cards, and is usually the first card of this group.

master clock [COMPUT SCI] The electronic or electric source of standard timing signals, often called clock pulses, required for sequencing the operation of a computer. Also known as main clock; master synchronizer; master timer.

master control [COMMUN] The control console that contains the main program controls for a radio or television transmitter or network. [COMPUT SCI] A computer program, oriented toward applications, which carries out the highest level of control in a hierarchy of programs, routines, and subroutines.

master cylinder [MECH ENG] The container for the fluid and the piston, forming part of a device such as a hydraulic brake or clutch.

master file [COMPUT SCI] 1. A computer file containing relatively permanent information, usually updated periodically, such as subscriber records or payroll data other than time worked. 2. A computer file that is used as an authoritative source of data in carrying out a particular job on the computer.

master frequency meter See integrating frequency meter.

master phonograph record [ENG ACOUS] The negative metal counterpart of a disk recording, produced by electroforming as one step in the production of phonograph records. Also known as master.

master program file [COMPUT SCI] The tape record of all programs for a system of runs.

master record [COMPUT SCI] The basic updated record which will be used for the next run.

master routine See executive routine.

master/slave manipulator [ENG] A mechanical, electromechanical, or hydromechanical device which reproduces the hand or arm motions of an operator, enabling the operator to perform manual motions while separated from the site of the work.

master/slave mode [COMPUT SCI] The feature ensuring the protection of each program when more than one program resides in memory.

master station [NAV] In a radio navigation system such as loran, the controlling station of two or more synchronized transmitting stations. Also known as master.

master switch [ELEC] 1. Switch that dominates the operation contactors, relays, or other magnetically operated devices. 2. Switch electrically ahead of a number of individual switches.

master synchronizer See master clock.

master system tape [COMPUT SCI] A monitor program centralizing the control of program operation by loading and executing any program on a system tape.

master tape [COMPUT SCI] A magnetic tape that contains data which must not be overwritten, such as an executive routine or master file; updating a master tape means generating a new master tape onto which supplementary data have been added.

master timer See master clock.

Mastigamoebidae [INV ZOO] A family of ameboid protozoans possessing one or two flagella, belonging to the order Rhizomastigida.

Mastigophora [INV ZOO] A superclass of the Protozoa characterized by possession of flagella.

mastodon [PALEON] A member of the Mastodontidae, especially the genus *Mammut.*

Mastodontidae [PALEON] An extinct family of elephantoid proboscideans that had low-crowned teeth with simple ridges and without cement.

mastoid [ANAT] 1. Breast-shaped. 2. The portion of the temporal bone where the mastoid process is located.

Mastotermitidae [INV ZOO] A family of lower termites in the order Isoptera with a single living species, in Australia.

match [COMPUT SCI] A data-processing operation similar to a merge, except that instead of producing a sequence of items made up from the input sequences, the sequences are matched against each other on the basis of some key. [ENG] 1. A charge of gunpowder put in a paper several inches long and used for igniting explosives. 2. A short flammable piece of wood, paper, or other material tipped with a combustible mixture that bursts into flame through friction. [IMMUNOL] To select blood donors whose erythrocytes are compatible with those of the recipient.

matched filter [COMPUT SCI] In character recognition, a method employed in character property detection in which a vertical projection of the input character produces an analog waveform which is then compared to a set of stored waveforms for the purpose of determining the character's identity. [ELECTR] A filter with the property that, when the input consists of noise in addition to a specified desired signal, the signal-to-noise ratio is the maximum which can be obtained in any linear filter.

matched impedance [ELEC] An impedance of a load which is equal to the impedance of a generator, so that maximum power is delivered to the load.

matched pairs [STAT] The design of an experiment for paired comparison in which the assignment of subjects to treatment or control is not completely at random, but the randomization is restricted to occur separately within each pair.

matched pulse intercepting [COMMUN] System for intercepting calls on party lines in a terminal-per-line office; operates on a ground pulse which is matched in time with the intercepted station's particular ringing frequency.

match gate *See* equivalence gate.

matching distribution [STAT] The distribution of number of matches obtained if N tickets labeled 1 to N are drawn at random one at a time and laid in a row, and a match is counted when a ticket's label matches its position.

material particle [MECH] An object which has rest-mass and an observable position in space, but has no geometrical extension, being confined to a single point. Also known as particle.

maternal effect [GEN] Determination of characters of the progeny by the maternal parent; mediated by the genetic constitution of the mother.

maternal inheritance [GEN] The acquisition of characters transmitted through the cytoplasm of the egg.

mathematical analysis *See* analysis.

mathematical biology [BIOL] A discipline that encompasses all applications of mathematics, computer technology, and quantitative theorizing to biological systems, and the underlying processes within the systems.

mathematical forecasting *See* numerical forecasting.

mathematical function program [COMPUT SCI] A set of routinely used mathematical functions, such as square root, which are efficiently coded and called for by special symbols.

mathematical geography [GEOGR] The branch of geography that deals with the features and processes of the earth, and their representations on maps and charts.

mathematical induction [MATH] A general method of proving statements concerning a positive integral variable: if a statement is proven true for $x = 1$, and if it is proven that, if the statement is true for $x = 1, \ldots, n$, then it is true for $x = n + 1$, it follows that the statement is true for any integer. Also known as complete induction.

mathematical logic [MATH] The study of mathematical theories from the viewpoint of model theory, recursive function theory, proof theory, and set theory.

mathematical model [MATH] 1. A mathematical representation of a process, device, or concept by means of a number of variables which are defined to represent the inputs, outputs, and internal states of the device or process, and a set of equations and inequalities describing the interaction of these variables. 2. A mathematical theory or system together with its axioms.

mathematical physics [PHYS] The study of the mathematical systems which represent physical phenomena; particular areas are, for example, quantum and statistical mechanics and field theory.

mathematical probability [MATH] The ratio of the number of mutually exclusive, equally likely outcomes of interest to the total number of such outcomes when the total is exhaustive. Also known as a priori probability.

mathematical programming *See* optimization theory.

mathematical subroutine [COMPUT SCI] A computer subroutine in which a well-defined mathematical function, such as exponential, logarithm, or sine, relates the output to the input.

mathematics [SCI TECH] The deductive study of shape, quantity, and dependence; the two main areas are applied mathematics and pure mathematics, the former arising from the study of physical phenomena, the latter the intrinsic study of mathematical structures.

matildite [MINERAL] $AgBiS_2$ An iron black to gray, orthorhombic mineral consisting of silver bismuth sulfide; occurrence is massive or granular.

matrix [COMPUT SCI] A latticework of input and output leads with logic elements connected at some of their intersections. [ELECTR] 1. The section of a color television transmitter that transforms the red, green, and blue camera signals into color-difference signals and combines them with the chrominance subcarrier. Also known as color coder; color encoder; encoder. 2. The section of a color television receiver that transforms the color-difference signals into the red, green, and blue signals needed to drive the color picture tube. Also known as color decoder; decoder. [ENG] A recessed mold in which something is formed or cast. [GRAPHICS] 1. In a type-casting machine, the portion of the mold that forms the letter face. 2. A heavy, unsized, unfinished paper that is used for molds for stereotype plates. [HISTOL] 1. The intercellular substance of a tissue. Also known as ground substance. 2. The epithelial tissue from which a toenail or fingernail develops. [MATER] A binding agent used to make an agglomerate mass. [MATH] A rectangular array of numbers or scalars from a vector space. [MET] 1. The principal component of an alloy. 2. The precisely shaped form used as the cathode in electroforming. [MYCOL] The substrate on or in which fungus grows. [PETR] The continuous, fine-grained material in which large grains of a sediment or sedimentary rock are embedded. Also known as groundmass.

matrix algebra [MATH] An algebra whose elements are matrices and whose operations are addition and multiplication of matrices.

matrix calculus [MATH] The treatment of matrices whose entries are functions as functions in their own right with a corresponding theory of differentiation; this has application to the study of multidimensional derivatives of functions of several variables.

matrix element [MATH] One of the set of numbers which form a matrix. [QUANT MECH] The scalar product of a member of a complete, orthogonal set of vectors, representing states, with a vector which results from applying a specified operator to another member of this set.

matrix isolation [SPECT] A spectroscopic technique in which reactive species can be characterized by maintaining them in a very cold, inert environment while they are examined by an absorption, electron-spin resonance, or laser excitation spectroscope.

matrix management [COMPUT SCI] An organized approach to administration of a program by defining and structuring all elements to form a single system with components united by interaction.

matrix mechanics [QUANT MECH] The theory of quantum mechanics developed by using the Heisenberg picture and representing operators by their matrix elements between eigenfunctions of the Hamiltonian operator; Heisenberg's original formulation of quantum mechanics.

matrix printing [COMPUT SCI] High-speed printing in which characterlike configurations of dots are printed through the proper selection of wire ends from a matrix of wire ends. Also known as stylus printing; wire printing.

matrix storage [COMPUT SCI] A computer storage in which coordinates are used to describe the locations or circuit elements. Also known as coordinate storage.

matter [PHYS] The substance composing bodies perceptible to the senses; includes any entity possessing mass when at rest.

matter wave *See* de Broglie wave.

Matthias' rules [SOLID STATE] Several empirical rules giving the dependence of the transition temperatures of superconducting metals and alloys on the position of the metals in the periodic table and in the composition of the alloys.

Matthiessen's rule [SOLID STATE] An empirical rule which states that the total resistivity of a crystalline metallic specimen is the sum of the resistivity due to thermal agitation of the metal ions of the lattice and the resistivity due to imperfections in the crystal.

mature [BIOL] **1.** Being fully grown and developed. **2.** Ripe. [FOOD ENG] Having attained the final state of processing, as certain wines. [GEOL] **1.** Pertaining to a topography or region, and to its landforms, having undergone maximum development and accentuation of form. **2.** Pertaining to the third stage of textural maturity of a clastic sediment. [PSYCH] Having the emotional qualities of a well-adjusted adult.

mature soil *See* zonal soil.

maturity [GEOL] **1.** The second stage of the erosion cycle in the topographic development of a landscape or region characterized by numerous and closely spaced mature streams, reduction of level surfaces to slopes, large well-defined drainage systems, and the absence of swamps or lakes on the uplands. Also known as topographic maturity. **2.** A stage in the development of a shore or coast that begins with the attainment of a profile of equilibrium. **3.** The extent to which the texture and composition of a clastic sediment approach the ultimate end product. **4.** The stage of stream development at which maximum vigor and efficiency has been reached.

maturity-onset diabetes [MED] A type of diabetes mellitus which develops later in life; characterized by more gradual development and less severe symptoms than juvenile-onset diabetes.

maucherite [MINERAL] $Ni_{11}As_8$ A reddish silver-white mineral composed of nickel arsenide.

maunder minimum [ASTRON] A period of time from about 1650 to 1710 when the sun did not appear to have sunspots.

mavar *See* parametric amplifier.

max *See* maximum.

maxilla [ANAT] **1.** The upper jawbone. **2.** The upper jaw. [INV ZOO] Either of the first two pairs of mouthparts posterior to the mandibles in certain arthropods.

maxillary artery [ANAT] A branch of the external carotid artery which supplies the deep structures of the face (internal maxillary) and the side of the face and nose (external maxillary).

maximax criterion [MATH] In decision theory, one of several possible prescriptions for making a decision under conditions of uncertainty; it prescribes the strategy which will maximize the maximum possible profit.

maximizing a function [MATH] Finding the largest value assumed by a function.

maximum [MATH] The maximum of a real-valued function is the greatest value it assumes. Abbreviated max.

maximum available gain [ELECTR] The theoretical maximum power gain available in a transistor stage; it is seldom achieved in practical circuits because it can be approached only when feedback is negligible. Abbreviated MAG.

maximum demand [ELEC] The greatest average value of the power, apparent power, or current consumed by a customer of an electric power system, the averages being taken over successive time periods, usually 15 or 30 minutes in length.

maximum likelihood method [STAT] A technique in statistics where the likelihood distribution is so maximized as to produce an estimate to the random variables involved.

maximum-minimum principle *See* min-max theorem.

maximum operating frequency [COMPUT SCI] The highest rate at which the modules perform iteratively and reliably.

maximum permissible dose [MED] The dose of ionizing radiation that a person may receive in his lifetime without appreciable bodily injury.

maximum usable frequency [COMMUN] The upper limit of the frequencies that can be used at a specified time for point-to-point radio transmission involving propagation by reflection from the regular ionized layers of the ionosphere. Abbreviated MUF.

maximum-wind level [METEOROL] The height at which the maximum wind speed occurs, determined in a winds-aloft observation. Also known as max-wind level.

maxwell [ELECTROMAG] A centimeter-gram-second electromagnetic unit of magnetic flux, equal to the magnetic flux which produces an electromotive force of 1 abvolt in a circuit of one turn linking the flux, as the flux is reduced to zero in 1 second at a uniform rate. Abbreviated Mx. Also known as abweber (abWb); line of magnetic induction.

Maxwell-Boltzmann density function *See* Maxwell-Boltzmann distribution.

Maxwell-Boltzmann distribution [STAT MECH] Any function giving the probability (or some function proportional to it) that a molecule of a gas in thermal equilibrium will have values of certain variables within given infinitesimal ranges, assuming that the gas molecules obey classical mechanics, and possibly making other assumptions; examples are the Maxwell distribution and the Boltzmann distribution. Also known as Maxwell-Boltzmann density function.

Maxwell-Boltzmann equation *See* Boltzmann transport equation.

Maxwell-Boltzmann statistics [STAT MECH] The classical statistics of identical particles, as opposed to the Bose-Einstein or Fermi-Dirac statistics. Also known as Boltzmann statistics.

Maxwell bridge [ELEC] A four-arm alternating-current bridge used to measure inductance (or capacitance) in terms of resistance and capacitance (or inductance); bridge balance is independent of frequency. Also known as Maxwell-Wien bridge; Wien-Maxwell bridge.

Maxwell distribution [STAT MECH] A function giving the number of molecules of a gas in thermal equilibrium whose velocities lie within a given, infinitesimal range of values, assuming that the molecules obey classical mechanics, and do not interact. Also known as Maxwellian distribution.

Maxwell equal area rule [THERMO] At temperatures for which the theoretical isothermal of a substance, on a graph of pressure against volume, has a portion with positive slope (as occurs in a substance with liquid and gas phases obeying the Van der Waals equation), a horizontal line drawn at the equilibrium vapor pressure and connecting two parts of the isothermal with negative slope has the property that the area between the horizontal and the part of the isothermal above it is equal to the area between the horizontal and the part of the isothermal below it.

Maxwell equations *See* Maxwell field equations.

Maxwell field equations [ELECTROMAG] Four differential equations which relate the electric and magnetic fields to electric charges and currents, and form the basis of the theory of electromagnetic waves. Also known as electromagnetic field equations; Maxwell equations.

Maxwellian distribution *See* Maxwell distribution.

Maxwell relation [ELECTROMAG] According to Maxwell's electromagnetic theory, that relation wherein the dielectric constant of a substance equals the square of its index of refraction. [THERMO] One of four equations for a system in thermal equilibrium, each of which equates two partial derivatives, involving the pressure, volume, temperature, and entropy of the system.

Maxwell's demon *See* demon of Maxwell.

Maxwell's displacement current *See* displacement current.

Maxwell's electromagnetic theory [ELECTROMAG] A mathematical theory of electric and magnetic fields which predicts the propagation of electromagnetic radiation, and is valid for electromagnetic phenomena where effects on an atomic scale can be neglected.

Maxwell's law [ELECTROMAG] A movable portion of a circuit will always move in such a direction as to give maximum magnetic flux linkages through the circuit.

Maxwell's theorem [MECH] If a load applied at one point *A* of an elastic structure results in a given deflection at another point *B*, then the same load applied at *B* will result in the same deflection at *A*.

Maxwell-Wien bridge *See* Maxwell bridge.

max-wind level *See* maximum-wind level.

mayfly [INV ZOO] The common name for insects composing the order Ephemeroptera.

MBE *See* molecular beam epitaxy.

Mc *See* megahertz.

McLeod gage [FL MECH] A type of instrument used to measure vacuum by measuring the height of a column of mercury supported by the gas whose pressure is to be measured, when this gas is trapped and compressed into a capillary tube.

Md *See* mendelevium.

M display [ELECTR] A modified radarscope A display in which target distance is determined by moving an adjustable pedestal signal along the baseline until it coincides with the horizontal position of the target deflection.

meadow [ECOL] A vegetation zone which is a low grassland, dense and continuous, variously interspersed with forbs but few if any shrubs. Also known

as pelouse; Wiesen. [ENG] Range of air-fuel ratio within which smooth combustion may be had.

meadow ore *See* bog iron ore.

mealybug [INV ZOO] Any of various scale insects of the family Pseudococcidae which have a powdery substance covering the dorsal surface; all are serious plant pests.

mean [MATH] A single number that typifies a set of numbers, such as the arithmetic mean, the geometric mean, or the expected value. Also known as mean value.

mean British thermal unit *See* British thermal unit.

mean carrier frequency [ELECTR] Average carrier frequency of a transmitter corresponding to the resting frequency in a frequency-modulated system.

meander [HYD] A sharp, sinuous loop or curve in a stream, usually part of a series. [OCEANOGR] A deviation of the flow pattern of a current.

meandering stream [HYD] A stream having a pattern of successive meanders. Also known as snaking stream.

meander plain [GEOL] A plain built by the meandering process, or a plain of lateral accretion.

mean deviation *See* average deviation.

mean difference [STAT] The average of the absolute values of the $n(n - 1)/2$ differences between pairs of elements in a statistical distribution that has n elements.

mean effective pressure [MECH ENG] A term commonly used in the evaluation for positive displacement machinery performance which expresses the average net pressure difference in pounds per square inch on the two sides of the piston in engines, pumps, and compressors. Abbreviated mep; mp. Also known as mean pressure.

mean free path [ACOUS] For sound waves in an enclosure, the average distance sound travels between successive reflections in the enclosure. [PHYS] The average distance traveled between two similar events, such as elastic collisions of molecules in a gas, of electrons or phonons in a crystal, or of neutrons in a moderator.

mean high water [OCEANOGR] The average height of all high waters recorded at a given place over a 19-year period.

mean latitude [GEOD] Half the arithmetical sum of the latitudes of two places on the same side of the equator; mean latitude is labeled N or S to indicate whether it is north or south of the equator.

mean low water [OCEANOGR] The average height of all low waters recorded at a given place over a 19-year period.

mean motion [ASTRON] The speed which a planet or its satellite would have if it were moving in a circular orbit with radius equal to its distance from the sun or a central planet with a period equal to its actual period.

mean neap range *See* neap range.

mean noon [ASTRON] Twelve o'clock mean time, or the instant the mean sun is over the upper branch of the meridian; it may be either local or Greenwich, depending upon the reference meridian.

mean pressure *See* mean effective pressure.

mean range [OCEANOGR] The difference in the height between mean high water and mean low water. [ORD] Average distance reached by a group of shots fired with the same firing data. [SCI TECH] The average difference in the extreme values of a variable quantity.

mean sea level [OCEANOGR] The average sea surface level for all stages of the tide over a 19-year period, usually determined from hourly height readings from a fixed reference level.

means-ends analysis   [COMPUT SCI] A method of problem solving in which the difference between the form of the data in the present and desired situations is determined, and an operator is then found to transform from one into the other, or, if this is not possible, objects between the present and desired objects are created, and the same procedure is then repeated on each of the gaps between them.

mean sidereal time   [ASTRON] Sidereal time adjusted for nutation, to eliminate slight irregularities in the rate.

mean solar day   [ASTRON] The duration of one rotation of the earth on its axis, with respect to the mean sun; the length of the mean solar day is 24 hours of mean solar time or $24^h\ 03^m\ 56.555^s$ of mean sidereal time.

mean spring range   *See* spring range.

mean square deviation   [STAT] A measure of the extent to which a collection $v_1, v_2, \ldots, v_n$ of numbers is unequal; it is given by the expression $(1/n) \cdot [(v_1 - \bar{v})^2 + \ldots + (v_n - \bar{v})^2]$, where $\bar{v}$ is the mean of the numbers.

mean stress   [MECH] **1.** The algebraic mean of the maximum and minimum values of a periodically varying stress. **2.** *See* octahedral normal stress.

mean temperature   [METEOROL] The average temperature of the air as indicated by a properly exposed thermometer during a given time period, usually a day, month, or year.

mean tide   *See* half tide.

mean tide level   [OCEANOGR] The tide level halfway between mean high water and mean low water.

mean time   [ASTRON] Time based on the rotation of the earth relative to the mean sun.

mean time between failures   [COMPUT SCI] A measure of the reliability of a computer system, equal to average operating time of equipment between failures, as calculated on a statistical basis from the known failure rates of various components of the system. Abbreviated MTBF.

mean value   [MATH] **1.** For a function $f(x)$ defined on an interval $\langle a,b \rangle$, the integral from a to b of $f(x)$ dx divided by $b - a$. **2.** *See* mean.

measles   [MED] An acute, highly infectious viral disease with cough, fever, and maculopapular rash; the appearance of Koplik spots on the oral mucous membranes marks the onset. Also known as rubeola.

measurable function   [MATH] **1.** A real valued function f defined on a measurable space X, where for every real number a all those points x in X for which $f(x) \geq a$ form a measurable set. **2.** A function on a measurable space to a measurable space such that the inverse image of a measurable set is a measurable set.

measure   [MATH] A nonnegative real valued function m defined on a sigma-algebra of subsets of a set S whose value is zero on the empty set, and whose value on a countable union of disjoint sets is the sum of its values on each set.

measured daywork   [IND ENG] Work done for an hourly wage on which specific productivity levels have been determined but which provides no incentive pay.

measured mile   [CIV ENG] The distance of 1 mile (1609.344 meters), the units of which have been accurately measured and marked. [NAV] A length of 1 nautical mile (1852 meters), the limits of which have been accurately measured and are indicated by ranges ashore; used by vessels to calibrate logs,

engine revolution counters, and such, and to determine speed.

measured ore   *See* developed reserves.

measurement ton   *See* ton.

mechanical   [ENG] Of, pertaining to, or concerned with machinery or tools. [GRAPHICS] A finished copy that usually contains hand lettering, type proofs, and art especially positioned and mounted so that a photochemical reproduction can be made on a letterpress, offset, or other printing plate. Also known as keyline layout; paste-up.

mechanical advantage   [MECH ENG] The ratio of the force produced by a machine such as a lever or pulley to the force applied to it.

mechanical analog computer   [COMPUT SCI] A machine aid to computation in which variables are represented as continuously variable displacements or motions of mechanical elements, such as gears and shafts.

mechanical bearing cursor   *See* bearing cursor.

mechanical classification   [MECH ENG] A sorting operation in which mixtures of particles of mixed sizes, and often of different specific gravities, are separated into fractions by the action of a stream of fluid, usually water.

mechanical damping   [ENG ACOUS] Mechanical resistance which is generally associated with the moving parts of an electromechanically transducer such as a cutter or a reproducer.

mechanical efficiency   [MECH ENG] In an engine, the ratio of brake horsepower to indicated horsepower.

mechanical engineering   [ENG] The branch of engineering that deals with the generation, transmission, and utilization of heat and mechanical power and with the production of tools, machines, and their products.

mechanical erosion   *See* corrasion.

mechanical expression   *See* expression.

mechanical filter   [ELECTR] Filter, used in intermediate-frequency amplifiers of highly selective superheterodyne receivers, consisting of shaped metal rods that act as coupled mechanical resonators when used with piezoelectric input and output transducers. Also known as mechanical wave filter. [PETRO ENG] Granule-packed steel shell used to filter suspended floc or undissolved solids out of treated waterflood water; granules can be graded sand and gravel, anthracite coal, graphite ore, or aluminum-oxide plates with granular filter medium.

mechanical impedance   [MECH] The complex ratio of a phasor representing a sinusoidally varying force applied to a system to a phasor representing the velocity of a point in the system.

mechanical integrator   [COMPUT SCI] A mechanical device which draws the graph of the integral of a function when a tracing point is passed over a graph of the function.

mechanical linkage   [MECH ENG] A set of rigid bodies, called links, joined together at pivots by means of pins or equivalent devices.

mechanical plotting board   *See* coordinate plotter.

mechanical pump   [MECH ENG] A pump through which fluid is conveyed by direct contact with a moving part of the pumping machinery.

mechanical rectifier   [ELEC] A rectifier in which rectification is accomplished by mechanical action, as in a synchronous vibrator.

mechanical refrigeration   [MECH ENG] The removal of heat by utilizing a refrigerant subjected to cycles of refrigerating thermodynamics and employing a mechanical compressor.

mechanical resistance *See* resistance.

mechanical rotational impedance *See* rotational impedance.

mechanical scanner [COMPUT SCI] In optical character recognition, a device that projects an input character into a rotating disk, on the periphery of which is a series of small, uniformly spaced apertures; as the disk rotates, a photocell collects the light passing through the apertures.

mechanical separation [MECH ENG] A group of industrial operations by means of which particles of solid or drops of liquid are removed from a gas or liquid, or are separated into individual fractions, or both, by gravity separation (settling), centrifugal action, and filtration.

mechanical spring *See* spring.

mechanical stepping motor [ELEC] A device in which a voltage pulse through a solenoid coil causes reciprocating motion by a solenoid plunger, and this is transformed into rotary motion through a definite angle by ratchet-and-pawl mechanisms or other mechanical linkages.

mechanical translation [COMPUT SCI] Automatic translation of one language into another by means of a computer or other machine that contains a dictionary look-up in its memory, along with the programs needed to make logical choices from synonyms, supply missing words, and rearrange word order as required for the new language. Also known as machine translation.

mechanical vibration [MECH] A motion, usually unintentional and often undesirable, of parts of machines and structures.

mechanical wave filter *See* mechanical filter.

mechanics [PHYS] 1. In the original sense, the study of the behavior of physical systems under the action of forces. 2. More broadly, the branch of physics which seeks to formulate general rules for predicting the behavior of a physical system under the influence of any type of interaction with its environment.

mechanism [MECH ENG] That part of a machine which contains two or more pieces so arranged that the motion of one compels the motion of the others.

mechanized *See* machine-sensible.

mechanocaloric effect [CRYO] An effect resulting from the fact that a temperature gradient in helium II is invariably accompanied by a pressure gradient, and conversely; examples are the fountain effect, and the heating of liquid helium left behind in a container when part of it leaks out through a small orifice.

mechanoreceptor [PHYSIO] A receptor that provides the organism with information about such mechanical changes in the environment as movement, tension, and pressure.

Mecoptera [INV ZOO] The scorpion flies, a small order of insects; adults are distinguished by the peculiar prolongation of the head into a beak, which bears chewing mouthparts.

media conversion [COMPUT SCI] The transfer of data from one storage type (such as punched cards) to another storage type (such as magnetic tape).

media conversion buffer [COMPUT SCI] Large storage area, such as a drum, on which data may be stored at low speed during nonexecution time, to be later transferred at high speed into core memory during execution time.

medial [ANAT] 1. Being internal as opposed to external (lateral). 2. Toward the midline of the body. [SCI TECH] Located in the middle.

medial moraine [GEOL] 1. An elongate moraine carried in or upon the middle of a glacier and parallel to its sides. 2. A moraine formed by glacial abrasion of a rocky protuberance near the middle of a glacier.

median [MATH] Any line in a triangle which joins a vertex to the midpoint of the opposite side. [SCI TECH] Located in the middle. [STAT] An average of a series of quantities or values; specifically, the quantity or value of that item which is so positioned in the series, when arranged in order of numerical quantity or value, that there are an equal number of items of greater magnitude and lesser magnitude.

median effective dose *See* effective dose 50.

median infective dose *See* infective dose 50.

mediastinum [ANAT] 1. A partition separating adjacent parts. 2. The space in the middle of the chest between the two pleurae.

medical bacteriology [MED] A branch of medical microbiology that deals with the study of bacteria which affect human health, especially those which produce disease.

medical electronics [ELECTR] A branch of electronics in which electronic instruments and equipment are used for such medical applications as diagnosis, therapy, research, anesthesia control, cardiac control, and surgery.

medical entomology [MED] The study of insects that are vectors for diseases and parasitic infestations in humans and domestic animals.

medical examiner [MED] A professionally qualified physician duly authorized and charged by a governmental unit to determine facts concerning causes of death, particularly deaths not occurring under natural circumstances, and to testify thereto in courts of law.

medical mycology [MED] A branch of medical microbiology that deals with fungi that are pathogenic to humans.

medical parasitology [MED] A branch of medical microbiology which deals with the relationship between humans and those animals which live in or on them.

medical protozoology [MED] A branch of medical microbiology that deals with the study of Protozoa which are parasites of humans.

Mediterranean fever *See* brucellosis.

medium [CHEM ENG] 1. The carrier in which a chemical reaction takes place. 2. Material of controlled pore size used to remove foreign particles or liquid droplets from fluid carriers. [COMPUT SCI] The material, or configuration thereof, on which data are recorded; usually not applied to disk, drum, or core, but to storable, removable media, such as paper tape, cards, and magnetic tape. [PHYS] That entity in which objects exist and phenomena take place; examples are free space and various fluids and solids.

medium artillery [ORD] Artillery which includes guns of caliber greater than 105 millimeters but less than 155 mm, and howitzers of caliber greater than 105 mm but not greater than 155 mm.

medium frequency [COMMUN] A Federal Communications Commission designation for the band from 300 to 3000 kilohertz in the radio spectrum. Abbreviated mf.

medium-range forecast [METEOROL] A forecast of weather conditions for a period of 48 hours to a week in advance. Also known as extended-range forecast.

medulla [ANAT] 1. The central part of certain organs and structures such as the adrenal glands and hair. 2. Marrow, as of bone or the spinal cord. 3. *See* medulla oblongata. [BOT] 1. Pith. 2. The central spongy portion of some fungi.

medulla oblongata [ANAT] The somewhat pyramidal, caudal portion of the vertebrate brain extending from the pons to the spinal cord. Also known as medulla.

Medullosaceae [PALEOBOT] A family of seed ferns; these extinct plants all have large spirally arranged petioles with numerous vascular bundles.

medusa See jellyfish.

meerschaum See sepiolite.

mega- [SCI TECH] A prefix representing 10^6, or one million. Abbreviated M.

megabit [COMPUT SCI] One million binary bits.

Megachilidae [INV ZOO] The leaf-cutting bees, a family of hymenopteran insects in the superfamily Apoidea.

Megachiroptera [VERT ZOO] The fruit bats, a group of Chiroptera restricted to the Old World; most species lack a tail, but when present it is free of the interfemoral membrane.

megacycle See megahertz.

Megadermatidae [VERT ZOO] The false vampires, a family of tailless bats with large ears and a nose leaf; found in Africa, Australia, and the Malay Archipelago.

megaelectronvolt [PHYS] A unit of energy commonly used in nuclear and particle physics, equal to the energy acquired by an electron in falling through a potential of 10^6 volts. Abbreviated MeV.

megahertz [PHYS] Unit of frequency, equal to 1,000,000 hertz. Abbreviated MHz. Also known as megacycle (Mc).

Megalodontoidea [INV ZOO] A superfamily of hymenopteran insects in the suborder Symphyta.

megalomania [PSYCH] The delusion of greatness and omnipotence characterizing certain psychotic reactions.

Megalomycteroidei [VERT ZOO] The mosaic-scaled fishes, a monofamilial suborder of the Cetomimiformes; members are rare species of small, elongate deep-sea fishes with degenerate eyes and irregularly disposed scales.

Megaloptera [INV ZOO] A suborder included in the order Neuroptera by some authorities.

Megalopygidae [INV ZOO] The flannel moths, a small family of lepidopteran insects in the suborder Heteroneura.

Megamerinidae [INV ZOO] A family of myodarian cyclorrhaphous dipteran insects in the subsection Acalypteratae.

Megapodiidae [VERT ZOO] The mound birds and brush turkeys, a family of birds in the order Galliformes; distinguished by their method of incubating eggs in mounds of dirt or in decomposing vegetation.

megaspore See macrospore.

Megathymiinae [INV ZOO] The giant skippers, a subfamily of lepidopteran insects in the family Hesperiidae.

megaton [PHYS] The energy released by 1,000,000 metric tons of chemical high explosive calculated at a rate of 1000 calories per gram, or a total of 4.18 $\times 10^{15}$ joules; used principally in expressing the energy released by a nuclear bomb. Abbreviated MT.

Meibomian cyst See chalazion.

Meibomian gland See tarsal gland.

Meinertellidae [INV ZOO] A family of wingless insects belonging to the Microcoryphia.

meionite [MINERAL] $3CaAl_2Si_2O_8 \cdot CaCO_3$ A scapolite mineral composed of calcium aluminosilicate and calcium carbonate; it is isomorphous with marialite.

meiosis [CYTOL] A type of cell division occurring in diploid or polyploid tissues that results in a reduction in chromosome number, usually by half.

Meissner effect [SOLID STATE] The expulsion of magnetic flux from the interior of a piece of superconducting material as the material undergoes the transition to the superconducting phase. Also known as flux jumping; Meissner-Ochsenfeld effect.

Meissner-Ochsenfeld effect See Meissner effect.

Meissner's corpuscle [ANAT] An ovoid, encapsulated cutaneous sense organ presumed to function in touch sensation in hairless portions of the skin.

mel [ACOUS] A unit of pitch, equal to one-thousandth of the pitch of a simple tone whose frequency is 1000 hertz and whose loudness is 40 decibels above a listener's threshold.

Melamphaidae [VERT ZOO] A family of bathypelagic fishes in the order Beryciformes.

Melampsoraceae [MYCOL] A family of parasitic fungi in the order Uredinales in which the teleutospores are laterally united to form crusts or columns.

Melanconiaceae [MYCOL] The single family of the order Melanconiales.

Melanconiales [MYCOL] An order of the class Fungi Imperfecti including many plant pathogens commonly causing anthracnose; characterized by densely aggregated cnidophores on an acervulus.

Melandryidae [INV ZOO] The false darkling beetles, a family of coleopteran insects in the superfamily Tenebrionoidea.

melanin [BIOCHEM] Any of a group of brown or black pigments occurring in plants and animals.

melanocerite [MINERAL] $(Ca,Ce,Y)_8(BO_3)(SiO_4)_4 \cdot (F,OH)_4$ A brown or black rhombohedral mineral composed of complex silicate, borate, fluoride, tantalate, or other anion of cerium, yttrium, calcium, and other metals; occurs as crystals.

melanocyte [HISTOL] A cell containing dark pigments.

melanocyte-stimulating hormone [BIOCHEM] A protein substance secreted by the intermediate lobe of the pituitary of man which causes dispersion of pigment granules in the skin; similar to intermedins in other vertebrates. Abbreviated MSH. Also known as melanophore-dilating principle; melanophore hormone.

melanoma [MED] **1.** A malignant tumor composed of anaplastic melanocytes. **2.** A benign or malignant tumor composed of melanocytes.

melanophore-dilating principle See melanocyte-stimulating hormone.

melanophore hormone See melanocyte-stimulating hormone.

melanose [PL PATH] **1.** A fungus disease of grapevine caused by *Septoria ampelina*; leaves are infected and fall off. **2.** A fungus disease of citrus trees and fruits caused by *Diaporthe citri*, characterized by hard, brown, usually gummy elevations on the rind, twigs, and leaves.

melanostibian [MINERAL] $Mn(Sb,Fe)O_3$ A black mineral consisting of iron and manganese antimonite; occurs as foliated masses and as striated crystals.

melanterite [MINERAL] $FeSO_4 \cdot 7H_2O$ A green mineral occurring mainly in fibrous or concretionary masses, or in short, monoclinic, prismatic crystals; hardness is 2 on Mohs scale, and specific gravity is 1.90.

Melasidae [INV ZOO] The equivalent name for Eucnemidae.

Melastomataceae [BOT] A large family of dicotyledonous plants in the order Myrtales characterized by an inferior ovary, axile placentation, up to twice as many stamens as petals (or sepals), anthers opening by terminal pores, and leaves with prominent, subparallel longitudinal ribs.

Meleagrididae [VERT ZOO] The turkeys, a family of birds in the order Galliformes characterized by a bare head and neck.

M electron [ATOM PHYS] An electron whose principal quantum number is 3.

Meliaceae [BOT] A family of dicotyledonous plants in the order Sapindales characterized by mostly exstipulate, alternate leaves, stamens mostly connate by their filaments, and syncarpous flowers.

melilite [MINERAL] A sorosilicate mineral group of complex composition [(Na, Ca)$_2$(Mg, Al)(Si, Al)$_2$O$_7$] crystallizing in the tetragonal system; luster is vitreous to resinous, and color is white, yellow, greenish, reddish, or brown; hardness is 5 on Mohs scale, and specific gravity varies from 2.95 to 3.04.

melilitite [PETR] An extrusive rock that is generally olivine-free and composed of more than 90% mafic mineral such as melilite and augite, with minor amounts of feldspathoids and sometimes plagioclase.

Melinae [VERT ZOO] The badgers, a subfamily of carnivorous mammals in the family Mustelidae.

Melinninae [INV ZOO] A subfamily of sedentary annelids belonging to the Ampharetidae which have a conspicuous dorsal membrane, with or without dorsal spines.

Meliolaceae [MYCOL] The sooty molds, a family of ascomycetous fungi in the order Erysiphales, with dark mycelia and conidia.

meliphane See meliphanite.

meliphanite [MINERAL] (Ca,Na)$_2$Be(Si,Al)$_2$(O,OH,F)$_7$ A yellow, red, or black mineral composed of sodium calcium beryllium fluosilicate. Also known as meliphane.

Melittidae [INV ZOO] A family of hymenopteran insects in the superfamily Apoidea.

mellite [MINERAL] Al$_2$[C$_6$(COO)$_6$]·18H$_2$O A honey-colored mineral with resinous luster composed of the hydrous aluminum salt of mellitic acid, occurring as nodules in brown coal; it is in part a product of vegetable decomposition.

Meloidae [INV ZOO] The blister beetles, a large cosmopolitan family of coleopteran insects in the superfamily Meloidea; characterized by soft, flexible elytra and the strongly vesicant properties of the body fluids.

Meloidea [INV ZOO] A superfamily of coleopteran insects in the suborder Polyphaga.

melon [BOT] Either of two soft-fleshed edible fruits, muskmelon or watermelon, or varieties of these.

melonite [MINERAL] NiTe$_2$ A reddish-white mineral composed of nickel telluride.

melt [CHEM] 1. To change a solid to a liquid by the application of heat. 2. A melted material. [MET] A charge of molten metal.

melting See fusion.

melting point [THERMO] 1. The temperature at which a solid of a pure substance changes to a liquid. Abbreviated mp. 2. For a solution of two or more components, the temperature at which the first trace of liquid appears as the solution is heated.

meltwater [HYD] Water derived from melting ice or snow, especially glacier ice.

Melusinidae [INV ZOO] A family of orthorrhaphous dipteran insects in the series Nematocera.

Melyridae [INV ZOO] The soft-winged flower beetles, a large family of cosmopolitan coleopteran insects in the superfamily Cleroidea.

member [CIV ENG] A structural unit such as a wall, column, beam, or tie, or a combination of any of

these. [GEOL] A rock stratigraphic unit of subordinate rank comprising a specially developed part of a varied formation. [MATH] An element of a set.

Membracidae [INV ZOO] The treehoppers, a family of homopteran insects included in the series Auchenorrhyncha having a pronotum that extends backward over the abdomen, and a vertical upper portion of the head.

membrane [BUILD] In built-up roofing, a weather-resistant (flexible or semiflexible) covering consisting of alternate layers of felt and bitumen, fabricated in a continuous covering and surfaced with aggregate or asphaltic material. [CHEM ENG] 1. The medium through which the fluid stream is passed for purposes of filtration. 2. The ion-exchange medium used in dialysis, diffusion, osmosis and reverse osmosis, and electrophoresis. [HISTOL] A thin layer of tissue surrounding a part of the body, separating adjacent cavities, lining cavities, or connecting adjacent structures.

membrane bone See dermal bone.

membrane potential [PHYSIO] A potential difference across a living cell membrane.

memistor [ELEC] Nonmagnetic memory device consisting of a resistive substrate in an electrolyte; when used in an adaptive system, a direct-current signal removes copper from an anode and deposits it on the substrate, thus lowering the resistance of the substrate; reversal of the current reverses the process, raising the resistance of the substrate.

memory [COMPUT SCI] Any apparatus in which data may be stored and from which the same data may be retrieved; especially, the internal, high-speed, large-capacity working storage of a computer, as opposed to external devices. Also known as computer memory. [PSYCH] The recollection of past events or sensations, or the performance of previously learned skills without practice.

memory address register [COMPUT SCI] A special register containing the address of a word currently required.

memory bank [COMPUT SCI] A physical section of a computer memory, which may be designed to handle information transfers independently of other such transfers in other such sections.

memory buffer register [COMPUT SCI] A special register in which a word is stored as it is read from memory or just prior to being written into memory.

memory capacity See storage capacity.

memory cell [COMPUT SCI] A single storage element of a memory, together with associated circuits for storing and reading out one bit of information.

memory core See magnetic core.

memory cycle See cycle time.

memory dump See storage dump.

memory dump routine [COMPUT SCI] A debugging routine which produces a listing of a consecutive section of memory, either numbers or instructions, at selected points in a program.

memory fill See storage fill.

memory hierarchy [COMPUT SCI] A ranking of computer memory devices, with devices having the fastest access time at the top of the hierarchy, and devices with slower access times but larger capacity and lower cost at lower levels.

memory lockout register [COMPUT SCI] A special register containing the limiting addresses of an area in memory which may not be accessed by the program.

memory map [COMPUT SCI] The list of variables, constants, identifiers, and their memory locations when a FORTRAN program is being run. Also known as memory map list.

memory map list *See* memory map.

memory overlay [COMPUT SCI] The efficient use of memory space by allowing for repeated use of the same areas of internal storage during the different stages of a program; for instance, when a subroutine is no longer required, another routine can replace all or part of it.

memory port [COMPUT SCI] A logical connection through which data are transferred in or out of main memory under control of the central processing unit.

memory power [COMPUT SCI] A relative characteristic pertaining to differences in access time speeds in different parts of memory; for instance, access time from the buffer may be a tenth of the access time from core.

memory print *See* storage dump.

memory-reference instruction [COMPUT SCI] A type of instruction usually requiring two machine cycles, one to fetch the instruction, the other to fetch the data at an address (part of the instruction itself) and to execute the instruction.

memory register *See* storage register.

memory search routine [COMPUT SCI] A debugging routine which has as an essential feature the scanning of memory in order to locate specified instructions.

memory storage [COMPUT SCI] The sum total of the computer's storage facilities, that is, core, drum, disk, cards, and paper tape.

memory switch *See* ovonic memory switch.

memory trace [PHYSIO] *See* engram. [PSYCH] An experience intentionally forgotten but not fully repressed, which may result in the development of a neurotic conflict.

memory tube *See* storage tube.

mendelevium [CHEM] Synthetic radioactive element, symbol Md, with atomic number 101; made by bombarding lighter elements with light nuclei accelerated in cyclotrons.

Mendelian genetics [GEN] Scientific study of heredity as related to or in accordance with Mendel's laws.

Mendel's laws [GEN] Two basic principles of genetics formulated by Mendel: the law of segregation and law of independent assortment.

Ménière's syndrome [MED] A disease of the inner ear characterized by deafness, vertigo, and tinnitus; possibly an allergic process. Also known as labyrinthine syndrome.

meninges [ANAT] The membranes that cover the brain and spinal cord; there are three in mammals and one or two in submammalian forms.

meningitis [MED] Inflammation of the meninges of the brain and spinal cord, caused by viral, bacterial, and protozoan agents.

meningocele [MED] Hernia of the meninges through a defect in the skull or vertebral column, forming a cyst filled with cerebrospinal fluid.

meninx [ANAT] Any one of the three membranes covering the brain and spinal cord.

menisc-, menisco- [SCI TECH] A combining form denoting crescentic, sickle-shaped, semilunar; denoting meniscus, semilunar cartilage.

Meniscotheriidae [PALEON] A family of extinct mammals of the order Condylarthra possessing selenodont teeth and molarized premolars.

meniscus [ANAT] A crescent-shaped body, especially an interarticular cartilage. [FL MECH] The free surface of a liquid which is near the walls of a vessel and which is curved because of surface tension.

[MET] In reference to a solder joint, the minimum angle at which the solder tapers from the joint to the flat area.

meniscus lens [OPTICS] A lens with one convex surface and one concave surface.

Menispermaceae [BOT] A family of dicotyledonous woody vines in the order Ranunculales distinguished by mostly alternate, simple leaves, unisexual flowers, and a dioecious habit.

menopause [PHYSIO] The natural physiologic cessation of menstruation, usually occurring in the last half of the fourth decade. Also known as climacteric.

Menoponidae [INV ZOO] A family of biting lice (Mallophaga) adapted to life only upon domestic and sea birds.

menses *See* menstruation.

menstruation [PHYSIO] The periodic discharge of sanguineous fluid and sloughing of the uterine lining in women from puberty to the menopause. Also known as menses.

mensuration [MATH] The measurement of geometric quantities; for example, length, area, and volume. [SCI TECH] The act or process of measuring.

mental age [PSYCH] The degree of mental development of an individual in terms of the chronological age of the average individual of equivalent mental ability; specifically, a score derived from intelligence tests.

mental deficiency [PSYCH] A condition characterized by intellectual retardation, social inadequacy, and persistent dependency.

mental health [PSYCH] A relatively enduring state of being in which an individual has effected an integration of his instinctual drives in a way that is reasonably satisfying to himself as reflected in his zest for living and his feeling of self-realization.

mental illness [PSYCH] Any form of mental aberration; usually refers to a chronic or prolonged disorder in which there are wide deviations from the normal.

mental retardation [PSYCH] An abnormal slowness of mental function and behavior patterns relative to age and development.

mental telepathy [PSYCH] A form of extrasensory perception in which one person is aware of an external event through direct sensory perception, and another person, not in the same place, also becomes aware of the event but not through direct sensory perception.

Menthaceae [BOT] An equivalent name for Labiatae.

mentum [ANAT] The chin. [BOT] A projection formed by union of the sepals at the base of the column in some orchids. [INV ZOO] 1. A projection between the mouth and foot in certain gastropods. 2. The median or basal portion of the labium in insects.

menu [COMPUT SCI] A list of computer functions appearing on a video display terminal which indicates the possible operations that a computer can perform next, only one of which can be selected by the operator.

Menurae [VERT ZOO] A small suborder of suboscine perching birds restricted to Australia, including the lyrebirds and scrubbirds.

Menuridae [VERT ZOO] The lyrebirds, a family of birds in the suborder Menurae notable for their vocal mimicry.

mep *See* mean effective pressure.

Meramecian [GEOL] A North American provincial series of geologic time: Upper Mississippian (above Osagian, below Chesterian).

mercallite [MINERAL] $KHSO_4$ A colorless or sky blue, orthorhombic mineral consisting of potassium acid sulfate; occurs as stalactites composed of minute crystals.

mercapt-, mercapto- [CHEM] A combining form denoting the presence of the thiol (SH) group.

mercaptal [ORG CHEM] A group of organosulfur compounds that contain the group $=C(SR)_2$.

mercaptan [ORG CHEM] A group of organosulfur compounds that are derivatives of hydrogen sulfide in the same way that alcohols are derivatives of water; have a characteristically disagreeable odor, and are found with other sulfur compounds in crude petroleum; an example is methyl mercaptan. Also known as thiol.

mercaptol [ORG CHEM] A compound formed by combining a mercaptal and a ketone.

Mercator bearing See rhumb bearing.

Mercator projection [MAP] A conformal cylindrical map projection in which the surface of a sphere or spheroid, such as the earth, is conceived as developed on a cylinder tangent along the Equator; meridians appear as equally spaced vertical lines, and parallels as horizontal lines drawn farther apart as the latitude increases, such that the correct relationship between latitude and longitude scales at any point is maintained.

Mercer engine [MECH ENG] A revolving-block engine in which two opposing pistons operate in a single cylinder with two rollers attached to each piston; intake ports are uncovered when the pistons are closest together, and exhaust ports are uncovered when they are farthest apart.

mercerization [TEXT] A technique used to increase luster, dye absorptivity, and strength in cotton and linen goods; the cloth is put into a heated solution of caustic soda at a controlled temperature, then washed, neutralized, and rinsed.

mercurial horn ore See calomel.

mercuric [INORG CHEM] The mercury ion with a 2+ oxidation state, for example $Hg(NO_3)_2$.

mercurous [INORG CHEM] Referring to mercury with a 1+ valence; for example, mercurous chloride, Hg_2Cl_2, where the mercury is covalently bonded, as Cl—Hg—Hg—Cl.

mercury [CHEM] A metallic element, symbol Hg, atomic number 80, atomic weight 200.59, existing at room temperature as a silvery, heavy liquid. Also known as quicksilver.

Mercury [ASTRON] The planet nearest to the sun; it is visible to the naked eye shortly after sunset or before sunrise when it is nearest to its greatest angular distance from the sun.

mercury arc [ELECTR] An electric discharge through ionized mercury vapor, giving off a brilliant bluish-green light containing strong ultraviolet radiation.

mercury barometer [ENG] An instrument which determines atmospheric pressure by measuring the height of a column of mercury which the atmosphere will support; the mercury is in a glass tube closed at one end and placed, open end down, in a well of mercury. Also known as Torricellian barometer.

mercury cell [ELEC] A primary dry cell that delivers an essentially constant output voltage throughout its useful life by means of a chemical reaction between zinc and mercury oxide; widely used in hearing aids. Also known as mercury oxide cell.

mercury lamp See mercury-vapor lamp.

mercury oxide cell See mercury cell.

mercury-vapor lamp [ELECTR] A lamp in which light is produced by an electric arc between two electrodes in an ionized mercury-vapor atmosphere; it gives off a bluish-green light rich in ultraviolet radiation. Also known as mercury lamp.

merganser [VERT ZOO] Any of several species of diving water fowl composing a distinct subfamily of Anatidae and characterized by a serrate bill adapted for catching fish.

merge [COMPUT SCI] To create an ordered set of data by combining properly the contents of two or more sets of data, each originally ordered in the same manner as the output data set. Also known as mesh.

merged-transistor logic See integrated injection logic.

merge sort [COMPUT SCI] To produce a single sequence of items ordered according to some rule, from two or more previously ordered or unordered sequences, without changing the items in size, structure, or total number; although more than one pass may be required for a complete sort, items are selected during each pass on the basis of the entire key.

merging routine [COMPUT SCI] A program that creates a single sequence of items, ordered according to some rule, out of two or more sequences of items, each sequence ordered according to the same rule.

meridian [ASTRON] 1. A great circle passing through the poles of the axis of rotation of a planet or satellite. 2. See celestial meridian. [GEOD] A north-south reference line, particularly a great circle through the geographical poles of the earth.

meridian angle [ASTRON] Angular distance east or west of the local celestial meridian; the arc of the celestial equator, or the angle at the celestial pole, between the upper branch of the local celestial meridian and the hour circle of a celestial body, measured eastward or westward from the local celestial meridian through 180°, and labeled E or W to indicate the direction of measurement.

meridian circle See transit circle.

meridian transit See transit; transit circle.

meridional circulation [METEOROL] An atmospheric circulation in a vertical plane oriented along a meridian; it consists, therefore, of the vertical and the meridional (north or south) components of motion only. [OCEANOGR] The exchange of water masses between northern and southern oceanic regions.

meridional ray [OPTICS] A ray that lies within a plane which also contains the axis of an optical system.

Meridosternata [INV ZOO] A suborder of echinoderms including various deep-sea forms of sea urchins.

meristem [BOT] Formative plant tissue composed of undifferentiated cells capable of dividing and giving rise to other meristematic cells as well as to specialized cell types; found in growth areas.

Mermithidae [INV ZOO] A family of filiform nematodes in the superfamily Mermithoidea; only juveniles are parasitic.

Mermithoidea [INV ZOO] A superfamily of nematodes composed of two families, both of which are invertebrate parasites.

merocrine [PHYSIO] Pertaining to glands in which the secretory cells undergo cytological changes without loss of cytoplasm during secretion.

merocrystalline See hypocrystalline.

merogony [EMBRYO] The normal or abnormal development of a part of an egg following cutting, shaking, or centrifugation of the egg before or after fertilization.

merohedral [CRYSTAL] Of a crystal class in a system, having a general form with only one-half, one-fourth, or one-eighth the number of equivalent faces of the corresponding form in the holohedral class of the same system. Also known as merosymmetric.

meromictic [HYD] Of or pertaining to a lake whose water is permanently stratified and therefore does not circulate completely throughout the basin at any time during the year.

Meropidae [VERT ZOO] The bee-eaters, a family of brightly colored Old World birds in the order Coraciiformes.

Merostomata [INV ZOO] A class of primitive arthropods of the subphylum Chelicerata distinguished by their aquatic mode of life and the possession of abdominal appendages which bear respiratory organs; only three living species are known.

merosymmetric See merohedral.

merwinite [MINERAL] $Ca_3MgSi_2O_8$ A rare colorless or pale-green neosilicate mineral crystallizing in the monoclinic system; occurs in granular aggregates showing polysynthetic twinning; hardness is 6 on Mohs scale, and specific gravity is 3.15.

Merycoidodontidae [PALEON] A family of extinct tylopod ruminants in the superfamily Merycoidodontoidea.

Merycoidodontoidea [PALEON] A superfamily of extinct ruminant mammals in the infraorder Tylopoda which were exceptionally successful in North America.

merzlota See frozen ground.

mesa [GEOGR] A broad, isolated, flat-topped hill bounded by a steep cliff or slope on at least one side; represents an erosion remnant.

Mesacanthidae [PALEON] An extinct family of primitive acanthodian fishes in the order Acanthodiformes distinguished by a pair of small intermediate spines, large scales, superficially placed fin spines, and a short branchial region.

mesa diode [ELECTR] A diode produced by diffusing the entire surface of a large germanium or silicon wafer and then delineating the individual diode areas by a photoresist-controlled etch that removes the entire diffused area except the island or mesa at each junction site.

mesa transistor [ELECTR] A transistor in which a germanium or silicon wafer is etched down in steps so the base and emitter regions appear as physical plateaus above the collector region.

mescaline [ORG CHEM] $C_{11}H_{17}NO_3$ The alkaloid 3,4,5-trimethoxyphenethylamine, found in mescal buttons; produces unusual psychic effects and visual hallucinations.

mesencephalon [EMBRYO] The middle portion of the embryonic vertebrate brain; gives rise to the cerebral peduncles and the tectum. Also known as midbrain.

mesenchyme [EMBRYO] That part of the mesoderm from which all connective tissues, blood vessels, blood, lymphatic system proper, and the heart are derived.

mesenteric artery [ANAT] Either of two main arterial branches arising from the abdominal aorta: the inferior, supplying the descending colon and the rectum, and the superior, supplying the small intestine, the cecum, and the ascending and transverse colon.

mesenteron [EMBRYO] See midgut. [INV ZOO] Central gastric cavity in an actinozoan.

MESFET See metal semiconductor field-effect transistor.

mesh [COMPUT SCI] See merge. [DES ENG] A size of screen or of particles passed by it in terms of the number of openings occurring per linear inch. Also known as mesh size. [ELEC] A set of branches

forming a closed path in a network so that if any one branch is omitted from the set, the remaining branches of the set do not form a closed path. Also known as loop. [MECH ENG] Engagement or working contact of teeth of gears or of a gear and a rack. [MIN ENG] 1. A closed path traversed through the network in ventilation surveys. 2. The size of diamonds as determined by sieves. [TEXT] Any fabric, knitted or woven, with an open, fine or coarse texture.

mesh connection See delta connection.

mesh impedance [ELEC] The ratio of the voltage to the current in a mesh when all other meshes are open. Also known as self-impedance.

mesh size See mesh.

mesic atom [PARTIC PHYS] An atom in which one of the electrons is replaced by a negative muon or meson orbiting close to or within the nucleus. Also known as mesonic atom.

mesobenthos [OCEANOGR] Of or pertaining to the sea bottom at depths of 180–900 meters (100-500 fathoms).

mesoblastema See mesoderm.

mesoclimate [CLIMATOL] 1. The climate of small areas of the earth's surface which may not be representative of the general climate of the district. 2. A climate characterized by moderate temperatures, that is, in the range 20–30°C. Also known as mesothermal climate.

mesoderm [EMBRYO] The third germ layer, lying between the ectoderm and endoderm; gives rise to the connective tissues, muscles, urogenital system, vascular system, and the epithelial lining of the coelom. Also known as mesoblastema.

Mesogastropoda [INV ZOO] The equivalent name for Pectinibranchia.

mesoglea [INV ZOO] The gelatinous layer between the ectoderm and endoderm in coelenterates and certain sponges.

Mesohippus [PALEON] An early ancestor of the modern horse; occurred during the Oligocene.

mesolite [MINERAL] $Na_2Ca_2Al_6Si_9O_{30} \cdot 8H_2O$ Zeolite mineral composed of hydrous sodium calcium aluminosilicate, usually found in white or colorless tufts of acicular crystals; used as cation exchangers or molecular sieves.

mesometeorology [METEOROL] That portion of the science of meteorology concerned with the study of atmospheric phenomena on a scale larger than that of micrometeorology, but smaller than the cyclonic scale.

mesomorphism [PHYS CHEM] A state of matter intermediate between a crystalline solid and a normal isotropic liquid, in which long rod-shaped organic molecules contain dipolar and polarizable groups.

meson [PARTIC PHYS] Any elementary (noncomposite) particle with strong nuclear interactions and baryon number equal to zero.

mesonephric duct [EMBRYO] The efferent duct of the mesonephros. Also known as Wolffian duct.

mesonic atom See mesic atom.

Mesonychidae [PALEON] A family of extinct mammals of the order Condylarthra.

mesopause [METEOROL] The top of the mesosphere; corresponds to the level of minimum temperature at 80 to 95 kilometers.

mesophyll [BOT] Parenchymatous tissue between the upper and lower epidermal layers in foliage leaves.

Mesosauria [PALEON] An order of extinct aquatic reptiles which is known from a single genus, *Mesosaurus*, characterized by a long snout, numerous slender teeth, small forelimbs, and webbed hind-feet.

mesoscale [METEOROL] A term used to classify meteorological phenomena extending approximately 0.6–60 miles or 1–100 kilometers (mesoscale cloud patterns, for example).

mesosiderite [GEOL] A stony-iron meteorite containing about equal amounts of silicates and nickel-iron, with considerable troilite. Also known as grahamite.

mesosphere [GEOL] *See* lower mantle. [METEOROL] The atmospheric shell between about 45–55 kilometers and 80–95 kilometers, extending from the top of the stratosphere to the mesopause; characterized by a temperature that generally decreases with altitude.

Mesostigmata [INV ZOO] The mites, a suborder of the Acarina characterized by a single pair of breathing pores (stigmata) located laterally in the middle of the idiosoma between the second and third, or third and fourth, legs.

Mesosuchia [PALEON] A suborder of extinct crocodiles of the Late Jurassic and Early Cretaceous.

Mesotaeniaceae [BOT] The saccoderm desmids, a family of fresh-water algae in the order Conjugales; cells are oval, cylindrical, or rectangular and have simple, undecorated walls in one piece.

Mesotardigrada [INV ZOO] An order of tardigrades which combines certain echiniscoidean features with eutardigradan characters.

mesothelium [ANAT] The simple squamous-cell epithelium lining the pleural, pericardial, peritoneal, and scrotal cavities. [EMBRYO] The lining of the wall of the primitive body cavity situated between the somatopleure and splanchnopleure.

mesothermal [MINERAL] Of a hydrothermal mineral deposit, formed at great depth at temperatures of 200–300°C.

mesothermal climate *See* mesoclimate.

Mesoveliidae [INV ZOO] The water treaders, a small family of hemipteran insects in the subdivision Amphibicorisae having well-developed ocelli.

mesoxalyurea *See* alloxan.

Mesozoa [INV ZOO] A division of the animal kingdom sometimes ranked intermediate between the Protozoa and the Metazoa; composed of two orders of small parasitic, wormlike organisms.

Mesozoic [GEOL] A geologic era from the end of the Paleozoic to the beginning of the Cenozoic; commonly referred to as the Age of Reptiles.

message [COMMUN] A series of words or symbols, transmitted with the intention of conveying information. [COMPUT SCI] An arbitrary amount of information with beginning and end defined or implied: usually, it originates in one place and is intended to be transmitted to another place.

message authentication [COMMUN] Security measure designed to establish the authenticity of a message by means of an authenticator within the transmission derived from certain predetermined elements of the message itself.

message processing [COMMUN] In communication operations, the acceptance, preparation for transmission, transmission, receipt, or delivery of a series of words or symbols intended for conveying information.

message switching [COMMUN] A system in which data transmitted between stations on different circuits within a network are routed through central points.

messenger ribonucleic acid [BIOCHEM] A linear sequence of nucleotides which is transcribed from and complementary to a single strand of deoxyribonucleic acid and which carries the information for protein synthesis to the ribosomes. Abbreviated m-RNA.

Messier number [ASTRON] A number by which star clusters and nebulae are listed in Messier's catalog; for example, the Andromeda Galaxy is M31.

metabiosis [ECOL] An ecological association in which one organism precedes and prepares a suitable environment for a second organism.

metabolism [PHYSIO] The physical and chemical processes by which foodstuffs are synthesized into complex elements (assimilation, anabolism), complex substances are transformed into simple ones (disassimilation, catabolism), and energy is made available for use by an organism.

metabolite [BIOCHEM] A product of intermediary metabolism.

metacarpus [ANAT] The portion of a hand or forefoot between the carpus and the phalanges.

metacenter [FL MECH] The intersection of a vertical line through the center of buoyancy of a floating body, slightly displaced from its equilibrium position, with a line connecting the center of gravity and the equilibrium center of buoyancy; the floating body is stable if the metacenter lies above the center of gravity.

metacercaria [INV ZOO] Encysted cercaria of digenetic trematodes; the infective form.

metacestode [INV ZOO] Encysted larva of a tapeworm; occurs in the intermediate host.

Metachlamydeae [BOT] An artificial group of flowering plants, division Magnoliophyta, recognized in the Englerian system of classification; consists of families of dicotyledons in which petals are characteristically fused, forming a sympetalous corolla.

metachromasia [CHEM] 1. The property exhibited by certain pure dyestuffs, chiefly basic dyes, of coloring certain tissue elements in a different color, usually of a shorter wavelength absorption maximum, than most other tissue elements. 2. The assumption of different colors or shades by different substances when stained by the same dye. Also known as metachromatism.

metachromatic granules [CYTOL] Granules which assume a color different from that of the dye used to stain them.

metachromatism *See* metachromasia.

metacinnabar [MINERAL] HgS A black isometric mineral that represents an ore of mercury. Also known as metacinnabarite.

metacinnabarite *See* metacinnabar.

Metacopina [PALEON] An extinct suborder of ostracods in the order Podocopida.

metacryst [PETR] A large crystal, such as garnet, formed in metamorphic rock by recrystallization. Also known as metacrystal.

metacrystal *See* metacryst.

metagenesis [BIOL] The phenomenon in which one generation of certain plants and animals reproduces asexually, followed by a sexually reproducing generation. Also known as alternation of generations.

metahewettite [MINERAL] $CaV_6O_{16} \cdot 9H_2O$ A deep red, probably orthorhombic mineral consisting of hydrated calcium vanadate; occurs as pulverulent masses.

metal [MATER] An opaque crystalline material usually of high strength with good electrical and thermal conductivities, ductility, and reflectivity; properties are related to the structure, in which the positively charged ions are bonded through a field of free electrons which surrounds them forming a close-packed structure.

metal ceramic   *See* cermet.

metal cluster compound   [CHEM] A compound in which two or more metal atoms aggregate so as to be within bonding distance of one another and each metal atom is bonded to at least two other metal atoms; some nonmetal atoms may be associated with the cluster.

metal forming   [MET] Any manufacturing process by which parts or components are fabricated by shaping or molding a piece of metal stock.

metalimnion   *See* thermocline.

metal-insulator semiconductor   [SOLID STATE] Semiconductor construction in which an insulating layer, generally a fraction of a micrometer thick, is deposited on the semiconducting substrate before the pattern of metal contacts is applied. Abbreviated MIS.

metallic-disk rectifier   *See* metallic rectifier.

metallic element   [CHEM] An element generally distinguished (from a nonmetallic one) by its luster, electrical conductivity, malleability, and ability to form positive ions.

metallic insulator   [ELECTROMAG] Section of transmission line used as a mechanical support device; the section is an odd number of quarter-wavelengths long at the frequency of interest, and the input impedance becomes high enough so that the section effectively acts as an insulator.

metallic paint   [MATER] 1. Paint used for covering metal surfaces; the pigment is commonly iron oxide. 2. Paint with a metal pigment.

metallic rectifier   [ELECTR] A rectifier consisting of one or more disks of metal under pressure-contact with semiconductor coatings or layers, such as a copper oxide, selenium, or silicon rectifier. Also known as contact rectifier; dry-disk rectifier; dry-plate rectifier; metallic-disk rectifier; semiconductor rectifier.

metalliferous   [MINERAL] Pertaining to mineral deposits from which metals can be extracted.

metallocene   [ORG CHEM] Organometallic coordination compound which is obtained as a cyclopentadienyl derivative of a transition metal or a metal halide.

metallography   [MET] The study of the structure of metals and alloys by various methods, especially by the optical and the electron microscope, and by x-ray diffraction.

metalloid   [CHEM] A nonmetallic element, such as carbon or nitrogen, which can combine with a metal to form an alloy.

metallurgical engineering   [ENG] Application of the principles of metallurgy to the engineering sciences.

metallurgical microscope   [ENG] A microscope used in the study of metals, usually optical.

metallurgy   [SCI TECH] The science and technology of metals and alloys.

metal-nitride-oxide semiconductor   [SOLID STATE] A semiconductor structure that has a double insulating layer; typically, a layer of silicon dioxide (SiO_2) is nearest the silicon substrate, with a layer of silicon nitride (Si_3N_4) over it. Abbreviated MNOS.

metal oxide semiconductor   [SOLID STATE] A metal insulator semiconductor structure in which the insulating layer is an oxide of the substrate material; for a silicon substrate, the insulating layer is silicon dioxide (SiO_2). Abbreviated MOS.

metal oxide semiconductor field-effect transistor   [ELECTR] A field-effect transistor having a gate that is insulated from the semiconductor substrate by a thin layer of silicon dioxide. Abbreviated MOSFET; MOST; MOS transistor. Formerly known as insulated-gate field-effect transistor.

metal plating   *See* plating.

metal powder   [MET] A finely divided metal or alloy.

metal replacement   *See* immersion plating.

metal rolling   *See* rolling.

metal semiconductor field-effect transistor   [ELECTR] A field-effect transistor that uses a thin film of gallium arsenide, with a Schottky barrier gate formed by depositing a layer of metal directly onto the surface of the film. Abbreviated MESFET.

metal spinning   *See* spinning.

metamerism   [ZOO] The condition of an animal body characterized by the repetition of similar segments (metameres), exhibited especially by arthropods, annelids, and vertebrates in early embryonic stages and in certain specialized adult structures. Also known as segmentation.

metamict   [MINERAL] Of a radioactive mineral, exhibiting lattice disruption due to radiation damage while the original external morphology is retained.

metamorphic aureole   *See* aureole.

metamorphic rock   [PETR] A rock formed from preexisting solid rocks by mineralogical, structural, and chemical changes, in response to extreme changes in temperature, pressure, and shearing stress.

metamorphic zone   *See* aureole.

metamorphism   [PETR] The mineralogical and structural changes of solid rock in response to environmental conditions at depth in the earth's crust.

metamorphosis   [BIOL] 1. A structural transformation. 2. A marked structural change in an animal during postembryonic development.   [MED] A degenerative change in tissue or organ structure.

metanauplius   [INV ZOO] A primitive larval stage of certain decapod crustaceans characterized by seven pairs of appendages; follows the nauplius stage.

metaphase   [CYTOL] 1. The phase of mitosis during which centromeres are arranged on the equator of the spindle. 2. The phase of the first meiotic division when centromeric regions of homologous chromosomes come to lie equidistant on either side of the equator.

metaphysis   *See* epiphyseal plate.

metaplasia   [PATH] Transformation of one form of tissue to another.

metarossite   [MINERAL] $CaV_2O_6 \cdot 2H_2O$ A light yellow mineral consisting of hydrated calcium vanadate; occurs as masses and veinlets.

metasomatism   [PETR] A variety of metamorphism in which one mineral or a mineral assemblage is replaced by another of different composition without melting.

metastable   *See* labile.

metastable equilibrium   [PHYS] A condition in which a system returns to equilibrium after small (but not large) displacements; it may be represented by a ball resting in a small depression on top of a hill.   [PHYS CHEM] A state of pseudoequilibrium having higher free energy than the true equilibrium state.

metastable ion   [ANALY CHEM] In mass spectroscopy, an ion formed by a secondary dissociation process in the analyzer tube (formed after the parent or initial ion has passed through the accelerating field).

metastable phase   [PHYS CHEM] Existence of a substance as either a liquid, solid, or vapor under conditions in which it is normally unstable in that state.

metastable state [QUANT MECH] An excited stationary energy state whose lifetime is unusually long.

metastasis [MED] Transfer of the causal agent (cell or microorganism) of a disease from a primary focus to a distant one through the blood or lymphatic vessels. [PHYS] A transition of an electron or nucleon from one bound state to another in an atom or molecule, or the capture of an electron by a nucleus.

Metastrongylidae [INV ZOO] A family of roundworms belonging to the Strongyloidea; species are parasitic in sheep, cattle, horses, dogs, and other domestic animals.

metatarsus [ANAT] The part of a foot or hindfoot between the tarsus and the phalanges.

Metatheria [VERT ZOO] An infraclass of therian mammals including a single order, the Marsupialia; distinguished by a small braincase, a total of 50 teeth, the inflected angular process of the mandible, and a pair of marsupial bones articulating with the pelvis.

metatorbernite [MINERAL] $Cu(UO_2)_2(PO_4)_2 \cdot 8H_2O$ A green secondary mineral composed of hydrous copper uranium phosphate; similar to torbernite, but with less water content.

metavariscite [MINERAL] $AlPO_6 \cdot 2H_2O$ A green monoclinic mineral composed of hydrous aluminum phosphate; it is isomorphous with phosphosiderite.

metavauxite [MINERAL] $FeAl_2(PO_4)_2(OH)_2 \cdot 8H_2O$ A colorless mineral composed of hydrous basic phosphate of iron and aluminum; similar to vauxite, but with more water.

metazeunerite [MINERAL] $Cu(UO_2)_2(AsO_4)_2 \cdot 8H_2O$ A grass to emerald green, tetragonal mineral consisting of a hydrated arsenate of copper and uranium; occurs in tabular form.

Metazoa [ZOO] The multicellular animals that make up the major portion of the animal kingdom; cells are organized in layers or groups as specialized tissues or organ systems.

metencephalon [EMBRYO] The cephalic portion of the rhombencephalon; gives rise to the cerebellum and pons.

meteor [ASTRON] The phenomena which accompany a body from space (a meteoroid) in its passage through the atmosphere, including the flash and streak of light and the ionized trail.

Meteoriaceae [BOT] A family of mosses in the order Isobryales in which the calyptra is frequently hairy.

meteoric stone *See* stony meteorite.

meteorite [GEOL] Any meteoroid that has fallen to the earth's surface.

meteoroid [ASTRON] Any solid object moving in interplanetary space that is smaller than a planet or asteroid but larger than a molecule.

meteorolite *See* stony meteorite.

meteorological chart [METEOROL] A weather map showing the spatial distribution, at an instant of time, of atmospheric highs and lows, rain clouds, and other phenomena.

meteorological equator *See* intertropical convergence zone.

meteorological optics [OPTICS] A branch of atmospheric physics or physical meteorology in which optical phenomena occurring in the atmosphere are described and explained. Also known as atmospheric optics.

meteorological radar [ENG] Radar which is used to study the scattering of radar waves by various types of atmospheric phenomena, for making weather observations and forecasts.

meteorological range [METEOROL] An empirically consistent measure of the visual range of a target; a concept developed to eliminate from consideration the threshold contrast and adaptation luminance, both of which vary from observer to observer. Also known as standard visibility; standard visual range.

meteorological rocket [ENG] Small rocket system used to extend observation of atmospheric character above feasible limits for balloon-borne observing and telemetering instruments. Also known as rocketsonde.

meteorological satellite [AERO ENG] Earth-orbiting spacecraft carrying a variety of instruments for measuring visible and invisible radiations from the earth and its atmosphere.

meteorological solenoid [METEOROL] A hypothetical tube formed in space by the intersection of a set of surfaces of constant pressure and a set of surfaces of constant specific volume of air. Also known as solenoid.

meteorological tide [OCEANOGR] A change in water level caused by local meteorological conditions, in contrast to an astronomical tide, caused by the attractions of the sun and moon.

meteorology [SCI TECH] The science concerned with the atmosphere and its phenomena; the meteorologist observes the atmosphere's temperature, density, winds, clouds, precipitation, and other characteristics and aims to account for its observed structure and evolution (weather, in part) in terms of external influence and the basic laws of physics.

meteor shower [ASTRON] A number of meteors with approximately parallel trajectories.

meteor trail *See* ion column.

meter [MECH] The international standard unit of length, equal to $1,650,763.73$ times the wavelength of the orange light emitted when a gas consisting of the pure krypton isotope of mass number 86 is excited in an electrical discharge. Abbreviated m. [ENG] A device for measuring the value of a quantity under observation; the term is usually applied to an indicating instrument alone.

meter-candle *See* lux.

metering pin *See* metering rod.

metering pump [CHEM ENG] Plunger-type pump designed to control accurately small-scale fluid-flow rates; used to inject small quantities of materials into continuous-flow liquid streams. Also known as proportioning pump.

metering rod [ENG] A device consisting of a long metallic pin of graduated diameters fitted to the main nozzle of a carburetor (on an internal combustion engine) or passage leading thereto in such a way that it measures or meters the amount of gasoline permitted to flow by it at various speeds. Also known as metering pin.

meter-kilogram [MECH] **1.** A unit of energy or work in a meter-kilogram-second gravitational system, equal to the work done by a kilogram-force when the point at which the force is applied is displaced 1 meter in the direction of the force; equal to 9.80665 joules. Abbreviated m-kgf. Also known as meter kilogram-force. **2.** A unit of torque, equal to the torque produced by a kilogram-force acting at a perpendicular distance of 1 meter from the axis of rotation. Also known as kilogram-meter (kgf-m).

meter kilogram-force *See* meter-kilogram.

meter-kilogram-second-ampere system [PHYS] A system of electrical and mechanical units in which length, mass, time, and electric current are the fundamental quantities, and the units of these quantities are the meter, the kilogram, the second, and the ampere respectively. Abbreviated mksa system. Also known as Giorgi system; practical system.

meter-kilogram-second system [MECH] A metric system of units in which length, mass, and time are fundamental quantities, and the units of these quantities are the meter, the kilogram, and the second respectively. Abbreviated mks system.

meter-proving tank See calibrating tank.

methanation [ORG CHEM] A chemical process for the conversion of organic compounds to methane.

methanogenesis [BIOCHEM] The biosynthesis of the hydrocarbon methane; common in certain bacteria. Also known as bacterial methanogenesis.

methemoglobin See ferrihemoglobin.

methionine [BIOCHEM] $C_5H_{11}O_2NS$ An essential amino acid; furnishes both labile methyl groups and sulfur necessary for normal metabolism.

methods engineering [IND ENG] A technique used by management to improve working methods and reduce labor costs in all areas where human effort is required.

methoxide [ORG CHEM] A compound formed from a metal and the methoxy radical; an example is sodium methoxide. Also known as methylate.

methoxy- [ORG CHEM] OCH_3- A combining form indicating the oxygen-containing methane radical, found in many organic solvents, insecticides, and plasticizer intermediates.

para-methoxybenzoic acid See anisic acid.

methyl [ORG CHEM] The alkyl group derived from methane and usually written CH_3-. Also known as carbinyl.

methylate See methoxide.

methylbenzoylecgonine See cocaine.

2-methyl-4-chlorophenoxyacetic acid See bromate.

methylene [ORG CHEM] $-CH_2-$ A radical that contains a bivalent carbon.

ortho-methylnitrobenzene See ortho-nitrotoluene.

metric [MATH] A real valued "distance" function on a topological space X satisfying four rules: for x, y, and z in X, the distance from x to itself is zero; the distance from x to y is positive if x and y are different; the distance from x to y is the same as the distance from y to x; and the distance from x to y is less than or equal to the distance from x to z plus the distance from z to y (triangle inequality).

metricate [SCI TECH] To use the metric system in expressing all physical quantities.

metric carat See carat.

metric centner [MECH] 1. A unit of mass equal to 50 kilograms. 2. A unit of mass equal to 100 kilograms. Also known as quintal.

metric line See millimeter.

metric system [MECH] A system of units used in scientific work throughout the world and employed in general commercial transactions and engineering applications; its units of length, time, and mass are the meter, second, and kilogram respectively, or decimal multiples and submultiples thereof.

metric tensor [MATH] A second rank tensor of a Riemannian space whose components are functions which help define magnitude and direction of vectors about a point. Also known as fundamental tensor.

metric ton See tonne.

Metridiidae [INV ZOO] A family of zoantharian coelenterates in the order Actiniaria.

metritis [MED] Inflammation of the uterus, usually involving both the endometrium and myometrium.

metrorrhagia [MED] Uterine bleeding during the intermenstrual cycle. Also known as intermenstrual flow; polymenorrhea.

MeV See megaelectronvolt.

Mexican onyx See onyx marble.

Meyliidae [INV ZOO] A family of free-living nematodes in the superfamily Desmoscolecoidea.

Meziridae [INV ZOO] A family of hemipteran insects in the superfamily Aradoidea.

mf See medium frequency.

Mg See magnesium.

MHD See magnetohydrodynamics.

MHD generator See magnetohydrodynamic generator.

mho See siemens.

MHz See megahertz.

mi See mile.

Miacidae [PALEON] The single, extinct family of the carnivoran superfamily Miacoidea.

Miacoidea [PALEON] A monofamilial superfamily of extinct carnivoran mammals; a stem group thought to represent the progenitors of the earliest member of modern carnivoran families.

miargyrite [MINERAL] $AgSbS_2$ An iron-black to steel-gray mineral that crystallizes in the monoclinic system.

miarolitic [PETR] Of igneous rock, characterized by small irregular cavities into which well-formed crystals of the rock-forming mineral protrude.

MIC See microwave integrated circuit.

mica [MINERAL] A group of phyllosilicate minerals (with sheetlike structures) of general formula $(K,Na,Ca)(Mg,Fe,Li,Al)_{2-3}(Al,Si)_4O_{10}(OH,F)_2$ characterized by low hardness (2–2½) and perfect basal cleavage.

mica capacitor [ELEC] A capacitor whose dielectric consists of thin rectangular sheets of mica and whose electrodes are either thin sheets of metal foil stacked alternately with mica sheets, or thin deposits of silver applied to one surface of each mica sheet.

micellar flooding [PETRO ENG] A two-step enhanced oil recovery process in which a surfactant slug is injected into the well followed by a larger slug of water containing a high-molecular-weight polymer which pushes the chemicals through the field and improves mobility and sweep efficiency. Also known as microemulsion flooding; surfactant flooding.

micelle [MOL BIO] A submicroscopic structural unit of protoplasm built up from polymeric molecules. [PHYS CHEM] A colloidal aggregate of a unique number (between 50 and 100) of amphipathic molecules, which occurs at a well-defined concentration known as the critical micelle concentration.

Michelson interferometer [OPTICS] An interferometer in which light strikes a partially reflecting plate at an angle of 45°, the light beams reflected and transmitted by the plate are both reflected back to the plate by mirrors, and the beams are recombined at the plate, interfering constructively or destructively depending on the distances from the plate to the two mirrors.

Michelson stellar interferometer [OPTICS] An instrument for measuring angular diameters of astronomical objects, in which a system of mirrors directs two parallel beams of light into a telescope, and angular diameter is determined from the maximum distance between the beams at which interference fringes are observable.

MICR See magnetic-ink character recognition.

micrencephaly [MED] The condition of having an abnormally small brain. Also spelled microencephaly.

micrinite [PETR] An opaque granular variety of inertinite of medium hardness showing no plant-cell structure.

micrite [PETR] A semiopaque crystalline limestone matrix that consists of chemically precipitated calcite mud, whose crystals are generally 1–4 micrometers in diameter.

micro- [MATH] A prefix representing 10^{-6}, or one-millionth. [SCI TECH] 1. A prefix indicating smallness, as in microwave. 2. A prefix indicating extreme sensitivity, as in microradiometer and microphone.

microaerophilic [MICROBIO] Pertaining to those microorganisms requiring free oxygen but in very low concentration for optimum growth.

microanalysis [ANALY CHEM] Identification and chemical analysis of material on a small scale so that specialized instruments such as the microscope are needed; the material analyzed may be on the scale of 1 microgram.

microanatomy [ANAT] Anatomical study of microscopic tissue structures.

microbalance [ENG] A small, light type of analytical balance that can weigh loads of up to 0.1 gram to the nearest microgram.

microbe [MICROBIO] A microorganism, especially a bacterium of a pathogenic nature.

microbiology [BIOL] The science and study of microorganisms, including protozoans, algae, fungi, bacteria, viruses, and rickettsiae.

microbit [COMPUT SCI] A unit of information equal to one-millionth of a bit.

microbreccia [GEOL] A poorly sorted sandstone containing large, angular sand particles in a fine silty or clayey matrix.

microcephaly [MED] The condition of having an abnormally small head, with a circumference less than two standard deviations below the mean.

microcircuitry [ELECTR] Electronic circuit structures that are orders of magnitude smaller and lighter than circuit structures produced by the most compact combinations of discrete components. Also known as microelectronic circuitry; microminiature circuitry.

microclimate [CLIMATOL] The local, rather uniform climate of a specific place or habitat, compared with the climate of the entire area of which it is a part.

microclimatology [CLIMATOL] The study of a microclimate, including the study of profiles of temperature, moisture and wind in the lowest stratum of air, the effect of the vegetation and of shelterbelts, and the modifying effect of towns and buildings.

microcline [MINERAL] $KAlSi_3O_8$ A triclinic potassium-rich feldspar, usually containing minor amounts of sodium; may be clear, white, pale-yellow, brick-red, or green, and is generally characterized by crosshatch twinning.

microcomputer [ELECTR] A microprocessor combined with input/output interface devices, some type of external memory, and the other elements required to form a working computer system; it is smaller, lower in cost, and usually slower than a minicomputer. Also known as micro.

microconsumer *See* decomposer.

microcontroller [ELECTR] A microcomputer, microprocessor, or other equipment used for precise process control in data handling, communication, and manufacturing.

Microcotyloidea [INV ZOO] A superfamily of ecto-parasitic trematodes in the subclass Monogenea.

Microcyprini [VERT ZOO] The equivalent name for Cyprinodontiformes.

Microdomatacea [PALEON] An extinct superfamily of gastropod mollusks in the order Aspidobranchia.

microelectronic circuitry *See* microcircuitry.

microelectronics [ELECTR] The technology of constructing circuits and devices in extremely small packages by various techniques. Also known as microminiaturization; microsystem electronics.

microelement [ELECTR] Resistor, capacitor, transistor, diode, inductor, transformer, or other electronic element or combination of elements mounted on a ceramic wafer 0.025 centimeter thick and about 0.75 centimeter square; individual microelements are stacked, interconnected, and potted to form micromodules.

microemulsion flooding *See* micellar flooding.

microencapsulation [CHEM ENG] Enclosing of materials in capsules from well below 1 micrometer to over 2000 micrometers in diameter.

microencephaly *See* micrencephaly.

microevolution [EVOL] 1. Evolutionary processes resulting from the accumulation of minor changes over a relatively short period of time; evolutionary changes due to gene mutation. 2. Evolution of species.

microfibril [MOL BIOL] The submicroscopic unit of a microscopic cellular fiber.

microfiche [GRAPHICS] A microfilm card or sheet used in some information storage systems; consists of a film format about 4 by 6 inches (10 by 15 centimeters) containing microimages of type and other information, with a title heading large enough to be read by the unaided eye.

microfilament [CYTOL] One of the cytoplasmic fibrous structures, about 5 nanometers in diameter, virtually identical to actin; thought to be important in the processes of phagocytosis and pinocytosis.

microfilaria [INV ZOO] Slender, motile prelarval forms of filarial nematodes measuring 150–300 micrometers in length; adult filaria are mammalian parasites.

microfilm [GRAPHICS] Greatly reduced film records of such things as books, newspapers, engineering drawings, reports, and manuscripts; copies are made on fine-grain film of 16, 35, 70, and 105-millimeter size, permitting easy storage and handling.

microfossil [PALEON] A small fossil which is studied and identified by means of the microscope.

microgamete [BIOL] The smaller, or male gamete produced by heterogametic species.

microgram [MECH] A unit of mass equal to one-millionth of a gram. Abbreviated μg.

microgroove record *See* long-playing record.

microhabitat [ECOL] A small, specialized, and effectively isolated location.

microhardness [MET] Hardness of microscopic areas of a metal or alloy.

microhm [ELEC] A unit of resistance, reactance, and impedance, equal to 10^{-6} ohm.

Microhylidae [VERT ZOO] A family of anuran amphibians in the suborder Diplasiocoela including many heavy-bodied forms with a pointed head and tiny mouth.

microinstruction [COMPUT SCI] The portion of a microprogram that specifies the operation of individual computing elements and such related subunits as the main memory and the input/output interfaces; usually includes a next-address field that eliminates the need for a program counter.

microlite [CRYSTAL] A microscopic crystal which polarizes light. Also known as microlith. [MINERAL] $(Na,Ca)_2(Ta, Nb)_2O_6(O,OH,F)$ A pale-yellow, reddish, brown, or black isometric mineral composed of sodium calcium tantalum oxide with a small amount of fluorine; it is isomorphous with pyrochlore. Also known as djalmaite.

microlith [CRYSTAL] See microlite. [MED] A calculus of microscopic size.

microlitic [PETR] Of the texture of a porphyritic igneous rock, having a groundmass composed of an aggregate of microlites in a generally glassy base.

Micromalthidae [INV ZOO] A family of coleopteran insects in the superfamily Cantharoidea; the single species is the telephone pole beetle.

micromanipulation [BIOL] The techniques and practice of microdissection, microvivisection, microisolation, and microinjection.

micromere [EMBRYO] A small blastomere of the upper or animal hemisphere in eggs that undergo uneven cleavage.

micrometeorite [ASTRON] A very small meteorite or meteoritic particle with a diameter generally less than a millimeter.

micrometeorology [METEOROL] That portion of the science of meteorology that deals with the observation and explanation of the smallest-scale physical and dynamic occurrences within the atmosphere; studies are confined to the surface boundary layer of the atmosphere, that is, from the earth's surface to an altitude where the effects of the immediate underlying surface upon air motion and composition become negligible.

micrometer [ENG] 1. An instrument attached to a telescope or microscope for measuring small distances or angles. 2. A caliper for making precise measurements; a spindle is moved by a screw thread so that it touches the object to be measured; the dimension can then be read on a scale. Also known as micrometer caliper. [MECH] A unit of length equal to one-millionth of a meter. Abbreviated μm. Also known as micron (μ).

micrometer caliper See micrometer.

micrometer of mercury See micron.

micromicro- See pico-.

micromini [COMPUT SCI] The central processing unit of a minicomputer placed on one of more integrated circuit chips.

microminiature circuitry See microcircuitry.

microminiaturization See microelectronics.

micron [MECH] 1. A unit of pressure equal to the pressure exerted by a column of mercury 1 micrometer high, having a density of 13.5951 grams per cubic centimeter, under the standard acceleration of gravity; equal to 0.133322387415 pascal; it differs from the millitorr by less than one part in seven million. Also known as micrometer of mercury. 2. See micrometer.

micronekton [ECOL] Active pelagic crustaceans and other forms intermediate between thrusting nekton and feebler-swimming plankton.

micronucleus [INV ZOO] The smaller, reproductive nucleus in multinucleate protozoans.

micronutrient [BIOCHEM] Trace elements and compounds required by living systems only in minute amounts.

microorganism [MICROBIO] A microscopic organism, including bacteria, protozoans, yeast, viruses, and algae.

Micropezidae [INV ZOO] A family of myodarian cyclorrhaphous dipteran insects in the subsection Acalypteratae.

microphone [ENG ACOUS] An electroacoustic device containing a transducer which is actuated by sound waves and delivers essentially equivalent electric waves.

microphonics [ELECTR] Noise caused by mechanical vibration of the elements of an electron tube, component, or system. Also known as microphonism.

microphonism See microphonics.

Microphysidae [INV ZOO] A palearctic family of hemipteran insects in the subfamily Cimicimorpha.

microphyte [ECOL] 1. A microscopic plant. 2. A dwarfed plant due to unfavorable environmental conditions.

micropipet [ENG] 1. A pipet with capacity of 0.5 milliliter or less, to measure small volumes of liquids with a high degree of accuracy; types include lambda, straight-bore, and Lang-Levy. 2. A fine-pointed pipette used for microinjection.

microprobe [SPECT] An instrument for chemical microanalysis of a sample, in which a beam of electrons is focused on an area less than a micrometer in diameter, and the characteristic x-rays emitted as a result are dispersed and analyzed in a crystal spectrometer to provide a qualitative and quantitative evaluation of chemical composition.

microprocessor [ELECTR] A single silicon chip on which the arithmetic and logic functions of a computer are placed.

microprogram [COMPUT SCI] A computer program that consists only of basic elemental commands which directly control the operation of each functional element in a microprocessor.

microprogramming [COMPUT SCI] Transformation of a computer instruction into a sequence of elementary steps (microinstructions) by which the computer hardware carries out the instruction.

Micropterygidae [INV ZOO] The single family of the lepidopteran superfamily Micropterygoidea; members are minute moths possessing toothed, functional mandibles and lacking a proboscis.

Micropterygoidea [INV ZOO] A monofamilial superfamily of lepidopteran insects in the suborder Homoneura.

micropump See electroosmotic driver.

Micropygidae [INV ZOO] A family of echinoderms in the order Diadematoida that includes only one genus, *Micropyga*, which has noncrenulate tubercles and umbrellalike outer tube feet.

micropyle [BOT] A minute opening in the integument at the tip of an ovule through which the pollen tube commonly enters; persists in the seed as an opening or a scar between the hilum and point of radicle.

microradiography [ANALY CHEM] Technique for the study of surfaces of solids by monochromatic-radiation (such as x-ray) contrast effects shown via projection or enlargement of a contact radiograph. [GRAPHICS] The radiography of small objects having details too fine to be seen by the unaided eye, with optical enlargement of the resulting negative.

Microsauria [PALEON] An order of Carboniferous and early Permian lepospondylous amphibians.

microscope [OPTICS] An instrument through which minute objects are enlarged by means of a lens or lens system; principal types include optical, electron, and x-ray.

microseism [GEOPHYS] A weak, continuous, oscillatory motion in the earth having a period of 1–9 seconds and caused by a variety of agents, especially atmospheric agents; not related to an earthquake.

microsome [CYTOL] 1. A fragment of the endoplasmic reticulum. 2. A minute granule of protoplasm.

microspore [BOT] The smaller spore of heterosporous plants; gives rise to the male gametophyte.

Microsporida [INV ZOO] The single order of the class Microsporidea.

Microsporidae [INV ZOO] The equivalent name for Sphaeriidae.

Microsporidea [INV ZOO] A class of Cnidospora characterized by the production of minute spores with a single intrasporal filament or one or two intracapsular filaments and a single sporoplasm; mainly intracellular parasites of arthropods and fishes.

microsystem electronics *See* microelectronics.

microtectonics *See* structural petrology.

microtome [ENG] An instrument for cutting thin sections of tissues or other materials for microscopical examination.

Microtragulidae [PALEON] A group of saltatorial caenolistoid marsupials that appeared late in the Cenozoic and paralleled the small kangaroos of Australia.

microtubule [CYTOL] One of the hollow tubelike filaments found in certain cell components, such as cilia and the mitotic spindle, and composed of repeating subunits of the protein tubulin.

microwave [ELECTROMAG] An electromagnetic wave which has a wavelength between about 0.3 and 30 centimeters, corresponding to frequencies of 1–100 gigahertz; however, there are no sharp boundaries distinguishing microwaves from infrared and radio waves.

microwave amplification by stimulated emission of radiation *See* maser.

microwave background *See* cosmic microwave radiation.

microwave bridge [ELECTROMAG] A microwave circuit equivalent to an ordinary electrical bridge and used to measure impedance; consists of six waveguide sections arranged to form a multiple junction.

microwave cavity *See* cavity resonator.

microwave circuit [ELECTROMAG] Any particular grouping of physical elements, including waveguides, attenuators, phase changers, detectors, wavemeters, and various types of junctions, which are arranged or connected together to produce certain desired effects on the behavior of microwaves.

microwave circulator *See* circulator.

microwave gyrator *See* gyrator.

microwave integrated circuit [ELECTR] A microwave circuit that uses integrated-circuit production techniques involving such feature as thin or thick films, substrates, dielectrics, conductors, resistors, and microstrip lines, to build passive assemblies on a dielectric. Abbreviated MIC.

microwave landing system [NAV] A system of ground equipment which generates guidance beams at microwave frequencies for guiding aircraft to landings; it is intended to replace the present lower-frequency instrument landing system. Abbreviated MLS.

microwave optics [ELECTROMAG] The study of those properties of microwaves which are analogous to the properties of light waves in optics.

microwave radiometer *See* radiometer.

microwave reflectometer [ELECTROMAG] A pair of single-detector couplers on opposite sides of a waveguide, one of which is positioned to monitor transmitted power, and the other to measure power reflected from a single discontinuity in the line.

microwave resonance cavity *See* cavity resonator.

microwave solid-state device [ELECTR] A semiconductor device for the generation or amplification of electromagnetic energy at microwave frequencies.

microwave spectrometer [SPECT] An instrument which makes a graphical record of the intensity of microwave radiation emitted or absorbed by a substance as a function of frequency, wavelength, or some related variable.

microwave spectroscope [SPECT] An instrument used to observe the intensity of microwave radiation emitted or absorbed by a substance as a function of frequency, wavelength, or some related variable.

microwave spectrum [ELECTROMAG] The range of wavelengths or frequencies of electromagnetic radiation that are designated microwaves. [SPECT] A display, photograph, or plot of the intensity of microwave radiation emitted or absorbed by a substance as a function of frequency, wavelength, or some related variable.

microwave waveguide *See* waveguide.

microyield strength [MECH] Stress at which a microstructure (single crystal, for example) exhibits a specified deviation in its stress-strain curve.

mictic [BIOL] 1. Requiring or produced by sexual reproduction. 2. Of or pertaining to eggs which without fertilization develop into males and with fertilization develop into amictic females, as occurs in rotifers.

midbrain [ANAT] Those portions of the adult brain derived from the embryonic midbrain. [EMBRYO] The middle portion of the embryonic vertebrate brain. Also known as mesencephalon.

Middle Cambrian [GEOL] The geologic epoch occurring between Upper and Lower Cambrian, beginning approximately 540,000,000 years ago.

Middle Cretaceous [GEOL] The geologic epoch between the Upper and Lower Cretaceous, beginning approximately 120,000,000 years ago.

Middle Devonian [GEOL] The geologic epoch occurring between the Upper and Lower Devonian, beginning approximately 385,000,000 years ago.

middle ear [ANAT] The middle portion of the ear in higher vertebrates; in mammals it contains three ossicles and is separated from the external ear by the tympanic membrane and from the inner ear by the oval and round windows.

Middle Jurassic [GEOL] The geologic epoch occurring between the Upper and Lower Jurassic, beginning approximately 170,000,000 years ago.

middle latitude Also known as mid-latitude. [GEOGR] A point of latitude that is midway on a north-and-south line between two parallels. [NAV] The latitude at which the arc length of the parallel separating the meridians passing through two specific points is exactly equal to the departure in proceeding from one point to the other by middle-latitude sailing.

middle-latitude westerlies *See* westerlies.

midgut [EMBRYO] The middle portion of the digestive tube in vertebrate embryos. Also known as mesenteron. [INV ZOO] The mesodermal intermediate part of an invertebrate intestine.

mid-latitude *See* middle latitude.

mid-latitude westerlies *See* westerlies.

mid-ocean canyon *See* deep-sea channel.

mid-oceanic ridge [GEOL] A continuous, median, seismic mountain range on the floor of the ocean, extending through the North and South Atlantic oceans, the Indian Ocean, and the South Pacific Ocean; the topography is rugged, elevation is 1–3 kilometers (km), width is about 1500 km, and length is over 84,000 km. Also known as mid-ocean ridge; mid-ocean rise; oceanic ridge.

mid-ocean ridge *See* mid-oceanic ridge.

mid-ocean rift *See* rift valley.

mid-ocean rise *See* mid-oceanic ridge.

midpoint [MATH] The midpoint of a line segment is the point which separates the segment into two equal parts.

migmatite [PETR] A mixed rock exhibiting crystalline textures in which a truly metamorphic component is streaked and mixed with obviously once-molten material of a more or less granitic character.

migration [COMPUT SCI] Movement of frequently used data items to more accessible storage locations, and of infrequently used data items to less accessible locations. [GEOL] 1. Movement of a topographic feature from one place to another, especially movement of a dune by wind action. 2. Movement of liquid or gaseous hydrocarbons from their source into reservoir rocks. [HYD] Slow, downstream movement of a system of meanders. [MET] The uncontrolled movement of certain metals, particularly silver, from one location to another, usually associated with undesirable effects such as oxidation or corrosion. [SOLID STATE] 1. The movement of charges through a semiconductor material by diffusion or drift of charge carriers or ionized atoms. 2. The movement of crystal defects through a semiconductor crystal under the influence of high temperature, strain, or a continuously applied electric field. [VERT ZOO] Periodic movement of animals to new areas or habitats.

migratory dune *See* wandering dune.

mil [MATH] A unit of angular measure which, due to nonuniformity of usage, may have any one of three values: 0.001 radian or approximately $0.0572958°$; 1/6400 of a full revolution or $0.5625°$; 1/1000 of a right angle or $0.09°$. [MECH] 1. A unit of length, equal to 0.001 inch, or to 2.54×10^{-5} meter. Also known as milli-inch; thou. 2. *See* milliliter.

milarite [MINERAL] $K_2Ca_4Be_4Al_2Si_{24}O_{62} \cdot H_2O$ A colorless to greenish, glassy, hexagonal mineral composed of a hydrous silicate of potassium, calcium, beryllium, and aluminum, occurring in crystals.

mildew [MYCOL] 1. A whitish growth on plants, organic matter, and other materials caused by a parasitic fungus. 2. Any fungus producing such growth.

mile [MECH] A unit of length in common use in the United States, equal to 5280 feet, or 1609.344 meters. Abbreviated mi. Also known as land mile; statute mile.

Milichiidae [INV ZOO] A family of myodarian cyclorrhaphous dipteran insects in the subsection Acalypteratae.

Miliolacea [INV ZOO] A superfamily of marine or brackish foraminiferans in the suborder Miliolina characterized by an imperforate test wall of tiny, disordered calcite rhombs.

Miliolidae [INV ZOO] A family of foraminiferans in the superfamily Miliolacea.

Miliolina [INV ZOO] A suborder of the Foraminiferida characterized by a porcelaneous, imperforate calcite wall.

military engineering [ENG] Science, art, and practice involved in design and construction of defensive and offensive military works as well as construction and maintenance of transportation systems.

military satellite [AERO ENG] An artificial earth satellite used for military purposes; the six mission categories are communication, navigation, geodesy, nuclear test detection, surveillance, and research and technology.

military technology [ENG] The technology needed to develop and support the armament used by the military.

milk [CHEM] A suspension of certain metallic oxides, as milk of magnesia, iron, or bismuth. [PHYSIO] 1. The whitish fluid secreted by the mammary gland for the nourishment of the young; composed of carbohydrates, proteins, fats, mineral salts, vitamins, and antibodies. 2. Any whitish fluid in nature resembling milk, as coconut milk.

milk glass [MATER] A white, and sometimes colored, opaque glass made by adding calcium fluoride and alumina to soda-lime glass.

milk sugar *See* lactose.

milkweed [BOT] Any of several latex-secreting plants of the genus *Asclepias* in the family Asclepiadaceae.

milky quartz [MINERAL] An opaque, milk-white variety of crystalline quartz, often with a greasy luster; milkiness is due to the presence of air-filled cavities. Also known as greasy quartz.

Milky Way [ASTRON] The faint band of light which encircles the sky and results from the combined light of the many stars near the plane of our galaxy.

Milky Way Galaxy [ASTRON] The large aggregation of stars and interstellar gas and dust of which the sun is a member. Also known as Galaxy.

milky weather *See* whiteout.

mill [FOOD ENG] 1. A machine for grinding grain into flour. 2. A building that houses milling machines. [IND ENG] 1. A machine that manufactures paper, textiles, or other products by the continuous repetition of some simple process or action. 2. A building that houses machinery for manufacturing processes. [MIN ENG] 1. An excavation made in the country rock, by a crosscut from the workings on a vein, to obtain waste for filling; it is left without timber so that the roof can fall in and furnish the required rock. 2. A passage connecting a stope or upper level with a lower level intended to be filled with broken ore that can then be drawn out at the bottom as desired for further transportation.

Milleporina [INV ZOO] An order of the class Hydrozoa known as the stinging corals; they resemble true corals because of a calcareous exoskeleton.

miller *See* milling machine.

Miller code [COMPUT SCI] A code used internally in some computers, in which a binary 1 is represented by a transition (either up or down), and a binary 0 is represented by no transition following a binary 1; a transition between bits represents successive 0's; in this code, the longest period possible without a transition is two bit times.

Miller generator *See* bootstrap integrator.

Miller indices [CRYSTAL] Three integers identifying a type of crystal plane; the intercepts of a plane on the three crystallographic axes are expressed as fractions of the crystal parameters; the reciprocals of these fractions, reduced to integral proportions, are the Miller indices. Also known as crystal indices.

millerite [MINERAL] NiS A brass to bronze-yellow mineral that crystallizes in the hexagonal system and usually contains trace amounts of cobalt, copper, and iron; hardness is 3–3.5 on Mohs scale, and specific gravity is 5.5; it generally occurs in fine crystals, chiefly as nodules in clay ironstone. Also known as capillary pyrites; hair pyrites; nickel pyrites.

millet [BOT] A common name applied to at least five related members of the grass family grown for their edible seeds.

milli- [MATH] A prefix representing 10^{-3}, or one-thousandth. Abbreviated m.

milliampere-second [NUCLEO] A unit of radiation dose resulting from exposure to x-rays, equal to the dose produced by an electron beam, carrying a current of 1 milliampere, bombarding the target of an x-ray tube for 1 second. Abbreviated mA s.

millier See tonne.

millihg See millimeter of mercury.

milli-inch See mil.

milliliter [MECH] A unit of volume equal to 10^{-3} liter or 10^{-6} cubic meter. Abbreviated ml. Also known as mil.

millimeter [MECH] A unit of length equal to one-thousandth of a meter. Abbreviated mm. Also known as metric line; strich.

millimeter of mercury [MECH] A unit of pressure, equal to the pressure exerted by a column of mercury 1 millimeter high with a density of 13.5951 grams per cubic centimeter under the standard acceleration of gravity; equal to 133.322387415 pascals; it differs from the torr by less than 1 part in 7,000,000. Abbreviated mmHg. Also known as millihg.

milli-micro- See nano-.

milling [MECH ENG] Mechanical treatment of materials to produce a powder, to change the size or shape of metal powder particles, or to coat one powder mixture with another. [MIN ENG] A combination of open-cut and underground mining, wherein the ore is mined in open cut and handled underground.

milling machine [MECH ENG] A machine for the removal of metal by feeding a workpiece through the periphery of a rotating circular cutter. Also known as miller.

million electron volts See megaelectronvolt.

millipede [INV ZOO] The common name for members of the arthropod class Diplopoda.

millisite [MINERAL] $(Na,K)CaAl_6(PO_4)_4(OH)_9 \cdot 3H_2O$ White mineral composed of a basic hydrous phosphate of sodium, potassium, calcium, and aluminum.

mill ore [MIN ENG] An ore that must be given some preliminary treatment before a marketable grade or a grade suitable for further treatment can be obtained.

millstone See buhrstone.

Mimas [ASTRON] A satellite of Saturn orbiting at a mean distance of 186,000 kilometers.

mimetene See mimetite.

mimetesite See mimetite.

mimetic [CRYSTAL] Pertaining to a crystal that is twinned or malformed but whose crystal symmetry appears to be of a higher grade than it actually is. [PETR] Of a tectonite, having a deformation fabric, formed by mimetic crystallization, that reflects and is influenced by preexisting anisotropic structure. [ZOO] Pertaining to or exhibiting mimicry.

mimetite [MINERAL] $Pb_5(AsO_4)_3Cl$ A yellow to yellowish-brown mineral of the apatite group, commonly containing calcium or phosphate; a minor ore of lead. Also known as mimetene; mimetesite.

mimicry [ZOO] Assumption of color, form, or behavior patterns by one species of another species, for camouflage and protection.

Mimidae [VERT ZOO] The mockingbirds, a family of the Oscines in the order Passeriformes.

Mimosoideae [BOT] A subfamily of the legume family, Leguminosae; members are largely woody and tropical or subtropical with regular flowers and usually numerous stamens.

min See minim.

minasragrite [MINERAL] $(VO)_2H_2(SO_4)_3 \cdot 15H_2O$ A blue, monoclinic mineral consisting of hydrated acid vanadyl sulfate; occurs in efflorescences and as aggregates or masses.

M indicator See M scope.

mine [MIN ENG] An opening or excavation in the earth for extracting minerals. [ORD] An encased explosive or chemical charge designed to be positioned so that it detonates when the target touches or moves near it or when it is fired by remote control; general types are land mines and underwater mines.

mineral [GEOL] A naturally occurring substance with a characteristic chemical composition expressed by a chemical formula; may occur as individual crystals or may be disseminated in some other mineral or rock; most mineralogists include the requirements of inorganic origin and internal crystalline structure.

mineral dressing See beneficiation.

mineral engineering See mining engineering.

mineral facies [PETR] Rocks of any origin whose components have been formed within certain temperature-pressure limits characterized by the stability of certain index minerals.

mineralization [GEOL] **1.** The process of fossilization whereby inorganic materials replace the organic constituents of an organism. **2.** The introduction of minerals into a rock, resulting in a mineral deposit.

mineralocorticoid [BIOCHEM] A steroid hormone secreted by the adrenal cortex that regulates mineral metabolism and, secondarily, fluid balance.

mineralogy [INORG CHEM] The science which concerns the study of natural inorganic substances called minerals.

mineral sequence See paragenesis.

mineral spring [HYD] A spring whose water has a definite taste due to a high mineral content.

mineral suite [MINERAL] **1.** A group of associated minerals in one deposit. **2.** A representative group of minerals from a certain locality. **3.** A group of specimens showing variations, as in color or form, in a single mineral species.

mineral tallow See hatchettite.

mineral water [HYD] Water containing naturally or artificially supplied minerals or gases.

mineral wax See ozocerite.

minesweeper [NAV ARCH] A ship specially designed and equipped to remove or destroy underwater mines. [ORD] Heavy road roller pushed in front of a tank, to destroy land mines by exploding them.

miniaturization [ELECTR] Reduction in the size and weight of a system, package, or component by using small parts arranged for maximum utilization of space.

minicomputer [COMPUT SCI] A small computer which in its basic configuration has at least 4096 words of memory, employs words between 8 and 16 bits in length.

minim [MECH] A unit of volume in the apothecaries' measure; equals 1/60 fluidram (approximately 0.061612 cubic centimeter) or about 1 drop (of water). Abbreviated min.

minimal-latency coding See minimum-access coding.

minimal polynomial [MATH] The polynomial of least degree which both divides the characteristic polynomial of a matrix and has the same roots.

minimal surface [MATH] A surface whose mean curvature is identically zero.

mini-maxi regret [CONT SYS] In decision theory, a criterion which selects that strategy which has the smallest maximum difference between its payoff and that of the best hindsight choice.

minimax technique _See_ min-max technique.

minimum [MATH] The least value that a real valued function assumes.

minimum-access coding [COMPUT SCI] Coding in such a way that a minimum time is required to transfer words to and from storage, for a computer in which this time depends on the location in storage. Also known as minimal-latency coding; minimum-delay coding; minimum-latency coding.

minimum-access programming [COMPUT SCI] The programming of a digital computer in such a way that minimum waiting time is required to obtain information out of the memory. Also known as forced programming; minimum-latency programming.

minimum-access routine _See_ minimum-latency routine.

minimum-delay coding _See_ minimum-access coding.

minimum detectable signal _See_ threshold signal.

minimum-latency coding _See_ minimum-access coding.

minimum-latency programming _See_ minimum-access programming.

minimum-latency routine [COMPUT SCI] A computer routine that is constructed so that the latency in serial-access storage is less than the random latency that would be expected if storage locations were chosen without regard for latency. Also known as minimum-access routine.

minimum-loss attenuator [ELECTR] A section linking two unequal resistive impedances which is designed to introduce the smallest attenuation possible. Also known as minimum-loss pad.

minimum-loss pad _See_ minimum-loss attenuator.

minimum thermometer [ENG] A thermometer that automatically registers the lowest temperature attained during an interval of time.

minimum turning circle [ENG] The diameter of the circle described by the outermost projection of a vehicle when the vehicle is making its shortest possible turn.

mining engineering [ENG] Engineering concerned with the discovery, development, and exploitation of coal, ores, and minerals, as well as the cleaning, sizing, and dressing of the product. Also known as mineral engineering.

minium [MINERAL] Pb_3O_4 A scarlet or orange-red mineral consisting of an oxide of lead; found in Wisconsin and the western United States. Also known as red lead.

mink [VERT ZOO] Any of three species of slender-bodied aquatic carnivorous mammals in the genus _Mustela_ of the family Mustelidae.

Minkowski space-time [RELAT] The space-time of special relativity; it is completely flat and contains no gravitating matter. Also known as Minkowski universe.

Minkowski universe _See_ Minkowski space-time.

min-max technique [MATH] A method of approximation of a function f by a function g from some class where the maximum of the modulus of $f - g$ is minimized over this class. Also known as Chebyshev approximation; minimax technique.

min-max theorem [MATH] The theorem that provides information concerning the nth eigenvalue of a symmetric operator on an inner product space without necessitating knowledge of the other eigenvalues. Also known as maximum-minimum principle.

minor control data [COMPUT SCI] Control data which are at the least significant level used, or which are used to sort records into the smallest groups used; for example, if control data are used to specify state, town, and street, then the data specifying street would be minor control data.

minor cycle [COMPUT SCI] The time required for the transmission or transfer of one machine word, including the space between words, in a digital computer using serial transmission. Also known as word time.

minor diatonic scale [ACOUS] A diatonic scale in which the relative sizes of the sequence of intervals are approximately 2,1,2,2,2,2,1.

minority carrier [SOLID STATE] The type of carrier, electron, or hole that constitutes less than half the total number of carriers in a semiconductor.

minority emitter [ELECTR] Of a transistor, an electrode from which a flow of minority carriers enters the interelectrode region.

minor loop [CONT SYS] A portion of a feedback control system that consists of a continuous network containing both forward elements and feedback elements.

minor planet [ASTRON] 1. Those planets smaller than the earth, specifically Mercury, Venus, Mars, and Pluto. 2. _See_ asteroid.

minuend [MATH] The quantity from which another quantity is to be subtracted.

minus [MATH] A minus B means that the quantity B is to be subtracted from the quantity A.

minus angle _See_ angle of depression.

minute [MATH] A unit of measurement of angle that is equal to $\frac{1}{60}$ of a degree. Symbolized ′. Also known as arcmin. [MECH] A unit of time, equal to 60 seconds.

Miocene [GEOL] A geologic epoch of the Tertiary period, extending from the end of the Oligocene to the beginning of the Pliocene.

miocrystalline _See_ hypocrystalline.

Miosireninae [PALEON] A subfamily of extinct sirenian mammals in the family Dugongidae.

mirabilite [MINERAL] $Na_2SO_4 \cdot 10H_2O$ A yellow or white monoclinic mineral consisting of hydrous sodium sulfate, occurring as a deposit from saline lakes, playas, and springs, and as an efflorescence; the pure crystals are known as Glauber's salt.

mirage [OPTICS] Any one of a variety of unusual images of distant objects seen as a result of the bending of light rays in the atmosphere during abnormal vertical distribution of air density.

Mirapinnatoidei [VERT ZOO] A suborder of tiny oceanic fishes in the order Cetomimiformes.

Miridae [INV ZOO] The largest family of the Hemiptera; included in the Cimicomorpha, it contains herbivorous and predacious plant bugs which lack ocelli and have a cuneus and four-segmented antennae.

Miripinnati [VERT ZOO] The equivalent name for Marapinnatoidei.

mirror [OPTICS] A surface which specularly reflects a large fraction of incident light.

mirror fusion [PL PHYS] An open-ended configuration which traps low-beta plasmas; it is realized by associating two identical magnetic mirrors having the same axis.

mirror galvanometer [ELEC] A galvanometer having a small mirror attached to the moving element, to permit use of a beam of light as an indicating pointer. Also known as reflecting galvanometer.

mirror glance _See_ wehrlite.

mirror interference [OPTICS] Interference occurring between two beams, one or both of which are reflected from a mirror at a small angle.

mirror interferometer [ENG] An interferometer used in radio astronomy, in which the sea surface acts as a mirror to reflect radio waves up to a single antenna, where the reflected waves interfere with the waves arriving directly from the source. [OPTICS] Any interferometer which makes use of mirror interference.

mirror machine [PL PHYS] A device which confines plasma in a tube with magnetic mirrors at each end to prevent it from escaping.

mirror optics [OPTICS] The science and technology of mirrors which, by means of reflecting rays of light, either revert optical bundles or focus them to form images.

mirror plane of symmetry See plane of mirror symmetry.

mirror reflection See specular reflection.

mirror stone See muscovite.

MIRV See multiple independently targeted reentry vehicle.

MIS See metal-insulator semiconductor.

miscibility [CHEM] The tendency or capacity of two or more liquids to form a uniform blend, that is, to dissolve in each other; degrees are total miscibility, partial miscibility, and immiscibility.

misfire [CHEM] Failure of fuel or an explosive charge to ignite properly. [ELECTR] Failure to establish an arc between the main anode and the cathode of an ignitron or other mercury-arc rectifier during a scheduled conducting period.

mismatch factor See reflection factor.

mispickel See arsenopyrite.

misregistration [COMPUT SCI] In character recognition, the improper state of appearance of a character, line, or document, on site in a character reader, with respect to a real or imaginary horizontal baseline. [GRAPHICS] Improper alignment of register marks in matching overlays on a page or in matching the front of a type page to the back.

missense codon [GEN] A mutant codon that directs the incorporation of a different amino acid and results in the synthesis of a protein with a sequence in which one amino acid has been replaced by a different one; in some cases the mutant protein may be unstable or less active.

missile [ORD] Any object that is, or is designed to be, thrown, dropped, projected, or propelled, for the purpose of making it strike a target; examples are guided missile and ballistic missile.

Mississippian [GEOL] A large division of late Paleozoic geologic time, after the Devonian and before the Pennsylvanian, named for a succession of highly fossiliferous marine strata consisting largely of limestones found along the Mississippi River between southeastern Iowa and southern Illinois; approximately equivalent to the European Lower Carboniferous.

Missourian [GEOL] A North American provincial series of geologic time: lower Upper Pennsylvanian (above Desmoinesian, below Virgilian).

mist [FL MECH] Fine liquid droplets suspended in or falling through a moving or stationary gas atmosphere. [METEOROL] A hydrometeor consisting of an aggregate of microscopic and more or less hygroscopic water droplets suspended in the atmosphere; it produces, generally, a thin, grayish veil over the landscape; it reduces visibility to a lesser extent than fog; the relative humidity with mist is often less than 95%.

mistbow See fogbow.

mistral [METEOROL] A north wind which blows down the Rhone Valley south of Valence, France, and into the Gulf of Lions. Strong, squally, cold, and dry, it is the combined result of the basic circulation, a fall wind, and jet-effect wind.

mite [INV ZOO] The common name for the acarine arthropods composing the diverse suborders Onychopalpida, Mesostigmata, Trombidiformes, and Sarcoptiformes.

miter box [ENG] A troughlike device of metal or wood with vertical slots set at various angles in the upright sides, for guiding a handsaw in making a miter joint.

mitochondria [CYTOL] Minute cytoplasmic organelles in the form of spherical granules, short rods, or long filaments found in almost all living cells; submicroscopic structure consists of an external membrane system.

mitosis [CYTOL] Nuclear division involving exact duplication and separation of the chromosome threads so that each of the two daughter nuclei carries a chromosome complement identical to that of the parent nucleus.

mitscherlichite [MINERAL] $K_2CuCl_4 \cdot 2H_2O$ A greenish-blue, tetragonal mineral consisting of potassium copper chloride dihydrate.

mix crystal See mixed crystal.

mixed crystal [CRYSTAL] A crystal whose lattice sites are occupied at random by different ions or molecules of two different compounds. Also known as mix crystal.

mixed gland [PHYSIO] A gland that secretes more than one substance, especially a gland containing both mucous and serous components.

mixed layer [OCEANOGR] The layer of water which is mixed through wave action or thermohaline convection. Also known as surface water.

mixed nerve [PHYSIO] A nerve containing both sensory and motor components.

mixer [ELECTR] 1. A device having two or more inputs, usually adjustable, and a common output; used to combine separate audio or video signals linearly in desired proportions to produce an output signal. 2. The stage in a superheterodyne receiver in which the incoming modulated radio-frequency signal is combined with the signal of a local r-f oscillator to produce a modulated intermediate-frequency signal. Also known as first detector; heterodyne modulator; mixer-first detector. [OPTICS] A nonlinear device in which two light beams are combined to form new beams having frequencies equal to the sum or the difference of the input wavelengths.

mixer-first detector See mixer.

mixer tube [ELECTR] A multigrid electron tube, used in a superheterodyne receiver, in which control voltages of different frequencies are impressed upon different control grids, and the nonlinear properties of the tube cause the generation of new frequencies equal to the sum and difference of the impressed frequencies.

mixing [CHEM ENG] The intermingling of different materials (liquid, gas, solid) to produce a homogeneous mixture. [ELECTR] Combining two or more signals, such as the outputs of several microphones. [SCI TECH] The thorough intermingling of two or more different materials.

Mixodectidae [PALEON] A family of extinct insectivores assigned to the Proteutheria; a superficially rodentlike group confined to the Paleocene of North America.

mixolimnion [HYD] The upper layer of a meromictic lake, characterized by low density and free circulation; this layer is mixed by the wind.

mixture [PHARM] A liquid medicine prepared by adding insoluble substances to a liquid medium, usually with a suspending agent. [SCI TECH] The product of mixing; components are not in a fixed proportion to each other.

m-kgf *See* meter-kilogram.

mksa system *See* meter-kilogram-second-ampere system.

mks system *See* meter-kilogram-second system.

ml *See* milliliter.

MLS *See* microwave landing system.

mm *See* millimeter.

mmf *See* magnetomotive force.

mmHg *See* millimeter of mercury.

Mn *See* manganese.

mnemonic code [COMPUT SCI] A programming code that is easy to remember because the codes resemble the original words, such as MPY for multiply and ACC for accumulator.

Mnesarchaeidae [INV ZOO] A family of lepidopteran insects in the suborder Homoneura; members are confined to New Zealand.

MNOS *See* metal-nitride-oxide semiconductor.

Mo *See* molybdenum.

moat [GEOL] **1.** A ringlike depression around the base of a seamount. **2.** A valleylike depression around the inner side of a volcanic cone, between the rim and the lava dome. [HYD] **1.** A glacial channel in the form of a deep, wide trench. **2.** *See* oxbow lake.

mobile radio [COMMUN] Radio communication in which the transmitter is installed in a vessel, vehicle, or airplane and can be operated while in motion.

mobile-relay station [COMMUN] Base station in which the base receiver automatically tunes on the base station transmitter and which retransmits all signals received by the base station receiver; used to extend the range of mobile units, and requires two frequencies for operation.

mobile station [COMMUN] Station in the mobile service intended to be used while in motion or during halts at unspecified points.

Mobilina [INV ZOO] A suborder of ciliophoran protozoans in the order Peritrichida.

mobility [PHYS] Freedom of particles to move, either in random motion or under the influence of fields or forces. [SOLID STATE] *See* drift mobility.

Möbius band [MATH] The nonorientable surface obtained from a rectangular strip by twisting it once and then gluing the two ends. Also known as Möbius strip.

Möbius function [MATH] The function μ of the positive integers where $\mu(1) = 1$, $\mu(n) = (-1)^r$ if n factors into r distinct primes, and $\mu(n) = 0$ otherwise; also, $\mu(n)$ is the sum of the primitive nth roots of unity.

Möbius strip *See* Möbius band.

Mobulidae [VERT ZOO] The devil rays, a family of batoids that are surface feeders and live mostly on plankton.

mock lead *See* sphalerite.

mock ore *See* sphalerite.

mock sun *See* paranthelion; parhelion.

mode [COMMUN] Form of the information in a communication such as literal language, digital data, and video. [COMPUT SCI] One of several alternative conditions or methods of operation of a device. [ELECTROMAG] A form of propagation of guided waves that is characterized by a particular field pattern in a plane transverse to the direction of propagation. Also known as transmission mode. [PETR] The mineral composition of a rock, usually expressed as percentages of total weight or volume. [PHYS] A state of an oscillating system that corresponds to a particular field pattern and one of the possible resonant frequencies of the system. [STAT] The most frequently occurring member of a set of numbers.

mode converter *See* mode transducer.

model [COMPUT SCI] *See* macroskeleton. [SCI TECH] A mathematical or physical system, obeying certain specified conditions, whose behavior is used to understand a physical, biological, or social system to which it is analogous in some way.

mode-locked laser [OPTICS] A laser designed so that several modes of oscillation with closely spaced wavelengths, in which the laser would normally oscillate, are synchronized so that a pulse of light, lasting for as little as a picosecond, is generated.

modem [ELECTR] A combination modulator and demodulator at each end of a telephone line to convert binary digital information to audio tone signals suitable for transmission over the line, and vice versa. Also known as dataset. Derived from modulator-demodulator.

mode of oscillation *See* mode of vibration.

mode of vibration [MECH] A characteristic manner in which a system which does not dissipate energy and whose motions are restricted by boundary conditions can oscillate, having a characteristic pattern of motion and one of a discrete set of frequencies. Also known as mode of oscillation.

moder *See* coder.

moderate breeze [METEOROL] In the Beaufort wind scale, a wind whose speed is from 11 to 16 knots (13 to 18 miles per hour or 20 to 30 kilometers per hour).

moderator [NUCLEO] The material used in a nuclear reactor to moderate or slow down neutrons from the high velocities at which they are created in the fission process.

mode switch [COMPUT SCI] A preset control which affects the normal response of various components of a mechanical desk calculator. [ELECTR] A microwave control device, often consisting of a waveguide section of special cross section, which is used to change the mode of microwave power transmission in the waveguide.

mode transducer [ELECTR] Device for transforming an electromagnetic wave from one mode of propagation to another. Also known as mode converter; mode transformer.

mode transformer *See* mode transducer.

modification [ENG] A major or minor change in the design of an item, effected in order to correct a deficiency, to facilitate production, or to improve operational effectiveness. [MET] Treatment of molten aluminum alloys containing 8–13% silicon with small amounts of a sodium fluoride or sodium chloride mixture; improves mechanical properties. [SCI TECH] Any change brought about by external or internal factors.

modifier [COMPUT SCI] A quantity used to alter the address of an operand in a computer, such as the cycle index. Also known as index word. [MATER] In flotation, any of the chemicals which increase the specific attraction between collector agents and particle surfaces or which increase the wettability of those surfaces.

modifier gene [GEN] A gene that alters the phenotypic expression of a nonallelic gene.

modifier register *See* index register.

modify [COMPUT SCI] **1.** To alter a portion of an instruction so its interpretation and execution will be other than normal; the modification may permanently change the instruction or leave it unchanged and affect only the current execution; the most frequent modification is that of the effective address through the use of index registers. **2.** To alter a subroutine according to a defined parameter.

modiolus [ANAT] The central axis of the cochlea.

modularity [COMPUT SCI] The property of functional flexibility built into a computer system by assembling discrete units which can be easily joined to or arranged with other parts or units.

modular programming [COMPUT SCI] The construction of a computer program from a collection of modules, each of workable size, whose interactions are rigidly restricted.

modular structure [BUILD] A building that is constructed of preassembled or presized units of standard sizes; uses a 4-inch (10.16-centimeter) cubical module as a reference. [ELECTR] **1.** An assembly involving the use of integral multiples of a given length for the dimensions of electronic components and electronic equipment, as well as for spacings of holes in a chassis or printed wiring board. **2.** An assembly made from modules.

modulate [ELECTR] To vary the amplitude, frequency, or phase of a wave, or vary the velocity of the electrons in an electron beam in some characteristic manner.

modulated amplifier [ELECTR] Amplifier stage in a transmitter in which the modulating signal is introduced and modulates the carrier.

modulation [COMMUN] The process or the result of the process by which some parameter of one wave is varied in accordance with some parameter of another wave. [MECH ENG] Regulation of the fuel-air mixture to a burner in response to fluctuations of load on a boiler.

modulation spectroscopy [SPECT] A branch of spectroscopy concerned with the measurement and interpretation of changes in transmission or reflection spectra induced (usually) by externally applied perturbation, such as temperature or pressure change, or an electric or magnetic field.

modulation transformer [ENG ACOUS] An audio-frequency transformer which matches impedances and transmits audio frequencies between one or more plates of an audio output stage and the grid or plate of a modulated amplifier.

modulator [ELECTR] **1.** The transmitter stage that supplies the modulating signal to the modulated amplifier stage or that triggers the modulated amplifier stage to produce pulses at desired instants as in radar. **2.** A device that produces modulation by any means, such as by virtue of a nonlinear characteristic or by controlling some circuit quantity in accordance with the waveform of a modulating signal. **3.** One of the electrodes of a spacistor.

modulator-demodulator *See* modem.

module [AERO ENG] A self-contained unit which serves as a building block for the overall structure in space technology; usually designated by its primary function, such as command module or lunar landing module. [COMPUT SCI] **1.** A distinct and identifiable unit of computer program for such purposes as compiling, loading, and linkage editing. **2.** One memory bank and associated electronics in a computer. [ELECTR] A packaged assembly of wired components, built in a standardized size and having standardized plug-in or solderable terminations. [ENG] A unit of size used as a basic component for standardizing the design and construction of buildings, building parts, and furniture. [MATH] A vector space in which the scalars are a ring rather than a field.

modulo [MATH] **1.** A group G modulo a subgroup H is the quotient group G/H of cosets of H in G. **2.** A technique of identifying elements in an algebraic structure in such a manner that the resulting collection of identified objects is the same type of structure.

modulus of a complex number *See* absolute value of a complex number.

modulus of a logarithm [MATH] The modulus of a logarithm with a given base is the factor by which a logarithm with a second base must be multiplied to give the first logarithm.

modulus of compression *See* bulk modulus of elasticity.

modulus of continuity [MATH] For a real valued continuous function f, this is the function whose value at a real number r is the maximum of the modulus of $f(x) - f(y)$ where the modulus of $x - y$ is less than r; this function is useful in approximation theory.

modulus of elasticity [MECH] The ratio of the increment of some specified form of stress to the increment of some specified form of strain, such as Young's modulus, the bulk modulus, or the shear modulus. Also known as coefficient of elasticity; elasticity modulus; elastic modulus.

modulus of elasticity in shear [MECH] A measure of a material's resistance to shearing stress, equal to the shearing stress divided by the resultant angle of deformation expressed in radians. Also known as coefficient of rigidity; modulus of rigidity; shear modulus.

modulus of rigidity *See* modulus of elasticity in shear.

modulus of torsion *See* torsional modulus.

modulus of volume elasticity *See* bulk modulus of elasticity.

Moeritherioidea [PALEON] A suborder of extinct sirenian mammals considered as primitive proboscideans by some authorities and as a sirenian offshoot by others.

mohavite *See* tincalconite.

Mohnian [GEOL] A North American stage of geologic time: Miocene (above Luisian, below Delmontian).

Moho *See* Mohorovičić discontinuity.

Mohorovičić discontinuity [GEOPHYS] A seismic discontinuity that separates the earth's crust from the subjacent mantle, inferred from travel time curves indicating that seismic waves undergo a sudden increase in velocity. Also known as Moho.

Mohr's circle [MECH] A graphical construction making it possible to determine the stresses in a cross section if the principal stresses are known.

mohsite *See* ilmenite.

Mohs scale [MINERAL] An empirical scale consisting of 10 minerals with reference to which the hardness of all other minerals is measured; it includes, from softest (designated 1) to hardest (10): talc, gypsum, calcite, fluorite, apatite, orthoclase, quartz, topaz, corundum, and diamond.

moiety [CHEM] A part or portion of a molecule, generally complex, having a characteristic chemical or pharmacological property.

moiré [COMMUN] In television, the spurious pattern in the reproduced picture resulting from interference beats between two sets of periodic structures in the image. [GRAPHICS] Undesirable patterns that occur when a halftone is made from a previously printed halftone or steel engraving; they are caused by the conflict between the dot arrangement produced by the halftone screen and the dots or lines of the original halftone or engraving; careful rotation of the halftone screen by the photographer or engraver may minimize moiré. [TEXT] A fabric finish in which the warp has yarn of harder twist than the filling, with a surface pattern resembling water ripples that is produced by engraved rollers, heat, pressure, steam, and chemicals.

moiré effect [OPTICS] The effect whereby, when one family of curves is superposed on another family of curves so that the curves cross at angles of less than about 45°, a new family of curves appears which pass through intersections of the original curves.

moist air [METEOROL] 1. In atmospheric thermodynamics, air that is a mixture of dry air and any amount of water vapor. 2. Generally, air with a high relative humidity.

moist-heat sterilization [ENG] Sterilization with steam under pressure, as in an autoclave, pressure cooker, or retort; most bacteriological media are sterilized by autoclaving at 121°C, with 15 pounds (103 kilopascals) of pressure, for 20 minutes or more.

moisture [CLIMATOL] The quantity of precipitation or the precipitation effectiveness. [METEOROL] The water vapor content of the atmosphere, or the total water substance (gaseous, liquid, and solid) present in a given volume of air. [PHYS CHEM] Water that is dispersed through a gas in the form of water vapor or small droplets, dispersed through a solid, or condensed on the surface of a solid.

moisture barrier [MATER] A material that retards the passage of moisture into walls.

moisture flux *See* eddy flux.

mol *See* mole.

molality [CHEM] Concentration given as moles per 1000 grams of solvent.

molal solution [CHEM] Concentration of a solution expressed in moles of solute divided by 1000 grams of solvent.

molal specific heat *See* molar specific heat.

molarity [CHEM] Measure of the number of gram-molecular weights of a compound present (dissolved) in 1 liter of solution; it is indicated by M, preceded by a number to show solute concentration.

molar refraction [OPTICS] Equation for the refractive index of a compound modified by the compound's molecular weight and density. Also known as the Lorentz-Lorenz molar refraction.

molar solution [CHEM] Aqueous solution that contains 1 mole (gram-molecular weight) of solute in 1 liter of the solution.

molar specific heat [PHYS CHEM] The ratio of the amount of heat required to raise the temperature of 1 mole of a compound 1°, to the amount of heat required to raise the temperature of 1 mole of a reference substance, such as water, 1° at a specified temperature. Also known as molal specific heat; molecular specific heat.

molasse [GEOL] A paralic sedimentary facies consisting mainly of shale, subgraywacke sandstone, and conglomerate; it is more clastic and less rhythmic than the preceding flysch and is generally postorogenic.

mold [ENG] 1. A pattern or template used as a guide in construction. 2. A cavity which imparts its form to a fluid or malleable substance. [ENG ACOUS] The metal part derived from the master by electroforming in reproducing disk recordings; has grooves similar to those of the recording. [GEOL] Soft, crumbling friable earth. [GRAPHICS] 1. To form a plastic substance by placing it in a matrix or form. 2. The form or matrix for shaping a plastic substance. [MYCOL] Any of various woolly fungus growths. [PALEON] An impression made in rock or earth material by an inner or outer surface of a fossil shell or other organic structure; a complete mold would be the hollow space.

molded capacitor [ELEC] Capacitor, usually mica, that has been encased in a molded plastic insulating material.

molding time *See* curing time.

mole [CHEM] An amount of substance of a system which contains as many elementary units as there are atoms of carbon in 0.012 kilogram of the pure nuclide carbon-12; the elementary unit must be specified and may be an atom, molecule, ion, electron, photon, or even a specified group of such units. Symbolized mol. [CIV ENG] A breakwater or berthing facility, extending from shore to deep water, with a core of stone or earth. [MECH ENG] A mechanical tunnel excavator. [MED] 1. A mass formed in the uterus by the maldevelopment of all or part of the embryo or of the placenta and membranes. 2. A fleshy, pigmented nevus. [VERT ZOO] Any of 19 species of insectivorous mammals composing the family Talpidae; the body is stout and cylindrical, with a short neck, small or vestigial eyes and ears, a long naked muzzle, and forelimbs adapted for digging.

molectronics *See* molecular electronics.

molecular adhesion [PHYS CHEM] A particular manifestation of intermolecular forces which causes solids or liquids to adhere to each other; usually used with reference to adhesion of two different materials, in contrast to cohesion.

molecular association [PHYS CHEM] The formation of double molecules or polymolecules from a single species as a result of specific and moderately strong intermolecular forces.

molecular beam [PHYS] A beam of neutral molecules whose directions of motion lie within a very small solid angle.

molecular beam epitaxy [SOLID STATE] A technique of growing single crystals in which beams of atoms or molecules are made to strike a single-crystalline substrate in a vacuum, giving rise to crystals whose crystallographic orientation is related to that of the substrate. Abbreviated MBE.

molecular binding [SOLID STATE] The force which holds a molecule at some site on the surface of a crystal.

molecular biology [BIOL] That part of biology which attempts to interpret biological events in terms of the physicochemical properties of molecules in a cell.

molecular cloud [ASTRON] A thick, dense interstellar cloud consisting mainly of molecular hydrogen but also a high concentration of dust grains.

molecular crystal [CRYSTAL] A solid consisting of a lattice array of molecules such as hydrogen, methane, or more complex organic compounds, bound by weak van der Waals forces, and therefore retaining much of their individuality.

molecular diffusion [FL MECH] The transfer of mass between adjacent layers of fluid in laminar flow.

molecular dipole [PHYS CHEM] A molecule having an electric dipole moment, whether it is permanent or produced by an external field.

molecular distillation [CHEM] A process by which substances are distilled in high vacuum at the lowest possible temperature and with least damage to their composition.

molecular electronics [ELECTR] The branch of electronics that deals with the production of complex electronic circuits in microminiature form by producing semiconductor devices and circuit elements integrally while growing multizoned crystals in a furnace. Also known as molectronics.

molecular energy level [PHYS CHEM] One of the states of motion of nuclei and electrons in a molecule, having a definite energy, which is allowed by quantum mechanics.

molecular engineering [ELECTR] The use of solid-state techniques to build, in extremely small volumes, the components necessary to provide the functional requirements of overall equipments, which when handled in more conventional ways are vastly bulkier.

molecular exclusion chromatography *See* gel filtration.

molecular field theory *See* Weiss theory.

molecular genetics [MOL BIO] The approach which deals with the physics and chemistry of the processes of inheritance.

molecular orbital [PHYS CHEM] A wave function describing an electron in a molecule.

molecular pathology [PATH] The study of the bases and mechanisms of disease on a molecular or chemical level.

molecular physics [PHYS] The study of the behavior and structure of molecules, including the quantum-mechanical explanation of several kinds of chemical binding between atoms in a molecule, directed valence, the polarizability of molecules, the quantization of vibrational, rotational, and electronic motions of molecules, and the phenomena arising from intermolecular forces.

molecular polarizability [PHYS CHEM] The electric dipole moment induced in a molecule by an external electric field, divided by the magnitude of the field.

molecular pump [MECH ENG] A vacuum pump in which the molecules of the gas to be exhausted are carried away by the friction between them and a rapidly revolving disk or drum.

molecular relaxation [PHYS CHEM] Transition of a molecule from an excited energy level to another excited level of lower energy or to the ground state.

molecular rotation [OPTICS] In a solution of an optically active compound, the specific rotation (angular rotation of polarized light) multiplied by the compound's molecular weight.

molecular sieve chromatography *See* gel filtration.

molecular specific heat *See* molar specific heat.

molecular spectrum [SPECT] The intensity of electromagnetic radiation emitted or absorbed by a collection of molecules as a function of frequency, wave number, or some related quantity.

molecular theory *See* kinetic theory.

molecular vibration [PHYS CHEM] The theory that all atoms within a molecule are in continuous motion, vibrating at definite frequencies specific to the molecular structure as a whole as well as to groups of atoms within the molecule; the basis of spectroscopic analysis.

molecular volume [CHEM] The volume that is occupied by 1 mole (gram-molecular weight) of an element or compound; equals the molecular weight divided by the density.

molecular weight [CHEM] The sum of the atomic weights of all the atoms in a molecule.

molecule [CHEM] A group of atoms held together by chemical forces; the atoms in the molecule may be identical as in H_2, S_2, and S_8, or different as in H_2O and CO_2; a molecule is the smallest unit of matter which can exist by itself and retain all its chemical properties.

mole fraction [CHEM] The ratio of the number of moles of a substance in a mixture or solution to the total number of moles of all the components in the mixture or solution.

mole percent [CHEM] Percentage calculation expressed in terms of moles rather than weight.

Molidae [VERT ZOO] A family of marine fishes, including some species of sunfishes, in the order Perciformes.

Mollicutes [MICROBIO] The mycoplasmas, a class of prokaryotic organisms lacking a true cell wall; cells are very small to submicroscopic.

Mollisol [GEOL] An order of soils having dark or very dark, friable, thick A horizons high in humus and bases such as calcium and magnesium; most have lighter-colored or browner B horizons that are less friable and about as thick as the A horizons; all but a few have paler C horizons, many of which are calcareous.

Mollusca [INV ZOO] One of the divisions of phyla of the animal kingdom containing snails, slugs, octopuses, squids, clams, mussels, and oysters; characterized by a shell-secreting organ, the mantle, and a radula, a food-rasping organ located in the forward area of the mouth.

mollusk [INV ZOO] Any member of the Mollusca.

Molossidae [VERT ZOO] The free-tailed bats, a family of tropical and subtropical insectivorous mammals in the order Chiroptera.

Molpadida [INV ZOO] An order of sea cucumbers belonging to the Apodacea and characterized by a short, plump body bearing a taillike prolongation.

Molpadidae [INV ZOO] The single family of the echinoderm order Molpadida.

molten-salt reactor [NUCLEO] A nuclear reactor in which the fissile and fertile material, in the form of fluoride salts, is dissolved in the coolant, which is a molten mixture of salts such as lithium fluoride and beryllium fluoride. Abbreviated MSR. Also known as fused-salt reactor.

molybdate [INORG CHEM] A salt derived from a molybdic acid.

molybdenite [MINERAL] MoS_2 A metallic, lead-gray mineral that crystallizes in the hexagonal system and is commonly found in scales or foliated masses; hardness is 1.5 on Mohs scale, and specific gravity is 4.7; it is chief ore of molybdenum.

molybdenum [CHEM] A chemical element, symbol Mo, atomic number 42, and atomic weight 95.95. [MET] A silvery-gray metal used in iron-based alloys.

molybdic ocher *See* molybdite.

molybdine *See* molybdite.

molybdite [MINERAL] MoO_3 A mineral, much of which is actually ferrimolybdite. Also known as molybdic ocher; molybdine.

molybdophyllite [MINERAL] $(Pb,Mg)_2SiO_4 \cdot H_2O$ A colorless, white, or pale-green mineral composed of a silicate of lead and magnesium.

molysite [MINERAL] $FeCl_3$ A brownish-red or yellow mineral composed of native ferric chloride, occurring in lava at Vesuvius.

moment [MECH] Static moment of some quantity, except in the term "moment of inertia." [STAT] The nth moment of a distribution $f(x)$ about a point x_0 is the expected value of $(x-x_0)^n$, that is, the integral of $(x-x_0)^n df(x)$, where $df(x)$ is the probability of some quantity's occurrence; the first moment is the mean of the distribution, while the variance may be found in terms of the first and second moments.

momental ellipsoid *See* inertia ellipsoid.

moment of force *See* torque.

moment of inertia [MECH] The sum of the products formed by multiplying the mass (or sometimes, the area) of each element of a figure by the square of its distance from a specified line. Also known as rotational inertia.

moment of momentum *See* angular momentum.

momentum [MECH] Also known as linear momentum; vector momentum. **1.** For a single nonrelativistic particle, the product of the mass and the velocity of a particle. **2.** For a single relativistic particle, $mv/(1 - v^2/c^2)^{1/2}$, where m is the rest-mass, v the velocity, and c the speed of light. **3.** For a system of particles, the vector sum of the momenta (as in the first or second definition) of the particles.

momentum conservation *See* conservation of momentum.

momentum wave function [QUANT MECH] A function of the momenta of a system of particles and of time which results from taking Fourier transforms, over the coordinates of all the particles, of the Schrödinger wave function; the absolute value squared is proportional to the probability that the particles will have given momenta at a given time.

Momotidae [VERT ZOO] The motmots, a family of colorful New World birds in the order Coraciiformes.

monadelphous [BOT] Having the filaments of the stamens united into one set.

Monadidae [INV ZOO] A family of flagellated protozoans in the order Kinetoplastida having two flagella of uneven length.

monadnock [GEOL] A remnant hill of resistant rock rising abruptly from the level of a peneplain; commonly represents an outcrop of rock that has withstood erosion. Also known as torso mountain.

monalbite [MINERAL] A modification of albite with monoclinic symmetry that is stable under equilibrium conditions at temperatures (about 1000°C) near the melting point.

monandrous [BOT] Having one stamen.

monatomic gas [CHEM] A gas whose molecules have only one atom; the inert gases are examples.

monaural sound [ENG ACOUS] Sound produced by a system in which one or more microphones are connected to a single transducing channel which is coupled to one or two earphones worn by the listener.

monazite [MINERAL] A yellow or brown rare-earth phosphate monoclinic mineral with appreciable substitution of thorium for rare-earths and silicon for phosphorus; the principal ore of the rare earths and of thorium. Also known as cryptolite.

Monera [BIOL] A kingdom that includes the bacteria and blue-green algae in some classification schemes.

monetite [MINERAL] $CaHPO_4$ A yellowish-white mineral consisting of an acid calcium hydrogen phosphate, occurring in crystals.

mongolism *See* Down's syndrome.

mongoose [VERT ZOO] The common name for 39 species of carnivorous mammals which are members of the family Viverridae; they are plantigrade animals and have a long slender body, short legs, nonretractile claws, and scent glands.

Monhysterida [INV ZOO] An order of aquatic nematodes in the subclass Chromadoria.

Monhysteroidea [INV ZOO] A superfamily of free-living nematodes in the order Monhysterida characterized by single or paired outstretched ovaries, circular to cryptospiral amphids, and a stoma which is usually shallow and unarmed.

Moniliaceae [MYCOL] A family of fungi in the order Moniliales; sporophores are usually lacking, but when present they are aggregated into fascicles, and hyphae and spores are hyaline or brightly colored.

Moniliales [MYCOL] An order of fungi of the Fungi Imperfecti containing many plant pathogens; asexual spores are always formed free on the surface of the material on which the organism is living, and never occur in either pycnidia or acervuli.

moniliform [BIOL] Constructed with contractions and expansions at regular alternating intervals, giving the appearance of a string of beads.

monimolimnion [HYD] The dense bottom stratum of a meromictic lake; it is stagnant and does not mix with the water above.

monimolite [MINERAL] $(Pb,Ca)_3Sb_2O_8$ Yellowish to brownish or greenish mineral composed of lead calcium antimony oxide; it may contain ferrous iron.

monitor [COMPUT SCI] To supervise a program, and check that it is operating correctly during its execution, usually by means of a diagnostic routine. [ENG] **1.** An instrument used to measure continuously or at intervals a condition that must be kept within prescribed limits, such as radioactivity at some point in a nuclear reactor, a variable quantity in an automatic process control system, the transmissions in a communication channel or bank, or the position of an aircraft in flight. **2.** To use meters or special techniques to measure such a condition. **3.** A person who watches a monitor. [MIN ENG] *See* hydraulic monitor. [VERT ZOO] Any of 27 carnivorous, voracious species of the reptilian family Varanidae characterized by a long, slender forked tongue and a dorsal covering of small, rounded scales containing pointed granules.

monitor display [COMPUT SCI] The facility of stopping the central processing unit and displaying information of main storage and internal registers; after manual intervention, normal instruction execution can be initiated.

monitoring key [ELECTR] Key which, when operated, makes it possible for an attendant or operator to listen on a telephone circuit without appreciably impairing transmission on the circuit.

monitor printer [COMMUN] A teleprinter used in a technical control facility or communications center for checking incoming teletypewriter signals. [COMPUT SCI] Input-output device, capable of receiving coded signals from the computer, which automatically operates the keyboard to print a hard copy and, when desired, to punch paper tape.

monitor routine *See* executive routine.

monitor system *See* executive system.

monkey [MIN ENG] **1.** An appliance for mechanically gripping or releasing the rope in rope haulage. **2.** An airway in an anthracite mine. [VERT ZOO] Any of several species of frugivorous and carnivorous primates which compose the families Cercopithecidae and Cebidae in the suborder Anthropoidea; the face is typically flattened and hairless, all species are pentadactyl, and the mammary glands are always in the pectoral region.

mono- [CHEM] A prefix for chemical compounds to show a single radical; for example, monoglyceride, a glycol ester on which a single acid group is attached to the glycerol group.

monobasic [CHEM] Pertaining to an acid with one displaceable hydrogen atom, such as hydrochloric acid, HCl.

Monoblepharidales [MYCOL] An order of aquatic fungi in the class Phycomycetes; distinguished by a mostly hyphal thallus and zoospores with one posterior flagellum.

Monobothrida [PALEON] An extinct order of monocyclic camerate crinoids.

monobrid circuit [ELECTR] Integrated circuit using a combination of monolithic and multichip techniques by means of which a number of monolithic circuits, or a monolithic device in combination with separate diffused or thin-film components, are interconnected in a single package.

monocarpic [BOT] Bearing fruit once and then dying.

monocarpous [BOT] Having a single ovary.

Monoceros [ASTRON] A constellation, right ascension 7 hours, declination 5° south; it has mostly faint stars.

monochlamydous [BOT] Referring to flowers having only one set of floral envelopes, that is, either a calyx or a corolla.

monochromasia [MED] Complete color blindness in which all colors appear as shades of gray. Also known as monochromatism.

monochromatic [OPTICS] Pertaining to the color of a surface which radiates light having an extremely small range of wavelengths. [PHYS] Consisting of electromagnetic radiation having an extremely small range of wavelengths, or particles having an extremely small range of energies.

monochromatic filter See birefringent filter.

monochromatic light [OPTICS] Light of one color, having wavelengths confined to an extremely narrow range.

monochromatic temperature scale [THERMO] A temperature scale based upon the amount of power radiated from a blackbody at a single wavelength.

monochromatism See monochromasia.

monochromator [SPECT] A spectrograph in which a detector is replaced by a second slit, placed in the focal plane, to isolate a particular narrow band of wavelengths for refocusing on a detector or experimental object.

monochrome [OPTICS] Having only one chromaticity.

monochrome signal [ELECTR] 1. A signal wave used for controlling luminance values in monochrome television. 2. The portion of a signal wave that has major control of the luminance values in a color television system, regardless of whether the picture is displayed in color or in monochrome. Also known as M signal.

monocline [GEOL] A stratigraphic unit that dips from the horizontal in one direction only, not as part of an anticline or syncline.

monoclinic [BOT] Having both stamens and pistils in the same flower.

monoclinic system [CRYSTAL] One of the six crystal systems characterized by a single, two-fold symmetry axis or a single symmetry plane.

monocotyledon [BOT] Any plant of the class Liliopsida; all have a single cotyledon.

Monocotyledoneae [BOT] The equivalent name for Liliopsida.

Monocyathea [PALEON] A class of extinct parazoans in the phylum Archaeocyatha containing single-walled forms.

monocyte [HISTOL] A large (about 12 micrometers), agranular leukocyte with a relatively small, eccentric, oval or kidney-shaped nucleus.

monocytic angina See infectious mononucleosis.

Monodellidae [INV ZOO] A monogeneric family of crustaceans in the order Thermosbaenacea distinguished by seven pairs of biramous pereiopods on thoracomeres 2–8, and by not having the telson united to the last pleonite.

monodelphic [VERT ZOO] 1. Having a single genital tract, in the female. 2. Having a single uterus.

monoecious [BOT] 1. Having both staminate and pistillate flowers on the same plant. 2. Having archegonia and antheridia on different branches. [ZOO] Having male and female reproductive organs in the same individual. Also known as hermaphroditic.

Monoedidae [INV ZOO] An equivalent name for Colydiidae.

Monogenea [INV ZOO] A diverse subclass of the Trematoda which are principally ectoparasites of fishes; individuals have enlarged anterior and posterior holdfasts with paired suckers anteriorly and opisthaptors posteriorly.

Monogenoidea [INV ZOO] A class of the Trematoda in some systems of classification; equivalent to the Monogenea of other systems.

monogeosyncline [GEOL] A primary geosyncline that is long, narrow, and deeply subsided; composed of the sediments of shallow water and situated along the inner margin of the borderlands.

monoglyceride [ORG CHEM] Any of the fatty-acid glycerol esters where only one acid group is attached to the glycerol group, for example, $RCOOCH_2$-$CHOHCH_2OH$; examples are glyceryl monostearate and monolaurate; used as emulsifiers in cosmetics and lubricants.

Monogonota [INV ZOO] An order of the class Rotifera, characterized by the presence of a single gonad in both males and females.

monogynous [BOT] Having only one pistil. [VERT ZOO] 1. Having only one female in a colony. 2. Consorting with only one female.

monolayer See monomolecular film.

monolithic [CIV ENG] Pertaining to concrete construction which is cast in one jointless piece. [SCI TECH] Constructed from a single crystal or other single piece of material.

monolithic filter [COMMUN] A device used to separate telephone communications sent simultaneously over the transmission line, consisting of a series of electrodes vacuum-deposited on a crystal plate so that the plated sections are resonant with ultrasonic sound waves, and the effect of the device is similar to that of an electric filter.

monolithic integrated circuit [ELECTR] An integrated circuit having elements formed in place on or within a semiconductor substrate, with at least one element being formed within the substrate.

monomer [CHEM] A simple molecule which is capable of combining with a number of like or unlike molecules to form a polymer; it is a repeating structure unit within a polymer.

Monommidae [INV ZOO] A family of coleopteran insects in the superfamily Tenebrionoidea.

monomolecular film [PHYS CHEM] A film one molecule thick. Also known as monolayer.

mononucleosis [MED] Any of various conditions marked by an abnormal increase in monocytes in the peripheral blood.

Monophisthocotylea [INV ZOO] An order of the Monogenea in which the posthaptor is without discrete multiple suckers or clamps.

Monophlebinae [INV ZOO] A subfamily of the homopteran superfamily Coccoidea distinguished by a dorsal anus.

monophonic sound [ENG ACOUS] Sound produced by a system in which one or more microphones feed a single transducing channel which is coupled to one or more loudspeakers.

Monopisthocotylea [INV ZOO] An order of trematode worms in the subclass Pectobothridia.

Monoplacophora [INV ZOO] A group of shell-bearing mollusks represented by few living forms; considered to be a sixth class of mollusks.

monoploid [GEN] 1. Having only one set of chromosomes. 2. Having the haploid number of chromosomes.

monopole *See* magnetic monopole.

monopole antenna [ELECTROMAG] An antenna, usually in the form of a vertical tube or helical whip, on which the current distribution forms a standing wave, and which acts as one part of a dipole whose other part is formed by its electrical image in the ground or in an effective ground plane. Also known as spike antenna.

monopropellant [MATER] A rocket propellant consisting of a single substance, especially a liquid, capable of creating rocket thrust without the addition of a second substance.

monopulse radar [ENG] Radar in which directional information is obtained with high precision by using a receiving antenna system having two or more partially overlapping lobes in the radiation patterns.

Monopylina [INV ZOO] A suborder of radiolarian protozoans in the order Oculosida in which pores lie at one pole of a single-layered capsule.

monopyroxene clinoaugite *See* clinopyroxene.

Monorhina [VERT ZOO] The subclass of Agnatha that includes the jawless vertebrates with a single median nostril.

monosaccharide [BIOCHEM] A carbohydrate which cannot be hydrolyzed to a simpler carbohydrate; a polyhedric alcohol having reducing properties associated with an actual or potential aldehyde or ketone group; classified on the basis of the number of carbon atoms, as triose (3C), tetrose (4C), pentose (5C), and so on.

Monosigales [BOT] A botanical order equivalent to the Choanoflagellida in some systems of classification.

monostable [ELECTR] Having only one stable state.

monostable blocking oscillator [ELECTR] A blocking oscillator in which the electron tube or other active device carries no current unless positive voltage is applied to the grid. Also known as driven blocking oscillator.

monostable circuit [ELECTR] A circuit having only one stable condition, to which it returns in a predetermined time interval after being triggered.

monostable multivibrator [ELECTR] A multivibrator with one stable state and one unstable state; a trigger signal is required to drive the unit into the unstable state, where it remains for a predetermined time before returning to the stable state. Also known as one-shot multivibrator; single-shot multivibrator; start-stop multivibrator; univibrator.

Monostylifera [INV ZOO] A suborder of the Hoplonemertini characterized by a single stylet.

monotectic alloy [MET] A metallic composite material having a dispersed phase of solidification products distributed within a matrix.

Monotomidae [INV ZOO] The equivalent name for Rhizophagidae.

monotone [SCI TECH] A quantity which never increases (or which never decreases) as a function of some other quantity. Also known as monotonic.

monotonic *See* monotone.

Monotremata [VERT ZOO] The single order of the mammalian subclass Prototheria containing unusual mammallike reptiles, or quasi-mammals.

monotrichous [MICROBIO] Of bacteria, having an individual flagellum at one pole.

Monotropaceae [BOT] A family of dicotyledonous herbs or half shrubs in the order Ericales distinguished by a small, scarcely differentiated embryo without cotyledons, lack of chlorophyll, leaves reduced to scales, and anthers opening by longitudinal slits.

monotropic [PHYS] Pertaining to an element which may exist in two or more forms, but in which one form is the stable modification at all temperatures and pressures.

monotype [BIOL] A single type of organism that constitutes a species or genus. [GRAPHICS] A printing technique in which a picture is painted on a sheet of glass or metal; the picture is transferred to a sheet of paper by pressure; additional copies require that the subject be repainted on the plate.

monovalent [CHEM] A radical or atom whose valency is 1.

monoxide [CHEM] A compound that contains a single oxygen atom, such as carbon monoxide, CO.

monozygotic twins [BIOL] Twins which develop from a single fertilized ovum. Also known as identical twins.

mons [ANAT] An eminence.

monsoon [METEOROL] A large-scale wind system which predominates or strongly influences the climate of large regions, and in which the direction of the wind flow reverses from winter to summer; an example is the wind system over the Asian continent.

Monstrilloida [INV ZOO] A suborder or order of microscopic crustaceans in the subclass Copepoda; adults lack a second antenna and mouthparts, and the digestive tract is vestigial.

montanite [MINERAL] $Bi_2O_3 \cdot TeO_3 \cdot 2H_2O$ A yellowish mineral consisting of a hydrated tellurate of bismuth; occurs in soft and earthy to compact form.

montebrasite [MINERAL] $LiAlPO_4(OH)$ A mineral composed of basic lithium aluminum phosphate; it is isomorphous with amblygonite and natromontebrasite.

Monte Carlo method [STAT] A technique which obtains a probabilistic approximation to the solution of a problem by using statistical sampling techniques.

montgomeryite [MINERAL] $Ca_2Al_2(PO_4)_3(OH) \cdot 7H_2O$ A green to colorless mineral composed of hydrous basic calcium aluminum phosphate.

month [ASTRON] 1. The period of the revolution of the moon around the earth (sidereal month). 2. The period of the phases of the moon (synodic month). 3. The month of the calendar (calendar month).

Montian [GEOL] A European stage of geologic time: Paleocene (above Danian, below Thanetian).

Monticellidae [INV ZOO] A family of tapeworms in the order Proteocephaloidea, in which some or all of the organs are in the cortical mesenchyme; catfish parasites.

monticellite [MINERAL] $CaMgSiO_4$ A colorless or gray mineral of the olivine structure type; isomorphous with kirsch steinite.

montmorillonite [MINERAL] 1. A group name for all clay minerals with an expanding structure, except vermiculite. 2. The high-alumina end member of the montmorillonite group; it is grayish, pale red, or blue and has some replacement of aluminum ion by magnesium ion. 3. Any mineral of the montmorillonite group.

monzonite [PETR] A phaneritic (visibly crystalline) plutonic rock composed chiefly of sodic plagioclase and alkali feldspar, with subordinate amounts of dark-colored minerals, intermediate between syenite and dorite.

moon [ASTRON] **1.** The natural satellite of the earth. **2.** A natural satellite of any planet.

moonstone [MINERAL] An alkali feldspar or cryptoperthite that is semitransparent to translucent and exhibits a bluish to milky-white, pearly, or opaline luster; used as a gemstone if flawless. Also known as hecatolite.

moor [ECOL] *See* bog. [ENG] Securing a ship or aircraft by attaching it to a fixed object or a mooring buoy with chains or lines, or with anchors or other devices.

moose [VERT ZOO] An even-toed ungulate of the genus *Alces* in the family Cervidae; characterized by spatulate antlers, long legs, a short tail, and a large head with prominent overhanging snout.

mor *See* ectohumus.

Moraceae [BOT] A family of dicotyledonous woody plants in the order Urticales characterized by two styles or style branches, anthers inflexed in the bud, and secretion of a milky juice.

morainal apron *See* outwash plain.

morainal plain *See* outwash plain.

moraine [GEOL] An accumulation of glacial drift deposited chiefly by direct glacial action and possessing initial constructional form independent of the floor beneath it.

morass ore *See* bog iron ore.

moravite [MINERAL] $Fe_2(N,Fe)_4Si_7O_{20}(OH)_4$ A black mineral of the chlorite group, composed of basic iron aluminum silicate, occurring as fine scales.

morbidity [MED] **1.** The quantity or state of being diseased. **2.** The conditions inducing disease. **3.** The ratio of the number of sick individuals to the total population of a community.

mordant [CHEM] An agent, such as alum, phenol, or aniline, that fixes dyes to tissues, cells, textiles, and other materials by combining with the dye to form an insoluble compound. Also known as dye mordant.

Mordellidae [INV ZOO] The tumbling flower beetles, a family of coleopteran insects in the superfamily Meloidea.

mordenite [MINERAL] $(Ca,Na_2,K_2)_4Al_8Si_{40}O_{96}\cdot28H_2O$ A zeolite mineral crystallizing in the orthorhombic system and found in minute crystals or fibrous concretions. Also known as arduinite; ashtonite; flokite; ptilolite.

morencite *See* nontronite.

morenosite [MINERAL] $NiSO_4\cdot7H_2O$ An apple-green or light-green mineral composed of hydrous nickel sulfate, occurring in crystals or fibrous crusts. Also known as nickel vitriol.

Moridae [VERT ZOO] A family of actinopterygian fishes in the order Gadiformes.

Morinae [VERT ZOO] The deep-sea cods, a subfamily of the Moridae.

morinite [MINERAL] $Na_2Ca_3Al_3H(PO_4)_4F_6\cdot8H_2O$ A mineral composed of hydrous acid phosphate of sodium, calcium, and aluminum. Also known as jezekite.

Mormyridae [VERT ZOO] A large family of electrogenic fishes belonging to the Osteoglossiformes; African river and lake fishes characterized by small eyes, a slim caudal peduncle, and approximately equal dorsal and anal fins in most.

Mormyriformes [VERT ZOO] Formerly an order of fishes which are now assigned to the Osteoglossiformes.

morning star [ASTRON] A misnomer given to a planet visible to the naked eye, when it rises before the sun.

Morphinae [INV ZOO] A subfamily of large tropical butterflies in the family Nymphalidae.

morphine [PHARM] $C_{17}H_{19}NO_3\cdot H_2O$ A white crystalline narcotic powder, melting point 254°C, an alkaloid obtained from opium; used in medicine in the form of a hydrochloride or sulfate salt.

morphogenesis [EMBRYO] The transformation involved in the growth and differentiation of cells and tissue. Also known as topogenesis.

morphology [BIOL] A branch of biology that deals with structure and form of an organism at any stage of its life history.

Morse code [COMMUN] **1.** A telegraph code for manual operating, consisting of short (dot) and long (dash) signals and various-length spaces; now used only for wire telegraphy. Also known as American Morse code. **2.** Collective term for Morse code (American Morse code) and continental code (International Morse code).

mortality rate [MED] For a given period of time, the ratio of the number of deaths occurring per 1000 population. Also known as death rate.

mortar [MATER] A mixture of cement, lime, and sand used for laying bricks or masonry. [ORD] A complete projectile-firing weapon, with rifled or smooth bore, characterized by a shorter barrel, lower velocity, shorter range, and higher angle of fire than a howitzer or a gun; most present-day mortars are muzzle-loaded and of simple construction for lightness and mobility. [SCI TECH] A bowl-shaped vessel made of hard material in which solids are crushed by hand with a pestle.

mortar structure [PETR] A cataclastic structure produced by dynamic metamorphism of crystalline rocks and characterized by a mica-free aggregate of finely crushed grains of quartz and feldspar filling the interstices between or forming borders on the edges of larger, rounded relics. Also known as cataclastic structure; murbruk structure; porphyroclastic structure.

mortise and tenon [DES ENG] A type of joint, principally used for wood, in which a hole, slot, or groove (mortise) in one member is fitted with a projection (tenon) from the second member.

mortlake *See* oxbow lake.

morula [EMBRYO] A solid mass of blastomeres formed by cleavage of the eggs of many animals; precedes the blastula or gastrula, depending on the type of egg. [INV ZOO] A cluster of immature male gametes in which differentiation occurs outside the gonad; common in certain annelids.

Moruloidea [INV ZOO] The only class of the phylum Mesozoa; embryonic development in the organisms proceeds as far as the morula or stereoblastula stage.

MOS *See* metal oxide semiconductor.

mosaic [BIOL] An organism or part made up of tissues or cells exhibiting mosaicism. [ELECTR] A light-sensitive surface used in television camera tubes, consisting of a thin mica sheet coated on one side with a large number of tiny photosensitive silver-cesium globules, insulated from each other. [EMBRYO] An egg in which the cytoplasm of early cleavage cells is of the type which determines its later fate. [PETR] **1.** Pertaining to a granoblastic texture in a rock formed by dynamic metamorphism in which the boundaries between individual grains are straight or slightly curved. Also known as cyclopean. **2.** Pertaining to a texture in a crystalline sedimentary rock in which contacts at grain boundaries are more or less regular.   [SCI TECH] A surface pattern made by the assembly and arrangement of many small pieces.

mosaicism   [GEN] The coexistence in an individual of somatic cells of genetically different types; it is caused by gene or chromosome mutations, especially nondisjunction, after fertilization, by double fertilization, or by fusion of embryos.

mosandrite   [MINERAL] A reddish-brown or yellowish-brown mineral composed of a silicate of sodium, calcium, titanium, zirconium, and cerium. Also known as khibinite; lovchorrite; rinkite; rinkolite.

moschellandsbergite   [MINERAL] Ag_2Hg_3 A silver-white mineral consisting of a silver and mercury compound; occurs in dodecahedral crystals and in massive and granular forms.

moscovite   See muscovite.

Moseley's law   [SPECT] The law that the square-root of the frequency of an x-ray spectral line belonging to a particular series is proportional to the difference between the atomic number and a constant which depends only on the series.

MOSFET   See metal oxide semiconductor field-effect transistor.

mosquito   [INV ZOO] Any member of the dipterous subfamily Culicinae; a slender fragile insect, with long legs, a long slender abdomen, and narrow wings.

moss   [BOT] Any plant of the class Bryatae, occurring in nearly all damp habitats except the ocean.

moss agate   [MINERAL] A milky or almost transparent chalcedony containing dark inclusions in a dendritic pattern.

Mössbauer effect   [NUC PHYS] The emission and absorption of gamma rays by certain nuclei, bound in crystals, without loss of energy through nuclear recoil, with the result that radiation emitted by one such nucleus can be absorbed by another.

Mössbauer spectroscopy   [SPECT] The study of Mössbauer spectra, for example, for nuclear hyperfine structure, chemical shifts, and chemical analysis.

Mössbauer spectrum   [SPECT] A plot of the absorption, by nuclei bound in a crystal lattice, of gamma rays emitted by similar nuclei in a second crystal, as a function of the relative velocity of the two crystals.

MOST   See metal oxide semiconductor field-effect transistor.

MOS transistor   See metal oxide semiconductor field-effect transistor.

most significant bit   [COMPUT SCI] The left-most bit in a word. Abbreviated msb.

most significant character   [COMPUT SCI] The character in the leftmost position in a number or word.

Motacillidae   [VERT ZOO] The pipits, a family of passeriform birds in the suborder Oscines.

moth   [INV ZOO] Any of various nocturnal or crepuscular insects belonging to the lepidopteran suborder Heteroneura; typically they differ from butterflies in having the antennae feathery and rarely clubbed, a stouter body, less brilliant coloration, and proportionately smaller wings.

mother liquor   See discharge liquor.

mother lode   [GEOL] A main unit of mineralized matter that may not have economic value but to which workable veins are related.

mother map   See base map.

mother rock   See source rock.

motion   [MECH] A continuous change of position of a body.

motion register   [COMPUT SCI] The register which controls the go/stop, forward/reverse motion of a tape drive.

motivation   [PSYCH] The comparatively spontaneous drive, force, or incentive, which partly determines the direction and strength of the response of a higher organism to a given situation; it arises out of the internal state of the organism.

motoneuron   See motor neuron.

motor   [ELEC] A machine that converts electric energy into mechanical energy by utilizing forces produced by magnetic fields on current-carrying conductors. Also known as electric motor.   [PHYSIO] 1. That which causes action or movement. 2. Pertaining to efferent nerves which innervate muscles and glands.

motorboating   [ELECTR] Undesired oscillation in an amplifying system or transducer, usually of a pulse type, occurring at a subaudio or low-audio frequency.

motor end plate   [ANAT] A specialized area beneath the sarcolemma where functional contact is made between motor nerve fibers and muscle fibers.

motor-generator set   [ELEC] A motor and one or more generators that are coupled mechanically for use in changing one power-source voltage to other desired voltages or frequencies.

motor neuron   [PHYSIO] An efferent nerve cell. Also known as motoneuron.

motor unit   [ANAT] The axon of an anterior horn cell, or the motor fiber of a cranial nerve, together with the striated muscle fibers innervated by its terminal branches.

mottle-leaf   [PL PATH] 1. A virus disease characterized by chlorotic mottling and wrinkling of leaves. 2. A zinc-deficiency disease characterized by partial chlorosis of the leaves and stunting of the plant.

mottramite   [MINERAL] $(Cu,Zn)Pb(VO_4)(OH)$ A mineral composed of a basic lead copper zinc vanadate; it is isomorphous with descloizite. Also known as cuprodescloizite; psittacinite.

moulin   [HYD] A shaft or hole in the ice of a glacier which is roughly cylindrical and nearly vertical, formed by swirling meltwater pouring down from the surface. Also known as glacial mill; glacier mill; glacier pothole; glacier well; pothole.

mound   [GEOL] 1. A low, isolated, rounded natural hill, usually of earth. Also known as tuft. 2. A structure built by fossil colonial organisms.

mount   [ELECTROMAG] The flange or other means by which a switching tube, or tube and cavity, is connected to a waveguide.   [ENG] 1. Structure supporting any apparatus, as a gun, searchlight, telescope, or surveying instrument. 2. To fasten an apparatus in position, such as a gun on its support.   [ORD] To equip; to put into operation; to go into operation, as to mount an offensive.

mountain   [GEOGR] A feature of the earth's surface that rises high above the base and has generally steep slopes and a relatively small summit area.

mountain butter   See halotrichite.

mountain chain   See mountain system.

mountain crystal   See rock crystal.

mountain glacier   See alpine glacier.

mountain lion   See puma.

mountain mahogany   See obsidian.

mountain range   [GEOGR] A succession of mountains or narrowly spaced mountain ridges closely related in position, direction, and geologic features.

mountain soap   See saponite.

mountain system   [GEOGR] A group of mountain ranges tied together by common geological features. Also known as mountain chain.

mountain tallow   See hatchettite.

mouse [VERT ZOO] Any of various rodents which are members of the families Muridae, Heteromyidae, Cricetidae, and Zapodidae; characterized by a pointed snout, short ears, and an elongated body with a long, slender, sparsely haired tail.

mouse deer See chevrotain.

mouth [ANAT] The oral or buccal cavity and its related structures. [ENG ACOUS] The end of a horn that has the larger cross-sectional area. [GEOGR] 1. The place where one body of water discharges into another. Also known as influx. 2. The entrance or exit of a geomorphic feature, such as of a cave or valley. [MIN ENG] 1. The end of a shaft, adit, drift, entry, or tunnel emerging at the surface. 2. The collar of a borehole. [SCI TECH] Something resembling a mouth, that is, a place where one thing enters another or an opening at the receiving end of a container or enclosure.

movable bridge [CIV ENG] A bridge in which either the horizontal or vertical alignment can be readily changed to permit the passage of traffic beneath it. Often called drawbridge (an anachronism).

move mode [COMPUT SCI] A method of communicating between an operating program and an input/output control system in which the data records to be read or written are actually moved into and out of program-designated memory areas; in contrast to locate mode.

moving bed [CHEM ENG] Granulated solids in a process vessel that are circulated (moved) either mechanically or by gravity flow; used in catalytic and absorption processes.

moving-boundary electrophoresis [ANALY CHEM] A U-tube variation of electrophoresis analysis that uses buffered solution so that all ions of a given species move at the same rate to maintain a sharp, moving front (boundary).

moving-coil galvanometer [ENG] Any galvanometer, such as the d'Arsonval galvanometer, in which the current to be measured is sent through a coil suspended or pivoted in a fixed magnetic field, and the current is determined by measuring the resulting motion of the coil.

moving-coil instrument [ELEC] Any instrument in which current is sent through one or more coils suspended or pivoted in a magnetic field, and the motion of the coils is used to measure either the current in the coils or the strength of the field.

moving-coil loudspeaker See dynamic loudspeaker.

moving-coil microphone See dynamic microphone.

moving-coil pickup See dynamic pickup.

moving-coil wattmeter See electrodynamic wattmeter.

moving-head disk [COMPUT SCI] A disk-storage device in which one or more read-write heads are attached to a movable arm which allows each head to cover many tracks of information.

moving-target indicator [ELECTR] A device that limits the display of radar information primarily to moving targets; signals due to reflections from stationary objects are canceled by a memory circuit. Abbreviated MTI.

Mozambique Current [OCEANOGR] The portion of the South Equatorial Current that turns and flows along the coast of Africa in the Mozambique Channel, forming one of the western boundary currents in the Indian Ocean.

mp See mean effective pressure; melting point.

m-RNA See messenger ribonucleic acid.

msb See most significant bit.

M scan See M scope.

M scope [ELECTR] A modified form of A scope on which part of the time base is slightly displaced in a vertical direction by insertion of an adjustable step which serves as a range marker. Also known as M indicator; M scan.

MSH See melanocyte-stimulating hormone.

M shell [ATOM PHYS] The third layer of electrons about the nucleus of an atom, having electrons characterized by the principal quantum number 3.

M signal See monochrome signal.

MSR See molten salt reactor.

M star [ASTRON] A spectral classification for a star whose spectrum is characterized by the presence of titanium oxide bands; M stars have surface temperatures of 3000 K for giants and 3400 K for dwarfs.

MT See megaton.

MTBF See mean time between failures.

MTI See moving-target indicator.

M-type backward-wave oscillator [ELECTR] A backward-wave oscillator in which focusing and interaction are through magnetic fields, as in a magnetron. Also known as M-type carcinotron; type-M carcinotron.

M-type carcinotron See M-type backward-wave oscillator.

muc-, muci-, muco- [ZOO] A combining form denoting pertaining to mucus, mucin, mucosa.

Mucedinaceae [MYCOL] The equivalent name for Moniliaceae.

mucking [ENG] Clearing and loading broken rock and other excavated materials, as in tunnels or mines.

mucoid [BIOCHEM] Any of various glycoproteins, similar to mucins but differing in solubilities and precipitation properties and found in cartilage, in the crystalline lens, and in white of egg. 2. Resembling mucus. [MICROBIO] Pertaining to large colonies of bacteria characterized by being moist and sticky.

mucopolysaccharide [BIOCHEM] Any of a group of polysaccharides containing an amino sugar and uronic acid; a constituent of mucoproteins, glycoproteins, and blood-group substances.

mucoprotein [BIOCHEM] Any of a group of glycoproteins containing a sugar, usually chondroitinsulfuric or mucoitinsulfuric acid, combined with amino acids or polypeptides.

Mucorales [MYCOL] An order of terrestrial fungi in the class Phycomycetes, characterized by a hyphal thallus and nonmotile sporangiospores, or conidiospores.

mucous [PHYSIO] Of or pertaining to mucus; secreting mucus.

mucous colitis See irritable colon.

mucous membrane [HISTOL] The type of membrane lining cavities and canals which have communication with air; it is kept moist by glandular secretions. Also known as tunica mucosa.

mucoviscidosis See cystic fibrosis.

mucus [PHYSIO] A viscid fluid secreted by mucous glands, consisting of mucin, water, inorganic salts, epithelial cells, and leukocytes, held in suspension.

mud [ENG] See slime. [GEOL] An indurated mixture of clay and silt with water; it is slimy with a consistency varying from that of a semifluid to that of a soft and plastic sediment. [PETR] The silt plus clay portion of a sedimentary rock.

mud cone [GEOL] A cone of sulfurous mud built around the opening of a mud volcano or mud geyser, with slopes as steep as 40° and diameters ranging upward to several hundred yards. Also known as puff cone.

mud crack [GEOL] An irregular fracture formed by shrinkage of clay, silt, or mud under the drying effects of atmospheric conditions at the surface. Also known as desiccation crack; sun crack.

mudfish *See* bowfin.

mud flat [GEOL] A relatively level, sandy or muddy coastal strip along a shore or around an island; may be alternately covered and uncovered by the tide or may be covered by shallow water. Also known as flat.

mudflow [GEOL] A flowing mass of fine-grained earth material having a high degree of fluidity during movement.

mudslide [GEOL] A slow-moving mudflow in which movement is mainly by sliding upon a discrete boundary shear surface.

mud volcano [GEOL] A conical accumulation of variable admixtures of sand and rock fragments, the whole resulting from eruption of wet mud and impelled upward by fluid or gas pressure. Also known as hervidero; macaluba.

MUF *See* maximum usable frequency.

muffler [ENG] A device to deaden the noise produced by escaping gases or vapors.

Mugilidae [VERT ZOO] The mullets, a family of perciform fishes in the suborder Mugiloidei.

Mugiloidei [VERT ZOO] A suborder of fishes in the order Perciformes; individuals are rather elongate, terete fishes with a short spinous dorsal fin that is well separated from the soft dorsal fin.

mulberry [BOT] Any of various trees of the genus *Morus* (family Moraceae), characterized by milky sap and simple, often lobed alternate leaves.

mulch [MATER] A mixture of organic material, such as straw, peat moss, or leaves, that is spread over soil to prevent evaporation, maintain an even soil temperature, prevent erosion, control weeds, and enrich soil.

mule [MIN ENG] *See* barney. [VERT ZOO] The sterile hybrid offspring of the male ass and the mare, or female horse.

mull [ENG] To mix thoroughly or grind. [GEOGR] *See* headland. [GEOL] Granular forest humus that is incorporated with mineral matter. [TEXT] A thin, sheer cotton or cotton and polyester fabric.

Müller's glass *See* hyalite.

mullion [BUILD] A vertical bar separating two windows in a multiple window. [GEOL] In folded sedimentary and metamorphic rocks, a columnar structure in which the rock columns seem to intersect.

mullite [MINERAL] $Al_6Si_2O_{13}$ An orthorhombic mineral consisting of an aluminum silicate that is resistant to corrosion and heat; used as a refractory. Also known as porcelainite.

multiaccess computer [COMPUT SCI] A computer system in which computational and data resources are made available simultaneously to a number of users who access the system through terminal devices, normally on an interactive or conversational basis.

multiaddress [COMPUT SCI] Referring to an instruction that has more than one address part.

multiaspect [COMPUT SCI] Pertaining to searches or systems which permit more than one aspect, or facet, of information to be used in combination, one with the other to effect identifying or selecting operations.

multicellular horn [ELECTROMAG] A cluster of horn antennas having mouths that lie in a common surface and that are fed from openings spaced one wavelength apart in one face of a common waveguide.

[ENG ACOUS] A combination of individual horn loudspeakers having individual driver units or joined in groups to a common driver unit. Also known as cellular horn.

multichannel communication [COMMUN] Communication in which there are two or more communication channels over the same path, such as a communication cable, or a radio transmitter which can broadcast on two different frequencies, either individually or simultaneously.

multicomputer system [COMPUT SCI] A system consisting of more than one computer, usually under the supervision of a master computer, in which smaller computers handle input/output and routine jobs while the large computer carries out the more complex computations.

multicycle feeding *See* multiread feeding.

multidimensional derivative [MATH] The generalized derivative of a function of several variables which is usually represented as a matrix involving the various partial derivatives of the function.

multilevel address *See* indirect address.

multilevel indirect addressing [COMPUT SCI] A programming device whereby the address retrieved in the memory word may itself be an indirect address that points to another memory location, which in turn may be another indirect address, and so forth.

multilevel transmission [COMMUN] Transmission of digital information in which three or more levels of voltage are recognized as meaningful, as 0,1,2 instead of simply 0,1.

multilinear algebra [MATH] The study of functions of several variables which are linear relative to each variable.

Multillidae [INV ZOO] An economically important family of Hymenoptera; includes the cow killer, a parasite of bumblebee pupae.

multilocular [BIOL] Having many small chambers or vesicles.

multinomial [MATH] An algebraic expression which involves the sum of at least two terms.

multinomial trials [STAT] Unrelated trials with more than two possible outcomes the probabilities of which do not change from trial to trial.

multipass sort [COMPUT SCI] Computer program designed to sort more data than can be contained within the internal storage of a computer; intermediate storage, such as disk, tape, or drum, is required.

multipath *See* multipath transmission.

multipath transmission [ELECTROMAG] The propagation phenomenon that results in signals reaching a radio receiving antenna by two or more paths, causing distortion in radio and ghost images in television. Also known as multipath.

multiple [ELEC] 1. Group of terminals arranged to make a circuit or group of circuits accessible at a number of points at any one of which connection can be made. 2. To connect in parallel. 3. *See* parallel. [MET] A piece of stock cut from bar for use in a forging which provides the exact length needed for a single workpiece.

multiple-access computer [COMPUT SCI] A computer system whose facilities can be made available to a number of users at essentially the same time, normally through terminals, which are often physically far removed from the central computer and which typically communicate with it over telephone lines.

multiple accumulating registers [COMPUT SCI] Special registers capable of handling factors larger than one computer word in length.

multiple-address code [COMPUT SCI] A computer instruction code in which more than one address or storage location is specified; the instruction may give the locations of the operands, the destination of the result, and the location of the next instruction.

multiple-address computer [COMPUT SCI] A computer whose instruction contains more than one address, for example, an operation code and three addresses A, B, C, such that the content of A is multiplied by the content of B and the product stored in location C.

multiple-address instruction [COMPUT SCI] An instruction which has more than one address in a computer; the addresses give locations of other instructions, or of data or instructions that are to be operated upon.

multiple-contact switch See selector switch.

multiple decay See branching.

multiple discharge See composite flash.

multiple disintegration See branching.

multiple-entry system [MIN ENG] A system of access or development openings generally in bituminous coal mines involving more than one pair of parallel entries, one for haulage and fresh-air intake and the other for return air.

multiple fault See step fault.

multiple-function chip See large-scale integrated circuit.

multiple independently targeted reentry vehicle [ORD] A type of intercontinental ballistics missile which carries several nuclear warheads; before the missile reenters the atmosphere, the warheads separate and follow different trajectories to various targets. Abbreviated MIRV.

multiple integral [MATH] An integral over a subset of n-dimensional space.

multiple-length arithmetic [COMPUT SCI] Arithmetic performed by a computer in which two or more machine words are used to represent each number in the calculations, usually to achieve higher precision in the result.

multiple-loop system [CONT SYS] A system whose block diagram has at least two closed paths, along each of which all arrows point in the same direction.

multiple modulation [COMMUN] A succession of modulating processes in which the modulated wave from one process becomes the modulating wave for the next. Also known as compound modulation.

multiple module access [COMPUT SCI] Device which establishes priorities in storage access in a multiple computer environment.

multiple precision arithmetic [COMPUT SCI] Method of increasing the precision of a result by increasing the length of the number to encompass two or more computer words in length.

multiple reflection [GEOPHYS] A seismic wave which has more than one reflection. Also known as repeated reflection; secondary reflection. [OPTICS] Reflection of light back and forth several times between a pair of strongly reflecting surfaces.

multiple root [MATH] A polynomial $f(x)$ has c as a multiple root if $(x - c)^n$ is a factor for some $n > 1$. Also known as repeated root.

multiple sclerosis [MED] A degenerative disease of the nervous system of unknown cause in which there is demyelination followed by gliosis.

multiple-stage rocket See multistage rocket.

multiplet [QUANT MECH] A collection of relatively closely spaced energy levels which result from the splitting of a single energy level by an interaction which is relatively weak; examples are spin-orbit multiplets and isospin multiplets. [SPECT] A collection of relatively closely spaced spectral lines resulting from transitions to or from the members of a multiplet (as in the quantum-mechanics definition).

multiplexer [ELECTR] A device for combining two or more signals, as for multiplex, or for creating the composite color video signal from its components in color television. Also spelled multiplexor.

multiplex mode [COMPUT SCI] The utilization of differences in operating speeds between a computer and transmission lines; the multiplexor channel scans each line in sequence, and any transmitted pulse on a line is assembled in an area reserved for this line; consequently, a number of users can be handled by the computer simultaneously. Also known as multiplexor channel operation.

multiplex operation [COMMUN] Simultaneous transmission of two or more messages in either or both directions over a carrier channel.

multiplexor See multiplexer.

multiplexor channel operation See multiplex mode.

multiplex transmission [COMMUN] The simultaneous transmission of two or more programs or signals over a single radio-frequency channel, such as by time division, frequency division, or phase division.

multiplication [ELECTR] An increase in current flow through a semiconductor because of increased carrier activity. [MATH] Any algebraic operation analogous to multiplication of real numbers. [NUCLEO] The ratio of neutron flux in a subcritical reactor to that supplied by a neutron source; it is the factor by which, in effect, the reactor multiplies the source strength.

multiplication table [COMPUT SCI] In certain computers, a part of memory holding a table of numbers in which the computer looks up values in order to perform the multiplication operation.

multiplication time [COMPUT SCI] The time required for a computer to perform a multiplication; for a binary number it will be equal to the total of all the addition times and all the shift times involved in the multiplication.

multiplicity [MATH] A root of a polynomial $f(x)$ has multiplicity n if $(x - a)^n$ is a factor of $f(x)$ and n is the largest possible integer for which this is true. [PHYS] In a system having Russell-Saunders coupling, the quantity $2S + 1$, where S is the total spin quantum number.

multiplier [ELEC] A resistor used in series with a voltmeter to increase the voltage range. Also known as multiplier resistor. [ELECTR] **1.** A device that has two or more inputs and an output that is a representation of the product of the quantities represented by the input signals; voltages are the quantities commonly multiplied. **2.** See electron multiplier. **3.** See frequency multiplier. [MATH] If a number x is to be multiplied by a number y, then y is called the multiplier.

multiplier phototube [ELECTR] A phototube with one or more dynodes between its photocathode and the output electrode; the electron stream from the photocathode is reflected off each dynode in turn, with secondary emission adding electrons to the stream at each reflection. Also known as electron-multiplier phototube; photoelectric electron-multiplier tube; photomultiplier; photomultiplier tube.

multiplier resistor See multiplier.

multiplier tube [ELECTR] Vacuum tube using secondary emission from a number of electrodes in sequence to obtain increased output current; the electron stream is reflected, in turn, from one electrode of the multiplier to the next.

multipolar [ELECTROMAG] Having more than one pair of magnetic poles.

multipole [ELECTROMAG] One of a series of types of static or oscillating distributions of charge or magnetization; namely, an electric multipole or a magnetic multipole.

multipole radiation [PHYS] **1.** Electromagnetic radiation which has characteristics equivalent to those of radiation generated by an oscillating electric or magnetic multipole, and is made up of photons of well-defined angular momentum and parity. **2.** Internal conversion electrons, or positron-electron pairs having similar characteristics, emitted from an atom when the nucleus makes a transition between two energy states.

multiport memory [COMPUT SCI] A memory shared by many processors to communicate among themselves.

multiprecision arithmetic [COMPUT SCI] A form of arithmetic similar to double precision arithmetic except that two or more words may be used to represent each number.

multiprocessing [COMPUT SCI] Carrying out of two or more sequences of instructions at the same time in a computer.

multiprocessing system See multiprocessor.

multiprocessor [COMPUT SCI] A data-processing system that can carry out more than one program, or more than one arithmetic operation, at the same time. Also known as multiprocessing system.

multiprogramming [COMPUT SCI] The interleaved execution of two or more programs by a computer, in which the central processing unit executes a few instructions from each program in succession.

multiread feeding [COMPUT SCI] A system of reading punched cards in which the card passes a sensing station several times and successive fields of the card are read on consecutive machine cycles, enabling several lines to be printed from a single card. Also known as multicycle feeding.

multistable circuit [ELECTR] A circuit having two or more stable operating conditions.

multistage amplifier See cascade amplifier.

multistage compressor [MECH ENG] A machine for compressing a gaseous fluid in a sequence of stages, with or without intercooling between stages.

multistage pump [MECH ENG] A pump in which the head is developed by multiple impellers operating in series.

multistage rocket [AERO ENG] A vehicle having two or more rocket units, each unit firing after the one in back of it has exhausted its propellant; normally, each unit, or stage, is jettisoned after completing its firing. Also known as multiple-stage rocket; step rocket.

multistatic radar [ENG] Radar in which successive antenna lobes are sequentially engaged to provide a tracking capability without physical movement of the antenna.

multistrip coupler [ELECTR] A series of parallel metallic strips placed on a surface acoustic wave filter between identical apodized interdigital transducers; it converts the spatially nonuniform surface acoustic wave generated by one transducer into a spatially uniform wave received at the other transducer, and helps to reject spurious bulk acoustic modes.

multitask operation [COMPUT SCI] A sophisticated form of multijob operation in a computer which allows a single copy of a program module to be used for more than one task.

multitrack operation [COMPUT SCI] The selection of the next read/write head in a cylinder, usually indicated by bit zero of the operation code in the channel command word.

Multituberculata [PALEON] The single order of the nominally mammalian suborder Allotheria; multituberculates had enlarged incisors, the coracoid bones were fused to the scapula, and the lower jaw consisted of the dentary bone alone.

multiuser system [COMPUT SCI] A computer system with multiple terminals, enabling several users, each at their own terminal, to use the computer.

multivalent See polyvalent.

multivariate analysis [STAT] The study of random variables which are multidimensional.

multivibrator [ELECTR] A relaxation oscillator using two tubes, transistors, or other electron devices, with the output of each coupled to the input of the other through resistance-capacitance elements or other elements to obtain in-phase feedback voltage.

mu meson See muon.

mundic See pyrite.

Munsell chroma See chroma.

muon [PARTIC PHYS] Collective name for two semistable elementary particles with positive and negative charge, designated μ^+ and μ^- respectively, which are leptons and have a spin of $1/2$ and a mass of approximately 105.7 MeV. Also known as mu meson.

muonic atom [PARTIC PHYS] An atom in which an electron is replaced by a negatively charged muon orbiting close to or within the nucleus.

muonium [PARTIC PHYS] An atom consisting of an electron bound to a positively charged muon by their mutual Coulomb attraction, just as an electron is bound to a proton in the hydrogen atom.

muramidase [BIOCHEM] Lysozyme when acting as an enzyme on the hydrolysis of the muramic acid-containing mucopeptide in the cell walls of some bacteria.

murbruk structure See mortar structure.

Murchisoniacea [PALEON] An extinct superfamily of gastropod mollusks in the order Prosobranchia.

Muricacea [INV ZOO] A superfamily of gastropod mollusks in the order Prosobranchia.

Muricidae [INV ZOO] A family of predatory gastropod mollusks in the order Neogastropoda; contains the rock snails.

Muridae [VERT ZOO] A large diverse family of relatively small cosmopolitan rodents; distinguished from closely related forms by the absence of cheek pouches.

muriform [BIOL] **1.** Resembling the arrangement of courses in a brick wall, especially having both horizontal and vertical septa. **2.** Pertaining to or resembling a rat or mouse.

Murinae [VERT ZOO] A subfamily of the Muridae which contains such forms as the striped mouse, house mouse, harvest mouse, and field mouse.

Murray loop test [ELEC] A method of localizing a fault in a cable by replacing two arms of a Wheatstone bridge with a loop formed by the cable under test and a good cable connected to the far end of the defective cable.

Musaceae [BOT] A family of monocotyledonous plants in the order Zingiberales characterized by five functional stamens, unisexual flowers, spirally arranged leaves and bracts, and fleshy, indehiscent fruit.

Musca [ASTRON] A southern constellation, right ascension 12 hours, declination 70°S. Also known as Fly.

Muschelkalk [GEOL] A European stage of geologic time equivalent to the Middle Triassic, above Bunter and below Keuper.

Musci *See* Bryopsida.

Muscicapidae [VERT ZOO] A family of passeriform birds assigned to the Oscines; includes the Old World flycatchers or fantails.

Muscidae [INV ZOO] A family of myodarian cyclorrhaphous dipteran insects in the subsection Calyptratae; includes the houseflies, stable flies, and allies.

muscle [ANAT] A contractile organ composed of muscle tissue that changes in length and effects movement when stimulated. [HISTOL] A tissue composed of cells containing contractile fibers; three types are smooth, cardiac, and skeletal.

muscle hemoglobin *See* myoglobin.

muscovite [MINERAL] $KAl_2(AlSi_3)O_{10}(OH)_2$ One of the mica group of minerals, occurring in some granites and abundant in pegmatities; it is colorless, whitish, or pale brown, and the crystals are tabular sheets with prominent base and hexagonal or rhomboid outline; hardness is 2–2.5 on Mohs scale, and specific gravity is 2.7–3.1. Also known as common mica; mirror stone; moscovite; Muscovy glass; potash mica; white mica.

Muscovy glass *See* muscovite.

muscul-, musculo- [ZOO] A combining form denoting muscle, muscular.

mushroom [MYCOL] 1. A fungus belonging to the basidiomycetous order Agaricales. 2. The fruiting body (basidiocarp) of such a fungus.

mushroom anchor [NAV ARCH] A large anchor having the approximate shape of a mushroom, capable of grasping the ground whichever way it falls; used mainly where the bottom is sandy or muddy for permanent moorings of light ships and moorings of modern submarines.

musical quality *See* timbre.

music wire [MET] High-quality, high-carbon steel wire used for making mechanical springs.

Musidoridae [INV ZOO] A family of orthorrhaphous dipteran insects in the series Brachycera distinguished by spear-shaped wings.

musk [PHYSIO] Any of various strong-smelling substances obtained from the musk glands of musk deer or similar animals; used in the form of a tincture as a fixative for perfume.

muskeg [ECOL] A peat bog or tussock meadow, with variably woody vegetation.

muskmelon [BOT] *Cucumis melo.* The edible, fleshy, globular to long-tapered fruit of a trailing annual plant of the order Violales; surface is uniform to broadly sutured to wrinkled, and smooth to heavily netted, and flesh is pale green to orange; varieties include cantaloupe, Honey Dew, Casaba, and Persian melons.

musk-ox [VERT ZOO] *Ovibos moschatus.* An eventoed ungulate which is a member of the mammalian family Bovidae; a heavy-set animal with a shag pilage, splayed feet, and flattened horns set low on the head.

muskrat [VERT ZOO] *Ondatra zibethica.* The largest member of the rodent subfamily Microtinae; essentially a water rat with a laterally flattened, long, naked tail, a broad blunt head with short ears, and short limbs.

Musophagidae [VERT ZOO] The turacos, an African family of birds of uncertain affinities usually included in the order Cuculiformes; resemble the cuckoos anatomically but have two unique pigments, turacin and turacoverdin.

mustard [BOT] Any of several annual crucifers belonging to the genus *Brassica* of the order Capparales; leaves are lyrately lobed, flowers are yellow, and pods have linear beaks; the mustards are cultivated for their pungent seed and edible foliage, and the seeds of *B. niger* are used as a condiment, prepared as a powder, paste, or oil.

Mustilidae [VERT ZOO] A large, diverse family of low-slung, long-bodied carnivorous mammals including minks, weasels, and badgers; distinguished by having only one molar in each upper jaw, and two at the most in the lower jaw.

mutagen [GEN] An agent that raises the frequency of mutation above the spontaneous rate.

mutant [GEN] An individual bearing an allele that has undergone mutation and is expressed in the phenotype.

mutarotation [CHEM] A change in the optical rotation of light that takes place in the solutions of freshly prepared sugars.

mutase [BIOCHEM] An enzyme able to catalyze a dismutation or a molecular rearrangement.

mutation [GEN] An abrupt change in the genotype of an organism, not resulting from recombination; genetic material may undergo qualitative or quantitative alteration, or rearrangement.

muthmannite [MINERAL] $(Ag,Au)Te$ A bright brass yellow mineral consisting of silver-gold telluride; occurs as tabular crystals.

Mutillidae [INV ZOO] The velvet ants, a family of hymenopteran insects in the superfamily Scolioidea.

muting circuit [ELECTR] 1. Circuit which cuts off the output of a receiver when no radio-frequency carrier greater than a predetermined intensity is reaching the first detector. 2. Circuit for making a receiver insensitive during operation of its associated transmitter.

muting switch [ELEC] 1. A switch used in connection with automatic tuning systems to silence the receiver while tuning from one station to another. 2. A switch used to ground the output of a phonograph pickup automatically while a record changer is in its change cycle.

mutual branch *See* common branch.

mutual capacitance [ELEC] The accumulation of charge on the surfaces of conductors of each of two circuits per unit of potential difference between the circuits.

mutual conductance *See* transconductance.

mutual impedance [ELEC] For two meshes of a network carrying alternating current, the ratio of the complex voltage in one mesh to the complex current in the other, when all meshes besides the latter one carry no current.

mutual inductance [ELECTROMAG] Property of two neighboring circuits, equal to the ratio of the electromotive force induced in one circuit to the rate of change of current in the other circuit.

mutualism [ECOL] Mutual interactions between two species that are beneficial to both species.

Mx *See* maxwell.

myasthenia gravis [MED] A muscle disorder of unknown etiology characterized by varying degrees of weakness and excessive fatigability of voluntary muscle.

myatonia [MED] Lack of muscle tone.

Mycelia Sterilia [MYCOL] An order of fungi of the class Fungi Imperfecti distinguished by the lack of spores; certain members are plant pathogens.

mycelium [BIO] A mass of filaments, or hyphae, composing the vegetative body of many fungi and some bacteria.

Mycetaeidae [INV ZOO] The equivalent name for Endomychidae.

Mycetophagidae [INV ZOO] The hairy fungus beetles, a cosmopolitan family of coleopteran insects in the superfamily Cucujoidea.

Mycetozoa [BIOL] A zoological designation for organisms that exhibit both plant and animal characters during their life history (Myxomycetes); equivalent to the botanical Myxomycophyta.

Mycetozoia [INV ZOO] A subclass of the protozoan class Rhizopodea.

Mycobacteriaceae [MICROBIO] A family of bacteria in the order Actinomycetales; acid-fast, aerobic rods form a filamentous or myceliumlike growth.

mycology [BOT] The branch of botany that deals with the study of fungi.

mycophagous [ZOO] Feeding on fungi.

Mycophiformes [VERT ZOO] An equivalent name for Salmoniformes.

Mycoplasmataceae [MICROBIO] A family of the order Mycoplasmatales; distinguished by sterol requirement for growth.

Mycoplasmatales [MICROBIO] The single order of the class Mollicutes; organisms are gram-negative, generally nonmotile, nonsporing bacteria which lack a true cell wall.

mycorrhiza [BOT] A mutual association in which the mycelium of a fungus invades the roots of a seed plant.

mycosis [MED] An infection with or a disease caused by a fungus.

Mycota [MYCOL] An equivalent name for Eumycetes.

Myctophidae [VERT ZOO] The lantern fishes, a family of deep-sea forms of the suborder Myctophoidei.

Myctophoidei [VERT ZOO] A large suborder of marine salmoniform fishes characterized by having the upper jaw bordered only by premaxillae, and lacking a mesocoracoid arch in the pectoral girdle.

Mydaidae [INV ZOO] The mydas flies, a family of orthorrhaphous dipteran insects in the series Brachycera.

myel-, myelo- [ANAT] A combining form for bone marrow, spinal cord.

myelin [BIOCHEM] A soft, white fatty substance that forms a sheath around certain nerve fibers.

myelitis [MED] 1. Inflammation of the spinal cord. 2. Inflammation of the bone marrow.

myelography [MED] Roentgenographic visualization of the subarachnoid space, after the injection of air or an opaque medium.

myeloid [ANAT] 1. Of or pertaining to bone marrow. 2. Of or pertaining to the spinal cord.

myeloma [MED] A primary tumor of the bone marrow composed of any of the bone marrow cell types.

Mygalomorphae [INV ZOO] A suborder of spiders (Araneida) including American tarantulas, trap-door spiders, and purse-web spiders; the tarantulas may attain a leg span of 10 inches (25 centimeters).

myiasis [MED] Infestation of vertebrates by the larvae, or maggots, of flies.

Mylabridae [INV ZOO] The equivalent name for Bruchidae.

Myliogatidae [VERT ZOO] The eagle rays, a family of batoids which may reach a length of 15 feet (4.6 meters).

mylonite [PETR] A hard, coherent, often glassy-looking rock that has suffered extreme mechanical deformation and granulation but has remained chemically unaltered; appearance is flinty, banded, or streaked, but the nature of the parent rock is easily recognized.

Mymaridae [INV ZOO] The fairy flies, a family of hymenopteran insects in the superfamily Chalcidoidea.

myocardium [HISTOL] The muscular tissue of the heart wall.

myoclonus [MED] 1. Clonic muscle spasm. 2. Any disorder characterized by scattered, irregular, arrhythmic muscle spasms.

Myodaria [INV ZOO] A section of the Schizophora series of cyclorrhaphous dipterans; in this group adult antennae consist of three segments, and all families except the Conopidae have the second cubitus and the second anal veins united for almost their entire length.

Myodocopa [INV ZOO] A suborder of the order Myodocopida; includes exclusively marine ostracods distinguished by possession of a heart.

Myodocopida [INV ZOO] An order of the subclass Ostracoda.

Myodopina [INV ZOO] The equivalent name for Myodocopa.

myofibril [CYTOL] A contractile fibril in a muscle cell. [INV ZOO] See myoneme.

myofilament [CYTOL] The structural unit of muscle proteins in a muscle cell.

myoglobin [BIOCHEM] A hemoglobinlike iron-containing protein pigment occurring in muscle fibers. Also known as muscle hemoglobin; myohemoglobin.

myohemoglobin See myoglobin.

myoma [MED] 1. A benign uterine tumor composed principally of smooth muscle cells. 2. Any neoplasm originating in muscle.

Myomorpha [VERT ZOO] A suborder of rodents recognized in some systems of classification.

myoneme [INV ZOO] A contractile fibril in a protozoan. Also known as myofibril.

myoneural junction [ANAT] The point of junction of a motor nerve with the muscle which it innervates. Also known as neuromuscular junction.

myopathia See myopathy.

myopathy [MED] Any disease of the muscles. Also known as myopathia.

myopia [MED] A condition in which the focal image is formed in front of the retina of the eye. Also known as nearsightedness.

Myopsida [INV ZOO] A natural assemblage of cephalopod mollusks considered as a suborder in the order Teuthoida according to some systems of classification, and a group of the Decapoda according to other systems; the eye is covered by the skin of the head in all species.

myosin [BIOCHEM] A muscle protein, comprising up to 50% of the total muscle proteins; combines with actin to form actomycin.

myostatic reflex See stretch reflex.

myotonia [MED] Tonic muscular spasm occurring after injury or infection.

Myriangiales [MYCOL] An order of parasitic fungi of the class Ascomycetes which produce asci at various levels in uniascal locules within stromata.

Myriapoda [INV ZOO] Informal designation for those mandibulate arthropods having two body tagmata, one pair of antennae, and more than three pairs of adult ambulatory appendages.

Myricaceae [BOT] The single family of the plant order Myricales.

Myricales [BOT] An order of dicotyledonous plants in the subclass Hamamelidae, marked by its simple, resinous-dotted, aromatic leaves, and a unilocular ovary with two styles and a single ovule.

Myrientomata [INV ZOO] The equivalent name for the Protura.

Myriotrochidae [INV ZOO] A family of holothurian echinoderms in the order Apodida, distinguished by eight or more spokes in each wheel-shaped spicule.

Myrmecophagidae [VERT ZOO] A small family of arboreal anteaters in the order Edentata.

myrmecophile [ECOL] An organism, usually a beetle, that habitually inhabits the nest of ants.

myrmekitic [PETR] 1. Pertaining to the texture of an igneous rock marked by intergrowths of feldspar and vermicular quartz. 2. Having characteristic properties of myrmekite.

Myrmeleontidae [INV ZOO] The ant lions, a family of insects in the order Neuroptera; larvae are commonly known as doodlebugs.

Myrmicinae [INV ZOO] A large diverse subfamily of ants (Formicidae); some members are inquilines and have no worker caste.

myrrh [MATER] A gum resin of species of myrrh (*Commiphora*); partially soluble in water, alcohol, and ether; used in dentifrices, perfumery, and pharmaceuticals.

Myrsinaceae [BOT] A family of mostly woody dicotyledonous plants in the order Primulales characterized by flowers without staminodes, a schizogenous secretory system, and gland-dotted leaves.

Myrtaceae [BOT] A family of dicotyledonous plants in the order Myrtales characterized by an inferior ovary, numerous stamens, anthers usually opening by slits, and fruit in the form of a berry, drupe, or capsule.

Myrtales [BOT] An order of dicotyledonous plants in the subclass Rosidae characterized by opposite, simple, entire leaves and perigynous to epigynous flowers with a compound pistil.

Mysida [INV ZOO] A suborder of the crustacean order Mysidacea characterized by fusion of the sixth and seventh abdominal somites in the adult, lack of gills, and other specializations.

Mysidacea [INV ZOO] An order of free-swimming Crustacea included in the division Pericarida; adult consists of 19 somites, each bearing one pair of functionally modified, biramous appendages, and the carapace envelops most of the thorax and is fused dorsally with up to four of the anterior thoracic segments.

Mystacinidae [VERT ZOO] A monospecific family of insectivorous bats (Chiroptera) containing the New Zealand short-tailed bat; hindlegs and body are stout, and fur is thick.

Mystacocarida [INV ZOO] An order of primitive Crustacea; the body is wormlike and the cephalothorax bears first and second antennae, mandibles, and first and second maxillae.

Mysticeti [VERT ZOO] The whalebone whales, a suborder of the mammalian order Cetacea, distinguished by horny filter plates of suspended from the upper jaws.

Mytilacea [INV ZOO] A suborder of bivalve mollusks in the order Filibranchia.

Mytilidae [INV ZOO] A family of mussels in the bivalve order Anisomyaria.

myx-, myxo- [ZOO] A combining form denoting mucus, mucous, mucin, mucinous.

myxedema [MED] A condition caused by hypothyroidism characterized by a subnormal basal metabolic rate, dry coarse hair, loss of hair, mental dullness, anemia, and slowed reflexes.

Myxicolinae [INV ZOO] A subfamily of sedentary polychaete annelids in the family Sabellidae.

Myxiniformes [VERT ZOO] The equivalent name for the Myxinoidea.

Myxinoidea [VERT ZOO] The hagfishes, an order of eellike, jawless vertebrates (Agnatha) distinguished by having the nasal opening at the tip of the snout and leading to the pharynx, with barbels around the mouth and 6–15 pairs of gill pouches.

Myxobacterales [MICROBIO] An order of gliding bacteria; unicellular, gram-negative rods embedded in a layer of slime and capable of gliding movement; form fruiting bodies containing resting cells (myxospores) under certain environmental conditions.

Myxococcaceae [MICROBIO] A family of the order Myxobacterales; vegetative cells are straight to slightly tapered, and spherical to ovoid microcysts (myxospores) are produced.

myxofibroma of nerve sheath *See* neurofibroma.

Myxogastromycetidae [MYCOL] A large subclass of plasmodial slime molds (Myxomycetes).

myxolipoma *See* liposarcoma.

myxoma lipomatodes *See* liposarcoma.

Myxomycetes [BIOL] Plasmodial (acellular or true) slime molds, a class of microorganisms of the division Mycota; they are on the borderline of the plant and animal kingdoms and have a noncellular, multinucleate, jellylike, creeping, assimilative stage (the plasmodium) which alternates with a myxameba stage.

Myxomycophyta [BOT] An order of microorganisms, equivalent to the Mycetozoia of zoological classification.

Myxophaga [INV ZOO] A suborder of the Coleoptera.

Myxophyceae [BOT] An equivalent name for the Cyanophyceae.

Myxosporida [INV ZOO] An order of the protozoan class Myxosporidea characterized by the production of spores with one or more valves and polar capsules, and by possession of a single sporoplasm.

Myxosporidea [INV ZOO] A class of the protozoan subphylum Cnidospora; members are parasitic in some fish, a few amphibians, and certain invertebrates.

myxovirus [VIROL] A group of ribonucleic-acid animal viruses characterized by hemagglutination and hemadsorption; includes influenza and fowl plague viruses and the paramyxoviruses.

Myzopodidae [VERT ZOO] A monospecific order of insectivorous bats (Chiroptera) containing the Old World disk-winged bat of Madagascar; characterized by long ears and by a vestigial thumb with a monostalked sucking disk.

Myzostomaria [INV ZOO] An aberrant group of Polychaeta; most are greatly depressed, broad, and very small, and true segmentation is delayed or absent in the adult; all are parasites of echinoderms.

Myzostomidae [INV ZOO] A monogeneric family of the Myzostomaria.

N

n- [ORG CHEM] Chemical prefix for "normal" (straight-carbon-chain) hydrocarbon compounds.

N See newton; nitrogen; normality.

Na See sodium.

N.A. See numerical aperture.

nabam [ORG CHEM] $NaSSCNHCH_2CH_2NHCSSNa$ Water-soluble, colorless crystals that will irritate skin and eyes; used as a pesticide and pesticides intermediate. Also known as disodium ethylene-bis-dithiocarbamate.

Nabidae [INV ZOO] The damsel bugs, a family of hemipteran insects in the superfamily Cimicimorpha.

nabla See del operator.

nacelle [AERO ENG] A separate streamlined enclosure on an airplane for sheltering or housing something, as the crew or an engine.

nacre [INV ZOO] An iridescent inner layer of many mollusk shells.

nacreous [OPTICS] Having an iridescent luster resembling that of mother-of-pearl. Also known as pearly.

nacrite [MINERAL] $Al_2Si_2O_5(OH)_4$ A crystallized clay mineral of the kaolinite group; structurally distinct in being the most closely stacked in the *c*-axis direction.

NAD See diphosphopyridine nucleotide.

nadir [ASTRON] That point on the celestial sphere vertically below the observer, or 180° from the zenith.

nadorite [MINERAL] $PbSbO_2Cl$ A smoky brown or brownish-yellow to yellow, orthorhombic mineral consisting of an oxychloride of lead and antimony.

nagyagite [MINERAL] $Pb_5Au(Te,Sb)_4S_{5-8}$ A lead-gray mineral consisting of a sulfide of lead, gold, tellurium, and antimony. Also known as black tellurium; tellurium glance.

nail [ANAT] The horny epidermal derivative covering the dorsal aspect of the terminal phalanx of each finger and toe. [DES ENG] A slender, usually pointed fastener with a head, designed for insertion by impact. [ENG] To drive nails in a manner that will position and hold two or more members, usually of wood, in a desired relationship. [MED] A metallic rod with one blunt end and one sharp end, used surgically to anchor bone fragments.

nailhead spot [PL PATH] A fungus rot of tomato caused by *Alternaria tomato* and marked by small brown to black sunken spots on the fruit.

Najadaceae [BOT] A family of monocotyledonous, submerged aquatic plants in the order Najadales distinguished by branching stems and opposite or whorled leaves.

Najadales [BOT] An order of aquatic and semiaquatic flowering plants in the subclass Alismatidae; the perianth, when present, is not differentiated into sepals and petals, and the flowers are usually not individually subtended by bracts.

naked bud [BOT] A bud covered only by rudimentary foliage leaves.

Namanereinae [INV ZOO] A subfamily of largely freshwater errantian annelids in the family Nereidae.

Namurian [GEOL] A European stage of geologic time; divided into a lower stage (Lower Carboniferous or Upper Mississippian) and an upper stage (Upper Carboniferous or Lower Pennsylvanian).

NAND [MATH] A logic operator having the characteristic that if P, Q, R, … are statements, then the NAND of P, Q, R, … is true if at least one statement is false, false if all statements are true. Derived from NOT-AND. Also known as sheffer stroke.

NAND circuit [ELECTR] A logic circuit whose output signal is a logical 1 if any of its inputs is a logical 0, and whose output signal is a logical 0 if all of its inputs are logical 1.

nannoplankton [BIOL] Minute plankton; the smallest plankton, including algae, bacteria, and protozoans.

nano- [BIOL] A prefix meaning dwarfed. [MATH] A prefix representing 10^{-9}, which is 0.000000001 or one-billionth of the unit adjoined. Also known as milli-micro- (deprecated usage).

naphthine See hatchettite.

napier See neper.

Napierian logarithm See logarithm.

Napier's rules [MATH] Two rules which give the formulas necessary in the solution of right spherical triangles.

nappe [GEOL] A sheetlike, allochthonous rock unit that is formed by thrust faulting or recumbent folding or both. [MATH] One of the two parts of a conical surface defined by the vertex.

narco- [MED] Combining form meaning numbness, narcosis, or stupor.

narcolepsy [MED] A disorder of sleep mechanism characterized by two or more of four distinct symptoms: uncontrollable periods of daytime drowsiness, cataleptic attacks of muscular weakness, sleep paralysis, and vivid nocturnal or hypnogogic hallucinations.

Narcomedusae [INV ZOO] A suborder of hydrozoan coelenterates in the order Trachylina; the hydroid generation is represented by an actinula larva.

narcotic [PHARM] A drug which in therapeutic doses diminishes awareness of sensory impulses, especially pain, by the brain; in large doses, it causes stupor, coma, or convulsions.

nari See caliche.

narrow-band amplifier [ELECTR] An amplifier which increases the magnitude of signals over a band of frequencies whose bandwidth is small compared to the average frequency of the band.

narrow-band frequency modulation [COMMUN] Frequency-modulation broadcasting system used primarily for two-way voice communication, having a maximum permissible deviation of 15 kilohertz or less.

narrow-band-pass filter [ELECTR] A band-pass filter in which the band of frequencies transmitted by the filter has a bandwidth which is small compared to the average frequency of the band.

narrow gage [CIV ENG] A railway gage narrower than the standard gage of 4 feet 8½ inches (143.51 centimeters).

narrows [GEOGR] A navigable narrow part of a bay, strait, or river.

narwhal [VERT ZOO] *Monodon monoceros*. An arctic whale characterized by lack of a dorsal fin, and by possession in the male of a long, twisted, pointed tusk (or rarely, two tusks) which is a source of ivory.

nasal [ANAT] Of or pertaining to the nose.

nascent [CHEM] Pertaining to an atom or simple compound at the moment of its liberation from chemical combination, when it may have greater activity than in its usual state.

nasonite [MINERAL] $Ca_4Pb_6Si_6O_{21}Cl_2$ A white mineral composed of silicate and chloride of calcium and lead and occurring in granular masses.

nasopharynx [ANAT] The space behind the posterior nasal orifices, above a horizontal plane through the lower margin of the palate.

nastic movement [BOT] Movement of a flat plant part, oriented relative to the plant body and produced by diffuse stimuli causing disproportionate growth or increased turgor pressure in the tissues of one surface.

nasturan *See* pitchblende.

Nasutitermitinae [INV ZOO] A subfamily of termites in the family Termitidae, characterized by having the cephalic glands open at the tip of an elongated tube which projects anteriorly.

Natalidae [VERT ZOO] The funnel-eared bats, a monogeneric family of small, tropical American insectivorous bats (Chiroptera) with large, funnellike ears.

Natantia [INV ZOO] A suborder of decapod crustaceans comprising shrimp and related forms characterized by a long rostrum and a ventrally flexed abdomen.

Naticacea [INV ZOO] A superfamily of gastropod mollusks in the order Prosobranchia.

Naticidae [INV ZOO] A family of gastropod mollusks in the order Pectinibranchia comprising the moonshell snails.

native element [GEOL] Any of 20 elements, such as copper, gold, and silver, which occur naturally uncombined in a nongaseous state; there are three groups—metals, semimetals, and nonmetals.

native language [COMPUT SCI] Machine language that is executed by the computer for which it is specifically designed, in contrast to a computer using an emulator.

native metal [GEOCHEM] A metallic native element; includes silver, gold, copper, iron, mercury, iridium, lead, palladium, and platinum.

native paraffin *See* ozocerite.

natrolite [MINERAL] $Na_2Al_2Si_3O_{10}\cdot2H_2O$ A zeolite mineral composed of hydrous silicate of sodium and aluminum; usually occurs in slender acicular or prismatic crystals.

natromontebrasite [MINERAL] $(Na,Li)Al(PO_4)$-(OH,F) Mineral composed of hydrous basic phosphate of sodium, lithium, and aluminum; it is iso-

morphous with montebrasite and amblygonite. Also known as fremontite.

natron [MINERAL] $Na_2CO_3\cdot10H_{20}$ A white, yellow, or gray mineral that crystallizes in the monoclinic system, is soluble in water, and generally occurs in solution or in saline residues.

natronborocalcite *See* ulexite.

natural boundary [MATH] Those points of the boundary of a region where an analytic function is defined through which the function cannot be continued analytically.

natural convection [THERMO] Convection in which fluid motion results entirely from the presence of a hot body in the fluid, causing temperature and hence density gradients to develop, so that the fluid moves under the influence of gravity. Also known as free convection.

natural coordinates [FL MECH] An orthogonal, or mutually perpendicular, system of curvilinear coordinates for the description of fluid motion, consisting of an axis t tangent to the instantaneous velocity vector and an axis n normal to this velocity vector to the left in the horizontal plane, to which a vertically directed axis z may be added for the description of three-dimensional flow; such a coordinate system often permits a concise formulation of atmospheric dynamical problems, especially in the Lagrangian system of hydrodynamics.

natural draft [FL MECH] Unforced gas flow through a chimney or vertical duct, directly related to chimney height and the temperature difference between the ascending gases and the atmosphere, and not dependent upon the use of fans or other mechanical devices.

natural fiber [TEXT] A textile fiber of mineral, plant, or animal origin.

natural frequency [ELECTR] The lowest resonant frequency of an antenna, circuit, or component. [PHYS] The frequency with which a system oscillates in the absence of external forces; or, for a system with more than one degree of freedom, the frequency of one of the normal modes of vibration.

natural function [MATH] A trigonometric function, as opposed to its logarithm.

natural function generator *See* analytical function generator.

natural gas [MATER] A combustible, gaseous mixture of low-molecular-weight paraffin hydrocarbons, generated below the surface of the earth; contains mostly methane and ethane with small amounts of propane, butane, and higher hydrocarbons, and sometimes nitrogen, carbon dioxide, hydrogen sulfide, and helium.

natural glass [GEOL] An amorphous, vitreous inorganic material that has solidified from magma too quickly to crystallize.

natural immunity [IMMUNOL] Native immunity possessed by the individuals of a race, strain, or species.

natural language [COMPUT SCI] A computer language whose rules reflect and describe current rather than prescribed usage; it is often loose and ambiguous in interpretation, meaning different things to different hearers.

natural logarithm *See* logarithm.

natural period [PHYS] Period of the free oscillation of a body or system; when the period varies with amplitude, the natural period is the period when the amplitude approaches zero.

natural radiation *See* background radiation.

natural remanent magnetization [GEOPHYS] The magnetization of rock which exists in the absence of a magnetic field and has been acquired from the influence of the earth's magnetic field at the time of their formation or, in certain cases, at later times. Abbreviated NRM.

natural selection [EVOL] Darwin's theory of evolution, according to which organisms tend to produce progeny far above the means of subsistence; in the struggle for existence that ensues, only those progeny with favorable variations survive; the favorable variations accumulate through subsequent generations, and descendants diverge from their ancestors.

Naucoridae [INV ZOO] A family of hemipteran insects in the superfamily Naucoroidea.

Naucoroidea [INV ZOO] The creeping water bugs, a superfamily of hemipteran insects in the subdivision Hydrocorisae; they are suboval in form, with chelate front legs.

Naumanniella [MICROBIO] A genus of bacteria in the family Siderocapsaceae; rod-shaped cells surrounded by a delicate sheath in which iron and manganese oxides are deposited.

naumannite [MINERAL] Ag_2Se An iron-black mineral that crystallizes in the isometric system; consists of silver selenide, and occurs massive or in crystals; specific gravity is 8.

nauplius [INV ZOO] A larval stage characteristic of many groups of Crustacea; the oval, unsegmented body has three pairs of appendages: uniramous antennules, biramous antennae, and mandibles.

nautical almanac [NAV] A book published annually by the governments of the principal maritime nations which contains the astronomical data required for navigation by observations of celestial objects; an abridged version is known as the abridged nautical almanac.

nautical chain [MECH] A unit of length equal to 15 feet or 4.572 meters.

nautical mile [NAV] A unit of distance used principally in navigation; for practical consideration it is usually considered the length of 1 minute of any great circle of the earth, the meridian being the great circle most commonly used; the International Hydrographic Bureau in 1929 proposed a standard length of 1852 meters, which is known as the international nautical mile.

nautical twilight [ASTRON] The interval of incomplete darkness between sunrise or sunset and the time at which the center of the sun's disk is 12° below the celestial horizon.

Nautilidae [INV ZOO] A monogeneric family of cephalopod mollusks in the order Nautiloidea; *Nautilus pompilius* is the only well-known living species.

Nautiloidea [INV ZOO] A primitive order of tetrabranchiate cephalopods; shells are external and smooth, being straight or coiled and chambered with curved transverse septa.

naval architecture [ENG] The study of the physical characteristics and the design and construction of buoyant structures, such as ships, boats, barges, submarines, and floats, which operate in water; includes the construction and operation of the power plant and other mechanical equipment of these structures.

naval meteorology [METEOROL] The branch of meteorology which studies the interaction between the ocean and the overlying air mass, and which is concerned with atmospheric phenomena over the oceans, the effect of the ocean surface on these phenomena, and the influence of such phenomena on shallow and deep seawater.

naval stores [MATER] **1.** Pitch and rosin formerly used in the construction of wooden ships. **2.** All pine wood products, including rosin, turpentine, and pine oils.

navarho [NAV] A long-distance, low-frequency continuous-wave navigation system providing simultaneous bearing and distance information; the portion providing bearing is termed navaglobe.

Navier-Stokes equations [FL MECH] The equations of motion for a viscous fluid which may be written $d\mathbf{V}/dt = -(1/\rho)\nabla p + F + \nu\nabla^2\mathbf{V} + (1/3)\nu\nabla(\nabla\cdot\mathbf{V})$, where p is the pressure, ρ the density, F the total external force per unit mass, $\mathbf{V}$ the fluid velocity, and ν the kinematic viscosity; for an incompressible fluid, the term in $\nabla\cdot\mathbf{V}$ (divergence) vanishes, and the effects of viscosity then play a role analogous to that of temperature in thermal conduction and to that of density in simple diffusion.

navigating bridge *See* flying bridge.

navigation [ENG] The process of directing the movement of a craft so that it will reach its intended destination; subprocesses are position fixing, dead reckoning, pilotage, and homing.

Nb *See* niobium.

n-body problem *See* many-body problem.

NC *See* numerical control.

N/C *See* numerical control.

n-channel [ELECTR] A conduction channel formed by electrons in an n-type semiconductor, as in an n-type field-effect transistor.

Nd *See* neodymium.

N display [ELECTR] Radar display in which the target appears as a pair of vertical deflections from a horizontal time base; direction is indicated by relative amplitude of the blips; target distance is determined by moving an adjustable pedestal signal along the base line until it coincides with the horizontal position of the blips; the pedestal control is calibrated in distance.

NDRO *See* nondestructive readout.

Ne *See* neon.

neap range [OCEANOGR] The mean semidiurnal range of tide when neap tides are occurring; the mean difference in height between neap high water and neap low water. Also known as mean neap range.

neaps *See* neap tide.

neap tide [OCEANOGR] Tide of decreased range occurring about every 2 weeks when the moon is in quadrature, that is, during its first and last quarter. Also known as neaps.

near-end crosstalk [COMMUN] A type of interference that may occur at carrier telephone repeater stations when output signals of one repeater leak into the same end of the other repeater.

near field [ACOUS] The acoustic radiation field that is close to an acoustic source such as a loudspeaker. [ELECTROMAG] The electromagnetic field that exists within one wavelength of a source of electromagnetic radiation, such as a transmitting antenna.

near-infrared radiation [ELECTROMAG] Infrared radiation having a relatively short wavelength, between 0.75 and about 2.5 micrometers (some scientists place the upper limit from 1.5 to 3 micrometers), at which radiation can be detected by photoelectric cells, and which corresponds in frequency range to the lower electronic energy levels of molecules and semiconductors. Also known as photoelectric infrared radiation.

nearsightedness *See* myopia.

near-ultraviolet radiation [ELECTROMAG] Ultraviolet radiation having relatively long wavelength, in the approximate range from 300 to 400 nanometers.

near wilt [PL PATH] A fungus disease of peas caused by *Fusarium oxysporum pisi;* affects scattered plants and develops more slowly than true wilt.

neat line [CIV ENG] The line to which a masonry wall should generally conform. [MAP] That border line which indicates the limits of an area shown on a map or chart.

Nebaliacea [INV ZOO] A small, marine order of Crustacea in the subclass Leptostraca distinguished by a large bivalve shell, without a definite hinge line, an anterior articulated rostrum, eight thoracic and seven abdominal somites, a pair of articulated furcal rami, and the telson.

Nebraskan glaciation [GEOL] The first glacial stage of the Pleistocene epoch in North America, beginning about 1,000,000 years ago, and preceding the Aftonian interglacial stage.

nebula [ASTRON] Interstellar clouds of gas or small particles; an example is the Horsehead Nebula in Orion.

neck [ANAT] The usually constricted communicating column between the head and trunk of the vertebrate body. g] The part of a furnace where the flame is contracted before reaching the stack. [GEOGR] A narrow strip of land, especially one connecting two larger areas. [GEOL] *See* pipe. [MET] In a tensile test, that portion of the metal at which fracture is imminent during the later stages of plastic deformation in a tensile test. [OCEANOGR] The narrow band of water forming the part of a rip current where feeder currents converge and flow swiftly through the incoming breakers and out to the head.

Neckeraceae [BOT] A family of mosses in the order Isobryales distinguished by undulate leaves.

neck rot [PL PATH] A fungus disease of onions caused by species of *Botrytis* and characterized by rotting of the leaves just above the bulb.

necr-, necro- [MED] Combining form denoting death.

Necrolestidae [PALEON] An extinct family of insectivorous marsupials.

necrosis [MED] Death of a cell or group of cells as a result of injury, disease, or other pathologic state.

necrotic ring spot [PL PATH] A virus leaf spot of cherries marked by small, dark water-soaked rings which may drop out, giving the leaf a tattered appearance.

nectar [BOT] A sugar-containing liquid secretion of the nectaries of many flowers.

nectarine [BOT] A smooth-skinned, fuzzless fruit originating as a spontaneous somatic mutation of the peach, *Prunus persica* and *P. persica* var. *nectarina*.

nectary [BOT] A secretory organ or surface modification of a floral organ in many flowers, occurring on the receptacle, in and around ovaries, on stamens, or on the perianth; secretes nectar.

Nectonematoidea [INV ZOO] A monogeneric order of worms belonging to the class Nematomorpha, characterized by dorsal and ventral epidermal chords, a pseudocoele, and dorsal and ventral rows of bristles; adults are parasites of true crabs and hermit crabs.

Nectridea [PALEON] An order of extinct lepospondylous amphibians characterized by vertebrae in which large fan-shaped hemal arches grow directly downward from the middle of each caudal centrum.

Nectrioidaceae [MYCOL] The equivalent name for Zythiaceae.

needle [BOT] A slender-pointed leaf, as of the firs and other evergreens. [COMPUT SCI] A slender rod or probe used to sort decks of edge-punched cards by inserting it through holes along the margin of the deck and vibrating the deck so that cards having that particular hole are retained, but those having a notch cut at that hole position drop out. [DES ENG] **1.** A device made of steel pointed at one end with a hole at the other; used for sewing. **2.** A device made of steel with a hook at one end; used for knitting. [ENG] **1.** A piece of copper or brass about ½ inch (13 millimeters) in diameter and 3 or 4 feet (90 or 120 centimeters) long, pointed at one end, thrust into a charge of blasting powder in a borehole and then withdrawn, leaving a hole for the priming, fuse, or squib. Also known as pricker. **2.** A thin pointed indicator on an instrument dial. [ENG ACOUS] *See* stylus. [GEOL] A pointed, elevated, and detached mass of rock formed by erosion, such as an aiguille. [HYD] A long, slender snow crystal that is at least five times as long as it is broad. [MINERAL] A needle-shaped or acicular mineral crystal.

needle nozzle [MECH ENG] A streamlined hydraulic turbine nozzle with a movable element for converting the pressure and kinetic energy in the pipe leading from the reservoir to the turbine into a smooth jet of variable diameter and discharge but practically constant velocity.

needle valve [DES ENG] A slender, pointed rod fitting in a hole or circular or conoidal seat; used in hydraulic turbines and hydroelectric systems.

needle weir [CIV ENG] A type of frame weir in which the wooden barrier is constructed of vertical square-section timbers placed side by side against the iron frames.

Néel point *See* Néel temperature.

Néel's theory [SOLID STATE] A theory of the behavior of antiferromagnetic and other ferrimagnetic materials in which the crystal lattice is divided into two or more sublattices; each atom in one sublattice responds to the magnetic field generated by nearest neighbors in other sublattices, with the result that magnetic moments of all the atoms in any sublattice are parallel, but magnetic moments of two different sublattices can be different.

Néel temperature [SOLID STATE] A temperature, characteristic of certain metals, alloys, and salts, below which spontaneous nonparalleled magnetic ordering takes place so that they become antiferromagnetic, and above which they are paramagnetic. Also known as Néel point.

negative [ELEC] Having a negative charge. [GRAPHICS] The image on film in which the dark tones of the original appear transparent, and the light tones appear black and opaque. Also known as reversed image.

negative acceleration [MECH] Acceleration in a direction opposite to the velocity, or in the direction of the negative axis of a coordinate system.

negative afterimage [PHYSIO] An afterimage that is seen on a bright background and is complementary in color to the initial stimulus.

negative angle [MATH] The angle subtended by moving a ray in the clockwise direction.

negative charge [ELEC] The type of charge which is possessed by electrons in ordinary matter, and which may be produced in a resin object by rubbing with wool. Also known as negative electricity.

negative electricity *See* negative charge.

negative electrode *See* cathode; negative plate.

negative electron *See* electron.

negative feedback [CONT SYS] Feedback in which a portion of the output of a circuit, device, or machine is fed back 180° out of phase with the input signal, resulting in a decrease of amplification so as to stabilize the amplification with respect to time or frequency, and a reduction in distortion and noise. Also known as inverse feedback; reverse feedback; stabilized feedback. [SCI TECH] Feedback which tends to reduce the output in a system.

negative g [MECH] In designating the direction of acceleration on a body, the opposite of positive *g*; for example, the effect of flying an outside loop in the upright seated position.

negative glow [ELECTR] The luminous flow in a glow-discharge cold-cathode tube occurring between the cathode dark space and the Faraday dark space.

negative impedance [ELECTR] An impedance such that when the current through it increases, the voltage drop across the impedance decreases.

negative indication [COMPUT SCI] A hole punched in a specified column and specified punch position on a punch card to indicate that a number represented in a particular field of the card has a negative sign.

negative integer [MATH] The additive inverse of a positive integer relative to the additive group structure of the integers.

negative ion [CHEM] An atom or group of atoms which by gain of one or more electrons has acquired a negative electric charge. [PHYS] An electron or negatively charged subatomic particle.

negative lens *See* diverging lens.

negative meniscus lens [OPTICS] A lens having one convex and one concave surface, with the radius of curvature of the convex surface greater than that of the concave surface. Also known as diverging meniscus lens.

negative plate [ELEC] The internal plate structure that is connected to the negative terminal of a storage battery. Also known as negative electrode.

negative pressure [PHYS] A way of expressing vacuum; a pressure less than atmospheric or the standard 760 mmHg (101,325 newtons per square meter).

negative-resistance device [ELECTR] A device having a range of applied voltages within which an increase in this voltage produces a decrease in the current.

negative skewness [MATH] Skewness in which the mean is smaller than the mode.

negative terminal [ELEC] The terminal of a battery or other voltage source that has more electrons than normal; electrons flow from the negative terminal through the external circuit to the positive terminal.

negatron *See* dynatron; electron.

Negri bodies [PATH] Acidophil cytoplasmic inclusion bodies in neurons, considered diagnostic of rabies.

Neididae [INV ZOO] A small family of thread-legged hemipteran insects in the superfamily Lygaeoidea.

neighbor [CRYSTAL] One of a pair of atoms or ions in a crystal which are close enough to each other for their interaction to be of significance in the physical problem being studied.

neighborhood of a point [MATH] A set in a topological space which contains an open set which contains the point; in euclidean space, an example of a neighborhood of a point is an open (without boundary) ball centered at that point.

Neisseriaceae [MICROBIO] The single family of gram-negative aerobic cocci and coccobacilli; some species are human parasites and pathogens.

N electron [ATOM PHYS] An electron in the fourth (N) shell of electrons surrounding the atomic nucleus, having the principal quantum number 4.

Nelumbonaceae [BOT] A family of flowering aquatic herbs in the order Nymphaeales characterized by having roots, perfect flowers, alternate leaves, and triaperturate pollen.

Nemata [INV ZOO] A proposed equivalent name for Nematoda.

Nemataceae [BOT] A family of mosses in the order Hookeriales distinguished by having perichaetial leaves only.

Nemathelminthes [INV ZOO] A subdivision of the Amera which comprised the classes Rotatoria, Gastrotrichia, Kinorhyncha, Nematoda, Nematomorpha, and Acanthocephala.

nematic phase [PHYS CHEM] A phase of a liquid crystal in the mesomorphic state, in which the liquid has a single optical axis in the direction of the applied magnetic field, appears to be turbid and to have mobile threadlike structures, can flow readily, has low viscosity, and lacks a diffraction pattern.

nematoblastic [PETR] Pertaining to a metamorphic rock with a homeoblastic texture due to development during recrystallization of slender prismatic crystals.

Nematocera [INV ZOO] A series of dipteran insects in the suborder Orthorrhapha; adults have antennae that are usually longer than the head, and the flagellum consists of 10–65 similar segments.

nematocyst [INV ZOO] An intracellular effector organelle in the form of a coiled tube which may be rapidly everted in food gathering or defense by coelenterates.

Nematoda [INV ZOO] A group of unsegmented worms which have been variously recognized as an order, class, and phylum.

nematode [INV ZOO] **1.** Any member of the Nematoda. **2.** Of or pertaining to the Nematoda.

Nematodonteae [BOT] A group of mosses included in the subclass Eubrya in which there may be faint transverse bars on the peristome teeth.

Nematognathi [VERT ZOO] The equivalent name for Siluriformes.

Nematoidea [INV ZOO] An equivalent name for Nematoda.

nematology [INV ZOO] The study of nematodes.

Nematomorpha [INV ZOO] A group of the Aschelminthes or a separate phylum that includes the horsehair worms.

Nematophytales [PALEOBOT] A group of fossil plants from the Silurian and Devonian periods that bear some resemblance to the brown seaweeds (Phaeophyta).

Nematosporoideae [BOT] A subfamily of the Saccharomycetaceae containing parasitic yeasts; two genera have been studied in culture: *Nematospora* with asci that contain eight spindle-shaped ascospores, and *Metschnikowia* whose asci contain one or two needle-shaped ascospores.

Nemertea [INV ZOO] An equivalent name for Rhynchocoela.

Nemertina [INV ZOO] An equivalent name for Rhynchocoela.

Nemertinea [INV ZOO] An equivalent name for Rhynchocoela.

Nemestrinidae [INV ZOO] The hairy flies, a family of dipteran insects in the series Brachycera of the suborder Orthorrhapha.

Nemichthyidae [VERT ZOO] A family of bathypelagic, eellike amphibians in the order Apoda.

Nemognathinae [INV ZOO] A subfamily of the coleopteran family Meloidae; members have greatly elongate maxillae that form a poorly fitted tube.

neo-, ne- [ORG CHEM] Prefix indicating hydrocarbons where a carbon is bonded directly to at least four other carbon atoms, such as neopentane. [SCI TECH] Prefix meaning new, or different in form;

indicating a compound related to an older one, or a precursor.

Neoanthropinae [PALEON] A subfamily of the Hominidae in some systems of classification, set up to include *Homo sapiens* and direct ancestors of *H. sapiens.*

Neocathartidae [PALEON] An extinct family of vulturelike diurnal birds of prey (Falconiformes) from the Upper Eocene.

Neocomian [GEOL] A European stage of Lower Cretaceous geologic time; includes Berriasian, Valanginian, Hauterivian, and Barremian.

neodymium [CHEM] A metallic element, symbol Nd, with atomic weight 144.24, atomic number 60; a member of the rare-earth group of elements.

neodymium glass laser [OPTICS] An amorphous solid laser in which glass is doped with neodymium; characteristics are comparable with those of a pulsed ruby laser, but the wavelength of radiation is outside the visible range.

Neogastropoda [INV ZOO] An order of gastropods which contains the most highly developed snails; respiration is by means of ctenidia, the nervous system is concentrated, an operculum is present, and the sexes are separate.

Neogene [GEOL] An interval of geologic time incorporating the Miocene and Pliocene of the Tertiary period; the Upper Tertiary.

Neognathae [VERT ZOO] A superorder of the avian order Neornithes, characterized as flying birds with fully developed wings and sternum with a keel, fused caudal vertebrae, and absence of teeth.

Neogregarinida [INV ZOO] An order of sporozoan protozoans in the subclass Gregarinia which are insect parasites.

neon [CHEM] A gaseous element, symbol Ne, atomic number 10, atomic weight 20.183; a member of the family of noble gases in the zero group of the periodic table.

neonatal [MED] Pertaining to a newborn infant.

neon glow lamp [ELECTR] A glow lamp containing neon gas, usually rated between 1/25 and 3 watts, and producing a characteristic red glow; used as an indicator light and electronic circuit component.

neon-helium laser [OPTICS] A continuous-wave gas laser using a combination of neon and helium gases to obtain a 632.8-nanometer visible red beam.

neon tube [ELECTR] An electron tube in which neon gas is ionized by the flow of electric current through long lengths of gas tubing, to produce a luminous red glow discharge; used chiefly in outdoor advertising signs.

Neopseustidae [INV ZOO] A family of Lepidoptera in the superfamily Eriocranioidea.

Neoptera [INV ZOO] A section of the insect subclass Pterygota; members have a muscular and articular mechanism allowing the wings to be flexed over the abdomen when at rest.

Neopterygii [VERT ZOO] An equivalent name for Actinopterygii.

Neorhabdocoela [INV ZOO] A group of the Rhabdocoela comprising fresh-water, marine, or terrestrial forms, with a bulbous pharynx, paired protonephridia, sexual reproduction, and ventral gonopores.

Neornithes [VERT ZOO] A subclass of the class Aves containing all known birds except the fossil *Archaeopteryx.*

neosilicate [MINERAL] A structural type of silicate mineral characterized by linkage of isolated SiO_4 tetrahedra by ionic bonding only; an example is olivine.

neotenin [BIOCHEM] A hormone secreted by cells of the corpus allatum in arthropod larvae and nymphs; inhibits the development of adult characters. Also known as juvenile hormone.

neoteny [VERT ZOO] A phenomenon peculiar to some salamanders, in which large larvae become sexually mature while still retaining gills and other larval features.

neotype [SYST] A specimen selected as type subsequent to the original description when the primary types are known to be destroyed; a nomenclatural type.

Nepenthaceae [BOT] A family of dicotyledonous plants in the order Sarraceniales; includes many of the pitcher plants.

neper [PHYS] Abbreviated Np. Also known as napier. **1.** A unit used for expressing the ratio of two currents, voltages, or analogous quantities; the number of nepers is the natural logarithm of this ratio. **2.** A unit used for expressing the ratio of two powers (even when this ratio is not the square of the corresponding current or voltage ratio); the number of nepers is the natural logarithm of the square root of this ratio; to avoid confusion, this usage should be accompanied by a specific statement.

nepheline [MINERAL] $(Na,K)AlSiO_4$ A mineral of the feldspathoid group crystallizing in the hexagonal system and occurring as glassy or coarse crystals or colorless grains or green to brown masses of greasy luster in alkalic igneous rocks; hardness is 5.5–6 on Mohs scale. Also known as eleolite; nephelite.

nepheline basalt *See* olivine nephelinite.

nepheline syenite [PETR] A phaneritic plutonic rock with granular texture, composed largely of alkali feldspar, nepheline, and dark-colored materials.

nephelinite [PETR] A dark-colored, aphanitic rock of volcanic origin, composed essentially of nepheline and pyroxene; texture is usually porphyritic with large crystals of augite and nepheline in a very-fine-grained matrix.

nephelite *See* nepheline.

nephelometer [OPTICS] A type of instrument that measures, at more than one angle, the scattering function of particles suspended in a medium; information obtained may be used to determine the size of the suspended particles and the visual range through the medium.

nephric tubule *See* uriniferous tubule.

nephridium [INV ZOO] Any of various paired excretory structures present in the Platyhelminthes, Rotifera, Rhynchocoela, Acanthocephala, Priapuloidea, Entoprocta, Gastrotricha, Kinorhyncha, Cephalochorda, and some Archiannelida and Polychaeta.

nephrite [MINERAL] An exceptionally tough, compact, fine-grained, greenish or bluish amphibole constituting the less valuable type of jade; formerly worn as a remedy for kidney diseases. Also known as greenstone; kidney stone.

nephritis [MED] Inflammation of the kidney.

nephrodystrophy *See* nephrosis.

nephron [ANAT] The functional unit of a kidney, consisting of the glomerulus with its capsule and attached uriniferous tubule.

nephropathy [MED] **1.** Any disease of the kidney. **2.** *See* nephrosis.

Nephropidae [INV ZOO] The true lobsters, a family of decapod crustaceans in the superfamily Nephropidea.

Nephropidea [INV ZOO] A superfamily of the decapod section Macrura including the true lobsters and crayfishes, characterized by a rostrum and by chelae on the first three pairs of pereiopods, with the first pair being noticeably larger.

nephrosis [PATH] Degenerative or retrogressive renal lesions, distinct from inflammation (nephritis) or vascular involvement (nephrosclerosis), especially as applied to tubular lesions (tubular nephritis). Also known as nephrodystrophy; nephropathy.

nephsystem See cloud system.

Nephtyidae [INV ZOO] A family of errantian annelids of highly opalescent colors, distinguished by an eversible pharynx.

Nepidae [INV ZOO] The water scorpions, a family of hemipteran insects in the superfamily Nepoidea, characterized by a long breathing tube at the tip of the abdomen, chelate front legs, and a short stout beak.

Nepoidea [INV ZOO] A superfamily of hemipteran insects in the subdivision Hydrocorisae.

Nepticulidae [INV ZOO] The single family of the lepidopteran superfamily Nepticuloidea.

Nepticuloidea [INV ZOO] A monofamilial superfamily of heteroneuran Lepidoptera; members are tiny moths with wing spines, and the females have a single genital opening.

Neptune [ASTRON] The outermost of the four giant planets, and the next to last planet, from the sun; it is 30 astronomical units from the sun, and the sidereal revolution period is 164.8 years.

neptunite [MINERAL] $(Na, K)_2(Fe, Mn)TiSi_4O_{12}$ Black mineral composed of silicate of sodium, potassium, iron, manganese, and titanium.

neptunium [CHEM] A chemical element, symbol Np, atomic number 93, atomic weight 237.0482; a member of the actinide series of elements.

nepuite See garnierite.

Nereid [ASTRON] One of the two satellites of the planet Neptune; it is the smaller, with a diameter of about 322 kilometers.

Nereidae [INV ZOO] A large family of mostly marine errantian annelids that have a well-defined head, elongated body with many segments, and large complex parapodia on most segments.

Nerillidae [INV ZOO] A family of archiannelids characterized by well-developed parapodia and setae.

Neritacea [INV ZOO] A superfamily of gastropod mollusks in the order Aspidobranchia.

neritic [OCEANOGR] Of or pertaining to the region of shallow water adjoining the seacoast and extending from low-tide mark to a depth of about 200 meters.

Neritidae [INV ZOO] A family of primitive marine, fresh-water, and terrestrial snails in the order Archaeogastropoda.

Nernst bridge [ELEC] A four-arm bridge containing capacitors instead of resistors, used for measuring capacitance values at high frequencies.

Nernst effect See Ettingshausen-Nernst effect.

Nernst equation [PHYS CHEM] The relationship showing that the electromotive force developed by a dry cell is determined by the activities of the reacting species, the temperature of the reaction, and the standard free-energy change of the overall reaction.

Nernst heat theorem [THERMO] The theorem expressing that the rate of change of free energy of a homogeneous system with temperature, and also the rate of change of enthalpy with temperature, approaches zero as the temperature approaches absolute zero.

nerve [ANAT] A bundle of nerve fibers or processes held together by connective tissue.

nerve cord [INV ZOO] Paired, ventral cords of nervous tissue in certain invertebrates, such as insects or the earthworm. [ZOO] Dorsal, hollow tubular cord of nervous tissue in chordates.

nerve fiber [CYTOL] The long process of a neuron, usually the axon.

nerve gas [MATER] Chemical agent which is absorbed into the body by breathing, by ingestion, or through the skin, and affects the nervous and respiratory systems and various body functions; an example is isopropylphosphonofluoridate.

nervous system [ANAT] A coordinating and integrating system which functions in the adaptation of an organism to its environment; in vertebrates, the system consists of the brain, brainstem, spinal cord, cranial and peripheral nerves, and ganglia.

Nesiotinidae [INV ZOO] A family of bird-infesting biting lice (Mallophaga) that are restricted to penguins.

Nesophontidae [PALEON] An extinct family of large, shrewlike lipotyphlans from the Cenozoic found in the West Indies.

nest [COMPUT SCI] To include data or subroutines in other items of a similar nature with a higher hierarchical level so that it is possible to access or execute various levels of data or routines recursively. [GEOL] A concentration of some relatively conspicuous element of a geologic feature, such as pebbles or inclusions, within a sand layer or igneous rock. [VERT ZOO] A bed, receptacle, or location in which the eggs of animals are laid and hatched.

nesting [COMPUT SCI] 1. Inclusion of a routine wholly within another routine. 2. Inclusion of a DO statement within a DO statement in FORTRAN.

nesting storage See push-down storage.

net [COMMUN] A number of communication stations equipped for communicating with each other, often on a definite time schedule and in a definite sequence. [ENG] 1. Threads or cords tied together at regular intervals to form a mesh. 2. A series of surveying or leveling stations that have been interconnected in such a manner that closed loops or circuits have been formed, or that are arranged so as to provide a check on the consistency of the measured values. Also known as network. [GEOL] 1. In structural petrology, coordinate network of meridians and parallels, projected from a sphere at intervals of 2°; used to plot points whose spherical coordinates are known and to study the distribution and orientation of planes and points. Also known as projection net; stereographic net. 2. A form of horizontal patterned ground whose mesh is intermediate between a circle and a polygon. [MATH] A set whose members are indexed by elements from a directed set; this is a generalization of a sequence. [TEXT] Any fabric made in open hexagonal mesh.

Net See Reticulum.

net blotch [PL PATH] A fungus disease of barley caused by *Helminthosporium teres* and marked by spotting of the foliage.

net head [FL MECH] The difference in elevation between the last free water surface in a power conduit above the waterwheel and the first free water surface in the conduit below the waterwheel, less the friction losses in the conduit.

net loss [COMMUN] The ratio of the power at the input of a transmission system to the power at the output; expressed in nepers, it is one-half the natural logarithm of this ratio, and in decibels it is 10 times the common logarithm of the ratio.

nettle cell *See* cnidoblast.

net ton *See* ton.

network [COMMUN] A number of radio or television broadcast stations connected by coaxial cable, radio, or wire lines, so all stations can broadcast the same program simultaneously. [ELEC] A collection of electric elements, such as resistors, coils, capacitors, and sources of energy, connected together to form several interrelated circuits. Also known as electric network. [ENG] *See* net. [MATH] The name given to a graph in applications in management and the engineering sciences; to each segment linking points in the graph, there is usually associated a direction and a capacity on the flow of some quantity.

network analyzer [COMPUT SCI] An analog computer in which networks are used to simulate power line systems or physical systems and obtain solutions to various problems before the systems are actually built.

network filter [ELEC] A combination of electrical elements (for example, interconnected resistors, coils, and capacitors) that represents relatively small attenuation to signals of a certain frequency, and great attenuation to all other frequencies.

network theory [ELEC] The systematizing and generalizing of the relations between the currents, voltages, and impedances associated with the elements of an electrical network.

Neumann function [MATH] 1. One of a class of Bessel functions arising in the study of the solutions to Bessel's differential equation. 2. A harmonic potential function in potential theory occurring in the study of Neumann's problem.

Neumann problem [MATH] The determination of a harmonic function within a finite region of three-dimensional space enclosed by a closed surface when the normal derivatives of the function on the surface are specified.

neuralgia [MED] Pain in or along the course of one or more nerves. Also known as neurodynia.

neurilemma [HISTOL] A thin tissue covering the axon directly, or covering the myelin sheath when present, of peripheral nerve fibers.

neuritis [MED] Degenerative or inflammatory nerve lesions associated with pain, hypersensitivity, anesthesia or paresthesia, paralysis, muscular atrophy, and loss of reflexes in the innervated part of the body.

neuroanatomy [ANAT] The study of the anatomy of the nervous system and nerve tissue.

Neurodontiformes [PALEON] A suborder of Conodontophoridia having a lamellar internal structure.

neuroelectricity [PHYSIO] A current or voltage generated in the nervous system.

neuroendocrinology [BIOL] The study of the structural and functional interrelationships between the nervous and endocrine systems.

neuroepithelioma [MED] A tumor resembling primitive medullary epithelium, containing cells of small cuboidal or columnar form with a tendency to form true rosettes, occurring in the retina, central nervous system, and occasionally in peripheral nerves. Also known as diktoma; esthesioneuroblastoma; esthesioneuroepithelioma.

neurofibroma [MED] A tumor characterized by the diffuse proliferation of peripheral nerve elements. Also known as endoneural fibroma; myxofibroma of nerve sheath; neurofibromyxoma; perineural fibroblastoma; perineural fibroma.

neurofibromyxoma *See* neurofibroma.

neuroglia [HISTOL] The nonnervous, supporting elements of the nervous system.

neurohormone [BIOCHEM] A hormone produced by nervous tissue.

neurohypophysis [ANAT] The neural portion or posterior lobe of the hypophysis.

neuroleptic [PHARM] 1. A drug that is useful in the treatment of mental disorders, especially psychoses. 2. Pertaining to the actions of such a drug.

neurologist [MED] A person versed in neurology, usually a physician who specializes in the diagnosis and treatment of disorders of the nervous system and the study of its functioning.

neurology [MED] The study of the anatomy, physiology, and disorders of the nervous system.

neuromuscular junction *See* myoneural junction.

neuromyasthenia [MED] Fatigue, headache, intense muscle pain, slight or transient muscle weakness, mental disturbances, objective signs in neurologic examination but usually normal cerebrospinal fluid findings, occurring in epidemics and thought to be viral in origin. Also known as benign myalgic encephalomyelitis.

neuron [HISTOL] A nerve cell, including the cell body, axon, and dendrites.

Neuroptera [INV ZOO] An order of delicate insects having endopterygote development, chewing mouthparts, and soft bodies.

neurosecretion [PHYSIO] The synthesis and release of hormones by nerve cells.

neurosis [PSYCH] A category of emotional maladjustments characterized by some impairment of thinking and judgment, with anxiety as the chief symptom.

neurotoxin [BIOCHEM] A poisonous substance in snake venom that acts as a nervous system depressant by blocking neuromuscular transmission by binding acetylcholine receptors on motor end plates, or on the innervated face of an electroplax.

neutral atom [ATOM PHYS] An atom in which the number of electrons that surround the nucleus is equal to the number of protons in the nucleus, so that there is no net electric charge.

neutral axis [MECH] In a beam bent downward, the line of zero stress below which all fibers are in tension and above which are in compression.

neutral conductor [ELEC] A conductor of a polyphase circuit or of a single-phase, three-wire circuit which is intended to have a potential such that the potential differences between it and each of the other conductors are approximately equal in magnitude and are also equally spaced in phase.

neutral ground [ELEC] Ground connected to the neutral point or points of an electric circuit, transformer, rotating machine, or system.

neutralize [CHEM] To make a solution neutral (neither acidic nor basic, pH of 7) by adding a base to an acidic solution, or an acid to a basic solution. [ELECTR] To nullify oscillation-producing voltage feedback from the output to the input of an amplifier through tube interelectrode capacitances; an external feedback path is used to produce at the input a voltage that is equal in magnitude but opposite in phase to that fed back through the interelectrode capacitance. [ORD] 1. To destroy or reduce the effectiveness of enemy personnel and materiel by gunfire, bombing, or any other means. 2. To make a toxic chemical agent harmless by chemical action. 3. To disarm or otherwise render safe a mine, bomb, missile, or booby trap.

neutralizing antibody [IMMUNOL] An antibody that reduces or abolishes some biological activity of a soluble antigen or of a living microorganism.

neutralizing capacitor [ELECTR] Capacitor, usually variable, employed in a radio receiving or transmitting circuit to feed a portion of the signal voltage from the plate circuit of a stage back to the grid circuit.

neutralizing circuit [ELECTR] Portion of an amplifier circuit which provides an intentional feedback path from plate to grid to prevent regeneration.

neutral molecule [PHYS CHEM] A molecule in which the number of electrons surrounding the nuclei is the same as the total number of protons in the nuclei, so that there is no net electric charge.

neutral point [ELEC] Point which has the same potential as the point of junction of a group of equal nonreactive resistances connected at their free ends to the appropriate main terminals or lines of the system. [FL MECH] *See* hyperbolic point. [MET] In rolling mills, the point at which the speed of the work is equal to the peripheral speed of the rolls. [METEOROL] *See* col. [OPTICS] In atmospheric optics, one of several points in the sky for which the degree of polarization of diffuse sky radiation is zero.

neutral relay [ELEC] Relay in which the movement of the armature does not depend upon the direction of the current in the circuit controlling the armature. Also known as nonpolarized relay.

neutral temperature [ELECTR] The temperature of the hot junction of a thermocouple at which the electromotive force of the thermocouple attains its maximum value, when the cold junction is maintained at a constant temperature of 0°C.

neutral wave [PHYS] Any wave whose amplitude does not change with time; in most contexts the wave is referred to as a stable wave, the term "neutral wave" being used when it is important to emphasize that the wave is neither damped nor amplified.

neutral zone *See* dead band.

neutrino [PHYS] A neutral particle having zero rest mass and spin ½ $(h/2\pi)$, where h is Planck's constant; experimentally, there are two such particles known as the e-neutrino (ν_e) and the μ-neutrino (μ_μ).

neutron [PHYS] An elementary particle which has approximately the same mass as the proton but lacks electric charge, and is a constituent of all nuclei having mass number greater than 1.

neutron absorption *See* neutron capture.

neutron activation analysis [NUCLEO] Activation analysis in which the specimen is bombarded with neutrons; identification is made by measuring the resulting radio isotopes.

neutron age *See* Fermi age.

neutron capture [NUC PHYS] A process in which the collision of a neutron with a nucleus results in the absorption of the neutron into the nucleus with the emission of one or more prompt gamma rays; in certain cases, beta decay or fission of the nucleus results. Also known as neutron absorption; neutron radiative capture.

neutron cross section [NUC PHYS] A measure of the probability that an interaction of a given kind will take place between a nucleus and an incident neutron; it is an area such that the number of interactions which occur in a sample exposed to a beam of neutrons is equal to the product of the number of nuclei in the sample and the number of neutrons in the beam that would pass through this area if their velocities were perpendicular to it.

neutron cycle [NUCLEO] The life history of the neutrons in a nuclear reactor, extending from the initial fission process until all the neutrons have been absorbed or have leaked out.

neutron diffraction [PHYS] The phenomenon associated with the interference processes which occur when neutrons are scattered by the atoms within solids, liquids, and gases.

neutron-gamma well logging [ENG] Neutron well logging in which the varying intensity of gamma rays produced artificially by neutron bombardment is recorded.

neutron logging *See* neutron well logging.

neutron magnetic moment [NUC PHYS] A vector whose scalar product with the magnetic flux density gives the negative of the energy of interaction of a neutron with a magnetic field.

neutron optics [PHYS] The study of certain phenomena, for example, crystal diffraction, in which the wave character of neutrons dominates and leads to behavior similar to that of light.

neutron radiative capture *See* neutron capture.

neutron radiography [NUCLEO] Radiography that uses a neutron beam generated by a nuclear reactor; the neutrons are detected by placing a conventional x-ray film next to a converter screen composed of potentially radioactive materials or prompt emission materials which convert the neutron radiation to other types of radiation more easily detected by the film.

neutron spectrometer [NUCLEO] An instrument used to determine the energies of neutrons and the relative intensities of neutrons of different energies in a neutron beam.

neutron spectrometry [NUC PHYS] A method of observing excited states of nuclei in which neutrons are used to bombard a target, causing nuclei to be transmuted into excited states by various nuclear reactions; the resultant excited states are determined by observing resonances in the reaction cross sections or by observing spectra of emitted particles or gamma rays. Also known as neutron spectroscopy.

neutron spectroscopy *See* neutron spectrometry.

neutron spectrum [NUC PHYS] A plot or display of the number of neutrons at various energies, such as the neutrons emitted in a nuclear reaction, or the neutrons in a nuclear reactor.

neutron star [ASTRON] A star that is supposed to occur in the final stage of stellar evolution; it consists of a superdense mass mainly of neutrons, and has a strong gravitational attraction from which only neutrinos and high-energy photons could escape so that the star is invisible.

neutron well logging [ENG] Study of formation fluid-content properties down a wellhole by neutron bombardment and detection of resultant radiation (neutrons or gamma rays). Also known as neutron logging.

neutrophil [HISTOL] A large granular leukocyte with a highly variable nucleus, consisting of three to five lobes, and cytoplasmic granules which stain with neutral dyes and eosin.

Nevadan orogeny [GEOL] Orogenic episode during Jurassic and Early Cretaceous geologic time in the western part of the North American Cordillera. Also known as Nevadian orogeny; Nevadic orogeny.

Nevadian orogeny *See* Nevadan orogeny.

Nevadic orogeny *See* Nevadan orogeny.

névé [GEOGR] A geographic area of perennial snow. [HYD] An accumulation of compacted, granular snow in transition from soft snow to ice; it contains much air; the upper portions of most glaciers and ice shelves are usually composed of névé.

nevus [MED] A lesion containing melanocytes.

newberyite [MINERAL] $MgH(PO_4)\cdot 3H_2O$ A white, orthorhombic member of the brushite mineral group; it is isostructural with gypsum.

new candle *See* candela.

newel post [CIV ENG] **1.** A pillar at the end of an oblique retaining wall of a bridge. **2.** The post about which a circular staircase winds. **3.** A large post at the foot of a straight stairway or on a landing.

new moon [ASTRON] The moon at conjunction, when little or none of it is visible to an observer on the earth because the illuminated side is turned away.

newt [VERT ZOO] Any of the small, semiaquatic salamanders of the genus *Triturus* in the family Salamandridae; all have an aquatic larval stage.

newton [MECH] The unit of force in the meter-kilogram-second system, equal to the force which will impart an acceleration of 1 meter per second squared to the International Prototype Kilogram mass. Symbolized N. Formerly known as large dyne.

Newton formula for the stress *See* Newtonian friction law.

Newtonian-Cassegrain telescope [OPTICS] A modification of a Cassegrain telescope in which the light reflected from the hyperboloidal secondary mirror is again reflected from a diagonal plane mirror and focused at a point on the side of the telescope, avoiding the need to pierce the primary mirror and making the eyepiece more accessible. Also known as Cassegrain-Newtonian telescope.

Newtonian flow [FL MECH] Flow system in which the fluid performs as a Newtonian fluid, that is, shear stress is proportional to shear rate.

Newtonian fluid [FL MECH] A simple fluid in which the state of stress at any point is proportional to the time rate of strain at that point; the proportionality factor is the viscosity coefficient.

Newtonian friction law [FL MECH] The law that shear stress in a fluid is proportional to the shear rate; it holds only for some fluids, which are then called Newtonian. Also known as Newton formula for the stress.

Newtonian mechanics [MECH] The system of mechanics based upon Newton's laws of motion in which mass and energy are considered as separate, conservative, mechanical properties, in contrast to their treatment in relativistic mechanics.

Newtonian reference frame [MECH] One of a set of reference frames with constant relative velocity and within which Newton's laws hold; the frames have a common time, and coordinates are related by the Galilean transformation rule.

Newtonian telescope [OPTICS] A reflecting telescope in which the light reflected from a concave mirror is reflected again by a plane mirror making an angle of 45° with the telescope axis, so that it passes through a hole in the side of the telescope containing the eyepiece.

newton-meter of energy *See* joule.

Newton's first law [MECH] The law that a particle not subjected to external forces remains at rest or moves with constant speed in a straight line. Also known as first law of motion; Galileo's law of inertia.

Newton's law of gravitation [MECH] The law that every two particles of matter in the universe attract each other with a force that acts along the line joining them, and has a magnitude proportional to the product of their masses and inversely proportional to the square of the distance between them. Also known as law of gravitation.

Newton's laws of motion [MECH] Three fundamental principles (called Newton's first, second, and third laws) which form the basis of classical, or Newtonian, mechanics, and have proved valid for all mechanical problems not involving speeds comparable with the speed of light and not involving atomic or subatomic particles.

Newton's second law [MECH] The law that the acceleration of a particle is directly proportional to the resultant external force acting on the particle and is inversely proportional to the mass of the particle. Also known as second law of motion.

Newton's theory of light *See* corpuscular theory of light.

Newton's third law [MECH] The law that, if two particles interact, the force exerted by the first particle on the second particle (called the action force) is equal in magnitude and opposite in direction to the force exerted by the second particle on the first particle (called the reaction force). Also known as law of action and reaction; third law of motion.

NGU *See* nongonococcal urethritis.

Ni *See* nickel.

niacin *See* nicotinic acid.

niacinamide *See* nicotinamide.

Niagaran [GEOL] A North American provincial geologic series, in the Middle Silurian.

nibbling [MECH ENG] Contour cutting of material by the action of a reciprocating punch that takes repeated small bites as the work is passed beneath it.

nicad battery *See* nickel-cadmium battery.

nicarbing *See* carbonitriding.

niccolite [MINERAL] NiAs A pale-copper-red, hexagonal mineral with metallic luster; an important ore of nickel; hardness is 5–5.5 on Mohs scale. Also known as arsenical nickel; copper nickel; nickeline.

nickel [CHEM] A chemical element, symbol Ni, atomic number 28, atomic weight 58.71. [MET] A silver-gray, ductile, malleable, tough metal; used in alloys, plating, coins (to replace silver), ceramics, and electronic circuits.

nickel-cadmium battery [ELEC] A sealed storage battery having a nickel anode, a cadmium cathode, and an alkaline electrolyte; widely used in cordless appliances; without recharging, it can serve as a primary battery. Also known as cadmium-nickel storage cell; nicad battery.

nickel glance *See* gersdorffite.

nickeline *See* niccolite.

nickel-iron battery *See* Edison battery.

nickel pyrites *See* millerite.

nickel vitriol *See* morenosite.

nickpoint *See* knickpoint.

Nicoletiidae [INV ZOO] A family of the insect order Thysanura proper.

Nicol prism [OPTICS] A device for producing plane-polarized light, consisting of two pieces of transparent calcite (a birefringent crystal) which together form a parallelogram and are cemented together with Canada balsam.

Nicomachinae [INV ZOO] A subfamily of the limnivorous sedentary annelids in the family Maldanidae.

nicotinamide [BIOCHEM] $C_6H_6ON_2$ Crystalline basic amide of the vitamin B complex that is interconvertible with nicotinic acid in the living organism; amide of nicotinic acid. Also known as niacinamide.

nicotinamide adenine dinucleotide *See* diphosphopyridine nucleotide.

nicotinic acid [BIOCHEM] $C_6H_5NO_2$ A component of the vitamin B complex; a white, water-soluble powder stable to heat, acid, and alkali; used for the treatment of pellagra. Also known as niacin.

nictitating membrane [VERT ZOO] A membrane of the inner angle of the eye or below the eyelid in many vertebrates, and capable of extending over the eyeball.

night-sky light   See airglow.

night-sky luminescence   See airglow.

nighttime visual range   See night visual range.

night-vision telescope   [OPTICS] A telescope that has sufficient electronic amplification of images to be used at night without artificial illumination; may have television, optoelectronic, or other means of providing the necessary image amplification.

night visual range   [OPTICS] The greatest distance at which a point source of light of a given candlepower can be perceived at night by an observer under given atmospheric conditions. Also known as nighttime visual range; penetration range; transmission range.

Nilionidae   [INV ZOO] The false ladybird beetles, a family of coleopteran insects in the superfamily Tenebrionoidea.

nimbostratus   [METEOROL] A principal cloud type, or cloud genus, gray-colored and often dark, rendered diffuse by more or less continuously falling rain, snow, or sleet of the ordinary varieties, and not accompanied by lightning, thunder, or hail; in most cases the precipitation reaches the ground.

nimbus   [ASTRON] See halo.   [METEOROL] A characteristic rain cloud; the term is not used in the international cloud classification except as a combining term, as cumulonimbus.

N indicator   See N scope.

ninety-column card   [COMPUT SCI] A card, punched or to be punched, divided in half horizontally, each half containing 45 columns, and each column containing six punch positions.

niobium   [CHEM] A chemical element, symbol Nb, atomic number 41, atomic weight 92.906.   [MET] A platinum-gray, ductile metal with brilliant luster; used in alloys, especially stainless steels. Also known as columbium.

nip   [GEOL] 1. A small, low cliff or break in slope which is produced by wavelets at the high-water mark. 2. The point on the bank of a meander lake where erosion takes place due to crowding of the stream current toward the lake. 3. Thinning of a coal seam, particularly if caused by tectonic movements. Also known as want.   [MET] See angle of nip.   [MIN ENG] See squeeze.

Nipkow disk   [COMPUT SCI] In optical character recognition, a disk having one or more spirals of holes around the outer edge, with successive openings positioned so that rotation of the disk provides mechanical scanning, as of a document.

nipple   [ANAT] The conical projection in the center of the mamma, containing the outlets of the milk ducts.   [DES ENG] A short piece of tubing, usually with an internal or external thread at each end, used to couple pipes.

Nippotaeniidea   [INV ZOO] An order of tapeworms of the subclass Cestoda including some internal parasites of certain fresh-water fishes; the head bears a single terminal sucker.

Nitelleae   [BOT] A tribe of stoneworts, order Charales, characterized by 10 cells in two tiers of five each composing the apical crown.

Nitidulidae   [INV ZOO] The sap-feeding beetles, a large family of coleopteran insects in the superfamily Cucujoidea; individuals have five-jointed tarsi and antennae with a terminal three-jointed clavate expansion.

nitrate   [CHEM] 1. A salt or ester of nitric acid. 2. Any compound containing the NO_3^- radical.

nitrate mineral   [MINERAL] Any of several generally rare minerals characterized by a fundamental ionic structure of NO_3^-; examples are soder niter, niter, and nitrocalcite.

nitratine   See soda niter.

nitration   [ORG CHEM] Introduction of an NO_2^- group into an organic compound.

nitride   [INORG CHEM] Compound of nitrogen and a metal, such as Mg_3N_2.

nitriding   [MET] Surface hardening of steel by formation of nitrides; nitrogen is introduced into the steel usually by heating in gaseous ammonia.

nitrification   [MICROBIO] Formation of nitrous and nitric acids or salts by oxidation of the nitrogen in ammonia; specifically, oxidation of ammonium salts to nitrites and oxidation of nitrites to nitrates by certain bacteria.

nitrite   [CHEM] A compound containing the radical NO_2^-; can be organic or inorganic.

nitro-   [CHEM] Chemical prefix showing the presence of the NO_2^- radical.

Nitrobacteraceae   [MICROBIO] The nitrifying bacteria, a family of gram-negative, chemolithotrophic bacteria; autotrophs which derive energy from nitrification of ammonia or nitrite, and obtain carbon for growth by fixation of carbon dioxide.

nitrogelatin   See gelatin dynamite.

nitrogen   [CHEM] A chemical element, symbol N, atomic number 1, atomic weight 14.0067; it is a gas, diatomic (N_2) under normal conditions; about 78% of the atmosphere is N_2; in the combined form the element is a constituent of all proteins.

nitrogen cycle   See carbon-nitrogen cycle.

nitrogen fixation   [CHEM ENG] Conversion of atmospheric nitrogen into compounds such as ammonia, calcium cyanamide, or nitrogen oxides by chemical or electric-arc processes.   [MICROBIO] Assimilation of atmospheric nitrogen by heterotrophic bacteria. Also known as dinitrogen fixation.

nitrohydrochloric acid   See aqua regia.

nitromuriatic acid   See aqua regia.

nitroso   [CHEM] The radical NO^- with trivalent nitrogen. Also known as hydroximino; oximido.

nivation   [GEOL] Rock or soil erosion beneath a snowbank or snow patch, due mainly to frost action but also involving chemical weathering, solifluction, and meltwater transport of weathering products. Also known as snow patch erosion.

nivenite   [MINERAL] UO_2 A velvet-black member of the uranite group; contains rare-earth metals cerium and yttrium; a source of uranium.

NMR   See nuclear magnetic resonance.

nn junction   [ELECTR] In a semiconductor, a region of transition between two regions having different properties in n-type semiconducting material.

No   See nobelium.

no-address instruction   [COMPUT SCI] An instruction which a computer can carry out without using an operand from storage.

nobelium   [CHEM] A chemical element, symbol No, atomic number 102, atomic weight 254 when the element is produced in the laboratory; a synthetic element, in the actinium series.

noble gas   [CHEM] A gas in group 0 of the periodic table of the elements; it is monatomic and, with limited exceptions, chemically inert. Also known as inert gas.

noble metal   [MET] A metal, or alloy, such as gold, silver, or platinum having high resistance to corrosion and oxidation; used in the construction of thin-film circuits, metal-film resistors, and other metal-film devices.

Nocardiaceae [MICROBIO] A family of aerobic bacteria in the order Actinomycetales; mycelium and spore production is variable.

nocerite *See* fluoborite.

nocioceptive reflex *See* flexion reflex.

noct-, nocti-, nocto-, noctu- [SCI TECH] Combining form meaning night.

Noctilionidae [VERT ZOO] The fish-eating bats, a tropical American monogeneric family of the Chiroptera having small eyes and long, narrow wings.

noctilucent cloud [METEOROL] A cloud of unknown composition which occurs at great heights and high altitudes; photometric measurements have located such clouds between 75 and 90 kilometers; they resemble thin cirrus, but usually with a bluish or silverish color, although sometimes orange to red, standing out against a dark night sky.

Noctuidae [INV ZOO] A large family of dull-colored, medium-sized moths in the superfamily Noctuoidea; larva are mostly exposed foliage feeders, representing an important group of agricultural pests.

Noctuoidea [INV ZOO] A large superfamily of lepidopteran insects in the suborder Heteroneura; most are moderately large moths with reduced maxillary palpi.

nodal line [ASTRON] The line passing through the ascending and descending nodes of the orbit of a celestial body. [PHYS] **1.** A line or curve in a two-dimensional standing-wave system, such as a vibrating diaphragm, where some specified characteristic of the wave, such as velocity of pressure, does not oscillate. **2.** A line which remains fixed during some deformation or rotation of a body or coordinate system.

nodal points [ELEC] Junction points in a transmission system; the automatic switches and switching centers are the nodal points in automated systems. [OCEANOGR] The no-tide points in amphidromic regions. [OPTICS] A pair of points on the axis of an optical system such that an incident ray passing through one of them results in a parallel emergent ray passing through the other.

node [ANAT] **1.** A knob or protuberance. **2.** A small, rounded mass of tissue, such as a lymph node. **3.** A point of constriction along a nerve. [ASTRON] **1.** One of two points at which the orbit of a planet, planetoid, or comet crosses the plane of the ecliptic. **2.** One of two points at which a satellite crosses the equatorial plane of its primary. [ELEC] *See* branch point. [ELECTR] A junction point within a network. [GEOL] That point along a fault at which the direction of apparent displacement changes. [MATH] *See* crunode. [PHYS] A point, line, or surface in a standing-wave system where some characteristic of the wave has essentially zero amplitude.

Nodosariacea [INV ZOO] A superfamily of Foraminiferida in the suborder Rotaliina characterized by a radial calcite test wall with monolamellar septa, and a test that is coiled, uncoiled, or spiral about the long axis.

nodular cast iron [MET] Cast iron treated in the molten state with a master alloy containing an element such as magnesium which favors formation of spheroidal graphite. Also known as ductile iron; spheroidal graphite cast iron.

nodule [ANAT] **1.** A small node. **2.** A small aggregation of cells. [GEOL] A small, hard mass or lump of a mineral or mineral aggregate characterized by a contrasting composition from and a greater hardness than the surrounding sediment or rock matrix

in which it is embedded. [MED] A primary skin lesion, seen as a circumscribed solid elevation.

Noeggerathiales [PALEOBOT] A poorly defined group of fossil plants whose geologic range extends from Upper Carboniferous to Triassic.

no-go gage [ENG] A limit gage designed not to fit a part being tested; usually employed with a go gage to set the acceptable maximum and minimum dimension limits of the part.

noise [ACOUS] Sound which is unwanted, either because of its effect on humans, its effect on fatigue or malfunction of physical equipment, or its interference with the perception or detection of other sounds. [ELEC] Interfering and unwanted currents or voltages in an electrical device or system.

noise-canceling microphone *See* close-talking microphone.

noise digit [COMPUT SCI] A digit, usually 0, inserted into the rightmost position of the mantissa of a floating point number during a left-shift operation associated with normalization. Also known as noisy digit.

noise factor [ELECTR] The ratio of the total noise power per unit bandwidth at the output of a system to the portion of the noise power that is due to the input termination, at the standard noise temperature of 290 K. Also known as noise figure.

noise figure *See* noise factor.

noise filter [ELECTR] **1.** A filter that is inserted in an alternating-current power line to block noise interference that would otherwise travel through the line in either direction and affect the operation of receivers. **2.** A filter used in a radio receiver to reduce noise, usually an auxiliary low-pass filter which can be switched in or out of the audio system.

noise generator [ELECTR] A device which produces (usually random) electrical noise, for use in tests of the response of electrical systems to noise, and in measurements of noise intensity. Also known as noise source.

noise level [PHYS] The intensity of unwanted sound, or the magnitude of unwanted currents or voltages, averaged over a specified frequency range and time interval, and weighted with frequency in a specified manner; usually expressed in decibels relative to a specified reference.

noise limiter [ELECTR] A limiter circuit that cuts off all noise peaks that are stronger than the highest peak in the desired signal being received, thereby reducing the effects of atmospheric or man-made interference. Also known as noise silencer; noise suppressor.

noise silencer *See* noise limiter.

noise source *See* noise generator.

noise suppressor [ELECTR] **1.** A circuit that blocks the audio-frequency amplifier of a radio receiver automatically when no carrier is being received, to eliminate background noise. Also known as squelch circuit. **2.** A circuit that reduces record surface noise when playing phonograph records, generally by means of a filter that blocks out the higher frequencies where such noise predominates. **3.** *See* noise limiter.

noisy digit *See* noise digit.

noisy mode [COMPUT SCI] A floating-point arithmetic procedure associated with normalization in which "1" bits, rather than "0" bits, are introduced in the low-order bit position during the left shift.

no-load current [ELEC] The current which flows in a network when the output is open-circuited.

no-load loss [ELEC] The power loss of a device that is operated at rated voltage and frequency but is not supplying power to a load.

nomen dubium [SYST] A proposed taxonomic name invalid because it is not accompanied by a definition or description of the taxon to which it applies.

nomen nudum [SYST] A proposed taxonomic name invalid because the accompanying definition or description of the taxon cannot be interpreted satisfactorily.

nomogram *See* nomograph.

nomograph [MATH] A chart which represents an equation containing three variables by means of three scales so that a straight line cuts the three scales in values of the three variables satisfying the equation. Also known as abac; alignment chart; nomogram.

nonarithmetic shift *See* cyclic shift.

nonbearing wall [CIV ENG] A wall that bears no vertical weight other than its own.

nonconformity [GEOL] A type of unconformity in which rocks below the surface of unconformity are either igneous or metamorphic.

nonconsumable electrode [MET] An electrode, such as of carbon or tungsten, that is not consumed during a welding or melting operation.

noncontacting piston *See* choke piston.

noncontacting plunger *See* choke piston.

noncoring bit [ENG] A general type of bit made in many shapes which does not produce a core and with which all the rock cut in a borehole is ejected as sludge; used mostly for blasthole drilling and in the unmineralized zones in a borehole where a core sample is not wanted. Also known as borehole bit; plug bit.

noncrossing rule [PHYS CHEM] The rule that when the potential energies of two electronic states of a diatomic molecule are plotted as a function of distance between the nuclei, the resulting curves do not cross, unless the states have different symmetry.

nondepositional unconformity *See* paraconformity.

nondestructive breakdown [ELECTR] Breakdown of the barrier between the gate and channel of a field-effect transistor without causing failure of the device; in a junction field-effect transistor, avalanche breakdown occurs at the *pn* junction.

nondestructive read [COMPUT SCI] A reading process that does not erase the data in memory; the term sometimes includes a destructive read immediately followed by a restorative write-back. Also known as nondestructive readout (NDRO).

nondestructive readout *See* nondestructive read.

nondestructive testing [ENG] Any testing method which does not involve damaging or destroying the test sample; includes use of x-rays, ultrasonics, radiography, magnetic flux, and so on.

nondeterministic [SCI TECH] Unpredictable in terms of observable antecedents and known laws; this is a relative term pertaining to a given state of knowledge but not necessarily implying ultimate unpredictability.

nondimensional parameter *See* dimensionless number.

nondirectional antenna *See* omnidirectional antenna.

nondisjunction mosaic [GEN] A population of cells with different chromosome numbers produced when one chromosome is lost during mitosis or when both members of a pair of chromosomes are included in the same daughter nucleus; can occur during embryogenesis or adulthood.

nondissipative muffler *See* reactive muffler.

nonequilibrium thermodynamics [THERMO] A quantitative treatment of irreversible processes and of rates at which they occur. Also known as irreversible thermodynamics.

nonerasable storage [COMPUT SCI] A device that permits a nondestructive read, such as punched cards, electrically conductive sheets, or paper tape.

nonexecutable statement [COMPUT SCI] A statement in a higher-level programming language which cannot be related to the instructions in the machine language program ultimately produced, but which provides the compiler with essential information from which it may determine the allocation of storage and other organizational characteristics of the final program.

nonferrous metal [MET] Any metal other than iron and its alloys.

nonfunctional packages software [COMPUT SCI] General-purpose software which permits the user to handle his particular applications requirements with little or no additional program or systems design work, or to perform certain specialized computational functions.

nongonococcal urethritis [MED] Human urethral inflammation not associated with common bacterial pathogens; thought to be caused by bacteria of the Bedsonia group. Abbreviated NGU.

nonholonomic constraints [MATH] A nonintegrable set of differential equations which describe the restrictions on the motion of a system or in optimization.

nonintelligible crosstalk [COMMUN] Crosstalk which cannot be understood regardless of its received volume, but which because of its syllabic nature is more annoying subjectively than thermal-type noise.

noninverting amplifier [ELECTR] An operational amplifier in which the input signal is applied to the ungrounded positive input terminal to give a gain greater than unity and make the output voltage change in phase with the input voltage.

Nonionacea [INV ZOO] A superfamily of Foraminiferida in the suborder Orbitoidacea, characterized by a granular calcite test wall with monolamellar septa, and a planispiral to trochospiral test.

nonionic detergent [MATER] A detergent with molecules that do not ionize in aqueous solution, for example, detergents derived from condensation products of long-chain glycols and octyl or nonyl phenols.

nonlinear [PHYS] Pertaining to a response which is other than directly or inversely proportional to a given variable.

nonlinear distortion [ELECTR] Distortion in which the output of a system or component does not have the desired linear relation to the input. [ENG ACOUS] The ratio of the total root-mean-square (rms) harmonic distortion output of a microphone to the rms value of the fundamental component of the output.

nonlinear equation [MATH] An equation in variables $x_1, \ldots, x_n, y$ which cannot be put into the form $a_1x_1 + \ldots + a_nx_n = y$.

nonlinear optics [OPTICS] The study of the interaction of radiation with matter in which certain variables describing the response of the matter (such as electric polarization or power absorption) are not proportional to variables describing the radiation (such as electric field strength or energy flux).

nonlinear programming [MATH] A branch of applied mathematics concerned with finding the maximum or minimum of a function of several variables, when the variables are constrained to yield values of other functions lying in a certain range, and either the function to be maximized or minimized, or at least one of the functions whose value is constrained, is nonlinear.

non-Newtonian fluid [FL MECH] A fluid whose flow behavior departs from that of a Newtonian fluid, so that the rate of shear is not proportional to the corresponding stress. Also known as non-Newtonian system.

non-Newtonian system *See* non-Newtonian fluid.

non-Newtonian viscosity [FL MECH] The behavior of a fluid which, when subjected to a constant rate of shear, develops a stress which is not proportional to the shear. Also known as anomalous viscosity.

nonpolarized relay *See* neutral relay.

nonprint code [COMPUT SCI] A bit combination which is interpreted as no printing, no spacing.

nonpriority interrupt [COMPUT SCI] Any one of a group of interrupts which may be disregarded by the central processing unit.

nonprocedural language [COMPUT SCI] A programming language in which the program does not follow the actual steps a computer follows in executing a program.

nonreactive load *See* resistive load.

nonredundant system [COMPUT SCI] A computer system designed in such a way that only the absolute minimum amount of hardware is utilized to implement its function.

nonreproducing code [COMPUT SCI] A code which normally does not appear as such in a generated output but will result in a function such as paging or spacing.

nonresonant antenna [ELECTROMAG] A long-wire or traveling-wave antenna which does not have natural frequencies of oscillation, and responds equally well to radiation over a broad range of frequencies.

non-return-to-zero [COMPUT SCI] A mode of recording and readout in which it is not necessary for the signal to return to zero after each item of recorded data.

nonrotational strain *See* irrotational strain.

nonsaccharine sorghum *See* grain sorghum.

nonselective radiator *See* graybody.

nonsense mutation [MOL BIO] A mutation that changes a codon that codes for one amino acid into a codon that does not specify any amino acid (a nonsense codon).

nonskid [CIV ENG] Pertaining to a surface that is roughened to reduce slipping, as a concrete floor treated with iron filings or carborundum powder, or indented while wet.

nonsynchronous [ELEC] Not related in phase, frequency, or speed to other quantities in a device or circuit.

nonthermal radiation [PHYS] Electromagnetic radiation emitted by accelerated charged particles that are not in thermal equilibrium; aurora light and fluorescent-lamp light are examples.

nontronite [MINERAL] $Na(Al,Fe,Si)O_{10}(OH)_2$ An iron-rich clay mineral of the montmorillonite group that represents the end member in which the replacement of aluminum by ferric ion is essentially complete. Also known as chloropal; gramenite; morencite; pinguite.

nonviscous flow *See* inviscid flow.

nonviscous fluid *See* inviscid fluid.

nonvolatile memory *See* nonvolatile storage.

nonvolatile storage [COMPUT SCI] A computer storage medium that retains information in the absence of power, such as a magnetic tape, drum, or core. Also known as nonvolatile memory.

nonwoven fabric [TEXT] Cloth produced from a random arrangement or matting of natural or synthetic fibers held together by adhesives, heat and pressure, or needling; felt is an example.

noon [ASTRON] The instant at which a time reference is over the upper branch of the reference meridian.

NO OP [COMPUT SCI] An instruction telling the computer to do nothing, except to proceed to the next instruction in sequence. Also known as no-operation instruction.

no-operation instruction *See* NO OP.

nor- [CHEM] Chemical formula prefix for normal; indicates a parent for another compound to be formed by removal of one or more carbons and associated hydrogens.

NOR [MATH] A logic operator having the property that if P, Q, R, ... are statements, then the NOR of P, Q, R, ... is true if all statements are false, false if at least one statement is true. Derived from NOT-OR. Also known as Peirce stroke relationship.

noradrenaline *See* norepinephrine.

norbergite [MINERAL] $Mg_3SiO_4(F,OH)_2$ A yellow or pink orthorhombic mineral composed of magnesium silicate with fluoride and hydroxyl; it is a member of the humite group.

NOR circuit [ELECTR] A circuit in which output voltage appears only when signal is absent from all of its input terminals.

Nordheim's rule [SOLID STATE] The rule that the residual resistivity of a binary alloy that contains mole fraction x of one element and $1 - x$ of the other is proportional to $x(1 - x)$.

nor'easter *See* northeaster.

norepinephrine [BIOCHEM] $C_8H_{11}O_3N$ A hormone produced by chromaffin cells of the adrenal medulla; acts as a vasoconstrictor and mediates transmission of sympathetic nerve impulses. Also known as noradrenaline.

norite [PETR] A coarse-grained plutonic rock composed principally of basic plagioclase with orthopyroxene (hypersthene) as the dominant mafic material. Also known as hypersthenfels.

norm [MATH] A scalar valued function on a vector space with properties analogous to those of the modulus of a complex number; namely: the norm of the zero vector is zero, all other vectors have positive norm, the norm of a scalar times a vector equals the absolute value of the scalar times the norm of the vector, and the norm of a sum is less than or equal to the sum of the norms. [PETR] The theoretical mineral composition of a rock expressed in terms of standard mineral molecules as determined by means of chemical analyses. [QUANT MECH] **1.** The square of the modulus of a Schrödinger-Pauli wave function, integrated over the space coordinates and summed over the spin coordinates of the particles it describes. **2.** The square root of this quantity.

normal curve *See* Gaussian curve.

normal cycle [GEOL] A cycle of erosion whereby a region is reduced to base level by running water, especially by the action of rivers. Also known as fluvial cycle of erosion.

normal density function [STAT] A normally distributed frequency distribution of a random variable x with mean e and variance σ is given by $(1/\sigma\sqrt{2\pi}) \exp[-(x-e)^2/2\sigma^2]$.

normal derivative [MATH] The directional derivative of a function at a point on a given curve or surface in the direction of the normal to the curve or surface.

normal dip *See* regional dip.

normal displacement *See* dip slip.

normal distribution [STAT] The most commonly occurring probability distributions have the form

$$(1/\sigma\sqrt{2\pi}) \int_{-\infty}^{u} \exp(-u^2/2)du, \ u = (x - e)/\sigma$$

where e is the mean and σ is the variance. Also known as Gauss' error curve; Gaussian distribution.

normal fault [GEOL] A fault, usually of 45–90°, in which the hanging wall appears to have shifted downward in relation to the footwall. Also known as gravity fault; normal slip fault; slump fault.

normal fold *See* symmetrical fold.

normal horizontal separation *See* offset.

normality [CHEM] Measure of the number of gram-equivalent weights of a compound per liter of solution. Abbreviated N.

normalization [COMPUT SCI] Breaking down of complex data structures into flat files.

normalize [COMPUT SCI] 1. To adjust the representation of a quantity so that this representation lies within a prescribed range. 2. In particular, to adjust the exponent and mantissa of a floating point number so that the mantissa falls within a prescribed range. [MATH] To multiply a quantity by a suitable constant or scalar so that it then has norm one; that is, its norm is then equal to one. [MET] To heat a ferrous alloy to some temperature above the transformation range, followed by air cooling. [QUANT MECH] To multiply a wave function by a constant so that its norm is equal to unity. [STAT] To carry out a normal transformation on a variate.

normal magnetization curve [ELECTROMAG] Curve traced on a graph of magnetic induction versus magnetic field strength in an originally unmagnetized specimen, as the magnetic field strength is increased from zero. Also known as magnetization curve.

normal mode [COMPUT SCI] Operation of a computer in which it executes its own instructions rather than those of a different computer.

normal operator [MATH] A linear operator where composing it with its adjoint operator in either order gives the same result. Also known as normal transformation.

normal pitch [MECH ENG] The distance between working faces of two adjacent gear teeth, measured between the intersections of the line of action with the faces.

normal pressure *See* standard pressure.

normal salt [CHEM] A salt in which all of the acid hydrogen atoms have been replaced by a metal, or the hydroxide radicals of a base are replaced by an acid radical; for example, Na_2CO_3.

normal slip fault *See* normal fault.

normal solution [CHEM] An aqueous solution containing one equivalent of the active reagent in grams in 1 liter of the solution.

normal space [MATH] A topological space in which any two disjoint closed sets may be covered respectively by two disjoint open sets.

normal surface [OPTICS] The surface that is generated by taking, at each point of the ray surface, the intersection of the tangent plane to the ray surface at that point with the perpendicular from the origin to this plane.

normal temperature and pressure *See* standard conditions.

normal transformation [MATH] *See* normal operator. [STAT] A transformation on a variate that converts it into a variate which has a normal distribution.

normal volume *See* standard volume.

normative mineral *See* standard mineral.

north [GEOD] The direction of the north terrestrial pole; the primary reference direction on the earth; the direction indicated by 000° in any system other than relative.

North Atlantic Current [OCEANOGR] A wide, slow-moving continuation of the Gulf Stream originating in the region east of the Grand Banks of Newfoundland.

northbound node *See* ascending node.

North Cape Current [OCEANOGR] A warm current flowing northeastward and eastward around northern Norway, and curving into the Barents Sea.

Northeast Drift Current [OCEANOGR] A North Atlantic Ocean current flowing northeastward toward the Norwegian Sea, gradually widening and, south of Iceland, branching and continuing as the Irminger Current and the Norwegian Current; it is the northern branch of the North Atlantic Current.

northeaster [METEOROL] A northeast wind, particularly a strong wind or gale. Also spelled nor'easter.

North Equatorial Current [OCEANOGR] Westward ocean currents driven by the northeast trade winds blowing over tropical oceans of the Northern Hemisphere. Also known as Equatorial Current.

northern anthracnose [PL PATH] A fungus disease of red and crimson clovers in North America, Asia, and Europe caused by *Kabatulla caulivora*; depressed, linear brown lesions form on the stems and petioles.

Northern Cross *See* Cygnus.

Northern Crown *See* Corona Borealis.

north geographic pole *See* North Pole.

north geomagnetic pole *See* north pole.

north magnetic pole *See* north pole.

North Pacific Current [OCEANOGR] The warm branch of the Kuroshio Extension flowing eastward across the Pacific Ocean.

north point [ASTRON] The point on the celestial sphere, due north of the observer, at which the celestial meridian intersects the celestial horizon.

North Polar Spur [ASTRON] One of the largest sources of diffuse radio emission outside the galactic plane; the Spur, a ridge of enhanced emission, is possibly the remnant of the shells of supernovae which exploded over 100,000 years ago.

north pole [ASTRON] The north celestial pole that indicates the zenith of the heavens when viewed from the north geographic pole. [ELECTROMAG] The pole of a magnet at which magnetic lines of force are considered as leaving the magnet; the lines enter the south pole; if the magnet is freely suspended, its north pole points toward the north geomagnetic pole. [GEOPHYS] The geomagnetic pole in the Northern Hemisphere, at approximately latitude 78.5°N, longitude 69°W. Also known as north geomagnetic pole; north magnetic pole.

North Pole [GEOGR] The geographic pole located at latitude 90°N in the Northern Hemisphere of the earth; it is the northernmost point of the earth, and the northern extremity of the earth's axis of rotation. Also known as north geographic pole.

North Star *See* Polaris.

northwester [METEOROL] A northwest wind. Also spelled nor'wester.

Norton's theorem [ELEC] The theorem that the voltage across an element that is connected to two terminals of a linear network is equal to the short-circuit current between these terminals in the absence of the element, divided by the sum of the admittances between the terminals associated with the element and the network respectively.

Norway Current [OCEANOGR] A continuation of the North Atlantic Current, which flows northward along the coast of Norway. Also known as Norwegian Current.

Norwegian Current *See* Norway Current.

nor'wester *See* northwester.

nose [ANAT] The nasal cavities and the structures surrounding and associated with them in all vertebrates. [ENG] The foremost point or section of a bomb, missile, or something similar. [FL MECH] The dense, forward part of a turbidity current. [GEOL] 1. A plunging anticline that is short and without closure. 2. A projecting and generally overhanging buttress of rock. 3. The projecting end of a hill, spur, ridge, or mountain. 4. The central forward part of a parabolic dune.

nosean *See* noselite.

nose cone [AERO ENG] A protective cone-shaped case for the nose section of a missile or rocket; may include the warhead, fusing system, stabilization system, heat shield, and supporting structure and equipment.

noselite [MINERAL] $Na_4Al_3Si_3O_{12} \cdot SO_4$ A gray, blue or brown mineral of the sodalite group; similar to haüynite; hardness is 5.5 on Mohs scale. Also known as nosean.

nosing [BUILD] Projection of a tread of a stair beyond the riser below it.

Notacanthidae [VERT ZOO] A family of benthic, deep-sea teleosts in the order Notacanthiformes, including the spiny eel.

Notacanthiformes [VERT ZOO] An order of actinopterygian fishes whose body is elongated, tapers posteriorly, and has no caudal fin.

NOT-AND *See* NAND.

notation [COMPUT SCI] *See* positional notation. [MATH] 1. The use of symbols to denote quantities or operations. 2. *See* positional notation.

notch [ELECTR] Rectangular depression extending below the sweep line of the radar indicator in some types of equipment. [ENG] A V-shaped indentation or cut in a surface or edge. [GEOL] A deep, narrow cut near the high-water mark at the base of a sea cliff. [GEOGR] A narrow passage between mountains or through a ridge, hill, or mountain.

notch filter [ELECTR] A band-rejection filter that produces a sharp notch in the frequency response curve of a system; used in television transmitters to provide attenuation at the low-frequency end of the channel, to prevent possible interference with the sound carrier of the next lower channel.

notch test [MET] A tensile or creep test of a metal to determine the effect of a surface notch.

NOT circuit [ELECTR] A logic circuit with one input and one output that inverts the input signal at the output; that is, the output signal is a logical 1 if the input signal is a logical 0, and vice versa. Also known as inverter circuit.

Noteridae [INV ZOO] The burrowing water beetles, a small family of coleopteran insects in the suborder Adephaga.

NOT function [MATH] A logical operator having the property that if P is a statement, then the NOT of P is true if P is false, and false if P is true.

Nothosauria [PALEON] A suborder of chiefly marine Triassic reptiles in the order Sauropterygia.

Notidanoidea [VERT ZOO] A suborder of rare sharks in the order Selachii; all retain the primitive jaw suspension of the order.

Notiomastodontinae [PALEON] A subfamily of extinct elephantoid proboscidean mammals in the family Gomphotheriidae.

Notioprogonia [PALEON] A suborder of extinct mammals comprising a diversified archaic stock of Notoungulata.

notochord [VERT ZOO] An elongated dorsal cord of cells which is the primitive axial skeleton in all chordates; persists in adults in the lowest forms (*Branchiostoma* and lampreys) and as the nuclei pulposi of the intervertebral disks in adult vertebrates.

Notodelphyidiformes [INV ZOO] A tribe of the Gnathostoma in some systems of classification.

Notodelphyoida [INV ZOO] A small group of crustaceans bearing a superficial resemblance to many insect larvae as a result of uniform segmentation, comparatively small trunk appendages, and crowding of inconspicuous oral appendages into the anterior portion of the head.

Notodontidae [INV ZOO] The puss moths, a family of lepidopteran insects in the superfamily Noctuoidea, distinguished by the apparently three-branched cubitus.

Notommatidae [INV ZOO] A family of rotifers in the order Monogonota including forms with a cylindrical body covered by a nonchitinous cuticle and with a slender posterior foot.

Notomyotina [INV ZOO] A suborder of echinoderms in the order Phanerozonida in which the upper marginals alternate in position with the lower marginals, and each tube foot has a terminal sucking disk.

Notonectidae [INV ZOO] The backswimmers, a family of aquatic, carnivorous hemipteran insects in the superfamily Notonectoidea; individuals swim ventral side up, aided in breathing by an air bubble.

Notonectoidea [INV ZOO] A superfamily of Hemiptera in the subdivision Hydrocorisae.

Notopteridae [VERT ZOO] The featherbacks, a family of actinopterygian fishes in the order Osteoglossiformes; bodies are tapered and compressed, with long anal fins that are continuous with the caudal fin.

NOT-OR *See* NOR.

Notoryctidae [PALEON] An extinct family of Australian insectivorous mammals in the order Marsupialia.

Notostigmata [INV ZOO] The single suborder of the Opilioacriformes, an order of mites.

Notostigmophora [INV ZOO] A subclass or suborder of the Chilopoda, including those centipedes embodying primitive as well as highly advanced characters, distinguished by dorsal respiratory openings.

Notostraca [INV ZOO] The tadpole shrimps, an order of crustaceans generally referred to the Branchiopoda, having a cylindrical trunk that consists of 25–44 segments, a dorsoventrally flattened dorsal shield, and two narrow, cylindrical cercopods on the telson.

Nototheniidae [VERT ZOO] A family of perciform fishes in the suborder Blennioidei, including most of the fishes of the permanently frigid waters surrounding Antarctica.

Notoungulata [PALEON] An extinct order of hoofed herbivorous mammals, characterized by a skull with an expanded temporal region, primitive dentition, and primitive feet with five toes, the weight borne mainly by the third digit.

nought state *See* zero condition.

noumeite *See* garnierite.

nova [ASTRON] A star that suddenly becomes explosively bright, the term is a misnomer because it does not denote a new star but the brightening of an existing faint star.

novaculite [GEOL] A siliceous sedimentary rock that is dense, hard, even-textured, light-colored, and characterized by dominance of microcrystalline quartz over chalcedony. Also known as razor stone.

nozzle [DES ENG] A tubelike device, usually streamlined, for accelerating and directing a fluid, whose pressure decreases as it leaves the nozzle.

nozzle efficiency [MECH ENG] The efficiency with which a nozzle converts potential energy into kinetic energy, commonly expressed as the ratio of the actual change in kinetic energy to the ideal change at the given pressure ratio.

Np See neper; neptunium.

npin transistor [ELECTR] An *npn* transistor which has a layer of high-purity germanium between the base and collector to extend the frequency range.

N-plus-one address instruction [COMPUT SCI] An instruction with N + 1 address parts, one of which gives the location of the next instruction to be carried out.

npnp diode See pnpn diode.

npnp transistor [ELECTR] An *npn*-junction transistor having a transition or floating layer between *p* and *n* regions, to which no ohmic connection is made. Also known as *pnpn* transistor.

npn semiconductor [ELECTR] Double junction formed by sandwiching a thin layer of *p*-type material between two layers of *n*-type material of a semiconductor.

npn transistor [ELECTR] A junction transistor having a *p*-type base between an *n*-type emitter and an *n*-type collector; the emitter should then be negative with respect to the base, and the collector should be positive with respect to the base.

np semiconductor [ELECTR] Region of transition between *n*- and *p*-type material.

NRM See natural remanent magnetization.

N scan See N scope.

N scope [ELECTR] A cathode-ray scope combining the features of K and M scopes. Also known as N indicator; N scan.

N shell [ATOM PHYS] The fourth layer of electrons about the nucleus of an atom, having electrons characterized by the principal quantum number 4.

NTP See standard conditions.

n-type conduction [ELECTR] The electrical conduction associated with electrons, as opposed to holes, in a semiconductor.

n-type semiconductor [ELECTR] An extrinsic semiconductor in which the conduction electron density exceeds the hole density.

Nubeculae See Magellanic Clouds.

Nubecula Major See Large Magellanic Cloud.

Nubecula Minor See Small Magellanic Cloud.

nucellus [BOT] The oval central mass of tissue in the ovule; contains the embryo sac.

nuchal ligament [ANAT] An elastic ligament extending from the external occipital protuberance and middle nuchal line to the spinous process of the seventh cervical vertebra. Also known as ligamentum nuchae.

nuclear [CHEM] Pertaining to a group of atoms joined directly to the central group of atoms or central ring of a molecule. [NUCLEO] Pertaining to nuclear energy. [NUC PHYS] Pertaining to the atomic nucleus.

nuclear age determination See radiometric dating.

nuclear angular momentum See nuclear spin.

nuclear atom [CHEM] An atomic structure consisting of dense, positively charged nucleus (neutrons and protons) surrounded by a corresponding set of negatively charged electrons.

nuclear battery [NUCLEO] A primary battery in which the energy of radioactive material is converted into electric energy by solar cells or other energy converters. Also known as atomic battery; radioisotope battery; radioisotopic generator.

nuclear binding energy [NUC PHYS] The energy required to separate an atom into its constituent protons, neutrons, and electrons.

nuclear capture [NUC PHYS] Any process in which a particle, such as a neutron, proton, electron, muon, or alpha particle, combines with a nucleus.

nuclear chain reaction [NUCLEO] A succession of generation after generation of acts of nuclear division such that the neutrons set free in the nuclear disruptions of the *n*th generation split the fissile nuclei (U^{233}, U^{235}, Pu^{239}) of the $(n + 1)$st generation. Also known as chain reaction.

nuclear chemistry [ATOM PHYS] Study of the atomic nucleus, including fission and fusion reactions and their products.

nuclear converter See converter.

nuclear cross section [NUC PHYS] A measure of the probability for a reaction to occur between a nucleus and a particle; it is an area such that the number of reactions which occur in a sample exposed to a beam of particles equals the product of the number of nuclei in the sample and the number of incident particles which would pass through this area if their velocities were perpendicular to it.

nuclear decay mode [NUC PHYS] One of the ways in which a nucleus can undergo radioactive decay, distinguished from other decay modes by the resulting isotope and the particles emitted.

nuclear energy [NUCLEO] Energy released by nuclear fission or nuclear fusion. Also known as atomic energy.

nuclear engine [NUCLEO] A type of thermal engine utilizing nuclear fission or fusion reactions to heat a working fluid for propulsive purposes.

nuclear engineering [NUCLEO] The branch of technology that deals with the utilization of the nuclear fission process, and is concerned with the design and construction of nuclear reactors and auxiliary facilities, the development and fabrication of special materials, and the handling and processing of reactor products.

nuclear fission See fission.

nuclear force [NUC PHYS] That part of the force between nucleons which is not electromagnetic; it is much stronger than electromagnetic forces, but drops off very rapidly at distances greater than about 10^{-13} centimeter; it is responsible for holding the nucleus together.

nuclear fuel [NUCLEO] A fissionable or fertile isotope with a reasonably long half-life, used as a source of energy in a nuclear reactor. Also known as fission fuel; reactor fuel.

nuclear fuel cycle See reactor fuel cycle.

nuclear fuel pebble See nuclear fuel pellet.

nuclear fuel pellet [NUCLEO] A piece of nuclear fuel usually in the shape of a sphere or cylinder, used in pebble-bed reactors, inserted in graphite blocks, or used in metallic tubular fuel elements. Also known as fuel ball; nuclear fuel pebble; reactor fuel pellet.

nuclear fusion See fusion.

nuclear induction [PHYS] Magnetic induction originating in the magnetic moments of nuclei; the effect depends on the unequal population of energy states available when the material is placed in a magnetic field.

nuclear magnetic moment [NUC PHYS] The magnetic dipole moment of an atomic nucleus; a vector whose scalar product with the magnetic flux density gives the negative of the energy of interaction of a nucleus with a magnetic field.

nuclear magnetic resonance   [PHYS] A phenomenon exhibited by a large number of atomic nuclei, in which nuclei in a static magnetic field absorb energy from a radio-frequency field at certain characteristic frequencies. Abbreviated NMR. Also known as magnetic nuclear resonance.

nuclear magnetism   [PHYS] The phenomena associated with the magnetic dipole, octupole, and higher moments of a nucleus, including the magnetic field generated by the nucleus, the force on the nucleus in an inhomogeneous magnetic field, and the splitting of nuclear energy levels in a magnetic field.

nuclear magneton   [NUC PHYS] A unit of magnetic dipole moment used to express magnetic moments of nuclei and baryons; equal to the electron charge times Planck's constant divided by the product of 4π, the proton mass, and the speed of light.

nuclear mass   [NUC PHYS] The mass of an atomic nucleus, which is usually measured in atomic mass units; it is less than the sum of the masses of its constituent protons and neutrons by the binding energy of the nucleus divided by the square of the speed of light.

nuclear moment   [NUC PHYS] One of the various static electric or magnetic multipole moments of a nucleus.

nuclear number   See mass number.

nuclear paramagnetism   [PHYS] Paramagnetism in which a substance develops a net magnetic moment because the magnetic moments of nuclei tend to point in the direction of the field.

nuclear physics   [PHYS] The study of the characteristics, behavior, and internal structures of the atomic nucleus.

nuclear pile   See nuclear reactor.

nuclear polarization   [NUC PHYS] For a nucleus in a mixed state, with spin I and probability $p(I_z)$ that the I_z substate is populated, the polarization is the sum over allowed values of I_z of $I_z p(I_z)/I$.

nuclear power   [NUCLEO] Power whose source is nuclear fission or fusion

nuclear quadrupole moment   [NUC PHYS] The electric quadrupole moment of an atomic nucleus.

nuclear radiation   [NUC PHYS] A term used to denote alpha particles, neutrons, electrons, photons, and other particles which emanate from the atomic nucleus as a result of radioactive decay and nuclear reactions.

nuclear reaction   [NUC PHYS] A reaction involving a change in an atomic nucleus, such as fission, fusion, neutron capture, or radioactive decay, as distinct from a chemical reaction, which is limited to changes in the electron structure surrounding the nucleus. Also known as reaction.

nuclear reactor   [NUCLEO] A device containing fissionable material in sufficient quantity and so arranged as to be capable of maintaining a controlled, self-sustaining nuclear fission chain reaction. Also known as atomic pile (deprecated usage); atomic reactor; fission reactor; nuclear pile (deprecated); pile (deprecated); reactor.

nuclear relaxation   [PHYS] The approach of a system of nuclear spins to a steady-state or equilibrium condition over a period of time, following a change in the applied magnetic field.

nuclear resonance   [NUC PHYS] 1. An unstable excited state formed in the collision of a nucleus and a bombarding particle, and associated with a peak in a plot of cross section versus energy. 2. The absorption of energy by nuclei from radio-frequency fields at certain frequencies when these nuclei are also subjected to certain types of static fields, as in magnetic resonance and nuclear quadrupole resonance.

nuclear spallation   See spallation.

nuclear species   See nuclide.

nuclear spectrum   [NUC PHYS] 1. The relative number of particles emitted by atomic nuclei as a function of energy or momenta of these particles. 2. The graphical display of data from devices used to measure these quantities.

nuclear spin   [NUC PHYS] The total angular momentum of an atomic nucleus, resulting from the coupled spin and orbital angular momenta of its constituent nuclei. Also known as nuclear angular momentum. Symbolized I.

nuclear spontaneous reaction   See radioactive decay.

nuclear star   See star.

nuclear transformation   See transmutation.

nuclear weapon   See atomic weapon.

nuclease   [BIOCHEM] An enzyme that catalyzes the splitting of nucleic acids to nucleotides, nucleosides, or the components of the latter.

nucleation   [CHEM] In crystallization processes, the formation of new crystal nuclei in supersaturated solutions.

nucleic acid   [BIOCHEM] A large, acidic, chainlike molecule containing phosphoric acid, sugar, and purine and pyrimidine bases; two types are ribonucleic acid and deoxyribonucleic acid.

nucleolus   [CYTOL] A small, spherical body composed principally of protein and located in the metabolic nucleus. Also known as plasmosome.

nucleon   [PHYS] A collective name for a proton or a neutron; these particles are the main constituents of atomic nuclei, have approximately the same mass, have a spin of $\frac{1}{2}$, and can transform into each other through the process of beta decay.

nucleon number   See mass number.

nucleoprotein   [BIOCHEM] Any member of a class of conjugated proteins in which molecules of nucleic acid are closely associated with molecules of protein.

nucleoside   [BIOCHEM] The glycoside resulting from removal of the phosphate group from a nucleotide; consists of a pentose sugar linked to a purine or pyrimidine base.

nucleosynthesis   [ASTROPHYS] The formation of the various nuclides present in the universe by various nuclear reactions, occurring chiefly in the early universe following the big bang, in the interiors of stars, and in supernovae.

nucleotide   [BIOCHEM] An ester of a nucleoside and phosphoric acid; the structural unit of a nucleic acid.

nucleus   [COMPUT SCI] 1. That portion of the control program that must always be present in main storage. 2. The main storage area used in the nucleus (first definition) and other transient control program routines.   [ASTRON] The small permanent body of a comet, believed to have a diameter between one and a few tens of kilometers, and to be composed of water and volatile hydrocarbons.   [CYTOL] A small mass of differentiated protoplasm rich in nucleoproteins and surrounded by a membrane; found in most animal and plant cells, contains chromosomes, and functions in metabolism, growth, and reproduction.   [HISTOL] A mass of nerve cells in the central nervous system.   [HYD] A particle of any nature upon which, or a locus at which, molecules of water or ice accumulate as a result of a phase change to a more condensed state.   [NUC PHYS] The central, positively charged, dense portion of an atom. Also known as atomic nucleus. [SCI TECH] A central mass about which accretion takes place.

nuclide [NUC PHYS] A species of atom characterized by the number of protons, number of neutrons, and energy content in the nucleus, or alternatively by the atomic number, mass number, and atomic mass; to be regarded as a distinct nuclide, the atom must be capable of existing for a measurable lifetime, generally greater than 10^{-10} second. Also known as nuclear species; species.

Nuda [INV ZOO] A class of the phylum Ctenophora distinguished by the lack of tentacles.

Nudechiniscidae [INV ZOO] A family of heterotardigrades in the suborder Echiniscoidea characterized by a uniform cuticle.

Nudibranchia [INV ZOO] A suborder of the Opisthobranchia containing the sea slugs; these mollusks lack a shell and a mantle cavity, and the gills are variable in size and shape.

null [MATH] A term meaning that an object is nonexistent or a quantity is zero. [NAV] 1. The azimuth or elevation reading on a navigational device indicated by minimum signal output. 2. Any of the nodal points on the radiation patterns of some antennas.

null character [COMPUT SCI] A control character used as a filler in data processing; may be inserted or removed from a sequence of characters without affecting the meaning of the sequence, but may affect format or control equipment.

null-current circuit [ELECTR] A circuit used to measure current, in which the unknown current is opposed by a current resulting from applying a voltage controlled by a slide wire across a series resistor, and the slide wire is continuously adjusted so that the resulting current, as measured by a direct-current detector amplifier, is equal to zero.

null detection [ELEC] Altering of adjustable bridge circuit components, to obtain zero current. [NAV] Method of determining the radio direction by altering the antenna position to obtain minimum signal strength.

null detector *See* null indicator.

null hypothesis [STAT] The hypothesis that there is no validity to the specific claim that two variations (treatments) of the same thing can be distinguished by a specific procedure.

null indicator [ENG] A galvanometer or other device that indicates when voltage or current is zero; used chiefly to determine when a bridge circuit is in balance. Also known as null detector.

null space [MATH] For a linear transformation, the vector subspace of all vectors which the transformation sends to the zero vector. Also known as kernel.

null vector [MATH] A vector whose invariant length, that is, the sum over the coordinates of the vector space of the product of its covariant component and contravariant component, is equal to zero. [RELAT] In special relativity, a four vector whose spatial part in any Lorentz frame has a magnitude equal to the speed of light multiplied by its time part in that frame; a special case of the mathematics definition.

number [MATH] 1. Any real or complex number. 2. The number of elements in a set is the cardinality of the set.

number cruncher [COMPUT SCI] A computer with great power to carry out computations, designed to maximize this ability rather than to process large amounts of data.

number scale [MATH] Representation of points on a line with numbers arranged in some order.

numeration [MATH] The listing of numbers in their natural order.

numerator [MATH] In a fraction a/b, the numerator is the quantity a.

numeric [COMPUT SCI] In computers, pertaining to data composed wholly or partly of digits, as distinct from alphabetic.

numerical analysis [MATH] The study of approximation techniques using arithmetic for solutions of mathematical problems.

numerical aperture [OPTICS] A measure of the resolving power of a microscope objective, equal to the product of the refractive index of the medium in front of the objective and the sine of the angle between the outermost ray entering the objective and the optical axis. Abbreviated N.A.

numerical control [CONT SYS] A control system for machine tools and some industrial processes, in which numerical values corresponding to desired positions of tools or controls are recorded on punched paper tapes, punched cards, or magnetic tapes so that they can be used to control the operation automatically. Abbreviated NC; N/C.

numerical decrement *See* decrement.

numerical display device [ELECTR] Any device for visually displaying numerical figures, such as a numerical indicator tube, a device utilizing electroluminescence, or a device in which any one of a stack of transparent plastic strips engraved with digits can be illuminated by a small light at the edge of the strip.

numerical forecasting [METEOROL] The forecasting of the behavior of atmospheric disturbances by the numerical solution of the governing fundamental equations of hydrodynamics, subject to observed initial conditions. Also known as dynamic forecasting; mathematical forecasting; numerical weather prediction; physical forecasting.

numerical taxonomy [SYST] The numerical evaluation of the affinity or similarity between taxonomic units and the ordering of these units into taxa on the basis of their affinities.

numerical weather prediction *See* numerical forecasting.

numeric character *See* digit.

numeric coding [COMPUT SCI] Code in which only digits are used, usually binary or octal.

numeric control [COMPUT SCI] The action of programs written for specialized computers which operate machine tools.

Numididae [VERT ZOO] A family of birds in the order Galliformes commonly known as guinea fowl; there are few if any feathers on the neck or head, but there may be a crest of feathers and various fleshy appendages.

nun buoy [NAV] A red buoy made of two conical or truncated cone-shaped sections joined at the base; marks the starboard side when entering a channel from the sea.

nut [BOT] 1. A fruit which has at maturity a hard, dry shell enclosing a kernel consisting of an embryo and nutritive tissue. 2. An indehiscent, one-celled, one-seeded, hard fruit derived from a single, simple, or compound ovary. [DES ENG] An internally threaded fastener for bolts and screws.

nutating antenna [ENG] An antenna system used in conical scan radar, in which a dipole or feed horn moves in a small circular orbit about the axis of a paraboloidal reflector without changing its polarization.

nutating-disk meter [ENG] An instrument for measuring flow of a liquid in which liquid passing through a chamber causes a disk to nutate, or roll back and forth, and the total number of rolls is mechanically counted.

nutation [ASTRON] A slight, slow, nodding motion of the earth's axis of rotation which is superimposed on the precession of the equinoxes; it is the combination of a number of perturbations (lunar, solar, and fortnightly nutation). [BOT] The rhythmic change in the position of growing plant organs caused by variation in the growth rates on different sides of the growing apex. [MECH] A bobbing or nodding up-and-down motion of a spinning rigid body, such as a top, as it precesses about its vertical axis.

nutrition [BIOL] The science of nourishment, including the study of nutrients that each organism must obtain from its environment in order to maintain life and reproduce.

nutritional dystrophy *See* kwashiorkor.

nutritional edema [MED] Edema resulting from starvation or malnutrition.

nutritional hypochromic anemia *See* iron-deficiency anemia.

Nuttalliellidae [INV ZOO] A family of ticks (Ixodides) containing one rare African species, *Nuttalliella namaqua*, morphologically intermediate between the families Argasidae and Ixodidae.

Nyctaginaceae [BOT] A family of dicotyledonous plants in the order Caryophyllales characterized by an apocarpous, monocarpous, or syncarpous gynoecium, sepals joined to a tube, a single carpel, and a cymose inflorescence.

Nycteridae [VERT ZOO] The slit-faced bats, a monogeneric family of insectivorous chiropterans having a simple, well-developed nose leaf, and large ears joined together across the forehead.

Nyctibiidae [VERT ZOO] A family of birds in the order Caprimulgiformes including the neotropical potoos.

Nyctribiidae [INV ZOO] The bat tick flies, a family of myodarian cyclorrhaphous dipteran insects in the subsection Acalyptratae.

nylon [MATER] Generic name for long-chain polymeric amide molecules in which recurring amide groups are part of the main polymer chain; used to make fibers, fabrics, sheeting, and extruded forms.

nymph [INV ZOO] Any immature larval stage of various hemimetabolic insects.

Nymphaeaceae [BOT] A family of dicotyledonous plants in the order Nymphaeales distinguished by the presence of roots, perfect flowers, alternate leaves, and uniaperturate pollen.

Nymphaeales [BOT] An order of flowering aquatic herbs in the subclass Magnoliidae; all lack cambium and vessels and have laminar placentation.

Nymphalidae [INV ZOO] The four-footed butterflies, a family of lepidopteran insects in the superfamily Papilionoidea; prothoracic legs are atrophied, and the well-developed patagia are heavily sclerotized.

Nymphalinae [INV ZOO] A subfamily of the lepidopteran family Nymphalidae.

Nymphonidae [INV ZOO] A family of marine arthropods in the subphylum Pycnogonida; members have chelifores, five-jointed palpi, and ten-jointed ovigers.

Nymphulinae [INV ZOO] A subfamily of the lepidopteran family Pyralididae which is notable because some species are aquatic.

Nyquist rate [COMMUN] The maximum rate at which code elements can be unambiguously resolved in a communications channel with a limited range of frequencies; equal to twice the frequency range.

Nyquist stability criterion *See* Nyquist stability theorem.

Nyquist stability theorem [CONT SYS] The theorem that the net number of counterclockwise rotations about the origin of the complex plane carried out by the value of an analytic function of a complex variable, as its argument is varied around the Nyquist contour, is equal to the number of poles of the variable in the right half-plane minus the number of zeros in the right half-plane. Also known as Nyquist stability criterion.

Nyquist's theorem [ELECTR] The mean square noise voltage across a resistance in thermal equilibrium is four times the product of the resistance, Boltzmann's constant, the absolute temperature, and the frequency range within which the voltage is measured.

Nyssaceae [BOT] A family of dicotyledonous plants in the order Cornales characterized by perfect or unisexual flowers with imbricate petals, a solitary ovule in each locule, a unilocular ovary, and more stamens than petals.

nystagmus [MED] Involuntary oscillatory movement of the eyeballs.

O *See* oxygen.

oak [BOT] Any tree of the genus *Quercus* in the order Fagales, characterized by simple, usually lobed leaves, scaly winter buds, a star-shaped pith, and its fruit, the acorn, which is a nut; the wood is tough, hard, and durable, generally having a distinct pattern.

oak wilt [PL PATH] A fungus disease of oak trees caused by *Chalara quercina*, characterized by wilting and yellow and red discoloration of the leaves progressing from the top downward and inward.

oasis [GEOGR] An isolated fertile area, usually limited in extent and surrounded by desert, and marked by vegetation and a water supply.

oat [BOT] Any plant of the genus *Avena* in the family Graminae, cultivated as an agricultural crop for its seed, a cereal grain, and for straw.

obelisk [ARCH] A four-sided pillar, tapering toward the top. [MATH] A frustrum of a regular, rectangular pyramid.

Oberon [ASTRON] One of the five satellites of Uranus; diameter about 1400 kilometers.

object computer [COMPUT SCI] The computer processing an object program; the same computer compiling the source program could, therefore, be called the source computer; such terminology is seldom used in practice.

object deck [COMPUT SCI] The set of machine-readable computer instructions produced by a compiler, either in absolute format (that is, containing only fixed addresses) or, more frequently, in relocatable format.

object glass *See* objective.

objective [OPTICS] The first lens, lens system, or mirror through which light passes or from which it is reflected in an optical system; many scientists exclude mirrors from the definition. Also known as object glass.

objective prism [OPTICS] A large prism, usually having a small angle, which is placed in front of the objective of a photographic telescope to make spectroscopic observations.

object language [COMPUT SCI] The intended and desired output language in the translation or conversion of information from one language to another.

object library *See* object program library.

object module [COMPUT SCI] The computer language program prepared by an assembler or a compiler after acting on a programmer-written source program.

object program [COMPUT SCI] The computer language program prepared by an assembler or a compiler after acting on a programmer-written source program. Also known as object routine; target program; target routine.

object program library [COMPUT SCI] A collection of computer programs in the form of relocatable instructions, which reside on, and may be read from, a mass storage device. Also known as object library.

object routine *See* object program.

object tape [COMPUT SCI] A tape, paper or magnetic, containing the machine language instructions resulting from a compiler or assembler, often found in minicomputer environments.

oblate spheroid [MATH] The surface or ellipsoid generated by rotating an ellipse about one of its axes so that the diameter of its equatorial circle exceeds the length of the axis of revolution. Also known as oblate ellipsoid.

oblique angle [MATH] An angle that is neither a right angle nor a multiple of a right angle.

oblique ascension [ASTRON] The arc of the celestial equator, or the angle at the celestial pole, between the hour circle of the vernal equinox and the hour circle through the intersection of the celestial equator and the eastern horizon at the instant a point on the oblique sphere rises, measured eastward from the hour circle of the vernal equinox through 24 hours.

oblique coordinates [MATH] Magnitudes defining a point relative to two intersecting nonperpendicular lines, called axes; the magnitudes indicate the distance from each axis, measured along a parallel to the other axis; oblique coordinates are a form of cartesian coordinates.

oblique cylindrical orthomorphic projection *See* oblique Mercator projection.

oblique extinction *See* inclined extinction.

oblique fault *See* diagonal fault.

oblique joint *See* diagonal joint.

oblique Mercator projection [MAP] A conformal cylindrical map projection in which points on the surface of a sphere or spheroid, such as the earth, are conceived as developed by Mercator principles on a cylinder tangent along an oblique great circle. Also known as oblique cylindrical orthomorphic projection.

Obolellida [PALEON] A small order of Early and Middle Cambrian inarticulate brachiopods, distinguished by a shell of calcium carbonate.

obscuration [METEOROL] In United States weather observing practice, the designation for the sky cover when the sky is completely hidden by surface-based obscuring phenomena, such as fog. Also known as obscured sky cover.

obscured sky cover *See* obscuration.

obsequent [GEOL] Of a stream, valley, or drainage system, being in a direction opposite to that of the original consequent drainage.

observable operator [QUANT MECH] A Hermitian operator with a complete, orthonormal set of eigenfunctions on the Hilbert space representing the states of a physical system; such operators are postulated to represent the observable quantities of the system.

obsidian [GEOL] A jet-black volcanic glass, usually of rhyolitic composition, formed by rapid cooling of viscous lava; generally forms the upper parts of lava flows. Also known as hyalopsite; Iceland agate; mountain mahogany.

obsidianite *See* tektite.

obstetrics [MED] The branch of medicine that deals with pregnancy, labor, and the puerperium.

obturator [ANAT] **1.** Pertaining to that which closes or stops up, as an obturator membrane. **2.** Either of two muscles, originating at the pubis and ischium, which rotate the femur laterally. [MED] A solid wire or rod contained within a hollow needle or cannula. [ORD] **1.** Assembly of steel spindle, mushroom head, obturator rings, and a gas-check or obturator pad of tough plastic material used as a seal to prevent the escape of propellant gases around the breechblock of guns using separate-loading ammunition, and therefore not having the obturation provided by a cartridge case. **2.** A device incorporated in a projectile to make the tube of a weapon gastight, preventing escape of gas until the projectile has left the muzzle of the weapon.

obtuse angle [MATH] An angle of more than 90° and less than 180°.

occipital artery [ANAT] A branch of the external carotid which branches into the mastoid, auricular, sternocleidomastoid, and meningeal arteries.

occipital bone [ANAT] The bone which forms the posterior portion of the skull, surrounding the foramen magnum.

occipital condyle [ANAT] An articular surface on the occipital bone which articulates with the atlas. [INV ZOO] A projection on the posterior border of an insect head which articulates with the lateral neck plates.

occipital lobe [ANAT] The posterior lobe of the cerebrum having the form of a three-sided pyramid.

occluded front [METEOROL] A composite of two fronts, formed as a cold front overtakes a warm front or quasi-stationary front. Also known as frontal occlusion; occlusion.

occlusion [ANAT] The relationship of the masticatory surfaces of the maxillary teeth to the masticatory surfaces of the mandibular teeth when the jaws are closed. [ENG] The retention of undissolved gas in a solid during solidification. [MED] A closing or shutting up. [METEOROL] *See* occluded front. [PHYS] Adhesion of gas or liquid on a solid mass, or the trapping of a gas or liquid within a mass. [PHYSIO] The deficit in muscular tension when two afferent nerves that share certain motor neurons in the central nervous system are stimulated simultaneously, as compared to the sum of tensions when the two nerves are stimulated separately.

occultation [ASTRON] The disappearance of the light of a celestial body by intervention of another body of larger apparent size; especially, a lunar eclipse of a star or planet.

occult blood [PATH] Blood in body products such as feces, not detectable on gross examination.

occupied bandwidth [COMMUN] Frequency bandwidth such that, below its lower and above its upper frequency limits, the mean powers radiated are each equal to 0.5% of the total mean power radiated by a given emission.

OC curve *See* operating characteristic curve.

ocean [GEOGR] A major primary subdivision of the intercommunicating body of salt water occupying the depressions of the earth's surface; bounded by continents and imaginary lines. Also known as sea.

oceanic climate *See* marine climate.

oceanic crust [GEOL] A thick mass of igneous rock which lies under the ocean floor.

oceanic island [GEOL] Any island which rises from the deep-sea floor rather than from shallow continental shelves.

oceanic ridge *See* mid-oceanic ridge.

oceanic rise [GEOL] A long, broad elevation of the bottom of the ocean.

oceanographic platform [ENG] A man-made structure with a flat horizontal surface higher than the water, on which oceanographic equipment is suspended or installed.

oceanographic submersible [NAV ARCH] Any small research vessel designed for undersea operations.

oceanographic vessel [NAV ARCH] A research ship or other manned vehicle used in oceanography.

oceanography [GEOPHYS] The scientific study and exploration of the oceans and seas in all their aspects. Also known as oceanology.

oceanology *See* oceanography.

ocellus [INV ZOO] A small, simple invertebrate eye composed of photoreceptor cells and pigment cells. [PETR] A phenocryst in an ocellar rock.

ocelot [VERT ZOO] *Felis pardalis.* A small arboreal wild cat, of the family Felidae, characterized by a golden head and back, silvery flanks, and rows of somewhat metallic spots on the body.

Ochnaceae [BOT] A family of dicotyledonous plants in the order Theales, characterized by simple, stipulate leaves, a mostly gynobasic style, and anthers that generally open by terminal pores.

Ochoan [GEOL] A North American provincial series that is uppermost in the Permian, lying above the Guadalupian and below the lower Triassic.

Ochotonidae [VERT ZOO] A family of the mammalian order Lagomorpha; members are relatively small, and all four legs are about equally long.

Ochrept [GEOL] A suborder of the soil order Inceptisol, with horizon below the surface, lacking clay, sesquioxides, or humus; widely distributed, occurring from the margins of the tundra region through the temperate zone, but not into the tropics.

Ochteridae [INV ZOO] The velvety shorebugs, the single family of the hemipteran superfamily Ochteroidea.

Ochteroidea [INV ZOO] A monofamilial tropical and subtropical superfamily of hemipteran insects in the subdivision Hydrocorisae; individuals are black with a silky sheen, and the antennae are visible from above.

OCR *See* optical character recognition.

octagon [MATH] A polygon with eight sides.

octahedral borax *See* tincalconite.

octahedral cleavage [CRYSTAL] Crystal cleavage in the four planes parallel to the face of the octahedron.

octahedral iron ore *See* magnetite.

octahedral normal stress [MECH] The normal component of stress across the faces of a regular octahedron whose vertices lie on the principal axes of stress; it is equal in magnitude to the spherical stress across any surface. Also known as mean stress.

octahedron [MATH] A polyhedron having eight faces, each of which is an equilateral triangle.

octal loading program [COMPUT SCI] Computer utility program providing a method for making changes in programs and tables existing in core memory or drum storage, by reading in words coded in octal notation on punched cards or tape.

octal number system [MATH] A number system in which a number r is written as $n_k n_{k-1} \ldots n_1$ where $r = n_1 8^0 + n_2 8^1 + \ldots + n_k 8^{k-1}$.

octane number [ENG] A rating that indicates the tendency to knock when a fuel is used in a standard internal combustion engine under standard conditions; n-heptane is 0, isooctane is 100; different test methods give research octane, motor octane, and road octane.

octave [ACOUS] The interval in pitch between two tones such that one tone may be regarded as duplicating at the next higher pitch the basic musical import of the other tone; the sounds producing these tones then have a frequency ratio of 2 to 1. [PHYS] The interval between any two frequencies having a ratio of 2 to 1.

octet [ATOM PHYS] A collection of eight valence electrons in an atom or ion, which form the most stable configuration of the outermost, or valence, electron shell. [PARTIC PHYS] A multiplet of eight elementary particles, corresponding to a representation of the approximate unitary symmetry (SU_3) of the strong interactions.

octillion [MATH] **1.** The number 10^{27}. **2.** In British and German usage, the number 10^{48}.

octode [ELECTR] An eight-electrode electron tube containing an anode, a cathode, a control electrode, and five additional electrodes that are ordinarily grids.

Octopoda [INV ZOO] An order of the dibranchiate cephalopods, characterized by having eight arms equipped with one to three rows of suckers.

Octopodidae [INV ZOO] The octopuses, in family of cephalopod mollusks in the order Octopoda.

octopus [INV ZOO] Any member of the genus *Octopus* in the family Octopodidae; the body is round with a large head and eight partially webbed arms, each bearing two rows of suckers, and there is no shell.

octupole [PHYS] **1.** Two electric or magnetic quadrupoles having charge distributions of opposite signs and separated from each other by a small distance. **2.** Any device for controlling beams of electrons or other charged particles, consisting of eight electrodes or magnetic poles arranged in a circular pattern, with alternating polarities; commonly used to correct aberrations of quadrupole systems.

octyl- [ORG CHEM] Prefix indicating the eight-carbon hydrocarbon radical ($C_8 H_{17} -$).

ocular [BIOL] Of or pertaining to the eye. [OPTICS] *See* eyepiece.

ocular prism [OPTICS] The prism employed in a range finder to bend the line of sight through the instrument into the eyepiece.

oculomotor nerve [ANAT] The third cranial nerve; a paired somatic motor nerve arising in the floor of the midbrain, which innervates all extrinsic eye muscles except the lateral rectus and superior oblique, and furnishes autonomic fibers to the ciliary and pupillary sphincter muscles within the eye.

Oculosida [INV ZOO] An order of the protozoan subclass Radiolaria; pores are restricted to certain areas in the central capsule, and an olive-colored material is always present near the astropyle.

OD *See* optical density; outside diameter.

odd-even check [COMPUT SCI] A means of detecting certain kinds of errors in which an extra bit, carried along with each word, is set to zero or one so that the total number of zeros or ones in each word is always made even or always made odd. Also known as parity check.

odd number [MATH] A natural number not divisible by 2.

odd parity [COMPUT SCI] Property of an expression in binary code which has an odd number of ones. [QUANT MECH] Property of a system whose state vector is multiplied by -1 under the operation of space inversion, that is, the simultaneous reflection of all spatial coordinates through the origin.

odd parity check [COMPUT SCI] A parity check in which the number of 0's or 1's in each word is expected to be odd; if the number is even, the check bit is 1, and if the number is odd, the check bit is 0.

Odiniidae [INV ZOO] A family of cyclorrhaphous myodarian dipteran insects in the subsection Acalyptratae.

Odobenidae [VERT ZOO] A family of carnivorous mammals in the suborder Pinnipedia; contains a single species, the walrus (*Odobenus rosmarus*).

odometer [ENG] **1.** An instrument for measuring distance traversed, as of a vehicle. **2.** The indicating gage of such an instrument. **3.** A wheel pulled by surveyors to measure distance traveled.

Odonata [INV ZOO] The dragonflies, an order of the class Insecta, characterized by a head with large compound eyes, and wings with clear or transparent membranes traversed by networks of veins.

odont-, odonto- [VERT ZOO] A combining form meaning tooth.

Odontognathae [PALEON] An extinct superorder of the avian subclass Neornithes, including all large, flightless aquatic forms and other members of the single order Hesperornithiformes.

Odontostomatida [INV ZOO] An order of the protozoan subclass Spirotrichia; individuals are compressed laterally and possess very little ciliature.

Oecophoridae [INV ZOO] A family of small to moderately small moths in the lepidopteran superfamily Tineoidea, characterized by a comb of bristles, the pecten, on the scape of the antennae.

Oedemeridae [INV ZOO] The false blister beetles, a large family of coleopteran insects in the superfamily Tenebrionoidea.

Oedogoniales [BOT] An order of fresh-water algae in the division Chlorophyta; characterized as branched or unbranched microscopic filaments with a basal holdfast cell.

Oegophiurida [INV ZOO] An order of echinoderms in the subclass Ophiuroidea, represented by a single living genus; members have few external skeletal plates and lack genital bursae, dorsal and ventral arm plates, and certain jaw plates.

Oegopsida [INV ZOO] A suborder of cephalopod mollusks in the order Decapoda of one classification system, and in the order Teuthoidea of another system.

O electron [ATOM PHYS] An electron in the fifth (O) shell of electrons surrounding the atomic nucleus, having the principal quantum number 5.

Oepikellacea [PALEON] A dimorphic superfamily of extinct ostracods in the order Paleocopa, distinguished by convex valves and the absence of any trace of a major sulcus in the external configuration.

oersted [ELECTROMAG] The unit of magnetic field strength in the centimeter-gram-second electromagnetic system of units, equal to the field strength at the center of a plane circular coil of one turn and 1-centimeter radius, when there is a current of $1/2 \pi$ abamp in the coil.

Oestridae [INV ZOO] A family of cyclorrhaphous myodarian dipteran insects in the subsection Calyptratae.

offgassing [MATER] The relative high mass loss characteristic of many nonmetallic materials upon initial vacuum exposure.

offlap [GEOL] The successive lateral contraction extent of strata (in an upward sequence) due to their deposition in a shrinking sea or on the margin of a rising landmass. Also known as regressive overlap.

off-line [COMPUT SCI] Describing equipment not connected to a computer, or temporarily disconnected from one. [ENG] 1. A condition existing when the drive rod of the drill swivel head is not centered and parallel with the borehole being drilled. 2. A borehole that has deviated from its intended course. 3. A condition existing wherein any linear excavation (shaft, drift, borehole) deviates from a previously determined or intended survey line or course. [IND ENG] State in which an equipment or subsystem is in standby, maintenance, or mode of operation other than on-line.

off-lining [COMPUT SCI] The process of separating card reading and printing from the actual running of jobs, by transcribing data to and from magnetic tape or other high-speed input/output devices.

offset {CONT SYS} The steady-state difference between the desired control point and that actually obtained in a process control system. [ENG] 1. A short perpendicular distance measured to a traverse course or a surveyed line or principal line of measurement in order to locate a point with respect to a point on the course or line. 2. In seismic prospecting, the horizontal distance between a shothole and the line of profile, measured perpendicular to the line. 3. In seismic refraction prospecting, the horizontal displacement, measured from the detector, of a point for which a calculated depth is relevant. 4. In seismic reflection prospecting, the correction of a reflecting element from its position on a preliminary working profile to its actual position in space. [GEOL] 1. The movement of an upcurrent part of a shore to a more seaward position than a downcurrent part. 2. A spur from a mountain range. 3. A level terrace on the side of a hill. 4. The horizontal displacement component in a fault, measured parallel to the stroke of the fault. Also known as normal horizontal separation. [MAP] During construction of a map projection, the small distance added to the length of meridians on either side of the central meridian in order to determine the chart's top latitude. [MECH] The value of strain between the initial linear portion of the stress-strain curve and a parallel line that intersects the stress-strain curve of an arbitrary value of strain; used as an index of yield stress; a value of 0.2% is common. [MIN ENG] 1. A short drift or crosscut driven from a main gangway or level. 2. The horizontal distance between the outcrops of a dislocated bed. [NAV ARCH] One of a series of measurements of the perpendicular distance of various points on a ship's hull from the centerline and above the molded baseline; used in ship construction. [ORD] The horizontal distance of forward travel covered by the missile after it strikes the ground; this distance is measured from the center of the hole of entry to the most forward part of the missile.

offset lithography [GRAPHICS] A system of printing that depends on the principle that the printing area accepts greasy ink while the nonprinting area is dampened with water and repels the ink; in practice, the image from a plate is offset onto the rubber blanket of an impression cylinder, and transferred to a sheet of paper.

offshore bar *See* longshore bar.

offshore current [OCEANOGR] 1. A prevailing nontidal current usually setting parallel to the shore outside the surf zone. 2. Any current flowing away from shore.

ogee [ARCH] A reverse curve, shaped like an elongated letter S, as the outline of an ogee molding.

ogive [ARCH] 1. An arch or rib placed diagonally across a Gothic vault. 2. A pointed arch. [GEOL] One of a periodically repeated series of dark, curved structures occurring down a glacier that resemble a pointed arch. [ORD] The curved or tapered front of a projectile.

ohm [ELEC] The unit of electrical resistance in the rationalized meter-kilogram-second system of units, equal to the resistance through which a current of 1 ampere will flow when there is a potential difference of 1 volt across it. Symbolized Ω.

ohmmeter [ENG] An instrument for measuring electric resistance; scale may be graduated in ohms or megohms.

Ohm's law [ELEC] The law that the direct current flowing in an electric circuit is directly proportional to the voltage applied to the circuit; it is valid for metallic circuits and many circuits containing an electrolytic resistance.

OHV engine *See* overhead-valve engine.

Oikomonadidae [INV ZOO] A family of protozoans in the order Kinetoplastida containing organisms that have a single flagellum.

oil [GEOL] *See* petroleum. [MATER] Any of various viscous, combustible, water-immiscible liquids that are soluble in certain organic solvents, as ether and naphtha; may be of animal, vegetable, mineral, or synthetic origin; examples are fixed oils, volatile or essential oils, and mineral oils.

oil accumulation *See* oil pool.

oil circuit breaker [ELECTR] A high-voltage circuit breaker in which the arc is drawn in oil to dissipate the heat and extinguish the arc; the intense heat of the arc decomposes the oil, generating a gas whose high pressure produces a flow of fresh fluid through the arc that furnishes the necessary insulation to prevent a restrike of the arc.

oil cup [ENG] A permanently mounted cup used to feed lubricant to a gear, usually with some means of regulating the flow.

oil field [PETRO ENG] The surface boundaries of an area from which petroleum is obtained; may correspond to an oil pool or may be circumscribed by political or legal limits.

oil-field brine [HYD] Connate waters, usually containing a high concentration of calcium and sodium salts and found during deep rock penetration by the drill.

oil-field separator *See* gas-oil separator.

oil filter [ENG] Cartridge-type filter used in automotive oil-lubrication systems to remove metal particles and products of heat decomposition from the circulating oil.

oil-gas separator *See* gas-oil separator.

oil gland *See* uropygial gland.

oil pool [GEOL] An accumulation of petroleum locally confined by subsurface geologic features. Also known as oil accumulation; oil reservoir.

oil reservoir *See* oil pool.

oil-reservoir water *See* formation water.

oil ring [MECH ENG] 1. A ring located at the lower part of a piston to prevent an excess amount of oil from being drawn up onto the piston during the suction stroke. 2. A ring on a journal, dipping into an oil bath for lubrication.

oil sand [GEOL] An unconsolidated, porous sand formation or sandstone containing or impregnated with petroleum or hydrocarbons.

oil separator See gas-oil separator.

oil shale [GEOL] A finely layered brown or black shale that contains kerogen and from which liquid or gaseous hydrocarbons can be distilled. Also known as kerogen shale.

oil trap [GEOL] An accumulation of petroleum which, by a combination of physical conditions, is prevented from escaping laterally or vertically. Also known as trap.

Oiluvium See Pleistocene.

oil well [PETRO ENG] A hole drilled (usually vertically) into an oil reservoir for the purpose of recovering the oil trapped in porous formations.

O indicator See O scope.

ointment [PHARM] A semisolid preparation used for a protective and emollient effect or as a vehicle for the local or endermic administration of medicaments; ointment bases are composed of various mixtures of fats, waxes, animal and vegetable oils, and solid and liquid hydrocarbons.

okapi [VERT ZOO] Okapia johnstoni. An artiodactylous mammal in the family Giraffidae; has a hazel coat with striped hindquarters, and the head shape, lips, and tongue are the same as those of the giraffe, but the neck is not elongate.

okra [BOT] Hibiscus esculentus. A tall annual plant grown for its edible immature pods. Also known as gumbo.

OL See only loadable.

Olacaceae [BOT] A family of dicotyledonous plants in the order Santalales characterized by dry or fleshy indehiscent fruit, the presence of petals, stamens, and chlorophyll, and a 2–5-celled ovary.

old age [GEOL] The last stage of the erosion cycle in the development of the topography of a region in which erosion has reduced the surface almost to base level and the land forms are marked by simplicity of form and subdued relief. Also known as topographic old age.

Oldham coupling See slider coupling.

Oldhaminidina [PALEON] A suborder of extinct articulate brachiopods in the order Strophomenida distinguished by a highly lobate brachial valve seated within an irregular convex pedicle valve.

oldhamite [MINERAL] CaS A pale-brown mineral known only from meteorites; unstable under earth conditions; member of the galena group with face-centered isometric structure.

old snow [HYD] Deposited snow in which the original crystalline forms are no longer recognizable, such as firn or spring snow. Also known as firn snow.

Oleaceae [BOT] A family of dicotyledonous plants in the order Scrophulariales characterized generally by perfect flowers, two stamens, axile to parietal or apical placentation, a four-lobed corolla, and two ovules in each locule.

oleate [ORG CHEM] Salt made up of a metal or alkaloid with oleic acid; used for external medicines and in soaps and paints.

olecranon [ANAT] The large process at the distal end of the ulna that forms the bony prominence of the elbow and receives the insertion of the triceps muscle.

olefin [ORG CHEM] C_nH_{2n} A family of unsaturated, chemically active hydrocarbons with one carbon-carbon double bond; includes ethylene and propylene. [TEXT] A manufactured fiber in which the fiber-forming substance is any long-chain synthetic polymer composed of at least 85% by weight of ethylene, propylene, or other olefin units except amorphous (noncrystalline) polyolefins qualifying as rubber.

olefin resin [ORG CHEM] Long-chain polymeric material produced by the chain reaction of olefinic monomers, such as polyethylene from ethylene, or polypropylene from propylene.

Olenellidae [PALEON] A family of extinct arthropods in the class Trilobita.

oleoresin [MATER] A resin-essential oil mixture with pungent taste; extracted from various plants; used in pharmaceutical preparations; examples are Peru, tulu, and styrax balsams.

Olethreutidae [INV ZOO] A family of moths in the superfamily Tortricoidea whose hindwings usually have a fringe of long hairs along the basal part of the cubitus.

olfaction [PHYSIO] **1.** The function of smelling. **2.** The sense of smell.

olfactory bulb [VERT ZOO] The bulbous distal end of the olfactory tract located beneath each anterior lobe of the cerebrum; well developed in lower vertebrates.

olfactory lobe [VERT ZOO] A lobe projecting forward from the inferior surface of the frontal lobe of each cerebral hemisphere, including the olfactory bulb, tracts, and trigone; well developed in most vertebrates, but reduced in man.

olfactory nerve [ANAT] The first cranial nerve; a paired sensory nerve with its origin in the olfactory lobe and formed by processes of the olfactory cells which lie in the nasal mucosa; greatly reduced in man.

olig-, oligo- [SCI TECH] A combining form denoting few, scant, or deficiency.

Oligobrachiidae [INV ZOO] A monotypic family of the order Athecanephria.

Oligocene [GEOL] The third of the five major worldwide divisions (epochs) of the Tertiary period (Cenozoic era), extending from the end of the Eocene to the beginning of the Miocene.

Oligochaeta [INV ZOO] A class of the phylum Annelida including worms that exhibit both external and internal segmentation, and setae which are not borne on parapodia.

oligochromemia See anemia.

oligoclase [MINERAL] A plagioclase feldspar mineral with a composition ranging from $Ab_{90}An_{10}$ to $Ab_{70}An_{30}$ ($Ab = NaAlSi_3O_8$ and $An = CaAl_2O_8$).

oligomer [ORG CHEM] A polymer made up of two, three, or four monomer units.

Oligomera [INV ZOO] A subphylum of the phylum Vermes comprising groups with two or three coelomic divisions.

oligomerous [BOT] Having one or more whorls with fewer members than other whorls of the flower.

Oligopygidae [PALEON] An extinct family of exocyclic Euechinoidia in the order Holectypoida which were small ovoid forms of the Early Tertiary.

oligosaccharide [BIOCHEM] A sugar composed of two to eight monosaccharide units joined by glycosidic bonds. Also known as compound sugar.

Oligotrichida [INV ZOO] A minor order of the Spirotrichia; the body is round in cross section, and the adoral zone of membranelles is often highly developed at the oral end of the organism.

oligotrophic [HYD] Of a lake, lacking plant nutrients and usually containing plentiful amounts of dissolved oxygen without marked stratification.

olive [BOT] Any plant of the genus Olea in the order Schrophulariales, especially the evergreen olive tree (O. europeae) cultivated for its drupaceous fruit, which is eaten ripe (black olives) and unripe (green), and is of high oil content.

olivenite [MINERAL] $Cu_2(AsO_4)(OH)$ An olive-green, dull-brown, gray, or yellow mineral crystallizing in the orthorhombic system and consisting of a basic arsenate of copper. Also known as leucochalcite; wood copper.

Olividae [INV ZOO] A family of snails in the gastropod order Neogastropoda.

olivine [MINERAL] $(Mg,Fe_2)SiO_4$ A neosilicate group of olive-green magnesium-iron silicate minerals crystallizing in the orthorhombic system and having a vitreous luster; hardness is 6½–7 on Mohs scale; specific gravity is 3.27–3.37.

olivine diabase [PETR] An igneous rock composed principally of olivine and formed from tholeiitic magmas by differentiation in thick sills.

olivine nephelinite [PETR] An extrusive igneous rock differing in composition from nephelinite only by the presence of olivine. Also known as ankaratrite; nepheline basalt.

omasum [VERT ZOO] The third chamber of the ruminant stomach where the contents are mixed to a more or less homogeneous state. Also known as manyplies; psalterium.

ombrometer *See* rain gage.

ombroscope [ENG] An instrument consisting of a heated, water-sensitive surface which indicates by mechanical or electrical techniques the occurrence of precipitation; the output of the instrument may be arranged to trip an alarm or to record on a time chart.

omega hyperon [PARTIC PHYS] A semistable baryon with a mass of approximately 1672 MeV, negative charge, spin of ³⁄₂, and positive parity; constitutes an isoptic spin singlet. Also known as omega particle. Symbolized Ω^1.

omega meson [PARTIC PHYS] An unstable, neutral vector meson having a mass of about 783 MeV, a width of about 12 MeV, and negative charge parity and G parity. Symbolized $\omega(783)$.

omega particle *See* omega hyperon.

omentum [ANAT] A fold of the peritoneum connecting or supporting abdominal viscera.

omission factor [COMPUT SCI] In information retrieval, the ratio obtained in dividing the number of nonretrieved relevant documents by the total number of relevant documents in the file.

ommatidium [INV ZOO] The structural unit of a compound eye, composed of a cornea, a crystalline cone, and a receptor element connected to the optic nerve.

omnibearing [NAV] The magnetic bearing of an omnidirectional radio range.

omnidirectional antenna [ELECTROMAG] An antenna that has an essentially circular radiation pattern in azimuth and a directional pattern in elevation. Also known as nondirectional antenna.

omnidirectional range *See* omnirange.

omnirange [NAV] A radio aid to navigation providing direct indication of the magnetic bearing (omnibearing) of that station from any direction. Also known as omnidirectional range.

omnivore [ZOO] An organism that eats both animal and vegetable matter.

Omophronidae [INV ZOO] The savage beetles, a small family of coleopteran insects in the suborder Adephaga.

omphacite [MINERAL] A grassy- to pale-green, granular or foliated, high-temperature aluminous clinopyroxene mineral with a vitreous luster that commonly occurs in the rock eclogite; a variety of augite.

OMS *See* ovonic memory switch.

Onagraceae [BOT] A family of dicotyledonous plants in the order Myrtales characterized by an inferior ovary, axile placentation, twice as many stamens as petals, a four-nucleate embryo sac, and many ovules.

on-call circuit [COMMUN] A permanently designated circuit that is activated only upon request of the user; this type of circuit is usually provided when a full-period circuit cannot be justified and the duration of use cannot be anticipated; during unactivated periods, the communications facilities required for the circuit are available for other requirements.

on center [BUILD] The measurement made between the centers of two adjacent members.

Onchidiidae [INV ZOO] An intertidal family of sluglike pulmonate mollusks in the order Systellommatophora in which the body is oval or lengthened, with the convex dorsal integument lacking a mantle cavity or shell.

oncogene [GEN] A gene that causes cancer in an animal; the gene specifies the structure of an enzyme that catalyzes events that can induce cancerous growth.

oncology [MED] The study of the causes, development, characteristics, and treatment of tumors.

Oncopoda [INV ZOO] A phylum of the superphylum Articulata.

one-address code [COMPUT SCI] In computers, a code using one-address instructions.

one-address instruction [COMPUT SCI] A digital computer programming instruction that explicitly describes one operation and one storage location. Also known as single-address instruction.

one condition [COMPUT SCI] The state of a magnetic core or other computer memory element in which it represents the value 1. Also known as one state.

one-dimensional flow [FL MECH] Fluid flow in which all flow is parallel to some straight line, and characteristics of flow do not change in moving perpendicular to this line.

one-level address [COMPUT SCI] In digital computers, an address that directly indicates the location of an instruction or some data.

one-level code [COMPUT SCI] Any code using absolute addresses and absolute operation codes.

one-level subroutine [COMPUT SCI] A subroutine that does not use other subroutines during its execution.

one-pass operation [COMPUT SCI] An operating method, now standard, which produces an object program from a source program in one pass.

one-plus-one address instruction [COMPUT SCI] A digital computer instruction whose format contains two address parts; one address designates the operand to be involved in the operation; the other indicates the location of the next instruction to be executed.

one's complement [COMPUT SCI] A numeral in binary notation, derived from another binary number by simply changing the sense of every digit.

ones-complement code [COMPUT SCI] A number coding system used in some computers, where, for any number x, $x = (1 - 2^{n-1}) \cdot a_0 + 2^{n-2} a_1 + \ldots + a_{n-1}$, where $a_i = 1$ or 0.

one-shot multivibrator *See* monostable multivibrator.

one-shot operation *See* single-step operation.

one-sided limit [MATH] Either a limit on the left or a limit on the right.

one-sided test [STAT] A test statistic T which rejects a hypothesis only for $T \geq d$ or $T \leq c$ but not for both (here d and c are critical values).

one state *See* one condition.

one-step operation *See* single-step operation.

one-to-one assembler [COMPUT SCI] An assembly program which produces a single instruction in machine language for each statement in the source language. Also known as one-to-one translater.

one-to-one mapping *See* injection.

one-to-one translater *See* one-to-one assembler.

one-way trunk [ELEC] Trunk between two central offices, used for calls that originate at one of those offices, but not for calls that originate at the other. Also known as outgoing trunk.

on grade [CIV ENG] **1.** At ground level. **2.** Supported directly on the ground.

onion [BOT] **1.** *Allium cepa.* A biennial plant in the order Liliales cultivated for its edible bulb. **2.** Any plant of the genus *Allium.*

onion scab *See* onion smudge.

onion smudge [PL PATH] A fungus disease of the onion caused by *Colletotrichum circinans* and characterized by black concentric integral rings or smutty spots on the bulb scales. Also known as onion scab.

onion smut [PL PATH] A fungus disease of onion, especially seedlings, caused by *Urocystis cepulae* and characterized by elongate black blisters on the scales and foliage.

Oniscoidea [INV ZOO] A terrestrial suborder of the Isopoda; the body is either dorsoventrally flattened or highly vaulted, and the head, thorax, and abdomen are broadly joined.

-onium [INORG CHEM] Chemical suffix indicating a complex cation, as for oxonium, $(H_3O)^+$.

on-line [COMPUT SCI] Pertaining to equipment capable of interacting with a computer. [ELECTR] The state in which a piece of equipment or a subsystem is connected and powered to deliver its proper output to the system.

on-line central file [COMPUT SCI] An organized collection of data, such as an on-line disk file, in a storage device under direct control of a central processing unit, that serves as a continually available source of data in applications where real-time or direct-access capabilities are required.

on-line cipher [COMMUN] A method of encryption directly associated with a particular transmission system, whereby messages may be encrypted and simultaneously transmitted from one station to one or more stations where reciprocal equipment is automatically operated.

on-line cryptographic operation *See* on-line operation.

on-line inquiry [COMPUT SCI] A level of computer processing that results from adding to an expanded batch system the capability to immediately access, from any terminal, any record that is stored in the disk files attached to the computer.

on-line operation [COMPUT SCI] Computer operation in which input data are fed into the computer directly from observing instruments or other input equipment, and computer results are obtained during the progress of the event. [COMMUN] A method of operation whereby messages are encrypted and simultaneously transmitted from one station to one or more other stations where reciprocal equipment is automatically operated to permit reception and simultaneous decryptment of the message. Also known as on-line cryptographic operation.

only loadable [COMPUT SCI] Attribute of a load module which can be brought into main memory only by a LOAD macroinstruction given from another module. Abbreviated OL.

on-off control [CONT SYS] A simple control system in which the device being controlled is either full on or full off, with no intermediate operating positions. Also known as on-off system.

on-off keying [COMMUN] Binary form of amplitude modulation in which one of the states of the modulated wave is the absence of energy in the keying interval.

on-off system *See* on-off control.

Onsager equation [PHYS CHEM] An equation which relates the measured equivalent conductance of a solution at a certain concentration to that of the pure solvent.

Onsager reciprocal relations [THERMO] A set of conditions which state that the matrix, whose elements express various fluxes of a system (such as diffusion and heat conduction) as linear functions of the various conjugate affinities (such as mass and temperature gradients) for systems close to equilibrium, is symmetric when certain definitions are chosen for these fluxes and affinities.

Onsager theory of dielectrics [ELEC] A theory for calculating the dielectric constant of a material with polar molecules in which the local field at a molecule is calculated for an actual spherical cavity of molecular size in the dielectric using Laplace's equation, and the polarization catastrophe of the Lorentz field theory is thereby avoided.

onshore wind [METEOROL] Wind blowing from the sea toward the land.

on stream [CHEM ENG] Of plant or process-operations, unit, being in operation.

on-the-fly printer [COMPUT SCI] A high-speed line printer using continuously rotating print wheels and fast-acting hammers to print the letters contained on one line of text so rapidly that the characters appear to be printed simultaneously.

ontogeny [EMBRYO] The origin and development of an organism from conception to adulthood.

Onuphidae [INV ZOO] A family of tubicolous, herbivorous, scavenging errantian annelids in the superfamily Eunicea.

onych-, onycho- [ZOO] A combining form denoting claw or nail.

Onychodontidae [PALEON] A family of Lower Devonian lobefin fishes in the order Osteolepiformes.

Onychopalpida [INV ZOO] A suborder of mites in the order Acarina.

Onychophora [INV ZOO] A phylum of wormlike animals that combine features of both the annelids and the arthropods.

Onygenaceae [MYCOL] A family of ascomycetous fungi in the order Eurotiales comprising forms that inhabit various animal substrata, such as horns and hoofs.

onyx [MINERAL] **1.** Banded chalcedonic quartz, in which the bands are straight and parallel; natural colors are usually red or brown with white, although black is occasionally encountered. **2.** *See* onyx marble.

onyx marble [MINERAL] A hard, compact, dense, generally translucent variety of calcite resembling true onyx and usually banded. Also known as alabaster; Algerian onyx; Gibraltar stone; Mexican onyx; onyx; oriental alabaster.

oocyte [HISTOL] An egg before the completion of maturation.

oogenesis [PHYSIO] Processes involved in the growth and maturation of the ovum in preparation for fertilization.

oogonium [BOT] The unisexual female sex organ in oogamous algae and fungi. [HISTOL] A descendant of a primary germ cell which develops into an oocyte.

oolite [PETR] A sedimentary rock, usually a limestone, composed principally of cemented ooliths. Also known as eggstone; roestone.

oolith [PETR] A small (0.25–2.0 millimeters), rounded accretionary body in a sedimentary rock; generally formed of calcium carbonate by inorganic precipitation or by replacement; ooliths generally exhibit concentric or radial internal structure.

Oomycetes [MYCOL] A class of the Phycomycetes comprising the biflagellate water molds and downy mildews.

ooze [GEOL] 1. A soft, muddy piece of ground, such as a bog, usually resulting from the flow of a spring or brook. 2. A marine pelagic sediment composed of at least 30% skeletal remains of pelagic organisms, the rest being clay minerals. 3. Soft mud or slime, typically covering the bottom of a lake or river.

opacity [OPTICS] The light flux incident upon a medium divided by the light flux transmitted by it.

opal [MINERAL] A natural hydrated form of silica; it is amorphous, usually occurs in botryoidal or stalactic masses, has a hardness of 5–6 on Mohs scale, and specific gravity is 1.9–2.2.

opalescence [OPTICS] The milky, iridescent appearance of a dense, transparent medium or colloidal system when it is illuminated by polychromatic radiation in the visible range, such as sunlight.

opal glass [MATER] Translucent or opaque glass, often milky white, made by adding impurities such as fluorine compounds to the melt; it appears white by reflected light but shows color images through thin sections; used for ornamental glass and as an efficient light diffuser.

Opalinata [INV ZOO] A superclass of the subphylum Sarcomastigophora containing highly specialized forms which resemble ciliates.

opaque medium [OPTICS] A medium impervious to rays of light, that is, not transparent to the human eye. [PHYS] 1. A medium which does not transmit electromagnetic radiation of a specified type, such as that in the infrared, x-ray, ultraviolet, and microwave regions. 2. A medium which prevents the passage of particles of a specified type.

opaque sky cover [METEOROL] In United States weather observing practice, the amount (in tenths) of sky cover that completely hides all that might be above it; opposed to transparent sky cover.

Opegraphaceae [BOT] A family of the Hysteriales characterized by elongated ascocarps; members are crustose on bark and rocks.

open-arc furnace [MET] An electrosmelting furnace in which the arc is generated above the level of the furnace feed.

open-belt drive [DES ENG] A belt drive having both shafts parallel and rotating in the same direction.

open caisson [CIV ENG] A caisson in the form of a cylinder or shaft that is open at both ends; it is set in place, pumped dry, and filled with concrete.

open-cast mining See open-pit mining.

open circuit [ELEC] An electric circuit that has been broken, so that there is no complete path for current flow.

open cluster [ASTRON] One of the groupings of stars that are concentrated along the central plane of the Milky Way; most have an asymmetrical appearance and are loosely assembled, and the stars are concentrated in their central region; they may contain from a dozen to many hundreds of stars. Also known as galactic cluster.

opencut mining See open-pit mining.

open cycle [THERMO] A thermodynamic cycle in which new mass enters the boundaries of the system and spent exhaust leaves it; the automotive engine and the gas turbine illustrate this process.

open-delta connection [ELEC] An unsymmetrical transformer connection which is employed when one transformer of a bank of three single-phase delta-connected units must be cut out, because of failure. Also known as V connection.

open die [MET] A forming or forging die in which there is little or no restriction to the lateral flow of metal within the die set.

open-ended [COMPUT SCI] Of techniques, designed to facilitate or permit expansion, extension, or increase in capability; the opposite of closed-in and artificially constrained.

open-ended system [COMPUT SCI] In character recognition, a system in which the input data to be read are derived from sources other than the computer with which the character reader is associated.

open file [COMPUT SCI] A file that can be accessed for reading, writing, or both.

open-hearth furnace [MET] A reverberatory melting furnace with a shallow hearth and a low roof, in which the charge is heated both by direct flame and by radiation from the roof and walls of the furnace.

open-hearth process [MET] A steel-making process carried out in an open-hearth furnace in which selected pig iron and malleable scrap iron are melted, with the addition of pure iron ore.

open interval [MATH] An open interval of real numbers, denoted by (a,b), consists of all numbers strictly greater than a and strictly less than b.

open-loop control system [CONT SYS] A control system in which the system outputs are controlled by system inputs only, and no account is taken of actual system output.

open-phase relay [ELEC] Relay which functions by reason of the opening of one or more phases of a polyphase circuit, when sufficient current is flowing in the remaining phase or phases.

open-pit mining [MIN ENG] Extracting metal ores and minerals that lie near the surface by removing the overlying material and breaking and loading the ore. Also known as open-cast mining; opencut mining.

open routine [COMPUT SCI] 1. A routine which can be inserted directly into a larger routine without a linkage or calling sequence. 2. A computer program that changes the state of a file from closed to open.

open sea [GEOGR] 1. That part of the ocean not enclosed by headlands, not within narrow straits, and so on. 2. That part of the ocean outside the territorial jurisdiction of any country.

open set [MATH] A set included in a topology; equivalently, a set which is a neighborhood of each of its points; a topology on a space is determined by a collection of subsets which are called open.

open stope [MIN ENG] Underground working place that is unsupported, or supported by timbers or pillars of rock.

open subroutine [COMPUT SCI] A set of computer instructions that collectively perform some particular function and are inserted directly into the program each and every time that particular function is required.

open system [HYD] A condition of freezing of the ground in which additional groundwater is available either through free percolation or through capillary movement. [THERMO] A system across whose boundaries both matter and energy may pass.

open-web girder See lattice girder.

open-window unit *See* sabin.

open-wire feeder *See* open-wire transmission line.

open-wire transmission line [ELEC] A transmission line consisting of two spaced parallel wires supported by insulators, at the proper distance to give a desired value of surge impedance. Also known as open-wire feeder.

operand [COMPUT SCI] Any one of the quantities entering into or arising from an operation.

operand-precision register [COMPUT SCI] A special register found in some minicomputers which can be programmed from 8- to 32-bit precision.

operate time [COMPUT SCI] The phase of computer operation when an instruction is being carried out. [ELEC] Total elapsed time from application of energizing current to a relay coil to the time the contacts have opened or closed.

operating characteristic curve [STAT] In hypothesis testing, a plot of the probability of accepting the hypothesis against the true state of nature. Abbreviated OC curve.

operating instructions [COMPUT SCI] A detailed description of the actions that must be carried out by a computer operator in running a program or group of interrelated programs, usually included in the documentation of a program supplied by a programmer or systems analyst, along with the source program and flow charts.

operating ratio [COMPUT SCI] The time during which computer hardware operates and gives reliable results divided by the total time scheduled for computer operation.

operating system *See* executive system.

operation [COMPUT SCI] 1. A process or procedure that obtains a unique result from any permissible combination of operands. 2. The sequence of actions resulting from the execution of one digital computer instruction. [IND ENG] A job, usually performed in one location, and consisting of one or more work elements.

operational amplifier [ELECTR] An amplifier having high direct-current stability and high immunity to oscillation, generally achieved by using a large amount of negative feedback; used to perform analog-computer functions such as summing and integrating.

operational analysis *See* operational calculus.

operational calculus [MATH] A technique by which problems in analysis, in particular differential equations, are transformed into algebraic problems, usually the problem of solving a polynomial equation. Also known as operational analysis.

operation code [COMPUT SCI] A field or portion of a digital computer instruction that indicates which action is to be performed by the computer. Also known as command code.

operation cycle [COMPUT SCI] The portion of a memory cycle required to perform an operation; division and multiplication usually require more than one memory cycle to be completed.

operation decoder [COMPUT SCI] A device that examines the operation contained in an instruction of a computer program and sends signals to the circuits required to carry out the operation.

operation number [COMPUT SCI] 1. Number designating the position of an operation, or its equivalent subroutine, in the sequence of operations composing a routine. 2. Number identifying each step in a program stated in symbolic code.

operation part [COMPUT SCI] That portion of a digital computer instruction which is reserved for the operation code.

operation register [COMPUT SCI] A register used to store and decode the operation code for the next instruction to be carried out by a computer.

operations research [MATH] The mathematical study of systems with input and output from the viewpoint of optimization subject to given constraints. [SCI TECH] The application of objective and quantitative criteria to decision making previously undertaken by empirical methods.

operative ankylosis *See* arthrodesis.

operator [COMPUT SCI] Anything that designates an action to be performed, especially the operation code of a computer instruction. [ENG] A person whose duties include the operation, adjustment, and maintenance of a piece of equipment. [GEN] A sequence at one end of an operon on which a repressor acts, thus regulating the transcription of the operon. [MATH] A function between vector spaces.

operator hierarchy [COMPUT SCI] A sequence of mathematical operators which designates the order in which these operators are to be applied to any mathematical expression in a given programming language.

operator interrupt [COMPUT SCI] A step whereby control is passed to the monitor, and a message, usually requiring a typed answer, is printed on the console typewriter.

operator theory [MATH] The general qualitative study of operators in terms of such concepts as eigenvalues, range, domain, and continuity.

operculum [ANAT] 1. The soft tissue partially covering the crown of an erupting tooth. 2. That part of the cerebrum which borders the lateral fissure. [BIOL] 1. A lid, flap, or valve. 2. A lidlike body process.

operon [GEN] A functional unit composed of a number of adjacent cistrons on the chromosome; its transcription is regulated by a receptor sequence, the operator, and a repressor.

Opheliidae [INV ZOO] A family of limivorous worms belonging to the annelid group Sedentaria.

Ophiacodonta [VERT ZOO] A suborder of extinct reptiles in the order Pelycosauria, including primitive, partially aquatic carnivores.

Ophidiidae [VERT ZOO] A family of small actinopterygian fishes in the order Gadiformes, comprising the cusk eels and brotulas.

Ophiocanopidae [INV ZOO] A family of asterozoan echinoderms in the subclass Ophiuroidea.

Ophiocistioidea [PALEON] A small class of extinct Echinozoa in which the domed aboral surface of the test was roofed by a polygonal plates and carried an anal pyramid.

Ophioglossales [BOT] An order of ferns in the subclass Ophioglossidae.

Ophioglossidae [BOT] The adder's-tongue ferns, a small subclass of the class Polypodiopsida; the plants are homosporous and eusporangiate and are distinguished by the arrangement of the sporogenous tissue in the characteristic fertile spike of the sporophyte.

ophiolite [PETR] A group of mafic and ultramafic igneous rocks, including spilite, basalt, gabbro, peridotite, and their metamorphic alternation products such as serpentine.

Ophiomyxidae [INV ZOO] The single family of the echinoderm suborder Ophiomyxina distinguished by a soft, unprotected integument.

Ophiomyxina [INV ZOO] A monofamilial suborder of ophiuroid echinoderms in the order Phrynophiurida.

ophitic [PETR] Of the holocrystalline, hypidiomorphic-granular texture of an igneous rock, exhibiting lath-shaped plagioclase crystals partly or wholly included within pyroxene crystals.

Ophiurida [INV ZOO] An order of echinoderms in the subclass Ophiuroidea in which the vertebrae articulate by means of ball-and-socket joints, and the arms, which do not branch, move mainly from side to side.

Ophiuroidea [INV ZOO] The brittle stars, a subclass of the Asterozoa in which the arms are usually clearly demarcated from the central disk and perform whiplike locomotor movements.

ophthalmia [MED] Inflammation of the eye, especially involving the conjunctiva.

ophthalmic nerve [ANAT] A sensory branch of the trigeminal nerve which supplies the lacrimal glands, upper eyelids, skin of the forehead, and anterior portion of the scalp, meninges, nasal mucosa, and frontal, ethmoid, and sphenoid air sinuses.

ophthalmology [MED] The study of the anatomy, physiology, and diseases of the eye.

opiate [PHARM] **1.** A sleep-inducing drug. **2.** Any narcotic. **3.** An opium preparation. **4.** Any tranquilizing agent.

Opilioacaridae [INV ZOO] The single family of moderately large mites of the suborder Notostigmata which comprises the Opilioacariformes.

Opilioacariformes [INV ZOO] A small monofamilial order of the Acari comprising large mites characterized by long legs and by the possession of a pretarsus on the pedipalp, with prominent claws.

Opisthobranchia [INV ZOO] A subclass of the class Gastropoda containing the sea hares, sea butterflies, and sea slugs; generally characterized by having gills, a small external or internal shell, and two pairs of tentacles.

Opisthocoela [VERT ZOO] A suborder of the order Anura; members have opisthocoelous trunk vertebrae, and the adults typically have free ribs.

Opisthocomidae [VERT ZOO] A family of birds in the order Galliformes, including the hoatzins.

opisthognathous [INV ZOO] Having the mouthparts ventral and posterior to the cranium. [VERT ZOO] Having retreating jaws.

opisthonephros [VERT ZOO] The fundamental adult kidney in amphibians and fishes.

Opisthopora [INV ZOO] An order of the class Oligochaeta distinguished by meganephridiostomal, male pores opening posteriorly to the last testicular segment.

Opomyzidae [INV ZOO] A family of cyclorrhaphous myodarian dipteran insects in the subsection Acalypteratae.

opossum [VERT ZOO] Any member of the family Didelphidae in the order Marsupialia; these mammals are arboreal and mainly omnivorous, and have many incisors, with all teeth pointed and sharp.

opposed engine [MECH ENG] A reciprocating engine having the pistons on opposite sides of the crankshaft, with the piston strokes on each side working in a direction opposite to the direction of the strokes on the other side.

opposition [ASTRON] The situation of two celestial bodies having either celestial longitudes or sidereal hour angles differing by 180°; the term is usually used only in relation to the position of a superior planet or the moon with reference to the sun. [PHYS] The condition in which the phase difference between two periodic quantities having the same frequency is 180°, corresponding to one half-cycle.

opsonin [IMMUNOL] A substance in blood serum that renders bacteria more susceptible to phagocytosis by leukocytes.

optic [BIOL] Pertaining to the eye. [OPTICS] Pertaining to the lenses, prisms, and mirrors of a camera, microscope, or other conventional optical instrument.

optical [OPTICS] Pertaining to or utilizing visible or near-visible light; the extreme limits of the optical spectrum are about 100 nanometers (0.1 micrometer or 3×10^{15} hertz) in the far ultraviolet and 30,000 nanometers (30 micrometers or 10^{13} hertz) in the far infrared.

optical aberration [OPTICS] Deviation from perfect image formation by an optical system; examples are spherical aberration, coma, astigmatism, curvature of field, distortion, and chromatic aberration. Also known as aberration.

optical achromatism *See* visual achromatism.

optical activity [OPTICS] The behavior of substances which rotate the plane of polarization of plane-polarized light, as it passes through them. Also known as rotary polarization.

optical amplifier [ENG] An optoelectronic amplifier in which the electric input signal is converted to light, amplified as light, then converted back to an electric signal for the output.

optical antipode *See* enantiomorph.

optical axis [ANAT] An imaginary straight line passing through the midpoint of the cornea (anterior pole) and the midpoint of the retina (posterior pole). [OPTICS] **1.** A line passing through a radially symmetrical optical system such that rotation of the system about this line does not alter it in any detectable way. **2.** *See* optic axis.

optical bar-code reader [COMPUT SCI] A device which uses any of various photoelectric methods to read information which has been coded by placing marks in prescribed boxes on documents with ink, pencil, or other means.

optical character recognition [COMPUT SCI] That branch of character recognition concerned with the automatic identification of handwritten or printed characters by any of various photoelectric methods. Abbreviated OCR. Also known as electrooptical character recognition.

optical communication [COMMUN] The use of electromagnetic waves in the region of the spectrum near visible light for the transmission of signals representing speech, pictures, data pulses, or other information, usually in the form of a laser beam modulated by the information signal.

optical comparator [ENG] Any comparator in which movement of a measuring plunger tilts a small mirror which reflects light in an optical system. Also known as visual comparator.

optical computer [COMPUT SCI] A computer that uses various combinations of holography, lasers, and mass-storage memories for such applications as ultra-high-speed signal processing, image deblurring, and character recognition.

optical coupler *See* optoisolator.

optical coupling [ELECTR] Coupling between two circuits by means of a light beam or light pipe having transducers at opposite ends, to isolate the circuits electrically.

optical density [OPTICS] The degree of opacity of a translucent medium expressed by $\log I_0/I$, where I_0 is the intensity of the incident ray, and I is the intensity of the transmitted ray. Abbreviated OD.

optical diffraction velocimeter *See* diffraction velocimeter.

optical disk [COMPUT SCI] A type of video disk storage device consisting of a pressed disk with a spiral groove at the bottom of which are submicrometer-sized depressions that are sensed by a laser beam.

optical disk storage [COMPUT SCI] A computer storage technology in which information is stored in submicrometer-sized holes on a rotating disk, and is recorded and read by laser beams focused on the disk. Also known as video disk storage.

optical dispersion [OPTICS] Separation of different colors of light such as occurs when it passes from one medium to another or is reflected from a diffraction grating.

optical distance *See* optical path.

optical double star [ASTRON] Two stars not formally a physical system but that appear to be a typical double star; a false binary star whose components happen to lie nearby in the same line of sight.

optical electronic reproducer *See* optical sound head.

optical element [OPTICS] A part of an optical instrument which acts upon the light passing through the instrument, such as a lens, prism, or mirror.

optical encoder [ELECTR] An encoder that converts positional information into corresponding digital data by interrupting light beams directed on photoelectric devices.

optical fiber [OPTICS] A long, thin thread of fused silica, or other transparent substance, used to transmit light. Also known as light guide.

optical-fiber cable *See* optical waveguide.

optical filter *See* filter.

optical indicatrix *See* index ellipsoid.

optical isolator *See* optoisolator.

optical isomer *See* enantiomorph.

optical isomerism [PHYS CHEM] Existence of two forms of a molecule such that one is a mirror image of the other; the two molecules differ in that rotation of light is equal but in opposite directions.

optical length *See* optical path.

optical lens *See* lens.

optically coupled isolator *See* optoisolator.

optical mask [ELECTR] A thin sheet of metal or other substance containing an open pattern, used to suitably expose to light a photoresistive substance overlaid on a semiconductor or other surface to form an integrated circuit.

optical memory [COMPUT SCI] A computer memory that uses optical techniques which generally involve an addressable laser beam, a storage medium which responds to the beam for writing and sometimes for erasing, and a detector which reacts to the altered character of the medium when it uses the beam to read out stored data.

optical microscope [OPTICS] An instrument used to obtain an enlarged image of a small object, utilizing visible light; in general it consists of a tube, a condenser, an objective lens, and an ocular or eyepiece, which can be replaced by a recording device. Also known as light microscope; photon microscope.

optical modulator [COMMUN] A device used for impressing information on a light beam, such as a blinker system or a device which electrically changes the properties of a material through which light is being transmitted.

optical moment [OPTICS] For a ray of light passing through an optical system, the triple product of a vector from an arbitrary origin on the optical axis to a point on the ray, a vector tangent to the ray at that point whose length equals the refractive index, and a unit vector along the optical axis; it does not depend on the point on the ray.

optical parallax [OPTICS] A fault in an optical measuring instrument in which the image being observed does not lie in the plane of the wires or marks used to make the measurement, so that motion of the observer's eye causes displacement of the image relative to these wires or marks.

optical path [OPTICS] For a ray of light traveling along a path between two points, the optical path is the integral, over elements of length along the path, of the refractive index. Also known as optical distance; optical length.

optical-path difference *See* retardation.

optical phase conjugation [OPTICS] The use of nonlinear optical effects to precisely reverse the direction of propagation of each plane wave in an arbitrary beam of light, thereby causing the return beam to exactly retrace the path of the incident beam. Also known as time-reversal reflection; wavefront reversal.

optical prism *See* prism.

optical projection system [OPTICS] An optical system which forms a real image of a suitably illuminated object so that it can be viewed, photographed, or otherwise observed. Also known as optical projector; projector.

optical projector *See* optical projection system.

optical pumping [OPTICS] The process of causing strong deviations from thermal equilibrium populations of selected quantized states of different energy in atomic or molecular systems by the use of electromagnetic radiation in or near the visible region.

optical rangefinder [ENG] An optical instrument for measuring distance, usually from its position to a target point, by measuring the angle between rays of light from the target, which enter the rangefinder through the windows spaced apart, the distance between the windows being termed the baselength of the rangefinder; the two types are coincidence and stereoscopic.

optical reader [COMPUT SCI] A computer data-entry machine that converts printed characters, bar or line codes, and pencil-shaded areas into a computer-input code format.

optical recording [ENG] Production of a record by focusing on photographic paper a beam of light whose position on the paper depends on the quantity to be measured, as in a light-beam galvanometer.

optical relay [ELECTR] An optoisolator in which the output device is a light-sensitive switch that provides the same on and off operations as the contacts of a relay.

optical rotary dispersion [OPTICS] Specific rotation, considered as a function of wavelength. Abbreviated ORD.

optical rotation [OPTICS] Rotation of the plane of polarization of plane-polarized light, or of the major axis of the polarization ellipse of elliptically polarized light by transmission through a substance or medium.

optical scanner *See* flying-spot scanner.

optical sound head [ELECTR] The assembly in motion picture projection which reproduces photographically recorded sound; light from an incandescent lamp is focused on a slit, light from the slit is in turn focused on the optical sound track of a film, and the light passing through the film is detected by a photoelectric cell. Also known as optical electronic reproducer.

optical spectra [SPECT] Electromagnetic spectra for wavelengths in the ultraviolet, visible and infrared regions, ranging from about 10 nanometers to 1 millimeter, associated with excitations of valence electrons of atoms and molecules, and vibrations and rotations of molecules.

optical spectrometer [SPECT] An optical spectroscope that is provided with a calibrated scale either for measurement of wavelength or for measurement of refractive indices of transparent prism materials.

optical spectroscopy [SPECT] The production, measurement, and interpretation of optical spectra arising from either emission or absorption of radiant energy by various substances.

optical storage [COMPUT SCI] Storage of large amounts of data in permanent form on photographic film or its equivalent, for nondestructive readout by means of a light source and photodetector.

optical surface [OPTICS] An interface between two media, such as between air and glass, which is used to reflect or refract light.

optical system [OPTICS] A collection comprising mirrors, lens, prisms, and other devices, placed in some specified configuration, which reflect, refract, disperse, absorb, polarize, or otherwise act on light.

optical twinning [CRYSTAL] Growing together of two crystals which are the same except that the structure of one is the mirror image of the structure of the other. Also known as chiral twinning.

optical type font [COMPUT SCI] A special type font whose characters are designed to be easily read by both people and optical character recognition machines.

optical waveguide [ELECTROMAG] A waveguide in which a light-transmitting material such as a glass or plastic fiber is used for transmitting information from point to point at wavelengths somewhere in the ultraviolet, visible-light, or infrared portions of the spectrum. Also known as fiber waveguide; optical-fiber cable.

optical window [OPTICS] The spectral region between 300 and 2000 nanometers (0.3 and 2 micrometers in wavelength), in which visible and near-visible radiation will pass through the earth's atmosphere.

optic angle See axial angle.

optic-axial angle See axial angle.

optic axis [OPTICS] The axis in a doubly refracting medium in which the ordinary and extraordinary waves propagate with the same velocity, and double refraction vanishes. Also known as optical axis; principal axis.

optic chiasma [ANAT] The partial decussation of the optic nerves on the undersurface of the hypothalamus.

optic disk [ANAT] The circular area in the retina that is the site of the convergence of fibers from the ganglion cells of the retina to form the optic nerve.

optic lobe [ANAT] One of the anterior pair of colliculi of the mammalian corpora quadrigemina. [INV ZOO] A lateral lobe of the forebrain in certain arthropods. [VERT ZOO] Either of the corpora bigemina of lower vertebrates.

optic nerve [ANAT] The second cranial nerve; a paired sensory nerve technically consisting of three layers of special nerve cells in the retina of the eye; fibers converge to form the optic tracts.

optics [PHYS] 1. Narrowly, the science of light and vision. 2. Broadly, the study of the phenomena associated with the generation, transmission, and detection of electromagnetic radiation in the spectral range extending from the long-wave edge of the x-ray region to the short-wave edge of the radio region, or in wavelength from about 1 nanometer to about 1 millimeter.

optimal control theory [MATH] A generalized calculus of variations dealing with the analysis of dynamical systems with the viewpoint of finding optimizing conditions.

optimization [MATH] The maximizing or minimizing of a given function possibly subject to some type of constraints. [SYS ENG] 1. Broadly, the efforts and processes of making a decision, a design, or a system as perfect, effective, or functional as possible. 2. Narrowly, the specific methodology, techniques, and procedures used to decide on the one specific solution in a defined set of possible alternatives that will best satisfy a selected criterion. Also known as system optimization.

optimization theory [MATH] The specific methodology, techniques, and procedures used to decide on the one specific solution in a defined set of possible alternatives that will best satisfy a selected criterion; includes linear and nonlinear programming, stochastic programming, and control theory. Also known as mathematical programming.

optimize [COMPUT SCI] To rearrange the instructions or data in storage so that a minimum number of time-consuming jumps or transfers are required in the running of a program.

optimum code [COMPUT SCI] A computer code which is particularly efficient with regard to a particular aspect; for example, minimum time of execution, minimum or efficient use of storage space, and minimum coding time.

optimum coupling See critical coupling.

optimum programming [COMPUT SCI] Production of computer programs that maximize efficiency with respect to some criteria such as least cost, least use of storage, least time, or least use of time-sharing peripheral equipment.

optimum reverberation time [ACOUS] The reverberation time which is most desirable for a given room size and a given use, such as speech, chamber music, or symphony orchestra.

optimum traffic frequency See optimum working frequency.

optimum working frequency [COMMUN] The most effective frequency at a specified time for ionospheric propagation of radio waves between two specified points. Also known as frequency optimum traffic; optimum traffic frequency.

optional halt instruction [COMPUT SCI] A halt instruction that can cause a computer program to stop either before or after the instruction is obeyed if certain criteria are met. Also known as optional stop instruction.

optional stop instruction See optional halt instruction.

optoacoustic effect [PHYS] A phenomenon in which a periodically interrupted beam of light generates sound in a gas through which it is passing; this results from energy in the light beam being transformed first into internal motions of the gas molecules, then into random translational motions of these molecules, or heat, and finally into periodic pressure fluctuations or sound. Also known as thermoacoustic effect.

optocoupler See optoisolator.

optoelectronic amplifier [ENG] An amplifier in which the input and output signals and the method of amplification may be either electronic or optical.

optoelectronic isolator See optoisolator.

optoelectronics [ELECTR] The branch of electronics that deals with solid-state and other electronic devices for generating, modulating, transmitting, and sensing electromagnetic radiation in the ultraviolet, visible-light, and infrared portions of the spectrum.

optogalvanic spectroscopy [ANALY CHEM] A method of obtaining absorption spectra of atomic and molecular species in flames and electrical discharges by measuring voltage and current changes upon irradiation.

optoisolator [ELECTR] A coupling device in which a light-emitting diode, energized by the input signal, is optically coupled to a photodetector such as a light-sensitive output diode, transistor, or silicon controlled rectifier. Also known as optical coupler; optical isolator; optically coupled isolator; optocoupler; optoelectronic isolator; photocoupler; photoisolator.

optometry [MED] Measurement of visual powers.

or [COMPUT SCI] An instruction which performs the logical operation "or" on a bit-by-bit basis for its two or more operand words, usually storing the result in one of the operand locations. Also known as OR function. [MATH] A logical operation whose result is false (or zero) only if every one of its operands is false, and true (or one) otherwise. Also known as inclusive or.

oral [ANAT] Of or pertaining to the mouth.

oral contraceptive [PHARM] Any medication taken by mouth that renders a woman nonfertile as long as the medication is continued.

oral disc [INV ZOO] The flattened upper or free end of the body of a polyp that has the mouth in the center and tentacles around the margin.

orange [BOT] Any of various evergreen trees of the genus *Citrus*, cultivated for the edible fruit, a berry with an aromatic, leathery rind containing numerous oil glands. [OPTICS] The hue evoked in an average observer by monochromatic radiation having a wavelength in the approximate range from 597 to 622 nanometers; however, the same sensation can be produced in a variety of other ways.

orange-peel bucket [DES ENG] A type of grab bucket that is multileaved and generally round in configuration.

orangutan [VERT ZOO] *Pongo pygmaeus*. The largest of the great apes, a long-armed primate distinguished by long sparse reddish-brown hair, naked face and hands and feet, and a large laryngeal cavity which appears as a pouch below the chin.

O ray *See* ordinary ray.

orbicular [PETR] Of the structure of a rock, containing large quantities of orbicules. [SCI TECH] Having the form of a sphere or orb.

Orbiniidae [INV ZOO] A family of polychaete annelids belonging to the Sedentaria; the prostomium is exposed, and the thorax and abdomen are weakly separated.

Orbiniinae [INV ZOO] A subfamily of sedentary polychaete annelids in the family Orbiniidae.

orbit [ANAT] The bony cavity in the lateral front of the skull beneath the frontal bone which contains the eyeball. Also known as eye socket. [OCEANOGR] The path of a water particle affected by wave motion; it is almost circular in deep-water waves and almost elliptical in shallow-water waves. [PHYS] **1.** Any closed path followed by a particle or body, such as the orbit of a celestial body under the influence of gravity, the elliptical path followed by electrons in the Bohr theory, or the paths followed by particles in a circular particle accelerator. **2.** More generally, any path followed by a particle, such as helical paths of particles in a magnetic field, or the parabolic path of a comet.

orbital [ATOM PHYS] The space-dependent part of the Schrödinger wave function of an electron in an atom or molecule in an approximation such that each electron has a definite wave function, independent of the other electrons.

orbital angular momentum [MECH] The angular momentum associated with the motion of a particle about an origin, equal to the cross product of the position vector with the linear momentum. Also known as orbital momentum. [QUANT MECH] The angular momentum operator associated with the motion of a particle about an origin, equal to the cross product of the position vector with the linear momentum, as opposed to the intrinsic spin angular momentum. Also known as orbital moment.

orbital electron [ATOM PHYS] An electron which has a high probability of being in the vicinity (at distances on the order of 10^{-10} meter or less) of a particular nucleus, but has only a very small probability of being within the nucleus itself. Also known as planetary electron.

orbital magnetic moment [QUANT MECH] The magnetic dipole moment associated with the motion of a charged particle about an origin, rather than with its intrinsic spin.

orbital moment *See* orbital angular momentum.

orbital momentum *See* orbital angular momentum.

orbital plane [MECH] The plane which contains the orbit of a body or particle in a central force field; it passes through the center of force.

orbital symmetry [PHYS CHEM] The property of certain molecular orbitals of being carried into themselves or into the negative of themselves by certain geometrical operations, such as a rotation of 180° about an axis in the plane of the molecule, or reflection through this plane.

orbital velocity [ASTRON] The instantaneous velocity at which an earth satellite or other orbiting body travels around the origin of its central force field.

Orbitoidacea [INV ZOO] A superfamily of foraminiferan protozoans in the suborder Rotaliina characterized by a low trochospire or a planispiral, uncoiled or branching test composed of radial calcite with bilamellar septa.

orbit transfer vehicles [AERO ENG] A propulsive (velocity-producing) rocket or stage for use with crew transfer modules, crewed sortie modules, or other payloads. Also know as OTV.

orch-, orchi-, orchid-, orchido-, orchio- [ZOO] A combining form denoting testis.

Orchidaceae [BOT] A family of monocotyledonous plants in the order Orchidales characterized by irregular flowers with only one or two stamens which are adnate to the style, and pollen grains which cohere in large masses called pollinia.

Orchidales [BOT] An order of monocotyledonous plants in the subclass Liliidae; plants are mycotropic and sometimes nongreen with numerous tiny seeds that have an undifferentiated embryo and little or no endosperm.

OR circuit *See* OR gate.

ORD *See* optical rotary dispersion.

order [CHEM] A classification of chemical reactions, in which the order is described as first, second, third, or higher, according to the number of molecules (one, two, three, or more) which appear to enter into the reaction; decomposition of H_2O_2 to form water and oxygen is a first-order reaction. [PHYS] A range of magnitudes of a quantity (and of all other quantities having the same physical dimensions) extending from some value of the quantity to some small multiple of the quantity (usually 10). Also known as order of magnitude. [SYST] A taxonomic category ranked below the class and above the family, made up either of families, subfamilies, or suborders.

order-disorder transition [SOLID STATE] The transition of an alloy or other solid solution between a state in which atoms of one element occupy certain regular positions in the lattice of another element, and a state in which this regularity is not present.

ordered field [MATH] A field with an ordering as a set analogous to the properties of less than or equal for real numbers relative to addition and multiplication.

ordered pair [MATH] A pair of elements x and y from a set, written (x,y), where x is distinguished as first and y as second.

ordering [MATH] A binary relation, denoted $\leq$, among the elements of a set such that $a \leq b$ and $b \leq c$ implies $a \leq c$, and $a \leq b$, $b \leq a$ implies $a = b$; it need not be the case that either $a \leq b$ or $b \leq a$. Also known as order relation; partial ordering. [SOLID STATE] A solid-state transformation in certain solid solutions, in which a random arrangement in the lattice is transformed into a regular ordered arrangement of the atoms with respect to one another; a so-called superlattice is formed.

order of magnitude *See* order.

order point [IND ENG] The inventory level at which a replenishment order must be placed.

order relation *See* ordering.

ordinal number [MATH] A generalized number which expresses the size of a set, in the sense of "how many" elements.

ordinary differential equation [MATH] An equation involving functions of one variable and their derivatives.

ordinary gear train [MECH ENG] A gear train in which all axes remain stationary relative to the frame.

ordinary ray [OPTICS] One of two rays into which a ray incident on an anisotropic uniaxial crystal is split; it obeys the ordinary laws of refraction, in contrast to the extraordinary ray. Also known as O ray.

ordinary-wave component [GEOPHYS] One of the two components into which an electromagnetic wave entering the ionosphere is divided under the influence of the earth's magnetic field; it has characteristics more nearly like those expected in the absence of a magnetic field. Also known as O-wave component. [OPTICS] The component of electromagnetic radiation propagating in an anisotropic uniaxial crystal whose electric displacement vector is perpendicular to the optical axis and the direction normal to the wavefront; gives rise to the ordinary ray.

ordinate [MATH] The perpendicular distance of a point (x,y) of the plane from the x-axis.

Ordovician [GEOL] The second period of the Paleozoic era, above the Cambrian and below the Silurian, from approximately 500 million to 440 million years ago.

ore [GEOL] **1.** The naturally occurring material from which economically valuable minerals can be extracted. **2.** Specifically, a natural mineral compound of the elements, of which one element at least is a metal. **3.** More loosely, all metalliferous rock, though it contains the metal in a free state. **4.** Occasionally, a compound of nonmetallic substances, as sulfur ore.

orebody [GEOL] Generally, a solid and fairly continuous mass of ore, which may include low-grade ore and waste as well as pay ore, but is individualized by form or character from adjoining country rock.

ore chimney *See* pipe.

Orectolobidae [VERT ZOO] An ancient isurid family of galeoid sharks, including the carpet and nurse sharks, which are primarily bottom feeders with small teeth and a blunt rostrum with barbels near the mouth.

ore deposit [GEOL] Rocks containing minerals of economic value in such amount that they can be profitably exploited.

ore dressing [MIN ENG] The cleaning of ore by the removal of certain valueless portions, as by jigging, cobbing, or vanning.

oregano [FOOD ENG] A spice prepared from leaves of various aromatic mints, especially wild marjoram.

ore in sight [MIN ENG] **1.** Ore exposed on at least three sides within reasonable distance of each other. **2.** Ore which may be reasonably assumed to exist, though not actually blocked out. **3.** *See* developed reserves.

ore-lead age [GEOL] Measurement of the age of the earth by comparing the relative progress of the two radioactive decay schemes $^{235}U-^{207}Pb$ and $^{238}U-^{206}Pb$.

ore of sedimentation *See* placer.

ore pipe *See* pipe.

ore pocket [MIN ENG] **1.** Excavation near the hoisting shaft into which ore from stopes is moved, preliminary to hoisting. **2.** An unusual concentration of ore in the lode.

ore shoot [GEOL] **1.** A large, generally vertical, pipelike ore body that is economically valuable. Also known as shoot. **2.** A large and usually rich aggregation of mineral in a vein.

OR function *See* or.

organ [ANAT] A differentiated structure of an organism composed of various cells or tissues and adapted for a specific function.

organelle [CYTOL] A specialized subcellular structure, such as a mitochondrion, having a special function; a condensed system showing a high degree of internal order and definite limits of size and shape.

organic [ORG CHEM] Of chemical compounds, based on carbon chains or rings and also containing hydrogen with or without oxygen, nitrogen, or other elements.

organic acid [ORG CHEM] A chemical compound with one or more carboxyl radicals (COOH) in its structure; examples are butyric acid, $CH_3(CH_2)_2COOH$, maleic acid, HOOCCHCHCOOH, and benzoic acid, C_6H_5COOH.

organic chemistry [CHEM] The study of the composition, reactions, and properties of carbon-chain or carbon-ring compounds or mixtures thereof.

organic evolution [EVOL] The processes of change in organisms by which descendants come to differ from their ancestors, and a history of the sequence of such changes.

organicism *See* holism.

organic reef [GEOL] A sedimentary rock structure of significant dimensions erected by, and composed almost exclusively of the remains of, corals, algae, bryozoans, sponges, and other sedentary or colonial organisms.

organic semiconductor [MATER] An organic material having unusually high conductivity, often enhanced by the presence of certain gases, and other properties commonly associated with semiconductors; an example is anthracene.

organic soil [GEOL] Any soil or soil horizon consisting chiefly of, or containing at least 30% of, organic matter; examples are peat soils and muck soils.

organization chart [IND ENG] Graphic representation of the interrelationships within an organization, depicting lines of authority and responsibility and provisions for control.

organ of Corti [ANAT] A specialized structure located on the basilar membrane of the mammalian cochlea, which contains rods of Corti and hair cells connected to ganglia of the cochlear nerve. Also known as spiral organ.

organometallic compound [ORG CHEM] Molecules containing carbon-metal linkage; a compound containing an alkyl or aryl radical bonded to a metal, such as tetraethyllead, $Pb(C_2H_5)_4$.

organophosphate [ORG CHEM] A soluble fertilizer material made up of organic phosphate esters such as glucose, glycol, or sorbitol; useful for providing phosphorus to deep-root systems.

organosulfur compound [ORG CHEM] One of a group of substances which contain both carbon and sulfur.

organs of Zuckerkandl *See* aortic paraganglion.

orgasm [PHYSIO] The intense, diffuse, and subjectively pleasurable sensation experienced during sexual intercourse or genital manipulation, culminating in the male with seminal ejaculation and in the female with uterine contractions, warm suffusion, and pelvic throbbing sensations.

OR gate [ELECTR] A multiple-input gate circuit whose output is energized when any one or more of the inputs is in a prescribed state; performs the function of the logical inclusive-or; used in digital computers. Also known as OR circuit.

Oribatei [INV ZOO] A heavily sclerotized group of free-living mites in the suborder Sarcoptiformes which serve as intermediate hosts of tapeworms.

Oribatulidae [INV ZOO] A family of oribatid mites in the suborder Sarcoptiformes.

orient [COMPUT SCI] To change relative and symbolic addresses to absolute form. [ENG] **1.** To place or set a map so that the map symbols are parallel with their corresponding ground features. **2.** To turn a transit so that the direction of the 0° line of its horizontal circle is parallel to the direction it had in the preceding or initial setup, or parallel to a standard reference line. [OPTICS] The play of color upon or just below the surface of a gem-quality pearl.

oriental alabaster *See* onyx marble.

orientation [CRYSTAL] The directions of the axes of a crystal lattice relative to the surfaces of the crystal, to applied fields, or to some other planes or directions of interest. [ELECTROMAG] The physical positioning of a directional antenna or other device having directional characteristics. [ENG] Establishment of the correct relationship in direction with reference to the points of the compass. [MATH] A choice of sense or direction in a topological space. [PHYS] **1.** The direction of some vector or set of vectors, such as the direction of the electric vector and the propagation direction of plane polarized light, or the direction of a preponderance of nuclear spins in a crystal near absolute zero, relative to some other directions of interest. **2.** Any process in which vectors associated with atoms or molecules in the substance are organized relative to some direction, rather than pointed at random; examples include dipole moments of polar molecules in an electric field, and nuclear spins in a crystal in a magnetic field at temperatures near absolute zero. [PHYS CHEM] The arrangement of radicals in an organic compound in relation to each other and to the parent compound. [PSYCH] Determination of one's relation to the environment.

orifice [ELECTROMAG] Opening or window in a side or end wall of a waveguide or cavity resonator through which energy is transmitted. [SCI TECH] An aperture or hole.

orifice meter [ENG] An instrument that measures fluid flow by recording differential pressure across a restriction placed in the flow stream and the static or actual pressure acting on the system.

origin [COMPUT SCI] Absolute storage address in relative coding to which addresses in a region are referenced. [MATH] The center of a coordinate system, where all coordinate axes meet, usually denoted by $(O, O, \ldots, O)$.

original dip *See* primary dip.

original document *See* source document.

Orion [ASTRON] A northern constellation near the celestial equator, right ascension 5 hours, declination 5° north. Also known as Warrior.

Orion Nebula [ASTRON] A luminous cloud surrounding Ori, the northern star in Orion's dagger; visible to the naked eye as a hazy object. Also known as Great Nebula of Orion.

ormer *See* abalone.

Ormyridae [INV ZOO] A small family of hemipteran insects in the superfamily Chalcidoidea.

Orneodidae [INV ZOO] A small family of lepidopteran insects in the superfamily Tineoidea; adults have each wing divided into six featherlike plumes.

Ornithischia [PALEON] An order of extinct terrestrial reptiles, popularly known as dinosaurs; distinguished by a four-pronged pelvis, and a median, toothless predentary bone at the front of the lower jaw.

ornithology [VERT ZOO] The study of birds.

Ornithopoda [PALEON] A suborder of extinct reptiles in the order Ornithischia including all bipedal forms in the order.

Ornithorhynchidae [VERT ZOO] A monospecific order of monotremes containing the semiaquatic platypus; characterized by a duck-billed snout, horny plates instead of teeth in the adult, and a flattened, well-developed tail.

ornithosis [MED] Any form of psittacosis originating in birds other than psittacines.

orogen *See* orogenic belt.

orogene *See* orogenic belt.

orogenesis *See* orogeny.

orogenic belt [GEOL] A linear region that has undergone folding or other deformation during the orogenic cycle. Also known as fold belt; orogen; orogene.

orogeny [GEOL] The process or processes of mountain formation, especially the intense deformation of rocks by folding and faulting which, in many mountainous regions, has been accompanied by metamorphism, invasion of molten rock, and volcanic eruption; in modern usage, orogeny produces the internal structure of mountains, and epeirogeny produces the mountainous topography. Also known as orogenesis; tectogenesis.

orographic precipitation [METEOROL] Precipitation which results from the lifting of moist air over an orographic barrier such as a mountain range; strictly, the amount so designated should not include that part of the precipitation which would be expected from the dynamics of the associated weather disturbance, if the disturbance were over flat terrain.

Oromericidae [VERT ZOO] An extinct family of camellike tylopod ruminants in the superfamily Cameloidea.

orpiment [MINERAL] As_2S_3 A lemon-yellow mineral, crystallizing in the monoclinic system, and generally occurring in foliated or columnar masses; luster is resinous and pearly on the cleavage surface, hardness is 1.5–2 on Mohs scale, and specific gravity is 3.49. Also known as yellow arsenic.

orris [MATER] The fragrant powder from the root of the plants *Iris florentinea, I. germanica,* and *I. pallida;* used in perfume, medicine, and tooth powder. Also known as orrisroot.

orrisroot *See* orris.

Orthacea [PALEON] An extinct group of articulate brachiopods in the suborder Orthidina in which the delthyrium is open.

Orthent [GEOL] A suborder of the soil order Entisol, well drained and of medium or fine texture, usually shallow to bedrock and lacking evidence of horizonation; occurs mostly on strong slopes.

Ortheziinae [INV ZOO] A subfamily of homopteran insects in the superfamily Coccoidea having abdominal spiracles present in all stages and a flat anal ring bearing pores and setae in immature forms and adult females.

orthicon [ELECTR] A camera tube in which a beam of low-velocity electrons scans a photoemissive mosaic that is capable of storing a pattern of electric charges; has higher sensitivity than the iconoscope.

Orthida [PALEON] An order of extinct articulate brachiopods which includes the oldest known representatives of the class.

Orthidina [PALEON] The principal suborder of the extinct Orthida, including those articulate brachiopods characterized by biconvex, finely ribbed shells with a straight hinge line and well-developed interareas on both valves.

orthoaxis [CRYSTAL] The diagonal or lateral axis perpendicular to the vertical axis in the monoclinic system.

orthoboric acid *See* boric acid.

orthocenter [MATH] The point at which the altitudes of a triangle intersect.

orthochromatic [BIOL] Having normal staining characteristics. [GRAPHICS] Pertaining to sensitized materials that can be exposed by ultraviolet, blue, and green light, but not deep orange or red.

orthoclase [MINERAL] $KAlSi_3O_8$ A colorless, white, cream-yellow, flesh-reddish, or gray potassium feldspar that usually contains some sodium feldspar, either as albite or analbite or in some intermediate state; it is or appears to be monoclinic. Also known as common feldspar; orthose; pegmatolite.

Orthod [GEOL] A suborder of the soil order Spodosol having accumulations of humus, aluminum, and iron; widespread in Canada and the Soviet Union.

orthodontics [MED] A branch of dentistry that deals with the prevention and treatment of malocclusion.

orthogeosyncline [GEOL] A linear geosynclinal belt lying between continental and oceanic cratons, and having internal volcanic belts (eugeosynclinal) and external nonvolcanic belts (miogeosynclinal). Also known as geosynclinal couple; primary geosyncline.

orthogonal [MATH] Perpendicular, or some concept analogous to it.

orthogonal antennas [ELECTROMAG] In radar, a pair of transmitting and receiving antennas, or a single transmitting-receiving antenna, designed for the detection of a difference in polarization between the transmitted energy and the energy returned from the target.

orthogonal family *See* orthogonal system.

orthogonalization [MATH] A procedure in which, given a set of linearly independent vectors in an inner product space, a set of orthogonal vectors is recursively obtained so that each set spans the same subspace.

orthogonal parity check [COMPUT SCI] A parity checking system involving both a lateral and a longitudinal parity check.

orthogonal polynomial [MATH] Orthogonal polynomials are various families of polynomials, which arise as solutions to differential equations related to the hypergeometric equation, and which are mutually orthogonal as functions.

orthogonal projection [MATH] A continuous linear map P of a Hilbert space H onto a subspace M such that if $\mathbf{h}$ is any vector in H, $\mathbf{h} = P\mathbf{h} + \mathbf{w}$, where $\mathbf{w}$ is in the orthogonal complement of M. Also known as orthographic projection.

orthogonal system [MATH] **1.** A system made up of n families of curves on an n-dimensional manifold in an $n + 1$ dimensional euclidean space, such that exactly one curve from each family passes through every point in the manifold, and, at each point, the tangents to the n curves that pass through that point are mutually perpendicular. **2.** A set of real-valued functions, the inner products of any two of which vanish. Also known as orthogonal family.

orthogonal transformation [MATH] A linear transformation between real inner product spaces which preserves the length of vectors.

orthogonal vectors [MATH] In an inner product space, two vectors are orthogonal if their inner product vanishes.

orthographic projection [CRYSTAL] A projection for displaying the poles of a crystal in which the poles are projected from a reference sphere onto an equatorial plane by dropping perpendiculars from the poles to the plane. [MAP] A perspective azimuthal projection of one hemisphere produced by straight parallel lines from any point desired from an infinite distance; it is true to scale at the center only. [MATH] *See* orthogonal projection.

orthomorphic map projection *See* conformal map projection.

Orthonectida [INV ZOO] An order of Mesozoa; orthonectids parasitize various marine invertebrates as multinucleate plasmodia, and sexually mature forms are ciliated organisms.

orthonormal coordinates [MATH] In an inner product space, the coordinates for a vector expressed relative to an orthonormal basis.

orthonormal functions [MATH] Orthogonal functions $f_1, f_2 \ldots$ with the additional property that the inner product of $f_n (x)$ with itself is 1.

orthonormal vectors [MATH] A collection of mutually orthogonal vectors, each having length 1.

orthopedics [MED] The branch of surgery concerned with corrective treatment of musculoskeletal deformities, diseases, and ailments by manual and instrumental measures.

Orthoperidae [INV ZOO] The minute fungus beetles, a family of coleopteran insects in the superfamily Cucujoidea.

orthophosphate [INORG CHEM] One of the possible salts of orthophosphoric acid; the general formula is M_3PO_4, where M may be potassium as in potassium orthophosphate, K_3PO_4.

Orthopsida [INV ZOO] An order of echinoderms in the subclass Euechinoidea.

Orthopsidae [PALEON] A family of extinct echinoderms in the order Hemicidaroida distinguished by a camarodont lantern.

Orthoptera [INV ZOO] A heterogeneous order of generalized insects with gradual metamorphosis, chewing mouthparts, and four wings.

orthopyroxene [MINERAL] A series of pyroxene minerals crystallizing in the orthorhombic system;

members include enstatite, bronzite, hypersthene, ferrohypersthene, eulite, and orthoferrosilite.

orthoquartzite [PETR] A clastic sedimentary rock composed almost entirely of detrital quartz grains; a quartzite of sedimentary origin. Also known as orthoquartzitic sandstone; sedimentary quartzite.

orthoquartzitic sandstone *See* orthoquartzite.

orthorhombic lattice [CRYSTAL] A crystal lattice in which the three axes of a unit cell are mutually perpendicular, and no two have the same length. Also known as rhombic lattice.

orthorhombic system [CRYSTAL] A crystal system characterized by three axes of symmetry that are mutually perpendicular and of unequal length. Also known as rhombic system.

Orthorrhapha [INV ZOO] A suborder of the Diptera; in this group of flies, the adult escapes from the puparium through a T-shaped opening.

orthoscopic system [OPTICS] An optical system that has been corrected so that distortion and spherical aberration are eliminated. Also known as rectilinear system.

orthose *See* orthoclase.

orthotropous [BOT] Having a straight ovule with the micropyle at the end opposite the stalk.

Orthox [GEOL] A suborder of the soil order Oxisol that is moderate to low in organic matter, well drained, and moist all or nearly all year; believed to be extensive at low altitudes in the heart of the humid tropics.

Orussidae [INV ZOO] A small family of hymenopteran insects in the superfamily Siricoidea.

Os *See* osmium.

Osagean [GEOL] A provincial series of geologic time in North America; Lower Mississippian (above Kinderhookian, below Meramecian).

osage orange [BOT] *Maclura pomifera.* A tree in the mulberry family of the Urticales characterized by yellowish bark, milky sap, simple entire leaves, strong axillary thorns, and aggregate green fruit about the size and shape of an orange.

osar *See* esker.

O scan *See* O scope.

oscillation [CONT SYS] *See* cycling. [PHYS] Any effect that varies periodically back and forth between two values.

oscillator [ELECTR] **1.** An electronic circuit that converts energy from a direct-current source to a periodically varying electric output. **2.** The stage of a superheterodyne receiver that generates a radio-frequency signal of the correct frequency to mix with the incoming signal and produce the intermediate-frequency value of the receiver. **3.** The stage of a transmitter that generates the carrier frequency of the station or some fraction of the carrier frequency. [PHYS] Any device (mechanical or electrical), which, in the absence of external forces, can have a periodic back-and-forth motion, the frequency determined by the properties of the oscillator.

Oscillatoriales [BOT] An order of blue-green algae (Cyanophyceae) which are filamentous and truly multicellular. ·

oscillator-mixer-first detector *See* converter.

oscillatory circuit [ELEC] Circuit containing inductance or capacitance, or both, and resistance, connected so that a voltage impulse will produce an output current which periodically reverses or oscillates.

oscillatory discharge [ELEC] Alternating current of gradually decreasing amplitude which, under certain conditions, flows through a circuit containing inductance, capacitance, and resistance when a voltage is applied.

oscillatory extinction *See* undulatory extinction.

oscillistor [ELECTR] A bar of semiconductor material, such as germanium, that will oscillate much like a quartz crystal when it is placed in a magnetic field and is carrying direct current that flows parallel to the magnetic field.

oscillograph [ENG] A measurement device for determining waveform by recording the instantaneous values of a quantity such as voltage as a function of time.

oscilloscope *See* cathode-ray oscilloscope.

Oscines [VERT ZOO] The songbirds, a suborder of the order Passeriformes.

O scope [ELECTR] An A scope modified by the inclusion of an adjustable notch for measuring range. Also known as O indicator; O scan.

osculating orbit [ASTRON] The orbit which would be followed by a body such as an asteroid or comet if, at a given time, all the planets suddenly disappeared, and it then moved under the gravitational force of the sun alone.

O shell [ATOM PHYS] The fifth layer of electrons about the nucleus of an atom, having electrons characterized by the principal quantum number 5.

osmate [INORG CHEM] A salt or ester of osmic acid, containing the osmate radical, OsO_4^{--}; for example, potassium osmate (K_2OsO_4).

osmium [CHEM] A chemical element, symbol Os, atomic number 76, atomic weight 190.2. [MET] A hard white metal of rare natural occurrence.

osmolality [CHEM] The molality of an ideal solution of a nondissociating substance that exerts the same osmotic pressure as the solution being considered.

osmolarity [CHEM] The molarity of an ideal solution of a nondissociating substance that exerts the same osmotic pressure as the solution being considered.

osmole [CHEM] **1.** The unit of osmolarity equal to the osmolarity of a solution that exerts an osmotic pressure equal to that of an ideal solution of a non-dissociating substance that has a concentration of 1 mole of solute per liter of solution. **2.** The unit of osmolality equal to the osmolality of a solution that exerts an osmotic pressure equal to that of an ideal solution of a nondissociating substance that has a concentration of 1 mole of solute per kilogram of solvent.

osmoregulatory mechanism [PHYSIO] Any physiological mechanism for the maintenance of an optimal and constant level of osmotic activity of the fluid in and around the cells.

osmosis [PHYS CHEM] The transport of a solvent through a semipermeable membrane separating two solutions of different solute concentration, from the solution that is dilute in solute to the solution that is concentrated.

osmotic gradient *See* osmotic pressure.

osmotic pressure [PHYS CHEM] **1.** The applied pressure required to prevent the flow of a solvent across a membrane which offers no obstruction to passage of the solvent, but does not allow passage of the solute, and which separates a solution from the pure solvent. **2.** The applied pressure required to prevent passage of a solvent across a membrane which separates solutions of different concentration, and which allows passage of the solute, but may also allow limited passage of the solvent. Also known as osmotic gradient.

osseous [ANAT] Bony; composed of or resembling bone.

ossicle [ANAT] Any of certain small bones, as those of the middle ear. [INV ZOO] Any of various calcareous bodies.

ossify [PHYSIO] To form or turn into bone.

ost-, oste-, osteo- [ANAT] A combining form meaning bone.

Ostariophysi [VERT ZOO] A superorder of actinopterygian fishes distinguished by the structure of the anterior four or five vertebrae which are modified as an encasement for the bony ossicles connecting the inner ear and swim bladder.

Osteichthyes [VERT ZOO] The bony fishes, a class of fishlike vertebrates distinguished by having a bony skeleton, a swim bladder, a true gill cover, and mesodermal ganoid, cycloid, or ctenoid scales.

osteitis [MED] Inflammation of bone.

osteoarthritis See degenerative joint disease.

osteocyte [HISTOL] A bone cell.

osteogenic sarcoma See osteosarcoma.

Osteoglossidae [VERT ZOO] The bony tongues, a family of actinopterygian fishes in the order Osteoglossiformes.

Osteoglossiformes [VERT ZOO] An order of soft-rayed, actinopterygian fishes distinguished by paired, usually bony rods at the base of the second gill arch, a single dorsal fin, no adipose fin, and a usually abdominal pelvic fin.

Osteolepidae [PALEON] A family of extinct fishes in the order Osteolepiformes.

Osteolepiformes [PALEON] A primitive order of fusiform lobefin fishes, subclass Crossopterygii, generally characterized by rhombic bony scales, two dorsal fins placed well back on the body, and a well-ossified head covered with large dermal plating bones.

osteoma [MED] A benign bone tumor, especially in membrane bones of the skull.

osteomyelitis [MED] Inflammation of bone tissue and bone marrow.

osteon [HISTOL] A microscopic unit of mature bone composed of layers of osteocytes and bone surrounding a central canal. Also known as Haversian system.

osteoporosis [MED] Deossification with absolute decrease in bone tissue, resulting in enlargement of marrow and Haversian spaces, decreased thickness of cortex and trabeculae, and structural weakness.

osteosarcoma [MED] A malignant tumor principally composed of anaplastic cells of mesenchymal derivation. Also known as osteogenic sarcoma.

Ostomidae [INV ZOO] The bark-gnawing beetles, a family of coleopteran insects in the superfamily Cleroidea.

Ostracoda [INV ZOO] A subclass of the class Crustacea containing small, bivalved aquatic forms; the body is unsegmented and there is no true abdominal region.

ostracoderm [PALEON] Any of various extinct jawless vertebrates covered with an external skeleton of bone which together with the Cyclostomata make up the class Agnatha.

Ostreidae [INV ZOO] A family of bivalve mollusks in the order Anisomyaria containing the oysters.

ostrich [VERT ZOO] Struthio camelus. A large running bird with soft plumage, naked head, neck and legs, small wings, and thick powerful legs with two toes on each leg; the only living species of the Struthioniformes.

Ostwald process [CHEM ENG] An industrial preparation of nitric acid by the oxidation of ammonia; the oxidation takes place in successive stages to nitric oxide, nitrogen dioxide, and nitric acid; a catalyst of platinum gauze is used and high temperatures are needed.

Ostwald viscometer [ENG] A viscometer in which liquid is drawn into the higher of two glass bulbs joined by a length of capillary tubing, and the time for its meniscus to fall between calibration marks above and below the upper bulb is compared with that for a liquid of known viscosity.

Otariidae [VERT ZOO] The sea lions, a family of carnivorous mammals in the superfamily Canoidea.

Othniidae [INV ZOO] The false tiger beetles, a small family of coleopteran insects in the superfamily Tenebrionoidea.

otic [ANAT] Of or pertaining to the ear or a part thereof.

-otic [SCI TECH] A suffix meaning of, pertaining to, characterized by, or causing the process.

Otitidae [INV ZOO] A family of cyclorrhaphous myodarian dipteran insects in the subsection Acalyptratae.

otitis [MED] Inflammation of the ear.

otolaryngology [MED] A branch of medicine that deals with the ear, nose, and throat. Also known as otorhinolaryngology.

otolith [ANAT] A calcareous concretion on the end of a sensory hair cell in the vertebrate ear and in some invertebrates.

otorhinolaryngology See otolaryngology.

otoscope [MED] An apparatus designed for examination of the ear and for rendering the tympanic membrane visible.

OTS See ovonic threshold switch.

otter [ENG] See paravane. [VERT ZOO] Any of various members of the family Mustelidae, having a long thin body, short legs, a somewhat flattened head, webbed toes, and a broad flattened tail; all are adapted to aquatic life.

Otto cycle [THERMO] A thermodynamic cycle for the conversion of heat into work, consisting of two isentropic phases interspersed between two constant-volume phases. Also known as spark-ignition combustion cycle.

Otto engine [MECH ENG] An internal combustion engine that operates on the Otto cycle, where the phases of suction, compression, expansion, and exhaust occur sequentially in a four-stroke-cycle or two-stroke-cycle reciprocating mechanism.

OTV See orbital transfer vehicles.

O-type backward-wave oscillator [ELECTR] A backward-wave tube in which an electron gun produces an electron beam focused longitudinally throughout the length of the tube, a slow-wave circuit interacts with the beam, and at the end of the tube a collector terminates the beam. Also known as O-type carcinotron; type-O carcinotron.

O-type carcinotron See O-type backward-wave oscillator.

O-type star [ASTRON] A spectral-type classification in the Draper catalog of stars; a star having spectral type O; a very hot, blue star in which the spectral lines of ionized helium are prominent.

ouachitite [PETR] A biotite monchiquite with no olivine and a glassy or analcime groundmass.

ounce [MECH] 1. A unit of mass in avoirdupois measure equal to 1/16 pound or to approximately 0.0283495 kilogram. Abbreviated oz. 2. A unit of mass in either troy or apothecaries' measure equal to 480 grains or exactly 0.0311034768 kilogram. Also known as apothecaries' ounce or troy ounce (abbreviations are oz ap and oz t in the United States, and oz apoth and oz tr in the United Kingdom).

outbreeding See exogamy.

outburst [MIN ENG] The sudden issue of gases, chiefly methane (sometimes accompanied by coal dust), from the working face of a coal mine.

outcrop [GEOL] Exposed stratum or body of ore at the surface of the earth. Also known as cropout.

outer atmosphere [METEOROL] Very generally, the atmosphere at a great distance from the earth's surface; possibly best usage of the term is as an approximate synonym for exosphere.

outer mantle *See* upper mantle.

outer planets [ASTRON] The planets with orbits larger than that of Mars: Jupiter, Saturn, Uranus, Neptune, and Pluto.

outer-shell electron *See* conduction electron.

outer space [ASTRON] A general term for any region that is beyond the earth's atmosphere.

outface *See* dip slope.

outfall [CIV ENG] The point at which a sewer or drainage channel discharges to the sea or to a river. [HYD] The narrow part of a stream, lake, or other body of water where it drops away into a larger body.

outgoing trunk *See* one-way trunk.

outlet [ELEC] A power line termination from which electric power can be obtained by inserting the plug of a line cord. Also known as convenience receptacle; electric outlet; receptacle.

outlier [GEOL] A group of rocks separated from the main mass and surrounded by outcrops of older rocks. [STAT] In a set of data, a value so far removed from other values in the distribution that its presence cannot be attributed to the random combination of chance causes.

output [COMPUT SCI] 1. The data produced by a data-processing operation, or the information that is the objective or goal in data processing. 2. The data actively transmitted from within the computer to an external device, or onto a permanent recording medium (paper, microfilm). 3. The activity of transmitting the generated information. 4. The readable storage medium upon which generated data are written, as in hard-copy output. [ELECTR] 1. The current, voltage, power, driving force, or information which a circuit or device delivers. 2. Terminals or other places where a circuit or device can deliver current, voltage, power, driving force, or information. [SCI TECH] The product of a system.

output area [COMPUT SCI] A part of storage that has been reserved for output data. Also known as output block.

output block [COMPUT SCI] 1. A portion of the internal storage of a computer that is reserved for receiving, processing, and transmitting data to be transferred out. 2. *See* output area.

output monitor interrupt [COMPUT SCI] A data-processing step in which control is passed to the monitor to determine the precedence order for two requests having the same priority level.

output program *See* output routine.

output record [COMPUT SCI] 1. A unit of data that has been transcribed from a computer to an external medium or device. 2. The unit of data that is currently held in the output area of a computer before being transcribed to an external medium or device.

output routine [COMPUT SCI] A series of computer instructions which organizes and directs all operations associated with the transcription of data from a computer to various media and external devices by various types of output equipment. Also known as output program.

output shaft [MECH ENG] The shaft that transfers motion from the prime mover to the driven machines.

output standard *See* standard time.

output transformer [ELECTR] The iron-core audio-frequency transformer used to match the output stage of a radio receiver or an amplifier to its loudspeaker or other load.

output unit [COMPUT SCI] In computers, a unit which delivers information from the computer to an external device or from internal storage to external storage.

outside diameter [DES ENG] The outer diameter of a pipe, including the wall thickness; usually measured with calipers. Abbreviated OD.

outwash [GEOL] 1. Sand and gravel transported away from a glacier by streams of meltwater and either deposited as a floodplain along a preexisting valley bottom or broadcast over a preexisting plain in a form similar to an alluvial fan. Also known as glacial outwash; outwash drift; overwash. 2. Soil material washed down a hillside by rainwater and deposited on more gently sloping land.

outwash apron *See* outwash plain.

outwash cone [GEOL] A cone-shaped deposit consisting chiefly of sand and gravel found at the edge of shrinking glaciers and ice sheets.

outwash drift *See* outwash.

outwash plain [GEOL] A broad, outspread flat or gently sloping alluvial deposit of outwash in front of or beyond the terminal moraine of a glacier. Also known as apron; frontal apron; frontal plain; marginal plain; morainal apron; morainal plain; outwash apron; overwash plain; sandur; wash plain.

outwash train *See* valley train.

ouvarovite *See* uvarovite.

ovalbumin [BIOCHEM] The major, conjugated protein of eggwhite.

oval window [ANAT] The membrane-covered opening into the inner ear of tetrapods, to which the ossicles of the middle ear are connected.

ovarian follicle [HISTOL] An ovum and its surrounding follicular cells, found in the ovarian cortex.

ovary [ANAT] A glandular organ that produces hormones and gives rise to ova in female vertebrates. [BOT] The enlarged basal portion of a pistil that bears the ovules in angiosperms.

oven [ENG] A heated enclosure for baking, heating, or drying. [GEOL] 1. A rounded, saclike, chemically weathered pit or hollow in a rock (especially a granitic rock) which has an arched roof and resembles an oven. 2. *See* spouting horn.

overburden [GEOL] 1. Rock material overlying a mineral deposit or coal seam. Also known as baring; top. 2. Material of any nature, consolidated or unconsolidated, that overlies a deposit of useful materials, ores, or coal, especially those deposits that are mined from the surface by open cuts. 3. Loose soil, sand, or gravel that lies above the bedrock. [MIN ENG] To charge in a furnace too much ore and flux in proportion to the amount of fuel.

overcast [METEOROL] 1. Pertaining to a sky cover of 1.0 (95% or more) when at least a portion of this amount is attributable to clouds or obscuring phenomena aloft, that is, when the total sky cover is not due entirely to surface-based obscuring phenomena. 2. Cloud layer that covers most or all of the sky; generally, a widespread layer of clouds such as that which is considered typical of a warm front. [MIN ENG] 1. An enclosed airway to permit one air current to pass over another without interruption. 2. To move overburden removed from coal mined from surface mines to an area from which the coal has been mined.

overcritical electric field   [ATOM PHYS] An electric field so strong that an electron-positron pair is created spontaneously; quantum electrodynamics predicts that this will happen near a nucleus having more than approximately 173 protons.

overcurrent protection   *See* overload protection.

overdrive   [MECH ENG] An automobile engine device that lowers the gear ratio, thereby reducing fuel consumption.

overflow   [CIV ENG] Any device or structure that conducts excess water or sewage from a conduit or container.   [COMPUT SCI] **1.** The condition that arises when the result of an arithmetic operation exeeds the storage capacity of the indicated result-holding storage. **2.** That part of the result which exceeds the storage capacity.   [SCI TECH] Excess liquid which overflows its given limits.

overflow check indicator   *See* overflow indicator.

overflow error   [COMPUT SCI] The condition in which the numerical result of an operation exceeds the capacity of the register.

overflow indicator   [COMPUT SCI] A bistable device which changes state when an overflow occurs in the register associated with it, and which is designed so that its condition can be determined, and its original condition restored. Also known as overflow check indicator.

overflow record   [COMPUT SCI] A unit of data whose length is too great for it to be stored in an assigned section of a direct-access storage, and which must be stored in another area from which it may be retrieved by means of a reference stored in the original assigned area in place of the record.

overflow storage   [COMMUN] Additional storage provided in a store-and-forward-switching center to prevent the loss of messages (or parts of messages) offered to a completely filled line store.

overgear   [MECH ENG] A gear train in which the angular velocity ratio of the driven shaft to driving shaft is greater than unity, as when the propelling shaft of an automobile revolves faster than the engine shaft.

overgrowth   [CRYSTAL] A crystal growth in optical and crystallographic continuity around another crystal of different composition.   [MINERAL] A mineral deposited on and growing in oriented, crystallographic directions on the surface of another mineral.

overhand stoping   [MIN ENG] A method of mining in which the ore is blasted from a series of ascending stepped benches; both horizontal and vertical holes may be employed.

overhang   [BUILD] The distance measured horizontally that a roof projects beyond a wall.   [GEOL] The part of a salt plug that projects from the top.

Overhauser effect   [ATOM PHYS] The effect whereby, if a radio frequency field is applied to a substance in an external magnetic field, whose nuclei have spin ½ and which has unpaired electrons, at the electron spin resonance frequency, the resulting polarization of the nuclei is as great as if the nuclei had the much larger electron magnetic moment.

overhead   [CHEM ENG] Pertaining to fluid (gas or liquid) effluent from the top of a process vessel, such as a distillation column.   [COMPUT SCI] The time a computer system spends doing computations that do not contribute directly to the progress of any user tasks in the system, such as allocation of resources, responding to exceptional conditions, providing protection and reliability, and accounting.

overhead camshaft   [MECH ENG] A camshaft mounted above the cylinder head.

overhead-valve engine   [MECH ENG] A four-stroke-cycle internal combustion engine having its valves located in the cylinder head, operated by pushrods that actuate rocker arms. Abbreviated OHV engine. Also known as valve-in-head engine.

overlap   [COMMUN] **1.** In teletypewriter practice, the selecting of another code group while the printing of a previously selected code group is taking place. **2.** Amount by which the effective height of the scanning facsimile spot exceeds the nominal width of the scanning line.   [COMPUT SCI] To perform some or all of an operation concurrently with one or more other operations.   [GEOL] **1.** Movement of an upcurrent part of a shore to a position extending seaward beyond a downcurrent part. **2.** Extension of strata over or beyond older underlying rocks. **3.** The horizontal component of separation measured parallel to the strike of a fault.   [MET] **1.** Projection of the weld metal beyond the bond at the toe of the weld. **2.** Extension of one sheet over another in spot, seam, or projection welding.

overlapping   [COMPUT SCI] An operation whereby, if the processor determines that the current instruction and the next instruction lie in different storage modules, the two words may be retrieved in parallel.

overlay   [CIV ENG] A repair topping of asphalt or concrete placed on a worn roadway.   [COMPUT SCI] A technique for bringing routines into high-speed storage from some other form of storage during processing, so that several routines will occupy the same storage locations at different times; overlay is used when the total storage requirements for instructions exceed the available main storage.   [ENG] **1.** Nonwoven fibrous mat (glass or other fiber) used as the top layer in a cloth or mat layup to give smooth finish to plastic products or to minimize the fibrous pattern on the surface. Also known as surfacing mat. **2.** An ornamental covering, as of wood or metal.   [GRAPHICS] **1.** A sheet attached to copy or artwork and containing special instructions about reproduction or arrangement. **2.** A transparent or translucent film attached to artwork and carrying additional detail to be reproduced; in multicolor printing these overlays may represent the separation of the colors to be printed.

overload   [ELECTR] A load greater than that which a device is designed to handle; may cause overheating of power-handling components and distortion in signal circuits.   [GEOL] The amount of sediment that exceeds the ability of a stream to transport it and is therefore deposited.

overload current   [ELECTR] A current greater than that which a circuit is designed to carry; may melt wires or damage elements of the circuit.

overload protection   [ELEC] Effect of a device operative on excessive current, but not necessarily on short circuit, to cause and maintain the interruption of current flow to the device governed. Also known as overcurrent protection.

overload relay   [ELEC] A relay that opens a circuit when the load in the circuit exceeds a preset value, in order to provide overload protection; usually responds to excessive current, but may respond to excessive values of power, temperature, or other quantities. Also known as overload release.

overload release   *See* overload relay.

overpotential   *See* overvoltage.

overprint   [GEOCHEM] A complete or partial disturbance of an isolated radioactive system by thermal, igneous, or tectonic activities which results in loss or gain of radioactive or radiogenic isotopes and, hence, a change in the radiometric age that will be given the disturbed system. [GEOL] The development or superposition of metamorphic structures on original structures. Also known as imprint; metamorphic overprint; superprint. [GRAPHICS] **1.** To imprint over something that has been printed. **2.** To apply a varnish or lacquer to printed matter from a type or litho process, by means of a brush, spray, or roller coating. **3.** To print in a primary color over an existing color print to obtain a compound shade.

overrunning clutch [MECH ENG] A clutch that allows the driven shaft to turn freely only under certain conditions; for example, a clutch in an engine starter that allows the crank to turn freely when the engine attempts to run.

overshoot [ELECTROMAG] The reception of microwave signals where they were not intended, due to an unusual atmospheric condition that sets up variations in the index of refraction. [ENG] **1.** An initial transient response to a unidirectional change in input which exceeds the steady-state response. **2.** The maximum amount by which this transient response exceeds the steady-state response.

overshot [ENG] **1.** A fishing tool for recovering lost drill pipe or casing. **2.** *See* bullet.

over-the-horizon radar [ELECTROMAG] Long-range radar in which the transmitted and reflected beams are bounced off the ionosphere layers to achieve ranges far beyond line of sight.

overthrust [GEOL] **1.** A thrust fault that has a low dip or a net slip that is large. Also known as low-angle thrust; overthrust fault. **2.** A thrust fault with the active element being the hanging wall.

overthrust fault *See* overthrust.

overtone [ACOUS] **1.** A component of a complex sound whose frequency is an integral multiple, greater than 1, of the fundamental frequency. **2.** A component of a complex tone having a pitch higher than that of the fundamental pitch. [MECH] One of the normal modes of vibration of a vibrating system whose frequency is greater than that of the fundamental mode. [PHYS] A harmonic other than the fundamental component.

overvoltage [ELEC] A voltage greater than that at which a device or circuit is designed to operate. Also known as overpotential. [ELECTR] The amount by which the applied voltage exceeds the Geiger threshold in a radiation counter tube. [PHYS CHEM] The difference between electrode potential under electrolysis conditions and the thermodynamic value of the electrode potential in the absence of electrolysis for the same experimental conditions. Also known as overpotential.

overwash [GEOL] **1.** A mass of water representing the part of the wave advancing up a beach that runs over the highest part of the berm (or other structure) and that does not flow directly back to the sea or lake. **2.** *See* outwash.

overwash plain *See* outwash plain.

overwash pool [OCEANOGR] A tidal pool between a berm and a beach scarp which water enters only at high tide.

oviduct [ANAT] A tube that serves to conduct ova from the ovary to the exterior or to an intermediate organ such as the uterus. Also known in mammals as Fallopian tube; uterine tube.

ovine encephalomyelitis *See* louping ill.

oviparous [VERT ZOO] Producing eggs that develop and hatch externally.

ovipositor [INV ZOO] A specialized structure in many insects for depositing eggs. [VERT ZOO] A tubular extension of the genital orifice in most fishes.

ovonic device *See* glass switch.

ovonic memory switch [ELECTR] A glass switch which, after being brought from the highly resistive state to the conducting state, remains in the conducting state until a current pulse returns it to its highly resistive state. Abbreviated OMS. Also known as memory switch.

ovonic threshold switch [ELECTR] A glass switch which, after being brought from the highly resistive state to the conducting state, returns to the highly resistive state when the current falls below a holding current value. Abbreviated OTS.

ovoviviparous [VERT ZOO] Producing eggs that develop internally and hatch before or soon after extrusion.

ovulation [PHYSIO] Discharge of an ovum or ovule from the ovary.

ovule [BOT] A structure in the ovary of a seed plant that develops into a seed following fertilization.

ovum [CYTOL] A female gamete. Also known as egg.

O-wave component *See* ordinary-wave component.

Oweniidae [INV ZOO] A family of limivorous polychaete annelids of the Sedentaria.

owl [VERT ZOO] Any of a number of diurnal and nocturnal birds of prey composing the order Strigiformes; characterized by a large head, more or less forward-placed large eyes, a short hooked bill, and strong talons.

own coding [COMPUT SCI] A series of instructions added to a standard software routine to change or extend the routine so that it can carry out special tasks.

OW unit *See* sabin.

Oxalidaceae [BOT] A family of dicotyledonous plants in the order Geraniales, generally characterized by regular flowers, two or three times as many stamens as sepals or petals, a style which is not gynobasic, and the fruit which is a beakless, loculicidal capsule.

oxalite *See* humboldtine.

oxammite [MINERAL] $(NH_4)_2C_2O_4 \cdot H_2O$ A yellowish-white, orthorhombic mineral consisting of ammonium oxalate monohydrate; occurs as lamellar masses.

oxatyl *See* carboxyl.

oxbow [HYD] **1.** A closely looping, U-shaped stream meander whose curvature is so extreme that only a neck of land remains between the two parts of the stream. Also known as horseshoe bend. **2.** *See* oxbow lake. [GEOL] The abandoned, horseshoe-shaped channel of a former stream meander after the stream formed a neck cutoff. Also known as abandoned channel.

oxbow lake [HYD] The crescent-shaped body of water located alongside a stream in an abandoned oxbow after a neck cutoff is formed and the ends of the original bends are silted up. Also known as crescentic lake; cutoff lake; horseshoe lake; loop lake; moat; mortlake; oxbow.

Oxfordian [GEOL] A European stage of geologic time, in the Upper Jurassic (above Callovian, below Kimmeridgean). Also known as Divesian.

oxidase [BIOCHEM] An enzyme that catalyzes oxidation reactions by the utilization of molecular oxygen as an electron acceptor.

oxidation [CHEM] **1.** A chemical reaction that increases the oxygen content of a compound. **2.** A chemical reaction in which a compound or radical loses electrons, that is in which the positive valence is increased.

oxidation number [CHEM] 1. Numerical charge on the ions of an element. 2. *See* oxidation state.

oxidation potential [PHYS CHEM] The difference in potential between an atom or ion and the state in which an electron has been removed to an infinite distance from this atom or ion.

oxidation-reduction potential *See* redox potential.

oxidation-reduction reaction [CHEM] An oxidizing chemical change, where an element's positive valence is increased (electron loss), accompanied by a simultaneous reduction of an associated element (electron gain).

oxidation state [CHEM] The number of electrons to be added (or subtracted) from an atom in a combined state to convert it to elemental form. Also known as oxidation number.

oxide [CHEM] Binary chemical compound in which oxygen is combined with a metal (such as Na_2O; basic) or nonmetal (such as NO_2; acidic).

oxidite *See* shale ball.

oxidizing agent [CHEM] Compound that gives up oxygen easily, removes hydrogen from another compound, or attracts negative electrons.

oximido *See* nitroso.

Oxisol [GEOL] A soil order characterized by residual accumulations of inactive clays, free oxides, kaolin, and quartz; mostly tropical.

oxonium ion *See* hydronium ion.

oxo process [CHEM ENG] Catalytic process for production of alcohols, aldehydes, and other oxygenated organic compounds by reaction of olefin vapors with carbon monoxide and hydrogen.

oxy- [CHEM] 1. Prefix indicating the oxygen radical (—O—) in a chemical compound. 2. Prefix incorrectly used as a substitute for hydroxy-.

oxyacetylene torch [ENG] An acetylene gas-mixing and burning tool that produces a hot flame for the welding or cutting of metal. Also known as acetylene torch.

Oxyaenidae [PALEON] An extinct family of mammals in the order Deltatheridea; members were short-faced carnivores with powerful jaws.

oxygen [CHEM] A gaseous chemical element, symbol O, atomic number 8, and atomic weight 15.9994; an essential element in cellular respiration and in combustion processes; the most abundant element in the earth's crust, and about 20% of the air by volume.

oxygen-18 [NUC PHYS] Oxygen isotope with atomic weight 18; found 8 parts to 10,000 of oxygen-16 in water, air, and rocks; used in tracer experiments. Also known as heavy oxygen.

oxygen point [THERMO] The temperature at which liquid oxygen and its vapor are in equilibrium, that is, the boiling point of oxygen, at standard atmospheric pressure; it is taken as a fixed point on the International Practical Temperature Scale of 1968, at $-182.962°C$.

oxyheeite [MINERAL] $Pb_5Ag_2Sb_6S_{15}$ A light steel gray to silver white mineral consisting of lead and silver antimony sulfide; occurs as acicular needles or in massive form.

oxyhemoglobin [BIOCHEM] The red crystalline pigment formed in blood by the combination of oxygen and hemoglobin, without the oxidation of iron.

Oxymonadida [INV ZOO] An order of xylophagous protozoans in the class Zoomastigophorea; colorless flagellate symbionts in the digestive tract of the roach *Cryptocercus* and of certain termites.

Oxystomata [INV ZOO] A subsection of the Brachyura, including those true crabs in which the first pair of pereiopods is chelate, and the mouth frame is triangular and forward.

Oxystomatidae [INV ZOO] A family of free-living marine nematodes in the superfamily Enoploidea, distinguished by amphids that are elongated longitudinally.

Oxyurata [INV ZOO] The equivalent name for Oxyurina.

Oxyuridae [INV ZOO] A family of the nematode superfamily Oxyuroidea.

Oxyurina [INV ZOO] A suborder of nematodes in the order Ascaridida.

Oxyuroidea [INV ZOO] A superfamily of the class Nematoda.

Oyashio [OCEANOGR] A cold current flowing from the Bering Sea southwest along the coast of Kamchatka, past the Kuril Islands, continuing close to the northeast coast of Japan, and reaching nearly 35°N.

oyster [INV ZOO] Any of various bivalve mollusks of the family Ostreidae; the irregular shell is closed by a single adductor muscle, the foot is small or absent, and there is no siphon.

oz *See* ounce.

ozalid [GRAPHICS] A print on light-sensitized material produced directly from a positive transparency and developed via a dry process with ammonia vapor.

oz ap *See* ounce.

oz apoth *See* ounce.

Ozawainellidae [PALEON] A family of extinct protozoans in the superfamily Fusulinacea.

ozocerite [GEOL] A natural, brown to jet black paraffin wax occurring in irregular veins; consists principally of hydrocarbons, is soluble in water, and has a variable melting point. Also known as ader wax; earth wax; fossil wax; mineral wax; native paraffin; ozokerite. [MATER] *See* ceresin.

ozokerite *See* ozocerite.

ozone [CHEM] O_3 Unstable blue gas with pungent odor; an allotropic form of oxygen; a powerful oxidant boiling at $-112°C$; used as an oxidant, bleach, and water purifier, and to treat industrial wastes.

ozone layer *See* ozonosphere.

ozonosphere [METEOROL] The general stratum of the upper atmosphere in which there is an appreciable ozone concentration and in which ozone plays an important part in the radiative balance of the atmosphere; lies roughly between 10 and 50 kilometers, with maximum ozone concentration at about 20 to 25 kilometers. Also known as ozone layer.

oz t *See* ounce.

oz tr *See* ounce.

P

p- *See* para-; peta-.

P *See* phosphorus; poise.

Pa *See* pascal; protactinium.

PABA *See* para-aminobenzoic acid.

PABX *See* private automatic branch exchange.

paca [VERT ZOO] Any of several rodents of the genus *Cuniculus*, especially *C. paca*, with a white-spotted brown coat, found in South and Central America.

pacemaker [MED] A pulsed battery-operated oscillator implanted in the body to deliver electric impulses to the muscles of the lower heart, either at a fixed rate or in response to a sensor that detects when the patient's pulse rate slows or ceases. Also known as cardiac pacemaker; heart pacer.

pachnolite [MINERAL] $NaCaAlF_6 \cdot H_2O$ Colorless to white mineral composed of hydrous sodium calcium aluminum fluoride, occurring in monoclinic crystals.

pachoidal structure *See* flaser structure.

Pachycephalosauridae [PALEON] A family of ornithischian dinosaurs characterized by a skull with a solid rounded mass of bone 10 centimeters thick above the minute brain cavity.

pachyderm [VERT ZOO] Any of various nonruminant hooved mammals characterized by thick skin, including the elephants, hippopotamuses, rhinoceroses, and others.

pachynema *See* pachytene.

pachytene [CYTOL] The third stage of meiotic prophase during which paired chromosomes thicken, each chromosome splits into chromatids, and breakage and crossing over between nonsister chromatids occur. Also known as pachynema.

Pacific anticyclone *See* Pacific high.

Pacific Equatorial Countercurrent [OCEANOGR] The Equatorial Countercurrent flowing east across the Pacific Ocean between 3° and 10°N.

Pacific high [METEOROL] The nearly permanent subtropical high of the North Pacific Ocean, centered, in the mean, at 30–40°N and 140–150°W. Also known as Pacific anticyclone.

Pacific North Equatorial Current [OCEANOGR] The North Equatorial Current which flows westward between 10° and 20°N in the Pacific Ocean.

Pacific South Equatorial Current [OCEANOGR] The South Equatorial Current flowing westward between 3°N and 10°S in the Pacific Ocean.

Pacific Standard Time *See* Pacific time.

Pacific time [ASTRON] The time for a given time zone that is based on the 120th meridian and is the eighth zone west of Greenwich. Also known as Pacific Standard Time.

Pacinian corpuscle [ANAT] An encapsulated lamellar sensory nerve ending that functions as a kinesthetic receptor.

pack [COMPUT SCI] To reduce the amount of storage required to hold information by changing the method of encoding the data. [IND ENG] To provide protection for an article or group of articles against physical damage during shipment; packing is accomplished by placing articles in a shipping container, and blocking, bracing, and cushioning them when necessary, or by strapping the articles or containers on a pallet or skid. [ORD] Part of a parachute assembly in which the canopy and shroud lines are folded and carried. Also known as pack assembly. [MIN ENG] **1.** A pillar built in the waste area or roadside within a mine to support the mine roof; constructed from loose stones and dirt. **2.** Waste rock or timber used to support the roof or underground workings or used to fill excavations. Also known as fill. [OCEANOGR] *See* pack ice.

package [COMPUT SCI] A program that is written for a general and widely used application in such a way that its usefulness is not impaired by the problems of data or organization of a particular user.

pack assembly *See* pack.

pack carburizing [MET] A method of surface hardening of steel in which parts are packed in a steel box with the carburizing compound and heated to elevated temperatures.

packed bed [CHEM ENG] A fixed layer of small particles or objects arranged in a vessel to promote intimate contact between gases, vapors, liquids, solids, or various combinations thereof; used in catalysis, ion exchange, sand filtration, distillation, absorption, and mixing.

packed decimal [COMPUT SCI] A means of representing two digits per character, to reduce space and increase transmission speed.

packed tower [CHEM ENG] A fractionating or absorber tower filled with small objects (packing) to bring about intimate contact between rising fluid (vapor or liquid) and falling liquid.

packed tube [CHEM ENG] A pipe or tube filled with high-heat-capacity granular material; used to heat gases when tubes are externally heated.

packet [COMMUN] A digital data message which is usually preceded by headers (containing address information and other control characters) and followed by control characters which signify the end of a message.

packet transmission [COMMUN] Transmission of standardized packets of data over transmission lines in a fraction of a second by networks of high-speed switching computers that have the message packets stored in fast-access core memory.

pack hardening [MET] A process of heat treating in which the workpiece is packed in a metal box together with carbonaceous material; carbon penetration is proportional to the length of heating; after treatment the workpiece is reheated and quenched.

pack ice [OCEANOGR] Any area of sea ice, except fast ice, composed of a heterogeneous mixture of ice of varying ages and sizes, and formed by the packing together of pieces of floating ice. Also known as ice canopy; ice pack; pack.

packing [CRYSTAL] Arrangement of atoms or ions in a crystal lattice. [ENG] *See* stuffing. [ENG ACOUS] Excessive crowding of carbon particles in a carbon microphone, produced by excessive pressure or by fusion particles due to excessive current, and causing lowered resistance and sensitivity. [GEOL] The arrangement of solid particles in a sediment or in sedimentary rock. [GRAPHICS] Paper used as a layer under the image or impression cylinder in letterpress printing or under the plate or blanket in lithographic printing in order to produce suitable pressure. [MET] In powder metallurgy, a material in which compacts are embedded during presintering or sintering operations.

packing density [COMPUT SCI] The amount of information per unit of storage medium, as characters per inch on tape, bits per inch or drum, or bits per square inch in photographic storage. [ELECTR] The number of devices or gates per unit area of an integrated circuit. [GEOL] A measure of the extent to which the grains of a sedimentary rock occupy the gross volume of the rock in contrast to the spaces between the grains; equal to the cumulative grain-intercept length along a traverse in a thin section.

packing fraction [NUC PHYS] The quantity $(M-A)/A$, where M is the mass of an atom in atomic mass units and A is its atomic number.

packing routine [COMPUT SCI] A subprogram which compresses data so as to eliminate blanks and reduce the storage needed for a file.

pad [AERO ENG] *See* launch pad. [ANAT] A small circumscribed mass of fatty tissue, as in terminal phalanges of the fingers or the underside of the toes of an animal, such as a dog. [ELECTR] **1.** An arrangement of fixed resistors used to reduce the strength of a radio-frequency or audio-frequency signal by a desired fixed amount without introducing appreciable distortion. Also known as fixed attenuator. **2.** *See* terminal area. [ENG] **1.** A layer of material used as a cushion or for protection. **2.** A projection of excess metal on a casting forging, or welded part. **3.** A takeoff or landing point for a helicopter or space vehicle. [MET] The brickwork that is beneath the molten iron at the base of a blast furnace.

Paenungulata [VERT ZOO] A superorder of mammals, including proboscideans, xenungulates, and others.

Paeoniaceae [BOT] A monogeneric family of dicotyledonous plants in the order Dilleniales; members are mesophyllic shrubs characterized by cleft leaves, flowers with an intrastaminal disk, and seeds having copious endosperm.

page [COMPUT SCI] **1.** A standard quantity of main-memory capacity, usually 512 to 4096 bytes or words, used for memory allocation and for partitioning programs into control sections. **2.** A standard quantity of source program coding, usually 8 to 64 lines, used for displaying the coding on a cathode-ray tube.

page printer [COMMUN] A high-speed printer used to transpose messages received on paper tape to full-page format, by printing characters one at a time. [COMPUT SCI] A computer output device which composes a full page of characters before printing the page.

page proof [GRAPHICS] A proof received from a compositor after the galley, and having the form of the final page, usually including any illustrations.

page reader [COMPUT SCI] In character recognition, a character reader capable of processing cut-form documents of varying sizes; sometimes capable of reading information in reel forms.

Paget's disease [MED] **1.** A type of carcinoma of the breast that involves the nipple or areola and the larger ducts, characterized by the presence of Paget's cells. **2.** Osseous hyperplasia simultaneous with accelerated deossification. **3.** An apocrine gland skin cancer, composed principally of Paget's cells.

page turning [COMPUT SCI] **1.** The process of moving entire pages of information between main memory and auxiliary storage, usually to allow several concurrently executing programs to share a main memory of inadequate capacity. **2.** In conversational time-sharing systems, the moving of programs in and out of memory on a round-robin, cyclic schedule so that each program may use its allotted share of computer time.

paging [COMPUT SCI] The scheme used to locate pages, to move them between main storage and auxiliary storage, or to exchange them with pages of the same or other computer programs; used in computers with virtual memories.

paging system [COMMUN] A system which gives an indication to a particular individual that he is wanted at the telephone, such as by sounding a number on musical gongs, calling by name over a loudspeaker, or producing an audible signal in a radio receiver carried in the individual's pocket.

Paguridae [INV ZOO] The hermit crabs, a family of decapod crustaceans belonging to the Paguridea.

Paguridea [INV ZOO] A group of anomuran decapod crustaceans in which the abdomen is nearly always asymmetrical, being either soft and twisted or bent under the thorax.

pahoehoe [GEOL] A type of lava flow whose surface is glassy, smooth, and undulating; the lava is basaltic, glassy, and porous. Also known as ropy lava.

pair production [PHYS] The conversion of a photon into an electron and a positron when the photon traverses a strong electric field, such as that surrounding a nucleus or an electron.

Palaeacanthaspidoidei [PALEON] A suborder of extinct, placoderm fishes in the order Rhenanida; members were primitive, arthrodire-like species.

Palaeacanthocephala [INV ZOO] An order of the Acanthocephala including parasitic worms characterized by fragmented nuclei in the hypodermis, lateral placement of the chief lacunar vessels, and proboscis hooks arranged in long rows.

Palaechinoida [PALEON] An extinct order of echinoderms in the subclass Perischoechinoidea with a rigid test in which the ambulacra bevel over the adjoining interambulacra.

Palaemonidae [INV ZOO] A family of decapod crustaceans in the group Caridea.

Palaeocaridacea [INV ZOO] An order of crustaceans in the superorder Syncarida.

Palaeocaridae [INV ZOO] A family of the crustacean order Palaeocaridacea.

Palaeoconcha [PALEON] An extinct order of simple, smooth-hinged bivalve mollusks.

Palaeocopida [PALEON] An extinct order of crustaceans in the subclass Ostracoda characterized by a straight hinge and by the anterior location for greatest height of the valve.

Palaeodonta [VERT ZOO] A suborder of artiodactylous mammals including piglike forms such as the extinct "giant pigs" and the hippopotami.

Palaeognathae [VERT ZOO] The ratites, making up a superorder of birds in the subclass Neornithes; merged with the Neognathae in some systems of classification.

Palaeoisopus [PALEON] A singular, monospecific, extinct arthropod genus related to the pycnogonida, but distinguished by flattened anterior appendages.

Palaeomastodontinae [PALEON] An extinct subfamily of elaphantoid proboscidean mammals in the family Mastodontidae.

Palaeomerycidae [PALEON] An extinct family of pecoran ruminants in the superfamily Cervoidea.

Palaeonemertini [INV ZOO] A family of the class Anopla distinguished by the two- or three-layered nature of the body-wall musculature.

Palaeonisciformes [PALEON] A large extinct order of chondrostean fishes including the earliest known and most primitive ray-finned forms.

Palaeoniscoidei [PALEON] A suborder of extinct fusiform fishes in the order Palaeonisciformes with a heavily ossified exoskeleton and thick rhombic scales on the body surface.

Palaeopantopoda [PALEON] A monogeneric order of extinct marine arthropods in the subphylum Pycnogonida.

Palaeopneustidae [INV ZOO] A family of deep-sea echinoderms in the order Spatangoida characterized by an oval test, long spines, and weakly developed fascioles and petals.

Palaeopterygii [VERT ZOO] An equivalent name for the Actinopterygii.

Palaeoryctidae [PALEON] A family of extinct insectivorous mammals in the order Deltatheridia.

Palaeospondyloidea [PALEON] An ordinal name assigned to the single, tiny fish *Palaeospondylus*, known only from Middle Devonian shales in Caithness, Scotland.

Palaeotheriidae [PALEON] An extinct family of perissodactylous mammals in the superfamily Equoidea.

palatine bone [ANAT] Either of a pair of irregularly L-shaped bones forming portions of the hard palate, orbits, and nasal cavities.

palatine tonsil [ANAT] Either of a pair of almond-shaped aggregations of lymphoid tissue embedded between folds of tissue connecting the pharynx and posterior part of the tongue with the soft palate. Also known as faucial tonsil; tonsil.

paleoanthropology [ANTHRO] A branch of anthropology concerned with the study of fossil man.

paleobiochemistry [PALEON] The study of chemical processes used by organisms that lived in the geologic past.

paleobiology [PALEON] The study of life and organisms that existed in the geologic past.

paleobotany [PALEON] The branch of paleontology concerned with the study of ancient and fossil plants and vegetation of the geologic past.

Paleocene [GEOL] A major worldwide division (epoch) of geologic time of the Tertiary period; extends from the end of the Cretaceous period to the Eocene epoch.

Paleocharaceae [PALEOBOT] An extinct group of fossil plants belonging to the Charophyta distinguished by sinistrally spiraled gyrogonites.

paleoclimatology [GEOL] The study of climates in the geologic past, involving the interpretation of glacial deposits, fossils, and paleogeographic, isotopic, and sedimentologic data.

Paleocopa [PALEON] An order of extinct ostracods distinguished by a long, straight hinge.

paleoecology [PALEON] The ecology of the geologic past.

Paleogene [GEOL] A geologic time interval comprising the Oligocene, Eocene, and Paleocene of the lower Tertiary period. Also known as Eogene.

paleogeographic stage *See* palstage.

paleogeography [GEOL] The geography of the geologic past; concerns all physical aspects of an area that can be determined from the study of the rocks.

paleogeology [GEOL] The geology of the past, applied particularly to the interpretation of the rocks at a surface of unconformity.

paleogeomorphology [GEOL] A branch of geomorphology concerned with the recognition of ancient erosion surfaces and the study of ancient topographies and topographic features that are now concealed beneath the surface and have been removed by erosion. Also known as paleophysiography.

paleomagnetics [GEOPHYS] The study of the direction and intensity of the earth's magnetic field throughout geologic time.

Paleonthropinae [PALEON] A former subfamily of fossil man in the family Hominidae; set up to include the Neanderthalers together with Rhodesian man.

paleontology [BIOL] The study of life of the past as recorded by fossil remains.

Paleoparadoxidae [PALEON] A family of extinct hippopotamuslike animals in the order Desmostylia.

paleophysiography *See* paleogeomorphology.

Paleoptera [INV ZOO] A section of the insect subclass Pterygota including primitive forms that are unable to flex their wings over the abdomen when at rest.

paleosol [GEOL] A soil horizon that formed on the surface during the geologic past, that is, an ancient soil. Also known as buried soil; fossil soil.

Paleozoic [GEOL] The era of geologic time from the end of the Precambrian (600 million years before present) until the beginning of the Mesozoic era (225 million years before present).

palette [GEOL] A broad sheet of calcite representing a solutional remnant in a cave. Also known as shield.

Palinuridae [INV ZOO] The spiny lobsters or langoustes, a family of macruran decapod crustaceans belonging to the Scyllaridea.

palisade mesophyll [BOT] A tissue system of the chlorenchyma in well-differentiated broad leaves composed of closely spaced palisade cells oriented parallel to one another, but with their long axes perpendicular to the surface of the blade.

palladium [CHEM] A chemical element, symbol Pd, atomic number 46, atomic weight 106.4. [MET] A white, ductile malleable metal that resembles platinum and follows it in abundance and importance of applications; does not tarnish at normal temperatures.

pallasite [GEOL] 1. A stony-iron meteorite composed essentially of large single glassy crystals of olivine embedded in a network of nickel-iron. 2. An ultramafic rock, of either meteoric or terrestrial origin, which contains more than 60% iron in the former, or more iron oxides than silica in the latter.

pallasite shell *See* lower mantle.

pallet [BUILD] A flat piece of wood laid in a wall to which woodwork may be securely fastened. [ENG] 1. A lever that regulates or drives a ratchet wheel. 2. A hinged valve on a pipe organ. 3. A tray or platform used in conjunction with a fork lift for lifting and moving materials. [INV ZOO] One of a pair of plates on the siphon tubes of certain Bivalvia. [GRAPHICS] An instrument consisting of a flat blade with a handle, used in clay work. [MECH ENG] One of the disks or pistons in a chain pump.

palletize [IND ENG] To package material for convenient handling on a pallet or lift truck.

Pallopteridae [INV ZOO] A family of myodarian cyclorrhaphous dipteran insects in the subsection Acalypteratae.

Palmales [BOT] An equivalent name for Arecales.

palmate [BOT] Having lobes, such as on leaves, that radiate from a common point. [VERT ZOO] Having webbed toes. [ZOO] Having the distal portion broad and lobed, resembling a hand with the fingers spread.

Palmyridae [INV ZOO] A mongeneric family of errantian polychaete annelids.

palp [INV ZOO] Any of various sensory, usually fleshy appendages near the oral aperture of certain invertebrates.

Palpicornia [INV ZOO] The equivalent name for Hydrophiloidea.

Palpigradida [INV ZOO] An order of rare tropical and warm-temperate arachnids; all are minute, whitish, eyeless animals with an elongate body that terminates in a slender, multisegmented flagellum set with setae.

palpitate [MED] To flutter, or beat abnormally fast; applied especially to the rate of the heartbeat.

palstage [GEOL] A period of time when paleogeographic conditions were relatively static or were changing gradually and progressively with relation to such factors as sea level, surface relief, or the distance of the shoreline from the region in question. Also known as paleogeographic stage.

paludal [ECOL] Relating to swamps or marshes and to material that is deposited in a swamp environment.

paludification [ECOL] Bog expansion resulting from the gradual rising of the water table as accumulation of peat impedes water drainage.

palygorskite [MINERAL] 1. A chain-structure type of clay mineral. 2. A group of lightweight, tough, fibrous clay minerals showing extensive substitution of aluminum for magnesium.

palynofacies [PALEON] An assemblage of palynomorphs in a portion of a sediment, representing local environmental conditions, but not representing the regional palynoflora.

palynology [PALEON] The study of spores, pollen, microorganisms, and microscopic fragments of megaorganisms that occur in sediments.

palynomorph [PALEON] A microscopic feature such as a spore or pollen that is of interest in palynological studies.

palynostratigraphy [PALEON] The stratigraphic application of palynologic methods.

palytoxin [BIOCHEM] A water-soluble toxin produced by several species of *Palythoa*; considered to be one of the most poisonous substances known.

PAM See pulse-amplitude modulation.

Pamphiliidae [INV ZOO] The web-spinning sawflies, a family of hymenopteran insects in the superfamily Megalodontoidea.

pan [COMMUN] To tilt or otherwise move a television or movie camera vertically and horizontally to keep it trained on a moving object or to secure a panoramic effect. [GEOL] 1. A shallow, natural depression or basin containing a body of standing water. 2. A hard, cementlike layer, crust, or horizon of soil within or just beneath the surface; may be compacted, indurated, or very high in clay content. [MIN ENG] 1. A shallow, circular, concave steel or porcelain dish in which drillers or sam-

plers wash the drill sludge to gravity-separate the particles of heavy, dense minerals from the lighter rock powder as a quick visual means of ascertaining if the rocks traversed by the borehole contain minerals of value. 2. The act or process of performing the above operation. [OCEANOGR] See pancake ice.

panabase See tetrahedrite.

panama [TEXT] A plain woven hopsacking of coarse-yarn basket weave, plain or in two colors, producing an effect similar to the texture of panama hats.

panautomorphic rock See panidiomorphic rock.

pancake [MIN ENG] A concrete disk employed in stope support. [OCEANOGR] See pancake ice.

pancake ice [OCEANOGR] One or more small, newly formed pieces of sea ice, generally circular with slightly raised edges and about 1 to 10 feet (0.3 to 3 meters) across. Also known as lily-pad ice; pan; pancake; pan ice; plate ice.

Pancarida [INV ZOO] A superorder of the subclass Malacostraca; the cylindrical, cruciform body lacks an external division between the thorax and pleon and has the cephalon united with the first thoracomere.

panchromatic [GRAPHICS] Of a photographic emulsion, film, or plate, sensitive to all wavelengths within the visible spectrum, though not uniformly so.

panclimax [ECOL] Two or more related climax communities or formations having similar climate, life forms, and genera or dominants. Also known as panformation.

pancreas [ANAT] A composite gland in most vertebrates that produces and secretes digestive enzymes, as well as at least two hormones, insulin and glucagon.

pancreozymin [BIOCHEM] A crude extract of the intestinal mucosa that stimulates secretion of pancreatic juice.

panda [VERT ZOO] Either of two Asian species of carnivores in the family Procyonidae; the red panda (*Ailurus fulgens*) has long, thick, red fur, with black legs; the giant panda (*Ailuropoda melanoleuca*) is white, with black legs and black patches around the eyes.

Pandanaceae [BOT] The single, pantropical family of the plant order Pandanales.

Pandanales [BOT] A monofamilial order of monocotyledonous plants; members are more or less arborescent and sparingly branched, with numerous long, firm, narrow, parallel-veined leaves that usually have spiny margins.

Pandaridae [INV ZOO] A family of dimorphic crustaceans in the suborder Caligoida; members are external parasites of sharks.

pandemic [MED] Epidemic occurring over a widespread geographic area.

pandermite See priceite.

Pandionidae [VERT ZOO] A monospecific family of birds in the order Falconiformes; includes the osprey (*Pandion haliaetus*), characterized by a reversible hindtoe, well-developed claws, and spicules on the scales of the feet.

panel [CIV ENG] 1. One of the divisions of a lattice girder. 2. A sheet of material held in a frame. 3. A distinct, usually rectangular, raised or sunken part of a construction surface or a material. [COMPUT SCI] The face of the console, which is normally equipped with lights, switches, and buttons to control the machine, correct errors, determine the status of the various CPU (central processing unit) parts, and determine and revise the contents of various locations. Also known as control panel;

patch panel. [ENG] A metallic or nonmetallic sheet on which operating controls and dials of an electronic unit or other equipment are mounted. [MIN ENG] 1. A system of coal extraction in which the ground is laid off in separate districts or panels, pillars of extra-large size being left between. 2. A large rectangular block or pillar of coal.

panel board [ELECTR] *See* control board. [ENG] A drawing board with an adjustable outer frame that is forced over the drawing paper to hold and strain it. [MATER] A rigid paperboard used for paneling in buildings and automobile bodies.

panel cooling [CIV ENG] A system in which the heat-absorbing units are in the ceiling, floor, or wall panels of the space which is to be cooled.

panel display [ELECTR] An unconventional method of displaying color television pictures in which luminescent conversion devices, such as light-emitting diodes or electroluminescent devices, are arranged in a matrix array, forming a flat-panel screen, and are controlled by signals sent over vertical and horizontal wires connected to both electrodes of the devices.

panel heating [CIV ENG] A system in which the heat-emitting units are in the ceiling, floor, or wall panels of the space which is to be heated.

panformation *See* panclimax.

Pangea [GEOL] Postulated former supercontinent supposedly composed of all the continental crust of the earth, and later fragmented by drift into Laurasia and Gondwana.

pangolin [VERT ZOO] Any of seven species composing the mammalian family Manidae; the entire dorsal surface of the body is covered with broad, horny scales, the small head is elongate, and the mouth is terminal in the snout.

panhypopituitarism *See* hypopituitarism.

pan ice *See* pancake ice.

panidiomorphic rock [GEOL] An igneous rock that is completely or predominantly idiomorphic. Also known as panautomorphic rock.

panmixis [BIOL] Random mating within a breeding population; in a closed population this results in a high degree of uniformity.

Pannonian [GEOL] A European stage of geologic time comprisong the lower Pliocene.

panplanation [GEOL] The action or process of formation or development of a panplain.

Panthalassa [GEOL] The hypothetical proto-ocean surrounding Pangea, supposed by some geologists to have combined all the oceans or areas of oceanic crust of the earth at an early time in the geologic past.

Pantodonta [PALEON] An extinct order of mammals which included the first large land animals of the Tertiary.

Pantodontidae [VERT ZOO] A family of fishes in the order Osteoglossiformes; the single, small species is known as African butterflyfish because of its expansive pectoral fins.

pantograph [ENG] A device that sits on the top of an electric locomotive or cars in an electric train and picks up electricity from overhead wires to run the train. [GRAPHICS] A drawing instrument used for copying and consisting of four rigid bars linked together in a parallelogram form; one arm, equipped with a pencil, is connected through the bars to a pointer that is used to trace the original drawing.

Pantolambdidae [PALEON] A family of middle to late Paleocene mammals of North America in the superfamily Pantolambdoidea.

Pantolambdodontidae [PALEON] A family of late Eocene mammals of Asia in the superfamily Pantolambdoidea.

Pantolambdoidea [PALEON] A superfamily of extinct mammals in the order Pantodonta.

Pantolestidae [PALEON] An extinct family of large aquatic insectivores referred to the Proteutheria.

Pantophthalmidae [INV ZOO] The wood-boring flies, a family of orthorrhaphous dipteran insects in the series Brachycera.

Pantopoda [INV ZOO] The equivalent name for Pycnogonida.

pantothenic acid [BIOCHEM] $C_9H_{17}O_5N$ A member of the vitamin B complex that is essential for nutrition of some animal species. Also known as vitamin B_3.

Pantotheria [PALEON] An infraclass of carnivorous and insectivorous Jurassic mammals; early members retained many reptilian features of the jaws.

Papanicolaou test [PATH] A technique for the detection of precancerous and early noninvasive cancer by the staining and examination of exfoliated cells; used especially in the diagnosis of uterine cervical and endometrial cancer. Also known as Pap test.

Papaveraceae [BOT] A family of dicotyledonous plants in the order Papaverales, with regular flowers, numerous stamens, and a well-developed latex system.

Papaverales [BOT] An order of dicotyledonous plants in the subclass Magnoliidae, marked by a syncarpous gynoecium, parietal placentation, and only two sepals.

paper capacitor [ELEC] A capacitor whose dielectric material consists of oiled paper sandwiched between two layers of metallic foil.

paper chromatography [ANALY CHEM] Procedure for analysis of complex chemical mixtures by the progressive absorption of the components of the unknown sample (in a solvent) on a special grade of paper.

paper master [GRAPHICS] A paper printing plate that is used on an offset duplicator.

paper shale [GEOL] A shale that easily separates on weathering into very thin, tough, uniform, and somewhat flexible layers or laminae suggesting sheets of paper.

paper spar [GEOL] A crystallized variety of calcite occurring in thin lamellae or paperlike plates.

paper tape [COMPUT SCI] A paper ribbon in which data may be represented by means of partially or completely punched holes.

paper-tape Turing machine [COMPUT SCI] A variation of a Turing machine in which a blank square can have a nonblank symbol written on it, but this symbol cannot be changed thereafter.

Papilionidae [INV ZOO] A family of lepidopteran insects in the superfamily Papilionoidea; members are the only butterflies with fully developed forelegs bearing an epiphysis.

Papilionoidea [INV ZOO] A superfamily of diurnal butterflies (Lepidoptera) with clubbed antennae, which are rounded at the tip, and forewings that always have two or more veins.

Papilionoideae [BOT] A subfamily of the family Leguminosae with characteristic irregular flowers that have a banner, two wing petals, and two lower petals united to form a boat-shaped keel.

papilla [BIOL] A small, nipplelike eminence.

papilloma [MED] A growth pattern of epithelial tumors in which the proliferating epithelial cells grow outward from a surface, accompanied by vascularized cores of connective tissue, to form a branching structure.

Pappotheriidae [PALEON] A family of primitive, ten-reclike Cretaceous insectivores assigned to the Proteutheria.

paprika [BOT] *Capsicum annuum*. A type of pepper with nonpungent flesh, grown for its long red fruit from which a dried, ground condiment is prepared.

Pap test *See* Papanicolaou test.

PAR *See* precision approach radar.

para- [ORG CHEM] Chemical prefix designating the positions of substituting radicals on the opposite ends of a benzene nucleus, for example, paraxylene, $CH_3C_6H_4CH_3$. Abbreviated *p*-.

parabasal body *See* kinetoplast.

parabituminous coal [GEOL] Bituminous coal that contains 84–87% carbon, analyzed on a dry, ash-free basis.

parabola [MATH] The U-shaped curve in the plane given by the equation $y = ax^2 + bx + c$.

parabolic antenna [ELECTROMAG] Antenna with a radiating element and a parabolic reflector that concentrates the radiated power into a beam.

parabolic microphone [ENG ACOUS] A microphone used at the focal point of a parabolic sound reflector to give improved sensitivity and directivity, as required for picking up a band marching down a football field.

parabolic reflector [ELECTROMAG] An antenna having a concave surface which is generated either by translating a parabola perpendicular to the plane in which it lies (in a cylindrical parabolic reflector), or rotating it about its axis of symmetry (in a paraboloidal reflector). Also known as dish. [OPTICS] *See* paraboloidal reflector.

paraboloid [ENG] A reflecting surface which is a paraboloid of revolution and is used as a reflector for sound waves and microwave radiation. [MATH] A surface where sections through one of its axes are ellipses or hyperbolas, and sections through the other are parabolas.

paraboloidal antenna *See* paraboloidal reflector.

paraboloidal reflector [ELECTROMAG] An antenna having a concave surface which is a paraboloid of revolution; it concentrates radiation from a source at its focal point into a beam. Also known as paraboloidal antenna. [OPTICS] A concave mirror which is a paraboloid of revolution and produces parallel rays of light from a source located at the focus of the parabola. Also known as parabolic reflector.

Paracanthopterygii [VERT ZOO] A superorder of teleost fishes, including the codfishes and allied groups.

paracentesis [MED] Puncture of the wall of a fluid-filled cavity by means of a hollow needle to draw off the contents.

parachronology [GEOL] 1. Practical dating and correlation of stratigraphic units. 2. Geochronology based on fossils that supplement, or replace, biostratigraphically significant fossils.

paraclinal [GEOL] Referring to a stream or valley that is oriented in a direction parallel to the fold axes of a region.

paraconformity [GEOL] A type of unconformity in which strata are parallel; there is little apparent erosion and the unconformity surface resembles a simple bedding plane. Also known as nondepositional unconformity; pseudoconformity.

paraconglomerate [GEOL] A conglomerate that is not a product of normal aqueous flow but is deposited by such modes of mass transport as subaqueous turbidity currents and glacier ice; characterized by a disrupted gravel framework, often unstratified, and notable for a matrix of greater than gravel-sized fragments.

Paracrinoidea [PALEON] A class of extinct Crinozoa characterized by the numerous, irregularly arranged plates, uniserial armlike appendages, and no clear distinction between adoral and aboral surfaces.

Paracucumidae [INV ZOO] A family of holothurian echinoderms in the order Dendrochirotida; the body is invested with plates and has a simplified calcareous ring.

paracyanogen [INORG CHEM] $(CN)_x$ A white solid produced by polymerization of cyanogen gas when heated to 400°C.

paraffin coal [GEOL] A type of light-colored bituminous coal from which oil and paraffin are produced.

paraffin dirt [GEOL] A clay soil appearing rubbery or curdy and occurring in the upper several inches of a soil profile near gas seeps; probably formed by biodegradation of natural gas.

paraffin press [ENG] A filter press used during petroleum refining for the separation of paraffin oil and crystallizable paraffin wax from distillates.

paraffin scale [MATER] Unrefined paraffin wax remaining in the chamber after oil has been removed from a mixture of oil and paraffin by sweating.

paraganglion [ANAT] Any of various isolated chromaffin bodies associated with structures such as the abdominal aorta, heart, kidney, and gonads. Also known as chromaffin body.

paragenesis [MINERAL] 1. The association and order of crystallization of minerals in a rock or vein. 2. The effect of one mineral on the development of another. Also known as mineral sequence; paragenetic sequence.

paragenetic sequence *See* paragenesis.

paragneiss [GEOL] A gneiss showing a sedimentary parentage.

paragon [LAP] A perfect diamond whose mass is equal to or greater than 100 carats (20 grams).

paragonite [MINERAL] $NaAl_2(AlSi_3)O_{10}(OH)_2$ A yellowish or greenish monoclinic mica species that contains sodium and usually occurs in metamorphic rock. Also known as soda mica.

paralgesia [MED] 1. Paresthesia characterized by pain. 2. Any perverted and disagreeable cutaneous sensation, as of formication, cold, or burning.

paralimnion [HYD] The littoral part of a lake, extending from the margin to the deepest limit of rooted vegetation.

parallactic angle *See* position angle.

parallactic motion [ASTRON] An apparent motion of stars away from the point in the celestial sphere toward which the sun is moving.

parallax [OPTICS] The change in the apparent relative orientations of objects when viewed from different positions.

parallax-second *See* parsec.

parallel [COMPUT SCI] Simultaneous transmission of, storage of, or logical operations on the parts of a word, character, or other subdivision of a word in a computer, using separate facilities for the various parts. [ELEC] Connected to the same pair of terminals. Also known as multiple; shunt. [GEOD] A circle on the surface of the earth, parallel to the plane of the equator and connecting all points of equal latitude. Also known as circle of longitude; parallel of latitude. [MATH] 1. Lines are parallel in a euclidean space if they lie in a common plane and do not intersect. 2. Planes are parallel in a euclidean three-dimensional space if they do not intersect. 3. A circle parallel to the primary great circle of a sphere or spheroid. [PHYS] Of two or more displacements or other vectors, having the same direction.

parallel access [COMPUT SCI] Transferral of information to or from a storage device in which all elements in a unit of information are transferred simultaneously. Also known as simultaneous access.

parallel addition [COMPUT SCI] A method of addition by a computer in which all the corresponding pairs of digits of the addends are processed at the same time during one cycle, and one or more subsequent cycles are used for propagation and adjustment of any carries that may have been generated.

parallel algorithm [COMPUT SCI] An algorithm in which several computations are carried on simultaneously.

parallel circuit [ELEC] An electric circuit in which the elements, branches (having elements in series), or components are connected between two points, with one of the two ends of each component connected to each point.

parallel compensation See feedback compensation.

parallel computation [COMPUT SCI] The simultaneous computation of several parts of a problem.

parallel computer [COMPUT SCI] 1. A computer that can carry out more than one logic or arithmetic operation at one time. 2. See parallel digital computer.

parallel course computer See course-line computer.

parallel digital computer [COMPUT SCI] Computer in which the digits are handled in parallel; mixed serial and parallel machines are frequently called serial or parallel, according to the way arithmetic processes are performed; an example of a parallel digital computer is one which handles decimal digits in parallel, although it might handle the bits constituting a digit either serially or in parallel.

parallel dot character printer See line dot matrix.

parallelepiped [MATH] A polyhedron all of whose faces are parallelograms.

parallel feed [COMPUT SCI] See sideways feed. [ELECTR] Application of a direct-current voltage to the plate or grid of a tube in parallel with an alternating-current circuit, so that the direct-current and the alternating-current components flow in separate paths. Also known as shunt feed.

parallel fold See concentric fold.

parallel impedance [ELEC] One of two or more impedances that are connected to the same pair of terminals.

parallel linkage [MECH ENG] A linkage system in which reciprocating motion is amplified.

parallel of altitude [ASTRON] A circle on the celestial sphere parallel to the horizon connecting all points of equal altitude. Also known as almucantar; altitude circle.

parallel of latitude [ASTRON] See circle of longitude. [GEOD] See parallel.

parallelogram [MATH] A four-sided polygon with each pair of opposite sides parallel.

parallel operation [COMPUT SCI] Performance of several actions, usually of a similar nature, by a computer system simultaneously through provision of individual similar or identical devices. [ELECTR] The connecting together of the outputs of two or more batteries or other power supplies so that the sum of their output currents flows to a common load.

parallel-plate capacitor [ELEC] A capacitor consisting of two parallel metal plates, with a dielectric filling the space between them.

parallel-plate laser [OPTICS] A laser which has two small parallel plates facing each other at a distance which is large compared with their diameters; one

of them reflects light and the other is partially reflecting, so that light can bounce back and forth between the plates enough to build up a strong pulse.

parallel programming [COMPUT SCI] A method for performing simultaneously the normally sequential steps of a computer program, using two or more processors. [ELECTR] Method of parallel operation for two or more power supplies in which their feedback terminals (voltage control terminals) are also paralleled; these terminals are often connected to a separate programming source.

parallel rectifier [ELECTR] One of two or more rectifiers that are connected to the same pair of terminals, generally in series with small resistors or inductors, when greater current is desired than can be obtained with a single rectifier.

parallel reliability [SYS ENG] Property of a system composed of functionally parallel elements in such a way that if one of the elements fails, the parallel units will continue to carry out the system function.

parallel resonance Also known as antiresonance. [ELEC] 1. The frequency at which the inductive and capacitive reactances of a parallel resonant circuit are equal. 2. The frequency at which the parallel impedance of a parallel resonant circuit is a maximum. 3. The frequency at which the parallel impedance of a parallel resonant circuit has a power factor of unity.

parallel roads [GEOL] A series of horizontal beaches or wave-cut terraces occurring parallel to each other at different levels on each side of a glacial valley.

parallel running [COMPUT SCI] 1. The running of a newly developed system in a data-processing area in conjunction with the continued operation of the current system. 2. The final step in the debugging of a system; this step follows a system test.

parallel shot [ENG] In seismic prospecting, a test shot which is made with all the amplifiers connected in parallel and activated by a single geophone so that lead, lag, polarity, and phasing in the amplifier-to-oscillograph circuits can be checked.

parallel texture [PETR] A rock texture characterized by tabular-to-prismatic crystals oriented parallel to a plane or line.

parallel-T network [ELEC] A network used in capacitance measurements at radio frequencies, having two sets of three impedances, each in the form of the letter T, with the arms of the two Ts joined to common terminals, and the source and detector each connected between two of these terminals. Also known as twin-T network.

parallel twin [CRYSTAL] A twinned crystal whose twin axis is parallel to the composition surface.

parallochthon [GEOL] Rocks that were brought from intermediate distances and deposited near an allochthonous mass during transit.

paralysis [MED] Complete or partial loss of motor or sensory function.

paralysis agitans See parkinsonism.

paramagnetic amplifier See maser.

paramagnetic cooling See adiabatic demagnetization.

paramagnetic material [ELECTROMAG] A material within which an applied magnetic field is increased by the alignment of electron orbits.

paramagnetic resonance See electron paramagnetic resonance.

paramagnetism [ELECTROMAG] A property exhibited by substances which, when placed in a magnetic field, are magnetized parallel to the field to an extent proportional to the field (except at very low temperatures or in extremely large magnetic fields).

Parameciidae [INV ZOO] A family of ciliated protozoans in the order Holotrichia; the body has differentiated anterior and posterior ends and is bounded by a hard but elastic pellicle.

Paramelina [VERT ZOO] An order of marsupials that includes the bandicoots in some systems of classification.

parameter [CRYSTAL] Any of the axial lengths or interaxial angles that define a unit cell. [ELEC] 1. The resistance, capacitance, inductance, or impedance of a circuit element. 2. The value of a transistor or tube characteristic. [MATH] An arbitrary constant or variable so appearing in a mathematical expression that changing it gives various cases of the phenomenon represented. [PHYS] A quantity which is constant under a given set of conditions, but may be different under other conditions.

parameter word [COMPUT SCI] A word in a computer storage containing one or more parameters that specify the action of a routine or subroutine.

parametric amplifier [ELECTR] A highly sensitive ultra-high-frequency or microwave amplifier having as its basic element an electron tube or solid-state device whose reactance can be varied periodically by an alternating-current voltage at a pumping frequency. Also known as mavar; paramp; reactance amplifier. [OPTICS] A device consisting of an optically nonlinear crystal in which an optical or infrared beam draws power from a laser beam at a higher frequency and is amplified.

parametric converter [ELECTR] Inverting or noninverting parametric device used to convert an input signal at one frequency into an output signal at a different frequency.

parametric equation [MATH] An equation where coordinates of points appear dependent on parameters such as the parametric equation of a curve or a surface.

parametric generation [OPTICS] A process in which a single electromagnetic wave propagating in a nonlinear medium is converted to two lower-frequency waves, the sum of whose frequencies equals the frequency of the original wave.

parametric hydrology [HYD] That branch of hydrology dealing with the development and analysis of relationships among the physical parameters involved in hydrologic events and the use of these relationships to generate, or synthesize, hydrologic events.

parametric oscillator [ELECTR] An oscillator in which the reactance parameter of an energy-storage device is varied to obtain oscillation. [OPTICS] A device consisting of an optically nonlinear crystal surrounded by a pair of mirrors to which is applied a relatively high-frequency laser beam and a relatively low-frequency signal, resulting in a low-frequency output whose frequency can be varied, usually by varying the indices of refraction.

parametrized voice response system [ENG ACOUS] A voice response system which first extracts informative parameters from human speech, such as natural resonant frequencies (formants) of the speaker's vocal tract and the fundamental frequency (pitch) of the voice, and which later reconstructs speech from such stored parameters.

paramp See parametric amplifier.

paramyosin [BIOCHEM] A type of fibrous protein found in the adductor muscles of bivalves and thought to form the core of a filament with myosin molecules at the surface.

paranoia [PSYCH] A rare form of paranoid psychosis characterized by the slow development of a complex, internally logical system of persecutory or grandiose delusions.

paranoid schizophrenia [PSYCH] A form of schizophrenia in which delusions of persecution or grandeur (or both), hallucinations, and ideas of reference predominate and sometimes are systematized.

paranthelion [ASTRON] A refraction phenomenon similar to a parhelion, but occurring generally at a distance of 120° (occasionally 90° and 140°) from the sun, on the parhelic circle. Also known as mock sun.

Paranyrocidae [PALEON] An extinct family of birds in the order Anseriformes, restricted to the Miocene of South Dakota.

Paraonidae [INV ZOO] A family of small, slender polychaete annelids belonging to the Sedentaria.

Paraparchitacea [PALEON] A superfamily of extinct ostracods in the suborder Kloedenellocopina including nonsulcate, nondimorphic forms.

parapatric [ECOL] Referring to populations or species that occupy nonoverlapping but adjacent geographical areas without interbreeding.

paraphase amplifier [ELECTR] An amplifier that provides two equal output signals 180° out of phase.

Parasaleniidae [INV ZOO] A family of echinacean echinoderms in the order Echinoida composed of oblong forms with trigeminate ambulacral plates.

Paraselloidea [INV ZOO] A group of the Asellota that contains forms in which the first pleopods of the male are coupled along the midline, and are lacking in the female.

Paraseminotidae [PALEON] A family of Lower Triassic fishes in the order Palaeonisciformes.

parasite [BIOL] An organism that lives in or on another organism of different species from which it derives nutrients and shelter. [ELEC] Current in a circuit, due to some unintentional cause, such as inequalities of temperature or of composition; particularly troublesome in electrical measurements.

parasitic absorption See parasitic capture.

parasitic antenna See parasitic element.

parasitic capture [NUCLEO] Any absorption of a neutron that does not result in a fission or the production of a desired element. Also known as parasitic absorption.

parasitic castration [BIOL] Destruction of the reproductive organs by parasites.

parasitic cone See adventive cone.

parasitic element [ELECTROMAG] An antenna element that serves as part of a directional antenna array but has no direct connection to the receiver or transmitter and reflects or reradiates the energy that reaches it, in a phase relationship such as to give the desired radiation pattern. Also known as parasitic antenna; parasitic reflector; passive element.

parasitic reflector See parasitic element.

parasitism [ECOL] A symbiotic relationship in which the host is harmed, but not killed immediately, and the species feeding on it is benefited.

parasitology [BIOL] A branch of biology which deals with those organisms, plant or animal, which have become dependent on other living creatures.

parastratigraphy [GEOL] 1. Supplemental stratigraphy based on fossils other than those governing the prevalent orthostratigraphy. 2. Stratigraphy based on operational units.

parastratotype [GEOL] Another section in the original locality where a stratotype was defined.

Parasuchia [PALEON] The equivalent name for Phytosauria.

parasympathetic nervous system [ANAT] The craniosacral portion of the autonomic nervous system, consisting of preganglionic nerve fibers in certain sacral and cranial nerves, outlying ganglia, and postganglionic fibers.

paratacamite [MINERAL] $Cu_2(OH)_3Cl$ Rhombohedral mineral composed of basic copper chloride; it is dimorphous with tacamite.

parathene [MATER] Any of a group of high-grade hydrocarbons that are extracted from lubricating oil stocks by the solvent process or refining.

parathormone [BIOCHEM] A polypeptide hormone that functions in regulating calcium and phosphate metabolism. Also known as parathyroid hormone.

parathyroid gland [ANAT] A paired endocrine organ located within, on, or near the thyroid gland in the neck region of all vertebrates except fishes.

parathyroid hormone See parathormone.

paratill [GEOL] A till formed by ice-rafting in a marine or lacustrine environment; includes deposits from ice floes and icebergs.

paratrachoma See inclusion conjunctivitis.

paratype [SYST] A specimen other than the holotype which is before the author at the time of original description and which is designated as such or is clearly indicated as being one of the specimens upon which the original description was based.

parautochthonous [GEOL] Pertaining to a mobilized part of an autochthonous granite moved higher in the crust or into a tectonic area of lower pressure and characterized by variable and diffuse contacts with country rocks. [PETR] Pertaining to a rock that is intermediate in tectonic character between autochthonous and allochthonous.

paravane [ENG] A torpedo-shaped device with sawlike teeth along its forward end, towed with a wire rope underwater from either side of the bow of a ship to cut the cables of anchored mines. Also known as otter.

paravauxite [MINERAL] $FeAl_2(PO_4)_2(OH)_2 \cdot 8H_2O$ A colorless mineral composed of hydrous basic iron aluminum phosphate; contains more water than vauxite.

parawollastonite [MINERAL] $CaSiO_3$ A monoclinic mineral composed of silicate of calcium; it is dimorphous with wollastonite.

paraxial rays [OPTICS] Rays which are close enough to the optical axis of a system, and thus whose directions are sufficiently close to being parallel to it, so that sines of angles between the rays and the optical axis may be replaced by the angles themselves in calculations.

Parazoa [INV ZOO] A name proposed for a subkingdom of animals which includes the sponges (Porifera).

Pareiasauridae [PALEON] A family of large, heavy-boned terrestrial reptiles of the late Permian, assigned to the order Cotylosauria.

parenchyma [BOT] A tissue of higher plants consisting of living cells with thin walls that are agents of photosynthesis and storage; abundant in leaves, roots, and the pulp of fruit, and found also in leaves and stems. [HISTOL] The specialized epithelial portion of an organ, as contrasted with the supporting connective tissue and nutritive framework.

parent [NUC PHYS] A radionuclide that upon disintegration yields a specified nuclide, the daughter, either directly, or indirectly as a later member of a radioactive series.

parent compound [CHEM] A chemical compound that is the basis for one or more derivatives; for example, ethane is the parent compound for ethyl alcohol and ethyl acetate.

parenthesis-free notation See Polish notation.

parent rock [GEOL] 1. The rock mass from which parent material is derived. 2. See source rock.

paresthesia [MED] Tingling, crawling, or burning sensation of the skin.

Pareulepidae [INV ZOO] A monogeneric family of errantian polychaete annelids.

parging [CIV ENG] A thin coating of mortar or plaster on a brick or stone surface.

parhelion [ASTRON] Either of two colored luminous spots that appear at points 22° (or somewhat more) on both sides of the sun and at the same elevation as the sun; the solar counterpart of the lunar paraselene. Also known as mock sun; sun dog.

parietal [ANAT] Of or situated on the wall of an organ or other body structure. [BOT] Of a plant part, having a peripheral location or orientation; in particular, attached to the main wall of an ovary.

Parietales [BOT] An order of plants in the Englerian system; families are placed in the order Violales in other systems.

parietal lobe [ANAT] The cerebral lobe of the brain above the lateral cerebral sulcus and behind the central sulcus.

parity [COMPUT SCI] The use of a self-checking code in a computer employing binary digits in which the total number of 1's or 0's in each permissible code expression is always even or always odd. [MATH] Two integers have the same parity if they are both even or both odd. [QUANT MECH] A physical property of a wave function which specifies its behavior under an inversion, that is, under simultaneous reflection of all three spatial coordinates through the origin; if the wave function is unchanged by inversion, its parity is 1 (or even); if the function is changed only in sign, its parity is -1 (or odd). Also known as space reflection symmetry.

parity check See odd-even check.

parity conservation See conservation of parity.

parity error [COMPUT SCI] A machine error in which an odd number of bits are accidentally changed, so that the error can be detected by a parity check.

parity selection rules [QUANT MECH] Rules which specify whether or not a change in parity occurs during a given type of transition of an atom, molecule, or nucleus; for example, the Laporte selection rule, or the rule that there is no parity change in an allowed β-decay transition of a nucleus.

parker See rep.

parking orbit [AERO ENG] A temporary earth orbit during which the space vehicle is checked out and its trajectory carefully measured to determine the amount and time of increase in velocity required to send it into a final orbit or into space in the desired direction.

parkinsonism [MED] A clinical state characterized by tremor at a rate of three to eight tremors per second, with "pill-rolling" movements of the thumb common, muscular rigidity, dyskinesia, hypokinesia, and reduction in number of spontaneous and autonomic movements; produces a masked facies, disturbances of posture, gait, balance, speech, swallowing, and muscular strength. Also known as paralysis agitans; Parkinson's disease.

Parkinson's disease See parkinsonism.

Parmeliaceae [BOT] The foliose shield lichens, a family of the order Lecanorales.

Parnidae [INV ZOO] The equivalent name for Dryopidae.

parogenetic [GEOL] Formed previously to the enclosing rock; especially said of a concretion formed in a different (older) rock from its present (younger) host.

parotid gland [ANAT] The salivary gland in front of and below the external ear; the largest salivary gland in humans; a compound racemose serous gland that communicates with the mouth by Steno's duct.

parrot [VERT ZOO] Any member of the avian family Psittacidae, distinguished by the short, stout, strongly hooked beak.

parsec [ASTRON] The distance at which a star would have a parallax equal to 1 second of arc; 1 parsec equals 3.258 light-years or 3.08572×10^{13} kilometers. Derived from parallax-second.

parsing [COMPUT SCI] A process whereby phrases in a string of characters in a computer language are associated with the component names of the grammar that generated the string.

pars intermedia [ANAT] The intermediate lobe of the adenohypophysis.

pars nervosa [ANAT] The inferior subdivision of the neurohypophysis. Also known as pars neuralis.

pars neuralis *See* pars nervosa.

parsnip [BOT] *Pastinaca sativa*. A biennial herb of Mediterranean origin belonging to the order Umbellales; grown for its edible thickened taproot.

parthenocarpy [BOT] Production of fruit without fertilization.

parthenogenesis [INV ZOO] A special type of sexual reproduction in which an egg develops without entrance of a sperm; common among rotifers, aphids, thrips, ants, bees, and wasps.

partial coherence [PHYS] Property of two waves whose relative phase undergoes random fluctuations which are not, however, sufficient to make the wave completely incoherent.

partial correlation coefficient [STAT] A measure of the strength of association between a dependent variable and one independent variable when the effect of all other independent variables is removed; equal to the square root of the partial coefficient of determination.

partial dislocation [CRYSTAL] The line at the edge of an extended dislocation where a slip through a fraction of a lattice constant has occurred.

partial fractions [MATH] A collection of fractions which when added are a given fraction whose numerator and denominator are usually polynomials; the partial fractions are usually constants or linear polynomials divided by factors of the denominator of the given fraction.

partial function [COMPUT SCI] A partial function from a set A to a set B is a correspondence between some subset of A and B which associates with each element of the subset of A a unique element of B.

partial node [PHYS] That part (a point, line, or surface) of a standing wave where some characteristic of the wave field has a minimum amplitude other than zero.

partial ordering *See* ordering.

partial pediment [GEOL] **1.** A broadly planate, gravel-capped, interstream bench or terrace. **2.** A broad, planate erosion surface which is formed by the coalescence of contemporaneous, valley-restricted benches developed at the same elevation in proximate valleys, and which would produce a pediment if uninterrupted planation were to continue at this level.

partial pressure [PHYS] The pressure that would be exerted by one component of a mixture of gases if it were present alone in a container.

partial thermoremanent magnetization [GEOPHYS] The thermoremanent magnetization acquired by cooling in an ambient field over only a restricted temperature interval, as opposed to the entire temperature range from Curie point to room temperature. Abbreviated PTRM.

particle [MECH] *See* material particle. [PARTIC PHYS] *See* elementary particle. [PHYS] **1.** Any very small part of matter, such as a molecule, atom, or electron. Also known as fundamental particle. **2.** Any relatively small subdivision of matter, ranging in diameter from a few angstroms (as with gas molecules) to a few millimeters (as with large raindrops).

particle accelerator [NUCLEO] A device which accelerates electrically charged atomic or subatomic particles, such as electrons, protons, or ions, to high energies. Also known as accelerator; atom smasher.

particle board [MATER] Construction board made with wood particles impregnated with low-molecular-weight resin and then cured.

particle detector [NUCLEO] A device used to indicate the presence of fast-moving charged atomic or nuclear particles by observation of the electrical disturbance created by a particle as it passes through the device. Also known as radiation detector.

particle electrophoresis [PHYS CHEM] Electrophoresis in which the particles undergoing analysis are of sufficient size to be viewed either with the naked eye or with the assistance of an optical microscope.

particle multiplet *See* isospin multiplet.

particle physics [PHYS] The branch of physics concerned with understanding the properties and behavior of elementary particles, especially through study of collisions or decays involving energies of hundreds of Mev (million electron volts) or more. Also known as high-energy physics.

particle spectrum *See* mass spectrum.

particle track [PHYS] Any visible phenomenon along the path of an ionizing particle, such as a trail of bubbles, water droplets, or sparks in a bubble chamber, cloud chamber, or spark chamber respectively, or of altered material in an emulsion or in glass.

parting cast [GEOL] A sand-filled tension crack produced by creep along the sea floor.

parting plane lineation [GEOL] A parting lineation on a laminated surface, consisting of subparallel, linear, shallow grooves and ridges of low relief, generally less than 1 millimeter.

parting-step lineation [GEOL] A parting lineation characterized by subparallel, steplike ridges where the parting surface cuts across several adjacent laminae.

partition chromatography [ANALY CHEM] Chromatographic procedure in which the stationary phase is a high-boiling liquid spread as a thin film on an inert support, and the mobile phase is a vaporous mixture of the components to be separated in an inert carrier gas.

partition coefficient [ANALY CHEM] In the equilibrium distribution of a solute between two liquid phases, the constant ratio of the solute's concentration in the upper phase to its concentration in the lower phase. Symbolized K.

partitioned data set [COMPUT SCI] A single data set, divided internally into a directory and one or more sequentially organized subsections called members, residing on a direct access for each device, and commonly used for storage or program libraries.

partition function [STAT MECH] **1.** The integral, over the phase space of a system, of the exponential of $(-E/kT)$, where E is the energy of the system, k is

Boltzmann's constant, and T is the temperature; from this function all the thermodynamic properties of the system can be derived. **2.** In quantum statistical mechanics, the sum over allowed states of the exponential of $(-E/kT)$. Also known as sum of states; sum over states.

parton [PARTIC PHYS] One of the very singular (or hard), small charged particles of which hadrons are proposed to be constructed, according to a theory developed to account for the scattering of very-high-energy electrons from protons at large angles and with large momentum transfers.

party wall [BUILD] A wall providing joint service between two buildings.

PAS *See para*-aminosalicylic acid; photoacoustic spectroscopy.

PASCAL [COMPUT SCI] A procedure-oriented programming language whose highly structured design facilitates the rapid location and correction of coding errors.

pascal [MECH] A unit of pressure equal to the pressure resulting from a force of 1 newton acting uniformly over an area of 1 square meter. Symbolized Pa.

Pascal's law [FL MECH] The law that a confined fluid transmits externally applied pressure uniformly in all directions, without change in magnitude.

Pascal's theorem [MATH] The theorem that when one inscribes a simple hexagon in a conic, the three pairs of opposite sides meet in collinear points.

Paschen-Back effect [SPECT] An effect on spectral lines obtained when the light source is placed in a very strong magnetic field; the anomalous Zeeman effect obtained with weaker fields changes over to what is, in a first approximation, the normal Zeeman effect.

Paschen's law [ELECTR] The law that the sparking potential between two parallel plate electrodes in a gas is a function of the product of the gas density and the distance between the electrodes. Also known as Paschen's rule.

Paschen's rule *See* Paschen's law.

Pasiphae [ASTRON] A small satellite of Jupiter with a diameter of about 35 miles (56 kilometers), orbiting with retrograde motion at a mean distance of about 1.46×10^7 miles (2.35×10^7 kilometers). Also known as Jupiter VIII.

pass [AERO ENG] **1.** A single circuit of the earth made by a satellite; it starts at the time the satellite crosses the equator from the Southern Hemisphere into the Northern Hemisphere. **2.** The period of time in which a satellite is within telemetry range of a data acquisition station. [COMPUT SCI] A complete cycle of reading, processing, and writing in a computer. [GEOGR] **1.** A natural break, depression, or other low place providing a passage through high terrain, such as a mountain range. **2.** A navigable channel leading to a harbor or river. **3.** A narrow opening through a barrier reef, atoll, or sand bar. [MECH ENG] **1.** The number of times that combustion gases are exposed to heat transfer surfaces in boilers (that is, single-pass, double-pass, and so on). **2.** In metal rolling, the passage in one direction of metal deformed between rolls. **3.** In metal cutting, transit of a metal cutting tool past the workpiece with a fixed tool setting. [MET] **1.** Passage of a metal bar between rolls. **2.** Open space between two grooved rolls through which metal is processed. **3.** Weld metal deposited in one trip along the axis of a weld. [MIN ENG] **1.** A mine opening through which coal or ore is delivered from a higher to a lower level. **2.** A passage left

in old workings for workers to travel as they move from one level to another. **3.** A treatment of the whole ore sample in a sample divider. **4.** A passage of an excavation or grading machine. **5.** In surface mining, a complete excavator cycle in removing overburden.

Passalidae [INV ZOO] The peg beetles, a family of tropical coleopteran insects in the superfamily Scarabaeoidea.

passband [ELECTR] A frequency band in which the attenuation of a filter is essentially zero.

pass element [ELECTR] Controlled variable resistance device, either a vacuum tube or power transistor, in series with the source of direct-current power; the pass element is driven by the amplified error signal to increase its resistance when the output needs to be lowered or to decrease its resistance when the output must be raised.

Passeres [VERT ZOO] The equivalent name for Oscines.

Passeriformes [VERT ZOO] A large order of perching birds comprising two major divisions: Suboscines and Oscines.

Passifloraceae [BOT] A family of dicotyledonous, often climbing plants in the order Violales; flowers are polypetalous and hypogynous with a corona, and seeds are arillate with an oily endosperm.

passivation [ELECTR] Growth of an oxide layer on the surface of a semiconductor to provide electrical stability by isolating the transistor surface from electrical and chemical conditions in the environment; this reduces reverse-current leakage, increases breakdown voltage, and raises power dissipation rating. [MET] To render passive; to reduce the reactivity of a chemically active metal surface by electrochemical polarization or by immersion in a passivating solution.

passive AND gate [ELECTR] *See* AND gate. [ENG] A fluidic device which achieves an output signal, by stream interaction, only when both of two control signals appear simultaneously.

passive antenna [ELECTROMAG] An antenna which influences the directivity of an antenna system but is not directly connected to a transmitter or receiver.

passive component *See* passive element.

passive device [COMPUT SCI] A unit of a computer which cannot itself initiate a request for communication with another device, but which honors such a request from another device.

passive earth pressure [CIV ENG] The maximum value of lateral earth pressure exerted by soil on a structure, occurring when the soil is compressed sufficiently to cause its internal shearing resistance along a potential failure surface to be completely mobilized.

passive element [ELEC] An element of an electric circuit that is not a source of energy, such as a resistor, inductor, or capacitor. Also known as passive component. [ELECTROMAG] *See* parasitic element.

passive filter [ELEC] An electric filter composed of passive elements, such as resistors, inductors, or capacitors, without any active elements, such as vacuum tubes or transistors.

passive fold [GEOL] A fold in which the mechanism of folding, either flow or slip, crosses the boundaries of the strata at random.

passive glacier [HYD] A glacier with sluggish movement, generally occurring in a continental environment at a high latitude, where both accumulation and ablation are minimal.

passive immunity [IMMUNOL] **1.** Immunity acquired by injection of antibodies in another individual or in an animal. **2.** Immunity acquired by the fetus by the transfer of maternal antibodies through the placenta.

passive method [CIV ENG] A construction method in permafrost areas in which the frozen ground near the structure is not disturbed or altered, and the foundations are provided with additional insulation to prevent thawing of the underlying ground.

passive reflector [ELECTROMAG] A flat reflector used to change the direction of a microwave or radar beam; often used on microwave relay towers to permit placement of the transmitter, repeater, and receiver equipment on the ground, rather than at the tops of towers. Also known as plane reflector.

passive system [ELECTR] Electronic system which emits no energy, and does not give away its position or existence.

pass point [GRAPHICS] A point whose horizontal or vertical position is determined from photographs by photogrammetric methods and which is intended for use as a supplemental control point in the orientation of other photographs.

pasteboard [MATER] A type of thin cardboard made from gluing together two or more sheets of paper.

paste-up *See* mechanical.

pasteurization [FOOD ENG] The application of heat for a specified time to a liquid food or beverage to enhance its keeping properties by destroying harmful microorganisms.

patch [COMPUT SCI] **1.** To modify a program or routine by inserting a machine language correction in an object deck, or by inserting it directly into the computer through the console. **2.** The section of coding inserted in this way. [ELEC] A temporary connection between jacks or other terminations on a patch board.

patch board [ELEC] A board or panel having a number of jacks at which circuits are terminated; patch cords are plugged into the jacks to connect various circuits temporarily as required in broadcast, communication, and computer work.

patch panel *See* control panel; panel.

patella [ANAT] A sesamoid bone in front of the knee, developed in the tendon of the quadriceps femoris muscle. Also known as kneecap.

Patellacea [PALEON] An extinct superfamily of gastropod mollusks in the order Aspidobranchia which developed a cap-shaped shell and were specialized for clinging to rock.

Patellidae [INV ZOO] The true limpets, a family of gastropod mollusks in the order Archeogastropoda.

patent base [GRAPHICS] In letterpress technology, a metal base with slots that is used to hold unmounted electrotypes.

Paterinida [PALEON] A small extinct order of inarticulated brachiopods, characterized by a thin shell of calcium phosphate and convex valves.

paternoite *See* kaliborite.

paternoster lake [HYD] One of a linear chain or series of small circular lakes, usually at different levels, which occupy rock basins in a glacial valley and are separated by morainal dams or riegels, but connected by streams, rapids, or waterfalls to resemble a rosary or string of beads. Also known as beaded lake; rock-basin lake; step lake.

path [COMPUT SCI] The logical sequence of instructions followed by a computer in carrying out a routine. [MATH] **1.** In a topological space, a path is a continuous curve joining two points. **2.** In graph theory, a walk whose vertices are all distinct. Also known as simple path. **3.** *See* walk. [NAV] A line

connecting a series of points and constituting a proposed or traveled route.

path coefficient [COMMUN] The ratio of the power transmitted over some designated path to that transmitted over the most direct path.

pathogen [MED] A disease-producing agent; usually refers to living organisms.

pathology [MED] The study of the causes, nature, and effects of diseases and other abnormalities.

pathotoxin [PL PATH] A chemical of biological origin, other than an enzyme, that plays an important causal role in a plant disease.

patina [GEOL] A thin, colored film produced on a rock surface by weathering. [MET] The greenish product, usually basic copper sulfate, formed on copper and copper-rich alloys as a result of prolonged atmospheric corrosion.

pattern analysis [COMPUT SCI] The phase of pattern recognition that consists of using whatever is known about the problem at hand to guide the gathering of data about the patterns and pattern classes, and then applying techniques of data analysis to help uncover the structure present in the data.

patterned ground [GEOL] Any of several well-defined, generally symmetrical forms, such as circles, polygons, and steps, that are characteristic of surficial material subject to intensive frost action.

pattern generator [ELECTR] A signal generator used to generate a test signal that can be fed into a television receiver to produce on the screen a pattern of lines having usefulness for servicing purposes.

pattern recognition [COMPUT SCI] The automatic identification of figures, characters, shapes, forms, and patterns without active human participation in the decision process.

Patterson projection [SOLID STATE] A projection of the Patterson function on a section through a crystal.

Patterson vectors [SOLID STATE] In analysis of crystal structure, the vectors of peaks relative to the origin in a Patterson function or Patterson projection.

paua *See* abalone.

Paucituberculata [VERT ZOO] An order of marsupial mammals in some systems of classification, including the opossum, rats, and polydolopids.

Pauli exclusion principle *See* exclusion principle.

Pauli g-sum rule [ATOM PHYS] For all the states arising from a given electron configuration, the sum of the g-factors for levels with the same J value is a constant, independent of the coupling scheme.

Pauling rule [SOLID STATE] A rule governing the number of ions of opposite charge in the neighborhood of a given ion in an ionic crystal, in accordance with the requirement of local electrical neutrality of the structure.

paulopost *See* deuteric.

Pauropoda [INV ZOO] A class of the Myriapoda distinguished by bifurcate antennae, 12 trunk segments with 9 pairs of functional legs, and the lack of eyes, spiracles, tracheae, and a circulatory system.

Paussidae [INV ZOO] The flat-horned beetles, a family of coleopteran insects in the suborder Adephaga.

pavilion [LAP] The portion of a faceted gemstone below the girdle. Also known as base.

pavilion facet [LAP] A main facet on the pavilion of any fashioned gemstone.

Pavo [ASTRON] A southern constellation; right ascension 20 hours, declination 65°S. Also known as Peacock.

pawdite [PETR] A dark-colored, fine-grained, granular hypabyssal rock composed of magnetite, titanite, biotite, hornblende, calcic plagioclase, and traces of quartz.

pawl [MECH ENG] The driving link or holding link of a ratchet mechanism, permits motion in one direction only.

PAX *See* private automatic exchange.

Paxillosida [INV ZOO] An order of the Asteroidea in some systems of classification, equivalent to the Paxillosina.

Paxillosina [INV ZOO] A suborder of the Phanerozonida with pointed tube feet which lack suckers, and with paxillae covering the upper body surface.

pay streak [MIN ENG] A layer of oil, ore, or other mineral that can be mined profitably.

Pb *See* lead.

P band [COMMUN] A band of radio frequencies extending from 225 to 390 megahertz, corresponding to wavelengths of 133.3 to 76.9 centimeters.

PBI test *See* protein-bound iodine test.

PBX *See* private branch exchange.

PCM *See* pulse-code modulation.

PCP *See* primary control program.

Pd *See* palladium.

PDA *See* postacceleration.

P display *See* plan position indicator.

pdl-ft *See* foot-poundal.

PDM *See* pulse-duration modulation.

4PDT *See* four-pole double-throw.

pea [BOT] **1.** *Pisum sativum.* The garden pea, an annual leafy leguminous vine cultivated for its smooth or wrinkled, round edible seeds which are borne in dehiscent pods. **2.** Any of several related or similar cultivated plants.

peach [BOT] *Prunus persica.* A low, spreading, freely branching tree of the order Rosales, cultivated in less rigorous parts of the temperate zone for its edible fruit, a juicy drupe with a single large seed, a pulpy yellow or white mesocarp, and a thin firm epicarp.

peachblossom ore *See* erythrite.

pea coal [GEOL] A size of anthracite that will pass through a $13/16$-inch (20.6-millimeter) round mesh but not through a $9/16$-inch (14.3-millimeter) round mesh.

Peacock *See* Pavo.

peak [GEOL] **1.** The conical or pointed top of a hill or mountain. **2.** An individual mountain or hill taken as a whole, used especially when it is isolated or has a pointed, conspicuous summit. [METEOROL] The point of intersection of the cold and warm fronts of a mature extra-tropical cyclone. [SCI TECH] The maximum instantaneous value of a quantity.

peaking circuit [ELECTR] A circuit used to improve the high-frequency response of a broad-band amplifier; in shunt peaking, a small coil is placed in series with the anode load; in series peaking, the coil is placed in series with the grid of the following stage.

peaking network [ELECTR] Type of interstage coupling network in which an inductance is effectively in series (series-peaking network), or in shunt (shunt-peaking network), with the parasitic capacitance to increase the amplification at the upper end of the frequency range.

peaking transformer [ELEC] A transformer in which the number of ampere-turns in the primary is high enough to produce many times the normal flux density values in the core; the flux changes rapidly from one direction of saturation to the other twice per cycle, inducing a highly peaked voltage pulse in a secondary winding.

peak limiter *See* limiter.

peak load [ELEC] The maximum instantaneous load or the maximum average load over a designated in-

terval of time. Also known as peak power. [ENG] The maximum quantity of a specified material to be carried by a conveyor per minute in a specified period of time.

peak plain [GEOL] A high-level plain formed by a series of summits of approximately the same elevation, often described as an uplifted and fully dissected peneplain. Also known as summit plain.

peak power [ELEC] *See* peak load. [ELECTROMAG] The maximum instantaneous power of a transmitted radar pulse.

peak zone [PALEON] An informal biostratigraphic zone consisting of a body of strata characterized by the exceptional abundance of some taxon (or taxa) or representing the maximum development of some taxon.

peanut [BOT] *Arachis hypogaea.* A low, branching, self-pollinated annual legume cultivated for its edible seed, which is a one-loculed legume formed beneath the soil in a pod.

pea ore [MINERAL] A variety of pisolitic limonite or bean ore occurring in small, rounded grains or masses about the size of a pea.

pear [BOT] Any of several tree species of the genus *Pyrus* in the order Rosales, cultivated for their fruit, a pome that is wider at the apical end and has stone cells throughout the flesh.

pearl [MATER] A dense, more or less round, white or light-colored concretion having various degrees of luster formed within or beneath the mantle of various mollusks by deposition of thin concentric layers of nacre about a foreign particle. [PATH] **1.** Rounded masses of concentrically arranged squamous epithelial cells, seen in some carcinomas. **2.** Mucous casts of the bronchi or bronchioles found in the sputum of asthmatic persons.

pearlite [GEOL] *See* perlite. [MET] A lamellar aggregate of ferrite (almost pure iron) and cementite (Fe_3C) often occurring in carbon steels and in cast iron.

pearl sinter *See* siliceous sinter.

pearlstone *See* perlite.

pearly *See* nacreous.

pear-shape cut [LAP] A variation of the brilliant cut, generally having 58 facets, and a pear-shaped girdle outline.

Pearson Type I distribution *See* beta distribution.

peat [GEOL] A dark-brown or black residuum produced by the partial decomposition and disintegration of mosses, sedges, trees, and other plants that grow in marshes and other wet places.

peat ball [ECOL] A lake ball containing an abundance of peaty fragments.

peat breccia [GEOL] Peat that has been broken up and then redeposited in water. Also known as peat slime.

peat coal [GEOL] A coal transitional between peat and lignite. [MATER] Artificially carbonized peat that is used as a fuel.

peat flow [ECOL] A mudflow of peat produced in a peat bog by a bog burst.

peat moss [ECOL] Moss, especially sphagnum moss, from which peat has been produced.

peat slime *See* peat breccia.

peat tar [MATER] A peat distillate containing 2–6% tar.

pebble [GEOL] A clast, larger than a granule and smaller than a cobble having a diameter in the range of 4–64 millimeters. Also known as pebblestone. [MINERAL] *See* rock crystal.

pebble bed [GEOL] Any pebble conglomerate, especially one in which the pebbles weather conspicuously and become loose. Also known as popple rock.

pebble-bed reactor [NUCLEO] A nuclear reactor in which the fuel consists of small spheres or pellets stacked in the core; the reaction rate is controlled by coolant flow and by loading and unloading pellets.

pebble conglomerate [PETR] A consolidated rock consisting mainly of pebbles.

pebble dike [GEOL] 1. A clastic dike composed largely of pebbles. 2. A tabular body containing sedimentary fragments in an igneous matrix.

pebble mill [MECH ENG] A solids size-reduction device with a cylindrical or conical shell rotating on a horizontal axis, and with a grinding medium such as balls of flint, steel, or porcelain.

pebble peat [GEOL] Peat this is formed in a semiarid climate by the accumulation of moss and algae, no more than ¼ inch (6 millimeters) in thickness, under the surface pebbles of well-drained soils.

pebble phosphate [GEOL] A secondary phosphorite of either residual or transported origin, consisting of pebbles or concretions of phosphatic material.

pebblestone *See* pebble.

pebbly sandstone [GEOL] A sandstone that contains 10–20% pebbles.

pecan [BOT] *Carya illinoensis.* A large deciduous hickory tree in the order Fagales which produces an edible, oblong, thin-shelled nut.

peccary [VERT ZOO] Either of two species of small piglike mammals in the genus *Tayassu,* composing the family Tayassuidae.

peck [MECH] Abbreviated pk. 1. A unit of volume used in the United States for measurement of solid substances, equal to 8 dry quarts, or ¼ bushel, or 537.605 cubic inches, or 0.0088097674172 cubic meter. 2. A unit of volume used in the United Kingdom for measurement of solid and liquid substances, although usually the former, equal to 2 gallons, or approximately 0.00909218 cubic meter.

Pecora [VERT ZOO] An infraorder of the Artiodactyla; includes those ruminants with a reduced ulna and usually with antlers, horns, or deciduous horns.

Pectenidae [INV ZOO] A family of bivalve mollusks in the order Anisomyaria; contains the scallops.

pectin [BIOCHEM] A purified carbohydrate obtained from the inner portion of the rind of citrus fruits, or from apple pomace; consists chiefly of partially methoxylated polygalacturonic acids.

Pectinariidae [INV ZOO] The cone worms, a family of polychaete annelids belonging to the Sedentaria.

Pectinibranchia [INV ZOO] An order of gastropod mollusks which contains many families of snails; respiration is by means of ctenidia, the nervous system is not concentrated, and sexes are separate.

Pectobothridia [INV ZOO] A subclass of parasitic worms in the class Trematoda, characterized by caudal hooks or hard posterior suckers or both.

pectolite [MINERAL] $NaCa_2Si_3O_8(OH)$ A colorless, white, or gray inosilicate, crystallizing in the monoclinic system and having a vitreous to silky luster; hardness is 5 on Mohs scale, and specific gravity is 2.75.

pectoral fin [VERT ZOO] One of the pair of fins of fishes corresponding to forelimbs of a quadruped.

pectoral girdle [ANAT] The system of bones supporting the upper or anterior limbs in vertebrates. Also known as shoulder girdle.

pectoralis major [ANAT] The large muscle connecting the anterior aspect of the chest with the shoulder and upper arm.

pectoralis minor [ANAT] The small, deep muscle connecting the third to fifth ribs with the scapula.

ped [GEOL] A naturally formed unit of soil structure.

pedal [BIOL] Of or pertaining to the foot. [DES ENG] A lever operated by foot.

pedalfer [GEOL] A soil in which there is an accumulation of sesquioxides; it is characteristic of a humid region.

pedality [GEOL] The physical nature of a soil as expressed by the features of its constituent peds.

pedestal [CIV ENG] 1. The support for a column. 2. A metal support carrying one end of a bridge truss or girder and transmitting any load to the top of a pier or abutment. [ELECTR] *See* blanking level. [ENG] A supporting part or the base of an upright structure, such as a radar antenna. [GEOL] A relatively slender column of rock supporting a wider rock mass and formed by undercutting as a result of wind abrasion or differential weathering. Also known as rock pedestal.

pedestal level *See* blanking level.

pedial class [CRYSTAL] That class in the triclinic system which has no symmetry.

pediatrics [MED] The branch of medicine that deals with the growth and development of the child through adolescence, and with the care, treatment, and prevention of diseases, injuries, and defects of children.

pedicel [BOT] 1. The stem of a fruiting or spore-bearing organ. 2. The stem of a single flower. [ZOO] A short stalk in an animal body.

pedicellaria [INV ZOO] In echinoids and starfishes, any of various small grasping organs in the form of a beak carried on a stalk.

Pedicellinea [INV ZOO] The single order of the class Calyssozoa, including all entoproct bryozoans.

pediculosis [MED] Infestation with lice, especially of the genus *Pediculus.*

pedigree mud [MATER] A high-chemical-content drilling mud that includes barium sulfate, caustic soda, soda ash, sodium bicarbonate, and phosphates.

Pedilidae [INV ZOO] The false ant-loving flower beetles, a family of coleopteran insects in the superfamily Tenebrionoidea.

pediment [ARCH] A triangular face forming the gable of a two-pitched roof. [GEOL] A piedmont slope formed from a combination of processes which are mainly erosional; the surface is chiefly bare rock but may have a covering veneer of alluvium or gravel. Also known as conoplain; piedmont interstream flat.

pediment gap [GEOL] A broad opening formed by the enlargement of a pediment pass.

Pedinidae [INV ZOO] The single family of the order Pedinoida.

Pedinoida [INV ZOO] An order of Diadematacea making up those forms of echinoderms which possess solid spines and a rigid test.

pediocratic [GEOL] Pertaining to a period of time in which there is little diastrophism.

Pedionomidae [VERT ZOO] A family of quaillike birds in the order Gruiformes.

Pedipalpida [INV ZOO] Former order of the Arachnida; these animals are now placed in the orders Uropygi and Amblypygi.

pedogenics [GEOL] The study of the origin and development of soil.

pedogeography [GEOL] The study of the geographic distribution of soils.

pedography [GEOL] The systematic description of soils; an aspect of soil science.

pedolith [GEOL] A surface formation that has undergone one or more pedogenic processes.

pedologic age [GEOL] The relative maturity of a soil profile.

pedologic unit [GEOL] A soil considered without regard to its stratigraphic relations.

pedology [GEOL] *See* soil science. [MED] The science of the study of the physiological as well as the psychological aspects of childhood.

pedon [GEOL] The smallest unit or volume of soil that represents or exemplifies all the horizons of a soil profile; it is usually a horizontal, hexagonal area of about 1 square meter, or possibly larger.

pedorelic [GEOL] Referring to a soil feature that is derived from a preexisting soil horizon.

pedosphere [GEOL] That shell or layer of the earth in which soil-forming processes occur.

pedotubule [GEOL] A soil feature consisting of skeleton grains, or skeleton grains plus plasma, and having a tubular external form (either single tubes or branching systems of tubes) characterized by relatively sharp boundaries and relatively uniform cross-sectional size and shape (circular or elliptical).

peduncle [ANAT] A band of white fibers joining different portions of the brain. [BOT] **1.** A flower-bearing stalk. **2.** A stalk supporting the fruiting body of certain thallophytes. [INV ZOO] The stalk supporting the whole or a large part of the body of certain crinoids, brachiopods, and barnacles.

peen [DES ENG] The end of a hammer head with a hemispherical, wedge, or other shape; used to bend, indent, or cut.

peg [ENG] **1.** A small pointed or tapered piece, often cylindrical, used to pin down or fasten parts. **2.** A projection used to hang or support objects. [MET] *See* plug.

Pegasidae [VERT ZOO] The single family of the order Pegasiformes.

Pegasiformes [VERT ZOO] The sea moths or sea dragons, a small order of actinopterygian fishes; the anterior of the body is encased in bone, and the nasal bones are enlarged to form a rostrum that projects well forward of the mouth.

Pegasus [ASTRON] A northern constellation; right ascension 22 hours, declination 20°N. Also known as Winged Horse.

pegmatite [PETR] Any extremely coarse-grained, igneous rock with interlocking crystals; pegmatites are relatively small, are relatively light colored, and range widely in composition, but most are of granitic composition; they are principal sources for feldspar, mica, gemstones, and rare elements. Also known as giant granite; granite pegmatite.

pegmatolite *See* orthoclase.

Peirce stroke relationship *See* NOR.

Peisidicidae [INV ZOO] A monogeneric family of polychaete annelids belonging to the Errantia.

Peking man [PALEON] *Sinanthropus pekinensis*. An extinct human type; the braincase was thick, with a massive basal and occipital torus structure and heavy browridges.

pekovskite [MINERAL] $CaTiO_3$ Lustrous, yellowish to gray-black, isometric or orthorhombic crystals with a hardness of 5.5; decomposes in hot sulfuric acid.

pel *See* pixel.

pelagic [GEOL] Pertaining to regions of a lake at depths of 10–20 meters or more, characterized by deposits of mud or ooze and by the absence of vegetation. Also known as eupelagic. [OCEANOGR] Pertaining to water of the open portion of an ocean, above the abyssal zone and beyond the outer limits of the littoral zone.

Pelecanidae [VERT ZOO] The pelicans, a family of aquatic birds in the order Pelecaniformes.

Pelecaniformes [VERT ZOO] An order of aquatic, fish-eating birds characterized by having all four toes joined by webs.

Pelecanoididae [VERT ZOO] The diving petrels, a family of oceanic birds in the order Procellariiformes.

Pelecinidae [INV ZOO] The pelecinid wasps, a monospecific family of hymenopteran insects in the superfamily Proctotrupoidea.

p electron [ATOM PHYS] In the approximation that each electron has a definite central-field wave function, an atomic electron that has an orbital angular momentum quantum number of unity.

Pele's hair [GEOL] A spun volcanic glass formed naturally by blowing out during quiet fountaining of fluid lava. Also known as capillary ejecta; filiform lapilli; lauoho o pele.

Pele's tears [GEOL] Volcanic glass in the form of small, solidified drops which precede pendants of Pele's hair.

pelican [VERT ZOO] Any of several species of birds composing the family Pelecanidae, distinguished by the extremely large bill which has a distensible pouch under the lower mandible.

pelite [GEOL] A sediment or sedimentary rock, such as mudstone, composed of fine, clay- or mud-size particles. Also spelled pelyte.

pellagra [MED] A disease caused by nicotinic acid deficiency characterized by skin lesions, inflammation of the soft tissues of the mouth, diarrhea, and central nervous system disorders.

pellet [AGR] A small, cylindrical, compressed mass of livestock feed. [GEOL] A fine-grained, sand-size, spherical to elliptical aggregate of clay-sized calcareous material, devoid of internal structure, and contained in the body of a well-sorted carbonate rock. [ORD] A small stone or metal ball used as a missile in firearms. [PHARM] A small pill. [SCI TECH] A small spherical or cylindrical body. [VERT ZOO] A mass of undigestible material regurgitated by a carnivorous bird.

pelletron [NUCLEO] A type of electrostatic accelerator that utilizes a charging system consisting of steel cylinders joined by links of solid insulating material such as nylon to form a chain; the metal cylinders are charged as they leave a pulley at ground potential, and the charge is removed as they pass over a pulley in the high-potential terminal.

pellicle [CYTOL] A plasma membrane. [INV ZOO] A thin protective membrane, as on certain protozoans.

Pelmatozoa [INV ZOO] A division of the Echinodermata made up of those forms which are anchored to the substrate during at least part of their life history.

pelmicrite [GEOL] A limestone containing less than 25% each of intraclasts and ooliths, having a volume ratio of pellets to fossils greater than 3 to 1, and with the micrite matrix more abundant than the sparry-calcite cement.

Pelobatidae [VERT ZOO] A family of frogs in the suborder Anomocoela, including the spadefoot toads.

Pelodytidae [VERT ZOO] A family of frogs in the suborder Anomocoela.

Pelogonidae [INV ZOO] The equivalent name for Ochteridae.

Pelomedusidae [VERT ZOO] The side-necked or hidden-necked turtles, a family of the order Chelonia.

Pelonemataceae [MICROBIO] A family of gliding bacteria of uncertain affiliation; straight, flexuous, or spiral, unbranched filaments containing colorless, cylindrical cells.

Pelopidae [INV ZOO] A family of oribatid mites, order Sarcoptiformes.

Peloridiidae [INV ZOO] The single family of the homopteran series Coleorrhyncha.

pelouse *See* meadow.

pelsparite [PETR] A limestone containing less than 25% each of intraclasts and ooliths, having a volume ratio of pellets to fossils greater than 3 to 1, and with the sparry-calcite cement more abundant than the micrite matrix.

Peltier coefficient [PHYS] The ratio of the rate at which heat is evolved or absorbed at a junction of two metals in the Peltier effect to the current passing through the junction.

Peltier effect [PHYS] Heat is evolved or absorbed at the junction of two dissimilar metals carrying a small current, depending upon the direction of the current.

Pelton turbine *See* Pelton wheel.

Pelton wheel [MECH ENG] An impulse hydraulic turbine in which pressure of the water supply is converted into velocity by a few stationary nozzles, and the water jets then impinge on the buckets mounted on the rim of a wheel; usually limited to high head installations, exceeding 500 feet (150 meters). Also known as Pelton turbine.

pelvic cavity *See* pelvis.

pelvic fin [VERT ZOO] One of the pair of fins of fishes corresponding to the hindlimbs of a quadruped.

pelvic girdle [ANAT] The system of bones supporting the lower limbs, or the hindlimbs, of vertebrates.

pelvis [ANAT] **1.** The main, basin-shaped cavity of the kidney into which urine is discharged by nephrons. **2.** The basin-shaped structure formed by the hipbones together with the sacrum and coccyx, or caudal vertebrae. **3.** The cavity of the bony pelvis. Also known as pelvic cavity.

Pelycosauria [PALEON] An extinct order of primitive, mammallike reptiles of the subclass Synapsida, characterized by a temporal fossa that lies low on the side of the skull.

pelyte *See* pelite.

Penaeidea [INV ZOO] A primitive section of the Decapoda in the suborder Natantia; in these forms, the pleurae of the first abdominal somite overlap those of the second, the third legs are chelate, and the gills are dendrobranchiate.

pencil beam [ELECTROMAG] A beam of radiant energy concentrated in an approximately conical or cylindrical portion of space of relatively small diameter; this type of beam is used for many revolving navigational lights and radar beams.

pencil follower [COMPUT SCI] A device for converting graphic images to digital form; the information to be analyzed appears on a reading table where a reading pencil is made to follow the trace, and a mechanism beneath the table surface transmits position signals from the pencil to an electronic console for conversion to digital form.

pencil ore [GEOL] Hard, fibrous masses of hematite that can be broken up into splinters.

pencil stone *See* pyrophyllite.

pendant *See* roof pendant.

pendant atomizer *See* hanging-drop atomizer.

pendant-drop method [PHYS] Method for the measurement of liquid surface tension by the elongation of a hanging drop of the liquid.

pendulum [PHYS] A rigid body mounted on a fixed horizontal axis, about which it is free to rotate under the influence of gravity. Also known as gravity pendulum.

penecontemporaneous [GEOL] Of a geologic process or the structure or mineral that is formed by the process, occurring immediately following deposition but before consolidation of the enclosing rock.

penetrance [GEN] The proportion of individuals carrying a dominant gene in the heterozygous condition or a recessive gene in the homozygous condition in which the specific phenotypic effect is apparent. Also known as gene penetrance.

penetration hardness *See* indentation hardness.

penetration probability [QUANT MECH] The probability that a particle will pass through a potential barrier, that is, through a finite region in which the particle's potential energy is greater than its total energy. Also known as transmission coefficient.

penetration range *See* night visual range.

penetration twin *See* interpenetration twin.

penetrometer [ENG] **1.** An instrument that measures the penetrating power of a beam of x-rays or other penetrating radiation. **2.** An instrument used to determine the consistency of a material by measurement of the depth to which a standard needle penetrates into it under standard conditions.

penfieldite [MINERAL] $Pb_2(OH)Cl_3$ A white hexagonal mineral composed of basic chloride of lead, occurring in hexagonal prisms.

penguin [VERT ZOO] Any member of the avian order Sphenisciformes; structurally modified wings do not fold and they function like flippers, the tail is short, feet are short and webbed, and the legs are set far back on the body.

peninsula [GEOGR] A body of land extending into water from the mainland, sometimes almost entirely separated from the mainland except for an isthmus.

penis [ANAT] The male organ of copulation in humans and certain other vertebrates. Also known as phallus.

penitent ice [HYD] A jagged spike or pillar of compacted firn caused by differential melting and evaporation; necessary for this formation are air temperature near freezing, dew point much below freezing, and strong insolation.

penitent snow [HYD] A jagged spike or pillar of compacted snow caused by differential melting and evaporation.

penna *See* contour feather.

Pennales [BOT] An order of diatoms (Bacillariophyceae) in which the form is often circular, and the markings on the valves are radial.

pennantite [MINERAL] $Mn_9Al_6Si_5O_{20}(OH)_{16}$ Orange mineral composed of basic manganese aluminum silicate; member of the chlorite group; it is isomorphous with thuringite.

Pennatulacea [INV ZOO] The sea pens, an order of the subclass Alcyonaria; individuals lack stolons and live with their bases embedded in the soft substratum of the sea.

Pennellidae [INV ZOO] A family of copepod crustaceans in the suborder Caligoida; skin-penetrating external parasites of various marine fishes and whales.

Penning gage *See* Philips ionization gage.

Penning ion source [NUCLEO] A source of positively charged heavy ions used in accelerators, consisting of an anode chamber with cathodes at each end, into which the desired element is introduced; an arc discharge is generated in the chamber, creating a plasma, and electrons confined by the cathodes and an axial magnetic field make many ionizing collisions with the gas in the chamber.

penninite [MINERAL] $(Mg,Fe,Al)_6(Si,Al)_4O_{11}(OH)_8$ An emerald-green, olive-green, pale-green, or bluish mineral of the chlorite group crystallizing in the monoclinic system, with a hardness of 2–2.5 on Mohs scale, and specific gravity of 2.6–2.85.

Pennsylvanian [GEOL] A division of late Paleozoic geologic time, extending from 320 to 280 million years ago, varyingly considered to rank as an independent period or as an epoch of the Carboniferous period; named for outcrops of coal-bearing rock formations in Pennsylvania.

pennyweight [MECH] A unit of mass equal to $\frac{1}{20}$ troy ounce or to 1.55517384 grams; the term is employed in the United States and in England for the valuation of silver, gold, and jewels. Abbreviated dwt; pwt.

penstock [CIV ENG] A valve or sluice gate for regulating water or sewage flow. [ENG] A closed water conduit controlled by valves and located between the intake and the turbine in a hydroelectric plant.

pentadactyl [VERT ZOO] Having five digits on the hand or foot.

pentagon [MATH] A polygon with five sides.

pentagonal dodecahedron See pyritohedron.

pentagrid See heptode.

pentahydrite [MINERAL] $MgSO_4 \cdot 5H_2O$ A triclinic mineral composed of hydrous magnesium sulfate; it is isostructural with chalcanthite.

Pentamerida [PALEON] An extinct order of articulate brachiopods.

Pentameridina [PALEON] A suborder of extinct brachiopods in the order Pentamerida; dental plates associated with the brachiophores were well developed, and their bases enclosed the dorsal adductor muscle field.

pentamerous [BOT] Having each whorl of the flower consisting of five members, or a multiple of five.

Pentastomida [INV ZOO] A class of bloodsucking parasitic arthropods; the adult is vermiform, and there are two pairs of hooklike, retractile claws on the cephalothorax.

pentastyle [ARCH] Having five columns across the front.

Pentatomidae [INV ZOO] The true stink bugs, a family of hemipteran insects in the superfamily Pentatomoidea.

Pentatomoidea [INV ZOO] A subfamily of the hemipteran group Pentatomorpha distinguished by marginal trichobothria and by antennae which are usually five-segmented.

Pentatomorpha [INV ZOO] A large group of hemipteran insects in the subdivision Geocorisae in which the eggs are nonoperculate, a median spermatheca is present, accessory salivary glands are tubular, and the abdomen has trichobothria.

pentavalent [CHEM] An atom or radical that exhibits a valency of 5.

pentlandite [MINERAL] $(Fe,Ni)_9S_8$ A yellowish-bronze mineral having a metallic luster and crystallizing in the isometric system; hardness is 3.5–4 on Mohs scale, and specific gravity is 4.6–5.0; the major ore of nickel.

pentobarbital sodium [PHARM] $C_{11}H_{17}N_2NaO_3$ A short- to intermediate-acting barbiturate; used as a hypnotic and sedative drug. Also known as sodium pentobarbitone.

pentode [ELECTR] A five-electrode electron tube containing an anode, a cathode, a control electrode, and two additional electrodes that are ordinarily grids.

pentose [BIOCHEM] Any one of a class of carbohydrates containing five atoms of carbon.

pentose phosphate pathway [BIOCHEM] A pathway by which glucose is metabolized or transformed in plants and microorganisms; glucose-6-phosphate is oxidized to 6-phosphogluconic acid, which then undergoes oxidative decarboxylation to form ribulose-5-phosphate, which is ultimately transformed to fructose-6-phosphate.

penumbra [ASTRON] The outer, relatively light part of a sunspot. [OPTICS] That portion of a shadow illuminated by only part of a radiating source.

penumbral eclipse See lunar appulse.

Penutian [GEOL] A North American stage of geologic time: lower Eocene (above Bulitian, below Ulatasian).

pepo [BOT] A fleshy indehiscent berry with many seeds and a hard rind; characteristic of the Cucurbitaceae.

pepper [BOT] Any of several warm-season perennials of the genus *Capsicum* in the order Polemoniales, especially *C. annum* which is cultivated for its fruit, a many-seeded berry with a thickened integument. [FOOD ENG] Any of various spices and condiments obtained from the fruits of plants of the genus *Piper*.

peppermint [BOT] Any of various aromatic herbs of the genus *Mentha* in the family Labiatae, especially *M. piperita*.

PeP reaction See proton-electron-proton reaction.

pepsinogen [BIOCHEM] The precursor of pepsin, found in the stomach mucosa.

peptic ulcer [MED] An ulcer involving the mucosa, submucosa, and muscular layer on the lower esophagus, stomach, or duodenum, due in part at least to the action of acid-pepsin gastric juice.

peptide [BIOCHEM] A compound of two or more amino acids joined by peptide bonds.

peptide bond [ORG CHEM] A bond in which the carboxyl group of one amino acid is condensed with the amino group of another to form a —CO·NH— linkage. Also known as peptide linkage.

peptide linkage See peptide bond.

Peptococcaceae [MICROBIO] A family of gram-positive cocci; organisms can use either amino acids or carbohydrates for growth and energy.

per- [CHEM] Chemical prefix meaning: 1. Complete, as in hydrogen peroxide. 2. Extreme, or the presence of the peroxy (—O—O—) group. 3. Exhaustive (complete) substitution, as in perchloroethylene.

Peracarida [INV ZOO] A superorder of the Eumalacostraca; these crustaceans have the first thoracic segment united with the head, the cephalothorax usually larger than the abdomen, and some thoracic segments free from the carapace.

peracid [CHEM] Acid containing the peroxy (—O—O—) group, such as peracetic acid or perchloric acid.

peralcohol [ORG CHEM] Chemical compound containing the peroxy group (—O—O—), such as peracetic acid and perchromic acid.

peralkaline [PETR] Of igneous rock, having a molecular proportion of aluminum lower than that of sodium oxide and potassium oxide combined.

Peramelidae [VERT ZOO] The bandicoots, a family of insectivorous mammals in the order Marsupialia.

perceived noise level [ACOUS] In perceived noise decibels, the noise level numerically equal to the sound pressure level, in decibels, of a band of random noise of width one-third to one octave centered on a frequency of 1000 hertz which is judged by listeners to be equally noisy.

percent [MATH] A quantitative term whereby n-percent of a number is n one-hundredths of the number. Symbolized %.

percentile [STAT] A value in the range of a set of data which separates the range into two groups so that a given percentage of the measures lies below this value.

perception [PHYSIO] Recognition in response to sensory stimuli; the act or process by which the memory of certain qualities of an object is associated with other qualities impressing the senses, thereby making possible recognition of the object.

perceptron [COMPUT SCI] A pattern recognition machine, based on an analogy to the human nervous system, capable of learning by means of a feedback system which reinforces correct answers and discourages wrong ones.

perch [MECH] Also known as pole; rod. **1.** A unit of length, equal to 5.5 yards, or 16.5 feet, or 5.0292 meters. **2.** A unit of area, equal to 30.25 square yards, or 272.25 square feet, or 25.29285264 square meters. [NAV] A staff placed on top of a buoy, rock, or shoal as a mark for navigators; a ball or cage is sometimes placed at the top of the perch, as an identifying mark. [VERT ZOO] **1.** Any member of the family Percidae. **2.** The common name for a number of unrelated species of fish belonging to the Centrarchidae, Anabantoidei, and Percopsiformes.

perched groundwater *See* perched water.

perched lake [HYD] A perennial lake whose surface level lies at a considerably higher elevation than those of other bodies of water, including aquifers, directly or closely associated with the lake.

perched stream [HYD] A stream whose surface level is above that of the water table and that is separated from underlying groundwater by an impermeable bed in the zone of aeration.

perched water [HYD] Groundwater that is unconfined and separated from an underlying main body of groundwater by an unsaturated zone. Also known as perched groundwater.

Percidae [VERT ZOO] A family of fresh-water actinopterygian fishes in the suborder Percoidei; comprises the true perches.

Perciformes [VERT ZOO] The typical spiny-rayed fishes, comprising the largest order of vertebrates; characterized by fin spines, a swim bladder without a duct, usually ctenoid scales, and 17 or fewer caudal fin rays.

Percoidei [VERT ZOO] A large, typical suborder of the order Perciformes; includes over 50% of the species in this order.

percolation [COMPUT SCI] The transfer of needed data back from secondary storage devices to main storage. [HYD] Gravity flow of groundwater through the pore spaces in rock or soil. [MIN ENG] Gentle movement of a solvent through an ore bed in order to extract a mineral. [SCI TECH] Slow movement of a liquid through a porous material.

percolation filtration [CHEM ENG] A continuous petroleum-refinery process in which lubricating oils and waxes are percolated through a clay bed to improve color, odor, and stability.

percolation test [CIV ENG] A test to determine the suitability of a soil for the installation of a domestic sewage-disposal system, in which a hole is dug and filled with water and the rate of water-level decline is measured.

Percomorphi [VERT ZOO] An equivalent, ordinal name for the Perciformes.

Percopsidae [VERT ZOO] A family of fishes in the order Percopsiformes.

Percopsiformes [VERT ZOO] A small order of actinopterygian fishes characterized by single, ray-supported dorsal and anal fins and a subabdominal pelvic fin with three to eight soft rays.

percussion drill [MECH ENG] A drilling machine usually using compressed air to drive a piston that delivers a series of impacts to the shank end of a drill rod or steel and attached bit.

percussion welding [MET] Resistance welding with arc heat and simultaneously applied pressure from a hammerlike blow.

percylite [MINERAL] $PbCuCl_2(OH)_2$ Mineral made up of a basic chloride of copper and lead and occurring as cubic blue crystals, with a hardness of 2.5.

perennial [BOT] A plant that lives for an indefinite period, dying back seasonally and producing new growth from a perennating part.

perfect dielectric *See* ideal dielectric.

perfect flower [BOT] A flower having both stamens and pistils.

perfect fluid *See* inviscid fluid.

perfect gas *See* ideal gas.

perfectly inelastic collision [PHYS] A collision in which as much translational kinetic energy is converted into internal energy of the colliding systems as is consistent with the conservation of momentum. Also known as completely inelastic collision.

perfect number [MATH] An integer which equals the sum of all its factors other than itself.

perfect set [MATH] A set in a topological space which equals its set of accumulation points.

perfoliate [BOT] Pertaining to the form of a leaf having its base united around the stem. [INV ZOO] Pertaining to the form of certain insect antennae having the terminal joints expanded and flattened to form plates which encircle the stalk.

perforatorium *See* acrosome.

Pergidae [INV ZOO] A small family of hymenopteran insects in the superfamily Tenthredinoidea.

perhydro- [ORG CHEM] Prefix designating a completely saturated aromatic compound, as for decalin $(C_{10}H_{18})$, also known as perhydronaphthalene.

perianth [BOT] The calyx and corolla considered together.

periapsis [ASTRON] The orbital point nearest the center of attraction of an orbiting body.

pericarditis [MED] Inflammation of the pericardium.

pericardium [ANAT] The membranous sac that envelops the heart; it contains 5–20 grams of clear serous fluid.

pericarp [BOT] The wall of a fruit, developed by ripening and modification of the ovarian wall.

pericentric inversion [GEN] A type of chromosome aberration in which chromosome material involving both arms of the chromosome is inverted, thus spanning the centromere.

periclase [MINERAL] MgO Native magnesia; a mineral occurring in granular forms or isometric crystals, with hardness of 6 on Mohs scale, and specific gravity of 3.67–3.90. Also known as periclasite.

periclasite *See* periclase.

periclinal [GEOL] Referring to strata and structures that dip radially outward from, or inward toward, a center, forming a dome or a basin.

pericline [GEOL] A fold characterized by central orientation of the dip of the beds. [MINERAL] A variety of albite elongated, and often twinned, along the *b*-axis.

pericycle [BOT] The outer boundary of the stele of plants; may not be present as a distinct layer of cells.

periderm [BOT] A group of secondary tissues forming a protective layer which replaces the epidermis of many plant stems, roots, and other parts; composed of cork cambium, phelloderm, and cork. [EMBRYO] The superficial transient layer of epithelial cells of the embryonic epidermis.

peridot [MINERAL] 1. A gem variety of olivine that is transparent to translucent and pale-, clear-, or yellowish-green in color. 2. A variety of tourmaline approaching olivine in color.

peridotite [PETR] A dark-colored, ultrabasic phaneritic igneous rock composed largely of olivine, with smaller amounts of pyroxene or hornblende.

peridotite shell See upper mantle.

perigee [ASTRON] The point in the orbit of the moon or other satellite when it is nearest the earth.

perigee-to-perigee period See anomalistic period.

perihelion [ASTRON] That orbital point nearest the sun when he sun is the center of attraction.

Perilampidae [INV ZOO] A family of hymenopteran insects in the superfamily Chalcidoidea.

perilymph [PHYSIO] The fluid separating the membranous from the osseous labyrinth of the internal ear.

perimeter [MATH] The total length of a closed curve; for example, the perimeter of a polygon is the total length of its sides.

perineum [ANAT] 1. The portion of the body included in the outlet of the pelvis, bounded in front by the pubic arch, behind by the coccyx and sacrotuberous ligaments, and at the sides by the tuberosities of the ischium. 2. The region between the anus and the scrotum in the male, between the anus and the posterior commissure of the vulva in the female.

perineural fibroblastoma See neurofibroma.

perineural fibroma See neurofibroma.

period [CHEM] A family of elements with consecutive atomic numbers in the periodic table and with closely related properties; for example, chromium through copper. [GEOL] A unit of geologic time constituting a subdivision of an era; the fundamental unit of the standard geologic time scale. [NUCLEO] The time required for exponentially rising or falling neutron flux in a nuclear reactor to change by a factor of e (2.71828). [PHYS] The duration of a single repetition of a cyclic phenomenon.

periodic function [MATH] A function $f(x)$ of a real or complex variable is periodic with period T if $f(x + T) = f(x)$ for every value of x.

periodic lattice See lattice.

periodic law [CHEM] The law that the properties of the chemical elements and their compounds are a periodic function of their atomic weights.

periodic motion [MECH] Any motion that repeats itself identically at regular intervals.

periodic perturbation [ASTRON] Small deviations from the computed orbit of a planet or satellite; the deviations extend through cycles that generally do not exceed a century. [MATH] A perturbation which is periodic as a function.

periodic table [CHEM] A table of the elements, written in sequence in the order of atomic number or atomic weight and arranged in horizontal rows (periods) and vertical columns (groups) to illustrate the occurrence of similarities in the properties of the elements as a periodic function of the sequence.

period-luminosity relation [ASTRON] Relation between the periods of Cepheid variable stars and their absolute magnitude; the absolutely brighter the star, the longer the period.

periosteum [ANAT] The fibrous membrane enveloping bones, except at joints and the points of tendonous and ligamentous attachment.

peripheral [ANAT] Pertaining to or located at or near the surface of a body or an organ. [SCI TECH] Remote from the center; marginal; on the periphery.

peripheral buffer [COMPUT SCI] A device acting as a temporary storage when transmission occurs between two devices operating at different transmission speeds.

peripheral control unit [COMPUT SCI] A device which connects a unit of peripheral equipment with the central processing unit of a computer and which interprets and responds to instructions from the central processing unit.

peripheral equipment [COMPUT SCI] Equipment that works in conjunction with a computer but is not part of the computer itself, such as a card or paper-tape reader or punch, magnetic-tape handler, or line printer.

peripheral-limited [COMPUT SCI] Property of a computer whose processing time is determined by the speed of its peripheral equipment rather than by the speed of its central processing unit. Also known as I/O-bound.

peripheral nervous system [ANAT] The autonomic nervous system, the cranial nerves, and the spinal nerves including their associated sensory receptors.

peripheral speed See cutting speed.

peripheral vision [PHYSIO] The act of seeing images that fall upon parts of the retina outside the macula lutea. Also known as indirect vision.

Periptychidae [PALEON] A family of extinct herbivorous mammals in the order Condylartha distinguished by specialized, fluted teeth.

Peripylina [INV ZOO] An equivalent name for Porulosida.

Periscelidae [INV ZOO] A family of myodarian cyclorrhaphous dipteran insects in the subsection Acalypteratae.

Perischoechinoidea [INV ZOO] A subclass of principally extinct echinoderms belonging to the Echinoidea and lacking stability in the number of columns of plates that make up the ambulacra and interambulacra.

periscope [OPTICS] An optical instrument used to provide a raised line of vision where it may not be practical or possible, as in entrenchments, tanks, or submarines; the raised line of vision is obtained by the use of mirrors or prisms within the structure of the item; it may have single or dual optical systems.

perisperm [BOT] In a seed, the nutritive tissue that is derived from the nucellus and deposited on the outside of the embryo sac.

Perissodactyla [VERT ZOO] An order of exclusively herbivorous mammals distinguished by an odd number of toes and mesaxonic feet, that is, with the axis going through the third toe.

peristalsis [PHYSIO] The rhythmic progressive wave of muscular contraction in tubes, such as the intestine, provided with both longitudinal and transverse muscular fibers.

peristome [BOT] The fringe around the opening of a moss capsule. [INV ZOO] The area surrounding the mouth of various invertebrates.

peritectic [PHYS CHEM] An isothermal reversible reaction in which a liquid phase reacts with a solid phase during cooling to produce a second solid phase.

peritectic point [PHYS CHEM] In a binary two-phase heteroazeotropic system at constant pressure, that point up to which the boiling point has remained constant until one of the phases has boiled away.

peritectoid [PHYS CHEM] An isothermal reversible reaction in which a solid phase on cooling reacts with another solid phase to form a third solid phase.

peritoneal cavity [ANAT] The potential space between the visceral and parietal layers of the peritoneum.

peritoneoscopy   *See* laparoscopy.

peritoneum   [ANAT] The serous membrane enveloping the abdominal viscera and lining the abdominal cavity.

Peritrichia   [INV ZOO] A specialized subclass of the class Ciliatea comprising both sessile and mobile forms.

Peritrichida   [INV ZOO] The single order of the protozoan subclass Peritrichia.

perlite   [GEOL] A rhyolitic glass with abundant spherical or convolute cracks that cause it to break into small pearllike masses or pebbles, usually less than a centimeter across; it is commonly gray or green with a pearly luster and has the composition of rhyolite. Also known as pearlite; pearlstone.

perlitic   [PETR] **1.** Of the texture of a glassy igneous rock, exhibiting small spheruloids formed from cracks due to contraction during cooling. **2.** Pertaining to or characteristic of perlite.

permafrost   [GEOL] Perennially frozen ground, occurring wherever the temperature remains below 0°C for several years, whether the ground is actually consolidated by ice or not and regardless of the nature of the rock and soil particles of which the earth is composed.

permanent aurora   *See* airglow.

permanent magnet   [ELECTROMAG] A piece of hardened steel or other magnetic material that has been strongly magnetized and retains its magnetism indefinitely. Abbreviated PM.

permanent set   [MECH] Permanent plastic deformation of a structure or a test piece after removal of the applied load. Also known as set.

permanent storage   [COMPUT SCI] A means of storing data for rapid retrieval by a computer; does not permit changing the stored data.

permanent teeth   [ANAT] The second set of teeth of a mammal, following the milk teeth; in humans, the set of 32 teeth consists of 8 incisors, 4 canines, 8 premolars, and 12 molars.

permeability   [ELECTROMAG] A factor, characteristic of a material, that is proportional to the magnetic induction produced in a material divided by the magnetic field strength; it is a tensor when these quantities are not parallel. [FL MECH] **1.** The ability of a membrane or other material to permit a substance to pass through it. **2.** Quantitatively, the amount of substance which passes through the material under given conditions. [GEOL] The capacity of a porous rock, soil, or sediment for transmitting a fluid without damage to the structure of the medium. Also known as perviousness. [NAV ARCH] The percentage of a given space in a ship that can be occupied by water.

permeable membrane   [CHEM] A thin sheet or membrane of material through which selected liquid or gas molecules or ions will pass, either through capillary pores in the membrane or by ion exchange; used in dialysis, electrodialysis, and reverse osmosis.

permeameter   [ENG] **1.** Device for measurement of the average size or surface area of small particles; consists of a powder bed of known dimension and degree of packing through which the particles are forced; pressure drop and rate of flow are related to particle size, and pressure drop is related to surface area. **2.** A device for measuring the coefficient of permeability by measuring the flow of fluid through a sample across which there is a pressure drop produced by gravity. **3.** An instrument for measuring the magnetic flux or flux density produced in a test specimen of ferromagnetic material by a given magnetic intensity, to permit computation of the magnetic permeability of the material.

permeance   [ELECTROMAG] A characteristic of a portion of a magnetic circuit, equal to magnetic flux divided by magnetomotive force; the reciprocal of reluctance. Symbolized P.

permeation   [CHEM] The movement of atoms, molecules, or ions into or through a porous or permeable substance (such as zeolite or a membrane).

Permian   [GEOL] The last period of geologic time in the Paleozoic era, from 280 to 225 million years ago.

permineralization   [GEOL] A fossilization process whereby additional minerals are deposited in the pore spaces of originally hard animal parts.

permittivity   [ELEC] The dielectric constant multiplied by the permittivity of empty space, where the permittivity of empty space (ϵ_0) is a constant appearing in Coulomb's law, having the value of 1 in centimeter-gram-second electrostatic units, and of 8.854×10^{-12} farad/meter in rationalized meter-kilogram-second units. Symbolized ϵ.

permutation   [MATH] A function which rearranges a finite number of symbols; more precisely, a one-to-one function of a finite set onto itself.

permutation table   [COMMUN] In computers, a table designed for the systematic construction of code groups; it may also be used to correct garbles in groups of code text.

pernicious anemia   [MED] A megaloblastic macrocytic anemia resulting from lack of vitamin B_{12}, secondary to gastric atrophy and loss of intrinsic factor necessary for vitamin B_{12} absorption, and accompanied by degeneration of the posterior and lateral columns of the spinal cord.

Perognathinae   [VERT ZOO] A subfamily of the rodent family Heteromyidae, including the pocket and kangaroo mice.

Peronosporales   [MYCOL] An order of aquatic and terrestrial phycomycetous fungi with a hyphal thallus and zoospores with two flagella.

Perothopidae   [INV ZOO] A small family of coleopteran insects in the superfamily Elateroidea found only in the United States.

perovskite   [MINERAL] Ca[TiO$_3$] A natural, yellow, brownish-yellow, reddish, brown, or black mineral and a structure type which includes no less than 150 synthetic compounds; the crystal structure is ideally cubic, it occurs as rounded cubes modified by the octahedral and dodecahedral forms, luster is subadamantine to submetallic, hardness is 5.5 on Mohs scale, and specific gravity is 4.0.

peroxidase   [BIOCHEM] An enzyme that catlyzes reactions in which hydrogen peroxide is an electron acceptor.

peroxide   [CHEM] **1.** A compound containing the peroxy (—O—O—) group, as in hydrogen peroxide. **2.** *See* hydrogen peroxide.

perpendicular   [MATH] Geometric objects are perpendicular if they intersect in an angle of 90°.

Perseus   [ASTRON] A northern constellation; right ascension 3 hours; declination 45°N.

Persian ammoniac   *See* ammoniac.

Persian melon   [BOT] A variety of muskmelon (*Cucumis melo*) in the order Violales; the fruit is globular and without sutures, and has dark-green skin, thin abundant netting, and firm, thick, orange flesh.

persistence   [ELECTR] A measure of the length of time that the screen of a cathode-ray tube remains luminescent after excitation is removed; ranges from 1 for short persistence to 7 for long persistence. [METEOROL] With respect to the long-term nature of the wind at a given location, the ratio of the magnitude of the mean wind vector to the average speed of the wind without regard to direction. Also known as constancy; steadiness.

persistent [BOT] Of a leaf, withering but remaining attached to the plant during the winter.

persistron [ELECTR] A device in which electroluminescence and photoconductivity are used in a single panel capable of producing a steady or persistent display with pulsed signal input.

personal computer [COMPUT SCI] A computer for home or personal use.

perspective [GRAPHICS] The technique representing a figure or the space relationships of natural objects, on either a plane or curved surface, by means of projecting lines emanating from a single point, which may be infinity.

perspective projection [GRAPHICS] A projection of points by straight lines drawn through them from some given point to an intersection with the plane of projection.

perspiration [PHYSIO] 1. The secretion of sweat. 2. See sweat.

PERT [SYS ENG] A management control tool for defining, integrating, and interrelating what must be done to accomplish a desired objective on time; a computer is used to compare current progress against planned objectives and give management the information needed for planning and decision making. Derived from program evaluation and review technique.

perthite [GEOL] A parallel to subparallel intergrowth of potassium and sodium feldspar; the potassium-rich phase is usually the host from which the sodium-rich phase evolves.

perturbation [ASTRON] A deviation of an astronomical body from its computed orbit because of the attraction of another body or bodies. [MATH] A function which produces a small change in the values of some given function. [PHYS] Any effect which makes a small modification in a physical system, especially in case the equations of motion could be solved exactly in the absence of this effect.

perturbation theory [MATH] The study of the solutions of differential and partial differential equations from the viewpoint of perturbation of solutions. [PHYS] The theory of obtaining approximate solutions to the equations of motion of a physical system when these equations differ by a small amount from equations which can be solved exactly.

pertussis [MED] An infectious inflammatory bacterial disease of the air passages, caused by *Hemophilus pertussis* and characterized by explosive coughing ending in a whooping inspiration. Also known as whooping cough.

Peru Current [OCEANOGR] The cold ocean current flowing north along the coasts of Chile and Peru. Also known as Humboldt Current.

Peru saltpeter See soda niter.

perviousness See permeability.

peta- [SCI TECH] A prefix that represents 10^{15}. Abbreviated P.

petal [BOT] One of the sterile, leaf-shaped flower parts that make up the corolla.

Petalichthyida [PALEON] A small order of extinct dorsoventrally flattened fishes belonging to the class Placodermi; the external armor is in two shields of large plates.

petalite [MINERAL] $LiAlSi_4O_{10}$ A white, gray, or colorless monoclinic mineral composed of silicate of lithium and aluminum, occurring in foliated masses or as crystals.

Petalodontidae [PALEON] A family of extinct cartilaginous fishes in the order Bradyodonti distinguished by teeth with deep roots and flattened diamond-shaped crowns.

Petaluridae [INV ZOO] A family of dragonflies in the suborder Anisoptera.

petit mal [MED] A generalized epileptic seizure of the absence type, that is, characterized by different degrees of impaired consciousness.

petrel [VERT ZOO] A sea bird of the families Procellariidae and Hydrobatidae, generally small to medium-sized with long wings and dark plumage with white areas near the rump.

petrifaction [GEOL] A fossilization process whereby inorganic matter dissolved in water replaces the original organic materials, converting them to a stony substance.

Petriidae [INV ZOO] A small family of coleopteran insects in the superfamily Tenebrionoidea.

petrochemicals [ORG CHEM] Chemicals made from feedstocks derived from petroleum or natural gas; examples are ethylene, butadiene, most large-scale plastics and resins, and petrochemical sulfur. Also known as petroleum chemicals.

petrofabric See fabric.

petrofabric analysis See structural petrology.

petrofabric diagram See fabric diagram.

petrofabrics See structural petrology.

petrogenesis [PETR] That branch of petrology dealing with the origin of ocks, particularly igneous rocks. Also known as petrogeny.

petrogeny See petrogenesis.

petrogeometry See structural petrology.

petrographic province [GEOL] A broad area in which similar igneous rocks are formed during the same period of igneous activity. Also known as comagmatic region; igneous province; magma province.

petrography [GEOL] The branch of geology that deals with the description and systematic classification of rocks, especially by means of microscopic examination.

petrol See gasoline.

petroleum [GEOL] A naturally occurring complex liquid hydrocarbon which after distillation yields combustible fuels, petrochemicals, and lubricants; can be gaseous (natural gas), liquid (crude oil, crude petroleum), solid (asphalt, tar, bitumen), or a combination of states.

petroleum chemicals See petrochemicals.

petroleum engineering [ENG] The application of almost all types of engineering to the drilling for and production of oil, gas, and liquefiable hydrocarbons.

petroleum geology [GEOL] The branch of economic geology dealing with the origin, occurrence, movement, accumulation, and exploration of hydrocarbon fuels.

petroleum secondary engineering [PETRO ENG] The process of removing oil from its native reservoirs by the use of supplemental energies after the natural energies causing oil production have been depleted.

petroleum trap [GEOL] Stable underground formation (geological or physical) of such nature as to trap and hold liquid or gaseous hydrocarbons; usually consists of sand or porous rock surrounded by impervious rock or clay formations.

petrologen See kerogen.

petrology [GEOL] The branch of geology concerned with the origin, occurrence, structure, and history of rocks, principally igneous and metamorphic rock.

petromorphology *See* structural petrology.

Petromyzonida [VERT ZOO] The lampreys, an order of eellike, jawless vertebrates (Agnatha) distinguished by a single, dorsal nasal opening, and the mouth surrounded by an oral disk and provided with a rasping tongue.

Petromyzontiformes [VERT ZOO] The equivalent name for Petromyzonida.

Petrosaviaceae [BOT] A small family of monocotyledonous plants in the order Triuridales characterized by perfect flowers, three carpels, and numerous seeds.

Pettit truss [CIV ENG] A bridge truss in which the panel is subdivided by a short diagonal and a short vertical member, both intersecting the main diagonal at its midpoint.

Petzval lens [OPTICS] A photographic objective which consists of four lenses ordered in two pairs widely separated from each other, with the first pair cemented together and the second usually having a small air space.

pewter [MET] An alloy that typically contained tin as the principal component and some antimony and copper; older produced pewter typically contains lead along with the other components.

pf *See* power factor.

PFE *See* photoferroelectric effect.

Pfeiffer's disease *See* infectious mononucleosis.

PGA *See* folic acid.

pH [CHEM] A term used to describe the hydrogen-ion activity of a system; it is equal to $-\log a_H^+$; here a_H^+ is the activity of the hydrogen ion; in dilute solution, activity is essentially equal to concentration and pH is defined as $-\log_{10}[H^+]$, where $[H^+]$ is hydrogen-ion concentration in moles per liter; a solution of pH 0 to 7 is acid, pH of 7 is neutral, pH over 7 to 14 is alkaline.

phacellite *See* kaliophilite.

phacolith [GEOL] A minor, concordant, lens-shaped, and usually granitic intrusion into folded sedimentary strata.

Phaenocephalidae [INV ZOO] A monospecific family of coleopteran insects in the superfamily Cucujoidea, found only in Japan.

Phaenothontidae [VERT ZOO] The tropic birds, a family of fish-eating aquatic forms in the order Pelecaniformes.

Phaeocoleosporae [MYCOL] A spore group of the Fungi Imperfecti with dark filiform spores.

Phaeodictyosporae [MYCOL] A spore group of the Fungi Imperfecti with dark muriform spores.

Phaeodidymae [MYCOL] A spore group of the Fungi Imperfecti with dark two-celled spores.

Phaeodorina [INV ZOO] The equivalent name for Tripylina.

Phaeohelicosporae [MYCOL] A spore group of the Fungi Imperfecti with dark, spirally coiled, septate spores.

Phaeophragmiae [MYCOL] A spore group of the Fungi Imperfecti with dark three- to many-celled spores.

Phaeophyta [BOT] The brown algae, constituting a division of plants; the plant body is multicellular, varying from a simple filamentous form to a complex, sometimes branched body having a basal attachment.

Phaeosporae [MYCOL] A spore group of Fungi Imperfecti characterized by dark one-celled, nonfiliform spores.

Phaeostaurosporae [MYCOL] A spore group of the Fungi Imperfecti with dark star-shaped or forked spores.

phage *See* bacteriophage.

phagocyte [CYTOL] An ameboid cell that engulfs foreign material.

phagocytic vacuole *See* food vacuole.

phagocytosis [CYTOL] A specialized form of macropinocytosis in which cells engulf large solid objects such as bacteria and deliver the internalized objects to special digesting vacuoles; exists in certain cell types, such as macrophages and neutrophils.

Phalacridae [INV ZOO] The shining flower beetles, a family of coleopteran insects in the superfamily Cucujoidea.

Phalacrocoracidae [VERT ZOO] The cormorants, a family of aquatic birds in the order Pelecaniformes.

Phalaenidae [INV ZOO] The equivalent name for Noctuidae.

Phalangeridae [VERT ZOO] A family of marsupial mammals in which the marsupium is well developed and opens anteriorly, the hindfeet are syndactylous, and the hallux is opposable and lacks a claw.

Phalangida [INV ZOO] An order of the class Arachnida characterized by an unsegmented cephalothorax broadly joined to a segmented abdomen, paired chelate chelicerae, and paired palpi.

phalanx [ANAT] One of the bones of the fingers or toes.

Phalaropodidae [VERT ZOO] The phalaropes, a family of migratory shore birds characterized by lobate toes and by reversal of the sex roles with respect to dimorphism and care of the young.

Phallostethidae [VERT ZOO] A family of actinopterygian fishes in the order Atheriniformes.

Phallostethiformes [VERT ZOO] An equivalent name for Atheriniformes.

phallus [ANAT] *See* penis. [EMBRYO] An undifferentiated embryonic structure derived from the genital tubercle that differentiates into the penis in males and the clitoris in females.

phaneritic [PETR] Of the texture of an igneous rock, being visibly crystalline. Also known as coarse-grained; phanerocrystalline; phenocrystalline.

phanerocrystalline *See* phaneritic.

Phanerorhynchidae [PALEON] A family of extinct chondrostean fishes in the order Palaeonisciformes having vertical jaw suspension.

Phanerozoic [GEOL] The part of geologic time for which there is abundant evidence of life, especially higher forms, in the corresponding rock, essentially post-Precambrian.

Phanerozonida [INV ZOO] An order of the Asteroidea in which the body margins are defined by two conspicuous series of plates and in which pentamerous symmetry is generally constant.

Phanodermatidae [INV ZOO] A family of free-living nematodes in the superfamily Enoploidea.

phanotron [ELECTR] A hot-filament diode rectifier tube utilizing an arc discharge in mercury vapor or an inert gas, usually xenon.

phantastron [ELECTR] A monostable pentode circuit used to generate sharp pulses at an adjustable and accurately timed interval after receipt of a triggering signal.

phantom bottom [OCEANOGR] A false bottom indicated by an echo sounder, some distance above the actual bottom; such an indication, quite common in the deeper parts of the ocean, is due to large quantities of small organisms.

phantom crystal [CRYSTAL] A crystal containing an earlier stage of crystallization outlined by dust, minute inclusions, or bubbles. Also known as ghost crystal.

phantom signals [ELECTR] Signals appearing on the screen of a cathode-ray-tube indicator, the cause of which cannot be readily be determined and which may be caused by circuit fault, interference, propagation anomalies, jamming, and so on.

Pharetronida [INV ZOO] An order of calcareous sponges in the subclass Calcinea characterized by a leuconoid structure.

pharmacognosy [PHARM] The science of crude drugs.

pharmacolite [MINERAL] $CaH(AsO_4)\cdot2H_2O$ A white to grayish monoclinic mineral composed of hydrous acid arsenate of calcium, occurring in fibrous form.

pharmacology [CHEM] The science dealing with the nature and properties of drugs, particularly their actions.

pharmacosiderite [MINERAL] $Fe_3(AsO_4)_2(OH)_3\cdot5H_2O$ Green or yellowish-green mineral composed of a hydrous basic iron arsenate and commonly found in cubic crystals. Also known as cube ore.

pharmacy [MED] 1. The art and science of the preparation and dispensation of drugs. 2. A place where drugs are dispensed.

pharyngeal tonsil *See* adenoid.

pharynx [ANAT] A chamber at the oral end of the vertebrate alimentary canal, leading to the esophagus.

phase [ASTRON] One of the cyclically repeating appearances of the moon or other orbiting body as seen from earth. [CHEM] Portion of a physical system (liquid, gas, solid) that is homogeneous throughout, has definable boundaries, and can be separated physically from other phases. [MATH] An additive constant in the argument of a trigonometric function. [MET] A constituent of an alloy that is physically distinct and is homogeneous in chemical composition. [PHYS] 1. The fractional part of a period through which the time variable of a periodic quantity (alternating electric current, vibration) has moved, as measured at any point in time from an arbitrary time origin; usually expressed in terms of angular measure, with one period being equal to 360° or 2π radians. 2. For a sinusoidally varying quantity, the phase (first definition) with the time origin located at the last point at which the quantity passed through a zero from a negative to a positive direction. 3. The argument of the trigonometric function describing the space and time variation of a sinusoidal disturbance, $y = A \cos [(2\pi/\lambda) (x - vt)]$, where x and t are the space and time coordinates, v is the velocity of propagation, and λ is the wavelength. [THERMO] The type of state of a system, such as solid, liquid, or gas.

phase angle [PHYS] The difference between the phase of a sinusoidally varying quantity and the phase of a second quantity which varies sinusoidally at the same frequency. Also known as phase difference.

phase-angle meter *See* phase meter.

phase boundary [PHYS] The interface between two or more separate phases, such as liquid-gas, liquid-solid, gas-solid, or, for immiscible materials, liquid-liquid or solid-solid.

phase change [PHYS] 1. The metamorphosis of a material or mixture from one phase to another, such as gas to liquid, solid to gas. 2. *See* phase shift.

phase change material [MATER] A material which is undergoing solid-liquid phase transformations and whose latent heat of fusion properties are used to store and deliver thermal energy, usually solar energy.

phase comparator [COMPUT SCI] A comparator that accepts two radio-frequency input signals of the same frequency and provides two video outputs which are proportional, respectively, to the sine and cosine of the phase difference between the two inputs.

phase-contrast microscope [OPTICS] A compound microscope that has an annular diaphragm in the front focal plane of the substage condenser and a phase plate at the rear focal plane of the objective, to make visible differences in phase or optical path in transparent or reflecting media.

phase control [ELECTR] 1. A control that changes the phase angle at which the alternating-current line voltage fires a thyratron, ignitron, or other controllable gas tube. Also known as phase-shift control. 2. *See* hue control.

phase converter [ELEC] A converter that changes the number of phases in an alternating-current power source without changing the frequency.

phase-correcting network *See* phase equalizer.

phased array [ELECTROMAG] An array of dipoles on a radar antenna in which the signal feeding each dipole is varied so that antenna beams can be formed in space and scanned very rapidly in azimuth and elevation.

phase delay [COMMUN] Ratio of the total phase shift (radians) of a sinusoidal signal in transmission through a system or transducer, to the frequency (radians/second) of the signal.

phase detector [ELECTR] A circuit that provides a direct-current output voltage which is related to the phase difference between an oscillator signal and a reference signal, for use in controlling the oscillator to keep it in synchronism with the reference signal. Also known as phase discriminator.

phase diagram [MET] *See* constitution diagram. [PHYS CHEM] A graphical representation of the equilibrium relationships between phases (such as vapor-liquid, liquid-solid) of a chemical compound, mixture of compounds, or solution. [THERMO] 1. A graph showing the pressures at which phase transitions between different states of a pure compound occur, as a function of temperature. 2. A graph showing the temperatures at which transitions between different phases of a binary system occur, as a function of the relative concentrations of its components.

phase difference *See* phase angle.

phase discriminator *See* phase detector.

phase distortion [COMMUN] 1. The distortion which occurs in an instrument when the relative phases of the input signal differ from those of the output signal. 2. *See* phase-frequency distortion.

phase encoding [COMPUT SCI] A method of recording data on magnetic tape in which a logical 1 is defined as the transition from one magnetic polarity to another positioned at the center of the bit cell, and 0 is defined as the transition in the opposite direction, also at the center of the cell.

phase equalizer [ELECTR] A network designed to compensate for phase-frequency distortion within a specified frequency band. Also known as phase-correcting network.

phase equilibria [PHYS CHEM] The equilibrium relationships between phases (such as vapor, liquid, solid) of a chemical compound or mixture under various conditions of temperature, pressure, and composition.

phase factor [ELEC] *See* power factor. [SOLID STATE] The argument (phase) of a structure factor; it cannot be directly observed.

phase-frequency distortion [COMMUN] Distortion occurring because phase shift is not proportional to frequency over the frequency range required for transmission. Also known as phase distortion.

phase integral *See* action.

phase integral method *See* Wentzel-Kramers-Brillouin method.

phase inversion [ELECTR] Production of a phase difference of 180° between two similar wave shapes of the same frequency.

phase inverter [ELECTR] A circuit or device that changes the phase of a signal by 180°, as required for feeding a push-pull amplifier stage without using a coupling transformer, or for changing the polarity of a pulse; a triode is commonly used as a phase inverter. Also known as inverter.

phase lead *See* lead angle.

phase-locked loop [ELECTR] A circuit that consists essentially of a phase detector which compares the frequency of a voltage-controlled oscillator with that of an incoming carrier signal or reference-frequency generator; the output of the phase detector, after passing through a loop filter, is fed back to the voltage-controlled oscillator to keep it exactly in phase with the incoming or reference frequency. Abbreviated PLL.

phase matching [OPTICS] A condition in which the polarization wave produced by two or more beams of incident radiation in a nonlinear medium has the same phase velocity as a freely propagating wave of the same frequency; the amplitude of the polarization wave is then greatly enhanced.

phase meter [ENG] An instrument for the measurement of electrical phase angles. Also known as phase-angle meter.

phase modulation [COMMUN] A special kind of modulation in which the linearly increasing angle of a sine wave has added to it a phase angle that is proportional to the instantaneous value of the modulating wave (message to be communicated). Abbreviated PM.

phase modulator [ELECTR] An electronic circuit that causes the phase angle of a modulated wave to vary (with respect to an unmodulated carrier) in accordance with a modulating signal.

phase quadrature *See* quadrature.

phaser [COMMUN] Facsimile device for adjusting equipment so the recorded elemental area bears the same relation to the record sheet as the corresponding transmitted elemental area bears to the subject copy in the direction of the scanning line. [ELECTROMAG] Microwave ferrite phase shifter employing a longitudinal magnetic field along one or more rods of ferrite in a waveguide.

phase shift [ELECTR] The phase angle between the input and output signals of a network or system. [PHYS] **1.** A change in the phase of a periodic quantity. Also known as phase change. **2.** A change in the phase angle between two periodic quantities. [QUANT MECH] For a partial wave of a particle scattered by a spherically symmetric potential, the phase shift is the difference between the phase of the wave function far from the scatterer and the corresponding phase of a free particle.

phase-shift control *See* phase control.

phase-shift keying [COMMUN] A form of phase modulation in which the modulating function shifts the instantaneous phase of the modulated wave between predetermined discrete values. Abbreviated PSK.

phase space [MATH] In a dynamical system or transformation group, this is the topological space whose points are being moved about by the given transformations. [STAT MECH] For a system with n degrees of freedom, a euclidean space with $2n$ dimensions, one dimension for each of the generalized coordinates and one for each of the corresponding momenta.

phase splitter [ELECTR] A circuit that takes a single input signal voltage and produces two output signal voltages 180° apart in phase.

phase stability [NUCLEO] A principle governing the stability of motion of particles in a synchrotron; the charged particle must be accelerated in each cycle at a time slightly earlier than the peak value of the accelerating potential.

phase transformation [ELEC] A change of polyphase power from three-phase to six-phase, from three-phase to twelve-phase, and so forth, by use of transformers. [PHYS] *See* phase transition.

phase transition [PHYS] A change of a substance from one phase to another. Also known as phase transformation.

Phasianidae [VERT ZOO] A family of game birds in the order Galliformes; typically, members are ground feeders, have bare tarsi and copious plumage, and lack feathers around the nostrils.

phasing *See* framing.

phasitron [ELECTR] An electron tube used to frequency-modulate a radio-frequency carrier; internal electrodes are designed to produce a rotating diskshaped corrugated sheet of electrons; audio input is applied to a coil surrounding the glass envelope of the tube, to produce a varying axial magnetic field that gives the desired phase or frequency modulation of the rf carrier input to the tube.

Phasmidae [INV ZOO] A family of the insect order Orthoptera including the walking sticks and leaf insects.

Phasmidea [INV ZOO] An equivalent name for Secernentea.

Phasmidia [INV ZOO] An equivalent name for Secernentea.

phasor [PHYS] **1.** A rotating line used to represent a sinusoidally varying quantity; the length of the line represents the magnitude of the quantity, and its angle with the x-axis at any instant represents the phase. **2.** Any quantity (such as impedance or admittance) which is a complex number. [SOLID STATE] A low-energy collective excitation of the conduction electrons in a metal, corresponding to a slowly varying phase modulation of a charge-density wave.

phasotron *See* cyclotron.

pH electrode [ANALY CHEM] Membrane-type glass electrode used as the hydrogen-ion sensor of most pH meters; the pH-response electrode surface is a thin membrane made of a special glass.

phenacite *See* phenakite.

Phenacodontidae [PALEON] An extinct family of large herbivorous mammals in the order Condylarthra.

phenakite [MINERAL] Be_2SiO_4 A colorless, white, wine-yellow, pink, blue, or brown glassy mineral that crystallizes in the rhombohedral system; used as a minor gemstone. Also spelled phenacite.

Phengodidae [INV ZOO] The fire beetles, a New World family of coleopteran insects in the superfamily Cantharoidea.

phenicochroite *See* phoenicochroite.

phenoclasts [PETR] The larger, conspicuous fragments in a sediment or sedimentary rock, such as cobbles in a conglomerate.

phenocrystalline *See* phaneritic.

phenotype [GEN] The observable characters of an organism.

phenylalanine [BIOCHEM] $C_9H_{11}O_2N$ An essential amino acid, obtained in the levo form by hydrolysis of proteins (as lactalbumin); converted to tyrosine in the normal body. Also known as α-aminohydrocinnamic acid; α-amino-β-phenylpropionic acid; β-phenylalanine.

β-phenylalanine *See* phenylalanine.

phenylketonuria [MED] A hereditary disorder of metabolism, transmitted as an autosomal recessive, in which there is a lack of the enzyme phenylalanine hydroxylase, resulting in excess amounts of phenylalanine in the blood and of excess phenylpyruvic and other acids in the urine. Abbreviated PKU. Also known as phenylpyruvic oligophrenia.

α-phenyl phenacyl *See* desyl.

phenylpyruvic oligophrenia *See* phenylketonuria.

Philips ionization gage [ELECTR] An ionization gage in which a high voltage is applied between two electrodes, and a strong magnetic field deflects the resulting electron stream, increasing the length of the electron path and thus increasing the chance for ionizing collisions of electrons with gas molecules. Abbreviated pig. Also known as cold-cathode ionization gage; Penning gage.

philipstadite [MINERAL] $Ca_2(Fe,Mg)_5(Si,Al)_8O_{22}(OH)_2$ Monoclinic mineral composed of basic silicate of calcium, iron, magnesium, and aluminum; member of the amphibole group.

phillipsite [MINERAL] $(K_2,Na_2CA)Al_2Si_4O_{12}\cdot H_2O$ A white or reddish zeolite mineral crystallizing in the orthorhombic system; occurs in complex fibrous crystals, which make up a large part of the red-clay sediments in the Pacific Ocean.

Phillips screw [DES ENG] A screw having in its head a recess in the shape of a cross; it is inserted or removed with a Phillips screwdriver that automatically centers itself in the screw.

Philomycidae [INV ZOO] A family of pulmonate gastropods composed of slugs.

Philopteridae [INV ZOO] A family of biting lice (Mallophaga) that are parasitic on most land birds and water birds.

phlebitis [MED] Inflammation of a vein.

phleger corer [ENG] A device for obtaining ocean bottom cores up to about 4 feet (1.2 meters) in length; consists of an upper tube, main body weight, and tailfin assembly with a check valve that prevents the flow of water into the upper section and a consequent washing out of the core sample while hoisting the corer.

Phloeidae [INV ZOO] The bark bugs, a small neotropical family of hemipteran insects in the superfamily Pentatomoidea.

phloem [BOT] A complex, food-conducting vascular tissue in higher plants; principal conducting cells are sieve elements. Also known as bast; sieve tissue.

phlogopite [MINERAL] $K_2[Mg,Fe(II)]_6(Si_6,Al_2)O_{20}(OH)_4$ A yellow-brown to copper mineral of the mica group occurring in disseminated flakes, foliated masses, or large crystals; hardness is 2.5–3.0 on Mohs scale, and specific gravity is 2.8–3.0. Also known as bronze mica; brown mica.

Phobos [ASTRON] A satellite of Mars; it is the larger of the two satellites, with a diameter of about 24 kilometers.

Phocaenidae [VERT ZOO] The porpoises, a family of marine mammals in the order Cetacea.

Phocidae [VERT ZOO] The seals, a pinniped family of carnivoran mammals in the superfamily Canoidea.

Phodilidae [VERT ZOO] A family of birds in the order Strigiformes; the bay owl (*Pholidus bodius*) is the single species.

Phoebe [ASTRON] A satellite of the planet Saturn; its diameter is judged to be about 320 kilometers; it has an eccentric orbit and retrograde revolution.

phoenicite *See* phoenicochroite.

phoenicochroite [MINERAL] Pb_2CrO_5 A red mineral composed of basic chromate of lead, occurring in crystals and masses. Also known as beresovite; phenicochroite; phoenicite.

Phoenicopteridae [VERT ZOO] The flamingos, a family of long-legged, long-necked birds in the order Ciconiiformes.

Phoenicopteriformes [VERT ZOO] An order comprising the flamingos in some systems of classification.

Phoeniculidae [VERT ZOO] The African wood hoopoes, a family of birds in the order Coraciiformes.

Phoenix [ASTRON] A southern constellation; right ascension 1 hour, declination 50°S. [ORD] An air-to-air solid-propellant guided missile that has both radar and infrared acquisition, a speed of about Mach 5, and a range of about 400 nautical miles (740 kilometers).

Pholadidae [INV ZOO] A family of bivalve mollusks in the subclass Eulamellibranchia; individuals may have one or more dorsal accessory plates, and the visceral mass is attached to the valves in the dorsal portion of the body.

Pholidophoridae [PALEON] A generalized family of extinct fishes belonging to the Pholidophoriformes.

Pholidophoriformes [PALEON] An extinct actinopterygian group composed of mostly small fusiform marine and fresh-water fishes of an advanced holostean level.

Pholidota [VERT ZOO] An order of mammals comprising the living pangolins and their fossil predecessors; characterized by an elongate tubular skull with no teeth, a long protrusive tongue, strong legs, and five-toed feet with large claws.

Phomaceae [MYCOL] The equivalent name for Sphaerioidaceae.

Phomales [MYCOL] The equivalent name for Sphaeropsidales.

phon [ACOUS] A unit of loudness level; the loudness level, in phons, of a sound is numerically equal to the sound pressure level, in decibels, of a 1000-hertz reference tone which is judged by listeners to be equally loud to the sound under evaluation.

phonation [PHYSIO] Production of speech sounds.

phone *See* headphone.

phonemic synthesizer [ENG ACOUS] A voice response system in which each word is abstractly represented as a sequence of expected vowels and consonants, and speech is composed by juxtaposing the expected phonemic sequence for each word with the sequences for the preceding and following words.

phonocardiograph [MED] An instrument that provides a graphic record of heart murmurs and other sounds.

phonograph cartridge *See* phonograph pickup.

phonograph cutter *See* cutter.

phonograph needle *See* stylus.

phonograph pickup [ENG ACOUS] A pickup that converts variations in the grooves of a phonograph record into corresponding electric signals. Also known as cartridge; phonograph cartridge.

phonograph record [ENG ACOUS] A shellac-composition or vinyl-plastic disk, usually 7 or 12 inches (18 or 30 centimeters) in diameter, on which sounds have been recorded as modulations in grooves. Also known as disk; disk recording.

phono jack [ELECTR] A jack designed to accept a phono plug and provide a ground connection for the shield of the conductor connected to the plug.

phonon [SOLID STATE] A quantum of an acoustic mode of thermal vibration in a crystal lattice.

phonon wind [SOLID STATE] A stream of nonthermal phonons that is effective in propelling electron-hole droplets through a crystal.

phono plug [ELECTR] A plug designed for attaching to the end of a shielded conductor, for feeding audio-frequency signals from a phonograph or other a-f source to a mating phono jack on a preamplifier or amplifier.

phonoreception [PHYSIO] The perception of sound through specialized sense organs.

Phoridae [INV ZOO] The hump-backed flies, a family of cyclorrhaphous dipteran insects in the series Aschiza.

phorogenesis [GEOL] The shifting or slipping of the earth's crust relative to the mantle.

Phoronida [INV ZOO] A small, homogeneous group, or phylum, of animals having an elongate body, a crown of tentacles surrounding the mouth, and the anus occurring at the level of the mouth.

phosphatase [BIOCHEM] An enzyme that catalyzes the hydrolysis and synthesis of phosphoric acid esters and the transfer of phosphate groups from phosphoric acid to other compounds.

phosphate [CHEM] 1. Generic term for any compound containing a phosphate group (PO_4^{3-}), such as potassium phosphate, K_3PO_4. 2. Generic term for a phosphate-containing fertilizer material. [MINERAL] A mineral compound characterized by a tetrahedral ionic group of phosphate and oxygen, PO_4^{3-}.

phosphate rock [MATER] 1. A rock that is naturally high enough in phosphorus to be used directly in fertilizer manufacturing. 2. The beneficiated concentrate of a phosphate deposit.

phosphatide See phospholipid.

phosphatization [GEOCHEM] Conversion to a phosphate or phosphates; for example, the diagenetic replacement of limestone, mudstone, or shale by phosphate-bearing solutions, producing phosphates of calcium, aluminum, or iron.

phosphazene [ORG CHEM] A ring or chain polymer that contains alternating phosphorus and nitrogen atoms, with two substituents on each phosphorus atom.

phospholipid [BIOCHEM] Any of a class of esters of phosphoric acid containing one or two molecules of fatty acid, an alcohol, and a nitrogenous base. Also known as phosphatide.

phosphor See luminophor.

phosphor dot [ELECTR] One of the tiny dots of phosphor material that are used in groups of three, one group for each primary color, on the screen of a color television picture tube.

phosphorescence [ATOM PHYS] 1. Luminescence that persists after removal of the exciting source. Also known as afterglow. 2. Luminescence whose decay, upon removal of the exciting source, is temperature-dependent.

phosphorite [PETR] A sedimentary rock composed chiefly of phosphate minerals.

phosphorization [GEOCHEM] Impregnation or combination with phosphorus or a compound of phosphorus; for example, the diagenetic process of phosphatization.

phosphorus [CHEM] A nonmetallic element, symbol P, atomic number 15, atomic weight 30.98; used to manufacture phosphoric acid, in phosphor bronzes, incendiaries, pyrotechnics, matches, and rat poisons; the white (or yellow) allotrope is a soft waxy solid melting at 44.5°C, is soluble in carbon disulfide, insoluble in water and alcohol, and is poisonous and self-igniting in air; the red allotrope is an amorphous powder subliming at 416°C, igniting at 260°C, is insoluble in all solvents, and is nonpoisonous; the black allotrope comprises lustrous crystals similar to graphite, and is insoluble in most solvents.

phosphorylation [ORG CHEM] The esterification of compounds with phosphoric acid.

phot [OPTICS] A unit of illumination equal to the illumination of a surface, 1 square centimeter in area, on which there is a luminous flux of 1 lumen, or the illumination on a surface all points of which are at a distance of 1 centimeter from a uniform point source of 1 candela. Also known as centimeter-candle (deprecated usage).

photic zone [ECOL] The uppermost layer of a body of water (approximately the upper 100 meters) that receives enough sunlight to permit the occurrence of photosynthesis.

Photidae [INV ZOO] A family of amphipod crustaceans in the suborder Gammaridea.

photoacoustic spectroscopy [SPECT] A spectroscopic technique for investigating solid and semisolid materials, in which the sample is placed in a closed chamber filled with a gas such as air and illuminated with monochromatic radiation of any desired wavelength, with intensity modulated at some suitable acoustic frequency; absorption of radiation results in a periodic heat flow from the sample, which generates sound that is detected by a sensitive microphone attached to the chamber. Abbreviated PAS.

photocathode [ELECTR] A photosensitive surface that emits electrons when exposed to light or other suitable radiation; used in phototubes, television camera tubes, and other light-sensitive devices.

photocell [ELECTR] A solid-state photosensitive electron device whose current-voltage characteristic is a function of incident radiation. Also known as electric eye; photoelectric cell.

photochemical smog [METEOROL] Chemical pollutants in the atmosphere resulting from chemical reactions involving hydrocarbons and nitrogen oxides in the presence of sunlight.

photochemistry [PHYS CHEM] The study of the effects of light on chemical reactions.

photochromism [CHEM] The ability of a chemically treated plastic or other transparent material to darken reversibly in strong light.

photoclinometer [ENG] A directional surveying instrument which records photographically the direction and magnitude of well deviations from the vertical.

photoconduction [SOLID STATE] An increase in conduction of electricity resulting from absorption of electromagnetic radiation.

photoconductive cell [ELECTR] A device for detecting or measuring electromagnetic radiation by variation of the conductivity of a substance (called a photoconductor) upon absorption of the radiation by this substance. Also known as photoresistive cell; photoresistor.

photoconductivity [SOLID STATE] The increase in electrical conductivity displayed by many nonmetallic solids when they absorb electromagnetic radiation.

photoconductor   [SOLID STATE] A nonmetallic solid whose conductivity increases when it is exposed to electromagnetic radiation.

photoconductor diode   *See* photodiode.

photocoupler   *See* optoisolator.

photodetector   [ELECTR] A detector that responds to radiant energy; examples include photoconductive cells, photodiodes, photoresistors, photoswitches, phototransistors, phototubes, and photovoltaic cells. Also known as light-sensitive cell; light-sensitive detector; light sensor photodevice; photoelectric detector; photosensor.

photodiode   [ELECTR] A semiconductor diode in which the reverse current varies with illumination; examples include the alloy-junction photocell and the grown-junction photocell. Also known as photoconductor diode.

photodissociation   [PHYS CHEM] The removal of one or more atoms from a molecule by the absorption of a quantum of electromagnetic energy.

photoelastic effect   [OPTICS] Changes in optical properties of a transparent dielectric when it is subjected to mechanical stress, such as mechanical birefringence. Also known as photoelasticity.

photoelasticity   [OPTICS] **1.** An experimental technique for the measurement of stresses and strains in material objects by means of the phenomenon of mechanical birefringence. **2.** *See* photoelastic effect.

photoelectret   [SOLID STATE] An electret produced by the removal of light from an illuminated photoconductor in an electric field.

photoelectric cell   *See* photocell.

photoelectric color comparator   *See* color comparator.

photoelectric constant   [ELECTR] The ratio of the frequency of radiation causing emission of photoelectrons to the voltage corresponding to the energy absorbed by a photoelectron; equal to Planck's constant divided by the electron charge.

photoelectric detector   *See* photodetector.

photoelectric effect   *See* photoelectricity.

photoelectric electron-multiplier tube   *See* multiplier phototube.

photoelectric infrared radiation   *See* near-infrared radiation.

photoelectricity   [ELECTR] The liberation of an electric charge by electromagnetic radiation incident on a substance; includes photoemission, photoionization, photoconduction, the photovoltaic effect, and the Auger effect (an internal photoelectric process). Also known as photoelectric effect; photoelectric process.

photoelectric process   *See* photoelectricity.

photoelectric relay   [ELECTR] A relay combined with a phototube and amplifier, arranged so changes in incident light on the phototube make the relay contacts open or close. Also known as light relay.

photoelectric tube   *See* phototube.

photoelectrochemistry   [PHYS CHEM] The study of the interaction between impinging light energy and the electropotential of the chemical changes in the electrode, electrolytic solution, or a photosensitive membrane.

photoelectromagnetic effect   [ELECTR] The effect whereby, when light falls on a flat surface of an intermetallic semiconductor located in a magnetic field that is parallel to the surface, excess hole-electron pairs are created, and these carriers diffuse in the direction of the light but are deflected by the magnetic field to give a current flow through the semiconductor that is at right angles to both the light rays and the magnetic field.

photoelectron spectroscopy   [SPECT] The branch of electron spectroscopy concerned with the energy analysis of photoelectrons ejected from a substance as the direct result of bombardment by ultraviolet radiation or x-radiation.

photoemission   [ELECTR] The ejection of electrons from a solid (or less commonly, a liquid) by incident electromagnetic radiation. Also known as external photoelectric effect.

photoemissivity   [ELECTR] The property of a substance that emits electrons when struck by light.

photoemitter   [SOLID STATE] A material that emits electrons when sufficiently illuminated.

photoengraving   [GRAPHICS] The technique of producing relief plates such as halftones or zinc etchings by photography; a metal plate is coated with a photosensitive emulsion and exposed to light under a reversed positive; the picture is developed by dissolving away the portion of the emulsion not acted upon by the light, and the plate is etched.

photoenlarger   *See* enlarger.

photoferroelectric effect   [SOLID STATE] An effect observed in ferroelectric ceramics such as PLZT materials, in which light at or near the band-gap energy of the material has an effect on the electric field in the material created by an applied voltage, and, at a certain value of the voltage, also influences the degree of ferroelectric remanent polarization. Abbreviated PFE.

photoflash lamp   [ELEC] A lamp consisting of a glass bulb filled with finely shredded aluminum foil in an atmosphere of oxygen; when the foil is ignited by a low-voltage dry cell, it burns with a burst of high-intensity light of short time duration and with definitely regulated time characteristics.

photoflood lamp   [ELEC] An incandescent lamp used in photography which has a high-temperature filament, so that it gives high illumination and high color temperature for a short lifetime.

photofluorography   *See* fluorography.

photogrammetry   [ENG] **1.** The science of making accurate measurements and maps from aerial photographs. **2.** The practice of obtaining surveys by means of photography.

photographic emulsion   [GRAPHICS] Microscopic grains of light-sensitive silver halide suspended in a gelatin surface on paper, plastic, metal, or glass; used to coat photographic film.

photography   [GRAPHICS] The process of forming visible images directly or indirectly by the action of light or other forms of radiation on sensitive surfaces.

photoionization   [PHYS CHEM] The removal of one or more electrons from an atom or molecule by absorption of a photon of visible or ultraviolet light. Also known as atomic photoelectric effect.

photoisolator   *See* optoisolator.

photolysis   [PHYS CHEM] The use of radiant energy to produce chemical changes.

photometer   [ENG] An instrument used for making measurements of light or electromagnetic radiation, in the visible range.

photometric parallax   [ASTRON] The annual parallax of a star too far away for its parallax to be measured directly, as calculated from its apparent magnitude and its absolute magnitude inferred from its spectral type.

photometric titration   [ANALY CHEM] A titration in which the titrant and solution cause the formation of a metal complex accompanied by an observable change in light absorbance by the titrated solution.

photometry [OPTICS] The calculation and measurement of quantities describing light, such as luminous intensity, luminous flux, luminous flux density, light distribution, color, absorption factor, spectral distribution, and the reflectance and transmittance of light; sometimes taken to include measurement of near-infrared and near-ultraviolet radiation as well as visible light.

photomicrography [GRAPHICS] The photography of the image formed by the microscope.

photomorphogenesis [BOT] The control exerted by light over growth, development, and differentiation of plants that is independent of photosynthesis.

photomultiplier *See* multiplier phototube.

photomultiplier tube *See* multiplier phototube.

photon [OPTICS] *See* troland. [QUANT MECH] A massless particle, the quantum of the electromagnetic field, carrying energy, momentum, and angular momentum. Also known as light quantum.

photon emission spectrum [PHYS] The relative numbers of optical photons emitted by a scintillator material per unit wavelength as a function of wavelength; the emission spectrum may also be given in alternative units such as wave number, photon energy, or frequency.

photon microscope *See* optical microscope.

photophore gland [VERT ZOO] A highly modified integumentary gland which develops into a luminous organ composed of a lens and a light-emitting gland; occurs in deep-sea teleosts and elasmobranchs.

photopic vision *See* foveal vision.

photopolymer [PHYS CHEM] Any of a class of light-sensitive polymers which undergo a spontaneous and permanent change in physical properties on exposure to light.

photoreception [PHYSIO] The process of absorption of light energy by plants and animals and its utilization for biological functions, such as photosynthesis and vision.

photoresist [GRAPHICS] A light-sensitive coating that is applied to a substrate or board, exposed, and developed prior to chemical etching; the exposed areas serve as a mask for selective etching.

photoresistive cell *See* photoconductive cell.

photoresistor *See* photoconductive cell.

photorespiration [BIOCHEM] Respiratory activity taking place in plants during the light period; CO_2 is released and O_2 is taken up, but no useful form of energy, such as adenosinetriphosphate, is derived.

photosensor *See* photodetector.

photosphere [ASTRON] The intensely bright portion of the sun visible to the unaided eye; it is a shell a few hundred miles in thickness marking the boundary between the dense interior gases of the sun and the more diffuse cooler gases in the outer portions of the sun.

photospheric granulation *See* granulation.

photosynthesis [BIOCHEM] Synthesis of chemical compounds in light, especially the manufacture of organic compounds (primarily carbohydrates) from carbon dioxide and a hydrogen source (such as water), with simultaneous liberation of oxygen, by chlorophyll-containing plant cells.

phototaxis [BIOL] Movement of a motile organism or free plant part in response to light stimulation.

phototelegraphy *See* facsimile.

photothermal conversion [ENG] Conversion of optical radiation into thermal energy by a photoabsorptive or photoselective material.

phototransistor [ELECTR] A junction transistor that may have only collector and emitter leads or also a base lead, with the base exposed to light through a tiny lens in the housing; collector current increases with light intensity, as a result of amplification of base current by the transistor structure.

phototronic photocell *See* photovoltaic cell.

phototroph [BIOL] An organism that utilizes light as a source of metabolic energy.

phototropism [BOT] A growth-mediated response of a plant to stimulation by visible light. [SOLID STATE] A reversible change in the structure of a solid exposed to light or other radiant energy, accompanied by a change in color. Also known as phototropy.

phototropy *See* phototropism.

phototube [ELECTR] An electron tube containing a photocathode from which electrons are emitted when it is exposed to light or other electromagnetic radiation. Also known as electric eye; light-sensitive tube; photoelectric tube.

phototypesetter [GRAPHICS] Machine for placing individual characters on photographic film.

photovoltaic cell [ELECTR] A device that detects or measures electromagnetic radiation by generating a potential at a junction (barrier layer) between two types of material, upon absorption of radiant energy. Also known as barrier-layer cell; barrier-layer photocell; boundary-layer photocell; photronic photocell.

photovoltaic effect [ELECTR] The production of a voltage in a nonhomogeneous semiconductor, such as silicon, or at a junction between two types of material, by the absorption of light or other electromagnetic radiation.

photovoltaic meter [ELECTR] An exposure cell in which a photovoltaic cell produces a current proportional to the light falling on the cell, and this current is measured by a sensitive microammeter.

Phoxichilidiidae [INV ZOO] A family of marine arthropods in the subphylum Pycnogonida; typically, chelifores are present, palpi are lacking, and ovigers have five to nine joints in males only.

Phoxocephalidae [INV ZOO] A family of amphipod crustaceans in the suborder Gammaridea.

Phractolaemidae [VERT ZOO] A family of tropical African fresh-water fishes in the order Gonorynchiformes.

Phragmidiothrix [MICROBIO] A genus of sheathed bacteria; cells are nonmotile, and sheaths are attached and not encrusted with iron or manganese oxides.

Phragmobasidiomycetes [MYCOL] An equivalent name for Heterobasidiomycetidae.

Phragmosporae [MYCOL] A spore group of the Fungi Imperfecti with three- to many-celled spores.

phreatic [GEOL] Of a volcanic explosion of material such as steam or mud, not being incandescent.

phreatic surface *See* water table.

phreatic water [HYD] Groundwater in the zone of saturation.

phreatic-water discharge *See* groundwater discharge.

phreatic zone *See* zone of saturation.

Phreatoicidae [INV ZOO] A family of isopod crustaceans in the suborder Phreatoicoidea in which only the left mandible retains a lacinia mobilis.

Phreatoicoidea [INV ZOO] A suborder of the Isopoda having a subcylindrical body that appears laterally compressed, antennules shorter than the antennae, and the first thoracic segment fused with the head.

Phrynophiurida [INV ZOO] An order of the Ophiuroidea in which the vertebrae usually articulate by means of hourglass-shaped surfaces, and the arms are able to coil upward or downward in the vertical plane.

phthisis [MED] **1.** Any disease characterized by emaciation and loss of strength. **2.** *See* tuberculosis.

phycite *See* erythritol.

Phycitinae [INV ZOO] A large subfamily of moths in the family Pyralididae in which the frenulum of the female is a simple spine rather than a bundle of bristles.

Phycomycetes [MYCOL] A primitive class of true fungi belonging to the Eumycetes; they lack regularly spaced septa in the actively growing portions of the plant body, and have the sporangiospore, produced in the sporangium by cleavage, as the fundamental, asexual reproductive unit.

Phycosecidae [INV ZOO] A small family of coleopteran insects of the superfamily Cucujoidea, including five species found in New Zealand, Australia, and Egypt.

Phylactolaemata [INV ZOO] A class of fresh-water ectoproct bryozoans; individuals have lophophores which are U-shaped in basal outline, and relatively short, wide zooecia.

phyletic evolution [EVOL] The gradual evolution of population without separation into isolated parts.

phyllite [PETR] A metamorphic rock intermediate in grade between slate and schist, and derived from argillaceous sediments; has a silky sheen on the cleavage surface.

Phyllobothrioidea [INV ZOO] The equivalent name for Tetraphyllidea.

Phyllodocidae [INV ZOO] A leaf-bearing family of errantian annelids in which the species are often brilliantly iridescent and are highly motile.

Phyllogoniaceae [BOT] A family of mosses in the order Isobryales in which the leaves are equitant.

Phyllolepida [PALEON] A monogeneric order of placoderms from the late Upper Devonian in which the armor is broad and low with a characteristic ornament of concentric and transverse ridges on the component plates.

Phyllophoridae [INV ZOO] A family of dendrochirotacean holothurians in the order Dendrochirotida having a rather naked skin and a complex calcareous ring.

phyllosilicate [MINERAL] A structural type of silicate mineral in which flat sheets are formed by the sharing of three of the four oxygen atoms in each tetrahedron with neighboring tetrahedrons. Also known as layer silicate; sheet mineral; sheet silicate.

Phyllosticales [MYCOL] An equivalent name for Sphaeropsidales.

Phyllostomatidae [VERT ZOO] The New World leafnosed bats (Chiroptera), a large tropical and subtropical family of insect- and fruit-eating forms with narrow, pointed ears.

Phylloxerinae [INV ZOO] A subfamily of homopteran insects in the family Chermidae in which the sexual forms lack mouthparts, and the parthenogenetic females have a beak but the digestive system is closed, and no honeydew is produced.

phylogeny [EVOL] The evolutionary or ancestral history of organisms.

phylum [SYST] A major taxonomic category in classifying animals (and plants in some systems), composed of groups of related classes.

Phymatidae [INV ZOO] A family of carnivorous hemipteran insects characterized by strong, thick forelegs.

Phymosomatidae [INV ZOO] A family of echinacean echinoderms in the order Phymosomatoida with imperforate crenulate tubercles; one surviving genus is known.

Phymosomatoida [INV ZOO] An order of Echinacea with a stirodont lantern and diademoid ambulacral plates.

Physalopteridae [INV ZOO] A family of parasitic nematodes in the superfamily Spiruroidea.

Physaraceae [MYCOL] A family of slime molds in the order Physarales.

Physarales [MYCOL] An order of Myxomycetes in the subclass Myxogastromycetidae.

physical anthropology [ANTHRO] The science that deals with the biological aspects of humankind and their relation to historical or cultural aspects.

physical chemistry [CHEM] The branch of chemistry that deals with the interpretation of chemical phenomena and properties in terms of the underlying physical processes, and with the development of techniques for their investigation.

physical constant [PHYS] A physical quantity which has a fixed and unchanging numerical value.

physical electronics [ELECTR] The study of physical phenomena basic to electronics, such as discharges, thermionic and field emission, and conduction in semiconductors and metals.

physical forecasting *See* numerical forecasting.

physical geography [GEOGR] The branch of geography which deals with the description, analysis, classification, and genetic interpretation of the natural features and phenomena of the earth's surface.

physical geology [GEOL] That branch of geology concerned with understanding the composition of the earth and the physical changes occurring in it, based on the study of rocks, minerals, and sediments, their structures and formations, and their processes of origin and alteration.

physical law [PHYS] A property of a physical phenomenon, or a relationship between the various quantities or qualities which may be used to describe the phenomenon, that applies to all members of a broad class of such phenomena, without exception.

physical libration of the moon *See* lunar libration.

physical meteorology [METEOROL] That branch of meteorology which deals with optical, electrical, acoustical, and thermodynamic phenomena of the atmosphere, its chemical composition, the laws of radiation, and the explanation of clouds and precipitation.

physical oceanography [OCEANOGR] The study of the physical aspects of the ocean, the movements of the sea, and the variability of these factors in relationship to the atmosphere and the ocean bottom.

physical property [CHEM] Property of a compound that can change without involving a change in chemical composition; examples are the melting point and boiling point.

physical record [COMPUT SCI] A set of adjacent data characters recorded on some storage medium, physically separated from other physical records that may be on the same medium by means of some indication that can be recognized by a simple hardware test. Also known as record block.

physics [SCI TECH] The study of those aspects of nature which can be understood in a fundamental way in terms of elementary principles and laws.

physiographic province [GEOL] A region having a pattern of relief features or landforms that differs significantly from that of adjacent regions.

physiological dead space See dead space.

physiology [BIOL] The study of the basic activities that occur in cells and tissues of living organisms by using physical and chemical methods.

physisorption [PHYS CHEM] A physical adsorption process in which there are van der Waals forces of interaction between gas or liquid molecules and a solid surface.

Physopoda [INV ZOO] The equivalent name for Thysanoptera.

Physosomata [INV ZOO] A superfamily of amphipod crustaceans in the suborder Hyperiidea; the eyes are small or rarely absent, and the inner plates of the maxillipeds are free at the apex.

Phytalmiidae [INV ZOO] A family of myodarian cyclorrhaphous dipteran insects in the subsection Acalypteratae.

Phytamastigophorea [INV ZOO] A class of the subphylum Sarcomastigophora, including green and colorless phytoflagellates.

phytochrome [BIOCHEM] A protein plant pigment which serves to direct the course of plant growth and development in response variously to the presence or absence of light, to photoperiod, and to light quality.

phytohormone See plant hormone.

Phytomastigina [INV ZOO] The equivalent name for Phytamastigophorea.

Phytomonadida [INV ZOO] The equivalent name for Volvocida.

phytoplankton [ECOL] Planktonic plant life.

Phytosauria [PALEON] A suborder of Late Triassic long-snouted aquatic thecodonts resembling crocodiles but with posteriorly located external nostrils, absence of a secondary palate, and a different structure of the pelvic and pectoral girdles.

Phytoseiidae [INV ZOO] A family of the suborder Mesostigmata.

phytotoxin [BIOCHEM] 1. A substance toxic to plants. 2. A toxin produced by plants.

phytotron [BOT] An apparatus for the growth of plants under a variety of controlled environmental conditions.

pi [MATH] The irrational number which is the ratio of the circumference of any circle to its radius; an approximation is 3.14159. Symbolized π.

Piacention See Plaisancian.

pia mater [ANAT] The vascular membrane covering the surface of the brain and spinal cord.

pi bonding [PHYS CHEM] Covalent bonding in which the greatest overlap between atomic orbitals is along a plane perpendicular to the line joining the nuclei of the two atoms.

Picidae [VERT ZOO] The woodpeckers, a large family of birds in the order Piciformes; adaptive modifications include a long tongue and hyoid mechanism, and stiffened tail feathers.

Piciformes [VERT ZOO] An order of birds characterized by the peculiar arrangement of the tendons of the toes.

Picinae [VERT ZOO] The true woodpeckers, a subfamily of the Picidae.

pickeringite [MINERAL] $MgAl_2(SO_4)_4 \cdot 22H_2O$ A white or faintly colored mineral composed of hydrous sulfate of magnesium and aluminum, occurring in fibrous masses.

pick hammer [DES ENG] A hammer with a point at one end of the head and a blunt surface at the other end.

pickling [CHEM ENG] A method of preparing hides for tanning by immersion in a salt solution with a pH of 2.5 or less. [FOOD ENG] A method of preserving food by using salt, sugar, spices, and acetic acid. [MET] Preferential removal of oxide or mill scale from the surface of a metal by immersion usually in an acidic or alkaline solution.

pickoff [ELECTR] A device used to convert mechanical motion into a proportional electric signal. [MECH ENG] A mechanical device for automatic removal of the finished part from a press die.

pickup [AERO ENG] A potentiometer used in an automatic pilot to detect the motion of the airplane around the gyro and initiate corrective adjustments. [ELEC] 1. A device that converts a sound, scene, measurable quantity, or other form of intelligence into corresponding electric signals, as in a microphone, phonograph pickup, or television camera. 2. The minimum current, voltage, power, or other value at which a relay will complete its intended function. 3. Interference from a nearby circuit or system. [MET] Transfer of metal from the work to the tool, or from the tool to the work, during a forming operation. [NUC PHYS] A type of nuclear reaction in which the incident particle takes a nucleon from the target nucleus and proceeds with this nucleon bound to itself.

pickup tube See camera tube.

pico- [MATH] A prefix meaning 10^{-12}; used with metric units. Also known as micromicro-.

picornavirus [VIROL] A viral group made up of small (18–30 nanometers), ether-sensitive viruses that lack an envelope and have a ribonucleic acid genome; among subgroups included are enteroviruses and rhinoviruses, both of human origin.

picotite [MINERAL] A dark-brown variety of hercynite that contains chromium and is commonly found in dunites. Also known as chrome spinel.

picrite [PETR] A medium- to fine-grained igneous rock composed chiefly of olivine, with smaller amounts of pyroxene, hornblende, and plagioclase felspar.

Picrodendraceae [BOT] A small family of dicotyledonous plants in the order Juglandales characterized by unisexual flowers borne in catkins, four apical ovules in a superior ovary, and trifoliate leaves.

picromerite [MINERAL] $K_2Mg(SO_4)_2 \cdot 6H_2O$ A white mineral composed of hydrous sulfate of magnesium and potassium, occurring as crystalline encrustations.

Pictor [ASTRON] A southern constellation; right ascension 6 hours, declination 55°S. Also known as Easel.

picture black See black signal.

picture compression [COMPUT SCI] The elimination of redundant information from a digital picture through the use of efficient encoding techniques in which frequently occurring gray levels or blocks of gray levels are represented by short codes and infrequently occurring ones by longer codes.

picture element [ELECTR] 1. That portion, in facsimile, of the subject copy which is seen by the scanner at any instant; it can be considered a square area having dimensions equal to the width of the scanning line. 2. In television, any segment of a scanning line, the dimension of which along the line is exactly equal to the nominal line width; the area which is being explored at any instant in the scanning process. Also known as critical area; elemental area; recording spot; scanning spot.

picture processing See image processing.

picture segmentation [COMPUT SCI] The division of a complex picture into parts corresponding to regions or objects, so that the picture can then be described in terms of the parts, their properties, and their spatial relationships. Also known as scene analysis.

picture transmitter See visual transmitter.

picture tube [ELECTR] A cathode-ray tube used in television receivers to produce an image by varying the electron-beam intensity as the beam is deflected from side to side and up and down to scan a raster on the fluorescent screen at the large end of the tube. Also known as kinescope; television picture tube.

Picumninae [VERT ZOO] The piculets, a subfamily of the avian family Picidae.

piecewise linear topology See combinatorial topology.

piedmont [GEOL] Lying or formed at the base of a mountain or mountain range, as a piedmont terrace or a piedmont pediment.

piedmont glacier [HYD] A thick, continuous ice sheet formed at the base of a mountain range by the spreading out and coalescing of valley glaciers from higher mountain elevations.

piedmont ice [HYD] An ice sheet formed by the joining of two or more glaciers on a comparatively level plain at the base of the mountains down which the glaciers descended; it may be partly afloat.

piedmont interstream flat See pediment.

piedmontite See piemontite.

pi electron [PHYS CHEM] An electron which participates in pi bonding.

piemontite [MINERAL] $Ca_2(Al, Mn^{3+}, Fe)_3Si_3O_{12}(OH)$ Reddish-brown epidote mineral that contains manganese. Also known as manganese epidote; piedmontite.

pier [BUILD] A concrete block that supports the floor of a building. [CIV ENG] 1. A vertical, rectangular or circular support for concentrated loads from an arch or bridge superstructure. 2. A structure with a platform projecting from the shore into navigable waters for mooring vessels.

piercement dome See diapir.

Pierce oscillator [ELECTR] Oscillator in which a piezoelectric crystal unit is connected between the grid and the plate of an electron tube, in what is basically a Colpitts oscillator, with voltage division provided by the grid-cathode and plate-cathode capacitances of the circuit.

piercing See fusion piercing.

piercing fold See diapir.

pier foundation See caisson foundation.

Pieridae [INV ZOO] A family of lepidopteran insects in the superfamily Papilionoidea including white, sulfur, and orange-tip butterflies; characterized by the lack of a prespiracular bar at the base of the abdomen.

Piesmatidae [INV ZOO] The ash-gray leaf bugs, a family of hemipteran insects belonging to the Pentatomorpha.

Piesmidae [INV ZOO] A small family of hemipteran insects in the superfamily Lygaeoidea.

piezoelectric effect [SOLID STATE] 1. The generation of electric polarization in certain dielectric crystals as a result of the application of mechanical stress. 2. The reverse effect, in which application of a voltage between certain faces of the crystal produces a mechanical distortion of the material.

piezoelectric hysteresis [SOLID STATE] Behavior of a piezoelectric crystal whose electric polarization depends not only on the mechanical stress to which the crystal is subjected, but also on the previous history of this stress.

piezoelectricity [SOLID STATE] Electricity or electric polarization resulting from the piezoelectric effect.

piezoelectric loudspeaker See crystal loudspeaker.

piezoelectric microphone See crystal microphone.

piezoelectric oscillator See crystal oscillator.

piezoelectric pickup See crystal pickup.

piezoelectric semiconductor [SOLID STATE] A semiconductor exhibiting the piezoelectric effect, such as quartz, Rochelle salt, and barium titanate.

piezoelectric transducer [ELECTR] A piezoelectric crystal used as a transducer, either to convert mechanical or acoustical signals to electric signals, as in a microphone, or vice versa, as in ultrasonic metal inspection.

piezometer [ENG] 1. An instrument for measuring fluid pressure, such as a gage attached to a pipe containing a gas or liquid. 2. An instrument for measuring the compressibility of materials, such as a vessel that determines the change in volume of a substance in response to hydrostatic pressure.

piezooptical effect [OPTICS] The change produced in the index of refraction of a light-transmitting material by externally applied stress.

pig [ELECTR] 1. An ion source based on the same principle as the Philips ionization gage. 2. See Philips ionization gage. [ENG] In-line scraper (brush, blade cutter, or swab) forced through pipelines by fluid pressure; used to remove scale, sand, water, and other foreign matter from the interior surfaces of the pipe. [MET] A crude metal casting prepared for storage, transportation, or remelting. [NUCLEO] A heavily shielded container, usually lead, used to ship or store radioisotopes and other radioactive materials. [VERT ZOO] Any wild or domestic mammal of the superfamily Suoidea in the order Artiodactyla; toes terminate in nails which are modified into hooves, the tail is short, and the body is covered sparsely with hair which is frequently bristlelike.

pigeonite [MINERAL] $(Mg, Fe^{2+}, Ca)(MgFe^{2+})Si_2O_6$ Clinopyroxene mineral species intermediate in composition between clinoenstatite and diopside, found in basic igneous rocks.

pig iron [MET] 1. Crude, high-carbon iron produced by reduction of iron ore in a blast furnace. 2. Cast iron in the form of pigs.

pigment [BIOCHEM] Any coloring matter in plant or animal cells. [MATER] A solid that reflects light of certain wavelengths while absorbing light of other wavelengths, without producing appreciable luminescence; used to impart color to other materials.

pigmentation [PHYSIO] The normal color of the body and its organs, resulting from a summation of the natural color of the tissue, the pigments deposited therein, and the pigments carried through the blood bathing the tissue.

pika [VERT ZOO] Any member of the family Ochotonidae, which includes 14 species of lagomorphs resembling rabbits but having a vestigial tail and short, rounded ears.

pike [GEOL] A mountain or hill which has a peaked summit. [VERT ZOO] Any of about five species of predatory fish which compose the family Esocidae in the order Clupeiformes; the body is cylindrical and compressed, with cycloid scales that have deeply scalloped edges.

Pilacraceae [MYCOL] A family of Basidiomycetes.

Pilargidae [INV ZOO] A family of small, short, depressed errantian polychaete annelids.

pilaster [CIV ENG] A vertical rectangular architectural member that is structurally a pier and architecturally a column.

pile [ENG] A long, heavy timber, steel, or reinforced concrete post that has been driven, jacked, jetted, or cast vertically into the ground to support a load. [NUCLEO] *See* nuclear reactor. [TEXT] Loops on a fabric surface.

pile hammer [MECH ENG] The heavy weight of a pile driver that depends on gravity for its striking power and is used to drive piles into the ground. Also known as drop hammer.

pileup [ELECTR] A set of moving and fixed contacts, insulated from each other, formed as a unit for incorporation in a relay or switch. Also known as stack.

Pilidae [INV ZOO] A family of fresh-water snails in the order Pectinibranchia.

pillar [CIV ENG] A column for supporting part of a structure. [GEOL] 1. A natural formation shaped like a pillar. 2. A joint block produced by columnar jointing. [MIN ENG] An area of coal or ore left to support the overlying strata or hanging wall in a mine.

pillow lava [GEOL] Any lava characterized by pillow structure and presumed to have formed in a subaqueous environment. Also known as ellipsoidal lava.

pillow structure [GEOL] A primary sedimentary structure that resembles a pillow in size and shape. Also known as mammillary structure. [PETR] A pillow-shaped structure visible in some extrusive lavas attributed to the congealment of lava under water.

pilot [AERO ENG] 1. A person who handles the controls of an aircraft or spacecraft from within the craft, and guides or controls the craft in flight. 2. A mechanical system designed to exercise control functions in an aircraft or spacecraft. [COMMUN] 1. In a transmission system, a signal wave, usually single frequency, transmitted over the system to indicate or control its characteristics. 2. Instructions, in tape relay, appearing in routing line, relative to the transmission or handling of that message. [MECH ENG] A cylindrical steel bar extending through, and about 8 inches (20 centimeters) beyond the face of, a reaming bit; it acts as a guide that follows the original unreamed part of the borehole and hence forces the reaming bit to follow, and be concentric with, the smaller-diameter, unreamed portion of the original borehole. [NAV] 1. A person who directs the movements of a vessel through pilot waters, usually a person who has demonstrated extensive knowledge of channels, aids to navigation, dangers to navigation, and so on, in a particular area and is licensed for that area. 2. A book of sailing directions; for waters of the United States and its possessions, the books are prepared by the U.S. Coast and Geodetic Survey, and are called coast pilots. 3. The person who flies aircraft.

pilotage [NAV] 1. A marine pilot's fee. 2. The process of directing a vehicle by reference to recognizable landmarks or soundings; these observations may be made by optical, aural, mechanical, or electronic means.

pilot hole [ENG] A small hole drilled ahead of a larger borehole.

pilot lamp [ELEC] A small lamp used to indicate that a circuit is energized. Also known as pilot light.

pilot light [ELEC] *See* pilot lamp. [ENG] A small, constantly burning flame used to ignite a gas burner.

pilot system [COMPUT SCI] A system for evaluating new procedures for handling data in which a sample that is representative of the data to be handled is processed.

Piltdown man [PALEON] An alleged fossil man based on fragments of a skull and mandible that were eventually discovered to constitute a skillful hoax.

pilus [ANAT] A hair. [BIOL] A fine, slender, hairlike body. [MICROBIO] Any filamentous appendage other than flagella on certain gram-negative bacteria. Also known as fimbria.

pimento [BOT] *Capsicum annuum.* A type of pepper in the order Polemoniales grown for its thick, sweet-fleshed red fruit.

pi meson [PARTIC PHYS] 1. Collective name for three semistable mesons which have charges of $+1$, 0, and -1 times the proton charge, and form a charge multiplet, with an approximate mass of 138 MeV (million electron volts), spin 0, negative parity, negative G parity, and positive charge parity (for the neutral meson). Also known as pion. Symbolized π. 2. Any meson belonging to an isospin triplet with hypercharge 0, negative G parity, and positive charge parity (for the neutral meson).

pi-mu atom *See* pionium.

pin [DES ENG] 1. A cylindrical fastener made of wood, metal, or other material used to join two members or parts with freedom of angular movement at the joint. 2. A short, pointed wire with a head used for fastening fabrics, paper, or similar materials. [ELECTR] A terminal on an electron tube, semiconductor, integrated circuit, plug, or connector. Also known as base pin; prong.

piña [BOT] A fiber obtained from the large leaves of the pineapple plant. Also known as pineapple fiber.

pinacoid [CRYSTAL] An open crystal form that comprises two parallel faces.

Pinales [BOT] An order of gymnospermous woody trees and shrubs in the class Pinopsida, including pine, spruce, fir, cypress, yew, and redwood; the largest plants are the conifers.

Pinatae *See* Pinopsida.

pinch [ENG] The closing-in of borehole walls before casing is emplaced, resulting from rock failure when drilling in formations having a low compressional strength. [GEOL] Thinning of a rock layer, as where a vein narrows. [MIN ENG] *See* horseback; squeeze.

pinch effect [ELEC] Manifestation of the magnetic self-attraction of parallel electric currents, such as constriction of ionized gas in a discharge tube, or constriction of molten metal through which a large current is flowing. Also known as cylindrical pinch; magnetic pinch; rheostriction.

pinch-off *See* cutoff.

pinch resistor [ELECTR] A silicon integrated-circuit resistor produced by diffusing an *n*-type layer over a *p*-type resistor; this narrows or pinches the resistive channel, thereby increasing the resistance value.

pin diode [ELECTR] A diode consisting of a silicon wafer containing nearly equal *p*-type and *n*-type impurities, with additional *p*-type impurities diffused from one side and additional *n*-type impurities from the other side; this leaves a lightly doped intrinsic layer in the middle, to act as a dielectric barrier between the *n*-type and *p*-type regions. Also known as power diode.

pine [BOT] Any of the cone-bearing trees composing the genus *Pinus*; characterized by evergreen leaves (needles), usually in tight clusters of two to five.

pineal body [ANAT] An unpaired, elongated, club-shaped, knoblike or threadlike organ attached by a stalk to the roof of the vertebrate forebrain. Also known as conarium; epiphysis.

pineapple [BOT] *Ananas sativus.* A perennial plant of the order Bromeliales with long, swordlike, usually rough-edged leaves and a dense head of small abortive flowers; the fruit is a sorosis that develops from the fleshy inflorescence and ripens into a solid mass, covered by the persistent bracts and crowned by a tuft of leaves.

pineapple fiber *See* piña.

pine nut [BOT] The edible seed borne in the cone of various species of pine (*Pinus*), such as stone pine (*P. pinea*) and piñon pine (*P. cembroides* var. *edulis*).

pinguite *See* nontronite.

Pinicae [BOT] A large subdivision of the Pinophyta, comprising woody plants with a simple trunk and excurrent branches, simple, usually alternate, needlelike or scalelike leaves, and wood that lacks vessels and usually has resin canals.

pinion [MECH ENG] The smaller of a pair of gear wheels or the smallest wheel of a gear train. [VERT ZOO] The distal portion of a bird's wing.

pin junction [ELECTR] A semiconductor device having three regions: *p*-type impurity, intrinsic (electrically pure), and *n*-type impurity.

pinkeye [MED] 1. A contagious, mucopurulent conjunctivitis. 2. *See* catarrhal conjunctivitis.

pink root [PL PATH] A fungus disease of onion and garlic caused by various organisms, especially species of *Phoma* and *Fusarium;* marked by red discoloration of the roots.

pink rot [PL PATH] 1. A fungus disease of potato tubers caused by *Phytophtora erythroseptica* and characterized by wet rot and pink color of the cut surfaces of the tuber upon exposure to air. 2. A rot disease of apples caused by the fungus *Tricothecium roseum.* 3. A watery soft rot of celery caused by the fungus *Sclerotinia sclerotiorum.*

pinna [ANAT] The cartilaginous, projecting flap of the external ear of vertebrates. Also known as auricle.

pinnacle [ARCH] A projection on the highest point of the roof of a building. [GEOL] 1. A sharp-pointed rock rising from the bottom, which may extend above the surface of the water, and may be a hazard to surface navigation; due to the sheer rise from the sea floor, no warning is given by sounding. 2. Any high tower or spire-shaped pillar of rock, alone or cresting a summit.

pinnate [BOT] Having parts arranged like a feather, branching from a central axis.

pinnate joint *See* feather joint.

Pinnipedia [VERT ZOO] A suborder of aquatic mammals in the order Carnivora, including walruses and seals.

Pinnotheridae [INV ZOO] The pea crabs, a family of decapod crustaceans belonging to the Brachygnatha.

pinocytosis [CYTOL] Deprecated term formerly used to describe the process of uptake or internalization of particles, macromolecules, and fluid droplets by living cells; the process is now termed endocytosis.

Pinophyta [BOT] The gymnosperms, a division of seed plants characterized as vascular plants with roots, stems, and leaves, and with seeds that are not enclosed in an ovary but are borne on cone scales or exposed at the end of a stalk.

Pinopsida [BOT] A class of gymnospermous plants in the subdivision Pinicae characterized by entire-margined or slightly toothed, narrow leaves.

pint [MECH] Abbreviated pt. 1. A unit of volume, used in the United States for measurement of liquid substances, equal to ⅛ U.S. gallon, or 29⅞ cubic inches, or $4.73176473 \times 10^{-4}$ cubic meters. Also known as liquid pint (liq pt). 2. A unit of volume used in the United States for measurement of solid substances, equal to ¼64 U.S. bushel, or 107,521/3200 cubic inches, or approximately 5.50610×10^{-4} cubic meters. Also known as dry pint (dry pt). 3. A unit of volume, used in the United Kingdom for measurement of liquid and solid substances, although usually the former, equal to ⅛ imperial gallon, or approximately 5.68261×10^{-4} cubic meters. Also known as imperial pint.

Piobert lines *See* Lüders lines.

pion *See* pi meson.

pionium [PARTIC PHYS] 1. An exotic atom consisting of a muon orbiting about an oppositely charged pion. Also known as pi-mu atom. 2. An exotic atom consisting of an electron in orbit about an oppositely charged pion.

Piophilidae [INV ZOO] The skipper flies, a family of myodarian cyclorrhaphous dipteran insects in the subsection Acalyptratae.

piotine *See* saponite.

pip *See* blip.

pipe [DES ENG] A tube made of metal, clay, plastic, wood, or concrete and used to conduct a fluid, gas, or finely divided solid. [GEOL] 1. A vertical, cylindrical ore body. Also known as chimney; neck; ore chimney; ore pipe; stock. 2. A tubular cavity of varying depth in calcareous rocks, often filled with sand and gravel. 3. A vertical conduit through the crust of the earth below a volcano, through which magmatic materials have passed. Also known as breccia pipe. [MET] 1. The central cavity in an ingot or casting formed by contraction of the metal during solidification. 2. An extrusion defect caused by the oxidized surface of the billet flowing toward the center of the rod at the back end.

Piperaceae [BOT] A family of dicotyledonous plants in the order Piperales characterized by alternate leaves, a solitary ovule, copious perisperm, and scanty endosperm.

Piperales [BOT] An order of dicotyledonous herbaceous plants marked by ethereal oil cells, uniaperturate pollen, and reduced crowded flowers with orthotropous ovules.

piperidic acid *See* γ-aminobutyric acid.

pipe still [CHEM ENG] A petroleum-refinery still in which heat is applied to the oil while it is being pumped through a coil or pipe arranged in a firebox, the oil then running to a fractionator with continuous removal of overhead vapor and liquid bottoms.

pipet [CHEM] Graduated or calibrated tube which may have a center reservoir (bulb); used to transfer known volumes of liquids from one vessel to another; types are volumetric or transfer, graduated, and micro.

Pipidea [VERT ZOO] A family of frogs sometimes included in the suborder Opisthocoela, but more commonly placed in its own suborder, Aglossa; a definitive tongue is lacking, and free ribs are present in the tadpole but they fuse to the vertebrae in the adult.

Pipridae [VERT ZOO] The manakins, a family of colorful, neotropical suboscine birds in the order Passeriformes.

Pirani gage [PHYS] A thermal conductivity gage (where the thermal conductivity of a gas heated by a hot wire varies with pressure) connected to a Wheatstone bridge to measure the resistance of the hot wire, thus the gas pressure; used to measure pressure from 1 to 10^{-3} mmHg (133.32 to 0.13332 newtons per square meter).

Piroplasmea [INV ZOO] A class of parasitic protozoans in the superclass Sarcodina; includes the single genus *Babesia.*

Pisces [ASTRON] A northern constellation; right ascension 1 hour, declination 15°N. Also know as Fishes. [VERT ZOO] The fish and fishlike vertebrates, including the classes Agnatha, Placodermi, Chondrichthyes, and Osteichthyes.

Pisces Volan *See* Volan.

Piscis Australis [ASTRON] A southern constellation; right ascension 22 hours, declination 30°S. Also known as Southern Fish.

Pisionidae [INV ZOO] A small family of errantian polychaete annelids; allies of the scale bearers.

pisolite [PETR] A sedimentary rock composed principally of pisoliths.

pisolith [GEOL] Small, more or less spherical particles found in limestones and dolomites, having a diameter of 2–10 millimeters and often formed of calcium carbonate.

pistachio [BOT] *Pistacia vera.* A small, spreading dioecious evergreen tree with leaves that have three to five broad leaflets, and with large drupaceous fruit; the edible seed consists of a single green kernel covered by a brown coat and enclosed in a tough shell.

pistil [BOT] The ovule-bearing organ of angiosperms; consists of an ovary, a style, and a stigma.

pistillate [BOT] **1.** Having a pistil. **2.** Having pistils but no stamens.

piston [ELECTROMAG] A sliding metal cylinder used in waveguides and cavities for tuning purposes or for reflecting essentially all of the incident energy. Also known as plunger; waveguide plunger. [ENG] *See* force plug. [MECH ENG] A sliding metal cylinder that reciprocates in a tubular housing, either moving against or moved by fluid pressure.

piston attenuator [ELECTROMAG] A microwave attenuator inserted in a waveguide to introduce an amount of attenuation that can be varied by moving an output coupling device along its longitudinal axis.

piston corer [MECH ENG] A steel tube which is driven into the sediment by a free fall and by lead attached to the upper end, and which is capable of recovering undistorted vertical sections of sediment.

piston drill [MECH ENG] A heavy percussion-type rock drill mounted either on a horizontal bar or on a short horizontal arm fastened to a vertical column; drills holes to 6 inches (15 centimeters) in diameter. Also known as reciprocating drill.

piston engine [MECH ENG] A type of engine characterized by reciprocating motion of pistons in a cylinder. Also known as displacement engine; reciprocating engine.

piston meter [ENG] A variable-area, constant-head fluid-flow meter in which the position of the piston, moved by the buoyant force of the liquid, indicates the flow rate. Also known as piston-type area meter.

piston pump [MECH ENG] A pump in which motion and pressure are applied to the fluid by a reciprocating piston in a cylinder. Also known as reciprocating pump.

piston rod [MECH ENG] The rod which is connected to the piston, and moves or is moved by the piston.

piston-type area meter *See* piston meter.

pit [BOT] **1.** A cavity in the secondary wall of a plant cell, formed where secondary deposition has failed to occur, and the primary wall remains uncovered; two main types are simple pits and bordered pits. **2.** The stone of a drupaceous fruit. [MET] A small hole in the surface of a metal; usually caused by corrosion or formed during electroplating operations. [MIN ENG] **1.** A coal mine; the term is not commonly used by the coal industry, except in reference to surface mining where the workings may be known as a strip pit. **2.** Any quarry, mine, or excavation area worked by the open-cut method to obtain material of value.

pitch [ACOUS] That psychological property of sound characterized by highness or lowness, depending primarily upon frequency of the sound stimulus, but also upon its sound pressure and waveform. [ARCH] The ratio of the rise of a roof to its span. [COMPUT SCI] The distance between the centerlines of adjacent rows of hole positions in punched paper tape. [DES ENG] The distance between similar elements arranged in a pattern or between two points of a mechanical part, as the distance between the peaks of two successive grooves on a disk recording or on a screw. [GEOL] *See* plunge. [MATER] A dark heavy liquid or solid substance obtained as a residue after distillation of tar, oil, and such materials; occurs naturally as asphalt. [MECH] **1.** Of an aerospace vehicle, an angular displacement about an axis parallel to the lateral axis of the vehicle. **2.** The rising and falling motion of the bow of a ship or the tail of an airplane as the craft oscillates about a transverse axis. [MOL BIO] The distance between two adjacent turns of double-stranded deoxyribonucleic acid. [SCI TECH] The inclination or degree of slope of an object or structure.

pitch axis [MECH] A lateral axis through an aircraft, missile, or similar body, about which the body pitches. Also known as pitching axis.

pitchblende [MINERAL] A massive, brown to black, and fine-grained, amorphous, or microcrystalline variety of uraninite which has a pitchy to dull luster and contains small quantities of uranium. Also known as nasturan; pitch ore.

pitching axis *See* pitch axis.

pitch ore *See* pitchblende.

pitchstone [GEOL] A type of volcanic glass distinguished by a waxy, dull, resinous, pitchy luster. Also known as fluolite.

pith [BOT] A central zone of parenchymatous tissue that occurs in most vascular plants and is surrounded by vascular tissue.

pitometer [ENG] Reversed pitot-tube-type flow-measurement device with one pressure opening facing upstream and the other facing downstream.

pitot tube [ENG] An instrument that measures the stagnation pressure of a flowing fluid, consisting of an open tube pointing into the fluid and connected to a pressure-indicating device. Also known as impact tube.

Pittidae [VERT ZOO] The pittas, a homogeneous family of brightly colored suboscine birds with an erectile crown of feathers, in the suborder Tyranni.

pitting [MED] **1.** The formation of pits; in the fingernails, a consequence and sign of psoriasis. **2.** The preservation for a short time of indentations on the skin made by pressing with the finger; seen in pitting edema. [MET] Selective localized formation of rounded cavities in a metal surface due to corrosion or to nonuniform electroplating. [MIN ENG] The act of digging or sinking a pit.

pi-T transformation *See* Y-delta transformation.

pituitary gland *See* hypophysis.

Pityaceae [PALEOBOT] A family of fossil plants in the order Cordaitales known only as petrifactions of branches and wood.

pivot [MECH] A short, pointed shaft forming the center and fulcrum on which something turns, balances, or oscillates.

pivot joint [ANAT] A diarthrosis that permits a rotation of one bone around another; an example is the articulation of the atlas with the axis. Also known as trochoid.

pixel [COMPUT SCI] The smallest part of an electronically coded picture image. [ELECTR] The smallest addressable element in an electronic display; a short form for picture element. Also known as pel.

pk *See* peck.

pK [CHEM] The logarithm (to base 10) of the reciprocal of the equilibrium constant for a specified reaction under specified conditions.

PKU *See* phenylketonuria.

Pl *See* poiseuille.

PLA *See* programmed logic array.

place [MATH] A position corresponding to a given power of the base in positional notation. Also known as column.

placenta [BOT] A plant surface bearing a sporangium. [EMBRYO] A vascular organ that unites the fetus to the wall of the uterus in all mammals except marsupials and monotremes.

placer [GEOL] A mineral deposit at or near the surface of the earth, formed by mechanical concentration of mineral particles from weathered debris. Also known as ore of sedimentation.

placer mining [MIN ENG] 1. The extraction and concentration of heavy metals from placers. 2. Mining of gold by washing the sand, gravel, or talus.

placode [EMBRYO] A platelike epithelial thickening, frequently marking, in the embryo, the anlage of an organ or part.

Placodermi [PALEON] A large and varied class of Paleozoic fishes characterized by a complex bony armor covering the head and the front portion of the trunk.

Placodontia [PALEON] A small order of Triassic marine reptiles of the subclass Euryapsida characterized by flat-crowned teeth in both the upper and lower jaws and on the palate.

Placothuriidae [INV ZOO] A family of holothurian echinoderms in the order Dendrochirotida; individuals are invested in plates and have a complex calcareous ring mechanism.

plage [ASTRON] One of the luminous areas that appear in the vicinity of sunspots or disturbed areas on the sun; they may be seen distinctively in spectroheliograms taken in the calcium K line.

Plaggept [GEOL] A suborder of the soil order Inceptisol, with very thick surface horizons of mixed mineral and organic materials resulting from manure or human wastes added over long periods of time.

Plagiaulacida [PALEON] A primitive, monofamilial suborder of multituberculate mammals distinguished by their dentition (dental formula I 3/0 C 0/0 Pm 5/4 M 2/2), having cutting premolars and two rows of cusps on the upper molars.

Plagiaulacidae [PALEON] The single family of the extinct mammalian suborder Plagiaulacida.

plagioclase [MINERAL] 1. A type of triclinic feldspars having the general formula $(Na,Ca)Al(Si,Al)Si_2O_8$; they are common rock-forming minerals. 2. A series in the plagioclase group which can be divided into a number of varieties based on the relative proportion of the solid solution end members, albite and anorthite (An): albite (An 0–10) oligoclase (An 10–30), andesine (An 30–50), labradorite (An 50–70), bytownite (An 70–90), and anorthite (An 90–100). Also known as sodium-calcium feldspar.

plagiohedral [CRYSTAL] Pertaining to obliquely arranged spiral faces; in particular, to a member of a group in the isometric system with 13 axes but no center or planes.

Plagiosauria [PALEON] An aberrant Triassic group of labyrinthodont amphibians.

plagiosere [ECOL] A plant succession deflected from its normal course by biotic factors.

plain [GEOGR] An extensive, broad tract of level or rolling, almost treeless land with a shrubby vegetation, usually at a low elevation. [GEOL] A flat, gently sloping region of the sea floor. Also known as submarine plain.

plaintext [COMMUN] The form of a message in which it can be generally understood, before it has been transformed by a code or cipher into a form in which it can be read only by those privy to the secrets of the form.

Plaisancian [GEOL] A European stage of geologic time: lower Pliocene (above Pontian of Miocene, below Astian). Also known as Piacention; Plaisanzian.

Plaisanzian *See* Plaisancian.

plan [GRAPHICS] 1. An orthographic drawing on a horizontal plane, as of an instrument, a horizontal section, or a layout. 2. A large-scale map or chart of a small area.

planar array [ELECTR] An array of ultrasonic transducers that can be mounted in a single plane or sheet, to permit closer conformation with the hull design of a sonar-carrying ship.

planar device [ELECTR] A semiconductor device having planar electrodes in parallel planes, made by alternate diffusion of *p*- and *n*-type impurities into a substrate.

planar flow structure *See* platy flow structure.

planaria [INV ZOO] Any flatworm of the turbellarian order Tricladida; the body is broad and dorsoventrally flattened, with anterior lateral projections, the auricles, and a pair of eyespots on the anterior dorsal surface.

planar transistor [ELECTR] A transistor constructed by an etching and diffusion technique in which the junction is never exposed during processing, and the junctions reach the surface in one plane; characterized by very low leakage current and relatively high gain.

Planck distribution law *See* Planck radiation formula.

Planck oscillator [QUANT MECH] An oscillator which can absorb or emit energy only in amounts which are integral multiples of Planck's constant times the frequency of the oscillator. Also known as radiation oscillator.

Planck radiation formula [STAT MECH] A formula for the intensity of radiation emitted by a blackbody within a narrow band of frequencies (or wavelengths), as a function of frequency, and of the body's temperature. Also known as Planck distribution law; Planck's law.

Planck's constant [QUANT MECH] A fundamental physical constant, the elementary quantum of action; the ratio of the energy of a photon to its frequency, it is equal to $6.62620 \pm 0.00005 \times 10^{-34}$ joule-second. Symbolized h. Also known as quantum of action.

Planck's law [QUANT MECH] A fundamental law of quantum theory stating that energy associated with electromagnetic radiation is emitted or absorbed in discrete amounts which are proportional to the frequency of radiation. [STAT MECH] *See* Planck radiation formula.

plane [ELECTR] Screen of magnetic cores; planes are combined to form stacks. [DES ENG] A tool consisting of a smooth-soled stock from the face of which extends a wide-edged cutting blade for smoothing and shaping wood. [MATH] A surface containing any straight line through any two of its points.

plane cyclic curve *See* cyclic curve.

plane geometry [MATH] The geometric study of the figures in the euclidean plane such as lines, triangles, and polygons.

plane mirror [OPTICS] A mirror whose surface lies in a plane; it forms an image of an object such that the mirror surface is perpendicular to and bisects the line joining all corresponding object-image points.

plane of mirror symmetry Also known as mirror plane of symmetry; plane of symmetry; reflection plane; symmetry plane. [CRYSTAL] In certain crystals, a symmetry element whereby reflection of the crystal through a certain plane leaves the crystal unchanged. [MATH] An imaginary plane which divides an object into two halves, each of which is the mirror image of the other in this plane.

plane of polarization [ELECTROMAG] Plane containing the electric vector and the direction of propagation of electromagnetic wave.

plane of reflection [CRYSTAL] *See* plane of mirror symmetry. [MATH] *See* plane of mirror symmetry. [OPTICS] A plane containing the direction of propagation of radiation reflected from a surface, and the normal to the surface. Also known as reflection plane.

plane of saturation *See* water table.

plane of symmetry *See* plane of mirror symmetry.

plane polarization *See* linear polarization.

plane-polarized wave [ELECTROMAG] An electromagnetic wave whose electric field vector at all times lies in a fixed plane that contains the direction of propagation through a homogeneous isotropic medium.

plane polygon [MATH] A polygon lying in the euclidean plane.

plan equation [MECH ENG] The mathematical statement that horsepower = $plan/33,000$, where p = mean effective pressure (pounds per square inch), l = length of piston stroke (feet), a = net area of piston (square inches), and n = number of cycles completed per minute.

plane reflector *See* passive reflector.

planet [ASTRON] A relatively small, solid celestial body circulating around a star, in particular the star known as the sun (which has nine planets).

plane table [ENG] A surveying instrument consisting of a drawing board mounted on a tripod and fitted with a compass and a straight-edge ruler; used to graphically plot survey lines directly from field observations.

planetary electron *See* orbital electron.

planetary gear train [MECH ENG] An assembly of meshed gears consisting of a central gear, a coaxial internal or ring gear, and one or more intermediate pinions supported on a revolving carrier.

planetary nebula [ASTRON] An oval or round nebula of expanding concentric rings of gas associated with a hot central star.

planetary physics [ASTROPHYS] The study of the structure, composition, and physical and chemical properties of the planets of the solar system, including their atmospheres and immediate cosmic environment.

planetary wave *See* long wave; Rossby wave.

planetoid *See* asteroid.

plane trigonometry [MATH] The study of triangles in the euclidean plane with the use of functions defined by the ratios of sides of right triangles.

plane wave [PHYS] Wave in which the wavefront is a plane surface; a wave whose equiphase surfaces form a family of parallel planes.

planform [AERO ENG] The shape or form of an object, such as an airfoil, as seen from above, as in a plan view. [GEOGR] A body of water's outline or morphology as defined by the still water line.

planigraphy *See* sectional radiography.

planimetric map [MAP] A map indicating only the horizontal positions of features, without regard to elevation, in contrast with a topographic map, which indicates both horizontal and vertical positions. Also known as line map.

Planipennia [INV ZOO] A suborder of insects in the order Neuroptera in which the larval mandibles are modified for piercing and for sucking.

plankton [ECOL] Passively floating or weakly motile aquatic plants and animals.

planoconcave lens [OPTICS] A lens for which one surface is plane and the other is concave.

planoconvex lens [OPTICS] A lens for which one surface is plane and the other is convex.

Planosol [GEOL] An intrazonal, hydromorphic soil having a clay pan or hardpan covered with a leached surface layer; developed in a humid to subhumid climate.

plan position indicator [ELECTR] A radarscope display in which echoes from various targets appear as bright spots at the same locations as they would on a circular map of the area being scanned, the radar antenna being at the center of the map. Abbreviated PPI. Also known as P display.

plant [BOT] Any organism belonging to the kingdom Plantae, generally distinguished by the presence of chlorophyll, a rigid cell wall, and abundant, persistent, active embryonic tissue, and by the absence of the power of locomotion. [COMPUT SCI] To place a number or instruction that has been generated in the course of a computer program in a storage location where it will be used or obeyed at a later stage of the program. [IND ENG] The land, buildings, and equipment used in an industry.

Plantae [BOT] The plant kingdom.

Plantaginaceae [BOT] The single family of the plant order Plantaginales.

Plantaginales [BOT] An order of dicotyledonous herbaceous plants in the subclass Asteridae, marked by small hypogynous flowers with a persistent regular corolla and four petals.

plantar reflex [PHYSIO] Flexion of the toes in response to stroking of the outer surface of the sole, from heel to little toe.

plant hormone [BIOCHEM] An organic compound that is synthesized in minute quantities by one part of a plant and translocated to another part, where it influences physiological processes. Also known as phytohormone.

plant pathology [BOT] The branch of botany concerned with diseases of plants.

plant physiology [BOT] The branch of botany concerned with the processes which occur in plants.

plaque [MED] **1.** A patch, or an abnormal flat area on any internal or external body surface. **2.** A localized area of atherosclerosis. [VIROL] A clear area representing a colony of viruses on a plate culture formed by lysis of the host cell.

plasma [GEOL] The part of a soil material that can be, or has been, moved, reorganized, or concentrated by soil-forming processes. [HISTOL] The fluid portion of blood or lymph. [MINERAL] A faintly translucent or semitranslucent and bright green, leek green, or nearly emerald green variety of chalcedony, sometimes having white or yellowish spots. [PL PHYS] **1.** A highly ionized gas which contains equal numbers of ions and electrons in sufficient density so that the Debye shielding length is much smaller than the dimensions of the gas. **2.** A completely ionized gas, composed entirely of a nearly equal number of positive and negative free charges (positive ions and electrons).

plasma accelerator [PL PHYS] An accelerator that forms a high-velocity jet of plasma by using a magnetic field, an electric arc, a traveling wave, or other similar means.

plasma-arc cutting [ENG] Metal cutting by melting a localized area with an arc followed by removal of metal by high-velocity, high-temperature ionized gas.

plasma-arc welding [MET] Welding metal in a gas stream heated by a tungsten arc to temperatures approaching 60,000°F (33,315°C).

plasma core reactor [NUC ENG] A nuclear reactor that utilizes fissionable plasmas (such as uranium fluoride) for the fuel.

plasma current [PL PHYS] An electric current induced in plasma by injection of fast ion beams or some other means.

plasma diode [ELECTR] A diode used for converting heat directly into electricity; it consists of two closely spaced electrodes serving as cathode and anode, mounted in an envelope in which a low-pressure cesium vapor fills the interelectrode space; heat is applied to the cathode, causing emission of electrons.

plasma drift [GEOPHYS] Movement of ion and plasma concentration in the ionosphere by electric field variations in the upper atmosphere.

plasma etching [ENG] The removal of material by use of a focused plasma beam.

plasma instability [PL PHYS] A sudden change in the quasistatic distribution of positions or velocities of particles constituting a plasma, and a sudden change in the accompanying electromagnetic field.

plasmalemma *See* cell membrane.

plasma membrane *See* cell membrane.

plasma physics [PHYS] The study of highly ionized gases.

plasma pumping [PL PHYS] The application of radiation of appropriate frequencies to a plasma to increase the population of atoms or molecules in the higher energy states.

plasmasphere [GEOPHYS] A region of relatively dense, cold plasma surrounding the earth and extending out to altitudes of approximately 2 to 6 earth radii, composed predominantly of electrons and protons, with thermal energies not exceeding several electronvolts.

plasma thromboplastin component *See* Christmas factor.

plasmatron [ELECTR] A gas-discharge tube in which independently generated plasma serves as a conductor between a hot cathode and an anode; the anode current is modulated by varying either the conductivity or the effective cross section of the plasma.

plasmin [BIOCHEM] A proteolytic enzyme in plasma which can digest many proteins through the process of hydrolysis. Also known as fibrinolysin.

Plasmodiidae [INV ZOO] A family of parasitic protozoans in the suborder Haemosporina inhabiting the erythrocytes of the vertebrate host.

Plasmodiophorida [INV ZOO] An order of the protozoan subclass Mycetozoia occurring as endoparasites of plants.

Plasmodiophoromycetes [MYCOL] A class of the Fungi.

Plasmodroma [INV ZOO] A subphylum of the Protozoa, including Mastigophora, Sarcodina, and Sporozoa, in some taxonomic systems.

plasmogamy [INV ZOO] Fusion of protoplasts, without nuclear fusion, to form a multinucleate mass; occurs in certain protozoans.

plasmolysis [PHYSIO] Shrinking of the cytoplasm away from the cell wall due to exosmosis by immersion of a plant cell in a solution of higher osmotic activity.

plasmosome *See* nucleolus.

plaster [MATER] A plastic mixture of various materials, such as lime or gypsum, and water which sets to a hard, coherent solid.

plasterboard [MATER] A large, thin sheet of pulpboard, paper, or felt bonded to a hardened gypsum plaster core and used as a wall backing or as a substitute for plaster.

plaster of paris [INORG CHEM] White powder consisting essentially of the hemihydrate of calcium sulfate ($CaSO_4 \cdot \frac{1}{2}H_2O$ or $2CaSO_4 \cdot H_2O$), produced by calcining gypsum until it is partially dehydrated; forms with water a paste that quickly sets; used for casts and molds, building materials, and surgical bandages. Also known as calcined gypsum.

plastic [MATER] A polymeric material (usually organic) of large molecular weight which can be shaped by flow; usually refers to the final product with fillers, plasticizers, pigments, and stabilizers included (versus the resin, the homogeneous polymeric starting material); examples are polyvinyl chloride, polyethylene, and urea-formaldehyde. [MECH] Displaying, or associated with, plasticity.

plastic deformation [MECH] Permanent change in shape or size of a solid body without fracture resulting from the application of sustained stress beyond the elastic limit.

plastic design *See* ultimate-load design.

plastic flow [PHYS] Rheological phenomenon in which flowing behavior of the material occurs after the applied stress reaches a critical (yield) value, such as with putty.

plastic foam *See* expanded plastic.

plasticity [MECH] The property of a solid body whereby it undergoes a permanent change in shape or size when subjected to a stress exceeding a particular value, called the yield value.

plastid [CYTOL] One of the specialized cell organelles containing pigments or protein materials, often serving as centers of special metabolic activities; examples are chloroplasts and leukoplasts.

plastron [INV ZOO] The ventral plate of the cephalothorax of spiders. [VERT ZOO] The ventral portion of the shell of tortoises and turtles.

plat [MAP] A plan that shows land ownership, boundaries, and subdivisions together with data for description and identification of various parts.

Platanaceae [BOT] A small family of monoecious dicotyledonous plants in which flowers have several carpels which are separate, three or four stamens, and more or less orthotropous ovules, and leaves are stipulate.

Plataspidae [INV ZOO] A family of shining, oval hemipteran insects in the superfamily Pentatomoidea.

plate [BUILD] **1.** A shoe or base member, such as of a partition or other kind of frame. **2.** The top horizontal member of a row of studs used in a frame wall. [DES ENG] A rolled, flat piece of metal of some arbitrary minimum thickness and width depending on the type of metal. [ELEC] **1.** One of the conducting surfaces in a capacitor. **2.** One of the electrodes in a storage battery. [ELECTR] *See* anode. [GEOL] **1.** A smooth, thin, flat fragment of rock, such as a flagstone. **2.** A large rigid, but mobile, block involved in plate tectonics; thickness ranges from 50 to 250 kilometers and includes both crust and a portion of the upper mantle. [GRAPHICS] **1.** In etching, the piece of copper, zinc, or other metal that constitutes the base from which prints are made. **2.** In photography, a sheet of glass coated with a sensitized emulsion. **3.** In printing, the reproduction of type or cuts in metal or other material; a plate may bear a relief, intaglio, or planographic printing surface. [MET] A thick flat particle of metal powder.

plate anemometer *See* pressure-plate anemometer.

plateau [ELECTR] The portion of the plateau characteristic of a counter tube in which the counting rate is substantially independent of the applied voltage. [GEOGR] An extensive, flat-surfaced upland region, usually more than 150–300 meters in elevation and considerably elevated above the adjacent country and limited by an abrupt descent on at least one side. [GEOL] A broad, comparatively flat and poorly defined elevation of the sea floor, commonly over 200 meters in elevation.

plate circuit *See* anode circuit.

plate current *See* anode current.

plate girder [CIV ENG] A riveted or welded steel girder having a deep vertical web plate with a pair of angles along each edge to act as compression and tension flanges.

plate ice *See* pancake ice.

platelet [HISTOL] *See* thrombocyte. [HYD] A small ice crystal which, when united with other such crystals, forms a layer of floating ice, especially sea ice, and serves as seed crystals for further thickening of the ice cover.

platen [ENG] **1.** A flat plate against which something rests or is pressed. **2.** The rubber-covered roller of a typewriter against which paper is pressed when struck by the typebars. [MECH ENG] A flat surface for exchanging heat in a boiler or heat exchanger which may have extended heat transfer surfaces.

plate resistance *See* anode resistance.

plate saturation *See* anode saturation.

plate tectonics [GEOL] Global tectonics based on a model of the earth characterized by a small number (10–25) of semirigid plates which float on some viscous underlayer in the mantle; each plate moves more or less independently and grinds against the others, concentrating most deformation, volcanism, and seismic activity along the periphery. Also known as raft tectonics.

plate tower [CHEM ENG] A distillation tower along the internal height of which is a series of transverse plates (bubble-cap or sieve) to force intimate contact between downward flowing liquid and upward flowing vapor.

plating [MET] Forming a thin, adherent layer of metal on an object. Also known as metal plating.

platinite *See* platynite.

platinum [CHEM] A chemical element, symbol Pt, atomic number 78, atomic weight 195.09. [MET] A soft, ductile, malleable, grayish white noble metal with relatively high electric resistance; used in alloys, in electrical and electronic devices, and in jewelry.

platinum resistance thermometer [ENG] The basis of the International Practical Temperature Scale of 1968 from 259.35° to 630.74°C; used in industrial thermometers in the range 0 to 650°C; capable of high accuracy because platinum is noncorrosive,

ductile, and nonvolatile, and can be obtained in a very pure state. Also known as Callendar's thermometer.

Platyasterida [INV ZOO] An order of Asteroidea in which traces of metapinnules persist, the ossicles of the arm skeleton being arranged in two growth gradient systems.

Platybelondoninae [PALEON] A subfamily of extinct elephantoid mammals in the family Gomphotheriidae consisting of species with digging specializations of the lower tusks.

Platycephalidae [VERT ZOO] The flatheads, a family of perciform fishes in the suborder Cottoidei.

Platyceratacea [PALEON] A specialized superfamily of extinct gastropod mollusks which adapted to a coprophagous life on crinoid calices.

Platycopa [INV ZOO] A suborder of ostracod crustaceans in the order Podocopida including marine forms with two pairs of thoracic legs.

Platycopina [INV ZOO] The equivalent name for Platycopa.

Platyctenea [INV ZOO] An order of the ctenophores whose members are sedentary or parasitic; adults often lack ribs and are flattened due to shortening of the main axis.

platy flow structure [PETR] Structure of an igneous rock characterized by tabular sheets which suggest stratification, and formation by contraction during cooling. Also known as linear flow structure; planar flow structure.

Platygasteridae [INV ZOO] A family of hymenopteran insects in the superfamily Proctotrupoidea.

Platyhelminthes [INV ZOO] A phylum of invertebrates composed of bilaterally symmetrical, nonsegmented, dorsoventrally flattened worms characterized by lack of coelom, anus, circulatory and respiratory systems, and skeleton.

platynite [MINERAL] $PbBi_2(Se,S)_3$ An iron-black mineral composed of selenide and sulfide of lead and bismuth; occurs in thin metallic plates resembling graphite. Also spelled platinite.

Platypodidae [INV ZOO] The ambrosia beetles, a family of coleopteran insects in the superfamily Curculionoidea.

Platypsyllidae [INV ZOO] The equivalent name for Leptinidae.

platypus [VERT ZOO] *Ornithorhynchus anatinus.* A monotreme, making up the family Ornithorhynchidae, which lays and incubates eggs in a manner similar to birds, and retains some reptilian characters; the female lacks a marsupium. Also known as duckbill platypus.

Platysomidae [PALEON] A family of extinct palaeonisciform fishes in the suborder Platysomoidei; typically, the body is laterally compressed and rhombic-shaped, with long dorsal and anal fins.

Platysomoidei [PALEON] A suborder of extinct deep-bodied marine and fresh-water fishes in the order Palaeonisciformes.

Platysternidae [VERT ZOO] The big-headed turtles, a family of Asiatic fresh-water Chelonia with a single species (*Platysternon megacephalum*), characterized by a large head, hooked mandibles, and a long tail.

playa [GEOL] **1.** A low, essentially flat part of a basin or other undrained area in an arid region. **2.** A small, generally sandy land area at the mouth of a stream or along the shore of a bay. **3.** A flat, alluvial coastland, as distinguished from a beach.

play of color [OPTICS] An optical phenomenon consisting of a rapid succession of flashes of a variety of prismatic colors as certain minerals or cabochon-cut gems are moved about; caused by diffraction of light from spherical particles of amorphous silica stacked in an orderly three-dimensional pattern. Also known as schiller.

Plecoptera [INV ZOO] The stoneflies, an order of primitive insects in which adults differ only slightly from immature stages, except for wings and tracheal gills.

Plectascales [MYCOL] An equivalent name for Eurotiales.

Plectognathi [VERT ZOO] The equivalent name for Tetraodontiformes.

Plectoidea [INV ZOO] A superfamily of small, free-living nematodes characterized by simple spiral amphids or variants thereof, elongate cylindroconoid stoma, and reflexed ovaries.

Pleiades [ASTRON] An open cluster of a few hundred stars in the constellation Taurus; six of the stars are easily visible to the naked eye.

Pleidae [INV ZOO] A family of hemipteran insects in the superfamily Pleoidea.

Pleistocene [GEOL] An epoch of geologic time of the Quaternary period, following the Tertiary and before the Holocene. Also known as Ice Age; Oiluvium.

plenum [ENG] A condition in which air pressure within an enclosed space is greater than that in the outside atmosphere.

pleochroic halos [OPTICS] Halos of color or color differences that are sometimes observed around inclusions in minerals, resulting from irradiation by alpha particles.

pleochroism [OPTICS] Phenomenon exhibited by certain transparent crystals in which light viewed through the crystal has different colors when it passes through the crystal in different directions. Also known as polychroism.

Pleoidea [INV ZOO] A superfamily of suboval hemipteran insects belonging to the subdivision Hydrocoriseae.

pleomorphism [BIOL] The occurrence of more than one distinct form of an organism in a single life cycle. [CRYSTAL] *See* polymorphism.

pleonaste *See* ceylonite.

Pleosporales [BOT] The equivalent name for the lichenized Pseudophaeriales.

Plesiocidaroida [PALEON] An extinct order of echinoderms assigned to the Euechinoidea.

Plesiosauria [PALEON] A group of extinct reptiles in the order Sauropterygia constituting a highly specialized offshoot of the nothosaurs.

plesiotype [SYST] A specimen or specimens on which subsequent descriptions are based.

Plethodontidae [VERT ZOO] A large family of salamanders in the suborder Salamandroidea characterized by the absence of lungs and the presence of a fine groove from nostril to upper lip.

pleura [ANAT] The serous membrane covering the lung and lining the thoracic cavity.

Pleuracanthodii [PALEON] An order of Paleozoic sharklike fishes distinguished by two-pronged teeth, a long spine projecting from the posterior braincase, and direct backward extension of the tail.

pleural cavity [ANAT] The potential space included between the parietal and visceral layers of the pleura.

pleurisy [MED] Inflammation of the pleura. Also known as pleuritis.

pleuritis *See* pleurisy.

Pleuroceridae [INV ZOO] A family of fresh-water snails in the order Pectinibranchia.

Pleurocoelea [INV ZOO] An extinct superfamily of gastropod mollusks of the order Opisthobranchia in which the shell, mantle cavity, and gills were present.

Pleurodira [VERT ZOO] A suborder of turtles (Chelonia) distinguished by spines on the posterior cervical vertebrae so that the head is retractile laterally.

pleurodontia [VERT ZOO] Attachment of the teeth to the inner surface of the jawbone.

Pleuromeiaceae [PALEOBOT] A family of plants in the order Pleuromiales, but often included in the Isoetales due to a phylogenetic link.

Pleuromeiales [PALEOBOT] An order of Early Triassic lycopods consisting of the genus *Pleuromeia*; the upright branched stem had grasslike leaves and a single terminal strobilus.

Pleuronectiformes [VERT ZOO] The flatfishes, an order of actinopterygian fishes distinguished by the loss of bilateral symmetry.

Pleuronematina [INV ZOO] A suborder of the Hymenostomatida.

Pleurostigmophora [INV ZOO] A subclass of the centipedes, in some taxonomic systems, distinguished by lateral spiracles.

Pleurotomariacea [PALEON] An extinct superfamily of gastropod mollusks in the order Aspidobranchia.

plexus [ANAT] A network of interlacing nerves or anastomosing vessels. [GEOL] An area on a subglacial deposit that encloses a giant's kettle.

Pliensbachian [GEOL] A European stage of geologic time: Lower Jurassic (above Sinemurian, below Toarcian).

plinth [ARCH] 1. The block forming the lowest member or base of a column or pedestal. 2. A slight widening at the base of a column or wall. [GEOL] The lower and outer part of a seif dune, beyond the slip-face boundaries, that has never been subjected to sand avalanches.

Pliocene [GEOL] A worldwide epoch of geologic time of the Tertiary period, extending from the end of the Miocene to the beginning of the Pleistocene.

Pliohyracinae [PALEON] An extinct subfamily of ungulate mammals in the family Procaviidae.

PLL *See* phase-locked loop.

ploidy [GEN] Number of complete chromosome sets in a nucleus.

Plokiophilidae [INV ZOO] A small family of predacious hemipteran insects in the superfamily Cimicoidea; individuals live in the webs of spiders and embiids.

Plotosidae [VERT ZOO] A family of Indo-Pacific saltwater catfishes (Siluriformes).

plotter [ENG] A visual display or board on which a dependent variable is graphed by an automatically controlled pen or pencil as a function of one or more variables.

plotting board [ENG] The surface portion of a plotter, on which graphs are recorded. Also known as plotting table.

plotting table *See* plotting board.

ploughed-and-tongued joint *See* feather joint.

plug [ELEC] The half of a connector that is normally movable and is generally attached to a cable or removable subassembly; inserted in a jack, outlet, receptacle, or socket. [GEOL] 1. A vertical pipelike magmatic body representing the conduit to a former volcanic vent. 2. A crater filling of lava, the surrounding material of which has been removed by erosion. 3. A mass of clay, sand, or other sediment filling the part of a stream channel abandoned by the formation of a cutoff. [MET] 1. A rod or mandrel over which a pierced tube is forced, or that fills a tube as it is drawn through a die. 2. A punch or mandrel over which a cup is drawn. 3. A protruding portion of a die impression for forming a corresponding recess in the forging. 4.

A false bottom in a die. Also known as peg. [MIN ENG] A watertight seal in a shaft formed by removing the lining and inserting a concrete dam, or by placing a plug of clay over ordinary debris used to fill the shaft up to the location of the plug. [SCI TECH] **1.** A piece of material used to fill a hole. **2.** A small segment of material removed from a larger object.

plug bit *See* noncoring bit.

plugboard *See* control panel.

plugging [ELEC] Braking an electric motor by reversing its connections, so it tends to turn in the opposite direction; the circuit is opened automatically when the motor stops, so the motor does not actually reverse. [ENG] The formation of a barrier (plug) of solid material in a process flow system, such as a pipe or reactor. [MIN ENG] *See* blinding. [PETRO ENG] The act or process of stopping the flow of water, oil, or gas in strata penetrated by a borehole or well so that fluid from one stratum will not escape into another or to the surface; especially the sealing up of a well that is dry and is to be abandoned.

plum [BOT] Any of various shrubs or small trees of the genus *Prunus* that bear smooth-skinned, globular to oval, drupaceous stone fruit. [GEOL] A clast embedded in a matrix of a different kind, especially a pebble in a conglomerate.

Plumatellina [INV ZOO] The single order of the ectoproct bryozoan class Phylactolaemata.

plumb [ENG] Pertaining to an object or structure in true vertical position as determined by a plumb bob.

Plumbaginaceae [BOT] The leadworts, the single family of the order Plumbaginales.

Plumbaginales [BOT] An order of dicotyledonous plants in the subclass Caryophyllidae; flowers are pentamerous with fused petals, trinucleate pollen, and a compound ovary containing a single basal ovule.

plumbago *See* graphite.

plumb bob [ENG] A weight suspended on a string to indicate the direction of the vertical.

plume *See* column.

plummet [ENG] A loose-fitting metal plug in a tapered rotameter tube which moves upward (or downward) with an increase (or decrease) in fluid flow rate upward through the tube. Also known as float.

plunge [ENG] **1.** To set the horizontal cross hair of a theodolite in the direction of a grade when establishing a grade between two points of known level. **2.** *See* transit. [GEOL] The inclination of a geologic structure, especially a fold axis, measured by its departure from the horizontal. Also known as pitch; rake.

plunger [DES ENG] A wooden shaft with a large rubber suction cup at the end, used to clear plumbing traps and waste outlets. [ELECTROMAG] *See* piston. [ENG] *See* force plug. [MECH ENG] The long rod or piston of a reciprocating pump.

plus [MATH] A mathematical symbol; *A* plus *B*, where *A* and *B* are mathematical quantities, denotes the quantity obtained by taking their sum in an appropriate context.

Pluto [ASTRON] The most distant planet in the solar system; mean distance to the sun is about 5.6×10^9 kilometers; it has no known satellite, and its sidereal revolution period is 248.4 years.

pluton [GEOL] **1.** An igneous intrusion. **2.** A body of rock formed by metasomatic replacement.

plutonian *See* plutonic.

plutonic [GEOL] Pertaining to rocks formed at a great depth. Also known as abyssal; deep-seated; plutonian.

plutonium [CHEM] A reactive metallic element, symbol Pu, atomic number 94, in the transuranium series of elements; the first isotope to be identified was plutonium-239; used as a nuclear fuel, to produce radioactive isotopes for research, and as the fissile agent in nuclear weapons.

plutonium-238 [NUC PHYS] The first synthetic isomer made of plutonium; similar chemically to uranium and neptunium; atomic number 94; formed by bombardment of uranium with deuterons.

plutonium-239 [NUC PHYS] A synthetic element chemically similar to uranium and neptunium; atomic number 94; made by bombardment of uranium-238 with slow electrons in a nuclear reactor; used as nuclear reactor fuel and an ingredient for nuclear weapons.

pluvial [GEOL] Of a geologic process or feature, effected by rain action. [METEOROL] Pertaining to rain, or more broadly, to precipitation, particularly to an abundant amount thereof.

pluviometer *See* rain gage.

plywood [MATER] A material composed of thin sheets of wood glued together, with the grains of adjacent sheets oriented at right angles to each other.

Pm *See* promethium.

PM *See* permanent magnet; phase modulation.

PMR *See* projection microradiography.

pneumatic [ENG] Pertaining to or operated by air or other gas.

pneumatics [FL MECH] Fluid statics and behavior in closed systems when the fluid is a gas.

pneumatolysis [GEOL] Rock alteration or mineral crystallization effected by gaseous emanations from solidifying magma.

pneumonia [MED] An acute or chronic inflammation of the lungs caused by numerous microbial, immunological, physical, or chemical agents, and associated with exudate in the alveolar lumens.

pneumothorax [MED] The presence of air or gas in the pleural cavity.

pn hook transistor *See* hook collector transistor.

pnicogen [CHEM] Any member of the nitrogen family of elements, group V in the periodic table.

pnictide [CHEM] A simple compound of a pnicogen and an electropositive element.

pnip transistor [ELECTR] An intrinsic junction transistor in which the intrinsic region is sandwiched between the *n*-type base and the *p*-type collector.

pn junction [ELECTR] The interface between two regions in a semiconductor crystal which have been treated so that one is a *p*-type semiconductor and the other is an *n*-type semiconductor; it contains a permanent dipole charge layer.

pnpn diode [ELECTR] A semiconductor device consisting of four alternate layers of *p*-type and *n*-type semiconductor material, with terminal connections to the two outer layers. Also known as *npnp* diode.

pnpn transistor *See* npnp transistor.

pnp transistor [ELECTR] A junction transistor having an *n*-type base between a *p*-type emitter and a *p*-type collector.

Po *See* polonium.

Poaceae [BOT] The equivalent name for Gramineae.

Poales [BOT] The equivalent name for Cyperales.

Pockels cell [OPTICS] A crystal that exhibits the Pockels effect, such as potassium dihydrogen phosphate, which is placed between crossed polarizers and has ring electrodes bonded to two faces to allow application of an electric field; used to modulate light beams, especially laser beams.

Pockels effect [OPTICS] Changes in the refractive properties of certain crystals in an applied electric field, which are proportional to the first power of the electric field strength.

Pockels readout optical modulator [ELECTR] A device for storing data in the form of images; it consists of bismuth silicon oxide crystal coated with an insulating layer of parylene and transparent electrodes evaporated on the surfaces; a blue laser is used for writing and a red laser is used for nondestructive readout or processing. Abbreviated PROM.

pocket gopher See gopher.

pod [AERO ENG] An enclosure, housing, or detachable container of some kind on an airplane or space vehicle, as an engine pod. [BOT] A dry dehiscent fruit; a legume. [DES ENG] **1.** The socket for a bit in a brace. **2.** A straight groove in the barrel of a pod auger. [GEOL] An orebody of elongate, lenticular shape. Also known as podiform orebody.

Podargidae [VERT ZOO] The heavy-billed frogmouths, a family of Asian and Australian birds in the order Caprimulgiformes.

pod blight [PL PATH] A fungus disease of legumes caused by *Diaporthe* species.

podiatrist See chiropodist.

Podicipedide [VERT ZOO] The single family of the avian order Podicipediformes.

Podicipediformes [VERT ZOO] The grebes, an order of swimming and diving birds distinguished by dense, silky plumage, a rudimentary tail, and toes that are individually broadened and lobed.

Podicipitiformes [VERT ZOO] The equivalent name for Podicipediformes.

Podocopa [INV ZOO] A suborder of fresh-water ostracod crustaceans in the order Podocopida in which the inner lamella has a calcified rim joining the outer lamella along a chitinous zone of concrescence, and the two valves fit together firmly.

Podocopida [INV ZOO] An order of the Ostracoda; contains all fresh-water ostracods and is divided into the suborders Podocopa, Metacopina, and Platycopina.

Podocopina [INV ZOO] The equivalent name for Podocopa.

Podostemaceae [BOT] The single family of the order Podostemales.

Podostemales [BOT] An order of dicotyledonous plants in the subclass Rosidae; plants are submerged aquatics with modified, branching shoots and small, perfect flowers having a superior ovary and united carpels.

Podzol [GEOL] A soil group characterized by mats of organic matter in the surface layer and thin horizons of organic minerals overlying gray, leached horizons and dark-brown illuvial horizons; found in cool forests to temperate coniferous or mixed forests.

Poeciliidae [VERT ZOO] A family of fishes in the order Atheriniformes including the live-bearers, such as guppies, swordtails, and mollies.

Poecilosclerida [INV ZOO] An order of sponges of the class Demospongiae in which the skeleton includes two or more types of megascleres.

Poeobiidae [INV ZOO] A monotypic family of spioniform worms (*Poeobius meseres*) belonging to the Sedentaria and found in the North Pacific Ocean.

pogonip See ice fog.

Pogonophora [INV ZOO] The single class of the phylum Brachiata; the elongate body consists of three segments, each with a separate coelom; there is no mouth, anus, or digestive canal, and sexes are separate.

poikilitic [PETR] Of the texture of an igneous rock, having small crystals of one mineral randomly scattered without common orientation in larger crystals of another mineral.

poikiloblastic [PETR] Of a metamorphic texture, simulating the poikilitic texture of igneous rocks in having small idioblasts of one constituent lying within larger xenoblasts. Also known as sieve texture.

poikilotherm [ZOO] An animal, such as reptiles, fishes, and invertebrates, whose body temperature varies with and is usually higher than the temperature of the environment; a cold-blooded animal.

Poinsot ellipsoid See inertia ellipsoid.

Poinsot motion [MECH] The motion of a rigid body with a point fixed in space and with zero torque or moment acting on the body about the fixed point.

Poinsot's method [MECH] A method of describing Poinsot motion, by means of a geometrical construction in which the inertia ellipsoid rolls on the invariable plane without slipping.

point [GEOGR] A tapering piece of land projecting into a body of water; it is generally less prominent than a cape. [GRAPHICS] A printer's unit of measurement, equivalent to 1/72 or 0.013837 inch (0.3514598 millimeter) or 1/12 pica. Also known as printer's point; typography point. [LAP] A unit of mass, used in measuring precious stones, equal to 0.01 metric carat, or to 2 milligrams. [MATH] **1.** An element in a topological space. **2.** One of the basic undefined elements of geometry possessing position but no nonzero dimension. **3.** In positional notation, the character or the location of an implied symbol that separates the integral part of a numerical expression from its fractional part; for example, it is called the binary point in binary notation and the decimal point in decimal notation. [NAV] In marine operation, one thirty-second of a circle, or $11\frac{1}{4}$ degrees.

point at infinity See ideal point.

point-bearing pile See end-bearing pile.

point contact [ELECTR] A contact between a specially prepared semiconductor surface and a metal point, usually maintained by mechanical pressure but sometimes welded or bonded.

point-contact diode [ELECTR] A semiconductor rectifier that uses the barrier formed between a specially prepared semiconductor surface and a metal point to produce the rectifying action.

point-contact transistor [ELECTR] A transistor having a base electrode and two or more point contacts located near each other on the surface of an *n*-type semiconductor.

point defect [CRYSTAL] A departure from crystal symmetry which affects only one, or, in some cases, two lattice sites.

point diagram [PETR] A fabric diagram in which a point represents the preferred orientation of each individual fabric element. Also known as scatter diagram.

pointed bracket See angle bracket.

pointer [COMPUT SCI] The part of an instruction which contains the address of the next record to be accessed. [ENG] The needle-shaped rod that moves over the scale of a meter.

point-junction transistor [ELECTR] Transistor having a base electrode and both point-contact and junction electrodes.

point mutation [GEN] Mutation of a single gene due to addition, loss, replacement, or change of sequence in one or more base pairs of the deoxyribo-nucleic acid of that gene.

point of contraflexure [MECH] A point at which the direction of bending changes. Also known as point of inflection.

point of inflection [MATH] A point where a plane curve changes from the concave to the convex relative to some fixed line; equivalently, if the function determining the curve has a second derivative, this derivative changes sign at this point. Also known as inflection point. [MECH] See point of contraflexure.

points of the compass See compass points.

point source [PHYS] A source of radiation having definite position but no extension in space; this is an ideal which is a good approximation for distances from the source sufficiently large compared to the dimensions of the source.

poise [FL MECH] A unit of dynamic viscosity equal to the dynamic viscosity of a fluid in which there is a tangential force 1 dyne per square centimeter resisting the flow of two parallel fluid layers past each other when their differential velocity is 1 centimeter per second per centimeter of separation. Abbreviated P.

poiseuille [FL MECH] A unit of dynamic viscosity of a fluid in which there is a tangential force of 1 newton per square meter resisting the flow of two parallel layers past each other when their differential velocity is 1 meter per second per meter of separation; equal to 10 poise; used chiefly in France. Abbreviated Pl.

Poiseuille's law [FL MECH] The law that the volume flow of an incompressible fluid through a circular tube is equal to $\pi/8$ times the pressure differences between the ends of the tube, times the fourth power of the tube's radius divided by the product of the tube's length and the dynamic viscosity of the fluid.

poison [ATOM PHYS] A substance which reduces the phosphorescence of a luminescent material. [CHEM] A substance that exerts inhibitive effects on catalysts, even when present only in small amounts; for example, traces of sulfur or lead will poison platinum-based catalysts. [ELECTR] A material which reduces the emission of electrons from the surface of a cathode. [MATER] A substance that in relatively small doses has an action that either destroys life or impairs seriously the functions of organs or tissues. [NUCLEO] A substance that absorbs neutrons without any fission resulting, and thereby lowers the reactivity of a nuclear reactor.

Poisson distribution [STAT] A probability distribution whose mean and variance have a common value k, and whose frequency is $f(x) = k^x e^{-k}/x!$, for $x = 0, 1, 2, \ldots$.

Poisson ratio [MECH] The ratio of the transverse contracting strain to the elongation strain when a rod is stretched by forces which are applied at its ends and which are parallel to the rod's axis.

polar [MATH] 1. For a conic section, the polar of a point is the line that passes through the points of contact of the two tangents drawn to the conic from the point. 2. For a quadric surface, the polar of a point is the plane that passes through the curve which is the locus of the points of contact of the tangents drawn to the surface from the point. 3. For a quadric surface, the polar of a line is the line of intersection of the planes which are tangent to the surface at its points of intersection with the original line.

polar air [METEOROL] A type of air whose characteristics are developed over high latitudes; there are two types: continental polar air and maritime polar air.

polar anticyclone See arctic high; subpolar high.

polar axis [CRYSTAL] An axis of crystal symmetry which does not have a plane of symmetry perpendicular to it. [MATH] The directed straight line relative to which the angle is measured for a representation of a point in the plane by polar coordinates.

polar cap [ASTRON] Any of the bright areas covering the poles of Mars, believed to be composed of frozen carbon dioxide and water-ice. [HYD] An ice sheet centered at one of the poles of the earth.

polar circle [GEOD] A parallel of latitude whose distance from the pole is equal to the obliquity of the ecliptic (approximately 23° 27′).

polar compound [CHEM] Molecules which contain polar covalent bonds; they can ionize when dissolved or fused; polar compounds include inorganic acids, bases, and salts.

polar-coordinate navigation system [NAV] A system in which one or more signals are emitted from a facility (or co-located facilities) to produce simultaneous indication of bearing and distance.

polar coordinates [MATH] A point in the plane may be represented by coordinates (r,θ), where θ is the angle between the positive x-axis and the ray from the origin to the point, and r the length of that ray.

polar covalent bond [PHYS CHEM] A bond in which a pair of electrons is shared in common between two atoms, but the pair is held more closely by one of the atoms.

polar diagram [PHYS] A diagram employing polar coordinates to show the magnitude of a quantity in some or all directions from a point; examples include directivity patterns and radiation patterns.

polar high See arctic high; subpolar high.

polarimeter [OPTICS] An instrument used to determine the rotation of the plane of polarization of plane polarized light when it passes through a substance; the light is linearly polarized by a polarizer (such as a Nicol prism), passes through the material being analyzed, and then passes through an analyzer (such as another Nicol prism).

polarimetry [OPTICS] Determination of the rotation of the plane of polarization of plane polarized light when it passes through a substance, using a polarimeter.

Polaris [ASTRON] A creamy supergiant star of stellar magnitude 2.0, spectral classification F8, in the constellation Ursa Minor; marks the north celestial pole, being about 1° from this point; the star Ursae Minoris. Also known as North Star; Pole Star.

polariscope [OPTICS] Any of several instruments used to determine the effects of substances on polarized light, in which linearly or elliptically polarized light passes through the substance being studied, and then through an analyzer.

polarity [COMMUN] 1. The direction in which a direct current flows, in a teletypewriter system. 2. The sense of the potential of a portion of a television picture signal representing a dark area of a scene relative to the potential of a portion of the signal representing a light area. [MATH] Property of a line segment whose two ends are distinguishable. [MOL BIO] The orientation of a strand of polynucleotide with respect to its partner, expressed in terms of nucleotide linkages. [PHYS] Property of a physical system which has two points with different (usually opposite) characteristics, such as one which has opposite charges or electric potentials, or opposite magnetic poles.

polarizability [ELEC] The electric dipole moment induced in a system, such as an atom or molecule, by an electric field of unit strength.

polarizability ellipsoid *See* index ellipsoid.

polarization [ELEC] **1.** The process of producing a relative displacement of positive and negative bound charges in a body by applying an electric field. **2.** A vector quantity equal to the electric dipole moment per unit volume of a material. Also known as dielectric polarization; electric polarization. **3.** A chemical change occurring in dry cells during use, increasing the internal resistance of the cell and shortening its useful life. [PHYS] **1.** Phenomenon exhibited by certain electromagnetic waves and other transverse waves in which the direction of the electric field or the displacement direction of the vibrations is constant or varies in some definite way. **2.** The direction of the electric field or the displacement vector of a wave exhibiting polarization (first definition). **3.** The process of bringing about polarization (first definition) in a transverse wave. **4.** Property of a collection of particles with spin, in which the majority have spin components pointing in one direction, rather than at random.

polarization potential [ELECTROMAG] One of two vectors from which can be derived, by differentiation, an electric scalar potential and magnetic vector potential satisfying the Lorentz condition. Also known as Hertz vector. [PHYS CHEM] The reverse potential of an electrolytic cell which opposes the direct electrolytic potential of the cell.

polarized ceramics [MATER] A substance, such as lead zirconate and barium titanate, having high electromechanical conversion efficiency and used as a transducer element in an ultrasonic system.

polarized electrolytic capacitor [ELEC] An electrolytic capacitor in which the dielectric film is formed adjacent to only one metal electrode; the impedance to the flow of current is then greater in one direction than in the other.

polarized light [OPTICS] Polarized electromagnetic radiation whose frequency is in the optical region.

polarized plug [ELEC] A plug that can be inserted in its receptacle only when in a predetermined position.

polarized receptacle [ELEC] A receptacle designed for use with a polarized plug, to ensure that the grounded side of an alternating-current line or the positive side of a direct-current line is always connected to the same terminal on a piece of equipment.

polarizing filter [OPTICS] A device which selectively absorbs components of electromagnetic radiation passing through it, so that light emerging from it is plane-polarized.

polar migration *See* polar wandering.

polar molecule [PHYS CHEM] A molecule having a permanent electric dipole moment.

polarographic analysis [ANALY CHEM] An electroanalytical technique in which the current through an electrolysis cell is measured as a function of the applied potential; the apparatus consists of a potentiometer for adjusting the potential, a galvanometer for measuring current, and a cell which contains two electrodes, a reference electrode whose potential is constant and an indicator electrode which is commonly the dropping mercury electrode. Also known as polarography.

polarography *See* polarographic analysis.

polaron [SOLID STATE] An electron in a crystal lattice together with a cloud of phonons that result from the deformation of the lattice produced by the interaction of the electron with ions or atoms in the lattice.

polar triangle [MATH] A triangle associated to a given spherical triangle obtained from three directed lines perpendicular to the planes associated with the sides of the original triangle.

polar vector *See* vector.

polar wandering [GEOL] Migration during geologic time of the earth's poles of rotation and magnetic poles. Also known as Chandler motion; polar migration.

polar westerlies *See* westerlies.

pole [CRYSTAL] A direction perpendicular to one of the faces of a crystal. [ELEC] **1.** One of the electrodes in an electric cell. **2.** An output terminal on a switch; a double-pole switch has two output terminals. [MATH] **1.** An isolated singular point z_0 of a complex function whose Laurent series expansion about z_0 will include finitely many terms of form $a_n(z-z_0)^{-n}$. **2.** For a great circle on a sphere, the pole of the circle is a point of intersection of the sphere and a line that passes through the center of the sphere and is perpendicular to the plane of the circle. **3.** For a conic section, the pole of a line is the intersection of the tangents to the conic at the points of intersection of the conic with the line. **4.** For a quadric surface, the pole of a plane is the vertex of the cone which is tangent to the surface along the curve where the plane intersects the surface. [MECH] **1.** A point at which an axis of rotation or of symmetry passes through the surface of a body. **2.** *See* perch.

pole blight [PL PATH] A destructive disease of white pines characterized by shortening of the needle-bearing stems, yellowing and shortening of needles, and copious flow of resin.

Polemoniaceae [BOT] A family of autotrophic dicotyledonous plants in the order Polemoniales distinguished by lack of internal phloem, corolla lobes that are convolute in the bud, three carpels, and axile placentation.

Polemoniales [BOT] An order of dicotyledonous plants in the subclass Asteridae, characterized by sympetalous flowers, a regular, usually five-lobed corolla, and stamens equal in number and alternate with the petals.

pole-positioning [CONT SYS] A design technique used in linear control theory in which many or all of a system's closed-loop poles are positioned as required, by proper choice of a linear state feedback law; if the system is controllable, all of the closed-loop poles can be arbitrarily positioned by this technique.

Pole Star *See* Polaris.

poliomyelitis [MED] An acute infectious viral disease which in its most serious form involves the central nervous system and, by destruction of motor neurons in the spinal cord, produces flaccid paralysis. Also known as Heine-Medin disease; infantile paralysis.

Polish notation [COMPUT SCI] **1.** A notation system for digital-computer or calculator logic in which there are no parenthetical expressions and each operator is a binary or unary operator in the sense that it operates on not more than two operands. Also known as Lukasiewicz notation; parenthesis-free notation. **2.** The version of this notation in which operators precede the operands with which they are associated. Also known as prefix notation.

pollen [BOT] The small male reproductive bodies produced in pollen sacs of the seed plants.

pollination [BOT] The transfer of pollen from a stamen to a pistil; fertilization in flowering plants.

pollucite [MINERAL] $(Cs,Na)_2Al_2Si_4O_{12} \cdot H_2O$ A colorless, transparent zoolite mineral composed of hydrous silicate of cesium, sodium, and aluminum, occurring massive or in cubes; used as a gemstone. Also known as pollux.

pollux *See* pollucite.

Pollux [ASTRON] A giant orange-yellow star with visual brightness of 1.16, a little less than 35 light-years from the sun, spectral classification K0-III, in the constellation Gemini; the star β Geminorum.

polonium [CHEM] A chemical element, symbol Po, atomic number 84; all polonium isotopes are radioactive; polonium-210 is the naturally occurring isotope found in pitchblende.

polonium-210 [NUC PHYS] Radioactive isotope of polonium; mass 210, half-life 140 days, α-radiation; used to calibrate radiation counters, and in oil well logging and atomic batteries. Also known as radium F.

poly- [ORG CHEM] A chemical prefix meaning many; for example, a polymer is made of a number of single molecules known as monomers, as polyethylene is made from ethylene.

polyacetals *See* acetal resins.

polyalgorithm [MATH] A set of algorithms together with a strategy for choosing and switching among them.

Polyangiaceae [MICROBIO] A family of bacteria in the order Myxobacterales; vegetative cells and myxospores are cylindrical with blunt, rounded ends; the slime capsule is lacking; sporangia are sessile or stalked.

polyargyrite [MINERAL] $Ag_{24}Sb_2S_{15}$ A gray to black mineral composed of antimony silver sulfide.

polyatomic molecule [CHEM] A chemical molecule with three or more atoms.

polybasic [CHEM] A chemical compound in solution that yields two or more H^- ions per molecule, such as sulfuric acid, H_2SO_4.

Polybrachiidae [INV ZOO] A family of sedentary marine animals in the order Thecanephria.

Polychaeta [INV ZOO] The largest class of the phylum Annelida, distinguished by paired, lateral, fleshy appendages (parapodia) provided with setae, on most segments.

polychroism *See* pleochroism.

Polycirrinae [INV ZOO] A subfamily of polychaete annelids in the family Terebellidae.

Polycladida [INV ZOO] A class of marine Turbellaria whose leaflike bodies have a central intestine with radiating branches, many eyes, and tentacles in most species.

polyclinal fold [GEOL] One of a group of adjacent folds, the axial surfaces of which are oriented randomly, but which have similar surface axes.

polycondensation [ORG CHEM] A chemical condensation leading to the formation of a polymer by the linking together of molecules of a monomer and the releasing of water or a similar simple substance.

polycrase [MINERAL] $(Y,Ca,Ce,U,Th)(Ti,Cb,Ta)_2O_6$ Black mineral composed of titanate, columbate, and tantalate of yttrium-group metals; it is isomorphous with euxenite and occurs in granite pegmatites.

polycrystalline [MATER] 1. Pertaining to a material composed of aggregates of individual crystals. 2. Characterized by variously oriented crystals.

Polyctenidae [INV ZOO] A family of hemipteran insects in the superfamily Cimicoidea; the individuals are bat ectoparasites which resemble bedbugs but lack eyes and have ctenidia and strong claws.

polycyclic [ORG CHEM] A molecule that contains two or more closed atomic rings; can be aromatic (such as DDT), aliphatic (bianthryl), or mixed (dicarbazyl).

polycyclic hydrocarbon *See* polynuclear hydrocarbon.

polycystic kidney [MED] A usually hereditary, congenital, and bilateral disease in which a large number of cysts are present on the kidney.

polycythemia [MED] A condition characterized by an increased number of erythrocytes in the circulation.

polycythemia vera [MED] An absolute increase in all blood cells derived from bone marrow, especially erythrocytes.

Polydolopidae [PALEON] A Cenozoic family of rodentlike marsupial mammals.

polyethylene *See* ethylene resin.

polyethylene resin *See* ethylene resin.

Polygalaceae [BOT] A family of dicotyledonous plants in the order Polygalales distinguished by having a bicarpellate pistil and monadelphous stamens.

Polygalales [BOT] An order of dicotyledonous plants in the subclass Rosidae characterized by its simple leaves and usually irregular, hypogynous flowers.

polygamous [BOT] Having both perfect and imperfect flowers on the same plant. [VERT ZOO] Having more than one mate at one time.

polygen *See* polyvalent.

Polygnathidae [PALEON] A family of Middle Silurian to Cretaceous conodonts in the suborder Conodontiformes, having platforms with small pitlike attachment scars.

polygon [MATH] A figure in the plane given by points $p_1, p_2, ..., p_n$ and line segments $p_1p_2, p_2p_3, ..., p_{n-1}p_n, p_np_1$.

Polygonaceae [BOT] The single family of the order Polygonales.

Polygonales [BOT] An order of dicotyledonous plants in the subclass Caryophyllidae characterized by well-developed endosperm, a unilocular ovary, and often trimerous flowers.

polygraph *See* lie detector.

polyhedral angle [MATH] The shape formed by the lateral faces of a polyhedron which have a common vertex.

polyhedron [MATH] A solid bounded by planar polygons.

polymenorrhea *See* metrorrhagia.

polymer [ORG CHEM] Substance made of giant molecules formed by the union of simple molecules (monomers); for example polymerization of ethylene forms a polyethylene chain, or condensation of phenol and formaldehyde (with production of water) forms phenol-formaldehyde resins.

Polymera [INV ZOO] Formerly a subphylum of the Vermes; equivalent to the phylum Annelida.

polymerase [BIOCHEM] An enzyme that links nucleotides together to form polynucleotide chains.

polymerization [CHEM] 1. The bonding of two or more monomers to produce a polymer. 2. Any chemical reaction that produces such a bonding.

polymictic [HYD] Pertaining to or characteristic of a lake having no stabile thermal stratification. [PETR] Of a clastic sedimentary rock, being made up of many rock types or of more than one mineral species.

polymignite *See* polymignyte.

polymignyte [MINERAL] $(Ca,Fe,Y,Zr,Th)(Nb,Ti,Ta)O_4$ A black mineral composed of niobate, titanate, and tantalate of cerium-group metals, with calcium and iron. Also spelled polymignite.

polymorph [BIOL] An organism that exhibits polymorphism. [CRYSTAL] One of the crystal forms of a substance displaying polymorphism. Also known as polymorphic modification. [HISTOL] *See* granulocyte.

polymorphic modification *See* polymorph.

polymorphism [BIOL] **1.** Occurrence of different forms of individual in a single species. **2.** Occurrence of different structural forms in a single individual at different periods in the life cycle. [CRYSTAL] The property of a chemical substance crystallizing into two or more forms having different structures, such as diamond and graphite. Also known as pleomorphism. [GEN] The coexistence of genetically determined distinct forms in the same population, even the rarest of them being too common to be maintained solely by mutation; human blood groups are an example.

polymorphonuclear leukocyte *See* granulocyte.

Polynemidae [VERT ZOO] A family of perciform shore fishes in the suborder Mugiloidei.

Polynoidae [INV ZOO] The largest family of polychaetes, included in the Errantia and having a body of varying size and shape that is covered with elytra.

polynomial [MATH] A polynomial in the quantities $x_1, x_2,...,x_n$ is an expression involving a finite sum of terms of form $bx_1{}^{p1}x_2{}^{p2}... x_n{}^{pn}$, where b is some number, and $p_1,..., p_n$ are integers.

polynuclear hydrocarbon [ORG CHEM] Hydrocarbon molecule with two or more closed rings; examples are naphthalene, $C_{10}H_8$, with two benzene rings side by side, or diphenyl, $(C_6H_5)_2$, with two bond-connected benzene rings. Also known as polycyclic hydrocarbon.

Polyodontidae [INV ZOO] A family of tubicolous, often large-bodied errantian polychaetes with characteristic cephalic and parapodial structures.

polyoma virus [VIROL] A small deoxyribonucleic acid virus normally causing inapparent infection in mice, but experimentally capable of producing parotid tumors and a wide variety of other tumors.

Polyopisthocotylea [INV ZOO] An order of the trematode subclass Monogenea having a solid posterior holdfast bearing suckers or clamps.

polyp [INV ZOO] A sessile coelenterate individual having a hollow, somewhat cylindrical body, attached at one end, with a mouth surrounded by tentacles at the free end; may be solitary (hydra) or colonial (coral). [MED] A smooth, rounded or oval mass projecting from a membrane-covered surface.

polypeptide [BIOCHEM] A chain of amino acids linked together by peptide bonds but with a lower molecular weight than a protein; obtained by synthesis, or by partial hydrolysis of protein.

polypetalous [BOT] Having distinct petals, in reference to a flower or a corolla. Also known as choripetalous.

Polyphaga [INV ZOO] A suborder of the order Coleoptera; members are distinguished by not having the hind coxae fused to the metasternum and by lacking notopleural sutures.

polyphase circuit [ELEC] Group of alternating-current circuits (usually interconnected) which enter (or leave) a delimited region at more than two points of entry; they are intended to be so energized that, in the steady state, the alternating currents through the points of entry, and the alternating potential differences between them, all have exactly equal periods, but have differences in phase, and may have differences in waveform.

polyphase rectifier [ELECTR] A rectifier which utilizes two or more diodes (usually three), each of which operates during an equal fraction of an alternating-current cycle to achieve an output current which varies less than that in an ordinary half-wave or full-wave rectifier.

Polyplacophora [INV ZOO] The chitons, an order of mollusks in the class Amphineura distinguished by an elliptical body with a dorsal shell that comprises eight calcareous plates overlapping posteriorly.

polyploidy [GEN] The occurrence of related forms possessing chromosome numbers which are three or more times the haploid number.

Polypodiales [BOT] The true ferns; the largest order of modern ferns, distinguished by being leptosporangiate and by having small sporangia with a definite number of spores.

Polypodiatae *See* Polypodiopsida.

Polypodiophyta [BOT] The ferns, a division of the plant kingdom having well-developed roots, stems, and leaves that contain xylem and phloem and show well-developed alternation of generations.

Polypodiopsida [BOT] A class of the division Polypodiophyta; stems of these ferns bear several large, spirally arranged, compound leaves with sporangia grouped in sori on their undermargins.

Polypteridae [VERT ZOO] The single family of the order Polypteriformes.

Polypteriformes [VERT ZOO] An ancient order of actinopterygian fishes distinguished by thick, rhombic, ganoid scales with an enamellike covering, a slitlike spiracle behind the eye, a symmetrical caudal fin, and a dorsal series of free, spinelike finlets.

polyribosome *See* polysome.

polysaccharide [BIOCHEM] A carbohydrate composed of many monosaccharides.

polysepalous [BOT] Having separate sepals. Also known as chorisepalous.

polysome [CYTOL] A complex of ribosomes bound together by a single messenger ribonucleic acid molecule. Also known as polyribosome.

Polystomatoidea [INV ZOO] A superfamily of monogeneid trematodes characterized by strong suckers and hooks on the posterior end.

Polystylifera [INV ZOO] A suborder of the Hoplonemertini distinguished by many stylets.

polytene chromosome [GEN] A giant, multistranded, cablelike chromosome composed of many identical chromosomes having their chromomeres in register and produced by polyteny.

Polytrichales [BOT] An order of ascocarpous perennial mosses; rigid, simple stems are highly developed and arise from a prostrate subterranean rhizome.

polytrifluorochloroethylene resin *See* chlorotrifluoroethylene polymer.

polytropic process [THERMO] An expansion or compression of a gas in which the quantity pV^n is held constant, where p and V are the pressure and volume of the gas, and n is some constant.

polyunsaturated fat [MATER] A fat or oil based on fatty acids such as linoleic or linolenic acids which have two or more double bonds in each molecule; corn oil and safflower oil are examples.

polyvalent [CHEM] An ion or radical with more than one valency, such as the sulfate ion, $SO_4{}^{2-}$. Also known as multivalent; polygen. [IMMUNOL] **1.** Of antigens, having many combining sites or determinants. **2.** Pertaining to vaccines composed of mixtures of different organisms, and to the resulting mixed antiserum.

polyvalent number [COMPUT SCI] A number, consisting of several figures, used for description, wherein each figure represents one of the characteristics being described.

Polyzoa [INV ZOO] The equivalent name for Bryozoa.

Pomacentridae [VERT ZOO] The damselfishes, a family of perciform fishes in the suborder Percoidei.

Pomadasyidae [VERT ZOO] The grunts and sweetlips, a family of perciform fishes in the suborder Percoidei.

Pomatiasidae [INV ZOO] A family of land snails in the order Pectinibranchia.

Pomatomidae [VERT ZOO] A monotypic family of the Perciformes containing the bluefish (*Pomatomus saltatrix*).

pomegranate [BOT] *Punica granatum*. A small, deciduous ornamental tree of the order Myrtales cultivated for its fruit, which is a reddish, pomelike berry containing numerous seeds embedded in crimson pulp.

Pompilidae [INV ZOO] The spider wasps, the single family of the superfamily Pompiloidea.

Pompiloidea [INV ZOO] A monofamilial superfamily of hymenopteran insects in the suborder Apocrita with oval abdomen and strong spinose legs.

pond [GEOGR] A small natural body of standing fresh water filling a surface depression, usually smaller than a lake. [MECH] *See* gram-force.

ponderosa pine [BOT] *Pinus ponderosa*. A hard pine tree of western North America; attains a height of 150–225 feet (46–69 meters) and has long, dark-green leaves in bundles of two to five and tawny, yellowish bark.

Ponerinae [INV ZOO] A subfamily of tropical carnivorous ants (Formicidae) in which pupae characteristically form in cocoons.

Pongidae [VERT ZOO] A family of anthropoid primates in the superfamily Hominoidea; includes the chimpanzee, gorilla, and orangutan.

pons [ANAT] 1. A process or bridge of tissue connecting two parts of an organ. 2. A convex white eminence located at the base of the brain; consists of fibers receiving impulses from the cerebral cortex and sending fibers to the contralateral side of the cerebellum.

Pontodoridae [INV ZOO] A monotypic family of pelagic polychaetes assigned to the Errantia.

poplar [BOT] Any tree of the genus *Populus*, family Salicaceae, marked by simple, alternate leaves, scaly buds, bitter bark, and flowers and fruits in catkins.

popliteal artery [ANAT] A continuation of the femoral artery in the posterior portion of the thigh above the popliteal space and below the buttock.

popliteal nerve [ANAT] Either of two branches of the sciatic nerve in the lower part of the thigh; the larger branch continues as the tibial nerve, and the smaller branch continues as the peroneal nerve.

popliteal vein [ANAT] A vein passing through the popliteal space, formed by merging of the tibial veins and continuing to become the femoral vein.

poppet valve [MECH ENG] A cam-operated or spring-loaded reciprocating-engine mushroom-type valve used for control of admission and exhaust of working fluid; the direction of movement is at right angles to the plane of its seat.

popple rock *See* pebble bed.

population covariance [STAT] The number $(1/N)\cdot[(v_1 - \bar{v})(w_1 - \bar{w}) + \ldots + (v_N - \bar{v})(w_N - \bar{w})]$, where v_i and w_i, $i = 1, 2, \ldots, N$, are the values

obtained from two populations, and $\bar{v}$ and $\bar{w}$ are the respective means.

population inversion [ATOM PHYS] The condition in which a higher energy state in an atomic system is more heavily populated with electrons than a lower energy state of the same system.

population mean [STAT] The average of the numbers obtained for all members in a population by measuring some quantity associated with each member.

population variance [STAT] The arithmetic average of the numbers $(v_1 - \bar{v})^2, \ldots, (v_N - \bar{v})^2$, where v_i are numbers obtained from a population with N members, one for each member, and $\bar{v}$ is the population mean.

p orbital [ATOM PHYS] The orbital of an atomic electron with an orbital angular momentum quantum number of unity.

porcelain enamel *See* vitreous enamel.

porcelainite *See* mullite.

Porcellanasteridae [INV ZOO] A family of essentially deep-water forms in the suborder Paxillosina.

Porcellanidae [INV ZOO] The rock sliders, a family of decapod crustaceans of the group Anomura which resemble true crabs but are distinguished by the reduced, chelate fifth pereiopods and the well-developed tail fan.

porcupine [VERT ZOO] Any of about 26 species of rodents in two families (Hystricidae and Erethizontidae) which have spines or quills in addition to regular hair.

pore [ASTRON] A very small, dark area on the sun formed by the separation of adjacent flocculi. [BIOL] Any minute opening by which matter passes through a wall or membrane. [GEOL] An opening or channelway in rock or soil. [MET] A minute cavity in a powder compact, metal casting, or electroplated coating.

pore fungus [MYCOL] The common name for members of the families Boletaceae and Polyporaceae in the group Hymenomycetes; sporebearing surfaces are characteristically within tubes or pores.

pore space [GEOL] The pores in a rock or soil considered collectively. Also known as pore volume.

pore volume *See* pore space.

Porifera [INV ZOO] The sponges, a phylum of the animal kingdom characterized by the presence of canal systems and chambers through which water is drawn in and released; tissues and organs are absent.

porosity [PHYS] 1. Property of a solid which contains many minute channels or open spaces. 2. The fraction as a percent of the total volume occupied by these channels or spaces; for example, in petroleum engineering the ratio (expressed in percent) of the void space in a rock to the bulk volume of that rock.

porosity trap *See* stratigraphic trap.

porous [MATER] 1. Filled with pores. 2. Capable of absorbing liquids.

Poroxylaceae [PALEOBOT] A monogeneric family of extinct plants included in the Cordaitales.

porphrite *See* porphyry.

porphyroclastic structure *See* mortar structure.

porphyry [PETR] An igneous rock in which large phenocrysts are enclosed in a very-fine-grained to aphanitic matrix. Formerly known as porphrite.

porpoise [VERT ZOO] Any of several species of marine mammals of the family Phocaenidae which have small flippers, a highly developed sonar system, and smooth, thick, hairless skin.

port [COMPUT SCI] An interface between a communications channel and a unit of computer hardware. [ELEC] An entrance or exit for a network.

[ELECTROMAG] An opening in a waveguide component, through which energy may be fed or withdrawn, or measurements made. [ENG] The side of a ship or airplane on the left of a person facing forward. [ENG ACOUS] An opening in a bass-reflex enclosure for a loudspeaker, designed and positioned to improve bass response. [GEOGR] *See* harbor. [NAV ARCH] An opening in a vessel to provide access for passengers, cargo handling, discharging water, and so forth. [NUCLEO] An opening in a research reactor through which objects are inserted for irradiation or from which beams of radiation emerge for experimental use.

portability [COMPUT SCI] Property of a computer program that is sufficiently flexible to be easily transferred to run on a computer of a type different from the one for which it was designed.

portland cement [MATER] A hydraulic cement resembling portland stone when hardened; made of pulverized, calcined argillaceous and calcareous materials; the proper name for ordinary cement.

Portlandian [GEOL] A European geologic stage of the Upper Jurassic, above Kimmeridgian, below Berriasian of Cretaceous.

portlandite [MINERAL] $Ca(OH)_2$ A colorless, hexagonal mineral consisting of calcium hydroxide; occurs as minute plates.

Portulacaceae [BOT] A family of dicotyledonous plants in the order Caryophyllales distinguished by a syncarpous gynoecium, few, cyclic tepals and stamens, two sepals, and two to many ovules.

Portunidae [INV ZOO] The swimming crabs, a family of the Brachyura having the last pereiopods modified as swimming paddles.

Porulosida [INV ZOO] An order of the protozoan subclass Radiolaria in which the central capsule shows many pores.

positional astronomy [ASTRON] The branch of astronomy that deals with the determination of the positions of celestial objects.

positional notation [MATH] Any of several numeration systems in which a number is represented by a sequence of digits in such a way that the significance of each digit depends on its position in the sequence as well as its numeric value. Also known as notation.

position angle [ASTRON] 1. The angle formed by the great circle running through two celestial objects and the hour circle running through one of the objects. 2. In measuring double stars, the angle formed between the great circle running through both components and the hour circle going through the primary measured from the north through the east from 0 to 360°. [NAV] That angle of the navigational triangle at the celestial body having the hour circle and the vertical circle at its sides. Also known as parallactic angle.

position line *See* line of position.

position pulse *See* commutator pulse.

position vector [MATH] The position vector of a point in euclidean space is a vector whose length is the distance from the origin to the point and whose direction is the direction from the origin to the point. Also known as radius vector.

positive [ELEC] Having fewer electrons than normal, and hence having ability to attract electrons. [GRAPHICS] Having the same rendition of light and shade as in the original scene. [MATH] Having value greater than zero.

positive acceleration [MECH] 1. Accelerating force in an upward sense or direction, such as from bottom to top, or from seat to head; 2. The acceleration in the direction that this force is applied.

positive charge [ELEC] The type of charge which is possessed by protons in ordinary matter, and which may be produced in a glass object by rubbing with silk.

positive-displacement compressor [MECH ENG] A compressor that confines successive volumes of fluid within a closed space in which the pressure of the fluid is increased as the volume of the closed space is decreased.

positive-displacement pump [MECH ENG] A pump in which a measured quantity of liquid is entrapped in a space, its pressure is raised, and then it is delivered; for example, a reciprocating piston-cylinder or rotary-vane, gear, or lobe mechanism.

positive drive belt *See* timing belt.

positive electrode *See* anode.

positive electron *See* positron.

positive feedback [CONT SYS] Feedback in which a portion of the output of a circuit or device is fed back in phase with the input so as to increase the total amplification. Also known as reaction (British usage); regeneration; regenerative feedback; retroaction (British usage).

positive-grid oscillator *See* retarding-field oscillator.

positive image [GRAPHICS] A picture as normally seen on a television picture tube or in a photograph, having the same rendition of light and shade as in the original scene.

positive ion [CHEM] An atom or group of atoms which by loss of one or more electrons has acquired a positive electric charge; occurs on ionization of chemical compounds as H^+ from ionization of hydrochloric acid, HCl.

positive logic [ELECTR] Pertaining to a logic circuit in which the more positive voltage (or current level) represents the 1 state; the less positive level represents the 0 state.

positive meniscus lens [OPTICS] A lens having one convex (bulging) and one concave (depressed) surface, with the radius of curvature of the convex surface smaller than that of the concave surface.

positive real function [MATH] An analytic function whose value is real when the independent variable is real, and whose real part is positive or zero when the real part of the independent variable is positive or zero.

positive terminal [ELEC] The terminal of a battery or other voltage source toward which electrons flow through the external circuit.

positron [PARTIC PHYS] An elementary particle having mass equal to that of the electron, and having the same spin and statistics as the electron, but a positive charge equal in magnitude to the electron's negative charge. Also known as positive electron.

positronium [PARTIC PHYS] The bound state of an electron and a positron.

postacceleration [ELECTR] Acceleration of beam electrons after deflection in an electron-beam tube. Also known as postdeflection acceleration (PDA).

post-and-beam construction [BUILD] A type of wall construction using posts instead of studs.

post-and-lintel [ARCH] Pertaining to construction employing vertical supports and horizontal beams instead of arches or vaults.

postdeflection acceleration *See* postacceleration.

posterior [ZOO] 1. The hind end of an organism. 2. Toward the back, or hinder end, of the body.

postfix notation *See* reverse Polish notation.

postglacial [GEOL] Referring to the interval of geologic time since the total disappearance of continental glaciers in middle latitudes or from a particular area.

postmeridian [SCI TECH] After noon, or the period of time between noon (1200) and midnight (2400).

postmortem dump [COMPUT SCI] 1. The printout showing the state of all registers and the contents of main memory, taken after a computer run terminates normally or terminates owing to fault. 2. The program which generates this printout.

postmortem program *See* postmortem routine.

postmortem routine [COMPUT SCI] A computer routine designed to provide information about the operation of a program after the program is completed. Also known as postmortem program.

pot *See* potentiometer; pothole.

Potamogalinae [VERT ZOO] An aberrant subfamily of West African tenrecs (Tenrecidae).

Potamogetonaceae [BOT] A large family of monocotyledonous plants in the order Najadales characterized by a solitary, apical or lateral ovule, usually two or more carpels, flowers in spikes or racemes, and four each of tepals and stamens.

Potamogetonales [BOT] The equivalent name for Najadales.

Potamonidae [INV ZOO] A family of fresh-water crabs included in the Brachyura.

potash alum *See* kalinite.

potash feldspar *See* potassium feldspar.

potash mica *See* muscovite.

potassium [CHEM] A chemical element, symbol K, atomic number 19, atomic weight 39.102; an alkali metal. Also known as kalium.

potassium-42 [NUC PHYS] Radioactive isotope with mass number of 42; half-life is 12.4 hours, with β- and γ-radiation; radiotoxic; used as radiotracer in medicine.

potassium alum *See* potassium aluminum sulfate.

potassium aluminum sulfate [INORG CHEM] $KAl(SO_4)_2 \cdot 12H_2O$ White, odorless crystals that are soluble in water; used in medicines and baking powder, in dyeing, papermaking, and tanning. Also known as alum; aluminum potassium sulfate; potassium alum.

potassium-argon dating [GEOL] Dating of archeological, geological, or organic specimens by measuring the amount of argon accumulated in the matrix rock through decay of radioactive potassium.

potassium feldspar [MINERAL] Any alkali feldspar (orthoclase, microcline, sonidine, adularia) containing the molecule $KAlSi_3O_8$. Incorrectly known as K feldspar; potash feldspar.

potato [BOT] *Solanum tuberosum.* An erect herbaceous annual that has a round or angular aerial stem, underground lateral stems, pinnately compound leaves, and white, pink, yellow, or purple flowers occurring in cymose inflorescences; produces an edible tuber which is a shortened, thickened underground stem having nodes (eyes) and internodes. Also known as Irish potato; white potato.

potential [ELEC] *See* electric potential. [PHYS] A function or set of functions of position in space, from whose first derivatives a vector can be formed, such as that of a static field intensity.

potential barrier [PHYS] The potential in a region in a field of force where the force exerted on a particle is such as to oppose the passage of the particle through the region. Also known as barrier; potential hill.

potential divider *See* voltage divider.

potential drop [ELEC] The potential difference between two points in an electric circuit. [FL MECH]

The difference in pressure head between one equipotential line and another.

potential flow analyzer *See* electrolytic tank.

potential gradient [ELEC] Difference in the values of the voltage per unit length along a conductor or through a dielectric.

potential hill *See* potential barrier.

potential instability *See* convective instability.

potential transformer *See* voltage transformer.

potentiometer [ELEC] A resistor having a continuously adjustable sliding contact that is generally mounted on a rotating shaft; used chiefly as a voltage divider. Also known as pot (slang). [ENG] A device for the measurement of an electromotive force by comparison with a known potential difference.

potentiometric titration [ANALY CHEM] Solution titration in which the end point is read from the electrode-potential variations with the concentrations of potential-determining ions, following the Nernst concept. Also known as constant-current titration.

potentiometry [ELEC] Use of a potentiometer to measure electromotive forces, and the applications of such measurements.

pothole [CIV ENG] A pot-shaped hole in a pavement surface. [GEOL] 1. A shaftlike cave opening upward to the surface. 2. Any bowl-shaped, cylindrical, or circular hole formed by the grinding action of a stone in the rocky bed of a river or stream. Also known as churn hole; colk; eddy mill; evorsion hollow; kettle; pot. 3. A vertical, or nearly vertical shaft in limestone. Also known as aven; cenote. 4. A small depression with steep sides in a coastal marsh; contains water at or below low-tide level. Also known as rotten spot. [HYD] *See* moulin.

potomology [CIV ENG] The systematic study of the factors affecting river channels to provide the basis for predictions of the effects of proposed engineering works on channel characteristics.

Potter-Bucky grid [MED] An assembly of lead strips resembling an open venetian blind, placed between a patient being x-rayed and the screen or film, to reduce the effects of scattered radiation. Also known as Bucky diaphragm; grid.

Pottiales [BOT] An order of mosses distinguished by erect stems, lanceolate to broadly ovate or obovate leaves, a strong, mostly percurrent or excurrent costa, and a cucullate calyptra.

potting [ELECTR] Process of filling a complete electronic assembly with a thermosetting compound for resistance to shock and vibration, and for exclusion of moisture and corrosive agents.

pound [MECH] 1. A unit of mass in the English absolute system of units, equal to 0.45359237 kilogram. Abbreviated lb. Also known as avoirdupois pound; pound mass. 2. A unit of force in the English gravitational system of units, equal to the gravitational force experienced by a pound mass when the acceleration of gravity has its standard value of 9.80665 meters per second per second (approximately 32.1740 ft/sec²) equal to 4.4482216152605 newtons. Abbreviated lb. Also spelled Pound (Lb). Also known as pound force (lbf). 3. A unit of mass in the troy and apothecaries' systems, equal to 12 troy or apothecaries' ounces, or 5760 grains, or 5760/7000 avoirdupois pound, or 0.3732417216 kilogram. Also known as apothecaries' pound (abbreviated lb ap in the US or lb apoth in the UK), troy pound (abbreviated lb UK).

poundal [MECH] A unit of force in the British absolute system of units equal to the force which will impart an acceleration of 1 ft/sec² to a pound mass, or to 0.138254954376 newton.

poundal-foot *See* foot-poundal.

pound-foot *See* foot-pound.

pound force *See* pound.

pound mass *See* pound.

pour-plate culture [MICROBIO] A technique for pure-culture isolation of bacteria; liquid, cooled agar in a test tube is inoculated with one loopful of bacterial suspension and mixed by rolling the tube between the hands; subsequent transfers are made from this to a second test tube, and from the second to a third; contents of each tube are poured into separate petri dishes; pure cultures can be isolated from isolated colonies appearing on the plates after incubation.

pour point [FL MECH] Lowest test temperature at which a liquid will flow. [MET] Temperature at which a molten alloy is cast.

Pourtalesiidae [INV ZOO] A family of exocyclic Euechinoidea in the order Holasteroida, including those forms with a bottle-shaped test.

powder clutch [MECH ENG] A type of electromagnetic disk clutch in which the space between the clutch members is filled with dry, finely divided magnetic particles; application of a magnetic field coalesces the particles, creating friction forces between clutch members.

powder metallurgy [MET] The production of massive materials and shaped objects by pressing, binding, and sintering powdered metal.

powder method [SOLID STATE] A method of x-ray diffraction analysis in which a collimated, monochromatic beam of x-rays is directed at a sample consisting of an enormous number of tiny crystals having random orientation, producing a diffraction pattern that is recorded on film or with a counter tube. Also known as x-ray powder method.

powder molding [ENG] Generic term for plastics-molding techniques to produce objects of varying sizes and shapes by melting polyethylene powder, usually against the heated inside of a mold.

powder snow [HYD] A cover of dry snow that has not been compacted in any way.

powdery mildew [MYCOL] A fungus characterized by production of abundant powdery conidia on the host; a member of the family Erysiphaceae or the genus *Oidium*. [PL PATH] A plant disease caused by a powdery mildew fungus.

powdery scab [PL PATH] A fungus disease of potato tubers caused by *Spongospora subterranea* and characterized by nodular discolored lesions, which burst and expose masses of powdery fungus spores.

powellite [MINERAL] Ca(WMo)O₄ A commercially important tungsten mineral, crystallizing in the tetragonal system; isomorphous with scheelite (CaWO₄).

power [MATH] 1. The value that is assigned to a mathematical expression and its exponent. 2. The power of a set is its cardinality. [PHYS] The time rate of doing work.

power amplification *See* power gain.

power amplifier [ELECTR] The final stage in multistage amplifiers, such as audio amplifiers and radio transmitters, designed to deliver maximum power to the load, rather than maximum voltage gain, for a given percent of distortion.

power amplifier tube *See* power tube.

power attenuation *See* power loss.

power bandwidth [COMMUN] The frequency range for which half the rated power of an audio amplifier is available at rated distortion.

power brake [MECH ENG] An automotive brake with engine-intake-manifold vacuum used to amplify the atmospheric pressure on a piston operated by movement of the brake pedal.

power component *See* active component.

power cord *See* line cord.

power density [ELECTROMAG] The amount of power per unit area in a radiated microwave or other electromagnetic field, usually expressed in units of watts per square centimeter. [NUCLEO] The power generation per unit volume of a nuclear-reactor core.

power-density spectrum *See* frequency spectrum.

power diode *See* pin-diode.

power factor [ELEC] The ratio of the average (or active) power to the apparent power (root-mean-square voltage times rms current) of an alternating-current circuit. Abbreviated pf. Also known as phase factor.

power gain [ELECTR] The ratio of the power delivered by a transducer to the power absorbed by the input circuit of the transducer. Also known as power amplification. [ELECTROMAG] An antenna ratio equal to 4π (12.57) times the ratio of the radiation intensity in a given direction to the total power delivered to the antenna.

power generator [ELEC] A device for producing electric energy, such as an ordinary electric generator or a magnetohydrodynamic, thermionic, or thermoelectric power generator.

power level [ELEC] The ratio of the amount of power being transmitted past any point in an electric system to a reference power value; usually expressed in decibels. [NUCLEO] The power production of a nuclear reactor in watts.

power line [ELEC] Two or more wires conducting electric power from one location to another. Also known as electric power line.

power-line filter *See* line filter.

power loss [ELECTR] The ratio of the power absorbed by the input circuit of a transducer to the power delivered to a specified load; usually expressed in decibels. Also known as power attenuation.

power meter *See* electric power meter.

power of a test [STAT] One minus the probability that a given test causes the acceptance of the null hypothesis when it is false due to the validity of an alternative hypothesis; this is the same as the probability of rejecting the null hypothesis by the test when the alternative is true.

power output [ELECTR] The alternating-current power in watts delivered by an amplifier to a load.

power output tube *See* power tube.

power pack [ELECTR] Unit for converting power from an alternating- or direct-current supply into an alternating- or direct-current power at voltages suitable for supplying an electronic device.

power package [MECH ENG] A complete engine and its accessories, designed as a single unit for quick installation or removal.

power plant [MECH ENG] Any unit that converts some form of energy into electrical energy, such as a hydroelectric or steam-generating station, a diesel-electric engine in a locomotive, or a nuclear power plant. Also known as electric power plant.

power rating [ELEC] The power available at the output terminals of a component or piece of equipment that is operated according to the manufacturer's specifications.

power reactor [NUCLEO] A nuclear reactor designed to provide useful power, as for submarines, aircraft, ships, vehicles, and power plants.

power semiconductor [ELECTR] A semiconductor device capable of dissipating appreciable power (generally over 1 watt) in normal operation; may handle currents of thousands of amperes or voltages up into thousands of volts, at frequencies up to 10 kilohertz.

power series [MATH] An infinite series composed of functions having nth term of the form $a_n(x - x_0)^n$, where x_0 is some point and a_n some constant.

power spectrum *See* frequency spectrum.

power steering [MECH ENG] A steering control system for a propelled vehicle in which an auxiliary power source assists the driver by providing the major force required to direct the road wheels.

power stroke [MECH ENG] The stroke in an engine during which pressure is applied to the piston by expanding steam or gases.

power supply [ELECTR] A source of electrical energy, such as a battery or power line, employed to furnish the tubes and semiconductor devices of an electronic circuit with the proper electric voltages and currents for their operation. Also known as electronic power supply.

power train [MECH ENG] The part of a vehicle connecting the engine to propeller or driven axle; may include drive shaft, clutch, transmission, and differential gear.

power transformer [ELEC] An iron-core transformer having a primary winding that is connected to an alternating-current power line and one or more secondary windings that provide different alternating voltage values.

power transistor [ELECTR] A junction transistor designed to handle high current and power; used chiefly in audio and switching circuits.

power transmission line [ELEC] The facility in an electric power system used to transfer large amounts of power from one location to a distant location; distinguished from a subtransmission or distribution line by higher voltage, greater power capability, and greater length. Also known as electric main; main (both British usages).

power tube [ELECTR] An electron tube capable of handling more current and power than an ordinary voltage-amplifier tube; used in the last stage of an audio-frequency amplifier or in high-power stages of a radio-frequency amplifier. Also known as power amplifier tube; power output tube.

power winding [ELEC] In a saturable reactor, a winding to which is supplied the power to be controlled; commonly the functions of the output and power windings are accomplished by the same winding, which is then termed the output winding.

poxvirus [VIROL] A deoxyribonucleic acid–containing animal virus group including the viruses of smallpox, molluscum contagiosum, and various animal pox and fibromas.

Poynting vector [ELECTROMAG] A vector, equal to the cross product of the electric-field strength and the magnetic-field strength (mks units) whose outward normal component, when integrated over a closed surface, gives the outward flow of electromagnetic energy through that surface.

pozzolan [GEOL] A finely ground burnt clay or shale resembling volcanic dust, found near Pozzuoli, Italy; used in cement because it hardens underwater. [MATER] Cement made by mixing and grinding together slaked lime and pozzolan without burning; sometimes used for concrete not exposed to the air.

PPI *See* plan position indicator.

PPM *See* pulse-position modulation.

P pulse *See* commutator pulse.

Pr *See* praseodymium.

practical entropy *See* virtual entropy.

practical system *See* meter-kilogram-second-ampere system.

practical units [ELECTROMAG] The units of the meter-kilogram-second-ampere system.

Praesepe [ASTRON] A cluster of faint stars in the center of the constellation Cancer. Also known as Beehive; Manger.

prairie [GEOGR] An extensive level-to-rolling treeless tract of land in the temperate latitudes of central North America, characterized by deep, fertile soil and a cover of coarse grass and herbaceous plants.

prairie dog [VERT ZOO] The common name for three species of stout, fossorial rodents belonging to the genus *Cynomys* in the family Sciuridae; all have a short, flat tail, small ears, and short limbs terminating in long claws.

prairie wolf *See* coyote.

praseodymium [CHEM] A chemical element, symbol Pr, atomic number 59, atomic weight 140.91; a metallic element of the rare-earth group.

Prasinovolvocales [BOT] An order of green algae in which there are lateral appendages in the flagellum.

Pratt truss [CIV ENG] A truss having both vertical and diagonal members between the upper and lower chords, with the diagonals sloped toward the center.

preamplifier [ELECTR] An amplifier whose primary function is boosting the output of a low-level audio-frequency, radio-frequency, or microwave source to an intermediate level so that the signal may be further processed without appreciable degradation of the signal-to-noise ratio of the system. Also known as preliminary amplifier.

Precambrian [GEOL] All geologic time prior to the beginning of the Paleozoic era (before 600,000,000 years ago); equivalent to about 90% of all geologic time.

precast concrete [MATER] Concrete components which are cast and partly matured in a factory or on the site before being lifted into their final position on a structure.

precession [MECH] The angular velocity of the axis of spin of a spinning rigid body, which arises as a result of external torques acting on the body.

precession of the equinoxes [ASTRON] A slow conical motion of the earth's axis about the vertical to the plane of the ecliptic, having a period of 26,000 years, caused by the attractive force of the sun, moon, and other planets on the equatorial protuberance of the earth; it results in a gradual westward motion of the equinoxes.

precipitable water [METEOROL] The total atmospheric water vapor contained in a vertical column of unit cross-sectional area extending between any two specified levels, commonly expressed in terms of the height to which that water substance would stand if completely condensed and collected in a vessel of the same unit cross section. Also known as precipitable water vapor.

precipitable water vapor *See* precipitable water.

precipitant [CHEM] A chemical or chemicals that cause a precipitate to form when added to a solution.

precipitate [CHEM] 1. A substance separating, in solid particles, from a liquid as the result of a chemical or physical change. 2. To form a precipitate.

precipitation [CHEM] The process of producing a separable solid phase within a liquid medium; represents the formation of a new condensed phase, such as a vapor or gas condensing to liquid droplets; a new solid phase gradually precipitates within a solid alloy as a result of slow, inner chemical reaction; in analytical chemistry, precipitation is used to separate a solid phase in an aqueous solution. [IMMUNOL] Aggregation of soluble antigen by an antibody. [METEOROL] **1.** Any or all of the forms of water particles, whether liquid or solid, that fall from the atmosphere and reach the ground. **2.** The amount, usually expressed in inches of liquid water depth, of the water substance that has fallen at a given point over a specified period of time.

precipitation titration [ANALY CHEM] Amperometric titration in which the potential of a suitable indicator electrode is measured during the titration.

precipitator *See* electrostatic precipitator.

precipitin test [IMMUNOL] An immunologic test in which a specific reaction between antigen and antibody results in a visible precipitate.

precision approach radar [NAV] A radar system located on an airfield for observation of the position of an aircraft with respect to an approach path, and specifically intended to provide guidance to the aircraft during its approach to the field; the system consists of a ground radar equipment which is alternately connected to two antenna systems; one antenna system sweeps a narrow beam over a 20° sector in the horizontal plane; the second sweeps a narrow beam over a 7° sector in the vertical plane; course correction is transmitted to the aircraft from the ground. Abbreviated PAR.

precision block *See* gage block.

precombustion chamber [MECH ENG] A small chamber before the main combustion space of a turbine or reciprocating engine in which combustion is initiated.

precompact set [MATH] A set in a metric space which can always be covered by open balls of any diameter about some finite number of its points. Also known as totally bounded set.

preconsolidation pressure [GEOL] The greatest effective stress exerted on a soil; result of this pressure from overlying materials is compaction. Also known as prestress.

predation [BIOL] The killing and eating of an individual of one species by an individual of another species.

preen gland *See* uropygial gland.

preferred orientation [PETR] The nonrandom orientation of planar or linear fabric elements in structural petrology.

prefix notation *See* Polish notation.

preform [ENG ACOUS] The small slab of record stock material that is loaded into a press to be formed into a disk recording. Also known as biscuit (deprecated usage).

prefrontal [ANAT] Situated in the anterior part of the frontal lobe of the brain. [VERT ZOO] **1.** Of or pertaining to a bone of some vertebrate skulls, located anterior and lateral to the frontal bone. **2.** Of, pertaining to, or being a scale or plate in front of the frontal scale on the head of some reptiles and fishes.

prefrontal lobotomy *See* lobotomy.

prehensile [VERT ZOO] Adapted for seizing, grasping, or plucking, especially by wrapping around some object.

prehnite [MINERAL] $Ca_2Al_2Si_3O_{10}(OH)_2$ A light-green to white mineral sorosilicate crystallizing in the orthorhombic system and generally found in reniform and stalactitic aggregates with crystalline surface; it has a vitreous luster, hardness is 6–6.5 on Mohs scale, and specific gravity is 2.8–2.9.

preignition [MECH ENG] Ignition of the charge in the cylinder of an internal combustion engine before ignition by the spark.

preliminary amplifier *See* preamplifier.

premaxilla [ANAT] Either of two bones of the upper jaw of vertebrates located in front of and between the maxillae.

preprogramming [COMPUT SCI] The prerecording of instructions or commands for a machine, such as an automated tool in a factory.

prepuce [ANAT] **1.** The foreskin of the penis, a fold of skin covering the glans penis. **2.** A similar fold over the glans clitoridis.

preread head [COMPUT SCI] A read head that is placed near another read head in such a way that it can read data stored on a moving medium such as a tape or disk before these data reach the second head.

presbyopia [MED] Diminished ability to focus the eye on near objects due to gradual loss of elasticity of the crystalline lens with age.

preselection [COMPUT SCI] A technique for saving time available in buffered computers by which a block of data is read into computer storage from the next input tape to be called upon before the data are required in the computer; the selection of the next input tape is determined by instructions to the computer.

presentation *See* radar display.

preservative [MATER] A chemical added to foodstuffs to prevent oxidation, fermentation, or other deterioration, usually by inhibiting the growth of bacteria.

preset [COMPUT SCI] **1.** Of a variable, having a value established before the first time it is used. **2.** To initialize a value of a variable before the value of the variable is used or tested.

presort [COMPUT SCI] **1.** The first part of a sort program in which data items are arranged into strings that are equal to or greater than some prescribed length. **2.** The sorting of data on off-line equipment before it is processed by a computer.

press [MECH ENG] Any of various machines by which pressure is applied to a workpiece, by which a material is cut or shaped under pressure, by which a substance is compressed, or by which liquid is expressed.

press fit [ENG] An interference or force fit assembled through the use of a press. Also known as force fit.

press forging [MET] Forging hot metal between dies in a press.

press proof [GRAPHICS] Proof removed from the printing press to inspect line and color values and overall quality; this is the last proof before the complete printing is done.

pressure [MECH] A type of stress which is exerted uniformly in all directions; its measure is the force exerted per unit area.

pressure altimeter [ENG] A highly refined aneroid barometer that precisely measures the pressure of the air at the altitude an aircraft is flying, and converts the pressure measurement to an indication of height above sea level according to a standard pressure-altitude relationship. Also known as barometric altimeter.

pressure breccia *See* tectonic breccia.

pressure carburetor *See* injection carburetor.

pressure center [METEOROL] **1.** On a synoptic chart (or on a mean chart of atmospheric pressure), a point of local minimum or maximum pressure; the center of a low or high. **2.** A center of cyclonic or anticyclonic circulation.

pressure chamber [ENG] A chamber in which an artificial environment is established at low or high pressures to test equipment under simulated conditions of operation. [MIN ENG] An enclosed space that seals off a part of a mine and in which the air pressure can be raised or lowered.

pressure coefficient [THERMO] The ratio of the fractional change in pressure to the change in temperature under specified conditions, usually constant volume.

pressure drawdown [PETRO ENG] The drop in reservoir pressure related to the withdrawal of gas from a producing well; for low-permeability formations, pressures near the wellbores can be much lower than in the main part of the reservoir; leads to pressure decline in the reservoir and ultimate pressure depletion.

pressure front See shock front.

pressure gradient [FL MECH] The rate of decrease (that is, the gradient) of pressure in space at a fixed time; sometimes loosely used to denote simply the magnitude of the gradient of the pressure field. Also known as barometric gradient. [METEOROL] The change in atmospheric pressure per unit horizontal distance, usually measured along a line perpendicular to the isobars.

pressure head [FL MECH] Also known as head. **1.** The height of a column of fluid necessary to develop a specific pressure. **2.** The pressure of water at a given point in a pipe arising from the pressure in it.

pressure ice [OCEANOGR] Ice, especially sea ice, which has been deformed or altered by the lateral stresses of any combination of wind, water currents, tides, waves, and surf; may include ice pressed against the shore, or one piece of ice upon another.

pressure microphone [ENG ACOUS] A microphone whose output varies with the instantaneous pressure produced by a sound wave acting on a diaphragm; examples are capacitor, carbon, crystal, and dynamic microphones.

pressure pad [ENG] A steel reinforcement in the face of a plastics mold to help the land absorb the closing pressure. [ENG ACOUS] A felt pad mounted on a spring arm, used to hold magnetic tape in close contact with the head on some tape recorders.

pressure plate [MECH ENG] The part of an automobile disk clutch that presses against the flywheel.

pressure-plate anemometer [ENG] An anemometer which measures wind speed in terms of the drag which the wind exerts on a solid body; may be classified according to the means by which the wind drag is measured. Also known as plate anemometer.

pressure point [PHYSIO] A point of marked sensibility to pressure or weight, arranged like the temperature spots, and showing a specific end apparatus arranged in a punctate manner and connected with the pressure sense.

pressure rating [ENG] The operating (allowable) internal pressure of a vessel, tank, or piping used to hold or transport liquids or gases.

pressure ridge [GEOL] **1.** A seismic feature resulting from transverse pressure and shortening of the land surface. **2.** An elongate upward movement of the congealing crust of a lava flow. **3.** A ridge of glacier ice. [OCEANOGR] A ridge or wall of hummocks where one ice floe has been pressed against another.

pressure seal [ENG] A seal used to make pressure-proof the interface (contacting surfaces) between two parts that have frequent or continual relative rotational or translational motion.

pressure still [CHEM ENG] A continuous-flow, petroleum-refinery still in which heated oil (liquid and vapor) is kept under pressure so that it will crack (decompose into smaller molecules) to produce lower-boiling products (pressure distillate or pressure naphtha).

pressure system [ENG] Any system of pipes, vessels, tanks, reactors, and other equipment, or interconnections thereof, operating with an internal pressure greater than atmospheric. [METEOROL] An individual cyclonic-scale feature of atmospheric circulation, commonly used to denote either a high or a low, less frequently a ridge or a trough.

pressure transducer [ENG] An instrument component that detects a fluid pressure and produces an electrical signal related to the pressure. Also known as electrical pressure transducer.

pressure-tube anemometer [ENG] An anemometer which derives wind speed from measurements of the dynamic wind pressures; wind blowing into a tube develops a pressure greater than the static pressure, while wind blowing across a tube develops a pressure less than the static; this pressure difference, which is proportional to the square of the wind speed, is measured by a suitable manometer.

pressure-tube reactor [NUCLEO] A nuclear reactor in which the fuel elements are located inside numerous tubes containing coolant circulating at high pressure; the tube assembly is surrounded by a tank containing the moderator at low pressure.

pressure ulcer See decubitus ulcer.

pressure wave [METEOROL] A wave or periodicity which exists in the variation of atmospheric pressure on any time scale, usually excluding normal diurnal or seasonal trends. [PHYS] See compressional wave.

pressure welding [MET] Welding of metal surfaces by the application of pressure; examples are percussion welding, resistance welding, seam welding, and spot welding.

pressurization [ENG] **1.** Use of an inert gas or dry air, at several pounds above atmospheric pressure, inside the components of a radar system or in a sealed coaxial line, to prevent corrosion by keeping out moisture, and to minimize high-voltage breakdown at high altitudes. **2.** The act of maintaining normal atmospheric pressure in a chamber subjected to high or low external pressure.

pressurized blast furnace [ENG] A blast furnace operated under pressure above the ambient; pressure is obtained by throttling the off-gas line, which permits a greater volume of air to be passed through the furnace at a lower velocity, and results in increase in smelting rate.

pressurized water reactor [NUCLEO] A nuclear reactor in which water is circulated under enough pressure to prevent it from boiling, while serving as moderator and coolant for the uranium fuel; the heated water is then used to produce steam for a power plant. Abbreviated PWR.

presswork [GRAPHICS] In printing, the actual operation of putting ink on paper; this activity is preceded by composition and perhaps platemaking, and is followed by binding.

prestore [COMPUT SCI] To store a quantity in an available computer location before it is required in a routine.

prestress [ENG] To apply a force to a structure to condition it to withstand its working load more effectively or with less deflection. [GEOL] See preconsolidation pressure.

prestressed concrete [MATER] Concrete compressed with heavily loaded wires or bars to reduce or eliminate cracking and tensile forces.

presumptive address See address constant.

presumptive instruction See basic instruction.

pretensioning [ENG] Process of precasting concrete beams with tensioned wires embedded in them. Also known as Hoyer method of prestressing.

pretersonics *See* acoustoelectronics.

prevailing wind *See* prevailing wind direction.

prevailing wind direction [METEOROL] The wind direction most frequently observed during a given period; the periods most often used are the observational day, month, season, and year. Also known as prevailing wind.

prevaporization [PHYS CHEM] The phase transformations of liquids to gases prior to some physical or chemical reactions.

previewing [COMPUT SCI] In character recognition, a process of attempting to gain prior information about the characters that appear on an incoming source document; this information, which may include the range of ink density, relative positions, and so forth, is used as an aid in the normalization phase of character recognition.

PRF *See* pulse repetition rate.

pri *See* primary winding.

Priapulida [INV ZOO] A minor phylum of wormlike marine animals; the body is made up of three distinct portions (proboscis, trunk, and caudal appendage) and is often covered with spines and tubercles, and the mouth is surrounded by concentric rows of teeth.

Priapuloidea [INV ZOO] An equivalent name for Priapulida.

priceite [MINERAL] $Ca_4B_{10}O_{19} \cdot 7H_2O$ A snow-white earthy mineral composed of hydrous calcium borate, occurring as a massive. Also known as pandermite.

pricker *See* needle.

prill [CHEM ENG] To form pellet-sized crystals or agglomerates of material by the action of upward-blowing air on falling hot solution; used in the manufacture of ammonium nitrate and urea fertilizers. [MATER] Spherical particles about the size of buckshot. [MIN ENG] 1. The best ore after cobbing. 2. A circular particle about the size of buckshot. 3. Compressed and sized explosives such as ammonium nitrate.

primary [ASTRON] 1. A planet with reference to its satellites, or the sun with reference to its planets. 2. The brighter star of a double star system. [CHEM] A term used to distinguish basic compounds from similar or isomeric forms; in organic compounds, for example, RCH_2OH is a primary alcohol, R_1R_2CHOH is a secondary alcohol, and $R_1R_2R_3COH$ is a tertiary alcohol; in inorganic compounds, for example, NaH_2PO_4 is primary sodium phosphate, Na_2HPO_4 is the secondary form, and Na_3PO_4 is the tertiary form. [ELEC] 1. *See* primary winding. 2. One of the high-voltage conductors of a power distribution system. [GEOL] 1. A young shoreline whose features are produced chiefly by nonmarine agencies. 2. Of a mineral deposit, unaffected by supergene enrichment. [MET] Of a metal, obtained directly from ore. [VERT ZOO] Of or pertaining to quills on the distal joint of a bird wing.

primary battery [ELEC] A battery consisting of one or more primary cells.

primary cell [ELEC] A cell that delivers electric current as a result of an electrochemical reaction that is not efficiently reversible, so that the cell cannot be recharged efficiently.

primary circle *See* primary great circle.

primary colors [OPTICS] Three colors, red, yellow, and blue, which can be combined in various proportions to produce any other color.

primary constriction *See* centromere.

primary control program [COMPUT SCI] The program which provides the sequential scheduling of jobs and basic operating systems functions. Abbreviated PCP.

primary cosmic rays *See* cosmic rays.

primary detector *See* sensor.

primary dip [GEOL] The slight dip assumed by a bedded deposit at its moment of deposition. Also known as depositional dip; initial dip; original dip.

primary extinction [SOLID STATE] A weakening of the stronger beams produced in x-ray diffraction by a very perfect crystal, as compared with the weaker.

primary fabric *See* apposition fabric.

primary fault [ELEC] In an electric circuit, the initial breakdown of the insulation of a conductor, usually followed by a flow of power current.

primary frequency [COMMUN] Frequency assigned for normal use on a particular circuit or communications channel.

primary geosyncline *See* orthogeosyncline.

primary great circle [GEOD] A great circle used as the origin of measurement of a coordinate; particularly, such a circle 90° from the poles of a system of spherical coordinates, as the equator. Also known as fundamental circle; primary circle.

primary hypertension *See* essential hypertension.

primary lateral sclerosis [MED] A sclerotic disease of the crossed pyramidal tracts of the spinal cord, characterized by paralysis of the limbs, with rigidity, increased tendon reflexes, and absence of sensory and nutritive disorders. Also known as lateral sclerosis.

primary mineral [MINERAL] A mineral that is formed at the same time as the rock in which it is contained, and that retains its original form and composition.

primary radar [ENG] Radar in which the incident beam is reflected from the target to form the return signal. Also known as primary surveillance radar (PSR).

primary reserve [PETRO ENG] Petroleum reserve recoverable commercially at current prices and costs by conventional methods and equipment as a result of the natural energy inherent in the reservoir.

primary root [BOT] The first plant root to develop; derived from the radicle.

primary scattering [PHYS] Any scattering process in which radiation is received at a detector, such as the eye, after having been scattered just once; distinguished from multiple scattering.

primary sedimentary structure [GEOL] A sedimentary structure produced during deposition, such as ripple marks and graded bedding.

primary standard [SCI TECH] A unit directly defined and established by some authority, against which all secondary standards are calibrated; for example, in analytical chemistry, reference substances or solutions of known chemical purity and concentration are used to standardize laboratory solutions prior to volumetric analysis or titration.

primary storage [COMPUT SCI] Main internal storage of a computer.

primary stress [MECH] A normal or shear stress component in a solid material which results from an imposed loading and which is under a condition of equilibrium and is not self-limiting.

primary surveillance radar *See* primary radar.

primary tissue [BOT] Plant tissue formed during primary growth. [HISTOL] Any of the four fundamental tissues composing the vertebrate body.

primary wave [COMMUN] A radio wave traveling by a direct path, as contrasted with skips. [GEOPHYS]

The first seismic wave that reaches a station from an earthquake.

primary winding [ELEC] The transformer winding that receives signal energy or alternating-current power from a source. Also known as primary. Abbreviated pri. Symbolized P.

Primates [VERT ZOO] The order of mammals to which man belongs; characterized in terms of evolutionary trends by retention of a generalized limb structure and dentition, increasing digital mobility, replacement of claws by flat nails, development of stereoscopic vision, and progressive development of the cerebral cortex.

prime [ENG] **1.** Main or primary, as in prime contractor. **2.** In blasting, to place a detonator in a cartridge or charge of explosive. **3.** To treat wood with a primer or penetrant primer.

prime coat See primer.

prime meridian [GEOD] The meridian of longitude 0°, used as the origin for measurement of longitude; the meridian of Greenwich, England, is almost universally used for this purpose.

prime mover [MECH ENG] The component of a power plant that transforms energy from the thermal or the pressure form to the mechanical form.

prime number [MATH] A positive integer having no divisors except itself and the integer 1.

prime polynomial [MATH] A polynomial whose only factors are itself and constants.

primer [ENG] In general, a small, sensitive initial explosive train component which on being actuated initiates functioning of the explosive train, and will not reliably initiate high explosive charge; classified according to the method of initiation, for example, percussion primer, electric primer, or friction primer. [MATER] A prefinishing coat applied to a wood surface that is to be painted or otherwise finished. Also known as prime coat.

priming pump [MECH ENG] A device on motor vehicles and tanks, providing a means of injecting a spray of fuel into the engine to facilitate starting.

Primitiopsacea [PALEON] A small dimorphic superfamily of extinct ostracods in the suborder Beyrichicopina; the velum of the male was narrow and uniform, but that of the female was greatly expanded posteriorly.

primitive circle [MATH] The stereographic projection of the great circle whose plane is perpendicular to the diameter of the projected sphere that passes through the point of projection.

primitive polynomial [MATH] A polynomial with integer coefficients which have 1 as their greatest common divisor.

primitive root [MATH] An nth root of unity that is not an mth root of unity for any m less than n.

primitive water [HYD] Water that has been imprisoned in the earth's interior, in either molecular or dissociated form, since the formation of the earth.

primordium See anlage.

Primulaceae [BOT] A family of dicotyledonous plants in the order Primulales characterized by a herbaceous habit and capsular fruit with two to many seeds.

Primulales [BOT] An order of dicotyledonous plants in the subclass Dilleniidae distinguished by sympetalous flowers, stamens located opposite the corolla lobes, and a compound ovary with a single style.

principal axis [CRYSTAL] The longest axis in a crystal. [ENG ACOUS] A reference direction for angular coordinates used in describing the directional characteristics of a transducer; it is usually an axis of structural symmetry or the direction of maximum response. [MECH] One of three perpen-

dicular axes in a rigid body such that the products of inertia about any two of them vanish. [OPTICS] See optic axis.

principal branch [MATH] For complex valued functions such as the logarithm which are multiple-valued, a selection of values so as to obtain a genuine single-valued function.

principal curvatures [MATH] For a point on a surface, the absolute maximum and absolute minimum values attained by the normal curvature.

principal focus See focal point.

principal lobe [PHYS] The lobe of a radiation pattern or directivity pattern that lies on the axis of symmetry of an acoustic or electromagnetic transmitter or receptor.

principal plane [OPTICS] **1.** Two planes perpendicular to the optical axis such that objects in one plane form images in the other with a lateral magnification of unity. **2.** See principal section. **3.** The vertical plane passing through the internal perspective center and containing the perpendicular from that center to the plane of a tilted photograph.

principal quantum number [ATOM PHYS] A quantum number for orbital electrons, which, together with the orbital angular momentum and spin quantum numbers, labels the electron wave function; the energy level and the average distance of an electron from the nucleus depend mainly upon this quantum number.

principal root [MATH] The positive real root of a positive number, or the negative real root in the case of odd roots of negative numbers.

principal section [OPTICS] A plane in a crystal that contains the crystal's optic axis and the ray of light under consideration. Also known as principal plane.

principle of covariance [RELAT] **1.** In classical physics and in special relativity, the principle that the laws of physics take the same mathematical form in all inertial reference frames. **2.** In general relativity, the principle that the laws of physics take the same mathematical form in all conceivable curvilinear coordinate systems.

principle of duality See duality principle.

principle of equivalence See equivalence principle.

principle of inaccessibility See Carathéodory's principle.

principle of least action [MECH] The principle that, for a system whose total mechanical energy is conserved, the trajectory of the system in configuration space is that path which makes the value of the action stationary relative to nearby paths between the same configurations and for which the energy has the same constant value. Also known as least-action principle.

principle of reciprocity See reciprocity theorem.

principle of superposition [ELEC] **1.** The principle that the total electric field at a point due to the combined influence of a distribution of point charges is the vector sum of the electric field intensities which the individual point charges would produce at that point if each acted alone. **2.** The principle that, in a linear electrical network, the voltage or current in any element resulting from several sources acting together is the sum of the voltages or currents resulting from each source acting alone. Also known as superposition theorem. [MECH] The principle that when two or more forces act on a particle at the same time, the resultant force is the vector sum of the two. [PHYS] Also known as superposition principle. **1.** A general principle applying to many physical systems which states that if a number of independent influences act on the system, the resultant influence is the sum of the individual influences acting separately. **2.** In all theories characterized by linear homogeneous differential equations, such as optics, acoustics, and quantum theory, the principle that the sum of any number of solutions to the equations is another solution.

principle of the maximum [MATH] The principle that for a nonconstant complex analytic function defined in a domain, the absolute value of the function cannot attain its maximum at any interior point of the domain.

principle of the minimum [MATH] The principle that for a nonvanishing nonconstant complex analytic function defined in a domain, the absolute value of the function cannot attain its minimum at any interior point of the domain.

principle of virtual work [MECH] The principle that the total work done by all forces acting on a system in static equilibrium is zero for any infinitesimal displacement from equilibrium which is consistent with the constraints of the system. Also known as virtual work principle.

printed circuit [ELECTR] A conductive pattern that may or may not include printed components, formed in a predetermined design on the surface of an insulating base in an accurately repeatable manner.

printed circuit board [ELECTR] A flat board whose front contains slots for integrated circuit chips and connections for a variety of electronic components, and whose back is printed with electrically conductive pathways between the components.

print hammer [GRAPHICS] A device on certain kinds of printers which, upon receiving the proper signal, strikes the paper, bringing it in contact with the character to be printed.

printout [COMPUT SCI] A printed output of a data-processing machine or system.

printthrough [ELECTR] Transfer of signals from one recorded layer of magnetic tape to the next on a reel.

print wheel [COMPUT SCI] A disk which has around its rim the letters, numerals, and other characters that are used in printing in a wheel printer.

Prioniodidae [PALEON] A family of conodonts in the suborder Conodontiformes having denticulated bars with a large denticle at one end.

Prioniodinidae [PALEON] A family of conodonts in the suborder Conodontiformes characterized by denticulated bars or blades with a large denticle in the middle third of the specimen.

priorite [MINERAL] $(Y, Ca, Th)(Ti, Nb)_2O_6$ A mineral composed of titanoniobate of rare-earth metals; it is isomorphous with eschynite. Also known as blomstrandine.

priority indicator [COMMUN] Data attached to a message to indicate its relative priority and hence the order in which it will be transmitted. [COMPUT SCI] Data attached to a computer program or job which are used to determine the order in which it will be processed by the computer.

priority interrupt [COMPUT SCI] An interrupt procedure in which control is passed to the monitor, the required operation is initiated, and then control returns to the running program, which never knows that it has been interrupted.

priority phase [COMPUT SCI] Phase consisting of execution of operations in response to instruments or process interrupts other than clock interrupts.

prism [CRYSTAL] A crystal which has three, four, six, eight, or twelve faces, with the face intersection edges parallel, and which is open only at the two ends of

the axis parallel to the intersection edges. [GEOL] A long, narrow, wedge-shaped sedimentary body with a width-thickness ratio greater than 5 to 1 but less than 50 to 1. [MATH] A polyhedron with two parallel, congruent faces and all other faces parallelograms. [OPTICS] An optical system consisting of two or more usually plane surfaces of a transparent solid or embedded liquid at an angle with each other. Also known as optical prism.

prismatic astrolabe [ENG] A surveying instrument that makes use of a pan of mercury forming an artificial horizon, and a prism mounted in front of a horizontal telescope to determine the exact times at which stars reach a fixed altitude, and thereby to establish an astronomical position.

prismatic binoculars See prism binoculars.

prismatic cleavage [CRYSTAL] A type of crystal cleavage that occurs parallel to the faces of a prism.

prismatic jointing See columnar jointing.

prismatic structure See columnar jointing.

prismatoid [MATH] A polyhedron whose vertices all are in one or the other of two parallel planes.

prism binoculars [OPTICS] A type of binoculars, each half of which is a Kepler telescope that employs a Porro prism erecting system both to erect the image and to reduce the length of the instrument. Also known as prismatic binoculars.

prism level [ENG] A surveyor's level with prisms that allow the levelman to view the level bubble without moving his eye from the telescope.

prismoid [MATH] A prismatoid whose two parallel faces are polygons having the same number of sides while the other faces are trapezoids or parallelograms.

prism spectrograph [SPECT] Analysis device in which a prism is used to give two different but simultaneous light wavelengths derived from a common light source; used for the analysis of materials by flame photometry.

Pristidae [VERT ZOO] The sawfishes, a family of modern sharks belonging to the batoid group.

Pristiophoridae [VERT ZOO] The saw sharks, a family of modern sharks often grouped with the squaloids which have a greatly extended rostrum with enlarged denticles along the margins.

privacy system [COMMUN] A device or method for scrambling overseas telephone conversations handled by radio links in order to make them unintelligible to outside listeners. Also known as privacy transformation; secrecy system.

privacy transformation See privacy system.

private automatic branch exchange [COMMUN] A private automatic branch exchange in which connections are made by remote-controlled switches. Abbreviated PABX.

private automatic exchange [COMMUN] A private telephone exchange in which connections are made by remote-controlled switches. Abbreviated PAX.

private branch exchange [COMMUN] A telephone exchange serving a single organization, having a switchboard and associated equipment, usually located on the customer's premises; provides for switching calls between any two extensions served by the exchange or between any extension and the national telephone system via a trunk to a central office. Abbreviated PBX.

privileged instruction [COMPUT SCI] A class of instructions, usually including storage protection setting, interrupt handling, timer control, input/output, and special processor status-setting instructions, that can be executed only when the computer is in a special privileged mode that is generally available to an operating or executive system, but not to user programs.

probability [STAT] The probability of an event is the ratio of the number of times it occurs to the large number of trials that take place; the mathematical model of probability is a positive measure which gives the measure of the space the value 1.

probability amplitude *See* Schrödinger wave function.

probability density function [STAT] A real-valued function whose integral over any set gives the probability that a random variable has values in this set. Also known as density function; frequency function.

probability deviation *See* probable error.

probability distribution *See* distribution of a random variable.

probability ratio test [STAT] Testing a simple hypothesis against a simple alternative by using the ratio of the probability of each simple event under the alternative to the probability of the event under the hypothesis.

probability theory [MATH] The study of the mathematical structures and constructions used to analyze the probability of a given set of events from a family of outcomes.

probable error [STAT] The error that is exceeded by a variable with a probability of ½. Also known as probability deviation.

probable reserves [PETRO ENG] Primary petroleum reserves based on limited evidence, but not proved by a commercial oil-production rate.

probe [AERO ENG] An instrumented vehicle moving through the upper atmosphere or space or landing upon another celestial body in order to obtain information about the specific environment. [COMMUN] To determine a radio interference by obtaining the relative interference level in the immediate area of a source by the use of a small, insensitive antenna in conjunction with a receiving device. [ELECTROMAG] A metal rod that projects into but is insulated from a waveguide or resonant cavity; used to provide coupling to an external circuit for injection or extraction of energy or to measure the standing-wave ratio. Also known as waveguide probe. [ENG] A small tube containing the sensing element of electronic equipment, which can be lowered into a borehole to obtain measurements and data. [PHYS] A small device which can be brought into contact with or inserted into a system in order to make measurements on the system; ordinarily it is designed so that it does not significantly disturb the system.

probertite [MINERAL] $NaCaB_5O_9 \cdot 5H_2O$ A colorless mineral crystallizing in the monoclinic system, consisting of hydrous sodium calcium borate.

problem definition [COMPUT SCI] The art of compiling logic in the form of general flow charts and logic diagrams which clearly explain and present the problem to the programmer in such a way that all requirements involved in the run are presented.

problem-defining language [COMPUT SCI] A programming language that literally defines a problem and may specifically define the input and output, but does not define the method of transforming one to the other. Also known as problem specification language.

problem-describing language [COMPUT SCI] A programming language that describes, in the most general way, the problem to be solved, but gives no indication of the problem's detailed characteristics or its solution.

problem file *See* run book.

problem folder *See* run book.

problem mode [COMPUT SCI] A condition of computer operation in which, in contrast to supervisor mode, the privileged instructions cannot be executed, preventing the program from upsetting the supervisor program or any other program.

problem-oriented language [COMPUT SCI] A language designed to facilitate the accurate expression of problems belonging to specific sets of problem types.

problem-solving language [COMPUT SCI] A programming language that can be used to specify a complete solution to a problem.

problem-specification language *See* problem-defining language.

Proboscidea [VERT ZOO] An order of herbivorous placental mammals characterized by having a proboscis, incisors enlarged to become tusks, and pillarlike legs with five toes bound together on a broad pad.

proboscis [INV ZOO] A tubular organ of varying form and function on a large number of invertebrates, such as insects, annelids, and tapeworms. [VERT ZOO] The flexible, elongated snout of certain mammals.

Procampodeidae [INV ZOO] A family of the insect order Diplura.

Procaviidae [VERT ZOO] A family of mammals in the order Hyracoidea including the hyraxes.

Procaviinae [VERT ZOO] A subfamily of ungulate mammals in the family Procaviidae.

procedural language [COMPUT SCI] A programming language made up of macroinstructions, each macroinstruction usually written in assembly language.

procedure [COMPUT SCI] A sequence of actions (or computer instructions) which collectively accomplish some desired task.

procedure-oriented language [COMPUT SCI] A language designed to facilitate the accurate description of procedures, algorithms, or routines belonging to a certain set of procedures.

Procellariidae [VERT ZOO] A family of birds in the order Procellariiformes comprising the petrels, fulmars, and shearwaters.

Procellariiformes [VERT ZOO] An order of oceanic birds characterized by tubelike nostril openings, webbed feet, dense plumage, compound horny sheath of the bill, and, often, a peculiar musky odor.

process [COMPUT SCI] To assemble, compile, generate, interpret, compute, and otherwise act on information in a computer. [ENG] A system or series of continuous or regularly occurring actions taking place in a predetermined or planned manner; for example, as oil refining process or chemicals manufacturing process.

process annealing [MET] Softening a ferrous alloy by heating to a temperature close to but below the lower limit of the transformation range and then cooling.

process color [GRAPHICS] Method of reproducing full-color originals such as paintings and color photographs; four-color process plates print in yellow, magenta, cyan, and black.

process engineering [ENG] A service function of production engineering that involves selection of the processes to be used, determination of the sequence of all operations, and requisition of special tools to make a product.

process furnace [CHEM ENG] Furnace used to heat process-stream materials (liquids, gases, or solids) in a chemical-plant operation; types are direct-fired, indirect-fired, and pebble heaters.

process heat [PHYS CHEM] Increase in enthalpy accompanying chemical reactions or phase transformations at constant pressure; heat of crystallization and heat of sublimination are examples.

processing [COMMUN] Further handling, manipulation, consolidation, compositing, and so on, of information to convert it from one format to another or to reduce it to manageable or intelligible information. [ENG] The act of converting material from one form into another desired form.

processing interrupt [COMPUT SCI] The interruption of the batch processing mode in a real-time system when live data are entered into the system.

processing section [COMPUT SCI] The computer unit that does the actual changing of input into output; includes the arithmetic unit and intermediate storage.

process-limited *See* processor-limited.

process metallurgy [MET] The branch of metallurgy concerned with the extraction of metals from ore, and with the refining of metals; usually synonymous with extractive metallurgy.

processor [COMPUT SCI] **1.** A device that performs one or many functions, usually a central processing unit. **2.** A program that transforms some input into some output, such as an assembler, compiler, or linkage editor.

processor error interrupt [COMPUT SCI] The interruption of a computer program because a parity check indicates an error in a word that has been transferred to or within the central processing unit.

processor-limited [COMPUT SCI] Property of a computer system whose processing time is determined by the speed of its central processing unit rather than by the speed of its peripheral equipment. Also known as process-limited.

processor stack pointer [COMPUT SCI] A programmable register used to access all temporary-storage words related to an interrupt-service routine which was halted when a new service routine was called in.

process variable [CHEM ENG] Any of those varying operational and physical conditions associated with a chemical processing operation, such as temperature, pressure, flowrate, density, pH, viscosity, or chemical composition.

prochirality [ORG CHEM] The property displayed by a molecule or atom which contains (or is bonded to) two constitutionally identical ligands; Also known as prostereoisomerism.

Procoela [VERT ZOO] A suborder of the Anura characterized by a procoelous vertebral column and a free coccyx articulating with a double condyle.

Procolophonia [PALEON] A subclass of extinct cotylosaurian reptiles.

proctology [MED] A branch of medicine concerned with the structure and disease of the anus, rectum, and sigmoid colon.

Proctotrupidae [INV ZOO] A family of hymenopteran insects in the superfamily Proctotrupoidea.

Proctotrupoidea [INV ZOO] A superfamily of parasitic Hymenoptera in the suborder Apocrita.

procumbent [BOT] Having stems that lie flat on the ground but do not root at the nodes. [SCI TECH] **1.** Lying stretched out. **2.** Slanting forward. **3.** Lying face down.

Procyon [ASTRON] A star of magnitude 0.3, of spectral type F5, and 11 light-years (1.041×10^{17} meters) from earth; one of a binary. Also known as α Canis Minoris.

Procyonidae [VERT ZOO] A family of carnivoran mammals in the superfamily Canoidea, including raccoons and their allies.

Prodinoceratinae [PALEON] A subfamily of extinct herbivorous mammals in the family Untatheriidae; animals possessed a carnivorelike body of moderate size.

producer gas [MATER] Fuel gas high in carbon monoxide and hydrogen, produced by burning a solid fuel with a deficiency of air or by passing a mixture of air and steam through a bed of incandescent fuel; used as a cheap, low-Btu industrial fuel.

producing reserves [PETRO ENG] Developed (proved) petroleum reserves to be produced by existing wells in that portion of a reservoir subjected to full-scale secondary-recovery operations.

product [CHEM ENG] *See* discharge liquor. [IND ENG] **1.** An item or goods made by an industrial firm. **2.** The total of such items or goods. [MATH] **1.** The product of two algebraic quantities is the result of their multiplication relative to an operation analogous to multiplication of real numbers. **2.** The product of a collection of sets $A_1, A_2, ..., A_n$ is the set of all elements of the form $(a_1, a_2, ..., a_n)$ where each a_i is an element of A_i for each $i = 1, 2, ..., n$.

Productinida [PALEON] A suborder of extinct articulate brachiopods in the order Strophomenida characterized by the development of spines.

production engineering [IND ENG] The planning and control of the mechanical means of changing the shape, condition, and relationship of materials within industry toward greater effectiveness and value.

production model [IND ENG] A model in its final mechanical and electrical form of final production design made by production tools, jigs, fixtures, and methods.

production program [COMPUT SCI] A proprietary program used primarily for internal processing in a business and not generally made available to third parties for profit.

production reactor [NUCLEO] A nuclear reactor designed primarily for large-scale production of transmutation products, such as plutonium.

production standard *See* standard time.

productivity [AGR] The yield of a given crop per unit of land. [IND ENG] **1.** The effectiveness with which labor and equipment are utilized. **2.** The production output per unit of effort. [PETRO ENG] Measure of an oil well's ability to produce liquid or gaseous hydrocarbons; categories include relative, specific, ultimate, and fractured-well productivity.

proenzyme *See* zymogen.

profile [GEOL] **1.** The outline formed by the intersection of the plane of a vertical section and the ground surface. Also known as topographic profile. **2.** Data recorded by a single line of receivers from one shot point in seismic prospecting. [GEOPHYS] A graphic representation of the variation of one property, such as gravity, usually as ordinate, with respect to another property, usually linear, such as distance. [HYD] A vertical section of a potentiometric surface, such as a water table. [PETR] In structural petrology, a cross section of a homoaxial structure.

Proganosauria [PALEON] The equivalent name for Mesosauria.

progesterone [BIOCHEM] $C_{21}H_{30}O_2$ A steroid hormone produced in the corpus luteum, placenta, testes, and adrenals; plays an important physiological role in the luteal phase of the menstrual cycle and in the maintenance of pregnancy; it is an intermediate in the biosynthesis of androgens, estrogens, and the corticoids.

progradation [GEOL] Seaward buildup of a beach, delta, or fan by nearshore deposition of sediments transported by a river, by accumulation of material thrown up by waves, or by material moved by longshore drifting.

program [AERO ENG] In missile guidance, the planned flight path events to be followed by a missile in flight, including all the critical functions, preset in a program device, which control the behavior of the missile. [COMMUN] A sequence of audio signals alone, or audio and video signals, transmitted for entertainment or information. [COMPUT SCI] A detailed and explicit set of directions for accomplishing some purpose, the set being expressed in some language suitable for input to a computer, or in machine language.

program card [COMPUT SCI] A punched card containing one or more instructions in a computer program in either source language or machine language, in contrast to a card that contains data to be processed according to the instructions.

program check [COMPUT SCI] A built-in check system in a program to determine that the program is running correctly.

program compatibility [COMPUT SCI] The type of compatibility shared by two computers that can process the identical program or programs written in the same source language or machine language.

program counter *See* instruction counter.

program editor [COMPUT SCI] A computer routine used in time-sharing systems for on-line modification of computer programs.

program element [COMPUT SCI] Part of a central computer system that carries out the instruction sequence scheduled by the programmer.

program evaluation and review technique *See* PERT.

program generator [COMPUT SCI] A program that permits a computer to write other programs automatically.

program library [COMPUT SCI] An organized set of computer routines and programs.

programmable calculator [COMPUT SCI] An electronic calculator that has some provision for changing its internal program, usually by inserting a new magnetic card on which the desired calculating program has been stored.

programmable controller [CONT SYS] A control device, normally used in industrial control applications, that employs the hardware architecture of a computer and a relay ladder diagram language. Also known as programmable logic controller.

programmable logic array *See* field-programmable logic array.

programmable read-only memory [ELECTR] A large-scale integrated-circuit chip for storing digital data; it can be erased with ultraviolet light and reprogrammed, or it can be programmed only once either at the factory or in the field. Abbreviated PROM.

programmed check [COMPUT SCI] 1. An error-detecting operation programmed by instructions rather than built into the hardware. 2. A computer check in which a sample problem with known answer, selected for having a program similar to that of the next problem to be run, is put through the computer.

programmed dump [COMPUT SCI] A storage dump which results from an instruction in a computer program at a particular point in the program.

programmed halt [COMPUT SCI] A halt that occurs deliberately as the result of an instruction in the program. Also known as programmed stop.

programmed logic array [ELECTR] An array of AND/OR logic gates that provides logic functions for a given set of inputs programmed during manufacture and serves as a read-only memory. Abbreviated PLA.

programmed operators [COMPUT SCI] Computer instructions which enable subroutines to be accessed with a single programmed instruction.

programmed stop *See* programmed halt.

programmer [COMPUT SCI] A person who prepares sequences of instructions for a computer, without necessarily converting them into the detailed codes. [CONT SYS] A device used to control the motion of a missile in accordance with a predetermined plan.

programming [COMPUT SCI] Preparing a detailed sequence of operating instructions for a particular problem to be run on a digital computer. Also known as computer programming. [ENG] In a plastics process, extruding a parison whose thickness differs longitudinally in order to equalize wall thickness of the blown container.

programming language [COMPUT SCI] The language used by a programmer to write a program for a computer.

program module [COMPUT SCI] A logically self-contained and discrete part of a larger computer program, for example, a subroutine or a coroutine.

program parameter [COMPUT SCI] In computers, an adjustable parameter in a subroutine which can be given a different value each time the subroutine is used.

program register [COMPUT SCI] The register in the control unit of a digital computer that stores the current instruction of the program and controls the operation of the computer during the execution of that instruction. Also known as computer control register.

program specification [COMPUT SCI] A statement of the precise functions which are to be carried out by a computer program, including descriptions of the input to be processed by the program, the processing needed, and the output from the program.

program step [COMPUT SCI] In computers, some part of a program, usually one instruction.

program stop [COMPUT SCI] An instruction built into a computer program that will automatically stop the machine under certain conditions, or upon reaching the end of processing or completing the solution of a program. Also known as halt instruction; stop instruction.

program storage [COMPUT SCI] Portion of the internal storage reserved for the storage of programs, routines, and subroutines; in many systems, protection devices are used to prevent inadvertent alteration of the contents of the program storage; contrasted with temporary storage.

program test [COMPUT SCI] A system of checking before running any problem in which a sample problem of the same type with a known answer is run.

progression [MATH] A sequence or series of mathematical objects or quantities, each entry determined from its predecessors by some algorithm.

progressive-wave antenna *See* traveling-wave antenna.

Progymnospermopsida [PALEON] A class of plants intermediate between ferns and gymnosperms; comprises the Denovian genus *Archaeopteris*.

Projapygidae [INV ZOO] A family of wingless insects in the order Diplura.

project engineering [ENG] 1. The engineering design and supervision (coordination) aspects of building a manufacturing facility. 2. The engineering aspects of a specific project, such as development of a product or solution to a problem.

projection [MAP] A system for presenting on a plane surface the spherical surface of the earth or the celestial sphere; some of these systems are conic, cylindrical, gnomonic, Mercator, orthographic, and stereographic. Also known as map projection. [MATH] 1. The continuous map for a fiber bundle. 2. Geometrically, the image of a geometric object or vector superimposed on some other. 3. A linear map P from a linear space to itself such that $P \cdot P$ is equal to P. [PSYCH] Ascribing one's motives to someone else to disguise a source of conflict in oneself.

projection microradiography [PHYS] Microradiography in which an electron beam, focused into an extremely fine pencil, generates a point source of x-rays, and enlargement is achieved by placing the sample very near this source, and several centimeters from the recording material. Abbreviated PMR. Also known as shadow microscopy; x-ray projection microscopy.

projection microscope [PHYS] An x-ray microscope which magnifies by image projection, either in contact microradiography or in projection microradiography.

projection net *See* net.

projection optics *See* Schmidt system.

projection welding [MET] Resistance welding in which the welds are localized at projections, intersections, and overlaps on the parts.

projective geometry [MATH] The study of those properties of geometric objects which are invariant under projection.

projector [ENG ACOUS] 1. A horn designed to project sound chiefly in one direction from a loudspeaker. 2. An underwater acoustic transmitter. [OPTICS] *See* optical projection system. [ORD] 1. Any apparatus for launching a projectile, such as a gun or rocket launcher. 2. Smooth-bore-type barrel or other unrifled weapon from which pyrotechnic signals, grenades, and certain mortar projectiles are fired. 3. A rack for launching target rockets. 4. Special type of gun for projecting antisubmarine projectiles.

prokaryote [CYTOL] 1. A primitive nucleus, where the deoxyribonucleic acid–containing region lacks a limiting membrane. 2. Any cell containing such a nucleus, such as the bacteria and the blue-green algae.

Prolacertiformes [PALEON] A suborder of extinct terrestrial reptiles in the order Eosuchia distinguished by reduction of the lower temporal arcade.

prolactin [BIOCHEM] A protein hormone produced by the adenohypophysis; stimulates lactation and promotes functional activity of the corpus luteum. Also known as lactogenic hormone; luteotropic hormone; mammary-stimulating hormone; mammogen; mammogenic hormone; mammotropin.

proliferative arthritis *See* rheumatoid arthritis.

proline [BIOCHEM] $C_5H_9O_2$ A heterocyclic amino acid occurring in essentially all proteins, and as a major constituent in collagen proteins.

PROM *See* Pockels readout optical modulator; programmable read-only memory.

promethium [CHEM] A chemical element, symbol Pm, atomic number 61; atomic weight of the most abundant isotope is 147; a member of the rare-earth group of metals.

promethium-147 [NUC PHYS] Artificially produced rare-earth element with atomic number 61 and mass 147; produced during fission of U^{235}. Also known as florentium; illinium.

prominence [ASTROPHYS] A volume of luminous, predominantly hydrogen gas that appears on the sun above the chromosphere; occurs only in the region of horizontal magnetic fields because these fields support the prominences against solar gravity.

promoter [CHEM] A chemical which itself is a feeble catalyst, but greatly increases the activity of a given catalyst. [GEN] The site on deoxyribonucleic acid to which ribonucleic acid polymerase binds preparatory to initiating transcription of an operon.

PROM programmer [COMPUT SCI] A hardware device that writes programmable read-only memory chips in a permanent form.

pronate [ANAT] 1. To turn the forearm so that the palm of the hand is down or toward the back. 2. To turn the sole of the foot outward with the lateral margin of the foot elevated; to evert.

prong *See* pin.

pronghorn [VERT ZOO] *Antilocapra americana*. An antelopelike artiodactyl composing the family Antilocapridae; the only hollow-horned ungulate with branched horns present in both sexes.

prony brake [MECH ENG] An absorption dynamometer that applies a friction load to the output shaft by means of wood blocks, a flexible band, or other friction surface.

proof [ENG] Reproduction of a die impression by means of a cast. [FOOD ENG] The strength of the ethyl alcohol in distilled spirits; in the United States, each degree of proof is equal to ½% of alcohol by volume. [GRAPH] The inked impression of composed type or a plate; used for inspection purposes or for pasting up with other artwork. [MATH] A deductive demonstration of a mathematical statement.

propagation *See* wave motion.

propagation constant [ELECTROMAG] A rating for a line or medium along or through which a wave of a given frequency is being transmitted; it is a complex quantity; the real part is the attenuation constant in nepers per unit length, and the imaginary part is the phase constant in radians per unit length.

propagation delay [ELECTR] The time required for a signal to pass through a given complete operating circuit; it is generally of the order of nanoseconds, and is of extreme importance in computer circuits.

Propalticidae [INV ZOO] A family of coleopteran insects of the superfamily Cucujoidea found in Old World tropics and Pacific islands.

propellant [MATER] A combustible substance that produces heat and supplies ejection particles as in a rocket engine.

propeller [MECH ENG] A bladed device that rotates on a shaft to produce a useful thrust in the direction of the shaft axis.

propeller strut *See* strut.

proper motion [ASTRON] That component of the space motion of a celestial body perpendicular to the line of sight, resulting in the change of a star's apparent position relative to that of other stars; expressed in angular units.

proper subset [MATH] A set X is a proper subset of a set Y if there is an element of Y which is not in X while X is a subset of Y.

property detector [COMPUT SCI] In character recognition, that electronic component of a character reader which processes the normalized signal for the purpose of extracting from it a set of characteristic properties on the basis of which the character can be subsequently identified.

proper value See eigenvalue.

prophage [VIROL] Integrated unit formed by union of the provirus into the bacterial genome.

prophase [CYTOL] The initial stage of mitotic or meiotic cell division in which chromosomes are condensed from the nuclear material and split logitudinally to form pairs.

Propionibacteriaceae [MICROBIO] A family of bacteria related to the actinomycetes; gram-positive, anaerobic to aerotolerant rods or filaments; ferment carbohydrates, with propionic acid as the principal product.

proportion [MATH] The proportion of two quantities is their ratio.

proportional control [CONT SYS] Control in which the amount of corrective action is proportional to the amount of error; used, for example, in chemical engineering to control pressure, flow rate, or temperature in a process system.

proportional counter [NUCLEO] A radiation counter consisting of a proportional counter tube and its associated circuits; resembles a Geiger-Müller counter, but with a different counting gas (argon methane) and a lower voltage on the tube; used to measure α, β, and x-rays; has low sensitivity for γ-radiation.

proportional limit [MECH] The greatest stress a material can sustain without departure from linear proportionality of stress and strain.

proportional parts [MATH] Numbers in the same proportion as a set of given numbers; such numbers are used in an auxiliary interpolation table based on the assumption that the tabulated quantity and entering arguments differ in the same proportion.

proportioning pump See metering pump.

proportioning reactor [ELECTROMAG] A saturable-core reactor used for regulation and control; increasing the input control current from zero to rated value makes output current increase in proportion from cutoff up to full load value.

proprioceptor [PHYSIO] A sense receptor that signals spatial position and movements of the body and its members, as well as muscular tension.

propulsion [MECH] The process of causing a body to move by exerting a force against it.

propyl- [ORG CHEM] The $CH_3CH_2CH_2-$ radical, derived from propane; found, for example, in 1-propanol.

propylite [PETR] A modified andesite, altered by hydrothermal processes, resembling a greenstone and consisting of calcite, epidote, serpentine, quartz, pyrite, and iron ore.

Prorastominae [PALEON] A subfamily of extinct dugongs (Dugongidae) which occur in the Eocene of Jamaica.

Prorhynchidae [INV ZOO] A family of turbellarians in the order Alloeocoela.

Prosauropoda [PALEON] A division of the extinct reptilian suborder Sauropodomorpha; they possessed blunt teeth, long forelimbs, and extremely large claws on the first finger of the forefoot.

Prosobranchia [INV ZOO] The largest subclass of the Gastropoda; generally, respiration is by means of

ctenidia, an operculum is present, there is one pair of tentacles, and the sexes are separate.

prosopite [MINERAL] $CaAl_2(F,OH)_8$ A colorless mineral composed of basic calcium aluminum fluoride.

Prosopora [INV ZOO] An order of the class Oligochaeta comprising mesonephridiostomal forms in which there are male pores in the segment of the posterior testes.

prostaglandin [BIOCHEM] Any of various physiologically active compounds containing 20 carbon atoms and formed from essential fatty acids; found in highest concentrations in normal human semen; activities affect the nervous system, circulation, female reproductive organs, and metabolism.

prostate [ANAT] A glandular organ that surrounds the urethra at the neck of the urinary bladder in the male.

prostereoisomerism See prochirality.

prosthesis [MED] An artificial substitute for a missing part of the body, such as a substitute hand, leg, eye, or denture.

prosthodontics [MED] The science and practice of replacement of missing dental and oral structures.

Prostigmata [INV ZOO] The equivalent name for Trombidiformes.

protactinium [CHEM] A chemical element, symbol Pa, atomic number 91; the third member of the actinide group of elements; all the isotopes are radioactive; the longest-lived isotope is protactinium-231.

protandry [PHYSIO] That condition in which an animal is first a male and then becomes a female.

Proteaceae [BOT] A large family of dicotyledonous plants in the order Proteales, notable for often having a large cluster of small or reduced flowers.

Proteales [BOT] An order of dicotyledonous plants in the subclass Rosidae marked by its strongly perigynous flowers, a four-lobed, often corollalike calyx, and reduced or absent true petals.

protease [BIOCHEM] An enzyme that digests proteins.

protected location [COMPUT SCI] A storage cell arranged so that access to its contents is denied under certain circumstances, in order to prevent programming accidents from destroying essential programs and data.

protection code [COMPUT SCI] A component of a task descriptor that specifies the protection domain of the task, that is, the authorizations it has to perform certain actions.

protective coloration [ZOO] A color pattern that blends with the environment and increases the animal's probability of survival.

protective device See electric protective device.

protective relay [ELEC] A relay whose principal function is to protect service from interruption or to prevent or limit damage to apparatus.

protector gap [ELEC] A device designed to limit or equalize voltage in order to protect telephone and telegraph equipment; consists of two carbon blocks with an air gap between them, which are brought into contact when there is a steady-state discharge across the gap. Also known as gap.

Proteida [VERT ZOO] A suborder coextensive with Proteidae in some classification systems.

Proteidae [VERT ZOO] A family of the amphibian suborder Salamandroidea; includes the neotenic, aquatic *Necturus* and *Proteus* species.

protein [BIOCHEM] Any of a class of high-molecular-weight polymer compounds composed of a variety of α-amino acids joined by peptide linkages.

protein-bound iodine test [PATH] A test of thyroid function that reflects the level of circulating thyroid hormone by determination of the level of protein-bound iodine in the blood. Abbreviated PBI test.

Proteocephalidae [INV ZOO] A family of tapeworms in the order Proteocephaloidea in which the reproductive organs are within the central mesenchyme of the segment.

Proteocephaloidea [INV ZOO] An order of tapeworms of the subclass Cestoda in which the holdfast organ bears four suckers and, frequently, a sucker-like apical organ.

proteolytic enzyme [BIOCHEM] Any enzyme that catalyzes the breakdown of protein.

Proteomyxida [INV ZOO] The single order of the Proteomyxidia.

Proteomyxidia [INV ZOO] A subclass of Actinopodea including protozoan organisms which lack protective coverings or skeletal elements and have reticulopodia, or filopodia.

Proterostomia [ZOO] That part of the animal kingdom in which cleavage of the egg is of the determinate type; includes all bilateral phyla except Echinodermata, Chaetognatha, Pogonophora, Hemichordata, and Chordata.

Proterosuchia [PALEON] A suborder of moderate-sized thecodont reptiles with lightly built triangular skulls, downturned snouts, and palatal teeth.

Proterotheriidae [PALEON] A group of extinct herbivorous mammals in the order Litopterna which displayed an evolutionary convergence with the horses in their dentition and in reduction of the lateral digits of their feet.

Proteutheria [VERT ZOO] A group of primatelike insectivores that contains the living tree shrews.

prothorax [INV ZOO] The first thoracic segment of an insect; bears the first pair of legs.

prothrombin [BIOCHEM] An inactive plasma protein precursor of thrombin. Also known as factor II; thrombinogen.

prothrombin factor See vitamin K.

prothrombin time [PATH] A one-stage clotting test based on the time required for clotting to occur after the addition of tissue thromboplastin and calcium to decalcified plasma.

Protista [BIOL] A proposed kingdom to include all unicellular organisms lacking a definite cellular arrangement, such as bacteria, algae, diatoms, and fungi.

Protoariciinae [INV ZOO] A subfamily of polychaete annelids in the family Orbiniidae.

Protobranchia [INV ZOO] A small and primitive order in the class Bivalvia; the hinge is taxodont in all but one family, there is a central ligament pit, and the anterior and posterior adductor muscles are nearly equal in size.

Protoceratidae [PALEON] An extinct family of pecoran ruminants in the superfamily Traguloidea.

Protochordata [INV ZOO] The equivalent name for Hemichordata.

Protococcaceae [BOT] A monogeneric family of green algae in the suborder Ulotrichineae in which reproduction is entirely vegetative.

Protococcida [INV ZOO] A small order of the protozoan subclass Coccidia; all are invertebrate parasites, and only sexual reproduction is known.

Protocucujidae [INV ZOO] A small family of coleopteran insects in the superfamily Cucujoidea found in Chile and Australia.

Protodonata [PALEON] An extinct order of huge dragonflylike insects found in Permian rocks.

Protodrilidae [INV ZOO] A family of annelids belonging to the Archiannelida.

protoenstatite [MINERAL] An artificial, unstable, altered form of $MgSiO_3$ produced by thermal decomposition of talc; convertible to enstatite by grinding or heating to a high temperature.

Protoeumalacostraca [PALEON] The stem group of the crustacean series Eumalacostraca.

protogyny [PHYSIO] A condition in hermaphroditic or dioecious organisms in which the female reproductive structures mature before the male structures.

Protomastigida [INV ZOO] The equivalent name for Kinetoplastida.

Protomonadina [INV ZOO] An order of flagellates, subclass Mastigophora, with one or two flagella, including many species showing protoplasmic collars ringing the base of the flagellum.

Protomonida [INV ZOO] The equivalent name for Protomonadina.

Protomyzostomidae [INV ZOO] A family of parasitic polychaetes belonging to the Myzostomaria and known for three species from Japan and the Murman Sea.

proton [PHYS] An elementary particle that is the positively charged constituent of ordinary matter and, together with the neutron, is a building stone of all atomic nuclei; its mass is approximately 938 MeV (million electron volts) and spin ½.

protonate [CHEM] To add protons to a base by a proton source.

proton-electron-proton reaction [NUC PHYS] A nuclear reaction in which two protons and an electron react to form a deuteron and a neutrino; it is an important source of detectable neutrinos from the sun. Abbreviated PeP reaction.

protonium [ATOM PHYS] A bound state of a proton and an antiproton.

proton number See atomic number.

proton-proton chain [NUC PHYS] An energy-releasing nuclear reaction chain which is believed to be of major importance in energy production in hydrogen-rich stars. Also known as deuterium cycle.

proton resonance [SPECT] A phenomenon in which protons absorb energy from an alternating magnetic field at certain characteristic frequencies when they are also subjected to a static magnetic field; this phenomenon is used in nuclear magnetic resonance quantitative analysis technique.

proton scattering microscope [SOLID STATE] A microscope in which protons produced in a cold-cathode discharge are accelerated and focused on a crystal in a vacuum chamber; protons reflected from the crystal strike a fluorescent screen to give a visual and photographable display that is related to the structure of the target crystal.

proton storage ring [NUCLEO] A machine consisting of magnets and vacuum chambers in which beams of high-energy protons can be stored.

proton synchrotron [NUCLEO] A device for accelerating protons in circular orbits in a time-varying magnetic field, in which the orbit radius is kept constant.

Protophyta [BOT] A division of the plant kingdom, according to one system of classification, set up to include the bacteria, the blue-green algae, and the viruses.

protoplasm [CYTOL] The colloidal complex of protein that composes the living material of a cell.

protoplast [CYTOL] The living portion of a cell considered as a unit; includes the cytoplasm, the nucleus, and the plasma membrane.

Protopteridales [PALEOBOT] An extinct order of ferns, class Polypodiatae.

Protosireninae [PALEON] An extinct superfamily of sirenian mammals in the family Dugongidae found in the middle Eocene of Egypt.

Protospondyli [VERT ZOO] An equivalent name for Semionotiformes.

Protostomia [INV ZOO] A major division of bilateral animals; includes most worms, arthropods, and mollusks.

Protosuchia [PALEON] A suborder of extinct crocodilians from the Late Triassic and Early Jurassic.

Prototheria [VERT ZOO] A small subclass of Mammalia represented by a single order, the Monotremata.

Prototrupoidea [INV ZOO] A superfamily of the Hymenoptera.

prototype [ENG] A model suitable for use in complete evaluation of form, design, and performance.

Protozoa [INV ZOO] A diverse phylum of eukaryotic microorganisms; the structure varies from a simple uninucleate protoplast to colonial forms, the body is either naked or covered by a test, locomotion is by means of pseudopodia or cilia or flagella, there is a tendency toward universal symmetry in floating species and radial symmetry in sessile types, and nutrition may be phagotrophic or autotrophic or saprozoic.

protozoology [INV ZOO] That branch of biology which deals with the Protozoa.

Protrachaeta [INV ZOO] The equivalent name for Onychophora.

protractor [ENG] An instrument used to construct and measure angles formed by lines of a plane; the midpoint of the diameter of the semicircle is marked and serves as the vertex of angles contructed or measured.

Protura [INV ZOO] An order of primitive wingless insects belonging to the subclass Apterygota; individuals are elongate and eyeless, lack antennae, and are from pale amber to white in color; anamorphosis is characteristic of the group.

proustite [MINERAL] Ag_3AsS_3 A cochineal-red mineral that crystallizes in the rhombohedral system, consists of silver arsenic sulfide, is isomorphous with pyrargyrite, and occurs massively and in crystals. Also known as light-red silver ore; light-ruby silver.

proved reserves [PETRO ENG] Reserves (primary or secondary) that have been proved by production at commercial flow rates.

proventriculus [INV ZOO] 1. A sac anterior to the gizzard in earthworms. 2. A dilation of the foregut anterior to the midgut of Mandibulata. [VERT ZOO] The true stomach of a bird, usually separated from the gizzard by a constriction.

proving [COMPUT SCI] Testing whether a computer is free of faults and capable of functioning normally, usually by having it carry out a check routine or diagnostic routine.

Proxima Centauri [ASTRON] The star that is the sun's nearest neighbor; stellar magnitude is ll, and it is 2° from the bright star α Centauri.

proximity warning indicator [NAV] An airborne instrument which produces a warning signal indicating the approach of an aircraft on a possible collision course. Abbreviated PWI.

PRR See pulse repetition rate.

pruritus [MED] Localized or generalized itch due to irritation of sensory nerve endings.

psalterium See omasum.

Psamment [GEOL] A suborder of the soil order Entisol, characterized by a texture of loamy fine sand or coarser sand, and by a coarse fragment content of less than 35%.

psammite See arenite.

Psammodontidae [PALEON] A family of extinct cartilaginous fishes in the order Bradyodonti in which the upper and lower dentitions consisted of a few large quadrilateral plates arranged in two rows meeting in the midline.

Psammodrilidae [INV ZOO] A small family of spioniform worms belonging to the Sedentaria.

Pselaphidae [INV ZOO] The ant-loving beetles, a large family of coleopteran insects in the superfamily Staphylinoidea.

Psephenidae [INV ZOO] The water penny beetles, a small family of coleopteran insects in the superfamily Dryopoidea.

Pseudaliidae [INV ZOO] A family of roundworms belonging to the Strongyloidea which occur as parasites of whales and porpoises.

pseudoadiabatic chart See Stuve chart.

pseudoalleles [GEN] Closely linked genes that behave as alleles and can be separated by crossing over.

Pseudoborniales [PALEOBOT] An order of fossil plants found in Middle and Upper Devonian rocks.

pseudobreccia [PETR] Limestone that is partially and irregularly dolomitized and is characterized by a mottled, breccialike appearance. Also known as recrystallization breccia.

pseudobrookite [MINERAL] Fe_2TiO_5 A brown or black mineral consisting of iron titanium oxide and occurring in orthorhombic crystals; specific gravity is 4.4–4.98.

pseudocarburizing See blank carburizing.

pseudocode See interpretive language.

Pseudocoelomata [INV ZOO] A group comprising the animal phyla Entoprocta, Aschelminthes, and Acanthocephala; characterized by a pseudocoelom.

pseudoconformity See paraconformity.

Pseudocycnidae [INV ZOO] A family of the Caligoida which comprises external parasites on the gills or various fishes.

Pseudodiadematidae [INV ZOO] A family of Jurassic and Cretaceous echinoderms in the order Phymosomatoida which had perforate crenulate tubercles.

pseudoequivalent temperature See equivalent temperature.

pseudogalena See sphalerite.

pseudoinstruction [COMPUT SCI] 1. A symbolic representation in a compiler or interpreter. 2. See quasi-instruction.

pseudomalachite [MINERAL] $Cu_5(PO_4)_2(OH)_4 \cdot H_2O$ An emerald green to dark green and blackish-green, monoclinic mineral consisting of a hydrated basic copper phosphate. Also known as tagilite.

Pseudomonadaceae [MICROBIO] A family of gram-negative, aerobic, rod-shaped bacteria; cells are straight or curved and motile by polar flagella.

Pseudomonadales [MICROBIO] Formerly an order of ovoid, rod-shaped, comma-shaped, or spiral bacteria in the class Schizomycetes; cells characterized as rigid and motile by means of polar flagella.

pseudomorph [MINERAL] An altered mineral whose crystal form has the outward appearance of another mineral species. Also known as false form.

Pseudophoracea [INV ZOO] An extinct superfamily of gastropod mollusks in the order Aspidobranchia.

Pseudophyllidea [INV ZOO] An order of tapeworms of the subclass Cestoda, parasitic principally in the intestine of cold-blooded vertebrates.

pseudopodium [BOT] A slender, leafless branch of the gametophyte in certain Bryatae. [CYTOL] Temporary projection of the protoplast of ameboid cells in which cytoplasm streams actively during extension and withdrawal. [INV ZOO] Foot of a rotifer.

pseudorandom numbers [COMPUT SCI] Numbers produced by a definite arithmetic process, but satisfying one or more of the standard tests for randomness.

pseudoscalar meson [PARTIC PHYS] A meson, such as the pion, which has spin 0 and negative parity, and may be described by a field which is a pseudoscalar quantity. Also known as pseudoscalar particle.

pseudoscalar particle See pseudoscalar meson.

Pseudoscorpionida [INV ZOO] An order of terrestrial Arachnida having the general appearance of miniature scorpions without the postabdomen and sting.

Pseudosphaeriales [BOT] An order of the class Ascolichenes, shared by the class Ascomycetes; the ascocarp is flask-shaped and lined with a layer of interwoven, branched pseudoparaphyses.

Pseudosporidae [INV ZOO] A family of the protozoan subclass Proteomyxidia; flagellated stages invade Volvocidae and filamentous algae and become amebas.

pseudostratification See sheeting structure.

pseudostratified epithelium [HISTOL] A type of epithelium in which all cells reach to the basement membrane but some extend toward the surface only part way, while others reach the surface.

Pseudosuchia [PALEON] A suborder of extinct reptiles of the order Thecodontia comprising bipedal, unarmored or feebly armored forms which resemble dinosaurs in many skull features but retain a primitive pelvis.

pseudotensor [PHYS] **1.** A quantity which transforms as a tensor under space rotations, but which transforms as a tensor, together with a change in sign, under space inversion. **2.** A quantity which transforms as a tensor under Lorentz transformations, but with an additional sign change under space reflection or time reflection or both.

Pseudothelphusidae [INV ZOO] A family of freshwater crabs belonging to the Brachyura.

Pseudotriakidae [VERT ZOO] The false catsharks, a family of galeoids in the carcharinid line.

pseudovector [PHYS] **1.** A quantity which transforms as a vector under space rotations but which transforms as a vector, together with a change in sign, under a space inversion. Also known as axial vector. **2.** A quantity which transforms as a four-vector under Lorentz transformations, but with an additional sign change under space reflection or time reflection or both.

P shell [ATOM PHYS] The sixth layer of electrons about the nucleus of an atom, having electrons whose principal quantum number is 6.

psi function [MATH] The special function of a complex variable which is obtained from differentiating the logarithm of the gamma function. [QUANT MECH] See Schrödinger wave function.

Psilidae [INV ZOO] The rust flies, a family of myodarian cyclorrhaphous dipteran insects in the subsection Acalyptratae.

psilomelane [MINERAL] $BaMn_9O_{16}(OH)_4$ A massive, hard, black, botryoidal manganese oxide mineral mixture with a specific gravity ranging from 3.7 to 4.7.

Psilophytales [PALEOBOT] A group formerly recognized as an order of fossil plants.

Psilophytineae [PALEON] The equivalent name for Rhyniopsida.

Psilopsida [BOT] A subdivision of the Tracheophyta.

Psilorhynchidae [VERT ZOO] A small family of actinopterygian fishes belonging to the Cyprinoidei.

Psilotales [BOT] The equivalent name for Psilotophyta.

Psilotatae [BOT] A class of the Psilotophyta.

Psilotophyta [BOT] A division of the plant kingdom represented by three living species; the life cycle is typical of the vascular cryptogams.

psi particle See J particle.

Psittacidae [VERT ZOO] The single family of the Psittaciformes.

Psittaciformes [VERT] The parrots, a monofamilial order of birds that exhibit zygodactylism and have a strong hooked bill.

psittacinite See mottramite.

psittacosis [MED] Pneumonia and generalized infection of man and of birds caused by agents of the PLT-Bedsonia group; transmitted to man by psittacine birds.

PSK See phase-shift keying.

psoas [ANAT] Either of two muscles: psoas major which arises from the bodies and transverse processes of the lumbar vertebrae and is inserted into the lesser trochanter of the femur, and psoas minor which arises from the bodies and transverse processes of the lumbar vertebrae and is inserted on the pubis.

Psocoptera [INV ZOO] An order of small insects in which wings may be present or absent, tarsi are two- or three-segmented, cerci are absent, and metamorphosis is gradual.

Psolidae [INV ZOO] A family of echinoderms in the order Dendrochirotida characterized by a ventral adhesive sucker and a U-shaped gut, with the mouth and anus opening upward on the adoral surface.

Psophiidae [VERT ZOO] The trumpeters, a family of birds in the order Gruiformes.

psoriasis [MED] A usually chronic, often acute inflammatory skin disease of unknown cause; characterized by dull red, well-defined lesions covered by silvery scales which when removed disclose tiny capillary bleeding points.

psorosis [PL PATH] A virus disease of tangerine, grapefruit, and sweet orange trees characterized by scaly bark, a gummy exudate, retarded growth, small yellow leaves, and dieback of twigs. Also known as scaly bark.

PSR See primary radar.

psychiatry [MED] The medical science that deals with the origins, diagnosis, and treatment of mental and emotional disorders.

Psychidae [INV ZOO] The bagworms, a family of lepidopteran insects in the superfamily Tineoidea; males are large, hairy moths, but females are degenerate, wingless, and legless and live in bag-shaped cases.

psychointegroammeter See lie detector.

psycholinguistics [PSYCH] The study of linguistic behavior such as conditioning by psychological factors, including the speaker's and listener's culturally determined categories of expression and comprehension.

psychology [BIOL] **1.** The science that deals with the functions of the mind and the behavior of an organism in relation to its environment. **2.** The mental activity characteristic of a person or a situation.

psychosis [PSYCH] An impairment of mental functioning to the extent that it interferes grossly with an individual's ability to meet the ordinary demands of life, characterized generally by severe affective disturbance, profound introspection, and withdrawal from reality, formation of delusions or hallucinations, and regression presenting the appearance of personality disintegration.

psychosurgery [MED] The branch of medicine that deals with the treatment of various psychoses, severe neuroses, and chronic painful conditions by means of operative procedures on the brain.

psychrometer [ENG] A device comprising two thermometers, one a dry bulb, the other a wet or wick-covered bulb, used in determining the moisture content or relative humidity of air or other gases. Also known as wet and dry bulb thermometer.

psychrometric chart [THERMO] A graph each point of which represents a specific condition of a gas-vapor system (such as air and water vapor) with regard to temperature (horizontal scale) and absolute humidity (vertical scale); other characteristics of the system, such as relative humidity, wet-bulb temperature, and latent heat of vaporization, are indicated by lines on the chart.

Psyllidae [INV ZOO] The jumping plant lice, a family of the Homoptera in the series Sternorrhyncha in which adults have a transverse head with protuberant eyes and three ocelli, 6- to 10-segmented antennae, and wings with reduced but conspicuous venation.

pt *See* pint.

Pt *See* platinum.

PTC *See* Christmas factor.

Pteraspidomorphi [VERT ZOO] The equivalent name for Diplorhina.

Pterasteridae [INV ZOO] A family of deep-water echinoderms in the order Spinulosida distinguished by having webbed spine fins.

Pteridophyta [BOT] The equivalent name for Polypodiophyta.

Pteridospermae [PALEOBOT] Seed ferns, a class of the Cycadicae comprising extinct plants characterized by naked seeds borne on large fernlike fronds.

Pteridospermophyta [PALEOBOT] The equivalent name for Pteridospermae.

Pteriidae [INV ZOO] Pearl oysters, a family of bivalve mollusks with have nacreous shells.

Pterobranchia [INV ZOO] A group of small or microscopic marine animals regarded as a class of the Hemichordata; all are sessile, tubicolous organisms with a U-shaped gut and three body segments.

Pteroclidae [VERT ZOO] The sandgrouse, a family of gramnivorous birds in the order Columbiformes; mainly an Afro-Asian group resembling pigeons and characterized by cryptic coloration, usually corresponding with the soil color of the habitat.

pterodactyl [PALEON] The common name for members of the extinct reptilian order Pterosauria.

Pterodactyloidea [PALEON] A suborder of Late Jurassic and Cretaceous reptiles in the order Pterosauria distinguished by lacking tails and having increased functional wing length due to elongation of the metacarpels.

Pteromalidae [INV ZOO] A family of hymenopteran insects in the superfamily Chalcidoidea.

Pteromedusae [INV ZOO] A suborder of hydrozoan coelenterates in the order Trachylina characterized by a modified, bipyramidal medusae.

Pterophoridae [INV ZOO] The plume moths, a family of the lepidopteran superfamily Pyralidoidea in which the wings are divided into featherlike plumes, maxillary palpi are lacking, and the legs are long.

Pteropidae [INV ZOO] The fruit bats, a large family of the Chiroptera found in Asia, Australia, and Africa.

Pteropoda [INV ZOO] The sea butterflies, an order of pelagic gastropod mollusks in the subclass Opisthobranchia in which the foot is modified into a pair of large fins and the shell, when present, is thin and glasslike.

Pteropodidae [VERT ZOO] A family of fruit-eating bats in the suborder Megachiroptera, characterized by primitive ears and by shoulder joints.

Pteropsida [BOT] A large group of vascular plants characterized by having parenchymatous leaf gaps in the stele and by having leaves which are thought to have originated in the distant past as branched stem systems.

Pterosauria [PALEON] An extinct order of flying reptiles of the Mesozoic era belonging to the subclass Archosauria; the wing resembled that of a bat, and a large heeled sternum supported strong wing muscles.

pteroylglutamic acid *See* folic acid.

Ptiliidae [INV ZOO] The feather-winged beetles, a family of coleopteran insects in the superfamily Staphylinoidea.

Ptilodactylidae [INV ZOO] The toed-winged beetles, a family of the Coleoptera in the superfamily Dryopoidea.

Ptilodontoidea [PALEON] A suborder of extinct mammals in the order Multituberculata.

ptilolite *See* mordenite.

Ptinidae [INV ZOO] The spider beetles, a family of coleopteran insects in the superfamily Bostrichoidea.

PTM *See* pulse-time modulation.

ptosis [MED] Prolapse, abnormal depression, or falling down of an organ or part; applied especially to drooping of the upper eyelid, from paralysis of the third cranial nerve.

ptyalase *See* ptyalin.

ptyalin [BIOCHEM] A diastatic enzyme found in saliva which catalyzes the hydrolysis of starch to dextrin, maltose, and glucose, and the hydrolysis of sucrose to glucose and fructose. Also known as ptyalase; salivary amylase; salivary diastase.

Ptychodactiaria [INV ZOO] An order of the zoantharian anthozoans of the phylum Coelenterata known only from two genera, *Ptychodactis* and *Dactylanthus.*

Ptychomniaceae [BOT] A family of mosses in the order Isobryales distinguished by an eight-ribbed capsule.

Ptyctodontida [PALEON] An order of Middle and Upper Devonian fishes of the class Placodermi in which both the head and trunk shields are present, and the joint between them is a well-differentiated and variable structure.

p-type semiconductor [ELECTR] An extrinsic semiconductor in which the hole density exeeds the conduction electron density.

Pu *See* plutonium.

pubic symphysis [ANAT] The fibrocartilaginous union of the pubic bones. Also known as symphysis pubis.

pubis [ANAT] The pubic bone, the portion of the hipbone forming the front of the pelvis.

public address system *See* sound reinforcement system.

public-key algorithm [COMMUN] A cryptographic algorithm in which one key (usually the enciphering key) is made public and a different key (usually the deciphering key) is kept secret; it must not be possible to deduce the private key from the public key.

pucherite [MINERAL] $BiVO_4$ A reddish-brown orthorhombic mineral composed of bismuth vanadate, occurring as small crystals.

puddling [MET] A process for the production of wrought iron by agitation of a bath of molten pig iron with iron oxide in order to reduce the carbon, silicon, phosphorus, and manganese content.

puff cone *See* mud cone.

pulley [DES ENG] A wheel with a flat, round, or grooved rim that rotates on a shaft and carries a flat belt, V-belt, rope, or chain to transmit motion and energy.

pullshovel *See* backhoe.

pulmonary artery [ANAT] A large artery that conducts venous blood from the heart to the lungs of tetrapods.

pulmonary stenosis [MED] Narrowing of the orifice of the pulmonary artery.

pulmonary valve [ANAT] A valve consisting of three semilunar cusps situated between the right ventricle and the pulmonary trunk.

pulmonary vein [ANAT] A large vein that conducts oxygenated blood from the lungs to the heart in tetrapods.

Pulmonata [INV ZOO] A subclass of the gastropod mollusks which contains the "lung"-bearing snails; the gills have been lost and in their place the mantle cavity has become a pulmonary sac.

pulp [ANAT] A mass of soft spongy tissue in the interior of an organ. [BOT] The soft succulent portion of a fruit. [ENG] *See* slime. [MATER] The cellulosic material produced by reducing wood mechanically or chemically and used in making paper and cellulose products. Also known as wood pulp.

pulsar [ASTROPHYS] A celestial radio source, emitting intense short bursts of radio emission; the periods of known pulsars range between 33 milliseconds and 3.75 seconds, and pulse durations range from 2 to about 150 milliseconds with longer-period pulsars generally having a longer pulse duration.

pulsating star [ASTRON] Variable star whose luminosity fluctuates as the star expands and contracts; the variation in brightness is thought to come from the periodic change of radiant energy to gravitational energy and back.

pulse [PHYS] A variation in a quantity which is normally constant; has a finite duration and is usually brief compared to the time scale of interest. [PHYSIO] 1. The regular, recurrent, palpable wave of arterial distention due to the pressure of the blood ejected with each contraction of the heart. 2. A single wave.

pulse amplifier [ELEC] An amplifier designed specifically to amplify electric pulses without appreciably changing their waveforms.

pulse amplitude [PHYS] The peak, average, effective, instantaneous, or other magnitude of a pulse, usually with respect to the normal constant value; the exact meaning should be specified when giving a numerical value.

pulse-amplitude modulation [COMMUN] Amplitude modulation of a pulse carrier. Abbreviated PAM.

pulse bandwidth [COMMUN] The bandwidth outside of which the amplitude of a pulse-frequency spectrum is below a prescribed fraction of the peak amplitude.

pulse carrier [COMMUN] A pulse train used as a carrier.

pulse code [COMMUN] A code consisting of various combinations of pulses, such as the Morse code, Baudot code, and the binary code used in computers.

pulse-code modulation [COMMUN] Modulation in which the peak-to-peak amplitude range of the signal to be transmitted is divided into a number of standard values, each having its own three-place code; each sample of the signal is then transmitted as the

code for the nearest standard amplitude. Abbreviated PCM.

pulse coder *See* coder.

pulse-compression radar [ENG] A radar system in which the transmitted signal is linearly frequency-modulated or otherwise spread out in time to reduce the peak power that must be handled by the transmitter; signal amplitude is kept constant; the receiver uses a linear filter to compress the signal and thereby reconstitute a short pulse for the radar display.

pulsed laser [OPTICS] A laser in which a pulse of coherent light is produced at fixed time intervals, as required for ranging and tracking applications or to permit higher output power than can be obtained with continuous operation.

pulsed light [OPTICS] A beam of light whose intensity is modulated in some prescribed manner; analogous to a radar pulse.

pulsed oscillator [ELECTR] An oscillator that generates a carrier-frequency pulse or a train of carrier-frequency pulses as the result of self-generated or externally applied pulses.

pulsed reactor [NUCLEO] A research nuclear reactor in which continual short, intense surges of power and radiation can be produced; the neutron flux during the surge is much higher than could be tolerated during steady-state operation.

pulse-duration coder *See* coder.

pulse-duration modulation [COMMUN] Modulation of a pulse carrier wherein the value of each instantaneous sample of a modulating wave produces a pulse of proportional duration by varying the leading, trailing, or both edges of a pulse. Abbreviated PDM. Also known as pulse-length modulation; pulse-width modulation.

pulse generator [ELEC] *See* impulse generator. [ELECTR] A generator that produces repetitive pulses or signal-initiated pulses.

pulse group *See* pulse train.

pulse height [ELECTR] The strength or amplitude of a pulse, measured in volts.

pulsejet engine [AERO ENG] A type of compressorless jet engine in which combustion occurs intermittently so that the engine is characterized by periodic surges of thrust; the inlet end of the engine is provided with a grid to which are attached flap valves; these can be sucked inward by a negative differential pressure to allow a regulated amount of air to flow inward to mix with the fuel. Also known as aeropulse engine.

pulse-length modulation *See* pulse-duration modulation.

pulse-mode multiplexing [COMMUN] A type of time-division multiplexing employing pulse-amplitude modulation in which a sequence of pulses is repeatedly transmitted, and the amplitude of each pulse in the sequence is modulated by a different communication channel.

pulse modulation [COMMUN] A system of modulation in which the amplitude, duration, position, or mere presence of discrete pulses may be so controlled as to represent the message to be communicated.

pulse-phase modulation *See* pulse-position modulation.

pulse-position modulation [COMMUN] Modulation of a pulse carrier wherein the value of each instantaneous sample of a modulating wave varies the position in time of a pulse relative to its unmodulated time of occurrence. Abbreviated PPM. Also known as pulse-phase modulation.

pulse radar [ENG] Radar in which the transmitter sends out high-power pulses that are spaced far apart in comparison with the duration of each pulse; the receiver is active for reception of echoes in the interval following each pulse.

pulse recurrence rate *See* pulse repetition rate.

pulse repetition frequency *See* pulse repetition rate.

pulse repetition rate [ELECTR] The number of times per second that a pulse is transmitted. Abbreviated PRR. Also known as pulse recurrence rate; pulse repetition frequency (PRF).

pulse shaper [ELECTR] A transducer used for changing one or more characteristics of a pulse, such as a pulse regenerator or pulse stretcher.

pulse-time modulation [COMMUN] Modulation in which the time of occurrence of some characteristic of a pulse carrier is varied from the unmodulated value; examples include pulse-duration, pulse-interval, and pulse-position modulation. Abbreviated PTM.

pulse train [PHYS] A series of regularly recurrent pulses having similar characteristics. Also known as pulse group.

pulse transformer [ELECTR] A transformer capable of operating over a wide range of frequencies, used to transfer nonsinusoidal pulses without materially changing their waveforms.

pulse-type altimeter *See* radar altimeter.

pulse voltage *See* impulse voltage.

pulse-width modulated static inverter [ELEC] A variation of the quasi-square-wave static inverter, operating at high frequency, in which the pulse width, and not the amplitude, of the square wave is adjusted to approximate the sine wave.

pulse-width modulation *See* pulse-duration modulation.

pulverization *See* comminution.

puma [VERT ZOO] *Felis concolor*. A large, tawny brown wild cat (family Felidae) once widespread over most of the Americas. Also known as American lion; catamount; cougar; mountain lion.

pumice [GEOL] A rock froth, formed by the extreme puffing up of liquid lava by expanding gases liberated from solution in the lava prior to and during solidification. Also known as foam; pumice stone; pumicite; volcanic foam.

pumice stone *See* pumice.

pumicite *See* pumice.

pump [ELECTR] Of a parametric device, the source of alternating-current power which causes the nonlinear reactor to behave as a time-varying reactance. [MECH ENG] A machine that draws a fluid into itself through an entrance port and forces the fluid out through an exhaust port.

pumped hydroelectric storage [ELEC] A method of energy storage in which excess electrical energy produced at times of low demand is used to pump water into a reservoir, and this water is released at times of high demand to operate hydroelectric generators.

pumpellyite [MINERAL] $Ca_2Al_3Si_3O_{12}(OH)$ A greenish epidotelike mineral that is probably related to clinozoisite. Also known as lotrite; zonochlorite.

pumping [FL MECH] Unsteadiness of the mercury in the barometer, caused by fluctuations of the air pressure produced by a gusty wind or due to the motion of a vessel. [PHYS] **1.** The application of optical, infrared, or microwave radiation of appropriate frequency to a laser or maser medium so that absorption of the radiation increases the population of atoms or molecules in higher energy states. Also known as electronic pumping. **2.** The removal of gases and vapors from a vacuum system.

pumpkin [BOT] Any of several prickly vines with large lobed leaves and yellow flowers in the genus Cucurbita of the order Violales; the fruit is orange-colored and large, with a firm rind.

pump oscillator [ELECTR] Alternating-current generator that supplies pumping energy for maser and parametric amplifiers; operates at twice or some higher multiple of the signal frequency.

puna [ECOL] An alpine biological community in the central portion of the Andes Mountains of South America characterized by low-growing, widely spaced plants that lack much green color most of the year.

punch card [COMPUT SCI] A medium by means of which data are fed into a computer in the form of rectangular holes punched in the card. Also known as punched card.

punched card *See* punch card.

punched-card interpreter *See* interpreter.

punched-card reader *See* card reader.

punched-card sorter *See* card sorter.

punched tape *See* punch tape.

punching rate [COMPUT SCI] The number of cards, characters, blocks, fields, or words of information placed in the form of holes distributed on cards, or paper tape per unit of time.

punch position [COMPUT SCI] The location of the row in a columnated card; for example, in an 80-column card the rows or punch position may be 0 to 9 or X and Y corresponding to positions 11 and 12.

punch press [MECH ENG] **1.** A press consisting of a frame in which slides or rams move up and down, of a bed to which the die shoe or bolster plate is attached, and of a source of power to move the slide. Also known as drop press. **2.** Any mechanical press.

punch tape [COMPUT SCI] A paper or plastic ribbon in which data may be represented by means of partially or completely punched holes; it generally has one row of small sprocket-feed holes and five, seven, or eight rows of larger data-representing holes. Also known as punched tape.

pupa [INV ZOO] The quiescent, intermediate form assumed by an insect that undergoes complete metamorphosis; it follows the larva and precedes the adult stages and is enclosed in a hardened cuticle or a cocoon.

pupil [ANAT] The contractile opening in the iris of the vertebrate eye.

Pupipara [INV ZOO] A section of cyclorrhaphous dipteran insects in the Schizophora series in which the young are born as mature maggots ready to become pupae.

Puppis [ASTRON] A southern constellation; right ascension 8 hours, declination 40° south. Also known as Stern.

Purbeckian [GEOL] A stage of geologic time in Great Britain; uppermost Jurassic (above Bononian, below Cretaceous).

pure coal *See* vitrain.

pure culture [MICROBIO] A culture that contains cells of one kind, all progeny of a single cell.

pure imaginary number [MATH] A complex number $z = x + iy$, where $x = 0$.

pure mathematics [MATH] The intrinsic study of mathematical structures, with no consideration given as to the utility of the results for practical purposes.

pure projective geometry [MATH] The axiomatic study of geometric systems which exhibit invariance relative to a notion of projection.

pure tone *See* simple tone.

purine [BIOCHEM] A heterocyclic compound containing fused pyrimidine and imidazole rings; adenine and guanine are the purine components of nucleic acids and coenzymes.

purity [CHEM] The state of a chemical compound when no impurity can be detected by any experimental method; absolute purity is never reached in practice. [OPTICS] The degree to which a primary color is pure and not mixed with the other two primary colors.

Purkinje fibers [HISTOL] Modified cardiac muscle fibers composing the terminal portion of the conducting system of the heart.

purple blende *See* kermesite.

purple blotch [PL PATH] A fungus disease of onions, garlic, and shallots caused by *Alternaria porri* and characterized by small white spots which become large purplish blotches.

purple-top [PL PATH] A virus disease of potato plants characterized by purplish or chlorotic discoloration of the top shoots, swelling of axillary branches, and severe wilting.

purpurite [MINERAL] $(Mn,Fe)PO_4$ A dark-red or purple mineral composed of ferric-manganic phosphate; it is isomorphous with heterosite.

push-bar conveyor [MECH ENG] A type of chain conveyor in which two endless chains are cross-connected at intervals by push bars which propel the load along a stationary bed or trough of the conveyor.

pushbroom sensor modes [AERO ENG] Spacecraft instrument arrangements in which large numbers of detectors composing linear arrays are swept by the forward motion of the spacecraft to attain increased fidelity and high sensitivity in the data captured.

push-button dialing [ELECTR] Dialing a number by pushing buttons on the telephone rather than turning a circular wheel; each depressed button causes a transistor oscillator to oscillate simultaneously at two different frequencies, generating a pair of audio tones which are recognized by central-office (or PBX) switching equipment as digits of a telephone number. Also known as tone dialing; touch call.

push-down list [COMPUT SCI] An ordered set of data items so constructed that the next item to be retrieved is the item most recently stored; in other words, last-in, first-out (LIFO).

push-down storage [COMPUT SCI] A computer storage in which each new item is placed in the first location in the storage and all the other items are moved back one location; it thus follows the principle of a push-down list. Also known as cellar; nesting storage; running accumulator.

push fit [DES ENG] A hand-tight sliding fit between a shaft and a hole.

push-pull amplifier [ELECTR] A balanced amplifier employing two similar electron tubes or equivalent amplifying devices working in phase opposition.

push-pull currents *See* balanced currents.

push-pull oscillator [ELECTR] A balanced oscillator employing two similar electron tubes or equivalent amplifying devices in phase opposition.

push-pull transformer [ELECTR] An audio-frequency transformer having a center-tapped winding and designed for use in a push-pull amplifier.

push-push amplifier [ELECTR] An amplifier employing two similar electron tubes with grids connected in phase opposition and with anodes connected in parallel to a common load; usually used as a frequency multiplier to emphasize even-order harmonics; transistors may be used in place of tubes.

push rod [MECH ENG] A rod, as in an internal combustion engine, which is actuated by the cam to open and close the valves.

push-up list [COMPUT SCI] An ordered set of data items so constructed that the next item to be retrieved will be the item that was inserted earliest in the list, resulting in a first-in, first-out (FIFO) structure.

Pustulosa [PALEON] An extinct suborder of echinoderms in the order Phanerozonida found in the Paleozoic.

putrefaction [BIOCHEM] Decomposition of organic matter, particularly the anaerobic breakdown of proteins by bacteria, with the production of foul-smelling compounds.

P wave *See* compressional wave.

PWI *See* proximity warning indicator.

PWR *See* pressurized water reactor.

pwt *See* pennyweight.

Pycnodontiformes [PALEON] An extinct order of specialized fishes characterized by a laterally compressed, disk-shaped body, long dorsal and anal fins, and an externally symmetrical tail.

Pycnogonida [INV ZOO] The sea spiders, a subphylum of marine arthropods in which the body is reduced to a series of cylindrical trunk somites supporting the appendage.

Pycnogonidae [INV ZOO] A family of the Pycnogonida lacking both chelifores and palpi and having six to nine jointed ovigers in the male only.

pycnocline [GEOPHYS] A change in density of ocean or lake water or rock with displacement in some direction, especially a rapid change in density with vertical displacement. [OCEAN] A region in the ocean where water density increases relatively rapidly with depth.

pycnometer [ENG] A container whose volume is precisely known, used to determine the density of a liquid by filling the container with the liquid and then weighing it. Also spelled pyknometer.

Pygasteridae [PALEON] The single family of the extinct order Pygasteroida.

Pygasteroida [PALEON] An order of extinct echinoderms in the superorder Diadematacea having four genital pores, noncrenulate tubercles, and simple ambulacral plates.

Pygopodidae [VERT ZOO] The flap-footed lizards, a family of the suborder Sauria.

pyknometer *See* pycnometer.

pylon [AERO ENG] A suspension device externally installed under the wing or fuselage of an aircraft; it is aerodynamically designed to fit the configuration of specific aircraft, thereby creating an insignificant amount of drag; it includes means of attaching to accommodate fuel tanks, bombs, rockets, torpedoes, rocket motors, or the like. [CIV ENG] **1.** A massive structure, such as a truncated pyramid, on either side of an entrance. **2.** A tower supporting a wire over a long span. **3.** A tower or other structure marking a route for an airplane.

pylorus [ANAT] The orifice of the stomach communicating with the small intestine.

pyoderma [MED] Any pus-producing skin lesion or lesions, used in reference to groups of furuncles, pustules, or even carbuncles.

Pyralidae [INV ZOO] The equivalent name for Pyralididae.

Pyralididae [INV ZOO] A large family of moths in the lepidopteran superfamily Pyralidoidea; the labial palpi are well developed, and the legs are usually long and slender.

Pyralidinae [INV ZOO] A subfamily of the Pyralididae.

Pyralidoidea [INV ZOO] A superfamily of the Lepidoptera belonging to the Heteroneura and including long-legged, slender-bodied moths with well-developed maxillary palpi.

pyramid [CRYSTAL] An open crystal having three, four, six, eight, or twelve nonparallel faces that meet at a point. [MATH] A polyhedron with one face a polygon and all other faces triangles with a common vertex.

Pyramidellidae [INV ZOO] A family of gastropod mollusks in the order Tectibranchia; the operculum is present in this group.

pyranometer [ENG] An instrument used to measure the combined intensity of incoming direct solar radiation and diffuse sky radiation; compares heating produced by the radiation on blackened metal strips with that produced by an electric current. Also known as solarimeter.

pyrargyrite [MINERAL] Ag_3SbS_3 A deep ruby-red to black mineral, crystallizing in the hexagonal system, occurring in massive form and in disseminated grains, and having an adamantine luster; hardness is 2.5 on Mohs scale, and specific gravity is 5.85; an important silver ore. Also known as dark-red silver ore; dark ruby silver.

Pyraustinae [INV ZOO] A large subfamily of the Pyralididae containing relatively large, economically important moths.

Pyrenolichenes [BOT] The equivalent name for Pyrenulales.

Pyrenulaceae [BOT] A family of the Pyrenulales; all species are crustose and most common on tree bark in the tropics.

Pyrenulales [BOT] An order of the class Ascolichenes including only those lichens with perithecia that contain true paraphyses and unitunicate asci.

Pyrgotidae [INV ZOO] A family of myodarian cyclorrhaphous dipteran insects in the subsection Acalyptratae.

pyrheliometer [ENG] An instrument for measuring the total intensity of direct solar radiation received at the earth.

pyrimidine [BIOCHEM] $C_4H_4N_2$ A heterocyclic organic compound containing nitrogen atoms at positions 1 and 3; naturally occurring derivatives are components of nucleic acids and coenzymes.

pyrite [MINERAL] FeS_2 A hard, brittle, brass-yellow mineral with metallic luster, crystallizing in the isometric system; hardness is 6–6.5 on Mohs scale, and specific gravity is 5.02. Also known as common pyrite; fool's gold; iron pyrite; mundic.

pyritohedron [CRYSTAL] A dodecahedral crystal with 12 irregular pentagonal faces; it is characteristic of pyrite. Also known as pentagonal dodecahedron; pyritoid; regular dodecahedron.

pyritoid See pyritohedron.

pyro- [CHEM] A chemical prefix for compounds formed by heat, such as pyrophosphoric acid, an inorganic acid formed by the loss of one water molecule from two molecules of an ortho acid.

pyroaurite [MINERAL] $Mg_6Fe_2(OH)_{16}\cdot CO_34H_2O$ A goldlike or brownish rhombohedral mineral composed of hydrous basic magnesium iron carbonate.

pyrobelonite [MINERAL] $PbMn(VO_4)(OH)$ A fire-red to deep brilliant-red mineral composed of basic vanadate of manganese and lead, occurring as crystal needles.

pyroborate See borax.

pyrochlore [MINERAL] $(Na,Ca)_2(Nb,Ta)_2\!-\!O_6(OH,F)$ Pale-yellow, reddish, brown, or black mineral, crystallizing in the isometric system, and occurring in pegmatites derived from alkalic igneous rocks. Also known as pyrrhite.

pyroclastic flow [GEOL] Ash flow not involving high-temperature conditions.

pyroclastic rock [PETR] A rock that is composed of fragmented volcanic products ejected from volcanoes in explosive events.

pyroelectricity [SOLID STATE] The property of certain crystals to produce a state of electrical polarity by a change of temperature.

pyrogen [BIOCHEM] A group of substances thought to be polysaccharides of microbial origin that produce an increase in body temperature when injected into humans and some animals.

pyrolusite [MINERAL] MnO_2 An iron-black mineral that crystallizes in the tetragonal system and is the most important ore of manganese; hardness is 1-2 on Mohs scale, and specific gravity is 4.75.

pyrolysis [CHEM] The breaking apart of complex molecules into simpler units by the use of heat, as in the pyrolysis of heavy oil to make gasoline.

pyrometallurgy [MET] High-temperature process metallurgy.

pyrometer [ENG] Any of a broad class of temperature-measuring devices; they were originally designed to measure high temperatures, but some are now used in any temperature range; includes radiation pyrometers, thermocouples, resistance pyrometers, and thermistors.

pyromorphite [MINERAL] $Pb_5(PO_4)_3Cl$ A green, yellow, brown, gray, or white mineral of the apatite group, crystallizing in the hexagonal system; a minor ore of lead. Also known as green lead ore.

pyrope [MINERAL] $Mg_3Al_2(SiO_4)_3$ A mineral species of the garnet group characterized by a deep fiery-red color and occurring in basic and ultrabasic igneous rocks.

pyrophane See fire opal.

pyrophanite [MINERAL] $MnTiO_3$ A blood-red rhombohedral mineral consisting of manganese titanate; it is isomorphous with ilmenite.

pyrophyllite [MINERAL] $AlSi_2O_5(OH)$ A white, greenish, gray, or brown phyllosilicate mineral that resembles talc and occurs in a foliated form or in compact masses in quartz veins, granites, and metamorphic rocks. Also known as pencil stone.

pyrosmalite [MINERAL] $(Mn,Fe)_4Si_3O_7(OH,Cl)_6$ A colorless, pale-brown, gray, or gray-green mineral composed mainly of basic iron manganese silicate with chlorine.

Pyrosomida [INV ZOO] An order of pelagic tunicates in the class Thaliacea in which species form tubular swimming colonies and are often highly luminescent.

pyrostibite See kermesite.

pyrotechnics [ENG] Art and science of preparing and using fireworks. [MATER] Items which are used for both military and nonmilitary purposes to produce a bright light for illumination, or colored lights or smoke for signaling, and which are consumed in the process.

Pyrotheria [PALEON] An extinct monofamilial order of primitive, mastodonlike, herbivorous, hoofed mammals restricted to the Eocene and Oligocene deposits of South America.

Pyrotheriidae [PALEON] The single family of the Pyrotheria.

pyroxene [MINERAL] A family of diverse and important rock-forming minerals having infinite (Si_2O_6) single inosilicate chains as their principal motif; colors range from white through yellow and green to brown and greenish black; hardness is 5.5-6 on Mohs scale, and specific gravity is 3.2–4.0.

pyroxenite [PETR] A heavy, dark-colored, phaneritic igneous rock composed largely of pyroxene with smaller amounts of olivine and hornblende, and formed by crystallization of gabbraic magma.

pyroxenoids [MINERAL] A mineral group (including wollastonite and rhodonite) compositionally similar to pyroxene, but SiO_4 tetrahedrons are connected in rings rather than chains.

pyrrhite _See_ pyrochlore.

Pyrrhocoridae [INV ZOO] A family of hemipteran insects belonging to the superfamily Pyrrhocoroidea.

Pyrrhocoroidea [INV ZOO] A superfamily of the Pentatomorpha.

Pyrrhophyta [BOT] A small division of motile, generally unicellular flagellate algae characterized by the presence of yellowish-green to golden-brown plastids and by the general absence of cell walls.

pyrrhotite [MINERAL] $Fe_{1-x}S$ A common reddish-brown to brownish-bronze mineral that occurs as rounded grains to large masses, more rarely as tabular pseudohexagonal crystals and rosettes; hardness is 4 on Mohs scale, and specific gravity is 4.6 (for the composition Fe_7S_8).

Pythagorean theorem [MATH] In a right triangle the square of the length of the hypotenuse equals the sum of the squares of the lengths of the other two sides.

python [VERT ZOO] The common name for members of the reptilian subfamily Pythoninae.

Pythoninae [VERT ZOO] A subfamily of the reptilian family Boidae distinguished anatomically by the skull structure and the presence of a pair of vestigial hindlegs in the form of stout, movable spurs.

Pyxis [ASTRON] A southern constellation; right ascension 9 hours, declination 30° south. Also known as Malus.

Q

Q [NUC PHYS] *See* disintegration energy. [PHYS] A measure of the ability of a system with periodic behavior to store energy equal to 2π times the average energy stored in the system divided by the energy dissipated per cycle. Also known as *Q*factor; quality factor; storage factor. [THERMO] A unit of heat energy, equal to 10^{18} British thermal units, or approximately 1.055×10^{21} joules.

QCD *See* quantum chromodynamics.

Q factor [ORD] A correction factor applied to a bombsight setting to help account for the differing winds between flight altitude and the ground; this factor corrects only for that component of the differential ballistic wind that is parallel to the actual wind at flight level. [PHYS] *See* Q.

Q meter [ENG] A direct-reading instrument which measures the Q of an electric circuit at radio frequencies by determining the ratio of inductance to resistance, and which has also been developed to measure many other quantities. Also known as quality-factor meter.

Q multiplier [ELECTR] A filter that gives a sharp response peak or a deep rejection notch at a particular frequency, equivalent to boosting the Q of a tuned circuit at that frequency.

QPRK *See* quadrature partial-response keying.

Q signal [COMMUN] A three-letter abbreviation starting with Q, used in the International List of Abbreviations for radiotelegraphy to represent complete sentences. [ELECTR] The quadrature component of the chrominance signal in color television, having a bandwidth of 0 to 0.5 megahertz; it consists of $+0.48(R-Y)$ and $+0.41(B-Y)$, where Y is the luminance signal, R is the red camera signal, and B is the blue camera signal.

QSO *See* quasar.

Q-switched laser [OPTICS] A laser whose Q factor is kept at a low value while an ion population inversion is built up, and then is suddenly switched to a high value just before instability occurs, resulting in a very high rate of stimulated emission. Also known as giant pulse laser.

qt *See* quart.

quad [ELEC] A series of four separately insulated conductors, generally twisted together in pairs. [ELECTR] A series-parallel combination of transistors; used to obtain increased reliability through double redundancy, because the failure of one transistor will not disable the entire circuit. [GRAPHICS] One of the small pieces of metal used in typesetting to space or to fill out a line of characters; used mostly to fill space when indenting the first line and to fill out the last line of type. [THERMO] A unit of heat energy, equal to 10^{15} British thermal units, or approximately 1.055×10^{18} joules.

quadrangle [CIV ENG] **1.** A four-cornered, four-sided courtyard, usually surrounded by buildings. **2.** The buildings surrounding such a courtyard. **3.** A four-cornered, four-sided building. [MAP] A four-cornered, four-sided tract of land, defined by parallels of latitude and meridians of longitude, used as an area unit in systematic mapping. [MATH] A geometric figure bounded by four straight-line segments called sides, each of which intersects each of two adjacent sides in points called vertices, but fails to intersect the opposite sides. Also known as quadrilateral.

quadrant [ANAT] One of the four regions into which the abdomen may be divided for purposes of physical diagnosis. [ELECTROMAG] *See* international henry. [ENG] **1.** An instrument for measuring altitudes, used, for example, in astronomy, surveying, and gunnery; employs a sight that can be moved through a graduated $90°$ arc. **2.** A lever that can move through a $90°$ arc. [MATH] **1.** A quarter of a circle; either an arc of $90°$ or the area bounded by such an arc and the two radii. **2.** Any of the four regions into which the plane is divided by a pair of coordinate axes. [MECH ENG] A device for converting horizontal reciprocating motion to vertical reciprocating motion. [NAV] One of the four areas between consecutive equisignal zones of a four course radio range station. [NAV ARCH] A casting, forging, or built-up frame in the shape of a sector of a circle attached to the rudder stock and through which the steering gear leads turn the rudder. [OPTICS] A double-reflecting instrument for measuring angles, used primarily for measuring altitudes of celestial bodies; the instrument was replaced by the sextant. [PHYSIO] A sector of one-fourth of the field of vision of one or both eyes.

quadrantal point *See* intercardinal point.

quadraphonic sound system [ENG ACOUS] A system for reproducing sound by means of four loudspeakers properly situated in the listening room, usually at the four corners of a square, with each loudspeaker being fed its own identifiable segment of the program signal. Also known as four-channel sound system.

quadrate bone [VERT ZOO] A small element forming part of the upper jaw joint on each side of the head in vertebrates below mammals.

quadratic equation [MATH] Any second-degree polynomial equation.

quadratic formula [MATH] A formula giving the roots of a quadratic equation in terms of the coefficients; for the equation $ax^2 + bx + c = 0$, the roots are $x = (-b \pm \sqrt{b^2 - 4ac})/2a$.

quadratic Stark effect [ATOM PHYS] A splitting of spectral lines of atoms in an electric field in which the energy levels shift by an amount proportional to the square of the electric field, and all levels shift to lower energies; observed in lines resulting from the lower energy states of many-electron atoms.

quadratic Zeeman effect [ATOM PHYS] A splitting of spectral lines of atoms in a magnetic field in which the energy levels shift by an amount proportional to the square of the magnetic field.

quadrature [ASTRON] The right-angle physical alignment of the sun, moon, and earth. [MATH] The construction of a square whose area is equal to that of a given surface. [PHYS] State of being separated in phase by 90°, or one quarter-cycle. Also known as phase quadrature.

quadrature current See reactive current.

quadriceps [ANAT] Four-headed, as a muscle.

quadric surface [MATH] A surface whose equation is a second-degree algebraic equation.

quadrigeminal body See corpora quadrigemina.

Quadrijugatoridae [PALEON] A monomorphic family of extinct ostracods in the superfamily Hollinacea.

quadrilateral See quadrangle.

quadruple point [PHYS CHEM] Temperature at which four phases are in equilibrium, such as a saturated solution containing an excess of solute.

quadrupole [ELECTROMAG] A distribution of charge or magnetization which produces an electric or magnetic field equivalent to that produced by two electric or magnetic dipoles whose dipole moments have the same magnitude but point in opposite directions, and which are separated from each other by a small distance.

quadrupole amplifier [ELECTR] A low-noise parametric amplifier consisting of an electron-beam tube in which quadrupole fields act on the fast cyclotron wave of the electron beam to produce high amplification at frequencies in the range of 400–800 megahertz.

quadrupole lens [ELECTROMAG] A device for focusing beams of charged particles which has four electrodes or magnetic poles of alternating sign arranged in a circle about the beam; used in instruments such as electron microscopes and particle accelerators.

quadrupole moment [ELECTROMAG] A quantity characterizing a distribution of charge or magnetization; it is given by integrating the product of the charge density or divergence of magnetization density, the second power of the distance from the origin, and a spherical harmonic $Y*_{2m}$ over the charge or magnetization distribution.

quagmire See bog.

qualifier [COMPUT SCI] A name that is associated with another name to give additional information about the latter and distinguish it from other things having the same name.

qualitative analysis [ANALY CHEM] The analysis of a gas, liquid, or solid sample or mixture to identify the elements, radicals, or compounds composing the sample.

quality assurance [IND ENG] Testing and inspecting all of a portion of the final product to ensure that the desired quality level of product reaches the consumer.

quality control [IND ENG] Inspection, analysis, and action applied to a portion of the product in a manufacturing operation to estimate overall quality of the product and determine what, if any, changes must be made to achieve or maintain the required level of quality.

quality factor [NUCLEO] The factor by which absorbed dose is to be multiplied to obtain a quantity that expresses on a common scale, for all ionizing radiations, the irradiation incurred by exposed persons. [PHYS] See Q.

quality-factor meter See Q meter.

quantification [SCI TECH] The act of quantifying, that is, of giving a numerical value to a measurement of something, as in computer applications, psychology, or market research.

quantifier [MATH] Either of the phrases "for all" and "there exists"; these are symbolized respectively by an inverted A and a backward E.

quantile [MATH] The arrangement of a set of N observations forming a frequency distribution in order of magnitude with every (N/P)-th observation marked off.

quantitative analysis [ANALY CHEM] The analysis of a gas, liquid, or solid sample or mixture to determine the precise percentage composition of the sample in terms of elements, radicals, or compounds.

quantization [COMMUN] Division of the range of values of a wave into a finite number of subranges, each of which is represented by an assigned or quantized value within the subrange. [QUANT MECH] 1. The restriction of an observable quantity, such as energy or angular momentum, associated with a physical system, such as an atom, molecule, or elementary particle, to a discrete set of values. 2. The transition from a description of a system of particles or fields in the classical approximation where canonically conjugate variables commute, to a description where these variables are treated as noncommuting operators; quantization (first definition) is a result of this procedure. [SCI TECH] The restriction of a variable to a discrete number of possible values; thus the age of a person is usually quantized as a whole number of years.

quantized spin wave See magnon.

quantized vortex [CRYO] A circular flow pattern observed in superfluid helium and type II superconductors, in which a superfluid flows about a normal (nonsuperfluid) cylindrical region or core which has the form of a thin line, and either the circulation or the magnetic flux is quantized.

quantum [COMMUN] One of the subranges of possible values of a wave which is specified by quantization and represented by a particular value within the subrange. [QUANT MECH] 1. For certain physical quantities, a unit such that the values of the quantity are restricted to integral multiples of this unit; for example, the quantum of angular momentum is Planck's constant divided by 2π. 2. An entity resulting from quantization of a field or wave, having particlelike properties such as energy, mass, momentum and angular momentum; for example, the photon is the quantum of an electromagnetic field, and the phonon is the quantum of a lattice vibration.

quantum chemistry [PHYS CHEM] A branch of physical chemistry concerned with the explanation of chemical phenomena by means of the laws of quantum mechanics.

quantum chromodynamics [PARTIC PHYS] A gauge theory of the strong interactions among quarks; the mathematical structure of the theory resembles that of quantum electrodynamics, with color as the conserved charge. Abbreviated QCD.

quantum electrodynamics [QUANT MECH] The quantum theory of electromagnetic radiation, synthesizing the wave and corpuscular pictures, and of the interaction of radiation with electrically charged matter, in particular with atoms and their constituent electrons. Also known as quantum theory of light; quantum theory of radiation.

quantum electronics [ELECTR] The branch of electronics associated with the various energy states of matter, motions within atoms or groups of atoms, and various phenomena in crystals; examples of practical applications include the atomic hydrogen maser and the cesium atomic-beam resonator.

quantum field theory [QUANT MECH] Quantum theory of physical systems possessing an infinite number of degrees of freedom, such as the electromagnetic field, gravitation field, or wave fields in a medium.

quantum mechanics [PHYS] The modern theory of matter, of electromagnetic radiation, and of the interaction between matter and radiation; it differs from classical physics, which it generalizes and supersedes, mainly in the realm of atomic and subatomic phenomena. Also known as quantum theory.

quantum number [QUANT MECH] One of the quantities, usually discrete with integer or half-integer values, needed to characterize a quantum state of a physical system; they are usually eigenvalues of quantum-mechanical operators or integers sequentially assigned to these eigenvalues.

quantum of action *See* Planck's constant.

quantum state [QUANT MECH] 1. The condition of a physical system as described by a wave function; the function may be simultaneously an eigenfunction of one or more quantum-mechanical operators; the eigenvalues are then the quantum numbers that label the state. 2. *See* energy state.

quantum statistics [STAT MECH] The statistical description of particles or systems of particles whose behavior must be described by quantum mechanics rather than classical mechanics.

quantum theory *See* quantum mechanics.

quantum theory of light *See* quantum electrodynamics.

quantum theory of radiation [QUANT MECH] 1. The theory of heat radiation based on Planck's law; its principal result is the Planck radiation formula. 2. *See* quantum electrodynamics.

quark [PARTIC PHYS] One of the hypothetical basic particles, having charges whose magnitudes are $\frac{1}{3}$ or $\frac{2}{3}$ of the electron charge, from which many of the elementary particles may, in theory, be built up; for example, nucleons may be formed from three quarks and mesons from quark-antiquark combinations; no experimental evidence for the actual existence of free quarks has been found.

quart [MECH] Abbreviated qt. 1. A unit of volume used for measurement of liquid substances in the United States, equal to 2 pints, or $\frac{1}{4}$ gallon, or $57\frac{3}{4}$ cubic inches, or $9.46352946 \times 10^{-4}$ cubic meter. 2. A unit of volume used for measurement of solid substances in the United States, equal to 2 dry pints, or $\frac{1}{32}$ bushel, or $107,521/1,600$ cubic inches, or approximately 1.10122×10^{-3} cubic meter. 3. A unit of volume used for measurement of both liquid and solid substances, although mainly the former, in the United Kingdom, equal to 2 U.K. pints, or $\frac{1}{4}$ U.K. gallon, or approximately 1.13652×10^{-3} cubic meter.

quarternary phase-shift keying [ELECTR] Modulation of a microwave carrier with two parallel streams of nonreturn-to-zero data in such a way that the data is transmitted as 90° phase shifts of the carrier; this gives twice the message channel capacity of binary phase-shift keying in the same bandwidth. Abbreviated QPSK.

quarter-phase *See* two-phase.

quarter-square multiplier [COMPUT SCI] A device used to carry out function multiplication in an analog computer by implementing the algebraic identity $xy = \frac{1}{4}[(x+y)^2 - (x-y)^2]$.

quarter-wave antenna [ELECTROMAG] An antenna whose electrical length is equal to one quarter-wavelength of the signal to be transmitted or received.

quarter-wave plate [OPTICS] A thin sheet of mica or other doubly refracting crystal material of such thickness as to introduce a phase difference of one quarter-cycle between the ordinary and the extraordinary components of light passing through; such a plate converts circularly polarized light into plane-polarized light.

quartic *See* biquadratic.

quartic equation [MATH] Any fourth-degree polynomial equation. Also known as biquadratic equation.

quartz [MINERAL] SiO_2 A colorless, transparent rock-forming mineral with vitreous luster, crystallizing in the trigonal trapezohedral class of the rhombohedral subsystem; hardness is 7 on Mohs scale, and specific gravity is 2.65; the most abundant and widespread of all minerals.

quartz clock [HOROL] A clock using the piezoelectric property of a quartz crystal, in which the crystal is introduced into an oscillating electric circuit having a frequency nearly equal to the natural frequency of vibration of the crystal.

quartz crystal [ELECTR] A natural or artificially grown piezoelectric crystal composed of silicon dioxide, from which thin slabs or plates are carefully cut and ground to serve as a crystal plate. [MINERAL] *See* rock crystal.

quartzite [PETR] A granoblastic metamorphic rock consisting largely or entirely of quartz; most quartzites are formed by metamorphism of sandstone.

quartz lamp [ELECTR] A mercury-vapor lamp having a transparent envelope made from quartz instead of glass; quartz resists heat, permitting higher currents, and passes ultraviolet rays that are absorbed by ordinary glass.

quartz lattice *See* rhyodacite.

quartz monzonite [PETR] Granitic rock in which 10–50% of the felsic constituents are quartz, and in which the ratio of alkali feldspar to total feldspar is between 35% and 65%. Also known as adamellite.

quartz oscillator [ELECTR] An oscillator in which the frequency of the output is determined by the natural frequency of vibration of a quartz crystal.

quartz porphyry [PETR] A porphyritic extrusive or hypabyssal rock containing quartz and alkali feldspar phenocrysts embedded in a microcrystalline or cryptocrystalline matrix. Also known as granite porphyry.

quartz wedge [OPTICS] A very thin wedge of quartz cut parallel to an optic axis; used to determine the sign of double refraction of biaxial crystals, and in other applications involving polarized light and its interaction with matter.

quasar [ASTRON] Quasi-stellar astronomical object, often a radio source; all quasars have large red shifts; they have small optical diameter, but may have large radio diameter. Also known as quasi-stellar object (QSO).

quasi-atom [ATOM PHYS] A system formed by two colliding atoms whose nuclei approach each other so closely that, for a very short time, the atomic electrons arrange themselves as if they belonged to a single atom whose atomic number equals the sum of the atomic numbers of the colliding atoms.

quasi-fission [NUC PHYS] A nuclear reaction induced by heavy ions in which the two product nuclei have kinetic energies typical of fission products, but have masses close to those of the target and projectile, individually. Also known as deep inelastic transfer; incomplete fusion; relaxed peak process; strongly damped collision.

quasi-instruction [COMPUT SCI] An expression in a source program which resembles an instruction in form, but which does not have a corresponding machine instruction in the object program, and is directed to the assembler or compiler. Also known as pseudoinstruction.

quasi-molecule [ATOM PHYS] The structure formed by two colliding atoms when their nuclei are close enough for the atoms to interact, but not so close as to form a quasi-atom.

quasi-particle [PHYS] An entity used in the description of a system of many interacting particles which has particlelike properties such as mass, energy, and momentum, but which does not exist as a free particle; examples are phonons and other elementary excitations in solids, and "dressed" helium-3 atoms in Landau's theory of liquid helium-3.

quasi-reflection [OPTICS] A term applied to the very strong return of light produced by dust particles and other suspensoids whose diameters are large compared to the wavelength of the incident radiation.

quasi-stable elementary particle [PARTIC PHYS] A term formerly (before the discovery of charmed particles) used for elementary particles that cannot decay into other particles through strong interactions and that have lifetimes longer than 10^{-20} second. Also known as semistable elementary particle.

quasi-static process *See* reversible process.

quasi-stationary front [METEOROL] A front which is stationary or nearly so; conventionally, a front which is moving at a speed less than about 5 knots (0.26 meter per second) is generally considered to be quasi-stationary. Commonly known as stationary front.

quasi-stellar object *See* quasar.

Quaternary [GEOL] The second period of the Cenozoic geologic era, following the Tertiary, and including the last 2–3 million years.

quaternary system [PHYS CHEM] An equilibrium relationship between a mixture of four (four phases, four components, and so on).

quaternion [MATH] The division algebra over the real numbers generated by elements i, j, k subject to the relations $i^2 = j^2 = k^2 = -1$ and $ij = -ji = k$, $jk = -kj = i$, and $ki = -ik = j$. Also known as hypercomplex number.

quay [CIV ENG] A solid embankment or structure parallel to a waterway; used for loading and unloading ships.

quebracho [BOT] Any of a number of South American trees in different genera in the order Sapindales, but all being a valuable source of wood, bark, and tannin. [MATER] A drilling-fluid additive used for thinning or dispersing in order to control viscosity and thixotropy; made from an extract of the quebracho tree and consisting essentially of tannic acid.

quench annealing [MET] Annealing an austenitic ferrous alloy by heating followed by quenching from solution temperatures.

quenched spark gap [ELEC] A spark gap having provisions for rapid deionization; one form consists of many small gaps between electrodes that have relatively large mass and are good radiators of heat; the electrodes serve to cool the gaps rapidly and thereby stop conduction.

quench hardening [MET] The hardening of a ferrous alloy by quenching from a temperature above the transformation range.

quenching [ATOM PHYS] Phenomenon in which a very strong electric field, such as a crystal field, causes the orbit of an electron in an atom to precess rapidly so that the average magnetic moment associated with its orbital angular momentum is reduced to zero. [ELECTR] **1.** The process of terminating a discharge in a gas-filled radiation-counter tube by inhibiting reignition. **2.** Reduction of the intensity of resonance radiation resulting from deexcitation of atoms, which would otherwise have emitted this radiation, in collisions with electrons or other atoms in a gas. [ENG] Shock cooling by immersing liquid or molten material into a cooling medium (liquid or gas); used in metallurgy, plastics forming, and petroleum refining. [IMMUNOL] An adaptation of immunofluorescence that uses two fluorochromes, one of which absorbs light emitted by the other; one fluorochrome labels that antigen, another the antibody, and the antigen-antibody complexes retain both; the initialy emitted light is absorbed and so quenched by the second compound. [MET] Rapid cooling from solution temperatures. [SOLID STATE] Reduction in the intensity of sensitized luminescence radiation when energy migrating through a crystal by resonant transfer is dissipated in crystal defects or impurities rather than being reemitted as radiation.

query [COMPUT SCI] A computer instruction to interrogate a data base.

query language [COMPUT SCI] A generalized computer language that is used to interrogate a data base.

query program [COMPUT SCI] A computer program that allows a user to retrieve information from a data base and have it displayed on a terminal or printed out.

queuing [ENG] The movement of discrete units through channels, such as programs or data arriving at a computer, or movement on a highway of heavy traffic.

queuing theory [MATH] The area of stochastic processes emphasizing those processes modeled on the situation of individuals lining up for service.

quibinary [COMPUT SCI] A numeration system, used in data processing, in which each decimal digit is represented by seven binary digits, a group of five which are coefficients of 8, 6, 4, 2, and 0, and a group of two which are coefficients of 1 and 0.

quick-break fuse [ELEC] A fuse designed to draw out the arc and break the circuit rapidly when the fuse wire melts, generally by separating the broken ends with a spring.

quick-break switch [ELEC] A switch that breaks a circuit rapidly, independently of the rate at which the switch handle is moved, to minimize arcing.

quicksand [GEOL] A highly mobile mass of fine sand consisting of smooth, rounded grains with little tendency to mutual adherence, usually thoroughly saturated with upward-flowing water; tends to yield under pressure and to readily swallow heavy objects on the surface. Also known as running sand. [MATER] A loose sand mixture with a high proportion of water, thus having a low bearing pressure.

quicksilver *See* mercury.

quiescent [ELECTR] Condition of a circuit element which has no input signal, so that it does not perform its active function. [ENG] State of a body at rest, or inactive, such as an undisturbed liquid in a storage or process vessel. [MED] Inactive, latent, or dormant, referring to a disease or pathological process.

quiet automatic volume control   *See* delayed automatic gain control.

quiet sun   [ASTROPHYS] The sun when it is free from unusual radio wave or thermal radiation such as that associated with sunspots.

quill   [DES ENG] A hollow shaft into which another shaft is inserted in mechanical devices.   [TEXT] A shaft or spool on which filling yarn is wound before insertion into a shuttle.   [VERT ZOO] The hollow, horny shaft of a large stiff wing or tail feather.

quill drive   [MECH ENG] A drive in which the motor is mounted on a nonrotating hollow shaft surrounding the driving-wheel axle; pins on the armature mesh with spokes on the driving wheels, thereby transmitting motion to the wheels; used on electric locomotives.

quillwort   [BOT] The common name for plants of the genus *Isoetes*.

quinary code   [COMPUT SCI] A code based on five possible combinations for representing digits.

quince   [BOT] *Cydonia oblonga*. A deciduous tree of the order Rosales characterized by crooked branching, leaves that are densely hairy on the underside and solitary white or pale-pink flowers; fruit is an edible pear- or apple-shaped tomentose pome.

quinoa   [BOT] *Chenopodium quinoa*. An annual herb of the family Chenopodiaceae grown at high altitudes in South America for the highly nutritious seeds.

quintal   *See* hundredweight; metric centner.

quoin   [BUILD] One of the members forming an outside corner or exterior angle of a building, and differentiated from the wall by color, texture, size, or projection.   [GRAPHICS] One of the wedge-shaped devices made of steel, generally triangular, and less than type-high; used to lock type and plates in chases for the press.

quotient   [MATH] The result of dividing one quantity by another.

Q value   *See* disintegration energy.

R

r *See* roentgen.

R *See* roentgen.

Ra *See* radium.

R.A. *See* right ascension.

rabbet [ENG] 1. A groove cut into a part. 2. A strip applied to a part as, for example, a stop or seal. 3. A joint formed by fitting one member into a groove, channel, or recess in the face or edge of a second member.

rabbit [NUCLEO] A small container that is propelled, usually pneumatically or hydraulically, through a tube into a nuclear reactor; used to expose samples to the radiation, especially neutron flux, then remove them rapidly for measurements of radioactive atoms having short half-lives. Also known as shuttle. [PETRO ENG] A small plug driven by pressure through a flow line to clean the line or to check that it is unobstructed. [VERT ZOO] Any of a large number of burrowing mammals in the family Leporidae.

rabies [VET MED] An acute, encephalitic viral infection transmitted to humans by the bite of a rabid animal. Also known as hydrophobia.

raccoon [VERT ZOO] Any of 16 species of carnivorous nocturnal mammals belonging to the family Procyonidae; all are arboreal or semiarboreal and have a bushy, long ringed tail.

race [ANTHRO] 1. A distinctive human type possessing characteristic traits that are transmissible by descent. 2. Descendants of a common ancestor. [BIOL] 1. An infraspecific taxonomic group of organisms, such as subspecies or microspecies. 2. A fixed variety or breed. [DES ENG] Either of the concentric pair of steel rings of a ball bearing or roller bearing. [ENG] A channel transporting water to or away from hydraulic machinery, as in a power house. [OCEANOGR] A rapid current, or a constricted channel in which such a current flows; the term is usually used only in connection with a tidal current, which may be called a tide race.

racemase [BIOCHEM] Any of a group of enzymes that catalyze racemization reactions.

raceme [BOT] An inflorescence on which flowers are borne on stalks of equal length on an unbranched main stalk that continues to grow during flowering.

racemic mixture [CHEM] A compound which is a mixture of equal quantities of dextrorotatory and levorotatory isomers of the same compound, and therefore is optically inactive.

racemose [ANAT] Of a gland, compound and shaped like a bunch of grapes, with freely branching ducts that terminate in acini. [BOT] Bearing, or occurring in the form of, a raceme.

race track [NUCLEO] An assembly of several Calutron isotope separators in the shape of a race track, having a common magnetic field. Also known as track.

raceway [ELEC] A channel used to hold and protect wires, cables, or busbars. Also known as electric raceway.

rack [AERO ENG] A suspension device permanently fixed to an aircraft; it is designed for attaching, arming, and releasing one or more bombs; it may also be utilized to accommodate other items such as mines, rockets, torpedoes, fuel tanks, rescue equipment, sonobuoys, and flares. [CIV ENG] A fixed screen composed of parallel bars placed in a waterway to catch debris. [DES ENG] *See* relay rack. [ENG] A frame for holding or displaying articles. [MECH ENG] A bar containing teeth on one face for meshing with a gear. [MIN ENG] An inclined trough or table for washing or separating ore.

racon *See* radar beacon.

rad [NUCLEO] The standard unit of absorbed dose, equal to energy absorption of 100 ergs per gram (0.01 joule per kilogram); supersedes the roentgen as the unit of dosage.

radar [ENG] 1. A system using beamed and reflected radio-frequency energy for detecting and locating objects, measuring distance or altitude, navigating, homing, bombing, and other purposes; in detecting and ranging, the time interval between transmission of the energy and reception of the reflected energy establishes the range of an object in the beam's path. Derived from radio detection and ranging. 2. *See* radar set.

radar altimeter [NAV] A radio altimeter, useful at altitudes much greater than the 5000-foot (1500 meter) limit of frequency-modulated radio altimeters, in which simple pulse-type radar equipment is used to send a pulse straight down from an aircraft and to measure its total time of travel to the surface and back to the aircraft. Also known as high-altitude radio altimeter; pulse-type altimeter.

radar astronomy [ASTRON] The study of astronomical bodies and the earth's atmosphere by means of radar pulse techniques, including tracking of meteors and the reflection of radar pulses from the moon and the planets.

radar attenuation [ELECTROMAG] Ratio of the power delivered by the transmitter to the transmission line connecting it with the transmitting antenna, to the power reflected from the target which is delivered to the receiver by the transmission line connecting it with the receiving antenna.

radar beacon [NAV] A radar receiver-transmitter that transmits a strong coded radar signal whenever its radar receiver is triggered by an interrogating radar on an aircraft or ship; the coded beacon reply can be used by the navigator to determine his own position in terms of bearing and range from the beacon. Also known as racon; radar transponder.

radar beam [ELECTROMAG] The movable beam of radio-frequency energy produced by a radar transmitting antenna; its shape is commonly defined as the loci of all points at which the power has decreased to one-half of that at the center of the beam.

radar clutter *See* clutter.

radar display [ELECTR] The pattern representing the output data of a radar set, generally produced on the screen of a cathode-ray tube. Also known as presentation; radar presentation.

radar echo *See* echo.

radar horizon [NAV] The distance to which a radar's operation is limited by the quasi-optical characteristics of the radio waves employed.

radar image [ELECTR] The image of an object which is produced on a radar screen.

radar indicator [ELECTR] A cathode-ray tube and associated equipment used to provide a visual indication of the echo signals picked up by a radar set.

radar meteorological observation [METEOROL] Evaluation of the echoes appearing on the indicator of a weather radar, in terms of orientation, coverage, intensity, tendency of intensity, height, movement, and unique characteristics of echoes, that may be indicative of certain types of severe storms (such as hurricanes, tornadoes, or thunderstorms) and of anomalous propagation. Also known as radar weather observation.

radar meteorology [METEOROL] The study of the scattering of radar waves by all types of atmospheric phenomena and the use of radar for making weather observations and forecasts.

radar presentation *See* radar display.

radar pulse [ELECTROMAG] Radio-frequency radiation emitted with high power by a pulse radar installation for a period of time which is brief compared to the interval between such pulses.

radar range [ELECTROMAG] The maximum distance at which a radar set is ordinarily effective in detecting objects.

radar range marker *See* distance marker.

radar reflection [ELECTROMAG] The return of electromagnetic waves, generated by a radar installation, from an object on which the waves are incident.

radar reflectivity [ELECTROMAG] The fraction of electromagnetic energy generated by a radar installation which is reflected by an object.

radar relay [ENG] **1.** Equipment for relaying the radar video and appropriate synchronizing signal to a remote location. **2.** Process or system by which radar echoes and synchronization data are transmitted from a search radar installation to a receiver at a remote point.

radar repeater [ELECTR] A cathode-ray indicator used to reproduce the visible intelligence of a radar display at a remote position; when used with a selector switch, the visible intelligence of any one of several radar systems can be reproduced.

radar return [NAV] The signal indication of an object which has reflected energy that was transmitted by a primary radar. Also known as radio echo.

radarscope [ELECTR] Cathode-ray tube, serving as an oscilloscope, the face of which is the radar viewing screen. Also known as scope.

radar set [ENG] A complete assembly of radar equipment for detecting and ranging, consisting essentially of a transmitter, antenna, receiver, and indicator. Also known as radar.

radar transponder *See* radar beacon.

radar triangulation [ENG] A radar system of locating targets, usually aircraft, in which two or more separate radars are employed to measure range only; the target is located by automatic trigonometric solution of the triangle composed of a pair of radars and the target in which all three sides are known.

radar weather observation *See* radar meteorological observation.

radechon [ELECTR] A storage tube having a single electron gun and a dielectric storage medium consisting of a sheet of mica sandwiched between a continuous metal backing plate and a fine-mesh screen; used in simple delay schemes, signal-to-noise improvement, signal comparison, and conversion of signal-time bases. Also known as barrier-grid storage tube.

radial artery [ANAT] A branch of the brachial artery in the forearm; principal branches are the radial recurrent and the main artery of the thumb.

radial bearing [MECH ENG] A bearing with rolling contact in which the direction of action of the load transmitted is radial to the axis of the shaft.

radial chromatography [ANALY CHEM] A circular disk of absorbent paper which has a strip (wick) cut from edge to center to dip into a solvent; the solvent climbs the wick, touches the sample, and resolves it into concentric rings (the chromatogram). Also known as circular chromatography; radial-paper chromatography.

radial cleavage [EMBRYO] A cleavage pattern characterized by formation of a mass of cells that show radial symmetry.

radial distribution function [MATH] A function $F(r)$ equal to the average of a given function of the three coordinates over a sphere of radius r centered at the origin of the coordinate system. [PHYS CHEM] A function $\rho(r)$ equal to the average over all directions of the number density of molecules at distance r from a given molecule in a liquid.

radial drill [MECH ENG] A drilling machine in which the drill spindle can be moved along a horizontal arm which itself can be rotated about a vertical pillar.

radial gate *See* tainter gate.

radial motion [MECH] Motion in which a body moves along a line connecting it with an observer or reference point; for example, the motion of stars which move toward or away from the earth without a change in apparent position.

radial nerve [ANAT] A large nerve that arises in the brachial plexus and branches to enervate the extensor muscles and skin of the posterior aspect of the arm, forearm, and hand.

radial-paper chromatography *See* radial chromatography.

radial-ply [DES ENG] Pertaining to the construction of a tire in which the cords run straight across the tire, and an additional layered belt of fabric is placed around the circumference between the plies and the tread.

radial velocity [MECH] The component of the velocity of a body that is parallel to a line from an observer or reference point to the body; the radial velocities of stars are valuable in determining the structure and dynamics of the Galaxy. Also known as line-of-sight velocity.

radial wave equation [MECH] Solutions to wave equations with spherical symmetry can be found by separation of variables; the ordinary differential equation for the radial part of the wave function is called the radial wave equation.

radian [MATH] The central angle of a circle determined by two radii and an arc joining them, all of the same length.

radiance [OPTICS] The radiant flux per unit solid angle per unit of projected area of the source; the usual unit is the watt per steradian per square meter. Also known as steradiancy.

radian frequency *See* angular frequency.

radiant [ASTRON] A point on the celestial sphere through which pass the backward extensions of the trail of a meteor as observed at various locations, or the backward extensions of trails of a number of meteors traveling parallel to each other. [PHYS] 1. Pertaining to motion of particles or radiation along radii from a common point or a small region. 2. A point, region, substance, or entity from which particles or radiations are emitted.

radiant energy *See* radiation.

radiant-energy thermometer *See* radiation pyrometer.

radiant exposure [OPTICS] A measure of the total radiant energy incident on a surface per unit area; equal to the integral over time of the radiant flux density. Also known as exposure.

radiant flux density [ELECTROMAG] The amount of radiant power per unit area that flows across or onto a surface. Also known as irradiance.

radiant heating [ENG] Any system of space heating in which the heat-producing means is a surface that emits heat to the surroundings by radiation rather than by conduction or convection.

Radiata [INV ZOO] Members of the Eumetazoa which have a primary radial symmetry; includes the Coelenterata and Ctenophora.

radiating power *See* emittance.

radiation [ENG] A method of surveying in which points are located by knowledge of their distances and directions from a central point. [PHYS] 1. The emission and propagation of waves transmitting energy through space or through some medium; for example, the emission and propagation of electromagnetic, sound, or elastic waves. 2. The energy transmitted by waves through space or some medium; when unqualified, usually refers to electromagnetic radiation. Also known as radiant energy. 3. A stream of particles, such as electrons, neutrons, protons, α-particles, or high-energy photons, or a mixture of these.

radiational cooling [METEOROL] The cooling of the earth's surface and adjacent air, accomplished (mainly at night) whenever the earth's surface suffers a net loss of heat due to terrestrial radiation.

radiation angle [ELECTROMAG] The vertical angle between the line of radiation emitted by a directional antenna and the horizon.

radiation biology *See* radiobiology.

radiation chemistry [NUCLEO] The branch of chemistry that is concerned with the chemical effects, including decomposition, of energetic radiation or particles on matter.

radiation counter [NUCLEO] An instrument used for detecting or measuring nuclear radiation by counting the resultant ionizing events; examples include Geiger counters and scintillation counters. Also known as counter.

radiation counter tube *See* counter tube.

radiation damping [ELECTROMAG] Damping of a system which loses energy by electromagnetic radiation. [QUANT MECH] Damping which arises in quantum electrodynamics from the virtual interaction of a particle with its zero point field.

radiation detector *See* particle detector.

radiation dose [NUCLEO] The total amount of ionizing radiation absorbed by material or tissues, in the sense of absorbed dose (expressed in rads), exposure dose (expressed in roentgens), or dose equivalent (expressed in rems).

radiation field [ELECTROMAG] The electromagnetic field that breaks away from a transmitting antenna and radiates outward into space as electromagnetic waves; the other type of electromagnetic field associated with an energized antenna is the induction field.

radiation hardening [ENG] Improving the ability of a device or piece of equipment to withstand nuclear or other radiation; applies chiefly to dielectric and semiconductor materials.

radiation intensity [ELECTROMAG] The power radiated from an antenna per unit solid angle in a given direction. [NUCLEO] The quantity of radiant energy passing perpendicularly through a specified location of unit area in unit time; reported as a number of particles or photons per square centimeter per second, or in energy units such as ergs per square centimeter per second.

radiation ionization [PHYS] Ionization of the atoms or molecules of a gas or vapor by the action of electromagnetic radiation.

radiation laws [PHYS] 1. The four physical laws which, together, fundamentally describe the behavior of blackbody radiation, Kirchhoff's law, Planck's law, Stefan-Boltzmann law, and Wien's law. 2. All of the more inclusive assemblage of empirical and theoretical laws describing all manifestations of radiative phenomena.

radiation lobe *See* lobe.

radiation oscillator *See* Planck oscillator.

radiation oven [ENG] Heating chamber relying on tungsten-filament infrared lamps with reflectors to create temperatures up to 600°F (315°C); used to dry sheet and granular material and to bake surface coatings.

radiation pattern [ELECTROMAG] Directional dependence of the radiation of an antenna. Also known as antenna pattern; directional pattern; field pattern.

radiation pressure [ACOUS] The average pressure exerted on a surface or interface between two media by a sound wave. [ELECTROMAG] The pressure exerted by electromagnetic radiation on objects on which it impinges.

radiation pyrometer [ENG] An instrument which measures the temperature of a hot object by focusing the thermal radiation emitted by the object and making some observation on it; examples include the total-radiation, optical, and ratio pyrometers. Also known as radiant-energy thermometer; radiation thermometer.

radiation sickness [MED] 1. Illness, usually manifested by nausea and vomiting, resulting from the effects of therapeutic doses of radiation. 2. Radiation injury following exposure to excessive doses of radiation, such as the explosion of an atomic bomb.

radiation standards [NUCLEO] Exposure standards, permissible concentrations, rules for safe handling, regulations for transportation, regulations for industrial control of radiation, and control of radiation exposure by legislative means.

radiation therapy [MED] The use of ionizing radiation or radioactive substances to treat disease. Also known as actinotherapy; radiotherapy.

radiation thermometer *See* radiation pyrometer.

radiation well logging *See* radioactive well logging.

radiative capture [NUC PHYS] A nuclear capture process whose prompt result is the emission of electromagnetic radiation only.

radiative transfer [THERMO] The transmission of heat by electromagnetic radiation.

radiative transition [QUANT MECH] A change of a quantum-mechanical system from one energy state to another in which electromagnetic radiation is emitted.

radiator [ACOUS] A vibrating element of a transducer which radiates sound waves. [ELECTROMAG] **1.** The part of an antenna or transmission line that radiates electromagnetic waves either directly into space or against a reflector for focusing or directing. **2.** A body that emits radiant energy. [ENG] Any of numerous devices, units, or surfaces that emit heat, mainly by radiation, to objects in the space in which they are installed. [PHYS] **1.** In general, a body which emits particles or radiation in any form. **2.** A body placed in a beam of ionizing radiation which, as a result, emits radiation of another kind.

radical [BOT] **1.** Of, pertaining to, or proceeding from the root. **2.** Arising from the base of a stem or from an underground stem. [MATH] **1.** In a ring, the intersection of all maximal ideals. Also known as Jacobson radical. **2.** An indicated root of a quantity. Symbolized $\sqrt{}$ [ORG CHEM] See free radical.

radio [COMMUN] The transmission of signals through space by means of electromagnetic waves; usually applied to the transmission of sound and code signals, although television and radar also depend on electromagnetic waves. [ELECTR] See radio receiver.

radio- [ELECTROMAG] A prefix denoting the use of radiant energy, particularly radio waves. [NUCLEO] Chemical prefix designating radiation or radioactivity; used to designate radioactive elements (such as radiocarbon) and substances containing them (such as radiochemicals, radiocolloids, or radio compounds).

radioacoustic position finding See radioacoustic ranging.

radioacoustic ranging [ENG] A method for finding the position of a vessel at sea; a bomb is exploded in the water, and the sound of the explosion transmitted through water is picked up by the vessel and by shore stations, other vessels, or buoys whose positions are known; the received sounds are transmitted instantaneously by radio to the surveying vessel, and the elapsed times are proportional to the distances to the known positions. Abbreviated RAR. Also known as radioacoustic position finding; radioacoustic sound ranging.

radioacoustic sound ranging See radioacoustic ranging.

radioactive [NUC PHYS] Exhibiting radioactivity or pertaining to radioactivity.

radioactive age determination See radiometric dating.

radioactive carbon dating See carbon-14 dating.

radioactive chain See radioactive series.

radioactive clock [NUC PHYS] A radioactive isotope such as potassium-40 which spontaneously decays to a stable end product at a constant rate, allowing absolute geologic age to be determined.

radioactive cloud [NUCLEO] A mass of air and vapor in the atmosphere carrying radioactive debris from a nuclear explosion.

radioactive dating See radiometric dating.

radioactive decay [NUC PHYS] The spontaneous transformation of a nuclide into one or more different nuclides, accompanied by either the emission of particles from the nucleus, nuclear capture or ejection of orbital electrons, or fission. Also known as decay; nuclear spontaneous reaction; radioactive dis-

integration; radioactive transformation; radioactivity.

radioactive decay constant See decay constant.

radioactive decay product See daughter.

radioactive decay series See radioactive series.

radioactive disintegration See radioactive decay.

radioactive displacement law [NUC PHYS] The statement of the changes in mass number A and atomic number Z that take place during various nuclear transformations. Also known as displacement law.

radioactive element [NUC PHYS] An element all of whose isotopes spontaneously transform into one or more different nuclides, giving off various types of radiation; examples include promethium, radium, thorium, and uranium.

radioactive fallout See fallout.

radioactive half-life See half-life.

radioactive isotope See radioisotope.

radioactive series [NUC PHYS] A succession of nuclides, each of which transforms by radioactive disintegration into the next until a stable nuclide results. Also known as decay chain; decay family; decay series; disintegration chain; disintegration family; disintegration series; radioactive chain; radioactive decay series; series decay; transformation series.

radioactive standard [NUCLEO] A sample of radioactive material which contains a known number and type of radioactive atoms at some definite time; used to calibrate radiation measuring instruments.

radioactive tracer [NUCLEO] A radioactive isotope which, when attached to a chemically similar substance or injected into a biological or physical system, can be traced by radiation detection devices, permitting determination of the distribution or location of the substance to which it is attached. Also known as radiotracer.

radioactive transformation See radioactive decay.

radioactive waste [NUCLEO] Liquid, solid, or gaseous waste resulting from mining of radioactive ore, production of reactor fuel materials, reactor operation, processing of irradiated reactor fuels, and related operations, and from use of radioactive materials in research, industry, and medicine.

radioactive well logging [ENG] The recording of the differences in radioactive content (natural or neutron-induced) of the various rock layers found down an oil well borehole; types include γ-ray, neutron, and photon logging. Also known as radiation well logging; radioactivity prospecting.

radioactivity [NUC PHYS] **1.** A particular type of radiation emitted by a radioactive substance, such as α-radioactivity. **2.** See radioactive decay. **3.** See activity.

radioactivity analysis See activation analysis.

radioactivity equilibrium [NUC PHYS] A condition which may arise in the decay of a radioactive parent with short-lived descendants, in which the ratio of the activity of a parent to that of a descendant remains constant.

radioactivity prospecting See radioactive well logging.

radio altimeter [ENG] An absolute altimeter that depends on the reflection of radio waves from the earth for the determination of altitude, as in a frequency-modulated radio altimeter and a radar altimeter. Also known as electronic altimeter; reflection altimeter.

radio antenna See antenna.

radio astronomy [ASTRON] The study of celestial objects by measurement and analysis of their emitted electromagnetic radiation in the wavelength range from roughly 1 millimeter to 30 millimeters.

radio attenuation [ELECTROMAG] For one-way propagation, the ratio of the power delivered by the transmitter to the transmission line connecting it with the transmitting antenna to the power delivered to the receiver by the transmission line connecting it with the receiving antenna.

radio beam [ELECTROMAG] A concentrated stream of radio-frequency energy as used in radio ranges, microwave relays, and radar.

radiobiology [BIOL] Study of the scientific principles, mechanisms, and effects of the interaction of ionizing radiation with living matter. Also known as radiation biology.

radio blackout [COMMUN] A fadeout that may last several hours or more at a particular frequency. Also known as blackout.

radio broadcasting [COMMUN] Radio transmission intended for general reception.

radiocarbon See carbon-14.

radiocarbon dating See carbon-14 dating.

radiocesium See cesium-137.

radiochemistry [CHEM] That area of chemistry concerned with the study of radioactive substances.

radiochronology [GEOL] An absolute-age dating method based on the existing ratio between radioactive parent elements (such as uranium-238) and their radiogenic daughter isotopes (such as lead-206).

radio compass See automatic direction finder.

radio detection and ranging See radar.

radio direction finder [NAV] A radio aid to navigation that uses a rotatable loop or other highly directional antenna arrangement to determine the direction of arrival of a radio signal. Abbreviated RDF. Also known as direction finder.

radio duct [GEOPHYS] An atmospheric layer, typically shallow and almost horizontal, in which radio waves propagate in an anomalous fashion; ducts occur when, due to sharp inversions of temperature or humidity, the vertical gradient of the radio index of refraction exceeds a critical value.

radio echo See radar return.

radioecology [ECOL] The interdisciplinary study of organisms, radionuclides, ionizing radiation, and the environment.

radio engineering [ENG] The field of engineering that deals with the generation, transmission, and reception of radio waves and with the design, manufacture, and testing of associated equipment.

radio fadeout [COMMUN] Increased absorption of radio waves passing through the lower layers of the ionosphere due to a sudden and abnormal increase in ionization in these regions; signals as receivers then fade out or disappear. Also known as fadeout.

radio frequency [ELECTROMAG] A frequency at which coherent electromagnetic radiation of energy is useful for communication purposes; roughly the range from 10 kilohertz to 100 gigahertz. Abbreviated rf.

radio-frequency amplifier [ELECTR] An amplifier that amplifies the high-frequency signals commonly used in radio communications.

radio-frequency bandwidth [COMMUN] Band of frequencies comprising 99% of the total radiated power extended to include any discrete frequency on which the power is at least 0.25% of the total radiated power.

radio-frequency choke [ELEC] A coil designed and used specifically to block the flow of radio-frequency current while passing lower frequencies or direct current.

radio-frequency filter [ELECTR] An electric filter which enhances signals at certain radio frequencies or attenuates signals at undesired radio frequencies.

radio-frequency generator [ELECTR] A generator capable of supplying sufficient radio-frequency energy at the required frequency for induction or dielectric heating.

radio-frequency heating See electronic heating.

radio-frequency interference [COMMUN] Interference from sources of energy outside a system or systems, as contrasted to electromagnetic interference generated inside systems. Abbreviated RFI.

radio-frequency line See radio-frequency transmission line.

radio-frequency pulse [COMMUN] A radio-frequency carrier that is amplitude-modulated by a pulse; the amplitude of the modulated carrier is zero before and after the pulse. Also known as radio pulse.

radio-frequency resistance See high-frequency resistance.

radio-frequency shift See frequency shift.

radio-frequency spectrometer [SPECT] An instrument which measures the intensity of radiation emitted or absorbed by atoms or molecules as a function of frequency at frequencies from 10^5 to 10^9 hertz; examples include the atomic-beam apparatus, and instruments for detecting magnetic resonance.

radio-frequency spectrum See radio spectrum.

radio-frequency transformer [ELECTROMAG] A transformer having a tapped winding or two or more windings designed to furnish inductive reactance or to transfer radio-frequency energy from one circuit to another by means of a magnetic field; may have an air core or some form of ferrite core. Also known as radio transformer.

radio-frequency transmission line [ELECTROMAG] A transmission line designed primarily to conduct radio-frequency energy, consisting of two or more conductors supported in a fixed spatial relationship along their own length. Also known as radio-frequency line.

radio-frequency welding See high-frequency welding.

radio galaxy [ASTROPHYS] A galaxy that is emitting much energy in radio frequencies often from regions devoid of visible matter.

radiogenic [NUC PHYS] Pertaining to a material produced by radioactive decay, as the production of lead from uranium decay.

radiogenic age determination See radiometric dating.

radiogenic dating See radiometric dating.

radiograph [GRAPHICS] The photographic image produced in radiography. Also known as shadowgraph.

radiography [GRAPHICS] The technique of producing a photographic image of an opaque specimen by transmitting a beam of x-rays or γ-rays through it onto an adjacent photographic film; the image results from variations in thickness, density, and chemical composition of the specimen; used in medicine and industry.

radio hole [GEOPHYS] Strong fading of the radio signal at some position in space along an air-to-air or air-to-ground path; the effect is caused by the abnormal refraction of radio waves.

radio homing beacon See homing beacon.

radio horizon [COMMUN] The locus of points at which direct rays from a transmitter become tangential to the surface of the earth; the distance to the radio horizon is affected by atmospheric refraction.

radioimmunoassay [IMMUNOL] A sensitive method for determining the concentration of an antigenic substance in a sample by comparing its inhibitory effect on the binding of a radioactivity-labeled antigen to a limited amount of a specific antibody with the inhibitory effect of known standards.

radio interference See interference.

radioisotope [NUC PHYS] An isotope which exhibits radioactivity. Also known as radioactive isotope; unstable isotope.

radioisotope battery See nuclear battery.

radioisotopic generator See nuclear battery.

Radiolaria [INV ZOO] A subclass of the protozoan class Actinopodea whose members are noted for their siliceous skeletons and characterized by a membranous capsule which separates the outer from the inner cytoplasm.

radiolocation [ENG] Determination of relative position of an object by means of equipment operating on the principle that propagation of radio waves is at a constant velocity and rectilinear.

radiology [MED] The medical science concerned with radioactive substances, x-rays, and other ionizing radiations, and the application of the principles of this science to diagnosis and treatment of disease.

radioluminescence [PHYS] Luminescence produced by x-rays or γ-rays, or by particles emitted in radioactive decay.

radiolysis [PHYS CHEM] The dissociation of molecules by radiation; for example, a small amount of water in a reactor core dissociates into hydrogen and oxygen during operation.

radiometer [ELECTR] A receiver for detecting microwave thermal radiation and similar weak wideband signals that resemble noise and are obscured by receiver noise; examples include the Dicke radiometer, subtraction-type radiometer, and two-receiver radiometer. Also known as microwave radiometer; radiometer-type receiver. [ENG] An instrument for measuring radiant energy; examples include the bolometer, microradiometer, and thermopile.

radiometer-type receiver See radiometer.

radiometric age [GEOL] Geologic age expressed in years determined by quantitatively measuring radioactive elements and their decay products.

radiometric analysis [ANALY CHEM] Quantitative chemical analysis that is based on measurement of the absolute disintegration rate of a radioactive component having a known specific activity.

radiometric dating [NUCLEO] A technique for measuring the age of an object or sample of material by determining the ratio of the concentration of a radioisotope to that of a stable isotope in it; for example, the ratio of carbon-14 to carbon-12 reveals the approximate age of bones, pieces of wood, and other archeological specimens. Also known as isotopic age determination; nuclear age determination; radioactive age determination; radioactive dating; radiogenic age determination; radiogenic dating.

radiometric titration [ANALY CHEM] Use of radioactive indicator to track the transfer of material between two liquid phases in equilibrium, such as titration of $Ag^{110}NO_3$ (silver nitrate, with the silver atom having mass number 110) against potassium chloride.

radiometry [PHYS] The detection and measurement of radiant electromagnetic energy, especially that associated with infrared radiation.

radio navigation [NAV] The use of apparatus operating at radio frequencies to determine parameters useful for navigation; it includes radio direction finding, radio ranges, radio compasses, radio homing beacons, and loran.

radionuclide [NUC PHYS] A nuclide that exhibits radioactivity.

radiopaque [ELECTROMAG] Not appreciably penetrable by x-rays or other forms of radiation.

radiophoto See facsimile.

radio pulse [COMMUN] See radio-frequency pulse. [ELECTROMAG] An intense burst of radio-frequency energy lasting for a fraction of a second.

radio range [COMMUN] A radio facility, usually landbased, that emits signals which when received by appropriate equipment provide navigational information. Also known as range.

radio receiver [ELECTR] A device that converts radio waves into intelligible sounds or other perceptible signals. Also known as radio; radio set; receiving set.

radio relay satellite See communications satellite.

radio repeater [COMMUN] A repeater that acts as an intermediate station in transmitting radio communications signals or radio programs from one fixed station to another; serves to extend the reliable range of the originating station; a microwave repeater is an example.

radio scattering See scattering.

radio set See radio receiver; radio transmitter.

radiosonde [ENG] A balloon-borne instrument for the simultaneous measurement and transmission of meteorological data; the instrument consists of transducers for the measurement of pressure, temperature, and humidity, a modulator for the conversion of the output of the transducers to a quantity which controls a property of the radio-frequency signal, a selector switch which determines the sequence in which the parameters are to be transmitted, and a transmitter which generates the radio-frequency carrier.

radio sonobuoy See sonobuoy.

radio spectrum [COMMUN] The entire range of frequencies in which useful radio waves can be produced, extending from the audio range to about 300,000 megahertz. Also known as radio-frequency spectrum.

radio telemetry [COMMUN] The presentation of data at a location remote from the source of the data, using radio-frequency electromagnetic radiation as the means of transmission.

radio telescope [ENG] An astronomical instrument used to measure the amount of radio energy coming from various directions in the sky, consisting of a highly directional antenna and associated electronic equipment.

radiotherapy See radiation therapy.

radio tower [COMMUN] A tower, usually several hundred meters tall, either guyed or freestanding, on which a transmitting antenna is mounted to increase the range of radio transmission; in some cases, the tower itself may be the antenna.

radiotracer See radioactive tracer.

radio transmitter [ELECTR] The equipment used for generating and amplifying a radio-frequency carrier signal, modulating the carrier signal with intelligence, and feeding the modulated carrier to an antenna for radiation into space as electromagnetic waves. Also known as radio set; transmitter.

radio tube See electron tube.

radio wave [ELECTROMAG] An electromagnetic wave produced by reversal of current in a conductor at a frequency in the range from about 10 kilohertz to about 300,000 megahertz.

radio window [GEOPHYS] A band of frequencies extending from about 6 to 30,000 megahertz, in which radiation from the outer universe can enter and travel through the atmosphere of the earth.

radish [BOT] *Raphanus sativus.* **1.** An annual or biennial crucifer belonging to the order Capparales. **2.** The edible, thickened hypocotyl of the plant.

radium [CHEM] A radioactive member of group IIA, symbol Ra, atomic number 88; the most abundant naturally occurring isotope has mass number 226 and a half-life of 1620 years. A highly toxic solid that forms water-soluble compounds; decays by emission of α, β, and γ-radiation; melts at 700°C, boils at 1140°C; turns black in air; used in medicine, in industrial radiography, and as a source of neutrons and radon.

radium age [NUCLEO] The age of a mineral as calculated from the numbers of radium atoms present originally, now, and when equilibrium is established with ionium.

radium F *See* polonium-210.

radius [ANAT] The outer of the two bones of the human forearm or of the corresponding part in vertebrates other than fish. [MATH] **1.** A line segment joining the center and a point of a circle or sphere. **2.** The length of such a line segment.

radius of convergence [MATH] The positive real number corresponding to a power series expansion about some number a with the property that if $x - a$ has absolute value less than this number the power series converges at x, and if $x - a$ has absolute value greater than this number the power series diverges at x.

radius of curvature [MATH] The radius of the circle of curvature at a point of a curve.

radius of gyration [MATH] The square root of the ratio of the moment of inertia of a plane figure about a given axis to its area. [MECH] The square root of the ratio of the moment of inertia of a body about a given axis to its mass.

radius vector [ASTRON] A line joining the center of an orbiting body with the focus of its orbit located near its primary.

radix *See* base of a number system; root.

radix complement [MATH] A numeral in positional notation that can be derived from another by subtracting the original numeral from the numeral of highest value with the same number of digits, and adding 1 to the difference. Also known as complement; true complement.

radix notation [MATH] A positional notation in which the successive digits are interpreted as coefficients of successive integral powers of a number called the radix or base; the represented number is equal to the sum of this power series. Also known as base notation.

radix point [MATH] A dot written either on or slightly above the line, used to mark the point at which place values change from positive to negative powers of the radix in a number system; a decimal point is a radix point for radix 10.

radome [ELECTROMAG] A strong, thin shell, made from a dielectric material that is transparent to radio-frequency radiation, and used to house a radar antenna, or a space communications antenna of similar structure.

radon [CHEM] A chemical element, symbol Rn, atomic number 86; all isotopes are radioactive, the longest half-life being 3.82 days for mass number 222; it is the heaviest element of the noble-gas group, produced as a gaseous emanation from the radioac-

tive decay of radium. [NUC PHYS] The conventional name for radon-222. Symbolized Rn.

radon-220 [NUC PHYS] The isotope of radon having mass number 220, symbol ^{220}Ra, which is a radioactive member of the thorium series with a half-life of 56 seconds.

radon-222 [NUC PHYS] The isotope of radon having mass number 222, symbol ^{222}Ra, which is a radioactive member of the uranium series with a half-life of 3.82 days.

radula [INV ZOO] A filelike ribbon studded with horny or chitinous toothlike structures, found in the mouth of all classes of mollusks except Bivalvia.

Rafflesiales [BOT] A small order of dicotyledonous plants; members are highly specialized, nongreen, rootless parasites which grow from the roots of the host.

rafted ice [OCEANOGR] A form of pressure ice composed of overlying pieces of ice floe.

raft foundation [CIV ENG] A continuous footing that supports an entire structure, such as a floor. Also known as foundation mat.

raft tectonics *See* plate tectonics.

rail [ENG] **1.** A bar extending between posts or other supports as a barrier or guard. **2.** A steel bar resting on the crossties to provide track for railroad cars and other vehicles with flanged wheels. [MECH ENG] A high-pressure manifold in some fuel injection systems.

railhead [CIV ENG] **1.** The topmost part of a rail, supporting the wheels of railway vehicles. **2.** A point at which railroad traffic originates and terminates. **3.** The temporary ends of a railroad line under construction.

Raillietiellidae [INV ZOO] A small family of parasitic arthropods in the order Cephalobaenida.

railroad engineering [CIV ENG] That part of transportation engineering involved in the planning, design, development, operation, construction, maintenance, use, or economics of facilities for transportation of goods and people in wheeled units of rolling stock running on, and guided by, rails normally supported on crossties and held to fixed alignment. Also known as railway engineering.

railway engineering *See* railroad engineering.

rain [METEOROL] Precipitation in the form of liquid water drops with diameters greater than 0.5 millimeter, or if widely scattered the drops may be smaller; the only other form of liquid precipitation is drizzle.

rainbow [ELECTR] Technique which applies pulse-to-pulse frequency changing to identifying and discriminating against decoys and chaff. [OPTICS] Colored arc seen in the sky when the sun or moon is illuminating large numbers of falling raindrops. [PETRO ENG] Chromatic iridescence observed in drilling fluid that has been circulated in a well, indicating contamination or contact with fresh hydrocarbons.

rainfall [METEOROL] The amount of precipitation of any type; usually taken as that amount which is measured by means of a rain gage (thus a small, varying amount of direct condensation is included).

rainforest [ECOL] A forest of broad-leaved, mainly evergreen, trees found in continually moist climates in the tropics, subtropics, and some parts of the temperate zones.

rain gage [ENG] An instrument designed to collect and measure the amount of rain that has fallen. Also known as ombrometer; pluviometer; udometer.

rain gush *See* cloudburst.

rain gust *See* cloudburst.

rain shadow [METEOROL] An area of diminished precipitation on the lee side of mountains or other topographic obstacles.

rainsquall [METEOROL] A squall associated with heavy convective clouds, frequently the cumulonimbus type; usually sets in shortly before the thunderstorm rain, blowing outward from the storm and generally lasting only a short time. Also known as thundersquall.

raise [MIN ENG] A shaftlike mine opening, driven upward from a level to connect with a level above, or to the surface.

Rajidae [VERT ZOO] The skates, a family of elasmobranchs included in the batoid group.

Rajiformes [VERT ZOO] The equivalent name for Batoidea.

rake [BUILD] The exterior finish and trim applied parallel to the sloping end walls of a gabled roof. [DES ENG] A hand tool consisting of a long handle with a row of projecting prongs at one end; for example, the tool used for gathering leaves or grass on the ground. [ENG] The angle between an inclined plane and the vertical. [GEOL] See plunge. [MECH ENG] The angle between the tooth face or a tangent to the tooth face of a cutting tool at a given point and a reference plane or line. [NAV ARCH] The angle between the vertical direction and a part of a ship, such as a mast, funnel, bow, stern, rudder, or sternpost. [ORD] To sweep a target, especially a ship or a column of troops, with gun or cannon fire.

Rallidae [VERT ZOO] A large family of birds in the order Gruiformes comprising rails, gallinules, and coots.

ralstonite [MINERAL] $NaMgAl_5F_{12}(OH)_6 \cdot 3H_2O$ A colorless, white, or yellowish mineral composed of hydrous basic sodium magnesium aluminum fluoride, occurring in octahedral crystals.

ram [AERO ENG] The forward motion of an air scoop or air inlet through the air. [HYD] An underwater ledge or projection from an ice wall, ice front, iceberg, or floe, usually caused by the more intensive melting and erosion of the unsubmerged part. Also known as apron; spur. [MECH ENG] A plunger, weight, or other guided structure for exerting pressure or drawing something by impact. [MIN ENG] See barney. [VERT ZOO] A male sheep or goat.

Ram See Aries.

RAM See random-access memory.

Raman effect [OPTICS] A phenomenon observed in the scattering of light as it passes through a transparent medium; the light undergoes a change in frequency and a random alteration in phase due to a change in rotational or vibrational energy of the scattering molecules. Also known as Raman scattering.

Raman scattering See Raman effect.

Raman spectroscopy [SPECT] Analysis of the intensity of Raman scattering of monochromatic light as a function of frequency of the scattered light.

Ramapithecinae [PALEON] A subfamily of Hominidae including the protohominids of the Miocene and Pliocene.

ramie [BOT] *Boehmeria nivea*. A shrub or half-shrub of the nettle family (Urticaceae) cultivated as a source of a tough, strong, durable, lustrous natural woody fiber resembling flax, obtained from the phloem of the plant; used for high-quality papers and fabrics. Also known as China grass; rhea.

ramjet engine [AERO ENG] A type of jet engine with no mechanical compressor, consisting of a specially shaped tube or duct open at both ends, the air necessary for combustion being shoved into the duct and compressed by the forward motion of the engine; the air passes through a diffuser and is mixed with fuel and burned, the exhaust gases issuing in a jet from the rear opening.

rammelsbergite [MINERAL] $NiAs_2$ A gray mineral composed of nickel diarsenide; it is dimorphous with pararammelsbergite. Also known as white nickel.

ramp [ENG] **1.** A uniformly sloping platform, walkway, or driveway. **2.** A stairway which gives access to the main door of an airplane. [HYD] An accumulation of snow forming an inclined plane between land or land ice and sea ice or shelf ice. Also known as drift ice foot. [MIN ENG] A slope between levels in open-pit mining.

Ramphastidae [VERT ZOO] The toucans, a family of birds with large, often colorful bills in the order Piciformes.

ram recovery See recovery.

ramsdellite [MINERAL] MnO_2 An orthorhombic mineral composed of manganese dioxide; it is dimorphous with pyrolusite.

Ramsden circle [OPTICS] A sharp, bright circle of light which appears on a sheet of white paper held near the eyepiece of a telescope focused for infinity and pointed at a bright sky. Also known as Ramsden disk.

Ramsden disk See Ramsden circle.

Ramsden eyepiece [OPTICS] An eyepiece consisting of two planoconvex lenses with their plane sides facing outward, having the same power and focal length, and separated by a distance equal to their common focal length.

ramus [ANAT] A slender bone process branching from a large bone. [VERT ZOO] The barb of a feather. [ZOO] The branch of a structure such as a blood vessel, nerve, arthropod appendage, and so on.

random access [COMPUT SCI] **1.** The ability to read or write information anywhere within a storage device in an amount of time that is constant regardless of the location of the information accessed and of the location of the information previously accessed. Also known as direct access. **2.** A process in which data are accessed in nonsequential order and possibly at irregular intervals of time. Also known as single reference.

random-access disk file [COMPUT SCI] A file which is contained on a disk having one head per track and in which consecutive records are not necessarily in consecutive locations.

random-access memory [COMPUT SCI] A data storage device having the property that the time required to access a randomly selected datum does not depend on the time of the last access or the location of the most recently accessed datum. Abbreviated RAM. Also known as direct-access memory; direct-access storage; random-access storage; random storage; uniformly accessible storage.

random-access programming [COMPUT SCI] Programming without regard for the time required for access to the storage positions called for in the program, in contrast to minimum-access programming.

random-access storage See random-access memory.

random diffusion chamber See reverberation chamber.

random error [STAT] An error that can be predicted only on a statistical basis.

random function [MATH] A function whose domain is an interval of the extended real numbers and has range in the set of random variables on some probability space; more precisely, a mapping of the cartesian product of an interval in the extended reals with a probability space to the extended reals so that each section is a random variable.

randomization [STAT] Assigning subjects to treatment groups by use of tables of random numbers.

random noise [MATH] A form of random stochastic process arising in control theory. [PHYS] Noise characterized by a large number of overlapping transient disturbances occurring at random, such as thermal noise and shot noise. Also known as fluctuation noise.

random number generator [COMPUT SCI] **1.** A mathematical program which generates a set of numbers which pass a randomness test. **2.** An analog device that generates a randomly fluctuating variable, and usually operates from an electrical noise source.

random ordered sample [STAT] An ordered sample of size s drawn from a population of size N such that the probability of any particular ordered sample is the reciprocal of the number of permutations of N things taken s at a time.

random sampling [STAT] A sampling from some population where each entry has an equal chance of being drawn.

random storage *See* random-access memory.

random variable [MATH] A measurable function on a probability space; usually real valued, but possibly with values in a general measurable space.

range [CIV ENG] Any series of contiguous townships of the U.S. Public Land Survey system. [COMMUN] **1.** In printing telegraphy, that fraction of a perfect signal element through which the time of selection may be varied to occur earlier or later than the normal time of selection without causing errors while signals are being received; the range of a printing telegraph receiving device is commonly measured in percent of a perfect signal element by adjusting the indicator. **2.** Upper and lower limits through which the index arm of the rangefinder mechanism of a teletypewriter may be moved and still receive correct copy. [ENG] **1.** The distance capability of an aircraft, missile, gun, radar, or radio transmitter. **2.** A line defined by two fixed landmarks, used for missile or vehicle testing and other test purposes. [MATH] The range of a function f from a set X to a set Y consists of those elements y in Y for which there is an x in X with $f(x) = y$. [MECH] The horizontal component of a projectile displacement at the instant it strikes the ground. [NAV] **1.** A line of bearing defined by a radio range. **2.** *See* radio range. [NUCLEO] The distance that a given ionizing particle can penetrate a given medium before its energy drops to the point that the particle no longer produces ionization. [PHYS] The greatest distance between two particles at which a given force between them is appreciable. [STAT] The difference between the maximums and minimums of a variable quantity.

range attenuation [ELECTROMAG] In radar terminology, the decrease in power density (flux density) caused by the divergence of the flux lines with distance, this decrease being in accordance with the inverse-square law.

range-bearing display *See* B display.

range delay [ELECTROMAG] A control used in radars which permits the operator to present on the radarscope only those echoes from targets which lie beyond a certain distance from the radar; by using range delay, undesired echoes from nearby targets may be eliminated while the indicator range is increased.

rangefinder [COMMUN] A movable, calibrated unit of the receiving mechanism of a teletypewriter by means of which the selecting interval may be moved with respect to the start signal. [ELECTR] A device which determines the distance to an object by measuring the time it takes for a radio wave to travel to the object and return. [ENG] *See* optical rangefinder.

range-height indicator [ENG] A scope which simultaneously indicates range and height of a radar target; this presentation is commonly used by height finders.

range marker *See* distance marker.

range of tide [OCEANOGR] The difference in height between consecutive high and low tides at a place.

range recorder [ENG] An item which makes a permanent representation of distance, expressed as range, versus time. [ENG ACOUS] A display used in sonar in which a stylus sweeps across a paper moving at a constant rate and chemically treated so that it is darkened by an electrical signal from the stylus; the stylus starts each sweep as a sound pulse is emitted so that the distance along the trace at which the echo signal appears is a measure of the range to the target.

Ranidae [VERT ZOO] A family of frogs in the suborder Diplasiocoela including the large, widespread genus *Rana*.

rank [GEOL] **1.** A coal classification based on degree of metamorphism. **2.** *See* stack. [MATH] **1.** The rank of a matrix is its maximum number of linearly independent rows. **2.** The rank of a system of homogeneous linear equations equals the rank of the matrix of its coefficients. **3.** A tensor in an n-dimensional space is of rank r if it has n^r components.

Rankine cycle [THERMO] An ideal thermodynamic cycle consisting of heat addition at constant pressure, isentropic expansion, heat rejection at constant pressure, and isentropic compression; used as an ideal standard for the performance of heat-engine and heat-pump installations operating with a condensable vapor as the working fluid, such as a steam power plant. Also known as steam cycle.

Rankine temperature scale [THERMO] A scale of absolute temperature; the temperature in degrees Rankine (°R) is equal to $9/5$ of the temperature in kelvins and is equal to the temperature in degrees Fahrenheit plus 459.67.

rank of an observation [STAT] The number assigned to an observation if a collection of observations is ordered from smallest to largest and each observation is given the number corresponding to its place in the order.

ransomite [MINERAL] $Cu(Fe,Al)_2(SO_4)_4 \cdot 7H_2O$ A sky-blue mineral composed of hydrous copper iron aluminum sulfate.

Ranunculaceae [BOT] A family of dicotyledonous herbs in the order Ranunculales distinguished by alternate leaves with net venation, two or more distinct carpels, and numerous stamens.

Ranunculales [BOT] An order of dicotyledons in the subclass Magnoliidae characterized by its mostly separate carpels, triaperturate pollen, herbaceous or only secondarily woody habit, and frequently numerous stamens.

Raoult's law [PHYS CHEM] The law that the vapor pressure of a solution equals the product of the vapor pressure of the pure solvent and the mole fraction of solvent.

rapakivi [PETR] Granite or quartz monzonite characterized by orthoclase phenocrysts mantled with plagioclase. Also known as wiborgite.

rape [BOT] *Brassica napus.* A plant of the cabbage family in the order Capparales; the plant does not form a compact head, the leaves are bluish-green, deeply lobed, and curled, and the small flowers produce black seeds; grown for forage.

raphe [ANAT] A broad seamlike junction between two lateral halves of an organ or other body part. [BOT] 1. The part of the funiculus attached along its full length to the integument of an anatropous ovule, between the chalaza and the attachment to the placenta. 2. The longitudinal median line or slit on a diatom valve.

Raphidae [VERT ZOO] A family of birds in the order Columbiformes that included the dodo (*Raphus calcullatus*); completely extirpated during the 17th and early 18th centuries.

rapid access loop [COMPUT SCI] A small section of storage, particularly in drum, tape, or disk storage units, which has much faster access than the remainder of the storage.

rapid-eye-movement sleep [PSYCH] That part of the sleep cycle during which the eyes move rapidly, accompanied by a loss of muscle tone and a low-amplitude encephalogram recording; most dreaming occurs during this stage of sleep. Abbreviated REM sleep.

rapid memory *See* rapid storage.

rapid sand filter [CIV ENG] A system for purifying water, which is forced through layers of sand and gravel under pressure.

rapid selector [COMPUT SCI] A device which scans codes recorded on microfilm; microimages of the documents associated with the codes may also be recorded on the film.

rapid sequence camera [OPTICS] A conventional camera in most respects except that it is designed to permit a number of photographs to be obtained in rapid succession with one winding of the shutter.

rapid storage [COMPUT SCI] In computers, storage with a very short access time; rapid access is generally gained by limiting storage capacity. Also known as high-speed storage; rapid memory.

rare-earth element [CHEM] The name given to any of the group of chemical elements with atomic numbers 58 to 71; the name is a misnomer since they are neither rare nor earths; examples are cerium, erbium, and gadolinium.

rarefaction [ACOUS] The instantaneous, local reduction in density of a gas resulting from passage of a sound wave, or the region in which the density is reduced at some instant.

rarefaction wave [FL MECH] A pressure wave or rush of air or water induced by rarefaction; it travels in the opposite direction to that of a shock wave directly following an explosion. Also known as suction wave.

rarefied gas [FL MECH] A gas whose pressure is much less than atmospheric pressure.

Raschig process [CHEM ENG] A method for production of phenol that begins with a first-stage chlorination of benzene, using an air–hydrochloric acid mixture.

rasorite *See* kernite.

raspberry [BOT] Any of several species of upright shrubs of the genus *Rubus*, with perennial roots and prickly biennial stems, in the order Rosales; the edible black or red juicy berries are aggregate fruits, and when ripe they are easily separated from the fleshy receptacle.

raspite [MINERAL] PbWO$_4$ A yellow or brownish-yellow mineral composed of lead tungstate, occurring as monoclinic crystals.

raster [ELECTR] A predetermined pattern of scanning lines that provides substantially uniform coverage of an area; in television the raster is seen as closely spaced parallel lines, most evident when there is no picture.

raster scanning [ELECTR] Radar scan very similar to electron-beam scanning in an ordinary television set; horizontal sector scan that changes in elevation.

rat [VERT ZOO] The name applied to over 650 species of mammals in several families of the order Rodentia; they differ from mice in being larger and in having teeth modified for gnawing.

ratchet [DES ENG] A wheel, usually toothed, operating with a catch or a pawl so as to rotate in only a single direction.

rate action *See* derivative action.

rate constant [PHYS CHEM] Numerical constant in a rate-of-reaction equation; for example, $r_A = kC_A{}^aC_B{}^bC_C{}^c$, where C_A, C_B, and C_C are reactant concentrations, k is the rate constant (specific reaction rate constant), and a, b, and c are empirical constants.

rated capacity [MECH ENG] The maximum capacity for which a boiler is designed, measured in pounds of steam per hour delivered at specified conditions of pressure and temperature.

rated engine speed [MECH ENG] The rotative speed of an engine specified as the allowable maximum for continuous reliable performance.

rated horsepower [MECH ENG] The normal maximum, allowable, continuous power output of an engine, turbine motor, or other prime mover.

rated load [MECH ENG] The maximum load a machine is designed to carry.

rate-grown transistor [ELECTR] A junction transistor in which both impurities (such as gallium and antimony) are placed in the melt at the same time and the temperature is suddenly raised and lowered to produce the alternate *p*-type and *n*-type layers of rate-grown junctions. Also known as graded-junction transistor.

rate multiplier [COMPUT SCI] An integrator in which the quantity to be integrated is held in a register and is added to the number standing in an accumulator in response to pulses which arrive at a constant rate.

rate of change *See* derivative.

rate of climb [AERO ENG] Ascent of aircraft per unit time, usually expressed as feet per minute.

rate of flow *See* flow rate.

rate-of-flow control valve *See* flow control valve.

rate test [COMPUT SCI] A test that verifies that the time constants of the integrators are correct; used in analog computers.

rathite [MINERAL] Pb$_{13}$As$_{18}$S$_{40}$ A dark-gray mineral with metallic luster composed of sulfide of lead and arsenic; occurs as orthorhombic crystals.

rating [ENG] A designation of an operating limit for a machine, apparatus, or device used under specified conditions.

ratio [MATH] A ratio of two quantities or mathematical objects A and B is their quotient or fraction A/B.

ratio arm circuit [ELEC] Two adjacent arms of a Wheatstone bridge, designed so they can be set to provide a variety of indicated resistance ratios.

ratio control system [CONT SYS] Control system in which two process variables are kept at a fixed ratio, regardless of the variation of either of the variables, as when flow rates in two separate fluid conduits are held at a fixed ratio.

ratio detector [ELECTR] A frequency-modulation detector circuit that uses two diodes and requires no limiter at its input; the audio output is determined by the ratio of two developed intermediate-frequency voltages whose relative amplitudes are a function of frequency.

ratio meter [ENG] A meter that measures the quotient of two electrical quantities; the deflection of the meter pointer is proportional to the ratio of the currents flowing through two coils.

rational horizon *See* celestial horizon.

rational number [MATH] A number which is the quotient of two integers.

ratites [VERT ZOO] A group of flightless, mostly large, running birds comprising several orders and including the emus, cassowaries, kiwis, and ostriches.

rattlesnake [VERT ZOO] Any of a number of species of the genera *Sistrurus* or *Crotalus* distinguished by the characteristic rattle on the end of the tail.

ravine [GEOGR] A small and narrow valley with steeply sloping sides.

raw [METEOROL] Colloquially descriptive of uncomfortably cold weather, usually meaning cold and damp, but sometimes cold and windy.

raw data [SCI TECH] Data that have not been processed; may be in machine-readable form.

raw humus *See* ectohumus.

rawin [METEOROL] A method of winds-aloft observation, that is, the determination of wind speeds and directions in the atmosphere above a station; accomplished by tracking a balloon-borne radar target, responder, or radiosonde transmitter with either radar or a radio direction finder.

raw score [STAT] Any number as it originally appears in an experiment; for example, in evaluating test results the raw scores express the number of correct answers, uncorrected for position in the reference population.

ray [ASTRON] One of the broad streaks that radiate from some craters on the moon, especially Copernicus and Tycho; they consist of material of high reflectivity and are seen from earth best at full moon. [MATH] A straight-line segment emanating from a point. Also known as half line. [OPTICS] A curve whose tangent at any point lies in the direction of propagation of a light wave. [PHYS] A moving particle or photon of ionizing radiation. [VERT ZOO] Any of about 350 species of the elasmobranch order Batoidea having flattened bodies with large pectoral fins attached to the side of the head, ventral gill slits, and long, spikelike tails.

ray acoustics [ACOUS] The study of the behavior of sound under the assumption that sound traversing a homogeneous medium travels along straight lines or rays. Also known as geometrical acoustics.

rayfin fish [VERT ZOO] The common name for members of the Actinopterygii.

rayleigh [OPTICS] A unit of brightness, used to measure the brightness of the night sky and aurorae, equal to $10^{10}/4\pi$ quanta per square meter per second per steradian.

Rayleigh balance [ELECTROMAG] An apparatus for assigning the value of the ampere in which the force exerted on a movable circular coil by larger circular coils above and below, but coaxial with, the movable coil is compared with the gravitational force on a known mass.

Rayleigh criterion [OPTICS] A criterion for the resolving power of an optical instrument which states that the images of two point objects are resolved

when the principal maximum of the diffraction pattern of one falls exactly on the first minimum of the diffraction pattern of the other.

Rayleigh disk [ACOUS] An acoustic radiometer, used to measure particle velocity, consisting of a thin disk set at an angle of 45° to a sound beam; the particle velocity is calculated from the resulting torque on the disk.

Rayleigh distribution [STAT] A normal distribution of two uncorrelated variates with the same variance.

Rayleigh flow [FL MECH] An idealized type of gas flow in which heat transfer may occur, satisfying the assumptions that the flow takes place in constant-area cross section and is frictionless and steady, that the gas is perfect and has constant specific heat, that the composition of the gas does not change, and that there are no devices in the system which deliver or receive mechanical work.

Rayleigh interferometer [OPTICS] An optical interferometer in which two rays of light, emanating from a single slit, are collimated by a lens, pass through separate slits and cells, and are brought to focus by a second lens so that interference fringes become visible. Also known as Rayleigh refractometer.

Rayleigh-Jeans law [STAT MECH] A law giving the intensity of radiation emitted by a blackbody within a narrow band of wavelengths; it states that this intensity is proportional to the temperature divided by the fourth power of the wavelength; it is a good approximation to the experimentally verified Planck radiation formula only at long wavelengths.

Rayleigh prism [OPTICS] A system of prisms used to produce greater dispersion of light than would be produced by a single prism.

Rayleigh ratio [OPTICS] Light-scattering relationship defined by the ratio of intensities of incident and scattered light at a specified distance; used in photometric and refractometric analyses.

Rayleigh refractometer *See* Rayleigh interferometer.

Rayleigh-Ritz method [MATH] An approximation method for finding solutions of functional equations in terms of finite systems of equations.

Rayleigh scattering [ELECTROMAG] Scattering of electromagnetic radiation by independent particles which are much smaller than the wavelength of the radiation.

Rayleigh wave [GEOPHYS] In seismology, a surface wave with a retrograde, elliptical motion at the free surface. Also known as R wave. [MECH] A wave which propagates on the surface of a solid; particle trajectories are ellipses in planes normal to the surface and parallel to the direction of propagation. Also known as surface wave.

rayon [TEXT] A fiber made from regenerated cellulose by the viscose or cuprammonium process.

Rb *See* rubidium.

R-C amplifier *See* resistance-capacitance coupled amplifer.

R-C circuit *See* resistance-capacitance circuit.

R-C constant *See* resistance-capacitance constant.

R-C coupled amplifier *See* resistance-capacitance coupled amplifier.

R-C coupling *See* resistance coupling.

R-C network *See* resistance-capacitance network.

R-C oscillator *See* resistance-capacitance oscillator.

RDF *See* radio direction finder.

R display [ELECTR] Radar display, essentially an expanded A display, in which an echo can be expanded for more detailed examination.

Re *See* rhenium.

reach [CIV ENG] A portion of a waterway between two locks or gages. [ENG] The length of a channel, uniform with respect to discharge, depth, area, and slope. [GEOGR] **1.** A continuous, unbroken surface of land or water. Also known as stretch. **2.** A bay, estuary, or other arm of the sea extending up into the land. [HYD] A straight, continuous, or extended part of a river, stream, or restricted waterway.

reach rod [MECH ENG] A rod motion in a link used to transmit motion from the reversing rod to the lifting shaft.

reactance [ELEC] The imaginary part of the impedance of an alternating-current circuit.

reactance amplifier *See* parametric amplifier.

reactance drop [ELEC] The component of the phasor representing the voltage drop across a component or conductor of an alternating-current circuit which is perpendicular to the current.

reactants [CHEM] The molecules that act upon one another to produce a new set of molecules (products); for example, in the reaction HCl + NaOH → NaCl + H_2O, the HCl and NaOH are the reactants.

reaction [CONT SYS] *See* positive feedback. [MECH] The equal and opposite force which results when a force is exerted on a body, according to Newton's third law of motion. [NUC PHYS] *See* nuclear reaction.

reaction border *See* reaction rim.

reaction energy *See* disintegration energy.

reaction engine [AERO ENG] An engine that develops thrust by its reaction to a substance ejected from it; specifically, such an engine that ejects a jet or stream of gases created by the addition of energy to the gases in the engine. Also known as reaction motor.

reactions 31 kinetics *See* chemical kinetics.

reaction motor [AERO ENG] *See* reaction engine. [ELEC] A synchronous motor whose rotor contains salient poles but which has no windings and no permanent magnets.

reaction pair [MINERAL] Any two minerals, one of which is formed at the expense of the other by reaction with liquid.

reaction rim [PETR] A surficial rim around one mineral produced by the reaction of the core mineral with the surrounding magma. Also known as reaction border.

reaction series [MINERAL] Any series of minerals in which early formed varieties react with the melt to yield new minerals; two distinct types of reaction series exist, continuous and discontinuous.

reactive bond [CHEM] A bond between atoms that is easily invaded (reacted to) by another atom or radical; for example, the double bond in CH_2═CH_2 (ethylene) is highly reactive to other ethylene molecules in the reaction known as polymerization to form polyethylene.

reactive current [ELEC] In the phasor representation of alternating current, the component of the current perpendicular to the voltage, which contributes no power but increases the power losses of the system. Also known as idle current; quadrature current; wattless current.

reactive dye [MATER] Dye that reacts with the textile fiber to produce both a hydroxyl and an oxygen linkage, the chlorine combining with the hydroxyl to form a fast ether linkage; gives fast, brilliant colors.

reactive muffler [ENG] A muffler that attenuates by reflecting sound back to the source. Also known as nondissipative muffler.

reactive volt-ampere hour *See* var hour.

reactive volt-ampere meter *See* varmeter.

reactivity [CHEM] The relative capacity of an atom, molecule, or radical to combine chemically with another atom, molecule, or radical. [NUCLEO] A measure of the deviation of a nuclear reactor from the critical state at any instant of time such that positive and negative values correspond to reactors above and below critical, respectively; measured in percent k, millikays, dollars, or inhours.

reactor [CHEM ENG] Device or process vessel in which chemical reactions (catalyzed or noncatalyzed) take place during a chemical conversion type of process. [ELEC] A device that introduces either inductive or capacitive reactance into a circuit, such as a coil or capacitor. Also known as electric reactor. [NUC PHYS] *See* nuclear reactor.

reactor fuel *See* nuclear fuel.

reactor fuel cycle [NUCLEO] The processes of preparing fuel elements and assemblies for use in a reactor, using these elements in reactor operation, recovering radioactive by-products from spent fuel, and reprocessing remaining fissionable material into new fuel elements. Also known as fuel cycle; nuclear fuel cycle.

reactor fuel pellet *See* nuclear fuel pellet.

reactor physics [NUCLEO] The science of the interaction of the elementary particles and radiations characteristic of nuclear reactors with matter in bulk.

read [COMPUT SCI] **1.** To acquire information, usually from some form of storage in a computer. **2.** To convert magnetic spots, characters, or punched holes into electrical impulses. [COMMUN] To understand clearly, as in radio communication. [ELECTR] To generate an output corresponding to the pattern stored in a charge storage tube.

read-around number *See* read-around ratio.

read-around ratio [COMPUT SCI] The number of times that a particular bit in electrostatic storage may be read without seriously affecting nearby bits. Also known as read-around number.

read-back check *See* echo check.

reader [COMPUT SCI] A device that converts information from one form to another, as from punched paper tape to magnetic tape. [GRAPHICS] A projection device for viewing an enlarged microimage with the unaided eye.

reader-interpreter [COMPUT SCI] A service routine that reads an input string, stores programs and data on random-access storage for later processing, identifies the control information contained in the input string, and stores this control information separately in the appropriate control lists.

read error [COMPUT SCI] A condition in which the content of a storage device cannot be electronically identified.

read head [COMPUT SCI] A device that converts digital information stored on a magnetic tape, drum, or disk into electrical signals usable by the computer arithmetic unit.

read-in [COMPUT SCI] To sense information contained in some source and transmit this information to an internal storage.

reading rate [COMPUT SCI] Number of characters, words, fields, blocks, or cards sensed by an input sensing device per unit of time.

reading station [COMPUT SCI] The position in a punched-card machine at which the data on the card are read, by sensing the positions of the holes, and converted into electrical impulses. Also known as sensing station.

read-in program [COMPUT SCI] Computer program that can be put into a computer in a simple binary form and allows other programs to be read into the computer in more complex forms.

read-only memory [COMPUT SCI] A device for storing data in permanent, or nonerasable, form; usually a static electronic or magnetic device allowing extremely rapid access to data. Abbreviated ROM. Also known as read-only storage.

read-only storage See read-only memory.

readout [COMPUT SCI] 1. The presentation of output information by means of lights, printed or punched tape or cards, or other methods. 2. To sense information contained in some computer internal storage and transmit this information to a storage external to the computer.

readout station [COMMUN] A recording or receiving radio station at which data are received, as the transmitter in a missile, probe, satellite, or other spacecraft reads the data out.

read time [COMPUT SCI] The time interval between the instant at which information is called for from storage and the instant at which delivery is completed in a computer.

read/write channel [COMPUT SCI] A path along which information is transmitted between the central processing unit of a computer and an input, output, or storage unit under the control of the computer.

read/write head [COMPUT SCI] A magnetic head that both senses and records data. Also known as combined head.

read/write memory [COMPUT SCI] A computer storage in which data may be stored or retrieved at comparable intervals.

read/write random-access memory [COMPUT SCI] A random access memory in which data can be written into memory as well as read out of memory.

reagent [ANALY CHEM] A substance, chemical, or solution used in the laboratory to detect, measure, or otherwise examine other substances, chemicals, or solutions; grades include ACS (American Chemical Society standards), reagent (for analytical reagents), CP (chemically pure), USP (U.S. Pharmacopeia standards), NF (National Formulary standards), and purified, technical (for industrial use). [CHEM] The compound that supplies the molecule, ion, or free radical which is arbitrarily considered as the attacking species in a chemical reaction.

reagin [IMMUNOL] 1. An antibody which occurs in human atopy, such as hay fever and asthma, and which readily sensitizes the skin. 2. An antibody which reacts in various serologic tests for syphilis.

real axis [MATH] The horizontal axis of the cartesian coordinate system for the euclidean or complex plane.

real data type [COMPUT SCI] A scalar data type which contains a normalized fraction (mantissa) and an exponent (characteristic) and is used to represent floating-point data, usually decimal.

real fluid flow [FL MECH] The flow in which effects of tangential or shearing forces are taken into account; these forces give rise to fluid friction, because they oppose the sliding of one particle past another.

realgar [MINERAL] AsS A red to orange mineral crystallizing in the monoclinic system, having a resinous luster and found in short, vertical striated crystals; specific gravity is 3.48, and hardness is 1.5–2 on Mohs scale. Also known as red arsenic; red orpiment; sandarac.

real gas [THERMO] A gas, as considered from the viewpoint in which deviations from the ideal gas law, resulting from interactions of gas molecules, are taken into account. Also known as imperfect gas.

real image [OPTICS] An optical image such that all the light from a point on an object that passes through an optical system actually passes close to or through a point on the image.

real number [MATH] Any member of the unique (to within isomorphism) complete ordered field.

real-time [COMPUT SCI] Pertaining to a data-processing system that controls an ongoing process and delivers its outputs (or controls its inputs) not later than the time when these are needed for effective control; for instance, airline reservations booking and chemical processes control.

real-time operation [COMPUT SCI] 1. Of a computer or system, an operation or other response in which programmed responses to an event are essentially simultaneous with the event itself. 2. An operation in which information obtained from a physical process is processed to influence or control the physical process.

real-time processing [COMPUT SCI] The handling of input data at a rate sufficient to ensure that the instructions generated by the computer will influence the operation under control at the required time.

real-time programming [COMPUT SCI] Programming for a situation in which results of computations will be used immediately to influence the course of ongoing physical events.

ream [ENG] To enlarge or clean out a hole. [MATER] 1. A layer of nonhomogeneous material in flat glass. 2. Five hundred sheets of paper; a printer's ream consists of 516 sheets.

reboiler [CHEM ENG] An auxiliary heating unit for a fractionating tower designed to supply additional heat to the lower portion of the tower; liquid withdrawn from the side or bottom of the tower is reheated by heat exchange, then reintroduced into the tower.

rebreather [ENG] A closed-loop oxygen supply system consisting of gas supply and face mask.

received power [ELECTROMAG] 1. The total power received at an antenna from a signal, such as a radar target signal. 2. In a mobile communications system, the root-mean-square value of power delivered to a load which properly terminates an isotropic reference antenna.

receiver [CHEM ENG] Vessel, container, or tank used to receive and collect liquid material from a process unit, such as the distillate receiver from the overhead condenser of a distillation column. [ELECTR] The complete equipment required for receiving modulated radio waves and converting them into the original intelligence, such as into sounds or pictures, or converting to desired useful information as in a radar receiver. [MECH ENG] An apparatus placed near the compressor to equalize the pulsations of the air as it comes from the compressor to cause a more uniform flow of air through the pipeline and to collect moisture and oil carried in the air.

receiver bandwidth [ELECTR] Spread, in frequency, between the halfpower points on the receiver response curve.

receiver lockout system See lockout.

receiver primary See display primary.

receiver synchro See synchro receiver.

receiving set See radio receiver.

receiving tube [ELECTR] A low-voltage and low-power vacuum tube used in radio receivers, computers, and sensitive control and measuring equipment.

Recent [GEOL] An epoch of geologic time (late Quaternary) following the Pleistocene; referred to as Holocene in several European countries.

receptacle   *See* outlet.

reception   [COMMUN] The conversion of modulated electromagnetic waves or electric signals, transmitted through the air or over wires or cables, into the original intelligence, or into desired useful information (as in radar), by means of antennas and electronic equipment.

receptor   [BIOCHEM] A site or structure in a cell which combines with a drug or other biological to produce a specific alteration of cell function.   [PHYSIO] A sense organ.

recessional moraine   [GEOL] 1. An end moraine formed during a temporary halt in the final retreat of a glacier. 2. A moraine formed during a minor readvance of the ice front during a period of glacial recession. Also known as stadial moraine.

recessive   [GEN] 1. An allele that is not expressed phenotypically when present in the heterozygous condition. 2. An organism homozygous for a recessive gene.

recharge   [HYD] 1. The processes involved in the replenishment of water to the zone of saturation. 2. The amount of water added or absorbed. Also known as groundwater increment; groundwater recharge; groundwater replenishment; increment; intake.

rechargeable battery   *See* storage battery.

recharge well   [HYD] A well used as a source of water in the process of artificial recharge. Also known as injection well.

recharging   [ELEC] The restoring of discharged electric storage batteries to a charged condition by passing direct current through them in a direction opposite to that of the discharging current.

reciprocal   [MATH] The reciprocal of a number A is the number $1/A$.

reciprocal bearing   *See* back bearing.

reciprocal ellipsoid   *See* index ellipsoid.

reciprocal impedance   [ELEC] Two impedances Z_1 and Z_2 are said to be reciprocal impedances with respect to an impedance Z (invariably a resistance) if they are so related as to satisfy the equation $Z_1 Z_2 = Z^2$.

reciprocal lattice   [CRYSTAL] A lattice array of points formed by drawing perpendiculars to each plane (hkl) in a crystal lattice through a common point as origin; the distance from each point to the origin is inversely proportional to spacing of the specific lattice planes; the axes of the reciprocal lattice are perpendicular to those of the crystal lattice.

reciprocal ohm   *See* siemens.

reciprocal translocation   [CYTOL] The special case of translocation in which two segments exchange positions.

reciprocal vectors   [CRYSTAL] For a set of three vectors forming the primitive translations of a lattice, the vectors that form the primitive translations of the reciprocal lattice.

reciprocating compressor   [MECH ENG] A positive-displacement compressor having one or more cylinders, each fitted with a piston driven by a crankshaft through a connecting rod.

reciprocating drill   *See* piston drill.

reciprocating engine   *See* piston engine.

reciprocating-plate column   *See* reciprocating-plate extractor.

reciprocating-plate extractor   [CHEM ENG] A liquid-liquid contactor in which equally spaced perforated plates (as in a distillation column) move up and down rapidly over a short distance to cause liquid agitation and mixing. Also known as reciprocating-plate column.

reciprocating pump   *See* piston pump.

reciprocity theorem   Also known as principle of reciprocity.   [ACOUS] The theorem that, in an acoustic system consisting of a fluid medium with boundary surfaces and subject to no impressed body forces, if p_1 and p_2 are the pressure fields produced respectively by the components of the fluid velocities V_1 and V_2 normal to the boundary surfaces, then the integral over the boundary surfaces of $p_1 V_2 - p_2 V_1$ vanishes.   [ELEC] 1. The electric potentials V_1 and V_2 produced at some arbitrary point, due to charge distributions having total charges of q_1 and q_2 respectively, are such that $q_1 V_2 = q_2 V_1$. 2. In an electric network consisting of linear passive impedances, the ratio of the electromotive force introduced in any branch to the current in any other branch is equal in magnitude and phase to the ratio that results if the positions of electromotive force and current are exchanged.   [ELECTROMAG] Given two loop antennas, a and b, then $I_{ab}/V_a = I_{ba}/V_b$, where I_{ab} denotes the current received in b when a is used as transmitter, and V_a denotes the voltage applied in a; I_{ba} and V_b are the corresponding quantities when b is the transmitter, a the receiver; it is assumed that the frequency and impedances remain unchanged.   [ENG ACOUS] The sensitivity of a reversible electroacoustic transducer when used as a microphone divided by the sensitivity when used as a source of sound is independent of the type and construction of the transducer.   [PHYS] In general, any theorem that expresses various reciprocal relations for the behavior of some physical systems, in which input and output can be interchanged without altering the response of the system to a given excitation.

recirculator   [ENG] A self-contained underwater breathing apparatus that recirculates an oxygen supply (mix-gas or pure) to the diver until the oxygen is depleted.

reclamation   [CIV ENG] 1. The recovery of land or other natural resource that has been abandoned because of fire, water, or other cause. 2. Reclaiming dry land by irrigation.

reclined fold   *See* recumbent fold.

recognition   [COMPUT SCI] The act or process of identifying (or associating) an input with one of a set of possible known alternatives, as in character recognition and pattern recognition.

recognition gate   [COMPUT SCI] A logic circuit used to select devices identified by a binary address code. Also known as decoding gate.

recoil electron   [PHYS] An electron that has been set into motion by a collision.

recoil escapement   *See* anchor escapement.

recoilless   [ORD] Built so as to eliminate or cancel out recoil; most recoilless guns are designed to let part of the propellant gases escape to the rear.

recombination   [GEN] 1. The occurrence of gene combinations in the progeny that differ from those of the parents as a result of independent assortment, linkage, and crossing-over. 2. The production of genetic information in which there are elements of one line of descent replaced by those of another line, or additional elements.   [PHYS] The combination and resultant neutralization of particles or objects having unlike charges, such as a hole and an electron or a positive ion and a negative ion.

record   [COMPUT SCI] A group of adjacent data items in a computer system, manipulated as a unit.   [SCI TECH] 1. To preserve for later reproduction or reference. 2. *See* recording.

record block   *See* physical record.

record density   *See* bit density; character density.

recorder   *See* recording instrument.

record gap   [COMPUT SCI] An area in a storage medium, such as magnetic tape or disk, which is devoid of information; it delimits records, and, on tape, allows the tape to stop and start between records without loss of data. Also known as interrecord gap (IRG).

recording   [SCI TECH] **1.** Any process for preserving signals, sounds, data, or other information for future reference or reproduction, such as disk recording, facsimile recording, ink-vapor recording, magnetic tape or wire recording, and photographic recording. **2.** The end product of a recording process, such as the recorded magnetic tape, disk, or record sheet. Also known as record.

recording head   [ELECTR] A magnetic head used only for recording. Also known as record head.   [ENG ACOUS] *See* cutter.

recording instrument   [ENG] An instrument that makes a graphic or acoustic record of one or more variable quantities. Also known as recorder.

recording level   [ELECTR] Amplifier output level required to secure a satisfactory recording.

recording spot   *See* picture element.

recording thermometer   *See* thermograph.

record length   [COMPUT SCI] The number of characters required for all the information in a record.

record mark   [COMPUT SCI] A symbol that signals a record's beginning or end.

record observation   [METEOROL] A type of aviation weather observation; the most complete of all such observations and usually taken at regularly specified and equal intervals (hourly, usually on the hour). Also known as hourly observation.

record storage mark   [COMPUT SCI] A special character which appears only in the record storage unit of the card reader to limit the length of the record read into storage.

recoupling   [QUANT MECH] A transformation between eigenfunctions of total angular momentum resulting from coupling eigenfunctions of three or more angular momenta in some order, and eigenfunctions of total angular momentum resulting from coupling of the same eigenfunctions in a different order.

recoverable shear   [FL MECH] Measure of the elastic content of a fluid, related to elastic recovery (mechanicallike property of elastic recoil); found in unvulcanized, unfilled natural rubber, and certain polymer solutions, soap gels, and biological fluids.

recovery   [AERO ENG] **1.** The procedure or action that obtains when the whole of a satellite, or a section, instrumentation package, or other part of a rocket vehicle, is retrieved after a launch. **2.** The conversion of kinetic energy to potential energy, such as in the deceleration of air in the duct of a ramjet engine. Also known as ram recovery. **3.** In flying, the action of a lifting vehicle returning to an equilibrium attitude after a nonequilibrium maneuver.   [HYD] The rise in static water level in a well, occurring upon the cessation of discharge from that well or a nearby well.   [MET] **1.** The percentage of valuable material obtained from a processed ore. **2.** Reduction or elimination of work-hardening effects, usually by heat treatment.   [MIN ENG] The proportion or percentage of coal or ore mined from the original seam or deposit.   [PETRO ENG] The removal (recovery) of oil or gas from reservoir formations.

recovery interrupt   [COMPUT SCI] A type of interruption of program execution which provides the computer with access to subroutines to handle an error and, if successful, to continue with the program execution.

recovery time   [ELECTR] **1.** The time required for the control electrode of a gas tube to regain control after anode-current interruption. **2.** The time required for a fired TR (transmit-receive) or pre-TR tube to deionize to such a level that the attenuation of a low-level radio-frequency signal transmitted through the tube is decreased to a specified value. **3.** The time required for a fired ATR (anti-transmit-receive) tube to deionize to such a level that the normalized conductance and susceptance of the tube in its mount are within specified ranges. **4.** The interval required, after a sudden decrease in input signal amplitude to a system or component, to attain a specified percentage (usually 63%) of the ultimate change in amplification or attenuation due to this decrease. **5.** The time required for a radar receiver to recover to half sensitivity after the end of the transmitted pulse, so it can receive a return echo.   [NUCLEO] The minimum time from the start of a counted pulse to the instant a succeeding pulse can attain a specific percentage of the maximum value of the counted pulse in a Geiger counter.

recrystallization   [CHEM] Repeated crystallization of a material from fresh solvent to obtain an increasingly pure product.   [CRYSTAL] A change in the structure of a crystal without a chemical alteration.   [MET] A process which takes place in metals and alloys following distortion and fragmentation of constituent crystals by severe mechanical deformation, in which some fragments grow at the expense of others, so that larger, strain-free grains are formed; it progresses slowly at room temperature, but is greatly speeded by annealing.   [PETR] The formation of new mineral grains in crystalline form in a rock under the influence of metamorphic processes.

recrystallization breccia   *See* pseudobreccia.

rectangle   [MATH] A plane quadrilateral having four interior right angles and opposite sides of equal length.

rectangular cartesian coordinate system   *See* cartesian coordinate system.

rectangular distribution   *See* uniform distribution.

rectangular projection   [MAP] A cylindrical map projection with uniform spacing of the parallels; used for the star chart in the Air Almanac.

rectangular pulse   [ELECTR] A pulse in which the wave amplitude suddenly changes from zero to another value at which it remains constant for a short period of time, and then suddenly changes back to zero.

rectangular wave   [ELECTR] A periodic wave that alternately and suddenly changes from one to the other of two fixed values. Also known as rectangular wave train.

rectangular wave train   *See* rectangular wave.

Recticornia   [INV ZOO] A family of amphipod crustaceans in the superfamily Genuina containing forms in which the first antennae are straight, arise from the anterior margin of the head, and have few flagellar segments.

rectification   [CIV ENG] A new alignment to correct a deviation of a stream channel or bank.   [ELEC] The process of converting an alternating current to a unidirectional current.   [GEOL] The simplification and straightening of the outline of an initially irregular and crenulate shoreline through the cutting back of headlands and offshore islands by marine erosion, and through deposition of waste from erosion or of sediment brought down by neighboring rivers.   [GRAPHICS] The transformation of a photograph onto a horizontal plane so as to remove or correct displacements (distortions in perspective) by tilt.

rectifier [ELEC] A nonlinear circuit component that allows more current to flow in one direction than the other; ideally, it allows current to flow in one direction unimpeded but allows no current to flow in the other direction.

rectifying column [CHEM ENG] Portion of a distillation column above the feed tray in which rising vapor is enriched by interaction with a countercurrent falling stream of condensed vapor; contrasted to the stripping column section below the column feed tray.

rectilinear [MATH] Consisting of or bounded by lines.

rectilinear motion [MECH] A continuous change of position of a body so that every particle of the body follows a straight-line path. Also known as linear motion.

rectilinear system See orthoscopic system.

rectus [ANAT] Having a straight course, as certain muscles.

recumbent [BOT] Of or pertaining to a plant or plant part that tends to rest on the surface of the soil.

recumbent fold [GEOL] An overturned fold with a nearly horizontal axial surface. Also known as reclined fold.

recuperative air heater [ENG] An air heater in which the heat-transferring metal parts are stationary and form a separating boundary between the heating and cooling fluids.

recurrence rate See repetition rate.

recurring decimal See repeating decimal.

recursion [COMPUT SCI] A technique in which an apparently circular process is used to perform an iterative process.

recursion formula [MATH] An algorithm allowing computation of a succession of quantities. Also known as recursion relation.

recursion relation See recursion formula.

recursive functions [MATH] Functions that can be obtained by a finite number of operations, computations, or algorithms.

recursive procedure [COMPUT SCI] A method of calculating a function by deriving values of it which become more elementary at each step; recursive procedures are explicitly outlawed in most systems with the exception of a few which use languages such as ALGOL and LISP.

recursive subroutine [COMPUT SCI] A reentrant subroutine whose partial results are stacked, with a processor stack pointer advancing and retracting as the subroutine is called and completed.

recurved [SCI TECH] Curving inward to backward.

recurved spit [GEOGR] See hook.

red [OPTICS] The hue evoked in an average observer by monochromatic radiation having a wavelength in the approximate range from 622 to 770 nanometers; however, the same sensation can be produced in a variety of other ways.

red algae [BOT] The common name for members of the phylum Rhodophyta.

red antimony See kermesite.

red arsenic See realgar.

redbed [GEOL] Continentally deposited sediment composed principally of sandstone, siltsone, and shale; red in color due to the presence of ferric oxide (hematite). Also known as red rock.

red blood cell See erythrocyte.

red clay [GEOL] A fine-grained, reddish-brown pelagic deposit consisting of relatively large proportions of windblown particles, meteoric and volcanic

dust, pumice, shark teeth, manganese nodules, and debris transported by ice. Also known as brown clay.

red cobalt See erythrite.

red dwarf star [ASTRON] A red star of low luminosity, so designated by E. Hertzsprung; dwarf stars are commonly those main-sequence stars fainter than an absolute magnitude of +1, and red dwarfs are the faintest and coolest of the dwarfs.

red giant star [ASTRON] A star whose evolution has progressed to the point where hydrogen core burning has been completed, the helium core has become denser and hotter than originally, and the envelope has expanded to perhaps 100 times its initial size.

red hematite See hematite.

redingtonite [MINERAL] $(Fe,Mg,Ni)(Cr,Al)_2(SO_4)_4 \cdot 22H_2O$ A pale-purple mineral composed of a hydrous sulfate of iron, magnesium, nickel, chromium, and aluminum.

red iron ore See hematite.

red leaf [PL PATH] Any of various nonparasitic plant diseases marked by red discoloration of the foliage.

red ocher See ferric oxide; hematite.

red orpiment See realgar.

redox cell [ELEC] Cell designed to convert the energy of reactants to electrical energy; an intermediate reductant, in the form of liquid electrolyte, reacts at the anode in a conventional manner; i is then regenerated by reaction with a primary fuel.

red oxide of zinc See zincite.

redox potential [PHYS CHEM] Voltage difference at an inert electrode immersed in a reversible oxidation-reduction system; measurement of the state of oxidation of the system. Also known as oxidation-reduction potential.

redox titration [ANALY CHEM] A titration characterized by the transfer of electrons from one substance to another (from the reductant to the oxidant) with the end point determined colorimetrically or potentiometrically.

red rock See redbed.

red rot [PL PATH] Any of several fungus diseases of plants characterized by red patches on stems or leaves; common in sugarcane, sisal, and various evergreen and deciduous trees.

red rust [PL PATH] An algal disease of certain subtropical plants, such as tea and citrus, caused by the green alga *Cephaleuros virescens* and characterized by a rusty appearance of the leaves or twigs.

redshift [ASTROPHYS] A systematic displacement toward longer wavelengths of lines in the spectra of distant galaxies and also of the continuous portion of the spectrum; increases with distance from the observer. Also known as Hubble effect.

red stele [PL PATH] A fatal fungus disease of strawberries caused by *Phytophthora fragariae* invading the roots, producing redness and rotting with consequent dwarfing and wilting of the plant.

red stripe [PL PATH] 1. A fungus decay of timber caused by *Polyporus vaporarius* and characterized by reddish streaks. 2. A bacterial disease of sugarcane caused by *Xanthomonas rubrilineans* and characterized by red streaks in the young leaves, followed by invasion of the vascular system.

red-tape operation See bookkeeping operation.

red thread [PL PATH] A fungus disease of turf grasses caused by *Corticium fuciforme* and characterized by the appearance of red stromata on the pinkish hyphal threads.

reduced-pressure distillation See vacuum distillation.

reducer [BIOL] *See* decomposer. [CHEM] *See* reducing agent. [DES ENG] A fitting having a larger size at one end than at the other and threaded inside, unless specifically flanged or for some special joint. [GRAPHICS] A solution capable of dissolving silver; used to cut down the contrast or density of a negative or positive image.

reducing agent [CHEM] Also known as reducer. **1.** A material that adds hydrogen to an element or compound. **2.** A material that adds an electron to an element or compound, that is, decreases the positiveness of its valence.

reducing flame [CHEM] A flame having excess fuel and being capable of chemical reduction, such as extracting oxygen from a metallic oxide.

reducing sugar [ORG CHEM] Any of the sugars that because of their free or potentially free aldehyde or ketone groups, possess the property of readily reducing alkaline solutions of many metallic salts such as copper, silver, or bismuth; examples are the monosaccharides and most of the disaccharides, including maltose and lactose.

reduction [CHEM] **1.** Reaction of hydrogen with another substance. **2.** Chemical reaction in which an element gains an electron (has a decrease in positive valence). [COMPUT SCI] Any process by which data are condensed, such as changing the encoding to eliminate redundancy, extracting significant details from the data and eliminating the rest, or choosing every second or third out of the totality of available points. [GEOL] The lowering of a land surface by erosion. [NAV] The process of substituting for an observed value one derived therefrom, as the calculation of the corresponding meridian altitude from an observation of a celestial body near the meridian, or the derivation from a celestial observation of the information needed for establishing a line of position.

reduction gear [MECH ENG] A gear train which lowers the output speed.

reduction potential [PHYS CHEM] The potential drop involved in the reduction of a positively charged ion to a neutral form or to a less highly charged ion, or of a neutral atom to a negatively charged ion.

redundancy [COMPUT SCI] Any deliberate duplication or partial duplication of circuitry or information to decrease the probability of a system or communication failure. [COMMUN] In the transmission of information, the fraction of the gross information content of a message which can be eliminated without loss of essential information. [GEN] **1.** Repetition of a specified deoxyribonucleic acid sequence in a nucleus. **2.** Multiplicity of codons for individual amino acids. [MATH] A repetitive statement. [MECH] A statically indeterminate structure.

redundancy check [COMPUT SCI] A forbidden-combination check that uses redundant digits called check digits to detect errors made by a computer.

redundant character [COMPUT SCI] A character specifically added to a group of characters to ensure conformity with certain rules which can be used to detect computer malfunction.

redundant digit [COMPUT SCI] Digit that is not necessary for an actual computation but serves to reveal a malfunction in a digital computer.

Reduviidae [INV ZOO] The single family of the hemipteran group Reduvioidea; nearly all have a stridulatory furrow on the prosternum, ocelli are generally present, and the beak is three-segmented.

Reduvioidea [INV ZOO] The assassin bugs or cone-nose bugs, a monofamilial group of hemipteran insects in the subdivision Geocorisae.

redwood [BOT] *Sequoia sempervirens.* An evergreen tree of the pine family; it is the tallest tree in the Americas, attaining 350 feet (107 meters); its soft heartwood is a valuable building material.

red zinc ore *See* zincite.

reed frequency meter *See* vibrating-reed frequency meter.

reef [GEOL] **1.** A ridge- or moundlike layered sedimentary rock structure built almost exclusively by organisms. **2.** An offshore chain or range of rock or sand at or near the surface of the water. [MIN ENG] A major ore trend or ore body.

reef limestone [PETR] Limestone composed of the remains of sedentary organisms such as sponges, and of sediment-binding organic constituents such as calcareous algae. Also known as coral rock.

reef rock [PETR] A hard, unstratified rock composed of sand, shale, and the calcareous remains of sedentary organisms, cemented by calcium carbonate.

reenterable [COMPUT SCI] The attribute that describes a program or routine which can be shared by several tasks concurrently.

reentrant program [COMPUT SCI] A subprogram in a time-sharing or multiprogramming system that can be shared by a number of users, and can therefore be applied to a given user program, interrupted and applied to some other user program, and then reentered at the point of interruption of the original user program.

reentry [AERO ENG] The event when a spacecraft or other object comes back into the sensible atmosphere after being in space.

reference acoustic pressure [ACOUS] Magnitude of any complex sound that will produce a sound-level meter reading equal to that produced by a sound pressure of 0.0002 dyne per square centimeter at 1000 hertz. Also known as reference sound level.

reference address *See* address constant.

reference angle [ELECTROMAG] Angle formed between the center line of a radar beam as it strikes a reflecting surface and the perpendicular drawn to that reflecting surface.

reference block [COMPUT SCI] A block within a computer program governing a numerically controlled machine which has enough data to allow resumption of the program following an interruption.

reference burst *See* color burst.

reference dipole [ELECTROMAG] Straight half-wave dipole tuned and matched for a given frequency, and used as a unit of comparison in antenna measurement work.

reference electrode [PHYS CHEM] A nonpolarizable electrode that generates highly reproducible potentials; used for pH measurements and polarographic analyses; examples are the calomel electrode, silver-silver chloride electrode, and mercury pool.

reference level [ENG] *See* datum plane. [ENG ACOUS] The level used as a basis of comparison when designating the level of an audio-frequency signal in decibels or volume units. Also known as reference signal level. [OCEANOGR] **1.** Level of no motion. **2.** A level for which current is known; allows determination of absolute current from relative current.

reference listing [COMPUT SCI] A list printed by a compiler showing the instructions in the machine language program which it generates.

reference plane [ENG] *See* datum plane. [MECH ENG] The plane containing the axis and the cutting point of a cutter.

reference record [COMPUT SCI] Output of a compiler that lists the operations and their positions in the final specific routine and contains information describing the segmentation and storage allocation of the routine.

reference signal level *See* reference level.

reference sound level *See* reference acoustic pressure. .

reference voltage [ELEC] An alternating-current voltage used for comparison, usually to identify an in-phase or out-of-phase condition in an ac circuit.

referred pain [MED] Pain felt in one area but originating in another area.

refine [ENG] To free from impurities, as the separation of petroleum, ores, or chemical mixtures into their component parts.

refined lecithin *See* lecithin.

refinery [CHEM ENG] System of process units used to convert crude petroleum into fuels, lubricants, and other petroleum-derived products. [MET] System of process units used to convert nonferrous-metal ores into pure metals, such as copper or zinc.

refinery gas [MATER] Gas produced in petroleum refineries by cracking, reforming, and other processes; principally methane, ethane, ethylene, butanes, and butylenes.

reflectance [COMPUT SCI] In optical character recognition, the relative brightness of the inked area that forms the printed or handwritten character; distinguished from background reflectance and brightness. [ELEC] *See* reflection factor. [PHYS] *See* reflectivity.

reflected binary [COMPUT SCI] A particular form of gray code which is constructed according to the following rule: Let the first 2^N code patterns be given, for any N greater than 1; the next 2^n code patterns are derived by changing the $(N + 1)$-th bit from the right from 0 to 1 and repeating the original 2^n patterns in reverse order in the N rightmost positions. Also known as reflected code.

reflected code *See* reflected binary.

reflected impedance [ELEC] 1. Impedance value that appears to exist across the primary of a transformer due to current flowing in the secondary. 2. Impedance which appears at the input terminals as a result of the characteristics of the impedance at the output terminals.

reflected resistance [ELEC] Resistance value that appears to exist across the primary of a transformer when a resistive load is across the secondary.

reflected wave [PHYS] A wave reflected from a surface, discontinuity, or junction of two different media, such as the sky wave in radio, the echo wave from a target in radar, or the wave that travels back to the source end of a mismatched transmission line.

reflecting antenna [ELECTROMAG] An antenna used to achieve greater directivity or desired radiation patterns, in which a dipole, slot, or horn radiates toward a larger reflector which shapes the radiated wave to produce the desired pattern; the reflector may consist of one or two plane sheets, a parabolic or paraboloidal sheet, or a paraboloidal horn.

reflecting galvanometer *See* mirror galvanometer.

reflecting microscope [OPTICS] A microscope whose objective is composed of two mirrors, one convex and the other concave; its imaging properties are independent of the wavelength of light, allowing it to be used even for infrared and ultraviolet radiation.

reflecting prism [OPTICS] A prism used in place of a mirror for deviating light, usually designed so that there is no dispersion of light; the light undergoes at least one internal reflection.

reflecting telescope [OPTICS] A telescope in which a concave parabolic mirror gathers light and forms a real image of an object. Also known as reflector telescope.

reflection altimeter *See* radio altimeter.

reflection angle *See* angle of reflection.

reflection coefficient [PHYS] The ratio of the amplitude of a wave reflected from a surface to the amplitude of the incident wave. Also known as coefficient of reflection.

reflection factor [ELEC] Ratio of the load current that is delivered to a particular load when the impedances are mismatched to that delivered under conditions of matched impedances. Also known as mismatch factor; reflectance; transition factor.

reflection law [PHYS] When a wave, such as electromagnetic radiation or sound, is reflected from a surface in a sharply defined direction, the reflected and incident waves travel in directions that make the same angle with a perpendicular to the surface and lie in a common plane with it. Also known as law of reflection.

reflection loss [ELEC] 1. Reciprocal of the ratio, expressed in decibels, of the scalar values of the volt-amperes delivered to the load to the volt-amperes that would be delivered to a load of the same impedance as the source. 2. Apparent transmission loss of a line which results from a portion of the energy being reflected toward the source due to a discontinuity in the transmission line.

reflection plane *See* plane of mirror symmetry; plane of reflection.

reflection survey [ENG] Study of the presence, depth, and configuration of underground formations; a ground-level explosive charge (shot) generates vibratory energy (seismic rays) that strike formation interfaces and are reflected back to ground-level sensors. Also known as seismic survey.

reflectivity [PHYS] The ratio of the energy carried by a wave which is reflected from a surface to the energy carried by the wave which is incident on the surface. Also known as reflectance.

reflectometer [ELECTROMAG] *See* microwave reflectometer. [ENG] A photoelectric instrument for measuring the optical reflectance of a reflecting surface.

reflector [ELECTR] *See* repeller. [ELECTROMAG] 1. A single rod, system of rods, metal screen, or metal sheet used behind an antenna to increase its directivity. 2. A metal sheet or screen used as a mirror to change the direction of a microwave radio beam. [GEOPHYS] A layer or horizon that reflects seismic waves. [NUCLEO] A layer of water, graphite, beryllium, or other scattering material placed around the core of a nuclear reactor to reduce the loss of neutrons. Also known as tamper.

reflector microphone [ENG ACOUS] A highly directional microphone which has a surface that reflects the rays of impinging sound from a given direction to a common point at which a microphone is located, and the sound waves in the speech-frequency range are in phase at the microphone.

reflector telescope *See* reflecting telescope.

reflex [PHYSIO] An automatic response mediated by the nervous system.

reflex angle [MATH] An angle greater than 180° and less then 360°.

reflex camera [OPTICS] A camera in which a mirror is used to reflect a full-size image of a scene on a ground glass so that the composition and focus may be judged.

reflex klystron [ELECTR] A single-cavity klystron in which the electron beam is reflected back through the cavity resonator by a repelling electrode having a negative voltage; used as a microwave oscillator. Also known as reflex oscillator.

reflex oscillator *See* reflex klystron.

reflux [CHEM ENG] In a chemical process, that part of the product stream that may be returned to the process to assist in giving increased conversion or recovery, as in distillation or liquid-liquid extraction.

Reformatsky reaction [ORG CHEM] A condensation-type reaction between ketones and α-bromoaliphatic acids in the presence of zinc or magnesium, such as $R_2CO + BrCH_2\cdot COOR + Zn \rightarrow (ZnO\cdot HBr) + R_2C(OH)CH_2COOR.$

reforming [CHEM ENG] The thermal or catalytic conversion of petroleum naphtha into more volatile products of higher octane number; represents the total effect of numerous simultaneous reactions, such as cracking, polymerization, dehydrogenation and isomerization.

refracting telescope [OPTICS] A telescope in which a lens gathers light and forms a real image of an object. Also known as refractor telescope.

refraction [PHYS] The change of direction of propagation of any wave, such as an electromagnetic or sound wave, when it passes from one medium to another in which the wave velocity is different, or when there is a spatial variation in a medium's wave velocity.

refraction loss [ELECTROMAG] Portion of the transmission loss that is due to refraction resulting from nonuniformity of the medium.

refractive index *See* index of refraction.

refractivity [ELECTROMAG] 1. Some quantitative measure of refraction, usually a measure of the index of refraction. 2. The index of refraction minus 1.

refractometer [ENG] An instrument used to measure the index of refraction of a substance in any one of several ways, such as measurement of the refraction produced by a prism, measurement of the critical angle, observation of an interference pattern produced by passing light through the substance, and measurement of the substance's dielectric constant.

refractor telescope *See* refracting telescope.

refractory [MATER] 1. A material of high melting point. 2. The property of resisting heat. [MED] Not readily yielding to treatment.

refracture index *See* index of refraction.

refrigerant [MATER] A substance that by undergoing a change in phase (liquid to gas, gas to liquid) releases or absorbs a large latent heat in relation to its volume, and thus effects a considerable cooling effect; examples are ammonia, sulfur dioxide, ethyl or methyl chloride (these are no longer widely used), and the fluorocarbons, such as Freon, Ucon, and Genetron.

refrigeration [MECH ENG] The cooling of a space or substance below the environmental temperature.

refrigeration cycle [THERMO] A sequence of thermodynamic processes whereby heat is withdrawn from a cold body and expelled to a hot body.

regelation [HYD] Phenomenon in which ice melts at the bottom of droplets of highly concentrated saline solution that are trapped in ice which has frozen over

polar waters, and freezes at the top of these droplets, so that the droplets move downward through the ice, leaving it hard and clear. [THERMO] Phenomenon in which ice (or any substance which expands upon freezing) melts under intense pressure and freezes again when this pressure is removed; accounts for phenomena such as the slippery nature of ice and the motion of glaciers.

regenerate [CHEM ENG] To clean of impurities and make reusable as in regeneration of a catalytic cracking catalyst by burning off carbon residue, regeneration of clay adsorbent by washing free of adherents, or regeneration of a filtration system by cleaning off the filter media. [ELECTR] 1. To restore pulses to their original shape. 2. To restore stored information to its original form in a storage tube in order to counteract fading and disturbances.

regeneration [BIOL] The replacement by an organism of tissues or organs which have been lost or severely injured. [CONT SYS] *See* positive feedback. [ELECTR] Replacement or restoration of charges in a charge storage tube to overcome decay effects, including loss of charge by reading. [NUCLEO] Restoration of contaminated nuclear fuel to a usable condition.

regenerative air heater [MECH ENG] An air heater in which the heat-transferring members are alternately exposed to heat-surrendering gases and to air.

regenerative braking [ELEC] A system of dynamic braking in which the electric drive motors are used as generators and return the kinetic energy of the motor armature and load to the electric supply system.

regenerative cycle [MECH ENG] *See* bleeding cycle. [THERMO] An engine cycle in which low-grade heat that would ordinarily be lost is used to improve the cyclic efficiency.

regenerative feedback *See* positive feedback.

regenerative pump [MECH ENG] Rotating-vane device that uses a combination of mechanical impulse and centrifugal force to produce high liquid heads at low volumes. Also known as turbine pump.

regenerative reactor [NUCLEO] A nuclear reactor that produces fissionable material in addition to energy; when loaded with fissionable uranium-235, and nonfissionable uranium-238 or thorium, it converts the uranium-238 or thorium into fissionable materials which are then used as fuel in the core of the reactor.

regenerative read [COMPUT SCI] A read operation in which the data are automatically written back into the locations from which they are taken.

regenerative storage [COMPUT SCI] A storage unit, such as a delay line or storage tube, in which the stored data must be constantly read and restored to prevent decay or loss.

regenerative track [COMPUT SCI] Track on a magnetic drum with interconnected reading and writing heads; information stored on these tracks is continuously read from the drum, transmitted round a closed circuit, and recorded back on the drum; consequently, access times to these data are short; operation is analogous to that of the acoustic delay line.

regenerator [CHEM ENG] Device or system used to return a system or a component of it to full strength in a chemical process; examples are a furnace to burn carbon from a catalyst, a tower to wash impurities from clay, and a flush system to clean off the surface of filter media. [MECH ENG] A device used with hot-air engines and gas-burning furnaces which transfers heat from effluent gases to incoming air or gas.

Regge pole [PARTIC PHYS] A pole singularity of a scattering amplitude in the complex angular momentum plane; the scattering amplitude is formed by continuing partial wave amplitudes from positive integer values of the angular momentum to the complex plane.

Regge recurrence [PARTIC PHYS] One of a sequence of hadrons, with successive hadrons increasing by one in spin and also increasing in mass, but with the same values of other quantum numbers, except for parity, charge parity and G parity, which alternate in sign; it is believed that they are rotationally excited states of a particle, and that they alternate between two Regge trajectories.

Regge trajectory [PARTIC PHYS] **1.** The path followed by a Regge pole in the complex angular momentum plane as the center-of-mass energy is varied. **2.** The relationship between the spin and mass of a sequence of hadrons, with successive hadrons increasing by 2 in spin and also increasing in mass, but with the same values of other quantum numbers; the hadrons are thought to correspond to energies at which a Regge pole passes near positive integers (or half integers).

regime [GEOL] The existence in a stream channel of a balance between erosion and deposition over a period of years.

region [COMPUT SCI] A group of machine addresses which refer to a base address. [MATH] *See* domain.

regional dip [GEOL] The nearly uniform and generally low-angle inclination of strata over a wide area. Also known as normal dip.

region of escape *See* exosphere.

register [COMPUT SCI] The computer hardware for storing one machine word. [COMMUN] Part of an automatic switching telephone system which receives and stores the dialing pulses which control the further operations necessary in establishing a telephone connection. [ENG] Also known as registration. **1.** The accurate matching or superimposition of two or more images, such as the three color images on the screen of a color television receiver, or the patterns on opposite sides of a printed circuit board, or the colors of a design on a printed sheet. **2.** The alignment of positions relative to a specified reference or coordinate, such as hole alignments in punched cards, or positioning of images in an optical character recognition device. [GRAPHICS] **1.** Exact agreement in the position of printed material on both sides of a sheet or on all pages of a book or pamphlet. **2.** Exact overprinting of colorplates, or other subsequent plates, so that all printed detail is correctly combined; proper color overprinting is checked by the exact superimposition of the register marks that are printed with each color run. **3.** In flat preparation, the exact agreement between color or complementary flats. [MECH ENG] The portion of a burner which directs the flow of air used in the combustion process. [ORD] **1.** To adjust fire on a visible point, called a check point, and compute accurate adjusted data so that firing data for later targets may be computed with reference to that check point. **2.** To adjust fire on several selected points in order that they may serve later as auxiliary targets.

register circuit [ELECTR] A switching circuit with memory elements that can store from a few to millions of bits of coded information; when needed, the information can be taken from the circuit in the same code as the input, or in a different code.

register mark [ENG] A mark or line printed or otherwise impressed on a web of material for use as a reference to maintain register. [GRAPHICS] One of a set of marks, usually cross-shaped, placed on the margin of colored originals to be photographed for colorplates; the marks are used to register the images when the plates are printed.

register ton *See* ton.

registration *See* register.

registration mark [COMPUT SCI] In character recognition, a preprinted indication of the relative position and direction of various elements of the source document to be recognized.

regolith [GEOL] The layer rock or blanket of unconsolidated rocky debris of any thickness that overlies bedrock and forms the surface of the land. Also known as mantle rock.

regression [GEOL] The theory that some rivers have sources on the rainier sides of mountain ranges and gradually erode backward until the ranges are cut through. [OCEANOGR] Retreat of the sea from land areas, and the consequent evidence of such withdrawal. [PSYCH] A mental state and a mode of adjustment to difficult and unpleasant situations, characterized by behavior of a type that had been satisfying and appropriate at an earlier stage of development but which no longer befits the age and social status of the individual. [STAT] Given two stochastically dependent random variables, regression functions measure the mean expectation of one relative to the other.

regressive overlap *See* offlap.

regular dodecahedron [CRYSTAL] *See* pyritohedron. [MATH] A regular polyhedron of twelve faces.

Regularia [INV ZOO] An assemblage of echinoids in which the anus and periproct lie within the apical system; not considered a valid taxon.

regularization [QUANT MECH] A formal procedure used to eliminate ambiguities which arise in evaluating certain integrals in a quantized field theory; corresponds to adding extra fields whose masses are allowed to approach infinity.

regular polygon [MATH] A polygon with congruent sides and congruent interior angles.

regular polyhedron [MATH] A polyhedron all of whose faces are regular polygons, and whose polyhedral angles are congruent.

regular reflection *See* specular reflection.

regular reflector *See* specular reflector.

regular topological space [MATH] A topological space where any point and a closed set not containing it can be enclosed in disjoint open sets.

regulating rod [NUCLEO] A control rod intended to accomplish rapid, fine, and sometimes continuous adjustment of the reactivity of a nuclear reactor; it usually can move much more rapidly than a shim rod but makes a smaller change in reactivity.

regulating system *See* automatic control system.

regulating transformer [ELEC] Transformer having one or more windings excited from the system circuit or a separate source and one or more windings connected in series with the system circuit for adjusting the voltage or the phase relation or both in steps, usually without interrupting the load.

regulation [CONT SYS] The process of holding constant a quantity such as speed, temperature, voltage, or position by means of an electronic or other system that automatically corrects errors by feeding back into the system the condition being regulated; regulation thus is based on feedback, whereas control is not. [ELEC] The change in output voltage that occurs between no load and full load in a transformer, generator, or other source. [ELECTR] The difference between the maximum and minimum tube voltage drops within a specified range of anode current in a gas tube.

regulator [CONT SYS] A device that maintains a desired quantity at a predetermined value or varies it according to a predetermined plan. [MIN ENG] An opening in a wall or door in the return airway of a district to increase its resistance and reduce the volume of air flowing.

regulator gene [GEN] A gene that controls the rate of transcription of one or more other genes.

Reighardiidae [INV ZOO] A monotypic family of arthropods in the order Cephalobaenida; the posterior end of the organism is rounded, without lobes, and the cuticula is covered with minute spines.

reindeer [VERT ZOO] *Rangifer tarandus.* A migratory ruminant of the deer family (Cervidae) which inhabits the Arctic region and has a circumpolar distribution; characteristically, both sexes have antlers and are brown with yellow-white areas on the neck and chest.

reinforced beam [CIV ENG] A concrete beam provided with steel bars for longitudinal tension reinforcement and sometimes compression reinforcement and reinforcement against diagonal tension.

reinforced column [CIV ENG] **1.** A long concrete column reinforced with longitudinal bars with ties or circular spirals. **2.** A composite column. **3.** A combination column.

reinforced concrete [CIV ENG] Concrete containing reinforcing steel rods or wire mesh.

reinforced plastic [MATER] High-strength filled plastic product used for mechanical, construction, and electrical products, automotive components, and ablative coatings; filling can be whiskers of glass, metal, boron, or other materials.

reinserter *See* direct-current restorer.

rejection band Also known as stop band. [ELECTROMAG] The band of frequencies below the cutoff frequency in a uniconductor waveguide. [PHYS] A frequency band within which electrical or electromagnetic signals are reduced or eliminated.

rejector *See* trap.

rejector impedance *See* dynamic impedance.

rejuvenation [GEOL] The restoration of youthful features to fluvial landscapes; the renewal of youthful vigor to low-gradient streams is usually caused by regional upwarping of broad areas formerly at or near base level. [HYD] **1.** The stimulation of a stream to renew erosive activity. **2.** The renewal of youthful vigor in a mature stream.

rejuvenation head *See* knickpoint.

relapsing fever [MED] An acute infectious disease caused by various species of the spirochete *Borrelia*, characterized by episodes of fever which subside spontaneously and recur over a period of weeks.

relational operator [COMPUT SCI] An operator that indicates whether one quantity is equal to, greater than, or less than another.

relational system [COMPUT SCI] A type of data-base management system in which data are represented as tables in which no entry contains more that one value.

relative address [COMPUT SCI] The numerical difference between a desired address and a known reference address.

relative coding [COMPUT SCI] A form of computer programming in which the address part of an instruction indicates not the desired address but the difference between the location of the instruction and the desired address.

relative damping ratio *See* damping ratio.

relative deflection *See* astrogeodetic deflection.

relative density *See* specific gravity.

relative dielectric constant *See* dielectric constant.

relative divergence *See* development index.

relative frequency [STAT] The ratio of the number of occurrences of a given type of event or the number of members of a population in a given class to the total number of events or the total number of members of the population.

relative humidity [METEOROL] The (dimensionless) ratio of the actual vapor pressure of the air to the saturation vapor pressure. Abbreviated RH.

relative luminosity factor *See* luminosity function.

relatively compact set *See* conditionally compact set.

relative Mach number *See* Mach number.

relative motion [MECH] The continuous change of position of a body with respect to a second body or to a reference point that is fixed. Also known as apparent motion.

relative orbit [ASTRON] The closed path described by the apparent position of the fainter member of a binary system relative to the brighter member.

relative permittivity *See* dielectric constant.

relative power gain [ELECTROMAG] Of one transmitting or receiving antenna over another, the measured ratio of the signal power one produces at the receiver input terminals to that produced by the other, the transmitting power level remaining fixed.

relative topology [MATH] In a topological space X any subset A has a topology on it relative to the given one by intersecting the open sets of X with A to obtain open sets in A.

relativistic electrodynamics [ELECTROMAG] The study of the interaction between charged particles and electric and magnetic fields when the velocities of the particles are comparable with that of light.

relativistic kinematics [RELAT] A description of the motion of particles compatible with the special theory of relativity, without reference to the causes of motion.

relativistic mass [RELAT] The mass of a particle moving at a velocity exceeding about one-tenth the velocity of light; it is significantly larger than the rest mass.

relativistic mechanics [RELAT] **1.** Any form of mechanics compatible with either the special or the general theory of relativity. **2.** The nonquantum mechanics of a system of particles or of a fluid interacting with an electromagnetic field, in the case when some of the velocities are comparable with the speed of light.

relativistic particle [RELAT] A particle moving at a speed comparable with the speed of light.

relativistic quantum theory [QUANT MECH] The quantum theory of particles which is consistent with the special theory of relativity, and thus can describe particles moving arbitrarily close to the speed of light.

relativistic theory [PHYS] Any theory which is consistent with the special or general theory of relativity.

relativity [PHYS] Theory of physics which recognizes the universal character of the propagation speed of light and the consequent dependence of space, time, and other mechanical measurements on the motion of the observer performing the measurements; it has two main divisions, the special theory and the general theory.

relaxation [GEOL] In experimental structural geology, the diminution of applied stress with time, as the result of any of various creep processes. [MATH] *See* relaxation method. [MECH] **1.** Relief of stress in a strained material due to creep. **2.** The lessening of elastic resistance in an elastic medium under an applied stress resulting in permanent deformation. [PHYS] A process in which a physical system approaches a steady state after conditions affecting it have been suddenly changed, and in which the presence of dissipative agents prevents the system from overshooting and then oscillating about this state.

relaxation circuit [ELECTR] Circuit arrangement, usually of vacuum tubes, reactances, and resistances, which has two states or conditions, one, both, or neither of which may be stable; the transient voltage produced by passing from one to the other, or the voltage in a state of rest, can be used in other circuits.

relaxation method [MATH] A successive approximation method for solving systems of equations where the errors from an initial approximation are viewed as constraints to be minimized or relaxed within a toleration limit. Also known as relaxation.

relaxation oscillations [PHYS] Oscillations having a sawtooth waveform in which the displacement increases to a certain value and then drops back to zero, after which the cycle is repeated.

relaxation oscillator [ELECTR] An oscillator whose fundamental frequency is determined by the time of charging or discharging a capacitor or coil through a resistor, producing waveforms that may be rectangular or sawtooth.

relaxation test [ENG] A creep test in which the decrease of stress with time is measured while the total strain (elastic and plastic) is maintained constant.

relaxation time [PHYS] For many physical systems undergoing relaxation, a time τ such that the displacement of a quantity from its equilibrium value at any instant of time t is the exponential of $-t/\tau$. [SOLID STATE] The travel time of an electron in a metal before it is scattered and loses its momentum.

relaxed peak process *See* quasi-fission.

relay [COMMUN] A microwave or other radio system used for passing on a signal from one radio communication link to another. [ELEC] A device that is operated by a variation in the conditions in one electric circuit and serves to make or break one or more connections in the same or another electric circuit. Also known as electric relay.

relay rack [DES ENG] A standardized steel rack designed to hold 19-inch (48.26-centimeter) panels of various heights, on which are mounted radio receivers, amplifiers, and other units of electronic equipment. Also known as rack.

relay satellite *See* communications satellite.

release joint *See* sheeting structure.

reliability [ENG] The probability that a component part, equipment, or system will satisfactorily perform its intended function under given circumstances, such as environmental conditions, limitations as to operating time, and frequency and thoroughness of maintenance for a specified period of time. [STAT] **1.** The amount of credence placed in a result. **2.** The precision of a measurement, as measured by the variance of repeated measurements of the same object.

relic [GEOL] **1.** A landform that remains intact after decay or disintegration or that remains after the disappearance of the major portion of its substance. **2.** A vestige of a particle in a sedimentary rock, such as a trace of a fossil fragment.

relict dike [GEOL] In a granitized mass, a tabular, crystalloblastic body that represents a dike which

was emplaced prior to, and which was relatively resistant to, the granitization process.

relict mineral [MINERAL] A mineral of a rock that persists from an earlier rock.

relict sediment [GEOL] A sediment which was in equilibrium with its environment when first deposited but which is unrelated to its present environment even though it is not buried by later sediments, such as a shallow-marine sediment on the deep ocean floor.

relict soil [GEOL] A soil formed on a preexisting landscape but not subsequently buried under younger sediments.

relief [CRYSTAL] The apparent topography exhibited by minerals in thin section as a consequence of refractive index. [GEOD] The configuration of a part of the earth's surface, with reference to altitude and slope variations and to irregularities of the land surface. [MECH ENG] **1.** A passage made by cutting away one side of a tailstock center so that the facing or parting tool may be advanced to or almost to the center of the work. **2.** Clearance provided around the cutting edge by removal of tool material.

relief angle [MECH ENG] The angle between a relieved surface and a tangential plane at a cutting edge.

relief map [MAP] A map of an area showing the topographic relief.

relief press *See* letterpress.

relief printing [GRAPHICS] A method of printing in which the type or other images stand above the printing surface; even though the lower areas may have ink in them, they do not print because the paper does not contact them.

relocatable code [COMPUT SCI] A code generated by an assembler or compiler, and in which all memory references needing relocation are either specially marked or relative to the current program-counter reading.

relocatable program [COMPUT SCI] A program coded in such a way that it may be located and executed in any part of memory.

relocate [COMPUT SCI] To establish or change the location of a program routine while adjusting or modifying the address references within the instructions to correctly indicate the new locations.

relocation register [COMPUT SCI] A hardware element that holds a constant to be added to the address of each memory location in a computer program running in a multiprogramming system, as determined by the location of the area in memory assigned to the program.

reluctance [ELECTROMAG] A measure of the opposition presented to magnetic flux in a magnetic circuit, analogous to resistance in an electric circuit; it is equal to magnetomotive force divided by magnetic flux. Also known as magnetic reluctance.

reluctance microphone *See* magnetic microphone.

reluctance motor [ELEC] A synchronous motor, similar in construction to an induction motor, in which the member carrying the secondary circuit has salient poles but no direct-current excitation; it starts as an induction motor but operates normally at synchronous speed.

reluctivity [PHYS] The reciprocal of magnetic permeability; the reluctivity of empty space is unity. Also known as magnetic reluctivity; specific reluctance.

rem [NUCLEO] A unit of ionizing radiation, equal to the amount that produces the same damage to humans as 1 roentgen of high-voltage x-rays. Derived from roentgen equivalent man.

remainder [MATH] **1.** The remaining integer when a division of an integer by another is performed; if $l = m \cdot p + r$, where l, m, p, and r are integers and r is less than p, then r is the remainder when l is divided by p. **2.** The remaining polynomial when division of a polynomial is performed; if $l = m \cdot p + r$, where l, m, p, and r are polynomials, and the degree of r is less than that of p, then r is the remainder when l is divided by p. **3.** The remaining part of a convergent infinite series after a computation, for some n, of the first n terms.

remanence [ELECTROMAG] The magnetic flux density that remains in a magnetic circuit after the removal of an applied magnetomotive force; if the magnetic circuit has an air gap, the remanence will be less than the residual flux density.

remanent magnetization [GEOPHYS] That component of a rock's magnetization whose direction is fixed relative to the rock and which is independent of moderate, applied magnetic fields.

remex See flight feather.

remodulator [ELECTR] A circuit that converts amplitude modulation to audio frequency-shift modulation for transmission of facsimile signals over a voice-frequency radio channel. Also known as converter.

remote batch computing [COMPUT SCI] The running of programs, usually during nonprime hours, or whenever the demands of real-time or time-sharing computing slacken sufficiently to allow less pressing programs to be run.

remote batch processing [COMPUT SCI] Batch processing in which an input device is located at a distance from the main installation and has access to a computer through a communication link.

remote computing system [COMPUT SCI] A data-processing system that has terminals distant from the central processing unit, from which users can communicate with the central processing unit and compile, debug, test, and execute programs.

remote control [CONT SYS] Control of a quantity which is separated by an appreciable distance from the controlling quantity; examples include master-slave manipulators, telemetering, telephone, and television.

remote data station [COMPUT SCI] A terminal in a data-processing system at which data can be sent to or received from a central computer over telephone or telegraph circuits, but which exerts no direct operating control over the central computer. Also known as remote data terminal.

remote data terminal See remote data station.

remote indicator [ELECTR] **1.** An indicator located at a distance from the data-gathering sensing element, with data being transmitted to the indicator mechanically, electrically over wires, or by means of light, radio, or sound waves. **2.** See repeater.

remote job entry [COMPUT SCI] The submission of jobs to a central computer from a location at least a few hundred meters and sometimes many kilometers distant from the computer, requiring the use of a telephone or other common-carrier communications link. Abbreviated RJE.

remotely piloted vehicle [AERO ENG] A robot aircraft, controlled over a two-wave radio link from a ground station or mother aircraft that can be hundreds of miles away; electronic guidance is generally supplemented by remote control television cameras feeding monitor receivers at the control station. Abbreviated RPV.

remote manipulation [ENG] Use of mechanical equipment controlled from a distance to handle materials, such as radioactive materials.

remote manipulator [ENG] A mechanical, electromechanical, or hydromechanical device which enables a person to perform manual operations while separated from the site of the work.

remote metering See telemetering.

remote sensing [ELEC] Sensing, by a power supply, of voltage directly at the load, so that variations in the load lead drop do not affect load regulation. [ENG] The gathering and recording of information without actual contact with the object or area being investigated.

remote terminal [COMPUT SCI] A computer terminal which is located away from the central processing unit of a data-processing system, at a location convenient to a user of the system.

removable plugboard See detachable plugboard.

REM sleep See rapid-eye-movement sleep.

renal artery [ANAT] A branch of the abdominal or ventral aorta supplying the kidneys in vertebrates.

renal corpuscle [ANAT] The glomerulus together with its Bowman's capsule in the renal cortex. Also known as Malpighian corpuscle.

renal tubule [ANAT] One of the glandular tubules which elaborate urine in the kidneys.

renal vein [ANAT] A vein which returns blood from the kidney to the vena cava.

renardite [MINERAL] $Pb(UO_2)_4(PO_4)_2(OH)_4 \cdot 7H_2O$ A yellow mineral composed of hydrous basic lead uranyl phosphate.

Rendoll [GEOL] A suborder of the soil order Mollisol, formed in highly calcareous parent materials, mostly restricted to humid, temperate regions; the soil profile consists of a dark upper horizon grading to a pale lower horizon.

reniform [SCI TECH] Bean- or kidney-shaped, as describing the structure of a crystal in which rounded masses occur at the ends of radiating crystals, or certain structures in animals and plants.

renin [BIOCHEM] A proteolytic enzyme produced in the afferent glomerular arteriole which reacts with the plasma component hypertensinogen to produce angiotensin II.

rennin [BIOCHEM] An enzyme found in the gastric juice of the fourth stomach of calves; used for coagulating milk casein in cheesemaking. Also known as chymosin.

renormalization [QUANT MECH] In certain quantum field theories, a procedure in which nonphysical bare values of certain quantities such as mass and charge are eliminated and the corresponding physically observable quantities are introduced.

reorder point [IND ENG] An arbitrary level of stock on hand plus stock due in, at or below which routine requisitions for replenishment purposes are submitted in accordance with established requisitioning schedules.

rep [NUCLEO] A unit of ionizing radiation, equal to the amount that causes absorption of 93 ergs per gram of soft tissue. Derived from roentgen equivalent physical. Also known as parker; tissue roentgen. [TEXT] A type of fabric characterized by distinct, round, padded ribs running from selvage to selvage.

repeated reflection See multiple reflection.

repeated root See multiple root.

repeater [ELEC] See repeating coil. [ELECTR] **1.** An amplifier or other device that receives weak signals and delivers corresponding stronger signals with or without reshaping of waveforms; may be either a one-way or two-way repeater. **2.** An indicator that shows the same information as is shown on a master indicator. Also known as remote indicator.

repeating coil [ELEC] A transformer used to provide inductive coupling between two sections of a telephone line when a direct connection is undesirable. Also known as repeater.

repeating decimal [MATH] A decimal that is either finite or infinite with a finite block of digits repeating indefinitely. Also known as recurring decimal.

repeat operator [COMPUT SCI] A pseudo instruction using two arguments, a count p and an increment n: the word immediately following the instruction is repeated p times, with the values $0, n, 2n, ..., (p-1)n$ added to the successive words.

repeller [ELECTR] An electrode whose primary function is to reverse the direction of an electron stream in an electron tube. Also known as reflector.

repent [BOT] Of a stem, creeping along the ground and rooting at the nodes.

repetition frequency *See* repetition rate.

repetition rate [COMMUN] The rate at which recurrent signals are produced or transmitted by radar. Also known as recurrence rate; repetition frequency.

repetitive addressing [COMPUT SCI] A system used on some computers in which, under certain conditions, an instruction is written without giving the address of the operand, and the operand address is automatically that of the location addressed by the last previous instruction.

repetitive analog computer [COMPUT SCI] An analog computer which repeatedly carries out the solution of a problem at a rapid rate (10 to 60 times a second) while an operator may vary parameters in the problem.

replacement [GEOL] Growth of a new or chemically different mineral in the body of an old mineral by simultaneous capillary solution and deposition. [PALEON] Substitution of inorganic matter for the original organic constituents of an organism during fossilization.

replication [ANALY CHEM] The formation of a faithful mold or replica of a solid that is thin enough for penetration by an electron microscope beam; can use plastic (such as collodion) or vacuum deposition (such as of carbon or metals) to make the mold. [MOL BIO] Duplication, as of a nucleic acid, by copying from a molecular template. [VIROL] Multiplication of phage in a bacterial cell.

reply [COMMUN] A radio-frequency signal or combination of signals transmitted by a transponder in response to an interrogation. Also known as response.

representation theory [MATH] **1.** The study of groups by the use of their representations. **2.** The determination of representations of specific groups. [QUANT MECH] Quantum-mechanical device in which one selects the common eigenfunctions of a complete set of quantum-mechanical operators as a basis of vectors in a Hilbert space, and expresses wave functions and operators in terms of column matrices and square matrices, respectively, which correspond to this basis.

repressor [BIOCHEM] An end product of metabolism which represses the synthesis of enzymes in the metabolic pathway. [GEN] The product of a regulator gene that acts to repress the transcription of another gene.

repro *See* reproduction proof.

reproducing stylus *See* stylus.

reproducing system *See* sound-reproducing system.

reproduction proof [GRAPHICS] A page proof made with great care on smooth paper and used for photographic reproduction; it smudges easily and must be handled with care. Also known as repro; slick paper proof.

Reptantia [INV ZOO] A suborder of the crustacean order Decapoda including all decapods other than shrimp.

reptile [VERT ZOO] Any member of the class Reptilia.

Reptilia [VERT ZOO] A class of terrestrial vertebrates composed of turtles, tuatara, lizards, snakes, and crocodileans; characteristically they lack hair, feathers, and mammary glands, the skin is covered with scales, they have a three-chambered heart, and the pleural and peritoneal cavities are continuous.

repulsion [MECH] A force which tends to increase the distance between two bodies having like electric charges, or the force between atoms or molecules at very short distances which keeps them apart. Also known as repulsive force.

repulsion motor [ELEC] An alternating-current motor having stator windings connected directly to the source of ac power and rotor windings connected to a commutator; brushes on the commutator are short-circuited and are positioned to produce the rotating magnetic field required for starting and running.

repulsive force *See* repulsion.

rerun routine [COMPUT SCI] A routine designed to be used in the wake of a computer malfunction or a coding or operating mistake to reconstitute a routine from the last previous rerun point.

RES *See* reticuloendothelial system.

rescue dump [COMPUT SCI] The copying of the entire contents of a computer memory into auxiliary storage devices, carried out periodically during the course of a computer program so that in case of a machine failure the program can be reconstituted at the last point at which this operation was executed.

resequent stream [HYD] A stream whose direction follows an original consequent stream but is generally lower; resequent streams are generally tributary to a subsequent stream.

reserve [COMPUT SCI] To assign portions of a computer memory and of input/output and storage devices to a specific computer program in a multiprogramming system.

reserve battery [ELEC] A battery which is inert until an operation is performed which brings all the cell components into the proper state and location to become active.

reserved minerals [MIN ENG] Economic minerals that belong to the state, which confers the right to prospect for and to mine them on any applicant.

reserved word [COMPUT SCI] A word which cannot be used in a programming language to represent an item of data because it has some particular significance to the compiler, or which can be used only in a particular context.

reserves [MIN ENG] The quantity of workable mineral or of gas or oil which is calculated to lie within given boundaries.

reservoir [CIV ENG] A pond or lake built for storage of water, usually by the construction of a dam across a river. [GEOL] A subsurface accumulation of crude oil or natural gas under adequate trap conditions.

reset *See* clear.

reset mode [COMPUT SCI] The phase of operation of an analog computer during which the required initial conditions are entered into the system and the computing units are inoperative. Also known as initial condition mode.

reset pulse [ELECTR] **1.** A drive pulse that tends to reset a magnetic cell in the storage section of a digital computer. **2.** A pulse used to reset an electronic counter to zero or to some predetermined position.

residence time [CHEM ENG] The average length of time a particle of reactant spends within a process vessel or in contact with a catalyst. [NUCLEO] The time during which radioactive material remains in the atmosphere following the detonation of a nuclear explosive; it is usually expressed as a half-time, since the time for all material to leave the atmosphere is not well known.

resident executive [COMPUT SCI] The portion of the executive program (sometimes called monitor system) which is permanently stored in core. Also known as resident monitor.

resident monitor *See* resident executive.

resident routine [COMPUT SCI] Any computer routine which is stored permanently in the memory, such as the resident executive.

residual [GEOL] **1.** Of a mineral deposit, formed by either mechanical or chemical concentration. **2.** Pertaining to a residue left in place after weathering of rock. **3.** Of a topographic feature, representing the remains of a formerly great mass or area and rising above the surrounding surface.

residual air *See* residual volume.

residual charge [ELEC] The charge remaining on the plates of a capacitor after initial discharge.

residual current [ELECTR] Current flowing through a thermionic diode when there is no anode voltage, due to the velocity of the electrons emitted by the heated cathode.

residual elements [MET] Elements present in small amounts in a metal or alloy, not added intentionally.

residual flux density [ELECTROMAG] The magnetic flux density at which the magnetizing force is zero when the material is in a symmetrically and cyclically magnetized condition. Also known as residual induction; residual magnetic induction; residual magnetism.

residual induction *See* residual flux density.

residual ionization [PHYS] Ionization of air or other gas in a closed chamber, not accounted for by recognizable neighboring agencies; now attributed to cosmic rays.

residual magnetic induction *See* residual flux density.

residual magnetism *See* residual flux density.

residual modulation *See* carrier noise.

residual resistance [SOLID STATE] The value to which the electrical resistance of a metal drops as the temperature is lowered to near absolute zero, caused by imperfections and impurities in the metal rather than by lattice vibrations.

residual stress *See* internal stress.

residual vibration *See* zero-point vibration.

residual voltage [ELEC] Vector sum of the voltages to ground of the several phase wires of an electric supply circuit.

residual volume [PHYSIO] Air remaining in the lungs after the most complete expiration possible; it is elevated in diffuse obstructive emphysema and during an attack of asthma. Also known as residual air.

residue [CHEM ENG] **1.** The substance left after distilling off all but the heaviest components from crude oil in petroleum refinery operations. Also known as bottoms; residuum. **2.** Solids deposited onto the filter medium during filtration. Also known as cake; discharged solids. [GEOL] The in-place accumulation of rock debris which remains after weathering has removed all but the least soluble constituent.

[MATH] The residue of a complex function $f(z)$ at an isolated singularity z_0 is given by $(1/2\pi i) \int f(z)dz$ along a simple closed curve interior to an annulus about z_0; equivalently, the coefficient of the term $(z - z_0)^{-1}$ in the Laurent series expansion of $f(z)$ about z_0.

resilience [MECH] **1.** Ability of a strained body, by virtue of high yield strength and low elastic modulus, to recover its size and form following deformation. **2.** The work done in deforming a body to some predetermined limit, such as its elastic limit or breaking point, divided by the body's volume.

resin [ORG CHEM] Any of a class of solid or semisolid organic products of natural or synthetic origin with no definite melting point, generally of high molecular weight; most resins are polymers.

resist [GRAPHICS] A protective layer applied to the image, or other parts of a plate, to protect that portion of the metal from the action of an etching bath or a sandblasting operation. [MATER] An acid-resistant nonconducting coating used to protect desired portions of a wiring pattern from the action of the etchant during manufacture of printed wiring boards. [MET] An insulating material, for example lacquer, applied to the surface of work to prevent electroplating or electrolytic action at the coated area. Also known as stopoff.

resistance [ACOUS] *See* acoustic resistance. [FL MECH] *See* fluid resistance. [ELEC] **1.** The opposition that a device or material offers to the flow of direct current, equal to the voltage drop across the element divided by the current through the element. Also known as electrical resistance. **2.** In an alternating-current circuit, the real part of the complex impedance. [MECH] In damped harmonic motion, the ratio of the frictional resistive force to the speed. Also known as damping coefficient; damping constant; mechanical resistance.

resistance box [ELEC] A box containing a number of precision resistors connected to panel terminals or contacts so that a desired resistance value can be obtained by withdrawing plugs (as in a post-office bridge) or by setting multicontact switches.

resistance bridge *See* Wheatstone bridge.

resistance-capacitance circuit [ELEC] A circuit which has a resistance and a capacitance in series, and in which inductance is negligible. Abbreviated R-C circuit.

resistance-capacitance constant [ELEC] Time constant of a resistive-capacitive circuit; equal in seconds to the resistance value in ohms multiplied by the capacitance value in farads. Abbreviated R-C constant.

resistance-capacitance coupled amplifier [ELECTR] An amplifier in which a capacitor provides a path for signal currents from one stage to the next, with resistors connected from each side of the capacitor to the power supply or to ground; it can amplify alternating-current signals but cannot handle small changes in direct currents. Also known as R-C amplifier; R-C coupled amplifier; resistance-coupled amplifier.

resistance-capacitance network [ELEC] Circuit containing resistances and capacitances arranged in a particular manner to perform a specific function. Abbreviated R-C network.

resistance-capacitance oscillator [ELECTR] Oscillator in which the frequency is determined by resistance and capacitance elements. Abbreviated R-C oscillator.

resistance-coupled amplifier *See* resistance-capacitance coupled amplifier.

resistance coupling [ELECTR] Coupling in which resistors are used as the input and output impedances of the circuits being coupled; a coupling capacitor is generally used between the resistors to transfer the signal from one stage to the next. Also known as *R-C* coupling; resistance-capacitance coupling; resistive coupling.

resistance drop [ELEC] The voltage drop occurring between two points on a conductor due to the flow of current through the resistance of the conductor; multiplying the resistance in ohms by the current in amperes gives the voltage drop in volts. Also known as *IR* drop.

resistance furnace [ENG] An electric furnace in which the heat is developed by the passage of current through a suitable internal resistance that may be the charge itself, a resistor embedded in the charge, or a resistor surrounding the charge. Also known as electric resistance furnace.

resistance grounding [ELEC] Electrical grounding in which lines are connected to ground by a resistive (totally dissipative) impedance.

resistance heating [ELEC] The generation of heat by electric conductors carrying current; degree of heating is proportional to the electrical resistance of the conductor; used in electrical home appliances, home or space heating, and heating ovens and furnaces.

resistance noise *See* thermal noise.

resistance pyrometer *See* resistance thermometer.

resistance seam welding [MET] Resistance welding process which produces a series of individual spot welds, overlapping spot welds, or a continuous nugget weld made by circular or wheel-type electrodes.

resistance spot welding [MET] Resistance welding process in which the parts are lapped and held in place under pressure; the size and shape of the electrodes (usually circular) control the size and shape of the welds.

resistance thermometer [ENG] A thermometer in which the sensing element is a resistor whose resistance is an accurately known function of temperature. Also known as electrical resistance thermometer; resistance pyrometer.

resist-dyeing [TEXT] A cross-dyeing method in which a chemical is applied to certain yarns before weaving so that, when the material is dyed, only the untreated yarns take the dye, producing a colorful pattern; the chemical is later removed from the fabric.

resisting moment [MECH] A moment produced by internal tensile and compressive forces that balances the external bending moment on a beam.

resistive coupling *See* resistance coupling.

resistive load [ELEC] A load whose total reactance is zero, so that the alternating current is in phase with the terminal voltage. Also known as nonreactive load.

resistivity *See* electrical resistivity.

resistivity index [PETRO ENG] Ratio of the true electrical resistivity of a rock system at a specified water saturation, to the resistivity of the rock itself; used for calculation of electrical well-logging data.

resistivity well logging [PETRO ENG] The measurement of subsurface electrical resistivities (normal and lateral to the borehole) during electrical logging of oil wells.

resistor [ELEC] A device designed to have a definite amount of resistance; used in circuits to limit current flow or to provide a voltage drop. Also known as electrical resistor.

resistor-transistor logic [ELECTR] One of the simplest logic circuits, having several resistors, a transistor, and a diode. Abbreviated RTL.

resolution [ELECTR] In television, the maximum number of lines that can be discerned on the screen at a distance equal to tube height; this ranges from 350 to 400 for most receivers. [ELECTROMAG] In radar, the minimum separation between two targets, in angle or range, at which they can be distinguished on a radar screen. Also known as resolving power. [PHYS] **1.** For a measurement of energy or momentum of a collection of particles, the difference between the highest and lowest energies at which the response of an instrument to a beam of monoenergetic particles is at least half its maximum value, divided by the energy of the particles. **2.** The procedure of breaking up a vectorial quantity into its components. [OPTICS] *See* resolving power. [SPECT] *See* resolving power.

resolution chart [COMMUN] *See* test pattern. [OPTICS] A device to test resolving power; usually alternate black and white lines of equal width arranged in groups of decreasing line width, identified as the number of line pairs per millimeter.

resolution error [COMPUT SCI] An error of an analog computing unit that results from its inability to respond to changes of less than a given magnitude.

resolution factor [COMPUT SCI] In information retrieval, the ratio obtained in dividing the total number of documents retrieved (whether relevant or not to the user's needs) by the total number of documents available in the file.

resolving power [ELECTROMAG] **1.** The reciprocal of the beam width of a unidirectional antenna, measured in degrees. **2.** *See* resolution. [OPTICS] A quantitative measure of the ability of an optical instrument to produce separable images of different points on an object; usually, the smallest angular or linear separation of two object points for which they may be resolved according to the Rayleigh criterion. Also known as resolution. [PHYS] A measure of the ability of a mass spectroscope to separate particles of different masses, equal to the ratio of the average mass of two particles whose mass spectrum lines can just be completely separated, to the difference in their masses. [SPECT] A measure of the ability of a spectroscope or interferometer to separate spectral lines of nearly equal wavelength, equal to the average wavelength of two equally strong spectral lines whose images can barely be separated, divided by the difference in wavelengths; for spectroscopes, the lines must be resolved according to the Rayleigh criterion; for interferometers, the wavelengths at which the lines have half of maximum intensity must be equal. Also known as resolution.

resonance [ELEC] A phenomenon exhibited by an alternating-current circuit in which there are relatively large currents near certain frequencies, and a relatively unimpeded oscillation of energy from a potential to a kinetic form; a special case of the physics definition. [PHYS] **1.** A phenomenon exhibited by a physical system acted upon by an external periodic driving force, in which the resulting amplitude of oscillation of the system becomes large when the frequency of the driving force approaches a natural free oscillation frequency of the system. **2.** In general, any phenomenon which is greatly enhanced at frequencies or energies that are at or very close to a given characteristic value. [QUANT MECH] *See* resonance absorption; resonance level.

resonance absorption [NUCLEO] The absorption of neutrons having a narrow range of energies corresponding to a nuclear resonance level of the absorber in a nuclear reactor. [QUANT MECH] The absorption of electromagnetic radiation by a quantum-mechanical system at a characteristic frequency satisfying the Bohr frequency condition. Also known as resonance.

resonance bridge [ELEC] A four-arm alternating-current bridge used to measure inductance, capacitance, or frequency; the inductor and the capacitor, which may be either in series or in parallel, are tuned to resonance at the frequency of the source before the bridge is balanced.

resonance capture [NUC PHYS] The combination of an incident particle and a nucleus in a resonance level of the resulting compound nucleus, characterized by having a large cross section at and very near the corresponding resonance energy.

resonance frequency [PHYS] A frequency at which some measure of the response of a physical system to an external periodic driving force is a maximum; three types are defined, namely, phase resonance, amplitude resonance, and natural resonance, but they are nearly equal when dissipative effects are small. Also known as resonant frequency. [QUANT MECH] A characteristic frequency, satisfying the Bohr frequency condition, at which a quantum-mechanical system absorbs radiation.

resonance hybrid [CHEM] A molecule that may be considered an intermediate between two or more valence bond structures.

resonance level [QUANT MECH] An unstable state of a compound system capable of being formed in a collision between two particles, and associated with a peak in a graph of cross section versus energy for the scattering of the particles. Also known as resonance.

resonance scattering [NUC PHYS] A peak in the cross section of a nucleus for elastic scattering of neutrons at energies near a resonance level, accompanied by an anomalous phase shift in the scattered neutrons.

resonance spectrum [SPECT] An emission spectrum resulting from illumination of a substance (usually a molecular gas) by radiation of a definite frequency or definite frequencies.

resonance transformer [ELEC] A high-voltage transformer in which the secondary circuit is tuned to the frequency of the power supply. [ELECTR] An electrostatic particle accelerator, used principally for acceleration of electrons, in which the high-voltage terminal oscillates between voltages which are equal in magnitude and opposite in sign.

resonance vibration [MECH] Forced vibration in which the frequency of the disturbing force is very close to the natural frequency of the system, so that the amplitude of vibration is very large.

resonant antenna [ELECTROMAG] An antenna for which there is a sharp peak in the power radiated or intercepted by the antenna at a certain frequency, at which electric currents in the antenna form a standing-wave pattern.

resonant cavity *See* cavity resonator.

resonant chamber *See* cavity resonator.

resonant-chamber switch [ELECTROMAG] Waveguide switch in which a tuned cavity in each waveguide branch serves the functions of switch contacts; detuning of a cavity blocks the flow of energy in the associated waveguide.

resonant circuit [ELEC] A circuit that contains inductance, capacitance, and resistance of such values as to give resonance at an operating frequency.

resonant coupling [ELEC] Coupling between two circuits that reaches a sharp peak at a certain frequency.

resonant element *See* cavity resonator.

resonant frequency *See* resonance frequency.

resonant line [ELECTROMAG] A transmission line having values of distributed inductance and distributed capacitance so as to make the line resonant at the frequency it is handling.

resonant-line tuner [ELECTR] A television tuner in which resonant lines are used to tune the antenna, radio-frequency amplifier, and radio-frequency oscillator circuits; tuning is achieved by moving shorting contacts that change the electrical lengths of the lines.

resonant scattering [QUANT MECH] Scattering of a photon by a quantum-mechanical system (usually an atom or nucleus) in which the system first absorbs the photon by undergoing a transition from one of its energy states to one of higher energy, and subsequently reemits the photon by the exact inverse transition.

resonating cavity [ELECTROMAG] Short piece of waveguide of adjustable length, terminated at either or both ends by a metal piston, an iris diaphragm, or some other wave-reflecting device; it is used as a filter, as a means of coupling between guides of different diameters, and as impedance networks corresponding to those used in radio circuits.

resonator [PHYS] A device that exhibits resonance at a particular frequency, such as an acoustic resonator or cavity resonator.

resorption [PETR] The process by which a magma redissolves previously crystallized minerals. [PHYS] Absorption or, less commonly, adsorption of material by a body or system from which the material was previously released.

respirator [ENG] A device for maintaining artificial respiration to protect the respiratory tract against irritating and poisonous gases, fumes, smoke, and dusts, with or without equipment supplying oxygen or air; some types have a fitting which covers the nose and mouth.

respiratory quotient [PHYSIO] The ratio of volumes of carbon dioxide evolved and oxygen consumed during a given period of respiration. Abbreviated RQ.

respiratory system [ANAT] The structures and passages involved with the intake, expulsion, and exchange of oxygen and carbon dioxide in the vertebrate body.

respiratory tree [ANAT] The trachea, bronchi, and bronchioles. [INV ZOO] Either of a pair of branched tubular appendages of the cloaca in certain holothurians that is thought to have a respiratory function.

responder [ELECTR] The transmitter section of a radar beacon.

response [COMMUN] *See* reply. [CONT SYS] A quantitative expression of the output of a device or system as a function of the input. Also known as system response. [STAT] The value of some measurable quantity after a treatment has been applied.

response time [COMPUT SCI] The delay experienced in time sharing between request and answer, a delay which increases when the number of users on the system increases. [CONT SYS] The time required for the output of a control system or element to reach a specified fraction of its new value after application of a step input or disturbance. [ELEC] The time it takes for the pointer of an electrical or electronic instrument to come to rest at a new value, after the quantity it measures has been abruptly changed.

rest energy [RELAT] The energy equivalent to the rest mass m_0 of a particle or body; that is, the quantity of m_0c^2, where c is the speed of light; often expressed in electronvolts.

resting frequency *See* carrier frequency.

resting potential [PHYSIO] The potential difference between the interior cytoplasm and the external aqueous medium of the living cell.

Restionaceae [BOT] A large family of monocotyledonous plants in the order Restionales characterized by unisexual flowers, wholly cauline leaves, unilocular anthers, and a more or less open inflorescence.

Restionales [BOT] An order of monocotyledonous plants in the subclass Commelinidae having reduced flowers and a single, pendulous, orthotropous ovule in each of the one to three locules of the ovary.

rest mass [RELAT] The mass of a particle in a Lorentz reference frame in which it is at rest.

restore [COMPUT SCI] To regenerate, to return a cycle index or variable address to its initial value, or to store again. [ELECTR] Periodic charge regeneration of volatile computer storage systems.

restorer pulses [ELECTR] In computers, pairs of complement pulses, applied to restore the coupling-capacitor charge in an alternating-current flip-flop.

restricted basin [GEOL] A depression in the floor of the ocean in which the water circulation is topographically restricted and therefore generally is oxygen-depleted. Also known as barred basin; silled basin.

restricted waters [NAV] Areas which for navigational reasons, such as the presence of shoals or other dangers, confine the movements of shipping within narrow limits.

resultant of forces [MECH] A system of at most a single force and a single couple whose external effects on a rigid body are identical with the effects of the several actual forces that act on that body.

resuscitation [MED] Restoration of consciousness or life functions after apparent death.

retainer wall [ENG] A wall, usually earthen, around a storage tank or an area of storage tanks (tank farm); used to hold (retain) liquid in place if one or more tanks begin to leak.

retaining wall [CIV ENG] A wall designed to maintain differences in ground elevations by holding back a bank of material.

retardant *See* retarder.

retardation [MED] Slow mental or physical functioning. [NAV] The amount of delay in time or phase angle introduced by the resistivity of the surface over which the radio wave in radio navigation is passing. [OCEANOGR] The amount of time by which corresponding tidal phases grow later day by day, averaging approximately 50 minutes. [OPTICS] In interference microscopy, the difference in optical path between the light passing through the specimen and the light bypassing the specimen. Also known as optical-path difference.

retardation coil [ELECTROMAG] A high-inductance coil used in telephone circuits to permit passage of direct current or low-frequency ringing current while blocking the flow of audio-frequency currents.

retarder [MATER] A material that inhibits the action of another substance, such as flameproofing agents or substances added to cement to retard setting time. Also known as retardant. [MECH ENG] 1. A braking device used to control the speed of railroad cars moving along the classification tracks in a hump yard. 2. A strip inserted in a tube of a fire-tube boiler to increase agitation of the hot gases flowing therein.

retarding-field oscillator [ELECTR] An oscillator employing an electron tube in which the electrons oscillate back and forth through a grid that is maintained positive with respect to both the cathode and anode; the field in the region of the grid exerts a retarding effect through the grid in either direction. Also known as positive-grid oscillator.

retention index [ANALY CHEM] In gas chromatography, the relationship of retention volume with arbitrarily assigned numbers to the compound being analyzed; used to indicate the volume retention behavior during analysis.

retention period [COMPUT SCI] The length of time that data must be kept on a reel of magnetic tape before it can be destroyed.

retention time [ANALY CHEM] In gas chromatography, the time at which the center, or maximum, of a symmetrical peak occurs on a gas chromatogram. [ELECTR] The maximum time between writing into a storage tube and obtaining an acceptable output by reading. Also known as storage time.

reticle [OPTICS] A series of intersecting fine lines, wires, or the like which are placed in the focus of the objective of an optical instrument to aid in measurement of angles or distances.

reticle image [OPTICS] A light image of the reticle in a computing gunsight or in certain types of optical gunsights and bombsights, cast on a reflector plate and superimposed on the target.

reticular cell *See* reticulocyte.

reticular formation [ANAT] The portion of the central nervous system which consists of small islands of gray matter separated by fine bundles of nerve fibers running in every direction.

Reticulariaceae [MYCOL] A family of plasmodial slime molds in the order Liceales.

reticular system *See* reticuloendothelial system.

reticulate [BIOL] Having or resembling a network of fibers, veins, or lines. [GEOL] 1. Referring to a vein or lode with netlike texture. 2. Referring to rock texture in which crystals are partly altered to a secondary material, forming a network that encloses the remnants of the original mineral. Also known as mesh texture; reticular; reticulated. [GEN] Of or relating to evolutionary change resulting from genetic recombination between strains in an interbreeding population.

reticulocyte [HISTOL] Also known as reticular cell. 1. A large, immature red blood cell, having a reticular appearance when stained due to retention of portions of the nucleus. 2. A cell of reticular tissue.

reticuloendothelial system [ANAT] The macrophage system, including all phagocytic cells such as histiocytes, macrophages, reticular cells, monocytes, and microglia, except the granular white blood cells. Abbreviated RES. Also known as hematopoietic system; reticular system.

reticulopodia [INV ZOO] Pseudopodia in the form of a branching network.

Reticulosa [PALEON] An order of Paleozoic hexactinellid sponges with a branching form in the subclass Hexasterophora.

Reticulum [ASTRON] A southern constellation, right ascension 4 hours, declination 60° south. Also known as Net.

retina [COMPUT SCI] In optical character recognition, a scanning device. [ANAT] The photoreceptive layer and terminal expansion of the optic nerve in the dorsal aspect of the vertebrate eye.

retinal pigment *See* rhodopsin.

retinene [BIOCHEM] A pigment extracted from the retina, which turns yellow by the action of light; the chief carotenoid of the retina.

retinite [MINERAL] A fossil resin, such as glessite, krantzite, muckite, and ambrite, composed of 6–15% oxygen, lacking succinic acid, and found in brown coals and peat.

retinitis pigmentosa [MED] A hereditary affection inherited as a sex-linked recessive and characterized by slowly progressing atrophy of the retinal nerve layers, and clumping of retinal pigment, followed by attenuation of the retinal arterioles and waxy atrophy of the optic disks.

retinol *See* vitamin A.

retort [CHEM ENG] 1. A closed refractory chamber in which coal is carbonized for manufacture of coal gas. 2. A vessel for the distillation or decomposition of a substance.

Retortamonadida [INV ZOO] An order of parasitic flagellate protozoans belonging to class Zoomastigophorea, having two or four flagella and a complex blepharoplast-centrosome-axostyle apparatus.

retrace *See* flyback.

retractor [ANAT] A muscle that draws a limb or other body part toward the body. [MED] A clawlike instrument for holding tissues away from the surgical field.

retreat [MIN ENG] Workings in the opposite direction of advance work which, when completed, will permit the area to be abandoned as finished.

retroaction *See* positive feedback.

retroactive refit *See* retrofit.

retrofit [ENG] A modification of equipment to incorporate changes made in later production of similar equipment; it may be done in the factory or field. Derived from retroactive refit.

retroflexion [ANAT] The state of being bent backward. [MED] A condition in which the uterus is bent backward on itself, producing a sharp angle in its longitudinal axis at the junction of the cervix and the fundus.

retrogradation [CHEM] 1. Generally, a process of deterioration; a reversal or retrogression to a simpler physical form. 2. A chemical reaction involving vegetable adhesives, which revert to a simpler molecular structure.

retrograde amnesia [MED] Loss of memory for events occurring prior to, but not after, the onset of a current disease or trauma.

retrograde metamorphism [PETR] Formation of metamorphic minerals of a lower grade of metamorphism at the expense of minerals which are characteristic of a higher grade. Also known as diaphthoresis; retrogressive metamorphism.

retrograde motion [ASTRON] 1. An apparent backward motion of a planet among the stars resulting from the observation of the planet from the planet earth which is also revolving about the sun at a different velocity. Also known as retrogression. 2. *See* retrograde orbit.

retrograde orbit [ASTRON] Motion in an orbit opposite to the usual orbital direction of celestial bodies within a given system; specifically, of a satellite, motion in a direction opposite to the direction of rotation of the primary. Also known as retrograde motion.

retrogression [ASTRON] *See* retrograde motion. [GEOL] *See* recession. [MED] Going backward, as in degeneration or atrophy of tissues. [METEOROL] The motion of an atmospheric wave or pressure system in a direction opposite to that of

the basic flow in which it is embedded. [PSYCH] Return to infantile behavior.

retrogressive metamorphism *See* retrograde metamorphism.

retroreflection [PHYS] Reflection wherein the reflected rays of radiation return along paths parallel to those of their corresponding incident rays.

retrorocket [AERO ENG] A rocket fitted on or in a spacecraft, satellite, or the like to produce thrust opposed to forward motion. Also known as braking rocket.

retting [CHEM ENG] Soaking vegetable stalks to decompose the gummy material and release the fibers.

return [BUILD] The continuation of a molding, projection, member, cornice, or the like, in a different direction, usually at a right angle. [COMPUT SCI] 1. To return control from a subroutine to the calling program. 2. To go back to a planned point in a computer program and rerun a portion of the program, usually when an error is detected; rerun points are usually not more than 5 minutes apart. [ELECTR] *See* echo. [GEOPHYS] Any of those surface waves on the record of a large earthquake which have traveled around the earth's surface by the long (greater than 180°) arc between epicenter and station, or which have passed the station and returned after traveling the entire circumference of the earth.

return jump [COMPUT SCI] A jump instruction in a subroutine which passes control to the first statement in the program which follows the instruction called the subroutine.

return loss [COMMUN] 1. The difference between the power incident upon a discontinuity in a transmission system and the power reflected from the discontinuity. 2. The ratio in decibels of the power incident upon a discontinuity to the power reflected from the discontinuity.

return streamer [GEOPHYS] The intensely luminous streamer which propagates upward from earth to cloud base in the last phase of each lightning stroke of a cloud-to-ground discharge. Also known as main stroke; return stroke.

return stroke *See* return streamer.

return trace *See* flyback.

retzian [MINERAL] $Mn_2Y(AsO_4)(OH)_4$ A chocolate brown to chestnut brown, orthorhombic mineral consisting of a basic arsenate of calcium, rare earths, and manganese.

reverberation [ACOUS] The prolongation of sound at a given point after direct reception from the source has ceased, due to such causes as reflections from bounding surfaces, scattering from inhomogeneities in a medium, and vibrations excited by the original sound.

reverberation chamber [ACOUS] An enclosure with heavy surfaces which randomly reflect as great an amount of sound as possible; used in acoustic measurements. Also known as random diffusion chamber.

reverberation time [ACOUS] The time in seconds required for the average sound-energy density at a given frequency to reduce to one-millionth of its initial steady-state value after the sound source has been stopped; this corresponds to a decrease of 60 decibels.

reverberatory furnace [ENG] A furnace in which heat is supplied by burning of fuel in a space between the charge and the low roof.

reversal film [GRAPHICS] A type of film designed to yield a positive image directly when reversal processed.

reversal of dip [GEOL] Change in the dip direction of bedding near a fault such that the beds curve toward the fault surface in a direction exactly opposite that of the drag folds. Also known as dip reversal.

reverse bias [ELECTR] A bias voltage applied to a diode or a semiconductor junction with polarity such that little or no current flows; the opposite of forward bias.

reverse-blocking tetrode thyristor *See* silicon controlled switch.

reverse-blocking triode thyristor *See* silicon controlled rectifier.

reverse bonded-phase chromatography [ANALY CHEM] A technique of bonded-phase chromatography in which the stationary phase is nonpolar and the mobile phase is polar.

reverse Brayton cycle [THERMO] A refrigeration cycle using air as the refrigerant but with all system pressures above the ambient. Also known as dense-air refrigeration cycle.

reverse Carnot cycle [THERMO] An ideal thermodynamic cycle consisting of the processes of the Carnot cycle reversed and in reverse order, namely, isentropic expansion, isothermal expansion, isentropic compression, and isothermal compression.

reverse code dictionary [COMPUT SCI] Alphabetic or alphanumeric arrangement of codes associated with their corresponding English words or terms.

reverse-current cleaning *See* anodic cleaning.

reverse-current protection [ELEC] A device which senses when there is a reversal in the normal direction of current in an electric power system, indicating an abnormal condition of the system, and which initiates appropriate action to prevent damage to the system.

reverse curve [MATH] An S-shaped curve, that is, one having two arcs with their centers on opposite sides of the curve. Also known as S curve.

reversed image [GRAPHICS] 1. A mirror image in which the right and left sides of the picture are interchanged. 2. *See* negative. [OPTICS] *See* inverted image.

reverse direction *See* inverse direction.

reversed-phase partition chromatography [ANALY CHEM] Paper chromatography in which the low-polarity phase (such as paraffin, paraffin jelly, or grease) is put onto the support (paper) and the high-polarity phase (such as water, acids, or organic solvents) is allowed to flow over it.

reverse fault *See* thrust fault.

reverse feedback *See* negative feedback.

reverse mutation [GEN] A mutation in a mutant allele which makes it capable of producing the nonmutant phenotype; may actually restore the original deoxyribonucleic-acid sequence of the gene or produce a new one which has a similar effect. Also known as back mutation.

reverse osmosis [CHEM ENG] A technique used in desalination and waste-water treatment; pressure is applied to the surface of a saline (or waste) solution, forcing pure water to pass from the solution through a membrane (hollow fibers of cellulose acetate or nylon) that will not pass sodium or chloride ions.

reverse Polish notation [COMPUT SCI] The version of Polish notation, used in some calculators, in which operators follow the operators with which they are associated. Abbreviated RPN. Also known as postfix notation; suffix notation.

reverse slip fault *See* thrust fault.

reversible chemical reaction [CHEM] A chemical reaction that can be made to proceed in either direction by suitable variations in the temperature, volume, pressure, or quantities of reactants or products.

reversible electrode [PHYS CHEM] An electrode that owes its potential to unit charges of a reversible nature, in contrast to electrodes used in electroplating and destroyed during their use.

reversible engine [THERMO] An ideal engine which carries out a cycle of reversible processes.

reversible process [THERMO] An ideal thermodynamic process which can be exactly reversed by making an indefinitely small change in the external conditions. Also known as quasi-static process.

reversing current [OCEANOGR] Any current that changes direction, with a period of slack water at each reversal of direction.

reversing motor [ELEC] A motor for which the direction of rotation can be reversed by changing electric connections or by other means while the motor is running at full speed; the motor will then come to a stop, reverse, and attain full speed in the opposite direction.

reversing thermometer [ENG] A mercury-in-glass thermometer which records temperature upon being inverted and thereafter retains its reading until returned to the first position.

revetment [CIV ENG] A facing made on a soil or rock embankment to prevent scour by weather or water. [ORD] A retaining wall with a facing such as concrete or stone, commonly used for fortifications or to protect against explosions.

revolution [GEOL] A little-used term to describe a time of profound crustal movements, on a continentwide or worldwide scale, which led to abrupt geographic, climatic, and environmental changes that were related to changes in forms of life. [MECH] The motion of a body around a closed orbit.

revolution per minute [MECH] A unit of angular velocity equal to the uniform angular velocity of a body which rotates through an angle of 360° (2π radians), so that every point in the body returns to its original position, in 1 minute. Abbreviated rpm.

Reynolds criterion [FL MECH] The principle that the type of fluid motion, that is, laminar flow or turbulent flow, in geometrically similar flow systems depends only on the Reynolds number; for example, in a pipe, laminar flow exists at Reynolds numbers less than 2000, turbulent flow at numbers above about 3000.

Reynolds stress [FL MECH] The net transfer of momentum across a surface in a turbulent fluid because of fluctuations in fluid velocity. Also known as eddy stress.

rf *See* radio frequency.

RFI *See* radio-frequency interference.

Rh *See* rhodium.

RH *See* relative humidity.

Rhabdiasoidea [INV ZOO] An order or superfamily of parasitic nematodes.

rhabdite [INV ZOO] A small rodlike or fusiform body secreted by epidermal or parenchymal cells of certain turbellarians and trematodes. [MINERAL] *See* schreibersite.

Rhabditia [INV ZOO] A subclass of nematodes in the class Secernentea.

Rhabditidia [INV ZOO] An order of nematodes in the subclass Rhabditia including parasites of man and domestic animals.

Rhabditoidea [INV ZOO] A superfamily of small to moderate-sized nematodes in the order Rhabditidia with small, porelike, anteriorly located amphids, and esophagus with corpus, isthmus, and valvulated basal bulb.

Rhabdocoela [INV ZOO] Formerly an order of the Turbellaria, and now divided into three orders, Catenulida, Macrostomida, and Neorhabdocoela.

rhabdome [INV ZOO] The central translucent cylinder in the retinula of a compound eye.

rhabdophane [MINERAL] $(Ce,Y,La,Di)(PO_4) \cdot H_2O$ A brown, pinkish, or yellowish-white mineral consisting of a hydrated phosphate of cerium, yttrium, and rare earths.

Rhabdophorina [INV ZOO] A suborder of ciliates in the order Gymnostomatida.

rhabdovirus [VIROL] A group of ribonucleic acid–containing animal viruses, including rabies virus and certain infective agents of fish and insects.

Rhachitomi [PALEON] A group of extinct amphibians in the order Temnospondyli in which pleurocentra were retained.

Rhacophoridae [VERT ZOO] A family of arboreal frogs in the suborder Diplasiocoela.

Rhacopilaceae [BOT] A family of mosses in the order Isobryales generally having dimorphous leaves with smaller dorsal leaves and a capsule that is plicate when dry.

Rhaetian [GEOL] A European stage of geologic time; the uppermost Triassic (above Norian, below Hettangian of Jurassic). Also known as Rhaetic.

Rhaetic See Rhaetian.

Rhagionidae [INV ZOO] The snipe flies, a family of predatory orthorrhaphous dipteran insects in the series Brachycera that are brownish or gray with spotted wings.

Rhamnaceae [BOT] A family of dicotyledonous plants in the order Rhamnales characterized by a solitary ovule in each locule, free stamens, simple leaves, and flowers that are hypogynous to perigynous or epigynous.

Rhamnales [BOT] An order of dicotyledonous plants in the subclass Rosidae having a single set of stamens, opposite the petals, usually a well-developed intrastamenal disk, and two or more locules in the ovary.

Rhamphorhynchoidea [PALEON] A Jurassic suborder of the Pterosauria characterized by long, slender tails with an expanded tip.

Rh antigen See Rh factor.

Rh blood group [IMMUNOL] The extensive, genetically determined system of red blood cell antigens defined by the immune serum of rabbits injected with rhesus monkey erythrocytes, or by human antiserums. Also known as rhesus blood group.

rhea [BOT] See ramie. [VERT ZOO] The common name for members of the avian order Rheiformes.

Rhea [ASTRON] A satellite of Saturn, estimated diameter is 1200 kilometers.

Rheidae [VERT ZOO] The single family of the avian order Rheiformes.

Rheiformes [VERT ZOO] The rheas, an order of South American running birds; called American ostriches, they differ from the true ostrich in their smaller size, feathered head and neck, three-toed feet, and other features.

Rhenanida [PALEON] An order of extinct marine fishes in the class Placodermi distinguished by mosaics of small bones between the large plates in the head shield.

rhenium [CHEM] A metallic element, symbol Re, atomic number 75, atomic weight 186.2; a transition element.

rheocasting [MET] A process in which a liquid metal is vigorously agitated during initial stages of solidification to produce a globular semisolid structure which remains highly fluid when more than 60% solidification has occurred.

rheology [MECH] The study of the deformation and flow of matter, especially non-Newtonian flow of liquids and plastic flow of solids.

rheopectic fluid [FL MECH] A fluid for which the structure builds up on shearing; this phenomenon is regarded as the reverse of thixotropy.

rheopexy [PHYS CHEM] A property of certain sols, having particles shaped like rods or plates, which set to gel form more quickly when mechanical means are used to hasten the orientation of the particles.

rheostat [ELEC] A resistor constructed so that its resistance value may be changed without interrupting the circuit to which it is connected. Also known as variable resistor.

rheostatic braking [ENG] A system of dynamic braking in which direct-current drive motors are used as generators and convert the kinetic energy of the motor rotor and connected load to electrical energy, which in turn is dissipated as heat in a braking rheostat connected to the armature.

rheostatic control [ELEC] A method of controlling the speed of electric motors that involves varying the resistance or reactance in the armature or field circuit; used in motors that drive elevators.

rheostriction See pinch effect.

rheotaxis [BIOL] Movement of a motile cell or organism in response to the direction of water currents.

rheotron See betatron.

rheotropism [BIOL] Orientation response of an organism to the stimulus of a flowing fluid, as water.

rhesus blood group See Rh blood group.

rhesus factor See Rh factor.

rhesus macaque See rhesus monkey.

rhesus monkey [VERT ZOO] *Macaque mulatta*. An agile, gregarious primate found in southern Asia and having a short tail, short limbs of almost equal length, and a stocky build. Also known as rhesus macaque.

rheumatic fever [MED] A febrile disease occurring in childhood as a delayed sequel of infection by *Streptococcus hemolyticus*, group A; characterized by arthritis, carditis, nosebleeds, and chorea.

rheumatism [MED] Any combination of muscle or joint pain, stiffness, or discomfort arising from nonspecific disorders.

rheumatoid arthritis [MED] A chronic systemic inflammatory disease of connective tissue in which symptoms and changes predominate in articular and related structures. Also known as atrophic arthritis; chronic infectious arthritis; proliferative arthritis.

Rh factor [IMMUNOL] Any of several red blood cell antigens originally identified in the blood of rhesus monkeys. Also known as Rh antigen; rhesus factor.

Rhigonematidae [INV ZOO] A family of nematodes in the superfamily Oxyuroidea.

Rhincodontidae [VERT ZOO] The whale sharks, a family of essentially tropical galeoid elasmobranchs in the isurid line.

rhinencephalon [ANAT] The anterior olfactory portion of the vertebrate brain.

rhinitis [MED] Inflammation of the mucous membranes in the nose.

Rhinobatidae [VERT ZOO] The guitarfishes, a family of elasmobranchs in the batoid group.

Rhinoceratidae [VERT ZOO] A family of perissodactyl mammals in the superfamily Rhinoceratoidea, comprising the living rhinoceroses.

Rhinoceratoidea [VERT ZOO] A superfamily of perissodactyl mammals in the suborder Ceratomorpha including living and extinct rhinoceroses.

rhinoceros [VERT ZOO] The common name for the odd-toed ungulates composing the family Rhinoceratidae, characterized by massive, thick-skinned limbs and bodies, and one or two horns which are composed of a solid mass of hairs attached to the bony prominence of the skull.

Rhinochimaeridae [VERT ZOO] A family of ratfishes, order Chimaeriformes, distinguished by an extremely elongate rostrum.

Rhinocryptidae [VERT ZOO] The tapaculos, a family of ground-inhabiting suboscine birds in the suborder Tyranni characterized by a large, movable flap which covers the nostrils.

Rhinolophidae [VERT ZOO] The horseshoe bats, a family of insect-eating chiropterans widely distributed in the Eastern Hemisphere and distinguished by extremely complex, horseshoe-shaped nose leaves.

Rhinopomatidae [VERT ZOO] The mouse-tailed bats, a small family of insectivorous chiropterans found chiefly in arid regions of northern Africa and southern Asia and characterized by long, wirelike tails and rudimentary nose leaves.

Rhinopteridae [VERT ZOO] The cow-nosed rays, a family of batoid sharks having a fleshy pad at the front end of the head and a well-developed poison spine.

Rhinotermitidae [INV ZOO] A family of lower termites of the order Isoptera.

rhinovirus [VIROL] A subgroup of the picornavirus group including small, ribonucleic acid–containing forms which are not inactivated by ether.

Rhipiceridae [INV ZOO] The cedar beetles, a family of coleopteran insects in the superfamily Elateroidea.

Rhipidistia [VERT ZOO] The equivalent name for Osteolepiformes.

Rhipiphoridae [INV ZOO] The wedge-shaped beetles, a family of coleopteran insects in the superfamily Meloidea.

rhizic water See soil water.

Rhizocephala [INV ZOO] An order of crustaceans which parasitize other crustaceans; adults have a thin-walled sac enclosing the visceral mass and show no trace of segmentation, appendages, or sense organs.

Rhizochloridina [INV ZOO] A suborder of flagellate protozoans in the order Heterochlorida.

Rhizodontidae [PALEON] An extinct family of lobe-fin fishes in the order Osteolepiformes.

rhizoid [BOT] A rootlike structure which helps to hold the plant to a substrate; found on fungi, liverworts, lichens, mosses, and ferns.

Rhizomastigida [INV ZOO] An order of the protozoan class Zoomastigophorea; all species are microscopic and ameboid, and have one or two flagella.

Rhizomastigina [INV ZOO] The equivalent name for Rhizomastigida.

rhizome [BOT] An underground horizontal stem, often thickened and tuber-shaped, and possessing buds, nodes, and scalelike leaves.

Rhizophagidae [INV ZOO] The root-eating beetles, a family of minute coleopteran insects in the superfamily Cucujoidea.

Rhizophoraceae [BOT] A family of dicolyledonous plants in the order Cornales distinguished by opposite, stipulate leaves, two ovules per locule, folded or convolute bud petals, and a berry fruit.

rhizopod [INV ZOO] An anastomosing rootlike pseudopodium.

Rhizopodea [INV ZOO] A class of the protozoan superclass Sarcodina in which pseudopodia may be filopodia, lobopodia, or reticulopodia, or may be absent.

rhizosphere [GEOL] The soil region subject to the influence of plant roots and characterized by a zone of increased microbiological activity.

Rhizostomeae [INV ZOO] An order of the class Scyphozoa having the umbrella generally higher than it is wide with the margin divided into many lappets but not provided with tentacles.

Rhodesian man [PALEON] A type of fossil man inhabiting southern and central Africa during the late Pleistocene; the skull was large and low, marked by massive browridges, with a cranial capacity of 1300 cubic centimeters or less.

Rhodininae [INV ZOO] A subfamily of limivorous worms in the family Maldanidae.

rhodium [CHEM] A chemical element, symbol Rh, atomic number 45, atomic weight 102.905. [MET] A silver-white metal in the platinum family; sometimes alloyed with platinum for thermocouples or used as a tarnish-resistant electrode posit.

rhodizite [MINERAL] $CsAl_4Be_4B_{11}O_{25}(OH)_4$ A white mineral composed of a basic borate of cesium, aluminum, and beryllium, occurring as isometric crystals.

rhodochrosite [MINERAL] $MnCO_3$ A rose-red to pink or gray mineral form of manganese carbonate with hexagonal symmetry but occurring in massive or columnar form; isomorphous with calcite and siderite, has a hardness of 3.5–4 on Mohs scale, and a specific gravity of 3.7; a minor ore of manganese.

rhodolite [MINERAL] A violet-red garnet species composed of a mixture of almandite and pyrope in about a 3:1 ratio.

rhodonite [MINERAL] $MnSiO_3$ A pink or brown mineral inosilicate crystallizing in the triclinic system and commonly found in cleavable to compact masses or in embedded grains; luster is vitreous, hardness is 5.5–6 on Mohs scale, and specific gravity is 3.4–3.7.

Rhodophyceae [BOT] A class of algae belonging to the division or subphylum Rhodophyta.

Rhodophyta [BOT] The red algae, a large diverse phylum or division of plants distinguished by having an abundance of the pigment phycoerythrin.

rhodopsin [BIOCHEM] A deep-red photosensitive pigment contained in the rods of the retina of marine fishes and most higher vertebrates. Also known as retinal pigment; visual purple.

Rhoipteleaceae [BOT] A monotypic family of dicotyledonous plants in the order Juglandales having pinnately compound leaves, and flowers in triplets with four sepals and six stamens, and the lateral flowers female but sterile.

rhomb See rhombohedron.

rhombencephalon [EMBRYO] The most caudal of the primary brain vesicles in the vertebrate embryo. Also known as hindbrain.

rhombic antenna [ELECTROMAG] A horizontal antenna having four conductors forming a diamond or rhombus; usually fed at one apex and terminated with a resistance or impedance at the opposite apex. Also known as diamond antenna.

rhombic lattice See orthorhombic lattice.

rhombic sulfur [CHEM] Crystalline sulfur with three unequal axes, all at right angles.

rhombic system See orthorhombic system.

Rhombifera [PALEON] An extinct order of Cystoidea in which the thecal canals crossed the sutures at the edges of the plates, so that one-half of any canal lay in one plate and the other half on an adjoining plate.

rhomboclase [MINERAL] $HFe^{3+}(SO_4)_2 \cdot 4H_2O$ A colorless mineral composed of hydrous acid ferric sulfate, occurring in rhombic plates.

rhombohedral [CRYSTAL] **1.** Of or pertaining to the rhombohedral system. **2.** Of or pertaining to crystal cleavage in or a centered lattice of the hexagonal system.

rhombohedral iron ore *See* hematite; siderite.

rhombohedral system [CRYSTAL] A division of the trigonal crystal system in which the rhombohedron is the basic unit cell.

rhombohedron [CRYSTAL] A trigonal crystal form that is a parallelepiped, the six identical faces being rhombs. Also known as rhomb. [MATH] A prism with six parallelogram faces.

rhomboid [MATH] A parallelogram whose adjacent sides are not equal.

rho meson [PARTIC PHYS] Collective name for vector meson resonances belonging to a charge multiplet with total isospin 1, hypercharge 0, negative charge conjugation parity, positive g-parity, mass of about 770 MeV, and width of about 146 MeV. Designated $\rho(770)$.

Rhopalidae [INV ZOO] A family of pentatomorphan hemipteran insects in the superfamily Coreoidea.

Rhopalodinidae [INV ZOO] A family of holothurian echinoderms in the order Dactylochirotida in which the body is flask-shaped, the mouth and anus lying together.

Rhopalosomatidae [INV ZOO] A family of hymenopteran insects in the superfamily Scolioidea.

rhubarb [BOT] *Rheum rhaponticum.* A herbaceous perennial of the order Polygoniales grown for its thick, edible petioles.

rhumbatron *See* cavity resonator.

rhumb bearing [NAV] The direction of a rhumb line through two terrestrial points, expressed as angular distance from a reference direction; usually measured from 000° at the reference direction clockwise through 360°. Also known as Mercator bearing.

rhumb line [MAP] A line on the surface of the earth making the same oblique angle with all meridians. Also known as loxodrome.

rhyacolite *See* sanidine.

Rhynchobdellae [INV ZOO] An order of the class Hirudinea comprising leeches that possess an eversible proboscis and lack hemoglobin in the blood.

Rhynchocephalia [VERT ZOO] An order of lepidosaurian reptiles represented by a single living species, *Sphenodon punctatus,* and characterized by a diapsid skull, teeth fused to the edges of the jaws, and an overhanging beak formed by the upper jaw.

rhynchocoel [INV ZOO] A cavity that holds the inverted proboscis in nemertinean worms.

Rhynchocoela [INV ZOO] A phylum of bilaterally symmetrical, unsegmented, ribbonlike worms having an eversible proboscis and a complete digestive tract with an anus.

Rhynchodina [INV ZOO] A suborder of ciliate protozoans in the order Thigmotrichida.

Rhynchonellida [INV ZOO] An order of articulate brachiopods; typical forms are dorsibiconvex, the posterior margin is curved, the dorsal interarea is absent, and the ventral one greatly reduced.

Rhynchosauridae [PALEON] An extinct family of generally large, stout, herbivorous lepidosaurian reptiles in the order Rhynchocephalea.

Rhynchotheriinae [PALEON] A subfamily of extinct elaphantoid mammals in the family Gomphotheriidae comprising the beak-jawed mastodonts.

Rhyniatae *See* Rhyniopsida

Rhyniophyta [PALEOBOT] A subkingdom of the Embryobionta including the relatively simple, uppermost Silurian-Devonian vascular plants.

Rhyniopsida [PALEOBOT] A class of extinct plants in the subkingdom Rhyniophyta characterized by leafless, usually dichotomously branched stems that bore terminal sporangia.

Rhynochetidae [VERT ZOO] A monotypic family of gruiform birds containing only the kagu of New Caledonia.

rhyodacite [PETR] A group of extrusive porphyritic igneous rocks containing quartz, plagioclase, and biotite phenocrysts in a fine-grained to glassy groundmass composed of alkali feldspar and silica minerals. Also known as dellenite; quartz lattite.

rhyolite [PETR] A light-colored, aphanitic volcanic rock composed largely of alkali feldspar and free silica with minor amounts of mafic minerals; the extrusive equivalent of granite.

rhyolitic magma [PETR] A type of magma formed by differentiation from basaltic magma in combination with assimilation of siliceous material, or by melting of portions of the earth's sialic layer.

Rhysodidae [INV ZOO] The wrinkled bark beetles, a family of coleopteran insects in the suborder Adephaga.

ria [GEOGR] **1.** Any broad, estuarine river mouth. **2.** A long, narrow coastal inlet, except a fjord, whose depth and width gradually and uniformly diminish inland.

rib [AERO ENG] A transverse structural member that gives cross-sectional shape and strength to a portion of an airfoil. [ANAT] One of the long curved bones forming the wall of the thorax in vertebrates. [BOT] A primary vein in a leaf. [GEOL] A layer or dike of rock forming a small ridge on a steep mountainside. [MIN ENG] **1.** A solid pillar of coal or ore left for support. **2.** A thin stratum in a seam of coal. [TEXT] A straight, raised cord in a fabric, formed by a heavy thread in any direction.

rib arch [CIV ENG] An arch consisting of ribs placed side by side and extending from the springings on one end to those on the other end.

ribbon cable [ELEC] A cable made of normal, round, insulated wires arranged side by side and bonded together by a cohesion process to form a flexible ribbon.

ribbon conveyor [MECH ENG] A type of screw conveyor which has an open space between the shaft and a ribbon-shaped flight, used for wet or sticky materials which would otherwise build up on the spindle.

ribbon lightning [GEOPHYS] Ordinary streak lightning that appears to be spread horizontally into a ribbon of parallel luminous streaks when a very strong wind is blowing at right angles to the observer's line of sight; successive strokes of the lightning flash are then displaced by small angular amounts and may appear to the eye or camera as distinct paths. Also known as band lightning; fillet lightning.

ribbon microphone [ENG ACOUS] A microphone whose electric output results from the motion of a thin metal ribbon mounted between the poles of a permanent magnet and driven directly by sound waves; it is velocity-actuated if open to sound waves on both sides, and pressure-actuated if open to sound waves on only one side.

riboflavin [BIOCHEM] $C_{17}H_{20}N_4O_6$ A water-soluble, yellow-orange fluorescent pigment that is essential to human nutrition as a component of the coenzymes flavin mononucleotide and flavin adenine dinucleotide. Also known as lactoflavin; vitamin B_2; vitamin G.

ribonuclease [BIOCHEM] $C_{587}H_{909}N_{171}O_{197}S_{12}$ An enzyme that catalyzes the depolymerization of ribonucleic acid.

ribonucleic acid [BIOCHEM] A long-chain, usually single-stranded nucleic acid consisting of repeating nucleotide units containing four kinds of heterocyclic, organic bases: adenine, cytosine, quanine, and uracil; they are conjugated to the pentose sugar ribose and held in sequence by phosphodiester bonds; involved intracellularly in protein synthesis. Abbreviated RNA.

ribonucleoprotein [BIOCHEM] Any of a large group of conjugated proteins in which molecules of ribonucleic acid are closely associated with molecules of protein.

ribose [BIOCHEM] $C_5H_{10}O_5$ A pentose sugar occurring as a component of various nucleotides, including ribonucleic acid.

ribosome [CYTOL] One of the small, complex particles composed of various proteins and three molecules of ribonucleic acid which synthesize proteins within the living cell.

Ricci tensor See contracted curvature tensor.

rice [BOT] *Oryza sativa.* An annual cereal grass plant of the order Cyperales, cultivated as a source of human food for its carbohydrate-rich grain.

Richardson effect See thermionic emission.

richellite [MINERAL] $Ca_3Fe_{10}(PO_4)_8(OH,F)_{12} \cdot nH_2O$ A yellow mineral composed of hydrous basic iron calcium fluophosphate; occurs in masses.

Richmondian [GEOL] A North American stage of geologic time: Upper Ordovician (above Maysvillian, below Lower Silurian).

richterite [MINERAL] $(Na,K)_2(Mg,Mn,Ca)_6Si_8O_{22}$-$(OH)_2$ A brown, yellow, or rose-red monoclinic mineral composed of basic silicate of sodium, potassium, magnesium, manganese, and calcium; a member of the amphibole group.

Richter scale [GEOPHYS] A scale of numerical values of earthquake magnitude ranging from 1 to 9.

Ricinidae [INV ZOO] A family of bird lice, order Mallophaga, which occur on numerous land and water birds.

Ricinuleida [INV ZOO] An order of rare, ticklike arachnids in which the two anterior pairs of appendages are chelate, and the terminal segments of the third legs of the male are modified as copulatory structures.

rickardite [MINERAL] Cu_4Te_3 A deep-purple mineral composed of copper telluride, occurring in masses.

rickets [MED] A disorder of calcium and phosphorus metabolism affecting bony structures, due to vitamin D deficiency.

Rickettsiaceae [MICROBIO] A family of the order Rickettsiales; small, rod-shaped, coccoid, or diplococcoid cells often found in arthropods; includes human and animal parasites and pathogens.

Rickettsiales [MICROBIO] An order of prokaryotic microorganisms; gram-negative, obligate, intracellular animal parasites (may be grown in tissue cultures); many cause disease in humans and animals.

rickettsiosis [MED] Any disease caused by rickettsiae.

ridge [ARCH] The line on which the sides of a sloping roof meet. [GEOL] An elongate, narrow, steep-sided elevation of the earth's surface or the ocean floor. [METEOROL] An elongated area of relatively high atmospheric pressure, almost always associated with, and most clearly identified as, an area of

maximum anticyclonic curvature of wind flow. Also known as wedge.

ridge board [BUILD] A horizontal board placed on edge at the apex of the roof.

ridge pole [BUILD] The horizontal supporting member placed along the ridge of a roof.

riebeckite [MINERAL] $Na_2(Fe,Mg)_5Si_8O_{22}(OH)_2$ A blue or black monoclinic amphibole occurring as a primary constituent in some acid- or sodium-rich igneous rocks.

riegel [GEOL] A low, traverse ridge of bedrock on the floor of a glacial valley. Also known as rock bar; threshold; verrou.

Riemann-Christoffel tensor [MATH] The basic tensor used for the study of curvature of a Riemann space; it is a fourth-rank tensor, formed from Christoffel symbols and their derivatives, and its vanishing is a necessary condition for the space to be flat. Also known as curvature tensor.

Riemannian geometry See elliptic geometry.

Riemannian manifold [MATH] A differentiable manifold where the tangent vectors about each point have an inner product so defined as to allow a generalized study of distance and orthogonality.

Riemann integral [MATH] The Riemann integral of a real function $f(x)$ on an interval (a,b) is the unique limit (when it exists) of the sum of $f(a_i)(x_i - x_{i-1})$, $i = 1, ..., n$, taken over all partitions of (a,b), $a = x_0 < a_1 < x_1 < ... < a_n < x_n = b$, as the maximum distance between x_i and x_{i-1} tends to zero.

Riemann space [MATH] A Riemannian manifold or subset of a euclidean space where tensors can be defined to allow a general study of distance, angle, and curvature.

Riemann surfaces [MATH] Sheets or surfaces obtained by analyzing multiple-valued complex functions and the various choices of principal branches.

Riemann tensors [MATH] Various types of tensors used in the study of curvature for a Riemann space.

rift [GEOL] **1.** A narrow opening in a rock caused by cracking or splitting. **2.** A high, narrow passage in a cave.

rift valley [GEOL] A deep, central cleft with a mountainous floor in the crest of a midoceanic ridge. Also known as central valley; midocean rift.

right ascension [ASTRON] A celestial coordinate; the angular distance taken along the celestial equator from the vernal equinox eastward to the hour circle of a given celestial body. Abbreviated R.A.

right circular cylinder [MATH] A solid bounded by two parallel planes and by a cylindrical surface consisting of the straight lines perpendicular to the planes and passing through a circle in one of them.

right-handed [CRYSTAL] Having a crystal structure with a mirror-image relationship to a left-handed structure. [DES ENG] **1.** Pertaining to screw threads that allow coupling only by turning in a clockwise direction. **2.** See right-laid.

right-hand limit See limit from the right.

right-hand rule [ELECTROMAG] **1.** For a current-carrying wire, the rule that if the fingers of the right hand are placed around the wire so that the thumb points in the direction of current flow, the fingers will be pointing in the direction of the magnetic field produced by the wire. **2.** For a moving wire in a magnetic field, such as the wire on the armature of a generator, if the thumb, first, and second fingers of the right hand are extended at right angles to one another, with the first finger representing the direction of magnetic lines of force and the second finger representing the direction of current flow induced by the wire's motion, the thumb will be pointing in the direction of motion of the wire. Also known as Fleming's rule.

right-hand screw [DES ENG] A screw that advances when turned clockwise.

right-laid [DES ENG] Rope or cable construction in which strands are twisted counterclockwise. Also known as right-handed.

right triangle [MATH] A triangle one of whose angles is a right angle.

rigid body [MECH] An idealized extended solid whose size and shape are definitely fixed and remain unaltered when forces are applied.

rigid coupling [MECH ENG] A mechanical fastening of shafts connected with the axes directly in line.

rigidity [MECH] The quality or state of resisting change in form.

rigid pavement [CIV ENG] A thick portland cement pavement on a gravel base and subbase, with steel reinforcement and often with transverse joints.

rigid rotor [PL PHYS] An ensemble of electrons moving in a circular or nearly circular orbit at a constant angular frequency.

Rigil Kent *See* α Centauri.

rill [ASTRON] A crooked, narrow crack on the moon's surface; may be a kilometer or more in width and a few to several hundred kilometers in length. Also spelled rille. [GEOL] A small, transient runnel. [HYD] A small brook or stream.

rille *See* rill.

rill erosion [GEOL] The formation of numerous, closely spaced rills due to the uneven removal of surface soil by streamlets of running water. Also known as rilling; rill wash; rillwork.

rilling *See* rill erosion.

rill wash *See* rill erosion.

rillwork *See* rill erosion.

rim blight [PL PATH] A fungus disease of tea caused by members of the genus *Cladosporium* and characterized by yellow discoloration of the leaf margins followed by browning.

rim clutch [MECH ENG] A frictional contact clutch having surface elements that apply pressure to the rim either externally or internally.

rime [HYD] A white or milky and opaque granular deposit of ice formed by the rapid freezing of supercooled water drops as they impinge upon an exposed object; composed essentially of discrete ice granules, and has densities as low as 0.2–0.3 gram per cubic centimeter.

rime fog *See* ice fog.

rimmed steel [MET] Low-carbon steel, partially deoxidized, which on cooling continuously, evolves sufficient carbon monoxide to form a case or rim of metal virtually free of voids.

rimrock [GEOL] A top layer of resistant rock on a plateau outcropping with vertical or near vertical walls.

R indicator *See* R scope.

ring [COMPUT SCI] A cyclic arrangement of data elements, usually including a specified entry pointer. [DES ENG] A tie member or chain link; tension or compression applied through the center of the ring produces bending moment, shear, and normal force on radial sections. [MATH] **1.** An algebraic system with two operations called multiplication and addition; the system is a commutative group relative to addition, and multiplication is associative, and is distributive with respect to addition. **2.** A ring of sets is a collection of sets where the union and difference of any two members is also a member.

ring canal [INV ZOO] In echinoderms, the circular tube of the water-vascular system that surrounds the esophagus.

ring current [GEOPHYS] A westward electric current which is believed to circle the earth at an altitude of several earth radii during the main phase of geomagnetic storms, resulting in a large worldwide decrease in the geomagnetic field horizontal component at low latitudes.

ring deoxyribonucleic acid *See* circular deoxyribonucleic acid.

Ringer's solution [CHEM] A solution of 0.86 gram sodium chloride, 0.03 gram potassium chloride, and 0.033 gram calcium chloride in boiled, purified water, used topically as a physiological salt solution.

ring gage [DES ENG] A cylindrical ring of steel whose inside diameter is finished to gage tolerance and is used for checking the external diameter of a cylindrical object.

ring galaxy [ASTRON] A class of galaxy whose ringlike structure has clumps of ionized hydrogen clouds on its periphery, may have a nucleus of stars, and is usually accompanied by a small galaxy; probably formed when a small galaxy crashes through the disk of a spiral galaxy.

ring gate [CIV ENG] A type of gate used to regulate and control the discharge of a morning-glory spillway; like a drum gate, it offers a minimum of interference to the passage of ice or drift over the gate and requires no external power for operation.

ring gear [MECH ENG] The ring-shaped gear in an automobile differential that is driven by the propeller shaft pinion and transmits power through the differential to the line axle.

ring isomerism [ORG CHEM] A type of geometrical isomerism in which bond lengths and bond angles prevent the existence of the trans structure if substituents are attached to alkenic carbons which are part of a cyclic system, the ring of which contains fewer than eight members; for example, 1,2-dichlorocyclohexene.

ring network [COMMUN] A communications network in which the nodes can be considered to be on a circle, about which messages must be routed. Also known as loop network.

ring-roller mill [MECH ENG] A grinding mill in which material is fed past spring-loaded rollers that apply force against the sides of a revolving bowl. Also known as roller mill.

ring rot [PL PATH] **1.** A fungus disease of the sweet potato root caused by *Rhizopus stolonifer* and marked by rings of dry rot. **2.** A bacterial disease of potatoes caused by *Corynebacterium sepedonicum* and characterized by brown discoloration of the annular vascular tissue.

ring shift *See* cyclic shift.

ring silicate *See* cyclosilicate.

ring spot [PL PATH] Any of various virus and fungus diseases of plants characterized by the appearance of a discolored, annular lesion.

ring structure [COMPUT SCI] A chained file organization such that the end of the chain points to its beginning. [GEOL] A formation on the surface of the earth, moon, or a planet, having a ring-shaped trace in plan.

ring system [ORG CHEM] Arbitrary designation of certain compounds as close, circular structures, as in the six-carbon benzene ring; common rings have four, five, and six members, either carbon or some combination of carbon, nitrogen, oxygen, sulfur, or other elements.

ringtail *See* cacomistle.

ring vessel [INV ZOO] A part of the water-vascular system in echinoderms; it is the circular canal around the mouth into which the stone canal empties, and from which a radial water vessel traverses to each of five radii.

ringworm [MED] A fungus infection of skin, hair, or nails producing annular lesions with elevated margins. Also known as tinea.

rinkite *See* mosandrite.

rinkolite *See* mosandrite.

rinneite [MINERAL] NaK_3FeCl_6 A colorless, pink, violet, or yellow mineral composed of sodium potassium iron chloride, occurring in granular masses.

Riodininae [INV ZOO] A subfamily of the lepidopteran family Lycaenidae in which prothoracic legs are nonfunctional in the male.

rip current [OCEANOGR] The return flow of water piled up on shore by incoming waves and wind.

ripe [BOT] Of fruit, fully developed, having mature seed and so usable as food. [FOR] Of timber or a forest, ready to be cut. [GEOL] Referring to peat, in an advanced state of decay. [HYD] Descriptive of snow that is in a condition to discharge meltwater; ripe snow usually has a coarse crystalline structure, a snow density near 0.5, and a temperature near 32°F (0°C).

ripidolite [MINERAL] $(Mg,Fe^{2+})_9Al_6Si_5O_{20}(OH)_{16}$ A mineral of the chlorite group; consists of basic magnesium iron aluminum silicate. Also known as aphrosiderite.

ripple [ELEC] The alternating-current component in the output of a direct-current power supply, arising within the power supply from incomplete filtering or from commutator action in a dc generator. [FL MECH] *See* capillary wave. [GEOL] A very small ridge of sand resembling or suggesting a ripple of water and formed on the bedding surface of a sediment. [OCEANOGR] A small curling or undulating wave controlled to a significant degree by both surface tension and gravity.

ripple filter [ELECTR] A low-pass filter designed to reduce ripple while freely passing the direct current obtained from a rectifier or direct-current generator. Also known as smoothing circuit; smoothing filter.

ripple mark [GEOL] 1. A surface pattern on incoherent sedimentary material, especially loose sand, consisting of alternating ridges and hollows formed by wind or water action. 2. One of the ridges on a ripple-marked surface.

ripple tank [PHYS] A shallow tray containing a liquid and equipped with means for generating surface waves; used to illustrate several types of wave phenomena, such as interference and diffraction.

ripple voltage [ELEC] The alternating component of the unidirectional voltage from a rectifier or generator used as a source of direct-current power.

rips [OCEANOGR] A turbulent agitation of water, generally caused by the interaction of currents and wind; in nearshore regions they may be currents flowing swiftly over an irregular bottom; sometimes referred to erroneously as tide rips.

ripsaw [MECH ENG] A heavy-tooth power saw used for cutting wood with the grain.

rise [ASTRON] Of a celestial body, to cross the visible horizon while ascending. [GEOL] A long, broad elevation which rises gently from its surroundings, such as the sea floor. [HYD] *See* resurgence. [SCI TECH] The increase in the height or the value of something, such as a rise of tide or a rise of temperature.

riser [CHEM ENG] That portion of a bubble-cap assembly in a distillation tower that channels the rising vapor and causes it to flow downward to pass through the liquid held on the bubble plate. [CIV ENG] **1.** A board placed vertically beneath the tread of a step in a staircase. **2.** A vertical steam, water, or gas pipe. [GEOL] A steplike topographic feature, such as a steep slope between terraces. [MET] *See* feedhead. [PETRO ENG] In an offshore drilling facility, a system of piping extending from the hole and terminating at the rig. [TEXT] A raised spot in a weaving pattern where the warp traverses the weft or the filling.

rising tide [OCEANOGR] The portion of the tide cycle between low water and the following high water.

Riss [GEOL] **1.** A European stage of geologic time: Pleistocene (above Mindel, below Würm). **2.** The third stage of glaciation of the Pleistocene in the Alps.

Rissoacea [PALEON] An extinct superfamily of gastropod mollusks.

Riss-Würm [GEOL] The third interglacial stage of the Pleistocene in the Alps, following the Riss glaciation and preceding the Würm glaciation.

Ritchie wedge [OPTICS] A photometer in which a test source and a standard source of light illuminate two perpendicular white, diffusing surfaces which intersect in a movable wedge, and these surfaces are viewed from a direction perpendicular to a line connecting the sources.

Ritz's combination principle [SPECT] The empirical rule that sums and differences of the frequencies of spectral lines often equal other observed frequencies. Also known as combination principle.

river [HYD] A large, natural freshwater surface stream having a permanent or seasonal flow and moving toward a sea, lake, or another river in a definite channel. [LAP] A pure-white diamond of very high grade.

riverbed [GEOL] The channel which contains, or formerly contained, a river.

river engineering [CIV ENG] A branch of transportation engineering consisting of the physical measures which are taken to improve a river and its banks.

river gage [ENG] A device for measuring the river stage; types in common use include the staff gage, the water-stage recorder, and wire-weight gage. Also known as stream gage.

River Po *See* Eridanus.

river terrace *See* stream terrace.

rivet [DES ENG] A short rod with a head formed on one end; it is inserted through aligned holes in parts to be joined, and the protruding end is pressed or hammered to form a second head.

RJE *See* remote job entry.

R meson [PARTIC PHYS] A meson resonance observed in missing mass experiments involving scattering of negative pions by protons; it has been resolved into three peaks labeled R_1, R_2, and R_3, having masses of 1640, 1700, and 1750 MeV respectively, and is believed to have a spin of 3.

rms value *See* root-mean-square value.

Rn *See* radon.

RNA *See* ribonucleic acid.

roach [INV ZOO] An insect of the family Blattidae; the body is wide and flat, the anterior part of the thorax projects over the head, and antennae are long and filiform, with many segments. Also known as cockroach.

roadbed [CIV ENG] The earth foundation of a highway or a railroad.

roast [MET] To heat ore to effect some chemical change that will facilitate smelting.

roast sintering *See* blast roasting.

Robertinacea [INV ZOO] A superfamily of marine, benthic foraminiferans in the suborder Rotaliina characterized by a trochospiral or modified test with a wall of radial aragonite, and having bilamellar septa.

Roberts evaporator *See* short-tube vertical evaporator.

robot [CONT SYS] A completely self-controlled electronic, electric, or mechanical device.

Roccilaceae [BOT] A family of fruticose species of Hysterales that grow profusely on trees and rocks along the coastlines of Portugal, California, and western South America.

Rochalimaea [MICROBIO] A genus of the tribe Rickettsieae; short rods usually found as extracellular parasites in arthropods.

Roche lobes [MECH] **1.** Regions of space surrounding two massive bodies revolving around each other under their mutual gravitational attraction, such that the gravitational attraction of each body dominates the lobe surrounding it. **2.** In particular, the effective potential energy (referred to a system of coordinates rotating with the bodies) is equal to a constant V_0 over the surface of the lobes, and if a particle is inside one of the lobes and if the sum of its effective potential energy and its kinetic energy is less than V_0, it will remain inside the lobe.

Roche's limak [ASTROPHYS] The limiting distance below which a satellite orbiting a celestial body would be disrupted by the tidal forces generated by the gravitational attraction of the primary; the distance depends on the relative densities of the bodies and on the orbit of the satellite; it is computed by $R = 2.45(Lr)$, where L is a factor that depends on the relative densities of the satellite and the body, R is the radius of the satellite's orbit measured from the center of the primary body, and r is the radius of the primary body; if the satellite and the body have the same density, the relationship is $R = 2.45r$.

rock [PETR] **1.** A consolidated or unconsolidated aggregate of mineral grains consisting of one or more mineral species and having some degree of chemical and mineralogic constancy. **2.** In the popular sense, a hard, compact material with some coherence, derived from the earth.

rock bar *See* riegel.

rock-basin lake *See* paternoster lake.

rockbolt [ENG] A bar, usually constructed of steel, which is inserted into predrilled holes in rock and secured for the purpose of ground control.

rockburst [MIN ENG] A sudden and violent rock failure around a mining excavation on a sufficiently large scale to be considered a hazard.

rock cleavage [PETR] The capacity of a rock to split along certain parallel surfaces more easily than along others.

rock creep [GEOL] A form of slow flowage in rock materials evident in the downhill bending of layers of bedded or foliated rock and in the slow downslope migration of large blocks of rock away from their parent outcrop.

rock crystal [MINERAL] A transparent, colorless form of quartz with low brilliance; used for lenses, wedges, and prisms in optical instruments. Also known as berg crystal; crystal; mountain crystal; pebble; quartz crystal.

rock cycle [GEOL] The interrelated sequence of events by which rocks are initially formed, altered, destroyed, and reformed as a result of magmatism, erosion, sedimentation, and metamorphism.

rocker [CIV ENG] A support at the end of a truss or girder which permits rotation and horizontal movement to allow for expansion and contraction.

[GRAPHICS] A type of chisel, with a sharp beveled edge, used in mezzotint engraving; it is set on the surface of a copper plate and rocked back and forth to produce a rough grain, which, when printed, produces a velvety black. [MIN ENG] A small digging bucket mounted on two rocker arms in which auriferous alluvial sands are agitated by oscillation, in water, to collect gold. [ORD] Movable, built-in support in a field gun carriage, between the trail and the cradle, that allows changes in elevation to be made without disturbing the angle of position setting.

rocker arm [MECH ENG] In an internal combustion engine, a lever that is pivoted near its center and operated by a pushrod at one end to raise and depress the valve stem at the other end.

rocket [AERO ENG] **1.** Any kind of jet propulsion capable of operating independently of the atmosphere. **2.** A complete vehicle driven by such a propulsive system.

rocket astronomy [ASTRON] The discipline comprising measurements of the electromagnetic radiation from the sun, planets, stars, and other bodies, of wavelengths that are almost completely absorbed below the 250-kilometer level, by using a rocket to carry instruments above 250 kilometers to measure these phenomena.

rocket engine [AERO ENG] A reaction engine that contains within itself, or carries along with itself, all the substances necessary for its operation or for the consumption or combustion of its fuel, not requiring intake of any outside substance and hence capable of operation in outer space. Also known as rocket motor.

rocket motor *See* rocket engine.

rocket ramjet [AERO ENG] A ramjet engine having a rocket mounted within the ramjet duct, the rocket being used to bring the ramjet up to the necessary operating speed. Also known as ducted rocket.

rocketry [AERO ENG] **1.** The science or study of rockets, embracing theory, research, development, and experimentation. **2.** The art and science of using rockets, especially rocket ammunition.

rocket-sled testing [AERO ENG] A method of subjecting structures and devices to high accelerations or decelerations and aerodynamic flow phenomena under controlled conditions; the test object is mounted on a sled chassis running on precision steel rails and accelerated by rockets or decelerated by water scoops.

rocketsonde *See* meteorological rocket.

rocket staging [AERO ENG] The use of successive rocket sections or stages, each having its own engine or engines; each stage is a complete rocket vehicle in itself.

rock fabric *See* fabric.

rock failure [GEOL] Fracture of a rock that has been stressed beyond its ultimate strength.

rockfall [GEOL] **1.** The fastest-moving landslide; free fall of newly detached bedrock segments from a cliff or other steep slope; usually occurs during spring thaw. **2.** The rock material moving in or moved by a rockfall.

rocking furnace [MECH ENG] A horizonal, cylindrical melting furnace that is rolled back and forth on a geared cradle.

rocking pier [CIV ENG] A pier that is hinged to allow for longitudinal expansion or contraction of the bridge.

rocking valve [MECH ENG] An engine valve in which a disk or cylinder turns in its seat to permit fluid flow.

rock magnetism [GEOPHYS] The natural remanent magnetization of igneous, metamorphic, and sedimentary rocks resulting from the presence of iron oxide minerals.

rock mechanics [GEOPHYS] Application of the principles of mechanics and geology to quantify the response of rock when it is acted upon by environmental forces, particularly when human-induced factors alter the original ambient forces.

rock pedestal *See* pedestal.

rock pressure [GEOPHYS] **1.** Stress in underground geologic material due to weight of overlying material, residual stresses, and pressures resulting from swelling clays. **2.** *See* ground pressure.

rock salt *See* halite.

rock shell [INV ZOO] The common name for a large number of gastropod mollusks composing the family Muridae and characterized by having conical shells with various sculpturing.

rockslide [GEOL] The sudden, rapid downward movement of newly detached bedrock segments over a surface of weakness, such as of bedding, jointing, or faulting. Also known as rock slip.

rock slip *See* rockslide.

rock step *See* knickpoint.

rock-stratigraphic unit [GEOL] A lithologically homogeneous body of strata characterized by certain observable physical features, or by the dominance of a certain rock type or combination of rock types; rock-stratigraphic units include groups, formations, members, and beds. Also known as geolith; lithologic unit; lithostratic unit; lithostratigraphic unit; rock unit.

rock terrace [GEOL] A stream terrace on the side of a valley composed of resistant bedrock which remains during erosion of weaker overlying and underlying beds.

rock unit *See* rock-stratigraphic unit.

Rockwell hardness test [ENG] One of the arbitrarily defined measures of resistance of a material to indentation under static or dynamic load; depth of indentation of either a steel ball or a 120° conical diamond with rounded point, 1/16, 1/8, 1/4, or 1/2 inch (1.5875, 3.175, 6.35, 12.7 millimeters) in diameter, called a brale, under prescribed load is the basis for Rockwell hardness; 60, 100, 150 kilogram load is applied with a special machine, and depth of impression under initial minor load is indicated on a dial whose graduations represent hardness number.

Rocky Mountain spotted fever [MED] An acute, infectious, typhuslike disease of man caused by the rickettsial organism *Rickettsia rickettsi* and transmitted by species of hard-shelled ticks; characterized by sudden onset of chills, headache, fever, and an exanthem on the extremities. Also known as American spotted fever; tick fever; tick typhus.

rod [DES ENG] **1.** A bar whose end is slotted, tapered, or screwed for the attachment of a drill bit. **2.** A thin, round bar of metal or wood. [GEOL] A rodlike sedimentary particle characterized by a width-length ratio less than 2/3 and a thickness-width ratio more than 2/3. Also known as roller. [HISTOL] One of the rod-shaped sensory bodies in the retina which are sensitive to dim light. [MECH] *See* perch. [NUCLEO] A relatively long and slender body of material used in, or in conjunction with, a nuclear reactor; may contain fuel, absorber, or fertile material or other material in which activation or transmutation is desired.

rodent [VERT ZOO] The common name for members of the order Rodentia.

Rodentia [VERT ZOO] An order of mammals characterized by a single pair of ever-growing upper and lower incisors, a maximum of five upper and four

lower cheek teeth on each side, and free movement of the lower jaw in an anteroposterior direction.

rod mill [MECH ENG] A pulverizer operated by the impact of heavy metal rods. [MET] A mill for making metal rods.

rod pump [PETRO ENG] Type of oil well sucker-rod pump that can be inserted into or removed from oil well tubing without moving or disturbing the tubing itself. Also known as insert pump.

rod string [MECH ENG] Drill rods coupled to form the connecting link between the core barrel and bit in the borehole and the drill machine at the collar of the borehole.

roentgen [NUCLEO] An exposure dose of x- or γ-radiation such that the electrons and positrons liberated by this radiation produce, in air, when stopped completely, ions carrying positive and negative charges of 2.58×10^{-4} coulomb per kilogram of air. Abbreviated R (formerly r). Also spelled röntgen.

roentgen diffractometry *See* x-ray crystallography.

roentgen equivalent man *See* rem.

roentgen equivalent physical *See* rep.

roentgen optics *See* x-ray optics.

roentgen rays *See* x-rays.

roentgen spectrometry *See* x-ray spectrometry.

roesslerite [MINERAL] $MgH(AsO_4)\cdot 7H_2O$ A monoclinic mineral composed of hydrous acid magnesium arsenate; it is isomorphous with phosphorroesslerite.

roestone *See* oolite.

roll [GEOL] A primary sedimentary structure produced by deformation involving subaqueous slump or vertical foundering. [MECH] Rotational or oscillatory movement of an aircraft or similar body about a longitudinal axis through the body; it is called roll for any degree of such rotation. [MECH ENG] A cylinder mounted in bearings; used for such functions as shaping, crushing, moving, or printing work passing by it. [MIN ENG] *See* horseback. [TEXT] A continuous strand made by rolling, rubbing, or twisting fibers.

roll axis [MECH] A longitudinal axis through an aircraft, rocket, or similar body, about which the body rolls.

roll control [ENG] The exercise of control over a missile so as to make it roll to a programmed degree, usually just before pitchover.

roll crusher [MECH ENG] A crusher having one or two toothed rollers to reduce the material.

rolled gold [MET] Same as gold-filled except that the proportion of gold alloy to total weight of the article may be less than 1/20; fineness of the gold alloy may not be less than 10 karat.

roller [DES ENG] A cylindrical device for transmitting motion and force by rotation. [GEOL] *See* rod. [OCEANOGR] A long, massive wave which usually retains its form without breaking until it reaches the beach or a shoal.

roller bearing [MECH ENG] A shaft bearing characterized by parallel or tapered steel rollers confined between outer and inner rings.

roller conveyor [MECH ENG] A gravity conveyor having a track of parallel tubular rollers set at a definite grade, usually on antifriction bearings, at fixed locations, over which package goods which are sufficiently rigid to prevent sagging between rollers are moved by gravity or propulsion.

roller mill *See* ring-roller mill.

roll flattening *See* flattening.

rolling [MECH] Motion of a body across a surface combined with rotational motion of the body so that the point on the body in contact with the surface is instantaneously at rest. [MET] Reducing or changing the cross-sectional area of a workpiece by the compressive forces exerted by rotating rolls. Also known as metal rolling. [NAV ARCH] The oscillating motion of a vessel from side to side due to ground swell, heavy sea, or other causes.

rolling contact [MECH] Contact between bodies such that the relative velocity of the two contacting surfaces at the point of contact is zero.

roll mill [MECH ENG] A series of rolls operating at different speeds for grinding and crushing.

roll-off [ELECTR] Gradually increasing loss or attenuation with increase or decrease of frequency beyond the substantially flat portion of the amplitude-frequency response characteristic of a system or transducer.

roll-out [COMPUT SCI] **1.** To make available additional main memory for one task by copying another task onto auxiliary storage. **2.** To read a computer register or counter by adding a one to each digit column simultaneously until all have returned to zero, with a signal being generated at the instant a column returns to zero.

ROM *See* read-only memory.

romeite [MINERAL] $(Ca, Fe, Mn, Na)_2(Sb, Ti)_2O_6\cdot(O, OH, F)$ A honey-yellow to yellowish-brown mineral composed of oxide of calcium, iron, manganese, sodium, antimony, and titanium, occurring in minute octahedrons.

röntgen *See* roentgen.

roof pendant [GEOL] Downward projection or sag into an igneous intrusion of the country rock of the roof. Also known as pendant.

roof stringer [MIN ENG] A lagging bar running parallel with the working place above the header in a weak or scaly top in narrow rooms or entries which have short life.

roof truss [BUILD] A truss used in roof construction; it carries the weight of roof deck and framing and of wind loads on the upper chord; an example is a Fink truss.

room-and-pillar [MIN ENG] A system of mining in which the coal or ore is mined in rooms separated by narrow ribs or pillars; pillars are subsequently worked.

room crosscut *See* breakthrough.

rooseveltite [MINERAL] $BiAsO_4$ A gray mineral consisting of bismuth arsenate; occurs as thin botryoidal crusts.

root [COMPUT SCI] The origin or most fundamental point of a tree diagram. Also known as base. [BOT] The absorbing and anchoring organ of a vascular plant; it bears neither leaves nor flowers and is usually subterranean. [CIV ENG] The portion of a dam which penetrates into the ground where the dam joins the hillside. [DES ENG] The bottom of a screw thread. [GEOL] **1.** The lower limit of an ore body. Also known as bottom. **2.** The part of a fold nappe that was originally linked to its root zone. [MATH] **1.** A root of a given real or complex number is a number which when raised to some exponent equals that number. Also known as radix. **2.** A root of a polynomial $p(x)$ is a number a such that $p(a) = 0$.

root canal [ANAT] The cavity within the root of a tooth, occupied by pulp, nerves, and vessels.

root cap [BOT] A thick, protective mass of parenchymal cells covering the meristematic tip of the root.

root hair [BOT] One of the hairlike outgrowths of the root epidermis that function in absorption.

root knot [PL PATH] Any of various plant diseases caused by root-knot nematodes which produce gall-like enlargements on the roots.

root-mean-square current *See* effective current.

root-mean-square error [STAT] The square root of the second moment corresponding to the frequency function of a random variable.

root-mean-square sound pressure *See* effective sound pressure.

root-mean-square value Abbreviated rms value. [PHYS] The square root of the time average of the square of a quantity; for a periodic quantity the average is taken over one complete cycle. Also known as effective value. [STAT] The square root of the average of the squares of a series of related values.

root of an equation [MATH] A number or quantity which satisfies a given equation.

root of weld [MET] The points at which the bottom of the weld and the base metal surfaces intersect.

root rot [PL PATH] Any of various plant diseases characterized by decay of the roots.

root segment [COMPUT SCI] The master or controlling segment of an overlay structure which always resides in the main memory of a computer.

rootstock [BOT] A root or part of a root used as the stock for grafting.

Roproniidae [INV ZOO] A small family of hymenopteran insects in the superfamily Proctotrupoidea.

ropy lava *See* pahoehoe.

Rosaceae [BOT] A family of dicotyledonous plants in the order Rosales typically having stipulate leaves and hypogynous, slightly perigynous, or epigynous flowers, numerous stamens, and several or many separate carpels.

Rosales [BOT] A morphologically diffuse order of dicotyledonous plants in the subclass Rosidae.

rosasite [MINERAL] $(Cu, Zn)_2(OH)_2(CO_3)$ A green to bluish-green and sky blue mineral consisting of a carbonate-hydroxide of copper and zinc.

roscherite [MINERAL] $(Ca, Mn, Fe)_2Al(PO_4)(OH)\cdot 2H_2O$ A dark-brown mineral composed of hydrous basic phosphate of aluminum, calcium, manganese, and iron, occurring as monoclinic crystals.

roscoelite [MINERAL] $K(V, Al, Mg)_3Si_3O_{10}(OH)_2$ Tan, grayish-brown, or greenish-brown vanadium-bearing mica mineral occurring in minute scales or flakes.

rose [BOT] A member of the genus *Rosa* in the rose family (Rosaceae); plants are erect, climbing, or trailing shrubs, generally prickly stemmed, and bear alternate, odd-pinnate single leaves. [MATH] A graph consisting of loops shaped like rose petals arising from the equations in polar coordinates $r = a \sin n\theta$ or $r = a \cos n\theta$.

rosebloom *See* false blossom.

roselite [MINERAL] $(Ca, Co)_2(Co, Mg)(AsO_4)_2\cdot 2H_2O$ A pink or rose-colored, monoclinic mineral consisting of a hydrated arsenate of calcium, cobalt, and magnesium.

rosemary [BOT] *Rosmarinus officinalis.* A fragrant evergreen of the mint family from France, Spain, and Portugal; leaves have a pungent bitter taste and are used as an herb and in perfumes.

roseola infantum *See* exanthem subitum.

rose quartz [MINERAL] A pink variety of crystalline quartz; commonly massive and used as a gemstone. Also known as Bohemian ruby.

rosette [BIOL] Any structure or marking resembling a rose. [MET] **1.** Rounded constituents in a microstructure arranged in whorls. **2.** Strain gages arranged to indicate at a single position the strains in three different directions. [MINERAL] Rose-shaped, crystalline aggregates of barite, marcasite, or pyrite formed in sedimentary rock. [PL PATH] Any of various plant diseases in which the leaves become clustered in the form of a rosette.

Rosidae [BOT] A large subclass of the class Magnoliatae; most have a well-developed corolla with petals separate from each other, binucleate pollen, and ovules usually with two integuments.

rosin [MATER] A translucent yellow, umber, or reddish resinous residue from the distillation of crude turpentine from the sap of pine trees (gum rosin) or from an extract of the stumps and other parts of the tree (wood rosin); used in varnishes, lacquers, printing inks, adhesives, and soldering fluxes, in medical ointments, and as a preservative.

Rossby diagram [THERMO] A thermodynamic diagram, named after its designer, with mixing ratio as abscissa and potential temperature as ordinate; lines of constant equivalent potential temperature are added.

Rossby wave [METEOROL] A wave on a uniform current in a two-dimensional nondivergent fluid system, rotating with varying angular speed about the local vertical (beta plane); this is a special case of a barotropic disturbance, conserving absolute vorticity; applied to atmospheric flow, it takes into account the variability of the Coriolis parameter while assuming the motion to be two-dimensional. Also known as planetary wave.

Rossel Current [OCEANOGR] A seasonal Pacific Ocean current flowing westward and northwestward along both the southern and northeastern coasts of New Guinea, the southern part flowing through Torres Strait and losing its identity in the Arafura Sea, and the northern part curving northeastward to join the equatorial countercurrent of the Pacific Ocean.

rossite [MINERAL] $CaV_2O_6 \cdot 4H_2O$ A yellow, triclinic mineral consisting of a hydrated calcium vanadate.

Rostratulidae [VERT ZOO] A small family of birds in the order Charadriiformes containing the painted snipe; females are more brightly colored than males.

rostrum [BIOL] A beak or beaklike process.

rot [MATER] See curl. [PL PATH] Any plant disease characterized by breakdown and decay of plant tissue.

Rotaliacea [INV ZOO] A superfamily of foraminiferans in the suborder Rotaliina characterized by a planispiral or trochospiral test having apertural pores and composed of radial calcite, with secondarily bilamellar septa.

rotameter [ENG] A variable-area, constant-head, rate-of-flow volume meter in which the fluid flows upward through a tapered tube, lifting a shaped weight to a position where upward fluid force just balances its weight.

rotary [MECH ENG] 1. A rotary machine, such as a rotary printing press or a rotary well-drilling machine. 2. The turntable and its supporting and rotating assembly in a well-drilling machine.

rotary actuator [MECH ENG] A device that converts electric energy into controlled rotary force; usually consists of an electric motor, gear box, and limit switches.

rotary beam [ELECTROMAG] Short-wave antenna system highly directional in azimuth and altitude, mounted in such a manner that it can be rotated to any desired position, either manually or by an electric motor drive.

rotary calculator [COMPUT SCI] A type of mechanical desk calculator that is distinguished from key-driven calculators by virtue of a latching selection keyboard in which the multiplicand or divisor could be indexed and then repeatedly transferred to the accumulator, positively or negatively by turning a crank; the accumulator and a cycle-counting register were mounted in a carriage that could be shifted right or left, relative to the selection keyboard, to accommodate the successive digits of the multiplier or quotient.

rotary compressor [MECH ENG] A positive-displacement machine in which compression of the fluid is effected directly by a rotor and without the usual piston, connecting rod, and crank mechanism of the reciprocating compressor.

rotary converter See dynamotor.

rotary crane [MECH ENG] A crane consisting of a boom pivoted to a fixed or movable structure.

rotary crusher [MECH ENG] Solids-reduction device in which a high-speed rotating cone on a vertical shaft forces solids against a surrounding shell.

rotary dispersion [OPTICS] The change in the angle through which an optically active substance rotates the plane of polarization of plane polarized light as the wavelength of the light is varied. Also known as rotatory dispersion.

rotary drill [MECH ENG] Any of various drill machines that rotate a rigid, tubular string of rods to which is attached a rock cutting bit, such as an oil well drilling apparatus.

rotary engine [MECH ENG] A positive displacement engine (such as a steam or internal combustion type) in which the thermodynamic cycle is carried out in a mechanism that is entirely rotary and without the more customary structural elements of a reciprocating piston, connecting rods, and crankshaft.

rotary filter See drum filter.

rotary furnace [MECH ENG] A heat-treating furnace of circular construction which rotates the workpiece around the axis of the furnace during heat treatment; workpieces are transported through the furnace along a circular path.

rotary gap See rotary spark gap.

rotary kiln [ENG] A long cylindrical kiln lined with refractory, inclined at a slight angle, and rotated at a slow speed.

rotary polarization See optical activity.

rotary pump [MECH ENG] A displacement pump that delivers a steady flow by the action of two members in rotational contact.

rotary spark gap [ELEC] A spark gap in which sparks occur between one or more fixed electrodes and a number of electrodes projecting outward from the circumference of a motor-driven metal disk. Also known as rotary gap.

rotary stepping relay See stepping relay.

rotary stepping switch See stepping relay.

rotary switch [ELEC] A switch that is operated by rotating its shaft.

rotary transformer [ELEC] A rotating machine used to transform direct-current power from one voltage to another.

rotary vacuum filter See drum filter.

rotary valve [MECH ENG] A valve for the admission or release of working fluid to or from an engine cylinder where the valve member is a ported piston that turns on its axis.

rotary-vane attenuator [ELECTROMAG] Device designed to introduce attenuation into a waveguide circuit by varying the angular position of a resistive material in the guide.

rotary-vane meter [ENG] A type of positive-displacement rate-of-flow meter having spring-loaded vanes mounted on an eccentric drum in a circular cavity; each time the drum rotates, a fixed volume of fluid passes through the meter.

rotation [MATH] *See* curl. [MECH] Also known as rotational motion. **1.** Motion of a rigid body in which either one point is fixed, or all the points on a straight line are fixed. **2.** Angular displacement of a rigid body. **3.** The motion of a particle about a fixed point.

rotational energy [MECH] The kinetic energy of a rigid body due to rotation. [PHYS CHEM] For a diatomic molecule, the difference between the energy of the actual molecule and that of an idealized molecule which is obtained by the hypothetical process of gradually stopping the relative rotation of the nuclei without placing any new constraint on their vibration, or on motions of electrons.

rotational flow [FL MECH] Flow of a fluid in which the curl of the fluid velocity is not zero, so that each minute particle of fluid rotates about its own axis. Also known as rotational motion.

rotational impedance [MECH] A complex quantity, equal to the phasor representing the alternating torque acting on a system divided by the phasor representing the resulting angular velocity in the direction of the torque at its point of application. Also known as mechanical rotational impedance.

rotational inertia *See* moment of inertia.

rotational motion *See* rotation; rotational flow.

rotational position sensing [ADP] A fast disk search method whereby the control unit looks for a specified sector, and then receives the sector number required to access the record.

rotational stability [MECH] Property of a body for which a small angular displacement sets up a restoring torque that tends to return the body to its original position.

rotational transform [PL PHYS] Property possessed by a magnetic field, in a system used to confine a plasma, in which magnetic lines of force do not close in on themselves after making a circuit around the system, but are rotationally displaced.

rotational wave *See* shear wave; S wave.

rotation axis [CRYSTAL] A symmetry element of certain crystals in which the crystal can be brought into a position physically indistinguishable from its original position by a rotation through an angle of $360°/n$ about the axis, where n is the multiplicity of the axis, equal to 2, 3, 4, or 6. Also known as symmetry axis.

rotation coefficients [MECH] Factors employed in computing the effects on range and deflection which are caused by the rotation of the earth; they are published only in firing tables involving comparatively long ranges.

rotation-inversion axis [CRYSTAL] A symmetry element of certain crystals in which a crystal can be brought into a position physically indistinguishable from its original position by a rotation through an angle of $360°/n$ about the axis followed by an inversion, where n is the multiplicity of the axis, equal to 1, 2, 3, 4, or 6. Also known as inversion axis.

rotation moment *See* torque.

rotation-reflection axis [CRYSTAL] A symmetry element of certain crystals in which a crystal can be brought into a position physically indistinguishable from its original position by a rotation through an angle of $360°/n$ about the axis followed by a reflection in the plane perpendicular to the axis, where n is the multiplicity of the axis, equal to 1, 2, 3, 4, or 6.

rotation twin [CRYSTAL] A twin crystal in which the parts will coincide if one part is rotated 180° (sometimes 30, 60, or 120°).

rotator [ELECTROMAG] A device that rotates the plane of polarization of a plane-polarized electromagnetic wave, such as a twist in a waveguide. [MECH] A rotating rigid body. [QUANT MECH] A molecule or other quantum-mechanical system which behaves as the quantum-mechanical analog of a rotating rigid body. Also known as top.

rotatory [OPTICS] Having the capability to rotate the plane of polarization of polarized electromagnetic radiation.

rotatory dispersion *See* rotary dispersion.

röteln *See* rubella.

Rotifera [INV ZOO] A class of the phylum Aschelminthes distinguished by the corona, a retractile trochal disk provided with several groups of cilia and located on the head.

Rotliegende [GEOL] A European series of geologic time: Lower and Middle Permian.

rotor [AERO ENG] An assembly of blades designed as airfoils that are attached to a helicopter or similar aircraft and rapidly rotated to provide both lift and thrust. [COMMUN] **1.** Disk with a set of input contacts and a set of output contacts, connected by any prearranged scheme designed to rotate within an electrical cipher machine. **2.** Disk whose rotation produces a variation of some cryptographic element in a cipher machine usually by means of lugs (or pins) in or on its periphery. [ELEC] The rotating member of an electrical machine or device, such as the rotating armature of a motor or generator, or the rotating plates of a variable capacitor. [MECH ENG] *See* impeller.

rotten spot *See* pothole.

rough bark [PL PATH] **1.** Any of various virus diseases of woody plants characterized by roughening and often splitting of the bark. **2.** A fungus disease of apples caused by *Phomopsis mali* and characterized by rough cankers on the bark.

roughcast [CIV ENG] A rough finish on a surface; in particular, a plaster made of lime and shells or pebbles, applied by throwing it against a wall with a trowel.

rounding [MATH] Dropping or neglecting decimals after some significant place. Also known as truncation.

round ligament [ANAT] **1.** A flattened band extending from the fovea on the head of the femur to attach on either side of the acetabular notch between which it blends with the transverse ligament. **2.** A fibrous cord running from the umbilicus to the notch in the anterior border of the liver; represents the remains of the obliterated umbilical vein.

round of beam [NAV ARCH] The arch or slope from side to side of a vessel's weather deck for water drainage. Also known as camber.

round off [MATH] To truncate the least significant digit or digits of a numeral, and adjust the remaining numeral to be as close as possible to the original number.

round window [ANAT] A membrane-covered opening between the middle and inner ears in amphibians and mammals through which energy is dissipated after traveling in the membranous labyrinth.

routine [COMPUT SCI] A set of digital computer instructions designed and constructed so as to accomplish a specified function.

routine library [COMPUT SCI] Ordered set of standard and proven computer routines by which problems or parts of problems may be solved.

rouvite [MINERAL] $CaU_2V_{12}O_{36}\cdot20H_2O$ A purplish- to bluish-black mineral consisting of a hydrated vanadate of calcium and uranium; occurs as dense masses, crusts, and coatings.

row [COMPUT SCI] **1.** The characters, or corresponding bits of binary-coded characters, in a computer word. **2.** Equipment which simultaneously processes the bits of a character, the characters of a word, or corresponding bits of binary-coded characters in a word. **3.** Corresponding positions in a group of columns.

row address [COMPUT SCI] An index array entry field which contains the main storage address of a data block.

row binary [COMPUT SCI] A method of encoding binary information onto punched cards in which successive bits are punched row-wise onto the card.

roweite [MINERAL] $(Mn,Mg,Zn)Ca(BO_2)_2(OH)_2$ A light-brown mineral composed of basic borate of calcium, manganese, magnesium, and zinc.

Rowland circle [SPECT] A circle drawn tangent to the face of a concave diffraction grating at its midpoint, having a diameter equal to the radius of curvature of a grating surface; the slit and camera for the grating should lie on this circle.

Rowland grating *See* concave grating.

Rowland mounting [SPECT] A mounting for a concave grating spectrograph in which camera and grating are connected by a bar forming a diameter of the Rowland circle, and the two run on perpendicular tracks with the slit placed at their junction.

royal jelly [MATER] A protein complex high in vitamin B secreted by bees to nourish the egg of the queen bee; used in face creams.

rpm *See* revolution per minute.

RPN *See* reverse Polish notation.

RPV *See* remotely piloted vehicle.

RQ *See* respiratory quotient.

RR Lyrae stars [ASTRON] Pulsating variable stars with a period of 0.05–1.2 days in the halo population of the Milky Way Galaxy; color is white, and they are mostly stars of spectral class A. Also known as cluster cepheids; cluster variables.

R scan *See* R scope.

R scope [ELECTR] An A scope presentation with a segment of the horizontal trace expanded near the target spot (pip) for greater accuracy in range measurement. Also known as R indicator; R scan.

RTL *See* resistor-transistor logic.

Ru *See* ruthenium.

rubber [ORG CHEM] A natural, synthetic, or modified high polymer with elastic properties and, after vulcanization, elastic recovery; the generic term is elastomer.

rubber tree [BOT] *Hevea brasiliensis*. A tall tree of the spurge family (Euphorbiaceae) from which latex is collected and coagulated to produce rubber.

rubble [CIV ENG] **1.** Rough, broken stones and other debris resulting from the deterioration and destruction of a building. **2.** Rough stone or brick used in coarse masonry or to fill the space in a wall between the facing courses. [GEOL] **1.** A loose mass of rough, angular rock fragments, coarser than sand. **2.** *See* talus. [HYD] Fragments of floating or grounded sea ice in hard, roughly spherical blocks measuring 0.5–1.5 meters (1.5–4.5 feet) in diameter, and resulting from the breakup of larger ice formations. Also known as rubble ice.

rubella [MED] An infectious virus disease of humans characterized by coldlike symptoms, fever, and transient, generalized pale-pink rash; its occurrence in early pregnancy is associated with congenital abnormalities. Also known as epidemic roseola; French measles; German measles; röteln.

rubellite [MINERAL] The red to red-violet variety of the gem mineral tourmaline; hardness is 7–7.5 on Mohs scale, and specific gravity is near 3.04.

rubeola *See* measles.

Rubiaceae [BOT] The single family of the plant order Rubiales.

Rubiales [BOT] An order of dicotyledonous plants marked by their inferior ovary, regular or nearly regular corolla, and opposite leaves with interpetiolar stipules or whorled leaves without stipules.

rubidium [CHEM] A chemical element, symbol Rb, atomic number 37, atomic weight 85.47; a reactive alkali metal; salts of the metal may be used in glass and ceramic manufacture.

ruby [MINERAL] The red variety of the mineral corundum; in its finest quality, the most valuable of gemstones.

ruby laser [OPTICS] An optically pumped solid-state laser that uses a ruby crystal to produce an intense and extremely narrow beam of coherent red light.

ruby maser [PHYS] A maser that uses a ruby crystal in the cavity resonator.

ruby zinc *See* zincite.

rudder [ENG] **1.** A flat, usually foil-shaped movable control surface attached upright to the stern of a boat, ship, or aircraft, and used to steer the craft. **2.** *See* rudder angle.

rudder angle [ENG] The acute angle between a ship or plane's rudder and its fore-and-aft line. Also known as rudder.

Ruffini cylinder [ANAT] A cutaneous nerve ending suspected as the mediator of warmth.

Rugosa [PALEON] An order of extinct corals having either simple or compound skeletons with internal skeletal structures consisting mainly of three elements, the septa, tabulae, and dissepiments.

rugose [BIOL] Having a wrinkled surface.

ruminant [PHYSIO] Characterized by the act of regurgitation and rechewing of food. [VERT ZOO] A mammal belonging to the Ruminantia.

Ruminantia [VERT ZOO] A suborder of the Artiodactyla including sheep, goats, camels, and other forms which have a complex stomach and ruminate their food.

run [COMPUT SCI] A single, complete execution of a computer program, or one continuous segment of computer processing, used to complete one or more tasks for a single customer or application. Also known as machine run. [CHEM ENG] **1.** The amount of feedstock processed by a petroleum refinery unit during a given time; often used colloquially in relation to the type of stock being processed, as in crude run or naphtha run. **2.** A processing-cycle or batch-treatment operation. [ENG] A portion of pipe or fitting lying in a straight line in the same direction of flow as the pipe to which it is connected. [GEOL] **1.** A ribbonlike, flat-lying, irregular orebody following the stratification of the host rock. **2.** A branching or fingerlike extension of the feeder of an igneous intrusion. [MIN ENG] *See* slant. [NAV] The distance traveled by a craft during any given time interval, or since leaving a designated place. [NAV ARCH] The underwater portion of that part of the aft end of a ship where it curves inward and upward to the stern. [ORD] **1.** Steady, level flight of an aircraft across a target to enable bombs to be dropped accurately in horizontal bombing. **2.** Passing of a moving target once across the range.

runback [CHEM ENG] A pipe through which all or part of a distillation column's overhead condensate can be run back into the column, instead of being drawn off as product. [ENG] **1.** To retract the drill feed mechanism to its starting position. **2.** To drill slowly downward toward the bottom of the hole when the drill string has been lifted off-bottom for rechucking.

run book [COMPUT SCI] The collection of materials necessary to document a program run on a computer. Also known as problem file; problem folder.

run chart [COMPUT SCI] A flow chart for one or more computer runs which shows input, output, and the use of peripheral units, but no details of the execution of the run. Also known as run diagram.

run diagram *See* run chart.

Runge vector [MECH] A vector which describes certain unchanging features of a nonrelativistic two-body interaction obeying an inverse-square law, either in classical or quantum mechanics; its constancy is a reflection of the symmetry inherent in the inverse-square interaction.

runnel [GEOL] A troughlike hollow on a tidal sand beach which carries water drainage off the beach as the tide retreats.

runner [BOT] A horizontally growing, sympodial stem system; adventitious roots form near the apex, and a new runner emerges from the axil of a reduced leaf. Also known as stolon. [ENG] In a plastics injection or transfer mold, the channel (usually circular) that connects the sprue with the gate to the mold cavity. [MET] **1.** The part of a casting between itself and the gate assembly of the mold. **2.** A channel through which molten metal flows from one receptacle to another. [MIN ENG] A vertical timber sheet pile used to prevent collapse of an excavation.

running accumulator *See* push-down storage.

running fit [DES ENG] The intentional difference in dimensions of mating mechanical parts that permits them to move relative to each other.

running sand *See* quicksand.

runoff [HYD] **1.** Surface streams that appear after precipitation. **2.** The flow of water in a stream, usually expressed in cubic feet per second; the net effect of storms, accumulation, transpiration, meltage, seepage, evaporation, and percolation. [MIN ENG] Collapse of a coal pillar in a mine.

runway [CIV ENG] A straight path, often hard-surfaced, within a landing strip, normally used for landing and takeoff of aircraft. [GEOL] The channel of a stream.

rupture *See* fracture; hernia.

russellite [MINERAL] Bi_2WO_6 A pale yellow to greenish, tetragonal mineral consisting of an oxide of bismuth and tungsten; occurs as fine-grained compact masses.

rust [MET] The iron oxides formed on corroded ferrous metals and alloys. [PL PATH] Any plant disease caused by rust fungi (Uredinales) and characterized by reddish-brown lesions on the plant parts.

rusty blotch [PL PATH] A fungus disease of barley caused by *Helminthosporium californicum* and characterized by brown blotches on the foliage.

rusty mottle [PL PATH] A virus disease of cherry characterized by retarded development of blossoms and leaves in the spring, followed by necrotic spotting and shot-holing of the foliage with considerable defoliation.

rut [PHYSIO] The period during which the male animal has a heightened mating drive.

rutabaga [BOT] *Brassica napobrassica.* A biennial crucifer of the order Capparales probably resulting from the natural crossing of cabbage and turnip and characterized by a large, edible, yellowish fleshy root.

Rutaceae [BOT] A family of dicotyledonous plants in the order Sapindales distinguished by mostly free stamens and glandular-punctate leaves.

ruthenium [CHEM] A chemical element, symbol Ru, atomic number 44, atomic weight 101.07. [MET] A hard, brittle, grayish-white metal used as a catalyst; workable slightly at high temperatures.

rutile [MINERAL] TiO_2 A reddish-brown tetragonal mineral common in acid igneous rocks, in metamorphic rocks, and as residual grain in beach sand.

R wave *See* Rayleigh wave.

ry *See* rydberg.

rydberg [ATOM PHYS] A unit of energy used in atomic physics, equal to the square of the charge of the electron divided by twice the Bohr radius; equal to 13.60583 ± 0.00004 electron-volts. Symbolized ry. [SPECT] *See* kayser.

Rydberg constant [ATOM PHYS] **1.** An atomic constant which enters into the formulas for wave numbers of atomic spectra, equal to $2\pi^2 me^4/ch^3$, where m and e are the rest mass and charge of the electron, c is the speed of light, and h is Planck's constant; equal to $109,737.31 \pm 0.01$ inverse centimeters. Symbolized R_∞. **2.** For an atom, the Rydberg constant (first definition) divided by $1 + m/M$, where m and M are the masses of an electron and of the nucleus.

rye [BOT] *Secale cereale.* A cereal plant of the order Cyperales cultivated for its grain, which contains the most desirable gluten, next to wheat.

Rynchopidae [VERT ZOO] The skimmers, a family of birds in the order Charadriiformes distinguished by a knifelike lower beak that is longer and narrower than the upper one.

Rytiodinae [VERT ZOO] A subfamily of trichechiform sirenians in the family Dugongidae.

S

s *See* second; strange quark.
S *See* secondary winding; siemens; stoke; sulfur.
sabach *See* caliche.
Sabellariidae [INV ZOO] The sand-cementing worms, a family of polychaete annelids belonging to the Sedentaria and characterized by a compact operculum formed of setae of the first several segments.
Sabellidae [INV ZOO] A family of sedentary polychaete annelids often occurring in intertidal depths but descending to great abyssal depths; one of two families that make up the feather-duster worms.
Sabellinae [INV ZOO] A subfamily of the Sabellidae including the most numerous and largest members.
saber saw [MECH ENG] A portable saw consisting of an electric motor, a straight saw blade with reciprocating mechanism, a handle, baseplate, and other essential parts.
sabin [ACOUS] A unit of sound absorption for a surface, equivalent to 1 square foot (0.09290304 square meter) of perfectly absorbing surface. Also known as absorption unit; open window unit (OW unit); square-foot unit of absorption.
Sabin vaccine [IMMUNOL] A live-poliovirus vaccine that is administered orally.
sable [VERT ZOO] *Martes zibellina*. A carnivore of the family Mustelidae; a valuable fur-bearing animal, quite similar to the American marten.
saccate [BOT] Having a saclike or pouchlike form.
saccharase [BIOCHEM] An enzyme that catalyzes the hydrolysis of disaccharide to monosaccharides, specifically of sucrose to dextrose and levulose. Also known as invertase; invertin; sucrase.
saccharose *See* sucrose.
Saccoglossa [INV ZOO] An order of gastropod mollusks belonging to the Opisthobranchia.
Saccopharyngiformes [VERT ZOO] Formerly an order of actinopterygian fishes, the gulpers, now included in the Anguilliformes.
Saccopharyngoidei [VERT ZOO] The gulpers, a suborder of actinopterygian fishes in the order Anguilliformes having degenerative adaptations, including loss of swim bladder, opercle, branchiostegal ray, caudal fin, scales, and ribs.
sacculus [ANAT] The smaller, lower saclike chamber of the membranous labyrinth of the vertebrate ear.
saccus *See* vesicle.
sac fungus [MYCOL] The common name for members of the class Ascomycetes.
sacral nerve [ANAT] Any of five pairs of spinal nerves in the sacral region which innervate muscles and skin of the lower back, lower extremities, and perineum, and branches to the hypogastric and pelvic plexuses.
sacral vertebrae [ANAT] Three to five fused vertebrae that form the sacrum in most mammals; am-

phibians have one sacral vertebra, reptiles usually have two, and birds have 10–23 fused in the synsacrum.
saddle [DES ENG] A support shaped to fit the object being held. [GEOL] **1.** A gap that is broad and gently sloping on both sides. **2.** A relatively flat ridge that connects the peaks of two higher elevations. **3.** That part along the surface axis or axial trend of an anticline that is a low point or depression.
saddle point [GEOL] *See* col. [MATH] A point where all the first partial derivatives of a function vanish but which is not a local maximum or minimum.
safe load [MECH] The stress, usually expressed in tons per square foot, which a soil or foundation can safely support.
safety engineering [IND ENG] The testing and evaluating of equipment and procedures to prevent accidents.
safety explosive [MATER] An explosive which may be handled safely under ordinary conditions; it requires a powerful detonating force.
safety glass [MATER] **1.** A glass that resists shattering (such as a glass containing a net of wire or constructed of sheets separated by plastic film). **2.** A glass that has been tempered so that when it shatters, it breaks up into grains instead of jagged fragments.
safety lamp [MIN ENG] In coal mining, a lamp that is relatively safe to use in atmospheres which may contain flammable gas.
safety pin [ORD] **1.** A device designed to fit the mechanism of a fuse in order to prevent accidental arming or functioning of the fuse and to ensure transport safety; the device is removed just before employment of the fuse. **2.** *See* safety wire.
safety rail *See* guard rail.
safety relief valve *See* safety valve.
safety rod [NUCLEO] A control rod capable of shutting down a reactor quickly in case of failure of the ordinary control system using regulating rods and shim rods; a safety rod may be suspended above the core by a magnetic coupling and allowed to fall in when power reaches a predetermined level. Also known as scram rod.
safety valve [ENG] A spring-loaded, pressure-actuated valve that allows steam to escape from a boiler at a pressure slightly above the safe working level of the boiler; fitted by law to all boilers. Also known as safety relief valve.
safety wire [ORD] Wire set into the body of a fuse to lock all movable parts into safe positions so that the fuse will not be set off accidentally; it is pulled out just before firing. Also known as safety pin.
safflorite [MINERAL] CoAs$_2$ A cobalt arsenide mineral that occurs in tin-white masses, and is dimor-

phous with smaltite; found in Canada, Morocco, and the United States.

safflower [BOT] *Carthamnus tinctorius*. An annual thistlelike herb belonging to the composite family (Compositae); the leaves are edible, flowers yield dye, and seeds yield a cooking oil.

saffron [BOT] *Crocus sativus*. A crocus of the iris family (Iridaceae); the source of a yellow dye used for coloring food and medicine.

sage [BOT] *Salvia officinalis*. A half-shrub of the mint family (Labiatae); the leaves are used as a spice.

sagebrush [BOT] Any of various hoary undershrubs of the genus *Artemisia* found on the alkaline plains of the western United States.

Sagenocrinida [PALEON] A large order of extinct, flexible crinoids that occurred from the Silurian to the Permian.

Saghathiinae [PALEON] An extinct subfamily of hyracoids in the family Procaviidae.

Sagitta [ASTRON] A small constellation; right ascension 20 hours, declination 10° north. Also known as Arrow.

sagittal [ZOO] In the median longitudinal plane of the body, or parallel to it.

Sagittariidae [VERT ZOO] A family of birds in the order Falconiformes comprising a single species, the secretary bird, noted for its nuchal plumes resembling quill pens stuck behind an ear.

Sagittarius [ASTRON] A constellation whose major portion lies in the Milky Way; right ascension 19 hours, declination 25° south. Also known as Archer.

sahlinite [MINERAL] $Pb_{14}(AsO_4)_2O_9Cl_4$ A pale sulfur yellow, monoclinic mineral consisting of a basic chloride-arsenate of lead; occurs in aggregates of small scales.

sahlite *See* salite.

Sail *See* Vela.

sailfish [VERT ZOO] Any of several large fishes of the genus *Istiophorus* characterized by a very large dorsal fin that is highest behind its middle.

Saint Elmo's fire [ELEC] A visible electric discharge, sometimes seen on the mast of a ship, on metal towers, and on projecting parts of aircraft, due to concentration of the atmospheric electric field at such projecting parts.

Saint Vitus dance [MED] Chorea associated with rheumatic fever. Also known as Sydenham's chorea.

Sakmarian [GEOL] A European stage of geologic time; the lowermost Permian, above Stephanian of Carboniferous and below Artinskian.

sal *See* sial.

salamander [VERT ZOO] The common name for members of the order Urodela.

Salamandridae [VERT ZOO] A family of urodele amphibians in the suborder Salamandroidea characterized by a long row of prevomerine teeth.

Salamandroidea [VERT ZOO] The largest suborder of the Urodela characterized by teeth on the roof of the mouth posterior to the openings of the nostrils.

salammoniac [MINERAL] NH_4Cl A white, isometric, crystalline mineral composed of native ammonium chloride.

Salam-Weinberg theory *See* Weinberg-Salam theory.

Salangidae [VERT ZOO] A family of soft-rayed fishes, in the suborder Galaxioidei, which live in estuaries of eastern Asia.

Saldidae [INV ZOO] The shore bugs, a family of predacious hemipteran insects in the superfamily Saldoidea.

Saldoidea [INV ZOO] A superfamily of the hemipteran group Leptopodoidea.

Saleniidae [INV ZOO] A family of echinoderms in the order Salenioida distinguished by imperforate tubercles.

Salenioida [INV ZOO] An order of the Echinacea in which the apical system includes one or several large angular plates covering the periproct.

salesite [MINERAL] $Cu(IO_3)(OH)$ A bluish-green mineral composed of basic iodate of copper.

salic [GEOL] A soil horizon enriched with secondary salts, at least 2 percent, and measuring at least 15 centimeters in thickness. [MINERAL] Pertaining to certain light-colored minerals, such as quartz and feldspars, that are rich in silica or magnesium and commonly occur in igneous rock.

Salicaceae [BOT] The single family of the order Salicales.

Salicales [BOT] A monofamilial order of dicotyledonous plants in the subclass Dilleniidae; members are dioecious, woody plants, with alternate, simple, stipulate leaves and plumose-hairy mature seeds.

salient pole [ELECTROMAG] A structure of magnetic material on which is mounted a field coil of a generator, motor, or similar device.

salina [ENG] *See* saltworks. [GEOL] An area, such as a salt flat, in which deposits of crystalline salts are formed or found. [HYD] A body of water containing high concentrations of salt.

salinity [OCEANOGR] The total quantity of dissolved salts in sea water, measured by weight in parts per thousand.

salinity logging [PETRO ENG] Technique for measurement and recording of saltwater-bearing zones in an oil or gas reservoir; uses a combination of neutron logging with a chlorine curve.

salite [MINERAL] $(Mg,Fe)_2Si_2O_6$ A grayish-green to black mineral variety of diopside containing more magnesium than iron; member of the clinopyroxene group. Also spelled sahlite.

salivary amylase *See* ptyalin.

salivary diastase *See* ptyalin.

Salk vaccine [IMMUNOL] A killed-virus vaccine administered for active immunization against poliomyelitis.

salmon [VERT ZOO] The common name for a number of fish in the family Salmonidae which live in coastal waters of the North Atlantic and North Pacific and breed in rivers tributary to the oceans.

Salmonidae [VERT ZOO] A family of soft-rayed fishes in the suborder Salmonoidei including the trouts, salmons, whitefishes, and graylings.

Salmoniformes [VERT ZOO] An order of soft-rayed fishes comprising salmon and their allies; the stem group from which most higher teleostean fishes evolved.

Salmonoidei [VERT ZOO] A suborder of the Salmoniformes comprising forms having an adipose fin.

salmonsite [MINERAL] A buff-colored mineral composed of hydrous phosphate of manganese and iron occurring in cleavable masses.

Salmopercae [VERT ZOO] An equivalent name for Percopsiformes.

Salpida [INV ZOO] An order of tunicates in the class Thaliacea including transparent forms ringed by muscular bands.

Salpingidae [INV ZOO] The narrow-waisted bark beetles, a family of coleopteran insects in the superfamily Tenebrionoidea.

salpingitis [MED] 1. Inflammation of the fallopian tube. 2. Inflammation of the eustachian tube.

salt [CHEM] The reaction product when a metal displaces the hydrogen of an acid; for example, $H_2SO_4 + 2NaOH \rightarrow Na_2SO_4$ (a salt) $+ 2H_2O$. [ENG]

To add an accelerator or retardant to cement. [MIN ENG] **1.** To introduce extra amounts of a valuable or waste mineral into a sample to be assayed. **2.** To artificially enrich, as a mine, usually with fraudulent intent.

saltation [GEOL] Transport of a sediment in which the particles are moved forward in a series of short intermittent bounces from a bottom surface.

salt bed [GEOL] Deposit of sodium chloride and other salts resulting from the evaporation or precipitation of ancient oceans.

salt bridge [PHYS CHEM] A bridge of a salt solution, usually potassium chloride, placed between the two half-cells of a galvanic cell, either to reduce to a minimum the potential of the liquid junction between the solutions of the two half-cells or to isolate a solution under study from a reference half-cell and prevent chemical precipitations.

salt dome [GEOL] A diapiric or piercement structure in which there is a central, equidimensional salt plug.

salt flowers *See* ice flowers.

salt gland [VERT ZOO] A compound tubular gland, located around the eyes and nasal passages in certain marine turtles, snakes, and birds, which copiously secretes a watery fluid containing a high percentage of salt.

Salticidea [INV ZOO] The jumping spiders, a family of predacious arachnids in the suborder Dipneumonomorphae having keen vision and rapid movements.

salting-out effect [CHEM ENG] The growth of crystals of a substance on heated, liquid-holding surfaces of a crystallizing evaporator as a result of the decrease in solubility of the substance with increase in temperature.

salt lake [HYD] A confined inland body of water having a high concentration of salts, principally sodium chloride.

saltmarsh [ECOL] A maritime habitat found in temperate regions, but typically associated with tropical and subtropical mangrove swamps, in which excess sodium chloride is the predominant environmental feature.

salt water *See* seawater.

Salviniales [BOT] A small order of heterosporous, leptosporangiate ferns (division Polypodiophyta) which float on the surface of the water.

SAM *See* surface-to-air missile.

samarium [CHEM] Group III rare-earth metal, atomic number 62, symbol Sm; melts at 1350°C, tarnishes in air, ignites at 200–400°C.

samarskite [MINERAL] $(Y,Ce,U,Ca,Fe,Pb,Th)(Nb,Ta,Ti,Sn)_2O_6$ A velvet-black to brown metamict orthorhombic mineral with splendent vitreous to resinous luster occurring in granite pegmatites. Also known as ampangabeite; uranotantalite.

Sambonidae [INV ZOO] A family of pentastomid arthropods in the suborder Porocephaloidea of the order Porocephalida.

sample [SCI TECH] Representative fraction of material tested or analyzed in order to determine the nature, composition, and percentage of specified constituents, and possibly their reactivity. [STAT] A selection of a certain collection from a larger collection.

sampled data [STAT] Data that are obtained at discrete rather than continuous intervals.

sample function [STAT] A function or procedure which, when applied repeatedly to a given population, produces a collection of samples.

sampleite [MINERAL] $NaCaCu_5(PO_4)_4Cl \cdot 5H_2O$ A blue mineral composed of hydrous phosphate and chloride of sodium, calcium, and copper.

sampling [ENG] Process of obtaining a sequence of instantaneous values of a wave. [SCI TECH] The obtaining of small representative quantities of materials (gas, liquid, solid) for the purpose of analysis. [STAT] A drawing of a collection from a given population.

sampling switch *See* commutator switch.

sampling time [ENG] The time between successive measurements of a physical quantity.

sampling voltmeter [ENG] A special type of voltmeter that detects the instantaneous value of an input signal at prescribed times by means of an electronic switch connecting the signal to a memory capacitor; it is particularly effective in detecting high-frequency signals (up to 12 gigahertz) or signals mixed with noise.

samsonite [MINERAL] $Ag_4MnSb_2S_6$ A black mineral composed of sulfide of silver, manganese, and antimony occurring in monoclinic prismatic crystals.

samson post *See* kingpost.

Samythinae [INV ZOO] A subfamily of sedentary polychaete annelids in the family Ampharetidae having a conspicuous dorsal membrane.

sand [GEOL] A loose material consisting of small mineral particles, or rock and mineral particles, distinguishable by the naked eye; grains vary from almost spherical to angular, with a diameter range from $\frac{1}{16}$ to 2 millimeters.

Sandalidae [INV ZOO] The equivalent name for Rhipiceridae.

sandalwood [BOT] **1.** Any species of the genus *Santalum* of the sandalwood family (Santalaceae) characterized by a fragrant wood. **2.** *S. album*. A parasitic tree with hard, close-grained, aromatic heartwood used in ornamental carving and cabinetwork.

sandarac *See* realgar.

sandbar [GEOL] A bar or low ridge of sand bordering the shore and built up, or near, to the surface of the water by currents or wave action. Also known as sand reef.

sandblasting [ENG] *See* grit blasting. [GEOL] Abrasion affected by the action of hard, windblown mineral grains.

sand boil *See* blowout.

sand-cast [MET] Made by pouring molten metal into a mold made of sand.

sand dollar [INV ZOO] The common name for the flat, disk-shaped echinoderms belonging to the order Clypeasteroida.

sand dune [GEOL] A mound of loose windblown sand commonly found along low-lying seashores above high-tide level.

sandfly [INV ZOO] Any of various small biting Diptera, especially of the genus *Phlebotomus*, which are vectors for phlebotomus (sandfly) fever.

sand levee *See* whaleback dune.

sandpiper [VERT ZOO] Any of various small birds that are related to plovers and that frequent sandy and muddy shores in temperate latitudes; bill is moderately long with a soft, sensitive tip, legs and neck are moderately long, and plumage is streaked brown, gray, or black above and is white below.

sand pump [MECH ENG] A pump, usually a centrifugal type, capable of handling sand- and gravel-laden liquids without clogging or wearing unduly; used to extract mud and cuttings from a borehole. Also known as sludge pump.

sand reef *See* sandbar.

sand ridge [GEOL] **1.** Any low ridge of sand formed at some distance from the shore, and either submerged or emergent, such as a longshore bar or a barrier beach. **2.** One of a series of long, wide, extremely low, parallel ridges believed to represent the eroded stumps of former longitudinal sand dunes. **3.** A crescent-shaped landform found on a sandy beach, such as a beach cusp. **4.** *See* sand wave.

sand shark [VERT ZOO] Any of various shallow-water predatory elasmobranchs of the family Carchariidae. Also known as tiger shark.

sandstone [PETR] A detrital sedimentary rock consisting of individual grains of sand-size particles 0.06 to 2 millimeters in diameter either set in a fine-grained matrix (silt or clay) or bonded by chemical cement.

sandstorm [METEOROL] A strong wind carrying sand through the air, the diameter of most particles ranging from 0.08 to 1 millimeter; in contrast to a duststorm, the sand particles are mostly confined to the lowest 2 meters above ground, rarely rising more than 11 meters.

sandur *See* outwash plain.

sand wave [GEOL] A large, ridgelike primary structure resembling a water wave on the upper surface of a sedimentary bed that is formed by high-velocity air or water currents. Also known as sand ridge.

Sangamon [GEOL] The third interglacial stage of the Pleistocene epoch in North America, following the Illinoian glacial and preceding the Wisconsin.

sanidine [MINERAL] $KAlSi_3O_8$ An alkali feldspar mineral occurring in clear, glassy crystals embedded in unaltered acid volcanic rocks; a high-temperature, disordered form. Also known as glassy feldspar; ice spar; rhyacolite.

sanitary engineering [CIV ENG] A field of civil engineering concerned with works and projects for the protection and promotion of public health.

sanitary landfill [CIV ENG] The disposal of garbage by spreading it in layers covered with soil or ashes to a depth sufficient to control rats, flies, and odors.

sanitary sewer [CIV ENG] A sewer which is restricted to carrying sewage and to which storm and surface waters are not admitted.

sanitation [CIV ENG] The act or process of making healthy environmental conditions.

San Joaquin Valley fever *See* coccidioidomycosis.

SA node *See* sinoauricular node.

Santalaceae [BOT] A family of parasitic dicotyledonous plants in the order Santalales characterized by dry or fleshy indehiscent fruit, plants with chlorophyll, petals absent, and ovules without integument.

Santalales [BOT] An order of dicotyledonous plants in the subclass Rosidae characterized by progressive adaptation to parasitism, accompanied by progressive simplification of the ovules.

sap [BOT] The fluid part of a plant which circulates through the vascular system and is composed of water, gases, salts, and organic products of metabolism.

saphenous nerve [ANAT] A somatic sensory nerve arising from the femoral nerve and innervating the skin of the medial aspect of the leg, foot, and knee joint.

Sapindaceae [BOT] A family of dicotyledonous plants in the order Sapindales distinguished by mostly alternate leaves, usually one and less often two ovules per locule, and seeds lacking endosperm.

Sapindales [BOT] An order of mostly woody dicotyledonous plants in the subclass Rosidae with compound or lobed leaves and polypetalous, hypogynous to perigynous flowers with one or two sets of stamens.

saponification [CHEM] The process of converting chemicals into soap; involves the alkaline hydrolysis of a fat or oil, or the neutralization of a fatty acid.

saponite [MINERAL] A soft, soapy, white or light-buff to bluish or reddish trioctahedral montmorillonitic clay mineral consisting of hydrous magnesium aluminosilicate and occurring in masses in serpentine and basaltic rocks. Also known as bowlingite; mountain soap; piotine; soapstone.

Sapotaceae [BOT] A family of dicotyledonous plants in the order Ebenales characterized by a well-developed latex system.

sappare *See* kyanite.

sapphire [MINERAL] Any of the gem varieties of the mineral corundum, especially the blue variety, except those that have medium to dark tones of red that characterize ruby; hardness is 9 on Mohs scale, and specific gravity is near 4.00.

Saprist [GEOL] A suborder of the soil order Histosol consisting of residues in which plant structures have been largely obliterated by decay; saturated with water most of the time.

saprobic [BOT] Living on decaying organic matter; applied to plants and microorganisms.

Saprolegniales [MYCOL] An order of aquatic fungi belonging to the class Phycomycetes, having a mostly hyphal thallus and zoospores with two flagella.

saprolite [GEOL] A soft, earthy red or brown, decomposed igneous or metamorphic rock that is rich in clay and formed in place by chemical weathering. Also known as saprolith; sathrolith.

saprolith *See* saprolite.

sapropel [GEOL] A mud, slime, or ooze deposited in more or less open water.

saprophage [BIOL] An organism that lives on decaying organic matter.

sapwood [BOT] The younger, softer, outer layers of a woody stem, between the cambium and heartwood. Also known as alburnum.

Sapygidae [INV ZOO] A family of hymenopteran insects in the superfamily Scolioidea.

Sarcodina [INV ZOO] A superclass of Protozoa in the subphylum Sarcomastigophora in which movement involves protoplasmic flow, often with recognizable pseudopodia.

sarcolemma [HISTOL] The thin connective tissue sheath enveloping a muscle fiber.

sarcoma [MED] A malignant tumor arising in connective tissue and composed principally of anaplastic cells that resemble those of supportive tissues.

Sarcomastigophora [INV ZOO] A subphylum of Protozoa comprising forms that possess flagella or pseudopodia or both.

sarcomere [HISTOL] One of the segments defined by Z disks in a skeletal muscle fibril.

Sarcophagidae [INV ZOO] A family of the myodarian orthorrhaphous dipteran insects in the subsection Calyptratae comprising flesh flies, blowflies, and scavenger flies.

sarcopside [MINERAL] $(Fe,Mn,Mg)_3(PO_4)_2$ A mineral composed of a phosphate of manganese, magnesium, and iron.

Sarcopterygii [VERT ZOO] A subclass of Osteichthyes, including Crossopterygii and Dipnoi in some systems of classification.

Sarcoptiformes [INV ZOO] A suborder of the Acarina including minute globular mites without stigmata.

Sarcosporida [INV ZOO] An order of Protozoa of the class Haplosporea which comprises parasites in skeletal and cardiac muscle of vertebrates.

sard [MINERAL] A translucent brown, reddish-brown, or deep orange-red variety of chalcedony. Also known as sardine; sardius.

sardine [MINERAL] *See* sard. [VERT ZOO] **1.** *Sardina pilchardus*. The young of the pilchard, a herringlike fish in the family Clupeidae found in the Atlantic along the European coasts. **2.** The young of any of various similar and related forms which are processed and eaten as sardines.

sardius *See* sard.

sarkinite [MINERAL] $Mn_2(AsO_4)(OH)$ A flesh-red monoclinic mineral composed of hydrous manganese arsenate, occurring in crystals.

Sarmatian [GEOL] A European stage of geologic time: the upper Miocene, above Tortonian, below Pontian.

sarmientite [MINERAL] $Fe_2(AsO_4)(SO_4)(OH)\cdot5H_2O$ A yellow mineral composed of basic hydrous arsenate and sulfate of iron; it is isomorphous with diadochite.

Sarothriidae [INV ZOO] The equivalent name for Jacobsoniidae.

Sarraceniaceae [BOT] A small family of dicotyledonous plants in the order Sarraceniales in which leaves are modified to form pitchers, placentation is axile, and flowers are perfect with distinct filaments.

Sarraceniales [BOT] An order of dicotyledonous herbs or shrubs in the subclass Dilleniidae; plants characteristically have alternate, simple leaves that are modified for catching insects, and grow in waterlogged soils.

sarsaparilla [BOT] Any of various tropical American vines of the genus *Smilax* (family Liliaceae) found in dense, moist jungles; a flavoring material used in medicine and soft drinks is obtained from the dried roots of at least four species.

sartorius [ANAT] A large muscle originating in the anterior superior iliac spine and inserting in the tibia; flexes the hip and knee joints, and rotates the femur laterally.

sassafras [BOT] *Sassafras albidum*. A medium-sized tree of the order Magnoliales recognized by the bright-green color and aromatic odor of the leaves and twigs.

sassoline *See* sassolite.

sassolite [MINERAL] H_3BO_3 A white or gray mineral consisting of native boric acid usually occurring in small pearly scales as an incrustation or as tabular triclinic crystals. Also known as sassoline.

satellite [AERO ENG] *See* artificial satellite. [ASTRON] A small, solid body moving in an orbit around a planet; the moon is a satellite of earth. [CYTOL] A chromosome segment distant from but attached to the rest of the chromosome by an achromatic filament.

satellite computer [COMPUT SCI] A computer which, under control of the main computer, handles the input and output routines, thereby allowing the main computer to be fully dedicated to computations.

sathrolith *See* saprolite.

satin spar [MINERAL] A white, translucent, fine fibrous variety of gypsum having a silky luster. Also known as satin stone.

satin stone *See* satin spar.

saturable-core reactor *See* saturable reactor.

saturable reactor [ELECTROMAG] An iron-core reactor having an additional control winding that carries direct current whose value is adjusted to change the degree of saturation of the core, thereby changing the reactance that the alternating-current winding offers to the flow of alternating current; with appropriate external circuits, a saturable reactor can serve

as a magnetic amplifier. Also known as saturable-core reactor; transductor.

saturable transformer [ELECTROMAG] A saturable reactor having additional windings to provide voltage transformation or isolation from the alternating-current supply.

saturated air [METEOROL] Moist air in a state of equilibrium with a plane surface of pure water or ice at the same temperature and pressure; that is, air whose vapor pressure is the saturation vapor pressure and whose relative humidity is 100%.

saturated compound [ORG CHEM] An organic compound with all carbon bonds satisfied; it does not contain double or triple bonds and thus cannot add elements or compounds.

saturated diode [ELECTR] A diode that is passing the maximum possible current, so further increases in applied voltage have no effect on current.

saturated interference spectroscopy [SPECT] A version of saturation spectroscopy in which the gas sample is placed inside an interferometer that splits a probe laser beam into parallel components in such a way that they cancel on recombination; intensity changes in the recombined probe beam resulting from changes in absorption or refractive index induced by a laser saturating beam are then measured.

saturated liquid [CHEM] A solution that contains enough of a dissolved solid, liquid, or gas so that no more will dissolve into the solution at a given temperature and pressure.

saturated surface *See* water table.

saturated vapor [THERMO] A vapor whose temperature equals the temperature of boiling at the pressure existing on it.

saturated zone *See* zone of saturation.

saturation [ELECTR] **1.** The condition that occurs when a transistor is driven so that it becomes biased in the forward direction (the collector becomes positive with respect to the base, for example, in a *pnp* type of transistor). **2.** *See* anode saturation. **3.** *See* temperature saturation. [ELECTROMAG] *See* magnetic saturation. [NUCLEO] **1.** The condition in which the decay rate of a given radionuclide is equal to its rate of production in an induced nuclear reaction. **2.** The condition in which the voltage applied to an ionization chamber is high enough to collect all the ions formed by radiation but not high enough to produce ionization by collision. [OPTICS] *See* color saturation. [ORD] The striking of a target area with such numbers of missiles that no place in it remains untouched by destruction. [PHYS] The condition in which a further increase in some cause produces no further increase in the resultant effect. [PHYS CHEM] The condition in which the partial pressure of any fluid constituent is equal to its maximum possible partial pressure under the existing environmental conditions, such that any increase in the amount of that constituent will initiate within it a change to a more condensed state.

saturation current [ELECTR] **1.** In general, the maximum current which can be obtained under certain conditions. **2.** In a vacuum tube, the space-charge-limited current, such that further increase in filament temperature produces no specific increase in anode current. **3.** In a vacuum tube, the temperature-limited current, such that a further increase in anode-cathode potential difference produces only a relatively small increase in current. **4.** In a gaseous-discharge device, the maximum current which can be obtained for a given mode of discharge. **5.** In a semiconductor, the maximum current which just precedes a change in conduction mode. [NUCLEO] The ionization current in a gas tube when the applied potential is large enough to collect all ions produced by ionizing radiation.

saturation diving [PHYSIO] Diving in which the tissues exposed to high pressure at great ocean depths for 24 hours become saturated with gases, especially inert gases, thereby reaching a new equilibrium state.

saturation limiting [ELECTR] Limiting the minimum output voltage of a vacuum-tube circuit by operating the tube in the region of plate-current saturation (not to be confused with emission saturation).

saturation mixing ratio [METEOROL] A thermodynamic function of state; the value of the mixing ratio of saturated air at the given temperature and pressure; this value may be read directly from a thermodynamic diagram.

saturation of forces [PHYS] Property exhibited by forces between particles wherein each particle can interact strongly with only a limited number of other particles, as in the forces between atoms in a molecule, and between nucleons in a nucleus.

saturation spectroscopy [SPECT] A branch of spectroscopy in which the intense, monochromatic beam produced by a laser is used to alter the energy-level populations of a resonant medium over a narrow range of particle velocities, giving rise to extremely narrow spectral lines that are free from Doppler broadening; used to study atomic, molecular, and nuclear structure, and to establish accurate values for fundamental physical constants.

Saturn [AERO ENG] One of the very large launch vehicles built primarily for the Apollo program; begun by Army Ordnance but turned over to the National Aeronautics and Space Administration for the manned space flight program to the moon. [ASTRON] The second largest planet in the solar system (mass is 95.3 compared to earth's 1) and the sixth in the order of distance to the sun; it is visible to the naked eye as a yellowish first-magnitude star except during short periods near its conjunction with the sun; it is surrounded by a series of rings.

Saturniidae [INV ZOO] A family of medium- to large-sized moths in the superfamily Saturnioidea including the giant silkworms, characterized by reduced, often vestigial, mouthparts and strongly bipectinate antennae.

Saturnioidea [INV ZOO] A superfamily of medium- to very-large-sized moths in the suborder Heteroneura having the frenulum reduced or absent, reduced mouthparts, no tympanum, and pectinate antennae.

Satyrinae [INV ZOO] A large, cosmopolitan subfamily of lepidopterans in the family Nymphalidae, containing the wood nymphs, meadow browns, graylings, and arctics, characterized by bladderlike swellings at the bases of the forewing veins.

Sauria [VERT ZOO] The lizards, a suborder of the Squamata, characterized generally by two or four limbs but sometimes none, movable eyelids, external ear openings, and a pectoral girdle.

Saurichthyidae [PALEON] A family of extinct chondrostean fishes bearing a superficial resemblance to the Aspidorhynchiformes.

Saurischia [PALEON] The lizard-hipped dinosaurs, an order of extinct reptiles in the subclass Archosauria characterized by an unspecialized, three-pronged pelvis.

Sauropoda [PALEON] A group of fully quadrupedal, seemingly herbivorous dinosaurs from the Jurassic and Cretaceous periods in the suborder Sauropodomorpha; members had small heads, spoon-shaped teeth, long necks and tails, and columnar legs.

Sauropodomorpha [PALEON] A suborder of extinct reptiles in the order Saurischia, including large, solid-limbed forms.

Sauropterygia [PALEON] An order of Mesozoic marine reptiles in the subclass Euryapsida.

Saururaceae [BOT] A family of dicotyledonous plants in the order Piperales distinguished by mostly alternate leaves, two to ten ovules per carpel, and carpels distinct or united into a compound ovary.

sausage instability See kink instability.

savanna [ECOL] Any of a variety of physiognomically or environmentally similar vegetation types in tropical and extratropical regions; all contain grasses and one or more species of trees of the families Leguminosae, Bombacaceae, Bignoniaceae, or Dilleniaceae.

SAW See surface acoustic wave.

sawfish [VERT ZOO] Any of several elongate viviparous fishes of the family Pristidae distinguished by a dorsoventrally flattened elongated snout with stout toothlike projections along each edge.

sawtooth generator [ELECTR] A generator whose output voltage has a sawtooth waveform; used to produce sweep voltages for cathode-ray tubes.

sawtooth pulse [ELECTR] An electric pulse having a linear rise and a virtually instantaneous fall, or conversely, a virtually instantaneous rise and a linear fall.

sawtooth waveform [ELECTR] A waveform characterized by a slow rise time and a sharp fall, resembling a tooth of a saw.

saxicolous [ECOL] Living or growing among rocks.

Saxifragaceae [BOT] A family of dicotyledonous plants in the order Rosales which are scarcely or not at all succulent and have two to five carpels usually more or less united, and leaves not modified into pitchers.

sb See stilb.

Sb See antimony.

Sc See scandium.

scabies [MED] A contagious skin disorder caused by the mite Sarcoptes scabiei burrowing beneath the skin, causing the formation of multiform lesions with intense itching.

scabrous [BIOL] Having a rough surface covered with stiff hairs or scales.

scacchite [MINERAL] $MnCl_2$ A mineral composed of native manganese chloride, found in volcanic regions.

scalar [MATH] One of the algebraic quantities which form a field, usually the real or complex numbers, by which the vectors of a vector space are multiplied. [PHYS] **1.** A quantity which has magnitude only and no direction, in contrast to a vector. **2.** A quantity which has magnitude only, and has the same value in every coordinate system. Also known as scalar invariant.

scalar field [MATH] **1.** The field consisting of the scalars of a vector space. **2.** A function on a vector space into the scalars of the vector space. [PHYS] A field which is characterized by a function of position and time whose value at each point is a scalar.

scalar function [MATH] A function from a vector space to its scalar field. [PHYS] A function of position and time whose value at each point is a scalar.

scalar invariant See scalar.

scalar potential [PHYS] A scalar function whose negative gradient is equal to some vector field, at least when this field is time-independent; for example, the potential energy of a particle in a conservative force field, and the electrostatic potential.

scale [ACOUS] A series of musical notes arranged from low to high by a specified scheme of intervals suitable for musical purposes. [BOT] The bract of a catkin. [ENG] **1.** A series of markings used for reading the value of a quantity or setting. **2.** To change the magnitude of a variable in a uniform way, as by multiplying or dividing by a constant factor, or the ratio of the real thing's magnitude to the magnitude of the model or analog of the model. **3.** A weighing device. **4.** A ruler or other measuring stick. An indication of represented to actual distances on a map, chart or drawing. [MET] A thick metallic oxide coating formed usually by heating metals in air. [PHYS] **1.** A one-to-one correspondence between numbers and the value of some physical quantity, such as the centigrade or Kelvin temperature scales on the API or Baumé scales of specific gravity. **2.** To determine a quantity at some order of magnitude by using data or relationships which are known to be valid at other (usually lower) orders of magnitude. [VERT ZOO] A flat calcified or cornified platelike structure on the skin of most fishes and of some tetrapods.

scale factor [ENG] The factor by which the reading of an instrument or the solution of a problem should be multiplied to give the true final value when a corresponding scale factor is used initially to bring the magnitude within the range of the instrument or computer.

scale insect [INV ZOO] Any of various small, structurally degenerate homopteran insects in the superfamily Coccoidea which resemble scales on the surface of a host plant.

scalene triangle [MATH] A triangle where no two angles are equal.

scaler [ELECTR] A circuit that produces an output pulse when a prescribed number of input pulses is received. Also known as counter; scaling circuit.

scale scar [BOT] A mark left on a stem after bud scales have fallen off.

Scalibregmidae [INV ZOO] A family of mud-swallowing worms belonging to the Sedentaria and found chiefly in sublittoral and great depths.

scaling [BIOL] The removing of scales from fishes. [ELECTR] Counting pulses with a scaler when the pulses occur too fast for direct counting by conventional means. [ENG] Removing scale (rust or salt) from a metal or other surface. [GRAPHICS] Using a scale to measure dimensions in a scale drawing. [MECH] Expressing the terms in an equation of motion in powers of nondimensional quantities (such as a Reynolds number), so that terms of significant magnitude under conditions specified in the problem can be identified, and terms of insignificant magnitude can be dropped. [MET] **1.** Forming of a thick layer of metallic oxide on metals at high temperatures. **2.** Depositing of solid inorganic solutes from water on a metal surface, such as a cooling tube or boiler. [MIN ENG] Removing loose rocks and coal from the roof, walls, or face after blasting.

scaling circuit *See* scaler.

scaling factor [ELECTR] The number of input pulses per output pulse of a scaling circuit. Also known as scaling ratio. [ENG] Factor used in heat-exchange calculations to allow for the loss in heat conductivity of a material because of the development of surface scale, as inside pipelines and heat-exchanger tubes. [PHYS] A constant of proportionality which appears in a scaling law.

scaling law [PHYS] A law, stating that two quantities are proportional, which is known to be valid at certain orders of magnitude and is used to calculate the value of one of the quantities at another order of magnitude.

scaling ratio [ELECTR] *See* scaling factor. [ENG] The ratio of a certain property of a laboratory model to the same property in the natural prototype.

scallop [GEOL] *See* scalloping. [INV ZOO] Any of various bivalve mollusks in the family Pectinidae distinguished by radially ribbed valves with undulated margins.

scalloped upland [GEOL] The region near or at the divide of an upland into which glacial cirques have cut from opposite sides.

scaly bark *See* psorosis.

scan [ELECTR] The motion, usually periodic, given to the major lobe of an antenna; the process of directing the radio-frequency beam successively over all points in a given region of space. [ENG] **1.** To examine an area, a region in space, or a portion of the radio spectrum point by point in an ordered sequence; for example, conversion of a scene or image to an electric signal or use of radar to monitor an airspace for detection, navigation, or traffic control purposes. **2.** One complete circular, up-and-down, or left-to-right sweep of the radar, light, or other beam or device used in making a scan.

scandium [CHEM] A metallic group III element, symbol Sc, atomic number 21; melts at 1200°C; found associated with rare-earth elements.

scanner [COMPUT SCI] In character recognition, a magnetic or photoelectric device which converts the input character into corresponding electric signals for processing by electronic apparatus. [COMMUN] That part of a facsimile transmitter which systematically translates the densities of the elemental areas of the subject copy into corresponding electric signals. [ENG] **1.** Any device that examines an area or region point by point in a continuous systematic manner, repeatedly sweeping across until the entire area or region is covered; for example, a flying-spot scanner. **2.** A device that automatically samples, measures, or checks a number of quantities or conditions in sequence, as in process control.

scanning circuit *See* sweep circuit.

scanning electron microscope [ELECTR] A type of electron microscope in which a beam of electrons, a few hundred angstroms in diameter, systematically sweeps over the specimen; the intensity of secondary electrons generated at the point of impact of the beam on the specimen is measured, and the resulting signal is fed into a cathode-ray-tube display which is scanned in synchronism with the scanning of the specimen.

scanning HEED [PHYS] A form of HEED in which the diffracted electrons are directly measured electronically with a sensitive detector, and the diffraction pattern is recorded either by moving the detector across it or by deflecting the diffracted electrons across a stationary detector. Abbreviated SHEED.

scanning line [COMMUN] **1.** In television, a single, continuous, narrow strip which is determined by the process of scanning. **2.** Path traced by the scanning or recording spot in one sweep across the subject copy or record sheet.

scanning loss [ELECTROMAG] In a radar system employing a scanning antenna, the reduction in sensitivity (usually expressed in decibels) due to scanning across the target, compared with that obtained when the beam is directed constantly at the target.

scanning proton microprobe [ENG] An instrument used for determining the spatial distribution of trace elements in samples, in which a beam of energetic protons is focused on a narrow spot which is swept over the sample, and the characteristic x-rays emitted from the target are measured.

scanning radiometer [ENG] An image-forming system consisting of a radiometer which, by the use of a plane mirror rotating at 45° to the optical axis, can see a circular path normal to the instrument.

scanning spot *See* picture element.

scanning switch *See* commutator switch.

scanning yoke *See* deflection yoke.

Scapanorhychidae [VERT ZOO] The goblin sharks, a family of deep-sea galeoids in the isurid line having long, sharp teeth and a long, pointed rostrum.

Scaphidiidae [INV ZOO] The shining fungus beetles, a family of coleopteran insects in the superfamily Staphylinoidea.

Scaphopoda [INV ZOO] A class of the phylum Mollusca in which the soft body fits the external, curved and tapering, nonchambered, aragonitic shell which is open at both ends.

scapolite [MINERAL] A white, gray, or pale-green complex aluminosilicate of sodium and calcium belonging to the tectosilicate group of silicate minerals; crystallizes in the tetragonal system and is vitreous; hardness is 5–6 on Mohs scale, and specific gravity is 2.65–2.74. Also known as wernerite.

scapula [ANAT] The large, flat, triangular bone forming the back of the shoulder. Also known as shoulder blade.

scar [GEOL] **1.** A steep, rocky eminence, such as a cliff or precipice, where bare rock is well exposed. Also known as scaur; scaw. **2.** *See* shore platform. [MED] A permanent mark on the skin or other tissue, formed from connective-tissue replacement of tissue destroyed by a wound or disease process.

Scarabaeidae [INV ZOO] The lamellicorn beetles, a large cosmopolitan family of coleopteran insects in the superfamily Scarabaeoidea including the Japanese beetle and other agricultural pests.

Scarabaeoidea [INV ZOO] A superfamily of Coleoptera belonging to the suborder Polyphaga.

scarfing [MET] **1.** Cutting away of surface defects in metals by use of a gas torch. **2.** A forging process in which the ends of two pieces to be joined are tapered to avoid an enlarged joint.

scarf joint [DES ENG] A joint made by the cutting of overlapping mating parts so that the joint is not enlarged and the patterns are complementary, and securing them by glue, fasteners, welding, or other joining method.

Scaridae [VERT ZOO] The parrotfishes, a family of perciform fishes in the suborder Percoidei which have the teeth of the jaw generally coalescent.

scarlet fever [MED] An acute, contagious bacterial disease caused by *Streptococcus hemolyticus;* characterized by a papular, or rough, bright-red rash over the body, with fever, sore throat, headache, and vomiting occurring 2–3 days after contact with a carrier.

scarp *See* escarpment.

scarpline [GEOL] A relatively straight line of cliffs of considerable extent, produced by faulting or erosion along a fault.

Scatopsidae [INV ZOO] The minute black scavenger flies, a family of orthorrhaphous dipteran insects in the series Nematocera.

scatter diagram [PETR] *See* point diagram. [STAT] A plot of the pairs of values of two variates in rectangular coordinates.

scattering [ELECTROMAG] Diffusion of electromagnetic waves in a random manner by air masses in the upper atmosphere, permitting long-range reception, as in scatter propagation. Also known as radio scattering. [PHYS] **1.** The change in direction of a particle or photon because of a collision with another particle or a system. **2.** Diffusion of acoustic or electromagnetic waves caused by inhomogeneity or anisotropy of the transmitting medium. **3.** In general, causing a collection of entities to assume a less orderly arrangement.

scattering amplitude [QUANT MECH] A quantity, depending in general on the energy and scattering angle, which specifies the wave function of particles scattered in a collision, and whose squared modulus is proportional to the number of particles scattered in a given direction.

scattering coefficient [ELECTROMAG] One of the elements of the scattering matrix of a waveguide junction; that is, a transmission or reflection coefficient of the junction. [PHYS] The fractional decrease in intensity of a beam of electromagnetic radiation or particles per unit distance traversed, which results from scattering rather than absorption. Also known as dissipation coefficient.

scattering cross section [ELECTROMAG] The power of electromagnetic radiation scattered by an antenna divided by the incident power. [PHYS] The sum of the cross sections for elastic and inelastic scattering.

scattering function [ELECTROMAG] The intensity of scattered radiation in a given direction per lumen of flux incident upon the scattering material.

scattering layer [OCEANOGR] A layer of organisms in the sea which causes sound to scatter and to return echoes.

scattering loss [ELECTROMAG] The portion of the transmission loss that is due to scattering within the medium or roughness of the reflecting surface.

scattering matrix [ELECTROMAG] A square array of complex numbers consisting of the transmission and reflection coefficients of a waveguide junction. [QUANT MECH] A matrix which expresses the initial state in a scattering experiment in terms of the possible final states. Also known as collision matrix; S matrix.

scattering-matrix theory *See* S-matrix theory.

scatter loading [COMPUT SCI] The process of loading a program into main memory such that each section or segment of the program occupies a single, connected memory area but the several sections of the program need not be adjacent to each other.

scaur *See* scar.

scaw *See* scar.

Scelionidae [INV ZOO] A family of small, shining wasps in the superfamily Proctotrupoidea, characterized by elbowed, 11- or 12-segmented antennae.

scene analysis *See* picture segmentation.

scent gland [VERT ZOO] A specialized skin gland of the tubuloalveolar or acinous variety which produces substances having peculiar odors; found in many mammals.

schach effect [AERO ENG] When a slowly rotating or nonrotating satellite is heated on its sunward side, the photons of thermal radiation carry away more momentum from the hot sunward side than the cold shadowed side, thereby giving the satellite a certain net acceleration in the direction away from the sun.

schafarzikite [MINERAL] $Fe_5Sb_4O_{11}$ A red to brown mineral composed of iron antimony oxide.

schairerite [MINERAL] $Na_3(SO_4)(F,Cl)$ A colorless rhombohedral mineral composed of sodium sulfate with fluorine and chlorine, occurring in crystals.

scheelite [MINERAL] $CaWO_4$ A yellowish-white mineral crystallizing in the tetragonal system and occurring in tabular or massive form in pneumatolytic veins associated with quartz; an ore of tungsten.

schefflerite [MINERAL] $(Ca,Mn)(Mg,Fe,Mn)Si_2O_6$ Brown to black variety of pyroxene that crystallizes in the monoclinic system and contains manganese and frequently iron.

schematic circuit diagram *See* circuit diagram.

schematic diagram [GRAPHICS] A presentation of the element-by-element relationship of all parts of a system.

Schering bridge [ELEC] A four-arm alternating-current bridge used to measure capacitance and dissipation factor; bridge balance is independent of frequency.

Schick test [IMMUNOL] A skin test for determining susceptibility to diphtheria performed by the intradermal injection of diluted diphtheria toxin; a positive reaction, showing edema and scaling after 5 to 7 days, indicates lack of immunity.

Schiff base [ORG CHEM] $RR'C\!=\!NR''$ Any of a class of derivatives of the condensation of aldehydes or ketones with primary amines; colorless crystals, weakly basic; hydrolyzed by water and strong acids to form carbonyl compounds and amines; used as chemical intermediates and perfume bases, in dyes and rubber accelerators, and in liquid crystals for electronics.

schiller *See* play of color.

Schindleriidae [VERT ZOO] The single family of the order Schindlerioidei.

Schindlerioidei [VERT ZOO] A suborder of fishes in the order Perciformes composed of one monogeneric family comprising two tiny oceanic species that are transparent and neotenic.

schist [GEOL] A large group of coarse-grained metamorphic rocks which readily split into thin plates or slabs as a result of the alignment of lamellar or prismatic minerals.

schistosity [GEOL] A type of cleavage characteristic of metamorphic rocks, notably schists and phyllites, in which the rocks tend to split along parallel planes defined by the distribution and parallel arrangement of platy mineral crystals.

Schistostegiales [BOT] A monospecific order of mosses; the small, slender, glaucous plants are distinguished by the luminous protonema.

schizogamy [BIOL] A form of reproduction involving division of an organism into a sexual and an asexual individual.

Schizogoniales [BOT] A small order of the Chlorophyta containing algae that are either submicroscopic filaments or macroscopic ribbons or sheets a few centimeters wide and attached by rhizoids to rocks.

schizogony [INV ZOO] Asexual reproduction by multiple fission of a trophozoite; a characteristic of certain Sporozoa.

Schizomeridaceae [BOT] A family of green algae in the order Ulvales.

Schizomycetes [MICROBIO] Formerly a class of the division Protophyta which included the bacteria.

Schizomycophyta [BOT] The designation for bacteria in those taxonomic systems that consider bacteria as plants.

Schizopathidae [INV ZOO] A family of dimorphic zoantharians in the order Antipatharia.

Schizophora [INV ZOO] A series of the dipteran suborder Cyclorrhapha in which adults possess a frontal suture through which a distensible sac, or ptilinum, is pushed to help the organism escape from its pupal case.

schizophrenia [PSYCH] A group of mental disorders characterized by withdrawal from reality and by alterations in thinking, feeling, and concept formations. Also known as dementia praecox.

Schizophyceae [MICROBIO] The blue-green algae, a class of the division Protophyta.

Schizophyta [BOT] The prokaryotes, a division of the plant subkingdom Thallobionta; includes the bacteria and blue-green algae.

Schizopteridae [INV ZOO] A family of minute ground-inhabiting hemipterans in the group Dipsocoeoidea; individuals characteristically live in leaf mold.

schlieren [OPTICS] In atmospheric optics, parcels or strata of air having densities sufficiently different from that of their surroundings so that they may be discerned by means of refraction anomalies in transmitted light. [PETR] Irregular streaks with shaded borders in some igneous rocks, representing the segregation of light and dark minerals or altered inclusions, elongated by flow.

Schmidt camera *See* Schmidt system.

Schmidt-Cassegrain telescope [OPTICS] A variant of the Schmidt system which uses a Schmidt corrector plate together with a pair of spheroidal or slightly aspherical mirrors arranged as in a Cassegrain telescope.

Schmidt objective *See* Schmidt system.

Schmidt optics *See* Schmidt system.

Schmidt system [OPTICS] An optical system designed to eliminate spherical aberration and coma, which, in its original form, consists of a spherical mirror, a Schmidt correction plate near the focus of the mirror, and usually a curved reflecting plate at the focus of the mirror; used in astronomical telescopes with unusually wide fields of view and in spectroscopes, and to project a television image from a cathode-ray tube onto a screen. Also known as projection optics; Schmidt camera; Schmidt objective; Schmidt optics.

Schoenbiinae [INV ZOO] A subfamily of moths in the family Pyralididae, including the genus *Acentropus*, the most completely aquatic Lepidoptera.

schorl *See* schorlite.

schorlite [MINERAL] The black, iron-rich, opaque variety of tourmaline. Also known as schorl.

schorlomite [MINERAL] $Ca_3(Fe,Ti)_2(Si,Ti)_3O_{12}$ Black mineral of the garnet group that has a vitreous luster and usually occurs in masses; hardness is 7–7.5 on Mohs scale, and specific gravity is 3.81–3.88.

Schottky barrier [ELECTR] A transition region formed within a semiconductor surface to serve as a rectifying barrier at a junction with a layer of metal.

Schottky barrier diode [ELECTR] A semiconductor diode formed by contact between a semiconductor layer and a metal coating; it has a nonlinear rectifying characteristic; hot carriers (electrons for n-type material or holes for p-type material) are emitted from the Schottky barrier of the semiconductor and move to the metal coating that is the diode base; since majority carriers predominate, there is essentially no injection or storage of minority carriers to limit switching speeds. Also known as hot-carrier diode; Schottky diode.

Schottky diode *See* Schottky barrier diode.

Schottky line [SOLID STATE] A graph of the logarithm of the saturation current from a thermionic cathode as a function of the square root of anode voltage; it is a straight line according to the Schottky theory.

Schottky noise See shot noise.

Schottky theory [SOLID STATE] A theory describing the rectification properties of the junction between a semiconductor and a metal that result from formation of a depletion layer at the surface of contact.

schreibersite [MINERAL] $(Fe,Ni)_3P$ A silver-white to tin-white magnetic meteorite mineral crystallizing in the tetragonal system and occurring in tables or plates as oriented inclusions in iron meteorites. Also known as rhabdite.

Schrödinger equation [QUANT MECH] A partial differential equation governing the Schrödinger wave function ψ of a system of one or more nonrelativistic particles; $\hbar(\partial\psi/\partial t) = H\psi$, where H is a linear operator, the Hamiltonian, which depends on the dynamics of the system, and $\hbar$ is Planck's constant divided by 2π.

Schrödinger-Klein-Gordon equation See Klein-Gordon equation.

Schrödinger's wave mechanics [QUANT MECH] The version of nonrelativistic quantum mechanics in which a system is characterized by a wave function which is a function of the coordinates of all the particles of the system and time, and obeys a differential equation, the Schrödinger equation; physical quantities are represented by differential operators which may act on the wave function and expectation values of measurements are equal to integrals involving the corresponding operator and the wave function. Also known as wave mechanics.

Schrödinger wave function [QUANT MECH] A function of the coordinates of the particles of a system and of time which is a solution of the Schrödinger equation and which determines the average result of every conceivable experiment on the system. Also known as probability amplitude; psi function; wave function.

Schubertellidae [PALEON] An extinct family of marine protozoans in the superfamily Fusulinacea.

Schuler pendulum [MECH] Any apparatus which swings, because of gravity, with a natural period of 84.4 minutes, that is, with the same period as a hypothetical simple pendulum whose length is the earth's radius; the pendulum arm remains vertical despite any motion of its pivot, and the apparatus is therefore useful in navigation.

schultenite [MINERAL] $PbHAsO_4$ A colorless mineral composed of lead hydrogen arsenate occurring in tabular orthorhombic crystals.

schuppen structure See imbricate structure.

Schwagerinidae [PALEON] A family of fusulinacean protozoans that flourished during the Early and Middle Pennsylvanian and became extinct during the Late Permian.

Schwann cell [HISTOL] One of the cells that surround peripheral axons forming sheaths of the neurilemma.

Schwarz-Christoffel transformations [MATH] Complex transformations which conformally map the interior of a given polygon onto the portion of the complex plane above the real axis.

Schwarz' inequality See Cauchy-Schwarz inequality.

Schwarzschild radius [RELAT] Conventionally taken to be twice the black hole mass appearing in the general relativistic Schwarzschild solution times the gravitational constant divided by the square of the

speed of light; the event horizon in a Schwarzschild solution is at the Schwarzschild radius.

Schwarzschild solution [RELAT] The unique solution of general relativity theory describing a nonrotating black hole in empty space.

Sciaenidae [VERT ZOO] A family of perciform fishes in the suborder Percoidei, which includes the drums.

sciatica [MED] Neuralgic pain in the lower extremities, hips, and back caused by inflammation or injury to the sciatic nerve.

sciatic nerve [ANAT] Either of a pair of long nerves that originate in the lower spinal cord and send fibers to the upper thigh muscles and the joints, skin, and muscles of the leg.

Scincidae [VERT ZOO] The skinks, a family of the reptilian suborder Sauria which have reduced limbs and snakelike bodies.

Scinidae [INV ZOO] A family of bathypelagic, amphipod crustaceans in the suborder Hyperiidea.

scintillation [ELECTROMAG] 1. A rapid apparent displacement of a target indication from its mean position on a radar display; one cause is shifting of the effective reflection point on the target. Also known as target glint; target scintillation; wander. 2. Random fluctuation, in radio propagation, of the received field about its mean value, the deviations usually being relatively small. [LAP] The flashing, twinkling, or sparkling of light, or the alternating display of reflections, from the polished facets of a gemstone. [NUCLEO] A flash of light produced in a phosphor by an ionizing particle or photon. [OPTICS] Rapid changes of brightness of stars or other distant, celestial objects caused by variations in the density of the air through which the light passes.

scintillation counter [NUCLEO] A device in which the scintillations produced in a fluorescent material by an ionizing radiation are detected and counted by a multiplier phototube and associated circuits; used in medical and nuclear research and in prospecting for radioactive ores. Also known as scintillation detector; scintillometer.

scintillation detector See scintillation counter.

scintillometer See scintillation counter.

Sciomyzidae [INV ZOO] A family of myodarian cyclorrhaphous dipteran insects in the subsection Acalypteratae.

scion [BOT] A section of a plant, usually a stem or bud, which is attached to the stock in grafting.

scissor engine See cat-and-mouse engine.

scissors fault [GEOL] A fault on which the offset or separation along the strike increases in one direction from an initial point and decreases in the other direction. Also known as differential fault.

Scitaminales [BOT] An equivalent name for Zingiberales.

Scitamineae [BOT] An equivalent name for Zingiberales.

Sciuridae [VERT ZOO] A family of rodents including squirrels, chipmunks, marmots, and related forms.

Sciuromorpha [VERT ZOO] A suborder of Rodentia according to the classical arrangement of the order.

sclera [ANAT] The hard outer coat of the eye, continuous with the cornea in front and with the sheath of the optic nerve behind.

Scleractinia [INV ZOO] An order of the subclass Zoantharia which comprises the true, or stony, corals; these are solitary or colonial anthozoans which attach to a firm substrate.

Scleraxonia [INV ZOO] A suborder of coelenterates in the order Gorgonacea in which the axial skeleton has calcareous spicules.

sclerenchyma [BOT] A supporting plant tissue composed principally of sclereids whose walls are often mineralized.

scleriasis *See* scleroderma.

sclerite [INV ZOO] One of the sclerotized plates of the integument of an arthropod.

Sclerodactylidae [INV ZOO] A family of echinoderms in the order Dendrochirotida having a complex calcareous ring.

scleroderma [MED] An abnormal increase in collagenous connective tissue in the skin. Also known as chorionitis; dermatosclerosis; scleriasis.

Sclerogibbidae [INV ZOO] A monospecific family of the hymenopteran superfamily Bethyloidea.

sclerophyllous [BOT] Characterized by thick, hard foliage due to well-developed sclerenchymatous tissue.

scleroprotein [BIOCHEM] Any one of a class of proteins, such as keratin, fibroin, and the collagens, which occur in hard parts of the animal body and serve to support or protect. Also known as albuminoid.

scleroscope hardness test *See* Shore scleroscope hardness test.

sclerosis [PATH] Hardening of a tissue, especially by proliferation of fibrous connective tissue.

scolecite [MINERAL] $CaAl_2Si_3O_{10}$ A zeolite mineral that occurs in delicate, radiating groups of white fibrous or acicular crystals; sometimes shows wormlike motion upon heating.

Scolecosporae [MYCOL] A spore group of Fungi Imperfecti characterized by filiform spores.

scolex [INV ZOO] The head of certain tapeworms, typically having a muscular pad with hooks, and two pairs of lateral suckers.

Scoliidae [INV ZOO] A family of the Hymenoptera in the superfamily Scolioidea.

Scolioidea [INV ZOO] A superfamily of hymenopteran insects in the suborder Apocrita.

Scolopacidea [VERT ZOO] A large, cosmopolitan family of birds of the order Charadriiformes including snipes, sandpipers, curlews, and godwits.

Scolopendridae [INV ZOO] A family of centipedes in the order Scolopendromorpha which characteristically possess eyes.

Scolopendromorpha [INV ZOO] An order of the chilopod subclass Pleurostigmophora containing the dominant tropical forms, and also the largest of the centipedes.

Scolytidae [INV ZOO] The bark beetles, a large family of coleopteran insects in the superfamily Curculionoidea characterized by a short beak and clubbed antennae.

Scombridae [VERT ZOO] A family of perciform fishes in the suborder Scombroidei including the mackerels and tunas.

Scombroidei [VERT ZOO] A suborder of fishes in the order Perciformes; all are moderate- to large-sized shore and oceanic fishes having fixed premaxillae.

scope *See* cathode-ray oscilloscope; radarscope.

Scopeumatidae [INV ZOO] The dung flies, a family of myodarian cyclorrhaphous dipteran insects in the subsection Calyptratae.

Scopidae [VERT ZOO] A family of birds in the order Ciconiiformes containing a single species, the hammerhead (*Scopus umbretta*) of tropical Africa.

scoria [GEOL] Vesicular, cindery, dark lava formed by the escape and expansion of gases in basaltic or andesitic magma; generally denser and darker than pumice. [MATER] Refuse after melting metals or reducing ore.

scorification [MET] Concentration of precious metals, such as gold and silver, in molten lead by oxidation employing appropriate fluxes.

scorodite [MINERAL] $FeAsO_4 \cdot 2H_2O$ A pale leek-green or liver-brown orthorhombic mineral consisting of ferric arsenate; isomorphous with mansfieldite and represents a minor ore of arsenic.

Scorpaenidae [VERT ZOO] The scorpion fishes, a family of Perciformes in the suborder Cottoidei, including many tropical shorefishes, some of which are venomous.

Scorpaeniformes [VERT ZOO] An order of fishes coextensive with the perciform suborder Cottoidei in some systems of classification.

scorpion [INV ZOO] The common name for arachnids constituting the order Scorpionida.

Scorpion [AERO ENG] An all-weather interceptor aircraft with twin turbojet engines; its armament consists of air-to-air rockets with nuclear or nonnuclear warheads. Designated F-89. [ASTRON] *See* Scorpius.

Scorpionida [VERT ZOO] The scorpions, an order of arachnids characterized by a shieldlike carapace covering the cephalothorax and by large pedipalps armed with chelae.

Scorpius [ASTRON] A southern constellation, right ascension 16 hours, declination 40° south; the bright-red star Antares is located in it. Also known as Scorpion.

scotch boiler [MECH ENG] A fire-tube boiler with one or more cylindrical internal furnaces enveloped by a boiler shell equipped with fire tubes in its upper part; heat is transferred to water partly in the furnace area and partly in passage of hot gases through the tubes. Also known as dry-back boiler; scotch marine boiler (marine usage).

Scotch bond *See* American bond.

scotch marine boiler *See* scotch boiler.

Scotch mist [METEOROL] A combination of thick mist (or fog) and heavy drizzle occurring frequently in Scotland and in parts of England.

scotopic vision [PHYSIO] Vision that is due to the activity of the rods of the retina only; it is the type of vision that occurs at very low levels of illumination, and it can detect differences of brightness but not of hue.

scouring [ENG] Physical or chemical attack on process equipment surfaces, as in a furnace or fluid catalytic cracker. [GEOL] 1. An erosion process resulting from the action of the flow of air, ice, or water. 2. *See* glacial scour. [MATER] *See* attrition. [MECH ENG] Mechanical finishing or cleaning of a hard surface by using an abrasive and low pressure. [TEXT] 1. Removal of grease and dirt from wool. 2. The cleaning of fabric before the dyeing step.

scouring basin [CIV ENG] A basin containing impounded water which is released at about low water in order to maintain the desired depth in the entrance channel. Also known as sluicing pond.

scout hole [MIN ENG] 1. A borehole penetrating only the uppermost part of an ore body in order to delineate the surface configuration. 2. A shallow borehole used to ascertain the presence of ore or to explore an area in a preliminary manner.

SCR *See* silicon controlled rectifier.

scram [NUCLEO] 1. A sudden shutting down of a nuclear reactor, usually by dropping safety rods, when a predetermined neutron flux or other dangerous condition occurs. 2. To close down a reactor by bringing about a scram.

scrambler [ELECTR] A circuit that divides speech frequencies into several ranges by means of filters, then inverts and displaces the frequencies in each range so that the resulting reproduced sounds are unintelligible; the process is reversed at the receiving apparatus to restore intelligible speech. Also known as speech inverter; speech scrambler.

scramjet [AERO ENG] Essentially a ramjet engine, intended for flight at hypersonic speeds. Derived from supersonic combustion ramjet.

scram rod *See* safety rod.

Scraptiidae [INV ZOO] An equivalent name for Melandryidae.

scratch filter [ENG ACOUS] A low-pass filter circuit inserted in the circuit of a phonograph pickup to suppress higher audio frequencies and thereby minimize needle-scratch noise.

scratch hardness [MATER] A measure of the resistance of minerals or metals to scratching; for minerals it is defined by comparison with 10 selected minerals which are numbered in order of increasing hardness according to the Mohs scale.

scree [GEOL] **1.** A mound of loose, angular material, less than 10 centimeters. **2.** *See* talus.

screed wire *See* ground wire.

screen [COMPUT SCI] To make a preliminary selection from a set of entities, selection criteria being based on a given set of rules or conditions. [ELECTR] **1.** The surface on which a television, radar, x-ray, or cathode-ray oscilloscope image is made visible for viewing; it may be a fluorescent screen with a phosphor layer that converts the energy of an electron beam to visible light, or a translucent or opaque screen on which the optical image is projected. Also known as viewing screen. **2.** *See* screen grid. [ELECTROMAG] Metal partition or shield which isolates a device from external magnetic or electric fields. [ENG] **1.** A large sieve of suitably mounted wire cloth, grate bars, or perforated sheet iron used to sort rock, ore, or aggregate according to size. **2.** A covering to give physical protection from light, noise, heat, or flying particles. **3.** A filter medium for liquid-solid separation.

screen grid [ELECTR] A grid placed between a control grid and an anode of an electron tube, and usually maintained at a fixed positive potential, to reduce the electrostatic influence of the anode in the space between the screen grid and the cathode. Also known as screen.

screening [ATOM PHYS] The reduction of the electric field about a nucleus by the space charge of the surrounding electrons. [ELECTROMAG] *See* electric shielding. [ENG] **1.** The separation of a mixture of grains of various sizes into two or more size-range portions by means of a porous or woven-mesh screening media. **2.** The removal of solid particles from a liquid-solid mixture by means of a screen. **3.** The material that has passed through a screen. [IND ENG] The elimination of defective pieces from a lot by inspection for specified defects. Also known as detailing.

screen printing [GRAPHICS] A method of printing in which ink is forced by a rubber squeegee through a silk, paint, or stencil screen (as through a sieve or strainer) onto a paper or a fabric. Also known as stencil printing.

screw [DES ENG] **1.** A cylindrical body with a helical groove cut into its surface. **2.** A fastener with continuous ribs on a cylindrical or conical shank and a slotted, recessed, flat, or rounded head. Also known as screw fastener.

screw axis [CRYSTAL] A symmetry element of some crystal lattices, in which the lattice is unaltered by a rotation about the axis combined with a translation parallel to the axis and equal to a fraction of the unit lattice distance in this direction.

screw conveyor [MECH ENG] A conveyor consisting of a helical screw that rotates upon a single shaft within a stationary trough or casing, and which can move bulk material along a horizontal, inclined, or vertical plane. Also known as auger conveyor; spiral conveyor; worm conveyor.

screw dislocation [CRYSTAL] A dislocation in which atomic planes form a spiral ramp winding around the line of the dislocation.

screw fastener *See* screw.

screwfeed [MECH ENG] A system or combination of gears, ratchets, and friction devices in the swivel head of a diamond drill, which controls the rate at which a bit penetrates a rock formation.

screw jack *See* jackscrew.

screw propeller [MECH ENG] A marine and airplane propeller consisting of a streamlined hub attached outboard to a rotating engine shaft on which are mounted two to six blades; the blades form helicoidal surfaces in such a way as to advance along the axis about which they revolve.

screw thread [DES ENG] A helical ridge formed on a cylindrical core, as on fasteners and pipes.

scrod [VERT ZOO] A young fish, especially a cod.

scrolling [COMPUT SCI] The continuous movement of information either vertically or horizontally on a video screen.

Scrophulariaceae [BOT] A large family of dicotyledonous plants in the order Scrophulariales, characterized by a usually herbaceous habit, irregular flowers, axile placentation, and dry, dehiscent fruit.

Scrophulariales [BOT] An order of flowering plants in the subclass Asteridae distinguished by a usually superior ovary and, generally, either by an irregular corolla or by fewer stamens than corolla lobes, or commonly both.

scrubber [ENG] A device for the removal, or washing out, of entrained liquid droplets or dust, or for the removal of an undesired gas component from process gas streams. Also known as wet collector. [MIN ENG] A device, such as a wash screen, wash trommel, log washer, and hydraulic jet or monitor, in which a coarse and sticky material, for example, ore or clay, is either washed free of adherents or mildly disintegrated.

SCS *See* silicon controlled switch.

scuba diving [ENG] Any of various diving techniques using self-contained underwater breathing apparatus.

sculpin [VERT ZOO] Any of several species of small fishes in the family Cottidae characterized by a large head that sometimes has spines, spiny fins, broad mouth, and smooth, scaleless skin.

Sculptor [ASTRON] A southern constellation, right ascension 0 hours, declination 30° south. Also known as Sculptor's Apparatus; Workshop.

Sculptor's Apparatus *See* Sculptor.

S curve *See* reverse curve.

scurvy [MED] An acute or chronic nutritional disorder due to vitamin C deficiency; characterized by weakness, subcutaneous hemorrhages, and alterations of any tissue containing collagen, ground substance, dentine, intercellular cement, or osteoid.

Scutechiniscidae [INV ZOO] A family of heterotardigrades in the suborder Echiniscoidea, with segmental and intersegmental thickenings of cuticle.

Scutelleridae [INV ZOO] The shield bugs, a family of Hemiptera in the superfamily Pentatomoidea.

Scutigeromorpha [INV ZOO] The single order of notostigmophorous centipedes; members are distinguished by a dorsal respiratory opening, compound-type eyes, long flagellate multisegmental antennae, and long thin legs with multisegmental tarsi.

scutum [INV ZOO] 1. A bony, horny, or chitinous plate. 2. The second of four pieces forming the upper part of the thoracic segment in certain insects. 3. One or two lower opercular valves in certain barnacles.

Scutum [ASTRON] A southern constellation, right ascension 19 hours, declination 10° south. Also known as Shield.

Scydmaenidae [INV ZOO] The antlike stone beetles, a large cosmopolitan family of the Coleoptera in the superfamily Staphylinoidea.

Scyllaridae [INV ZOO] The Spanish, or shovel-nosed, lobsters, a family of the Scyllaridea.

Scyllaridea [INV ZOO] A superfamily of decapod crustaceans in the section Macrura including the heavily armored spiny lobsters and the Spanish lobsters, distinguished by the absence of a rostrum and chelae.

Scylliorhinidae [VERT ZOO] The catsharks, a family of the cacharinid group of galeoids; members exhibit the most exotic color patterns of all sharks.

Scyphomedusae [INV ZOO] A subclass of the class Scyphozoa characterized by reduced marginal tentacles, tetramerous medusae, and medusalike polyploids.

Scyphozoa [INV ZOO] A class of the phylum Coelenterata; all members are marine and are characterized by large, well-developed medusae and by small, fairly well-organized polyps.

Scythian stage [GEOL] A stage in the lesser Triassic series of the alpine facies. Also known as Werfenian stage.

Se *See* selenium.

sea [GEOGR] A usually salty lake lacking an outlet to the ocean. [OCEANOGR] 1. A major subdivision of the ocean. 2. A heavy swell or ocean wave still under the influence of the wind that produced it. 3. *See* ocean.

sea anemone [INV ZOO] Any of the 1000 marine coelenterates that constitute the order Actiniaria; the adult is a cylindrical polyp or hydroid stage with the free end bearing tentacles that surround the mouth.

sea bank *See* seawall.

seabed *See* sea floor.

sea bottom *See* sea floor.

sea breeze [METEOROL] A coastal, local wind that blows from sea to land, caused by the temperature difference when the sea surface is colder than the adjacent land; it usually blows on relatively calm, sunny summer days, and alternates with the oppositely directed, usually weaker, nighttime land breeze.

sea cucumber [INV ZOO] The common name for the echinoderms that make up the class Holothuroidea.

sea fan [GEOL] *See* submarine fan. [INV ZOO] A form of horny coral that branches like a fan.

sea floor [GEOL] The bottom of the ocean. Also known as seabed; sea bottom.

sea-foam *See* sepiolite.

sea fog [METEOROL] A type of advection fog formed over the ocean as a result of any of a variety of processes, as when air that has been lying over a warm water surface is transported over a colder water surface, resulting in a cooling of the lower layer of air below its dew point.

sea gate [CIV ENG] A gate which serves to protect a harbor or tidal basin from the sea, such as one of a pair of supplementary gates at the entrance to a tidal basin exposed to the sea. [GEOGR] A way giving access to the sea such as a gate, channel, or beach.

sea horse [INV ZOO] Any of about 50 species of tropical and subtropical marine fishes constituting the genus *Hippocampus* in the family Syngnathidae; the body is compressed, the head is bent ventrally and has a tubiform snout, and the tail is tapering and prehensile.

sea ice [OCEANOGR] 1. Ice formed from seawater. 2. Any ice floating in the sea.

seal [ENG] 1. Any device or system that creates a nonleaking union between two mechanical or process-system elements; for example, gaskets for pipe connection seals, mechanical seals for rotating members such as pump shafts, and liquid seals to prevent gas entry to or loss from a gas-liquid processing sequence. 2. A tight, perfect closure or joint. [VERT ZOO] Any of various carnivorous mammals of the suborder Pinnipedia, especially the families Phoridae, containing true seals, and Otariidae, containing the eared and fur seals.

sea-lane *See* seaway.

sea level [GEOL] The level of the surface of the ocean; especially, the mean level halfway between high and low tide, used as a standard in reckoning land elevation or sea depths.

sea lion [VERT ZOO] Any of several large, eared seals of the Pacific Ocean; related to fur seals but lack a valuable coat.

seam [ENG] 1. A mechanical or welded joint. 2. A mark on ceramic or glassware where matching mold parts join. [GEOL] 1. A stratum or bed of coal or other mineral. 2. A thin layer or stratum of rock. 3. A very narrow coal vein. [MET] An unwelded fold or lap which appears as a crack on the surface of a casting or wrought product.

seamanite [MINERAL] $Mn_3(PO_4)(BO_3)\cdot3H_2O$ A pale-to wine-yellow orthorhombic mineral that is a phosphate and borate of manganese; occurs in crystals.

sea mist *See* steam fog.

seamount [GEOL] An elevation of the sea floor that is either flat-topped or peaked, rising to about 3000–1000 feet (900–300 meters) or more, with the summit approximately 1000–6000 feet (300–1800 meters) below sea level.

seam weld [MET] 1. A longitudinal weld joining of sheet-metal parts or in making tubing. 2. Arc or resistance welding in which a series of overlapping spot welds is produced.

sea otter [VERT ZOO] *Enhydra lutris.* A large marine otter found close to the shoreline in the North Pacific; these animals are diurnally active and live in herds.

sea pen [INV ZOO] The common name for coelenterates constituting the order Pennatulacea.

seaquake [GEOPHYS] An earth tremor whose epicenter is beneath the ocean and can be felt only by ships in the vicinity of the epicenter. Also known as submarine earthquake.

search [COMPUT SCI] To seek a desired item or condition in a set of related or similar items or conditions, especially a sequentially organized or nonorganized set, rather than a multidimensional set. [ENG] To explore a region in space with radar. [NAV] An orderly arrangement of course lines used in searching an area.

search coil *See* exploring coil.

searching lighting *See* horizontal scanning.

search radar [ENG] A radar intended primarily to cover a large region of space and to display targets as soon as possible after they enter the region; used for early warning, in connection with ground-controlled approach and interception, and in air-traffic control.

search time [COMPUT SCI] Time required to locate a particular field of data in a computer storage device; requires a comparison of each field with a predetermined standard until an identity is obtained.

sea rim [ASTRON] The apparent horizon as actually observed at sea.

searlesite [MINERAL] $NaB(SiO_3)_2 \cdot H_2O$ A white mineral composed of hydrous sodium borosilicate occurring as spherulites.

sea salt [OCEANOGR] The salt remaining after the evaporation of seawater, containing sodium and magnesium chlorides and magnesium and calcium sulfates.

seashore [GEOL] 1. The strip of land that borders a sea or ocean. Also known as seaside; shore. 2. The ground between the usual tide levels. Also known as seastrand.

seaside *See* seashore.

sea slug [INV ZOO] The common name for the naked gastropods composing the suborder Nudibranchia.

sea smoke *See* steam fog.

season [CLIMATOL] A division of the year according to some regularly recurrent phenomena, usually astronomical or climatic.

seasonally frozen ground [GEOL] Ground that is frozen during low temperatures and remains so only during the winter season. Also known as frost zone.

seasonal variation [GEOPHYS] The variation of any parameter of the upper atmosphere with season; for example, the variation of ion densities of different parts of the ionosphere, and the resulting variation in transmission of radio signals over large distances.

seasoned lumber [MATER] Lumber which has been cured by drying to ensure a uniform moisture content.

sea spider [INV ZOO] The common name for arthropods in the subphylum Pycnogonida.

sea squirt [INV ZOO] A sessile, marine tunicate of the class Ascidiacea; it squirts water from two openings in the unattached end when touched or disturbed.

sea state [OCEANOGR] The numerical or written description of ocean-surface roughness.

seastrand *See* seashore.

seat [MECH ENG] The fixed, pressure-containing portion of a valve which comes into contact with the moving portions of that valve. [ORD] 1. Support or holder for a mechanism, or for a part of one. 2. To fit correctly in or on a holder, or prepared position, such as to seat a fuse in a bomb, a projectile in the bore of a gun, or a cartridge in a chamber.

sea terrace *See* marine terrace.

sea turtle [VERT ZOO] Any of various marine turtles, principally of the families Cheloniidae and Dermochelidae, having paddle-shaped feet.

sea urchin [INV ZOO] A marine echinoderm of the class Echinoidea; the soft internal organs are enclosed in and protected by a test or shell consisting of a number of close-fitting plates beneath the skin.

seawall [CIV ENG] A concrete, stone, or metal wall or embankment constructed along a shore to reduce wave erosion and encroachment by the sea. Also known as sea bank. [GEOL] A steep-faced, long embankment situated by powerful storm waves along a seacoast at high-water mark.

seawater [OCEANOGR] Water of the seas, distinguished by high salinity. Also known as salt water.

seaway [NAV] 1. The motion or rate of motion of a vessel. 2. Headway of a vessel. 3. The sea as a route of travel from one place to another; a shipping lane. Also known as sea-lane.

seaweed [BOT] A marine plant, especially algae.

sebaceous gland [PHYSIO] A gland, arising in association with a hair follicle, which produces and liberates sebum.

Sebekidae [INV ZOO] A family of pentastomid arthropods in the suborder Porocephaloidea.

sebum [PHYSIO] The secretion of sebaceous glands, composed of fat, cellular debris, and keratin.

sec *See* secant; second; secondary winding.

secant [MATH] 1. The function given by the reciprocal of the cosine function. Abreviated sec. 2. The secant of an angle A is $1/\cos A$.

Secernentea [INV ZOO] A class of the phylum Nematoda in which the primary excretory system consists of intracellular tubular canals joined anteriorly and ventrally in an excretory sinus, into which two ventral excretory gland cells may also open.

secohm *See* international henry.

second [MATH] A unit of plane angle, equal to 1/60 minute, or 1/3,600 degree, or $\pi/648,000$ radian. [MECH] The fundamental unit of time equal to 9,192,631,770 periods of the radiation corresponding to the transition between the two hyperfine levels of the ground state of an atom of cesium-133. Abbreviated s; sec.

secondary [ELEC] Low-voltage conductors of a power distributing system. [ELECTROMAG] *See* secondary winding. [GEOL] A term with meanings that changed from early to late in the 19th century, when the term was confined to the entire Mesozoic era; it was finally replaced by Mesozoic era.

secondary battery *See* storage battery.

secondary bow *See* secondary rainbow.

secondary cell *See* storage cell.

secondary circle *See* secondary great circle.

secondary cold front [METEOROL] A front which forms behind a frontal cyclone and within a cold air mass, characterized by an appreciable horizontal temperature gradient.

secondary creep [MECH] The change in shape of a substance under a minimum and almost constant differential stress, with the strain-time relationship a constant. Also known as steady-state creep.

secondary cyclone [METEOROL] A cyclone which forms near or in association with a primary cyclone. Also known as secondary low.

secondary electron [ELECTR] 1. An electron emitted as a result of bombardment of a material by an incident electron. 2. An electron whose motion is due to a transfer of momentum from primary radiation.

secondary emission [ELECTR] The emission of electrons from the surface of a solid or liquid into a vacuum as a result of bombardment by electrons or other charged particles.

secondary enrichment [GEOL] The addition to a vein or ore body of material that originated later in time from the oxidation of decomposed ore masses that overlie the vein.

secondary front [METEOROL] A front which may form within a baroclinic cold air mass which itself is separated from a warm air mass by a primary frontal system; the most common type is the secondary cold front.

secondary great circle [GEOD] A great circle perpendicular to a primary great circle, as a meridian. Also known as secondary circle.

secondary hardening [MET] The hardening of certain alloy steels at moderate temperatures (250–650°C) by the precipitation of carbides; the resultant hardness is greater than that obtained by tempering the steel at some lower temperature for the same time.

secondary ion mass spectrometer [ENG] An instrument for microscopic chemical analysis, in which a beam of primary ions with an energy in the range 5–20 kiloelectronvolts bombards a small spot on the surface of a sample, and positive and negative secondary ions sputtered from the surface are analyzed in a mass spectrometer. Abbreviated SIMS. Also known as ion microprobe; ion probe.

secondary low *See* secondary cyclone.

secondary mineral [MINERAL] A mineral produced in an enclosing rock after the rock was formed as a result of weathering or metamorphic or solution activity, and usually at the expense of a primary material that came into existence earlier.

secondary oil recovery [PETRO ENG] Procedures used to increase the flow of oil from depleted or nearly depleted wells; includes fracturing, acidizing, waterflood, and gas injection.

secondary phloem [BOT] Phloem produced by the cambium, consisting of two interpenetrating systems, the vertical or axial and the horizontal or ray.

secondary radar [ELECTR] Radar which receives pulses transmitted by an interrogator and makes a return transmission (usually on a different frequency) by its transponder, as opposed to a primary radar which receives pulses returned from illuminated objects.

secondary radiation [PHYS] Particles or photons produced by the action of primary radiation on matter, such as Compton recoil electrons, delta rays, secondary cosmic rays, and secondary electrons.

secondary rainbow [OPTICS] A faint rainbow of angular radius about 50° which may appear outside the primary rainbow of 42° radius, and which has its colors in reverse order to those of the primary. Also known as secondary bow.

secondary reflection *See* multiple reflection; shoot.

secondary reserves [PETRO ENG] Reserves recoverable commercially at current prices and costs as a result of artificial supplementation of the reserve's natural (gas-drive) energy.

secondary storage [COMPUT SCI] Any means of storing and retrieving data external to the main computer itself but accessible to the program.

secondary stratification [GEOL] The layering that occurs when sediments that were at one time deposited are resuspended and redeposited. Also known as indirect stratification.

secondary stratigraphic trap *See* stratigraphic trap.

secondary stress [MECH] A self-limiting normal or shear stress which is caused by the constraint of a structure and which is expected to cause minor distortions that would not result in a failure of the structure.

secondary structure [GEOL] A structure such as a fault, fold, or joint resulting from tectonic movement that started after the rock in which it is found was emplaced. [PALEON] A coarse structure usually between the thin sheets in the protective wall of a tintinnid.

secondary wave [GEOPHYS] *See* S wave. [OPTICS] One of the waves that radiate from each point on a wavefront, according to Huygens' principle.

secondary winding [ELECTROMAG] A transformer winding that receives energy by electromagnetic induction from the primary winding; a transformer may have several secondary windings, and they may provide alternating-current voltages that are higher, lower, or the same as that applied to the primary winding. Abbreviated sec. Also known as secondary.

secondary xylem [BOT] Xylem produced by cambium, composed of two interpenetrating systems, the horizontal (ray) and vertical (axial).

second-channel interference *See* alternate-channel interference.

second generation *See* F_2.

second-generation computer [COMPUT SCI] A computer characterized by the use of transistors rather than vacuum tubes, the execution of input/output operations simultaneously with calculations, and the use of operating systems.

second law of motion *See* Newton's second law.

second law of thermodynamics [THERMO] A general statement of the idea that there is a preferred direction for any process; there are many equivalent statements of the law, the best known being those of Clausius and of Kelvin.

second-level controller [CONT SYS] A controller which influences the actions of first-level controllers, in a large-scale control system partitioned by plant decomposition, to compensate for subsystem interactions so that overall objectives and constraints of the system are satisfied. Also known as coordinator.

second-order equation [MATH] A differential equation where some term includes the second derivative of the unknown function and no derivative of higher order is present.

second-order leveling [ENG] Spirit leveling that has less stringent requirements than those of first-order leveling, in which lines between benchmarks established by first-order leveling are run in only one direction.

second-order reaction [PHYS CHEM] A reaction whose rate of reaction is determined by the concentration of two chemical species.

second-order subroutine [COMPUT SCI] A subroutine that is entered from another subroutine, in contrast to a first-order subroutine; it constitutes the second level of a two-level or higher-level routine. Also known as second-remove subroutine.

second-order transition [THERMO] A change of state through which the free energy of a substance and its first derivatives are continuous functions of temperature and pressure, or other corresponding variables.

second-remove subroutine *See* second-order subroutine.

second sound [ACOUS] A transverse sound wave which propagates in smectic liquid crystals, and whose behavior resembles mathematically that of second sound in superfluid helium. [CRYO] A type of wave propagated in the superfluid phase of liquid helium (helium II), in which temperature and entropy variations propagate with no appreciable variation in density or pressure.

secrecy system *See* privacy system.

secretion [GEOL] A secondary structure formed of material deposited (from solution) within an empty cavity in any rock, especially a deposit formed on or parallel to the walls of the cavity, the first layer being the outer one. [PHYSIO] **1.** The act or process of producing a substance which is specialized to perform a certain function within the organism or is excreted from the body. **2.** The material produced by such a process.

sectional header boiler [MECH ENG] A horizontal boiler in which tubes are assembled in sections into front and rear headers; the latter, in turn, are connected to the boiler drum by vertical tubes.

sectional radiography [ELECTR] The technique of making radiographs of plane sections of a body or an object; its purpose is to show detail in a predetermined plane of the body, while blurring the images of structures in other planes. Also known as laminography; planigraphy; tomography.

sector [COMPUT SCI] 1. A portion of a track on a magnetic disk or a band on a magnetic drum. 2. A unit of data stored in such a portion. [CIV ENG] A clearly defined area or airspace designated for a particular purpose. [ELECTROMAG] Coverage of a radar as measured in azimuth. [MATH] A portion of a circle bounded by two radii and an arc joining their end points. [METEOROL] Something resembling the sector of a circle, as a warm sector between the warm and cold fronts of a cyclone.

sector display [ELECTR] A display in which only a sector of the total service area of a radar system is shown; usually the sector is selectable.

sector gate [CIV ENG] A horizontal gate with a pie-slice cross section used to regulate the level of water at the crest of a dam; it is raised and lowered by a rack and pinion mechanism.

secular equilibrium [NUCLEO] Radioactive equilibrium in which the parent has such a small decay constant that there has been no appreciable change in the quantity of parent present by the time the decay products have reached radioactive equilibrium.

secular parallax [ASTRON] An apparent angular displacement of a star, resulting from the sun's motion.

secular perturbations [ASTROPHYS] Changes in the orbit of a planet, or of a satellite, that operates in extremely long cycles.

secular variation [ASTRON] A perturbation of the moon's motion caused by variations in the effect of the sun's gravitational attraction on the earth and moon as their relative distances from the sun vary during the synodic month. [GEOPHYS] The changes, measured in hundreds of years, in the magnetic field of the earth. Also known as geomagnetic secular variation.

sedative [PHARM] An agent or drug that has a quieting effect on the central nervous system.

Sedentaria [INV ZOO] A group of families of polychaete annelids in which the anterior, or cephalic, region is more or less completely concealed by overhanging peristomial structures, or the body is divided into an anterior thoracic and a posterior abdominal region.

sediment [GEOL] 1. A mass of organic or inorganic solid fragmented material, or the solid fragment itself, that comes from weathering of rock and is carried by, suspended in, or dropped by air, water, or ice; or a mass that is accumulated by any other natural agent and that forms in layers on the earth's surface such as sand, gravel, silt, mud, fill, or loess. 2. A solid material that is not in solution and either is distributed through the liquid or has settled out of the liquid.

sedimentary breccia [PETR] A rock composed of fragments that are larger than 2 millimeters in diameter and are the result of sedimentary processes; characterized by imperfect mechanical sorting of its materials and by a higher concentration of fragments from one local source or by a wide variety of materials mixed together in no particular pattern. Also known as sharpstone conglomerate.

sedimentary cycle See cycle of sedimentation.

sedimentary facies [GEOL] A stratigraphic facies differing from another part or parts of the same unit in both lithologic and paleontologic characters.

sedimentary injection See injection.

sedimentary intrusion See intrusion.

sedimentary petrology [PETR] The study of the composition, characteristics, and origin of sediments and sedimentary rocks.

sedimentary quartzite See orthoquartzite.

sedimentary rock [PETR] A rock formed by consolidated sediment deposited in layers. Also known as derivative rock; neptunic rock; stratified rock.

sedimentary structure [GEOL] A structure in sedimentary rocks, such as cross-bedding, ripple marks, and sandstone dikes, produced either contemporaneously with deposition (primary sedimentary structures) or shortly after deposition (secondary sedimentary structures).

sedimentation [GEOL] 1. The act or process of accumulating sediment in layers. 2. The process of deposition of sediment. [MET] Classification of metal powders by the rate of settling in a fluid.

sedimentation equilibrium [ANALY CHEM] The equilibrium between the forward movement of a sample's liquid-sediment boundary and reverse diffusion during centrifugation; used in molecular-weight determinations.

sedimentation tank [ENG] A tank in which suspended matter is removed either by quiescent settlement or by continuous flow at high velocity and extended retention time to allow deposition.

sedimentation velocity [ANALY CHEM] The rate of movement of the liquid-sediment boundary in the sample holder during centrifugation; used in molecular-weight determinations.

sedimentology [GEOL] The science concerned with the description, classification, origin, and interpretation of sediments and sedimentary rock.

Seebeck effect [ELECTR] The development of a voltage due to differences in temperature between two junctions of dissimilar metals in the same circuit. [GRAPHICS] A photographic emulsion that is exposed until a faint visible image appears, and is then exposed to colored light and takes on the color of the light to which it is exposed.

seed [BOT] A fertilized ovule containing an embryo which forms a new plant upon germination. [CHEM] A small, single crystal of a desired substance added to a solution to induce crystallization. [SOLID STATE] A small, single crystal of semiconductor material used to start the growth of a large, single crystal for use in cutting semiconductor wafers.

seed fern [PALEOBOT] The common name for the extinct plants classified as Pteridospermae, characterized by naked seeds borne on large, fernlike fronds.

seek [COMPUT SCI] 1. To position the access mechanism of a random-access storage device at a designated location or position. 2. The command that directs the positioning to take place. [ORD] To go toward a target or other object in reaction to some influence such as heat, light, sound, or other radiation emitted by the target or object.

seek area [COMPUT SCI] An area of a direct-access storage device, such as a magnetic disk file, assigned to hold records to which rapid access is needed, and located so that the physical characteristics of the device permit such access. Also known as cylinder.

seepage [FL MECH] The slow movement of water or other fluid through a porous medium. [HYD] The slow movement of water through small openings and spaces in the surface of unsaturated soil into or out of a body of surface or subsurface water.

segment [COMPUT SCI] 1. A single section of an overlay program structure, which can be loaded into the main memory when and as needed. 2. In some direct-access storage devices, a hardware-defined portion of a track having fixed data capacity. [MATH]

1. A segment of a line or curve is any connected piece. 2. A segment of a circle is a portion of the circle bounded by a chord and an arc subtended by the chord. [NAV] In air operations, a basic functional division of an instrument approach procedure; it bears a fixed orientation with respect to the course to be flown; it is assigned specific geometric coordinates which uniquely determine its position; the location of the segment is assigned with respect to the obstacles in the operations area.

segmental meter [ENG] A variable head meter whose orifice plate has an opening in the shape of a half circle.

segmental reflex [PHYSIO] A reflex arc having afferent inputs by way of the spinal dorsal roots, and efferent outputs over spinal ventral roots of the same or adjacent segments.

segmentation [COMPUT SCI] The division of virtual storage into identifiable functional regions, each having enough addresses so that programs or data stored in them will not assign the same addresses more than once. [ZOO] *See* metamerism.

segmentation cavity *See* blastocoele.

segregation [ENG] 1. The keeping apart of process streams. 2. In plastics molding, a close succession of parallel, relatively narrow, and sharply defined wavy lines of color on the surface of a plastic that differ in shade from surrounding areas and create the impression that the components have separated. [GEN] The separation of alleles and homologous chromosomes during meiosis in the formation of gametes. [GEOL] The formation of a secondary feature within a sediment after deposition due to chemical rearrangement of minor constituents. [MET] The nonuniform distribution of alloying elements, impurities, or microphases, resulting in localized concentrations.

seiche [FL MECH] An oscillation of a fluid body in response to the disturbing force having the same frequency as the natural frequency of the fluid system. [OCEANOGR] A standing-wave oscillation of an enclosed or semienclosed water body, continuing pendulum-fashion after cessation of the originating force, which is usually considered to be strong winds or barometric pressure changes.

seignette-electric *See* ferroelectric.

seismic activity *See* seismicity.

seismic area *See* earthquake zone.

seismic discontinuity [GEOPHYS] 1. A surface at which velocities of seismic waves change abruptly. 2. A boundary between seismic layers of the earth. Also known as interface; velocity discontinuity.

seismic event [GEOPHYS] An earthquake or a somewhat similar transient earth motion caused by an explosion.

seismicity [GEOPHYS] The phenomena of earth movements. Also known as seismic activity.

seismic prospecting [GEOPHYS] Geophysical prospecting based on the analysis of elastic waves generated in the earth by artificial means.

seismic risk [GEOPHYS] 1. An assortment of earthquake effects that range from ground shaking, surface faulting, and landsliding to economic loss and casualties. 2. The probability that social or economic consequences of earthquakes will equal or exceed specified values at a site, at several sites, or in an area, during a specified exposure time.

seismic sea wave *See* tsunami.

seismic survey *See* reflection survey.

seismograph [ENG] An instrument that records vibrations in the earth, especially earthquakes.

seismology [GEOPHYS] 1. The study of earthquakes. 2. The science of strain-wave propagation ins24the earth.

Seisonacea [INV ZOO] A monofamiliar order of the class Rotifera characterized by an elongated jointed body with a small head, a long slender neck region, a thick fusiform trunk, and an elongated foot terminating in a perforated disk.

Seisonidea [INV ZOO] The equivalent name for Seisonacea.

seizure [MED] 1. The sudden onset or recurrence of a disease or an attack. 2. Specifically, an epileptic attack, fit, or convulsion.

Selachii [VERT ZOO] An order of elasmobranchs including all fossil sharks, except Cladoselachii and Pleuracanthodii.

Selaginellales [BOT] The plant order of small club mosses, containing one living genus, *Selaginella*; distinguished from other lycopods in being heterosporous and in having a ligule borne on the upper base of the leaf.

select [COMPUT SCI] 1. To choose a needed subroutine from a file of subroutines. 2. To take one alternative if the report on a condition is of one state, and another alternative if the report on the condition is of another state. 3. To pull from a mass of data certain items that require special attention; selection of individual cards is accomplished automatically by either the sorter or collator, according to the type of selection.

select bit [COMPUT SCI] The bit (or bits) in an input/output instruction word which selects the function of a specified device. Also known as subdevice bit.

selecting circuit [ELEC] A simple switching circuit that receives the identity (the address) of a particular item and selects that item from among a number of similar ones.

selection [COMMUN] The process of addressing a call to a specific station in a selective calling system. [GEN] Any natural or artificial process which favors the survival and propagation of individuals of a given genotype in a population.

selection bias [STAT] A bias built into an experiment by the method used to select the subjects which are to undergo treatment.

selection rules [PHYS] Rules summarizing the changes that must take place in the quantum numbers of a quantum-mechanical system for a transition between two states to take place with appreciable probability; transitions that do not agree with the selection rules are called forbidden and have considerably lower probability.

selective absorption [ELECTROMAG] A greater absorption of electromagnetic radiation at some wavelengths (or frequencies) than at others.

selective dump [COMPUT SCI] An edited or nonedited listing of the contents of selected areas of memory or auxiliary storage.

selective permeability [PHYS] The property of a membrane or other material that allows some substances to pass through it more easily than others.

selectivity [ELECTR] The ability of a radio receiver to separate a desired signal frequency from other signal frequencies, some of which may differ only slightly from the desired value.

selector [COMPUT SCI] Computer device which interrogates a condition and initiates a particular operation dependent upon the report. [CIV ENG] A device that automatically connects the appropriate railroad signal to control the track selected. [ELEC] An automatic or other device for making connections to any one of a number of circuits, such as a selector relay or selector switch. [ENG] 1. A device for selecting objects or materials according to predetermined properties. **2.** A device for starting or stopping at predetermined positions. [MECH ENG] **1.** The part of the gearshift in an automotive transmission that selects the required gearshift bar. **2.** The lever with which a driver operates an automatic gearshift. [MET] A converter that separates purified copper from residue in a single operation.

selector channel [COMPUT SCI] A unit which connects high-speed input/output devices, such as magnetic tapes, disks, and drums, to a computer memory.

selector switch [ELEC] A manually operated multiposition switch. Also called multiple-contact switch.

s-electron [ATOM PHYS] An atomic electron that is described by a wave function with orbital angular momentum quantum number of zero in the independent particle approximation.

selenite [MINERAL] The clear, colorless variety of gypsum crystallizing in the monoclinic system and occurring in crystals or in crystal mass. Also known as spectacle stone.

selenium [CHEM] A highly toxic, nonmetallic element in group VI, symbol Se, atomic number 34; steel-gray color; soluble in carbon disulfide, insoluble in water and alcohol; melts at 217°C; and boils at 690°C; used in analytical chemistry, metallurgy, and photoelectric cells, and as a lube-oil stabilizer and chemicals intermediate.

selenodesy [ASTRON] The branch of applied mathematics that determines, by observation and measurement, the exact positions of points on the moon's surface, as well as the shape and size of the moon.

selenodont [VERT ZOO] **1.** Being or pertaining to molars having crescentic ridges on the crown. **2.** A mammal with selenodont dentition.

selenology [ASTRON] A branch of astronomy that treats of the moon, including such attributes as magnitude, motion and constitution. Also known as lunar geology.

self-bias [ELECTR] A grid bias provided automatically by the resistor in the cathode or grid circuit of an electron tube; the resulting voltage drop across the resistor serves as the grid bias. Also known as automatic C bias; automatic grid bias.

self-bias transistor circuit [ELECTR] A transistor with a resistance in the emitter lead that gives rise to a voltage drop which is in the direction to reverse-bias the emitter junction; the circuit can be used even if there is zero direct-current resistance in series with the collector terminal.

self-checking code [COMPUT SCI] An encoding of data so designed and constructed that an invalid code can be rapidly detected; this permits the detection, but not the correction, of almost all errors. Also known as error-checking code; error-detecting code.

self-checking number [COMPUT SCI] A number with a suffix figure related to the figure of the number, used to check the number after it has been transferred from one medium or device to another.

self-complementing code [COMPUT SCI] A binary-coded-decimal code in which the combination for the complement of a digit is the complement of the combination for that digit.

self-consistent field method *See* Hartree method.

self-contained breathing apparatus [ENG] A portable breathing unit which permits freedom of movement.

self-contained data-base management system [COMPUT SCI] A data-base management system that is in no way an extension of any programming language, and is usually quite independent of any language.

self-excited [ELEC] Operating without an external source of alternating-current power.

self-excited vibration *See* self-induced vibration.

self-hardening steel *See* air-hardening steel.

self-impedance *See* mesh impedance.

self-induced vibration [MECH] The vibration of a mechanical system resulting from conversion, within the system, of nonoscillatory excitation to oscillatory excitation. Also known as self-excited vibration.

self-inductance [ELECTROMAG] **1.** The property of an electric circuit whereby an electromotive force is produced in the circuit by a change of current in the circuit itself. **2.** Quantitatively, the ratio of the electromotive force produced to the rate of change of current in the circuit.

self-induction [ELECTROMAG] The production of a voltage in a circuit by a varying current in that same circuit.

self-propelled [MECH ENG] Pertaining to a vehicle given motion by means of a self-contained motor. [ORD] **1.** Pertaining to a gun mounted on a vehicle that has its own motive power. **2.** Pertaining to a missile that is propelled by fuel carried by the missile itself, as in the case of a rocket. **3.** Pertaining to a military unit having self-propelled guns.

self-reset [ELEC] Automatically returning to the original position when normal conditions are resumed; applied chiefly to relays and circuit breakers.

self-selection bias [STAT] Bias introduced into an experiment by having the subjects decide themselves whether or not they will receive treatment.

self-synchronous device *See* synchro.

self-synchronous repeater *See* synchro.

self-tapping screw [DES ENG] A screw with a specially hardened thread that makes it possible for the screw to form its own internal thread in sheet metal and soft materials when driven into a hole. Also known as sheet-metal screw; tapping screw.

self-triggering program [COMPUT SCI] A computer program which automatically commences execution as soon as it is fed into the central processing unit.

seligmannite [MINERAL] PbCuAsS$_3$ A metallic gray orthorhombic mineral, occurring in crystals.

sellaite [MINERAL] MgF$_2$ A colorless mineral composed of magnesium fluoride occurring in tetragonal prismatic crystals.

selsyn *See* synchro.

selsyn generator *See* synchro transmitter.

selsyn motor *See* synchro receiver.

selsyn receiver *See* synchro receiver.

selsyn transmitter *See* synchro transmitter.

Selwood engine [MECH ENG] A revolving-block engine in which two curved pistons opposed 180° run in toroidal tracks, forcing the entire engine block to rotate.

Semaeostomeae [INV ZOO] An order of the class Scyphozoa including most of the common medusae, characterized by a flat, domelike umbrella whose margin is divided into many lappets.

semantics [COMMUN] The branch of semiotics that deals with the relations between symbols and what they stand for, and defines the meaning that is prescribed for a statement by its originator.

semaphore [COMPUT SCI] A memory cell that is shared by two parallel processes which rely on each other for their continued operation, and that provides an elementary form of communication between them by indicating when significant events have taken place.

semialgorithm [COMPUT SCI] A procedure for solving a problem that will continue endlessly if the problem has no solution.

semiautomatic [ORD] Pertaining to a firearm or gun that utilizes a part of the force of an exploding cartridge to extract the empty case and to chamber the next round, but requires a separate pull on the trigger to fire each round; examples are the semiautomatic rifle and the semiautomatic pistol.

semiautomatic telephone system [COMMUN] Telephone system that limits automatic dialing to only those subscribers who are served by the same exchange as the calling subscriber.

semiautomatic welding [MET] An arc-welding method in which the electrode, a long length of small-diameter bare wire, usually in coil form, is positioned and advanced by the operator from a hand-held welding gun which feeds the electrode through the nozzle.

semibituminous coal [GEOL] Coal that is harder and more brittle than bituminous coal, has a high fuel ratio, contains 10–20% volatile matter, and burns without smoke; ranks between bituminous and semianthracite coals.

semichemical pulping [CHEM ENG] A method of producing wood-fiber products in which the wood chips are merely softened by chemical treatment (neutral sodium sulfite solution), while the remainder of the pulping action is supplied by a disk attrition mill or by some similar mechanical device for separating the fibers.

semicircular canal [ANAT] Any of three loop-shaped tubular structures of the vertebrate labyrinth; they are arranged in three different spatial planes at right angles to each other, and function in the maintenance of body equilibrium.

semiconductor [SOLID STATE] A solid crystalline material whose electrical conductivity is intermediate between that of a conductor and an insulator, ranging from about 10^5 mhos to 10^{-7} mho per meter, and is usually strongly temperature-dependent.

semiconductor detector [NUCLEO] A particle detector which detects ionization produced by energetic charged particles in the depletion layer of a reverse-biased pn junction in a semiconductor, usually a very pure single crystal of silicon or germanium.

semiconductor device [ELECTR] Electronic device in which the characteristic distinguishing electronic conduction takes place within a semiconductor.

semiconductor diode [ELECTR] Also known as crystal diode; crystal rectifier; diode. **1.** A two-electrode semiconductor device that utilizes the rectifying properties of a pn junction or a point contact. **2.** More generally, any two-terminal electronic device that utilizes the properties of the semiconductor from which it is constructed.

semiconductor doping See doping.

semiconductor heterostructure [ELECTR] A structure of two different semiconductors in junction contact having useful electrical or electrooptical characteristics not achievable in either conductor separately; used in certain types of lasers and solar cells.

semiconductor-insulator-semiconductor [SOLID STATE] A semiconductor device consisting of an electrically insulating layer sandwiched between two semiconducting materials.

semiconductor junction [ELECTR] Region of transition between semiconducting regions of different electrical properties, usually between p-type and n-type material.

semiconductor laser [OPTICS] A laser in which stimulated emission of coherent light occurs at a pn junction when electrons and holes are driven into the junction by carrier injection, electron-beam excitation, impact ionization, optical excitation, or other means. Also known as diode laser; laser diode.

semiconductor memory [COMPUT SCI] A device for storing digital information that is fabricated by using integrated circuit technology. Also known as integrated circuit memory; large-scale integrated memory; transistor memory.

semiconductor rectifier See metallic rectifier.

semiconductor thermocouple [ELECTR] A thermocouple made of a semiconductor, which offers the prospect of operation with high-temperature gradients, because semiconductors are good electrical conductors but poor heat conductors.

semiconductor trap See trap.

semiconservative replication [MOL BIO] Replication of deoxyribonucleic acid by longitudinal separation of the two complementary strands of the molecule, each being conserved and acting as a template for synthesis of a new complementary strand.

semicrystalline See hypocrystalline.

semidiurnal [ASTRON] Having a period of, occurring in, or related to approximately half a day.

semifloating axle [MECH ENG] A supporting member in motor vehicles which carries torque and wheel loads at its outer end.

semigroup [MATH] A set which is closed with respect to a given associative binary operation.

semi-invariants See cumulants.

semikilled steel [MET] Incompletely deoxidized steel containing enough dissolved oxygen to react with the carbon it contains to form carbon monoxide, the latter offsetting solidification shrinkage.

semilethal gene [GEN] A mutant causing the death of some of the individuals of the relevant genotype, but never 100%. Also known as sublethal gene.

semilunar valve [ANAT] Either of two tricuspid valves in the heart, one at the orifice of the pulmonary artery and the other at the orifice of the aorta.

seminal receptacle See spermatheca.

seminal vesicle [ANAT] A saclike, glandular diverticulum on each ductus deferens in male vertebrates; it is united with the excretory duct and serves for temporary storage of semen.

semiochemical [PHYSIO] Any of a class of substances produced by organisms, especially insects, that participate in regulation of their behavior in such activities as aggregation of both sexes, sexual stimulation, and trail following.

semipermeable membrane [PHYS] A membrane which allows a solvent to pass through it, but not certain dissolved or colloidal substances.

semiprecious [LAP] Pertaining to gemstones whose value is lower than that of precious stones; in particular, stones whose hardness is less than 8.

semistable elementary particle See quasistable elementary particle.

semisteel [MET] Low-carbon steel made by replacing about one-fourth of the pig iron in the cupola with steel scrap.

semitrailer [ENG] A cargo-carrying piece of equipment that has one or two axles at the rear; the load is carried on these axles and on the fifth wheel of the tractor that supplies motive power to the semitrailer.

senaite [MINERAL] $(Fe,Mn,Pb)TiO_3$ A black mineral consisting of a lead- and manganese-bearing ilmenite; occurs as rough crystals and rounded fragments.

senarmontite [MINERAL] Sb_2O_3 A colorless or grayish mineral composed of native antimony trioxide occurring in masses or as octahedral crystals.

sender [COMMUN] Part of an automatic-switching telephone system that receives pulses from a dial or other source and, in accordance with them, controls the further operations necessary in establishing a telephone connection.

senescence [BIOL] The study of the biological changes related to aging, with special emphasis on plant, animal, and clinical observations which may apply to man. [GEOL] The part of the erosion cycle at which the stage of old age begins.

senescent arthritis See degenerative joint disease.

senile [GEOL] Pertaining to the stage of senility of the cycle or erosion. [MED] Of, pertaining to, or caused by the aging process or by the infirmities of old age.

senility [GEOL] The stage of the cycle of erosion in which erosion of a land surface has reached a minimum, most of the hills have disappeared, and base level has been approached.

sensation [PHYSIO] The subjective experience that results from the stimulation of a sense organ.

sense [COMPUT SCI] To read punched holes in tape or cards. [ENG] To determine the arrangement or position of a device or the value of a quantity. [NAV] The general direction from which a radio signal arrives; if a radio bearing is received by a simple loop antenna, there are two possible readings approximately 180° apart; the resolving of this ambiguity is called sensing of the bearing.

sense indicator See to-from indicator.

sense organ [PHYSIO] A structure which is a receptor for external or internal stimulation.

sense switch See alteration switch.

sensible heat See enthalpy.

sensible-heat factor [THERMO] The ratio of space sensible heat to space total heat; used for air-conditioning calculations. Abbreviated SHF.

sensible heat flow [METEOROL] In the atmosphere, the poleward transport of sensible heat (enthalpy) across a given latitude belt by fluid flow. [THERMO] The heat given up or absorbed by a body upon being cooled or heated, as the result of the body's ability to hold heat; excludes latent heats of fusion and vaporization.

sensible horizon [ASTRON] That circle of the celestial sphere formed by the intersection of the celestial sphere and a plane through any point, such as the eye of an observer, and perpendicular to the zenith-nadir line.

sensible temperature [METEOROL] The temperature at which air with some standard humidity, motion, and radiation would provide the same sensation of human comfort as existing atmospheric conditions.

sensing element See sensor.

sensing station See reading station.

sensitivity [ELECTR] 1. The minimum input signal required to produce a specified output signal, for a radio receiver or similar device. 2. Of a camera tube, the signal current developed per unit incident radiation, that is, per watt per unit area. [ENG] 1. A measure of the ease with which a substance can be caused to explode. 2. A measure of the effect of a change in severity of engine-operating conditions on the antiknock performance of a fuel; expressed as the difference between research and motor octane numbers. Also known as spread.

[GEOL] The effect of remolding on the consistency of a clay or cohesive soil, regardless of the physical nature of the causes of the change. [PHYSIO] The capacity for receiving sensory impressions from the environment. [SCI TECH] 1. The ability of the output of a device, system, or organism to respond to an input stimulus. 2. Mathematically, the ratio of the response or change induced in the output to a stimulus or change in the input.

sensitivity function [CONT SYS] The ratio of the fractional change in the system response of a feedback-compensated feedback control system to the fractional change in an open-loop parameter, for some specified parameter variation.

sensitization See activation.

sensor [ENG] The generic name for a device that senses either the absolute value or a change in a physical quantity such as temperature, pressure, flow rate, or pH, or the intensity of light, sound, or radio waves and converts that change into a useful input signal for an information-gathering system; a television camera is therefore a sensor, and a transducer is a special type of sensor. Also known as primary detector; sensing element.

sensory nerve [PHYSIO] A nerve that conducts afferent impulses from the periphery to the central nervous system.

sepal [BOT] One of the leaves composing the calyx.

separated sets [MATH] Sets A and B in a topological space are separated if both the closure of A intersected with B and the closure of B intersected with A are disjoint.

separation [AERO ENG] The action of a fallaway section or companion body as it casts off from the remaining body of a vehicle, or the action of the remaining body as it leaves a fallaway section behind it. [CHEM ENG] The separation of liquids or gases in a mixture, as by distillation or extraction. [ENG] 1. The action segregating phases, such as gas-liquid, gas-solid, liquid-solid. 2. The segregation of solid particles by size range, as in screening. [ENG ACOUS] The degree, expressed in decibels, to which left and right stereo channels are isolated from each other. [GEOL] The apparent relative displacement on a fault, measured in any given direction. [MIN ENG] The removal of gangue from raw ores, as in frothing.

separation energy [NUC PHYS] The energy needed to remove a proton, neutron, or alpha particle from a nucleus.

separation negatives [GRAPHICS] The negatives made from full-color originals and used in the preparation of colorplates; four negatives are made, for yellow, magenta, cyan, and black printing plates.

separation of variables [MATH] A technique where certain differential equations are rewritten in the form $f(x)dx = g(y)dy$ which is then solvable by integrating both sides of the equation.

separator [COMPUT SCI] A datum or character that denotes the beginning or ending of a unit of data. [ELEC] A porous insulating sheet used between the plates of a storage battery. [ELECTR] A circuit that separates one type of signal from another by clipping, differentiating, or integrating action. [ENG] 1. A machine for separating materials of different specific gravity by means of water or air. 2. Any machine for separating materials, as the magnetic separator. [MECH ENG] See cage. [PETRO ENG] See gas-oil separator.

separatory funnel [CHEM] A funnel-shaped device used for the careful and accurate separation of two immiscible liquids; a stopcock on the funnel stem controls the rate and amount of outflow of the lower liquid.

Sepioidea [INV ZOO] An order of the molluscan subclass Coleoidea having a well-developed eye, an internal shell, fins separated posteriorly, and chromatophores in the dermis.

sepiolite [MINERAL] $Mg_4(Si_2O_5)_3(OH)_2 \cdot 6H_2O$ A soft, lightweight, absorbent, white to light-gray or light-yellow clay mineral, found principally in Asia Minor; used for tobacco pipe bowls and ornamental carvings. Also known as meerschaum; sea-foam.

Sepsidae [INV ZOO] The spiny-legged flies, a family of myodarian cyclorrhaphous dipteran insects in the subsection Acalypteratae; development takes place in decaying organic matter.

sepsis [MED] **1.** Poisoning by products of putrefaction. **2.** The severe toxic, febrile state resulting from infection with pyogenic microorganisms, with or without associated septicemia.

Septibranchia [INV ZOO] A small order of bivalve mollusks in which the anterior and posterior abductor muscles are about equal in size, the foot is long and slender, and the gills have been transformed into a muscular septum.

septicemia [MED] A clinical syndrome in which infection is disseminated through the body in the bloodstream. Also known as blood poisoning.

septic tank [CIV ENG] A settling tank in which settled sludge is in immediate contact with sewage flowing through the tank while solids are decomposed by anaerobic bacterial action.

septum [BIOL] A partition or dividing wall between two cavities. [ELECTROMAG] A metal plate placed across a waveguide and attached to the walls by highly conducting joints; the plate usually has one or more windows, or irises, designed to give inductive, capacitive, or resistive characteristics.

Sequanian [GEOL] Upper Lower Jurassic (Upper Lusitanian) geologic time. Also known as Astartian.

sequence [COMPUT SCI] To put a set of symbols into an arbitrarily defined order; that is, to select A if A is greater than or equal to B, or to select B if A is less than B. [ENG] An orderly progression of items of information or of operations in accordance with some rule. [GEOL] **1.** A sequence of geologic events, processes, or rocks, arranged in chronological order. **2.** A geographically discrete, major informal rock-stratigraphic unit of greater than group or supergroup rank. Also known as stratigraphic sequence. [MATH] A listing of mathematical entities $x_1, x_2 \ldots$ which is indexed by the positive integers; more precisely, a function whose domain is an infinite subset of the positive integers. Also known as infinite sequence. [METEOROL] See collective.

sequence calling [COMPUT SCI] The instructions used for linking a closed subroutine with a main routine; that is, standard linkage and a list of the parameters.

sequence check [COMPUT SCI] To verify that correct precedence relationships are obeyed, usually by checking for ascending sequence numbers.

sequence counter See instruction counter.

sequence number [COMPUT SCI] A number assigned to an item to indicate its relative position in a series of related items.

sequencer [COMPUT SCI] A machine which puts items of information into a particular order, for example, it will determine whether A is greater than, equal to, or less than B, and sort or order accordingly. Also known as sorter. [ENG] A mechanical or electronic device that may be set to initiate a series of events and to make the events follow in a given sequence.

sequence register [COMPUT SCI] A counter which contains the address of the next instruction to be carried out.

sequential access [COMPUT SCI] A process that involves reading or writing data serially and, by extension, a data-recording medium that must be read serially, as a magnetic tape.

sequential circuit [ELEC] A switching circuit whose output depends not only upon the present state of its input, but also on what its input conditions have been in the past.

sequential control [COMPUT SCI] Manner of operating a computer by feeding orders into the computer in a given order during the solution of a problem.

sequential logic element [ELECTR] A circuit element having at least one input channel, at least one output channel, and at least one internal state variable, so designed and constructed that the output signals depend on the past and present states of the inputs.

sequential operation [COMPUT SCI] The consecutive or serial execution of operations, without any simultaneity or overlap.

sequential organization [COMPUT SCI] The write and read of records in a physical rather than a logical sequence.

sequential processing [COMPUT SCI] Processing items in a collection of data according to some specified sequence of keys, in contrast to serial processing.

sequential search [COMPUT SCI] A procedure for searching a table that consists of starting at some table position (usually the beginning) and comparing the file-record key in hand with each table-record key, one at a time, until either a match is found or all sequential positions have been searched.

sequential trials [STAT] The outcome of each trial is known before the next trial is performed.

Sequoia [BOT] A genus of conifers having overlapping, scalelike evergreen leaves and vertical grooves in the trunk; the giant sequoia (*Sequoia gigantea*) is the largest and oldest of all living things.

serandite [MINERAL] $Na(Mn,Ca)_2Si_3O_8(OH)$ A rosered mineral composed of a basic silicate of manganese, lime, potash, and soda occurring in monoclinic crystals.

serial-access [COMPUT SCI] **1.** Pertaining to memory devices having structures such that data storage sites become accessible for read/write in time-sequential order; circulating memories and magnetic tapes are examples of serial-access memories. **2.** Pertaining to a particular process or program that accesses data items sequentially, without regard to the capability of the memory hardware. **3.** Pertaining to character-by-character transmission from an on-line real-time keyboard.

serial bit [COMPUT SCI] Digital computer storage in which the individual bits that make up a computer word appear in time sequence.

serial dot character printer [COMPUT SCI] A computer printer in which the dot matrix technique is used to print characters, one at a time, with a movable print head that is driven back and forth across the page.

serial feed [COMPUT SCI] The method of placing cards in the feed hopper of a punched-card machine in which one of the short edges of the card enters the machine first, so that the columns of the card are read sequentially.

serial file [COMPUT SCI] The simplest type of file organization, in which no subsets are defined, no directories are provided, no particular file order is specified, and a search is performed by sequential comparison of the query with identifiers of all stored items.

serial memory [COMPUT SCI] A computer memory in which data are available only in the same sequence as originally stored.

serial operation [COMPUT SCI] The flow of information through a computer in time sequence, using only one digit, word, line, or channel at a time.

serial parallel [COMPUT SCI] **1.** A combination of serial and parallel; for example, serial by character, parallel by bits comprising the character. **2.** Descriptive of a device which converts a serial input into a parallel output.

serial-parallel conversion [COMPUT SCI] The transformation of a serial data representation as found on a disk or drum into the parallel data representation as exists in core.

serial processing [COMPUT SCI] Processing items in a collection of data in the order that they appear in a storage device, in contrast to sequential processing.

serial processor [COMPUT SCI] A computer in which data are handled sequentially by separate units of the system.

serial programming [COMPUT SCI] In computers, programming in which only one operation is executed at one time.

serial storage [COMPUT SCI] Computer storage in which time is one of the coordinates used to locate any given bit, character, or word; access time, therefore, includes a variable waiting time, ranging from zero to many word times.

serial transfer [COMPUT SCI] Transfer of the characters of an element of information in sequence over a single path in a digital computer.

series [ELEC] An arrangement of circuit components end to end to form a single path for current. [GEOL] **1.** A number of rocks, minerals, or fossils that can be arranged in a natural sequence due to certain characteristics, such as succession, composition, or occurrence. **2.** A time-stratigraphic unit, below system and above stage, composed of rocks formed during an epoch of geologic time. [MATH] An expression of the form $x_1 + x_2 + x_3 + \ldots$, where x_i are real or complex numbers.

series circuit [ELEC] A circuit in which all parts are connected end to end to provide a single path for current.

series compensation *See* cascade compensation.

series decay *See* radioactive series.

series generator [ELEC] A generator whose armature winding and field winding are connected in series. Also known as series-wound generator.

series motor [ELEC] A commutator-type motor having armature and field windings in series; characteristics are high starting torque, variation of speed with load, and dangerously high speed on no-load. Also known as series-wound motor.

series-parallel circuit [ELEC] A circuit in which some of the components or elements are connected in parallel, and one or more of these parallel combinations are in series with other components of the circuit.

series reactor [ELEC] A reactor used in alternating-current power systems for protection against excessively large currents under short-circuit or transient conditions; it consists of coils of heavy insulated cable either cast in concrete columns or supported in rigid frames and mounted on insulators. Also known as current-limiting reactor.

series regulator [ELEC] A regulator that controls output voltage or current by automatically varying a resistance in series with the voltage source.

series repeater [ELEC] A type of negative impedance telephone repeater which is stable when terminated in an open circuit and oscillates when it is connected to a low impedance, in contrast to a shunt repeater.

series resonance [ELEC] Resonance in a series resonant circuit, wherein the inductive and capacitive reactances are equal at the frequency of the applied voltage; the reactances then cancel each other, reducing the impedance of the circuit to a minimum, purely resistive value.

series resonant circuit [ELEC] A resonant circuit in which the capacitor and coil are in series with the applied alternating-current voltage.

series-shunt network *See* ladder network.

series transistor regulator [ELECTR] A voltage regulator whose circuit has a transistor in series with the output voltage, a Zener diode, and a resistor chosen so that the Zener diode is approximately in the middle of its operating range.

series-tuned circuit [ELEC] A simple resonant circuit consisting of an inductance and a capacitance connected in series.

series winding [ELEC] A winding in which the armature circuit and the field circuit are connected in series with the external circuit.

series-wound generator *See* series generator.

series-wound motor *See* series motor.

serine [BIOCHEM] $C_3H_7O_3N$ An amino acid obtained by hydrolysis of many proteins; a biosynthetic precursor of several metabolites, including cysteine, glycine, and choline.

Serolidae [INV ZOO] A family of isopod crustaceans which contains greatly flattened forms that live partially buried on sandy bottoms.

serology [BIOL] The branch of science dealing with the properties and reactions of blood sera.

serosa [ANAT] The serous membrane lining the pleural, peritoneal, and pericardial cavities. [EMBRYO] The chorion of reptile and bird embryos.

serotonin [BIOCHEM] $C_{10}H_{12}ON_2$ A compound derived from tryptophan which functions as a local vasoconstrictor, plays a role in neurotransmission, and has pharmacologic properties. Also known as 5-hydroxytryptamine.

serous gland [PHYSIO] A structure that secretes a watery, albuminous fluid.

serous membrane [HISTOL] A delicate membrane covered with flat, mesothelial cells lining closed cavities of the body.

Serpens [ASTRON] A constellation, right ascension 17 hours, declination 0°. Also known as Serpent.

Serpent *See* Serpens.

Serpentes [VERT ZOO] The snakes, a suborder of the Squamata characterized by the lack of limbs and pectoral girdle and external ear openings, immovable eyelids, and a braincase that is completely bony anteriorly.

serpentine [MINERAL] $(Mg,Fe)_3Si_2O_5(OH)_4$ A group of green, greenish-yellow, or greenish-gray ferromagnesian hydrous silicate rock-forming minerals having greasy or silky luster and a slightly soapy feel; translucent varieties are used for gemstones as substitutes for jade.

serpentine cooler *See* cascade cooler.

serpentine curve [MATH] The curve given by the equation $x^2y + b^2y - a^2x = 0$, passing through and having symmetry about the origin while being asymptotic to the x axis in both directions.

serpentine rock *See* serpentinite.

serpentinite [PETR] A rock composed almost entirely of serpentine minerals. Also known as serpentine rock.

serpent kame *See* esker.

serpierite [MINERAL] $(Cu,Zn,Ca)_5(SO_4)_2(OH)_6 \cdot 3H_2O$ A bluish-green mineral composed of hydrous basic sulfate of copper, zinc, and calcium; occurs in tabular crystals and tufts.

Serpulidae [INV ZOO] A family of polychaete annelids belonging to the Sedentaria including many of the feather-duster worms which construct calcareous tubes in the earth, sometimes in such abundance as to clog drains and waterways.

Serranidae [VERT ZOO] A family of perciform fishes in the suborder Percoidei including the sea basses and groupers.

serrate [BIOL] Possessing a notched or toothed edge. [GEOL] Pertaining to topographic features having a notched or toothed edge, or a saw-edge profile.

serrated pulse [ELECTR] Vertical and horizontal synchronizing pulse divided into a number of small pulses, each of which acts for the duration of half a line in a television system.

serrate ridge *See* arête.

Serridentinae [PALEON] An extinct subfamily of elephantoids in the family Gomphotheriidae.

Serritermitidae [INV ZOO] A family of the Isoptera which contains the single monotypic genus *Serritermes*.

Serropalpidae [INV ZOO] An equivalent name for Melandryidae.

serum [PHYSIO] The liquid portion that remains when blood clots spontaneously and the formed and clotting elements are removed by centrifugation; it differs from plasma by the absence of fibrinogen.

serum hepatitis [MED] A form of viral hepatitis transmitted by parenteral injection of human blood or blood products contaminated with the type B virus.

serum shock [MED] An anaphylactic reaction following the injection of foreign serum into a sensitized individual.

serum sickness [MED] A syndrome manifested in 8–12 days after the administration of serum by an urticarial rash, edema, enlargement of lymph nodes, arthralgia, and fever.

service bit [COMMUN] A bit used in data transmission to monitor the transmission rather than to convey information, such as a request that part of a message be repeated.

service life [ENG] The length of time during which a machine, tool, or other apparatus or device can be operated or used economically or before breakdown.

service routine [COMPUT SCI] A section of a computer code that is used in so many different jobs that it cannot belong to any one job.

servo *See* servomotor.

servo brake [MECH ENG] **1.** A brake in which the motion of the vehicle is used to increase the pressure on one of the shoes. **2.** A brake in which the force applied by the operator is augmented by a power-driven mechanism.

servomechanism [CONT SYS] An automatic feedback control system for mechanical motion; it applies only to those systems in which the controlled quantity or output is mechanical position or one of its derivatives (velocity, acceleration, and so on). Also known as servo system.

servomotor [CONT SYS] The electric, hydraulic, or other type of motor that serves as the final control element in a servomechanism; it receives power from the amplifier element and drives the load with a linear or rotary motion. Also known as servo.

servomultiplier [ELECTR] An electromechanical multiplier in which one variable is used to position one or more ganged potentiometers across which the other variable voltages are applied.

servo system *See* servomechanism.

Sessilina [INV ZOO] A suborder of ciliates in the order Peritrichida.

set [ASTRON] Of a celestial body, to cross the visible western horizon while descending. [CHEM] The hardening or solidifying of a plastic or liquid substance. [COMMUN] A radio or television receiver. [COMPUT SCI] A collection of record types. [ELECTR] The placement of a storage device in a prescribed state, for example, a binary storage cell in the high or 1 state. [ENG] **1.** A combination of units, assemblies, and parts connected or otherwise used together to perform an operational function, such as a radar set. **2.** In plastics processing, the conversion of a liquid resin or adhesive into a solid state by curing or evaporation of solvent or suspending medium, or by gelling. **3.** Saw teeth bent out of the plane of the saw body, resulting in a wide cut in the workpiece. [GEOL] A group of essentially conformable strata or cross-strata, separated from other sedimentary units by surfaces of erosion, nondeposition, or abrupt change in character. [GRAPHICS] The fixing or drying of a printing ink on a printed sheet, so that, though not completely dry, the sheet can be handled without smudging. [MATER] **1.** The hardening or firmness displayed by some materials when left undisturbed. **2.** Permanent deformation of a material, such as metal or plastic, when stressed beyond the elastic limit. [MATH] A collection of objects which has the property that, given any thing, it can be determined whether or not the thing is in the collection. [MECH] *See* permanent set. [MIN ENG] *See* frame set. [NAV] The establishment of a course. [OCEANOGR] The direction toward which an oceanic current flows.

seta [BIOL] A slender, usually rigid bristle or hair. Also known as chaeta.

set analyzer *See* analyzer.

set screw [DES ENG] A small headless machine screw, usually having a point at one end and a recessed hexagonal socket or a slot at the other end, used for such purposes as holding a knob or gear on a shaft.

set theory [MATH] The study of the structure and size of sets from the viewpoint of the axioms imposed.

settling [ENG] The gravity separation of heavy from light materials; for example, the settling out of dense solids or heavy liquid droplets from a liquid carrier, or the settling out of heavy solid grains from a mixture of solid grains of different densities. [GEOL] The sag in outcrops of layered strata, caused by rock creep. Also known as outcrop curvature.

settling basin [CIV ENG] An artificial trap designed to collect suspended stream sediment before discharge of the stream into a reservoir. [IND ENG] A sedimentation area designed to remove pollutants from factory effluents.

settling chamber [ENG] A vessel in which solids or heavy liquid droplets settle out of a liquid carrier by gravity during processing or storage.

settling tank [ENG] A tank into which a two-phase mixture is fed and the entrained solids settle by gravity during storage.

settling velocity [FL MECH] The rate at which suspended solids subside and are deposited. Also known as fall velocity. [MECH] The velocity reached by a particle as it falls through a fluid, dependent on its size and shape, and the difference between its specific gravity and that of the settling medium; used to sort particles by grain size.

severe storm [METEOROL] In general, any destructive storm, but usually applied to a severe local storm, that is, an intense thunderstorm, hail storm, or tornado.

sewage [CIV ENG] The fluid discharge from medical, domestic, and industrial sanitary appliances.

sewage sludge [CIV ENG] A semiliquid waste with a solid concentration in excess of 2500 parts per million, obtained from the purification of municipal sewage. Also known as sludge.

sewage system [CIV ENG] Any of several drainage systems for carrying surface water and sewage for disposal.

sewer [CIV ENG] An underground pipe or open channel in a sewage system for carrying water or sewage to a disposal area.

sewer gas [MATER] The gas evolved from the decomposition of municipal sewage; it has a high content of methane and hydrogen sulfide, and can be used as a fuel gas.

sexadecimal See hexadecimal.

sexadecimal number system See hexadecimal number system.

sex chromosome [GEN] Either member of a pair of chromosomes responsible for sex determination of an organism.

sex factor See fertility factor.

sex hormone [BIOCHEM] Any hormone secreted by a gonad, but also found in other tissues.

sex-influenced inheritance [GEN] That part of the inheritance pattern on which sex differences operate to promote character differences.

sexless connector See hermaphroditic connector.

sex-limited inheritance [GEN] Expression of a phenotype in only one sex; may be due to either a sex-linked or autosomal gene.

sex-linked inheritance [GEN] The transmission to successive generations of differences that are due to genes located in the sex chromosomes.

sex organs [ANAT] The organs pertaining entirely to the sex of an individual, both physiologically and anatomically.

Sextans [ASTRON] A constellation in the southern hemisphere, right ascension 10 hours, declination 0°. Also known as Sextant.

sextant [MATH] A unit of plane angle, equal to 60° or $\pi/3$ radians. [NAV] An optical instrument; a double reflecting instrument used in navigation for measuring angles, primarily altitudes of celestial bodies.

Sextant See Sextans.

sextillion [MATH] 1. The number 10^{21}. 2. In British and German usage, the number 10^{36}.

sexual dimorphism [BIOL] Diagnostic morphological differences between the sexes.

seybertite See clintonite.

Seymouriamorpha [PALEON] An extinct group of labyrinthodont Amphibia of the Upper Carboniferous and Permian in which the intercentra were reduced.

sferics See atmospheric interference.

shaded-pole motor [ELEC] A single-phase induction motor having one or more auxiliary short-circuited windings acting on only a portion of the magnetic circuit; generally, the winding is a closed copper ring embedded in the face of a pole; the shaded pole provides the required rotating field for starting purposes.

shaded-relief map [MAP] A map of an area whose relief is made to appear three-dimensional by the hill-shading method.

shadow [OPTICS] A region of darkness caused by the presence of an opaque object interposed between such a region and a source of light. [PHYS] A region which some type of radiation, such as sound or x-rays, does not reach because of the presence of an object, which the radiation cannot penetrate, interposed between the region and the source of radiation.

shadow bands [ASTRON] Rippling bands of shadow that appear on every white surface of flat terrestrial objects a few minutes before the total eclipse of the sun.

shadow effect [COMMUN] Reduction in the strength of an ultra-high-frequency signal caused by some object (such as a mountain or a tall building) between the points of transmission and reception.

shadowgraph [GRAPHICS] 1. A photographic image in the form of a shadow. 2. See radiograph. [OPTICS] A simple method of making visible the disturbances that occur in fluid flow at high velocity, in which light passing through a flowing fluid is refracted by density gradients in the fluid, resulting in bright and dark areas on a screen placed behind the fluid.

shadow mask [ELECTR] A thin, perforated metal mask mounted just back of the phosphor-dot faceplate in a three-gun color picture tube; the holes in the mask are positioned to ensure that each of the three electron beams strikes only its intended color phosphor dot. Also known as aperture mask.

shadow microscopy See projection microradiography.

shaft [GEOL] A passage in a cave that is vertical or nearly vertical. [MECH ENG] A cylindrical piece of metal used to carry rotating machine parts, such as pulleys and gears, to transmit power or motion. [MIN ENG] An excavation of limited area compared with its depth, made for finding or mining ore or coal, raising water, ore, rock, or coal, hoisting and lowering men and material, or ventilating underground workings; the term is often specifically applied to approximately vertical shafts as distinguished from an incline or an inclined shaft. [SCI TECH] A long, slender, usually cylindrical part.

shaft coupling See coupling.

shaft furnace [ENG] A vertical, refractory-lined cylinder in which a fixed bed (or descending column) of solids is maintained, and through which an ascending stream of hot gas is forced; for example, the pig-iron blast furnace and the phosphors-from-phosphate-rock furnace.

shaft kiln [ENG] A kiln in which raw material fed into the top, moves down through hot gases flowing up from burners on either side at the bottom, and emerges as a product from the bottom; used for calcining operations.

shaker [ELECTROMAG] An electromanetic device capable of imparting known and usually controlled vibratory acceleration to a given object. Also known as electrodynamic shaker; shake table.

shaker conveyor [MIN ENG] A conveyor consisting of a length of metal troughs, with suitable supports, to which a reciprocating motion is imparted by drives.

shake table See shaker; vibration machine.

shake wave See S wave.

shaking-out [CHEM ENG] A procedure in which a sample of crude oil is centrifuged at high speed to separate its components; used to determine sediment and water content.

shaking table See Wilfley table.

shale [PETR] A fine-grained laminated or fissile sedimentary rock made up of silt- or clay-size particles; generally consists of about one-third quartz, one-third clay materials, and one-third miscellaneous minerals, including carbonates, iron oxides, feldspars, and organic matter.

shale ball [GEOL] A meteorite partly or wholly converted to iron oxides by weathering. Also known as oxidite.

shale break [GEOL] A thin layer or parting of shale between harder strata or within a bed of sandstone or limestone.

shale crescent [GEOL] A crescent formed by the filling of a ripple-mark trough by shale.

shale ice [HYD] A mass of thin and brittle plates of river or lake ice formed when sheets of skim ice break up into small pieces.

shale naphtha [MATER] Naphtha derived from shale oil, usually containing 60–70% olefins and other hydrocarbons.

shale oil [MATER] Liquid obtained from the destructive distillation of kerogens in oil shale; further processing is needed to convert shale oil into products similar to petroleum oils.

shalification [GEOL] The formation of shale.

shallow-focus earthquake [GEOPHYS] An earthquake whose focus is located within 70 kilometers of the earth's surface.

shallow water [HYD] Water of such a depth that bottom topography affects surface waves.

shandite [MINERAL] $Ni_3Pb_2S_2$ A rhombohedral mineral composed of nickel lead sulfide, occurring in crystals.

shank [DES ENG] **1.** The end of a tool which fits into a drawing holder, as on a drill. **2.** See bit blank.

Shannon formula [COMMUN] A theorem in information theory which states that the highest number of binary digits per second which can be transmitted with arbitrarily small frequency of error is equal to the product of the bandwidth and $\log_2 (1 + R)$, where R is the signal-to-noise ratio.

shaped-beam antenna [ELECTROMAG] Antenna with a directional pattern which, over a certain angular range, is of special shape for some particular use.

shaped charge [ORD] An explosive charge with a shaped cavity that forces the impact of the explosion to the front so that there is an armor-piercing force. Also known as cavity charge.

shape factor [ELEC] See form factor. [FL MECH] The quotient of the area of a sphere equivalent to the volume of a solid particle divided by the actual surface of the particle; used in calculations of gas flow through beds of granular solids. [OPTICS] For a lens, the quantity $(R_2 + R_1)/(R_2 − R_1)$, where R_1 and R_2 are the radii of the first and second surface of the lens. Also known as Coddington shape factor.

shape memory alloy [MET] A martensitic alloy which exhibits shape recovery characteristics by stress-induced transformation and reorientation; reverse transformation during heating restores the original grain structure of the high-temperature phase.

shaping circuit See corrective network.

shaping network See corrective network.

shared logic [COMPUT SCI] The simultaneous use of a single computer by multiple users.

shark [VERT ZOO] Any of about 225 species of carnivorous elasmobranchs which occur principally in tropical and subtropical oceans; the body is fusiform with a heterocercal tail and a tough, usually gray, skin roughened by tubercles, and the snout extends beyond the mouth.

sharpstone conglomerate See sedimentary breccia.

shattercrack See flake.

shattuckite [MINERAL] $Cu_5(SiO_3)_4H_2O$ A blue mineral composed of basic copper silicate, occurring in fibrous masses.

shear [DES ENG] A cutting tool having two opposing blades between which a material is cut. [MECH] See shear strain. [MIN ENG] To make vertical cuts in a coal seam that has been undercut.

shear center See center of twist.

shear fold [GEOL] A similar fold whose mechanism is shearing or slipping along closely spaced planes that are parallel to the fold's axial surface. Also known as glide fold; slip fold.

shearing instability See Helmholtz instability.

shearing stress [MECH] A stress in which the material on one side of a surface pushes on the material on the other side of the surface with a force which is parallel to the surface. Also known as shear stress; tangential stress.

shear joint [GEOL] A joint that is a shear fracture; it is a potential plane of shear. Also known as slip joint.

shear modulus See modulus of elasticity in shear.

shear rate [FL MECH] The relative velocities in laminar flow of parallel adjacent layers of a fluid body under shear force.

shear spinning [MECH ENG] A sheet-metal-forming process which forms parts with rotational symmetry over a mandrel with the use of a tool or roller in which deformation is carried out with a roller in such a manner that the diameter of the original blank does not change but the thickness of the part decreases by an amount dependent on the mandrel angle.

shear strain [MECH] Also known as shear. **1.** A deformation of a solid body in which a plane in the body is displaced parallel to itself relative to parallel planes in the body; quantitatively, it is the displacement of any plane relative to a second plane, divided by the perpendicular distance between planes. **2.** The force causing such deformation.

shear strength [MECH] **1.** The maximum shear stress which a material can withstand without rupture. **2.** The ability of a material to withstand shear stress.

shear stress See shearing stress.

shear structure [GEOL] A local structure in which earth stresses have been relieved by many small, closely spaced fractures.

shearwater [VERT ZOO] Any of various species of oceanic birds of the genus *Puffinus* having tubular nostrils and long wings.

shear wave [GEOPHYS] See S wave. [MECH] A wave that causes an element of an elastic medium to change its shape without changing its volume. Also known as rotational wave.

shear zone [GEOL] A tabular area of rock that has been crushed and brecciated by many parallel fractures resulting from shear strain; often becomes a channel for underground solutions and the seat of ore deposition.

sheath [ELEC] A protective outside covering on a cable. [ELECTR] A space charge formed by ions near an electrode in a gas tube. [ELECTROMAG] The metal wall of a waveguide. [SCI TECH] A protective case or cover.

SHEED *See* scanning HEED.

sheep [VERT ZOO] Any of various mammals of the genus *Ovis* in the family Bovidae characterized by a stocky build and horns, when present, which tend to curl in a spiral.

sheer line [NAV ARCH] The longitudinal curve of the rail or decks, which shows the variation in height above water or freeboard throughout the vessel's entire length.

sheer strake [NAV ARCH] The uppermost line of planking of a wooden ship, or the upper strake of plating of a steel ship's main deck.

sheet [GEOL] 1. A thin flowstone coating of calcite in a cave. 2. A tabular igneous intrusion, especially when concordant or only slightly discordant. [HYD] *See* sheetflood. [MATER] A material in a configuration similar to a film except that its thickness is greater than 0.25 millimeter. [NAV ARCH] A rope or chain used to haul the clew of a sail out toward the yard arm or downward toward the deck and aft.

sheet cavitation [FL MECH] A type of cavitation in which cavities form on a solid boundary and remain attached as long as the conditions that led to their formation remain unaltered. Also known as steady-state cavitation.

sheet erosion [GEOL] Erosion of thin layers of surface materials by continuous sheets of running water. Also known as sheetflood erosion; sheetwash; surface wash; unconcentrated wash.

sheetflood [HYD] A broad expanse of moving, storm-borne water that spreads as a thin, continuous, relatively uniform film over a large area for a short distance and duration. Also known as sheet; sheetwash.

sheetflood erosion *See* sheet erosion.

sheet ice [HYD] A smooth, thin layer of ice formed by rapid freezing of the surface layer of a body of water.

sheeting [GEOL] The process by which thin sheets, slabs, scales, plates, or flakes of rock are successively broken loose or stripped from the outer surface of a large rock mass in response to release of load. Also known as exfoliation. [MATER] 1. A continuous film of a material such as plastic. 2. Steel or wood members used to face the walls of an excavation such as a basement or a trench.

sheeting structure [GEOL] A fracture or joint formed by pressure-release jointing or exfoliation. Also known as exfoliation joint; expansion joint; pseudostratification; release joint; sheet joint; sheet structure.

sheet joint *See* sheeting structure.

sheet lightning [GEOPHYS] A diffuse, but sometimes fairly bright, illumination of those parts of a thundercloud that surround the path of a lightning flash, particularly a cloud discharge or cloud-to-cloud discharge. Also known as luminous cloud.

sheet metal [MET] Thin sections of metal formed by rolling hot metal and usually less than 0.25 inch (6.35 millimeters) thick; when thicker than 0.25 inch, called plate.

sheet-metal screw *See* self-tapping screw.

sheet mineral *See* phyllosilicate.

sheet silicate *See* phyllosilicate.

sheet structure *See* sheeting structure.

sheetwash [GEOL] 1. The detritus deposited by a sheetflood. 2. *See* sheet erosion. [HYD] 1. A wide, moving expanse of water on an arid plain; the com-

bined result of many streams issuing from the mountains. 2. *See* sheetflood.

sheffer stroke *See* NAND.

shelf [GEOL] 1. Solid rock beneath alluvial deposits. 2. A flat, projecting ledge of rock. 3. *See* continental shelf.

shelf ice [HYD] The ice of an ice shelf. Also known as barrier ice.

shelf sea [OCEANOGR] A shallow marginal sea located on the continental shelf, usually less than 150 fathoms (275 meters) in depth; an example is the North Sea.

shell [ARCH] A reinforced concrete arched or domed roof used over unpartitioned areas. [BUILD] A building without internal partitions or furnishings. [DES ENG] 1. The case of a pulley block. 2. A thin hollow cylinder. 3. A hollow hemispherical structure. 4. The outer wall of a vessel or tank. [GEOL] 1. The crust of the earth. 2. A thin hard layer of rock. [GRAPHICS] An engraved roller made of copper and used in calico printing. [MET] 1. The outer wall of a metal mold. 2. The hard layer of sand and thermosetting plastic formed over a pattern and used as a mold wall in shell molding. 3. The metal sleeve remaining when a billet is extruded with a dummy block at smaller diameter. 4. A tubular casting used in preparing seamless drawn tubes. 5. A pierced forging. [ORD] 1. A hollow metal projectile designed to be projected from a gun, containing or intended to contain, a high-explosive, chemical, atomic, or other charge. 2. A shotgun cartridge or a cartridge for artillery of small arms. [ZOO] 1. A hard, usually calcareous, outer covering on an animal body, as of bivalves and turtles. 2. The hard covering of an egg. 3. Chitinous exoskeleton of certain arthropods.

shellac [MATER] A natural, alcohol-soluble, water-insoluble, flammable resin; made from lac resin deposited on tree twigs in India by the lac insect (*Laccifer lecca*) used as an ingredient of wood coatings.

shell-and-tube exchanger [ENG] A device for the transfer of heat from a hot fluid to a cooler fluid; one fluid passes through a group (bundle) of tubes, the other passes around the tubes, through a surrounding shell.

shellfish [INV ZOO] An aquatic invertebrate, such as a mollusk or crustacean, that has a shell or exoskeleton.

shell gland [INV ZOO] An organ that secretes the embryonic shell in many mollusks. [VERT ZOO] A specialized structure attached to the oviduct in certain animals that secretes the eggshell material.

shell model [NUC PHYS] A model of the nucleus in which the shell structure is either postulated or is a consequence of other postulates; especially the model in which the nucleons act as independent particles filling a preassigned set of energy levels as permitted by the quantum numbers and Pauli principle.

shell structure [NUC PHYS] Structure of the nucleus in which nucleons of each kind occupy quantum states which are in groups of approximately the same energy, called shells, the number of nucleons in each shell being limited by the Pauli exclusion principle.

sherardizing [MET] Coating iron with zinc by tumbling the article in powdered zinc at about 250–375°C.

sheridanite [MINERAL] $(Mg,Al)_6(Al,Si)_4O_{10}(OH)_8$ A pale-green to colorless talclike mineral composed of basic magnesium aluminum silicate.

SHF *See* sensible-heat factor; superhigh frequency.

shield [ENG] An iron, steel, or wood framework used to support the ground ahead of the lining in tunneling and mining. [GEOL] *See* palette. [NUCLEO]

The material placed around a nuclear reactor, or other source of radiation, to reduce escaping radiation or particles to a permissible level. Also known as shielding. [ORD] Armor plate mounted on a gun carriage to protect the operating mechanism and gun crew from enemy fire.

Shield See Scutum.

shield basalt [GEOL] A basaltic lava flow from a group of small, close-spaced shield-volcano vents that coalesced to form a single unit.

shield cone [GEOL] A cone or dome-shaped volcano built up by successive outpourings of lava.

shielded arc welding [MET] Arc welding in which the electric arc and the weld metal are protected by gas, decomposition products of the electrode covering, or a blanket of fusible flux.

shielded joint [ELEC] Cable joint having its insulation so enveloped by a conducting shield that substantially every point on the surface of the insulation is at ground potential, or at some predetermined potential with respect to ground.

shielded line [ELECTROMAG] Transmission line, the elements of which confine the propagated waves to an essentially finite space; the external conducting surface is called the sheath.

shielded wire [ELEC] Insulated wire covered with a metal shield, usually of tinned braided copper wire.

shield grid [ELECTR] A grid that shields the control grid of a gas tube from electrostatic fields, thermal radiation, and deposition of thermionic emissive material; it may also be used as an additional control electrode.

shielding [ELECTROMAG] See electric shielding. [MET] Placing a nonconducting object in an electrolytic bath during plating to alter the current distribution. [NUCLEO] 1. Reducing the ionizing radiation reaching one region of space from another region by using a shield or other device. 2. See shield.

shield volcano [GEOL] A broad, low volcano shaped like a flattened dome and built of basaltic lava. Also known as basaltic dome; lava dome.

shift [COMPUT SCI] A movement of data to the right or left, in a digital-computer location, usually with the loss of characters shifted beyond a boundary. [GEOL] The relative displacement of the units affected by a fault but outside the fault zone itself. [IND ENG] The number of hours or the part of any day worked. Also known as tour. [MECH ENG] To change the ratio of the driving to the driven gears to obtain the desired rotational speed or to avoid overloading and stalling an engine or a motor. [MET] A casting defect caused by malalignment of the mold parts.

shift register [COMPUT SCI] A computer hardware element constructed to perform shifting of its contained data.

shift-register generator [COMPUT SCI] A random-number generator which consists of a sequence of shift operations and other operations, such as no-carry addition.

shingle-block structure See imbricate structure.

shingles See herpes zoster.

shingle structure See imbricate structure.

Ship See Argo.

shipping ton See ton.

ship-tended acoustic relay [NAV] An acoustic navigation system that employs ocean bottom transponders and determines the distance of the ship from these transponders by measuring the time required for a signal broadcast by an acoustic transducer below the hull of the ship to travel to the transponder,

be rebroadcast there, and travel back to the ship. Abbreviated STAR.

shipworm [INV ZOO] Any of several bivalve mollusk species belonging to the family Teredinidae and which superficially resemble earthworms because the two valves are reduced to a pair of plates at the anterior of the animal or are used for boring into wood.

shipwright [CIV ENG] A worker whose responsibility is to ensure that the structure of a ship is straight and true and to the designed dimensions; the work starts with the laying down of the keel blocks and continues throughout the steelwork; applicable also to wood ship builders.

shipyard eye See keratoconjunctivitis.

SHM See harmonic motion.

shoal [GEOL] A submerged elevation rising from the bed of a shallow body of water and consisting of, or covered by, unconsolidated material, and may be exposed at low water.

shock [MECH] A pulse or transient motion or force lasting thousandths to tenths of a second which is capable of exciting mechanical resonances; for example, a blast produced by explosives. [MED] Clinical manifestations of circulatory insufficiency, including hypotension, weak pulse, tachycardia, pallor, and diminished urinary output.

shock absorber [MECH ENG] A spring, a dashpot, or a combination of the two, arranged to minimize the acceleration of the mass of a mechanism or portion thereof with respect to its frame or support.

shock excitation [ELEC] Excitation produced by a voltage or current variation of relatively short duration; used to initiate oscillation in the resonant circuit of an oscillator. Also known as impulse excitation.

shock front [PHYS] The outer side of a shock wave whose pressure rises from zero up to its peak value. Also known as pressure front.

shock isolation [MECH ENG] The application of isolators to alleviate the effects of shock on a mechanical device or system.

Shockley diode [ELECTR] A *pnpn* silicon controlled switch having characteristics that permit operation as a unidirectional diode switch.

shock tube [FL MECH] A long tube divided into two parts by a diaphragm; the volume on one side of the diaphragm constitutes the compression chamber, the other side is the expansion chamber; a high pressure is developed by suitable means in the compression chamber, and the diaphragm ruptured; the shock wave produced in the expansion chamber can be used for the calibration of air blast gages, or the chamber can be instrumented for the study of the characteristics of the shock wave.

shock wave [PHYS] A fully developed compression wave of large amplitude, across which density, pressure, and particle velocity change drastically.

shoe [ENG] In glassmaking, an open-ended crucible placed in a furnace for heating the blowing irons. [MECH ENG] 1. A metal block used as a form or support in various bending operations. 2. A replaceable piece used to break rock in certain crushing machines. 3. See brake shoe. [MIN ENG] 1. Pieces of steel fastened to a mine cage and formed to fit over the guides to guide it when it is in motion. 2. The bottom wedge-shaped piece attached to tubbing when sinking through quicksand. 3. A trough to convey ore to a crusher. 4. A coupling of rolled, cast, or forged steel to protect the lower end of the casting or drivepipe in overburden, or the bottom end of a sampler when pressed into a formation being sampled.

shoe brake [MECH ENG] A type of brake in which friction is applied by a long shoe, extending over a large portion of the rotating drum; the shoe may be external or internal to the drum.

shoofly *See* slant.

shoot [BOT] 1. The aerial portion of a plant, including stem, branches, and leaves. 2. A new, immature growth on a plant. [ENG] To detonate an explosive, used to break coal loose from a seam or in blasting operation or in a borehole. [GEOL] *See* ore shoot. [GEOPHYS] The energy that goes up through the strata from a seismic profiling shot and is reflected downward at the surface or at the base of the weathering; appears either as a single wave or unites with a wave train that is traveling downward. Also known as secondary reflection. [HYD] 1. A place where a stream flows or descends swiftly. 2. A natural or artificial channel, passage, or trough through which water is moved to a lower level. 3. A rush of water down a steep place or a rapids. [ORD] To project a missile with force; to fire a weapon, as a gun or cannon; to strike or hit something with a missile.

shooting star [ASTRON] A small meteor that has the brief appearance of a darting, starlike object.

shore [ENG] Timber or other material used as a temporary prop for excavations or buildings; may be sloping, vertical, or horizontal. [GEOL] 1. The narrow strip of land immediately bordering a body of water. 2. *See* seashore.

shore current [HYD] A water current near a shoreline, often flowing parallel to the shore.

shoreface [GEOL] The narrow, steeply sloping zone between the seaward limit of the shore at low water and the nearly horizontal offshore zone.

shore ice [OCEANOGR] Sea ice that has been beached by wind, tides, currents, or ice pressure; it is a type of fast ice, and may sometimes be rafted ice.

shoreline [GEOL] The intersection of a specified plane of water, especially mean high water, with the shore; a limit which changes with the tide or water level. Also known as strandline; waterline.

shore platform [GEOL] The horizontal or gently sloping surface produced along a shore by wave erosion. Also known as scar.

Shore scleroscope hardness test [MET] A rebound hardness test in which a metal body is dropped vertically down a glass tube onto the surface of the material being tested; the height of the rebound is a measure of the hardness. Also known as scleroscope hardness test.

shore terrace [GEOL] 1. A terrace produced along the shore by wave and current action. 2. *See* marine terrace.

short [ELEC] *See* short circuit. [ENG] In plastics injection molding, the failure to fill the mold completely. Also known as short shot. [ORD] A bomb or projectile hit short of the target.

short circuit [ELEC] A low-resistance connection across a voltage source or between both sides of a circuit or line, usually accidental and usually resulting in excessive current flow that may cause damage. Also known as short.

short-contact switch [ELEC] Selector switch in which the width of the movable contact is greater than the distance between contact clips, so that the new circuit is contacted before the old one is broken; this avoids noise during switching.

short fuse [ENG] 1. Any fuse that is cut too short. 2. The practice of firing a blast, the fuse on the primer of which is not sufficiently long to reach from the top of the charge to the collar of the borehole; the primer, with fuse attached, is dropped into the charge while burning.

short hundredweight *See* hundredweight.

shortite [MINERAL] $Na_2Ca_2(CO_3)_3$ A mineral composed of sodium and calcium carbonate.

short-path principle *See* Hittorf principle.

short-period comet [ASTRON] A comet whose period is short enough for observations at two or more apparitions to be interrelated; usually taken to be a comet whose period is shorter than 200 years.

short-pulse laser [OPTICS] A laser designed to generate a pulse of light lasting on the order of nanoseconds or less, and having very high power, such as by Q switching or mode-locking.

short-range attack missile [ORD] An air-to-surface and air-to-air solid-fuel guided missile that uses inertial guidance for complete immunity to jamming; target coordinates are fed into the missile by a computer on the aircraft just before launch. Abbreviated SRAM.

short-range forecast [METEOROL] A weather forecast made for a time period generally not greater than 48 hours in advance.

short-range radar [ENG] Radar whose maximum line-of-sight range, for a reflecting target having 1 square meter of area perpendicular to the beam, is between 50 and 150 miles (80 and 240 kilometers).

short shot *See* short.

short takeoff and landing [AERO ENG] The ability of an aircraft to clear a 50-foot (15-meter) obstacle within 1500 feet (450 meters) of commencing takeoff, or in landing, to stop within 1500 feet after passing over a 50-foot obstacle. Abbreviated STOL.

short ton *See* ton.

short-tube vertical evaporator [CHEM ENG] A liquid evaporation process unit with a vertical bundle of tubes 2–3 inches (5–8 centimeters) in diameter and 4–6 feet (1.2–1.8 meters) long; the heating fluid is inside the tubes, and the liquid to be evaporated is in the shell area outside the tubes; used mainly to evaporate cane-sugar juice. Also known as calandria evaporator; Roberts evaporator; standard evaporator.

shortwall [MIN ENG] 1. A method of mining in which comparatively small areas are worked separately. 2. A length of coal face between about 5 and 30 yards (4.6 and 27 meters), generally employed in pillar methods of working.

shortwave broadcasting [COMMUN] Radio broadcasting at frequencies in the range from about 1600 to 30,000 kilohertz, above the standard broadcast band.

shortwave radiation [ELECTROMAG] A term used loosely to distinguish radiation in the visible and near-visible portions of the electromagnetic spectrum (roughly 0.4 to 1.0 micrometer in wavelength) from long-wave radiation (infrared radiation).

short word [COMPUT SCI] The fixed word of lesser length in computers capable of handling words of two different lengths; in many computers this is referred to as a half-word because the length is exactly the half-length of the full word.

shot [AERO ENG] An act or instance of firing a rocket, especially from the earth's surface. [ENG] 1. A charge of some kind of explosive. 2. Small spherical particles of steel. 3. Small steel balls used as the cutting agent of a shot drill. 4. The firing of a blast. 5. In plastics molding, the yield from one complete molding cycle, including scrap. [MIN ENG] Coal broken by blasting or other methods. [ORD] 1. A solid projectile for cannon, without a bursting charge; the term projectile is preferred for uniformity in nomenclature. **2.** A mass or load of numerous, relatively small, lead pellets used in a shotgun, as birdshot or buckshot.

shot blasting [MET] Cleaning and descaling metal by shot peening or by means of a stream of abrasive powder blown through a nozzle under air pressure in the range 30–150 pounds per square inch (2×10^5 to 1.0×10^6 newtons per square meter).

shot effect *See* shot noise.

shot feed [MECH ENG] A device to introduce chilled-steel shot, at a uniform rate and in the proper quantities, into the circulating fluid flowing downward through the rods or pipe connected to the core barrel and bit of a shot drill.

shothole [ENG] The borehole in which an explosive is placed for blasting.

shot noise [ELECTR] Noise voltage developed in a thermionic tube because of the random variations in the number and the velocity of electrons emitted by the heated cathode; the effect causes sputtering or popping sounds in radio receivers and snow effects in television pictures. Also known as Schottky noise; shot effect.

shot peening [MET] Shot blasting with small steel balls driven by a blast of air.

shot point [ENG] The point at which an explosion (such as in seismic prospecting) originates, generating vibrations in the ground.

shoulder [ANAT] **1.** The area of union between the upper limb and the trunk in humans. **2.** The corresponding region in other vertebrates. [DES ENG] The portion of a shaft, a stepped object, or a flanged object that shows an increase of diameter. [ENG] A projection made on a piece of shaped wood, metal, or stone, where its width or thickness is suddenly changed. [GEOL] **1.** A short, rounded spur protruding laterally from the slope of a mountain or hill. **2.** The sloping segment below the summit of a mountain or hill. **3.** A bench on the flanks of a glaciated valley, located at the sharp change of slope where the steep sides of the inner glaciated valley meet the more gradual slope above the level of glaciation. **4.** A joint structure on a joint face produced by the intersection of plume-structure ridges with fringe joints.

shoulder blade *See* scapula.

shoulder girdle *See* pectoral girdle.

shoulder screw [DES ENG] A screw with an unthreaded cylindrical section, or shoulder, between threads and screwhead; the shoulder is larger in diameter than the threaded section and provides an axis around which close-fitting moving parts operate.

shovel dozer *See* tractor loader.

shovel loader [MECH ENG] A loading machine mounted on wheels, with a bucket hinged to the chassis which scoops up loose material, elevates it, and discharges it behind the machine.

shower [METEOROL] Precipitation from a convective cloud; characterized by the suddenness with which it starts and stops, by the rapid changes of intensity, and usually by rapid changes in the appearance of the sky. [NUC PHYS] *See* cosmic-ray shower.

shrew [VERT ZOO] Any of more than 250 species of insectivorous mammals of the family Soricidae; individuals are small with a moderately long tail, minute eyes, a sharp-pointed snout, and small ears.

shrimp [INV ZOO] The common name for a number of crustaceans, principally in the decapod suborder Natantia, characterized by having well-developed pleopods and by having the abdomen sharply bent in most species, producing a humped appearance.

shrinkage stoping [MIN ENG] A modification of overhead stoping, involving the use of a part of the ore for the purpose of support and as a working platform. Also known as back stoping.

shrink fit [DES ENG] A tight interference fit between mating parts made by shrinking-on, that is, by heating the outer member to expand the bore for easy assembly and then cooling so that the outer member contracts.

shrink forming [DES ENG] Forming metal wherein the piece undergoes shrinkage during cooling following the application of heat, cold upset, or pressure.

shrouded propeller *See* ducted fan.

shunt [ELEC] **1.** A precision low-value resistor placed across the terminals of an ammeter to increase its range by allowing a known fraction of the circuit current to go around the meter. Also known as electric shunt. **2.** To place one part in parallel with another. **3.** *See* parallel. [ELECTROMAG] A piece of iron that provides a parallel path for magnetic flux around an air gap in a magnetic circuit. [MED] A vascular passage by which blood is diverted from its normal circulatory path; frequently it is a surgical passage created between two blood vessels, but it may also be an anatomical feature. [MIN ENG] To shove or turn off to one side, as a car or train from one track to another.

shunt-excited [ELECTROMAG] Having field windings connected across the armature terminals, as in a direct-current generator.

shunt-excited antenna [ELECTROMAG] A tower antenna, not insulated from the ground at the base, whose feeder is connected at a point about one-fifth of the way up the antenna and usually slopes up to this point from a point some distance from the antenna's base.

shunt feed *See* parallel feed.

shunt generator [ELEC] A generator whose field winding and armature winding are connected in parallel, and in which the armature supplies both the load current and the field current.

shunting [ELEC] The act of connecting one device to the terminals of another so that the current is divided between the two devices in proportion to their respective admittances.

shunt motor [ELEC] A direct-current motor whose field circuit and armature circuit are connected in parallel.

shunt regulator [ELEC] A regulator that maintains a constant output voltage by controlling the current through a dropping resistance in series with the load.

shunt repeater [ELEC] A type of negative impedance telephone repeater which is stable when it is short-circuited, but oscillates when terminated by a high impedance, in contrast to a series repeater; it can be thought of as a negative admittance.

shunt valve [ENG] A valve that gives a fluid under pressure a more readily available escape route than the normal route.

shunt wound [ELEC] Having armature and field windings in parallel, as in a direct-current generator or motor.

shutter [NUCLEO] A movable plate of absorbing material used to cover a window or a beam hole in a reactor when radiation is not desired, or used to shut off a flow of slow neutrons. [OPTICS] A mechanical device that cuts off a beam of light by opening and closing at different rates of speed to expose film or plates; used in cameras and motion picture projectors. [ORD] A barrier in an explosive train used to stop a detonation wave; an interrupter which opens or closes as a shutter; often used to obtain fuse safety.

shuttering See formwork.

shuttle [MECH ENG] A back-and-forth motion of a machine which continues to face in one direction. [NUCLEO] See rabbit. [TEXT] A device on a loom that moves filling yarns between the warp yarns during weaving.

shuttle car [MIN ENG] An electrically propelled vehicle on rubber tires or caterpillar treads used to transfer raw materials, such as coal and ore, from loading machines in trackless areas of a mine to the main transportation system.

shuttle conveyor [MECH ENG] Any conveyor in a self-contained structure movable in a defined path parallel to the flow of the material.

Si See silicon.

SI See International System of units.

sial [PETR] A petrologic term for the silica- and alumina-rich upper rock layers of the earth's crust; gives rise to granite magma; the bulk of the continental blocks is sialic. Also known as granitic layer; sal.

Siboglinidae [INV ZOO] A family of pogonophores in the order Athecanephria.

SIC See dielectric constant.

sickle-cell anemia [MED] A chronic, hereditary hemolytic and thrombotic disorder in which hypoxia causes the erythrocyte to assume a sickle shape; occurs in individuals homozygous for sickle-cell hemoglobin trait.

sicklerite [MINERAL] $(Li,Mn)(PO_4)$ A dark-brown mineral composed of hydrous lithium manganese phosphate occurring in cleavable masses.

SID See sudden ionospheric disturbance.

sideband [ELECTROMAG] 1. The frequency band located either above or below the carrier frequency, within which fall the frequency components of the wave produced by the process of modulation. 2. The wave components lying within such bands.

side chain [ORG CHEM] A grouping of similar atoms (two or more, generally carbons, as in the ethyl radical, C_2H_5—) that branches off from a straight-chain or cyclic (for example, benzene) molecule.

side drift See adit.

side-looking radar [ENG] A high-resolution airborne radar having antennas aimed to the right and left of the flight path; used to provide high-resolution strip maps with photographlike detail, to map unfriendly territory while flying along its perimeter, and to detect submarine snorkels against a background of sea clutter.

sidereal [ASTRON] Referring to a quantity, such as time, to indicate that it is measured in relation to the apparent motion or position of the stars.

sidereal day [ASTRON] The time between two successive upper transits of the vernal equinox; this period measures one sidereal day.

sidereal hour angle [ASTRON] The angle along the celestial equator formed between the hour circle of a celestial body and the hour circle of the vernal equinox, measuring westward from the vernal equinox through 360°.

sidereal month [ASTRON] The time period of one revolution of the moon about the earth relative to the stars; this period varies because of perturbations, but it is a little less than 27 1/3 days.

sidereal noon [ASTRON] The instant in time that the vernal equinox is on the meridian.

sidereal period [ASTRON] The length of time required for one revolution of a celestial body about its primary, with respect to the stars.

sidereal time [ASTRON] Time based on diurnal motion of stars; it is used by astronomers but is not convenient for ordinary purposes.

sidereal year [ASTRON] The time period relative to the stars of one revolution of the earth around the sun; it is about 365.2564 mean solar days.

siderite [MINERAL] $FeCO_3$ A brownish, gray, or greenish rhombohedral mineral composed of ferrous carbonate; hardness is 4 on Mohs scale, and specific gravity is 3.9. Also known as chalybite; iron spar; rhombohedral iron ore; siderose; sparry iron; spathic iron; white iron ore.

side rod [MECH ENG] 1. A rod linking the crankpins of two adjoining driving wheels on the same side of a locomotive; distributes power from the main rod to the driving wheels. 2. One of the rods linking the piston-rod crossheads and the side levers of a side-lever engine.

sideronatrite [MINERAL] $Na_2Fe(SO_4)(OH)\cdot3H_2O$ A yellow mineral composed of basic hydrous sodium iron sulfate occurring in fibrous masses.

siderose See siderite.

siderotil [MINERAL] $(Cu,Fe)SO_4\cdot5H_2O$ A white to yellowish or pale greenish-white mineral consisting of ferrous sulfate pentahydrate; occurs as fibrous crusts and groups of needlelike crystals.

sidestream [CHEM ENG] A liquid stream taken from an intermediate point of a liquids-processing unit, for example, a distillation or extraction tower.

side stream See tributary.

sidetone [COMMUN] The sound of the speaker's own voice as heard in his telephone receiver; the effect is undesirable and is usually suppressed by special circuits.

sidetrack [CIV ENG] 1. To move railroad cars onto a siding. 2. See siding.

sideways feed [COMPUT SCI] The method of placing cards in the feed hopper of a punched-card machine in which one of the long edges of the card enters the machine first, so that the columns of the card are read simultaneously. Also known as parallel feed.

siding [CIV ENG] A short railroad track connected to the main track at one or more points and used to move railroad cars in order to free traffic on the main line or for temporary storage of cars. Also known as sidetrack. [MATER] Any wall cladding, except masonry or brick.

siegbahn [SPECT] A unit of length, formerly used to express wavelengths of x-rays, equal to exactly 1/3029.45 of the spacing of the (200) planes of calcite at 18°C, or to $(1.00202 \pm 0.00003) \times 10^{-13}$ meter. Also known as x-ray unit; X-unit. Symbolized X; XU.

siegenite [MINERAL] $(Co,Ni)_3S_4$ A mineral composed of nickel cobalt sulfide.

siemens [ELEC] A unit of conductance, admittance, and susceptance, equal to the conductance between two points of a conductor such that a potential difference of 1 volt between these points produces a current of 1 ampere; the conductance of a conductor in siemens is the reciprocal of its resistance in ohms. Formerly known as mho (Ω); reciprocal ohm. Symbolized S.

Sierolomorphidae [INV ZOO] A small family of hymenopteran insects in the superfamily Scolioidea.

sierra [GEOGR] A high range of hills or mountains with irregular peaks that give a sawtooth profile.

sieve [ENG] 1. A meshed or perforated device or sheet through which dry loose material is refined, liquid is strained, and soft solids are comminuted. 2. A meshed sheet with apertures of uniform size used for sizing granular materials.

sieve cell [BOT] A long, tapering cell that is characteristic of phloem in gymnosperms and lower vascular plants, in which all the sieve areas are of equal specialization.

sieve plate [CHEM ENG] A distillation-tower tray that is perforated so that the vapor emerges vertically through the tray, passing through the liquid holdup on top of the tray; used as a replacement for bubblecap trays in distillation. Also known as sieve tray.

sieve texture See poikiloblastic.

sieve tissue See phloem.

sieve tray See sieve plate.

sieve tube [BOT] A phloem element consisting of a series of thin-walled cells arranged end to end, in which some sieve areas are more specialized than others.

Sigalionidae [INV ZOO] A family of scale-bearing polychaete annelids belonging to the Errantia.

Siganidae [VERT ZOO] A small family of herbivorous perciform fishes in the suborder Acanthuroidei having minute concealed scales embedded in the skin and strong, sharp fin spines.

sighting [OPTICS] 1. The act or procedure of aiming with the aid of a sight. 2. The action of bringing something into view; the action of seeing something.

sighting tube [ENG] A tube, usually ceramic, inserted into a hot chamber whose temperature is to be measured; an optical pyrometer is sighted into the tube to observe the interior end of the tube to give a temperature reading.

sight reduction [NAV] The information needed for establishing a line of position, derived from the use of a sextant or an octant.

sigma hyperon [PARTIC PHYS] 1. The collective name for three semistable baryons with charges of $+1$, 0, and -1 times the proton charge, designated Σ^+, Σ^0, Σ^-, all having masses of approximately 1193 megaelectronvolts, spin of $\frac{1}{2}$, and positive parity; they form an isotopic spin multiplet with a total isotopic spin of 1 and a hypercharge of 0. 2. Any baryon belonging to an isotopic spin multiplet having a total isotopic spin of 1 and a hypercharge of 0; designated $\Sigma_J^P(m)$, where m is the mass of the baryon in million electron volts, and J and P are its spin and parity; the $\Sigma = _{\Sigma \times +}(1385)$ is sometimes designated Σ^*.

sigma pile [NUCLEO] An assembly of moderating material containing a neutron source, used to study the absorption cross sections and other neutron properties of the material.

sigmoid colon [ANAT] The S-shaped portion of the colon between the descending colon and the rectum.

sigmoidoscope [MED] An appliance for the inspection, by artificial light, of the sigmoid colon; it differs from the proctoscope in its greater length and diameter.

sign [COMMUN] In semiotics, an entity that signifies some other thing, and may be interpreted. [MATH] 1. A symbol which indicates whether a quantity is greater than zero or less than zero; the signs are often the marks $+$ and $-$ respectively, but other arbitrarily selected symbols are used, especially in automatic data processing. 2. A unit of plane angle, equal to $30°$ or $\pi/6$ radians.

signage [GRAPHICS] Environmental graphic communications whose functions include direction, identification, information or orientation, regulation, warning, or restriction.

signal [COMMUN] 1. A visual, aural, or other indication used to convey information. 2. The intelligence, message, or effect to be conveyed over a communication system. 3. See signal wave.

signal channel [COMMUN] A signal path for transmitting electric signals; such paths may be separated by frequency division or time division.

signal distance [COMPUT SCI] The number of bits that are not the same in two binary words of equal length. Also known as hamming distance.

signal generator [ENG] An electronic test instrument that delivers a sinusoidal output at an accurately calibrated frequency that may be anywhere from the audio to the microwave range; the frequency and amplitude are adjustable over a wide range, and the output usually may be amplitude- or frequency-modulated. Also known as test oscillator.

signaling key See key.

signal-shaping network [ELECTR] Network inserted in a telegraph circuit, usually at the receiving end, to improve the waveform of the code signals.

signal strength [ELECTROMAG] The strength of the signal produced by a radio transmitter at a particular location, usually expressed as microvolts or millivolts per meter of effective receiving antenna height.

signal-to-noise ratio [ELECTR] The ratio of the amplitude of a desired signal at any point to the amplitude of noise signals at that same point; often expressed in decibels; the peak value is usually used for pulse noise, while the root-mean-square (rms) value is used for random noise. Abbreviated S/N; SNR.

signal wave [COMMUN] A wave whose characteristics permit some intelligence, message, or effect to be conveyed. Also known as signal.

signal winding [ELEC] Control winding, of a saturable reactor, to which the independent variable (signal wave) is applied.

signature [ELECTR] The characteristic pattern of a target as displayed by detection and classification equipment. [GRAPHICS] A folded, printed sheet, usually consisting of 16 or 32 pages, that forms a section of a book or a pamphlet; the sheet may have fewer pages, but is always in multiples of four. [NAV ARCH] The graphic record of the magnetic properties of a vessel automatically traced as the vessel passes over the sensitive element of a recording instrument; more accurately called magnetic signature. [ORD] The identifying characteristics peculiar to each type of target which enable detecting apparatus, such as certain fuses, to sense and differentiate targets.

sign bit [COMPUT SCI] A sign digit consisting of one bit.

sign check indicator [COMPUT SCI] An error checking device, indicating no sign or improper signing of a field used for arithmetic processes; the machine can, upon interrogation, be made to stop or enter into a correction routine.

sign-control flip-flop [COMPUT SCI] In computers, a flip-flop in the arithmetic unit used for storing the sign of the result of an operation.

sign digit [COMPUT SCI] A digit containing one to four binary bits, associated with a data item and used to denote an algebraic sign.

signed field [COMPUT SCI] A field of data that contains a number which includes a sign digit indicating the number's sign.

significance level See level of significance of a test.

significance probability [STAT] The probability of observing a value of a test statistic as significant as, or even more significant than, the value actually observed.

significant digit See significant figure.

significant figure [MATH] A prescribed decimal place which determines the amount of rounding off to be done; this is usually based upon the degree of accuracy in measurement. Also known as significant digit.

sign position [COMPUT SCI] That position, always at or near the left or right end of a numeral, in which the algebraic sign of the number is represented.

sign test [STAT] A test which can be used whenever an experiment is conducted to compare a treatment with a control on a number of matched pairs, provided the two treatments are assigned to the members of each pair at random.

silica [MINERAL] SiO_2 Naturally occurring silicon dioxide; occurs in five crystalline polymorphs (quartz, tridymite, cristobalite, coesite, and stishovite), in cryptocrystalline form (as chalcedony), in amorphous and hydrated forms (as opal), and combined in silicates.

silicate [INORG CHEM] The generic term for a compound that contains silicon, oxygen, and one or more metals, and may contain hydrogen. [MINERAL] Any of a large group of minerals whose crystal lattice contains SiO_4 tetrahedra, either isolated or joined through one or more of the oxygen atoms.

siliceous [PETR] Describing a rock containing abundant silica, especially free silica.

siliceous sinter [MINERAL] A white, lightweight, porous, opaline variety of silica, deposited by a geyser or hot spring. Also known as fiorite; geyserite; pearl sinter; sinter.

silicification [GEOL] Introduction of or replacement by silica. Also known as silification.

Silicoflagellata [BOT] A class of unicellular flagellates of the plant division Chrysophyta represented by a single living genus, *Dictyocha*.

Silicoflagellida [INV ZOO] An order of marine flagellates in the class Phytamastigophorea which have an internal, siliceous, tubular skeleton, numerous yellow chromatophores, and a single flagellum.

silicon [CHEM] A group IV nonmetallic element, symbol Si, with atomic number 14, atomic weight 28.086; dark-brown crystals that burn in air when ignited; soluble in hydrofluoric acid and alkalies; melts at 1410°C; used to make silicon-containing alloys, as an intermediate for silicon-containing compounds, and in rectifiers and transistors.

silicon controlled rectifier [ELECTR] A semiconductor rectifier that can be controlled; it is a *pnpn* four-layer semiconductor device that normally acts as an open circuit, but switches rapidly to a conducting state when an appropriate gate signal is applied to the gate terminal. Abbreviated SCR. Also known as reverse-blocking triode thyristor.

silicon controlled switch [ELECTR] A four-terminal switching device having four semiconductor layers, all of which are accessible; it can be used as a silicon controlled rectifier, gate-turnoff switch, complementary silicon controlled rectifier, or conventional silicon transistor. Abbreviated SCS. Also known as reverse-blocking tetrode thyristor.

silicon detector See silicon diode.

silicon diode [ELECTR] A crystal diode that uses silicon as a semiconductor; used as a detector in ultrahigh- and super-high-frequency circuits. Also known as silicon detector.

silicone [MATER] A fluid, resin, or elastomer; can be a grease, a rubber, or a foamable powder; the group name for heat-stable, water repellent, semiorganic polymers of organic radicals attached to the silicones, for example, dimethyl silicone; used in adhesives, cosmetics, and elastomers.

silicon rectifier [ELECTR] A metallic rectifier in which rectifying action is provided by an alloy junction formed in a high-purity silicon slab.

silicon resistor [ELECTR] A resistor using silicon semiconductor material as a resistance element, to obtain a positive temperature coefficient of resistance that does not appreciably change with temperature; used as a temperature-sensing element.

silicon solar cell [ELECTR] A solar cell consisting of *p* and *n* silicon layers placed one above the other to form a *pn* junction at which radiant energy is converted into electricity.

silification See silicification.

silk [GEOL] Microscopic needle-shaped crystalline inclusions of rutile in a natural gem from which subsurface reflections produce a whitish sheen resembling that of a silk fabric. [INV ZOO] A continuous protein fiber consisting principally of fibroin and secreted by various insects and arachnids, especially the silkworm, for use in spinning cocoons, webs, egg cases, and other structures. [TEXT] A thread or fabric made from silk secretions of the silkworm.

silk cotton tree See kapok tree.

silk gland [INV ZOO] A gland in certain insects which secretes a viscous fluid in the form of filaments known as silk; it is a salivary gland in insects and an abdominal gland in spiders.

silk-screen printing [GRAPHICS] The process of printing a flat color design through a piece of silk; the silk is tightly stretched on a wooden frame and a design is transferred to the silk; the parts of the design not to be printed are stopped out with a resist medium; the ink is pushed through the open mesh of the design area with a rubber-edged squeegee; only one color can be printed at a time.

silkworm [INV ZOO] The larva of various moths, especially *Bombyx mori*, that produces a large amount of silk for building its cocoon.

sill [BUILD] The lowest horizontal member of a framed partition or of a window or door frame. [CIV ENG] 1. A timber laid across the foot of a trench or a heading under the side truss. 2. The horizontal overflow line of a dam spillway or other weir structure. 3. A horizontal member on which a lift gate rests when closed. 4. A low concrete or masonry dam in a small stream to retard bottom erosion. [GEOL] 1. Submarine ridge in relatively shallow water that separates a partly closed basin from another basin or from an adjacent sea. 2. A tabular igneous intrusion that is oriented parallel to the planar structure of surrounding rock. [MIN ENG] 1. A piece of wood laid across a drift to constitute a frame to support uprights of timber sets and to carry the track of the tramway. 2. The floor of a gallery or passage in a mine.

silled basin See restricted basin.

sillenite [MINERAL] Bi_2O_3 A mineral composed of native bismuth oxide, is polymorphous with bismite, and occurs as earthy masses.

sillimanite [MINERAL] Al_2SiO_5 A brown, pale-green, or white neosilicate mineral with vitreous luster crystallizing in the orthorhombic system; commonly occurs in slender crystals, often in fibrous aggregates; hardness is 6–7 on Mohs scale, and specific gravity is 3.23. Also known as fibrolite.

silo [AERO ENG] A missile shelter that consists of a hardened vertical hole in the ground with facilities either for lifting the missile to a launch position, or for direct launch from the shelter. [CIV ENG] A large vertical, cylindrical structure, made of reinforced concrete, steel, or timber, and used for storing grain, cement, or other materials.

Silphidae [INV ZOO] The carrion beetles, a family of coleopteran insects in the superfamily Staphylinoidea.

silt [GEOL] 1. A rock fragment or a mineral or detrital particle in the soil having a diameter of 0.002–0.05 millimeter that is, smaller than fine sand and larger than coarse clay. 2. Sediment carried or deposited by water. 3. Soil containing at least 80% silt and less than 12% clay.

silting [CIV ENG] The filling up or raising of the bed of a body of water by depositing silt. [GEOL] The deposition or accumulation of stream-deposited silt that is suspended in a body of standing water.

siltite *See* siltstone.

siltstone [GEOL] Indurated silt having a shalelike texture and composition. Also known as siltite.

Silurian [GEOL] 1. A period of geologic time of the Paleozoic era, covering a time span of between 430–440 and 395 million years ago. 2. The rock system of this period.

Siluridae [VERT ZOO] A family of European catfish in the suborder Siluroidei in which the adipose dorsal fin is rudimentary or lacking.

Siluriformes [VERT ZOO] The catfishes, a distinctive order of actinopterygian fishes in the superorder Ostariophysi, distinguished by a complex Weberian apparatus that involves the fifth vertebrae and one to four pair of barbels.

Siluroidei [VERT ZOO] A suborder of the Siluriformes.

Silvanidae [INV ZOO] An equivalent name for Cucujidae.

silver [CHEM] A white metallic element in group I, symbol Ag, with atomic number 47; soluble in acids and alkalies, insoluble in water; melts at 961°C, boils at 2212°C; used in photographic chemicals, alloys, conductors, and plating. [MET] A sonorous, ductile, malleable metal that is capable of a high degree of polish and that has high thermal and electric conductivity.

silver-brazing alloy *See* silver solder.

silverfish [INV ZOO] Any of over 350 species of insects of the order Thysanura; they are small, wingless forms with biting mouthparts.

silver frost [METEOROL] A deposit of glaze built up on trees, shrubs, and other exposed objects during a fall of freezing precipitation; the product of an ice storm. Also known as silver thaw.

silver solder [MET] A solder composed of silver, copper, and zinc, having a melting point lower than silver but higher than lead-tin solder. Also known as silver-brazing alloy.

silver storm *See* ice storm.

silver thaw *See* silver frost.

silviculture [FOR] The theory and practice of controlling the establishment, composition, and growth of stands of trees for any of the goods and benefits that they may be called upon to produce.

sima [PETR] A petrologic term for the lower layer of the earth's crust, composed of silica- and magnesia-rich rocks; source of basaltic magma; sima is equivalent to the lower part of the continental crust and the bulk of the oceanic crust. Also known as intermediate layer.

similar figures [MATH] Two figures or bodies that are identical except for size; similar figures can be placed in perspective, so that straight lines joining corresponding parts of the two figures will pass through a common point.

similar fold [GEOL] A fold in deformed beds in which the successive folds resemble each other.

similitude [ENG] A likeness or resemblance; for example, the scale-up of a chemical process from a laboratory or pilot-plant scale to a commercial scale. [PHYS] The use in scientific studies and engineering designs of the corresponding behavior between large and small objects or systems which are of similar nature and, more precisely, have geometrical, kinematic, and dynamical similarity.

simple balance [ENG] An instrument for measuring weight in which a beam can rotate about a knife-edge or other point of support, the unknown weight is placed in one of two pans suspended from the ends of the beam and the known weights are placed in the other pan, and a small weight is slid along the beam until the beam is horizontal.

simple buffering [COMPUT SCI] A technique for obtaining simultaneous performance of input/output operations and computing; it involves associating a buffer with only one input or output file (or data set) for the entire duration of the activity on that file (or data set).

simple closed curve [MATH] A closed curve which never crosses itself.

simple conic projection [MAP] A conic map projection in which the surface of a sphere or spheroid, such as the earth, is developed on a tangent cone which is then spread out to form a plane.

simple continuous distillation *See* equilibrium flash vaporization.

simple cubic lattice [CRYSTAL] A crystal lattice whose unit cell is a cube, and whose lattice points are located at the vertices of the cube.

simple engine [MECH ENG] An engine (such as a steam engine) in which expansion occurs in a single phase, after which the working fluid is exhausted.

simple harmonic current [ELEC] Alternating current, the instantaneous value of which is equal to the product of a constant, and the cosine of an angle varying linearly with time. Also known as sinusoidal current.

simple harmonic motion *See* harmonic motion.

simple leaf [BOT] A leaf having one blade, or a lobed leaf in which the separate parts do not reach down to the midrib.

simple lens [OPTICS] A lens consisting of a single element. Also known as single lens.

simple machine [MECH ENG] Any of several elementary machines, one or more being incorporated in every mechanical machine; usually, only the lever, wheel and axle, pulley (or block and tackle), inclined plane, and screw are included, although the gear drive and hydraulic press may also be considered simple machines.

simple microscope [OPTICS] A diverging lens system, which can form an enlarged image of a small object. Also known as hand lens; magnifier; magnifying glass.

simple ore [GEOL] An ore of a single metal.

simple oscillator *See* harmonic oscillator.

simple path *See* path.

simple pendulum [MECH] A device consisting of a small, massive body suspended by an inextensible object of negligible mass from a fixed horizontal axis about which the body and suspension are free to rotate.

simple protein [BIOCHEM] One of a group of proteins which, upon hydrolysis, yield exclusively amino acids; included are globulins, glutelins, histones, prolamines, and protamines.

simple salt [CHEM] One of four classes of salts in a classification system that depends on the character of completeness of the ionization; examples are NaCl, $NaHCO_3$, and $Pb(OH)Cl$.

simple tone [ACOUS] Also known as pure tone. **1.** A sound wave whose instantaneous sound pressure is a simple sinusoidal function of time. **2.** A sound sensation characterized by singleness of pitch.

simplex [MATH] An n-dimensional simplex in a euclidean space consists of $n + 1$ linearly independent points $p_0, p_1, \ldots, p_n$ together with all line segments $a_0 p_0 + a_1 p_1 + \ldots + a_n p_n$ where the $a_i \geq 0$ and $a_0 + a_1 + \ldots + a_n = 1$; a triangle with its interior and a tetrahedron with its interior are examples.

simplex channel [COMMUN] A channel which permits transmission in one direction only.

simplex method [MATH] A finite iterative algorithm used in linear programming whereby successive solutions are obtained and tested for optimality.

simplex structure [COMPUT SCI] The structure of an information processing system designed in such a way that only the minimum amount of hardware is utilized to implement its function.

simplex uterus [ANAT] A uterus consisting of a single cavity, representing the greatest degree of fusion of the Müllerian ducts; found in man and apes.

simpsonite [MINERAL] $AlTaO_4$ A hexagonal mineral composed of aluminum tantalum oxide and occurring in short crystals.

SIMS See secondary ion mass spectrometer.

simulate [ENG] To mimic some or all of the behavior of one system with a different, dissimilar system, particularly with computers, models, or other equipment.

simulator [COMPUT SCI] A routine which is executed by one computer but which imitates the operations of another computer. [ENG] A computer or other piece of equipment that simulates a desired system or condition and shows the effects of various applied changes, such as a flight simulator.

Simuliidae [INV ZOO] The black flies, a family of orthorrhaphous dipteran insects in the series Nematocera.

simultaneous access See parallel access.

simultaneous computer [COMPUT SCI] A computer, usually of the analog or hybrid type, in which separate units of hardware are used to carry out the various parts of a computation, the execution of different parts usually overlap in time, and the various hardware units are interconnected in a manner determined by the computation.

simultaneous equations [MATH] A collection of equations considered to be a set of joint conditions imposed on the variables involved.

simultaneous lobing [ELECTR] A radar direction-finding technique in which the signals received by two partly overlapping antenna lobes are compared in phase or power to obtain a measure of the angular displacement of a target from the equisignal direction.

sin A See sine.

sincosite [MINERAL] $Ca(VO)_2(PO_4)_2 \cdot 5H_2O$ A leek-green mineral composed of hydrous calcium vanadyl phosphate and occurring in tetragonal scales or plates.

sine [MATH] The sine of an angle A in a right triangle with hypotenuse of length c given by the ratio a/c, where a is the length of the side opposite A; more generally, the sine function assigns to any real number A the ordinate of the point on the unit circle obtained by moving from $(1,0)$ counterclockwise A units along the circle, or clockwise $|A|$ units if A is less than 0. Denoted sin A.

sine wave [PHYS] A wave whose amplitude varies as the sine of a linear function of time. Also known as sinusoidal wave.

sine-wave oscillator See sinusoidal oscillator.

sine-wave response See frequency response.

singeing [TEXT] Passing a fabric over heated plates or gas flames during finishing to remove lint or loose threads from the surface. Also known as gassing.

single acting [MECH ENG] Acting in one direction only, as a single-acting plunger, or a single-acting engine (admitting the working fluid on one side of the piston only).

single-address instruction See one-address instruction.

single-block brake [MECH ENG] A friction brake consisting of a short block fitted to the contour of a wheel or drum and pressed up against the surface by means of a lever on a fulcrum; used on railroad cars.

single-carrier theory [SOLID STATE] A theory of the behavior of a rectifying barrier which assumes that conduction is due to the motion of carriers of only one type; it can be applied to the contact between a metal and a semiconductor.

single crystal [CRYSTAL] A crystal, usually grown artificially, in which all parts have the same crystallographic orientation.

single-current transmission [COMMUN] Telegraph transmission in which a current flows, in only one direction, during marking intervals, and no current flows during spacing intervals.

single-cut file [DES ENG] A file with one set of parallel teeth, extending diagonally across the face of the file.

single-degree-of-freedom gyro [MECH] A gyro the spin reference axis of which is free to rotate about only one of the orthogonal axes, such as the input or output axis.

single-end amplifier [ELECTR] Amplifier stage which normally employs only one tube or semiconductor or, if more than one tube or semiconductor is used, they are connected in parallel so that operation is asymmetric with respect to ground. Also known as single-sided amplifier.

single-gun color tube [ELECTR] A color television picture tube having only one electron gun and one electron beam; the beam is sequentially deflected across phosphors for the three primary colors to form each color picture element, as in the chromatron.

single-length [COMPUT SCI] Pertaining to the expression of numbers in binary form in such a way that they can be included in a single computer word.

single lens See simple lens.

single-loop feedback [CONT SYS] A system in which feedback may occur through only one electrical path.

single-phase circuit [ELEC] Either an alternating-current circuit which has only two points of entry, or one which, having more than two points of entry, is intended to be so energized that the potential differences between all pairs of points of entry are either in phase or differ in phase by 180°.

single-phase meter [ENG] A type of power-factor meter that contains a fixed coil that carries the load current, and crossed coils that are connected to the load voltage; there is no spring to restrain the moving system, which takes a position to indicate the angle between the current and voltage.

single-phase motor [ELEC] A motor energized by a single alternating voltage.

single-phase rectifier [ELECTR] A rectifier whose input voltage is a single sinusoidal voltage, in contrast to a polyphase rectifier.

single-point grounding [ELEC] Grounding system that attempts to confine all return currents to a network that serves as the circuit reference; to be effective, no appreciable current is allowed to flow in the circuit reference, that is, the sum of the return currents is zero.

single-pole double-throw [ELEC] A three-terminal switch or relay contact arrangement that connects one terminal to either of two other terminals. Abbreviated SPDT.

single-pole single-throw [ELEC] A two-terminal switch or relay contact arrangement that opens or closes one circuit. Abbreviated SPST.

single-precision number [COMPUT SCI] A number having as many digits as are ordinarily used in a given computer, in contrast to a double-precision number.

single reference *See* random access.

single refraction [OPTICS] Any refraction that occurs in an isotropic crystal.

single-shot multivibrator *See* monostable multivibrator.

single-shot operation *See* single-step operation.

single-sideband communication [COMMUN] A communication system in which one of the two sidebands used in amplitude-modulation is suppressed; the carrier wave may be either transmitted or suppressed.

single-sideband modulation [COMMUN] Modulation resulting from elimination of all components of one sideband from an amplitude-modulated wave.

single-sideband transmission [COMMUN] Transmission of a carrier and substantially only one sideband of modulation frequencies, as in television where only the upper sideband is transmitted completely for the picture signal; the carrier wave may be either transmitted or suppressed.

single-sided amplifier *See* single-end amplifier.

single-signal receiver [ELECTR] A highly selective superheterodyne receiver for code reception, having a crystal filter in the intermediate-frequency amplifier.

single-stage compressor [MECH ENG] A machine that effects overall compression of a gas or vapor from suction to discharge conditions without any sequential multiplicity of elements, such as cylinders or rotors.

single-stage pump [MECH ENG] A pump in which the head is developed by a single impeller.

single-stage rocket [AERO ENG] A rocket or rocket missile to which the total thrust is imparted in a single phase, by either a single or multiple thrust unit.

single-step operation [COMPUT SCI] A method of computer operation, used in debugging or detecting computer malfunctions, in which a program is carried out one instruction at a time, each instruction being performed in response to a manual control device such as a switch or button. Also known as one-shot operation; one-step operation; single-shot operation; step-by-step operation.

single-throw switch [ELEC] A switch in which the same pair of contacts is always opened or closed.

single tuned circuit [ELEC] A circuit whose behavior is the same as that of a circuit with a single inductance and a single capacitance, together with associated resistances.

singular integral equation [MATH] An integral equation where the integral appearing either has infinite limits of integration or the kernel function has points where it is infinite.

singularity [MATH] A point where a function of real or complex variables is not differentiable or analytic. Also known as singular point of a function. [METEOROL] A characteristic meteorological condition which tends to occur on or near a specific calendar date more frequently than chance would indicate; an example is the January thaw.

singular point of a function *See* singularity.

singular transformation [MATH] A linear transformation which has no corresponding inverse transformation.

sinhalite [MINERAL] $MgAl(BO_4)$ A mineral composed of magnesium aluminum borate; sometimes used as a gem.

sink [ELECTROMAG] The region of a Rieke diagram where the rate of change of frequency with respect to phase of the reflection coefficient is maximum for an oscillator; operation in this region may lead to unsatisfactory performance by reason of cessation or instability of oscillations. [GEOL] **1.** A circular or ellipsoidal depression formed by collapse on the flank of or near to a volcano. **2.** A slight, low-lying desert depression containing a central playa or saline lake with no outlet, as where a desert stream comes to an end or disappears by evaporation. [MIN ENG] **1.** To excavate strata downward in a vertical line for the purpose of winning and working minerals. **2.** To drill or put down a shaft or borehole. [PHYS] A device or system where some extensive entity is absorbed, such as a heat sink, a sink flow, a load in an electrical circuit, or a region in a nuclear reactor where neutrons are strongly absorbed.

sink flow [FL MECH] **1.** In three-dimensional flow, a point into which fluid flows uniformly from all directions. **2.** In two-dimensional flow, a straight line into which fluid flows uniformly from all directions at right angles to the line.

sinkhead *See* feedhead.

sinkhole [GEOL] Closed surface depressions in regions of karst topography produced by solution of surface limestone or the collapse of cavern roofs.

sinoatrial node [ANAT] A bundle of Purkinje fibers located near the junction of the superior vena cava with the right atrium which acts as a pacemaker for cardiac excitation. Abbreviated SA node. Also known as sinoauricular node.

sinoauricular node *See* sinoatrial node.

Sinope [ASTRON] A small satellite of Jupiter with a diameter of about 17 miles (27 kilometers), orbiting with retrograde motion at a mean distance of about 1.47×10^7 miles (2.37×10^7 kilometers). Also known as Jupiter IX.

sinter [MET] **1.** The product of a sintering operation. **2.** A shaped body composed of metal powders and produced by sintering with or without previous compacting. [MINERAL] *See* siliceous sinter. [PETR] A chemical sedimentary rock deposited by precipitation from mineral waters, especially siliceous sinter and calcareous sinter.

sintering [MET] Forming a coherent bonded mass by heating metal powders without melting; used mostly in powder metallurgy.

sinus [BIOL] A cavity, recess, or depression in an organ, tissue, or other part of an animal body.

sinus hairs *See* vibrissae.

sinusoidal angular modulation *See* angle modulation.

sinusoidal current *See* simple harmonic current.

sinusoidal function [MATH] The real or complex function $\sin(u)$ or any function with analogous continuous periodic behavior.

sinusoidal oscillator [ELECTR] An oscillator circuit whose output voltage is a sine-wave function of time. Also known as harmonic oscillator; sine-wave oscillator.

sinusoidal wave *See* sine wave.

Siphinodentallidae [INV ZOO] A family of mollusks in the class Scaphopoda characterized by a subterminal epipodial ridge which is not slit dorsally and which terminates with a crenulated disk.

siphon [BOT] A tubular element in various algae. [ENG] A tube, pipe, or hose through which a liquid can be moved from a higher to a lower level by atmospheric pressure forcing it up the shorter leg while the weight of the liquid in the longer leg causes continuous downward flow. [GEOL] A passage in a cave system that connects with a water trap. [INV ZOO] **1.** A tubular structure for intake or output of water in bivalves and other mollusks. **2.** The sucking-type of proboscis in many arthropods.

Siphonales [BOT] A large order of green algae (Chlorophyta) which are coenocytic, nonseptate, and mostly marine.

Siphonaptera [INV ZOO] The fleas, an order of insects characterized by a small, laterally compressed, oval body armed with spines and setae, three pairs of legs modified for jumping, and sucking mouthparts.

Siphonocladaceae [BOT] A family of green algae in the order Siphonocladales.

Siphonocladales [BOT] An order of green algae in the division Chlorophyta including marine, mostly tropical forms.

Siphonolaimidae [INV ZOO] A family of nematodes in the superfamily Monhysteroidea in which the stoma is modified into a narrow, elongate, hollow, spearlike apparatus.

Siphonophora [INV ZOO] An order of the coelenterate class Hydrozoa characterized by the complex organization of components which may be connected by a stemlike region or may be more closely united into a compact organism.

Siphonotretacea [PALEON] A superfamily of extinct, inarticulate brachiopods in the suborder Acrotretidina of the order Acrotretida having an enlarged, tear-shaped, apical pedicle valve.

Siphunculata [INV ZOO] The equivalent name for Anoplura.

Sipunculida [INV ZOO] A phylum of marine worms which dwell in burrows, secreted tubes, or adopted shells; the mouth and anus occur close together at one end of the elongated body, and the jawless mouth, surrounded by tentacles, is situated in an eversible proboscis.

Sipunculoidea [INV ZOO] An equivalent name for Sipunculida.

Sirenia [VERT ZOO] An order of aquatic placental mammals which include the living manatees and dugongs; these are nearly hairless, thick-skinned mammals without hindlimbs and with paddlelike forelimbs.

Siricidae [INV ZOO] The horntails, a family of the Hymenoptera in the superfamily Siricoidea; females use a stout, hornlike ovipositor to deposit eggs in wood.

Siricoidea [INV ZOO] A superfamily of wasps of the suborder Symphala in the order Hymenoptera.

sirocco [METEOROL] A warm south or southeast wind in advance of a depression moving eastward across the southern Mediterranean Sea or North Africa.

SIS *See* semiconductor-insulator-semiconductor.

sisal [BOT] *Agave sisalina.* An agave of the family Amaryllidaceae indigenous to Mexico and Central America; a coarse, stiff yellow fiber produced from the leaves is used for making twine and brush bristles.

sister chromatids [CYTOL] The two daughter strands of a chromosome after it has duplicated.

site [COMPUT SCI] **1.** A position available for the symbols of information, for example, a digital place. **2.** A location on a tally that can bear either a mark or a blank; for example, a location that can be punched or left unpunched on a card. [ENG] Position of anything; for example, the position of a gun emplacement.

Sitka cypress *See* Alaska cedar.

six-j-symbol [QUANT MECH] A coefficient that appears in the transformation between various modes of coupling eigenfunctions of three angular momenta; it is equal to the Racah coefficient, except perhaps in sign, and has greater symmetry than the Racah coefficient.

sixty degrees Fahrenheit British thermal unit *See* British thermal unit.

size block *See* gage block.

sizing [ENG] **1.** Separating an aggregate of mixed particles into groups according to size, using a series of screens. Also known as size classification. **2.** *See* sizing treatment. [MECH ENG] A finishing operation to correct surfaces and shapes to meet specified dimensions and tolerances. [MET] Final pressing of a metal powder compact after sintering.

sizing treatment [ENG] Also known as sizing; surface sizing. **1.** Application of material to a surface to fill pores and thus reduce the absorption of subsequently applied adhesive or coating; used for textiles, paper, and other porous materials. **2.** Surface-treatment applied to glass fibers used in reinforced plastics.

sjogrenite [MINERAL] $Mg_6Fe_2(OH)_{16}(CO_3) \cdot 4H_2O$ A hexagonal mineral composed of hydrous basic magnesium iron carbonate.

skarn [GEOL] A lime-bearing silicate derived from nearly pure limestone and dolomite with the introduction of large amounts of silicon, aluminum, iron, and magnesium.

skate [VERT ZOO] Any of various batoid elasmobranchs in the family Rajidae which have flat bodies with winglike pectoral fins and a slender tail with two small dorsal fins.

skeletal coding [COMPUT SCI] A set of incomplete instructions in symbolic form, intended to be completed and specialized by a processing program written for that purpose.

skeletal muscle [ANAT] A striated, voluntary muscle attached to a bone and concerned with body movements.

skeletal system [ANAT] Structures composed of bone or cartilage or a combination of both which provide a framework for the vertebrate body and serve as attachment for muscles.

skeleton framing [BUILD] Framing in which steel framework supports all the gravity loading of the structure; this system is used for skyscrapers.

skew [COMPUT SCI] In character recognition, a condition arising at the read station whereby a character or a line of characters appears in a "twisted" manner in relation to a real or imaginary horizontal baseline. [ELECTR] **1.** The deviation of a received facsimile frame from rectangularity due to lack of synchronism between scanner and recorder; expressed numerically as the tangent of the angle of this deviation. **2.** The degree of nonsynchronism of supposedly parallel bits when bit-coded characters are read from magnetic tape. [MECH ENG] Gearing whose shafts are neither interesecting nor parallel. [SCI TECH] Deviating from rectangularity or a straight line.

skew bridge [CIV ENG] A bridge which spans a gap obliquely and is therefore longer than the width of the gap.

skewed density function [STAT] A density function which is not symmetrical, and which depends not only on the magnitude of the difference between the average value and the variate, but also on the sign of this difference.

skew failure [COMPUT SCI] In character recognition, the condition that exists during document alignment whereby the document reference edge is not parallel to that of the read station.

skew field [MATH] A ring whose nonzero elements form a non-Abelian group with respect to the multiplicative operation.

skew lines [MATH] Lines which do not lie in the same plane in euclidean three-dimensional space.

skew product [MATH] A multiplicative operation or structure induced upon a cartesian product of sets, where each has some algebraic structure.

skew surface [MATH] A ruled surface whose total curvature vanishes everywhere.

skid [AERO ENG] The metal bar or runner used as part of the landing gear of helicopters and planes. [ENG] **1.** A device attached to a chain and placed under a wheel to prevent its turning when descending a steep hill. **2.** A timber, bar, rail, or log placed under a heavy object when it is being moved over bare ground. **3.** A wood or metal platform support on wheels, legs, or runners used for handling and moving material. Also known as skid platform. [MECH ENG] A brake for a power machine. [MIN ENG] An arrangement upon which certain coal-cutting machines travel along the working faces.

skimmer [VERT ZOO] Any of various ternlike birds, members of the Rhynchopidae, that inhabit tropical waters around the world and are unique in having the knifelike lower mandible substantially longer than the wider upper mandible.

skimming plant [CHEM ENG] A petroleum refinery designed to remove and finish only the lighter constituents of crude oil, such as gasoline and kerosine; the heavy ends are sold as fuel oil or for further processing elsewhere.

skin [AERO ENG] The covering of a body, such as the covering of a fuselage, a wing, a hull, or an entire aircraft. [ANAT] The external covering of the vertebrate body, consisting of two layers, the outer epidermis and the inner dermis. [BUILD] The exterior wall of a building. [ENG] In flexible bag molding, a protective covering for the mold; it may consist of a thin piece of plywood or a thin hardwood. [MET] A thin outside layer of metal differing in composition, structure, or other characteristics from the main mass of metal but not formed by bonding or electroplating.

skin depth [ELECTROMAG] The depth beneath the surface of a conductor, which is carrying current at a given frequency due to electromagnetic waves incident on its surface, at which the current density drops to one neper below the current density at the surface.

skin effect [ELEC] The tendency of alternating currents to flow near the surface of a conductor thus being restricted to a small part of the total sectional area and producing the effect of increasing the resistance. Also known as conductor skin effect; Kelvin skin effect. [PETRO ENG] The restriction to fluid flow through a reservoir adjacent to the borehole; calculated as a factor of reservoir pressure, product rate, formation volume and thickness, porosity, and other related parameters.

skin friction [FL MECH] A type of friction force which exists at the surface of a solid body immersed in a much larger volume of fluid which is in motion relative to the body.

skink [VERT ZOO] Any of numerous small- to medium-sized lizards comprising the family Scincidae with a cylindrical body; short, sometimes vestigial, legs; cores of bone in the body scales; and pleurodont dentition.

skin test [IMMUNOL] A procedure for evaluating immunity status involving the introduction of a reagent into or under the skin.

skip [COMPUT SCI] **1.** In fixed-instruction-length digital computers, to bypass or ignore one or more instructions in an otherwise sequential process. **2.** A device on a card punch that causes columns on a punch in fields where no punching is desired to move rapidly past the punching station. [MECH ENG] *See* skip hoist.

skip chain [COMPUT SCI] A programming technique which matches a word against a set of test words; if there is a match, control is transferred (skipped) to a routine, otherwise the word is matched with the next test word in sequence.

skip effect [COMMUN] The existence of a ring-shaped area around a radio transmitter within which no radio signals are received, because ground signals are received only inside the ring and sky-wave signals are received only outside the ring.

skip hoist [MECH ENG] A basket, bucket, or open car mounted vertically or on an incline on wheels, rails, or shafts and hoisted by a cable; used to raise materials. Also known as skip.

skipjack *See* bluefish.

skip mark [GEOL] A crescent-shaped mark that is one of a linear pattern of regularly spaced marks made by an object that skipped along the bottom of a stream.

skip zone [ACOUS] A region in the air surrounding a source of sound in which no sound is heard, although the sound becomes audible at greater distances. Also known as zone of silence. [COMMUN] The area between the outer limit of reception of radio high-frequency ground waves and the inner limit of reception of sky waves, where no signal is received.

skull [ANAT] The bones and cartilages of the vertebrate head which form the cranium and the face. [MET] A layer of solidified metal or dross left in the pouring vessel after the molten metal has been poured.

skull cracker [ENG] A heavy iron or steel ball that can be swung freely or dropped by a derrick to raze buildings or to compress bulky scrap. Also known as wrecking ball.

skunk [VERT ZOO] Any one of a group of carnivores in the family Mustelidae characterized by a glossy black and white coat and two musk glands at the base of the tail.

skutterudite [MINERAL] $(Co,Ni)As_3$ A tin-white mineral with metallic luster composed of cobalt and nickel arsenides; crystallizes in the isometric system but commonly is massive; hardness is 5.5–6 on Mohs scale, and specific gravity is 6.6; it is a minor ore of cobalt and nickel.

sky cover [METEOROL] In surface weather observations, the amount of sky covered but not necessarily concealed by clouds or by obscuring phenomena aloft, the amount of sky concealed by obscuring phenomena that reach the ground, or the amount of sky covered or concealed by a combination of the two phenomena.

skylight [ASTROPHYS] *See* diffuse sky radiation. [ENG] An opening in a roof or ship deck that is covered with glass or plastic and designed to admit daylight.

sky map [ASTRON] A planar representation of areas of the sky showing positions of celestial bodies. [METEOROL] A pattern of variable brightness observable on the underside of a cloud layer, and caused by the different reflectivities of material on the earth's surface immediately beneath the clouds; this term is used mainly in polar regions.

sky radiation *See* diffuse sky radiation.

slab [CIV ENG] That part of a reinforced concrete floor, roof, or platform which spans beams, columns, walls, or piers. [ELECTR] A relatively thick-cut crystal from which blanks are obtained by subsequent transverse cutting. [ENG] The outside piece cut from a log when sawing it into boards. [GEOL] A cleaved or finely parallel jointed rock, which splits into tabular plates from 1 to 4 inches (2.5 to 10 centimeters) thick. Also known as slabstone. [HYD] A layer in, or the whole-thickness of, a snowpack that is very hard and has the ability to sustain elastic deformation under stress. [MATER] A thin piece of concrete or stone. [MET] A piece of metal, intermediate between ingot and plate, with the width at least twice the thickness. [MIN ENG] A slice taken off the rib of an entry or room in a mine.

slabbing machine [MIN ENG] A coal-cutting machine designed to make cuts in the side of a room or entry pillar preparatory to slabbing.

slabstone *See* slab.

slack water [OCEANOGR] The interval when the speed of the tidal current is very weak or zero; usually refers to the period of reversal between ebb and flood currents.

slag [MET] A nonmetallic product resulting from the interaction of flux and impurities in the smelting and refining of metals.

slaking [GEOL] 1. Crumbling and disintegration of earth materials when exposed to air or moisture. 2. The breaking up of dried clay when saturated with water.

slant [MIN ENG] 1. Any short, inclined crosscut connecting the entry with its air course to facilitate the hauling of coal. Also known as shoofly. 2. A heading driven diagonally between the dip and the strike of a coal seam. Also known as run.

slant culture [MICROBIO] A method for maintaining bacteria in which the inoculum is streaked on the surface of agar that has solidified in inclined glass tubes.

slate [PETR] A group name for various very-fine-grained rocks derived from mudstone, siltstone, and other clayey sediment as a result of low-degree regional metamorphism; characterized by perfect fissility or slaty cleavage which is a regular or perfect planar schistosity.

slave antenna [ELECTROMAG] A directional antenna positioned in azimuth and elevation by a servo system; the information controlling the servo system is supplied by a tracking or positioning system.

slave mode *See* user mode.

slavikite [MINERAL] $MgFe_3^{3+}(SO_4)_4(OH)_3 \cdot 18H_2O$ A greenish-yellow mineral composed of hydrous basic magnesium ferric sulfate and occurring as rhombohedral crystals.

sleeping sickness *See* African sleeping sickness; encephalitis lethargica.

sleet [METEOROL] Colloquially in some parts of the United States, precipitation in the form of a mixture of snow and rain.

sleeve [ELEC] 1. The cylindrical contact that is farthest from the tip of a phone plug. 2. Insulating tubing used over wires or components. Also known as sleeving. [ENG] A cylindrical part designed to fit over another part.

sleeve antenna [ELECTROMAG] A single vertical half-wave radiator, the lower half of which is a metallic sleeve through which the concentric feed line runs; the upper radiating portion, one quarter-wavelength long, connects to the center of the line.

sleeve coupling [DES ENG] A hollow cylinder which fits over the ends of two shafts or pipes, thereby joining them.

sleeve valve [MECH ENG] An admission and exhaust valve on an internal-combustion engine consisting of one or two hollow sleeves that fit around the inside of the cylinder and move with the piston so that their openings align with the inlet and exhaust ports in the cylinder at proper stages in the cycle.

sleeving *See* sleeve.

slicer *See* amplitude gate; slitting mill.

slicer amplifier *See* amplitude gate.

slicing [ELECTR] Transmission of only those portions of a waveform lying between two amplitude values.

slick paper proof *See* reproduction proof.

slide [ENG] 1. A sloping chute with a flat bed. 2. A sliding mechanism. [GEOL] 1. A vein of clay intersecting and dislocating a vein vertically, or the vertical dislocation itself. 2. A rotational or planar mass movement of earth, snow, or rock resulting from failure under shear stress along one or more surfaces. [MECH ENG] The main reciprocating member of a mechanical press, guided in a press frame, to which the punch or upper die is fastened. [MIN ENG] 1. An upright rail fixed in a shaft with corresponding grooves for steadying the cages. 2. A trough used to guide and to support rods in a tripod when drilling an angle hole. Also known as rod slide. [ORD] 1. Sliding part of the receiver of certain automatic weapons. 2. Sliding catch on the breech mechanism of certain weapons.

slide conveyor [ENG] A slanted, gravity slide for the forward downward movement of flowable solids, slurries, liquids, or small objects.

slide rail *See* guard rail.

slider coupling [MECH ENG] A device for connecting shafts that are laterally misaligned. Also known as double-slider coupling; Oldham coupling.

slide rule [MATH] A mechanical device, composed of a ruler with sliding insert, marked with various number scales, which facilitates such calculations as division, multiplication, finding roots, and finding logarithms.

slide valve [MECH ENG] A sliding mechanism to cover and uncover ports for the admission of fluid, as in some steam engines.

slide-wire bridge [ELEC] A bridge circuit in which the resistance in one or more branches is controlled by the position of a sliding contact on a length of resistance wire stretched along a linear scale.

sliding gear [DES ENG] A change gear in which speed changes are made by sliding gears along their axes, so as to place them in or out of mesh.

sliding pair [MECH ENG] Two adjacent links, one of which is constrained to move in a particular path with respect to the other; the lower, or closed, pair is completely constrained by the design of the links of the pair.

slime [ENG] Liquid slurry of very fine solids with slime- or mudlike appearance. Also known as mud; pulp; sludge.

slime fungus See slime mold.

slime mold [MYCOL] The common name for members of the Myxomycetes. Also known as slime fungus.

sling psychrometer [ENG] A psychrometer in which the wet- and dry-bulb thermometers are mounted upon a frame connected to a handle at one end by means of a bearing or a length of chain; the psychrometer may be whirled in the air for the simultaneous measurement of wet- and dry-bulb temperatures.

slip [CIV ENG] A narrow body of water between two piers. [CRYSTAL] The movement of one atomic plane over another in a crystal; it is one of the ways that plastic deformation occurs in a solid. Also known as glide. [ELEC] **1.** The difference between synchronous and operating speeds of an induction machine. Also known as slip speed. **2.** Method of interconnecting multiple wiring between switching units by which trunk number 1 becomes the first choice for the first switch, trunk number 2 first choice for the second switch, and trunk number 3 first choice for the third switch, and so on. [ELECTR] Distortion produced in the recorded facsimile image which is similar to that produced by skew but is caused by slippage in the mechanical drive system. [FL MECH] The difference between the velocity of a solid surface and the mean velocity of a fluid at a point just outside the surface. [GEOL] The actual relative displacement along a fault plane of two points which were formerly adjacent on either side of the fault. Also known as actual relative movement; total displacement. [MATER] A suspension of fine clay in water with a creamy consistency, used in the casting process and in decorating ceramic ware. Also known as slurry. [NAV ARCH] **1.** To part from an anchor by releasing the shackles from the anchor chain. **2.** The reduction in the distance a propeller advances, per unit time, due to yielding of the fluid.

slipband [CRYSTAL] One of the microscopic parallel lines (Lüders' lines) on the surface of a crystalline material stretched beyond its elastic limit, located at the intersection of the surface with intracrystalline slip planes in the grains of the material. Also known as slip line.

slip bedding [GEOL] Convolute bedding formed as the result of subaqueous sliding.

slip casting [ENG] A process in the manufacture of shaped refractories, cermets, and other materials in which the slip is poured into porous plaster molds.

slip cleavage [GEOL] Cleavage that is superposed on slaty cleavage or schistosity, characterized by spaced cleavage with thin tabular bodies of rock between the cleavage planes. Also known as close-joints cleavage; crenulation cleavage; shear cleavage; strainslip cleavage.

slip fold See shear fold.

slip joint [CIV ENG] **1.** Contraction joint between two adjoining wall sections, or at the horizontal bearing of beams, slabs, or precast units, consisting of a vertical tongue fitted into a groove which allows independent movement of the two sections. **2.** A telescoping joint between two parts. [ENG] **1.** A method of laying-up plastic veneers in flexible-bag molding, wherein edges are beveled and allowed to overlap part or all of the scarfed area. **2.** A mechanical union that allows limited endwise movement of two solid items for example, pipe, rod, or duct with relation to each other. [GEOL] See shear joint.

slip line See slipband.

slipped disk See herniated disk.

slip plane [CRYSTAL] See glide plane. [GEOL] A planar slip surface. [ENG] A plane visible by reflected light in a transparent material; caused by poor welding and shrinkage during cooling.

slip ring [ELEC] A conductive rotating ring which, in combination with a stationary brush, provides a continuous electrical connection between rotating and stationary conductors; used in electric rotating machinery, synchros, gyroscopes, and scanning radar antennas.

slip speed See slip.

slit scan [COMPUT SCI] In character recognition, a magnetic or photoelectric device that obtains the horizontal structure of an inputted character by vertically projecting its component elements at given intervals.

slitting mill [LAP] A rotating disk used by gem cutters in slitting. Also known as slicer.

slope [GEOL] The inclined surface of any part of the earth's surface. [MATH] **1.** The slope of a line through the points (x_1, y_1) and (x_2, y_2) is number $(y_2 - y_1)/(x_2 - x_1)$. **2.** The slope of a curve at a point p is the slope of the tangent line to the curve at p. [NAV] The projection of a flight path in the vertical plane.

slope angle [MATH] The angle of inclination of a line in the plane, where this angle is measured from the positive x-axis to the line in the counterclockwise direction.

slope current See gradient current.

slope failure [GEOL] The downward and outward movement of a mass of soil beneath a natural slope or other inclined surface; four types of slope failure are rockfall, rock flow, plane shear, and rotational shear.

slot [AERO ENG] **1.** An air gap between a wing and the length of a slat or certain other auxiliary airfoils, the gap providing space for airflow or room for the auxiliary airfoil to be depressed in such a manner as to make for smooth air passage on the surface. **2.** Any of certain narrow apertures made through a wing to improve aerodynamic characteristics. [COMPUT SCI] A punched-out area of a hand-sorted card to connect two or more guide holes; slots can be extended to the outside edge with notches. [DES ENG] A narrow, vertical opening. [ELEC] One of the conductor-holding grooves in the face of the rotor or stator of an electric rotating machine. [MIN ENG] To hole; to undercut or channel.

slot antenna [ELECTROMAG] An antenna formed by cutting one or more narrow slots in a large metal surface fed by a coaxial line or waveguide.

sloth [VERT ZOO] Any of several edentate mammals in the family Bradypodidae found exclusively in Central and South America; all are slow-moving, arboreal species that cling to branches upside down by means of long, curved claws.

slotted line See slotted section.

slotted nut [DES ENG] A regular hexagon nut with slots cut across the flats of the hexagon so that a cotter pin or safety wire can hold it in place.

slotted section [ELECTROMAG] A section of waveguide or shielded transmission line in which the shield is slotted to permit the use of a traveling probe for examination of standing waves. Also known as slotted line; slotted waveguide.

slotted waveguide See slotted section.

slow-blow fuse [ELEC] A fuse that can withstand up to 10 times its normal operating current for a brief period, as required for circuits and devices which draw a very heavy starting current.

slow memory See slow storage.

slow neutron [NUC PHYS] **1.** A neutron having low-kinetic energy, up to about 100 electron volts. **2.** See thermal neutron.

slow storage [COMPUT SCI] In computers, storage with a relatively long access time. Also known as slow memory.

slow virus [VIROL] Any member of a group of animal viruses characterized by prolonged periods of incubation and an extended clinical course lasting months or years.

slow wave [ELECTROMAG] A wave having a phase velocity less than the velocity of light, as in a ridge waveguide.

slow wave sleep See deep sleep.

sludge [CHEM ENG] **1.** Residue left after acid treatment of petroleum oils. **2.** Any semisolid waste from a chemical process. [CIV ENG] See sewage sludge. [ENG] **1.** Mud from a drill hole in boring. **2.** Sediment in a steam boiler. **3.** Precipitate from oils, such as the products from crankcase oils in engines. **4.** See slime. [GEOL] A soft or muddy bottom deposit as on tideland or in a stream bed. [OCEANOGR] A dense, soupy accumulation of new sea ice consisting of incoherent floating frazil crystals. Also known as cream ice; sludge ice; slush.

sludge barrel See calyx.

sludge bucket See calyx.

sludge ice See sludge.

sludge pump [MECH ENG] See sand pump. [MIN ENG] A short iron pipe or tube fitted with a valve at the lower end, with which the sludge is extracted from a borehole.

sludging See solifluction.

slug [ELECTROMAG] **1.** A heavy copper ring placed on the core of a relay to delay operation of the relay. **2.** A movable iron core for a coil. **3.** A movable piece of metal or dielectric material used in a wave guide for tuning or impedance-matching purposes. [GRAPHICS] A strip of metal used to space between lines of type. [INV ZOO] Any of a number of pulmonate gastropods which have a rudimentary shell and the body elevated toward the middle and front end where the mantle covers the lung region. [MECH] A unit of mass in the British gravitational system of units, equal to the mass which experiences an acceleration of 1 foot per second per second when a force of 1 pound acts on it; equal to approximately 32.1740 pound mass or 14.5939 kilograms. Also known as geepound. [MET] **1.** A small, roughly shaped piece of metal for subsequent processing, as by forging or extruding. **2.** The piece of material produced by piercing a hole in a sheet. [MIN ENG] To inject a borehole with cement, slurry, or various liquids containing shredded materials in an attempt to restore lost circulation by sealing off the openings in the borehole-wall rocks. [NUCLEO] A short fuel rod inserted in a hole or channel in the active lattice of

a nuclear reactor. [ORD] **1.** As pertains to shaped charge ammunition, massive and relatively slow-moving remnant of the collapsed metal liner, as distinguished from the jet. **2.** A solid cast iron projectile used in test firing.

sluice [CIV ENG] **1.** A passage fitted with a vertical sliding gate or valve to regulate the flow of water in a channel or lock. **2.** A body of water retained by a floodgate. **3.** A channel serving to drain surplus water.

sluice box [MIN ENG] A long, inclined trough or launder with riffles in the bottom that provide a lodging place for heavy minerals in ore concentration.

sluicing pond See scouring basin.

slump [GEOL] A type of landslide characterized by the downward slipping of a mass of rock or unconsolidated debris, moving as a unit or several subsidiary units, characteristically with backward rotation on a horizontal axis parallel to the slope; common on natural cliffs and banks and on the sides of artificial cuts and fills.

slump fault See normal fault.

slurry [MATER] **1.** A semiliquid refractory material, such as clay, used to repair furnace refractories. **2.** A free-flowing, pumpable suspension of fine solid material in liquid. **3.** An emulsion of a sulfonated soluble oil in water used to cool and lubricate metal during cutting operations. **4.** A plastic mixture of portland cement and water pumped into an oil well; after hardening, it provides support for the casing and a seal for the well bore. **5.** See slip.

slurry mining [MIN ENG] The hydraulic breakdown of a subsurface ore matrix with drill-hole equipment, and the eduction of the resulting slurry to the surface for processing.

slush [HYD] Snow or ice on the ground that has been reduced to a soft, watery mixture by rain, warm temperature, or chemical treatment. [OCEANOGR] See sludge.

slush molding [ENG] A thermoplastic casting in which a liquid resin is poured into a hot, hollow mold where a viscous skin forms; excess slush is drained off, the mold is cooled, and the molded product is stripped out.

Sm See samarium.

small calorie See calorie.

small intestine [ANAT] The anterior portion of the intestine in man and other mammals; it is divided into three parts, the duodenum, the jejunum, and the ileum.

Small Magellanic Cloud [ASTRON] The smaller of the two star clouds near the south celestial pole; it is about 170,000 light-years away and contains a wide assortment of giant and variable stars, star clusters, and nebulae. Also known as Nubecula Minor.

smallpox [MED] An acute, infectious, viral disease characterized by severe systemic involvement and a single crop of skin lesions which proceeds through macular, papular, vesicular, and pustular stages. Also known as variola.

small-scale integration [ELECTR] Integration in which a complete major subsystem or system is fabricated on a single integrated-circuit chip that contains integrated circuits which have appreciably less complexity than for medium-scale integration. Abbreviated SSI.

smaragd See emerald.

smart terminal See intelligent terminal.

S matrix See scattering matrix.

S-matrix theory [PARTIC PHYS] A theory of elementary particles based on the scattering matrix, and on its properties such as unitarity and analyticity. Also known as scattering-matrix theory.

smectic phase [PHYS CHEM] A form of the liquid crystal (mesomorphic) state in which molecules are arranged in layers that are free to glide over each other with relatively small viscosity.

smectite [MINERAL] Dioctahedral (montmorillonite) and trioctahedral (saponite) clay minerals, and their chemical varieties characterized by swelling properties and high cation-exchange capacities.

smegma [PHYSIO] The sebaceous secretion that accumulates around the glans penis and the clitoris.

smelting [MET] The heating of ore mixtures accompanied by a chemical change resulting in liquid metal.

Smilacaceae [BOT] A family of monocotyledonous plants in the order Liliales; members are usually climbing, leafy-stemmed plants with tendrils, trimerous flowers, and a superior ovary.

Sminthuridae [INV ZOO] A family of insects in the order Collembola which have simple tracheal systems.

smithite [MINERAL] $AgAsS_2$ A red monoclinic mineral composed of silver arsenic sulfide and occurring as small crystals.

smithsonite [MINERAL] $ZnCO_3$ White, yellow, gray, brown, or green secondary carbonate mineral associated with sphalerite and commonly reniform, botryoidal, stalactitic, or granular; hardness is 5 on Mohs scale, and specific gravity is 4.30–4.45; it is an ore of zinc. Also known as calamine; dry-bone ore; szaskaite; zinc spar.

smog [METEOROL] Air pollution consisting of smoke and fog.

smoke [ENG] Dispersions of finely divided (0.01–5.0 micrometers) solids or liquids in a gaseous medium.

smokebox [MECH ENG] A chamber external to a boiler for trapping the unburned products of combustion.

smokestone *See* smoky quartz.

smoky quartz [MINERAL] A smoky-yellow, smoky-brown, or brownish-gray, often transparent variety of crystalline quartz containing inclusions of carbon dioxide; may be used as a semiprecious stone. Also known as cairngorm; smokestone.

smooth [OCEANOGR] Comparatively calm water between heavy seas. [STAT] To modify a sequential set of numerical data items in a manner designed to reduce the differences in value between adjacent items.

smoothing circuit *See* ripple filter.

smoothing filter *See* ripple filter.

smooth manifold [MATH] A differentiable manifold whose local coordinate systems depend upon those of euclidean space in an infinitely differentiable manner.

smooth muscle [ANAT] The involuntary muscle tissue found in the walls of viscera and blood vessels, consisting of smooth muscle fibers.

smudge [PL PATH] Any of several fungus diseases of cereals and other plants characterized by dark, sooty discolorations.

smut [MET] A reaction product left on the surface of a metal after pickling. [PL PATH] Any of various destructive fungus diseases of cereals and other plants characterized by large dusty masses of dark spores on the plant organs.

smut fungus [MYCOL] The common name for members of the Ustilaginales.

Sn *See* tin.

S/N *See* signal-to-noise ratio.

snail [INV ZOO] Any of a large number of gastropod mollusks distinguished by a spiral shell that encloses the body, a head, a foot, and a mantle.

snake [MET] **1.** A twisted and bent hod rod formed before subsequent rolling operations. **2.** A flexible mandrel used to prevent collapse of a shaped piece during bending operations. [VERT ZOO] Any of about 3000 species of reptiles which belong to the 13 living families composing the suborder Serpentes in the order Squamata.

snaking stream *See* meandering stream.

snap ring [DES ENG] A form of spring used as a fastener; the ring is elastically deformed, put in place, and allowed to snap back toward its unstressed position into a groove or recess.

snapshot dump [COMPUT SCI] An edited printout of selected parts of the contents of main memory, performed at one or more times during the execution of a program without materially affecting the operation of the program.

sneak circuit *See* sneak path.

Snell laws of refraction [OPTICS] When light travels from one medium into another the incident and refracted rays lie in one plane with the normal to the surface; are on opposite sides of the normal; and make angles with the normal whose sines have a constant ratio to one another. Also known as Descartes laws of refraction; laws of refraction.

snow [ELECTR] Small, random, white spots produced on a television or radar screen by inherent noise signals originating in the receiver. [METEOROL] The most common form of frozen precipitation, usually flakes of starlike crystals, matted ice needles, or combinations, and often rime-coated.

snow blindness [MED] A transient visual impairment and actinic keratoconjunctivitis caused by exposure of the eyes to ultraviolet rays reflected from snow. Also known as solar photophthalmia.

snow blink [METEOROL] A bright, white glare on the underside of clouds, produced by the reflection of light from a snow-covered surface; this term is used in polar regions with reference to the sky map. Also known as snow sky.

snowbreak [CIV ENG] Any barrier designed to shelter an object or area from snow.

snow cover [HYD] **1.** All accumulated snow on the ground, including that derived from snowfall, snowslides, and drifting snow. Also known as snow mantle. **2.** The extent, expressed as a percentage, of snow cover in a particular area.

snow crystal [METEOROL] Any of several types of ice crystal found in snow; a snow crystal is a single crystal, in contrast to a snowflake which is usually an aggregate of many single snow crystals.

snowfield [HYD] **1.** A broad, level, relatively smooth and uniform snow cover on ground or ice at high altitudes or in mountainous regions above the snow line. **2.** The accumulation area of a glacier. **3.** A small glacier or accumulation of perennial ice and snow too small to be designated a glacier.

snowflake [MET] *See* flake. [METEOROL] An ice crystal or, much more commonly, an aggregation of many crystals which falls from a cloud; simple snowflakes (single crystal) exhibit beautiful variety of form, but the symmetrical shapes reproduced so often in photomicrographs are not actually found frequently in snowfalls; broken single crystals, fragments, or clusters of such elements are much more typical of actual snows.

snow flurry [METEOROL] Popular term for snow shower, particularly of a very light and brief nature.

snow ice [HYD] Ice crust formed from snow, either by compaction or by the refreezing of partially thawed snow.

snow line [GEOGR] **1.** A transient line delineating a snow-covered area or altitude. **2.** An area with more than 50% snow cover. **3.** The altitude or geographic line separating areas in which snow melts in summer from areas having perennial ice and snow.

snow mantle *See* snow cover.

snowpack [HYD] The amount of annual accumulation of snow at higher elevations in the western United States, usually expressed in terms of average water equivalent.

snow patch erosion *See* nivation.

snow sky *See* snow blink.

SNR *See* signal-to-noise ratio.

soaking [MET] Heating an alloy, usually an ingot, to a temperature not far below its melting temperature and holding it there for a long time to eliminate segregation that occurred on solidification.

soap [MATER] **1.** A particular type of detergent, in which the water-solubilizing group is a carboxylate, COO—, and the positive ion is usually sodium, Na^+, or potassium, K^+. **2.** A soap compound mixed with a fragrance and other ingredients and then cast into soap bars of different shapes.

soaprock *See* soapstone.

soapstone [MINERAL] **1.** A mineral name applied to steatite or to massive talc. Also known as soaprock. **2.** *See* saponite. [PETR] A metamorphic rock characterized by massive, schistose, or interlaced fibrous texture and a soft unctuous feel.

socket-head screw [DES ENG] A screw fastener with a geometric recess in the head into which an appropriate wrench is inserted for driving and turning, with consequent improved nontamperability.

soda-acid extinguisher [ENG] A fire-extinguisher from which water is expelled at a high rate by the generation of carbon dioxide, the result of mixing (when the extinguisher is tilted) of sulfuric acid and sodium bicarbonate.

sodaclase *See* albite.

soda lime [MATER] A mixture of sodium or potassium hydroxide with calcium oxide; granules are used to absorb water vapor and carbon dioxide gas.

sodalite [MINERAL] $Na_2Al_3Si_3O_{12}Cl$ A blue or sometimes white, gray, or green mineral tectosilicate of the feldspathoid group, crystallizing in the isometric system, with vitreous luster, hardness of 5 on Mohs scale, and specific gravity of 2.2–2.4; used as an ornamental stone.

soda mica *See* paragonite.

soda niter [MINERAL] $NaNO_3$ A colorless to white mineral composed of sodium nitrate, crystallizing in the rhombohedral division of the hexagonal system; hardness is 1½ to 2 on Mohs scale and specific gravity is 2.266. Also known as nitratine; Peru saltpeter.

sodium [CHEM] A metallic element of group I, symbol Na, with atomic number 11, atomic weight 22.9898; silver-white, soft, and malleable; oxidizes in air; melts at 97.6°C; used as a chemical intermediate and in pharmaceuticals, petroleum refining, and metallurgy; the source of the symbol Na is natrium.

sodium-24 [NUC PHYS] A radioactive isotope of sodium, mass 24, half-life 15.5 hours; formed by deuteron bombardment of sodium; decomposes to magnesium with emission of beta rays.

sodium (1:2) borate *See* borax.

sodium-calcium feldspar *See* plagioclase.

sodium chloride [INORG CHEM] NaCl Colorless or white crystals; soluble in water and glycerol, slightly soluble in alcohol; melts at 804°C; used in foods and as a chemical intermediate and an analytical reagent. Also known as common salt; table salt.

sodium feldspar *See* albite.

sodium sulfite process [CHEM ENG] A process for the digestion of wood chips in a solution of magnesium, ammonium, or calcium disulfite containing free sulfur dioxide; used in papermaking.

sodium-vapor lamp [ELECTR] A discharge lamp containing sodium vapor, used chiefly for outdoor illumination.

sofar [NAV] A system of fixing a position at sea by exploding a charge under water, measuring the time for the shock waves to travel through water to three widely separated shore stations, and calculating the position of the explosive by triangulation; the explosive can be dropped from a lifeboat by survivors of air or sea disasters. Derived from sound fixing and ranging.

soffit [CIV ENG] The underside of a horizontal structural member, such as a beam or a slab.

soffosian knob *See* frost mound.

soft chancre *See* chancroid.

soft coal *See* bituminous coal.

soft copy [COMPUT SCI] Information that is displayed on a screen, given by voice, or stored in a form that cannot be read directly by a person, as on magnetic tape, disk, or microfilm.

soft coral [INV ZOO] The common name for coelenterates composing the order Alcyonacea; the colony is supple and leathery.

softening point [PHYS] For a substance which does not have a definite melting point, the temperature at which viscous flow changes to plastic flow.

soft palate [ANAT] The posterior part of the palate which consists of an aggregation of muscles, the tensor veli palatini, levator veli palatini, azygos uvulae, palatoglossus, and palatopharyngeus, and their covering mucous membrane.

soft radiation [PHYS] Radiation whose particles or photons have a low energy, and, as a result, do not penetrate any type of material readily.

soft rime [HYD] A white, opaque coating of fine rime deposited chiefly on vertical surfaces, especially on points and edges of objects, generally in supercooled fog.

soft rock [MIN ENG] Rock that can be removed by air-operated hammers, but cannot be handled economically by a pick. [PETR] **1.** A broad designation for sedimentary rock. **2.** A rock that is relatively nonresistant to erosion.

soft rot [PL PATH] A mushy, watery, or slimy disintegration of plant parts caused by either fungi or bacteria.

software [COMPUT SCI] The totality of programs usable on a particular kind of computer, together with the documentation associated with a computer or program, such as manuals, diagrams, and operating instructions.

software compatibility [COMPUT SCI] Property of two computers, with respect to a particular programming language, in which a source program from one machine in that language will compile and execute to produce acceptably similar results in the other.

software engineering [COMPUT SCI] The systematic application of scientific and technological knowledge, through the medium of sound engineering principles, to the production of computer programs, and to the requirements definition, functional specification, design description, program implementation, and test methods that lead up to this code.

software flexibility [COMPUT SCI] The ability of software to change easily in response to different user and system requirements.

software interface [COMPUT SCI] A computer language whereby computer programs can communicate with each other, and one language can call upon another for assistance.

software monitor [COMPUT SCI] A system, used to evaluate the performance of computer software, that is similar to accounting packages, but can collect more data concerning usage of various components of a computer system and is usually part of the control program.

software package [COMPUT SCI] A program for performing some specific function or calculation which is useful to more than one computer user and is sufficiently well documented to be used without modification on a defined configuration of some computer system.

software piracy [COMPUT SCI] The process of copying commercial software without the permission of the originator.

soft water [CHEM] Water that is free of magnesium or calcium salts.

soft x-ray [ELECTROMAG] An x-ray having a comparatively long wavelength and poor penetrating power.

soil [GEOL] **1.** Unconsolidated rock material over bedrock. **2.** Freely divided rock-derived material containing an admixture of organic matter and capable of supporting vegetation.

soil blister *See* frost mound.

soil chemistry [GEOCHEM] The study and analysis of the inorganic and organic components and the life cycles within soils.

soil flow *See* solifluction.

soil fluction *See* solifluction.

soil mechanics [ENG] The application of the laws of solid and fluid mechanics to soils and similar granular materials as a basis for design, construction, and maintenance of stable foundations and earth structures.

soil moisture *See* soil water.

soil physics [GEOPHYS] The study of the physical characteristics of soils; concerned also with the methods and instruments used to determine these characteristics.

soil profile [GEOL] A vertical section of a soil, showing horizons and parent material.

soil science [GEOL] The study of the formation, properties, and classification of soil; includes mapping. Also known as pedology.

soil series [GEOL] A family of soils having similar profiles, and developing from similar original materials under the influence of similar climate and vegetation.

soil structure [GEOL] Arrangement of soil into various aggregates, each differing in the characteristics of its particles.

soil survey [GEOL] The systematic examination of soils, their description and classification, mapping of soil types, and the assessment of soils for various agricultural and engineering uses.

soil water [HYD] Water in the belt of soil water. Also known as rhizic water; soil moisture.

sol [CHEM] A colloidal solution consisting of a suitable dispersion medium, which may be gas, liquid, or solid, and the colloidal substance, the disperse phase, which is distributed throughout the dispersion medium.

Sol *See* sun.

Solanaceae [BOT] A family of dicotyledonous plants in the order Polemoniales having internal phloem,

mostly numerous ovules and seeds on axile placentae, and mostly cellular endosperm.

solar activity [ASTRON] Disturbances on the surface of the sun; examples are sunspots, prominences, and solar flares.

solar battery [ELECTR] An array of solar cells, usually connected in parallel and series.

solar blanket [ELECTR] A large, high-temperture, low-mass solar array consisting of ultrathin silicon solar cells interconnected, welded, and bonded to flexible substances.

solar cell [ELECTR] A *pn*-junction device which converts the radiant energy of sunlight directly and efficiently into electrical energy.

solar constant [METEOROL] The rate at which energy from the sun is received just outside the earth's atmosphere on a surface normal to the incident radiation and at the earth's mean distance from the sun; it is 0.140 watt per square centimeter.

solar corona [ASTRON] The upper, rarefied solar atmosphere which becomes visible around the darkened sun during a total solar eclipse. Also known as corona.

solar cycle [ASTRON] The periodic change in the number of sunspots; the cycle is taken as the interval between successive minima and is about 11.1 years.

solar day [ASTRON] A time measurement, the duration of one rotation of the earth on its axis with respect to the sun; this may be a mean solar day or an apparent solar day as the reference is to the mean sun or apparent sun.

solar eclipse [ASTRON] An eclipse that takes place when the new moon passes between the earth and the sun and the shadow formed reaches the earth; may be classified as total, partial, or annular.

solar energy [ASTROPHYS] The energy transmitted from the sun in the form of electromagnetic radiation.

solar-excited laser *See* sun-pumped laser.

solar flare [ASTROPHYS] An abrupt increase in the intensity of the H-α and other emission near a sunspot region; the brightness may be many times that of the associated plage.

solar heating [MECH ENG] The conversion of solar radiation into heat for technological, comfort-heating, and cooking purposes.

solarimeter [ENG] **1.** A type of pyranometer consisting of a Moll thermopile shielded from the wind by a bell glass. **2.** *See* pyranometer.

solar month [ASTRON] A time interval equal to one-twelfth of the solar year.

solar nutation [ASTRON] Nutation caused by the change in declination of the sun.

solar parallax [ASTRON] The sun's mean equatorial horizontal parallax p, which is the angle subtended by the equatorial radius r of the earth at mean distance a of the sun.

solar photophthalmia *See* snow blindness.

solar physics [ASTROPHYS] The scientific study of all physical phenomena connected with the sun; it overlaps with geophysics in the consideration of solar-terrestrial relationships, such as the connection between solar activity and auroras.

solar pond [ENG] A large, shallow pond covered with a thin, transparent plastic shield and used for collecting and storing solar heat for conversion to electric power.

solar power [MECH ENG] The conversion of the energy of the sun's radiation to useful work.

solar prominence [ASTRON] Sheets of luminous gas emanating from the sun's surface; they appear dark against the sun's disk but bright against the dark sky, and occur only in regions of horizontal magnetic fields.

solar radiation [ASTROPHYS] The electromagnetic radiation and particles (electrons, protons, and rarer heavy atomic nuclei) emitted by the sun.

solar spicule See spicule.

solar still [CHEM ENG] A device for evaporating seawater, in which water is confined in one or more shallow pools, over which is placed a roof-shaped transparent cover made of glass or plastic film; the sun's heat evaporates the water, leaving behind a residue of salt; the vapor from the evaporated water condenses on the surface of the cover and trickles down into gutters, which thus collect fresh water.

solar system [ASTRON] The sun and the celestial bodies moving about it; the bodies are planets, satellites of the planets, asteroids, comets, and meteor swarms.

solar telescope [OPTICS] An observational instrument of the solar astronomer; it is designed so that heating effects produced by the sun do not distort the images; the two classes consist of those designed for observations of the brilliant solar disk, and those designed for the study of the much fainter prominences and the still fainter corona through the relatively bright, scattered light of the sky.

solar tide [OCEANOGR] The tide caused solely by the tide-producing forces of the sun.

solar time [ASTRON] Time based on the rotation of the earth relative to the sun.

solar wind [GEOPHYS] The supersonic flow of gas, composed of ionized hydrogen and helium, which continuously flows from the sun out through the solar system with velocities of 300 to 1000 kilometers per second; it carries magnetic fields from the sun.

Solasteridae [INV ZOO] The sun stars, a family of asteroid echinoderms in the order Spinulosida.

solation [PHYS CHEM] The change of a substance from a gel to a sol.

solder [MET] 1. To join by means of solder. 2. An alloy, such as of zinc and copper, or of tin and lead, used when melted to join metallic surfaces.

solderless contact See crimp contact.

sole [BUILD] The horizontal member beneath the studs in a framed building. [ELECTR] Electrode used in magnetrons and backward-wave oscillators to carry a current that generates a magnetic field in the direction wanted. [GEOGR] The lowest part of a valley. [GEOL] 1. The bottom of a sedimentary stratum. 2. The middle and lower portion of the shear surface of a landslide. 3. The underlying fault plane of a thrust nappe. Also known as sole plane. [HYD] The basal ice of a glacier, often dirty in appearance due to contained rock fragments.

Solemyidae [INV ZOO] A family of bivalve mollusks in the order Protobranchia.

Solenichthyes [VERT ZOO] An equivalent name for Gasterosteiformes.

solenodon [VERT ZOO] Either of two species of insectivorous mammals comprising the family Solenodontidae; the almique (*Atopogale cubana*) is found only in Cuba, while the white agouta (*Solenodon paradoxus*) is found in Haiti.

Solenodontidae [VERT ZOO] The solenodons, a family of insectivores belonging to the group Lipotyphla.

solenoid [ELECTROMAG] Also known as electric solenoid. 1. An electrically energized coil of insulated wire which produces a magnetic field within the coil. 2. In particular, a coil that surrounds a movable iron core which is pulled to a central position with re-

spect to the coil when the coil is energized by sending current through it. [METEOROL] See meteorological solenoid.

solenoid brake [MECH ENG] A device that retards or arrests rotational motion by means of the magnetic resistance of a solenoid.

Solenopora [PALEOBOT] A genus of extinct calcareous red algae in the family Solenoporaceae that appeared in the Late Cambrian and lasted until the Early Tertiary.

Solenoporaceae [PALEOBOT] A family of extinct red algae having compact tissue and the ability to deposit calcium carbonate within and between the cell walls.

sole plane See sole.

soletta [AERO ENG] An orbiting solar mirror (reflector).

solid [PHYS] 1. A substance that has a definite volume and shape and resists forces that tend to alter its volume or shape. 2. A crystalline material, that is, one in which the constituent atoms are arranged in a three-dimensional lattice, periodic in three independent directions.

solid angle [MATH] A surface formed by all rays joining a point to a closed curve.

solid coupling [MECH ENG] A flanged-face or a compression-type coupling used to connect two shafts to make a permanent joint and usually designed to be capable of transmitting the full load capacity of the shaft; a solid coupling has no flexibility.

solid-dielectric capacitor [ELEC] A capacitor whose dielectric is one of several solid materials such as ceramic, mica, glass, plastic film, or paper.

solid-electrolyte battery [ELEC] A primary battery whose electrolyte is either a solid crystalline salt, such as silver iodide or lead chloride, or an ion-exchange membrane; in either case, conductivity is almost entirely ionic.

solid helium [CRYO] A certain state which is not attained by helium under its own vapor pressure down to absolute zero, but which requires an external pressure of 25 atmospheres at absolute zero.

solid laser [OPTICS] A laser in which either a crystalline or amorphous solid material, usually in the form of a rod, is excited by optical pumping; the most common crystalline materials are ruby, neodymium-doped ruby, and neodymium-doped yttrium aluminum garnet.

solid propellant [MATER] A rocket propellant in solid form, usually containing both fuel and oxidizer combined or mixed, and formed into a monolithic (not powdered or granulated) grain. Also known as solid rocket fuel; solid rocket propellant.

solid rocket fuel See solid propellant.

solid rocket propellant See solid propellant.

solid solution [PHYS] A homogeneous crystalline phase composed of several distinct chemical species, occupying the lattice points at random and existing in a range of concentrations.

solid state [ENG] Pertaining to a circuit, device, or system that depends on some combination of electrical, magnetic, and optical phenomena within a solid that is usually a crystalline semiconductor material. [PHYS] The condition of a substance in which it is a solid.

solid-state battery [ELEC] A battery in which both the electrodes and the electrolyte are solid-state materials.

solid-state circuit [ELECTR] Complete circuit formed from a single block of semiconductor material.

solid-state computer   [COMPUT SCI] A digital computer which uses diodes and transistors instead of vacuum tubes.

solid-state lamp   *See* light-emitting diode.

solid-state laser   [OPTICS] A laser in which a semiconductor material produces the coherent output beam.

solid-state maser   [PHYSICS] A maser in which a semiconductor material produces the coherent output beam; two input waves are required: one wave, called the pumping source, induces upward energy transitions in the active material, and the second wave, of lower frequency, causes downward transitions and undergoes amplification as it absorbs photons from the active material.

solid-state memory   [COMPUT SCI] A computer memory whose elements consist of integrated-circuit bistable multivibrators in which bits of information are stored as one of two states.

solid-state physics   [PHYS] The branch of physics centering about the physical properties of solid materials.

solid-state switch   [ELECTR] A microwave switch in which a semiconductor material serves as the switching element; a zero or negative potential applied to the control electrode will reverse-bias the switch and turn it off, and a slight positive voltage will turn it on.

solidus   [PHYS CHEM] In a constitution or equilibrium diagram, the locus of points representing the temperature below which the various compositions finish freezing on cooling, or begin to melt on heating.

solifluction   [GEOL] A rapid soil creep, especially referring to downslope soil movement in periglacial areas. Also known as sludging; soil flow; soil fluction.

solion   [ELEC] An electrochemical device in which amplification is obtained by controlling and monitoring a reversible electrochemical reaction.

solitary wave   [PHYS] A traveling wave in which a single disturbance is neither preceded by nor followed by other such disturbances, but which does not involve unusually large amplitudes or rapid changes in variables, in contrast to a shock wave.

soliton   [MATH] A solution of a nonlinear differential equation that propagates with a characteristic constant shape.

Solo man   [PALEON] A relative but primitive form of fossil man from Java; this form had a small brain, heavy horizontal browridges, and a massive cranial base.

Solpugida   [INV ZOO] The sun spiders, an order of nonvenomous, spiderlike, predatory arachnids having large chelicerae for holding and crushing prey.

solstice   [ASTRON] The two days (actually, instants) during the year when the earth is so located in its orbit that the inclination (about $23\frac{1}{2}°$) of the polar axis is toward the sun; the days are June 21 for the North Pole and December 22 for the South Pole; because of leap years, the dates vary a little.

solubility   [PHYS CHEM] The ability of a substance to form a solution with another substance.

solubility coefficient   [PHYS CHEM] The volume of a gas that can be dissolved by a unit volume of solvent at a specified pressure and temperature.

solubility product constant   [PHYS CHEM] A type of simplified equilibrium constant, K_{sp}, defined for and useful for equilibria between solids and their respective ions in solution; for example, the equilibrium

$$AgCl(s) \rightleftarrows Ag^+ + Cl^-, \quad [Ag^+][Cl^-] \cong K_{sp},$$

where $[Ag^+]$ and $[Cl^-]$ are molar concentrations of silver ions and chloride ions.

solum   [GEOL] The upper part of a soil profile, composed of A and B horizons in mature soil. Also known as true soil.

solute   [CHEM] The substance dissolved in a solvent.

solution   [CHEM] A single, homogeneous liquid, solid, or gas phase that is a mixture in which the components (liquid, gas, solid, or combinations thereof) are uniformly distributed throughout the mixture.

solution gas   [PETRO ENG] Gaseous reservoir hydrocarbons dissolved in liquid reservoir hydrocarbons because of the prevailing pressures in the reservoir. Also known as dissolved gas.

solution pressure   [PHYS CHEM] **1.** A measure of the tendency of molecules or atoms to cross a bounding surface between phases and to enter into a solution. **2.** A measure of the tendency of hydrogen, metals, and certain nonmetals to pass into solution as ions.

solvable group   [MATH] A group G which has subgroups $G_0, G_1, \ldots, G_n$, where $G_0 = G$, $G_n =$ the identity element alone, and each G_i is a normal subgroup of G_{i-1} with the quotient group G_{i-1}/G_i Abelian.

solvation   [CHEM] The process of swelling, gelling, or dissolving of a material by a solvent; for resins, the solvent can be a plasticizer.

solvent   [CHEM] That part of a solution that is present in the largest amount, or the compound that is normally liquid in the pure state (as for solutions of solids or gases in liquids).

solvent extraction   [CHEM ENG] The separation of materials of different chemical types and solubilities by selective solvent action; that is, some materials are more soluble in one solvent than in another, hence there is a preferential extractive action; used to refine petroleum products, chemicals, vegetable oils, and vitamins.   [NUCLEO] A process for removing uranium fuel residue from used fuel elements of a reactor; it generally involves decay cooling under water for up to 6 months, removal of cladding, dissolution, separation of reusable fuel, decontamination, and disposal of radioactive wastes. Also known as liquid extraction.

solvent-refined coal   [MATER] Low-sulfur distillate fuels from coal, plus the by-products of methane, light hydrocarbons, and naphtha, all useful for making pipeline gas, ethylene, and high-octane unleaded gasoline.

solvent refining   [CHEM ENG] The process of treating a mixed material with a solvent that preferentially dissolves and removes certain minor constituents (usually the undesired ones); common in the petroleum refining industry.

solvus   [PHYS CHEM] In a phase or equilibrium diagram, the locus of points representing the solid-solubility temperatures of various compositions of the solid phase.

Somali Current   *See* East Africa Coast Current.

Somasteroidea   [INV ZOO] A subclass of Asterozoa comprising sea stars of generalized structure, the jaws often only partly developed, and the skeletal elements of the arm arranged in a double series of transverse rows termed metapinnules.

somatic cell   [BIOL] Any cell of the body of an organism except the germ cells.

somatization   [PSYCH] A type of neurosis manifested in neurasthenias, hypochondriacal symptoms, and conversion hysterias.

somatotropin [BIOCHEM] The growth hormone of the pituitary gland.

somesthesis [PHYSIO] The general name for all systems of sensitivity present in the skin, muscles and their attachments, visceral organs, and nonauditory labyrinth of the ear.

Sommerfeld model *See* free-electron theory of metals.

Sommerfeld theory *See* free-electron theory of metals.

sonar [ENG] **1.** A system that uses underwater sound, at sonic or ultrasonic frequencies, to detect and locate objects in the sea, or for communication; the commonest type is echo-ranging sonar, other versions are passive sonar, scanning sonar, and searchlight sonar. Derived from sound navigation and ranging. **2.** *See* sonar set.

sonar boomer transducer [ENG ACOUS] A sonar transducer that generates a large pressure wave in the surrounding water when a capacitor bank discharges into a flat, epoxy-encapsulated coil, creating opposed magnetic fields from the coil and from eddy currents in an adjacent aluminum disk, which cause the disk to be driven away from the coils with great force.

sonar set [ENG] A complete assembly of sonar equipment for detecting and ranging or for communication. Also known as sonar.

sonde [ENG] An instrument used to obtain weather data during ascent and descent through the atmosphere, in a form suitable for telemetering to a ground station by radio, as in a radiosonde.

sone [ACOUS] A unit of loudness, equal to the loudness of a simple 1000-hertz tone with a sound pressure level 40 decibels above 0.0002 microbar; a sound that is judged by listeners to be n times as loud as this tone has a loudness of n sones.

sonic [ACOUS] **1.** Of or pertaining to the speed of sound. **2.** Pertaining to that which moves at acoustic velocity, as in sonic flow. **3.** Designed to operate or perform at the speed of sound, as in sonic leading edge.

sonic barrier [AERO ENG] A popular term for the large increase in drag that acts upon an aircraft approaching acoustic velocity; the point at which the speed of sound is attained and existing subsonic and supersonic flow theories are rather indefinite. Also known as sound barrier.

sonic boom [ACOUS] A noise caused by a shock wave that emanates from an aircraft or other object traveling at or above sonic velocity.

sonic depth finder [ENG] A sonar-type instrument used to measure ocean depth and to locate underwater objects; a sound pulse is transmitted vertically downward by a piezoelectric or magnetostriction transducer mounted on the hull of the ship; the time required for the pulse to return after reflection is measured electronically. Also known as echo sounder.

sonics [ACOUS] The technology of sound, or elastic wave motion, as applied to problems of measurement, control, and processing.

sonic speed *See* speed of sound.

sonic velocity *See* speed of sound.

sonic well logging [ENG] A well logging technique that uses a pulse-echo system to measure the distance between the instrument and a sound-reflecting surface; used to measure the size of cavities around brine wells, and capacities of underground liquefied petroleum gas storage chambers.

sonobuoy [ENG] An acoustic receiver and radio transmitter mounted in a buoy that can be dropped from an aircraft by parachute to pick up underwater sounds of a submarine and transmit them to the aircraft; to track a submarine, several buoys are dropped in a pattern that includes the known or suspected location of the submarine, with each buoy transmitting an identifiable signal; an electronic computer then determines the location of the submarine by comparison of the received signals and triangulation of the resulting time-delay data. Also known as radio sonobuoy.

sonoencephalograph *See* echoencephalograph.

sonoscan [ENG] A type of acoustic microscope in which an unfocused acoustic beam passes through the object and produces deformations in a liquid-solid interface that are sensed by a laser beam reflected from the surface.

sordawalite *See* tachylite.

sore shin [PL PATH] A fungus disease of cowpea, cotton, tobacco, and other plants, beyond the seedling stage, marked by annular growth of the pathogen on the stem at the groundline.

sorghum [BOT] Any of a variety of widely cultivated grasses, especially *Sorghum bicolor* in the United States, grown for grain and herbage; growth habit and stem form are similar to Indian corn, but leaf margins are serrate and spikelets occur in pairs on a hairy rachis.

Soricidae [VERT ZOO] The shrews, a family of insectivorous mammals belonging to the Lipotyphla.

sorosilicate [MINERAL] A structural type of silicate whose crystal lattice has two SiO_4 tetrahedra sharing one oxygen atom.

sorption [PHYS CHEM] A general term used to encompass the processes of adsorption, absorption, desorption, ion exchange, ion exclusion, ion retardation, chemisorption, and dialysis.

sorption pumping [ENG] A technique used to reduce the pressure of gas in an atmosphere; the gas is adsorbed on a granular sorbent material such as a molecular sieve in a metal container; when this sorbent-filled container is immersed in liquid nitrogen, the gas is sorbed.

sorrel tree *See* sourwood.

sort [COMPUT SCI] **1.** To rearrange a set of data items into a new sequence, governed by specific rules of precedence. **2.** The program designed to perform this activity.

sorter *See* card sorter; sequencer.

sort generator [COMPUT SCI] A computer program that produces other programs which arrange collections of items into sequences as specified by parameters in the original program.

sorus [BOT] **1.** A cluster of sporangia on the lower surface of a fertile fern leaf. **2.** A clump of reproductive bodies or spores in lower plants.

sou'easter *See* southeaster.

sound [ACOUS] **1.** An alteration of properties of an elastic medium, such as pressure, particle displacement, or density, that propagates through the medium, or a superposition of such alterations; sound waves having frequencies above the audible (sonic) range are termed ultrasonic waves; those with frequencies below the sonic range are called infrasonic waves. Also known as acoustic wave; sound wave. **2.** The auditory sensation which is produced by these alterations. Also known as sound sensation.

sound absorption coefficient [ACOUS] The ratio of sound energy absorbed to that arriving at a surface or medium. Also known as acoustic absorption coefficient; acoustic absorptivity.

sound attenuation [ACOUS] Diminution of the intensity of sound energy propagating in a medium; caused by absorption, spreading, and scattering.

sound barrier *See* sonic barrier.

sound carrier [COMMUN] The television carrier that is frequency-modulated by the sound portion of a television program; the unmodulated center frequency of the sound carrier is 4.5 megahertz higher than the video carrier frequency for the same channel.

sound channel [ACOUS] A layer of seawater extending from about 700 meters down to about 1500 meters, in which sound travels at about 450 meters per second, the slowest it can travel in seawater; below 1500 meters the speed of sound increases as a result of pressure. [ELECTR] The series of stages that handles only the sound signal in a television receiver.

sound energy [ACOUS] The difference between the total energy and the energy which would exist if no sound waves were present. Also known as acoustic energy.

sound fixing and ranging *See* sofar.

sounding [ENG] 1. Determining the depth of a body of water by an echo sounder or sounding line. 2. Measuring the depth of bedrock by driving a steel rod into the soil. 3. Any penetration of the natural environment for scientific observation. [METEOROL] *See* upper-air observation. [MIN ENG] 1. Knocking on a mine roof to see whether it is sound or safe to work under. 2. Subsurface investigation by observing the penetration resistance of the subsurface material without drilling holes, by driving a rod into the ground or by using a penetrometer.

sounding velocity [ACOUS] The vertical velocity of sound in water, usually assumed to be constant at 800 to 820 fathoms (1464 to 1501 meters) per second for sounding measurements.

sound intensity [ACOUS] For a specified direction and point in space, the average rate at which sound energy is transmitted through a unit area perpendicular to the specified direction.

sound level [ACOUS] The sound pressure level (in decibels) at a point in a sound field, averaged over the audible frequency range and over a time interval, with a frequency weighting and the time interval as specified by the American National Standards Association.

sound navigation and ranging *See* sonar.

sound pressure *See* effective sound pressure.

sound pressure level [ACOUS] A value in decibels equal to 20 times the logarithm to the base 10 of the ratio of the pressure of the sound under consideration to a reference pressure; reference pressures in common use are 0.0002 microbar and 1 microbar. Abbreviated SPL.

soundproofing *See* damping.

sound ranging [ENG ACOUS] Determining the location of a gun or other sound source by measuring the travel time of the sound wave to microphones at three or more different known positions.

sound recording [ENG ACOUS] The process of recording sound signals so they may be reproduced at any subsequent time, as on a phonograph disk, motion picture sound track, or magnetic tape.

sound reflection coefficient *See* acoustic reflectivity.

sound-reinforcement system [ENG ACOUS] An electronic means for augmenting the sound output of a speaker, singer, or musical instrument in cases where it is either too weak to be heard above the general noise or too reverberant; basic elements of

such a system are microphones, amplifiers, volume controls, and loudspeakers. Also known as public address system.

sound-reproducing system [ENG ACOUS] A combination of transducing devices and associated equipment for picking up sound at one location and time and reproducing it at the same or some other location and at the same or some later time. Also known as audio system; reproducing system; sound system.

sound sensation *See* sound.

sound system *See* sound-reproducing system.

sound track [ENG ACOUS] A narrow band, usually along the margin of a sound film, that carries the sound record; it may be a variable-width or variable-density optical track or a magnetic track.

sound transducer *See* electroacoustic transducer.

sound transmission coefficient [ACOUS] The ratio of transmitted to incident sound energy at an interface in a sound medium; the value depends on the angle of incidence of the sound. Also known as acoustic transmission coefficient; acoustic transmissivity.

sound wave *See* sound.

source [ELEC] The circuit or device that supplies signal power or electric energy or charge to a transducer or load circuit. [ELECTR] The terminal in a field-effect transistor from which majority carriers flow into the conducting channel in the semiconductor material. [NUCLEO] A radioactive material packaged so as to produce radiation for experimental or industrial use. [PHYS] 1. In general, a device that supplies some extensive entity, such as energy, matter, particles, or electric charge. 2. A point, line, or area at which mass or energy is added to a system, either instantaneously or continuously. 3. A point at which lines of force in a vector field originate, such as a point in an electrostatic field where there is positive charge. [SPECT] The arc or spark that supplies light for a spectroscope. [THERMO] A device that supplies heat.

source data automation *See* automation source data.

source data entry [COMPUT SCI] Entry of data into a computer system directly from its source, without transcription.

source document [COMPUT SCI] The original medium containing the basic data to be used by a data-processing system, from which the data are converted into a form which can be read into a computer. Also known as original document.

source flow [FL MECH] 1. In three-dimensional flow, a point from which fluid issues at a uniform rate in all directions. 2. In two-dimensional flow, a line normal to the planes of flow, from which fluid flows uniformly in all directions at right angles to the line.

source-follower amplifier *See* common-drain amplifier.

source language [COMPUT SCI] The language in which a program (or other text) is originally expressed.

source program [COMPUT SCI] The form of a program just as the programmer has written it, often on coding forms or machine-readable media; a program expressed in a source-language form.

source region [METEOROL] An extensive area of the earth's surface characterized by essentially uniform surface conditions and so situated with respect to the general atmospheric circulation that an air mass may remain over it long enough to acquire its characteristic properties.

source rock [GEOL] 1. Rock from which fragments have been derived which form a later, usually sedimentary rock. Also known as mother rock; parent rock. 2. Sedimentary rock, usually shale and limestone, deposited together with organic matter which was subsequently transformed to liquid or gaseous hydrocarbons.

sour gas [MATER] Natural gas that contains corrosive, sulfur-bearing compounds, such as hydrogen sulfide and mercaptans.

sourwood [BOT] *Oxydendrum arboreum*. A deciduous tree of the heath family (Ericaceae) indigenous along the Alleghenies and having long, simple, finely toothed, long-pointed leaves that have an acid taste, and white, urn-shaped flowers. Also known as sorrel tree.

South American trypanosomiasis *See* Chagas' disease.

South Atlantic Current [OCEANOGR] An eastward-flowing current of the South Atlantic Ocean that is continuous with the northern edge of the West Wind Drift.

southbound node *See* descending node.

Southeast Drift Current [OCEANOGR] A North Atlantic Ocean current flowing southeastward and southward from a point west of the Bay of Biscay toward southwestern Europe and the Canary Islands, where it continues as the Canary Current.

southeaster [METEOROL] A southeasterly wind, particularly a strong wind or gale; for example, the winter southeast storms of the Bay of San Francisco. Also spelled sou'easter.

South Equatorial Current [OCEANOGR] Any of several ocean currents, flowing westward, driven by the southeast trade winds blowing over the tropical oceans of the Southern Hemisphere and extending slightly north of the equator. Also known as Equatorial Current.

Southern Cross *See* Crux.

Southern Crown *See* Corona Australis.

Southern Fish *See* Piscis Australis.

Southern Polar Front *See* Antarctic Convergence.

Southern Triangle *See* Triangulum Australe.

south geographical pole [GEOGR] The geographical pole in the Southern Hemisphere, at latitude 90°S. Also known as South Pole.

south geomagnetic pole [GEOPHYS] The geomagnetic pole in the Southern Hemisphere at approximately 78.5°S, longitude 111°E, 180° from the north geomagnetic pole. Also known as south magnetic pole.

South Pacific Current [OCEANOGR] An eastward-flowing current of the South Pacific Ocean that is continuous with the northern edge of the West Wind Drift.

south pole [ELECTROMAG] The pole of a magnet at which magnetic lines of force are assumed to enter. [GEOPHYS] *See* south geomagnetic pole.

South Pole *See* south geographical pole.

southwester [METEOROL] A southwest wind, particularly a strong wind or gale. Also spelled sou'wester.

sou'wester *See* southwester.

soybean [BOT] *Glycine max*. An erect annual legume native to China and Manchuria and widely cultivated for forage and for its seed.

soybean lecithin *See* lecithin.

soy lecithin *See* lecithin.

space [ASTRON] 1. Specifically, the part of the universe lying outside the limits of the earth's atmosphere. 2. More generally, the volume in which all celestial bodies, including the earth, move. [COMMUN] The open-circuit condition or the signal causing the open-circuit condition in telegraphic communication; the closed-circuit condition is called the mark. [MATH] In context, usually a set with a topology on it or some other type of structure.

space biology [BIOL] A term for the various biological sciences and disciplines that are concerned with the study of living things in the space environment.

space capsule [AERO ENG] A container, manned or unmanned, used for carrying out an experiment or operation in space.

space character *See* blank character.

space charge [ELEC] The net electric charge within a given volume. [GEOPHYS] In atmospheric electricity, space charge refers to a preponderance of either negative or positive ions within any given portion of the atmosphere.

space-charge effect [ELECTR] Repulsion of electrons emitted from the cathode of a thermionic vacuum tube by electrons accumulated in the space charge near the cathode.

space-charge layer *See* depletion layer.

space-charge limitation [ELECTR] The current flowing through a vacuum between a cathode and an anode cannot exceed a certain maximum value, as a result of modification of the electric field near the cathode due to space charge in this region.

space-charge polarization [ELEC] Polarization of a dielectric which occurs when charge carriers are present which can migrate an appreciable distance through the dielectric but which become trapped or cannot discharge at an electrode. Also known as interfacial polarization.

space-charge region [ELECTR] Of a semiconductor device, a region in which the net charge density is significantly different from zero.

space communication [COMMUN] Communication between a vehicle in outer space and the earth, using high-frequency electromagnetic radiation.

space coordinates [MATH] A three-dimensional system of cartesian coordinates by which a point is located by three magnitudes indicating distance from three planes which intersect at a point.

spacecraft [AERO ENG] Devices, manned and unmanned, which are designed to be placed into an orbit about the earth or into a trajectory to another celestial body. Also known as space ship; space vehicle.

spaced antenna [ELECTROMAG] Antenna system consisting of a number of separate antennas spaced a considerable distance apart, used to minimize local effects of fading at short-wave receiving stations.

space flight [AERO ENG] Travel beyond the earth's sensible atmosphere; space flight may be an orbital flight about the earth or it may be a more extended flight beyond the earth into space.

space frame [BUILD] A three-dimensional steel building frame which is stable against wind loads.

space group [CRYSTAL] A group of operations which leave the infinitely extended, regularly repeating pattern of a crystal unchanged; there are 230 such groups.

space inversion *See* inversion.

space lattice [BUILD] A space frame built of lattice girders. [CRYSTAL] *See* crystal lattice; lattice.

space medicine [MED] A branch of medicine that deals with the physiologic disturbances and diseases produced in man by high-velocity projection through and beyond the earth's atmosphere, flight through interplanetary space, and return to earth.

space perception [PHYSIO] The awareness of the spatial properties and relations of an object, or of one's own body, in space; especially, the sensory appreciation of position, size, form, distance, and direction of an object, or of the observer himself, in space.

space platform [AERO ENG] A gimbal-mounted platform equipped with gyros and accelerometers for maintaining a desired orientation in inertial space independent of spacecraft motion.

space probe [AERO ENG] An instrumented vehicle, the payload of a rocket-launching system designed specifically for flight missions to other planets or the moon and into deep space, as distinguished from earth-orbiting satellites.

space processing [ENG] Forming and fabrication techniques employed aboard a spacecraft in a weightless or low-gravity environment and involving improved chemical or physical procedures for the creation of new or better products.

space reflection symmetry See parity.

space satellite [AERO ENG] A vehicle, manned or unmanned, for orbiting the earth.

space ship See spacecraft.

space shuttle [AERO ENG] A spacecraft designed to travel from the earth to a space station and to return to earth.

space station [AERO ENG] An autonomous, permanent facility in space for the conduct of scientific and technological research, earth-oriented applications, and astronomical observations.

space technology [AERO ENG] The systematic application of engineering and scientific disciplines to the exploration and utilization of outer space.

spacetenna [AERO ENG] The transmitting antenna of a solar power satellite transmission system which directs the high-power beam from space to a focus on the rectennas on earth.

space vehicle See spacecraft.

spacing bias See bias telegraph distortion.

spallation [NUC PHYS] A nuclear reaction in which the energy of each incident particle is so high that more than two or three particles are ejected from the target nucleus and both its mass number and atomic number are changed. Also known as nuclear spallation.

spandrel [BUILD] The part of a wall between the sill of a window and the head of the window below it.

spangolite [MINERAL] $Cu_6Al(SO_4)(OH)_{12}Cl\cdot3H_2O$ A dark-green hexagonal mineral composed of hydrous basic sulfate and chloride of aluminum and copper and occurring as crystals.

spanloader aircraft [AERO ENG] An advanced distributed-load cargo aircraft configuration in which the payload is distributed across the span of the wing for a close match between aerodynamic and inertial loading for minimal bending stress.

spanned record [COMPUT SCI] A logical record which covers more than one block, used when the size of a data buffer is fixed or limited.

spar [AERO ENG] A principal spanwise member of the structural framework of an airplane wing, aileron, stabilizer, and such; it may be of one-piece design or a fabricated section. [MIN ENG] A small clay vein in a coal seam. [MINERAL] Any transparent or translucent, nonmetallic, light-colored, readily cleavable, crystalline mineral; examples are calcspar and fluorspar. [NAV ARCH] A long, round stick of steel or wood, often tapered at one or both ends, and usually a part of a ship's masts or rigging.

Sparganiaceae [BOT] A family of monocotyledonous plants in the order Typhales distinguished by the inflorescence of globose heads, a vestigial perianth, and achenes that are sessile or nearly sessile.

Sparidae [VERT ZOO] A family of perciform fishes in the suborder Percoidei, including the porgies.

spark [ELEC] A short-duration electric discharge due to a sudden breakdown of air or some other dielectric material separating two terminals, accompanied by a momentary flash of light. Also known as electric spark; spark discharge; sparkover.

spark chamber [NUCLEO] A particle-detecting device in which the trajectory of a charged particle is made visible by a series of sparks that are triggered by the particle as it passes through an array of spark gaps.

spark counter [NUCLEO] A particle detector in which high-speed charged particles ionize a gas, consisting of argon mixed with an organic gas, triggering a spark between two plane parallel metal electrodes.

spark discharge See spark.

spark gap [ELEC] An arrangement of two electrodes between which a spark may occur; the insulation (usually air) between the electrodes is self-restoring after passage of the spark; used as a switching device, for example, to protect equipment against lightning or to switch a radar antenna from receiver to transmitter and vice versa.

spark-ignition combustion cycle See Otto cycle.

spark-ignition engine [MECH ENG] An internal combustion engine in which an electrical discharge ignites the explosive mixture of fuel and air.

sparkover See spark.

sparkover voltage See flashover voltage.

spark plug [ELEC] A device that screws into the cylinder of an internal combustion engine to provide a pair of electrodes between which an electrical discharge is passed to ignite the explosive mixture.

Sparnacean [GEOL] A European stage of geologic time; upper upper Paleocene, above Thanetian, below Ypresian of Eocene.

sparry iron See siderite.

spartalite See zincite.

spastic colon See irritable colon.

spatial resolution [OPTICS] The precision with which an optical instrument can produce separable images of different points on an object.

spatter cone [GEOL] A low, steep-sided cone of small pyroclastic fragments built up on a fissure or vent. Also known as agglutinate cone; volcanello.

spawn [ZOO] **1.** The collection of eggs deposited by aquatic animals, such as fish. **2.** To produce or deposit eggs or discharge sperm; applied to aquatic animals.

spay [VET MED] To remove the ovaries.

SPDT See single-pole double-throw.

speaker See loudspeaker.

spearmint [BOT] Mentha spicata. An aromatic plant of the mint family, Labiatae; the leaves are used as a flavoring in foods.

special functions [MATH] The various families of solution functions corresponding to cases of the hypergeometric equation or functions used in the equation's study, such as the gamma function.

special-purpose language [COMPUT SCI] A programming language designed to solve a particular type of problem.

special relativity [RELAT] The division of relativity theory which relates the observations of observers moving with constant relative velocities and postulates that natural laws are the same for all such observers.

species [CHEM] A chemical entity or molecular particle, such as a radical, ion, molecule, or atom. [NUC PHYS] See nuclide. [SYST] A taxonomic category ranking immediately below a genus and including closely related, morphologically similar individuals which actually or potentially interbreed.

specific conductance   *See* conductivity.

specific gravity   [MECH] The ratio of the density of a material to the density of some standard material, such as water at a specified temperature, for example, 4°C or 60°F, or (for gases) air at standard conditions of pressure and temperature. Abbreviated sp gr. Also known as relative density.

specific heat   [THERMO] **1.** The ratio of the amount of heat required to raise a mass of material 1 degree in temperature to the amount of heat required to raise an equal mass of a reference substance, usually water, 1 degree in temperature; both measurements are made at a reference temperature, usually at constant pressure or constant volume. **2.** The quantity of heat required to raise a unit mass of homogeneous material one degree in temperature in a specified way; it is assumed that during the process no phase or chemical change occurs.

specific humidity   [METEOROL] In a system of moist air, the (dimensionless) ratio of the mass of water vapor to the total mass of the system.

specific impulse   [AERO ENG] A performance parameter of a rocket propellant, expressed in seconds, equal to the thrust in pounds divided by the weight flow rate in pounds per second. Also known as specific thrust.

specific inductive capacity   *See* dielectric constant.

specific power   [NUCLEO] The power produced per unit mass of fuel present in a nuclear reactor.

specific reluctance   *See* reluctivity.

specific resistance   *See* electrical resistivity.

specific rotation   [OPTICS] The calculated rotation of light passing through a solution as related to the solution volume and depth, the amount of solute, and the observed optical rotation at a given wavelength and temperature.

specific routine   [COMPUT SCI] Computer routine to solve a particular data-handling problem in which each address refers to explicitly stated registers and locations.

specific thrust   *See* specific impulse.

specific viscosity   [FL MECH] The specific viscosity of a polymer is the relative viscosity of a polymer solution of known concentration minus 1; usually determined at low concentration of the polymer; for example, 0.5 gram per 100 milliliters of solution, or less.

specific volume   [MECH] The volume of a substance per unit mass; it is the reciprocal of the density. Abbreviated sp vol.

specific weight   [MECH] The weight per unit volume of a substance.

speckle   [OPTICS] A phenomenon in which the scattering of light from a highly coherent source, such as a laser, by a rough surface or inhomogeneous medium generates a random-intensity distribution of light that gives the surface or medium a granular appearance.

speckle interferometry   [OPTICS] The use of speckle patterns in the study of object displacements, vibration, and distortion, and in obtaining diffraction-limited images of stellar objects.

spectacle stone   *See* selenite.

spectral classification   [ASTRON] A classification of stars by characteristics revealed by study of their spectra; the six classes B, A, F, G, K, and M include 99% of all known stars.

spectral color   [OPTICS] **1.** A color corresponding to light of a pure frequency; the basic spectral colors are violet, blue-green, yellow, orange, and red. **2.**

A color that is represented by a point on the chromaticity diagram that lies on a straight line between some point on the spectral color (first definition) locus and the achromatic points; purple, for example, is not a spectral color.

spectral density   [ELECTROMAG] *See* spectral energy distribution.   [MATH] The density function for the spectral measure of a linear transformation on a Hilbert space.   [SYS ENG] *See* frequency spectrum.

spectral energy distribution   [ELECTROMAG] The power carried by electromagnetic radiation within some small interval of wavelength (of frequency) of fixed amount as a function of wavelength (of frequency). Also known as spectral density.

spectral line   [SPECT] A discrete value of a quantity, such as frequency, wavelength, energy, or mass, whose spectrum is being investigated; one may observe a finite spread of values resulting from such factors as level width, Doppler broadening, and instrument imperfections. Also known as spectrum line.

spectral luminous efficiency   *See* luminosity function.

spectral reflectance   [ANALY CHEM] Situation when the desired directions for analysis of energy from (reflected from) an object under spectrophotometric colorimetric analysis is diffused in all directions (not directed as a single beam).

spectral regions   [SPECT] Arbitrary ranges of wavelength, some of them overlapping, into which the electromagnetic spectrum is divided, according to the types of sources that are required to produce and detect the various wavelengths, such as x-ray, ultraviolet, visible, infrared, or radio-frequency.

spectral response   *See* spectral sensitivity.

spectral sensitivity   [ELECTR] Radiant sensitivity, considered as a function of wavelength.   [PHYS] The response of a device or material to monochromatic light as a function of wavelength. Also known as spectral response.

spectral transmission   [OPTICS] The radiant flux which passes through a filter divided by the radiant flux incident upon it, for monochromatic light of a specified wavelength.

spectral type   [ASTRON] A label used to indicate the physical and chemical characteristics of a star as indicated by study of the star's spectra; for example, the stars in the spectral type known as class B are blue-white, and are referred to as helium stars because the dominant lines in their spectra are the lines in helium spectra.

spectroheliograph   [OPTICS] An instrument used to photograph the sun in one spectral band.

spectrometer   [SPECT] **1.** A spectroscope that is provided with a calibrated scale either for measurement of wavelength or for measurement of refractive indices of transparent prism materials. **2.** A spectroscope equipped with a photoelectric photometer to measure radiant intensities at various wavelengths.

spectrophotometer   [SPECT] An instrument that measures transmission or apparent reflectance of visible light as a function of wavelength, permitting accurate analysis of color or accurate comparison of luminous intensities of two sources or specific wavelengths.

spectrophotometric titration   [ANALY CHEM] An analytical method in which the radiant-energy absorption of a solution is measured spectrophotometrically after each increment of titrant is added.

spectrophotometry   [SPECT] A procedure to measure photometrically the wavelength range of radiant energy absorbed by a sample under analysis; can be by visible light, ultraviolet light, or x-rays.

spectropyrheliometer [SPECT] An astronomical instrument used to measure distribution of radiant energy from the sun in the ultraviolet and visible wavelengths.

spectroscope [SPECT] An optical instrument consisting of a slit, collimator lens, prism or grating, and a telescope or objective lens which produces a spectrum for visual observation.

spectroscopic binary star [ASTRON] A binary star that may be distinguished from a single star only by noting the Doppler shift of the spectral lines of one or both stars as they revolve about their common center of mass.

spectroscopic parallax [ASTRON] Parallax as determined from examination of a stellar spectrum; critical spectral lines indicate the star's absolute magnitude, from which the star's distance, or parallax, can be deduced.

spectroscopic splitting factor *See* Landé g factor.

spectroscopy [PHYS] The branch of physics concerned with the production, measurement, and interpretation of electromagnetic spectra arising from either emission or absorption of radiant energy by various substances.

spectrum [MATH] If T is a linear operator of a normed space X to itself and I is the identity transformation $(\text{I }(x)\equiv x)$, the spectrum of T consists of all scalars λ for which either $T - \lambda\text{I}$ has no inverse or the range of $T - \lambda\text{I}$ is not dense in X. [PHYS] **1.** A display or plot of intensity of radiation (particles, photons, or acoustic radiation) as a function of mass, momentum, wavelength, frequency, or some related quantity. **2.** The set of frequencies, wavelengths, or related quantities, involved in some process; for example, each element has a characteristic discrete spectrum for emission and absorption of light. **3.** A range of frequencies within which radiation has some specified characteristic, such as audio-frequency spectrum, ultraviolet spectrum, or radio spectrum.

spectrum line *See* spectral line.

specular iron *See* specularite.

specularite [MINERAL] A black or gray variety of hematite with brilliant metallic luster, occurring in micaceous or foliated masses, or in tabular or disklike crystals. Also known as gray hematite; iron glance; specular iron.

specular reflection [PHYS] Reflection of electromagnetic, acoustic, or water waves in which the reflected waves travel in a definite direction, and the directions of the incident and reflected waves make equal angles with a line perpendicular to the reflecting surface, and lie in the same plane with it. Also known as direct reflection; mirror reflection; regular reflection.

specular reflector [OPTICS] A reflecting surface (polished metal or silvered glass) that gives a direct image of the source, with the angle of reflection equal to the angle of incidence. Also known as regular reflector.

speech amplifier [ENG ACOUS] An audio-frequency amplifier designed specifically for amplification of speech frequencies, as for public-address equipment and radiotelephone systems.

speech compression [COMMUN] Modulation technique that takes advantage of certain properties of the speech signal to permit adequate information quality, characteristics, and the sequential pattern of a speaker's voice to be transmitted over a nar-

rower frequency band than would otherwise be necessary.

speech frequency *See* voice frequency.

speech intelligibility *See* intelligibility.

speech inverter *See* scrambler.

speech scrambler *See* scrambler.

speed [GRAPHICS] The sensitivity of a photographic film, expressed according to one of several scales. [MECH] The time rate of change of position of a body without regard to direction; in other words, the magnitude of the velocity vector. [OPTICS] **1.** The light-gathering power of a lens, expressed as the reciprocal of the f number. **2.** The time that a camera shutter is open. [PHYS] In general, the rapidity with which a process takes place.

speed of light [ELECTROMAG] The speed of propagation of electromagnetic waves in a vacuum, which is a physical constant equal to $299,792.4580 \pm 0.0012$ kilometers per second. Also known as electromagnetic constant; velocity of light.

speed of response [PHYS] The time required for a system to react to some signal; for example, the delay time for a photon detector to react to a radiation pulse, or the time needed for a current or voltage in a circuit to reach a definite fraction of its final value as a result of an abrupt change in the electromotive force.

speed of sound [ACOUS] The phase velocity of a sound wave. Also known as sonic speed; sonic velocity; velocity of sound.

Spelaeogriphacea [INV ZOO] A peracaridan order of the Malacostraca comprised of the single species *Spelaeogriphus lepidops*, a small, blind, transparent, shrimplike crustacean with a short carapace that coalesces dorsally with the first thoracic somite.

speleothem [GEOL] A secondary mineral deposited in a cave by the action of water. Also known as cave formation.

spencerite [MINERAL] $Zn_4(PO_4)_2(OH)_2 \cdot 3H_2O$ A pearly white monoclinic mineral composed of hydrous basic zinc phosphate and occurring in scaly masses and small crystals.

spent fuel [NUCLEO] Nuclear reactor fuel that has been irradiated to the extent that it can no longer effectively sustain a chain reaction because its fissionable isotopes have been partially consumed and fission-product poisons have accumulated in it.

spent liquor [MATER] The liquid effluent from the digestion of wood during pulping; contains wood chemicals (for example, lignin) and spent digestant (caustic, sulfite, or sulfate, depending on the process used).

spergenite [GEOL] A biocalcarenite which contains ooliths and fossil debris and has a maximum quartz content of 10. Also known as Bedford limestone; Indiana limestone.

sperm *See* spermatozoon.

spermaceti [MATER] A white, crystalline, oily (waxy) solid that separates from sperm oil; soluble in ether, chloroform, and carbon disulfide, insoluble in water; melts at 42 to 50°C; used in ointments, emulsions, candles, soaps, and cosmetics; and for linen finishing. Also known as spermaceti wax.

spermaceti wax *See* spermaceti.

spermatheca [ZOO] A sac in the female for receiving and storing sperm until fertilization; found in many invertebrates and certain vertebrates. Also known as seminal receptacle.

spermatid [HISTOL] A male germ cell immediately before assuming its final typical form.

spermatogenesis [PHYSIO] The process by which spermatogonia undergo meiosis and transform into spermatozoa.

spermatozoon [HISTOL] A mature male germ cell. Also known as sperm.

sperm nucleus [BOT] One of the two nuclei in a pollen grain that function in double fertilization in seed plants.

sperm whale [VERT ZOO] *Physeter catadon*. An aggressive toothed whale belonging to the group Odontoceti of the order Cetacea; it produces ambergris and contains a mixture of spermaceti and oil in a cavity of the nasal passage.

sperrylite [MINERAL] PtAs$_2$ A tin-white isometric mineral composed of platinum arsenide; the only platinum compound known to occur in nature; hardness is 6–7 on Mohs scale, and specific gravity is 10.60.

spessartite [MINERAL] Mn$_3$Al$_2$(SiO$_4$)$_3$ A mineral composed of manganese aluminum silicate with small amounts of iron, magnesium, or other elements. [PETR] A lamprophyre composed of a sodic plagioclase groundmass in which green hornblende phenocrysts are embedded; also contains accessory olivine, biotite, apatite, and opaque oxides.

sp gr *See* specific gravity.

Sphaeractinoidea [PALEON] An extinct group of fossil marine hydrozoans distinguished in part by the relative prominence of either vertical or horizontal trabeculae and by the presence of long, tabulate tubes called autotubes.

Sphaeriales [MYCOL] An order of fungi in the subclass Euascomycetes characterized by hard, dark perithecia with definite ostioles.

Sphaeriidae [INV ZOO] The minute bog beetles, a small family of coleopteran insects in the suborder Myxophaga.

Sphaerioidaceae [MYCOL] A family of fungi of the order Sphaeropsidales in which the pycnidia are black or dark-colored and are flask-, cone-, or lens-shaped with thin walls and a round, relatively small pore.

sphaerite [MINERAL] Light-gray or bluish mineral composed of hydrous aluminum phosphate and occurring in global concretions.

Sphaeroceridae [INV ZOO] A family of myodarian cyclorrhaphous dipteran insects in the subsection Acalypteratae.

Sphaerodoridae [INV ZOO] A family of polychaete annelids belonging to the Errantia in which species are characterized by small bodies, and are usually papillated.

Sphaerolaimidae [INV ZOO] A family of free-living nematodes in the superfamily Monhysteroidea characterized by a spacious and deep stoma.

Sphaeromatidae [INV ZOO] A family of isopod crustaceans in the suborder Flabellifera in which the body is broad and oval and the inner branch of the uropod is immovable.

Sphaerophoraceae [BOT] A family of the Ascolichenes in the order Caliciales which are fruticose with a solid thallus.

Sphaeropleineae [BOT] A suborder of green algae in the order Ulotrichales distinguished by long, coenocytic cells, numerous bandlike chloroplasts, and heterogametes produced in undifferentiated vegetative cells.

Sphaeropsidaceae [MYCOL] An equivalent name for Sphaerioidaceae.

Sphaeropsidales [MYCOL] An order of fungi of the class Fungi Imperfecti in which asexual spores are formed in pycnidia, which may be separate or joined

to vegetative hyphae, conidiophores are short or absent, and conidia are always slime spores.

Sphagnaceae [BOT] The single monogeneric family of the order Sphagnales.

Sphagnales [BOT] The single order of mosses in the subclass Sphagnobrya containing the single family Sphagnaceae.

Sphagnobrya [BOT] A subclass of the Bryopsida; plants are grayish-green with numerous, spirally arranged branches and grow in deep tufts or mats, commonly in bogs and in other wet habitats.

sphalerite [MINERAL] (Zn,Fe)S The low-temperature form and common polymorph of zinc sulfide; a usually brown or black mineral that crystallizes in the hextetrahedral class of the isometric system, occurs most commonly in coarse to fine, granular, cleanable masses, has resinous luster, hardness of 3.5 on Mohs scale, and specific gravity of 4.1. Also known as blende; false galena; jack; lead marcasite; mock lead; mock ore; pseudogalena; steel jack.

Sphecidae [INV ZOO] A large family of hymenopteran insects in the superfamily Sphecoidea.

Sphecoidea [INV ZOO] A superfamily of wasps belonging to the suborder Apocrita.

Sphenacodontia [PALEON] A suborder of extinct reptiles in the order Pelycosauria which were advanced, active carnivores.

sphene [MINERAL] CaTiSiO$_5$ A brown, green, yellow, gray, or black neosilicate mineral common as an accessory mineral in igneous rocks; it is monoclinic and has resinous luster; hardness is 5–5.5 on Mohs scale; specific gravity is 3.4–3.5. Also known as grothite; titanite.

Spheniscidae [VERT ZOO] The single family of the avian order Sphenisciformes.

Sphenisciformes [VERT ZOO] The penguins, an order of aquatic birds found only in the Southern Hemisphere and characterized by paddlelike wings, erect posture, and scalelike feathers.

Sphenodontidae [VERT ZOO] A family of lepidosaurian reptiles in the order Rhynchocephalia represented by a single living species, *Sphenodon punctatus*, a lizardlike form distinguished by lack of a penis.

sphenoid [CRYSTAL] An open crystal, occurring in monoclinic crystals of the sphenoidal class, and characterized by two nonparallel faces symmetrical with an axis of twofold symmetry. [SCI TECH] Wedge-shaped.

sphenoid bone [ANAT] The butterfly-shaped bone forming the anterior part of the base of the skull and portions of the cranial, orbital, and nasal cavities.

Sphenophyllatae *See* Sphenophyllopsida.

Sphenophyllopsida [PALEOBOT] An extinct class of embryophytes in the division Equisetophyta.

Sphenopsida [BOT] A group of vascular cryptogams characterized by whorled, often very small leaves and by the absence of true leaf gaps in the stele; essentially equivalent to the division Equisetophyta.

sphere [MATH] 1. The set of all points in a euclidean space which are a fixed common distance from some given point; in euclidean three-dimensional space the Riemann sphere consists of all points (x,y,z) which satisfy the equation $x^2 + y^2 + z^2 = 1$. 2. The set of points in a metric space whose distance from a fixed point is constant.

spherical aberration [OPTICS] Aberration arising from the fact that rays which are initially at different distances from the optical axis come to a focus at different distances along the axis when they are reflected from a spherical mirror or refracted by a lens with spherical surfaces.

spherical coordinates [MATH] A system of curvilinear coordinates in which the position of a point in space is designated by its distance r from the origin or pole, called the radius vector, the angle ϕ between the radius vector and a vertically directed polar axis, called the cone angle or colatitude, and the angle θ between the plane of ϕ and a fixed meridian plane through the polar axis, called the polar angle or longitude. Also known as spherical polar coordinates.

spherical cyclic curve See cyclic curve.

spherical polar coordinates See spherical coordinates.

spherical sector [MATH] The cap and cone formed by the intersection of a plane with a sphere, the cone extending from the plane to the center of the sphere and the cap extending from the plane to the surface of the sphere.

spherical segment [MATH] A solid that is bounded by a sphere and two parallel planes which intersect the sphere or are tangent to it.

spherical triangle [MATH] A three-sided surface on a sphere the sides of which are arcs of great circles.

spherical trigonometry [MATH] The study of spherical triangles from the viewpoint of angle, length, and area.

spherical wave [PHYS] A wave whose equiphase surfaces form a family of concentric spheres; the direction of travel is always perpendicular to the surfaces of the spheres.

spheroid See ellipsoid of revolution.

spheroidal galaxy See elliptical galaxy.

spheroidal graphite cast iron See nodular cast iron.

spheroidizing [MET] Heating steels just below Ae_1 until the shape of cementite particles becomes relatively spherical.

spherulite [GEOL] A spherical body or coarsely crystalline aggregate having a radial internal structure arranged about one or more centers.

sphincter [ANAT] A muscle that surrounds and functions to close an orifice.

Sphinctozoa [PALEON] A group of fossil sponges in the class Calcarea which have a skeleton of massive calcium carbonate organized in the form of hollow chambers.

Sphindidae [INV ZOO] The dry fungus beetles, a family of coleopteran insects in the superfamily Cucujoidea.

Sphingidae [INV ZOO] The single family of the lepidopteran superfamily Sphingoidea.

Sphingoidea [INV ZOO] A superfamily of Lepidoptera in the suborder Heteroneura consisting of the sphinx, hawk, or hummingbird moths; these are heavy-bodied forms with antennae that are thickened with a pointed apex, a well-developed proboscis, and narrow wings.

sphingolipid [BIOCHEM] Any lipid, such as a sphingomyelin, that yields sphingosine or one of its derivatives as a product of hydrolysis.

Sphyraenidae [VERT ZOO] A family of shore fishes in the suborder Mugiloidei of the order Perciformes comprising the barracudas.

Sphyriidae [INV ZOO] A family of ectoparasitic Crustacea belonging to the group Lernaeopodoida; the parasite embeds its head and part of its thorax into the host.

spicule [ASTRON] One of an irregular distribution of jets shooting up from the sun's chromosphere. Also known as solar spicule. [BOT] An empty diatom shell. [INV ZOO] A calcareous or siliceous, usually spikelike supporting structure in many invertebrates, particularly in sponges and alcyonarians.

spider [ELEC] A structure on the shaft of an electric rotating machine that supports the core or poles of the rotor, consisting of a hub, spokes, and rim, or some similar arrangement. [ENG] 1. The part of an ejector mechanism which operates ejector pins in a molding press. 2. In extrusion, the membranes which support a mandrel within the head-die assembly. [ENG ACOUS] A highly flexible perforated or corrugated disk used to center the voice coil of a dynamic loudspeaker with respect to the pole piece without appreciably hindering in-and-out motion of the voice coil and its attached diaphragm. [INV ZOO] The common name for arachnids comprising the order Araneida. [PETRO ENG] A steel block with a tapered opening which permits passage of pipe during movement into or from a well; designed to hold pipe suspended in the well when the slips are placed in the tapered opening and in contact with the pipe.

spike [BOT] An indeterminate inflorescence with sessile flowers. [DES ENG] A large nail, especially one longer than 3 inches (7.6 centimeters), and often of square section. [PHYS] A short-duration transient whose amplitude considerably exceeds the average amplitude of the associated pulse or signal. [SOLID STATE] A sputtering event in which the process from impact of a bombarding projectile to the ejection of target atoms involves motion of a large number of particles in the target, so that collisions between particles become significant.

spike antenna See monopole antenna.

spilite [PETR] An altered basalt containing albitized feldspar accompanied by low-temperature, hydrous crystallization products such as chlorite, calcite, and epidote.

spill [ENG] The accidental release of some material, such as nuclear material or oil, from a container. [NUCLEO] The accidental release of radioactive material.

spillover [COMMUN] The receiving of a radio signal of a different frequency from that to which the receiver is tuned, due to broad tuning characteristics. [METEOROL] That part of orographic precipitation which is carried along by the wind so that it reaches the ground in the nominal rain shadow on the lee side of the barrier.

spillway [CIV ENG] A passage in or about a dam or other hydraulic structure for escape of surplus water.

spin [AERO ENG] The descent of an aircraft with the nose down and the tail spinning in circles overhead. [MECH] Rotation of a body about its axis. [QUANT MECH] The intrinsic angular momentum of an elementary particle or nucleus, which exists even when the particle is at rest, as distinguished from orbital angular momentum.

spina bifida [MED] A congenital anomaly characterized by defective closure of the vertebral canal with herniation of the spinal cord meninges.

spinach [BOT] *Spinacia oleracea*. An annual potherb of Asiatic origin belonging to the order Caryophyllales and grown for its edible foliage.

spinal column See spine.

spinal cord [ANAT] The cordlike posterior portion of the central nervous system contained within the spinal canal of the vertebral column of all vertebrates.

spinal nerve [ANAT] Any of the paired nerves arising from the spinal cord.

spinal reflex [PHYSIO] A reflex mediated through the spinal cord without the participation of the more cephalad structures of the brain or spinal cord.

spin axis [PHYS] The axis of rotation of a gyroscope.

spin-density wave [SOLID STATE] The ground state of a metal in which the conduction–electron spin density has a sinusoidal variation in space.

spine [ANAT] An articulated series of vertebrae forming the axial skeleton of the trunk and tail, and being a characteristic structure of vertebrates. Also known as backbone; spinal column; vertebral column. [BOT] A rigid sharp-pointed process in plants; many are modified leaves. [INV ZOO] One of the processes covering the surface of a sea urchin. [VERT ZOO] **1.** One of the spiny rays supporting the fins of most fishes. **2.** A sharp-pointed modified hair on certain mammals, such as the porcupine.

spinel [MINERAL] **1.** $MgAl_2O_4$ A colorless, purplish-red, greenish, yellow, or black mineral, usually forming octahedral crystals, and characterized by great hardness; used as a gemstone. **2.** A group of minerals of general formula AB_2O_4, where A is magnesium, ferrous iron, zinc, or manganese, or a combination of them, and B is aluminum, ferric iron, or chromium.

spin-flip laser [OPTICS] A semiconductor laser in which the output wavelength is continuously tunable by a magnetic field; operation is based on exciting conduction-band electrons to a higher energy level by reversing the direction of the electrons as they spin about an axis in the direction of the magnetic field.

spin glass [SOLID STATE] A substance in which the atomic spins are oriented in random but fixed directions.

spin label [PHYS CHEM] A molecule which contains an atom or group of atoms exhibiting an unpaired electron spin that can be detected by electron spin resonance (ESR) spectroscopy and can be bonded to another molecule.

spin-lattice relaxation [SOLID STATE] Magnetic relaxation in which the excess potential energy associated with electron spins in a magnetic field is transferred to the lattice.

spinneret [ENG] An extrusion die with many holes through which plastic melt is forced to form filaments. [INV ZOO] An organ that spins fiber from the secretion of silk glands. [TEXT] A metal device with tiny holes through which a solution is forced at high speeds to make fine textile filaments.

spinning [ENG] The extrusion of a spinning solution (such as molten plastic) through a spinneret. [MECH ENG] Shaping and finishing sheet metal by rotating the workpiece over a mandrel and working it with a round-ended tool. Also known as metal spinning. [TEXT] Converting fibers or filaments into thread or yarn by drawing and twisting.

spinode See cusp.

spinor [MATH] **1.** A vector with two complex components, which undergoes a unitary unimodular transformation when the three-dimensional coordinate system is rotated; it can represent the spin state of a particle of spin ½. **2.** More generally, a spinor of order (or rank) n is an object with 2^n components which transform as products of components of n spinors of rank one. **3.** A quantity with four complex components which transforms linearly under a Lorentz transformation in such a way that if it is a solution of the Dirac equation in the original Lorentz frame it remains a solution of the Dirac equation in the transformed frame; it is formed from two spinors (definition 1). Also known as Dirac spinor.

spin-orbit coupling [QUANT MECH] The interaction between a particle's spin and its orbital angular momentum.

spin-orbit multiplet [PHYS] A collection of atomic or nuclear states which differ in energy only on account of spin-orbit coupling; the total spin angular momentum quantum number S and total orbital angular momentum quantum number L are the same for all states; the energy levels are labeled by the total angular momentum quantum number J.

spin-parity [PARTIC PHYS] A combined symbol J^P for an elementary particle's spin J, and its intrinsic parity P.

spin quantum number [QUANT MECH] The ratio of the maximum observable component of a system's spin to Planck's constant divided by 2π; it is an integer or a half-integer.

spin state [QUANT MECH] Condition of a particle in which its total spin, and the component of its spin along some specified axis, have definite values; more precisely, the particle's wave function is an eigenfunction of the operators corresponding to these quantities.

Spintheridae [INV ZOO] An amphinomorphan family of small polychaete annelids included in the Errantia.

Spinulosida [INV ZOO] An order of Asteroidea in which pedicellariae rarely occur, marginal plates bounding the arms and disk are small and inconspicuous, and spines occur in groups on the upper surface.

spin wave [SOLID STATE] A sinusoidal variation, propagating through a crystal lattice, of that angular momentum which is associated with magnetism (mostly spin angular momentum of the electrons).

spiny-rayed fish [VERT ZOO] The common designation for actinopterygian fishes, so named for the presence of stiff, unbranched, pointed fin rays, known as spiny rays.

Spionidae [INV ZOO] A family of spioniform annelid worms belonging to the Sedentaria.

spiracle [INV ZOO] An external breathing orifice of the tracheal system in insects and certain arachnids. [VERT ZOO] **1.** The external respiratory orifice in cetaceous and amphibian larvae. **2.** The first visceral cleft in fishes.

spiral [MATH] A simple curve in the plane which continuously winds about itself either into some point or out from some point.

spiral arms [ASTRON] The shape of sections of certain galaxies called spirals; these sections are two so-called arms composed of stars, dust, and gas extending from the center of the galaxy and coiled about it.

spiral cleavage [EMBRYO] A cleavage pattern characterized by formation of a cell mass showing spiral symmetry; occurs in mollusks.

spiral conveyor See screw conveyor.

spiral gage See spiral pressure gage.

spiral galaxy [ASTRON] A type of galaxy classified on the basis of appearance of its photographic image; this type includes two main groups: normal spirals with circular symmetry of the nucleus and of the spiral arms, and barred spirals in which the dominant form is a luminous bar crossing the nucleus with spiral arms starting at the ends of the bar or tangent to a luminous rim on which the bar terminates.

spiral gear [MECH ENG] A helical gear that transmits power from one shaft to another, nonparallel shaft.

spiral layer See Ekman layer.

spiral organ See organ of Corti.

spiral pressure gage [ENG] A device for measurement of pressures; a hollow tube spiral receives the system pressure which deforms (unwinds) the spiral in direct relation to the pressure in the tube. Also known as spiral gage.

spiral spring [DES ENG] A spring bar or wire wound in an Archimedes spiral in a plane; each end is fastened to the force-applying link of the mechanism.

spiral valve [VERT ZOO] A spiral fold of mucous membrane in the small intestine of elasmobranchs and some primitive fishes which increases the surface area for absorption.

Spiriferida [PALEON] An order of fossil articulate brachiopods distinguished by the spiralium, a pair of spirally coiled ribbons of calcite supported by the crura.

Spiriferidina [PALEON] A suborder of the extinct brachiopod order Spiriferida including mainly ribbed forms having laterally or ventrally directed spires, well-developed interareas, and a straight hinge line.

Spirillaceae [MICROBIO] A family of bacteria; motile, helically curved rods that move with a characteristic corkscrewlike motion.

Spirillinacea [INV ZOO] A superfamily of foraminiferan protozoans in the suborder Rotaliina characterized by a planispiral or low conical test with a wall composed of radial calcite.

spirit level See level.

spirit varnish [MATER] An artificial varnish consisting of resin, asphalt, or a cellulose ester dissolved in a volatile solvent.

Spirobrachiidae [INV ZOO] A family of the Brachiata in the order Thecanephria.

Spirochaetaceae [MICROBIO] The single family of the order Spirochaetales.

Spirochaetales [MICROBIO] An order of bacteria characterized by slender, helically coiled cells sometimes occurring in chains.

spirometry [PHYSIO] The measurement, by a form of gas meter (spirometer), of volumes of air that can be moved in or out of the lungs.

Spirulidae [INV ZOO] A family of cephalopod mollusks containing several species of squids.

Spiruria [INV ZOO] A subclass of nematodes in the class Secernentea.

Spirurida [INV ZOO] An order of phasmid nematodes in the subclass Spiruria.

Spiruroidea [INV ZOO] A superfamily of spirurid nematodes which are parasitic in the respiratory and digestive systems of vertebrates.

spit [ENG] To light a fuse. [GEOGR] A small point of land commonly consisting of sand or gravel and which terminates in open water.

Spitsbergen Current [OCEANOGR] An ocean current flowing northward and westward from a point south of Spitsbergen, and gradually merging with the East Greenland Current in the Greenland Sea; the Spitsbergen Current is the continuation of the northwestern branch of the Norwegian Current.

SPL See sound pressure level.

spleen [ANAT] A blood-forming lymphoid organ of the circulatory system, present in most vertebrates.

splenic fever See anthrax.

spline [DES ENG] One of a number of equally spaced keys cut integral with a shaft, or similarly, keyways in a hubbed part; the mated pair permits the transmission of rotation or translatory motion along the axis of the shaft. [ENG] A strip of wood, metal, or plastic. [GRAPHICS] A flexible strip used in drawing curves.

split-phase motor [ELEC] A single-phase induction motor having an auxiliary winding connected in parallel with the main winding, but displaced in magnetic position from the main winding so as to produce the required rotating magnetic field for starting;

the auxiliary circuit is generally opened when the motor has attained a predetermined speed.

split-word operation [COMPUT SCI] A computer operation performed with portions of computer words rather than whole words as is normally done.

Spodosol [GEOL] A soil order characterized by accumulations of amorphous materials in subsurface horizons.

spodumene [MINERAL] $LiAlSi_2O_6$ A white to yellowish-, purplish-, or emerald-green clinopyroxene mineral occurring in prismatic crystals; hardness is 6.5–7 on Mohs scale, and specific gravity 3.13–3.20; an ore of lithium. Also known as triphane.

spoil [MIN ENG] 1. The overburden or nonore material from a coal mine. 2. A stratum of coal and dirt mixed.

spoiler [AERO ENG] A plate, series of plates, comb, tube, bar, or other device that projects into the airstream about a body to break up or spoil the smoothness of the flow, especially such a device that projects from the upper surface of an airfoil, giving an increased drag and a decreased lift. [ELECTROMAG] Rod grating mounted on a parabolic reflector to change the pencil-beam pattern of the reflector to a cosecant-squared pattern; rotating the reflector and grating 90° with respect to the feed antenna changes one pattern to the other.

sponge metal [MET] Any porous metal made by decomposition or reduction of a compound without melting.

Spongiidae [INV ZOO] A family of sponges of the order Dictyoceratida; members are encrusting, massive, or branching in form and have small spherical flagellated chambers which characteristically join the exhalant canals by way of narrow channels.

Spongillidae [INV ZOO] A family of fresh- and brackish water sponges in the order Haplosclerida which are chiefly gray, brown, or white in color, and encrusting, massive, or branching in form.

Spongiomorphida [PALEON] A small, extinct Mesozoic order of fossil colonial Hydrozoa in which the skeleton is a reticulum composed of perforate lamellae parallel to the upper surface and of regularly spaced vertical elements in the form of pillars.

Spongiomorphidae [PALEON] The single family of extinct hydrozoans comprising the order Spongiomorphida.

spontaneous combustion See spontaneous ignition.

spontaneous ignition [CHEM] Ignition which can occur when certain materials such as tung oil are stored in bulk, resulting from the generation of heat, which cannot be readily dissipated; often heat is generated by microbial action.

spore [BIOL] A uni- or multicellular, asexual, reproductive or resting body that is resistant to unfavorable environmental conditions and produces a new vegetative individual when the environment is favorable.

Sporobolomycetaceae [MYCOL] The single family of the order Sporobolomycetales.

Sporobolomycetales [MYCOL] An order of yeastlike and moldlike fungi assigned to the class Basidiomycetes characterized by the formation of sterigmata, upon which the asexual ballistospores are formed.

Sporocytophaga [MICROBIO] A genus of bacteria in the family Cytophagaceae; cells are motile rods with rounded ends; microcysts are formed.

sporophyte [BOT] 1. An individual of the spore-bearing generation in plants exhibiting alternation of generation. 2. The spore-producing generation. 3. The diplophase in a plant life cycle.

Sporozoa [INV ZOO] A subphylum of parasitic Protozoa, typically producing spores during the asexual stages of the life cycle.

spot blight *See* grease spot.

spot blotch [PL PATH] A fungus disease of barley caused by *Helminthosporium sativum* and characterized by the appearance of dark, elongated spots on the foliage.

spotted wilt [PL PATH] A virus disease of various crop and wild plants, especially tomato, characterized by bronzing and downward curling of the leaves.

spot test [ANALY CHEM] The addition of a drop of reagent to a drop or two of sample solution to obtain distinctive colors or precipitates; used in qualitative analysis.

spot welding [MET] Resistance welding in which fusion is localized in small circular areas; sometimes also accomplished by various arc-welding processes.

spray dryer [MECH ENG] A machine for drying an atomized mist by direct contact with hot gases.

spray nozzle [MECH ENG] A device in which a liquid is subdivided into a stream (mist) of small drops.

spray tower [CHEM ENG] A vertical column, at the top of which is a liquid spray device; used to contact liquids with gas streams for absorption, humidification, or drying.

spread [ENG] **1.** The layout of geophone groups from which data from a single shot are recorded simultaneously. **2.** *See* sensitivity. [STAT] The range within which the values of a variable quantity occur.

spreader stoker [MECH ENG] A coal-burning system where mechanical feeders and distributing devices form a thin fuel bed on a traveling grate, intermittent-cleaning dump grate, or reciprocating continuous-cleaning grate.

spreading coefficient [THERMO] The work done in spreading one liquid over a unit area of another, equal to the surface tension of the stationary liquid, minus the surface tension of the spreading liquid, minus the interfacial tension between the liquids.

spreading factor *See* hyaluronidase.

spread spectrum transmission [ELECTR] Communications technique in which many different signal waveforms are transmitted in a wide band; power is spread thinly over the band so narrow-band radios can operate within the wide-band without interference; used to achieve security and privacy, prevent jamming, and utilize signals buried in noise.

spring [ASTRON] The period extending from the vernal equinox to the summer solstice; comprises the transition period from winter to summer. [ENG] To enlarge the bottom of a drill hole by small charges of a high explosive in order to make room for the full charge; to chamber a drill hole. [HYD] A general name for any discharge of deep-seated, hot or cold, pure or mineralized water. [MECH ENG] An elastic, stressed, stored-energy machine element that, when released, will recover its basic form or position. Also known as mechanical spring.

springback [MET] **1.** Return of a metal part to its original shape after release of stress. **2.** The degree to which a metal returns to its original shape after forming operations. **3.** In flash, upset or pressure welding, the deflection in the welding machine caused by the upset pressure.

spring balance [ENG] An instrument which measures force by determining the extension of a helical spring.

spring bolt [DES ENG] A bolt which must be retracted by pressure and which is shot into place by a spring when the pressure is released.

spring equinox *See* vernal equinox.

spring gravimeter [ENG] An instrument for making relative measurements of gravity; the elongation *s* of the spring may be considered proportional to gravity *g*, $s = (1/k)g$, and the basic formula for relative measurements is

$$g_2 - g_1 = k(s_2 - s_1).$$

spring modulus [MECH] The additional force necessary to deflect a spring an additional unit distance; if a certain spring has a modulus of 100 newtons per centimeter, a 100-newton weight will compress it 1 centimeter, a 200-newton weight 2 centimeters, and so on.

spring range [OCEANOGR] The mean semidiurnal range of tide when spring tides are occurring; the mean difference in height between spring high water and spring low water. Also known as mean spring range.

sprocket [DES ENG] A tooth on the periphery of a wheel or cylinder to engage in the links of a chain, the perforations of a motion picture film, or other similar device.

sprocket chain [MECH ENG] A continuous chain which meshes with the teeth of a sprocket and thus can transmit mechanical power from one sprocket to another.

sprocket pulse [COMPUT SCI] **1.** A pulse generated by a magnetized spot which accompanies every character recorded on magnetic tape; this pulse is used during read operations to regulate the timing of the read circuits, and also to provide a count on the number of characters read from the tape. **2.** A pulse generated by the sprocket or driving hole in paper tape which serves as the timing pulse for reading or punching the paper tape.

sprocket wheel [DES ENG] A wheel with teeth or cogs, used for a chain drive or to engage the blocks on a cable.

spruce [BOT] An evergreen tree belonging to the genus *Picea* characterized by single, four-sided needles borne on small peglike projections, pendulous cones, and resinous wood.

sprue [ENG] **1.** A feed opening or vertical channel through which molten material, such as metal or plastic, is poured in an injection or transfer mold. **2.** A slug of material that solidifies in the channel. [MED] A syndrome characterized by impaired absorption of food, water, and minerals by the small intestine; symptoms are the result of nutritional deficiencies.

sprung axle [MECH ENG] A supporting member for carrying the rear wheels of an automobile.

sprung weight [MECH ENG] The weight of a vehicle which is carried by the springs, including the frame, radiator, engine, clutch, transmission, body, load, and so forth.

SPST *See* single-pole single-throw.

spud [DES ENG] **1.** A diamond-point drill bit. **2.** An offset type of fishing tool used to clear a space around tools stuck in a borehole. **3.** Any of various spade- or chisel-shaped tools or mechanical devices. **4.** *See* grouser. [MIN ENG] A nail, resembling a horseshoe nail, with a hole in the head, driven into mine timbering or into a wooden plug inserted in the rock to mark a surveying station. [NAV ARCH] A foot piece to provide support for the legs of the A frame of a floating dipper dredge.

spur [BOT] **1.** A hollow process at the base of a petal or sepal. **2.** A short fruit-bearing tree branch. **3.** A short projecting root. [GEOL] A ridge or rise projecting from a larger elevational feature. [HYD] *See* ram. [MATH] *See* trace. [PHYS] A cluster of ionized molecules near the path of an energetic charged particle, consisting of the molecule ionized directly by the charged particle, and secondary ionizations produced by electrons released in the primary ionization; it usually forms a side track from the path of the particle. [ZOO] A stiff, sharp outgrowth, as on the legs of certain birds and insects.

spur blight [PL PATH] A fungus disease of raspberries and blackberries caused by *Didymella applanata* which kills the fruit spurs and causes dark spotting of the cane.

spur dike *See* groin.

spurious emission *See* spurious radiation.

spurious radiation [ELECTROMAG] Any emission from a radio transmitter at frequencies outside its frequency band. Also known as spurious emission.

spurious response [ELECTR] **1.** Response of a radio receiver to a frequency different from that to which the receiver is tuned. **2.** In electronic warfare, the undesirable signal images in the intercept receiver resulting from the mixing of the intercepted signal with harmonics of the local oscillators in the receiver.

spurrite [MINERAL] $Ca_5(SiO_4)_2(CO_3)$ A light-gray mineral occurring in granular masses.

sputtering [ELECTR] Also known as cathode sputtering. **1.** The ejection of atoms or groups of atoms from the surface of the cathode of a vacuum tube as the result of heavy-ion impact. **2.** The use of this process to deposit a thin layer of metal on a glass, plastic, metal, or other surface in vacuum.

sputter-ion pump *See* getter-ion pump.

sp vol *See* specific volume.

sq *See* square.

Squalidae [VERT ZOO] The spiny dogfishes, a family of squaloid elasmobranchs recognized by their well-developed fin spines.

squall [METEOROL] A strong wind with sudden onset and more gradual decline, lasting for several minutes; in the United States observational practice, a squall is reported only if a wind speed of 16 knots or higher (8.23 meters per second) is sustained for at least 2 minutes.

squall line [METEOROL] A line of thunderstorms near whose advancing edge squalls occur along an extensive front; the region of thunderstorms is typically 20 to 50 kilometers wide and a few hundred to 2000 kilometers long.

Squamata [VERT ZOO] An order of reptiles, composed of the lizards and snakes, distinguished by a highly modified skull that has only a single temporal opening, or none, by the lack of shells or secondary palates, and by possession of paired penes on the males.

squamosal bone [ANAT] The part of the temporal bone in man corresponding with the squamosal bone in lower vertebrates. [VERT ZOO] A membrane bone lying external and dorsal to the auditory capsule of many vertebrate skulls.

squamous epithelium [HISTOL] A single-layered epithelium consisting of thin, flat cells.

square [MATH] **1.** The square of a number r is the number r^2, that is, r times r. **2.** The plane figure with four equal sides and four interior right angles. [MECH] Denotes a unit of area; if x is a unit of length, a square x is the area of a square whose sides have a length of $1x$: for example, a square meter, or a meter squared, is the area of a square whose sides have a length of 1 meter. Abbreviated sq.

square degree [MATH] A unit of a solid angle equal to $(\pi/180)^2$ steradian, or approximately 3.04617×10^{-4} steradian.

square-foot unit of absorption *See* sabin.

square-jaw clutch [MECH ENG] A type of positive clutch consisting of two or more jaws of square section which mesh together when they are aligned.

squareness ratio [ELECTROMAG] **1.** The magnetic induction at zero magnetizing force divided by the maximum magnetic induction, in a symmetric cyclic magnetization of a material. **2.** The magnetic induction when the magnetizing force has changed halfway from zero toward its negative limiting value divided by the maximum magnetic induction in a symmetric cyclic magnetization of a material.

square-nose bit *See* flat-face bit.

square root [MATH] A square root of a real or complex number s is a number t for which $t^2 = s$.

square-root law [STAT] The standard deviation of the ratio of the number of successes to number of trials is inversely proportional to the square root of the number of trials.

square set [MIN ENG] A set of timbers composed of a cap, girt, and post which meet so as to form a solid 90° angle; they are so framed at the intersection as to form a compression joint, and join with three other similar sets.

square thread [DES ENG] A screw thread having a square cross section; the width of the thread is equal to the pitch or distance between threads.

square wave [ELEC] An oscillation the amplitude of which shows periodic discontinuities between two values, remaining constant between jumps.

squaring circuit [ELECTR] **1.** A circuit that reshapes a sine or other wave into a square wave. **2.** A circuit that contains nonlinear elements proportional to the square of the input voltage.

squash [BOT] Either of two plants of the genus *Cucurbita*, order Violales, cultivated for its fruit; some types are known as pumpkins.

Squatinidae [VERT ZOO] A group of squaloid elasmobranchs of uncertain affinity characterized by a greatly extended rostrum with enlarged denticles along the margins; maximum length is under 4 feet (1.2 meters).

squealing [ELECTR] A condition in which a radio receiver produces a high-pitched note or squeal along with the desired radio program, due to interference between stations or to oscillation in some receiver circuit.

squeeze [ENG] **1.** To inject a grout into a borehole under high pressure. **2.** The plastic movement of a soft rock in the walls of a borehole or mine working that reduced the diameter of the opening. [MIN ENG] **1.** The settling, without breaking, of the roof over a considerable area of working. Also known as creep; nip; pinch. **2.** The gradual upheaval of the floor of a mine due to the weight of the overlying strata. **3.** The sections in coal seams that have become constricted by the squeezing in of the overlying or underlying rock. [PHYS] Increasing external pressure upon the ears and sinuses in diving.

squeeze film [MATER] A thin viscoelastic fluid film squeezed between two usually planar structures to serve as a sealant, load damper, lubricant, and so on.

squegger *See* blocking oscillator.

squegging [ELECTR] Condition of self-blocking in an electron-tube-oscillator circuit.

squegging oscillator See blocking oscillator.

squelch [ELECTR] To automatically quiet a receiver by reducing its gain in response to a specified characteristic of the input.

squelch circuit See noise suppressor.

squid [INV ZOO] Any of a number of marine cephalopod mollusks characterized by a reduced internal shell, ten tentacles, an ink sac, and chromatophores.

SQUID See superconducting quantum interference device.

Squillidae [INV ZOO] The single family of the eumalacostracan order Stomatopoda, the mantis shrimp.

squint [ELECTROMAG] 1. The angle between the two major lobe axes in a radar lobe-switching antenna. 2. The angular difference between the axis of radar antenna radiation and a selected geometric axis, such as the axis of the reflector. 3. The angle between the full-right and full-left positions of the beam of a conical-scan radar antenna. [MED] See strabismus.

squirrel [VERT ZOO] Any of over 200 species of arboreal rodents of the families Sciuridae and Anomaluridae having a bushy tail and long, strong hind limbs.

squirrel-cage motor [ELEC] An induction motor in which the secondary circuit consists of a squirrel-cage winding arranged in slots in the iron core.

squirrel-cage rotor See squirrel-cage winding.

squirrel-cage winding [ELEC] A permanently short-circuited winding, usually uninsulated, around the periphery of the rotor and joined by continuous end rings. Also known as squirrel-cage rotor.

sr See steradian.

Sr See strontium.

SRAM See short-range attack missile.

SS See stainless steel.

SS Cygni stars See U Geminorum stars.

SSD See steady-state distribution.

SSI See small-scale integration.

SSM See surface-to-surface missile.

S star [ASTRON] A spectral classification of stars, comprising red stars with surface temperature of about 2200 K; prominent in the spectra is zirconium oxide.

s-state [QUANT MECH] A single-particle state whose orbital angular momentum quantum number is zero.

St See stoke.

stab culture [MICROBIO] A culture of anaerobic bacteria made by piercing a solid agar medium in a test tube with an inoculating needle covered with the bacterial inoculum.

stability [CHEM] The property of a chemical compound which is not readily decomposed and does not react with other compounds. [CONT SYS] The property of a system for which any bounded input signal results in a bounded output signal. [ENG] The property of a body, as an aircraft, rocket, or ship, to maintain its attitude or to resist displacement, and, if displaced, to develop forces and moments tending to restore the original condition. [FL MECH] The resistance to overturning or mixing in the water column, resulting from the presence of a positive (increasing downward) density gradient. [GEOL] 1. The resistance of a structure, spoil heap, or clay bank to sliding, overturning, or collapsing. 2. Chemical durability, resistance to weathering. [MATER] Of a fuel, the capability to retain its characteristics in an adverse environment, for example, extreme temperature. [MECH] See dynamic stability. [PHYS] 1. The property of a system which does not undergo any change without the application of an external agency. 2. The property of a system in which any departure from an equilibrium state gives rise to

forces or influences which tend to return the system to equilibrium. Also known as static stability. [PL PHYS] The property of a plasma which maintains its shape against externally applied forces (usually pressure of magnetic fields) and whose constituents can pass through confining fields only by diffusion of individual particles.

stabilization [CHEM ENG] A petroleum-refinery process for separating light gases from petroleum or gasoline, thus leaving a stable (less volatile) liquid so that it can be handled or stored with less change in composition. [CONT SYS] See compensation. [ELECTR] Feedback introduced into vacuum tube or transistor amplifier stages to reduce distortion by making the amplification substantially independent of electrode voltages and tube constants. [ELECTROMAG] Treatment of a magnetic material to improve the stability of its magnetic properties. [ENG] Maintenance of a desired orientation independent of the roll and pitch of a ship or aircraft.

stabilized feedback See negative feedback.

stabilizer [AERO ENG] Any airfoil or any combination of airfoils considered as a single unit, the primary function of which is to give stability to an aircraft or missile. [CHEM ENG] The fractionation column in a petroleum refinery used to stabilize (remove fractions from) hydrocarbon mixtures. [ENG] 1. A hardened, splined bushing, sometimes freely rotating, slightly larger than the outer diameter of a core barrel and mounted directly above the core barrel back head. Also known as ferrule; fluted coupling. 2. A tool located near the bit in the drilling assembly to modify the deviation angle in a well by controlling the location of the contact point between the hole and the drill collars. [MATER] 1. Any powdered or liquid additive used as an agent in soil stabilization. 2. An ingredient used in the formulation of some compounded plastics to maintain the physical and chemical properties at their initial values throughout the processing and service life of the material, for example, heat and RV stabilizers. [NAV ARCH] Any of the submerged fins used on ships to prevent rolling.

stabistor [ELECTR] A diode component having closely controlled conductance, controlled storage charge, and low leakage, as required for clippers, clamping circuits, bias regulators, and other logic circuits that require tight voltage-level tolerances.

stable equilibrium [SCI TECH] Equilibrium in which any departure from the equilibrium state gives rise to forces or influences which tend to return the system to equilibrium.

stable isotope [NUC PHYS] An isotope which does not spontaneously undergo radioactive decay.

stable nucleus [NUC PHYS] A nucleus which does not spontaneously undergo radioactive decay.

stack [BUILD] The portion of a chimney rising above the roof. [CHEM ENG] In gas works, a row of benches containing retorts. [COMPUT SCI] A portion of a computer memory used to temporarily hold information, organized as a linear list for which all insertions and deletions, and usually all accesses, are made at one end of the list. [ELECTR] See pileup. [ENG] 1. To stand and rack drill rods in a drill tripod or derrick. 2. Any structure or part thereof that contains a flue or flues for the discharge of gases. 3. One or more filter cartridges mounted on a single column. 4. Tall, vertical conduit (such as smokestack, flue) for venting of combustion or evaporation products or gaseous process wastes. 5. The exhaust pipe of an internal combustion engine. [GEOL] An erosional, coastal landform that is a steep-sided, pillarlike rocky island or mass that has been detached by wave action from a shore made up of cliffs; applies particularly to a stack that is columnar in structure and has horizontal stratifications. Also known as marine stack; rank. [MET] The cone-shaped section of a blast furnace or cupola above the hearth and melting zone and extending to the throat. [NAV] To assign different altitudes by radio to aircraft awaiting their turns to land at an airport.

stacked antennas [ELECTROMAG] Two or more identical antennas arranged above each other on a vertical supporting structure and connected in phase to increase the gain.

stacked array [ELECTROMAG] An array in which the antenna elements are stacked one above the other and connected in phase to increase the gain.

stacked-beam radar [ENG] Three-dimensional radar system that derives elevation by emitting narrow beams stacked vertically to cover a vertical segment, azimuth information from horizontal scanning of the beam, and range information from echo-return time.

stacked-dipole antenna [ELECTROMAG] Antenna in which directivity is increased by providing a number of identical dipole elements, excited either directly or parasitically; the resultant radiation pattern depends on the number of dipole elements used, the spacing and phase difference between the elements, and the relative magnitudes of the currents.

stacked loops [ELECTROMAG] Two or more loop antennas arranged above each other on a vertical supporting structure and connected in phase to increase the gain. Also known as vertically stacked loops.

stacker [COMPUT SCI] That part (or parts) of a punched-card handling device which arranges the processed cards into an orderly stack and holds them until they are removed by the operator. [MECH ENG] A machine for lifting merchandise on a platform or fork and arranging it in tiers; operated by hand, or electric or hydraulic mechanisms.

stacking fault [CRYSTAL] A defect in a face-centered cubic or hexagonal close-packed crystal in which there is a change from the regular sequence of positions of atomic planes.

stack operation [COMPUT SCI] A computer system in which flags, return address, and all temporary addresses are saved in the core in sequential order for any interrupted routine so that a new routine (including the interrupted routine) may be called in.

stadia [ENG] A surveying instrument consisting of a telescope with special horizontal parallel lines or wires, used in connection with a vertical graduated rod.

stadial moraine See recessional moraine.

Staffellidae [PALEON] An extinct family of marine protozoans (superfamily Fusulinacea) that persisted during the Pennsylvanian and Early Permian.

stage [AERO ENG] A self-propelled separable element of a rocket vehicle or spacecraft. [ELECTR] A circuit containing a single section of an electron tube or equivalent device or two or more similar sections connected in parallel, push-pull, or push-push; it includes all parts connected between the control-grid input terminal of the device and the input terminal of the next adjacent stage. [GEOL] 1. A developmental phase of an erosion cycle in which landscape features have distinctive characteristic forms. 2. A phase in the historical development of a geologic feature. 3. A major subdivision of a glacial epoch. 4. A time-stratigraphic unit ranking below series and above substage.

[HYD] The elevation of the water surface in a stream as measured by a river gage with reference to some arbitrarily selected zero datum. [MIN ENG] 1. A certain length of underground roadway worked by one horse. 2. A narrow thin dike, especially one where the material of which the dike is composed is soft. 3. A platform on which mine cars stand.

staggered tuning [ELECTR] Alignment of successive tuned circuits to slightly different frequencies in order to widen the overall amplitude-frequency response curve.

staggering [COMMUN] Offsetting of two channels of different carrier systems from exact side-band frequency coincidence to avoid mutual interference.

staghead See witches'-broom disease.

stagnation point [FL MECH] A point in a field of flow about a body where the fluid particles have zero velocity with respect to the body.

stagnation pressure See dynamic pressure.

stainierite See heterogenite.

stainless alloy [MET] Any of a large and complex group of corrosion-resistant iron-chromium alloys (containing 10% or more chromium), sometimes containing other elements, such as nickel, silicon, molybdenum, tungsten, and niobium. Commonly known as stainless steel (SS).

stainless iron See ferritic stainless steel.

stainless steel See stainless alloy.

stalactite [GEOL] A conical or roughly cylindrical speleothem formed by dripping water and hanging from the roof of a cave; usually composed of calcium carbonate.

stalagmite [GEOL] A conical speleothem formed upward from the floor of a cave by the action of dripping water; usually composed of calcium carbonate.

stall [AERO ENG] 1. The action or behavior of an airplane (or one of its airfoils) when by the separation of the airflow, as in the case of insufficient airspeed or of an excessive angle of attack, the airplane or airfoil tends to drop; the condition existing during this behavior. 2. A flight performance in which an airplane is made to lose flying speed and to drop by pointing the nose steeply upward. 3. An act or instance of stalling.

stamen [BOT] The male reproductive structure of a flower, consisting of an anther and a filament.

stamping mill [MIN ENG] A machine in which ore is finely crushed by descending pestles (stamps), usually operated by hydraulic power. Also known as crushing mill.

stand [ECOL] A group of plants, distinguishable from adjacent vegetation, which is generally uniform in species composition, age, and condition. [FOR] The amount of standing timber per unit area; usually expressed in terms of volume. [OCEANOGR] The interval at high or low water when there is no appreciable change in the height of the tide. Also known as tidal stand.

stand-alone machine [COMPUT SCI] A machine capable of functioning independently of a master computer, either part of the time or all of the time.

standard [PHYS] An accepted reference sample used for establishing a unit for the measurement of a physical quantity.

standard atmosphere [METEOROL] A hypothetical vertical distribution of atmospheric temperature, pressure, and density which is taken to be representative of the atmosphere for purposes of pressure altimeter calibrations, aircraft performance calculations, aircraft and missile design, and ballistic tables; the air is assumed to obey the perfect gas law and hydrostatic equation, which, taken together, relate temperature, pressure, and density variations in the vertical; it is further assumed that the air contains no water vapor, and that the acceleration of gravity does not change with height. [PHYS] *See* atmosphere.

standard broadcast band *See* broadcast band.

standard broadcasting [COMMUN] Broadcasting using amplitude modulation in the band of frequencies from 535 to 1605 kilohertz; carrier frequencies are placed 10 kilohertz apart.

standard calomel electrode [PHYS CHEM] A mercury-mercurous chloride electrode used as a reference (standard) measurement in polarographic determinations.

standard candle *See* international candle.

standard capacitor [ELEC] A capacitor constructed in such a manner that its capacitance value is not likely to vary with temperature and is known to a high degree of accuracy. Also known as capacitance standard.

standard cell [ELEC] A primary cell whose voltage is accurately known and remains sufficiently constant for instrument calibration purposes; the Weston standard cell has a voltage of 1.018636 volts at 20°C.

standard conditions [PHYS] 1. A temperature of 0°C and a pressure of 1 atmosphere (760 torr). Also known as normal temperature and pressure (NTP); standard temperature and pressure (STP). 2. According to the American Gas Association, a temperature of 60°F (15 5⁄9°C) and a pressure of 762 millimeters (30 inches) of mercury. 3. According to the Compressed Gas Institute, a temperature of 20°C (68°F) and a pressure of 1 atmosphere. [SOLID STATE] The allotropic form in which a substance most commonly occurs.

standard deviation [STAT] The positive square root of the expected value of the square of the difference between a random variable and its mean.

standard electrode potential [PHYS CHEM] The reversible or equilibrium potential of an electrode in an environment where reactants and products are at unit activity.

standard error [STAT] A measure of the variability any statistical constant would be expected to show in taking repeated random samples of a given size from the same universe of observations.

standard evaporator *See* short-tube vertical evaporator.

standard fit [DES ENG] A fit whose allowance and tolerance are standardized.

standard form [COMPUT SCI] The form of a floating point number whose mantissa lies within a standard specified range of values.

standard-frequency signal [COMMUN] One of the highly accurate signals broadcast by government radio stations and used for testing and calibrating radio equipment all over the world; in the United States signals are broadcast by the National Bureau of Standards' radio stations WWV, WWVH, WWVB, and WWVL.

standard interface [COMPUT SCI] 1. A joining place of two systems or subsystems that has a previously agreed-upon form, so that two systems may be readily connected together. 2. In particular, a system of uniform circuits and input/output channels connecting the central processing unit of a computer with various units of peripheral equipment.

standardization [DES ENG] The adoption of generally accepted uniform procedures, dimensions, materials, or parts that directly affect the design of a product or a facility. [ENG] The process of establishing by common agreement engineering criteria,

terms, principles, practices, materials, items, processes, and equipment parts and components.

standardize [COMPUT SCI] To replace any given floating point representation of a number with its representation in standard form; that is, to adjust the exponent and fixed-point part so that the new fixed-point part lies within a prescribed standard range.

standardized units [STAT] A random variable Z has been reduced to standardized units when it has zero expected value and standard deviation 1; this is accomplished by dividing the difference of Z and the expected value of Z by the standard deviation of Z.

standard load [DES ENG] A load which has been preplanned as to dimensions, weight, and balance, and designated by a number or some classification.

standard meridian [GEOD] The meridian used for reckoning standard time; throughout most of the world the standard meridians are those whose longitudes are exactly divisible by 15°. [MAP] A meridian of a map projection along which the scale is as stated.

standard mineral [MINERAL] A mineral that, on the basis of chemical analyses, is theoretically capable of being present in a rock. Also known as normative mineral.

standard parallel [MAP] A parallel on a map or chart along which the scale is as stated for that map or chart. [GEOD] The parallel or parallels of latitude used as control lines in the computation of a map projection.

standard pitch [ACOUS] A musical pitch based on 440 hertz for tone A; with this standard, the frequency of middle C is 261 hertz.

standard pressure [METEOROL] The arbitrarily selected atmospheric pressure of 1000 millibars to which adiabatic processes are referred for definitions of potential temperature, equivalent potential temperature, and so on. [PHYS] A pressure of 1 atmosphere (101,325 newtons per square meter), to which measurements of quantities dependent on pressure, such as the volume of a gas, are often referred. Also known as normal pressure.

standard state [PHYS] The stable and pure form of a substance at standard pressure and ordinary temperature.

standard temperature and pressure *See* standard conditions.

standard time [ASTRON] The mean solar time, based on the transit of the sun over a specified meridian, called the time meridian, and adopted for use over an area that is called a time zone. [IND ENG] A unit time value for completion of a work task as determined by the proper application of the appropriate work-measurement techniques. Also known as direct labor standard; output standard; production standard; time standard.

standard ton *See* ton.

standard visibility *See* meteorological range.

standard visual range *See* meteorological range.

standard volume [PHYS] The volume of 1 mole of a gas at a pressure of 1 atmosphere and a temperature of 0°C. Also known as normal volume.

standard waveguide [ELECTROMAG] Any one of several rectangular waveguides whose dimensions have been specified by various agencies and which are in general use.

standby computer [COMPUT SCI] A computer in a duplex system that takes over when the need arises.

standby register [COMPUT SCI] In computers, a register into which information can be copied to be available in case the original information is lost or mutilated in processing.

standby time [COMPUT SCI] 1. The time during which two or more computers are tied together and available to answer inquiries or process intermittent actions on stored data. 2. The elapsed time between inquiries when the equipment is operating on an inquiry application.

standing-on-nines carry [COMPUT SCI] In high-speed parallel addition of decimal numbers, an arrangement that causes carry digits to pass through one or more nine digits, while signaling that the skipped nines are to be reset to zero.

standing rigging [NAV ARCH] Rigging that is permanently secured such as shrouds, stays, bob-stays, martingales, and mast pendants.

standing valve [PETRO ENG] A sucker-rod-pump (oil well) discharge valve that remains stationary during the pumping cycle, in contrast to a traveling valve.

standing wave [PHYS] A wave in which the ratio of an instantaneous value at one point to that at any other point does not vary with time. Also known as stationary wave.

standing-wave detector [ELECTROMAG] An electric indicating instrument used for detecting a standing electromagnetic wave along a transmission line or in a waveguide and measuring the resulting standing-wave ratio; it can also be used to measure the wavelength, and hence the frequency, of the wave. Also known as standing-wave indicator; standing-wave meter; standing-wave-ratio meter.

standing-wave indicator *See* standing-wave detector.

standing-wave meter *See* standing-wave detector.

standing-wave-ratio meter *See* standing-wave detector.

standpipe [ENG] A vertical tube filled with a material, for example, water, or in petroleum refinery catalytic cracking, a catalyst to serve as a seal between high- and low-pressure parts of the process equipment.

stannite [MINERAL] Cu_2FeSnS_4 A steel-gray or iron-black mineral crystallizing in the tetragonal system and occurring in granular masses; luster is metallic, hardness is 4 on Mohs scale, and specific gravity is 4.3–4.53. Also known as bell-metal ore; tin pyrites.

stapes [ANAT] The stirrup-shaped middle-ear ossicle, articulating with the incus and the oval window. Also known as columella.

Staphylinidae [INV ZOO] The rove beetles, a very large family of coleopteran insects in the superfamily Staphylinoidea.

Staphylinoidea [INV ZOO] A superfamily of Coleoptera in the suborder Polyphaga.

star [ASTRON] A celestial body consisting of a large, self-luminous mass of hot gas held together by its own gravity; the sun is a typical star. [NUCLEO] A star-shaped group of tracks made by ionizing particles originating at a common point in a nuclear emulsion, cloud chamber, or bubble chamber; some stars are produced by successive disintegrations of an atom in a radioactive series, and others by nuclear reactions of the spallation type, such as those due to cosmic-ray particles. Also known as nuclear star.

star catalog [ASTRON] A comprehensive tabulation of data concerning the stars listed; the data may include, for example, apparent positions, brightness, motions, parallaxes, and other properties of stars.

starch [BIOCHEM] Any one of a group of carbohydrates or polysaccharides, of the general composition $(C_6H_{10}O_5)_n$, occurring as organized or structural granules of varying size and markings in many plant cells; it hydrolyzes to several forms of dextrin and glucose; its chemical structure is not completely known, but the granules consist of concentric shells containing at least two fractions: an inner portion called amylose, and an outer portion called amylopectin.

star cluster [ASTRON] A group of stars held together by gravitational attraction; the two chief types are open clusters (composed of from 12 to hundreds of stars) and globular clusters (composed of thousands to hundreds of thousands of stars).

star drift [ASTRON] A description of two star groups in the Milky Way traveling through each other in opposite directions; individual stars have movements that are relative to each other. Also known as star stream.

starfish [INV ZOO] The common name for echinoderms belonging to the subclass Asteroidea.

Stark effect [SPECT] The effect on spectrum lines of an electric field which is either externally applied or is an internal field caused by the presence of neighboring ions or atoms in a gas, liquid, or solid. Also known as electric field effect.

star stream *See* star drift.

start bit [COMPUT SCI] The first bit transmitted in asynchronous data transmission to unequivocally indicate the start of the word.

star telescope [OPTICS] An accessory of the marine navigational sextant designed primarily for star observations; it has a large objective to give a greater field of view and increased illumination; it is an erect telescope, that is, the object viewed is seen erect as opposed to the inverting telescope in which the object viewed is inverted.

starter [ELEC] 1. A device used to start an electric motor and to accelerate the motor to normal speed. 2. *See* engine starter. [ELECTR] An auxiliary control electrode used in a gas tube to establish sufficient ionization to reduce the anode breakdown voltage. Also known as trigger electrode. [ENG] A drill used for making the upper part of a hole, the remainder of the hole being made with a drill of smaller gage, known as a follower. [MICROBIO] A culture of microorganisms, either pure or mixed, used to commence a process, for example, cheese manufacture.

starting friction *See* static friction.

starting motor *See* engine starter.

star tracker *See* astrotracker.

start-stop multivibrator *See* monostable multivibrator.

stat- [ELEC] A prefix indicating an electrical unit in the electrostatic centimeter-gram-second system of units; it is attached to the corresponding SI unit.

statC *See* statcoulomb.

statcoulomb [ELEC] The unit of charge in the electrostatic centimeter-gram-second system of units, equal to the charge which exerts a force of 1 dyne on an equal charge at a distance of 1 centimeter in a vacuum; equal to approximately 3.3356×10^{-10} coulomb. Abbreviated statC. Also known as franklin (Fr); unit charge.

state [CONT SYS] A minimum set of numbers which contain enough information bout a system's history to enable its future behavior to be computed. [PHYS] The condition of a system which is specified as completely as possible by observations of a specified nature, for example, thermodynamic state, energy state. [QUANT MECH] The condition in which a system exists; the state may be pure and describable by a wave function or mixed and describable by a density matrix.

state equations [CONT SYS] Equations which express the state of a system and the output of a system at any time as a single valued function of the system's input at the same time and the state of the system at some fixed initial time.

state feedback [CONT SYS] A class of feedback control laws in which the control inputs are explicit memoryless functions of the dynamical system state, that is, the control inputs at a given time t_a are determined by the values of the state variables at t_a and do not depend on the values of these variables at earlier times $t \geqslant t_a$.

statement [COMPUT SCI] An elementary specification of a computer action or process, complete and not divisible into smaller meaningful units; it is analogous to the simple sentence of a natural language.

statement editor [COMPUT SCI] A text editor in which the text is divided into superlines, that is, units greater than ordinary lines, resulting in easier editing and freedom from truncation problems.

state parameter *See* thermodynamic function of state.

state variable [CONT SYS] One of a minimum set of numbers which contain enough information about a system's history to enable computation of its future behavior. [THERMO] *See* thermodynamic function of state.

statF *See* statfarad.

statfarad [ELEC] Unit of capacitance in the electrostatic centimeter-gram-second system of units, equal to the capacitance of a capacitor having a charge of 1 statcoulomb, across the plates of which the charge is 1 statvolt; equal to approximately 1.1126×10^{-12} farad. Abbreviated statF. Also known as centimeter.

static [COMMUN] A hissing, crackling, or other sudden sharp sound that tends to interfere with the reception, utilization, or enjoyment of desired signals or sounds. [PHYS] Without motion or change.

static bed [CHEM ENG] Refers to a layer of solids in a process vessel (absorber, catalytic reactor, packed distillation column, or granular filter bed) in which the particles rest upon one another at essentially the settled bulk density of the solids phase; contrasted to moving-solids or fluidized-solids beds.

static breeze *See* convective discharge.

static charge [ELEC] An electric charge accumulated on an object.

static check [COMPUT SCI] Of a computer, one or more tests of computing elements, their interconnections, or both, performed under static conditions.

static debugging routine [COMPUT SCI] A debugging routine which is used after the program being checked has been run and has stopped.

static dump [COMPUT SCI] An edited printout of the contents of main memory or of the auxiliary storage, performed in a fixed way; it is usually taken at the end of a program run either automatically or by operator intervention.

static electricity [ELEC] 1. The study of the effects of macroscopic charges, including the transfer of a static charge from one object to another by actual contact or by means of a spark that bridges an air gap between the objects. 2. *See* electrostatics.

static equilibrium *See* equilibrium.

static friction [MECH] 1. The force that resists the initiation of sliding motion of one body over the other with which it is in contact. 2. The force required to move one of the bodies when they are at rest. Also known as starting friction.

static induction transistor [ELECTR] A type of transistor capable of operating at high current and voltage, whose current-voltage characteristics do not saturate, and are similar in form to those of a vacuum triode. Abbreviated SIT.

staticize [COMPUT SCI] 1. To capture transient data in stable form, thus converting fleeting events into examinable information. 2. To extract an instruction from the main computer memory and store the various component parts of it in the appropriate registers, preparatory to interpreting and executing it.

static load [MECH] A nonvarying load; the basal pressure exerted by the weight of a mass at rest, such as the load imposed on a drill bit by the weight of the drill-stem equipment or the pressure exerted on the rocks around an underground opening by the weight of the superimposed rocks. Also known as dead load.

static magnetism *See* magnetostatics.

static moment [MECH] 1. A scalar quantity (such as area or mass) multiplied by the perpendicular distance from a point connected with the quantity (such as the centroid of the area or the center of mass) to a reference axis. 2. The magnitude of some vector (such as force, momentum, or a directed line segment) multiplied by the length of a perpendicular dropped from the line of action of the vector to a reference point.

static pressure [ACOUS] The pressure that would exist at a point in a medium if no sound waves were present. [FL MECH] 1. The normal component of stress, the force per unit area, exerted across a surface moving with a fluid, especially across a surface which lies in the direction of fluid flow. 2. The average of the normal components of stress exerted across three mutually perpendicular surfaces moving with a fluid.

static random-access memory [COMPUT SCI] A read-write random-access memory whose storage cells are made up of four or six transistors forming flip-flop elements that indefinitely remain in a given state until the information is intentionally changed, or the power to the memory circuit is shut off.

statics [MECH] The branch of mechanics which treats of force and force systems abstracted from matter, and of forces which act on bodies in equilibrium.

static seal *See* gasket.

static stability [METEOROL] The stability of an atmosphere in hydrostatic equilibrium with respect to vertical displacements, usually considered by the parcel method. Also known as convectional stability; hydrostatic stability; vertical stability. [PHYS] *See* stability.

static storage [COMPUT SCI] Computer storage such that information is fixed in space and available at any time, as in flip-flop circuits, electrostatic memories, and coincident-current magnetic-core storage.

static subroutine [COMPUT SCI] In computers, a subroutine which involves no parameters other than the addresses of the operands.

static switching [ELEC] Switching of circuits by means of magnetic amplifiers, semiconductors, and other devices that have no moving parts.

station [COMPUT SCI] One of a series of essentially similar positions or facilities occurring in a data-processing system. [COMMUN] *See* broadcast station. [ELEC] An assembly line or assembly machine location at which a wiring board or chassis is stopped for insertion of one or more parts. [ELECTR] A location at which radio, television, radar, or other electric equipment is installed. [ENG] Any predetermined point or area on the seas or oceans which is patrolled by naval vessels. [MIN ENG] 1. An enlargement in a mining shaft or gallery on any level used for a landing at any desired place and also for receiving loaded mine cars that are to be sent to the surface. **2.** An opening into a level which heads out of the side of an inclined plane; the point at which a surveying instrument is planted or observations are made. [SCI TECH] A geographic location at which scientific observations and measurements are made.

stationary field [PHYS] Field which does not change during the time interval under consideration. Also known as constant field.

stationary front *See* quasi-stationary front.

stationary noise [ELECTR] A random noise for which the probability that the noise voltage lies within any given interval does not change with time.

stationary orbit [AERO ENG] A circular, equatorial orbit in which the satellite revolves about the primary body at the angular rate at which the primary body rotates on its axis; from the primary body, the satellite thus appears to be stationary over a point on the primary body; a stationary orbit must be synchronous, but the reverse need not be true.

stationary point [ASTRON] A point at which a planet's apparent motion changes from direct to retrograde motion, or vice versa.

stationary satellite [AERO ENG] A satellite in a stationary orbit.

stationary state *See* energy state.

stationary time principle *See* Fermat's principle.

stationary wave *See* standing wave.

station buoy [NAV] A buoy used to mark the approximate station (position) of an important buoy or lightship should it be carried away or temporarily removed. Also known as marker buoy; watch buoy.

statistic [STAT] An estimate or piece of data, concerning some parameter, obtained from a sampling.

statistical analysis [STAT] The body of techniques used in statistical inference concerning a population.

statistical distribution *See* distribution of a random variable.

statistical forecast [METEOROL] A weather forecast based upon a systematic statistical examination of the past behavior of the atmosphere, as distinguished from a forecast based upon thermodynamic and hydrodynamic considerations.

statistical hypothesis [STAT] A statement about the way a random variable is distributed.

statistical mechanics [PHYS] That branch of physics which endeavors to explain and predict the macroscopic properties and behavior of a system on the basis of the known characteristics and interactions of the microscopic constituents of the system, usually when the number of such constituents is very large. Also known as statistical thermodynamics.

statistical thermodynamics *See* statistical mechanics.

statistics [MATH] A discipline dealing with methods of obtaining data, analyzing and summarizing it, and drawing inferences from data samples by the use of probability theory.

statolith [BOT] A sand grain or other solid inclusion which moves readily in the fluid contents of a statocyst, comes to rest on the lower surface of the cell, and is believed to function in gravity perception. [INV ZOO] A secreted calcareous body, a sand grain, or other solid inclusion contained in a statocyst.

stator [ELEC] The portion of a rotating machine that contains the stationary parts of the magnetic circuit and their associated windings. [MECH ENG] A stationary machine part in or about which a rotor turns.

stator armature [ELEC] A stator which includes the main current-carrying winding in which electromotive force produced by magnetic flux rotation is induced; it is found in most alternating-current machines.

statoreceptor [PHYSIO] A sense organ concerned primarily with equilibrium.

stator plate [ELEC] One of the fixed plates in a variable capacitor; stator plates are generally insulated from the frame of the capacitor.

status word [COMPUT SCI] A word indicating the state of the system or the diagnosis of a state into which the system has entered.

statute mile *See* mile.

staurolite [MINERAL] $FeAl_4(SiO_4)_2(OH)_2$ A reddish-brown to black neosilicate mineral that crystallizes in the orthorhombic system, has resinous to vitreous luster, hardness is 7–7.5 on Mohs scale, and specific gravity is 3.7. Also known as cross-stone; fairy stone; grenatite; staurotide.

Stauromedusae [INV ZOO] An order of the class Scyphozoa in which the medusa is composed of a cuplike bell called a calyx and a stem that terminates in a pedal disk.

Staurosporae [MYCOL] A spore group of the Fungi Imperfecti characterized by star-shaped or forked spores.

staurotide *See* staurolite.

stayed-cable bridge [CIV ENG] A modified cantilever bridge consisting of girders or trusses cantilevered both ways from a central tower and supported by inclined cables attached to the tower at the top or sometimes at several levels.

steadiness *See* persistence.

steady flow [FL MECH] Fluid flow in which all the conditions at any one point are constant with respect to time.

steady state [PHYS] The condition of a body or system in which the conditions at each point do not change with time, that is after initial transients or fluctuations have disappeared.

steady-state cavitation *See* sheet cavitation.

steady-state conduction [THERMO] Heat conduction in which the temperature and heat flow at each point does not change with time.

steady-state creep *See* secondary creep.

steady-state distribution [ANALY CHEM] The equilibrium condition between phases in each step of a multistage, countercurrent liquid-liquid extraction. Abbreviated SSD.

steady-state error [CONT SYS] The error that remains after transient conditions have disappeared in a control system.

steady-state reactor [NUCLEO] A reactor in which conditions such as temperature, reaction rate, and neutron flux do not change appreciably with time.

steady-state wave motion [PHYS] Wave motion in which the wave quantities at each point in the region through which the wave is passing repeat themselves periodically.

steam accumulator [MECH ENG] A pressure vessel in which water is heated by steam during off-peak demand periods and regenerated as steam when needed.

steam boiler [MECH ENG] A pressurized system in which water is vaporized to steam by heat transferred from a source of higher temperature, usually the products of combustion from burning fuels. Also known as steam generator.

steam condenser [MECH ENG] A device to maintain vacuum conditions on the exhaust of a steam prime mover by transfer of heat to circulating water or air at the lowest ambient temperature.

steam cycle *See* Rankine cycle.

steam distillation [CHEM ENG] A distillation in which vaporization of the volatile constituents of a liquid mixture takes place at a lower temperature by the introduction of steam directly into the charge; steam used in this manner is known as open steam. Also known as steam stripping.

steam engine [MECH ENG] A thermodynamic device for the conversion of heat in steam into work, generally in the form of a positive displacement, piston and cylinder mechanism.

steam fog [METEOROL] Fog formed when water vapor is added to air which is much colder than the vapor's source; most commonly, when very cold air drifts across relatively warm water. Also known as frost smoke; sea mist; sea smoke; steam mist; water smoke.

steam generator *See* steam boiler.

steam jacket [MECH ENG] A casing applied to the cylinders and heads of a steam engine, or other space, to keep the surfaces hot and dry.

steam-jet cycle [MECH ENG] A refrigeration cycle in which water is used as the refrigerant; high-velocity steam jets provide a high vacuum in the evaporator, causing the water to boil at low temperature and at the same time compressing the flashed vapor up to the condenser pressure level.

steam mist *See* steam fog.

steam point [THERMO] The boiling point of pure water whose isotopic composition is the same as that of sea water at standard atmospheric pressure; it is assigned a value of 100°C on the International Practical Temperature Scale of 1968.

steam refining [CHEM ENG] A petroleum refinery distillation process, in which the only heat used comes from steam in open and closed coils near the bottom of the still; used to produce gasoline and naphthas where odor and color are of prime importance; where open steam is used, it is known as steam distillation.

steam stripping *See* steam distillation.

steam trap [MECH ENG] A device which drains and removes condensate automatically from steam lines.

steam turbine [MECH ENG] A prime mover for the conversion of heat energy of steam into work on a revolving shaft, utilizing fluid acceleration principles in jet and vane machinery.

steatite [PETR] A compact, massive, fine-ground rock composed principally of talc, but with much other material.

Steatornithidae [VERT ZOO] A family of birds in the order Caprimulgiformes which contains a single, South American species, the oilbird or guacharo (*Steatornis caripensis*).

steel [MET] An iron base alloy, malleable under proper conditions, containing up to about 2% carbon.

steel jack [MIN ENG] A screw jack suitable in mechanical mining; used for legs or upright timbers. [MINERAL] *See* sphalerite.

Stefan-Boltzmann constant [STAT MECH] The energy radiated by a blackbody per unit area per unit time divided by the fourth power of the body's temperature; equal to $(5.6696 \pm 0.0010) \times 10^{-8}$ (watts)(meter)$^{-2}$(degrees Kelvin)$^{-4}$.

Stefan-Boltzmann law [STAT MECH] The total energy radiated from a blackbody is proportional to the fourth power of the temperature of the body. Also known as fourth-power law; Stefan's law of radiation.

Stefan's law of radiation *See* Stefan-Boltzmann law.

Steganopodes [VERT ZOO] Formerly, an order of birds that included the totipalmate swimming birds.

Stegodontinae [PALEON] An extinct subfamily of elephantoid proboscideans in the family Elephantidae.

Stegosauria [PALEON] A suborder of extinct reptiles of the order Ornithischia comprising the plated dinosaurs of the Jurassic which had tiny heads, great triangular plates arranged on the back in two alternating rows, and long spikes near the end of the tail.

steigerite [MINERAL] $4AlVO_4 \cdot 13H_2O$ A canary-yellow mineral composed of hydrous aluminum vanadate and occurring in masses.

Steinheim man [PALEON] A prehistoric man represented by a skull, without mandible, found near Stuttgart, Germany; the browridges are massive, the face is relatively small, and the braincase is similar in shape to that of *Homo sapiens*.

Steinmetz's law [ELECTROMAG] The energy converted into heat per unit volume per cycle during a cyclic change of magnetization is proportional to the maximum magnetic induction raised to the 1.6 power, the constant of proportionality depending only on the material.

stele [BOT] The part of a plant stem including all tissues and regions of plants from the cortex inward, including the pericycle, phloem, cambium, xylem, and pith.

Stelenchopidae [INV ZOO] A family of polychaete annelids belonging to the Myzostomaria, represented by a single species from Crozet Island in the Antarctic Ocean.

stellar [ASTRON] Relating to or consisting of stars.

stellarator [PL PHYS] A device for confining a high-temperature plasma, consisting of a tube, which closes in on itself in a figure-eight or race-track configuration, and external coils which generate magnetic fields whose lines of force run parallel to the walls of the tube and prevent the plasma from touching the walls.

stellar corona [ASTRON] An ionized region about a star formed by x-rays emitted during stellar flares.

stellar interferometer [OPTICS] An optical interferometer for measuring angular diameters of stars; it is attached to a telescope and measures interference rings at the telescope's focus.

stellar magnitude *See* magnitude.

stellar parallax [ASTRON] The subtended angle at a star formed by the mean radius of the earth's orbit; it indicates distance to the star.

stellar population [ASTRON] Either of two classes of stars, termed population I and population II; population I are relatively young stars, found in the arms of spiral galaxies, especially the blue stars of high luminosity; population II stars are the much older, more evolved stars of lower metallic content; many high luminosity red giants and many variable stars are members of population II.

stellar pulsation [ASTROPHYS] Expansion of a star followed by contraction so that its surface temperature and intrinsic brightness undergo periodic variation.

stellar spectrum [ASTRON] The spectrum of a star normally obtained with a slit spectrograph by black-and-white photography; the spectrum of a star in a large majority of cases shows absorption lines superposed on a continuous background.

Stelleroidea [INV ZOO] The single class of echinoderms in the subphylum Asterozoa; characters coincide with those of the subphylum.

stem [BOT] The organ of vascular plants that usually develops branches and bears leaves and flowers. [ENG] **1.** The heavy iron rod acting as the connecting link between the bit and the balance of the string of tools on a churn rod. **2.** To insert packing or tamping material in a shothole. [NAV] To make headway against an obstacle, as a current. [NAV ARCH] The foremost part of a ship's hull.

stem blight [PL PATH] Any of various fungus blights that affect the plant stem.

stem canker [PL PATH] Any canker disease affecting the stem.

stem correction [THERMO] A correction which must be made in reading a thermometer in which part of the stem, and the thermometric fluid within it, is at a temperature which differs from the temperature being measured.

Stemonitaceae [MYCOL] The single family of the order Stemonitales.

Stemonitales [MYCOL] An order of fungi in the subclass Myxogastromycetidae of the class Myxomycetes.

stem rust [PL PATH] Any of several fungus diseases, especially of grasses, affecting the stem and marked by black or reddish-brown lesions.

stencil printing *See* screen printing.

Stenetrioidea [INV ZOO] A group of isopod crustaceans in the suborder Asellota consisting mostly of tropical marine forms in which the first pleopods are fused.

Stenocephalidae [INV ZOO] A family of Old World, neotropical Hemiptera included in the Pentatomorpha.

Stenolaemata [INV ZOO] A class of marine ectoproct bryozoans having lophophores which are circular in basal outline and zooecia which are long, slender, tubular or prismatic, and gradually tapering to their proximal ends.

Stenomasteridae [PALEON] An extinct family of Euchinoidea in the order Holasteroida comprising oval and heart-shaped forms with fully developed pore pairs.

Stenopodidea [INV ZOO] A section of decapod crustaceans in the suborder Natantia which includes shrimps having the third pereiopods chelate and much longer and stouter than the first pair.

stenosis [MED] Constriction or narrowing, as of the heart or blood vessels.

Stenostomata [INV ZOO] The equivalent name for Cyclostomata.

Stenothoidae [INV ZOO] A family of amphipod crustaceans in the suborder Gammaridea containing semiparasitic and commensal species.

Stensioellidae [PALEON] A family of Lower Devonian placoderms of the order Petalichthyida having large pectoral fins and a broad subterminal mouth.

Stenurida [PALEON] An order of Ophiuroidea, comprising the most primitive brittlestars, known only from Paleozoic sediments.

step [COMPUT SCI] A single computer instruction or operation. [ENG] A small offset on a piece of core or in a drill hole resulting from a sudden sidewise deviation of the bit as it enters a hard, tilted stratum or rock underlying a softer rock. [GEOL] A hitch or dislocation of the strata. [MIN ENG] The portion of a longwall face at right angles to the line of the face formed when a place is worked in front of or behind an adjoining place. [ORG CHEM] *See* elementary reaction.

step aeration [CIV ENG] An activated sludge process in which the settled sewage is introduced into the aeration tank at more than one point.

step attenuator [ELECTR] An attenuator in which the attenuation can be varied in precisely known steps by means of switches.

step bearing [MECH ENG] A device which supports the bottom end of a vertical shaft. Also known as pivot bearing.

step block [ENG] A metal block, usually of steel or cast iron, with integral stepped sections to allow application of clamps when securing a workpiece to a machine tool table.

step-by-step operation *See* single-step operation.

step counter [COMPUT SCI] In computers, a counter in the arithmetic unit used to count the steps in multiplication, division, and shift operations.

step-down transformer [ELEC] A transformer in which the alternating-current voltages of the secondary windings are lower than those applied to the primary winding.

step fault [GEOL] One of a set of closely spaced, parallel faults. Also known as distributive fault; multiple fault.

step-function generator [ELECTR] A function generator whose output waveform increases and decreases suddenly in steps that may or may not be equal in amplitude.

Stephanidae [INV ZOO] A small family of the Hymenoptera in the superfamily Ichneumonoidea characterized by many-segmented filamentous antennae.

stephanite [MINERAL] Ag_5SbS_4 An iron-black mineral crystallizing in the orthorhombic system and having a metallic luster; an ore of silver. Also known as black silver; brittle silver ore; goldschmidtine.

step lake *See* paternoster lake.

steppe [GEOGR] An extensive grassland in the semiarid climates of southeastern Europe and Asia; it is similar to but more arid than the prairie of the United States.

stepper motor [ELEC] A motor that rotates in short and essentially uniform angular movements rather than continuously; typical steps are 30, 45, and 90°; the angular steps are obtained electromagnetically rather than by the ratchet and pawl mechanisms of stepping relays. Also known as magnetic stepping motor; stepping motor; step-servo motor.

stepping *See* zoning.

stepping motor *See* stepper motor.

stepping relay [ELEC] A relay whose contact arm may rotate through 360° but not in one operation. Also known as rotary stepping relay; rotary stepping switch; stepping switch.

stepping switch *See* stepping relay.

step rocket *See* multistage rocket.

step-servo motor *See* stepper motor.

step-up transformer [ELEC] Transformer in which the energy transfer is from a low-voltage winding to a high-voltage winding or windings.

step voltage regulator [ELEC] A type of voltage regulator used on distribution feeder lines; it provides increments or steps of voltage change.

sterad *See* steradian.

steradian [MATH] The unit of measurement for solid angles; it is equal to the solid angle subtended at the center of a sphere by a portion of the surface of the sphere whose area equals the square of the sphere's radius. Abbreviated sr; sterad.

steradiancy *See* radiance.

Stercorariidae [VERT ZOO] A family of predatory birds of the order Charadriiformes including the skuas and jaegers.

stercorite [MINERAL] $Na(NH_4)H(PO_4)\cdot4H_2O$ A white to yellowish and brown, triclinic mineral consisting of a hydrated acid phosphate of sodium and ammonium.

Sterculiaceae [BOT] A family of dicotyledonous trees and shrubs of the order Malvales distinguished by imbricate or contorted petals, bilocular anthers, and ten to numerous stamens arranged in two or more whorls.

stereo *See* stereophonic; stereo sound system.

stereo- [PHYS] A prefix used to designate a three-dimensional characteristic.

stereochemistry [PHYS CHEM] The study of the spatial arrangement of atoms in molecules and the chemical and physical consequences of such arrangement.

stereographic net *See* net.

stereographic projection [MAP] A perspective conformal, azimuthal map projection in which points on the surface of a sphere or spheroid, such as the earth, are conceived as projected by radial lines from any point on the surface to a plane tangent to the antipode of the point of projection; circles project as circles through the point of tangency, except for great circles which project as straight lines; the principal navigational use of the projection is for charts of the polar regions. Also known as azimuthal orthomorphic projection. [MATH] The projection of the Riemann sphere onto the euclidean plane performed by emanating rays from the north pole of the sphere through a point on the sphere.

stereoisomers [ORG CHEM] Compounds whose molecules have the same number and kind of atoms and the same atomic arrangement, but differ in their spatial relationship.

stereo multiplex [COMMUN] Stereo broadcasting by a frequency-modulation station, in which the output of two microphones is transmitted on the same carrier by frequency-division multiplexing.

stereophonic [ENG ACOUS] Pertaining to three-dimensional pickup or reproduction of sound, as achieved by using two or more separate audio channels. Also known as stereo.

stereophonic sound system *See* stereo sound system.

stereopsis *See* stereoscopy.

stereoscope [OPTICS] An optical instrument in which each eye views one of two photographs taken with the camera or object of study displaced, or simultaneously with two cameras, or with a stereoscopic camera, so that a sensation of depth is produced.

stereoscopic parallax *See* absolute stereoscopic parallax.

stereoscopic vision *See* stereoscopy.

stereoscopy [PHYSIO] The phenomenon of simultaneous vision with two eyes in which there is a vivid perception of the distances of objects from the viewer; it is present because the two eyes view objects in space from two points, so that the retinal image patterns of the same object are slightly different in the two eyes. Also known as stereopsis; stereoscopic vision.

stereo sound system [ENG ACOUS] A sound reproducing system in which a stereo pickup, stereo tape recorder, stereo tuner, or stereo microphone system feeds two independent audio channels, each of which terminates in one or more loudspeakers arranged to give listeners the same audio perspective that they would get at the original sound source. Also known as stereo; stereophonic sound system.

stereospecificity [ORG CHEM] The condition of a polymer whose molecular structure has a fixed spatial (geometric) arrangement of its constituent atoms, thus having crystalline properties; for example, synthetic natural rubber, *cis*-polyisoprene.

Stereospondyli [PALEON] A group of labyrinthodont amphibians from the Triassic characterized by a flat body without pleurocentra and with highly developed intercentra.

stereotaxis [BIOL] An orientation movement in response to stimulation by contact with a solid body. Also known as thigmotaxis.

stereotropism [BIOL] Growth or orientation of a sessile organism or part of an organism in response to the stimulus of a solid body. Also known as thigmotropism.

stereotype [GRAPHICS] A duplicate printing plate made from type and cuts; a paper matrix, or mat, is forced down over the type and cuts to form a mold, into which molten metal is poured, resulting in a new metal printing surface that exactly duplicates the original.

steric effect [PHYS CHEM] The influence of the spatial configuration of reacting substances upon the rate, nature, and extent of reaction.

sterling silver [MET] A silver alloy having a defined standard of purity of 92.5% silver and the remaining 7.5% usually of copper.

Sternaspidae [INV ZOO] A monogeneric family of polychaete annelids belonging to the Sedentaria.

sternbergite [MINERAL] $AgFe_2S_3$ A dark-brown or black mineral composed of silver iron sulfide and occurring as tabular crystals or flexible laminae.

Sterninae [VERT ZOO] A subfamily of birds in the family Laridae, including the Arctic tern.

Sternorrhyncha [INV ZOO] A series of the insect order Homoptera in which the beak appears to arise either between or behind the fore coxae, and the antennae are long and filamentous with no well-differentiated terminal setae.

sternum [ANAT] The bone, cartilage, or series of bony or cartilaginous segments in the median line of the anteroventral part of the body of vertebrates above fishes, connecting with the ribs or pectoral girdle.

steroid [BIOCHEM] A member of a group of compounds, occurring in plants and animals, that are considered to be derivatives of a fused, reduced ring system, cyclopenta[α]-phenanthrene, which consists of three fused cyclohexane rings in a nonlinear or phenanthrene arrangement.

sterol [BIOCHEM] Any of the natural products derived from the steroid nucleus; all are waxy, colorless solids soluble in most organic solvents but not in water, and contain one alcohol functional group.

sterrettite *See* kolbeckite.

Sthenurinae [PALEON] An extinct subfamily of marsupials of the family Diprotodontidae, including the giant kangaroos.

stibiconite [MINERAL] $Sb_3O_6(OH)$ A pale yellow to yellowish- or reddish-white mineral consisting of a basic or hydrated oxide of antimony; occurs in massive form, as a powder, and in crusts.

stibiocolumbite [MINERAL] $Sb(Nb,Ta,Cb)O_4$ A dark brown to light yellowish- or reddish-brown, orthorhombic mineral consisting of an oxide of antimony and tantalum-columbium.

stibnite *See* antimonite.

Stichaeidae [VERT ZOO] The pricklebacks, a family of perciform fishes in the suborder Blennioidei.

Stichocotylidae [INV ZOO] A family of trematodes in the subclass Aspidogastrea in which adults are elongate and have a single row of alveoli.

Stichopodidae [INV ZOO] A family of the echinoderm order Aspidochirotida characterized by tentacle ampullae and by left and right gonads.

stichtite [MINERAL] $Mg_6Cr_2(CO_3)(OH)_{16}\cdot4H_2O$ A lilac-colored rhombohedral mineral composed of hydrous basic carbonate of magnesium and chromium.

stickleback [VERT ZOO] Any fish which is a member of the family Gasterosteidae, so named for the variable number of free spines in front of the dorsal fin.

stiffness [ACOUS] *See* acoustic stiffness. [MECH] The ratio of a steady force acting on a deformable elastic medium to the resulting displacement.

stiffness constant [MECH] Any one of the coefficients of the relations in the generalized Hooke's law used to express stress components as linear functions of the strain components. Also known as elastic constant.

stigma [BOT] The rough or sticky apical surface of the pistil for reception of the pollen. [INV ZOO] **1.** The eyespot of certain protozoans, such as *Euglena*. **2.** The spiracle of an insect or arthropod. **3.** A colored spot on many lepidopteran wings. [MECH] A unit of length used mainly in nuclear measurements, equal to 10^{-12} meter. Also known as bicron.

stigmatic [OPTICS] **1.** Property of an optical system whose focal power is the same in all meridians. **2.** *See* homocentric.

stigmatism [PHYSIO] A condition of the refractive media of the eye in which rays of light from a point are accurately brought to a focus on the retina.

stilb [OPTICS] A unit of luminance, equal to 1 candela per square centimeter. Abbreviated sb.

Stilbellaceae [MYCOL] A family of fungi of the order Moniliales in which conidiophores are aggregated in long bundles or fascicles, forming synnemata or coremia, generally having the conidia in a head at the top.

stilbite [MINERAL] $Ca(Al_2Si_7O_{18})\cdot7H_2O$ A white, brown, or yellow mineral belonging to the zeolite family of silicates; crystallizes in the monoclinic system, occurs in sheaflike aggregates of tabular crystals, and has pearly luster; hardness is 3.5–4 on Mohs scale, and specific gravity is 2.1–2.2. Also known as desmine.

still [CHEM ENG] A device used to evaporate liquids; heat is applied to the liquid, and the resulting vapor is condensed to a liquid state.

stilpnomelane [MINERAL] $K(Fe,Mg,Al)_3Si_4O_{10}(OH)_2\cdot H_2O$ A black or greenish-black mineral composed of basic hydrous potassium iron magnesium aluminum silicate; occurs as fibers, iron magnesium aluminum silicate; occurs as fibers, incrustations, and foliated plates.

stimulated emission device [ELECTR] A device that uses the principle of amplification of electromagnetic waves by stimulated emission, namely, a maser or a laser.

stinging cell *See* cnidoblast.

stingray [VERT ZOO] Any of various rays having a whiplike tail armed with a long serrated spine, at the base of which is a poison gland.

stipe [BOT] **1.** The petiole of a fern frond. **2.** The stemlike portion of the thallus in certain algae. [MYCOL] The short stalk or stem of the fruit body of a fungus, such as a mushroom.

stipoverite *See* stishovite.

stipule [BOT] Either of a pair of appendages that are often present at the base of the petiole of a leaf.

Stirling cycle [THERMO] A regenerative thermodynamic power cycle using two isothermal and two constant volume phases.

Stirling engine [MECH ENG] An engine in which work is performed by the expansion of a gas at high tem-

perature; heat for the expansion is supplied through the wall of the piston cylinder.

Stirodonta [INV ZOO] Formerly, an order of Euechinoidea that included forms with stirodont dentition.

stishovite [MINERAL] SiO_2 A polymorph of quartz, a dense, fine-grained mineral formed under very high pressure (about 1,000,000 pounds per square inch or 7×10^9 newtons per square meter); it is the only mineral in which the silicon atom has a coordination number of six; specific gravity is 4.28. Also known as stipoverite.

stochastic [MATH] Pertaining to random variables.

stock [GEOL] *See* pipe. [IND ENG] **1.** A product or material kept in storage until needed for use or transferred to some ultimate point for use, for example, crude oil tankage or paper-pulp feed. **2.** Designation of a particular material, such as bright stock or naphtha stock. [PETR] A usually discordant, batholithlike body of intrusive igneous rock not exceeding 40 square miles (103.6 square kilometers) in surface exposure and usually discordant.

stoichiometry [PHYS CHEM] The numerical relationship of elements and compounds as reactants and products in chemical reactions.

stoke [FL MECH] A unit of kinematic viscosity, equal to the kinematic viscosity of a fluid with a dynamic viscosity of 1 poise and a density of 1 gram per cubic centimeter. Symbol St (formerly S). Also known as lentor (deprecated usage); stokes.

stokes *See* stoke.

stokesite [MINERAL] $CaSnSi_3O_9\cdot2H_2O$ A colorless orthorhombic mineral composed of hydrous calcium tin silicate occurring in crystals.

Stokes' law [FL MECH] At low velocities, the frictional force on a spherical body moving through a fluid at constant velocity is equal to 6π times the product of the velocity, the fluid viscosity, and the radius of the sphere. [SPECT] The wavelength of luminescence excited by radiation is always greater than that of the exciting radiation.

Stokes stream function [FL MECH] A one-component vector potential function used in analyzing and describing a steady, axially symmetric fluid flow; at any point it is equal to $\frac{1}{2}\pi$ times the mass rate of flow inside the surface generated by rotating the streamline on which the point is located about the axis of symmetry.

STOL *See* short takeoff and landing.

stolon [BOT] *See* runner. [INV ZOO] An elongated projection of the body wall from which buds are formed giving rise to new zooids in Anthozoa, Hydrozoa, Bryozoa, and Ascidiacea. [MYCOL] A hypha produced above the surface and connecting a group of conidiophores.

Stolonifera [INV ZOO] An order of the Alcyonaria, lacking a coenenchyme; they form either simple (*Clavularia*) or rather complex colonies (*Tubipora*).

stolzite [MINERAL] $PbWO_4$ A tetragonal mineral composed of native lead tungstate; it is isomorphous with wulfenite and dimorphous with raspite.

stoma [BIOL] A small opening or pore in a surface. [BOT] One of the minute openings in the epidermis of higher plants which are regulated by guard cells and through which gases and water vapor are exchanged between internal spaces and the external atmosphere.

stomach [ANAT] The tubular or saccular organ of the vertebrate digestive system located between the esophagus and the intestine and adapted for temporary food storage and for the preliminary stages of food breakdown.

Stomatopoda [INV ZOO] The single order of the Eumalacostraca in the superorder Hoplocarida distinguished by raptorial arms, especially the second pair of maxillipeds.

Stomiatoidei [VERT ZOO] A suborder of fishes of the order Salmoniformes including the lightfishes and allies, which are of small size and often grotesque form and are equipped with photophores.

stone [GEOL] 1. A small fragment of rock or mineral. 2. *See* stony meteorite. [LAP] A cut and polished natural gemstone. [MECH] A unit of mass in common use in the United Kingdom, equal to 14 pounds or 6.35029318 kilograms.

stone coal *See* anthracite.

stone fruit *See* drupe.

stone ice *See* ground ice.

stonewort [BOT] The common name for algae comprising the class Charophyceae, so named because most species are lime-encrusted.

stony coral [INV ZOO] Any coral characterized by a calcareous skeleton.

stony meteorite [GEOL] Any meteorite composed principally of silicate minerals, especially olivine, pyroxene, and plagioclase. Also known as aerolite; asiderite; meteoric stone; meteorolite; stone.

stop [OPTICS] The aperture or useful opening of a lens, usually adjustable by means of a diaphragm.

stop band *See* rejection band.

stop bits [COMPUT SCI] The last two bits transmitted in asynchronous data transmission to unequivocally indicate the end of a word.

stope [MIN ENG] 1. To excavate ore in a vein by driving horizontally upon it a series of workings, one immediately over the other, or vice versa; each horizontal working is called a stope because when a number of them are in progress, each working face under attack assumes the shape of a flight of stairs. 2. Any subterranean extraction of ore except that which is incidentally performed in sinking shafts or driving levels for the purpose of opening the mine.

stop element [COMMUN] The last element of a character in certain serial transmissions, used to ensure the recognition of the next start element.

stop loop *See* loop stop.

stop number *See* f number.

stop nut [DES ENG] 1. An adjustable nut that restricts the travel of an adjusting screw. 2. A nut with a compressible insert that binds it so that a lock washer is not needed.

stopoff *See* resist.

stopping capacitor *See* coupling capacitor.

storage area [COMPUT SCI] A specified set of locations in a storage unit. Also known as zone.

storage battery [ELEC] A connected group of two or more storage cells or a single storage cell. Also known as accumulator; accumulator battery; rechargeable battery; secondary battery.

storage calorifier *See* cylinder.

storage camera *See* iconoscope.

storage capacity [COMPUT SCI] The quantity of data that can be retained simultaneously in a storage device; usually measured in bits, digits, characters, bytes, or words. Also known as capacity; memory capacity.

storage cell [COMPUT SCI] An elementary (logically indivisible) unit of storage; the storage cell can contain one bit, character, byte, digit (or sometimes word) of data. [ELEC] An electrolytic cell for generating electric energy, in which the cell after being discharged may be restored to a charged condition

by sending a current through it in a direction opposite to that of the discharging current. Also known as secondary cell.

storage compacting [COMPUT SCI] The practice, followed on multiprogramming computers which use dynamic allocation, of assigning and reassigning programs so that the largest possible area of adjacent locations remains available for new programs.

storage cycle [COMPUT SCI] 1. Periodic sequence of events occurring when information is transferred to or from the storage device of a computer. 2. Storing, sensing, and regeneration from parts of the storage sequence.

storage density [COMPUT SCI] The number of characters stored per unit-length of area of storage medium (for example, number of characters per inch of magnetic tape).

storage device [COMPUT SCI] A mechanism for performing the function of data storage: accepting, retaining, and emitting (unchanged) data items. Also known as computer storage device.

storage dump [COMPUT SCI] A printout of the contents of all or part of a computer storage. Also known as memory dump; memory print.

storage element [COMPUT SCI] Smallest part of a digital computer storage used for storing a single bit.

storage factor *See* Q.

storage fill [COMPUT SCI] Storing a pattern of characters in areas of a computer storage that are not intended for use in a particular machine run; these characters cause the machine to stop if one of these areas is erroneously referred to. Also known as memory fill.

storage hierarchy [COMPUT SCI] The sequence of storage devices, characterized by speed, type of access, and size for the various functions of a computer; for example, core storage from programs and data, disks or drums for temporary storage of massive amounts of data, tapes and cards for back up storage.

storage integrator [COMPUT SCI] In an analog computer, an integrator used to store a voltage in the hold condition for future use while the rest of the computer assumes another computer control state.

storage key [COMPUT SCI] A special set of bits associated with every word or character in some block of storage, which allows tasks having a matching set of protection key bits to use that block of storage.

storage location [COMPUT SCI] A digital-computer storage position holding one machine word and usually having a specific address.

storage medium [COMPUT SCI] Any device or recording medium into which data can be copied and held until some later time, and from which the entire original data can be obtained.

storage register [COMPUT SCI] A register in the main internal memory of a digital computer storing one computer word. Also known as memory register.

storage reservoir *See* impounding reservoir.

storage rings [NUCLEO] Annular vacuum chambers in which charged particles can be stored, without acceleration, by a magnetic field of suitable focusing properties; they are used to stretch effectively the duty cycle of a particle accelerator or to produce colliding beams of particles, resulting in a greater possible center of mass energy.

storage time [ELECTR] 1. The time required for excess minority carriers stored in a forward-biased *pn* junction to be removed after the junction is switched to reverse bias, and hence the time interval between the application of reverse bias and the cessation of forward current. 2. The time required for excess charge carriers in the collector region of a saturated transistor to be removed when the base signal is changed to cut-off level, and hence for the collector current to cease. [PHYS] See decay time.

storage tube [ELECTR] An electron tube employing cathode-ray beam scanning and charge storage for the introduction, storage, and removal of information. Also known as electrostatic storage tube; memory tube (deprecated usage).

storage-type camera tube See iconoscope.

store [COMPUT SCI] 1. To record data into a (static) data storage device. 2. To preserve data in a storage device.

store and forward [COMMUN] A procedure in data communications in which data are stored at some point between the sender and the receiver and are later forwarded to the receiver.

stored-program computer [COMPUT SCI] A digital computer which executes instructions that are stored in main memory as patterns of data.

stork [VERT ZOO] Any of several species of long-legged wading birds in the family Ciconiidae.

storm [METEOROL] An atmospheric disturbance involving perturbations of the prevailing pressure and wind fields on scales ranging from tornadoes (1 kilometer across) to extratropical cyclones (2–3000 km across); also the associated weather (rain storm or blizzard) and the like.

storm center [METEOROL] The area of lowest atmospheric pressure of a cyclone; this is a more general expression than eye of the storm, which refers only to the center of a well-developed tropical cyclone, in which there is a tendency of the skies to clear.

storm drain [CIV ENG] A drain which conducts storm surface, or wash water, or drainage after a heavy rain from a building to a storm or a combined sewer. Also known as storm sewer.

storm sewer See storm drain.

storm surge [OCEANOGR] A rise above normal water level on the open coast due only to the action of wind stress on the water surface; includes the rise in level due to atmospheric pressure reduction as well as that due to wind stress. Also known as storm wave; surge.

storm wave See storm surge.

stove bolt [DES ENG] A coarsely-threaded bolt with a slotted head, which with a square nut is used to join metal parts.

STP See standard conditions.

strabismus [MED] Incoordinate action of the extrinsic ocular muscles resulting in failure of the visual axes to meet at the desired objective point. Also known as cast; heterotropia; squint.

straight-line coding [COMPUT SCI] A digital computer program or routine (section of program) in which instructions are executed sequentially, without branching, looping, or testing.

straight-line mechanism [MECH ENG] A linkage so proportioned and constrained that some point on it describes over part of its motion a straight or nearly straight line.

straight-run [CHEM ENG] Petroleum fractions derived from the straight distillation of crude oil without chemical reaction or molecular modification. Also known as virgin.

straight-tube boiler [MECH ENG] A water-tube boiler in which all the tubes are devoid of curvature and therefore require suitable connecting devices to complete the circulatory system. Also known as header-type boiler.

strain [MECH] Change in length of an object in some direction per unit undistorted length in some direction, not necessarily the same; the nine possible strains form a second-rank tensor.

strain bursts [MIN ENG] Rock bursts in which there is spitting, flaking, and sudden fracturing at the face, indicating increased pressure at the site.

strain ellipsoid [MECH] A mathematical representation of the strain of a homogeneous body by a strain that is the same at all points or of unequal stress at a particular point. Also known as deformation ellipsoid.

strain gage [ENG] A device which uses the change of electrical resistance of a wire under strain to measure pressure.

strain hardening [MET] Increasing the hardness and tensile strength of a metal by cold plastic deformation.

strain rosette [MECH] A pattern of intersecting lines on a surface along which linear strains are measured to find stresses at a point.

strain seismometer [ENG] A seismometer that measures relative displacement of two points in order to detect deformation of the ground.

strain shadow See pressure shadow.

strain-slip cleavage See slip cleavage.

strake [MIN ENG] A relatively wide trough set at a slope and covered with a blanket or corduroy for catching comparatively coarse gold and any valuable mineral. [NAV ARCH] A continuous band of planking or plating running fore and aft along the hull of a ship.

strand [ENG] 1. One of a number of steel wires twisted together to form a wire rope or cable or an electrical conductor. 2. A thread, yarn, string, rope, wire, or cable of specified length. 3. One of the fibers or filaments twisted or laid together into yarn, thread, rope, or cordage. [GEOL] A beach bordering a sea or an arm of an ocean. [NAV] To run aground; term strand usually refers to a serious grounding, while the term "ground" refers to any grounding, however slight. [TEXT] An element of a woven material.

stranded caisson See box caisson.

stranded ice [OCEANOGR] Ice held in place by virtue of being grounded. Also known as grounded ice.

strandline [GEOL] 1. A beach raised above the present sea level. 2. The level at which a body of standing water meets the land. See shoreline.

strangeness number [PARTIC PHYS] A quantum number carried by hadrons, equal to the hypercharge minus the baryon number. Symbol S.

strange particle [PARTIC PHYS] A hadron whose strangeness number is not zero, for example, a K-meson or a Σ-hyperon.

strange quark [PARTIC PHYS] A quark with an electric charge of $-\frac{1}{3}$, baryon number of $\frac{1}{3}$, strangeness of -1, and 0 charm. Symbolized s.

strapped magnetron [ELECTR] A multicavity magnetron in which resonator segments having the same polarity are connected together by small conducting strips to suppress undesired modes of oscillation.

stratification [GEOL] An arrangement or deposition of sedimentary material in layers, or of sedimentary rock in strata. [HYD] 1. The arrangement of a body of water, as a lake, into two or more horizontal layers of differing characteristics, especially densities. 2. The formation of layers in a mass of snow, ice, or firn.

stratified rock See sedimentary rock.

stratiform [GEOL] 1. Descriptive of a layered mineral deposit of either igneous or sedimentary origin.

2. Consisting of parallel bands, layers, or sheets. [METEOROL] Description of clouds of extensive horizontal development, as contrasted to the vertically developed cumuliform types.

stratigraphic geology *See* stratigraphy.

stratigraphic sequence *See* sequence.

stratigraphic trap [GEOL] Sealing of a reservoir bed due to lithologic changes rather than geologic structure. Also known as porosity trap; secondary stratigraphic trap.

stratigraphic unit [GEOL] A stratum of rock or a body of strata classified as a unit on the basis of character, property, or attribute.

stratigraphy [GEOL] A branch of geology concerned with the form, arrangement, geographic distribution, chronologic succession, classification, correlation, and mutual relationships of rock strata, especially sedimentary. Also known as stratigraphic geology.

stratocumulus [METEOROL] A principal cloud type predominantly stratiform, in the form of a gray or whitish layer of patch, which nearly always has dark parts.

stratopause [METEOROL] The boundary or zone of transition separating the stratosphere and the mesosphere; it marks a reversal of temperature change with altitude.

stratosphere [METEOROL] The atmospheric shell above the troposphere and below the mesosphere; it extends, therefore, from the tropopause to about 55 kilometers, where the temperature begins again to increase with altitude.

stratus [METEOROL] A principal cloud type in the form of a gray layer with a rather uniform base; a stratus does not usually produce precipitation, but when it does occur it is in the form of minute particles, such as drizzle, ice crystals, or snow grains.

strawberry [BOT] A low-growing perennial of the genus *Fragaria*, order Rosales, that spreads by stolons; the juicy, usually red, edible fruit consists of a fleshy receptacle with numerous seeds in pits or nearly superficial on the receptacle.

stray capacitance [ELECTR] Undesirable capacitance between circuit wires, between wires and the chassis, or between components and the chassis of electronic equipment.

stray current [ELEC] **1.** A portion of a current that flows over a path other than the intended path, and may cause electrochemical corrosion of metals in contact with electrolytes. **2.** An undesirable current generated by discharge of static electricity; it commonly arises in loading and unloading petroleum fuels and some chemicals, and can initiate explosions.

strays *See* atmospheric interference.

streak [MINERAL] The color of a powdered mineral, obtained by rubbing the mineral on a streak plate.

streak lightning [GEOPHYS] Ordinary lightning, of a cloud-to-ground discharge, that appears to be entirely concentrated in a single, relatively straight lightning channel.

streak photography [GRAPHICS] The process of taking a time exposure photograph of a tracer particle in a fluid so that the motion of each tracer particle represented as a streak may be interpreted as a velocity vector.

stream [HYD] A body of running water moving under the influence of gravity to lower levels in a narrow, clearly defined natural channel.

stream capacity [GEOL] The ability of a stream to carry detritus, measured at a given point per unit of time.

stream cipher [COMMUN] A cipher that makes use of an algorithmic procedure to produce an unending sequence of binary digits which is then combined either with plaintext to produce ciphertext or with ciphertext to recover plaintext.

stream current [HYD] A steady current in a stream or river. [OCEANOGR] A deep, narrow, well-defined fast-moving ocean current.

stream editor [COMPUT SCI] A modification of a statement editor to allow superlines that expand and contract as necessary; the most powerful type of text editor. Also known as string editor.

streamer [GEOPHYS] A sinuous channel of very high ion-density which propagates itself through a gas by continual establishment of an electron avalanche just ahead of its advancing tip; in lightning discharges, the stepped leader, and return streamer all constitute special types of streamers.

stream function *See* Lagrange stream function.

stream gage *See* river gage.

stream gradient [GEOL] The angle, measured in the direction of flow, between the water surface (for large streams) or the channel flow (for small streams) and the horizontal. Also known as stream slope.

streaming current [ELEC] The electric current which is produced when a liquid is forced to flow through a diaphragm, capillary, or porous solid.

streaming potential [ELEC] The difference in electric potential between a diaphragm, capillary, or porous solid and a liquid that is forced to flow through it.

streamline flow [FL MECH] Flow of a fluid in which there is no turbulence: particles of the fluid follow well-defined continuous paths, and the flow velocity at a fixed point either remains constant or varies in a regular fashion with time.

streamlining [DES ENG] The contouring of a body to reduce its resistance to motion through a fluid.

stream slope *See* stream gradient.

stream terrace [GEOL] One of a series of level surfaces on a stream valley flanking and parallel to a stream channel and above the stream level, representing the uneroded remnant of an abandoned floodplain or stream bed. Also known as river terrace.

Streblidae [INV ZOO] The bat flies, a family of cyclorrhaphous dipteran insects in the section Pupipara; adults are ectoparasites on bats.

Strelitziaceae [BOT] A family of monocotyledonous plants in the order Zingiberales distinguished by perfect flowers with five functional stamens and without an evident hypanthium, penniveined leaves, and symmetrical guard cells.

strengite [MINERAL] $FePO_4 \cdot 2H_2O$ A pale-red mineral crystallizing in the orthorhombic system, isomorphous with variscite and dimorphous with phosphosiderite, and specific gravity 2.87.

strength [ACOUS] The maximum instantaneous rate of volume displacement produced by a sound source when emitting a wave with sinusoidal time variation. [MECH] The stress at which material ruptures or fails.

Strepsiptera [INV ZOO] An order of the Coleoptera that is coextensive with the family Stylopidae.

Streptococcaceae [MICROBIO] A family of gram-positive cocci; chemoorganotrophs with fermentative metabolism.

Streptomycetaceae [MICROBIO] A family of soil-inhabiting bacteria in the order Actinomycetales; branched mycelia are produced by vegetative hyphae; spores are produced on aerial hyphae.

stress [BIOL] A stimulus or succession of stimuli of such magnitude as to tend to disrupt the homeostasis of the organism. [MECH] The force acting across a unit area in a solid material in resisting the separation, compacting, or sliding that tends to be induced by external forces.

stress corrosion [MET] Corrosion that is accelerated by stress, applied or residual, in a metal.

stress crack [MECH] An external or internal crack in a solid body (metal or plastic) caused by tensile, compressive, or shear forces.

stress ellipsoid [MECH] A mathematical representation of the state of stress at a point that is defined by the minimum, intermediate, and maximum stresses and their intensities.

stress lines *See* isostatics.

stress ratio [MECH] The ratio of minimum to maximum stress in fatigue testing, considering tensile stresses as positive and compressive stresses as negative.

stress-strain curve *See* deformation curve.

stress trajectories *See* isostatics.

stress-wave emission *See* acoustic emission.

stretcher [CIV ENG] A brick or block that is laid with its length paralleling the wall. [MED] A litter usually made of canvas stretched on a frame for carrying injured, disabled, or dead persons. [MIN ENG] A bar used for roof support on roadways and which is either wedged against or pocketed into the sides of the roadway without support of legs or struts.

stretcher strains *See* Lüders' lines.

stretch forming [MECH ENG] Shaping metals and plastics by applying tension to stretch the heated sheet or part, wrapping it around a die, and then cooling it. Also known as wrap forming.

stretch reflex [PHYSIO] Contraction of a muscle in response to a sudden, brisk, longitudinal stretching of the same muscle. Also known as myostatic reflex.

striation [ELECTR] A succession of alternately luminous and dark regions sometimes observed in the positive column of a glow-discharge tube near the anode. [GEOL] One of a series of parallel or subparallel scratches, small furrows, or lines on the surface of a rock or rock fragment; usually inscribed by rock fragments embedded at the base of a moving glacier. [MINERAL] One of a series of parallel, shallow depressions or narrow bands on the cleavage face of a mineral caused either by growth twinning or oscillatory growth of different crystal faces.

strich *See* millimeter.

stridulation [INV ZOO] Creaking and other audible sounds made by certain insects, produced by rubbing various parts of the body together.

Strigidae [VERT ZOO] A family of birds of the order Strigiformes containing the true owls.

Strigiformes [VERT ZOO] The order of birds containing the owls.

strigovite [MINERAL] $Fe_3(Al,Fe)_3Si_3O_{11}(OH)_7$ A dark-green mineral of the chlorite group, composed of basic aluminum iron silicate; occurs as crystalline incrustations.

Strigulaceae [BOT] A family of Ascolichenes in the order Pyrenulales comprising crustose species confined to tropical evergreens, and which form extensive crusts on or under the cuticle of leaves.

strike [GEOL] The direction taken by a structural surface, such as a fault plane, as it intersects the horizontal. Also known as line of strike. [MET] **1.** A very thin, initially electroplated film or the plating solution with which to deposit such a film. **2.**

A local crater in a metal surface due to accidental contact with the welding electrode. [ORD] Concerted air attack on a single objective.

strike fault [GEOL] A fault whose strike is parallel with that of the strata involved.

strike joint [GEOL] A joint that strikes parallel to the bedding or cleavage of the constituent rock.

strike-shift fault *See* strike-slip fault.

strike-slip fault [GEOL] A fault whose direction of movement is parallel to the strike of the fault. Also known as strike-shift fault.

string [COMPUT SCI] A set of consecutive, adjacent items of similar type; normally a bit string or a character string. [ENG] A piece of pipe, casing, or other down-hole drilling equipment coupled together and lowered into a borehole. [GEOL] A very small vein, either independent or occurring as a branch of a larger vein. Also known as stringer. [MECH] A solid body whose length is many times as large as any of its cross-sectional dimensions, and which has no stiffness.

string break [COMPUT SCI] In the sorting of records, the situation that arises when there are no records having keys with values greater than the highest key already written in the sequence of records currently being processed.

string constant [COMPUT SCI] An arbitrary combination of letters, digits, and other symbols that is treated in a manner completely analogous to numeric constants.

string editor *See* stream editor.

stringer [CIV ENG] **1.** A long horizontal member used to support a floor or to connect uprights in a frame. **2.** An inclined member supporting the treads and risers of a staircase. [GEOL] *See* string. [MET] An elongated mass of microconstituents or foreign material in wrought metal oriented in the direction of working.

string galvanometer [ENG] A galvanometer consisting of a silver-plated quartz fiber under tension in a magnetic field, used to measure oscillating currents. Also known as Einthoven galvanometer.

stringing [PETRO ENG] The connecting of lengths of pipe end to end (tubing or casing) to make a string long enough to reach to the desired depth in a well bore.

stringy floppy [COMPUT SCI] A peripheral storage device for microcomputers that uses a removable magnetic tape cartridge with a 1/16-inch wide (1.5875-millimeter) loop of magnetic tape.

strip-cropping [AGR] Growing separate crops in adjacent strips that follow the contour of the land as a method of reducing soil erosion.

strip line [ELECTROMAG] A strip transmission line that consists of a flat metal-strip center conductor which is separated from flat metal-strip outer conductors by dielectric strips.

strip mining [MIN ENG] The mining of coal by surface mining methods.

stripping [CHEM ENG] In petroleum refining, the removal (by flash evaporation or steam-induced vaporation) of the more volatile components from a cut or fraction; used to raise the flash point of kerosine, gas oil, or lubricating oil. [GRAPHICS] In offset lithography, the process of positioning negatives or positives on a flat before making plates. [MET] Removing a coating from the surface of a metal.

stripping reaction [NUC PHYS] A nuclear reaction in which part of the incident nucleus combines with the target nucleus, and the other part proceeds with most of its original momentum in practically its original direction; especially the reaction in which the incident nucleus is a deuteron and only a proton emerges from the target.

strip transmission line [ELECTROMAG] A microwave transmission line consisting of a thin, narrow, rectangular metal strip that is supported above a ground-plane conductor or between two wide ground-plane conductors and is usually separated from them by a dielectric material.

strobe [ELECTR] 1. Intensified spot in the sweep of a deflection-type indicator, used as a reference mark for ranging or expanding the presentation. 2. Intensified sweep on a plan-position indicator or B-scope; such a strobe may result from certain types of interference, or it may be purposely applied as a bearing or heading marker. 3. Line on a console oscilloscope representing the azimuth data generated by a jammed radar site.

strobilation [INV ZOO] Asexual reproduction by segmentation of the body into zooids, proglottids, or separate individuals.

strobilus [BOT] 1. A conelike structure made up of sporophylls, or spore-bearing leaves, as in Equisetales. 2. The cone of members of the Pinophyta.

stroboscope [ENG] An instrument for making moving bodies visible intermittently, either by illuminating the object with brilliant flashes of light or by imposing an intermittent shutter between the viewer and the object; a high-speed vibration can be made visible by adjusting the strobe frequency close to the vibration frequency.

stroboscopic lamp *See* flash lamp.

stroke [COMPUT SCI] 1. A key-depressing operation in keypunching. 2. In optical character recognition, straight or curved portion of a letter, such as is commonly made with one smooth motion of a pen. Also known as character stroke. 3. That segment of a printed or handwritten character which has been temporarily isolated from other segments for the purpose of analyzing it, particularly with regard to its dimensions and relative reflectance. Also known as character stroke. [ELECTR] The penlike motion of a focused electron beam in cathode-ray-tube displays. [MECH ENG] The linear movement, in either direction, of a reciprocating mechanical part. [MED] A sudden cerebrovascular accident.

stroma [ANAT] The supporting tissues of an organ, including connective and nervous tissues and blood vessels.

Stromateidae [INV ZOO] A family of perciform fishes in the suborder Stromateoidei containing the butterfishes.

Stromateoidei [VERT ZOO] A suborder of fishes of the order Perciformes in which most species have teeth in pockets behind the pharyngeal bone.

stromatolite [GEOL] A structure in calcareous rocks consisting of concentrically laminated masses of calcium carbonate and calcium-magnesium carbonate which are believed to be of calcareous algal origin; these structures are irregular to columnar and hemispheroidal in shape, and range from 1 millimeter to many meters in thickness. Also known as callenia.

Strombacea [PALEON] An extinct superfamily of gastropod mollusks in the order Prosobranchia.

Strombidae [INV ZOO] A family of gastropod mollusks comprising tropical conchs.

stromeyerite [MIN] CuAgS A metallic-gray orthorhombic mineral with a blue tarnish composed of silver copper sulfide occurring in compact masses.

strong interaction [PARTIC PHYS] One of the fundamental interactions of elementary particles, primarily responsible for nuclear forces and other interactions among hadrons.

strongly damped collision *See* quasi-fission.

strong topology [MATH] The topology on a normed space obtained from the given norm; the basic open neighborhoods of a vector x are sets consisting of all those vectors y where the norm of $x - y$ is less than some number.

Strongyloidea [INV ZOO] The hookworms, an order or superfamily of roundworms which, as adults, are endoparasites of most vertebrates, including man.

strontianite [MINERAL] $SrCO_3$ A pale-green, white, gray, or yellowish mineral of the aragonite group having orthorhombic symmetry and occurring in veins or as masses; hardness is 3.5 on Mohs scale, and specific gravity is 3.76.

strontium [CHEM] A metallic element in group IIA, symbol Sr, with atomic number 38, atomic weight 87.62; flammable, soft, pale-yellow solid; soluble in alcohol and acids, decomposes in water; melts at 770°C, boils at 1380°C; chemistry is similar to that of calcium; used as electron-tube getter.

strontium-90 [NUC PHYS] A poisonous, radioactive isotope of strontium; 28-year half life with β radiation; derived from reactor-fuel fission products; used in thickness gages, medical treatment, phosphor activation, and atomic batteries.

strophoid [MATH] The curve traced in the plane by a point P moving along a varying line L passing through a fixed point, where the distance of P to L's intersection with the y-axis always is equal to the y-intercept value.

Strophomenida [PALEON] A large diverse order of articulate brachiopods which first appeared in Lower Ordovician times and became extinct in the Late Triassic.

Strophomenidina [PALEON] A suborder of extinct, articulate brachiopods in the order Strophomenida characterized by a concavo-convex shell, the pseudodeltidium and socket plates disposed subparallel to the hinge.

structural analysis [ENG] The determination of stresses and strains in a given structure. [PETR] *See* structural petrology.

structural drawings [GRAPHICS] The design and working drawings for structures such as buildings, bridges, dams, tanks, and highways.

structural engineering [CIV ENG] A branch of civil engineering dealing with the design of structures such as buildings, dams, and bridges.

structural fabric *See* fabric.

structural formula [CHEM] A system of notation used for organic compounds in which the exact structure, if it is known, is given in schematic representation.

structural gene *See* cistron.

structural geology [GEOL] A branch of geology concerned with the form, arrangement, and internal structure of the rocks.

structural information [COMPUT SCI] Information specifying the number of independently variable features or degrees of freedom of a pattern.

structural petrology [PETR] The study of the internal structure of a rock to determine its deformational history. Also known as fabric analysis; microtectonics; petrofabric analysis; petrofabrics; petrogeometry; petromorphology; structural analysis.

structural wall *See* bearing wall.

structure [AERO ENG] The construction or makeup of an airplane, spacecraft, or missile, including that of the fuselage, wings, empennage, nacelles, and landing gear, but not that of the power plant, furnishings, or equipment. [CIV ENG] Something, as a bridge or a building, that is built or constructed and designed to sustain a load. [GEOL] 1. An assemblage of rocks upon which erosive agents have been or are acting. **2.** The sum total of the structural features of an area. [MINERAL] The form taken by a mineral, such as tabular or fibrous. [PETR] A macroscopic feature of a rock mass or rock unit, best seen in an outcrop. [SCI TECH] The arrangement and interrelation of the parts of an object.

structure cell *See* unit cell.

structure contour [GEOL] A contour that portrays a structural surface, such as a fault. Also known as subsurface contour.

structured data type [COMPUT SCI] The manner in which a collection of data items, which may have the same or different scalar data types, is represented in a computer program.

structured programming [COMPUT SCI] The use of program design and documentation techniques that impose a uniform structure on all computer programs.

structure type [CRYSTAL] The structural arrangement of a crystal, regardless of the atomic elements present; it corresponds to the crystal's space group.

strut [AERO ENG] A bar supporting the wing or landing gear of an airplane. [CIV ENG] A long structural member of timber or metal, or a bar designed to resist pressure in the direction of its length. [ENG] **1.** A brace or supporting piece. **2.** A diagonal brace between two legs of a drill tripod or derrick. [MIN ENG] A vertical-compression member in a structure or in an underground timber set. [NAV ARCH] A bracket outside the hull of a ship, supporting the propeller shaft. Also known as propeller strut.

Struthionidae [VERT ZOO] The single family of the avian order Struthioniformes.

Struthioniformes [VERT ZOO] A monofamilial order of ratite birds comprising the single living species of ostrich (*Struthio camelus*).

struvite [MINERAL] $Mg(NH_4)PO_4 \cdot 6H_2O$ A colorless to yellow or pale-brown mineral consisting of a hydrous ammonium magnesium phosphate, and occurring in orthorhombic crystals; hardness is 2 on Mohs scale, and specific gravity is 1.7.

stub [COMPUT SCI] The left-hand portion of a decision table, consisting of a single column, and comprising the condition stub and the action stub. [ELECTROMAG] **1.** A short section of transmission line, open or shorted at the far end, connected in parallel with a transmission line to match the impedance of the line to that of an antenna or transmitter. **2.** A solid projection one-quarter-wavelength long, used as an insulating support in a waveguide or cavity.

stubborn disease [PL PATH] A virus disease of citrus trees characterized by short internodes resulting in stiff brushy growth and chlorotic leaves.

stub entry [MIN ENG] A short, narrow entry turned from another entry and driven into the solid coal, but not connected with other mine workings.

stud [BUILD] One of the vertical members in the walls of a framed building to which wallboards, lathing, or paneling is nailed or fastened. [DES ENG] **1.** A rivet, boss, or nail with a large, ornamental head. **2.** A short rod or bolt threaded at both ends without a head.

Student's distribution [STAT] The probability distribution used to test the hypothesis that a random sample of *n* observations comes from a normal population with a given mean.

Student's t-statistic [STAT] A one-sample test statistic computed by $T = \sqrt{n}(\overline{X} - \mu_h)/S$, where $\overline{X}$ is the

mean of a collection of *n* observations, *S* is the square root of the mean square deviation, and μ_h is the hypothesized mean.

Student's t-test [STAT] A test in a one-sample problem which uses Student's t-statistic.

stuffing [ENG] A method of sealing the mechanical joint between two metal surfaces; packing (stuffing) material is inserted within the seal area container (the stuffing or packing box), and compressed to a liquid-proof seal by a threaded packing ring follower. Also known as packing.

stull [MIN ENG] A platform laid on timbers, braced across a working from side to side, to support workers or to carry ore or waste.

stunt [PL PATH] Any of several plant diseases marked by reduction in size of the plant.

sturgeon [VERT ZOO] Any of 10 species of large bottom-living fish which comprise the family Acipenseridae; the body has five rows of bony plates, and the snout is elongate with four barbels on its lower surface.

Sturm-Liouville problem [MATH] The general problem of solving a given linear differential equation of order $2n$ together with $2n$-boundary conditions. Also known as eigenvalue problem.

Sturm's theorem [MATH] This gives a method to determine the number of real roots of a polynomial $p(x)$ which lie between two given values of x; the Sturm sequence of $p(x)$ provides the necessary information.

sturtite [MINERAL] A black mineral composed of hydrous silicate of iron, manganese, calcium, and magnesium; occurs in compact masses.

Stuve chart [METEOROL] A thermodynamic diagram with atmospheric temperature as the *x* axis and atmospheric pressure to the power 0.286 as the *y* ordinate, increasing downward; named after G. Stuve. Also known as adiabatic chart; pseudoadiabatic chart.

sty *See* hordeolum.

Styginae [INV ZOO] A subfamily of butterflies in the family Lycaenidae in which the prothoracic legs in the male are nonfunctional.

Stygocaridacea [INV ZOO] An order of crustaceans in the superorder Syncarida characterized by having a furca.

Stylasterina [INV ZOO] An order of the class Hydrozoa, including several brightly colored branching or encrusting coral-like coelenterates of warm seas.

style [BOT] The portion of a pistil connecting the stigma and ovary. [ENG] *See* gnomon. [ZOO] A slender elongated process on an animal.

styloglossus [ANAT] A muscle arising from the styloid process of the temporal bone, and inserted into the tongue.

stylolite [GEOL] An irregular surface, generally parallel to a bedding plane, in which small toothlike projections on one side of the surface fit into cavities of complementary shape on the other surface; interpreted to result diagenetically by pressure solution.

Stylommatophora [INV ZOO] A large order of the molluscan subclass Pulmonata characterized by having two pairs of retractile tentacles with eyes located on the tips of the large tentacles.

stylotypite *See* tetrahedrite.

stylus [ENG ACOUS] The portion of a phonograph pickup that follows the modulations of a record groove and transmits the resulting mechanical motions to the transducer element of the pickup for conversion to corresponding audio-frequency signals. Also known as needle; phonograph needle; reproducing stylus. [GRAPHICS] A rather blunt metal point sometimes used in painting to make lightly ruled lines.

stylus printing *See* matrix printing.

Stypocapitellidae [INV ZOO] A family of polychaete annelids belonging to the Sedentaria and consisting of a monotypic genus found in western Germany.

subadditive function [MATH] A function F is subadditive if $f(x + y)$ is less than or equal to $f(x) + f(y)$ for all x and y in its domain.

subalgebra [MATH] **1.** A subset of an algebra which itself forms an algebra relative to the same operations. **2.** A subalgebra (of sets) is any algebra (of sets) contained in some given algebra.

subangular [SCI TECH] Somewhat angular but free from sharp edges and corners.

Subantarctic Intermediate Water [OCEANOGR] A layer of water above the deep-water layer in the South Atlantic.

subaqueous [HYD] Pertaining to conditions and processes occurring in, under, or beneath the surface of water, especially fresh water.

subarachnoid space [ANAT] The space between the pia mater and the arachnoid of the brain.

subarctic [GEOGR] Pertaining to regions adjacent to the Arctic Circle or having characteristics somewhat similar to those of these regions.

subarid [CLIMATOL] Pertaining to regions that are moderately or slightly arid.

subarkose [GEOL] Sandstone that is intermediate in composition between arkose and pure quartz sandstone; it contains less feldspar than arkose.

subassembly [ELECTR] Two or more components combined into a unit for convenience in assembling or servicing equipment; an intermediate-frequency strip for a receiver is an example. [ENG] A structural unit, which, though manufactured separately, was designed for incorporation with other parts in the final assembly of a finished product.

subatomic particle [PHYS] A particle which is smaller than an atom, namely, an elementary particle or an atomic nucleus.

subbituminous coal [GEOL] Black coal intermediate in rank between lignite and bituminous coal; has more carbon and less moisture than lignite.

subbottom depth recorder [ENG] A compact seismic instrument which can provide continuous soundings of strata beneath the ocean bottom utilizing the low-frequency output of an intense electrical spark discharge source in water.

subcarrier [COMMUN] **1.** A carrier that is applied as a modulating wave to modulate another carrier. **2.** *See* chrominance subcarrier.

subchannel [COMPUT SCI] The portion of an input/output channel associated with a specific input/output operation. [COMMUN] In a telemetry system, the route followed to convey the magnitude of a single subcommutated measurand.

subclavian artery [ANAT] The proximal part of the principal artery in the arm or forelimb.

subcontinent [GEOGR] **1.** A landmass such as Greenland that is large but not as large as the generally recognized continents. **2.** A large subdivision of a continent (for example, the Indian subcontinent) distinguished geologically or geomorphically from the rest of the continent.

subcritical [NUCLEO] Having an effective multiplication constant less than one, so that a self-supporting chain reaction cannot be maintained in a nuclear reactor.

subdevice bit *See* select bit.

subduction [GEOL] The process by which one crustal block descends beneath another, such as the descent of the Pacific plate beneath the Andean plate along the Andean Trench.

suberin [BIOCHEM] A fatty substance found in many plant cell walls, especially cork.

subfield [MATH] **1.** A subset of a field which itself forms a field relative to the same operations. **2.** A subfield (of sets) is any field (of sets) contained in some given field of sets.

subgiant star [ASTRON] A member of the family of stars whose luminosity is intermediate between giants and the main sequence in the Hertzsprung-Russell diagram; spectral classes G and K are most frequent.

subgrade [CIV ENG] The soil or rock leveled off to support the foundation of a structure.

subgraywacke [PETR] An argillaceous sandstone with a composition intermediate between graywacke and orthoquartzite; a clay matrix is usually present but it amounts to less than 15%.

subgroup [MATH] A subset N of a group G which is itself a group relative to the same operation.

subharmonic [PHYS] A sinusoidal quantity having a frequency that is an integral submultiple of the frequency of some other sinusoidal quantity to which it is referred; a third subharmonic would be one-third the fundamental or reference frequency.

subhedral [MINERAL] **1.** Pertaining to an individual mineral crystal that is partly bounded by its own crystal faces and partly bounded by surfaces formed against preexisting crystals. **2.** Descriptive of a crystal having partially developed crystal faces.

subjacent [GEOL] Being lower than but not directly underneath.

subject copy [GRAPHICS] Material that is to be transmitted for facsimile reproduction. Also known as copy.

sublacustrine [GEOL] Existing or formed on the bottom of a lake.

sublethal gene *See* semilethal gene.

sublevel [ATOM PHYS] *See* subshell. [MIN ENG] An intermediate level opened a short distance below the main level; or in the caving system of mining, a 15–20-foot (4.6–6.1-meter) level below the top of the ore body, preliminary to caving the ore between it and the level above.

sublimation [PSYCH] A defense mechanism whereby the energies of undesirable instinctual cravings and impulses are converted into socially acceptable activities. [THERMO] The process by which solids are transformed directly to the vapor state or vice versa without passing through the liquid phase.

sublimation point [THERMO] The temperature at which the vapor pressure of the solid phase of a compound is equal to the total pressure of the gas phase in contact with it; analogous to the boiling point of a liquid.

sublingual gland [ANAT] A complex of salivary glands located in the sublingual fold on each side of the floor of the mouth.

sublittoral zone [OCEANOGR] The benthic region extending from mean low water (or 40–60 meters, according to some authorities) to a depth of about 100 fathoms (200 meters), or the edge of a continental shelf, beyond which most abundant attached plants do not grow.

submandibular gland [ANAT] A large seromucous or mixed salivary gland located below the mandible on each side of the jaw. Also known as mandibular gland; submaxillary gland.

submarine cable [ELEC] A cable designed for service under water; usually a lead-covered cable with steel armor applied between layers of jute.

submarine canyon [GEOL] Steep-sided valleys winding across the continental shelf or continental slope, probably originally produced by Pleistocene stream erosion, but presently the site of turbidity flows.

submarine cave   *See* submarine fan.

submarine delta   *See* submarine fan.

submarine earthquake   *See* seaquake.

submarine fan   [GEOL] A shallow marine sediment that is fan- or cone-shaped and lies off the seaward opening of large rivers and submarine canyons. Also known as abyssal cave; abyssal fan; sea fan; submarine cave; submarine delta; subsea apron.

submarine plain   *See* plain.

submarine trough   *See* trough.

submarine valley   *See* valley.

submaxillary gland   *See* submandibular gland.

submerged-arc furnace   [MET] An arc-heating furnace in which the arcs may be completely submerged under the charge or in the molten bath under the charge.

submerged-arc welding   [MET] Arc welding with a bare metal electrode, the arc and tip of the electrode being shielded by a blanket of granular, fusible material.

submerged-combustion heater   [ENG] A combustion device in which fuel and combustion air are mixed and ignited below the surface of a liquid; used in heaters and evaporators where absorption of the combustion products will not be detrimental.

submergence   [GEOL] A change in the relative levels of water and land either from a sinking of the land or a rise of the water level.

submersible pump   [MECH ENG] A pump and its electric motor together in a protective housing which permits the unit to operate under water.

submucosa   [HISTOL] The layer of fibrous connective tissue that attaches a mucous membrane to its subadjacent parts.

Suboscines   [VERT ZOO] A major division of the order Passeriformes, usually divided into the suborders Eurylaimi, Tyranni, and Memirae.

subpolar anticyclone   *See* subpolar high.

subpolar high   [METEOROL] A high that forms over the cold continental surfaces of subpolar latitudes, principally in Northern Hemisphere winters; these highs typically migrate eastward and southward. Also known as polar anticyclone; polar high; subpolar anticyclone.

subpolar westerlies   *See* westerlies.

subprogram   [COMPUT SCI] A part of a larger program which can be converted independently into machine language.

subroutine   [COMPUT SCI] **1.** A body of computer instruction (and the associated constants and working-storage areas, if any) designed to be used by other routines to accomplish some particular purpose. **2.** A statement in FORTRAN used to define the beginning of a closed subroutine (first definition).

subscriber line   [ELEC] A telephone line between a central office and a telephone station, private branch exchange, or other end equipment. Also known as central office line; subscriber loop.

subscriber loop   *See* subscriber line.

subsea apron   *See* submarine fan.

subsequence   [MATH] A subsequence of a given sequence is any sequence all of whose entries appear in the original sequence and in the same manner of succession.

subsequent   [GEOL] Referring to a geologic feature that followed in time the development of a consequent feature of which it is a part.

subset   [COMMUN] A telephone or other subscriber equipment connected to a communication system,

such as a modem. Derived from subscriber set. [MATH] A subset of a set B is a set all of whose elements are included in B.

subshell   [ATOM PHYS] Electrons of an atom within the same shell (energy level) and having the same azimuthal quantum numbers. Also known as sublevel.

subsidence   [METEOROL] A descending motion of air in the atmosphere, usually with the implication that the condition extends over a rather broad area. [MIN ENG] A sinking down of a part of the earth's crust due to underground excavations.

subsidiary fracture   *See* tension fracture.

subsoil ice   *See* ground ice.

subsonic   [ACOUS] *See* infrasonic. [PHYS] Of, pertaining to, or dealing with speeds less than acoustic velocity, as in subsonic aerodynamics.

subspace   [MATH] A subset of a space which, in the appropriate context, is a space in its own right.

subspecies   [SYST] A geographically defined grouping of local populations which differs taxonomically from similar subdivisions of species.

substantive dye   *See* direct dye.

substation   [ELEC] *See* electric power substation. [ENG] An intermediate compression station to repressure a fluid being transported by pipeline over a long distance. [MIN ENG] A subsidiary station for the conversion of power to the type, usually direct current, and voltage needed for mining equipment and fed into the mine power system.

substitute mode   [COMPUT SCI] One method of exchange buffering, in which segments of storage function alternately as buffer and as program work area.

substitution reaction   [CHEM] Replacement of an atom or radical by another one in a chemical compound.

substrate   [BIOCHEM] The substance with which an enzyme reacts. [ELECTR] The physical material on which a microcircuit is fabricated; used primarily for mechanical support and insulating purposes, as with ceramic, plastic, and glass substrates; however, semiconductor and ferrite substrates may also provide useful electrical functions. [ENG] Basic surface on which a material adheres, for example, paint or laminate. [ORG CHEM] A compound with which a reagent reacts.

substructure   [CIV ENG] The part of a structure which is below ground.

subsurface contour   *See* structure contour.

subsystem   [ENG] A major part of a system which itself has the characteristics of a system, usually consisting of several components.

subtense bar   [ENG] The horizontal bar of fixed length in the subtense technique of distance measurement method.

subtense technique   [CIV ENG] A distance measuring technique in which the transit angle subtended by the subtense bar enables the computation of the transit-to-bar distance.

subterranean ice   *See* ground ice.

subtraction   [MATH] The addition of one quantity with the negative of another; in a system with an additive operation this is formally the sum of one element with the additive inverse of another.

subtractive primaries   [OPTICS] The three colors, usually yellow, magenta, and cyan (greenish-blue), which are mixed together in a subtractive process.

subtractive process   [OPTICS] The process of producing colors by mixing absorbing media or filters of subtractive primary colors.

subtrahend   [MATH] A quantity which is to be subtracted from another given quantity.

Subtriquetridae [INV ZOO] A family of arthropods in the suborder Porocephaloidea.

subtropic [METEOROL] An indefinite belt in each hemisphere between the tropic and temperate regions; the polar boundaries are considered to be roughly 35–40° northern and southern latitudes, but vary greatly according to continental influence, being farther poleward on the western coasts of continents and farther equatorward on the eastern coasts.

Subtropical Convergence [OCEANOGR] The zone of converging currents, generally located in midlatitudes.

subtropical westerlies See westerlies.

Subulitacea [PALEON] An extinct superfamily of gastropod mollusks in the order Prosobranchia which possessed a basal fold but lacked an apertural sinus.

Subuluridae [INV ZOO] The equivalent name for Heterakidae.

succession [ECOL] A gradual process brought about by the change in the number of individuals of each species of a community and by the establishment of new species populations which may gradually replace the original inhabitants.

sucker rod [PETRO ENG] A connecting rod between a down-hole oil-well pump and the lifting or pumping device on the surface.

sucker-rod pump [PETRO ENG] A cylinder-piston-type pump used to displace oil into the oil-well tubing string, and to the surface.

sucking louse [INV ZOO] The common name for insects of the order Anoplura, so named for the slender, tubular mouthparts.

sucrase See saccharase.

sucrose [ORG CHEM] $C_{12}H_{22}O_{11}$ Combustible, white crystals soluble in water, decomposes at 160 to 186°C; derived from sugarcane or sugarbeet; used as a sweetener in drinks and foods and to make syrups, preserves, and jams. Also known as saccharose; table sugar.

suction pump [MECH ENG] A pump that raises water by the force of atmospheric pressure pushing it into a partial vacuum under the valved piston, which retreats on the upstroke.

suction stroke [MECH ENG] The piston stroke that draws a fresh charge into the cylinder of a pump, compressor, or internal combustion engine.

suction wave See rarefaction wave.

Suctoria [INV ZOO] A small subclass of the protozoan class Ciliatea, distinguished by having tentacles which serve as mouths.

Suctorida [INV ZOO] The single order of the protozoan subclass Suctoria.

sudden death syndrome See crib death.

sudden ionospheric disturbance [GEOPHYS] A complex combination of sudden changes in the condition of the ionosphere following the appearance of solar flares, and the effects of these changes. Abbreviated SID.

suffix notation See reverse Polish notation.

suffrutescent [BOT] Of or pertaining to a stem intermediate between herbaceous and shrubby, becoming partly woody and perennial at the base.

sugarbeet [BOT] *Beta vulgaris.* A beet characterized by a white root and cultivated for the high sugar content of the roots.

sugarcane [BOT] *Saccharum officinarum.* A stout, perennial grass plant characterized by two-ranked leaves, and a many-jointed stalk with a terminal inflorescence in the form of a silky panicle; the source of more than 50% of the world's annual sugar production.

sugarcane gummosis See Cobb's disease.

sugar maple [BOT] *Acer saccharum.* A commercially important species of maple tree recognized by its gray furrowed bark, sharp-pointed scaly winter buds, and symmetrical oval outline of the crown.

sugar snow See depth hoar.

Suidae [VERT ZOO] A family of paleodont artiodactyls in the superfamily Suoidae including wild and domestic pigs.

suite [COMPUT SCI] A collection of related computer programs run one after another.

sulcus [ZOO] A furrow or groove, especially one on the surface of the cerebrum.

sulfate [CHEM] **1.** A compound containing the —SO_4 group, as in sodium sulfate Na_2SO_4. **2.** A salt of sulfuric acid.

sulfate mineral [MINERAL] A mineral compound characterized by the sulfate radical SO_4.

sulfation [CHEM] The conversion of a compound into a sulfate by the oxidation of sulfur, as in sodium sulfide, Na_2S, oxidized to sodium sulfate, Na_2SO_4; or the addition of a sulfate group, as in the reaction of sodium and sulfuric acid to form Na_2SO_4.

sulfide [CHEM] Any compound with one or more sulfur atoms in which the sulfur is connected directly to a carbon, metal, or other nonoxygen atom; for example, sodium sulfide, Na_2S.

sulfide mineral [MINERAL] A mineral compound characterized by the linkage of sulfur with a metal or semimetal.

sulfo- [CHEM] Prefix for a compound with either a divalent sulfur atom or the presence of —SO_3H, the sulfo group. Also spelled sulpho-.

sulfonation [CHEM] Substitution of —SO_3H groups (from sulfuric acid) for hydrogen atoms, for example, conversion of benzene, C_6H_6, into benzenesulfonic acid, $C_6H_5SO_3H$.

sulfoxide [ORG CHEM] R_2SO A compound with the radical $=SO$; derived from oxidation of sulfides, the proportion of oxidant, such as hydrogen peroxide, and temperature being set to avoid excessive oxidation; an example is dimethyl sulfoxide, $(CH_3)_2SO$.

sulfur [CHEM] A nonmetallic element in group VIa, symbol S, atomic number 16, atomic weight 32.064, existing in a crystalline or amorphous form and in four stable isotopes; used as a chemical intermediate and fungicide, and in rubber vulcanization. [MINERAL] A yellow orthorhombic mineral occurring in crystals, masses, or layers, and existing in several allotropic forms; the native form of the element.

sulfur-35 [NUC PHYS] Radioactive sulfur with mass number 35; radiotoxic, with 87.1-day half-life, β radiation; derived from pile irradiation; used as a tracer to study chemical reactions, engine wear, and protein metabolism.

sulfur test [ANALY CHEM] **1.** Method to determine the sulfur content of a petroleum material by combustion in a bomb. **2.** Analysis of sulfur in petroleum products by lamp combustion in which combustion of the sample is controlled by varying the flow of carbon dioxide and oxygen to the burner.

Sulidae [VERT ZOO] A family of aquatic birds in the order Pelecaniformes including the gannets and boobies.

sulpho- See sulfo-.

sulvanite [MINERAL] Cu_3VS_4 A bronze-yellow mineral composed of copper vanadium sulfide occurring in masses.

sum [MATH] **1.** The addition of numbers or mathematical objects in context. **2.** The sum of an infinite series is the limit of the sequence consisting of all partial sums of the series.

summation check [COMPUT SCI] An error-detecting procedure involving adding together all the digits of some number and comparing this sum to a previously computed value of the same sum.

summation convention [MATH] An abbreviated notation used particularly in tensor analysis and relativity theory, in which a product of tensors is to be summed over all possible values of any index which appears twice in the expression.

summation network *See* summing network.

summer [ASTRON] The period from the summer solstice to the autumnal equinox; popularly and for most meteorological purposes, it is taken to include June through August in the Northern Hemisphere, and December through February in the Southern Hemisphere.

summer solstice [ASTRON] **1.** The sun's position on the ecliptic when it reaches its greatest northern declination. Also known as first point of Cancer. **2.** The date, about June 21, on which the sun has its greatest northern declination.

summer time *See* daylight saving time.

summing amplifier [ELECTR] An amplifier that delivers an output voltage which is proportional to the sum of two or more input voltages or currents.

summing network [ELEC] A passive electric network whose output voltage is proportional to the sum of two or more input voltages. Also known as summation network.

summit plain *See* peak plain.

sum of states *See* partition function.

sum over states *See* partition function.

sump [ENG] A pit or tank which receives and temporarily stores drainage at the lowest point of a circulating or drainage system.

sum rule [QUANT MECH] A formula which sets some quantity equal to the sum, over all the states of a system, of another quantity, usually involving the square of the magnitude of the matrix element of some operator between a given state and the state being summed over.

sun [ASTRON] The star about which the earth revolves; it is a globe of gas 1.4×10^6 kilometers in diameter, held together by its own gravity; thermonuclear reactions take place in the deep interior of the sun converting hydrogen into helium releasing energy which streams out. Also known as Sol.

sun crack *See* mud crack.

sun cross [METEOROL] A rare halo phenomenon in which bands of white light intersect over the sun at right angles; it appears probable that most of such observed crosses appear merely as a result of the superposition of a parhelic circle and a sun pillar.

sun crust [HYD] A type of snow crust, formed by refreezing of surface snow crystals after having been melted by the sun.

sundew [BOT] Any plant of the genus *Drosera* of the family Droseraceae; the genus comprises small, herbaceous, insectivorous plants that grow on all continents, especially Australia.

sun dog *See* parhelion.

sunfish [VERT ZOO] Any of several species of marine and freshwater fishes in the families Centrarchidae and Molidae characterized by brilliant metallic skin coloration.

sunflower [BOT] *Helianthus annuus.* An annual plant native to the United States characterized by broad, ovate leaves growing from a single, usually long (3–20 feet or 1–6 meters) stem, and large, composite flowers with yellow petals.

sunlamp [ELEC] A mercury-vapor gas-discharge tube used to produce ultraviolet radiation for therapeutic or cosmetic purposes.

sun opal *See* fire opal.

sun pillar [METEOROL] A luminous streak of light, white or slightly reddened, extending above and below the sun, most frequently observed near sunrise or sunset; it may extend to about 20° above the sun, and generally ends in a point. Also known as light pillar.

sun-pumped laser [OPTICS] A continuous-wave laser in which pumping is achieved by concentrating the energy of the sun on the laser crystal with a parabolic mirror. Also known as solar-excited laser.

sunrise [ASTRON] The exact moment the upper limb of the sun appears above the horizon.

sunscald [PL PATH] An injury to woody plants which results in local death of the plant tissues; in summer it is caused by excessive action of the sun's rays, in winter, by the great variation of temperature on the side of trees that is exposed to the sun in cold weather.

sunset [ASTRON] The exact moment the upper limb of the sun disappears below the horizon.

sunspot [ASTRON] A dark area in the photosphere of the sun caused by a lowered surface temperature.

sunspot cycle [ASTRON] Variation of the size and number of sunspots in an 11-year cycle which is shared by all other forms of solar activity.

sunstone [MINERAL] An aventurine feldspar containing minute flakes of hematite; usually brilliant and translucent, it emits reddish or golden billowy reflection. Also known as heliolite.

sun-synchronous orbit [AERO ENG] An earth orbit of a spacecraft so that the craft is always in the same direction relative to that of the sun; as a result, the spacecraft passes over the equator at the same spots at the same times.

Suoidea [VERT ZOO] A superfamily of artiodactyls of the suborder Paleodonta which comprises the pigs and peccaries.

superacid [CHEM] **1.** An acidic medium that has a proton-donating ability equal to or greater than 100% sulfuric acid. **2.** A solution of acetic or phosphoric acid.

superaerodynamics [FL MECH] That branch of gas dynamics dealing with the flow of gases at such low density that the molecular mean free path is not negligibly small; under these conditions the gas no longer behaves as a continuous fluid.

superalloy [MET] A thermally resistant alloy for use at elevated temperatures where high stresses and oxidation are encountered.

supercavitating propeller [NAV ARCH] A marine propeller which has special blade sections so that at sufficiently high speed the whole back of each blade becomes enveloped by a smooth sheet of cavitation.

supercharger [MECH ENG] An air pump or blower in the intake system of an internal combustion engine used to increase the weight of air charge and consequent power output from a given engine size.

supercompressibility factor *See* compressibility factor.

supercomputer [COMPUT SCI] A computer which is among those with the highest speed, largest functional size, biggest physical dimensions, or greatest monetary cost in any given period of time.

superconducting computer [COMPUT SCI] A high-performance computer whose circuits employ superconductivity and the Josephson effect to reduce computer cycle time.

superconducting device *See* cryogenic device.

superconducting material *See* superconductor.

superconducting memory [CRYO] A computer memory made up of a number of cryotrons, thin-film cryotrons, superconducting thin films, or other superconducting storage devices; these operate only under cryogenic conditions and dissipate power only during the read or write operation, which permits construction of large, dense memories.

superconducting quantum interference device [ELECTR] A superconducting ring that couples with one or two Josephson junctions; applications include high-sensitivity magnetometers, near-magnetic-field antennas, and measurement of very small currents or voltages. Abbreviated SQUID.

superconducting thin film [CRYO] A thin film of indium, tin, or other superconducting element, used as a cryogenic switching or storage device, as in a thin-film cryotron.

superconductivity [SOLID STATE] A property of many metals, alloys, and chemical compounds at temperatures near absolute zero by virtue of which their electrical resistivity vanishes and they become strongly diamagnetic.

superconductor [SOLID STATE] Any material capable of exhibiting superconductivity; examples include iridium, lead, mercury, niobium, tin, tantalum, vanadium, and many alloys. Also known as cryogenic conductor; superconducting material.

supercooling [THERMO] Cooling of a substance below the temperature at which a change of state would ordinarily take place without such a change of state occurring, for example, the cooling of a liquid below its freezing point without freezing taking place; this results in a metastable state.

supercritical [NUCLEO] Having an effective multiplication constant greater than 1, so that the rate of reaction rises rapidly in a nuclear reactor. [THERMO] Property of a gas which is above its critical pressure and temperature.

supercritical wing [AERO ENG] A wing developed to permit subsonic aircraft to maintain an efficient cruise speed very close to the speed of sound; the middle portion of the wing is relatively flat with substantial downward curvature near the trailing edge; in addition, the forward region of the lower surface is more curved than that of the upper surface with a substantial cusp of the rearward portion of the lower surface.

superdense theory *See* big bang theory.

superficial deposit *See* surficial deposit.

superfines [MET] The portion of a metal powder composed of particles smaller than 10 micrometers.

superfluidity [CRYO] The frictionless flow of liquid helium at temperatures very close to absolute zero through holes as small as 10^{-7} centimeter in diameter, and for particle velocities below a few centimeters per second.

superfluorescence [ATOM PHYS] The process of spontaneous emission of electromagnetic radiation from a collection of excited atoms.

supergene [MINERAL] Referring to mineral deposits or enrichments formed by descending solutions. Also known as hypergene.

supergiant star [ASTRON] A member of the family containing the intrinsically brightest stars, populating the top of the Hertzsprung-Russell diagram; supergiant stars occur at all temperatures from 30,000 to 3000 K and have luminosities from 10^4 to 10^6 times that of the sun; the star Betelgeuse is an example.

supergravity [PHYS] A supersymmetry which is used to unify general relativity and quantum theory; it is formed by adding to the Poincaré group, as a symmetry of space-time, four new generators that behave as spinors and vary as the square root of the translations.

superheater [MECH ENG] A component of a steam-generating unit in which steam, after it has left the boiler drum, is heated above its saturation temperature.

superheating [THERMO] Heating of a substance above the temperature at which a change of state would ordinarily take place without such a change of state occurring, for example, the heating of a liquid above its boiling point without boiling taking place; this results in a metastable state.

superhet *See* superheterodyne receiver.

superheterodyne receiver [ELECTR] A receiver in which all incoming modulated radio-frequency carrier signals are converted to a common intermediate-frequency carrier value for additional amplification and selectivity prior to demodulation, using heterodyne action; the output of the intermediate-frequency amplifier is then demodulated in the second detector to give the desired audio-frequency signal. Also known as superhet.

superhigh frequency [COMMUN] A frequency band from 3,000 to 30,000 megahertz, corresponding to wavelengths from 1 to 10 centimeters. Abbreviated SHF.

superimposed glacier [GEOL] A glacier whose course is maintained despite different preexisting structures and lithologies as the glacier erodes downward.

superimposed stream [HYD] A stream, started on a new surface, that kept its course through the different preexisting lithologies and structures encountered as it eroded downward into the underlying rock. Also known as superinduced stream.

superincumbent [GEOL] Pertaining to a superjacent layer, especially one that is situated so as to exert pressure.

superinduced stream *See* superimposed stream.

superior [BOT] **1.** Positioned above another organ or structure. **2.** Referring to a calyx that is attached to the ovary. **3.** Referring to an ovary that is above the insertion of the floral parts.

superior conjunction [ASTRON] A conjunction when an astronomical body is opposite the earth on the other side of the sun.

superior planet [ASTRON] Any of the planets that are farther than the earth from the sun; includes Mars, Jupiter, Saturn, Uranus, Neptune, and Pluto.

superior transit *See* upper transit.

superior vena cava [ANAT] The principal vein collecting blood from the head, chest wall, and upper extremities and draining into the right atrium.

superjacent [GEOL] Pertaining to a stratum situated immediately upon or over a particular lower stratum or above an unconformity.

superlattice [ELECTR] A structure consisting of alternating layers of two different semiconductor materials, each several nanometers thick. [SOLID STATE] An ordered arrangement of atoms in a solid solution which forms a lattice superimposed on the normal solid solution lattice. Also known as superstructure.

superline [COMPUT SCI] A unit of text longer than an ordinary line, used in some of the more powerful text editors.

supermassive star [ASTRON] A star with a mass exceeding about 50 times that of the sun.

supermini [COMPUT SCI] A large minicomputer.

supermolecule [PHYS CHEM] A single quantum-mechanical entity presumably formed by two reacting molecules and in existence only during the collision process; a concept in the hard-sphere collision theory of chemical kinetics.

supermultiplet [QUANT MECH] A set of quantum-mechanical states each of which has the same value of some fundamental quantum numbers and differs from the other members of the set by other quantum numbers, which take values from a range of numbers dictated by the fundamental quantum numbers.

supernova [ASTRON] A star that suddenly bursts into very great brilliance as a result of its blowing up; it is orders of magnitude brighter than a nova.

supernumerary bud See accessory bud.

supernumerary chromosome [CYTOL] A chromosome present in addition to the normal chromosome complement. Also known as accessory chromosome.

superphosphate [MATER] The most important phosphorus fertilizer, derived by action of sulfuric acid on phosphate rock (mostly tribasic calcium phosphate) to produce a mix of gypsum and monobasic calcium phosphate.

superplasticity [MET] The unusual ability of some metals and alloys to elongate uniformly by thousands of percent at elevated temperatures, much like hot polymers or glasses.

superposed stream See consequent stream.

superposition [GEOL] 1. The order in which sedimentary layers are deposited, the highest being the youngest. 2. The process by which the layering occurs. [MATH] The principle of superposition states that any given geometric figure in a euclidean space can be so moved about as not to change its size or shape. [PHYS] Addition of phenomena when the sum of two physically realizable disturbances is also physically realizable; for example, sound waves are superposable in this sense, but shock waves are not.

superposition principle See principle of superposition.

superposition theorem See principle of superposition.

superpressure balloon [METEOROL] A meteorological balloon consisting of a nonextensible envelope designed to withstand higher internal pressure differentials than external ones; it will maintain constant elevations until sufficient gas diffuses from it to cause a change in buoyancy.

superrotation [GEOPHYS] The generally more rapid relative motion of the atmosphere found in the very tenuous regions at heights about 185 mi (300 km).

supersaturation [METEOROL] The condition existing in a given portion of the atmosphere when the relative humidity is greater than 100%, in respect to a plane surface of pure water or pure ice. [PHYS CHEM] The condition existing in a solution when it contains more solute than is needed to cause saturation. Also known as supersolubility.

supersolubility See supersaturation.

supersonic [ACOUS] See ultrasonic. [PHYS] Of, pertaining to, or dealing with speeds greater than the speed of sound.

supersonic combustion ramjet See scramjet.

superstructure [CIV ENG] The part of a structure that is raised on the foundation. [NAV ARCH] The entire structure of a ship above the main deck. [SOLID STATE] See superlattice.

supersymmetry [PARTIC PHYS] A generalization of previously known symmetries of elementary particles to new kinds of supermultiplets that include both bosons and fermions; it is based on graded Lie algebras rather than on Lie algebras.

superthermal source [NUCLEO] A source of ultracold neutrons consisting of a container of liquid helium, cooled to about 1 K or less, which has walls that are good reflectors of ultracold neutrons but are transparent to electrons with energies of about 1 millielectronvolt, and which is placed in a beam of such electrons.

supervisor call [COMPUT SCI] A mechanism whereby a computer program can interrupt the normal flow of processing and ask the supervisor to perform a function for the program that the program cannot or is not permitted to perform for itself.

supervisor interrupt [COMPUT SCI] An interruption caused by the program being executed which issues an instruction to the master control program.

supervisor mode [COMPUT SCI] A method of computer operation in which the computer can execute all its own instructions, including the privileged instruction not normally allowed to the programmer, in contrast to problem mode.

supervisory controlled manipulation [ENG] A form of remote manipulation in which a computer enables the operator to teach the manipulator motion patterns to be remembered and repeated later.

supervisory program [COMPUT SCI] A program that organizes and regulates the flow of work in a computer system, for example, it may automatically change over from one run to another and record the time of the run.

supervisory routine [COMPUT SCI] A program or routine that initiates and guides the execution of several (or all) other routines and programs; it usually forms part of (or is) the operating system.

supervisory signal [ELEC] A signal which indicates the operating condition of a circuit or a combination of circuits in a switching apparatus or other electrical equipment to an attendant.

supplementary angle [MATH] One angle is supplementary to another angle if their sum is 180°.

suppressed carrier [COMMUN] A carrier that is suppressed at the transmitter; the chrominance subcarrier in a color television transmitter is an example.

suppression [COMPUT SCI] 1. Removal or deletion usually of insignificant digits in a number, especially zero suppression. 2. Optional function in either on-line or off-line printing devices that permits them to ignore certain characters or groups of characters which may be transmitted through them. [ELECTR] Elimination of any component of an emission, as a particular frequency or group of frequencies in an audio-frequency of a radio-frequency signal.

suppressor [ELEC] 1. In general, a device used to reduce or eliminate noise or other signals that interfere with the operation of a communication system, usually at the noise source. 2. Specifically, a resistor used in series with a spark plug or distributor of an automobile engine or other internal combustion engine to suppress spark noise that might otherwise interfere with radio reception. [ELECTR] See suppressor grid.

suppressor gene [GEN] A gene that reverses the effect of a mutation in another gene.

suppressor grid [ELECTR] A grid placed between two positive electrodes in an electron tube primarily to reduce the flow of secondary electrons from one electrode to the other; it is usually used between the screen grid and the anode. Also known as suppressor.

suppressor mutation [GEN] A mutation that restores functional loss of a primary mutation and is located at a different genetic site from the primary mutation.

suprarenal gland   See adrenal gland.

surface acoustic wave   [ACOUS] A sound wave that propagates along and is bound to the surface of a solid; ordinarily it contains both compressional and shear components. Abbreviated SAW.

surface acoustic wave device   [ELECTR] Any device, such as a filter, resonator, or oscillator, which employs surface acoustic waves with frequencies in the range 10^7–10^9 hertz, traveling on the optically polished surface of a piezoelectric substrate, to process electronic signals.

surface barrier   [ELECTR] A potential barrier formed at a surface of a semiconductor by the trapping of carriers at the surface.

surface-barrier diode   [ELECTR] A diode utilizing thin-surface layers, formed either by deposition of metal films or by surface diffusion, to serve as a rectifying junction.

surface-barrier transistor   [ELECTR] A transistor in which the emitter and collector are formed on opposite sides of a semiconductor wafer, usually made of n-type germanium, by training two jets of electrolyte against its opposite surfaces to etch and then electroplate the surfaces.

surface boundary layer   [METEOROL] That thin layer of air adjacent to the earth's surface, extending up to the so-called anemometer level (the base of the Ekman layer); within this layer the wind distribution is determined largely by the vertical temperature gradient and the nature and contours of the underlying surface, and shearing stresses are approximately constant. Also known as atmospheric boundary layer; friction layer; ground layer; surface layer.

surface chemistry   [PHYS CHEM] The study and measurement of the forces and processes that act on the surfaces of fluids (gases and liquids) and solids, or at an interface separating two phases; for example, surface tension.

surface condenser   [MECH ENG] A heat-transfer device used to condense a vapor, usually steam under vacuum, by absorbing its latent heat in cooling fluid, ordinarily water.

surface current   [OCEANOGR] **1.** Water movement which extends to depths of 3–10 feet (1–3 meters) below the surface in nearshore areas, and to about 33 feet (10 meters) in deep-ocean areas. **2.** Any current whose maximum velocity core is at or near the surface.

surface deposit   See surficial deposit.

surface energy   [FL MECH] The energy per unit area of an exposed surface of a liquid; generally greater than the surface tension, which equals the free energy per unit surface.

surface hardening   [MET] Hardening the surface of steel by one of several processes, such as carburizing, carbonitriding, nitriding, flame or induction hardening, and surface working.

surface layer   See surface boundary layer.

surface mining   [MIN ENG] Mining at or near the surface; includes placer mining, mining in open gloryhole or milling pits, mining and removing ore from opencuts by hand or with mechanical excavating and transportation equipment, and the removal of capping or overburden to uncover the ores.

surface of revolution   [MATH] A surface realized by rotating a planar curve about some axis in its plane.

surface passivation   [ELECTR] A method of coating the surface of a p-type wafer for a diffused junction transistor with an oxide compound, such as silicon oxide, to prevent penetration of the impurity in undesired regions.

surface pipe   [PETRO ENG] The string of casing first set in a well, usually to shut off shallow, fresh-water sands from contamination by deeper, saline waters.

surface pressure   [METEOROL] The atmospheric pressure at a given location on the earth's surface; the expression is applied loosely and about equally to the more specific terms: station pressure and sea-level pressure. [PHYS] See film pressure.

surface resistivity   [ELEC] The electric resistance of the surface of an insulator, measured between the opposite sides of a square on the surface; the value in ohms is independent of the size of the square and the thickness of the surface film.

surface retention   See surface storage.

surface sizing   See sizing treatment.

surface soil   [GEOL] The soil extending 5 to 8 inches (13 to 20 centimeters) below the surface.

surface state   [SOLID STATE] An electron state in a semiconductor whose wave function is restricted to a layer near the surface.

surface storage   [HYD] The part of precipitation retained temporarily at the ground surface as interception or depression storage so that it does not appear as infiltration or surface runoff either during the rainfall period or shortly thereafter. Also known as initial detention; surface retention.

surface temperature   [METEOROL] Temperature of the air near the surface of the earth. [OCEANOGR] Temperature of the layer of seawater nearest the atmosphere.

surface tension   [FL MECH] The force acting on the surface of a liquid, tending to minimize the area of the surface; quantitatively, the force that appears to act across a line of unit length on the surface. Also known as interfacial force; interfacial tension; surface tensity.

surface tensity   See surface tension.

surface-to-air missile   [ORD] A guided missile designed to be fired at an airborne target from the ground or from the deck of a surface ship; examples include Bomarc, Hawk, Nike, Talos, Tartar, Terrier, and Wizard. Abbreviated SAM.

surface-to-surface missile   [ORD] A guided missile designed to be fired at a surface target from a surface position on land or water; examples include Atlas, Corporal, Dart, Jupiter, Lacross, Mace, Matador, Navaho, Pershing, Polaris, Redstone, Regulus, Sergeant, Snark, Thor, and Titan. Abbreviated SSM.

surface visibility   [METEOROL] The visibility determined from a point on the ground, as opposed to control-tower visibility.

surface wash   See sheet erosion.

surface water   [HYD] All bodies of water on the surface of the earth. [OCEANOGR] See mixed layer.

surface wave   [COMMUN] See ground wave. [ELECTROMAG] A wave that can travel along an interface between two different mediums without radiation; the interface must be essentially straight in the direction of propagation; the commonest interface used is that between air and the surface of a circular wire. [FL MECH] A wave that distorts the free surface that separates two fluid phases, usually a liquid and a gas. [MECH] See Rayleigh wave. [OCEANOGR] A progressive gravity wave in which the disturbance is of greatest amplitude at the air-water interface.

surface wind   [METEOROL] The wind measured at a surface observing station; customarily, it is measured at some distance above the ground itself to minimize the distorting effects of local obstacles and terrain.

surfacing mat   See overlay.

surfactant flooding   See micellar flooding.

surficial deposit [GEOL] Unconsolidated alluvial, residual, or glacial deposits overlying bedrock or occurring on or near the surface of the earth. Also known as superficial deposit; surface deposit.

surf zone [OCEANOGR] The area between the landward limit of wave uprush and the farthest seaward breaker.

surge [ASTROPHYS] An unusually violent solar prominence that usually accompanies a smaller flare, consisting of a brilliant jet of gas which shoots out into the solar corona with a speed on the order of 300 kilometers per second and reaches a height on the order of 100,000 kilometers. [ELEC] A momentary large increase in the current or voltage in an electric circuit. [ENG] 1. An upheaval of fluid in a processing system, frequently causing a carryover (puking) of liquid through the vapor lines. 2. The peak system pressure. 3. An unstable pressure buildup in a plastic extruder leading to variable throughput and waviness of the hollow plastic tube. [FL MECH] A wave at the free surface of a liquid generated by the motion of a vertical wall, having a change in the height of the surface across the wavefront and violent eddy motion at the wavefront. [OCEANOGR] 1. Wave motion of low height and short period, from about ½ to 60 minutes. 2. See storm surge.

surge arrester [ELEC] A protective device designed primarily for connection between a conductor of an electrical system and ground to limit the magnitude of transient overvoltages on equipment. Also known as arrester; lightning arrester.

surge current [ELEC] A short-duration, high-amperage electric current wave that may sweep through an electrical network, as a power transmission network, when some portion of it is strongly influenced by the electrical activity of a thunderstorm.

surge electrode current See fault electrode current.

surge generator [ELEC] A device for producing high-voltage pulses, usually by charging capacitors in parallel and discharging them in series.

surging glacier [HYD] A glacier that alternates periodically between surges (brief periods of rapid flow) and stagnation.

surveillance radar [NAV] Ground radar used for traffic-control purposes in the approach and landing zone; it is used to assist controllers in converting random arrivals to regular landings and in positioning such aircraft so that they may make low approaches by the use of a fixed-beam, low-approach system or by a precision radar low-approach system.

survey [ENG] 1. The process of determining accurately the position, extent, contour, and so on, of an area, usually for the purpose of preparing a chart. 2. The information so obtained. [NUCLEO] Measurement of radiation in the vicinity of a nuclear reactor or other source.

surveyor's level [ENG] A telescope and spirit level mounted on a tripod, rotating vertically and having leveling screws for adjustment.

surveyor's measure [ENG] A system of measurement used in surveying having the engineer's, or Gunter's, chain as a unit.

suscept [PL PATH] A plant that is susceptible to disease caused by either parasitic or nonparasitic plant pathogens.

susceptance [ELEC] The imaginary component of admittance.

susceptibility [ELEC] See electric susceptibility. [ELECTROMAG] See magnetic susceptibility. [ORD] The degree to which a device, equipment, or

weapons system is open to effective attack due to one or more inherent weaknesses.

suspended acoustical ceiling [BUILD] An acoustical ceiling which is suspended from either the roof or a higher ceiling.

suspended solids See suspension.

suspension [CHEM] A mixture of fine, nonsettling particles of any solid within a liquid or gas, the particles being the dispersed phase, while the suspending medium is the continuous phase. Also known as suspended solids. [ENG] A fine wire or coil spring that supports the moving element of a meter. [MIN ENG] The bolting of rock to secure fragments or sections, such as small slabs barred down after blasting blocks of rock broken by fracture or joint patterns, which may subsequently loosen and fall.

suspension bridge [CIV ENG] A fixed bridge consisting of either a roadway or a truss suspended from two cables which pass over two towers and are anchored by backstays to a firm foundation.

suspension current See turbidity current.

suspension feeder [ZOO] An animal that feeds on small particles suspended in water; particles may be minute living plants or animals, or products of excretion or decay from these or larger organisms.

suspension system [MECH ENG] A system of springs, shock absorbers, and other devices supporting the upper part of a motor vehicle on its running gear.

sustained yield [BIOL] In a biological resource such as timber or grain, the replacement of a harvest yield by growth or reproduction before another harvest occurs.

sutured [PETR] Referring to rock texture in which mineral grains or irregularly shaped crystals interfere with their neighbors, producing interlocking, irregular contacts without interstitial spaces.

Svedberg equation [STAT MECH] An equation which states that the amplitude of vibration of a particle which exhibits Brownian motion is proportional to its period.

SW See switch.

swab [MIN ENG] 1. A pistonlike device provided with a rubber cap ring that is used to clean out debris inside a borehole or casing. 2. The act of cleaning the inside of a tubular object with a swab. [PETRO ENG] In petroleum drilling, to pull the drill string so rapidly that the drill mud is sucked up and overflows the collar of the borehole, thus leaving an undesirably empty borehole.

swage bolt [DES ENG] A bolt having indentations with which it can be gripped in masonry.

swaging [MET] Tapering a rod or tube or reducing its diameter by any of several methods, such as forging, squeezing, or hammering.

swale [GEOL] 1. A slight depression, sometimes swampy, in the midst of generally level land. 2. A shallow depression in an undulating ground moraine due to uneven glacial deposition. 3. A long, narrow, generally shallow, troughlike depression which lies between two beach ridges and is aligned roughly parallel to the coastline.

swan [VERT ZOO] Any of several species of large waterfowl comprising the subfamily Anatinae; they are herbivorous stout-bodied forms with long necks and spatulate bills.

Swan See Cygnus.

Swanscombe man [PALEON] A partial skull recovered in Swanscombe, Kent, England, which represents an early stage of *Homo sapiens* but differing in having a vertical temporal region and a rounded occipital profile.

swartzite [MINERAL] $CaMg(UO_2)(CO_3)_3 \cdot 12H_2O$ A green monoclinic mineral composed of hydrous carbonate of calcium, magnesium, and uranium.

S wave [GEOPHYS] A seismic body wave propagated in the crust or mantle of the earth by a shearing motion of material; speed is 3.0–4.0 kilometers per second in the crust and 4.4–4.6 in the mantle. Also known as distortional wave; equivoluminal wave; rotational wave; secondary wave; shake wave; shear wave; tangential wave; transverse wave.

sway brace [CIV ENG] One or a pair of diagonal members designed to resist horizontal forces, such as wind.

sway frame [CIV ENG] A unit in the system of members of a bridge that provides bracing against side sway; consists of two diagonals, the verticals, the floor beam, and the bottom strut.

sweat [CHEM] Exudation of nitroglycerin from dynamite due to separation of nitroglycerin from its adsorbent. [MET] Exudate of low-melting-point constituents from a metal on solidification. [PHYSIO] The secretion of the sweat glands. Also known as perspiration. [SCI TECH] Formation of moisture beads on a surface as a result of concentration.

sweep [ELECTR] 1. The steady movement of the electron beam across the screen of a cathode-ray tube, producing a steady bright line when no signal is present; the line is straight for a linear sweep and circular for a circular sweep. 2. The steady change in the output frequency of a signal generator from one limit of its range to the other. [MET] A profile pattern used to form molds for symmetrical articles made by sweep casting. [ORD] 1. Swift flight of a formation of combat airplanes over enemy territory. 2. To cover a wide area with gunfire.

sweep amplifier [ELECTR] An amplifier used with a cathode-ray tube, such as in a television receiver or cathode-ray oscilloscope, to amplify the sawtooth output voltage of the sweep oscillator, to shape the waveform for the deflection circuits of a television picture tube, or to provide balanced signals to the deflection plates.

sweep circuit [ELECTR] The sweep oscillator, sweep amplifier, and any other stage used to produce the deflection voltage or current for a cathode-ray tube. Also known as scanning circuit.

sweep generator Also known as sweep oscillator. [ELECTR] 1. An electronic circuit that generates a voltage or current, usually recurrent, as a prescribed function of time; the resulting waveform is used as a time base to be applied to the deflection system of an electron-beam device, such as a cathode-ray tube. Also known as time-base generator; timing-axis oscillator. 2. A test instrument that generates a radio-frequency voltage whose frequency varies back and forth through a given frequency range at a rapid constant rate; used to produce an input signal for circuits or devices whose frequency response is to be observed on an oscilloscope.

sweep oscillator See sweep generator.

sweep rate [ELECTR] The number of times a radar radiation pattern rotates during 1 minute; sometimes expressed as the duration of one complete rotation in seconds.

sweep voltage [ELECTR] Periodically varying voltage applied to the deflection plates of a cathode-ray tube to give a beam displacement that is a function of time, frequency, or other data base.

sweet corrosion [PETRO ENG] Corrosion occurring in oil or gas wells where there is no iron-sulfide corrosion product, and there is no odor of hydrogen sulfide in the produced reservoir fluid.

sweet crudes [MATER] Crude petroleum oil containing little sulfur.

sweetening [CHEM ENG] Improvement of a petroleum-product color and odor by converting sulfur compounds into disulfides with sodium plumbite (doctor treating), or by removing them by contacting the petroleum stream with alkalies or other sweetening agents.

sweetgum [BOT] *Liquidambar styraciflua.* A deciduous tree of the order Hamamelidales found in the southeastern United States, and distinguished by its five-lobed, or star-shaped, leaves, and by the corky ridges developed on the twigs.

sweet potato [BOT] *Ipomoea batatas.* A tropical vine having variously shaped leaves, purplish flowers, and a sweet, fleshy, edible tuberous root.

swell [GEOL] 1. The volumetric increase of soils on being removed from their compacted beds due to an increase in void ratio. 2. A local enlargement or thickening in a vein or ore deposit. 3. A low dome or quaquaversal anticline of considerable areal extent; long and generally symmetrical waves contribute to the mixing processes in the surface layer and thus to its sound transmission properties. 4. Gently rising ground, or a rounded hill above the surrounding ground or ocean floor. [MIN ENG] See horseback. [OCEANOGR] Ocean waves which have traveled away from their generating area; these waves are of relatively long length and period, and regular in character.

swim bladder [VERT ZOO] A gas-filled cavity found in the body cavities of most bony fishes; has various functions in different fishes, acting as a float, a lung, a hearing aid, and a sound-producing organ.

swimmer's conjunctivitis See inclusion conjunctivitis.

swimming pool conjunctivitis See inclusion conjunctivitis.

swing bridge [CIV ENG] A movable bridge that pivots in a horizontal plane about a center pier.

swing joint [DES ENG] A pipe joint in which the parts may be rotated relative to each other.

switch [COMPUT SCI] 1. A hardware or programmed device for indicating that one of several alternative states or conditions have been chosen, or to interchange or exchange two data items. 2. A symbol used to indicate a branch point, or a set of instructions to condition a branch. [CIV ENG] 1. A device for enabling a railway car to pass from one track to another. 2. The junction of two tracks. [ELEC] A manual or mechanically actuated device for making, breaking, or changing the connections in an electric circuit. Also known as electric switch. Symbolized SW.

switchboard [COMMUN] A manually operated apparatus at a telephone exchange, on which the various circuits from subscribers and other exchanges are terminated to enable operators to establish communications either between two subscribers on the same exchange, or between subscribers on different exchanges. Also known as telephone switchboard. [ELEC] A single large panel or assembly of panels on which are mounted switches, circuit breakers, meters, fuses, and terminals essential to the operation of electric equipment. Also known as electric switchboard.

switched capacitor [ELECTR] An integrated circuit element, consisting of a capacitor with two metal oxide semiconductor (MOS) switches, whose function is approximately equivalent to that of a resistor.

switched-message network [COMPUT SCI] A data transmission system in which a user can communicate with any other user of the network.

switch function [ELECTR] A circuit having a fixed number of inputs and outputs designed such that the output information is a function of the input information, each expressed in a certain code or signal configuration or pattern.

switchgear [ELEC] The aggregate of switching devices for a power or transforming station, or for electric motor control.

switching [ELEC] Making, breaking, or changing the connections in an electrical circuit.

switching circuit [ELEC] A constituent electric circuit of a switching or digital processing system which receives, stores, or manipulates information in coded form to accomplish the specified objectives of the system.

switching device [ENG] An electrical or mechanical device or mechanism, which can bring another device or circuit into an operating or nonoperating state. Also known as switching mechanism.

switching diode [ELECTR] A crystal diode that provides essentially the same function as a switch; below a specified applied voltage it has high resistance corresponding to an open switch, while above that voltage it suddenly changes to the low resistance of a closed switch.

switching gate [ELECTR] An electronic circuit in which an output having constant amplitude is registered if a particular combination of input signals exists; examples are the OR, AND, NOT, and IN-HIBIT circuits. Also known as logical gate.

switching key See key.

switching mechanism See switching device.

switching system [COMMUN] An assembly of switching and control devices provided so that any station in a communications system may be connected as desired with any other station.

switching theory [ELECTR] The theory of circuits made up of ideal digital devices; included are the theory of circuits and networks for telephone switching, digital computing, digital control, and data processing.

switching trunk [ELEC] Trunk from a long-distance office to a local exchange office used for completing a long-distance call.

switch register [COMPUT SCI] A manual switch on the control panel by means of which a bit may be entered in a processor register.

Swordfish See Dorado.

SWS See deep sleep.

sycamore [BOT] 1. Any of several species of deciduous trees of the genus *Platanus*, especially *P. occidentalis* of eastern and central North America, distinguished by simple, large, three- to five-lobed leaves and spherical fruit heads. 2. The Eurasian maple (*Acer pseudoplatanus*).

Sycettida [INV ZOO] An order of calcareous sponges of the subclass Calcaronea in which choanocytes occur in flagellated chambers, and the spongocoel is not lined with these cells.

Sycettidae [INV ZOO] A family of sponges in the order Sycettida.

Sycidales [PALEON] A group of fossil aquatic plants assigned to the Charophyta, characterized by vertically ribbed gyrogonites.

Sydenham's chorea See St. Vitus dance.

syenite [PETR] A visibly crystalline plutonic rock with granular texture composed largely of alkali feldspar, with subordinate plagioclose and mafic minerals; the intrusive equivalent of trachyte.

Syllidae [INV ZOO] A large family of polychaete annelids belonging to the Errantia; identified by their long, linear, translucent bodies with articulated cirri; size ranges from minute *Exogone* to *Trypanosyllis*, which may be 100 millimeters long.

Syllinae [INV ZOO] A subfamily of polychaete annelids of the family Syllidae.

Sylonidae [INV ZOO] A family of parasitic crustaceans in the order Rhizocephala.

Sylopidae [INV ZOO] A family of coleopteran insects in the superfamily Meloidea in which the elytra in males are reduced to small leathery flaps while the hindwings are large and fan-shaped.

sylvanite [MINERAL] $(Au,Ag)Te_2$ A steel-gray, silver-white, or brass-yellow mineral that crystallizes in the monoclinic system and often occurs in implanted crystals. Also known as goldschmidtite; graphic tellurium; white tellurium; yellow tellurium.

Sylvicolidae [INV ZOO] A family of orthorrhaphous dipteran insects in the series Nematocera.

sylvine See sylvite.

sylvite [MINERAL] KCl A salty-tasting, white or colorless isometric mineral, occurring in cubes or crystalline masses or as a saline residue; the chief ore of potassium. Also known as leopoldite; sylvine.

sym- [ORG CHEM] A chemical prefix; denotes structure of a compound in which substituents are symmetrical with respect to a functional group or to the carbon skeleton.

symbiosis [ECOL] 1. An interrelationship between two different species. 2. An interrelationship between two different organisms in which the effects of that relationship is expressed as being harmful or beneficial. Also known as consortism.

symbol [CHEM] Letter or combination of letters and numbers that represent various conditions or properties of an element, for example, a normal atom, O (oxygen); with its atomic weight, ^{16}O; its atomic number, $_8^{16}O$; as a molecule, O_2; as an ion, O^{2+}; in excited state, O^*; or as an isotope, ^{18}O. [SCI TECH] 1. A design used on a diagram to represent a component or to identify specific characteristics, quantities, or objects. 2. A sign letter or abbreviation used on a diagram or in an equation to represent a quantity or to identify an object.

symbolic address [COMPUT SCI] In coding, a programmer-defined symbol that represents the location of a particular datum item, instruction, or routine. Also known as symbolic number.

symbolic age of neutrons See Fermi age.

symbolic assembly language listing [COMPUT SCI] A list that may be produced by a computer during the compilation of a program showing the source language statements together with the corresponding machine language instructions generated by them.

symbolic number See symbolic address.

symmetrical alternating quantity [PHYS] Alternating quantity of which all values separated by a half period have the same magnitude but opposite sign.

symmetrical avalanche rectifier [ELECTR] Avalanche rectifier that can be triggered in either direction, after which it has a low impedance in the triggered direction.

symmetrical band-pass filter [ELECTR] A band-pass filter whose attenuation as a function of frequency is symmetrical about a frequency at the center of the pass band.

symmetrical band-reject filter [ELECTR] A band-rejection filter whose attenuation as a function of frequency is symmetrical about a frequency at the center of the rejection band.

symmetrical fold [GEOL] A fold whose limbs have approximately the same angle of dip relative to the axial surface. Also known as normal fold.

symmetrical lens [OPTICS] A lens system consisting of two parts, each of which is the mirror image of the other.

symmetrical transducer [ELECTR] A transducer is symmetrical with respect to a specified pair of terminations when the interchange of that pair of terminations will not affect the transmission.

symmetric difference [MATH] The symmetric difference of two sets consists of all points in one or the other of the sets but not in both.

symmetric function [MATH] A function whose value is unchanged for any permutation of its variables.

symmetric space [MATH] A differentiable manifold which has a differentiable multiplication operation that behaves similarly to the multiplication of a complex number and its conjugate.

Symmetrodonta [PALEON] An order of the extinct mammalian infraclass Pantotheria distinguished by the central high cusp, flanked by two smaller cusps and several low minor cusps, on the upper and lower molars.

symmetry [BIOL] The disposition of organs and other constituent parts of the body of living organisms with respect to imaginary axes. [MATH] A geometric object G has this property relative to some configuration S of its points if S determines two pieces of G which can be reflected onto each other through S. [PHYS] *See* invariance.

symmetry axis *See* axis of symmetry; rotation axis.

symmetry center *See* center of symmetry.

symmetry class *See* crystal class.

symmetry element [CRYSTAL] 1. Some combination of rotations and reflections and translations which brings a crystal into a position that cannot be distinguished from its original position. Also known as symmetry operation; symmetry transformation. 2. The rotational axes, mirror planes, and center of symmetry characteristic of a given crystal.

symmetry function *See* symmetry transformation.

symmetry group [MATH] A group composed of all rigid motions or similarity transformations of some geometric object onto itself.

symmetry law *See* invariance principle.

symmetry operation *See* symmetry element.

symmetry plane *See* plane of mirror symmetry.

symmetry principle [MATH] The centroid of a geometrical figure (line, area, or volume) is at a point on a line or plane of symmetry of the figure. [PHYS] *See* invariance principle.

symmetry transformation [CRYSTAL] *See* symmetry element. [MATH] A rigid motion sending a geometric object onto itself; examples are rotations and, for the case of a polygon, permutations of the vertices. Also known as symmetry function.

symon fault *See* horseback.

sympathetic nervous system [ANAT] The portion of the autonomic nervous system, innervating smooth muscle and glands of the body, which upon stimulation produces a functional state of preparation for flight or combat.

sympatric [ECOL] Of a species, occupying the same range as another species but maintaining identity by not interbreeding.

sympetalous *See* gamopetalous.

symphile [ECOL] An organism, usually a beetle, living as a guest in the nest of a social insect, such as an ant, where it is reared and bred in exchange for its exudates.

Symphyla [INV ZOO] A class of the Myriapoda comprising tiny, pale, centipedelike creatures which inhabit humus or soil.

symphysis [ANAT] An immovable articulation of bones connected by fibrocartilaginous pads.

symphysis pubis *See* pubic symphysis.

Symphyta [INV ZOO] A suborder of the Hymenoptera including the sawflies and horntails characterized by a broad base attaching the abdomen to the thorax.

sympodium [BOT] A branching system in trees in which the main axis is comprised of successive secondary branches, each representing the dominant fork of a dichotomy.

Synallactidae [INV ZOO] A family of echinoderms of the order Aspidochirotida comprising mainly deep-sea forms which lack tentacle ampullae.

synantectic [MINERAL] Refers to a mineral that was formed by the reaction of two other minerals.

Synanthae [BOT] An equivalent name for Cyclanthales.

Synanthales [BOT] An equivalent name for Cyclanthales.

synapse [ANAT] A site where the axon of one neuron comes into contact with and influences the dendrites of another neuron or a cell body.

Synaptidae [INV ZOO] A family of large sea cucumbers of the order Apodida having a respiratory tree and having a reduced water-vascular system.

synarthrosis [ANAT] An articulation in which the connecting material (fibrous connective tissue) is continuous, immovably binding the bones.

Synbranchiformes [VERT ZOO] An order of eellike actinopterygian fishes that, unlike true eels, have the premaxillae present as distinct bones.

Synbranchii [VERT ZOO] The equivalent name for Synbranchiformes.

Syncarida [INV ZOO] A superorder of crustaceans of the subclass Melacostraca lacking a carapace and oostegites and having exopodites on all thoracic limbs.

syncarpous [BOT] Descriptive of a gynoecium having the carpels united in a compound ovary.

sync generator *See* synchronizing generator.

synchro [ELEC] Any of several devices used for transmitting and receiving angular position or angular motion over wires, such as a synchro transmitter or synchro receiver. Also known as mag-slip (British usage); self-synchronous device; self-synchronous repeater; selsyn.

synchro- [SCI TECH] Occurring at the same time or made to occur at the same time.

synchrocyclotron [NUCLEO] A circular particle accelerator for accelerating protons, deuterons, or alpha particles, in which the frequency of the accelerating voltage is modulated to maintain synchronism with the frequency of the particles which spiral outward and attain energies at which the relativistic mass increase becomes significant. Also known as frequency-modulated cyclotron; synchrophasotron.

synchro differential receiver [ELEC] A synchro receiver that subtracts one electrical angle from another and delivers the difference as a mechanical angle. Also known as differential synchro.

synchro differential transmitter [ELEC] A synchro transmitter that adds a mechanical angle to an electrical angle and delivers the sum as an electrical angle. Also known as differential synchro.

synchro generator *See* synchro transmitter.

synchromesh [MECH ENG] An automobile transmission device that minimizes clashing; acts as a friction clutch, bringing gears approximately to correct speed just before meshing.

synchro motor   *See* synchro receiver.

synchronism   [ELEC] Of a synchronous motor, the condition under which the motor runs at a speed which is directly related to the frequency of the power applied to the motor and is not dependent upon variables.   [PHYS] Condition of two periodic quantities which have the same frequency, and whose phase difference is either constant or varies around a constant average value.

synchronizer   [COMPUT SCI] A computer storage device used to compensate for a difference in rate of flow of information or time of occurrence of events when transmitting information from one device to another.   [ELECTR] The component of a radar set which generates the timing voltage for the complete set.

synchronizing generator   [ELECTR] An electronic generator that supplies synchronizing pulses to television studio and transmitter equipment. Also known as sync generator; sync-signal generator.

synchronizing pulse   [COMMUN] In pulse modulation, a pulse which is transmitted to synchronize the transmitter and the receiver; it is usually distinguished from signal-carrying pulses by some special characteristic.

synchronizing reactor   [ELEC] Current-limiting reactor for connecting momentarily across the open contacts of a circuit-interrupting device for synchronizing purposes.

synchronizing relay   [ELEC] Relay which functions when two alternating-current sources are in agreement within predetermined limits of phase angle and frequency.

synchronizing signal   *See* sync signal.

synchronous   [ENG] In step or in phase, as applied to two or more circuits, devices, or machines.   [GEOL] Geological rock units or features formed at the same time.

synchronous capacitor   [ELEC] A synchronous motor running without mechanical load and drawing a large leading current, like a capacitor; used to improve the power factor and voltage regulation of an alternating-current power system.

synchronous clamp circuit   *See* keyed clamp circuit.

synchronous computer   [COMPUT SCI] A digital computer designed to operate in sequential elementary steps, each step requiring a constant amount of time to complete, and being initiated by a timing pulse from a uniformly running clock.

synchronous converter   [ELEC] A converter in which motor and generator windings are combined on one armature and excited by one magnetic field; normally used to change alternating to direct current. Also known as converter; electric converter.

synchronous data transmission   [COMMUN] Data transmission in which a clock defines transmission times for data; since start and stop bits for each character are not needed, more of the transmission bandwidth is available for message bits.

synchronous demodulator   *See* synchronous detector.

synchronous detector   [ELECTR] **1.** A detector that inserts a missing carrier signal in exact synchronism with the original carrier at the transmitter; when the input to the detector consists of two suppressed-carrier signals in phase quadrature, as in the chrominance signal of a color television receiver, the phase of the reinserted carrier can be adjusted to recover either one of the signals. Also known as synchronous demodulator. **2.** *See* cross-correlator.

synchronous gate   [ELECTR] A time gate in which the output intervals are synchronized with an incoming signal.

synchronous generator   [ELEC] A machine that generates an alternating voltage when its armature or field is rotated by a motor, an engine, or other means. The output frequency is exactly proportional to the speed at which the generator is driven.

synchronous inverter   *See* dynamotor.

synchronous machine   [ELEC] An alternating-current machine whose average speed is proportional to the frequency of the applied or generated voltage.

synchronous motor   [ELEC] A synchronous machine that transforms alternating-current electric power into mechanical power, using field magnets excited with direct current.

synchronous orbit   [AERO ENG] **1.** An orbit in which a satellite makes a limited number of equatorial crossing points which are then repeated in synchronism with some defined reference (usually earth or sun). **2.** Commonly, the equatorial, circular, 24-hour case in which the satellite appears to hover over a specific point of the earth.

synchronous rectifier   [ELECTR] A rectifier in which contacts are opened and closed at correct instants of time for rectification by a synchronous vibrator or by a commutator driven by a synchronous motor.

synchronous satellite   *See* geostationary satellite.

synchronous switch   [ELECTR] A thyratron circuit used to control the operation of ignitrons in such applications as resistance welding.

synchronous system   [COMMUN] A telecommunication system in which transmitting and receiving apparatus operate continuously at substantially the same rate, and correction devices are used, if necessary, to maintain them in a fixed relationship.

synchronous vibrator   [ELECTROMAG] An electromagnetic vibrator that simultaneously converts a low direct voltage to a low alternating voltage and rectifies a high alternating voltage obtained from a power transformer to which the low alternating voltage is applied; in power packs, it eliminates the need for a rectifier tube.

synchrophasotron   *See* synchrocyclotron.

synchro receiver   [ELEC] A synchro that provides an angular position related to the applied angle-defining voltages; when two of its input leads are excited by an alternating-current voltage and the other three input leads are excited by the angle-defining voltages, the rotor rotates to the corresponding angular position; the torque of rotation is proportional to the sine of the difference between the mechanical and electrical angles. Also known as receiver synchro; selsyn motor; selsyn receiver; synchro motor.

synchroscope   [ELECTR] A cathode-ray oscilloscope designed to show a short-duration pulse by using a fast sweep that is synchronized with the pulse signal to be observed.   [ENG] An instrument for indicating whether two periodic quantities are synchronous; the indicator may be a rotating-pointer device or a cathode-ray oscilloscope providing a rotating pattern; the position of the rotating pointer is a measure of the instantaneous phase difference between the quantities.

synchro transmitter   [ELEC] A synchro that provides voltages related to the angular position of its rotor; when its two input leads are excited by an alternating-current voltage, the magnitudes and polarities of the voltages at the three output leads define the rotor position. Also known as selsyn generator; selsyn transmitter; synchro generator; transmitter; transmitter synchro.

synchrotron [NUCLEO] A device for accelerating electrons or protons in closed orbits in which the frequency of the accelerating voltage is varied (or held constant in the case of electrons) and the strength of the magnetic field is varied so as to keep the orbit radius constant.

synchrotron radiation [ELECTROMAG] Electromagnetic radiation generated by the acceleration of charged relativistic particles, usually electrons, in a magnetic field.

syncline [GEOL] A fold having stratigraphically younger rock material in its core; it is concave upward.

synclinorium [GEOL] A composite synclinal structure in a region of lesser folds.

sync signal [COMMUN] A signal transmitted after each line and field to synchronize the scanning process in a television or facsimile receiver with that of the receiver. Also known as synchronizing signal.

sync-signal generator See synchronizing generator.

syncytium [CYTOL] A mass of multinucleated cytoplasm without division into separate cells.

syndesmosis [ANAT] An articulation in which the bones are joined by collagen fibers.

syndet See synthetic detergent.

syndrome [MED] A group of signs and symptoms which together characterize a disease. Also known as complex.

syneresis [CHEM] Spontaneous separation of a liquid from a gel or colloidal suspension due to contraction of the gel.

synergid [BOT] Either of two small cells lying in the embryo sac in seed plants adjacent to the egg cell toward the micropylar end.

synergism [ECOL] An ecological association in which the physiological processes or behavior of an individual are enhanced by the nearby presence of another organism. [MATER] An action where the total effect of two active components in a mixture is greater than the sum of their individual effects, for example, a mixture volume that is greater than the sum of the individual volumes, or in resin formulation, the use of two or more stabilizers, where the combination improves polymer stability more than expected from the additive effect of the stabilizers, a material that causes such an effect is known as a synergist.

Syngamidae [INV ZOO] A family of roundworms belonging to the Strongyloidea and including parasites of birds and mammals.

syngamy [BIOL] Sexual reproduction involving union of gametes.

syngenetic [GEOL] 1. Pertaining to a primary sedimentary structure formed contemporaneously with sediment deposition. 2. Pertaining to a mineral deposit formed contemporaneously with the enclosing rock. Also known as ideogenous.

Syngnathidae [VERT ZOO] A family of fishes in the order Gasterosteiformes including the seahorses and pipefishes.

synodic period [ASTRON] The time period between two successive astronomical conjunctions of the same celestial objects.

synopsis [SYST] In taxonomy, a brief summary of current knowledge about a taxon.

synoptic [METEOROL] Refers to the use of meteorological data obtained simultaneously over a wide area for the purpose of presenting a comprehensive and nearly instantaneous picture of the state of the atmosphere.

synoptic oceanography [OCEANOGR] The study of the physical spatial parameters of the ocean through analysis of simultaneous observations from many stations.

synovia See synovial fluid.

synovial fluid [PHYSIO] A transparent viscid fluid secreted by synovial membranes. Also known as synovia.

synovial membrane [HISTOL] A layer of connective tissue which lines sheaths of tendons at freely moving articulations, ligamentous surfaces of articular capsules, and bursae.

syntax [COMPUT SCI] The set of rules needed to construct valid expressions or sentences in a language.

syntaxial overgrowth [MINERAL] A crystallographically oriented overgrowth of two alternating, chemically identical substances.

syntectic alloy [MET] A metallic composite material characterized by a reversible convertibility of its solid phase into two liquid phases by the application of heat.

Synteliidae [INV ZOO] The sap-flow beetles, a small family of coleopteran insects in the superfamily Histeroidea.

Syntexidae [INV ZOO] A family of the Hymenoptera in the superfamily Siricoidea.

synthesis [CHEM] Any process or reaction for building up a complex compound by the union of simpler compounds or elements. [CONT SYS] See system design.

synthesis gas [CHEM ENG] A mixture of gases prepared as feedstock for a chemical reaction, for example, carbon monoxide and hydrogen to make hydrocarbons or organic chemicals, or hydrogen and nitrogen to make ammonia.

synthesizer [ELECTR] An electronic instrument which combines simple elements to generate more complex entities; examples are frequency synthesizer and sound synthesizer.

synthetase See ligase.

synthetic address See generated address.

synthetic aperture [ENG] A method of increasing the ability of an imaging system, such as radar or acoustical holography, to resolve small details of an object, in which a receiver of large size (or aperture) is in effect synthesized by the motion of a smaller receiver and the proper correlation of the detected signals.

synthetic-aperture radar [ENG] A radar system in which an aircraft moving along a very straight path emits microwave pulses continuously at a frequency constant enough to be coherent for a period during which the aircraft may have traveled about 1 kilometer; all echoes returned during this period can then be processed as if a single antenna as long as the flight path had been used.

synthetic crude [MATER] The total liquid, multicomponent hydrocarbon mixture resulting from a process involving molecular rearrangement of charge stock, as from oil shale or synthesis gas. Also known as synthetic oil.

synthetic detergent [MATER] A liquid or solid material able to dissolve oily materials and disperse them (or emulsify them) in water. Also known as syndet.

synthetic oil See synthetic crude.

Syntrophiidina [PALEON] A suborder of extinct articulate brachiopods of the order Pentamerida characterized by a strong dorsal median fold.

syntype [SYST] Any specimen of a series when no specimen is designated as the holotype. Also known as cotype.

Synxiphosura [PALEON] An extinct heterogeneous order of arthropods in the subclass Merostomata possibly representing an explosive proliferation of aberrant, terminal, and apparently blind forms.

syphilis [MED] An infectious disease caused by the spirochete *Treponema pallidum*, transmitted principally by sexual intercourse.

syphilitic meningoencephalitis See general paresis.

Syringophyllidae [PALEON] A family of extinct corals in the order Tabulata.

syrinx [PALEON] A tube surrounding the pedicle in certain fossil brachiopods. [VERT ZOO] The vocal organ in birds.

Syrphidae [INV ZOO] The flower flies, a family of cyclorrhaphous dipteran insects in the series Aschiza.

Systellomatophora [INV ZOO] An order of the subclass Pulmonata in which the eyes are contractile but stalks are not retractile, the body is sluglike, oval, or lengthened, and the lung is posterior.

system [ELECTR] A combination of two or more sets generally physically separated when in operation, and such other assemblies, subassemblies, and parts necessary to perform an operational function or functions. [ENG] A combination of several pieces of equipment integrated to perform a specific function; thus a fire control system may include a tracking radar, computer, and gun. [GEOL] **1.** A major time-stratigraphic unit of worldwide significance, representing the basic unit of Phanerozic rocks. **2.** A group of related structures, such as joints. [PHYS] A region in space or a portion of matter that has a certain amount of one or more substances, ordered in one or more phases. [SCI TECH] A method of organizing entities or terms; in particular, organizing such entities into a larger aggregate.

systematic error [ENG] An error due to some known physical law by which it might be predicted; these errors produced by the same cause affect the mean in the same sense, and do not tend to balance each other but rather give a definite bias to the mean. [STAT] An error which results from some bias in the measurement process and is not due to chance, in contrast to random error.

systematic error checking code [COMPUT SCI] A type of self-checking code in which a valid character consists of the minimum number of digits needed to identify the character and distinguish it from any other valid character, and a set of check digits which maintain a minimum specified signal distance between any two valid characters. Also known as group code.

systematic joints [GEOL] Joints occurring in patterns or sets and oriented perpendicular to the boundaries of the constituent rock unit.

systematics [BIOL] The science of animal and plant classification.

system chart [COMPUT SCI] A flowchart that emphasizes the component operations which make up a system.

system check [COMPUT SCI] A check on the overall performance of the system, usually not made by built-in computer check circuits; for example, control total, hash totals, and record counts.

system design [COMPUT SCI] Determination in detail of the exact operational requirements of a system, resolution of these into file structures and in-

put/output formats, and relation of each to management tasks and information requirements. [CONT SYS] A technique of constructing a system that performs in a specified manner, making use of available components. Also known as synthesis.

Système International d'Unités See International System of Units.

system engineering See systems engineering.

system generation [COMPUT SCI] A process that creates a particular and uniquely specified operating system; it combines user-specified options and parameters with manufacturer-supplied general-purpose or nonspecialized program subsections to produce an operating system (or other complex software) of the desired form and capacity.

system optimization See optimization.

system response See response.

systems definition [COMPUT SCI] A document describing a computer-based system for processing data or solving a problem, including a general description of the aims and benefits of the system and clerical procedures employed, and detailed program specification. Also known as systems specification.

systems ecology [ECOL] The combined approaches of systems analysis and the ecology of whole ecosystems and subsystems.

systems engineering [ENG] The design of a complex interrelation of many elements (a system) to maximize an agreed-upon measure of system performance, taking into consideration all of the elements related in any way to the system, including utilization of manpower as well as the characteristics of each of the system's components. Also known as system engineering.

system software [COMPUT SCI] Computer software involved with data and program management, including operating systems, control programs, and data-base management systems.

systems programming [COMPUT SCI] The development and production of programs that have to do with translation, loading, supervision, maintenance, control, and running of computers and computer programs.

systems specification See systems definition.

systems test [COMPUT SCI] The running of the whole system against test data, a complete simulation of the actual running system for purposes of testing out the adequacy of the system. [ENG] A test of an entire interconnected set of components for the purpose of determining proper functions and interconnections.

system supervisor [COMPUT SCI] A control program which ensures an efficient transition in running program after program and accomplishing setups and control functions.

syzygy [ASTRON] **1.** One of the two points in a celestial object's orbit where it is in conjunction with or opposition to the sun. **2.** Those points in the moon's orbit where the moon, earth, and sun are in a straight line. [INV ZOO] End-to-end union of the sporonts of certain gregarine protozoans.

szaibelyite [MINERAL] $(Mn, Mg)(BO_2)(OH)$ A white to buff or straw yellow, orthorhombic mineral consisting of a basic borate of manganese and magnesium; occurs as veinlets, masses, or embedded nodules.

szaskaite See smithsonite.

T

t *See* troy system.

T *See* tera-; tesla.

Ta *See* tantalum.

Tabanidae [INV ZOO] The deer and horse flies, a family of orthorrhaphous dipteran insects in the series Brachycera.

table [COMPUT SCI] A set of contiguous, related items, each uniquely identified either by its relative position in the set or by some label. [LAP] The flat face forming the top of a brilliant-cut stone. [MATH] An array or listing of computed quantities. [MECH ENG] That part of a grinding machine which directly or indirectly supports the work being ground. [MIN ENG] **1.** In placer mining, a wide, shallow sluice box designed to recover gold or other valuable material from screened gravel. **2.** A platform or plate on which coal is screened and picked.

table iceberg *See* tabular iceberg.

tableland [GEOGR] A broad, elevated, nearly level, and extensive region of land that has been deeply cut at intervals by valleys or broken by escarpments. Also known as continental plateau.

table look-up [COMPUT SCI] A procedure for calculating the location of an item in a table by means of an algorithm, rather than by conducting a search for the item.

tablemount *See* guyot.

table salt *See* sodium chloride.

table sugar *See* sucrose.

tabular [GEOL] Referring to a sedimentary particle whose length is two to three times its thickness.

tabular berg *See* tabular iceberg.

tabular iceberg [OCEANOGR] An iceberg with cliff-like sides and a flat top; usually arises by detachment from an ice shelf. Also known as table iceberg; tabular berg.

tabular language [COMPUT SCI] A part of a program which represents the composition of a decision table required by the problem considered.

tabular spar *See* wollastonite.

Tabulata [PALEON] An extinct Paleozoic order of corals of the subclass Zoantharia characterized by an exclusively colonial mode of growth and by secretion of a calcareous exoskeleton of slender tubes.

tabulating system [COMPUT SCI] Any group of machines which is capable of entering, converting, receiving, classifying, computing, and recording data by means of tabulating cards, and in which tabulating cards are used for storing data and for communicating with the system.

tabulation character [COMPUT SCI] A character that controls the action of a computer printer and is not itself printed, although it forms part of the data to be printed.

tabulator [COMPUT SCI] A machine that reads information from punched cards and produces lists, tables, and totals on separate forms or continuous paper.

Tachinidae [INV ZOO] The tachina flies, a family of bristly, grayish or black Diptera whose larvae are parasitic in caterpillars and other insects.

tachometer [ENG] An instrument that measures the revolutions per minute or the angular speed of a rotating shaft.

tachycardia [MED] Excessive rapidity of the heart's action.

Tachyglossidae [VERT ZOO] A family of monotreme mammals having relatively large brains with convoluted cerebral hemispheres; comprises the echidnas or spiny anteaters.

tachyhydrite [MINERAL] $CaMg_2Cl_6 \cdot 12H_2O$ A honey yellow, hexagonal mineral consisting of a hydrated chloride of calcium and magnesium; occurs in massive form.

tachylite [GEOL] A black, green, or brown volcanic glass formed from basaltic magma. Also known as basalt glass; basalt obsidian; hyalobasalt; jaspoid; sordawalite; wichtisite.

Tachyniscidae [INV ZOO] A family of myodarian cyclorrhaphous dipteran insects in the subsection Acalypteratae.

tachyon [PARTIC PHYS] A hypothetical particle that travels faster than light, consistent with the theory of relativity.

Taconian orogeny [GEOL] A process of formation of mountains in the latter part of the Ordovician period, particularly in the northern Appalachians. Also known as Taconic orogeny.

Taconic orogeny *See* Taconian orogeny.

taconite [GEOL] The siliceous iron formation from which high-grade iron ores of the Lake Superior district have been derived; consists chiefly of fine-grained silica mixed with magnetite and hematite.

tactile [PHYSIO] Pertaining to the sense of touch.

tactile hairs *See* vibrissae.

tadpole [VERT ZOO] The larva of a frog or toad; at hatching it has a rounded body with a long fin-bordered tail, and the gills are external but shortly become enclosed.

taele *See* frozen ground.

taenia [ANAT] A ribbon-shaped band of nerve fibers or muscle.

Taeniodidea [INV ZOO] An equivalent name for Cyclophyllidea.

Taeniodonta [PALEON] An order of extinct quadrupedal land mammals, known from early Cenozoic deposits in North America.

Taenioidea [INV ZOO] An equivalent name for Cyclophyllidea.

Taeniolabidoidea [PALEON] An advanced suborder of the extinct mammalian order Multituberculata having incisors that were self-sharpening in a limited way.

taenite [MINERAL] A meteoritic mineral consisting of a nickel-iron alloy, with a nickel content varying from about 27 to 65.

tag [COMPUT SCI] 1. A unit of information used as a label or marker. 2. The symbol written in the location field of an assembly-language coding form, and used to define the symbolic address of the data or instruction written on that line. [NUCLEO] *See* isotopic tracer.

tagilite *See* pseudomalachite.

Tahuian [GEOL] A local Eocene time subdivision in Australia whose identification is based on foraminiferans.

taiga [ECOL] A zone of forest vegetation encircling the Northern Hemisphere between the arctic-subarctic tundras in the north and the steppes, hardwood forests, and prairies in the south.

tail [AERO ENG] 1. The rear part of a body, as of an aircraft or a rocket. 2. The tail surfaces of an aircraft or a rocket. [ASTRON] The part of a comet that extends from the coma in a direction opposite to the sun; it consists of dust and gas that have been blown away from the coma by the solar wind and the sun's radiation pressure. [ELECTR] 1. A small pulse that follows the main pulse of a radar set and rises in the same direction. 2. The trailing edge of a pulse. [VERT ZOO] 1. The usually slender appendage that arises immediately above the anus in many vertebrates and contains the caudal vertebrae. 2. The uropygium, and its feathers, of a bird. 3. The caudal fin of a fish or aquatic mammal.

tail assembly *See* empennage.

tailings [ENG] The lighter particles which pass over a sieve in milling, crushing, or purifying operations. [MIN ENG] 1. The parts, or a part, of any incoherent or fluid material separated as refuse, or separately treated as inferior in quality or value. 2. The decomposed outcrop of a vein or bed. 3. The refuse material resulting from processing ground ore.

tailstock [MECH ENG] A part of a lathe that holds the end of the work not being shaped, allowing it to rotate freely.

tailwind [METEOROL] A wind which assists the intended progress of an exposed, moving object, for example, rendering an airborne object's ground speed greater than its airspeed; the opposite of a headwind. Also known as following wind.

tainter gate [CIV ENG] A spillway gate whose face is a section of a cylinder; rotates about a horizontal axis on the downstream end of the gate and can be closed under its own weight. Also known as radial gate.

takedown [COMPUT SCI] The actions performed at the end of an equipment operating cycle to prepare the equipment for the next setup; for example, to remove the tapes from the tape handlers at the end of a computer run is a takedown operation.

takeup [MECH ENG] A tensioning device in a belt-conveyor system for taking up slack of loose parts.

talc [MINERAL] $Mg_3Si_4O_{10}(OH)_2$ A whitish, greenish, or grayish hydrated magnesium silicate mineral crystallizing in the monoclinic system; it is extremely soft (hardness is 1 on Mohs scale) and has a characteristic soapy or greasy feel.

Talitridae [INV ZOO] A family of terrestrial amphipod crustaceans in the suborder Gammaridea.

talk-back circuit *See* interphone.

Talpidae [VERT ZOO] The moles, a family of insectivoran mammals; distinguished by the forelimbs which are adapted for digging, having powerful muscles and a spadelike bony structure.

talus [ANAT] *See* astragalus. [GEOL] Also known as rubble; scree. 1. Coarse and angular rock fragments derived from and accumulated at the base of a cliff or steep, rocky slope. 2. The accumulated heap of such fragments.

tamp [ENG] To tightly pack a drilled hole with clay or other stemming material after the charge has been placed.

tamper [NUCLEO] *See* reflector. [ORD] In a weapon, any substance that resists movement for a split microsecond, used so that the active materials may build up greater pressure behind the substance.

tan *See* tangent.

Tanaidacea [INV ZOO] An order of eumalacostracans of the crustacean superorder Peracarida; the body is linear, more or less cylindrical or dorsoventrally depressed, and the first and second thoracic segments are fused with the head, forming a carapace.

Tanaostigmatidae [INV ZOO] A small family of hymenopteran insects in the superfamily Chalcidoidea.

tandem accelerator [NUCLEO] An electrostatic accelerator in which negative hydrogen ions generated in a special ion source are accelerated as they pass from ground potential up to a high-voltage terminal, both electrons are then stripped from the negative ion by passage through a very thin foil or gas cell, and the proton is again accelerated as it passes to ground potential.

tandem central office [COMMUN] A telephone office that makes connections between local offices in an area where there is such a high density of local offices that it would be uneconomical to make direct connections between them. Also known as tandem office.

tandem compensation *See* cascade compensation.

tandem connection *See* cascade connection.

tandem mill [MET] A rolling mill consisting of two or more stands in succession, synchronized so that the metal passes directly from one to another.

tandem office *See* tandem central office.

tandem switching [COMMUN] System of routing telephone calls in which calls do not travel directly between local offices, but rather through a central office.

tandem system [COMPUT SCI] A computing system in which there are two central processing units, usually with one controlling the other, and with data proceeding from one processing unit into the other.

tangeite *See* calciovolborthite.

tangelo [BOT] A tree that is hybrid between a tangerine or other mandarin and a grapefruit or shaddock; produces an edible fruit.

tangent [MATH] 1. A line is tangent to a curve at a fixed point P if it is the limiting position of a line passing through P and a variable point on the curve Q, as Q approaches P. 2. The function which is the quotient of the sine function by the cosine function. Abbreviated tan. 3. The tangent of an angle is the ratio of its sine and cosine. Abbreviated tan.

tangential acceleration [MECH] The component of linear acceleration tangent to the path of a particle moving in a circular path.

tangential stress *See* shearing stress.

tangential velocity [MECH] 1. The instantaneous linear velocity of a body moving in a circular path; its direction is tangential to the circular path at the point in question. 2. The component of the velocity of a body that is perpendicular to a line from an observer or reference point to the body.

tangential wave *See* S wave.

tangential wave path [ELECTROMAG] In radio propagation over the earth, a path of propagation of a direct wave which is tangential to the surface of the earth; the tangential wave path is curved by atmospheric refraction.

tangent vector [MATH] A tangent vector at a point of a differentiable manifold is any vector tangent to a differentiable curve in the manifold at this point; alternatively, a member of the tangent plane to the manifold at the point.

tangerine [BOT] Any of several trees of the species *Citrus reticulata*; the fruit is a loose-skinned mandarin with a deep-orange or scarlet rind.

tank [ELECTR] 1. A unit of acoustic delay-line storage containing a set of channels, each forming a separate recirculation path. 2. The heavy metal envelope of a large mercury-arc rectifier or other gas tube having a mercury-pool cathode. 3. *See* tank circuit. [ENG] A large container for holding, storing, or transporting a liquid.

tank circuit [ELECTR] A circuit which exhibits resonance at one or more frequencies, and which is capable of storing electric energy over a band of frequencies continuously distributed about the resonant frequency, such as a coil and capacitor in parallel. Also known as electrical resonator; tank.

tank farm [PETRO ENG] An area in which a number of large-capacity storage tanks are located, generally used for crude oil or petroleum products.

tank reactor [NUCLEO] A nuclear reactor in which the core is suspended in a closed tank, as distinct from an open-pool reactor.

tanning [ENG] A process of preserving animal hides by chemical treatment (using vegetable tannins, metallic sulfates, and sulfurized phenol compounds, or syntans) to make them immune to bacterial attack, and subsequent treatment with fats and greases to make them pliable.

tantalite [MINERAL] (Fe,Mn)Ta_2O_6 An iron-black mineral that crystallizes in the orthorhombic system and commonly occurs in short prismatic crystals; luster is submetallic, hardness is 6 on Mohs scale, and specific gravity is 7.95; principal ore of tantalum.

tantalum [CHEM] Metallic element in group V, symbol Ta, atomic number 73, atomic weight 180.948; black powder or steel-blue solid soluble in fused alkalies, insoluble in acids (except hydrofluoric and fuming sulfuric); melts about 3000°C. [MET] A lustrous, platinum-gray ductile metal used in making dental and surgical tools, pen points, and electronic equipment.

T antenna [ELECTROMAG] An antenna consisting of one or more horizontal wires, with a lead-in connection being made at the approximate center of each wire.

Tanyderidae [INV ZOO] The primitive crane flies, a family of orthorrhaphous dipteran insects in the series Nematocera.

Tanypezidae [INV ZOO] A family of myodarian cyclorrhaphous dipteran insects in the subsection Acalyptratae.

tap [DES ENG] 1. A plug of accurate thread, form, and dimensions on which cutting edges are formed; it is screwed into a hole to cut an internal thread. 2. A threaded cone-shaped fishing tool. [ELEC] A connection made at some point other than the ends of a resistor or coil. [ENG] A small, threaded hole drilled into a pipe or process vessel; used as connection points for sampling devices, instruments, or controls. [MET] 1. A quantity of molten metal run out from a furnace at one time. 2. To remove excess slag from the floor of a pot furnace. [MIN

ENG] To intersect with a borehole and withdraw or drain the contained liquid, as water from a water-bearing formation or from underground workings.

tap bolt [DES ENG] A bolt with a head that can be screwed into a hole and held in place without a nut. Also known as tap screw.

tap changer [ELEC] A device which is used to change the ratio of the input and output voltages of a transformer over any one of a definite number of steps.

tap crystal [ELECTR] Compound semiconductor that stores current when stimulated by light and then gives up energy as flashes of light when it is physically tapped.

tape [COMPUT SCI] A ribbonlike material used to store data in lengthwise sequential position. [ENG] A graduated steel ribbon used, instead of a chain, in surveying.

tape alternation [COMPUT SCI] The switching of a computer program back and forth between two tape units in order to avoid interruption of the program during mounting and removal of tape reels.

tape cartridge [ENG ACOUS] A cartridge that holds a length of magnetic tape in such a way that the cartridge can be slipped into a tape recorder and played without threading the tape; in stereophonic usage, usually refers to an eight-track continuous-loop cartridge, which is larger than a cassette. Also known as cartridge.

tape control unit [COMPUT SCI] A device which senses which tape unit is to be accessed for read or write purpose and opens up the necessary electronic paths. Formerly known as hypertape control unit.

tape drive [COMPUT SCI] A tape reading or writing device consisting of a tape transport, electronics, and controls; it usually refers to magnetic tape exclusively. [ENG ACOUS] *See* tape transport.

tape grass [BOT] *Vallisnerida spiralis*. An aquatic flowering plant belonging to the family Hydrocharitaceae. Also known as eel grass.

tape-limited [COMPUT SCI] Pertaining to a computer operation in which the time required to read and write tapes exceeds the time required for computation.

tape mark [COMPUT SCI] 1. A special character or coding, an attached piece of reflective material, or other device that indicates the physical end of recording on a magnetic tape. Also known as destination warning mark; end-of-tape mark. 2. A special character that divides a file of magnetic tape into sections, usually followed by a record with data describing the particular section of the file. Also known as control mark.

tape operating system [COMPUT SCI] A computer operating system in which source programs and sometimes incoming data are stored on magnetic tape, rather than in the computer memory. Abbreviated TOS.

tapered transmission line *See* tapered waveguide.

tapered waveguide [ELECTROMAG] A waveguide in which a physical or electrical characteristic changes continuously with distance along the axis of the waveguide. Also known as tapered transmission line.

tape search unit [COMPUT SCI] Small, fully transistorized, special-purpose, digital data-processing system using a stored program to perform logical functions necessary to search a magnetic tape in off-line mode, in response to a specific request.

tape skip [COMPUT SCI] A machine instruction to space forward and erase a portion of tape when a defect on the tape surface causes a write error to persist.

tape station [COMPUT SCI] A tape reading or writing device consisting of a tape transport, electronics, and controls; it may use either magnetic tape or paper tape.

tape transport [COMPUT SCI] The mechanism that physically moves a tape past a stationary head. Also known as transport. [ENG ACOUS] The mechanism of a tape recorder that holds the tape reels, drives the tape past the heads, and controls various modes of operation. Also known as tape drive.

tape unit [COMPUT SCI] A tape reading or writing device consisting of a tape transport, electronics, controls, and possibly a cabinet; the cabinet may contain one or more magnetic tape stations.

tape verifier [COMPUT SCI] A verifier for checking the accuracy of a punched paper tape by comparing it with a second manual punch of the same data; the machine stops whenever a character being punched the second time differs from that on the first tape.

tapeworm [INV ZOO] Any member of the class Cestoidea; all are vertebrate endoparasites, characterized by a ribbonlike body divided into proglottids, and the anterior end modified into a holdfast organ.

taphole [MET] A hole in a furnace or ladle through which molten metal is tapped.

taphonomy [PALEON] The study of fossil preservation, including all events during the transition of organisms from the biosphere to the lithosphere.

tapir [VERT ZOO] Any of several large odd-toed ungulates of the family Tapiridae that have a heavy, sparsely hairy body, stout legs, a prehensile muzzle, a short tail, and small eyes.

Tapiridae [VERT ZOO] The tapirs, a family of perissodactyl mammals in the superfamily Tapiroidea.

Tapiroidea [VERT ZOO] A superfamily of the mammalian order Perissodactyla in the suborder Ceratomorpha.

tapped control [ELECTR] A rheostat or potentiometer having one or more fixed taps along the resistance element, usually to provide a fixed grid bias or for automatic bass compensation.

tappet [MECH ENG] A lever or oscillating member moved by a cam and intended to tap or touch another part, such as a push rod or valve system.

tapping screw See self-tapping screw.

taproot [BOT] A root system in which the primary root forms a dominant central axis that penetrates vertically and rather deeply into the soil; it is generally larger in diameter than its branches.

tap screw See tap bolt.

tap switch [ELEC] Multicontact switch used chiefly for connecting a load to any one of a number of taps on a resistor or coil.

tarantula [INV ZOO] 1. Any of various large hairy spiders of the araneid suborder Mygalomorphae. 2. Any of the wolf spiders comprising the family Lycosidae.

Tardigrada [INV ZOO] A class of microscopic, bilaterally symmetrical invertebrates in the subphylum Malacopoda; the body consists of an anterior prostomium and five segments surrounded by a soft, nonchitinous cuticle, with four pairs of ventrolateral legs.

tare [MECH] The weight of an empty vehicle or container; subtracted from gross weight to ascertain net weight.

target [COMPUT SCI] An index card or test document used to assist, reference, or calibrate equipment. [ATOM PHYS] The atom or nucleus in an atomic or nuclear reaction which is initially stationary. [ELECTR] 1. In an x-ray tube, the anode or anticathode which emits x-rays when bombarded with electrons. 2. In a television camera tube, the storage surface that is scanned by an electron beam to generate an output signal current corresponding to the charge-density pattern stored there. 3. In a cathode-ray tuning indicator tube, one of the electrodes that is coated with a material that fluoresces under electron bombardment. [ENG] 1. The sliding weight on a leveling rod used in surveying to enable the staffman to read the line of collimation. 2. The point that a borehole or an exploratory work is intended to reach. 3. In radar and sonar, any object capable of reflecting the transmitted beam. [ORD] 1. A geographical area, complex, or installation planned for capture or destruction by military forces. 2. A paper or pasteboard item of square or rectangular shape designed to be fired upon from a specified range during practice or while testing an automatic firearm such as an automatic rifle, machine gun, or submachine gun; it is used to establish a degree of accuracy; it usually consists of a series of geometric patterns of various shapes on a common background. [PHYS] An object or substance subjected to bombardment or irradiation by particles or electromagnetic radiation.

target acquisition [AERO ENG] The process of optically, manually, mechanically, or electronically orienting a tracking system in direction and range to lock on a target. [ELECTR] 1. The first appearance of a recognizable and useful echo signal from a new target in radar and sonar. 2. See acquire.

target acquisition radar [ENG] An antiaircraft artillery radar, normally of lesser range capabilities but of greater inherent accuracy than that of surveillance radar, whose normal function is to acquire aerial targets either by independent search or on direction of the surveillance radar, and to transfer these targets to tracking radars.

target configuration [COMPUT SCI] The combination of input, output, and storage units and the amount of computer memory required to carry out an object program.

target discrimination [ELECTR] The ability of a detection or guidance system to distinguish a target from its background or to discriminate between two or more targets that are close together.

target echo [ELECTROMAG] A radio signal reflected by an airborne or other target and received by the radar station which transmitted the original signal.

target glint See scintillation.

target language [COMPUT SCI] The language into which a program (or text) is to be converted.

target phase [COMPUT SCI] The stage of handling a computer program at which the object program is first carried out after it has been compiled.

target program See object program.

target routine See object program.

target scintillation See scintillation.

target signal [ELECTROMAG] The radio energy returned to a radar by a target. Also known as echo signal; video signal.

target spot [PL PATH] Any plant disease characterized by lesions in the form of concentric markings.

tarn [GEOGR] A landlocked pool or small lake that may occur in a marsh or swamp, or that may occupy a basin amid mountain ranges.

tarpon [VERT ZOO] Megalops atlantica. A herring-like fish of the family Elopidae weighing up to 300 pounds (136 kilograms) and reaching a length of 8 feet (2.4 meters); it has a single soft, rayed dorsal fin, strong jaws, a bony plate under the mouth, numerous small teeth, and coarse. bony flesh covered with large scales.

tarragon [FOOD ENG] A herb prepared from the pungent leaves of the tarragon tree (*Artemisia dracunculus*).

tarsal gland [ANAT] Any of the sebaceous glands in the tarsal plates of the eyelids. Also known as Meibomian gland.

tar sand [GEOL] A type of oil sand; a sand whose interstices are filled with asphalt that remained after the escape of the lighter fractions of crude oil.

tarsier [VERT ZOO] Any of several species of primates comprising the genus *Tarsius* of the family Tarsiidae characterized by a round skull, a flattened face, and large eyes that are separated from the temporal fossae in the orbital depression, and by adhesive pads on the expanded ends of the fingers and toes.

Tarsiidae [VERT ZOO] The tarsiers, a family of prosimian primates distinguished by incomplete postorbital closure and a greatly elongated ankle region.

Tarsonemidae [INV ZOO] A small family of phytophagous mites in the suborder Trombidiformes.

tarsus [ANAT] **1.** The instep of the foot consisting of the calcaneus, talus, cuboid, navicular, medial, intermediate, and lateral cuneiform bones. **2.** The dense connective tissues supporting an eyelid.

tartar *See* dental calculi.

task management [COMPUT SCI] The functions, assumed by the operating system, of switching the processor among tasks, scheduling, sending messages or timing signals between tasks, and creating or removing tasks.

taste bud [ANAT] An end organ consisting of goblet-shaped clusters of elongate cells with microvilli on the distal end to mediate the sense of taste.

T attenuator [ELEC] **1.** A resistive attenuator with three resistors forming a T network. **2.** A power-tap type of attenuator which removes part of the power from a main line through a T connection and dissipates the power, without reflection into the main line.

tau meson [PARTIC PHYS] Former name for the *K* meson, especially one which decays into three pions.

Taurus [ASTRON] A northern constellation; right ascension 4 hours, declination 15° north; it includes the star Aldebaran, useful in navigation. Also known as Bull.

tautomerism [CHEM] The reversible interconversion of structural isomers of organic chemical compounds; such interconversions usually involve transfer of a proton.

Taxales [BOT] A small order of gymnosperms in the class Pinatae; members are trees or shrubs with evergreen, often needlelike leaves, with a well-developed fleshy covering surrounding the individual seeds, which are terminal or subterminal on short shoots.

taxis [PHYSIO] A mechanism of orientation by means of which an animal moves in a direction related to a source of stimulation.

Taxocrinida [PALEON] An order of flexible crinoids distributed from Ordovician to Mississippian.

Taxodonta [INV ZOO] A subclass of pelecypod mollusks in which the hinge is of the taxodont type, that is, the dentition is a series of similar alternating teeth and sockets along the hinge margin.

taxon [SYST] A taxonomic group or entity.

taxonomic category [SYST] One of a hierarchy of levels in the biological classification of organisms; the seven major categories are kingdom, phylum, class, order, family, genus, species.

taxonomy [SYST] A study aimed at producing a hierarchical system of classification of organisms which best reflects the totality of similarities and differences.

Tayassuidae [VERT ZOO] The peccaries, a family of artiodactyl mammals in the superfamily Suoidae.

taylorite *See* bentonite.

Taylor-Orowan dislocation *See* edge dislocation.

Tay-Sachs disease [MED] A form of sphingolipidosis, transmitted as an autosomal recessive, in which there is an accumulation in neuronal cells of the neuraminic fraction of gangliosides; manifested clinically within the first few months of life by hypotonia progressing to spasticity, convulsions, and visual loss accompanied by the appearance of a cherry-red spot at the macula lutea. Also known as infantile amaurotic familial idiocy.

Tb *See* terbium.

TBE *See* binding energy.

T beam [CIV ENG] A metal beam or bar with a T-shaped cross section.

Tc *See* technetium.

T cell [IMMUNOL] One of a heterogeneous population of thymus-derived lymphocytes which participates in the immune responses.

Tchernozem *See* Chernozem.

T connector [ELEC] A type of electric connector that joins a through conductor to another conductor at right angles to it.

Te *See* tellurium.

tea [BOT] *Thea sinensis.* A small tree of the family Theaceae having lanceolate leaves and fragrant white flowers; a caffeine beverage is made from the leaves of the plant.

tear gland *See* lacrimal gland.

technetium [CHEM] A member of group VII, symbol Tc, atomic number 43; derived from uranium and plutonium fission products; chemically similar to rhenium and manganese; isotope Tc^{99} has a half-life of 2×10^5 years; used to absorb slow neutrons in reactor technology. [MET] Silver-gray metal with a high melting point, slightly magnetic.

technology [SCI TECH] Systematic knowledge of and its application to industrial processes; closely related to engineering and science.

Tectibranchia [INV ZOO] An order of mollusks in the subclass Opisthobranchia containing the sea hares and the bubble shells; the shell may be present, rudimentary, or absent.

tectite *See* tektite.

tectogenesis *See* orogeny.

tectonic breccia [PETR] A breccia developed from brittle rocks, formed as a result of crustal movements and produced by lateral or vertical pressure. Also known as dynamic breccia; pressure breccia.

tectonic conglomerate *See* crush conglomerate.

tectonic cycle [GEOL] The orogenic cycle which relates larger crustal features, such as mountain belts, to a series of stages of development. Also known as geosynclinal cycle.

tectonic patterns [GEOL] The arrangement of the large structural units of the earth's crust, such as mountain systems, shields or stable areas, basins, arches, and volcanic archipelagoes.

tectonics [CIV ENG] **1.** The science and art of construction with regard to use and design. **2.** Design relating to crustal deformations of the earth. [GEOL] A branch of geology that deals with regional structural and deformational features of the earth's crust, including the mutual relations, origin, and historical evolution of the features. Also known as geotectonics.

tectonophysics [GEOPHYS] A branch of geophysics dealing with the physical processes involved in forming geological structures.

tectosilicate [MINERAL] A structural type of silicate in which all four oxygen atoms of the silicate tetrahedra are shared with neighboring silicate tetrahedra; tectosilicates include quartz, the feldspars, the feldspathoids, and zeolites. Also known as framework silicate.

tectum [ANAT] A rooflike structure of the body, especially the roof of the midbrain including the corpora quadrigemina.

tegmentum [ANAT] A mass of white fibers with gray matter in the cerebral peduncles of higher vertebrates. [BOT] The outer layer, or scales, of a leaf bud. [INV ZOO] The upper layer of a shell plate in Amphineura.

Teiidae [VERT ZOO] The tegus lizards, a diverse family of the suborder Sauria that is especially abundant and widespread in South America.

teineite [MINERAL] $CuTeO_3 \cdot 2H_2O$ A greenish to yellowish, probably triclinic mineral consisting of a hydrated sulfate-tellurate of copper; occurs as crystals.

tektite [GEOL] A collective term applied to certain objects of natural glass of debatable origin that are widely strewn over the land and in sediments under the oceans; composition and size vary, and overall shapes resemble splash forms; most tektites are believed to be of extraterrestrial origin. Also known as obsidianite; tectite.

telecast [COMMUN] A television broadcast intended for reception by the general public, involving the transmission of the picture and sound portions of the program.

telecommunications [COMMUN] Communication over long distances.

teleconference [COMMUN] A conference between persons remote from one another but linked by a telecommunications system.

Telegeusidae [INV ZOO] The long-lipped beetles, a small family of colepteran insects in the superfamily Cantharoidea confined to the western United States.

telegraph alphabet *See* telegraph code.

telegraph circuit [COMMUN] The complete wire or radio circuit over which signal currents flow between transmitting and receiving apparatus in a telegraph system.

telegraph code [COMMUN] A system of symbols for transmitting telegraph messages in which each letter or other character is represented by a set of long and short electrical pulses, or by pulses of opposite polarity, or by time intervals of equal length in which a signal is present or absent. Also known as telegraph alphabet.

telegraph receiver [ELEC] A tape reperforator, teletypewriter, or other equipment which converts telegraph signals into a pattern of holes on a tape, printed letters, or other forms of information.

telegraph transmitter [ELEC] A device that controls an electric power source in order to form telegraph signals.

telegraphy [COMMUN] Communication at a distance by means of code signals consisting of current pulses sent over wires or by radio.

telemeter [ENG] 1. The complete measuring, transmitting, and receiving apparatus for indicating or recording the value of a quantity at a distance. Also known as telemetering system. 2. To transmit the value of a measured quantity to a remote point.

telemetering [ENG] Transmitting the readings of instruments to a remote location by means of wires, radio waves, or other means. Also known as remote metering; telemetry.

telemetering system *See* telemeter.

telemetry *See* telemetering.

telencephalon [EMBRYO] The anterior subdivision of the forebrain in a vertebrate embryo; gives rise to the olfactory lobes, cerebral cortex, and corpora striata.

teleology [SCI TECH] The doctrine that explanations of phenomena are to be sought in terms of final causes, purpose, or design in nature.

Teleosauridae [PALEON] A family of Jurassic reptiles in the order Crocodilia characterized by a long snout and heavy armor.

Teleostei [VERT ZOO] An infraclass of the subclass Actinopterygii, or rayfin fishes; distinguished by paired bracing bones in the supporting skeleton of the caudal fin, a homocercal caudal fin, thin cycloid scales, and a swim bladder with a hydrostatic function.

telephone [COMMUN] A system of converting sound waves into variations in electric current that can be sent over wires and reconverted into sound waves at a distant point, used primarily for voice communication; it consists essentially of a telephone transmitter and receiver at each station, interconnecting wires, signaling devices, a central power supply, and switching facilities. Also known as telephone system. [ENG ACOUS] *See* telephone set.

telephone central office *See* central office.

telephone channel [COMMUN] A one-way or two-way path suitable for the transmission of audio signals between two stations.

telephone circuit [ELEC] The complete circuit over which audio and signaling currents travel in a telephone system between the two telephone subscribers in communication with each other; the circuit usually consists of insulated conductors, as ground returns are now rarely used in telephony.

telephone data set [COMPUT SCI] Equipment interfacing a data terminal with a telephone circuit.

telephone modem [ELECTR] A piece of equipment that modulates and demodulates one or more separate telephone circuits, each containing two or more telephone channels; it may include multiplexing and demultiplexing circuits, individual amplifiers, and carrier-frequency sources.

telephone pickup [ELEC] A large flat coil placed under a telephone set to pick up both voices during a telephone conversation for recording purposes.

telephone receiver [ENG ACOUS] The portion of a telephone set that converts the audio-frequency current variations of a telephone line into sound waves, by the motion of a diaphragm activated by a magnet whose field is varied by the electrical impulses that come over the telephone wire.

telephone relay [ELEC] A relay having a multiplicity of contacts on long spring strips mounted parallel to the coil, actuated by a lever arm or other projection of the hinged armature; used chiefly for switching in telephone circuits.

telephone set [ENG ACOUS] An assembly including a telephone transmitter, a telephone receiver, and associated switching and signaling devices. Also known as telephone.

telephone switchboard *See* switchboard.

telephone system *See* telephone.

telephone transmitter [ENG ACOUS] The microphone used in a telephone set to convert speech into audio-frequency electric signals.

telephoto *See* facsimile.

telephotography *See* facsimile.

telephoto lens [OPTICS] A lens for photographing distant objects; it is designed in a compact manner so that the distance from the front of the lens to the film plane is less than the focal length of the lens.

teleprinter [COMMUN] A device that responds to teletype signals and prints the corresponding characters on paper tape. [COMPUT SCI] Any typewriter-type device capable of being connected to a computer and of printing out a set of messages under computer control.

teleprocessing [COMPUT SCI] 1. The use of telecommunications equipment and systems by a computer. 2. A computer service involving input/output at locations remote from the computer itself.

teleprocessing monitor [COMPUT SCI] A computer program that manages the transfer of information between local and remote terminals. Abbreviated TP monitor.

telescope [ENG] Any device that collects radiation, which may be in the form of electromagnetic or particle radiation, from a limited direction in space. [OPTICS] An assemblage of lenses or mirrors, or both, that enhances the ability of the eye either to see objects with greater resolution or to see fainter objects.

Telescope *See* Telescopium.

Telescopium [ASTRON] A constellation, right ascension 19 hours, declination 50° south. known as Telescope.

Telestacea [INV ZOO] An order of the subclass Alcyonaria comprised of individuals which form erect branching colonies by lateral budding from the body wall of an elongated axial polyp.

teletext [COMMUN] A data broadcasting service in which preprogrammed sequences of frames of data are broadcast cyclically, and a user equipped with a standard television receiver and a special decoder selects the desired frames of information for viewing.

teletypesetter [GRAPHICS] A system automatically operating a linecasting machine (Linotype or Intertype) to produce lines of type at high speed under the control of perforated tape or equivalent electrical signals.

teletypewriter [COMMUN] A special electric typewriter that produces coded electric signals corresponding to manually typed characters, and automatically types messages when fed with similarly coded signals produced by another machine. Also known as TWX machine.

teletypewriter code [COMMUN] Special code in which each code group is made up of five units, or elements, of equal length which are known as marking or spacing impulses; the five-unit start-stop code consists of five signal impulses preceded by a start impulse and followed by a stop impulse.

television [COMMUN] A system for converting a succession of visual images into corresponding electric signals and transmitting these signals by radio or over wires to distant receivers at which the signals can be used to reproduce the original images. Abbreviated TV.

television bandwidth [COMMUN] The difference between the limiting frequencies of a television channel; in the United States, this is 6 megahertz.

television broadcasting [COMMUN] Transmission of television programs by means of radio waves, for reception by the public.

television camera [ELECTR] The pickup unit used to convert a scene into corresponding electric signals; optical lenses focus the scene to be televised on the photosensitive surface of a camera tube, and

the tube breaks down the visual image into small picture elements and converts the light intensity of each element in turn into a corresponding electric signal. Also known as camera.

television camera tube *See* camera tube.

television channel [COMMUN] A band of frequencies 6 megahertz wide in the television broadcast band, available for assignment to a television broadcast station.

television network [COMMUN] An arrangement of communication channels, suitable for transmission of video and accompanying audio signals, which link together groups of television broadcasting stations or closed-circuit television users in different cities so that programs originating at one point can be fed simultaneously to all others.

television picture tube *See* picture tube.

television receiver [ELECTR] A receiver that converts incoming television signals into the original scenes along with the associated sounds. Also known as television set.

television set *See* television receiver.

television tower [ENG] A tall metal structure used as a television transmitting antenna, or used with another such structure to support a television transmitting antenna wire.

television transmitter [ELECTR] An electronic device that converts the audio and video signals of a television program into modulated radio-frequency energy that can be radiated from an antenna and received on a television receiver.

telltale [ENG] A marker on the outside of a tank that indicates on an exterior scale the amount of fluid inside the tank.

telluric current *See* earth current.

tellurite [MINERAL] TeO$_2$ A white or yellowish orthorhombic mineral consisting of tellurium dioxide, and occurring in crystals; it is dimorphous with paratellurite.

tellurium [CHEM] A member of group VI, symbol Te, atomic number 52, atomic weight 127.60; dark-gray crystals, insoluble in water, soluble in nitric and sulfuric acids and potassium hydroxide; melts at 452°C, boils at 1390°C; used in alloys (with lead or steel), glass, and ceramics.

tellurium glance *See* nagyagite.

tellurobismuthite [MINERAL] Bi$_2$Te$_3$ A pale lead gray, hexagonal mineral consisting of a bismuth and tellurium compound; occurs as irregular plates or foliated masses.

tellurometer [ENG] A microwave instrument used in surveying to measure distance; the time for a radio wave to travel from one observation point to the other and return is measured and converted into distance by phase comparison, much as in radar.

telocentric [CYTOL] Pertaining to a chromosome with a terminal centromere.

telolecithal [CYTOL] Of an ovum, having a large, evenly dispersed volume of yolk and a small amount of cytoplasm concentrated at one pole.

telophase [CYTOL] The phase of meiosis or mitosis at which the chromosomes, having reached the poles, reorganize into interphase nuclei with the disappearance of the spindle and the reappearance of the nuclear membrane; in many organisms telophase does not occur at the end of the first meiotic division.

Telosporea [INV ZOO] A class of the protozoan subphylum Sporozoa in which the spores lack a polar capsule and develop from an oocyst.

telson [INV ZOO] The postabdominal segment in lobsters, amphipods, and certain other invertebrates.

TEM mode *See* transverse electromagnetic mode.

Temnocephalida [INV ZOO] A group of rhabdocoeles sometimes considered a distinct order but usually classified under the Neorhabdocoela; members are characterized by the possession of tentacles and adhesive organs.

Temnochilidae [INV ZOO] The equivalent name for Ostomidae.

Temnopleuridae [INV ZOO] A family of echinoderms in the order Temnopleuroida whose tubercles are imperforate, though usually crenulate.

Temnopleuroida [INV ZOO] An order of echinoderms in the superorder Echinacea with a camarodont lantern, smooth or sculptured test, imperforate or perforate tubercles, and bronchial slits which are usually shallow.

Temnospondyli [PALEON] An order of extinct amphibians in the subclass Labyrinthodontia having vertebrae with reduced pleurocentra and large intercentra.

TE mode *See* transverse electric mode.

temperate belt [CLIMATOL] A belt around the earth within which the annual mean temperature is less than 20°C (68°F) and the mean temperature of the warmest month is higher than 10°C (50°F).

temperate climate [CLIMATOL] The climate of the middle latitudes; the climate between the extremes of tropical climate and polar climate.

temperate phage [VIROL] A deoxyribonucleic acid phage, the genome (DNA) of which can under certain circumstances become integrated with the genome of the host.

temperate westerlies *See* westerlies.

Temperate Zone [CLIMATOL] Either of the two latitudinal zones on the earth's surface which lie between 23°27′ and 66°32′ N and S (the North Temperate Zone and South Temperate Zone, respectively).

temperature [THERMO] A property of an object which determines the direction of heat flow when the object is placed in thermal contact with another object: heat flows from a region of higher temperature to one of lower temperature; it is measured either by an empirical temperature scale, based on some convenient property of a material or instrument, or by a scale of absolute temperature, for example, the Kelvin scale.

13.0 temperature *See* annealing point.

temperature coefficient [PHYS] The rate of change of some physical quantity (such as resistance of a conductor or voltage drop across a vacuum tube) with respect to temperature.

temperature compensation [ELECTR] The process of making some characteristic of a circuit or device independent of changes in ambient temperature.

temperature-humidity index [METEOROL] An index which gives a numerical value, in the general range of 70–80, reflecting outdoor atmospheric conditions of temperature and humidity as a measure of comfort (or discomfort) during the warm season of the year; equal to 15 plus 0.4 times the sum of the dry-bulb and wet-bulb temperatures in degrees Fahrenheit. Also known as comfort index; discomfort index. Abbreviated CI; DI; THI.

temperature inversion [METEOROL] A layer in the atmosphere in which temperature increases with altitude; the principal characteristic of an inversion layer is its marked static stability, so that very little turbulent exchange can occur within it; strong wind shears often occur across inversion layers, and abrupt changes in concentrations of atmospheric particulates and atmospheric water vapor may be en-

countered on ascending through the inversion layer.

[OCEANOGR] A layer of a large body of water in which temperature increases with depth.

temperature saturation [ELECTR] The condition in which the anode current of a thermionic vacuum tube cannot be further increased by increasing the cathode temperature at a given value of anode voltage; the effect is due to the space charge formed near the cathode. Also known as filament saturation; saturation.

temperature scale [THERMO] An assignment of numbers to temperatures in a continuous manner, such that the resulting function is single valued; it is either an empirical temperature scale, based on some convenient property of a substance or object, or it measures the absolute temperature.

temperature transducer [ENG] A device in an automatic temperature-control system that converts the temperature into some other quantity such as mechanical movement, pressure, or electric voltage; this signal is processed in a controller, and is applied to an actuator which controls the heat of the system.

temperature wave [CRYO] A disturbance in which a variation in temperature propagates through a medium; the chief example of this is second sound. Also known as thermal wave.

tempering [MET] Heat treatment of hardened steels to temperatures below the transformation temperature range, usually to improve toughness.

temporal bone [ANAT] The bone forming a portion of the lateral aspect of the skull and part of the base of the cranium in vertebrates.

TEM wave *See* transverse electromagnetic wave.

Tendipedidae [INV ZOO] The midges, a family of orthorrhaphous dipteran insects in the series Nematocera whose larvae occupy intertidal wave-swept rocks on the seacoasts.

tendon [ANAT] A white, glistening, fibrous cord which joins a muscle to some movable structure such as a bone or cartilage; tendons permit concentration of muscle force into a small area and allow the muscle to act at a distance. [CIV ENG] A steel bar or wire that is tensioned, anchored to formed concrete, and allowed to regain its initial length to induce compressive stress in the concrete before use.

tendril [BOT] A stem modification in the form of a slender coiling structure capable of twining about a support to which the plant is then attached.

Tenebrionidae [INV ZOO] The darkling beetles, a large cosmopolitan family of coleopteran insects in the superfamily Tenebrionoidea; members are common pests of grains, dried fruits, beans, and other food products.

Tenebrionoidea [INV ZOO] A superfamily of the Coleoptera in the suborder Polyphaga.

tennantite [MINERAL] $(Cu,Fe)_{12}As_4S_{13}$ A lead-gray mineral crystallizing in the isometric system; it is isomorphous with tetrahedrite; an important ore of copper.

tenorite [MINERAL] CuO A triclinic mineral that occurs in small, shining, steel-gray scales, in black powder, or in black earthy masses; an ore of copper.

tenrec [VERT ZOO] Any of about 30 species of unspecialized, insectivorous mammals indigenous to Madagascar, and which have poor vision and clawed digits.

Tenrecidae [VERT ZOO] The tenrecs, a family of insectivores in the group Lipotyphla.

ten's complement [MATH] In decimal arithmetic, the unique numeral that can be added to a given N-digit numeral to form a sum equal to 10^N (that is, a one followed by N zeros).

tensile modulus [MECH] The tangent or secant modulus of elasticity of a material in tension.

tensile strength [MECH] The maximum stress a material subjected to a stretching load can withstand without tearing. Also known as hot strength.

tensile test [ENG] A test in which a specimen is subjected to increasing longitudinal pulling stress until fracture occurs.

tension [MECH] **1.** The condition of a string, wire, or rod that is stretched between two points. **2.** The force exerted by the stretched object on a support.

tension fault [GEOL] A fault in which crustal tension is a factor, such as a normal fault. Also known as extensional fault.

tension fracture [GEOL] A minor rock fracture developed at right angles to the direction of maximum tension. Also known as subsidiary fracture.

tension pulley [MECH ENG] A pulley around which an endless rope passes mounted on a trolley or other movable bearing so that the slack of the rope can be readily taken up by the pull of the weights.

tensor [MATH] An object relative to a locally euclidean space which possesses a specified system of components for every coordinate system and which changes under a transformation of coordinates.

tensor analysis [MATH] The abstract study of mathematical objects having components which express properties similar to those of a geometric tensor; this study is fundamental to Riemannian geometry and the structure of euclidean spaces. Also known as tensor calculus.

tensor calculus See tensor analysis.

tensor field [MATH] A tensor or collection of tensors defined in some open subset of a Riemann space.

tensor force [NUC PHYS] A spin-dependent force between nucleons, having the same form as the interaction between magnetic dipoles; it is introduced to account for the observed values of the magnetic dipole moment and electric quadrupole moment of the deuteron.

tensor muscle [PHYSIO] A muscle that stretches a part or makes it tense.

tentacle [INV ZOO] Any of various elongate, flexible processes with tactile, prehensile, and sometimes other functions, and which are borne on the head or about the mouth of many animals.

Tentaculata [INV ZOO] A class of the phylum Ctenophora whose members are characterized by having variously modified tentacles.

tenthmeter See angstrom.

Tenthredinidae [INV ZOO] A family of hymenopteran insects in the superfamily Tenthredinoidea including economically important species whose larvae are plant pests.

Tenthredinoidea [INV ZOO] A superfamily of Hymenoptera in the suborder Symphyla.

Tenuipalpidae [INV ZOO] A small family of mites in the suborder Trombidiformes.

tepetate See caliche.

tephrite [PETR] A group of basaltic extrusive rocks composed chiefly of calcic plagioclase, augite, and nepheline or leucite, with some sodic sanidine.

Tephritidae [INV ZOO] The fruit flies, a family of myodarian cyclorrhaphous dipteran insects in the subsection Acalyptratae.

tera- [MATH] A prefix representing 10^{12}, which is 1,000,000,000,000 or a million million. Abbreviated T.

teratogen [MED] An agent causing formation of a congenital anomaly or monstrosity.

teratology [MED] The science of fetal malformations and monstrosities.

teratoma [MED] A true neoplasm composed of bizarre and chaotically arranged tissues that are foreign embryologically as well as histologically to the area in which the tumor is found.

Teratornithidae [PALEON] An extinct family of vulturelike birds of the Pleistocene of western North America included in the order Falconiformes.

terbium [CHEM] A rare-earth element, symbol Tb, in the yttrium subgroup of group III, atomic number 65, atomic weight 158.924.

tercentesimal thermometric scale See approximate absolute temperature.

Terebellidae [INV ZOO] A family of polychaete annelids belonging to the Sedentaria which are chiefly large, thick-bodied, tubicolous forms with the anterior end covered by a matted mass of tentacular cirri.

Terebratellidina [PALEON] An extinct suborder of articulate brachiopods in the order Terebratulida in which the loop is long and offers substantial support to the side arms of the lophophore.

Terebratulida [INV ZOO] An order of articulate brachiopods that has a punctate shell structure and is characterized by the possession of a loop extending anteriorly from the crural bases, providing some degree of support for the lophophore.

Terebratulidina [INV ZOO] A suborder of articulate brachiopods in the order Terebratulida distinguished by a short V- or W-shaped loop.

Teredinidae [INV ZOO] The pileworms or shipworms, a family of bivalve mollusks in the subclass Eulamellibranchia distinguished by having the two valves reduced to a pair of small plates at the anterior end of the animal.

tergite [INV ZOO] The dorsal plate covering a somite in arthropods and certain other articulate animals.

tergum [INV ZOO] A dorsal plate of the operculum in barnacles.

terminal [ARCH] The ornamental finish, decorative element, or termination of an object, item of construction, or structural part. [COMPUT SCI] A site or location at which data can leave or enter a system. [ELEC] **1.** A screw, soldering lug, or other point to which electric connections can be made. Also known as electric terminal. **2.** The equipment at the end of a microwave relay system or other communication channel. **3.** One of the electric input or output points of a circuit or component.

terminal area [ELECTR] The enlarged portion of conductor material surrounding a hole for a lead on a printed circuit. Also known as land; pad.

terminal board [ELEC] An insulating mounting for terminal connections. Also known as terminal strip.

terminal box [ELEC] An enclosure which includes, mounts, and protects one or more terminals or terminal boards; it may include a cover and such accessories as mounting hardware, brackets, locks, and conduit fittings.

terminal bud [BOT] A bud that develops at the apex of a stem. Also known as apical bud.

terminal equipment [COMMUN] **1.** Assemblage of communications-type equipment required to transmit or receive a signal on a channel or circuit, whether it be for delivery or relay. **2.** In radio relay systems, equipment used at points where intelligence is inserted or derived, as distinct from equipment used to relay a reconstituted signal. **3.** Telephone and teletypewriter switchboards and other centrally located equipment at which wire circuits are terminated.

terminal leg See terminal stub.

terminal moraine [GEOL] An end moraine that extends as an arcuate or crescentic ridge across a glacial valley; marks the farthest advance of a glacier. Also known as marginal moraine.

terminal pair [ELEC] An associated pair of accessible terminals, such as the input or output terminals of a device or network.

terminal strip *See* terminal board.

terminal stub [ELEC] Piece of cable that comes with a cable terminal for splicing into the main cable. Also known as terminal leg.

terminal velocity [FL MECH] The velocity with which a body moves relative to a fluid when the resultant force acting on it (due to friction, gravity, and so forth) is zero. [PHYS] The maximum velocity attainable, especially by a freely falling body, under given conditions.

termite [INV ZOO] A soft-bodied insect of the order Isoptera; individuals feed on cellulose and live in colonies with a caste system comprising three types of functional individuals: sterile workers and soldiers, and the reproductives. Also known as white ant.

Termitidae [INV ZOO] A large family of the order Isoptera which contains the higher termites, representing 80% of the species.

Termopsidae [INV ZOO] A family of insects in the order Isoptera composed of damp wood-dwelling forms.

ternary [SCI TECH] Consisting of three, as in a three-phase (that is, ternary) liquid system.

Ternifine man [PALEON] The name for a fossil human type, represented by three lower jaws and a parietal bone discovered in France and thought to be from the upper part of the middle Pleistocene.

terrace [BUILD] **1.** A flat roof. **2.** A colonnaded promenade. **3.** An open platform extending from a building, usually at ground level. [GEOL] **1.** A horizontal or gently sloping embankment of earth along the contours of a slope to reduce erosion, control runoff, or conserve moisture. **2.** A narrow coastal strip sloping gently toward the water. **3.** A long, narrow, nearly level surface bounded by a steeper descending slope on one side and by a steeper ascending slope on the other side. **4.** A benchlike structure bordering an undersea feature.

terracing *See* contour plowing.

terrain-avoidance radar [NAV] Airborne radar which provides a display of terrain ahead of a low-flying aircraft to permit horizontal avoidance of obstacles.

terrain echoes *See* ground clutter.

terrain-following radar [NAV] Airborne radar which provides a display of terrain ahead of a low-flying aircraft to permit manual control, or signals for automatic control, to maintain constant altitude above the ground.

terrain sensing [ENG] The gathering and recording of information about terrain surfaces without actual contact with the object or area being investigated; in particular, the use of photography, radar, and infrared sensing in airplanes and artificial satellites.

terrestrial coordinates *See* geographical coordinates.

terrestrial electricity [GEOPHYS] Electric phenomena and properties of the earth; used in a broad sense to include atmospheric electricity. Also known as geoelectricity.

terrestrial equator *See* astronomical equator.

terrestrial frozen water [HYD] Seasonally or perennially frozen waters of the earth, exclusive of the atmosphere.

terrestrial magnetism *See* geomagnetism.

terrestrial meridian *See* astronomical meridian.

terrestrial planet [ASTRON] One of the four small planets near the sun (Earth, Mercury, Venus, and Mars).

terrestrial radiation [GEOPHYS] Electromagnetic radiation originating from the earth and its atmosphere at wavelengths determined by their temperature. Also known as earth radiation; eradiation.

terrigenous sediment [GEOL] Shallow marine sedimentary deposits composed of eroded terrestrial material.

tert- [ORG CHEM] $(R_1R_2R_3C—)$ Abbreviation for tertiary; trisubstituted methyl radical with the central carbon attached to three other carbons.

Tertiary [GEOL] The older major subdivision (period) of the Cenozoic era, extending from the end of the Cretaceous to the beginning of the Quaternary, from 70,000,000 to 2,000,000 years ago.

tertiary storage [COMPUT SCI] Any of several types of computer storage devices, usually consisting of magnetic tape transports and mass storage tape systems, which have slower access times, larger capacity, and lower cost than main storage or secondary storage.

tesla [ELECTROMAG] The International System unit of magnetic flux density, equal to one weber per square meter. Symbolized T.

Tesla coil [ELECTROMAG] An air-core transformer used with a spark gap and capacitor to produce a high voltage at a high frequency.

Tessaratomidae [INV ZOO] A family of large tropical Hemiptera in the superfamily Pentatomoidea.

Testacellidae [INV ZOO] A family of pulmonate gastropods that includes some species of slugs.

test data [COMPUT SCI] A set of data developed specifically to test the adequacy of a computer run or system; the data may be actual data that has been taken from previous operations, or artificial data created for this purpose.

test function [MATH] An infinitely differentiable function of several real variables used in studying solutions of partial differential equations.

testing level [ELEC] Value of power used for reference represented by 0.001 watt working in 600 ohms.

testis [ANAT] One of a pair of male reproductive glands in vertebrates; after sexual maturity, the source of sperm and hormones.

test of hypothesis [STAT] A rule for rejecting or accepting a hypothesis concerning a population which is based upon a given sample of data.

test of significance [STAT] A test of a hypothetical population property against a sample property where an acceptance interval is used as the rule for rejection.

test oscillator *See* signal generator.

testosterone [BIOCHEM] $C_{19}H_{28}O_2$ The principal androgenic hormone released by the human testis; may be synthesized from cholesterol and certain other sterols.

test pattern [COMMUN] A chart having various combinations of lines, squares, circles, and graduated shading, transmitted from time to time by a television station to check definition, linearity, and contrast for the complete system from camera to receiver. Also known as resolution chart.

test point [ELEC] A terminal or plug-in connector provided in a circuit to facilitate monitoring, calibration, or trouble-shooting.

test program *See* check routine.

test reactor [NUCLEO] A nuclear reactor designed to test the behavior of materials and components under the neutron and gamma fluxes and temperature conditions of an operating reactor.

test routine *See* check routine.

test run [COMPUT SCI] The performance of a computer program to check that it is operating correctly, by using test data to generate results that can be compared with expected answers.

Testudinata [VERT ZOO] The equivalent name for Chelonia.

Testudinellidae [INV ZOO] A family of free-swimming rotifers in the suborder Flosculariacea.

Testudinidae [VERT ZOO] A family of tortoises in the suborder Cryptodira; there are about 30 species found on all continents except Australia.

tetanus [MED] An infectious disease of humans and animals caused by the toxin of *Clostridium tetani* and characterized by convulsive tonic contractions of voluntary muscles; infection commonly follows dirt contamination of deep wounds or other injured tissue. Also known as lockjaw.

Tethinidae [INV ZOO] A family of myodarian cyclorrhaphous dipteran insects in the subsection Acalyptratae.

Tethys [ASTRON] A satellite of the planet Saturn having a diameter of about 1300 kilometers. [GEOL] **1.** A sea which existed for extensive periods of geologic time between the northern and southern continents of the Eastern Hemisphere. **2.** A composite geosyncline from which many structures of the present Alpine-Himalayan orogenic belt were formed.

Tetrabranchia [INV ZOO] A subclass of primitive mollusks of the class Cephalopoda; *Nautilus* is the only living form and is characterized by having four gills.

Tetracentraceae [BOT] A family of dicotyledonous trees in the order Trochodendrales distinguished by possession of a perianth, four stamens, palmately veined leaves, and secretory idioblasts.

Tetracorallia [PALEON] The equivalent name for Rugosa.

Tetractinomorpha [INV ZOO] A heterogeneous subclass of Porifera in the class Demospongiae.

tetrad [CYTOL] A group of four chromatids lying parallel to each other as a result of the longitudinal division of each of a pair of homologous chromosomes during the pachytene and later stages of the prophase of meiosis.

tetrad of Fallot *See* tetralogy of Fallot.

tetradymite [MINERAL] Bi_2Te_2S A pale steel-gray mineral that usually occurs in foliated masses in auriferous veins; has metallic luster, hardness of 1.5–2 on Mohs scale, and specific gravity of 7.2–7.6.

tetragonal lattice [CRYSTAL] A crystal lattice in which the axes of a unit cell are perpendicular, and two of them are equal in length to each other, but not to the third axis.

tetragonal trisoctahedron *See* trapezohedron.

tetrahedrite [MINERAL] $(Cu,Fe,Zn,Ag)_{12}Sb_4S_{13}$ A grayish-black mineral crystallizing in the isometric system as tetrahedrons and occurring in massive or granular form; luster is metallic, hardness is 3.5–4 on Mohs scale, and specific gravity is 4.6–5.1; an important ore of copper. Also known as fahlore; gray copper ore; panabase; stylotypite.

tetrahedron [CRYSTAL] An isometric crystal form in cubic crystals, in the shape of a four-faced polyhedron, each face of which is a triangle. [MATH] A four-sided polyhedron.

tetrahexahedron [CRYSTAL] A form of regular crystal system with four triangular isosceles faces on each

side of a cube; there are altogether 24 congruent faces.

tetrahydroxy butane *See* erythritol.

tetralogy of Fallot [MED] A congenital abnormality of the heart consisting of pulmonary stenosis, defect of the interventricular septum, hypertrophy of the right ventricle, and overriding or dextroposition of the aorta. Also known as tetrad of Fallot.

Tetralophodontinae [PALEON] An extinct subfamily of proboscidean mammals in the family Gomphotheridae.

tetramer [ORG CHEM] A polymer that results from the union of four identical monomers; for example, the tetramer C_8H_8 forms from union of four molecules of C_2H_2.

Tetranychidae [INV ZOO] The spider mites, a family of phytophagous trombidiform mites.

†72Tetraodontiformes [VERT ZOO] An order of specialized teleost fishes that includes the triggerfishes, puffers, trunkfishes, and ocean sunfishes.

Tetraonidae [VERT ZOO] The ptarmigans and grouse, a family of upland game birds in the order Galliformes characterized by rounded tails and wings and feathered nostrils.

Tetraphidaceae [BOT] The single family of the plant order Tetraphidales.

Tetraphidales [BOT] A monofamilial order of mosses distinguished by scalelike protonema and the peristomes of four rigid, nonsegmented teeth.

Tetraphyllidea [INV ZOO] An order of small tapeworms of the subclass Cestoda characterized by the variation in the structure of the scolex; all species are intestinal parasites of elasmobranch fishes.

tetraploidy [CYTOL] The occurrence of related forms possessing in the somatic cells chromosome numbers four times the haploid number.

tetrapod [VERT ZOO] A four-footed animal.

Tetrapoda [VERT ZOO] The superclass of the subphylum Vertebrata whose members typically possess four limbs; includes all forms above fishes.

Tetrarhynchoidea [INV ZOO] The equivalent name for Trypanorhyncha.

Tetrasporales [BOT] A heterogeneous and artificial assemblage of colonial fresh-water and marine algae in the division Chlorophyta.

Tetrigidae [INV ZOO] The grouse locusts or pygmy grasshoppers in the order Orthoptera in which the front wings are reduced to small scalelike structures.

tetrode [ELECTR] A four-electrode electron tube containing an anode, a cathode, a control electrode, and one additional electrode that is ordinarily a grid.

tetrode transistor [ELECTR] A four-electrode transistor, such as a tetrode point-contact transistor or double-base junction transistor.

Tettigoniidae [INV ZOO] A family of insects in the order Orthoptera which have long antennae, hindlegs fitted for jumping, and an elongate, vertically flattened ovipositor; consists of the longhorn or green grasshopper.

Teuthidae [VERT ZOO] The rabbitfishes, a family of perciform fishes in the suborder Acanthuroidei.

Teuthoidea [INV ZOO] An order of the molluscan subclass Coleoidea in which the rostrum is not developed, the proostracum is represented by the elongated pen or gladus, and ten arms are present.

TE wave *See* transverse electric wave.

text-editing system [COMPUT SCI] A computer program, together with associated hardware, for the online creation and modification of computer programs and ordinary text.

textile [MATER] A material made of natural or man-made fibers and used for the manufacture of items such as clothing and furniture fittings.

Textulariina [INV ZOO] A suborder of foraminiferan protozoans characterized by an agglutinated wall.

Th *See* thorium.

thalamus [ANAT] Either one of two masses of gray matter located on the sides of the third ventricle and forming part of the lateral wall of that cavity.

Thalassinidea [INV ZOO] The mud shrimps, a group of thin-shelled, burrowing decapod crustaceans belonging to the Macrura; individuals have large chelate or subchelate first pereiopods, and no chelae on the third pereiopods.

Thalattosauria [PALEON] A suborder of extinct reptiles in the order Eosuchia from the Middle Triassic.

Thaliacea [INV ZOO] A small class of pelagic Tunicata in which oral and atrial apertures occur at opposite ends of the body.

thallium [CHEM] A metallic element in group III, symbol Tl, atomic number 81, atomic weight 204.37; insoluble in water, soluble in nitric and sulfuric acids, melts at 302°C, boils at 1457°C. [MET] Bluish-white metal with tinlike malleability, but a little softer; used in alloys.

Thallobionta [BOT] One of the two subkingdoms of plants, characterized by the absence of specialized tissues or organs and multicellular sex organs.

Thallophyta [BOT] The equivalent name for Thallobionta.

thallus [BOT] A plant body that is not differentiated into special tissue systems or organs and may vary from a single cell to a complex, branching multicellular structure.

Thanetian [GEOL] A European stage of geologic time; uppermost Paleocene, above Montian, below Ypresian of Eocene.

Thaumaleidae [INV ZOO] A family of orthorrhaphous dipteran insects in the series Nematocera.

Thaumastellidae [INV ZOO] A monospecific family of the Hemiptera assigned to the Pentatomorpha found only in Ethiopia.

Thaumastocoridae [INV ZOO] The single family of the hemipteran superfamily Thaumastocoroidea.

Thaumastocoroidea [INV ZOO] A monofamilial superfamily of the Hemiptera in the subdivision Geocorisae which occurs in Australia and the New World tropics.

Thaumatoxenidae [INV ZOO] A family of cyclorrhaphous dipteran insects in the series Aschiza.

thaw pipe [MIN ENG] A string of pipe drilled into a string of drill rods that is frozen in a borehole in permafrost, through which water is circulated to thaw the ice and free the drill rods.

Theaceae [BOT] A family of dicotyledonous erect trees or shrubs in the order Theales characterized by alternate, exstipulate leaves, usually five petals, and mostly numerous stamens.

Theales [BOT] An order of dicotyledonous mostly woody plants in the subclass Dilleniidae with simple or occasionally compound leaves, petals usually separate, numerous stamens, and the calyx arranged in a tight spiral.

theca [ANAT] The sheath of dura mater which covers the spinal cord. [BOT] **1.** A moss capsule. **2.** A pollen sac. [HISTOL] The layer of stroma surrounding a Graafian follicle. [INV ZOO] The test of a testate protozoan or a rotifer.

Thecanephria [INV ZOO] An order of the phylum Brachiata containing a group of elongate, tube-dwelling tentaculate, deep-sea animals of bizarre structure.

Thecideidina [PALEON] An extinct suborder of articulate brachiopods doubtfully included in the order Terebratulida.

Thecodontia [PALEON] An order of archosaurian reptiles, confined to the Triassic and distinguished by the absence of a supratemporal bone, parietal foramen, and palatal teeth, and by the presence of an antorbital fenestra.

Thelastomidae [INV ZOO] A family of nematode worms in the superfamily Oxyuroidea.

Themis [ASTRON] An asteroid with a diameter of roughly 225 kilometers, mean distance from the sun of 3.138 astronomical units, and C-type surface composition.

thenardite [MINERAL] Na_2SO_4 A colorless, grayish-white, yellowish, yellow-brown, or reddish, orthorhombic mineral consisting of sodium sulfate.

theodolite [OPTICS] An optical instrument used in surveying which consists of a sighting telescope mounted so that it is free to rotate around horizontal and vertical axes, and graduated scales so that the angles of rotation may be measured; the telescope is usually fitted with a right-angle prism so that the observer continues to look horizontally into the eyepiece, whatever the variation of the elevation angle; in meteorology, it is used principally to observe the motion of a pilot balloon.

Theophrastaceae [BOT] A family of tropical and subtropical dicotyledonous woody plants in the order Primulales characterized by flowers having staminodes alternate with the corolla lobes.

theorem [MATH] A proven mathematical statement.

theoretical physics [PHYS] The description of natural phenomena in mathematical form.

theory [MATH] The collection of theorems and principles associated with some mathematical object or concept. [SCI TECH] An attempt to explain a certain class of phenomena by deducing them as necessary consequences of other phenomena regarded as more primitive and less in need of explanation.

theory of equations [MATH] The study of polynomial equations from the viewpoint of solution methods, relations among roots, and connections between coefficients and roots.

theory of games *See* game theory.

theralite [PETR] A dark-colored, visibly crystalline rock composed chiefly of pyroxene with smaller amounts of calcic plagioclase and nepheline.

Therapsida [PALEON] An order of mammallike reptiles of the subclass Synapsida which first appeared in mid-Permian times and persisted until the end of the Triassic.

Therevidae [INV ZOO] The stiletto flies, a family of orthorrhaphous dipteran insects in the series Brachycera.

Theria [VERT ZOO] A subclass of the class Mammalia including all living mammals except the monotremes.

Theridiidae [INV ZOO] The comb-footed spiders, a family of the suborder Dipneumonomorphae.

thermal [METEOROL] A relatively small-scale, rising current of air produced when the atmosphere is heated enough locally by the earth's surface to produce absolute instability in its lower layers. [THERMO] Of or concerning heat.

thermal ammeter *See* hot-wire ammeter.

thermal analysis [ANALY CHEM] Any analysis of physical or thermodynamic properties of materials in which heat (or its removal) is directly involved; for example, boiling, freezing, solidification-point determinations, heat of fusion and heat of vaporization measurements, distillation, calorimetry, and differential thermal, thermogravimetric, thermometric, and thermometric titration analyses. Also known as thermoanalysis. [MET] Determining transformations in a metal by observing the temperature-time relationship during uniform cooling or heating; phase tranformations are indicated by irregularities in a smooth curve.

thermal aureole   *See* aureole.

thermal barrier   [AERO ENG] A limit to the speed of airplanes and rockets in the atmosphere imposed by heat from friction between the aircraft and the air, which weakens and eventually melts the surface of the aircraft. Also known as heat barrier.

thermal battery   [ELEC] **1.** A combination of thermal cells. Also known as fused-electrolyte battery; heat-activated battery. **2.** A voltage source consisting of a number of bimetallic junctions connected to produce a voltage when heated by a flame.

thermal breeder reactor   [NUCLEO] A breeder reactor in which the fission chain reaction is sustained by thermal neutrons.

thermal capacity   *See* heat capacity.

thermal charge   *See* entropy.

thermal conductance   [THERMO] The amount of heat transmitted by a material divided by the difference in temperature of the surfaces of the material. Also known as conductance.

thermal conductivity   [THERMO] The heat flow across a surface per unit area per unit time, divided by the negative of the rate of change of temperature with distance in a direction perpendicular to the surface. Also known as heat conductivity.

thermal convection   [METEOROL] Atmospheric currents, predominantly vertical, arising from the release of gravitational visibility; commonly produced by solar heating of the ground; the cause of convective (cumulus) clouds. Also known as free convection; gravitational convection. [THERMO] *See* heat convection.

thermal converter   [ELECTR] A device that converts heat energy directly into electric energy by using the Seebeck effect; it is composed of at least two dissimilar materials, one junction of which is in contact with a heat source and the other junction of which is in contact with a heat sink. Also known as thermocouple converter; thermoelectric generator; thermoelectric power generator; thermoelement.

thermal cracking   [CHEM ENG] A petroleum refining process that decomposes, rearranges, or combines hydrocarbon molecules by the application of heat, without the aid of catalysts.

thermal detector   *See* bolometer.

thermal diffusion   [PHYS CHEM] A phenomenon in which a temperature gradient in a mixture of fluids gives rise to a flow of one constituent relative to the mixture as a whole. Also known as thermodiffusion.

thermal drift   [ELECTR] Drift caused by internal heating of equipment during normal operation or by changes in external ambient temperature.

thermal efficiency   [CHEM ENG] In a tube-and-shell heat-exchange system, the ratio of the actual temperature range of the tube-side fluid (inlet versus outlet temperature) to the maximum possible temperature range. [THERMO] *See* efficiency.

thermal emissivity   *See* emissivity.

thermal energy   [NUCLEO] Energy which is characteristic for thermal neutrons at room temperature, about 0.025 electron volt.

thermal equilibrium   [THERMO] Property of a system all parts of which have attained a uniform temperature which is the same as that of the system's surroundings.

thermal excitation   [ATOM PHYS] The process in which atoms or molecules acquire internal energy in collisions with other particles.

thermal expansion   [PHYS] The dimensional changes exhibited by solids, liquids, and gases for changes in temperature while pressure is held constant.

thermal high   [METEOROL] A high resulting from the cooling of air by a cold underlying surface, and remaining relatively stationary over the cold surface.

thermal hysteresis   [THERMO] A phenomenon sometimes observed in the behavior of a temperature-dependent property of a body; it is said to occur if the behavior of such a property is different when the body is heated through a given temperature range from when it is cooled through the same temperature range.

thermal inductance   [THERMO] The product of temperature difference and time divided by entropy flow.

thermal instability   [FL MECH] The instability resulting in free convection in a fluid heated at a boundary.

thermal low   [METEOROL] An area of low atmospheric pressure due to high temperatures caused by intensive heating at the earth's surface; common to the continental subtropics in summer, thermal lows remain stationary over the area that produces them, their cyclonic circulation is generally weak and diffuse, and they are nonfrontal. Also known as heat low.

thermal microphone   [ENG ACOUS] Microphone depending for its action on the variation in the resistance of an electrically heated conductor that is being alternately increased and decreased in temperature by sound waves.

thermal neutron   [NUCLEO] One of a collection of neutrons whose energy distribution is identical with or similar to the Maxwellian distribution in the material in which they are found; the average kinetic energy of such neutrons at room temperature is about 0.025 electron volt. Also known as slow neutron.

thermal noise   [ELECTR] Electric noise produced by thermal agitation of electrons in conductors and semiconductors. Also known as Johnson noise; resistance noise.

thermal photography   *See* thermography.

thermal process   [CHEM ENG] Any process that utilizes heat, without the aid of a catalyst, to accomplish chemical change; for example, thermal cracking, thermal reforming, or thermal polymerization.

thermal radiation   *See* heat radiation.

thermal reactor   [CHEM ENG] A device, system, or vessel in which chemical reactions take place because of heat (no catalysis); for example, thermal cracking, thermal reforming, or thermal polymerization. [NUCLEO] A nuclear reactor in which fission is induced primarily by neutrons of such low energy that they are in substantial thermal equilibrium with the material of the core.

thermal reforming   [CHEM ENG] A petroleum refining process using heat (but no catalyst) to effect molecular rearrangement of a low-octane naphtha to form high-octane motor gasoline.

thermal resistance   [ELECTR] *See* effective thermal resistance. [THERMO] A measure of a body's ability to prevent heat from flowing through it, equal to the difference between the temperatures of opposite faces of the body divided by the rate of heat flow. Also known as heat resistance.

thermal resistivity   [THERMO] The reciprocal of the thermal conductivity.

thermal resistor [ELEC] A resistor designed so its resistance varies in a known manner with changes in ambient temperature.

thermal scattering [SOLID STATE] Scattering of electrons, neutrons, or x-rays passing through a solid due to thermal motion of the atoms in the crystal lattice.

thermal spring [HYD] A spring whose water temperature is higher than the local mean annual temperature of the atmosphere.

thermal stress [MECH] Mechanical stress induced in a body when some or all of its parts are not free to expand or contract in response to changes in temperature.

thermal titration *See* thermometric titration.

thermal transducer [ENG] Any device which converts energy from some form other than heat energy into heat energy; an example is the absorbing film used in the thermal pulse method.

thermal value [THERMO] Heat produced by combustion, usually expressed in calories per gram or British thermal units per pound.

thermal volt *See* kelvin.

thermal wave [CRYO] *See* temperature wave. [SOLID STATE] A sound wave in a solid which has a short wavelength.

thermal wind [METEOROL] The mean wind-shear vector in geostrophic balance with the gradient of mean temperature of a layer bounded by two isobaric surfaces.

thermion [ELECTR] A charged particle, either negative or positive, emitted by a heated body, as by the hot cathode of a thermionic tube.

thermionic cathode *See* hot cathode.

thermionic converter [ELECTR] A device in which heat energy is directly converted to electric energy; it has two electrodes, one of which is raised to a sufficiently high temperature to become a thermionic electron emitter, while the other, serving as an electron collector, is operated at a significantly lower temperature. Also known as thermionic generator; thermionic power generator; thermoelectric engine.

thermionic current [ELECTR] Current due to directed movements of thermions, such as the flow of emitted electrons from the cathode to the plate in a thermionic vacuum tube.

thermionic emission [ELECTR] 1. The outflow of electrons into vacuum from a heated electric conductor. Also known as Edison effect; Richardson effect. 2. More broadly, the liberation of electrons or ions from a substance as a result of heat.

thermionic generator *See* thermionic converter.

thermionic power generator *See* thermionic converter.

thermionic tube [ELECTR] An electron tube that relies upon thermally emitted electrons from a heated cathode for tube current. Also known as hot-cathode tube.

thermionic work function [ELECTR] Energy required to transfer an electron from the fermi energy in a given metal through the surface to the vacuum just outside the metal.

thermistor [ELECTR] A resistive circuit component, having a high negative temperature coefficient of resistance, so that its resistance decreases as the temperature increases; it is a stable, compact, and rugged two-terminal ceramiclike semiconductor bead, rod, or disk. Derived from thermal resistor.

thermit welding [MET] Welding with molten iron which is obtained by igniting aluminum and an iron

oxide in a crucible, whereby the aluminum floats to the top of the molten metal and is poured off.

thermoacoustic effect *See* optoacoustic effect.

thermoammeter [ENG] An ammeter that is actuated by the voltage generated in a thermocouple through which is sent the current to be measured; used chiefly for measuring radio-frequency currents. Also known as electrothermal ammeter; thermocouple ammeter.

thermoanalysis *See* thermal analysis.

thermobalance [ANALY CHEM] An analytical balance modified for thermogravimetric analysis, involving the measurement of weight changes associated with the transformations of matter when heated.

thermochemical calorie *See* calorie.

thermochemistry [PHYS CHEM] The measurement, interpretation, and analysis of heat changes accompanying chemical reactions and changes in state.

thermocline [GEOPHYS] 1. A temperature gradient as in a layer of sea water, in which the temperature decrease with depth is greater than that of the overlying and underlying water. Also known as metalimnion. 2. A layer in a thermally stratified body of water in which such a gradient occurs.

thermocouple [ENG] A device consisting basically of two dissimilar conductors joined together at their ends; the thermoelectric voltage developed between the two junctions is proportional to the temperature difference between the junctions, so the device can be used to measure the temperature of one of the junctions when the other is held at a fixed, known temperature, or to convert radiant energy into electric energy.

thermocouple ammeter *See* thermoammeter.

thermocouple converter *See* thermal converter.

thermodiffusion *See* thermal diffusion.

thermodynamic cycle [THERMO] A procedure or arrangement in which some material goes through a cyclic process and one form of energy, such as heat at an elevated temperature from combustion of a fuel, is in part converted to another form, such as mechanical energy of a shaft, the remainder being rejected to a lower temperature sink. Also known as heat cycle.

thermodynamic equilibrium [THERMO] Property of a system which is in mechanical, chemical, and thermal equilibrium.

thermodynamic function of state [THERMO] Any of the quantities defining the thermodynamic state of a substance in thermodynamic equilibrium; for a perfect gas, the pressure, temperature, and density are the fundamental thermodynamic variables, any two of which are, by the equation of state, sufficient to specify the state. Also known as state parameter; state variable; thermodynamic variable.

thermodynamic potential at constant volume *See* free energy.

thermodynamics [PHYS] The branch of physics which seeks to derive, from a few basic postulates, relationships between properties of matter, especially those which are affected by changes in temperature, and a description of the conversion of energy from one form to another.

thermodynamic variable *See* thermodynamic function of state.

thermoelasticity [PHYS] Dependence of the stress distribution of an elastic solid on its thermal state, or of its thermal conductivity on the stress distribution.

thermoelectric converter [ELECTR] A converter that changes solar or other heat energy to electric energy; used as a power source on spacecraft.

thermoelectric cooling   [ENG] Cooling of a chamber based on the Peltier effect; an electric current is sent through a thermocouple whose cold junction is thermally coupled to the cooled chamber, while the hot junction dissipates heat to the surroundings. Also known as thermoelectric refrigeration.

thermoelectric effect   *See* thermoelectricity.

thermoelectric engine   *See* thermionic converter.

thermoelectric generator   *See* thermal converter.

thermoelectric heating   [ENG] Heating based on the Peltier effect, involving a device which is in principle the same as that used in thermoelectric cooling except that the current is reversed.

thermoelectricity   [PHYS] The direct conversion of heat into electrical energy, or the reverse; it encompasses the Seebeck, Peltier, and Thomson effects but, by convention, excludes other electrothermal phenomena, such as thermionic emission. Also known as thermoelectric effect.

thermoelectric nuclear battery   [NUCLEO] A low-voltage battery in which a heat source, consisting of a radioactive isotope such as polonium-210, is hermetically sealed in a strong, dense capsule, and a series of thermocouples are alternately connected thermally, but not electrically, to the heat source and the outer surface of the capsule.

thermoelectric power generator   *See* thermal converter.

thermoelectric refrigeration   *See* thermoelectric cooling.

thermoelectric series   [MET] A series of metals arranged in order of their thermoelectric voltage-generating ratings with respect to some reference metal, such as lead.

thermoelectric solar cell   [ELECTR] A solar cell in which the sun's energy is first converted into heat by a sheet of metal, and the heat is converted into electricity by a semiconductor material sandwiched between the first metal sheet and a metal collector sheet.

thermoelement   *See* thermal converter.

thermoforming   [ENG] Forming of thermoplastic sheet by heating it and then pulling it down onto a mold surface to shape it.

thermograph   [ENG] An instrument that senses, measures, and records the temperature of the atmosphere. Also known as recording thermometer. [OPTICS] A far-infrared image-forming device that provides a thermal photograph by scanning a far-infrared image of an object or scene.

thermography   [ENG] A method of measuring surface temperature by using luminescent materials: the two main types are contact thermography and projection thermography. [GRAPHICS] **1.** A photocopying process in which the original copy is placed in contact with a transparent sheet and is exposed to infrared rays; heat from carbon or a metallic compound in the text ink then causes a chemical change in a substance laminated between the transparent sheet of paper and a white waxy back. **2.** Photography that uses radiation in the long-wavelength far-infrared region, emitted by objects at temperatures ranging from $-170°F$ $(-112°C)$ to over 300°F (149°C). Also known as thermal photography.

thermoluminescence   [ATOM PHYS] **1.** Broadly, any luminescence appearing in a material due to application of heat. **2.** Specifically, the luminescence appearing as the temperature of a material is steadily increased; it is usually caused by a process in which electrons receiving increasing amounts of thermal energy escape from a center in a solid where they have been trapped and go over to a luminescent center, giving it energy and causing it to luminesce.

thermomagnetic effect   [PHYS] An electrical or thermal phenomenon occurring when a conductor or semiconductor is placed simultaneously in a temperature gradient and a magnetic field; examples are the Ettingshausen-Nernst effect and the Righi-Leduc effect.

thermometer   [ENG] An instrument that measures temperature.

thermometer bird   [VERT ZOO] The name applied to the brush turkey, native to Australia, because it lays its eggs in holes in mounds of earth and vegetation, with the heat from the decaying vegetation serving to incubate the eggs.

thermometric analysis   [PHYS CHEM] A method for determination of the transformations a substance undergoes while being heated or cooled at an essentially constant rate, for example, freezing-point determinations.

thermometric conductivity   *See* diffusivity.

thermometric titration   [ANALY CHEM] A titration in an adiabatic system, yielding a plot of temperature versus volume of titrant; used for neutralization, precipitation, redox, organic condensation, and complex-formation reactions. Also known as calorimetric titration; enthalpy titration; thermal titration.

thermometry   [THERMO] The science and technology of measuring temperature, and the establishment of standards of temperature measurement.

thermonatrite   [MINERAL] $Na_2CO_3 \cdot H_2O$ A colorless to white, grayish, or yellowish, orthorhombic mineral consisting of sodium carbonate monohydrate; occurs as a crust or efflorescence.

thermonuclear   [NUCLEO] Referring to any process in which a very high temperature is used to bring about the fusion of light nuclei, with the accompanying liberation of energy.

thermopile   [ENG] An array of thermocouples connected either in series to give higher voltage output or in parallel to give higher current output, used for measuring temperature or radiant energy or for converting radiant energy into electric power.

thermoplastic resin   [MATER] A material with a linear macromolecular structure that will repeatedly soften when heated and harden when cooled; for example, styrene, acrylics, cellulosics, polyethylenes, vinyls, nylons, and fluorocarbons.

thermoreceptor   [PHYSIO] A sense receptor that responds to stimulation by heat and cold.

thermoregulation   [PHYSIO] A mechanism by which mammals and birds attempt to balance heat gain and heat loss in order to maintain a constant body temperature when exposed to variations in cooling power of the external medium.

thermoregulator   [ENG] A high-accuracy or high-sensitivity thermostat; one type consists of a mercury-in-glass thermometer with sealed-in electrodes, in which the rising and falling column of mercury makes and breaks an electric circuit.

thermorelay   *See* thermostat.

thermoremanent magnetization   [GEOPHYS] The permanent magnetization of igneous rocks, acquired at the time of cooling from the molten state.

Thermosbaenacea   [INV ZOO] An order of small crustaceans in the superorder Pancarida.

Thermosbaenidae   [INV ZOO] A family of the crustacean order Thermosbaenacea.

thermosetting resin   [MATER] A plastic that solidifies when first heated under pressure, and which cannot be remelted or remolded without destroying its original characteristics; examples are epoxies, malamines, phenolics, and ureas.

thermostat [ENG] An instrument which measures changes in temperature and directly or indirectly controls sources of heating and cooling to maintain a desired temperature. Also known as thermorelay.

Theropoda [PALEON] A suborder of carnivorous bipedal saurischian reptiles which first appeared in the Upper Triassic and culminated in the uppermost Cretaceous.

Theropsida [PALEON] An order of extinct mammal-like reptiles in the subclass Synapsida.

theta pinch [PL PHYS] A device for producing a controlled nuclear fusion reaction, in which plasma in a long torus or skinny tube is confined by a magnetic field produced by current-carrying coils, and is shock-heated and compressed by pulses in this field to produce the high temperatures at which fusion reactions take place; the magnetic field is then sustained in order to maintain the plasma confinement.

theta rhythm [PSYCH] A brain rhythm having a frequency of about 4–7 hertz, and somewhat greater voltage than the alpha rhythm; thought to originate in the hippocampus.

Thévenin's theorem [ELEC] A valuable theorem in network problems which allows calculation of the performance of a device from its terminal properties only: the theorem states that at any given frequency the current flowing in any impedance, connected to two terminals of a linear bilateral network containing generators of the same frequency, is equal to the current flowing in the same impedance when it is connected to a voltage generator whose generated voltage is the voltage at the terminals in question with the impedance removed, and whose series impedance is the impedance of the network looking back from the terminals into the network with all generators replaced by their internal impedances. Also known as Helmholtz's theorem.

thiamine [BIOCHEM] $C_{12}H_{17}ClN_4OS$ A member of the vitamin B complex that occurs in many natural sources, frequently in the form of cocarboxylase. Also known as aneurine; vitamin B_1.

Thiaridae [INV ZOO] A family of freshwater gastropod mollusks in the order Pectinibranchia.

thicket *See* tropical scrub.

thick-film circuit [ELECTR] A microcircuit in which passive components, of a ceramic-metal composition, are formed on a ceramic substrate by successive screen-printing and firing processes, and discrete active elements are attached separately.

thickness gage [ENG] A gage for measuring the thickness of a sheet of material, the thickness of an object, or the thickness of a coating; examples include penetration-type and backscattering radioactive thickness gages and ultrasonic thickness gages.

thigh [ANAT] The upper part of the leg, from the pelvis to the knee.

thigmotaxis *See* stereotaxis.

Thigmotrichida [INV ZOO] An order of ciliated protozoans in the subclass Holotrichia.

thigmotropism *See* stereotropism.

thin film [ELECTR] A film a few molecules thick deposited on a glass, ceramic, or semiconductor substrate to form a capacitor, resistor, coil, cryotron, or other circuit component.

thin-film circuit [ELECTR] A circuit in which the passive components and conductors are produced as films on a substrate by evaporation or sputtering; active components may be similarly produced or mounted separately.

thin-film memory *See* thin-film storage.

thin-film storage [COMPUT SCI] A high-speed storage device that is fabricated by depositing layers, one molecule thick, of various materials which, after etching, provide microscopic circuits which can move and store data in small amounts of time. Also known as thin-film memory.

thin-layer chromatography [ANALY CHEM] Chromatographing on thin layers of adsorbents rather than in columns; adsorbent can be alumina, silica gel, silicates, charcoals, or cellulose.

thin lens [OPTICS] A lens whose thickness is small enough to be neglected in calculations of such quantities as object distance, image distance, and magnification.

Thinocoridae [VERT ZOO] The seed snipes, family of South American birds in the order Charadriiformes.

thio- [CHEM] A chemical prefix derived from the Greek *theion*, meaning sulfur; indicates the replacement of an oxygen in an acid radical by sulfur with a negative valence of 2.

thiol *See* mercaptan.

third-generation computer [COMPUT SCI] One of the general purpose digital computers introduced in the late 1960s; it is characterized by integrated circuits and has logical organization and software which permit the computer to handle many programs at the same time, allow one to add or remove units from the computer, permit some or all input/output operations to occur at sites remote from the main processor, and allow conversational programming techniques.

third harmonic [PHYS] A sine-wave component having three times the fundamental frequency of a complex signal.

third law of motion *See* Newton's third law.

third law of thermodynamics [THERMO] The entropy of all perfect crystalline solids is zero at absolute zero temperature.

third-order reaction [PHYS CHEM] A chemical reaction in which the rate of reaction is determined by the concentration of three reactants.

thixotropy [PHYS CHEM] Property of certain gels which liquefy when subjected to vibratory forces, such as ultrasonic waves or even simple shaking, and then solidify again when left standing.

Thlipsuridae [PALEON] A Paleozoic family of ostracod crustaceans in the suborder Platycopa.

tholeiite [PETR] 1. A group of basalts composed principally of plagioclase, pyroxene, and iron oxide minerals as phenocrysts in a glassy groundmass. 2. Any rock in the group.

thomsenolite [MINERAL] $NaCaAlF_6 \cdot H_2O$ A colorless to white, monoclinic mineral consisting of a hydrated aluminofluoride of sodium and calcium; it is dimorphous with pachnolite.

Thomson-Berthelot principle [PHYS CHEM] The assumption that the heat released in a chemical reaction is directly related to the chemical affinity, and that, in the absence of the application of external energy, that chemical reaction which releases the greatest heat is favored over others; the principle is in general incorrect, but applies in certain special cases.

Thomson bridge *See* Kelvin bridge.

Thomson coefficient [PHYS] The ratio of the voltage existing between two points on a metallic conductor to the difference in temperature of those points.

Thomson effect [PHYS] A thermoelectric effect in which heat flows into or out of a homogeneous conductor when an electric current flows between two points in the conductor at different temperatures, the direction of heat flow depending upon whether the current flows from colder to warmer metal or from warmer to colder.

thomsonite [MINERAL] $NaCa_2Al_5Si_5O_{20}\cdot 6H_2O$ Snow-white zeolite mineral forming orthorhombic crystals and occurring in masses of radiating crystals; hardness is 5–5.5 on Mohs scale.

Thomson relations [PHYS] Equations in the study of thermoelectricity, relating the Peltier coefficient and the Thomson coefficient to the Seebeck voltage; they are derived by thermodynamics. Also known as Kelvin relations.

Thomson scattering [ELECTROMAG] Scattering of electromagnetic radiation by free (or very loosely bound) charged particles, computed according to a classical nonrelativistic theory: energy is taken away from the primary radiation as the charged particles accelerated by the transverse electric field of the radiation, radiate in all directions.

Thoracica [INV ZOO] An order of the subclass Cirripedia; individuals are permanently attached in the adult stage, the mantle is usually protected by calcareous plates, and six pairs of biramous thoracic appendages are present.

thoracic cavity *See* thorax.

thoracic duct [ANAT] The common lymph trunk beginning in the crura of the diaphragm at the level of the last thoracic vertebra, passing upward, and emptying into the left subclavian vein at its junction with the left internal jugular vein.

thoracic vertebrae [ANAT] The vertebrae associated with the chest and ribs in vertebrates; there are 12 in humans.

Thoracostomopsidae [INV ZOO] A family of marine nematodes in the superfamily Enoploidea, which have the stomatal armature modified to form a hollow tube.

thorax [ANAT] The chest; the cavity of the mammalian body between the neck and the diaphragm, containing the heart, lungs, and mediastinal structures. Also known as thoracic cavity. [INV ZOO] The middle of three principal divisions of the body of certain classes of arthropods.

thorianite [MINERAL] ThO_2 A radioactive mineral that crystallizes in the isometric system, occurs in worn cubic crystals, is brownish black to reddish brown in color, and has resinous luster; hardness is 7 on the Mohs scale, and specific gravity is 9.7–9.8.

Thorictidae [INV ZOO] The ant blood beetles, a family of coleopteran insects in the superfamily Dermestoidea.

thorite [MINERAL] $ThSiO_4$ A brownish-yellow to brownish-black and black radioactive mineral that is tetragonal in crystallization; hardness is about 4.5 on Mohs scale, and specific gravity is 4.3–5.4.

thorium [CHEM] An element of the actinium series, symbol Th, atomic number 90, atomic weight 232; soft, radioactive, insoluble in water and alkalies, soluble in acids, melts at 1750°C, boils at 4500°C. [MET] A heavy malleable metal that changes from silvery-white to dark gray or black in air; potential source of nuclear energy; used in manufacture of sunlamps.

thou *See* mil.

thread [DES ENG] A continuous helical rib, as on a screw or pipe. [GEOL] An extremely small vein, even thinner than a stringer. [MIN ENG] A more or less straight line of stall faces, having no cuttings, loose ends, fast ends, or steps. [TEXT] A continuous strand formed by spinning and twisting together short strands of textile fibers.

thread blight [PL PATH] A fungus disease of a number of tropical and semitropical woody plants, including cocoa and tea, caused by species of *Pellicu-*

laria and *Marasmius* which form filamentous mycelia on the surface of twigs and leaves.

thread cutter [MECH ENG] A tool used to cut screw threads on a pipe, screw, or bolt.

threading machine [MECH ENG] A tool used to cut or form threads inside or outside a cylinder or cone.

three-address code [COMPUT SCI] In computers, a multiple-address code which includes three addresses, usually two addresses from which data are taken and one address where the result is entered; location of the next instruction is not specified, and instructions are taken from storage in preassigned order.

three-address instruction [COMPUT SCI] In computers, an instruction which includes an operation and specifies the location of three registers.

three-body problem [MECH] The problem of predicting the motions of three objects obeying Newton's laws of motion and attracting each other according to Newton's law of gravitation.

three-junction transistor [ELECTR] A *pnpn* transistor having three junctions and four regions of alternating conductivity; the emitter connection may be made to the *p* region at the left, the base connection to the adjacent *n* region, and the collector connection to the *n* region at the right, while the remaining *p* region is allowed to float.

three-phase circuit [ELEC] A circuit energized by alternating-current voltages that differ in phase by one-third of a cycle or 120°.

three-phase current [ELEC] Current delivered through three wires, with each wire serving as the return for the other two and with the three current components differing in phase successively by one-third cycle, or 120 electrical degrees.

three-phase system [PHYS] Any physical system in which three distinct phases coexist; phases can be liquid, solid, vapor (gas), or three mutually insoluble liquids, or any combination thereof.

three-plus-one address [COMPUT SCI] An instruction format containing an operation code, three operand address parts, and a control address.

three-wire generator [ELEC] Electric generator with a balance coil connected across the armature, the midpoint of the coil providing the potential of the neutral wire in a three-wire system.

thresher shark [VERT ZOO] Common name for fishes in the family Alopiidae; pelagic predacious sharks of generally wide distribution that have an extremely long, whiplike tail with which they thrash the water, destroying schools of small fishes.

threshold [BUILD] A piece of stone, wood, or metal that lies under an outside door. [ELECTR] In a modulation system, the smallest value of carrier-to-noise ratio at the input to the demodulator for all values above which a small percentage change in the input carrier-to-noise ratio produces a substantially equal or smaller percentage change in the output signal-to-noise ratio. [ENG] The least value of a current, voltage, or other quantity that produces the minimum detectable response in an instrument or system. [GEOL] *See* riegel. [MATH] A logic operator such that, if P, Q, R, S, ... are statements, then the threshold will be true if at least N statements are true, false otherwise. [PHYS] The minimum level of some input quantity needed for some process to take place, such as a threshold energy for a reaction, or the minimum level of pumping at which a laser can go into self-excited oscillation. [PHYSIO] The minimum level of a stimulus that will evoke a response in an irritable tissue.

threshold contrast [OPTICS] The smallest contrast of luminance (or brightness) that is perceptible to the human eye under specified conditions of adaptation luminance and target visual angle. Also known as contrast sensitivity; contrast threshold; liminal contrast.

threshold detector [NUC PHYS] An element or isotope in which radioactivity is induced only by the capture of neutrons having energies in excess of a certain characteristic threshold value; used to determine the neutron spectrum from a nuclear explosion.

threshold element [COMPUT SCI] A logic circuit which has one output and several weighted inputs, and whose output is energized if and only if the sum of the weights of the energized inputs exceeds a prescribed threshold value.

threshold signal [ELECTROMAG] A received radio signal (or radar echo) whose power is just above the noise level of the receiver. Also known as minimum detectable signal.

threshold voltage [ELECTR] 1. In general, the voltage at which a particular characteristic of an electronic device first appears. 2. The voltage at which conduction of current begins in a *pn* junction. 3. The voltage at which channel formation occurs in a metal oxide semiconductor field-effect transistor. 4. The voltage at which a solid-state lamp begins to emit light. [NUCLEO] The lowest voltage at which all pulses produced in a Geiger counter by any ionizing event are of the same size, regardless of the size of the initial ionizing event.

Threskiornithidae [VERT ZOO] The ibises, a family of long-legged birds in the order Ciconiiformes.

thrip [INV ZOO] A small, slender-bodied phytophagous insect of the order Thysanoptera with suctorial mouthparts, a stout proboscis, a vestigial right mandible, and a fully developed left mandible, while wings may be present or absent.

Thripidae [INV ZOO] A large family of thrips, order Thysanoptera, which includes the most common species.

throat [ANAT] The region of the vertebrate body that includes the pharynx, the larynx, and related structures. [BOT] The upper, spreading part of the tube of a gamopetalous calyx or corolla. [DES ENG] The narrowest portion of a constricted duct, as in a diffuser or a venturi tube; specifically, a nozzle throat. [ENG] 1. The smaller end of a horn or tapered waveguide. 2. The area in a fireplace that forms the passageway from the firebox to the smoke chamber.

throat velocity *See* critical velocity.

thrombin [BIOCHEM] An enzyme elaborated from prothrombin in shed blood which induces clotting by converting fibrinogen to fibrin.

thrombinogen *See* prothrombin.

thrombocyte [HISTOL] One of the minute protoplasmic disks found in vertebrate blood; thought to be fragments of megakaryocytes. Also known as blood platelet; platelet.

thromboplastin [BIOCHEM] Any of a group of lipid and protein complexes in blood that accelerate the conversion of prothrombin to thrombin. Also known as factor III.

thrombus [MED] A blood clot occurring on the wall of a blood vessel where the endothelium is damaged.

Throscidae [INV ZOO] The false metallic wood-boring beetles, a cosmopolitan family of the Coleoptera in the superfamily Elateroidea.

throttled flow [FL MECH] Flow which is forced to pass through a restricted area, where the velocity must increase.

throttle valve [MECH ENG] A choking device to regulate flow of a liquid, for example, in a pipeline, to an engine or turbine, from a pump or compressor.

throttling [AERO ENG] The varying of the thrust of a rocket engine during powered flight. [CONT SYS] Control by means of intermediate steps between full on and full off. [THERMO] An adiabatic, irreversible process in which a gas expands by passing from one chamber to another chamber which is at a lower pressure than the first chamber.

through arch [CIV ENG] An arch bridge from which the roadway is suspended as distinct from one which carries the roadway on top.

throughput [CHEM ENG] The volume of feedstock charged to a process equipment unit during a specified time. [COMMUN] A measure of the effective rate of transmission of data by a communications system. [COMPUT SCI] The productivity of a data-processing system, as expressed in computing work per minute or hour. [MIN ENG] The quantity of ore or other material passed through a mill or a section of a mill in a given time or at a given rate.

throw [GEOL] The vertical component of dip separation on a fault, or generally the amount of vertical displacement on any fault.

thrust [GEOL] Overriding movement of one crystal unit over another. [MECH] The force exerted in any direction by a fluid jet or by a powered screw. [MECH ENG] The weight or pressure applied to a bit to make it cut. [MIN ENG] 1. A crushing of coal pillars caused by excess weight of the superincumbent rocks, the floor being harder than the roof. 2. The ruins of the fallen roof, after pillars and stalls have been removed.

thrust axis [AERO ENG] A line or axis through an aircraft or a rocket, along which the thrust acts; an axis through the longitudinal center of a jet or rocket engine, along which the thrust of the engine acts. Also known as axis of thrust; center of thrust.

thrust bearing [MECH ENG] A bearing which sustains axial loads and prevents axial movement of a loaded shaft.

thruster [AERO ENG] A control jet employed in spacecraft; an example would be one utilizing hydrogen peroxide.

thrust fault [GEOL] A low-angle (less than a 45° dip) fault along which the hanging wall has moved up relative to the footwall. Also known as reverse fault; reverse slip fault; thrust slip fault.

thrust slip fault *See* thrust fault.

thulium [CHEM] A rare-earth element, symbol Tm, group IIIB, of the lanthanide group, atomic number 69, atomic weight 168.934; reacts slowly with water, soluble in dilute acids, melts at 1550°C, boils at 1727°C; the dust is a fire hazard; used as x-ray source and to make ferrites.

thulium-170 [NUC PHYS] The radioactive isotope of thulium, with mass number 170; used as a portable x-ray source.

thunder [GEOPHYS] The sound emitted by rapidly expanding gases along the channel of a lightning discharge.

thundercloud [METEOROL] A convenient and often used term for the cloud mass of a thunderstorm, that is, a cumulonimbus.

thunderhead *See* incus.

thundersquall *See* rainsquall.

thunderstorm [METEOROL] A convective storm accompanied by lightning and thunder and rain, rarely snow showers but often hail, and gusty squall winds at the onset of precipitation; the characteristic cloud is the cumulonimbus.

Thunnidae [VERT ZOO] The tunas, a family of perciform fishes; there are no scales on the posterior part of the body, and those on the anterior are fused to form an armored covering, the body is streamlined, and the tail is crescent-shaped.

Thurniaceae [BOT] A small family of monocotyledonous plants in the order Juncales distinguished by an inflorescence of one or more dense heads, vascular bundles of the leaf in vertical pairs, and silica bodies in the leaf epidermis.

Thylacinidae [VERT ZOO] A family of Australian carnivorous marsupials in the superfamily Dasyuroidea.

Thylacoleonidae [PALEON] An extinct family of carnivorous marsupials in the superfamily Phalangeroidea.

thyme [BOT] A perennial mint plant of the genus *Thymus*; pungent aromatic herb is made from the leaves.

Thymelaeaceae [BOT] A family of dicotyledonous woody plants in the order Myrtales characterized by a superior ovary with a solitary ovule, and petals, if present, are scalelike.

thymus gland [ANAT] A lymphoid organ in the neck or upper thorax of all vertebrates; it is most prominent in early life and is essential for normal development of the circulating pool of lymphocytes.

thyratron [ELECTR] A hot-cathode gas tube in which one or more control electrodes initiate but do not limit the anode current except under certain operating conditions. Also known as hot-cathode gas-filled tube.

Thyrididae [INV ZOO] The window-winged moths, a small tropical family of lepidopteran insects in the suborder Heteroneura.

thyristor [ELECTR] A transistor having a thyratron-like characteristic; as collector current is increased to a critical value, the alpha of the unit rises above unity to give high-speed triggering action.

thyrocalcitonin *See* calcitonin.

thyroglobulin [BIOCHEM] An iodinated protein found as the storage form of the iodinated hormones in the thyroid follicular lumen and epithelial cells.

thyroid cartilage [ANAT] The largest of the laryngeal cartilages in humans and most other mammals, located anterior to the cricoid; in man, forms the Adam's apple.

thyroid gland [ANAT] An endocrine gland found in all vertebrates that produces, stores, and secretes the thyroid hormones.

thyroid hormone [BIOCHEM] Commonly, thyroxine or triiodothyronine, or both; a metabolically active compound formed and stored in the thyroid gland which functions to regulate the rate of metabolism.

thyroid-stimulating hormone *See* thyrotropic hormone.

Thyropteridae [VERT ZOO] The New-World disk-winged bats, a family of the Chiroptera found in Central and South America, characterized by a stalked sucking disk and a well-developed claw on the thumb.

thyrotoxicosis *See* hyperthyroidism.

thyrotropic hormone [BIOCHEM] A hormone produced by the adenohypophysis which regulates thyroid gland function. Also known as thyroid-stimulating hormone (TSH).

thyroxine [BIOCHEM] $C_{15}H_{11}I_4NO_4$ The active physiologic principle of the thyroid gland; used in the form of the sodium salt for replacement therapy in states of hypothyroidism or absent thyroid function.

Thysanidae [INV ZOO] A family of hymenopteran insects in the superfamily Chalcidoidea.

Thysanoptera [INV ZOO] The thrips, an order of small, slender insects having exopterygote development, sucking mouthparts, and exceptionally narrow wings with few or no veins and bordered by long hairs.

Thysanura [INV ZOO] The silverfish, machilids, and allies, an order of primarily wingless insects with soft, fusiform bodies.

Ti *See* titanium.

tibia [ANAT] The larger of the two leg bones, articulating with the femur, fibula, and talus.

tibialis [ANAT] **1.** A muscle of the leg arising from the proximal end of the tibia and inserted into the first cuneiform and first metatarsal bones. **2.** A deep muscle of the leg arising proximally from the tibia and fibula and inserted into the navicular and first cuneiform bones.

tick [COMMUN] A pulse broadcast at 1-second intervals by standard frequency and time broadcasting stations to indicate the exact time. [COMPUT SCI] A time interval equal to 1/60 second, used primarily in discussing computer operations. [INV ZOO] Any arachnid comprising Ixodoidea; a blood-sucking parasite and important vector of various infectious diseases of humans and lower animals.

tick fever *See* Rocky Mountain spotted fever.

tick typhus *See* Rocky Mountain spotted fever.

tidal bore *See* bore.

tidal channel [OCEANOGR] A major channel followed by tidal currents, extending from the ocean into a tidal marsh or tidal flat.

tidal constants [OCEANOGR] Tidal relations that remain essentially constant for any particular locality.

tidal current [OCEANOGR] The alternating horizontal movement of water associated with the rise and fall of the tide caused by the astronomical tide-producing forces.

tidal cycle *See* tide cycle.

tidal datum [OCEANOGR] A level of the sea, defined by some phase of the tide, from which water depths and heights of tide are reckoned. Also known as tidal datum plane.

tidal datum plane *See* tidal datum.

tidal day [OCEANOGR] The interval between two consecutive high waters of the tide at a given place, averaging 24 hours and 51 minutes.

tidal flat [GEOL] A marshy, sandy, or muddy nearly horizontal coastal flatland which is alternately covered and exposed as the tide rises and falls.

tidal frequency [OCEANOGR] The rate of travel, in degrees per day, of a component of a tide, the component being created by a particular juxtaposition of forces in the sun-earth-moon system.

tidalite [GEOL] Any sediment transported and deposited by tidal currents.

tidal pool [OCEANOGR] An accumulation of sea water remaining in a depression on a beach or reef after the tide recedes.

tidal stand *See* stand.

tidal wave [OCEANOGR] **1.** Any unusually high and generally destructive sea wave or water level along a shore. **2.** *See* tide wave.

tide [OCEANOGR] The periodic rising and falling of the oceans resulting from lunar and solar tide-producing forces acting upon the rotating earth.

tide bulge *See* tide wave.

tide cycle [OCEANOGR] A period which includes a complete set of tide conditions or characteristics, such as a tidal day or a lunar month. Also known as tidal cycle.

tide gate [CIV ENG] **1.** A restricted passage through which water runs with great speed due to tidal action. **2.** An opening through which water may flow freely when the tide sets in one direction, but which closes automatically and prevents the water from flowing in the other direction when the direction of flow is reversed.

tidemark [OCEANOGR] **1.** A high-water mark left by tidal water. **2.** The highest point reached by a high tide.

tidewater [OCEANOGR] **1.** A body of water, such as a river, affected by tides. **2.** Water inundating land at flood tide.

tide wave [OCEANOGR] A long-period wave associated with the tide-producing forces of the moon and sun, and identified with the rising and falling of the tide. Also known as tidal wave; tide bulge.

tie [CIV ENG] One of the transverse supports to which railroad rails are fastened to keep them to line, gage, and grade. [ELEC] **1.** Electrical connection or strap. **2.** *See* tie wire. [ENG] A beam, post, rod, or angle to hold two pieces together; a tension member in a construction. [MIN ENG] A support for the roof in coal mines.

tie bar [CIV ENG] **1.** A bar used as a tie rod. **2.** A rod connecting two switch rails on a railway to hold them to gage. [GEOL] *See* tombolo.

tie line [COMMUN] **1.** A leased communication channel or circuit. **2.** *See* data link. [PHYS CHEM] A line on a phase diagram joining the two points which represent the composition of systems in equilibrium. Also known as conode.

tiemannite [MINERAL] HgSe A steel gray to blackish–lead gray mineral consisting of mercuric selenide; commonly occurs in massive form.

tie plate [CIV ENG] A metal plate between a rail and a tie to hold the rail in place and reduce wear on the tie. [MECH ENG] A plate used in a furnace to connect tie rods.

tie rod [CIV ENG] A structural member used as a brace to take tensile loads. [ENG] A round or square iron rod passing through or over a furnace and connected with buckstays to assist in binding the furnace together. [MECH ENG] A rod used as a mechanical or structural support between elements of a machine. [MIN ENG] Vertical rods mounted in overlying horizontal shaft timbers.

tie wire [ELEC] A short piece of wire used to tie an open-line wire to an insulator. Also known as tie.

tiger [VERT ZOO] *Felis tigris.* An Asiatic carnivorous mammal in the family Felidae characterized by a tawny coat with transverse black stripes and white underparts.

tiger beetle [INV ZOO] The common name for any of the bright-colored beetles in the family Cicindelidae; there are about 1300 species distributed all over the world.

tiger salamander [VERT ZOO] *Ambystoma tigrinum.* A salamander in the family Ambystomatidae, found in a variety of subspecific forms from Canada to Mexico and over most of the United States; lives in arid and humid regions, and is the only salamander in much of the Great Plains and Rocky Mountains.

tiger's-eye [MINERAL] A yellowish-brown crystalline variety of quartz; a translucent, fibrous, broadly chatoyant gemstone that may be dyed other colors.

tiger shark *See* sand shark.

tight coupling *See* close coupling.

tight fit [DES ENG] A fit between mating parts with slight negative allowance, requiring light to moderate force to assemble.

tight fold *See* closed fold.

TIG welding *See* tungsten–inert gas welding.

till [GEOL] Unsorted and unstratified drift consisting of a heterogeneous mixture of clay, sand, gravel, and boulders which is deposited by and underneath a glacier. Also known as boulder clay; glacial till; icelaid drift.

Tilletiaceae [MYCOL] A family of fungi in the order Ustilaginales in which basidiospores form at the tip of the apibasidium.

tillite [PETR] A sedimentary rock formed by lithification of till, especially pre-Pleistocene till.

Tillodontia [PALEON] An order of extinct quadrupedal land mammals known from early Cenozoic deposits in the Northern Hemisphere and distinguished by large, rodentlike incisors, blunt-cuspid cheek teeth, and five clawed toes.

tilt [AERO ENG] The inclination of an aircraft, winged missile, or the like from the horizontal, measured by reference to the lateral axis or to the longitudinal axis. [ELECTROMAG] **1.** Angle which an antenna forms with the horizontal. **2.** In radar, the angle between the axis of radiation in the vertical plane and a reference axis which is normally the horizontal. [METEOROL] The inclination to the vertical of a significant feature of the circulation (or pressure) pattern or of the field of temperature or moisture; for example, troughs in the westerlies usually display a westward tilt with altitude in the lower and middle troposphere. [OPTICS] The angle between the plane of a photograph from a downward-pointing camera and the horizontal plane.

tilt angle [ELECTROMAG] The angle between the axis of radiation of a radar beam in the vertical plane and a reference axis (normally the horizontal).

tiltmeter [ENG] An instrument used to measure small changes in the tilt of the earth's surface, usually in relation to a liquid-level surface or to the rest position of a pendulum.

tilt-rotor aircraft [AERO ENG] A type of convertible aircraft which takes off, hovers, and lands as a helicopter but is converted into a fixed-wing aircraft by the 90° tilting of its rotor or rotors for use as a propeller for forward flight.

timberline [ECOL] The elevation or latitudinal limits for arboreal growth. Also known as tree line.

timbre [ACOUS] That attribute of auditory sensation in terms of which a listener can judge that two sounds similarly presented and having the same loudness and pitch are dissimilar. Also known as musical quality.

time [PHYS] **1.** The dimension of the physical universe which, at a given place, orders the sequence of events. **2.** A designated instant in this sequence, as the time of day. Also known as epoch.

time-base generator *See* sweep generator.

time constant [PHYS] **1.** The time required for a physical quantity to rise from zero to $1 - 1/e$ (that is, 63.2%) of its final steady value when it varies with time t as $1 - e^{-kt}$. **2.** The time required for a physical quantity to fall to $1/e$ (that is, 36.8%) of its initial value when it varies with time t as e^{-kt}. **3.** Generally, the time required for an instrument to indicate a given percentage of the final reading resulting from an input signal. Also known as lag coefficient.

time-controlled system *See* clock control system.

time-correlation [GEOL] A correlation of age or mutual time relations between stratigraphic units in separated areas.

time-delay circuit [ELECTR] A circuit in which the output signal is delayed by a specified time interval with respect to the input signal. Also known as delay circuit.

time-division multiplexing [COMPUT SCI] The interleaving of bits or characters in time to compensate for the slowness of input devices as compared to data transmission lines. [COMMUN] A process for transmitting two or more signals over a common path by using successive time intervals for different signals. Also known as time multiplexing.

time-lapse photography [GRAPHICS] Motion picture photography in which a single frame is exposed at regular intervals; when the film is projected at normal speed, the action appears to be speeded up.

time line [GEOL] 1. A line that indicates equal geologic age in a correlation diagram. 2. A rock unit represented by a time line.

time-mark generator [ELECTR] A signal generator that produces highly accurate clock pulses which can be superimposed as pips on a cathode-ray screen for timing the events shown on the display.

time-multiplexing See multiprogramming; time-division multiplexing.

time of flight [MECH] Elapsed time in seconds from the instant a projectile or other missile leaves a gun or launcher until the instant it strikes or bursts. [PHYS] The elapsed time from the instant a particle leaves a source to the instant it reaches a detector.

time-of-flight mass spectrometer [SPECT] A mass spectrometer in which all the positive ions of the material being analyzed are ejected into the drift region of the spectrometer tube with essentially the same energies, and spread out in accordance with their masses as they reach the cathode of a magnetic electron multiplier at the other end of the tube.

time-phase [PHYS] Two disturbances are in time phase if they reach corresponding peak values at the same instants of time, though not necessarily at the same points in space.

time quadrature [PHYS] 1. Differing by a time interval corresponding to one-fourth the time of one cycle of the frequency in question. 2. An integration over time.

timer [ELECTR] A circuit used in radar and in electronic navigation systems to start pulse transmission and synchronize it with other actions, such as the start of a cathode-ray sweep. [ENG] 1. A device for automatically starting or stopping a machine or other device. 2. See interval timer. [MECH ENG] A device that controls timing of the ignition spark of an internal combustion engine at the correct time.

time reversal [PHYS] The replacement of the time coordinate t by its negative $-t$ in the equations of motion of a dynamical system; the time reversal operator, a symmetry operator for a quantum-mechanical system, contains also the complex conjugation operator and a matrix operating on the spin coordinate.

time-reversal reflection See optical phase conjugation.

time-rock unit See time-stratigraphic unit.

time-share [COMPUT SCI] To perform several independent processes almost simultaneously by interleaving the operations of the processes on a single high-speed processor.

time standard [HOROL] A recurring phenomenon, used as a reference for establishing a unit of time; the presently accepted standard is the second, defined to be 9,192,631,770 transitions between two specified hyperfine levels of the atom of cesium-133. [IND ENG] See standard time.

time-stratigraphic unit [GEOL] A stratigraphic unit based on geologic age or time of origin. Also known

as chronolith; chronolithologic unit; chronostratic unit; chronostratigraphic unit; time-rock unit.

time study [IND ENG] A work measurement technique, generally using a stopwatch or other timing device, to record the actual elapsed time for performance of a task, adjusted for any observed variance from normal effort or pace, unavoidable or machine delays, rest periods, and personal needs.

time zone [ASTRON] To avoid the inconvenience of the continuous change of mean solar time with longitude, the earth is divided into 24 time zones, each about 15° wide and centered on standard longitudes, 0°, 15°, 30°, and so on; within each zone the time kept is the mean solar time of the standard meridian.

timing-axis oscillator See sweep generator.

timing belt [DES ENG] A power transmission belt with evenly spaced teeth on the bottom side which mesh with grooves cut on the periphery of the pulley to produce a positive, no-slip, constant-speed drive. [MECH ENG] A positive drive belt that has axial cogs molded on the underside of the belt which fit into grooves on the pulley; prevents slip, and makes accurate timing possible; combines the advantages of belt drives with those of chains and gears. Also known as positive drive belt.

timing error [COMPUT SCI] An error made in planning or writing a computer program, usually in underestimating the time that will be taken by input/output or other operations, which causes unnecessary delays in the execution of the program.

timing gears [MECH ENG] The gear train of reciprocating engine mechanisms for relating camshaft speed to crankshaft speed.

timing motor [ELEC] A motor which operates from an alternating-current power system synchronously with the alternating-current frequency, used in timing and clock mechanisms. Also known as clock motor.

timing relay [ELEC] Form of auxiliary relay used to introduce a definite time delay in the performance of a function.

timing signal [ELECTR] Any signal recorded simultaneously with data on magnetic tape for use in identifying the exact time of each recorded event.

timothy [BOT] *Phleum pratense.* A perennial hay grass of the order Cyperales characterized by moderately leafy stems and a dense cylindrical inflorescence.

tin [CHEM] Metallic element in group IV, symbol Sn, atomic number 50, atomic weight 118.69; insoluble in water, soluble in acids and hot potassium hydroxide solution; melts at 232°C, boils at 2260°C. [MET] A lustrous silver-white ductile, malleable metal used in alloys, for solder, terneplate, and tinplate.

Tinamidae [VERT ZOO] The single family of the avian order Tinamiformes.

Tinamiformes [VERT ZOO] The tinamous, an order of South and Central American birds which are superficially fowllike but have fully developed wings and are weak fliers.

tincalconite [MINERAL] $Na_2B_4O_7 \cdot 5H_2O$ A colorless to dull-white mineral, crystallizing in the rhombohedral system; one of the principal ores of borax and boron compounds. Also known as mohavite; octahedral borax.

tincture [MATER] A dilute solution (aqueous or aqueous alcoholic) of a drug or chemical; more dilute than fluid extracts, less volatile than spirits.

tinea See ringworm.

Tineidae [INV ZOO] A family of small moths in the superfamily Tineoidea distinguished by an erect, bristling vestiture on the head.

Tineoidea [INV ZOO] A superfamily of heteroneuran Lepidoptera which includes small moths that usually have well-developed maxillary palpi.

Tingidae [INV ZOO] The lace bugs, the single family of the hemipteran superfamily Tingoidea.

Tingoidea [INV ZOO] A superfamily of the Hemiptera in the subdivision Geocorisae characterized by the wings with many lacelike areolae.

tinnitus [MED] A ringing, roaring, or hissing sound in one or both ears.

tin pyrites *See* stannite.

tin stone *See* cassiterite.

Tintinnida [INV ZOO] An order of ciliated protozoans in the subclass Spirotrichia whose members are conical or trumpet-shaped pelagic forms bearing shells.

tipburn [PL PATH] A disease of certain cultivated plants, such as potato and lettuce, characterized by browning of the leaf margins due to excessive loss of water.

tipper [MIN ENG] An apparatus for emptying coal or ore cars by turning them upside down and then righting them, with a minimum of manual labor.

tipple [MIN ENG] **1.** The place where the mine cars are tipped and emptied of their coal. **2.** The tracks, trestles, and screens at the entrance to a colliery, where coal is screened and loaded.

Tipulidae [INV ZOO] The crane flies, a family of orthorrhaphous dipteran insects in the series Nematocera.

tissue [HISTOL] An aggregation of cells more or less similar morphologically and functionally. [TEXT] A sheer woven fabric or gauze, usually of fine quality.

tissue roentgen *See* rep.

Titan [ASTRON] The largest satellite of Saturn, with a diameter estimated to be about 5800 kilometers. [ORD] A U.S. Air Force surface-to-surface intercontinental ballistic missile having a range of over 6000 miles (9700 kilometers) and a greater payload of nuclear or conventional weapons than Atlas; Titan uses inertial guidance alone or combined with radar guidance.

Titania [ASTRON] A satellite of Uranus, with a diameter estimated to be 1600 kilometers.

titanic iron ore *See* ilmenite.

titanite *See* sphene.

titanium [CHEM] A metallic element in group IV, symbol Ti, atomic number 22, atomic weight 47.90; ninth most abundant element in the earth's crust; insoluble in water, melts at 1660°C, boils above 3000°C. [MET] A lustrous, silvery-gray, strong, light metal that is hard and brittle when cold, malleable when heated, and ductile when pure; used in the pure state or in alloys for aircraft and chemical-plate metals, for surgical instruments, and in cermets, and metal-ceramic brazing.

Titanoideidae [PALEON] A family of extinct land mammals in the order Pantodonta.

titer [CHEM] **1.** The concentration in a solution of a dissolved substance as shown by titration. **2.** The least amount or volume needed to give a desired result in titration. **3.** The solidification point of hydrolyzed fatty acids. [TEXT] **1.** The weight per unit length of yarn. **2.** The number of filaments in reeled silk thread.

Titius Bode law *See* Bode's law.

titrant [ANALY CHEM] A standard solution of known concentration and composition used for analytical titrations.

titration [ANALY CHEM] A method of analyzing the composition of a solution by adding known amounts of a standardized solution until a given reaction (color change, precipitation, or conductivity change) is produced.

tjaele *See* frozen ground.

T junction [ELECTR] A network of waveguides with three waveguide terminals arranged in the form of a letter T; in a rectangular waveguide a symmetrical T junction is arranged by having either all three broadsides in one plane or two broadsides in one plane and the third in a perpendicular plane.

Tl *See* thallium.

T²L *See* transistor-transistor logic.

Tm *See* thulium.

TM mode *See* transverse magnetic mode.

TM wave *See* transverse magnetic wave.

T network [ELEC] A network composed of three branches, with one end of each branch connected to a common junction point, and with the three remaining ends connected to an input terminal, an output terminal, and a common input and output terminal, respectively.

toad [VERT ZOO] Any of several species of the amphibian order Anura, especially in the family Bufonidae; glandular structures in the skin secrete acrid, irritating substances of varying toxicity.

tobacco [BOT] **1.** Any plant of the genus *Nicotinia* cultivated for its leaves, which contain 1–3% of the alkaloid nicotine. **2.** The dried leaves of the plant.

tobacco jack *See* wolframite.

tobacco mosaic [PL PATH] Any of a complex of virus diseases of tobacco and other solanaceous plants in which the leaves are mottled with light- and dark-green patches, sometimes interspersed with yellow.

Todidae [VERT ZOO] The todies, a family of birds in the order Coraciiformes found in the West Indies.

toe [ANAT] One of the digits on the foot of man and other vertebrates. [CIV ENG] The part of a base of a dam or retaining wall on the side opposite to the retained material. [GEOL] The leading edge of a thrust nappe. [MET] The junction between the face of a weld and the base metal. [MIN ENG] **1.** The burden of material between the bottom of the borehole and the free face. **2.** The bottom of the borehole. **3.** A spurn, or small pillar of coal. **4.** The base of a bank in an open-pit mine.

toe hole [ENG] A blasting hole, usually drilled horizontally or at a slight inclination into the base of a bank, bench, or slope of a quarry or open-pit mine.

toe-in [MECH ENG] The degree (usually expressed in fractions of an inch) to which the forward part of the front wheels of an automobile are closer together than the rear part, measured at hub height with the wheels in the normal "straight ahead" position of the steering gear.

toe-out [MECH ENG] The outward inclination of the wheels of an automobile at the front on turns due to setting the steering arms at an angle.

to-from indicator [NAV] An indicator that shows whether an aircraft is flying toward or away from an omnirange station. Also known as sense indicator.

toggle [ELECTR] To switch over to an alternate state, as in a flip-flop. [MECH ENG] A form of jointed mechanism for the amplification of forces.

tokamak [PL PHYS] A device for confining a plasma within a toroidal chamber, which produces plasma temperatures, densities, and confinement times greater than that of any other such device; confinement is effected by a very strong externally applied toroidal field, plus a weaker poloidal field produced by a toroidally directed plasma current, and this current causes ohmic heating of the plasma.

tolerance [DES ENG] The permissible variations in the dimensions of machine parts. [ENG] A permissible deviation from a specified value, expressed in actual values or more often as a percentage of the nominal value. [PHARM] **1.** The ability of enduring or being less responsive to the influence of a drug or poison, particularly when acquired by continued use of the substance. **2.** The allowable deviation from a standard, as the range of variation permitted for the content of a drug in one of its dosage forms.

tomato [BOT] A plant of the genus *Lycopersicon*, especially *L. esculentum*, in the family Solanaceae cultivated for its fleshy edible fruit, which is red, pink, orange, yellow, white, or green, with fleshy placentas containing many small, oval seeds with short hairs and covered with a gelatinous matrix.

tombolo [GEOL] A sand or gravel bar or spit that connects an island with another island or an island with the mainland. Also known as connecting bar; tie bar; tying bar.

tomography *See* sectional radiography.

Tomopteridae [INV ZOO] The glass worms, a family of pelagic polychaete annelids belonging to the group Errantia.

ton [IND ENG] A unit of volume of sea freight, equal to 40 cubic feet. Also known as freight ton; measurement ton; shipping ton. [MECH] **1.** A unit of weight in common use in the United States, equal to 2000 pounds or 907.18474 kilogram-force. Also known as just ton; net ton; short ton. **2.** A unit of mass in common use in the United Kingdom equal to 2240 pounds, or to 1016.0469088 kilogram-force. Also known as gross ton; long ton. **3.** A unit of weight in troy measure, equal to 2000 troy pounds, or to 746.4834432 kilogram-force. **4.** *See* tonne. [MECH ENG] A unit of refrigerating capacity, that is, of rate of heat flow, equal to the rate of extraction of latent heat when one short ton of ice of specific latent heat 144 international table British thermal units per pound is produced from water at the same temperature in 24 hours; equal to 200 British thermal units per minute, or to approximately 3516.85 watts. Also known as standard ton. [NAV ARCH] A unit of internal capacity of ships, equal to 100 cubic feet. Also known as register ton. [NUCLEO] The energy released by one metric ton of chemical high explosives calculated at the rate of 1000 calories per gram; equal to 4.18 × 10⁹ joules; used principally in expressing the energy released by a nuclear bomb.

tone [ACOUS] **1.** A sound oscillation capable of exciting an auditory sensation having pitch. **2.** An auditory sensation having pitch. [GRAPHICS] Each distinguishable shade variation from black to white on photographs.

tone dialing *See* pushbutton dialing.

tone modulation [COMMUN] Type of code-signal transmission obtained by causing the radio-frequency carrier amplitude to vary at a fixed audio frequency.

Tongrian [GEOL] A European stage of geologic time; lower Oligocene (above Ludian of Eocene, below Rupelian). Also known as Lattorfian.

tongue [ANAT] A muscular organ located on the floor of the mouth in man and most vertebrates which may serve various functions, such as taking and swallowing food or tasting or as a tactile organ or sometimes a prehensile organ. [GEOL] **1.** A minor rock-stratigraphic unit of limited geographic extent; it

disappears laterally in one direction. **2.** A lava flow branching from a larger flow. [OCEANOGR] **1.** A protrusion of water into a region of different temperature, or salinity, or dissolved oxygen concentrating. **2.** A protrusion of one water mass into a region occupied by a different water mass.

tongue worm *See* acorn worm.

tonne [MECH] A unit of mass in the metric system, equal to 1000 kilograms or to approximately 2204.62 pound mass. Also known as metric ton; millier; ton; tonneau.

tonneau *See* tonne.

tonoplast [BOT] The membrane surrounding a plant-cell vacuole.

tonsil [ANAT] **1.** Localized aggregation of diffuse and nodular lymphoid tissue found in the throat where the nasal and oral cavities open into the pharynx. **2.** *See* palatine tonsil.

tonus [PHYSIO] The degree of muscular contraction when not undergoing shortening.

tool [ENG] Any device, instrument, or machine for the performance of an operation, for example, a hammer, saw, lathe, twist drill, drill press, grinder, planer, or screwdriver. [IND ENG] To equip a factory or industry for production by designing, making, and integrating machines, machine tools, and special dies, jigs, and instruments, so as to achieve manufacture and assembly of products on a volume basis at minimum cost.

tooth [ANAT] One of the hard bony structures supported by the jaws in mammals and by other bones of the mouth and pharynx in lower vertebrates serving principally for prehension and mastication. [DES ENG] **1.** One of the regular projections on the edge or face of a gear wheel. **2.** An angular projection on a tool or other implement, such as a rake, saw, or comb. [GRAPHICS] **1.** The coarse or abrasive quality of a paper or a painting ground that assists in the application of charcoal, pastels, or paint. **2.** A paper texture that holds ink more readily. [INV ZOO] Any of various sharp, horny, chitinous, or calcareous processes on or about any part of an invertebrate that functions like or resembles vertebrate jaws.

tooth shell [INV ZOO] A mollusk of the class Scaphopoda characterized by the elongate, tube-shaped, or cylindrical shell which is open at both ends and slightly curved.

top [GEOL] *See* overburden. [MECH] A rigid body, one point of which is held fixed in an inertial reference frame, and which usually has an axis of symmetry passing through this point; its motion is usually studied when it is spinning rapidly about the axis of symmetry. [QUANT MECH] *See* rotator.

topaz [MINERAL] Al₂SiO₄(F,OH) A red, yellow, green, blue, or brown neosilicate mineral that crystallizes in the orthorhombic system and commonly occurs in prismatic crystals with pyramidal terminations; hardness is 8 on Mohs scale, and specific gravity is 3.4–3.6; used as a gemstone.

top cut [MIN ENG] A machine cut made in the coal at or near the top of the working face in a mine.

tophus [MED] A localized swelling principally in cartilage and connective tissues in or adjacent to the small joints of the hands and feet; occurs specifically in gout.

topogenesis *See* morphogenesis.

topographical latitude *See* geodetic latitude.

topographic maturity *See* maturity.

topographic old age *See* old age.

topographic profile *See* profile.

topographic youth   See youth.

topography   [GEOGR] 1. The general configuration of a surface, including its relief; may be a land or water-bottom surface. 2. The natural surface features of a region, treated collectively as to form.

topological groups   [MATH] Groups which also have a topology with the property that the group operation and the inverse operation determine continuous functions.

topologically closed set   See closed set.

topological mapping   See homeomorphism.

topological space   [MATH] A set endowed with a topology.

topology   [MATH] 1. A collection of subsets of a set X, which includes X and the empty set, and has the property that any union or finite intersection of its members is also a member. 2. The generalized study of properties of spaces invariant under deformations and stretchings.

topotaxis   See tropism.

topotype   [SYST] A specimen of a species not of the original type series collected at the type locality.

topped crude   [MATER] A residual product remaining after the removal by distillation or other means of an appreciable quantity of the more volatile components of crude petroleum.

topping governor   See limit governor.

topset bed   [GEOL] One of the nearly horizontal sedimentary layers deposited on the top surface of an advancing delta.

top shell   [INV ZOO] Any of the marine snails of the family Trochidae characterized by a spiral conical shell with a flat base.

top slicing   [MIN ENG] A method of stoping in which the ore is extracted by excavating a series of horizontal (sometimes inclined) timbered slices alongside each other, beginning at the top of the ore body and working progressively downward.

topsoil   [GEOL] 1. Soil presumed to be fertile and used to cover areas of special planting. 2. Surface soil, usually corresponding with the A horizon, as distinguished from subsoil.

top steam   [CHEM ENG] Steam admitted near the top of a shell still to purge the still, and to prevent a vacuum from forming when pumping out the liquid contents.

torbanite   [GEOL] A variety of coal that resembles a carbonaceous shale in outward appearance; it is fine-grained, black to brown, and tough. Also known as bitumenite; kerosine shale.

torbernite   [MINERAL] $Cu(UO_2)_2(PO_4)_2 \cdot 8-12H_2O$ A green radioactive mineral crystallizing in the tetragonal system and occurring in tabular crystals or in foliated form. Also known as chalcolite; copper uranite; cuprouranite; uran-mica.

tornado   [METEOROL] An intense rotary storm of small diameter, the most violent of weather phenomena; tornadoes always extend downward from the base of a convective-type cloud, generally in the vicinity of a severe thunderstorm.

toroid   See doughnut; toroidal magnetic circuit.

toroidal coil   See toroidal magnetic circuit.

toroidal magnetic circuit   [ELECTROMAG] Doughnut-shaped piece of magnetic material, together with one or more coils of current-carrying wire wound about the doughnut, with the permeability of the magnetic material high enough so that the magnetic flux is almost completely confined within it. Also known as toroid; toroidal coil.

Torpedinidae   [VERT ZOO] The electric rays or torpedoes, a family of batoid sharks.

torpor   [PHYSIO] The condition in hibernating poikilotherms during winter when body temperature drops in a parallel relation to ambient environmental temperatures.

torque   [MECH] 1. For a single force, the cross product of a vector from some reference point to the point of application of the force with the force itself. Also known as moment of force; rotation moment. 2. For several forces, the vector sum of the torques (first definition) associated with each of the forces.

torque amplifier   [COMPUT SCI] An analog computer device having input and output shafts and supplying work to rotate the output shaft in positional correspondence with the input shaft without imposing any significant torque on the input shaft.

torque arm   [MECH ENG] In automotive vehicles, an arm to take the torque of the rear axle.

torque converter   [MECH ENG] A device for changing the torque speed or mechanical advantage between an input shaft and an output shaft.

torque motor   [ELECTROMAG] A motor designed primarily to exert torque while stalled or rotating slowly.

torr   [MECH] A unit of pressure, equal to 1/760 atmosphere; it differs from 1 millimeter of mercury by less than one part in seven million; approximately equal to 133.3224 pascals.

Torrert   [GEOL] A suborder of the soil order Vertisol; it is the driest soil of the order and forms cracks that tend to remain open; occurs in arid regions.

torreyite   [MINERAL] $(Mg,Mn,Zn)(SO_4)(OH)_{12} \cdot 4H_2O$ A bluish-white mineral consisting of a hydrated basic sulfate of magnesium, manganese, and zinc; occurs in massive form.

Torricellian barometer   See mercury barometer.

Torridincolidae   [INV ZOO] A small family of coleopteran insects in the suborder Myxophaga found only in Africa and Brazil.

torsion   [MECH] A twisting deformation of a solid body about an axis in which lines that were initially parallel to the axis become helices.

torsional modulus   [MECH] The ratio of the torsional rigidity of a bar to its length. Also known as modulus of torsion.

torsional pendulum   [MECH] A device consisting of a disk or other body of large moment of inertia mounted on one end of a torsionally flexible elastic rod whose other end is held fixed; if the disk is twisted and released, it will undergo simple harmonic motion, provided the torque in the rod is proportional to the angle of twist.

torsional rigidity   [MECH] The ratio of the torque applied about the centroidal axis of a bar at one end of the bar to the resulting torsional angle, when the other end is held fixed.

torsional vibration   [MECH] A periodic motion of a shaft in which the shaft is twisted about its axis first in one direction and then in the other; this motion may be superimposed on rotational or other motion.

torsion balance   [ENG] An instrument, consisting essentially of a straight vertical torsion wire whose upper end is fixed while a horizontal beam is suspended from the lower end; used to measure minute gravitational, electrostatic, or magnetic forces.

torsion bar   [MECH ENG] A spring flexed by twisting about its axis; found in the spring suspension of truck and passenger car wheels, in production machines where space limitations are critical, and in high-speed mechanisms where inertia forces must be minimized.

torsion fault   See wrench fault.

torso mountain   See monadnock.

tortoise [VERT ZOO] Any of various large terrestrial reptiles in the order Chelonia, especially the family Testudinidae.

Tortonian [GEOL] A European stage of geologic time: Miocene (above Helvetian, below Sarmatian).

Tortricidae [INV ZOO] A family of phytophagous moths in the superfamily Tortricoidea which have a stout body, lightly fringed wings, and threadlike antennae.

Tortricoidea [INV ZOO] A superfamily of small wide-winged moths in the suborder Heteroneura.

Torulopsidales [MYCOL] The equivalent name for Cryptococcales.

torus [ANAT] A rounded protruberance occurring on a body part. [BOT] The thickened membrane closing a bordered pit. [MATH] **1.** The surface of a doughnut. **2.** The topological space obtained by identifying the opposite sides of a rectangle. **3.** The group which is the product of two circles.

TOS See tape operating system.

total binding energy See binding energy.

total curvature See Gaussian curvature.

total differential [MATH] The total differential of a function of several variables, $f(x_1, x_2, \ldots, x_n)$, is the function given by the sum of terms $(\partial f/\partial x_i) dx_i$ as i runs from 1 to n. Also known as differential.

total displacement See slip.

total eclipse [ASTRON] An eclipse that obscures the entire surface of the moon or sun.

total evaporation See evapotranspiration.

total heat See enthalpy.

total heat of solution See heat of solution.

total internal reflection [OPTICS] A phenomenon in which electromagnetic radiation in a given medium which is incident on the boundary with a less-dense medium (one having a lower index of refraction) at an angle less than the critical angle is completely reflected from the boundary.

totally bounded set See precompact set.

total pressure [FL MECH] See dynamic pressure. [MECH] The gross load applied on a given surface. [MIN ENG] The total ventilating pressure in a mine, usually measured in the fan drift.

total radiation pyrometer [ENG] A pyrometer which focuses heat radiation emitted by a hot object on a detector (usually a thermopile or other thermal type detector), and which responds to a broad band of radiation, limited only by absorption of the focusing lens, or window and mirror.

totipalmate [VERT ZOO] Having all four toes connected by webs, as in the Pelecaniformes.

toucan [VERT ZOO] Any of numerous fruit-eating birds, of the family Ramphastidae, noted for their large and colorful bills.

Toucan See Tucana.

touch call See push-button dialing.

tour See shift.

tourmaline [MINERAL] $(Na,Ca)(Al,Fe,Li,Mg)_3Al_6\text{-}(BO_3)_3Si_6O_{18}(OH)_4$ Any of a group of cyclosilicate minerals with a complex chemical composition, vitreous to resinous luster, and variable color; crystallizes in the ditrigonal-pyramidal class of the hexagonal system, has piezoelectric properties, and is used as a gemstone.

Tournaisian [GEOL] European stage of lowermost Carboniferous time.

tower [CHEM ENG] A vertical, cylindrical vessel used in chemical and petroleum processing to increase the degree of separation of liquid mixtures by distillation or extraction. Also known as column. [ELECTROMAG] A tall metal structure used as a transmitting antenna, or used with another such

structure to support a transmitting antenna wire. [ENG] A concrete, metal, or timber structure that is relatively high for its length and width, and used for various purposes, including the support of electric power transmission lines, radio and television antennas, and rockets and missiles prior to launching.

tower crane [CIV ENG] A crane mounted on top of a tower which is sometimes incorporated in the frame of a building.

town plan See city plan.

Townsend avalanche See avalanche.

Townsend discharge [ELECTR] A discharge which occurs at voltages too low for it to be maintained by the electric field alone, and which must be initiated and sustained by ionization produced by other agents; it occurs at moderate pressures, above about 0.1 torr, and is free of space charges.

Townsend ionization See avalanche.

towrope horsepower See effective horsepower.

Toxasteridae [PALEON] A family of Cretaceous echinoderms in the order Spatangoida which lacked fascioles and petals.

toxemia [MED] A condition in which the blood contains toxic substances, either of microbial origin or as by-products of abnormal protein metabolism.

toxic goiter See hyperthyroidism.

toxicity [PHARM] **1.** The quality of being toxic. **2.** The kind and amount of poison or toxin produced by a microorganism, or possessed by a chemical substance not of biological origin.

toxicology [PHARM] The study of poisons, including their nature, effects, and detection, and methods of treatment.

toxin [BIOCHEM] Any of various poisonous substances produced by certain plant and animal cells, including bacterial toxins, phytotoxins, and zootoxins.

Toxodontia [PALEON] An extinct suborder of mammals representing a central stock of the order Notoungulata.

Toxoglossa [INV ZOO] A group of carnivorous marine gastropod mollusks distinguished by a highly modified radula (toxoglossate).

toxoid [IMMUNOL] Detoxified toxin, but with antigenic properties intact; toxoids of tetanus and diphtheria are used for immunization.

Toxoplasmea [INV ZOO] A class of the protozoan subphylum Sporozoa composed of small, crescent-shaped organisms that move by body flexion or gliding and are characterized by a two-layered pellicle with underlying microtubules, micropyle, paired organelles, and micronemes.

Toxoplasmida [INV ZOO] An order of the class Toxoplasmea; members are parasites of vertebrates.

Toxopneustidae [INV ZOO] A family of Tertiary and extant echinoderms of the order Temnopleuroida where the branchial slits are deep and the test tends to be absent.

Toxothrix [MICROBIO] A genus of gliding bacteria of uncertain affiliation; long, often U-shaped filaments contain colorless, cylindrical cells; move slowly with ends of filament trailing.

Toxotidae [VERT ZOO] The archerfishes, a family of small fresh-water forms in the order Perciformes.

TP monitor See teleprocessing monitor.

TPN See triphosphopyridine nucleotide.

Tr See trace.

trace [COMPUT SCI] To provide a record of every step, or selected steps, executed by a computer program, and by extension, the record produced by this operation. [ELECTR] The visible path of a moving spot on the screen of a cathode-ray tube. Also known as line. [ENG] The record made by a recording device, such as a seismometer or electrocardiograph. [GEOL] The intersection of two geological surfaces. [MATH] The trace of a matrix is the sum of the entries along its principal diagonal. Designated Tr. Also known as spur. [METEOROL] A precipitation of less than 0.005 inch (0.127 millimeter). [SCI TECH] An extremely small but detectable quantity of a substance.

trace analysis [ANALY CHEM] Analysis of a very small quantity of material of a sample by such techniques as polarography or spectroscopy.

trace element [GEOCHEM] A nonessential element found in small quantities (usually less than 1.0%) in a mineral. Also known as accessory element; guest element.

trace fossil [GEOL] A trail, track, or burrow made by an animal and found in ancient sediments such as sandstone, shale, or limestone. Also known as ichnofossil.

tracer [CHEM] A foreign substance, usually radioactive, that is mixed with or attached to a given substance so the distribution or location of the latter can later be determined; used to trace chemical behavior of a natural element in an organism. Also known as tracer element. [ENG] A thread of contrasting color woven into the insulation of a wire for identification purposes.

tracer element See tracer.

trace routine [COMPUT SCI] A routine which tracks the execution of a program, step by step, to locate a program malfunction. Also known as tracing routine.

trace statement [COMPUT SCI] A statement, included in certain programming languages, that causes certain error-checking procedures to be carried out on specified segments of a source program.

trachea [ANAT] The cartilaginous and membranous tube by which air passes to and from the lungs in humans and many vertebrates. [BOT] A xylem vessel resembling the trachea of vertebrates. [INV ZOO] One of the anastomosing air-conveying tubules composing the respiratory system in most insects.

tracheid [BOT] An elongate, spindle-shaped xylem cell, lacking protoplasm at maturity, and having secondary walls laid in various thicknesses and patterns over the primary wall.

Tracheophyta [BOT] A large group of plants characterized by the presence of specialized conducting tissues (xylem and phloem) in the roots, stems, and leaves.

trachoma [MED] An infectious disease of the conjunctiva and cornea caused by *Chlamydia trachomatis* producing photophobia, pain, and excessive lacrimation.

trachybasalt [PETR] An extrusive rock characterized by calcic plagioclase and sanidine, with augite, olivine, and possibly minor analcime or leucite.

Trachylina [INV ZOO] An order of moderate-sized jellyfish of the class Hydrozoa distinguished by having balancing organs and either a small polyp stage or none.

Trachymedusae [INV ZOO] A group of marine jellyfish, recognized as a separate order or as belonging to the order Trachylina whose tentacles have a solid core consisting of a single row of endodermal cells.

Trachypsammiacea [INV ZOO] An order of colonial anthozoan coelenterates characterized by a dendroid skeleton.

Trachystomata [VERT ZOO] The name given to the Meantes when the group is considered to be an order.

trachyte [PETR] The light-colored, aphanitic rock (the volcanic equivalent of syenite), composed largely of alkali feldspar with minor amounts of mafic minerals.

tracing routine See trace routine.

track [AERO ENG] The actual line of movement of an aircraft or a rocket over the surface of the earth; it is the projection of the history of the flight path on the surface. Also known as flight track. [COMPUT SCI] The recording path on a rotating surface. [DES ENG] As applied to a pattern of setting diamonds in a bit crown, an arrangement of diamonds in concentric circular rows in the bit crown, with the diamonds in a specific row following in the track cut by a preceding diamond. [ELECTR] 1. A path for recording one channel of information on a magnetic tape, drum, or other magnetic recording medium; the location of the track is determined by the recording equipment rather than by the medium. 2. The trace of a moving target on a plan-position-indicator radar screen or an equivalent plot. [ENG] 1. The groove cut in a rock by a diamond inset in the crown of a bit. 2. A pair of parallel metal rails for a railway, railroad, tramway, or for any wheeled vehicle. [MECH ENG] 1. The slide or rack on which a diamond-drill swivel head can be moved to positions above and clear of the collar of a borehole. 2. A crawler mechanism for earth-moving equipment. [NAV] 1. To follow the movements of an object by keeping the reticle of an optical system or a radar beam on the object, by plotting its bearing and distance at frequent intervals, or by a combination of the two. 2. To navigate by following the movements of a craft without regard for future positions; this is used when frequent changes of an unanticipated amount are expected in course or speed or both. 3. A recommended route on a nautical chart, such as a North Atlantic Track. [NUCLEO] 1. The visible path of an ionizing particle in a particle detector, such as a cloud chamber, bubble chamber, spark chamber, or nuclear photographic emulsion. 2. See race track.

track cable [ENG] Steel wire rope, usually a locked-coil rope which supports the wheels of the carriers of a cableway.

track gage [CIV ENG] The width between the rails of a railroad track; in the United States the standard gage is 4 feet 8½ inches.

tracking [ELEC] A leakage or fault path created across the surface of an insulating material when a high-voltage current slowly but steadily forms a carbonized path. [ELECTR] The condition in which all tuned circuits in a receiver accurately follow the frequency indicated by the tuning dial over the entire tuning range. [ENG] 1. A motion given to the major lobe of a radar or radio antenna such that some preassigned moving target in space is always within the major lobe. 2. The process of following the movements of an object; may be accomplished by keeping the reticle of an optical system or a radar beam on the object, by plotting its bearing and distance at frequent intervals, or by a combination of techniques. [ENG ACOUS] 1. The following of a groove by a phonograph needle. 2. Maintaining the same ratio of loudness in the two channels of a stereophonic sound system at all settings of the ganged volume control. [NAV] Navigation which follows the movements of a craft but does not anticipate future positions.

tracking error [ENG ACOUS] Deviation of the vibration axis of a phonograph pickup from tangency with a groove; true tangency is possible for only one groove when the pickup arm is pivoted; the longer the pickup arm, the less is the tracking error.

tracking filter [ELECTR] Electronic device for attenuating unwanted signals while passing desired signals, by phase-lock techniques that reduce the effective bandwidth of the circuit and eliminate amplitude variations.

tracking network [ENG] A group of tracking stations whose operations are coordinated in tracking objects through the atmosphere or space.

tracking station [ENG] A radio, radar, or other station set up to track an object moving through the atmosphere or space.

traction [GEOL] Transport of sedimentary particles along and parallel to a bottom surface of a stream channel by rolling, sliding, dragging, pushing, or saltation. [GRAPHICS] A defect in a paint coating in which the film cracks and wide fissures reveal the underlying surface. [MECH] Pulling friction of a moving body on the surface on which it moves.

tractional force [FL MECH] The force exerted on particles under flowing water by the current; it is proportional to the square of the velocity.

tractor gate [CIV ENG] A type of outlet control gate used to release water from a reservoir; there are two types, roller and wheel.

tractor loader [MECH ENG] A tractor equipped with a tipping bucket which can be used to dig and elevate soil and rock fragments to dump at truck height. Also known as shovel dozer; tractor shovel.

tractor shovel *See* tractor loader.

tractrix [MATH] A curve in the plane where every tangent to it has the same length.

trade wind [METEOROL] The wind system, occupying most of the tropics, which blows from the subtropical highs toward the equatorial trough; a major component of the general circulation of the atmosphere; the winds are northeasterly in the Northern Hemisphere and southeasterly in the Southern Hemisphere; hence they are known as the northeast trades and southeast trades, respectively.

traffic [COMMUN] The messages transmitted and received over a communication channel. [ENG] The passage or flow of vehicles, pedestrians, ships, or planes along defined routes such as highways, sidewalks, sea lanes, or air lanes.

Tragulidae [VERT ZOO] The chevrotains, a family of pecoran ruminants in the superfamily Traguloidea.

Traguloidea [VERT ZOO] A superfamily of pecoran ruminants, composed of the most primitive forms with large canines; the chevrotain is the only extant member. ·

trailer label [COMPUT SCI] A record appearing at the end of a magnetic tape that uniquely identifies the tape as one required by the system.

trailer record [COMPUT SCI] A record which contains data pertaining to an associated group of records immediately preceding it.

trailing antenna [ELECTROMAG] An aircraft radio antenna having one end weighted and trailing free from the aircraft when in flight.

trailing edge [AERO ENG] The rear section of a multipiece airfoil, usually that portion aft of the rear spar. [ELECTR] The major portion of the decay of a pulse.

train [ASTRON] The bright tail of a comet or meteor. [ENG] To aim or direct a radar antenna in azimuth.

trajectory [GEOPHYS] The path followed by a seismic wave. [MECH] The curve described by an object moving through space, as of a meteor through the atmosphere, a planet around the sun, a projectile fired from a gun, or a rocket in flight.

trammel [ENG] A device consisting of a bar, each of whose ends is constrained to move along one of two perpendicular lines; used in drawing ellipses and in the Rowland mounting.

transacter [COMPUT SCI] A system in which data from sources in a number of different locations, as in a factory, are transmitted to a data-processing center and immediately processed by a computer.

transaction [COMPUT SCI] General description of updating data relevant to any item.

transaction data [COMPUT SCI] A set of data in a data-processing area in which the incidence of the data is essentially random and unpredictable; hours worked, quantities shipped, and amounts invoiced are examples from, respectively, the areas of payroll, accounts receivable, and accounts payable.

transaction file *See* detail file.

transaction record *See* change record.

transaction tape *See* change tape.

transadmittance [ELECTR] A specific measure of transfer admittance under a given set of conditions, as in forward transadmittance, interelectrode transadmittance, short-circuit transadmittance, small-signal forward transadmittance, and transadmittance compression ratio.

transaminase [BIOCHEM] One of a group of enzymes that catalyze the transfer of the amino group of an amino acid to a keto acid to form another amino acid. Also known as aminotransferase.

transamination [CHEM] **1.** The transfer of one or more amino groups from one compound to another. **2.** The transposition of an amino group within a single compound.

transceiver [COMPUT SCI] A device which transmits and receives data from punch card to punch card; it is essentially a conversion device which at the sending end reads the card and transmits the data over the wire, and at the receiving end punches the data into a card. [ELECTR] A radio transmitter and receiver combined in one unit and having switching arrangements such as to permit use of one or more tubes for both transmitting and receiving. Also known as transmitter-receiver.

transcendental element [MATH] An element of a field K is transcendental relative to a subfield F if it satisfies no polynomial whose coefficients come from F.

transcendental functions [MATH] Functions which cannot be given by any algebraic expression involving only their variables and constants.

transcendental number [MATH] An irrational number that is the root of no polynomial with rational-number coefficients.

transconductance [ELECTR] An electron-tube rating, equal to the change in plate current divided by the change in control-grid voltage that causes it, when the plate voltage and all other voltages are maintained constant. Also known as grid-anode transconductance; grid-plate transconductance; mutual conductance. Symbolized G_m; g_m.

transcribe [COMPUT SCI] To copy, with or without translating, from one external computer storage medium to another. [ELECTR] To record, as to record a radio program by means of electric transcriptions or magnetic tape for future rebroadcasting.

transcription [ENG ACOUS] A 16-inch-diameter (40.6-centimeter-diameter), 33-1/3-rpm disk recording of a complete radio program, made especially for broadcast purposes. Also known as electrical transcription. [MOL BIO] The process by which ribonucleic acid is formed from deoxyribonucleic acid.

transducer [ENG] Any device or element which converts an input signal into an output signal of a different form; examples include the microphone, phonograph pickup, loudspeaker, barometer, photoelectric cell, automobile horn, doorbell, and underwater sound transducer.

transduction [MICROBIO] Transfer of genetic material between bacterial cells by bacteriophages.

transductor See magnetic amplifier; saturable reactor.

transesterification [ORG CHEM] Conversion of an organic acid ester into another ester of that same acid.

transfer [COMPUT SCI] See jump. [MIN ENG] A vertical or inclined connection between two or more levels, used as an ore pass. [NAV] **1.** The distance a vessel moves perpendicular to its initial direction in making a turn of 90° with a constant rudder angle. **2.** The distance a vessel moves perpendicular to its initial direction for turns of less than 90°.

transfer admittance [ELECTR] An admittance rating for electron tubes and other transducers or networks; it is equal to the complex alternating component of current flowing to one terminal from its external termination, divided by the complex alternating component of the voltage applied to the adjacent terminal on the cathode or reference side; all other terminals have arbitrary external terminations.

transferase [BIOCHEM] Any of various enzymes that catalyze the transfer of a chemical group from one molecule to another.

transfer card See transition card.

transfer conditionally [COMPUT SCI] To copy, exchange, read, record, store, transmit, or write data or to change control or jump to another location according to a certain specified rule or in accordance with a certain criterion.

transfer ellipse See transfer orbit.

transference [PSYCH] The unconscious transfer of the patient's feelings and reactions originally associated with important persons in the patient's life, usually father, mother, or siblings, toward others and in the analytic situation, toward the analyst.

transfer function [CONT SYS] The mathematical relationship between the output of a control system and its input: for a linear system, it is the Laplace transform of the output divided by the Laplace transform of the input under conditions of zero initial-energy storage.

transfer impedance [ELEC] The ratio of the voltage applied at one pair of terminals of a network to the resultant current at another pair of terminals, all terminals being terminated in a specified manner.

transfer-in-channel command [COMPUT SCI] A command used to direct channel control to a specified location in main storage when the next channel command word is not stored in the next location in sequence.

transfer instruction [COMPUT SCI] Step in computer operation specifying the next operation to be performed, which is not necessarily the next instruction in sequence.

transfer interpreter [COMPUT SCI] A variation of a punched-card interpreter that senses a punched card and prints the punched information on the following card. Also known as posting interpreter.

transfer molding [ENG] Molding of thermosetting materials in which the plastic is softened by heat and pressure in a transfer chamber, then forced at high pressure through suitable sprues, runners, and gates into a closed mold for final curing.

transfer orbit [AERO ENG] In interplanetary travel, an elliptical trajectory tangent to the orbits of both the departure planet and the target planet. Also known as transfer ellipse.

transfer rate [COMPUT SCI] The speed at which data are moved from a direct-access device to a central processing unit.

transferred-electron device [ELECTR] A semiconductor device, usually a diode, that depends on internal negative resistance caused by transferred electrons in gallium arsenide or indium phosphide at high electric fields; transit time is minimized, permitting oscillation at frequencies up to several hundred megahertz.

transferred-electron effect [SOLID STATE] The variation in the effective drift mobility of charge carriers in a semiconductor when significant numbers of electrons are transferred from a low-mobility valley of the conduction band in a zone to a high-mobility valley, or vice versa.

transfer ribonucleic acid [MOL BIO] The smallest ribonucleic acid molecule found in cells; its structure is complementary to messenger ribonucleic acid and it functions by transferring amino acids from the free state to the polymeric form of growing polypeptide chains. Abbreviated t-RNA.

transfer switch [ELEC] A switch for transferring one or more conductor connections from one circuit to another.

transfinite induction [MATH] A reasoning process by which if a theorem holds true for the first element of a well-ordered set N and is true for an element n whenever it holds for all predecessors of n, then the theorem is true for all members of N.

transform [COMPUT SCI] To change the form of digital-computer information without significantly altering its meaning. [MATH] **1.** A conjugate of an element of a group. **2.** An expression, commonly used in harmonic analysis, formed from a given function f by taking an integral of $f \cdot g$, where g is a member of an orthogonal family of functions. **3.** The value of a transformation at some point.

transformation [CRYSTAL] See inversion. [ELEC] For two networks which are equivalent as far as conditions at the terminals are concerned, a set of equations giving the admittances or impedances of the branches of one circuit in terms of the admittances or impedances of the other. [GRAPHICS] The process of projecting a photograph (mathematically, graphically, or photographically) from its plane onto another plane by translation, rotation, or scale change. [MATH] A function, usually between vector spaces.

transformation constant See decay constant.

transformation group [MATH] **1.** A collection of transformations which forms a group with composition as the operation. **2.** A dynamical system or, more generally, a topological group G together with a topological space X where each g in G gives rise to a homeomorphism of X in a continuous manner with respect to the algebraic structure of G.

transformation series See radioactive series.

transformation temperature [MET] **1.** The temperature at which a change in phase occurs in a metal during heating or cooling. **2.** The maximum or minimum temperature of a transformation temperature range.

transformation theory [QUANT MECH] The study of coordinate and other transformations in quantum mechanics, especially those which leave some properties of the system invariant.

transformer [ELECTROMAG] An electrical component consisting of two or more multiturn coils of wire placed in close proximity to cause the magnetic field of one to link the other; used to transfer electric energy from one or more alternating-current circuits to one or more other circuits by magnetic induction.

transformer bridge [ELEC] A network consisting of a transformer and two impedances, in which the input signal is applied to the transformer primary and the output is taken between the secondary center-tap and the junction of the impedances that connect to the outer leads of the secondary.

transformer coupling [ELEC] *See* inductive coupling. [ELECTR] Interconnection between stages of an amplifier which employs a transformer for connecting the plate circuit of one stage to the grid circuit of the following stage; a special case of inductive coupling.

transformer hybrid *See* hybrid set.

transformer loss [ELEC] Ratio of the signal power that an ideal transformer of the same impedance ratio would deliver to the load impedance, to the signal power that the actual transformer delivers to the load impedance; this ratio is usually expressed in decibels.

transform fault [GEOL] A strike-slip fault with offset ridges characteristic of a midoceanic ridge.

transgranular corrosion [MET] A slow mode of failure that requires the combined action of stress and aggressive environment where the path of failure runs through the grains, producing branched cracking.

transgression [GEOL] Geologic evidence of landward extension of the sea. Also known as invasion; marine transgression. [OCEANOGR] Extension of the sea over land areas.

transhybrid loss [ELEC] In a carrier telephone system, the transmission loss at a given frequency measured across a hybrid circuit joined to a given two-wire termination and balancing network.

transient [PHYS] A pulse, damped oscillation, or other temporary phenomenon occurring in a system prior to reaching a steady-state condition.

transient analyzer [ELECTR] An analyzer that generates transients in the form of a succession of equal electric surges of small amplitude and adjustable waveform, applies these transients to a circuit or device under test, and shows the resulting output waveforms on the screen of an oscilloscope.

transient motion [PHYS] An oscillatory or other irregular motion occurring while a quantity is changing to a new steady-state value.

transistance [ELECTR] The characteristic that makes possible the control of voltages or currents so as to accomplish gain or switching action in a circuit; examples of transistance occur in transistors, diodes, and saturable reactors.

transistor [ELECTR] An active component of an electronic circuit consisting of a small block of semiconducting material to which at least three electrical contacts are made, usually two closely spaced rectifying contacts and one ohmic (nonrectifying) contact; it may be used as an amplifier, detector, or switch.

transistor amplifier [ELECTR] An amplifier in which one or more transistors provide amplification comparable to that of electron tubes.

transistor biasing [ELECTR] Maintaining a direct-current voltage between the base and some other element of a transistor.

transistor chip [ELECTR] An unencapsulated transistor of very small size used in microcircuits.

transistor circuit [ELECTR] An electric circuit in which a transistor is connected.

transistor memory *See* semiconductor memory.

transistor radio [ELECTR] A radio receiver in which transistors are used in place of electron tubes.

transistor-transistor logic [ELECTR] A logic circuit containing two transistors, for driving large output capacitances at high speed. Abbreviated T^2L; TTL.

transit [ASTRON] **1.** A celestial body's movement across the meridian of a place. Also known as meridian transit. **2.** Passage of a smaller celestial body across a larger one. **3.** Passage of a satellite's shadow across the disk of its primary. [ENG] **1.** A surveying instrument with the telescope mounted so that it can measure horizontal and vertical angles. Also known as transit theodolite. **2.** To reverse the direction of the telescope of a transit by rotating 180° about its horizontal axis. Also known as plunge. [NAV] A positive-fixing system employing low-orbit satellites which constantly emit continuous-wave signals; on the surface vehicle, the signals are received and the Doppler shift is recorded; position is determined by computation based on the shift.

transit circle [ENG] A type of astronomical transit instrument having a micrometer eyepiece that has an extra pair of moving wires perpendicular to the vertical set to measure the zenith distance or declination of the celestial object in conjunction with readings taken from a large, accurately calibrated circle attached to the horizontal axis. Also known as meridian circle; meridian transit.

transit declinometer [ENG] A type of declinometer; a surveyor's transit, built to exacting specifications with respect to freedom from traces of magnetic impurities and quality of the compass needle, has a 17-power telescope for sighting on a mark and for making solar and stellar observations to determine true directions.

transit instrument *See* transit telescope.

transition [COMMUN] Change from one circuit condition to the other; for example, the change from mark to space or from space to mark. [MOL BIO] A mutation resulting from the substitution of deoxyribonucleic acid or ribonucleic acid of one purine or pyrimidine for another. [QUANT MECH] The change of a quantum-mechanical system from one energy state to another. [THERMO] A change of a substance from one of the three states of matter to another.

transitional epithelium [HISTOL] A form of stratified epithelium found in the urinary bladder; cells vary between squamous, when the tissue is stretched, and columnar, when not stretched.

transitional flow [FL MECH] A flow in which the viscous and Reynolds stresses are of approximately equal magnitude; it is transitional between laminar flow and turbulent flow.

transition card [COMPUT SCI] In reading a deck of punched cards by a computer, a card that causes the computer to stop reading cards and begin executing a program. Also known as transfer card.

transition element [CHEM] One of a group of metallic elements in which the members have the filling of the outermost shell to 8 electrons interrupted to bring the penultimate shell from 8 to 18 or 32 electrons; includes elements 21 through 29 (scandium through copper), 39 through 47 (yttrium through silver), 57 through 79 (lanthanum through gold), and all known elements from 89 (actinium) on. [ELECTROMAG] An element used to couple one type of transmission system to another, as for coupling a coaxial line to a waveguide.

transition factor *See* reflection factor.

transition frequency [ENG ACOUS] The frequency corresponding to the intersection of the asymptotes to the constant-amplitude and constant-velocity portions of the frequency-response curve for a disk recording; this curve is plotted with output-voltage ratio in decibels as the ordinate, and the logarithm of the frequency as the abscissa. Also known as crossover frequency; turnover frequency. [QUANT MECH] The characteristic frequency of radiation emitted or absorbed by a quantum-mechanical system as it changes from one energy state to another; equal to the energy difference between the states divided by Planck's constant.

transition loss [ELEC] At a junction between a source and a load, the ratio of the available power to the power delivered to the load.

transition moment [QUANT MECH] Any type of multipole moment which determines radiative transitions between states; it consists of an integral of the product of the conjugate of the final state wave function, a multipole moment operator, and the initial state wave function.

transition point [ELECTROMAG] A point at which the constants of a circuit change in such a way as to cause reflection of a wave being propagated along the circuit. [THERMO] Either the temperature at which a substance changes from one state of aggregation to another (a first-order transition), or the temperature of culmination of a gradual change, such as the lambda point, or Curie point (a second-order transition). Also known as transition temperature.

transition region [SOLID STATE] The region between two homogeneous semiconductors in which the impurity concentration changes.

transition temperature [MET] The temperature at which a fracture changes from tough to brittle in various tests, such as notched-bar impact test. [THERMO] *See* transition point.

transitron [ELECTR] Thermionic-tube circuit whose action depends on the negative transconductance of the suppressor grid of a pentode with respect to the screen grid.

transit telescope [OPTICS] A telescopic instrument adapted to the observation of the passage, or transit, of an astronomical object across the meridian of an observer; consists of a telescope mounted on a single fixed horizontal axis of rotation which has a central hollow cube (sometimes a sphere) and two conical semiaxes ending in cylindrical pivots; the objective and eyepiece halves of the instrument are also fastened to the cube of the instrument, perpendicular to the horizontal axis. Also known as transit instrument.

transit theodolite *See* transit.

transit time [ELECTR] The time required for an electron or other charge carrier to travel between two electrodes in an electron tube or transistor.

translate [COMPUT SCI] To convert computer information from one language to another, or to convert characters from one representation set to another, and by extension, the computer instruction which directs the latter conversion to be carried out.

translating circuit *See* translator.

translation [MATH] 1. A function changing the coordinates of a point in a euclidean space into new coordinates relative to axes parallel to the original. 2. A function on a group to itself given by operating on each element by some one fixed element. [MECH]

The linear movement of a point in space without any rotation. [MOL BIO] The process by which the linear sequence of nucleotides in a molecule of messenger ribonucleic acid directs the specific linear sequence of amino acids, as during protein synthesis.

translational fault [GEOL] A fault in which there has been uniform movement in one direction and no rotational component of movement. Also known as translatory fault.

translation algorithm [COMPUT SCI] A specific, effective, essentially computational method for obtaining a translation from one language to another.

translational motion [MECH] Motion of a rigid body in such a way that any line which is imagined rigidly attached to the body remains parallel to its original direction.

translator [COMPUT SCI] A computer network or system having a number of inputs and outputs, so connected that when signals representing information expressed in a certain code are applied to the inputs, the output signals will represent the same information in a different code. Also known as translating circuit. [ELECTR] A combination television receiver and low-power television transmitter, used to pick up television signals on one frequency and retransmit them on another frequency to provide reception in areas not served directly by television stations.

translator routine [COMPUT SCI] A program which accepts statements in one language and outputs them as statements in another language.

translatory fault *See* translational fault.

transliterate [COMPUT SCI] To represent the characters or words of one language by corresponding characters or words of another language.

translocation [BOT] Movement of water, mineral salts, and organic substances from one part of a plant to another. [CYTOL] The transfer of a chromosome segment from its usual position to a new position in the same or in a different chromosome.

translucent medium [OPTICS] A medium which transmits rays of light so diffused that objects cannot be seen distinctly; examples are various forms of glass which admit considerable light but impede vision.

transmethylation [BIOCHEM] A metabolic reaction in which a methyl group is transferred from one compound to another; methionine and choline are important donors of methyl groups.

transmission [ELECTR] 1. The process of transferring a signal, message, picture, or other form of intelligence from one location to another location by means of wire lines, radio, light beams, infrared beams, or other communication systems. 2. A message, signal, or other form of intelligence that is being transmitted. [ELECTROMAG] *See* transmittance. [MECH ENG] The gearing system by which power is transmitted from the engine to the live axle in an automobile. Also known as gearbox.

transmission band [ELECTROMAG] Frequency range above the cutoff frequency in a waveguide, or the comparable useful frequency range for any other transmission line, system, or device.

transmission coefficient [PHYS] 1. The value of some quantity associated with the resultant field produced by incident and reflected waves at a given point in a transmission medium divided by the corresponding quantity in the incident wave. 2. The ratio of transmitted to incident energy flux or flux of some other quantity at a discontinuity in a transmission medium; for sound waves, it is called the sound transmission coefficient. 3. The ratio of the transmitted flux of some quantity to the incident flux for a substance of unit thickness. [QUANT MECH] *See* penetration probability.

transmission diffraction [ANALY CHEM] A type of electron diffraction analysis in which the electron beam is transmitted through a thin film or powder whose smallest dimension is no greater than a few tenths of a micrometer.

transmission electron microscope [ELECTR] A type of electron microscope in which the specimen transmits an electron beam focused on it, image contrasts are formed by the scattering of electrons out of the beam, and various magnetic lenses perform functions analogous to those of ordinary lenses in a light microscope.

transmission gain *See* gain.

transmission gate [ELECTR] A gate circuit that delivers an output waveform that is a replica of a selected input during a specific time interval which is determined by a control signal.

transmission grating [OPTICS] A diffraction grating produced on a transparent base so radiation is transmitted through the grating instead of being reflected from it.

transmission interface converter [COMPUT SCI] A device that converts data to or from a form suitable for transfer over a channel connecting two computer systems or connecting a computer with its associated data terminals.

transmission level [COMMUN] The ratio of the signal power at any point in a transmission system to the signal power at some point in the system chosen as a reference point; usually expressed in decibels.

transmission line [ELEC] A system of conductors, such as wires, waveguides, or coaxial cables, suitable for conducting electric power or signals efficiently between two or more terminals.

transmission loss [COMMUN] 1. The ratio of the power at one point in a transmission system to the power at a point farther along the line; usually expressed in decibels. 2. The actual power that is lost in transmitting a signal from one point to another through a medium or along a line. Also known as loss.

transmission mode *See* mode.

transmission modulation [ELECTR] Amplitude modulation of the reading-beam current in a charge storage tube as the beam passes through apertures in the storage surface; the degree of modulation is controlled by the stored charge pattern.

transmission plane [OPTICS] The plane of vibration of polarized light that will pass through a Nicol prism or other polarizer.

transmission range *See* night visual range.

transmission time [COMMUN] Absolute time interval from transmission to reception of a signal.

transmissivity [ELECTROMAG] The ratio of the transmitted radiation to the radiation arriving perpendicular to the boundary between two mediums.

transmit [COMMUN] To send a message, program, or other information to a person or place by wire, radio, or other means. [COMPUT SCI] To move data from one location to another.

transmit-receive tube [ELECTR] A gas-filled radio-frequency switching tube used to disconnect a receiver from its antenna during the interval for pulse transmission in radar and other pulsed radio-frequency systems. Also known as TR box; TR cell (British usage); TR switch; TR tube.

transmittance [ANALY CHEM] During absorption spectroscopy, the amount of radiant energy transmitted by the solution under analysis. [ELECTRO-

MAG] The radiant power transmitted by a body divided by the total radiant power incident upon the body. Also known as transmission.

transmitter [COMMUN] 1. In telephony, the carbon microphone that converts sound waves into audio-frequency signals. 2. *See* radio transmitter. [ELEC] *See* synchro transmitter.

transmitter-receiver *See* transceiver.

transmitter synchro *See* synchro transmitter.

transmutation [NUC PHYS] A nuclear process in which one nuclide is transformed into the nuclide of a different element. Also known as nuclear transformation.

transonic [PHYS] That which occurs or is occurring within the range of speed in which flow patterns change from subsonic to supersonic (or vice versa), about Mach 0.8 to 1.2, as in transonic flight or transonic flutter.

transparent medium [OPTICS] 1. A medium which has the property of transmitting rays of light in such a way that the human eye may see through the medium distinctly. 2. A medium transparent to other regions of the electromagnetic spectrum, such as x-rays and microwaves.

transparent sky cover [METEOROL] In United States weather-observing practice, that portion of sky cover through which higher clouds and blue sky may be observed; opposed to opaque sky cover.

transpiration [BIOL] The passage of a gas or liquid (in the form of vapor) through the skin, a membrane, or other tissue.

transplantation [BIOL] 1. The artificial removal of part of an organism and its replacement in the body of the same or of a different individual. 2. To remove a plant from one location and replant it in another place.

transplutonium element [INORG CHEM] An element having an atomic number greater than that of plutonium (94).

transponder [COMMUN] A transmitter-receiver capable of accepting the challenge of an interrogator and automatically transmitting an appropriate reply.

transport [COMPUT SCI] 1. To convey as a whole from one storage device to another in a digital computer. 2. *See* tape transport. [ENG] Conveyance equipment such as vehicular transport, hydraulic transport, and conveyor-belt setups. [NAV ARCH] A ship designed to carry military personnel from one place to another. Also known as troop ship.

transport delay unit [COMPUT SCI] A device used in analog computers which produces an output signal as a delayed form of an input signal. Also known as delay unit; transport unit.

transport properties [PHYS] Properties of a compound or material associated with mass or heat transport; for example, viscosity, and thermal conductivity of liquids, gases, or solids.

transport unit *See* transport delay unit.

transposition [COMMUN] Interchanging the relative positions of conductors at regular intervals along a transmission line to reduce cross talk. [MATH] A permutation of a set of symbols which exchanges exactly two while leaving all others unaffected.

transrectification [ELEC] Rectification that occurs in one circuit when an alternating voltage is applied to another circuit.

transuranic elements [CHEM] Elements that have atomic numbers greater than 92; all are radioactive, are products of artificial nuclear changes, and are members of the actinide group. Also known as transuranium elements.

transuranium elements *See* transuranic elements.

transversal [MATH] 1. A line intersecting a given family of lines. 2. A curve orthogonal to a hypersurface. 3. If π is a given map of a set X onto a set Y, a transversal for π is a subset T of X with the property that T contains exactly one point of $\pi^{-1}(y)$ for each $y \in Y$.

transverse basin *See* exogeosyncline.

transverse cylindrical orthomorphic projection *See* transverse Mercator projection.

transverse Doppler effect [ELECTROMAG] An aspect of the optical Doppler effect, occurring when the direction of motion of the source relative to an observer is perpendicular to the direction of the light received by the observer; the observed frequency is smaller than the source frequency by the factor $[1 - (v/c)^2]^{1/2}$, where v is the speed of the source and c is the speed of light.

transverse dune [GEOL] A sand dune with a nearly straight ridge crest formed by the merger of crescentic dunes; elongated at right angles to the direction of prevailing winds, with a gentle windward slope and a steep leeward slope.

transverse electric mode [ELECTROMAG] A mode in which a particular transverse electric wave is propagated in a waveguide or cavity. Abbreviated TE mode. Also known as H mode (British usage).

transverse electric wave [ELECTROMAG] An electromagnetic wave in which the electric field vector is everywhere perpendicular to the direction of propagation. Abbreviated TE wave. Also known as H wave (British usage).

transverse electromagnetic mode [ELECTROMAG] A mode in which a particular transverse electromagnetic wave is propagated in a waveguide or cavity. Abbreviated TEM mode.

transverse electromagnetic wave [ELECTROMAG] An electromagnetic wave in which both the electric and magnetic field vectors are everywhere perpendicular to the direction of propagation. Abbreviated TEM wave.

transverse joint *See* cross joint.

transverse magnetic mode [ELECTROMAG] A mode in which a particular transverse magnetic wave is propagated in a waveguide or cavity. Abbreviated TM mode. Also known as E mode (British usage).

transverse magnetic wave [ELECTROMAG] An electromagnetic wave in which the magnetic field vector is everywhere perpendicular to the direction of propagation. Abbreviated TM wave. Also known as E wave (British usage).

transverse Mercator projection [MAP] A conformal map projection in which the regular Mercator projection is rotated (transversed) 90° in azimuth, the central meridian corresponding to the line which represents the equator on the regular Mercator; the characteristics as to scale are identical to those of the regular Mercator, except that the scale is dependent on distances east or west of the meridian instead of north or south of the equator. Also known as inverse cylindrical orthomorphic projection; inverse Mercator projection; transverse cylindrical orthomorphic projection.

transverse valley [GEOL] 1. A valley perpendicular to the general strike of the underlying strata. 2. A valley cutting perpendicularly across a ridge, range, or chain of mountains. Also known as cross valley.

transverse vibration [MECH] Vibration of a rod in which elements of the rod move at right angles to the axis of the rod.

transverse wave [GEOPHYS] *See* S wave. [PHYS] A wave in which the direction of the disturbance at each point of the medium is perpendicular to the wave vector and parallel to surfaces of constant phase.

trap [AERO ENG] That part of a rocket motor that keeps the propellant grain in place. [CIV ENG] A bend or dip in a soil drain which is always full of water, providing a water seal to prevent odors from entering the building. [COMPUT SCI] An automatic transfer of control of a computer to a known location, this transfer occurring when a specified condition is detected by hardware. [ELECTR] 1. A tuned circuit used in the radio-frequency or intermediate-frequency section of a receiver to reject undesired frequencies; traps in television receiver video circuits keep the sound signal out of the picture channel. Also known as rejector. 2. *See* wave trap. [GEOL] *See* oil trap. [MECH ENG] A device which reduces the effect of the vapor pressure of oil or mercury on the high-vacuum side of a diffusion pump. [PETR] Any dark-colored, fine-grained, nongranitic, hypabyssal or extrusive rock. Also known as trappide; trap rock. [SOLID STATE] Any irregularity, such as a vacancy, in a semiconductor at which an electron or hole in the conduction band can be caught and trapped until released by thermal agitation. Also known as semiconductor trap.

TRAPATT diode [ELECTR] A pn junction diode, similar to the IMPATT diode, but characterized by the formation of a trapped space-charge plasma within the junction region; used in the generation and amplification of microwave power. Derived from trapped plasma avalanche transit time diode.

trapezohedron [CRYSTAL] An isometric crystal form of 24 faces, each face of which is an irregular four-sided figure. Also known as icositetrahedron; leucitohedron; tetragonal trisoctahedron.

trapezoid [MATH] A quadrilateral having two parallel sides.

trapezoidal pulse [ELECTR] An electrical pulse in which the voltage rises linearly to some value, remains constant at this value for some time, and then drops linearly to the original value.

trapezoidal wave [ELECTR] A wave consisting of a series of trapezoidal pulses.

trapped vortex [FL MECH] Airflow in rotary motion but trapped relative to leading-edge vortex separation, which increases not only lift but also drag.

trappide *See* trap.

trapping [COMMUN] *See* guided propagation. [GRAPHICS] The process of an already printed ink film accepting a succeeding or overprinted ink film.

trapping mode [COMPUT SCI] A procedure by means of which the computer, upon encountering a predetermined set of conditions, saves the program in its present status, executes a diagnostic procedure, and then resumes the processing of the program as of the moment of interruption.

trap rock *See* trap.

traveling dune *See* wandering dune.

traveling-grate stoker [MECH ENG] A type of furnace stoker; coal feeds by gravity into a hopper located on top of one end of a moving (traveling) grate; as the grate passes under the hopper, it carries a bed of fresh coal toward the furnace.

traveling wave [PHYS] A wave in which energy is transported from one part of a medium to another, in contrast to a standing wave.

traveling-wave amplifier [ELECTR] An amplifier that uses one or more traveling-wave tubes to provide useful amplification of signals at frequencies of the order of thousands of megahertz.

traveling-wave antenna [ELECTROMAG] An antenna in which the current distributions are produced by waves of charges propagated in only one direction in the conductors. Also known as progressive-wave antenna.

traveling-wave maser [PHYS] A ruby maser used with a comblike slow-wave structure and a number of yttrium iron garnet isolators to give L-band amplification (390 to 1550 megahertz); operation is at the temperature of liquid helium (4.2 K).

traveling-wave tube [ELECTR] An electron tube in which a stream of electrons interacts continuously or repeatedly with a guided electromagnetic wave moving substantially in synchronism with it, in such a way that there is a net transfer of energy from the stream to the wave; the tube is used as an amplifier or oscillator at frequencies in the microwave region.

travel-time curve [GEOPHYS] A plot of P-, S-, and L-wave travel times used by seismologists to locate earthquakes.

traverse [ENG] 1. A survey consisting of a set of connecting lines of known length, meeting each other at measured angles. 2. Movement to right or left on a pivot or mount, as of a gun, launcher, or radar antenna. [GEOL] A line of survey or sampling across a thin section of geological region. [METEOROL] A westerly wind in central France; it is moderate to strong, generally squally, humid and thundery in summer, especially on slopes facing west; it is cold in winter and spring and brings snow or hail showers. [NAV] A series of directions and distances, as those involved when a sailing vessel beats into the wind or a steam vessel zigzags.

travertine [GEOL] Concretionary limestone deposited at the mouth of a hot spring.

tray tower [CHEM ENG] A vertical process tower for liquid-vapor contacting (as in distillation, absorption, stripping, evaporation, spray drying, dehumidification, humidification, flashing, rectification, dephlegmation), along the height of which is a series of trays designed to cause intimate contact between the falling liquid and the rising vapor.

TR box See transmit-receive tube.

TR cell See transmit-receive tube.

tree [BOT] A perennial woody plant at least 20 feet (6 meters) in height at maturity, having an erect stem or trunk and a well-developed crown or leaf canopy. [ELECTR] A set of connected circuit branches that includes no meshes; responds uniquely to each of the possible combinations of a number of simultaneous inputs. Also known as decoder. [MATH] A connected graph contained in a given connected graph having all the vertices of the original but without any closed circuit. [MET] A projecting treelike aggregate of crystals formed at areas of high local current density in electroplating.

tree diagram [COMPUT SCI] A flow diagram which has no closed paths.

tree fern [BOT] The common name for plants belonging to the families Cyatheaceae and Dicksoniaceae; all are ferns that exhibit an arborescent habit.

tree frog [VERT ZOO] Any of the arboreal frogs comprising the family Hylidae characterized by expanded digital adhesive disks.

tree line See timberline.

Trematoda [INV ZOO] A loose grouping of acoelomate, parasitic flatworms of the phylum Platyhelminthes; they exhibit cephalization, bilateral symmetry, and well-developed holdfast structures.

Trematosauria [PALEON] A group of Triassic amphibians in the order Temnospondyli.

trembling ill See louping ill.

Tremellales [MYCOL] An order of basidiomycetous fungi in the subclass Heterobasidiomycetidae in which basidia have longitudinal walls.

tremie [ENG] An apparatus for placing concrete underwater, consisting of a large metal tube with a hopper at the top end and a valve arrangement at the bottom, submerged end.

tremolite [MINERAL] $Ca_2Mg_5Si_8O_{22}(OH)_2$ Magnesium-rich monoclinic calcium amphibole that forms one end member of a group of solid-solution series with iron, sodium, and aluminum; occurs in long blade-shaped or short stout prismatic crystals and also in masses or compound aggregates.

tremor [GEOPHYS] A minor earthquake. Also known as earthquake tremor; earth tremor. [MED] Involuntary, rhythmic trembling of voluntary muscles resulting from alternate contraction and relaxation of opposing muscle groups.

trend [GEOL] The direction of an outcrop of a layer, vein, fold, or other kind of geologic feature. Also known as direction. [STAT] The general drift, tendency, or bent of a set of statistical data as related to time or another related set of statistical data.

Trentepohliaceae [BOT] A family of green algae belonging to the Ulotrichales having thick walls, bandlike or reticulate chloroplasts, and zoospores or isogametes produced in enlarged, specialized cells.

Trentonian [GEOL] A North American stage of geologic time; Middle Ordovician (above Wilderness, below Edenian); equivalent to the upper Mohawkian.

trepanning tool [MECH ENG] A cutting tool in the form of a circular tube, having teeth on the end; the workpiece or tube, or both, are rotated and the tube is fed axially into the workpiece, leaving behind a narrow grooved surface in the workpiece.

Trepostomata [PALEON] An extinct order of ectoproct bryozoans in the class Stenolaemata characterized by delicate to massive colonies composed of tightly packed zooecia with solid calcareous zooecial walls.

Treroninae [VERT ZOO] The fruit pigeons, a subfamily of the avian family Columbidae distinguished by the gaudy coloration of the feathers.

Tretothoracidae [INV ZOO] A family of the Coleoptera in the superfamily Tenebrionoidea which contains a single species found in Queensland, Australia.

triandrous [BOT] Possessing three stamens.

triangle [MATH] The figure realized by connecting three noncollinear points by line segments.

Triangle See Triangulum.

triangle equation See angle equation.

triangular pulse [ELECTR] An electrical pulse in which the voltage rises linearly to some value, and immediately falls linearly to the original value.

triangular wave [ELECTR] A wave consisting of a series of triangular pulses.

triangulation [ENG] A surveying method for measuring a large area of land by establishing a base line from which a network of triangles is built up; in a series, each triangle has at least one side common with each adjacent triangle. [MATH] A decomposition of a topological manifold into subsets homeomorphic with a polyhedron in some euclidean space. [NAV] Determination of the position of a ship or aircraft by obtaining bearings of the moving object with reference to two fixed radio stations a known distance apart; this gives the values of one side and all angles of a triangle, from which the position can be computed.

Triangulum [ASTRON] A northern constellation, right ascension 2 hours, declination 30°N. Also known as Triangle.

Triangulum Australe [ASTRON] A southern constellation, right ascension 16 hours, declination 65°S. Also known as Southern Triangle.

Triassic [GEOL] The first period of the Mesozoic era, lying above Permian and below Jurassic, 180–225 million years ago.

Triatominae [INV ZOO] The kissing bugs, a subfamily of hemipteran insects in the family Reduviidae, distinguished by a long, slender rostrum.

tribo- [PHYS] A prefix meaning pertaining to or resulting from friction.

tributary [HYD] A stream that feeds or flows into or joins a larger stream or a lake. Also known as contributory; feeder; side stream; tributary stream.

tributary stream *See* tributary.

tricarboxylic acid cycle *See* Krebs cycle.

trichalcite [MINERAL] $Cu_5Ca(AsO_4)_2(CO_3)(OH)_4 \cdot 6H_2O$ A verdigris green to blue-green, orthorhombic mineral consisting of hydrated copper arsenate. Also known as tyrolite.

Trichechiformes [VERT ZOO] A suborder of mammals in the order Sirenia which contains the manatees and dugongids.

Trichiaceae [MYCOL] A family of slime molds in the order Trichiales.

Trichiales [MYCOL] An order of Myxomycetes in the subclass Myxogastromycetidae.

trichinosis [MED] Infection by the nematode *Trichinella spiralis* following ingestion of encysted larvae in raw or partially cooked pork; characterized by eosinophilia, nausea, fever, diarrhea, stiffness, and painful swelling of muscles, and facial edema.

Trichiuridae [VERT ZOO] The cutlass-fishes, a family of the suborder Scombroidei.

trichloromethane *See* chloroform.

trichobothrium [INV ZOO] An erect, bristlelike sensory hair found on certain arthropods, insects, and other invertebrates.

Trichobranchidae [INV ZOO] A family of polychaete annelids belonging to the Sedentaria; most members are rare and live at great ocean depths.

Trichocomaceae [MYCOL] A small tropical family of ascomycetous fungi in the order Eurotiales with ascocarps from which a tuft of capillitial threads extrudes, releasing the ascospores after dissolution of the asci.

trichocyst [INV ZOO] A minute structure in the cortex of certain protozoans that releases filamentous or fibrillar threads when discharged.

Trichodactylidae [INV ZOO] A family of fresh-water crabs in the section Brachyura, found mainly in tropical regions.

Trichogrammatidae [INV ZOO] A family of the Hymenoptera in the superfamily Chalcidoidea whose larvae are parasitic in the eggs of other insects.

trichome [BOT] An appendage derived from the protoderm in plants, including hairs and scales. [INV ZOO] A brightly colored tuft of hairs on the body of a myrmecophile that releases an aromatic substance attractive to ants.

Trichomonadida [INV ZOO] An order of the protozoan class Zoomastigophorea which contains four families of uninucleate species.

Trichomonadidae [INV ZOO] A family of flagellate protozoans in the order Trichomonadida.

Trichoniscidae [INV ZOO] A primitive family of isopod crustaceans in the suborder Oniscoidea found in damp littoral, halophilic, or riparian habitats.

Trichophilopteridae [INV ZOO] A family of lice in the order Mallophaga adapted to life upon the lemurs of Madagascar.

Trichoptera [INV ZOO] The caddis flies, an aquatic order of the class Insecta; larvae are wormlike and adults have two pairs of well-veined hairy wings, long antennae, and mouthparts capable of lapping only liquids.

Trichopterygidae [INV ZOO] The equivalent name for Ptiliidae.

Trichostomatida [INV ZOO] An order of ciliated protozoans in the subclass Holotrichia in which no true buccal ciliature is present but there is a vestibulum.

Trichostrongylidae [INV ZOO] A family of parasitic roundworms belonging to the Strongyloidea; hosts are cattle, sheep, goats, swine, and cats.

trichroism [OPTICS] Phenomenon exhibited by certain optically anisotropic transparent crystals when subjected to white light, in which a cube of the material is found to transmit a different color through each of the three pairs of parallel faces.

Trichuroidea [INV ZOO] A group of nematodes parasitic in various vertebrates and characterized by a slender body sometimes having a thickened posterior portion.

trickle cooler *See* cascade cooler.

trickling filter [CIV ENG] A bed of broken rock or other coarse aggregate onto which sewage or industrial waste is sprayed intermittently and allowed to trickle through, leaving organic matter on the surface of the rocks, where it is oxidized and removed by biological growths.

Tricladida [INV ZOO] The planarians, an order of the Turbellaria distinguished by diverticulated intestines with a single anterior branch and two posterior branches separated by the pharynx.

triclinic crystal [CRYSTAL] A crystal whose unit cell has axes which are not at right angles, and are unequal. Also known as anorthic crystal.

triclinic system [CRYSTAL] The most general and least symmetric crystal system, referred to by three axes of different length which are not at right angles to one another.

tricolor picture tube *See* color picture tube.

Triconodonta [PALEON] An extinct mammalian order of small flesh-eating creatures of the Mesozoic era having no angle or a pseudoangle on the lower jaw and triconodont molars.

Trictenotomidae [INV ZOO] A small family of Indian and Malaysian beetles in the superfamily Tenebrionoidea.

tricuspid [ANAT] Having three cusps or points, as a tooth.

tricuspid valve [ANAT] A valve consisting of three flaps located between the right atrium and right ventricle of the heart.

Tridacnidae [INV ZOO] A family of bivalve mollusks in the subclass Eulamellibranchia which contains the giant clams of the tropical Pacific.

Tridactylidae [INV ZOO] The pygmy mole crickets, a family of insects in the order Orthoptera, highly specialized for fossorial existence.

tridymite [MINERAL] SiO_2 A white or colorless crystal occurring in minute, thin, tabular crystals or scales; a high-temperature polymorph of quartz.

trifoliate [BOT] Having three leaves or leaflets.

trigeminal nerve [ANAT] The fifth cranial nerve in vertebrates; either of a pair of composite nerves rising from the side of the medulla, and with three great branches: the ophthalmic, maxillary, and mandibular nerves.

trigger [COMPUT SCI] To execute a jump to the first instruction of a program after the program has been loaded into the computer. Also known as initiate. [ELECTR] **1.** To initiate an action, which then continues for a period of time, as by applying a pulse to a trigger circuit. **2.** The pulse used to initiate the action of a trigger circuit. **3.** *See* trigger circuit. [ORD] A metallic item, part of the firing mechanism of a firearm, designed to release a firing pin by the application of pressure by the finger.

trigger circuit [ELECTR] **1.** A circuit or network in which the output changes abruptly with an infinitesimal change in input at a predetermined operating point. Also known as trigger. **2.** A circuit in which an action is initiated by an input pulse, as in a radar modulator. **3.** *See* bistable multivibrator.

trigger electrode *See* starter.

triggering [ELECTR] Phenomenon observed in some high-performance magnetic amplifiers with very low leakage rectifiers; as the input current is decreased in magnitude, the amplifier remains at cutoff for some time, and the output then suddenly shoots upward.

trigger pulse [ELECTR] A pulse that starts a cycle of operation. Also known as tripping pulse.

Triglidae [VERT ZOO] The searobins, a family of perciform fishes in the suborder Cottoidei.

Trigonalidae [INV ZOO] A small family of hymenopteran insects in the superfamily Proctotrupoidea.

trigonal system [CRYSTAL] A crystal system which is characterized by threefold symmetry, and which is usually considered as part of the hexagonal system since the lattice may be either hexagonal or rhombohedral.

trigone [ANAT] A triangular area inside the bladder limited by the openings of the ureters and urethra. [BOT] A thickening of plant cell walls formed when three or more cells adjoin.

trigonite [MINERAL] $MnPb_3H(AsO_3)_3$ A sulfur yellow to yellowish-brown or dark brown, monoclinic mineral consisting of an acid arsenite of lead and manganese; occurs in domatic form.

trigonometric functions [MATH] The real-valued functions such as $\sin(x)$, $\tan(x)$, and $\cos(x)$ obtained from studying certain ratios of the sides of a right triangle. Also known as circular functions.

trigonometry [MATH] The study of triangles and the trigonometric functions.

trihedral [MATH] Any figure obtained from three noncoplanar lines intersecting in a common point.

trilateration [ENG] The measurement of a series of distances between points on the surface of the earth, for the purpose of establishing relative positions of the points in surveying.

Trilobita [PALEON] The trilobites, a class of extinct Cambrian-Permian arthropods characterized by an exoskeleton covering the dorsal surface, delicate biramous appendages, body segments divided by furrows on the dorsal surface, and a pygidium composed of fused segments.

Trilobitoidea [PALEON] A class of Cambrian arthropods that are closely related to the Trilobita.

Trilobitomorpha [INV ZOO] A subphylum of the Arthropoda including Trilobita.

trilocular [BIOL] Having three cavities or cells.

Trimenoponidae [INV ZOO] A family of lice in the order Mallophaga occurring as parasites on South American rodents.

trimer [CHEM] A condensation product of three monomer molecules; C_6H_6 is a trimer of C_2H_2.

Trimerellacea [PALEON] A superfamily of extinct inarticulate brachiopods in the order Lingulida; they have valves, usually consisting of calcium carbonate.

Trimerophytatae *See* Trimerophytopsida.

Trimerophytopsida [PALEOBOT] A group of extinct land vascular plants with leafless, dichotomously branched stems that bear terminal sporangia.

trimerous [BOT] Having parts in sets of three. [INV ZOO] In insects, having the tarsus divided or apparently divided into three segments.

trimetric drawing [GRAPHICS] A form of nonperspective pictorial drawing in which the object being drawn is turned so that three mutually perpendicular edges are unequally foreshortened.

trimmer capacitor [ELEC] A relatively small variable capacitor used in parallel with a larger variable or fixed capacitor to permit exact adjustment of the capacitance of the parallel combination.

trinomial [MATH] A polynomial comprising three terms. [SYST] A nomenclatural designation for an organism composed of three terms: genus, species, and subspecies or variety.

trinomial distribution [STAT] A multinomial distribution in which there are three distinct outcomes.

triode [ELECTR] A three-electrode electron tube containing an anode, a cathode, and a control electrode.

Trionychidae [VERT ZOO] The soft-shelled turtles, a family of reptiles in the order Chelonia.

triphane *See* spodumene.

triphosphopyridine dinucleotide *See* triphosphopyridine nucleotide.

triphosphopyridine nucleotide [BIOCHEM] C_{12}-$H_{28}N_7O_{17}P_3$ A grayish-white powder, soluble in methanol and in water; a coenzyme and an important component of enzymatic systems concerned with biological oxidation-reduction systems. Abbreviated TPN. Also known as codehydrogenase II; coenzyme II; triphosphopyridine dinucleotide.

triphylite [MINERAL] $Li(Fe^{2+},Mn^{2+})PO_4$ A grayish-green or bluish-gray mineral crystallizing in the orthorhombic system; it is isomorphous with lithiophilite.

triple point [PHYS CHEM] A particular temperature and pressure at which three different phases of one substance can coexist in equilibrium.

triplet state [ATOM PHYS] Electronic state of an atom or molecule whose total spin angular momentum quantum number is equal to 1. [QUANT MECH] Any multiplet having three states.

triplexer [ELECTR] Dual duplexer that permits the use of two receivers simultaneously and independently in a radar system by disconnecting the receivers from the transmitted pulse.

triplite [MINERAL] $(Mn,Fe,Mg,Ca)_2(PO)_4(F,OH)$ A dark brown, chestnut brown, reddish-brown, or salmon pink, monoclinic mineral consisting of a fluophosphate of iron, manganese, magnesium, and calcium; occurs in massive form.

triploidy [CYTOL] The occurrence of related forms possessing chromosome numbers three times the haploid number.

tripoli [GEOL] A lightweight, porous, friable, siliceous sedimentary rock that may have a white, gray, pink, red, or yellow color; used for polishing metals and stones.

tripper [CIV ENG] A device activated by a passing train to work a signal or switch or to apply brakes. [MECH ENG] A device that snubs a conveyor belt causing the load to be discharged.

tripping device [ELEC] Mechanical or electromagnetic device used to bring a circuit breaker or starter to its off or open position, either when certain abnormal electrical conditions occur or when a catch is actuated manually.

tripping pulse   See trigger pulse.

Tripylina   [INV ZOO] A subdivision of the protozoan order Oculosida in which the major opening (astropyle) usually contains a perforated plate.

trisaccharide   [BIOCHEM] A carbohydrate which, on hydrolysis, yields three molecules of monosaccharides.

trisomy   [CYTOL] The presence in triplicate of one of the chromosomes of the complement.

trisomy 21 syndrome   See Down's syndrome.

tristate logic   [ELECTR] A form of transistor-transistor logic in which the output stages or input and output stages can assume three states; two are the normal low-impedance 1 and 0 states, and the third is a high-impedance state that allows many tristate devices to time-share bus lines.

trisulfide   [CHEM] A binary chemical compound that contains three sulfur atoms in its molecule, for example, iron trisulfude, Fe_2S_3.

triterpene   [ORG CHEM] One of a class of compounds having molecular skeletons containing 30 carbon atoms, and theoretically composed of six isoprene units; numerous and widely distributed in nature, occurring principally in plant resins and sap; an example is ambrein.

tritium   [NUC PHYS] The hydrogen isotope having mass number 3; it is one form of heavy hydrogen, the other being deuterium. Symbolized H^3; T.

Triton   [ASTRON] One of the two satellites of the planet Neptune with a diameter of about 4800 kilometers.

Triuridaceae   [BOT] A family of monocotyledonous plants in the order Triuridales distinguished by unisexual flowers and several carpels with one seed per carpel.

Triuridales   [BOT] A small order of terrestrial, mycotrophic monocots in the subclass Alismatidae without chlorophyll, and with separate carpels, trinucleate pollen, and a well-developed endosperm.

trivial name   [ORG CHEM] Unsystematic nomenclature, being the name of a chemical compound derived from the names of the natural source of the compound at the time of its isolation and before anything is known about its molecular structure.

t-RNA   See transfer ribonucleic acid.

Trochacea   [PALEON] A recent subfamily of primitive gastropod mollusks in the order Aspidobranchia.

trochanter   [ANAT] A process on the proximal end of the femur in many vertebrates, which serves for muscle attachment and, in birds, for articulation with the ilium. [INV ZOO] The second segment of an insect leg, counting from the base.

Trochidae   [INV ZOO] A family of gastropod mollusks in the order Aspidobranchia, including many of the top shells.

Trochili   [VERT ZOO] A suborder of the avian order Apodiformes.

Trochilidae   [VERT ZOO] The hummingbirds, a tropical New World family of the suborder Trochili with tubular tongues modified for nectar feeding; slender bills and the ability to hover are further feeding adaptations.

Trochiliscales   [PALEOBOT] A group of extinct plants belonging to the Charophyta in which the gyrogonites are dextrally spiraled.

trochlear nerve   [ANAT] The fourth cranial nerve; either of a pair of somatic motor nerves which innervate the superior oblique muscle of the eyeball.

Trochodendraceae   [BOT] A family of dicotyledonous trees in the order Trochodendrales distin-

guished by the absence of a perianth and stipules, numerous stamens, and pinnately veined leaves.

Trochodendrales   [BOT] An order of dicotyledonous trees in the subclass Hamamelidae characterized by primitively vesselless wood and unique, elongate, often branched idioblasts in the leaves.

trochoid   [ANAT] See pivot joint. [MATH] The path in the plane obtained from a point on the radius of a circle as the circle rolls along a fixed straight line.

trochophore   [INV ZOO] A generalized but distinct free-swimming larva found in several invertebrate groups, having a pear-shaped form with an external circlet of cilia, apical ciliary tufts, a complete functional digestive tract, and paired nephridia with excretory tubules.   Also known as trochosphere.

trochosphere   See trochophore.

Troglodytidae   [VERT ZOO] The wrens, a family of songbirds in the order Passeriformes.

Trogonidae   [VERT ZOO] The trogons, the single, pantropical family of the avian order Trogoniformes.

Trogoniformes   [VERT ZOO] An order of brightly colored, slow-moving birds characterized by a unique foot structure with the first and second toes directed backward.

Trojan planet   [ASTRON] One of a group of asteroids whose periods of revolution are about equal to that of Jupiter, or about 12 years; these bodies move close to one or the other of two positions called Lagrangian points, 60° ahead of or 60° behind Jupiter; the asteroids near these positions are known as Greeks and Pure Trojans respectively.

troland   [OPTICS] A unit of retinal illuminance, equal to the retinal illuminance produced by a surface whose luminance is one nit when the apparent area of the entrance pupil of the eye is 1 square millimeter.   Also known as luxon; photon.

Trombiculidae   [INV ZOO] The chiggers, or red bugs, a family of mites in the suborder Trombidiformes whose larvae are parasites of vertebrates.

Trombidiformes   [INV ZOO] The trombidiform mites, a suborder of the Acarina distinguished by the presence of a respiratory system opening at or near the base of the chelicerae.

trommel   [MIN ENG] 1. A revolving cylindrical screen used to grade coarsely crushed ore: the ore is fed into the trommel at one end, the fine material drops through the holes, and the coarse is delivered at the other end.   Also known as trommel screen. 2. To separate coal into various sizes by passing it through a revolving screen.

trommel screen   See trommel.

trona   [MINERAL] $Na_2(CO_3)\cdot Na(HCO_3)\cdot 2H_2O$ A gray-white or yellowish-white mineral that crystallizes in the monoclinic system and occurs in fibrous or columnar layers or masses.   Also known as urao.

troop ship   See transport.

Tropaeolaceae   [BOT] A family of dicotyledonous plants in the order Geraniales characterized by strongly irregular flowers, simple peltate leaves, eight stamens, and schizocarpous fruit.

Tropept   [GEOL] A suborder of the order Inceptisol, characterized by moderately dark A horizons with modest additions of organic matter, B horizons with brown or reddish colors, and slightly pale C horizons; restricted to tropical regions with moderate or high rainfall.

trophic level   [ECOL] Any of the feeding levels through which the passage of energy through an ecosystem proceeds; examples are photosynthetic plants, herbivorous animals, and microorganisms of decay.

trophobiosis [ECOL] A nutritional relationship associated only with certain species of ants in which alien insects supply food to the ants and are milked by the ants for their secretions.

trophozoite [INV ZOO] A vegetative protozoan; used especially of a parasite.

tropical climate [CLIMATOL] A climate which is typical of equatorial and tropical regions, that is, one with continually high temperatures and with considerable precipitation, at least during part of the year.

tropical front *See* intertropical front.

tropical meteorology [METEOROL] The study of the tropical atmosphere; the dividing lines, in each hemisphere, between the tropical easterlies and the mid-latitude westerlies in the middle troposphere roughly define the poleward boundaries of this region.

tropical scrub [ECOL] A class of vegetation composed of low woody plants (shrubs), sometimes growing quite close together, but more often separated by large patches of bare ground, with clumps of herbs scattered throughout; an example is the Ghanaian evergreen coastal thicket. Also known as brush; bush; fourré; mallee; thicket.

Tropic of Cancer [ASTRON] A small circle on the celestial sphere connecting points with declination 23.45° north of the celestial equator, the northernmost declination of the sun. [GEOD] A parallel of latitude 23.45° north of the equator, marking ¡the northernmost latitude at which the sun reaches its zenith.

Tropic of Capricorn [ASTRON] A small circle on the celestial sphere connecting points with declination 23.45° south of the celestial equator, the southernmost declination of the sun. [GEOD] A parallel of latitude 23.45° south of the equator, marking the southernmost latitude at which the sun reaches its zenith.

tropics [CLIMATOL] Any portion of the earth characterized by a tropical climate.

Tropiometridae [INV ZOO] A family of feather stars in the class Crinoidea which are bottom crawlers.

tropism [BIOL] Orientation movement of a sessile organism in response to a stimulus. Also known as topotaxis.

tropomyosin [BIOCHEM] A muscle protein similar to myosin and implicated as being part of the structure of the Z bands separating sarcomeres from each other.

tropopause [METEOROL] The boundary between the troposphere and stratosphere, usually characterized by an abrupt change of lapse rate; the change is in the direction of increased atmospheric stability from regions below to regions above the tropopause; its height varies from 15 to 20 kilometers in the tropics to about 10 kilometers in polar regions.

troposphere [METEOROL] That portion of the atmosphere from the earth's surface to the tropopause, that is, the lowest 10 to 20 kilometers of the atmosphere.

tropospheric duct *See* duct.

tropospheric ducting *See* ducting.

trough [GEOL] **1.** A small, straight depression formed just offshore on the bottom of a sea or lake and on the landward side of a longshore bar. **2.** Any narrow, elongate depression in the surface of the earth. **3.** An elongate depression on the sea floor that is wider and shallower than a trench. Also known as submarine trough. **4.** The line connecting the lowest points of a fold. [METEOROL] An elongated area of relatively low atmospheric pressure; the opposite of a ridge.

trout [VERT ZOO] Any of various edible fresh-water fishes in the order Salmoniformes that are generally much smaller than the salmon.

troy ounce *See* ounce.

troy pound *See* pound.

troy system [MECH] A system of mass units used primarily to measure gold and silver; the ounce is the same as that in the apothecaries' system, being equal to 480 grains or 31.1034768 grams. Abbreviated t. Also known as troy weight.

troy weight *See* troy system.

TR switch *See* transmit-receive tube.

TR tube *See* transmit-receive tube.

Trucherognathidae [PALEON] A family of conodonts in the order Conodontophorida in which the attachment scar permits the conodont to rest on the jaw ramus.

truck [MECH ENG] A self-propelled wheeled vehicle, designed primarily to transport goods and heavy equipment; it may be used to tow trailers or other mobile equipment. [MIN ENG] *See* barney.

trudellite [MINERAL] $Al_{10}(SO_4)_3Cl_{12}(OH)_{12} \cdot 30H_2O$ An amber yellow, hexagonal mineral consisting of a hydrated basic sulfate-chloride of aluminum; occurs as compact masses.

true airspeed [AERO ENG] The actual speed of an aircraft relative to the air through which it flies, that is, the calibrated airspeed corrected for temperature, density, or compressibility.

true air temperature [METEOROL] Basic air temperature corrected for heat of compression error due to high-speed motion of the thermometer through the air, as on an aircraft.

true anomaly *See* anomaly.

true bearing [NAV] Bearing relative to true north; compass bearing corrected for magnetic deviation.

true complement *See* radix complement.

true dip *See* dip.

true freezing point [PHYS CHEM] The temperature at which the liquid and solid forms of a substance exist in equilibrium at a given pressure (usually 1 standard atmosphere, or 101,325 newtons per square meter).

true horizon [OPTICS] The boundary of a horizontal plane passing through a point of vision, or in photogrammetry, the perspective center of a lens system.

true north [NAV] **1.** The direction of the north geographical pole. **2.** The reference direction for measurement of true directions.

true soil *See* solum.

true solar day *See* apparent solar day.

true solar time *See* apparent solar time.

truffle [BOT] The edible underground fruiting body of various European fungi in the family Tuberaceae, especially the genus *Tuber*.

trumpeter [VERT ZOO] A bird belonging to the Psophiidae, a family with three South American species; the common trumpeter (*Psophia crepitans*) is the size of a pheasant and resembles a long-legged guinea fowl.

truncate [BIOL] Abbreviated at an end, as if cut off. [MATH] **1.** To drop digits at the end of a numerical value; the number 3.14159265 is truncated to five figures in 3.1415, whereas it would be 3.1416 if rounded off to five figures. **2.** To approximate the sum of an infinite series by the sum of a finite number of its terms. **3.** To terminate an infinite sequence of successively better approximations of a quantity after a finite number of such approximations.

truncation [MATH] 1. Approximating the sum of an infinite series by the sum of a finite number of its terms. 2. *See* rounding.

trunk [ANAT] The main mass of the human body, exclusive of the head, neck, and extremities; it is divided into thorax, abdomen, and pelvis. [BOT] The main stem of a tree. [COMMUN] 1. A path over which information is transferred in a computer. 2. A telephone line connecting two central offices. Also known as trunk circuit.

trunk circuit *See* trunk.

trunk feeder [ELEC] An electric power transmission line that connects two generating stations, or a generating station and an important substation, or two electrical distribution networks.

truss [CIV ENG] A frame, generally of steel, timber, concrete, or a light alloy, built from members in tension and compression.

truss bridge [CIV ENG] A fixed bridge consisting of members vertically arranged in a triangular pattern.

Tryblidiidae [PALEON] An extinct family of Paleozoic mollusks.

Trypanorhyncha [INV ZOO] An order of tapeworms of the subclass Cestoda; all are parasites in the intestine of elasmobranch fishes.

Trypanosomatidae [INV ZOO] A family of Protozoa, order Kinetoplastida, containing flagellated parasites which exhibit polymorphism during their life cycle.

trypsin [BIOCHEM] A proteolytic enzyme which catalyzes the hydrolysis of peptide linkages in proteins and partially hydrolyzed proteins; derived from trypsinogen by the action of enterokinase in intestinal juice.

tryptophan [BIOCHEM] $C_{11}H_{12}O_2N_2$ An amino acid obtained from casein, fibrin, and certain other proteins; it is a precursor of indoleacetic acid, serotonin, and nicotinic acid.

tsetse fly [INV ZOO] Any of various South African muscoid flies of the genus *Glossina*; medically important as vectors of sleeping sickness or trypanosomiasis.

TSH *See* thyrotropic hormone.

tsumebite [MINERAL] $Pb_2Cu(PO_4)(SO_4)(OH)$ An emerald green, monoclinic mineral consisting of a hydrated basic phosphate and sulfate of lead and copper.

tsunami [OCEANOGR] A long-period sea wave produced by a seaquake or volcanic eruption; it may travel for thousands of miles. Also known as seismic sea wave.

Tsushima Current [OCEANOGR] That part of the Kuroshio Current flowing northeastward through the Korea Strait and along the Japanese coast in the Sea of Japan.

t-test [STAT] A statistical test involving means of normal populations with unknown standard deviations; small samples are used, based on a variable t equal to the difference between the mean of the sample and the mean of the population divided by a result obtained by dividing the standard deviation of the sample by the square root of the number of individuals in the sample.

TTL *See* transistor-transistor logic.

tube [BIOL] A narrow channel within the body of an animal or plant. [ELECTR] *See* electron tube. [ENG] A long cylindrical body with a hollow center used especially to convey fluid. [GEOL] A passage in a cave having smooth sides and an elliptical to nearly circular cross section. [ORD] The main part of a gun, the cylindrical piece of metal surrounding the bore; tube is frequently used in referring

to artillery weapons, and barrel is more frequently used in referring to small arms.

tube foot [INV ZOO] One of the tentaclelike outpushings of the radial vessels of the water-vascular system in echinoderms; may be suctorial, or serve as stiltlike limbs or tentacles.

tube mill [MECH ENG] A revolving cylinder used for fine pulverization of ore, rock, and other such materials; the material, mixed with water, is fed into the chamber from one end, and passes out the other end as slime.

tuber [BOT] The enlarged end of a rhizome in which food accumulates, as in the potato.

tubercle [BIOL] A small knoblike prominence. [MET] A mound of corrosive products on the surface of a metal that is subjected to local corrosive attack.

Tuberculariaceae [MYCOL] A family of fungi of the order Moniliales having short conidia that form cushion-shaped, often waxy or gelatinous aggregates (sporodochia).

tuberculate [BIOL] Having or characterized by knoblike processes.

tuberculosis [MED] A chronic infectious disease of humans and animals primarily involving the lungs caused by the tubercle bacillus, *Mycobacterium tuberculosis*, or by *M. bovis*. Also known as consumption; phthisis.

tuberosity [ANAT] A large or obtuse prominence, especially as on bone for muscle attachment.

Tubicola [INV ZOO] An order of sedentary polychaete annelids that surround themselves with a calcareous tube or one which is composed of agglutinated foreign particles.

tubing hanger *See* hanger.

Tubulanidae [INV ZOO] A family of the order Palaeonemertini.

tubular gland [ANAT] A secreting structure whose secretory endpieces are tubelike or cylindrical in shape.

Tubulidentata [VERT ZOO] An order of mammals which contains a single living genus, the aardvark (*Orycteropus*) of Africa.

Tucana [ASTRON] A constellation in the southern hemisphere; right ascension 23 hours, declination 60° south. Also known as Toucan.

tufa [GEOL] A spongy, porous limestone formed by precipitation from evaporating spring and river waters, often onto leaves and stems of neighboring plants. Also known as calcareous sinter; calcareous tufa.

tuff [GEOL] Consolidated volcanic ash, composed largely of fragments (less than 4 millimeters) produced directly by volcanic eruption; much of the fragmented material represents finely comminuted crystals and rocks.

tuff lava *See* welded tuff.

tuft *See* mound.

tulip [BOT] Any of various plants with showy flowers constituting the genus *Tulipa* in the family Liliaceae; characterized by coated bulbs, lanceolate leaves, and a single flower with six equal perianth segments and six stamens.

tulip poplar *See* tulip tree.

tulip tree [BOT] *Liriodendron tulipifera*. A tree belonging to the magnolia family (Magnoliaceae) distinguished by leaves which are squarish at the tip, true terminal buds, cone-shaped fruit, and large greenish-yellow and orange-colored flowers. Also known as tulip poplar.

tumbleweed [BOT] Any of various plants that break loose from their roots in autumn and are driven by the wind in rolling masses over the ground.

tumbling [AERO ENG] An attitude situation in which the vehicle continues on its flight, but turns end over end about its center of mass. [ENG] A surface-finishing operation for small articles in which irregularities are removed or surfaces are polished by tumbling them together in a barrel, along with wooden pegs, sawdust, and polishing compounds. [MECH ENG] Loss of control in a two-frame free gyroscope, occurring when both frames of reference become coplanar.

tumbling mill [MECH ENG] A grinding and pulverizing machine consisting of a shell or drum rotating on a horizontal axis.

tumor [MED] Any abnormal mass of cells resulting from excessive cellular multiplication.

tuna [VERT ZOO] Any of the large, pelagic, cosmopolitan marine fishes which form the family Thunnidae including species that rank among the most valuable of food and game fish.

tunable filter [ELECTR] An electric filter in which the frequency of the passband or rejection band can be varied by adjusting its components.

tunable laser [OPTICS] A laser in which the frequency of the output radiation can be tuned over part or all of the ultraviolet, visible, and infrared regions of the spectrum.

tundra [ECOL] An area supporting some vegetation between the northern upper limit of trees and the lower limit of perennial snow on mountains, and on the fringes of the Antarctic continent and its neighboring islands.

tuned amplifier [ELECTR] An amplifier in which the load is a tuned circuit; load impedance and amplifier gain then vary with frequency.

tuned-anode oscillator [ELECTR] A vacuum-tube oscillator whose frequency is determined by a tank circuit in the anode circuit, coupled to the grid to provide the required feedback. Also known as tuned-plate oscillator.

tuned-base oscillator [ELECTR] Transistor oscillator in which the frequency-determining resonant circuit is located in the base circuit; comparable to a tuned-grid oscillator.

tuned cavity See cavity resonator.

tuned circuit [ELECTR] A circuit whose components can be adjusted to make the circuit responsive to a particular frequency in a tuning range. Also known as tuning circuit.

tuned-collector oscillator [ELECTR] A transistor oscillator in which the frequency-determining resonant circuit is located in the collector circuit; this is comparable to a tuned-anode electron-tube oscillator.

tuned-grid oscillator [ELECTR] Oscillator whose frequency is determined by a parallel-resonant circuit in the grid coupled to the plate to provide the required feedback.

tuned-plate oscillator See tuned-anode oscillator.

tuned-reed frequency meter See vibrating-reed frequency meter.

tuned relay [ELEC] A relay having mechanical or other resonating arrangements that limit response to currents at one particular frequency.

tuned transformer [ELEC] Transformer whose associated circuit elements are adjusted as a whole to be resonant at the frequency of the alternating current supplied to the primary, thereby causing the secondary voltage to build up to higher values than would otherwise be obtained.

tuner [ELECTR] The portion of a receiver that contains circuits which can be tuned to accept the carrier frequency of the alternating current supplied to

the primary, thereby causing the secondary voltage to build up to higher values than would otherwise be obtained.

tungsten [CHEM] Also known as wolfram. A metallic element in group VI, symbol W, atomic number 74, atomic weight 183.85; soluble in mixed nitric and hydrofluoric acids; melts at 3400°C. [MET] A hard, brittle, ductile, heavy gray-white metal used in the pure form chiefly for electrical purposes and with other substances in dentistry, pen points, x-ray-tube targets, phonograph needles, and high-speed tool metal, and as a radioactive shield.

tungsten–inert gas welding [MET] Welding in which an arc plasma from a nonconsumable tungsten electrode radiates heat onto the work surface, to create a weld puddle in a protective atmosphere provided by a flow of inert shielding gas; heat must then travel by conduction from this puddle to melt the desired depth of weld. Abbreviated TIG welding.

tungstenite [MINERAL] WS_2 A dark lead gray mineral consisting of tungsten disulfide; occurs in massive form, in scaly or feathery aggregates.

tungstite [MINERAL] $WO_3 \cdot H_2O$ A bright yellow, golden yellow, or yellowish-green mineral thought to consist of hydrated tungsten oxide; occurs in massive form and as platy crystals.

tung tree [BOT] *Aleurites fordii*. A plant of the spurge family in the order Euphorbiales, native to central and western China and grown in the southern United States.

tunica [BIOL] A membrane or layer of tissue that covers or envelops an organ or other anatomical structure.

tunica adventitia See adventitia.

tunica mucosa See mucous membrane.

Tunicata [INV ZOO] A subphylum of the Chordata characterized by restriction of the notochord to the tail and posterior body of the larva, absence of mesodermal segmentation, and secretion of an outer covering or tunic about the body.

tuning [ELECTR] The process of adjusting the inductance or the capacitance or both in a tuned circuit, for example, in a radio, television, or radar receiver or transmitter, so as to obtain optimum performance at a selected frequency.

tuning circuit See tuned circuit.

tunnel diode [ELECTR] A heavily doped junction diode that has a negative resistance at very low voltage in the forward bias direction, due to quantum-mechanical tunneling, and a short circuit in the negative bias direction. Also known as Esaki tunnel diode.

tunnel effect [QUANT MECH] The ability of a particle to pass through a region of finite extent in which the particle's potential energy is greater than its total energy; this is a quantum-mechanical phenomenon which would be impossible according to classical mechanics.

tunnel rectifier [ELECTR] Tunnel diode having a relatively low peak-current rating as compared with other tunnel diodes used in memory-circuit applications.

tunnel resistor [ELECTR] Resistor in which a thin layer of metal is plated across a tunneling junction, to give the combined characteristics of a tunnel diode and an ordinary resistor.

tunnel triode [ELECTR] Transistorlike device in which the emitter-base junction is a tunnel diode and the collector-base junction is a conventional diode.

Tupaiidae [VERT ZOO] The tree shrews, a family of mammals in the order Insectivora.

tupelo [BOT] Any of various trees belonging to the genus *Nyssa* of the sour gum family, Nyssaceae, distinguished by small, obovate, shiny leaves, a small blue-black drupaceous fruit, and branches growing at a wide angle from the axis.

turanite [MINERAL] $Cu_5(VO_4)_2(OH)_4$ An olive green, orthorhombic mineral consisting of basic copper vanadate; occurs as reniform crusts and spherical concretions.

Turbellaria [INV ZOO] A class of the phylum Platyhelminthes having bodies that are elongate and flat to oval or circular in cross section.

turbidimeter [OPTICS] A device that measures the loss in intensity of a light beam as it passes through a solution with particles large enough to scatter the light.

turbidimetric analysis [ANALY CHEM] A scattered-light procedure for the determination of the weight concentration of particles in cloudy, dull, or muddy solutions; uses a device that measures the loss in intensity of a light beam as it passes through the solution. Also known as turbidimetry.

turbidimetric titration [ANALY CHEM] Titration in which the end point is indicated by the developing turbidity of the titrated solution.

turbidimetry See tubidimetric analysis.

turbidite [GEOL] Any sediment or rock transported and deposited by a turbidity current, generally characterized by graded bedding, large amounts of matrix, and commonly exhibiting a Bouma sequence.

turbidity [METEOROL] Any condition of the atmosphere which reduces its transparency to radiation, especially to visible radiation. [ANALY CHEM] **1.** Measure of the clarity (using APHA or colorimetric scales) of an otherwise clear liquid. **2.** Cloudy or hazy appearance in a naturally clear liquid caused by a suspension of colloidal liquid droplets or fine solids.

turbidity current [OCEANOGR] A highly turbid, relatively dense current carrying large quantities of clay, silt, and sand in suspension which flows down a submarine slope through less dense sea water. Also known as density current; suspension current.

turbinate [BOT] Shaped like an inverted cone. [INV ZOO] Spiral with rapidly decreasing whorls from base to apex.

turbine [MECH ENG] A fluid acceleration machine for generating rotary mechanical power from the energy in a stream of fluid.

turbine pump See regenerative pump.

Turbinidae [INV ZOO] A family of gastropod mollusks including species of top shells.

turbodrill [PETRO ENG] A rotary tool used in drilling oil or gas wells in which the bit is rotated by a turbine motor inside the well.

turbofan [AERO ENG] An air-breathing jet engine in which additional propulsive thrust is gained by extending a portion of the compressor or turbine blades outside the inner engine case.

turbojet [AERO ENG] A jet engine incorporating a turbine-driven air compressor to take in and compress the air for the combustion of fuel (or for heating by a nuclear reactor), the gases of combustion (or the heated air) being used both to rotate the turbine and to create a thrust-producing jet.

turbulence See turbulent flow.

turbulent diffusion See eddy diffusion.

turbulent flow [FL MECH] Motion of fluids in which local velocities and pressures fluctuate irregularly, in a random manner. Also known as turbulence.

turbulent flux See eddy flux.

Turdidae [VERT ZOO] The thrushes, a family of passeriform birds in the suborder Oscines.

turgor pressure [BOT] The actual pressure developed by the fluid content of a turgid plant cell.

Turing machine [COMPUT SCI] A mathematical idealization of a computing automation similar in some ways to real computing machines; used by mathematicians to define the concept of computability.

turkey [VERT ZOO] Either of two species of wild birds, and any of various derived domestic breeds, in the family Meleagrididae characterized by a bare head and neck, and in the male a large pendant wattle which hangs on one side from the base of the bill.

Turkey stone See turquoise.

turn [ELEC] One complete loop of wire. [MATH] See circle.

turnaround [CHEM ENG] In petroleum refining, the shutdown of a unit after a normal run for maintenance and repair work, then putting the unit back into operation. [ENG] The length of time between arriving at a point and departing from that point; it is used in this sense for the turnaround of vehicles, ships in ports, and aircraft.

turnaround document [COMPUT SCI] A document, such as a punch card, that is produced by a computer, can be read by humans, and can be reread into the machine.

turnaround system [COMPUT SCI] In character recognition, a system in which the input data to be read have previously been printed by the computer with which the reader is associated; an application is invoice billing and the subsequent recording of payments.

turnbuckle [DES ENG] A sleeve with a thread at one end and a swivel at the other, or with threads of opposite hands at each end so that by turning the sleeve connected rods or wire rope will be drawn together and tightened.

Turnicidae [VERT ZOO] The button quails, a family of Old World birds in the order Gruiformes.

turnip [BOT] *Brassica rapa* or *B. campestris* var. *rapa*. An annual crucifer of Asiatic origin belonging to the family Brassiaceae in the order Capparales and grown for its foliage and edible root.

turnover frequency See transition frequency.

Turonian [GEOL] A European stage of geologic time: Upper or Middle Cretaceous (above Cenomanian, below Coniacian).

turquoise [MINERAL] $CuAl_6(PO_4)_4(OH)_8 \cdot 4H_2O$ A semitranslucent sky-blue, bluish-green, apple-green, or greenish-gray mineral that crystallizes in the triclinic system and occurs in veinlets or as crusts of massive, concretionary, and stalactite shapes; an important gem mineral. Also known as calaite; Turkey stone.

turret lathe [MECH ENG] A semiautomatic lathe differing from the engine lathe in having the tailstock replaced with a multisided, indexing tool holder or turret designed to hold several tools.

turtle [VERT ZOO] Any of about 240 species of reptiles which constitute the order Chelonia distinguished by the two bony shells enclosing the body.

tussock [ECOL] A small hummock of generally solid ground in a bog or marsh, usually covered with and bound together by the roots of low vegetation such as grasses, sedges, or ericaceous shrubs.

tuyere [MET] An opening in the shell and refractory lining of a furnace through which air is forced.

TV See television.

tweeter [ENG ACOUS] A loudspeaker designed to handle only the higher audio frequencies, usually those well above 3000 hertz; generally used in conjunction with a crossover network and a woofer.

twenty-nine feature [COMPUT SCI] A device used on some punched-card machines to represent values from 0 through 29 by a maximum of two punches on a single column; x and y punches represent 10 and 20, and these are added to punches in positions 0 through 9.

twilight [ASTRON] An intermediate period of illumination of the sky before sunrise and after sunset; the three forms are civil, nautical, and astronomical.

twin [BIOL] One of two individuals born at the same time. [CRYSTAL] *See* twin crystal.

twin crystal [CRYSTAL] A compound crystal which has one or more parts whose lattice structure is the mirror image of that in the other parts of the crystal. Also known as twin.

twinkling stars [ASTRON] Rapid fluctuations of the brightness and size of the images of stars caused by turbulence in the earth's atmosphere.

twin law [CRYSTAL] A statement relating two or more individuals of a twin to one another in terms of their crystallography (twin plane, twin axis, and so on).

twinning [CRYSTAL] The development of a twin crystal by growth, translation, or gliding.

twinning plane *See* twin plane.

twin paradox *See* clock paradox.

twin plane [CRYSTAL] The plane common to and across which the individual crystals or components of a crystal twin are symmetrically arranged or reflected. Also known as twinning plane.

Twins *See* Gemini.

twin-T filter [ELEC] An electric filter consisting of a parallel-T network with values of network elements chosen in such a way that the outputs due to each of the paths precisely cancel at a specified frequency.

twin-T network *See* parallel-T network.

twister [METEOROL] In the United States, a colloquial term for tornado. [SOLID STATE] A piezoelectric crystal that generates a voltage when twisted.

two-address code [COMPUT SCI] In computers, a code using two-address instructions.

two-address instruction [COMPUT SCI] In computers, an instruction which includes an operation and specifies the location of two registers.

two-body force [PHYS] A force between two particles which is not affected by the existence of other particles in the vicinity, such as a gravitational force or a Coulomb force between charged particles.

two-body problem [MECH] The problem of predicting the motions of two objects obeying Newton's laws of motion and exerting forces on each other according to some specified law such as Newton's law of gravitation, given their masses and their positions and velocities at some initial time.

two-carrier theory [SOLID STATE] A theory of the conduction properties of a material in bulk or in a rectifying barrier which takes into account the motion of both electrons and holes.

two-cycle engine [MECH ENG] A reciprocating internal combustion engine that requires two piston strokes or one revolution to complete a cycle.

two-dimensional chromatography [ANALY CHEM] A paper chromatography technique in which the sample is resolved by standard procedures (ascending, descending, or horizontal solvent movement) and then turned at right angles in a second solvent and re-resolved.

two-dimensional flow [FL MECH] Fluid flow in which all flow occurs in a set of parallel planes with no flow normal to them, and the flow is identical in each of these parallel planes.

two-dimensional storage [COMPUT SCI] A direct-access storage device in which the storage locations assigned to a particular file do not have to be physically adjacent, but instead may be taken from one or more seek areas.

two-gap head [COMPUT SCI] One of two separate magnetic tape heads, one for reading and the other for recording data.

two-level subroutine [COMPUT SCI] A subroutine in which entry is made to a second, lower-level subroutine.

two-part code [COMMUN] Randomized code consisting of an encoding section in which the plain text groups are arranged in alphabetical or other significant order accompanied by their code groups in nonalphabetical or random order, and a decoding section in which the code groups are arranged in alphabetical or numerical order and are accompanied by their meanings given in the encoding section.

two-phase [PHYS] Having a phase difference of one quarter-cycle or 90°. Also known as quarter-phase.

two-phase flow [CRYO] Flow of helium II, or of electrons in a superconductor, thought of as consisting of two interpenetrating, noninteracting fluids, a superfluid component which exhibits no resistance to flow and is responsible for superconducting properties, and a normal component, which behaves as does an ordinary fluid or as conduction electrons in a nonsuperconducting metal. [FL MECH] Concurrent movement of two phases (for example, gas and liquid) through a closed conduit or duct (for example, a pipe).

two-quadrant multiplier [COMPUT SCI] Of an analog computer, a multiplier in which operation is restricted to a single sign of one input variable only.

two's complement [MATH] A number derived from a given n-bit number by requiring the two numbers to sum to a value of 2^n.

two-sided test [STAT] A test which rejects the null hypothesis when the test statistic T is either less than or equal to c or greater than or equal to d, where c and d are critical values.

two-stroke cycle [MECH ENG] An internal combustion engine cycle completed in two strokes of the piston.

two-valued logic [MATH] A system of logic where each statement has two possible values or states, truth or falsehood.

two-way slab [CIV ENG] A concrete slab supported by beams along all four edges and reinforced with steel bars arranged perpendicularly.

TWX machine *See* teletypewriter.

Twyman-Green interferometer [OPTICS] An interferometer similar to the Michelson interferometer except that it is illuminated with a point source of light instead of an extended source.

tychite [MINERAL] $Na_6Mg_2(SO_4)(CO_3)_4$ A white, isometric mineral consisting of a sulfate-carbonate of sodium and magnesium.

Tychonoff space *See* completely regular space.

tying bar *See* tombolo.

Tylenchida [INV ZOO] An order of soil-dwelling or phytoparasitic nematodes in the subclass Rhabdita.

Tylenchoidea [INV ZOO] A superfamily of mainly soil and insect-associated nematodes in the order Tylenchida with a stylet for piercing live cells and sucking the juices.

Tylopoda [VERT ZOO] An infraorder of artiodactyls in the suborder Ruminantia that contains the camels and extinct related forms.

tympanic membrane [ANAT] The membrane separating the external from the middle ear. Also known as eardrum; tympanum.

tympanum [ANAT] *See* tympanic membrane. [INV ZOO] A thin membrane covering an organ of hearing in insects.

Tyndall effect [OPTICS] Visible scattering of light along the path of a beam of light as it passes through a system containing discontinuities, such as the surfaces of colloidal particles in a colloidal solution.

type [GRAPHICS] The relief or plane characters used to generate printed characters of various styles and sizes. [SYST] A specimen on which a species or subspecies is based.

type A wave *See* continuous wave.

type bar [GRAPHICS] A long, narrow box or magazine from which projects the type used at a particular print position of a printer, and which contains the entire collection of characters available to that print position.

type-bar printer [GRAPHICS] A serial printer in which two characters are mounted on a type bar, as in some electric typewriters and early teletypewriters, and desired characters are printed one at a time.

Type II Cepheids *See* W Virginis stars.

typeface *See* face.

type-M carcinotron *See* M-type backward-wave oscillator.

type-O carcinotron *See* O-type backward wave oscillator.

Typhaceae [BOT] A family of monocotyledonous plants in the order Typhales characterized by an inflorescence of dense, cylindrical spikes and absence of a perianth.

Typhales [BOT] An order of marsh or aquatic monocotyledons in the subclass Commelinidae with emergent or floating stems and leaves and reduced, unisexual flowers having a single ovule in an ovary composed of a single carpel.

Typhlopidae [VERT ZOO] A family of small, burrowing circumtropical snakes, suborder Serpentes, with vestigial eyes and toothless jaws.

Typhloscolecidae [INV ZOO] A family of pelagic polychaete annelids belonging to the Errantia.

typhoid fever [MED] A highly infectious, septicemic disease of humans caused by *Salmonella typhi* which enters the body by the oral route through ingestion of food or water contaminated by contact with fecal matter.

typhoon [METEOROL] A severe tropical cyclone in the western Pacific.

typhoon wind *See* hurricane wind.

typhus fever [MED] Any of three louse-borne human diseases caused by *Rickettsia prowazakii* characterized by fever, stupor, headaches, and a dark-red rash.

typography [GRAPHICS] The techniques involved in letterpress printing, including style, arrangements, and appearance of the printed matter.

Typotheria [PALEON] A suborder of extinct rodent-like herbivores in the order Notoungulata.

Tyranni [VERT ZOO] A suborder of suboscine Passeriformes containing birds with limited song power and having the tendon of the hind toe separate and the intrinsic muscles of the syrinx reduced to one pair.

Tyrannidae [VERT ZOO] The tyrant flycatchers, a family of passeriform birds in the suborder Tyranni confined to the Americas.

Tyrannoidea [VERT ZOO] The flycatchers, a superfamily of suboscine birds in the suborder Tyranni.

tyrolite *See* trichalcite.

tyrosine [BIOCHEM] $C_9H_{11}NO_3$ A phenolic alpha amino acid found in many proteins; a precursor of the hormones epinephrine, norepinephrine, thyroxine, and triiodothyronine, and of the black pigment melanin.

Tytonidae [VERT ZOO] The barn owls, a family of birds in the order Strigiformes distinguished by an unnotched sternum which is fused to large clavicles.

tyuyamunite [MINERAL] $Ca(UO_2)_2(VO_4)_2 \cdot 5-8H_2O$ A yellow orthorhombic mineral occurring in incrustations as a secondary mineral; an ore of uranium. Also known as calciocarnotite.

U

u *See* up quark.

U *See* uranium.

U center [CRYSTAL] The color-center type of point lattice defect in ionic crystals created by the incorporation of an impurity such as hydrogen into alkali halides.

Uda antenna *See* Yagi-Uda antenna.

Udalf [GEOL] A suborder of the soil order Alfisol; brown soil formed in a udic moisture regime and in a mesic or warmer temperature regime.

Udert [GEOL] A suborder of the soil order Vertisol; formed in a humid region so that surface cracks remain open only for 2–3 months.

Udoll [GEOL] A suborder of the Mollisol soil order; found in humid, temperate, and warm regions where maximum rainfall comes during growing season; has thick, very dark A horizons, brown B horizons, and paler C horizons.

udometer *See* rain gage.

Udult [GEOL] A suborder of the soil order Ultisol; organic-carbon content is low, argillic horizons are reddish or yellowish; formed in a udic moisture regime.

U format [COMPUT SCI] A record format which the input/output control system treats as completely unknown and unpredictable.

U Geminorum stars [ASTRON] A class of variable stars known as dwarf novae; their light curves resemble those of novae, with range brightness variations of about 4 magnitudes; examples are U Gemini and SS Cygni. Also known as SS Cygni stars.

UHF *See* ultra-high-frequency.

uhligite [MINERAL] A black, pseudoisometric mineral consisting of an oxide of titanium and calcium, with zirconium and aluminum replacing titanium.

Uintatheriidae [PALEON] The single family of the extinct mammalian order Dinocerata.

Uintatheriinae [PALEON] A subfamily of extinct herbivores in the family Uintatheriidae including all horned forms.

UJT *See* unijunction transistor.

Ulatisian [GEOL] A mammalian age in a local stage classification of the Eocene in use on the Pacific Coast based on foraminifers.

ulcer [MED] Localized interruption of the continuity of an epithelial surface, with an inflamed base.

ulexite [MINERAL] $NaCaB_5O_9 \cdot 8H_2O$ A white mineral that crystallizes in the triclinic system and forms rounded reniform masses of extremely fine acicular crystals. Also known as boronatrocalcite; cotton ball; natronborocalcite.

ullmannite [MINERAL] NiSbS A steel-gray to black mineral consisting of nickel antimonide and sulfide, usually with a little arsenic, occurring massive, and having a metallic luster. Also known as nickel-antimony glance.

Ulmaceae [BOT] A family of dicotyledonous trees in the order Urticales distinguished by alternate stipulate leaves, two styles, a pendulous ovule, and lack of a latex system.

ulmic acid *See* ulmin.

ulmin [GEOL] Alkali-soluble organic substances derived from decaying vegetable matter; occurs as amorphous brown to black gel material. Also known as carbohumin; fundamental jelly; fundamental substance; gelose; humin; humogelite; jelly; ulmic acid; vegetable jelly.

ulna [ANAT] The larger of the two bones of the forearm or forelimb in vertebrates; articulates proximally with the humerus and radius and distally with the radius.

Ulotrichaceae [BOT] A family of green algae in the suborder Ulotrichineae; contains both attached and floating filamentous species with cells having parietal, platelike or bandlike chloroplasts.

Ulotrichales [BOT] A large, artificial order of the Chlorophyta composed mostly of fresh-water, branched or unbranched filamentous species with mostly cylindrical, uninucleate cells having cellulose, but often mucilaginous walls.

Ulotrichineae [BOT] A suborder of the Ulotrichales characterized by short cylindrical cells.

ulrichite *See* uraninite.

ultimate-load design [DES ENG] Design of a beam that is proportioned to carry at ultimate capacity the design load multiplied by a safety factor. Also known as limit-load design; plastic design; ultimate-strength design.

ultimate recovery [PETRO ENG] Estimated total (ultimate) recovery of hydrocarbon fluids expected from a reservoir during its productive lifetime.

ultimate strength [MECH] The tensile stress, per unit of the original surface area, at which a body will fracture, or continue to deform under a decreasing load.

ultimate-strength design *See* ultimate-load design.

Ultisol [GEOL] A soil order characterized by typically moist soils, with horizons of clay accumulation and a low supply of bases.

ultrabasic [PETR] Of igneous rock, having a low silica content, as opposed to the higher silica contents of acidic, basic, and intermediate rocks.

ultrabasite *See* diaphorite.

ultracentrifuge [ENG] A laboratory centrifuge that develops centrifugal fields of more than 100,000 times gravity.

ultra-cold neutron [PHYS] A neutron whose energy is of the order of 10^{-7} electron volt or less, so that it is totally reflected from various materials and suitably constructed magnetic fields, regardless of the angle of incidence, and can be stored in suitably constructed bottles.

ultrafiltration [CHEM ENG] Separation of colloidal or very fine solid materials by filtration through microporous or semipermeable mediums.

ultra-high-frequency [COMMUN] The band of frequencies between 300 and 3000 megahertz in the radio spectrum, corresponding to wavelengths of 10 centimeters to 1 meter. Abbreviated UHF.

ultramafic [PETR] Referring to igneous rock composed principally of mafic minerals, such as olivine and pyroxene.

ultrasonic [ACOUS] Pertaining to signals, equipment, or phenomena involving frequencies just above the range of human hearing, hence above about 20,000 hertz. Also known as supersonic (deprecated usage).

ultrasonic imaging *See* acoustic imaging.

ultraviolet absorption spectrophotometry [SPECT] The study of the spectra produced by the absorption of ultraviolet radiant energy during the transformation of an electron from the ground state to an excited state as a function of the wavelength causing the transformation.

ultraviolet astronomy [ASTRON] The observation of astronomical phenomena in the ultraviolet spectrum.

ultraviolet lamp [ELECTR] A lamp providing a high proportion of ultraviolet radiation, such as various forms of mercury-vapor lamps.

ultraviolet light *See* ultraviolet radiation.

ultraviolet photoemission spectroscopy [SPECT] A spectroscopic technique in which photons in the energy range 10–200 electronvolts bombard a surface and the energy spectrum of the emitted electrons gives information about the states of electrons in atoms and chemical bonding. Abbreviated UPS.

ultraviolet radiation [ELECTROMAG] Electromagnetic radiation in the wavelength range 4–400 nanometers; this range begins at the short-wavelength limit of visible light and overlaps the wavelengths of long x-rays (some scientists place the lower limit at higher values, up to 40 nanometers). Also known as ultraviolet light.

ultraviolet telescope [OPTICS] An assemblage of mirrors, with special coatings imparting high ultraviolet reflectivity, which forms magnified ultraviolet images of objects in the same manner as an optical telescope forms images in visible light.

Ulvaceae [BOT] A large family of green algae in the order Ulvales.

Ulvales [BOT] An order of algae in the division Chlorophyta in which the thalli are macroscopic, attached tubes or sheets.

umangite [MINERAL] Cu_3Se_2 A dark cherry red mineral consisting of copper selenide; occurs in massive form, in small grains or fine granular aggregates.

umbel [BOT] An indeterminate inflorescence with the pedicels all arising at the top of the peduncle and radiating like umbrella ribs; there are two types, simple and compound.

Umbellales [BOT] An order of dicotyledonous herbs or woody plants in the subclass Rosidae with mostly compound or conspicuously lobed or dissected leaves, well-developed schizogenous secretory canals, separate petals, and an inferior ovary.

Umbelliferae [BOT] A large family of aromatic dicotyledonous herbs in the order Umbellales; flowers have an ovary of two carpels, ripening to form a dry fruit that splits into two halves, each containing a single seed.

umbilical artery [EMBRYO] Either of a pair of arteries passing through the umbilical cord to carry im-

pure blood from the mammalian fetus to the placenta.

umbilical cord [AERO ENG] Any of the servicing electrical or fluid lines between the ground or a tower and an uprighted rocket vehicle before the launch. Also known as umbilical. [EMBRYO] The long, cylindrical structure containing the umbilical arteries and vein, and connecting the fetus with the placenta.

umbilical vein [EMBRYO] A vein passing through the umbilical cord and conveying purified, nutrient-rich blood from placenta to fetus.

Umbilicariaceae [BOT] The rock tripes, a family of Ascolichens in the order Lecanorales having a large, circular, umbilicate thallus.

umbilicus [ANAT] The navel; the round, depressed cicatrix in the median line of the abdomen, marking the site of the aperture through which passed the fetal umbilical vessels.

umbo [ANAT] A rounded elevation of the surface of the tympanic membrane. [INV ZOO] A prominence above the hinge of a bivalve mollusk shell.

umbra [ASTRON] The dark, central region of a sunspot. [OPTICS] That portion of a shadow which is screened from light rays emanating from any part of an extended source.

umbrella antenna [ELECTROMAG] Antenna in which the wires are guyed downward in all directions from a central pole or tower to the ground, somewhat like the ribs of an open umbrella.

Umbrept [GEOL] A suborder of the Inceptisol soil order; has dark A horizon more than 10 inches (25 centimeters) thick, brown B horizons, and slightly paler C horizons; soil is strongly acid, and clay minerals are crystalline; occurs in cool or temperate climates.

Umbriel [ASTRON] One of the five satellites of the planet Uranus, with a diameter of about 400 kilometers.

Umklapp process [SOLID STATE] The interaction of three or more waves in a solid, such as lattice waves or electron waves, in which the sum of the wave vectors is not equal to zero but, rather, is equal to a vector in the reciprocal lattice. Also known as flipover process.

unbalanced output [ELEC] An output in which one of the two input terminals is substantially at ground potential.

unbounded wave [PHYS] A wave which propagates through a nondissipative, homogeneous medium which is infinite in extent, without any boundaries.

uncertainty principle [QUANT MECH] The precept that the accurate measurement of an observable quantity necessarily produces uncertainties in one's knowledge of the values of other observables. Also known as Heisenberg uncertainty principle; indeterminacy principle.

uncertainty relation [QUANT MECH] The relation whereby, if one simultaneously measures values of two canonically conjugate variables, such as position and momentum, the product of the uncertainties of their measured values cannot be less than approximately Planck's constant divided by 2π. Also known as Heisenberg uncertainty relation.

unconcentrated wash *See* sheet erosion.

unconditional convergence [MATH] A convergent series converges unconditionally if every series obtained by rearranging its terms also converges; equivalent to absolute convergence.

unconditional inequality [MATH] An inequality which holds true for all values of the variables involved, or which contains no variables; for example,

$y + 2 > y$, or $4 > 3$. Also known as absolute inequality.

unconditional jump [COMPUT SCI] A digital-computer instruction that interrupts the normal process of obtaining instructions in an ordered sequence, and specifies the address from which the next instruction must be taken. Also known as unconditional transfer.

unconditional transfer *See* unconditional jump.

unconformity [GEOL] The relation between adjacent rock strata whose time of deposition was separated by a period of nondeposition or of erosion; a break in a stratigraphic sequence.

unctuous [MATER] Greasy, oily, or soapy to the touch.

undamped wave [PHYS] A continuous wave produced by oscillations having constant amplitude.

undercurrent [OCEANOGR] A water current flowing beneath a surface current at a different speed or in a different direction.

undercut [ENG] Underside recess either cut or molded into an object so as to leave a topside lip or protuberance. [MET] An unfilled groove melted into the base metal at the toe of a weld. [MIN ENG] To cut below or in the lower part of a coal bed by chipping away the coal with a pick or mining machine; cutting is usually done on the level of the floor of the mine, extending laterally the entire face and 5 or 6 feet (1.5 or 1.8 meters) into the material.

undercutting [CHEM ENG] In distillation, the technique of taking the products coming off the distillation tower at a temperature below the desired ultimate boiling point range to prevent contaminating the products with the compound that would distill just beyond the ultimate boiling point range. [GEOL] Erosion of material at the base of a steep slope, cliff, or other exposed rock.

underflow [COMPUT SCI] The generation of a result whose value is smaller than the smallest quantity that can be represented or stored by a computer.

underground gasification *See* gasification.

underground ice *See* ground ice.

underhand stoping [MIN ENG] Mining downward or from upper to lower level; the stope may start below the floor of a level and be extended by successive horizontal slices, either worked sequentially or simultaneously in a series of steps; the stope may be left as an open stope or supported by stulls or pillars.

underhead crack [MET] A subsurface crack in the heat-affected zone of the base metal near a weld.

underhung crane [MECH ENG] An overhead traveling crane in which the end trucks carry the bridge suspended below the rails.

underpinning [CIV ENG] 1. Permanent supports replacing or reinforcing the older supports beneath a wall or a column. 2. Braced props temporarily supporting a structure. [MIN ENG] Building up the wall of a mine shaft to join that above it.

underthrust [GEOL] A thrust fault in which the lower, active rock mass has been moved under the upper, passive rock mass.

undertow [OCEANOGR] A subsurface seaward movement by gravity flow of water carried up on a sloping beach by waves or breakers.

undervoltage protection [ELEC] An undervoltage relay which removes a motor from service when a low-voltage condition develops, so that the motor will not draw excessive current, or which prevents a large induction or synchronous motor from starting under low-voltage conditions.

undistorted wave [COMMUN] Periodic wave in which both the attenuation and velocity of propagation are the same for all sinusoidal components, and in which no sinusoidal component is present at one point that is not present at all points.

undulant fever *See* brucellosis.

undulatory extinction [OPTICS] Extinction that occurs successively in adjacent areas as the microscope stage is turned. Also known as oscillatory extinction; strain shadow; wavy extinction.

undulatus *See* billow cloud.

ungemachite [MINERAL] $K_3Na_9Fe(SO_4)_6(OH)_3 \cdot 9H_2O$ A colorless to pale yellow, hexagonal mineral consisting of a hydrated basic sulfate of potassium, sodium, and iron; occurs in tabular form.

ungulate [VERT ZOO] Referring to an animal that has hoofs.

uniaxial crystal [OPTICS] A doubly refracting crystal which has a single axis along which light can propagate without exhibiting double refraction.

uniaxial stress [MECH] A state of stress in which two of the three principal stresses are zero.

unidirectional [PHYS] 1. Flowing in only one direction, such as direct current. 2. Radiating in only one direction.

unified field theory [RELAT] Any theory which attempts to express gravitational theory and electromagnetic theory within a single unified framework; usually, an attempt to generalize Einstein's general theory of relativity from a theory of gravitation alone to a theory of gravitation and classical electromagnetism.

uniform circular motion [MECH] Circular motion in which the angular velocity remains constant.

uniform distribution [STAT] The distribution of a random variable in which each value has the same probability of occurrence. Also known as rectangular distribution.

uniform field [PHYS] A field which, at the instant under consideration, has the same value at every point in the region under consideration.

uniform load [MECH] A load distributed uniformly over a portion or over the entire length of a beam; measured in pounds per foot.

uniformly accessible storage *See* random-access memory.

uniform plane wave [ELECTROMAG] Plane wave in which the electric and magnetic intensities have constant amplitude over the equiphase surfaces; such a wave can only be found in free space at an infinite distance from the source.

uniform space [MATH] A topological space X whose topology is derived from a family of subsets of $X \times X$, called a uniformity; intuitively, this gives a notion of "nearness" which is uniform throughout the space.

unijunction transistor [ELECTR] An n-type bar of semiconductor with a p-type alloy region on one side; connections are made to base contacts at either end of the bar and to the p-region. Abbreviated UJT. Formerly known as double-base diode; double-base junction diode.

unimolecular reaction [PHYS CHEM] A chemical reaction involving only one molecular species as a reactant; for example, $2H_2O \rightarrow 2H_2 + O_2$, as in the electrolytic dissociation of water.

uninterruptible power system [ELEC] A system that provides protection against primary alternating-current power failure and variations in power-line frequency and voltage. Abbreviated UPS.

union [DES ENG] A screwed or flanged pipe coupling usually in the form of a ring fitting around the outside of the joint. [MATH] A union of a given family of sets is a set consisting of those elements which are members of at least one set in the family.

Unionidae [INV ZOO] The fresh-water mussels, a family of bivalve mollusks in the subclass Eulamellibranchia; the larvae, known as glochidia, are parasitic on fish.

union joint [DES ENG] A threaded assembly used for the joining of ends of lengths of installed pipe or tubing where rotation of neither length is feasible.

unipolar [ELEC] Having but one pole, polarity, or direction; when applied to amplifiers or power supplies, it means that the output can vary in only one polarity from zero and, therefore, must always contain a direct-current component.

Unipolarina [INV ZOO] A suborder of the protozoan order Myxosporida characterized by spores with one to six (never five) polar capsules located at the anterior end.

unipolar machine See homopolar generator.

unipole [ELECTROMAG] A hypothetical antenna that radiates or receives signals equally well in all directions. Also known as isotropic antenna.

unitary spin [PARTIC PHYS] A quantum number associated with SU_3 symmetry and which determines the SU_3 supermultiplet to which a particle belongs, such as singlet, octet, or decuplet.

unitary symmetry [PARTIC PHYS] An approximate symmetry law obeyed by the strong interactions of elementary particles; it may be described as the equivalence of three fundamental particles, termed quarks, out of which all hadrons could be assumed to be composed. Also known as SU_3 symmetry.

unitary transformation [MATH] A linear transformation on a vector space which preserves inner products and norms; alternatively, a linear operator whose adjoint is equal to its inverse.

unit assembly [IND ENG] Assemblage of machine parts which constitutes a complete auxiliary part of an end item, and which performs a specific auxiliary function, and which may be removed from the parent item without itself being disassembled.

unit cell [CRYSTAL] A parallelepiped which will fill all space under the action of translations which leave the crystal lattice unchanged. Also known as structure cell. [MIN ENG] In flotation, a single cell.

unit charge See statcoulomb.

uniterm [COMPUT SCI] A word, symbol, or number used as a description for retrieval of information from a collection; especially, such a description used in a coordinate indexing system.

unitized body [ENG] An automotive body that has the body and frame in one unit; side members are designed on the principle of a bridge truss to gain stiffness, and sheet metal of the body is stressed so that it carries some of the load.

unit operations [CHEM ENG] The basic physical operations of chemical engineering in a chemical process plant, that is, distillation, fluid transport, heat and mass transfer, evaporation, extraction, drying, crystallization, filtration, mixing, size separation, crushing and grinding, and conveying.

unit operator [MATH] The identity operator.

unit pulse See baud.

unit record [COMPUT SCI] Any of a collection of records, all of which have the same form and the same data elements.

unit strain [MECH] 1. For tensile strain, the elongation per unit length. 2. For compressive strain, the shortening per unit length. 3. For shear strain, the change in angle between two lines originally perpendicular to each other.

unit systems [PHYS] Groups of units suitable for use in measurement of physical quantities and in the convenient statement of physical laws relating physical quantities.

universal algebra [MATH] The study of algebraic systems such as groups, rings, modules, and fields and the examination of what families of theorems are analogous in each system.

universal donor [IMMUNOL] An individual of O blood group; can give blood to persons of all blood types.

universal instrument See altazimuth.

universality [STAT MECH] The hypothesis that the critical exponents of a substance are the same within broad classes of substances of widely varying characteristics, and depend only on the microscopic symmetry properties of the substance.

universal joint [MECH ENG] A linkage that transmits rotation between two shafts whose axes are coplanar but not coinciding.

universal motor [ELEC] A motor that may be operated at approximately the same speed and output on either direct current or single-phase alternating current. Also known as ac/dc motor.

universal output transformer [ENG ACOUS] An output transformer having a number of taps on its winding, to permit its use between the audio-frequency output stage and the loudspeaker of practically any radio receiver by proper choice of connections.

universal product code [COMPUT SCI] 1. A 10-digit bar code on the outside of a package for electronic scanning at supermarket checkout counters; each digit is represented by the ratio of the widths of adjacent stripes and white areas. 2. The corresponding combinations of binary digits into which the scanned bars are converted for computer processing that provides continuously updated inventory data and printout of the register tape at the checkout counter.

universal recipient [IMMUNOL] An individual of AB blood group; can receive a blood transfusion of all blood types, A, AB, B, or O.

universal shunt See Ayrton shunt.

universal time See Greenwich mean time.

universal wavelength function [OPTICS] One of four functions which enable one to compute easily, with reasonable accuracy, the refractive index of glass or other transparent material when this index is known for four standard wavelengths.

universe [ASTRON] The totality of astronomical things, events, relations, and energies capable of being described objectively.

univibrator See monostable multivibrator.

unloading circuit [COMPUT SCI] In an analog computer, a computing element or combination of computing elements capable of reproducing or amplifying a given voltage signal while drawing negligible current from the voltage source.

unmodified instruction See basic instruction.

unrelated frequencies [STAT] The long run frequency of any result in one part of an experiment is approximately equal to the long run conditional frequency of that result, given that any specified result has occurred in the other part of the experiment.

unrestricted visibility [METEOROL] The visibility when no obstruction to vision exists in sufficient quantity to reduce the visibility to less than 7 miles (11.3 kilometers).

uns-, unsym- [ORG CHEM] A chemical prefix denoting that the substituents of an organic compound are structurally unsymmetrical with respect to the carbon skeleton, or with respect to a function group (for example, double or triple bond).

unsaturated compound [CHEM] Any chemical compound with more than one bond between adjacent atoms, usually carbon, and thus reactive toward the addition of other atoms at that point; for example, olefins, diolefins, and unsaturated fatty acids.

unsaturated hydrocarbon [ORG CHEM] One of a class of hydrocarbons that have at least one double or triple carbon-to-carbon bond that is not in an aromatic ring; examples are ethylene, propadiene, and acetylene.

unsaturated zone *See* zone of aeration.

unsprung axle [MECH ENG] A rear axle in an automobile in which the housing carries the right and left rear-axle shafts and the wheels are mounted at the outer end of each shaft.

unsprung weight [MECH ENG] The weight of the various parts of a vehicle that are not carried on the springs such as wheels, axles, brakes, and so forth.

unstable [PHYS] Capable of undergoing spontaneous change, as in a radioactive nuclide or an excited nuclear system.

unstable colon *See* irritable colon.

unstable isotope *See* radioisotope.

unsteady flow [FL MECH] Fluid flow in which properties of the flow change with respect to time.

untuned [ELEC] Not resonant at any of the frequencies being handled.

up-converter [ELECTR] Type of parametric amplifier which is characterized by the frequency of the output signal being greater than the frequency of the input signal.

updraft carburetor [MECH ENG] For a gasoline engine, a fuel-air mixing device in which both the fuel jet and the airflow are upward.

updraft furnace [MECH ENG] A furnace in which volumes of air are supplied from below the fuel bed or supply.

upper-air observation [METEOROL] A measurement of atmospheric conditions aloft, above the effective range of a surface weather observation. Also known as sounding; upper-air sounding.

upper-air sounding *See* upper-air observation.

upper atmosphere [METEOROL] The general term applied to the atmosphere above the troposphere.

Upper Cambrian [GEOL] The latest epoch of the Cambrian period of geologic time, beginning approximately 510 million years ago.

Upper Carboniferous [GEOL] The European epoch of geologic time equivalent to the Pennsylvanian of North America.

Upper Cretaceous [GEOL] The late epoch of the Cretaceous period of geologic time, beginning about 90 million years ago.

upper culmination *See* upper transit.

Upper Devonian [GEOL] The latest epoch of the Devonian period of geologic time, beginning about 365 million years ago.

Upper Jurassic [GEOL] The latest epoch of the Jurassic period of geologic time, beginning approximately 155 million years ago.

upper-level winds *See* winds aloft.

upper limb [ASTRON] That half of the outer edge of a celestial body having the greatest altitude.

upper mantle [GEOL] The portion of the mantle lying above a depth of about 1000 kilometers. Also known as outer mantle; peridotite shell.

Upper Mississippian [GEOL] The latest epoch of the Mississippian period of geologic time.

upper mixing layer [METEOROL] The region of the upper mesosphere between about 50 and 80 kilometers (that is, immediately above the mesopeak) through which there is a rapid decrease of temperature with height and where there appears to be considerable turbulence.

Upper Ordovician [GEOL] The latest epoch of the Ordovician period of geologic time, beginning approximately 440 million years ago.

Upper Pennsylvanian [GEOL] The latest epoch of the Pennsylvanian period of geologic time.

Upper Permian [GEOL] The latest epoch of the Permian period of geologic time, beginning about 245 million years ago.

Upper Silurian [GEOL] The latest epoch of the Silurian period of geologic time.

upper transit [ASTRON] The movement of a celestial body across a celestial meridian's upper branch. Also known as superior transit; upper culmination.

Upper Triassic [GEOL] The latest epoch of the Triassic period of geologic time, beginning about 200 million years ago.

upper winds *See* winds aloft.

up quark [PARTIC PHYS] A quark with an electric charge of $+2/3$, baryon number of $1/3$, and 0 strangeness and charm. Symbolized u.

UPS [ELEC] *See* uninterruptible power system. [SPECT] *See* ultraviolet photoemission spectroscopy.

upset welding [MET] Pressure butt welding in which heat is generated by resistance to the passage of current across the joint.

upstream [CHEM ENG] That portion of a process stream that has not yet entered the system or unit under consideration; for example, upstream to a refinery or to a distillation column. [HYD] Toward the source of a stream.

upthrow [GEOL] **1.** The fault side that has been thrown upward. **2.** The amount of vertical fault displacement.

Upupidae [VERT ZOO] The Old World hoopoes, a family of birds in the order Coraciiformes whose young are hatched with sparse down.

upward compatibility [COMPUT SCI] The ability of a newer or larger computer to accept programs from an older or smaller one.

upwelling [OCEANOGR] The process by which water rises from a deeper to a shallower depth, usually as a result of divergence of offshore currents.

upwind [METEOROL] In the direction from which the wind is flowing.

uracil [BIOCHEM] $C_4H_4N_2O_2$ A pyrimidine base important as a component of ribonucleic acid.

Uralean [GEOL] A stage of geologic time in Russia: uppermost Carboniferous (above Gzhelian, below Sakmarian of Permian).

uralite [MINERAL] A green variety of secondary amphibole; it is usually fibrous or acicular and is formed by alteration of pyroxene.

Uraniidae [INV ZOO] A tropical family of moths in the superfamily Geometroidea including some slender-bodied, brilliantly colored diurnal insects which lack a frenulum and are often mistaken for butterflies.

uraninite [MINERAL] UO_2 A black, brownish-black, or dark-brown radioactive mineral that is isometric in crystallization; often contains impurities such as thorium, radium, cerium, and yttrium metals, and lead; the chief ore of uranium; hardness is 5.5–6 on Mohs scale, and specific gravity of pure UO_2 is 10.9, but that of most natural material is 9.7–7.5. Also known as coracite; ulrichite.

uranium [CHEM] A metallic element in the actinide series, symbol U, atomic number 92, atomic weight 238.03; highly toxic and radioactive; ignites spontaneously in air and reacts with nearly all nonmetals;

melts at 1132°C, boils at 3818°C; used in nuclear fuel and as the source of uranium-235 and plutonium. [MET] A dense, silvery, ductile, strongly electropositive metal.

uranium age [GEOL] The age of a mineral as calculated from the numbers of ionium atoms present originally, now, and when equilibrium is established with uranium.

uranium enrichment [NUCLEO] A process carried out on uranium, in which the ratio of the abundance of the isotope uranium-235 to that of the isotope uranium-238 is increased above that found in natural uranium.

uranium-lead dating [GEOL] A method for calculating the geologic age of a material in years based on the radioactive decay rate of uranium-238 to lead-206 and of uranium-235 to lead-207.

uranium ocher *See* gummite.

uran-mica *See* torbernite.

uranopilite [MINERAL] $(UO_2)_6(SO_4)(OH)_{10} \cdot 12H_2O$ A bright yellow, lemon yellow, or golden yellow, monoclinic mineral consisting of a hydrated basic sulfate of uranium; occurs as encrustations and masses.

uranosphaerite [MINERAL] $Bi_2O_3 \cdot 2UO_3 \cdot 3H_2O$ An orange-yellow or brick red, orthorhombic mineral consisting of a hydrated oxide of bismuth and uranium.

uranospinite [MINERAL] $Ca(UO_2)_2(AsO_4)_2 \cdot 8H_2O$ A lemon yellow to siskin green, tetragonal mineral consisting of a hydrated arsenate of calcium and uranium; occurs in tabular form.

uranotantalite *See* samarskite.

Uranus [ASTRON] A planet, seventh in the order of distance from the sun; it has five known satellites, and its equatorial diameter is about four times that of the earth.

urao *See* trona.

urease [BIOCHEM] An enzyme that catalyzes the degradation of urea to ammonia and carbon dioxide; obtained from the seed of jack bean.

Urechinidae [INV ZOO] A family of echinoderms in the order Holasteroida which have an ovoid test lacking a marginal fasciole.

Uredinales [MYCOL] An order of parasitic fungi of the subclass Heterobasidiomycetidae characterized by the teleutospore, a spore with one or more cells, each of which is a modified hypobasidium; members cause plant diseases known as rusts.

uremia [MED] A condition resulting from kidney failure and characterized by azotemia, chronic acidosis, anemia, and a variety of systemic signs and symptoms.

ureter [ANAT] A long tube conveying urine from the renal pelvis to the urinary bladder or cloaca in vertebrates.

urethra [ANAT] The canal in most mammals through which urine is discharged from the urinary bladder to the outside.

urinary system [ANAT] The system which functions in the elaboration and excretion of urine in vertebrates; in man and most mammals, consists of the kidneys, ureters, urinary bladder, and urethra.

uriniferous tubule [ANAT] One of the numerous winding tubules of the kidney. Also known as nephric tubule.

urobilin [BIOCHEM] A bile pigment produced by reduction of bilirubin by intestinal bacteria and excreted by the kidneys or removed by the liver.

urobilinogen [BIOCHEM] A chromogen, formed in feces and present in urine, from which urobilin is formed by oxidation.

Urochordata [INV ZOO] The equivalent name for Tunicata.

Urodela [VERT ZOO] The tailed amphibians or salamanders, an order of the class Amphibia distinguished superficially from frogs and toads by the possession of a tail, and from caecilians by the possession of limbs.

urogenital system [ANAT] The combined urinary and genital system in vertebrates, which are intimately related embryologically and anatomically. Also known as genitourinary system.

urology [MED] The scientific study of urine and the diseases and abnormalities of the urinary and urogenital tracts.

Uropygi [INV ZOO] The tailed whip scorpions, an order of arachnids characterized by an elongate, flattened body which bears in front a pair of thickened, raptorial pedipalps set with sharp spines and used to hold and crush insect prey.

uropygial gland [VERT ZOO] A relatively large, compact, bilobed, secretory organ located at the base of the tail (uropygium) of most birds having a keeled sternum. Also known as oil gland; preen gland.

Urostylidae [INV ZOO] A family of hemipteran insects in the superfamily Pentatomoidea.

Ursa Major [ASTRON] A northern constellation, right ascension 11 hours, declination 50°N; it contains a group of seven stars known as the Big Dipper.

Ursa Minor [ASTRON] A northern constellation, right ascension 15 hours, declination 70°N; its brightest star, Polaris, is almost at the north celestial pole; seven of the eight stars form a dipper outline. Also known as Little Bear; Little Dipper.

Ursidae [VERT ZOO] A family of mammals in the order Carnivora including the bears and their allies.

Urticaceae [BOT] A family of dicotyledonous herbs in the order Urticales characterized by a single unbranched style, a straight embryo, and the lack of milky juice (latex).

Urticales [BOT] An order of dicotyledons in the subclass Hamamelidae; woody plants or herbs with simple, usually stipulate leaves, and reduced clustered flowers that usually have a vestigial perianth.

urticaria [MED] Hives or nettle rash; a skin condition characterized by the appearance of intensely itching wheals or welts with elevated, usually white centers and a surrounding area of erythema. Also known as hives.

user friendly [COMPUT SCI] Property of a computer system that is easy to use and sets up an easily understood dialog between the user and the computer.

user group [COMPUT SCI] An organization of users of the computers of a particular vendor, which shares information and ideas, and may develop system software and influence vendors to change their products.

user mode [COMPUT SCI] The mode of operation exercised by the user programs of a computer system in which there is a class of privileged instructions that is not permitted, since these can be executed only by the operating system or executive system. Also known as slave mode.

Usneaceae [BOT] The beard lichens, a family of Ascolichenes in the order Lecanorales distinguished by their conspicuous fruticose growth form.

Ustalf [GEOL] A suborder of the soil order Alfisol; red or brown soil formed in a ustic moisture regime and in a mesic or warmer temperature regime.

Ustert [GEOL] A suborder of the Vertisol soil order; has a faint horizon and is dry for an appreciable period or more than one period of the year.

Ustilaginaceae [MYCOL] A family of fungi in the order Ustilaginales in which basidiospores bud from the sides of the septate epibasidium.

Ustilaginales [MYCOL] An order of the subclass Heterobasidiomycetidae comprising the smut fungi which parasitize plants and cause diseases known as smut or bunt.

Ustoll [GEOL] A suborder of the soil order Mollisol; formed in a ustic moisture regime and in a mesic or warmer temperature regime; may have a calcic, petrocalcic, or gypsic horizon.

Ustox [GEOL] A suborder of the soil order Oxisol that is low to moderate in organic matter, well drained, and dry for at least 90 cumulative days each year.

Ustult [GEOL] A suborder of the soil order Ultisol; brownish or reddish, with low to moderate organic-carbon content; a well-drained soil of warm-temperate and tropical climates with moderate or low rainfall.

utahite *See* jarosite.

uterus [ANAT] The organ of gestation in mammals which receives and retains the fertilized ovum, holds the fetus during development, and becomes the principal agent of its expulsion at term.

utility routine [COMPUT SCI] A program or routine of general usefulness, usually not very complicated, and applicable to many jobs or purposes.

utricle *See* utriculus.

utriculus [ANAT] 1. That part of the membranous labyrinth of the ear into which the semicircular canals open. 2. A small, blind pouch extending from the urethra into the prostate. Also known as utricle.

U-tube heat exchanger [CHEM ENG] A heat-exchanger system consisting of a bundle of U tubes (hairpin tubes) surrounded by a shell (outer vessel); one fluid flows through the tubes, and the other fluid flows through the shell, around the tubes.

U-tube manometer [ENG] A manometer consisting of a U-shaped glass tube partly filled with a liquid of known specific gravity; when the legs of the manometer are connected to separate sources of pressure, the liquid rises in one leg and drops in the other; the difference between the levels is proportional to the difference in pressures and inversely proportional to the liquid's specific gravity. Also known as liquid-column gage.

uvanite [MINERAL] $U_2V_6O_{21} \cdot 15H_2O$ A brownish-yellow, orthorhombic mineral consisting of a hydrated uranium vanadate; occurs as crystalline masses and coatings.

uvarovite [MINERAL] $Ca_3Cr_2(SiO_4)_3$ The emerald-green, calcium-chromium end member of the garnet group. Also known as ouvarovite; uwarowite.

UV Ceti stars [ASTRON] A class of stars that have brief outbursts of energy over their surface areas; they may have an increase of about 1 magnitude for periods of 1 hour; the type star is UV Ceti. Also known as flare stars.

uvea [ANAT] The pigmented, vascular layer of the eye: the iris, ciliary body, and choroid.

uvula [ANAT] 1. A fingerlike projection in the midline of the posterior border of the soft palate. 2. A lobe of the vermiform process of the lower surface of the cerebellum.

uwarowite *See* uvarovite.

V

V *See* vanadium; volt.

VA *See* volt-ampere.

vacancy [SOLID STATE] A defect in the form of an unoccupied lattice position in a crystal.

vaccination [IMMUNOL] Inoculation of viral or bacterial organisms or antigens to produce immunity in the recipient.

vaccine [IMMUNOL] A suspension of killed or attenuated bacteria or viruses or fractions thereof, injected to produce active immunity.

vacuole [CYTOL] A membrane-bound cavity within a cell; may function in digestion, storage, secretion, or excretion. [GEOL] *See* vesicle.

vacuum [PHYS] **1.** Theoretically, a space in which there is no matter. **2.** Practically, a space in which the pressure is far below normal atmospheric pressure so that the remaining gases do not affect processes being carried on in the space.

vacuum brake [MECH ENG] A form of air brake which operates by maintaining low pressure in the actuating cylinder; braking action is produced by opening one side of the cylinder to the atmosphere so that atmospheric pressure, aided in some designs by gravity, applies the brake.

vacuum capacitor [ELEC] A capacitor with separated metal plates or cylinders mounted in an evacuated glass envelope to obtain a high breakdown voltage rating.

vacuum circuit breaker [ELEC] A circuit breaker in which a pair of contacts is hermetically sealed in a vacuum envelope; the contacts are separated by using a bellows to move one of them; an arc is produced by metallic vapor boiled from the electrodes, and is extinguished when the vapor particles condense on solid surfaces.

vacuum diffusion [ELECTR] Diffusion of impurities into a semiconductor material in a continuously pumped hard vacuum.

vacuum distillation [CHEM ENG] Liquid distillation under reduced (less than atmospheric) pressure; used to lower boiling temperatures and lessen the risk of thermal degradation during distillation. Also known as reduced-pressure distillation.

vacuum flashing [CHEM ENG] The heating of a liquid that, upon release to a lower pressure (vacuum), undergoes considerable vaporization (flashing). Also known as flash vaporization.

vacuum forming [ENG] Plastic-sheet forming in which the sheet is clamped to a stationary frame, then heated and drawn down into a mold by vacuum.

vacuum fusion [MET] A technique for determining the oxygen, hydrogen, and sometimes nitrogen content of metals; can be applied to a wide variety of metals with the exception of alkali and alkaline earth metals.

vacuum metallurgy [MET] The melting, shaping, and treating of metals and alloys under reduced pressure that ranges from subatmospheric pressure to ultrahigh vacuum.

vacuum plating [MET] Producing a surface film of metal on a heated surface, often in a vacuum, either by decomposition of the vapor of a compound at the work surface, or by direct reaction between the work surface and the vapor. Also known as vapor deposition.

vacuum polarization [QUANT MECH] A process in which an electromagnetic field gives rise to virtual electron-positron pairs that effectively alter the distribution of charges and currents that generated the original electromagnetic field.

vacuum pump [MECH ENG] A compressor for exhausting air and noncondensable gases from a space that is to be maintained at subatmospheric pressure.

vacuum relay [ELEC] A sensitive relay having its contacts mounted in a highly evacuated glass housing, to permit handling radio-frequency voltages as high as 20,000 volts without flashover between contacts even though contact spacing is but a few hundredths of an inch when open.

vacuum switch [ELEC] A switch having its contacts in an evacuated envelope to minimize sparking.

vacuum tube [ELECTR] An electron tube evacuated to such a degree that its electrical characteristics are essentially unaffected by the presence of residual gas or vapor.

vacuum ultraviolet radiation [ELECTROMAG] Ultraviolet radiation with a wavelength of less than 200 nanometers; absorption of radiation in this region by air and other gases requires the use of evacuated apparatus for transmission. Also known as extreme ultraviolet radiation.

vadose zone *See* zone of aeration.

vagina [ANAT] The canal from the vulvar opening to the cervix uteri.

vaginitis [MED] **1.** Inflammation of the vagina. **2.** Inflammation of a tendon sheath.

vagus [ANAT] The tenth cranial nerve; either of a pair of sensory and motor nerves forming an important par of the parasympathetic system in vertebrates.

valence [BIOCHEM] The relative ability of a biological substance to react or combine. [CHEM] A positive number that characterizes the combining power of an element for other elements, as measured by the number of bonds to other atoms which one atom of the given element forms upon chemical combination; hydrogen is assigned valence 1, and the valence is the number of hydrogen atoms, or their equivalent, with which an atom of the given element combines. [MATH] The number of lines incident on a specified point of a graph.

valence angle   See bond angle.

valence band   [SOLID STATE] The highest electronic energy band in a semiconductor or insulator which can be filled with electrons.

valence-bond method   [PHYS CHEM] A method of calculating binding energies and other parameters of molecules by taking linear combinations of electronic wave functions, some of which represent covalent structures, others ionic structures; the coefficients in the linear combination are calculated by the variational method. Also known as valence-bond resonance method.

valence-bond resonance method   See valence-bond method.

valence-bond theory   [CHEM] A theory of the structure of chemical compounds according to which the principal requirements for the formation of a covalent bond are a pair of electrons and suitably oriented electron orbitals on each of the atoms being bonded; the geometry of the atoms in the resulting coordination polyhedron is coordinated with the orientation of the orbitals on the central atom.

valence electron   [ATOM PHYS] An electron that belongs to the outermost shell of an atom.   [SOLID STATE] See conduction electron.

valence number   [CHEM] A number that is equal to the valence of an atom or ion multiplied by $+1$ or -1, depending on whether the ion is positive or negative, or equivalently on whether the atom in the molecule under consideration has lost or gained electrons from its free state.

valence shell   [ATOM PHYS] The electrons that form the outermost shell of an atom.

valid program   [COMPUT SCI] A computer program whose statements, individually and together, follow the syntactical rules of the programming language in which it is written, so that they are capable of being translated into a machine language program.

valine   [BIOCHEM] $C_5H_{11}NO_2$ An amino acid considered essential for normal growth of animals, and biosynthesized from pyruvic acid. Also known as 2-aminoisovaleric acid; α-aminoisovaleric acid; 2-amino-3-methylbutyric acid.

valley   [BUILD] An inside angle formed where two sloping sides intersect.   [GEOGR] A generally broad area of flat, low-lying land bordered by higher ground.   [GEOL] A relatively shallow, wide depression of the sea floor with gentle slopes. Also known as submarine valley.

valley glacier   [HYD] A glacier that flows down the walls of a mountain valley.

valley train   [GEOL] A long, narrow body of outwash, deposited by meltwater far beyond the margin of an active glacier and extending along the floor of a valley. Also known as outwash train.

valley wind   [METEOROL] A wind which ascends a mountain valley (up-valley wind) during the day; the daytime component of a mountain and valley wind system.

valuation   [MATH] A scalar function of a field which has properties similar to those of absolute value.

value   [SCI TECH] The magnitude of a quantity.

value-added network   [COMMUN] A communications network that provides not only communications channels but also other services such as automatic error detection and correction, protocol conversions, and store-and-forward message services.

value analysis   See value engineering.

value control   See value engineering.

value engineering   [IND ENG] The systematic application of recognized techniques which identify the function of a product or service, and provide the necessary function reliably at lowest overall cost. Also known as value analysis; value control.

value theory   [SYS ENG] A concept normally associated with decision theory; it strives to evaluate relative utilities of simple and mixed parameters which can be used to describe outcomes.

Valvatacea   [PALEON] A superfamily of extinct gastropod mollusks in the order Prosobranchia.

valvate   [BOT] Having valvelike parts, as those which meet edge to edge or which open as if by valves.

Valvatida   [INV ZOO] An order of echinoderms in the subclass Asteroidea.

Valvatina   [INV ZOO] A suborder of echinoderms in the order Phanerozonida in which the upper marginals lie directly over, and not alternate with, the corresponding lower marginals.

valve   [ANAT] A flat of tissue, as in the veins or between the chambers in the heart, which permits movement of fluid in one direction only.   [BOT] **1.** A segment of a dehiscing capsule or legume. **2.** The lidlike portion of certain anthers.   [ELECTR] See electron tube.   [INV ZOO] **1.** One of the distinct, articulated pieces composing the shell of certain animals, such as barnacles and brachiopods. **2.** One of two shells encasing the body of a bivalve mollusk or a diatom.   [MECH ENG] A device used to regulate the flow of fluids in piping systems and machinery.

valve arrester   [ELEC] A type of lightning arrester which consists of a single gap or multiple gaps in series with current-limiting elements; gaps between spaced electrodes prevent flow of current through the arrester except when the voltage across them exceeds the critical gap flashover.

valve follower   [MECH ENG] A linkage between the cam and the push rod of a valve train.

valve-in-head engine   See overhead-valve engine.

valve lifter   [MECH ENG] A device for opening the valve of a cylinder as in an internal combustion engine.

valve seat   [DES ENG] The circular metal ring on which the valve head of a poppet valve rests when closed.

valve train   [MECH ENG] The valves and valve-operating mechanism for the control of fluid flow to and from a piston-cylinder machine, for example, steam, diesel, or gasoline engine.

Valvifera   [INV ZOO] A suborder of isopod crustaceans distinguished by having a pair of flat, valvelike uropods which hinge laterally and fold inward beneath the rear part of the body.

Vampyrellidae   [INV ZOO] A family of protozoans in the order Proteomyxida including species which invade filamentous algae and sometimes higher plants.

Vampyromorpha   [INV ZOO] An order of dibranchiate cephalopod mollusks represented by *Vampyroteuthis infernalis*, an inhabitant of the deeper waters of tropical and temperate seas.

vanadate   [MINERAL] Any of several mineral compounds characterized by pentavalent vanadium and oxygen in the anion; an example is vanadinite.

vanadinite   [MINERAL] $Pb_5(VO_4)_3Cl$ A red, yellow, or brown opatite mineral often occurring as globular masses encrusting other minerals in lead mines; an ore of vanadium and lead hardness is 2.75–3 on Mohs scale, and specific gravity is 6.66–7.10.

vanadium   [CHEM] A metal in group Vb, symbol V, atomic number 23; soluble in strong acids and alkalies; melts at 1900°C, boils about 3000°C; used as a catalyst.   [MET] A silvery-white, ductile metal resistant to corrosion; used in alloy steels and as an x-ray target.

Van Allen radiation belt [GEOPHYS] One of the belts of intense ionizing radiation in space about the earth formed by high-energy charged particles which are trapped by the geomagnetic field.

Van de Graaff generator [ELECTR] A high-voltage electrostatic generator in which electrical charge is carried from ground to a high-voltage terminal by means of an insulating belt and is discharged onto a large, hollow metal electrode.

vandenbrandite [MINERAL] $CuO \cdot UO_3 \cdot 2H_2O$ A dark green to black mineral consisting of a hydrated oxide of copper and uranium; occurs in small crystals and massive form.

van der Waals attraction *See* van der Waals force.

van der Waals equation [PHYS CHEM] An empirical equation of state which takes into account the finite size of the molecules and the attractive forces between them: $p = [RT/(v - b)] - (a/v^2)$, where p is the pressure, v is the volume per mole, T is the absolute temperature, R is the gas constant, and a and b are constants.

van der Waals force [PHYS CHEM] An attractive force between two atoms or nonpolar molecules, which arises because a fluctuating dipole moment in one molecule induces a dipole moment in the other, and the two dipole moments then interact. Also known as dispersion force; London dispersion force; van der Waals attraction.

vandyke [GRAPHICS] A process used for photocopying; the material is paper-sensitized with ferric iron and silver salts and then exposed to strong light; upon processing, a negative print of white lines on a brown background results; from this a brown-line print can be made.

vane [AERO ENG] A device that projects ahead of an aircraft to sense gusts or other actions of the air so as to create impulses or signals that are transmitted to the control system to stabilize the aircraft. [MECH ENG] A flat or curved surface exposed to a flow of fluid so as to be forced to move or to rotate about an axis, to rechannel the flow, or to act as the impeller; for example, in a steam turbine, propeller fan, or hydraulic turbine. [NAV] A sight on an instrument used for observing bearings, such as on a pelorus or azimuth circle.

vane feather *See* contour feather.

Vaneyellidae [INV ZOO] A family of holothurian echinoderms in the order Dactylochirotida.

Vanhorniidae [INV ZOO] A monospecific family of the Hymenoptera in the superfamily Proctotrupoidea.

vanoxite [MINERAL] $V_4^4V_2^5O_{13} \cdot 8H_2O$ A black mineral consisting of a hydrous oxide of vanadium; occurs as microscopic crystals and in massive form.

V antenna [ELECTROMAG] An antenna having a V-shaped arrangement of conductors fed by a balanced line at the apex; the included angle, length, and elevation of the conductors are proportioned to give the desired directivity. Also spelled vee antenna.

van't Hoff equation [PHYS CHEM] An equation for the variation with temperature T of the equilibrium constant K of a gaseous reaction in terms of the heat of reaction at constant pressure, ΔH: $d(\ln K)/dT = \Delta H/RT^2$, where R is the gas constant. Also known as van't Hoff isochore.

van't Hoff isochore *See* van't Hoff equation.

vanthoffite [MINERAL] $Na_6Mg(SO_4)_4$ A colorless mineral consisting of a sulfate of sodium and magnesium; occurs in massive form.

van't Hoff's law [PHYS] The law that the osmotic pressure of a dissolved substance equals the gas pressure it would exert if it were an ideal gas that occupied the same volume as that of the solution.

Van Vleck equation [QUANT MECH] An equation based on quantum theory for the molar paramagnetism of a magnetically susceptible material from magnetic moment, absolute temperature, and various constants.

vapor [THERMO] A gas at a temperature below the critical temperature, so that it can be liquefied by compression, without lowering the temperature.

vapor barrier [CIV ENG] A layer of material applied to the inner (warm) surface of a concrete wall or floor to prevent absorption and condensation of moisture.

vapor-compression cycle [MECH ENG] A refrigeration cycle in which refrigerant is circulated through a machine which allows for successive boiling (or vaporization) of liquid refrigerant as it passes through an expansion valve, thereby producing a cooling effect in its surroundings, followed by compression of vapor to liquid.

vapor cycle [THERMO] A thermodynamic cycle, operating as a heat engine or a heat pump, during which the working substance is in, or passes through, the vapor state.

vapor deposition *See* vacuum plating.

vaporization *See* volatilization.

vaporization cooling [ENG] Cooling by volatilization of a nonflammable liquid having a low boiling point and high dielectric strength; the liquid is flowed or sprayed on hot electronic equipment in an enclosure where it vaporizes, carrying the heat to the enclosure walls, radiators, or heat exchanger. Also known as evaporative cooling.

vaporizer [CHEM ENG] A process vessel in which a liquid is heated until it vaporizes; heat can be indirect (steam or heat-transfer fluid) or direct (hot gases or submerged combustion).

vapor lamp *See* discharge lamp.

vapor lock [FL MECH] Interruption of the flow of fuel in a gasoline engine caused by formation of vapor or gas bubbles in the fuel-feeding system.

vapor-phase-epitaxy [CRYSTAL] A crystal growth process whereby an element or a compound is deposited as a thin layer on a slice of substrate single-crystal material by the vapor-phase technique.

vapor-phase reactor [CHEM ENG] A heavy steel vessel for carrying out chemical reactions on an industrial scale where efficient control over a vapor phase is needed, for example, in an oxidation process.

vapor pressure [METEOROL] The partial pressure of water vapor in the atmosphere. [THERMO] For a liquid or solid, the pressure of the vapor in equilibrium with the liquid or solid.

varactor [ELECTR] A semiconductor device characterized by a voltage-sensitive capacitance that resides in the space-charge region at the surface of a semiconductor bounded by an insulating layer. Also known as varactor diode; variable-capacitance diode; varicap; voltage-variable capacitor.

varactor diode *See* varactor.

Varanidae [VERT ZOO] The monitors, a family of reptiles in the suborder Sauria found in the hot regions of Africa, Asia, Australia, and Malaya.

var hour [ELEC] A unit of the integral of reactive power over time, equal to a reactive power of 1 var integrated over 1 hour; equal in magnitude to 3600 joules. Also known as reactive volt-ampere hour; volt-ampere-hour reactive.

variable [COMPUT SCI] A data item, or specific area in main memory, that can assume any of a set of values. [MATH] A symbol which is used to represent some undetermined element from a given set, usually the domain of a function.

variable attenuator [ELECTR] An attenuator for reducing the strength of an alternating-current signal either continuously or in steps, without causing appreciable signal distortion, by maintaining a substantially constant impedance match.

variable-bandwidth filter [ELECTR] An electric filter whose upper and lower cutoff frequencies may be independently selected, so that almost any bandwidth may be obtained; it usually consists of several stages of RC filters, each separated by buffer amplifiers; tuning is accomplished by varying the resistance and capacitance values.

variable-block [COMPUT SCI] Pertaining to an arrangement of data in which the number of words or characters in a block can vary, as determined by the programmer.

variable-capacitance diode See varactor.

variable capacitor [ELEC] A capacitor whose capacitance can be varied continuously by moving one set of metal plates with respect to another.

variable carrier modulation See controlled carrier modulation.

variable connector [COMPUT SCI] A flow chart symbol representing a sequence connection which is not fixed, but which can be varied by the flow-charted procedure itself; it corresponds to an assigned GO TO in a programming language such as FORTRAN.

variable-cycle operation [COMPUT SCI] An operation that requires a variable number of regularly timed execution cycles for its completion.

variable diode function generator [ELECTR] An improvement of a diode function generator in which fully adjustable potentiometers are used for breakpoint and slope resistances, permitting the programming of analytic, arbitrary, and empirical functions, including inflections. Abbreviated VDFG.

variable field [COMPUT SCI] A field of data whose length is allowed to vary within certain specified limits. [PHYS] Field which changes during the time under consideration.

variable flow [FL MECH] Fluid flow in which the velocity changes both with time and from point to point.

variable-focal-length lens See zoom lens.

variable-length field [COMPUT SCI] A data field in which the number of characters varies, the length of the field being stored within the field itself.

variable-length record [COMPUT SCI] A data or file format that allows each record to be exactly as long as needed.

variable-pitch propeller [ENG] A controllable-pitch propeller whose blade angle may be adjusted to any angle between the low and high pitch limits.

variable point [COMPUT SCI] A system of numeration in which the location of the decimal point is indicated by a special character at that position.

variable-reluctance microphone See magnetic microphone.

variable resistor See rheostat.

variable-speed drive [MECH ENG] A mechanism transmitting motion from one shaft to another that allows the velocity ratio of the shafts to be varied continuously.

variable star [ASTRON] A star that has a detectable change in its intensity which is often accompanied by other physical changes; changes in brightness may be a few thousandths of a magnitude to 20 magnitudes or even more.

variable transformer [ELEC] An iron-core transformer having provisions for varying its output voltage over a limited range or continuously from zero to maximum output voltage, generally by means of a contact arm moving along exposed turns of the secondary winding. Also known as adjustable transformer; continuously adjustable transformer.

variable word length [COMPUT SCI] A phrase referring to a computer in which the number of characters addressed is not a fixed number but is varied by the data or instruction.

variance [STAT] The square of the standard deviation.

variance ratio test [STAT] A technique for comparing the spreads or variabilities of two sets of figures to determine whether the two sets of figures were drawn from the same population. Also known as F test.

variation See declination.

variational principle [MATH] A technique for solving boundary value problems that is applicable when the given problem can be rephrased as a minimization problem.

varicap See varactor.

varicose vein [ANAT] An enlarged tortuous blood vessel that occurs chiefly in the superficial veins and their tributaries in the lower extremities. Also known as varicosity.

varicosity See varicose vein.

variegate [BIOL] Having irregular patches of diverse colors.

variety [SYS ENG] The logarithm (usually to base 2) of the number of discriminations that an observer or a sensing system can make relative to a system. [SYST] A taxonomic group or category inferior in rank to a subspecies.

varifocal lens See zoom lens.

variola See smallpox.

variometer [ELECTROMAG] A variable inductance having two coils in series, one mounted inside the other, with provisions for rotating the inner coil in order to vary the total inductance of the unit over a wide range. [ENG] A geomagnetic device for detecting and indicating changes in one of the components of the terrestrial magnetic field vector, usually magnetic declination, the horizontal intensity component, or the vertical intensity component.

Variscan orogeny [GEOL] The late Paleozoic orogenic era in Europe, extending through the Carboniferous and Permian. Also known as Hercynian orogeny.

varistor [ELECTR] A two-electrode semiconductor device having a voltage-dependent nonlinear resistance; its resistance drops as the applied voltage is increased. Also known as voltage-dependent resistor.

varmeter [ENG] An instrument for measuring reactive power in vars. Also known as reactive volt-ampere meter.

varnish tree [BOT] *Rhus vernicifera*. A member of the sumac family (Anacardiaceae) cultivated in Japan; the cut bark exudes a juicy milk which darkens and thickens on exposure and is applied as a thin film to become a varnish of extreme hardness. Also known as lacquer tree.

varulite [MINERAL] $(Na,Ca)(Mn,Fe)_2(PO_4)_2$ An olive green, orthorhombic mineral consisting of a phosphate of sodium, calcium, manganese, and iron; occurs in massive form.

varve [GEOL] A sedimentary bed, layer, or sequence of layers deposited in a body of still water within a year's time, and usually during a season. Also known as glacial varve.

vascular bundle [BOT] A strandlike part of the plant vascular system containing xylem and phloem.

vas deferens [ANAT] The portion of the excretory duct system of the testis which runs from the epididymal duct to the ejaculatory duct. Also known as ductus deferens.

vasectomy [MED] Cutting, or removing a section from, the ductus deferens.

vashegyite [MINERAL] $2Al_4(PO_4)_3(OH)_3 \cdot 27H_2O$ A white or pale green to yellow and brownish mineral consisting of a hydrous basic aluminum phosphate; occurs in massive and microcrystalline forms.

vasoconstrictor [PHYSIO] A nerve or an agent that causes blood vessel constriction.

vasodilator [PHYSIO] A nerve or an agent that causes blood vessel dilation.

vasopressin [BIOCHEM] A peptide hormone which is elaborated by the posterior pituitary and which has a pressor effect; used medicinally as an antidiuretic. Also known as antidiuretic hormone (ADH).

vaterite [MINERAL] $CaCO_3$ A rare hexagonal mineral consisting of unstable calcium carbonate; it is trimorphous with calcite and aragonite.

vauquelinite [MINERAL] $Pb_2Cu(CrO_4)PO_4(OH)$ A monoclinic mineral of varying color, consisting of a basic chromate-phosphate of lead and copper.

vauxite [MINERAL] $FeAl_2(PO_4)_2(OH)_2 \cdot 7H_2O$ A sky blue to Venetian blue, triclinic mineral consisting of a hydrated basic phosphate of iron and aluminum.

V band [ELECTROMAG] A radio-frequency band of 46.0 to 56.0 gigahertz. [SPECT] Absorption bands that appear in the ultraviolet part of the spectrum due to color centers produced in potassium bromide by exposure of the crystal at temperature of liquid nitrogen (81 K) to intense penetrating x-rays.

V belt [DES ENG] An endless power-transmission belt with a trapezoidal cross section which runs in a pulley with a V-shaped groove; it transmits higher torque at less width and tension than a flat belt. [MECH ENG] A belt, usually endless, with a trapezoidal cross section which runs in a pulley with a V-shaped groove, with the top surface of the belt approximately flush with the top of the pulley.

V connection *See* open-delta connection.

VD *See* venereal disease.

VDFG *See* variable diode function generator.

VDT *See* display terminal.

Vectian *See* Aptian.

vectograph [GRAPHICS] A picture or drawing having self-contained light polarization; at each point of such a picture, one can control the direction and magnitude of polarization, and the image can be expressed as a nonuniform vector field.

vector [MATH] An element of a vector space. [MED] An agent, such as an insect, capable of mechanically or biologically transferring a pathogen from one organism to another. [NAV] To guide a pilot, navigator, aircraft, or missile from one point to another within a given time by means of a direction communicated to the craft. [PHYS] A quantity which has both magnitude and direction, and whose components transform from one coordinate system to another in the same manner as the components of a displacement. Also known as polar vector.

vectorcardiography [PHYSIO] A method of recording the magnitude and direction of the instantaneous cardiac vectors.

vector coupling coefficient [QUANT MECH] One of the coefficients used to express an eigenfunction of the sum of two angular momenta in terms of sums of products of eigenfunctions of the original two angular momenta. Also known as Clebsch-Gordan coefficient; Wigner coefficient.

vector field [MATH] **1.** The field of vectors arising from considering a system of differential equations on a differentiable manifold. **2.** A function whose range is in a vector space. [PHYS] A field which is characterized by a vector function.

vector function [PHYS] A function of position and time whose value at each point is a vector. Also known as vector point function.

vectorial structure *See* directional structure.

vector meson [PARTIC PHYS] A meson which has spin quantum number 1 and negative parity, and may be described by a vector field; examples include the ω, ρ, ϕ, and K^* mesons.

vector momentum *See* momentum.

vector point function *See* vector function.

vector potential [ELECTROMAG] A vector function whose curl is equal to the magnetic induction. Symbolized **A**. Also known as magnetic vector potential. [PHYS] Any vector function whose curl is equal to some solenoidal vector field.

vector power [ELEC] Vector quantity equal in magnitude to the square root of the sum of the squares of the active power and the reactive power.

vector space [MATH] A system of mathematical objects which have an additive operation producing a group structure and which can be multiplied by elements from a field in a manner similar to contraction or magnification of directed line segments in euclidean space. Also known as linear space.

vee antenna *See* V antenna.

vegetable [AGR] The edible portion of a usually herbaceous plant; customarily served with the main course of a meal. [BOT] Resembling or relating to plants.

vegetable diastase *See* diastase.

vegetable jelly *See* ulmin.

vegetation [BOT] The total mass of plant life that occupies a given area.

Veillonellaceae [MICROBIO] The single family of gram-negative, anaerobic cocci; characteristically occur in pairs with adjacent sides flattened; parasites of homotherms, including humans, rodents, and pigs.

Veil Nebula [ASTRON] A nebula in the constellation Cygnus; speculation is that the nebula is the remnant of a gigantic supernova which occurred 30,000 years ago.

vein [ANAT] A relatively thin-walled blood vessel that carries blood from capillaries to the heart in vertebrates. [BOT] One of the vascular bundles in a leaf. [GEOL] A mineral deposit in tabular or shell-like form filling a fracture in a host rock. [INV ZOO] **1.** One of the thick, stiff ribs providing support for the wing of an insect. **2.** A venous sinus in invertebrates.

Vela [ASTRON] A southern constellation, right ascension 9 hours, declination 50°S. Also known as Sail.

Veliidae [INV ZOO] A family of the Hemiptera in the subdivision Amphibicorisae composed of small water striders which have short legs and a longitudinal groove between the eyes.

vellum [MATER] A high-grade paper made to resemble genuine parchment.

velocimeter [ENG] An instrument for measuring the speed of sound in water; two transducers transmit acoustic pulses back and forth over a path of fixed length, each transducer immediately initiating a pulse upon receiving the previous one; the number of pulses occurring in a unit time is measured.

Velocipedidae [INV ZOO] A tropical family of hemipteran insects in the superfamily Cimicoidea.

velocity [MECH] 1. The time rate of change of position of a body; it is a vector quantity having direction as well as magnitude. Also known as linear velocity. 2. The speed at which the detonating wave passes through a column of explosives, expressed in meters or feet per second.

velocity analysis [MECH] A graphical technique for the determination of the velocities of the parts of a mechanical device, especially those of a plane mechanism with rigid component links.

velocity coefficient [FL MECH] The ratio of the actual velocity of gas emerging from a nozzle to the velocity calculated under ideal conditions; it is less than 1 because of friction losses. Also known as coefficient of velocity.

velocity constant [CONT SYS] The ratio of the rate of change of the input command signal to the steady-state error, in a control system where these two quantities are proportional.

velocity discontinuity *See* seismic discontinuity.

velocity distribution [STAT MECH] For the molecules of a gas, a function of velocity whose value at any velocity v is proportional to the number of molecules with velocities in an infinitesimal range about v, per unit velocity range.

velocity-focusing mass spectrograph *See* velocity spectrograph.

velocity gradient [FL MECH] The rate of change of velocity of propagation with distance normal to the direction of flow. [GEOPHYS] *See* seismic gradient.

velocity head [FL MECH] The square of the speed of flow of a fluid divided by twice the acceleration of gravity; it is equal to the static pressure head corresponding to a pressure equal to the kinetic energy of the fluid per unit volume.

velocity level [ACOUS] A sound rating in decibels, equal to 20 times the logarithm to the base 10 of the ratio of the particle velocity of the sound to a specified reference particle velocity.

velocity microphone [ENG ACOUS] A microphone whose electric output depends on the velocity of the air particles that form a sound wave; examples are a hot-wire microphone and a ribbon microphone.

velocity-modulated oscillator [ELECTR] Oscillator which employs velocity modulation to produce radio-frequency power. Also known as klystron oscillator.

velocity of light *See* speed of light.

velocity of sound *See* speed of sound.

velocity pickup [ELEC] A device that generates a voltage proportional to the relative velocity between two principal elements of the pickup, the two elements usually being a coil of wire and a source of magnetic field.

velocity ratio [MECH ENG] The ratio of the velocity given to the effort or input of a machine to the velocity acquired by the load or output. [OCEANOGR] The ratio of the speed of tidal current at a subordinate station to the speed of the corresponding current at the reference station.

velocity spectrograph [PHYS] A mass spectrograph in which only positive ions having a certain velocity pass through all three slits and enter a chamber where they are deflected by a magnetic field in proportion to their charge-to-mass ratio. Also known as velocity-focusing mass spectrograph.

velum [BIOL] A veil- or curtainlike membrane. [INV ZOO] A swimming organ on the larva of certain

marine gastropod mollusks that develops as a contractile ciliated collar-shaped ridge. [METEOROL] An accessory cloud veil of great horizontal extent draped over or penetrated by cumuliform clouds; velum occurs with cumulus and cumulonimbus.

venereal bubo *See* lymphogranuloma venereum.

venereal disease [MED] Any of several contagious diseases generally acquired during sexual intercourse; includes gonorrhea, syphilis, chancroid, granuloma inguinale, and lymphogranuloma venereum. Abbreviated VD.

vent [ENG] 1. A small passage made with a needle through stemming, for admitting a squib to enable the charge to be lighted. 2. A hole, extending up through the bearing at the top of the core-barrel inner tube, which allows the water and air in the upper part of the inner tube to escape into the borehole. 3. A small hole in the upper end of a core-barrel inner tube that allows water and air in the inner tube to escape into the annular space between the inner and outer barrels. 4. An opening provided for the discharge of pressure or the release of pressure from tanks, vessels, reactors, processing equipment, and so on. [GEOL] The opening of a volcano on the surface of the earth. [MET] A small opening in a casting mold to allow for the escape of gases. [ZOO] The external opening of the cloaca or rectum, especially in fish, birds, and amphibians.

ventricle [ANAT] 1. A chamber, or one of two chambers, in the vertebrate heart which receives blood from the atrium and forces it into the arteries by contraction of the muscular wall. 2. One of the interconnecting, fluid-filled chambers of the vertebrate brain that are continuous with the canal of the spinal cord. [ZOO] A cavity in a body part or organ.

ventricose [BIOL] Swollen or distended, especially on one side.

ventricular septum *See* interventricular septum.

venturi meter [ENG] An instrument for efficiently measuring fluid flow rate in a piping system; a nozzle section increases velocity and is followed by an expanding section for recovery of kinetic energy.

venturi scrubber [CHEM ENG] A gas-cleaning device in which liquid injected at the throat of a venturi is used to scrub dust and mist from the gas flowing through the venturi.

venturi tube [ENG] A constriction that is placed in a pipe and causes a drop in pressure as fluid flows through it, consisting essentially of a short straight pipe section or throat between two tapered sections; it can be used to measure fluid flow rate (a venturi meter), or to draw fuel into the main flow stream, as in a carburetor.

Venus [ASTRON] The planet second in distance from the sun; the linear diameter, about 12,200 kilometers, includes the top of a cloud layer; the diameter of the solid globe is about 50 kilometers less; the mass is about 0.815 (earth = 1).

Venus' flytrap [BOT] *Dionaea muscipula.* An insectivorous plant (order Sarraceniales) of North and South Carolina; the two halves of a leaf blade can swing upward and inward as though hinged, thus trapping insects between the closing halves of the leaf blade.

Verbeekinidae [PALEON] A family of extinct marine protozoans in the superfamily Fusulinacea.

Verbenaceae [BOT] A family of variously woody or herbaceous dicotyledons in the order Lamiales characterized by opposite or whorled leaves and regular or irregular flowers, usually with four or two functional stamens.

verge [BUILD] The edge of a sloping roof which projects over a gable.

verglas *See* glaze.

verification [COMPUT SCI] The process of checking the results of one data transcription against the results of another data transcription; both transcriptions usually involve manual operations.

verifier [COMPUT SCI] A device for checking card punching semimechanically; it mimics keypunch machine operation, but reads prepunched cards without punching any new holes, and signals if the card does not agree with data entered through the verifier keyboard in some column.

verify [COMMUN] To ensure that the meaning and phraseology of the transmitted message convey the exact intention of the originator. [COMPUT SCI] To determine whether an operation has been completed correctly, and in particular, to check the accuracy of keypunching by using a verifier.

vermiculite [MINERAL] $(Mg,Fe,Al)_3(Al,Si)_4O_{10}(OH)_2 \cdot 4H_2O$ A clay mineral constituent similar to chlorite and montmorillonite, and consisting of trioctahedral mica sheets separated by double water layers; sometimes used as a textural material in painting, or as an aggregate in certain plaster formulations used in sculpture.

vermiform [BIOL] Wormlike; resembling a worm.

Vermilingua [VERT ZOO] An infraorder of the mammalian order Edentata distinguished by lack of teeth and in having a vermiform tongue; includes the South American true anteaters.

vernadskite *See* antlerite.

vernal equinox [ASTRON] The sun's position on the celestial sphere about March 21; at this time the sun's path on the ecliptic crosses the celestial equator. Also known as first point of Aries; March equinox; spring equinox.

vernalization [BOT] The induction in plants of the competence or ripeness to flower by the influence of cold, that is, at temperatures below the optimal temperature for growth.

vernier [ENG] A short, auxiliary scale which slides along the main instrument scale to permit accurate fractional reading of the least main division of the main scale.

vernier capacitor [ELEC] Variable capacitor placed in parallel with a larger tuning capacitor to provide a finer adjustment after the larger unit has been set approximately to the desired position.

vernier dial [ENG] A tuning dial in which each complete rotation of the control knob causes only a fraction of a revolution of the main shaft, permitting fine and accurate adjustment.

vernier engine [AERO ENG] A rocket engine of small thrust used primarily to obtain a final adjustment in the velocity and trajectory of a rocket vehicle just after the thrust cutoff of the last sustainer engine, and used secondarily to add thrust to a booster or sustainer engine. Also known as vernier rocket.

vernier rocket *See* vernier engine.

verrou *See* riegel.

verruca [BIOL] A wartlike elevation on the surface of a plant or animal.

Verrucariaceae [BOT] A family of crustose lichens in the order Pyrenulales typically found on rocks, especially in intertidal or salt-spray zones along rocky coastlines.

Verrucomorpha [INV ZOO] A suborder of the crustacean order Thoracica composed of sessile, asymmetrical barnacles.

verrucose [BIOL] Having the surface covered with wartlike protuberances.

versed sine [MATH] The versed sine of A is $1 - \cosine A$. Denoted vers. Also known as versine.

versine *See* versed sine.

vertebral column *See* spine.

Vertebrata [VERT ZOO] The major subphylum of the phylum Chordata including all animals with backbones, from fish to man.

vertebrate zoology [ZOO] That branch of zoology concerned with the study of members of the Vertebrata.

vertex [ASTRON] 1. The highest point that a celestial body attains. 2. On a great circle, that point that is closest to a pole. [MATH] For a polygon or polyhedron, any of those finitely many points which together with line segments or plane pieces determine the figure or solid. [OPTICS] One of the points where the surface of a lens intersects the optical axis.

vertical angles [MATH] The two angles produced by a pair of intersecting lines and lying on opposite sides of the point of intersection.

vertical circle [ASTRON] A great circle of the celestial sphere, through the zenith and nadir of the celestial sphere; vertical circles are perpendicular to the horizon.

vertical component effect *See* antenna effect.

vertical definition *See* vertical resolution.

vertical deflection oscillator [ELECTR] The oscillator that produces, under control of the vertical synchronizing signals, the sawtooth voltage waveform that is amplified to feed the vertical deflection coils on the picture tube of a television receiver. Also known as vertical oscillator.

vertical drop [MECH] The drop of an object in trajectory or along a plumb line, measured vertically from its line of departure to the object.

vertical instruction [COMPUT SCI] An instruction in machine language to carry out a single operation or a time-ordered series of a fixed number and type of operation on a single set of operands.

vertical intensity [GEOPHYS] The magnetic intensity of the vertical component of the earth's magnetic field, reckoned positive if downward, negative if upward.

vertically stacked loops *See* stacked loops.

vertical metal oxide semiconductor technology [ELECTR] For semiconductor devices, a technology that involves essentially the formation of four diffused layers in silicon and etching of a V-shaped groove to a precisely controlled depth in the layers, followed by deposition of metal over silicon dioxide in the groove to form the gate electrode. Abbreviated VMOS technology.

vertical oscillator *See* vertical deflection oscillator.

vertical resolution [ELECTR] The number of distinct horizontal lines, alternately black and white, that can be seen in the reproduced image of a television or facsimile test pattern; it is primarily fixed by the number of horizontal lines used in scanning. Also known as vertical definition.

vertical separation [AERO ENG] A specified vertical distance measured in terms of space between aircraft in flight at different altitudes or flight levels. [GEOL] The vertical component of the dip slip in a fault.

vertical stability *See* static stability.

vertical takeoff and landing [AERO ENG] A flight technique in which an aircraft rises directly into the air and settles vertically onto the ground. Abbreviated VTOL.

Vertisol [GEOL] A soil order formed in regoliths high in clay; subject to marked shrinking and swelling with changes in water content; low in organic content and high in bases.

very-high-frequency [COMMUN] The band of frequencies from 30 to 300 megahertz in the radio spectrum, corresponding to wavelengths of 1 meter to 10 meters. Abbreviated VHF.

very-large-scale integration [ELECTR] A very complex integrated circuit, which may contain 10,000 or more individual devices, such as basic logic gates and transistors, placed on a single semiconductor chip. Also known as VLSI.

very-long-baseline interferometry [ELECTR] A method of improving angular resolution in the observation of radio sources; these are simultaneously observed by two radio telescopes which are very far apart, and the signals are recorded on magnetic tapes which are combined electronically or on a computer. Abbreviated VLBI.

very-long-range radar [ELECTR] Equipment whose maximum range on a reflecting target of 1 square meter normal to the signal path exceeds 800 miles (1300 kilometers), provided line-of-sight exists between the target and the radar.

very-low-frequency [COMMUN] The band of frequencies from 3 to 30 kilohertz in the radio spectrum, corresponding to wavelengths of 10 to 100 kilometers. Abbreviated VLF.

very-short-range radar [ELECTR] Equipment whose range on a reflecting target of 1 square meter normal to the signal path is less than 50 miles (80 kilometers), provided line-of-sight exists between the target and the radar.

vesicant [PHARM] An agent that causes blistering.

vesicle [BIOL] A small, thin-walled bladderlike cavity, usually filled with fluid. [GEOL] A cavity in lava formed by entrapment of a gas bubble during solidification. Also known as air sac; bladder; saccus; vacuole; wing.

vesicular structure [PETR] A structure that is common in many volcanic rocks and which forms when magma is brought to or near the earth's surface; may form a structure with small cavities, or produce a pumiceous structure or a scoriaceous structure.

Vespertilionidae [VERT ZOO] The common bats, a large cosmopolitan family of the Chiroptera characterized by a long tail, extending to the edge of the uropatagium; almost all members are insect-eating.

Vespidae [INV ZOO] A widely distributed family of Hymenoptera in the superfamily Vespoidea including hornets, yellow jackets, and potter wasps.

Vespoidea [INV ZOO] A superfamily of wasps in the suborder Apocrita.

vestibular apparatus [ANAT] The anatomical structures concerned with the vestibular portion of the eighth cranial nerve; includes the saccule, utricle, semicircular canals, vestibular nerve, and vestibular nuclei of the ear.

vestibular nerve [ANAT] A somatic sensory branch of the auditory nerve, which is distributed about the ampullae of the semicircular canals, macula sacculi, and macula utriculi.

vestibule [ANAT] **1.** The central cavity of the bony labyrinth of the ear. **2.** The parts of the membranous labyrinth within the cavity of the bony labyrinth. **3.** The space between the labia minora. **4.** See buccal cavity. [BUILD] A hall or chamber between the outer door and the interior, or rooms, of a building.

vestibulocochlear nerve See auditory nerve.

vestigial sideband [COMMUN] The transmitted portion of an amplitude-modulated sideband that has been largely suppressed by a filter having a gradual cutoff in the neighborhood of the carrier frequency; the other sideband is transmitted without much suppression. Abbreviated VSB.

vestigial-sideband transmission [COMMUN] A type of radio signal transmission for amplitude modulation in which the normal complete sideband on one side of the carrier is transmitted, but only a part of the other sideband is transmitted. Also known as asymmetrical-sideband transmission.

vesuvian See leucite; vesuvianite.

Vesuvian garnet See leucite.

vesuvianite [MINERAL] $Ca_{10}Mg_2Al_4(SiO_4)_5(Si_2O_7)_2(OH)_4$ A brown, yellow, or green mineral found in contact-metamorphosed limestones. Also known as idocrase; vesuvian.

veszelyite [MINERAL] $(Cu,Zn)_3(PO_4)(OH)_3 \cdot 2H_2O$ A greenish-blue to dark blue, monoclinic mineral consisting of a hydrated basic phosphate of copper and zinc.

vetch [BOT] Any of a group of mostly annual legumes, especially of the genus *Vicia*, with weak viny stems terminating in tendrils and having compound leaves; some varieties are grown for their edible seed.

VF See voice frequency.

V format [COMPUT SCI] A data record format in which the logical records are of variable length and each record begins with a record length indication.

VFR See visual flight rules.

VHF See very-high-frequency.

viaduct [CIV ENG] A bridge structure supported on high towers with short masonry or reinforced concrete arched spans.

Vianaidae [INV ZOO] A small family of South American Hemiptera in the superfamily Tingordea.

vibrating capacitor [ELEC] A capacitor whose capacitance is varied in a cyclic manner to produce an alternating voltage proportional to the charge on the capacitor; used in a vibrating-reed electrometer.

vibrating needle [ENG] A magnetic needle used in compass adjustment to find the relative intensity of the horizontal components of the earth's magnetic field and the magnetic field at the compass location.

vibrating-reed frequency meter [ENG] A frequency meter consisting of steel reeds having different and known natural frequencies, all excited by an electromagnet carrying the alternating current whose frequency is to be measured. Also known as Frahm frequency meter; reed frequency meter; tuned-reed frequency meter.

vibration [MECH] A continuing periodic change in a displacement with respect to a fixed reference.

vibrational energy [PHYS CHEM] For a diatomic molecule, the difference between the energy of the molecule idealized by setting the rotational energy equal to zero, and that of a further idealized molecule which is obtained by gradually stopping the vibration of the nuclei without placing any new constraint on the motions of electrons.

vibration galvanometer [ENG] An alternating-current galvanometer in which the natural oscillation frequency of the moving element is equal to the frequency of the current being measured.

vibration isolation [ENG] The isolation, in structures, of those vibrations or motions that are classified as mechanical vibration; involves the control of the supporting structure, the placement and arrangement of isolators, and control of the internal construction of the equipment to be protected.

vibration machine [MECH ENG] A device for subjecting a system to controlled and reproducible mechanical vibration. Also known as shake table.

vibrator [ELEC] An electromechanical device used primarily to convert direct current to alternating current but also used as a synchronous rectifier; it contains a vibrating reed which has a set of contacts that alternately hit stationary contacts attached to the frame, reversing the direction of current flow; the reed is activated when a soft-iron slug at its tip is attracted to the pole piece of a driving coil. [MECH ENG] An instrument which produces mechanical oscillations.

Vibrionaceae [MICROBIO] A family of gram-negative, facultatively anaerobic rods; cells are straight or curved and usually motile by polar flagella; generally found in water.

vibrissae [VERT ZOO] Hairs with specialized erectile tissue; found on all mammals except man. Also known as sinus hairs; tactile hairs; whiskers.

Vickers hardness test *See* diamond pyramid hardness test.

vicuna [VERT ZOO] *Lama vicugna*. A rare, wild ruminant found in the Andes mountains; the fiber of the vicuna is strong, resilient, and elastic but is the softest and most delicate of animal fibers.

video [ELECTR] 1. Pertaining to picture signals or to the sections of a television system that carry these signals in either unmodulated or modulated form. 2. Pertaining to the demodulated radar receiver output that is applied to a radar indicator.

video amplifier [ELECTR] A low-pass amplifier having a band-width on the order of 2–10 megahertz, used in television and radar transmission and reception; it is a modification of an RC-coupled amplifier, such that the high-frequency half-power limit is determined essentially by the load resistance, the internal transistor capacitances, and the shunt capacitance in the circuit.

video disk recorder [ELECTR] A video recorder that records television visual signals and sometimes aural signals on a magnetic, optical, or other type of disk which is usually about the size of a long-playing phonograph record.

video disk storage *See* optical disk storage.

video display terminal *See* display terminal.

video frequency [COMMUN] One of the frequencies existing in the output of a television camera when an image is scanned; it may be any value from almost zero to well over 4 megahertz.

video integrator [ELECTR] 1. Electric counter-countermeasures device that is used to reduce the response to nonsynchronous signals such as noise, and is useful against random pulse signals and noise. 2. Device which uses the redundancy of repetitive signals to improve the output signal-to-noise ratio, by summing the successive video signals.

video recorder [ELECTR] A magnetic tape recorder capable of storing the video signals for a television program and feeding them back later to a television transmitter or directly to a receiver.

video sensing [COMPUT SCI] In optical character recognition, a scanning technique in which the document is flooded with light from an ordinary light source, and the image of the character is reflected onto the face of a cathode-ray tube, where it is scanned by an electron beam.

video signal [COMMUN] In television, the signal containing all of the visual information together with blanking and synchronizing pulses. [ELECTROMAG] *See* target signal.

video tape [ELECTR] A heavy-duty magnetic tape designed primarily for recording the video signals of television programs.

videotex [COMMUN] A computer communication service which uses standard television information from a data base, and which allows the user, equipped with a limited computer terminal, to interact with the service in selecting information to be displayed, so as to provide electronic mail, teleshopping, financial services, calculation services, and such.

video transformer [ELECTR] A transformer designed to transfer, from one circuit to another, the signals containing picture information in television.

video transmitter *See* visual transmitter.

viewdata [COMMUN] Alphanumeric characters or graphics which are produced on a home television screen by a special decoder in response to digital signals sent over a telephone line.

viewing screen *See* screen.

viewing storage tube *See* direct-view storage tube.

villous placenta *See* epitheliochorial placenta.

Vindobonian [GEOL] A European stage of geologic time, middle Miocene.

vinegar eel [INV ZOO] *Turbatrix aceti*. A very small nematode often found in large numbers in vinegar fermentation. Also known as vinegar worm.

vinegar worm *See* vinegar eel.

vinylation [CHEM] Formation of a vinyl-derived product by reaction with acetylene; for example, vinylation of alcohols gives vinyl ethers, such as vinyl ethyl ether.

Violaceae [BOT] A family of dicolyledonous plants in the order Violales characterized by polypetalous, mostly perfect, hypogynous flowers with a single style and five stamens.

Violales [BOT] A heterogeneous order of dicotyledons in the subclass Dilleniidae distinguished by a unilocular, compound ovary and mostly parietal placentation.

violarite [MINERAL] Ni_2FeS_4 A violet-gray mineral of the linnaeite group consisting of a sulfide of nickel and iron; found in meteorites.

violet [OPTICS] The hue evoked in an average observer by monochromatic radiation having a wavelength in the approximate range from 390 to 455 nanometers; however, the same sensation can be produced in a variety of other ways.

violle [OPTICS] A unit of luminous intensity, equal to the luminous intensity of 1 square centimeter of platinum at its temperature of solidification; it is found experimentally to be equal to 20.17 candelas.

viper [VERT ZOO] The common name for reptiles of the family Viperidae; thick-bodied poisonous snakes having a pair of long fangs, present on the anterior part of the upper jaw, which fold against the roof of the mouth when the jaws are closed.

Viperidae [VERT ZOO] A family of reptiles in the suborder Serpentes found in Eurasia and Africa; all species are proglyphodont.

viral pneumonia [MED] A form of pneumonia caused by a virus of various types, in which the inflammatory reaction predominates in the septa, and the alveoli contain fibrin, edema fluid, and some inflammatory cells.

Vireonidae [VERT ZOO] The vireos, a family of New World passeriform birds in the suborder Oscines.

virgin *See* straight-run.

Virgin *See* Virgo.

Virgo [ASTRON] A constellation, right ascension 13 hours, declination 0°. Also known as Virgin.

Virgo cluster [ASTRON] A cluster of galaxies which is the nearest to the galaxy that includes the sun; the cluster is centered in the constellation Virgo and is about 16,000,000 light-years (1.51×10^{23} m) from earth.

virial coefficients [THERMO] For a given temperature T, one of the coefficients in the expansion of P/RT in inverse powers of the molar volume, where P is the pressure and R is the gas constant.

virilism [MED] *See* gynandry. [PSYCH] **1.** Masculinity. **2.** Manifestation of male behavioral patterns in the female.

viroid [MICROBIO] The smallest known agents of infectious disease, characterized by the absence of encapsidated proteins.

virology [MICROBIO] The study of submicroscopic organisms known as viruses.

virtual address [COMPUT SCI] A symbol that can be used as a valid address part but does not necessarily designate an actual location.

virtual decimal point *See* assumed decimal point.

virtual direct-access storage [COMPUT SCI] A device used with mass-storage systems, whereby data are retrieved prior to usage by a batch-processing program and automatically transcribed onto disk storage.

virtual displacement [MECH] **1.** Any change in the positions of the particles forming a mechanical system. **2.** An infinitesimal change in the positions of the particles forming a mechanical system, which is consistent with the geometrical constraints on the system.

virtual entropy [THERMO] The entropy of a system, excluding that due to nuclear spin. Also known as practical entropy.

virtual height [GEOPHYS] The apparent height of a layer in the ionosphere, determined from the time required for a radio pulse to travel to the layer and return, assuming that the pulse propagates at the speed of light. Also known as equivalent height.

virtual image [OPTICS] An optical image from which rays of light only appear to diverge, without actually being focused there.

virtual memory [COMPUT SCI] A combination of primary and secondary memories that can be treated as a single memory by programmers because the computer itself translates a program or virtual address to the actual hardware address.

virtual state [NUC PHYS] An unstable state of a compound nucleus which has a lifetime many times longer than the time it takes a nucleon, with the same energy as it has in the virtual state, to cross the nucleus.

virtual work principle *See* principle of virtual work.

virulence [MICROBIO] The disease-producing power of a microorganism; infectiousness.

virus [VIROL] A large group of infectious agents ranging from 10 to 250 nanometers in diameter, composed of a protein sheath surrounding a nucleic acid core and capable of infecting all animals, plants, and bacteria; characterized by total dependence on living cells for reproduction and by lack of independent metabolism.

virus hepatitis *See* infectious hepatitis.

visbreaking *See* viscosity breaking.

visceral arch [ANAT] One of the series of mesodermal ridges covered by epithelium bounding the lateral wall of the oral and pharyngeal regions of vertebrates; embryonic in higher forms, they contribute to formation of the face and neck. [VERT ZOO] One of the first two arches of the series in gill-bearing forms.

viscid [BOT] Having a sticky surface, as certain leaves.

viscoelastic damping [MECH] The absorption of oscillatory motions by materials which are viscous while exhibiting certain elastic properties.

viscoelasticity [MECH] Property of a material which is viscous but which also exhibits certain elastic properties such as the ability to store energy of deformation, and in which the application of a stress gives rise to a strain that approaches its equilibrium value slowly.

viscometer [ENG] An instrument designed to measure the viscosity of a fluid.

viscometry [ENG] A branch of rheology; the study of the behavior of fluids under conditions of internal shear; the technology of measuring viscosities of fluids.

viscosity [FL MECH] The resistance that a gaseous or liquid system offers to flow when it is subjected to a shear stress. Also known as flow resistance.

viscosity breaking [CHEM ENG] A petroleum refinery process used to lower or break the viscosity of high-viscosity residuum by thermal cracking of molecules at relatively low temperatures. Also known as visbreaking.

viscosity curve [FL MECH] A graph showing the viscosity of a liquid or gaseous material as a function of temperature.

viscous damping [MECH ENG] A method of converting mechanical vibrational energy of a body into heat energy, in which a piston is attached to the body and is arranged to move through liquid or air in a cylinder or bellows that is attached to a support.

viscous drag [FL MECH] That part of the rearward force on an aircraft that results from the aircraft carrying air forward with it through viscous adherence.

viscous flow [FL MECH] **1.** The flow of a viscous fluid. **2.** The flow of a fluid through a duct under conditions such that the mean free path is small in comparison with the smallest, transverse section of the duct.

viscous fluid [FL MECH] A fluid whose viscosity is sufficiently large to make the viscous forces a significant part of the total force field in the fluid.

viscous force [FL MECH] The force per unit volume or per unit mass arising from viscous effects in fluid flow.

visibility [METEOROL] In weather observing practice, the greatest distance in a given direction at which it is just possible to see and identify with the unaided eye, in the daytime, a prominent dark object against the sky at the horizon and, at nighttime, a known, preferably unfocused, moderately intense light source.

visibility function *See* luminosity function.

visibility meter [ENG] An instrument for making direct measurements of visual range in the atmosphere or of the physical characteristics of the atmosphere which determine the visual range. [OPTICS] A type of photometer that operates on the principle of artificially reducing the visibility of objects to threshold values (borderline of seeing and not seeing) and measuring the amount of the reduction on an appropriate scale.

visible absorption spectrophotometry [SPECT] The study of spectra produced by the absorption of visible-light energy during the transformation of an electron from the ground state to an excited state as a function of the wavelength causing the transformation.

visible horizon [ASTRON] That line where earth and sky appear to meet, and the projection of this line upon the celestial sphere.

visible radiation *See* light.

visible spectrum [SPECT] **1.** The range of wavelengths of visible radiation. **2.** A display or graph of the intensity of visible radiation emitted or absorbed by a material as a function of wavelength or some related parameter.

visual achromatism [OPTICS] In an optical system, the removal of chromatic aberration or chromatic differences of magnification between light at the wavelength of the Fraunhofer C line at 656.3 nanometers and the F line at 486.1 nanometers in order to minimize these defects at wavelengths at which the human eye is most sensitive. Also known as optical achromatism.

visual acuity [PHYSIO] The ability to see fine details of an object; specifically, the ability to see an object whose angle subtended at the eye is 1 minute of arc.

visual binaries [ASTRON] Binary stars that to the naked eye seem to be single stars, but when viewed through the telescope, are separated into pairs. Also known as visual doubles.

visual comparator See optical comparator.

visual display unit See display tube.

visual doubles See visual binaries.

visual flight [AERO ENG] An aircraft flight occurring under conditions which allow navigation by visual reference to the earth's surface at a safe altitude and with sufficient horizontal visibility, and operating under visual flight rules. Also known as VFR flight.

visual flight rules [AERO ENG] A set of regulations set down by the U.S. Civil Aeronautics Board (in Civil Air Regulations) to govern the operational control of aircraft during visual flight. Abbreviated VFR.

visual purple See rhodopsin.

visual transmitter [ELECTR] Those parts of a television transmitter that act on picture signals, including parts that act on the audio signals as well. Also known as picture transmitter; video transmitter.

Vitaceae [BOT] A family of dicotyledonous plants in the order Rhamnales; mostly tendril-bearing climbers with compound or lobed leaves, as in grapes (*Vitis*).

vital capacity [PHYSIO] The volume of air that can be forcibly expelled from the lungs after the deepest inspiration.

vitamin [BIOCHEM] An organic compound present in variable, minute quantities in natural foodstuffs and essential for the normal processes of growth and maintenance of the body; vitamins do not furnish energy, but are essential for energy transformation and regulation of metabolism.

vitamin A [BIOCHEM] $C_{20}H_{29}OH$ A pale-yellow alcohol that is soluble in fat and insoluble in water; found in liver oils and carotenoids, and produced synthetically; it is a component of visual pigments and is essential for normal growth and maintenance of epithelial tissue. Also known as antiinfective vitamin; antixerophthalmic vitamin; retinol.

vitamin B₁ See thiamine.

vitamin B₂ See riboflavin.

vitamin B₃ See pantothenic acid.

vitamin B₆ [BIOCHEM] A vitamin which exists as three chemically related and water-soluble forms found in food: pyridoxine, pyridoxal, and pyridoxamine; dietary requirements and physiological activities are uncertain.

vitamin B₁₂ [BIOCHEM] A group of closely related polypyrrole compounds containing trivalent cobalt; the antipernicious anemia factor, essential for normal hemopoiesis. Also known as cobalamin; cyanocobalamin; extrinsic factor.

vitamin B complex [BIOCHEM] A group of water-soluble vitamins that include thiamine, riboflavin, nicotinic acid, pyridoxine, panthothenic acid, inositol, *p*-aminobenzoic acid, biotin, folic acid, and vitamin B₁₂.

vitamin C See ascorbic acid.

vitamin D [BIOCHEM] Either of two fat-soluble, sterol-like compounds, calciferol or ergocalciferol (vitamin D₂) and cholecalciferol (vitamin D₃); occurs in fish liver oils and is essential for normal calcium and phosphorus deposition in bones and teeth. Also known as antirachitic vitamin.

vitamin E [BIOCHEM] $C_{29}H_{50}O_2$ Any of a series of eight related compounds called tocopherols, α-tocopherol having the highest biological activity; occurs in wheat germ and other oils and is believed to be needed in certain human physiological processes.

vitamin G See riboflavin.

vitamin K [BIOCHEM] Any of three yellowish oils which are fat-soluble, nonsteroid, and nonsaponifiable; it is essential for formation of prothrombin. Also known as antihemorrhagic vitamin; prothrombin factor.

vitamin P [BIOCHEM] A substance, such as citrin or one or more of its components, believed to be concerned with maintenance of the normal state of the walls of small blood vessels.

vitamin P complex See bioflavanoid.

vitellogenesis [PHYSIO] The process by which yolk is formed in the ooplasm of an oocyte.

vitrain [GEOL] A brilliant black coal lithotype with vitreous luster and cubical cleavage. Also known as pure coal.

vitreous body See vitreous humor.

vitreous chamber [ANAT] A cavity of the eye posterior to the crystalline lens and anterior to the retina, which is filled with vitreous humor.

vitreous enamel [MATER] A glass coating applied to a metal by covering the surface with a powdered glass frit and heating until fusion occurs. Also known as porcelain enamel.

vitreous humor [PHYSIO] The transparent gel-like substance filling the greater part of the globe of the eye, the vitreous chamber. Also known as vitreous body.

vitreous state [SOLID STATE] A solid state in which the atoms or molecules are not arranged in any regular order, as in a crystal, and which crystallizes only after an extremely long time. Also known as glassy state.

vitrification [GEOL] Formation of a glassy or noncrystalline material.

vitrinite [GEOL] A maceral group that is rich in oxygen and composed of humic material associated with peat formation; characteristic of vitrain.

vitrophyre [PETR] Any porphyritic igneous rock whose groundmass is glassy. Also known as glass porphyry.

Viverridae [VERT ZOO] A family of carnivorous mammals in the superfamily Feloidea composed of the civets, genets, and mongooses.

vivianite [MINERAL] $Fe_3(PO_4)_2 \cdot 8H_2O$ A colorless, blue, or green mineral in the unaltered state (darkens upon oxidation); crystallizes in the monoclinic system and occurs in earth form and as globular and encrusting fibrous masses. Also known as blue iron earth; blue ocher.

Viviparidae [INV ZOO] A family of fresh-water gastropod mollusks in the order Pectinibranchia.

VLBI See very-long-baseline interferometry.

VLF See very-low-frequency.

VLSI See very-large-scale integration.

VMOS technology See vertical metal oxide semiconductor technology.

vocal cord See vocal fold.

vocal fold [ANAT] Either of a pair of folds of tissue covered by mucous membrane in the larynx. Also known as vocal cord.

Vochysiaceae [BOT] A family of dicotyledonous plants in the order Polygalales characterized by mostly three carpels, usually stipulate leaves, one fertile stamen, and capsular fruit.

vocoder [ELECTR] A system of electronic apparatus for synthesizing speech according to dynamic specifications derived from an analysis of that speech.

voglite [MINERAL] An emerald green to grass green, triclinic mineral consisting of a hydrated carbonate of calcium, copper, and uranium; occurs as coatings of scales.

voice channel [COMMUN] A communication channel having sufficient bandwidth to carry voice frequencies intelligibly; the minimum bandwidth for a voice channel is about 3000 hertz.

voice coder [ELECTR] Device that converts speech input into digital form prior to encipherment for secure transmission and converts the digital signals back to speech at the receiver.

voice control [ENG] The use of a voice to activate devices which respond or operate by means of speech recognition.

voice frequency [COMMUN] An audio frequency in the range essential for transmission of speech of commercial quality, from about 300 to 3400 hertz. Abbreviated VF. Also known as speech frequency.

voice print [ENG ACOUS] A voice spectrograph that has individually distinctive patterns of voice characteristics that can be used to identify one person's voice from other voice patterns.

voice response [COMPUT SCI] A computer-controlled recording system in which basic sounds, numerals, words, or phrases are individually stored for playback under computer control as the reply to a keyboarded query. [ENG ACOUS] The process of generating an acoustic speech signal that communicates an intended message, such that a machine can respond to a request for information by talking to a human user.

voice synthesizer [ELECTR] A synthesizer that simulates speech in any language by assembling a language's elements or phonemes under digital control, each with the correct inflection, duration, pause, and other speech characteristics.

vol See volume.

Volan [ASTRON] A southern constellation, right ascension 8 hours, declination 70°S. Also known as Flying Fish; Pisces Volan.

volar [ANAT] Pertaining to, or on the same side as, the palm of the hand or the sole of the foot.

volatile memory See volatile storage.

volatile storage [COMPUT SCI] A storage device that must be continuously supplied with energy, or it will lose its retained data. Also known as volatile memory.

volatility [THERMO] The quality of having a low boiling point or subliming temperature at ordinary pressure or, equivalently, of having a high vapor pressure at ordinary temperatures.

volatilization [THERMO] The conversion of a chemical substance from a liquid or solid state to a gaseous or vapor state by the application of heat, by reducing pressure, or by a combination of these processes. Also known as vaporization.

volcanello See spatter cone.

volcanic ash [GEOL] Fine pyroclastic material; particle diameter is less than 4 millimeters.

volcanic breccia [PETR] A pyroclastic rock that is composed of angular volcanic fragments having a diameter larger than 2 millimeters and that may or may not have a matrix.

volcanic foam See pumice.

volcanic glass [GEOL] Natural glass formed by the cooling of molten lava, or one of its liquid fractions, too rapidly to allow crystallization.

volcanicity See volcanism.

volcanic mud [GEOL] Sediment containing large quantities of ash from a volcanic eruption, mixed with water.

volcanic rock [GEOL] Finely crystalline or glassy igneous rock resulting from volcanic activity at or near the surface of the earth. Also known as extrusive rock.

volcanism [GEOL] The movement of magma and its associated gases from the interior into the crust and to the surface of the earth. Also known as volcanicity.

volcano [GEOL] 1. A mountain or hill, generally with steep sides, formed by the accumulation of magma extruded through openings or volcanic vents. 2. The vent itself.

volcanology [GEOL] The branch of geology that deals with volcanism.

vole [VERT ZOO] Any of about 79 species of rodent in the tribe Microtini of the family Cricetidae; individuals have a stout body with short legs, small ears, and a blunt nose.

volt [ELEC] The unit of potential difference or electromotive force in the meter-kilogram-second system, equal to the potential difference between two points for which 1 coulomb of electricity will do 1 joule of work in going from one point to the other. Symbolized V.

Volta effect See contact potential difference.

voltage [ELEC] Potential difference or electromotive force measured in volts.

voltage amplifier [ELECTR] An amplifier designed primarily to build up the voltage of a signal, without supplying appreciable power.

voltage coefficient [ELEC] For a resistor whose resistance varies with voltage, the ratio of the fractional change in resistance to the change in voltage.

voltage-dependent resistor See varistor.

voltage divider [ELEC] A tapped resistor, adjustable resistor, potentiometer, or a series arrangement of two or more fixed resistors connected across a voltage source; a desired fraction of the total voltage is obtained from the intermediate tap, movable contact, or resistor junction. Also known as potential divider.

voltage doubler [ELECTR] A transformerless rectifier circuit that gives approximately double the output voltage of a conventional half-wave vacuum-tube rectifier by charging a capacitor during the normally wasted half-cycle and discharging it in series with the output voltage during the next half-cycle. Also known as doubler.

voltage drop [ELEC] The voltage developed across a component or conductor by the flow of current through the resistance or impedance of that component or conductor.

voltage gain [ELECTR] The difference between the output signal voltage level in decibels and the input signal voltage level in decibels; this value is equal to 20 times the common logarithm of the ratio of the output voltage to the input voltage.

voltage generator [ELECTR] A two-terminal circuit element in which the terminal voltage is independent of the current through the element.

voltage gradient [ELEC] The voltage per unit length along a resistor or other conductive path.

voltage multiplier [ELEC] See instrument multiplier. [ELECTR] A rectifier circuit capable of supplying a direct-current output voltage that is two or more times the peak value of the alternating-current voltage.

voltage-range multiplier   *See* instrument multiplier.

voltage rating   [ELEC] The maximum sustained voltage that can safely be applied to an electric device without risking the possibility of electric breakdown. Also known as working voltage.

voltage regulator   [ELECTR] A device that maintains the terminal voltage of a generator or other voltage source within required limits despite variations in input voltage or load. Also known as automatic voltage regulator; voltage stabilizer.

voltage saturation   *See* anode saturation.

voltage stabilizer   *See* voltage regulator.

voltage transformer   [ELEC] An instrument transformer whose primary winding is connected in parallel with a circuit in which the voltage is to be measured or controlled. Also known as potential transformer.

voltage-tunable tube   [ELECTR] Oscillator tube whose operating frequency can be varied by changing one or more of the electrode voltages, as in a backward-wave magnetron.

voltage-variable capacitor   *See* varactor.

voltaic cell   [ELEC] A primary cell consisting of two dissimilar metal electrodes in a solution that acts chemically on one or both of them to produce a voltage.

voltametry   [PHYS CHEM] Any electrochemical technique in which a faradaic current passing through the electrolysis solution is measured while an appropriate potential is applied to the polarizable or indicator electrode; for example, polarography.

voltammeter   [ELEC] An instrument that may be used either as a voltmeter or ammeter.

volt-ampere   [ELEC] The unit of apparent power in the International System; it is equal to the apparent power in a circuit when the product of the root-mean-square value of the voltage, expressed in volts, and the root-mean-square value of the current, expressed in amperes, equals 1. Abbreviated VA.

volt-ampere hour   [ELEC] A unit for expressing the integral of apparent power over time, equal to the product of 1 volt-ampere and 1 hour, or to 3600 joules.

volt-ampere-hour reactive   *See* var hour.

Volta series   *See* displacement series.

Volterra dislocation   [SOLID STATE] A model of a dislocation which is formed in a ring of crystalline material by cutting the ring, moving the cut surfaces over each other, and then rejoining them.

voltmeter   [ENG] An instrument for the measurement of potential difference between two points, in volts or in related smaller or larger units.

voltzite   [MINERAL] Zn_5S_4O A rose red, yellowish, or brownish mineral consisting of an oxysulfide of zinc; occurs in implanted spherical globules and as a crust.

volume   [ACOUS] The intensity of a sound.   [COMPUT SCI] A single unit of external storage, all of which can be read or written by a single access mechanism or input/output device.   [ENG ACOUS] The magnitude of a complex audio-frequency current as measured in volume units on a standard volume indicator.   [MATH] A measure of the size of a body or definite region in three-dimensional space; it is equal to the least upper bound of the sum of the volumes of nonoverlapping cubes that can be fitted inside the body or region, where the volume of a cube is the cube of the length of one of its sides. Abbreviated vol.

volume acoustic wave   *See* bulk acoustic wave.

volume compressor   [ENG ACOUS] An audio-frequency circuit that limits the volume range of a radio program at the transmitter, to permit using a higher average percent modulation without risk of over-modulation; also used when making disk recordings, to permit a closer groove spacing without overcutting. Also known as automatic volume compressor.

volume dose   *See* integral dose.

volume expander   [ENG ACOUS] An audio-frequency control circuit sometimes used to increase the volume range of a radio program or recording by making weak sounds weaker and loud sounds louder; the expander counteracts volume compression at the transmitter or recording studio. Also known as automatic volume expander.

volume-limiting amplifier   [ELECTR] Amplifier containing an automatic device that functions only when the input signal exceeds a predetermined level, and then reduces the gain so the output volume stays substantially constant despite further increases in input volume; the normal gain of the amplifier is restored when the input volume returns below the predetermined limiting level.

volume meter   [ENG] Any flowmeter in which the actual flow is determined by the measurement of a phenomenon associated with the flow.

volume range   [ELEC] In a transmission system, the difference, expressed in decibels, between the maximum and minimum volumes that can be satisfactorily handled by the system.   [ENG ACOUS] The difference, expressed in decibels, between the maximum and minimum volumes of a complex audio-frequency signal occurring over a specified period of time.

volumetric analysis   [ANALY CHEM] Quantitative analysis of solutions of known volume but unknown strength by adding reagents of known concentration until a reaction end point (color change or precipitation) is reached; the most common technique is by titration.

volumetric efficiency   [MECH ENG] In describing an engine or gas compressor, the ratio of volume of working substance actually admitted, measured at a specified temperature and pressure, to the full piston displacement volume; for a liquid-fuel engine, such as a diesel engine, volumetric efficiency is the ratio of the volume of air drawn into a cylinder to the piston displacement.

volumetric flask   [ANALY CHEM] A laboratory flask primarily intended for the preparation of definite, fixed volumes of solutions, and therefore calibrated for a single volume only.

volumetric pipet   [ANALY CHEM] A graduated glass tubing used to measure quantities of a solution; the tube is open at the top and bottom, and a slight vacuum (suction) at the top pulls liquid into the calibrated section; breaking the vacuum allows liquid to leave the tube.

volume unit   [ENG ACOUS] A unit for expressing the audio-frequency power level of a complex electric wave, such as that corresponding to speech or music; the power level in volume units is equal to the number of decibels above a reference level of 1 milliwatt as measured with a standard volume indicator. Abbreviated VU.

voluntary muscle   [PHYSIO] A muscle directly under the control of the will of the organism.

Volvocales   [BOT] An order of one-celled or colonial green algae in the division Chlorophyta; individuals are motile with two, four, or rarely eight whiplike flagella.

Volvocida [INV ZOO] An order of the protozoan class Phytamastigophorea; individuals are grass-green or colorless, have one, two, four, or eight flagella, and thick cell walls of cellulose.

Vombatidae [VERT ZOO] A family of marsupial mammals in the order Diprotodonta in some classification systems.

vomer [ANAT] A skull bone below the ethmoid region constituting part of the nasal septum in most vertebrates.

vortex [FL MECH] 1. Any flow possessing vorticity; for example, an eddy, whirlpool, or other rotary motion. 2. A flow with closed streamlines, such as a free vortex or line vortex. 3. See vortex tube.

vortex alleviation [AERO ENG] The alteration of airfoil configurations to change the airflow patterns directly behind the wings to eliminate or inhibit the vortical motion which directly affects the aircraft immediately following, during closely spaced landings.

vortex filament [FL MECH] The line of concentrated vorticity in a line vortex. Also known as vortex line.

vortex flap [AERO ENG] A leading-edge flap design for highly swept wings, in which the leading edge tabs, which are counterdeflected, cause vortexes to form on the flap; the trapped vortexes cause significantly improved wind flow characteristics.

vortex line [FL MECH] 1. A line drawn through a fluid such that it is everywhere tangent to the vorticity. 2. See vortex filament.

vortex shedding [FL MECH] In the flow of fluids past objects, the shedding of fluid vortices periodically downstream from the restricting object (for example, smokestacks, pipelines, or orifices).

vortex sheet [FL MECH] A surface across which there is a discontinuity in fluid velocity, such as in slippage of one layer of fluid over another; the surface may be regarded as being composed of vortex filaments.

vortex street [FL MECH] A series of vortices which are systematically shed from the downstream side of a body around which fluid is flowing rapidly. Also known as vortex trail; vortex train.

vortex trail See vortex street.

vortex train See vortex street.

vortex tube [FL MECH] A tubular surface consisting of the collection of vortex lines which pass through a small closed curve. Also known as vortex.

vorticity [FL MECH] For a fluid flow, a vector equal to the curl of the velocity of flow.

VSB See vestigial sideband.

VTOL See vertical takeoff and landing.

VU See volume unit.

vulcanization [CHEM ENG] A chemical reaction of sulfur (or other vulcanizing agent) with rubber or plastic to cause cross-linking of the polymer chains; it increases strength and resiliency of the polymer.

vulture [VERT ZOO] The common name for any of various birds of prey in the families Cathartidae and Accipitridae of the order Falconiformes; the head of these birds is usually naked.

vulva [ANAT] The external genital organs of women.

vulval gland [ANAT] A scent gland in the vulval tissues of the human female.

W

W *See* tungsten; watt.

wacke [PETR] Sandstone composed of a mixture of angular and unsorted or poorly sorted fragments of minerals and rocks and an abundant matrix of clay and fine silt.

wafer [ELECTR] A thin semiconductor slice on which matrices of microcircuits can be fabricated, or which can be cut into individual dice for fabricating single transistors and diodes. [ENG] A flat element for a process unit, as in a series of stacked filter elements.

wagnerite [MINERAL] $Mg_2(PO_4)F$ A yellow, grayish, flash-red, or greenish, monoclinic mineral consisting of magnesium fluophosphate.

waiting time *See* idle time.

wake [FL MECH] The region behind a body moving relative to a fluid in which the effects of the body on the fluid's motion are concentrated.

wake flow [FL MECH] Turbulent eddying flow that occurs downstream from bluff bodies.

walk down [ELECTR] A malfunction in a magnetic core of a computer storage in which successive drive pulses or digit pulses cause charges in the magnetic flux in the core that persist after the magnetic fields associated with pulses have been removed. Also known as loss of information.

wallboard [MATER] Panels of various materials for surfacing ceilings and walls, including asbestos cement sheet, plywood, gypsum plasterboard, and laminated plastics.

wall energy [SOLID STATE] The energy per unit area of the boundary between two ferromagnetic domains which are oriented in different directions.

Walley engine [MECH ENG] A multirotor engine employing four approximately elliptical rotors that turn in the same clockwise sense, leading to excessively high rubbing velocities.

walnut [BOT] The common name for about a dozen species of deciduous trees in the genus *Juglans* characterized by pinnately compound, aromatic leaves and chambered or laminate pith; the edible nut of the tree is distinguished by a deeply furrowed or sculptured shell.

walpurgite [MINERAL] $Bi_4(UO_2)(AsO_4)_2O_4 \cdot 3H_2O$ A wax yellow to straw yellow, triclinic mineral consisting of a hydrated arsenate of bismuth and uranium. Also known as waltherite.

walrus [VERT ZOO] *Odobenus rosmarus.* The single species of the pinniped family Odobenidae distinguished by the upper canines in both sexes being prolonged as tusks.

waltherite *See* walpurgite.

wander *See* apparent wander; scintillation.

wandering dune [GEOL] A sand dune that has moved as a unit in the leeward direction of the prevailing winds, and that is characterized by the lack of vegetation to anchor it. Also known as migratory dune; traveling dune.

Wankel engine [MECH ENG] An eccentric-rotor-type internal combustion engine with only two primary moving parts, the rotor and the eccentric shaft; the rotor moves in one direction around the trochoidal chamber containing peripheral intake and exhaust ports.

want *See* nip.

wardite [MINERAL] $Na_4CaAl_{12}(PO_4)_8(OH)_{18} \cdot 6H_2O$ A blue-green to pale green, tetragonal mineral consisting of a hydrated basic phosphate of sodium, calcium, and aluminum.

warhead [ORD] An item which is designed to be mounted in or on a torpedo, guided missile, rocket, or bomb; it may contain high-explosive, nuclear, chemical, biological, or inert materials.

warm air mass [METEOROL] An air mass that is warmer than the surrounding air; an implication that the air mass is warmer than the surface over which it is moving.

warm anticyclone *See* warm high.

warm-core anticyclone *See* warm high.

warm-core cyclone *See* warm low.

warm-core high *See* warm high.

warm-core low *See* warm low.

warm cyclone *See* warm low.

warm front [METEOROL] Any nonoccluded front, or portion thereof, which moves in such a way that warmer air replaces colder air.

warm high [METEOROL] At a given level in the atmosphere, any high that is warmer at its center than at its periphery. Also known as warm anticyclone; warm-core anticyclone; warm-core high.

warm low [METEOROL] At a given level in the atmosphere, any low that is warmer at its center than at its periphery; the opposite of a cold low. Also known as warm-core cyclone; warm-core low; warm cyclone.

warp [GEOL] **1.** An upward or downward flexure of the earth's crust. **2.** A layer of sediment deposited by water. [NAV] To move a vessel or other waterborne object from one point to another by pulling on lines fastened to a fixed buoy, wharf, or such. [TEXT] Yarn extending lengthwise, under tension on a loom. Also known as end.

Warren truss [CIV ENG] A truss having only sloping members between the top and bottom horizontal members.

warringtonite *See* brochantite.

Warrior *See* Orion.

warwickite [MINERAL] $(Mg,Fe)_3Ti(BO_4)_2$ A hair brown to dull black, orthorhombic mineral consisting of a titanoborate of magnesium and iron; occurs as prismatic crystals.

wash [AERO ENG] The stream of air or other fluid sent backward by a jet engine or a propeller. [ENG] **1.** To clean cuttings or other fragmental rock materials out of a borehole by the jetting and buoyant action of a copious flow of water or a mud-laden liquid. **2.** The erosion of core or drill string equipment by the action of a rapidly flowing stream of water or mud-laden drill-circulation liquid. [FL MECH] The surge of disturbed air or other fluid resulting from the passage of something through the fluid. [GEOL] **1.** An alluvial placer. **2.** A piece of land washed by a sea or river. **3.** *See* alluvial cone. [GRAPHICS] To dip negatives and prints in water after fixing to remove the soluble silver halide–fixing agent complexes. [MET] **1.** A coating applied to the face of a mold prior to casting. **2.** A sand expansion defect on the surface finish of a casting due to radiation from the metal rising in the mold and causing increased volume and shear of the interface sand on the upper layers.

wash boring *See* jet drilling.

washout [ENG] **1.** An overlarge well bore caused by the solvent and erosional action of drilling fluid. **2.** A fluid-cut opening resulting from leaking fluid. [MIN ENG] *See* horseback.

wash plain *See* outwash plain to alluvial plain.

wasp [INV ZOO] The common name for members of 67 families of the order Hymenoptera; all are important as parasites or predators of injurious pests.

waste-heat boiler [CHEM ENG] A heat-retrieval unit using hot by-product gas or oil from chemical processes; used to produce steam in a boiler-type system. Also known as gas-tube boiler.

watch buoy *See* station buoy.

water-activated battery [ELEC] A primary battery that contains the electrolyte but requires the addition of or immersion in water before it is usable.

Water Bearer *See* Aquarius.

water block [PETRO ENG] The tendency of accumulated water-oil emulsion around the lower (producing) end of an oil well borehole to block the movement of formation fluids through the formation and toward the borehole.

water boiler *See* water-boiler reactor.

water-boiler reactor [NUCLEO] A homogeneous reactor that uses enriched uranium as fuel and ordinary water as moderator; the fuel is uranyl sulfate dissolved in water. Also known as water boiler.

water brake [ENG] An absorption dynamometer for measuring power output of an engine shaft; the mechanical energy is converted to heat in a centrifugal pump, with a free casing where turning moment is measured.

water cloud [METEOROL] Any cloud composed entirely of liquid water drops; to be distinguished from an ice-crystal cloud and from a mixed cloud.

water column [MECH ENG] A tubular column located at the steam and water space of a boiler to which protective devices such as gage cocks, water gage, and level alarms are attached.

water cooling [ELECTR] Cooling the electrodes of an electron tube by circulating water through or around them. [ENG] Cooling in which the primary coolant is water.

watercress [BOT] *Nasturtium officinale*. A perennial cress generally grown in flooded soil beds and used for salads and food garnishing.

water cycle *See* hydrologic cycle.

waterflooding *See* flooding.

water gas reaction [CHEM ENG] A method used to prepare carbon monoxide by passing steam over hot coke or coal at 600–1000°C.

water hammer [FL MECH] Pressure rise in a pipeline caused by a sudden change in the rate of flow or stoppage of flow in the line.

water level *See* water table.

waterline [GEOL] *See* shoreline. [HYD] *See* water table. [NAV ARCH] **1.** The intersection of the surface of the water with the side of a ship. **2.** A line painted on the hull of a ship showing the level of the water when the ship is properly trimmed.

water loss *See* evapotranspiration.

watermark [GRAPHICS] A localized modification of the structure and opacity of a sheet of paper so that a pattern or design can be seen when the sheet is held to the light.

water mass [OCEANOGR] A body of water identified by its temperature-salinity curve or chemical composition, and normally consisting of a mixture of two or more water types.

watermelon [BOT] *Citrullus vulgaris*. An annual trailing vine with light-yellow flowers and leaves having five to seven deep lobes; the edible, oblong or roundish fruit has a smooth, hard, green rind filled with sweet, tender, juicy, pink to red tissue containing many seeds.

water moccasin [VERT ZOO] *Agkistrodon piscivorus*. A semiaquatic venomous pit viper; skin is brownish or olive on the dorsal aspect, paler on the sides, and has indistinct black bars. Also known as cottonmouth.

Water Monster *See* Hydra.

water of hydration [CHEM] Water present in a definite amount and attached to a compound to form a hydrate; can be removed, as by heating, without altering the composition of the compound.

water opal *See* hyalite.

water pollution [ECOL] Contamination of water by materials such as sewage effluent, chemicals, detergents, and fertilizer runoff.

waterpower [MECH] Power, usually electric, generated from an elevated water supply by the use of hydraulic turbines.

waterproof [ENG] Impervious to water.

water repellent [MATER] Chemicals used to treat textiles, leather, paper, or wood to make them resistant (but not proof) to wetting by water; includes various types of resins, aluminum of zirconium acetates, or latexes.

water rheostat *See* electrolytic rheostat.

water scrubber [CHEM ENG] A device or system in which gases are contacted with water (either by spray or bubbling through) to wash out traces of water-soluble components of the gas stream.

water seal [ENG] A seal formed by water to prevent the passage of gas.

water smoke *See* steam fog.

Water Snake *See* Hydrus.

water softening [CHEM] Removal of scale-forming calcium and magnesium ions from hard water, or replacing them by the more soluble sodium ions; can be done by chemicals or ion exchange.

waterspout [ENG] A pipe or orifice through which water is discharged or by which it is conveyed. [METEOROL] A tornado occurring over water; rarely, a lesser whirlwind over water, comparable in intensity to a dust devil over land.

water table [HYD] The planar surface between the zone of saturation and the zone of aeration. Also known as free-water elevation; free-water surface; groundwater level; groundwater surface; groundwater table; level of saturation; phreatic surface; plane of saturation; saturated surface; water level; waterline.

water tower [CIV ENG] A tower or standpipe for storing water in areas where ordinary water pressure is inadequate for distribution to consumers.

water-tube boiler [MECH ENG] A steam boiler in which water circulates within tubes and heat is applied from outside the tubes to generate steam.

WATS *See* Wide Area Telephone Service.

watt [PHYS] The unit of power in the meter-kilogram-second system of units, equal to 1 joule per second. Symbolized W.

watt current *See* active current.

watt-hour [ELEC] A unit of energy used in electrical measurements, equal to the energy converted or consumed at a rate of 1 watt during a period of 1 hour, or to 3600 joules. Abbreviated Wh.

watt-hour capacity [ELEC] Number of watt-hours which can be delivered from a storage battery under specified conditions as to temperature, rate of discharge, and final voltage.

wattless current *See* reactive current.

watt-hour meter [ENG] A meter that measures and registers the integral, with respect to time, of the active power of the circuit in which it is connected; the unit of measurement is usually the kilowatt-hour.

wattmeter [ENG] An instrument that measures electric power in watts ordinarily.

Watt's law [THERMO] A law which states that the sum of the latent heat of steam at any temperature of generation and the heat required to raise water from 0°C to that temperature is constant; it has been shown to be substantially in error.

wave [FL MECH] A disturbance which moves through or over the surface of a liquid, as of a sea. [PHYS] A disturbance which propagates from one point in a medium to other points without giving the medium as a whole any permanent displacement.

wave acoustics [ACOUS] The study of the propagation of sound based on its wave properties.

wave angle [ELECTROMAG] The angle, either in bearing or elevation, at which a radio wave leaves a transmitting antenna or arrives at a receiving antenna.

wave antenna [ELECTROMAG] Directional antenna composed of a system of parallel, horizontal conductors, varying from a half to several wavelengths long, terminated to ground at the far end in its characteristic impedance.

wave-built platform *See* alluvial terrace.

wave-built terrace *See* alluvial terrace.

wave converter [ELECTROMAG] Device for changing a wave of a given pattern into a wave of another pattern, for example, baffle-plate converters, grating converters, and sheath-reshaping converters for waveguides.

wave-corpuscle duality *See* wave-particle duality.

wave-cut plain *See* wave-cut platform.

wave-cut platform [GEOL] A gently sloping surface which is produced by wave erosion and which extends into the sea for a considerable distance from the base of the wave-cut cliff. Also known as cut platform; erosion platform; strand flat; wave-cut plain; wave-cut terrace; wave platform.

wave-cut terrace *See* wave-cut platform.

wave duct [ELECTROMAG] 1. Waveguide, with tubular boundaries, capable of concentrating the propagation of waves within its boundaries. 2. Natural duct, formed in air by atmospheric conditions, through which waves of certain frequencies travel with more than average efficiency.

wave equation [PHYS] 1. In classical physics, a special equation governing waves that suffer no dissipative attenuation; it states that the second partial derivative with respect to time of the function characterizing the wave is equal to the square of the wave velocity times the Laplacian of this function. Also known as classical wave equation; d'Alembert's wave equation. 2. Any of several equations which relate the spatial and time dependence of a function characterizing some physical entity which can propagate as a wave, including quantum-wave equations for particles.

wave forecasting [OCEANOGR] The theoretical determination of future wave characteristics based on observed or forecasted meteorological phenomena.

waveform [PHYS] The pictorial representation of the form or shape of a wave, obtained by plotting the displacement of the wave as a function of time, at a fixed point in space.

waveform-amplitude distortion *See* frequency distortion.

wavefront [PHYS] 1. A surface of constant phase. 2. The portion of a wave envelope that is between the beginning zero point and the point at which the wave reaches its crest value, as measured either in time or distance.

wavefront reversal *See* optical phase conjugation.

wave function *See* Schrödinger wave function.

wave group [PHYS] A series of waves in which the wave direction, length, and height vary only slightly.

waveguide [ELECTROMAG] 1. Broadly, a device which constrains or guides the propagation of electromagnetic waves along a path defined by the physical construction of the waveguide; includes ducts, a pair of parallel wires, and a coaxial cable. Also known as microwave waveguide. 2. More specifically, a metallic tube which can confine and guide the propagation of electromagnetic waves in the lengthwise direction of the tube.

waveguide attenuation [ELECTROMAG] The decrease from one point of a waveguide to another, in the power carried by an electromagnetic wave in the waveguide.

waveguide cavity [ELECTROMAG] A cavity resonator formed by enclosing a section of waveguide between a pair of waveguide windows which form shunt susceptances.

waveguide filter [ELECTROMAG] A filter made up of waveguide components, used to change the amplitude-frequency response characteristic of a waveguide system.

waveguide junction *See* junction.

waveguide plunger *See* piston.

waveguide probe *See* probe.

waveguide resonator *See* cavity resonator.

waveguide window *See* iris.

wave height [OCEANOGR] The height of a water-surface wave is generally taken as the height difference between the wave crest and the preceding trough. [PHYS] Twice the wave amplitude.

wave impedance [ELECTROMAG] The ratio, at every point in a specified plane of a waveguide, of the transverse component of the electric field to the transverse component of the magnetic field.

wave intensity [PHYS] The average amount of energy transported by a wave in the direction of wave propagation, per unit area per unit time.

wave interference *See* interference.

wavelength [PHYS] The distance between two points having the same phase in two consecutive cycles of a periodic wave, along a line in the direction of propagation.

wavelength division multiplexing [COMMUN] In optical communications, the process in which each modulating wave modulates a separate subcarrier and the subcarriers are spaced in wavelengths.

wavelength standards [SPECT] Accurately measured lengths of waves emitted by specified light sources for the purpose of obtaining the wavelengths in other spectra by interpolating between the standards.

wavellite [MINERAL] $Al_3(PO_4)_2(OH)_3 \cdot 5H_2O$ A white to yellow, green, or black mineral crystallizing in the orthorhombic system and occurring in small hemispherical aggregates.

wave mechanics *See* Schrödinger's wave mechanics.

wavemeter [ENG] A device for measuring the geometrical spacing between successive surfaces of equal phase in an electromagnetic wave.

wave microphone [ENG ACOUS] Any microphone whose directivity depends upon some type of wave interference, such as a line microphone or a reflector microphone.

wave motion [PHYS] The process by which a disturbance at one point is propagated to another point more remote from the source with no net transport of the material of the medium itself; examples include the motion of electromagnetic waves, sound waves, hydrodynamic waves in liquids, and vibration waves in solids. Also known as propagation; wave propagation.

wave optics [OPTICS] The branch of optics which treats of light (or electromagnetic radiation in general) with explicit recognition of its wave nature.

wave packet [PHYS] In wave phenomena, a superposition of waves of differing lengths, so phased that the resultant amplitude is negligibly small except in a limited portion of space whose dimensions are the dimensions of the packet.

wave-particle duality [QUANT MECH] The principle that both matter and electromagnetic radiation exhibit phenomena in which they behave as waves and other phenomena in which they behave as particles, the two aspects being associated by the de Broglie relations. Also known as duality principle; wave-corpuscle duality.

wave platform *See* wave-cut platform.

wave propagation *See* wave motion.

wave-shaping circuit [ELECTR] An electronic circuit used to create or modify a specified time-varying electrical quantity, usually voltage or current, using combinations of electronic devices, such as vacuum tubes or transistors, and circuit elements, including resistors, capacitors, and inductors.

wave soldering *See* flow soldering.

wave theory of light [OPTICS] A theory which assumes that light is a wave motion, rather than a stream of particles.

wave train [PHYS] A series of waves produced by the same disturbance.

wave trap [CIV ENG] A device used to reduce the size of waves from sea or swell entering a harbor before they penetrate as far as the quayage; usually in the form of diverging breakwaters, or small projecting breakwaters situated close within the entrance. [ELECTR] A resonant circuit connected to the antenna system of a receiver to suppress signals at a particular frequency, such as that of a powerful local station that is interfering with reception of other stations. Also known as trap.

wavy extinction *See* undulatory extinction.

ways [CIV ENG] **1.** The tracks and sliding timbers used in launching a vessel. **2.** The building slip or space upon which the sliding timbers or ways, supporting a vessel to be launched, travel. [MECH ENG] Bearing surfaces used to guide and support moving parts of machine tools; may be flat, V-shaped, or dovetailed.

Wb *See* weber.

W boson [PARTIC PHYS] Collective name for two of the hypothetical intermediate vector bosons which carry electric charge; it is believed that charged current interactions are produced by the exhange of these bosons.

weak coupling [PARTIC PHYS] The coupling of four fermion fields in the weak interaction, having a strength many orders of magntiude weaker than that of the strong or electromagnetic interactions.

weak interaction [PARTIC PHYS] One of the fundamental interactions among elementary particles, responsible for beta decay of nuclei, and for the decay of elementary particles with lifetimes greater than about 10^{-10} second, such as muons, K mesons, and lambda hyperons; it is several orders of magnitude weaker than the strong and electromagnetic interactions, and fails to conserve strangeness or parity. Also known as beta interaction.

weak topology [MATH] A topology on a topological vector space X whose open neighborhoods around a point x are obtained from those points y of X for which every $f_i(x)$ is close to $f_i(y)$, f_i appearing in a finite list of linear functionals.

weapons system [ORD] Two or more instruments of combat operating as a single unit of striking power in military combat; specifically, a system in which two instruments of combat are required to perform a single mission.

wear [ENG] Deterioration of a surface due to material removal caused by relative motion between it and another part.

weasel [VERT ZOO] The common name for at least 12 species of small, slim carnivores which belong to the family Mustelidae and which have a reddish-brown coat with whitish underparts; species in the northern regions have white fur during the winter and are called ermine.

weather [METEOROL] **1.** The state of the atmosphere, mainly with respect to its effects upon life and human activities; as distinguished from climate, weather consists of the short-term (minutes to months) variations of the atmosphere. **2.** As used in the making of surface weather observations, a category of individual and combined atmospheric phenomena which must be drawn upon to describe the local atmospheric activity at the time of observation.

weathering [GEOL] Physical disintegration and chemical decomposition of earthy and rocky materials on exposure to atmospheric agents, producing an in-place mantle of waste. Also known as clastation; demorphism.

weather minimum [METEOROL] The worst weather conditions under which aviation operations may be conducted under either visual or instrument flight rules; usually prescribed by directives and standing operating procedures in terms of minimum ceiling, visibility, or specific hazards to flight.

weather station [METEOROL] A place and facility for the observation, measurement, and recording of the variable elements of weather; one of the most effective network facilities is that of the U.S. Weather Bureau.

weather strip [BUILD] A piece of material, such as wood or rubber, applied to the joints of a window or door to stop drafts.

web [ARCH] The portion of a ribbed vault between ribs. [CIV ENG] The vertical strip connecting the upper and lower flanges of a rail or girder. [GRAPHICS] The continuous length of paper formed when paper pulp moves through a papermaking machine; the web is then cut into sheets or wrapped onto rolls. [MATER] In a grain of propellant, the minimum thickness of the grain between any two adjacent surfaces. [MECH ENG] For twist drills and reamers, the central portion of the tool body that joins the loads. [MET] In forging, the thin section of metal remaining at the bottom of a depression or at the location of the punches. [OPTICS] *See* wire. [TEXT] A fabric as it is being woven on a loom. [VERT ZOO] The membrane between digits in many birds and amphibians.

weber [ELECTROMAG] The unit of magnetic flux in the meter-kilogram-second system, equal to the magnetic flux which, linking a circuit of one turn, produces in it an electromotive force of 1 volt as it is reduced to zero at a uniform rate in 1 second. Symbolized Wb.

Weberian apparatus [VERT ZOO] A series of bony ossicles which form a chain connecting the swim bladder with the inner ear in fishes of the superorder Ostariophysi.

weberite [MINERAL] Na_2MgAlF_7 A light gray, orthorhombic mineral consisting of an aluminofluoride of sodium and magnesium; occurs as grains and masses.

web-fed press [GRAPHICS] Printing press designed to accept paper from rolls instead of sheets; large web presses offer great speed and economy for long press runs, but high makeready and plate costs make them too costly for short runs.

websterite *See* aluminite.

Weddell Current [OCEANOGR] A surface current which flows in an easterly direction from the Weddell Sea outside the limit of the West Wind Drift.

weddellite [MINERAL] $CaC_2O_4 \cdot 2H_2O$ A colorless to white or yellowish-brown to brown, tetragonal mineral consisting of calcium oxalate dihydrate.

wedge [COMMUN] A convergent pattern of equally spaced black and white lines, used in a television test pattern to indicate resolution. [DES ENG] A piece of resistant material whose two major surfaces make an acute angle. [ELECTROMAG] A waveguide termination consisting of a tapered length of dissipative material introduced into the guide, such as carbon. [ENG] In ultrasonic testing, a device which directs waves of ultrasonic energy into the test piece at an angle. [METEOROL] *See* ridge. [OPTICS] 1. An optical filter in which the transmission decreases continuously or in steps from one end to the other. 2. A refracting prism of very small angle, inserted in an optical train to introduce a bend in the ray path.

wedge filter [NUCLEO] A radiation filter so constructed that its thickness or transmission characteristics vary continuously or in steps from one edge to the other; used to increase the uniformity of radiation in certain types of treatment.

wedge photometer [ENG] A photometer in which the luminous flux density of light from two sources is made equal by pushing into the beam from the brighter source a wedge of absorbing material; the wedge has a scale indicating how much it reduces the flux density, so that the luminous intensities of the sources may be compared.

week [ASTRON] A time period of 7 days which has been accepted from ancient Babylon; the 7 days of the week were first given names of the seven celestial bodies: the sun, moon, and five visible planets.

weep hole [CIV ENG] A hole in a wood sill, retaining wall, or other structure to allow accumulated water to escape.

weevil [INV ZOO] Any of various snout beetles whose larvae destroy crops by eating the interior of the fruit or grain, or bore through the bark into the pith of many trees.

wehrlite [MINERAL] BiTe A mineral that is a native alloy of bismuth and tellurium. Also known as mirror glance. [PETR] A peridotite composed principally of olivine and clinopyroxene with accessory opaque oxides.

weibullite [MINERAL] $Pb_4Bi_6S_9Se_4$ A steel gray mineral consisting of lead bismuth sulfide with selenium replacing the sulfide; occurs in indistinct prismatic crystals in massive form.

Weierstrass functions [MATH] Used in the calculus of variations, these determine functions satisfying the Euler-Lagrange equation and Jacobi's condition while maximizing a given definite integral.

Weierstrass M test [MATH] An infinite series of numbers will converge or functions will converge uniformly if each term is dominated in absolute value by a nonnegative constant M_n, where these M_n form a convergent series. Also known as Weierstrass' test for convergence.

Weierstrass' test for convergence *See* Weierstrass M test.

Weierstrass transform [MATH] This transform of a real function $f(y)$ is the function given by the integral from $-\infty$ to ∞ of $(4\pi t)^{-1/2} \exp[-(x-y)^2/4] f(y)dy$; this is used in studying the heat equation.

weight [MECH] 1. The gravitational force with which the earth attracts a body. 2. By extension, the gravitational force with which a star, planet, or satellite attracts a nearby body.

weighted area masks [COMPUT SCI] In character recognition, a set of characters (each character residing in the character reader in the form of weighted points) which theoretically render all input specimens unique, regardless of the size or style.

weighted average [STAT] The number obtained by adding the product of α_i times the ith number in a set of N numbers for $i = 1,2,...,N$, where α_i are numbers (weights) such that $\alpha_1 + \alpha_2 + ... + \alpha_N = 1$.

weighted code [COMPUT SCI] A method of representing a decimal digit by a combination of bits, in which each bit is assigned a weight, and the value of the decimal digit is found by multiplying each bit by its weight and then summing the results.

weightlessness [MECH] A condition in which no acceleration, whether of gravity or other force, can be detected by an observer within the system in question. Also known as zero gravity.

weightlessness switch *See* zero gravity switch.

Weinberg-Salam theory [PARTIC PHYS] A gage theory in which the electromagnetic and weak nuclear interactions are described by a single unifying framework in which both have a characteristic coupling parameter equal to the fine-structure constant; it predicts the existence of intermediate vector bosons and neutral current interactions. Also known as Salam-Weinberg theory.

weinschenkite [MINERAL] 1. $YPO_4 \cdot 2H_2O$ A white mineral consisting of a hydrous yttrium phosphate. Also known as churchite. 2. A dark-brown variety of hornblende high in ferric iron, aluminum, and water.

weir [CIV ENG] A dam in a waterway over which water flows, serving to regulate water level or measure flow.

weir tank [PETRO ENG] A type of oil-field storage tank with high- and low-level weir boxes and liquid-level controls for metering the liquid content of the tank.

weissite [MINERAL] Cu_5Te_3 A dark bluish-black mineral consisting of copper telluride; occurs in massive form.

Weiss magneton [ATOM PHYS] A unit of magnetic moment, equal to 1.853×10^{-21} erg/oersted, about one-fifth of the Bohr magneton; it is experimentally derived, the magnetic moments of certain molecules being close to integral multiples of this quantity.

Weiss molecular field [SOLID STATE] The effective magnetic field postulated in the Weiss theory of ferromagnetism, which acts on atomic magnetic moments within a domain, tending to align them, and is in turn generated by these magnetic moments.

Weiss theory [SOLID STATE] A theory of ferromagnetism based on the hypotheses that below the Curie point a ferromagnetic substance is composed of small, spontaneously magnetized regions called domains, and that each domain is spontaneously magnetized because a strong molecular magnetic field tends to align the individual atomic magnetic moments within the domain. Also known as molecular field theory.

weld [MET] A union made between two metals by welding.

weld bead [MET] A deposit of filler metal from a single welding pass. Also known as bead.

weldbonding [MET] A process for joining metals in which adhesive, typically an epoxy paste, is applied to the parts, which are then clamped together, spot-welded, and put into an oven (250°F, or 121°C, for 1 hour) to cure the adhesive.

welded tuff [PETR] A pyroclastic deposit hardened by the action of heat, pressure from overlying material, and hot gases. Also known as tuff lava.

welding [GEOL] Consolidation of sediments by pressure; water is squeezed out and cohering particles are brought within the limits of mutual molecular attraction. [MET] Joining two metals by applying heat to melt and fuse them, with or without filler metal.

welding electrode [MET] **1.** In arc welding, the current-carrying rod or rods used to strike an arc between rod and work. **2.** In resistance welding, the component of a machine through which current and pressure are applied to the work.

welding force See electrode force.

welding rod [MET] Filler metal in the form of a rod or heavy wire.

weld mark See flow line.

weldment [ENG] An assembly or structure whose component parts are joined by welding.

well [BUILD] An open shaft in a building, extending vertically through floors to accommodate stairs or an elevator. [ENG] A hole dug into the earth to reach a supply of water, oil, brine, or gas.

well completion [PETRO ENG] The final sealing off of a drilled well (after drilling apparatus is removed from the borehole) with valving, safety, and flow-control devices.

well core [ENG] A sample of rock penetrated in a well or other borehole obtained by use of a hollow bit that cuts a circular channel around a central column or core.

wellhead [CIV ENG] The top of a well. [HYD] The place where a stream emerges from the ground.

wellhole [MIN ENG] **1.** A large-diameter vertical hole used in quarries and opencast pits for taking heavy explosive charges in blasting. **2.** The sump, or portion of a shaft below the place where skips are caged at the bottom of the shaft, in which water collects.

well logging [ENG] The technique of analyzing and recording the character of a formation penetrated by a drill hole in mineral exploration and exploitation work.

well-ordered set [MATH] A linearly ordered set where every subset has a least element.

wellpoint [CIV ENG] A component of a wellpoint system consisting of a perforated pipe about 4 feet (1.2 meters) long and about 2 inches (5 centimeters) in diameter, equipped with a ball valve, a screen, and a jetting tip.

wellpoint system [CIV ENG] A method of keeping an excavated area dry by intercepting the flow of groundwater with pipe wells located around the excavation area.

well shooting [ENG] The firing of a charge of nitroglycerin, or other high explosive, in the bottom of a well for the purpose of increasing the flow of water, oil, or gas.

well-type manometer [ENG] A type of double-leg, glass-tube manometer; one leg has a relatively small diameter, and the second leg is a reservoir; the level of the liquid in the reservoir does not change appreciably with change of pressure; a mercury barometer is a common example.

Welwitschiales [BOT] An order of gymnosperms in the subdivision Geneticae represented by the single species *Welwitschia mirabilis* of southwestern Africa; distinguished by only two leaves and short, unbranched, cushion- or saucer-shaped woody main stem which tapers to a very long taproot.

Wenlockian [GEOL] A European stage of geologic time: Middle Silurian (above Tarannon, below Ludlovian).

Wentzel-Kramers-Brillouin method [QUANT MECH] Method of approximating quantum-mechanical wave functions and energy levels, in which the logarithm of the wave function is expanded in powers of Planck's constant, and all except the first two terms are neglected. Also known as phase integral method; WKB method.

Werfenian stage See Scythian stage.

Werner complex See coordination compound.

wernerite See scapolite.

West Australia Current [OCEANOGR] The complex current flowing northward along the west coast of Australia; it is strongest from November to January, and weakest and variable from May to July; it curves toward the west to join the South Equatorial Current.

westerlies [METEOROL] The dominant west-to-east motion of the atmosphere, centered over the middle latitudes of both hemispheres; at the earth's surface, the westerly belt (or west-wind belt) extends, on the average, from about 35 to 65° latitude. Also known as circumpolar westerlies; middle-latitude westerlies; mid-latitude westerlies; polar westerlies; subpolar westerlies; subtropical westerlies; temperate westerlies; zonal westerlies; zonal winds.

Western Equatorial Countercurrent [OCEANOGR] Weak, narrow bands of eastward-flowing water observed in some winter months in the western Atlantic near the equator.

Western equine encephalitis [MED] A type of equine encephalitis which occurs chiefly west of the Mississippi River; the chief vector is the culicine mosquito *Culex tarsalis.*

West Greenland Current [OCEANOGR] The current flowing northward along the west coast of Greenland into the Davis Strait; part of this current joins the Labrador Current, while the other part continues into Baffin Bay.

West Nile fever [MED] An acute, usually mild, mosquito-borne virus disease occurring in summer, chiefly in Egypt, Israel, Africa, India, and Korea; signs are fever and lymphadenopathy, sometimes with a rash.

Weston standard cell [ELEC] A standard cell used as a highly accurate voltage source for calibrating purposes; the positive electrode is mercury, the negative electrode is cadmium, and the electrolyte is a saturated cadmium sulfate solution; the Weston standard cell has a voltage of 1.018636 volts at 20°C.

West Wind Drift *See* Antarctic Circumpolar Current.

wet [PHYS] A liquid is said to wet a solid if the contact angle between the solid and the liquid, measured through the liquid, lies between 0 and 90°, and not to wet the solid if the contact angle lies between 90 and 180°.

wet and dry bulb thermometer *See* psychrometer.

wet assay [MIN ENG] The determination of the quantity of a desired constituent in ores, metallurgical residues, and alloys by the use of the processes of solution, flotation, or other liquid means.

wet-bulb depression [METEOROL] The difference in degrees between the dry-bulb temperature and the wet-bulb temperature.

wet-bulb temperature [METEOROL] **1.** Isobaric wet-bulb temperature, that is, the temperature an air parcel would have if cooled adiabatically to saturation at constant pressure by evaporation of water into it, all latent heat being supplied by the parcel. **2.** The temperature read from the wet-bulb thermometer; for practical purposes, the temperature so obtained is identified with the isobaric wet-bulb temperature.

wet-bulb thermometer [ENG] A thermometer having the bulb covered with a cloth, usually muslin or cambric, saturated with water.

wet cell [ELEC] A primary cell in which there is a substantial amount of free electrolyte in liquid form.

wet collector *See* scrubber.

wet cooling tower [MECH ENG] A structure in which water is cooled by atomization into a stream of air; heat is lost through evaporation. Also known as evaporative cooling tower.

wet drill [MECH ENG] A percussive drill with a water feed either through the machine or by means of a water swivel, to suppress the dust produced when drilling.

wet hole [ENG] A borehole that traverses a water-bearing formation from which the flow of water is great enough to keep the hole almost full of water.

wet mill [MECH ENG] **1.** A grinder in which the solid material to be ground is mixed with liquid. **2.** A mill in which the grinding energy is developed by a fast-flowing liquid stream; for example, a jet pulverizer.

wet snow [METEOROL] Deposited snow that contains a great deal of liquid water.

wet strength [MATER] **1.** The strength of a material saturated with water. **2.** The ability to withstand water (as for paper products) with a wet-strength additive or resin finish.

wettability [CHEM] The ability of any solid surface to be wetted when in contact with a liquid; that is, the surface tension of the liquid is reduced so that the liquid spreads over the surface.

wetted [CHEM] Pertaining to material that has accepted water or other liquid, either on its surface or within its pore structure.

wetted-wall column [CHEM ENG] A vertical column that operates with the inner walls wetted by the liquid being processed; used in theoretical studies of mass transfer rates and in analytical distillations; an example is a spinning-band column.

wetting [ELECTR] The coating of a contact surface with an adherent film of mercury. [MET] Spreading liquid filler metal or flux on a solid base metal.

wet well [MECH ENG] A chamber which is used for collecting liquid, and to which the suction pipe of a pump is attached.

Wh *See* watt-hour.

whale [VERT ZOO] A large marine mammal of the order Cetacea; the body is streamlined, the broad flat tail is used for propulsion, and the limbs are balancing structures.

Whale *See* Cetus.

whaleback dune [GEOL] A smooth, elongated mound or hill of desert sand shaped generally like a whale's back; formed by passage of a succession of longitudinal dunes along the same path. Also known as sand levee.

whalebone *See* baleen.

wharf [CIV ENG] A structure of open construction built parallel to the shoreline; used by vessels to receive and discharge passengers and cargo.

wheat [BOT] A food grain crop of the genus *Triticum*; plants are self-pollinating; the inflorescence is a spike bearing sessile spikelets arranged alternately on a zigzag rachis.

Wheatstone bridge [ELEC] A four-arm bridge circuit, all arms of which are predominately resistive; used to measure the electrical resistance of an unknown resistor by comparing it with a known standard resistance. Also known as resistance bridge; Wheatstone network.

Wheatstone network *See* Wheatstone bridge.

wheel base [DES ENG] The distance in the direction of travel from front to rear wheels of a vehicle, measured between centers of ground contact under each wheel.

wheel printer [COMPUT SCI] A line printer that prints its characters from the rim of a wheel around which is the type for the alphabet, numerals, and other characters.

whelk [INV ZOO] A gastropod mollusk belonging to the order Neogastropoda; species are carnivorous but also scavenge.

whewellite [MINERAL] $Ca(C_2O_4) \cdot H_2O$ A colorless or yellowish or brownish, monoclinic mineral consisting of calcium oxalate monohydrate; occurs as crystals.

whip antenna [ELECTROMAG] A flexible vertical rod antenna, used chiefly on vehicles. Also known as fishpole antenna.

whip grafting [BOT] A method of grafting by fitting a small tongue and notch cut in the base of the scion into corresponding cuts in the stock.

whipworm disease [MED] A chronic, wasting diarrhea produced by heavy parasitization of the large intestine by the nematode *Trichuris trichiura*, particularly in undernourished children in the tropics.

whiskers *See* vibrissae.

whiskey [FOOD ENG] A potable alcoholic beverage made by distilling fermented grain mashes and aging the distillate in wood, usually oak; principal sources of grain are barley, wheat, rye, oats, and corn.

whistler [GEOPHYS] An effect that occurs when a plasma disturbance, caused by a lightning discharge, travels out along lines of magnetic force of the earth's field and is reflected back to its origin from a magnetically conjugate point on the earth's surface; the disturbance may be picked up electromagnetically and converted directly to sound; the characteristic drawn-out descending pitch of the whistler is a dispersion effect due to the greater velocity of the higher-frequency components of the disturbance.

white ant *See* termite.

white blood cell *See* leukocyte.

white body [PHYS] A hypothetical substance whose surface absorbs no electromagnetic radiation of any wavelength, that is, one which exhibits zero absorptivity for all wavelengths.

whitecap [OCEANOGR] A cloud of bubbles at the sea surface caused by a breaking wave.

white clay *See* kaolin.

white compression [COMMUN] In facsimile or television the reduction in picture-signal gain at levels corresponding to light areas, with respect to the gain at the level for midrange light values; the overall effect of white compression is to reduce contrast in the highlights of the picture.

white corpuscle *See* leukocyte.

white dwarf star [ASTRON] An intrinsically faint star of very small radius and high density; the mass is about 0.6 that of the sun and the average radius is about 8000 kilometers; it is one final stage of stellar evolution with thermonuclear energy sources extinct.

white feldspar *See* albite.

whitefish [VERT ZOO] Any of various food fishes in the family Salmonidae, especially of the genus *Coregonus*, characterized by an adipose dorsal fin and nearly toothless mouth.

white frost *See* hoarfrost.

white garnet *See* leucite.

white iron ore *See* siderite.

white light [OPTICS] Any radiation producing the same color sensation as average noon sunlight.

white metal [MET] 1. Any of several white-colored metals and their alloys of relatively low melting points, such as lead, tin, antimony, and zinc. 2. A copper matte of about 77% copper, obtained from the smelting of sulfide copper ores.

white mica *See* muscovite.

white nickel *See* rammelsbergite.

white noise [PHYS] Random noise that has a constant energy per unit bandwidth at every frequency in the range of interest.

whiteout [METEOROL] An atmospheric optical phenomenon of the polar regions in which the observer appears to be engulfed in a uniformly white glow; shadows, horizon, and clouds are not discernible; sense of depth and orientation are lost; dark objects in the field of view appear to float at an indeterminable distance. Also known as milky weather.

white potato *See* potato.

white rainbow *See* fogbow.

white schorl *See* albite.

white tellurium *See* krannerite; sylvanite.

white transmission [COMMUN] 1. In an amplitude-modulated system, that form of transmission in which the maximum transmitted power corresponds to the minimum density of the subject copy. 2. In a frequency-modulation system, that form of transmission in which the lowest transmitted frequency corresponds to the minimum density of the subject copy.

white water [OCEANOGR] Frothy water, as in whitecaps or breakers.

whiting *See* chalk.

whole step *See* whole tone.

whole tone [ACOUS] The interval between two sounds whose basic frequency ratio is approximately equal to the sixth root of 2. Also known as whole step.

whooping cough *See* pertussis.

whooping crane [VERT ZOO] *Grus americana*. A member of a rare North American migratory species of wading birds; the entire species forms a single population.

whorl [ANAT] A fingerprint pattern in which at least two deltas are present with a recurve in front of each. [BOT] An arrangement of several identical anatomical parts, such as petals, in a circle around the same point.

wiborgite *See* rapakivi.

wichtisite *See* tachylite.

wicket dam [CIV ENG] A movable dam consisting of a number of rectangular panels of wood or iron hinged to a sill and propped vertically; the prop is hinged and can be tripped to drop the wickets flat on the sill.

Wide Area Telephone Service [COMMUN] A special telephone service that allows a customer to call anyone in one or more of six regions into which the continental United States has been divided, on a direct dialing basis, for a flat monthly charge related to the number of regions to be called. Abbreviated WATS.

wide band [ELECTR] Property of a tuner, amplifier, or other device that can pass a broad range of frequencies.

wide-open [ELECTR] Refers to the untuned characteristic or lack of frequency selectivity.

Wiedemann-Franz law [SOLID STATE] The law that the ratio of the thermal conductivity of a metal to its electrical conductivity is a constant, independent of the metal, times the absolute temperature. Also known as Lorentz relation.

Wien bridge oscillator [ELECTR] A phase-shift feedback oscillator that uses a Wien bridge as the frequency-determining element.

Wien capacitance bridge [ELEC] A four-arm alternating-current bridge used to measure capacitance in terms of resistance and frequency; two adjacent arms contain capacitors respectively in parallel and in series with resistors, while the other two arms are nonreactive resistors; bridge balance depends on frequency.

Wien constant [STAT MECH] The product of the temperature and the wavelength at which the intensity of radiation from a blackbody reaches its maximum; it is equal to approximately 2898 micrometer-kelvins.

Wien effect [PHYS CHEM] An increase in the conductance of an electrolyte at very high potential gradients.

Wiener process [MATH] A stochastic process with normal density at each stage, arising from the study of Brownian motion, which represents the limit of a sequence of experiments. Also known as Gaussian noise.

Wien frequency bridge [ELEC] A modification of the Wien capacitance bridge, used to measure frequencies.

Wien inductance bridge [ELEC] A four-arm alternating-current bridge used to measure inductance in terms of resistance and frequency; two adjacent arms contain inductors respectively in parallel and in series with resistors, while the other two arms are nonreactive resistors; bridge balance depends on frequency.

Wien-Maxwell bridge *See* Maxwell bridge.

Wien's displacement law [STAT MECH] A law for blackbody radiation which states that the wavelength at which the maximum amount of radiation occurs is a constant equal to approximately 2898 times the product of 1 micrometer and 1 kelvin. Also known as displacement law; Wien's radiation law.

Wien's distribution law [STAT MECH] A formula for the spectral distribution of radiation from a blackbody, which is a good approximation to the Planck radiation formula at sufficiently low temperatures or wavelengths, for example, in the visible region of the spectrum below 3000K. Also known as Wien's radiation law.

Wien's radiation law [STAT MECH] **1.** The law that the intensity of radiation emitted by a blackbody per unit wavelength, at that wavelength at which this intensity reaches a maximum, is proportional to the fifth power of the temperature. **2.** See Wien's displacement law. **3.** See Wien's distribution law.

Wiesen See meadow.

Wigner coefficient See vector coupling coefficient.

Wigner effect See discomposition effect.

Wigner force [NUCLEO] A short-range nonexchange force between nucleons, postulated to explain various phenomena.

Wigner nuclides [NUC PHYS] The most important class of mirror nuclides, comprising pairs of odd-mass-number isobars for which the atomic number and the neutron number differ by 1.

Wigner-Seitz method [SOLID STATE] A method of approximating the band structure of a solid: Wigner-Seitz cells surrounding atoms in the solid are approximated by spheres, and band solutions of the Schrödinger equation for one electron are estimated by using the assumption that an electronic wave function is the product of a plane wave function and a function whose gradient has a vanishing radial component at the sphere's surface.

Wigner's theorem [QUANT MECH] **1.** The theorem that, if ψ is an eigenfunction of the Hamiltonian operator and R is a symmetry element of the Hamiltonian, then $R\psi$ is an eigenfunction of the Hamiltonian having the same eigenvalue as ψ. **2.** Angular momentum of the electron spin is conserved in a collision of the second kind.

Wigner supermultiplet [NUC PHYS] A set of quantum-mechanical states of a collection of nucleons which form the basis of a representation of $SU(4)$, especially appropriate when spin and isospin dependence of the nuclear interaction may be disregarded; several combinations of spin and isospin multiplets may occur in a supermultiplet.

wildfire [PL PATH] A bacterial disease of tobacco caused by *Pseudomonas tabaci* and characterized by the appearance of brown spots surrounded by yellow rings, which turn dark, rot, and fall out.

Wilfley table [MIN ENG] A flat, rectangular surface that can be tilted and shaken about the long axis and has horizontal riffles for imposing restraint in removing minerals from classified sand. Also known as shaking table.

wilkeite [MINERAL] $Ca_5(SiO_4,PO_4,SO_4)_3(O,OH,F)$ A rose red or yellow, hexagonal mineral consisting of a basic sulfate-silicate-phosphate of calcium.

willemite [MINERAL] Zn_2SiO_4 A white, greenish-yellow, green, reddish, or brown mineral that forms rhombohedral crystals and exhibits intense bright-yellow fluorescence in ultraviolet light; a minor ore of zinc.

Williamsoniaceae [PALEOBOT] A family of extinct plants in the order Cycadeoidales distinguished by profuse branching.

williwaw [METEOROL] A very violent squall in the Straits of Magellan; it may occur in any month but occurs most frequently in winter.

willow [BOT] A deciduous tree and shrub of the genus *Salix*, order Salicales; twigs are often yellow-green and bear alternate leaves which are characteristically long, narrow, and pointed, usually with fine teeth along the margins.

Wilson cloud chamber [NUCLEO] A cloud chamber containing air supersaturated with water vapor by sudden expansion, in which rapidly moving nuclear particles such as alpha or beta rays produced ionization tracks by condensation of vapor on the ions produced by the rays.

wilt [PL PATH] Any of various plant diseases characterized by drooping and shriveling, following loss of turgidity.

winch [MECH ENG] A machine having a drum on which to coil a rope, cable, or chain for hauling, pulling, or hoisting.

Winchester disk [COMPUT SCI] A type of disk storage device characterized by nonremovable or sealed disk packs; extremely narrow tracks; a lubricated surface that allows the head to rest on the surface during start and stop operations; and servomechanisms which utilize a magnetic pattern, recorded in the medium itself, to position the head.

Winchester technology [COMPUT SCI] Innovations designed to achieve disks with up to 6×10^8 bytes per disk drive; the technology includes nonremovable or sealed disk packs, a read/write head that weighs only 0.25 gram and floats above the surface, magnetic orientation of iron oxide particles on the disk surface, and lubrication of the disk surface.

wind [ELECTR] The manner in which magnetic tape is wound onto a reel; in an A wind, the coated surface faces the hub; in a B wind, the coated surface faces away from the hub. [METEOROL] The motion of air relative to the earth's surface; usually means horizontal air motion, as distinguished from vertical motion, and air motion averaged over the response period of the particular anemometer.

wind-chill index [METEOROL] The cooling effect of any combination of temperature and wind, expressed as the loss of body heat in kilogram calories per hour per square meter of skin surface; it is only an approximation because of individual body variations in shape, size, and metabolic rate.

wind cone [ENG] A tapered fabric sleeve, shaped like a truncated cone and pivoted at its larger end on a standard, for the purpose of indicating wind direction; since the air enters the fixed end, the small end of the cone points away from the wind. Also known as wind sleeve; wind sock.

wind drift [ACOUS] Shift in the apparent position of a sound source or target observed by sound apparatus; it is caused by the effect of wind on sound waves, which changes their direction and increases or decreases sound lag. [OCEANOGR] See drift current.

wind-driven current See drift current.

wind gap [GEOL] A shallow, relatively high-level notch in the upper part of a mountain ridge, usually an abandoned water gap. Also known as air gap; wind valley.

winding [ELEC] **1.** One or more turns of wire forming a continuous coil for a transformer, relay, rotating machine, or other electric device. **2.** A conductive path, usually of wire, that is inductively coupled to a magnetic storage core or cell.

windmill anemometer [ENG] A rotation anemometer in which the axis of rotation is horizontal; the instrument has either flat vanes (as in the air meter) or helicoidal vanes (as in the propeller anemometer);

the relation between wind speed and angular rotation is almost linear.

window [AERO ENG] An interval of time during which conditions are favorable for launching a spacecraft on a specific mission. [BUILD] An opening in the wall of a building or the body of a vehicle to admit light and usually to permit vision through a transparent or translucent material, usually glass. [ELECTR] A material having minimum absorption and minimum reflection of radiant energy, sealed into the vacuum envelope of a microwave or other electron tube to permit passage of the desired radiation through the envelope to the output device. [ELECTROMAG] A hole in a partition between two cavities or waveguides, used for coupling. [GEOL] A break caused by erosion of a thrust sheet or a large recumbent anticline that exposes the rocks beneath the thrust sheet. Also known as fenster. [GEOPHYS] Any range of wavelengths in the electromagnetic spectrum to which the atmosphere is transparent. [HYD] The unfrozen part of a river surrounded by river ice during the winter. [MATER] A globular defect in a thermoplastic sheet or film caused by incomplete plasticization; similar to a fisheye. [NUCLEO] 1. An aperture for the passage of particles or radiation in a nuclear reactor. 2. An energy range of relatively high transparency in the total neutron cross section of a material; such windows arise from interference between potential and resonance scattering in elements of intermediate atomic weight, and can be of importance in neutron shielding. [ORD] A confusion reflector consisting of strips of chaff, wire, or bars cut to give resonance at expected enemy radar frequencies, and dropped in clusters from aircraft or expelled from shells or rockets as a radar countermeasure.

wind rose [METEOROL] A diagram in which statistical information concerning direction and speed of the wind at a location may be summarized; a line segment is drawn in each of perhaps eight compass directions from a common origin; the length of a particular segment is proportional to the frequency with which winds blow from that direction; thicknesses of a segment indicate frequencies of occurrence of various classes of wind speed.

windrow [GEOL] Any accumulation of material formed by wind or tide action.

winds aloft [METEOROL] Generally, the wind speeds and directions at various levels in the atmosphere above the domain of surface weather observations, as determined by any method of winds-aloft observation. Also known as upper-level winds; upper winds.

wind shear [METEOROL] The local variation of the wind vector or any of its components in a given direction.

wind-shift line [METEOROL] A line or narrow zone along which there is an abrupt change of wind direction.

wind sleeve See wind cone.

wind sock See wind cone.

windstorm [METEOROL] A storm in which strong wind is the most prominent characteristic.

wind stress [METEOROL] The drag or tangential force per unit area exerted on the surface of the earth by the adjacent layer of moving air.

wind tunnel [ENG] A duct in which the effects of airflow past objects can be determined.

wind valley See wind gap.

wine [FOOD ENG] An alcoholic beverage made by fermentation of the juice of fruits or berries, especially grapes; classified on the basis of color, sweetness, alcoholic content, variety of grape, presence of carbon dioxide, and region where the grapes are grown.

wing [AERO ENG] 1. A major airfoil. 2. An airfoil on the side of an airplane's fuselage or cockpit, paired off by one on the other side, the two providing the principal lift for the airplane. [GEOL] See vesicle. [ZOO] Any of the paired appendages serving as organs of flight on many animals.

wing dam See groin.

Winged Horse See Pegasus.

winglet [AERO ENG] A small, nearly vertical, winglike surface mounted rearward above the wing tip to reduce drag coefficients at lifting conditions.

wing nut [DES ENG] An internally threaded fastener with wings to permit it to be tightened or loosened by finger pressure only. Also known as butterfly nut.

wing section See airfoil profile.

winning [MIN ENG] 1. A new mine opening. 2. The portion of a coal field laid out for working. 3. Mining.

winter [ASTRON] The period from the winter solstice, about December 22, to the vernal equinox, about March 21; popularly and for most meteorological purposes, winter is taken to include December, January, and February in the Northern Hemisphere, and June, July, and August in the Southern Hemisphere.

Winteraceae [BOT] A family of dicotyledonous plants in the order Magnoliales distinguished by hypogynous flowers, exstipulate leaves, air vessels absent, and stamens usually laminar.

winter solstice [ASTRON] 1. The sun's position on the ecliptic (about December 22). Also known as first point of Capricorn. 2. The date (December 22) when the greatest southern declination of the sun occurs.

wiper [ELEC] That portion of the moving member of a selector, or other similar device, in communications practice, which makes contact with the terminals of a bank.

wire [ELEC] A single bare or insulated metallic conductor having solid, stranded, or tinsel construction, designed to carry current in an electric circuit. Also known as electric wire. [MET] A thin, flexible, continuous length of metal, usually of circular cross section. [OPTICS] A filament, usually consisting of a stretched strand of spider's web or a fine metal wire, mounted in the field of view of a telescope eyepiece to serve as a reference or for measurements. Also known as web.

wiregrating [ELECTROMAG] A series of wires placed in a waveguide that allow one or more types of waves to pass and block all others.

wirephoto [COMMUN] 1. A photograph transmitted over wires to a facsimile receiver. 2. See facsimile.

wire printing See matrix printing.

wire recorder [ENG ACOUS] A magnetic recorder that utilizes a round stainless steel wire about 0.004 inch (0.01 centimeter) in diameter instead of magnetic tape.

wire saw [MECH ENG] A machine employing one- or three-strand wire cable, up to 16,000 feet (4900 meters) long, running over a pulley as a belt; used in quarries to cut rock by abrasion.

wire-wound potentiometer [ELEC] A potentiometer which is similar to a slide-wire potentiometer, except that the resistance wire is wound on a form and contact is made by a slider which moves along an edge from turn to turn.

wire-wound rheostat [ELEC] A rheostat in which a sliding or rolling contact moves over resistance wire that has been wound on an insulating core.

wiring [ELEC] The installation and utilization of a system of wire for conduction of electricity. Also known as electric wiring. [SCI TECH] A system of wires.

wiring board *See* control panel.

wiring diagram *See* circuit diagram.

wiring harness [ELEC] An array of insulated conductors bound together by lacing cord, metal bands, or other binding, in an arrangement suitable for use only in specific equipment for which the harness was designed; it may include terminations.

Wisconsin false blossom *See* false blossom.

witches'-broom disease [PL PATH] An abnormal cluster of small branches or twigs that grow on a tree or shrub as a result of attack by fungi, viruses, dwarf mistletoes, or insect injury. Also known as hexenbesen; staghead.

witherite [MINERAL] $BaCO_3$ A yellowish- or grayish-white mineral of the aragonite group that has orthorhombic symmetry, hardness of $3\frac{1}{4}$ on Mohs scale, and specific gravity 4.3.

withertip [PL PATH] A blighting of the terminal shoots or the tips of leaves associated with certain plant diseases, such as anthracnose of citrus plants.

wittichenite [MINERAL] Cu_3BiS_3 A steel gray to tin white, orthorhombic mineral consisting of copper bismuth sulfide; occurs in tabular and massive form.

Wittig ether rearrangement [ORG CHEM] The rearrangement of benzyl and alkyl ethers when reacted with a methylating agent, producing secondary and tertiary alcohols.

WKB method *See* Wentzel-Kramers-Brillouin method.

wolf [VERT ZOO] Any of several wild species of the genus *Canis* in the family Canidae which are fierce and rapacious, sometimes attacking humans; includes the red wolf, gray wolf, and coyote.

wolfachite [MINERAL] $Ni(As,Sb)S$ A silver white to tin white mineral consisting of nickel, arsenic, and antimony sulfide; occurs in small crystals and in aggregates.

Wolfcampian [GEOL] A North American provincial series of geologic time; lowermost Permian (below Leonardian, above Virgilian of Pennsylvania).

wolfeite [MINERAL] $(Fe,Mn)_2(PO_4)(OH)$ A pinkish, wine-yellow to yellowish-brown or reddish-brown, monoclinic mineral consisting of a basic phosphate of iron and manganese.

Wolffian duct *See* mesonephric duct.

wolfram *See* tungsten; wolframite.

wolframine *See* wolframite.

wolframite [MINERAL] $(Fe,Mn)WO_4$ A brownish- or grayish-black mineral occurring in short monoclinic, prismatic, bladed crystals; the most important ore of tungsten. Also known as tobacco jack; wolfram; wolframine.

Wolf-Rayet star [ASTRON] A member of a class of very hot stars ($100,000–35,000$ K) which characteristically show broad bright emission lines in their spectra; luminosities are high, probably in the range 10^4–10^5 times that of the sun; these stars are probably very young and represent an early short-lived stage in stellar evolution.

wollastonite [MINERAL] $CaSiO_3$ A white to gray inosilicate mineral (a pyroxenoid) that crystallizes in the triclinic system in tabular crystals and has a pearly or silky luster on the cleavages; hardness is 5–5.5 on Mohs scale, and specific gravity is 2.85. Also known as tabular spar.

Wollaston polarizing prism [OPTICS] A device for producing linearly polarized beams of light, consisting of two adjacent quartz wedges with their optic axes perpendicular to each other and to the direction of incident light.

wolverine [VERT ZOO] *Gulo gulo.* A carnivorous mammal which is the largest and most vicious member of the family Mustelidae.

wood copper *See* olivenite.

woodhouseite [MINERAL] $CaAl_3(PO_4)(SO_4)(OH)_6$ A colorless to flesh-colored or white, hexagonal mineral consisting of a basic sulfate-phosphate of calcium and aluminum; occurs in small crystals and tabular form.

woodpecker [VERT ZOO] A bird of the family Picidae characterized by stiff tail feathers and zygodactyl feet which enable them to cling to a tree trunk while drilling into the bark for insects.

wood pulp *See* pulp.

wood screw [DES ENG] A threaded fastener with a pointed shank, a slotted or recessed head, and a sharp tapered thread of relatively coarse pitch for use only in wood.

wood sugar *See* xylose.

Woodward-Hoffmann rule [ORG CHEM] A concept which can predict or explain the stereochemistry of certain types of reactions in organic chemistry; it is also described as the conservation of orbital symmetry.

woodwardite [MINERAL] $Cu_4Al_2(SO_4)(OH)_{12} \cdot 2–4H_2O$ A greenish-blue to turquoise blue mineral consisting of a hydrated basic sulfate of copper and aluminum; occurs as botryoidal concretions and in spherulitic form.

woofer [ENG ACOUS] A large loudspeaker designed to reproduce low audio frequencies at relatively high power levels; usually used in combination with a crossover network and a high-frequency loudspeaker called a tweeter.

wool [TEXT] A textile fiber made from raw wool characterized by absorbency, resiliency, and insulation. [VERT ZOO] The soft undercoat of various animals such as sheep, angora, goat, camel, alpaca, llama, and vicuna.

wool-sorter's disease *See* anthrax.

word [COMPUT SCI] The fundamental unit of storage capacity for a digital computer, almost always considered to be more than eight bits in length. Also known as computer word.

word format [COMPUT SCI] Arrangement of characters in a word, with each position or group of positions in the word containing certain specified data.

word length [COMPUT SCI] The number of bits, digits, characters, or bytes in one word.

word mark [COMPUT SCI] A nondata punctuation bit used to delimit a word in a variable-word-length computer.

word processing [COMPUT SCI] The creation, dissemination, storage, and retrieval of the written word by typewriter terminals that use magnetic tape for storage, automatic control, editing, and retyping.

word rate [COMPUT SCI] In computer operations, the frequency derived from the elapsed period between the beginning of the transmission of one word and the beginning of the transmission of the next word.

word time *See* minor cycle.

work [ELEC] *See* load. [MECH] The transference of energy that occurs when a force is applied to a body that is moving in such a way that the force has a component in the direction of the body's motion; it is equal to the line integral of the force over the path taken by the body.

work angle [MET] In arc welding, the angle in a plane normal to the weld axis between the electrode and one member of the joint.

work cycle [IND ENG] A sequence of tasks, operations, and processes, or a pattern of manual motions, elements, and activities that is repeated for each unit of work.

work function [SOLID STATE] The minimum energy needed to remove an electron from the Fermi level of a metal to infinity; usually expressed in electron volts. [THERMO] *See* free energy.

work hardening [MET] Increased hardness accompanying plastic deformation of a metal below the recrystallization temperature range.

working load [ENG] The maximum load that any structural member is designed to support.

working pressure [ENG] The allowable operating pressure in a pressurized vessel or conduit, usually calculated by ASME (American Society of Mechanical Engineers) or API (American Petroleum Institute) codes.

working program [COMPUT SCI] A valid program which, when translated into machine language, can be executed on a computer.

working Q *See* loaded Q.

working set [COMPUT SCI] The smallest collection of instruction and data words of a given computer program which should be loaded into the main storage of a computer system so that efficient processing is possible.

working space *See* working storage.

working storage [COMPUT SCI] **1.** An area of main memory that is reserved by the programmer for storing temporary or intermediate values. Also known as working space. **2.** In COBOL (computer language), a section in the data division used for describing the name, structure, usage, and initial value of program variables that are neither constants nor records of input/output files.

working voltage *See* voltage rating.

Workshop *See* Sculptor.

work softening [MET] The phenomenon of a drop in the yield strength when a metal has been strained or cold-worked at low temperature and subsequently strained at an elevated temperature to cause the dislocations to become unstable.

worm [DES ENG] A shank having at least one complete tooth (thread) around the pitch surface; the driver of a worm gear. [INV ZOO] **1.** The common name for members of the Annelida. **2.** Any of various elongated, naked, soft-bodied animals resembling an earthworm. [MET] Sweat of molten metal which exudes through the crust of solidifying metal in a casting, and is caused by gas evolution.

worm conveyor *See* screw conveyor.

worm gear [DES ENG] A gear with teeth cut on an angle to be driven by a worm; used to connect nonparallel, nonintersecting shafts.

wound shock *See* hypovolemic shock.

wow [ENG ACOUS] A low-frequency flutter; when caused by an off-center hole in a disk record, occurs once per revolution of the turntable.

wrap [GRAPHICS] In a book or magazine, an insert, usually on a different stock, which is not tipped in but wrapped around one of the signatures, and which usually consists of at least two leaves (four pages) securely stitched into the binding.

wrap forming *See* stretch forming.

wrecking ball *See* skull cracker.

wren [VERT ZOO] Any of the various small brown singing birds in the family Troglodytidae; they are insectivorous and tend to inhabit dense, low vegetation.

wrench [ENG] A manual or power tool with adapted or adjustable jaws or sockets either at the end or between the ends of a lever for holding or turning a bolt, pipe, or other object. [MECH] The combination of a couple and a force which is parallel to the torque exerted by the couple.

wrench fault [GEOL] A lateral fault with a more or less vertical fault surface. Also known as basculating fault; torsion fault.

Wright system [PETRO ENG] A method for mining oil from partially drained sands that involves drilling a shaft through the productive strata, followed by long, slanting holes drilled radially in all directions from the shaft bottom into the oil sands.

wrist [ANAT] The part joining forearm and hand.

write [COMPUT SCI] **1.** To transmit data from any source onto an internal storage medium. **2.** A command directing that an output operation be performed.

write enable ring [COMPUT SCI] A file protection ring that must be attached to the hub of a reel of magnetic tape in order to physically allow data to be transcribed onto the reel. Also known as write ring.

write error [COMPUT SCI] **1.** A condition in which information cannot be written onto or into a storage device, due to dust, dirt, damage to the recording surface, or damaged electronic components. **2.** A condition in which there is an inconsistency between the pattern of bits transmitted to the write head of a magnetic tape drive and the pattern sensed immediately afterward by the read head.

write head [ELECTR] Device that stores digital information as coded electrical pulses on a magnetic drum, disk, or tape.

write inhibit ring [COMPUT SCI] A file protection ring that physically prevents data from being written on a reel of magnetic tape when it is attached to the hub of the reel.

write protection [COMPUT SCI] A form of memory protection in which a computer program can read from any area in memory but cannot write outside its own area.

write ring *See* write enable ring.

wrought alloy [MET] An alloy that has been mechanically worked after casting.

wulfenite [MINERAL] $PbMoO_4$ A yellow, orange, orange-yellow, or orange-red tetragonal mineral occurring in tabular crystals or granular masses; an ore of molybdenum. Also known as yellow lead ore.

Würm [GEOL] **1.** A European stage of geologic time: uppermost Pleistocene (above Riss, below Holocene). **2.** Pertaining to the fourth glaciation of the Pleistocene epoch in the Alps, equivalent to the Wisconsin glaciation in North America, following the Riss-Würm interglacial.

W Ursae Majoris stars [ASTRON] Eclipsing variable stars whose brightness is continuously varying in periods of a few hours; they are composed of two close stars that have a common gaseous envelope.

Wurtz-Fittig reaction [ORG CHEM] A modified Wurtz reaction in which an aromatic halide reacts with an aklyl halide in the presence of sodium and an anhydrous solvent to form alkylated aromatic hydrocarbons.

wurtzilite [GEOL] A black, massive, sectile, infusible, asphaltic pyrobitumen derived from the metamorphosis of petroleum.

wurtzite [MINERAL] (Zn,Fe)S A brownish-black hexagonal mineral consisting of zinc sulfide and occurring in hemimorphic pyramidal crystals, or in radiating needles and bundles.

Wurtz reaction [ORG CHEM] Synthesis of hydrocarbons by treating alkyl iodides in ethereal solution with sodium according to the reaction $2CH_3I + 2Na \rightarrow CH_3CH_3 + 2NaI$.

W Virginis stars [ASTRON] Periodic variable stars with periods of about 10 to 30 days; they exhibit two surges of activity from the same star so that there is a dou-

bling of their spectral lines. Also known as Type II Cepheids.

wye [ELEC] Polyphase circuit whose phase differences are 120° and which when drawn resembles the letter Y. [ENG] A pipe branching off a straight main run at an angle of 45°. Also known as Y; yoke.

Wynyardiidae [PALEON] An extinct family of herbivorous marsupial mammals in the order Diprotodonta.

Xanthidae [INV ZOO] The mud crabs, a family of decapod crustaceans in the section Brachyura.

xanthochroite *See* greenockite.

xanthoconite [MINERAL] Ag_3AsS_3 A dark red to dull orange to clove brown mineral consisting of silver arsenic sulfide.

Xanthophyceae [BOT] A class of yellow-green to green flagellate organisms of the division Chrysophyta; zoologists classify these organisms in the order Heterochlorida.

xanthophyllite *See* clintonite.

xanthosiderite *See* goethite.

xanthoxenite [MINERAL] $Ca_2Fe(PO_4)_2(OH)\cdot1\frac{1}{2}H_2O$ A pale yellow to brownish-yellow, monoclinic or triclinic mineral consisting of a hydrated basic phosphate of calcium and iron; occurs as masses and crusts.

Xantusiidae [VERT ZOO] The night lizards, a family of reptiles in the suborder Sauria.

x axis [CRYSTAL] A reference axis within a quartz crystal. [MATH] **1.** A horizontal axis in a system of rectangular coordinates. **2.** That line on which distances to the right or left (east or west) of the reference line are marked, especially on a map, chart, or graph.

X band [COMMUN] A radio-frequency band extending from 5200 to 10,900 megahertz, corresponding to wavelengths of 5.77 to 2.75 centimeters.

X chromosome [GEN] The sex chromosome occurring in double dose in the homogametic sex and in single dose in the heterogametic sex.

x coordinate [MATH] One of the coordinates of a point in a two- or three-dimensional cartesian coordinate system, equal to the directed distance of a point from the y axis in a two-dimensional system, or from the plane of the y and z axes in a three-dimensional system, measured along a line parallel to the x axis.

Xe *See* xenon.

Xenarthra [VERT ZOO] A suborder of mammals in the order Edentata including sloths, anteaters, and related forms; posterior vertebrae have extra articular facets and vertebrae in the hip, and shoulder regions tend to be fused.

X engine [MECH ENG] An in-line engine with the cylinder banks so arranged around the crankshaft that they resemble the letter X when the engine is viewed from the end.

xenoblast [MINERAL] A mineral which has grown during metamorphism without development of its characteristic crystal faces. Also known as allotrioblast.

xenocryst [CRYSTAL] A crystal in igneous rock that resembles a phenocryst and is foreign to the enclosing body of rock. Also known as chadacryst.

xenolith [PETR] An inclusion in an igneous rock which is not genetically related, such as an unmelted fragment of country rock. Also known as accidental inclusion; exogenous inclusion.

xenon [CHEM] An element, symbol Xe, member of the noble gas family, group O, atomic number 54, atomic weight 131.30; colorless, boiling point $-108°C$ (1 atmosphere, or 101,325 newtons per square meter), noncombustible, nontoxic, and nonreactive; used in photographic flash lamps, luminescent tubes, and lasers, and as an anesthetic.

xenon-135 [NUC PHYS] A radioactive isotope of xenon produced in nuclear reactors; readily absorbs neutrons; half-life is 9.2 hours.

Xenophyophorida [INV ZOO] An order of Protozoa in the subclass Granuloreticulosia; includes deep-sea forms that develop as discoid to fan-shaped branching forms which are multinucleate at maturity.

Xenopneusta [INV ZOO] A small order of wormlike animals belonging to the Echiurida.

Xenopterygii [VERT ZOO] The equivalent name for Gobiesociformes.

Xenosauridae [VERT ZOO] A family of four rare species of lizards in the suborder Sauria; composed of the Chinese lizard (*Shinisaurus crocodilurus*) and three Central American species of the genus *Xenosaurus*.

Xenungulata [PALEON] An order of large, digitigrade, extinct, tapirlike mammals with relatively short, slender limbs and five-toed feet with broad, flat phalanges; restricted to the Paleocene deposits of Brazil and Argentina.

Xeralf [GEOL] A suborder of the soil order Alfisol, having good drainage, and found in regions with rainy winters and dry summers in mediterranean climates; the surface horizons tend to become massive and hard during the dry seasons, with some soils having duripans that interfere with root growth.

Xerert [GEOL] A suborder of the soil order Vertisol, formed in a Mediterranean climate; wide surface cracks open and close once a year.

Xeroll [GEOL] A suborder of the soil order Mollisol, formed in a xeric moisture regime; may have a calcic, petrocalcic, or gypsic horizon, or a duripan.

xerophyte [ECOL] A plant adapted to life in areas where the water supply is limited.

Xerult [GEOL] A suborder of the soil order Ultisol, formed in a xeric moisture regime; brownish or reddish soil with a low to moderate organic-carbon content.

X frame [DES ENG] An automotive frame which either has side rails bent in at the center of the vehicle, making the overall form that of an X, or has an X-shaped member which joins the side rails with diagonals for added strength and resistance to torsional stresses.

xi hyperon [PARTIC PHYS] Also known as xi particle. **1.** Collective name for the xi-minus and xi-zero particles, which form an isotopic-spin multiplet of quasistable baryons, designated Ξ, having a hypercharge of -1, a total isotopic spin of $\frac{1}{2}$, a spin of $\frac{1}{2}$, positive parity, and an average mass of approximately 1318 MeV (million electron volts). Also known as cascade hyperon; cascade particle. **2.** A baryon belonging to any isotopic-spin multiplet having a hypercharge of -1 and a total isotopic spin of $\frac{1}{2}$; designated by $\Xi_{JP}(m)$, where m is the mass of the baryon in MeV, and J and P are its spin and parity (if known); the $\Xi_{3/2} + (1530)$ is sometimes designated Ξ^*.

xi-minus particle [PARTIC PHYS] A negatively charged xi hyperon, designated Ξ^-. Also known as cascade particle.

Xiphiidae [VERT ZOO] The swordfishes, a family of perciform fishes in the suborder Scombroidei characterized by a tremendously produced bill.

Xiphodontidae [PALEON] A family of primitive tylopod ruminants in the superfamily Anaplotherioidea from the late Eocene to the middle Oligocene of Europe.

Xiphosura [INV ZOO] The equivalent name for Xiphosurida.

Xiphosurida [INV ZOO] A subclass of primitive arthropods in the class Merostomata characterized by cephalothoracic appendages, ocelli, book lungs, a somewhat trilobed body, and freely articulating styliform telson.

Xiphydriidae [INV ZOO] A family of the Hymenoptera in the superfamily Siricoidea.

X organ [INV ZOO] A cluster of neurosecretory cells of the medulla terminales, a portion of the brain lying in the eyestalk in stalk-eyed crustaceans.

x-parallax See absolute stereoscopic parallax.

XPS See x-ray photoelectron spectroscopy.

X punch [COMPUT SCI] In an 80-column punched card, any hole in the second row from the top. Also known as eleven punch.

x-radiation See x-rays.

x-ray astronomy [ASTRON] The study of x-rays mainly from sources outside the solar system; it includes the study of novae and supernovae in the Milky Way Galaxy, together with extragalactic radio sources.

x-ray burster [ASTRON] One of a class of celestial x-ray sources which produce bursts of x-rays in the 1–20-kiloelectronvolt range and which are characterized by rise times of less than a few seconds and decay times of a few seconds to a few minutes; the peak luminosity is of the order of 10^{38} ergs/per second (10^{31} watts) and the sources have an average equivalent temperature of 10^8 K.

x-ray crystallography [CRYSTAL] The study of crystal structure by x-ray diffraction techniques. Also known as roentgen diffractometry.

x-ray diffraction [PHYS] The scattering of x-rays by matter, especially crystals, with accompanying variation in intensity due to interference effects. Also known as x-ray microdiffraction.

x-ray emission See x-ray fluorescence.

x-ray fluorescence [ATOM PHYS] Emission by a substance of its characteristic x-ray line spectrum upon exposure to x-rays. Also known as x-ray emission.

x-ray fluorescence analysis [ANALY CHEM] A nondestructive physical method used for chemical analyses of solids and liquids; the specimen is irradiated by an intense x-ray beam and the lines in the spectrum of the resulting x-ray fluorescence are diffracted at various angles by a crystal with known lattice spacing; the elements in the specimen are identified by the wavelengths of their spectral lines, and their concentrations are determined by the in-

tensities of these lines. Also known as x-ray fluorimetry.

x-ray fluorimetry See x-ray fluorescence analysis.

x-ray hardness [ELECTROMAG] The penetrating ability of x-rays; it is an inverse function of the wavelength.

x-ray image spectrography [SPECT] A modification of x-ray fluorescence analysis in which x-rays irradiate a cylindrically bent crystal, and Bragg diffraction of the resulting emissions produces a slightly enlarged image with a resolution of about 50 micrometers.

x-ray microdiffraction See x-ray diffraction.

x-ray microscope [ENG] **1.** A device in which an ultra-fine-focus x-ray tube and electron gun produces an electron beam focused to an extremely small image on a transmission-type x-ray target that serves as a vacuum seal; the magnification is by projection; specimens being examined can thus be in air, as also can the photographic film that records the magnified image. **2.** Any of several instruments which utilize x-radiation for chemical analysis and for magnification of 100–1000 diameters; it is based on contact or projection microradiography, reflection x-ray microscopy, or x-ray image spectrography.

x-ray optics [ELECTROMAG] A title-by-analogy of those phases of x-ray physics in which x-rays demonstrate properties similar to those of light waves. Also known as roentgen optics.

x-ray photoelectron spectroscopy [SPECT] A form of electron spectroscopy in which a sample is irradiated with a beam of monochromatic x-rays and the energies of the resulting photoelectrons are measured. Abbreviated XPS. Also known as electron spectroscopy for chemical analysis (ESCA).

x-ray powder method See powder method.

x-ray projection microscopy See projection microradiography.

x-rays [PHYS] A penetrating electromagnetic radiation, usually generated by accelerating electrons to high velocity and suddenly stopping them by collision with a solid body, or by inner-shell transitions of atoms with atomic number greater than 10; their wavelengths range from about 10^{-5} angstrom to 10^3 angstroms, the average wavelength used in research being about 1 angstrom. Also known as roentgen rays; x-radiation.

x-ray spectrometry [SPECT] The measure of wavelengths of x-rays by observing their diffraction by crystals of known lattice spacing. Also known as roentgen spectrometry; x-ray spectroscopy.

x-ray spectroscopy See x-ray spectrometry.

x-ray spectrum [SPECT] A display or graph of the intensity of x-rays, produced when electrons strike a solid object, as a function of wavelengths or some related parameter; it consists of a continuous bremsstrahlung spectrum on which are superimposed groups of sharp lines characteristic of the elements in the target.

x-ray star [ASTROPHYS] A source of x-rays from outside the solar system; examples are the point x-ray sources Scorpius X-1, Cygnus X-2, and the Crab x-ray source.

x-ray telescope [ENG] An instrument designed to detect x-rays emanating from a source outside the earth's atmosphere and to resolve the x-rays into an image; they are carried to high altitudes by balloons, rockets, or space vehicles; although several types of x-ray detector, involving gas counters, scintillation counters, and collimators, have been used, only one, making use of the phenomenon of total external reflection of x-rays from a surface at grazing incidence, is strictly an x-ray telescope.

x-ray tube [ELECTR] A vacuum tube designed to produce x-rays by accelerating electrons to a high velocity by means of an electrostatic field, then suddenly stopping them by collision with a target.

x-ray unit *See* siegbahn.

X test [STAT] A one-sample test which rejects the hypothesis $\mu = \mu_H$ in favor of the alternative $\mu > \mu_H$ if $X - \mu_H \geq c$ where c is an appropriate critical value, X is the arithmetic mean of observations, μ_H is a given number, and μ is the (unknown) expected value of the random variable X.

XU *See* siegbahn.

X unit *See* siegbahn.

X wave *See* extraordinary wave.

XY coordinate plotter *See* coordinate plotter.

Xyelidae [INV ZOO] A family of hymenopteran insects in the superfamily Megalodontoidea.

xylem [BOT] The principal water-conducting tissue and the chief supporting tissue of higher plants; composed of tracheids, vessel members, fibers, and parenchyma.

Xylocopidae [INV ZOO] A family of hairy tropical bees in the superfamily Apoidea.

Xylomyiidae [INV ZOO] A family of orthorrhaphous dipteran insects in the series Brachycera.

xylose [BIOCHEM] $C_5H_{10}O_5$ A pentose sugar found in many woody materials; combustible, white crystals with a sweet taste; soluble in water and alcohol; melts about 148°C; used as a nonnutritive sweetener and in dyeing and tanning. Also known as wood sugar.

XY plotter *See* coordinate plotter.

XY recorder [ENG] A recorder that traces on a chart the relation of two variables, neither of which is time.

Xyridaceae [BOT] A family of terrestrial monocotyledonous plants in the order Commelinales characterized by an open leaf sheath, three stamens, and a simple racemose head for the inflorescence.

XY switching system [ELECTR] A telephone switching system consisting of a series of flat bank and wiper switches in which the wipers move in a horizontal plane, first in one direction and then in another under the control of pulses from a subscriber's dial; the switches are stacked on frames, and are operated one after another.

Y

Y *See* wye; yttrium.

Yagi antenna *See* Yagi-Uda antenna.

Yagi-Uda antenna [ELECTROMAG] An end-fire antenna array having maximum radiation in the direction of the array line; it has one dipole connected to the transmission line and a number of equally spaced unconnected dipoles mounted parallel to the first in the same horizontal plane to serve as directors and reflectors. Also known as Uda antenna; Yagi antenna.

yag laser *See* yttrium-aluminum-garnet laser.

yak [VERT ZOO] *Poephagus grunniens.* A heavily built, long-haired mammal of the order Artiodactyla, with a shoulder hump; related to the bison, and resembles it in having 14 pairs of ribs.

yam [BOT] **1.** A plant of the genus *Dioscorea* grown for its edible fleshy root. **2.** An erroneous name for the Puerto Rico variety of sweet potato; the edible, starchy tuberous root of the plant.

yard [CIV ENG] A facility for building and repairing ships. [MECH] A unit of length in common use in the United States and United Kingdom, equal to 0.9144 meter, or 3 feet. Abbreviated yd. [NAV ARCH] A long spar, tapered at the ends, attached at its middle to a mast and running athwartships, and used to support a sail.

Yarmouth interglacial [GEOL] The second interglacial stage of the Pleistocene epoch in North America, following the Kansan glacial stage and before the Illinoian.

yaw [MECH **1.** The rotational or oscillatory movement of a ship, aircraft, rocket, or the like about a vertical axis. Also known as yawing. **2.** The amount of this movement, that is, the angle of yaw. **3.** To rotate or oscillate about a vertical axis.

yawing *See* yaw.

yaws [MED] An infectious tropical disease of humans caused by the spirochete *Treponema pertenue*; manifested by a primary cutaneous lesion followed by a granulomatous skin eruption.

y axis [CRYSTAL] A line perpendicular to two opposite parallel faces of a quartz crystal. [MATH] **1.** A vertical axis in a system of rectangular coordinates. **2.** That line on which distances above or below (north or south) the reference line are marked, especially on a map, chart, or graph.

Yb *See* ytterbium.

Y chromosome [GEN] The sex chromosome found only in the heterogametic sex.

Y connection *See* Y network.

y coordinate [MATH] One of the coordinates of a point in a two- or three-dimensional coordinate system, equal to the directed distance of a point from the *x* axis in a two dimensional system, or from the plane of the *x* and *z* axes in a three-dimensional co-ordinate system, measured along a line parallel to the *y* axis.

yd *See* yard.

Y-delta transformation [ELEC] One of two electrically equivalent networks with three terminals, one being connected internally by a Y configuration and the other being connected internally by a delta transformation. Also known as delta-Y transformation; pi-T transformation.

year [ASTRON] Any of several units of time based on the revolution of the earth about the sun; the tropical year to which the calendar is adjusted is the period required for the sun's longitude to increase 360°; it is about 365.24220 mean solar days. Abbreviated yr.

yellow [OPTICS] The hue evoked in an average observer by monochromatic radiation having a wavelength in the approximate range from 577 to 597 nanometers; however, the same sensation can be produced in a variety of other ways.

yellow arsenic *See* orpiment.

yellow cedar *See* Alaska cedar.

yellow copperas *See* copiapite.

yellow cypress *See* Alaska cedar.

yellow dwarf [PL PATH] Any of several plant viral diseases characterized by yellowing of the foliage and stunting of the plant.

yellow fat cell [HISTOL] A large, generally spherical fat cell with a thin shell of protoplasm and a single enlarged fat droplet which appears yellowish.

yellow fever [MED] An acute, febrile, mosquito-borne viral disease characterized in severe cases by jaundice, albuminuria, and hemorrhage.

yellow lead ore *See* wulfenite.

yellow leaf blotch [PL PATH] A fungus disease of alfalfa caused by *Pyrenopeziza medicaginis* characterized by the appearance of yellow or orange blotches with small black dots on the foliage.

yellows [PL PATH] Any of various fungus diseases of plants characterized by yellowing of the leaves which later turn brown, become brittle, and die; affects cabbage, lettuce, cauliflower, peach, sugarbeet, and other plants.

yellow tellurium *See* sylvanite.

yew [BOT] A genus of evergreen trees and shrubs, *Taxus*, with the fruit, an aril, containing a single seed surrounded by a scarlet, fleshy, cuplike envelope; the leaves are flat and acicular.

yield [ENG] Product of a reaction or process as in chemical reactions or food processing. [MECH] That stress in a material at which plastic deformation occurs. [ORD] The total effective energy released in a nuclear explosion; usually expressed in terms of the equivalent tonnage of trinitrotoluene (TNT) required to produce the same energy release.

yield-pillar system [MIN ENG] A method of roof control whereby the natural strength of the roof strata is maintained by the relief of pressure in working areas and the controlled transference of load to abutments which are clear of the workings and roadways.

yield point [MECH] The lowest stress at which strain increases without increase in stress.

yield strength [MECH] The stress at which a material exhibits a specified deviation from proportionality of stress and strain.

yield stress [MECH] The lowest stress at which extension of the tensile test piece increases without increase in load.

yield temperature [ENG] The temperature at which a fusible plug device melts and is dislodged by its holder and thus relieves pressure in a pressure vessel; it is caused by the melting of the fusible material, which is then forced from its holder.

yig device [ELECTR] A filter, oscillator, parametric amplifier, or other device that uses an yttrium-iron-garnet crystal in combination with a variable magnetic field to achieve wide-band tuning in microwave circuits. Derived from yttrium-iron-garnet device.

Y network [ELEC] A star network having three branches. Also known as Y connection.

yoke [ARCH] A horizontal member forming the head of a window frame. [DES ENG] A clamp or similar device to embrace and hold two other parts. [ELECTR] See deflection yoke. [ELECTROMAG] Piece of ferromagnetic material without windings, which permanently connects two or more magnet cores. [ENG] 1. A bar of wood used to join the necks of draft animals for working together. 2. See wye. [MECH ENG] A slotted crosshead used instead of a connecting rod in some steam engines.

yoked basin See zeugogeosyncline.

yolk [BIOCHEM] 1. Nutritive material stored in an ovum. 2. The yellow spherical mass of food material that makes up the central portion of the egg of a bird or reptile.

yolk sac [EMBRYO] A distended extraembryonic extension, heavy-laden with yolk, through the umbilicus of the midgut of the vertebrate embryo.

Y organ [INV ZOO] Either of a pair of nonneural structures found in the anterior portion of the crustacean body; source of the molting hormone, ecdysone.

Young-Helmholtz theory [PHYSIO] A theory of color vision according to which there are three types of color receptors that respond to short, medium, and long waves respectively; primary colors are those that stimulate most successfully the three types of receptors. Also known as Helmholtz theory.

Younginiformes [PALEON] A suborder of extinct small lizardlike reptiles in the order Eosuchia, ranging from the Middle Permian to the Lower Triassic in South Africa.

Young's modulus [MECH] The ratio of a simple tension stress applied to a material to the resulting strain parallel to the tension.

Young's two-slit interference [OPTICS] Interference of light from two parallel slits which are illuminated by light from a single slit, which in turn is illuminated by a source; the interference can be seen by letting the light fall on a screen, which then shows a series of parallel fringes.

youth [GEOL] The first stage of the cycle of erosion in which the original surface or structure is the dominant topographic feature; characterized by broad, flat-topped interstream divides, numerous swamps and shallow lakes, and progressive increase of local relief. Also known as topographic youth.

Yponomeutidae [INV ZOO] A heterogeneous family of small, often brightly colored moths in the superfamily Tineoidea; the head is usually smooth with reduced or absent ocelli.

Ypsilothuriidae [INV ZOO] A family of echinoderms in the order Dactylochirotida having 8–10 tentacles, a permanent spire on the plates of the test, and the body fusiform or U-shaped.

Y punch [COMPUT SCI] In a standard punched card, a hole in the topmost row.

yr See year.

ytterbium [CHEM] A rare-earth metal of the yttrium subgroup, symbol Yb, atomic number 70, atomic weight 173.04; lustrous, malleable, soluble in dilute acids and liquid ammonia, reacts slowly with water; melts at 824°C, boils at 1427°C; used in chemical research, lasers, garnet doping, and x-ray tubes.

yttrium [CHEM] A rare-earth metal, symbol Y, atomic number 39, atomic weight 88.905; dark-gray, flammable (as powder), soluble in dilute acids and potassium hydroxide solution, and decomposes in water; melts at 1500°C, boils at 2927°C; used in alloys and nuclear technology and as a metal deoxidizer.

yttrium-aluminum-garnet laser [OPTICS] A four-level infrared laser in which the active material is neodymium ions in an yttrium-aluminum-garnet crystal; it can provide a continuous output power of several watts. Abbreviated yag laser.

yttrium-iron-garnet device See yig device.

yttrocrasite [MINERAL] $(Y,Th,U,Ca)_2Ti_4O_{11}$ A black, orthorhombic mineral consisting of an oxide of rare earths and titanium.

yttrotantalite [MINERAL] $(Y,U,Fe)(Ta,Nb)O_4$ A black or brown, orthorhombic mineral consisting of an oxide of iron, yttrium, uranium, columbium, and tantalum; occurs in prismatic and tabular form.

Yucatán Current [OCEANOGR] A rapid northward flowing current along the western side of the Yucatán Strait; generally loops to the north and exits as the Florida Current.

yugawaralite [MINERAL] $CaAl_2Si_6O_{16} \cdot 4H_2O$ A zeolite mineral consisting of hydrous calcium aluminum silicate.

Yukawa force [NUC PHYS] The strong, short-range force between nucleons, as calculated on the assumption that this force is due to the exchange of a particle of finite mass (Yukawa meson), just as electrostatic forces are interpreted in quantum electrodynamics as being due to the exchange of photons.

Yukawa meson [PARTIC PHYS] A particle, having a finite rest mass, whose exchange between nucleons is postulated to account for the strong, short-range forces between them; such a contributor is the pi meson.

Z

Zalambdalestidae [PALEON] A family of extinct insectivorous mammals belonging to the group Proteutheria; they occur in the Late Cretaceous of Mongolia.

Zanclidae [VERT ZOO] The Moorish idols, a family of Indo-Pacific perciform fishes in the suborder Acanthuroidei.

Zapoididae [VERT ZOO] The Northern Hemisphere jumping mice, a family of the order Rodentia with long legs and large feet adapted for jumping.

zaratite [MINERAL] $Ni_3(CO_3)(OH)_4 \cdot 4H_2O$ An emerald-green mineral consisting of a hydrous basic nickel carbonate and occurring in incrustations or compact masses.

z axis [CRYSTAL] The optical axis of a quartz crystal, perpendicular to both the x and y axes. [MATH] One of the three axes in a three-dimensional cartesian coordinate system; in a rectangular coordinate system it is perpendicular to the x and y axes.

z coordinate [MATH] One of the coordinates of a point in a three-dimensional coordinate system, equal to the directed distance of a point from the plane of the x and y axes, measured along a line parallel to the z axis.

zebra [VERT ZOO] Any of three species of African mammals belonging to the family Equidae distinguished by a coat of black and white stripes.

zebu [VERT ZOO] A domestic breed of cattle, indigenous to India, belonging to the family Bovidae, distinguished by long drooping ears, a dorsal hump between the shoulders, and a dewlap under the neck; known as the Brahman in the United States.

Zechstein [GEOL] A European series of geologic time, especially in Germany: Upper Permian (above Rothliegende).

Zeeman effect [SPECT] A splitting of spectral lines in the radiation emitted by atoms or molecules in a static magnetic field.

Zeiformes [VERT ZOO] The dories, a small order of teleost fishes, distinguished by the absence of an orbitosphenoid bone, a spinous dorsal fin, and a pelvic fin with a spine and five to nine soft rays.

Zener breakdown [ELECTR] Nondestructive breakdown in a semiconductor, occurring when the electric field across the barrier region becomes high enough to produce a form of field emission that suddenly increases the number of carriers in this region. Also known as Zener effect.

zener diode [ELECTR] A semiconductor breakdown diode, usually constructed of silicon, in which reverse-voltage breakdown is based on the Zener effect.

Zener effect *See* Zener breakdown.

zenith [ASTRON] That point of the celestial sphere vertically overhead.

Zeoidea [VERT ZOO] An equivalent name for Zeiformes.

zeolite [MINERAL] **1.** A group of white or colorless, sometimes red or yellow, hydrous tectosilicate minerals characterized by an aluminosilicate tetrahedral framework, ion-exchangeable large cations, and loosely held water molecules permitting reversible dehydration. **2.** Any mineral of the zeolite group, such as analcime, chabazite, natrolite, and stilbite.

zeotrope [PHYS CHEM] A nonazeotropic liquid mixture which may be separated by distillation, and in which the components are miscible in all proportions (homogeneous zeotrope or homozeotrope) or not miscible in all proportions (heterogeneous zeotrope or heterozeotrope).

zephyr [METEOROL] Any soft, gentle breeze.

zero [MATH] The additive identity element of an algebraic system.

zero-access instruction [COMPUT SCI] An instruction consisting of an operation which does not require the designation of an address in the usual sense; for example, the instruction, "shift left 0003," has in its normal address position the amount of the shift desired.

zero-access storage [COMPUT SCI] Computer storage for which waiting time is negligible.

zero-address instruction format [COMPUT SCI] An instruction format in which the instruction contains no address, used when an address is not needed to specify the location of the operand, as in repetitive addressing. Also known as addressless instruction format.

zero bias [ELECTR] The condition in which the control grid and cathode of an electron tube are at the same direct-current voltage.

zero compression [COMPUT SCI] Any of a number of techniques used to eliminate the storage of nonsignificant leading zeros during data processing in a computer.

zero condition [COMPUT SCI] The state of a magnetic core or other computer memory element in which it represents the value 0. Also known as nought state; zero state.

zero error [ELECTR] Delay time occurring within the transmitter and receiver circuits of a radar system; for accurate range data, this delay time must be compensated for in the calibration of the range unit.

zero gravity *See* weightlessness.

zero-gravity switch [ELEC] A switch that closes as weightlessness or zero gravity is approached; in one version, conductive sphere of mercury encompasses two contacts at zero gravity but flattens away from the upper contact under the influence of gravity. Also known as weightlessness switch.

zero level [ENG ACOUS] Reference level used for comparing sound or signal intensities; in audio-frequency work, a power of 0.006 watt is generally used as zero level; in sound, the threshold of hearing is generally assumed as the zero level.

zero-level address [COMPUT SCI] The operand contained in an instruction so structured as to make immediate use of the operand.

zero-point vibration [STAT MECH] The vibrational motion which molecules in a crystal lattice, or particles in any oscillator potential, retain at a temperature of absolute zero; it is quantum-mechanical in origin. Also known as residual vibration.

zero state See zero condition.

zero-sum game [MATH] A two-person game where the sum of the payoffs to the two players is zero for each move.

zero suppression [COMPUT SCI] A process of replacing leading (nonsignificant) zeros in a numeral by blanks; it is an editing operation designed to make computable numerals easily readable to the human eye.

zeta pinch [PL PHYS] A type of plasma pinch produced by an electric current applied axially to a plasma cylinder in a controlled fusion reactor.

zeta potential [PHYS] The electrical potential that exists across the interface of all solids and liquids. Also known as electrokinetic potential.

zeugogeosyncline [GEOL] A geosyncline in a craton or stable area, within which is also an uplifted area, receiving clastic sediments. Also known as yoked basin.

zeunerite [MINERAL] $Cu(UO_2)_2(AsO_4)_2 \cdot 10-16H_2O$ A green secondary mineral of the autunite group consisting of a hydrous copper uranium arsenate; it is isomorphous with uranospinite.

zeylanite See ceylonite.

Ziegler catalyst [MATER] A special catalyst developed to produce stereospecific polymers, and derived from a transition-metal halide and a metal hydride or metal alkyl.

zigzag lightning [GEOPHYS] Ordinary lightning of a cloud-to-ground discharge that appears to have a single, but very irregular, lightning channel; viewed from the right angle, this may be observed as beaded lightning.

zinc [CHEM] A metal of group IIb, symbol Zn, atomic number 30, atomic weight 65.37; explosive as powder; soluble in acids and alkalies, insoluble in water; strongly electropositive; melts at 419°C, boils at 907°C. [MET] A shiny, bluish-white, lustrous metal that is ductile when pure; used in alloys, metal coatings, electrical fuses, anodes, and dry cells.

zinc-65 [NUC PHYS] A radioactive isotope of zinc, which has a 250-day half-life with beta and gamma radiation; used in alloy-wear tracer studies and body metabolism studies.

zincaluminite [MINERAL] $Zn_6Al_6(SO_4)_2(OH)_{26} \cdot 5H_2O$ A white to bluish-white and pale blue mineral consisting of a basic hydrated sulfate of zinc and aluminum; occurs in tufts and crusts.

zincite [MINERAL] $(Zn,Mn)O$ A deep-red to orange-yellow brittle mineral; an ore of zinc. Also known as red oxide of zinc; red zinc ore; ruby zinc; spartalite.

zinckenite See zinkenite.

zinc spar See smithsonite.

zinc spinel See gahnite.

Zingiberaceae [BOT] A family of aromatic monocotyledonous plants in the order Zingiberales characterized by one functional stamen with two pollen sacs, distichously arranged leaves and bracts, and abundant oil cells.

Zingiberales [BOT] An order of monocotyledonous herbs or scarcely branched shrubs in the subclass Commelinidae characterized by pinnately veined leaves and irregular flowers that have well-differentiated sepals and petals, an inferior ovary, and either one or five functional stamens.

zinkenite [MINERAL] $Pb_6Sb_{14}S_{27}$ A steel-gray orthorhombic mineral consisting of a lead antimony sulfide and occurring in crystals and in masses; has metallic luster, hardness of 3–3.5 on Mohs scale, and specific gravity of 5.30–5.35. Also spelled zincken-ite.

zinnwaldite [MINERAL] $K_2(Li,Fe,Al)_6(Si,Al)_8O_{20}-(OH,F)_4$ A pale-violet, yellowish, brown, or dark-gray mica mineral; an iron-bearing variety of lepidolite; the characteristic mica of greisens.

zircon [MINERAL] $ZrSiO_4$ A brown, green, pale-blue, red, orange, golden-yellow, grayish, or colorless neosilicate mineral occurring in tetragonal prisms; it is the chief source of zirconium; the colorless varieties provide brilliant gemstones. Also known as hyacinth; jacinth; zirconite.

zirconite See zircon.

zirconium [CHEM] A metallic element of group IVb, symbol Zr, atomic number 40, atomic weight 91.22; occurs as crystals, flammable as powder; insoluble in water, soluble in hot, concentrated acids; melts at 1850°C, boils at 4377°C. [MET] A hard, lustrous, grayish metal that is strong and ductile; used in alloys, pyrotechnics, welding fluxes, and explosives.

zirconium-95 [NUC PHYS] A radioactive isotope of zirconium; half-life of 63 days with beta and gamma radiation; used to trace petroleum-pipeline flows and in the circulation of a catalyst in a cracking plant.

zirkelite [MINERAL] A black mineral consisting of an oxide of zirconium, titanium, calcium, ferrous iron, thorium, uranium, and rare earths.

Z line [HISTOL] The line formed by attachment of the actin filaments between two sarcomeres.

Zn See zinc.

Zoantharia [INV ZOO] A subclass of the class Anthozoa; individuals are monomorphic and most have retractile, simple, tubular tentacles.

Zoanthidea [INV ZOO] An order of anthozoans in the subclass Zoantharia; these are mostly colonial, sedentary, skeletonless, anemonelike animals that live in warm, shallow waters and coral reefs.

Zoarcidae [VERT ZOO] The eelpouts, a family of actinopterygian fishes in the order Gadiformes which inhabit cold northern and far southern seas.

Zodiac [ASTRON] A band of the sky extending 8° on each side of the ecliptic, within which the moon and principal planets remain.

zodiacal counterglow See gegenschein.

zodiacal light [GEOPHYS] A diffuse band of luminosity occasionally visible on the ecliptic; it is sunlight diffracted and reflected by dust particles in the solar system within and beyond the orbit of the earth.

zoisite [MINERAL] $Ca_2Al_3Si_3O_{12}(OH)$ A white, gray, brown, green, or rose-red orthorhombic mineral of the epidote group consisting of a basic calcium aluminum silicate and occurring massive or in prismatic crystals.

zonal circulation See zonal flow.

zonal flow [METEOROL] The flow of air along a latitude circle; more specifically, the latitudinal (east or west) component of existing flow. Also known as zonal circulation.

zonal soil [GEOL] In early classification systems in the United States, a soil order including soils with well-developed characteristics that reflect the influence of agents of soil genesis. Also known as mature soil.

zonal westerlies *See* westerlies.

zonal winds *See* westerlies.

zone [ANALY CHEM] *See* band. [COMPUT SCI] **1.** One of the top three rows of a punched card, namely, the 11, 12, and zero rows. **2.** *See* storage area. [CRYSTAL] A set of crystal faces which intersect (or would intersect, if extended) along edges which are all parallel. [GEOGR] An area or region of latitudinal character. [GEOL] A belt, layer, band, or strip of earth material such as rock or soil. [MATH] The portion of a sphere lying between two parallel planes that intersect the sphere. [ORD] **1.** Any tactical area of importance, generally parallel to the front, such as a fortified area, a defensive position, a combat zone, or a traffic-control zone. **2.** An area in which projectiles will fall when a given propelling charge is used and the elevation is varied between the minimum and the maximum; in practice, generally limited to howitzer and mortar firings.

zone axis [CRYSTAL] A line through the center of a crystal which is parallel to all the faces of a zone.

zone bit [COMPUT SCI] A set of bits; for example, it may indicate whether the set of bits represents a numeric or alphabetic character.

zoned decimal [COMPUT SCI] A format for use with EBCDIC input and output permitting a sign overpunch in the low order position of the field; thus, + 1234 would be represented as: 1111/0001/1111/0010/1111/0011/1100/0100.

zone indices [CRYSTAL] Three integers identifying a zone of a crystal; they are the crystallographic coordinates of a point joined to the origin by a line parallel to the zone axis.

zone law [CRYSTAL] A law which states that the Miller indices (h, k, l) of any crystal plane lying in a zone with zone indices (u, v, w) satisfy the equation $hu + lv + kw = 0$.

zone of aeration [GEOL] A subsurface zone containing water below atmospheric pressure and air or gases at atmospheric pressure. Also known as unsaturated zone; vadose zone; zone of suspended water.

zone of cementation [GEOL] The layer of the earth's crust in which unconsolidated deposits are cemented by percolating water containing dissolved minerals from the overlying zone of weathering. Also known as belt of cementation.

zone of saturation [HYD] A subsurface zone in which water fills the interstices and is under pressure greater than atmospheric pressure. Also known as phreatic zone; saturated zone.

zone of silence *See* skip zone.

zone of suspended water *See* zone of aeration.

zone-position indicator [ENG] Auxiliary radar set for indicating the general position of an object to another radar set with a narrower field.

zone purification *See* zone refining.

zone refining [MET] A technique to purify materials in which a narrow molten zone is moved slowly along the complete length of the specimen to bring about impurity segregation, and which depends on differences in composition of the liquid and solid in equilibrium. Also known as zone purification.

zone time [ASTRON] The local mean time of a reference or zone meridian whose time is kept throughout a designated zone; the zone meridian is usually the nearest meridian whose longitude is exactly divisible by 15°.

zoning [CIV ENG] Designation and reservation under a master plan of land use for light and heavy industry, dwellings, offices, and other buildings; use is enforced by restrictions on types of buildings in each zone. [CRYSTAL] A variation in the composition of a crystal from core to margin due to a separation of the crystal phases during its growth by loss of equilibrium in a continuous reaction series. [ELECTROMAG] The displacement of various portions of the lens or surface of a microwave reflector so the resulting phase front in the near field remains unchanged. Also known as stepping.

zonochlorite *See* pumpellyite.

zoogeography [BIOL] The science that attempts to describe and explain the distribution of animals in space and time.

zooid [INV ZOO] A more or less independent individual of colonial animals such as bryozoans and coral.

zoology [BIOL] The science that deals with knowledge of animal life.

Zoomastigina [INV ZOO] The equivalent name for Zoomastigophorea.

Zoomastigophorea [INV ZOO] A class of flagellate protozoans in the subphylum Sarcomastigophora; some are simple, some are specialized, and all are colorless.

zoom lens [OPTICS] A system of lenses in which two or more parts are moved with respect to each other to obtain a continuously variable focal length and hence magnification, while the image is kept in the same image plane. Also known as variable-focal-length lens; varifocal lens.

zoonoses [BIOL] Diseases which are biologically adapted to and normally found in lower animals but which under some conditions also infect man.

zooplankton [ECOL] Microscopic animals which move passively in aquatic ecosystems.

Zoraptera [INV ZOO] An order of insects, related to termites and psocids, which live in decaying wood, sheltered from light; most individuals are wingless, pale in color, and blind.

Zoroasteridae [INV ZOO] A family of deep-water asteroid echinoderms in the order Forcipulatida.

Zorotypidae [INV ZOO] The single family, containing one genus, *Zorotypus*, in the order Zoraptera.

zorsite [MINERAL] $Ca_2Al_3Si_3O_{12}(OH)$ White, gray, brown, green, or rose-red orthorhombic mineral of the epidote group; an essential constituent of saussurite.

zoster *See* herpes zoster.

Zosteraceae [BOT] A family of monocotyledonous plants in the order Najadales; the group is unique among flowering plants in that they grow submerged in shallow ocean waters near the shore.

Zosterophyllatae [PALEOBOT] *See* Zosterophyllopsida.

Zosterophyllopsida [PALEOBOT] A group of early land vascular plants ranging from the Lower to the Upper Devonian; individuals were leafless and rootless.

Zr *See* zirconium.

Z score [STAT] A measure of how many standard deviations a raw score is from the mean.

Z time *See* Greenwich mean time.

z-transform [MATH] The z-transform of a sequence whose general term is f_n is the sum of a series whose general term is $f_n z^{-n}$, where z is a complex variable; n runs over the positive integers for a one-sided transform, over all the integers for a two-sided transform.

Zulu time *See* Greenwich mean time.

zwitterion *See* dipolar ion.

Zygaenidae [INV ZOO] A diverse family of small, often brightly colored African moths in the superfamily Zygaenoidea.

Zygaenoidea [INV ZOO] A superfamily of moths in the suborder Heteroneura characterized by complete venation, rudimentary palpi, and usually a rudimentary proboscis.

Zygnemataceae [BOT] A family of filamentous plants in the order Conjugales; they are differentiated into genera by chloroplast morphology, which may be spiral, bandlike, or cushionlike.

zygodactyl [VERT ZOO] Of birds, having a toe arrangement of two in front and two behind.

zygomatic bone [ANAT] A bone of the side of the face below the eye; forms part of the zygomatic arch and part of the orbit in mammals. Also known as malar bone.

zygomorphic [BIOL] Bilaterally symmetrical.

Zygomycetes [MYCOL] A class of fungi in the division Eumycetes.

Zygoptera [INV ZOO] The damsel flies, a suborder of insects in the order Odonata; individuals are slender, dainty creatures, often with bright-blue or orange coloring and usually with clear or transparent wings.

zygote [EMBRYO] 1. An organism produced by the union of two gametes. 2. The fertilized ovum before cleavage.

zygotene [CYTOL] The stage of meiotic prophase during which homologous chromosomes synapse; visible bodies in the nucleus are now bivalents. Also known as amphitene.

zymogen [BIOCHEM] The inactive precursor of an enzyme; liberates an active enzyme on reaction with an appropriate kinose. Also known as proenzyme.

zymosis *See* fermentation.

Zythiaceae [MYCOL] A family of fungi of the order Sphaeropsidales which contains many plant and insect pathogens.